PRINCIPLES OF ANATOMY AND PHYSIOLOGY

HPI — history of present illness
HR — heart rate
HSV — herpes simplex virus
HTLV — human T-cell leukemia-lymphoma virus
HTN — hypertension
Hx — history
I & D — incision and drainage
I & O — intake and output
IBD — inflammatory bowel disease
IBS — irritable bowel syndrome
ICC — intensive coronary care
ICF — intracellular fluid
ICN — intensive care nursery
ICU — intensive care unit
ID — intradermal
IF — intrinsic factor
IFN — interferon
Ig — immunoglobin
IM — intramuscular; infectious mononucleosis
IN — internist
IOP — intraocular pressure
IPPA — inspection, palpation, percussion, auscultation
IPSP — inhibitory postsynaptic potential
IUD — intrauterine device
IV — intravenous
IVC — inferior vena cava
IVF — in vitro fertilization
IVP — intravenous pyelogram
IVT — intravenous transfusion
JGA — juxtaglomerular apparatus
KS — Kaposi's sarcoma
KUB — kidneys, ureters, bladder
LBB — left breast biopsy
LDL — low-density lipoprotein
LFT — liver function test
LG — laryngectomy
LH — luteinizing hormone
LLQ — left lower quadrant
LMP — last menstrual period
LOC — loss of consciousness
LP — lumbar puncture
LPN — licensed practical nurse
LRI — lower respiratory infection
LUQ — left upper quadrant
LVAD — left ventricular assist device
MAb — monoclonal antibody
mEq/l — milliequivalents per liter
MG — myasthenia gravis
MI — myocardial infarction
MLT — medical laboratory technologist
mm^3 — cubic millimeter
mm Hg — millimeters of mercury
MOA — medical office assistant
MRI — magnetic resonance imaging
MS — multiple sclerosis
MSH — melanocyte-stimulating hormone
MSOF — multisystem organ failure
MVP — mitral valve prolapse
MVR — minute volume of respiration
NBM — nothing by mouth
ND — natural death
NE — norepinephrine
NGU — nongonococcal urethritis
NLMC — nocturnal leg muscle cramping
NMG — neuromuscular junction
NPN — nonprotein nitrogen
NREM — nonrapid eye movement
NSAID — nonsteroidal antiinflammatory drug
NSU — nonspecifric urethritis
NTG — nitroglycerin
NTP — normal temperature and pressure
OB/GYN — obstetrician-gynecologist; obstetrics-gynecology
OC — oral contraceptive
OD — overdose; right eye
OHS — open heart surgery
OI — opportunistic infection
OR — operating room
ORT — operating room technician
OT — oxytocin
OTC — over-the-counter
OV — office visit
P — pressure
PABA — para-aminobenzoic acid
PCP — *Pneumocystis carinii* pneumonia
PCV — packed cell volume
PD — Parkinson's disease
PE — pulmonary embolism; physical examination
PED — pediatrics; pediatrician
Peff — effective filtration pressure
PEG — pneumoencephalogram
PEMF — pulsating electromagnetic field
PET — positron emission tomography
PG — prostaglandin
pH — hydrogen-ion concentration
PID — pelvic inflammatory disease
PKU — phenylketonuria
PMH — past medical history
PMN — polymorphonuclear leucocyte
PMP — plasma membrane protein
PMS — premenstrual syndrome
PNS — peripheral nervous system
PRL — prolactin
PROG — progesterone
PT — prothrombin time; physical therapist
PTCA — percutaneous transluminal coronary angioplasty
PTD — permanent and total disability
PTH — parathyroid hormone
PTT — partial thromboplastin time
PTX — pneumothorax
PUBS — percutaneous umbilical blood sampling
PUL — percutaneous ultrasonic lithotripsy
Px — prognosis; pneumothorax
PX — physical examination
R — roentgen (unit of x radiation)
RA — rheumatoid arthritis
RAD — radiation absorbed dose
RAS — reticular activating system
RBB — right breast biopsy
RBC — red blood cell; red blood count
RBOW — rupture of bag of waters
RDA — recommended daily allowance
RDS — respiratory distress syndrome
REM — rapid eye movement
Rh — *Rhesus*
RHC — respirations have ceased
RIA — radioimmunoassay
RK — radial keratotomy
RLQ — right lower quadrant
RLX — relaxin
RM — radical mastectomy
RN — registered nurse
RNA — ribonucleic acid
ROS — review of symptoms
RR — respiratory rate
RRR — regular rate and rhythm (heart)
RS — Reye's syndrome
RT — radiotherapy; radiologic technologist
RUQ — right upper quadrant
Rx — prescription
SA — sinoatrial (sinuatrial)
SBP — systolic blood pressure
SC — subcutaneous
SCA — sickle-cell anemia
SCD — sudden cardiac death
SCID — severe combined immunodeficiency syndrome
SDS — same-day surgery
SF — synovial fluid
SG — skin graft; specific gravity
SH — social history
SIDS — sudden infant death syndrome
SIG — sigmoidoscopy; sigmoidoscope
SIW — self-inflicted wound
SLE — systemic lupus erythematosus
SMD — senile macular degeneration
SNS — somatic nervous system
SOB — shortness of breath
SPF — sun protection factor
S/S (Sx) — signs and symptoms
STD — sexually transmitted disease
SubQ or SQ — subcutaneous
SV — stroke volume
SVC — superior vena cava
T — temperature
TAH — total artificial heart
TB — tuberculosis
TBI — total body irradiation
TIA — transient ischemic attack
TLI — total lymphoid irradiation
Tm — tubular maximum
TM — transcendental meditation
TMJ — temporomandibular joint
TND — transient neurologic deficit
TOP — termination of pregnancy
t-PA — tissue plasminogen activator
TPE — therapeutic plasma exchange
TPN — total parenteral nutrition
TPR — temperature, pulse, and respiration
TSH — thyroid-stimulating hormone
TSS — toxic shock syndrome
Tx — treatment
UA — urinalysis
UDO — undetermined origin
UG — urogenital
URI — upper respiratory infection
US — ultrasound; ultrasonography
UTI — urinary tract infection
UV — ultraviolet
VD — venereal disease
VDRL — venereal disease research laboratory test (blood test for syphilis)
VF — ventricular fibrillation
VPC — ventricular premature contraction
VS — vital signs
VT — ventricular tachycardia
VV — varicose veins; vulva and vagina
WBC — white blood cell; white blood count
WNL — within normal limits
X-match — cross-match
XRT — x-ray therapy

Student:

To help you make the most of your study time and improve your grades, we have developed the following supplements designed to accompany Tortora/Anagnostakos: *Principles of Anatomy and Physiology,* 6/e:

- *Learning Guide,* by Kathleen Schmidt Prezbindowski and Gerard J. Tortora
  0–06–045374–5

- *Medical Terminology: An Illustrated Guide,* by Barbara J. Cohen
  0–06–04660–X

- *The Physiology Coloring Book,* by Wynn Kapit, Robert I. Macey, and Esmail Meisami
  0–06–043479–1

- *The Human Brain Coloring Book,* by Marian C. Diamond, Arnold B. Scheibel, and Lawrence M. Elson
  0–06–460306–7

- *The Anatomy Coloring Book,* by Wynn Kapit and Lawrence M. Elson
  0–06–453914–8

- *Atlas of Human Anatomy,* by J. Gosling et al.
  0–06–042425–7

You can order a copy at your local bookstore or call Harper & Row directly at 1-800-638-3030.

SIXTH EDITION

PRINCIPLES OF ANATOMY AND PHYSIOLOGY

Gerard J. Tortora
Bergen Community College

Nicholas P. Anagnostakos
Late, Bergen Community College

Medical Tests, Clinical Applications, and Disorders reviewed by
John G. Deaton, M.D.
University of Texas at Austin
Margaret B. Brenner, M.S.N., R.N.
Holy Family College
Jean Mueller, M.S.N., R.N., C.
Bellin College of Nursing

Art program reviewed by
Biagio John Melloni, Ph.D.
Michael C. Kennedy, Ph.D.
Hahnemann Medical College

Principal Medical Illustrators
Leonard Dank
Kevin A. Sommerville
Nadine Sokol

Sponsoring Editor: **Elizabeth A. Dollinger**
Development Editor: **Robert Ginsberg**
Project Editor: **Thomas R. Farrell**
Art Direction: **Teresa J. Delgado**
Art Coordinator: **Claudia DePolo Design**
Text Design: **Howard Petlack, AGT, Inc.**
Cover Coordinator: **Teresa J. Delgado**
Cover Design: **Circa '86, Inc.**
Cover Photo and Frontispiece: **From *The Dance Workshop.* Courtesy Gaia Books Ltd.**
Photo Research: **Mira Schachne**
Production: **Kewal K. Sharma**

Photo credits: Page 3, Focus on Sports; page 139, Loubat-Petit, Agence Vandystadt/Photo Researchers; page 329, Gerry Cranham/Photo Researchers; page 543, Comstock; page 875, Comstock.

Principles of Anatomy and Physiology, Sixth Edition

Library of Congress Cataloging-in-Publication Data

Tortora, Gerard J.
Principles of anatomy and physiology / Gerard J. Tortora, Nicholas P. Anagnostakos.—6th ed.
p. cm.
Includes bibliographies and index.
ISBN 0-06-046704-5
1. Human physiology. 2 Anatomy, Human. I. Anagnostakos, Nicholas Peter, 1924– . II. Title.
[DNLM: 1. Anatomy. 2. Homeostasis. 3. Physiology. QS 4 T712p]
QP34.5.T67 1990
612—dc20
DNLM/DLC
for Library of Congress **89-11035**
CIP

92 9 8 7 6

CONTENTS IN BRIEF

CONTENTS IN DETAIL

PREFACE

Principles of Anatomy and Physiology, Sixth Edition, is designed for use in an introductory course in anatomy and physiology and assumes no previous study of the human body. The text is geared to students in health-oriented, medical, and biological programs who are aiming for careers as nurses, medical assistants, physicians' assistants, medical laboratory technologists and technicians, perfusionists, radiation therapy technologists and radiographers, respiratory therapists, dental hygienists, physical and occupational therapists, surgical assistants and technologists, diagnostic medical sonographers, cytotechnologists and histologic technologists, electroencephalographic (EEG) technologists, emergency medical technicians-paramedics, nuclear medicine technologists, morticians, and medical record administrators and technicians. Because of its scope, the text is also useful for students in the biological sciences, science technology, liberal arts, and physical education and in premedical, predental, and prechiropractic programs.

OBJECTIVES

The objectives of the sixth edition are (1) to provide a basic understanding and working knowledge of the human body and (2) to present this essential material at a level that average students can handle.

Throughout, the goal has been to eliminate barriers to a ready comprehension of the structure and function of the human body. It is recognized, however, that some technical vocabulary and difficult concepts are vital to the course. Such material is developed in step-by-step, easy-to-understand explanations that avoid needlessly difficult technical vocabulary and syntax.

THEMES

As in previous editions, two major themes still dominate the book: ***homeostasis*** and ***pathology.*** Throughout, the book shows students how dynamic counterbalancing forces maintain normal anatomy and physiology. Pathology is viewed as a disruption in homeostasis. Accordingly, a number of clinical topics and disorders are presented and contrasted with previously learned normal processes.

ORGANIZATION

Based on comments by numerous reviewers and users, the book follows the same unit and topic sequence as its five earlier editions. It is divided into five principal areas of concentration. Unit 1, Organization of the Human Body, provides an understanding of the structural and functional levels of the body, from molecules to organ-systems. Unit 2, Principles of Support and Movement, analyzes the anatomy and physiology of the skeletal system, articulations, and the muscular system. Unit 3, Control Systems of the Human Body, emphasizes the importance of the nerve impulse in the immediate maintenance of homeostasis, the role of receptors in providing information about the internal and external environment, and the significance of hormones in maintaining long-range homeostasis. Unit 4, Maintenance of the Human Body, illustrates how the body maintains itself on a day-to-day basis through the mechanisms of circulation, respiration, digestion, cellular metabolism, urinary functions, and buffer systems. Unit 5, Continuity, covers the anatomy and physiology of the reproductive systems, development, and the basic concepts of genetics.

GENERAL CHANGES

1. Medical tests have been added to most chapters. The commonly employed tests provide information on diagnostic value, procedure, and normal values (when appropriate).

2. All of the art in Chapter 11 and much of the art in Chapter 28 has been redrawn using a new rendering technique.

3. A chapter outline has been added to each chapter to help students overview the sequence of topics.

4. Student objectives are now numbered.

5. Page numbers have been added to review questions to help students locate the answers.

6. Phonetic pronunciations have been added to medical terminology lists.

7. New line art has been placed adjacent to the epithelial tissue photomicrographs in Chapter 4.

8. Shock and homeostasis have been moved from Chapter 20 to Chapter 21, inflammation has been moved from Chapter 4 to Chapter 22, and regulation of food intake has been moved from Chapter 24 to Chapter 25.

9. Abbreviations (formerly Appendix C) and Eponyms Used in This Text (formerly Appendix D) are now found inside the covers of the text. Also added to the covers is a new section dealing with terms used in prescription writing.

10. Selected Readings now appear at the ends of each chapter rather than grouped at the end of the book.

11. In addition to updating and expanding the physiology throughout, several new pieces of physiology art have been added.

In the sixth edition, the strengths of the previous editions have been maintained. Revisions made for the new edition, based on extensive reviewer feedback, have focused on updating certain topics and strengthening the coverage of physiology. Among the specific changes made in topic coverage in the sixth edition are the following:

UNIT 1. ORGANIZATION OF THE HUMAN BODY

Chapter 1 has been expanded to include a new exhibit that summarizes representative structures found in the nine abdominopelvic regions. There is a new section on medical imaging that includes conventional radiography, computed tomography (CT) scanning, dynamic spatial reconstruction (DSR), magnetic resonance imaging (MRI), ultrasound (US), positron emission tomography (PET), and digital subtraction angiography (DSA). A new section on homeostasis and disease has also been included.

Chapter 2 contains new clinical applications on lasers. fat substitutes, and DNA fingerprinting; new sections on polar and nonpolar covalent bonds and the solvating property of water; and expanded coverage of levels of organization of proteins.

New to Chapter 3 are a clinical application on liposomes and drug therapy and discussions of the packaging of DNA in chromosomes, bulk flow, ligands, the microtrabecular lattice, DNA replication, and grading and staging tumors. Several new scanning electron micrographs (SEMs) of cell organelles have also been added. Among the topics revised are the structure, chemistry, and functions of the plasma membrane; lysosomal structure and functions; extracellular materials; cell division; and cells and cancer.

In Chapter 4 there is a new section on adhesion proteins and integrins and a new clinical application on Marfan syndrome. Membranes are now discussed before muscle and nervous tissue. In response to many requests by users, new line art has been placed adjacent to the epithelial tissue photomicrographs. Several new color photomicrographs have been added as replacements.

Changes in Chapter 5 include new clinical applications dealing with moles, liver spots, Retin-A and wrinkles, and skin grafts; new sections on scar formation and types of skin cancer; and a revised discussion of the histology of the epidermis.

UNIT 2. PRINCIPLES OF SUPPORT AND MOVEMENT

New to Chapter 6 are a medical test that deals with bone scanning; a clinical application on manganese and bone growth; and a section that discusses exercise and the skeletal system. The sections dealing with the histology of bone, homeostasis of bone, and osteoporosis have been revised.

In Chapter 7, there are new clinical applications on why scientists study bones and on temporomandibular joint (TMJ) syndrome; new sections on the bones of the orbits and treating herniated discs; and several new color photos of bones.

Chapter 9 contains a new medical test on arthrocentesis; a new discussion of the first transplant of an entire human knee; and a new exhibit on selected joints of the body according to definition, structural type, and movements permitted.

Chapter 10 has been revised considerably in response to user feedback. A new medical test on electromyography and new clinical applications on "second wind" and hypotonia and hypertonia have been added. There are new discussions of energy for muscular contraction, muscle action potential, factors that affect muscle tension, and regeneration of muscle tissue. Revised topics include oxygen debt, muscle fatigue, types of skeletal muscle fibers, muscular dystrophies, and abnormal contraction. Several new pieces of art have been added to demonstrate physiological concepts and several pieces have been redrawn.

Chapter 11 has been revised extensively. First, there are new exhibits and illustrations on muscles of the soft palate; muscles that move the wrist, hand, and fingers; intrinsic muscles of the hand; and intrinsic muscles of the foot. Also, every muscle exhibit now contains an overview section that focuses on the muscles under consideration. Several new cross-sectional illustrations have been added, as well as illustrations of the triangles of the neck and movements of the thumb. Another striking change is the use of a new technique for illustrating skeletal muscles. A new, detailed section on running injuries has been added.

UNIT 3. CONTROL SYSTEMS OF THE HUMAN BODY

Chapter 12 has been updated in response to user feedback to include new sections on electrical synapses, muscle action potentials, and presynaptic inhibition. Among the revised areas are regeneration of neurons, voltage-sensitive channels, the effect of "crack" on neurotransmitters, and impulse propagation.

In Chapter 13, new medical tests on spinal tap and myelography have been added. The section on peripheral nerve damage and repair has been revised.

Chapter 14 contains new sections on brain damage,

transient ischemic attack, and delirium. There is also a new medical test on electroencephalography, a new summary exhibit on neurotransmitters and neuropeptides, and new line art of the ventricles of the brain, brain stem, hypothalamus, limbic system, and origin of cranial nerves. Discussions of the blood–brain barrier, hypothalamus, limbic system, cerebellar structure and function, Parkinson's disease, cerebrovascular accident, multiple sclerosis, dyslexia, Alzheimer's disease, and Tay-Sachs disease have been revised.

Chapter 15 contains new clinical applications on anesthesia and coma, a new section on chronic and acute pain, and new line art of the reticular formation. The sections on generator and receptor potentials and memory have been revised.

In Chapter 17, a new clinical application on corneal surgery has been added. Also new are medical tests on tonometry, ophthalmoscopy, visual acuity, and audiometry. Several illustrations have been redrawn. The discussions of contact lenses, perforated eardrum, cataract, and conjunctivitis have been revised.

Chapter 18 has been updated by the inclusion of a new medical test on thyroid function and new sections on upregulation and down-regulation, hormonal interactions (permissive effect, synergism, antagonism), milk let-down reflex, growth factors, and the link between stress and immunity. There are also several new color photos of endocrine disorders.

UNIT 4. MAINTENANCE OF THE HUMAN BODY

Chapter 19 contains new medical tests dealing with erythrocyte sedimentation rate, reticulocyte count, hematocrit, histocompatibility testing, differential white blood cell count, complete blood count, blood typing, prothrombin time, and bleeding time. There are also new clinical applications related to taking blood samples, blood doping, and plasmapheresis. New material has been added on platelet structure and function, hemostatic control mechanisms, chronic fatigue syndrome, and colony-stimulating factors. Extrinsic and intrinsic clotting, polycythemia, and infectious mononucleosis have been revised.

In Chapter 20, new medical tests on serum enzyme studies; resting, stress, and ambulatory electrocardiograms; thallium imaging; and lipid profile have been added. Among the new sections are rheumatic fever, left ventricular assist device, hemopump, and stents. The revised topics include free radicals, artificial heart, heart murmur, coronary angiography, cardiac catheterization, development and treatment of atherosclerosis, and flutter and fibrillation. The section on shock has been moved to Chapter 21.

New topics in Chapter 21 are medical tests on digital subtraction angiography and Doppler ultrasound, a clinical application on edema, a section on shock and homeostasis, and illustrations of the branches of the celiac, superior mesenteric, and inferior mesenteric arteries. In response to user comments, the section on cardiovascular physiology (blood flow, blood pressure, resistance, vasomotor control, and exercise and the cardiovascular system) have been revised and new physiological illustrations have been added. Another notable change is the illustration of schemes of arterial distribution and venous drainage within exhibits, instead of as separate illustrations.

Chapter 22 contains a new medical test on lymphangiography, new clinical applications on metastasis and immunotherapy, and new summary exhibits on components of the lymphatic system and lymphokines. Among the revised topics are nonspecific resistance, phagocytosis, inflammation (moved from Chapter 4), complement, antibodies, T cells and cellular immunity, functions of macrophages, AIDS, and hypersensitivity.

New to Chapter 23 are medical tests dealing with bronchography, lung scans, pulmonary function tests, and arterial blood gas and a clinical application on rhinoplasty. The discussions of inspiration, expiration, tuberculosis, and pulmonary embolism have been revised.

Chapter 24 contains a new clinical application on root canal therapy, new medical tests on gastroscopy, liver function tests, oral cholecystogram, fecal occult blood, sigmoidoscopy and colonoscopy, and barium swallow and barium enema. There are also new sections on the functions of the liver, lipoproteins, dietary fiber, and colorectal cancer and new line art on the histology of the gastrointestinal tract. Revised sections include those dealing with jaundice, periodontal disease, anorexia nervosa, and bulimia. The section on regulation of food intake has been moved to Chapter 25.

New topics in Chapter 25 include a clinical application on carbohydrate loading and sections dealing with generation of ATP, chemiosmosis, hypothermia, malnutrition, and morbid obesity. The sections on metabolism, oxidation-reduction reactions, Krebs cycle, electron transport chain, regulation of body temperature, and fever have been revised.

Chapter 26 has new medical tests on cystoscopy, urinalysis, blood urea nitrogen, and creatinine; a new disorder, nephrotic syndrome; and several new physiology illustrations. The revised topics include glomerular filtration, tubular reabsorption, tubular secretion, renal clearance, renal failure, and extracorporeal shock wave lithotripsy.

In response to user feedback, the coverage of functions, regulation, and disorders related to electrolytes in Chapter 27 has been expanded greatly. The coverage of acid–base imbalances has also been increased along with a new summary exhibit.

UNIT 5. CONTINUITY

New topics in Chapter 28 include medical tests on semen analysis, Pap smear, colposcopy, cone biopsy, endocervical curettage, mammography, and laparoscopy; new clinical applications on cancer of the prostate gland and signs of ovulation; and sections on cervical mucus, mifepristone (RU 486), sympto-thermal method of birth control, genital warts, benign prostatic hyperplasia, transurethral sonography, and vulvovaginal candidiasis. Spermatogenesis, oogenesis, sexually transmitted diseases, impotence, dysmenorrhea, premenstrual syndrome, and toxic shock syndrome have been revised. Much of the art in this chapter has been redrawn using a new technique.

In Chapter 29, there are new medical tests on fetal ultrasonography and electronic fetal monitoring; new clinical applications on placenta previa, fetomaternal hemorrhage, umbilical cord accidents, fetal surgery, fetal-tissue transplantation, fertility and body fat, and karyotyping; and a new section dealing with diagnosis of pregnancy, exercise and pregnancy, Apgar scoring, aneuploidy, Klinefelter syndrome, Turner syndrome, metafemale syndrome, fragile X syndrome, separation of conjoined twins, and testis-determining factor. The sections on in vitro fertilization, the placenta, lactation, and genetics have been revised.

SPECIAL FEATURES

As in previous editions, the book contains numerous learning aids. Users of the book have cited the pedagogical aids as one of the book's many strengths. All of the tested and successful learning aids of previous editions have been retained in the sixth edition, and several new ones have been added. These special features are:

1. **Student Objectives.** Each chapter opens with a comprehensive list of Student Objectives. Each objective describes a knowledge or skill students should acquire while studying the chapter. (See ***Note to the Student*** for an explanation of how the objectives may be used.) In this edition, the Student Objectives are numbered.

2. **Chapter Outline.** New to the sixth edition, each chapter now contains an outline of its contents to help students preview the sequence of topics.

3. **Study Outline.** A study outline at the end of each chapter provides a brief summary of major topics. This section consolidates the essential points covered in the chapter so that students can recall and relate the points to one another. Page numbers given beside major headings in the outlines make it easy to refer to topics within chapters.

4. **Review Questions.** Review questions at the end of each chapter provide a check to see if the objectives stated at the beginning of the chapter have been mastered. Page numbers have been added to the questions to help students locate the answers. After answering the questions, students should reread the objectives to determine whether they have met the goals.

5. **Exhibits.** Health-science students are generally expected to learn a great deal about the anatomy of certain organ-systems, specifically, skeletal muscles, articulations, blood vessels, and nerves. To avoid interrupting the discussion of concepts and to organize the data, anatomical details have been presented in tabular form in exhibits, most of which are accompanied by illustrations. New summary exhibits have been added to the sixth edition, especially in relation to physiological principles.

6. **Disorders: Homeostatic Imbalances.** Abnormalities of structure or function are grouped at the end of appropriate chapters in sections titled "Disorders: Homeostatic Imbalances." These sections provide a review of normal body processes and demonstrate the importance of the study of anatomy and physiology to a career in any of the health fields. All disorders have been updated and several new ones have been added.

7. **Phonetic Pronunciations.** Throughout the text, phonetic pronunciations are provided in parentheses for selected anatomical and physiological terms. These pronunciations are given at the point where the terms are introduced and are repeated in the Glossary of Terms. Many new phonetic pronunciations have been added in the sixth edition. The ***Note to the Student*** explains the pronunciation key.

8. **Medical Terminology.** Glossaries of selected medical terms appear at the end of appropriate chapters; these listings are entitled "Medical Terminology." All of the glossaries have been revised for the sixth edition and phonetic pronunciations have been added.

9. **Clinical Applications.** Throughout the text, clinical applications are boxed off for greater emphasis. Many new ones have been added.

10. **Medical Tests.** New to the sixth edition are selected, commonly performed medical tests within appropriate chapters. Diagnostic value, procedure, and normal values, when applicable, are given for each medical test. It should be noted that the procedures described for the various medical tests are commonly used ones; other alternative procedures or variations also exist.

11. **Line Art.** The line drawings in the book are large so that details are easily seen. In the sixth edition, a large number have been redrawn (see especially Chapters 11 and 28) and many new ones have been added, many pertaining to physiology. Full color is used throughout to differentiate structures and regions.

12. **Photographs.** The photographs amplify the narrative and the line drawings. Numerous photomicrographs (in full color), scanning electron micrographs, and transmission electron micrographs enhance the histological discussions. Color photographs of specimens and regional dissections clarify gross anatomy discussions. New photographs have been added throughout.

13. **Appendixes.** Appendix A, Measurements, summarizes U.S., metric, and apothecary units of length, mass, volume, and time. Appendix B, Normal Values for Selected Blood and Urine Tests, contains a listing of normal values for the principal constituents of these fluids.

14. **Inside Front and Back Cover.** Three helpful listings have been placed on the inside of the front and back covers. Abbreviations (formerly Appendix C) is an alphabetical list of commonly encountered medical abbreviations, and it has been expanded for the sixth edition. Eponyms Used in This Text

(formerly Appendix D) is an alphabetical list of commonly encountered eponyms and the corresponding current terminology. (Eponyms are cited in the text in parentheses immediately following the preferred current terms.) Terms Used in Prescription Writing is entirely new, and consists of a listing of abbreviations, derivations, and English equivalents.

15. Glossaries. Two glossaries appear at the end of the book. The first deals with combining forms, prefixes, and suffixes. The second is a comprehensive glossary of terms. Both have been greatly updated and expanded.

16. Selected Readings. All new lists have been prepared for the sixth edition, and they now appear at the end of each chapter.

SUPPLEMENTS

The following supplementary items are available to accompany the sixth edition of *Principles of Anatomy and Physiology:*

1. Instructor's Manual. A new and considerably revised Instructor's Manual by Judith Lanum Mohan of Case Western Reserve University has been prepared for the sixth edition. Each chapter contains a chapter synopsis, a list of major concepts, a set of suggested activities, a discussion of clinical applications, suggestions for supplementary lecture material, a set of problem-solving essay questions for students, and lists of audiovisual aids. A set of student self-test questions is also included.

2. Transparencies. One hundred forty full-color transparencies will be available. The set contains illustrations that come from the text and were selected because they are frequently discussed in class.

3. Slides. Two sets totaling two hundred forty full-color slides are available for those professors who do not wish to use transparencies; 140 slides are from the text, and the remaining 100 are cadaver slides from Gosling et al., *Atlas of Human Anatomy.*

4. Test Bank. An entirely new Test Bank has been prepared for the sixth edition by John Dustman of Indiana University Northwest. This expanded Test Bank contains 3000 questions in a variety of formats (true–false, multiple-choice, completion, matching, and essay). The Test Bank is available in standard printed format as well as on Harper Test (a microcomputer-based test generator for the IBM PC, Apple II series, and Macintosh computers).

5. Learning Guide. By Kathleen S. Prezbindowski and Gerard J. Tortora, the *Learning Guide* is designed to help students *learn* anatomy and physiology. At the start of each chapter a *framework* permits students to visualize relationships among key concepts and terms. It is designed for both an introduction and a review of the chapter content. The framework provides a one-page synthesis of the chapter. It allows students to see "the forest," complete and organized, and also "the trees" (key terms). *Wordbytes* introduce prefixes, suffixes, and word roots of key terms. *Checkpoints* do not simply ask for repetition of text material, but challenge students to check progress as they handle new information in a variety of learning activities: labeling and coloring diagrams, placing physiological events in sequence, filling in paragraphs, matching, and multiple choice. *Clinical challenges* present students with opportunities for application of content. To enhance the effectiveness of the exercises, answers are provided for key-concept exercises so that the student has immediate feedback. A Mastery Test at the end of each chapter provides the student with a means of evaluating his or her learning of the chapter material and also gives practice for classroom testing situations.

6. Atlas of Human Anatomy with Integrated Text by J. A. Gosling et al. With a unique combination of photographs of cadaver dissections, accompanying diagrams and concise text, this volume provides the student with a better understanding of human anatomy.

7. Slide Atlas of Human Anatomy. Based on the material in the *Atlas of Human Anatomy with Integrated Text,* this collection of 506 slides includes every photograph from the text.

8. Laboratory Manuals. A laboratory manual with cat dissections by Patricia J. Donnelly and George A. Wistreich is available for students to purchase. Included in the manual are comparisons to the human body and full-color cadaver photographs. Another laboratory manual, written by Victor Eroschenko, features the cadaver. It is ideally suited for allied-health science students. Illustrations are taken from the text and show the student various structures in greater detail. An Instructor's Manual to accompany *Laboratory Manual for Anatomy and Physiology: With Cat Dissections* is available.

9. Medical Terminology: An Illustrated Guide by Barbara J. Cohen of Delaware County Community College. Provides students with a solid command of medical language. Suitable as a classroom text or self-study workbook, *Medical Terminology* is easy to read, inviting, concise, and attractive. Basic terminology is organized by body systems. Each chapter has a brief overview of the system, illustrations, an introduction of pertinent word parts with examples of their use, and numerous exercises. An Instructor's Manual to accompany Cohen's *Medical Terminology: An Illustrated Guide* is available. A printed Test Bank to accompany Cohen's *Medical Terminology: An Illustrated Guide* is also available.

10. Software. The following software packages are available for the IBM PC, Apple II series, Apple IIgs, and Macintosh computers. They include interactive tutorials, dissection simulations, and flashcards. Demonstration disks will be made available through your local representative should you wish to preview any of the programs.

BIOSOURCE. Four tutorial packages featuring a high degree of interaction, high-resolution graphics, and extensive multiple-choice testings on unprotected diskettes. These products include *Skeletal Muscle Anatomy and Physiology, Neuromuscular Concepts,* and *The Human Brain.* For the Apple II series and IBM.

HOMEOSTASIS. Focuses on the hypothalamus brain and the maintenance of the blood glucose level as it demonstrates visually how body temperature is maintained. A self-tutorial that tests the students' understanding of the material and leads them through simple experiments is included. For the Apple II series and IBM PC computers.

BODY LANGUAGE. A drill program designed to help students identify and name anatomical structures. Both matching and spelling drills are included for approximately 200 anatomical diagrams covering the major body systems. For the Apple II, Apple IIgs, and IBM PC computers.

ANATOMIST. Based on *The Anatomy Coloring Book,* this Hypercard tutorial incorporates audio pronunciations as well as in-depth discussions of each body region and system. The *Anatomist* also emphasizes the interrelation between parts and enables the students to "navigate" through various regions of the bodily functions. For the Apple Macintosh.

HYPERCARD STACKS FOR ANATOMY AND PHYSIOLOGY. Both a tutorial and a simulation, each organ stack includes textual material, graphics, and animation. By using the concept of hypertext, students may study the topics in any order they wish. A self-test of multiple-choice questions follows each topic. For the Apple Macintosh computer.

COMPUTER SIMULATION FOR THE LABORATORY. Also consisting of Hypercard stacks, this software is based on *Laboratory Manual for Anatomy and Physiology with Cat Dissections,* Third Edition, by Patricia J. Donnelly and George A. Wistreich. Each stack investigates anatomical structures and allows the students to explore the organization of the body through cat dissection simulations. For the Apple Macintosh computer.

FLASH!. Encourages students to learn vocabulary through comprehensive testing of both terms and definitions. The terms are organized around *Principles of Anatomy and Physiology,* Sixth Edition, so that students can quiz themselves after each chapter. For the Apple II series and IBM PC computers.

SENSORY AND MOTOR BRAIN. A tutorial dealing with the cerebral cortex. The cat brain is used to show the response of the animal to stimulation. For the Apple II series and IBM PC.

11. Coloring Books. For student purchase, these coloring books are available:

The Anatomy Coloring Book, by W. Kapit and L. Elson
The Physiology Coloring Book, by W. Kapit, R. Macey, and E. Meisami
The Human Brain Coloring Book, by M. Diamond, A. Scheibel, and L. Elson

12. Interactive Videos. An interactive video package, *Laser Touch Anatomy,* developed at Cuyahoga Community College, has been developed for student use. Five tutorials are provided for each section: general tutorial, fill-in-the-blank quiz, touch-screen quiz, vocabulary building exercises, and hear and spell exercises. Equipment requirements include an IBM PC/XT with 360-ICB floppy disk, 512-kB Memory, 20-MB hard-disk drive, serial port, laser disk player, and a touch-screen monitor

13. Cadaver Dissection Videos. A series of three videos introducing the student to the circulatory, muscular, and nervous systems of the body. Cadaver dissection illustrations from the book and animation are used to enhance the student's understanding of the human body.

14. Media Policy. Per Harper & Row's media policy for adopters, the following videos are available:

The New Womb
Windows on the Body
Generation to Generation: Genetic Screening, Counseling, and Therapy
Blood: The Vital Humor
To Hear a Pin Drop

15. Grades. A new program from Harper & Row that enables you to use your personal computer to enter and store students' quiz, test, and exam scores. You can also calculate intermediate and final grades using either point on percentage scores For the IBM PC or compatible personal computers.

16. Gross Anatomy Tutorial. This package of 9 disks provides a comprehensive review of anatomy for your students. EGA grahics are used, and many of the illustrations have been taken from *Principles of Anatomy and Physiology,* Sixth Edition. The software package will also drive some of the new videodisks available.

17. Integrator. This supplement has been designed to help professors organize assignments and lectures around the entire ancillary package. For every chapter in the book, the *Integrator* identifies similar materials that can be drawn from the supplements to enhance the student's understanding of the subject.

For further information on the supplement package to accompany *Principles of Anatomy and Physiology,* Sixth Edition, please contact your local representative or:

Harper & Row
College Marketing Group
10 East 53d Street
New York, NY 10022-5299

ACKNOWLEDGMENTS

I wish to thank the following reviewers for their outstanding contributions to the sixth edition of *Principles of Anatomy and Physiology.* They have generously shared their extensive knowledge, teaching experience, and concern for students with me, and their feedback has been invaluable.

Merlyn L. Anderberg
Spokane Falls Community College

Barry R. Anderson
University of Scranton

Rosemary L. Cannistraro
University of Missouri at St. Louis

John G. Deaton, MD
University of Texas at Austin

Sarah D. Gray
University of California, Davis

Ernest Joe Harber
San Antonio College

Charles W. Harnsberger
Macon College

Raymond Herndon
Danville Community College

James L. Larimer
University of Texas at Austin

Joseph A. Lipsky
The Ohio State University

Margaret H. Peaslee
Louisiana Tech University

E. Edward Sheeley
Valdosta State College

Ralph W. Stevens
Old Dominion University

David R. Wade
Southern Illinois University at Carbondale

In addition to the individuals who reviewed the entire manuscript for this edition, many people have been kind enough to take time to offer their comments and suggestions. I wish to publicly acknowledge their efforts and to thank them for encouraging me to write a better sixth edition. Among those to whom I express my deepest gratitude are Donna Alder of Roberts Wesleyan College, Edith Applegate of Kettering College of Medical Arts, Ian Barr of Camosun College, Bill Becker of Americal River College, Phil Bentley of Schenectady County Community College, Janene Blodgett of Black Hawk College, Colin Bond of College of Natural Therapies, Alan Bretag of South Australian Institute of Technology, Leigh Callan of Floyd College, Brian Case of St. Clair College, John Chilgren of Lewis & Clark College, M. James Cosentino of Millersville University, Laurie Cree of Kuring-gai College of Advanced Education, George Darnoff of Liberty University, Peter Dill of Okangan College, James Donney of Cape Cod Community College, Gerald R. Dotson of Front Range Community College, John Emes of British Columbia Institute of Technology, Jon R. Fortman of Mississippi University for Women, George Fortunato of Nassau Community College, John Galligan of Westchester Community College, Carol A. Gerding of Cuyahoga Community College, Bruce Graham of Riverina-Murray Institute of Higher Education, Robert Guimond of University of Massachusetts, John Gwinn of The University of Akron, James S. Hall of Our Lady of the Lake University, John Harling of Okangan College, Dean Hays of Auburn University, Aniko Hill of John Abbot College, Rick Hinterthuer of North Arkansas Community College, Julie Horsch of Casper College, B. R. Hudson of Glasgow School of Occupational Therapy, George Kahler of Cape Cod Community College, Frederick H, Karre of Carl Sandburg Community College, David Kaufman of Tasmanian State Institute of Technology, Bob Lenn of American River College, Sandy Lewis of Pierce College, Raymond Lo of College of New Caledonia, Patricia Mack of Clinton Community College, Zoriana K. Malseed of University of Pennsylvania, Christine Martin-Widder of Stark Technical College, Robert R. Montgomery of Oakland Community College, Phyllis Mooney of Tasmanian State Institute of Technology, John Natalini of Quincy College, George A. Newman of Hardin-Simmons University, Gregory A. Oseland of Illinois Valley Community College, Bruce Palmer of Assiniboine Community College, Izak Paul of Mount Royal College, Ron Paulson of Cariboo College, Ronald S. Payson of Columbia-Greene Community College, Deborah Perry of St. Lawrence College Saint Laurent, Clem Persaud of Canadore College, John A. Pitts of North Shore Community College, Emmanuel G. Pizania of Helene Fuld School of Nursing, Matthew Pravetz of Elizabeth Seton College, Keith Randolph of Ouachita Baptist University, Jerry W. Smith of St. Petersburg Community College, Wendy Smith of Northeastern University, John E. Stencel of Olney Central College, Analee G. Stone of Tunxis Community College, James F. Thompson of Huntingdon College, Gary Tieben of Saint Francis College, Robert Tiplady of Brainerd Community College, Thomas V. Tobin of King's College, Isabel Vance of Brevard Community College, Michael Venning of South Australian Institute of Technology, Frank V. Veselovsky of South Puget Sound Community College, Richard E. Welton of Southern Oregon State College, Paul F. White of Polk Community College, Richard White of Mendocino Academy of Science, Michael B. Wollam of Pasco-Hernando Community College, Florence Wonnacott of P.E.I. School of Nursing, and Fook Yong of University of New Mexico.

Gratitude is also extended for the contributions of the individuals and organizations whose names appear with the photographs in the text. Special thanks to Andrew J. Kuntzman of Wright State University School of Medicine for providing numerous excellent color photomicrographs and Dr. Michael C. Kennedy of Hahnemann Medical College for his help in reviewing the illustration program. I should especially like to thank Biagio John Melloni, who reviewed each piece of line art, every photo, and every caption for accuracy of drawing, labeling, and description. Thanks to his unremitting attention to detail, the art for the sixth edition is more accurate than ever before. His help is deeply appreciated. Finally, for typing drafts of the manuscript and numerous other duties associated with the task of putting together a textbook, thanks to Geraldine C. Tortora.

Finally, I would like to express my deep appreciation to all the many instructors and students who have used this text throughout its first five editions. It is deeply gratifying to know that over 1 million students have begun their careers in the allied health field by reading and studying this text. The responsibility placed upon me to maintain the quality of this text has been very great, but the knowledge that so many have benefitted from using it has provided an even greater reward. My deepest thanks go to all of you, past, present, and future.

As the acknowledgements indicate, the participation of many individuals of diverse talent and expertise is required in the production of a textbook of this scope and complexity. For this reason, readers and users of the sixth edition are invited to send their reactions and suggestions to me so that plans can be formulated for subsequent editions.

Gerard J. Tortora

Natural Sciences and Mathematics S229
Bergen Community College
400 Paramus Road
Paramus, NJ 07652

NOTE TO THE STUDENT

At the beginning of each chapter is a listing of ***Student Objectives.*** Before you read the chapter, please read the objectives carefully. Each objective is a statement of a skill or knowledge that you should acquire. To meet these objectives, you will have to perform several activities. Obviously, you must read the chapter carefully. If there are sections of the chapter that you do not understand after one reading, you should reread those sections before continuing. In conjunction with your reading, pay particular attention to the figures and exhibits; they have been carefully coordinated with the textual narrative.

At the end of each chapter are two and sometimes three other learning guides that you may find useful. The first, ***Study Outline,*** is a concise summary of important topics discussed in the chapter. This section is designed to consolidate the essential points covered in the chapter, so that you may recall and relate them to one another. The second guide, ***Review Questions,*** is a series of questions designed specifically to help you master the objectives. A third aid, ***Medical Terminology,*** appears in some chapters. This is a listing of terms designed to build your medical vocabulary. After you have answered the review questions, you should return to the beginning of the chapter and reread the objectives to determine whether you have achieved the goals.

As a further aid, we have included pronunciations for many terms that may be new to you. These appear in parentheses immediately following the new words, and they are repeated in the Glossary of Terms at the back of the book. (Of course, since there will always be some conflict among medical personnel and dictionaries about pronunciation, you will come across variations in different sources.) Look at the words carefully and say them out loud several times. Learning to pronounce a new word will help you remember it and make it a useful part of your medical vocabulary. Take a few minutes now to read the following pronunciation key, so it will be familiar as you encounter new words. The key is repeated at the beginning of the Glossary of Terms.

PRONUNCIATION KEY

1. The strongest accented syllable appears in capital letters, for example, bilateral (bī-LAT-er-al) and diagnosis (dī-ag-NŌ-sis).

2. If there is a secondary accent, it is noted by a single quote mark ('), for example, constitution (kon'-sti-TOO-shun) and physiology (fiz'-ē-OL-ō-jē). Any additional secondary accents are also noted by a single quote mark, for example, decarboxylation (dē'-kar-bok'-si-LĀ-shun).

3. Vowels marked with a line above the letter are pronounced with the long sound as in the following common words:

ā as in *māke*
ē as in *bē*
ī as in *īvy*
ō as in *pōle*

4. Vowels not so marked are pronounced with the short sound as in the following words:

e as in *bet*
i as in *sip*
o as in *not*
u as in *bud*

5. Other phonetic symbols are used to indicate the following sounds:

a as in *above*
oo as in *sue*
yoo as in *cute*
oy as in *oil*

PRINCIPLES OF ANATOMY AND PHYSIOLOGY

UNIT 1

ORGANIZATION OF THE HUMAN BODY

This unit is designed to show you how your body is organized at different levels. After you study the various regions and parts of your body, you will discover the importance of the chemicals that make it up. You will then find out how your cells, tissues, and organs form the systems that keep you alive and healthy.

Chapter 1

An Introduction to the Human Body

Chapter Contents at a Glance

Student Objectives

1. Define anatomy, with its subdivisions, and physiology.
2. Define each of the following levels of structural organization that make up the human body: chemical, cellular, tissue, organ, system, and organismic.
3. Identify the principal systems of the human body, list representative organs of each system, and describe the function of each system.
4. List and define several important life processes of humans.
5. Define the anatomical position and compare common and anatomical terms used to describe various regions of the human body.
6. Define several directional terms and anatomical planes used in association with the human body.
7. List by name and location the principal body cavities and the organs contained within them.
8. Describe the principle and importance of selected medical imaging techniques in the diagnosis of disease.
9. Define homeostasis and explain the effects of stress on homeostasis.
10. Define a feedback system and explain its role in homeostasis.

You are about to begin a study of the human body in order to learn how it is organized and how it functions. The study of the human body involves many branches of science. Each contributes to an understanding of how your body normally works and what happens when it is injured, diseased, or placed under stress.

ANATOMY AND PHYSIOLOGY DEFINED

Two branches of science that will help you understand your body parts and functions are anatomy and physiology. ***Anatomy*** (*anatome* = to dissect) refers to the study of *structure* and the relationships among structures. Anatomy is a broad science, and its study becomes more meaningful when specific aspects of the science are considered. Several subdivisions of anatomy are described in Exhibit 1-1.

Whereas anatomy and its branches deal with structures of the body, ***physiology*** (fiz'-ē-OL-ō-jē) deals with *functions* of the body parts, that is, how the body parts work. Since physiology cannot be completely separated from anatomy, you will learn about the human body by studying its structures and functions together, and you will see how each structure of the body is designed to carry out a particular function. The structure of a part often determines the functions it will perform. For example, the hairs lining the nose filter the air we breathe. The chambers of the heart that pump blood greater distances have thicker walls than those that pump blood shorter distances. The external ear is specifically shaped to efficiently collect sound waves that normally result in hearing. The air sacs of the lungs are so thin that oxygen and carbon dioxide readily pass between them and the blood. In turn, body functions often influence the size, shape, and health of the structures.

EXHIBIT 1-1 SUBDIVISIONS OF ANATOMY

Subdivision	Description
Surface anatomy	Study of the form (morphology) and markings of the surface of the body.
Gross (macroscopic) anatomy	Study of structures that can be examined without the use of a microscope.
Systemic (systematic) anatomy	Study of specific systems of the body such as the nervous system or respiratory system.
Regional anatomy	Study of a specific region of the body such as the head or chest.
Developmental anatomy	Study of development from the fertilized egg to adult form.
Embryology (em'-brē-OL-ō-jē; *logos* = study of)	Study of development from the fertilized egg through the eighth week in utero.
Pathological (path'-ō-LOJ-i-kal; *patho* = disease) **anatomy**	Study of structural changes associated with disease.
Histology (hiss'-TOL-ō-jē; *histio* = tissue)	Microscopic study of the structure of tissues.
Cytology (sī-TOL-ō-jē; *cyto* = cell)	Microscopic study of the structure of cells.
Radiographic (rā'-dē-ō-GRAF-ik; *radio* = ray; *graph* = to write) **anatomy**	Study of the structure of the body that includes the use of x-rays.

LEVELS OF STRUCTURAL ORGANIZATION

The human body consists of several levels of structural organization that are associated with one another in various ways. Here we will consider the principal levels that will help you to understand how your body is organized (Figure 1-1). The lowest level of organization, the ***chemical level,*** includes all chemical substances essential for maintaining life. All these chemicals are made up of atoms joined together in various ways.

The chemicals, in turn, are put together to form the next higher level of organization: the ***cellular level. Cells*** are the basic structural and functional units of an organism. Among the many kinds of cells in your body are muscle cells, nerve cells, and blood cells. Figure 1-1 shows several isolated cells from the lining of the stomach. Each has a different structure, and each performs a different function.

The next higher level of structural organization is the ***tissue level. Tissues*** are groups of similar cells that together with their intercellular material (substance between cells) usually have a similar embryological origin and perform special functions. When the cells shown in Figure 1-1 are joined together, they form a tissue called *epithelium,* which lines the stomach. Each cell in the tissue has a specific function. Mucous cells produce mucus, a secretion that lubricates food as it passes through the stomach. Parietal cells produce acid in the stomach. Zymogenic cells produce enzymes needed to digest proteins. Other examples of tissues in your body are muscle tissue, connective tissue, and nervous tissue.

In many places in the body, different kinds of tissues are joined together to form an even higher level of organization: the ***organ level. Organs*** are structures that are composed of two or more different tissues, have specific functions, and usually have recognizable shapes. Examples of organs are the heart, liver, lungs, brain, and stomach. Figure 1-1 shows several tissues that make up

FIGURE 1-1 Levels of structural organization that compose the human body.

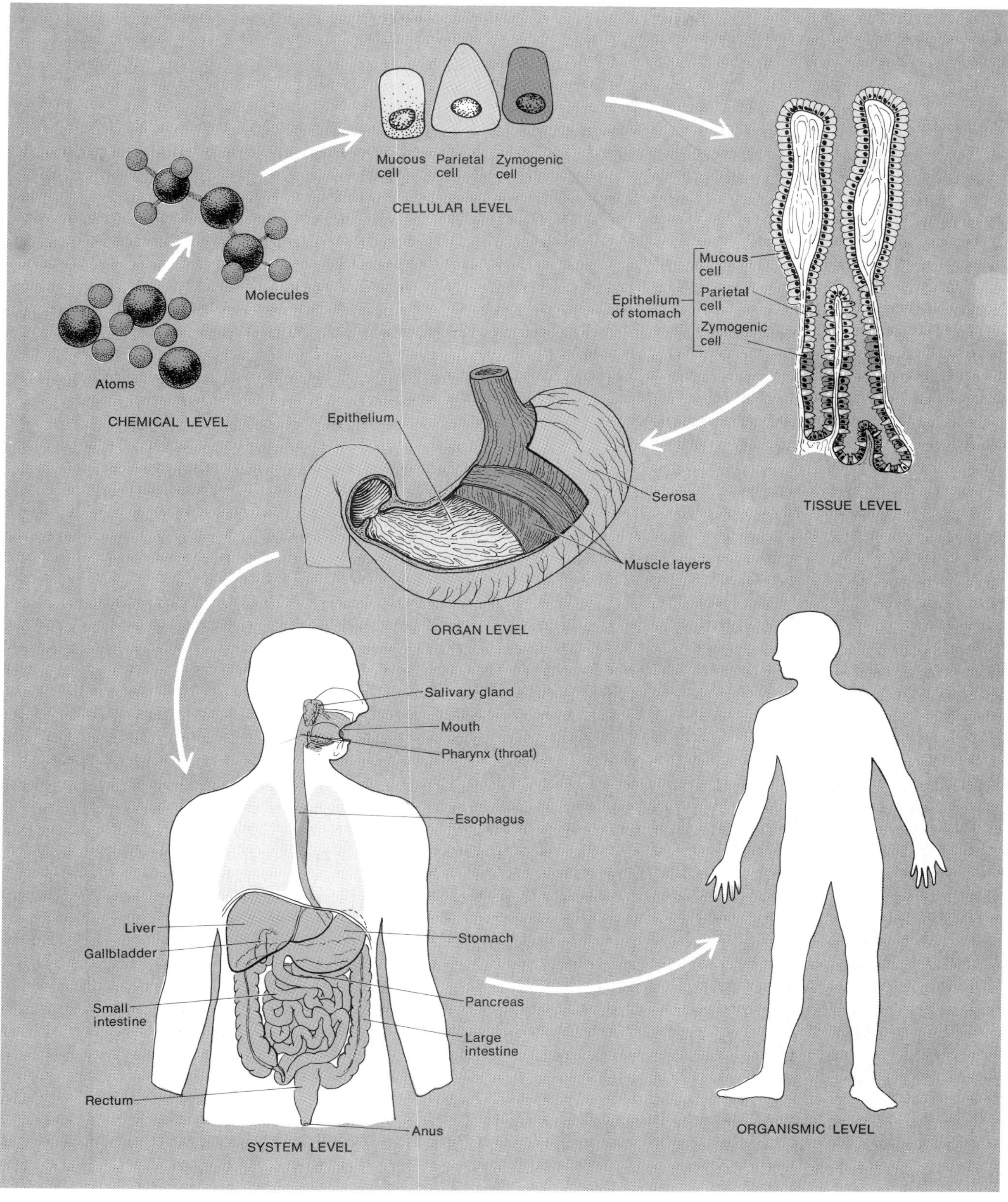

the stomach. The *serosa* is a layer of connective tissue and epithelium around the outside that protects the stomach and reduces friction when the stomach moves and rubs against other organs. The muscle tissue layers of the stomach contract to mix food and pass it on to the next digestive organ. The epithelial tissue layer lining the stomach produces mucus, acid, and enzymes.

The next higher level of structural organization in the body is the ***system level.*** A ***system*** consists of an association of organs that have a common function. The digestive system, which functions in the breakdown and absorption of food, is composed of the mouth, saliva-producing glands called salivary glands, pharynx (throat), esophagus, stomach, small intestine, large intestine, rectum, liver, gallbladder, and pancreas.

The highest level is the ***organismic level.*** All the parts of the body functioning with one another constitute the total ***organism***—one living individual.

In the chapters that follow, you will examine the anatomy and physiology of the major body systems. Exhibit 1-2 describes these systems in terms of their representative organs and their general functions. The systems are presented in the exhibit in the order in which they are discussed in later chapters.

EXHIBIT 1-2 PRINCIPAL SYSTEMS OF HUMAN BODY, REPRESENTATIVE ORGANS, AND FUNCTIONS

1. Integumentary
Definition: The skin and structures derived from it, such as hair, nails, and sweat and oil glands.
Function: Helps regulate body temperature, protects the body, eliminates wastes, synthesizes vitamin D, and receives certain stimuli such as temperature, pressure, and pain.
Reference: See Figure 5-1.

2. Skeletal
Definition: All the bones of the body, their associated cartilages, and the joints of the body.
Function: Supports and protects the body, provides leverage, houses cells that produce blood cells, and stores minerals.
Reference: See Figure 7-1.

3. Muscular
Definition: Specifically refers to skeletal muscle tissue; other muscle tissues include visceral and cardiac.
Function: Participates in bringing about movement, maintains posture, and produces heat.
Reference: See Figure 11-3.

4. Nervous
Definition: Brain, spinal cord, nerves, and sense organs, such as the eye and ear.
Function: Regulates body activities through nerve impulses.
Reference: See Figures 13-1 and 14-1.

5. Endocrine
Definition: All glands that produce hormones.
Function: Regulates body activities through hormones transported by the cardiovascular system.
Reference: See Figure 18-1.

6. Cardiovascular
Definition: Blood, heart, and blood vessels.
Function: Distributes oxygen and nutrients to cells, carries carbon dioxide and wastes from cells, maintains the acid–base balance of the body, protects against disease, prevents hemorrhage by forming blood clots, and helps regulate body temperature.
Reference: See Figures 20-1, 21-12, and 21-17.

7. Lymphatic
Definition: Lymph, lymphatic vessels, and structures or organs containing lymphatic tissue (large numbers of white blood cells called lymphocytes), such as the spleen, thymus gland, lymph nodes, and tonsils.
Function: Returns proteins and plasma to the cardiovascular system, transports fats from the gastrointestinal tract to the cardiovascular system, filters body fluid, produces white blood cells, and protects against disease.
Reference: See Figure 22-2.

8. Respiratory
Definition: The lungs and a series of associated passageways leading into and out of them.
Function: Supplies oxygen, eliminates carbon dioxide, and helps regulate the acid–base balance of the body.
Reference: See Figure 23-1.

9. Digestive
Definition: A long tube called the gastrointestinal (GI) tract and associated organs such as the salivary glands, liver, gallbladder, and pancreas.
Function: Performs the physical and chemical breakdown and absorption of food for use by cells and eliminates solid and other wastes.
Reference: See Figure 24-1.

10. Urinary
Definition: Organs that produce, collect, and eliminate urine.
Function: Regulates the chemical composition of blood, eliminates wastes, regulates fluid and electrolyte balance and volume, and helps maintain the acid–base balance of the body.
Reference: See Figure 26-1.

11. Reproductive
Definition: Organs (testes and ovaries) that produce reproductive cells (sperm and ova) and other organs that transport and store reproductive cells.
Function: Reproduces the organism.
Reference: See Figures 28-1 and 28-11.

LIFE PROCESSES

All living forms have certain characteristics that distinguish them from nonliving things. Following are several of the more important life processes of humans:

1. Metabolism. ***Metabolism*** is the sum of all the chemical processes that occur in the body. One phase of metabolism, called ***catabolism,*** provides us with energy needed to sustain life. The other phase, called ***anabolism,*** uses the energy from catabolism to make various substances that form the body's structural and functional components. Among the many processes contributing to metabolism are: ***ingestion,*** the taking in of foods; ***digestion,*** the breaking down of foods into simpler forms that can be used by cells; ***absorption,*** the uptake of substances by cells; ***assimilation,*** the buildup of absorbed substances into different materials required by cells; ***respiration,*** the generation of energy, usually in the presence of oxygen with the release of carbon dioxide; ***secretion,*** the production and release of a useful substance by cells; and ***excretion,*** the elimination of wastes produced as a result of metabolism.

2. Excitability. ***Excitability*** is our ability to sense changes within and around us. We do this by continually responding to stimuli (changes in the environment) such as light, pressure, heat, noises, chemicals, and pain to make adjustments that maintain health.

3. Conductivity. ***Conductivity*** refers to the ability of cells to carry the effect of a stimulus from one part of a cell to another. This characteristic is highly developed in nerve cells and is developed to a great extent in muscle fibers (cells).

4. Contractility. ***Contractility*** is the capacity of cells or parts of cells to actively generate force to undergo shortening and change form for purposeful movements. Muscle fibers exhibit a high degree of contractility.

5. Growth. ***Growth*** refers to an increase in size. It involves an increase in the number of cells or an increase in the size of existing cells or the substance surrounding cells as internal components increase in size or an increase in the size of the substance surrounding cells.

6. Differentiation. ***Differentiation*** is the process whereby unspecialized cells change to specialized cells. Specialized cells have structural and functional characteristics that differ from cells from which they originated. Through differentiation, a fertilized egg normally develops into an embryo, fetus, infant, child, and adult, each of which consists of a variety of diversified cells.

7. Reproduction. ***Reproduction*** refers to either the formation of new cells for growth, repair, or replacement, or the production of a new individual. Through reproduction, life is transmitted from one generation to the next.

STRUCTURAL PLAN

The human body has certain general ***anatomical characteristics*** that will help you to understand its overall structural plan. For example, humans have a ***backbone (vertebral column),*** a characteristic that places them in a large group of organisms called ***vertebrates.*** Another characteristic is the body's ***tube-within-a-tube*** construction. The outer tube is formed by the body wall; the inner tube is the gastrointestinal tract. Moreover, humans are for the most part ***bilaterally symmetrical;*** that is, essentially the left and right sides of the body are mirror images.

ANATOMICAL POSITION AND REGIONAL NAMES

In all anatomical texts and charts, descriptions of any region or part of the human body assume that the body is in a specific position, called the ***anatomical position,*** so that directional terms are clear and any part can be related to any other part. In the anatomical position, the subject is standing erect (upright position) facing the observer, the upper extremities (limbs) are placed at the sides, the palms of the hands are turned forward, and the feet are flat on the floor (Figure 1-2). Once the body is in the anatomical position, it is easier to visualize and understand how it is organized into various regions. The common and anatomical terms of the principal body regions are also presented in Figure 1-2.

DIRECTIONAL TERMS

In order to explain exactly where various body structures are located in relation to each other, anatomists use certain ***directional terms.*** Such terms are precise and avoid the use of unnecessary words. If you want to point out the sternum (breastbone) to someone who knows where the clavicle (collarbone) is, you can say that the sternum is inferior (farther away from the head) and medial (toward the middle of the body) to the clavicle. As you can see, using the terms *inferior* and *medial* avoids a great deal of complicated description. Many directional terms are defined in Exhibit 1-3, and the parts of the body referred to in the examples are labeled in Figure 1-3. Studying the exhibit and the figure together should make clear to you many of the directional relationships among various body parts.

PLANES AND SECTIONS

The structural plan of the human body may also be discussed with respect to ***planes*** (imaginary flat surfaces) that pass through it. Several of the commonly used planes are illustrated in Figure 1-4. A ***sagittal*** (SAJ-i-tal) ***plane*** is a vertical plane that divides the body or organs into right and left sides. Such a plane may be midsagittal or parasagittal. A ***midsagittal (median) plane*** is a vertical plane that passes through the midline of the body and divides the body or organs into *equal* right and left

FIGURE 1-2 Anatomical position. The common names and anatomical terms, in parentheses, are indicated for many of the regions of the body. (a) Anterior view. (b) Posterior view.

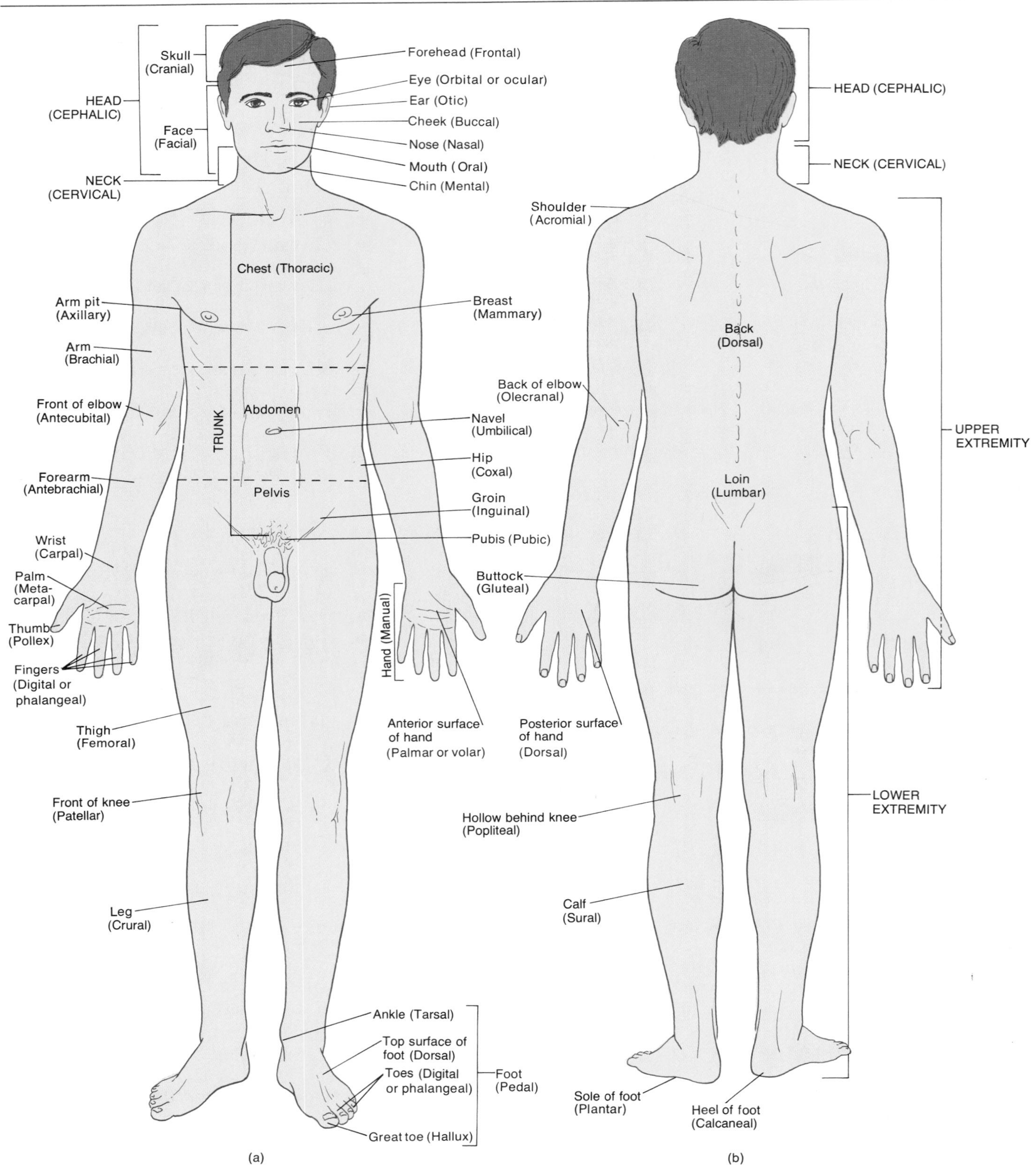

EXHIBIT 1-3 DIRECTIONAL TERMS[a]

Term	Definition	Example
Superior (cephalic or **cranial)**	Toward the head or the upper part of a structure; generally refers to structures in the trunk.	The heart is superior to the liver.
Inferior (caudal)	Away from the head or toward the lower part of a structure; generally refers to structures in the trunk.	The stomach is inferior to the lungs.
Anterior (ventral)	Nearer to or at the front of the body. In the ***prone position,*** the body lies anterior side down. In the ***supine position,*** the body lies anterior side up.	The sternum is anterior to the heart.
Posterior (dorsal)	Nearer to or at the back of the body.	The esophagus is posterior to the trachea.
Medial (mesial)	Nearer to the midline of the body or a structure. The ***midline*** is an imaginary vertical line that divides the body into equal left and right sides. The *anterior midline* is on the front surface of the body, and the *posterior midline* is on the back surface.	The ulna is on the medial side of the forearm.
Lateral	Farther from the midline of the body or a structure.	The lungs are lateral to the heart.
Intermediate	Between two structures.	The ring finger is intermediate between the little and middle fingers.
Ipsilateral	On the same side of the body.	The gallbladder and ascending colon of the large intestine are ipsilateral.
Contralateral	On the opposite side of the body.	The ascending and descending colons of the large intestine are contralateral.
Proximal	Nearer to the attachment of an extremity to the trunk or a structure; nearer to the point of origin.	The humerus is proximal to the radius.
Distal	Farther from the attachment of an extremity to the trunk or a structure; farther from the point of origin.	The phalanges are distal to the carpals (wrist bones).
Superficial	Toward or on the surface of the body.	The muscles of the thoracic wall are superficial to the viscera in the thoracic cavity. (See Figure 1-7.)
Deep	Away from the surface of the body.	The ribs are deep to the skin of the chest.
Parietal	Pertaining to or forming the outer wall of a body cavity.	The parietal pleura forms the outer layer of the pleural sacs that surround the lungs. (See Figure 1-7.)
Visceral	Pertaining to the covering of an organ (viscus) within the ventral body cavity.	The visceral pleura forms the inner layer of the pleural sacs and covers the external surface of the lungs. (See Figure 1-7.)

[a] Study this exhibit with Figures 1-3 and 1-7 in order to visualize the examples given.

sides. A ***parasagittal*** (*para* = near) ***plane*** is a vertical plane that does not pass through the midline of the body and divides the body or organs into *unequal* left and right portions. A ***frontal*** (***coronal;*** kō-RŌ-nal) ***plane*** is a plane at a right angle to a midsagittal (or parasagittal) plane that divides the body or organs into anterior and posterior portions. Finally, a ***horizontal*** (***transverse***) ***plane*** is a plane that is parallel to the ground, that is, at a right angle to the midsagittal, parasagittal, and frontal planes. It divides the body or organs into superior and inferior portions.

When you study a body structure, you will often view it in section, meaning that you look at the flat surface resulting from a cut made through the three-dimensional structure. It is important to know the plane of the section so that you can understand the anatomical relationship of one part to another. Figure 1-5 indicates how three different sections—a *cross* (*horizontal* or *transverse*) *sec-*

FIGURE 1-3 Anatomical and directional terms. By studying Exhibit 1-3 with this figure, you should gain an understanding of the meanings of the following directional terms: *superior, inferior, anterior, posterior, medial, lateral, intermediate, ipsilateral, contralateral, proximal,* and *distal.*

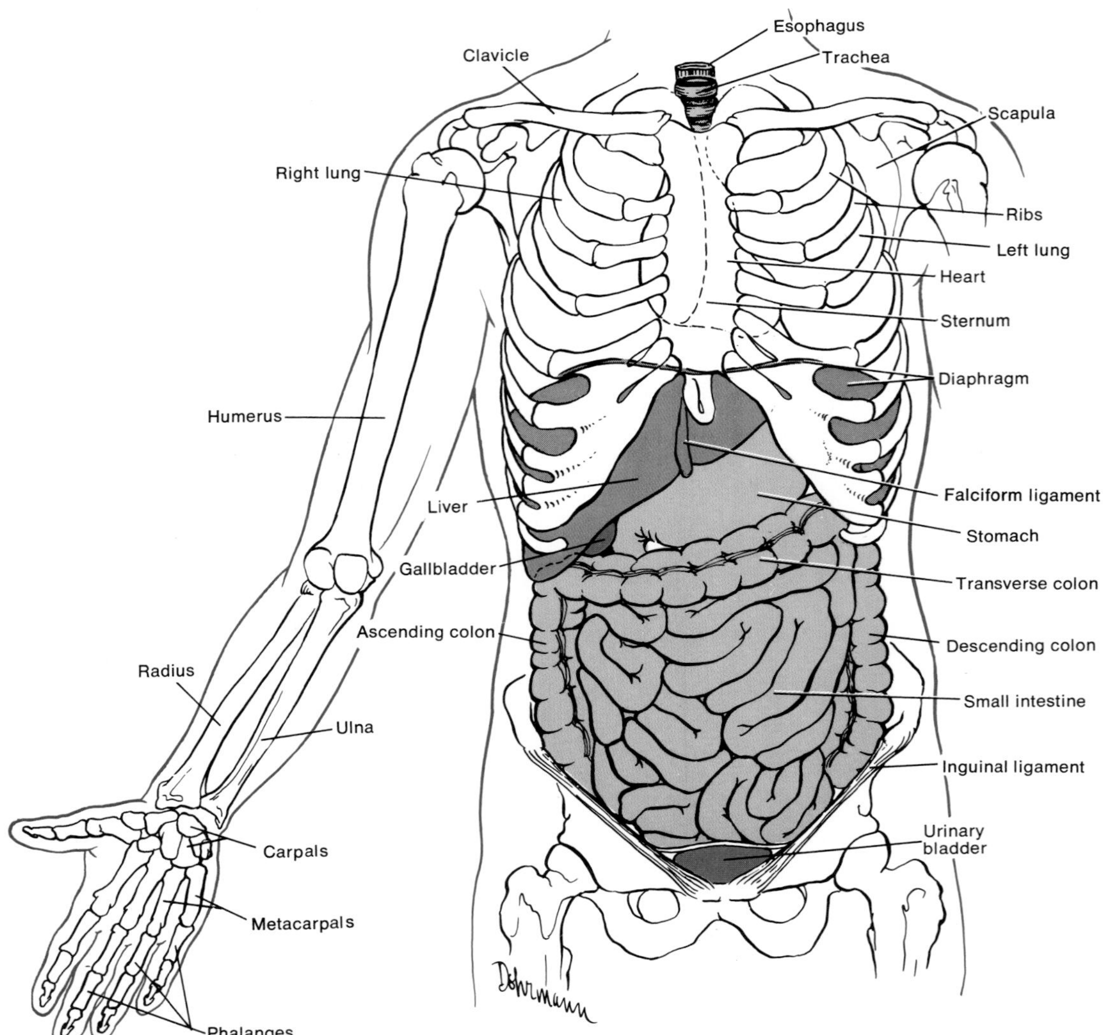

tion, a *frontal section,* and a *midsagittal section*—are made through different parts of the brain.

BODY CAVITIES

Spaces within the body that contain internal organs are called ***body cavities.*** The various body cavities may be separated by structures such as muscles, bones, or ligaments. Specific cavities may be distinguished if the body is divided into right and left halves. Figure 1-6 shows the two principal body cavities. The ***dorsal body cavity*** is located near the posterior (dorsal) surface of the body. It is further subdivided into a ***cranial cavity,*** which is a bony cavity that is formed by the cranial (skull) bones and contains the brain, and a ***vertebral (spinal) canal,*** which is a bony cavity that is formed by the vertebrae of the backbone and contains the spinal cord and the beginnings of spinal nerves.

The other principal body cavity is the ***ventral body cavity.*** This cavity is located on the anterior (ventral) aspect of the body. Its walls are composed of skin, connective tissue, bone, muscle, and a membrane called a *serous membrane.* The organs inside the ventral body cavity are called ***viscera*** (VIS-er-a). Like the dorsal body cavity, the ventral body cavity has two principal subdivisions—an upper portion, called the ***thoracic*** (thō-RAS-ik) ***cavity*** (or chest cavity), and a lower portion, called the ***abdominopelvic*** (ab-dom'-i-nō-PEL-vik) ***cavity.*** The anatomical landmark that divides the ventral body

FIGURE 1-4 Planes of the human body.

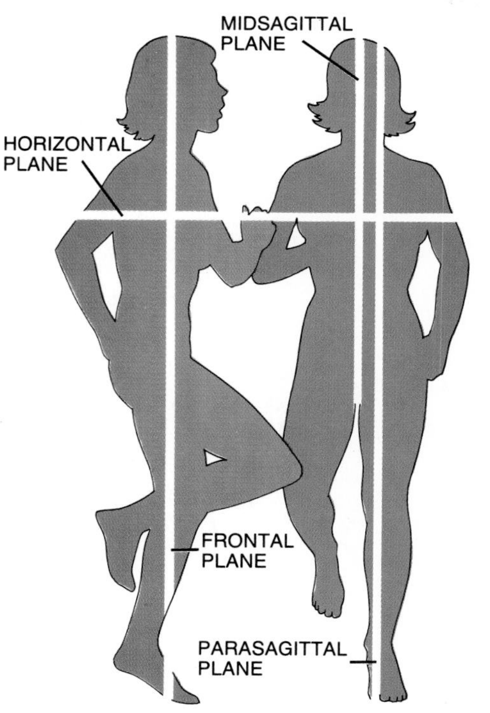

cavity into the thoracic and abdominopelvic cavities is the muscular diaphragm (*diaphragma* = partition or wall).

The thoracic cavity contains several divisions. There are two ***pleural cavities,*** each surrounding a lung (Figure 1-7). Each pleural cavity is a small potential space (not an actual natural space) between the parietal pleura, a membrane that lines the pleural cavities, and the visceral pleura, a membrane that covers the lungs. The pleural cavities contain a small amount of fluid. The ***mediastinum*** (mē'-dē-as-TĪ-num; *medias* = middle; *stare* = stand in) is a broad, median partition—actually a mass of tissues—between the lungs that extends from the sternum to the vertebral column (Figure 1-7). The mediastinum includes all the contents of the thoracic cavity except the lungs themselves. Among the structures in the mediastinum are the heart, thymus gland, esophagus, trachea, bronchi, and many large blood and lymphatic vessels. The ***pericardial*** (per'-i-KAR-dē-al; *peri* = around; *cardi* = heart) ***cavity*** is a small potential space between the visceral pericardium and parietal pericardium, the membranes covering the heart (Figure 1-7). It also contains a fluid.

The abdominopelvic cavity, as the name suggests, is divided into two portions, although no wall separates them (see Figure 1-6). The upper portion, the ***abdominal*** (*abdere* means to hide, because it hides the viscera) ***cavity,*** contains the stomach, spleen, liver, gallbladder, pancreas, small intestine, and most of the large intestine. The lower portion, the ***pelvic cavity,*** contains the urinary bladder, cecum, appendix, sigmoid colon, rectum, and the internal male or female reproductive organs. The pelvic cavity is the region between two imaginary planes. The top plane, the plane of the pelvic inlet, extends from the superior border of the sacrum (sacral promontory) to the upper margin of the symphysis pubis (anterior joint between hipbones). The bottom plane, the plane of the pelvic outlet, extends from the end of the coccyx (tailbone) to the pelvic arch (junction of bottom portions of symphysis pubis).

ABDOMINOPELVIC REGIONS

To describe the location of organs easily, the abdominopelvic cavity may be divided into the ***nine regions*** shown in Figure 1-8a. Note which organs and parts of organs are in the different regions by carefully examining Figure 1-8b–d and Exhibit 1-4. Although some parts of the body in the illustrations and exhibit may be unfamiliar to you at this point, they will be discussed in detail in later chapters.

ABDOMINOPELVIC QUADRANTS

The abdominopelvic cavity may be divided more simply into ***quadrants*** (*quad* = four). These are shown in Figure 1-9. In this method, frequently used by clinical

FIGURE 1-5 Sections through different parts of the brain. (a) Cross (horizontal or transverse) section; see also Figure 14-7d. (b) Frontal (coronal) section; see also Figure 14-5a. (c) Midsagittal section; see also Figure 14-1b.

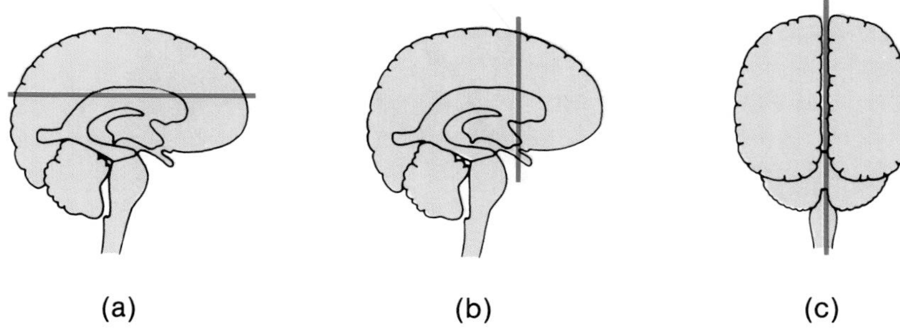

FIGURE 1-6 Body cavities. (a) Location of the dorsal and ventral body cavities seen in right lateral view. (b) Subdivisions of the ventral body cavity seen in anterior view. (c) Photograph of a frontal section of the chest, abdomen, and pelvis. (Courtesy of J. A. Gosling, P. F. Harris, et al., *Atlas of Human Anatomy*, Gower Medical Publishing Ltd., 1985.)

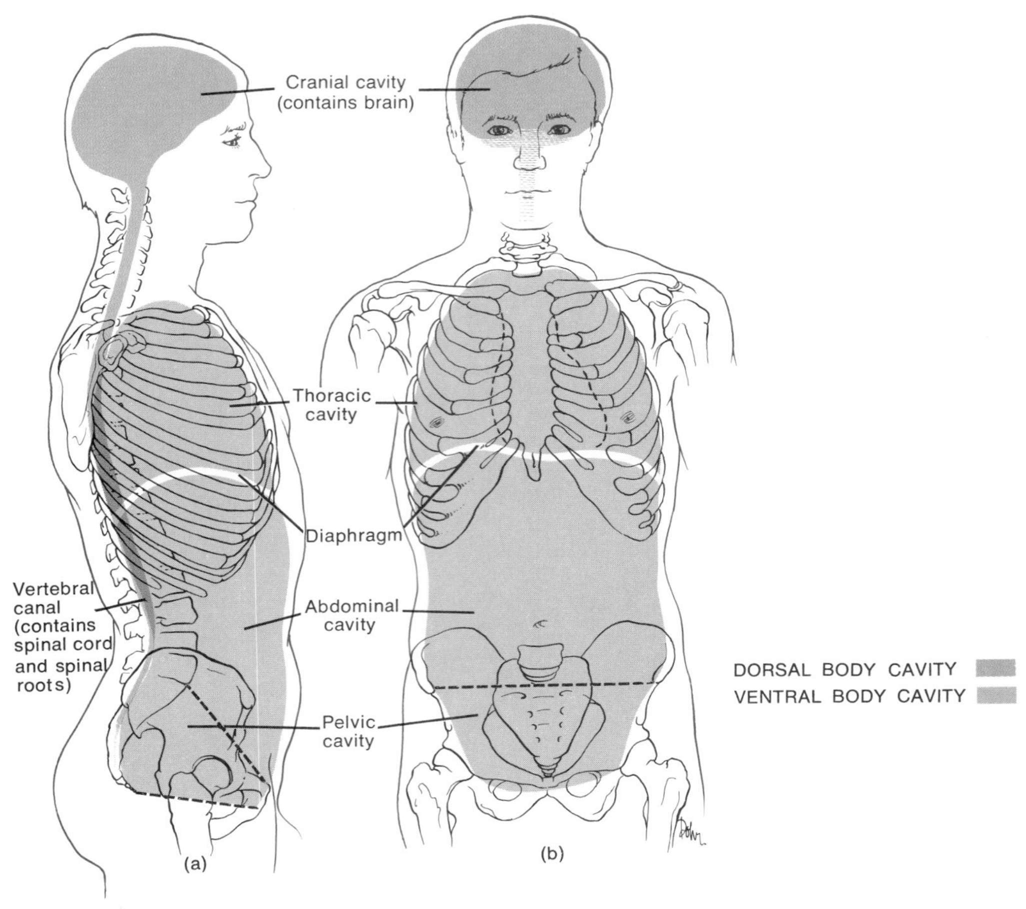

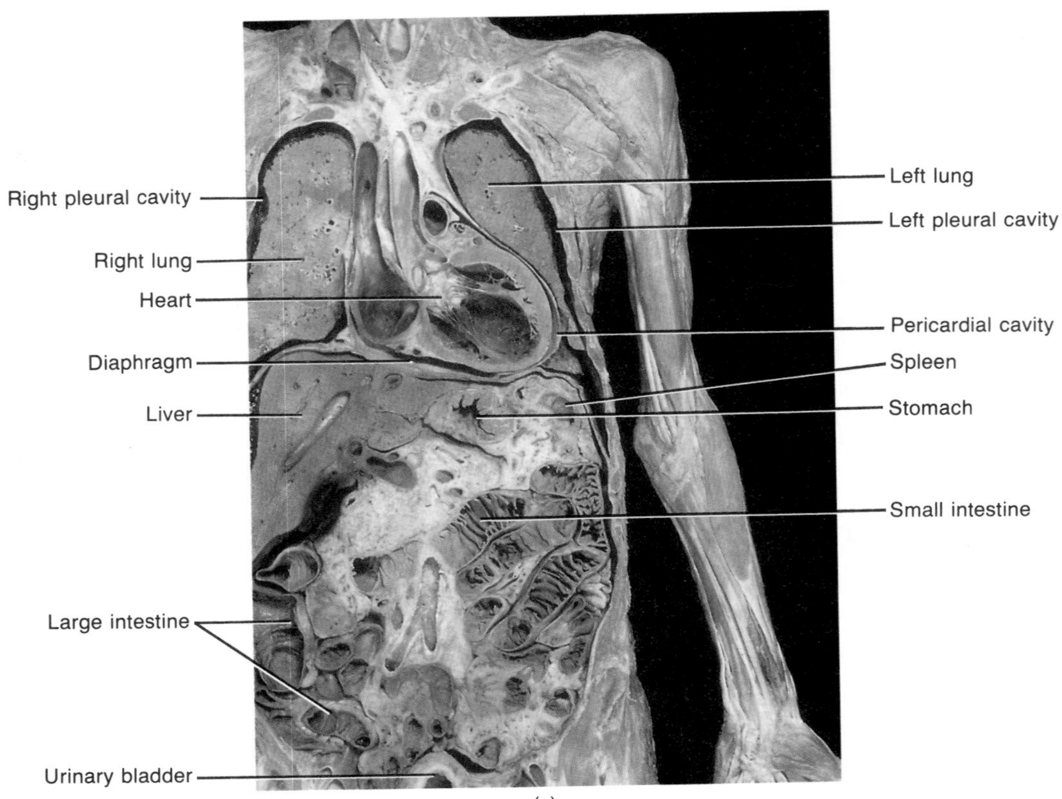

FIGURE 1-7 Mediastinum. (a) Diagram in a cross section of the thorax. Some of the structures shown and labeled may be unfamiliar to you now, but they will be discussed in detail in later chapters.

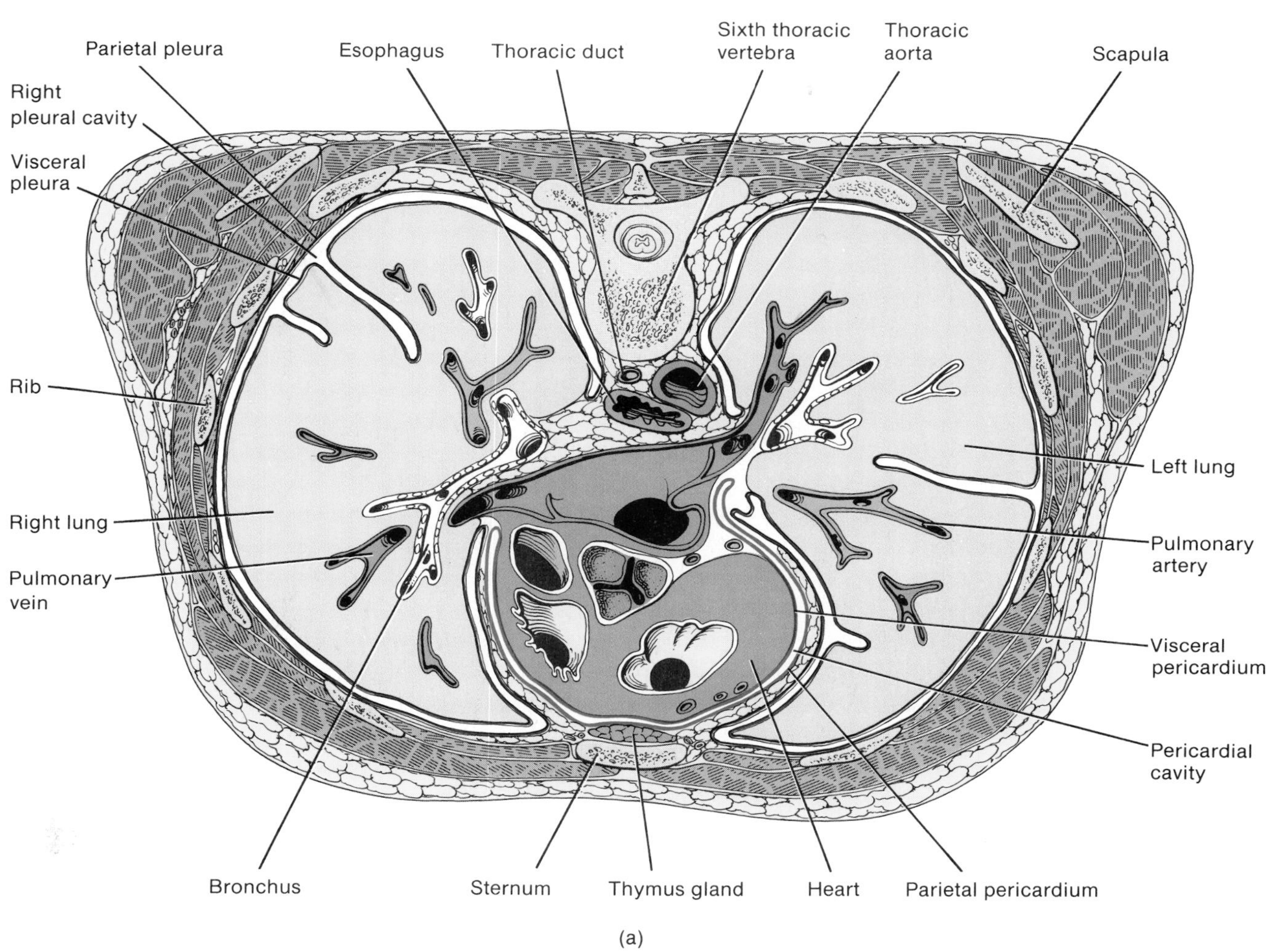

(a)

personnel, a horizontal line and a vertical line are passed through the umbilicus. These two lines divide the abdomen into a ***right upper quadrant (RUQ), left upper quadrant (LUQ), right lower quadrant (RLQ),*** and ***left lower quadrant (LLQ).*** Whereas the nine region division is more widely used for anatomical studies, the quadrant division is more commonly used for locating the site of an abdominopelvic pain, tumor, or other abnormality.

CLINICAL APPLICATION: AUTOPSY

To determine the cause of death accurately, it is necessary to perform an ***autopsy*** (AW-top-sē; *auto* = self; *opsis* = to see with one's own eyes), that is, a postmortem examination of the body. In addition, an autopsy can also be used to uncover the existence of diseases not detected during life, support the accuracy of diagnostic tests, determine the effectiveness and side effects of drugs, analyze the effects of environmental influences on the body, and educate health care students. Moreover, an autopsy can reveal conditions that may affect offspring or siblings (such as congenital heart defects) and can help resolve disputes and affect benefits that survivors receive as part of insurance settlements.

A typical autopsy consists of three principal phases of examination. The first phase is examination of the exterior of the body for the presence of wounds, scars, tumors, or other abnormalities. The second phase includes the dissection and gross examination of the major body organs. The third phase of autopsy consists of microscopic examination of tissues from organs to ascertain any pathology. Depending on the

FIGURE 1-7 (*Continued*) (b) Photograph of a cross section of the right side of the thorax, emphasizing the pleural and pericardial cavities. (Courtesy of J. A. Gosling, P. F. Harris, et al., *Atlas of Human Anatomy*, Gower Medical Publishing Ltd., 1985.)

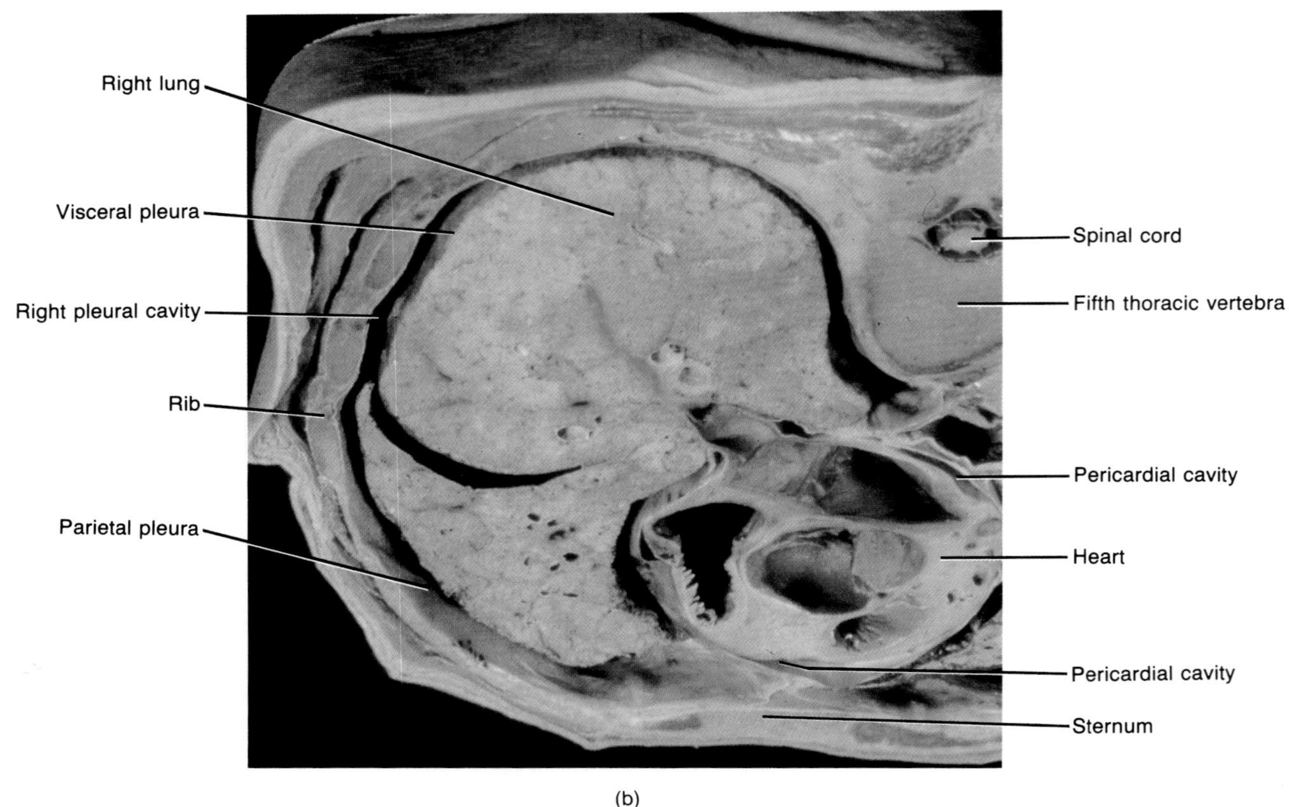

(b)

EXHIBIT 1-4 REPRESENTATIVE STRUCTURES FOUND IN THE ABDOMINOPELVIC REGIONS

Region	Representative Structures
Epigastric (ep-i-GAS-trik; *epi* = above; *gaster* = stomach)	Left lobe and medial part of right lobe of liver, pyloric portion and lesser curvature of stomach, superior and descending portions of duodenum, body and superior part of head of pancreas, and right and left adrenal (suprarenal) glands.
Right hypochondriac (hī-pō-KON-drē-ak; *hypo* = under; *chondro* = cartilage)	Right lobe of liver, gallbladder, and upper superior third of right kidney.
Left hypochondriac	Body and fundus of stomach, spleen, left colic (splenic) flexure, superior two-thirds of left kidney, and tail of pancreas.
Umbilical (um-BIL-i-kul)	Middle portion of transverse colon, inferior part of duodenum, jejunum, ileum, and bifurcations (branching) of abdominal aorta and inferior vena cava.
Right lumbar (*lumbus* = loin)	Superior part of cecum, ascending colon, right colic (hepatic) flexure, inferior lateral portion of right kidney, and small intestine.
Left lumbar	Descending colon, inferior third of left kidney, and small intestine.
Hypogastric (pubic)	Urinary bladder when full, small intestine, and part of sigmoid colon.
Right iliac (inguinal) (IL-ē-ak; iliac refers to superior part of hipbone)	Lower end of cecum, appendix, and small intestine.
Left iliac (inguinal)	Junction of descending and sigmoid parts of colon and small intestine.

FIGURE 1-8 Abdominopelvic cavity. (a) The nine regions. The top horizontal (subcostal) line is drawn just inferior to the bottom of the rib cage, across the lower portion (pylorus) of the stomach. The bottom horizontal (transtubercular) line is drawn just inferior to the tops of the hipbones. The two vertical (left and right midclavicular) lines are drawn through the midpoints of the clavicles, just medial to the nipples. The horizontal and vertical lines divide the abdominopelvic cavity into a larger middle section and smaller left and right sections. (b) Anterior view showing the most superficial organs. The greater omentum has been removed.

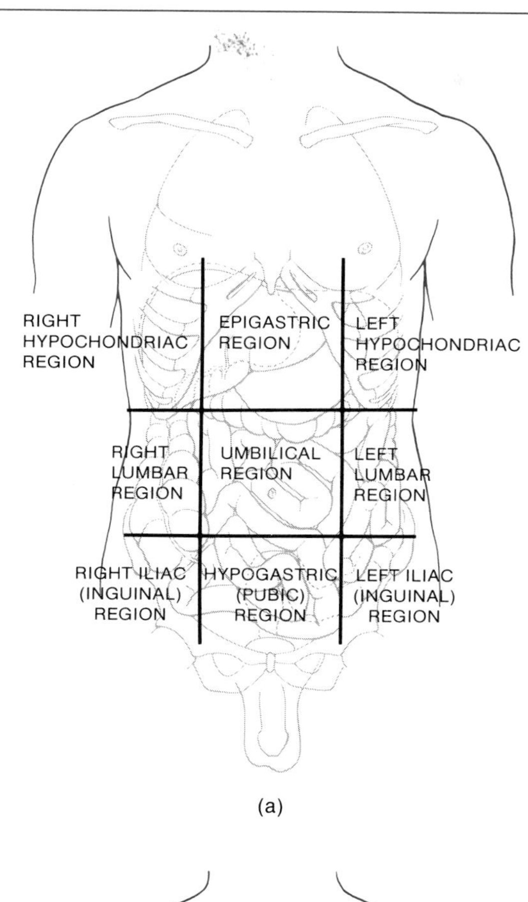

(a)

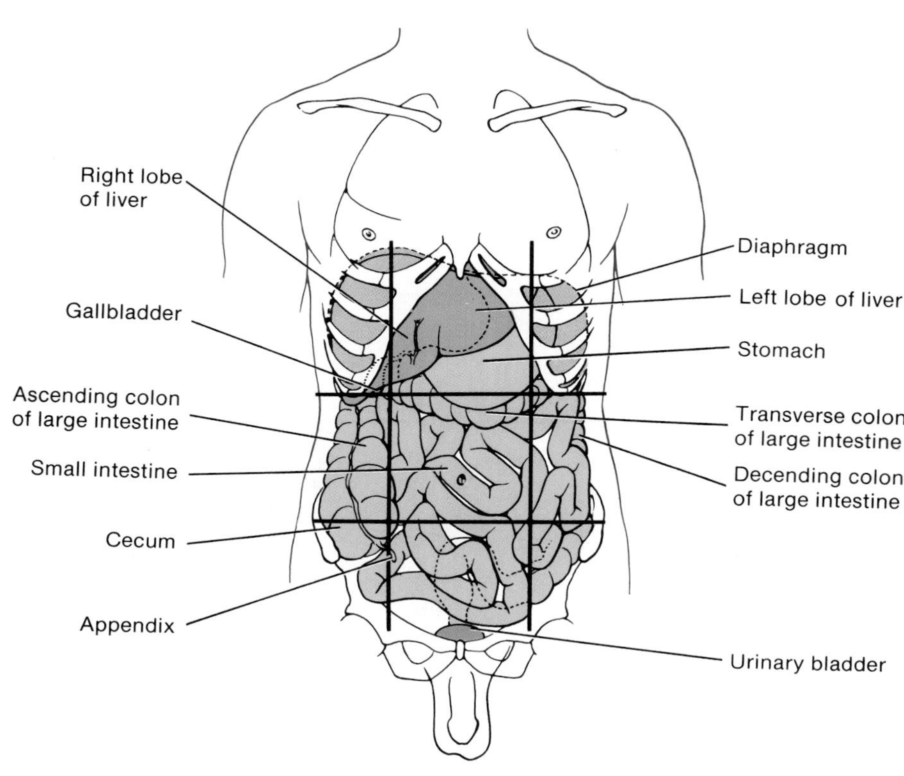

(b)

FIGURE 1-8 (*Continued*) (c) Anterior view in which most of the small intestine and transverse colon have been removed to expose deeper structures. (d) Anterior view in which many organs have been removed, exposing the posterior structures. The internal reproductive organs in the pelvic cavity are shown in Figures 28-1a and 28-11.

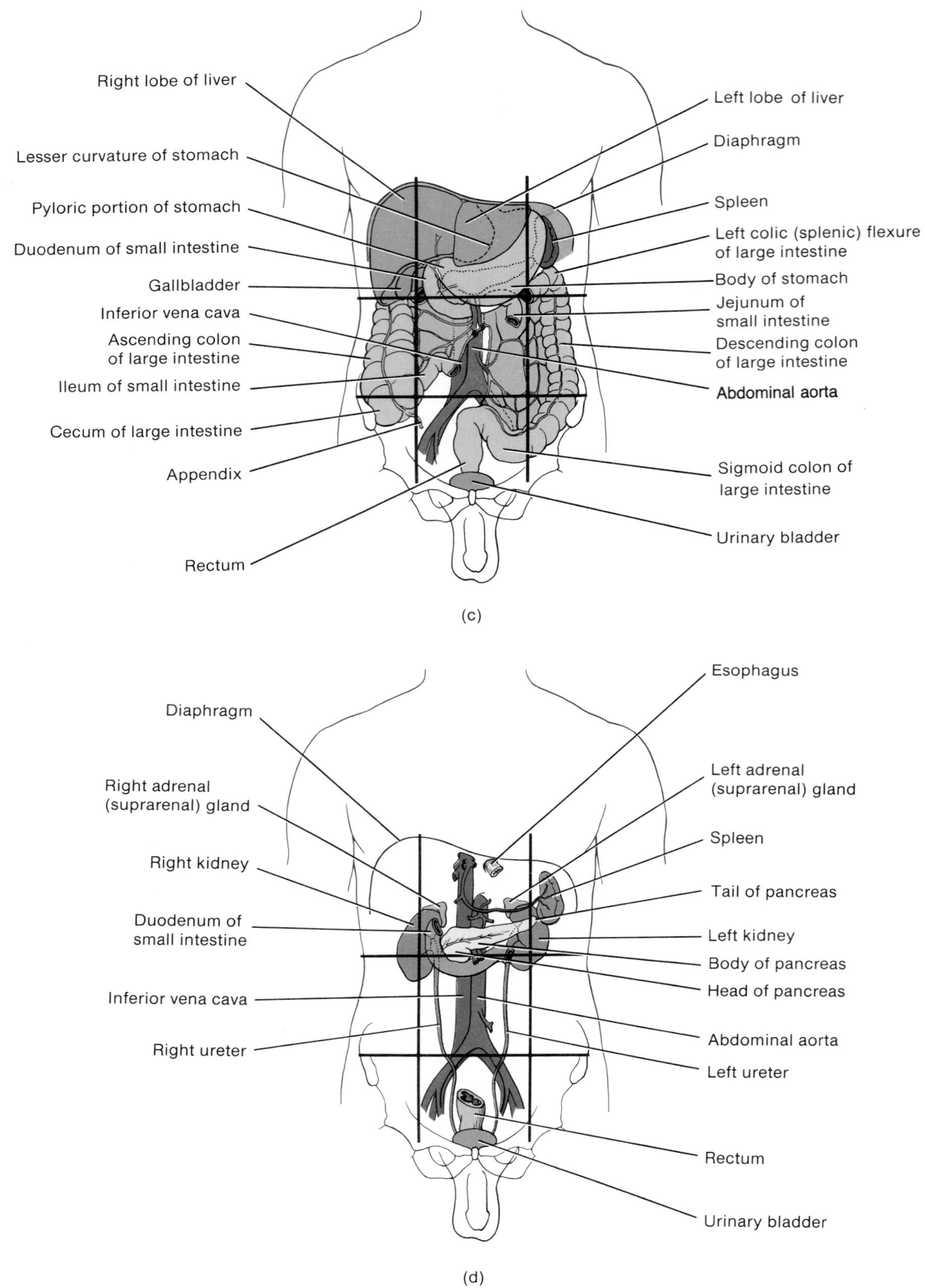

FIGURE 1-9 **Quadrants of the abdominopelvic cavity. The two lines intersect at right angles at the umbilicus.**

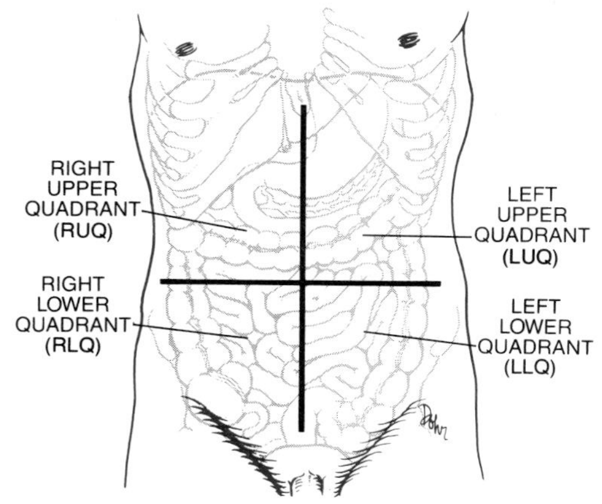

circumstances, techniques may also be used to detect and recover microbes and determine the presence of foreign substances in the body.

The first and second phases of an autopsy usually last two to four hours, depending on their comprehensiveness. When the procedure is completed, the organs are returned to the body, except for those that are donated or used to extract a particular pharmacologic substance, and all incisions are sutured.

MEDICAL IMAGING

In recent years, there has been an explosion in the development of various kinds of ***medical imaging*** techniques. These techniques are essential for diagnosing a wide range of disorders. Here we will briefly discuss conventional radiography, a procedure that has been used for many years, and several newer techniques such as computed tomography (CT) scanning, dynamic spatial reconstruction (DSR), magnetic resonance imaging (MRI), ultrasound (US), positron emission tomography (PET), and digital subtraction angiography (DSA).

Conventional Radiography

A specialized branch of anatomy that is essential for the diagnosis of many disorders is ***radiographic*** (*radius* = ray) ***anatomy,*** or ***radiology,*** which includes the use of x-rays. The most common and familiar type of radiographic anatomy employs the use of a single barrage of x-rays. The x-rays pass through the body and expose an x-ray film, producing a photographic image called a ***roentgenogram*** (RENT-gen-ō-gram). A roentgenogram provides a two-dimensional shadow image of the interior of the body (Figure 1-10a). As valuable as they are in diagnosis, conventional x-rays compress the body image onto a flat sheet of film, often resulting in an overlap of organs and tissues that could make diagnosis difficult. Moreover, x-rays do not always differentiate between subtle differences in tissue density.

Computed Tomography (CT) Scanning

These diagnostic difficulties have been virtually eliminated by the use of an x-ray technique called ***computed tomography (CT) scanning*** or ***computerized axial tomography (CAT) scanning.*** First introduced in 1971, CT scanning combines the principles of x-ray and advanced computer technologies. An x-ray source moves in an arc around the part of the body being scanned and repeatedly sends out x-ray beams. As the beams pass through the body, the tissues absorb small amounts of radiation, depending on their densities. Once the beams pass through, they are converted by light-sensitive crystal detectors to electronic signals that are transmitted to the scanner's computer. The computer then projects an image, called a ***CT scan,*** onto a television screen called the *physician's console.* The CT scan provides a very accurate cross-sectional picture of any area of the body (Figure 1-10b). A series of scans permits a physician to examine section after section of a patient's tissues. Additional scans taken from other angles can be used to pinpoint extremely small abnormalities in the organs. There are two types of CT scanners: one (*body scan*) permits examination of the whole body; the other (*head scan*) is a smaller model that is used to examine the head only.

The entire CT scanning process takes only seconds, it is completely painless, and the x-ray dose is equal to or less than that of many other diagnostic procedures. Moreover, the CT scanner produces images with 10 to 20 times the detail of conventional radiography. Among the clinical uses of CT scanning are the detection of tumors, aneurysms (bulges in blood vessels), kidney stones, gallstones, infections, tissue damage, and deformities.

Dynamic Spatial Reconstruction (DSR)

Another development in radiographic anatomy is ***dynamic spatial reconstruction (DSR)*** which uses a highly sophisticated x-ray machine that has the ability to construct *moving, three-dimensional,* life-size images of all or part of an internal organ from any view desired. The image produced can be rotated and tipped, and like an electronic knife, the dynamic spatial reconstructor can pictorially slice open an organ and expose its interior. The instrument also provides enlargements, stop-action, replay, high-speed, and slow-motion viewing. The level of radiation exposure of the reconstructor is about double that for a conventional chest roentgenogram.

FIGURE 1-10 Radiographic anatomy. (a) Roentgenogram of the abdomen. (b) CT scan of the abdomen. Note the greater detail in the CT scan. (Courtesy of Stephen A. Kieffer and E. Robert Heitzman, *An Atlas of Cross-Sectional Anatomy,* Harper & Row, Publishers, Inc., Hagerstown, MD, 1979.)

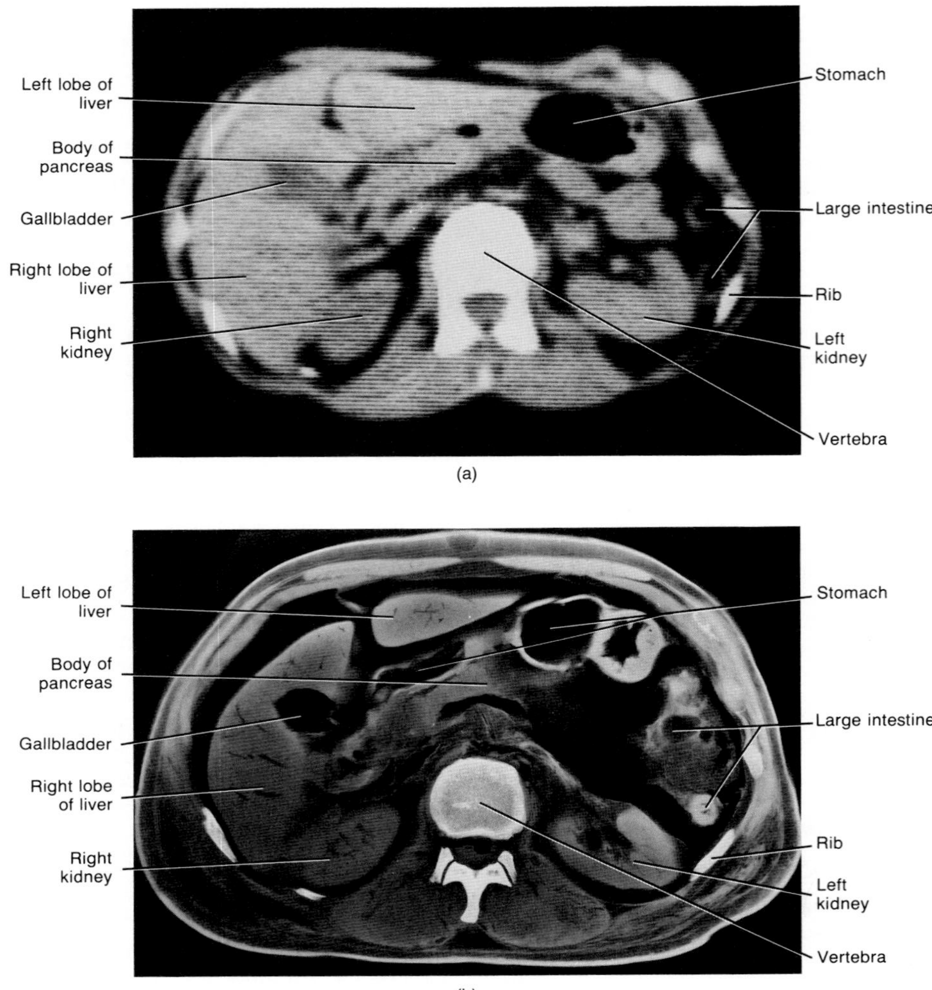

The dynamic spatial reconstructor has been designed to provide three-dimensional imaging of the heart, lungs, and blood vessels; measure the volumes and movements of the heart and lungs and other internal organs; detect cancer and heart defects; and measure tissue damage following a heart attack, stroke, or other disease.

Magnetic Resonance Imaging (MRI)

New in the arsenal for diagnosing disease is ***magnetic resonance imaging (MRI)***, formerly called ***nuclear magnetic resonance (NMR)***. MRI focuses on the nuclei of atoms of a single element in a tissue and determines their response to an external force such as magnetism. In most studies to date, MRI of hydrogen nuclei has been popular because of the body's large water content. The part of the body to be studied, ranging from a finger to the entire body, is placed in the scanner (magnet), exposing the nuclei to a uniform magnetic field (Figure 1-11a). The images produced, which indicate a biochemical blueprint of cellular activity, somewhat resemble a CT scan (Figure 1-11b) and can be displayed in color and as two- or three-dimensional images. The procedure typically takes about 30 minutes. MRI is not indicated for pregnant women, individuals who have artificial pacemakers or metal joints, or persons dependent on life-support equipment that has metal components.

The diagnostic advantage of MRI is that in addition to

providing images of diseased organs and tissues, it also provides information about what chemicals are present in an organ and tissue. MRI can be used to perform a "biopsy" on tumors without an operation, assess mental disorders, reveal brain changes from Parkinson's disease, diagnose spinal cord disorders, measure blood flow, study the evolution of hematomas and infarctions in the brain following a stroke, identify the potential for developing a stroke, assess treatment for conditions such as heart disease and stroke, monitor the progress and treatment of a disease, study the effects of toxic drugs on tissues, measure pH, and study metabolism. MRI offers the advantages of being noninvasive (not involving puncture or incision of the skin or insertion of an instrument or foreign material into the body), not utilizing radiation, and gathering biochemical information without time-consuming chemical analyses.

FIGURE 1-11 Magnetic resonance imaging (MRI). (a) MRI scanner. (b) Image of the human brain and surrounding structures. (Courtesy of Technicare Corporation, Cleveland, Ohio.)

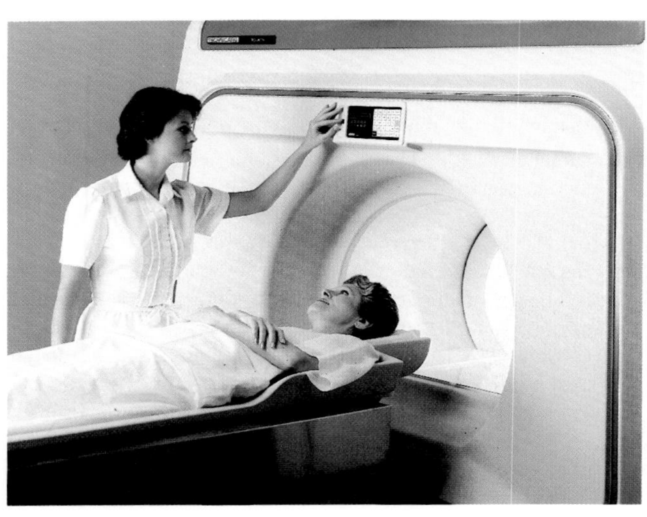

(a)

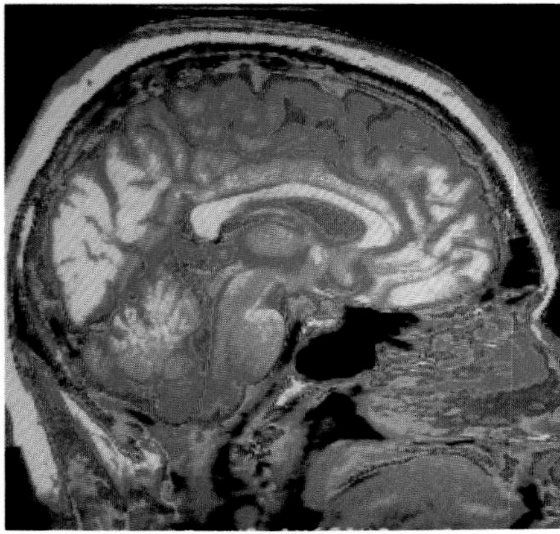

(b)

Ultrasound (US)

When high-frequency sound waves (sound waves that cannot be heard by the human ear) travel forward, they continue to move until they make contact with an object; then a certain amount of the sound bounces back. This is the principle of ***ultrasound (US).*** Submarines use the principle to detect the presence of other vessels and mines and to determine the depth of the ocean floor.

In medical practice, high-frequency sound waves are generated by a hand-held instrument called a ***transducer*** that is moved over the part of the body to be examined. The returning sound waves are also detected by the transducer. Since normal and abnormal body tissues have different densities, they reflect sound waves differently. This information is translated into an image that appears on a screen. The image, called a ***sonogram,*** can be photographed for future reference. Some images are still, two-dimensional cross sections; others, such as the heart or fetus, can be moving images.

Ultrasound may be used to study most abdominal organs, especially in diagnosing gallstones, pelvic organs, blood flow in arteries and veins (*Doppler ultrasound*), the heart (*echocardiography*), and a developing fetus (*fetal ultrasound*). A fetal sonogram is shown in Figure 29-14.

Among the advantages of ultrasound are the following: there is no exposure to radiation, it is noninvasive, it is painless, it does not require any dyes that might cause nausea or allergic reactions, and it is quick and readily available. Numerous experiments and years of experience have uncovered no clear evidence that US produces any harmful effects in humans. However, since ultrasound does not penetrate bone and air-filled spaces, it is not useful in diagnosing abnormalities in the skull, lungs, or intestines. Obesity and scars also interfere with the efficiency of the procedure.

Positron Emission Tomography (PET)

The branch of medicine that uses radioisotopes in the diagnosis of disease and therapy is called ***nuclear medicine.*** The principle behind ***positron emission tomography (PET) scanning*** is as follows. Short-lived radioisotopes such as ^{11}C, ^{13}N, or ^{15}O are produced and incorporated into a solution that can be injected into the body. As the radioisotope circulates through the body, it emits positively charged electronlike particles called *positrons.* Positrons collide with negatively charged electrons in body tissues, causing their annihilation and the release of gamma rays (bundles of energy similar to high-energy x-rays). Once released, gamma rays are detected and recorded by PET receptors. A computer then takes the information and constructs a colored ***PET scan*** that shows where the radioisotopes are being used in the body (Figure 1-12).

PET provides information that cannot be obtained by

any other technique by sensing *function* rather than structure. Using PET, physicians can study the effects of drugs in body organs, measure blood flow through organs such as the brain and heart, diagnose coronary artery disease, identify the extent of damage as a result of strokes or heart attacks, and detect cancers and measure the effects of treatment. PET studies are also used to examine chemical changes associated with schizophrenia, manic-depressive episodes, epilepsy, obsessive-compulsive disorder, and senile dementia. PET is also used to distinguish between the two major types of breast tumors (those with estrogen receptors and those without). This permits physicians to determine appropriate therapy very early in treatment.

Digital Subtraction Angiography (DSA)

A procedure that provides physicians with a much clearer look at diseased arteries is called ***digital subtraction angiography* (*DSA*).** The procedure employs a computer technique that compares an x-ray image of the same region of the body before and after a contrast substance containing iodine has been introduced intravenously. Any tissue or blood vessels that show up in the first image can be subtracted (eased) from the second image, leaving an unobstructed view of an artery (Figure 1-13). DSA figuratively lifts an artery out of the body so that it can be studied in isolation. DSA is frequently used to study the blood vessels of the brain and heart. DSA also helps in diagnosing lesions in the carotid arteries leading to the brain, a potential cause of strokes, and in evaluating patients before surgery and after coronary artery bypass grafting (CABG) and certain transplant operations.

HOMEOSTASIS: MAINTAINING PHYSIOLOGICAL LIMITS

Having considered some of the major anatomical features of the body, we will now turn our attention to an important physiological feature of the body. This is homeostasis, one of the major themes of this textbook.

Homeostasis (hō'-mē-ō-STĀ-sis) is a condition in which the body's internal environment remains within certain physiological limits (*homeo* = same; *stasis* = standing still). For the body's cells to survive, the composition of the surrounding fluids must be precisely maintained at all times. Fluid outside body cells is called ***extracellular*** (*extra* = outside) ***fluid* (*ECF*)** and is found in two principal places. The fluid filling the microscopic spaces between the cells of tissues is called ***interstitial*** (in'-ter-STISH-al) ***fluid*** (*inter* = between), ***intercellular fluid,*** or ***tissue fluid.*** The extracellular fluid in blood vessels is termed ***plasma*** (Figure 1-14). Fluid within cells is called ***intracellular*** (*intra* = within, inside) ***fluid* (*ICF*).** Among the substances in extracellular fluid are gases, nutrients, and electrically charged chemical particles called ions—all needed for the maintenance of life. Extracellular fluid circulates through the blood and lymphatic vessels and from there moves into the spaces between the tissue cells. Thus, it is in constant motion throughout the body. Essentially, all body cells are surrounded by the same fluid environment. For this reason, extracellular fluid is often called the body's ***internal environment.***

An organism is said to be in homeostasis when its internal environment (1) contains the optimum concentrations of gases, nutrients, ions, and water; (2) has an optimal temperature; and (3) has an optimal pressure for the health of the cells. When homeostasis is disturbed, ill health may result. If the body fluids are not eventually brought back into homeostasis, death may occur.

FIGURE 1-12 Positron emission tomography (PET) scans. (a) Oxygen metabolism in the brain. (b) Blood volume in the brain. (c) Blood flow through the brain. (Courtesy of Dr. Michel M. Ter-Rogossian, Washington University, School of Medicine.)

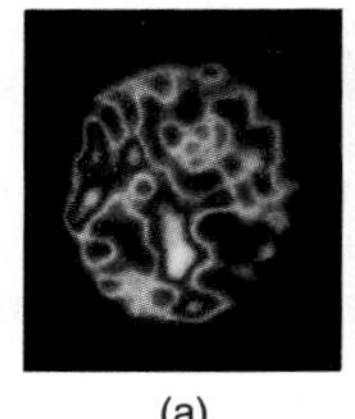

(a)

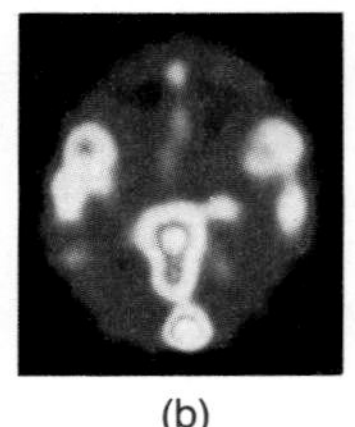

(b)

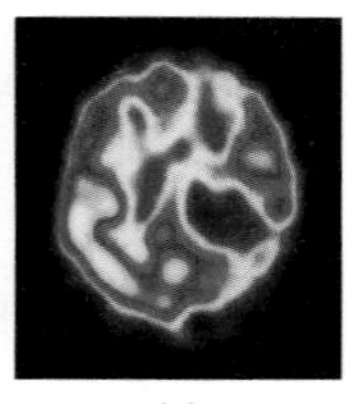

(c)

FIGURE 1-13 Digital subtraction angiography (DSA) image of the heart. The marker (triangle) indicates the point where a blood vessel of the heart (left coronary artery) is narrowed. (Courtesy of The Bayer Company.)

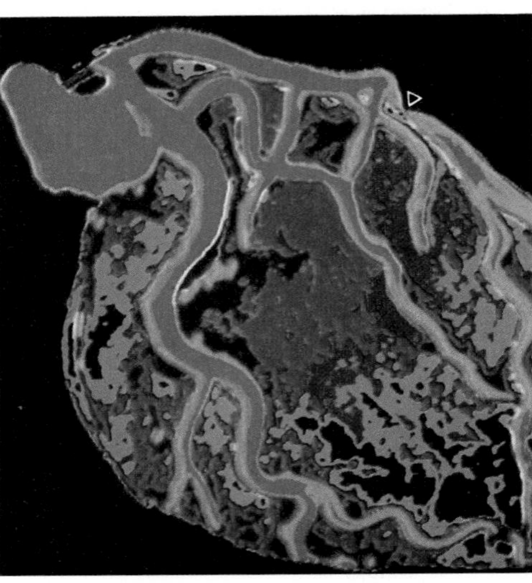

Stress and Homeostasis

Homeostasis in all organisms is continually disturbed by ***stress,*** which is any stimulus that creates an imbalance in the internal environment. The stress may come from

FIGURE 1-14 Internal environment of the body. (a) Extracellular fluid is found in two principal places: in blood vessels as plasma and between cells as interstitial fluid. Plasma circulates through arteries and arterioles and then into microscopic blood vessels called capillaries. From there, certain components of plasma move into the spaces between body cells, where it is called interstitial fluid. Some of this fluid then returns to capillaries as plasma and passes through the venules, then into the veins. (b) Enlarged detail.

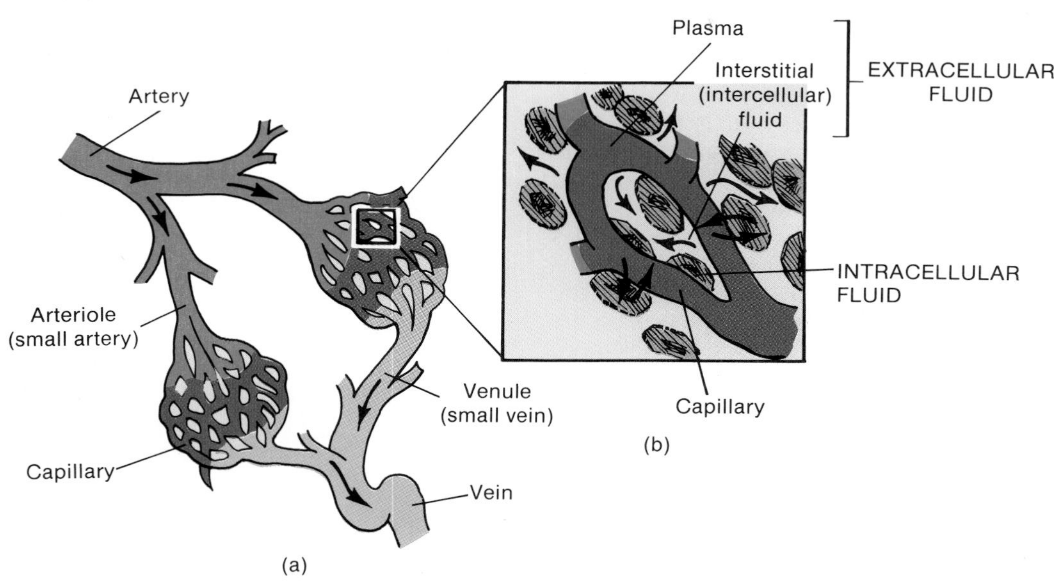

the external environment in the form of stimuli such as heat, cold, loud noises, or lack of oxygen. Or the stress may originate within the body in the form of stimuli such as high blood pressure, pain, tumors, or unpleasant thoughts. Most stresses are mild and routine. Poisoning, overexposure to temperature extremes, severe infection, and surgical operations are examples of extreme stress.

Fortunately, the body has many regulating (homeostatic) devices that oppose the forces of stress and bring the internal environment back into balance. High resistance to stress is a striking feature of all organisms. Some people live in deserts where the daytime temperatures easily reach 49°C (120°F). Others work outside all day in subzero weather. Yet everyone's internal body temperature remains near 37°C (98.6°F). Mountain climbers exercise strenuously at high altitudes, where the oxygen content of the air is low. But once they adjust to the new altitude, they do not suffer from oxygen shortage. The extremes in temperature and in oxygen content of the air are external stresses, and the exercise performed is an internal stress, yet the body compensates and remains in homeostasis. Walter B. Cannon (1871–1945), an American physiologist who coined the term *homeostasis,* noted that the heat produced by the muscles during strenuous exercise would curdle and inactivate the body's proteins if the body did not dissipate heat quickly. Muscles that are being exercised also produce, in addition to heat, a great deal of lactic acid. If the body did not have a homeostatic mechanism for reducing the amount of the acid, the extracellular fluid would become acidic and destroy the cells. Every body structure, from the cellular to the system level, contributes in some way to keeping the internal environment within normal limits.

The homeostatic responses of the body are themselves subject to control by the nervous system and the endocrine system. The nervous system regulates homeostasis by detecting when the body deviates from its balanced state and by sending messages, called ***nerve impulses,*** to the proper organs to counteract the stress. For instance, when muscle fibers (cells) are active, they take a great deal of oxygen from the blood. They also give off carbon dioxide, which is picked up by the blood. Certain nerve cells detect the chemical changes occurring in the blood and send a message to the brain. The brain then sends nerve impulses to the heart to pump blood more quickly and forcefully to the lungs so that the blood can give up its excess carbon dioxide and take on more oxygen. Simultaneously, the brain sends impulses to the muscles that control breathing to contract faster. As a result, more carbon dioxide can be exhaled and more oxygen can be inhaled.

Homeostasis is also controlled by the endocrine system—a series of glands that secrete chemical regulators, called ***hormones,*** into the blood. Whereas nerve impulses coordinate homeostasis rapidly, hormones work slowly. Both means of control are directed toward the same end. Here we will consider the control of homeostasis by the nervous system, using blood pressure as an example. The control of homeostasis by the endocrine system is considered in detail in Chapter 18.

Homeostasis of Blood Pressure (BP)

Blood pressure (BP) is the force exerted by blood as it presses against and attempts to stretch the walls of the blood vessels, especially the arteries. It is determined primarily by three factors: the rate and strength of the heartbeat, the amount of blood, and the resistance offered by the arteries as blood passes through them. The resistance of the arteries results from the physical properties of the blood and the size of the arteries.

If some stress, either internal or external, causes the heartbeat to speed up, the following sequence occurs (Figure 1-15). As the heart pumps faster, it pushes more blood into the arteries per minute, increasing pressure in the arteries. The higher pressure is detected by pressure-sensitive nerve cells in the walls of certain arteries, which send nerve impulses to the brain. The brain interprets the impulses and responds by sending impulses to the heart to slow the heart rate, thus decreasing blood pressure. The continual monitoring of blood pressure by the nervous system is an attempt to maintain a normal blood pressure and employs what is called a feedback system.

A ***feedback system*** is any circular situation in which information about the status of something is continually reported (fed back) to a central control region. The nervous control that results in a normal constant blood pressure is an example. In the case of regulating blood pressure, the ***input (stimulus)*** is the information picked up by the pressure-sensitive nerve cells (high blood pressure), and the ***output (response)*** is the return toward normal blood pressure, owing to decreased heartbeat. Figure 1-15 shows that the system runs in a circle. The pressure-sensitive nerve cells continue to monitor pressure and to feed this information back to the brain, even after the return to homeostasis has begun and blood pressure has begun to normalize. In other words, the pressure-sensitive nerve cells send nerve impulses to the brain about the changed blood pressure, and if the pressure is still too high, the brain continues to send out nerve impulses to slow the heartbeat.

This type of feedback system reverses the direction of the initial condition from a rising to a falling blood pressure and is called a ***negative feedback system.*** The reaction of the body (output) counteracts the stress (input) in order to restore homeostasis. Such a system is therefore a stimulatory-inhibitory one. If, instead, the brain had signaled the heart to beat even faster and the blood pressure had kept on rising, the system would have been a ***positive feedback system.*** In a positive feedback system the output ***intensifies*** the input. This system is therefore a stimulatory-stimulatory one. Although most positive feedback systems are destructive and result in various disorders, some are very beneficial as you will see in Chapter 18 (see Figure 18-9). Most of the feedback systems of the body are negative.

FIGURE 1-15 **Homeostasis of blood pressure. Note that the output is fed back into the system, and the system continues to lower blood pressure until there is a return to homeostasis.**

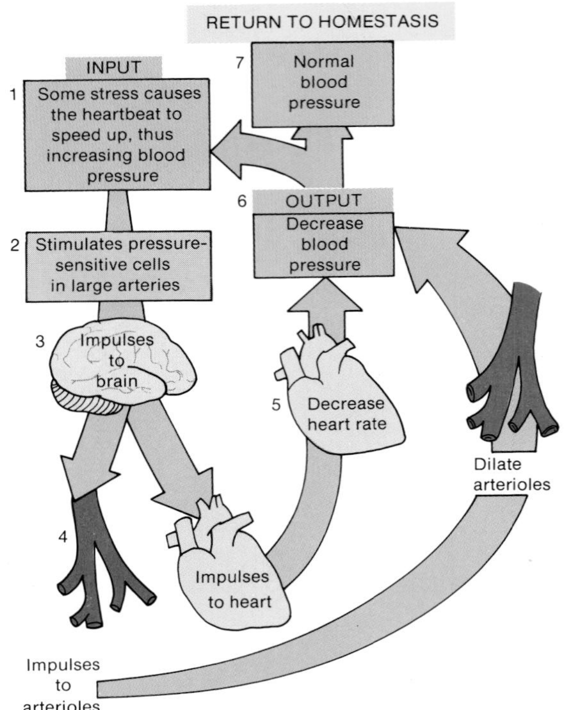

Figure 1-15 also shows that a second negative feedback control is involved in maintaining normal blood pressure. Small arteries, called *arterioles,* have muscular walls that can constrict or dilate (relax) upon receiving an appropriate nerve impulse from the brain. When the blood pressure increases, pressure-sensitive nerve cells in certain arteries send nerve impulses to the brain. The brain interprets the messages and responds by sending fewer nerve impulses to the arterioles, causing them to dilate. Thus, the blood flowing through the arterioles meets less resistance and blood pressure drops back to normal.

Homeostasis and Disease

As long as the various body processes remain within normal physiological limits, body cells function efficiently and homeostasis (health) is maintained. However, when one or more components of the body lose their ability to contribute to homeostasis, body processes do not function efficiently. If the dysfunction is moderate, disease may result. If the dysfunction is severe, death may result.

Disease is any change from a state of health in which part or all of the body is not carrying on its normal functions. A ***local disease*** is one that affects one part or a limited area of the body. A ***systemic disease*** affects either the entire body or several parts. Each disease affecting the body alters body structures and functions in particular ways, as indicated by several kinds of changes. For example, a patient may experience certain ***symptoms,*** that is, *subjective* changes in body functions not apparent to an observer. These might include pain or nausea. A patient can also exhibit ***signs,*** that is, *objective* changes that a clinician can observe and measure. Signs frequently evaluated include swelling, fever, rash, paralysis, and lesions. There are times when a specific group of symptoms and signs always accompanies a particular disease. Such a group is called a ***syndrome.***

The science that deals with why, when, and where diseases occur and how they are transmitted in a human community is known as ***epidemiology*** (ep'-i-dē'-mē-OL-ō-jē; *epidemios* = prevalent; *logos* = study of). The science that deals with the effects and uses of drugs in the treatment of disease is called ***pharmacology*** (far'-ma-KOL-ō-jē; *pharmakon* = medicine; *logos* = study of).

Diagnosis (dī'-ag-NŌ-sis; *dia* = through; *gnosis* = knowledge) is the art of distinguishing one disease from another or determining the nature of a disease. It is an early step in evaluating a disease, usually after a medical history is taken and a physical examination is given. A ***medical history*** consists of information that is collected concerning past events that might be related to a patient's illness. A medical history includes information such as the chief complaint (CC), history of present illness (HPI), past medical history (PMH), family medical history (FMH), social history (SH), and review of symptoms (ROS). A ***physical examination*** is a systematic evaluation that includes inspection (looking at or into a patient with various instruments); palpation (touching for irregularities); auscultation (listening); percussion (striking); measuring vital signs such as temperature (T), pulse (P), respiratory rate (RR), and blood pressure (BP); and laboratory tests.

MEASURING THE HUMAN BODY

An important aspect of describing the body and understanding how it works is ***measurement.*** Examples of measurements involve the size of an organ, the weight of an organ, the time it takes for a physiological reaction to occur, and the amount of a medication to be administered. Measurements involving time, weight, temperature, size, length, and volume are routine to a medical science program. The metric system of measurement is standardly used in sciences. In this book, measurements are given in metric units followed by the approximate U.S. equivalent in parentheses, where convenient. An example is: 2.54 cm (1 inch). If you are not familiar with U.S.–metric conversions, consult Appendix A.

STUDY OUTLINE

Anatomy and Physiology Defined (p. 6)

1. Anatomy is the study of structure and the relationship among structures.
2. Subdivisions of anatomy include surface anatomy (form and markings of surface features), gross (macroscopic) anatomy, systemic (systematic) anatomy (systems), regional anatomy (regions), developmental anatomy (development from fertilization to adulthood), embryology (development from fertilized egg through eighth week in utero), pathological anatomy (disease), histology (tissues), cytology (cells), and radiographic anatomy (x-rays).
3. Physiology is the study of how body structures function.

Levels of Structural Organization (p. 6)

1. The human body consists of several levels of structural organization; among these are the chemical, cellular, tissue, organ, system, and organismic levels.
2. Cells are the basic structural and functional units of an organism.
3. Tissues consist of groups of similarly specialized cells and their intercellular material that perform certain special functions.
4. Organs are structures of definite form that are composed of two or more different tissues and have specific functions.
5. Systems consist of associations of organs that have a common function.
6. The human organism is a collection of structurally and functionally integrated systems.
7. The systems of the human body are the integumentary, skeletal, muscular, nervous, endocrine, cardiovascular, lymphatic, respiratory, digestive, urinary, and reproductive (see Exhibit 1-2).

Life Processes (p. 9)

1. All living forms have certain characteristics that distinguish them from nonliving things.
2. Among the life processes in humans are metabolism, excitability, conductivity, contractility, growth, differentiation and reproduction.

Structural Plan (p. 9)

1. The human body has certain general characteristics.
2. Among the characteristics are a backbone, a tube-within-a tube organization, and bilateral symmetry.

Anatomical Position and Regional Names (p. 9)

1. When in the anatomical position, the subject stands erect facing the observer, the upper extremities are placed at the sides, the palms of the hands are turned forward, and the feet are flat on the floor.
2. Regional names are terms given to specific regions of the body for reference. Examples of regional names include cranial (skull), thoracic (chest), brachial (arm), patellar (knee), cephalic (head), and gluteal (buttock).

Directional Terms (p. 9)

1. Directional terms indicate the relationship of one part of the body to another.
2. Commonly used directional terms are superior (toward the head or upper part of a structure), inferior (away from the head or toward the lower part of a structure), anterior (near or at the front of the body), posterior (near or at the back of the body), medial (nearer the midline of the body or a structure), lateral (farther from the midline of the body or a structure), intermediate (between a medial and lateral structure), ipsilateral (on the same side of the body), contralateral (on the opposite side of the body), proximal (nearer the attachment of an extremity to the trunk or a structure), distal (farther from the attachment of an extremity to the trunk or a structure), superficial (toward or on the surface of the body), deep (away from the surface of the body), parietal (pertaining to the outer wall of a body cavity), and visceral (pertaining to the covering of an organ).

Planes and Sections (p. 9)

1. Planes are imaginary flat surfaces that are used to divide the body or organs into definite areas. A midsagittal (median) plane is a vertical plane through the midline of the body that divides the body or organs into equal right and left sides; a parasagittal plane is a plane that does not pass through the midline of the body and divides the body or organs into unequal right and left sides; a frontal (coronal) plane is a plane at a right angle to a midsagittal (or parasagittal) plane that divides the body or organs into anterior and posterior portions; and a horizontal (transverse) plane is a plane parallel to the ground and at a right angle to the midsagittal, parasagittal, and frontal planes that divides the body or organs into superior and inferior portions.
2. Sections are flat surfaces resulting from cuts through body structures. They are named according to the plane on which the cut is made and include cross sections, frontal sections, and midsagittal sections.

Body Cavities (p. 12)

1. Spaces in the body that contain internal organs are called cavities.
2. The dorsal and ventral cavities are the two principal body cavities. The dorsal cavity contains the brain and spinal cord. The organs of the ventral cavity are collectively called viscera.
3. The dorsal cavity is subdivided into the cranial cavity, which contains the brain, and the vertebral (spinal) canal, which contains the spinal cord and beginnings of spinal nerves.
4. The ventral body cavity is subdivided by the diaphragm into an upper thoracic cavity and a lower abdominopelvic cavity.
5. The thoracic cavity contains two pleural cavities and a mediastinum, which includes the pericardial cavity.
6. The mediastinum is a broad, median partition—actually, a mass of tissues—between the lungs that extends from the sternum to the vertebral column; it contains all contents of the thoracic cavity except the lungs.
7. The abdominopelvic cavity is divided into a superior abdominal and an inferior pelvic cavity.
8. The pelvic cavity is located between two imaginary planes: the superior plane of the pelvic inlet and inferior plane of the pelvic outlet.
9. Viscera of the abdominal cavity include the stomach, spleen, pancreas, liver, gallbladder, small intestine, and most of the large intestine.
10. Viscera of the pelvic cavity include the urinary bladder, sigmoid colon, rectum, and internal female and male reproductive structures.

Abdominopelvic Regions (p. 13)

1. To describe the location of organs easily, the abdominopelvic cavity may be divided into nine regions by drawing four imaginary lines (left midclavicular, right midclavicular, subcostal, and transtubercular).
2. The names of the nine abdominopelvic regions are epigastric, right hypochondriac, left hypochondriac, umbilical, right lumbar, left lumbar, hypogastric (pubic), right iliac (inguinal), and left iliac (inguinal).

Abdominopelvic Quadrants (p. 13)

1. To locate the site of an abdominopelvic abnormality in clinical studies, the abdominopelvic cavity may be divided into quadrants by passing imaginary horizontal and vertical lines through the umbilicus.
2. The names of the abdominopelvic quadrants are right upper quadrant (RUQ), left upper quadrant (LUQ), right lower quadrant (RLQ), and left lower quadrant (LLQ).

Medical Imaging (p. 19)

Conventional Radiography (p. 19)

1. Conventionl radiography uses a single barrage of x-rays.
2. The photographic two-dimensional image produced is called a roentgenogram.
3. Conventional radiography has several diagnostic drawbacks including overlapping of organs and tissues and inability to differentiate subtle differences in tissue density.

Computed Tomography (CT) Scanning (p. 19)

1. Computed tomography (CT) scanning combines the principles of x-ray and advanced computer technology.
2. The image produced, called a CT scan, provides a very accurate cross-sectional picture of any area of the body.

Dynamic Spatial Reconstruction (DSR) (p. 19)

1. The dynamic spatial reconstructor is a highly sophisticated x-ray machine that can produce moving, three-dimensional images of different organs of the body.
2. The reconstructor has been designed to provide imaging of the heart, lungs, and circulation.

Magnetic Resonance Imaging (MRI) (p. 20)

1. Magnetic resonance imaging (MRI) is based on the reaction of atomic nuclei to magnetism.
2. MRI can identify existing pathologies, evaluate drug therapy,

measure metabolism, and assess the potential for certain diseases.

Ultrasound (US) (p. 21)

1. Ultrasound (US) is based on high-frequency sound waves that are reflected and translated into images.
2. Applications of US include diagnosing diseases of abdominal and pelvic organs, analyzing blood flow, assessing heart disorders, and evaluating fetal development.

Positron Emission Tomography (PET) (p. 21)

1. Positron emission tomography (PET) is a form of radioisotope scanning based on the emission of positrons from radioisotopes.
2. PET not only helps in the diagnosis of disease but also analyzes a healthy brain.

Digital Subtraction Angiography (DSA) (p. 22)

1. This procedure compares a blood vessel before and after a contrast medium is introduced.
2. It is frequently used to study blood vessels of the heart.

Homeostasis: Maintaining Physiological Limits (p. 22)

1. Homeostasis is a condition in which the internal environment (extracellular fluid) of the body remains within physiological limits in terms of chemical composition, temperature, and pressure.
2. All body systems attempt to maintain homeostasis.
3. Homeostasis is controlled mainly by the nervous and endocrine systems.

Stress and Homeostasis (p. 22)

1. Stress is any external or internal stimulus that creates a change in the internal environment.
2. If a stress acts on the body, homeostatic mechanisms attempt to counteract the effects of the stress and bring the condition back to normal.

Homeostasis of Blood Pressure (BP) (p. 24)

1. Blood pressure (BP) is the force exerted by blood as it presses against and attempts to stretch the walls of arteries. It is determined by the rate and force of the heartbeat, the amount of blood, and arterial resistance.
2. If a stress causes the heartbeat to increase, blood pressure also increases; pressure-sensitive nerve cells in certain arteries send nerve impulses to the brain, and the brain responds by sending impulses that decrease heartbeat, thus decreasing blood pressure back to normal; a rise in blood pressure also causes the brain to send impulses that dilate arterioles, thereby also helping to decrease blood pressure back to normal.
3. Any circular situation in which information about the status of something is continually fed back to a control region is called a feedback system.
4. A negative feedback system is one in which the reaction of the body (output) counteracts the stress (input) in order to maintain homeostasis; most feedback systems of the body are negative. In a positive feedback system, the output intensifies the input; the system is usually destructive, but a few are very beneficial.

Homeostasis and Disease (p. 24)

1. Disruptions of homeostasis can lead to disease and death.
2. Disease is any change from a state of health, characterized by symptoms and signs.
3. The diagnosis of disease involves a medical history and physical examination.

Measuring the Human Body (p. 25)

1. Various kinds of measurements are important in understanding the human body.
2. Examples of such measurements include organ dimensions and weight, physiological response time, and amount of medication to be administered.
3. Measurements in this book are given in metric units followed by the approximate U.S. equivalents in parentheses, where convenient.
4. The principles and applications of the various metric and U.S. units of measurement are discussed in Appendix A.

REVIEW QUESTIONS

1. Define anatomy. List and define the various subdivisions of anatomy. Define physiology. (p. 6).
2. Give several examples of how structure and function are related. (p. 6).
3. Define each of the following terms: cell, tissue, organ, system, and organism. (p. 6).
4. Using Exhibit 1-2 as a guide, outline the functions of each system of the body, then list several organs that compose each system. (p. 8).
5. List and define the life processes of humans. (p. 9).
6. Define the anatomical position. Why is the anatomical position used? (p. 9).
7. Review Figure 1-2. See if you can locate each region on your own body and name each by its common and anatomical term. (p. 10).
8. What is a directional term? Why are these terms important? Use each of the directional terms listed in Exhibit 1-3 in a complete sentence. (p. 9, 11).
9. Define the various planes that may be passed through the body. Explain how each plane divides the body. Describe the meaning of cross section, frontal section, and midsagittal section. (p. 9).
10. Define a body cavity. List the body cavities discussed and tell which major organs are located in each. What landmarks separate the various body cavities from one another? What is the mediastinum? (p. 12).

11. Describe how the abdominopelvic area is subdivided into nine regions. Name and locate each region and list the organs, or parts of organs, in each. (p. 13).
12. Describe how the abdominopelvic cavity is divided into quadrants and name each quadrant. (p. 13).
13. Why is an autopsy performed? Describe the basic procedure. (p. 15).
14. Explain the principle of computed tomography (CT) scanning. Contrast it with conventional radiography in terms of principle and diagnostic value. (p. 19).
15. Explain the principle and clinical applications of the dynamic spatial reconstruction (DSR), magnetic resonance imaging (MRI), ultrasound (US), positron emission tomography (PET), and digital subtraction angiography (DSA). (p. 20).
16. Define homeostasis. What is extracellular fluid (ECF)? Why is it called the internal environment of the body? (p. 22).
17. Under what conditions is the internal environment said to be in homeostasis? (p. 22).
18. What is a stress? Give several examples. How is stress related to homeostasis? (p. 22).
19. How is homeostasis related to normal and abnormal conditions in the body? (p. 23).
20. What systems of the body control homeostasis? Explain. Discuss briefly how the regulation of blood pressure (BP) is an example of homeostasis. (p. 23).
21. Define a feedback system. Distinguish between a negative and a positive feedback system. (p. 24).
22. Define disease and distinguish between a symptom and a sign. (p. 24).
23. What is meant by diagnosis? On what basis is a diagnosis made? (p. 25).
24. Describe several situations that involve measurement of the human body. (p. 25).

SELECTED READINGS

Clemente, C. D. *Anatomy: A Regional Atlas of the Human Body*, 3rd ed. Baltimore: Urban & Schwarzenberg, 1987.

Geller, S. A. "Autopsy," *Scientific American*, March 1983.

Gosling, J. A., P. F. Harris, J. R. Humpherson, I. Whitmore, and P. L. T. Willan. *Atlas of Human Anatomy*. London: Gower Medical, 1985.

Keiffer, S. A., and E. R. Heitzman. *An Atlas of Cross-Sectional Anatomy*. New York: Harper & Row, 1979.

Netter, F. H. *CIBA Collection of Medical Illustrations*, Vols. 1–8. Summit, N.J.: CIBA, 1962–1987.

Rohen, J. W., and C. Yokochi. *Color Atlas of Anatomy*. Tokyo, New York: Igaku-Shoin, Ltd., 1983.

Sochurek, H. "Medicine's New Vision," *National Geographic*, January 1987.

Tortora, G. J. *Principles of Human Anatomy,* 5th ed. New York: Harper & Row, 1989.

Chapter 2

The Chemical Level of Organization

Chapter Contents at a Glance

Student Objectives

1. Identify by name and symbol the principal chemical elements of the human body.
2. Explain how ionic, covalent, and hydrogen bonds form.
3. Define a chemical reaction and explain the basic differences between synthesis, decomposition, exchange, and reversible chemical reactions.
4. Describe how chemical reactions occur and explain the relationship of energy to chemical reactions.
5. List and compare the properties of water and inorganic acids, bases, and salts.
6. Define pH and explain the role of a buffer system as a homeostatic mechanism that maintains the pH of a body fluid.
7. Compare the structure and functions of carbohydrates, lipids, proteins, deoxyribonucleic acid (DNA), ribonucleic acid (RNA), adenosine triphosphate (ATP), and cyclic AMP.

Many of the common substances we eat and drink—water, sugar, table salt, cooking oil—play vital roles in keeping us alive. In this chapter, you will learn something about how these substances function in your body. Fundamental to this study is a knowledge of basic chemistry and chemical processes, since your body is composed of chemicals and all body activities are chemical in nature. To understand the nature of the matter you are made from and the changes this matter goes through in your body, you will need to know which chemical elements are present in the human organism and how they interact.

INTRODUCTION TO BASIC CHEMISTRY

Chemical Elements

All living and nonliving things consist of ***matter,*** which is anything that occupies space and has mass. Matter may exist in a solid, liquid, or gaseous state. All forms of matter are made up of a limited number of building units called ***chemical elements,*** substances that cannot be decomposed into simpler substances by ordinary chemical reactions. At present, scientists recognize 106 different elements, of which 92 occur naturally. Elements are designated by letter abbreviations, usually derived from the first or first and second letters of the Latin or English name for the element. Such letter abbreviations are called ***chemical symbols.*** Examples of chemical symbols are H (hydrogen), C (carbon), O (oxygen), N (nitrogen), Na (sodium), K (potassium), Fe (iron), and Ca (calcium).

Approximately 26 elements are found in the human organism. Oxygen, carbon, hydrogen, and nitrogen make up about 96 percent of the body's weight. These four elements together with calcium and phosphorous constitute approximately 99 percent of the total body weight. Twenty other chemical elements, called ***trace elements,*** are found in low concentrations and compose the remaining 1 percent.

Structure of Atoms

Each element is made up of units of matter called ***atoms,*** the smallest units of matter that enter into chemical reactions. An ***element*** is simply a quantity of matter composed of atoms all of the same type. A handful of the element carbon, such as pure coal, contains only carbon atoms. A tank of oxygen contains only oxygen atoms. Measurements indicate that the smallest atoms are less than 0.00000001 cm (1/250,000,000 inch) in diameter, and the largest atoms are 0.00000005 cm (1/50,000,000 inch) in diameter. In other words, if 50 million of the largest atoms were placed end to end, they would measure approximately 2.5 cm (1 inch) in length.

An atom consists of two basic parts: the nucleus and electrons (Figure 2-1). The centrally located ***nucleus*** constitutes most of the atomic mass and contains positively charged particles called ***protons*** (p^+) and uncharged (neutral) particles called ***neutrons*** (n^0)**.** Because each proton has one positive charge, the nucleus itself is positively charged. Together, protons and neutrons are referred to as particles called ***nucleons.*** Various coordinated movements of nuclear particles cause the nucleus to vibrate in several distinctive patterns. ***Electrons*** (e^-) are negatively charged particles that move around the nucleus. (For reasons far beyond the scope of this textbook, the diagrams of atoms are very simplified, resembling planetary models of the solar system. (In actuality, electrons do not follow fixed paths around the nucleus but move about in probable locations.) The number of electrons in an atom of an element always equals the number of protons. Since each electron carries one negative charge, the negatively charged electrons and the positively charged protons balance each other, and the atom is electrically neutral.

FIGURE 2-1 Structure of an atom. In this highly simplified version of a carbon atom, note the centrally located nucleus. The nucleus contains six neutrons and six protons, although all are not visible in this view since some are behind others. The six electrons move about the nucleus at varying distances from its center.

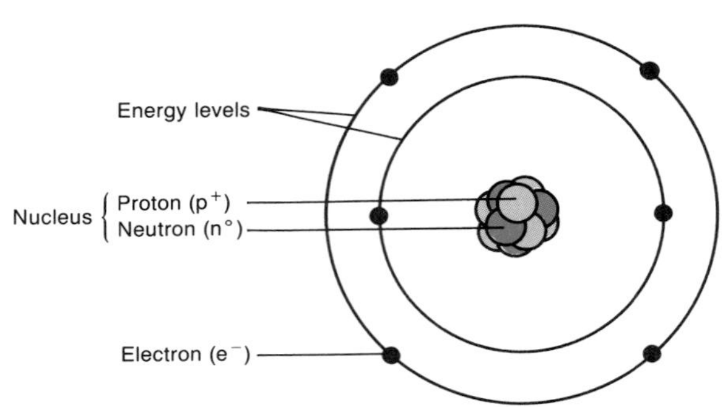

The standard relative weight unit for measuring subatomic particles and atoms is called an ***atomic mass unit (amu)*** or ***dalton.*** A neutron has a mass of 1.008 daltons, and a proton has a mass of 1.007 daltons. The mass of an electron is 0.0005 daltons, about 1/2000 the mass of a neutron or proton.

What makes the atoms of one element different from those of another? The answer lies in the number of protons. Figure 2-2 shows that the hydrogen atom contains one proton; the helium atom contains two; the carbon atom has six; and so on. Each different kind of atom has a different number of protons in its nucleus. The number of protons in an atom is called the atom's ***atomic number.*** Therefore, we can say that each kind of atom, or element, has a different atomic number. The total number of protons and neutrons in an atom is its approximate ***atomic weight.*** Each proton and neutron contributes one dalton unit of weight to the atom. Thus, an atom of sodium has an atomic weight of 23 because of the presence of 11 protons and 12 neutrons in its nucleus.

Atoms of an element, although chemically alike, may have different nuclear masses and thus different atomic

FIGURE 2-2 Atomic structures of some representative atoms.

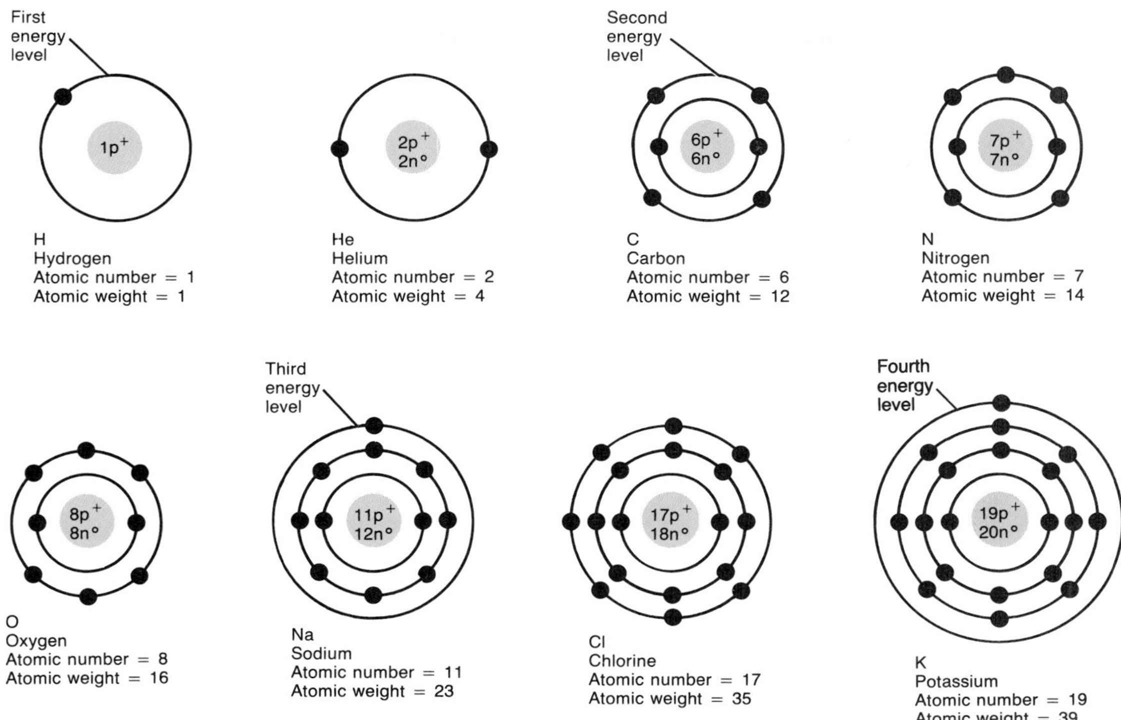

weights because of one or more extra neutrons in some of the atoms, so the atomic weight assigned to an element is only an average. Each of the chemically identical atoms of an element with a particular nuclear mass is an ***isotope*** of that element. All isotopes of an element have the same number of protons in their nuclei, but their atomic weights differ because of the difference in the number of neutrons. In a sample of oxygen, for example, most atoms have 8 neutrons, but a few have 9 or 10, even though all have 8 protons. The isotopes of oxygen are designated as ^{16}O, ^{17}O, and ^{18}O. The numbers indicate their atomic weights.

Certain isotopes called ***radioisotopes*** are unstable—they "decay" or change their nuclear structure to a more stable configuration. And in decaying they emit high-energy radiation (alpha, beta, or gamma particles) that can be detected by instruments. These instruments estimate the amount of radioisotope present in a part of the body or in a sample of material and form an image of its distribution. (Refer to the discussion of positron emission tomography, or PET, in Chapter 1.)

Atoms and Molecules

When atoms combine with or break apart from other atoms, a ***chemical reaction*** occurs. In the process, new products with different properties are formed. Chemical reactions are the foundation of all life processes.

The electrons of an atom actively participate in chemical reactions. The electrons move around the nucleus in regions, shown in Figure 2-2 as concentric circles lying at varying distances from the nucleus. We call these regions ***energy levels.*** Each energy level has a maximum number of electrons it can hold. For instance, the energy level nearest the nucleus never holds more than two electrons, no matter what the element. This energy level can be referred to as the first energy level. The second energy level holds a maximum of eight electrons. The third level of atoms whose atomic number is less than 20 also can hold a maximum of eight electrons. The third level of more complex atoms can hold a maximum of 18 electrons.

An atom always attempts to fill its outermost energy level with the maximum number of electrons it can hold. To do this, the atom may give up, take on, or share electrons with another atom—whichever is easiest. The ***valence*** (combining capacity) is the number of extra or deficient electrons in the outermost energy level. Take a look at the chlorine atom. Its outermost energy level, which happens to be the third level, has seven electrons. Since the third level of an atom can hold a maximum of eight electrons, chlorine can be described as having a shortage of one electron. In fact, chlorine usually does try to pick up an extra electron. Sodium, by contrast, has only one electron in its outer level. This again happens to be the third energy level. It is much easier for sodium to get rid of the one electron than to fill the third level by taking on seven more electrons. Atoms of a few elements, like helium, have completely filled outer energy levels and do not need to gain or lose electrons. These are called ***inert elements*** and are not chemically active.

Atoms with incompletely filled outer energy levels, like sodium and chlorine, tend to combine with other atoms in a chemical reaction. During the reaction, the atoms can trade off or share electrons and thereby fill their outer energy levels. Atoms that already have filled outer levels generally do not participate in chemical reactions for the simple reason that they do not need to gain or lose electrons. When two or more atoms combine in a chemical reaction, the resulting combination is called a ***molecule*** (MOL-e-kyool). A molecule may contain two atoms of the same kind, as in the hydrogen molecule: H_2. The subscript 2 indicates that there are two hydrogen atoms in the molecule. Molecules may also be formed by the reaction of two or more different kinds of atoms, as in the hydrochloric acid molecule: HCl. Here an atom of hydrogen is attached to an atom of chlorine. A ***compound*** is a substance that can be broken down into two or more other substances by chemical means. The molecules of a compound always contain atoms of two or more different elements. Hydrochloric acid, which is present in the digestive juices of the stomach, is a compound. A molecule of hydrogen is not.

CLINICAL APPLICATION: LASERS

The ***laser*** (acronym for ***light amplification*** by ***stimulated emission*** of ***radiation***) is based on a relatively simple principle. Atoms, molecules, or ions in a laser are excited by absorption of energy (thermal, electrical, or optical). After absorption, the atoms, molecules, or ions give off an equal amount of energy in the form of light. This energy, the ***laser beam,*** is an intense light that can produce surgical effects. These include coagulation to stop bleeding, incision making, and tissue removal.

Clinical applications have been found for lasers in several medical and surgical specialties. Among these are ophthalmology (treatment of glaucoma, tumors, cataracts), dermatology and plastic surgery (removal of lesions, port-wine stains, and tattoos), cardiovascular surgery (angioplasty to open blood vessels), gastrointestinal surgery (control of bleeding and tumor removal), general surgery (incisions and removal of decubitus ulcers and leukoplakia), gynecology (treatment of vulvar, vaginal, and cervical neoplasia and endometriosis), neurosurgery (incision or removal of tissue near sensitive neural and vascular structures), otolaryngology (tonsillectomy, treatment of tumors of the larynx, and removal of nasal polyps), and urology (removal of lesions and opening strictures).

The atoms in a molecule are held together by electrical forces of attraction called ***chemical bonds,*** a form of potential energy. As you will see, when chemical bonds are broken, energy is released. When chemical bonds are formed, energy is required. Here we will consider ionic bonds, covalent bonds, and hydrogen bonds.

Ionic Bonds

Atoms are electrically neutral because the number of positively charged protons equals the number of negatively charged electrons. But when an atom gains or loses electrons, this balance is upset. If the atom gains electrons, it acquires an overall negative charge. If the atom loses electrons, it acquires an overall positive charge. Such a negatively or positively charged particle is called an ***ion*** (Ī-on).

Consider the sodium ion (Figure 2-3a). The sodium atom (Na) has 11 protons and 11 electrons, with 1 electron in its outer energy level. When sodium gives up the single electron in its outer level, it is left with 11 protons and only 10 electrons. It is an ***electron donor.*** The atom now has an overall positive charge of one (+1). This positively charged sodium atom is called a sodium ion (written Na^+).

Another example is the formation of the chloride ion (Figure 2-3b). Chlorine has a total of 17 electrons, 7 of them in the outer energy level. Since this energy level can hold eight electrons, chlorine tends to pick up an electron that has been lost by another atom. Chlorine is an ***electron acceptor.*** By accepting an electron, chlorine acquires a total of 18 electrons. However, it still has only 17 protons in its nucleus. The chloride ion therefore has a negative charge of one (−1) and is written as Cl^-.

The positively charged sodium ion (Na^+) and the negatively charged chloride ion (Cl^-) attract each other—unlike charges attract each other. The attraction, called an ***ionic bond,*** holds the two ions together, and a molecule is formed (Figure 2-3c). The formation of this molecule, sodium chloride (NaCl), or table salt, is one of the most common examples of ionic bonding. Thus, an ionic bond is an attraction between ions formed when one atom loses electrons and another atom gains electrons. Generally, atoms whose outer energy level is less than half-filled lose electrons and form positively charged ions called ***cations*** (KAT-ī-ons). Examples of cations are potassium ion (K^+), calcium ion (Ca^{2+}), iron ion (Fe^{2+}), and sodium ion (Na^+). By contrast, atoms whose outer energy level is more than half-filled tend to gain electrons and form negatively charged ions called ***anions*** (AN-ī-ons). Examples of anions include iodide ion (I^-), chloride ion (Cl^-), and sulfur ion (S^{2-}).

Notice that an ion is always symbolized by writing the chemical abbreviation followed by the number of positive (+) or negative (−) charges the ion acquires.

Hydrogen is an example of an atom whose outer level is exactly half-filled. The first energy level can hold two electrons, but in hydrogen atoms, it contains only one. Hydrogen may lose its electron and become a positive ion (H^+). This is precisely what happens when hydrogen combines with chlorine to form hydrochloric acid (H^+Cl^-). However, hydrogen is equally capable of forming another kind of bond called a covalent bond.

FIGURE 2-3 Formation of an ionic bond. (a) An atom of sodium attains stability by passing a single electron to an electron acceptor. The loss of this single electron results in the formation of a sodium ion (Na^+). (b) An atom of chlorine attains stability by accepting a single electron from an electron donor. The gain of this single electron results in the formation of a chloride ion (Cl^-). (c) When the Na^+ and the Cl^- ions are combined, they are held together by the attraction of opposite charges, which is known as an ionic bond, and a molecule of sodium chloride (NaCl) is formed.

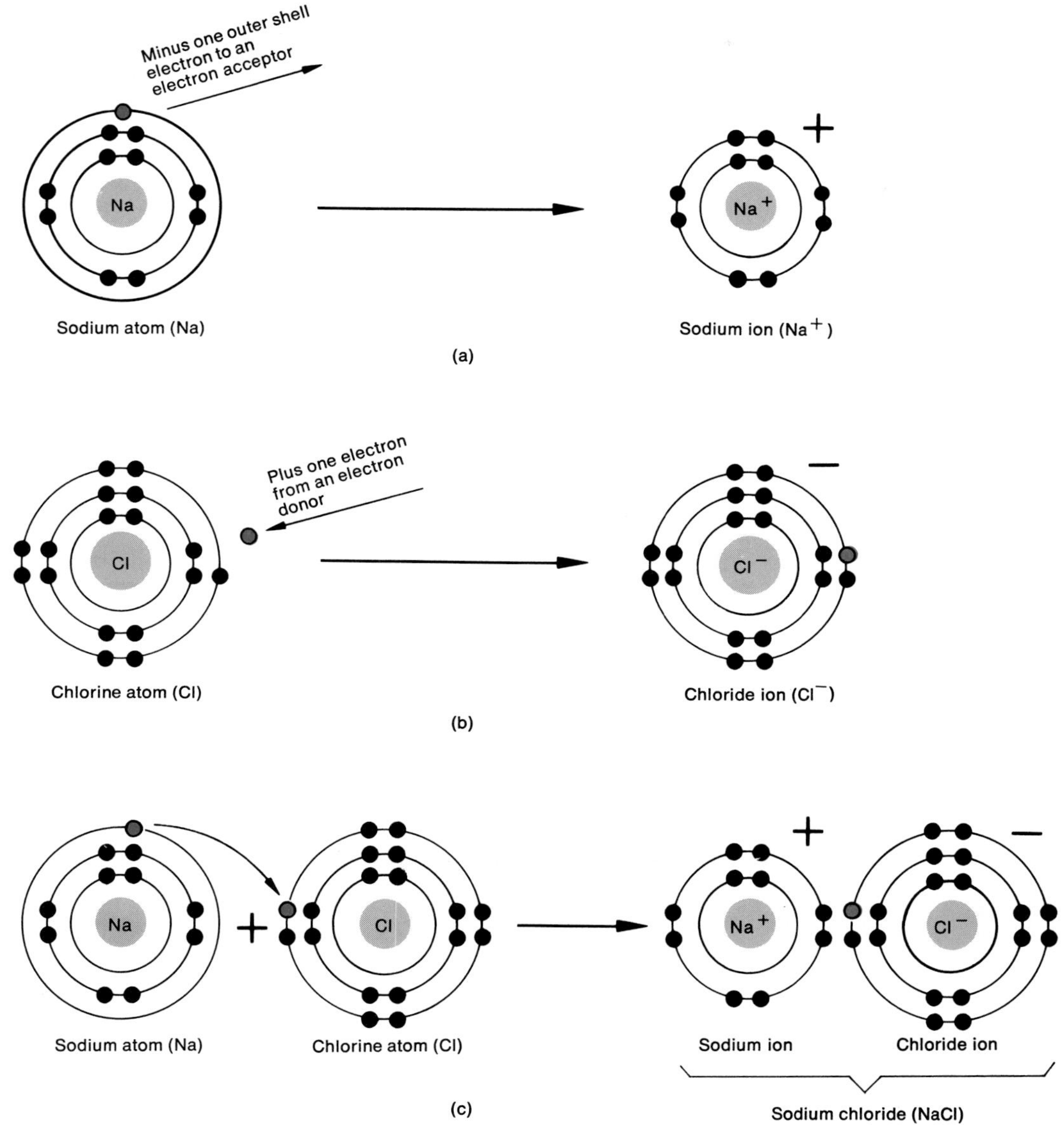

Covalent Bonds

The second chemical bond to be considered is the ***covalent bond.*** This bond is far more common in organisms than an ionic bond and is more stable. When a covalent bond is formed, neither of the combining atoms loses or gains an electron. Instead, the two atoms share one, two, or three electron pairs. Look at the hydrogen atom again. One way a hydrogen atom can fill its outer energy level is to combine with another hydrogen atom to form the molecule H_2 (Figure 2-4a). In the H_2 molecule, the two atoms share a pair of electrons. Each hydrogen atom has its own electron plus one electron from the other atom. The two electrons actually circle the nuclei of both atoms, spending their time equally between both atoms. Therefore, the outer energy levels of both atoms are filled half the time. When one pair of electrons is shared between atoms, as in the H_2 molecule, a *single covalent bond* is formed. A single covalent bond is expressed as a single line between the atoms (H—H). When two pairs of electrons are shared between two atoms, a *double covalent bond* is formed, which is expressed as

FIGURE 2-4 **Formation of a covalent bond between atoms of the same element and between atoms of different elements. In the symbols on the right, each covalent bond is represented by a straight line between atoms. (a) A single covalent bond between two hydrogen atoms. (b) A double covalent bond between two oxygen atoms. (c) A triple covalent bond between two nitrogen atoms. (d) Single covalent bonds between a carbon atom and four hydrogen atoms.**

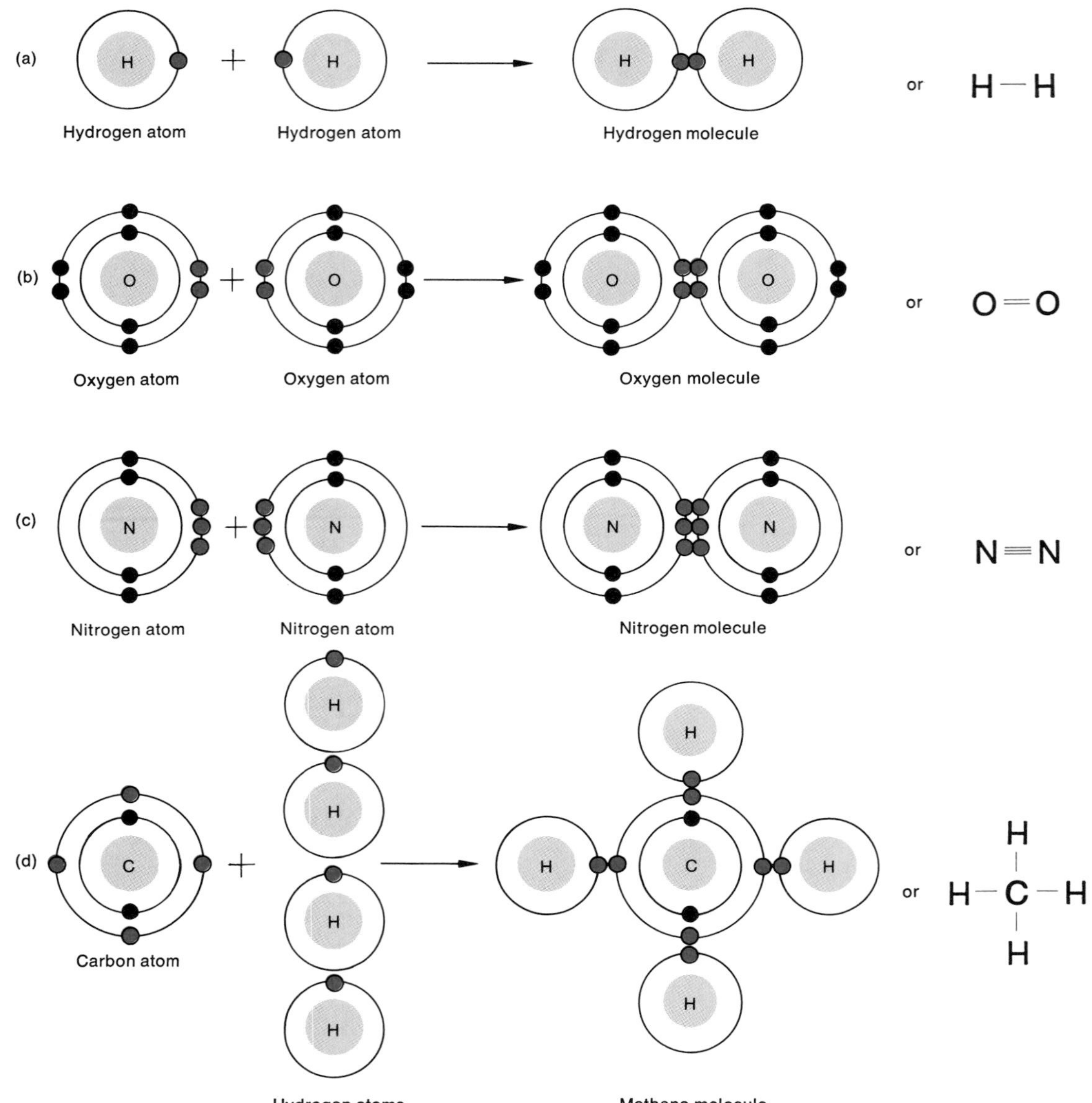

two parallel lines (=) (Figure 2-4b). A *triple covalent bond,* expressed by three parallel lines (≡), occurs when three pairs of electrons are shared (Figure 2-4c).

The same principles that apply to covalent bonding between atoms of the same element also apply to atoms of different elements. Methane (CH_4), also known as marsh gas, is an example of covalent bonding between atoms of different elements (Figure 2-4d). The outer energy level of the carbon atom can hold eight electrons but has only four of its own. Each hydrogen atom can hold two electrons but has only one of its own. In the methane molecule the carbon atom shares four pairs of electrons. One pair is shared with each hydrogen atom. Each of the four carbon electrons orbits around both the carbon nucleus and a hydrogen nucleus. Each hydrogen electron circles around its own nucleus and the carbon nucleus.

In some covalent bonds, the electrons are shared equally between atoms, that is, one atom does not attract the shared electrons more strongly than the other atom. Such covalent bonds are referred to as ***nonpolar covalent bonds.*** Examples include the bonds between hydrogen atoms and between oxygen atoms (see Figure 2-4a,b). In other covalent bonds, there is an unequal sharing of electrons between atoms, that is, one atom attracts the shared electrons more strongly than the other. This type of covalent bond is known as a ***polar covalent bond.*** An example of this type of bond occurs in a molecule of water (see Figure 2-6). This will be considered in more detail later in the chapter as part of the discussion of the function of water as a solvent.

Elements whose outer energy levels are half-filled, such as hydrogen and carbon, form covalent bonds quite easily. In fact, carbon always forms covalent bonds. It never becomes an ion. However, many atoms whose outer energy levels are more than half-filled also form covalent bonds. An example is oxygen. We won't go into the reasons why some atoms tend to form covalent bonds rather than ionic bonds.

Hydrogen Bonds

A ***hydrogen bond*** consists of a hydrogen atom covalently bonded to one oxygen atom or one nitrogen atom but attracted to another oxygen or nitrogen atom. Because hydrogen bonds are weak, only about 5 percent as strong as covalent bonds, they do not bind atoms into molecules. However, they do serve as bridges between different molecules or between various parts of the same molecule. The weak bonds may be formed and broken fairly easily. It is this property that accounts for the temporary bonding between certain atoms within large complex molecules such as proteins and nucleic acids (see Figure 2-12). It should be noted that even though hydrogen bonds are relatively weak, such large molecules may contain several hundred of these bonds, resulting in considerable strength and stability.

Chemical reactions are nothing more than the making or breaking of bonds between atoms. And these reactions occur continually in all the cells of your body. As you will see again and again, chemical reactions are the processes by which body structures are built and body functions carried out.

Chemical Reactions

As we said earlier, ***chemical reactions*** involve the making or breaking of bonds between atoms. After a chemical reaction, the total number of atoms remains the same, but because they are rearranged, there are new molecules with new properties. In this section, we will look at the basic chemical reactions common to all living cells. Once you have learned them, you will be able to understand the chemical reactions discussed later.

Synthesis Reactions—Anabolism

When two or more atoms, ions, or molecules combine to form new and larger molecules, the process is called a ***synthesis reaction.*** The word *synthesis* means "combination," and synthesis reactions involve the *forming of new bonds.* Synthesis reactions can be expressed in the following way:

$$\underset{\text{Atom, ion, or molecule A}}{A} + \underset{\text{Atom, ion, or molecule B}}{B} \xrightarrow{\text{Combine to form}} \underset{\text{New molecule AB}}{AB}$$

The combining substances, A and B, are called the ***reactants;*** the substance formed by the combination is the ***end product.*** The arrow indicates the direction in which the reaction is proceeding. An example of a synthesis reaction is:

$$\underset{\text{Nitrogen atom}}{N} + \underset{\text{Hydrogen atoms}}{3H} \rightarrow \underset{\text{Ammonia molecule}}{NH_3}$$

All the synthesis reactions that occur in your body are collectively called anabolic reactions, or simply ***anabolism*** (a-NAB-ō-lizm). Combining glucose molecules to form glycogen and combining amino acids to form proteins are two examples of anabolism. The importance of anabolism is considered in detail in Chapter 25.

Decomposition Reactions—Catabolism

The reverse of a synthesis reaction is a ***decomposition reaction.*** The word *decompose* means to break down into smaller parts. In a decomposition reaction, the *bonds are broken.* Large molecules are broken down into smaller molecules, ions, or atoms. A decomposition reaction occurs in this way:

$$\underset{\text{Molecule AB}}{AB} \xrightarrow{\text{Breaks down into}} \underset{\text{Atom, ion, or molecule A}}{A} + \underset{\text{Atom, ion, or molecule B}}{B}$$

Under the proper conditions, methane can decompose into carbon and hydrogen:

$$\underset{\text{Methane molecule}}{CH_4} \rightarrow \underset{\text{Carbon atom}}{C} + \underset{\text{Hydrogen atoms}}{4H}$$

The subscript 4 on the left-hand side of the reaction equation indicates that four atoms of hydrogen are bonded to one carbon atom in the methane molecule. The number 4 on the right-hand side of the equation shows that four single hydrogen atoms have been set free.

All the decomposition reactions that occur in your body are collectively called catabolic reactions, or simply ***catabolism*** (ka-TAB-ō-lizm). The digestion and oxidation of food molecules are examples. The importance of catabolism is also considered in detail in Chapter 25.

Exchange Reactions

All chemical reactions are based on synthesis or decomposition processes. In other words, chemical reactions are simply the making and/or breaking of ionic or covalent bonds. Many reactions, such as ***exchange reactions,*** are partly synthesis and partly decomposition. An exchange reaction works like this:

$$AB + CD \rightarrow AD + BC \quad \text{or} \quad AC + BD$$

The bonds between A and B and between C and D are broken in a decomposition process. New bonds are then formed between A and D and between B and C or between A and C and between B and D in a synthesis process.

Reversible Reactions

When chemical reactions are reversible, the end product can revert to the original combining molecules. A ***reversible reaction*** is indicated by two arrows:

$$A + B \underset{\text{Breaks down to}}{\overset{\text{Combines with}}{\rightleftharpoons}} AB$$

Some reversible reactions reverse themselves only under special conditions:

$$A + B \underset{\text{Water}}{\overset{\text{Heat}}{\rightleftharpoons}} AB$$

Whatever is written above or below the arrows indicates the special condition under which the reaction occurs. In this case, A and B react to produce AB only when heat is applied, and AB breaks down into A and B only when water is added. Figure 2-5 summarizes the basic chemical reactions that can occur.

How Chemical Reactions Occur

The ***collision theory*** explains how chemical reactions occur and how certain factors affect the rates of those reactions. According to this theory, all atoms, ions, and molecules are continuously moving and colliding with one another. The energy transferred by the particles in the collision might disrupt their electron structures enough that chemical bonds are broken or new ones are formed.

Several factors determine whether a collision will actually cause a chemical reaction. Among these are the velocities of the colliding particles, their energy, and their specific chemical configurations. Up to a point, the higher the particles' velocities, the greater the probability that their collision will result in a reaction. Also, each chemical reaction requires a specific level of energy. The collision energy required for a chemical reaction is its ***activation energy,*** which is the amount of energy needed to disrupt the stable electronic configuration of a specific molecule so that the electrons can be rearranged. (The relationship of enzymes to activation energy is considered in Chapter 25.) But even if colliding particles possess the minimum energy needed for reaction, no reaction will take place unless the particles are properly oriented toward each other.

Energy and Chemical Reactions

Energy is the capacity to do work. The two principal kinds of energy are ***potential*** (inactive or stored) and ***kinetic*** (energy of motion). Energy, whether potential or kinetic, exists in a number of different forms.

Chemical energy is the energy released or absorbed in the breaking or forming of chemical bonds. When a chemical bond is formed, energy is required. Such a chemical reaction is called an ***endergonic (energy inward) reaction.*** When a bond is broken, energy is released. Such a chemical reaction is called an ***exergonic (energy outward) reaction.*** This means that synthesis reactions are endergonic (need energy), whereas decomposition reactions are exergonic (give off energy). The building processes of the body—the construction of bones, the growth of hair and nails, the replacement of injured cells—occur basically through synthesis reactions. The breakdown of foods, on the other hand, occurs through decomposition reactions. When foods are decomposed, they release energy that can be used by the body for its building processes.

Radiant energy, such as heat and light, travels in waves. Some of the energy released during decomposition reactions is heat energy, which is used to help maintain normal body temperature, but does not accomplish cellular work.

Electrical energy is the result of the flow of charges, electrons, or charged particles called ions. As you will see later, electrical energy is essential for the conduction of action potentials by nerve and muscle cells.

The various forms of energy can be transduced (transformed) from one form into another. For example, potential energy can be transduced to kinetic energy and kinetic energy can be transduced to potential energy.

CHEMICAL COMPOUNDS AND LIFE PROCESSES

Most of the chemicals in the body exist in the form of compounds. Biologists and chemists divide these compounds into two principal classes: inorganic compounds and organic compounds. ***Inorganic compounds*** usually lack carbon. They are usually small, ionically bonded molecules that are vital to body functions. They include water and many salts, acids, and bases. ***Organic com-***

FIGURE 2-5 Kinds of chemical reactions. (a) Synthetic or anabolic reaction. When linked together as shown, molecules of glucose form a molecule of glycogen. Glucose is a sugar that is the primary source of energy. Glycogen is a stored form of that sugar found in the liver and skeletal muscles. (b) Decomposition or catabolic reaction. The example shows a molecule of fat breaking down into glycerol and fatty acids. This reaction occurs whenever a food that contains fat is digested. (c) Exchange reaction. In this reaction, atoms of different molecules are exchanged with each other. Shown is a buffer reaction in which the body eliminates strong acids to help maintain homeostasis. (d) Reversible reaction. ATP (adenosine triphosphate) is an important source of stored energy. When such energy is needed, the ATP breaks down into ADP (adenosine diphosphate) and PO_4^{3-} (phosphate group), releasing energy in the reaction. The phosphate group is symbolized as P. The cells of the body reconstruct ATP by using the energy of foods to attach ADP to PO_4^{3-}.

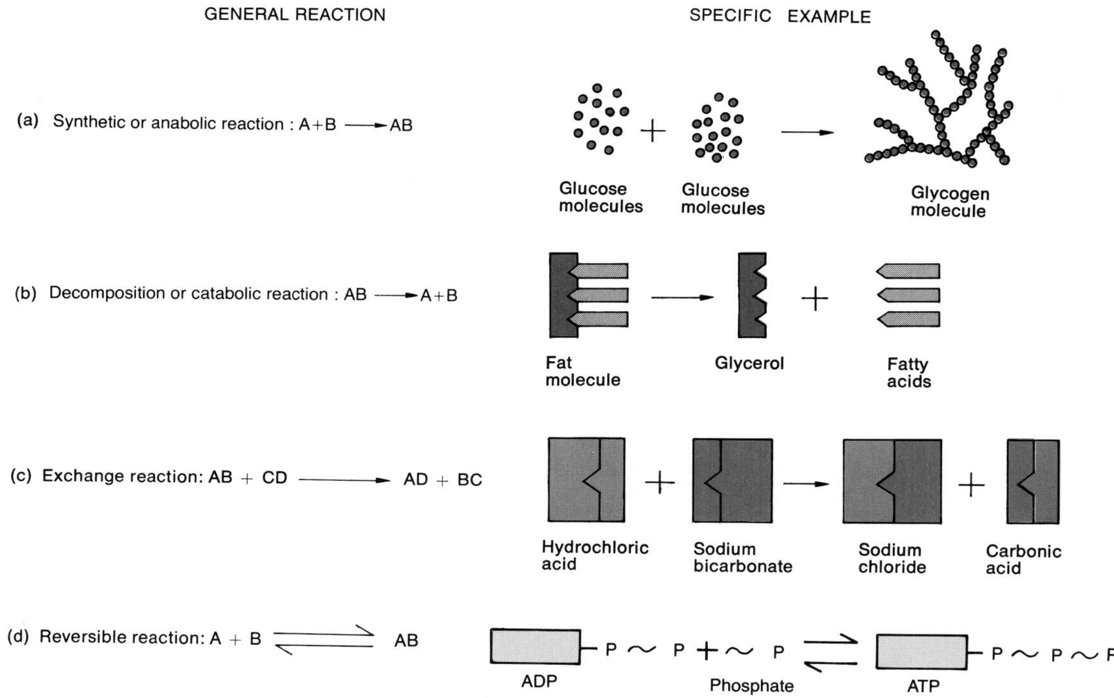

pounds always contain carbon and hydrogen. Carbon is a unique element in the chemistry of life. Because carbon has four electrons in its outer shell, it can combine with a variety of atoms, including other carbon atoms, to form straight or branched chains and rings. Carbon chains are the backbone for many substances of living cells. Organic compounds are held together mostly or entirely by covalent bonds. Organic compounds present in the body include carbohydrates, lipids, proteins, nucleic acids, and adenosine triphosphate (ATP).

Inorganic Compounds

Water

One of the most important, as well as the most abundant, inorganic substances in the human organism is ***water.*** In fact, with a few exceptions, such as tooth enamel and bone tissue, water is by far the most abundant material in tissues. About 60 percent of red blood cells, 75 percent of muscle tissue, and 92 percent of blood plasma is water. The following functions of water explain why it is such a vital compound in living systems:

1. Water is an excellent solvent and suspending medium. A ***solvent*** is a liquid or gas in which some other material (solid, liquid, or gas), called a ***solute,*** has been dissolved. The combination of solvent plus solute is called a ***solution,*** and one common example of a solution is salt water. A solute, such as salt when placed in water, typically does not settle out of its solution. The solute can be retrieved through a chemical reaction or, in some cases, by boiling off the solvent. In a ***suspension,*** by contrast, the suspended material mixes with the liquid or suspending medium, but it will eventually settle out of the mixture. An example of a suspension is cornstarch and water. If the two materials are shaken together, a milky mixture forms. After the mixture sits for a while, however, the water clears at the top

and the cornstarch settles to the bottom. Since water serves as a solvent for so many solutes, it is referred to as the ***universal solvent.***

The versatility of water as a solvent is related to its polar covalent bonds. Recall that a molecule of water contains polar covalent bonds, that is, bonds in which there is an unequal sharing of electrons between atoms. One atom attracts the shared electrons more strongly than others. In a molecule of water, there are both positive and negative areas (Figure 2-6a). When the two hydrogen atoms bond covalently to an oxygen atom, the shared electrons spend more time around oxygen than hydrogen. Since electrons have a negative charge, the unequal sharing causes the oxygen atom to have a slight negative charge and each hydrogen atom to have a slight positive charge.

In order to understand the solvating property of water, consider the following: If a crystal of an ionic compound, such as sodium chloride (NaCl), is placed in water, the sodium and chloride ions at the surface of the salt are exposed to the water molecules. The oxygen portions of the water molecule are negatively charged and are attracted to the sodium ion (Na^+) of the salt, while the hydrogen portions of the water molecule are positively charged and are attracted to the chloride ions (Cl^-) of the salt (Figure 2-6b). As the pattern repeats itself from the surface of the salt inward, water molecules surround the Na^+ and Cl^- ions and separate them from each other. In this way, the salt is taken apart by water molecules—it is dissolved in water.

The solvating property of water is essential to health and survival. For example, if the surfaces of the air sacs in your lungs are not moist, oxygen cannot dissolve and therefore cannot move into your blood to be distributed throughout your body. Water, moreover, is the solvent that carries nutrients into and wastes out of your body cells.

As a suspending medium, water is also vital to your survival. Many large organic molecules are suspended in the water of your body cells. These molecules are consequently able to come in contact with other chemicals, allowing various essential chemical reactions to occur.

2. Water can participate in chemical reactions. During digestion, for example, water can be added to large nutrient molecules in order to break them down into smaller molecules. This kind of breakdown is necessary if the body is to utilize the energy in nutrients. Water molecules are also used in synthesis reactions. Such reactions occur in the production of hormones and enzymes.

3. Water absorbs and releases heat very slowly. In comparison to other substances, water requires a large amount of heat to increase its temperature and a great loss of heat to decrease its temperature. Thus, the presence of a large amount of water moderates the effects of fluctuations in environmental temperature and thereby helps to maintain a homeostatic body temperature.

4. Water requires a large amount of heat to change from a liquid to a gas. When water (perspiration) evaporates from the skin, it takes with it large quantities of heat and provides an excellent cooling mechanism.

5. Water serves as a lubricant in various regions of the body. It is a major part of mucus and other lubricating fluids. Lubrication is especially necessary in the chest and abdomen, where internal organs touch and slide over each other. It is also needed at joints, where bones, ligaments, and tendons rub against each other. In the gastrointestinal tract, water in mucus moistens foods to ensure their smooth passage.

Inorganic Acids, Bases, and Salts

When molecules of inorganic acids, bases, or salts are dissolved in water in the body cells, they undergo ***ionization*** (ī'-on-i-ZĀ-shun) or ***dissociation*** (dis'-sō-sē-Ā-shun); that is, they dissociate (separate) into ions. Such particles are also called ***electrolytes*** (ē-LEK-trō-līts) be-

FIGURE 2-6 Solvating property of water. (a) Polar covalent bonds in a molecule of water. (b) The negative oxygen portions of the water molecules are attracted to the positive sodium ions (Na^+), while the positive hydrogen portions of the water molecules are attracted to the negative chloride ions (Cl^-).

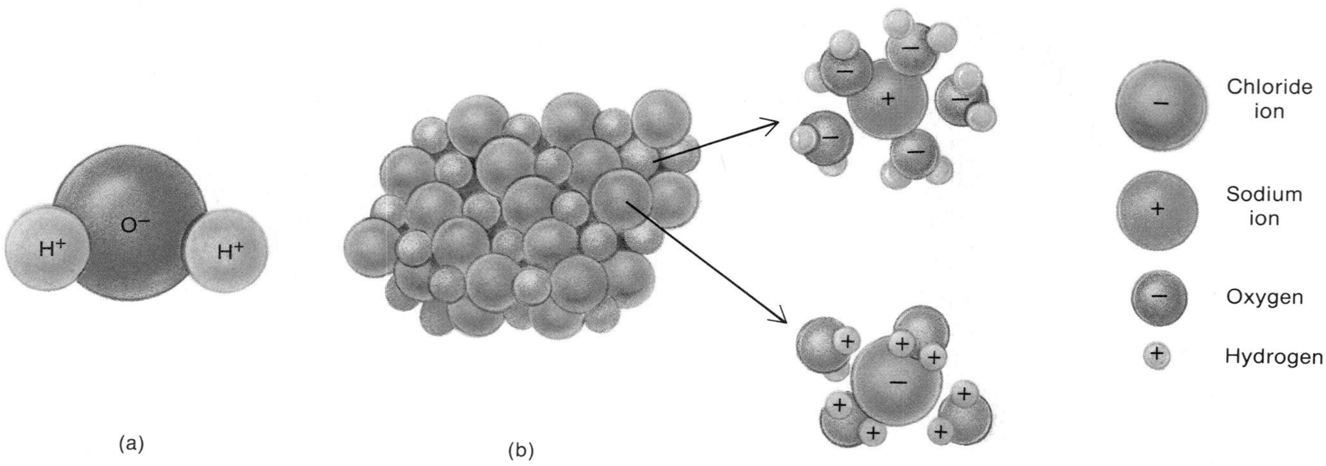

cause the solution will conduct an electric current (the chemistry and importance of electrolytes are discussed in detail in Chapter 27). An ***acid*** may be defined as a substance that dissociates into one or more ***hydrogen ions*** **(H^+)** and one or more negative ions **(*anions*).** Since an H^+ ion is a single proton with a charge of $+1$, an acid may also be defined as a proton donor. A ***base,*** by contrast, dissociates into one or more ***hydroxyl ions*** **(OH^-)** and one or more positive ions **(*cations*).** A base may also be viewed as a proton acceptor. Hydroxyl ions, as well as some other negative ions, have a strong attraction for protons. A ***salt,*** when dissolved in water, dissociates into cations and anions, neither of which is H^+ or OH^- (Figure 2-7). Acids and bases react with one another to form salts. For example, the combination of hydrochloric acid (HCl), an acid, and sodium hydroxide (NaOH), a base, produces sodium chloride (NaCl), a salt, and water (H_2O).

Many salts are found in the body. Some are in cells, whereas others are in the body fluids, such as lymph, blood, and the extracellular fluid of tissues. The ions of salts are the source of many essential chemical elements. Exhibit 2-1 shows how salts dissociate into ions that provide these elements. Chemical analyses reveal that sodium and chloride ions are present in higher concentrations than other ions in extracellular body fluids. Inside the cells, phosphate and potassium ions are more abundant than other ions. Chemical elements such as sodium, phosphorus, potassium, or iodine are present in the body only in chemical combination with other elements or as ions.

EXHIBIT 2-1 DISSOCIATION OF REPRESENTATIVE SALTS INTO IONS THAT PROVIDE ESSENTIAL CHEMICAL ELEMENTS

Salt	Dissociates into	Cation		Anion
NaCl Sodium chloride	→	Na^+ Sodium ion	+	Cl^- Chloride ion
KCl Potassium chloride	→	K^+ Potassium ion	+	Cl^- Chloride ion
$CaCl_2$ Calcium chloride	→	Ca^{2+} Calcium ion	+	$2Cl^-$ Chloride ions
$MgCl_2$ Magnesium chloride	→	Mg^{2+} Magnesium ion	+	$2Cl^-$ Chloride ions
$CaCO_3$ Calcium carbonate	→	Ca^{2-} Calcium ion	+	CO_3^{2-} Carbonate ion
$Ca_3(PO_4)_2$ Calcium phosphate	→	$3Ca^{2+}$ Calcium ions	+	$2PO_4^{3-}$ Phosphate ions
Na_2SO_4 Sodium sulfate	→	$2Na^+$ Sodium ions	+	SO_4^{2-} Sulfate ion

FIGURE 2-7 Ionization of inorganic acids, bases, and salts. (a) When placed in water, hydrochloric acid (HCl) dissociates into H^+ ions and Cl^- ions. Acids are proton donors. (b) When the base sodium hydroxide (NaOH) is placed in water, it dissociates into OH^- ions and Na^+ ions. Bases are proton acceptors. (c) When table salt (NaCl) is placed in water, it dissociates into positive and negative ions (Na^+ and Cl^-), neither of which is H^+ or OH^-.

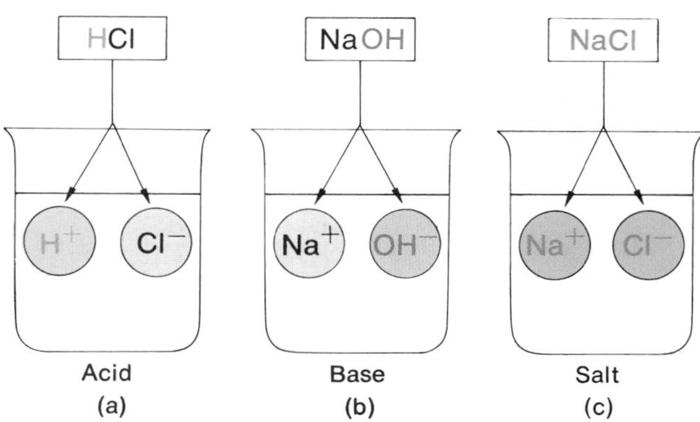

Their presence as free, un-ionized atoms could be instantly fatal. Exhibit 2-2 lists representative elements found in the body.

Acid–Base Balance: The Concept of pH

The fluids of your body must maintain a fairly constant balance of acids and bases. In solutions such as those found in body cells or in extracellular fluids, acids dissociate into hydrogen ions (H^+) and anions. Bases, on the other hand, dissociate into hydroxyl ions (OH^-) and cations. The more hydrogen ions that exist in a solution, the more acid the solution; conversely, the more hydroxyl ions, the more basic (alkaline) the solution. The term ***pH*** is used to describe the degree of ***acidity*** or ***alkalinity*** **(*basicity*)** of a solution.

Biochemical reactions—reactions that occur in living systems—are extremely sensitive to even small changes in the acidity or alkalinity of the environment in which they occur. In fact, H^+ and OH^- ions are involved in practically all biochemical processes, and the functions of cells are modified greatly by any departure from narrow limits of normal H^+ and OH^- concentrations. For this reason, the acids and bases that are constantly formed in the body must be kept in balance.

A solution's acidity or alkalinity is expressed on a ***pH scale*** that runs from 0 to 14 (Figure 2-8). The pH scale is based on the number of H^+ ions in a solution expressed in chemical units called ***moles per liter.*** A pH of 7 means that a solution contains one ten-millionth (0.0000001) of a mole* of H^+ ions per liter. The number 0.0000001 is written 10^{-7} in exponential form. To convert this value

* A ***mole*** of any substance is the weight, in grams, of the combined atomic weights of the atoms that make up a molecule of the substance. Example: A mole of H_2O weighs 18 grams (2 for the two hydrogen atoms + 16 for the oxygen atom).

EXHIBIT 2-2 REPRESENTATIVE CHEMICAL ELEMENTS FOUND IN THE BODY

Chemical Element	Comment
Oxygen (O)	Constituent of water and organic molecules; functions in cellular respiration.
Carbon (C)	Found in every organic molecule.
Hydrogen (H)	Constituent of water, all foods, and most organic molecules.
Nitrogen (N)	Component of all protein molecules and nucleic acid molecules.
Calcium (Ca)	Constituent of bone and teeth; required for blood clotting, intake (endocytosis) and output (exocytosis) of substances through plasma membranes, motility of cells, movement of chromosomes prior to cell division, glycogen metabolism, synthesis and release of neurotransmitters, and contraction of muscle.
Phosphorus (P)	Component of many proteins, nucleic acids, ATP, and cyclic AMP; required for normal bone and tooth structure; found in nerve tissue.
Chlorine (Cl)	Cl^- is an anion of NaCl, a salt important in water movement between cells.
Sulfur (S)	Component of many proteins, especially the contractile proteins of muscle.
Potassium (K)	Required for growth and important in conduction of nerve impulses and muscle contraction.
Sodium (Na)	Na^+ is a cation of NaCl; structural component of bone; essential in blood to maintain water balance; needed for conduction of nerve impulses.
Magnesium (Mg)	Component of many enzymes.
Iodine (I)	Vital to functioning of thyroid gland.
Iron (Fe)	Essential component of hemoglobin and respiratory enzymes.

to pH, the negative exponent (-7) is converted into the positive number 7. A solution with a concentration of 0.0001 (10^{-4}) of H^+ ions per liter has a pH of 4; a solution with a concentration of 0.000000001 (10^{-9}) has a pH of 9; and so on.

A solution that is zero on the pH scale has many H^+ ions and few OH^- ions. A solution that rates 14, by contrast, has many OH^- ions and few H^+ ions. The midpoint in the scale is 7, where the concentration of H^+ and OH^- ions is equal. A substance with a pH of 7, such as pure water, is neutral. A solution that has more H^+ ions than OH^- ions is an ***acid solution*** and has a pH below 7. A solution that has more OH^- ions than H^+ ions is a ***basic (alkaline) solution*** and has a pH above 7. A change of one whole number on the pH scale represents a 10-fold change from the previous concentration; that is, a pH of 2 indicates 10 times fewer H^+ ions than a pH of 1. A pH of 3 indicates 10 times fewer H^+ ions than a pH of 2 and 100 times fewer H^+ ions than a pH of 1.

Maintaining pH: Buffer Systems

Although the pH of body fluids may differ, the normal limits for the various fluids are generally quite specific and narrow. Exhibit 2-3 shows the pH values for certain body fluids compared with common substances. Even though strong acids and bases are continually taken into the body, the pH levels of these body fluids remain relatively constant. The mechanisms that maintain these homeostatic pH values in the body are called ***buffer systems.***

The essential function of a buffer system is to react with strong acids or bases in the body and replace them with weak acids or bases so that the strong acids or bases do not alter pH drastically. Strong acids (or bases) ionize easily and contribute many H^+ (or OH^-) ions to a solution. They therefore change the pH drastically. Weak acids (or bases) do not ionize so easily. They contribute fewer H^+ (or OH^-) ions and have little effect on the pH. The chemicals that replace strong acids or bases with weak ones are called ***buffers*** and are found in the body's fluids. Most buffers in the human body consist of a weak acid and the salt of that acid. It is very important to note that the salt of the acid functions as a weak base. At this point, we will examine the ***carbonic acid–bicarbonate buffer system***—the most important one found in extracellular fluid.

The carbonic acid–bicarbonate buffer system consists of a pair of chemicals: ***carbonic acid*** (***H_2CO_3***), which functions as the ***weak acid,*** and ***sodium bicarbonate*** (***$NaHCO_3$***), the salt of the weak acid, which functions as the ***weak base.*** (As you will see in Chapter 27, it is actually the anion of the salt that behaves like the weak base.) The carbonic acid is a proton donor, that is, a hydrogen ion (H^+) donor, and the bicarbonate ion (HCO_3^-) of the sodium bicarbonate is a proton acceptor, that is, a hydrogen ion (H^+) acceptor. In solution, the members of this buffer pair dissociate as follows:

FIGURE 2-8 **pH scale. At pH 7 (neutrality), the concentration of H^+ and OH^- ions is equal. A pH value below 7 indicates an acid solution; that is, there are more H^+ ions than OH^- ions. The lower the numerical value of the pH, the more acid the solution is because the H^+ ion concentration becomes progressively greater. A pH value above 7 indicates an alkaline (basic) solution; that is, there are more OH^- ions than H^+ ions. The higher the numerical value of the pH, the more alkaline the solution is because the OH^- ion concentration becomes progressively greater. A change in one whole number on the pH scale represents a 10-fold change from the previous concentration ($10^0 = 1.0$, $10^{-1} = 0.1$, $10^{-2} = 0.01$, $10^{-3} = 0.001$, and so on).**

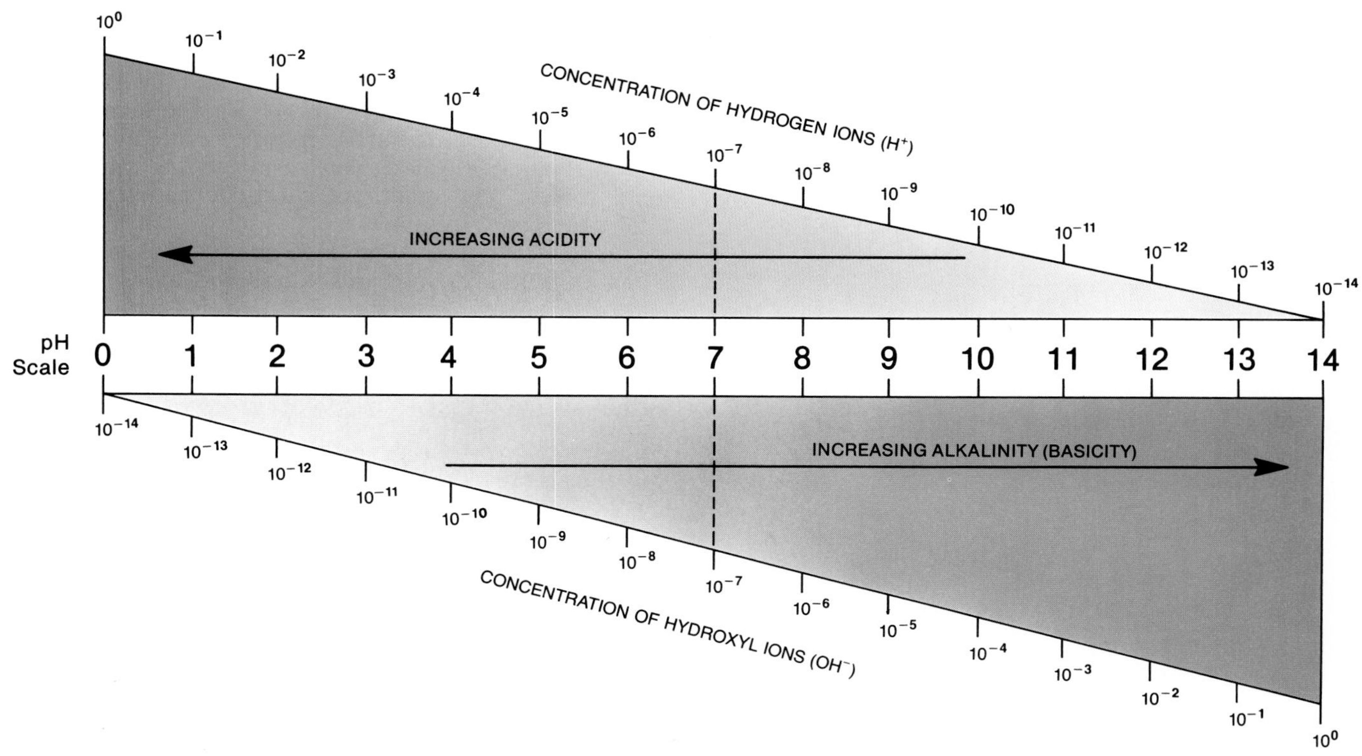

Weak acid component:	H_2CO_3	$\rightleftharpoons$	H^+	+	HCO_3^-
	Carbonic acid (weak acid)		Hydrogen ion		Bicarbonate ion
Salt of the acid component:	$NaHCO_3$	$\rightleftharpoons$	Na^+	+	HCO_3^-
	Sodium bicarbonate (weak base)		Sodium ion		Bicarbonate ion

Each member of the buffer pair has a specific role in helping the body maintain a constant pH. If the body's pH is threatened by the presence of a strong acid, the salt of the acid of the buffer pair behaves like a weak base and goes into operation. If the body's pH is threatened by a strong base, the weak acid goes into play.

Consider the following situation. If a strong acid, such as HCl, is added to extracellular fluid, the salt of the acid of the buffer system behaves like a weak base and goes to work, and the following acid-buffering reaction occurs:

HCl	+	$NaHCO_3$	$\rightleftharpoons$	NaCl	+	H_2CO_3
Hydrochloric acid (strong acid)		Sodium bicarbonate (weak base of buffer system)		Sodium chloride (salt)		Carbonic acid (weak acid)

The chloride ion (Cl^-) of HCl and the sodium ion (Na^+) of sodium bicarbonate combine to form NaCl, a substance that has no effect on pH. The hydrogen ion of the HCl could greatly lower pH by making the solution more acid, but this H^+ ion combines with the bicarbonate ion (HCO_3^-) of sodium bicarbonate to form carbonic acid, a weak acid that lowers pH only slightly. In other words, because of the action of the salt of the acid of the buffer system functioning as a weak base, the strong acid (HCl) has been replaced by a weak acid (H_2CO_3) and a salt (NaCl), and the pH remains relatively constant.

Now suppose a strong base, such as sodium hydroxide (NaOH), is added to the extracellular fluid. In this instance, carbonic acid, the weak acid of the buffer system, goes

EXHIBIT 2-3 NORMAL pH VALUES OF REPRESENTATIVE SUBSTANCES

Substance	pH Value
Gastric juice (digestive juice of the stomach)	1.2–3.0
Lemon juice	2.2–2.4
Grapefruit juice, vinegar, beer, wine	3.0
Cider	2.8–3.3
Carbonated soft drink	3.0–3.5
Pineapple juice, orange juice	3.5
Tomato juice	4.2
Coffee	5.0
Clam chowder	5.7
Urine	4.6–8.0
Saliva	6.35–6.85
Milk	6.6–6.9
Pure (distilled) water	7.0
Blood	7.35–7.45
Semen (fluid containing sperm)	7.20–7.60
Cerebrospinal fluid (fluid associated with nervous system)	7.4
Pancreatic juice (digestive juice of the pancreas)	7.1–8.2
Eggs	7.6–8.0
Bile (liver secretion that aids in fat digestion)	7.6–8.6
Milk of magnesia	10.0–11.0
Limewater	12.3

to work and the following base-buffering reaction takes place:

$$\underset{\text{Sodium hydroxide (strong base)}}{NaOH} + \underset{\text{Carbonic acid (weak acid of buffer system)}}{H_2CO_3} \rightleftharpoons \underset{\text{Water}}{H_2O} + \underset{\text{Sodium bicarbonate (weak base)}}{NaHCO_3}$$

In this reaction, the OH^- ion of sodium hydroxide could greatly raise the pH of the solution by making it more alkaline. However, the OH^- ion combines with an H^+ ion of carbonic acid and forms water, a substance that has no effect on pH. In addition, the Na^+ ion of sodium hydroxide combines with the bicarbonate ion (HCO_3^-) to form sodium bicarbonate, a salt of the acid that acts like a weak base and has little effect on pH. Thus, because of the action of the buffer system, the strong base (NaOH) is replaced by water and a salt of the acid ($NaHCO_3$), that functions like a weak base, and the pH remains relatively constant.

Whenever a buffering reaction occurs, the concentration of one member of the buffer pair is increased, whereas the concentration of the other decreases. When a strong acid is buffered, for example, the concentration of carbonic acid is increased, but the concentration of sodium bicarbonate is decreased. This happens because carbonic acid is produced and sodium bicarbonate is used up in the acid-buffering reaction. When a strong base is buffered, the concentration of sodium bicarbonate is increased, but the concentration of carbonic acid is decreased because sodium bicarbonate is produced and carbonic acid is used up in the base-buffering reaction. When the buffered substances—HCl and NaOH, in this case—are removed from the body via the kidneys, the carbonic acid and sodium bicarbonate formed as products of the reactions function again as components of the buffer pair. Do you now understand why buffers are sometimes called "chemical sponges"?

Although the preceding discussion of pH has focused on inorganic acids and bases, you should know that there are also organic acids and bases and that they, too, are involved in counteracting potential pH problems.

Organic Compounds

In addition to carbon, the most frequently found elements in organic compounds are hydrogen (which can form one bond), oxygen (two bonds), and nitrogen (three bonds). Sulfur (two bonds) and phosphorus (five bonds) appear less often. Other elements are found, but only in a relatively few organic compounds. Carbon has several properties that make it particularly useful to living organisms. For one thing, it can react with one to several hundred other carbon atoms to form large molecules of many different shapes. This means that the body can build many compounds out of carbon, hydrogen, and oxygen. Each compound can be especially suited for a particular structure or function. The relatively large size of most carbon-containing molecules and the fact that some do not dissolve easily in water make them useful materials for building body structures. Carbon compounds are mostly or entirely held together by covalent bonds and tend to decompose easily. This means that organic compounds are also a good source of energy. Ionic compounds are not good energy sources because they form new ionic bonds as soon as the old ones are broken.

Carbohydrates

A large and diverse group of organic compounds found in the body are the ***carbohydrates,*** also known as sugars and starches. The carbohydrates perform a number of major functions in living systems. A few even form structural units. For instance, one type of sugar (deoxyribose) is a building block of deoxyribonucleic acid (DNA), the molecule that carries hereditary information. Some carbohydrates are converted to other substances, which are used to build structures and provide an emergency source of energy. Other carbohydrates function as food reserves. One example is glycogen, which is stored in the liver and skeletal muscles. The principal function of carbohydrates, however, is to provide the most readily available source of energy to sustain life.

Carbon, hydrogen, and oxygen are the elements found in carbohydrates. The ratio of hydrogen to oxygen atoms is typically 2:1, the same as in water. This ratio can be seen in the formulas for carbohydrates such as ribose ($C_5H_{10}O_5$), glucose ($C_6H_{12}O_6$), and sucrose ($C_{12}H_{22}O_{11}$). Although there are exceptions, the general formula for carbohydrates is $(CH_2O)_n$, where n symbolizes three or more CH_2O units. Carbohydrates can be divided into three major groups on the basis of size: monosaccharides, disaccharides, and polysaccharides.

1. Monosaccharides. ***Monosaccharides*** (mon-ō-SAK-a-rīds), or simple sugars, are compounds containing from three to seven carbon atoms. Simple sugars with three carbons in the molecule are called trioses. The number of carbon atoms in the molecule is indicated by the prefix *tri.* There are also tetroses (four-carbon sugars), pentoses (five-carbon sugars), hexoses (six-carbon sugars), and heptoses (seven-carbon sugars). Pentoses and hexoses are exceedingly important to the human organism. The pentose called deoxyribose is a component of genes. The hexose called glucose is the main energy-supplying molecule of the body.

2. Disaccharides. A second group of carbohydrates, the ***disaccharides*** (dī-SAK-a-rīds), are also sugars and consist of two monosaccharides joined chemically. In the process of disaccharide formation, two monosaccharides combine to form a disaccharide molecule and a molecule of water is lost. This reaction is known as ***dehydration synthesis*** (*dehydration* = loss of water). The following reaction shows disaccharide formation. Molecules of the monosaccharides glucose and fructose combine to form a molecule of the disaccharide sucrose (table sugar):

$C_6H_{12}O_6$	+	$C_6H_{12}O_6$	→	$C_{12}H_{22}O_{11}$	+	H_2O
Glucose (monosaccharide)		Fructose (monosaccharide)		Sucrose (disaccharide)		Water

You may be puzzled to see that glucose and fructose have the same chemical formulas. Actually, they are different monosaccharides, since the relative positions of the oxygens and carbons vary in the two different molecules (see Figure 2-9). The formula for sucrose is $C_{12}H_{22}O_{11}$ and not $C_{12}H_{24}O_{12}$, since a molecule of H_2O is lost in the process of disaccharide formation. In every dehydration synthesis, a molecule of water is lost. Along with this water loss, there is the synthesis of two small molecules, such as glucose and fructose, into one large, more complex molecule, such as sucrose (Figure 2-9). Similarly, the dehydration synthesis of the two monosaccharides glucose and galactose forms the disaccharide lactose (milk sugar).

Disaccharides can also be broken down into smaller, simpler molecules by adding water. This reverse chemical reaction is called ***digestion (hydrolysis),*** which means to split by using water. A molecule of sucrose, for example, may be digested into its components of glucose and fructose by the addition of water. The mechanism of this reaction also is represented in Figure 2-9.

CLINICAL APPLICATION: ARTIFICIAL SWEETENERS

Artificial sweeteners **(saccharin, aspartame, and acesulfame potassium) are used in soft drinks and foods and for table use. Aspartame (NutraSweet or Equal) is now used in many such products as a replacement for saccharin. Aspartame is about 180 times sweeter than sucrose and reportedly lacks the bitter aftertaste of saccharin. Because of the small amounts needed, aspartame adds very few calories to drinks and foods—less than 0.5 calories for the equivalent of 1 teaspoon of sugar. It also poses no risk for the development of dental caries. However, since aspartame contains phenylalanine (an amino acid), its consumption should be restricted in children with phenylketonuria (PKU) since they cannot metabolize it, and in such persons it may result in neuronal damage during development. At present, acesulfame potassium (Surette) is used in dry food products and is for sale in powder form or tablets that can be applied directly by the consumer.**

FIGURE 2-9 Dehydration synthesis and hydrolysis of a molecule of sucrose. In the dehydration synthesis reaction (read from left to right), the two smaller molecules, glucose and fructose, are joined to form a larger molecule of sucrose. Note the loss of a water molecule. In hydrolysis (read from right to left), the larger sucrose molecule is broken down into the two smaller molecules, glucose and fructose. Here, a molecule of water is added to sucrose for the reaction to occur.

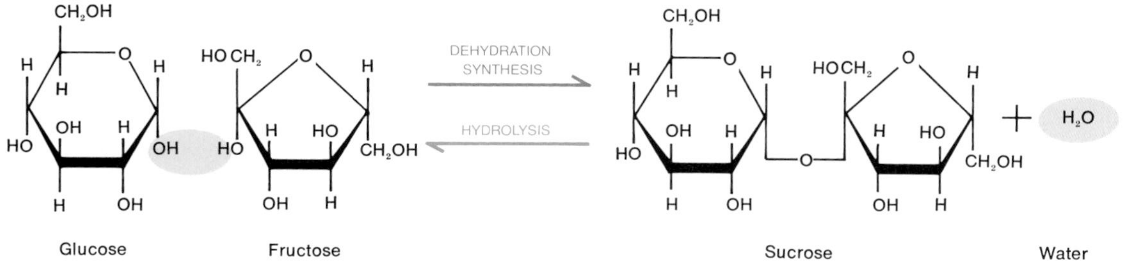

3. Polysaccharides. The third major group of carbohydrates, the ***polysaccharides*** (pol'-ē-SAK-a-rīds), consists of several monosaccharides joined together through dehydration synthesis. Polysaccharides have the formula $(C_6H_{10}O_5)_n$. Like disaccharides, polysaccharides can be broken down into their constituent sugars through hydrolysis reactions. Unlike monosaccharides or disaccharides, however, they usually lack the characteristic sweetness of sugars like fructose or sucrose and are usually not soluble in water. One of the chief polysaccharides is glycogen.

Lipids

A second group of organic compounds that is vital to the human organism is the ***lipids.*** Like carbohydrates, lipids are composed of carbon, hydrogen, and oxygen, but they do not have a 2:1 ratio of hydrogen to oxygen. In fact, the amount of oxygen in lipids is usually less than that in carbohydrates. Most lipids are insoluble in water, but they readily dissolve in solvents such as alcohol, chloroform, and ether. Among the groups of lipids are fats, phospholipids (lipids that contain phosphorous), steroids, carotenes, vitamins E and K, and prostaglandins (PGs). A listing of the various types of lipids is shown in Exhibit 2-4, along with their relationships to the human organism. Since lipids are a large and diverse group of compounds, we will discuss only two types in detail at this point: fats and prostaglandins.

A molecule of ***fat*** (triglyceride) consists of two basic components: ***glycerol*** and ***fatty acids*** (Figure 2-10). A single molecule of fat is formed when a molecule of glycerol combines with three molecules of fatty acids. This reaction, like the one described for disaccharide formation, is a dehydration synthesis reaction. During hydro-

EXHIBIT 2-4 RELATIONSHIPS OF REPRESENTATIVE LIPIDS TO THE HUMAN ORGANISM

Lipids	Relationship
FATS	Protection, insulation, source of energy.
PHOSPHOLIPIDS	
Lecithin	Major lipid component of cell membranes; constituent of plasma.
Cephalin and sphingomyelin	Found in high concentrations in nerves and brain tissue.
STEROIDS	
Cholesterol	Constituent of all animal cells, blood, and nervous tissue; suspected relationship to heart disease and atherosclerosis; precursor of bile salts, vitamin D, and steroid hormones.
Bile salts	Substances that emulsify or suspend fats before their digestion and absorption; needed for absorption of fat-soluble vitamins (A, D, E, K).
Vitamin D	Produced in skin on exposure to ultraviolet radiation; necessary for bone growth, development, and repair.
Estrogens	Sex hormones produced in large quantities by females.
Androgens	Sex hormones produced in large quantities by males.
OTHER LIPOID SUBSTANCES	
Carotenes	Pigment in egg yolk, carrots, and tomatoes; vitamin A is formed from carotenes; retinal, formed from vitamin A, is a photoreceptor in the retina of the eye.
Vitamin E	May promote wound healing, prevent scarring, and contribute to the normal structure and functioning of the nervous system; deficiency causes sterility in rats and muscular dystrophy in monkeys; deficiency in humans believed to cause the oxidation of certain fats (unsaturated), resulting in abnormal structure and function of certain parts of cells (mitochondria, lysosomes, and plasma membranes); may reduce the severity of visual loss associated with retrolental fibroplasia (eye disease in premature infants caused by too much oxygen in incubators) by functioning as an antioxidant.
Vitamin K	Needed for the formation of the vitamin K–dependent clotting factors: prothrombin and factors VII, IX, and X.
Prostaglandins	Membrane-associated lipids that stimulate uterine contractions, induce labor and abortions, regulate blood pressure, transmit nerve impulses, regulate metabolism, regulate stomach secretions, inhibit lipid breakdown, and regulate muscular contractions of the gastrointestinal tract; mediate hormones.

FIGURE 2-10 Structure and reactions of glycerol and fatty acids. (a) Structure of glycerol. (b) Structure of a fatty acid. The one shown here is palmitic acid. Note that when glycerol and the fatty acid are joined in dehydration synthesis, a molecule of water is lost. (c) Fats consist of one molecule of glycerol joined to three molecules of fatty acids, which vary in length and the number and location of double bonds between carbon atoms (C═C). Shown here is a molecule of a fat that contains three different fatty acids: palmitic acid, a saturated fatty acid; lineolic acid, a polyunsaturated fatty acid; and oleic acid, a monosaturated fatty acid. Note that when a molecule of glycerol combines with three fatty acids in dehydration synthesis to form a molecule of fat, three molecules of water are lost.

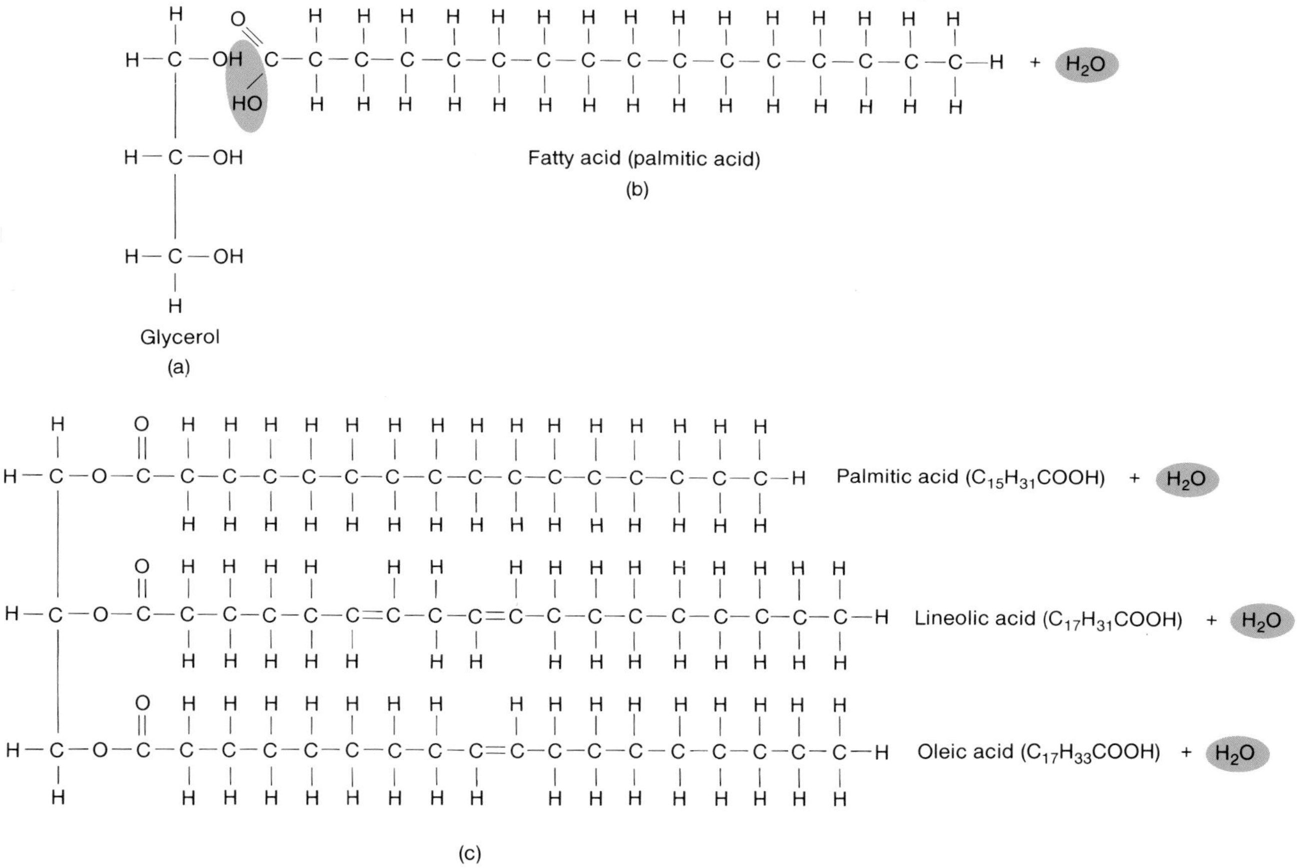

+ H_2O

Fatty acid (palmitic acid)
(b)

Glycerol
(a)

Palmitic acid ($C_{15}H_{31}COOH$) + H_2O

Lineolic acid ($C_{17}H_{31}COOH$) + H_2O

Oleic acid ($C_{17}H_{33}COOH$) + H_2O

(c)

lysis, a single molecule of fat is broken down into fatty acids and glycerol.

In subsequent chapters, we will be talking about saturated, monosaturated, and polyunsaturated fats. These terms have the following meanings. A ***saturated fat*** contains no double bonds between any of its carbon atoms. It contains only single covalent bonds between its carbon atoms, and all the carbon atoms are bonded to the maximum number of hydrogen atoms; thus, such a fat is saturated with hydrogen atoms (Figure 2-10c). Saturated fats (and some cholesterol) occur mostly in animal foods such as beef, pork, butter, whole milk, eggs, and cheese. They also occur in some plant products such as cocoa butter, palm oil, and coconut oil. Since the liver uses some breakdown products of saturated fats to produce cholesterol, consumption of these fats is discouraged for individuals with high cholesterol levels. A ***monosaturated fat*** contains one double covalent bond between its carbon atoms; it is not completely saturated with hydrogen atoms (Figure 2-10c). Examples are olive oil and peanut oil, which are believed to help reduce cholesterol levels. A ***polyunsaturated fat*** contains more than one double covalent bond between its carbon atoms (Figure 2-10c). Corn oil, safflower oil, sunflower oil, cottonseed oil, sesame oil, and soybean oil are examples of polyunsaturated fats, which researchers believe also help to reduce cholesterol in the blood.

Fats represent the body's most highly concentrated

source of energy. They provide more than twice as much energy per weight as either carbohydrates or proteins. In addition, fats do not attract water and thus do not result in excessive water retention in the body. In general, however, fats are about 10 to 12 percent less efficient as body fuels than are carbohydrates. A great amount of the fat calorie is wasted and thus not available for the body to use.

CLINICAL APPLICATIONL: FAT SUBSTITUTE

The Food and Drug Administration (FDA) is currently evaluating data related to a substance called *olestra.* It is a chemical that looks and tastes like vegetable oil but passes through the body without entering the bloodstream. Olestra is a cholesterol-free, calorie-free *fat substitute* that may even lower blood cholesterol levels. Olestra may be used in deep-fried foods, oils, shortenings, and salty snacks, if approved by the FDA.

Prostaglandins (pros'-ta-GLAN-dins), also called ***PGs,*** are a large group of membrane-associated lipids composed of 20-carbon fatty acids containing 5 carbon atoms joined to form a ring (cyclopentane ring). Prostaglandins are produced in all nucleated cells in the body and are able to influence the functioning of any type of cell.

Prostaglandins are produced in cell membranes and are rapidly decomposed by catabolic enzymes. Although synthesized in minute quantities, they are potent substances and exhibit a wide variety of effects on the body. Basically, prostaglandins mimic hormones. They are involved in modulating many hormonal responses (Chapter 18), inducing menstruation, and inducing second-trimester abortions. They are also involved in contributing to the inflammatory response (Chapter 22), preventing peptic ulcers, opening bronchial and nasal passages, platelet aggregation and inhibition of aggregation, and regulating body temperature.

Proteins

A third principal group of organic compounds is ***proteins.*** These compounds are much more complex in structure than the carbohydrates or lipids. They are also responsible for much of the structure of body cells and are related to many physiological activities. For example, proteins in the form of enzymes speed up most essential biochemical reactions. This is described in detail in Chapter 25. Other proteins assume a necessary role in muscular contraction. Antibodies are proteins that provide the human organism with defenses against invading microbes. And some hormones that regulate body functions are also proteins. A classification of proteins on the basis of function is shown in Exhibit 2-5.

Chemically, proteins always contain carbon, hydrogen, oxygen, and nitrogen. Many proteins also contain sulfur and phosphorus. Just as monosaccharides are the building units of sugars, and fatty acids and gycerol are the building units of fats, ***amino acids*** are the building blocks of proteins. In protein formation, amino acids combine to form more complex molecules, while water molecules are lost. The process is a dehydration synthesis reaction, and the bonds formed between amino acids are called ***peptide bonds*** (Figure 2-11).

When two amino acids combine, a ***dipeptide*** results. Adding another amino acid to a dipeptide produces a ***tripeptide.*** Further additions of amino acids result in the formation of chainlike ***polypeptides,*** which are large protein molecules that consist of 50 or more amino acids. At least 20 different amino acids are found in proteins. Although all amino acids have one component in common, each is distinguished on the basis of additional atoms or groups of atoms arranged in a specific way. A great variety of proteins is possible because each variation in the number or sequence of amino acids can produce a different protein. The situation is similar to using an alphabet of 20 letters to form words. Each letter could be

EXHIBIT 2-5 CLASSIFICATION OF PROTEINS BY FUNCTION

Type of Protein	Description
Structural	Form the structural framework of various parts of the body. Examples: keratin in the skin, hair, and fingernails (Chapter 5) and collagen in connective tissue (Chapter 4).
Regulatory	Function as hormones and regulate various physiological processes. Example: insulin, which regulates blood sugar level (Chapter 18).
Contractile	Serve as contractile elements in muscle tissue. Examples: myosin and actin (Chapter 10).
Immunological	Serve as antibodies to protect the body against invading microbes. Example: gamma globulin (Chapter 22).
Transport	Transport vital substances throughout the body. Example: hemoglobin, which transports oxygen and carbon dioxide in the blood (Chapter 19).
Catalytic	Act as enzymes and function in regulating biochemical reactions. Examples: salivary amylase, lipase, and lactase (Chapters 24 and 25).

FIGURE 2-11 Protein formation. When two or more amino acids are chemically united, the resulting bond between them is called a peptide bond. In the example shown here, the amino acids glycine and alanine are joined to form the dipeptide glycylalanine. The peptide bond is formed at the point where water is lost.

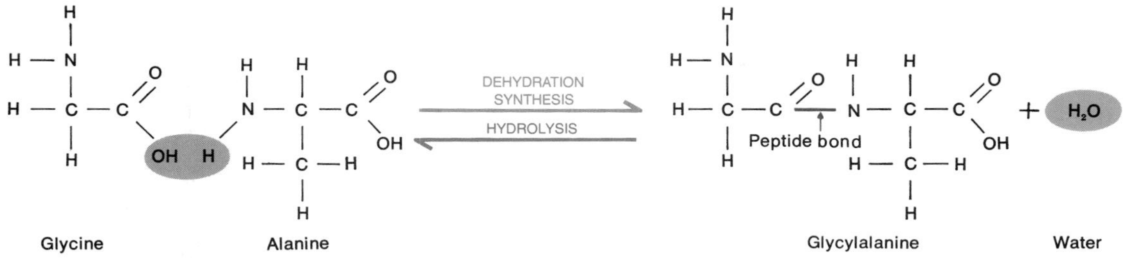

DEHYDRATION SYNTHESIS
HYDROLYSIS
Peptide bond
Glycine
Alanine
Glycylalanine
Water

compared to a different amino acid, and each word would be a different protein.

Proteins vary tremendously in structure. Different proteins have different architectures and different three-dimensional shapes. This variation in structure and shape is directly related to their diverse functions. When a cell makes a protein, the polypeptide chain folds spontaneously to assume a certain shape. One reason for folding of the polypeptide is that some parts of a protein are attracted to water, and other parts are repelled by it. In practically every case, the function of a protein depends on its ability to recognize and bind to some other molecule, a situation similar to the fit between a lock and key. As an example, an enzyme binds specifically with its substrate. A hormonal protein binds to a receptor on a cell whose function it will alter. An antibody binds to a foreign substance (antigen) that has invaded the body. The unique shape of a protein permits it to interact with other specific molecules in order to carry out specific functions.

Proteins exhibit four levels of structural organization. The ***primary structure*** is the unique order (sequence) of amino acids making up the protein. An alteration in primary structure can have serious consequences. For example, a single substitution of an amino acid in a blood protein can result in a deformed hemoglobin molecule that produces sickle-cell anemia. The ***secondary structure*** of a protein is the localized, repetitious twisting or folding of its polypeptide chain. Common secondary structures are clockwise spirals (helixes) and pleated sheets. The ***tertiary structure*** refers to the overall three-dimensional structure of a polypeptide chain. The folding is not repetitive or predictable as in secondary structures. The tertiary structure is very irregular, and it determines the particular function of a protein. The ***quaternary structure*** of a protein refers to two or more individual polypeptide chains bonded to each other that function as a single unit.

It was once thought that the atoms in proteins are in fixed positions and that proteins are therefore rigid molecules. However, based on computer simulations, it has been learned that the atoms in proteins are in constant motion, usually twisting movements. Such motion is due to the chemical bonds between atoms that act like springs. The internal motion of proteins results in their constantly changing structure and explains the many and varied functions of proteins. Scientists are now studying how proteins fold into three-dimensional structures that determine their biological functions.

If a protein encounters a hostile environment in terms of temperature, pH, or salt concentrations, it may unravel and lose its characteristic shape. This process is called ***denaturation.*** As a result of denaturation, the protein is no longer functional.

Nucleic Acids: Deoxyribonucleic Acid (DNA) and Ribonucleic Acid (RNA)

Nucleic (noo-KLĒ-ic) ***acids,*** compounds first discovered in the nuclei of cells, are exceedingly large organic molecules containing carbon, hydrogen, oxygen, nitrogen, and phosphorus. They are divided into two principal kinds: ***deoxyribonucleic*** (dē-ok′-sē-rī′-bō-noo-KLĒ-ik) ***acid (DNA)*** and ***ribonucleic acid (RNA).***

Whereas the basic structural units of proteins are amino acids, the basic units of nucleic acids are ***nucleotides.*** A molecule of DNA is a chain composed of repeating nucleotide units. Each nucleotide of DNA consists of three basic parts (Figure 2-12a):

1. It contains one of four possible *nitrogenous bases,* which are ring-shaped structures containing atoms of C, H, O, and N. The nitrogenous bases found in DNA are adenine, thymine, cytosine, and guanine. Adenine and guanine are double-ring structures, collectively referred to as *purines.* Thymine and cytosine are smaller, single-ring structures called *pyrimidines.*
2. It contains a pentose sugar called *deoxyribose.*
3. It also contains *phosphate groups.*

The nucleotides are named according to the nitrogenous base that is present. Thus, a nucleotide containing thymine is called a *thymine nucleotide.* One containing adenine is called an *adenine nucleotide,* and so on.

The chemical components of the DNA molecule were known before 1900, but it was not until 1953 that a model

FIGURE 2-12 DNA molecule. (a) Adenine nucleotide. (b) Portion of an assembled DNA molecule. (c) Strand of DNA viewed through a scanning tunneling microscope. This is the first direct image of chemically unaltered, uncoated, pure DNA. The looped DNA strand stretches across an area 400 Å wide. The circular structure is possibly an unresolved DNA fragment. (Courtesy of Lawrence Livermore National Laboratory.)

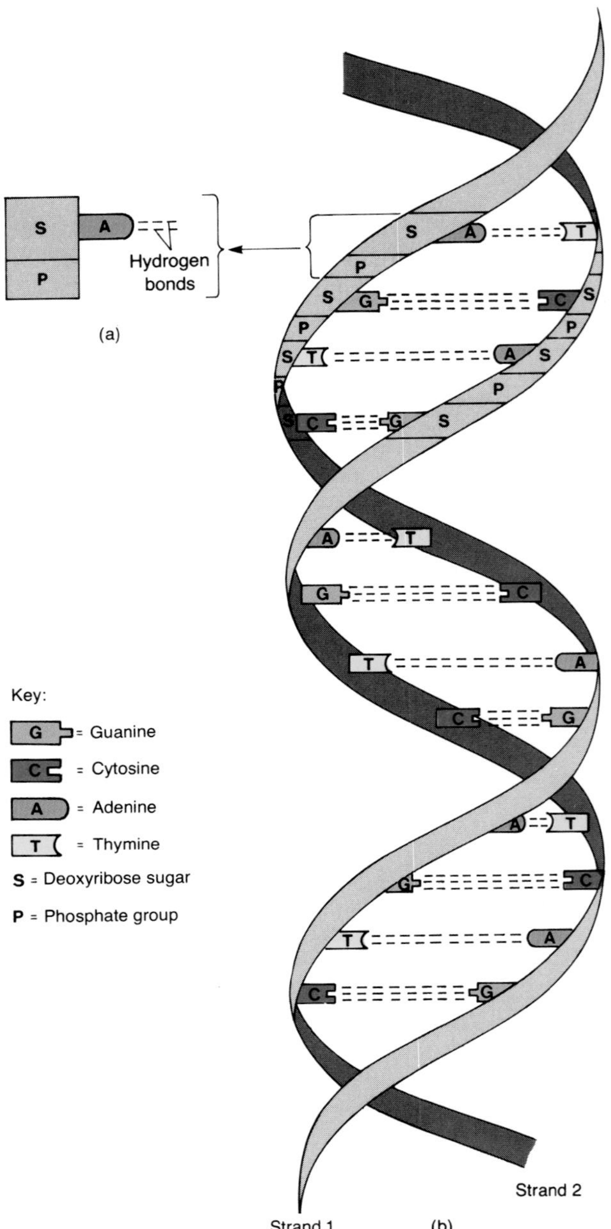

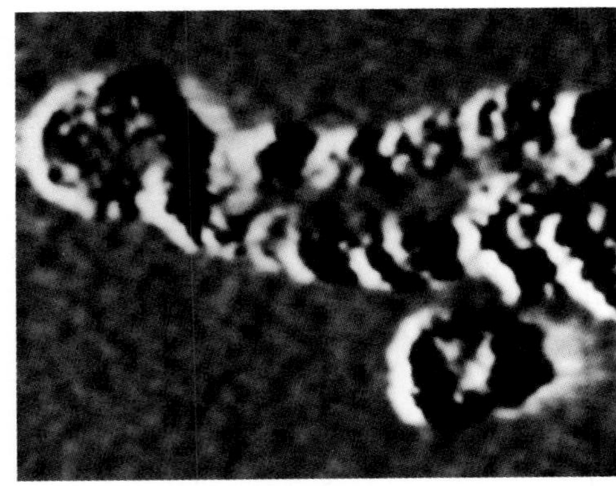

of the organization of the chemicals was constructed. This model was proposed by J. D. Watson and F. H. C. Crick on the basis of data from many investigations. Figure 2-12b shows the following structural characteristics of the DNA molecule.

1. The molecule consists of two strands with crossbars. The strands twist about each other in the form of a ***double helix,*** so that the shape resembles a twisted ladder. For many years, it was assumed that all DNA was in the form of a double helix that twisted smoothly to the right (right-handed DNA). There at least four varieties of right-handed DNA designated as A-DNA, B-DNA, C-DNA, and D-DNA. Watson and Crick's model, the one most commonly found in cells, is B-DNA. It was discovered, however, that DNA can also twist jaggedly to the left (Z-DNA or left-handed DNA). This alternate form of DNA is located near the ends of genes and may help to explain how genes turn on and off and how some cells may become malignant.

2. The uprights of the DNA ladder consist of alternating phosphate groups and the deoxyribose portions of the nucleotides.

3. The rungs of the ladder contain paired nitrogenous bases. As shown, adenine always pairs with thymine, and cytosine always pairs with guanine.

Cells contain hereditary material called ***genes,*** each of which is a segment of a DNA molecule. Our genes determine which traits we inherit, and they control all the activities that take place in our cells throughout a lifetime. When a cell divides, its hereditary information is passed on to the next generation of cells. The passing of information is possible because of DNA's unique structure.

CLINICAL APPLICATION: DNA FINGERPRINTING

A new technique called *DNA fingerprinting* is now used to help identify perpetrators of violent crimes, mainly murder and rape. It is based on the principle that DNA segments contain specific sequences of nitrogenous bases that are repeated several times and that the number of repeat regions are different from one person to another, a situation similar to regular fingerprints. Except for identical twins, the probability that two individuals would have the same DNA pattern is about 1 in 9.34 billion, about twice the current world population. DNA fingerprinting can be performed with minute quantities of DNA as might be found in a single strand of hair. Blood and semen samples can also be used. The FBI is planning to establish a data base of DNA fingerprints of convicted criminals.

RNA, the second principal kind of nucleic acid, differs from DNA in several respects. RNA is single-stranded; DNA is double-stranded. The sugar in the RNA nucleotide is the pentose ribose. And RNA does not contain the nitrogen base thymine. Instead of thymine, RNA has the nitrogen base uracil. At least three different kinds of RNA have been

FIGURE 2-13 **Structure of ATP and ADP. The high-energy bonds are indicated by ~.**

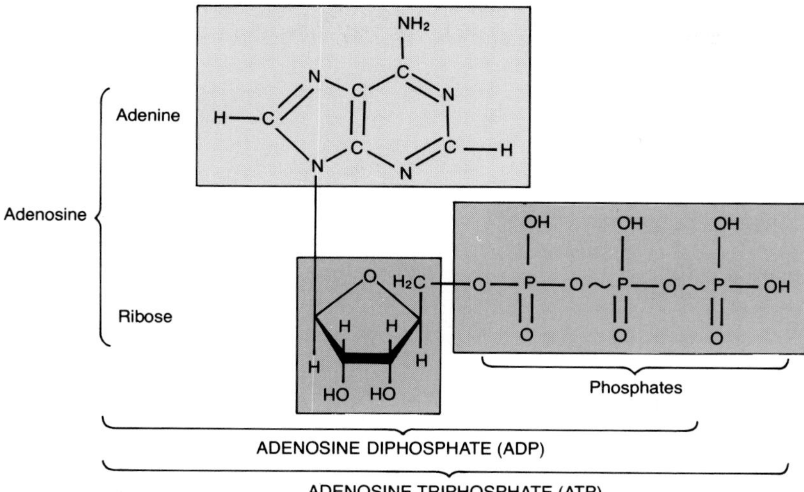

identified in cells. Each type has a specific role to perform with DNA in protein synthesis reactions (Chapter 3).

Adenosine Triphosphate (ATP)

A molecule that is indispensable to the life of the cell is ***adenosine*** (a-DEN-ō-sēn) ***triphosphate* (*ATP*).** This substance is found universally in living systems and performs the essential function of storing energy for various cellular activities. Structurally, ATP consists of three phosphate groups (PO_4^{3-}) and an adenosine unit composed of adenine and the five-carbon sugar ribose (Figure 2-13). ATP is regarded as a high-energy molecule because of the total amount of usable energy it releases when it is broken down by the addition of a water molecule (hydrolysis).

When the terminal phosphate group (here symbolized by *P*) is hydrolyzed, the reaction liberates a great deal of energy. This energy is used by the cell to perform its basic activities. Removal of the terminal phosphate group leaves a molecule called ***adenosine diphosphate* (*ADP*).** This reaction may be represented as follows:

ATP	⇌	ADP	+	P	+	E
Adenosine triphosphate		Adenosine diphosphate		Phosphate		Energy

The energy supplied by the catabolism of ATP into ADP is constantly being used by the cell. Since the supply of ATP at any given time is limited, a mechanism exists to replenish it—a phosphate group is added to ADP to manufacture more ATP. The reaction may be represented as follows:

ADP	+	P	+	E	⇌	ATP
Adenosine diphosphate		Phosphate		Energy		Adenosine triphosphate

Logically, energy is required to manufacture ATP. The energy required to attach a phosphate group to ADP is supplied by various decomposition reactions taking place in the cell, particularly by the decomposition of glucose. Glucose is completely metabolized in cells into carbon dioxide and water, and the energy released in this process is used to attach a phosphate group to ADP to resynthesize ATP. ATP can be stored in every cell, where it provides potential energy that is not released until needed.

Cyclic AMP

A chemical substance closely related to ATP is ***cyclic AMP,*** also known as ***adenosine-3′,5′-monophosphate.*** Essentially it is a molecule of adenosine monophosphate with the phosphate attached to the ribose sugar at two places (Figure 2-14). The attachment

FIGURE 2-14 **Structure of cyclic AMP.**

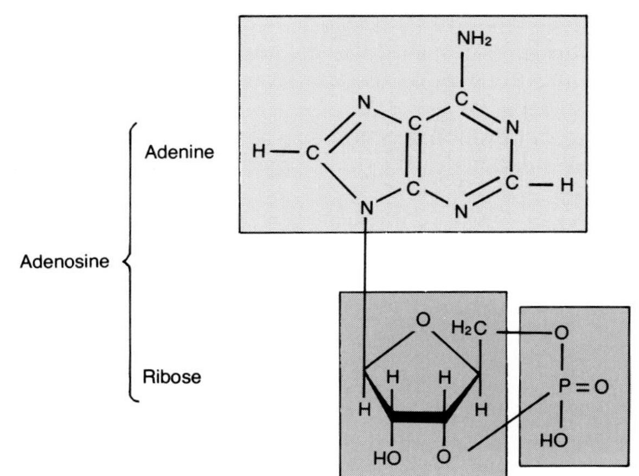

forms a ring-shaped structure and thus the name cyclic AMP.

Cyclic AMP is formed from ATP by the action of a special enzyme, called ***adenylate cyclase,*** located in the cell membrane. Although cyclic AMP was discovered in 1958, only recently has its function in cells become clear. One function is related to the action of hormones, a topic we explore in detail in Chapter 18.

STUDY OUTLINE

Introduction to Basic Chemistry (p. 30)

Chemical Elements (p. 30)

1. Matter is anything that occupies space and has mass. It is made up of building units called chemical elements.
2. Oxygen, carbon, hydrogen, and nitrogen make up 96 percent of body weight. These elements together with calcium and phosphorus make up 99 percent of total body weight.

Structure of Atoms (p. 30)

1. Units of matter of all chemical elements are called atoms.
2. Atoms consist of a nucleus, which contains protons and neutrons (nucleons), and electrons that move about the nucleus in energy levels.
3. The total number of protons of an atom is its atomic number. This number is equal to the number of electrons in the atom.

Atoms and Molecules (p. 31)

1. The electrons are the part of an atom that actively participate in chemical reactions.
2. A molecule is the smallest unit of two or more combined atoms. A molecule containing two or more different kinds of atoms is a compound.
3. In an ionic bond, outer-energy-level electrons are transferred from one atom to another. The transfer forms ions, whose unlike charges attract each other and form ionic bonds.
4. In a covalent bond, there is a sharing of pairs of outer-energy-level electrons.
5. Hydrogen bonding provides temporary bonding between certain atoms within large complex molecules such as proteins and nucleic acids.

Chemical Reactions (p. 35)

1. Synthesis reactions involve the combination of reactants to produce a new molecule. The reactions are anabolic: bonds are formed.
2. In decomposition reactions, a substance breaks down into other substances. The reactions are catabolic: bonds are broken.
3. Exchange reactions involve the replacement of one atom or atoms by another atom or atoms.
4. In reversible reactions, end products can revert to the original combining molecules.
5. When chemical bonds are formed (endergonic reaction), energy is needed. When bonds are broken (exergonic reaction), energy is released. This is known as chemical bond energy.
6. Other forms of energy include mechanical, radiant, and electrical.

Chemical Compounds and Life Processes (p. 36)

1. Inorganic substances usually lack carbon, contain ionic bonds, resist decomposition, and dissolve readily in water.
2. Organic substances always contain carbon and hydrogen. Most organic substances contain covalent bonds and many are insoluble in water.

Inorganic Compounds (p. 37)

1. Water is the most abundant substance in the body. It is an excellent solvent and suspending medium, participates in chemical reactions, absorbs and releases heat slowly, and lubricates.
2. Inorganic acids, bases, and salts dissociate into ions in water. An acid ionizes into H^+ ions; a base ionizes into OH^- ions. A salt ionizes into neither H^+ nor OH^- ions. Cations are positively charged ions; anions are negatively charged ions.
3. The pH of different parts of the body must remain fairly constant for the body to remain healthy. On the pH scale, 7 represents neutrality. Values below 7 indicate acid solutions, and values above 7 indicate alkaline solutions.
4. The pH values of different parts of the body are maintained by buffer systems, which usually consist of a weak acid and a weak base. Buffer systems eliminate excess H^+ ions and excess OH^- ions in order to maintain pH homeostasis.

Organic Compounds (p. 42)

1. Carbohydrates are sugars or starches that provide most of the energy needed for life. They may be monosaccharides, disaccharides, or polysaccharides. Carbohydrates, and other organic molecules, are joined together to form larger molecules with the loss of water by a process called dehydration synthesis. In the reverse process, called digestion (hydrolysis), large molecules are broken down into smaller ones upon the addition of water.
2. Lipids are a diverse group of compounds that includes fats, phospholipids, steroids, carotenes, vitamins E and K, and prostaglandins (PGs). Fats protect, insulate, provide energy, and are stored. Prostaglandins mimic the effects of hormones and are involved in the inflammatory response and the modulation of hormonal responses.
3. Proteins are constructed from amino acids. They give structure to the body, regulate processes, provide protection, help muscles to contract, transport substances, and serve as enzymes. Structural levels of organization among proteins include: primary, secondary, tertiary, and quaternary.
4. Deoxyribonucleic acid (DNA) and ribonucleic acid (RNA) are nucleic acids consisting of nitrogenous bases, sugar, and phosphate groups. DNA is a double helix and is the primary chemical in genes. RNA differs in structure and chemical composition from DNA and is mainly concerned with protein synthesis reactions.

5. The principal energy-storing molecule in the body is adenosine triphosphate (ATP). When its energy is liberated, it is decomposed to adenosine diphosphate (ADP) and P. ATP is manufactured from ADP and P using the energy supplied by various decomposition reactions, particularly of glucose.
6. Cyclic AMP is closely related to ATP and assumes a function in certain hormonal reactions.

REVIEW QUESTIONS

1. What is the relationship of matter and energy to the body? (p. 30, 36)
2. Define a chemical element. List the chemical symbols for 10 different chemical elements. Which chemical elements make up the bulk of the human organism? (p. 30)
3. What is an atom? Diagram the positions of the nucleus, protons, neutrons, and electrons in an atom of oxygen and in an atom of nitrogen. (p. 30)
4. What is an atomic number? Compare it to an atomic weight. (p. 30)
5. What is meant by an energy level? (p. 31)
6. How are chemical bonds formed? Distinguish between an ionic bond and a covalent bond. Give at least one example of each. (p. 32)
7. Can you determine how a molecule of $MgCl_2$ is ionically bonded? Magnesium has two electrons in its outer energy level. Construct a diagram to verify your answer.
8. Refer to Figure 2-4b and c. See if you can determine why there is a double covalent bond between atoms in an oxygen molecule (O_2) and a triple covalent bond between atoms in a nitrogen molecule (N_2). (p. 34)
9. Define a hydrogen bond. Why are hydrogen bonds important? (p. 35)
10. Define an isotope. What are radioisotopes? (p. 31)
11. What are the four principal kinds of chemical reactions? How are anabolism and catabolism related to synthesis and decomposition reactions, respectively? How is energy related to exergonic and endergonic chemical reactions? (p. 35)
12. Identify what kind of reaction each of the following represents:
    a. $H_2 + Cl_2 \rightarrow 2HCl$
    b. $3NaOH + H_3PO_4 \rightarrow Na_3PO_4 + 3H_2O$
    c. $CaCO_3 + CO_2 + H_2O \rightarrow Ca(HCO_3)_2$
    d. $HNO_3 \rightarrow H^+ + NO_3^-$
    e. $NH_3 + H_2O \rightleftharpoons NH_4^+ + OH^-$
13. How do inorganic compounds differ from organic compounds? List and define the principal inorganic and organic compounds that are important to the human body. (p. 36)
14. What are the essential functions of water in the body? Explain the solvating property of water. (p. 37)
15. Define an inorganic acid, a base, and a salt. How does the body acquire some of these substances? List some functions of the chemical elements furnished as ions of salts. (p. 39)
16. What is pH? Why is it important to maintain a relatively constant pH? What is the pH scale? If there are 100 OH^- ions at a pH of 8.5, how many OH^- ions are there at a pH of 9.5? (p. 39)
17. List the normal pH values of some common fluids, biological solutions, and foods. Refer to Exhibit 2-3 and select the two substances whose pH values are closest to neutrality. Is the pH of milk or of cerebrospinal fluid closer to 7? Is the pH of bile or of urine farther from neutrality? (p. 42)
18. What are the components of a buffer system? What is the function of a buffer pair? Diagram and explain how the carbonic acid–bicarbonate buffer system of extracellular fluid maintains a constant pH even in the presence of a strong acid or strong base. How is this an example of homeostasis? (p. 40)
19. Why are the reactions of buffer pairs more important with strong acids and bases than with weak acids and bases? (p. 40)
20. Define a carbohydrate. Why are carbohydrates essential to the body? How are carbohydrates classified? (p. 42)
21. Compare dehydration synthesis and hydrolysis. Why are they significant? (p. 43)
22. How do lipids differ from carbohydrates? What are some relationships of lipids to the body? (p. 44)
23. Define a prostaglandin (PG). List some physiological effects of prostaglandins. (p. 46)
24. Define a protein. What is a peptide bond? Discuss the classification of proteins on the basis of function and levels of structural organization. (p. 46)
25. What is a nucleic acid? How do deoxyribonucleic acid (DNA) and ribonucleic acid (RNA) differ with regard to chemical composition, structure, and function? (p. 47)
26. What is adenosine triphosphate (ATP)? What is the essential function of ATP in the human body? How is this function accomplished? (p. 49)
27. How is cyclic AMP related to ATP? What is the function of cyclic AMP? (p. 49)
28. What is a laser? Describe several clinical applications for lasers. (p. 32)
29. What are the advantages and disadvantages of using aspartame? Describe the potential benefits of olestra as a fat substitute. (p. 43, 46)
30. What is DNA fingerprinting? What is its value? (p. 48)

SELECTED READINGS

Darnell, J. E. "RNA," *Scientific American,* October 1985.

Dickerson, R. E., and I. Geis. *Chemistry, Matter, and the Universe.* Menlo Park, Calif.: Benjamin/Cummings, 1976.

Doolittle, R. F. "Proteins," *Scientific American,* October 1985.

Felsenfeld, G. "DNA," *Scientific American,* October 1985.

Frieden, E. "The Chemical Elements of Life," *Scientific American,* July 1972.

Karplus, M., and J. A. McCammon. "The Dynamics of Proteins," *Scientific American,* April 1986.

Sharon, N. "Carbohydrates," *Scientific American,* November 1980.

Weinberg, R. A. "The Molecular Basis of Life," *Scientific American,* October 1985.

———. "The Molecules of Life," *Scientific American,* October 1985.

Chapter 3

The Cellular Level of Organization

Chapter Contents at a Glance

Student Objectives

1. Define a cell and list its principal parts.
2. Explain the structure of the plasma (cell) membrane and describe how materials move across it.
3. Describe the structure and functions of the following cellular structures: cytoplasm, nucleus, ribosomes, agranular and granular endoplasmic reticulum (ER), Golgi complex, mitochondria, lysosomes, peroxisomes, cytoskeleton, centrioles, flagella, and cilia.
4. Distinguish between a cellular inclusion and extracellular material.
5. Define a gene and explain the sequence of events involved in protein synthesis.
6. Discuss the stages, events, and significance of somatic and reproductive cell division.
7. Describe cancer (CA) as a homeostatic imbalance of cells.
8. Explain the relationship of aging to cells.
9. Define medical terminology associated with cells.

The study of the body at the cellular level of organization is important because activities essential to life occur in cells and disease processes originate there. A ***cell*** may be defined as the basic, living, structural, and functional unit of the body and, in fact, of all organisms. ***Cytology*** (sī-TOL-ō-jē; *cyt* = cell; *logos* = study of) is the branch of science concerned with the study of cells. This chapter concentrates on the structure, functions, and reproduction of cells.

A series of illustrations accompanies each cell structure that you study. A diagram of a generalized animal cell shows the location of the structure within the cell. An electron micrograph shows the actual appearance of the structure.* A diagram of the electron micrograph clarifies some of the small details by exaggerating their outlines. Finally, an enlarged diagram of the structure shows its details.

* An ***electron micrograph (EM)*** is a photograph taken with an electron microscope. Some electron microscopes can magnify objects up to 1 million times. In comparison, the light microscope that you probably use in your laboratory magnifies objects up to 1000 times their size.

GENERALIZED ANIMAL CELL

A ***generalized animal cell*** is a composite of many different cells in the body. Examine the generalized cell illustrated in Figure 3-1, but keep in mind that no such single cell actually exists.

For convenience, we can divide the generalized cell into four principal parts:

1. **Plasma (cell) membrane.** The outer, limiting membrane separating the cell's internal parts from the extracellular materials and external environment.

2. **Cytoplasm.** The substance that surrounds organelles and is located between the nucleus and the plasma membrane.

3. **Organelles.** Permanent structures with characteristic morphology that are highly specialized for specific cellular activities.

4. **Inclusions.** The secretions and storage products of cells.

Extracellular materials, also referred to as the ***matrix,*** which are substances external to the cell surface, will also be examined in connection with cells.

FIGURE 3-1 Generalized animal cell based on electron microscope studies. The diagrams of the structures shown in external view represent their appearance as seen through the plasma membrane. They are not part of the plasma membrane.

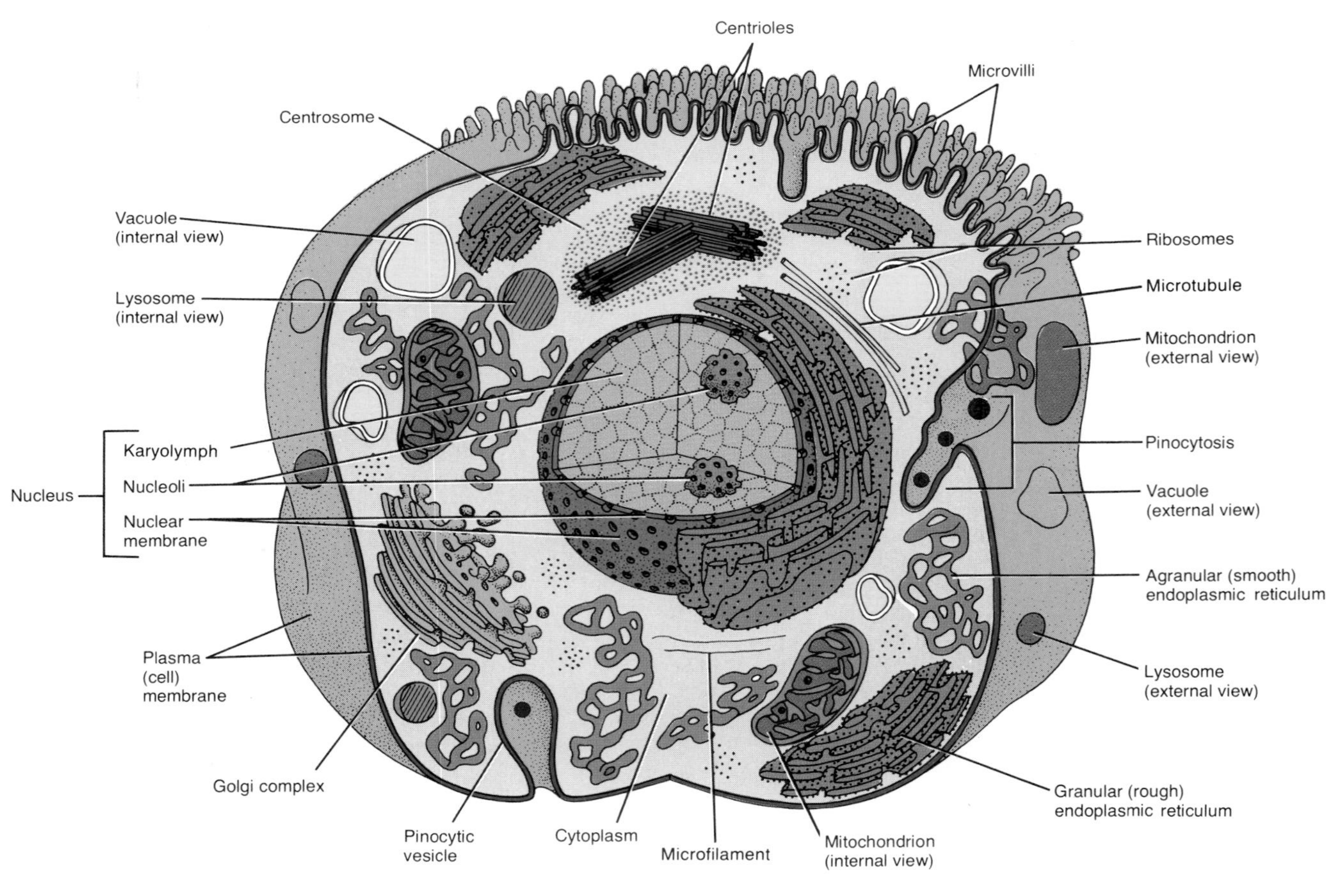

PLASMA (CELL) MEMBRANE

The exceedingly thin structure that separates the internal components of a cell from the external environment is called the ***plasma (cell) membrane*** (Figure 3-2a). The membrane measures from 4.5 nm at the phospholipid bilayer regions up to 10 nm in regions where membrane proteins are present. Formerly, dimensions of cells and parts of cells were given in angstroms (Å). 1 Å = 0.1 nm.

Chemistry and Structure

Plasma membranes consist primarily of phospholipids (lipids that contain phosphorous), the most abundant chemicals, and proteins. Other chemicals in lesser amounts include cholesterol (a lipid), glycolipids (combinations of carbohydrates and lipids), and carbohydrates called oligosaccharides (Figure 3-2b).

The phospholipid molecules are arranged in two parallel rows, forming a ***phospholipid bilayer.*** A phospholipid molecule consists of a polar, phosphate-containing "head" that mixes with water (hydrophilic) and nonpolar fatty acid "tails" that do not mix with water (hydrophobic). The molecules are oriented in the bilayer so that "heads" face outward on either side, and the "tails" face each other in the membrane's interior. The phospholipid bilayer is dynamic, since the phospholipid molecules can move sideways and exchange places in their own row; movement of phospholipid molecules between rows rarely occurs. The bilayer is also self-sealing; if a

FIGURE 3-2 Plasma membrane. (a) Electron micrograph of portions of two plasma membranes separated by an intercellular space at a magnification of 200,000×. (Copyright © Dr. Donald Fawcett, Science Source/Photo Researchers.) (b) Diagrammatic enlargement of the plasma membrane showing the relationship of the phospholipid bilayer and protein molecules. The separation of the bilayer is for illustrative purposes only.

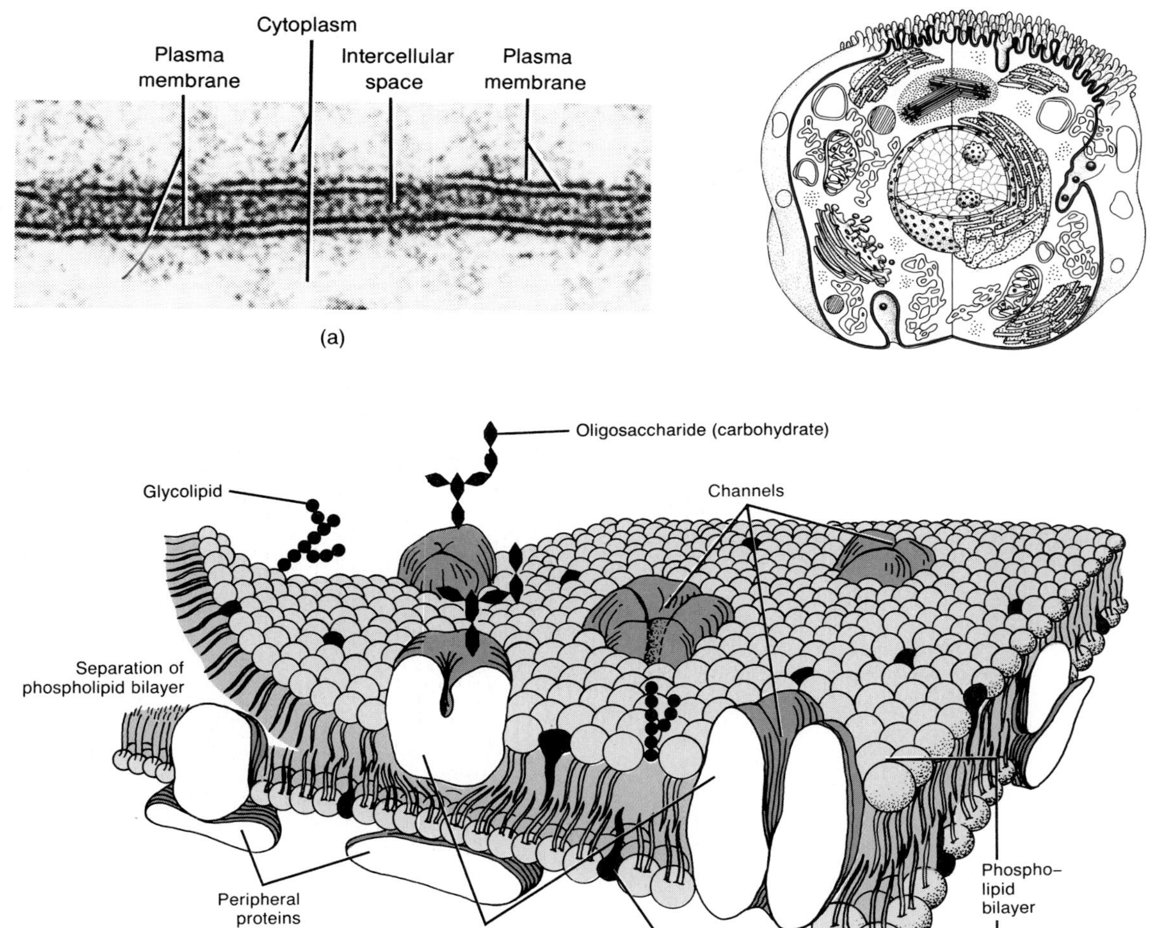

needle is pushed through it and pulled out, the puncture site will seal automatically. The phospholipid bilayer forms the basic framework of plasma membranes.

The plasma membrane proteins (PMPs) are classified into two categories: integral and peripheral. ***Integral proteins*** are embedded in the phospholipid bilayer among the fatty acid "tails." Some of the integral proteins lie at or near the inner and outer membrane surfaces; others penetrate the membrane completely. Since the phospholipid bilayer is somewhat fluid and flexible and the integral proteins have been observed moving from one location to another in the membrane, the relationship has been compared to icebergs (proteins) floating in the sea (phospholipid bilayer).

The subunits of some integral proteins form minute channels through which substances can be transported into and out of the cell (described shortly). Other integral proteins are bound to branching chains of oligosaccharides (carbohydrates). The oligosaccharides and integral proteins together provide receptor sites that enable a cell to recognize other cells of its own kind so that they can associate to form a tissue, to recognize and respond to foreign cells that might be potentially dangerous, and to recognize and attach to hormones, nutrients, and other chemicals. Red blood cells also have receptors that prevent them from clumping and producing unwanted clumps. Type II diabetes, one type of diabetes, is a disease believed to be caused by faulty cell receptor sites.

Peripheral proteins are loosely bound to the membrane surface and easily separated from it. Far less is known about them than integral proteins, and their functions are not yet completely understood. For example, some peripheral proteins are believed to serve as enzymes that catalyze cellular reactions. An example is cytochrome *c*, involved in cellular respiration. This enzyme is located in the inner membrane of a structure called a mitochondrion (discussed later). Other peripheral proteins, such as spectrin of red blood cells, are believed to have a mechanical function by serving as a scaffolding to support the plasma membrane. It is also believed that peripheral proteins may assume a role in changes in membrane shape during such processes as cell division, locomotion, and ingestion.

The currently accepted model of plasma membrane structure is known as the ***fluid mosaic model.*** A mosaic is a pattern of many small pieces fitted together. According to this concept, the membrane is a mosaic of proteins which move laterally in the phospholipid bilayer.

The presence of cholesterol molecules makes the membrane less flexible and less permeable. Glycolipids mediate cell-to-cell recognition and communication, participate in cellular growth and development, and may be infection sites for several kinds of viruses and bacteria.

Physiology

Based upon the discussion of the chemistry and structure of the plasma membrane, we can now describe its several important functions. First, the plasma membrane provides a flexible boundary that encloses the cellular contents and separates them from the extracellular fluid, and, in some cases, the external environment as well. Second, the membrane facilitates contact with other body cells or with foreign cells or substances. Third, the membrane provides receptors for chemicals such as hormones, neurotransmitters, enzymes, nutrients, and antibodies. Fourth, the plasma membrane mediates the entrance and exit of materials. The ability of a plasma membrane to permit certain substances to enter and exit, but to restrict the passage of others, is called ***selective permeability.*** Let us look at this mechanism in more detail.

A membrane is said to be *permeable* to a substance if it allows the free passage of that substance. Although plasma membranes are not actually freely permeable to any substance, they do permit some substances to pass more readily than others. For example, water passes more readily than most other substances. The permeability of a plasma membrane appears to be a function of several factors.

1. **Size of molecules.** Most large molecules cannot pass through the plasma membrane. Water and amino acids are small molecules and can enter and exit the cell easily. However, most proteins, which consist of many amino acids linked together, seem to be too large to pass through the membrane. Many scientists believe that the giant-sized molecules do not enter the cell because they are larger than the minute channels in the integral proteins and cannot dissolve in the phospholipid bilayer.
2. **Solubility in lipids.** Substances that dissolve easily in lipids pass through the membrane more readily than other substances, since a major part of the plasma membrane consists of phospholipid molecules. Examples of lipid-soluble substances are oxygen, carbon dioxide, and steroid hormones.
3. **Charge on ions.** The charge of an ion attempting to cross the plasma membrane can determine how easily the ion enters or leaves the cell. The protein portion of the membrane is capable of ionization. If an ion has a charge opposite that of the membrane, it is attracted to the membrane and passes through more readily. If the ion attempting to cross the membrane has the same charge as the membrane, it is repelled by the membrane and its passage is restricted. Opposite charges attract, whereas like charges repel each other.
4. **Presence of carrier molecules.** Some integral proteins called carriers are capable of attracting and transporting substances across the membrane regardless of size, ability to dissolve in lipids, or membrane charge. Their purpose is to modify membrane permeability. The mechanism by which carriers do this will be described shortly.

Movement of Materials Across Plasma Membranes

The mechanisms whereby substances move across a plasma membrane, or intracellular membrane, are essential to the life of the cell. Certain substances, for example, must move into the cell to support life, whereas waste materials or harmful substances must be moved out. Plasma membranes mediate the movements of such materials. The processes involved in these movements may be classified as either passive or active. In ***passive (physical) processes,*** substances move across plasma membranes without an expenditure of energy (from the splitting of ATP) by cells. Their movement is dependent on the kinetic energy (energy of motion) of individual molecules. The substances move on their own down a concentration gradient, that is, from an area where their concentration is high to an area where their concentration is low. The substances may also be forced across the plasma membrane by pressure from an area where the pressure is high to an area where it is low. In ***active (physiological) processes,*** the cell uses energy (from the splitting of ATP) in moving the substance across the membrane since the substance moves against a concentration gradient, that is, from an area where its concentration is low to an area where its concentration is high.

Passive Processes

▪ **Diffusion** A passive process called ***diffusion*** (*diffus* = spreading) occurs when there is a *net* (greater) movement of molecules or ions from a region of their higher concentration to a region of their lower concentration, that is, when the molecules move from an area where there are more of them to an area where there are fewer of them. The movement from high to low concentration continues until the molecules are evenly distributed. At this point, they move in both directions at an *equal* rate; there is no net diffusion. This point of even distribution is called equilibrium. The difference between high and low concentrations is called the ***concentration gradient.*** Molecules moving from the high-concentration area to the low-concentration area are said to move *down* or *with* the concentration gradient.

If a dye pellet is placed in a beaker filled with water, the color of the dye is seen immediately around the pellet. At increasing distances from the pellet, the color becomes lighter (Figure 3-3). Later, however, the water solution will be a uniform color. (Recall from Chapter 2 that a solution consists of a liquid or gas called a solvent in which another material, called the solute, is dissolved. In this case, water is the solvent and the dye pellet is the solute.) The dye molecules possess kinetic energy and move about at random. The dye molecules move down the concentration gradient from an area of high dye concentration to an area of low dye concentration. The water molecules also move from their high-concentration to their low-concentration area. When dye molecules and water molecules are evenly distributed among themselves, equilibrium is reached and diffusion ceases, even though molecular movements continue. As another example of diffusion, consider what would happen if you opened a bottle of perfume in a room. The perfume molecules would diffuse until an equilibrium is reached between the perfume molecules and the air molecules in the room.

In the examples cited, no membranes were involved. Diffusion may occur, however, through selectively permeable membranes in the body. Large and small lipid-soluble molecules pass through the phospholipid bilayer of the membrane. A good example of diffusion in the body is the movement of oxygen from the blood into the cells and the movement of carbon dioxide from the cells back into the blood. This movement is essential in order for body cells to maintain homeostasis. It ensures that cells receive adequate amounts of oxygen and eliminate carbon dioxide as part of their normal metabolism. Small molecules that are not lipid-soluble, such as certain ions (sodium, potassium, chloride), are able to diffuse through channels formed by integral proteins in the membrane.

FIGURE 3-3 **Principle of diffusion. Molecules of dye (solute) in a beaker of water (solvent) move down the concentration gradient from a region of high solute concentration to a region of low solute concentration. At the same time, solvent molecules are moving down their concentration gradient.**

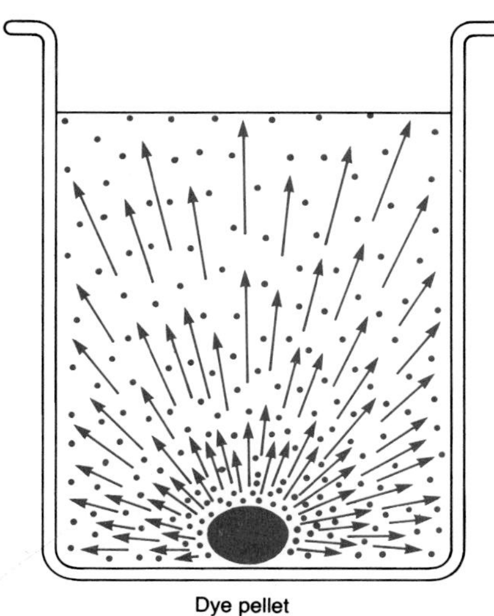

▪ **Facilitated Diffusion** Another type of diffusion through a selectively permeable membrane occurs by a process called ***facilitated diffusion.*** This process is accomplished with the assistance of integral proteins in the membrane that serve as carriers. Although some chemical substances are large molecules and insoluble in lipids, they can still pass through the plasma membrane. Among these are various sugars, especially glucose. In facilitated diffusion, it is believed that glucose is picked up by a carrier. The combined glucose–carrier is soluble in the phospholipid bilayer of the membrane and diffuses to the inside of the membrane where it releases glucose to the cytoplasm. The carrier makes the glucose soluble in the phospholipid bilayer of the membrane so that it can pass through the membrane. By itself, glucose is insoluble and cannot penetrate the membrane. In facilitated diffusion, the cell does not expend energy from the splitting of ATP, and the substance moves from a region of its higher concentration to a region of its lower concentration.

The rate of facilitated diffusion is considerably faster than that of simple diffusion and depends on (1) the difference in concentration of the substance on the two sides of the membrane, (2) the amount of carrier available to transport the substance, and (3) how quickly the carrier and substance combine. The process for glucose is greatly accelerated by insulin, a hormone produced by the pancreas. One of insulin's functions is to lower the blood glucose level by accelerating the transportation of glucose from the blood into body cells. This transportation, as we have just seen, is by facilitated diffusion.

▪ **Osmosis** Another passive process by which materials move across membranes is ***osmosis.*** It is the net movement of water molecules through a selectively permeable membrane from an area of higher water concentration to an area of lower water concentration. The water molecules pass through channels in integral proteins in the membrane. Once again, a simple apparatus may be used to demonstrate the process. The apparatus shown in Figure 3-4 consists of a sac constructed from cellophane, a selectively permeable membrane. The cellophane sac is filled with a colored, 20 percent sugar (sucrose) solution. Such a solution consists of 20 parts sugar and 80 parts water. The upper portion of the cellophane sac is plugged with a rubber stopper through which a glass tube is fitted. The cellophane sac is placed into a beaker containing distilled (pure) water. Pure water consists of 0 parts sugar and 100 parts water. Initially, the concentrations of water on either side of the selectively permeable membrane are different. There is a lower concentration of water inside the cellophane sac than outside. Because of this difference, water moves from the beaker into the cellophane sac. The passage of water through a selectively permeable membrane generates a pressure called osmotic pressure. ***Osmotic pressure*** is the pressure required to prevent the movement of pure water (no solutes) into a solution containing some solutes when the solutions are separated by a selectively permeable membrane. In other words, osmotic pressure is the pressure needed to stop the flow of water across the membrane. The greater the solute concentration of the solution, the greater its osmotic pressure. There is no movement of sugar from the cellophane sac inside the beaker, since the cellophane is impermeable to molecules of sugar—sugar molecules are too large to go through the pores of the membrane. As water moves into the cellophane sac, the sugar solution becomes increasingly diluted and the increased volume and pressure force the mixture up the glass tubing. In time, the weight of the water that has accumulated in the cellophane sac and the glass tube will be equal to the osmotic pressure, and this keeps the water

FIGURE 3-4 Principle of osmosis. (a) Apparatus at the start of the experiment. (b) Apparatus at equilibrium. In (a), the cellophane tube (selectively permeable membrane) contains a 20 percent sugar (sucrose) solution and is immersed in a beaker of distilled water. The 20 percent sugar solution contains 20 parts sugar and 80 parts water, whereas the distilled water contains 0 parts sugar and 100 parts water. The arrows indicate that water molecules can pass freely into the tube but that sugar (sucrose) molecules are held back by the selectively permeable membrane. As water moves into the tube by osmosis, the sugar solution is diluted and the volume of the solution in the cellophane tube increases. This increased volume is shown in (b), with the sugar solution moving up the glass tubing. The final height reached (FH) occurs at equilibrium and represents the osmotic pressure. At this point, the number of water molecules leaving the cellophane tube is equal to the number of water molecules entering the tube.

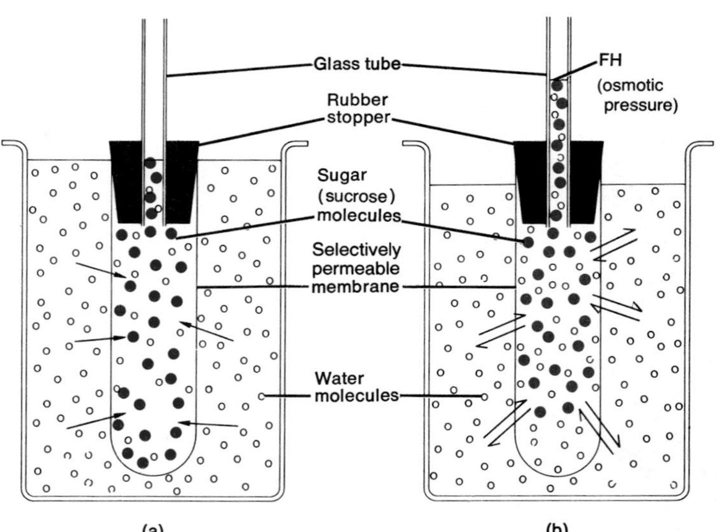

volume from increasing in the tube. When water molecules leave and enter the cellophane sac at the same rate, equilibrium is reached. Osmotic pressure is an important force in the movement of water between various compartments of the body.

▪ **Isotonic, Hypotonic, and Hypertonic Solutions** Osmosis may also be understood by considering the effects of different water concentrations on red blood cells. If the normal shape of a red blood cell is to be maintained, the cell must be placed in an ***isotonic solution.*** This is a solution in which the total concentrations of water molecules (solvent) and solute (solid) molecules are the same on both sides of the selectively permeable cell membrane. The concentrations of water and solute in the fluid outside the red blood cell must be the same as the concentration of the fluid inside the cell. Under ordinary circumstances, a 0.90 percent NaCl (salt) solution is isotonic for red blood cells. In this condition, water molecules enter and exit the cell at the same rate, allowing the cell to maintain its normal shape since the cell will neither increase nor decrease in volume and pressure.

A different situation results if red blood cells are placed in a solution that has a lower concentration of solutes and therefore a higher concentration of water. This is called a ***hypotonic solution.*** In this condition, water molecules enter the cells faster than they can leave, causing the red blood cells to swell and eventually burst. The rupture of red blood cells in this manner is called ***hemolysis*** (hē-MOL-i-sis) or ***laking.*** Distilled water is a strongly hypotonic solution.

A ***hypertonic solution*** has a higher concentration of solutes and a lower concentration of water than the red blood cells. One example of a hypertonic solution is a 10 percent NaCl solution. In such a solution, water molecules move out of the cells faster than they can enter. This situation causes the cells to shrink. The shrinkage of red blood cells in this manner is called ***crenation*** (kri-NĀ-shun). Red blood cells, as well as other body cells, may be greatly impaired or destroyed if placed in solutions that deviate significantly from the isotonic state.

▪ **Bulk Flow** The movement of *large* numbers of ions, molecules, or particles in the same direction as a result of forces that push them is referred to as ***bulk flow.*** The substances move in unison in response to forces such as osmotic or hydrostatic (water) pressure at rates far greater than can be accounted for by diffusion or osmosis alone. Examples of bulk flow in the body are movements of substances through blood capillary membranes, blood flow within vessels, and movement of air into and out of the lungs.

▪ **Filtration** Another passive process involved in moving materials in and out of cells is ***filtration.*** This process involves the movements of solvents such as water and dissolved substances such as sugar across a selectively permeable membrane by gravity or mechanical pressure, usually hydrostatic (water) pressure. Such a movement is always from an area of high pressure to an area of lower pressure and continues as long as a pressure difference exists. Most small- to medium-sized molecules can be forced through a cell membrane, while large molecules or aggregates cannot.

An example of filtration occurs in the kidneys, where the blood pressure supplied by the heart forces water and small molecules like urea through thin cell membranes of tiny blood vessels and into the kidney tubules. In this basic process, protein molecules are retained in the blood since they are too large to be forced through the cell membranes of the blood vessels. The molecules of harmful substances such as urea are small enough to be forced through and eliminated in the urine, however.

▪ **Dialysis** The final passive process to be considered is ***dialysis.*** Dialysis is the separation of small molecules from large molecules by diffusion across a selectively permeable membrane. Suppose a solution containing molecules of various sizes is placed in a sac that is permeable only to the smaller molecules. The sac is then placed in a beaker of distilled water. Eventually, half the smaller molecules will move from the sac into the water in the beaker and the larger molecules will be left behind. The fluid surrounding the sac has to be replaced on a regular and frequent basis in order to have most of the smaller molecules eventually removed from the sac.

The principle of dialysis is employed in artificial kidney machines. The patient's blood is exposed to a dialysis membrane outside the body. The dialysis membrane takes the place of the kidneys. As components of the blood move across the membrane, small-particle waste products pass from the blood into a solution surrounding the dialysis membrane. At the same time, certain nutrients can be passed from the solution into the blood. This solution is continuously replaced in order to maintain a high concentration gradient between the contents and blood. The blood is then returned to the body.

Active Processes

When cells actively participate in moving substances across membranes, they must expend energy by splitting of ATP. Cells can even move substances against a concentration gradient. The active processes considered here are active transport and endocytosis (phagocytosis, pinocytosis, and receptor-mediated endocytosis).

▪ **Active Transport** The process by which substances are transported across plasma membranes typically from an area of their low concentration to an area of their high concentration is called ***active transport.*** In order to transport a substance against a concentration gradient, the membrane uses energy supplied by the splitting of ATP. In fact, a typical body cell probably expends up to 40 percent of its ATP for active transport. Although the molecular events involved in active transport are not completely understood, integral proteins in the plasma membrane do assume a role by functioning as carrier molecules that attach to a substance before it can undergo active transport. Enzymes speed up the splitting of ATP which releases the energy needed for this process.

As just noted, glucose can be transported across cell membranes via facilitated diffusion from areas of high concentration to areas of low concentration. Glucose can also be moved by the cells lining the gastrointestinal tract from the cavity of the tract into the blood, even though blood concentration of glucose is higher. This movement involves active transport. One proposed mechanism is that glucose (or another substance) enters a channel in an integral membrane protein (Figure 3-5). When the glucose molecule makes contact with an active site in the channel, the energy from the splitting of ATP induces a change in the membrane protein that expels the glucose on the opposite side of the membrane. As you will see later, kidney cells also have the ability to actively transport glucose back into the blood so that it is not lost in the urine.

Active transport is also an important process in maintaining higher concentrations of some ions inside body cells and higher concentrations of other ions outside body cells. This mechanism is discussed in Chapter 12 in relation to nerve impulse conduction.

▪ **Endocytosis** Large molecules and particles pass through plasma membranes by a process called ***endocytosis,*** in which a segment of the plasma membrane surrounds the substance, encloses it, and brings it into the cell. The export of substances from the cell occurs by a reverse process called ***exocytosis,*** a very important mechanism for secretory cells. There are three basic kinds of endocytosis: phagocytosis, pinocytosis, and receptor-mediated endocytosis.

In ***phagocytosis*** (fag'-ō-sī-TŌ-sis) or "cell eating," projections of cytoplasm, called ***pseudopodia*** (soo'-dō-PŌ-dē-a), engulf large solid particles external to the cell (Figure 3-6a,b). Once the particle is surrounded, the membrane folds inwardly, forming a membrane sac around the particle. This newly formed sac, called a ***phagocytic vesicle,*** breaks off from the plasma membrane, and the solid material inside the vesicle is digested by enzymes provided by lysosomes. Indigestible particles and cell products are removed from the cell by exocytosis. Phagocytosis (and pinocytosis) are important because molecules and particles of material that would normally be restricted from crossing the plasma membrane because of their large size can be brought into or removed from the cell. Through phagocytosis, the white blood cells engulf and destroy bacteria and other foreign substances. The phagocytic white blood cells of the body constitute a vital defense mechanism that helps protect us against disease.

In ***pinocytosis*** (pi'-nō-sī-TŌ-sis), or "cell drinking," the engulfed material consists of an extracellular liquid rather than a solid (Figure 3-6c). Moreover, no cytoplasmic projections are formed. Instead, a minute droplet of liquid is attracted to the surface of the membrane. The membrane folds inwardly, forms a ***pinocytic vesicle*** that surrounds the liquid, and detaches from the rest of the intact membrane. Few cells, such as endothelial (lining) cells in the wall of blood vessels, especially capillaries, are capable of phagocytosis, but many cells carry on pinocytosis.

Receptor-mediated endocytosis is a highly selective process in which cells can take up large molecules or particles. The interstitial fluid that bathes cells contains a large number of chemicals, most of which are in concentrations lower than in the cells themselves. These chemicals, called ***ligands,*** serve a variety of functions, all of which are related to homeostasis. For example, some of the substances are nutrients—like amino acids, iron, and vitamins—needed for various chemical reactions that sustain life. Some are hormones that deliver messages to cells so that cells can carry on specific physiological responses. Other substances are waste products or poisonous materials that certain cells have the ability to break down so that they do not interfere with normal functioning of cells.

Receptor-mediated endocytosis is accomplished as follows. The plasma membrane contains protein receptors that have binding sites for ligands (Figure 3-6d). The binding between receptor and ligand causes the plasma membrane to fold inward, forming a ***vesicle*** around the ligand. As a vesicle moves inward from the plasma membrane, it fuses with another vesicle to form a larger structure called an ***endosome.*** Each endosome develops into an even larger structure called a ***CURL (compartment of uncoupling of receptor and ligand).*** Within the CURL, the ligands separate from the receptors. The ligands move into one part of the CURL (vesicular portion), and

FIGURE 3-5 **Proposed mechanism of active transport. The number ① indicates an earlier stage of the process; the number ② indicates a later stage.**

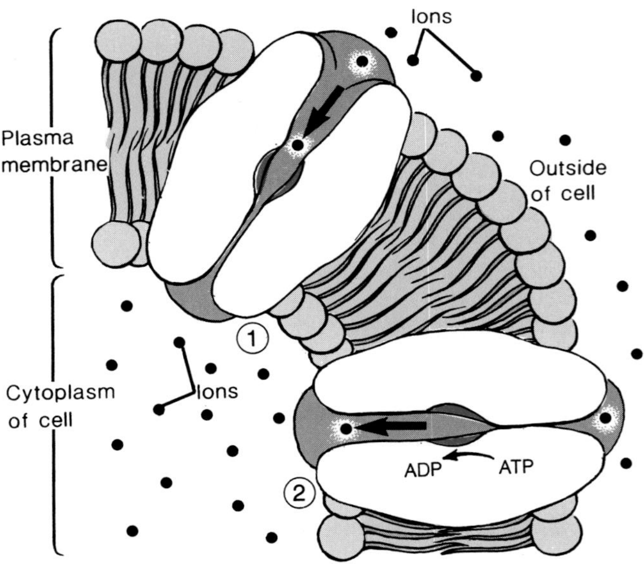

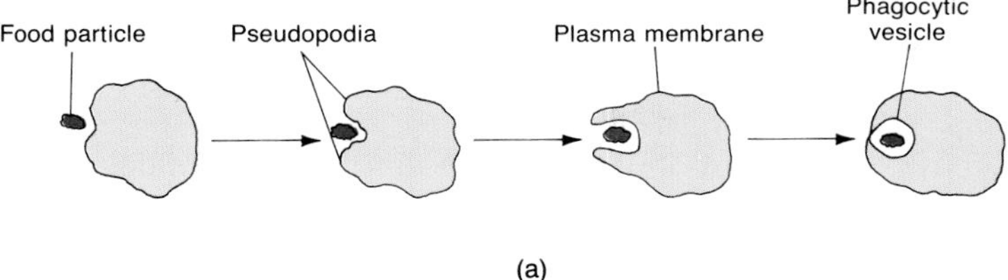

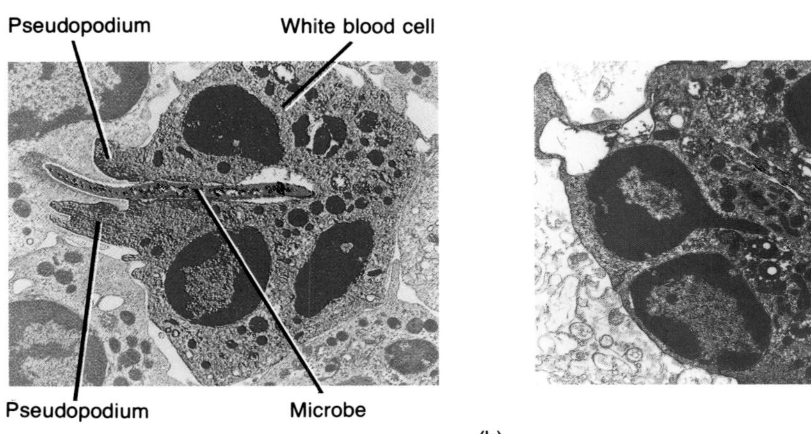

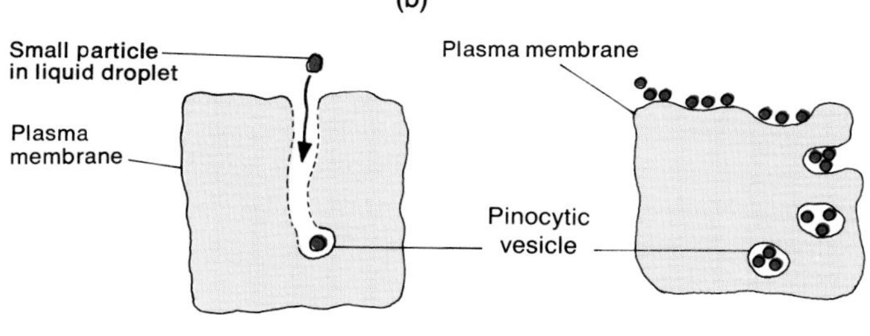

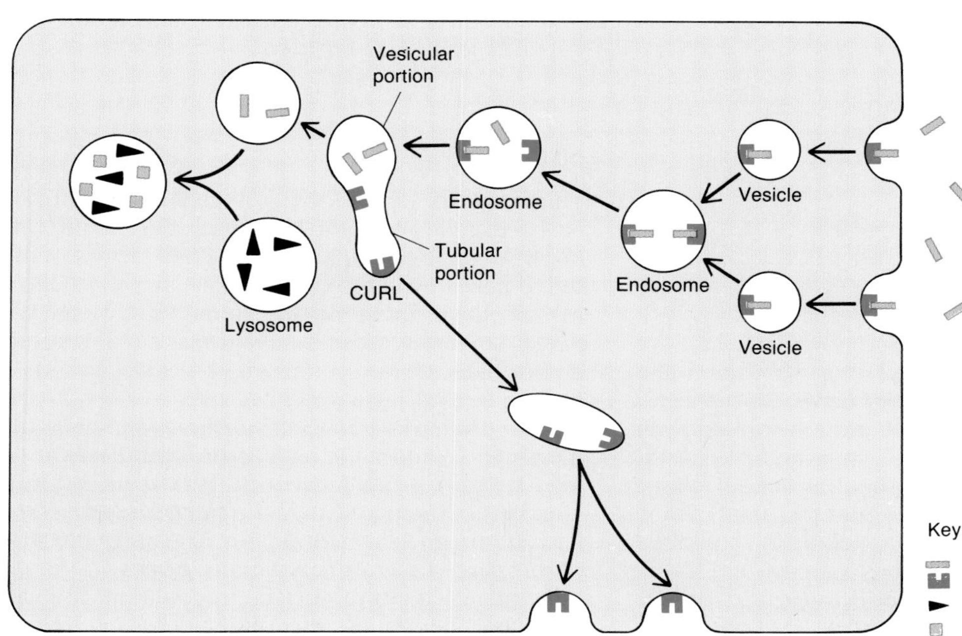

FIGURE 3-6 **Edocytosis.** (a) Diagram of phagocytosis. (b) Color-enhanced photomicrographs of phagocytosis. The photomicrograph on the left shows a human white blood cell (neutrophil) engulfing a microbe, and the photomicrograph on the right shows a later stage of phagocytosis in which the engulfed microbe is being destroyed. (Courtesy of Abbott Laboratories.) (c) Two variations of pinocytosis. In the variation on the left, the ingested substance enters a channel formed by the plasma membrane and becomes enclosed in a pinocytic vesicle at the base of the channel. In the variation on the right, the ingested substance becomes enclosed in a pinocytic vesicle that forms at the surface of the cell and detaches. (d) Receptor-mediated endocytosis.

the receptors accumulate in another portion of the CURL (tubular portion). Ligands have different destinations in the cell. However, in most instances following this separation, the vesicular portion fuses with a lysosome, where the ligands are broken down by powerful digestive enzymes. The tubular portion recycles receptors to the plasma membrane for reuse.

CLINICAL APPLICATION: LIPOSOMES AND DRUG THERAPY

Despite the impact of antibiotics and other drugs in the treatment of disease, they generally have disadvantages in that they are cleared from the blood rapidly and may be toxic to healthy cells as well as microorganisms and diseased cells. In an attempt to overcome these problems, scientists are experimenting with phospholipid sacs called ***liposomes*** (LĪP-ō-sōms). When mixed with a drug, the phospholipids enclose it and form a sac. Liposomes containing drugs can be injected, sprayed in aerosols, or applied to the skin. It is believed that liposomes release their drugs slowly to maintain a more even blood level and that the drugs are taken into cells by absorption or endocytosis.

Liposomes are also being tested to treat cancer by activating body cells called ***macrophages.*** These cells are scavenger cells that normally hunt down, ingest, and destroy foreign particles such as viruses and that, if properly activated, can attack tumor cells. One problem in the past with activating macrophages is that the activating chemicals also cause toxicity to other cells. In addition, the chemicals are degraded too quickly. By inserting the activating substances in liposomes, the sacs containing chemicals are ingested by macrophages, so the activators are released *within* the macrophages without affecting other cells and the rate of release is more controlled.

The various passive and active processes by which substances move across plasma membranes are summarized in Exhibit 3-1.

CYTOPLASM

Structure

The substance inside the cell's plasma membrane and external to the nucleus is called ***cytoplasm*** (SĪ-tō-plazm') (Figure 3-7a,b). It is the matrix, or ground substance, in which various cellular components are found. Physically, cytoplasm is described as a thick, semitransparent, elastic fluid containing suspended particles and a series of minute tubules and filaments that form a cytoskeleton (described shortly). The cytoskeleton provides support and shape and is involved in the movement of structures in the cytoplasm and even the entire cell, as occurs in phagocytosis. Chemically, cytoplasm is 75 to 90 percent water plus solid components. Proteins, carbohydrates, lipids, and inorganic substances compose the bulk of the solid components. The inorganic substances and most carbohydrates, amino acids, and peptides are soluble in water and are present as a solution. The majority of organic compounds, however, are found as ***colloids***—particles that remain suspended in the surrounding medium. Since the particles of a colloid bear electrical charges that repel each other, they remain suspended and separated from each other.

Physiology

Functionally, cytoplasm is the substance in which some chemical reactions occur. The cytoplasm receives raw materials from the external environment by way of extracellular fluid and converts them into usable energy by decomposition reactions. Cytoplasm is also the site where new substances are synthesized for cellular use. Also, chemicals are packaged for transport to other parts of the cell or other cells of the body and facilitate the excretion of waste materials.

ORGANELLES: STRUCTURE AND PHYSIOLOGY

Despite the numerous chemical activities occurring simultaneously in the cell, there is little interference of one reaction with another. This is because the cell has a system of compartmentalization provided by structures called ***organelles.*** These structures are specialized portions of the cell with characteristic shapes that assume specific roles in growth, maintenance, repair, and control. The number and types of organelles vary among different cells, depending on their functions.

Nucleus

The ***nucleus*** (NOO-klē-us) is generally a spherical or oval organelle and is the largest structure in the cell (Figure 3-7a–c). It contains the hereditary factors of the cell, called genes, which control cellular structure and direct many cellular activities. Most body cells contain a single nucleus, although some, such as mature red blood cells, do not. Skeletal muscle fibers (cells) and a few other cells contain several nuclei.

The nucleus is separated from the cytoplasm by a double membrane called the ***nuclear membrane (envelope)*** (Figure 3-7c). Between the two layers of the nuclear membrane is a space called the ***perinuclear cisterna.*** Each of the nuclear membranes resembles the structure of the plasma membrane. Minute ***pores*** in the nuclear membrane allow the nucleus to communicate with the cytoplasm. Substances entering and leaving the nucleus are believed to pass through the tiny pores. At many points the two nuclear membranes fuse with each other, and at the points of fusion, the membranes are permeable to practically all dissolved or suspended substances, including newly synthesized ribosomes.

Several structures are internal to the nuclear membrane. The first is a gellike fluid that fills the nucleus called ***karyolymph*** (*karyo* = nucleus) or ***nucleoplasm.*** One or more spherical bodies called ***nucleoli*** may also be present. These structures do not contain a membrane and are composed of protein, DNA, and RNA. Nucleoli disperse and disappear during cell division and re-form once new cells are formed. Nucleoli are the sites of the synthesis of a type of RNA called ribosomal RNA. The RNA is stored in nucleoli and, as you will see shortly, assumes a function in protein synthesis. Finally, there is the ***genetic material*** consisting principally of DNA. When the cell is not reproducing, the genetic material appears as a threadlike mass called ***chromatin.*** Prior to cellular reproduction the chromatin shortens and coils into rod-shaped bodies called ***chromosomes.*** The DNA is combined with protein.

Chromosomes contain a very large amount of DNA relative to their size. The DNA fits into chromosomes because of a very complex pattern of packing. In the first step of DNA packing, DNA is joined to proteins. The basic structural unit of a chromosome is called a ***nucleosome*** (Figure 3-7e). A nucleosome consists of DNA associated with one of five types of proteins called ***histones.*** Specifically, a nucleosome is composed of two molecules of each of four types of histone (called an ***octamer***) and a fixed length of DNA (about 146 nitrogenous base pairs). A fifth type of histone is associated with a length of ***linker DNA*** that holds adjacent nucleosomes together. The whole arrangement of nucleosomes and linker DNA resembles beads on a string. The functional significance of nucleosome structure is still speculative. It is believed that histones may facilitate changes in chromosomal structure that expose activated genes (DNA) to perform a specific task in the cell. The next level of DNA packing is called a ***chromatin fiber.*** It consists of regular, repeating units of six nucleosomes held together by a histone. Each chromatin fiber, under the right conditions, forms a coil called a ***solenoid,*** each of which consists of about six nucleosomes. In the next higher level of DNA packing, solenoids coil to form a hollow tube. Such structures, consisting of 20,000 to 100,000 nitrogenous base pairs, are called ***loop domains.*** In cells that are not dividing, this is the extent of DNA packing. As a cell enters its divisional cycle, the loop domains coil even more to form ***chromatids.*** Chromatids are organized into ***chromosomes.***

EXHIBIT 3-1 SUMMARY OF PROCESSES BY WHICH SUBSTANCES MOVE ACROSS MEMBRANES

Process	Description
PASSIVE PROCESSES	Substances move on their own down a concentration gradient from an area of higher to lower concentration or pressure; cell does not expend energy.
Diffusion	Net movement of molecules or ions due to their kinetic energy from an area of higher to lower concentration until an equilibrium is reached.
Facilitated diffusion	Diffusion of larger molecules across a selectively permeable membrane with the assistance of integral proteins in the membrane that serve as carriers.
Osmosis	Net movement of water molecules due to kinetic energy across a selectively permeable membrane from an area of higher to lower concentration of water until an equilibrium is reached.
Bulk flow	Movement of large numbers of ions, molecules, or particles in the same direction as a result of forces that push them.
Filtration	Movement of solvents (such as water) and solutes (such as glucose) across a selectively permeable membrane as a result of gravity or hydrostatic (water) pressure from an area of higher to lower pressure.
Dialysis	Diffusion of solute particles across a selectively permeable membrane in which small molecules are separated from large ones.
ACTIVE PROCESSES	Substances move against a concentration gradient from an area of lower to higher concentration; cell must expend energy released by the splitting of ATP.
Active transport	Movement of substances, usually ions, across a selectively permeable membrane from a region of lower to higher concentration by an interaction with integral proteins in the membrane; the process requires energy expenditure from the splitting of ATP.
Endocytosis	Movement of large molecules and particles through plasma membranes in which the membrane surrounds the substance, encloses it, and brings it into the cell. Examples include phagocytosis ("cell eating"), pinocytosis ("cell drinking"), and receptor-mediated endocytosis.
Exocytosis	Export of substances from the cell by reverse endocytosis.

FIGURE 3-7 Cytoplasm and nucleus. (a) Electron micrograph of cytoplasm and the nucleus at a magnification of 31,600×. (Copyright © Dr. Myron C. Ledbetter, Biophoto Associates/Photo Researchers.) (b) Diagram of the electron micrograph. (c) Diagram of a nucleus with two nucleoli. (d) Scanning electron micrograph at a magnification of 7800×. (Courtesy of L. Nilsson, *The Body Victorious,* Delacorte Press.)

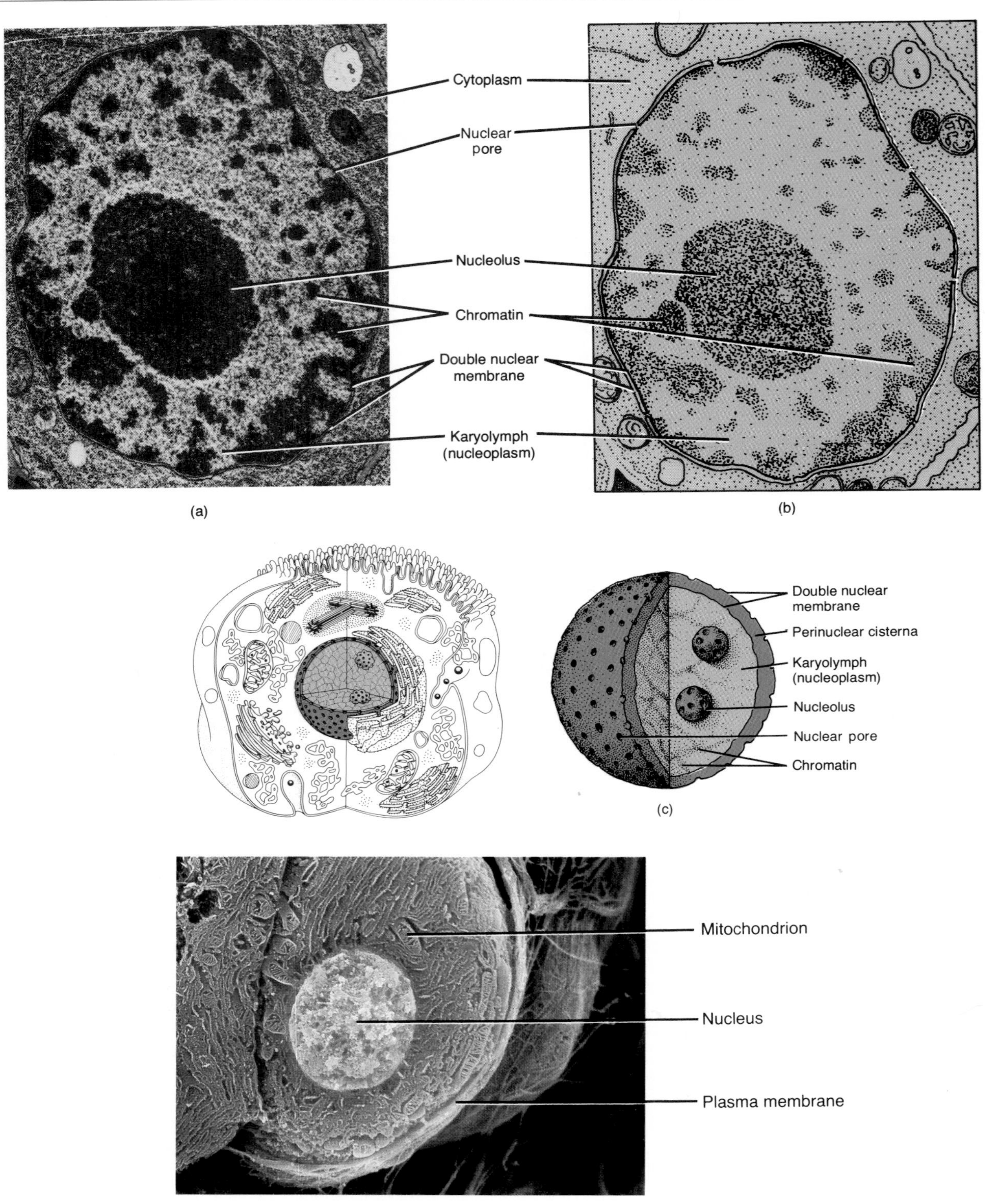

FIGURE 3-7 *(Continued)* (e) Packing of DNA into chromosomes.

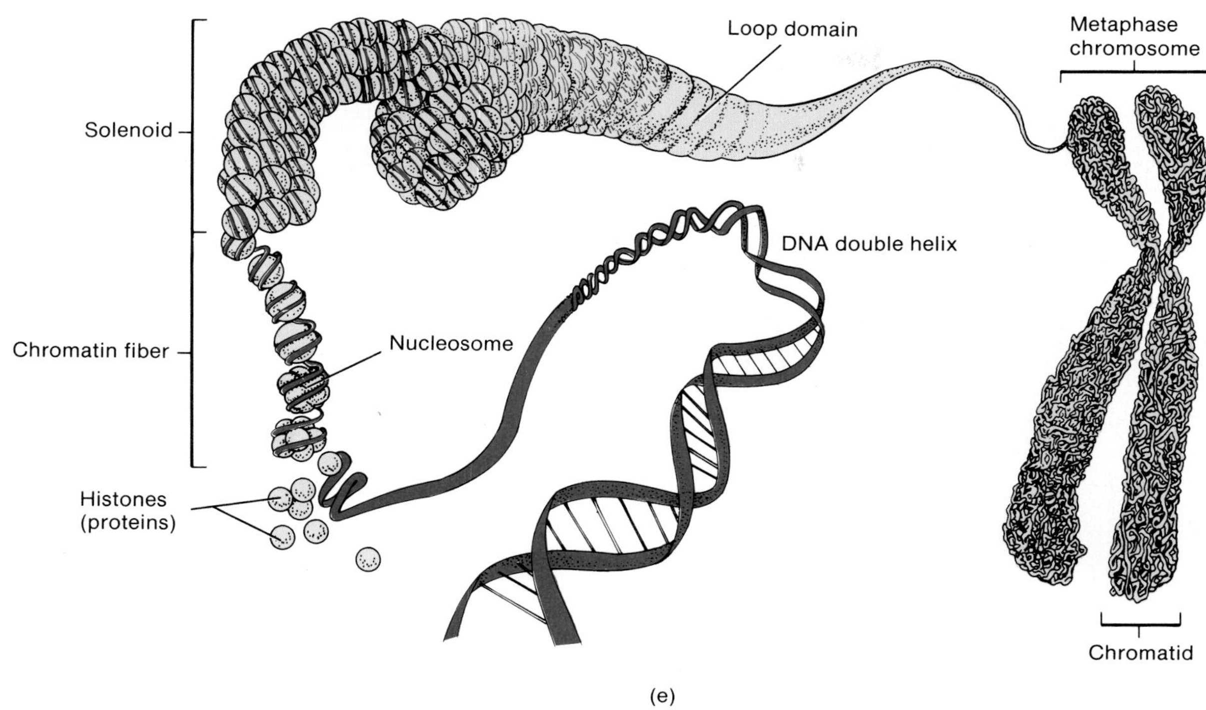

Ribosomes

Ribosomes (RĪ-bo-sōms) are tiny granules, 25 nm at their largest dimension, that are composed of a type of RNA called ***ribosomal RNA (rRNA)*** and a number of specific ribosomal proteins. The rRNA is manufactured by DNA in the nucleolus. Ribosomes were so named because of their high content of rRNA. Structurally, a ribosome consists of two subunits, one about half the size of the other. Scientists have provided three-dimensional models of the structure of a ribosome (Figure 3-8d). Functionally, ribosomes are the sites of protein synthesis: they receive genetic instructions and use them to produce protein. Amino acids are joined one at a time on ribosomes into a protein chain. The completed chain then may fold into a protein molecule that can serve as part of the cell's structure or as an enzyme. The mechanism of ribosomal function is discussed later in the chapter.

Some ribosomes, called ***free ribosomes,*** are scattered in the cytoplasm; they have no attachments to other parts of the cell. The free ribosomes occur singly or in clusters, and they are primarily concerned with synthesizing proteins for use inside the cell. Other ribosomes are attached to a cellular structure called the endoplasmic reticulum (ER). These ribosomes are concerned with the synthesis of proteins for export from the cell.

Endoplasmic Reticulum (ER)

Within the cytoplasm, there is a system of double membranous channels, called ***cisternae,*** of varying shapes. This system is known as the ***endoplasmic reticulum*** (en'-dō-PLAS-mik re-TIK-yoo-lum), or ***ER*** (Figure 3-8). The channels are continuous with the nuclear membrane.

On the basis of its association with ribosomes, the ER is divided into two types. ***Granular (rough) ER*** is studded with ribosomes; ***agranular (smooth) ER*** is free of ribosomes. Agranular ER is synthesized from granular ER.

Numerous functions related to homeostasis are attributed to the ER. It contributes to the mechanical support and distribution of the cytoplasm. The ER is also involved in the intracellular exchange of materials with the cytoplasm and provides a surface area for chemical reactions. Various products are transported from one portion of the cell to another via the ER, so the ER is considered an intracellular transportation system. The ER also serves as a storage area for synthesized molecules. Together with a cellular structure called the Golgi complex, the ER assumes a role in the synthesis and packaging of molecules. Ribosomes associated with granular ER synthesize proteins. Agranular ER is associated with lipid synthesis inactivation or detoxification of certain molecules, and release of calcium ions involved, with muscle contraction.

FIGURE 3-8 Endoplasmic reticulum and ribosomes. (a) Electron micrograph of the endoplasmic reticulum and ribosomes at a magnification of 76,000×. (Copyright © Dr. Myron C. Ledbetter, Biophoto Associates/Photo Researchers.) (b) Diagram of the electron micrograph. (c) Diagram of the endoplasmic reticulum and ribosomes. See if you can find the agranular (smooth) endoplasmic reticulum in Figure 3-9a. (d) Diagram of the three-dimensional structure of ribosomes. (e) Scanning electron micrograph at a magnification of 60,000×. (Courtesy of L. Nilsson, *The Body Victorious,* Delacorte Press.)

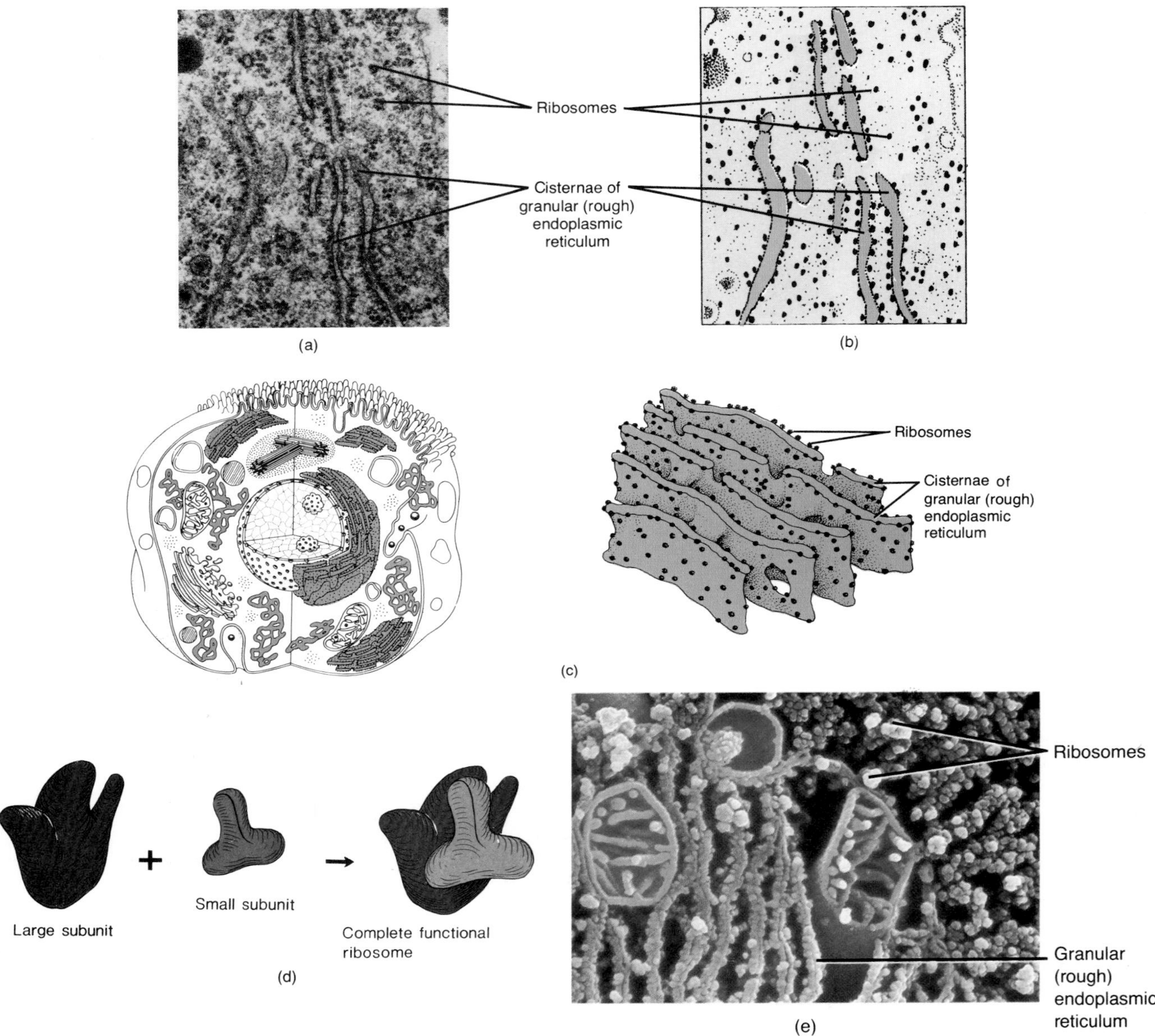

Golgi Complex

A cytoplasmic structure called the ***Golgi*** (GOL-jē) ***complex*** is generally near the nucleus. In cells with high secretory activity, the Golgi complex is extensive. It consists of four to eight flattened membranous sacs, stacked upon each other like a pile of dishes with expanded areas at their ends. Like those of the ER, the stacked elements are called ***cisternae*** (Figure 3-9). On the basis of function, cisternae are designated as *cis, medial,* and *trans* (described shortly).

The principal function of the Golgi complex is to process, sort, package, and deliver proteins to various parts of the cell. Proteins synthesized at ribosomes associated with granular ER are transported into the granular ER (Figure 3-10). While still in the granular ER, sugar molecules are added to the proteins if needed (glycoproteins). Next, the proteins become surrounded by a vesicle formed by a piece of the granular ER membrane, and the vesicle buds off and fuses with the *cis* cisternae of the Golgi complex. The *cis* cisternae are the ones closest to the agranular ER. As a result of this fusion, the proteins enter the Golgi complex. Once inside the Golgi complex, the proteins are transported by other vesicles formed by the Golgi complex from the *cis* cisternae to the *medial* cisternae to the *trans* cisternae, the ones farthest from the agranular ER. As the proteins pass in succession through the Golgi cisternae, they become modified in various ways depending on their function and destination. The proteins are sorted and packaged (in vesicles) in the *trans* cisternae. Some vesicles become ***secretory granules,*** which move toward the surface of the cell where the protein is released. The contents of the granules are discharged into the extracellular space, and the granule membrane is incorporated into the plasma membrane. Cells of the gastrointestinal tract that secrete enzymes utilize this mechanism. The secretory granule prevents "digestion" of the cytoplasm of the cells by the enzymes as it moves toward the cell surface. Other vesicles that pinch off from the Golgi complex are loaded with special digestive enzymes and remain within the cell. They become cellular structures called lysosomes.

The Golgi complex is also associated with lipid secretion. Lipids synthesized by the agranular ER pass through the ER into the Golgi complex. Packaged lipids are discharged at the surface of the cell and enter extracellular

FIGURE 3-9 Golgi complex. (a) Electron micrograph of the two Golgi complexes at a magnification of 78,000×. (Copyright © Dr. Myron C. Ledbetter, Biophoto Associates/Photo Researchers.) (b) Diagram of the electron micrograph. (c) Diagram of a Golgi complex. (d) Scanning electron micrograph at a magnification of 20,000×. (Courtesy of L. Nilsson, *The Body Victorious,* Delacorte Press.)

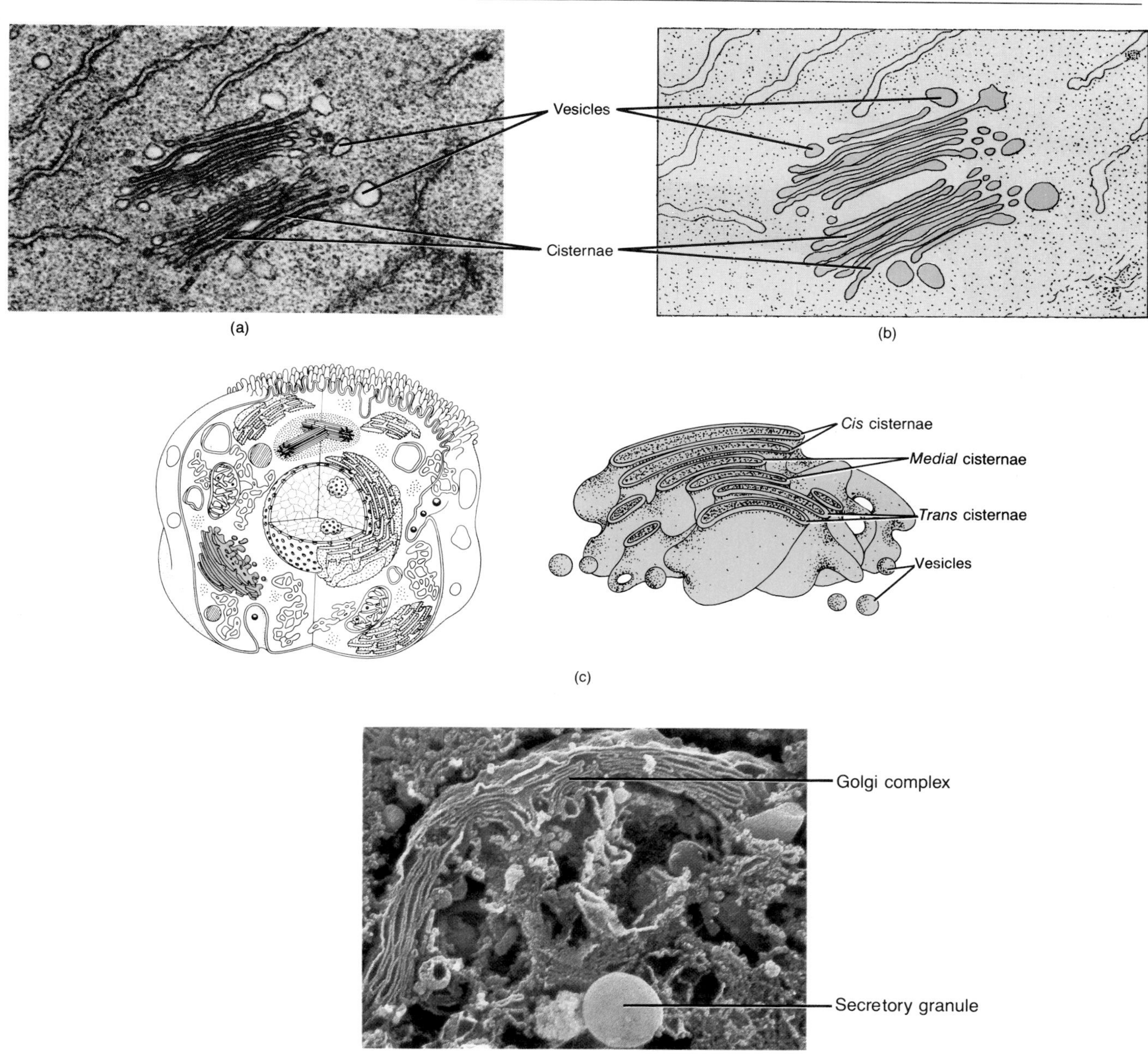

FIGURE 3-10 Packaging of synthesized protein for export from the cell.

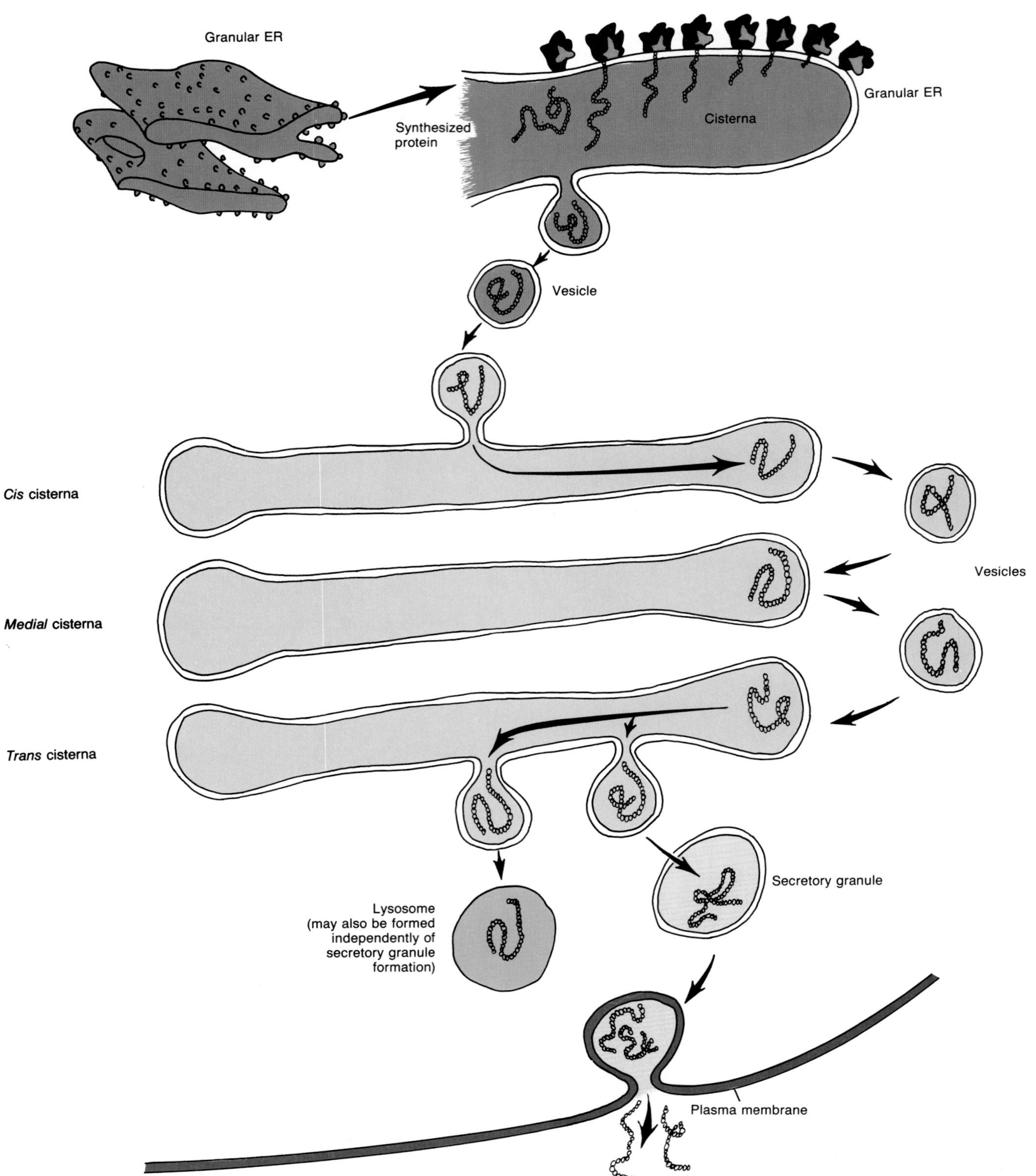

fluid. In the course of moving through the cytoplasm, the vesicle may release lipids into the cytoplasm before being discharged from the cell. These appear in the cytoplasm as lipid droplets. Among the lipids secreted in this manner are steroids (see Exhibit 2-4).

Mitochondria

Small, spherical, rod-shaped, or filamentous structures called ***mitochondria*** (mī-tō-KON-drē-a) appear throughout the cytoplasm. Because of their function in generating energy, they are referred to as the "powerhouses" of the cell. Within mitochondria, energy is transferred from carbon compounds, such as glucose, to ATP. When sectioned and viewed under an electron microscope, each reveals an elaborate internal organization (Figure 3-11). A mitochondrion consists of two membranes, each of which is similar in structure to the plasma membrane. The outer mitochondrial membrane is smooth, but the inner membrane is arranged in a series of folds called ***cristae.*** The central cavity of a mitochondrion enclosed by the inner membrane and cristae is called the ***matrix.***

Because of the nature and arrangement of the cristae, the inner membrane provides an enormous surface area for chemical reactions, collectively referred to as ***cellular respiration.*** Enzymes involved in energy-releasing reactions that form ATP are located on the cristae. Active cells, such as mucle, liver, and kidney tubule cells, have a large number of mitochondria because of their high energy expenditure.

Mitochondria are self-replicative—that is, they can divide to form new ones. The replication process is controlled by DNA that is incorporated into the mitochondrial structure. Self-replication usually occurs in response to increased cellular need for ATP. The energy releasing reactions only occur if oxygen (O_2) is present and result in the catabolism of nutrient molecules. Some of the energy released as these nutrients are broken down is used to re-form the high energy molecule ATP.

Lysosomes

When viewed under an electron microscope, ***lysosomes*** (LĪ-so-sōms; *lysis* = dissolution; *soma* = body) appear as membrane-enclosed spheres (Figure 3-12). They are formed from Golgi complexes and have a single membrane. They contain powerful digestive (hydrolytic) enzymes capable of breaking down many kinds of molecules. You will see in Chapter 14 that Tay-Sachs disease results from a deficiency of a lysosomal enzyme. These enzymes are also capable of digesting bacteria and other substances that enter the cell in phagocytic vesicles. White blood cells, which ingest bacteria by phagocytosis, contain large numbers of lysosomes.

Lysosomal enzymes are believed to be synthesized on the granular ER and then transported to the Golgi complex, probably in an inactive form to prevent the unwanted digestion of structures through which they pass. Lysosomes develop as buds that pinch off the ends of Golgi cisternae. When first formed, the lysosome is referred to as a ***primary lysosome,*** meaning that it contains enzymes but is not yet engaged in digestive activity. In order to participate in digestion, a primary lysosome must fuse with a membrane-bound vacuole or other organelles containing food particles. A ***vacuole*** is a membrane-bound organelle that, in animal cells, frequently functions in temporary storage or transportation. As a result of this fusion, the primary lysosome becomes known as a ***secondary lysosome,*** that is, a lysosome engaged in digestive activity. In the digestive process, the contents of the vacuole are broken down into smaller and smaller components. Eventually, the products of digestion are small enough to pass out of the lysosome into the cytoplasm of the cell to be recycled in the synthesis of various molecules needed by the cell.

Lysosomes function in intracellular digestion in a number of ways. For example, during phagocytosis lysosomal enzymes digest the solid material contained in phagocytic vesicles. A similar process occurs in pinocytosis and receptor-mediated endocytosis. Lysosomes also use their enzymes to recycle the cell's own molecules. A lysosome can engulf another organelle, digest it, and return the digested components to the cytoplasm for reuse. In this way, worn-out cellular structures are continually renewed. The process by which worn-out organelles are digested is called ***autophagy*** (aw-TOF-a-jē; *auto* = self; *phagio* = to eat). A human liver cell can recycle about half its contents in a week. Under normal conditions, the lysosomal membrane is impermeable to the passage of enzymes into cytoplasm. This prevents digestion of the cytoplasm. However, under certain conditions, ***autolysis*** or self-destruction by lysosomes occurs. For example, during human embryological development, the fingers and toes are webbed. The normal development of individual fingers and toes requires the selective removal of the cells of the web between the digits. This involves autolysis, in which lysosomal enzymes digest the tissue between the digits. It is actually a programmed destruction of cells. Because of this activity, lysosomes are sometimes referred to as "suicide packets."

Lysosomes also function in extracellular digestion. Lysosomal enzymes released at sites of injury help to digest away cellular debris which prepares the injured area for effective repair. Prior to fertilization, the head of a sperm cell releases lysosomal enzymes capable of digesting a barrier around the egg so that the sperm cell can penetrate it. Once this is accomplished, fertilization can take place. Also, release of lysosomal enzymes from a cell may be the process responsible for bone removal. In bone reshaping, especially during the growth process, special bone-destroying cells, called ***osteoclasts,*** secrete extracellular enzymes that dissolve bone. Bone tissue cultures given excess amounts of vitamin A remove bone apparently through an activation process involving lysosomes.

FIGURE 3-11 Mitochondria. (a) Electron micrograph of an entire mitochondrion (above) and a portion of another (below). (Courtesy of Lester V. Bergman & Associates, Inc.) (b) Diagram of the electron micrograph. (c) Diagram of a mitochondrion. (d) Scanning electron micrograph at a magnification of 30,000×. (Courtesy of L. Nilsson, *The Body Victorious,* Delacorte Press.)

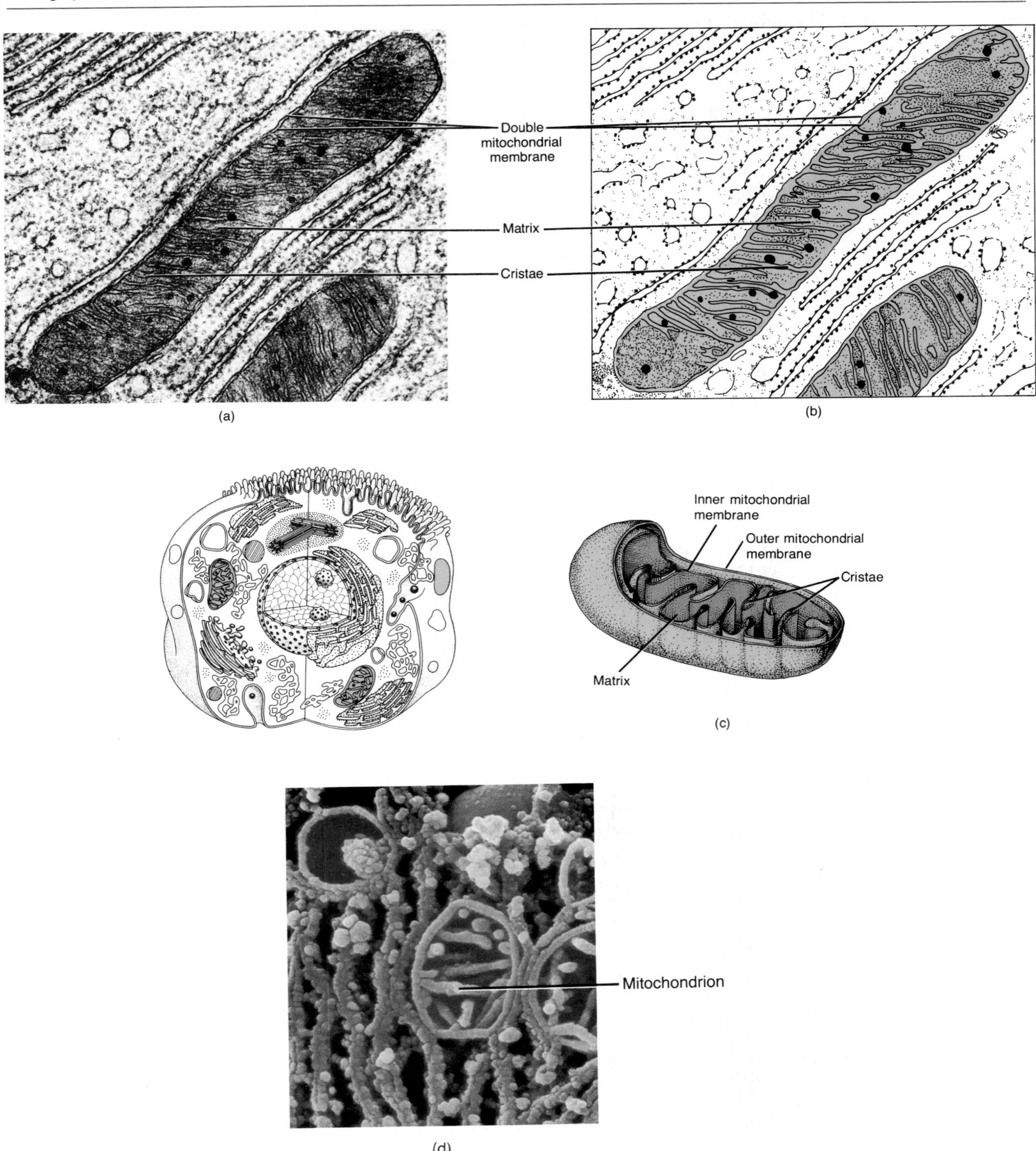

FIGURE 3-12 Lysosome. (a) Electron micrograph of a lysosome at a magnification of 55,000×. (Courtesy of F. Van Hoof, Université Catholique de Louvain.) (b) Diagram of the electron micrograph.

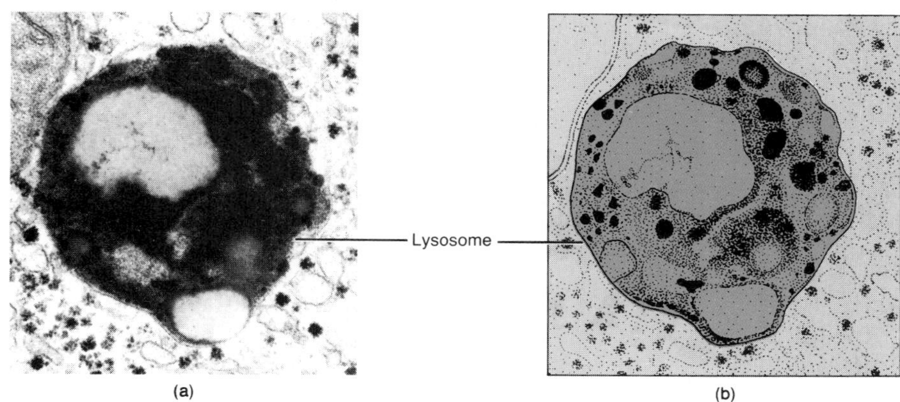

CLINICAL APPLICATION: LYSOSOMES AND STEROIDS

Although an adequate intake of vitamin A is required for lysosomal function, animals overfed on vitamin A develop spontaneous fractures, suggesting greatly increased lysosomal activity. On the other hand, cortisone and hydrocortisone, ***steroid hormones*** produced by the adrenal gland, ***have a stabilizing effect on lysosomal membranes.*** The steroid hormones are well known for their antiinflammatory properties, which suggests that they reduce destructive cellular activity by lysosomes.

Peroxisomes

Organelles similar in structure to lysosomes, but smaller, are called ***peroxisomes*** (pe-ROKS-i-sōms). They are abundant in liver cells and contain several enzymes related to the metabolism of hydrogen peroxide (H_2O_2), a substance that is toxic to body cells. One of the enzymes in peroxisomes, called *catalase,* immediately breaks down H_2O_2 into water and oxygen:

$$\underset{\text{Hydrogen peroxide}}{2H_2O_2} \xrightarrow{\text{catalase}} \underset{\text{Water}}{2H_2O} + \underset{\text{Oxygen}}{2O_2}$$

The Cytoskeleton

Cytoplasm has a complex internal structure, consisting of a series of exceedingly small microfilaments, microtubules, and intermediate filaments, together referred to as the ***cytoskeleton*** (see Figure 3-1).

Microfilaments are rodlike structures that are 6 nm in diameter and are of variable length. Microfilaments consist of a protein called ***actin.*** In muscle tissue, actin microfilaments (thin myofilaments) and myosin microfilaments (thick myofilaments) are involved in the contraction of muscle fibers (cells). This mechanism is described in Chapter 10. In nonmuscle cells, microfilaments help to provide support and shape and assist in the movement of entire cells (phagocytes and cells of developing embryos) and movements within cells (secretion, phagocytosis, pinocytosis).

Microtubules are relatively straight, slender, cylindrical structures that range in diameter from 18 to 30 nm, usually averaging about 24 nm. They consist of a protein called ***tubulin.*** Microtubules and microfilaments help to provide support and shape for cells. Microtubules may also form conducting channels through which various substances can move throughout the cytoplasm. This mechanism has been studied extensively in nerve cells. Microtubules also assist in the movement of pseudopodia that are characteristic of phagocytes. As you will see shortly, microtubules form the structure of flagella and cilia (cellular appendages involved in motility), centrioles (organelles that may direct the assembly of microtubules), and the mitotic spindle (structures that are involved in cell division).

Intermediate filaments range between 8 and 12 nm in diameter. An example is neurofilaments found in nerve cells (Chapter 12). Although the functions of intermediate filaments are not completely understood, it appears that they provide structural reinforcement in some cells and assist in contraction of others.

Some investigators believe that the microfilaments, microtubules, and intermediate filaments, as well as other cytoplasmic components, are held together by a three-dimensional meshwork of fine filaments called ***microtrabeculae*** (mī′-krō-tra-BEK-yoo-lē), which are about 10–15 nm thick. Together, all the microtrabeculae constitute the ***microtrabecular lattice*** (Figure 3-13). This lattice is believed to provide organization for chemical

FIGURE 3-13 Microtrabecular lattice.

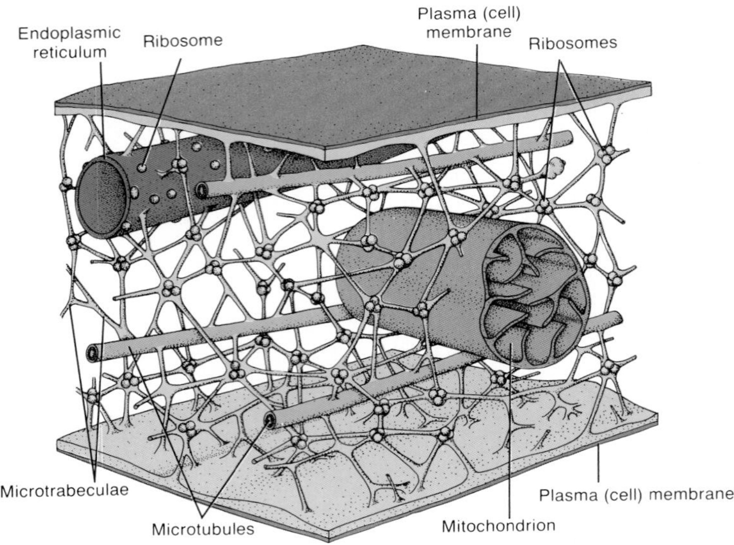

reactions that occur within the cytoplasm and to assist in the transport of substances through the cytoplasm.

Centrosome and Centrioles

A dense area of cytoplasm, generally spherical and located near the nucleus, is called the ***centrosome (centrosphere).*** Within the centrosome is a pair of cylindrical structures called ***centrioles*** (Figure 3-14). Each centriole is composed of nine triplet clusters of microtubules arranged in a circular pattern. Centrioles lack the two central single microtubules found in flagella and cilia. The two centrioles are situated so that the long axis of one is at right angles to the long axis of the other. Centrioles assume a role in cell reproduction by serving as centers about which microtubules involved in chromosome movement are organized. This role will be described shortly as part of cell division. Certain cells, such as most mature nerve cells, do not have a centrosome and so do not reproduce. This is why they cannot be replaced if they are destroyed. Like mitochondria, centrioles contain DNA that controls their self-replication.

FIGURE 3-14 Centrosome and centrioles. (a) Diagram of a centriole in longitudinal view. (b) Diagram of a centriole in cross section.

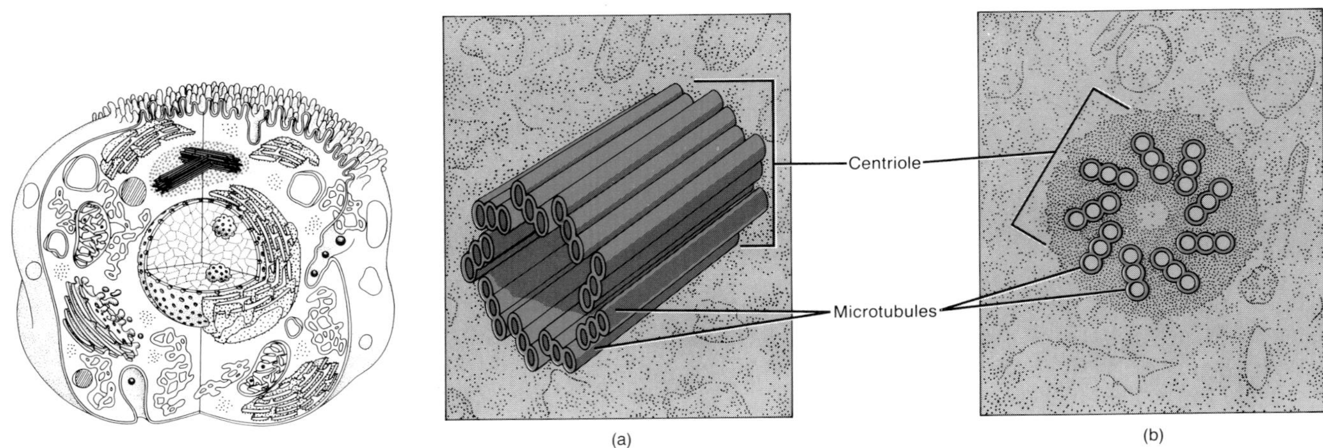

Flagella and Cilia

Some body cells possess projections for moving the entire cell or for moving substances along the surface of the cell. These projections contain cytoplasm and are bounded by the plasma membrane. If the projections are few and long in proportion to the size of the cell (typically occurring singly or in pairs), they are called ***flagella*** (fla-JEL-a). The only example of a flagellum in the human body is the tail of a sperm cell, used for locomotion (see Figure 28-5). If the projections are numerous and short, resembling many hairs, they are called ***cilia*** (SIL-ē-a). In humans, ciliated cells of the respiratory tract move mucus that has trapped foreign particles over the surface of the tissue (see Figure 23-5). Electron microscopy has revealed no fundamental structural difference between cilia and flagella. Both consist of nine pairs of microtubules that form a ring around two single microtubules in the center.

CELL INCLUSIONS

Cell inclusions are a large and diverse group of chemical substances produced by cells, some of which have recognizable shapes. These products are principally organic and may appear or disappear at various times in the life of the cell. Examples include melanin, glycogen, and lipids. ***Melanin*** is a pigment stored in certain cells of the skin, hair, and eyes. It protects the body by screening out harmful ultraviolet rays from the sun. ***Glycogen*** is a polysaccharide that is stored in the liver, skeletal muscle fibers (cells), and the vaginal mucosa. When the body requires quick energy, liver cells can break down the glycogen into glucose and release it. ***Lipids,*** which are stored in adipocytes (fat cells), may be decomposed for producing energy.

The major parts of the cell and their functions are summarized in Exhibit 3-2.

EXTRACELLULAR MATERIALS

The substances that lie outside cells are called ***extracellular materials.*** They include body fluids, which provide a medium for dissolving, mixing, transporting substances, and carry out chemical reactions. Among the body fluids are interstitial fluid, the fluid that fills the microscopic spaces (interstitial spaces), and plasma, the liquid portion of blood (see Figure 1-14). Extracellular materials also include secreted inclusions like mucus and special substances that form the matrix (substance between cells) in which some cells are embedded.

The matrix materials are produced by certain cells and deposited outside their plasma membranes. The matrix supports cells, binds them together, and gives strength and elasticity to the tissue. Some matrix materials are ***amorphous*** (*a* = without; *morpho* = shape); they have no specific shape. These include hyaluronic acid, chondroitin sulfate, dermatan sulfate, and keratan sulfate.

Hyaluronic (hī'-a-loo-RON-ik) ***acid*** is a viscous, fluidlike substance that binds cells together, lubricates joints, and maintains the shape of the eyeballs. ***Chondroitin*** (kon-DROY-tin) ***sulfate*** is a jellylike substance that provides support and adhesiveness in cartilage, bone, and blood vessels. ***Dermatan sulfate*** is found in the skin, tendons, and heart valves, and ***keratan sulfate*** is found in the cornea of the eye and bone.

Other matrix materials are ***fibrous,*** or threadlike. Fibrous materials provide strength and support for tissues. Among these are ***collagenous*** (*kolla* = glue) ***fibers*** consisting of the protein *collagen.* These fibers are found in all types of connective tissue, especially in bones, cartilage, tendons, and ligaments. ***Reticular*** (*rete* = net) ***fibers*** consisting of the protein collagen and a coating of glycoprotein, form a network around fat cells, nerve fibers, and skeletal and smooth muscle fibers (cells), and in the walls of blood vessels. They also are found in close association with the basal lamina of most epithelia and form the framework or stroma for many soft organs of the body such as the spleen. ***Elastic fibers,*** consisting of the protein *elastin,* give elasticity to skin and to tissues forming the walls of blood vessels.

GENE ACTION

Protein Synthesis

Although cells synthesize numerous chemicals in order to maintain homeostasis, much of the cellular machinery is concerned with producing proteins. Some of the proteins are structural, helping to form plasma membranes, microfilaments, microtubules, centrioles, flagella, cilia, the mitotic spindle, and other parts of cells. Other proteins serve as hormones, antibodies, and contractile elements in muscle tissue. Still other proteins serve as enzymes that regulate the rate of the myriad chemical reactions that occur in cells. Basically, cells are protein factories that constantly synthesize large numbers of diverse proteins that determine the physical and chemical characteristics of cells and, therefore, of organisms.

The genetic instructions for making proteins are found in DNA. Cells make proteins by translating the genetic information encoded in DNA. In the process, the genetic information in a region of DNA is copied to produce a specific molecule of RNA. Through a complex series of reactions, the information contained in RNA is translated into a corresponding specific sequence of amino acids in a newly produced protein molecule. Let us take a look at how DNA directs protein synthesis by considering the two principal steps in protein synthesis: transcription and translation.

EXHIBIT 3-2 CELL PARTS AND THEIR FUNCTIONS

Part	Functions
PLASMA MEMBRANE	Protects cellular contents; makes contact with other cells; provides receptors for hormones, enzymes, and antibodies; mediates the entrance and exit of materials.
CYTOPLASM	Serves as the ground substance in which chemical reactions occur.
ORGANELLES	
Nucleus	Contains genes and controls cellular activities.
Ribosomes	Sites of protein synthesis.
Endoplasmic reticulum (ER)	Contributes to mechanical support; facilitates intracellular exchange of materials with cytoplasm; provides a surface area for chemical reactions; provides a pathway for transporting chemicals; serves as a storage area; together with Golgi complex synthesizes and packages molecules for export; ribosomes associated with granular ER synthesize proteins; smooth ER synthesizes lipids, detoxifies certain molecules, and releases calcium ions involved with muscle contraction.
Golgi complex	Packages synthesized proteins for secretion in conjunction with endoplasmic reticulum; forms lysosomes; secretes lipids; synthesizes carbohydrates; combines carbohydrates with proteins to form glycoproteins for secretion.
Mitochondria	Sites of production of ATP.
Lysosomes	Digest substances and foreign microbes; may be involved in bone removal.
Peroxisomes	Contain several enzymes, such as catalase, related to hydrogen peroxide metabolism.
Microfilaments	Form part of cytoskeleton; involved in muscle fiber (cell) contraction; provide support and shape; assist in cellular and intracellular movement.
Microtubules	Form part of cytoskeleton; provide support and shape; form intracellular conducting channels; assist in cellular movement; form the structure of flagella, cilia, centrioles, and mitotic spindle.
Intermediate filaments	Form part of cytoskeleton; probably provide structural reinforcement in some cells.
Centrioles	Help organize mitotic spindle during cell division.
Flagella and cilia	Allow movement of entire cell (flagella) or movement of particles trapped in mucus along surface of cell (cilia).
INCLUSIONS	Melanin (pigment in skin, hair, eyes) screens out ultraviolet rays; glycogen (stored glucose) can be decomposed to provide energy; lipids (stored in fat cells) can be decomposed to produce energy.

Transcription

Transcription is the process by which genetic information encoded in DNA is copied by a strand of RNA called ***messenger RNA (mRNA).*** It is called transcription because it resembles the transcription of a sequence of words from one tape to another. By using a specific portion of the cell's DNA as a template, the genetic information stored in the sequence of nitrogenous bases of DNA is rewritten so that the same information appears in the nitrogenous bases of mRNA. As in DNA replication (see Figure 3-16), a cytosine (C) in the DNA template dictates a guanine (G) in the mRNA strand being made; a G in the DNA template dictates a C in the mRNA strand; and a thymine (T) in the DNA template dictates an adenine (A) in the mRNA. Since RNA contains uracil (U) instead of T, an A in the DNA template dictates a U in the mRNA. As an example, if the template portion of DNA has the base sequence ATGCAT, the transcribed mRNA strand would have the complementary base sequence UACGUA. Only one of the two DNA strands serves as a template for RNA synthesis. This strand is referred to as the ***sense strand.*** The other strand, the one not transcribed, is the complement of the sense strand and is called the ***antisense strand.***

One factor that complicates transcription is that there

are regions of DNA, called ***introns,*** that do not code for the synthesis of a protein. Introns are found between regions of DNA that do code for proteins, called ***exons.*** Since introns are transcribed on mRNA, they must be deleted from mRNA and the exons of mRNA must be rejoined before mRNA can leave the nucleus and enter the cytoplasm to participate further in protein synthesis. The function of cutting out introns and splicing together exons is performed by ***small nuclear ribonucleoproteins (snRNPs),*** pronounced ***"snurps."***

In addition to serving as the template for the synthesis of mRNA, DNA also synthesizes two other kinds of RNA. One is called ***ribosomal RNA (rRNA),*** which together with ribosomal proteins makes up ribosomes. The other is called ***transfer RNA (tRNA).*** Once synthesized, mRNA, rRNA, and tRNA leave the nucleus of the cell. In the cytoplasm, they participate in the next principal step in protein synthesis—translation.

Translation

The process by which information in the nitrogenous base sequence of mRNA is used to specify the amino acid sequence of a protein is called ***translation.*** The key events involved in translation are as follows (Figure 3-15).

1. In the cytoplasm, the small ribosomal subunit binds one end of the mRNA molecule (Figure 3-15a). In this respect, ribosomes are the sites of protein synthesis.

2. There are 20 different amino acids in the cytoplasm that may participate in protein synthesis. Whichever amino acids participate to form a particular protein are picked up by tRNAs (Figure 3-15b). For each different amino acid there is a different type of tRNA. One end of a tRNA molecule couples with a specific amino acid. This amino acid activation requires energy from ATP breakdown. The other end has a specific sequence of three nitrogenous bases (triplet) known as an ***anticodon.*** The complementary triplet on an mRNA strand is called a ***codon.***

3. By base pairing, the anticodon of a specific tRNA recognizes the corresponding codon of mRNA and attaches to it. If the tRNA anticodon is UAC, the mRNA codon would be AUG (Figure 3-15c). In the process, the tRNA also brings along the specific amino acid. The pairing of anticodon and codon occurs only where mRNA is attached to a ribosome.

4. Once the first tRNA attaches to mRNA, the ribosome moves along the mRNA, and the next tRNA with its amino acid moves into position (Figure 3-15d).

5. The two amino acids are joined by a peptide bond, and the first tRNA detaches itself from the mRNA strand. The larger ribosomal subunit contains the enzymes that join the amino acids together (Figure 3-15e). The released tRNA can now pick up another molecule of the same amino acid if necessary.

6. As the proper amino acids are brought into line, one by one, peptide bonds are formed between them, and the protein gets progressively longer (Figure 3-15f).

7. When the specified protein is completed, further synthesis is stopped by a special ***termination codon.*** The assembled protein is released from the ribosome, and the ribosome comes apart into its component subunits (Figure 3-15g,h).

As each ribosome moves along the mRNA strand, it "reads" the information coded in mRNA and synthesizes a protein according to that information; the ribosome synthesizes the protein by translating the codon sequences into an amino acid sequence.

Protein synthesis progresses at the rate of about 15 amino acids per second. As the ribosome moves along the mRNA and before it completes translation of that gene, another ribosome may attach and begin translation of the same mRNA strand, so that several ribosomes may be attached to the same mRNA strand. Such an mRNA strand with its several ribosomes attached is called a ***polyribosome.*** Several ribosomes moving simultaneously in tandem along the same mRNA molecule permit the translation of a single mRNA strand into several identical proteins simultaneously.

Based upon the description of protein synthesis just presented, we can appropriately define a ***gene*** as a group of nucleotides on a DNA molecule that serves as the master mold for manufacturing a specific protein. Genes average about 1000 pairs of nucleotides, which appear in a specific sequence on the DNA molecule. No two genes have exactly the same sequence of nucleotides, and this is the key to heredity.

Remember that the base sequence of the gene determines the sequence of nitrogenous bases in the mRNA. The sequence of the bases in the mRNA then determines the order and kind of amino acids that will form the protein. Thus, each gene is responsible for making a particular protein as follows:

$$\text{DNA} \xrightarrow{\text{Transcription}} \text{RNA} \xrightarrow{\text{Translation}} \text{Protein}$$

CLINICAL APPLICATION: GENETIC ENGINEERING

Different kinds of cells make different proteins following instructions encoded in the DNA of their genes. Since 1973, scientists have been able to alter those instructions in bacterial cells by adding genes from other organisms to the bacterial genes. This causes the bacterial cells to produce proteins that they normally do not synthesize. Bacteria so altered are called *recombinants,* and their DNA, a combination of DNA from different sources, is called *recombinant DNA.* When recombinant DNA is introduced into a bacterium, the bacterium will synthesize the proteins of whatever new gene it has acquired. The new technology that has arisen from manipulating the genetic material is called *genetic engineering.* Yeast cells are also being used in recombinant DNA research.

FIGURE 3-15 Translation. Note that during protein synthesis the ribosomal subunits join together. When protein synthesis ceases, they separate and are quickly metabolized. Ribosomes occurring singly in the cytoplasm are not active; only when they are linked by mRNA to form polyribosomes do they engage in protein synthesis.

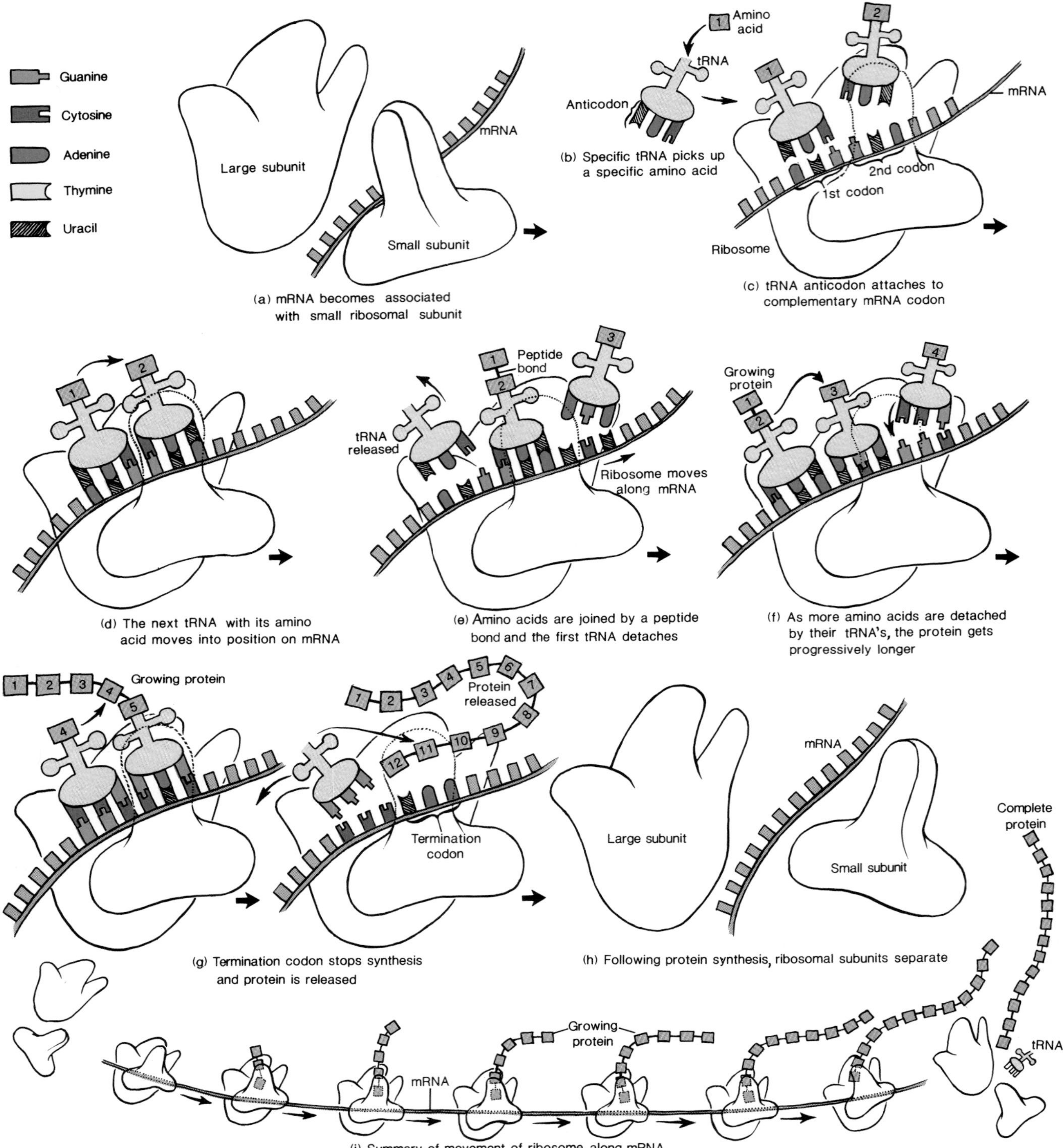
Guanine
Cytosine
Adenine
Thymine
Uracil
Large subunit
Small subunit
mRNA
(a) mRNA becomes associated with small ribosomal subunit
Amino acid
tRNA
Anticodon
(b) Specific tRNA picks up a specific amino acid
mRNA
2nd codon
1st codon
Ribosome
(c) tRNA anticodon attaches to complementary mRNA codon
(d) The next tRNA with its amino acid moves into position on mRNA
Peptide bond
tRNA released
Ribosome moves along mRNA
(e) Amino acids are joined by a peptide bond and the first tRNA detaches
Growing protein
(f) As more amino acids are detached by their tRNA's, the protein gets progressively longer
Growing protein
Protein released
Termination codon
(g) Termination codon stops synthesis and protein is released
mRNA
Large subunit
Small subunit
(h) Following protein synthesis, ribosomal subunits separate
Complete protein
Growing protein
mRNA
tRNA
(i) Summary of movement of ribosome along mRNA

An important goal of recombinant DNA research is to increase our understanding of how genes are organized, how they function, and how they are regulated. This information will enable scientists to identify each of the thousands of genes in a human cell and might be used to find a way to replace defective genes that are responsible for hemophilia, sickle-cell anemia, and other diseases. Recombinant DNA technology might also help scientists understand why normal cells turn malignant.

The practical applications of recombinant DNA technology are staggering. Strains of recombinant bacteria are presently producing several important therapeutic substances. These include *human growth hormone* (*hGH*), required for growth during childhood; *somatostatin,* a brain hormone that helps regulate growth; *insulin,* a hormone that helps regulate blood sugar level and is used by diabetics; *human chorionic gonadotropin* (*hCG*), a hormone required to maintain pregnancy; *interferon* (*INF*), an antiviral (and possibly anticancer) substance; *surfactant,* a phospholipid that lowers surface tension in the lungs; *factor VIII,* a clotting factor missing in people with hemophilia A, the main form of the inherited blood disorder; *tissue-plasminogen activator* (*t-PA*), a substance used to dissolve blood clots in coronary arteries; *hepatitis B protein,* a protein used in a vaccine against hepatitis B; *monoclonal antibodies* to diagnose and treat cancer and assist in AIDS research; and *beta-endorphin,* a neuropeptide that has morphinelike properties that suppress pain. Scientists are also using recombinant DNA techniques to develop vaccines against several viruses, including those that cause herpes, influenza, and malaria.

NORMAL CELL DIVISION

Most of the cell activities mentioned thus far maintain the life of the cell on a day-to-day basis. However, cells become damaged, diseased, or worn out and then die. New cells must be produced as replacements and for growth. In addition, sperm and egg cells must be produced by cell division.

Cell division is the process by which cells reproduce themselves. It consists of a nuclear division and a cytoplasmic division. Because nuclear division can be of two types, two kinds of cell division are recognized.

In the first kind of division, often called ***somatic cell division,*** a single starting cell called a ***parent cell*** divides to produce two identical cells called ***daughter cells.*** This process consists of a nuclear division called ***mitosis*** and a cytoplasmic division called ***cytokinesis*** (*kinesis* = motion). The process ensures that each daughter cell has the same *number* and *kind* of chromosomes as the original parent cell. After the process is complete, the two daughter cells have the same hereditary material and genetic potential as the parent cell. This kind of cell division results in an increase in the number of body cells. In a 24-hour period, the average adult loses billions of cells from different parts of the body. Obviously, these cells must be replaced. Cells that have a short life span—the cells of the outer layer of skin and those lining the gastrointestinal tract—are continually being replaced. Mitosis and cytokinesis are the means by which dead or injured cells are replaced and new cells are added for body growth.

The second type of cell division is called ***reproductive cell division*** and is the mechanism by which sperm and egg cells are produced, cells required to form a new organism. The process consists of a nuclear division called ***meiosis*** plus ***cytokinesis.*** We will first discuss somatic cell division.

Somatic Cell Division

When a cell reproduces, it must replicate (produce duplicates of) its chromosomes so that its hereditary traits may be passed on to succeeding generations of cells. A ***chromosome*** (*chromo* = colored) is a highly coiled DNA molecule that is partly covered by protein. The protein causes changes in the length and thickness of the chromosome. Hereditary information is contained in the DNA portion of the chromosome in units called ***genes.*** Humans have about 100,000 functional genes. However, at any given time, only a fraction of the genes are operating.

When a cell is between divisions it is said to be in ***interphase (metabolic phase).*** It is during this stage that the replication (synthesis) of chromosomes occurs and the RNA and protein needed to produce structures required for doubling all cellular components are manufactured. When DNA replicates, its helical structure partially uncoils (Figure 3-16). Those portions of DNA that remain coiled stain darker than the uncoiled portions. This unequal distribution of stain causes the DNA to appear as a granular mass called ***chromatin*** (Figure 3-17a). During uncoiling, DNA separates at the points where the nitrogenous bases are connected. In the presence of an enzyme, each exposed nitrogenous base then picks up a complementary nitrogenous base (with associated sugar and phosphate group). This uncoiling and complementary base pairing continues until each of the two original DNA strands is matched and joined with two newly formed DNA strands. The original DNA molecule has become two DNA molecules. Since each DNA molecule thus formed consists of one complete strand from the original (conserved) molecule and a newly produced strand, this DNA replication is called ***semiconservative.*** It is estimated that the nucleus of a single human diploid cell contains about 3 billion nitrogenous base pairs.

Generation after generation, as a result of DNA replication and cell division, DNA is copied with great accuracy. Some mistakes in copying are of no consequence and can even be beneficial. However, some are quite serious and

FIGURE 3-16 Replication of DNA. The two strands of the double helix separate by breaking the hydrogen bonds between nucleotides. New nucleotides attach at the proper sites, and a new strand of DNA is paired off with each of the original strands. After replication, the two DNA molecules, each consisting of a new and an old strand, return to their helical structure.

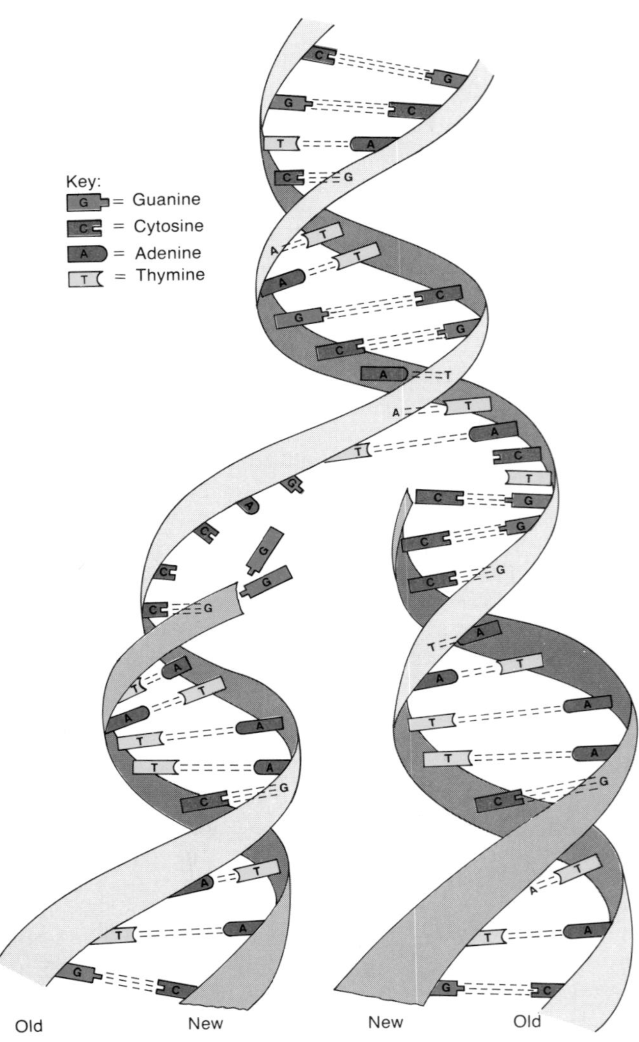

may result in sickle-cell anemia and several forms of cancer. The accuracy of DNA copying depends on three processes, all of which are enzymatically regulated. The first is selection of the proper complementary nitrogenous base following uncoiling. The second process involves removal of an uncomplementary nitrogenous base. And the third process, which takes place after replication, involves correcting errors in replication. These three processes are so effective in maintaining accurate DNA copying that it is estimated that only one mistake in 10 billion nitrogenous bases is made.

A microscopic view of a cell during interphase shows a clearly defined nuclear membrane, nucleoli, karyolymph, and chromatin. Once a cell completes its replication of DNA and centrioles and its production of RNA and proteins during interphase, mitosis begins.

Mitosis

The events that take place during mitosis and cytokinesis are plainly visible under a microscope after the cells have been stained in the laboratory.

The process called ***mitosis*** is the distribution of the two sets of chromosomes into two separate and equal nuclei following the replication of the chromosomes of the parent nucleus. For convenience, biologists divide the process into four stages: prophase, metaphase, anaphase, and telophase. These are arbitrary classifications. Mitosis is actually a continuous process, one stage merging imperceptibly into the next.

▪ **Prophase** During ***prophase*** (*pro* = before) (Figure 3-17b), the first stage of mitosis, chromatin shortens and coils into chromosomes. The nucleoli become less distinct and the nuclear membrane disappears. Each prophase "chromosome" is actually composed of a pair of structures called ***chromatids.*** A chromatid is a complete chromosome consisting of a double-stranded DNA molecule, and each is attached to its partner by a small spherical body called a ***centromere.*** During prophase, the chromatid pairs assemble near the center of the cell in a region called the ***equatorial plane region (equator)*** of the cell.

Also during prophase, the paired centrioles separate and each pair moves to an opposite pole (end) of the cell (the centrioles replicate during interphase). Between the centrioles, a series of microtubules is organized into two groups of fibers. The ***continuous (interpolar) microtubules*** originate from the vicinity of each pair of centrioles and grow toward each other. Thus, they extend from one pole of the cell to another. As they grow toward each other, the second group of microtubules develops. These are called ***chromosomal microtubules*** and apparently grow out of the centromeres; they extend from a centromere to a pole of the cell. Together, the continuous and chromosomal microtubules constitute the ***mitotic spindle*** and, with the centrioles, are referred to as the ***mitotic apparatus.***

▪ **Metaphase** During ***metaphase*** (*meta* = after) (Figure 3-17c), the second stage of mitosis, the centromeres of the chromatid pairs line up on the equatorial plane of the cell. The centromeres of each chromatid pair form a chromosomal microtubule that attaches the centromere to a pole of the cell.

▪ **Anaphase** The third stage of mitosis, ***anaphase*** (*ana* = upward) (Figure 3-17d), is characterized by the division of the centromeres and the movement of complete identical sets of chromatids, now called chromo-

FIGURE 3-17 Cell division: mitosis and cytokinesis. Photomicrographs and diagrammatic representations of the various stages of cell division in whitefish eggs. Read the sequence starting at (a) and move clockwise until you complete the cycle. (Photographs by Carolina Biological Supply Company.)

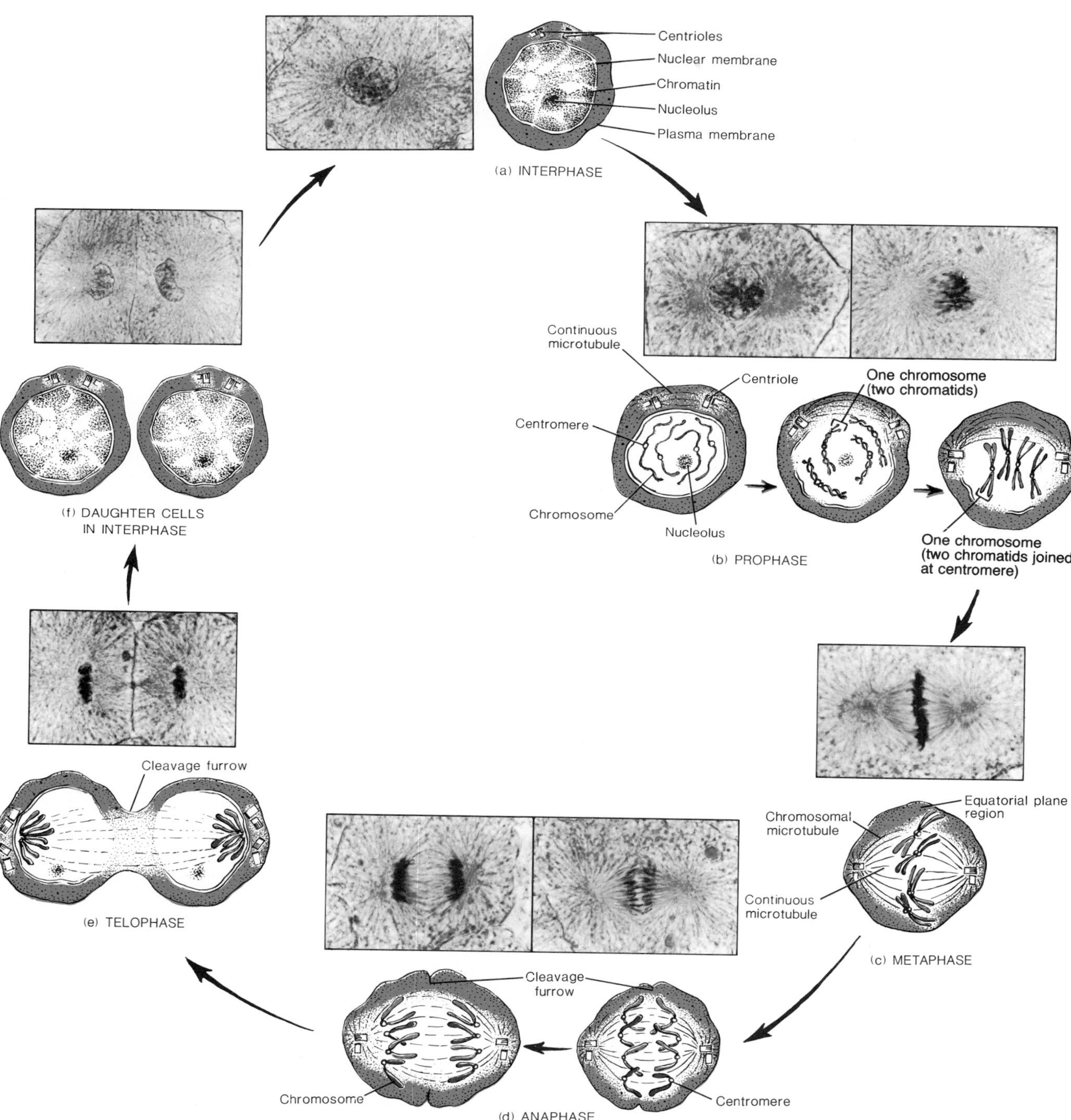

somes, to opposite poles of the cell. During this movement, the centromeres attached to the chromosomal microtubules seem to drag the trailing parts of the chromosomes toward opposite poles. Although several theories have been proposed, the mechanism by which the chromosomes move to opposite poles is not completely understood.

▪ Telophase ***Telophase*** (*telo* = far or end) (Figure 3-17e), the final stage of mitosis, consists of a series of events nearly the reverse of prophase. By this time, two identical sets of chromosomes have reached opposite poles. As telophase progresses, new nuclear membranes begin to enclose the chromosomes, the chromosomes start to assume their chromatin form, nucleoli reappear, and the mitotic spindles disappear. The formation of two nuclei identical to those of cells in interphase terminates telophase. A mitotic cycle has thus been completed (Figure 3-17f).

▪ Time Required The time required for mitosis varies with the kind of cell, its location, and the influence of factors such as temperature. Furthermore, the different stages of mitosis are not equal in duration. However, in order to give you some idea of the length of a cell cycle, mammalian cells in culture have been studied and often have the following time intervals. Interphase is highly variable, ranging from almost nonexistent in rapidly dividing cells to days, weeks, or years. However, it typically takes about 18–24 hours. Mitosis and cytokinesis require about 30–45 minutes. Within the mitosis and cytokinesis time interval, prophase takes longest and anaphase is shortest. As you can see, mitosis and cytokinesis represent only a small part of the life cycle of a cell. Together, the various phases of the cell cycle require about 18–24 hours in many cultured mammalian cells.

Cytokinesis

Division of the cytoplasm, a process called ***cytokinesis*** (sī′-tō-ki-NĒ-sis), often begins during late anaphase and terminates at the same time as telophase. Cytokinesis begins with the formation of a ***cleavage furrow*** that extends around the cell's equatorial plane region. The furrow progresses inward, resembling a constricting ring, and cuts completely through the cell to form two separate portions of cytoplasm (Figure 3-17e,f).

A summary of the events that occur during the interphase and cell division is presented in Exhibit 3-3.

EXHIBIT 3-3 SUMMARY OF EVENTS ASSOCIATED WITH INTERPHASE AND SOMATIC CELL DIVISION

Period or Stage	Activity
Interphase	Cell is between divisions. Cell engages in growth, metabolism, and production of substances required for division; chromosomal replication occurs.
Cell Division	Single parent cell produces two identical daughter cells.
Prophase	Chromatin shortens and coils into chromosomes (chromatids), nucleoli and nuclear membrane become less distinct, centrioles separate and move to opposite poles of cell, and mitotic spindle forms.
Metaphase	Centromeres of chromatid pairs line up on equatorial plane of cell and form chromosomal microtubules that attach centromeres to poles of cell.
Anaphase	Centromeres divide and identical sets of chromosomes move to opposite poles of cell.
Telophase	Nuclear membrane reappears and encloses chromosomes, chromosomes resume chromatin form, nucleoli reappear, mitotic spindle disappears, and centrioles duplicate.
Cytokinesis	Cleavage furrow forms around equatorial plane of cell, progresses inward, and separates cytoplasm into two separate and equal portions.

Reproductive Cell Division

In sexual reproduction, each new organism is produced by the union and fusion of two different sex cells, one produced by each parent. The sex cells, called ***gametes,*** are the ovum produced in the female gonads (ovaries) and the sperm produced in the male gonads (testes). The union and fusion of gametes is called ***fertilization,*** and the cell thus produced is known as a ***zygote.*** The zygote contains a mixture of chromosomes (DNA) from the two parents and, through its repeated mitotic division, develops into a new organism.

Gametes differ from all other body cells (somatic cells) with respect to the number of chromosomes in their nuclei. Somatic cells, such as brain cells, stomach cells, kidney cells, and all other uninucleated somatic cells, contain

46 chromosomes in their nuclei. Some somatic cells, such as skeletal muscle fibers (cells), are multinucleated and thus contain more than 46 chromosomes. However, since most somatic cells are uninucleated, these are the cells to which we will refer in the following discussion. Of the 46 chromosomes, 23 are a complete set that contain one copy of all the genes necessary for carrying out the activities of the cell. In a sense, the other 23 chromosomes are a duplicate set. The symbol n is used to designate the number of different chromosomes within the nucleus. Since somatic cells contain two sets of chromosomes, they are referred to as ***diploid*** (DIP-loyd; *di* = two) ***cells,*** symbolized as $2n$. In a diploid cell, two chromosomes that belong to a pair are called ***homologous*** (hō-MOL-ō-gus) ***chromosomes,*** or ***homologues.*** In human diploid cells, 22 of the 23 pairs of chromosomes are morphologically similar and are called ***autosomes.*** The other pair is called the ***sex chromosomes*** and designated as X and Y. In females, the homologous pair of sex chromosomes consists of two X chromosomes; in males, the pair consists of an X and a Y chromosome.

If gametes had the same number of chromosomes as somatic cells, the zygote formed from their fusion would have double the number. The somatic cells of the resulting individual would have twice the number of chromosomes ($4n$) as the somatic cells of the parents, and with every succeeding generation, the number of chromosomes would double. The chromosome number does not double with each generation because of a special nuclear division called ***meiosis.*** Meiosis occurs only in the development of gametes, and it results in the production of cells that contain only 23 chromosomes. Thus, gametes are ***haploid*** (HAP-loyd) ***cells,*** meaning "one-half," and are symbolized as n.

Meiosis

The formation of haploid sperm cells in the testes of the male consists of several phases and is called ***spermatogenesis.*** One of the phases involves meiosis. The formation of haploid ova (eggs) in the ovaries of the female also involves several phases and is referred to as ***oogenesis.*** It, too, involves meiosis. Both spermatogenesis and oogenesis are discussed in detail in Chapter 28. At this point, we will examine only the essentials of meiosis.

Meiosis occurs in two successive nuclear divisions referred to as ***reduction division (meiosis I)*** and ***equatorial division (meiosis II).*** During the interphase that precedes reduction division of meiosis, the chromosomes replicate themselves. This replication is similar to that in interphase preceding the mitosis of somatic cell division. Once chromosomal replication is complete, reduction division begins. It consists of four phases referred to as prophase I, metaphase I, anaphase I, and telophase I (Figure 3-18).

Prophase I is an extended phase in which the chromosomes shorten and thicken, the nuclear membrane and nucleoli disappear, the centrioles replicate, and the mitotic spindle appears. Unlike the prophase of mitosis, however, a unique event occurs in prophase I of meiosis. The chromosomes line up along the equatorial plane region in homologous pairs. The pairing is called ***synapsis.*** The four chromatids of each homologous pair are referred to as a ***tetrad.*** Another unique event of meiosis occurs within a tetrad. Portions of one chromatid may be exchanged with portions of another, a process called ***crossing-over*** (Figure 3-19). This process, among others, permits an exchange of genes among chromatids so that subsequent daughter cells produced are unlike each other genetically and unlike the parent cell that produced them. This phenomenon accounts for part of the great genetic variation among humans and other organisms that form gametes by meiosis. In metaphase I, the paired chromosomes line up along the equatorial plane of the cell, with one member of each pair on either side. Recall that there is no pairing of homologous chromosomes during the metaphase of mitosis. The centromeres of each chromatid pair form chromosomal microtubules that attach the centromeres to opposite poles of the cell. Anaphase I is characterized by separation of the members of each homologous pair, with one member of each pair moving to an opposite pole of the cell. During anaphase I, unlike mitotic anaphase, the centromeres do not split and the paired chromatids, held by a centromere, remain together. Telophase I and cytokinesis are similar to telophase and cytokinesis of mitosis. The net effect of reduction division is that each resulting daughter cell contains the haploid number of chromosomes; each cell contains only one member of each pair of the original homologous chromosomes in the starting parent cell.

The interphase between reduction division and equatorial division is either brief or lacking altogether. It does differ from the interphase preceding reduction division in that there is no replication of DNA between the reduction and equatorial divisions.

The equatorial division of meiosis consists of four phases referred to as prophase II, metaphase II, anaphase II, and telophase II. These phases are essentially similar to those that occur during mitosis since the centromeres divide and chromatids separate and move toward opposite poles of the cell.

In reviewing the overall process, note that during reduction division we start with a parent cell with the diploid number and end up with two daughter cells, each with the haploid number. During equatorial division, each haploid cell formed during reduction division divides, and the net result is four haploid cells. As you will see later, all four haploid cells develop into sperm cells in the testes of the male, but only one of the haploid cells has the potential to develop into an ovum in the female. The other three become structures called ***polar bodies*** that do not function as gametes.

A very simplified comparison of mitosis and meiosis is illustrated in Figure 3-20.

FIGURE 3-18 Meiosis. See text for details.

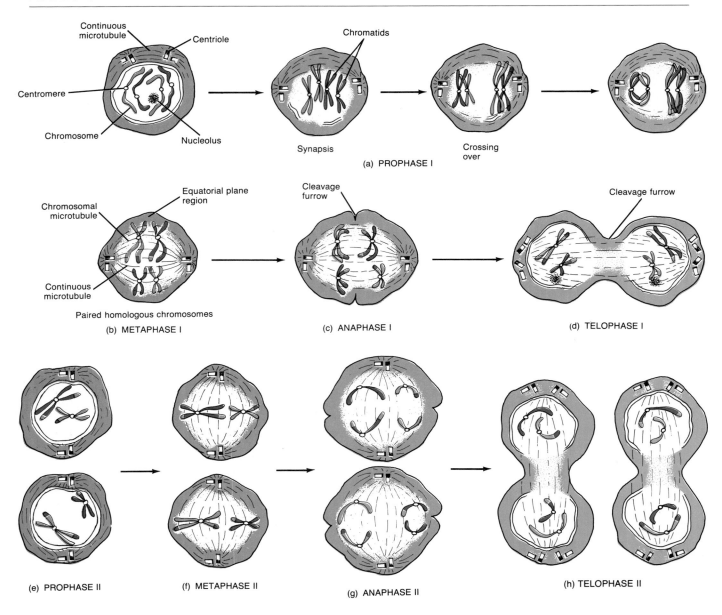

Continuous microtubule
Centriole
Chromatids
Centromere
Chromosome
Nucleolus
Synapsis
Crossing over
(a) PROPHASE I

Equatorial plane region
Chromosomal microtubule
Continuous microtubule
Paired homologous chromosomes
(b) METAPHASE I

Cleavage furrow
(c) ANAPHASE I

Cleavage furrow
(d) TELOPHASE I

(e) PROPHASE II

(f) METAPHASE II

(g) ANAPHASE II

(h) TELOPHASE II

FIGURE 3-19 Crossing-over within a tetrad.

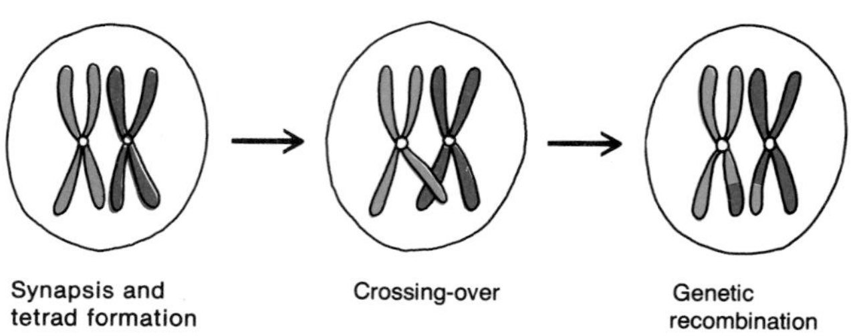

FIGURE 3-20 Very simplified comparison between (a) mitosis and (b) meiosis.

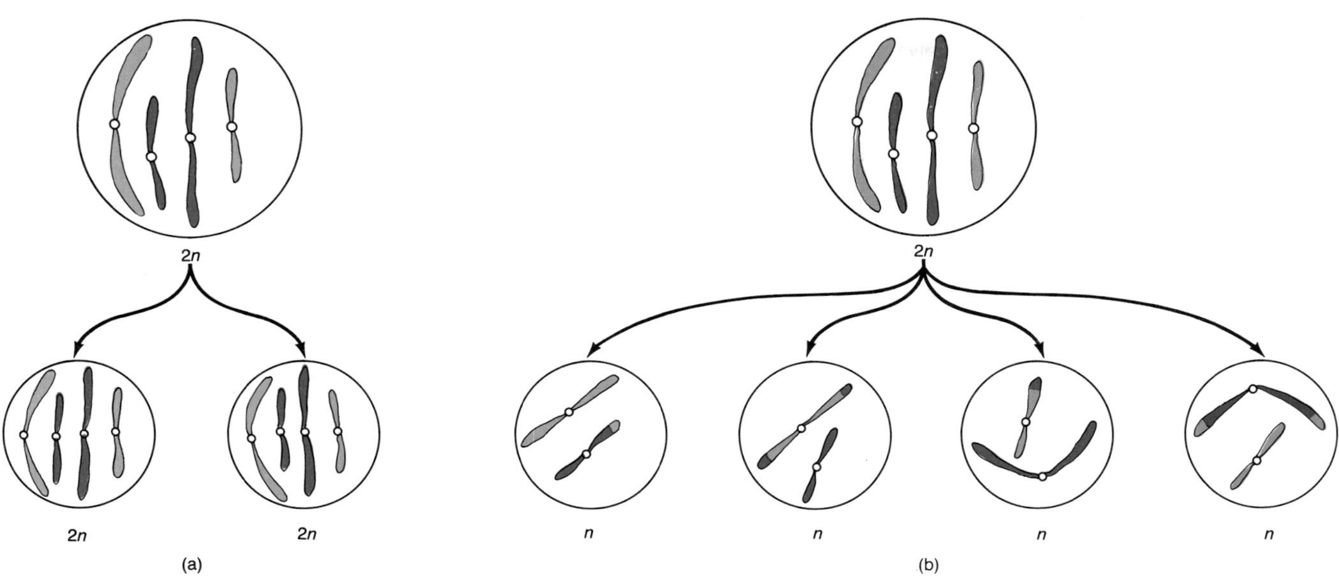

ABNORMAL CELL DIVISION: CANCER (CA)

Definition

When cells in some area of the body duplicate without control, the excess of tissue that develops is called a ***tumor, growth,*** or ***neoplasm.*** The study of tumors is called ***oncology*** (*onco* = swelling or mass; *logos* = study of), and a physician who specializes in this field is called an ***oncologist.*** Tumors may be cancerous and sometimes fatal, or they may be quite harmless. A cancerous growth is called a ***malignant tumor,*** or ***malignancy.*** A noncancerous growth is called a ***benign growth.*** Benign tumors are composed of cells that do not spread to other parts of the body, but they may be removed if they interfere with a normal body function or are disfiguring.

Spread

Cells of malignant growths duplicate continuously and very often quickly and without control. The majority of cancer patients are not killed by the ***primary tumor*** that develops but by secondary infections of bacteria and viruses due to lowered resistance as a result of ***metastasis*** (me-TAS-ta-sis), the spread of the disease to other parts of the body. One of the unique properties of a malignant tumor is its ability to metastasize. Tumor cells secrete a protein called ***autocrine motility factor (AMF)*** that enables them to metastasize. Metastatic groups of cells are more difficult to detect and eliminate than primary tumors.

In the process of metastasis, there is an initial invasion of the malignant cells into surrounding tissues. As the cancer grows, it expands and begins to compete with normal tissues for space and nutrients. Eventually, the normal tissue atrophies and dies. The invasiveness of the malignant cells may be related to mechanical pressure of the growing tumor, motility of the malignant cells, and enzymes produced by the malignant cells. Also, malignant cells lack what is called ***contact inhibition.*** When nonmalignant cells of the body divide and migrate (for example, skin cells that multiply to heal a superficial cut), their further migration is inhibited by contact on all sides with other skin cells. Unfortunately, malignant cells do not conform to the rules of contact inhibition; they have the ability to invade healthy body tissues with very few restrictions.

Following invasion, some of the malignant cells may detach from the primary tumor and invade a body cavity (abdominal or thoracic) or enter the blood or lymph. This latter condition can lead to widespread metastasis. In the next step in metastasis, those malignant cells that survive in the blood or lymph invade adjacent body tissues and establish ***secondary tumors.*** It is believed that some of the invading cells involved in metastasis have properties different from those of the primary tumor that enhance metastasis. These include appropriate mechanical, enzymatic, and surface properties. In the final stage of metastasis, the secondary tumors become vascularized; that is, they take on new networks of blood vessels that provide nutrients for their further growth. Any new tissue, whether it results from repair, normal growth, or tumors, requires a blood supply. Proteins that serve as chemical triggers for blood vessel growth are called ***tumor angiogenesis factors (TAFs)***, which have been isolated from human colon tumors. In all stages of metastasis, the malignant cells resist the antitumor defenses of the body. The pain associated with cancer develops when the growth puts pressure on nerves or blocks a passageway so that secretions build up pressure.

Types

At present, cancers are classified by their microscopic appearance and the body site from which they arise. At least 100 different cancers have been identified in this way. If finer details of appearance are taken into consideration, the number can be increased to 200 or more. The name of the cancer is derived from the type of tissue in which it develops. ***Carcinoma*** (*carc* = cancer; *oma* = tumor) refers to a malignant tumor consisting of epithelial cells. A tumor that develops from a gland is called an ***adenocarcinoma*** (*adeno* = gland). ***Sarcoma*** is a general term for any cancer arising from connective tissue. ***Osteogenic sarcomas*** (*osteo* = bone; *genic* = origin), the most frequent type of childhood cancer, destroy normal bone tissue and eventually spread to other areas of the body. ***Myelomas*** (*myelos* = marrow) are malignant tumors, usually occurring in middle-aged and older people, that interfere with the blood cell–producing function of bone marrow and cause anemia. ***Chondrosarcomas*** (*chondro* = cartilage) are cancerous growths of cartilage.

CLINICAL APPLICATION: GRADING AND STAGING TUMORS

Two methods are used to categorize malignant tumors: grading and staging. In *grading,* pathologists classify tumors into four categories based on the degree to which cells are altered in size, shape, and organization as seen microscopically. Grade I tumors closely resemble normal cells; grade IV tumors are altered dramatically from normal cells. Grades II and III tumors are intermediate in appearance between grades I and IV. Grading aids physicians in cancer prognosis in that grade I tumors have the highest survival rates, whereas grade IV tumors have the lowest. Grading is also used to evaluate precancerous cells such as those of the uterine cervix (Pap smear).

***Staging* is based on the progression of tumor growth and development. One example of a staging system is the most frequently used TNM system that is usually applied to breast and lung cancers as well as others. *T* stands for tumor (T1–T4 defines increasing extent of tumor size), *N* refers to regional lymph nodes affected by the malignancy (N1–N3 indicates advancing nodal disease), and *M* refers to metastasis (MO indicates no metastasis and M1–M3 indicates advancing degrees of metastasis).**

Possible Causes

What triggers a perfectly normal cell to lose control and become abnormal? Scientists are uncertain. First, there are environmental agents: substances in the air we breathe, the water we drink, the food we eat. A chemical or other environmental agent that produces cancer is called a ***carcinogen.*** The World Health Organization estimates that carcinogens may be associated with 60 to 90 percent of all human cancers. Examples of carcinogens are the hydrocarbons found in cigarette tar. Ninety percent of all lung cancer patients are smokers. Another environmental factor is radiation. Ultraviolet (UV) light from the sun, for example, may cause genetic mutations in exposed skin cells and lead to cancer, especially among light-skinned people.

Viruses are a second cause of cancer, at least in animals. These agents are tiny packages of nucleic acids, either DNA or RNA, that are capable of infecting cells and converting them to virus-producers. With over 100 separate viruses identified as carcinogens in many species and tissues of animals, it is also probable that at least some cancers in humans are due to viruses. For example, *human T-cell leukemia-lymphoma virus-1* (*HTLV-1*) is strongly associated with *leukemia,* a malignant disease of blood-forming tissues, and *lymphoma,* a cancer of lymphoid tissue. A variant of HTLV-1 known as *human immunodeficiency virus* (*HIV*) is the causative agent of acquired immune deficiency syndrome (AIDS). The *Epstein-Barr virus* (*EBV*), the causative agent of infectious mononucleosis, has been linked as the causative agent of several human cancers—*Burkitt's lymphoma* (a cancer of white blood cells called B-cells), *nasopharyngeal carcinoma* (common in Chinese males), and *Hodgkin's disease* (a cancer of the lymphatic system). The *hepatitis B virus* (*HBV*) has been associated with cancer of the liver. Also, *type 2 herpes simplex virus,* the causative agent of genital herpes, has been implicated in cancer of the cervix of the uterus, and the *papilloma virus,* a virus that causes warts, has been associated with cancer of the cervix, vagina, vulva, and penis, and cancer of the colon.

A great deal of cancer research is now directed toward studying ***oncogenes*** (ONG-kō-jēnz)—genes that have the ability to transform a normal cell into a cancerous cell. Oncogenes are derived from normal genes that regulate growth and development, called ***proto-oncogenes.*** These genes may undergo some change that either causes them to produce an abnormal product or disrupts their control so that they are expressed inappropriately, making their products in excessive amounts or at the wrong time. It is believed that some oncogenes cause extra production of growth factors, chemicals that stimulate cell growth. Other oncogenes may cause changes in a surface receptor, causing it to send signals as though it were being activated by a growth factor. As a result, the growth pattern of the cell becomes abnormal. Oncogenes are genes that can cause cancer when they are inappropriately activated.

Every human cell contains oncogenes. In fact, oncogenes apparently carry out normal cellular functions until a malignant change occurs. It appears that some proto-oncogenes are activated to oncogenes by various types of mutations in which the DNA of the proto-oncogenes is altered. Such mutations are induced by carcinogens. Other proto-oncogenes are activated by viruses. Some oncogenes can also be activated by a rearrangement of a cell's chromosomes in which segments of DNA are exchanged. This rearrangement is sufficient to activate oncogenes by placing them near genes that enhance their activity. Burkitt's lymphoma, malignant tumors of the colon and rectum, and one type of lung cancer are linked to oncogenes.

Researchers have also determined that some cancers are not caused by oncogenes but may be caused by genes called ***anti-oncogenes.*** These genes can cause cancer when they are inappropriately inactivated. The prototype for a cancer caused by an anti-oncogene is a rare, inherited childhood cancer of the eye called ***retinoblastoma.*** Such inactivation may also be involved in some types of breast cancer and one type of lung cancer.

Currently, scientists are also trying to establish a relationship between stress and cancer. Some believe that stress may play a role not only in the development but also in the metastasis of cancer.

There is also great interest in determining the effects of alterations of the immune system and nutrition in the development of cancer. In 1982 the National Research Council issued a series of guidelines related to diet and cancer. The main recommendations were: (1) to reduce fat intake; (2) to increase consumption of fiber, fruits, and vegetables; (3) to increase intake of complex carbohydrates (potatoes, pasta); and (4) to reduce consumption of salted, smoked, and pickled foods, and simple carbohydrates (refined sugars).

Treatment

Treating cancer is difficult because it is not a single disease and because all the cells in a single population (tumor) do not behave in the same way. The same cancer may contain a diverse population of cells by the time it reaches a clinically detectable size. Although tumor cells look alike when stained and viewed under the microscope, they do not necessarily behave in the same manner in the body. For example, some metastasize and others do not. Some divide and others do not. Some are sensitive to drugs and some are resistant. As a consequence of differences in drug resistance, a single chemotherapeutic drug may destroy susceptible cells but permit resistant cells to proliferate. This is probably one of the reasons that combination chemotherapy is usually more successful. In addition to chemotherapy, radiation therapy, surgery, hyperthermia (abnormally high temperatures), and immunotherapy (bolstering the body's own defenses) may be used alone or in combination.

Scientists are moving closer to developing a vaccine for cancer. What happens in cancer is that the immune system fails to protect the body. Accordingly, the goal of a cancer vaccine is to stimulate the immune system into marshaling a successful attack against the cancer cells.

There has been considerable debate over the use of ***Laetrile*** (LĀ-e-tril) in the treatment of human cancer. Laetrile is a naturally derived substance prepared from apricot pits. In response to public pressure, the National Cancer Institute (NCI) and the Food and Drug Administration (FDA) conducted a clinical trial to determine the effectivenes of Laetrile in the treatment of advanced cancer. The results of the trial were published in 1982 and indicated that Laetrile does not work and is a toxic drug.

CELLS AND AGING

Aging is a normal process accompanied by a progressive alteration of the body's homeostatic adaptive responses. It is a general response that produces observable changes in structure and function and increased vulnerability to environmental stress and disease. Disease and aging probably accelerate each other. The specialized branch of medicine that deals with the medical problems and care of elderly persons is called ***geriatrics*** (jer′-ē-AT-riks; *geras* = old age; *iatrike* = surgery, medicine).

The obvious characteristics of aging are well known: graying and loss of hair, loss of teeth, wrinkling of skin, decreased muscle mass, and increased fat deposits. The physiological signs of aging are gradual deterioration in function and capacity to respond to environmental stress. Thus, basic kidney and digestive metabolic rates decrease, as does the ability to respond effectively to changes in temperature, diet, and oxygen supply in order to maintain a constant internal environment. These manifestations of aging are related to a net decrease in the number of cells in the body (thousands of brain cells are lost each day) and to the dysfunctioning of the cells that remain.

The extracellular components of tissues also change with age. Collagen fibers, responsible for the strength in tendons, increase in number and change in quality with aging. These changes in the collagen of arterial walls are as much responsible for their loss of extensibility as are the deposits associated with atherosclerosis, the deposition of fatty materials in arterial walls. Elastin, another extracellular component, is responsible for the elasticity of blood vessels and skin. It thickens, fragments, and acquires a greater affinity for calcium with age—changes that may also be associated with the development of atherosclerosis.

Several kinds of cells in the body—heart cells, skeletal muscle fibers (cells), nerve cells—are incapable of replacement. Experiments have proved that many other cell types are limited when it comes to cell division. Cells grown outside the body divide only a certain number of times and then stop. The number of divisions correlates with the donor's age and with the normal life span of the different species from which the cells are obtained—strong evidence for the hypothesis that cessation of mitosis is a normal, genetically programmed event. According to this view, an "aging" gene is part of the genetic blueprint at birth, and it turns on at a preprogrammed time, slowing down or halting processes vital to life.

Another theory of aging is the ***free radical theory.*** Free radicals, or oxygen radicals, are oxygen molecules that bear free electrons, are highly reactive, and can easily tie up and weaken proteins. As a result, cells grow rigid as nutrients are excluded and wastes are locked in. Such effects are exhibited as wrinkled skin, stiff joints, and hardened arteries. Free radicals may also cause damage to DNA. Among the factors that produce free radicals are pollution, radiation, and certain foods we eat. Other substances in the diet such as vitamin E, vitamin C, beta-carotene, and selenium, are antioxidants and inhibit free radical formation.

Recently, it has been learned that glucose, the most abundant sugar in the body, may play a role in the aging process. According to one hypothesis, glucose is added, haphazardly, to proteins, forming irreversible cross-links between adjacent protein molecules. As a person ages, more cross-links are formed, and this probably contributes to the stiffening and loss of elasticity that occurs in aging tissues.

Whereas some theories of aging explain the process at the cellular level, others concentrate on regulatory mechanisms operating within the entire organism. For example, one such theory holds that the immune system, which manufactures antibodies against foreign invaders, turns on its own cells. This autoimmune response might be caused by changes in the surfaces of cells, causing antibodies to attack the body's own cells. As surface changes in cells increase, the autoimmune response intensifies, producing the well-known signs of aging. Another organismic theory suggests that aging is programmed in the pituitary gland, a gland that produces and stores hormones and is attached to the undersurface of the brain. Supposedly, at a set time in life, the gland releases a hormone that triggers age-associated disruptions.

The effects of aging on the various body systems are discussed in their respective chapters.

MEDICAL TERMINOLOGY ASSOCIATED WITH CELLS

NOTE TO THE STUDENT

Each chapter in this text that discusses a major system of the body is followed by a glossary of ***medical terminology.*** Both normal and pathological conditions of the system are included in these glossaries. You should familiarize yourself with the terms, since they will play an essential role in your medical vocabulary.

Atrophy (AT-rō-fē; *a* = without; *tropho* = nourish) A decrease in the size of cells with subsequent decrease in the size of the affected tissue or organ; wasting away.

Biopsy (BĪ-op-sē; *bio* = life; *opsis* = vision) The removal and microscopic examination of tissue from the living body for diagnosis.

Deterioration (de-te'-rē-ō-RĀ-shun; *deterior* = worse or poorer) The process or state of growing worse; disintegration or wearing away.

Dysplasia (dis-PLĀ-zē-a; *dys* = abnormal; *plas* = to grow) Alteration in the size, shape, and organization of cells owing to chronic irritation or inflammation; may progress to neoplasia (tumor formation, usually malignant) or revert to normal if the stress is removed.

Hyperplasia (hī'-per-PLĀ-zē-a; *hyper* = over) Increase in the number of cells owing to an increase in the frequency of cell division.

Hypertrophy (hī-PER-trō-fē) Increase in the size of cells without cell division.

Insidious (in-SID-ē-us) Hidden, not apparent, as a disease that does not exhibit distinct symptoms of its arrival.

Metaplasia (met'-a-PLĀ-zē-a; *meta* = change) The transformation of one cell into another.

Metastasis (me-TAS-ta-sis; *stasis* = standing still) The spread of cancer to surrounding tissues (*local metastasis*) or to other body sites (***distant metastasis***).

Necrosis (ne-KRŌ-sis; *necros* = death; *osis* = condition) Death of a group of cells.

Neoplasm (NĒ-ō-plazm; *neo* = new) A new growth that may be benign or malignant.

Progeny (PROJ-e-nē; *progignere* = to bring forth) Offspring or descendants.

Senescence (se-NES-ens) The process of growing old.

STUDY OUTLINE

Generalized Animal Cell (p. 54)

1. A cell is the basic, living, structural and functional unit of the body.
2. A generalized cell is a composite that represents various cells of the body.
3. Cytology is the science concerned with the study of cells.
4. The principal parts of a cell are the plasma (cell) membrane, cytoplasm, organelles, and inclusions. Extracellular materials are manufactured by the cell and deposited outside the plasma membrane.

Plasma (Cell) Membrane (p. 55)

Chemistry and Structure (p. 55)

1. The plasma (cell) membrane surrounds the cell and separates it from other cells and the external environment.
2. It is composed primarily of phospholipids and proteins. According to the fluid mosaic model, the membrane consists of a phospholipid bilayer with integral and peripheral proteins.

Physiology (p. 56)

1. Functionally, the plasma membrane facilitates contact with other cells, provides receptors, and regulates the passage of materials.
2. The membrane's selectively permeable nature restricts the passage of certain substances. Substances can pass through the membrane depending on their molecular size, lipid solubility, electrical charges, and the presence of carriers.

Movement of Materials Across Plasma Membranes (p. 56)

1. Passive (physical) processes depend on the kinetic energy of individual molecules.
2. Diffusion is the net movement of molecules or ions from an area of higher concentration to an area of lower concentration until an equilibrium is reached.
3. In facilitated diffusion, certain molecules, such as glucose, combine with a carrier to become soluble in the phospholipid portion of the membrane.
4. Osmosis is the movement of water through a selectively permeable membrane from an area of higher water concentration to an area of lower water concentration.
5. In an isotonic solution, red blood cells maintain their normal shape; in a hypotonic solution, they undergo hemolysis; in a hypertonic solution, they undergo crenation.
6. Bulk flow is the movement of large numbers of ions, molecules, or particles in the same direction as a result of forces (hydrostatic, osmotic) that push them.
7. Filtration is the movement of water and dissolved substances across a selectively permeable membrane by pressure.
8. Dialysis is the separation of small molecules from large molecules by diffusion across a selectively permeable membrane.
9. Active (physiological) processes depend on the use of ATP by the cell.
10. Active transport is the movement of ions across a cell membrane from lower to higher concentration.
11. Endocytosis is the movement of substances through plasma membranes in which the membrane surrounds the substance, encloses it, and brings it into the cell.
12. Phagocytosis is the ingestion of solid particles by pseudopodia. It is an important process used by white blood cells to destroy bacteria that enter the body.
13. Pinocytosis is the ingestion of a liquid by the plasma membrane. In this process, the liquid becomes surrounded by a vacuole.
14. Receptor-mediated endocytosis is the selective uptake of large molecules and particles by cells.

Cytoplasm (p. 62)

1. Cytoplasm is the substance inside the cell between the plasma membrane and nucleus that contains organelles and inclusions.
2. It is composed mostly of water plus proteins, carbohydrates, lipids, and inorganic substances. The chemicals in cytoplasm are either in solution or in a colloid (suspended) form.
3. Functionally, cytoplasm is the medium in which chemical reactions occur.

Organelles: Structure and Physiology (p. 62)

1. Organelles are specialized portions of the cell with characteristic morphology that carry on specific activities.
2. They assume specific roles in cellular growth, maintenance, repair, and control.

Nucleus (p. 62)

1. Usually the largest organelle, the nucleus controls cellular activities and contains the genetic information.
2. Most body cells have a single nucleus; some (red blood cells) have none, whereas others (skeletal muscle fibers) have several.
3. The parts of the nucleus include the double nuclear membrane, karyolymph, nucleoli, and genetic material (DNA), which comprises the chromosomes.
4. Chromosomes consist of subunits called nucleosomes that are composed of DNA and histones. The various levels of DNA packing are represented by nucleosomes, chromatin fibers, solenoids, loop domains, chromatids, and chromosomes.

Ribosomes (p. 65)

1. Ribosomes are granular structures consisting of ribosomal RNA and ribosomal proteins.
2. They occur free (singly or in clusters) or in conjunction with endoplasmic reticulum.
3. Functionally, ribosomes are the sites of protein synthesis.

Endoplasmic Reticulum (ER) (p. 65)

1. The ER is a network of parallel membranes continuous with the plasma membrane and nuclear membrane.
2. Granular or rough ER has ribosomes attached to it. Agranular or smooth ER does not contain ribosomes.
3. The ER provides mechanical support, releases calcium ions

involved in muscle contraction, conducts intracellular exchange of materials with cytoplasm, transports substances intracellularly, synthesizes lipids and proteins, stores synthesized molecules, and helps export chemicals from the cell.

Golgi Complex (p. 66)

1. The Golgi complex consists of four to eight stacked, flattened membranous sacs (cisternae) referred to as *cis, medial,* and *trans.*
2. The principal function of the Golgi complex is to process, sort, and deliver proteins within the cell. It also secretes proteins and lipids and forms lysosomes.

Mitochondria (p. 69)

1. Mitochondria consist of a smooth outer membrane and a folded inner membrane surrounding the interior matrix. The inner folds are called cristae.
2. The mitochondria are called "powerhouses" of the cell because ATP is produced in them.

Lysosomes (p. 69)

1. Lysosomes are spherical structures that contain digestive enzymes. They are formed from Golgi complexes.
2. They are found in large numbers in white blood cells, which carry on phagocytosis.
3. Lysosomes function in intracellular digestion.
4. If the cell is injured, lysosomes release enzymes and digest the cell. Thus, they are called "suicide packets," and the process is called autolysis.
5. Lysosomes may be involved in bone removal and play a role in embryonic development and the destruction of phagocytized microorganisms.

Peroxisomes (p. 71)

1. Peroxisomes are similar to lysosomes but smaller.
2. They contain enzymes (e.g., catalase) involved in the metabolism of hydrogen peroxide.

The Cytoskeleton (p. 71)

1. Together microfilaments, microtubules, and intermediate filaments form the cytoskeleton.
2. Microfilaments are rodlike structures consisting of the protein actin or myosin. They are involved in muscular contraction, support, and movement.
3. Microtubules are cylindrical structures consisting of the protein tubulin. They support, provide movement, and form the structure of flagella, cilia, centrioles, and the mitotic spindle.
4. Intermediate filaments appear to provide structural reinforcement in some cells.
5. The cytoskeleton and other cytoplasmic components are held together by micotrabeculae; interconnecting microtrabeculae form the microtrabecular lattice.

Centrosome and Centrioles (p. 72)

1. The dense area of cytoplasm containing the centrioles is called a centrosome.
2. Centrioles are paired cylinders arranged at right angles to one another. They assume an important role in cell reproduction by helping to organize the mitotic spindle.

Flagella and Cilia (p. 73)

1. These cellular projections have the same basic structure and are used in movement.
2. If projections are few (typically occurring singly or in pairs) and long, they are called flagella. If they are numerous and hairlike, they are called cilia.
3. The flagellum on a sperm cell moves the entire cell. The cilia on cells of the respiratory tract move foreign matter trapped in mucus along the cell surfaces toward the throat for elimination.

Cell Inclusions (p. 73)

1. Cell inclusions are chemical substances produced by cells. They are usually organic and may have recognizable shapes.
2. Examples are melanin, glycogen, and lipids.

Extracellular Materials (p. 73)

1. These are substances that lie outside the plasma membrane.
2. They provide support and a medium for the diffusion of nutrients and wastes.
3. Some, like hyaluronic acid and chondroitin sulfate, are amorphous. Others, like collagenous, reticular, and elastic fibers, are fibrous.

Gene Action (p. 73)

Protein Synthesis (p. 73)

1. Most of the cellular machinery is concerned with synthesizing proteins.
2. Cells make proteins by translating the genetic information encoded in DNA into specific proteins. This involves transcription and translation.
3. In transcription, genetic information encoded in DNA is copied by a strand of messenger RNA (mRNA); the DNA strand that serves as the template is called the sense strand.
4. DNA also synthesizes ribosomal RNA (rRNA) and transfer RNA (tRNA).
5. The process of using the information in the nitrogenous base sequence of mRNA to dictate the amino acid sequence of a protein is known as translation.
6. mRNA associates with ribosomes, which consist of rRNA and protein.
7. Specific amino acids are attached to molecules of tRNA. Another portion of the tRNA has a triplet of nitrogenous bases called an anticodon; a codon is a segment of three bases of mRNA.
8. tRNA delivers a specific amino acid to the codon; the ribosome moves along an mRNA strand as amino acids are joined to form a growing polypeptide.

Normal Cell Divison (p. 77)

1. Cell division is the process by which cells reproduce themselves. It consists of nuclear division and cytoplasmic division (cytokinesis).
2. Cell division that results in an increase in body cells is called somatic cell division and involves a nuclear division called mitosis and cytokinesis.
3. Cell division that results in the production of sperm and eggs is called reproductive cell division and consists of a nuclear division called meiosis and cytokinesis.

Somatic Cell Division (p. 77)

1. Prior to mitosis and cytokinesis, the DNA molecules, or chromosomes, replicate themselves so the same chromosomal complement can be passed on to future generations of cells.
2. A cell carrying on every life process except division is said to be in interphase (metabolic phase).
3. Mitosis is the distribution of two sets of chromosomes into separate and equal nuclei following their replication.
4. It consists of prophase, metaphase, anaphase, and telophase.
5. Cytokinesis usually begins in late anaphase and terminates in telophase.
6. A cleavage furrow forms at the cell's equatorial plane and progresses inward, cutting through the cell to form two separate portions of cytoplasms.

Reproductive Cell Division (p. 80)

1. Gametes contain the haploid (n) chromosome number, and uninucleated somatic cells contain the diploid ($2n$) chromosome number.
2. Meiosis is one of the processes that produces haploid gametes. It consists of two successive nuclear divisions called reduction division (meiosis I) and equatorial division (meiosis II).
3. During reduction division, homologous chromosomes undergo synapsis and crossing-over; the net result is two haploid daughter cells.
4. During equatorial division, the two haploid daughter cells undergo mitosis, and the net result is four haploid cells.

Abnormal Cell Division: Cancer (CA) (p. 83)

1. Cancerous tumors are referred to as malignant; noncancerous tumors are called benign; the study of tumors is called oncology.
2. The spread of cancer from its primary site is called metastasis.
3. Carcinogens are chemicals or environmental agents that can produce cancer.
4. Oncogenes are genes that can transform normal cells into cancerous cells; their normal counterparts are called protooncogenes.
5. Treating cancer is difficult because all the cells in a single population do not behave the same way.

Cells and Aging (p. 85)

1. Aging is a normal process accompanied by progressive alteration of the body's homeostatic adaptive responses.
2. Many theories of aging have been proposed, including genetically programmed cessation of cell division and excessive immune responses, but none successfully answers all the experimental objections.
3. All the various body systems exhibit definitive and sometimes extensive changes with aging.

REVIEW QUESTIONS

1. Define a cell. What are the four principal portions of a cell? What is meant by a generalized cell? (p. 54)
2. Discuss the chemistry and structure of the plasma membrane with respect to the fluid mosaic model. (p. 55)
3. How do integral and peripheral membrane proteins differ in function? (p. 56)
4. Describe the various functions of the plasma membrane. What determines selective permeability? (p. 56)
5. What are the major differences between passive (physical) processes and active (physiological) processes in moving substances across plasma membranes? (p. 57)
6. Define and give an example of each of the following: diffusion, facilitated diffusion, osmosis, bulk flow, filtration, dialysis, active transport, phagocytosis, pinocytosis, and receptor-mediated endocytosis. (p. 57)
7. Compare the effect on red blood cells of an isotonic, hypertonic, and hypotonic solution. What is osmotic pressure? (p. 59)
8. Discuss the chemical composition and physical nature of cytoplasm. What is its function? (p. 62)
9. What is an organelle? (p. 62) By means of a labeled diagram, indicate the parts of a generalized animal cell.
10. Describe the structure and functions of the nucleus of a cell. Describe how DNA is packed into chromosomes. (p. 62)
11. Discuss the distribution of ribosomes. What is their function? (p. 65)
12. Distinguish between granular (rough) and agranular (smooth) endoplasmic reticulum (ER). What are the functions of ER? (p. 65)
13. Describe the structure and functions of the Golgi complex. (p. 66)
14. Why are mitochondria referred to as "powerhouses" of the cell? (p. 69)
15. List and describe the various functions of lysosomes by contrasting autophagy and autolysis. (p. 69)
16. What is the importance of peroxisomes? (p. 71)
17. Contrast the structure and functions of microfilaments, microtubules, and intermediate filaments. What is the microtrabecular lattice? (p. 71)
18. Describe the structure and function of centrioles. (p. 72)
19. How are flagella and cilia distinguished on the basis of structure and function? (p. 73)
20. Define a cell inclusion. Provide examples and indicate their functions. (p. 73)
21. What are extracellular materials? Give examples and the functions of each. (p. 73)
22. Summarize the steps involving gene action in protein synthesis. (p. 73)
23. What is recombinant DNA? What is its clinical usefulness? (p. 75)
24. Distinguish between the two types of cell division. Why is each important? (p. 77)
25. Define interphase. How does DNA replicate itself? (p. 77)
26. Describe the principal events of each stage of mitosis. (p. 78)
27. Distinguish between haploid (n) and diploid ($2n$) cells. (p. 81)

28. Define meiosis and contrast the principal events of reduction division and equatorial division. (p. 81)
29. What is a tumor? Distinguish between malignant and benign tumors. Describe the principal types of malignant tumors. How are malignant tumors graded and staged? (p. 83)
30. Define metastasis. What factors contribute to metastasis? (p. 83)
31. Discuss several possible causes of cancer (CA). Distinguish oncogenes, proto-oncogenes, and anti-oncogenes. (p. 84)
32. How is cancer treated? What are some of the problems with respect to treating cancer? (p. 85)
33. What is aging? List some of the characteristics of aging. (p. 85)
34. Briefly describe the various theories regarding aging. (p. 86)
35. Define the following terms: liposome (p. 62) and genetic engineering (p. 75) List several substances produced by genetic engineering and explain their uses. (p. 75)
36. Refer to the glossary of medical terminology associated with cells. Be sure that you can define each term. (p. 86)

SELECTED READINGS

Becker, W. M. *The World of the Cell.* Menlo Park, Calif.: Benjamin/Cummings, 1986.

Begley, S., and M. Hager. "Brave New Gene Therapy," *Newsweek,* 13 February 1989.

Berridge, M. J. "The Molecular Basis of Communication within the Cell," *Scientific American,* October 1985.

Bretscher, M. S. "The Molecules of the Cell Membrane," *Scientific American,* October 1985.

Cairns, J. "The Treatment of Diseases and the War against Cancer," *Scientific American,* November 1985.

Dautry-Varsat, A., and H. F. Lodish. "How Receptors Bring Proteins and Particles into Cells," *Scientific American,* May 1984.

de Duve, C. *A Guided Tour of the Living Cell.* New York: Scientific American Books, 1984.

Feldman, M., and L. Eisenbach. "What Makes a Tumor Cell Metastatic?" *Scientific American,* November 1988.

Kartner, N. and V. Ling. "Multidrug Resistance in Cancer," *Scientific American,* March 1989.

Marx, J. L. "How DNA Viruses May Cause Cancer," *Science,* 24 February 1989.

Mazia, D. "The Cell Cycle," *Scientific American,* January 1974.

Porter, K. R., and J. B. Tucker. "The Ground Substance of the Living Cell," *Scientific American,* March 1981.

Ptashne, M. "How Gene Activators Work," *Scientific American,* January 1989.

Radman, M., and R. Wagner. "The High Fidelity of DNA Duplication," *Scientific American,* August 1988.

Rothman, J. E. "The Compartmental Organization of the Golgi Apparatus," *Scientific American,* September 1985.

Unwin, N., and R. Henderson, "The Structure of Proteins in Biological Membranes," *Scientific American,* February 1984.

Weinberg, R. A. "Finding the Anti-Oncogene," *Scientific American,* September 1988.

Chapter 4

The Tissue Level of Organization

Chapter Contents at a Glance

Student Objectives

1. Define a tissue and classify the tissues of the body into four major types.
2. Compare the distinguishing characteristics of epithelial and connective tissues.
3. List the structure, location, and function for the following types of epithelium: simple squamous, simple cuboidal, simple columnar (nonciliated and ciliated), stratified squamous, stratified cuboidal, stratified columnar, transitional, and pseudostratified columnar.
4. Define a gland and distinguish between exocrine and endocrine glands.
5. Discuss the intercellular substance, fibers, and cells that constitute connective tissue.
6. List the structure, function, and location of loose (areolar) connective tissue; adipose tissue; dense, elastic, and reticular connective tissue; cartilage; osseous (bone) tissue; and vascular (blood) tissue.
7. Define an epithelial membrane and list the location and function of mucous, serous, cutaneous, and synovial membranes.
8. Contrast the three types of muscle tissue with regard to structure, location, and modes of control.
9. Describe the structural features and functions of nervous tissue.
10. Describe the conditions necessary for tissue repair.

Cells are highly organized units, but, in multicellular organisms, they do not function in isolation. They work together in a group of similar cells called a tissue.

TYPES OF TISSUES

A ***tissue*** is a group of similar cells and their intercellular substance that have a similar origin in an embryo and function together to perform a specialized activity. The science that deals with the study of tissues is called ***histology*** (hiss'-TOL-ō-jē; *histio* = tissue; *logos* = study of). The various tissues of the body are classified into four principal types according to their function and structure:

1. Epithelial (ep'-i-THĒ-lē-al) tissue, which covers body surfaces, lines body cavities and ducts, and forms glands.
2. Connective tissue, which protects and supports the body and its organs, binds organs together, and stores energy reserves.
3. Muscular tissue, which is responsible for movement through the active generation of force.
4. Nervous tissue, which initiates, transmits, and interprets nerve impulses that coordinate body activities.

Epithelial tissue and connective tissue, except for bone and blood, will be discussed in detail in this chapter. The general features of bone tissue and blood will be introduced here, but their detailed discussion occurs later in the book. Similarly, the detailed discussion of muscle tissue and nervous tissue will be postponed.

EPITHELIAL TISSUE

Epithelial tissues perform many activities in the body, ranging from protection of underlying tissues against microbial invasion, drying out, and harmful environmental factors to secretion. ***Epithelial tissue,*** or more simply, ***epithelium,*** may be divided into two subtypes: (1) ***covering and lining epithelium*** and (2) ***glandular epithelium.*** Covering and lining epithelium forms the outer covering of external body surfaces and the outer covering of some internal organs. It lines body cavities and the interiors of the respiratory and gastrointestinal tracts, blood vessels, and ducts. It makes up, along with nervous tissue, the parts of the sense organs for smell, hearing, vision, and touch which respond to stimuli. And it is the tissue from which gametes (sperm and eggs) develop. Glandular epithelium constitutes the secreting portion of glands.

Both types of epithelium consist largely or entirely of closely packed cells with little or no intercellular material between adjacent cells. (Such intercellular material is also called the ***matrix.***) The points of attachment between adjacent plasma membranes of epithelial cells are called ***cell junctions.*** They not only provide cell-to-cell attachments but also inhibit the movement of materials into certain cells and provide channels for communication between other cells. Epithelial cells are arranged in continuous sheets that may be either single or multilayered. Nerves may extend through these sheets, but blood vessels do not. They are ***avascular*** (*a* = without; *vascular* = blood vessels). The vessels that supply nutrients and remove wastes are located in underlying connective tissue.

Both types of epithelium overlie and adhere firmly to the connective tissue, which holds the epithelium in position and prevents it from being torn. The attachment between the epithelium and the connective tissue is a thin extracellular layer called the ***basement membrane.*** With few exceptions, epithelial cells secrete along their basal surfaces a material consisting of a special type of collagen and glycoproteins. This structure averages 50 to 80 nm in thickness and is referred to as the ***basal lamina.*** Frequently, the basal lamina is reinforced by an underlying ***reticular lamina*** consisting of reticular fibers and glycoproteins. This lamina is produced by cells in the underlying connective tissue. The combination of the basal lamina and reticular lamina constitutes the basement membrane.

Within a tissue, most normal cells remain in place, anchored to basement membranes and connective tissues. In an adult, a few cells, such as phagocytes, routinely move through the intercellular material (matrix) during an infection. And, in an embryo, certain cells migrate extensively as part of the growth and development process. Various activities of cells depend on a variety of ***adhesion proteins*** found in the intercellular material and blood. These proteins interact with receptors on plasma membranes called ***integrins.*** Many adhesion proteins contain tripeptides (arginine, glycine, and aspartic acid) as the recognition site for integrins. Among the adhesion proteins are fibronectin, vitronectin, osteopontin, collagens, thrombospondin, and fibrinogen. Together, adhesion proteins and their integrins function in anchoring cells in position, providing traction for the movement of cells, differentiation, positioning of cells, and possibly growth.

Some tissues of the body are so highly differentiated (specialized) that they have lost their capacity for mitosis. Examples are muscle tissue and nervous tissue. Other tissues, such as epithelium, which are subjected to a certain amount of wear and tear and injury, have a continuous capacity for renewal because they contain stem cells that are capable of renewing the tissue. Examining cells that are sloughed off provides the basis for the Pap smear, a test for precancer and cancer diagnosis of the uterus, cervix, and vagina (Chapter 28).

Covering and Lining Epithelium

Arrangement of Layers

Covering and lining epithelium is arranged in several different ways related to location and function. If the epithelium is specialized for absorption or filtration and is in an area that has minimal wear and tear, the cells of the tissue are arranged in a single layer. Such an arrangement is called ***simple epithelium.*** If the epithelium is found in an area with a high degree of wear and tear, then the cells are stacked in several layers. This tissue is referred to as ***stratified epithelium.*** A third, less common arrangement of epithelium is called ***pseudostratified columnar.*** Like simple epithelium, pseudostratified epithelium has only one layer of cells. However, some of the cells do not reach the surface—an arrangement that gives the tissue a multilayered, or stratified, appearance. The cells in pseudostratified epithelium that do reach the surface either secrete mucus or contain cilia that move mucus and foreign particles for eventual elimination from the body.

Cell Shapes

In addition to classifying covering and lining epithelium according to the number of its layers, we may also categorize it by cell shape. The cells may be flat, cubelike, columnar or a combination of shapes. ***Squamous*** (SKWĀ-mus) cells are flattened and scalelike. They are attached to each other and form a mosaic pattern. ***Cuboidal*** cells are usually cube-shaped in cross section. They sometimes appear as hexagons. ***Columnar*** cells are tall and cylindrical, appearing as somewhat rectangular in shape when set on end. ***Transitional*** cells often have a combination of shapes and are found where there is a great degree of distention or expansion in the body. Transitional cells in the bottom layer of an epithelial tissue may range in shape from cuboidal to columnar. In the intermediate layer, they may be cuboidal or polyhedral. In the superficial layer, they may range from cuboidal to squamous, depending on how much they are pulled out of shape during certain body functions.

Classification

Considering layers and cell shapes in combination, we may classify covering and lining epithelium as follows:

Simple

1. Squamous
2. Cuboidal
3. Columnar

Stratified

1. Squamous
2. Cuboidal
3. Columnar
4. Transitional

Pseudostratified columnar

Each of the epithelial tissues described in the following sections is illustrated in Exhibit 4-1.

Simple Epithelium

▪ **Simple Squamous Epithelium** This type of simple epithelium consists of a single layer of flat, scalelike cells. Its surface resembles a tiled floor. The nucleus of each cell is centrally located and oval or spherical. Since simple squamous epithelium has only one layer of cells, it is highly adapted to diffusion, osmosis, and filtration. Thus, it lines the air sacs of the lungs, where respiratory gases (oxygen and carbon dioxide) are exchanged between air spaces and blood. It is present in the part of the kidney that filters blood. It also lines the inner surfaces of the membranous labyrinth and tympanic membrane of the ear. Simple squamous epithelium is found in body parts that have little wear and tear.

Simple squamous epithelium that lines the heart, blood vessels, and lymph vessels and forms the walls of capillaries is known as ***endothelium.*** Simple squamous epithelium that forms the epithelial layer of serous membranes is called ***mesothelium.*** Serous membranes line the thoracic and abdominopelvic cavities and cover the viscera within them.

▪ **Simple Cuboidal Epithelium** Viewed from above, the cells of simple cuboidal epithelium appear as closely fitted polygons. The cuboidal nature of the cells is obvious only when the tissue is sectioned at right angles. Like simple squamous epithelium, these cells possess a central nucleus that is usually round. Simple cuboidal epithelium covers the surface of the ovaries, lines the anterior surface of the capsule of the lens of the eye, and forms the pigmented epithelium of the retina of the eye. In the kidneys, where it forms the kidney tubules and contains microvilli, it functions in water reabsorption. It also lines the smaller ducts of some glands and the secreting units of glands, such as the thyroid.

Simple cuboidal epithelium performs the functions of secretion and absorption. ***Secretion*** is the production and release by cells of a fluid that may contain a variety of substances such as mucus, perspiration, or enzymes. ***Absorption*** is the intake of fluids or other substances by cells of the skin or mucous membranes.

▪ **Simple Columnar Epithelium** The surface view of simple columnar epithelium is similar to that of simple cuboidal tissue. When sectioned at right angles to the surface, however, the cells appear somewhat rectangular. The nuclei, usually located near the bases of the cells, are commonly oval.

The luminal surfaces (surfaces adjacent to the lumen, or cavity of a hollow organ, vessel, or duct) of simple columnar epithelial cells are modified in several ways, depending on location and function. Simple columnar epithelium lines the gastrointestinal tract from the cardia of the stomach to the anus, the gallbladder, and excretory ducts of many glands. In such sites, the cells protect the underlying tissues. Many of the cells are also modified to

EXHIBIT 4-1 EPITHELIAL TISSUES

COVERING AND LINING EPITHELIUM

Simple Squamous

Description: Single layer of flat, scalelike cells; centrally located nuclei.
Location: Lines air sacs of lungs, glomerular (Bowman's) capsule of kidneys, and inner surfaces of the membranous labyrinth and tympanic membrane of ear. Called endothelium when it lines heart, blood and lymphatic vessels, and forms capillaries. Called mesothelium when it lines the ventral body cavity and covers viscera as part of a serous membrane.
Function: Filtration, absorption, exchange, and secretion in serous membranes.

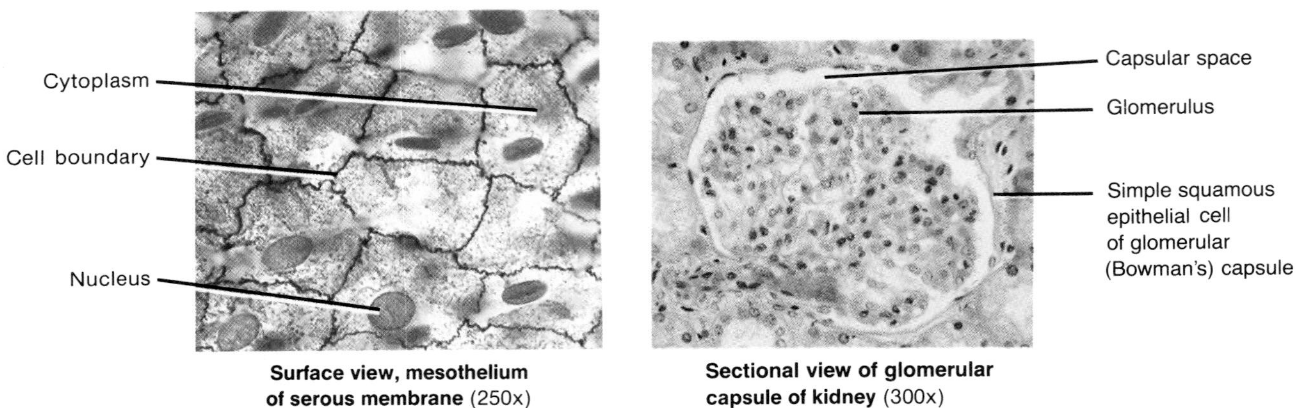

Surface view, mesothelium of serous membrane (250x)

Sectional view of glomerular capsule of kidney (300x)

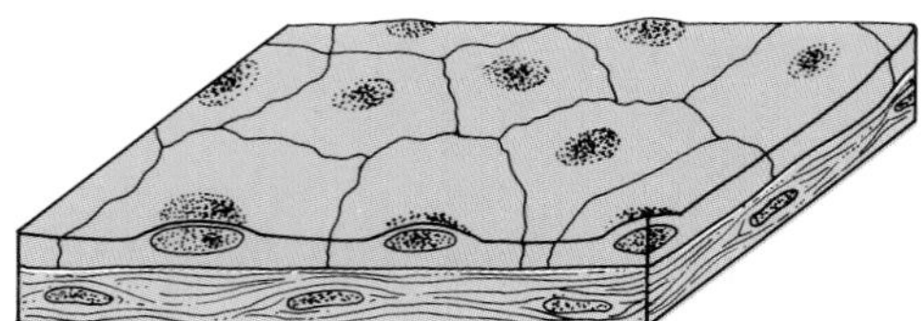

Simple Cuboidal

Description: Single layer of cube-shaped cells; centrally located nuclei.
Location: Covers surface of ovary, lines anterior surface of capsule of the lens of eyes, forms pigmented epithelium of retina of eye, and lines kidney tubules and smaller ducts of many glands.
Function: Secretion and absorption.

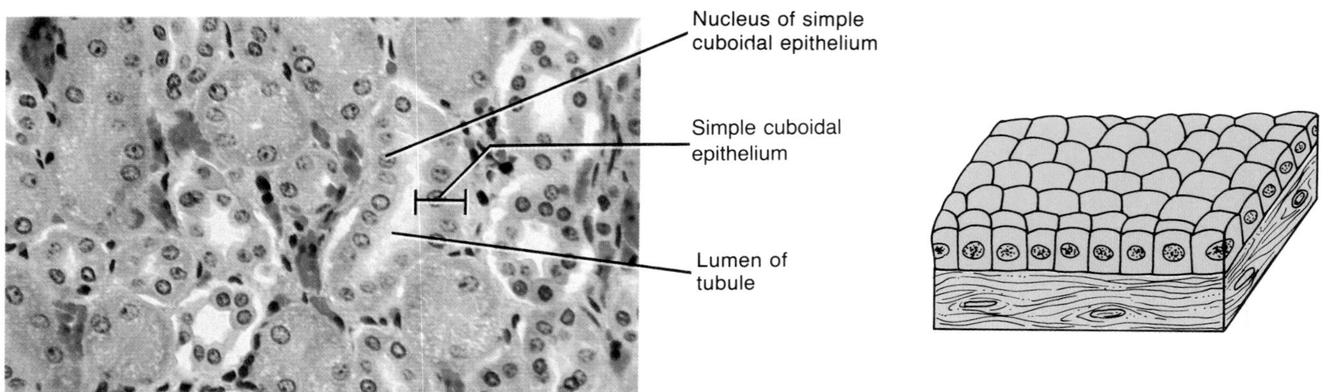

Sectional view of kidney (450x)

EXHIBIT 4-1 EPITHELIAL TISSUES (*Continued*)

Simple Columnar (nonciliated)
Description: Single layer of nonciliated rectangular cells; contains goblet cells in some locations; nuclei at bases of cells.
Location: Lines the gastrointestinal tract from the cardia of the stomach to the anus, excretory ducts of many glands, and gallbladder.
Function: Secretion and absorption.

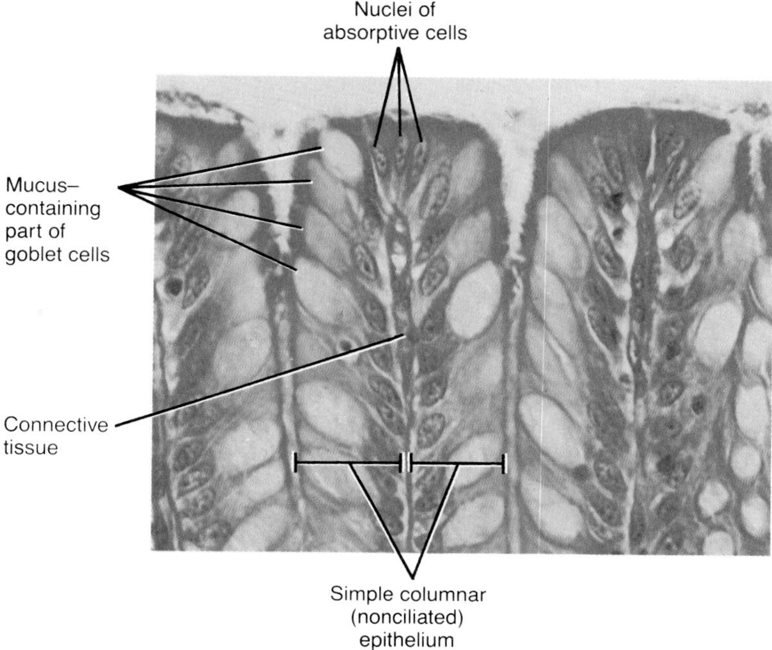

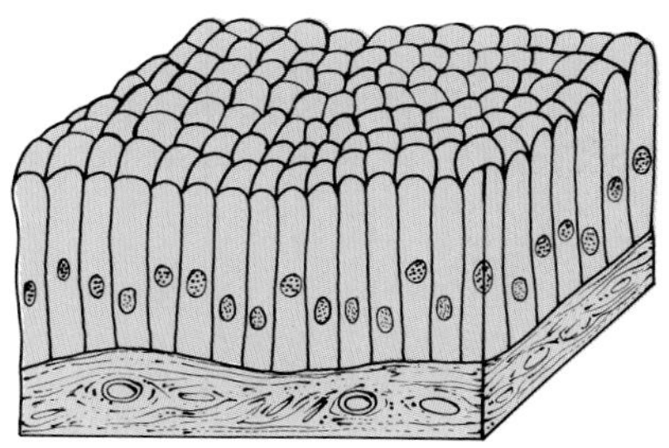

Sectional view of large intestine (600×)

Simple Columnar (ciliated)
Description: Single layer of ciliated columnar cells; contains goblet cells in some locations; nuclei at bases of cells.
Location: Lines a few portions of upper respiratory tract, uterine (Fallopian) tubes, uterus, some paranasal sinuses, and central canal of spinal cord.
Function: Moves mucus by ciliary action.

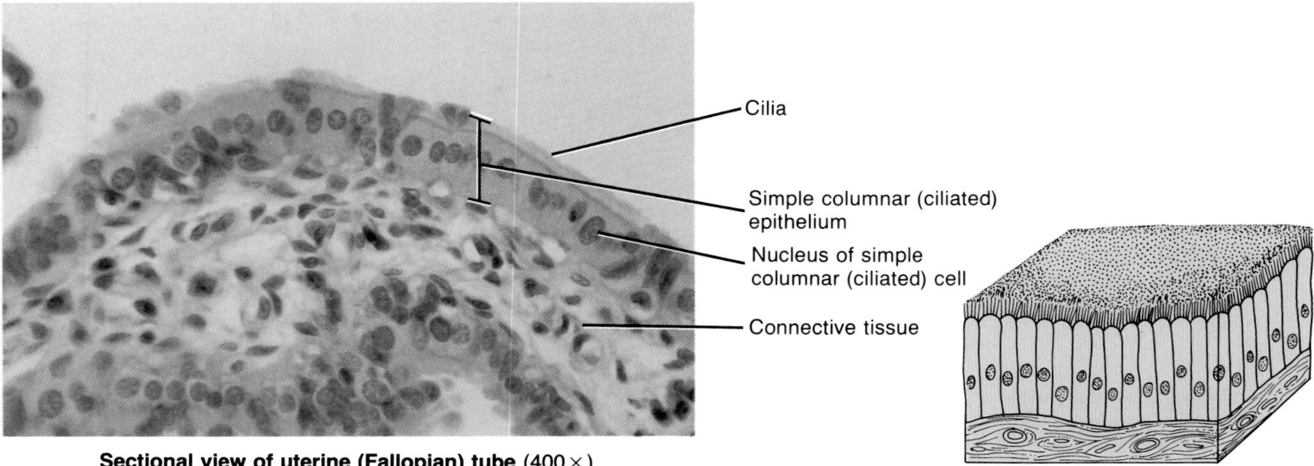

Sectional view of uterine (Fallopian) tube (400×)

EXHIBIT 4-1 EPITHELIAL TISSUES (*Continued*)

Stratified Squamous
Description: Several layers of cells; cuboidal to columnar shape in deep layers; squamous cells in superficial layers; basal cells replace surface cells as they are lost.
Location: Nonkeratinized variety lines wet surfaces such as lining of the mouth, tongue, esophagus, part of epiglottis, and vagina; keratinized variety forms outer layer of skin.
Function: Protection.

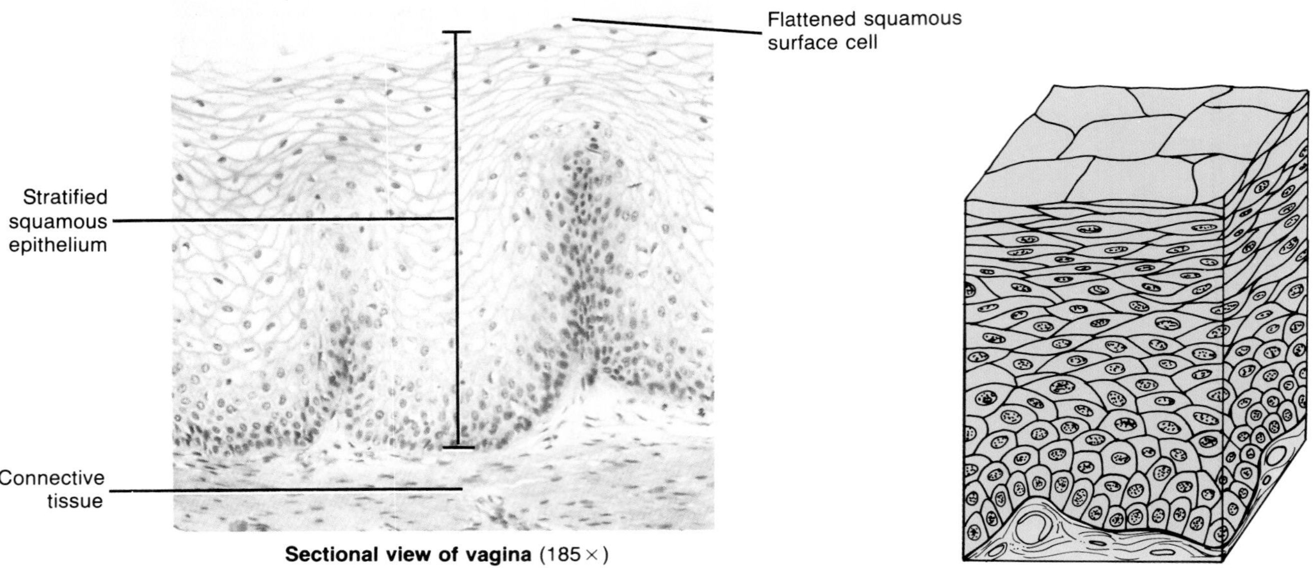

Sectional view of vagina (185 ×)

Stratified Cuboidal
Description: Two or more layers of cells in which the surface cells are cube-shaped.
Location: Ducts of adult sweat glands, fornix of conjunctiva of eye, cavernous urethra of male urogenital system, pharynx, and epiglottis.
Function: Protection.

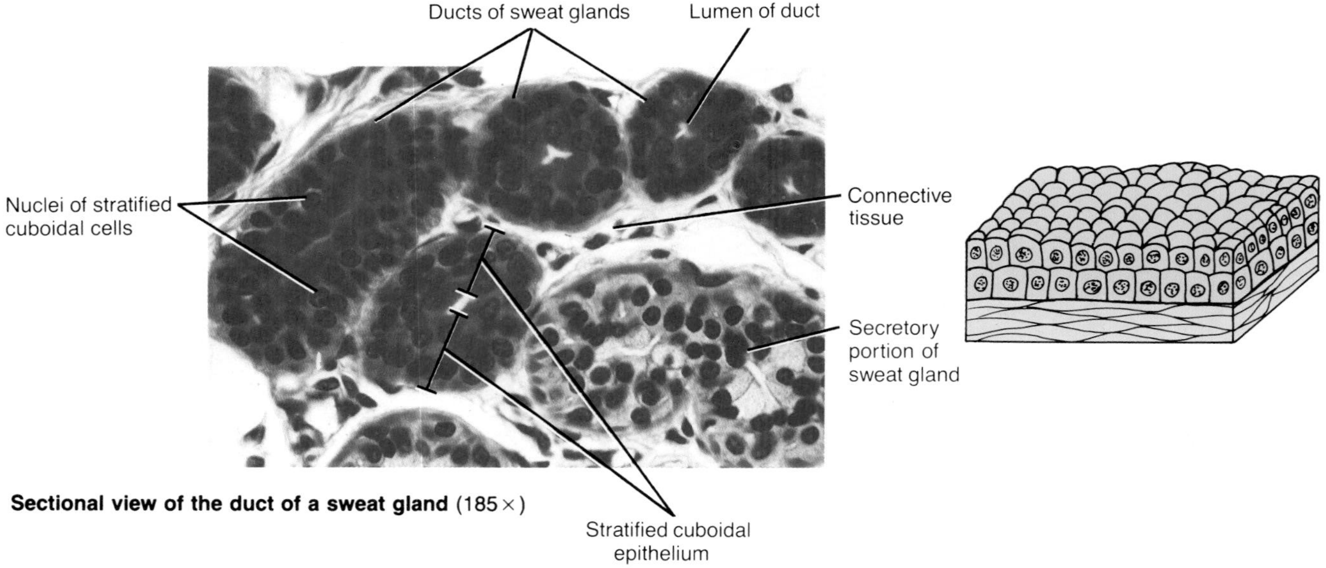

Sectional view of the duct of a sweat gland (185 ×)

EXHIBIT 4-1 EPITHELIAL TISSUES (*Continued*)

Stratified Columnar
Description: Several layers of polyhedral cells; columnar cells only in superficial layer.
Location: Lines part of male urethra, large excretory ducts of some glands, and small areas in anal mucous membrane.
Function: Protection and secretion.

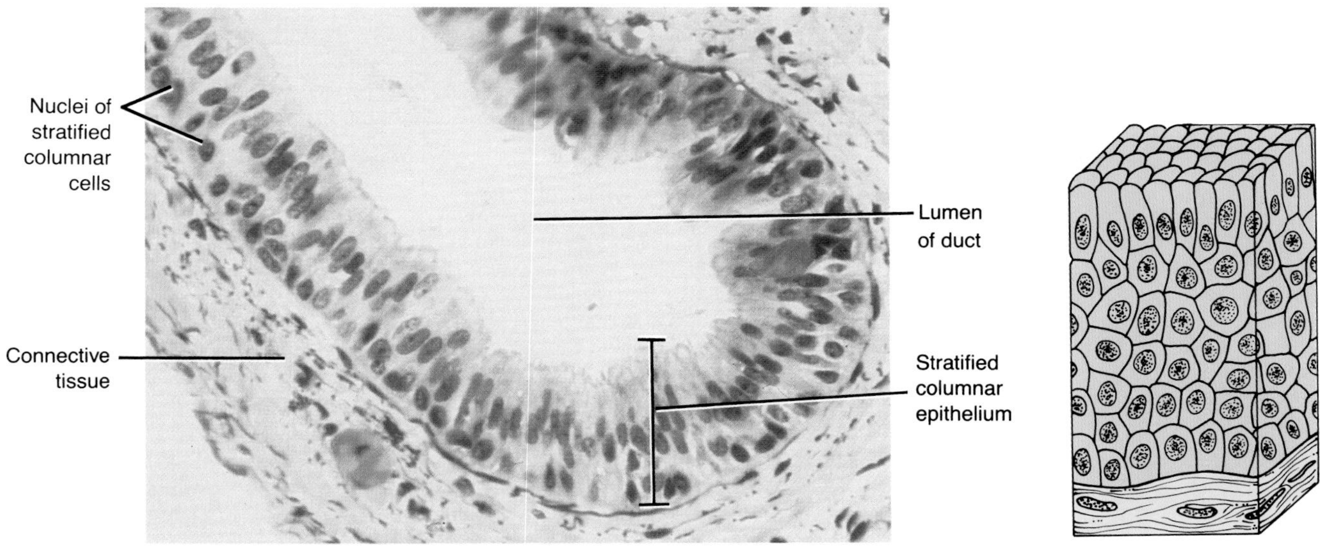

Sectional view of the duct of the submandibular salivary gland (375×)

Transitional
Description: Resembles nonkeratinized stratified squamous tissue, except that superficial cells are larger and have a rounded free surface.
Location: Lines urinary bladder and portions of ureters and urethra.
Function: Permits distention.

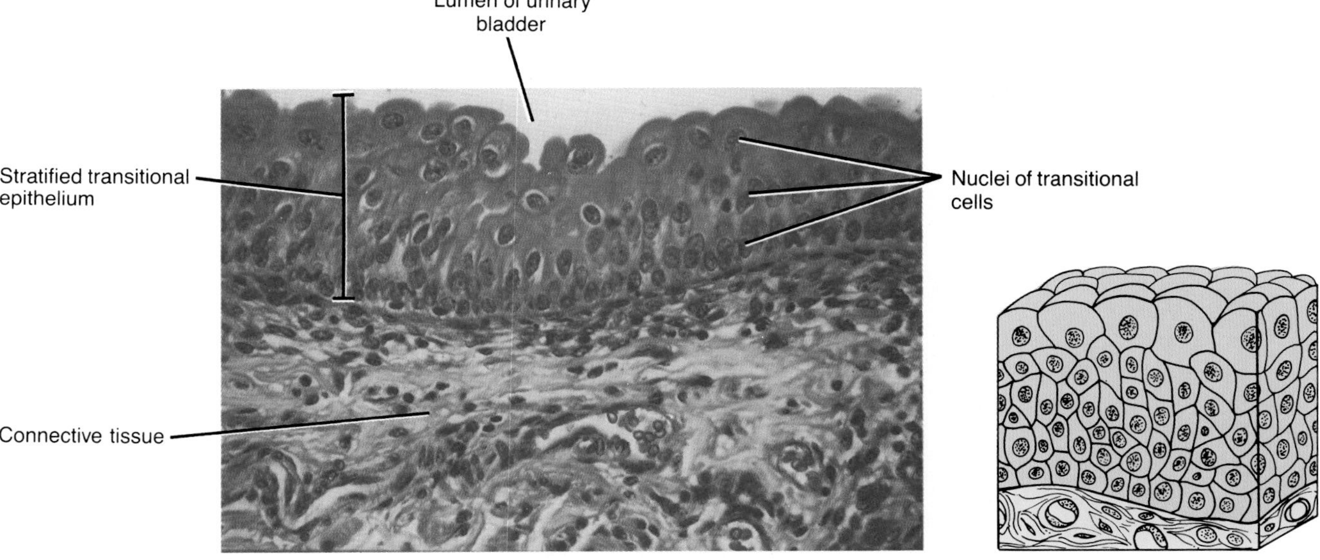

Section of urinary bladder in relaxed state (185×)

EXHIBIT 4-1 EPITHELIAL TISSUES (*Continued*)

Pseudostratified Columnar
Description: Not a true stratified tissue; nuclei of cells at different levels; all cells attached to basement membrane, but not all reach surface.
Location: Lines larger excretory ducts of many large glands, epididymis, male urethra, and auditory (Eustachian) tubes; ciliated variety with goblet cells lines most of the upper respiratory tract and some ducts of male reproductive system.
Function: Secretion and movement of mucus and sperm cells by ciliary action.

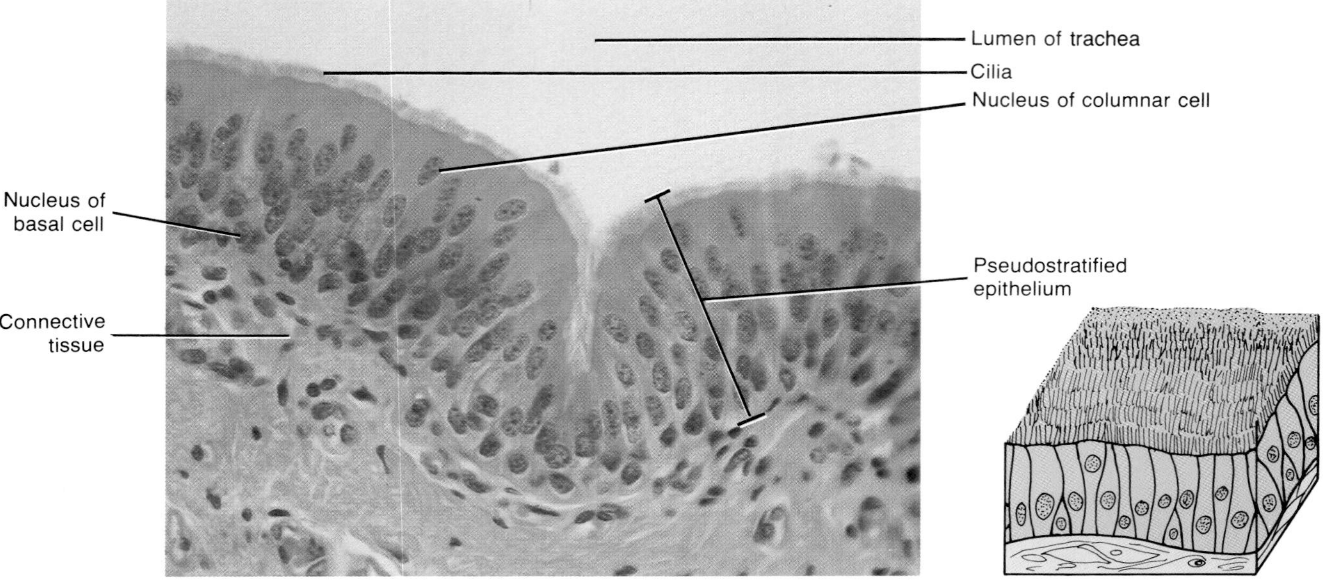

Sectional view of trachea (500x)

GLANDULAR EPITHELIUM

Exocrine Gland
Description: Secretes products into ducts.
Location: Sweat, oil, wax, and mammary glands of the skin; digestive glands such as salivary glands that secrete into mouth cavity and pancreas that secretes into the small intestine.
Function: Produces mucus, perspiration, oil, wax, milk, or digestive enzymes.

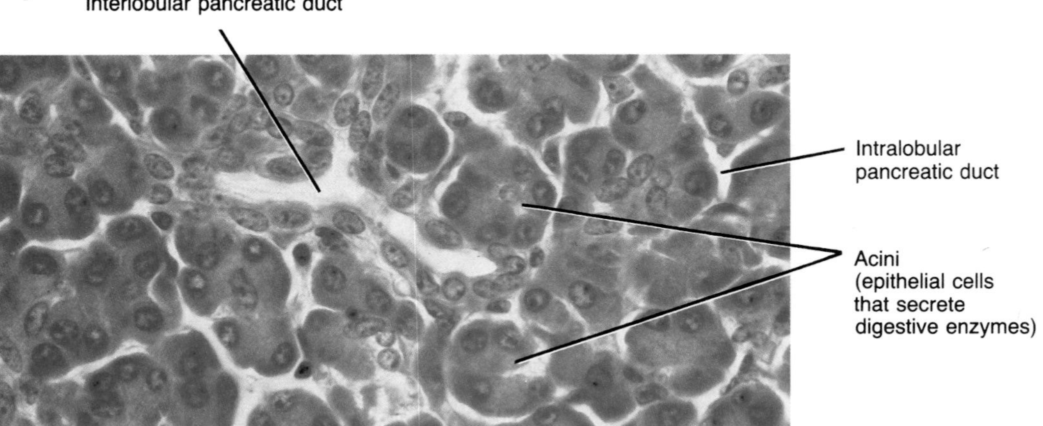

Sectional view of pancreas (300×)

EXHIBIT 4-1 EPITHELIAL TISSUES (*Continued*)

Endocrine Gland
Description: Secretes hormones into blood.
Location: Pituitary at base of brain, thyroid and parathyroids near larynx, adrenals (suprarenals) above kidneys, ovaries in pelvic cavity, testes in scrotum, pineal at base of brain, and thymus in the thoracic cavity.
Function: Produces hormones that regulate various body activities.

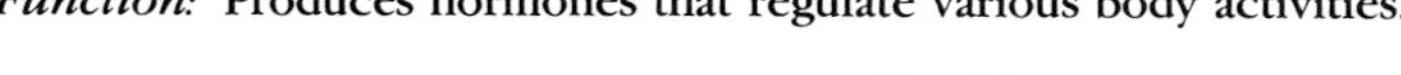

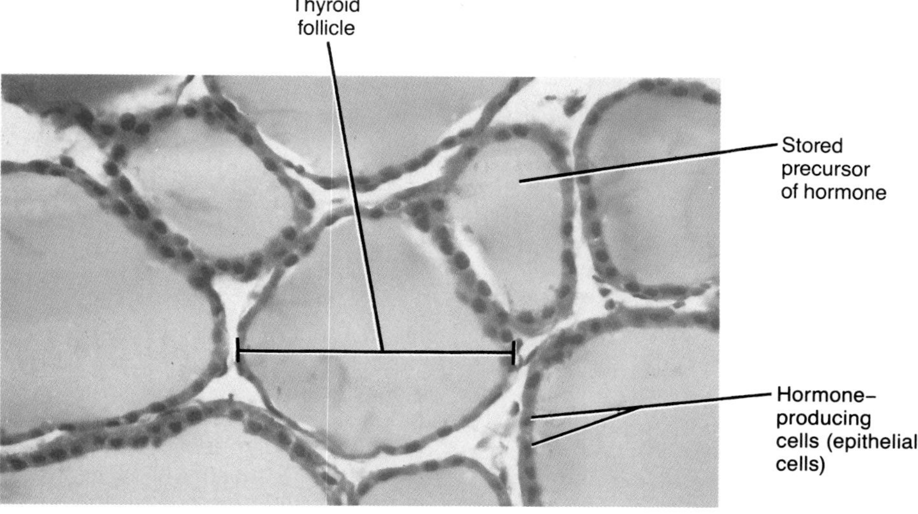

Sectional view of thyroid gland (180×)

Photomicrographs of mesothelium, simple columnar (nonciliated), stratified cuboidal, stratified transitional, exocrine gland, and endocrine gland. © 1983 by Michael H. Ross. Used by permission. Photomicrographs of glomerular capsule, simple cuboidal, simple columnar (ciliated), stratified squamous, stratified columnar, and pseudostratified columnar courtesy of Andrew Kuntzman.

aid in the digestive process. In the small intestine especially, the plasma membranes of the columnar cells have microscopic projections called ***microvilli*** (see Figure 3-1). These structures serve to increase the surface area of the plasma membrane and thereby allow larger amounts of digested nutrients and fluids to be absorbed into the body.

Interspersed among the typical columnar cells of the intestine are other modified columnar cells called ***goblet cells.*** These cells, which secrete mucus, are so named because the mucus accumulates in the upper half of the cell, causing the area to bulge out. The whole cell resembles a goblet or wine glass. The secreted mucus serves as a lubricant between the food and the walls of the gastrointestinal tract.

A third modification of columnar epithelium is found in cells with hairlike processes called ***cilia*** (*cillio* = to move). In a few portions of the upper respiratory tract, ciliated columnar cells are interspersed with goblet cells. Mucus secreted by the goblet cells forms a film over the respiratory surface. This film traps foreign particles that are inhaled. The cilia wave in unison and move the mucus, with any foreign particles, toward the throat, where it can be swallowed or eliminated. Ciliated columnar epithelium is also found in the uterus and uterine (Fallopian) tubes of the female reproductive system, some paranasal sinuses, and the central canal of the spinal cord.

Stratified Epithelium

In contrast to simple epithelium, stratified epithelium consists of at least two layers of cells. Thus, it is more durable and can protect underlying tissues from the external environment and from wear and tear. Some stratified epithelium cells also produce secretions. The name of the specific kind of stratified epithelium depends on the shape of the *surface* cells.

Stratified Squamous Epithelium In the more superficial layers of this type of epithelium, the cells are flat, whereas in the deep layers, cells vary in shape from cuboidal to columnar. The basal, or bottom, cells continually replicate by cell division. As new cells grow, they push the surface cells outward. The basal cells continually shift upward and outward. As they move farther from the deep layer and their blood supply, they become dehydrated, shrink, and become harder. At the surface, the cells are rubbed off. New cells continually emerge, are sloughed off, and replaced.

One form of stratified squamous epithelium is called ***nonkeratinized stratified squamous epithelium.*** This tissue is found on wet surfaces that are subjected to considerable wear and tear—such as the lining of the mouth, the tongue, the esophagus, and the vagina. Another form of stratified squamous epithelium is called ***keratinized stratified squamous epithelium.*** The surface cells of this type form a tough layer of material containing keratin. ***Keratin*** is a protein that is waterproof and resistant to friction and helps to resist bacterial invasion. The outer layer of skin, the epidermis, consists of keratinized stratified squamous epithelium.

▪ **Stratified Cuboidal Epithelium** This relatively rare type of epithelium is found in the ducts of the sweat glands of adults, fornix of the conjunctiva of the eye, cavernous urethra of the male urogenital system, pharynx, and epiglottis. It sometimes consists of more than two layers of cells. Its function is mainly protective.

▪ **Stratified Columnar Epithelium** Like stratified cuboidal epithelium, this type of tissue is also uncommon in the body. Usually the basal layer or layers consist of shortened, irregularly polyhedral cells. Only the superficial cells are columnar in form. This kind of epithelium lines part of the male urethra, some larger excretory ducts such as lactiferous (milk) ducts in the mammary glands, and small areas in the anal mucous membrane. It functions in protection and secretion.

▪ **Transitional Epithelium** This kind of epithelium is very much like nonkeratinized stratified squamous epithelium. The distinction is that cells of the outer layer in transitional epithelium tend to be large and rounded rather than flat. This feature allows the tissue to be stretched (distended) without the outer cells breaking apart from one another. When stretched, they are drawn out into squamouslike cells. Because of this arrangement, transitional epithelium lines hollow structures that are subjected to expansion from within, such as the urinary bladder, and parts of the ureters and urethra. Its function is to help prevent a rupture of the organ.

Pseudostratified Columnar Epithelium

The third category of covering and lining epithelium consists of columnar cells and is called pseudostratified columnar epithelium. The nuclei of the cells are at varying depths. Even though all the cells are attached to the basement membrane in a single layer, some do not reach the surface. This feature gives the impression of a multilayered tissue, the reason for the designation *pseudo*stratified epithelium. It lines the larger excretory ducts of many glands, epididymis, parts of the male urethra, and the auditory (Eustachian) tubes, the tubes that connect the middle ear cavity and upper part of the throat. Pseudostratified columnar epithelium may contain cilia and goblet cells. It lines most of the upper respiratory tract and certain ducts of the male reproductive system.

Glandular Epithelium

The function of glandular epithelium is secretion, accomplished by glandular cells that lie in clusters deep to the covering and lining epithelium. A ***gland*** may consist of one cell or a group of highly specialized epithelial cells that secrete substances into ducts, onto a surface, or into the blood. The production of such substances always requires active work by the cells and results in an expenditure of energy.

All glands of the body are classified as exocrine or endocrine according to whether they secrete substances into ducts (or directly onto a free surface) or into the blood. ***Exocrine glands*** secrete their products into ducts (tubes) that empty at the surface of covering and lining epithelium or directly onto a free surface. The product of an exocrine gland may be released at the skin surface or into the lumen of a hollow organ. The secretions of exocrine glands include mucus, perspiration, oil, wax, and digestive enzymes. Examples of exocrine glands are sweat glands, which eliminate perspiration to cool the skin, salivary glands, which secrete a digestive enzyme, and goblet cells, which produce mucus. ***Endocrine glands*** are ductless and ultimately secrete their products into the blood. The secretions of endocrine glands are always hormones, chemicals that regulate various physiological activities. The pituitary, thyroid, and adrenal (suprarenal) glands are examples of endocrine glands.

Structural Classification of Exocrine Glands

Exocrine glands are classified into two structural types: unicellular and multicellular. ***Unicellular glands*** are single-celled. A good example of a unicellular gland is a goblet cell (see Exhibit 4-1, simple columnar, nonciliated). Goblet cells are found in the epithelial lining of the digestive, respiratory, urinary, and reproductive systems. They produce mucus to lubricate the free surfaces of these membranes.

Multicellular glands occur in several different forms (Figure 4-1). If the secretory portions of a gland are tubular, it is referred to as a ***tubular gland.*** If they are flasklike, it is called an ***acinar*** (AS-i-nar) ***gland.*** If the gland contains both tubular and flasklike secretory portions, it is called a ***tubuloacinar gland.*** Further, if the

FIGURE 4-1 Structural types of multicellular exocrine glands. The secretory portions of the glands are indicated in gold. The blue areas represent the ducts of the glands.

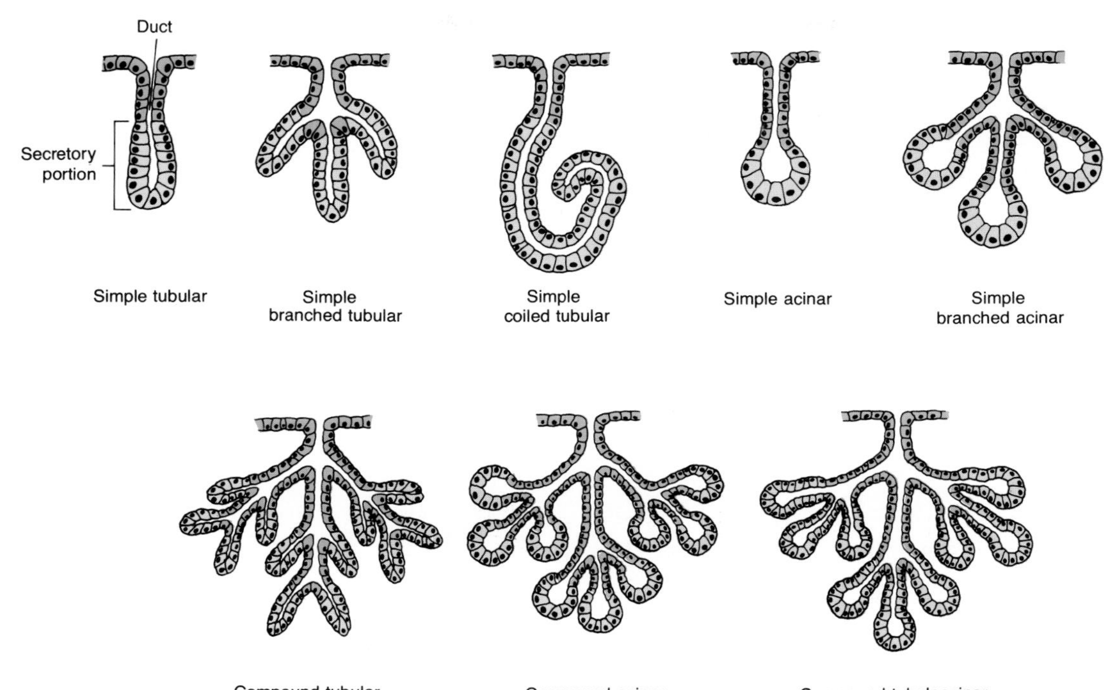

duct of the gland does not branch, it is referred to as a ***simple gland;*** if the duct does branch, it is called a ***compound gland.*** By combining the shape of the secretory portion with the degree of branching of the duct, we arrive at the following structural classification for exocrine glands:

I. Unicellular. Single-celled gland that secretes mucus. Example: goblet cell of the digestive and respiratory systems.

II. Multicellular. Many-celled glands.

- **A.** *Simple.* Single, nonbranched duct.
    1. **Tubular.** The secetory portion is straight and tubular. Example: intestinal glands.
    2. **Branched tubular.** The secretory portion is branched and tubular. Examples: gastric and uterine glands.
    3. **Coiled tubular.** The secretory portion is coiled. Example: eccrine sudoriferous (sweat) glands.
    4. **Acinar.** The secretory portion is flasklike. Example: seminal vesicle glands.
    5. **Branched acinar.** The secretory portion is branched and flasklike. Example: sebaceous (oil) glands.
- **B.** *Compound.* Branched duct.
    1. **Tubular.** The secretory portion is tubular. Examples: bulbourethral (Cowper's) glands, testes, and liver.
    2. **Acinar.** The secretory portion is flasklike. Examples: salivary glands (sublingual and submandibular).
    3. **Tubuloacinar.** The secretory portion is both tubular and flasklike. Examples: salivary glands (parotid) and pancreas.

Functional Classification of Exocrine Glands

The functional classification of exocrine glands is based on whether a secretion is a product of a cell or consists of entire or partial glandular cells themselves. The three recognized categories are holocrine, merocrine, and apocrine glands. ***Holocrine glands*** accumulate a secretory product in their cytoplasm. The cell then dies and is discharged with its contents as the glandular secretion (Figure 4-2a). The discharged cell is replaced by a new cell. One example of a holocrine gland is a sebaceous (oil) gland of the skin. ***Merocrine (eccrine) glands*** simply form the secretory product and discharge it from the cell (Figure 4-2b). Most exocrine glands of the body are merocrine. Examples of merocrine glands are the salivary glands and pancreas. ***Apocrine glands*** accumulate their secretory product at the apical (outer) margin of the secreting cell. That portion of the cell pinches off from the rest of the cell to form the secretion (Figure 4-2c). The remaining part of the cell repairs itself and repeats the process. Examples of apocrine glands are large, modified sweat glands found in the axillary, anal, and genital areas and in the mammary glands.

FIGURE 4-2 Functional classification of multicellular exocrine glands. (a) Holocrine gland. (b) Merocrine gland. (c) Apocrine gland.

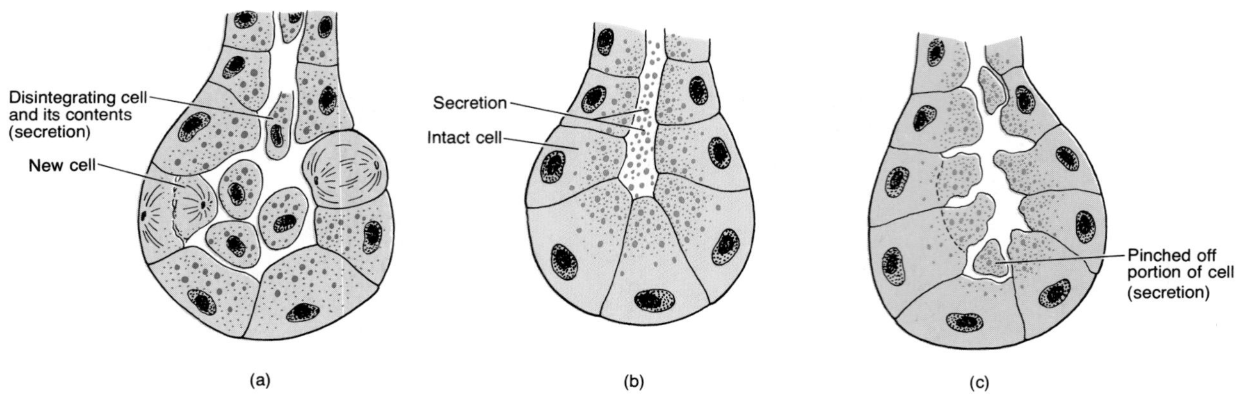

CONNECTIVE TISSUE

The most abundant tissue in the body is ***connective tissue.*** This binding and supporting tissue usually is highly vascular and thus has a rich blood supply. An exception is cartilage, which is avascular. The connective tissue cells are usually widely scattered, rather than closely packed, and there is considerable intercellular substance (matrix). In contrast to epithelium, connective tissues do not occur on free surfaces, such as the surfaces of a body cavity or the external surface of the body. The general functions of connective tissues are protection, support, binding together various organs, separating structures such as skeletal muscles, and storage of reserve energy.

The intercellular substance in a connective tissue largely determines the tissue's qualities. This substance is nonliving and may consist of fluid, semifluid, gellike, or fibrous material. In cartilage, the intercellular material is firm but pliable. In bone, it is considerably harder and not pliable. The cells of connective tissue produce the intercellular substances. The cells may also store fat, ingest bacteria and cell debris, form anticoagulants, or give rise to antibodies that protect against disease.

Classification

Connective tissues may be classified in several ways. We will classify them as follows:

I. Embryonic connective tissue
- **A.** Mesenchyme
- **B.** Mucous connective tissue

II. Adult connective tissue
- **A.** Connective tissue proper
  - **1.** Loose (areolar) connective tissue
  - **2.** Adipose tissue
  - **3.** Dense (collagenous) connective tissue
  - **4.** Elastic connective tissue
  - **5.** Reticular connective tissue
- **B.** Cartilage
  - **1.** Hyaline cartilage
  - **2.** Fibrocartilage
  - **3.** Elastic cartilage
- **C.** Osseous (bone) tissue
- **D.** Vascular (blood) tissue

Each of the connective tissues described in the following sections is illustrated in Exhibit 4-2.

Embryonic Connective Tissue

Connective tissue that is present primarily in the embryo or fetus is called ***embryonic connective tissue.*** The term *embryo* refers to a developing human from fertilization through the first two months of pregnancy; a *fetus* refers to a developing human from the third month of pregnancy to birth.

One example of embryonic connective tissue found almost exclusively in the embryo is ***mesenchyme*** (MEZ-en-kīm)—the tissue from which all other connective tissues eventually arise. Mesenchyme is located beneath the skin and along the developing bones of the embryo. Some mesenchymal cells are scattered irregularly throughout adult connective tissue, most frequently around blood vessels. Here mesenchymal cells differentiate into fibroblasts that assist in wound healing.

Another kind of embryonic connective tissue is ***mucous connective tissue* (*Wharton's jelly*),** found primarily in the fetus. This tissue is located in the umbilical cord of the fetus, where it supports the wall of the cord.

Adult Connective Tissue

Adult connective tissue is connective tissue that exists in the newborn that has differentiated from mesenchyme and that does not change after birth. It is subdivided into several kinds.

EXHIBIT 4-2 CONNECTIVE TISSUES

EMBRYONIC

Mesenchyme

Description: Consists of highly branched mesenchymal cells embedded in a fluid substance.
Location: Under skin and along developing bones of embryo; some mesenchymal cells found in adult connective tissue, especially along blood vessels.
Function: Forms all other kinds of connective tissue.

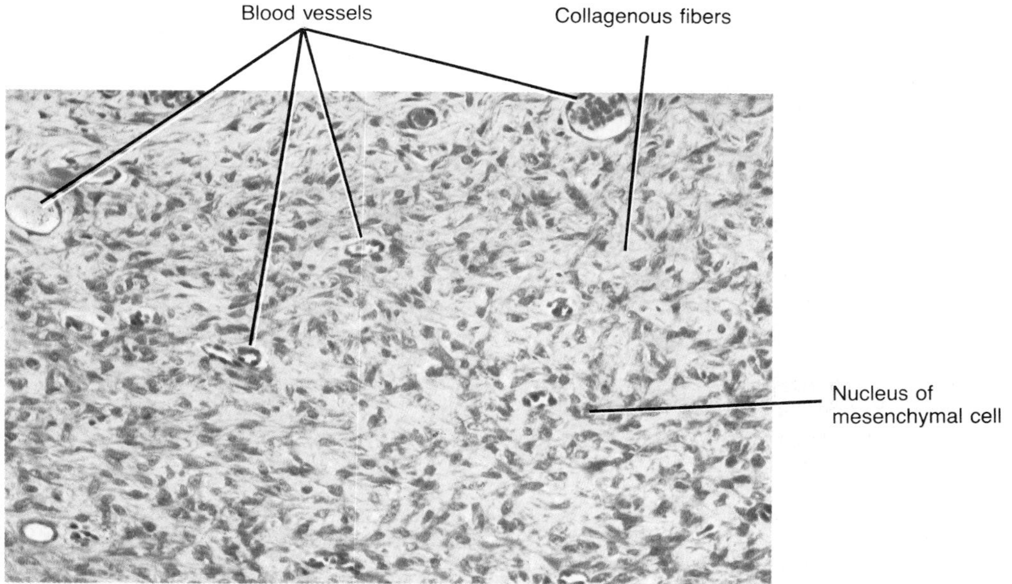

Sectional view of mesenchyme from a developing fetus (180 ×)

Mucous

Description: Consists of flattened or spindle-shaped cells embedded in a mucuslike substance containing fine collagenous fibers.
Location: Umbilical cord of fetus.
Function: Support.

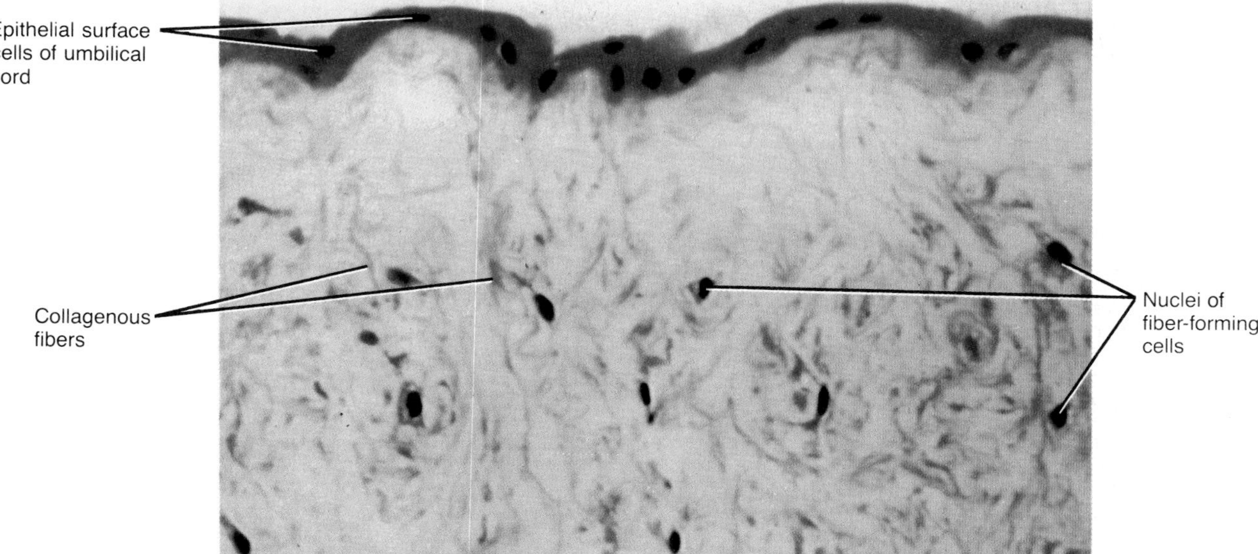

Sectional view of the umbilical cord (320 ×)

EXHIBIT 4-2 CONNECTIVE TISSUES (*Continued*)

ADULT

Loose or Areolar

Description: Consists of fibers (collagenous, elastic, and reticular) and several kinds of cells (fibroblasts, macrophages, plasma cells, adipocytes, and mast cells) embedded in a semifluid ground substance.

Location: Subcutaneous layer of skin, mucous membranes, blood vessels, nerves, and body organs.

Function: Strength, elasticity, and support.

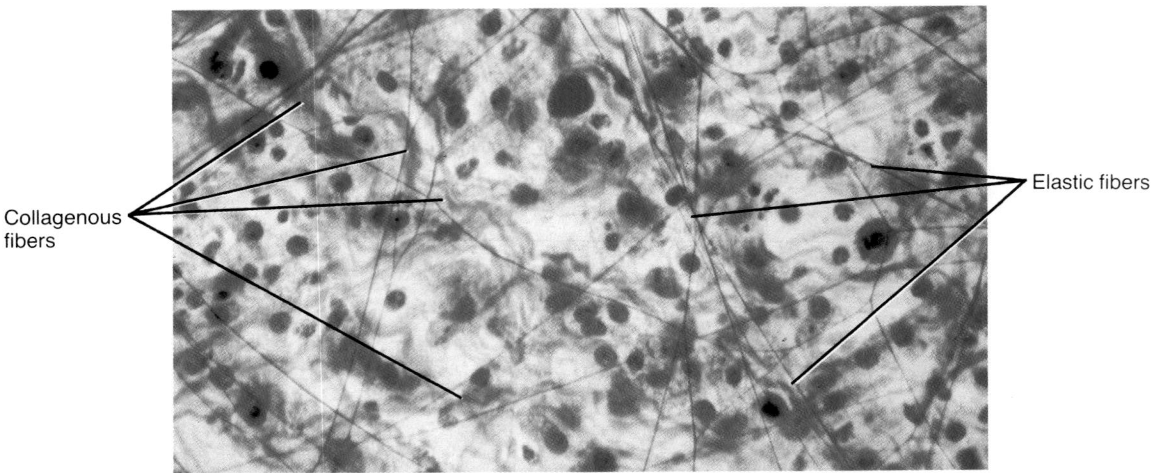

Sectional view of the mesentery of the small intestine of a rat (180×)

Adipose

Description: Consists of adipocytes, "signet ring"–shaped cells with peripheral nuclei, that are specialized for fat storage.

Location: Subcutaneous layer of skin, around heart and kidneys, marrow of long bones, and padding around joints.

Function: Reduces heat loss through skin, serves as an energy reserve, supports, and protects.

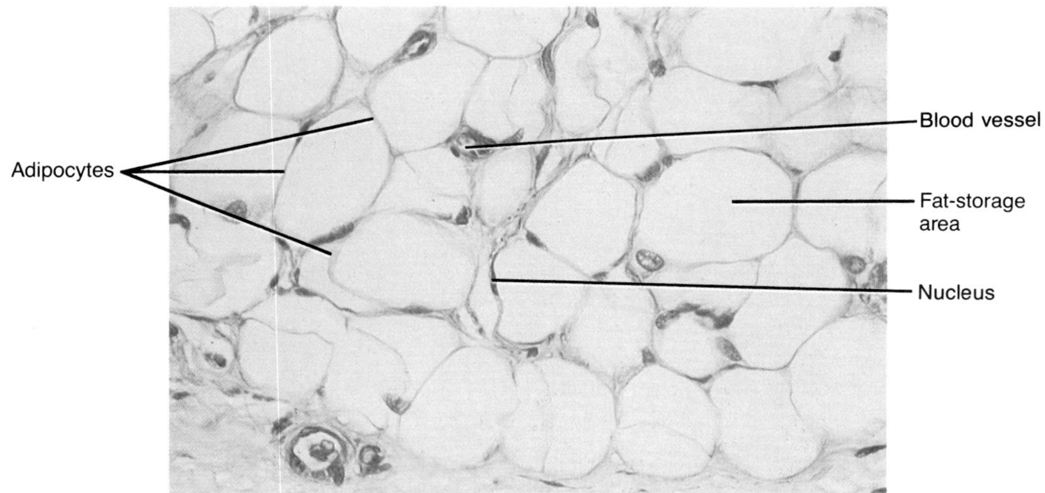

Sectional view of the adventia of the small intestine (adjacent to the mesentery) (250×)

EXHIBIT 4-2 CONNECTIVE TISSUES (*Continued*)

Dense or Collagenous
Description: Consists of predominately collagenous fibers arranged in bundles; fibroblasts present in rows between bundles.
Location: Forms tendons, ligaments, aponeuroses, membranes around various organs, and fasciae.
Function: Provides strong attachment between various structures.

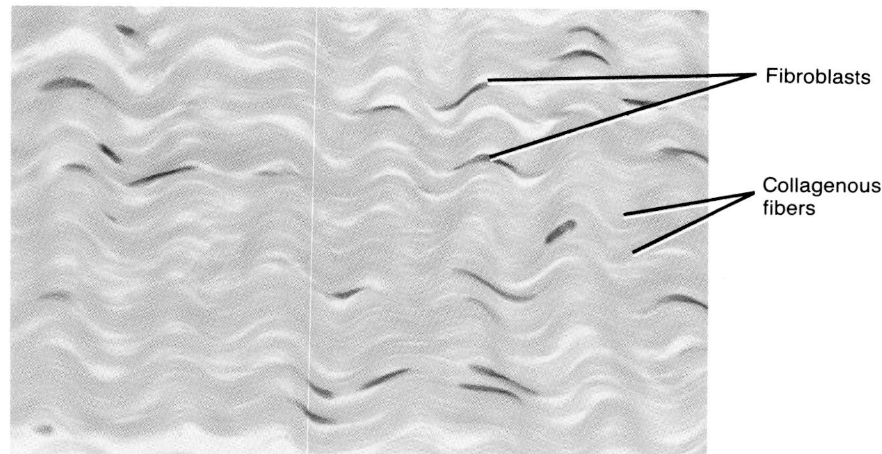

Sectional view of a tendon (250×)

Elastic
Description: Consists of predominantly freely branching elastic fibers; fibroblasts present in spaces between fibers.
Location: Lung tissue, wall of arteries, trachea, bronchial tubes, true vocal cords, and ligmenta flava of vertebrae.
Function: Allows stretching of various organs.

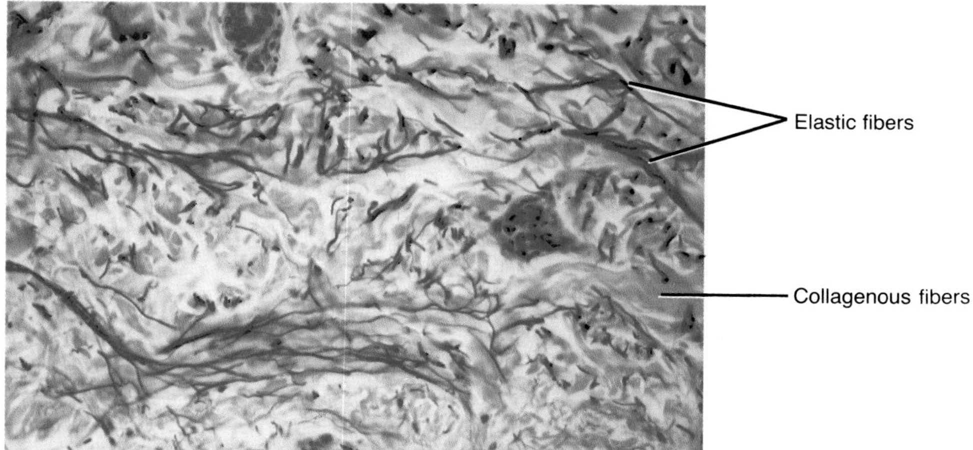

Sectional view of the dermis of the skin (180×)

EXHIBIT 4-2 CONNECTIVE TISSUES (*Continued*)

Reticular
Description: Consists of a network of interlacing reticular fibers with thin, flat cells wrapped around fibers.
Location: Liver, spleen, lymph nodes, and basal lamina underlying epithelia.
Function: Forms stroma of organs; binds together smooth muscle tissue cells.

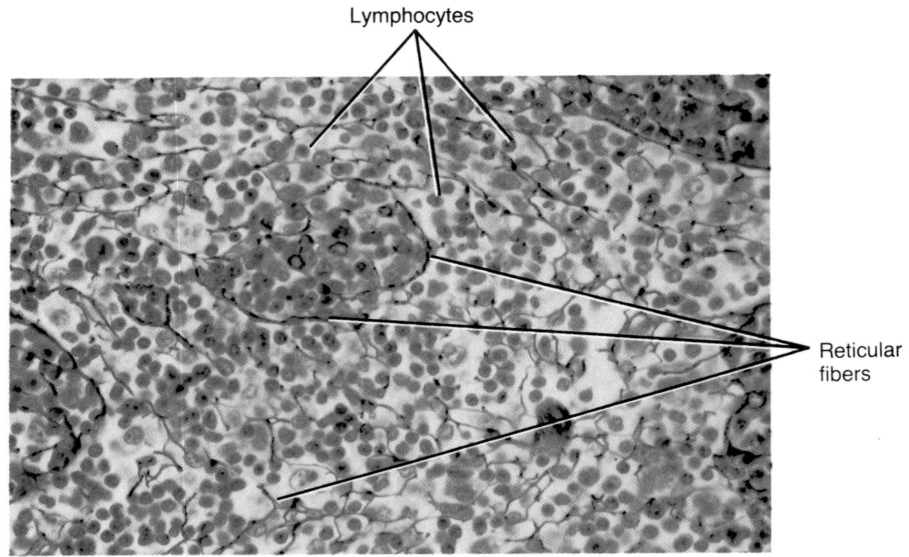

Sectional view of a lymph node (250 ×)

Hyaline Cartilage
Description: Also called gristle; appears as a bluish white, glossy mass; contains numerous chondrocytes; is the most abundant type of cartilage.
Location: Ends of long bones, ends of ribs, nose, parts of larynx, trachea, bronchi, bronchial tubes, and embryonic skeleton.
Function: Provides movement at joints, flexibility, and support.

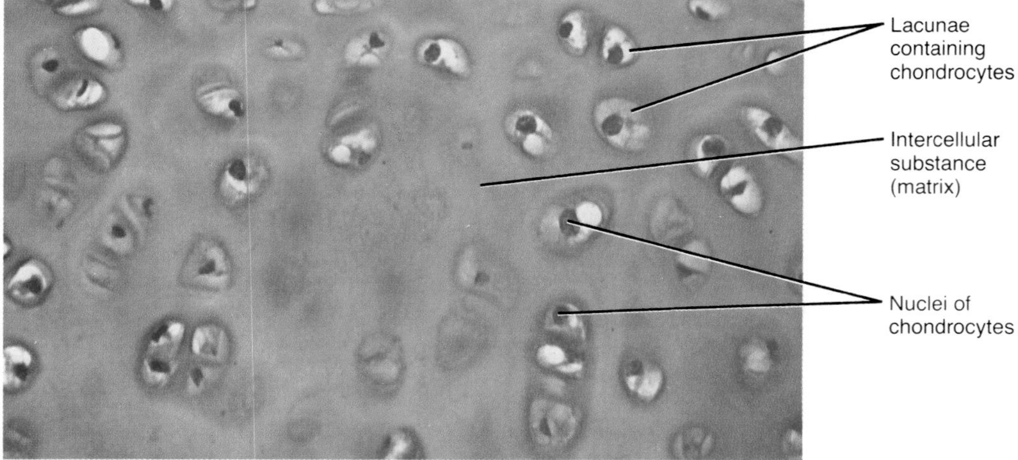

Sectional view of the trachea (350 ×)

EXHIBIT 4-2 CONNECTIVE TISSUES (*Continued*)

Fibrocartilage
Description: Consists of chondrocytes scattered among bundles of collagenous fibers.
Location: Symphysis pubis, intervertebral discs, and menisci of knee.
Function: Support and fusion.

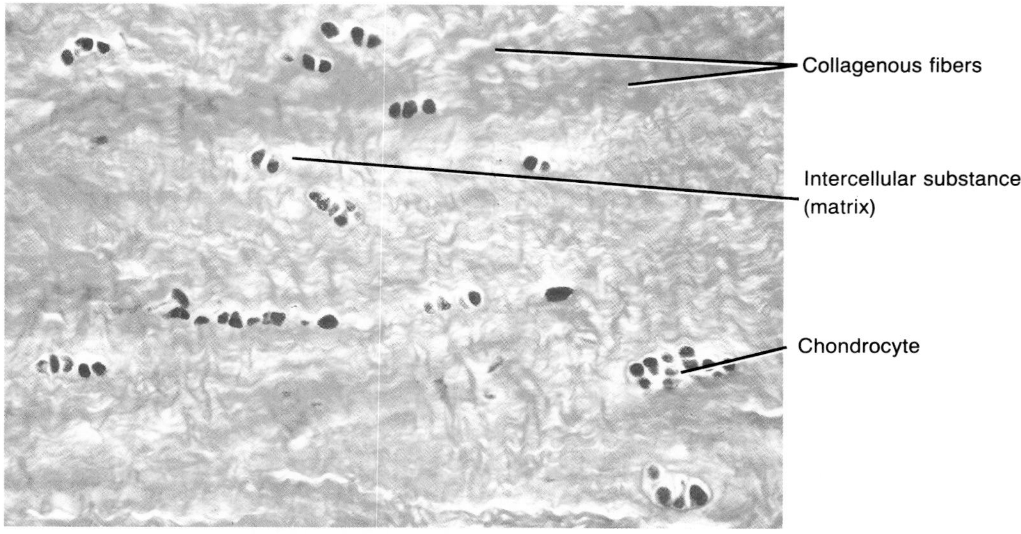

Sectional view of a meniscus (180 ×)

Elastic cartilage
Description: Consists of chondrocytes located in a threadlike network of elastic fibers.
Location: Epiglottis of larynx, external ear, and auditory (Eustachian) tubes.
Function: Gives support and maintains shape.

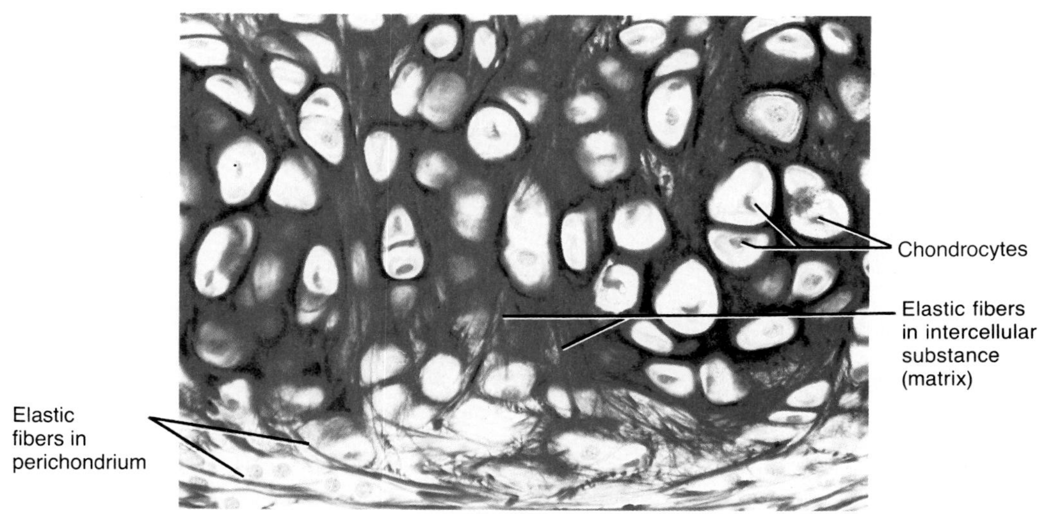

Sectional view of the epiglottis (350 ×)

EXHIBIT 4-2 CONNECTIVE TISSUES (*Continued*)

Osseous (bone)
Description: Compact bone consists of osteons (Haversian systems) that contain lamellae, lacunae, osteocytes, canaliculi, and central (Haversian) canals. See also Figure 6-3.
Location: Both compact and spongy bone comprise the various bones of the body.
Function: Support, protection, storage, houses blood forming tissue, and serves as levers that act in conjunction with muscle tissue to provide movement.

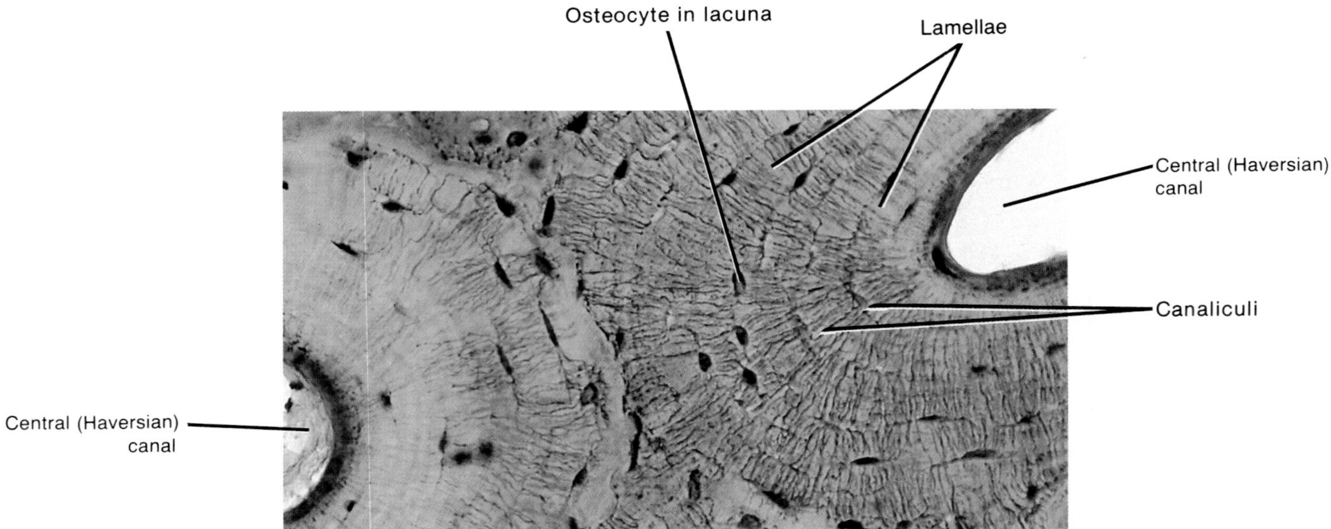

Vascular (blood)
Description: Consists of plasma (intercellular substance) and formed elements (erythrocytes, leucocytes, and thrombocytes).
Location: Within blood vessels (arteries, arterioles, capillaries, venules, and veins).
Function: Erythrocytes transport oxygen and carbon dioxide, leucocytes carry on phagocytosis and are involved in allergic reactions and immunity, and thrombocytes are essential to the clotting of blood.

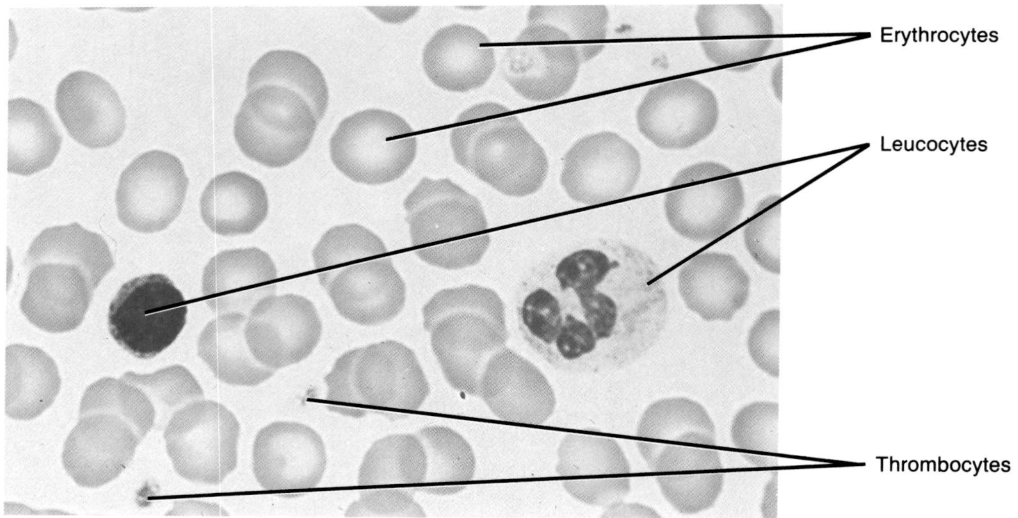

Blood smear (150×)

Connective Tissue Proper

Connective tissue that has a more or less fluid intercellular material and a fibroblast as the typical cell is termed ***connective tissue proper.*** Five examples of such tissues may be distinguished.

▪ **Loose (Areolar) Connective Tissue** Loose or areolar (a-RĒ-ō-lar) connective tissue is one of the most widely distributed connective tissues in the body. Structurally, it consists of fibers and several kinds of cells embedded in a semifluid intercellular substance. The term *loose* refers to the loosely woven arrangement of fibers in the intercellular substance. The fibers are neither abundant nor arranged to prevent stretching.

The intercellular substance consists primarily of hyaluronic acid, chondroitin sulfate, dermatan sulfate, and keratan sulfate. It is secreted by connective tissue cells, mostly fibroblasts. The intercellular substance normally facilitates the passage of nutrients from the blood vessels of the connective tissue into adjacent cells and tissues, although the thick consistency of this acid may impede the movement of some drugs. But if an enzyme called ***hyaluronidase*** is injected into the tissue, the intercellular substance changes to a watery consistency. This feature is of clinical importance because the reduced viscosity hastens the absorption and diffusion of injected drugs and fluids through the tissue and thus can lessen tension and pain. Some bacteria, white blood cells, and sperm cells produce hyaluronidase.

The three types of fibers embedded between the cells of loose connective tissue are collagenous, elastic, and reticular fibers. ***Collagenous fibers*** are very tough and resistant to a pulling force, yet allow some flexibility in the tissue because they are not taut. These fibers often occur in bundles. They are composed of many minute fibers called fibrils lying parallel to one another. The bundle arrangement affords a great deal of strength. Chemically, collagenous fibers consist of the protein collagen. ***Elastic fibers,*** by contrast, are smaller than collagenous fibers and freely branch and rejoin one another. Elastic fibers consist of a protein called elastin. These fibers also provide strength and have great elasticity, up to 50 percent of their length. ***Reticular fibers*** also consist of collagen, plus some glycoprotein. They are very thin fibers that form branching networks. Like collagenous fibers, reticular fibers provide support and strength and also form the ***stroma*** (framework) of many soft organs.

The cells in loose connective tissue are numerous and varied. Most are ***fibroblasts***—large, flat cells with branching processes. If the tissue is injured, fibroblasts are believed to form collagenous fibers, elastic fibers, and the intercellular substance. When a fibroblast becomes relatively inactive, it is sometimes referred to as a ***fibrocyte.***

Other cells found in loose connective tissue are called ***fixed macrophages*** (MAK-rō-fā-jēz; *macro* = large; *phagein* = to eat), or ***histiocytes,*** which are derived from ***monocytes,*** a type of white blood cell. Macrophages are irregular in form with short branching projections and are capable of engulfing bacteria and cellular debris by phagocytosis. Thus, they provide a vital defense for the body.

A third kind of cell in loose connective tissue is a ***plasma cell.*** It is small and either round or irregular and develops from a type of white blood cell called a ***lymphocyte (B cell).*** Plasma cells give rise to antibodies and, accordingly, provide a defense mechanism through immunity. They are found in many places in the body, but most are found in connective tissue, especially that of the gastrointestinal tract and the mammary glands.

Another cell in loose connective tissue is a ***mast cell.*** The mast cell is found in abundance along blood vessels. It forms heparin, an anticoagulant that prevents blood from clotting in the vessels. Mast cells are also believed to produce histamine and serotonin, chemicals that dilate small blood vessels.

Other cells in loose connective tissue include ***adipocytes (fat cells)*** and ***leucocytes (white blood cells).***

Loose connective tissue is continuous throughout the body. It is present in all mucous membranes and around all blood vessels and nerves. And it occurs around body organs and in the papillary (upper) region of the dermis of the skin. Combined with adipose tissue, it forms the ***subcutaneous*** (sub′-kyoo-TĀ-nē-us; *sub* = under; *cut* = skin) ***layer***—the layer of tissue that attaches the skin to underlying tissues and organs. The subcutaneous layer is also referred to as the ***superficial fascia*** (FASH-ē-a) or ***hypodermis.***

CLINICAL APPLICATION: MARFAN SYNDROME

Marfan (mar-FAN) syndrome is an inherited disorder that affects about 1 person in 10,000. It results in abnormalities of connective tissue, especially in the skeleton, eyes, and cardiovascular system. Marfan's victims tend to be tall, with disproportionately long arms, an unusually long lower half of the body, very long fingers and toes, an especially elongated thumb, an overly curved backbone, a malformed breastbone that either curves outward or inward, a backward curve of the legs, flat feet, and loose joints. Other common, visible signs of Marfan syndrome include leanness, small muscle mass, crowded teeth, and nearsightedness. Among the cardiovascular problems—the most serious ones—are a weakened area in the aorta (main artery that emerges from the heart), weakened heart valves, and heart murmurs. Flo Hyman, an outstanding American female volleyball player, died of Marfan syndrome in January 1986.

▪ **Adipose Tissue** Adipose tissue is basically a form of loose connective tissue in which the cells, called ***adipocytes,*** are specialized for fat storage. Adipocytes are derived from fibroblasts, and the cells have the shape of a "signet ring" because the cytoplasm and nucleus are pushed to the edge of the cell by a large droplet of fat. Adipose tissue is found wherever loose connective tissue is located. Specifically, it is in the subcutaneous layer below the skin, around the kidneys, at the base and on the surface of the heart, in the marrow of long bones, as a padding around joints, and behind the eyeball in the orbit. Adipose tissue is a poor conductor of heat and therefore reduces heat loss through the skin. It is a major energy reserve and generally supports and protects various organs.

CLINICAL APPLICATION: SUCTION LIPECTOMY

A surgical procedure, called *suction lipectomy* (*lipo* = fat; *ectomy* = removal of) involves suctioning out small amounts of fat from certain areas of the body. The technique can be used in a number of areas of the body in which fat accumulates, such as the inner and outer thighs, buttocks, area behind the knee, underside of the arms, breasts, and abdomen. However, suction lipectomy is not a treatment for obesity, since it will not result in a permanent reduction in fat in the area. Moreover, certain complications may develop, including fat emboli (clots), infection, fluid depletion, and injury to internal structures.

▪ **Dense (Collagenous) Connective Tissue** Dense (collagenous) connective tissue is characterized by a closer packing of fibers than in loose connective tissue. The fibers can be either irregularly arranged or regularly arranged. In areas of the body where tensions are exerted in various directions, the fiber bundles are interwoven and without regular orientation. Such dense connective tissue is referred to as ***irregularly arranged*** and occurs in sheets. It forms most fasciae, the reticular (deeper) region of the dermis of the skin, the periosteum of bone, the perichondrium of cartilage, and the ***membrane (fibrous) capsules*** around organs, such as the kidneys, liver, testes, and lymph nodes.

In other areas of the body, dense connective tissue is adapted for tension in one direction, and the fibers have an orderly, parallel arrangement. Such dense connective tissue is known as ***regularly arranged.*** The most common variety of dense regularly arranged connective tissue has a predominance of collagenous fibers arranged in bundles. Fibroblasts are placed in rows between the bundles. The tissue is silvery white, tough, yet somewhat pliable. Because of its great strength, it is the principal component of ***tendons,*** which attach muscles to bones; ***aponeuroses*** (ap'-ō-noo-RŌ-sēz), which are sheetlike tendons connecting one muscle with another or with bone; and many ***ligaments*** (collagenous ligaments), which hold bones together at joints.

▪ **Elastic Connective Tissue** Unlike collagenous connective tissue, elastic connective tissue has a predominance of freely branching elastic fibers. These fibers give the unstained tissue a yellowish color. Fibroblasts are present only in the spaces between the fibers. Elastic connective tissue can be stretched and will snap back into shape. It is a component of the walls of elastic arteries, trachea, bronchial tubes to the lungs, and the lungs themselves. Elastic connective tissue provides stretch and strength, allowing structures to perform their functions efficiently. Yellow elastic ligaments, as contrasted with collagenous ligaments, are composed mostly of elastic fibers; they form the ligamenta flava of the vertebrae (ligaments between successive vertebrae), the suspensory ligament of the penis, and the true vocal cords.

▪ **Reticular Connective Tissue** Reticular connective tissue consists of interlacing reticular fibers. It helps to form a delicate supporting stroma (framework) for many organs, including the liver, spleen, and lymph nodes. It is also found in the basal lamina underlying epithelia and around blood vessels and muscle. Reticular connective tissue also helps to bind together the fibers (cells) of smooth muscle tissue.

Cartilage

Cartilage is capable of enduring considerably more stress than the tissues just discussed. Unlike other connective tissues, cartilage has no blood vessels or nerves, except for those in the perichondrium (membranous covering). Cartilage consists of a dense network of collagenous fibers and elastic fibers firmly embedded in chondroitin sulfate, a jellylike intercellular substance (matrix). Whereas the strength of cartilage is due to its collagenous fibers, its resilience (ability to assume its original shape after deformation) is due to chondroitin sulfate. The cells of mature cartilage, called ***chondrocytes*** (KON-drō-sīts), occur singly or in groups within spaces called ***lacunae*** (la-KOO-nē) in the intercellular substance. The surface of cartilage is surrounded by irregularly arranged dense connective tissue called the ***perichondrium*** (per'-i-KON-drē-um; *peri* = around; *chondro* = cartilage). Three kinds of cartilage are recognized: hyaline cartilage, fibrocartilage, and elastic cartilage (Exhibit 4-2).

▪ **Hyaline Cartilage** This cartilage, also called ***gristle,*** appears in the body as a bluish-white, shiny substance. The collagenous fibers, although present, are not visible with ordinary staining techniques, and the prominent chondrocytes are found in lacunae. Hyaline cartilage is the most abundant kind of cartilage in the body. It is found at joints over the ends of the long bones (where it is called ***articular cartilage***) and forms the ***costal cartilages*** at the ventral ends of the ribs. Hyaline cartilage also helps to form the nose, larynx, trachea, bronchi, and bronchial tubes leading to the lungs. Most of the embryonic skeleton consists of hyaline cartilage. Hyaline cartilage affords flexibility and support and, as articular cartilage, reduces friction and absorbs shock.

▪ **Fibrocartilage** Chondrocytes scattered through many bundles of visible collagenous fibers are found in this type of cartilage. Fibrocartilage is found at the symphysis pubis, the point where the coxal (hip) bones fuse anteriorly at the midline. It is also found in the intervertebral discs between vertebrae and the menisci of the knee. This tissue combines strength and rigidity.

▪ **Elastic Cartilage** In this tissue, chondrocytes are located in a threadlike network of elastic fibers. Elastic cartilage provides strength and elasticity and maintains the shape of certain organs—the epiglottis of the larynx, the external part of the ear (pinna), and the auditory (Eustachian) tubes.

▪ **Growth of Cartilage** The growth of cartilage follows two basic patterns. In ***interstitial (endogenous) growth,*** the cartilage increases rapidly in size through the division of existing chondrocytes and continuous deposition of increasing amounts of intercellular matrix by the chondrocytes. The formation of new chondrocytes and their production of new intercellular matrix causes the cartilage to expand from within—thus, the term *interstitial growth.* This growth pattern occurs while the cartilage is young and pliable—during childhood and adolescence.

In ***appositional (exogenous) growth,*** the growth of cartilage occurs because of the activity of the inner chondrogenic layer of the perichondrium. The deeper cells of the perichondrium, the fibroblasts, divide. Some differentiate into chondroblasts (immature cells that develop into specialized cells) and then into chondrocytes. As differentiation occurs, the chondroblasts become surrounded with intercellular matrix and become chondrocytes. As a result, the matrix is deposited on the surface of the cartilage, increasing its size. The new layer of cartilage is added beneath the perichondrium on the surface of the cartilage, causing it to grow in width. Appositional growth starts later than interstitial growth and continues throughout life.

Osseous Tissue (Bone)

Together, cartilage, joints, and ***osseous*** (OS-ē-us) ***tissue (bone)*** comprise the skeletal system. Mature bone cells are called ***osteocytes.*** The intercellular substance consists of mineral salts, primarily calcium phosphate and calcium carbonate, and collagenous fibers. The salts are responsible for the hardness of bone.

Bone tissue is classified as either compact (dense) or spongy (cancellous), depending on how the intercellular substance and cells are organized. At this point, we will discuss compact bone only. The basic unit of compact bone is called an ***osteon (Haversian system).*** Each osteon consists of ***lamellae,*** concentric rings of hard, intercellular substance; ***lacunae,*** small spaces between lamellae that contain osteocytes; ***canaliculi,*** radiating minute canals that provide numerous routes so that nutrients can reach osteocytes and wastes can be removed from them; and a ***central (Haversian) canal*** that contains blood vessels and nerves. Notice that whereas osseous tissue is vascular, cartilage is not. Also, the lacunae of osseous tissue are interconnected by canaliculi; those of cartilage are not.

Functionally, the skeletal system supports soft tissues, protects delicate structures, works with skeletal muscles to facilitate movement, stores calcium and phosphorus, houses red marrow, which produces several kinds of blood cells, and houses yellow marrow, which contains lipids as an energy source.

The details of compact and spongy bone are discussed in Chapter 6.

Vascular Tissue (Blood)

Vascular tissue (blood) is a liquid connective tissue that consists of an intercellular substance called plasma and formed elements (cells and cell-like structures). ***Plasma*** is a straw-colored liquid that consists mostly of water plus some dissolved substances (nutrients, enzymes, hormones, respiratory gases, and ions). The formed elements are erythrocytes (red blood cells), leucocytes (white blood cells), and thrombocytes (platelets). The fibers characteristic of all connective tissues are present in blood only when it is clotted.

Erythrocytes function in transporting oxygen to body cells and removing carbon dioxide from them. ***Leucocytes*** are involved in phagocytosis, immunity, and allergic reactions. ***Thrombocytes*** function in blood clotting.

The details of blood are considered in Chapter 19.

MEMBRANES

The combination of an epithelial layer and an underlying connective tissue layer constitutes an ***epithelial membrane.*** The principal epithelial membranes of the body are mucous membranes, serous membranes, and the cutaneous membrane, or skin. Another kind of membrane, a ***synovial membrane,*** does not contain epithelium.

Mucous Membranes

A ***mucous membrane,*** or ***mucosa,*** lines a body cavity that opens directly to the exterior. Mucous membranes line the entire gastrointestinal, respiratory, excretory, and reproductive tracts (see Figure 24-2) and consist of a lining layer of epithelium and an underlying layer of connective tissue. In addition, most mucous membranes also contain a layer of smooth muscle called the ***muscularis mucosae.***

The epithelial layer of a mucous membrane secretes mucus, which prevents the cavities from drying out. It also traps particles in the respiratory passageways and lubricates food as it moves through the gastrointestinal tract. In addition, the epithelial layer is responsible for the secretion of digestive enzymes and the absorption of food.

The connective tissue layer of a mucous membrane is called the ***lamina propria.*** The lamina propria is so named because it belongs to the mucous membrane (*proprius* = one's own). It binds the epithelium to the underlying structures and allows some flexibility of the membrane. It also holds the blood vessels in place, protects underlying muscles from abrasion or puncture, and allows the diffusion of oxygen and nutrients to the epithelium covering it, and the diffusion of carbon dioxide and wastes from the epithelium covering it.

The muscularis mucosae contains smooth muscle fibers and separates the mucosa from the submucosa beneath. Its role in the gastrointestinal tract is considered in Chapter 24.

Serous Membranes

A ***serous membrane,*** or ***serosa,*** lines a body cavity that does not open directly to the exterior, and it covers the organs that lie within the cavity. Serous membranes consist of thin layers of loose connective tissue covered by a layer of mesothelium, and they are composed of two portions. The part attached to the cavity wall is called the ***parietal*** (pa-RĪ-e-tal) ***portion.*** The part that covers and attaches to the organs inside these cavities is the ***visceral portion.*** The serous membrane lining the thoracic cavity and covering the lungs is called the ***pleura*** (see Figure 1-7). The membrane lining the heart cavity and covering the heart is the ***pericardium*** (*cardio* = heart). The serous membrane lining the abdominal cavity and covering the abdominal organs and some pelvic organs is called the ***peritoneum.***

The epithelial layer of a serous membrane secretes a lubricating fluid, called ***serous fluid,*** that allows the organs to glide easily against one another or against the walls of the cavities. The connective tissue layer of the serous membrane consists of a relatively thin layer of loose connective tissue.

Cutaneous Membrane

The ***cutaneous membrane,*** or skin, constitutes an organ of the integumentary system and is discussed in the next chapter.

Synovial Membranes

Synovial membranes line the cavities of the freely movable joints (see Figure 9-1a). Like serous membranes, they line structures that do not open to the exterior. Unlike mucous, serous, and cutaneous membranes, they do not contain epithelium and are therefore not epithelial membranes. They are composed of loose connective tissue with elastic fibers and varying amounts of fat. Synovial membranes secrete ***synovial fluid,*** which lubricates the articular cartilage at the ends of bones as they move at joints and nourishes the articular cartilage covering the bones that form the joints. These are articular synovial membranes. Other synovial membranes line cushioning sacs, called bursae, and tendon sheaths in our hands and feet that facilitate the movement of muscle tendons.

MUSCLE TISSUE

Muscle tissue consists of fibers (cells) that are highly specialized for the active generation of force for contraction. As a result of this characteristic, muscle tissue provides motion, maintenance of posture, and heat production. On the basis of certain structural and functional characteristics, muscle tissue is classified into three types: skeletal, cardiac, and smooth (Exhibit 4-3).

Skeletal muscle tissue is named for its location—attached to bones. It is also ***striated;*** that is, the fibers (cells) contain alternating light and dark bands (striations) that are perpendicular to the long axes of the fibers. The striations are visible under a microscope. Skeletal muscle tissue is also ***voluntary*** because it can be made to contract or relax by conscious control. A single skeletal muscle fiber is cylindrical, and the fibers are parallel to each other in a tissue. Each muscle fiber contains a plasma membrane, the ***sarcolemma,*** surrounding the cytoplasm, or ***sarcoplasm.*** Skeletal muscle fibers are multinucleate (they have more than one nucleus), and the nuclei lie close to the sarcolemma. The contractile elements of skeletal muscle fibers are proteins called ***myofilaments.*** They contain wide, transverse, dark bands and narrow light ones that give the fibers the striated appearance.

Cardiac muscle tissue forms the bulk of the wall of the heart. Like skeletal muscle tissue, it is striated. However, unlike skeletal muscle tissue, it is usually involuntary; its contraction is usually not under conscious control. Cardiac muscle fibers are roughly quadrangular and branch to form networks throughout the tissue. The fibers usually have only one nucleus that is centrally located. Cardiac muscle fibers are bound to each other by transverse thickenings of the sarcolemma called ***intercalated discs.*** These are unique to cardiac muscle and serve to strengthen the tissue and aid muscle action potential conduction by way of channels called ***gap junctions.***

Smooth muscle tissue is located in the walls of hollow internal structures such as blood vessels, the stomach, intestines, and urinary bladder. Smooth muscle fibers are usually involuntary, and they are ***nonstriated (smooth).*** Each smooth muscle fiber is thickest in the midregion, with either end tapering to a point, and contains a single, centrally located nucleus.

A more detailed discussion of muscle tissue is considered in Chapter 10.

EXHIBIT 4-3 MUSCLE TISSUE

Skeletal mucle tissue
Description: Cylindrical, striated fibers with several peripheral nuclei; voluntary.
Location: Usually attached to bones.
Function: Motion, posture, heat production.

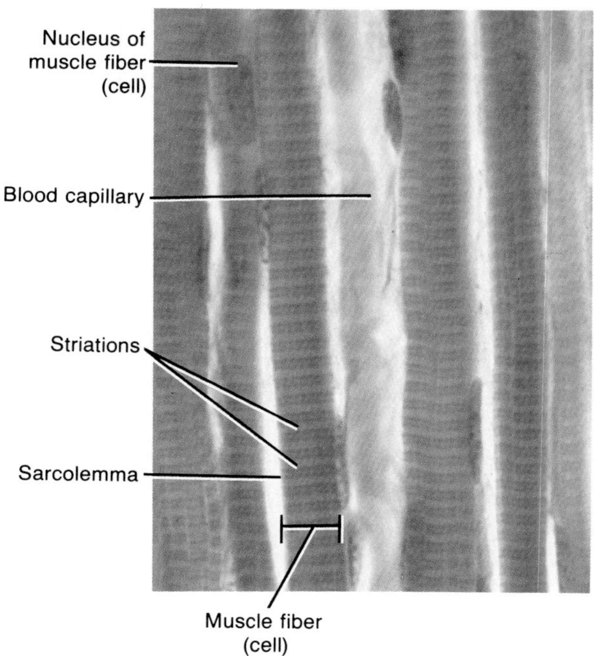

Section of skeletal muscle (800 ×)

Cardiac muscle tissue
Description: Quadrangular, branching, striated fibers with one centrally located nucleus; contains intercalated discs; usually involuntary.
Location: Heart wall.
Function: Motion (contraction of heart).

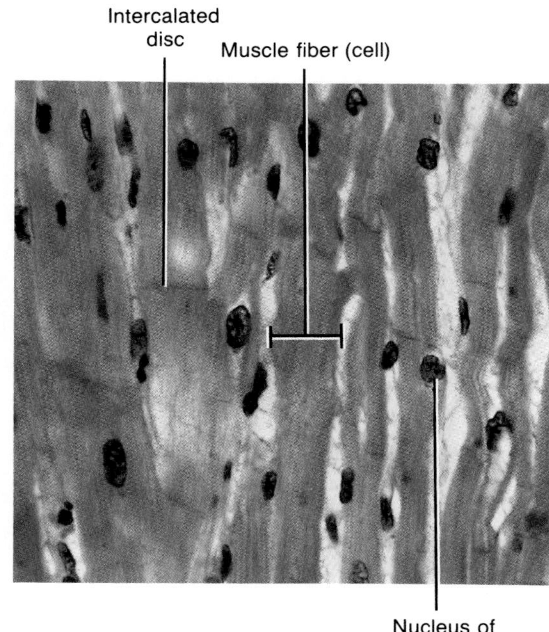

Section of cardiac muscle (400 ×)

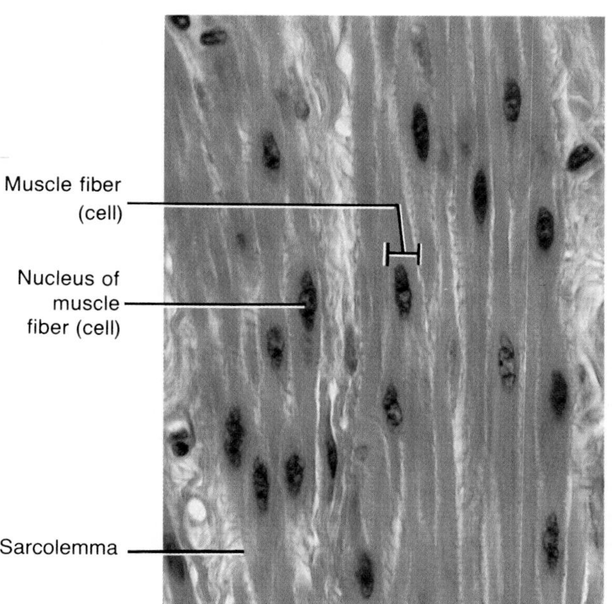

Section of smooth muscle (840x)

Smooth muscle tissue
Description: Spindle-shaped, nonstriated fibers with one centrally located nucleus; usually involuntary.
Location: Walls of hollow internal structures such as blood vessels, stomach, intestines, and urinary bladder.
Function: Motion (constriction of blood vessels, propulsion of foods through gastrointestinal tract; contraction of gallbladder).

Photomicrographs courtesy of Andrew Kuntzman.

NERVOUS TISSUE

Despite the tremendous complexity of the nervous system, it consists of only two principal kinds of cells: neurons and neuroglia. ***Neurons,*** or nerve cells, are highly specialized cells that are sensitive to various stimuli; converting stimuli to nerve impulses; and conducting nerve impulses to other neurons, muscle fibers, or glands. Neurons are the structural and functional units of the nervous system. Most consist of three basic portions: cell body and two kinds of processes called dendrites and axons (Exhibit 4-4). The ***cell body (perikaryon)*** contains the nucleus and other organelles. ***Dendrites*** are highly branched processes of the cell body that conduct nerve impulses toward the cell body. ***Axons*** are single, long processes of the cell body that conduct nerve impulses away from the cell body.

Neuroglia are cells that protect and support neurons. They are of clinical interest because they are frequently the sites of tumors of the nervous system.

The detailed structure and function of neurons and neuroglia are considered in Chapter 12.

TISSUE REPAIR: AN ATTEMPT TO RESTORE HOMEOSTASIS

Tissue repair is the process by which new tissues replace dead or damaged cells. New cells originate by cell duplication from the ***stroma,*** the supporting connective tissue, or from the ***parenchyma,*** cells that form the organ's functioning part. The epithelial cells that secrete and absorb are the parenchymal cells of the intestine, for example. The restoration of an injured organ or tissue to normal structure and function depends entirely on which type of cell—parenchymal or stromal—is active in the repair. If only parenchymal elements accomplish the repair, a perfect or near-perfect reconstruction of the injured tissue may occur. However, if the fibroblast cells of the stroma are active in the repair, the tissue will be replaced with new connective tissue called ***scar tissue.*** In this process, fibroblasts synthesize collagen and protein polysaccharides that aggregate to form scar tissue. The process of scar tissue formation is known as ***fibrosis.*** Since scar tissue is not specialized to perform the functions of the parenchymal tissue, the overall function of the tissue is impaired. If the rate of collagen breakdown in a scar exceeds production, the scar becomes softer and less bulky. If, on the other hand, the rate of collagen production exceeds breakdown, a ***keloid (hypertrophic) scar*** develops. Such a scar is sharply elevated, irregularly shaped, and unsightly.

CLINICAL APPLICATION: ADHESIONS

Scar tissue can cause the abnormal joining of tissues called *adhesions.* They are common in the abdomen and may occur around a site of previous inflammation such as an inflamed appendix, or they can follow surgery. Although adhesions do not always cause problems, they can affect tissue flexibility, cause obstruction (such as in the intestine), and make subsequent surgery more difficult. Surgery may be required to remove adhesions. Surgeons, aware of the causes of adhesions, are now more careful about methods of dissection (sharp as opposed to blunt is preferable), surgical bleeding (the less the better), and sutures used (those made of polyglycolic acid are less likely to form adhesions than silk or catgut sutures). In addition, talcum is completely rinsed from gloves before surgery to prevent talc granulomas, a kind of adhesion.

EXHIBIT 4-4 NERVOUS TISSUE

Description: **Neurons (nerve cells) consist of a cell body and processes extending from the cell body called dendrites (usually conduct impulses toward cell body) or axons (usually conduct impulses away from cell body).**
Location: **Nervous system.**
Function: **Exhibits sensitivity to various types of stimuli, converts stimuli to nerve impulses, and conducts nerve impulses to other neurons, muscle fibers, or glands.**

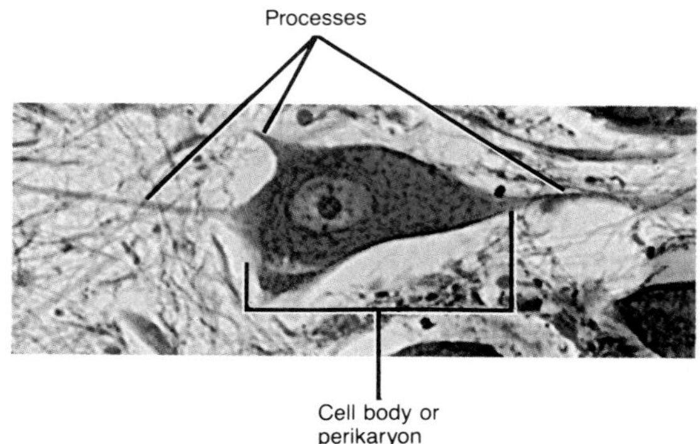

Efferent (motor) neuron (640 ×)

Courtesy of Biophoto Associates.

The cardinal factor in tissue repair lies in the capacity of parenchymal tissue to regenerate. This capacity, in turn, depends on the ability of the parenchymal cells to replicate quickly.

Repair Process

If injury to a tissue is slight, repair may sometimes be accomplished with the drainage and reabsorption of pus (accumulation of leucocytes and fluid resulting from inflammation), followed by parenchymal regeneration. When the area of skin loss is great, fluid moves out of the capillaries and the area becomes dry. Fibrin, an insoluble protein in a blood clot, seals the open tissue by hardening into a ***scab.***

When tissue and cell damage are extensive and severe, as in large, open wounds, both the connective tissue stroma and the parenchymal cells are active in repair. This repair involves the rapid cell division of many fibroblasts, the manufacture of new collagenous fibers to provide strength, and an increase by cell division of the number of small blood vessels in the area. All these processes create an actively growing connective tissue called ***granulation tissue.*** This new granulation tissue forms across a wound or surgical incision to provide a framework (stroma). The framework supports the epithelial cells that migrate into the open area and fill it. The newly formed granulation tissue also secretes a fluid that kills bacteria.

Conditions Affecting Repair

Three factors affect tissue repair: nutrition, blood circulation, and age. Nutrition is vital in the healing process since a great demand is placed on the body's store of nutrients. Protein-rich diets are important since most of the cell structure is made from proteins. Vitamins also play a direct role in wound healing. Among the vitamins involved and their roles in wound healing are the following:

1. Vitamin A is essential in the replacement of epithelial tissues, especially in the respiratory tract.

2. The B vitamins—thiamine, nicotinic acid, riboflavin—are coenzymes needed by many enzyme systems in cells. They are needed especially for the enzymes involved in decomposing glucose to CO_2 and H_2O, which is crucial to both heart and nervous tissue. These vitamins may relieve pain in some cases and are necessay for division of the cells that accomplish repair.

3. Vitamin C directly affects the normal production and maintenance of intercellular substances. It is required for the manufacture of cementing elements of connective tissues, especially collagen. Vitamin C also strengthens and promotes the formation of new blood vessels. With vitamin C deficiency, even superficial wounds fail to heal, and the walls of the blood vessels become fragile and are easily ruptured.

4. Vitamin D is necessary for the proper absorption of calcium from the intestine. Calcium gives bones their hardness and is necessary for the healing of fractures.

5. Vitamin E is believed to promote healing of injured tissues and may prevent scarring.

6. Vitamin K assists in the clotting of blood and thus prevents the injured person from bleeding to death. Although vitamin K is essential for the formation of certain clotting factors, it is not, itself, a coagulant. The body, with the aid of colon bacteria, normally synthesizes its own vitamin K. In humans, bacteria in the colon synthesize vitamin K which can be absorbed and utilized.

In tissue repair, proper blood circulation is indispensable. It is the blood that transports oxygen, nutrients, antibodies, and many defensive cells to the site of injury. The blood also plays an important role in the removal of tissue fluid, blood cells that have been depleted of oxygen, bacteria, foreign bodies, and debris. These elements would otherwise interfere with healing.

Generally, tissues heal faster and leave less obvious scars in the young than in the aged. The young body is generally in a much better nutritional state, its tissues have a better blood supply, and the cells of younger people have a faster metabolic rate. Thus, cells can duplicate their materials and divide more quickly.

STUDY OUTLINE

Types of Tissues (p. 92)

1. A tissue is a group of similar cells and their intercellular substance that have a similar embryological origin and are specialized for a particular function.
2. Depending on their function and structure, the various tissues of the body are classified into four principal types: epithelial, connective, muscular, and nervous.

Epithelial Tissue (p. 92)

1. Epithelium has many cells, little intercellular material, and no blood vessels (avascular). It is attached to connective tissue by a basement membrane. It can replace itself.
2. The subtypes of epithelium include covering and lining epithelium and glandular epithelium.

Covering and Lining Epithelium (p. 93)

1. Layers are arranged as simple (one layer), stratified (several layers), and pseudostratified (one layer that appears as several); cell shapes include squamous (flat), cuboidal (cubelike), columnar (rectangular), and transitional (variable).
2. Simple squamous epithelium is adapted for diffusion and filtration and is found in lungs and kidneys. Endothelium lines the heart and blood vessels. Mesothelium lines the thoracic and abdominopelvic cavities and covers the organs within them.
3. Simple cuboidal epithelium is adapted for secretion and absorption. It is found covering ovaries, in kidneys and eyes, and lining some glandular ducts.
4. Nonciliated simple columnar epithelium lines most of the gastrointestinal tract. Specialized cells containing microvilli perform absorption. Goblet cells secrete mucus. In a few portions of the respiratory tract, the cells are ciliated to move foreign particles trapped in mucus out of the body.
5. Stratified squamous epithelium is protective. It lines the upper gastrointestinal tract and vagina and forms the outer layer of skin.
6. Stratified cuboidal epithelium is found in adult sweat glands, pharynx, epiglottis, and portions of the urethra.
7. Stratified columnar epithelium protects and secretes. It is found in the male urethra and large excretory ducts.
8. Transitional epithelium lines the urinary bladder and is capable of stretching.
9. Pseudostratified columnar epithelium has only one layer but gives the appearance of many. It lines larger excretory ducts, parts of urethra, auditory (Eustachian) tubes, and most upper respiratory structures, where it protects and secretes.

Glandular Epithelium (p. 100)

1. A gland is a single cell or a mass of epithelial cells adapted for secretion.
2. Exocrine glands (sweat, oil, and digestive glands) secrete into ducts or directly onto a free surface.
3. Structural classification includes unicellular and multicellular glands; multicellular glands are further classified as tubular, acinar, tubuloacinar, simple, and compound.
4. Functional classification includes holocrine, merocrine, and apocrine glands.
5. Endocrine glands secrete hormones into the blood.

Connective Tissue (p. 102)

1. Connective tissue is the most abundant body tissue. It has few cells, an extensive intercellular substance, and a rich blood supply (vascular), except for cartilage. It does not occur on free surfaces.
2. The intercellular substance determines the tissue's qualities.
3. Connective tissue protects, supports, and binds organs together.
4. Connective tissue is classified into two principal types: embryonic and adult.

Embryonic Connective Tissue (p. 102)

1. Mesenchyme forms all other connective tissues.
2. Mucous connective tissue is found in the umbilical cord of the fetus, where it gives support.

Adult Connective Tissue (p. 102)

1. Adult connective tissue is connective tissue that differentiates from mesenchyme and exists in the newborn and does not change after birth. It is subdivided into several kinds: connective tissue proper, cartilage, bone tissue, and vascular tissue.
2. Connective tissue proper has a more or less fluid intercellular material, and a typical cell is the fibroblast. Five examples of such tissues may be distinguished.
3. Loose (areolar) connective tissue is one of the most widely distributed connective tissues in the body. Its intercellular substance (hyaluronic acid) contains fibers (collagenous, elastic, and reticular) and various cells (fibroblasts, macrophages, plasma, and mast). Loose connective tissue is found in all mucous membranes, around body organs, and in the subcutaneous layer.
4. Adipose tissue is a form of loose connective tissue in which the cells, called adipocytes, are specialized for fat storage. It is found in the subcutaneous layer and around various organs.
5. Dense (collagenous) connective tissue has a close packing of fibers (regularly or irregularly arranged). It is found as a component of fascia, membranes of organs, tendons, ligaments, and aponeuroses.
6. Elastic connective tissue has a predominance of freely branching elastic fibers that give it a yellow color. It is found in elastic arteries, trachea, bronchial tubes, and true vocal cords.
7. Reticular connective tissue consists of interlacing reticular fibers and forms the stroma of the liver, spleen, and lymph nodes.
8. Cartilage has a jellylike matrix containing collagenous and elastic fibers and chondrocytes.
9. Hyaline cartilage is found in the embryonic skeleton, at the ends of bones, in the nose, and in respiratory structures. It is flexible, allows movement, and provides support.
10. Fibrocartilage connects the pelvic bones and the vertebrae. It provides strength.
11. Elastic cartilage maintains the shape of organs such as the larynx, auditory (Eustachian) tubes, and external ear.
12. The growth of cartilage is accomplished by interstitial growth (from within) and appositional growth (from without).
13. Osseous tissue (bone) consists of mineral salts and collagenous fibers that contribute to the hardness of bone and cells called osteocytes. It supports, protects, helps provide movement, stores minerals, and houses red marrow.
14. Vascular tissue (blood) consists of plasma and formed elements (erythrocytes, leucocytes, and thrombocytes). Functionally, its cells transport, carry on phagocytosis, participate in allergic reactions, provide immunity, and bring about blood clotting.

Membranes (p. 111)

1. An epithelial membrane is an epithelial layer overlying a connective tissue layer. Examples are mucous, serous, and cutaneous membranes.
2. Mucous membranes line cavities that open to the exterior, such as the gastrointestinal tract.
3. Serous membranes (pleura, pericardium, peritoneum) line closed cavities and cover the organs in the cavities. These membranes consist of parietal and visceral portions.
4. The cutaneous membrane is the skin.
5. Synovial membranes line joint cavities, bursae, and tendon sheaths and do not contain epithelium.

Muscle Tissue (p. 112)

1. Muscle tissue is modified for contraction and thus provides motion, maintenance of posture, and heat production.
2. Skeletal muscle tissue is attached to bones, is striated, and is voluntary.
3. Cardiac muscle tissue forms most of the heart wall, is striated, and is usually involuntary.
4. Smooth muscle tissue is found in the walls of hollow internal structures (blood vessels and viscera), is nonstriated, and is usually involuntary.

Nervous Tissue (p. 114)

1. The nervous system is composed of neurons (nerve cells) and neuroglia (protective and supporting cells).
2. Most neurons consist of a cell body and two types of processes called dendrites and axons.
3. Neurons are sensitive to stimuli, convert stimuli into nerve impulses, and conduct nerve impulses.

Tissue Repair: An Attempt to Restore Homeostasis (p. 114)

1. Tissue repair is the replacement of damaged or destroyed cells by healthy ones.
2. It begins during the active phase of inflammation and is not completed until after harmful substances in the inflamed area have been neutralized or removed.

Repair Process (p. 114)

1. If the injury is superficial, tissue repair involves pus removal (if pus is present), scab formation, and parenchymal regeneration.
2. If damage is extensive, granulation tissue is involved.

Conditions Affecting Repair (p. 115)

1. Nutrition is important to tissue repair. Various vitamins (A, some B, D, C, E, and K) and a protein-rich diet are needed.
2. Adequate circulation of blood is needed.
3. The tissues of young people repair rapidly and efficiently; the process slows down with aging.

REVIEW QUESTIONS

1. Define a tissue. What are the four basic kinds of human tissue? (p. 92)
2. Distinguish covering and lining epithelium from glandular epithelium. What characteristics are common to all epithelium? Describe the structure of the basement membrane. (p. 92)
3. Describe the various layering arrangements and cell shapes of epithelium. (p. 92)
4. How is epithelium classified? List the various types. (p. 93)
5. For each of the following kinds of epithelium, briefly describe the microscopic appearance, location in the body, and functions: simple squamous, simple cuboidal, simple columnar (nonciliated and ciliated), stratified squamous (keratinized and nonkeratinized), stratified cuboidal, stratified columnar, transitional, and pseudostratified columnar. (pp. 94–98)
6. Define the following terms: endothelium, mesothelium, secretion, absorption, goblet cell, and keratin. (p. 93)
7. What is a gland? Distinguish between endocrine and exocrine glands. (p. 100)
8. Describe the classification of exocrine glands according to structure and function and give at least one example of each. (p. 100)
9. Enumerate the ways in which connective tissue differs from epithelium. (p. 102)
10. How are connective tissues classified? List the various types. (p. 102)
11. How are embryonic connective tissue and adult connective tissue distinguished? (p. 102)
12. Describe the following connective tissues with regard to microscopic appearance, location in the body, and function: loose (areolar), adipose, dense (collagenous), elastic, reticular, hyaline cartilage, fibrocartilage, elastic cartilage, osseous tissue (bone), and vascular tissue (blood). (p. 104–108)
13. Define the following terms: hyaluronic acid, collagenous fiber, elastic fiber, reticular fiber, fibroblast, macrophage, plasma cell, mast cell, melanocyte, adipocyte, chondrocyte, lacuna, osteocyte, lamella, and canaliculus. (p. 109)
14. Distinguish between the interstitial and appositional growth of cartilage. (p. 111)
15. Define the following kinds of membranes: mucous, serous, cutaneous, and synovial. Where is each located in the body? What are their functions? (p. 111)
16. Describe the histology of muscle tissue. How is it classified? What are its functions? (p. 112)
17. Distinguish between neurons and neuroglia. Describe the structure and function of neurons. (p. 114)
18. Following are some descriptions of various tissues of the body. For each description, name the tissue described.
    a. An epithelium that permits distention (stretching).
    b. A single layer of flat cells concerned with filtration and absorption.
    c. Forms all other kinds of connective tissue.
    d. Specialized for fat storage.
    e. An epithelium with waterproofing qualities.
    f. Forms the framework of many organs.
    g. Produces perspiration, wax, oil, or digestive enzymes.
    h. Cartilage that shapes the external ear.
    i. Contains goblet cells and lines the intestine.
    j. Most widely distributed connective tissue.
    k. Forms tendons, ligaments, and aponeuroses.
    l. Specialized for the secretion of hormones.
    m. Provides support in the umbilical cord.
    n. Lines kidney tubules and is specialized for absorption and secretion.
    o. Permits extensibility of lung tissue.
    p. Stores red marrow, protects, supports.
    q. Nonstriated, usually involuntary muscle tissue.
    r. Consists of granulocytes and agranulocytes.
    s. Composed of a cell body, dendrites, and axon.
19. What is meant by tissue repair? Distinguish between stromal and parenchymal repair. What is the importance of granulation tissue? (p. 114)
20. What conditions affect tissue repair? (p. 115)
21. Define the following: Marfan syndrome (p. 109), suction lipectomy (p. 110), and adhesions. (p. 114)

SELECTED READINGS

Cormack, D. H. *Ham's Histology,* 9th ed. Philadelphia: Lippincott, 1987.

Fawcett, D. W. *Bloom and Fawcett: A Textbook of Histology,* 11th ed. Philadelphia: Saunders, 1986.

Kelly, D. E., R. L. Wood, and A. C. Enders. *Bailey's Textbook of Microscopic Anatomy,* 18th ed. Baltimore: Williams and Wilkins, 1984.

Miller, J. "The Connective Tissue Perspective," *Science News,* 20 February 1982.

Ross, M. H., and E. J. Reith. *Histology: A Text and Atlas.* New York: Harper & Row, 1985.

Synder, S. H. "The Molecular Basis of Communication between Cells," *Scientific American,* October 1985.

Chapter 5

The Integumentary System

Chapter Contents at a Glance

Student Objectives

1. Describe the structure and functions of the skin.
2. Explain the basis for skin color.
3. Outline the steps involved in epidermal wound healing and deep wound healing.
4. Compare the structure, distribution, and functions of hair, sebaceous (oil), sudoriferous (sweat), and ceruminous glands.
5. Explain the role of the skin in helping to maintain the homeostasis of normal body temperature.
6. Describe the effects of aging on the integumentary system.
7. Describe the development of the epidermis, its derivatives, and the dermis.
8. Describe the causes and effects for the following skin disorders: acne, systemic lupus erythematosus (SLE), psoriasis, decubitus ulcers, sunburn, and skin cancer.
9. Define a burn; classify burns into first, second, and third degree; and describe how to estimate the extent of a burn.
10. Define medical terminology associated with the integumentary system.

An aggregation of tissues that performs a specific function is an ***organ.*** The next higher level of organization is a ***system***—a group of organs operating together to perform specialized functions. The skin and its derivatives, such as hair, nails, glands, and several specialized receptors, constitute the ***integumentary*** (in′-teg-yoo-MEN-tar-ē) ***system*** of the body.

The developmental anatomy of the integumentary system is considered at the end of the chapter.

SKIN

The ***skin*** is an organ because it consists of tissues structurally joined together to perform specific activities. It is one of the larger organs of the body in terms of surface area. For the average adult, the skin occupies a surface area of approximately 2 sq m (3,000 sq inches). The skin is not just a simple thin covering that keeps the body together and gives it protection. The skin is quite complex in structure and performs several functions essential for survival. ***Dermatology*** (der′-ma-TOL-ō-jē; *dermato* = skin; *logos* = study of) is the medical specialty that deals with the diagnosis and treatment of skin disorders.

Physiology

The numerous functions of the skin are as follows:

1. **Regulation of body temperature.** In response to high environmental temperature or strenuous exercise, the production of perspiration by sudoriferous (sweat) glands helps to lower body temperature back to normal. Changes in the flow of blood to the skin also alter its insulating properties and help to adjust body temperature. This is described in detail later in the chapter.

2. **Protection.** The skin covers the body and provides a physical barrier that protects underlying tissues from physical abrasion, bacterial invasion, dehydration, and ultraviolet (UV) radiation.

3. **Reception of stimuli.** The skin contains numerous nerve endings and receptors that detect stimuli related to temperature, touch, pressure, and pain (Chapter 15).

4. **Excretion.** Not only does perspiration assume a role in helping to regulate normal body temperature, it also assists in the excretion of small amounts of water, salts, and several organic compounds.

5. **Synthesis of vitamin D.** The term ***vitamin D*** actually refers to a group of closely related compounds synthesized naturally from a precursor molecule present in the skin upon exposure to ultraviolet (UV) radiation. In the skin, the precursor substance, 7-dehydrocholesterol, is converted to cholecalciferol (vitamin D_3) in the presence of UV radiation. In the liver, cholecalciferol is converted to 25-hydroxycholecalciferol. Then, in the kidneys, this substance is changed into 1,25-dihydroxycalciferol (calcitriol), the most active form of vitamin D which stimulates the absorption of calcium and phosphorous from dietary foods. In subsequent chapters, when we speak of vitamin D, we are really referring to 1,25-dihydroxycalciferol. Vitamin D is actually a hormone, since it is produced in one location in the body, transported by the blood, and then exerts its effect in another location.

6. **Immunity.** As you will see in Chapter 22, certain cells of the epidermis are important components of immunity, your ability to fight disease by producing antibodies.

Structure

Structurally, the skin consists of two principal parts (Figure 5-1). The outer, thinner portion, which is composed of *epithelium,* is called the ***epidermis.*** The epidermis is cemented to the inner, thicker, *connective tissue* part called the ***dermis.*** Thick skin has a relatively thick epidermis, whereas thin skin has a relatively thin epidermis. Beneath the dermis is a ***subcutaneous (SC) layer.*** This layer, also called the ***superficial fascia*** or ***hypodermis,*** consists of areolar and adipose tissues. Fibers from the dermis extend down into the subcutaneous layer and anchor the skin to it. The subcutaneous layer, in turn, is attached to underlying tissues and organs.

Epidermis

The ***epidermis*** is composed of stratified squamous epithelium and contains four distinct types of cells. The most numerous is known as a ***keratinocyte,*** a cell that undergoes keratinization. This process will be described shortly. The functions of these cells are to produce the protein keratin, which helps waterproof and protect the skin and underlying tissues, and to participate in immunity. A second type of cell in the epidermis is called a ***melanocyte.*** It is located at the base of the epidermis, and its role is to produce melanin, one of the pigments responsible for skin color and the absorption of ultraviolet (UV) light. The third and fourth types of cells in the epidermis are called ***nonpigmented granular dendrocytes,*** two distinct cell types, formerly known as ***Langerhans' cells*** and ***Granstein cells.*** These cells differ both in their sensitivity to damage by UV radiation and their functions in immunity. Langerhans' cells are a small population of dendrocytes that arise from bone marrow and migrate to the epidermis and other areas of the body that contain stratified squamous epithelial tissue. These cells are sensitive to UV radiation and lie above the basal layer of keratinocytes. Langerhans' cells interact with cells called helper T cells to assist in the immune response. Granstein cells are dendrocytes that are more resistant to UV radiation and interact with cells called suppressor T cells to assist in the immune response.

The keratinocytes of the epidermis are organized into four or five cell layers, depending on location in the body (see Figures 5-1 and 5-2). Where exposure to friction is greatest, such as in the palms and soles, the epidermis

FIGURE 5-1 **Skin. Structure of the skin and underlying subcutaneous layer.**

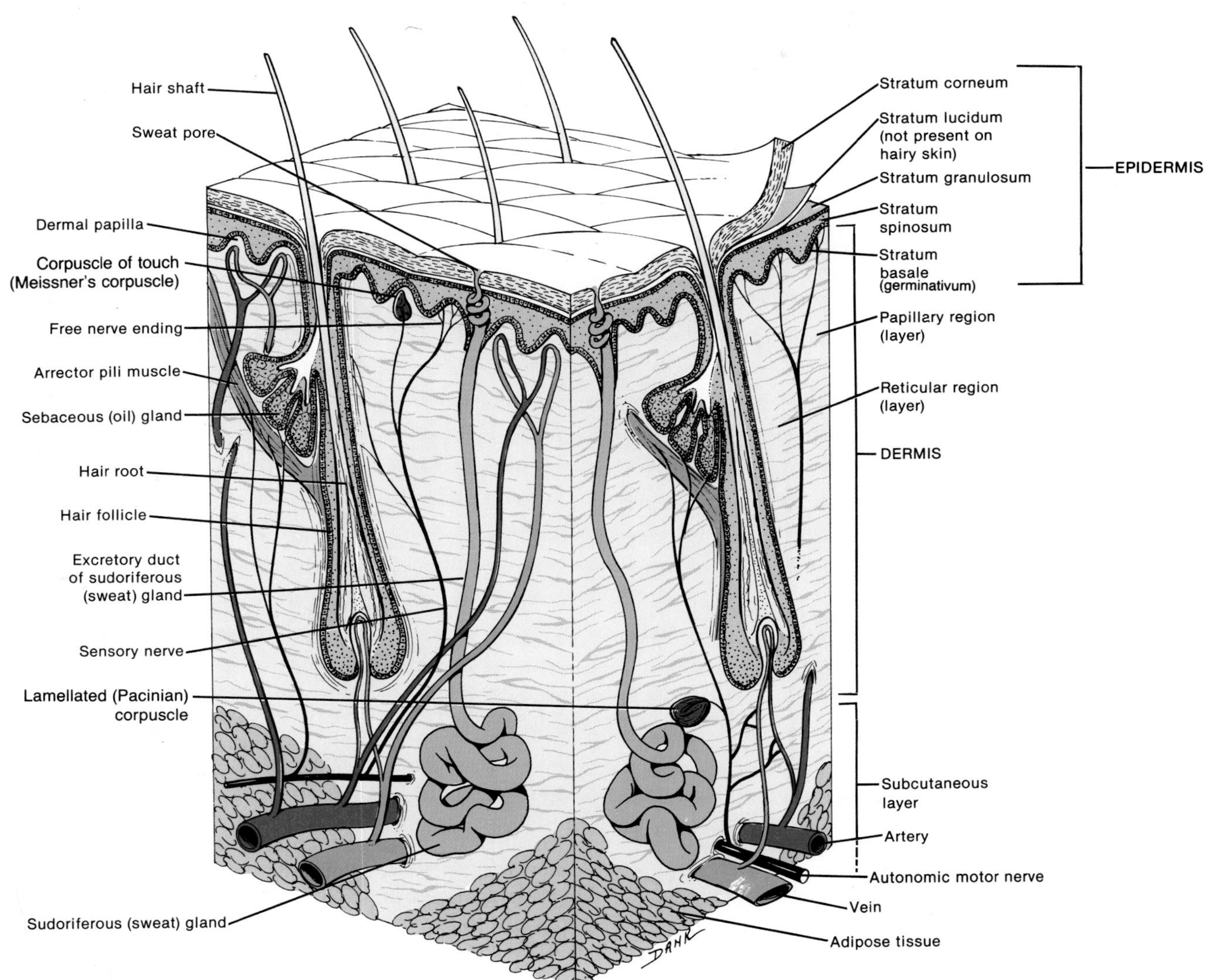

has five layers. In all other parts it has four layers. The names of the five layers from the deepest to the most superficial are as follows:

1. **Stratum basale.** This single layer of cuboidal to columnar cells is capable of continued cell division. As these cells multiply, they push up toward the surface and become part of the layers to be described next. Their nuclei degenerate, and the cells die. Eventually, the cells are shed from the top layer of the epidermis. Other cells in the stratum basale migrate into the dermis and give rise to sweat and oil glands and hair follicles. The stratum basale is sometimes referred to as the ***stratum germinativum*** (jer′-mi-na-TĒ-vum) to indicate its role in germinating new cells. The stratum basale of hairless skin contains nerve endings sensitive to touch called ***tactile (Merkel's) discs.***

2. **Stratum spinosum.** This layer of the epidermis contains 8 to 10 rows of polyhedral (many-sided) cells that fit closely together. The surfaces of these cells contain spinelike projections (*spinosum* = prickly) that join the cells together.

3. **Stratum granulosum.** The third layer of the epidermis consists of three to five rows of flattened cells that contain darkly staining granules of a substance called ***keratohyalin*** (ker'-a-tō-HĪ-a-lin). This compound is involved in the first step of keratin formation. ***Keratin*** is a waterproofing protein found in the top layer of the epidermis. The nuclei of the cells in the stratum granulosum are in various stages of degeneration. As these nuclei break down, the cells are no longer capable of carrying out vital metabolic reactions and die.

4. **Stratum lucidum.** This layer is normally found only in the thick skin of the palms and soles and is absent in thin skin. It consists of three to five rows of clear, flat, dead cells that

FIGURE 5-2 Photomicrograph of thick skin at a magnification of 155×. (Courtesy of Andrew Kuntzman).

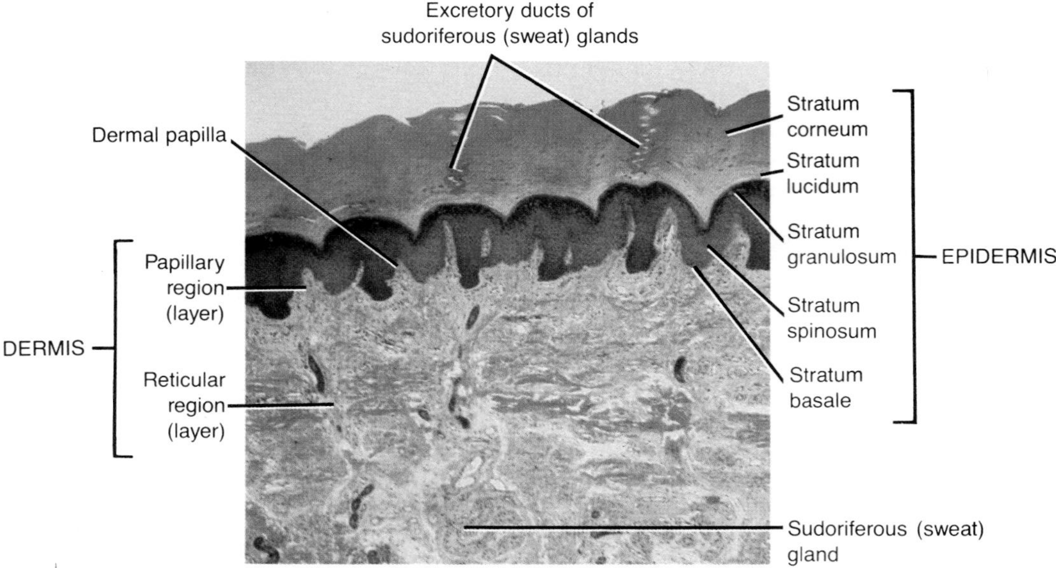

contain droplets of a substance called ***eleidin*** (el-Ē-i-din). The layer is so named because eleidin is translucent (*lucidum* = clear). Eleidin is formed from keratohyalin and is eventually transformed to keratin.

5. **Stratum corneum.** This layer consists of 25 to 30 rows of flat, dead cells completely filled with keratin. These cells are continuously shed and replaced. The stratum corneum serves as an effective barrier against light and heat waves, bacteria, and many chemicals.

In the process of ***keratinization,*** newly formed cells produced in the basal layers are pushed up to more superficial layers. As the cells move toward the surface, the cytoplasm, nucleus, and other organelles are replaced by keratohyalin, then eleidin, and finally keratin, and the cells die. Eventually, the keratinized cells are sloughed off and replaced by underlying cells that, in turn, become keratinized. The whole process by which a cell forms in the basal layers, rises to the surface, becomes keratinized, and sloughs off takes about two weeks.

Epidermal growth factor (EGF) is a protein hormone that functions as a growth factor. EGF stimulates the growth of epidermal cells and fibroblasts.

Dermis

The second principal part of the skin, the ***dermis,*** is composed of connective tissue containing collagenous and elastic fibers (see Figure 5-1). The dermis is very thick in the palms and soles and very thin in the eyelids, penis, and scrotum. It also tends to be thicker on the dorsal aspects of the body than the ventral and thicker on the lateral aspects of extremities than medial aspects. Numerous blood vessels, nerves, glands, and hair follicles are embedded in the dermis.

The upper region of the dermis, about one-fifth of the thickness of the total layer, is named the ***papillary region*** or ***layer.*** It consists of loose connective tissue containing fine elastic fibers. Its surface area is greatly increased by small, fingerlike projections called ***dermal papillae*** (pa-PIL-ē). These structures project into the epidermis, and many contain loops of capillaries. Some dermal papillae also contain ***corpuscles of touch,*** also called ***Meissner's*** (MĪS-nerz) ***corpuscles,*** nerve endings that are sensitive to touch. Dermal papillae cause ridges in the overlying epidermis. It is these ridges that leave fingerprints on objects that are handled.

The remaining portion of the dermis is called the ***reticular region*** or ***layer.*** It consists of dense, irregularly arranged connective tissue containing interlacing bundles of collagenous and coarse elastic fibers. It is named the reticular (*rete* = net) region because the bundles of collagenous fibers interlace in a netlike manner. Spaces between the fibers are occupied by a small quantity of adipose tissue, hair follicles, nerves, oil glands, and the ducts of sweat glands. Varying thicknesses of the reticular region, among other factors, are responsible for differences in the thickness of the skin.

The combination of collagenous and elastic fibers in the reticular region provides the skin with strength, extensibility, and elasticity. (***Extensibility*** is the ability to stretch; ***elasticity*** is the ability to return to original shape after extension or contraction.) The ability of the skin to stretch can readily be seen during conditions of pregnancy, obesity, and edema. The small tears that occur during extreme stretching are initially red and remain visible afterward as silvery white streaks called ***striae*** (STRĪ-ē).

The reticular region is attached to underlying organs, such as bone and muscle, by the subcutaneous layer. The subcutaneous layer also contains nerve endings called ***lamellated*** or ***Pacinian*** (pa-SIN-ē-an) ***corpuscles*** that are sensitive to pressure (see Figure 15-1).

CLINICAL APPLICATION: LINES OF CLEAVAGE AND SURGERY

The collagenous fibers in the dermis run in all directions, but in particular regions of the body they tend to run more in one direction than another. The predominant direction of the underlying collagenous fibers is indicated in the skin by ***lines of cleavage (tension lines).*** The lines are especially evident on the palmar surfaces of the fingers, where they run parallel to the long axis of the digit. Lines of cleavage are of particular interest to a surgeon because an incision running parallel to the collagen fibers will heal with only a fine scar. An incision made across the rows of fibers disrupts the collagen, and the wound tends to gape open and heal in a broad, thick scar.

Skin Color

The color of the skin is due to melanin, a pigment in the epidermis; carotene, a pigment mostly in the dermis; and blood in capillaries in the dermis. The amount of ***melanin*** varies the skin color from pale yellow to black. This pigment is found primarily in the basale and spinosum layers. Melanin is synthesized in cells called ***melanocytes*** (MEL-a-nō-sīts), located either just beneath or between cells of the stratum basale. Melanocytes are produced from ***melanoblasts*** (MEL-a-nō-blasts), precursor cells that vary in number from 800 to 2000 cells per cubic millimeter (mm^3). Maximum numbers of melanoblasts are found in mucous membranes, the penis, face, and extremities. Since the number of melanocytes is about the same in all races, differences in skin color are due to the amount of pigment the melanocytes produce and disperse. An inherited inability of an individual in any race to produce melanin results in ***albinism*** (AL-bin-izm). In albinism, the pigment is absent in the hair and eyes as well as the skin. An individual affected with albinism is called an ***albino*** (al-BĪ-no). The partial or complete loss of melanocytes from areas of skin produces patchy, white spots, and the condition is called ***vitiligo*** (vit-i-LĪ-gō). In some people, melanin tends to form in patches called ***freckles.***

Melanocytes synthesize melanin from the amino acid *tyrosine* in the presence of an enzyme called *tyrosinase.* Exposure to ultraviolet radiation increases the enzymatic activity of melanocytes and leads to increased melanin production. The cell bodies of melanocytes send out long processes between epidermal cells. Upon contact with the processes, epidermal cells take up the melanin by phagocytosis. When the skin is again exposed to ultraviolet radiation, both the amount and the darkness of melanin increase, tanning and further protecting the body against radiation. Thus, melanin serves a vital protective function. Melanocyte-stimulating hormone (MSH) produced by the anterior pituitary gland causes increased melanin synthesis and distribution through the epidermis.

CLINICAL APPLICATION: MALIGNANT MELANOMA

Overexposure of the skin to the ultraviolet light of the sun may lead to skin cancer. Among the most lethal skin cancers is ***malignant melanoma*** (*melano* = dark-colored; *oma* = tumor), cancer of the melanocytes. Fortunately, most skin cancers involve basal and squamous cells and can be treated by surgical excision.

It is a good practice to examine one's skin periodically for ***moles*** that may develop highly irregular borders, uneven surfaces, or a mixture of colors, especially red, white, and blue. These signs of changing appearance plus changes in size and bleeding may indicate that a melanoma is developing.

Some individuals, usually over 55 years of age, may develop ***liver spots (age spots).*** These are flat skin patches that look like freckles and range in color from light brown to black. Liver spots are clusters of melanocytes and are medically insignificant; they do not become cancerous.

Another pigment, called ***carotene*** (KAR-o-tēn), is found in the stratum corneum and fatty areas of the dermis in people of Asian origin. Together, carotene and melanin account for the yellowish hue of their skin.

The pink color of Caucasian skin is due to blood in capillaries in the dermis. The redness of the vessels is not heavily masked by pigment. The epidermis has no blood vessels, a characteristic of all epithelia.

Epidermal Ridges and Grooves

The outer surface of the skin of the palms and fingers and soles and toes is marked by a series of ridges and grooves that appear either as fairly straight lines or as a pattern of loops and whorls, as on the tips of the digits.

The ***epidermal ridges*** develop during the third and fourth fetal months as the epidermis conforms to the contours of the underlying dermal papillae (see Figure 5-1). The function of the ridges is to increase the grip of the hand or foot by increasing friction and acting like tiny suction cups. Since the ducts of sweat glands open on the summits of the epidermal ridges, fingerprints (or footprints) are left when a smooth object is touched. The ridge pattern, which is genetically determined, is unique for each individual. Normally, it does not change throughout life, except to enlarge, and thus can serve as the basis for identification through fingerprints or footprints.

Epidermal grooves on other parts of the skin divide the surface into a number of diamond-shaped areas. Examine the dorsum of the hand as an example. Note that hairs typically emerge at the points of intersection of the grooves. Note also that the grooves increase in frequency and depth in regions where your fingers bend.

CLINICAL APPLICATION: TREATING WRINKLES AND SCARS

It is possible to improve in many cases the appearance of scars from deep acne, chickenpox, burns, cleft lip, and other disorders and to make age-related wrinkles disappear. The procedure that makes this possible is called a *collagen implant.* Collagen is the body's principal structural material. It is found in skin, bone, cartilage, tendons, ligaments, and various viscera and accounts for almost one-third of the total protein of the body.

The collagen implant is prepared from cattle collagen that is suspended in a saline solution containing lidocaine, a local anesthetic. Once injected into the skin, the solution disappears, and the collagen, under the influence of body temperature, becomes a stationary fleshlike substance that is incorporated into the skin and smoothes it out. The collagen implant becomes colonized by blood vessels and cells and acts as a natural structural framework in the skin.

Another procedure, called *chemical exfoliation (skin peel)*, is used for more superficial problems such as frown lines, wrinkles, age spots, and superficial scars. It is performed under sedation or with light anesthesia. In this procedure, the natural oils in the skin are removed with solvents. Then, keratolytic agents (keratin-dissolving chemicals) are mixed with surgical soap or croton oil and applied thickly, and the area is wrapped in bandages for a few days. When the bandages are removed, the keratin-containing layers of the epidermis are peeled away. When the new surface layers grow back, the superficial problems are either greatly reduced or disappear completely.

The most recent treatment for wrinkles and brown spots related to sun-damaged (photo-aged) skin is a drug called *tretinoin (Retin-A)*, a derivative of vitamin A that has been used to treat acne since the 1960s. Among the reported benefits of using tretinoin are increased cell proliferation and thickening of the epidermis, smoother stratum corneum, diminished number and size of melanocytes, increased production of collagen and elastin, dilation of blood vessels in the dermis, and regression of precancerous lesions. Dermatologists caution that tretinoin does not reverse the aging process or improve advanced changes associated with aging.

SKIN WOUND HEALING: RESTORATION OF HOMEOSTASIS

Here we will discuss the basic mechanisms by which skin wounds are repaired. First we will consider wounds that affect primarily the epidermis. Then we will describe wounds that extend to tissues deep to the epithelium such as the dermis and subcutaneous layer.

Epidermal Wound Healing

The exposed location of the skin (and mucous membranes) causes it to be subjected to physical and chemical trauma. One common type of epidermal wound is an abrasion such as might be experienced in the form of a skinned knee or elbow. Another type is a first-degree or second-degree burn. In an epidermal wound, the central portion of the wound usually extends deep down to the dermis, whereas the edges of the wound usually involve only superficial damage to the epidermal cells.

In response to injury, basal epidermal cells in the area of the wound break their contacts with the basement membrane. These cells then enlarge and migrate across the wound (Figure 5-3). The cells appear to migrate as a sheet until advancing cells from opposite sides of the wound meet. When epidermal cells encounter each other, their continued migration is stopped by ***contact inhibition.*** According to this phenomenon, when one epidermal cell encounters another, its direction of movement is changed until it encounters another like cell, and so on. Continued migration of the epidermal cell is inhibited when it is finally in contact on all sides with other epidermal cells. Contact inhibition appears to occur only among like cells; in other words, contact inhibition does not occur between epidermal cells and other types of cells. Recall from Chapter 3 that malignant cells do not conform to the rules of contact inhibition, and so they have the ability to invade body tissues with few restrictions.

Simultaneous with the migration of some basal epidermal cells, stationary basal epidermal cells divide to replace the migrated cells. Migration continues until resurfacing of the wound is accomplished. Following this, the migrated cells themselves divide to form new strata, thus thickening the new epidermis. Once coverage of the wound surface beneath a scab is sufficient, the scab sloughs off and the surface epidermal cells become keratinized (see Figure 5-4b). The various events involved in epidermal wound healing occur within 24 to 48 hours after wounding.

Deep Wound Healing

When an injury extends to tissues deep to the epidermis, owing to accidental lacerations or surgical incisions, the repair process is more complex than epidermal healing, and scar formation results.

FIGURE 5-3 Epidermal wound healing. (a) Division of basal cells (germinal layer) and migration across wound. (b) Resurfacing of wound.

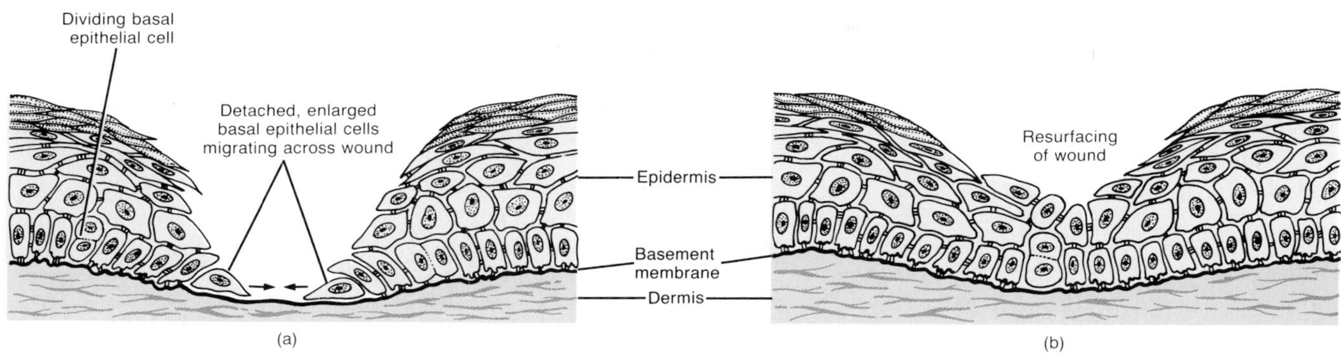

The first step in deep wound healing involves inflammation, a vascular and cellular response that serves to dispose of microbes, foreign material, and dying tissue in preparation for repair. During the ***inflammatory phase,*** a blood clot forms in the wound and loosely unites the wound edges, epithelial cells begin migrating across the wound (see Figure 5-3), vasodilation and increased permeability of blood vessels deliver neutrophils and monocytes that phagocytize microbes, and mesenchymal cells develop into fibroblasts (Figure 5-4a).

In the next phase, the ***migratory phase,*** the clot becomes a scab and epithelial cells migrate beneath the scab to bridge the wound, fibroblasts migrate along fibrin threads and begin synthesizing scar tissue (collagenous fibers and protein polysaccharides), and damaged blood vessels begin to regrow. During this phase, tissue filling the wound is called ***granulation tissue.***

The ***proliferative phase*** is characterized by an extensive growth of epithelial cells beneath the scab, the deposition of collagenous fibers in random patterns by fibroblasts, and the continued growth of blood vessels.

In the final phase, the ***maturation phase,*** the scab sloughs off once the epidermis is restored to normal thickness, collagenous fibers become more organized, fibroblasts begin to disappear, and blood vessels are restored to normal (Figure 5-4b).

The period of scar tissue formation is called ***fibroplasia*** (fī'-brō-PLĀ-zē-a; *fibro* = fibers; *plas* = to grow). In

FIGURE 5-4 Deep wound healing. (a) Inflammatory phase. (b) Maturation phase.

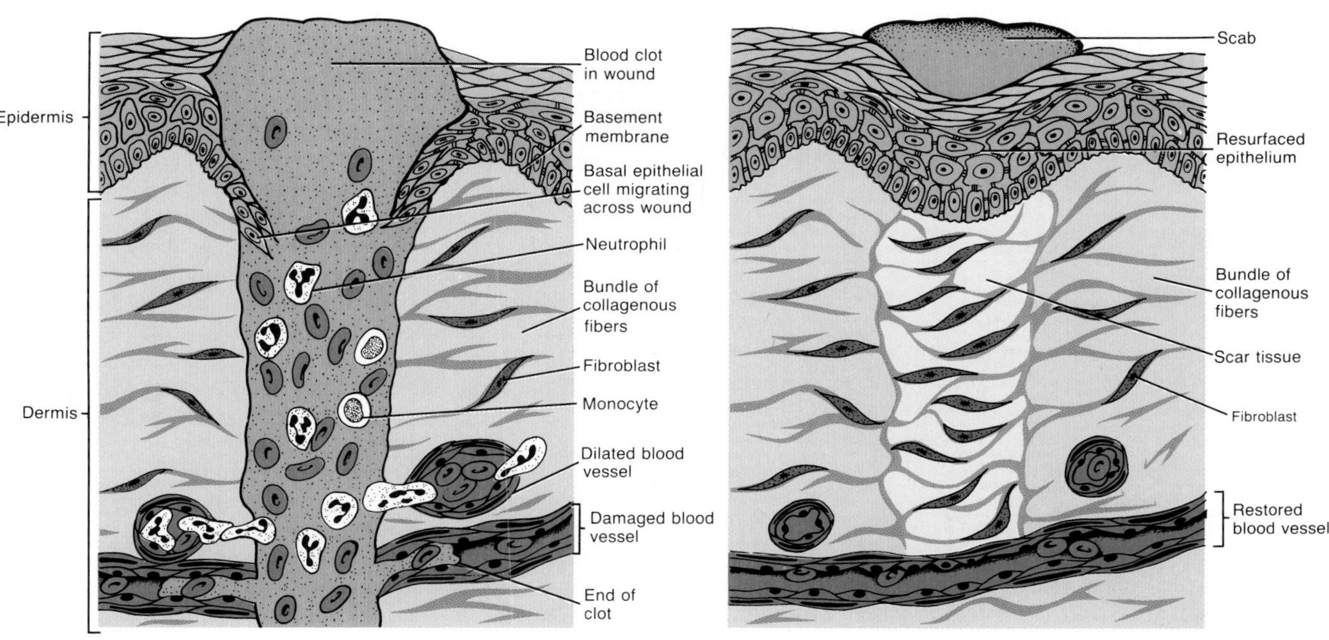

some cases, so much scar tissue is formed that a raised scar results, that is, one that is elevated above the normal epidermal surface. In some cases, such a scar remains within the boundaries of the original wound (***hypertrophic scar***), or it may extend beyond the boundaries of the original wound into normal surrounding tissues (***keloid scar***). Scar tissue differs from normal skin in that its collagenous fibers are more dense, and it has no epidermis, fewer blood vessels, and may not contain hair, skin glands, or sensory receptors.

CLINICAL APPLICATION: SKIN GRAFTS (SGs)

As a result of certain conditions, such as third-degree burns, the germinal portion of epidermis is destroyed and new skin cannot regenerate. Such wounds require *skin grafts (SGs)*. The most successful type of skin graft involves the transplantation of a segment of skin from a donor site to a recipient site of the same individual (autograft). Skin grafts vary in thickness from very thin, containing not much more than epidermis and part of the dermis (split-thickness), to full thickness, containing all of the dermis as well as epidermis.

If skin loss is so extensive that conventional grafting is impossible, a self-donation procedure called *autologous* (aw-TOL-o-gus) *skin transplantation* may be employed. In this procedure, small amounts of an individual's skin are removed, grown in the laboratory, and then transplanted back to the patient. The process normally takes three to four weeks and, during the waiting period, wound dressing or skin from another person (homograft) or animal (heterograft) help protect the patient from potentially fatal fluid loss and infection. Although homografts and heterografts are temporary because they are ultimately rejected by the body, they prevent loss of fluid, electrolytes, and protein from burn sites while also decreasing pain and increasing movement.

EPIDERMAL DERIVATIVES

Structures developed from the embryonic epidermis—hair, glands, nails—perform functions that are necessary and sometimes vital. Hair and nails protect the body. The sweat glands help regulate body temperature. The enamel of teeth is also an epidermal derivative and is discussed in Chapter 24.

Hair

Growths of the epidermis variously distributed over the body are ***hairs***, or ***pili*** (PI-lē). The primary function of hair is protection. Although the protection is limited, hair guards the scalp from injury and the sun's rays. Eyebrows and eyelashes protect the eyes from foreign particles. Hair in the external ear canal and nostrils protects these structures from insects and numerous inhaled particles. Touch receptors associated with hair follicles are activated when hair is even slightly touched.

Each hair consists of a shaft and a root (Figure 5-5a). The ***shaft*** is the superficial portion, most of which projects above the surface of the skin. The shaft of coarse hairs consists of three principal parts. The inner ***medulla*** is composed of rows of polyhedral cells containing granules of eleidin and air spaces. The middle ***cortex*** forms the major part of the shaft and consists of elongated cells that contain pigment granules in dark hair but mostly air in white hair. The ***cuticle of the hair***, the outermost layer, consists of a single layer of thin, flat, scalelike cells that are the most heavily keratinized. They are arranged like shingles on the side of a house, but the free edges of the cuticle cells point upward rather than downward like shingles.

The ***root*** is the portion below the surface that penetrates into the dermis and even into the subcutaneous layer and, like the shaft, contains a medulla, cortex, and cuticle (Figure 5-5a).

Surrounding the root is the ***hair follicle***, which is made up of an external root sheath and an internal root sheath. The ***external root sheath*** is a downward continuation of the basale and spinosum layers of the epidermis. Near the surface, it contains all the epidermal layers. At the bottom of the hair follicle, the external root sheath contains only the stratum basale. The ***internal root sheath*** is formed from proliferating cells of the matrix (described shortly) and takes the form of a cellular tubular sheath deep to the external root sheath.

The base of each follicle is enlarged into an onion-shaped structure, the ***bulb***. This structure contains an indentation, the ***papilla of the hair***, filled with loose connective tissue. The papilla of the hair contains many blood vessels and provides nourishment for the growing hair. The bulb also contains a region of cells called the ***matrix***, a germinal layer. The cells of the matrix produce new hairs by cell division when older hairs are shed. This replacement occurs within the same follicle.

Normal hair loss in an adult scalp is about 70 to 100 hairs per day. Both the rate of growth and the replacement cycle may be altered by illness, diet, and other factors. For example, high fever, major illness, major surgery, blood loss, or severe emotional stress may increase the rate of shedding. Rapid weight-loss diets involving severe restriction of calories or protein also increase hair loss. An increase in the rate of shedding can also occur for three to four months after childbirth. Certain drugs and radiation therapy are also factors in increasing hair loss.

A substance that removes superfluous hair is called a ***depilatory***. It dissolves the protein in the hair shaft, turning it into a gelatinous mass that can be wiped away. Since the hair root is not affected, regrowth of the hair occurs. In ***electrolysis***, the hair bulb is destroyed by an electric current so that the hair cannot regrow.

FIGURE 5-5 Hair. (a) Principal parts of a hair root and associated structures seen in longitudinal section. The relationship of a hair to the epidermis, sebaceous (oil) glands, and arrector pili muscle should be noted in Figure 5-1. (b) Color enhanced scanning electron micrograph of two hair shafts growing out of hair follicles at a magnification of 160×. Note the shinglelike cuticular scales of the hair shafts. (Courtesy of CNPI/Science Photo Library/Photo Researchers.)

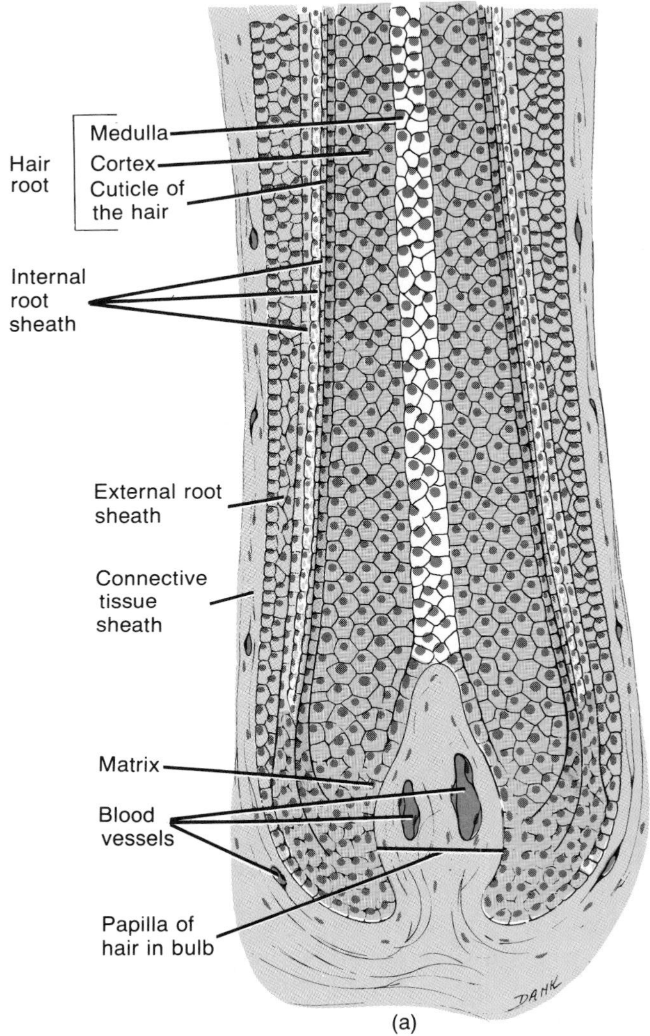

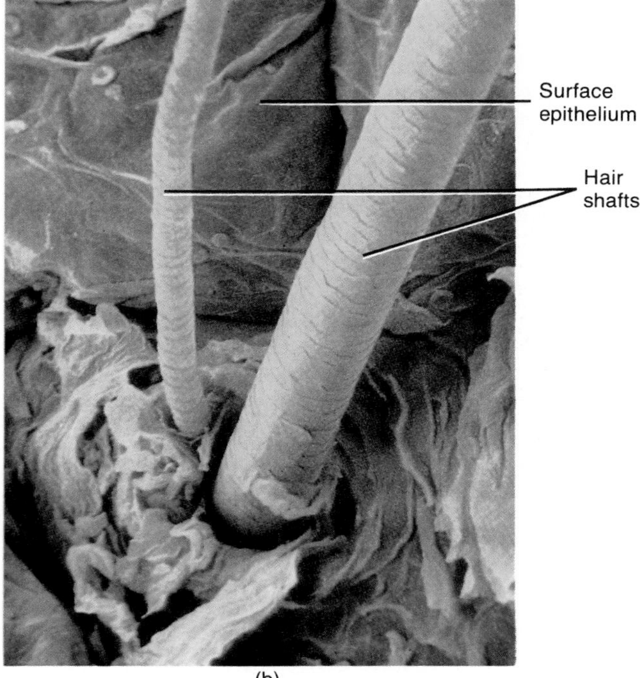

Sebaceous (oil) glands and a bundle of smooth muscle are also associated with hair. Details of the sebaceous glands will be discussed shortly. The smooth muscle is called ***arrector pili;*** it extends from the dermis of the skin to the side of the hair follicle (see Figure 5-1). In its normal position, hair is arranged at an angle to the surface of the skin. The arrectores pilorum muscles contract under stresses of fright, cold, and emotions, and pull the hairs into a vertical position. This contraction results in "goose bumps" or "gooseflesh" because the skin around the shaft forms slight elevations.

Around each hair follicle are nerve endings, called ***hair root plexuses,*** that are sensitive to touch (see Figure 15-1). They respond if a hair shaft is moved.

CLINICAL APPLICATION: COMMON BALDNESS

The age of onset, degree of thinning, and ultimate hair pattern associated with ***common baldness (male-pattern baldness)*** are determined by male

hormones, called androgens, and heredity. Androgens are also involved in promoting normal sexual development. In recent years, minoxidil (Rogaine), a potent vasodilator drug used to treat high blood pressure, has been applied topically to stimulate hair regrowth in *some* persons with thinning hair associated with common baldness. It will not help individuals who are bald. Since minoxidil increases blood flow in the skin, it is believed to stimulate existing hair follicles to become more active.

Glands

Three kinds of glands associated with the skin are sebaceous glands, sudoriferous glands, and ceruminous glands.

Sebaceous (Oil) Glands

Sebaceous (se-BĀ-shus), or ***oil glands,*** with few exceptions, are connected to hair follicles (see Figure 5-1). The secreting portions of the glands lie in the dermis, and those glands associated with hairs open into the necks of hair follicles. Sebaceous glands not associated with hair follicles open directly onto the surface of the skin (lips, glans penis, labia minora, and tarsal glands of the eyelids). Sebaceous glands are simple branched acinar glands. Absent in the palms and soles, they vary in size and shape in other regions of the body. For example, they are small in most areas of the trunk and extremities, but large in the skin of the breasts, face, neck, and upper chest.

The sebaceous glands secrete an oily substance called ***sebum*** (SĒ-bum), a mixture of fats, cholesterol, proteins, and inorganic salts. Sebum helps keep hair from drying and becoming brittle, forms a protective film that prevents excessive evaporation of water from the skin, keeps the skin soft and pliable, and inhibits the growth of certain bacteria.

CLINICAL APPLICATION: BLACKHEADS

When sebaceous glands of the face become enlarged because of accumulated sebum, acne lesions called *blackheads* develop. Since sebum is nutritive to certain bacteria, *pimples* or *boils* often result. The color of blackheads is due to melanin and oxidized oil, not dirt.

Sudoriferous (Sweat) Glands

Sudoriferous (soo'-dor-IF-er-us; *sudor* = sweat; *ferre* = to bear), or ***sweat glands*** are divided into two principal types on the basis of structure and location. ***Apocrine sweat glands*** are simple, branched tubular glands. Their distribution is limited primarily to the skin of the axilla, pubic region, and pigmented areas (areolae) of the breasts. The secretory portion of apocrine sweat glands is located in the dermis or subcutaneous layer and the excretory duct opens into hair follicles. Apocrine sweat glands begin to function at puberty and produce a more viscous secretion than eccrine sweat glands.

Eccrine sweat glands are much more common than apocrine sweat glands and are simple, coiled tubular glands. They are distributed throughout the skin except for the margins of the lips, nail beds of the fingers and toes, glans penis, glans clitoris, labia minora, and eardrums. Eccrine sweat glands are most numerous in the skin of the palms and the soles; their density can be as high as 3000 per square inch in the palms. The secretory portion of eccrine sweat glands is located in the subcutaneous layer, and the excretory duct projects upward through the dermis and epidermis to terminate at a pore at the surface of the epidermis (see Figure 5-1). Eccrine sweat glands function throughout life and produce a secretion that is more watery than that of apocrine sweat glands.

Perspiration, or ***sweat,*** is the substance produced by sudoriferous glands. It is a mixture of water, salts (mostly NaCl), urea, uric acid, amino acids, ammonia, sugar, lactic acid, and ascorbic acid. Its principal function is to help regulate body temperature by way of evaporation of water in perspiration which carries off large quantities of heat energy from the body surface. It also helps to eliminate wastes.

Since the mammary glands are actually modified sudoriferous glands, they could be discussed here. However, because of their relationship to the reproductive system, they will be considered in Chapter 28.

Ceruminous Glands

In certain parts of the skin, sudoriferous glands are modified as ***ceruminous*** (se-ROO-mi-nus) ***glands.*** Such modified glands are simple, coiled tubular glands present in the external auditory meatus (canal). The secretory portions of ceruminous glands lie in the subcutaneous layer, deep to sebaceous glands, and the excretory ducts open either directly onto the surface of the external auditory meatus or into ducts of sebaceous glands. The combined secretion of the ceruminous and sebaceous glands is called ***cerumen*** (*cera* = wax). Cerumen, together with hairs in the external auditory meatus, provides a sticky barrier that prevents the entrance of foreign bodies.

CLINICAL APPLICATION: IMPACTED CERUMEN (EARWAX)

Some people produce an abnormal amount of cerumen, or earwax, in the external auditory meatus. It then becomes impacted and prevents sound waves from reaching the tympanic membrane (eardrum). The treatment for *impacted cerumen* is usually periodic ear irrigation or removal of wax with a blunt instrument by trained medical personnel.

FIGURE 5-6 Structure of nails. (a) Fingernail viewed from above. (b) Sagittal section of a fingernail.

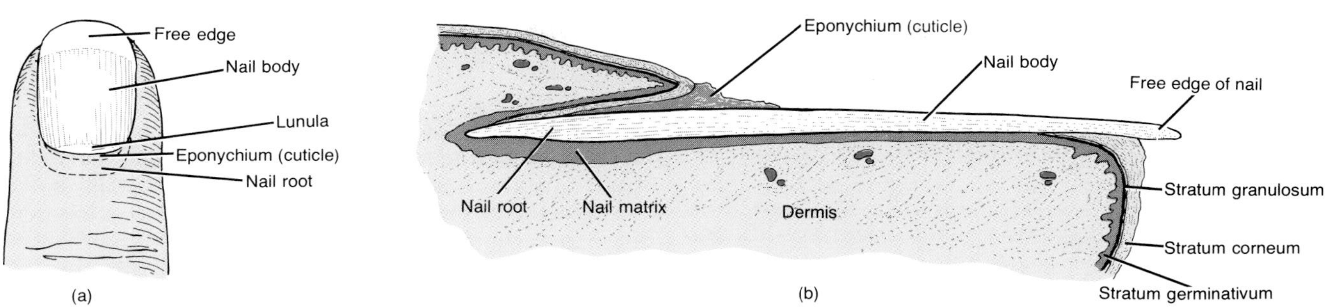

Nails

Hard, keratinized cells of the epidermis are referred to as ***nails.*** The cells form a clear, solid covering over the dorsal surfaces of the terminal portions of the fingers and toes. Each nail (Figure 5-6) consists of a nail body, a free edge, and a nail root. The ***nail body*** is the portion of the nail that is visible, the ***free edge*** is the part that may project beyond the distal end of the digit, and the ***nail root*** is the portion that is hidden in the proximal nail groove. Most of the nail body is pink because of the underlying vascular tissue. The whitish semilunar area of the proximal end of the body is called the ***lunula*** (LOO-nyoo-la). It appears whitish because the vascular tissue underneath does not show through owing to the thickened stratum basale in the area. The ***eponychium*** (ep'-ō-NIK-ē-um) or ***cuticle*** is a narrow band of epidermis that extends from the margin of the nail wall (lateral border), adhering to it. It occupies the proximal border of the nail and consists of stratum corneum.

The epithelium of the proximal part of the nail bed is known as the ***nail matrix.*** Its function is to bring about the growth of nails. Essentially, growth occurs by the transformation of superficial cells of the matrix into nail cells. In the process, the outer, harder layer is pushed forward over the stratum germinativum. The average growth in the length of fingernails is about 1 mm (0.04 inch) per week. The growth rate is somewhat slower in toenails. For some reason, the longer the digit, the faster the nail grows. Adding supplements such as gelatin to an otherwise healthy diet has no effect on making nails grow faster or stronger.

Functionally, nails help us to grasp and manipulate small objects in various ways and provide protection against trauma to the ends of the digits.

HOMEOSTASIS OF BODY TEMPERATURE

One of the best examples of homeostasis in humans is the regulation of body temperature by the skin. Humans, like other mammals, are ***homeotherms*** (*homeo* = same; *therm* = temperature)—warm-blooded organisms. This means that we are able to maintain a remarkably constant body temperature of about 37°C (98.6°F) even though the environmental temperature may vary over a broad range.

Suppose you are in an environment where the temperature is higher than normal body temperature at 37.8°C (100°F). A sequence of events is set into operation to counteract this above-normal temperature, which may be considered a stress. Sensing structures in the skin called ***thermoreceptors*** pick up the stimulus—in this case, heat—and activate neurons (nerve cells) that send the message (impulse) to your brain. A temperature-regulating area of the brain then sends nerve impulses to the sudoriferous (sweat) glands, which produce more perspiration. As the perspiration evaporates from the surface of your skin, it is cooled and your body temperature is lowered. This sequence of events is shown in Figure 5-7.

FIGURE 5-7 One role of the skin in regulating the homeostasis of body temperature.

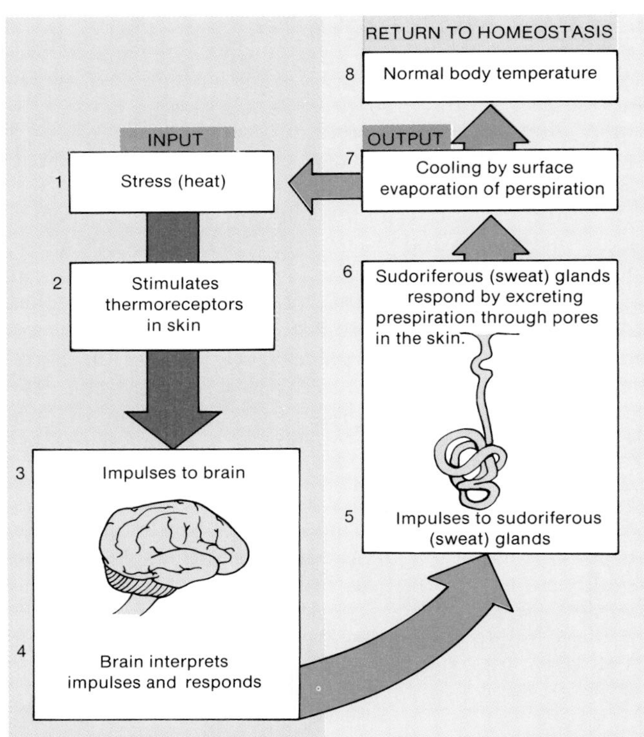

Note that temperature regulation by the skin involves a feedback system because the output (cooling of the skin) is fed back to the skin receptors and becomes part of a new stimulus–response cycle. In other words, after the sudoriferous glands are activated, the skin receptors keep the brain informed about the external temperature. The brain, in turn, continues to send impulses to the sudoriferous glands until the temperature returns to 37°C (98.6°F). Like most of the body's feedback systems, temperature regulation is a negative feedback system—the output, cooling, is the opposite of the original condition, overheating.

Temperature regulation by perspiration represents only one mechanism by which we can lower body temperature to normal. Other mechanisms include adjusting blood flow to the skin (dilation of blood vessels in the skin results in the release of more heat), regulating metabolic rate (a slower metabolic rate reduces heat production), and regulating skeletal musle contractions (decreased muscle tone results in less heat production). These mechanisms are discussed in detail in Chapter 25.

In response to a lower-than-normal body temperature, production of perspiration is decreased, blood vessels in the skin constrict so that less heat is released, metabolic rate is increased to produce more heat, and increased muscle tone and shivering of skeletal muscles results in increased heat production.

AGING AND THE INTEGUMENTARY SYSTEM

Although skin is constantly aging, the pronounced effects do not occur until a person reaches the late forties. Collagen fibers decrease in number, stiffen, break apart, and form into a shapeless, matted tangle. Elastic fibers lose some of their elasticity, thicken into clumps and fray, and as a result, the skin wrinkles. Fibroblasts, which produce both collagenous and elastic fibers, decrease in number, and macrophages become less efficient phagocytes. In addition, blood vessels in the dermis become thick walled and more permeable; there is a loss of subcutaneous fat; atrophy of sebaceous (oil) glands, producing dry and broken skin that is susceptible to infection; a decrease in the number of functioning melanocytes, resulting in gray hair and atypical skin pigmentation; and an increase in the size of some melanocytes that produces pigmented blotching (liver spots). In general, aged skin is thinner than young skin. Aged skin also heals poorly and becomes susceptible to pathological conditions such as skin cancer, senile pruritis (itching), decubitus ulcers (bedsores), and herpes zoster (shingles). Prolonged exposure to the UV rays of sunlight accelerates the aging of skin and results in considerable damage (photodamage).

DEVELOPMENTAL ANATOMY OF THE INTEGUMENTARY SYSTEM

In this and subsequent chapters, the developmental anatomy of the systems of the body will be discussed at the end of each appropriate chapter. Since the principal features of embryonic development are not treated in detail until Chapter 29, it will be necessary to explain a few terms at this point so that you can follow the development of organ systems.

As part of the early development of a fertilized egg, a portion of the developing embryo differentiates into three layers of tissue called ***primary germ layers.*** On the basis of position, the primary germ layers are referred to as ***ectoderm, mesoderm,*** and ***endoderm.*** They are the embryonic tissues from which all tissues and organs of the body will eventually develop (see Exhibit 29-1).

The *epidermis* is derived from the ***ectoderm,*** the outermost primary germ layer. At the beginning of the second month, the ectoderm consists of simple cuboidal epithelium. These cells become flattened and are known as the ***periderm.*** By the fourth month, all layers of the epidermis are formed and each layer assumes its characteristic structure.

Nails are developed during the third month. Initially, they consist of a thick layer of epithelium called the ***primary nail field.*** The nail itself is keratinized epithelium and grows forward from its base. It is not until the ninth month that the nails actually reach the tips of the digits.

Hair follicles develop between the third and fourth months as downgrowths of the stratum basale of the epidermis into the dermis below. The downgrowths soon differentiate into the bulb, papilla of the hair, beginnings of the epithelial portions of sebaceous glands, and other structures associated with hair follicles. By the fifth or sixth month, the follicles produce ***lanugo*** (delicate fetal hair), first on the head and then on other parts of the body. The lanugo is usually shed prior to birth.

The epithelial (secretory) portions of *sebaceous (oil) glands* develop from the sides of the hair follicles and remain connected to the follicles.

The epithelial portions of *sudoriferous (sweat) glands* are also derived from downgrowths of the stratum basale of the epidermis into the dermis. They appear during the fourth month on the palms and soles and a little later in other regions. The connective tissue and blood vessels associated with the glands develop from ***mesoderm,*** the middle primary germ layer.

The *dermis* is derived from wandering ***mesenchymal (mesodermal) cells.*** The mesenchyme becomes arranged in a zone beneath the ectoderm and there undergoes changes into the connective tissues that form the dermis.

DISORDERS: HOMEOSTATIC IMBALANCES

Acne

Acne is an inflammation of sebaceous (oil) glands and usually begins at puberty. At puberty, the sebaceous glands, under the influence of androgens (male hormones), grow in size and increase production of sebum. Although testosterone, a male hormone, appears to be the most potent circulating androgen for sebaceous cell stimulation, adrenal and ovarian androgens can stimulate sebaceous secretions as well.

Acne occurs predominantly in sebaceous follicles that are rapidly colonized by bacteria that thrive in the lipid-rich sebum. When this occurs, the cyst or sac of connective tissue cells can destroy and displace epidermal cells, resulting in permanent scarring, a condition called ***cystic acne.*** Care must be taken to avoid squeezing, pinching, or scratching the lesions.

Cystic acne may be treated by a synthetic form of vitamin A called Accutane (isotretinoin). However, it must not be used by females who are pregnant or who intend to become pregnant while undergoing treatment. Major fetal abnormalities have been traced to Accutane. This drug may also cause serious side effects in those who take it and should be used only if less toxic forms of therapy have failed.

Systemic Lupus Erythematosus (SLE)

Systemic lupus erythematosus (er-ē'-the-ma-TŌ-sus), ***SLE,*** or ***lupus*** is an autoimmune, inflammatory disease of connective tissue, occurring mostly in young women in their reproductive years. An ***autoimmune disease*** is one in which the body attacks its own tissues, failing to differentiate between what is foreign and what is not. In SLE, damage to blood vessel walls results in the release of chemicals that mediate the inflammatory response. The blood vessel damage can be associated with virtually every body system.

The cause of SLE is not known, and its onset may be abrupt or gradual. It is not contagious and is thought to be hereditary. There seems to be a strong incidence of other connective tissue disorders—especially rheumatoid arthritis (RA) and rheumatic fever—in relatives of SLE victims. The disease may be triggered by medication, such as penicillin, sulfa, or tetracycline, exposure to excessive sunlight, injury, emotional upset, infection, or other stress. These triggering factors, once recognized, are to be avoided by the patient in the future.

Symptoms include painful joints, low-grade fever, fatigue, mouth ulcers, weight loss, enlarged lymph nodes and spleen, photosensitivity, rapid loss of large amounts of scalp hair, and sometimes an eruption across the bridge of the nose and cheeks called a "butterfly rash." Other skin lesions may occur with blistering and ulceration. The erosive nature of some of the SLE skin lesions was thought to resemble the damage inflicted by the bite of a wolf—thus, the term ***lupus.*** The most serious complications of the disease involve inflammation of the kidneys, liver, spleen, lungs, heart, and the central nervous system.

Psoriasis

Psoriasis (sō-RĪ-a-sis) is a chronic, occasionally acute, noncontagious, relapsing skin disease. It is characterized by distinct, reddish, slightly raised plaques or small, round skin elevations covered with scales. Itching is seldom severe, and the lesions heal without scarring. Psoriasis ordinarily involves the scalp, the elbows and knees, the back, and the buttocks. Occasionally the disease is generalized.

The cause of psoriasis is an abnormally high rate of mitosis in epidermal cells that may be related to a substance carried in the blood, a defect in the immune system, or a virus. Triggering factors such as trauma, infections, certain drugs (beta blockers and lithium) seasonal and hormonal changes, and emotional stress can initiate and intensify the skin eruptions.

Psoriasis is treated by a number of methods including steroid ointments and creams; natural sunlight; tar preparations; retinoids (chemicals similar to vitamin A); PUVA, a therapy that combines psoralen (a chemical that increases the skin's reaction to light); and artificial UV light; and a newly approved drug called etretinate (Tegison). Some evidence suggests that PUVA may increase the long-term risk of developing two usually nonfatal skin cancers (squamous cell and basal cell).

Decubitus Ulcers

Decubitus (dē-KYOO-bi-tus) ***ulcers,*** also known as ***bedsores, pressure sores,*** or ***trophic ulcers,*** are caused by a constant deficiency of blood to tissues overlying a bony projection that has been subjected to prolonged pressure against an object such as a bed, cast, or splint. The deficiency results in tissue ulceration. Small breaks in the epidermis become infected, and the sensitive subcutaneous and deeper tissues are damaged. Eventually, the tissue is destroyed.

Decubitus ulcers are seen most frequently in patients who are bedridden for long periods of time. The most common areas involved are the skin over the sacrum, heels, ankles, buttocks, and the skin over other large bony projections. The chief causes are pressure from infrequent turning of the patient, trauma and maceration of the skin, and malnutrition. Maceration of the skin often follows soaking of bed and clothing by perspiration, urine, or feces.

Sunburn

Sunburn is injury to the skin as a result of acute, prolonged exposure to the UV rays of sunlight. The damage to skin cells caused by sunburn is due to inhibition of DNA and RNA synthesis, which leads to cell death. There can also be damage to blood vessels as well as other structures in the dermis. Overexposure over a period of years results in a leathery skin texture, wrinkles, skin folds, sagging skin, warty growths called keratoses, freckling, a yellow discoloration due to abnormal elastic tissue, premature aging of the skin, and skin cancer.

DISORDERS: HOMEOSTATIC IMBALANCES (*continued*)

Skin Cancer

Excessive sun exposure can result in ***skin cancer,*** the most common cancer in Caucasians. However, everyone, regardless of skin pigmentation, is a potential victim of skin cancer if exposure to sunlight is sufficiently intense and continuous. Natural skin pigment can never give complete protection.

The three most common forms of skin cancer are basal cell carcinoma (BCC), squamous cell carcinoma (SCC), and malignant melanoma. ***Basal cell carcinomas (BCCs)*** account for over 75 percent of all skin cancers. The tumors arise from the epidermis and rarely metastasize. They are believed to be caused by years of chronic sun exposure. ***Squamous cell carcinomas (SCCs)*** also arise from the epidermis, and although less common than BCCs, they have a variable tendency to metastasize. Most SCCs arise from preexisting lesions on sun-exposed skin. ***Malignant melanomas*** arise from melanocytes and are the leading cause of death from all diseases arising in the skin since they metastasize rapidly. Fortunately, malignant melanomas are the least common of the skin cancers, accounting for only about 3 percent of all skin cancers. The principal cause is chronic sun exposure. A clinical trial is under way to test an antimelanoma vaccine made from human melanoma cells. The vaccine apparently sensitizes the patients so that they are more receptive to other anticancer substances, such as interleukin-2, or chemotherapy. The antimelanoma vaccine is designed to stop continued tumor growth rather than prevent tumor formation.

The treatment of most skin cancers consists of curettage, excision, electrodesiccation, cryosurgery, laser surgery, and radiation therapy.

Among the risk factors for skin cancer are:

1. Skin type. Persons with fair skin who never tan but always burn are at high risk.
2. Geographic location. Areas with many days of sunlight per year and high-altitude locations show high incidences of skin cancer.
3. Age. Older people are more prone to skin cancer owing to greater exposure to sunlight.
4. Personal habits. Individuals engaged in outdoor occupations have a higher risk.
5. Immunologic status. Persons who are immunosuppressed have a higher incidence of skin cancer.

In recent years, ***suntanning salons*** have become very popular, and this has become a concern to physicians. Many salons claim to use "safe" wavelengths of UV light, that is, longer (A) portions of the UV spectrum (UVA). According to some medical authorities, the radiation emitted by lamps in many suntanning parlors is potentially dangerous. Among the harmful effects are increased risk for skin cancer, premature aging of the skin, supression of the immune system, and retinal damage or cataracts.

If you must be in direct sunlight for long periods of time, use a suitable sunscreen. One of the best agents for protection against overexposure to ultraviolet rays of the sun is para-aminobenzoic acid (PABA). The alcohol preparations of PABA are best because the active ingredient binds to the stratum corneum of the skin.

Burns

Tissues may be damaged by thermal (heat), electrical, radioactive, or chemical agents, all of which can destroy (denature) the proteins in the exposed cells and cause cell injury or death. Such damage is a ***burn.*** Burns disrupt homeostasis because they destroy the protection afforded by the skin, allowing microbial invasion and infection, loss of fluid, and loss of temperature control. The injury to tissues directly or indirectly in contact with the damaging agent, such as the skin or the linings of the respiratory and gastrointestinal tracts, is the local effect of a burn. Generally, however, the systemic effects of a burn are a greater threat to life than the local effects. The systemic effects of a burn may include (1) a large loss of water, plasma, and plasma proteins, which causes shock; (2) bacterial infection; (3) reduced circulation of blood; and (4) decreased production of urine.

Classification

A ***first-degree burn*** involves only the surface epidermis. It is characterized by mild pain, erythema (redness), dry skin, slight edema, and no blisters. Skin functions remain intact. The pain of a first-degree burn may be lessened by flushing the affected area with cold water. Generally, a first-degree burn will heal in about two to three days and may be accompanied by flaking or peeling. A typical sunburn is an example of a first-degree burn.

A ***second-degree burn*** involves the entire epidermis or varying portions of the dermis, and skin functions are lost. In a superficial second-degree burn, the deeper layers of the epidermis are injured, and there is characteristic redness, blister formation, marked edema, and pain. Such an injury usually heals within 7 to 10 days with only mild scarring. In a deep second-degree burn, there is destruction of the epidermis as well as the upper levels of the dermis. Epidermal derivatives, such as hair follicles, sebaceous glands, and sweat glands, are usually not injured. If there is no infection, deep second-degree burns heal without grafting in about three to four weeks. Scarring may result.

First- and second-degree burns are collectively referred to as ***partial-thickness burns.*** A ***third-degree burn*** or ***full-thickness burn*** involves destruction of the epidermis, dermis, and the epidermal derivatives, and skin functions are lost. Such burns vary in appearance from marble-white to mahogany colored to charred, dry wounds. There is marked edema, and such a burn is usually not painful to the touch owing to

DISORDERS: HOMEOSTATIC IMBALANCES (*continued*)

destruction of nerve endings. Regeneration is slow, and much granulation tissue forms before being covered by epithelium. Even if skin grafting is quickly begun, third-degree burns rapidly contract and produce scarring.

The seriousness of a burn is determined by depth, extent, area of involvement, age, and general health.

A fairly accurate method for estimating the amount of surface area affected by a burn is to apply the ***Lund-Browder method.*** This method estimates the extent by measuring the areas affected against the percentage of total surface area for body parts shown in Figure 5-8. For example, if the anterior of the head and neck of an adult is affected, the burn covers $4\frac{1}{2}$ percent of the body surface. Because the proportions of the body change with growth, the percentages vary for different ages. Thus, the extent of burn damage can be made fairly accurately for any age group.

A somewhat less accurate, although easy to apply, means for estimating the extent of a burn is the ***rule of nines.***

1. If the anterior and posterior surfaces of the head and neck are affected, the burn covers 9 percent of the body surface.
2. The anterior and posterior surfaces of each shoulder, arm, forearm, and hand constitute 9 percent of the body surface.
3. The anterior and posterior surfaces of the trunk, including the buttocks, constitute 36 percent.
4. The anterior and posterior surfaces of each foot, leg, and thigh as far up as the buttocks total 18 percent.
5. The perineum (per-i-NĒ-um) consists of 1 percent. The perineum includes the anal and urogenital regions.

FIGURE 5-8 The Lund-Browder method for estimating the extent of burns. Relative proportions of various body regions in (a) a young child compared with (b) an adult.

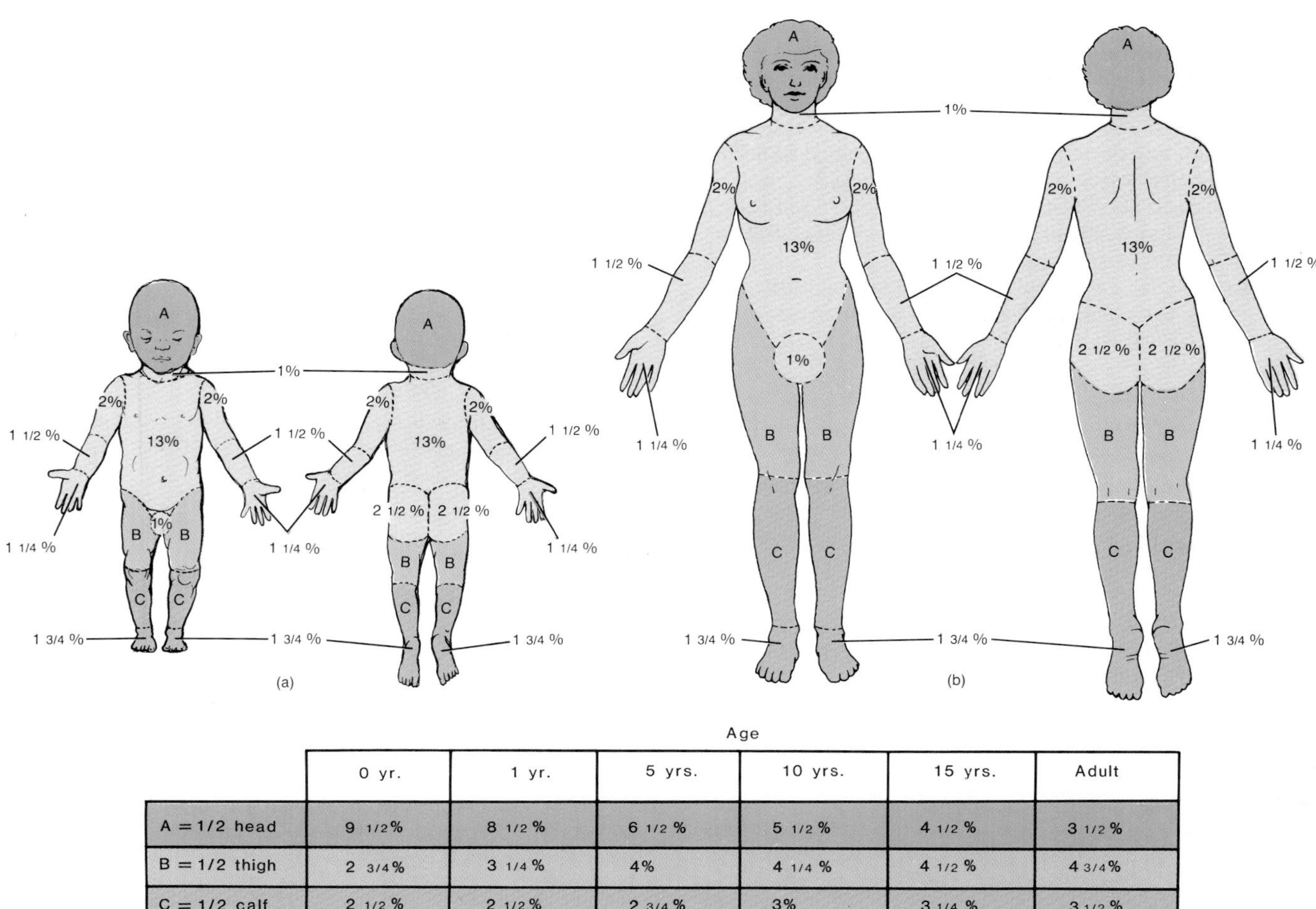

	0 yr.	1 yr.	5 yrs.	10 yrs.	15 yrs.	Adult
A = 1/2 head	9 1/2%	8 1/2 %	6 1/2 %	5 1/2 %	4 1/2 %	3 1/2 %
B = 1/2 thigh	2 3/4%	3 1/4 %	4%	4 1/4 %	4 1/2 %	4 3/4%
C = 1/2 calf	2 1/2 %	2 1/2%	2 3/4 %	3%	3 1/4 %	3 1/2 %

MEDICAL TERMINOLOGY ASSOCIATED WITH THE INTEGUMENTARY SYSTEM

Abrasion (a-BRĀ-shun; *ab* = away; *rasion* = scraped) A portion of the skin that has been scraped away.

Anhidrosis (an'-hī-DRŌ-sis; *an* = without; *hidr* = sweating; *osis* = condition) A rare genetic condition characterized by inability to sweat.

Athlete's (ATH-lēts) **foot** A superficial fungus infection of the skin of the foot.

Callus (KAL-lus) An area of hardened and thickened skin that is usually seen in palms and soles and is due to pressure and friction.

Carbuncle (KAR-bung-kl; *carbunculus* = little coal) A hard, round, deep, painful inflammation of the subcutaneous tissue that causes necrosis (death) and pus formation (abscess).

Chickenpox Highly contagious disease that initiates in the respiratory system and is caused by the varicella-zoster virus and characterized by vesicular eruptions on the skin that fill with pus, rupture, and form a scab before healing. Also called **varicella** (var'-i-SEL-a). Shingles is caused by the latent chickenpox virus.

Cold sore (KŌLD sor) A lesion, usually in oral mucous membrane, caused by type 1 herpes simplex virus (HSV), transmitted by oral or respiratory routes. Triggering factors include UV radiation, hormonal changes, and emotional stress. Also called a **fever blister.**

Comedo (KOM-ē-dō; *comedo* = to eat up) A collection of sebaceous material and dead cells in the hair follicle and excretory duct of the sebaceous (oil) gland. Usually found over the face, chest, and back, and more commonly during adolescence. Also called **blackhead** or **whitehead.**

Contusion (kon-TOO-shun; *contundere* = to bruise) Condition in which tissue below the skin is damaged, but the skin is not broken.

Corn (KORN) A painful conical thickening of the skin found principally over toe joints and between the toes. It may be hard or soft, depending on the location. Hard corns are usually found over toe joints, and soft corns are usually found between the fourth and fifth toes.

Cyst (SIST; *cyst* = sac containing fluid) A sac with a distinct connective tissue wall, containing a fluid or other material.

Detritus (de-TRĪ-tus; *deterere* = to rub away) Particulate matter produced by or remaining after the wearing away or disintegration of a substance or tissue; scales, crusts, and loosened skin.

Eczema (EK-ze-ma; *ekzein* = to boil out) An acute or chronic superficial inflammation of the skin, characterized by redness, oozing, crusting, and scaling. Also called **chronic dermatitis.**

Erythema (er'-e-THĒ-ma; *erythema* = redness) Redness of the skin caused by an engorgement of capillaries in lower layers of the skin. Erythema occurs with any skin injury, infection, or inflammation.

Furuncle (FYOOR-ung-kul) A boil; an abscess resulting from infection of a hair follicle.

German measles Highly contagious disease that initiates in the respiratory system and is caused by the rubella virus and characterized by a rash of small red spots on the skin. Also called **rubella** (roo-BEL-a).

Hemangioma (hē-man'-jē-Ō-ma; *hemo* = blood; *angio* = blood vessel; *oma* = tumor) Localized tumor of the skin and subcutaneous layer that results from an abnormal increase in blood vessels; one type is a **portwine stain,** a flat, pink, red, or purple lesion present at birth, usually at the nape of the neck.

Hives (HĪVZ) Condition of the skin marked by reddened elevated patches that are often itchy. Most commonly caused by infections, physical trauma, medications, emotional stress, food additives, and certain foods. Also called **urticaria** (yoor-ti-KAR-ē-a).

Impetigo (im'-pe-TĪ-go) Superficial skin infection caused by staphylococci or streptococci; most common in children.

Intradermal (in'-tra-DER-mal; *intra* = within) Within the skin. Also called **intracutaneous.**

Keratosis (ker'-a-TŌ-sis; *kera* = horn) Formation of a hardened growth of tissue.

Laceration (las'-er-Ā-shun; *lacerare* = to tear) Wound or irregular tear of the skin.

Measles Highly contagious disease that initiates in the respiratory system and is caused by the measles virus and characterized by a papular rash on the skin. Also called **rubeola** (roo-bē-Ō-la).

Nevus (NE-vus) A round, pigmented, flat, or raised skin area that may be present at birth or develop later. Varying in color from yellow-brown to black. Also called a **mole** or **birthmark.**

Papule (PAP-yool) A small, round skin elevation varying in size from a pinpoint to that of a split pea. One example is a pimple.

Polyp (POL-ip) A tumor on a stem found especially on mucous membranes.

Pruritus (proo-RĪ-tus; *pruire* = to itch) Itching, one of the most common dermatological disorders. It may be caused by skin disorders (infections), systemic disorders (cancer, kidney failure), or psychogenic factors (emotional stress).

Subcutaneous (sub'-kyoo-TĀ-nē-us; *sub* = under) Beneath the skin; also called **hypodermis.**

Topical (TOP-i-kal) Pertaining to a definite area; local. Also in reference to a medication, applied to the surface rather than ingested or injected.

Wart (WORT) Mass produced by uncontrolled growth of epithelial skin cells; caused by a virus (papillomavirus). Most warts are noncancerous. Treatment may involve cryosurgery with liquid nitrogen, electrosurgery, chemical destruction, injections, carbon dioxide laser surgery, surgical excision, and immunotherapy.

STUDY OUTLINE

Skin (p. 120)

1. The skin and its derivatives (hair, glands, and nails) constitute the integumentary system.
2. The skin is one of the larger organs of the body. It performs the functions of regulating body temperature; protection; receiving stimuli; excretion of water, salts, and several organic compounds; and synthesis of vitamin D.
3. The principal parts of the skin are the outer epidermis and inner dermis. The dermis overlies the subcutaneous layer.
4. The epidermal layers, from deepest to most superficial, are the strata basale, spinosum, granulosum, lucidum, and corneum. The basale layer undergoes continuous cell division and produces all other layers. Epidermal cells include keratinocytes, melanocytes, and nonpigmented granular dendrocytes (Langerhans' and Granstein cells).
5. The dermis consists of a papillary region and a reticular region. The papillary region is loose connective tissue containing blood vessels, nerves, hair follicles, dermal papillae, and corpuscles of touch (Meissner's). The reticular region is dense, irregularly arranged connective tissue containing adipose tissue, hair follicles, nerves, sebaceous (oil) glands, and ducts of sudoriferous (sweat) glands.
6. Lines of cleavage indicate the direction of collagenous fiber bundles in the dermis and are considered during surgery.
7. The color of skin is due to melanin, carotene, and blood in capillaries in the dermis.
8. Epidermal ridges increase friction for better grasping ability and provide the basis for fingerprints and footprints.

Skin Wound Healing: Restoration of Homeostasis (p. 124)

Epidermal Wound Healing (p. 124)

1. In an epidermal wound, the central portion of the wound usually extends deep down to the dermis, whereas the wound edges usually involve only superficial damage to the epidermal cells.
2. Epidermal wounds are repaired by enlargement and migration of basal cells, contact inhibition, and division of migrating and stationary basal cells.

Deep Wound Healing (p. 124)

1. During the inflammatory phase, a blood clot unites the wound edges, epithelial cells migrate across the wound, vasodilation and increased permeability of blood vessls deliver phagocytes, and fibroblasts form.
2. During the migratory phase, epithelial cells beneath the scab bridge the wound, fibroblasts begin to synthesize scar tissue, and damaged blood vessels begin to regrow.
3. During the proliferative phase, the events of the migratory phase intensify, and the open wound tissue is called granulation tissue.
4. During the maturation phase, the scab sloughs off, the epidermis is restored to normal thickness, collagenous fibers become more organized, fibroblasts begin to disappear, and blood vessels are restored to normal.

Epidermal Derivatives (p. 126)

1. Epidermal derivatives are structures developed from the embryonic epidermis.
2. Among the epidermal derivatives are hair, skin glands (sebaceous, sudoriferous, and ceruminous), and nails.

Hair (p. 126)

1. Hairs are epidermal growths that function in protection.
2. Hair consists of a shaft above the surface, a root that penetrates the dermis and subcutaneous layer, and a hair follicle.
3. Associated with hairs are sebaceous (oil) glands, arrectores pilorum muscles, and hair root plexuses.
4. New hairs develop from cell division of the matrix in the bulb; hair replacement and growth occur in a cyclic pattern. "Male-pattern" baldness is caused by androgens and heredity.

Glands (p. 129)

1. Sebaceous (oil) glands are usually connected to hair follicles; they are absent in the palms and soles. Sebaceous glands produce sebum, which moistens hairs and waterproofs the skin. Enlarged sebaceous glands may produce blackheads, pimples, and boils.
2. Sudoriferous (sweat) glands are divided into apocrine and eccrine. Apocrine sweat glands are limited in distribution to the skin of the axilla, pubis, and areolae; their ducts open into hair follicles. Eccrine sweat glands have an extensive distribution; their ducts terminate at pores at the surface of the epidermis. Sudoriferous glands produce perspiration, which carries small amounts of wastes to the surface and assists in maintaining body temperature.
3. Ceruminous glands are modified sudoriferous glands that secrete cerumen. They are found in the external auditory meatus.

Nails (p. 129)

1. Nails are hard, keratinized epidermal cells over the dorsal surfaces of the terminal portions of the fingers and toes.
2. The principal parts of a nail are the body, free edge, root, lunula, eponychium, and matrix. Cell division of the matrix cells produces new nails.

Homeostasis of Body Temperature (p. 129)

1. One of the functions of the skin is the maintenance of a normal body temperature of 37°C (98.6°F).
2. If environmental temperature is high, skin receptors sense the stimulus (heat) and generate impulses that are transmitted to the brain. The brain then causes the sweat glands to produce perspiration. As the perspiration evaporates, the skin is cooled.
3. The skin-cooling response is a negative feedback mechanism.
4. Temperature maintenance is also accomplished by adjusting blood flow to the skin, regulating metabolic rate, and regulating skeletal muscle contractions.

Aging and the Integumentary System (p. 130)

1. Most effects of aging occur when an individual reaches the late forties.
2. Among the effects of aging are wrinkling, loss of subcutaneous fat, atrophy of sebaceous glands, and decrease in the number of melanocytes.

Developmental Anatomy of the Integumentary System (p. 130)

1. The epidermis is derived from ectoderm. Hair, nails, and skin glands are epidermal derivatives.
2. The dermis is derived from wandering mesodermal cells.

Disorders: Homeostatic Imbalances (p. 131)

1. Acne is an inflammation of sebaceous (oil) glands.
2. Systemic lupus erythematosus (SLE) is an autoimmune inflammatory disease of connective tissue.
3. Psoriasis is a chronic skin disease characterized by reddish, raised plaques or papules.
4. Decubitus ulcers are caused by a chronic deficiency of blood to tissues subjected to prolonged pressure.
5. Sunburn is a skin injury resulting from prolonged exposure to the UV rays of sunlight.
6. Skin cancer can be caused by excessive exposure to sunlight. Types include basal cell carcinoma (BCC), squamous cell carcinoma (SCC), and malignant melanoma.
7. Tissue damage that destroys protein is called a burn. Depending on the depth of damage, skin burns are classified as first-degree and second-degree (partial-thickness) and third-degree (full-thickness). One method employed for determining the extent of a burn is the Lund-Browder method; another is the "rule of nines."

REVIEW QUESTIONS

1. What is the integumentary system? (p. 120)
2. List the six principal functions of the skin. (p. 120)
3. Compare the structure of epidermis and dermis. What is the subcutaneous (SC) layer? (p. 120)
4. List and describe the epidermal layers from the deepest outward. What is the importance of each layer? Describe the various cells that comprise the epidermis. (p. 121)
5. Explain the factors that produce skin color. What is an albino? (p. 123)
6. Describe how melanin is synthesized and distributed to epidermal cells. (p. 123)
7. How are epidermal ridges formed? Why are they important? (p. 123)
8. List the receptors in the epidermis, dermis, and subcutaneous (SC) layer and indicate the location and role of each. (pp. 121–123)
9. Outline the steps involved in epidermal wound healing and deep wound healing. (p. 124)
10. Describe the structure of a hair. How are hairs moistened? What produces "goose bumps" or "gooseflesh"? (p. 126)
11. Contrast the locations and functions of sebaceous (oil) glands, sudoriferous (sweat) glands, and ceruminous glands. (p. 128)
12. Distinguish between apocrine and eccrine sweat glands. (p. 128)
13. From what layer of the skin do nails form? Describe the principal parts of a nail. (p. 128)
14. Explain with a labeled diagram how the skin helps maintain normal body temperature. (p. 129)
15. Describe the effects of aging on the integumentary system. (p. 130)
16. Describe the origin of the epidermis, its derivatives, and the dermis. (p. 130)
17. Define each of the following disorders of the integumentary system: acne, systemic lupus erythematosus (SLE), psoriasis, decubitus ulcers, and sunburn. (p. 131)
18. Distinguish the three most common forms of skin cancer. (p. 132)
19. Define a burn. Classify burns according to degree. (p. 132)
20. Explain how the Lund-Browder method is used to estimate the extent of a burn. What is the "rule of nines"?
21. Define the following: lines of cleavage (p. 123), mole, liver (age) spot (123), collagen implant (p. 124), chemical exfoliation (p. 124), skin graft (SG) (p. 126), common baldness (p. 127), and blackhead (p. 128).
22. Refer to the glossary of medical terminology associated with the integumentary system. Be sure that you can define each term. (p. 134)

SELECTED READINGS

Dahl, M. V. "Acne: How It Happens and How It's Treated," *Modern Medicine,* September 1982.

Edelson, L. E., and J. M. Fink. "The Immunologic Function of Skin," *Scientific American,* June 1985.

Flanagan, B. P. "Skin Cancer: An Illustrated Guide to Early Diagnosis," *Modern Medicine,* October 1985.

Levine, N. "Sunburn: How to Stop the Pain, How to Prevent the Damage," *Modern Medicine,* July 1984.

Luterman, A., and P. W. Curreri. "Emergency Protocol for the Severely Burned Patient," *Hospital Medicine,* October 1984.

Montagna, W. "The Skin," *Scientific American,* June 1969.

Sober, A. "Malignant Melanoma: A Guide to Early, Accurate Diagnosis," *Modern Medicine,* July 1988.

UNIT 2

PRINCIPLES OF SUPPORT AND MOVEMENT

This unit considers two primary themes—support and movement. You will study the various ways in which the body is supported and the different movements it can perform. Both support and movement are made possible by the cooperative effort of bones, joints, and muscles.

Chapter 6

Skeletal Tissue

Chapter Contents at a Glance

Student Objectives

1. Discuss the components and functions of the skeletal system.
2. Describe the histological features of compact and spongy bone tissue.
3. Contrast the steps involved in intramembranous and endochondral ossification.
4. Describe the processes of bone construction and destruction involved in the homeostasis of bone remodeling.
5. Describe the conditions necessary for normal bone growth and replacement.
6. Explain the effects of exercise and aging on the skeletal system.
7. Describe the development of the skeletal system.
8. Contrast the causes and clinical symptoms associated with osteoporosis, rickets, osteomalacia, Paget's disease, and osteomyelitis.
9. Define a fracture (Fx), describe several common kinds of fractures, and describe the sequence of events involved in fracture repair.
10. Define medical terminology associated with skeletal tissue.

Without the skeletal system we would be unable to perform movements such as walking or grasping. The slightest jar to the head or chest could damage the brain or heart. It would even be impossible to chew food. The framework of bones and cartilage that protects our organs and allows us to move is called the ***skeletal*** (*skeletos* = dried up) ***system.*** The specialized branch of medicine that deals with the preservation and restoration of the skeletal system, articulations (joints), and associated structures is called ***orthopedics*** (or'-thō-PĒ-diks; *ortho* = correct or straighten; *pais* = child).

The developmental anatomy of the skeletal system is considered at the end of the chapter.

FUNCTIONS

The skeletal system performs several basic functions.

1. **Support.** The skeleton provides a framework for the body, and as such, it supports soft tissues and provides a point of attachment for many muscles.

2. **Protection.** Many internal organs are protected from injury by the skeleton. For example, the brain is protected by the cranial bones, the spinal cord by the vertebrae, the heart and lungs by the rib cage, and internal reproductive organs by the pelvic bones.

3. **Movement facilitation.** Bones serve as levers to which muscles are attached. When the muscles contract, bones acting as levers and movable joints acting as fulcrums produce movement.

4. **Mineral storage.** Bones store several minerals that can be distributed to other parts of the body upon demand. The principal stored minerals are calcium and phosphorus.

5. **Storage of blood cell–producing cells.** Red marrow in certain bones is capable of producing blood cells, a process called ***hematopoiesis*** (hēm'-a-tō-poy-Ē-sis) or ***hemopoiesis. Red marrow*** consists of blood cells in immature stages, fat cells, and macrophages. Red marrow produces red blood cells, some white blood cells, and platelets.

6. **Storage of energy.** Lipids stored in cells of yellow marrow are an important source of chemical energy.

HISTOLOGY

Structurally, the skeletal system consists of several types of connective tissue: cartilage, bone, and dense connective tissue. We described the microscopic structure of cartilage and dense connective tissue in Chapter 4. Here, our attention will be directed to a detailed discussion of the microscopic structure of bone tissue.

Like other connective tissues, ***bone,*** or ***osseous*** (OS-ē-us), ***tissue*** contains a great deal of intercellular substance surrounding widely separated cells. Four types of cells are characteristic of bone tissue: osteoprogenitor (osteogenic) cells, osteoblasts, osteocytes, and osteoclasts. ***Osteoprogenitor*** (os'-tē-ō-prō-JEN-i-tor; *osteo* = bone; *pro* = precursor; *gen* = to produce) ***cells*** are unspecialized cells derived from mesenchyme. They possess mitotic potential and have the ability to differentiate into osteoblasts (described shortly). Osteoprogenitor cells are found in the inner portion of the membrane (periosteum) around a bone, in the membrane (endosteum) that lines the medullary cavity, and in canals in bone (perforating and central) that contain blood vessels. ***Osteoblasts*** (OS-tē-ō-blasts'; *blast* = germ or bud) do not have mitotic potential and are associated with bone formation. They secrete some of the organic components and mineral salts involved in bone formation. Osteoblasts are found on the surfaces of bone. ***Osteocytes*** (OS-tē-ō-sīts'; *cyte* = cell), or mature bone cells, are the principal cells of bone tissue. Like osteoblasts, osteocytes have no mitotic potential. Osteocytes are actually osteoblasts that become isolated within the bony intercellular substance that they deposit around themselves and whose structure changes. Whereas osteoblasts initially form bone, osteocytes maintain daily cellular activities of bone tissue. ***Osteoclasts*** (OS-tē-o-clasts'; *clast* = to break) develop from circulating monocytes (one type of white blood cell). They are found around the surfaces of bone and function in bone resorption (degradation). This is important in the development, growth, maintenance, and repair of bone.

Unlike other connective tissues, the intercellular substance of bone contains abundant mineral salts, primarily calcium phosphate $[Ca_3(PO_4)_2 \cdot (OH)_2]$ and some calcium carbonate ($CaCO_3$). In addition, there are small amounts of magnesium hydroxide, fluoride, and sulfate. As these salts are deposited in the framework formed by the collagenous fibers of the intercellular substance, the tissue hardens, that is, becomes calcified. Mineral salts compose about 67 percent of the weight of bone, and collagenous fibers make up the remaining approximately 33 percent.

At one time, it was thought that calcification simply occurred when enough mineral salts were present to form crystals. Now, however, it is known that the process occurs only in the presence of collagen. Mineral salts accumulate in microscopic spaces between collagenous fibers. There the salts crystallize and become hardened. Then, after these spaces are filled, mineral salts are deposited around collagenous fibers, where the salts again crystallize and harden. The combination of crystallized salts and collagen produces the hardness characteristic of bone.

The microscopic structure of bone may be analyzed by first considering the anatomy of a long bone such as the humerus (arm bone) shown in Figure 6-1a. A typical long bone consists of the following parts:

1. **Diaphysis** (dī-AF-i-sis; *dia* = through; *physis* = growth). The shaft or long, main portion of the bone.

2. **Epiphyses** (ē-PIF-i-sēz; *epi* = above; *physis* = growth). The extremities or ends of the bone (singular is ***epiphysis***).

3. **Metaphysis** (me-TAF-i-sis). The region in a mature bone where the diaphysis joins the epiphysis. In a growing bone, it is the region including the epiphyseal plate where cartilage

FIGURE 6-1 Osseous tissue. (a) Macroscopic appearance of a long bone that has been partially sectioned. (b) Histological structure of bone.

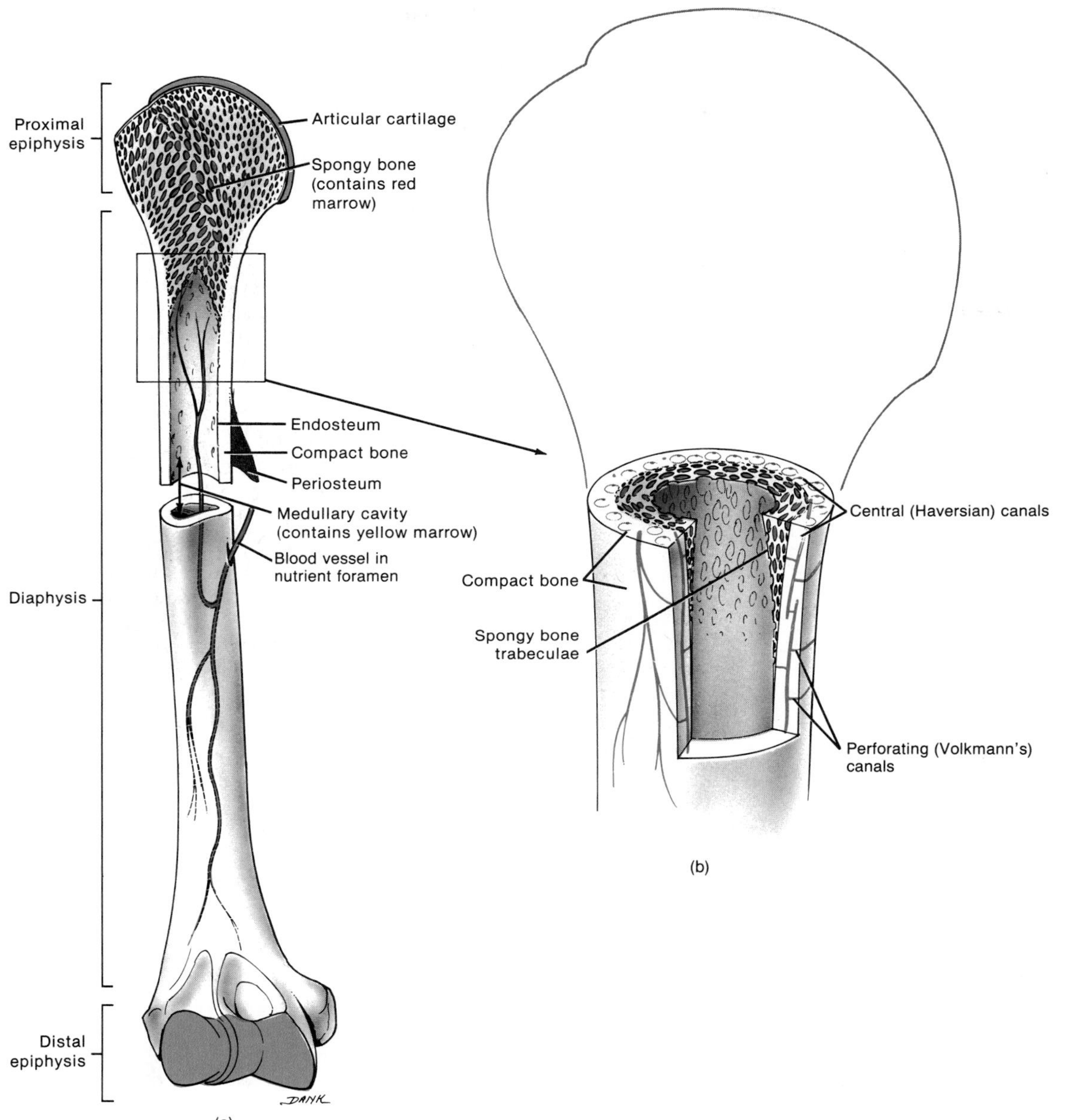

is reinforced and then replaced by bone (described later in the chapter).

4. Articular cartilage. A thin layer of hyaline cartilage covering the epiphysis where the bone forms a joint with another bone. The cartilage reduces friction and absorbs shock at freely movable joints.

5. Periosteum (per'-ē-OS-tē-um). A dense, white, fibrous covering around the surface of the bone not covered by articular cartilage. The periosteum (*peri* = around; *osteo* = bone) consists of two layers. The outer ***fibrous layer*** is composed of connective tissue containing blood vessels, lymphatic vessels, and nerves that pass into the bone. The inner ***osteogenic*** (os'-tē-ō-JEN-ik) ***layer*** contains elastic fibers, blood vessels, osteoprogenitor cells, osteoclasts, and osteoblasts. The periosteum is essential for bone growth, repair, and nutrition. It also serves as a point of attachment for ligaments and tendons.

6. Medullary (MED-yoo-lar'-ē) **or marrow cavity.** The space within the diaphysis that contains the fatty ***yellow marrow*** in adults. Yellow marrow consists primarily of fat cells and a few scattered blood cells. Thus, yellow marrow functions in fat storage.

7. **Endosteum** (end-OS-tē-um). A layer of osteoprogenitor cells and osteoblasts that lines the medullary cavity and also contains scattered osteoclasts (cells that assume a role in the removal of bone).

Bone is not completely solid. In fact, all bone has some spaces between its hard components. The spaces provide channels for blood vessels that supply bone cells with nutrients. The spaces also make bones lighter. Depending on the size and distribution of the spaces, the regions of a bone may be categorized as compact or spongy (Figure 6-2; see Figure 6-1a,b also).

Compact (dense) bone tissue contains few spaces. It is found in a layer over the spongy bone tissue. The layer of compact bone is thicker in the diaphysis than the epiphyses. Compact bone tissue provides protection and support and helps the long bones resist the stress of weight placed on them. ***Spongy (cancellous) bone tissue,*** by contrast, contains many larger spaces filled with red marrow. It makes up most of the bone tissue of short, flat, and irregularly shaped bones and most of the epiphyses of long bones. Spongy bone tissue in certain bones also provides a storage area for some red marrow.

FIGURE 6-2 **Spongy and compact bone. Photograph of a section through the femur to illustrate the positional and structural differences between spongy and compact bone. (Courtesy of Lester V. Bergman & Associates, Inc.)**

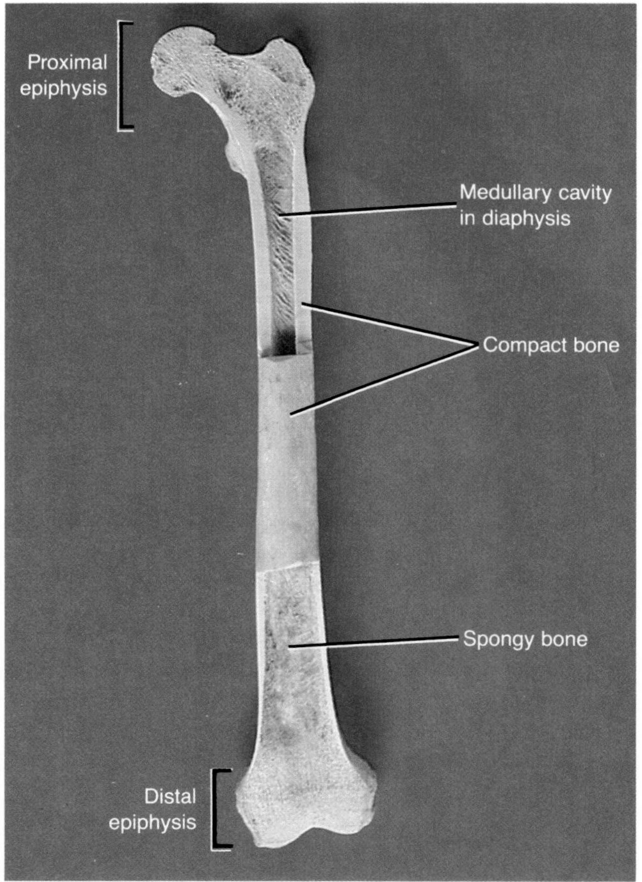

Compact Bone

We can compare the differences between spongy and compact bone tissues by looking at the highly magnified section in Figure 6-3a. (See Figure 6-2 also.) One main difference is that adult compact bone has a concentric-ring structure, whereas spongy bone does not. Blood vessels and nerves from the periosteum penetrate the compact bone through ***perforating (Volkmann's) canals.*** The blood vessels of these canals connect with blood vessels and nerves of the medullary cavity and those of the ***central (Haversian) canals.*** The central canals run longitudinally through the bone. Around the canals are ***concentric lamellae*** (la-MEL-ē)—rings of hard, calcified, intercellular substance. Between the lamellae are small spaces called ***lacunae*** (la-KOO-nē; *lacuna* = little lake) which contain osteocytes. ***Osteocytes,*** as noted earlier, are mature osteoblasts that no longer produce new bone tissue and function to support daily cellular activities of bone tissue. Radiating in all directions from the lacunae are minute canals called ***canaliculi*** (kan'-a-LIK-yoo-lī), which contain slender processes of osteocytes and extracellular fluid (Figure 6-3b). The canaliculi connect with those of other lacunae and, eventually, with the central canals. Thus an intricate network is formed throughout the bone. This branching network of canaliculi provides numerous routes so that nutrients can reach the osteocytes and wastes can be removed. Each central canal, with its surrounding lamellae, lacunae, osteocytes, and canaliculi, is called an ***osteon (Haversian system).*** Osteons are characteristic of adult compact bone. The areas between osteons contain ***interstitial lamellae.*** These also possess lacunae with osteocytes and canaliculi, but their lamellae are usually not connected to the osteons. Interstitial lamellae are fragments of older osteons that have been partially destroyed during bone replacement.

Spongy Bone

In contrast to compact bone, spongy bone does not contain true osteons (Figure 6-3a,c). It consists of an irregular latticework of thin plates of bone called ***trabeculae*** (tra-BEK-yoo-lē). See Figure 6-2 also. The spaces between the trabeculae of some bones are filled with red marrow. The cells of red marrow are responsible for producing blood cells. Within the trabeculae lie lacunae, which contain osteocytes. Blood vessels from the periosteum penetrate through to the spongy bone, and osteocytes in the trabeculae are nourished directly from the blood circulating through the marrow cavities.

FIGURE 6-3 **Histology of osseous tissue. (a) Enlarged aspect of several osteons (Haversian systems) in compact bone. A photomicrograph of compact bone is shown in Exhibit 4-2. (b) Electron micrograph of an osteocyte at a magnification of 10,000×. (Courtesy of Biophoto Associates/Photo Researchers.)**

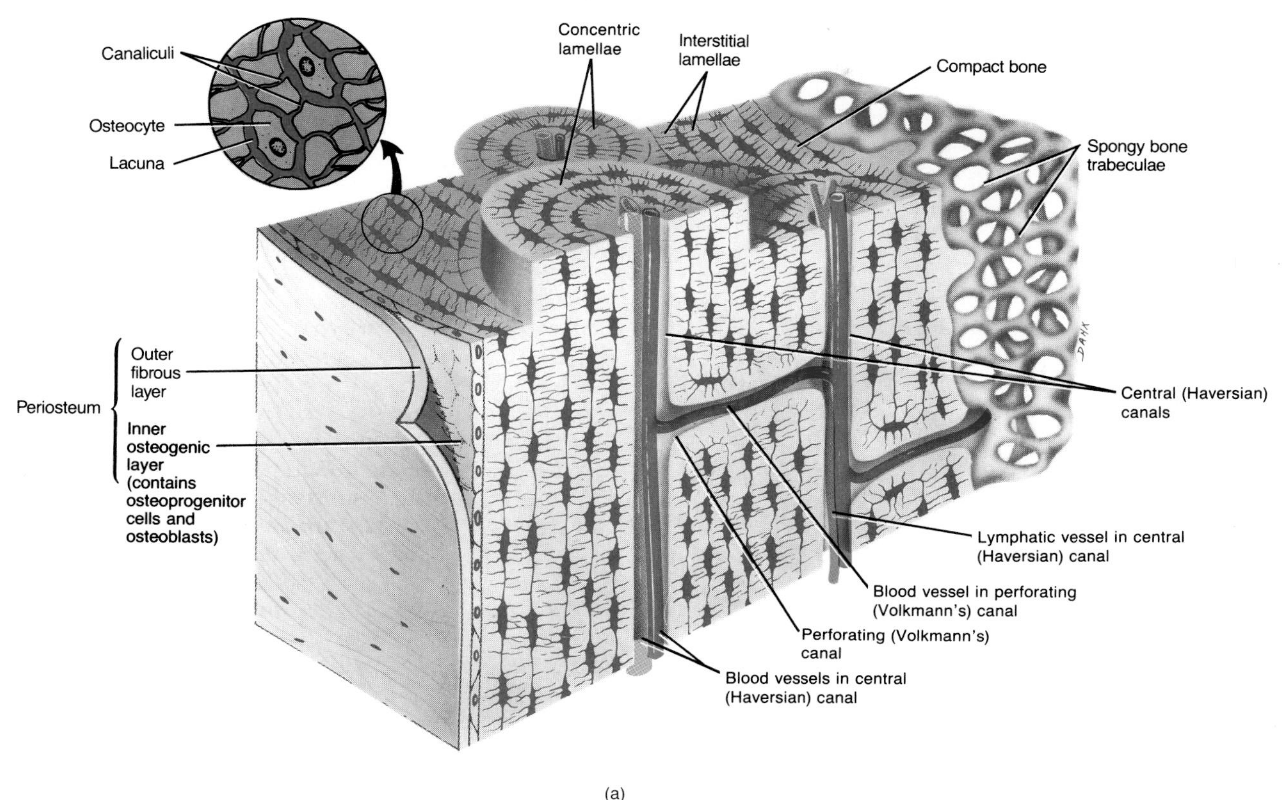

(a)

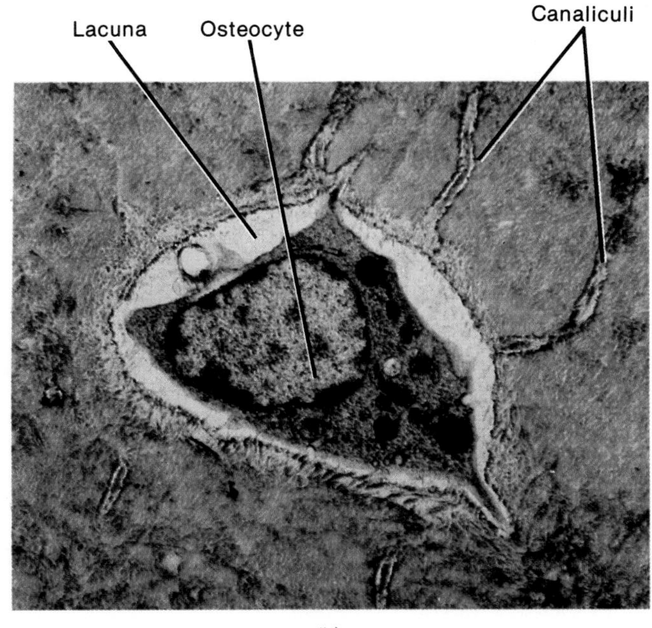

(b)

FIGURE 6-3 (*Continued*) (c) Enlarged aspect of several trabeculae of spongy bone. (d) Details of a section of a trabecula.

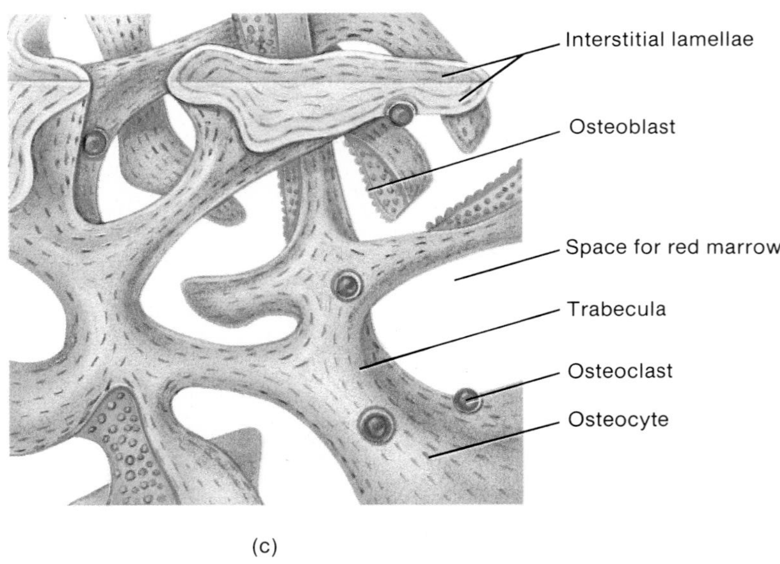

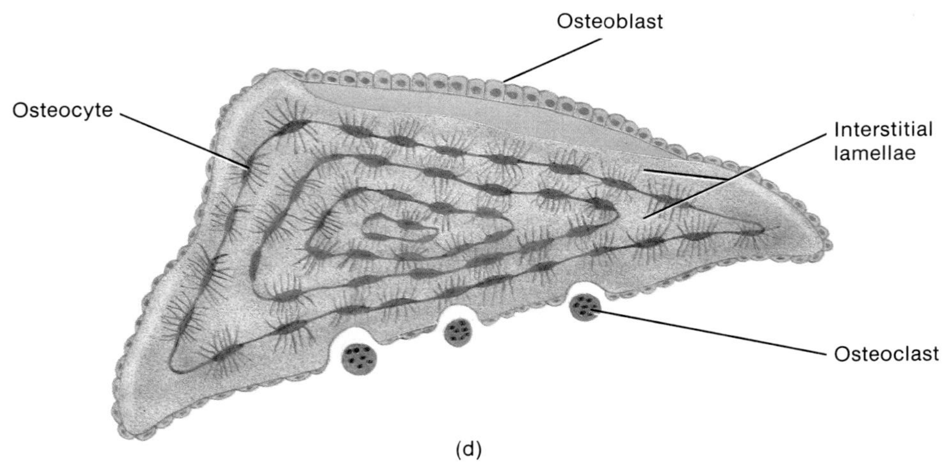

Most people think of all bone as a very hard, rigid material. Yet the bones of an infant are generally more pliable than those of an adult. The final shape and hardness of adult bones require many years to develop and depend on a complex series of chemical changes. Let us now see how bones are formed and how they grow.

PHYSIOLOGY OF OSSIFICATION: BONE FORMATION

The process by which bone forms in the body is called ***ossification*** (os'-i-fi-KĀ-shun) or ***osteogenesis.*** The "skeleton" of a human embryo is composed of fibrous membranes and hyaline cartilage. Both are shaped like bones and provide the medium for ossification. Ossification begins around the sixth or seventh week of embryonic life and continues throughout adulthood. Two kinds of bone formation occur. The first is called ***intramembranous*** (in'-tra-MEM-bra-nus; *intra* = within; *membranous* = membrane) ***ossification.*** This term refers to the formation of bone directly on or within the fibrous membranes. The second kind, ***endochondral*** (en'-dō-KON-dral; *endo* = within; *chondro* = cartilage) ***ossification,*** refers to the formation of bone in cartilage. These two kinds of ossification do *not* lead to differences in the structure of mature bones. They simply indicate different methods of bone formation. Both mechanisms involve the replacement of a preexisting connective tissue with bone.

The first stage in the development of bone is the migration of mesenchymal embryonic connective tissue cells into the area where bone formation is about to begin. These cells increase in number and size and become osteoprogenitor cells. In some skeletal structures where capillaries are lacking, they become chondroblasts; in others where capillaries are present, they become osteoblasts. The ***chondroblasts*** are responsible for cartilage formation. Osteoblasts form bone tissue by intramembranous or endochondral ossification.

Intramembranous Ossification

Of the two types of bone formation, the simpler and more direct is ***intramembranous ossification.*** Most of the surface skull bones and the clavicles (collarbones) are formed in this way. The essentials of this process are as follows.

Osteoblasts formed from osteoprogenitor cells cluster in a fibrous membrane. The site of such a cluster is called a ***center of ossification.*** The osteoblasts then secrete intercellular substances partly composed of collagenous fibers that form a framework, or matrix, in which calcium salts are quickly deposited. The deposition of calcium salts is called ***calcification.*** When a cluster of osteoblasts is completely surrounded by the calcified matrix, it is called a ***trabecula.*** As trabeculae form in nearby ossification centers, they fuse into the open latticework characteristic of spongy bone. With the formation of successive layers of bone, some osteoblasts become trapped in the lacunae. The entrapped osteoblasts lose their ability to form bone and are called osteocytes. The spaces between the trabeculae fill with red marrow. The original connective tissue that surrounds the growing mass of bone then becomes the periosteum. The ossified area has now become true spongy bone. Eventually, the surface layers of the spongy bone will be reconstructed into compact bone. Much of this newly formed bone will be destroyed and reformed so that the bone may reach its final adult size and shape.

Endochondral Ossification

The replacement of cartilage by bone is called ***endochondral (intracartilaginous) ossification.*** Most bones of the body, including bones of the base of the skull, are formed in this way. Since this type of ossification is best observed in a long bone, we will study the process in the tibia, or shinbone (Figure 6-4).

Early in embryonic life, a cartilage model or template of the future bone is laid down. This model is covered by a membrane called the ***perichondrium*** (per-i-KON-drē-um). Midway along the shaft of this model, a blood vessel penetrates the perichondrium, stimulating the osteoprogenitor cells in the internal layer of the perichondrium to enlarge and become osteoblasts. The cells begin to form a periosteal collar of compact bone around the middle of the diaphysis of the cartilage model. Once the perichondrium starts to form bone, it is called the ***periosteum.*** Simultaneously with the appearance of the periosteal collar and the penetration of blood vessels, changes occur in the cartilage in the center of the diaphysis. In this area, called the ***primary ossification center,*** cartilage cells hypertrophy (increase in size)—probably because they accumulate glycogen for energy and produce enzymes to catalyze future chemical reactions. These cells burst, resulting in a change in extracellular pH to a more alkaline pH, causing the intercellular substance to become ***calcified;*** that is, minerals are deposited within it. Once the cartilage becomes calcified, nutritive materials required by the cartilage cells can no longer diffuse through the intercellular substance, and this may cause the cartilage cells to die. Then the intercellular substance begins to degenerate, leaving large cavities in the cartilage model. Blood vessels grow along the spaces where cartilage cells were previously located and enlarge the cavities further. Gradually, these spaces in the middle of the shaft join with each other, and the marrow cavity is formed. Osteoprogenitor cells that may develop into osteoblasts are carried into the cartilage with the invading blood vessels.

As these developmental changes are occurring, the osteoblasts of the periosteum deposit successive layers of bone on the outer surface so that the periosteal collar thickens, becoming thickest in the diaphysis. The cartilage model continues to grow at its ends, steadily increasing in length. Eventually, blood vessels enter the epiphyses, and ***secondary ossification centers*** appear in the epiphyses and lay down spongy bone. In the tibia, one secondary ossification center develops in the proximal epiphysis soon after birth. The other center develops in the distal epiphysis during the child's second year.

After the two secondary ossification centers have formed, bone tissue completely replaces cartilage, except in two regions. Cartilage continues to cover the articular surfaces of the epiphyses, where it is called ***articular cartilage.*** It also remains as a region between the epiphysis and diaphysis, where it is called the ***epiphyseal plate,*** which is responsible for lengthwise growth of long bones.

MEDICAL TEST

Bone scan (bone scintigraphy) **(sin-TIG-ra-fē)**

Diagnostic Value: To determine if a cancer, such as breast or prostate cancer, has metastasized to bone; to evaluate an unexplained bone pain or possible fracture; to monitor response to therapy of a cancer that has spread to bone; to monitor the progress of bone grafts and degenerative bone disorders; and to identify bone infections as might occur following total hip replacement.

FIGURE 6-4 Endochondral ossification of the tibia. (a) Cartilage model. (b) Periosteal collar formation. (c) Development of primary ossification center. (d) Entrance of blood vessels (nutrient arteries). (e) Marrow-cavity formation. (f) Thickening and lengthening of collar. (g) Formation of secondary ossification centers. (h) Remains of cartilage as articular cartilage and epiphyseal plate. (i) Formation of epiphyseal lines.

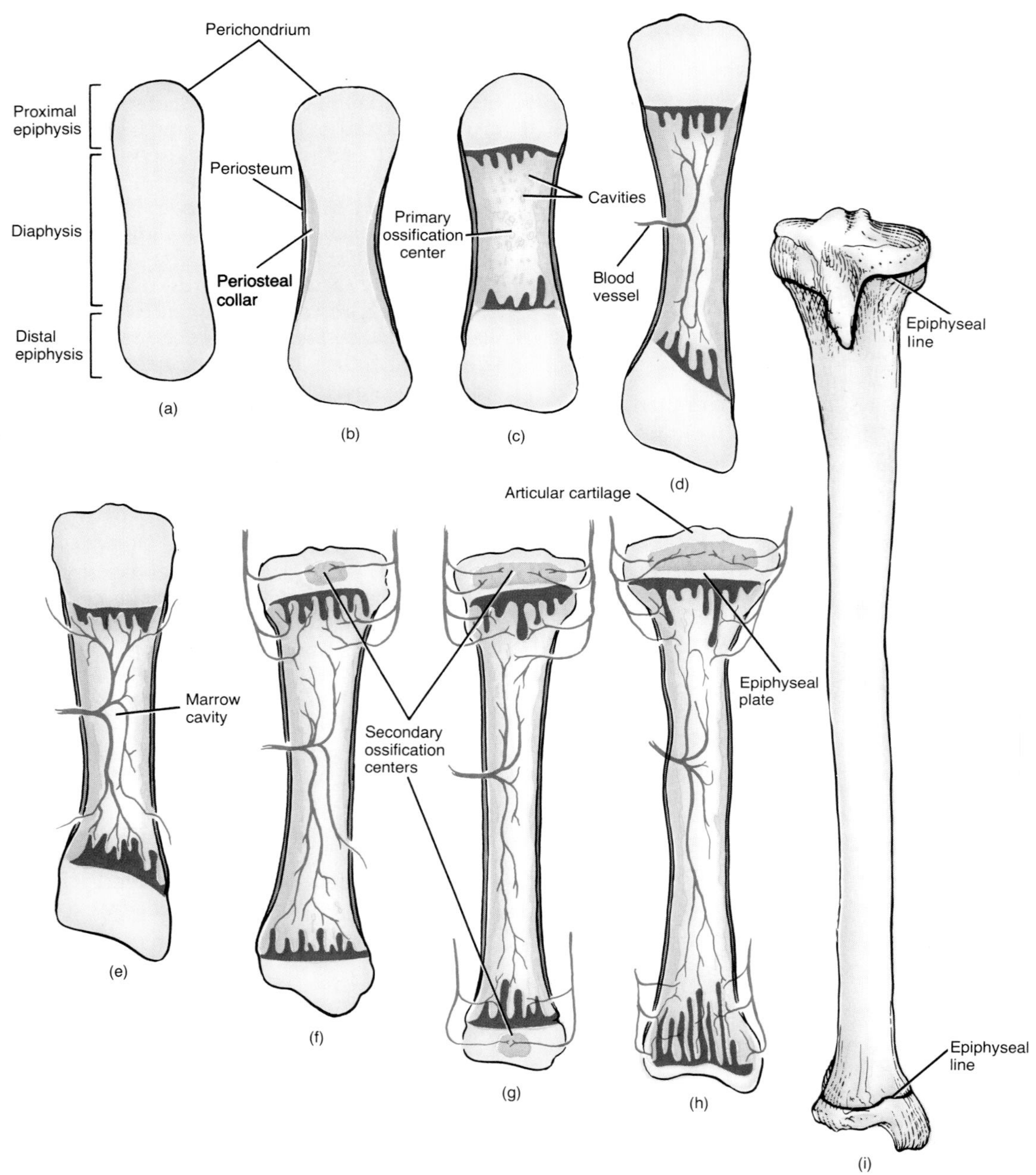

Procedure: A small amount of radioisotope, which is readily absorbed by bone, is injected intravenously. A scanning camera measures the radiation emitted subsequently from the bone. The information is translated into a photo or diagram that can be read like an x-ray. Areas of bone that absorb more of the isotope (owing to increased vascularity) are called ***hot spots***. These may indicate an abnormality such as a stress fracture, a joint infection (septic arthritis), an area of osteomyelitis (bone infection), or a primary or metastatic tumor. It is normal for the epiphyses of growing bones to show up as hot spots. A bone scan not only detects a pathology sooner than regular x-rays, but also uses less radiation.

PHYSIOLOGY OF BONE GROWTH

In order to understand how a bone grows in length, you will need to know some of the details of the structure of the epiphyseal plate.

The epiphyseal (ep'-i-FIZ-ē-al) plate consists of four zones (Figure 6-5). The ***zone of reserve cartilage*** is adjacent to the epiphysis and consists of small chondrocytes that are scattered irregularly throughout the intercellular matrix. The cells of this zone do not function in bone growth; they anchor the epiphyseal plate to the bone of the epiphysis.

The ***zone of proliferating cartilage*** consists of slightly larger chondrocytes arranged like stacks of coins. The function of this zone is to make new chondrocytes by cell division to replace those that die at the diaphyseal surface of the epiphyseal plate.

The ***zone of hypertrophic*** (hī-per-TRŌF-ik) ***cartilage*** consists of even larger chondrocytes that are also arranged in columns, with the more mature cells closer to the diaphysis. The lengthwise expansion of the epiphyseal plate is the result of cellular proliferation of the zone of proliferating cartilage and maturation of the cells in the zone of hypertrophic cartilage.

The ***zone of calcified matrix*** is only a few cells thick and consists mostly of dead cells because the intercellular matrix around them has calcified. The calcified matrix is taken up by osteoclasts, and the area is invaded by osteoblasts and capillaries from the bone in the diaphysis. These cells lay down bone on the calcified cartilage that persists. As a result, the diaphyseal border of the epiphyseal plate is firmly cemented to the bone of the diaphysis.

FIGURE 6-5 Epiphyseal plate. Photomicrograph of the epiphysis of a long bone showing the various zones of the epiphyseal plate at a magnification of 160×. (Copyright © 1983 by Michael H. Ross. Used by permission.)

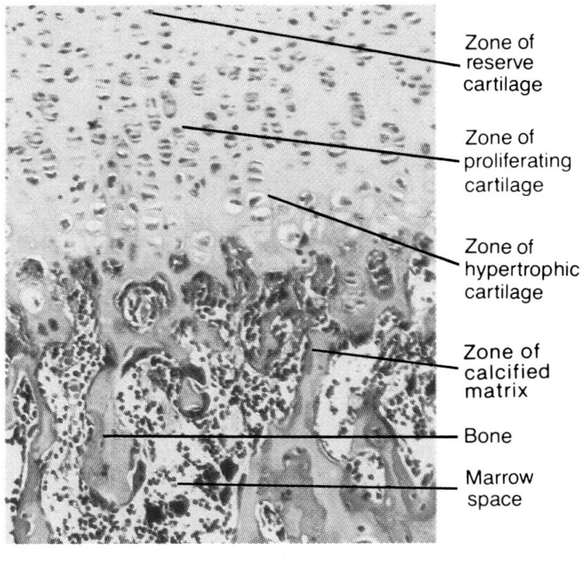

Epiphyseal side
Zone of reserve cartilage
Zone of proliferating cartilage
Zone of hypertrophic cartilage
Zone of calcified matrix
Bone
Marrow space
Diaphyseal side

The region between the diaphysis and epiphysis of a bone where the calcified matrix is replaced by bone is called the ***metaphysis*** (me-TAF-i-sis). The activity of the epiphyseal plate is the only mechanism by which the diaphysis can increase in length. Unlike cartilage, which can grow by both interstitial and appositional growth, bone can grow only by appositional growth.

The epiphyseal plate allows the diaphysis of the bone to increase in length until early adulthood. The rate of growth is controlled by hormones such as human growth hormone (hGH) produced by the pituitary gland and sex hormones produced by the ovaries and testes. As the child grows, cartilage cells are produced by mitosis on the epiphyseal side of the plate. Cartilage cells are then destroyed, and the cartilage is replaced by bone on the diaphyseal side of the plate. In this way, the thickness of the epiphyseal plate remains fairly constant, but the bone on the diaphyseal side increases in length.

Growth in diameter occurs along with growth in length. In this process, the bone lining the marrow cavity is destroyed so that the cavity increases in diameter. At the same time, osteoblasts from the periosteum add new osseous tissue around the outer surface of the bone. Initially, diaphyseal and epiphyseal ossification produce only spongy bone. Later, by reconstruction, the outer region of spongy bone is reorganized into compact bone.

Eventually, the epiphyseal cartilage cells stop dividing, and the cartilage is replaced by bone. The newly formed bony structure is called the ***epiphyseal line,*** a remnant of the once active epiphyseal plate. With the appearance of the epiphyseal line, bone growth in length stops. The clavicle is the last bone to stop growing. Ossification of most bones is usually completed by age 25. In general, lengthwise growth in bones in females is completed before that in males.

CLINICAL APPLICATION: OSTEOGENIC SARCOMA

Osteogenic sarcoma is a malignant bone tumor that primarily affects osteoblasts. It predominantly affects young people between the ages of 10 and 25 years and occurs slightly more often in males than females. Tumors often occur in the metaphyses of long bones such as the femur (thighbone), tibia (shinbone), and humerus (arm bone). Left untreated, osteogenic sarcoma metastasizes and leads to death quite rapidly. Metastases occur most frequently in the lungs. Treatment consists of multidrug chemotherapy following amputation or resection.

HOMEOSTASIS OF REMODELING

Bones undergoing ossification are continually remodeled from the time that initial calcification occurs until the final structure appears. ***Remodeling*** is the replacement of old bone tissue by new bone tissue. Compact bone is formed by the transformation of spongy bone. The diameter of a long bone is increased by the destruction of the bone closest to the marrow cavity and the construction of new bone around the outside of the diaphysis. However, even after bones have reached their adult shapes and sizes, old bone is perpetually destroyed and new bone tissue is formed in its place. Bone is never metabolically at rest; it constantly remodels and reappropriates its matrix and minerals along lines of mechanical stress.

Bone shares with skin the feature of replacing itself throughout adult life. Remodeling takes place at different rates in various body regions. The distal portion of the femur (thighbone) is replaced about every four months. By contrast, bone in certain areas of the shaft will not be completely replaced during the individual's life. Remodeling allows worn or injured bone to be removed and replaced with new tissue. It also allows bone to serve as the body's storage area for calcium. Many other tissues in the body need calcium in order to perform their functions. For example, a nerve cell needs calcium for nerve impulse conduction, muscle needs calcium to contract, and blood needs calcium to clot. The blood continually trades off calcium with the bones, removing calcium when it and other tissues are not receiving enough of this element and resupplying the bones with dietary calcium to keep them from losing too much bone mass.

The cells believed to be responsible for the resorption (loss of a substance through a physiological or pathological process) of bone tissue are osteoclasts. In the healthy adult, a delicate homeostasis is maintained between the action of the osteoclasts in removing calcium and collagen and the action of the bone-making osteoblasts in depositing calcium and collagen. Should too much new tissue be formed, the bones become abnormally thick and heavy. If too much calcium is deposited in the bone, the surplus may form thick bumps, or spurs, on the bone that interfere with movement at joints. A loss of too much tissue or calcium weakens the bones and allows them to break easily or to become very flexible. As you will see later, a greatly accelerated remodeling process results in a condition called Paget's diease.

In the process of resorption, it is believed that osteoclasts send out projections that secrete protein-digesting enzymes released from lysosomes and several acids (lactic and citric). The enzymes may function by digesting the collagen and other organic substances, whereas the acids may cause the bone salts (minerals) to dissolve. It is also presumed that the osteoclastic projections may phagocytose whole fragments of collagen and bone salts. Magnesium deficiency inhibits the activity to osteoblasts.

Normal bone growth in the young and bone replacement in the adult depend on several factors. First, sufficient quantities of calcium and phosphorus, components of the primary salt that makes bone hard, must be included in the diet. Recent studies also suggest that boron may be a factor in bone growth by inhibiting calcium loss and increasing levels of estrogen. Manganese may also be important in bone growth. It has been shown that manganese deficiency significantly inhibits laying down of new bone tissue.

CLINICAL APPLICATION: BONE GROWTH AND MANGANESE

In 1980, scientists became interested in the relationship between *bone growth* and *manganese* as it related to basketball player Bill Walton. His career was plagued by a broken ankle bone that failed to heal properly. Analysis of his blood indicated no manganese, below normal levels of copper and zinc, and high levels of calcium. Once his diet was altered by taking mineral supplements, including manganese, his condition improved dramatically.

Second, the individual must obtain sufficient amounts of vitamins, particularly vitamin D, which participates in the absorption of calcium from the gastrointestinal tract into the blood, calcium removal from bone, and kidney reabsorption of calcium that might otherwise be lost in urine. Vitamin C helps to maintain the intercellular substance of bone and other connective tissues. Vitamin C deficiency leads to decreased production of collagen and bone matrix that results in retardation of bone growth and delayed healing of fractures. Vitamin A helps to control the activity, distribution, and coordination of osteoblasts and osteoclasts during development. Its deficiency results in a decreased rate of growth in the skeleton. Vitamin B_{12} may play a role in osteoblast activity.

Third, the body must manufacture the proper amounts of the hormones responsible for bone tissue activity (Chapter 18). Human growth hormone (hGH), secreted by the pituitary gland, is responsible for the general growth of bones. Too much or too little of this hormone during childhood makes the adult abnormally tall or short. Other hormones specialize in regulating the osteoclasts. Calcitonin (CT), produced by the thyroid gland, inhibits osteoclastic activity and accelerates calcium absorption by bones. Parathormone (PTH), synthesized by the parathyroid glands, increases the number and activity of osteoclasts which release calcium and phosphate from bones into blood. PTH also causes transport of calcium from fluid that will become urine into blood, and transport of phosphate from blood into urine. And still others, especially the sex hormones (estrogen and testosterone), aid osteoblastic activity and thus promote the growth of new bone. (It has recently been shown that the nuclei of osteoblasts contain estrogen receptors.) The sex hormones act as a double-edged sword. They aid in the growth of new bone, but they also bring about the degeneration of all the cartilage cells in epiphyseal plates. Because of the sex hormones, the typical adolescent experiences a spurt of growth during puberty, when sex hormone levels start to increase. The individual then quickly completes the growth process as the epiphyseal cartilage disappears. Premature puberty can actually prevent one from reaching an average adult height because of the simultaneous premature degeneration of the plates. Insulin and thyroid hormones are also important for normal bone growth and maturity.

EXERCISE AND THE SKELETAL SYSTEM

Within limits, bone has the ability to alter its strength in response to mechanical stress. When placed under mechanical stress, bone tissue increases its deposition of mineral salts and production of collagen fibers. Removal of mechanical stress induces removal of mineral salts and collagen fibers. Among the mechanical stresses to which bone is subjected are those that result from the pull of skeletal muscles on bones and supporting the weight of the body against the pull of gravity. Bones of athletes, which are stressed to a high degree, become considerably thicker than those of nonathletes. In the absence of such stresses, bone does not develop normally. If a person has a fractured leg bone in a cast but continues to walk on the opposite leg, it can be noted that the fractured bone becomes decalcified within a few weeks, from lack of mechanical stress, whereas the opposite bone remains normally calcified.

In response to mechanical stress, bone (mostly crystals of calcium phosphate) produces very minute currents of electricity. This is called the ***piezoelectric*** (pē-e-zō-e-LEK-trik) ***effect,*** which is believed to stimulate formation of more osteoblasts which make additional bone matrix. One effect of regular exercise in which bones are stressed, such as walking or running, is to stimulate bone growth. Another effect is to increase the production of calcitonin (CT) by the thyroid gland, a hormone that inhibits the activity of bone-destroying osteoclasts. This inhibits bone resorption.

AGING AND THE SKELETAL SYSTEM

There are two principal effects of aging on the skeletal system. The first effect is the loss of calcium from bones. This loss usually begins after age 30 in females, accelerates greatly around age 40–45 as estrogen levels decrease, and continues until as much as 30 percent of the calcium in bones is lost by age 70. In males, calcium loss typically does not begin until after age 60. The loss of calcium from bones is one of the factors related to a condition called osteoporosis, which will be described shortly.

The second principal effect of aging on the skeletal system is a decrease in the rate of protein formation, which results in a decreased ability to produce the organic portion of bone matrix. As a consequence, bone matrix accumulates a lesser proportion of organic matrix and a greater proportion of inorganic matrix. In some elderly individuals, this process can cause their bones to become quite brittle and more susceptible to fracture.

DEVELOPMENTAL ANATOMY OF THE SKELETAL SYSTEM

Bone forms about the sixth or seventh week of embryonic development by either of two processes, ***intramembranous ossification*** or ***endochondral ossification.*** Both processes begin when ***mesenchymal (mesodermal) cells*** migrate into the area where bone formation will occur. In some skeletal structures, the mesenchymal cells develop into ***chondroblasts*** that form *cartilage.* In other skeletal structures, the mesenchymal cells develop into ***osteoblasts*** that form *bone tissue* by intramembranous or endochondral ossification. (The details of ossification have already been discussed earlier in the chapter.)

Discussion of the development of the skeletal system provides us with an excellent opportunity to trace the development of the extremities. The *extremities* make their appearance about the fifth week as small elevations at the sides of the trunk called ***limb buds*** (Figure 6-6). They consist of masses of general ***mesoderm*** surrounded by ***ectoderm.*** At this point, a mesenchymal skeleton exists in the limbs; some of the masses of mesoderm surrounding the developing bones will develop into the skeletal muscles of the extremities. By the sixth week, the limb buds develop a constriction around the middle portion. The constriction produces distal segments of the upper buds called ***hand plates*** and distal segments of the lower buds called ***foot plates.*** These plates represent the beginnings of the *hands* and *feet,* respectively. At this stage of limb development, a cartilaginous skeleton is present. By the seventh week, the *arm, forearm,* and *hand* are evident in the upper limb bud, and the *thigh, leg,* and *foot* appear in the lower limb bud. Endochondral ossification has begun. By the eighth week, the upper limb bud is appropriately called the *upper extremity* as the *shoulder, elbow,* and *wrist* areas become apparent, and the lower limb bud is referred to as the *lower extremity* with the appearance of the *knee* and *ankle* areas.

The ***notochord*** is a flexible rod of tissue that lies in a position where the future vertebral column will develop (see Figure 10-17b). As the vertebrae develop, the notochord becomes surrounded by the developing vertebral bodies, and the notochord eventually disappears except for remnants that persist as the *nucleus pulposus* of the intervertebral discs (see Figure 7-19).

FIGURE 6-6 External features of a developing human at various stages of development. Many of the labeled structures are discussed in later chapters. (a) Fifth week. (b) Sixth week. (c) Seventh week. (d) Eighth week.

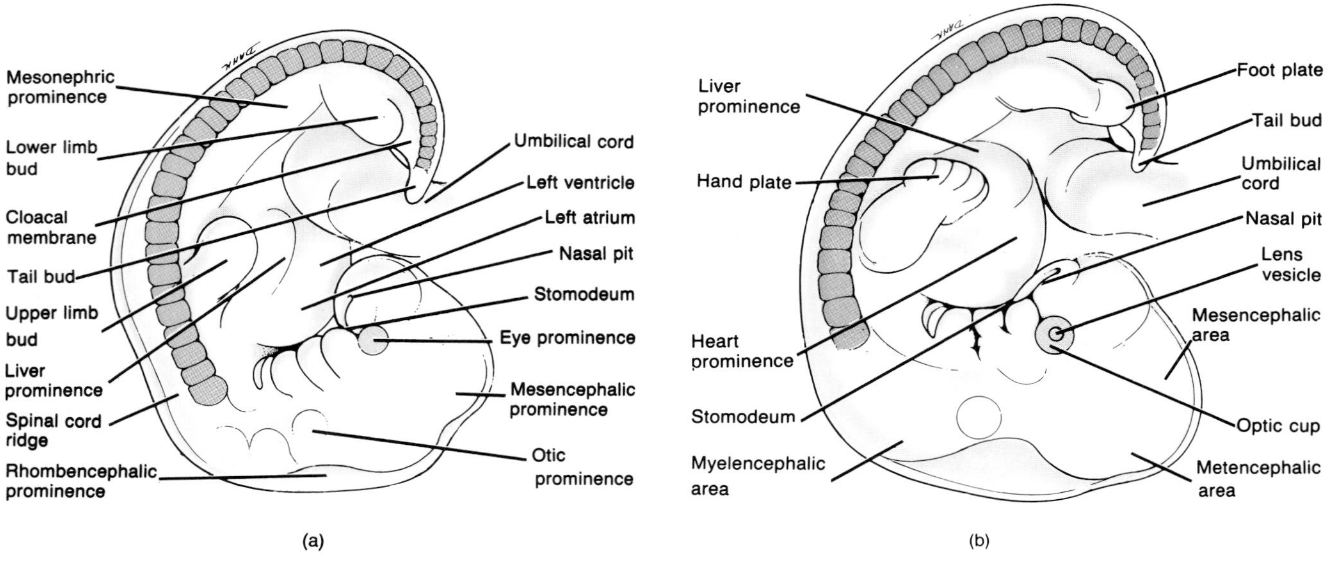

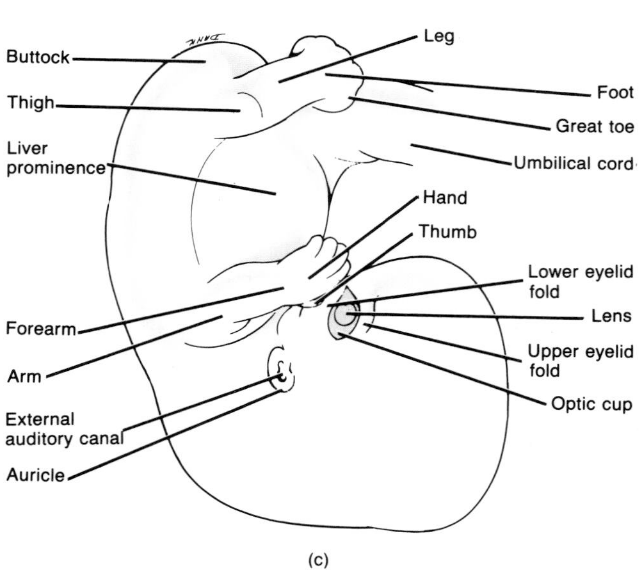

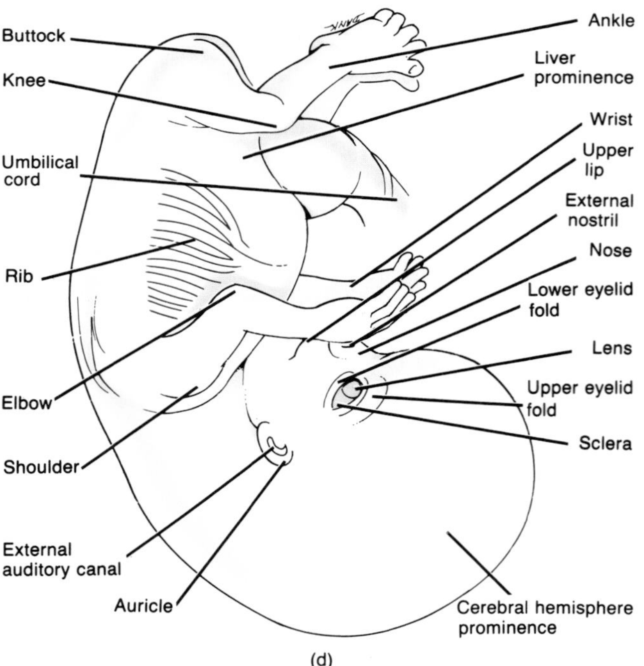

DISORDERS: HOMEOSTATIC IMBALANCES

Some bone disorders result from deficiencies in vitamins or minerals or from too much or too little of the hormones that regulate bone homeostasis. Infections and tumors are also responsible for certain bone disorders.

Osteoporosis

Osteoporosis (os'-tē-ō-pō-RŌ-sis) is an age-related disorder characterized by decreased bone mass and increased susceptibility to fractures as a result of decreased levels of estrogens. The disorder primarily affects middle-aged and elderly people—white women more than men and whites more than blacks. Between puberty and the middle years, sex hormones, especially estrogen, maintain osseous tissue by stimulating the osteoblasts to form new bone. (Recall that osteoblasts contain estrogen receptors in their nuclei.) Women produce smaller amounts of sex hormones after menopause, and both men and women produce smaller amounts during old age. As a result, the osteoblasts become less active, and there is a decrease in bone mass. Osteoporosis can also occur during nursing and pregnancy and in individuals exposed to prolonged treatment with cortisone or high levels of thyroid hormones. The first symptom of osteoporosis occurs when bone mass is so depleted that the skeleton can no longer withstand the mechanical stresses of everyday living, and fractures result. Osteoporosis is responsible for shrinkage of the backbone and height loss, hunched backs, hip fractures, and considerable pain. Osteoporosis affects the entire skeletal system, especially the vertebral bodies, ribs, proximal femur (hip), humerus, and distal radius.

Among the factors implicated in osteoporosis, besides race and sex, are: (1) body build (short females are at greater risk since they have less total bone mass, (2) weight (thin females are at greater risk since adipose tissue is a great source of estrone, an estrogen that retards bone loss, (3) smoking (smoking decreases blood estrogen levels), (4) calcium deficiency and malabsorption, (5) vitamin D deficiency, (6) exercise (sedentary people are more likely to develop bone loss), (7) certain drugs (alcohol, some diuretics, cortisone, and tetracycline promote bone loss), and (8) premature menopause. Estrogen replacement therapy (ERT), calcium supplements, and weight-bearing exercise are prescribed to prevent or retard the development of osteoporosis in postmenopausal women. Adequate diet and exercise are the mainstays of preventing osteoporosis in middle-aged and older men. Among treatments for established osteoporosis are the preventive measures given above as well as anabolic steroids, calcitonin (CT), and sodium fluoride. (A new treatment that halts the progression of spinal osteoporosis and reverses bone loss involves taking sodium fluoride along with a calcium supplement—fluoride stimulates osteoblasts). The most important aspect of treatment is prevention, and many physicians now urge women in their twenties and thirties to pay attention to exercise and adequate calcium intake in advance of the perimenopausal years when, without ERT and calcium supplements, a negative calcium balance favors the development of osteoporosis.

Vitamin Deficiencies

Rickets

A deficiency of vitamin D in children results in a condition called ***rickets.*** It is characterized by an inability of the body to transport calcium and phosphorus from the gastrointestinal tract into the blood for utilization by bones. As a result, epiphyseal cartilage cells cease to degenerate, and new cartilage continues to be produced. Epiphyseal cartilage thus becomes wider than normal. At the same time, the soft matrix laid down by the osteoblasts in the diaphysis fails to calcify. This causes bones to stay soft.

When the child walks, the weight of the body causes the bones in the legs to bow. Malformations of the head, chest, and pelvis also occur.

The cure and prevention of rickets consist of adding generous amounts of calcium, phosphorus, and vitamin D to the diet. Exposing the skin to the ultraviolet rays of sunlight also aids the body in manufacturing additional vitamin D.

Osteomalacia

Deficiency of vitamin D in adults causes the bones to give up excessive amounts of calcium and phosphorus. This loss, called ***demineralization,*** is especially heavy in the bones of the pelvis, legs, and spine. Demineralization caused by vitamin D deficiency is called ***osteomalacia*** (os'-tē-ō-ma-LĀ-shē-a; *malacia* = softness). After the bones demineralize, the weight of the body produces bowing of the leg bones, shortening of the backbone, and flattening of the pelvic bones. Osteomalacia mainly affects women who live on nutritionally poor cereal diets devoid of milk, are seldom exposed to the sun, and have repeated pregnancies that deplete the body of calcium. The condition responds to the same treatment as rickets; if the disease is severe enough and threatens life, large doses of vitamin D are given.

Osteomalacia may also result from failure to absorb fats (steatorrhea) because vitamin D is soluble in fats and calcium combines with fats. As a result, vitamin D and calcium remain with the unabsorbed fat and are lost in the feces.

Paget's Disease

Paget's disease is characterized by a greatly accelerated remodeling process in which osteoclastic resorption is massive and new bone formation by osteoblasts is extensive. As a result, there is an irregular thickening and softening of the bones and greatly increased vascularity, especially in bones of the skull, pelvis, and extremities. It rarely occurs in individuals under 50. The cause of the disease is unknown.

DISORDERS: HOMEOSTATIC IMBALANCES (*continued*)

Osteomyelitis

The term ***osteomyelitis*** (os'-tē-ō-mī-i-LĪ-tis) includes all infectious diseases of bone. These diseases may be localized or widespread and may also involve the periosteum, marrow, and cartilage. Various microorganisms may give rise to bone infection, but the most frequent are bacteria known as *Staphylococcus aureus,* commonly called "staph." These bacteria may reach the bone by various means: the bloodstream, an injury such as a fracture, or an infection such as a sinus infection or a tooth abscess. Antibiotics may be effective in treating the disease and in preventing it from spreading through extensive areas of bone, but complete recovery is usually a prolonged and painful process.

Fractures (Fxs)

A ***fracture (Fx)*** is any break in a bone. Usually, the fracture is restored to normal position by manipulation without surgery. This procedure of setting a fracture is called ***closed reduction.*** In other cases, the fracture must be exposed by surgery before the break is rejoined. This procedure is known as ***open reduction.***

Types

Although fractures may be classified in several different ways, the following scheme is useful (Figure 6-7):

1. Partial (incomplete). A fracture in which the break across the bone is incomplete.
2. Complete. A fracture in which the break across the bone is complete, so that the bone is broken into two or more pieces.
3. Closed (simple). A fracture in which the bone does not break through the skin.
4. Open (compound). A fracture in which the broken ends of the bone protrude through the skin.
5. Comminuted (KOM-i-nyoo'-ted). A fracture in which the bone is splintered at the site of impact, and smaller fragments of bone are found between the two main fragments.
6. Greenstick. A partial fracture in which one side of the bone is broken and the other side bends; occurs only in children.
7. Spiral. A fracture in which the bone is usually twisted apart.
8. Transverse. A fracture at right angles to the long axis of the bone.
9. Impacted. A fracture in which one fragment is firmly driven into the other.
10. Pott's. A fracture of the distal end of the fibula, with serious injury of the distal tibial articulation.
11. Colles' (KOL-ēz). A fracture of the distal end of the radius in which the distal fragment is displaced posteriorly.
12. Displaced. A fracture in which the anatomical alignment of the bone fragments is not preserved.
13. Nondisplaced. A fracture in which the anatomical alignment of the bone fragments is preserved.
14. Stress. A partial fracture resulting from inability to withstand repeated stress due to a change in training, harder surfaces, longer distances, and greater speed. About 25 percent of all stress fractures involve the fibula, typically the distal third.
15. Pathologic. A fracture due to weakening of a bone caused by disease processes such as neoplasia, osteomyelitis, osteoporosis, or osteomalacia.

Fracture Repair

Bone sometimes requires months to heal. A fractured femur, for example, may take six months to heal. Sufficient calcium to strengthen and harden new bone is deposited only gradually. Bone cells also grow and reproduce slowly. Moreover, in a fractured bone the blood supply is decreased, which helps to explain the difficulty in the healing of an infected bone.

The following steps occur in the repair of a fracture (Figure 6-8):

1. As a result of the fracture, blood vessels crossing the fracture line are broken. These vessels are found in the periosteum, osteons (Haversian systems), and marrow cavity. As blood pours from the torn ends of the vessels, it forms a clot in and about the site of the fracture. This clot, called a ***fracture hematoma*** (hē'-ma-TŌ-ma), usually occurs six to eight hours after the injury. Since the circulation of blood ceases when the fracture hematoma forms, bone cells and periosteal cells at the fracture site die. The hematoma serves as a focus for subsequent cellular invasion.
2. A growth of new bone tissue—a ***callus***—develops in and around the fractured area. It forms a bridge between separated areas of bone. The part of the callus that forms from the osteoblasts of the torn periosteum and develops around the outside of the fracture is called an ***external callus.*** The part of the callus that forms from the osteoblast cells of the endosteum and develops between the two ends of bone fragments and between the two marrow cavities is called the ***internal callus.***

   Approximately 48 hours after a fracture occurs, osteoblasts and osteoclasts that ultimately repair the fracture become actively mitotic. These cells come from the osteogenic layer of the periosteum, the endosteum of the marrow cavity, and bone marrow. As a result of their ac-

DISORDERS: HOMEOSTATIC IMBALANCES (*continued*)

FIGURE 6-7 **Types of fractures. (a) Comminuted. (b) Colles'. (c) Impacted. (d) Pott's. (e) Greenstick. (f) Compound.**

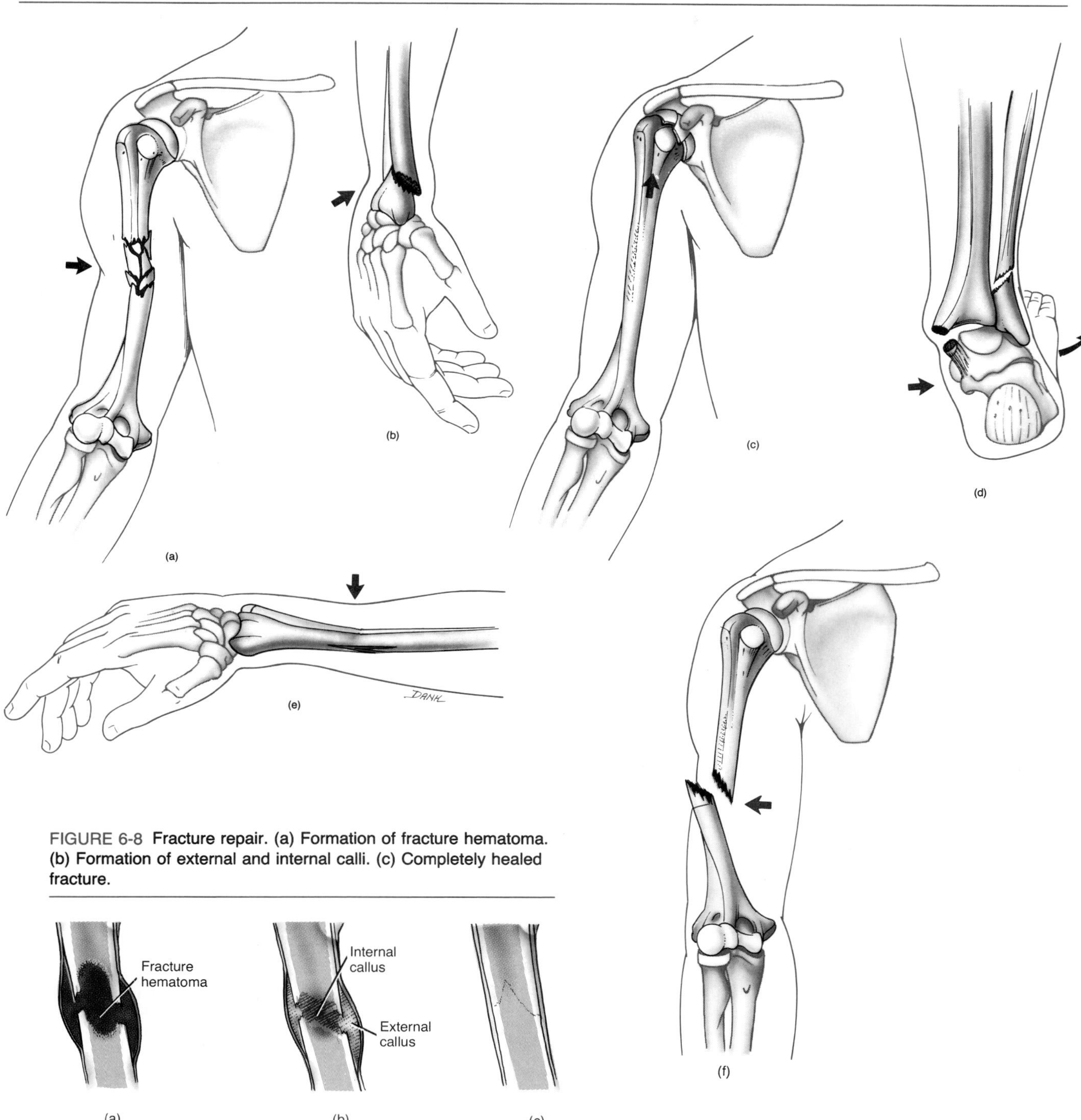

FIGURE 6-8 **Fracture repair. (a) Formation of fracture hematoma. (b) Formation of external and internal calli. (c) Completely healed fracture.**

DISORDERS: HOMEOSTATIC IMBALANCES (*continued*)

celerated mitotic activity, the cells of the three regions grow toward the fracture. During the first week following the fracture, the osteoblasts of the endosteum and bone marrow form new trabeculae in the marrow cavity near the line of fracture. This is the internal callus. During the next few days, osteoblasts of the periosteum form a collar around each bone fragment. The collar, or external callus, is replaced by trabeculae. The trabeculae of the calli are joined to living and dead portions of the original bone fragments.

3. The final phase of fracture repair is the ***remodeling*** of the calli. Dead portions of the original fragments are gradually resorbed by osteoclasts. Compact bone replaces spongy bone around the periphery of the fracture. In some cases, the healing is so complete that the fracture line is undetectable, even by x-ray. However, a thickened area on the surface of the bone usually remains as evidence of the fracture site.

CLINICAL APPLICATION: PULSATING ELECTROMAGNETIC FIELDS

In the past when a fracture failed to unite, the patient could either wait, hoping that nature and time would solve the problem, or choose surgery. Now there is another alternative, called ***pulsating electromagnetic fields (PEMFs)***, that involves electrotherapy to stimulate bone repair.

Essentially, the fracture is exposed to a weak electrical current generated from coils that are fastened around the cast. Osteoblasts adjacent to the fracture site become more active metabolically in response to the stimulation. The increased activity apparently causes an acceleration of calcification, vascularization, and endochondral ossification, resulting in acceleration of fracture repair. It was noted earlier that parathyroid hormone (PTH) increases osteoclastic activity and thus stimulates bone destruction. One hypothesis suggests that electricity also heals fractures by keeping PTH from acting on osteoclasts, thus increasing bone formation and repair.

MEDICAL TERMINOLOGY ASSOCIATED WITH SKELETAL TISSUE

Achondroplasia (a-kon'-drō-PLĀ-zē-a; *a* = without; *chondro* = cartilage; *plasia* = growth) Imperfect ossification within cartilage of long bones during fetal life; also called **fetal rickets.** It causes a form of dwarfism.

Craniotomy (krā'-nē-OT-ō-mē; *cranium* = skull; *tome* = a cutting) Any surgery that requires cutting through the bones surrounding the brain.

Necrosis (ne-KRŌ-sis; *necros* = death; *osis* = condition) Death of tissues or organs; in the case of bone, results from deprivation of blood supply resulting from fracture, extensive removal of periosteum in surgery, exposure to radioactive substances, or other causes.

Osteitis (os'-tē-Ī-tis; *osteo* = bone) Inflammation or infection of bone.

Osteoarthritis (os'-tē-ō-ar-THRĪ-tis; *arthro* = joint) The degeneration of cartilage, allowing the bony ends to touch and, from the friction of bone against bone, creating a bony reaction that worsens the friction and worsens the condition; usually associated with the elderly.

Osteoblastoma (os'-tē-ō-blas-TŌ-ma; *oma* = tumor) A benign tumor of osteoblasts.

Osteochondroma (os'-tē-ō-kon-DRŌ-ma; *chondro* = cartilage) A benign tumor of bone and cartilage.

Osteoma (os'-tē-Ō-ma) A benign bone tumor.

Osteosarcoma (os'-tē-ō-sar-KŌ-ma; *sarcoma* = connective tissue tumor) A malignant tumor composed of osseous tissue.

Pott's (POTS) **disease** Inflammation of the backbone, caused by the microorganism that produces tuberculosis.

STUDY OUTLINE

Functions (p. 142)

1. The skeletal system consists of all bones attached at joints and cartilage between joints.
2. The functions of the skeletal system include support, protection, leverage, mineral storage, and housing blood-forming tissue.

Histology (p. 142)

1. Osseous tissue consists of widely separated cells surrounded by large amounts of intercellular substance (bone matrix). The four principal types of cells are osteoprogenitor cells, osteoblasts, osteocytes, and osteoclasts. The intercellular substance contains collagenous fibers and abundant hydroxyapatites (mineral salts), consisting mainly of calcium phosphate salts.
2. Parts of a typical long bone are the diaphysis (shaft), epiphyses (ends), metaphysis, articular cartilage, periosteum, medullary (marrow) cavity, and endosteum.
3. Compact (dense) bone consists of osteons (Haversian systems) with little space between them. Compact bone lies over spongy bone and composes most of the bone tissue of the diaphysis. Functionally, compact bone protects, supports, and resists stress.
4. Spongy (cancellous) bone consists of trabeculae surrounding many red and yellow marrow–filled spaces. It forms most of the structure of short, flat, and irregular bones, and the epiphyses of long bones. Functionally, spongy bone stores some red and yellow marrow and provides some support.

Physiology of Ossification: Bone Formation (p. 146)

1. Bone forms by a process called ossification (osteogenesis), which begins when mesenchymal cells become transformed into osteoprogenitor cells, which undergo cell division giving rise to cells that differentiate into osteoblasts and osteoclasts.
2. The process begins during the sixth or seventh week of embryonic life and continues throughout adulthood. The two types of ossification, intramembranous and endochondral, involve the replacement of a preexisting connective tissue with bone.
3. Intramembranous ossification occurs within fibrous membranes of the embryo and the adult.
4. Endochondral ossification occurs within a cartilage model. The primary ossification center of a long bone is in the diaphysis. Cartilage degenerates, leaving cavities that merge to form the marrow cavity. Osteoblasts lay down bone. Next, ossification occurs in the epiphyses, where bone replaces cartilage, except for the epiphyseal plate.

Physiology of Bone Growth (p. 149)

1. The anatomical zones of the epiphyseal plate are the zones of reserve cartilage, proliferating cartilage, hypertrophic cartilage, and calcified matrix.
2. Because of the activity of the epiphyseal plate, the diaphysis of a bone increases in length by appositional growth.
3. Bone grows in diameter as a result of the addition of new bone tissue by periosteal osteoblasts around the outer surface of the bone.

Homeostasis of Remodeling (p. 150)

1. The homeostasis of bone growth and development depends on a balance between bone formation and resorption.
2. Old bone is constantly destroyed by osteoclasts, whereas new bone is constructed by osteoblasts. This process is called remodeling.
3. Normal growth depends on calcium, phosphorus, and vitamins, especially vitamin D, and is controlled by hormones that are responsible for bone mineralization and resorption.

Exercise and the Skeletal System (p. 151)

1. Bone can alter its strength in response to mechanical stress.
2. Bone that is stressed produces a minute electric current, by way of its mineral salt crystals (piezoelectric effect), that stimulates osteoblastic activity.
3. Regular exercise can stimulate osteoblasts and inhibit osteoclasts.

Aging and the Skeletal System (p. 151)

1. The principal effect of aging is a loss of calcium from bones, which may result in osteoporosis.
2. Another effect is a decreased production of organic matrix, which makes bones more susceptible to fracture.

Developmental Anatomy of the Skeletal System (p. 151)

1. Bone forms from mesoderm by intramembranous or endochondral ossification.
2. Extremities develop from limb buds, which consist of mesoderm and ectoderm.

Disorders: Homeostatic Imbalances (p. 153)

1. Osteoporosis is a decrease in the amount and strength of bone tissue owing to decreases in hormone output.
2. Rickets is a vitamin D deficiency in children in which the body does not absorb calcium and phosphorus. The bones soften and bend under the body's weight.
3. Demineralization caused by vitamin D deficiency in adults results in osteomalacia.
4. Paget's disease is the irregular thickening and softening of bones, related to a greatly accelerated remodeling process.
5. Osteomyelitis is a term for the infectious diseases of bones, marrow, and periosteum. It is frequently caused by "staph" bacteria.
6. A fracture (Fx) is any break in a bone.
7. The types of fractures include partial, complete, simple, compound, comminuted, greenstick, spiral, transverse, impacted, Pott's, Colles', displaced, nondisplaced, and stress.

8. Fracture repair consists of forming a fracture hematoma, forming a callus, and remodeling.
9. Treatment by pulsating electromagnetic fields (PEMFs) has provided dramatic results in healing fractures that would otherwise not have mended properly. Its application for limb regeneration and stopping the growth of tumor cells is being investigated.

REVIEW QUESTIONS

1. Define the skeletal system. What are its six principal functions? (p. 142)
2. Why is osseous tissue considered a connective tissue? Describe the cells present and the composition of the intercellular substance. (p. 142)
3. Diagram the parts of a long bone and list the functions of each part. (p. 142)
4. Distinguish between spongy and compact bone in terms of microscopic appearance, location, and function. (p. 144)
5. Diagram the microscopic appearance of compact bone and indicate the functions of the various components. (p. 145)
6. What is meant by ossification? Describe the initial events of ossification. (p. 146)
7. Outline the major events involved in intramembranous and endochondral ossification and explain the main differences. (p. 147)
8. Describe the histology of the various zones of the epiphyseal plate. How does the plate grow? What is the significance of the epiphyseal line? (p. 149)
9. Define remodeling. How does balance between osteoblast activity and osteoclast activity demonstrate the homeostasis of bone? (p. 150)
10. List the primary factors involved in bone growth and replacement. (p. 150)
11. Explain the effects of exercise and aging on the skeletal system. (p. 151)
12. Describe the development of the skeletal system. (p. 151)
13. What are the principal symptoms of osteoporosis, Paget's disease, and osteomyelitis? What is the etiology of each? (p. 153)
14. Distinguish between rickets and osteomalacia. What do the two diseases have in common? (p. 153)
15. What is a fracture (Fx)? Distinguish several principal kinds. Outline the three basic steps involved in fracture repair. (p. 154)
16. Explain the principle of pulsating electromagnetic fields (PEMFs) in the repair of fractures. (p. 156)
17. Define osteogenic carcinoma. How is manganese related to bone growth? (pp. 149, 150)
18. How is a bone scan performed? What is its diagnostic value? (p. 147)
19. Refer to the glossary of medical terminology associated with the skeletal system. Be sure that you can define each term. (p. 156)

SELECTED READINGS

Arehart-Treichel, J. "Boning Up on Osteoporosis," *Science News*, 27 August 1983.

Avioli, L. V. "Osteoporosis: A Guide to Detection," *Modern Medicine*, February 1986.

Bassett, C. A., S. N. Mitchell, and J. Gaston. "Pulsing Electromagnetic Field Treatment in Ununited Fractures and Failed Arthrodeses," *Journal of the American Medical Association*, 5 February 1982.

Cameron, N. U. "Electromagnetism—Quackery or Useful Therapy?" *Modern Medicine*, March 1984.

Gamble, J. G. *The Musculoskeletal System: Physiological Basis*. New York: Raven Press, 1988.

Raisz, L. G. "Local and Systemic Factors in the Pathogenesis of Osteoporosis," *New England Journal of Medicine*, 31 March 1988.

Ravnikav, V. "Clinical Considerations in the Diagnosis of Osteoporosis," *Modern Medicine*, May 1988.

Vaughan, J. M. *The Physiology of Bone*, 3rd ed. New York: Oxford University Press, 1981.

Wilson, F. C. *The Musculoskeletal System: Basic Processes and Disorders*, 2nd ed. Philadelphia: Lippincott, 1983.

Chapter 7

The Skeletal System: The Axial Skeleton

Chapter Contents at a Glance

Student Objectives

1. Classify the principal types of bones, on the basis of shape and location.
2. Describe the various markings on the surfaces of bones.
3. Identify the bones of the skull and the major markings associated with each.
4. Identify the principal sutures, fontanels, paranasal sinuses, and foramina of the skull.
5. Identify the bones of the vertebral column and their principal markings.
6. Identify the bones of the thorax and their principal markings.
7. Contrast herniated (slipped) disc, abnormal curves, spina bifida, and fractures of the vertebral column as disorders associated with the skeletal system.

The skeletal system forms the framework of the body. For this reason, a familiarity with the names, shapes, and positions of individual bones will help you understand some of the other organ-systems. For example, movements such as throwing a ball, typing, and walking require the coordinated use of bones and muscles. To understand how muscles produce different movements, you need to learn the parts of the bones to which the muscles attach and the types of joints acted upon by the contracting muscles. The respiratory system is also highly dependent on bone structure. The bones in the nasal cavity form a series of passageways that help clean, moisten, and warm inhaled air. Furthermore, the bones of the thorax are specially shaped and positioned so the chest can expand during inhalation. Many bones also serve as landmarks to students of anatomy as well as to surgeons. As you will see, bony landmarks can be used to locate the outlines of the lungs and heart, abdominal and pelvic viscera, and structures within the skull. Blood vessels and nerves often run parallel to bones. These structures can be located more easily if the bone is identified first.

CLINICAL APPLICATIONS: WHY STUDY BONES?

One of the reasons that scientists study *skeletal materials* is to obtain information about individuals and populations. Skeletal remains make it possible to trace patterns of disease and nutrition, evaluate the effects of certain social and economic changes, and deduce patterns of reproduction and mortality. Examination of skeletal materials relies on several points of information including sex, age, height, and race.

Many disorders can leave permanent effects on skeletal material. Three of the many common causes of bone pathologies are malnutrition, tumors, and infections. Each may cause specific alterations in bone that permit diagnosis from skeletal remains.

We will study bones by examining the various regions of the body. We will look at the skull first and see how the bones of the skull relate to each other. We will then move on to the vertebral column and the chest. This regional approach will allow you to see how all the many bones of the body relate to each other.

TYPES OF BONES

Almost all the bones of the body may be classified into four principal types on the basis of shape: long, short, flat, and irregular. ***Long bones*** have greater length than width and consist of a diaphysis and a variable number of epiphyses. For example, metacarpals, metatarsals, and phalanges have only one epiphysis. The femur actually has four. Other long bones have two. Long bones are slightly curved for strength. A curved bone is structurally designed to absorb the stress of the body weight at several different points so the stress is evenly distributed. If such bones were straight, the weight of the body would be unevenly distributed and the bone would easily fracture. Examples of long bones include bones of the thighs, legs, toes, arms, forearms, and fingers. Figure 6-1a shows the parts of a long bone.

Short bones are somewhat cube-shaped and nearly equal in length and width. Their texture is spongy except at the surface, where there is a thin layer of compact bone. Examples of short bones are the wrist and ankle bones.

Flat bones are generally thin and composed of two more or less parallel plates of compact bone enclosing a layer of spongy bone. Flat bones afford considerable protection and provide extensive areas for muscle attachment. Examples of flat bones include the cranial bones (which protect the brain), the sternum and ribs (which protect organs in the thorax), and the scapulas. The spongy bone found between plates of compact bone in skull bones is called the ***diploë*** (DIP-lō-ē).

Irregular bones have complex shapes and cannot be grouped into any of the three categories just described. They also vary in the amount of spongy and compact bone present. Such bones include the vertebrae and certain facial bones.

There are two additional types of bones that are not included in this classification by shape, but instead are classified by location. ***Sutural*** (SOO-chur-al) or ***Wormian bones*** are small bones between the joints of certain cranial bones (see Figure 7-2e). Their number varies greatly from person to person. ***Sesamoid bones*** are small bones in tendons where considerable pressure develops, for instance, in the wrist. These, like sutural bones, are also variable in number. Two sesamoid bones, the patellas (kneecaps), are present in all individuals.

SURFACE MARKINGS

The surfaces of bones reveal various structural features adapted to specific functions. These features are called ***surface markings.*** Long bones that bear a great deal of weight have large, rounded ends that can form sturdy joints and provide adequate surface area for the attachment of ligaments and muscles. Other bones have depressions that receive the rounded ends. Rough areas serve as points of attachment for muscles, tendons, and ligaments. Grooves in the surfaces of bones provide for the passage of blood vessels. Openings occur where blood vessels and nerves pass through the bone. Exhibit 7-1 describes the different markings and their functions.

EXHIBIT 7-1 BONE MARKINGS

Marking	Description	Example
DEPRESSIONS AND OPENINGS		
Fissure (FISH-ur)	A narrow, cleftlike opening between adjacent parts of bones through which blood vessels or nerves pass.	Superior orbital fissure of the sphenoid bone (Figure 7-2).
Foramen (fō-RĀ-men; *foramen* = hole)	An opening through which blood vessels, nerves, or ligaments pass.	Infraorbital foramen of the maxilla (Figure 7-2).
Meatus (mē-Ā-tus; *meatus* = canal)	A tubelike passageway running within a bone.	External auditory meatus of the temporal bone (Figure 7-2).
Paranasal sinus (*sin* = cavity)	An air-filled cavity within a bone connected to the nasal cavity.	Frontal sinus of the frontal bone (Figure 7-8).
Groove or *sulcus* (*sulcus* = ditchlike groove)	A furrow or depression that accommodates a soft structure such as a blood vessel, nerve, or tendon.	Intertubercular sulcus of the humerus (Figure 8-4).
Fossa (*fossa* = basinlike depression)	A depression in or on a bone.	Mandibular fossa of the temporal bone (Figure 7-4).
PROCESSES	Any prominent projection.	Mastoid process of the temporal bone (Figure 7-2).
Processes that form joints		
Condyle (KON-dīl; *condylus* = knuckle-like process	A large, rounded articular prominence.	Medial condyle of the femur (Figure 8-10).
Head	A rounded articular projection supported on the constricted portion (neck) of a bone.	Head of the femur (Figure 8-10).
Facet	A smooth, flat surface.	Articular facet for the tubercle of rib on a vertebra (Figure 7-18).
Processes to which tendons, ligaments, and other connective tissues attach		
Tubercle (TOO-ber-kul; *tube* = knob)	A small, rounded process.	Greater tubercle of the humerus (Figure 8-4).
Tuberosity	A large, rounded, usually roughened process.	Ischial tuberosity of the coxal (hip) bone (Figure 8-8).
Trochanter (trō-KAN-ter)	A large, blunt projection found only on the femur.	Greater trochanter of the femur (Figure 8-10).
Crest	A prominent border or ridge.	Iliac crest of the coxal (hip) bone (Figure 8-7).
Line	A less prominent ridge than a crest.	Linea aspera of the femur (Figure 8-10).
Spinous process (spine)	A sharp, slender process.	Spinous process of a vertebra (Figure 7-12).
Epicondyle (*epi* = above)	A prominence above a condyle.	Medial epicondyle of the femur (Figure 8-10).

DIVISIONS OF THE SKELETAL SYSTEM

The adult human skeleton usually consists of 206 named bones grouped in two principal divisions: the ***axial skeleton*** and the ***appendicular skeleton.*** The longitudinal ***axis,*** or center, of the human body is a straight line that runs through the body's center of gravity. This imaginary line extends through the head and down to the space between the feet. The midsagittal plane and midline are drawn through this line. The axial division of the skeleton consists of the bones that lie around the axis: ribs, breastbone, hyoid bone, bones of the skull, and backbone.

The appendicular division contains the bones of the

free ***appendages,*** which are the ***upper*** and ***lower extremities (limbs)***, plus the bones called ***girdles,*** which connect the extremities to the axial skeleton.

The 80 bones of the axial division and the 126 bones of the appendicular division are typically grouped as shown in Exhibit 7-2.

Now that you understand how the skeleton is organized into axial and appendicular divisions, refer to Figure 7-1 to see how the two divisions are joined to form the complete skeleton. The bones of the axial skeleton are shown in gold. Be certain to locate the following regions of the skeleton: skull, cranium, face, hyoid bone, vertebral column, thorax, shoulder girdle, upper extremity, pelvic girdle, and lower extremity.

SKULL

The ***skull,*** which contains 22 bones, rests on the superior end of the vertebral column and is composed of two sets of bones: cranial bones and facial bones. The ***cranial bones*** enclose and protect the brain. The 8 cranial bones are the frontal bone, parietal bones (2), temporal bones (2), occipital bone, sphenoid bone, and ethmoid bone. There are 14 ***facial bones:*** nasal bones (2), maxillae (2), zygomatic bones (2), mandible, lacrimal bones (2), palatine bones (2), inferior nasal conchae (2), and vomer. Be sure you can locate all the skull bones in the various views of the skull (Figure 7-2a–e).

Sutures

A ***suture*** (SOO-chur), meaning seam or stitch, is an immovable joint found only between skull bones. Very little connective tissue is found between the bones of a suture. Four prominent sutures are:

1. ***Coronal suture*** between the frontal bone and the two parietal bones.
2. ***Sagittal suture*** between the two parietal bones.
3. ***Lambdoidal*** (lam-DOY-dal) ***suture*** between the parietal bones and the occipital bone.
4. ***Squamosal*** (skwa-MŌ-sal) ***suture*** between the parietal bones and the temporal bones.

Refer to Figures 7-2 and 7-3 for the locations of these sutures. Several other sutures are also shown. Their names are descriptive of the bones they connect. For example, the frontonasal suture is between the frontal bone and the nasal bones. These sutures are indicated in Figures 7-2–7-5.

Fontanels

The "skeleton" of a newly formed embryo consists of cartilage or fibrous membrane structures shaped like bones. Gradually, the cartilage or fibrous membrane is

EXHIBIT 7-2 DIVISIONS OF THE SKELETAL SYSTEM

Regions of the Skeleton	Number of Bones
AXIAL SKELETON	
Skull	
Cranium	8
Face	14
Hyoid	1
Auditory ossicles* **(3 in each ear)**	6
Vertebral column	26
Thorax	
Sternum	1
Ribs	24
	80
APPENDICULAR SKELETON	
Pectoral (shoulder) girdles	
Clavicle	2
Scapula	2
Upper extremities	
Humerus	2
Ulna	2
Radius	2
Carpals	16
Metacarpals	10
Phalanges	28
Pelvic (hip) girdle	
Coxal, pelvic, or hip bone	2
Lower extremities	
Femur	2
Fibula	2
Tibia	2
Patella	2
Tarsals	14
Metatarsals	10
Phalanges	28
	126
Total =	206

* Although the auditory ossicles are not considered part of the axial or appendicular skeleton, but rather as a separate group of bones, they are placed with the axial skeleton for convenience. The auditory ossicles are exceedingly small bones named for their shapes. Their names are the malleus, incus, and stapes, commonly called the hammer, anvil, and stirrup, respectively. The middle portion of each ear contains three auditory ossicles held together by a series of ligaments. The auditory ossicles vibrate in response to sound waves that strike the eardrum and assume a key function in the mechanism involved in hearing. This is described in detail in Chapter 17.

FIGURE 7-1 **Divisions of the skeletal system. The axial skeleton is indicated in gold and the appendicular skeleton is indicated in light brown. (a) Anterior view. (b) Posterior view.**

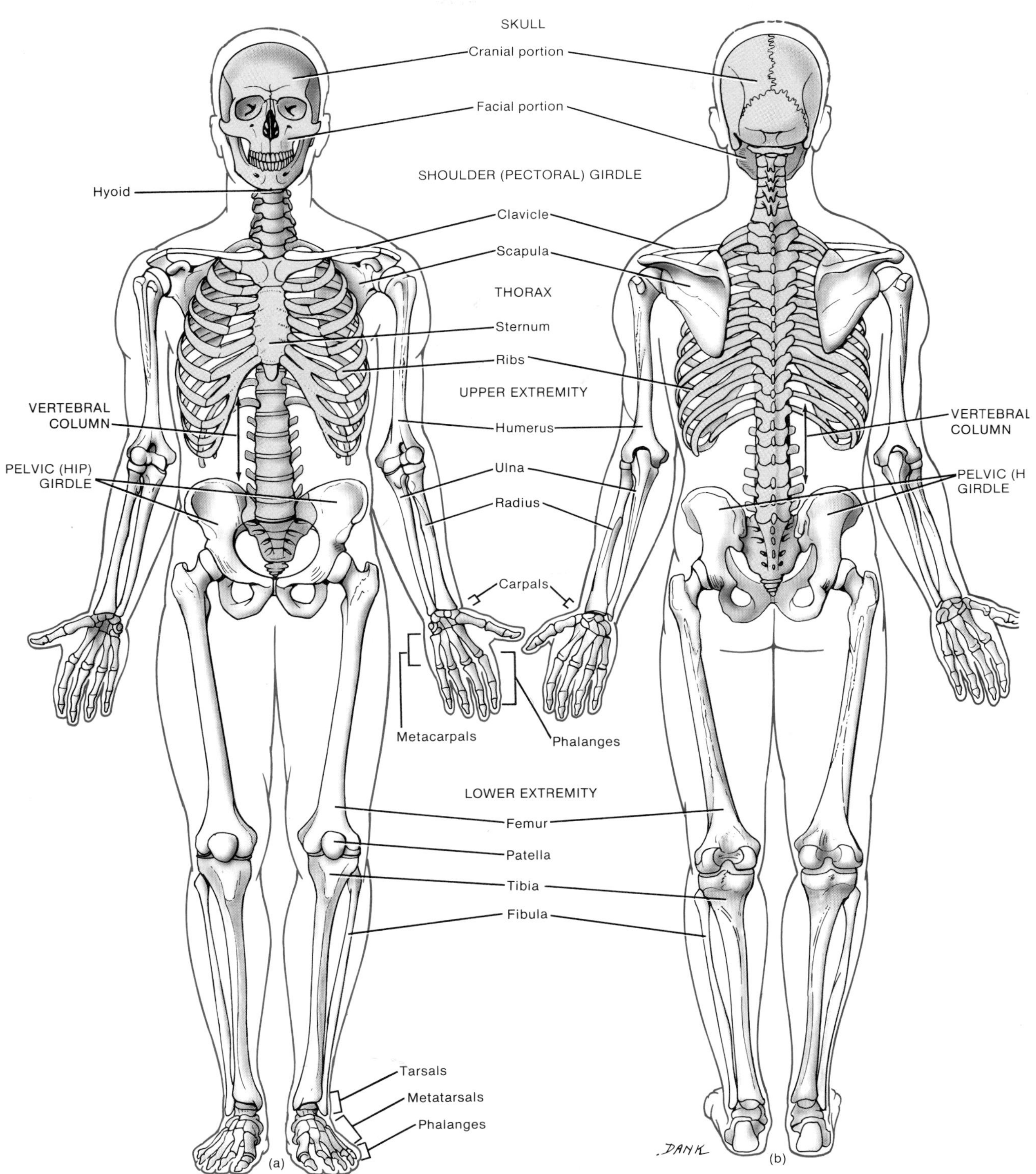

FIGURE 7-2 **Skull. (a) Anterior view.**

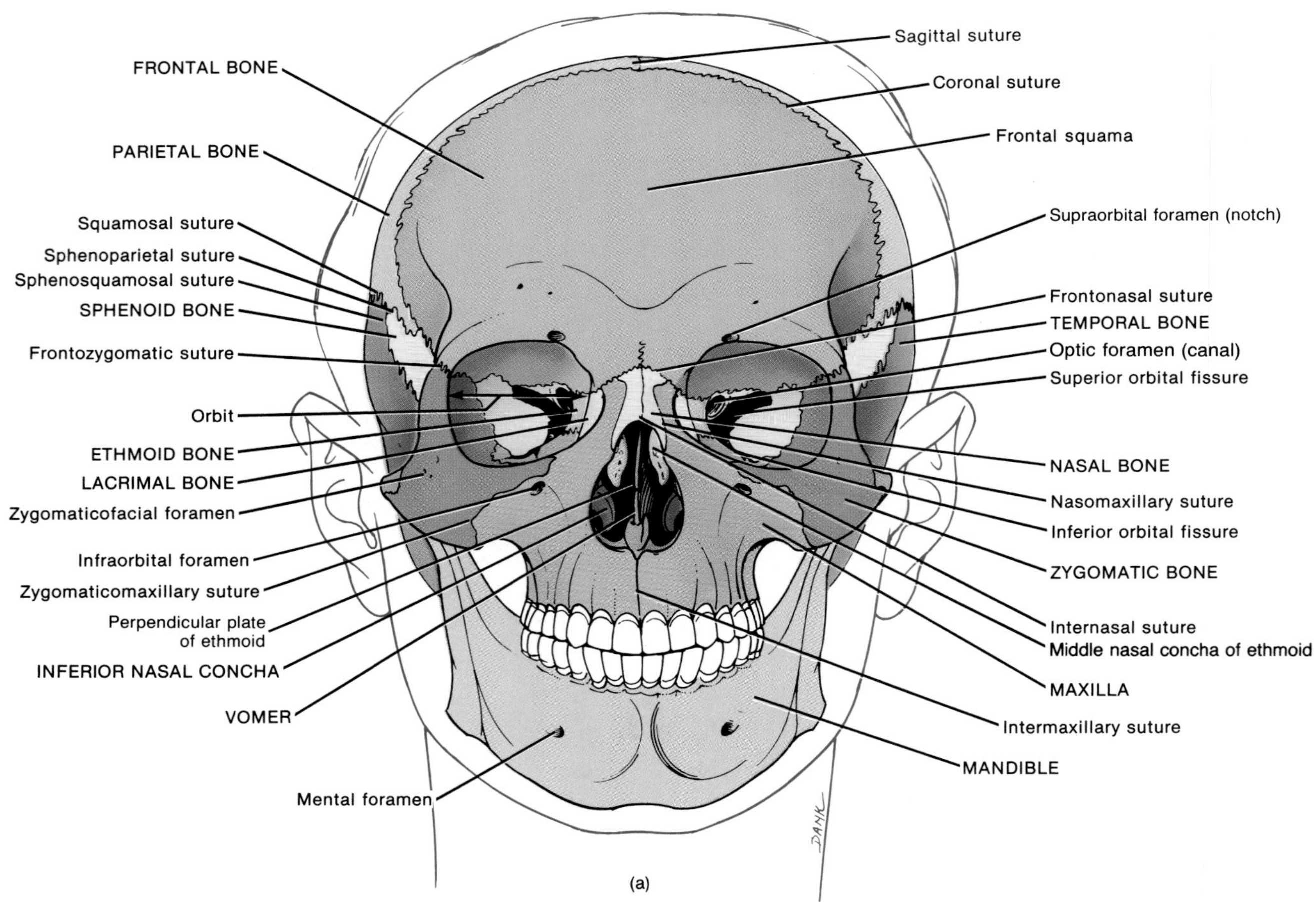

replaced by bone, a process called ossification. At birth, membrane-filled spaces called ***fontanels*** (fon'-ta-NELZ; = little fountains) are found between cranial bones (Figure 7-3). These "soft spots" are areas of dense connective tissue where intramembranous ossification will eventually replace the membrane-filled spaces with bone. They (1) enable the fetal skull to compress as it passes through the birth canal, (2) permit rapid growth of the brain during infancy, (3) facilitate determination of the degree of brain development by their state of closure, (4) serve as landmarks (anterior fontanel) for withdrawal of blood from the superior sagittal sinus, and (5) aid in determining the position of the fetal head prior to birth. Although an infant may have many fontanels at birth, the form and location of six are fairly constant.

The ***anterior (frontal) fontanel*** is located between the angles of the two parietal bones and the two segments of the frontal bone. This fontanel is roughly diamond-shaped, and it is the largest of the six fontanels. It usually closes 18 to 24 months after birth.

The ***posterior*** (***occipital***; ok-SIP-i-tal) ***fontanel*** is situated between the two parietal bones and the occipital bones. This diamond-shaped fontanel is considerably smaller than the anterior fontanel. It generally closes about two months after birth.

The ***anterolateral*** (***sphenoidal***; sfē-NOY-dal) ***fontanels*** are paired. One is located on each side of the skull at the junction of the frontal, parietal, temporal, and sphenoid bones. These fontanels are quite small and irregular in shape. They normally close about three months after birth.

The ***posterolateral*** (***mastoid***) ***fontanels*** are also paired. One is situated on each side of the skull at the junction of the parietal, occipital, and temporal bones. These fontanels are irregularly shaped. They begin to close 1 or 2 months after birth, but closure is not generally complete until 12 months.

Cranial Bones

Frontal Bone

The ***frontal bone*** forms the forehead (the anterior part of the cranium), the roofs of the ***orbits*** (eye sockets), and most of the anterior part of the cranial floor. Soon

FIGURE 7-2 (*Continued*) **(b) Details of the right orbit in anterior view. (c) Right lateral view. Although the hyoid bone is not part of the skull, it is included in the illustration for reference.**

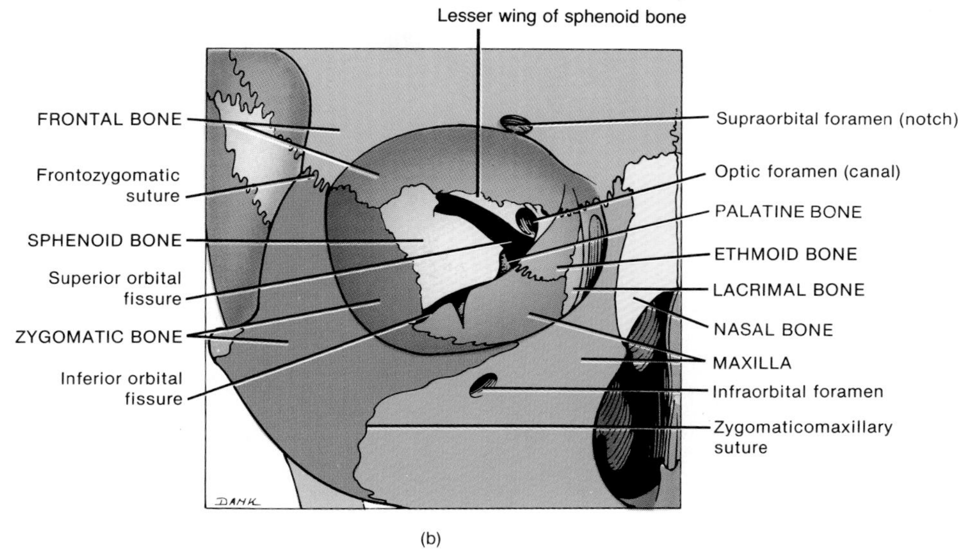

(b)

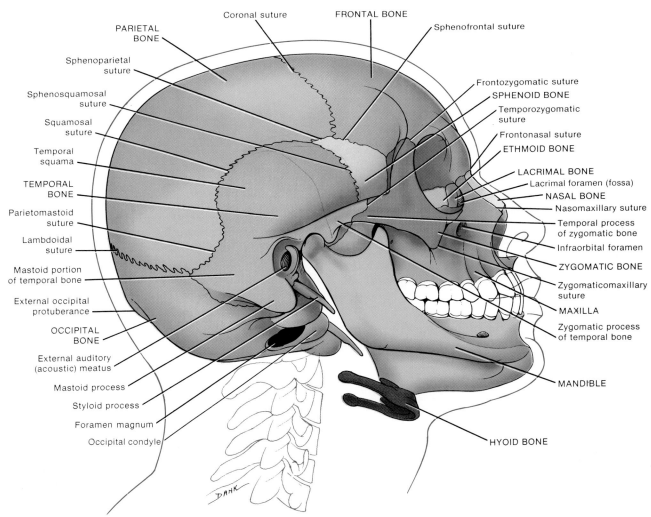

Coronal suture
FRONTAL BONE
PARIETAL BONE
Sphenofrontal suture
Sphenoparietal suture
Sphenosquamosal suture
Squamosal suture
Temporal squama
TEMPORAL BONE
Parietomastoid suture
Lambdoidal suture
Mastoid portion of temporal bone
External occipital protuberance
OCCIPITAL BONE
External auditory (acoustic) meatus
Mastoid process
Styloid process
Foramen magnum
Occipital condyle
Frontozygomatic suture
SPHENOID BONE
Temporozygomatic suture
Frontonasal suture
ETHMOID BONE
LACRIMAL BONE
Lacrimal foramen (fossa)
NASAL BONE
Nasomaxillary suture
Temporal process of zygomatic bone
Infraorbital foramen
ZYGOMATIC BONE
Zygomaticomaxillary suture
MAXILLA
Zygomatic process of temporal bone
MANDIBLE
HYOID BONE
DANK

(c)

FIGURE 7-2 (*Continued*) (d) Median view.

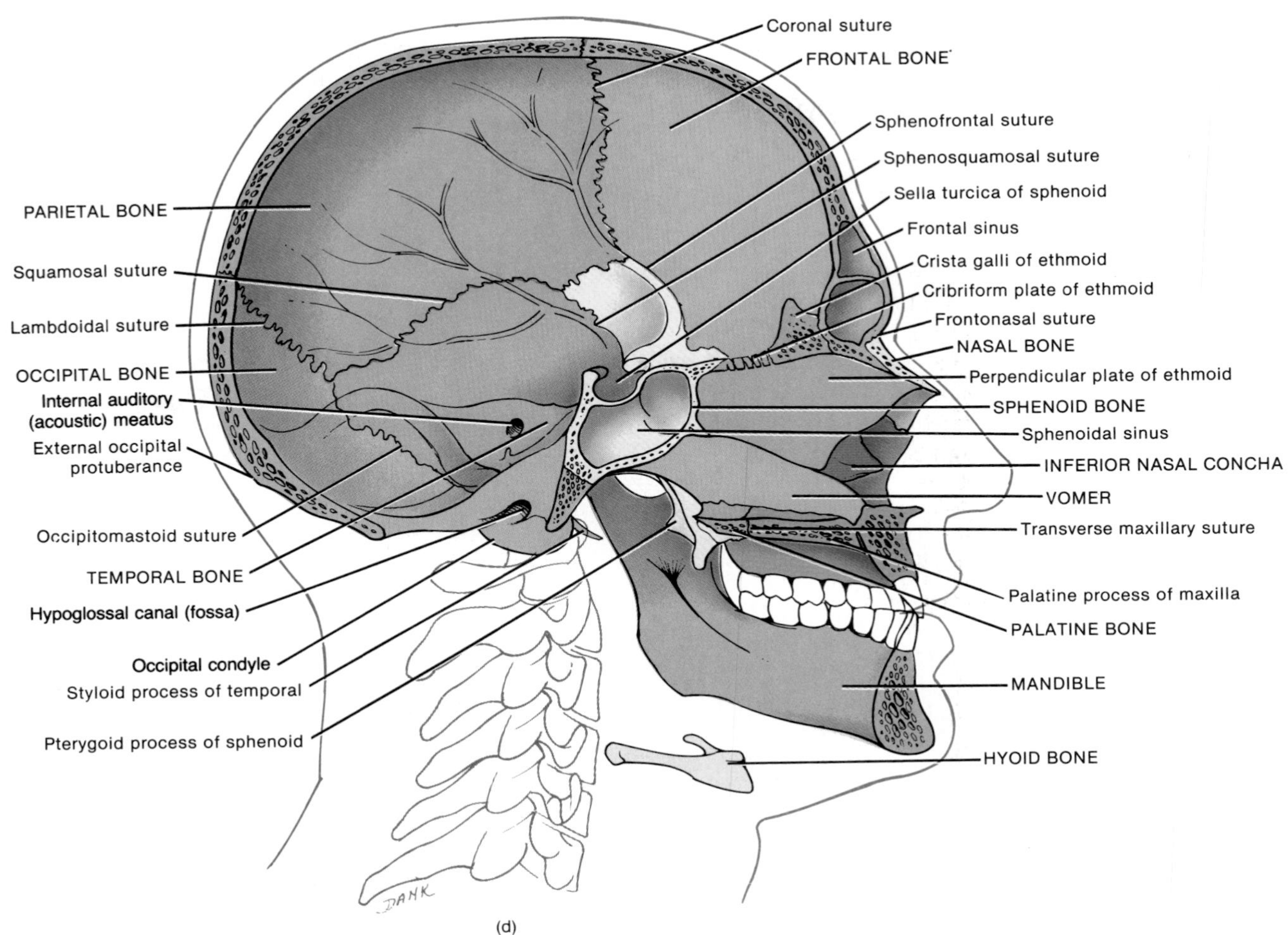

(d)

after birth, the left and right parts of the frontal bone are united by a suture. The suture usually disappears by age 6. If, however, the suture persists throughout life, it is referred to as the ***metopic suture.***

If you examine the anterior and lateral views of the skull in Figure 7-2, you will note the ***frontal squama*** (SKWĀ-ma; *squam* = scale), or ***vertical plate.*** This scalelike plate, which corresponds to the forehead, gradually slopes down from the coronal suture, then turns abruptly downward.

A thickening of the frontal bone is called the ***supraorbital margin.*** From this margin the frontal bone extends posteriorly to form the roof of the orbit and part of the floor of the cranial cavity.

CLINICAL APPLICATION: BLACK EYE

Just above the supraorbital margin is a relatively sharp ridge that overlies the frontal sinus. A blow to the ridge frequently lacerates the skin over it, resulting in bleeding. Bruising of the skin over the ridge causes tissue fluid and blood to accumulate in the surrounding connective tissue and gravitate into the upper eyelid. The resulting swelling and discoloration is called a *black eye.*

Within the supraorbital margin, slightly medial to its midpoint, is a hole (frequently a notch) called the ***supraorbital foramen (notch).*** The supraorbital nerve and artery pass through this foramen. The ***frontal sinuses*** lie deep to the frontal squama. These mucus-lined cavities act as sound chambers that give the voice resonance.

Parietal Bones

The two ***parietal*** (pa-RĪ-e-tal; *paries* = wall) ***bones*** form the greater portion of the sides and roof of the cranial cavity (see Figure 7-2). The internal surfaces of the bones

FIGURE 7-2 (*Continued*) (e) Posterior view. The sutural bones are exaggerated for emphasis.

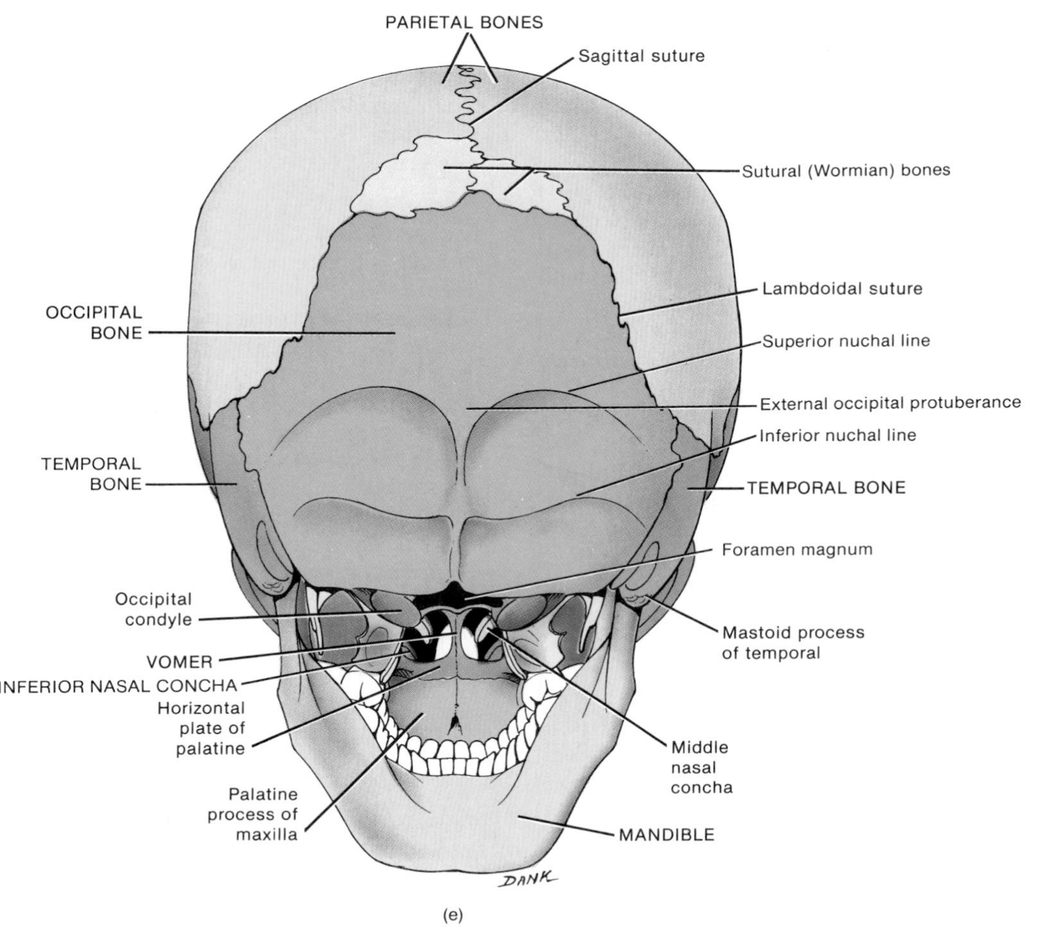

FIGURE 7-3 Fontanels of the skull at birth. (a) Superior view. (b) Right lateral view.

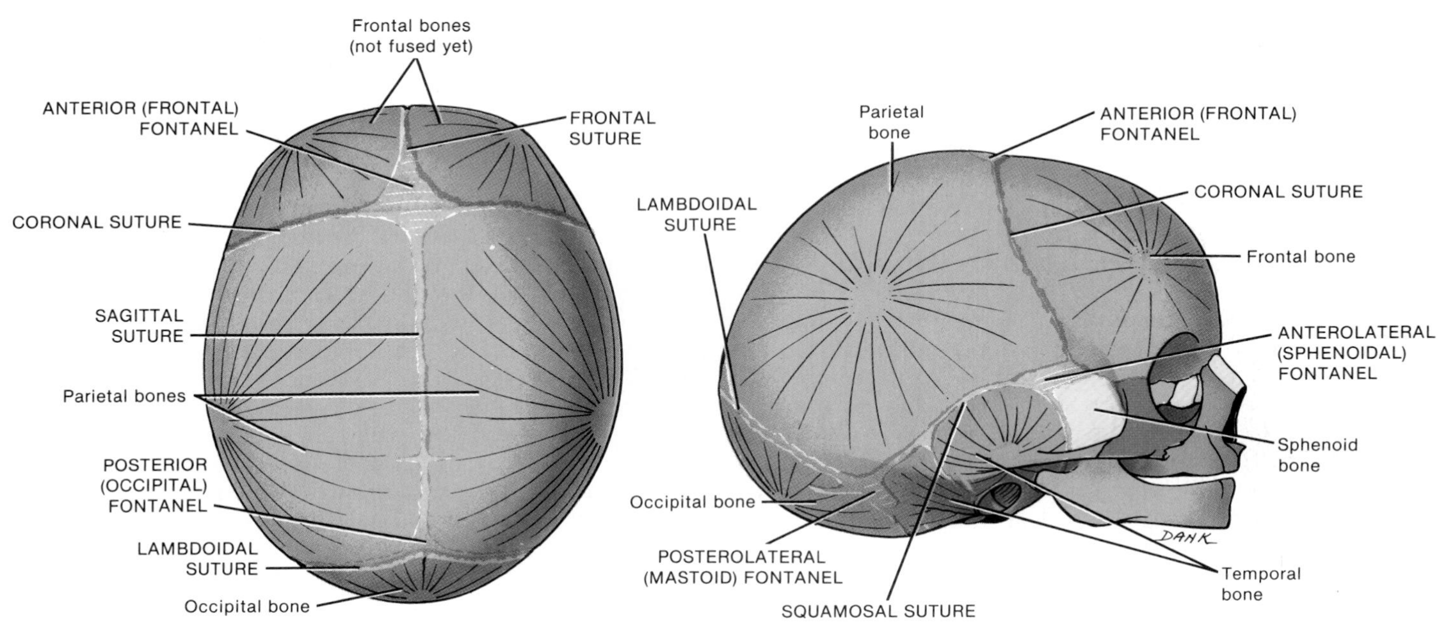

contain many eminences and depressions that accommodate the blood vessels supplying the outer meninx (covering) of the brain called the ***dura mater.***

Temporal Bones

The two ***temporal*** (*tempora* = temples) ***bones*** form the inferior sides of the cranium and part of the cranial floor.

In the lateral view of the skull in Figure 7-2c, notice the ***temporal squama***—a thin, large, expanded area that forms the anterior and superior part of the temple. Projecting from the inferior portion of the temporal squama is the ***zygomatic process,*** which articulates with the temporal process of the zygomatic bone. The zygomatic process of the temporal bone and the temporal process of the zygomatic bone constitute the ***zygomatic arch.***

At the floor of the cranial cavity, shown in Figure 7-5, is the ***petrous portion*** of the temporal bone. This portion is triangular and located at the base of the skull between the sphenoid and occipital bones. The petrous portion contains the internal ear in which are located the structures involved in hearing and equilibrium (balance). It also contains the ***carotid foramen (canal)*** through which the internal carotid artery passes (see Figure 7-4). Posterior to the carotid foramen and anterior to the occipital bone is the ***jugular foramen (fossa)*** through which the internal jugular vein and the glossopharyngeal (IX) nerve, vagus (X) nerve, and accessory (XI) nerve pass. (As you will see later, the Roman numerals associated with cranial nerves indicate the order in which the nerves arise from the brain, from front to back.)

Between the squamous and petrous portions is a socket called the ***mandibular fossa.*** Anterior to the mandibular fossa is a rounded eminence, the ***articular tubercle.*** The mandibular fossa and articular tubercle articulate with the condylar process of the mandible (lower jawbone) to form the temporomandibular joint (TMJ). The mandibular fossa and articular tubercle are seen best in Figure 7-4.

In the lateral view of the skull in Figure 7-2c, you will see the ***mastoid portion*** of the temporal bone, located posterior and inferior to the external auditory meatus, or ear canal. In the adult, this portion of the bone contains a number of ***mastoid air "cells."*** These air spaces are separated from the brain only by thin bony partitions.

If ***mastoiditis,*** the inflammation of these bony cells, occurs, the infection may spread to the brain or its outer covering. The mastoid air cells do not drain as do the paranasal sinuses.

The ***mastoid process*** is a rounded projection of the temporal bone posterior to the external auditory meatus. It serves as a point of attachment for several neck muscles. Near the posterior border of the mastoid process is the ***mastoid foramen*** through which a vein (emissary) to the transverse sinus and a small branch of the occipital artery to the dura mater pass. The ***external auditory (acoustic) meatus*** is the canal in the temporal bone that leads to the middle ear. The ***internal auditory (acoustic) meatus*** is superior to the jugular foramen (see Figure 7-5a). It transmits the facial (VII) and vestibulocochlear (VIII) nerves and the internal auditory artery. The ***styloid process*** projects downward from the undersurface of the temporal bone and serves as a point of attachment for muscles and ligaments of the tongue and neck.

Occipital Bone

The ***occipital*** (ok-SIP-i-tal) ***bone*** forms the posterior part and a prominent portion of the base of the cranium (Figure 7-4).

The ***foramen magnum*** is a large hole in the inferior part of the bone through which the medulla oblongata (part of the brain) and its membranes, the spinal portion of the accessory (XI) nerve, and the vertebral and spinal arteries pass.

The ***occipital condyles*** are oval processes with convex surfaces, one on either side of the foramen magnum, that articulate (form a joint) with depressions on the first cervical vertebra. At the base of each condyle is a ***hypoglossal canal (fossa)*** through which the hypoglossal (XII) nerve passes (see Figure 7-5).

The ***external occipital protuberance*** is a prominent projection on the posterior surface of the bone just superior to the foramen magnum. You can feel this structure as a definite bump on the back of your head, just above your neck. The protuberance is also visible in Figure 7-2d. Extending laterally from the protuberance are two curved lines, the ***superior nuchal lines,*** and below these two ***inferior nuchal lines,*** which are areas of muscle attachment (see Figure 7-2e).

Sphenoid Bone

The ***sphenoid*** (SFĒ-noyd) ***bone*** is situated at the middle part of the base of the skull (Figure 7-5). *Spheno* means "wedge." This bone is referred to as the keystone of the cranial floor because it articulates with all the other cranial bones. If you view the floor of the cranium from above, you will note that the sphenoid articulates with the temporal bones anteriorly and the occipital bone posteriorly. It lies posterior and slightly superior to the nasal cavities and forms part of the floor and sidewalls of the orbit (eye socket). The shape of the sphenoid is frequently described as a bat with outstretched wings.

The ***body*** of the sphenoid is the cubelike central portion between the ethmoid and occipital bones. It contains the ***sphenoidal sinuses,*** which drain into the nasal cavity (see Figure 7-8). On the superior surface of the body of the sphenoid is a depression called the ***sella turcica*** (SEL-a TUR-si-ka; = Turk's saddle). This depression houses the pituitary gland.

The ***greater wings*** of the sphenoid are lateral projections from the body and form the anterolateral floor of the cranium. The greater wings also form part of the lateral wall of the skull just anterior to the temporal bone.

FIGURE 7-4 **Skull in inferior view.**

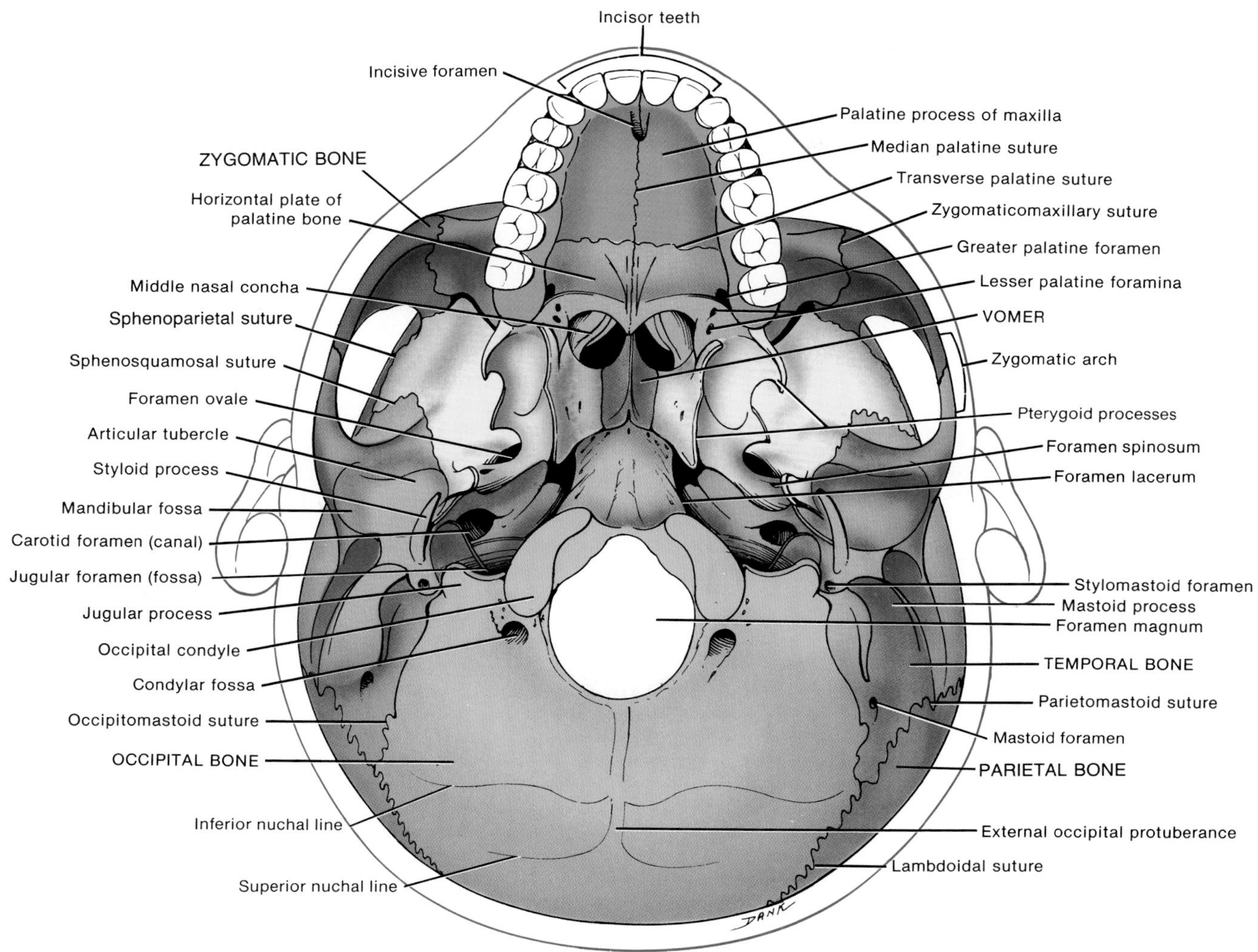

The ***lesser wings*** are anterior and superior to the greater wings. They form part of the floor of the cranium and the posterior part of the orbit.

Between the body and lesser wing, you can locate the ***optic foramen (canal)*** through which the optic (II) nerve and ophthalmic artery pass. Lateral to the body between the greater and lesser wings is a somewhat triangular slit called the ***superior orbital fissure.*** It is an opening for the oculomotor (III) nerve, trochlear (IV) nerve, ophthalmic branch of the trigeminal (V) nerve, and abducens (VI) nerve. This fissue may also be seen in the anterior view of the skull in Figure 7-2a.

On the inferior part of the sphenoid bone, you can see the ***pterygoid*** (TER-i-goyd) ***processes.*** These structures project inferiorly from the points where the body and greater wings unite. The pterygoid processes form part of the lateral walls of the nasal cavities.

Ethmoid Bone

The ***ethmoid bone*** is a light, spongy bone located in the anterior part of the floor of the cranium between the orbits. It is anterior to the sphenoid and posterior to the nasal bones (Figure 7-6). The ethmoid bone forms part of the anterior portion of the cranial floor, the medial wall of the orbits, the superior portions of the nasal septum, or partition, and most of the sidewalls of the nasal roof. The ethmoid is the principal supporting structure of the nasal cavities.

Its ***lateral masses (labyrinths)*** compose most of the wall between the nasal cavities and the orbits. They contain several air spaces, or "cells," ranging in number from 3 to 18. It is from these "cells" that the bone derives its name (*ethmos* = sieve). The ethmoid "cells" together form the ***ethmoidal sinuses.*** The sinuses are shown in

FIGURE 7-5 Sphenoid bone. (a) Viewed in the floor of the cranium from above. (b) Anterior view.

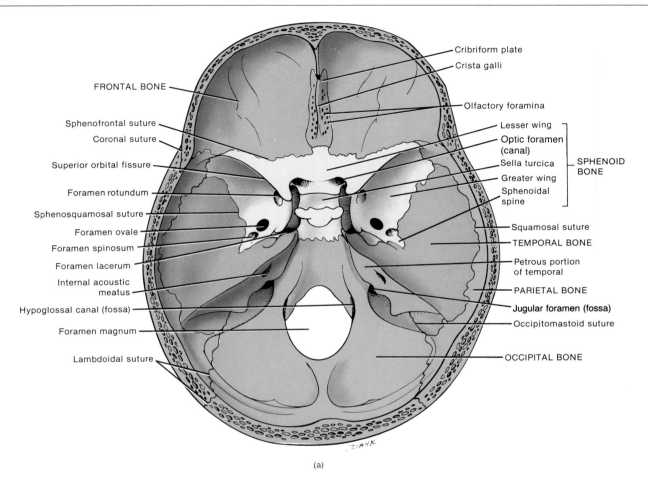

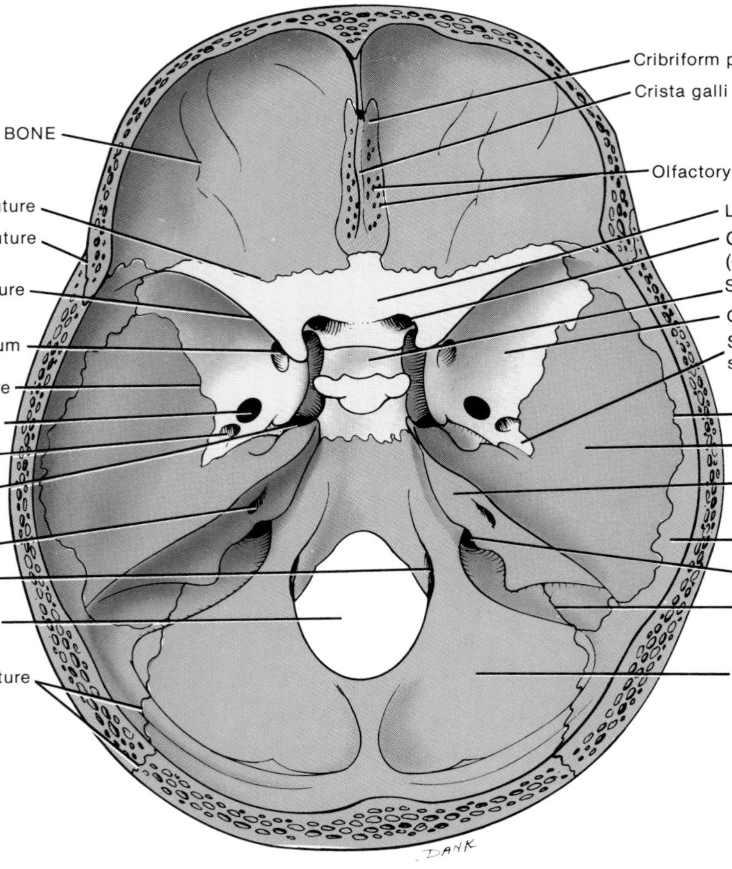

Cribriform plate
Crista galli
FRONTAL BONE
Olfactory foramina
Sphenofrontal suture
Lesser wing
Coronal suture
Optic foramen (canal)
Superior orbital fissure
Sella turcica
SPHENOID BONE
Greater wing
Foramen rotundum
Sphenoidal spine
Sphenosquamosal suture
Squamosal suture
Foramen ovale
TEMPORAL BONE
Foramen spinosum
Foramen lacerum
Petrous portion of temporal
Internal acoustic meatus
PARIETAL BONE
Hypoglossal canal (fossa)
Jugular foramen (fossa)
Foramen magnum
Occipitomastoid suture
Lambdoidal suture
OCCIPITAL BONE
DANK

(a)

Lesser wings
Optic foramen (canal)
Superior orbital fissure
Sphenoidal sinuses
Foramen rotundum
Pterygoid canal
Body
Sphenoidal spine
Greater wings
Pterygoid process
DANK

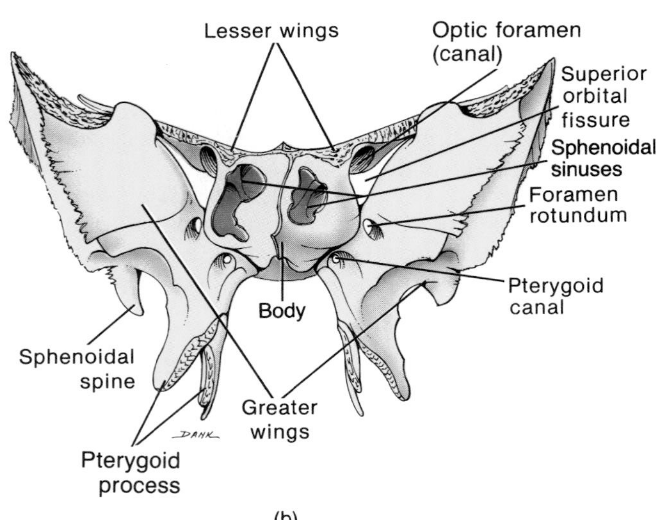

(b)

FIGURE 7-6 **Ethmoid bone. (a) Median view showing the ethmoid bone on the inner aspect of the left part of the skull. (b) Superior view. (c) Highly diagrammatic representation showing, in anterior view, the approximate position of the ethmoid bone in the skull.**

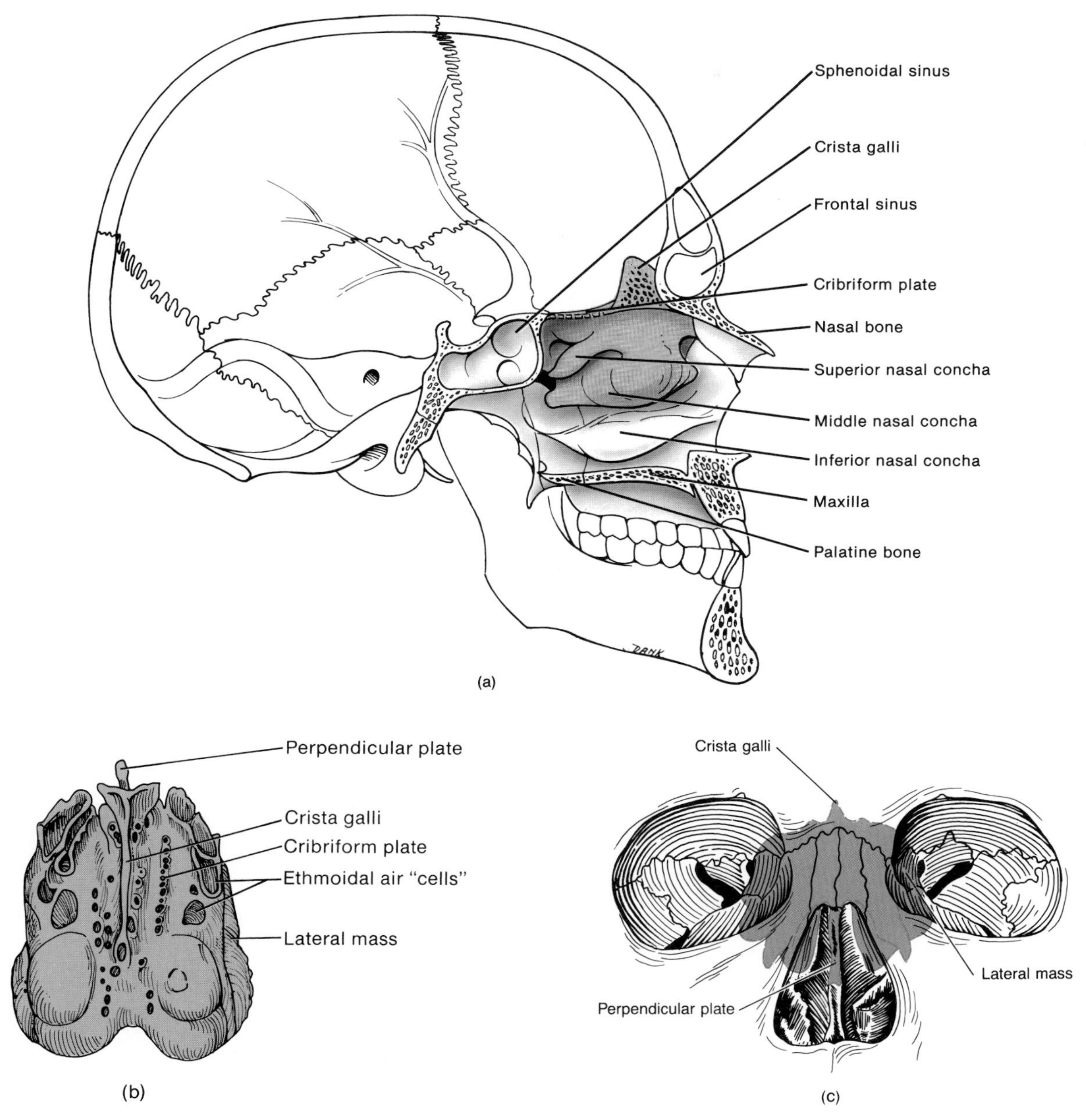

Figure 7-8. The ***perpendicular plate*** forms the superior portion of the nasal septum (see Figure 7-7). The ***cribriform (horizontal) plate*** lies in the anterior floor of the cranium and forms the roof of the nasal cavity. The cribriform plate contains the ***olfactory foramina*** through which the olfactory (I) nerves pass. These nerves function in smell. Projecting upward from the cribriform plate is a triangular process called the ***crista galli*** (= cock's comb). This structure serves as a point of attachment for the membranes (meninges) that cover the brain.

The labyrinths contain two thin, scroll-shaped bones on either side of the nasal septum. These are called the ***superior nasal concha*** (KONG-ka; *concha* = shell) and the ***middle nasal concha.*** The conchae allow for the turbulent circulation and filtration of inhaled air before it passes into the trachea, bronchi, and lungs by causing the air to whirl around, which results in many inhaled particles striking and becoming trapped in the mucus found on the lining of the nasal passageways.

Facial Bones

The shape of the face changes dramatically during the first two postnatal years, owing in part to the expanding brain and cranial bones, the formation and eruption of teeth, and the increase in size of the paranasal sinuses. Growth of the face ceases at approximately 16 years of age.

Nasal Bones

The paired ***nasal bones*** are small, oblong bones that meet at the middle and superior part of the face (see Figures 7-2 and 7-7). Their fusion forms part of the bridge of the nose. The inferior portion of the nose, indeed the major portion, consists of cartilage.

Maxillae

The paired ***maxillae*** (mak-SIL-ē; *macerae* = to chew) unite to form the upper jawbone (Figure 7-7) and articulate with every bone of the face except the mandible, or lower jawbone. They form part of the floors of the orbits, part of the roof of the mouth (most of the hard palate), and part of the lateral walls and floor of the nasal cavity.

Each maxillary bone contains a ***maxillary sinus*** that empties into the nasal cavity (see Figure 7-8). The ***alveolar*** (al-VĒ-ō-lar) ***process*** (*alveolus* = hollow) contains the ***alveoli*** (bony sockets) into which the maxillary (upper) teeth are set. The ***palatine process*** is a horizontal projection of the maxilla that forms the anterior three-fourths of the hard palate, or anterior portion of the roof of the oral cavity. The two portions of the maxillary bones unite, and the fusion is normally completed before birth.

CLINICAL APPLICATION: CLEFT PALATE AND CLEFT LIP

If the palatine processes of the maxillary bones do not unite before birth, a condition called *cleft palate* results. The condition may also involve incomplete fusion of the cribriform plates of the palatine bones (see Figure 7-4). Another form of this condition, called *cleft lip,* involves a split in the upper lip. Cleft lip is often associated with cleft palate. Depending on the extent and position of the cleft, speech and swallowing may be affected. Facial and oral surgeons recommend closure of cleft lip during the first year of life, and surgical results are excellent. Repair of cleft palate is done between the first and second year of life, before the child begins to

FIGURE 7-7 Maxillae. (a) Median view of the left maxilla. (b) Inferior view of the skull, showing the maxillae.

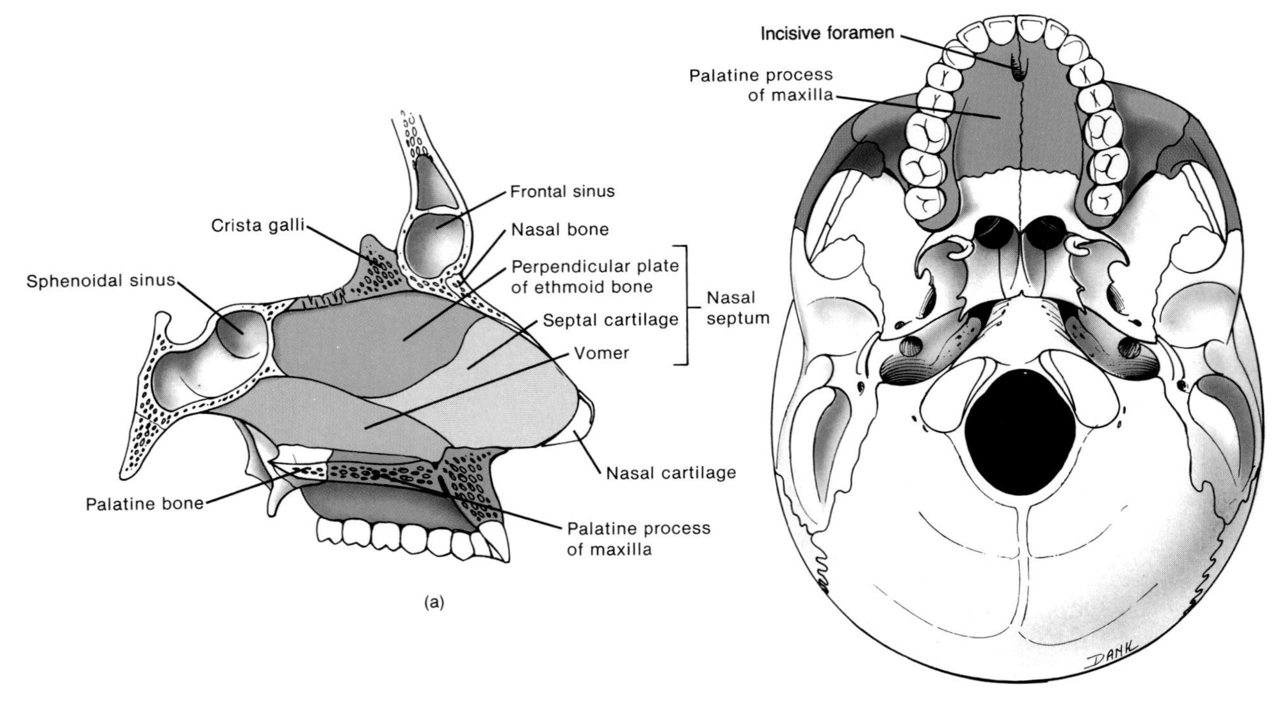

FIGURE 7-8 Paranasal sinuses seen in median view.

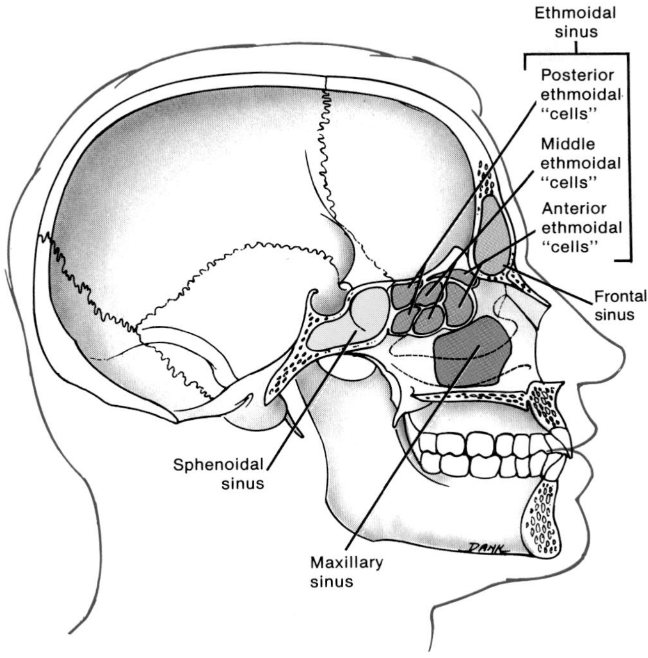

develop speech at about age 2. Here again, results are usually excellent, and orthodontic therapy may be valuable in aligning the teeth. Satisfactory speech will develop in three-fourths of such children, although a secondary surgical repair is sometimes necessary.

A fissure associated with the maxilla and sphenoid bone is the ***inferior orbital fissure.*** It is located between the greater wing of the sphenoid and the maxilla (see Figure 7-2b). It transmits the maxillary branch of the trigeminal (V) nerve, the infraorbital vessels, and the zygomatic nerve.

Paranasal Sinuses

Although not cranial or facial bones, this is an appropriate point to discuss ***paranasal*** (*para* = beside) ***sinuses,*** paired cavities in certain cranial and facial bones near the nasal cavity (Figure 7-8). The paranasal sinuses are lined with mucous membranes that are continuous with the lining of the nasal cavity. Skull bones containing paranasal sinuses are the frontal, sphenoid, ethmoid, and maxillae. (The paranasal sinuses were described in the discussion of each of these bones.) Besides producing mucus, the paranasal sinuses lighten the skull bones and serve as resonant chambers for sound as we speak or sing.

CLINICAL APPLICATION: SINUSITIS

Secretions produced by the mucous membranes of the paranasal sinuses drain into the nasal cavity. An inflammation of the membranes due to an allergic reaction or infection is called *sinusitis.* If the membranes swell enough to block drainage into the nasal cavity, fluid pressure builds up in the paranasal sinuses, and a sinus headache results.

Zygomatic Bones

The two ***zygomatic bones (malars),*** commonly referred to as the cheekbones, form the prominences of the cheeks and part of the outer wall and floor of the orbits (Figure 7-2b).

The ***temporal process*** of the zygomatic bone projects posteriorly and articulates with the zygomatic process of the temporal bone. These two processes form the ***zygomatic arch*** (Figure 7-4).

Mandible

The ***mandible*** (*mandere* = to chew), or lower jawbone, is the largest, strongest facial bone (Figure 7-9). It is the only movable skull bone (other than the auditory ossicles).

In the lateral view, you can see that the mandible consists of a curved, horizontal portion called the ***body*** and two perpendicular portions called the ***rami*** (singular = ***ramus***). The ***angle*** of the mandible is the area where each ramus meets the body. Each ramus has a ***condylar*** (KON-di-lar) ***process*** that articulates with the mandibular fossa and articular tubercle of the temporal bone to form the temporomandibular joint (TMJ). It also has a ***coronoid*** (KOR-ō-noyd) ***process*** to which the temporalis muscle attaches. The depression between the coronoid and condylar processes is called the ***mandibular notch.*** The ***alveolar process*** is an arch containing the ***alveoli*** (sockets) for the mandibular (lower) teeth.

The ***mental foramen*** (*mentum* = chin) is approximately below the first molar tooth. The mental nerve and

FIGURE 7-9 Mandible in right lateral view.

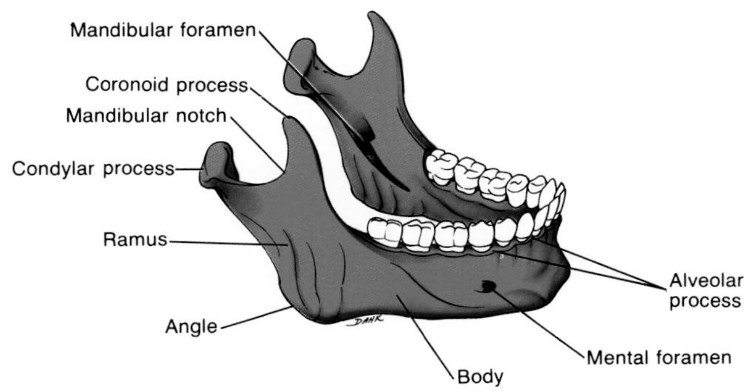

vessels pass through this opening. It is through this foramen that dentists sometimes reach the nerve when injecting anesthetics. Another foramen associated with the mandible is the ***mandibular foramen*** on the medial surface of the ramus, another site frequently used by dentists to inject anesthetics. It transmits the inferior alveolar nerve and vessels. The mandibular foramen is the beginning of the ***mandibular canal,*** which runs forward in the ramus deep to the roots of the teeth. The canal carries branches of the inferior alveolar nerve and vessels to the teeth. Parts of these nerves and vessels emerge through the mental foramen.

CLINICAL APPLICATION: TEMPOROMANDIBULAR JOINT (TMJ) SYNDROME

One problem associated with the temporomandibular joint (TMJ) is called *TMJ syndrome.* It is characterized by dull pain around the ear, tenderness of the jaw muscles, clicking or popping noise when opening or closing the mouth, limited or abnormal opening of the mouth, headache, tooth sensitivity, and abnormal wearing of the teeth. TMJ syndrome might be caused by improperly aligned teeth, grinding or clenching the teeth, trauma to the jaw, or arthritis. Treatment may consist of application of moist heat or ice, keeping lips together and teeth apart, a soft diet, taking aspirin, muscle retraining, use of an occlusal splint, adjusting or reshaping the teeth, orthodontic treatment, or surgery.

Lacrimal Bones

The paired ***lacrimal*** (LAK-ri-mal; *lacrima* = tear) ***bones*** are thin bones roughly resembling a fingernail in size and shape. They are the smallest bones of the face. These bones are posterior and lateral to the nasal bones in the medial wall of the orbit. They can be seen in the anterior and lateral views of the skull in Figure 7-2. The lacrimal bones form a part of the medial wall of the orbit.

Palatine Bones

The two ***palatine*** (PAL-a-tīn) ***bones*** are L-shaped and form the posterior portion of the hard palate, part of the floor and lateral wall of the nasal cavity, and a small portion of the floors of the orbits. The posterior portion of the hard palate, which separates the nasal cavity from the oral cavity, is formed by the ***horizontal plates*** of the palatine bones. These can be seen in Figure 7-4.

Inferior Nasal Conchae

Refer to the views of the skull in Figures 7-2a and 7-6a. The two ***inferior nasal conchae*** (KONG-kē) are scroll-like bones that form a part of the lateral wall of the nasal cavity and project into the nasal cavity inferior to the superior and middle nasal conchae of the ethmoid bone. They serve the same function as the superior and middle nasal conchae; that is, they allow for the turbulent circulation and filtration of air before it passes into the lungs. The inferior nasal conchae are separate bones and not part of the ethmoid.

Vomer

The ***vomer*** (= plowshare) is a roughly triangular bone that forms the inferior and posterior part of the nasal septum. It is clearly seen in the anterior view of the skull in Figure 7-2a and the inferior view in Figure 7-4.

The inferior border of the vomer articulates with the cartilage septum that divides the nose into a right and left nostril. Its superior border articulates with the perpendicular plate of the ethmoid bone. The structures that form the ***nasal septum,*** or partition, are the perpendicular plate of the ethmoid, septal cartilage, vomer, and parts of the palatine bones and maxillae (Figure 7-7a).

CLINICAL APPLICATION: DEVIATED NASAL SEPTUM (DNS)

A *deviated nasal septum (DNS)* is deflected laterally from the midline of the nose. The deviation usually occurs at the junction of bone with the septal cartilage. If the deviation is severe, it may entirely block the nasal passageway. Even though the blockage may not be complete, infection and inflammation may develop and cause nasal congestion, blockage of the paranasal sinus openings, chronic sinusitis, headache, and nosebleeds.

A summary of bones of the skull is presented in Exhibit 7-3.

Orbits

Each ***orbit*** (eye socket) is a pyramid-shaped space that contains the eyeball and associated structures. It is formed by seven bones of the skull (see Figure 7-2b) and has four walls and an apex (back end). The roof of the orbit consists of parts of the frontal and sphenoid bones. The lateral wall is formed by portions of the zygomatic and sphenoid bones. The floor of the orbit is formed by parts of the maxilla, zygomatic, and palatine bones. The medial wall of the orbit is formed from portions of the maxilla, lacrimal, ethmoid, and sphenoid bones.

The principal openings of each orbit are the following:

1. ***Optic foramen (canal)*** at the junction of the roof and medial wall.
2. ***Superior orbital fissure*** at the upper lateral angle of the apex.

EXHIBIT 7-3 SUMMARY OF BONES OF THE SKULL*

Cranial Bones	Facial Bones
Frontal (1)	Nasal (2)
Parietal (2)	Maxillae (2)
Temporal (2)	Zygomatic (2)
Occipital (1)	Mandible (1)
Sphenoid (1)	Lacrimal (2)
Ethmoid (1)	Palatine (2)
	Inferior nasal conchae (2)
	Vomer (1)

* The numbers in parentheses indicate how many of each bone are present.

3. ***Inferior orbital fissure*** at the junction of the lateral wall and floor.

4. ***Supraorbital foramen (notch)*** on the medial side of the supraorbital margins of the frontal bone.

5. ***Canal for nasolacrimal duct*** in the nasal bone.

Foramina

Some ***foramina*** (singular = foramen) of the skull were mentioned along with the descriptions of the cranial and facial bones with which they are associated. As preparation for studying other systems of the body, especially the nervous and cardiovascular systems, these foramina, as well as some additional ones, and the structures passing through them are listed in Exhibit 7-4. For your convenience and for future reference, the foramina are listed alphabetically.

HYOID BONE

The single ***hyoid bone*** (*hyoedes* = U-shaped) is a unique component of the axial skeleton because it does not articulate with any other bone. Rather, it is suspended from the styloid process of the temporal bone by ligaments and muscles. The hyoid is located in the neck between the mandible and larynx. It supports the tongue and provides attachment for some of its muscles. It also provides attachment for muscles of the neck and pharynx. Refer to the median and lateral views of the skull in Figure 7-2c,d to see the position of the hyoid bone.

The hyoid consists of a horizontal ***body*** and paired projections called the ***lesser cornu*** (*cornu* = horn) and the ***greater cornu*** (Figure 7-10). Muscles and ligaments attach to these paired projections.

The hyoid bone is frequently fractured during strangulation. As a result, it is carefully examined in an autopsy when strangulation is suspected.

FIGURE 7-10 Hyoid bone. (a) Anterior view. (b) Right lateral view.

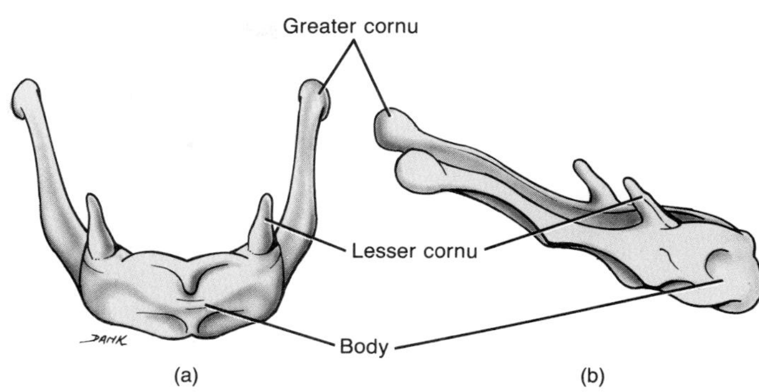

VERTEBRAL COLUMN

Divisions

The ***vertebral column (spine),*** together with the sternum and ribs, constitutes the skeleton of the ***trunk*** of the body. The vertebral column makes up about two-fifths of the total height of the body and is composed of a series of bones called ***vertebrae.*** In an average adult male, the column measures about 71 cm (28 inches) in length; in an average adult female, it measures about 61 cm (24 inches) in length. In effect, the vertebral column is a strong, flexible rod that moves anteriorly, posteriorly, and laterally and rotates. It encloses and protects the spinal cord, supports the head, and serves as a point of attachment for the ribs and the muscles of the back. Between vertebrae are openings called ***intervertebral foramina.*** The nerves that connect the spinal cord to various parts of the body pass through these openings.

The adult vertebral column typically contains 26 vertebrae (Figure 7-11a,b). These are distributed as follows: 7 ***cervical vertebrae*** (*cervix* = neck) in the neck region; 12 ***thoracic vertebrae*** (*thorax* = chest) posterior to the thoracic cavity; 5 ***lumbar vertebrae*** (*lumbus* = loin) supporting the lower back; 5 ***sacral vertebrae*** fused into one bone called the ***sacrum;*** and usually 4 ***coccygeal*** (kok-SIJ-ē-al) ***vertebrae*** fused into one or two bones called the ***coccyx*** (KOK-six). Prior to the fusion of the sacral and coccygeal vertebrae, the total number of vertebrae is 33.

Between adjacent vertebrae from the first vertebra (axis) to the sacrum are fibrocartilaginous ***intervertebral discs.*** Each disc is composed of an outer fibrous ring consisting of fibrocartilage called the ***annulus fibrosus*** and an inner soft, pulpy, highly elastic structure called the ***nucleus pulposus*** (see Figure 7-19). The discs form strong joints, permit various movements of the vertebral column, and absorb vertical shock. Under compression, they flatten, broaden, and bulge from their intervertebral spaces (Figure 7-11c).

EXHIBIT 7-4 SUMMARY OF FORAMINA OF THE SKULL

Foramen	Location	Structures Passing Through
Carotid (Figure 7-4)	Petrous portion of temporal.	Internal carotid artery.
Greater palatine (Figure 7-4)	Posterior angle of hard palate.	Greater palatine nerve and greater palatine vessels.
Hypoglossal (Figure 7-5)	Superior to base of occipital condyles.	Hypoglossal (XII) nerve and branch of ascending pharyngeal artery.
Incisive (Figure 7-7b)	Posterior to incisor teeth.	Branches of greater palatine vessels and nasopalatine nerve.
Inferior orbital (Figure 7-2b)	Between greater wing of sphenoid and maxilla.	Maxillary branch of trigeminal (V) nerve, zygomatic nerve, and infraorbital vessels.
Infraorbital (Figure 7-2a)	Inferior to orbit in maxilla.	Infraorbital nerve and artery.
Jugular (Figure 7-4)	Posterior to carotid canal between petrous portion of temporal and occipital.	Internal jugular vein, glossopharyngeal (IX) nerve, vagus (X) nerve, and accessory (XI) nerve.
Lacerum (Figure 7-5a)	Bounded anteriorly by sphenoid, posteriorly by petrous portion of temporal, and medially by the sphenoid and occipital.	Branch of ascending pharyngeal artery.
Lacrimal (Figure 7-2c)	Lacrimal bone.	Lacrimal (tear) duct.
Lesser palatine (Figure 7-4)	Posterior to greater palatine foramen.	Lesser palatine nerves and artery.
Magnum (Figure 7-4)	Occipital bone.	Medulla oblongata and its membranes, accessory (XI) nerve, and vertebral and spinal arteries and meninges.
Mandibular (Figure 7-9)	Medial surface of ramus of mandible.	Inferior alveolar nerve and vessels.
Mastoid (Figure 7-4)	Posterior border of mastoid process of temporal bone.	Emissary vein to transverse sinus and branch of occipital artery to dura mater.
Mental (Figure 7-9)	Inferior to second premolar tooth in mandible.	Mental nerve and vessels.
Olfactory (Figure 7-5a)	Cribriform plate of ethmoid.	Olfactory (I) nerve.
Optic (Figure 7-5a)	Between upper and lower portions of small wing of sphenoid.	Optic (II) nerve and ophthalmic artery.
Ovale (Figure 7-5a)	Greater wing of sphenoid.	Mandibular branch of trigeminal (V) nerve.
Rotundum (Figure 7-5a)	Junction of anterior and medial parts of sphenoid.	Maxillary branch of trigeminal (V) nerve.
Spinosum (Figure 7-5a)	Posterior angle of sphenoid.	Middle meningeal vessels.
Stylomastoid (Figure 7-4)	Between styloid and mastoid processes of temporal.	Facial (VII) nerve and stylomastoid artery.
Superior orbital (Figure 7-2b)	Between greater and lesser wings of sphenoid.	Oculomotor (III) nerve, trochlear (IV) nerve, ophthalmic branch of trigeminal (V) nerve, and abducens (VI) nerve.
Supraorbital (Figure 7-2a)	Supraorbital margin of orbit.	Supraorbital nerve and artery.
Zygomaticofacial (Figure 7-2a)	Zygomatic bone.	Zygomaticofacial nerve and vessels.

FIGURE 7-11 **Vertebral column. (a) Anterior view. (b) Right lateral view. (c) Intervertebral disc in its normal position (left) and under compression (right). The relative size of the disc has been enlarged for emphasis. A "window" has been cut in the annulus fibrosus so that the nucleus pulposus can be seen.**

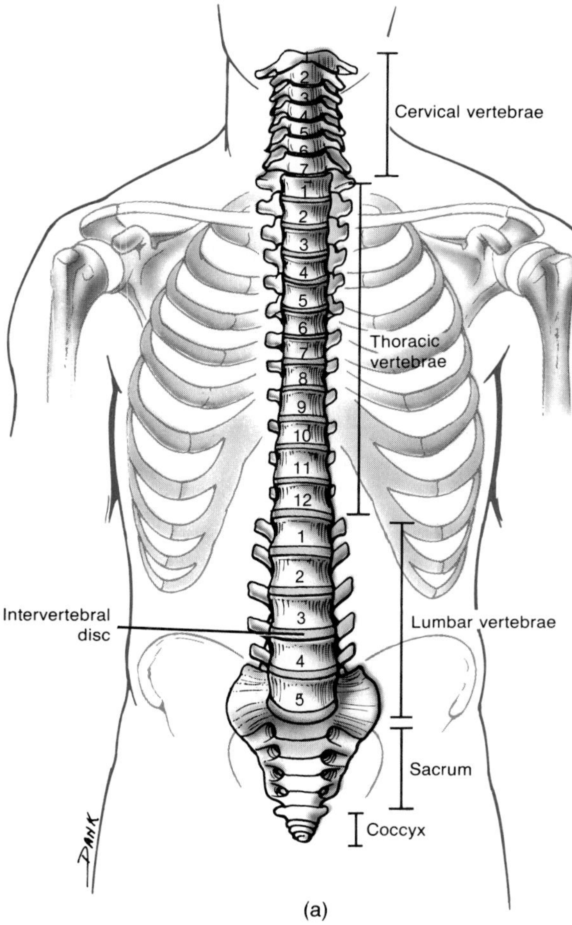

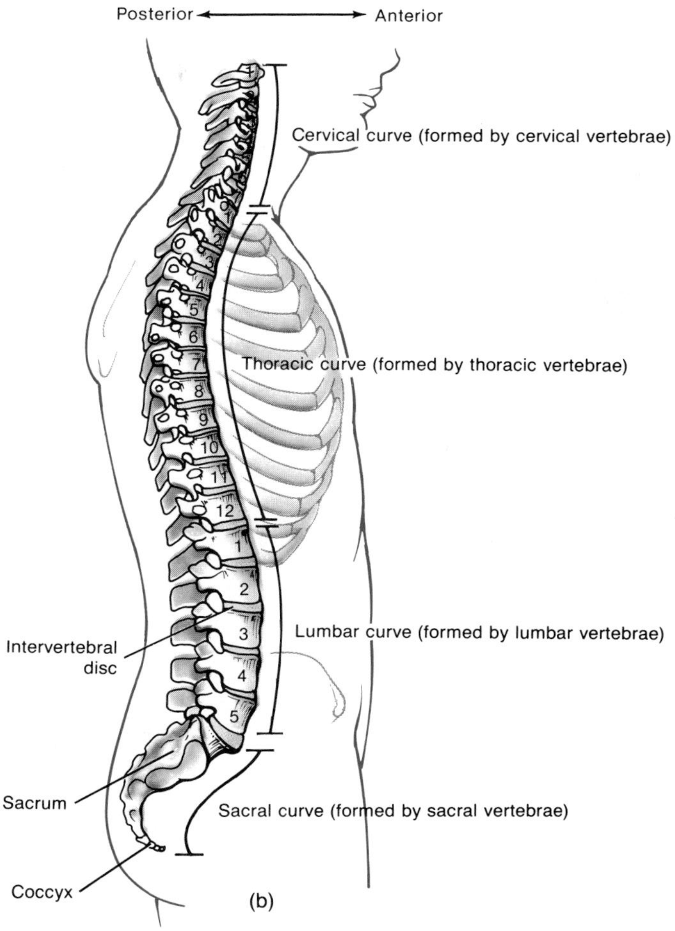

(a)

(b)

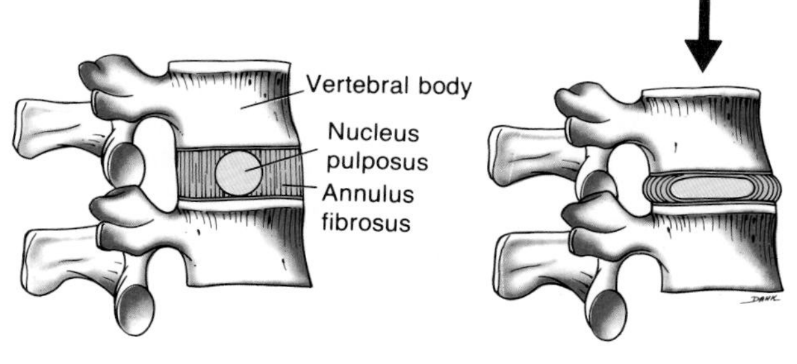

(c)

Normal Curves

When viewed from the side with the subject facing to the right, the vertebral column shows four ***normal curves*** (Figure 7-11b), two of which are convex), and two of which are concave (. The curves of the column, like the curves in a long bone, are important because they increase its strength. The curves also help maintain balance in the upright position, absorb shocks from walking, and help protect the column from fracture.

In the fetus, there is only a single anteriorly concave curve. At approximately the third postnatal month, when an infant begins to hold its head erect, the ***cervical curve*** develops. Later, when the child sits up, stands, and walks, the ***lumbar curve*** develops. The cervical and lumbar curves are anteriorly convex. Because they are modifications of the fetal positions, they are called ***secondary curves.*** The other two curves, the ***thoracic curve*** and the ***sacral curve,*** are anteriorly concave. Since they retain the anterior concavity of the fetus, they are referred to as ***primary curves.***

CLINICAL APPLICATION: GRAVITY INVERSION

In recent years, *gravity inversion* has been used by some people to decompress the backbone by using gravity and the body's own weight. This may be accomplished by hanging upside down by the ankles in gravity inversion boots suspended from a horizontal bar. Despite the popularity of this method of traction and exercise, there are some potentially harmful effects. For example, some healthy young individuals may experience increases in general blood pressure, pulse rate, intraocular pressure (IOP), and blood pressure in the eyes. Moreover, gravity inversion is not recommended for people with glaucoma, high blood pressure, heart disease, hiatal hernias, or disorders of the vertebral column.

Typical Vertebra

Although there are variations in size, shape, and detail in the vertebrae in different regions of the column, all the vertebrae are basically similar in structure (Figure 7-12). A typical vertebra consists of the following components.

1. The ***body (centrum)*** is the thick, disc-shaped anterior portion that is the weight-bearing part of a vertebra. Its superior and inferior surfaces are roughened for the attachment of intervertebral discs. The anterior and lateral surfaces contain nutrient foramina for blood vessels.

2. The ***vertebral (neural) arch*** extends posteriorly from the body of the vertebra. With the body of the vertebra, it surrounds the spinal cord. It is formed by two short, thick processes, the ***pedicles*** (PED-i-kuls), which project posteriorly from the body to unite with the laminae. The ***laminae*** (LAM-i-nē) are the flat parts that join to form the posterior portion of the vertebral arch. The space that lies between the vertebral arch and body contains the spinal cord. This space is known as the ***vertebral foramen.*** The vertebral foramina of all vertebrae together form the ***vertebral (spinal) canal.*** The pedicles are notched superiorly and inferiorly in such a way that when they are arranged in the column, there is an opening between ver-

FIGURE 7-12 Typical vertebra. (a) Superior view. (b) Right lateral view.

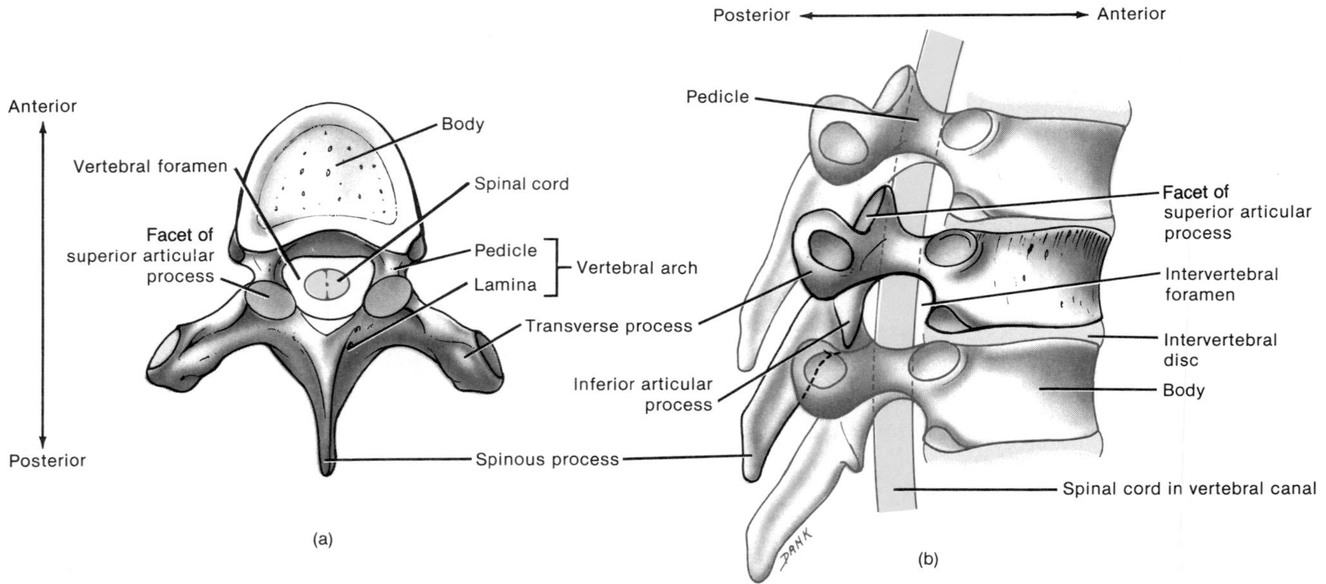

tebrae on each side of the column. This opening, the ***intervertebral foramen,*** permits the passage of a single spinal nerve.

3. Seven ***processes*** arise from the vertebral arch. At the point where a lamina and pedicle join, a ***transverse process*** extends laterally on each side. A single ***spinous process (spine)*** projects posteriorly and inferiorly from the junction of the laminae. These three processes serve as points of attachment for muscles. The remaining four processes form joints with other vertebrae. The two ***superior articular processes*** of a vertebra articulate with the vertebra immediately superior to them. The two ***inferior articular processes*** of a vertebra articulate with the vertebra inferior to them. The articulating surfaces of the articular processes are referred to as ***facets.***

Cervical Region

When viewed from above, it can be seen that the bodies of ***cervical vertebrae*** are smaller than those of thoracic vertebrae (Figure 7-13). The vertebral arches, however, are larger. The spinous processes of the second through sixth cervical vertebrae are often *bifid,* that is, with a cleft. All cervical vertebrae have three foramina: the vertebral foramen and two transverse foramina. Each cervical transverse process contains a ***transverse foramen*** through which the vertebral artery and its accompanying vein and nerve fibers pass.

The first two cervical vertebrae differ considerably from the others. The first cervical vertebra (C1), the ***atlas,*** is named for its support of the head. Essentially, the atlas is a ring of bone with ***anterior*** and ***posterior arches*** and large ***lateral masses.*** It lacks a body and a spinous process. The superior surfaces of the lateral masses, called ***superior articular facets,*** are concave and articulate with the occipital condyles of the occipital bone. This articulation permits the movement seen when nodding the head. The inferior surfaces of the lateral masses, the ***inferior articular facets,*** articulate with the second cervical vertebra. The transverse processes and transverse foramina of the atlas are quite large.

The second cervical vertebra (C2), the ***axis,*** does have a body. A peglike process called the ***dens*** (*dens* = tooth) projects up through the ring of the atlas. The dens makes a pivot on which the atlas and head rotate. This arrangement permits side-to-side rotation of the head. In various instances of trauma, the dens of the axis may be driven into the medulla oblongata of the brain. This injury is the usual cause of fatality in ***whiplash injuries*** that result in death.

The third through sixth cervical vertebrae (C3–C6) correspond to the structural pattern of the typical cervical vertebra previously described.

The seventh cervical vertebra (C7), called the ***vertebra prominens,*** is somewhat different. It is marked by a large, nonbifid spinous process that may be seen and felt at the base of the neck.

Thoracic Region

Viewing a typical ***thoracic vertebra*** from above, you can see that it is considerably larger and stronger than a vertebra of the cervical region (Figure 7-14). In addition, the spinous process on each vertebra is long, laterally flattened, and directed inferiorly. Thoracic vertebrae also have longer and heavier transverse processes than cervical vertebrae.

Except for the eleventh and twelfth thoracic vertebrae, the transverse processes have ***facets*** for articulating with the tubercles of the ribs. The bodies of thoracic vertebrae also have whole facets or half-facets, called ***demifacets,*** for articulation with the heads of the ribs. The first thoracic vertebra (T1) has, on either side of its body, a superior whole facet and an inferior demifacet. The superior facet articulates with the first rib, and the inferior demifacet, together with the superior demifacet of the second thoracic vertebra (T2), forms a facet for articulation with the second rib. The second through eighth thoracic vertebrae (T2–T8) have two demifacets on each side, a larger superior demifacet and a smaller inferior demifacet. When the vertebrae are articulated, they form whole facets for the heads of the ribs. The ninth thoracic vertebra (T9) has a single superior demifacet on either side of its body. The tenth through twelfth thoracic vertebrae (T10–T12) have whole facets on either side of their bodies.

Lumbar Region

The ***lumbar vertebrae*** (L1–L5) are the largest and strongest in the column (Figure 7-15). Their various projections are short and thick. The superior articular processes are directed medially instead of superiorly. The inferior articular processes are directed laterally instead of inferiorly. The spinous processes are quadrilateral in shape, thick, and broad and project nearly straight posteriorly. The spinous processes are well adapted for the attachment of the large back muscles.

Sacrum and Coccyx

The ***sacrum*** (= sacred or holy bone) is a triangular bone formed by the union of five sacral vertebrae. These are indicated in Figure 7-16 as S1–S5. Fusion begins between 16 and 18 years of age and is usually completed by the midtwenties. The sacrum serves as a strong foundation for the pelvic girdle. It is positioned at the posterior portion of the pelvic cavity between the two coxal (hip) bones.

The concave anterior side of the sacrum faces the pelvic cavity. It is smooth and contains four ***transverse lines (ridges)*** that mark the joining of the sacral vertebral bodies. At the ends of these lines are four pairs of ***anterior sacral (pelvic) foramina.*** The lateral portion of the superior surface contains a smooth surface called

FIGURE 7-13 Cervical vertebrae. (a) Photograph of a superior view of a typical cervical vertebrae. (b) Photograph of a superior view of the atlas. (c) Photograph of a superior view of the axis.

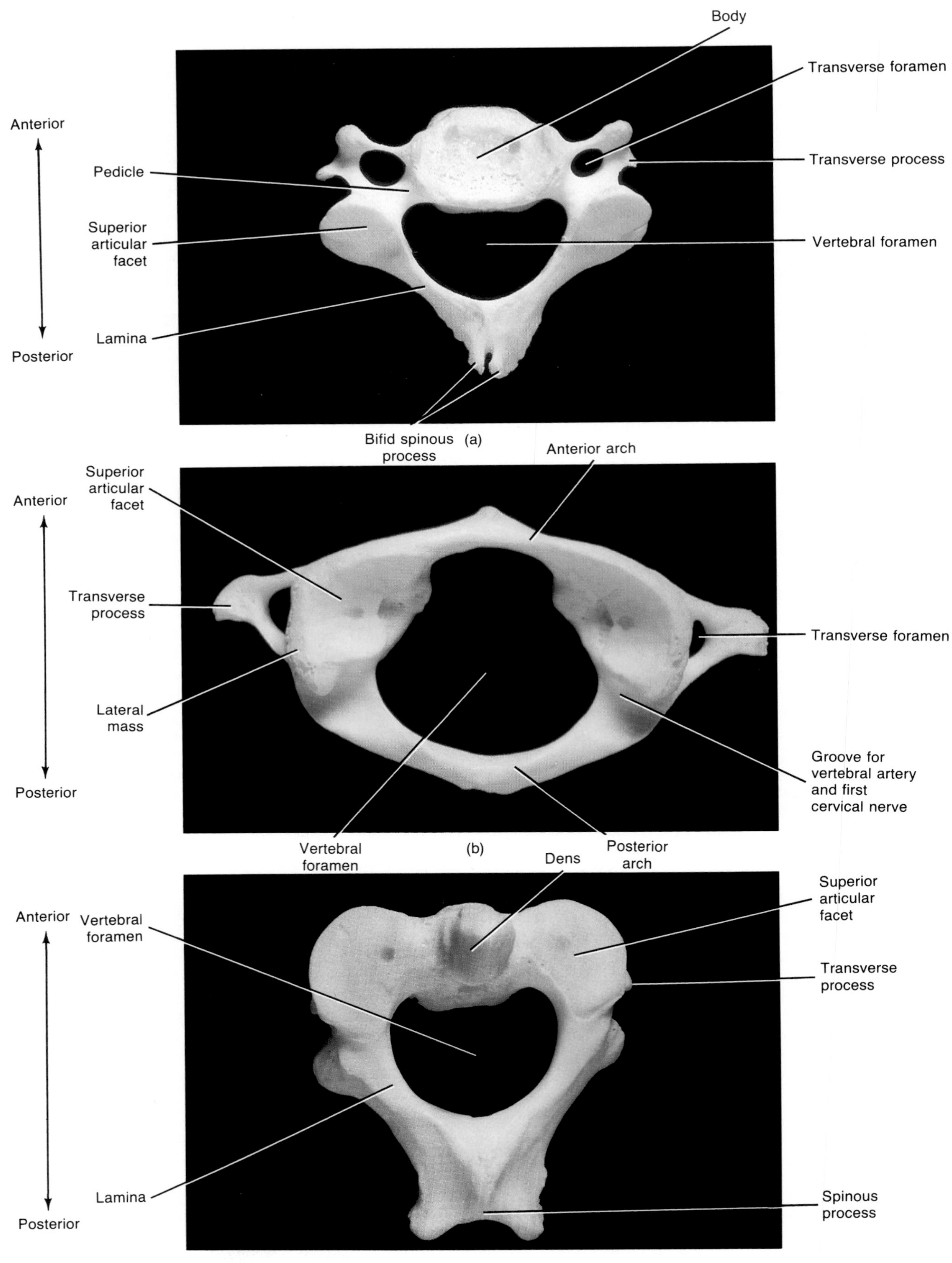

FIGURE 7-13 (*Continued*) (d) Photograph of the atlas and axis in posterior view. (e) Diagram of cervical vertebrae articulated in posterior view. (Photographs courtesy of J. A. Gosling, P. F. Harris, et al., *Atlas of Human Anatomy,* Gower Medical Publishing Ltd., 1985.)

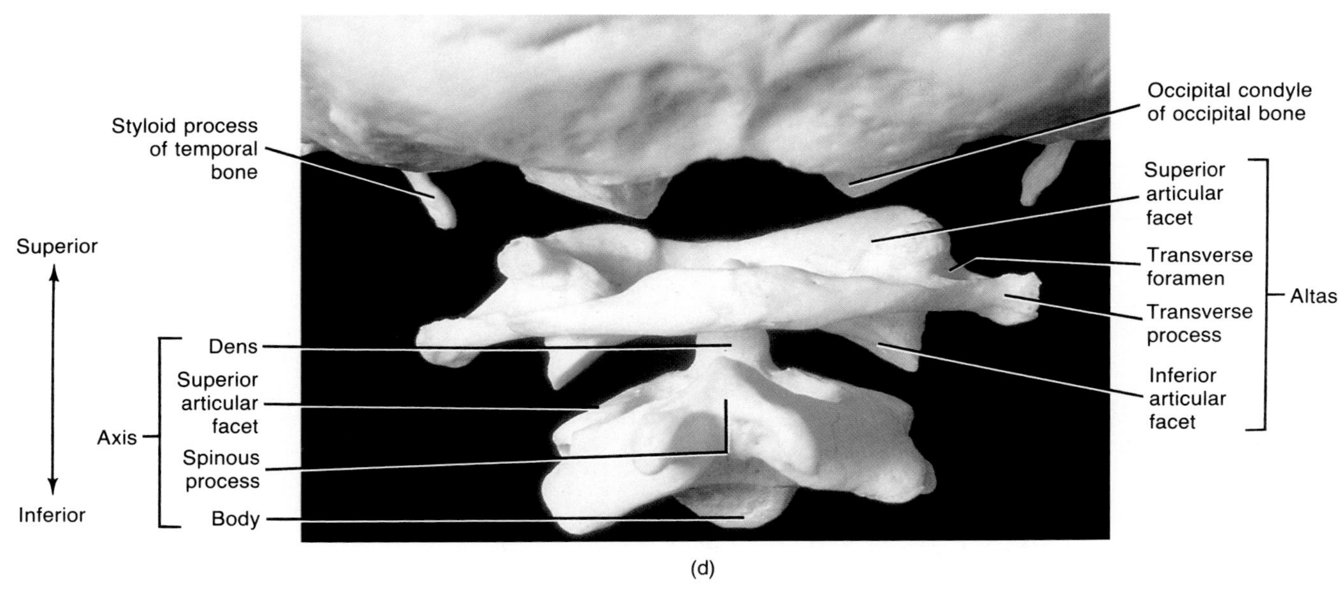

(d)

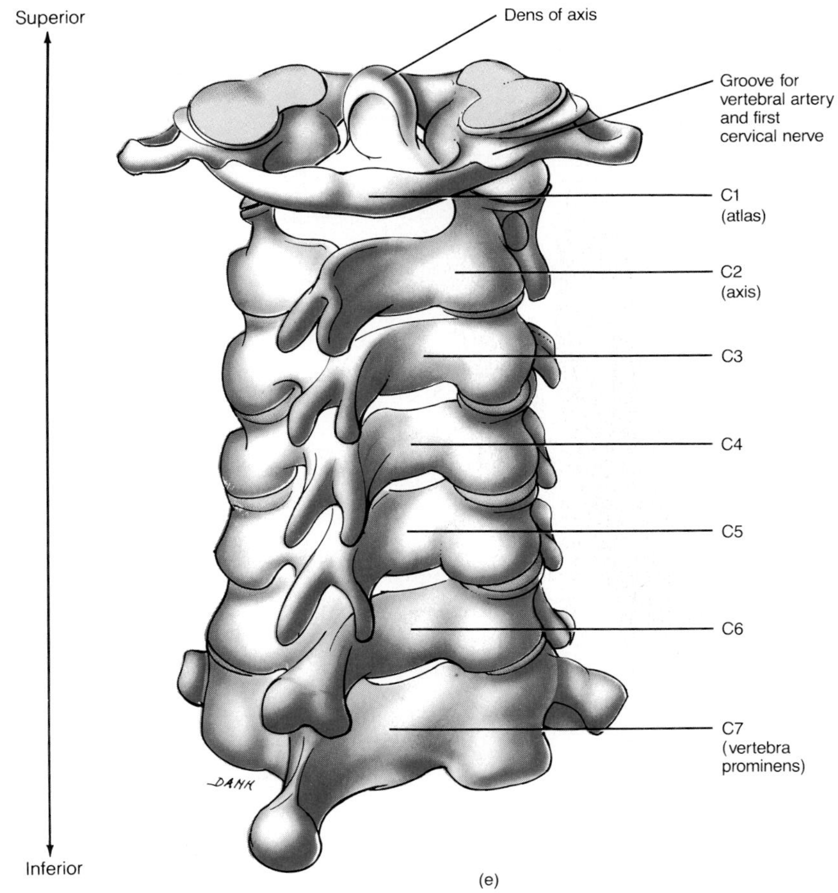

(e)

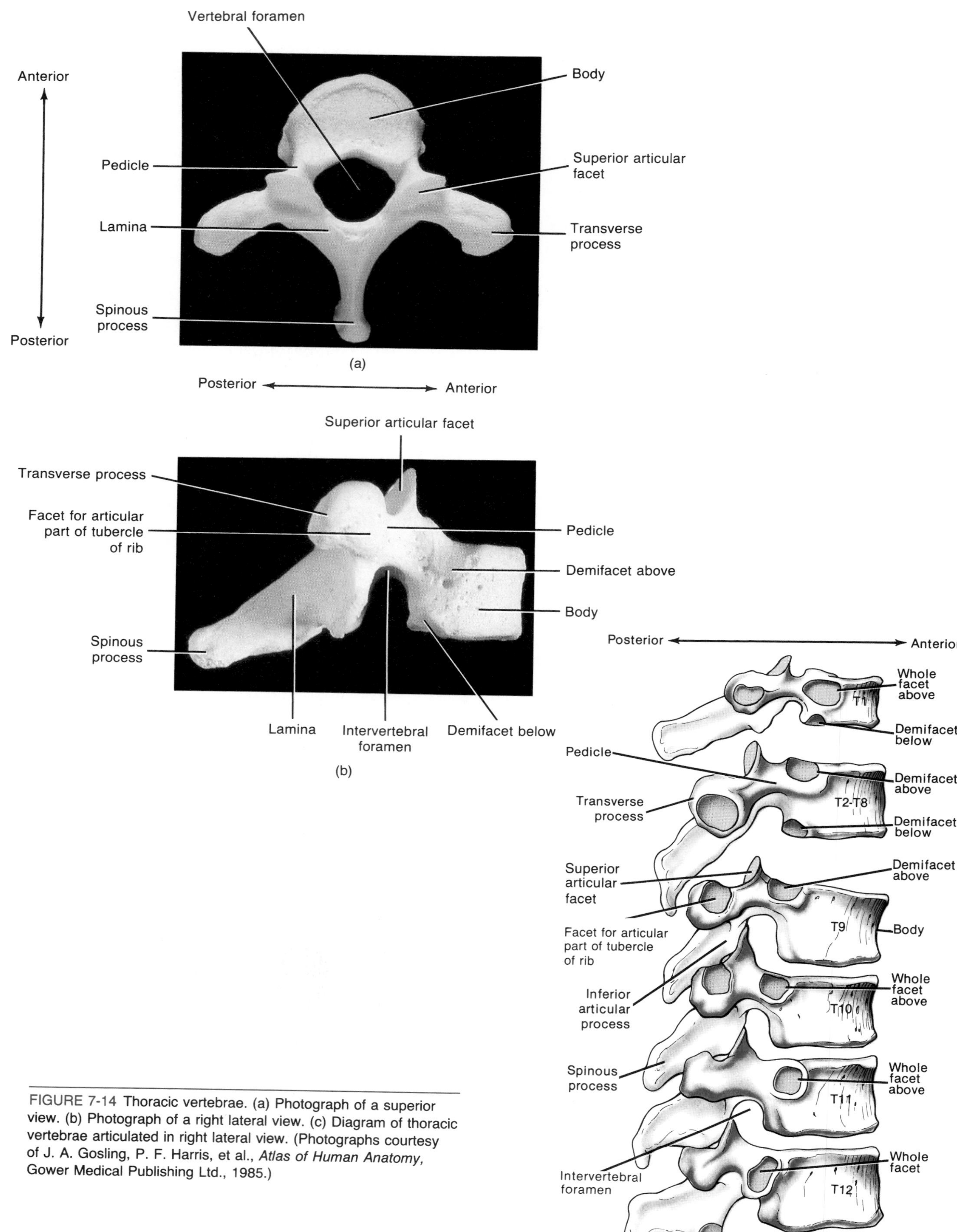

FIGURE 7-14 Thoracic vertebrae. (a) Photograph of a superior view. (b) Photograph of a right lateral view. (c) Diagram of thoracic vertebrae articulated in right lateral view. (Photographs courtesy of J. A. Gosling, P. F. Harris, et al., *Atlas of Human Anatomy,* Gower Medical Publishing Ltd., 1985.)

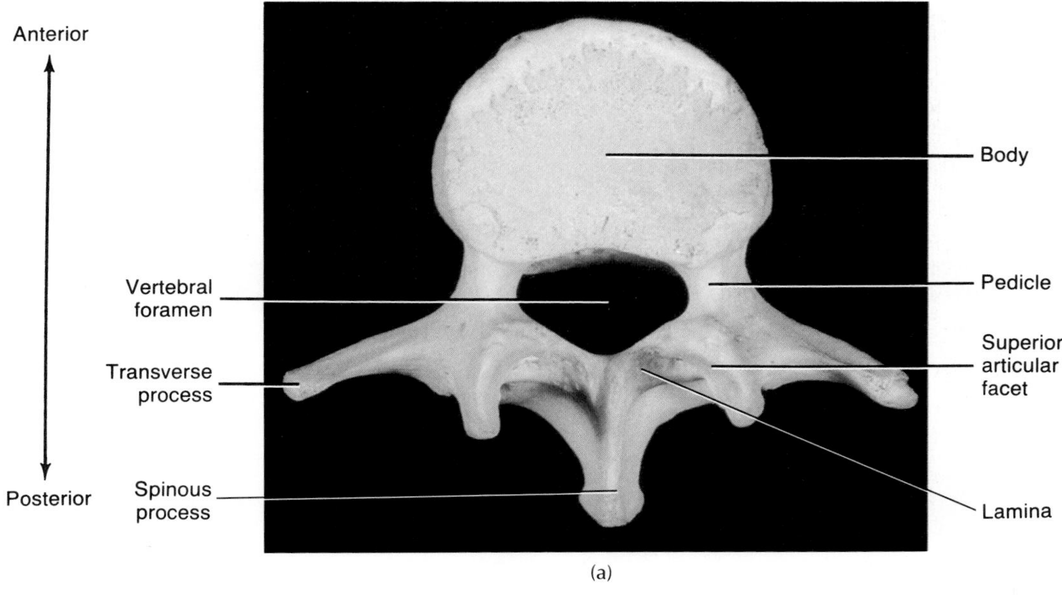

(a)

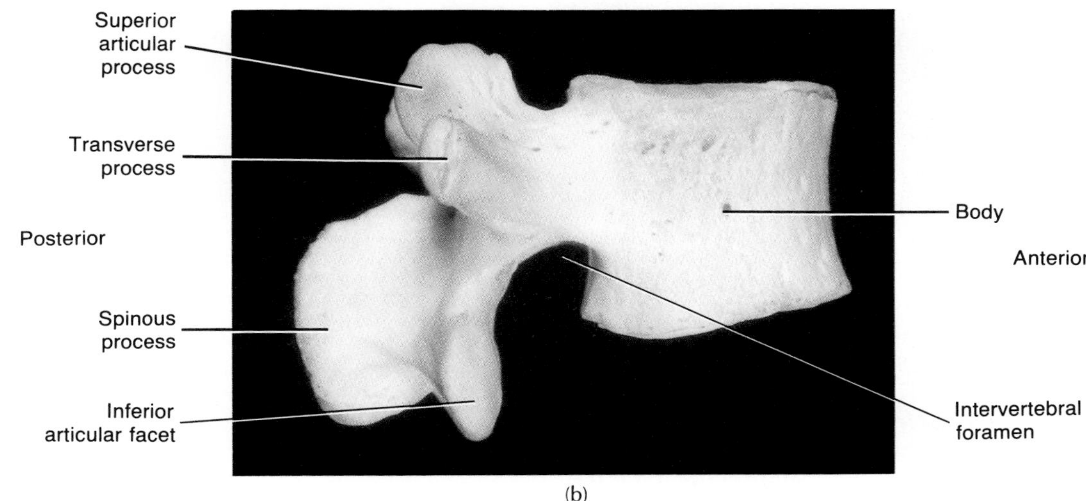

(b)

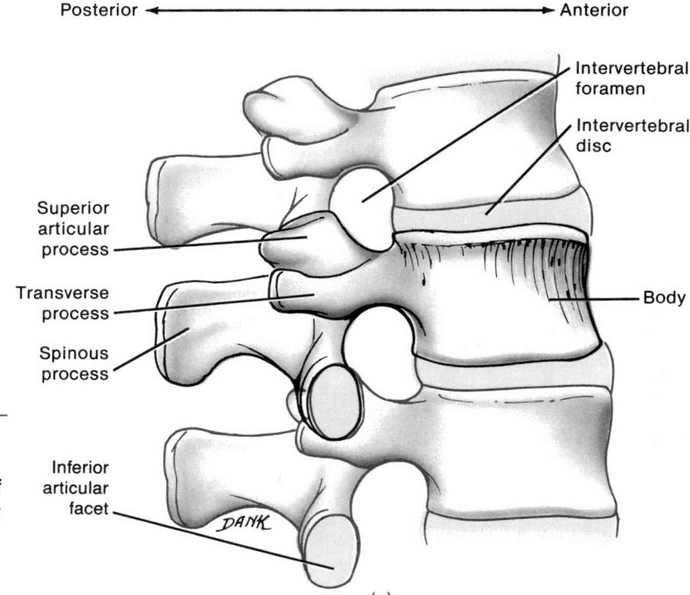

(c)

FIGURE 7-15 Lumbar vertebrae. (a) Photograph in superior view. (b) Photograph in right lateral view. (c) Diagram of lumbar vertebrae articulated in right lateral view. (Photographs courtesy of J. A. Gosling, P. F. Harris, et al., *Atlas of Human Anatomy*, Gower Medical Publishing Ltd., 1985.)

FIGURE 7-16 Sacrum and coccyx. (a) Anterior view. (b) Posterior view. (Photographs copyright © 1987 by Michael H. Ross. Used by permission.)

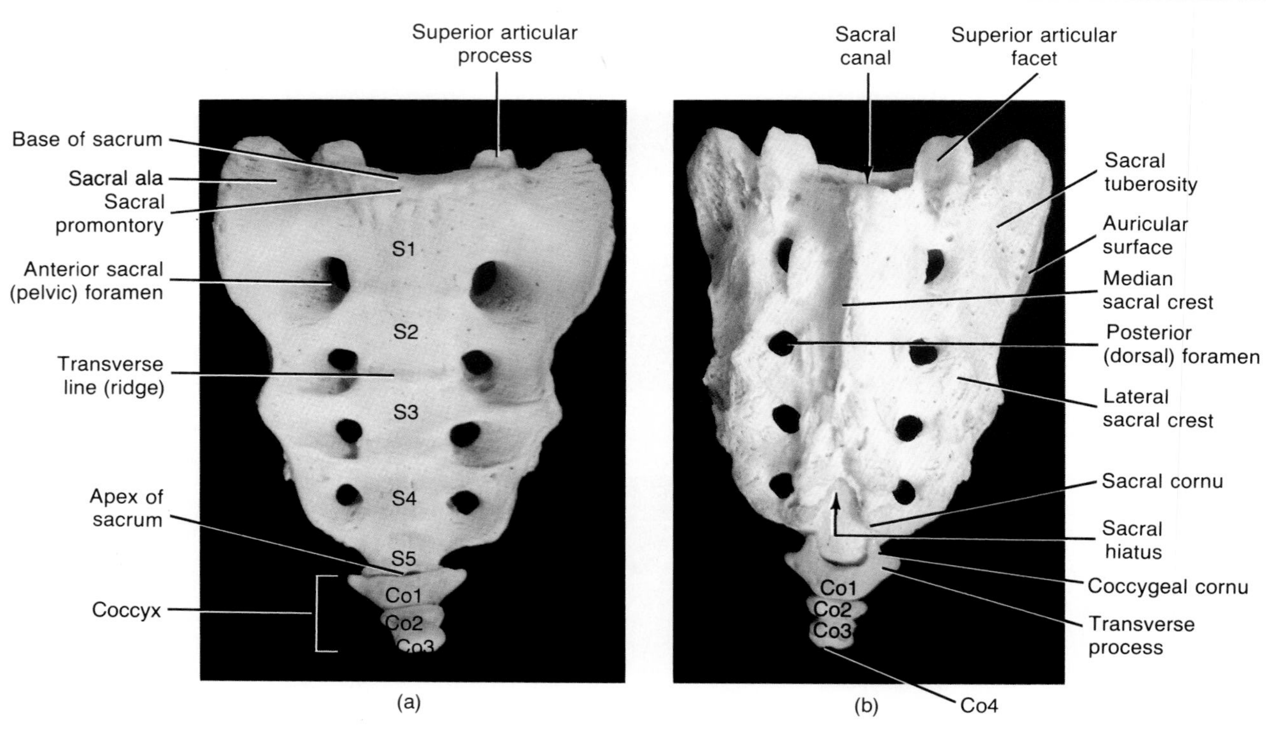

the ***ala*** (wing), which is formed by the transverse process of the first sacral vertebra (S1).

The convex, posterior surface of the sacrum is irregular. It contains a ***median sacral crest,*** the fused spinous processes of the upper sacral vertebrae; a ***lateral sacral crest,*** the transverse processes of the sacral vertebrae; and four pairs of ***posterior sacral (dorsal) foramina.*** These foramina communicate with the anterior sacral foramina through which nerves and blood vessels pass. The ***sacral canal*** is a continuation of the vertebral canal. The laminae of the fifth sacral vertebra, and sometimes the fourth, fail to meet. This leaves an inferior entrance to the vertebral canal called the ***sacral hiatus*** (hī-Ā-tus). On either side of the sacral hiatus are the ***sacral cornua,*** the inferior articular processes of the fifth sacral vertebra. They are connected by ligaments to the coccygeal cornua of the coccyx.

CLINICAL APPLICATION: CAUDAL ANESTHESIA

Anesthetic agents that act on the sacral and coccygeal nerves are sometimes injected through the sacral hiatus, a procedure called *caudal anesthesia* which is used most frequently in obstetrics. Since the sacral hiatus is between the sacral cornua, the cornua are important bony landmarks for locating the hiatus. Anesthetic agents may also be injected through the posterior sacral (dorsal) foramina.

The superior border of the sacrum exhibits an anteriorly projecting border, the ***sacral promontory*** (PROM-on-tō'-rē). It is an obstetrical landmark for measurements of the pelvis. Laterally, the sacrum has a large ***auricular surface*** for articulating with the ilium of the coxal (hip) bone. Posterior to the auricular surface is a roughened surface, the ***sacral tuberosity,*** that contains depressions for the attachment of ligaments. The sacral tuberosity is another surface of the sacrum that unites with the coxal (hip) bone to form the sacroiliac joint. The ***superior articular processes*** of the sacrum articulate with the fifth lumbar vertebra.

The ***coccyx*** is also triangular in shape and is formed by the fusion of the coccygeal vertebrae, usually the last four. These are indicated in Figure 7-16 as Co1–Co4. Fusion generally occurs between 20 and 30 years. The dorsal surface of the body of the coccyx contains two long ***coccygeal cornua*** that are connected by ligaments to the sacral cornua. The coccygeal cornua are the pedicles and superior articular processes of the first coccygeal vertebra. On the lateral surfaces of the body of the coccyx are a series of ***transverse processes,*** the first pair being the largest. The coccyx articulates superiorly with the sacrum.

THORAX

Anatomically, the term ***thorax*** refers to the chest. The skeletal portion of the thorax is a bony cage formed by the sternum, costal cartilage, ribs, and the bodies of the thoracic vertebrae (Figure 7-17).

The thoracic cage is roughly cone-shaped, the narrow portion being superior and the broad portion inferior. It is flattened from front to back. The thoracic cage encloses and protects the organs in the thoracic cavity and upper abdomen. It also provides support for the bones of the shoulder girdle and upper extremities.

Sternum

The ***sternum,*** or breastbone, is a flat, narrow bone measuring about 15 cm (6 inches) in length. It is located in the median line of the anterior thoracic wall.

The sternum (see Figure 7-17) consists of three basic portions: the ***manubrium*** (ma-NOO-brē-um), the superior portion; the ***body,*** the middle, largest portion; and the ***xiphoid*** (ZĪ-foyd) ***process,*** the inferior, smallest portion. The junction of the manubrium and body forms the ***sternal angle.*** The manubrium has a depression on its superior surface called the ***jugular (suprasternal) notch.*** On each side of the jugular notch are ***clavicular notches*** that articulate with the medial ends of the clavicles. The manubrium also articulates with the first and second ribs. The body of the sternum articulates directly or indirectly with the second through tenth ribs. The xiphoid process has no ribs attached to it but provides attachment for some abdominal muscles. The xiphoid process consists of hyaline cartilage during infancy and childhood and does not ossify completely until about age 40. If the hands of a rescuer are mispositioned during cardiopulmonary resuscitation (CPR), there is danger of fracturing the ossified xiphoid process, separating it from the body, and driving it into the liver.

FIGURE 7-17 Skeleton of the thorax in anterior view. (Courtesy of J. A. Gosling, P. F. Harris, et al., *Atlas of Human Anatomy,* Gower Medical Publishing Ltd., 1985.)

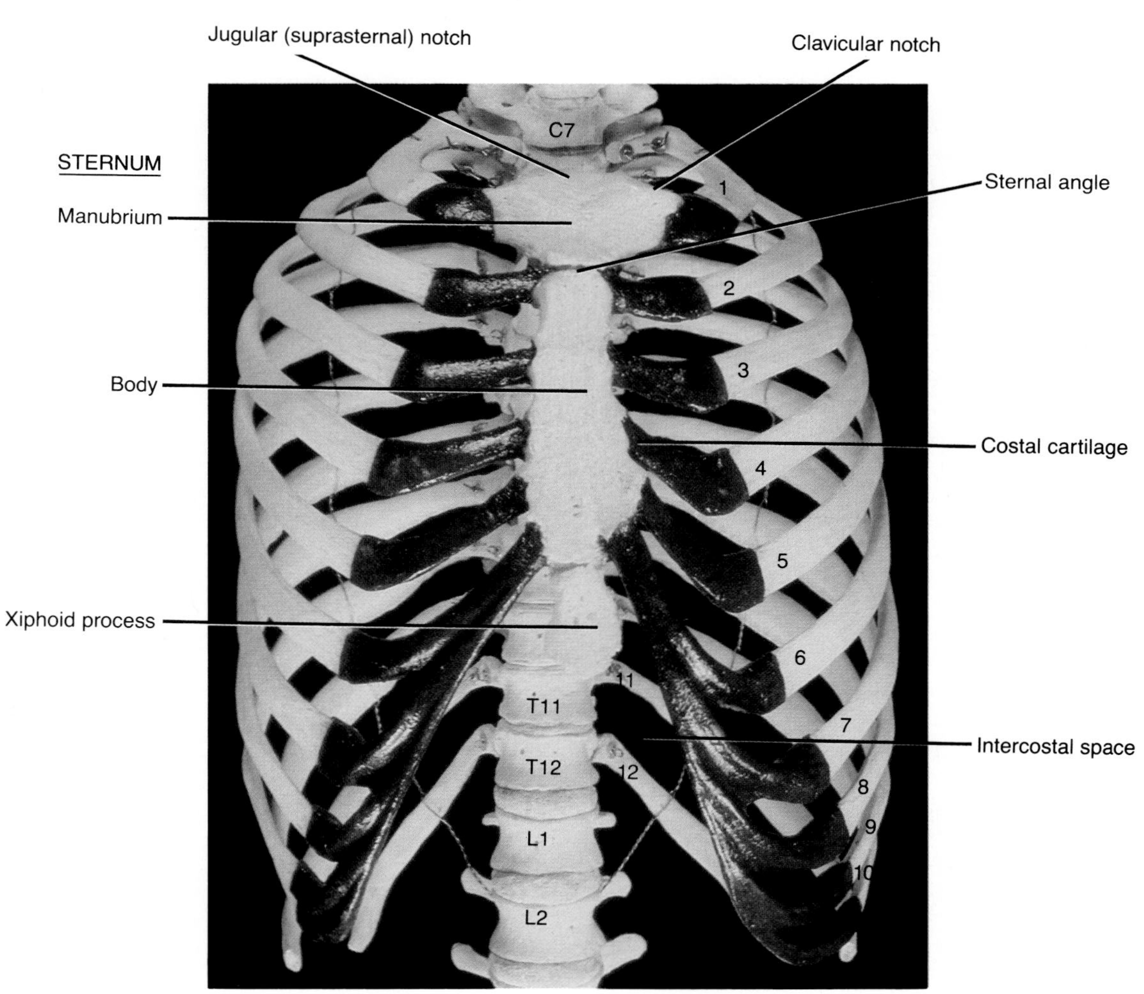

CLINICAL APPLICATION: STERNAL PUNCTURE

Since the sternum possesses red bone marrow throughout life and because it is readily accessible and has thin compact bone, it is a common site for withdrawal of marrow for biopsy (***marrow aspiration***). Under a local anesthetic, a wide-bore needle is introduced into the marrow cavity of the sternum for aspiration of a sample of red bone marrow. This procedure is called a ***sternal puncture.***

The sternum may also be split in the midsagittal plane to allow surgeons access to mediastinal structures such as the thymus gland, heart, and great vessels of the heart.

Ribs

Twelve pairs of ***ribs*** make up the sides of the thoracic cavity (Figure 7-17). The ribs increase in length from the first through seventh. Then they decrease in length to the twelfth rib. Each rib articulates posteriorly with its corresponding thoracic vertebra. In order to number the ribs anteriorly, count downward from the second costal cartilage, which articulates at the angle of the sternal angle.

The first through seventh ribs have a direct anterior attachment to the sternum by a strip of hyaline cartilage called ***costal cartilage*** (*costa* = rib). These ribs are called ***true (vertebrosternal) ribs.*** The remaining five pairs of ribs are referred to as ***false ribs*** because their costal cartilages do not attach directly to the sternum. The cartilages of the eighth, ninth, and tenth ribs attach to each other and then to the cartilage of the seventh rib. These false ribs are called ***vertebrochondral ribs.*** The eleventh and twelfth false ribs are designated as ***floating (vertebral) ribs*** because their anterior ends do not attach even indirectly to the sternum. They attach only posteriorly to the thoracic vertebrae.

Although there is some variation in rib structure, we will examine the parts of a typical (third through ninth) rib when viewed from the right side and from behind (Figure 7-18). The ***head*** of a typical rib is a projection at the posterior end of the rib. It is wedge-shaped and consists of one or two ***facets*** that articulate with facets on the bodies of adjacent thoracic vertebrae. The facets on the head of a rib are separated by a horizontal ***interarticular crest.*** The inferior facet on the head of a rib is larger than the superior facet. The ***neck*** is a constricted portion just lateral to the head. A knoblike structure on the posterior surface where the neck joins the body is called a ***tubercle*** (TOO-ber-kul). It consists of a ***nonarticular part*** that affords attachment to the ligament of the tubercle and an ***articular part*** that articulates with the facet of a transverse process of the inferior of the two vertebrae to which the head of the rib is connected. The ***body (shaft)*** is the main part of the rib. A short distance beyond the tubercle, there is an abrupt change in the curvature of the shaft. This point is called the ***costal angle.*** The inner surface of the rib has a ***costal groove*** that protects blood vessels and a small nerve.

CLINICAL APPLICATION: RIB FRACTURES

Rib fractures represent the most common chest injuries and usually result from direct blows, most commonly from steering wheel impact, falls, and crushing injuries to the chest. Ribs tend to break at the point where greatest force is applied, but may also break at the weakest point, that is, the site of greatest curvature which is just anterior to the costal angle. In children, the ribs are highly elastic, and fractures are less frequent than in adults. Since the first two ribs are protected by the clavicle and pectoralis major muscle, and the last two ribs are mobile, they are the least commonly injured. The middle ribs are the ones most commonly fractured. In some cases, fractured ribs may cause damage to the heart, great vessels of the heart, lungs, trachea, bronchi, esophagus, spleen, liver, and kidneys.

The posterior portion of the rib is connected to a thoracic vertebra by its head and articular part of a tubercle. The facet of the head fits into a facet on the body of a vertebra, and the articular part of the tubercle articulates with the facet of the transverse process of the vertebra. Each of the second through ninth ribs articulates with the bodies of two adjacent vertebrae. The first, tenth, eleventh, and twelfth ribs articulate with only one vertebra each. On the eleventh and twelfth ribs, there is no articulation between the tubercles and the transverse processes of their corresponding vertebrae.

Spaces between ribs, called ***intercostal spaces,*** are occupied by intercostal muscles, blood vessels, and nerves (see Figure 7-17).

Surgical access to the lungs or structures in the mediastinum is commonly undertaken through an intercostal space. Special rib retractors are used to create a wide separation between ribs. The costal cartilages are sufficiently elastic to permit considerable bending.

FIGURE 7-18 **Typical rib. A left rib viewed from below and behind.**

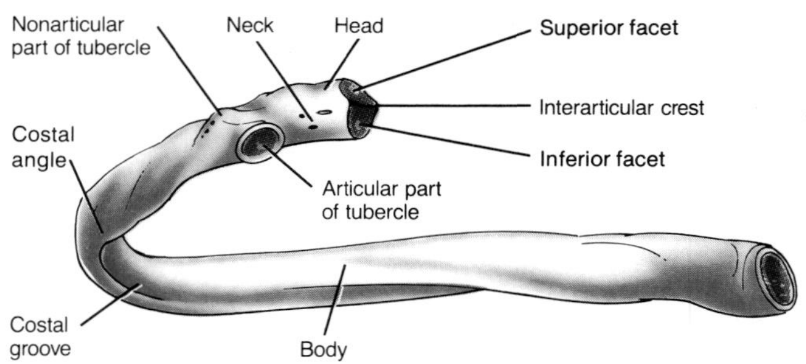

DISORDERS: HOMEOSTATIC IMBALANCES

Herniated (Slipped) Disc

In their function as shock absorbers, intervertebral discs are subject to compressional forces. The discs between the fourth and fifth lumbar vertebrae and between the fifth lumbar vertebra and sacrum usually are subject to more forces than other discs. If the anterior and posterior ligaments of the discs become injured or weakened, the pressure developed in the nucleus pulposus may be great enough to rupture the surrounding fibrocartilage. If this occurs, the nucleus pulposus may herniate (protrude) posteriorly or into one of the adjacent vertebral bodies. This condition is called a ***herniated (slipped) disc*** or ***herniated intervertebral disc (HIVD)*** or ***herniated nucleus pulposus (HNP).***

Most often the nucleus pulposus slips posteriorly toward the spinal cord and spinal nerves (Figure 7-19). This movement exerts pressure on the spinal nerves, causing considerable, sometimes very acute, pain. If the roots of the sciatic nerve, which passes from the spinal cord to the foot, are pressured, the pain radiates down the back of the thigh, through the calf, and occasionally into the foot. If pressure is exerted on the spinal cord itself, nervous tissue may be destroyed.

Traction, bed rest, and analgesia usually relieve the pain. If such treatment is ineffective, surgical decompression of the spinal nerves or removal of some of the nucleus pulposus may be necessary to relieve pain. Removal may involve conventional surgery or, for selected patients, ***chemonucleolysis*** (kē'-mō-noo'-klē-OL-i-sis), the use of a proteolytic enzyme called chymopapain (Chymodiactin), the active ingredient in meat tenderizer. This enzyme is injected into a herniated disc where it dissolves the nucleus pulposus, thus relieving pressure on spinal nerves and pain. Although chemonucleolysis provides good short-term results, long-term results are poor. Another procedure, called ***percutaneous lumbar diskectomy with aspiration probe,*** can also be used on selected patients. This procedure involves the use of an automated needlelike probe (nucleosome) that cuts away and removes disc material. The probe is inserted through the skin and guided by special x-ray equipment with an image projected on a TV screen. The nucleus pulposus is aspirated through the probe.

FIGURE 7-19 Superior view of a herniated (slipped) disc.

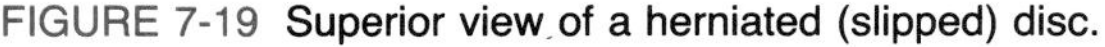

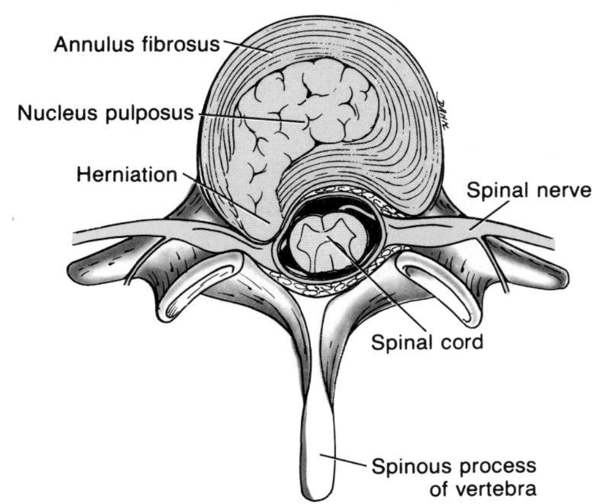

Abnormal Curves

As a result of various conditions, the normal curves of the vertebral column may become exaggerated, or the column may acquire a lateral bend, resulting in ***abnormal curves*** of the spine.

Scoliosis (skō'-lē-Ō-sis; *scolio* = bent) is a lateral bending of the vertebral column, usually in the thoracic region. This is the most common of the abnormal curves. It may be caused by a congenital condition in which vertebrae are malformed, chronic sciatica, paralysis of muscles on one side of the backbone, poor posture, and one leg being shorter than the other.

Until recently, a child with progressive scoliosis faced either years of treatment with a brace or major corrective surgery. Several research teams are experimenting with using electrical stimulation of muscles to limit the progression of scoliosis and even to reduce curvatures in some cases. The skeletal muscles on the convex side of the spinal curve are electrically stimulated.

Kyphosis (kī-FŌ-sis; *kypho* = hunchback) is an exaggeration of the thoracic curve of the vertebral column. In tuberculosis of the spine, vertebral bodies may partially collapse, causing an acute angular bending of the vertebral column. In the elderly, degeneration of the intervertebral discs leads to kyphosis. Kyphosis may also be caused by rickets and poor posture. The term *round-shouldered* is an expression for mild kyphosis. Electrical muscle stimulation is being studied to assess its effects on kyphosis as well as on scoliosis.

Lordosis (lor-DŌ-sis; *lordo* = swayback) is an exaggeration of the lumbar curve of the vertebral column. It may result from increased weight of abdominal contents as in pregnancy or extreme obesity, poor posture, rickets, and tuberculosis of the spine.

Spina Bifida

Spina bifida (SPĪ-na BIF-i-da) is a congenital defect of the vertebral column in which laminae fail to unite at the midline. The lumbar vertebrae are involved in about 50 percent of all cases. In less serious cases, the defect is small and the area is covered with skin. Only a dimple or tuft of hair may mark the site. In such cases, symptoms are mild, with perhaps intermittent urinary problems. An operation is not ordinarily required. Larger defects in the vertebral arches with protrusion of the membranes (meninges) around the spinal cord or the spinal cord itself produce serious problems, such as partial or complete paralysis, partial or complete loss of urinary bladder control, and the absence of reflexes. Spina bifida may be diagnosed prenatally by a test of the mother's blood, sonography, and/or amniocentesis.

DISORDERS: HOMEOSTATIC IMBALANCES (*continued*)

Fractures of the Vertebral Column

Fractures of the vertebral column most commonly involve T5, T6, and T9–L2. Thoracic fractures usually result from a flexion-compression type of injury such as might be sustained in landing on the feet or buttocks after a fall from a height or having a heavy weight fall on the shoulders. Cervical vertebrae may be fractured or, more commonly, dislodged by a fall on the head with acute flexion of the neck, as might happen on diving into shallow water. Dislocation may even result from the sudden forward jerk ("whiplash") that may occur when an automobile or airplane crashes. Spinal nerve damage may occur as a result of fractures of the vertebral column.

STUDY OUTLINE

Types of Bones (p. 160)

1. On the basis of shape, bones are classified as long, short, flat, or irregular.
2. Sutural (Wormian) bones are found between the sutures of certain cranial bones. Sesamoid bones develop in tendons or ligaments.

Surface Markings (p. 160)

1. Surface markings are structural features visible on the surfaces of bones.
2. Each marking is structured for a specific function—joint formation, muscle attachment, or passage of nerves and blood vessels.
3. Terms that describe markings include fissure, foramen, meatus, fossa, process, condyle, head, facet, tuberosity, crest, and spine.

Divisions of the Skeletal System (p. 161)

1. The axial skeleton consists of bones arranged along the longitudinal axis. The parts of the axial skeleton are the skull, hyoid bone, vertebral column, sternum, and ribs.
2. The appendicular skeleton consists of the bones of the girdles and the upper and lower extremities. The parts of the appendicular skeleton are the pectoral (shoulder) girdles, bones of the upper extremities, pelvic (hip) girdle, and bones of the lower extremities.

Skull (p. 162)

1. The skull consists of the cranium and the face. It is composed of 22 bones.
2. Sutures are immovable joints between bones of the skull. Examples are coronal, sagittal, lambdoidal, and squamosal sutures.
3. Fontanels are dense connective tissue membrane-filled spaces between the cranial bones of fetuses and infants. The major fontanels are the anterior, posterior, anterolaterals, and posterolaterals.
4. The 8 cranial bones include the frontal, parietal (2), temporal (2), occipital, sphenoid, and ethmoid.
5. The 14 facial bones are the nasal (2), maxillae (2), zygomatic (2), mandible, lacrimal (2), palatine (2), inferior nasal conchae (2), and vomer.
6. Paranasal sinuses are cavities in bones of the skull that communicate with the nasal cavity. They are lined by mucous membranes. The cranial bones containing the paranasal sinuses are the frontal, sphenoid, ethmoid, and maxillae.
7. The orbits (eye sockets) are formed by seven bones of the skull.
8. The foramina of the skull bones provide passages for nerves and blood vessels.

Hyoid Bone (p. 175)

1. The hyoid bone is a U-shaped bone that does not articulate with any other bone.
2. It supports the tongue and provides attachment for some of its muscles as well as some neck muscles and muscles of the pharynx.

Vertebral Column (p. 175)

1. The vertebral column, sternum, and ribs constitute the skeleton of the trunk.
2. The bones of the adult vertebral column are the cervical vertebrae (7), thoracic vertebrae (12), lumbar vertebrae (5), the sacrum (5, fused), and the coccyx (4, fused).
3. The vertebral column contains normal primary curves (thoracic and sacral) and normal secondary curves (cervical and lumbar). These curves give strength, support, and balance.
4. The vertebrae are similar in structure, each consisting of a body, vertebral arch, and seven processes. Vertebrae in the different regions of the column vary in size, shape, and detail.

Thorax (p. 185)

1. The thoracic skeleton consists of the sternum, ribs and costal cartilages, and thoracic vertebrae.
2. The thoracic cage protects vital organs in the chest area and upper abdomen.

Disorders: Homeostatic Imbalances (p. 187)

1. Protrusion of the nucleus pulposus of an intervertebral disc posteriorly or into an adjacent vertebral body is called a herniated (slipped) disc.

2. Exaggeration of a normal curve or lateral bending of the vertebral column results in an abnormal curve. Examples include scoliosis, kyphosis, and lordosis.
3. The imperfect union of the vertebral laminae at the midline, a congenital defect, is referred to as spina bifida.
4. Fractures of the vertebral column most often involve T5, T6, and T9–L2.

REVIEW QUESTIONS

1. What are the four principal types of bones? Give an example of each. Distinguish between a sutural (Wormian) and a sesamoid bone. (p. 160)
2. What are surface markings? Describe and give an example of each. (p. 160)
3. Distinguish between the axial and appendicular skeletons. What subdivisions and bones are contained in each? (p. 161)
4. What are the bones that compose the skull? The cranium? The face? (p. 162)
5. Define a suture. What are the four prominent sutures of the skull? Where are they located? (p. 162)
6. What is a fontanel? Describe the location of the six fairly constant fontanels. (p. 162)
7. What is a paranasal sinus? What cranial bones contain paranasal sinuses? (p. 173)
8. Describe the components of each orbit. (p. 174)
9. What bones form the skeleton of the trunk? Distinguish between the number of nonfused vertebrae found in the adult vertebral column and that of a child. (p. 175)
10. What are the normal curves in the vertebral column? How are primary and secondary curves differentiated? What are the functions of the curves? (p. 178)
11. What are the principal distinguishing characteristics of the bones of the various regions of the vertebral column? (p. 179)
12. What bones form the skeleton of the thorax? What are the functions of the thoracic skeleton? (p. 185)
13. How are ribs classified on the basis of their attachment to the sternum? (p. 186)
14. What is a herniated (slipped) disc? Why does it cause pain? How is it treated? (p. 187)
15. What is an abnormal curve? Describe the symptoms and causes of scoliosis, kyphosis, and lordosis. (p. 187)
16. Define spina bifida. (p. 187)
17. What are some common causes of fractures of the vertebral column? (p. 188)
18. Define the following: black eye (p. 166), cleft palate and cleft lip (p. 172), sinusitis (p. 173), temporomandibular joint (TMJ) syndrome (p. 174), deviated nasal septum (DNS) (p. 174), gravity inversion (p. 178), caudal anesthesia (p. 184), sternal puncture (p. 186), and fractured rib (p. 186).

SELECTED READINGS

Cristensen, J. B., and I. R. Telford. *Synopsis of Gross Anatomy*, 5th ed. Philadelphia: Lippincott, 1988.

Gosling, J. A., P. F. Harris, J. R. Humpherson, I. Whitmore, and P. L. T. Willon. *Atlas of Human Anatomy*. Philadelphia: Lippincott, 1985.

Netter, F. H. *Musculoskeletal System, Part I, Anatomy, Physiology, and Metabolic Disorders*. Summit, N.J.: CIBA, 1987.

Rohen, J. W., and C. Yokochi. *Color Atlas of Anatomy*. Tokyo, New York: Igaku-Shoin, Ltd., 1983.

Shipman, P., A. Walker, and D. Bichell. *The Human Skeleton*. Cambridge: Harvard University Press, 1985.

Steele, D. G., and C. A. Bramblett. *The Anatomy and Biology of the Human Skeleton*. College Station: Texas A&M University Press, 1988.

Chapter 8

The Skeletal System: The Appendicular Skeleton

Chapter Contents at a Glance

Student Objectives

1. Identify the bones of the pectoral (shoulder) girdle and their major markings.
2. Identify the upper extremity, its component bones, and their markings.
3. Identify the components of the pelvic (hip) girdle and their principal markings.
4. Identify the lower extremity, its component bones, and their markings.
5. Define the structural features and importance of the arches of the foot.
6. Compare the principal structural differences between female and male skeletons, especially those that pertain to the pelvis.

This chapter discusses the bones of the appendicular skeleton, that is, the bones of the pectoral (shoulder) and pelvic (hip) girdles and upper and lower extremities. The differences between female and male skeletons are also compared.

PECTORAL (SHOULDER) GIRDLE

The ***pectoral*** (PEK-tō-ral) or ***shoulder girdles*** attach the bones of the upper extremities to the axial skeleton (Figure 8-1). Each of the two pectoral girdles consists of two bones: a clavicle and a scapula. The clavicle is the anterior component and articulates with the sternum at the sternoclavicular joint. The posterior component, the scapula, which is positioned freely by complex muscle attachments, articulates with the clavicle and humerus. The pectoral girdles have no articulation with the vertebral column. Although the shoulder joints are not very stable, they are freely movable and thus allow movement in many directions.

Clavicle

The ***clavicles*** (KLAV-i-kuls), or ***collarbones,*** are long, slender bones with a double curvature (Figure 8-2). The medial one-third of the clavicle is convex anteriorly, whereas the lateral one-third is concave anteriorly. The clavicles lie horizontally in the superior and anterior part of the thorax superior to the first rib.

The medial end of the clavicle, the ***sternal extremity,*** is rounded and articulates with the sternum. The broad, flat, lateral end, the ***acromial*** (a-KRŌ-mē-al) ***extremity,*** articulates with the acromion of the scapula. This joint is called the ***acromioclavicular joint.*** (Refer to Figure 8-1 for a view of these articulations.) The ***conoid tubercle*** on the inferior surface of the lateral end of the bone serves as a point of attachment for a ligament. The ***costal tuberosity*** on the inferior surface of the medial end also serves as a point of attachment for a ligament.

CLINICAL APPLICATION: FRACTURED CLAVICLE

Because of its position, the clavicle transmits forces from the upper extremity to the trunk. If such forces are excessive, as in falling on one's outstretched arm, a *fractured clavicle* may result. In fact, it is the most frequently broken bone in the body. The junction of the two curves is the weakest point of the clavicle and the site of most fractures.

Scapula

The ***scapulae*** (SCAP-yoo-lē), or ***shoulder blades,*** are large, triangular, flat bones situated in the posterior part of the thorax between the levels of the second and seventh ribs (Figure 8-3). Their medial borders are located about 5 cm (2 inches) from the vertebral column.

A sharp ridge, the ***spine,*** runs diagonally across the posterior surface of the flattened, triangular ***body.*** The end of the spine projects as a flattened, expanded process called the ***acromion*** (a-KRŌ-mē-on), easily felt as the high point of the shoulder. This process articulates with the clavicle. Inferior to the acromion is a depression called the ***glenoid cavity (fossa).*** This cavity articulates with the head of the humerus to form the shoulder joint.

The thin edge of the body near the vertebral column is the ***medial (vertebral) border.*** The thick edge closer to the arm is the ***lateral (axillary) border.*** The medial and lateral borders join at the ***inferior angle.*** The superior edge of the scapular body, called the ***superior border,*** joins the vertebral border at the ***superior angle.*** The ***scapular notch*** is a prominent indentation along the superior border near the coracoid process; the notch permits passage of the suprascapular nerve.

At the lateral end of the superior border is a projection of the anterior surface called the ***coracoid*** (KOR-a-koyd) ***process*** to which muscles attach. Above and below the spine are two fossae: the ***supraspinous*** (soo'-pra-SPĪ-nus) ***fossa*** and the ***infraspinous fossa,*** respectively. Both serve as surfaces of attachment for shoulder muscles. On the ventral (costal) surface is a lightly hollowed-out area called the ***subscapular fossa,*** also a surface of attachment for shoulder muscles.

UPPER EXTREMITY

The ***upper extremities*** consist of 60 bones. The skeleton of the right upper extremity is shown in Figure 8-1. Each upper extremity includes a humerus in the arm, ulna and radius in the forearm, carpals (wrist bones), metacarpals (palm bones), and phalanges in the fingers of the hand.

Humerus

The ***humerus*** (HYOO-mer-us), or arm bone, is the longest and largest bone of the upper extremity (Figure 8-4). It articulates proximally with the scapula and distally at the elbow with both ulna and radius.

The proximal end of the humerus consists of a ***head*** that articulates with the glenoid cavity of the scapula. It also has an ***anatomical neck,*** which is an oblique groove just distal to the head and the site of the epiphyseal plate. The ***greater tubercle*** is a lateral projection distal to the neck. It is the most laterally palpable bony landmark of the shoulder region. The ***lesser tubercle*** is an anterior projection. Between these tubercles runs an ***intertubercular sulcus (bicipital groove).*** The ***surgical***

FIGURE 8-1 Right pectoral (shoulder) girdle and upper extremity. (a) Anterior view. (b) Posterior view.

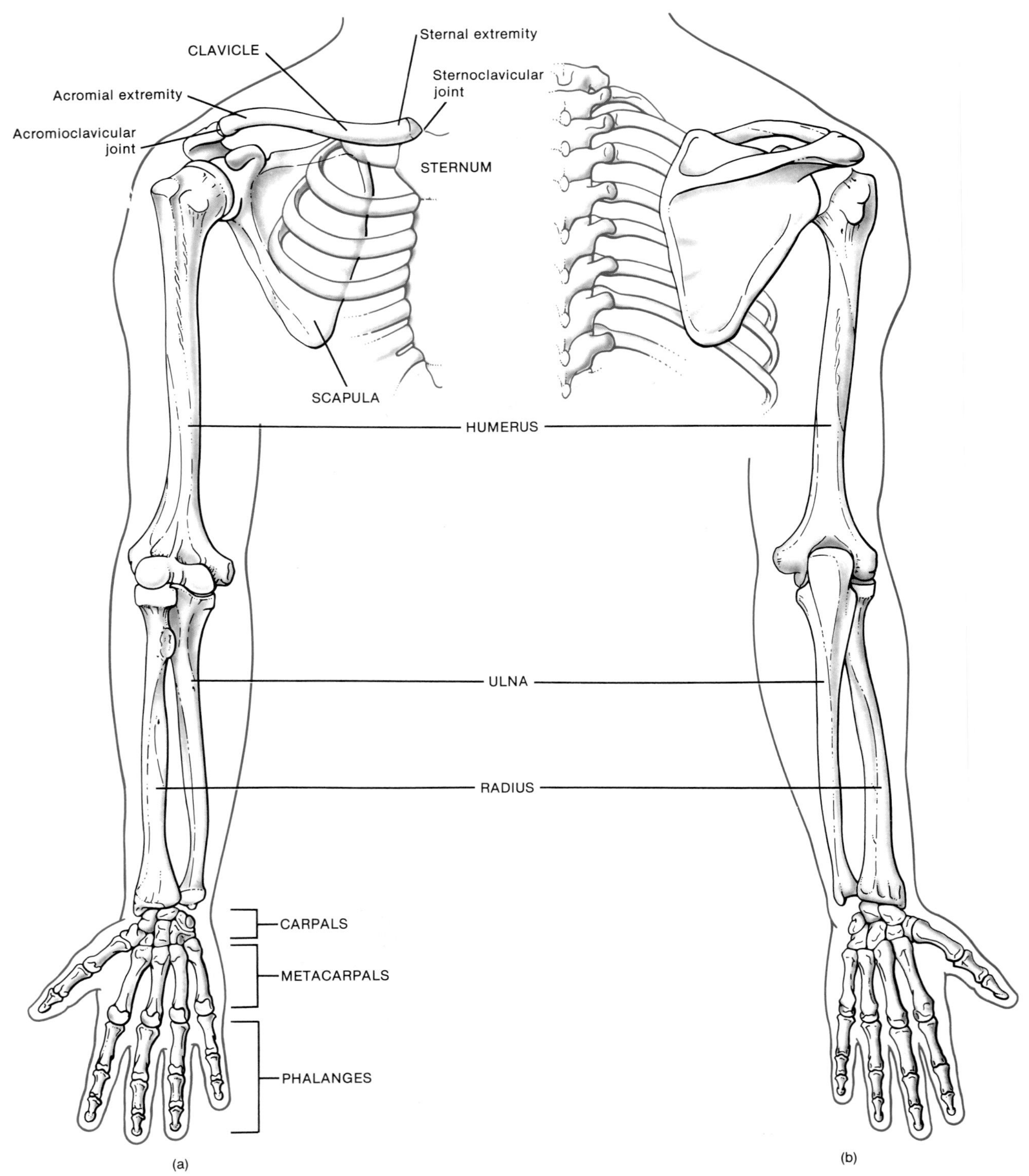

FIGURE 8-2 Photograph of right clavicle viewed from below. (Copyright © 1987 by Michael H. Ross. Reprinted by permission.)

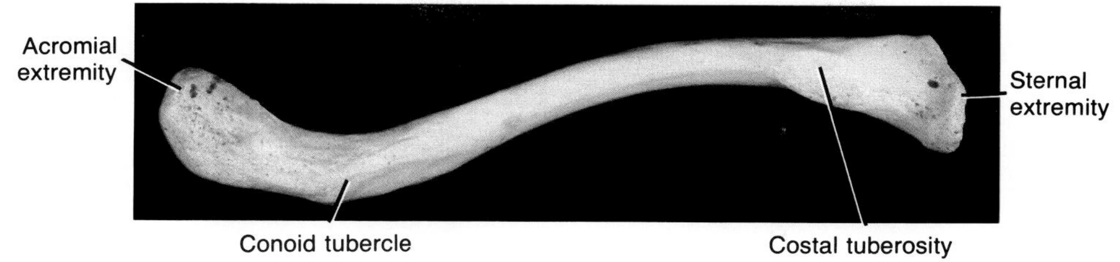

FIGURE 8-3 Photograph of right scapula. (a) Anterior view. (b) Posterior view. (c) Lateral border view. (Copyright © 1987 by Michael H. Ross. Reprinted by permission.)

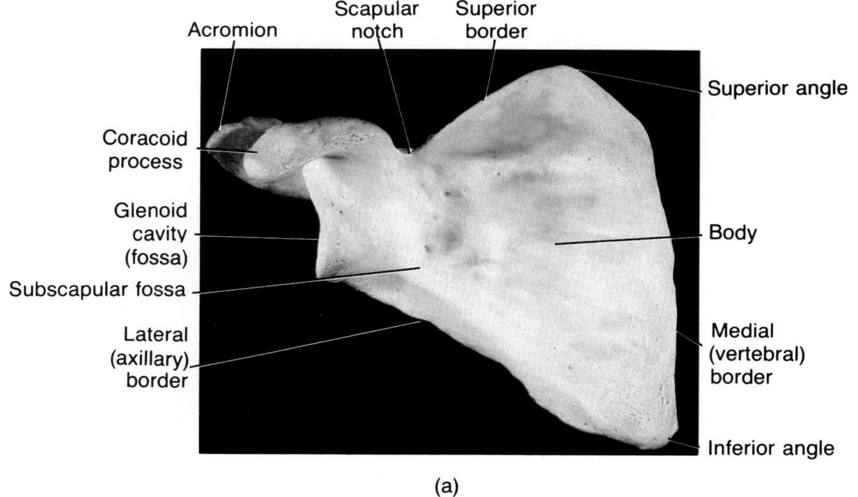

(a)

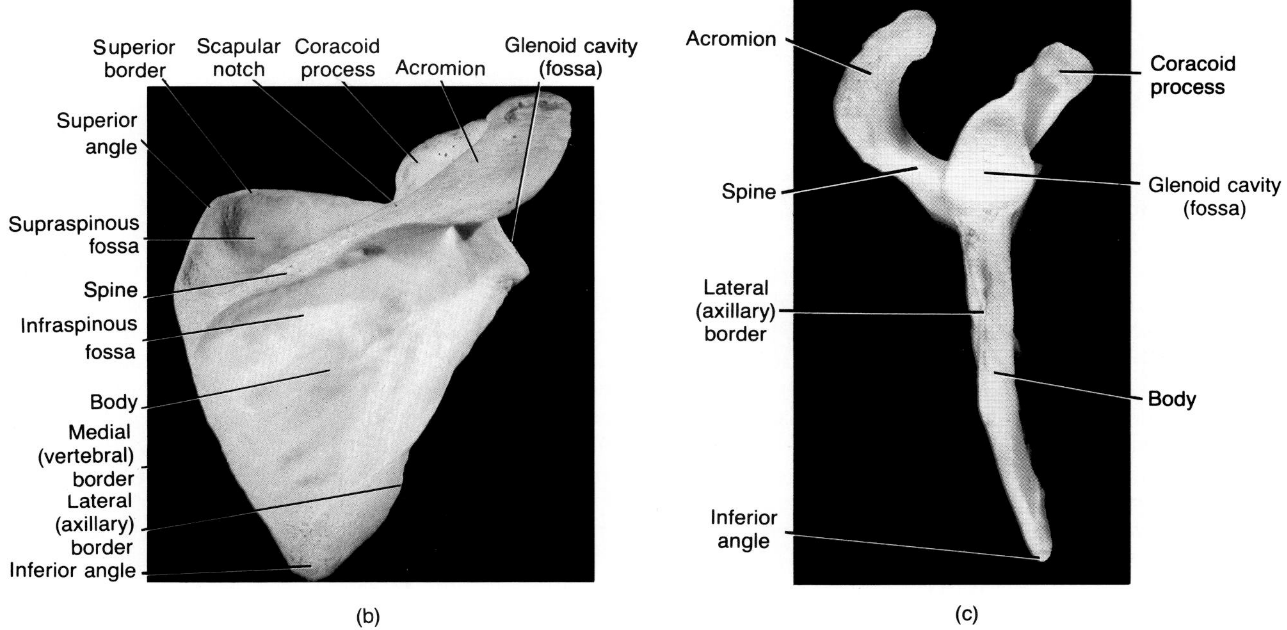

(b) (c)

FIGURE 8-4 Photograph of right humerus. (a) Anterior view. (b) Posterior view. (Copyright © 1987 by Michael H. Ross. Reprinted by permission.)

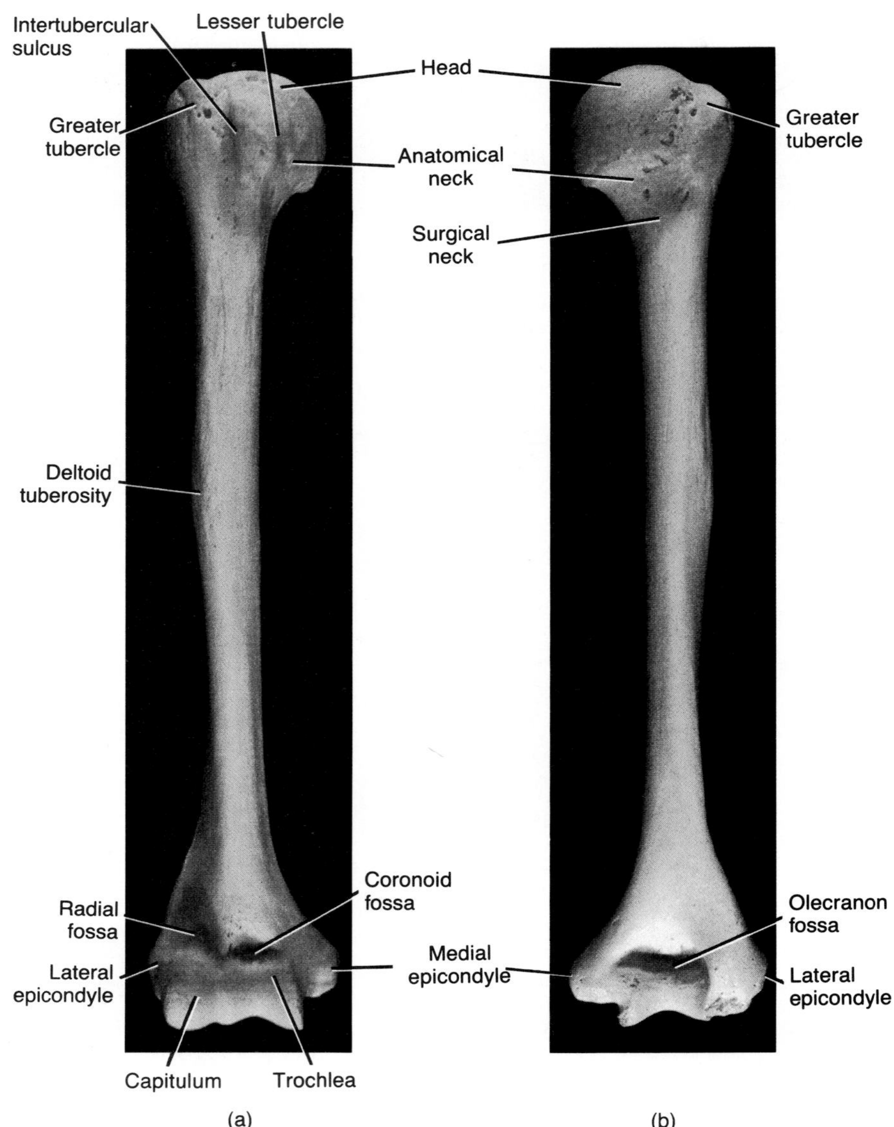

neck is a constricted portion just distal to the tubercles. It is so named because of its liability to fracture.

The ***body (shaft)*** of the humerus is cylindrical at its proximal end. It gradually becomes triangular and is flattened and broad at its distal end. Laterally, at the middle portion of the shaft, there is a roughened, V-shaped area called the ***deltoid tuberosity.*** This area serves as a point of attachment for the deltoid muscle.

The following parts are found at the distal end of the humerus. The ***capitulum*** (ka-PIT-yoo-lum) is a rounded knob that articulates with the head of the radius. The ***radial fossa*** is a depression that receives the head of the radius when the forearm is flexed. The ***trochlea*** (TRŌK-lē-a) is a pulleylike surface that articulates with the ulna. The ***coronoid fossa*** is an anterior depression that receives part of the ulna when the forearm is flexed. The ***olecranon*** (ō-LEK-ra-non) ***fossa*** is a posterior depression that receives the olecranon of the ulna when the forearm is extended. The ***medial epicondyle*** and ***lateral epicondyle*** are rough projections on either side of the distal end to which most muscles of the forearm are attached. The ulnar nerve lies on the posterior surface of the medial epicondyle and may easily be rolled between the finger and the medial epicondyle.

Ulna and Radius

The ***ulna*** is the medial bone of the forearm (Figure 8-5). In other words, it is located at the little finger side. The proximal end of the ulna presents an ***olecranon (olecranon process),*** which forms the prominence of the elbow. The ***coronoid process*** is an anterior projection that, together with the olecranon, receives the trochlea of the humerus. The ***trochlear (semilunar) notch*** is a curved area between the olecranon and the coronoid process. The trochlea of the humerus fits into this notch. The ***radial notch*** is a depression located laterally and inferiorly to the trochlear notch. It receives the head of the radius. The distal end of the ulna consists of a ***head*** that is separated from the wrist by a fibrocartilage disc. A ***styloid process*** is on the posterior side of the distal end.

The ***radius*** is the lateral bone of the forearm; that is, it is situated on the thumb side. The proximal end of the radius has a disc-shaped ***head*** that articulates with the capitulum of the humerus and radial notch of the ulna. It also has a raised, roughened area on the medial side called the ***radial tuberosity.*** This is a point of attachment for the biceps brachii muscle. The shaft of the radius widens distally to form a concave inferior surface that articulates with two bones of the wrist called the lunate and scaphoid bones. Also at the distal end is a ***styloid process*** on the lateral side and a medial, concave ***ulnar notch*** for articulation with the distal end of the ulna.

CLINICAL APPLICATION: COLLES' FRACTURE

When one falls on the outstretched arm, the radius bears the brunt of forces transmitted through the hand. If a fracture occurs in such a fall, it is usually a transverse break about 3 cm (1 inch) from the distal end of the bone. In this type of injury, called a *Colles' fracture,* the hand is displaced backward and upward (see Figure 6-7b).

Carpals, Metacarpals, and Phalanges

The ***carpus*** (or wrist) consists of eight small bones, the ***carpals,*** united to each other by ligaments (Figure 8-6). The bones are arranged in two transverse rows, with four bones in each row, and they are named for their shapes. In the anatomical position, the proximal row of carpals, from the lateral to medial position, consists of the ***scaphoid*** (resembles a boat), ***lunate*** (resembles a crescent moon in its anteroposterior aspect), ***triquetrum*** (has three articular surfaces), and ***pisiform*** (pea-shaped). In about 70 percent of cases involving carpal fractures, only the scaphoid is involved. The distal row of carpals, from the lateral to medial position, consists of the ***trapezium*** (four-sided), ***trapezoid*** (also four-sided), ***capitate*** (its rounded projection, the head, articulates with the lunate), and ***hamate*** (named for a large hook-shaped projection on its anterior surface).

The five bones of the ***metacarpus*** (*meta* = after) constitute the palm of the hand. Each metacarpal bone consists of a proximal ***base,*** a ***shaft,*** and a distal ***head.*** The metacarpal bones are numbered I to V, starting with the lateral bone. The bases articulate with the distal row of carpal bones. The heads articulate with the proximal phalanges of the fingers. The heads of the metacarpals are commonly called the "knuckles" and are readily visible when the fist is clenched.

The ***phalanges*** (fa-LAN-jēz), or bones of the fingers, number 14 in each hand. A single bone of the finger (or toe) is referred to as a ***phalanx*** (FĀ-lanks). Each phalanx consists of a proximal ***base,*** a ***shaft,*** and a distal ***head.*** There are two phalanges in the first digit, called the ***pollex (thumb),*** and three phalanges in each of the remaining four digits. These digits, moving medially from the thumb, are commonly referred to as the index finger, middle finger, ring finger, and little finger. The first row of phalanges, the ***proximal row,*** articulates with the metacarpal bones and second row of phalanges. The second row of phalanges, the ***middle row,*** articulates with the proximal row and the third row. The third row of phalanges, the ***distal row,*** articulates with the middle row. The thumb has no middle phalanx.

PELVIC (HIP) GIRDLE

The ***pelvic (hip) girdle*** consists of the two ***coxal*** (KOK-sal) ***bones*** or ***hipbones*** (Figure 8-7). The pelvic girdle provides a strong and stable support for the lower extremities on which the weight of the body is carried. The coxal bones are united to each other anteriorly at the ***symphysis*** (SIM-fi-sis) ***pubis.*** They unite posteriorly to the sacrum.

Together with the sacrum and coccyx, the two coxal (hip) bones of the pelvic girdle form the basinlike structure called the ***pelvis.*** The pelvis is divided into a greater pelvis and lesser pelvis by an oblique plane that passes through the sacral promontory (posterior), iliopectineal lines (laterally), and symphysis pubis (anteriorly). The circumference of this oblique plane is called the ***brim of the pelvis.*** The ***greater (false) pelvis*** represents the expanded portion situated superior to the brim of the pelvis. The greater pelvis consists laterally of the superior portions of the ilia and posteriorly of the superior portion of the sacrum. There is no bony component in the anterior aspect of the greater pelvis. Rather, the front is formed by the walls of the abdomen.

The ***lesser (true) pelvis*** is inferior and posterior to the brim of the pelvis. It is formed by the inferior portions of the ilia and sacrum, the coccyx, and the pubes. The

FIGURE 8-5 **Photograph of right ulna and radius. (a) Anterior view. (b) Posterior view. (c) Medial view of the right elbow. (Copyright © 1987 by Michael H. Ross. Reprinted by permission.)**

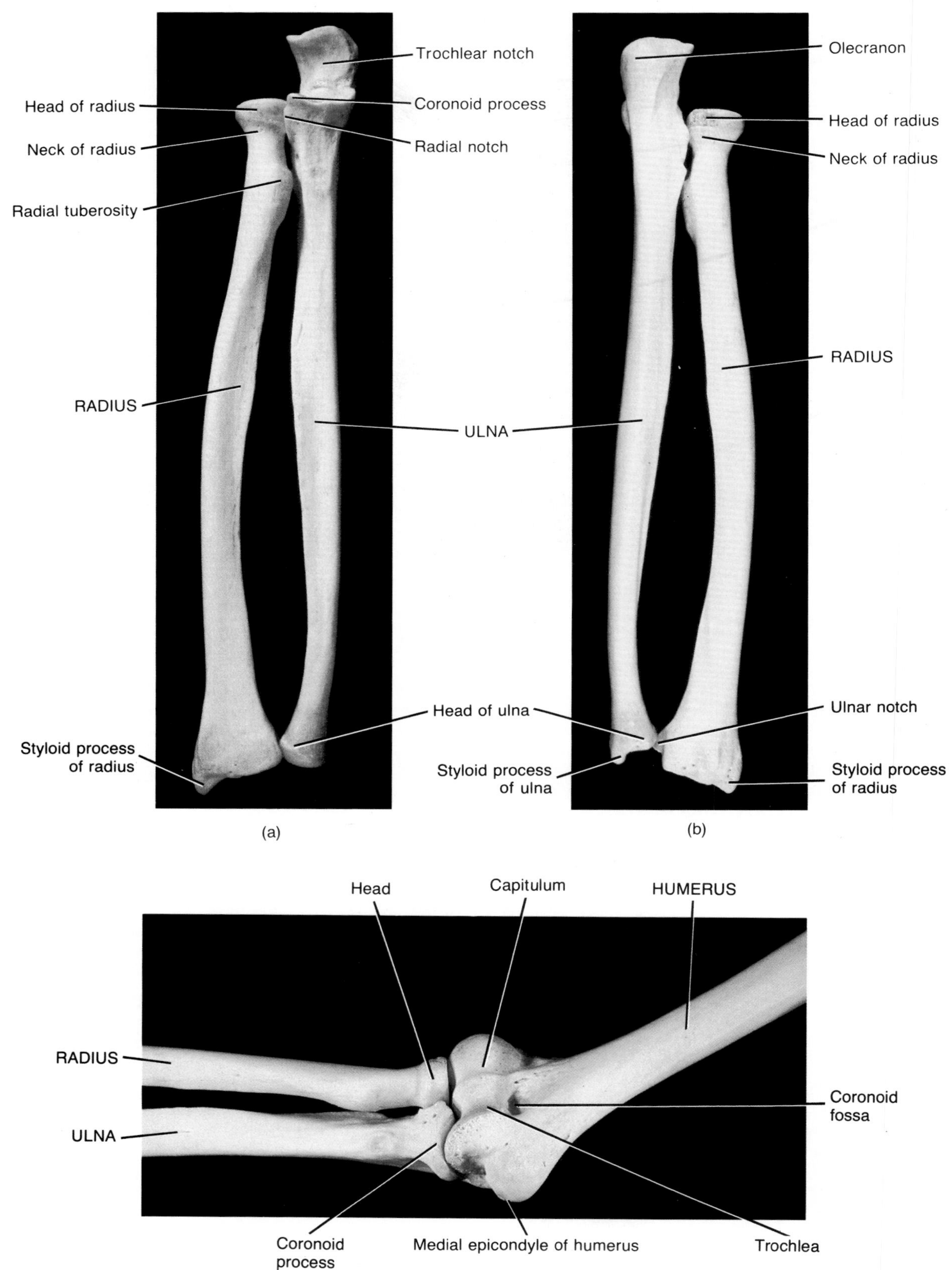

FIGURE 8-6 Photograph of right wrist and hand in anterior view. (Courtesy of J. A. Gosling et al., *Atlas of Human Anatomy with Integrated Text.* Copyright © 1985 by Gower Medical Publishing Ltd.)

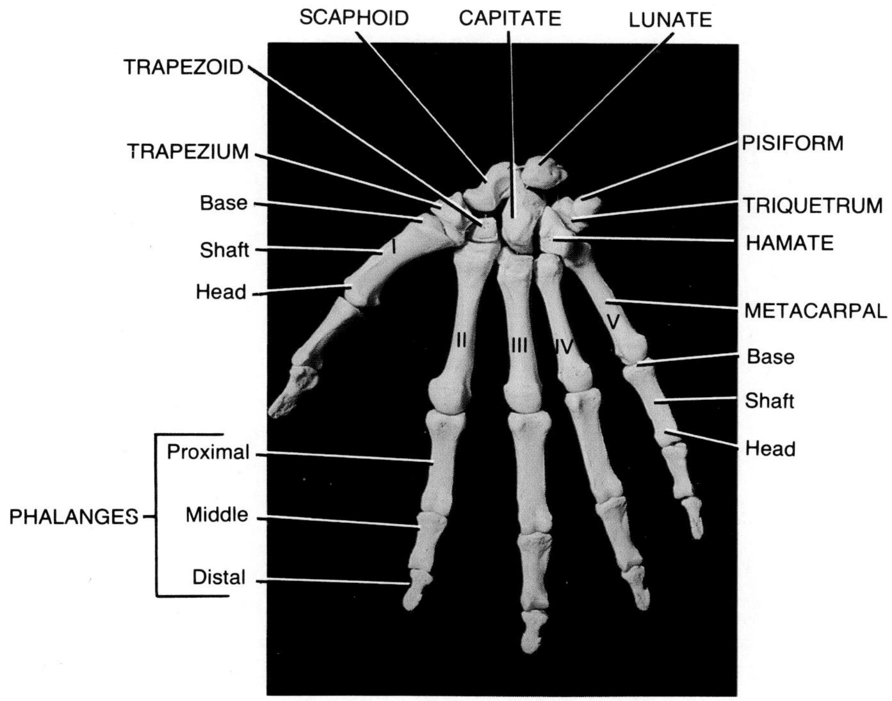

FIGURE 8-7 Pelvic (hip) girdle of a male in anterior view.

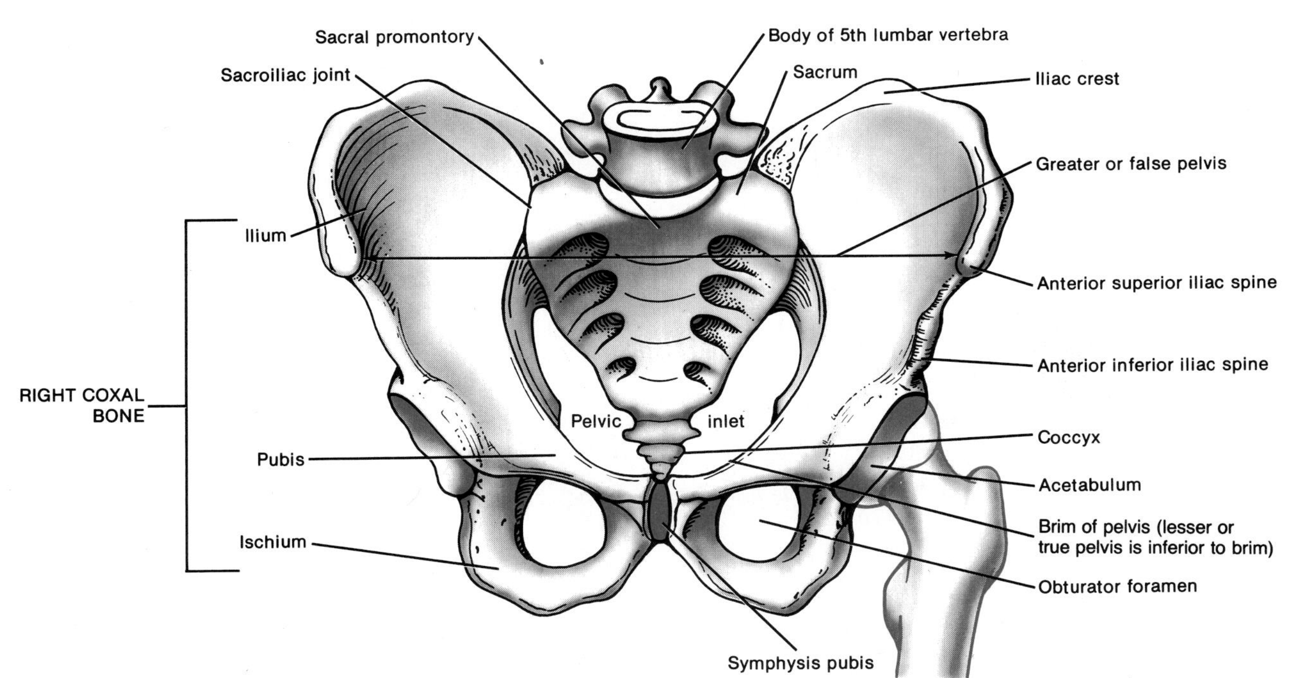

lesser pelvis contains a superior opening called the ***pelvic inlet*** and an inferior opening called the ***pelvic outlet.***

CLINICAL APPLICATON: PELVIMETRY

Pelvimetry **is the measurement (usually manually) of the size of the inlet and outlet of the birth canal. Measurement of the pelvic cavity in pregnant females is important to the physician, because the fetus must pass through the narrower opening of the lesser pelvis at birth.**

Each of the two ***coxal (hip) bones*** of a newborn consists of three components: a superior ***ilium,*** an inferior and anterior ***pubis,*** and an inferior and posterior ***ischium*** (Figure 8-8). Eventually, the three separate bones fuse into one. The area of fusion is a deep, lateral fossa called the ***acetabulum*** (as'-e-TAB-yoo-lum). Although the adult coxae are both single bones, it is common to discuss the bones as if they still consisted of three portions.

The ilium is the largest of the three subdivisions of the coxal bone. Its superior border, the ***iliac crest,*** ends anteriorly in the ***anterior superior iliac spine.*** Posteriorly, the iliac crest ends in the ***posterior superior iliac spine.*** The spines serve as points of attachment for muscles of the abdominal wall. Inferior to the posterior superior iliac spine is the ***greater sciatic*** (sī-AT-ik) ***notch.*** The internal surface of the ilium seen from the medial side is the ***iliac fossa.*** It is a concavity where the iliacus muscle attaches. Posterior to this fossa is the ***auricular surface,*** which articulates with the sacrum.

The ischium is the inferior, posterior portion of the coxal bone. It contains a prominent ***ischial spine,*** a ***lesser sciatic notch*** below the spine, and an ***ischial tuberosity.*** The rest of the ischium, the ***ramus,*** joins with the pubis, and together they surround the ***obturator*** (OB-too-rā'-ter) ***foramen.***

The pubis is the anterior and inferior part of the coxal bone. It consists of a ***superior ramus,*** an ***inferior ramus,*** and a ***body*** that contributes to the formation of the symphysis pubis.

The ***symphysis pubis*** is the joint between the two coxal bones (see Figure 8-7). It contains a pad of fibrocartilage. The acetabulum is the fossa formed by the ilium, ischium, and pubis. It is the socket for the head of the femur. Two-fifths of the acetabulum is formed by the ilium, two-fifths by the ischium, and one-fifth by the pubis.

LOWER EXTREMITY

The ***lower extremities*** are composed of 60 bones (Figure 8-9). Each extremity includes the femur in the thigh, patella (kneecap), fibula and tibia in the leg, tarsals (ankle bones), metatarsals, and phalanges in the toes.

Femur

The ***femur,*** or thighbone, is the longest and heaviest bone in the body (Figure 8-10). Its proximal end articulates with the coxal bone. Its distal end articulates with the tibia. The body (shaft) of the femur angles medially as it approaches the femur of the opposite thigh. As a result, the knee joints are brought nearer at the midline. The degree of convergence is greater in the female because the female pelvis is broader.

The proximal end of the femur consists of a rounded ***head*** that articulates with the acetabulum of the coxal bone. The ***neck*** of the femur is a constricted region distal to the head. A fairly common fracture in the elderly occurs at the neck of the femur. Apparently, the neck becomes so weak that it fails to support the body. The ***greater trochanter*** (trō-KAN-ter) and ***lesser trochanter*** are projections that serve as points of attachment for some of the thigh and buttock muscles. The greater trochanter is the prominence felt and seen anterior to the hollow on the side of the hip. Between the trochanters on the anterior surface is a narrow ***intertrochanteric line.*** Between the trochanters on the posterior surface is an ***intertrochanteric crest.***

The diaphysis (shaft) of the femur contains a rough vertical ridge on its posterior surface called the ***linea aspera.*** This ridge serves for the attachment of several thigh muscles.

The distal end of the femur is expanded and includes the ***medial condyle*** and ***lateral condyle.*** These articulate with the tibia. Superior to the condyles are the ***medial epicondyle*** and ***lateral epicondyle.*** A depressed area between the condyles on the posterior surface is called the ***intercondylar*** (in'-ter-KON-di-lar) ***fossa.*** The ***patellar surface*** is located between the condyles on the anterior surface.

Pathologic changes in the angle of the neck of the femur result in abnormal posture of the lower limbs. A decreased angle produces "knock-knee" condition. An abnormally large angle produces "bowleg" condition. Both conditions place an abnormal strain on the knee joints.

Patella

The ***patella,*** or kneecap, is a small, triangular bone anterior to the knee joint (Figure 8-11). It is a sesamoid bone that develops in the tendon of the quadriceps femoris muscle. The broad superior end of the patella is called the ***base.*** The pointed inferior end is the ***apex.*** The posterior surface contains two ***articular facets,*** one for the medial condyle and the other for the lateral condyle of the femur.

Tibia and Fibula

The ***tibia,*** or shinbone, is the larger, medial bone of the leg (Figure 8-12). It bears the major portion of the weight of the leg. The tibia articulates at its proximal end with

FIGURE 8-8 Right coxal bone. (a) Photograph in lateral view. (b) Photograph in medial view. (Copyright © 1987 by Michael H. Ross. Reprinted by permission.) (c) Three divisions of the coxal bone. The lines of fusion of the ilium, ischium, and pubis are not always visible in an adult bone.

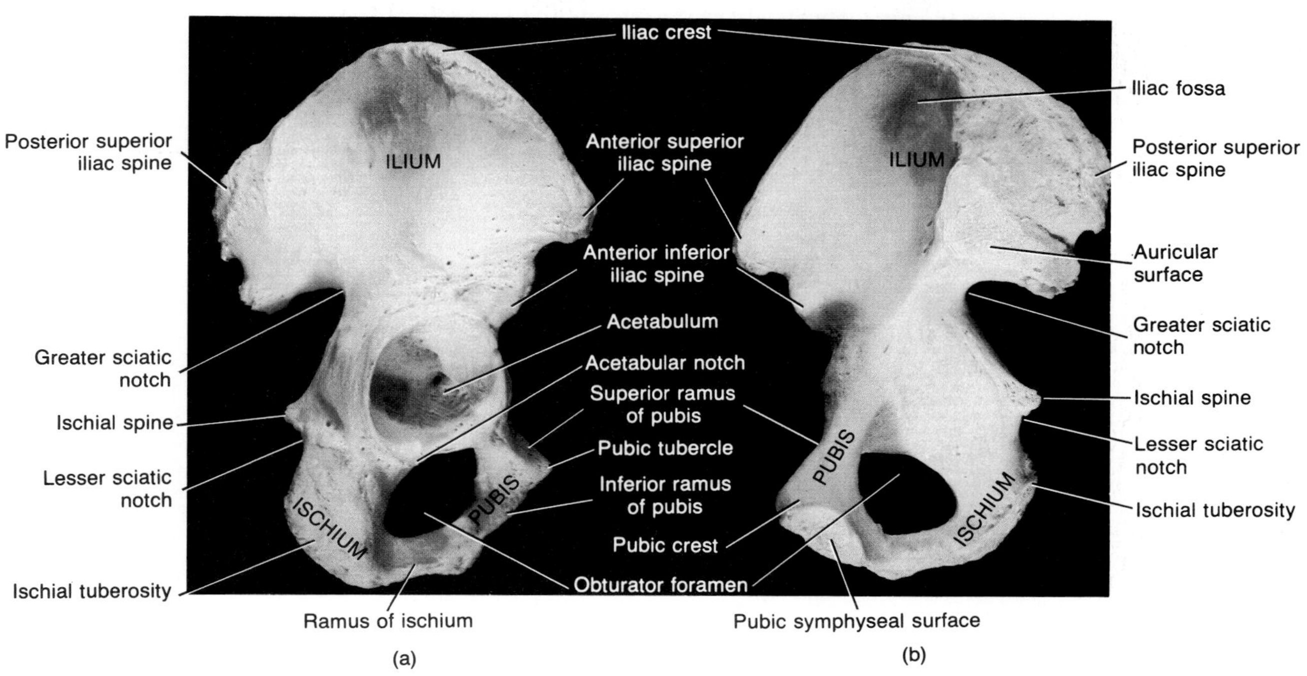

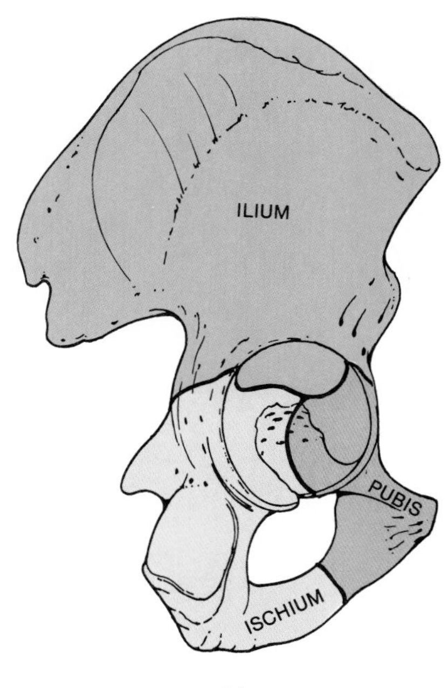

the femur and fibula, and at its distal end with the fibula of the leg and talus bone of the ankle.

The proximal end of the tibia is expanded into a ***lateral condyle*** and a ***medial condyle.*** These articulate with the condyles of the femur. The inferior surface of the lateral condyle articulates with the head of the fibula. The slightly concave condyles are separated by an upward projection called the ***intercondylar eminence.*** The ***tibial tuberosity*** on the anterior surface is a point of attachment for the patellar ligament.

The medial surface of the distal end of the tibia forms the ***medial malleolus*** (mal-LĒ-ō-lus). This structure articulates with the talus bone of the ankle and forms the prominence that can be felt on the medial surface of your ankle. The ***fibular notch*** articulates with the fibula.

The ***fibula*** is parallel and lateral to the tibia. It is considerably smaller than the tibia and is nonweight bearing. The ***head*** of the fibula, the proximal end, articulates with the inferior surface of the lateral condyle of the tibia below the level of the knee joint. The distal end has a projection called the ***lateral malleolus*** that articulates with the talus bone of the ankle. This forms the prominence on the lateral surface of the ankle. The inferior portion of the fibula also articulates with the tibia at the fibular notch. A fracture of the lower end of the fibula with injury to the tibial articulation is called a ***Pott's fracture.***

FIGURE 8-9 **Right pelvic (hip) girdle and lower extremity. (a) Anterior view. (b) Posterior view.**

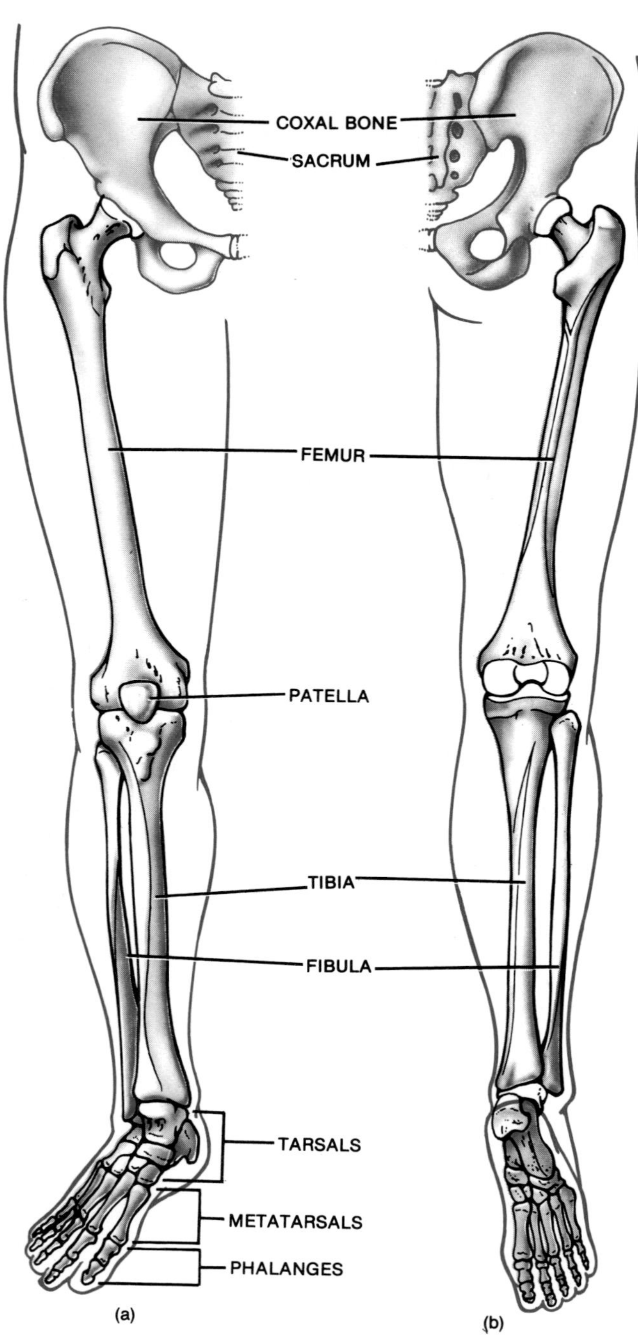

FIGURE 8-10 Photograph of right femur. (a) Anterior view. (b) Posterior view. (Copyright © 1987 by Michael H. Ross. Reprinted by permission.)

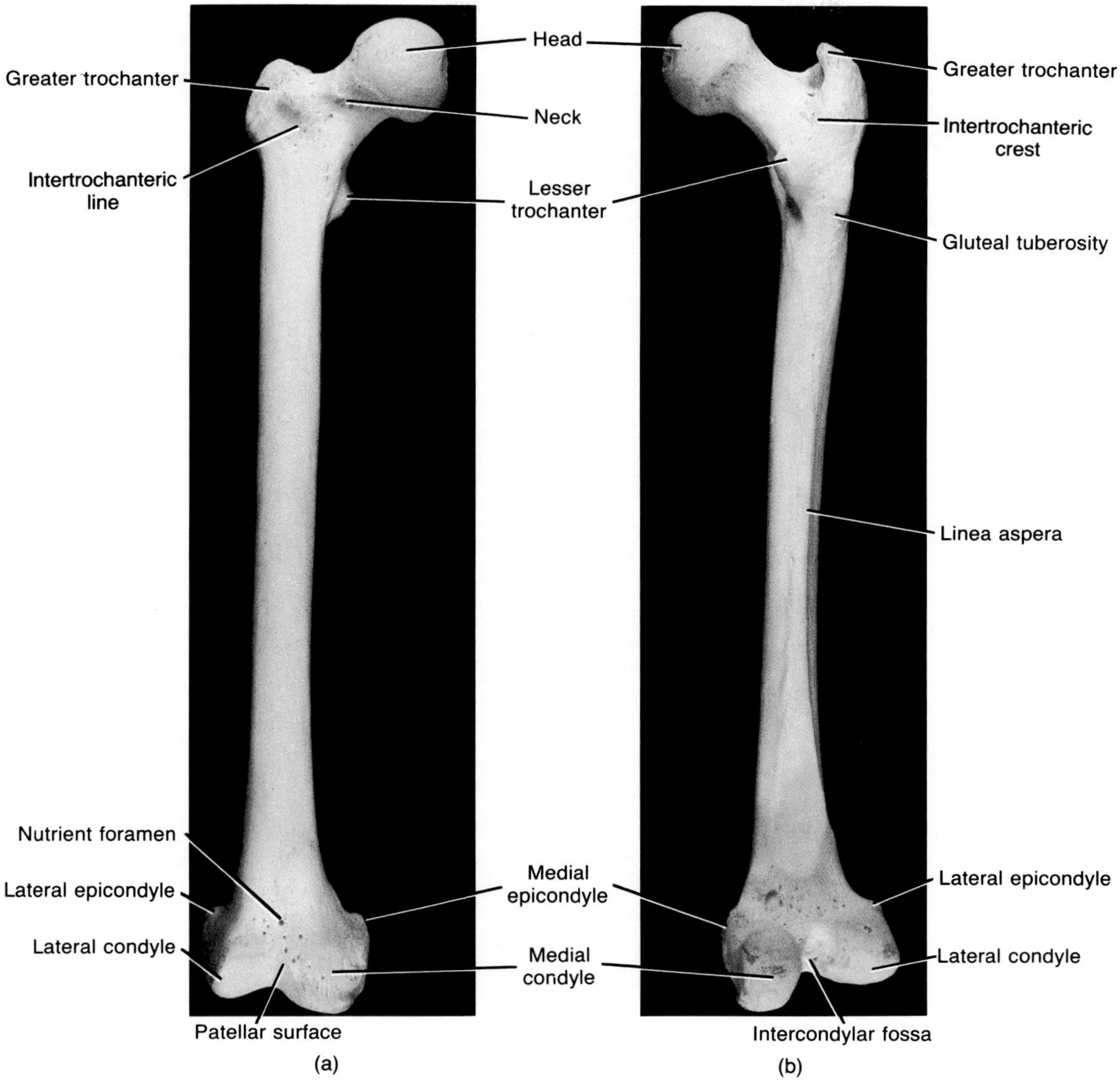

FIGURE 8-11 Photograph of right patella. (a) Anterior view. (b) Posterior view. (Copyright © 1987 by Michael H. Ross. Reprinted by permission.)

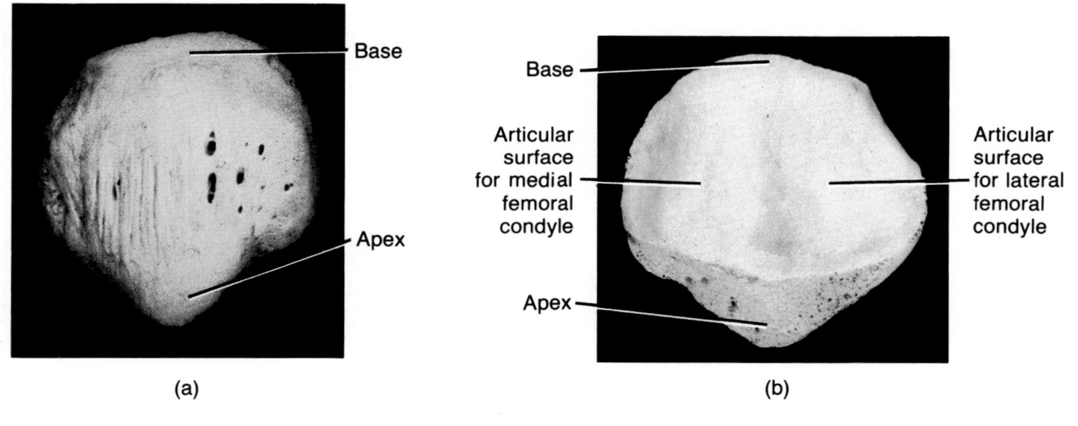

FIGURE 8-12 Photograph of right tibia and fibula. (a) Anterior view. (b) Posterior view. (Copyright © 1987 by Michael H. Ross. Reprinted by permission.)

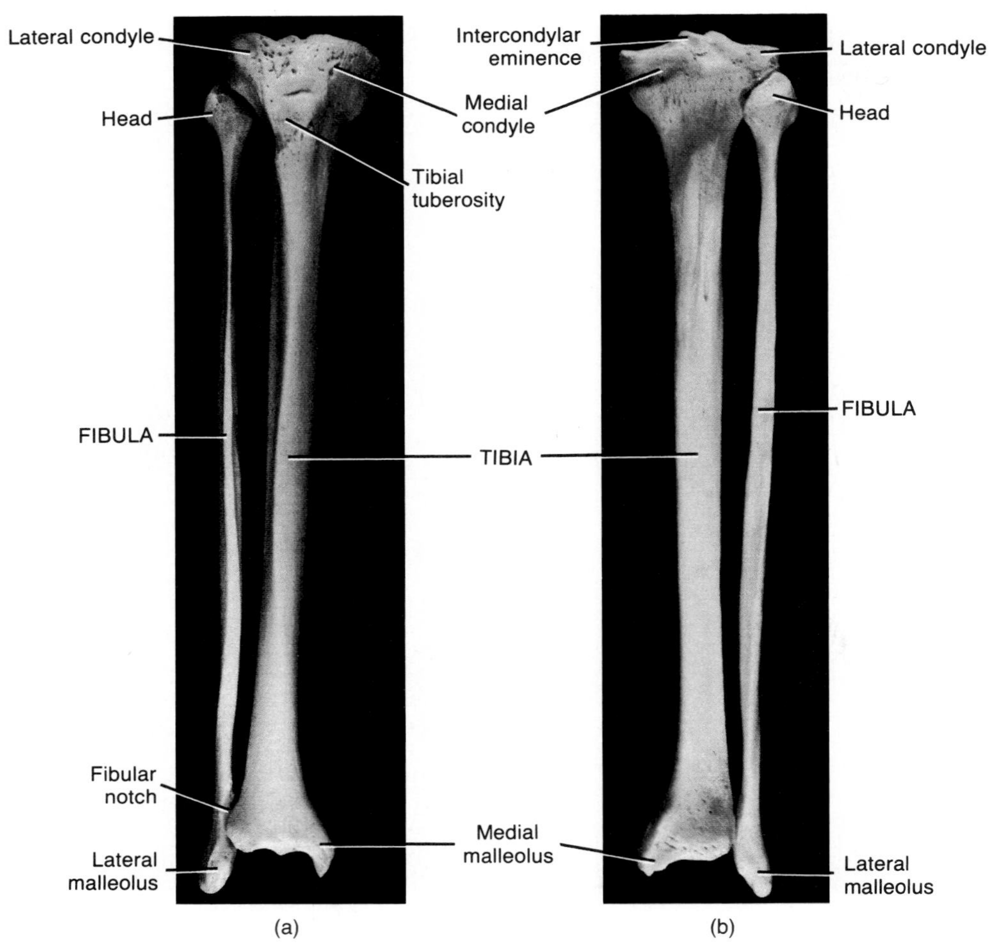

Tarsals, Metatarsals, and Phalanges

The ***tarsus*** is a collective designation for the seven bones of the ankle called ***tarsals*** (Figure 8-13). The term *tarsos* pertains to a broad, flat surface. The ***talus*** and ***calcaneus*** (kal-KĀ-nē-us) are located on the posterior part of the foot. The anterior part contains the ***cuboid, navicular,*** and three ***cuneiform*** (*cuneiform* = wedge-shaped) ***bones*** called the ***first (medial), second (intermediate),*** and ***third (lateral) cuneiform.*** The talus, the uppermost tarsal bone, is the only bone of the foot that articulates with the fibula and tibia. It is surrounded on one side by the medial malleolus of the tibia and on the other side by the lateral malleolus of the fibula. During walking, the talus initially bears the entire weight of the body. About half the weight is then transmitted to the calcaneus. The remainder is transmitted to the other tarsal bones. The calcaneus, or heel bone, is the largest and strongest tarsal bone.

The ***metatarsus*** consists of five metatarsal bones numbered I to V from the medial to lateral position. Like the metacarpals of the palm of the hand, each metatarsal consists of a proximal ***base,*** a ***shaft,*** with a distal ***head.*** The metatarsals articulate proximally with the first, second, and third cuneiform bones and with the cuboid. Distally, they articulate with the proximal row of phalanges. The first metatarsal is thicker than the others because it bears more weight.

The ***phalanges*** of the foot resemble those of the hand both in number and arrangement. Each also consists of a proximal ***base,*** a middle ***shaft,*** and a distal ***head.*** The ***hallux*** (great or big toe), has two large, heavy phalanges called proximal and distal phalanges. The other four toes each have three phalanages—proximal, middle, and distal.

Arches of the Foot

The bones of the foot are arranged in two ***arches*** (Figure 8-14). These arches enable the foot to support the weight of the body, provide an ideal distribution of body weight

FIGURE 8-13 **Photograph of right foot in superior view. (Copyright © by Michael H. Ross. Used by permission.)**

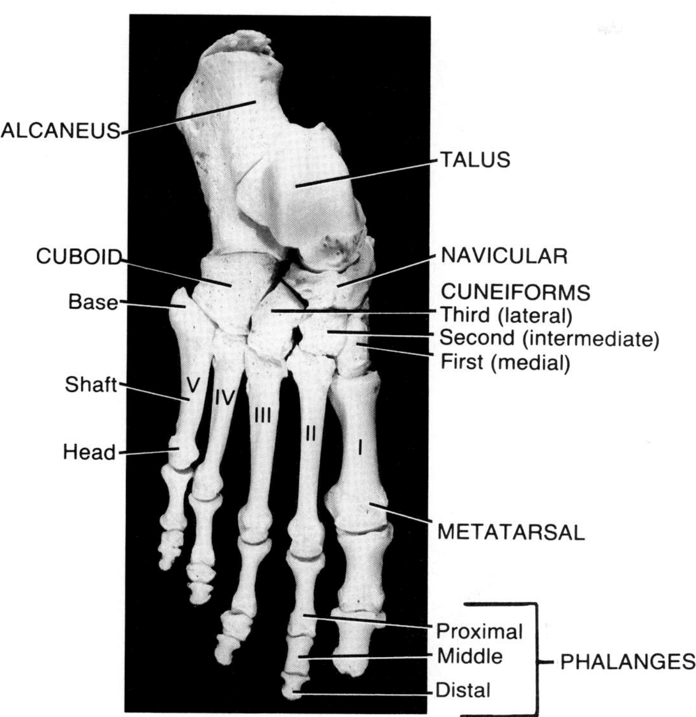

over the hard and soft tissues of the foot, and provide leverage while walking. The arches are not rigid. They yield as weight is applied and spring back when the weight is lifted, thus helping to absorb shocks.

The ***longitudinal arch*** has two parts. Both consist of tarsal and metatarsal bones arranged to form an arch from the anterior to the posterior part of the foot. The ***medial*** (inner) ***part*** of the longitudinal arch originates at the calcaneus. It rises to the talus and descends through the navicular, the three cuneiforms, and the heads of the three medial metatarsals. The talus is the keystone of this arch. The ***lateral*** (outer) ***part*** of the longitudinal arch also begins at the calcaneus. It rises at the cuboid and descends to the heads of the two lateral metatarsals. The cuboid is the keystone of this arch.

The ***transverse arch*** is formed by the navicular, three cuneiforms, cuboid, and the bases of the five metatarsals.

CLINICAL APPLICATION: FLATFOOT, CLAWFOOT, AND BUNIONS

The bones composing the arches are held in position by ligaments and tendons. If these ligaments and tendons are weakened, the height of the medial longitudinal arch may decrease or "fall." The result is *flatfoot.*

***Clawfoot* is a condition in which the medial longitudinal arch is abnormally elevated. It is frequently caused by muscle imbalance, such as may result from poliomyelitis.**

A *bunion (hallux valgus; valgus* = bent outward) is a deformity of the great toe. Although the condition may be inherited, it is typically caused by wearing tightly fitting shoes and is characterized by lateral deviation of the great toe and medial displacement of metatarsal I. Arthritis of the first metatarsophalangeal joint may also be a predisposing factor. The condition produces inflammation of bursae (fluid-filled sacs at the joint), bone spurs, and calluses.

FIGURE 8-14 **Arches of the right foot in lateral view.**

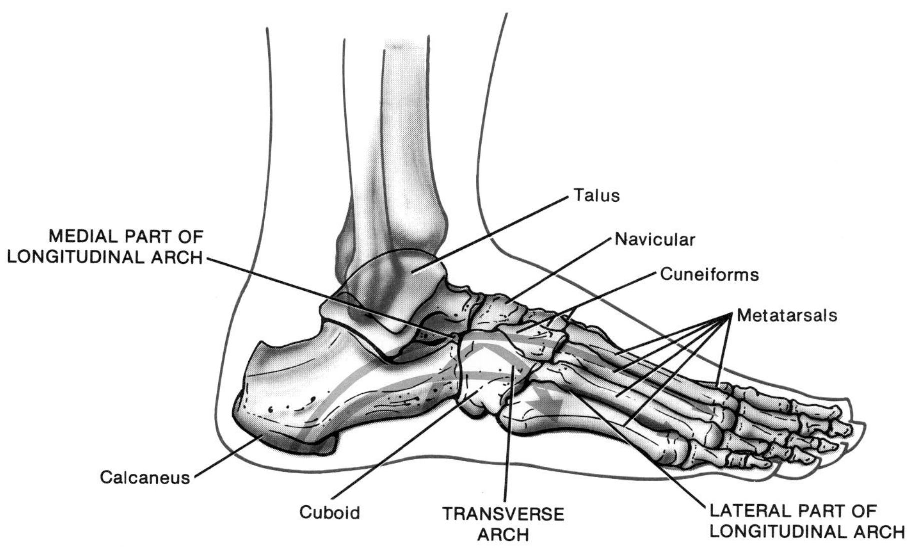

FEMALE AND MALE SKELETONS

The bones of the male are generally larger and heavier than those of the female. The articular ends are thicker in relation to the shafts. In addition, since certain muscles of the male are larger than those of the female, the points of attachment—tuberosities, lines, ridges—are larger in the male skeleton.

Many significant structural differences between female and male skeletons are noted in the pelvis; most are related to pregnancy and childbirth. The typical differences are listed in Exhibit 8-1 and illustrated in Figure 8-15.

FIGURE 8-15 **Comparison of (a) female and (b) male pelvis in anterior view.**

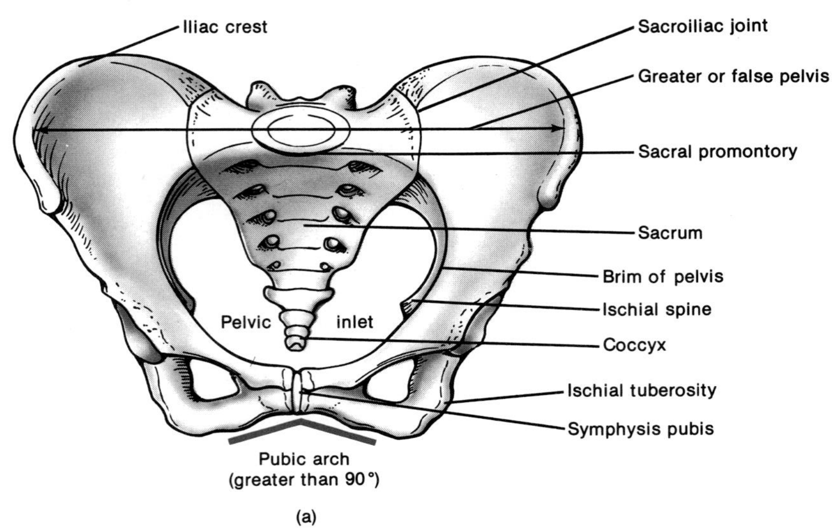

(a)

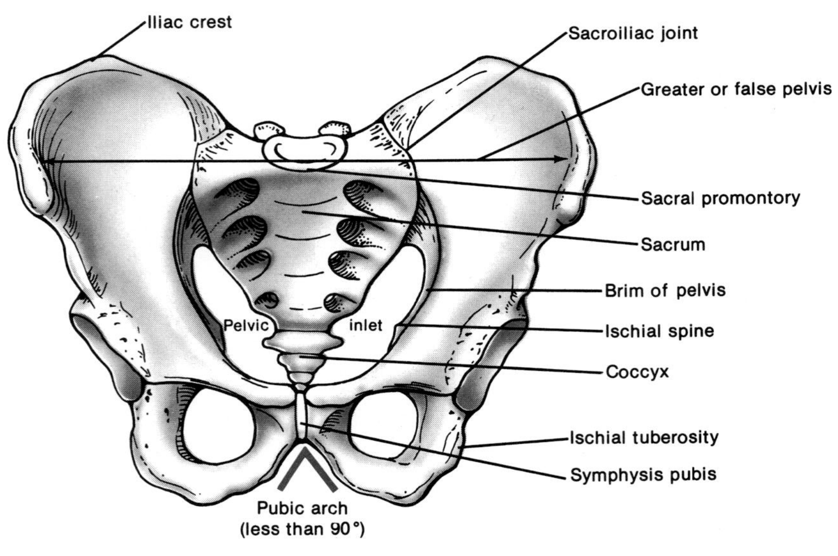

(b)

EXHIBIT 8-1 COMPARISON OF TYPICAL FEMALE AND MALE PELVIS

Point of Comparison	Female	Male
General structure	Light and thin.	Heavy and thick.
Joint surfaces	Small.	Large.
Muscle attachments	Rather indistinct.	Well marked.
Greater pelvis	Shallow.	Deep.
Pelvic inlet	Larger and more oval.	Heart-shaped.
Pelvic outlet	Comparatively large.	Comparatively small.
First piece of sacrum	Superior surface of the body spans about one-third the width of sacrum.	Superior surface of the body spans nearly one-half the width of the sacrum.
Sacrum	Short, wide, flat, curving forward in lower part.	Long, narrow, with smooth concavity.
Auricular surface	Extends only to upper border of third piece of the sacrum.	Extends well down the third piece of the sacrum.
Pubic arch	Greater than a 90° angle.	Less than a 90° angle.
Inferior ramus of pubis	Everted surface not present.	Presents strong everted surface for attachment of the crus of the penis.
Symphysis pubis	Less deep.	More deep.
Ischial spine	Turned inward less.	Turned inward more.
Ischial tuberosity	Turned outward.	Turned inward.
Ilium	Less vertical.	More vertical.
Iliac fossa	Shallow.	Deep.
Iliac crest	Less curved.	More curved.
Anterior superior iliac spine	Wider apart.	Closer.
Acetabulum	Small.	Large.
Obturator foramen	Oval.	Round.
Greater sciatic notch	Wide.	Narrow.

STUDY OUTLINE

Pectoral (Shoulder) Girdle (p. 191)

1. Each pectoral (shoulder) girdle consists of a clavicle and scapula.
2. Each attaches an upper extremity to the trunk.

Upper Extremity (p. 191)

1. The bones of each upper extremity include the humerus, ulna, radius, carpals, metacarpals, and phalanges.

Pelvic (Hip) Girdle (p. 195)

1. The pelvic (hip) girdle consists of two coxal (hip) bones.
2. It attaches the lower extremities to the trunk at the sacrum.
3. Each coxal bone consists of three fused components—ilium, pubis, and ischium.

Lower Extremity (p. 198)

1. The bones of each lower extremity include the femur, tibia, fibula, tarsals, metatarsals, and phalanges.
2. The bones of the foot are arranged in two arches, the longitudinal arch and the transverse arch, to provide support and leverage.

Female and Male Skeletons (p. 204)

1. The female pelvis is adapted for pregnancy and childbirth. Differences in pelvic structure are listed in Exhibit 8-1.
2. Male bones are generally larger and heavier than female bones and have more prominent markings for muscle attachment.

REVIEW QUESTIONS

1. What is the pectoral (shoulder) girdle? Why is it important? (p. 191)
2. What are the bones of the upper extremity? (p. 191)
3. What is the pelvic (hip) girdle? Why is it important? (p. 195)
4. What are the bones of the lower extremity? What is a Pott's fracture? (p. 198)
5. What is pelvimetry? What is the clinical importance of pelvimetry? (p. 198)
6. In what ways do the upper extremity and lower extremity differ structurally?
7. Describe the structure of the longitudinal and transverse arches of the foot. What is the function of an arch? (p. 202)
8. What are the principal structural differences between typical female and male skeletons? Use Exhibit 8-1 as a guide in formulating your response. (p. 205)
9. Define the following: fractured clavicle (p. 191), Colles' fracture (p. 195), and flatfoot, clawfoot, and bunion (p. 203).

SELECTED READINGS

Cristensen, J. B., and I. R. Telford. *Synopsis of Gross Anatomy,* 5th ed. Philadelphia: Lippincott, 1988.

Gosling, J. A., P. F. Harris, J. R. Humpherson, I. Whitmore, and P. L. T. Willon. *Atlas of Human Anatomy.* Philadelphia: Lippincott, 1985.

Netter, F. H. *Musculoskeletal System, Part I, Anatomy, Physiology, and Metabolic Disorders.* Summit, N.J.: CIBA, 1987.

Rohen, J. W., and C. Yokochi. *Color Atlas of Anatomy.* Tokyo, New York: Igaku-Shoin, Ltd., 1983.

Shipman, P., A. Walker, and D. Bichell. *The Human Skeleton.* Cambridge: Harvard University Press, 1985.

Steele, D. G., and C. A. Bramblett. *The Anatomy and Biology of the Human Skeleton.* College Station: Texas A&M University Press, 1988.

Chapter 9

Articulations

Chapter Contents at a Glance

Student Objectives

1. Define an articulation and identify the factors that determine the types and degree (range) of movement at a joint.
2. Contrast the structure, kind of movement, and location of fibrous, cartilaginous, and synovial joints.
3. Discuss and compare the movements possible at various synovial joints.
4. Describe selected articulations of the body with respect to the bones that enter into their formation, structural classification, and anatomical components.
5. Describe the causes and symptoms of common joint disorders, including rheumatism, rheumatoid arthritis (RA), osteoarthritis, gouty arthritis, bursitis, dislocation, and sprain.
6. Define medical terminology associated with articulations.

Bones are too rigid to bend without causing damage. Fortunately, the skeletal system consists of many separate bones, most of which are held together at joints by flexible connective tissue. All movements that change the positions of the bony parts of the body occur at joints. You can understand the importance of joints if you imagine how a cast over the knee joint prevents flexing the leg or how a splint on a finger limits the ability to manipulate small objects.

An ***articulation (joint)*** is a point of contact between bones, between cartilage and bones, or between teeth and bones. The scientific study of joints is referred to as ***arthrology*** (ar-THROL-ō-jē; *arthro* = joint; *logos* = study of). The joint's structure determines how it functions. Some joints permit no movement, others permit slight movement, and still others afford considerable movement. In general, the closer the fit at the point of contact, the stronger the joint. At tightly fitted joints, however, movement is restricted. The looser the fit, the greater the movement. Unfortunately, loosely fitted joints are prone to dislocation. Movement at joints is also determined by the structure (shape) of the articulating bones, the flexibility (tension or tautness) of the connective tissue ligaments and joint capsules that bind the bones together, and the position of ligaments, muscles, and tendons.

CLASSIFICATION

Functional

The functional classification of joints takes into account the degree of movement they permit. Functionally, joints are classified as ***synarthroses*** (sin'-ar-THRŌ-sēz), which are immovable joints; ***amphiarthroses*** (am'-fē-ar-THRŌ-sēz), which are slightly movable joints; and ***diarthroses*** (dī-ar-THRŌ-sēz), which are freely movable joints.

Structural

The structural classification of joints is based on the presence or absence of a synovial (joint) cavity (a space between the articulating bones) and the kind of connective tissue that binds the bones together. Structurally, joints are classified as ***fibrous,*** in which there is no synovial cavity and the bones are held together by fibrous connective tissue; ***cartilaginous,*** in which there is no synovial cavity and the bones are held together by cartilage; and ***synovial,*** in which there is a synovial cavity and the bones forming the joint are united by a surrounding articular capsule and frequently by accessory ligaments (described in detail later). We will discuss the joints of the body based upon their structural classification, but with reference to their functional classification as well.

FIBROUS JOINTS

Fibrous joints lack a synovial cavity, and the articulating bones are held very closely together by fibrous connective tissue. They permit little or no movement. The three types of fibrous joints are (1) sutures, (2) syndesmoses, and (3) gomphoses.

Suture

Sutures (SOO-cherz) are found between bones of the skull. In a suture, the bones are united by a thin layer of dense fibrous connective tissue. The irregular (interdigitated) structure of sutures gives them added strength and decreases their chance of fractures. Since sutures are immovable, they are functionally classified as synarthroses. Some sutures, present during growth, are replaced by bone in the adult. In this case, they are called ***synostoses*** (sin'-os-TŌ-sēz), or bony joints—joints in which there is a complete fusion of bone across the suture line. An example is the frontal suture between the left and right sides of the frontal bone that begins to fuse during infancy (see Figure 7-3a). Synostoses are also functionally classified as synarthroses.

Syndesmosis

A ***syndesmosis*** (sin'-dez-MŌ-sis) is a fibrous joint in which the uniting fibrous connective tissue is present in a much greater amount than in a suture, but the fit between the bones is not quite as tight. The fibrous connective tissue forms an interosseous membrane or ligament. A syndesmosis is slightly movable because the bones are separated more than in a suture and some flexibility is permitted by the interosseous membrane or ligament. Syndesmoses are functionally classified as amphiarthrotic and typically permit slight movement. An example of a syndesmosis is the distal articulation of the tibia and fibula (see Figure 8-12).

Gomphosis

A ***gomphosis*** (gom-FŌ-sis) is a type of fibrous joint in which a cone-shaped peg fits into a socket. The intervening substance is the periodontal ligament. A gomphosis is functionally classified as synarthrotic. Examples are the articulations of the roots of the teeth with the alveoli (sockets) of the maxillae and mandible.

CARTILAGINOUS JOINTS

Another joint that has no synovial cavity is a ***cartilaginous joint.*** Here the articulating bones are tightly connected by cartilage. Like fibrous joints, they allow little or no movement. The two types of cartilaginous joints are (1) synchondroses and (2) symphyses.

Synchondrosis

A ***synchondrosis*** (sin'-kon-DRŌ-sis) is a cartilaginous joint in which the connecting material is hyaline cartilage. The most common type of synchondrosis is the epiphyseal plate (see Figure 6-5). Such a joint is found between the epiphysis and diaphysis of a growing bone and is immovable. Thus, it is synarthrotic. Since the hyaline cartilage is eventually replaced by bone when growth ceases, the joint is temporary. It is replaced by a synostosis. Another example of a synchondrosis is the joint between the first rib and the sternum. The cartilage in this joint undergoes ossification during adult life.

Symphysis

A ***symphysis*** (SIM-fi-sis) is a cartilaginous joint in which the connecting material is a broad, flat disc of fibrocartilage. This joint is found between bodies of vertebrae (see Figure 7-11). A portion of the intervertebral disc is cartilaginous material. The symphysis pubis between the anterior surfaces of the coxal bones is another example (see Figure 8-15). These joints are slightly movable, or amphiarthrotic.

SYNOVIAL JOINTS

Structure

A joint in which there is a space between articulating bones is called a ***synovial*** (si-NŌ-vē-al) ***joint.*** The space is called a ***synovial (joint) cavity*** (Figure 9-1). Because of this cavity and because of the arrangement of the articular capsule and accessory ligaments, synovial joints are freely movable. Thus, synovial joints are functionally classified as diarthrotic.

Synovial joints are also characterized by the presence of ***articular cartilage.*** Articular cartilage covers the surfaces of the articulating bones but does not bind the bones together. The articular cartilage of synovial joints is hyaline cartilage.

Synovial joints are surrounded by a sleevelike ***articular capsule*** that encloses the synovial cavity and unites the articulating bones. The articular capsule is composed of two layers. The outer layer, the ***fibrous capsule,*** consists of dense connective (collagenous) tissue. It is attached to the periosteum of the articulating bones at a variable distance from the edge of the articular cartilage. The flexibility of the fibrous capsule permits movement at a joint, whereas its great tensile strength resists dislocation. The fibers of some fibrous capsules are arranged

FIGURE 9-1 **Generalized synovial joint in frontal section.**

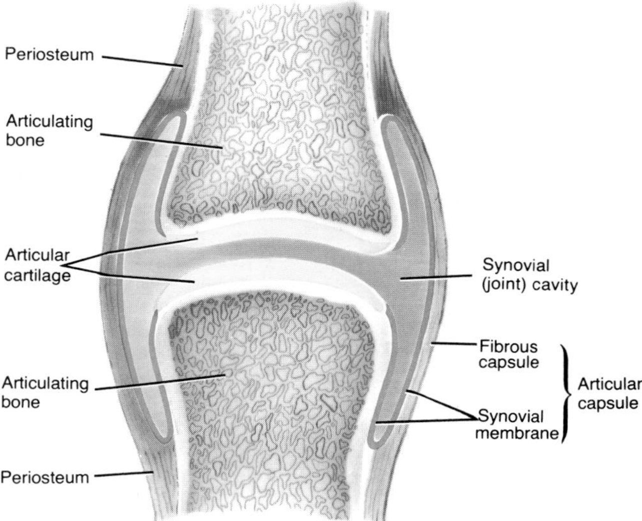

in parallel bundles and are therefore highly adapted to resist recurrent strain. Such fibers are called ***ligaments*** and are given special names. The strength of the ligaments is one of the principal factors in holding bone to bone.

The inner layer of the articular capsule is formed by a ***synovial membrane.*** The synovial membrane is composed of loose connective tissue with elastic fibers and a variable amount of adipose tissue. It secretes ***synovial fluid (SF),*** which lubricates the joint and provides nourishment for the articular cartilage. Synovial fluid also contains phagocytic cells that remove microbes and debris resulting from wear and tear in the joint. Synovial fluid consists of hyaluronic acid and an interstitial fluid formed from blood plasma and is similar in appearance and consistency to uncooked egg white. When there is no joint movement, the fluid is quite viscous, but as movement increases, the fluid becomes less viscous. The amount of synovial fluid varies in different joints of the body, ranging from a thin, viscous layer to about 3.5 ml (about 0.1 oz) of free fluid in a large joint such as the knee. The amount present in each joint is sufficient only to form a thin film over the surfaces within an articular capsule where it reduces friction and supplies nutrients to and removes metabolic wastes from the cartilage cells of the articular cartilage.

One interesting feature of some synovial joints is their ability to produce a ***cracking sound*** when pulled apart. When a synovial joint is first pulled on, there is little separation between the opposing articular surfaces. However, as the pull continues, a negative pressure develops in the synovial fluid, driving out carbon dioxide (CO_2). As a result, a bubble of gas forms in the fluid. Then, the opposing articular surfaces abruptly separate, until limited by the articular capsule. Once the surfaces are separated, the pressure within the joint exceeds that in the bubble and the bubble collapses, producing the cracking noise. The collapse of the large bubble creates a series of smaller bubbles, which gradually go back into solution. Until the small bubbles disappear and the gas is completely dissolved, the joint cannot be cracked again. Synovial joints whose surfaces are more congruent, such as those between phalanges and metacarpals, crack more easily than joints whose surfaces are less congruent.

MEDICAL TEST

Arthrocentesis (ar-thrō-sen-TĒ-sis; *arthro* = joint; *centesis* = to puncture) (***joint tap***)

Diagnostic Value: To determine the cause of joint swelling and inflammation or a painful joint by removing a sample of synovial fluid for analysis; to remove fluid (synovial fluid, blood, or pus) in order to decrease pressure and relieve pain; to inject local anesthesia or a steroid medication to relieve arthritic pain; or rarely, to evacuate abnormal joint fluid or blood and cleanse the joint by saline irrigations.

Procedure: The skin over the joint is cleansed with an antiseptic solution. A local anesthetic agent is usually given. Then a needle is inserted into the synovial (joint) cavity to remove a sample of synovial fluid. The fluid is analyzed for its cell content and is tested for urate and other crystals. Its color and viscosity are noted, and a sample is submitted for microbial culture and chemical analysis (glucose, protein, enzymes).

Many synovial joints also contain ***accessory ligaments*** (*ligare* = to bind), which are called extracapsular ligaments and intracapsular ligaments. ***Extracapsular ligaments*** are outside of the articular capsule. An example is the fibular collateral ligament of the knee joint (see Figure 9-8e). ***Intracapsular ligaments*** occur within the articular capsule but are excluded from the synovial cavity by reflections of the synovial membrane. Examples are the cruciate ligaments of the knee joint (see Figure 9-8e).

CLINICAL APPLICATION: ARTIFICIAL LIGAMENTS

Artificial ligaments are used to support or replace severely torn ligaments, especially in the knee. One such device is a carbon fiber implant. The implant consists of carbon fibers coated with a plastic called polylactic acid. The coated fibers are sewn in and around torn ligaments and tendons to reinforce them and to provide a scaffolding around which the body's own collagenous fibers grow. Within two weeks, the polylactic acid is absorbed by the body and the carbon fibers eventually fracture. By this time, the fibers are completely clad in collagen produced by fibroblasts.

Inside some synovial joints, there are pads of fibrocartilage that lie between the articular surfaces of the bones and are attached by their margins to the fibrous capsule. These pads are called ***articular discs (menisci)***. The discs usually subdivide the synovial cavity into two separate spaces. Articular discs allow two bones of different shapes to fit tightly; they modify the shape of the joint surfaces of the articulating bones. Articular discs also help to maintain the stability of the joint and direct the flow of synovial fluid to areas of greatest friction.

CLINICAL APPLICATION: TORN CARTILAGE

A tearing of articular discs in the knee, commonly called ***torn cartilage,*** occurs frequently among athletes. Such damaged cartilage requires surgical removal (meniscectomy), or it will begin to wear and cause arthritis. At one time, knee joint surgery for torn cartilage necessitated cutting through layers of healthy tissue and removing much, if not all, of the cartilage. This procedure is usually painful and expensive and does not always provide full recovery. These problems have been overcome by arthroscopy.

MEDICAL TEST

Arthroscopy (ar-THROS-kō-pē; *arthro* = joint; *skopein* = to view) Examination of the interior of a joint, usually the knee, using an arthroscope, a lighted instrument the diameter of a pencil.

Diagnostic Value: To determine the nature and extent of damage following knee injury; to remove torn cartilage and repair cruciate ligaments in the knee; to perform surgery on other joints and obtain tissue samples for analysis; to monitor the progression of disease and the effects of therapy; and to plan, if necessary, additional surgical procedures.

Procedure: After a local or general anesthetic is given, the arthroscope is inserted into the knee joint through an incision as small as one-quarter inch. A second small incision is made to insert a tube through which a saline (salt) solution is injected into the joint. (Alternately, the saline solution may be introduced through the arthroscope.) If necessary, another small incision is used for insertion of an instrument that shaves off and reshapes the damaged cartilage and then suctions out the shaved cartilage along with the salt solution. Some orthopedic surgeons attach a lightweight television camera to the arthroscope so that the image from inside the knee can be projected onto a screen. Since arthroscopy requires only small incisions, recovery is more

rapid than with conventional surgery, although a cast or splint may be worn for several days, depending on the extent of the procedure.

The various movements of the body create friction between moving parts. To reduce this friction, saclike structures called ***bursae*** are situated in the body tissues. These sacs resemble joint capsules in that their walls consist of connective tissue lined by a synovial membrane. They are also filled with a fluid similar to synovial fluid. Bursae are located between the skin and bone in places where skin rubs over bone. They are also found between tendons and bones, muscles and bones, and ligaments and bones. As fluid-filled sacs, they cushion the movement of one part of the body over another. An inflammation of a bursa is called ***bursitis.***

The articular surfaces of synovial joints are kept in contact with each other by several factors. One factor is the fit of the articulating bones. This interlocking is very obvious at the hip joint, where the head of the femur articulates with the acetabulum of the coxal (hip) bone. Another factor is the strength and tension (tautness) of the joint ligaments. This is especially important in the coxal (hip) joint. A third factor is the arrangement and tension of the muscles around the joint. For example, the fibrous capsule of the knee joint is formed principally from tendinous expansions by muscles acting on the joint.

Movements

The movements permitted at synovial joints are limited by several factors. The most important is the ***structure (shape) of the articulating bones,*** that is, the precise manner in which the articulating bones fit with respect to each other. A second factor is the ***tension of ligaments.*** The different components of a fibrous capsule are tense only when the joint is in certain positions. Tense ligaments not only restrict the range of movement but also direct the movement of the articulating bones with respect to each other. In the knee joint, for example, the major ligaments are lax when the knee is bent but tense when the knee is straightened. Also, when the knee is straightened, the surfaces of the articulating bones are in fullest contact with each other. A third factor that restricts movement at a synovial joint is ***muscle arrangement and tension,*** which reinforces the restraint placed on a joint by ligaments. A good example of the effect of muscle tension on a joint is seen at the coxal (hip) joint. When the thigh is raised with the knee straight, the movement is restricted by the tension of the hamstring muscles on the posterior surface of the thigh. But if the knee is bent, the tension on the hamstring muscles is lessened and the thigh can be raised further. Finally, in a few joints, the ***apposition of soft parts*** may limit mobility. For example, during bending at the elbow, the anterior surface of the forearm is pressed against the anterior surface of the arm.

Following is a description of the specific movements that occur at synovial joints.

Gliding

A ***gliding movement*** is the simplest kind that can occur at a joint. One surface moves back and forth and from side to side over another surface without angular or rotary motion. Some joints that glide are those between the carpals and between the tarsals. The heads and tubercles of ribs glide on the bodies and transverse processes of vertebrae.

Angular

Angular movements increase or decrease the angle between bones. Among the angular movements are flexion, extension, abduction, and adduction (Figure 9-2). ***Flexion*** involves a decrease in the angle between the surfaces of the articulating bones. Examples of flexion include bending the head forward (the joint is between the occipital bone and the atlas), bending the elbow, and bending the knee.

Extension involves an increase in the angle between the surfaces of the articulating bones. Extension restores a body part to its anatomical position after it has been flexed. Examples of extension are returning the head to the anatomical position after flexion, straightening the arm after flexion, and straightening the leg after flexion. Continuation of extension beyond the anatomical position, as in bending the head backward, is called ***hyperextension.***

Abduction usually means movement of a bone *away from* the midline of the body. An example of abduction is moving the arm upward and away from the body until it is held straight out at right angles to the chest. With the fingers and toes, however, the midline of the body is not used as the line of reference. Abduction of the fingers (not the thumb) is a movement away from an imaginary line drawn through the middle finger; in other words, it is spreading the fingers. Abduction of the thumb moves the thumb away from the plane of the palm at a right angle to the palm. Abduction of the toes is relative to an imaginary line drawn through the second toe.

Adduction is usually movement of a part *toward* the midline of the body. An example of adduction is returning the arm to the side after abduction. As in abduction, adduction of the fingers (not the thumb) is relative to the middle finger, and adduction of the toes is relative to the second toe. In adduction of the thumb, the thumb moves toward the plane of the palm at a right angle to the palm.

Rotation

Rotation is the movement of a bone around its own longitudinal axis. During rotation, no other motion is permitted. In ***medial rotation,*** the anterior surface of a bone or extremity moves toward the midline. In ***lateral rotation,*** the anterior surface moves away from the mid-

FIGURE 9-2 Angular movements at synovial joints. (Copyright © 1983 by Gerard J. Tortora. Courtesy of Lynne and James Borghesi.)

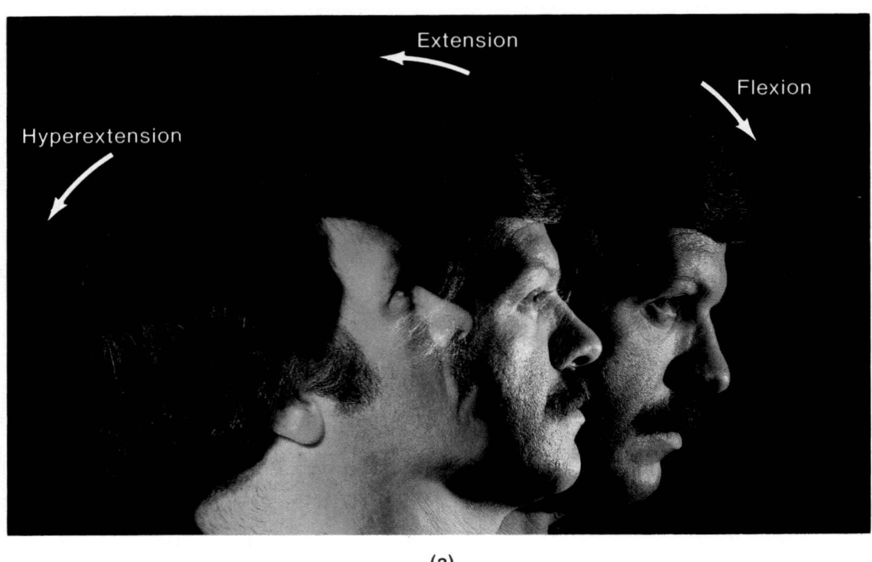

(a)

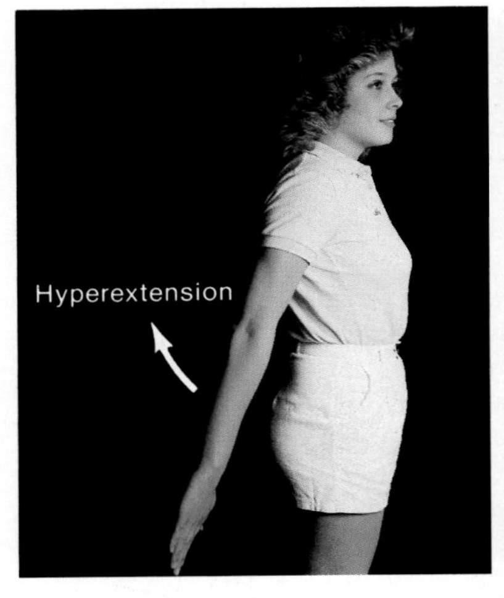

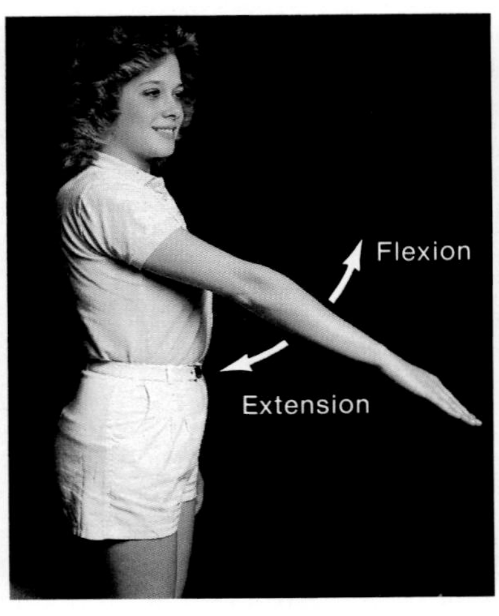

(b)

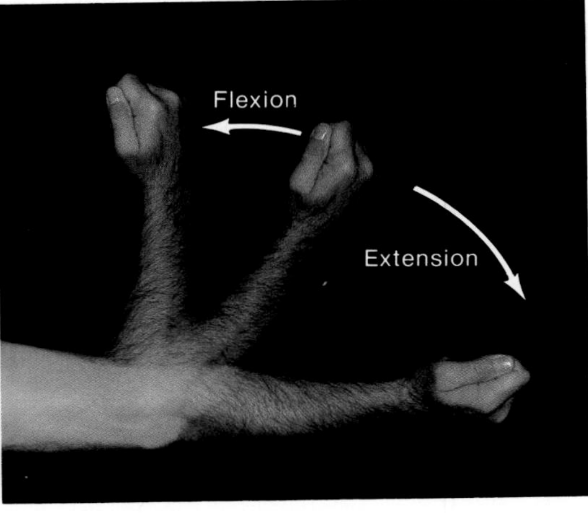

(c)

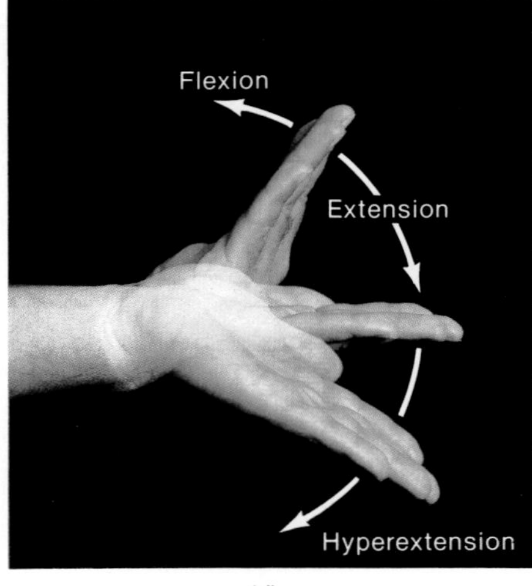

(d)

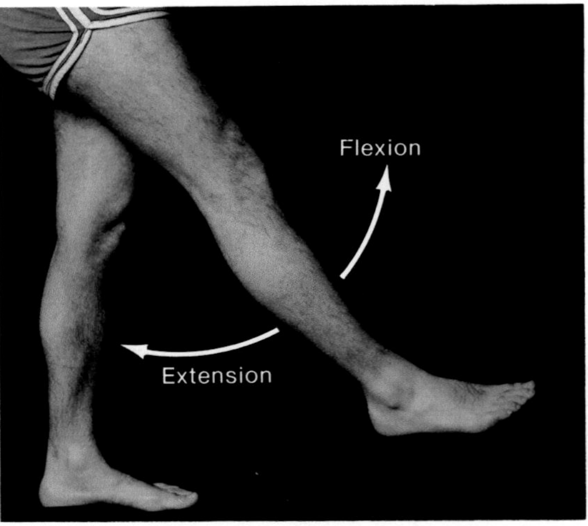

(e)

FIGURE 9-2 (*Continued*)

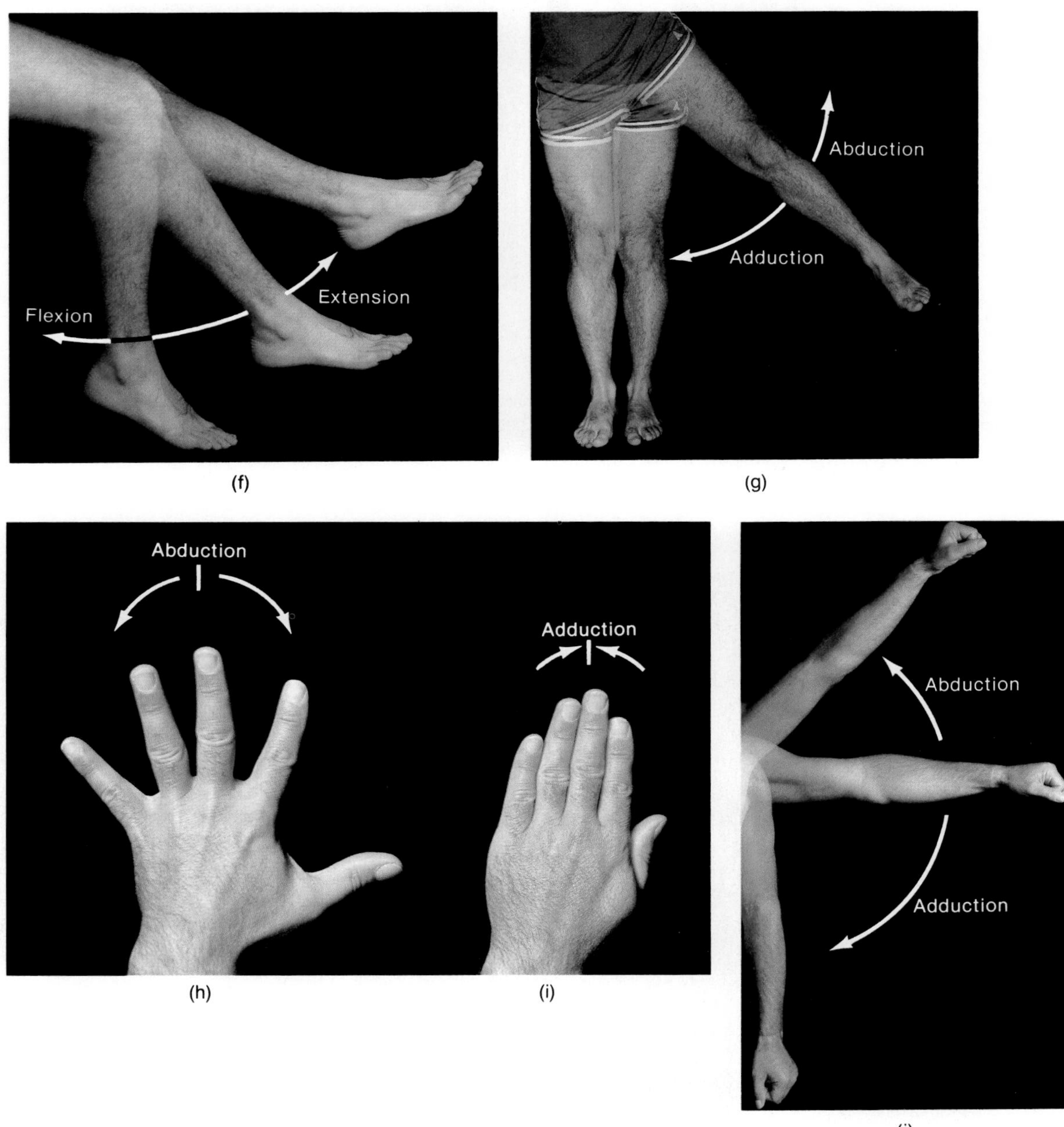

(f) (g) (h) (i) (j)

line. We rotate the atlas around the dens of the axis when we shake the head from side to side (Figure 9-3a). Rotation of the humerus turns the anterior surface of the humerus to face either medially or laterally.

Circumduction

Circumduction is a movement in which the distal end of a bone moves in a circle while the proximal end remains stable. The bone describes a cone in the air. Circumduction typically involves flexion, abduction, adduction, extension, and rotation. It involves a 360° rotation. An example is moving the outstretched arm in a circle to wind up to pitch a ball (Figure 9-3b).

Special

Special movements are those found only at the joints indicated in Figure 9-4. ***Inversion*** is the movement of the sole of the foot inward (medially) so that the soles face toward each other. ***Eversion*** is the movement of the sole outward (laterally) so that the soles face away

FIGURE 9-3 Rotation and circumduction. (a) Rotation at the atlantoaxial joint. (b) Circumduction of the humerus at the shoulder joint. (Copyright © 1983 by Gerard J. Tortora.)

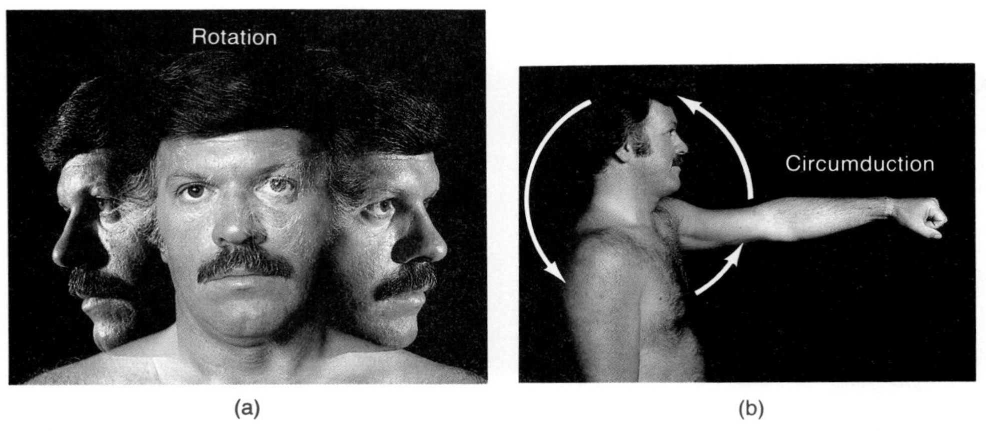

FIGURE 9-4 Special movements. (a) Inversion. (b) Eversion. (c) Dorsiflexion. (d) Plantar flexion. (e) Retraction. (f) Protraction. (g) Pronation. (h) Supination. (i) Elevation. (j) Depression. (Copyright © 1983 by Gerard J. Tortora. Courtesy of Lynne and James Borghesi.)

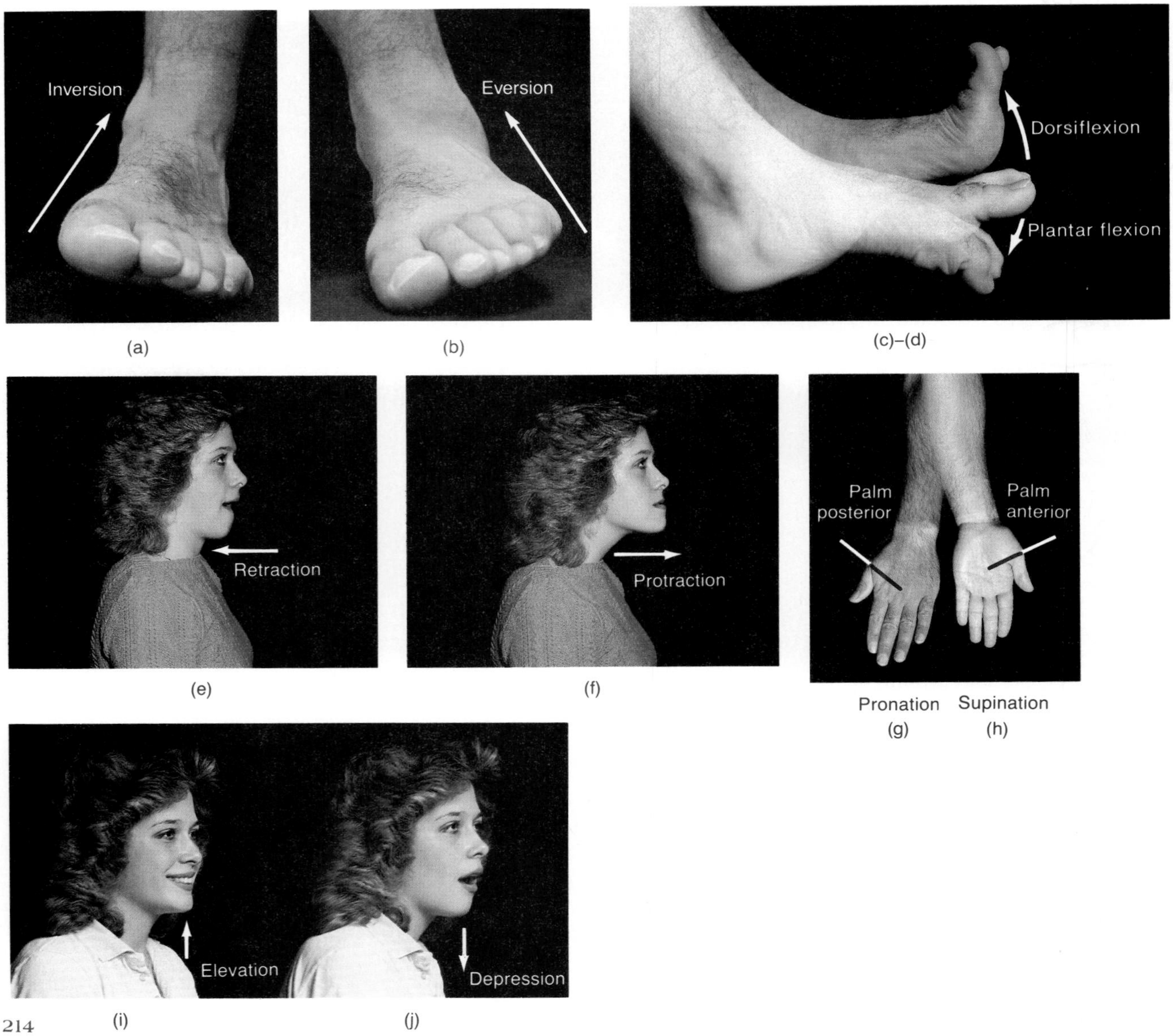

Inversion
(a)

Eversion
(b)

Dorsiflexion
Plantar flexion
(c)–(d)

Retraction
(e)

Protraction
(f)

Palm posterior
Palm anterior
Pronation (g)
Supination (h)

Elevation
(i)

Depression
(j)

from each other. ***Dorsiflexion*** involves bending of the foot in the direction of the dorsum (upper surface). ***Plantar flexion*** involves bending the foot in the direction of the plantar surface (sole).

Protraction is the movement of the mandible or shoulder girdle forward on a plane parallel to the ground. Thrusting the jaw outward is protraction of the mandible. Bringing your arms forward until the elbows touch requires protraction of the clavicle or shoulder girdle. ***Retraction*** is the movement of a protracted part of the body backward on a plane parallel to the ground. Pulling the lower jaw back in line with the upper jaw is retraction of the mandible.

Supination is a movement of the forearm in which the palm of the hand is turned anterior or superior. To demonstrate supination, flex your forearm at the elbow to prevent rotation of the humerus in the shoulder joint. ***Pronation*** is a movement of the forearm in which the palm is turned posterior or inferior.

Elevation is an upward movement of a part of the body. You elevate your mandible when you close your mouth. ***Depression*** is a downward movement of a part of the body. You depress your mandible when you open your mouth. The shoulders can also be elevated and depressed.

A summary of movements that occur at synovial joints is presented in Exhibit 9-1.

Types

Although all synovial joints are similar in structure, variations exist in the shape of the articulating surfaces. Accordingly, synovial joints are divided into six subtypes: gliding, hinge, pivot, ellipsoidal, saddle, and ball-and-socket joints.

Gliding

The articulating surfaces of bones in ***gliding joints*** or ***arthrodia*** (ar-THRŌ-dē-a) are usually flat. Only side-to-side and back-and-forth movements are permitted (Figure

EXHIBIT 9-1 SUMMARY OF MOVEMENTS AT SYNOVIAL JOINTS

Movement	Definition
GLIDING	One surface moves back and forth and from side to side over another surface without angular or rotary motion.
ANGULAR	There is an increase or decrease at the angle between bones.
Flexion	Involves a decrease in the angle between the surfaces of articulating bones.
Extension	Involves an increase in the angle between the surfaces of articulating bones.
Hyperextension	Continuation of extension beyond the anatomical position.
Abduction	Movement of a bone away from the midline.
Adduction	Movement of a bone toward the midline.
ROTATION	Movement of a bone around its longitudinal axis; may be medial or lateral.
CIRCUMDUCTION	A movement in which the distal end of a bone moves in a circle while the proximal end remains stable.
SPECIAL	Occurs at specific joints.
Inversion	Movement of the sole of the foot inward so that the soles face toward each other.
Eversion	Movement of the sole of the foot outward so that the soles face away from each other.
Dorsiflexion	Bending the foot in the direction of the dorsum (upper surface).
Plantar flexion	Bending the foot in the direction of the plantar surface (sole).
Protraction	Movement of the mandible or shoulder girdle forward on a plane parallel to the ground.
Retraction	Movement of a protracted part backward on a plane parallel to the ground.
Supination	Movement of the forearm in which the palm is turned anterior or superior.
Pronation	Movement of the flexed forearm in which the palm is turned posterior or inferior.
Elevation	Movement of a part of the body upward.
Depression	Movement of a part of the body downward.

9-5a). Twisting and rotation are inhibited at gliding joints, generally because ligaments or adjacent bones restrict the range of movement. Since gliding joints do not move around an axis, they are referred to as ***nonaxial.*** Examples are the joints between carpal bones, tarsal bones, the sternum and clavicle, and the scapula and clavicle.

Hinge

A ***hinge*** or ***ginglymus*** (JIN-gli-mus) ***joint*** is one in which the convex surface of one bone fits into the concave surface of another bone. Movement is primarily in a single plane, and the joint is therefore known as ***monaxial,*** or ***uniaxial*** (Figure 9-5b). The motion is similar to that of a hinged door. Movement is flexion and extension. Examples of hinge joints are the elbow, ankle, and interphalangeal joints. The movement allowed by a hinge joint is illustrated by flexion and extension at the elbow (see Figure 9-2c).

Pivot

In a ***pivot*** or ***trochoid*** (TRŌ-koyd) ***joint,*** a rounded, pointed, or conical surface of one bone articulates within

FIGURE 9-5 Subtypes of synovial joints. For each subtype shown, there is a simplified diagram and a photograph of the actual joint. (a) Gliding joint between the navicular and second and third cuneiforms of the tarsus. (b) Hinge joint at the elbow between the trochlea of the humerus and trochlear notch of the ulna. (c) Pivot joint between the head of the radius and the radial notch of the ulna. (d) Ellipsoidal joint at the wrist between the distal end of the radius and the scaphoid and lunate bones of the carpus. (e) Saddle joint between the trapezium of the carpus and the metacarpal of the thumb. (f) Ball-and-socket joint between the head of the femur and the acetabulum of the coxal (hip) bone. (Copyright © 1985 by Michael H. Ross. Used by permission.)

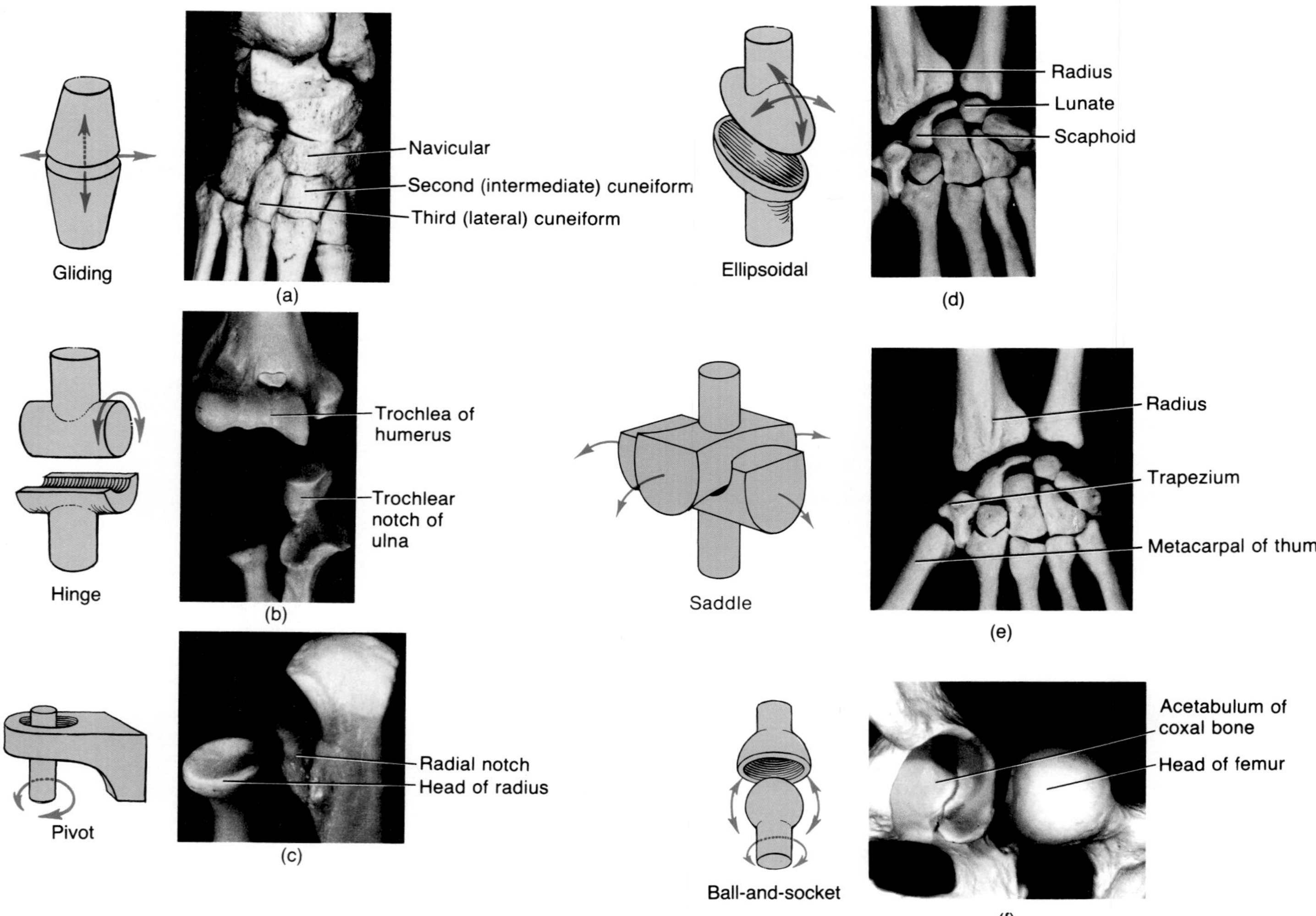

a ring formed partly by another bone and partly by a ligament. The primary movement permitted is rotation, and the joint is therefore ***monaxial*** (Figure 9-5c). Examples include the joint between the atlas and axis (atlantoaxial) and between the proximal ends of the radius and ulna. Movement at a pivot joint is illustrated by supination and pronation of the palms and rotation of the head from side to side (see Figure 9-3a).

Ellipsoidal

In an ***ellipsoidal*** or ***condyloid*** (KON-di-loyd) ***joint,*** an oval-shaped condyle of one bone fits into an elliptical cavity of another bone. Since the joint permits side-to-side and back-and-forth movements, it is ***biaxial*** (Figure 9-5d). The joint at the wrist between the radius and carpals is ellipsoidal. The movement permitted by such a joint is illustrated when you flex and extend (see Figure 9-2d) and abduct and adduct and circumduct the wrist.

Saddle

In a ***saddle*** or ***sellaris*** (sel-A-ris) ***joint,*** the articular surface of one bone is saddle-shaped and the articular surface of the other bone is shaped like a rider sitting in the saddle. Essentially, the saddle joint is a modified ellipsoidal joint in which the movement is somewhat freer. Movements at a saddle joint are side to side and back and forth. Thus, the joint is ***biaxial*** (Figure 9-5e). The joint between the trapezium of the carpus and metacarpal of the thumb is an example of a saddle joint.

Ball-and-Socket

A ***ball-and-socket*** or ***spheroid*** (SFĒ-royd) ***joint*** consists of a ball-like surface of one bone fitted into a cuplike depression of another bone. Such a joint permits ***triaxial*** movement, or movement in three planes of motion: flexion–extension, abduction–adduction, and rotation–circumduction (Figure 9-5f). Examples of ball-and-socket joints are the shoulder joint and coxal (hip) joint. The range of movement at a ball-and-socket joint is illustrated by circumduction of the arm (Figure 9-3b).

Summary of Joints

The summary of joints presented in Exhibit 9-2 is based on the anatomy of the joints. If we rearrange the types of joints into a classification based on movement, we arrive at the following:

Synarthroses: immovable joints
1. **Suture**
2. **Synchondrosis**
3. **Gomphosis**

Amphiarthroses: slightly movable joints
1. **Syndesmosis**
2. **Symphysis**

Diarthroses: freely movable joints
1. **Gliding**
2. **Hinge**
3. **Pivot**
4. **Ellipsoidal**
5. **Saddle**
6. **Ball-and-socket**

SELECTED ARTICULATIONS OF THE BODY

We will now examine in some detail the structure of three articulations of the body: humeroscapular (shoulder), coxal (hip), and tibiofemoral (knee) joint.

Humeroscapular (Shoulder) Joint

The ***humeroscapular (shoulder) joint*** is formed by the head of the humerus and the glenoid cavity of the scapula (Figure 9-6). It is a ball-and-socket (spheroid) joint.

Its anatomical components are as follows:

1. Articular capsule. Loose sac that completely envelops the joint, extending from the circumference of the glenoid cavity to the anatomical neck of the humerus.
2. Coracohumeral ligament. Strong, broad ligament that extends from the coracoid process of the scapula to the greater tubercle of the humerus.
3. Glenohumeral ligaments. Three thickenings of the articular capsule over the ventral surface of the joint.
4. Transverse humeral ligament. Narrow sheet extending from the greater tubercle to the lesser tubercle of the humerus.
5. Glenoid labrum. Narrow rim of fibrocartilage around the edge of the glenoid cavity.
6. Among the ***bursae*** associated with the shoulder joint are:
   a. **Subscapular bursa** between the tendon of the subscapularis muscle and the underlying joint capsule.
   b. **Subdeltoid bursa** between the deltoid muscle and joint capsule.
   c. **Subacromial bursa** between the acromion and joint capsule.
   d. **Subcoracoid bursa** either lies between the coracoid process and joint capsule or appears as an extension from the subacromial bursa.

CLINICAL APPLICATION: SEPARATED SHOULDER

Separation of the shoulder refers to dislocation of the acromioclavicular joint or displacement of the head of the humerus from the glenoid cavity.

EXHIBIT 9-2 SUMMARY OF JOINTS

Type	Description	Movement	Examples
FIBROUS	No synovial (joint) cavity; bones held together by a thin layer of fibrous tissue or dense fibrous tissue.		
Suture	Found only between bones of the skull; articulating bones separated by a thin layer of fibrous tissue.	None (synarthrotic).	Lambdoidal suture between occipital and parietal bones.
Syndesmosis	Articulating bones united by dense fibrous tissue.	Slight (amphiarthrotic).	Distal ends of tibia and fibula.
Gomphosis	Cone-shaped peg fits into a socket; articulating bones separated by periodontal ligament.	None (synarthrotic).	Roots of teeth in alveoli (sockets).
CARTILAGINOUS	No synovial cavity; articulating bones united by cartilage.		
Synchondrosis	Connecting material is hyaline cartilage.	None (synarthrotic).	Temporary joint between the diaphysis and epiphyses of a long bone and permanent joint between true ribs and sternum.
Symphysis	Connecting material is a broad, flat disc of fibrocartilage.	Slight (amphiarthrotic).	Intervertebral joints and symphysis pubis.
SYNOVIAL	Synovial cavity and articular cartilage present; articular capsule composed of an outer fibrous capsule and an inner synovial membrane; may contain accessory ligaments, articular discs (menisci), and bursae.	Freely movable (diarthrotic).	
Gliding	Articulating surfaces usually flat.	Nonaxial.	Intercarpal and intertarsal joints.
Hinge	Spoollike surface fits into a concave surface.	Monaxial (flexion–extension).	Elbow, ankle, and interphalangeal joints.
Pivot	Rounded, pointed, or concave surface fits into a ring formed partly by bone and partly by a ligament.	Monaxial (rotation).	Atlantoaxial and radioulnar joints.
Ellipsoidal	Oval-shaped condyle fits into an elliptical cavity.	Biaxial (flexion–extension, abduction–adduction).	Radiocarpal joint.
Saddle	Articular surface of one bone is saddle-shaped, and the articular surface of the other bone is shaped like a rider sitting in the saddle.	Biaxial (flexion–extension, abduction–adduction).	Carpometacarpal joint of thumb.
Ball-and-socket	Ball-like surface fits into a cuplike depression.	Triaxial (flexion–extension, abduction–adduction, rotation–circumduction).	Shoulder and coxal (hip) joints.

FIGURE 9-6 **Humeroscapular (shoulder) joint seen in anterior view.**

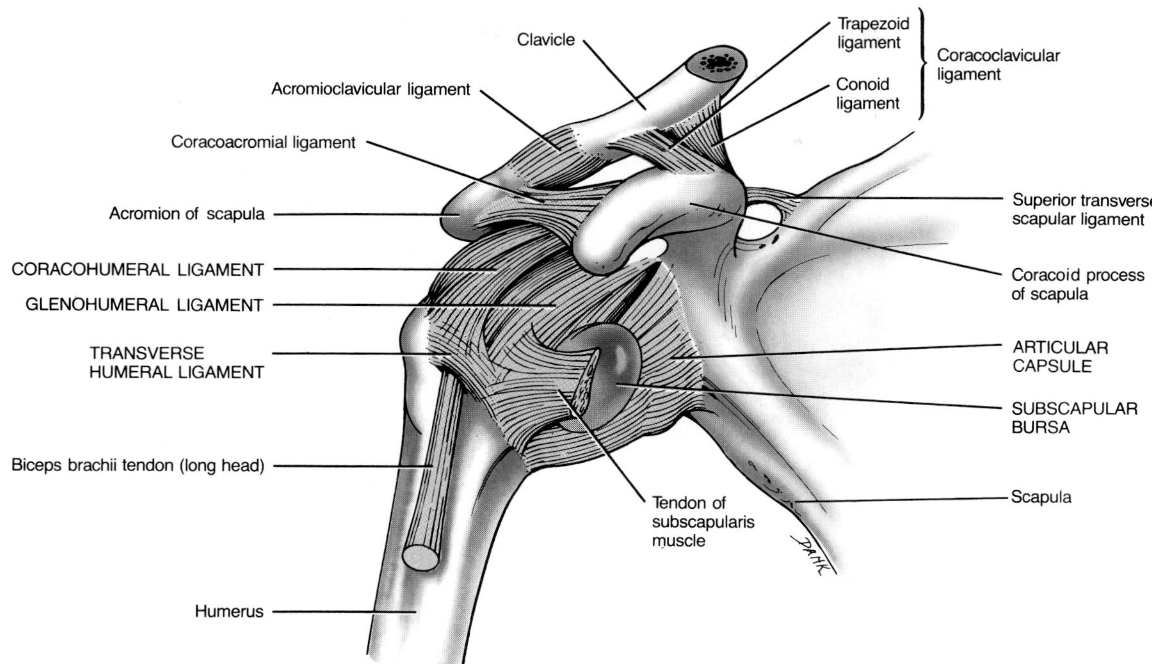

Coxal (Hip) Joint

The ***coxal (hip) joint*** is formed by the head of the femur and the acetabulum of the coxal bone (Figure 9-7). It is a ball-and-socket (spheroid) joint.

Its anatomical components are as follows:

1. **Articular capsule.** Extends from the rim of the acetabulum to the neck of the femur. One of the strongest ligaments of the body, the capsule consists of circular and longitudinal fibers. The circular fibers, called the ***zona orbicularis,*** form a collar around the neck of the femur. The longitudinal fibers are reinforced by accessory ligaments known as the iliofemoral ligament, pubofemoral ligament, and ischiofemoral ligament.
2. **Iliofemoral ligament.** Thickened portion of the articular capsule that extends from the anterior inferior iliac spine of the coxal bone to the intertrochanteric line of the femur.
3. **Pubofemoral ligament.** Thickened portion of the articular capsule that extends from the pubic part of the rim of the acetabulum to the neck of the femur.
4. **Ischiofemoral ligament.** Thickened portion of the articular capsule that extends from the ischial wall of the acetabulum to the neck of the femur.
5. **Ligament of the head of the femur (capitate ligament).** Flat, triangular band that extends from the fossa of the acetabulum to the head of the femur.
6. **Acetabular labrum.** Fibrocartilage rim attached to the margin of the acetabulum.
7. **Transverse ligament of the acetabulum.** Strong ligament that crosses over the acetabular notch, converting it to a foramen. It supports part of the acetabular labrum and is connected with the ligament of the head of the femur and the articular capsule.

CLINICAL APPLICATION: DISLOCATED HIP

Dislocation of the hip **among adults is quite rare because of (1) the stability of the ball-and-socket joint, (2) the strong, tough, articular capsule, (3) the strength of the intracapsular ligaments, and (4) the extensive musculature over the joint.**

Tibiofemoral (Knee) Joint

The ***tibiofemoral (knee) joint*** is the largest joint of the body, actually consisting of three joints: (1) an intermediate patellofemoral joint between the patella and the patellar surface of the femur, (2) a lateral tibiofemoral joint between the lateral condyle of the femur, lateral meniscus, and lateral condyle of the tibia, and (3) a medial tibiofemoral joint between the medial condyle of the femur, medial meniscus, and medial condyle of the tibia (Figure 9-8). The patellofemoral joint is a gliding (arthrodial) joint, and the lateral and medial tibiofemoral joints are hinge (ginglymus) joints.

The anatomical components of the tibiofemoral joint are as follows:

FIGURE 9-7 **Coxal (hip) joint. (a) Frontal section. (b) Anterior view. (c) Posterior view.**

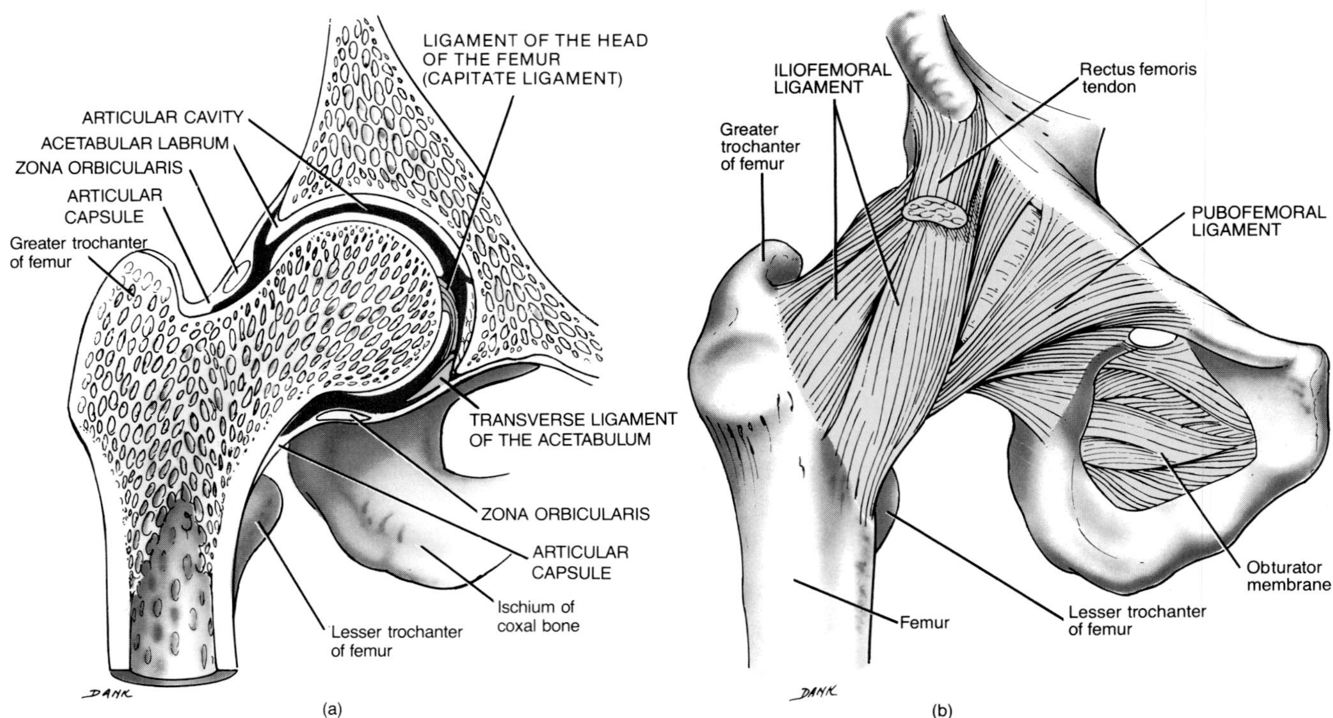

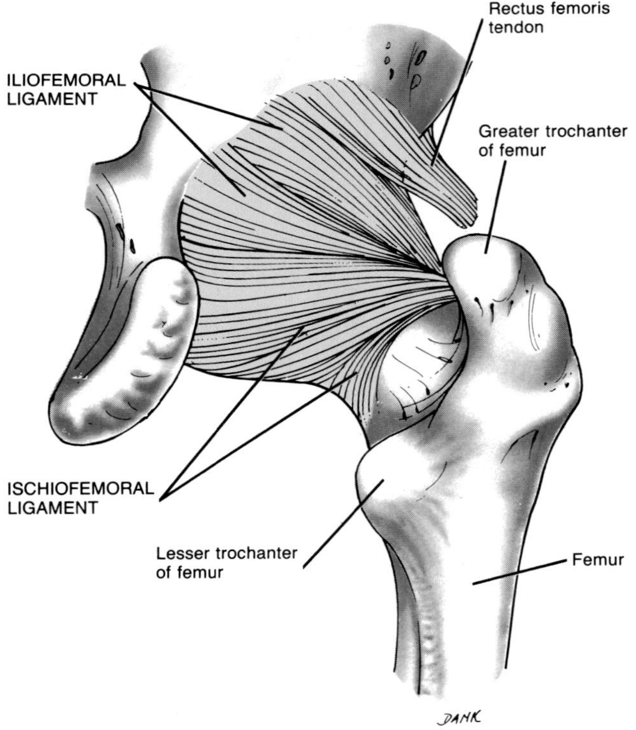

1. **Articular capsule.** No complete, independent capsule unites the bones. The ligamentous sheath surrounding the joint consists mostly of muscle tendons or expansions of them. There are, however, some capsular fibers connecting the articulating bones.

2. **Medial and lateral patellar retinacula.** Fused tendons of insertion of the quadriceps femoris muscle and the fascia lata that strengthen the anterior surface of the joint.

3. **Patellar ligament.** Central portion of the common tendon of insertion of the quadriceps femoris muscle that extends from the patella to the tibial tuberosity. This ligament also strengthens the anterior surface of the joint. The posterior surface of the ligament is separated from the synovial membrane of the joint by an ***infrapatellar fat pad.***

4. **Oblique popliteal ligament.** Broad, flat ligament that extends from the intercondylar fossa of the femur to the head of the tibia. The tendon of the semimembranous muscle is superficial to the ligament and passes from the medial condyle of the tibia to the lateral condyle of the femur. The ligament and tendon afford strength for the posterior surface of the joint.

5. **Arcuate popliteal ligament.** Extends from the lateral condyle of the femur to the styloid process of the head of the fibula. It strengthens the lower lateral part of the posterior surface of the joint.

6. **Tibial collateral ligament.** Broad, flat ligament on the medial surface of the joint that extends from the medial condyle of the femur to the medial condyle of the tibia. The ligament is crossed by tendons of the sartorius, gracilis, and

FIGURE 9-8 Tibiofemoral (knee) joint. (a) Diagram of anterior view. (b) Diagram of posterior view. (c) Diagram of sagittal section. (d) Photograph of sagittal section. (Courtesy of J. A. Gosling, P. F. Harris, et al., *Atlas of Human Anatomy,* Gower Medical Publishing Ltd., 1985.)

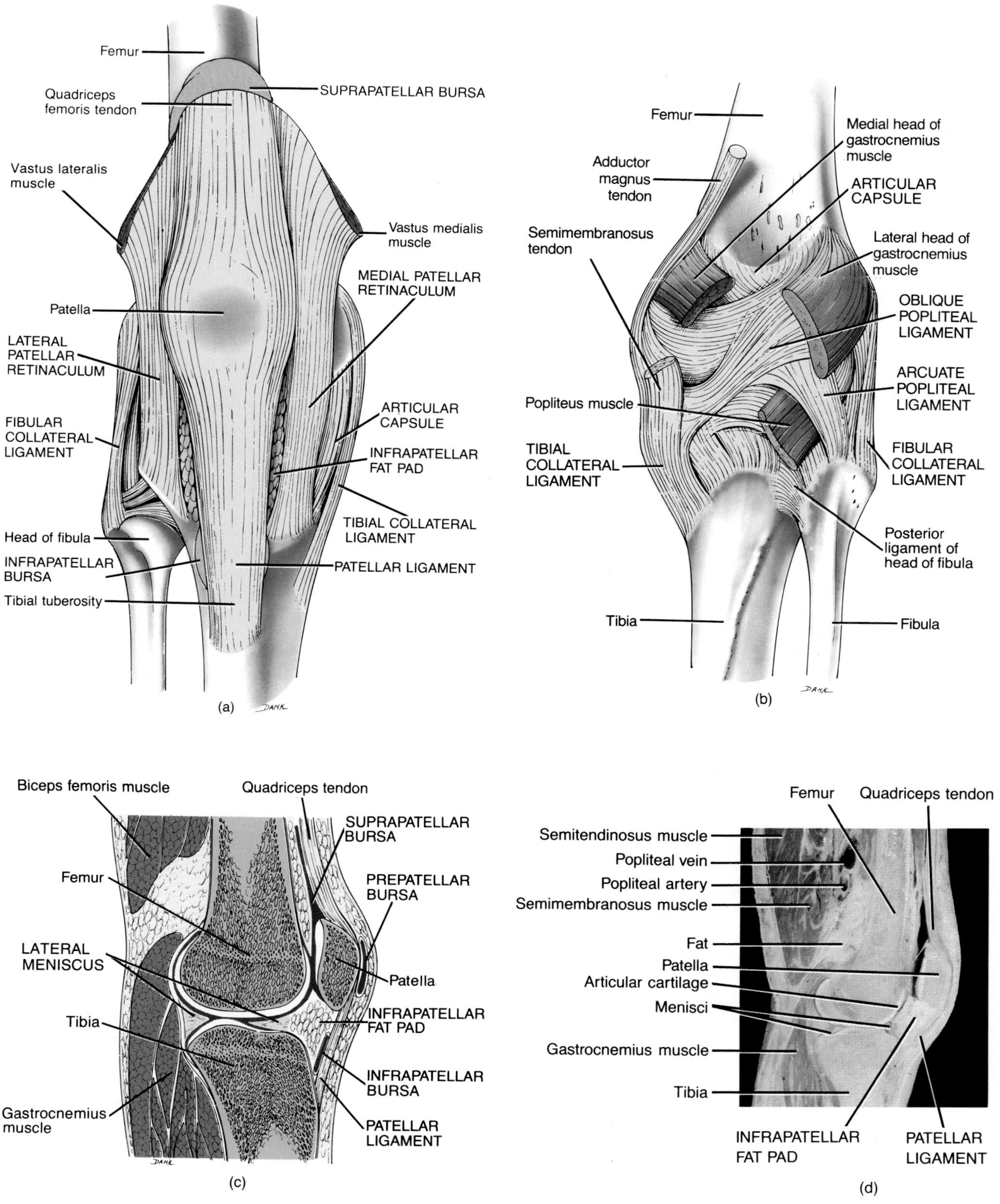

Femur
Quadriceps femoris tendon
SUPRAPATELLAR BURSA
Vastus lateralis muscle
Vastus medialis muscle
MEDIAL PATELLAR RETINACULUM
Patella
LATERAL PATELLAR RETINACULUM
FIBULAR COLLATERAL LIGAMENT
ARTICULAR CAPSULE
INFRAPATELLAR FAT PAD
TIBIAL COLLATERAL LIGAMENT
Head of fibula
INFRAPATELLAR BURSA
PATELLAR LIGAMENT
Tibial tuberosity
(a)
Femur
Adductor magnus tendon
Medial head of gastrocnemius muscle
ARTICULAR CAPSULE
Semimembranosus tendon
Lateral head of gastrocnemius muscle
OBLIQUE POPLITEAL LIGAMENT
ARCUATE POPLITEAL LIGAMENT
Popliteus muscle
TIBIAL COLLATERAL LIGAMENT
FIBULAR COLLATERAL LIGAMENT
Posterior ligament of head of fibula
Tibia
Fibula
(b)
Biceps femoris muscle
Quadriceps tendon
SUPRAPATELLAR BURSA
Femur
PREPATELLAR BURSA
LATERAL MENISCUS
Patella
INFRAPATELLAR FAT PAD
Tibia
INFRAPATELLAR BURSA
Gastrocnemius muscle
PATELLAR LIGAMENT
(c)
Femur
Quadriceps tendon
Semitendinosus muscle
Popliteal vein
Popliteal artery
Semimembranosus muscle
Fat
Patella
Articular cartilage
Menisci
Gastrocnemius muscle
Tibia
INFRAPATELLAR FAT PAD
PATELLAR LIGAMENT
(d)

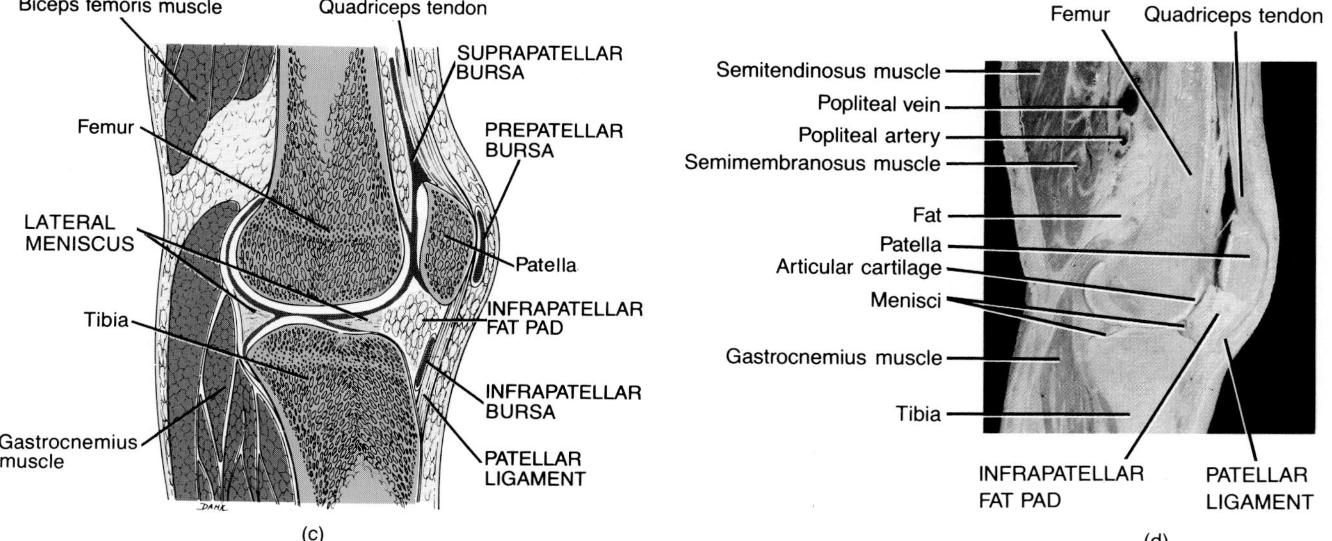

FIGURE 9-8 *(Continued)* (e) Diagram of anterior view (flexed).

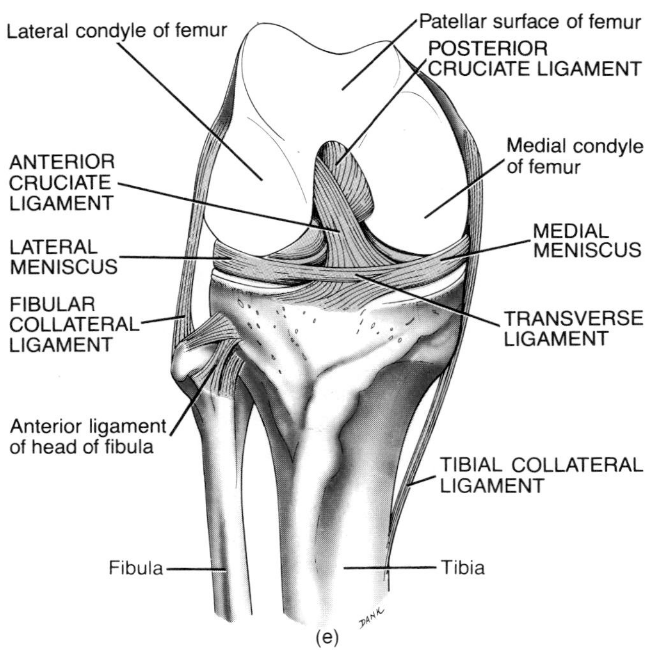

semitendinosus muscles, all of which strengthen the medial aspect of the joint.

7. Fibular collateral ligament. Strong, rounded ligament on the lateral surface of the joint that extends from the lateral condyle of the femur to the lateral side of the head of the fibula. The ligament is covered by the tendon of the biceps femoris muscle. The tendon of the popliteal muscle is deep to the tendon.

8. Intraarticular ligaments. Ligaments within the capsule that connect the tibia and femur.

a. **Anterior cruciate ligament.** Extends posteriorly and laterally from the area anterior to the intercondylar eminence of the tibia to the posterior part of the medial surface of the lateral condyle of the femur. This ligament is stretched or torn in about 70 percent of all serious knee injuries. Both O. J. Simpson and Gale Sayers had their careers in professional football ended by torn anterior cruciate ligaments.

b. **Posterior cruciate ligament.** Extends anteriorly and medially from the posterior intercondylar fossa of the tibia and lateral meniscus to the anterior part of the medial surface of the medial condyle of the femur.

9. Articular discs. Fibrocartilage discs between the tibial and femoral condyles. They help to compensate for the incongruence of the articulating bones.

a. **Medial meniscus.** Semicircular piece of fibrocartilage (C-shaped). Its anterior end is attached to the anterior intercondylar fossa of the tibia, in front of the anterior cruciate ligament. Its posterior end is attached to the posterior intercondylar fossa of the tibia between the attachments of the posterior cruciate ligament and lateral meniscus.

b. **Lateral meniscus.** Nearly circular piece of fibrocartilage (approaches an incomplete O in shape). Its anterior end is attached anterior to the intercondylar eminence of the tibia and lateral and posterior to the anterior cruciate ligament. Its posterior end is attached posterior to the intercondylar eminence of the tibia and anterior to the posterior end of the medial meniscus. The medial and lateral menisci are connected to each other by the ***transverse ligament*** and to the margins of the head of the tibia by the ***coronary ligaments.***

10. The principal ***bursae*** of the knee include:

a. **Anterior bursae.** (1) Between the patella and skin **(*prepatellar bursa*)**, (2) between upper part of tibia and patellar ligament **(*infrapatellar bursa*)**, (3) between lower part of tibial tuberosity and skin, and (4) between lower part of femur and deep surface of quadriceps femoris muscle **(*suprapatellar bursa*)**.

b. **Medial bursae.** (1) Between medial head of gastrocnemius muscle and the articular capsule; (2) superficial to the tibial collateral ligament between the ligament and tendons of the sartorius, gracilis, and semitendinosus muscles; (3) deep to the tibial collateral ligament between the ligament and the tendon of the semimembranosus muscle; (4) between the tendon of the semimembranosus muscle and the head of the tibia; and (5) between the tendons of the semimembranosus and semitendinosus muscles.

c. **Lateral bursae.** (1) Between the lateral head of the gastrocnemius muscle and articular capsule, (2) between the tendon of the biceps femoris muscle and fibular collateral ligament, (3) between the tendon of the popliteal muscle and fibular collateral ligament, and (4) between the lateral condyle of the femur and the popliteal muscle.

CLINICAL APPLICATION: KNEE PROBLEMS

The most common type of *knee injury* in football is rupture of the tibial collateral ligament, often associated with tearing of the anterior cruciate ligament and medial meniscus (torn cartilage). It is caused by a blow to the lateral side of the knee. When a knee is examined for such an injury, the three Cs are kept in mind: collateral ligament, cruciate ligament, and cartilage.

A *swollen knee* may occur immediately or be delayed. The immediate swelling is due to escape of blood from rupture of the anterior cruciate ligament, torn menisci, fractures, or collateral ligament sprains. Delayed swelling is due to an excessive pro-

duction of synovial fluid as a result of conditions that irritate the synovial membrane.

A ***dislocated knee*** refers to the displacement of the tibia relative to the femur. Accordingly, such dislocations are classified as anterior, posterior, medial, lateral, or rotatory. The most common type is anterior dislocation, resulting from hyperextension of the knee. A frequent consequence of a dislocated knee is damage to the popliteal artery.

In November 1987, a group of orthopedic surgeons at the Hospital of the University of Pennsylvania performed the ***first transplant of an entire human knee.*** The surgery was necessitated by a potentially malignant tumor on the knee of a 32-year-old female recipient. The donor was an 18-year-old male. Once the tumor was removed, the patient's knee joint was removed and replaced with the donor's joint. The donor knee joint, about 41 cm (16 inches) long, consisted of the lower portion of a femur that was connected to the recipient's femur by a rod; the upper portion of the tibia and head of the fibula that were connected by a metal plate; the patella; medial and lateral menisci; intracapsular and extracapsular ligaments; and certain tendons. The recipient's own muscles, nerves, and blood vessels were used. Since rejection occurs in only about 3 to 5 percent of bone transplant cases, no immunosuppressive drugs were needed. Although the recipient will not be permitted to participate in strenuous exercise and contact sports, she is expected to have near normal use of her leg.

Exhibit 9-3 contains a listing of selected joints of the body in terms of articulating components, classification, and movements.

EXHIBIT 9-3 REPRESENTATIVE JOINTS ACCORDING TO ARTICULAR COMPONENTS, CLASSIFICATION, AND MOVEMENTS

TEMPOROMANDIBULAR JOINT (TMJ)

Definition	Joint formed by mandibular condyle of the mandible and mandibular fossa and articular tubercle of temporal bone. The temporomandibular joint (TMJ) is the only movable joint between skull bones; all other skull joints are sutures and therefore immovable.
Type of joint	Synovial; combined hinge (ginglymus) and gliding (arthrodial) type.
Movements	Only the mandible moves since the maxilla is firmly anchored to other bones by sutures. Accordingly, the mandible may function in depression (jaw opening), elevation (jaw closing), protraction, retraction, lateral displacement, and slight rotation.

ATLANTO-OCCIPITAL JOINT

Definition	Joint formed by the superior articular surfaces of the atlas and the occipital condyles of the occipital bone.
Type of joint	Synovial; ellipsoidal (condyloid) type.
Movements	Flexion (forward movement of head), extension (backward movement of head), and slight lateral tilting of head to either side.

INTERVERTEBRAL JOINTS

Definition	Joints formed between (1) vertebral bodies and (2) vertebral arches.
Type of joint	Joints between vertebral bodies—cartilaginous (fibrocartilage), symphysis type. Joints between vertebral arches—synovial, gliding (arthrodial) type.
Movements	Flexion (bending the backbone forward), extension (bending the backbone backward), lateral displacement (bending the backbone to either side), and rotation.

LUMBOSACRAL JOINT

Definition	Joint formed by the body of the fifth lumbar vertebra and the superior surface of the first sacral vertebra of the sacrum.
Type of joint	Joint between the bodies of the fifth lumbar vertebra and the first sacral vertebra—cartilaginous joint, symphysis type. Joint between the articular processes—synovial joint, gliding (arthrodial) type.
Movements	Similar to those of intervertebral joints.

EXHIBIT 9-3 REPRESENTATIVE JOINTS ACCORDING TO ARTICULAR COMPONENTS, CLASSIFICATION, AND MOVEMENTS (*Continued*)

HUMEROSCAPULAR OR GLENOHUMERAL (SHOULDER) JOINT	
Definition	Joint formed by the head of the humerus and the glenoid cavity of the scapula.
Type of joint	Synovial joint, ball-and-socket (spheroid) type.
Movements	Flexion (humerus drawn forward), extension (humerus drawn backward), abduction (humerus drawn away from midline), adduction (humerus drawn toward midline), medial rotation, lateral rotation, and circumduction.
ELBOW JOINT	
Definition	Joint formed by the trochlea of the humerus, the trochlear notch of the ulna, and the head of the radius.
Type of joint	Synovial joint, hinge (ginglymus) type.
Movements	Flexion and extension of forearm.
RADIOCARPAL (WRIST) JOINT	
Definition	Joint formed by the distal end of the radius, the distal surface of the articular disc separating the carpal and distal radioulnar joint, and the scaphoid, lunate, and triquetral carpal bones.
Type of joint	Synovial joint, ellipsoidal (condyloid) type.
Movements	Flexion, extension, abduction, adduction, and circumduction.
COXAL (HIP) JOINT	
Definition	Joint formed by the head of the femur and the acetabulum of the coxal bone.
Type of joint	Synovial, ball-and-socket (spheroid) type.
Movements	Flexion, extension, abduction, adduction, circumduction, and rotation.
TIBIOFEMORAL (KNEE) JOINT	
Definition	The largest joint of the body, actually consisting of three joints: (1) an intermediate patellofemoral joint between the patella and the patellar surface of the femur; (2) a lateral tibiofemoral joint between the lateral condyle of the femur, lateral meniscus, and lateral condyle of the tibia; and (3) a medial tibiofemoral joint between the medial condyle of the femur, medial meniscus, and medial condyle of the tibia.
Type of joint	Patellofemoral joint—partly synovial, gliding (arthrodial) type. Lateral and medial tibiofemoral joints—synovial, hinge (ginglymus) type.
Movements	Flexion, extension, medial rotation, and lateral rotation.
TALOCRURAL (ANKLE) JOINTS	
Definition	Joints between (1) the distal end of the tibia and its medial malleolus and the talus and (2) the lateral malleolus of the fibula and the talus.
Type of joint	Both joints—synovial, hinge (ginglymus) type.
Movements	Dorsiflexion and plantar flexion.

DISORDERS: HOMEOSTATIC IMBALANCES

Rheumatism

Rheumatism (*rheumat* = subject to flux) refers to any painful state of the supporting structures of the body—its bones, ligaments, joints, tendons, or muscles. Arthritis is a form of rheumatism in which the joints have become inflamed.

Arthritis

The term ***arthritis*** refers to many different diseases, the most common of which are rheumatoid arthritis (RA), osteoarthritis, and gouty arthritis. All are characterized by inflammation of one or more joints. Inflammation, pain, and stiffness may also be present in adjacent parts of the body, such as the muscles near the joint.

Rheumatoid Arthritis (RA)

Rheumatoid (ROO-ma-toyd) ***arthritis (RA)*** is an autoimmune disease in which the body attacks its own tissues, in this case its own cartilage and joint linings. It is characterized by inflammation of the joint, swelling, pain, and loss of function. Usually, this form occurs bilaterally—if your left wrist is affected, your right wrist may also be affected, although usually not to the same degree.

The primary symptom of rheumatoid arthritis is inflammation of the synovial membrane. If untreated, the following sequential pathology may occur. The membrane thickens and synovial fluid accumulates. The resulting pressure causes pain and tenderness. The membrane then produces an abnormal granulation tissue called a ***pannus,*** which adheres to the surface of the articular cartilage. The pannus formation sometimes erodes the cartilage completely. When the cartilage is destroyed, fibrous tissue joins the exposed bone ends. The tissue ossifies and fuses the joint so that it is immovable—the ultimate crippling effect of rheumatoid arthritis. Most cases do not progress to this stage, but the range of motion of the joint is greatly inhibited by the severe inflammation and swelling. The growth of the pannus is what causes the distortion of the fingers that is so typical of the clinical appearance of hands that have been affected by rheumatoid arthritis. A plastic surgeon or orthopedic surgeon will remove the pannus growth before crippling and dysfunctional deformities have occurred—long before the joints are fused to the point of immobility.

Treatment is aimed at reducing pain and inflammation and preserving muscle strength and joint function. Therapies include adequate rest; antiinflammatory drugs, such as aspirin and in selected patients steroids; exercises to maintain full range of joint motion; heat and other forms of physical therapy; and weight loss to relieve pressure on weight-bearing joints. In severe cases, damaged joints may be surgically replaced, either partly or entirely, with artificial joints. The artificial parts are inserted after removal of the diseased portion of the articulating bone and its cartilage. The new metal or plastic joint is fixed in place with a special acrylic cement. When freshly mixed in the operating room, it hardens as strong as bone in minutes.

Osteoarthritis

A degenerative joint disease far more common than rheumatoid arthritis, and usually less damaging, is ***osteoarthritis*** (os'-tē-ō-ar-THRĪ-tis). It apparently results from a combination of aging, irritation of the joints, and wear and abrasion.

Degenerative joint disease is a noninflammatory, progressive disorder of movable joints, particularly weight-bearing joints. It is characterized by the deterioration of articular cartilage and by formation of new bone in the subchondral areas and at the margins of the joint. The cartilage slowly degenerates, and as the bone ends become exposed, small bumps, or ***spurs,*** of new osseous tissue are deposited on them. These spurs decrease the space of the joint cavity and restrict joint movement. Unlike rheumatoid arthritis, osteoarthritis usually affects only the articular cartilage. The synovial membrane is rarely destroyed, and other tissues are unaffected. The main distinction between osteoarthritis and rheumatoid arthritis is that the former strikes the big joints (knees, hips) first, whereas the latter strikes the small joints first. But osteoarthritis may affect the fingers, and when this is the case, the distal phalanges show the most prominent changes. The effect of rheumatoid arthritis on the fingers is most pronounced proximally, in the wrists and in the metacarpophalangeal, and proximal interphalangeal joints. Also, rheumatoid arthritis is more likely to be bilaterally symmetrical than is the case for osteoarthritis.

Gouty Arthritis

Uric acid (a substance that gives urine its name) is a waste product produced during the metabolism of nucleic acids. The person who suffers from ***gout*** either produces excessive amounts of uric acid or is not able to excrete normal amounts. The result is a buildup of uric acid in the blood. This excess acid then reacts with sodium to form a salt called sodium urate. Crystals of this salt are deposited in soft tissues. Typical sites are the kidneys and the cartilage of the ears and joints.

In ***gouty*** (GOW-tē) ***arthritis,*** sodium urate crystals are deposited in the soft tissues of the joints. The crystals irritate the cartilage, causing inflammation, swelling, and acute pain. Eventually, the crystals destroy all the joint tissues. If the disorder is not treated, the ends of the articulating bones fuse and the joint becomes immovable.

Gouty arthritis occurs primarily in middle-aged and older males. It is believed to be the cause of 2 to 5 percent of all chronic joint diseases. Numerous studies indicate that gouty arthritis is sometimes caused by an abnormal gene. As a result of this gene, the body manufactures unusually large amounts of uric acid. Diet and environmental factors such as stress and climate are also suspected causes of gouty arthritis.

DISORDERS: HOMEOSTATIC IMBALANCES (*continued*)

Although other forms of arthritis cannot be treated with complete success, the treatment of gouty arthritis with the use of various drugs has been quite effective. A chemical called colchicine has been utilized periodically since the sixth century to relieve the pain, swelling, and tissue destruction that occur during attacks of gouty arthritis. This chemical is derived from the variety of crocus plant from which the spice saffron is obtained. Other drugs, which either inhibit uric acid production or assist in the elimination of excess uric acid by the kidneys, are used to prevent further attacks. The drug allopurinal is used for the treatment of gouty arthritis because it prevents the formation of uric acid without interfering with nucleic acid synthesis. Relief from an attack of gouty arthritis may be obtained by rest and the use of antiinflammatory medication.

Bursitis

An acute chronic inflammation of a bursa is called ***bursitis.*** The condition may be caused by trauma, by an acute or chronic infection (including syphilis and tuberculosis), or by rheumatoid arthritis. Repeated excessive friction often results in a bursitis with local inflammation and the accumulation of fluid. Bunions are frequently associated with a friction bursitis over the head of the first metatarsal bone. Symptoms include pain, swelling, tenderness, and the limitation of motion involving the inflamed bursa. The prepatellar or subcutaneous infrapatellar bursae may become inflamed in individuals who spend a great deal of time kneeling. This bursitis is usually called ***"housemaid's knee" ("carpet layer's knee").***

Dislocation

A ***dislocation,*** or ***luxation*** (luks-Ā-shun), is the displacement of a bone from a joint with tearing of ligaments, tendons, and articular capsules. A partial or incomplete dislocation is called a ***subluxation.*** The most common dislocations are those involving a finger or shoulder. Those of the mandible, elbow, knee, or hip are less common. Symptoms include loss of motion, temporary paralysis of the involved joint, pain, swelling, and occasionally shock. A dislocation is usually caused by a blow or fall, although unusual physical effort may lead to this condition.

Sprain and Strain

A ***sprain*** is the forcible wrenching or twisting of a joint with partial rupture or other injury to its attachments without luxation. It occurs when the attachments are stressed beyond their normal capacity. There may be damage to the associated blood vessels, muscles, tendons, ligaments, or nerves. A sprain is more serious than a ***strain,*** which is the overstretching of a muscle. Severe sprains may be so painful that the joint cannot be moved. There is considerable swelling and pain may occur owing to underlying hemorrhage from ruptured blood vessels. The ankle joint is most often sprained; the low back area is another frequent location for sprains.

MEDICAL TERMINOLOGY ASSOCIATED WITH ARTICULATIONS

Ankylosis (ang'-ki-LŌ-sus; *ankyle* = stiff joint; *osis* = condition) Severe or complete loss of a movement at a joint.

Arthralgia (ar-THRAL-jē-a; *arth* = joint; *algia* = pain) Pain in a joint.

Arthrosis (ar-THRŌ-sis) Refers to an articulation; also a disease of a joint.

Bursectomy (bur-SEK-tō-mē; *ectomy* = removal of) Removal of a bursa.

Chondritis (kon-DRĪ-tis; *chondro* = cartilage) Inflammation of cartilage.

Rheumatology (roo'-ma-TOL-ō-jē; *rheumat* = subject to flux) The study of joints; the field of medicine devoted to joint diseases and related conditions.

Synovitis (sin'-ō-VĪ-tis; *synov* = joint) Inflammation of a synovial membrane in a joint.

STUDY OUTLINE

Classification (p. 208)

1. An articulation (joint) is a point of contact between two or more bones.
2. Functional classification of joints is based on the degree of movement permitted. Joints may be synarthroses, amphiarthroses, or diarthroses.

3. Structural classification is based on the presence of a synovial (joint) cavity and type of connecting tissue. Structurally, joints are classified as fibrous, cartilaginous, or synovial.

Fibrous Joints (p. 208)

1. Bones held by fibrous connective tissue, with no synovial cavity, are fibrous joints.
2. These joints include immovable sutures (found in the skull), slightly movable syndesmoses (such as the tibiofibular articulation), and immovable gomphoses (roots of teeth in alveoli of mandible and maxilla).

Cartilaginous Joints (p. 208)

1. Bones held together by cartilage, with no synovial cavity, are cartilaginous joints.
2. These joints include immovable synchondroses united by hyaline cartilage (temporary cartilage between diaphysis and epiphyses) and partially movable symphyses united by fibrocartilage (the symphysis pubis).

Synovial Joints (p. 209)

1. Synovial joints contain a synovial cavity, articular cartilage, and a synovial membrane; some also contain ligaments, articular discs, and bursae.
2. All synovial joints are freely movable.
3. Movements at synovial joints are limited by the structure (shape) of articulating bones, arrangement and tension of ligaments, muscle arrangement and tension, and apposition of soft parts.
4. Types of movements at synovial joints include gliding movements, angular movements, rotation, circumduction, and special movements.
5. Types of synovial joints include gliding joints (between wrist bones), hinge joints (elbow), pivot joints (between radius and ulna), ellipsoidal joints (between radius and wrist), saddle joints (between trapezium of wrist and metacarpal of thumb), and ball-and-socket joints (shoulder and coxal).
6. A joint may be described according to the number of planes of movement it allows as nonaxial, monaxial, biaxial, or triaxial.

Selected Articulations of the Body (p. 217)

1. The humeroscapular (shoulder joint) is formed by the humerus and scapula.
2. The coxal (hip) joint is formed by the femur and coxal bone.
3. The tibiofemoral (knee) joint is formed by the patella and femur and by the tibia and femur.

Disorders: Homeostatic Imbalances (p. 225)

1. Rheumatism is a painful state of supporting body structures such as bones, ligaments, tendons, joints, and muscles.
2. Arthritis refers to several disorders characterized by inflammation of joints, often accompanied by stiffness of adjacent structures.
3. Rheumatoid arthritis (RA) refers to inflammation of a joint accompanied by pain, swelling, and loss of function.
4. Osteoarthritis is a degenerative joint disease characterized by deterioration of articular cartilage and spur formation.
5. Gouty arthritis is a condition in which sodium urate crystals are deposited in the soft tissues of joints and eventually destroy the tissues.
6. Bursitis is an acute or chronic inflammation of bursae.
7. A dislocation (luxation) is a displacement of a bone from its joint; a partial dislocation is called subluxation.
8. A sprain is the forcible wrenching or twisting of a joint with partial rupture to its attachments without dislocation, whereas a strain is the stretching of a muscle.

REVIEW QUESTIONS

1. Define an articulation. What factors determine the degree of movement at joints? (p. 208)
2. Distinguish among the three kinds of joints on the basis of structure and function. List the subtypes. Be sure to include degree of movement and specific examples. (p. 208)
3. Explain the components of a synovial joint. Indicate the relationship of ligaments and tendons to the strength of the joint and restrictions on movement. (p. 209)
4. Explain how the articulating bones in a synovial joint are held together. (p. 209)
5. What is an accessory ligament? Define the two principal types. (p. 210)
6. What is an articular disc? Why are they important? (p. 210)
7. Describe the principle and importance of arthroscopy. (p. 210)
8. What are bursae? What is their function? (p. 211)
9. Define the following principal movements: gliding, angular, rotation, circumduction, and special. Name a joint where each occurs. (p. 211)
10. Have another person assume the anatomical position and execute each of the movements at joints discussed in the text. Reverse roles and see if you can execute the same movements. (p. 211)
11. Contrast nonaxial, monaxial, biaxial, and triaxial planes of movement. Give examples of each, then name a joint at which each occurs. (p. 218)
12. Be sure that you can name the bones that form the joint, identify the joint by type, and list the anatomical components of the joint for each of the following: humeroscapular (shoulder), coxal (hip), and tibiofemoral (knee) joints. (p. 217)
13. Distinguish among rheumatoid arthritis (RA), osteoarthritis, and gouty arthritis with respect to causes and symptoms. (p. 225)
14. Define bursitis. How is it caused? (p. 226)
15. Define dislocation. What are the symptoms of dislocation? (p. 226)
16. Distinguish between a sprain and a strain. (p. 226)
17. Define the following terms: torn cartilage (p. 210), separated shoulder (p. 217), dislocated hip (p. 219), swollen knee (p. 222), and dislocated knee. (p. 223)
18. Define arthrocentesis and arthroscopy. How is each performed? What are their diagnostic values? (p. 210)
19. Refer to the glossary of medical terminology associated with articulations at the end of the chapter. Be sure that you can define each term. (p. 226)

SELECTED READINGS

Allman, W. F. "The Knee," *Science 83,* November 1983.

Arehart-Trechel, J. "The Joint Destroyers," *Science News,* September 1982.

Bertram, Z., and M. Adams. "Knee Injuries in Sports," *New England Journal of Medicine,* 14 April 1988.

Bienenstock, H. "Diagnosis: Arthritis," *Hospital Medicine,* October 1984.

Edwards, D. D. "Severe Arthritis Under Attack," *Science News,* 19 October 1985.

Hogan, M. J. (ed.). "Arthroscopy," *Mayo Clinic Health Letter,* March 1989.

Hogan, M. J. (ed.). "Artificial Joints," *Mayo Clinic Health Letter,* Three parts. November and December 1988 and January 1989.

Kiley, J. M. (ed.). "Rheumatoid Arthritis," *Mayo Clinic Health Letter,* December 1986.

Chapter 10

Muscle Tissue

Chapter Contents at a Glance

Student Objectives

1. List the characteristics and functions of muscle tissue.
2. Compare the location, microscopic appearance, nervous control, functions, and regenerative capacities of the three kinds of muscle tissue.
3. Describe the principal events associated with the sliding-filament theory of muscle contraction.
4. Describe the structure and importance of a neuromuscular junction and a motor unit.
5. Identify the source of energy for muscular contraction.
6. Define the all-or-none principle of muscular contraction.
7. Describe different types of normal contractions performed by skeletal muscles.
8. Compare oxygen debt, fatigue, and heat production as examples of muscle homeostasis.
9. Explain the effects of aging on muscle tissue.
10. Describe the development of the muscular system.
11. Define such common muscular disorders as fibrosis, fibromyalgia, muscular dystrophies, myasthenia gravis (MG), spasm, tremor, fasciculation, fibrillation, and tic.
12. Define medical terminology associated with the muscular system.

Although bones and joints provide leverage and form the framework of the body, they are not capable of moving the body by themselves. Motion is an essential body function that results from the contraction and relaxation of muscles.

Muscle tissue is highly specialized to actively generate force and constitutes about 40 to 50 percent of total body weight. The scientific study of muscles is known as ***myology*** (mī-OL-ō-jē; *myo* = muscle; *logos* = study of).

The developmental anatomy of the muscular system is considered at the end of the chapter.

CHARACTERISTICS

Muscle tissue has four principal characteristics that assume key roles in maintaining homeostasis.

1. ***Excitability*** is the ability of muscle tissue to receive and respond to stimuli. A stimulus is a change in the internal or external environment strong enough to initiate an impulse (action potential).
2. ***Contractility*** is the ability of muscle tissue to actively generate force to shorten and thicken to do work (contract) when a sufficient stimulus is received.
3. ***Extensibility*** is the ability of muscle tissue to be stretched (extended). Many skeletal muscles are arranged in opposing pairs. While one is contracting, the other is relaxed and is undergoing extension.
4. ***Elasticity*** is the ability of muscle tissue to return to its original shape after contraction or extension.

FUNCTIONS

Through contraction, muscle performs three important functions:

1. Motion (both reflex and voluntary).
2. Maintenance of posture.
3. Heat production.

Motion is obvious in movements such as walking and running, and in localized movements, such as grasping a pencil, nodding the head, or chest movements involved with breathing. All these movements rely on the integrated functioning of the bones, joints, and skeletal muscles attached to the bones. Less noticeable kinds of motion produced by muscles are the beating of the heart, churning of food in the stomach, pushing of food through the intestines, contraction of the gallbladder to release bile, and contraction of the urinary bladder to expel urine.

In addition to the movement function, muscle tissue also enables the body to maintain posture. The contraction of skeletal muscles holds the body in stationary positions, such as standing and sitting.

The third function of muscle tissue is heat production. Skeletal muscle contractions produce most of the heat generated in the body and are thereby important in maintaining normal body temperature. It has been estimated that as much as 85 percent of all body heat is generated by muscle contractions.

TYPES

Types of muscle tissue are categorized by location, histology (microscopic structure), and nervous and other modes of control.

Skeletal muscle tissue, which is named for its location, is attached primarily to bones and moves parts of the skeleton. (Some skeletal muscles are also attached to skin, other muscles, or deep fascia.) Skeletal muscle tissue is ***striated*** because striations, or alternating light and dark bandlike structures, are visible when the tissue is examined under a microscope. It is a ***voluntary*** muscle tissue because it can be made to contract and relax by conscious control.

Cardiac muscle tissue forms the bulk of the wall of the heart. It is ***striated*** at the microscopic level and ***involuntary;*** that is, its contraction is usually not under conscious control. Cardiac muscle has built in controls (automaticity) and is also influenced by involuntary nerves and certain hormones.

Smooth muscle tissue is involved with processes related to maintaining the internal environment. It is located in the walls of hollow internal structures, such as blood vessels, the stomach, and the intestines. It is also found in the skin attached to hair follicles. It is referred to as ***nonstriated*** because it lacks striations at the microscopic level. It is ***involuntary*** muscle tissue which often has built-in controls (automaticity) and is also influenced by involuntary nerves and certain hormones.

Thus, all muscle tissues are classified in the following way: (1) skeletal, striated, voluntary muscle tissue; (2) cardiac, striated, involuntary muscle tissue; and (3) smooth, nonstriated, involuntary muscle tissue.

SKELETAL MUSCLE TISSUE

To understand the fundamental mechanisms of muscle movement, you will need some knowledge of the connective tissue components, nerve and blood supply, and histology of skeletal muscle.

Connective Tissue Components

The term ***fascia*** (FASH-ē-a; *fascia* = bandage) is applied to a sheet or broad band of fibrous connective tissue beneath the skin or around muscles and other organs of the body. ***Superficial fascia (subcutaneous layer)*** is immediately deep to the skin. It is composed of adipose tissue and loose connective tissue and has a number of important functions:

1. It serves as a storehouse for water and particularly for fat. Much of the fat of an overweight person is in the superficial fascia.

2. It forms a layer of insulation protecting the body from loss of heat.

3. It provides mechanical protection from blows.

4. It provides a pathway for nerves and vessels.

Deep fascia is a dense connective tissue that lines the body wall and extremities and holds muscles together, separating them into functioning groups. Functionally, deep fascia allows free movement of muscles, carries nerves and blood and lymph vessels, fills spaces between muscles, and sometimes provides the origin (one point of attachment to a bone) for muscles.

Skeletal muscles are further protected, strengthened, and attached to other structures by several other connective tissue coverings (Figure 10-1). The entire muscle is usually wrapped with a substantial quantity of fibrous connective tissue called the ***epimysium*** (ep'-i-MĪZ-ē-um). Bundles of muscle fibers (cells) called ***fasciculi*** (fa-SIK-yoo-lī) or ***fascicles*** (FAS-i-kuls) are covered by a fibrous connective tissue called the ***perimysium*** (per'-i-MĪZ-ē-um). ***Endomysium*** (en'-dō-MĪZ-ē-um) is a fibrous connective tissue that penetrates into the interior of each fascicle and surrounds and separates the muscle fibers. All three connective tissue coverings are extensions of deep fascia and contain collagenous fibers.

Epimysium, perimysium, and endomysium are all continuous with and contribute collagenous fibers to the connective tissue that attaches the muscle to another structure, such as bone or other muscle. All three elements may be extended beyond the muscle fibers as a ***tendon*** (*tendere* = to stretch out)—a cord of dense connective tissue that attaches a muscle to the periosteum of a bone. When the connective tissue elements extend as a broad, flat layer, the tendon is called an ***aponeurosis*** (*apo* = from; *neuron* = a tendon). This structure also attaches to the coverings of a bone, another muscle, or the skin. An example of an aponeurosis is the galea aponeurotica (see Figure 11-4). Certain tendons, especially those of the wrist and ankle, are enclosed by tubes of fibrous connective tissue called ***tendon (synovial) sheaths.*** They are similar in structure to bursae. The inner layer of a tendon sheath,

FIGURE 10-1 **Relationships of connective tissue to skeletal muscle depicted by a three-dimensional drawing indicating the relative positions of the epimysium, perimysium, and endomysium.**

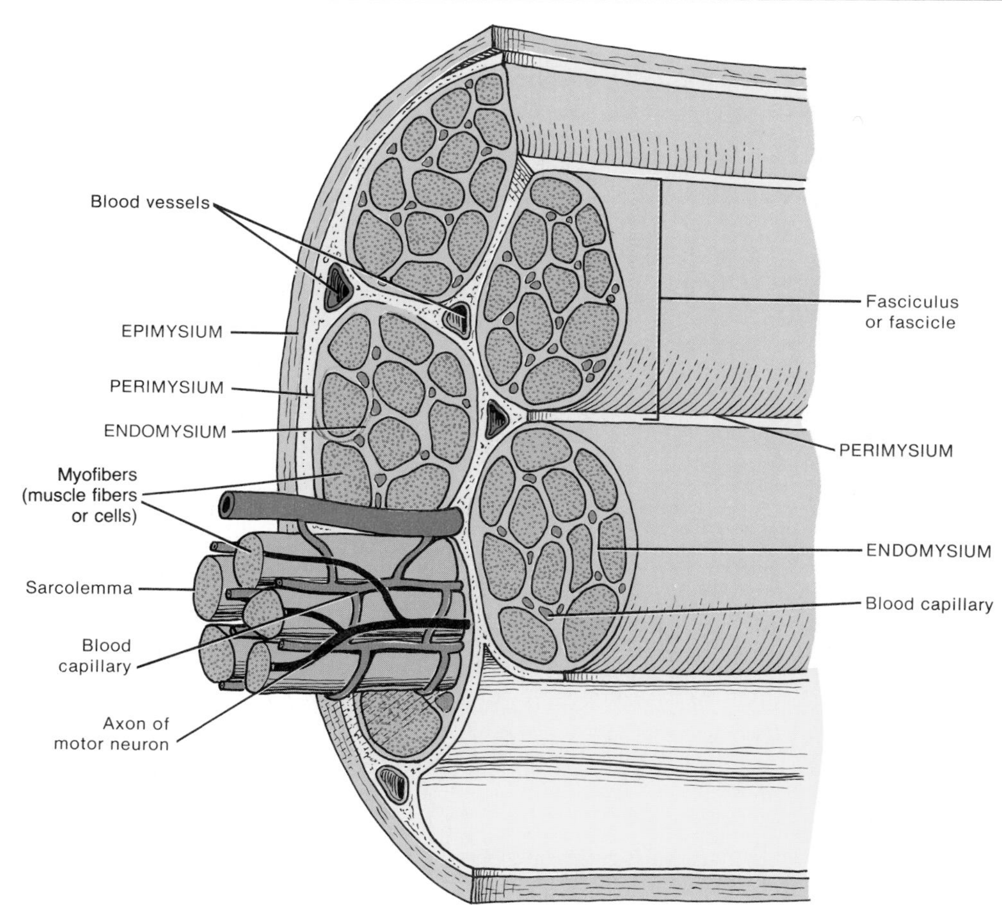

the visceral layer, is applied to the surface of the tendon. The outer layer is known as the parietal layer. Between the layers is a cavity that contains a film of synovial fluid. Tendon sheaths permit tendons to slide back and forth more easily.

CLINICAL APPLICATION: TENDINITIS

Tenosynovitis (ten'-ō-sin-ō-VĪ-tis) frequently occurs as inflammation involving the tendons, tendon sheaths, and synovial membrane surrounding certain joints. The wrists, shoulders, elbows (tennis elbow), finger joints (trigger finger), ankles, feet, and associated tendons are most often affected. The affected sheaths may become visibly swollen because of fluid accumulation, or they may remain dry. Local tenderness is variable, and there may be disabling pain with movement of the body part. The condition often follows some form of trauma, strain, or excessive exercise.

Nerve and Blood Supply

Skeletal muscles are well supplied with nerves and blood vessels. This innervation and vascularization is directly related to contraction, the chief characteristic of muscle. For a skeletal muscle fiber (cell) to contract, it must first be stimulated by an impulse from a nerve cell. Muscle contraction also requires a good deal of energy and therefore large amounts of nutrients and oxygen. Moreover, the waste products of these energy-producing reactions must be eliminated. Thus, prolonged muscle action depends on a rich blood supply to deliver nutrients and oxygen and remove wastes and heat.

Generally, an artery and one or two veins accompany each nerve that penetrates a skeletal muscle. The larger branches of the blood vessels accompany the nerve branches through the connective tissue of the muscle (Figure 10-2). Microscopic blood vessels called capillaries are distributed within the endomysium. Each muscle fiber is thus in close contact with one or more capillaries. Each skeletal muscle fiber usually makes contact with a portion of a nerve cell called a synaptic end bulb.

Histology

When a typical skeletal muscle is teased apart and viewed microscopically, it can be seen to consist of thousands of elongated cylindrical cells called ***muscle fibers*** or ***myofibers*** (see Exhibit 4-3). These fibers lie parallel to one another and range from 10 to 100 μm in diameter. Some fibers may reach lengths of 30 cm (12 inches) or

FIGURE 10-2 Relationship of blood vessels and nerves to skeletal muscles as seen in a photograph of a cross section through the mid thigh. (Courtesy of J. A. Gosling, P. F. Harris, et al., *Atlas of Human Anatomy,* Gower Medical Publishing Ltd., 1985.)

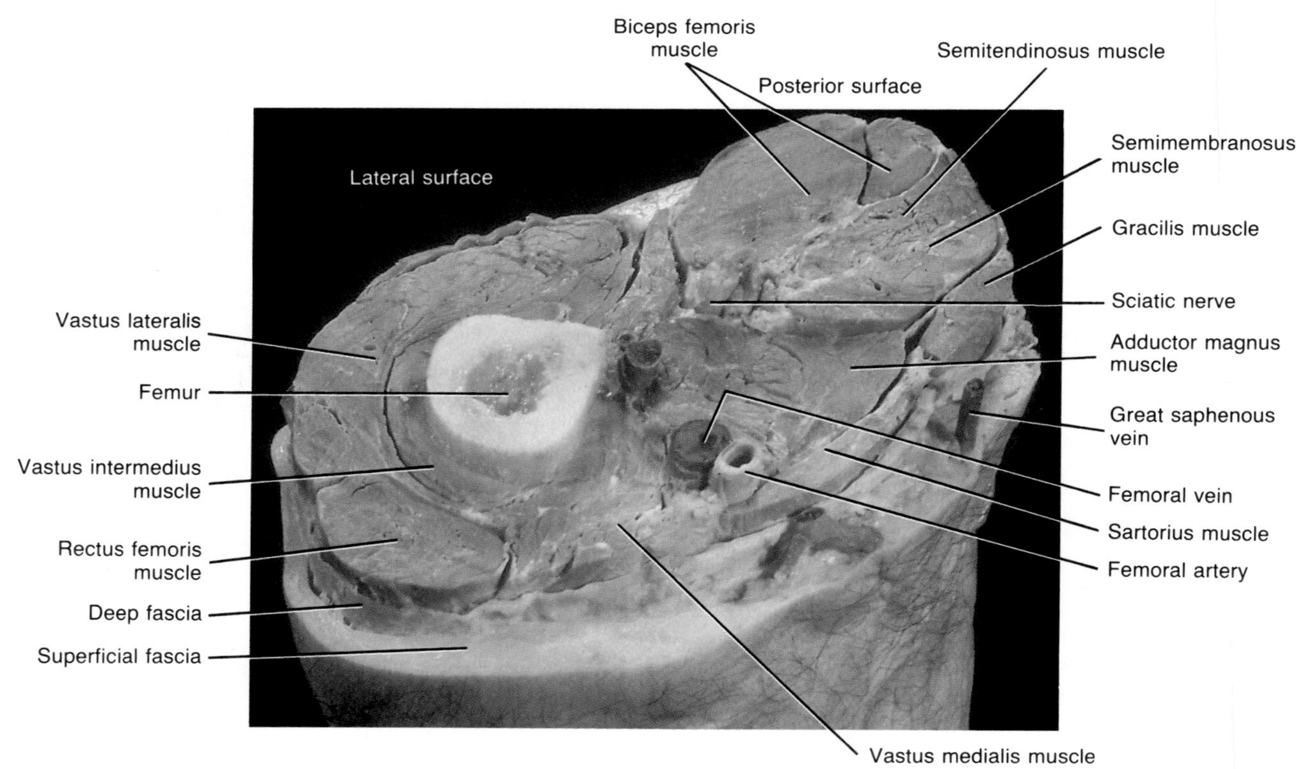

more. Each muscle fiber is enveloped by a plasma membrane called the ***sarcolemma*** (*sarco* = flesh; *lemma* = sheath). The sarcolemma surrounds a quantity of cytoplasm called ***sarcoplasm.*** Within the sarcoplasm of a muscle fiber and lying close to the sarcolemma are many nuclei. Skeletal muscle fibers are thus multinucleate. The sarcoplasm also contains myofibrils (described shortly), special high-energy molecules, enzymes, and ***sarcoplasmic reticulum*** (sar'-kō-PLAZ-mik re-TIK-yoo-lum), a network of membrane-enclosed tubules comparable to smooth endoplasmic reticulum (Figure 10-3a). Dilated sacs of sarcoplasmic reticulum, called ***terminal cisterns,*** form ringlike channels around myofibrils. Running transversely through the fiber and perpendicularly to the sarcoplasmic reticulum are ***transverse tubules (T tubules).*** The tubules are extensions of the sarcolemma that open to the outside of the fiber. A ***triad*** consists of a transverse tubule and the segments of sarcoplasmic reticulum (terminal cisterns) on either side.

A highly magnified view of skeletal muscle fibers reveals that they are composed of cylindrical structures, about 1 or 2 μm in diameter, called ***myofibrils*** (Figure 10-3a). Myofibrils, ranging in number from several hundred to several thousand, run longitudinally through the muscle fiber and consist of two kinds of even smaller structures called ***myofilaments.*** The ***thin myofilaments*** are about 6 nm in diameter. The ***thick myofilaments*** are about 16 nm in diameter.

The myofilaments of a myofibril do not extend the entire length of a muscle fiber—they are arranged in compartments called ***sarcomeres.*** Sarcomeres are separated from one another by narrow zones of dense material called ***Z lines.*** Within a sarcomere, certain areas can be distinguished (Figure 10-3b). A dark, dense area, called the ***A (anisotropic) band,*** represents the length of thick myofilaments. The sides of the A band are darkened by the overlapping of thick and thin myofilaments. The length of darkening depends on the extent of overlapping. As you will see later, the greater the degree of contraction, the greater the overlapping of thick and thin myofilaments. A light-colored, less dense area called the ***I (isotropic) band*** is composed of thin myofilaments only. This combination of alternating dark A bands and light I bands gives the muscle fiber its striated (striped) appearance. A narrow ***H zone*** is a region in the center of the A band that contains thick myofilaments only. In the center of the H zone is the ***M line,*** a series of fine threads that appear to connect the middle parts of adjacent thick myofilaments.

Thin myofilaments are anchored to the Z lines and project in both directions. They are composed mostly of the protein ***actin.*** The actin molecules are arranged in two single strands that entwine helically and give the thin myofilaments their characteristic shape (Figure 10-4a). Each actin molecule contains a ***myosin-binding site*** that interacts with a cross bridge of a myosin molecule (described shortly). Besides actin, the thin myofilaments contain two other protein molecules, ***tropomyosin*** and ***troponin,*** that are involved in the regulation of muscle contractions. Tropomyosin is arranged in strands that are loosely attached to the actin helices. Troponin is located at regular intervals on the surface of tropomyosin and is made up of three subunits: troponin I, which binds to actin; troponin C, which binds to calcium ions; and troponin T, which binds to tropomyosin. Together tropomyosin and troponin are referred to as the ***tropomyosin–troponin complex.***

Thick myofilaments overlap the free ends of the thin myofilaments and occupy the A band region of a sarcomere. These myofilaments are composed mostly of the protein ***myosin.*** A myosin molecule is shaped like a golf club. The tails (handles of the golf club) are arranged parallel to each other, forming the shaft of the thick myofilament. The heads of the golf clubs project outward from the shaft and are arranged spirally on the surface of the shaft. The projecting heads are referred to as ***cross bridges*** and contain an ***actin-binding site*** and an ***ATP-binding site*** (Figure 10-4b).

CLINICAL APPLICATION: MUSCLE-BUILDING ANABOLIC STEROIDS

Steroids are chemical substances derived from cholesterol and serve many useful functions in the body (discussed in later chapters). Most steroids are hormones. In recent years, attention has been focused on the use of ***muscle-building anabolic steroids*** by amateur athletes. These steroids, a derivative of the hormone testosterone, are used by the athletes supposedly to build muscle proteins and therefore increase strength and endurance during athletic events. However, physicians point out that use of anabolic steroids may cause a number of side effects, including liver cancer, kidney damage, increased risk of heart disease, muscle spasm, increased cholesterol, stunted growth in young people, increased irritability and aggressive behavior, psychotic symptoms (including hallucinations), manic episodes, major depression, and mood swings. In females, additional side effects include sterility, the development of facial hair, deepening of the voice, atrophy of the breasts and uterus, enlargement of the clitoris, and irregularities of menstruation. In males, additional side effects include testicular atrophy, baldness, excessive development of breast glands, and diminished hormone secretion and sperm production by the testes. Steroids may be addictive.

CONTRACTION

Sliding-Filament Theory

During muscle contraction, myosin cross bridges pull on thin myofilaments, causing them to slide inward toward the H zone. The sarcomere shortens, but the lengths of the thin and thick myofilaments do not change. The myosin

FIGURE 10-3 **Histology of skeletal muscle tissue. (a) Enlarged aspect of several myofibrils of a muscle fiber (cell) based on an electron micrograph. (b) Enlarged aspect of a sarcomere showing thin and thick myofilaments.**

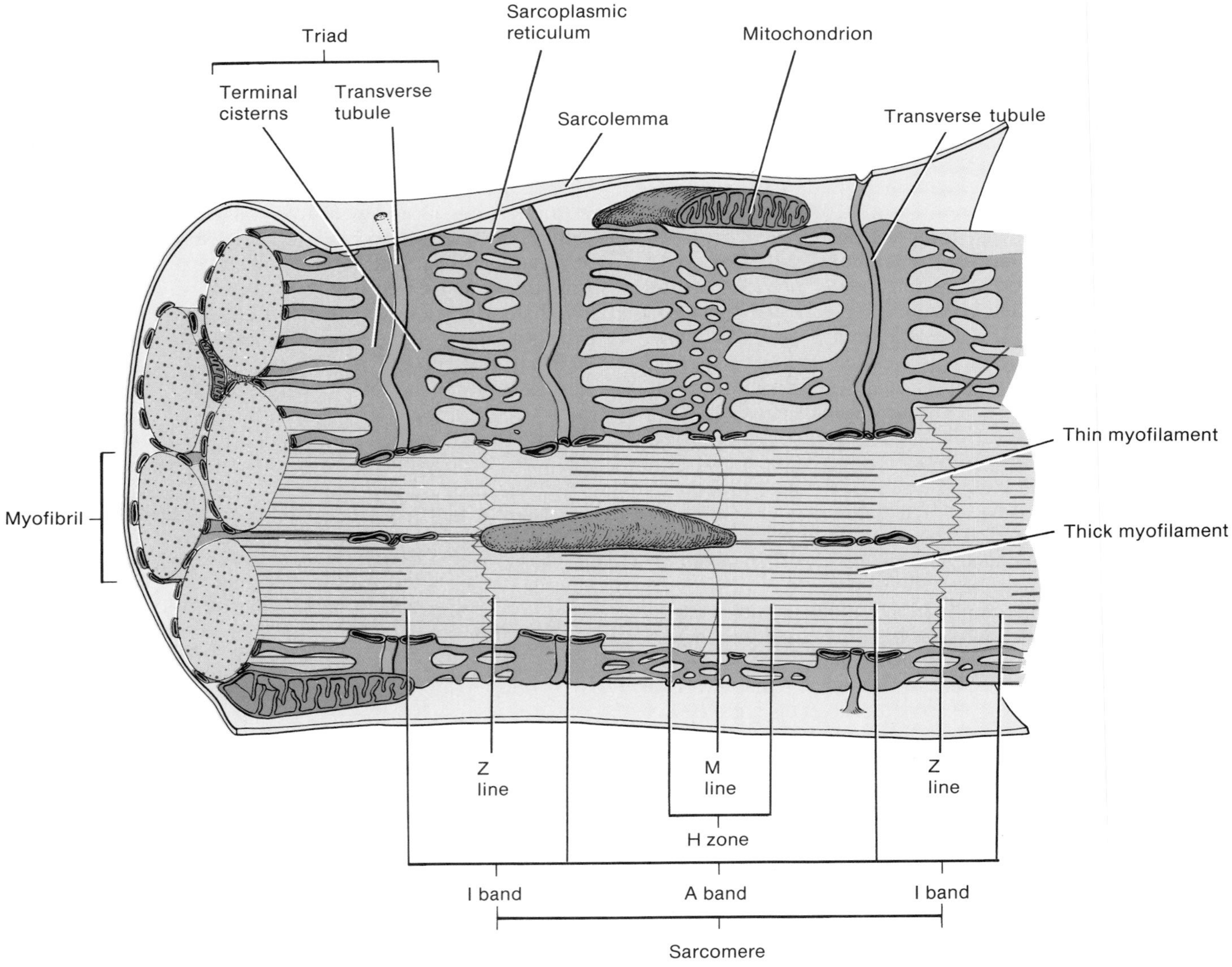

Triad
Terminal cisterns
Transverse tubule
Sarcoplasmic reticulum
Sarcolemma
Mitochondrion
Transverse tubule
Thin myofilament
Thick myofilament
Myofibril
Z line
M line
Z line
H zone
I band
A band
I band
Sarcomere

(a)

I band
A band
I band
M line
Z line
Thin myofilament
Thick myofilament
Cross bridge
H zone
Z line

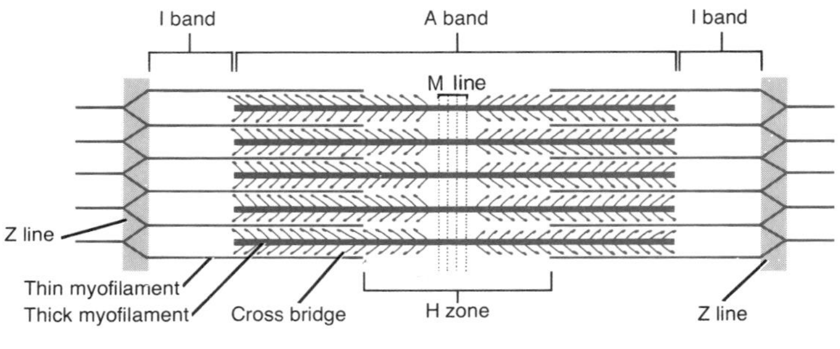

(b)

FIGURE 10-3 (*Continued*) Histology of skeletal muscle tissue. (c) Electron micrograph of several sarcomeres at a magnification of 35,000×. (Courtesy of D. E. Kelly, from *Introduction to the Musculoskeletal System* by Cornelius Rosse and D. Kay Clawson, Harper & Row, Publishers, Inc., New York, 1970.)

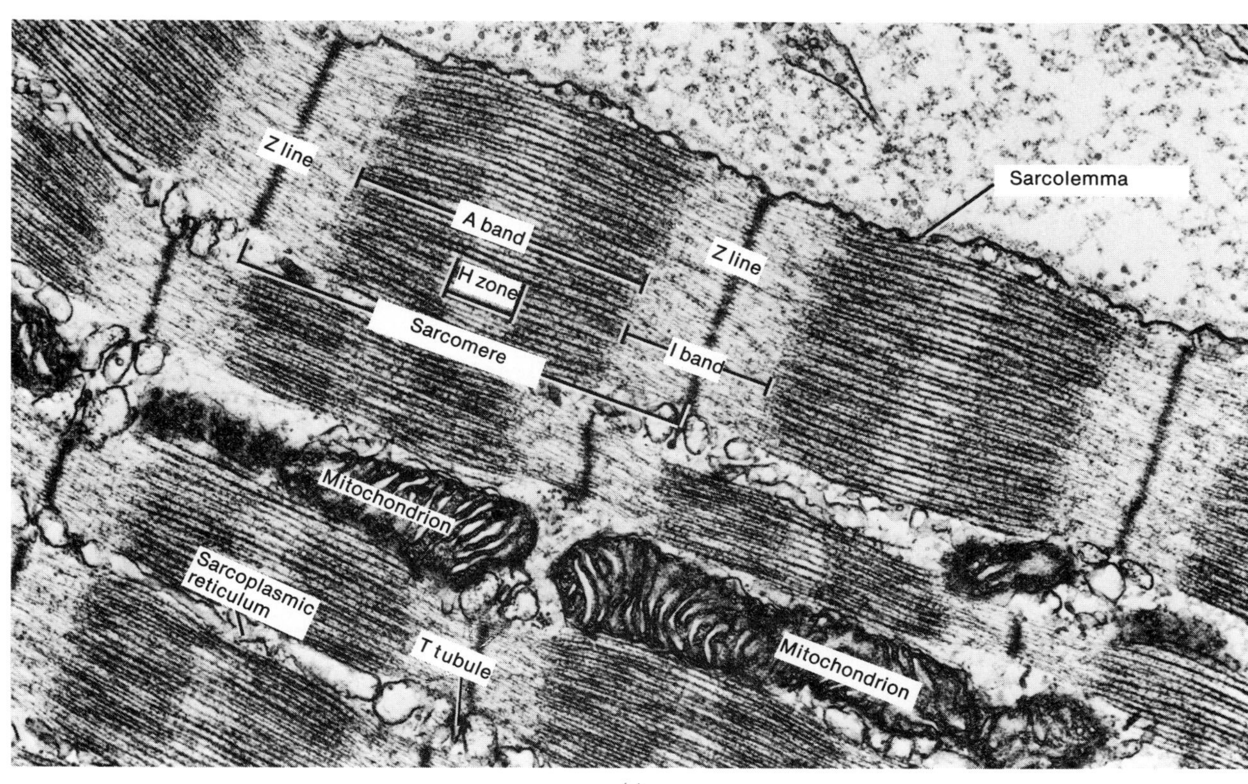

FIGURE 10-4 Detailed structure of portions of myofilaments. (a) Thin myofilament. (b) Thick myofilament.

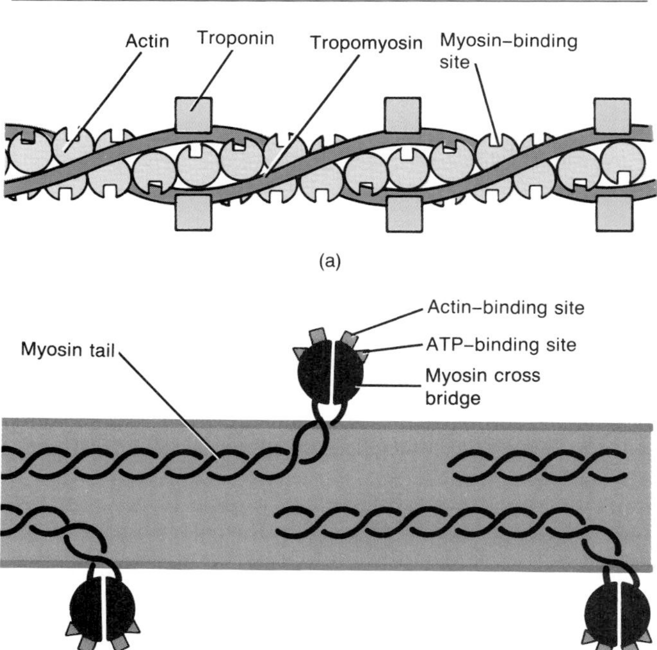

cross bridges of the thick myofilaments connect with portions of actin of the thin myofilaments. The myosin cross bridges move like the oars of a boat on the surface of the thin myofilaments, and the thin and thick myofilaments slide past each other as the cross bridges pull on, that is, apply force to, the thin myofilaments. As the thin myofilaments move past the thick myofilaments, the H zone narrows and even disappears when the thin myofilaments meet at the center of the sarcomere (Figure 10-5). In fact, the myosin cross bridges may pull the thin myofilaments of each sarcomere so far inward that their ends overlap. As the thin myofilaments slide inward, the Z lines are drawn toward each other and the sarcomere is shortened. The sliding of myofilaments and shortening of sarcomeres causes the shortening of the muscle fibers. All these events associated with the movement of myofilaments are known as the ***sliding-filament theory*** of muscle contraction.

Neuromuscular Junction

For a skeletal muscle fiber to contract, a stimulus must be applied to it. The stimulus is delivered by a nerve cell, or ***neuron.*** A neuron has a threadlike process called a

FIGURE 10-5 Sliding-filament theory of muscle contraction. Shown are the positions of the various parts of two sarcomeres in relaxed, contracting, and maximally contracted states. Note the movement of the thin myofilaments and the relative size of the I band and H zones.

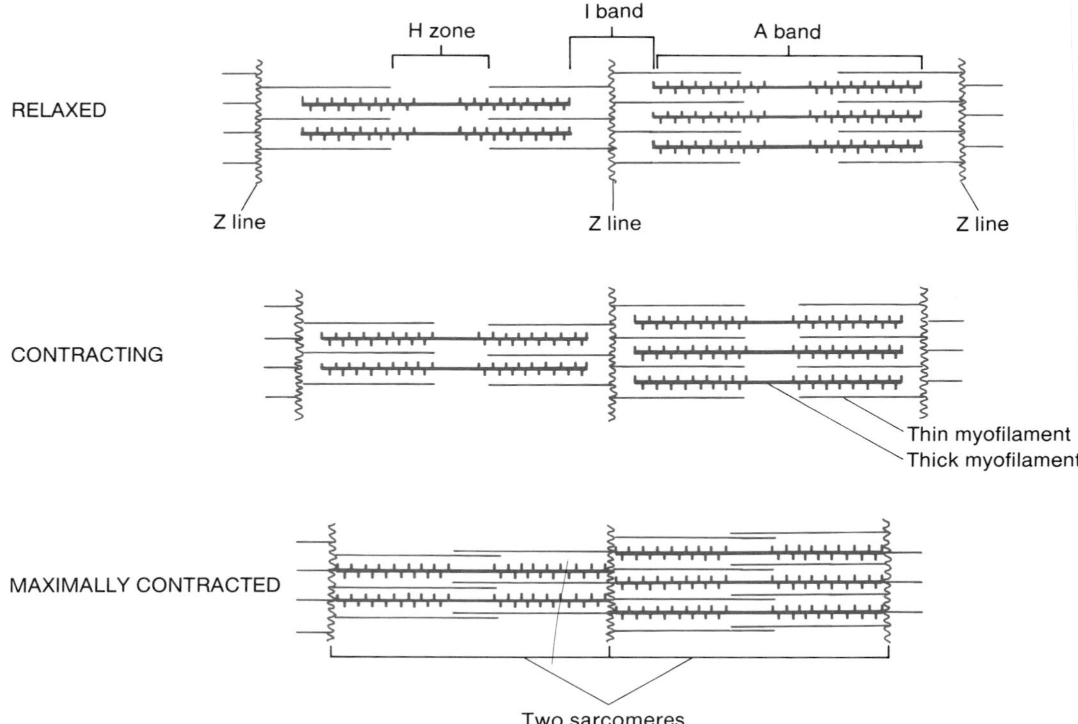

fiber, or axon, that may run 91 cm (3 ft) or more to a muscle. A bundle of such fibers from many different neurons composes a nerve. A neuron that stimulates muscle tissue is called a ***motor neuron.***

On entering a skeletal muscle, the axon of a motor neuron branches into axon terminals (telodendria) that come into close approximation with a portion of the sarcolemma of a muscle fiber. The region of the sarcolemma adjacent to the axon terminal is known as the ***motor end plate.*** The term ***neuromuscular junction*** or ***myoneural junction* (*MNJ*)** refers to the axon terminal of a motor neuron together with the motor end plate. (Figure 10-6). Close examination of a neuromuscular junction reveals that the distal ends of the axon terminals are expanded into bulblike structures called ***synaptic end bulbs*** (see Figure 12-3a). The bulbs contain membrane-enclosed sacs, the ***synaptic vesicles,*** that store chemicals called ***neurotransmitters.*** These chemicals determine whether an impulse is passed on to a muscle (or gland or another nerve cell). The invaginated area of the sarcolemma under the axon terminal is referred to as a ***synaptic gutter* (*trough*),** and the space between the axon terminal and sarcolemma is known as a ***synaptic cleft.*** There are numerous folds of the sarcolemma along the synaptic gutter, called ***subneural clefts,*** which greatly increase the surface area of the synaptic gutter. This allows for increased numbers of receptor site molecules that are able to bond to the neurotransmitter.

When a nerve impulse (nerve action potential) reaches an axon terminal, it initiates a sequence of reactions that liberates neurotransmitter molecules from synaptic vesicles (and possibly the cytoplasm as well). The neurotransmitter released at neuromuscular junctions in skeletal muscle is ***acetylcholine*** (as'-ē-til-KŌ-lēn), or **ACh** (see Figure 12-13). Upon its release, ACh diffuses across the synaptic cleft and combines with receptor sites on the sarcolemma of the muscle fiber. This combination alters the permeability of the sarcolemma to sodium (Na^+) and potassium (K^+) ions, and ultimately results in the development of a muscle action potential that travels along the sarcolemma, and is responsible for initiating the events leading to contraction. (The details of nerve impulse and muscle action potential generation are discussed in Chapter 12.)

In the vast majority of skeletal muscle fibers, there is only one neuromuscular junction for each fiber, and the junction is located in the middle of the fibers. Thus, the muscle action potential spreads from the center of the sarcolemma of the fiber to the ends and through the

FIGURE 10-6 Neuromuscular junction. (a) Photomicrograph at a magnification of 400×. (Courtesy of D. E. Kelly, from *Introduction to the Musculoskeletal System* by Cornelius Rosse and D. Kay Clawson, Harper & Row, Publishers, Inc., New York, 1970.) (b) Diagram based on a photomicrograph. (c) Enlarged aspect based on an electron micrograph.

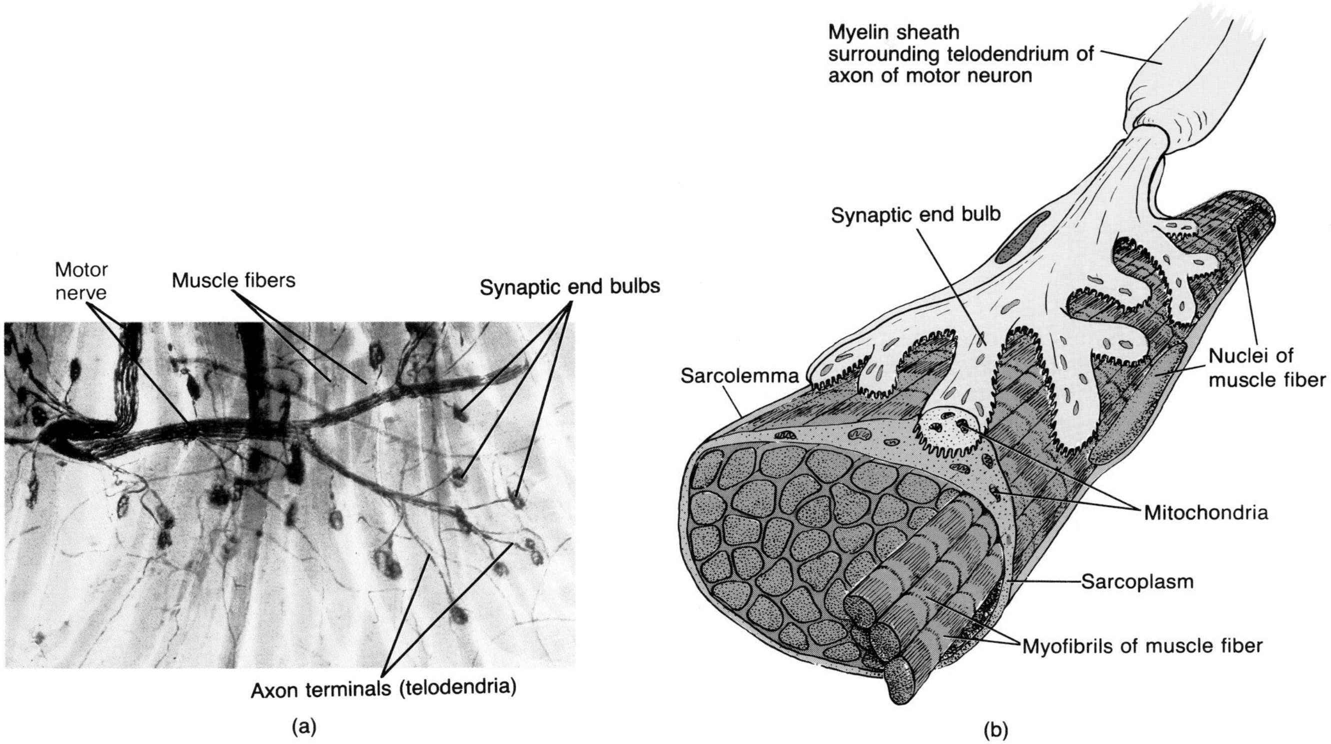

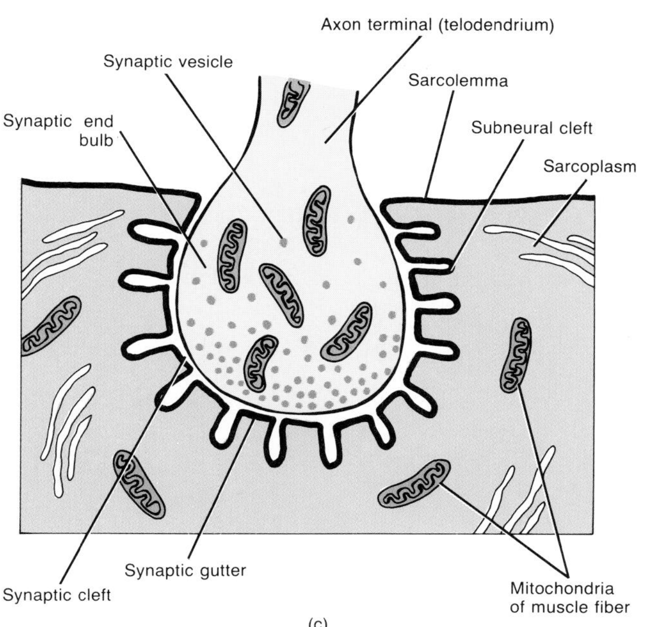

transverse tubules as well. This permits nearly simultaneous contraction of all sarcomeres in the fiber as the action potential spreads deeply into the sarcoplasm by way of the transverse tubules of the sarcolemma.

Motor Unit

A motor neuron, together with all the muscle fibers it stimulates, is referred to as a ***motor unit*** (Figure 10-7). A single motor neuron may innervate about 150 muscle fibers, depending on the region of the body. This means that stimulation of one neuron will tend to cause the simultaneous contraction of about 150 muscle fibers. In addition, all the muscle fibers of a motor unit that are sufficiently stimulated will contract and relax together. Muscles that control precise movements, such as the extrinsic (external) eye muscles, have fewer than 10 muscle fibers to each motor unit. Muscles of the body that are responsible for gross movements, such as the biceps brachii and gastrocnemius, may have as many as 2000 muscle fibers in each motor unit.

Stimulation of a motor neuron produces a contraction in all the muscle fibers in a particular motor unit. Accordingly, the total tension in a muscle can be varied by adjusting the number of motor units that are activated. The process of increasing the number of active motor units is called ***recruitment*** **(*motor unit summation*)**, and is determined by the needs of the body at a given time. The various motor neurons to a given muscle fire asynchronously; that is, while some are excited, others are inhibited. This means that while some motor units are active, others are inactive; all the motor units are not contracting at the same time. This pattern of firing of motor neurons prevents fatigue while maintaining contraction by allowing a brief rest for the inactive units. The alternating motor units relieve one another so smoothly that the contraction can be sustained for long periods. It also helps to maintain a state of partial contraction in a relaxed skeletal muscle, a phenomenon called muscle tone (described later in the chapter). Also, recruitment is one factor responsible for producing smooth movements during a muscle contraction, rather than a series of jerky movements.

Physiology of Contraction

When a muscle fiber is relaxed (not contracting), the concentration of calcium ions (Ca^{2+}) in the sarcoplasm is low; the ions are stored in the sarcoplasmic reticulum. Also, in a relaxed muscle fiber, ATP concentration is high and ATP is attached to the ATP-binding sites of the myosin cross bridges. The myosin cross bridges are prevented from combining with actin of the thin myofilaments by the binding of the tropomyosin–troponin complex to actin and the binding of ATP to the myosin cross bridges. In other words, the muscle fiber remains relaxed as long as there are few calcium ions in the sarcoplasm, the tropomyosin–troponin complex is attached to actin, and ATP is attached to the ATP-binding site of the myosin cross bridges (Figure 10-8a).

When a nerve impulse (nerve action potential) reaches the synaptic end bulb of an axon terminal, a small amount of calcium enters the synaptic end bulb, causing the synaptic vesicles (and possibly the cytoplasm as well) to release acetylcholine into the synaptic cleft. It diffuses across the synaptic cleft and combines with receptor sites on the sarcolemma of the muscle fiber directly beneath the synaptic end bulb. This combination causes a change in the permeability of the sarcolemma in which membrane proteins, called ***acetylcholine-gated ion channels,*** open and permit a rapid influx of positive ions, mainly Na^+. This initiates a ***muscle action potential,*** or impulse, that travels along the sarcolemma and then into the transverse tubules. When the muscle action potential is conveyed over the transverse tubules close to the sarcoplasmic reticulum, the action potential causes the reticulum to release some of the calcium ions from storage into the sarcoplasm surrounding the myofilaments. It has recently been shown that voltage-sensitive, tunnel-shaped proteins in the sarcoplasmic reticulum, called ***calcium release channels,*** regulate the release of calcium from the sarcoplasmic reticulum.

Calcium ions combine with a portion of troponin, a calcium-binding protein, causing it to undergo a structural change. This change moves the troponin and its attached tropomyosin out of the way. The tropomyosin–troponin complex moves into a groove between actin strands, thus exposing the myosin-binding sites on actin (Figure 10-8b).

FIGURE 10-7 **Motor Unit. Shown are two motor neurons, one in red and one in blue, each supplying its respective muscle fibers (cells).**

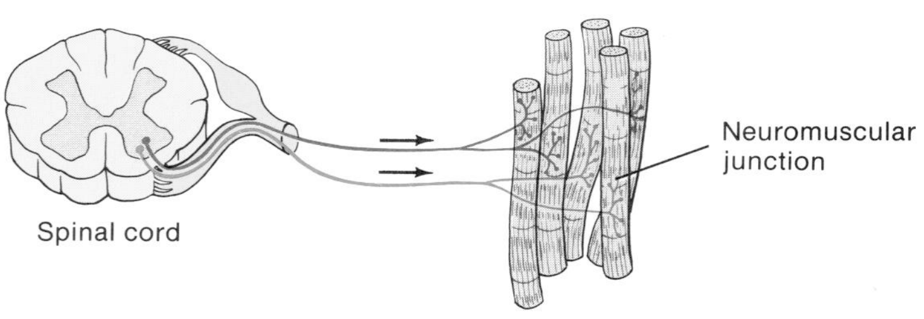

FIGURE 10-8 Mechanism of muscle contraction. (a) Relaxed state of a muscle fiber in which the tropomyosin–troponin complex covers the myosin-binding site on actin and the ATP-binding site of the myosin cross bridge is occupied. (b) Calcium ions combine with troponin, and the tropomyosin–troponin complex undergoes a structural change and moves, exposing the myosin-binding site. (c) A myosin cross bridge functioning as an ATPase enzyme, splits ATP into ADP + P. The energy from the splitting of ATP activates the myosin cross bridge and it combines with the myosin-binding site. The attachment produces a change in the orientation of the myosin cross bridge (power stroke) which slides the actin myofilament past the myosin myofilament.

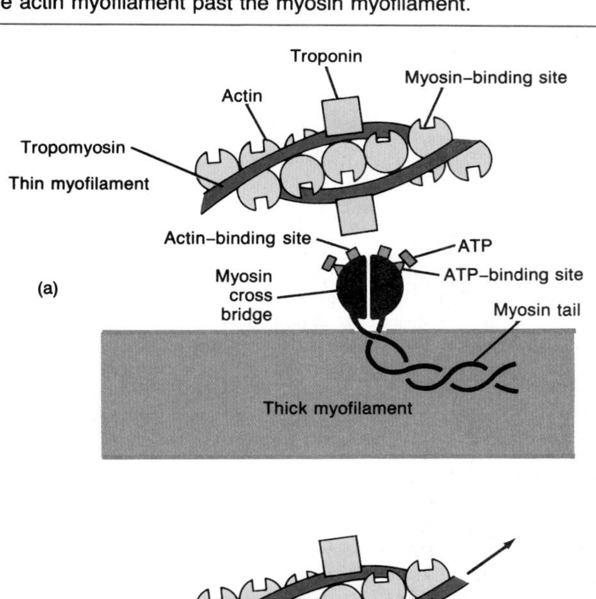

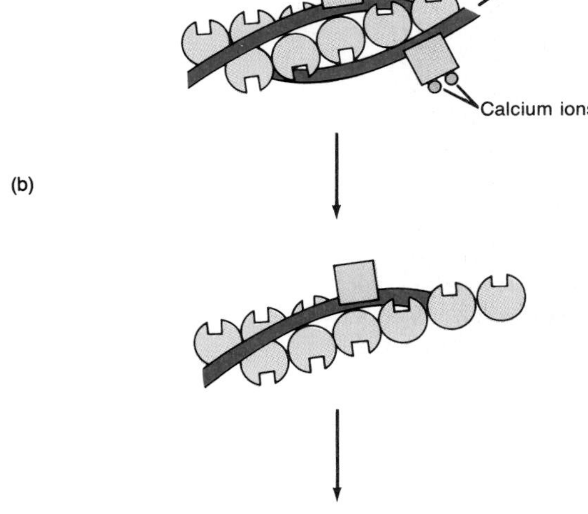

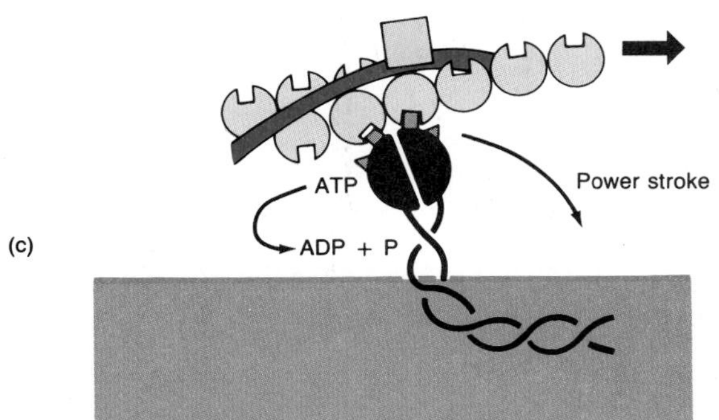

Muscle contraction requires energy, as well as calcium ions. This energy is supplied by ATP. ATP is found attached to ATP-binding sites on myosin cross bridges. Myosin cross bridges function as ATPase enzymes which can split ATP into ADP + P with the release of energy. When a muscle action potential stimulates a muscle fiber, the myosin cross bridges, functioning as ATPase enzymes, split ATP into ADP + P. The ADP + P remain attached to their binding sites and the myosin cross bridges, activated (energized) by the energy from the splitting of ATP, combine with myosin-binding sites on actin (Figure 10-8c).

The attachment of activated myosin cross bridges to myosin-binding sites on actin results in the release of ADP + P from their binding sites on the activated myosin cross bridges and a change in the orientation of the myosin cross bridges. The myosin cross bridge moves toward the H zone in the center of the sarcomere and in so doing applies force to the thin actin myofilaments. This movement of the myosin cross bridges, called the ***power stroke,*** is a force that causes the thin actin myofilaments to slide past the thick myosin myofilaments. Once the power stroke is complete, ATP combines with the ATP-binding sites on the myosin cross bridges, resulting in the detachment of the myosin cross bridge from actin. A given myosin cross bridge then combines with another myosin-binding site further along the actin strand. Again, ATP is split and the cycle repeats itself. The myosin cross bridges keep moving back and forth like the cogs of a ratchet with each power stroke, moving the thin actin myofilaments toward the H zone. This continual movement applies the force that causes the Z lines of a sarcomere to be drawn toward each other, and the sarcomere shortens. The muscle fibers thus contract and the muscle itself contracts. During maximal muscle contractions, the distance between Z lines can be shortened to 50 percent of the resting length. This cross bridge activity (contraction) does not always result in shortening of the muscle fibers and muscle. When the contraction does not result in muscle shortening, it is called isometric contraction. Using postural muscles to stand or sit is an example of an isometric contraction.

What happens when a muscle fiber goes from a contracted state back to a relaxed state? Acetylcholine, after it has caused the generation of a muscle action potential, is rapidly destroyed by an enzyme called ***acetylcholinesterase (AChE),*** which is found on the surfaces of the subneural clefts of the sarcolemma of muscle fibers. In the absence of action potentials, there will be no new release of ACh and this stops the generation of a muscle action potential. After the muscle action potential ends, the calcium ions are actively transported from the troponin of the thin filaments in the sarcoplasm back into the sarcoplasmic reticulum for storage. This is accomplished by two proteins called ***calsequestrin*** and ***calcium–ATPase*** and involves an expenditure of some ATP as the molecule is broken down to release energy used to actively transport calcium ions into storage in the sarcoplasmic reticulum. With the removal of calcium ions from the sarcoplasm, the tropomyosin–troponin complex is reattached to the actin strands, so that the myosin-binding sites of actin become covered and myosin cross bridges separate from actin and cannot reattach. ATP is also required to detach the cross bridges. Since the myosin cross bridges are broken, the thin myofilaments slip back to their relaxed position. After the muscle action potential ends, ADP is resynthesized into ATP, which again attaches to the ATP-binding site of the myosin cross bridge. The sarcomeres are thereby returned to their resting lengths, and the muscle fiber resumes its resting state. During extensibility, sarcomere length can be increased by about 20 percent of the resting length.

CLINICAL APPLICATION: RIGOR MORTIS

Following death, certain chemical changes occur in muscle tissue that affect the status of the muscles. Because of a lack of ATP, myosin cross bridges remain attached to the actin myofilaments, thus preventing relaxation. The resulting condition, in which muscles are in a state of rigidity (cannot contract or stretch), is called ***rigor mortis*** (rigidity of death). The time elapsing between death and the onset of rigor mortis varies greatly among individuals. Those who have had long, wasting illnesses undergo rigor mortis more quickly. Rigor mortis is not a permanent state that continues unabated after death. Depending on conditions, it lasts about 24 hours, then begins to abate, and disappears after another 12 hours when tissues begin to disintegrate, although a residual stiffness in the joints will remain.

A summary of the events associated with the contraction and relaxation of a muscle fiber is presented in Exhibit 10-1.

EXHIBIT 10-1 SUMMARY OF EVENTS INVOLVED IN CONTRACTION AND RELAXATION OF A SKELETAL MUSCLE FIBER

1. A nerve impulse (nerve action potential) causes synaptic vesicles in motor axon synaptic end bulbs to release acetylcholine (ACh).
2. Acetylcholine diffuses across the synaptic cleft within the neuromuscular junction and initiates a muscle action potential that spreads over the surface of the sarcolemma and transverse tubules.
3. The muscle action potential enters the transverse tubules and sarcoplasmic reticulum and stimulates the sarcoplasmic reticulum to release calcium ions from storage into the sarcoplasm.
4. Calcium ions combine with troponin, causing the tropomyosin–troponin complex to move, thus exposing the myosin-binding sites on actin.
5. When a muscle action potential stimulates a muscle fiber, ATPase splits ATP into ADP + P and energy is released. The released energy activates (energizes) myosin cross bridges, which combine with the exposed myosin-binding sites on actin, and apply force which causes the myosin cross bridges to move toward the H zone (power stroke). This movement results in the sliding of the thin myofilaments past thick myofilaments.
6. The sliding draws the Z lines toward each other, the sarcomere shortens, the muscle fibers contract, and the muscle contracts.
7. Acetylcholine is inactivated by acetylcholinesterase (AChE). As a result, ACh no longer has any effect at the neuromuscular junction.
8. Once no more muscle action potentials are generated, calcium ions are actively transported back into the sarcoplasmic reticulum by calsequestrin and calcium ATPase, using energy from ATP breakdown.
9. The low calcium concentration in the sarcoplasm permits the tropomyosin–troponin complex to reattach to actin. As a result, myosin-binding sites of actin become covered, myosin cross bridges separate from actin, ADP is resynthesized into ATP (which reattaches to the ATP-binding site of the myosin cross bridge), and the thin myofilaments return to their relaxed position.
10. Sarcomeres return to their resting lengths, muscle fibers relax, and the muscle relaxes.

Energy for Contraction and Relaxation

Phosphagen System

Contraction of a muscle requires energy. ATP is the immediate source of energy for contraction. When a muscle action potential stimulates a muscle fiber, ATP, in the presence of the enzyme ATPase, breaks down into ADP + P, and energy is released. Like other cells of the body, muscle fibers synthesize ATP as follows:

$$\text{ADP} + \text{P} + \text{Energy} \rightarrow \text{ATP}$$

Unfortunately, the amount of ATP present in skeletal muscle fibers is sufficient to maintain muscle contraction during vigorous exercise for only about five to six seconds.

Unlike most other cells of the body, skeletal muscle fibers function in a discontinuous manner; that is, they alternate between virtual inactivity and great activity. If strenuous exercise is to continue for more than a few seconds, additional ATP must be generated. Skeletal muscle fibers contain a high-energy molecule called ***phosphocreatine*** (fos′-fō-KRĒ-a-tin) that is used to generate ATP rapidly. The amount of phosphocreatine is about two to three times greater than ATP. Upon decomposition, phosphocreatine breaks down into creatine and phosphate, and in the process, large amounts of energy are released:

$$\text{Phosphocreatine} \rightarrow \text{Creatine} + \text{Phosphate} + \text{Energy}$$

Some of the released energy is used to convert ADP to ATP. The transfer of energy from phosphocreatine to ATP takes place in a fraction of a second. Together phosphocreatine and ATP constitute the ***phosphagen system*** and provide only enough energy for muscles to contract maximally for about 15 seconds. This energy system is used for maximal short bursts of energy and allows time for the rate of two metabolic processes, glycolysis and cellular respiration, to be increased. (Chapter 25).

Glycogen–Lactic Acid System

When muscle activity is continued so that even the supply of phosphocreatine is depleted, then the source of energy is glucose, which is derived from the breakdown of glycogen and also picked up from the blood. Glycogen, which is stored glucose, is always present in skeletal muscles and the liver. Its breakdown into glucose requires calcium ions and a calcium-binding protein called ***calmodulin,*** among other factors. Once stored glycogen is broken down into glucose, each molecule of glucose is split into two molecules of pyruvic acid, a process called ***glycolysis.*** In the process, energy is released and used to form ATP. Since glycolysis does not require oxygen, it is referred to an ***anaerobic process.*** Glycolysis may be summarized as follows:

$$1 \text{ Glucose} \rightarrow 2 \text{ Pyruvic acid} + \text{Energy (ATP)}$$

Ordinarily, because oxygen (O_2) is present, the pyruvic acid formed by glycolysis enters the mitochondria of muscle fibers, where it is completely catabolized to carbon dioxide and water in the metabolic process called cellular respiration. Since this process requires oxygen, it is referred to as ***aerobic.*** The complete catabolism of pyruvic acid also yields energy that is used to generate most of a muscle fiber's ATP:

$$\text{Pyruvic acid} + O_2 \rightarrow CO_2 + H_2O + \text{Energy (ATP)}$$

In some cases, as will be described shortly, there may not be sufficient oxygen for the complete catabolism of pyruvic acid. What happens then is that most of the pyruvic acid is converted to lactic acid, some of which then diffuses out of the muscle fibers and eventually into blood. The production of lactic acid in this way releases energy from glucose that can be used to produce ATP, and it occurs anaerobically. The ***glycogen–lactic acid system*** just described can provide sufficient energy for about 30 to 40 seconds of maximal muscle activity.

Aerobic System

As noted earlier, if sufficient oxygen is present, pyruvic acid can be completely catabolized to form carbon dioxide, water, and energy (ATP) in the mitochondria. This ***aerobic system,*** along with glycolysis, is used for prolonged muscular activity and will continue as long as nutrients and adequate oxygen last. These nutrients include fatty acids (from fats) and amino acids (from proteins), as well as glucose derived from glycogen breakdown or delivered to the muscle by way of blood.

In summary, the phosphagen system will provide enough ATP to sustain maximal muscular activity for about 15 seconds. A 100-meter dash is an example. The glycogen–lactic acid system will provide sufficient ATP to support about 30 to 40 seconds of maximal muscular activity, such as a 400-meter dash. The aerobic system will provide sufficient ATP for a prolonged activity as long as sufficient oxygen and nutrients are available. Jogging is an example.

All-or-None Principle

The weakest stimulus from a neuron that can still initiate a contraction is called a ***threshold (liminal) stimulus.*** A stimulus of lesser intensity which cannot initiate contraction, is referred to as a ***subthreshold (subliminal) stimulus.*** According to the ***all-or-none principle,*** once a threshold stimulus is applied, individual muscle fibers of a motor unit will contract to their fullest extent or will not contract at all, provided conditions remain constant. In other words *individual muscle fibers* do not partly contract. The principle does not mean that the entire muscle must be either fully relaxed or fully contracted because, of the many motor units that compose the entire muscle, some are contracting and some are relaxing. Thus, the muscle as a whole can have graded contractions (varying magnitudes) in order to perform a specific task. For example, the muscles of the arm do not contract to the same extent when you lift a 10-pound weight as compared with a 1-pound weight. But they do contract in a smooth, graded fashion to lift either weight. The strength of contraction may be decreased by fatigue, lack of nutrients, or lack of oxygen.

MEDICAL TEST

Electromyography (e-lek'-trō-mī-OG-ra-fē; *electro* = electricity; *myo* = muscle; *graph* = to write) The evaluation of electrical activity in resting and contracting muscles is called ***electromyography,*** or ***EMG.*** The record of the study is known as an ***electromyogram.***

Diagnostic Value: To help determine the cause of muscular weakness or paralysis, to evaluate involuntary muscle twitching, to determine why abnormal levels of muscle enzymes (e.g., creatine phosphokinase) appear in blood, and to serve as a component of biofeedback studies.

Procedure: A flat metal disc electrode is placed on the skin over the muscle to be tested. Then, a thin sterile needle attached by wires to a recording machine is inserted through the skin into the muscle. The electrical activity of the muscle is recorded at rest and during contraction and displayed as electrical waves on an oscilloscope at the same time that the activity is reproduced as sounds over a speaker.

During the test, the needle may be moved several times to evaluate different areas of one muscle, or different muscles. One condition that is diagnosed by EMG is ALS (amyotropic lateral sclerosis).

To determine if muscle weakness is due to a peripheral nerve or spinal cord disease, *nerve conduction studies* are sometimes done in conjunction with EMG testing. A nerve is stimulated electrically through the skin, while a recording device detects the speed of response through the nerve. Such studies help to determine the cause of numbness, tingling, or pain from nerve damage.

Kinds of Contractions

The various skeletal muscles are capable of producing different kinds of contractions, depending on the stimulation frequency.

Twitch

The ***twitch contraction*** is a rapid, jerky response to a single threshold or greater stimulus. Twitch contractions can be artificially brought about in muscles of an animal. A recording of a twitch contraction is the classic way to illustrate the different phases of one single contraction. Figure 10-9 is a graph of a twitch contraction. The record of a muscle contraction is called a ***myogram.*** Note that a brief period exists between application of the stimulus and the beginning of contraction: the ***latent period.*** During this time, it is believed that Ca^{2+} ions are released from the sarcoplasmic reticulum and the onset of myosin cross bridge activity occurs. In frog muscle, it lasts about 10 milliseconds (msec) (10 msec = 0.01 sec). The second phase, the ***contraction period,*** lasts about 40 msec (0.04 sec) and is indicated by the upward tracing that is caused by the cross bridge activity that brings about contraction. The third phase, the ***relaxation period,*** lasts about 50 msec (0.05 sec) and is indicated by the downward tracing. It is caused by the active transport of Ca^{2+} ions back into the sarcoplasmic reticulum which results in relaxation. The duration of these periods varies with the muscle involved. The latent period, contraction period, and relaxation period for muscles that move the eyes are very short. For muscles that move the leg, longer periods are required.

If additional stimuli are applied to the muscle after the initial stimulus, other responses may be noted. For example, if two stimuli are applied one immediately after the other, the muscle will respond to the first stimulus but not to the second. When a muscle fiber receives enough stimulation to contract, it temporarily loses its excitability and cannot contract again until its responsiveness is regained. This period of lost excitability is the ***refractory period.*** Its duration also varies with the muscle involved. Skeletal muscle has a short refractory period of 5 msec (0.005 sec). Cardiac muscle has a long refractory period of 300 msec (0.30 sec).

Tetanus

When two stimuli are applied and the second is delayed until the refractory period is over, the skeletal muscle will respond to both stimuli. In fact, if the second stimulus is applied after the refractory period, but before the muscle has finished relaxing, the second contraction will be stronger than the first. This phenomenon, in which stimuli arrive at different times, is called ***wave (temporal) summation*** (Figure 10-10). In another type of summation, called ***multiple motor unit (spatial) summation,*** the stimuli occur at the same time but at different locations (different motor units).

If a frog muscle is stimulated at a rate of 20 to 30 stimuli per second, the muscle can only partly relax between stimuli. As a result, the muscle maintains a sustained contraction called ***incomplete (unfused) tetanus*** (Figure

FIGURE 10-9 Myogram of a twitch contraction. The red arrow indicates the point at which the stimulus is applied.

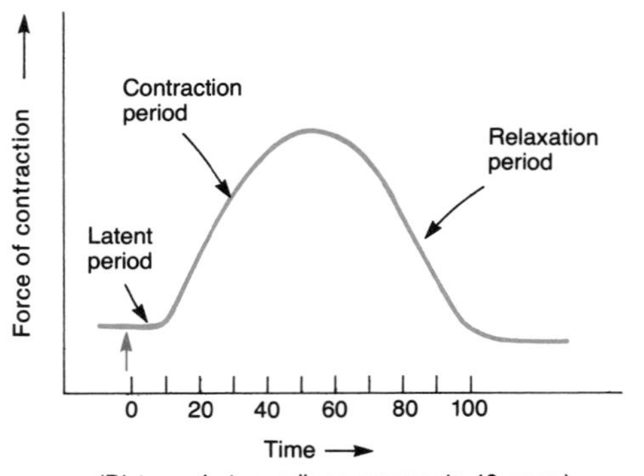

FIGURE 10-10 Myogram of wave summation. The second stimulus (long red arrow) is applied before the muscle has finished relaxing. The second contraction is even stronger than the first. The broken line represents the continuation of a twitch contraction.

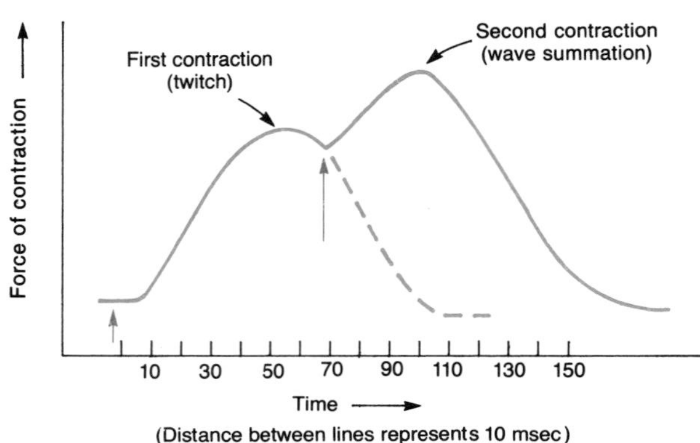

FIGURE 10-11 **Myograms of (a) incomplete and (b) complete tetanus.**

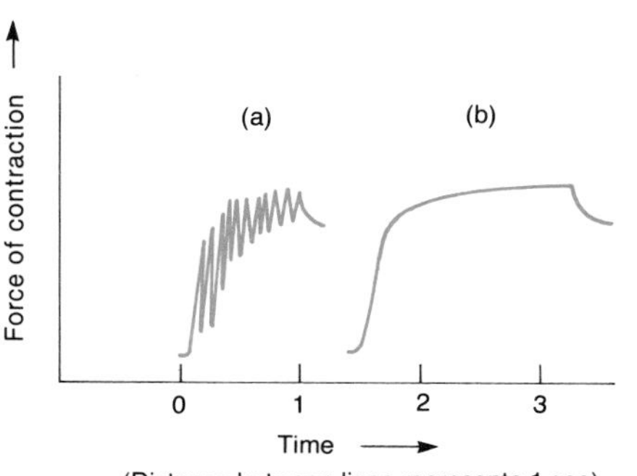

10-11a). Stimulation at an increased rate (35 to 50 stimuli per second) results in ***complete (fused) tetanus,*** a sustained contraction that lacks even partial relaxation between stimuli (Figure 10-11b). Essentially, both kinds of tetanus result from the addition of calcium ions released from the sarcoplasmic reticulum by the second, and subsequent stimuli, to the calcium ions still in the sarcoplasm from the first stimulus. This causes the rapid succession of separate twitches. Relaxation is either partial or does not occur at all. Voluntary contractions, such as contraction of the biceps brachii muscle in order to flex the forearm, are tetanic contractions. In fact, most of our muscular contractions involve short-term tetanic contractions and are thus smooth sustained contractions.

Treppe

Treppe is the condition in which a skeletal muscle contracts more forcefully in response to the same strength of stimulus after it has contracted several times. It is demonstrated by stimulating an isolated muscle with a series of stimuli at the same frequency and intensity but not at a rate fast enough to produce tetanus. Suppose a series of liminal stimuli are introduced into a muscle. Time must be allowed for the muscle to undergo its latent period, contract, and relax, but this takes only 0.1 sec in frog muscle. If stimuli are repeated at intervals of 0.5 sec, the first few tracings on the myogram will show an increasing height with each contraction. This is treppe—the staircase phenomenon (Figure 10-12). It is the principle athletes use when warming up. After the first few stimuli, the muscle reaches its peak of performance and undergoes its strongest contraction. Treppe is thought to result from increased availability of calcium ions that relieve tropomyosin–troponin inhibition after several contractions. Also, the internal conditions in the muscle, such as temperature, pH, and viscosity, have changed.

Isotonic and Isometric

Isotonic (*iso* = equal; *tonos* = tension) ***contractions*** are probably familiar to you. As the contraction occurs, the muscle shortens and pulls on another structure, such as a bone, to produce movement (Figure 10-13a). During such a contraction, the tension remains constant and energy is expended.

In an ***isometric contraction,*** there is minimal shortening of the muscle, but the *tension* on the muscle increases greatly (Figure 10-13b). Although isometric contractions do not result in body movement, energy is still expended. You can demonstrate such a contraction by carrying your books with your arm extended. The weight of the books pulls the arm downward, stretching the shoulder and arm muscles. The isometric contraction of the shoulder and arm muscles counteracts the stretch. The two forces—contraction and stretching—applied in opposite directions create the tension. The tension developed in a muscle for performing any kind of action depends on the total number of muscle fibers contracting at a time and the amount of tension each muscle fiber generates.

Although both isotonic and isometric training methods are able to increase muscular strength in relatively short periods, studies where direct comparisons are made tend to favor isotonic methods. The greatest advantage of isotonic exercise is that it works all the involved muscles over the entire range of a particular movement. Isometric exercise would require several separate and different maneuvers to work all the same muscles.

There is also the psychological advantage of performing isotonic exercises. The performer has some satisfaction in seeing the movements being accomplished, whereas isometrics are less satisfying and often considered boring since they are static.

Some experiments indicate that greater muscle enlargement (hypertrophy) and endurance result from isotonic exercise.

The main advantages of isometric exercises are their ease of implementation and time-saving features. Very little equipment is needed, and therefore larger groups of participants work out in a shorter period of time.

One caution needs to be added. It is well documented that the systolic and diastolic blood pressures increase considerably during isometric maneuvers. Therefore, they

FIGURE 10-12 **Myogram of treppe.**

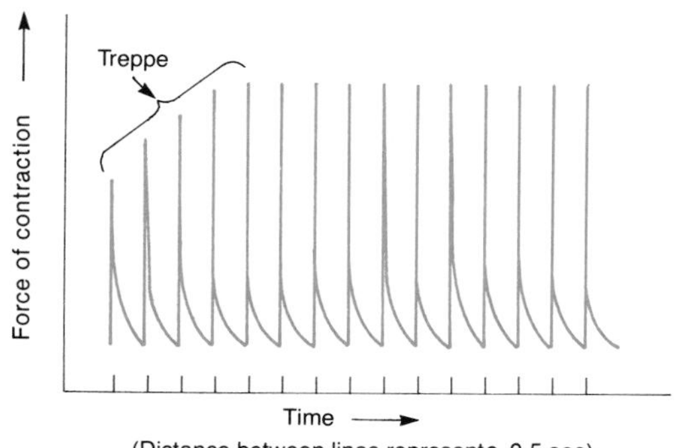

FIGURE 10-13 Comparison between an (a) isotonic contraction and an (b) isometric contraction. In an isotonic contraction, the muscle shortens. In an isometric contraction, tension develops, but the muscle does not shorten.

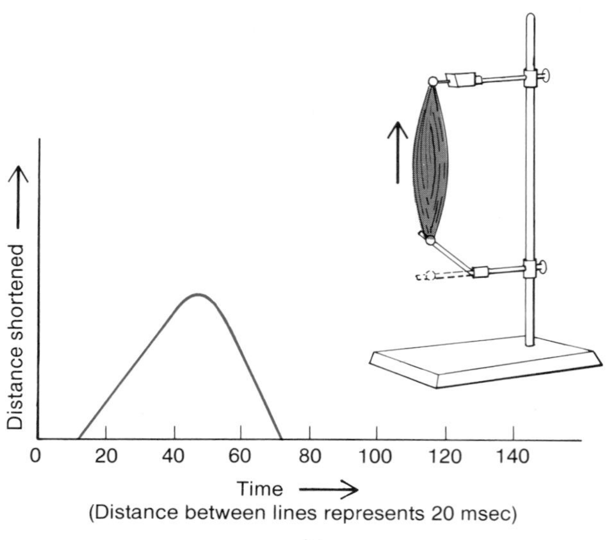

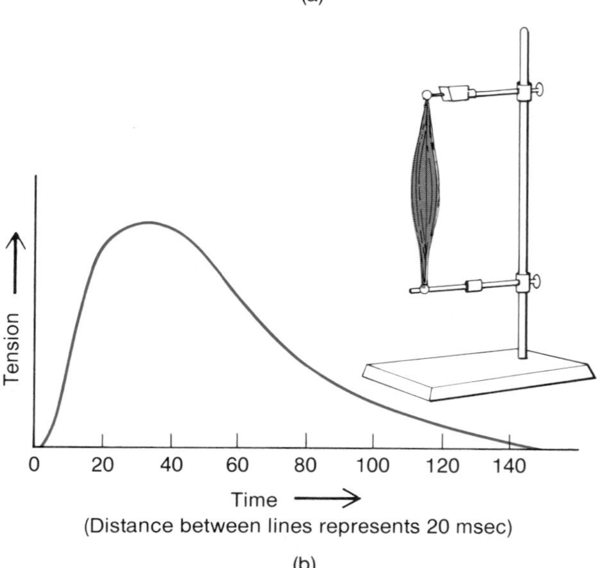

are a potentially dangerous form of exercise for rehabilitating cardiac patients and for older adults.

Muscle Tension

The amount of tension (force) that can be developed by a skeletal muscle depends on several factors. Among these are frequency of stimulation of muscle fibers by motor neurons, the number of muscle fibers contracting (number of active motor units) at a particular time, the components of the muscle fibers themselves, and the length of the muscle fibers at the time of contraction.

Frequency of Stimulation

The tension generated by a muscle fiber depends partially on the frequency of stimulation it receives from its motor neuron. Such neurons normally do not transmit just a single nerve impulse at a time. Rather, the impulses are transmitted in bursts, so that one impulse closely follows another. This results in a summation of individual contractions (wave summation). A muscle subjected to wave summation is in a state of sustained contraction called tetanus. Both wave summation and tetanus were described earlier.

Number of Muscle Fibers Contracting

As indicated before, the number of muscle fibers innervated by one motor neuron varies greatly. Precision movements require very subtle changes in muscle contraction. Therefore, in such muscles, the motor units are small. In this way, when a motor unit is recruited or turned off, only very slight but controlled changes in muscle contraction occur. Conversely, large motor units are employed where maintaining a constant position or posture is important and precision is not. Nerve fibers in a given motor nerve fire asynchronously. Thus, overall sustained muscle contraction can be maintained for long periods while the tension-generation process is shared at different moments in turn by the motor units.

Components of Muscle Fibers

Two components of muscle fibers are related to muscle tension: contractile elements and elastic elements. ***Contractile elements*** are those components that are actively involved in muscle contraction. Thin and thick myofilaments are examples. The tension generated by contractile elements is called ***active tension. Elastic elements,*** by contrast, are structures that are capable of being stretched. They include the connective tissue surrounding each muscle fiber and the tendon that attaches muscle to bone. The tension generated by elastic elements is called ***passive tension.*** Such tension is not dependent on muscular contraction, and it is proportional to the degree to which a skeletal muscle is stretched; within limits, the more the muscle is stretched, the greater its passive tension.

In order to understand how elastic elements are related to passive tension, consider the following. When a skeletal muscle contracts, it pulls on its connective tissue coverings and tendons. As a result, the coverings and tendons stretch, they become taut, and the tension passed through the tendons pulls on the bones to which they are attached. The result is movement of a part of the body. The role of elastic elements can be made clearer by imagining that you are trying to pull a heavy object along the ground by pulling on a spring attached to the object. When you pull on the spring, it stretches, but the object stays put until the tension in the stretched spring equals the weight of the object. When the tension exceeds the weight of the object, the object moves.

If the weight to be moved is light, only a slight tug on the spring is required to move it, and it will move quickly. Conversely, if the weight is heavy, a greater tug is required because the spring must stretch farther, and the weight moves slowly. A similar situation occurs in a skeletal muscle in that there is a latent period from the time of application of the stimulus to actual contraction. The latent period is longer when a heavier object has to be lifted.

The stretch of elastic elements is also related to wave summation and tetanus. During wave summation, elastic elements are not given much time to relax between contractions, and thus they remain taut. While in this state, the elastic elements do not require very much stretching before the beginning of the next muscular contraction. The combination of the tautness of the elastic elements and partially contracted state of myofilaments enables the force of one contraction to be added immediately to the one before.

Length of Muscle Fibers

Having already considered the sliding-filament theory of muscle contraction, we can now examine the relationship between muscle length and force of contraction (tension). As you already know, a skeletal muscle fiber contracts when myosin cross bridges of thick myofilaments connect with portions of thin myofilaments within a sarcomere. As it turns out, a muscle fiber develops its greatest tension when there is maximum overlap between thick and thin myofilaments (Figure 10-14). At this length, the optimal length, the maximum number of myosin cross bridges make contact with thin myofilaments to bring about the greatest force of contraction. As a muscle fiber is stretched, fewer and fewer myosin cross bridges make contact with thin myofilaments, and the force of contraction progressively decreases. In fact, if a muscle fiber is stretched to 175 percent its optimal length, no myosin cross bridges attach to thin myofilaments and no contraction occurs. At lengths less than the optimum length, the force of contraction also decreases. This is because extreme shortening of sarcomeres causes thin myofilaments to overlap and thick myofilaments to crumple as they run into Z lines, resulting in fewer myosin cross bridge contacts with thin myofilaments. In general, changes in resting muscle fiber length above or below the optimum length rarely exceed 30 percent.

FIGURE 10-14 Length–tension relationship in skeletal muscle fibers (cells). Maximum tension is produced at sarcomere length 100 percent.

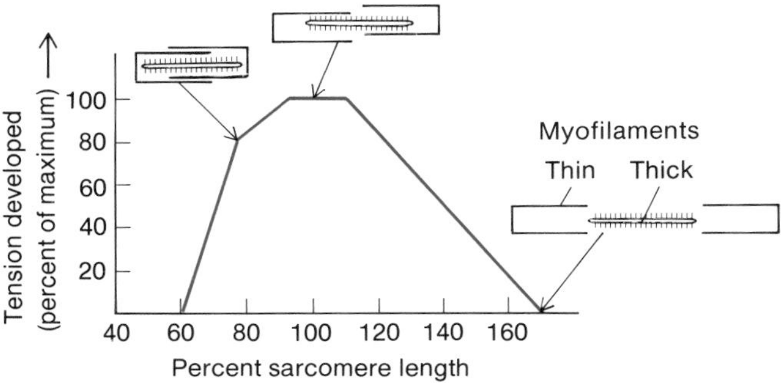

Muscle Tone

A muscle may be in a state of partial contraction even though the muscle fibers operate on an all-or-none basis. At any given time, some fibers in a muscle are contracted, whereas others are relaxed. This contraction tightens a muscle, but there may not be enough fibers contracting at the time to produce movement. Recruitment (asynchronous firing) allows the contraction to be sustained for long periods.

A sustained partial contraction of portions of a skeletal muscle in response to activation of stretch receptors results in ***muscle tone*** and occurs even in a relaxed muscle. Tone is essential for maintaining posture. For example, when the muscles in the back of the neck are in tonic contraction, they keep the head in the anatomical position and prevent it from slumping forward onto the chest, but they do not apply enough force to pull the head back into hyperextension. The degree of tone in a skeletal muscle is monitored by receptors in the muscle called ***muscle spindles.*** They provide feedback information on tone to the brain and spinal cord so that adjustments can be made (Chapter 15).

CLINICAL APPLICATION: HYPOTONIA AND HYPERTONIA

Abnormalities of muscle tone are expressed as hypotonia or hypertonia. ***Hypotonia*** refers to decreased or lost muscle tone. Such muscles are said to be ***flaccid*** (FLAK-sid or FLAS-sid). Flaccid muscles are loose, their normal rounded contour is replaced by a flattened appearance, and the affected limbs are hyperextended. As you will see later, certain disorders of the nervous system may result in ***flaccid paralysis,*** which is characterized by loss of muscle tone, loss or reduction of tendon reflexes, and atrophy (wasting away) and degeneration of muscles.

Hypertonia refers to increased muscle tone and is expressed in two ways: spasticity or rigidity. ***Spasticity*** is characterized by increased muscle tone (stiffness) associated with an increase in tendon reflexes and pathological reflexes (Babinski sign). Spastic muscles exhibit increased resistance to passive movement, followed by a sudden or gradual release of resistance (clasp-knife reaction). As you will see later, certain disorders of the nervous system may result in ***spastic paralysis,*** partial paralysis in which the muscles exhibit spasticity. ***Rigidity*** refers to increased muscle tone, although reflexes are not affected. If rigidity is uniform throughout a range of movement, it is called lead-pipe rigidity; if rigidity is interrupted by a series of jerks, it is called cog-wheel rigidity, as in Parkinson's disease.

Muscular Atrophy and Hypertrophy

Muscular atrophy (A-trō-fē) refers to a state of wasting away of muscles. Individual muscle fibers decrease in size owing to a progressive loss of myofibrils. Muscles may become atrophied if they are not used. This is termed ***disuse atrophy.*** Bedridden individuals and people with casts may experience atrophy because the flow of impulses to the inactive muscle is greatly reduced. If the nerve supply to a muscle is cut, it will undergo complete atrophy. This is termed ***denervation atrophy.*** In about six months to two years, the muscle will be one-quarter its original size and the muscle fibers will be replaced by fibrous tissue. The transition to fibrous tissue, when complete, cannot be reversed. Battery-operated transcutaneous muscle stimulators (TMSs) are used to maintain the strength of atrophied muscles immobilized by a cast by stimulating the muscles to contract. TMSs are also used to prevent atrophy of muscles whose motor nerves have been temporarily damaged by trauma or stroke. Some athletes even use TMSs to strengthen certain muscles.

Muscular hypertrophy (hī-PER-trō-fē) is the reverse of atrophy. It refers to an increase in the diameters of muscle fibers owing to the production of more myofibrils, mitochondria, sarcoplasmic reticulum, nutrients (glycogen and triglycerides), and energy-supplying molecules (ATP and phosphocreatine). Hypertrophic muscles are capable of more forceful contractions. Weak muscular activity does not produce significant hypertrophy. It results from very forceful muscular activity or repetitive muscular activity at moderate levels. There is some evidence that muscle growth during weight lifting may be partly due to a longitudinal splitting of muscle fibers, followed by hypertrophy. However, it is generally felt that the number of muscle fibers does not increase after birth. During childhood, the increase in the *size* of muscle fibers appears to be at least partially under the control of human growth hormone (hGH), which is produced by the anterior pituitary gland. A further increase in the size of muscle fibers appears to be due to the hormone testosterone, produced by the testes. The influence of testosterone probably accounts for the generally larger muscles in males than females. More forceful muscular contractions, as in weight lifting, also contribute to generally larger muscles in males.

Types of Skeletal Muscle Fibers

All skeletal muscle fibers are not alike in structure or function. For example, skeletal muscle fibers vary in color depending on their content of ***myoglobin,*** a reddish pigment similar to hemoglobin in blood. Myoglobin stores oxygen until needed by mitochondria, the organelles in which ATP generation occurs. Skeletal muscle fibers that have a high myoglobin content are referred to as ***red muscle fibers.*** Conversely, skeletal muscle fibers that have a low content of myoglobin are called ***white muscle fibers.*** Red muscle fibers are smaller in diameter than white muscle fibers, and red muscle fibers have more mitochondria and more blood capillaries. White muscle fibers have a more extensive sarcoplasmic reticulum than red muscle fibers.

Skeletal muscle fibers contract with different velocities, depending on their ability to split ATP. Faster-contracting fibers have greater ability to split ATP. In addition, skeletal muscle fibers vary with respect to the metabolic processes they use to generate ATP. They also differ in terms of the onset of fatigue, the mechanism of which is discussed at the end of the chapter.

On the basis of various structural and functional characteristics, skeletal muscle fibers are classified into three types:

1. **Type I fibers.** These fibers, also called ***slow twitch*** or ***slow oxidative fibers,*** contain large amounts of myoglobin, many mitochondria, and many blood capillaries and have a high capacity to generate ATP by oxidative metabolic processes (Chapter 25). Type I fibers are red. Such fibers also split ATP at a slow rate and, as a result, contraction velocity is slow. The fibers are very resistant to fatigue. Such fibers are found in large numbers in the postural muscles of the neck.
2. **Type II B fibers.** These fibers, also called ***fast twitch*** or ***fast glycolytic fibers,*** have a low content of myoglobin, relatively few mitochondria, and relatively few blood capillaries. They do, however, contain large amounts of glycogen. Type II B fibers are white and geared to generate ATP by anaerobic metabolic processes, which are not able to supply skeletal muscle fibers continuously with sufficient ATP. Accordingly, these fibers fatigue easily, but they split ATP at a fast rate so that contraction velocity is fast. Muscles of the arms contain many of these fibers.
3. **Type II A fibers.** These fibers, also called ***fast twitch*** or ***fast oxidative fibers,*** contain very large amounts of myoglobin, very many mitochondria, and very many blood capillaries. These fibers are red and have a very high capacity for generating ATP by oxidative metabolic processes. Such fibers also split ATP at a very rapid rate, and as a result, contraction velocity is fast. Type II A fibers are resistant to fatigue but not quite as much as type I fibers. In humans, these fibers are infrequent.

Most skeletal muscles of the body are a mixture of all three types of skeletal muscle fibers, but their proportion varies depending on the usual action of the muscle. For example, postural muscles of the neck, back, and legs have a higher proportion of type I fibers. Muscles of the shoulders and arms are not constantly active but are used intermittently, usually for short periods of time, to produce large amounts of tension such as in lifting and throwing. These muscles have a higher proportion of type II B fibers. Leg muscles not only support the body but are also used for walking and running. Such muscles have higher proportions of type I and type II B fibers.

Even though most skeletal muscles are a mixture of all three types of skeletal muscle fibers, all the skeletal muscle fibers of any one motor unit are all the same. In addition, the different skeletal muscle fibers in a muscle may be used in various ways, depending on need. For example, if only a weak contraction is needed to perform a task, only type I fibers are activated by their motor units. If a stronger contraction is needed, the motor units of type II A fibers are activated. And, if a maximal contraction is required, motor units of type II B fibers are activated as well. Activation of various motor units is determined in the brain and spinal cord.

Although the number of the different skeletal muscle fibers does not change, the characteristics of those present can be altered. Various types of exercises can bring about changes in the fibers in a skeletal muscle. Endurance-type exercises, such as running or swimming, cause a gradual transformation of type II B fibers into type II A fibers. The transformed muscle fibers show a slight increase in diameter, mitochondria, blood capillaries, and strength. Endurance exercises result in cardiovascular and respiratory changes that cause skeletal muscles to receive better supplies of oxygen and carbohydrates but do not contribute to muscle mass. On the other hand, exercises that require great strength for short periods of time, such as weight lifting, produce an increase in the size and strength of type II B fibers. The increase in size is due to increased synthesis of thin and thick myofilaments. The overall result is that the person develops large muscles.

CARDIAC MUSCLE TISSUE

Structure

The principal constituent of the heart wall is ***cardiac muscle tissue.*** Although it is striated in appearance like skeletal muscle, it is involuntary. The fibers of cardiac muscle tissue are roughly quadrangular and usually have only a single centrally located nucleus (see Exhibit 4-3). Skeletal muscle fibers contain several nuclei that are peripherally located. The thin sarcolemma of cardiac muscle fibers is similar to that of skeletal muscle, but the sarcoplasm is more abundant and the mitochondria are larger and more numerous. Cardiac muscle fibers have the same arrangement of actin and myosin and the same bands, zones, and lines as skeletal muscle fibers (Figure 10-15). Myofilaments in cardiac muscle fibers are not arranged in discrete myofibrils as in skeletal muscle. The transverse tubules of mammalian cardiac muscle are larger than those of skeletal muscle and are located at the Z lines rather than at the A–I band junctions as in skeletal muscle fibers. The sarcoplasmic reticulum of cardiac muscle is less well developed than that in skeletal muscle.

Note in Exhibit 4-3 that cardiac muscle fibers branch and interconnect with each other. Recall that skeletal muscle fibers are arranged in parallel fashion. Cardiac muscle fibers form two separate networks. The muscular walls and partition of the upper chambers of the heart (atria) compose one network. The muscular walls and partition of the lower chambers of the heart (ventricles) compose the other network. Each fiber in a network is separated from the next fiber by an irregular transverse thickening of the sarcolemma called an ***intercalated*** (in-TER-ka-lāt-ed) ***disc.*** These discs contain desmosomes, which hold fibers together, and gap junctions, which aid in conduction of muscle action potentials from one muscle fiber to another. When a single fiber of either network is stimulated, all the fibers in the network become stimulated as well. Thus, each network contracts as a functional unit. As you will see in Chapter 20, when the fibers of the atria contract as a unit, blood moves into the ventricles. Then, when the ventricular fibers contract as a unit, blood is pumped into arteries.

Physiology

Under normal resting conditions, cardiac muscle tissue contracts and relaxes rapidly, continuously, and rhythmically about 75 times a minute without stopping. This is a major physiological difference between cardiac and skeletal muscle tissue. Accordingly, cardiac muscle tissue requires a constant supply of oxygen. Energy generation occurs in large, numerous mitochondria. Another difference is the source of stimulation. Skeletal muscle tissue ordinarily contracts only when stimulated by a nerve impulse. In contrast, cardiac muscle tissue can contract without extrinsic (outside) nerve or hormonal stimulation. Its source of stimulation is a conducting tissue of specialized intrinsic (internal) cardiac muscle within the heart. Nerve stimulation merely causes the conducting tissue to increase or decrease its rate of discharge. Some types of smooth muscle fibers and nerve cells in the brain and spinal cord also possess spontaneous, rhythmical self-excitation, a phenomenon referred to as ***autorhythmicity.*** It is discussed in detail in Chapter 20.

Another difference between cardiac and skeletal muscle tissue is that cardiac muscle tissue remains contracted (depolarized) 10 to 15 times longer than skeletal muscle tissue. This is because there is a prolonged delivery of Ca^{2+} ions into the sarcoplasm. In cardiac muscle fibers, Ca^{2+} ions are derived from sarcoplasmic reticulum (as in skeletal muscle fibers) and from extracellular fluids. Although the release of Ca^{2+} ions from the sarcoplasmic reticulum is fast, passage of Ca^{2+} ions from extracellular fluids through the sarcolemma is much slower, thus accounting for the prolonged contraction of cardiac muscle fibers.

Cardiac muscle tissue also has an extra long refractory period, lasting several tenths of a second, that allows time for the heart to relax between beats. The long refractory period permits heart rate to be increased significantly but prevents the heart itself from undergoing incomplete or complete tetanus. Tetanus of heart muscle would stop blood flow through the body and result in death.

FIGURE 10-15 Histology of cardiac muscle tissue. (a) Diagram based on an electron micrograph showing several myofibrils.

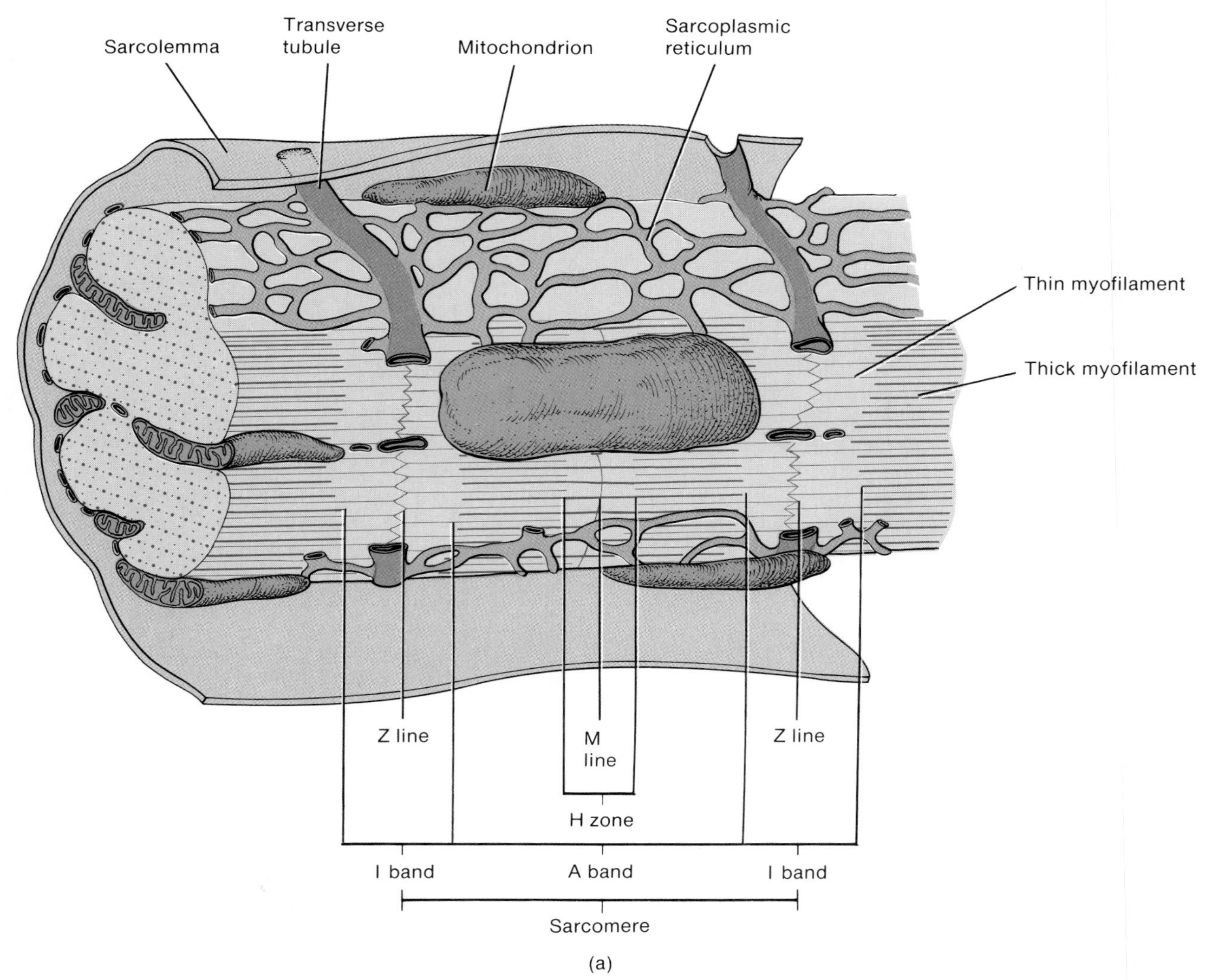

SMOOTH MUSCLE TISSUE

Structure

Like cardiac muscle tissue, ***smooth muscle tissue*** is usually involuntary. However, it is nonstriated. Smooth muscle fibers are considerably smaller than skeletal muscle fibers. A single fiber of smooth muscle tissue is about 5 to 10 μm in diameter and 30 to 200 μm long. Each fiber is thickest in the midregion and tapers at each end. Within the fiber is a single, oval, centrally located nucleus (Figure 10-16 and see Exhibit 4-3). The sarcoplasm of smooth muscle fibers contains thick myofilaments (longer than those of skeletal muscle fibers) and thin myofilaments, but not arranged as orderly sarcomeres as in striated muscle. In smooth muscle fibers, there are 10 to 15 thin myofilaments for each thick myofilament in the regions of myofilament overlap; in skeletal muscle fibers, the ratio is 2:1. Smooth muscle fibers also contain ***intermediate filaments.*** Since the various myofilaments have no regular pattern of organization and since there are no A or I bands of sarcomeres, smooth muscle fibers have no characteristic striations. Thus, the name *smooth.*

Intermediate filaments are attached to structures called ***dense bodies,*** which have characteristics similar to Z lines in striated muscle fibers. Some dense bodies are dispersed throughout the cytoplasm; others are attached to the sarcolemma. Bundles of intermediate filaments stretch from one dense body to another (Figure 10-16b). The sliding filament mechanism involving thick and thin myofilaments during contraction generates tension that is transmitted to intermediate filaments. These, in turn,

FIGURE 10-15 (*Continued*) (b) Diagram based on an electron micrograph showing intercalated discs and related structures.

Sarcoplasmic reticulum
Separation of intercalated discs
Nucleus
Mitochondrion
Thin myofilament
Thick myofilament
Intercalated discs
Z line
Z line
Sarcomere

(b)

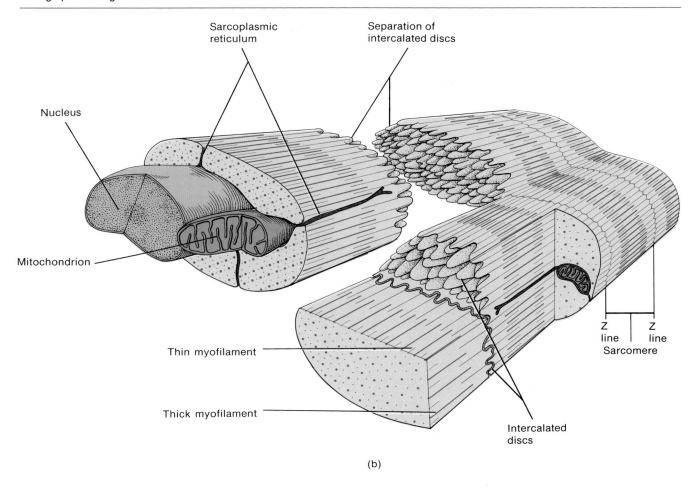

FIGURE 10-16 Histology of smooth muscle tissue. (a) Diagram of visceral (single-unit) smooth muscle tissue (left) and multiunit smooth muscle tissue (right). (b) Details of a fiber (cell) before contraction.

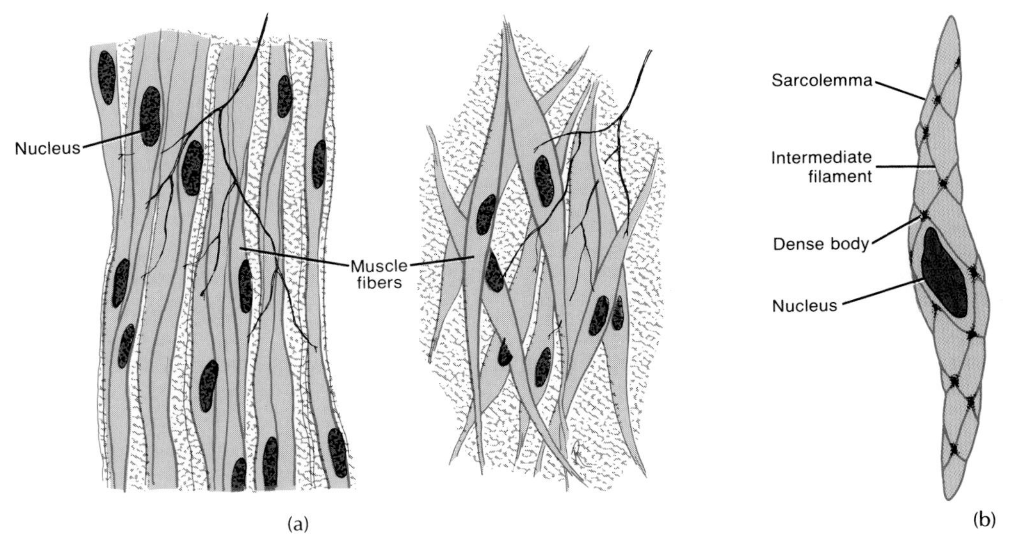

FIGURE 10-16 (*Continued*) (c) Sectional views showing relaxed and contracted fibers.

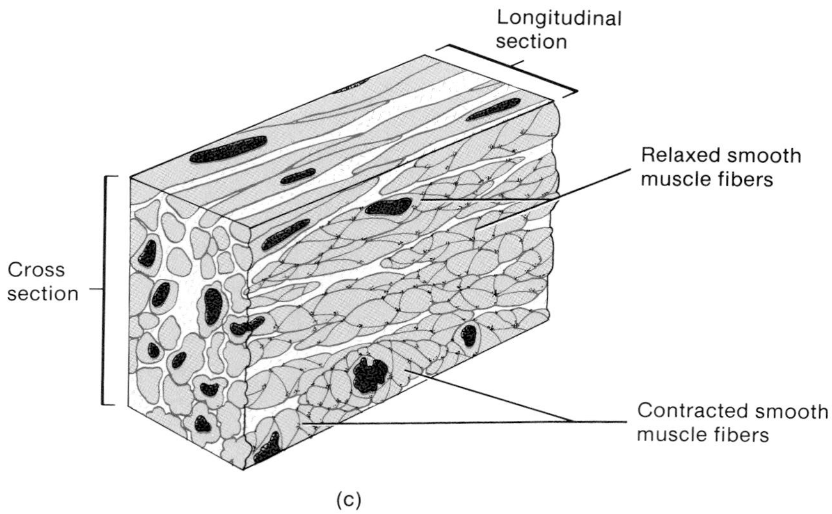

pull on the dense bodies attached to the sarcolemma, causing a lengthwise shortening of the muscle fiber. Note in Figure 10-16b that shortening of the muscle fiber produces a bubblelike expansion of the sarcolemma. Evidence suggests that a smooth muscle fiber contracts in a corkscrewlike manner; the fiber twists in a helix as it shortens and rotates in the opposite direction as it lengthens.

Smooth muscle fibers also possess a less well developed sarcoplasmic reticulum than skeletal muscle fibers. Small subsurface vesicles termed ***caveolae*** open onto the surface of the smooth muscle fiber and are thought to function in a way similar to the transverse tubules of striated fibers, namely, to carry muscle action potentials into the fibers. Tubules of the sarcoplasmic reticulum are associated with longitudinal rows of caveolae.

Two kinds of smooth muscle tissue, visceral and multiunit, are recognized (Figure 10-16a). The more common type is called ***visceral (single-unit) muscle tissue.*** It is found in wraparound sheets that form part of the walls of small arteries and veins and hollow viscera such as the stomach, intestines, uterus, and urinary bladder. The terms *smooth muscle tissue* and *visceral muscle tissue* are sometimes used interchangeably. The fibers in visceral muscle tissue are tightly bound together to form a continuous network. They contain gap junctions to facilitate conduction of muscle action potentials between fibers. When a neurotransmitter, hormone, or autorhythmic signal stimulates one fiber, the muscle action potential spreads to other fibers so that contraction occurs in a wave over many adjacent fibers. Thus, whereas skeletal muscle fibers contract as individual units, visceral muscle cells contract in sequence as the action potential spreads from one cell to another.

The second kind of smooth muscle tissue, ***multiunit smooth muscle tissue,*** consists of individual fibers, each with their own motor-nerve endings. Whereas stimulation of a single visceral muscle fiber causes contraction of many adjacent fibers, stimulation of a single multiunit fiber causes contraction of only that fiber. In this respect, multiunit muscle tissue is like skeletal muscle tissue. Multiunit smooth muscle tissue is found in the walls of large arteries, in large airways to the lungs, in the arrector pili muscles that attach to hair follicles, and in the intrinsic (internal) muscles of the eye, such as the iris.

Physiology

Although the principles of contraction are essentially the same in smooth and striated muscle tissue, smooth muscle tissue exhibits several important physiological differences. First of all, the duration of contraction and relaxation of smooth muscle fibers is about 5 to 500 times longer than in skeletal muscle fibers. Smooth muscle fibers derive their Ca^{2+} ions from sarcoplasmic reticulum and extracellular fluids. Since there are no transverse tubules in smooth muscle fibers, it takes longer for Ca^{2+} ions to reach the deep filaments in the center of the fiber to trigger the contractile process, and this accounts, in part, for the slow activation and prolonged contraction of smooth muscle. Moreover, smooth muscle fibers do not contain calcium-binding troponin to facilitate myosin cross bridge attachment to thin myofilaments. They use a different mechanism that takes more time and thus contributes to prolonged contraction.

Not only do Ca^{2+} ions move slowly into deep filaments, but they also move slowly out of the muscle fiber following contraction. This delays relaxation, and the prolonged stay of Ca^{2+} ions in the fibers provides for tone, the state of continued partial contraction. Smooth muscle tissue can undergo sustained, long-term tone, which is important in the gastrointestinal tract where the walls of the tract maintain a steady pressure on the contents of the tract, in the walls of blood vessels called arterioles that maintain a steady pressure on blood, and in the wall of the urinary bladder that maintains a steady pressure on urine.

Some smooth muscle fibers contract in response to nerve impulses from the autonomic (involuntary) nervous system. Thus, smooth muscle is normally not under voluntary control. Other smooth muscle fibers contract in response to hormones or local factors such as pH, oxygen and carbon dioxide levels, temperature, and ion concentrations.

Finally, unlike skeletal muscle fibers, smooth muscle fibers can stretch considerably without developing tension. When smooth muscle fibers are stretched, they initially develop increased tension. However, there is an almost immediate decrease in the tension. This phenomenon is referred to as ***stress–relaxation*** and is important because it permits smooth muscle to accommodate great changes in size while still retaining the ability to contract effectively. Thus, the smooth muscle in the wall of hollow organs such as the stomach, intestines, and urinary bladder can stretch as the viscera distend, while the pressure within them remains the same.

A summary of the principal characteristics of the three types of muscle tissue is presented in Exhibit 10-2.

REGENERATION OF MUSCLE TISSUE

Skeletal muscle fibers cannot divide. After the first year of life, all growth of skeletal muscle is due to enlargement of existing cells (hypertrophy), rather than increase in the number of fibers (hyperplasia). Skeletal muscle fibers, however, can be replaced on an individual basis by new fibers derived from ***satellite cells.*** These cells are dormant stem cells found in association with skeletal muscle fibers. During rapid postnatal growth, satellite cells lengthen existing skeletal muscle fibers by fusing with them. They also persist as a lifelong source of cells that can fuse with each other to form new skeletal muscle fibers. However, the number of new skeletal muscle fibers formed by this mechanism is not sufficient to compensate for any significant skeletal muscle damage. In cases of such damage, skeletal muscle tissue is replaced by fibrous scar tissue. For this reason, skeletal muscle tissue has only limited powers of regeneration.

Cardiac muscle fibers, like those of skeletal muscle tissue, have no capacity for division and increase in size by hypertrophy. In addition, cardiac muscle fibers do not associate with cells comparable to satellite cells. Healing of cardiac muscle tissue is by scar formation. Cardiac muscle tissue has no powers of regeneration.

EXHIBIT 10-2 SUMMARY OF THE PRINCIPAL CHARACTERISTICS OF MUSCLE TISSUE

Characteristic	Skeletal Muscle	Cardiac Muscle	Smooth Muscle
Location	Attached primarily to bones.	Heart.	Walls of hollow viscera, blood vessels, iris, arrector pili.
Microscopic appearance	Striated, multinucleated, unbranched fibers.	Striated uninucleated, branched fibers with intercalated discs.	Nonstriated (smooth) uninucleated, spindle-shaped fibers.
Nervous control	Voluntary.	Involuntary.	Involuntary.
Sarcomeres	Yes.	Yes.	No.
Transverse tubules	Yes.	Yes.	No.
Gap junctions	No.	Yes.	Yes, in visceral smooth muscle.
Cell size	Large.	Large.	Small.
Source of calcium	Sarcoplasmic reticulum.	Sarcoplasmic reticulum and extracellular fluids.	Sarcoplasmic reticulum and extracellular fluids.
Speed of contraction	Fast.	Moderate.	Slow.
Capacity for division	None.	None.	Limited.
Capacity for regeneration	Limited.	None.	Considerable compared with other muscle tissues but limited compared with tissues such as epithelium.

Smooth muscle tissue, like skeletal and cardiac muscle tissue, can undergo hypertrophy. In addition, certain smooth muscle fibers, such as those in the uterus, retain their capacity for division and thus can grow by hyperplasia. Also, new smooth muscle fibers can arise from cells called ***pericytes,*** stem cells found in association with the endothelium of blood capillaries and venules. It is also known that smooth muscle fibers can proliferate in certain pathological conditions such as occur in the development of atherosclerosis (Chapter 21). Compared with the other two types of muscle tissue, smooth muscle tissue has a considerably higher, though still limited, power of regeneration when compared with other tissues, such as epithelium.

HOMEOSTASIS

Muscle tissue plays a vital role in maintaining the body's homeostasis. Three examples are the relationship of muscle tissue to oxygen, to fatigue, and to heat production.

Oxygen Debt

Earlier in the chapter, it was noted that some of the energy used to generate ATP for muscle contractions is derived from the complete breakdown of pyruvic acid, in the presence of oxygen, to carbon dioxide and water by way of cellular respiration in the mitochondria of muscle fibers.

During muscular exercise, blood vessels in muscles dilate and blood flow is increased in order to increase the available oxygen supply. Up to a point, the available oxygen is sufficient to meet the energy needs of the body. But, when muscular exertion is very great, oxygen cannot be supplied to muscle fibers fast enough, and the aerobic breakdown of pyruvic acid cannot produce all the ATP required for further muscle contraction. During such periods, additional ATP is generated by anaerobic glycolysis. In the process, most of the pyruvic acid produced is converted to lactic acid. Although about 80 percent of the lactic acid diffuses from the skeletal muscles and is transported to the liver for conversion back to glucose or glycogen, some lactic acid accumulates in muscle tissue. Ultimately, once adequate oxygen is available, lactic acid must be catabolized completely into carbon dioxide and water. After exercise has stopped, extra oxygen is required to metabolize lactic acid; to replenish ATP, phosphocreatine, and glycogen; and to pay back any oxygen that has been borrowed from hemoglobin, myoglobin (an iron-containing substance similar to hemoglobin that is found in muscle fibers), air in the lungs, and body fluids. The additional oxygen that must be taken into the body after vigorous exercise to restore all systems to their normal states is called ***oxygen debt.*** The debt is paid back by labored breathing that continues after exercise has stopped. Thus, the accumulation of lactic acid causes hard breathing and sufficient discomfort to stop muscle activity until homeostasis is restored. Eventually, muscle glycogen must also be restored. This is accomplished through diet and may take several days, depending on the intensity of exercise.

The maximum rate of oxygen consumption during the aerobic catabolism of pyruvic acid is called ***maximal oxygen uptake.*** It is determined by sex (higher in males), age (highest at about age 20), and size (increases with body size). Highly trained athletes can have maximal oxygen uptakes that are twice that of average people, probably owing to a combination of genetics and training. As a result, they are capable of greater muscular activity without increasing their lactic acid production, and their oxygen debts are less. It is for these reasons that they do not become short of breath as readily as untrained individuals.

CLINICAL APPLICATION: "SECOND WIND"

Many athletes and individuals who participate in fitness exercises indicate that they experience less fatigue and less labored breathing after the first few minutes of continuous exercise. This phenomenon is called *"second wind."* Although the mechanism is not completely understood, it may actually be a warm-up effect that results from a leveling off of the rate of oxygen consumption after a few minutes of aerobic activity. At the beginning of vigorous physical activity, aerobic catabolism of pyruvic acid does not proceed fast enough to meet the full energy demands of the body. In this situation, anaerobic catabolism makes up the difference. The lactic acid produced during anaerobic catabolism is believed to be partially responsible for fatigue and stress during breathing. When oxygen debt is repaid, lactic acid buildup stops, fatigue is lessened, and breathing returns to normal.

Muscle Fatigue

If a skeletal muscle or group of skeletal muscles is continuously stimulated for an extended period of time, the strength of contraction becomes progressively weaker until the muscle no longer responds. The inability of a muscle to maintain its strength of contraction or tension is called ***muscle fatigue,*** which is related to an inability of muscle to produce sufficient energy to meet its needs. Although its exact mechanism is not completely understood, it may be related to insufficient oxygen, depletion of glycogen, and/or lactic acid buildup, which makes muscle tissue more acid. More acidic conditions could disrupt enzymes involved in ATP production and the contractile process. Recent evidence indicates that during fatigue, phosphate derived from the breakdown of creatine phosphate builds up and muscles lose strength. It is speculated that the excessive phosphate might inhibit effective cross bridge functioning.

Heat Production

The production of heat by skeletal muscles is one homeostatic mechanism for maintaining normal body temperature. Of the total energy released during muscular contraction, only a small amount is used for mechanical work (contraction). As much as 85 percent can be released as heat, some of which is utilized to help maintain a normal body temperature. The remainder is eliminated by the skin and lungs. Excessive heat loss by the body results in shivering, an increase in muscle tone, that can increase the rate of heat production by several hundred percent. Thus, shivering produces heat to raise body temperature back to normal.

Heat production by muscles may be divided into two phases: (1) ***initial heat,*** which is produced by the contraction and relaxation of a muscle; and (2) ***recovery heat,*** which is produced after relaxation. Initial heat is independent of O_2 and is associated with ATP breakdown. Recovery heat is associated with ATP restoration. It includes the anaerobic breakdown of glucose to pyruvic acid and pyruvic acid to lactic acid. It also includes the aerobic breakdown of pyruvic acid to CO_2 and H_2O and the aerobic conversion of lactic acid to CO_2 and H_2O.

AGING AND MUSCLE TISSUE

Beginning at about 30 years of age, there is a progressive loss of skeletal muscle mass that is largely replaced by fat. Accompanying the loss of muscle mass, there is a decrease in maximal strength and a diminishing of muscle reflexes.

DEVELOPMENTAL ANATOMY OF THE MUSCULAR SYSTEM

In this brief discussion of the development of the human muscular system, we will concentrate mostly on skeletal muscles. Except for the muscles of the iris of the eyes and the arrector pili muscles attached to hairs, all muscles of the body are derived from ***mesoderm.*** As the mesoderm develops, a portion of it becomes arranged in dense columns on either side of the developing nervous system. These columns of mesoderm undergo segmentation into a series of blocks of cells called ***somites*** (Figure 10-17a). The first pair of somites appears on the twentieth day of embryologic development. Eventually, 44 pairs of somites are formed by the thirtieth day.

With the exception of the skeletal muscles of the head and extremities, *skeletal muscles* develop from the ***mesoderm of somites.*** Since there are very few somites in the head region of the embryo, most of the skeletal muscles there develop from the ***general mesoderm*** in the head region. The skeletal muscles of the limbs develop from masses of general mesoderm around developing bones in embryonic limb buds (origins of future extremities: see Figure 6-6a).

The cells of a somite are differentiated into three regions: (1) ***myotome,*** which forms most of the skeletal muscles; (2) ***dermatome,*** which forms the connective tissues, including the dermis; and (3) ***sclerotome,*** which gives rise to the vertebrae (Figure 10-17b).

In the development of a skeletal muscle from a myotome of a somite, certain patterns emerge. For ex-

FIGURE 10-17 Development of the muscular system. (a) Dorsal aspect of an embryo indicating the location of somites. (b) Cross section of a portion of an embryo showing a somite.

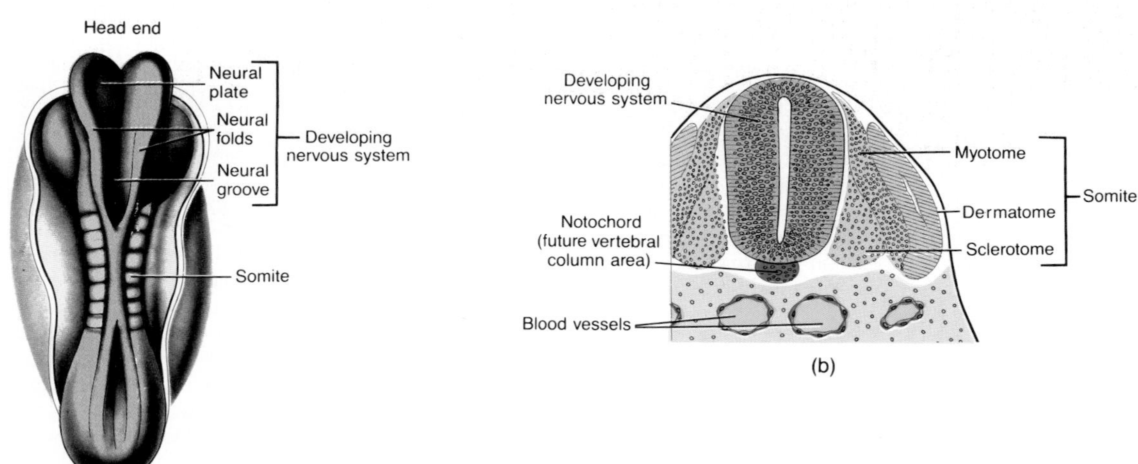

ample, a myotome may split longitudinally into two or more portions. This represents the manner in which the trapezius muscle forms. Other myotomes split into two or more layers. Such a pattern represents how the external oblique, internal oblique, and transversus abdominis muscles develop. In other instances, several myotomes fuse to form a single muscle. Such an example is the rectus abdominis muscle. Muscles may also migrate, wholly or in part, from their sites of origin. The latissimus dorsi, for example, originates from the cervical myotomes but extends all the way down to the thoracic and lumbar vertebrae and coxal (hip) bone. Another trend that may be observed is a change in the direction of muscle fibers. Initially, muscle fibers in a myotome are parallel to the long axis of the embryo. However, nearly all developed skeletal muscles do not have fibers that are parallel to the long axis. One example is the external oblique. Finally, *fasciae, ligaments,* and *aponeuroses* may form as a result of degeneration of all or parts of myotomes.

Smooth muscle develops from ***mesodermal cells*** that migrate to and envelop the developing gastrointestinal tract and viscera.

Cardiac muscle develops from ***mesodermal cells*** that migrate to and envelop the developing heart while it is still in the form of primitive heart tubes (see Figure 20-12).

DISORDERS: HOMEOSTATIC IMBALANCES

Disorders of the muscular system are related to disruptions of homeostasis. The disorders may involve a lack of nutrients, accumulation of toxic products, disease, injury, disuse, or faulty nervous connections (innervations).

Fibrosis

The formation of fibrous (containing fibers) connective tissue in locations where it normally does not exist is called ***fibrosis.*** Skeletal and cardiac muscle fibers cannot undergo mitosis, and dead muscle fibers are normally replaced with fibrous connective tissue. Fibrosis, then, is often a consequence of muscle injury or degeneration.

Fibromyalgia

Fibromyalgia (*algia* = painful condition) refers to a group of common nonarticular rheumatic disorders characterized by pain, tenderness, and stiffness of muscles, tendons, and surrounding soft tissues. The disorders specifically affect the fibrous connective tissue components of muscles, tendons, and ligaments. Fibromyalgia may be caused or aggravated by physical or mental stress, trauma, exposure to dampness or cold, poor sleep, or a rheumatic condition. Frequent sites at which fibromyalgia occur include the lumbar region (***lumbago***), neck, chest, and thighs (***charleyhorse***). Charleyhorse is a slang term that refers to a painful, localized cramp that sometimes is associated with pregnancy or may occur in otherwise healthy persons. In some cases, local muscle spasms are noted. The condition is relieved with heat, massage, and rest and completely disappears, although occasionally it may become chronic or recur at frequent intervals.

Muscular Dystrophies

Muscular dystrophies (*dystrophy* = degeneration) are muscle-destroying diseases. The diseases are characterized by degeneration of individual muscle fibers, which leads to a progressive atrophy of the skeletal muscle. Usually, the voluntary skeletal muscles are weakened equally on both sides of the body, whereas the internal muscles, such as the diaphragm, are not affected. Histologically, the changes that occur include variation in muscle fiber size, degeneration of fibers, and deposition of fat. Muscular dystrophies are classified by mode of inheritance, age of onset, and clinical features. The most common form of muscular dystrophies is called ***Duchenne*** (du-SHĀN) ***muscular dystrophy (DMD).*** The gene responsible for DMD has been identified, and its DNA code has been worked out. Hopefully, this information will lead to replacement therapy and a halt to muscle loss.

Muscular dystrophies are due to a genetic defect that may result in faulty metabolism of potassium, protein deficiency, or inability of the body to utilize creatine. Just recently, scientists discovered that a protein they named ***dystrophin*** is present in triads of normal muscle tissue but absent in persons with Duchenne muscular dystrophy. The function of dystrophin is unknown. According to one hypothesis, lack of dystrophin may result in leakage of calcium ions from the sarcoplasmic reticulum into sarcoplasm. This, in turn, may activate an enzyme (phospholipase A) that degenerates muscle fibers.

Muscular dystrophies are diagnosed on the basis of muscle enzyme studies, electromyography, and where indicated, muscle biopsy.

Myasthenia Gravis (MG)

Myasthenia (mī-as-THĒ-nē-a) ***gravis (MG)*** is a weakness of skeletal muscles. It is caused by an abnormality at the neuromuscular junction that prevents muscle fibers from contracting. Recall that motor neurons stimulate skeletal muscle fibers to contract by releasing acetylcholine (ACh). Myasthenia gravis is an autoimmune disorder caused by antibodies directed against ACh receptors of the muscle fiber sarcolemma. The antibodies bind to the receptors and hinder the attachment of ACh to the receptors (see Chapter 14). As the disease progresses, more neuromuscular junctions become affected.

DISORDERS: HOMEOSTATIC IMBALANCES (*continued*)

The muscle becomes increasingly weaker and may eventually cease to function altogether.

Myasthenia gravis is more common in females, occurring most frequently between the ages of 20 and 50. The muscles of the face and neck are most apt to be involved. Initial symptoms include a weakness of the eye muscles and difficulty in swallowing. Later, the individual has difficulty chewing and talking. Eventually, the muscles of the limbs may become involved. Death may result from paralysis of the respiratory muscles, but usually the disorder does not progress to this stage.

Anticholinesterase drugs such as neostigmine and pyridostigmine, derivatives of physostigmine, have been the primary treatment for the disease. They act as inhibitors of acetylcholinesterase, thus raising the level of ACh to bind with available receptors. More recently, steroid drugs, such as prednisone, have been used with great success to reduce antibody levels. Immunosuppressant drugs are also used to decrease the production of antibodies that interfere with normal muscle contraction. Another treatment involves *plasmapheresis,* a procedure that separates blood cells from the plasma that contains the unwanted antibodies. The blood cells are then mixed with a plasma substitute and pumped back into the individual. In some individuals, surgical removal of the thymus gland is indicated (thymectomy).

Abnormal Contractions

One kind of abnormal muscular contraction is a ***spasm,*** a sudden involuntary contraction of large groups of muscles. (Cerebral palsy is characterized by generalized spastic contractions.) ***Tremor*** is a rhythmic, involuntary, purposeless contraction of opposing muscle groups. (A resting tremor occurs in Parkinson's disease.) A ***fasciculation*** is an involuntary, brief twitch of a muscle visible under the skin. It occurs irregularly and is not associated with movement of the affected muscle. (Fasciculations may be seen in multiple sclerosis or amyotrophic lateral sclerosis, also called Lou Gehrig's disease, both of which are discussed in Chapter 14.) A ***fibrillation*** is similar to a fasciculation except that it is not visible under the skin. It is recorded by electromyography. A ***tic*** is a spasmodic twitching made involuntarily by muscles that are ordinarily under voluntary control. Twitching of the eyelid and face muscles are examples. In general, tics are of psychological origin.

MEDICAL TERMINOLOGY ASSOCIATED WITH THE MUSCULAR SYSTEM

Gangrene (GANG-rēn; *gangraena* = an eating sore) Death of a soft tissue, such as muscle, that results from interruption of its blood supply. One type is caused by various species of *Clostridium,* bacteria that live anaerobically in the soil.

Myalgia (mi-AL-jē-a; *algia* = painful condition) Pain in or associated with muscles.

Myoma (mī-Ō-ma; *oma* = tumor) A tumor consisting of muscle tissue.

Myomalacia (mī'-ō-ma-LĀ-shē-a; *malaco* = soft) Softening of a muscle.

Myopathy (mī-OP-a-thē; *pathos* = disease) Any disease of muscle tissue.

Myosclerosis (mī'-ō-skle-RŌ-sis; *scler* = hard) Hardening of a muscle.

Myositis (mī'-ō-SĪ-tis; *itis* = inflammation of) Inflammation of muscle fibers (cells).

Myospasm (MĪ-o-spazm) Spasm of a muscle.

Myotonia (mī-ō-TŌ-nē-a; *tonia* = tension) Increased muscular excitability and contractility with decreased power of relaxation; tonic spasm of the muscle.

Paralysis (pa-RAL-a-sis; *para* = beyond; *lyein* = to loosen) Loss or impairment of motor (muscular) function resulting from a lesion of nervous or muscular origin.

Trichinosis (trik'-i-NŌ-sis) A myositis caused by the parasitic worm *Trichinella spiralis,* which may be found in the muscles of humans, rats, and pigs. People contract the disease by eating insufficiently cooked infected pork.

Volkmann's contracture (FŌLK-manz kon-TRAK-tur; *contra* = against) Permanent contraction of a muscle due to replacement of destroyed muscle fibers (cells) with fibrous tissue that lacks ability to stretch. Destruction of muscle fibers may occur from interference with circulation caused by a tight bandage, a piece of elastic, or a cast.

Wryneck (RĪ-neck) or **torticollis** (*tortus* = twisted; *collum* = neck) Contracted state of several superficial and deep muscles of the neck that produces twisting of the neck and an unnatural position of the head; one of the common triggering factors is vigorous activity.

STUDY OUTLINE

Characteristics (p. 230)

1. Excitability is the property of receiving and responding to stimuli.
2. Contractility is the ability to shorten and thicken (contract).
3. Extensibility is the ability to be stretched (extended).
4. Elasticity is the ability to return to original shape after contraction or extension.

Functions (p. 230)

1. Through contraction, muscle tissue performs three important functions.
2. These functions are motion, maintenance of posture, and heat production.

Types (p. 230)

1. Skeletal muscle tissue is primarily attached to bones. It is striated and voluntary.
2. Cardiac muscle tissue forms the wall of the heart. It is striated and involuntary.
3. Visceral muscle tissue is located in viscera. It is nonstriated (smooth) and involuntary.

Skeletal Muscle Tissue (p. 230)

Connective Tissue Components (p. 230)

1. The term fascia is applied to a sheet or broad band of fibrous connective tissue underneath the skin or around muscles and organs of the body.
2. Other connective tissue components are epimysium, covering the entire muscle; perimysium, covering fasciculi; and endomysium, covering fibers; all are extensions of deep fascia.
3. Tendons and aponeuroses are extensions of connective tissue beyond muscle cells that attach the muscle to bone or other muscle.

Nerve and Blood Supply (p. 232)

1. Nerves convey impulses for muscular contraction.
2. Blood provides nutrients and oxygen for contraction.

Histology (p. 232)

1. Skeletal muscle consists of fibers (cells) covered by a sarcolemma. The fibers contain sarcoplasm, nuclei, sarcoplasmic reticulum, and transverse tubules.
2. Each fiber contains myofibrils that consist of thin and thick myofilaments. The myofilaments are compartmentalized into sarcomeres.
3. Thin myofilaments are composed of actin, tropomyosin, and troponin; thick myofilaments consist mostly of myosin.
4. Projecting myosin heads are called cross bridges and contain actin- and ATP-binding sites.

Contraction (p. 233)

Sliding-Filament Theory (p. 233)

1. A muscle action potential travels over the sarcolemma and enters the transverse tubules and affects the sarcoplasmic reticulum causing it to release Ca^{2+} into the sarcoplasm.
2. The muscle action potential leads to the release of calcium ions from the sarcoplasmic reticulum, triggering the contractile process.
3. Actual contraction is brought about when the thin myofilaments of a sarcomere slide toward each other as the myosin cross bridges pull on the actin myofilaments.

Neuromuscular Junction (p. 235)

1. A motor neuron transmits a nerve impulse (nerve action potential) to a skeletal muscle where it serves as a stimulus for contraction.
2. A neuromuscular junction refers to an axon terminal of a motor neuron and the portion of the muscle fiber sarcolemma in close approximation with it (motor end plate).

Motor Unit (p. 238)

1. A motor neuron and the muscle fibers it stimulates form a motor unit.
2. A single motor unit may innervate as few as 10 or as many as 2000 muscle fibers.

Physiology of Contraction (p. 238)

1. When a nerve impulse (nerve action potential) reaches an axon terminal, the synaptic vesicles of the terminal release acetylcholine (ACh), which ultimately initiates a muscle action potential in the muscle fiber sarcolemma that then travels into the transverse tubules and causes the sarcoplasmic reticulum to release some of its stored Ca^{2+} into the sarcoplasm.
2. The muscle action potential releases calcium ions that combine with troponin, causing it to pull on tropomyosin to change its orientation, thus exposing myosin-binding sites on actin.
3. ATPase splits ATP into ADP + P and the released energy activates (energizes) myosin cross bridges.
4. Activated cross bridges attach to actin and a change in the orientation of the cross bridge occurs (power stroke); their movement results in the sliding of thin myofilaments.

Energy for Contraction (p. 240)

1. The immediate, direct source of energy for muscle contraction is ATP.
2. Muscle fibers generate ATP continuously. This involves phosphocreatine and the metabolism of glycogen and fats.

All-or-None Principle (p. 241)

1. The weakest stimulus capable of causing contraction is a liminal (threshold) stimulus.
2. A stimulus not capable of inducing contraction is a subliminal (subthreshold) stimulus.
3. Muscle fibers of a motor unit contract to their fullest extent or not at all.

Kinds of Contractions (p. 242)

1. The various kinds of contractions are twitch, tetanus, treppe, isotonic, and isometric.
2. A record of a contraction is called a myogram. The refractory period is the time when a muscle has temporarily lost ex-

citability. Skeletal muscles have a short refractory period. Cardiac muscle has a long refractory period.
3. Wave summation is the increased strength of a contraction resulting from the application of a second stimulus before the muscle has completely relaxed after a previous stimulus.

Muscle Tension (p. 244)

1. Muscle tension (force) depends on several factors.
2. Among these are frequency of stimulation, number of contracting (active) fibers, components of muscle fiber (contractile and elastic), and length of muscle fibers.

Muscle Tone (p. 245)

1. A sustained partial contraction of portions of a skeletal muscle results in muscle tone.
2. Tone is essential for maintaining posture.
3. Flaccidity is a condition of less-than-normal tone. Atrophy is a wasting away or decrease in size; hypertrophy is an enlargement or overgrowth.

Muscular Atrophy and Hypertrophy (p. 246)

1. Muscular atrophy refers to a state of wasting away of muscles.
2. Muscular hypertrophy refers to an increase in the diameter of muscle fibers.

Types of Skeletal Muscle Fibers (p. 246)

1. On the basis of structure and function, skeletal muscle fibers are classified as type I, type II B, and type II A.
2. Most skeletal muscles contain a mixture of all three fiber types, their proportions varying with the usual action of the muscle.
3. Various exercises can modify the types of skeletal muscle fibers.

Cardiac Muscle Tissue (p. 247)

1. This muscle is found only in the heart. It is striated and involuntary.
2. The fibers are quadrangular and usually contain a single centrally placed nucleus.
3. Compared to skeletal muscle tissue, cardiac muscle tissue has more sarcoplasm, more mitochondria, less well-developed sarcoplasmic reticulum, and larger transverse tubules located at Z lines rather than at A–I band junctions. Myofilaments are not arranged in discrete myofibrils.
4. The fibers branch freely and are connected via gap junctions.
5. Intercalated discs provide strength and aid in conduction of muscle action potentials by way of gap junctions located in the discs.
6. Unlike skeletal muscle tissue, cardiac muscle tissue contracts and relaxes rapidly, continuously, and rhythmically. Energy is supplied by glycogen and fat in large, numerous mitochondria.
7. Cardiac muscle tissue can contract without extrinsic stimulation and can remain contracted longer than skeletal muscle tissue.
8. Cardiac muscle tissue has a long refractory period, which prevents tetanus.

Smooth Muscle Tissue (p. 248)

1. Smooth muscle is nonstriated and involuntary.
2. Smooth muscle fibers contain intermediate filaments, dense bodies (function as Z lines), and caveolae (function as transverse tubules).
3. Visceral (single-unit) smooth muscle is found in the walls of viscera. The fibers are arranged in a network.
4. Multiunit smooth muscle is found in blood vessels and the eye. The fibers operate singly rather than as a unit.
5. The duration of contraction and relaxation of smooth muscle is longer than in skeletal muscle.
6. Smooth muscle fibers contract in response to nerve impulses, hormones, and local factors.
7. Smooth muscle fibers can stretch considerably without developing tension.

Regeneration of Muscle Tissue (p. 251)

1. Skeletal muscle fibers cannot divide and have limited powers of regeneration.
2. Cardiac muscle fibers cannot divide or regenerate.
3. Smooth muscle fibers have limited capacity for division and regeneration.

Homeostasis (p. 252)

1. Oxygen debt is the amount of O_2 needed to convert accumulated lactic acid into CO_2 and H_2O. It occurs during strenuous exercise and is paid back by continuing to breathe rapidly after exercising. Until it is paid back, the homeostasis between muscular activity and oxygen requirements is not restored.
2. Muscle fatigue results from diminished availability of oxygen and toxic effects of carbon dioxide and lactic acid built up during exercise.
3. The heat given off during muscular contraction maintains the homeostasis of body temperature.

Aging and Muscle Tissue (p. 253)

1. Beginning at about 30 years of age, there is a progressive loss of skeletal muscle, which is replaced by fat.
2. There is also a decrease in muscle strength and diminished muscle reflexes.

Developmental Anatomy of the Muscular System (p. 253)

1. With few exceptions, muscles develop from mesoderm.
2. Skeletal muscles of the head and extremities develop from general mesoderm; the remainder of the skeletal muscles develop from the mesoderm of somites.

Disorders: Homeostatic Imbalances (p. 254)

1. Fibromyalgia refers to a group of nonarticular rheumatic disorders characterized by pain, tenderness, and stiffness of muscles, tendons, and ligaments. Frequent sites are the lower back (lumbago) and thigh (charleyhorse).
2. Muscular dystrophies refer to hereditary diseases of muscles characterized by degeneration of individual muscle fibers.
3. Myasthenia gravis (MG) is a disease characterized by great muscular weakness and fatigability resulting from improper neuromuscular transmission.
4. Abnormal contractions include spasm, tremor, fasciculation, fibrillation, and tic.

REVIEW QUESTIONS

1. How is the skeletal system related to the muscular system? What are the four characteristics of muscle tissue? (p. 230)
2. What are the three basic functions of the muscular system? (p. 230)
3. How can the three types of muscle tissue be distinguished on the basis of location, microscopic appearance, and nervous control? (p. 230)
4. What is fascia? Distinguish between superficial and deep fascia. (p. 230)
5. Define epimysium, perimysium, endomysium, tendon, and aponeurosis. Describe the nerve and blood supply to a skeletal muscle. (p. 231)
6. Describe the microscopic structure of skeletal muscle tissue. (p. 232)
7. In considering the contraction of skeletal muscle tissue, describe the following: neuromuscular junction (p. 235), motor unit (p. 238), role of calcium (p. 238), sources of energy (p. 240), and sliding-filament theory. (p. 233)
8. What is the all-or-none principle? Relate it to a threshold and subthreshold stimulus. (p. 241)
9. Define each of the following contractions and state the importance of each: twitch, tetanus, treppe, isotonic, and isometric. (p. 242)
10. Describe the various factors that affect muscle tension. (p. 244)
11. What is a myogram? Describe the latent period, contraction period, and relaxation period of a twitch muscle contraction. Construct a diagram to illustrate your answer. (p. 242)
12. Define the refractory period. How does it differ between skeletal and cardiac muscle? What is wave summation? (p. 242)
13. What is muscle tone? Why is it important? (p. 245)
14. Compare type I, type II B, and type II A fibers with respect to structure and function. (p. 246)
15. Compare skeletal, cardiac, and smooth muscle with regard to differences in structure, physiology, and capacity for division and regeneration. (p. 251)
16. Discuss each of the following as examples of muscle homeostasis: oxygen debt, fatigue, and heat production. (p. 252)
17. What do you think might be the relationship between shivering (uncontrolled muscular contractions) and body temperature? Can you relate sweating (cooling of the skin) after strenuous exercise to the homeostasis of body temperature?
18. Describe the effects of aging on muscle tissue. (p. 253)
19. Describe how the muscular system develops. (p. 253)
20. Define fibromyalgia, muscular dystrophies, myasthenia gravis (MG), spasm, tremor, fasciculation, fibrillation, and tic. (p. 254)
21. Define the following: tenosynovitis (p. 232), muscle-building anabolic steroid (p. 233), rigor mortis (p. 240), hypotonia (p. 245), flaccidity (p. 245), hypertonia (p. 245), spasticity (p. 245), rigidity (p. 245), muscular atrophy (p. 246), muscular hypertrophy (p. 246), and "second wind." (p. 252)
22. What is electromyography? Why is it performed? Briefly describe the procedure. (p. 241)
23. Refer to the glossary of medical terminology associated with the muscular system. Be sure that you can define each term. (p. 255)

SELECTED READINGS

Allman, W. F. "Weight-Lifting: Inside the Pumphouse," *Science 84,* September 1984.

Bohigian, G. M. "Drug Abuse in Athletes," *Journal of the American Medical Association,* 18 March 1988.

Gamble, J. G. *The Musculoskeletal System: Physiological Basics.* New York: Raven Press, 1988.

Hinson, M. M. *Kinesiology.* Dubuque, Iowa: Wm. C. Brown, 1981.

Huxley, H. E. "The Contraction of Muscle," *Scientific American,* November 1968.

Ravits, J. "Myasthenia Gravis," *Postgraduate Medicine,* January 1988.

Wilson, F. C. *The Musculoskeletal System: Basic Processes and Disorders,* 2nd ed. Philadelphia: Lippincott, 1983.

Windsor, R. E., and D. Dumitru. "Anabolic Steroid Use By Athletes," *Postgraduate Medicine,* September 1988.

Chapter 11

The Muscular System

Chapter Contents at a Glance

Student Objectives

1. Describe the relationship between bones and skeletal muscles in producing body movements.
2. Define a lever and fulcrum and compare the three classes of levers on the basis of placement of the fulcrum, effort, and resistance.
3. Identify the various arrangements of muscle fibers in a skeletal muscle and relate the arrangements to the strength of contraction and range of movement.
4. Discuss most body movements as activities of groups of muscles by explaining the roles of the prime mover, antagonist, synergist, and fixator.
5. Define the criteria employed in naming skeletal muscles.
6. Identify the principal skeletal muscles in different regions of the body by name, origin, insertion, action, and innervation.
7. Discuss the administration of drugs by intramuscular (IM) injection.
8. Describe several injuries related to running.

The term ***muscle tissue*** refers to all the contractile tissues of the body: skeletal, cardiac, and smooth muscle. The ***muscular system,*** however, refers to the *skeletal* muscle system: the skeletal muscle tissue and connective tissues that make up individual muscle organs, such as the biceps brachii muscle. Cardiac muscle tissue is located in the heart and is therefore considered part of the cardiovascular system. Smooth muscle tissue of the intestine is part of the digestive system, whereas smooth muscle tissue of the urinary bladder is part of the urinary system. In this chapter, we discuss only the muscular system. We will see how skeletal muscles produce movement, and we will describe the principal skeletal muscles.

HOW SKELETAL MUSCLES PRODUCE MOVEMENT

Origin and Insertion

Skeletal muscles produce movements by exerting force on tendons, which in turn pull on bones or other structures, such as skin. Most muscles cross at least one joint and are attached to the articulating bones that form the joint (Figure 11-1). When such a muscle contracts, it draws one articulating bone toward the other. The two articulating bones usually do not move equally in response to

FIGURE 11-1 **Relationship of skeletal muscles to bones. (a) Skeletal muscles produce movements by pulling on bones. (b) Bones serve as levers, and joints act as fulcrums for the levers. Here the lever–fulcrum principle is illustrated by the movement of the forearm lifting a weight. Note where the resistance and effort are applied in this example.**

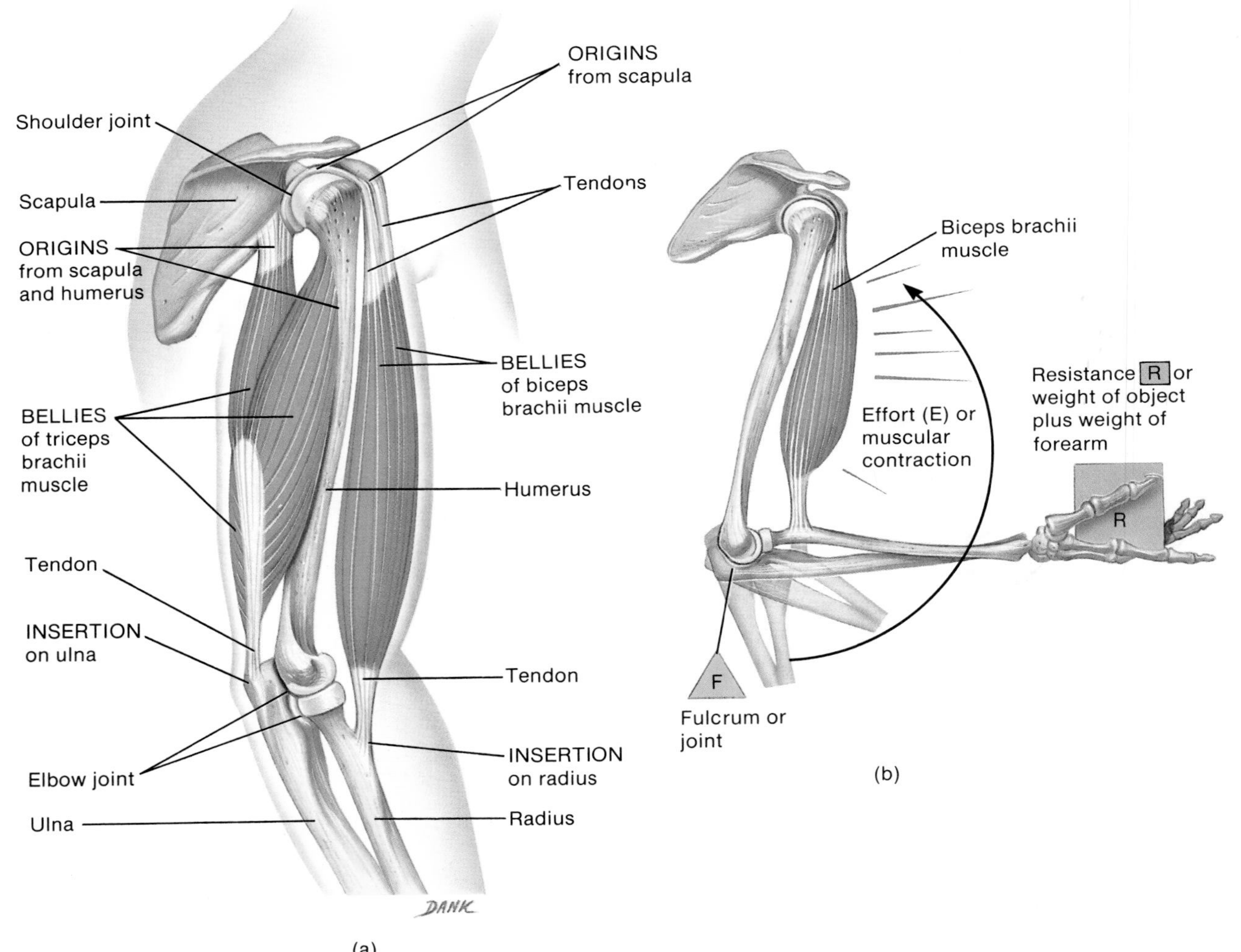

the contraction. One is held nearly in its original position because other muscles contract to pull it in the opposite direction or because its structure makes it less movable. Ordinarily, the attachment of a muscle tendon to the stationary bone is called the ***origin.*** The attachment of the other muscle tendon to the movable bone is the ***insertion.*** A good analogy is a spring on a door. The part of the spring attached to the door represents the insertion; the part attached to the frame is the origin. The fleshy portion of the muscle between the tendons of the origin and insertion is called the ***belly (gaster).*** The origin is usually proximal and the insertion distal, especially in the extremities. In addition, muscles that move a body part generally do not cover the moving part. Figure 11-1a shows that although one of the functions of the biceps brachii muscle is to move the forearm, the belly of the muscle lies over the humerus, not the forearm.

Lever Systems and Leverage

In producing a body movement, bones act as levers and joints function as fulcrums of these levers. A ***lever*** may be defined as a rigid rod that moves about on some fixed point called a ***fulcrum.*** A fulcrum may be symbolized as △F. A lever is acted on at two different points by two different forces: the ***resistance*** [R] and the ***effort*** (E). The resistance may be regarded as a force (load) to be overcome, whereas the effort is the force exerted to overcome the resistance. The resistance may be the weight of a part of the body that is to be moved. The muscular effort (contraction) is applied to the bone at the insertion of the muscle and produces motion if the effort exceeds the resistance (load). Consider the biceps brachii flexing the forearm at the elbow as a weight is lifted (Figure 11-1b). When the forearm is raised, the elbow is the fulcrum. The weight of the forearm plus the weight in the hand is the resistance. The shortening due to the force of contraction of the biceps brachii pulling the forearm up is the effort.

Levers are categorized into three types according to the positions of the fulcrum, the effort, and the resistance.

1. In ***first-class levers,*** the fulcrum is between the effort and resistance (Figure 11-2a). This is symbolized EFR. An example of a first-class lever is a seesaw. There are not many first-class levers in the body. One example is the head resting on the vertebral column. When the head is raised, the facial portion of the skull is the resistance. The joint between the atlas and occipital bone (atlanto-occipital joint) is the fulcrum. The contraction of the muscles of the back is the effort.

2. ***Second-class levers*** have the fulcrum at one end, the effort at the opposite end, and the resistance between them (Figure 11-2b). This is symbolized FRE. They operate like a wheelbarrow. Most authorities agree that there are very few examples of second-class levers in the body. One example is raising the body on the toes. The body is the resistance, the ball of the foot is the fulcrum, and the contraction of the calf muscles to pull the heel upward is the effort.

3. ***Third-class levers*** consist of the fulcrum at one end, the resistance at the opposite end, and the effort between them (Figure 11-2c). This is symbolized FER. They are the most common levers in the body. One example is adduction of the thigh, in which the weight of the thigh is the resistance, the hip joint is the fulcrum, and contraction of the adductor muscles is the effort. Another example is flexing the forearm at the elbow. As we have seen, the weight of the forearm is the resistance, the contraction of the biceps brachii is the effort, and the elbow joint is the fulcrum (see Figure 11-1b).

Leverage, the mechanical advantage gained by a lever, is largely responsible for a muscle's strength and range of movement. Consider strength first. Suppose we have two muscles of the same strength crossing and acting on the same joint. Assume also that one is attached farther from the joint and one is nearer. The muscle attached farther will produce the more powerful movement. Thus, strength of movement depends on the placement of muscle attachments.

In considering range of movement, again assume that we have two muscles of the same strength crossing and acting on the same joint and that one is attached farther from the joint than the other. The muscle inserting closer to the joint will produce the greater range and speed of movement. Thus, range of movement also depends on the placement of muscle attachments. Since strength increases with distance from the joint and range of movement decreases, maximal strength and maximal range are incompatible; strength and range vary inversely.

Arrangement of Fasciculi

Recall from Chapter 10 that skeletal muscle fibers (cells) are arranged within the muscle in bundles called fasciculi (fascicles). The muscle fibers are arranged in a parallel fashion within each bundle, but the arrangement of the fasciculi with respect to the tendons may take one of four characteristic patterns.

The first pattern is called ***parallel.*** The fasciculi are parallel with the longitudinal axis and terminate at either end in flat tendons. The muscle is typically quadrilateral in shape. An example is the stylohyoid muscle (see Figure 11-7). In a modification of the parallel arrangement, called ***fusiform,*** the fasciculi are nearly parallel with the longitudinal axis and terminate at either end in flat tendons, but the muscle tapers toward the tendons, where the diameter is less than that of the belly. An example is the biceps brachii muscle (see Figure 11-18).

The second distinct pattern is called ***convergent.*** A broad origin of fasciculi converges to a narrow, restricted insertion. Such a pattern gives the muscle a triangular shape. An example is the deltoid muscle (see Figure 11-16c).

The third distinct pattern is referred to as ***pennate.*** The fasciculi are short in relation to the entire length of the muscle, and the tendon extends nearly the entire

FIGURE 11-2 Classes of levers. Each is defined on the basis of the placement of the fulcrum, effort, and resistance. (a) First-class lever. (b) Second-class lever. (c) Third-class lever. A third-class lever is also shown in Figure 11-1b.

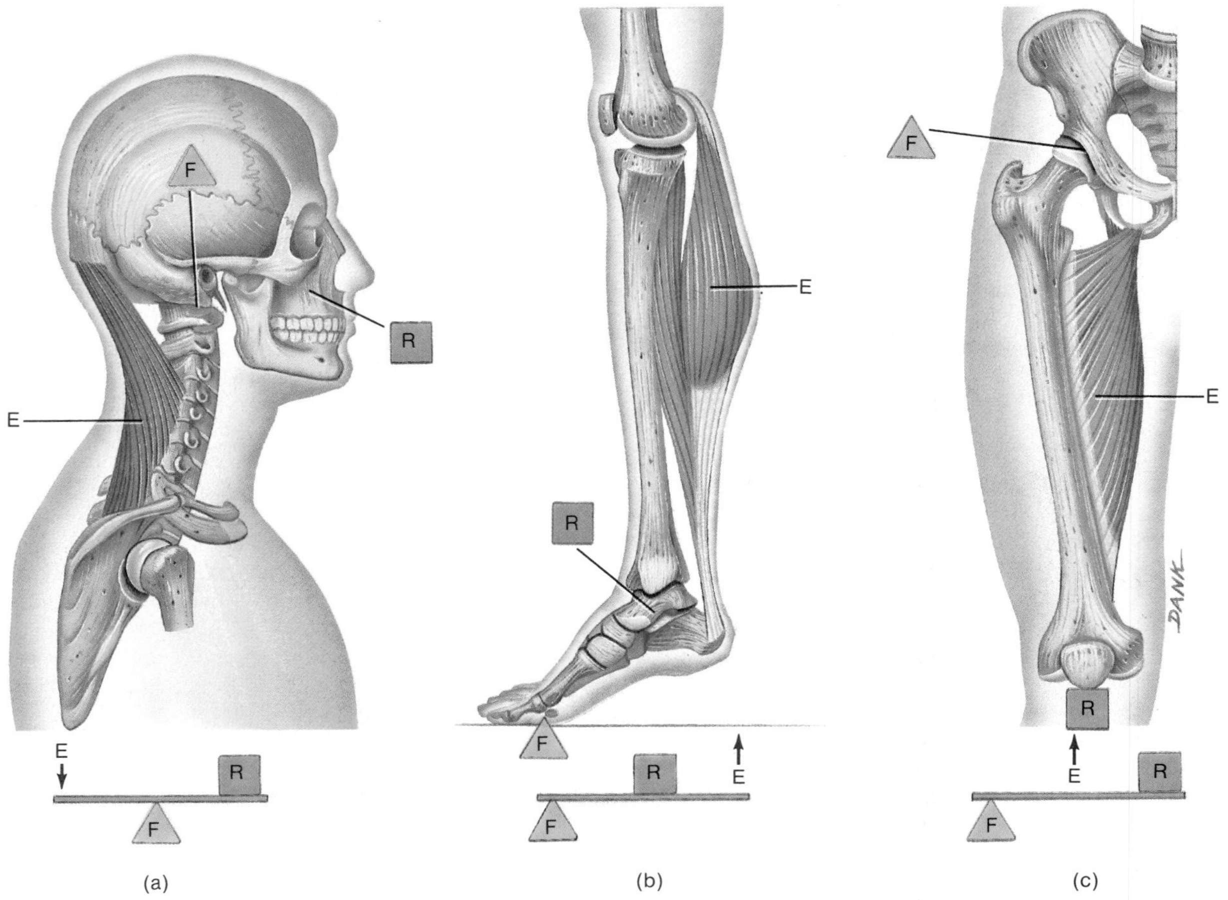

length of the muscle. The fasciculi are directed obliquely toward the tendon like the plumes of a feather. If the fasciculi are arranged on only one side of a tendon, as in the extensor digitorum longus muscle, the muscle is referred to as ***unipennate*** (see Figure 11-24). If the fasciculi are arranged on both sides of a centrally positioned tendon, as in the rectus femoris muscle, the muscle is referred to as ***bipennate*** (see Figure 11-22a).

The final distinct pattern is referred to as ***circular.*** The fasciculi are arranged in a circular pattern and enclose an orifice. An example is the orbicularis oris muscle (see Figure 11-4).

Fascicular arrangement is correlated with the power of a muscle and range of movement. When a muscle fiber contracts, it shortens to a length just slightly greater than half of its resting length. Thus, the longer the fibers in a muscle, the greater the range of movement it can produce. By contrast, the strength of a muscle depends on the total number of fibers it contains, since a short fiber can contract as forcefully as a long one. Because a given muscle can contain either a small number of long fibers or a large number of short fibers, fascicular arrangement represents a compromise between power and range of movement. Pennate muscles, for example, have a large number of fasciculi distributed over their tendons, giving them greater power, but a smaller range of movement. Parallel muscles, on the other hand, have comparatively few fasciculi that extend the length of the muscle. Thus, they have a greater range of movement but less power.

Group Actions

Most movements are coordinated by several skeletal muscles acting in groups rather than individually, and most skeletal muscles are arranged in opposing pairs at joints, that is, flexors–extensors, abductors–adductors, and so on. Consider flexing the forearm at the elbow, for example.

A muscle that causes a desired action is referred to as the ***prime mover (agonist).*** In this instance, the biceps brachii is the prime mover (see Figure 11-18). Simultaneously with the contraction of the biceps brachii, another muscle, called the ***antagonist,*** is relaxing. In this movement, the triceps brachii serves as the antagonist (see Figure 11-18). The antagonist has an effect opposite to that of the prime mover; that is, the antagonist relaxes and yields to the movement of the prime mover. You should not assume, however, that the biceps brachii is always the prime mover and the triceps brachii is always the antagonist. For example, when extending the forearm at the elbow, the triceps brachii serves as the prime mover and the biceps brachii functions as the antagonist; their roles are reversed. Note that if the prime mover and antagonist contracted simultaneously with equal force, there would be no movement, as in an isometric contraction.

In addition to prime movers and antagonists, most movements also involve muscles called ***synergists,*** which serve to steady a movement, thus preventing unwanted movements and helping the prime mover function more efficiently. For example, flex your hand at the wrist and then make a fist. Note how difficult this is to do. Now, extend your hand at the wrist and then make a fist. Note how much easier it is to clench your fist. In this case, the extensor muscles of the wrist act as synergists in cooperation with the flexor muscles of the fingers acting as prime movers. The extensor muscles of the fingers serve as antagonists (see Figure 11-19).

Some synergist muscles in a group also act as ***fixators,*** which stabilize the origin of the prime mover so that the prime mover can act more efficiently. For example, the scapula is a freely movable bone in the pectoral (shoulder) girdle that serves as a firm origin for several muscles that move the arm. However, for the scapula to do this, it must be held steady. This is accomplished by fixator muscles that hold the scapula firmly against the back of the chest. In abduction of the arm, the deltoid muscle serves as the prime mover, whereas fixators (pectoralis minor, rhomboideus major, rhomboideus minor, trapezius, subclavius, and serratus anterior muscles) hold the scapula firmly (see Figure 11-16). These fixators stabilize the scapula that serves as the attachment site for the origin of the deltoid muscle, whereas the insertion of the muscle pulls on the humerus to abduct the arm. Under different conditions and depending on the movement and which point is fixed, many muscles act, at various times, as prime movers, antagonists, synergists, or fixators.

NAMING SKELETAL MUSCLES

The names of most of the nearly 700 skeletal muscles are based on several types of characteristics. Learning the terms used to indicate specific characteristics will help you remember the names of muscles.

1. Muscle names may indicate the ***direction of the muscle fibers.*** *Rectus* fibers usually run parallel to the midline of the body. *Transverse* fibers run perpendicular to the midline. *Oblique* fibers are diagonal to the midline. Muscles named according to directions of fibers include the rectus abdominis, transversus abdominis, and external oblique.

2. A muscle may be named according to ***location.*** The temporalis is near the temporal bone. The tibialis anterior is near the front of the tibia.

3. ***Size*** is another characteristic. The term *maximus* means largest; *minimus,* smallest; *longus,* long; and *brevis,* short. Examples include the gluteus maximus, gluteus minimus, adductor longus, and peroneus brevis.

4. Some muscles are named for their ***number of origins.*** The biceps brachii has two origins; the triceps brachii, three; and the quadriceps femoris, four.

5. Other muscles are named on the basis of ***shape.*** Common examples include the deltoid (meaning triangular), trapezius (meaning trapezoid), serratus anterior (meaning saw-toothed), and rhomboideus major (meaning rhomboid or diamond shaped).

6. Muscles may be named after their ***origin*** and ***insertion.*** The sternocleidomastoid originates on the sternum and clavicle and inserts at the mastoid process of the temporal bone; the stylohyoid originates on the styloid process of the temporal bone and inserts at the hyoid bone.

7. Still another characteristic of muscles used for naming is ***action.*** Exhibit 11-1 lists the principal actions of muscles, their definitions, and examples of muscles that perform the actions. For convenience, the actions are grouped as antagonistic pairs where possible.

PRINCIPAL SKELETAL MUSCLES

Exhibits 11-2 through 11-24 list the principal muscles of the body with their origins, insertions, actions, and innervations. (By no means have all the muscles of the body been included.) An **overview** section in each Exhibit provides a general orientation to the muscles under consideration. Refer to Chapters 7 and 8 to review bone markings, since they serve as points of origin and insertion for muscles. The muscles are divided into groups according to the part of the body on which they act. If you have mastered the naming of the muscles, their actions will have more meaning. Figure 11-3 shows general anterior and posterior views of the muscular system. Do not try to memorize all these muscles yet. As you study groups of muscles in the following exhibits, refer to Figure 11-3 to see how each group is related to all others.

The figures that accompany the exhibits contain superficial and deep, anterior and posterior, or medial and lateral views to show each muscle's position as clearly as possible. An attempt has been made to show the relationship of the muscles under consideration to other muscles in the area you are studying.

EXHIBIT 11-1 PRINCIPAL ACTIONS OF MUSCLES

Action	Definition	Example
Flexor	Decreases the angle at a joint.	Flexor carpi radialis (see Exhibit 11-18).
Extensor	Increases the angle at a joint.	Extensor carpi ulnaris (see Exhibit 11-18).
Abductor	Moves a bone away from the midline.	Abductor pollicis brevis (see Exhibit 11-19).
Adductor	Moves a bone closer to the midline.	Adductor longus (see Exhibit 11-21).
Levator	Produces an upward movement.	Levator scapulae (see Exhibit 11-15).
Depressor	Produces a downward movement.	Depressor labii inferioris (see Exhibit 11-2).
Supinator	Turns the palm upward or anteriorly.	Supinator (see Exhibit 11-17).
Pronator	Turns the palm downward or posteriorly.	Pronator teres (see Exhibit 11-17).
Sphincter	Decreases the size of an opening.	External anal sphincter (see Exhibit 11-14).
Tensor	Makes a body part more rigid.	Tensor fasciae latae (see Exhibit 11-21).
Rotator	Moves a bone around its longitudinal axis.	Obturator externus (see Exhibit 11-21).

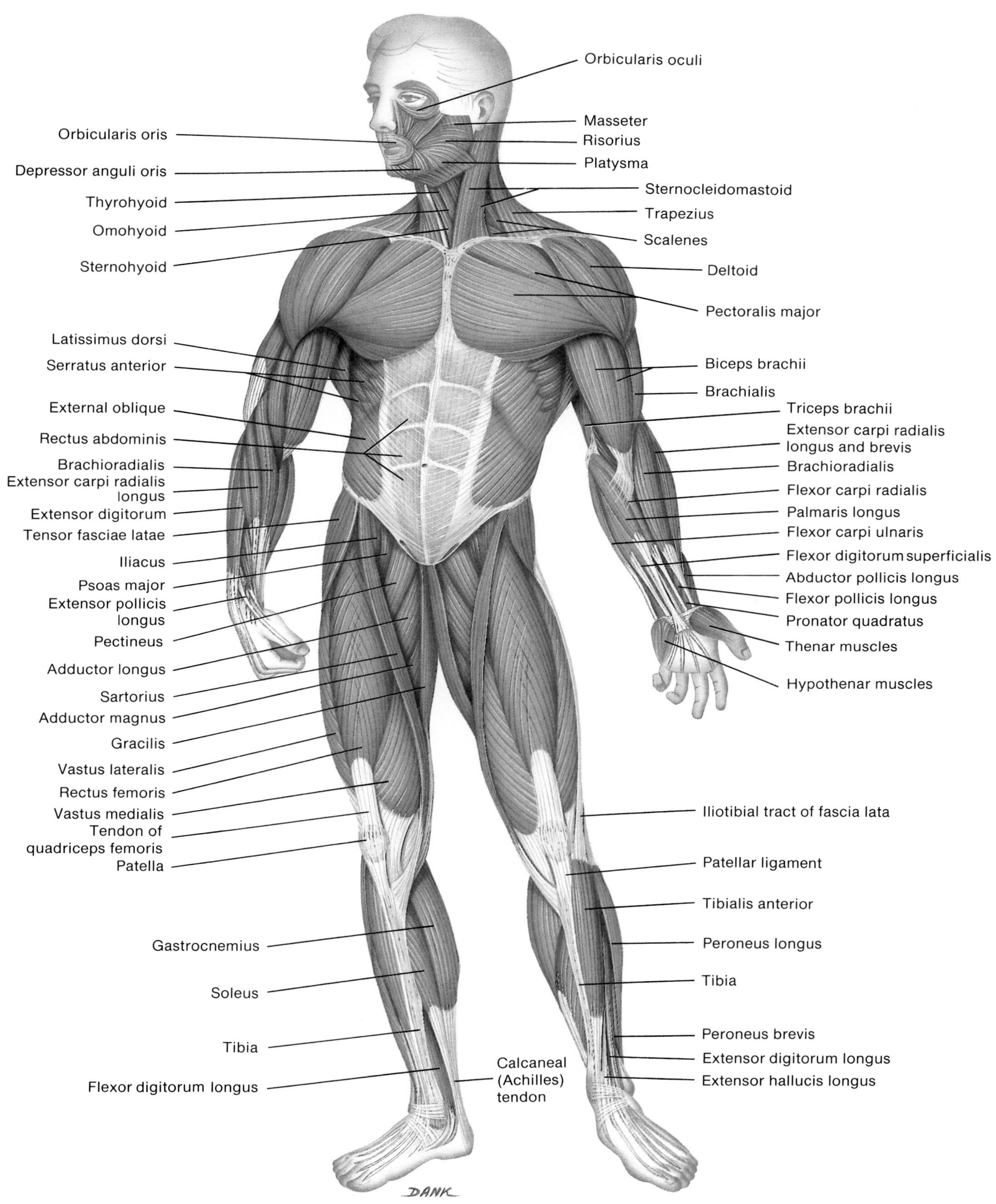
Orbicularis oculi
Orbicularis oris
Masseter
Risorius
Depressor anguli oris
Platysma
Thyrohyoid
Sternocleidomastoid
Trapezius
Omohyoid
Scalenes
Sternohyoid
Deltoid
Pectoralis major
Latissimus dorsi
Serratus anterior
Biceps brachii
Brachialis
External oblique
Triceps brachii
Rectus abdominis
Extensor carpi radialis longus and brevis
Brachioradialis
Brachioradialis
Extensor carpi radialis longus
Flexor carpi radialis
Extensor digitorum
Palmaris longus
Tensor fasciae latae
Flexor carpi ulnaris
Iliacus
Flexor digitorum superficialis
Abductor pollicis longus
Psoas major
Flexor pollicis longus
Extensor pollicis longus
Pronator quadratus
Pectineus
Thenar muscles
Adductor longus
Hypothenar muscles
Sartorius
Adductor magnus
Gracilis
Vastus lateralis
Rectus femoris
Vastus medialis
Iliotibial tract of fascia lata
Tendon of quadriceps femoris
Patella
Patellar ligament
Tibialis anterior
Gastrocnemius
Peroneus longus
Soleus
Tibia
Peroneus brevis
Tibia
Calcaneal (Achilles) tendon
Extensor digitorum longus
Flexor digitorum longus
Extensor hallucis longus
DANK

(a)

FIGURE 11-3 (*Continued*) (b) Posterior view.

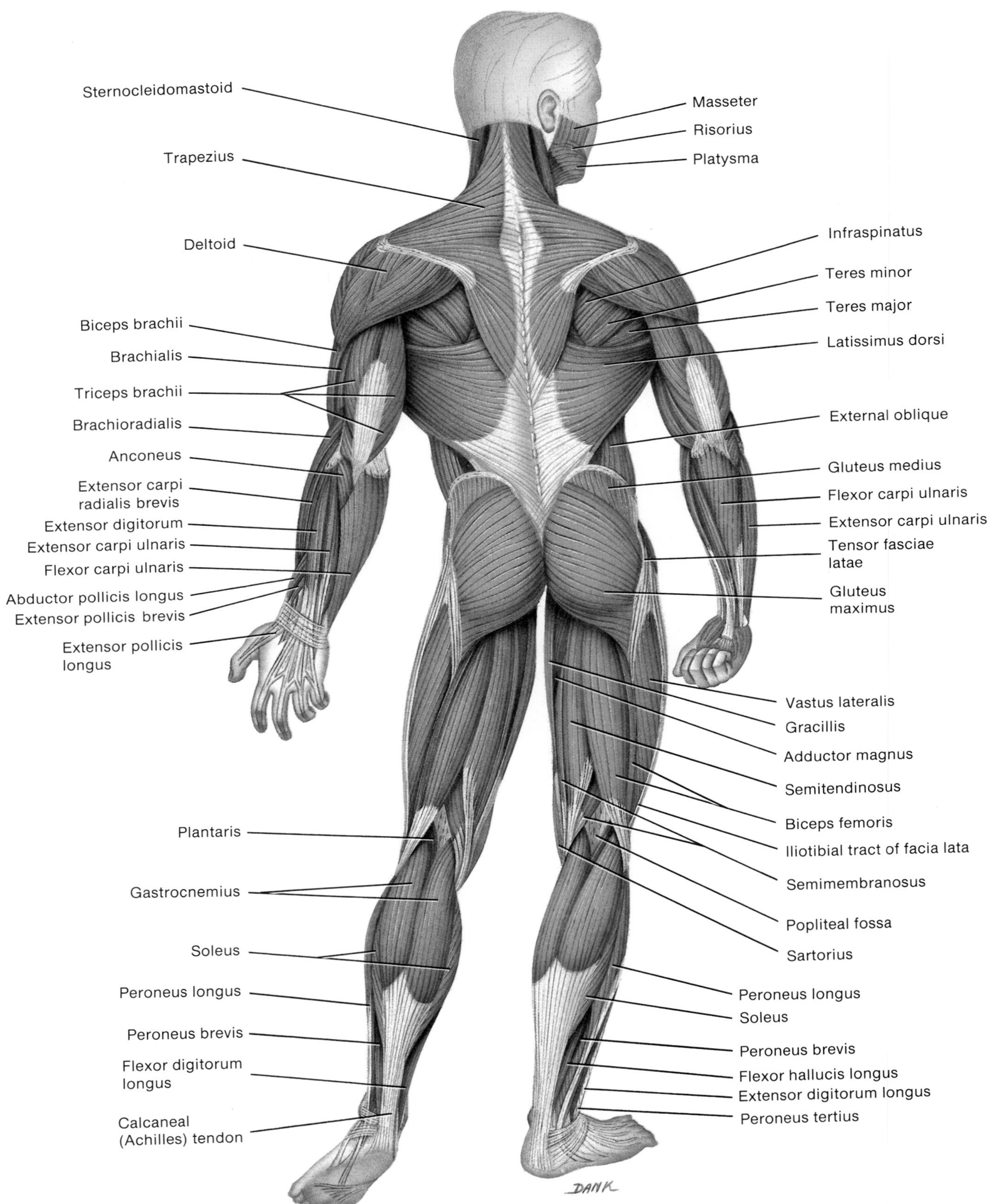
Sternocleidomastoid
Trapezius
Deltoid
Biceps brachii
Brachialis
Triceps brachii
Brachioradialis
Anconeus
Extensor carpi radialis brevis
Extensor digitorum
Extensor carpi ulnaris
Flexor carpi ulnaris
Abductor pollicis longus
Extensor pollicis brevis
Extensor pollicis longus
Plantaris
Gastrocnemius
Soleus
Peroneus longus
Peroneus brevis
Flexor digitorum longus
Calcaneal (Achilles) tendon
Masseter
Risorius
Platysma
Infraspinatus
Teres minor
Teres major
Latissimus dorsi
External oblique
Gluteus medius
Flexor carpi ulnaris
Extensor carpi ulnaris
Tensor fasciae latae
Gluteus maximus
Vastus lateralis
Gracillis
Adductor magnus
Semitendinosus
Biceps femoris
Iliotibial tract of facia lata
Semimembranosus
Popliteal fossa
Sartorius
Peroneus longus
Soleus
Peroneus brevis
Flexor hallucis longus
Extensor digitorum longus
Peroneus tertius
DANK

(b)

EXHIBIT 11-2 MUSCLES OF FACIAL EXPRESSION (Figure 11-4)

Overview: The muscles in this group provide humans with the ability to express a wide variety of emotions, including frowning, surprise, fear, and happiness. The muscles themselves lie within the layers of superficial fascia. As a rule, they arise from the fascia or bones of the skull and insert into the skin. Because of their insertions, the muscles of facial expression move the skin rather than a joint when they contract.

Muscle	Origin	Insertion	Action	Innervation
Epicranius (*epi* = over; *crani* = skull)	This muscle is divisible into two portions: the frontalis over the frontal bone and the occipitalis over the occipital bone. The two muscles are united by a strong aponeurosis, the galea aponeurotica, which covers the superior and lateral surfaces of the skull.			
Frontalis (*front* = forehead)	Galea aponeurotica.	Skin superior to supraorbital line.	Draws scalp forward, raises eyebrows, and wrinkles skin of forehead horizontally.	Facial (VII) nerve.
Occipitalis (*occipito* = base of skull)	Occipital bone and mastoid process of temporal bone.	Galea aponeurotica.	Draws scalp backward.	Facial (VII) nerve.
Orbicularis oris (*orb* = circular; *or* = mouth)	Muscle fibers surrounding opening of mouth.	Skin at corner of mouth.	Closes lips, compresses lips against teeth, protrudes lips, and shapes lips during speech.	Facial (VII) nerve.
Zygomaticus major (*zygomatic* = cheek bone; *major* = greater)	Zygomatic bone.	Skin at angle of mouth and orbicularis oris.	Draws angle of mouth upward and outward as in smiling or laughing.	Facial (VII) nerve.
Levator labii superioris (*levator* = raises or elevates; *labii* = lip; *superioris* = upper)	Superior to infraorbital foramen of maxilla.	Skin at angle of mouth and orbicularis oris.	Elevates (raises) upper lip.	Facial (VII) nerve.
Depressor labii inferioris (*depressor* = depresses or lowers; *inferioris* = lower)	Mandible.	Skin of lower lip.	Depresses (lowers) lower lip.	Facial (VII) nerve.
Buccinator (*bucc* = cheek)	Alveolar processes of maxilla and mandible and pterygomandibular raphe (fibrous band extending from the pterygoid hamulus to the mandible).	Orbicularis oris.	Major cheek muscle; compresses cheek as in blowing air out of mouth and causes cheeks to cave in, producing the action of sucking.	Facial (VII) nerve.
Mentalis (*mentum* = chin)	Mandible.	Skin of chin.	Elevates and protrudes lower lip and pulls skin of chin up as in pouting.	Facial (VII) nerve.
Platysma (*platy* = flat, broad)	Fascia over deltoid and pectoralis major muscles.	Mandible, muscles around angle of mouth, and skin of lower face.	Draws outer part of lower lip downward and backward as in pouting; depresses mandible.	Facial (VII) nerve.
Risorius (*risor* = laughter)	Fascia over parotid (salivary) gland.	Skin at angle of mouth.	Draws angle of mouth laterally as in tenseness.	Facial (VII) nerve.
Orbicularis oculi (*ocul* = eye)	Medial wall of orbit.	Circular path around orbit.	Closes eye.	Facial (VII) nerve.

EXHIBIT 11-2 MUSCLES OF FACIAL EXPRESSION (*Continued*)

Muscle	Origin	Insertion	Action	Innervation
Corrugator supercilli (*corrugo* = wrinkle; *supercilium* = eyebrow)	Medial end of superciliary arch of frontal bone.	Skin of eyebrow.	Draws eyebrow downward as in frowning.	Facial (VII) nerve.
Levator palpebrae superioris (*palpebrae* = eyelids) (see also Figure 11-6a)	Roof of orbit (lesser wing of sphenoid bone).	Skin of upper eyelid.	Elevates upper eyelid.	Oculomotor (III) nerve.

FIGURE 11-4 Muscles of facial expression. (a) Anterior superficial view. (b) Anterior deep view.

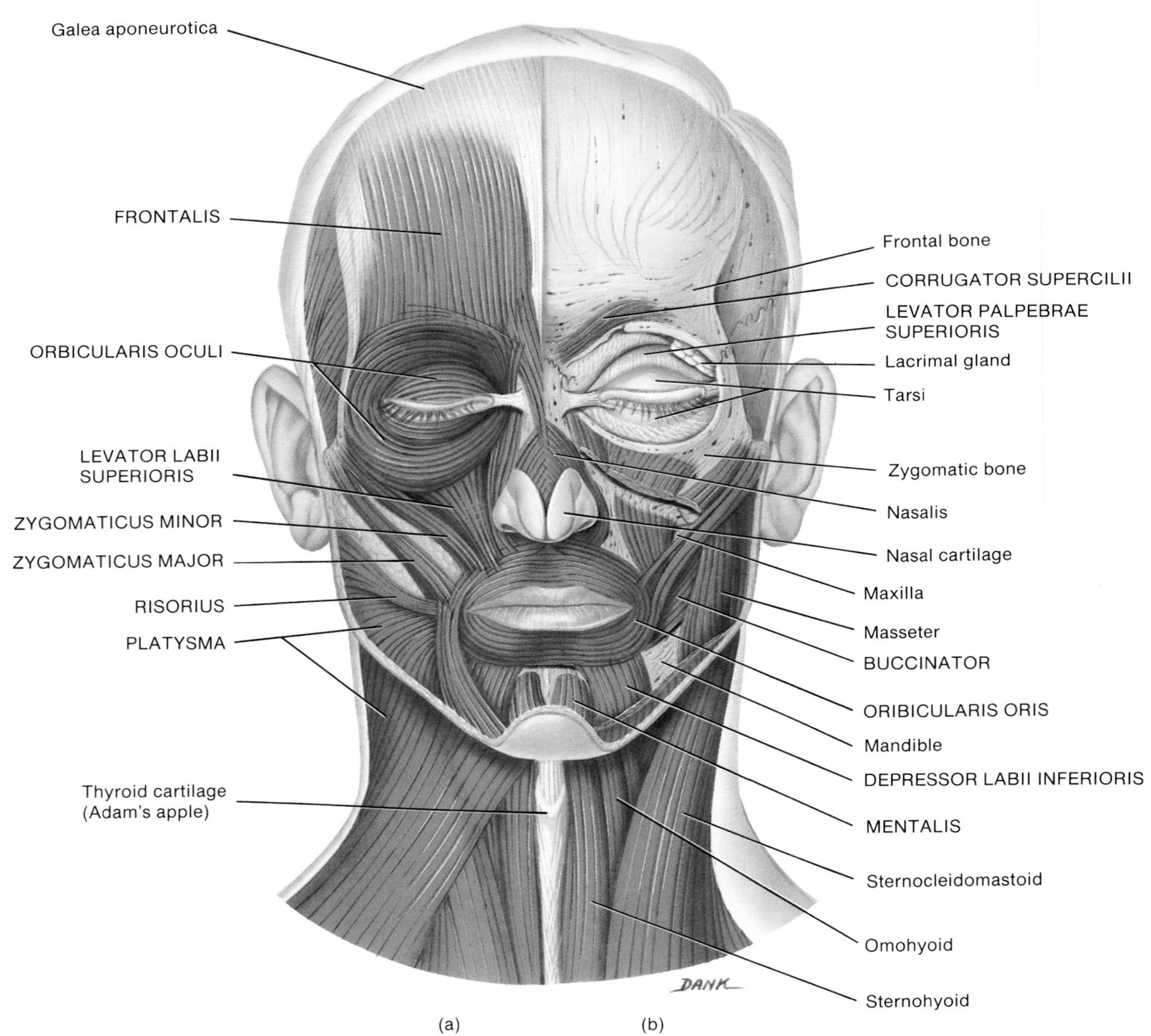

FIGURE 11-4 (*Continued*) (c) Right lateral superficial view.

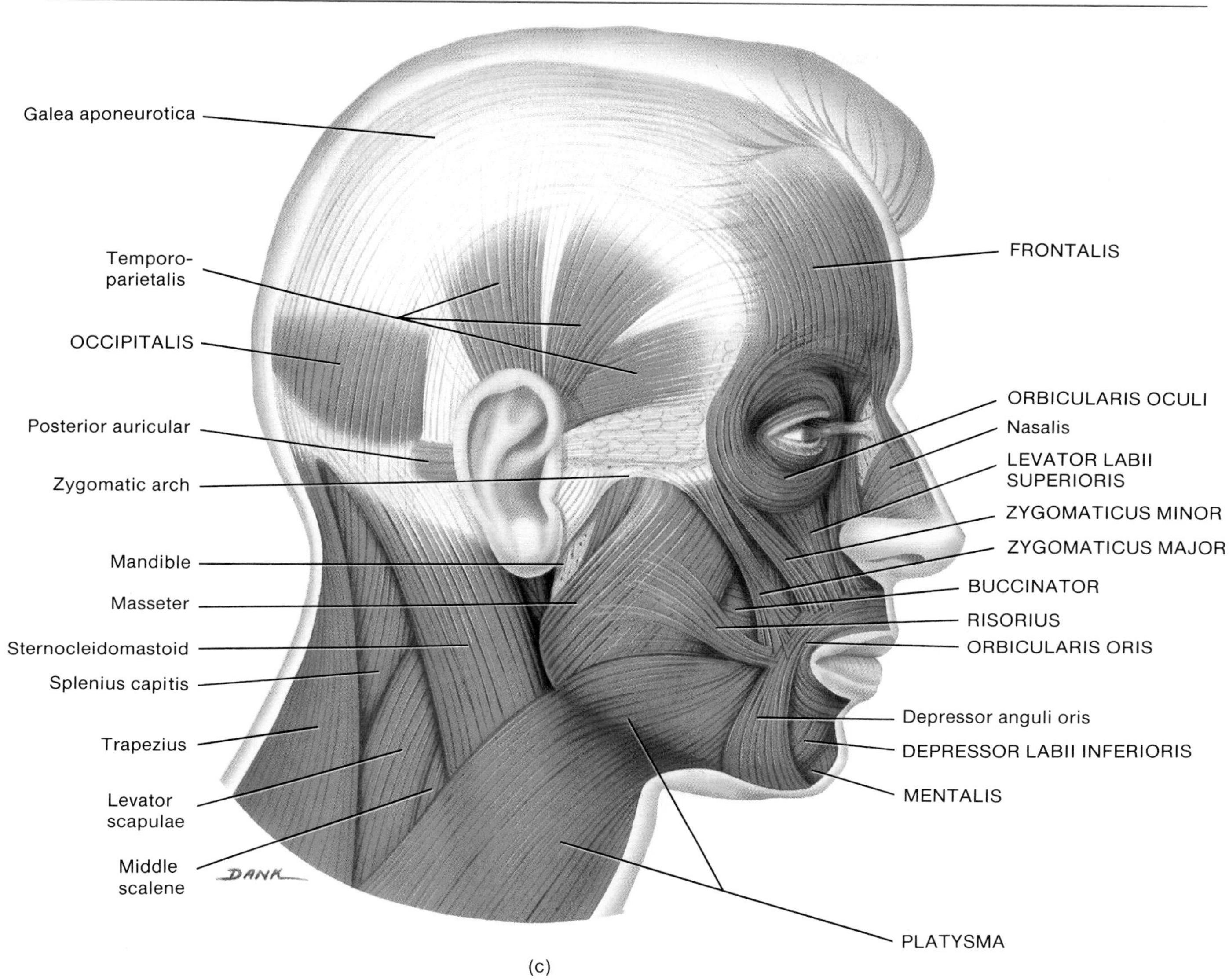

(c)

EXHIBIT 11-3 MUSCLES THAT MOVE THE LOWER JAW (Figure 11-5)

Overview: Muscles that move the lower jaw are also known as muscles of mastication because they are involved in biting and chewing. These muscles also assist in speech.

Muscle	Origin	Insertion	Action	Innervation
Masseter (*maseter* = chewer)	Maxilla and zygomatic arch.	Angle and ramus of mandible.	Elevates mandible as in closing mouth, assists in side to side movement of mandible, and protracts (protrudes) mandible.	Mandibular division of trigeminal (V) nerve.
Temporalis (*tempora* = temples)	Parietal bone.	Coronoid process of mandible.	Elevates and retracts mandible and assists in side to side movement of mandible.	Mandibular division of trigeminal (V) nerve.

FIGURE 11-5 Muscles that move the lower jaw. (a) Superficial right lateral view.

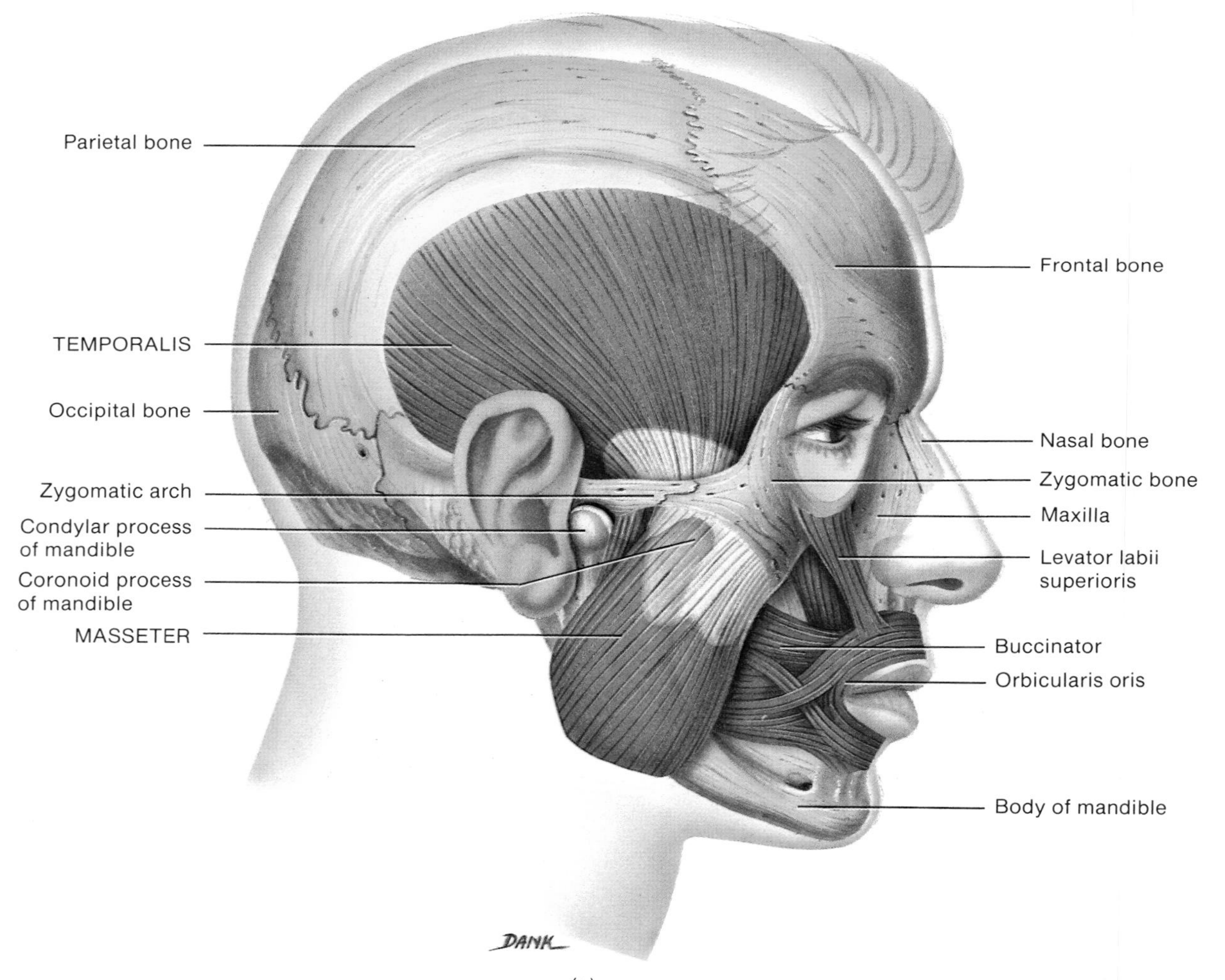

(a)

EXHIBIT 11-3 MUSCLES THAT MOVE THE LOWER JAW (*Continued*)

Muscle	Origin	Insertion	Action	Innervation
Medial pterygoid (*medial* = closer to midline; *pterygoid* = like a wing; pterygoid plate of sphenoid bone)	Medial surface of lateral pterygoid plate of sphenoid; maxilla.	Angle and ramus of mandible.	Elevates and protracts mandible and moves mandible from side to side.	Mandibular division of trigeminal (V) nerve.
Lateral pterygoid (*lateral* = farther from midline)	Greater wing and lateral surface of lateral pterygoid plate of sphenoid.	Condyle of mandible; temporomandibular articulation.	Protracts mandible, opens mouth, and moves mandible from side to side.	Mandibular division of trigeminal (V) nerve.

FIGURE 11-5 (*Continued*) (b) Deep right lateral view.

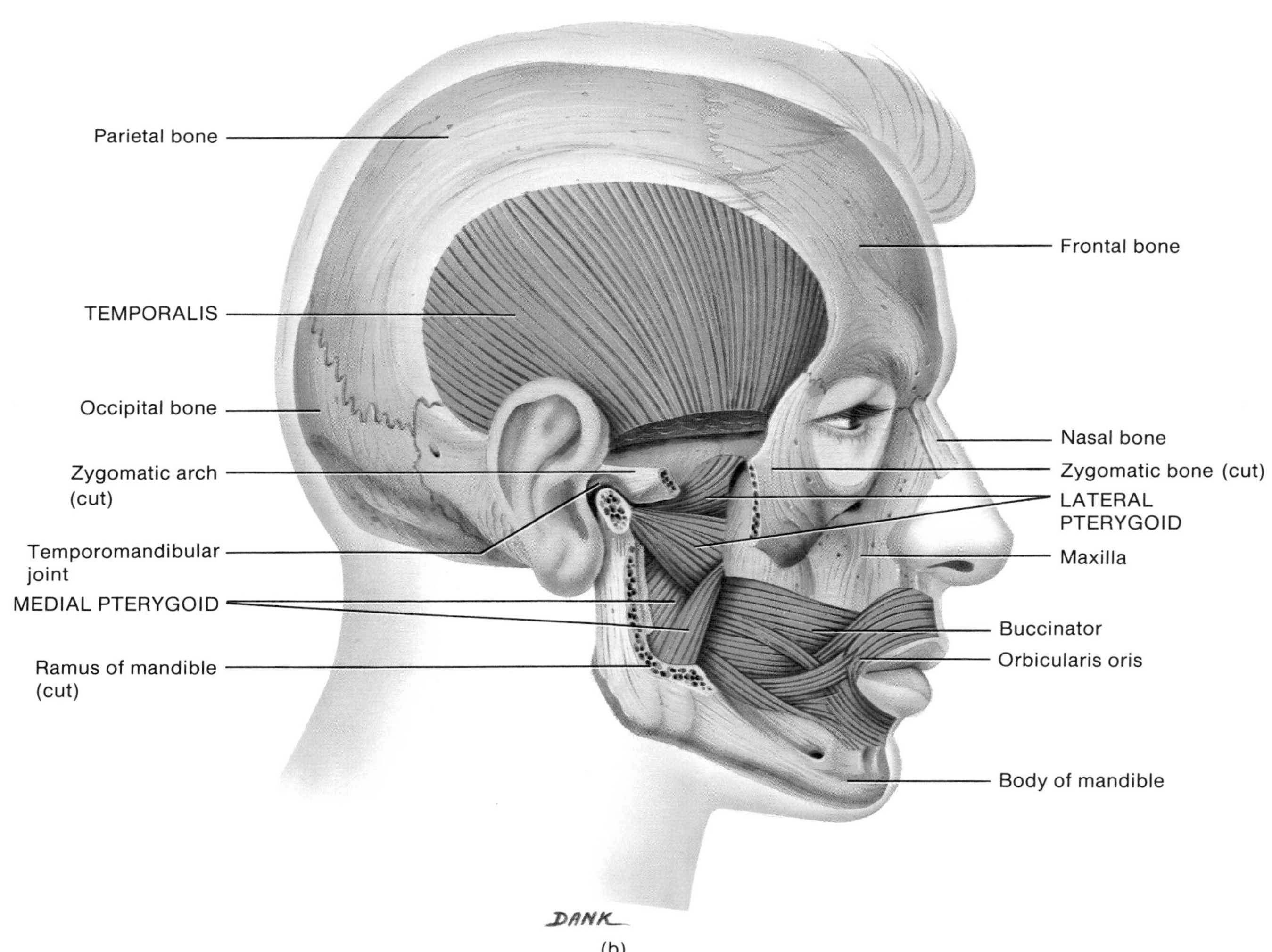

EXHIBIT 11-4 MUSCLES THAT MOVE THE EYEBALLS—EXTRINSIC MUSCLES (Figure 11-6)

Overview: Muscles associated with the eyeball are of two principal types: extrinsic and intrinsic. ***Extrinsic muscles*** originate outside the eyeball and are inserted on its outer surface (sclera). ***Intrinsic muscles*** originate and insert entirely within the eyeball.

Movements of the eyeballs are controlled by three pairs of extrinsic muscles. Two pairs of rectus muscles move the eyeball in the direction indicated by their respective names—superior, inferior, lateral, and medial. One pair of muscles, the oblique muscles—superior and inferior—rotate the eyeball on its axis. The extrinsic muscles of the eyeballs are among the fastest contracting and most precisely controlled skeletal muscles of the body.

Muscle	Origin	Insertion	Action	Innervation
Superior rectus (*superior* = above; *rectus* = in this case, muscle fibers running parallel to long axis of eyeball)	Tendinous ring attached to bony orbit around optic foramen.	Superior and central part of eyeball.	Rolls eyeball upward.	Oculomotor (III) nerve.
Inferior rectus (*inferior* = below)	Same as above.	Inferior and central part of eyeball.	Rolls eyeball downward.	Oculomotor (III) nerve.
Lateral rectus	Same as above.	Lateral side of eyeball.	Rolls eyeball laterally.	Abducens (VI) nerve.
Medial rectus	Same as above.	Medial side of eyeball.	Rolls eyeball medially.	Oculomotor (III) nerve.
Superior oblique (*oblique* = in this case, muscle fibers running diagonally to long axis of eye ball)	Same as above.	Eyeball between superior and lateral recti.	Rotates eyeball on its axis; directs cornea downward and laterally; note that it moves through a ring of fibrocartilaginous tissue called the trochlea (*trochlea* = pulley).	Trochlear (IV) nerve.
Inferior oblique	Maxilla (front of orbital cavity).	Eyeball between inferior and lateral recti.	Rotates eyeball on its axis; directs cornea upward and laterally.	Oculomotor (III) nerve.

FIGURE 11-6 **Extrinsic muscles of the eyeball. (a) Lateral view of the right eyeball. (b) Movements of the right eyeball in response to contraction of the various extrinsic muscles.**

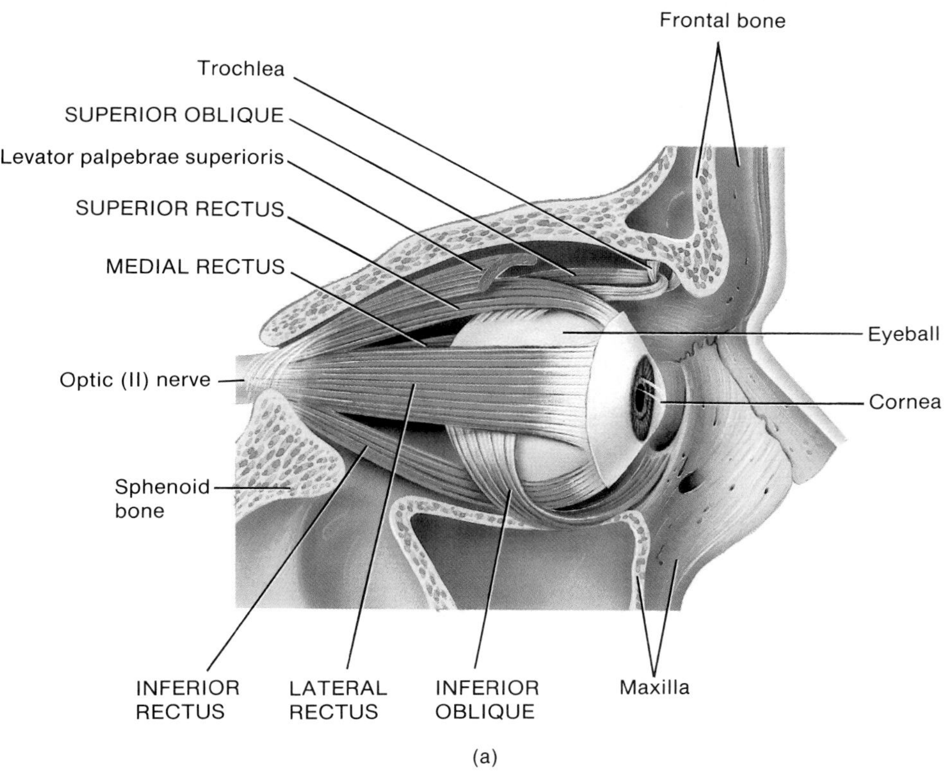

(a)

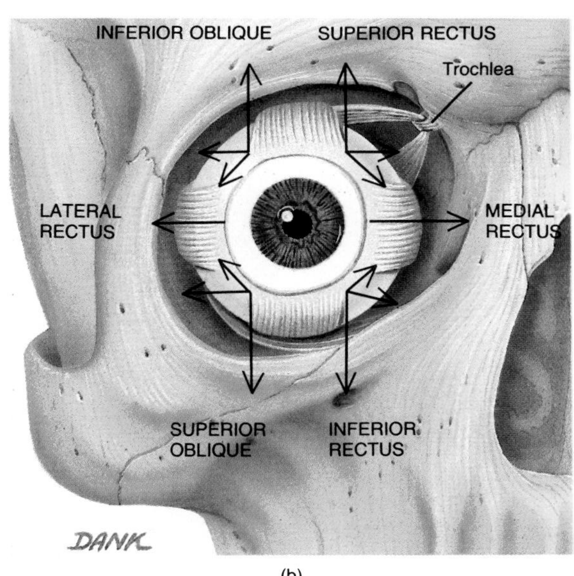

(b)

EXHIBIT 11-5 MUSCLES THAT MOVE THE TONGUE—EXTRINSIC MUSCLES (Figure 11-7)

Overview: The tongue is divided into lateral halves by a median fibrous septum. The septum extends throughout the length of the tongue and is attached inferiorly to the hyoid bone. Like the muscles of the eyeball, muscles of the tongue are of two principal types—extrinsic and intrinsic. ***Extrinsic muscles*** originate outside the tongue and insert into it. ***Intrinsic muscles*** originate and insert within the tongue. The extrinsic and intrinsic muscles of the tongue are arranged in both lateral halves of the tongue.

Muscle	Origin	Insertion	Action	Innervation
Genioglossus (*geneion* = chin; *glossus* = tongue)	Mandible.	Undersurface of tongue and hyoid bone.	Depresses tongue and thrusts it forward (protraction).	Hypoglossal (XII) nerve.
Styloglossus (*stylo* = stake or pole; styloid process of temporal bone)	Styloid process of temporal bone.	Side and undersurface of tongue.	Elevates tongue and draws it backward (retraction).	Hypoglossal (XII) nerve.
Palatoglossus (*palato* = palate)	Anterior surface of soft palate.	Side of tongue.	Elevates posterior portion of tongue and draws soft palate down on tongue.	Pharyngeal plexus.
Hyoglossus	Body of hyoid bone.	Side of tongue.	Depresses tongue and draws down its sides.	Hypoglossal (XII) nerve.

FIGURE 11-7 Muscles that move the tongue as viewed from the right side.

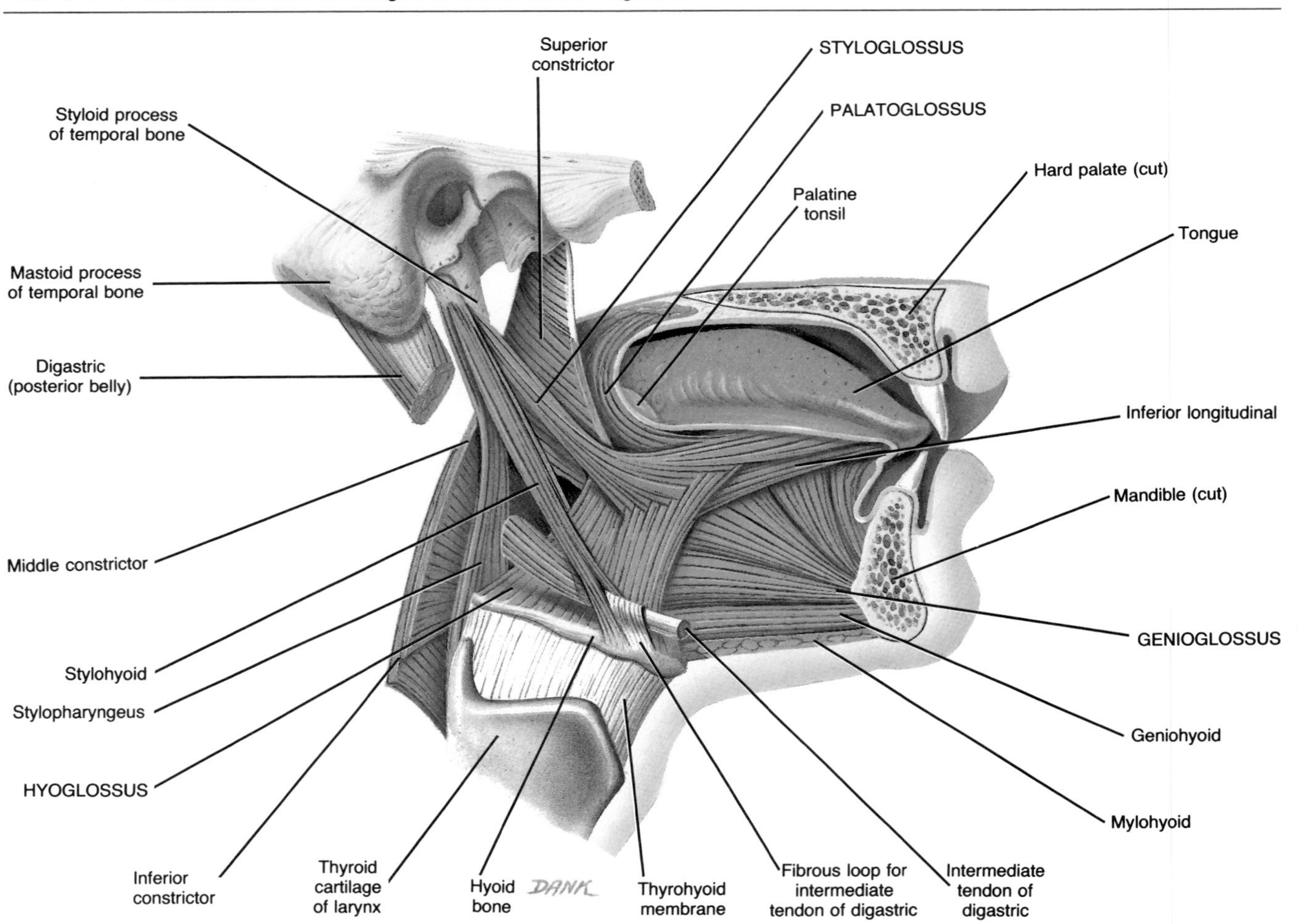

EXHIBIT 11-6 MUSCLES OF THE SOFT PALATE (Figure 11-8)

Overview: The soft palate is mainly muscular in structure and attaches anteriorly to the hard palate and blends posteriorly with the pharynx. The soft palate forms the posterior aspect of the partition that separates the oral (mouth) cavity from the pharynx. Hanging down from the free edge of the soft palate is a nipplelike structure, the uvula. During swallowing, the muscles of the soft palate tighten and elevate it, thus preventing food from entering the nasal cavities and openings of the auditory (Eustachian) tubes.

Muscle	Origin	Insertion	Action	Innervation
Levator veli palatini (*levator* = raises; *velum* = veil; *palato* = palate) (see also Figure 11-9)	Petrous portion of temporal bone and medial wall of auditory (Eustachian) tube.	Blends with corresponding muscle of opposite side.	Elevates soft palate during swallowing.	Pharyngeal plexus.
Tensor veli palatini (*tensor* = makes tense) (see also Figure 11-9)	Medial pterygoid plate of sphenoid bone, spine of sphenoid, lateral wall of auditory (Eustachian) tube.	Palatine aponeurosis and palatine bone.	Tenses (tightens) soft palate during swallowing.	Mandibular branch of trigeminal (V) nerve.
Musculus uvulae (*uvulae* = uvula)	Posterior border of hard palate and palatine aponeurosis.	Uvula.	Tenses (tightens) and raises uvula.	Pharyngeal plexus.

FIGURE 11-8 Muscles of the soft palate in posterior view.

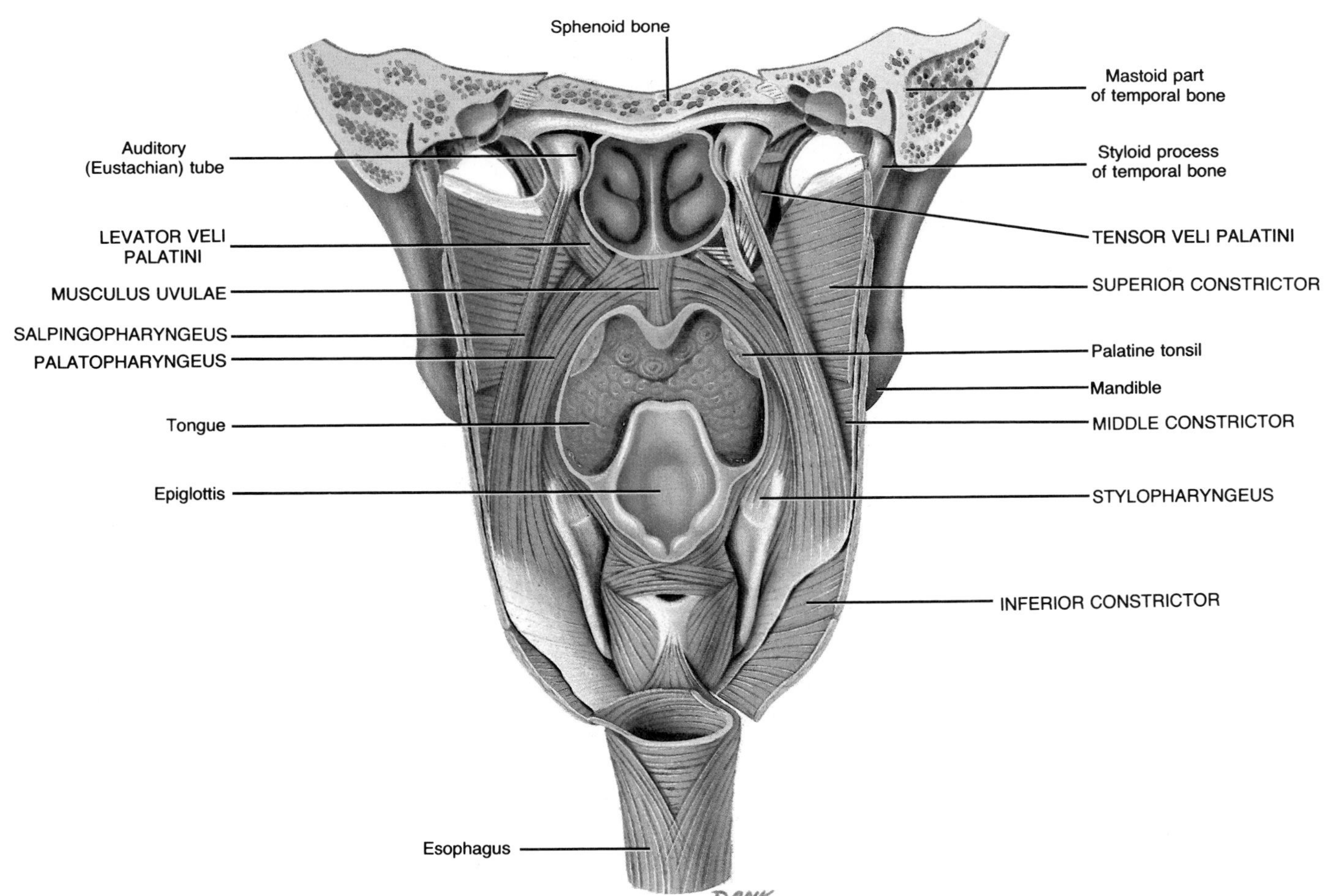

EXHIBIT 11-6 MUSCLES OF THE SOFT PALATE (*Continued*)

Muscle	Origin	Insertion	Action	Innervation
Palatoglossus (*palato* = palate; *glossus* = tongue) (see also Figure 11-7)	Anterior surface of soft palate.	Side of tongue.	Elevates posterior portion of tongue and draws soft palate down on tongue.	Pharyngeal plexus.
Palatopharyngeus (*pharyngo* = pharynx)	Posterior border of hard palate and palatine aponeurosis.	Posterior border of thyroid cartilage and lateral and posterior wall of pharynx.	Elevates larynx and pharynx and helps close nasopharynx during swallowing.	Pharyngeal plexus.

EXHIBIT 11-7 MUSCLES OF THE PHARYNX (Figure 11-9)

Overview: The pharynx (throat) is a funnel-shaped, muscular tube located posterior to the nasal cavities, mouth, and larynx (voice box). The muscles are arranged in two layers—an outer circular layer and an inner longitudinal layer. The ***circular layer*** is composed of the three constrictors, each overlapping the one above. The remaining muscles constitute the ***longitudinal layer.***

Muscle	Origin	Insertion	Action	Innervation
CIRCULAR LAYER				
Inferior constrictor (*inferior* = below; *constrictor* = decreases diameter of a lumen)	Cricoid and thyroid cartilages of larynx.	Posterior median raphe of pharynx.	Constricts inferior portion of pharynx to propel a bolus into esophagus.	Pharyngeal plexus.
Middle constrictor	Greater and lesser cornu of hyoid bone and stylohyoid ligament.	Posterior median raphe of pharynx.	Constricts middle portion of pharynx to propel a bolus into esophagus.	Pharyngeal plexus.
Superior constrictor (*superior* = above)	Pterygoid process, pterygomandibular raphe, and mylohyoid line of mandible.	Posterior median raphe of pharynx.	Constricts superior portion of pharynx to propel a bolus into esophagus.	Pharyngeal plexus.
LONGITUDINAL LAYER				
Stylopharyngeus (*stylo* = stake or pole; styloid process of temporal bone; *pharyngo* = pharynx) (see also Figure 11-7)	Medial side of base of styloid process.	Lateral aspects of pharynx and thyroid cartilage.	Elevates larynx and dilates pharynx to help bolus descend.	Glossopharyngeal (IX) nerve.
Salpingopharyngeus (*salping* = pertaining to the auditory or uterine tube) (see also Figure 11-8)	Inferior portion of auditory (Eustachian) tube.	Posterior fibers of palatopharyngeus muscle.	Elevates superior portion of lateral wall of pharynx during swallowing and opens orifice of auditory (Eustachian) tube.	Pharyngeal plexus.
Palatopharyngeus (*palato* = palate)	Soft palate.	Posterior border of thyroid cartilage and lateral and posterior wall of pharynx.	Elevates larynx and pharynx and helps close nasopharynx during swallowing.	Pharyngeal plexus.

FIGURE 11-9 **Muscles of the pharynx as seen in right lateral view.**

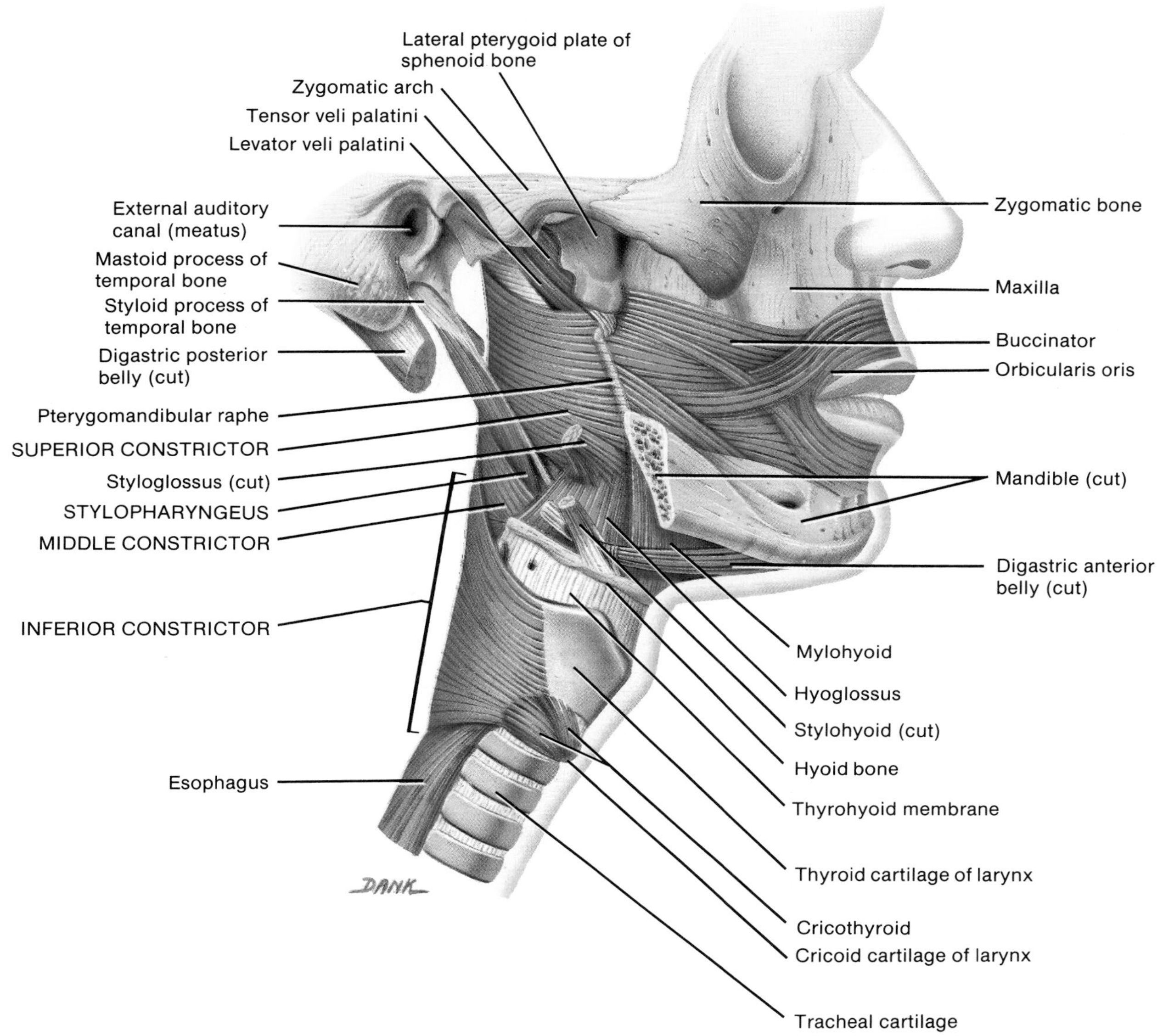
Lateral pterygoid plate of sphenoid bone
Zygomatic arch
Tensor veli palatini
Levator veli palatini
External auditory canal (meatus)
Mastoid process of temporal bone
Styloid process of temporal bone
Digastric posterior belly (cut)
Pterygomandibular raphe
SUPERIOR CONSTRICTOR
Styloglossus (cut)
STYLOPHARYNGEUS
MIDDLE CONSTRICTOR
INFERIOR CONSTRICTOR
Esophagus
Zygomatic bone
Maxilla
Buccinator
Orbicularis oris
Mandible (cut)
Digastric anterior belly (cut)
Mylohyoid
Hyoglossus
Stylohyoid (cut)
Hyoid bone
Thyrohyoid membrane
Thyroid cartilage of larynx
Cricothyroid
Cricoid cartilage of larynx
Tracheal cartilage
DANK

EXHIBIT 11-8 MUSCLES OF THE FLOOR OF THE ORAL CAVITY (Figure 11-10)

Overview: As a group, these muscles are referred to as ***suprahyoid muscles.*** They lie superior to the hyoid bone, and all insert into it. The digastric muscle consists of an anterior belly and a posterior belly united by an intermediate tendon that is held in position by a fibrous loop (see also Figure 11-7).

Muscle	Origin	Insertion	Action	Innervation
Digastric (*di* = two; *gaster* = belly)	Anterior belly from inner side of lower border of mandible; posterior belly from mastoid process of temporal bone.	Body of hyoid bone via an intermediate tendon.	Elevates hyoid bone and depresses mandible as in opening the mouth.	Anterior belly from mandibular division of trigeminal (V) nerve; posterior belly from facial (VII) nerve.
Stylohyoid (*stylo* = stake or pole, styloid process of temporal bone; *hyoedes* = U-shaped, pertaining to hyoid bone) (see also Figure 11-7)	Styloid process of temporal bone.	Body of hyoid bone.	Elevates hyoid bone and draws it posteriorly.	Facial (VII) nerve.
Mylohyoid	Inner surface of mandible.	Body of hyoid bone.	Elevates hyoid bone and floor of mouth and depresses mandible.	Mandibular division of trigeminal (V) nerve.
Geniohyoid (*geneion* = chin) (see also Figure 11-7)	Inner surface of mandible.	Body of hyoid bone.	Elevates hyoid bone, draws hyoid bone and tongue anteriorly, and depresses mandible.	Cervical nerve C1.

FIGURE 11-10 Muscles of the floor of the oral cavity and front of the neck. Superficial muscles are shown on the left side of the illustration; deep muscles are shown on the right side of the illustration.

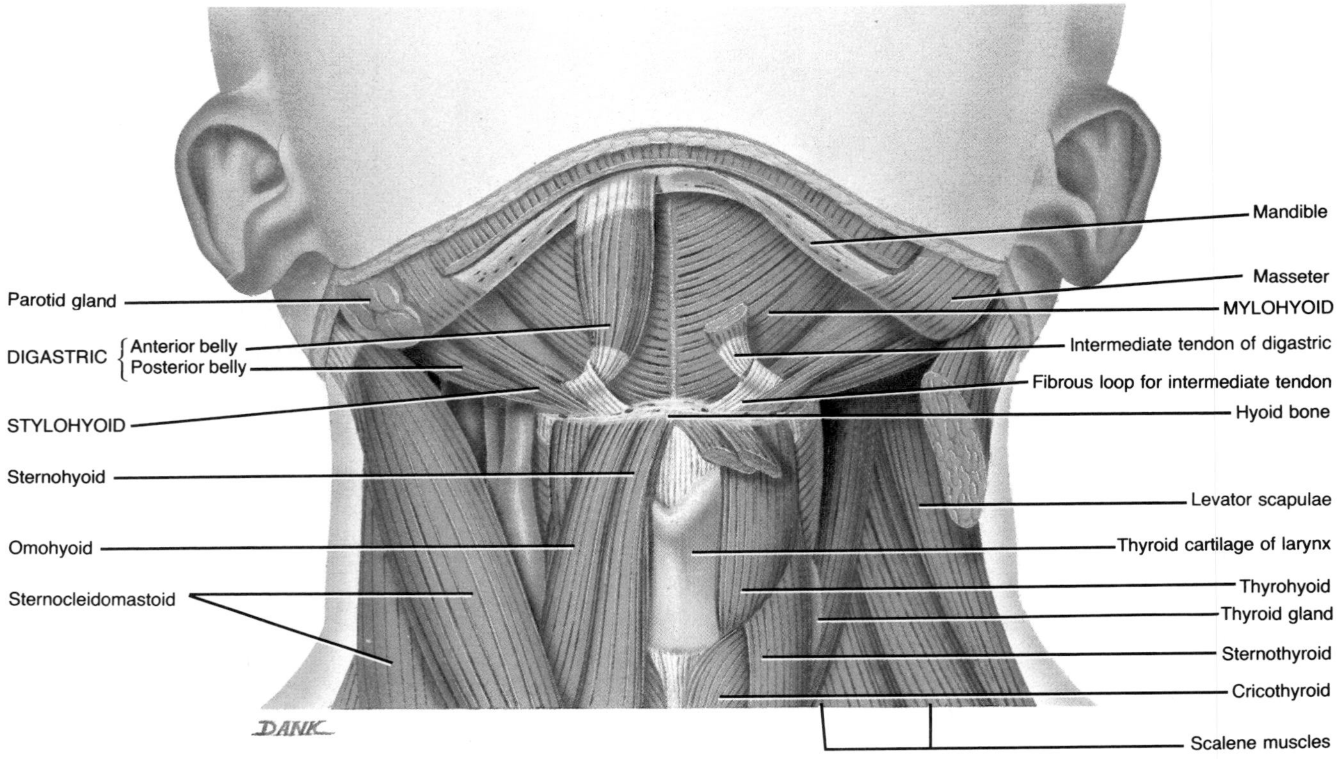

EXHIBIT 11-9 MUSCLES OF THE LARYNX (Figure 11-11)

Overview: The muscles of the larynx, like those of the eyeballs and tongue, are grouped into extrinsic and intrinsic. The extrinsic muscles of the larynx marked (*) are together referred to as ***infrahyoid (strap) muscles.*** They lie inferior to the hyoid bone. The omohyoid muscle, like the digastric muscle, is composed of two bellies and an intermediate tendon. In this case, however, the two bellies are referred to as superior and inferior, rather than anterior and posterior.

Muscle	Origin	Insertion	Action	Innervation
EXTRINSIC				
Omohyoid* (*omo* = relationship to the shoulder; *hyoedes* = U-shaped; pertaining to hyoid bone)	Superior border of scapula and superior transverse ligament.	Body of hyoid bone.	Depresses hyoid bone.	Branches of ansa cervicalis nerve (C1–C3).
Sternohyoid* (*sterno* = sternum)	Medial end of clavicle and manubrium of sternum.	Body of hyoid bone.	Depresses hyoid bone.	Branches of ansa cervicalis nerve (C1–C3).
Sternothyroid* (*thyro* = thyroid gland)	Manubrium of sternum.	Thyroid cartilage of larynx.	Depresses thyroid cartilage.	Branches of ansa cervicalis nerve (C1–C3).
Thyrohyoid*	Thyroid cartilage of larynx.	Greater cornu of hyoid bone.	Elevates thyroid cartilage and depresses hyoid bone.	Branches of ansa cervicalis nerve (C1–C2) and descending hypoglossal (XII) nerve.
Stylopharyngeus	See Exhibit 11-7.			
Palatopharyngeus	See Exhibit 11-7.			
Inferior constrictor	See Exhibit 11-7.			
Middle constrictor	See Exhibit 11-7.			
INTRINSIC				
Cricothyroid (*crico* = cricoid cartilage of larynx)	Anterior and lateral portion of cricoid cartilage of larynx.	Anterior border of inferior cornu of thyroid cartilage of larynx and posterior part of inferior border of lamina of thyroid cartilage.	Produces tension and elongation of vocal folds.	External laryngeal branch of vagus (X) nerve.
Posterior cricoarytenoid (*arytaina* = shaped like a jug)	Posterior surface of cricoid cartilage.	Posterior surface of muscular process of arytenoid cartilage of larynx.	Opens glottis.	Recurrent laryngeal branch of vagus (X) nerve.
Lateral cricoarytenoid	Superior border of cricoid cartilage.	Anterior surface of muscular process of arytenoid cartilage.	Closes glottis.	Recurrent laryngeal branch of vagus (X) nerve.
Arytenoid	Posterior surface and lateral border of one arytenoid cartilage.	Corresponding parts of opposite arytenoid cartilage.	Closes glottis.	Recurrent laryngeal branch of vagus (X) nerve.
Thyroarytenoid	Inferior portion of angle of thyroid cartilage and middle of cricothyroid ligament.	Base and anterior surface of arytenoid cartilage.	Shortens and relaxes vocal folds.	Recurrent laryngeal branch of vagus (X) nerve.

FIGURE 11-11 **Muscles of the larynx. (a) Anterior view. (b) Right posterolateral view.**

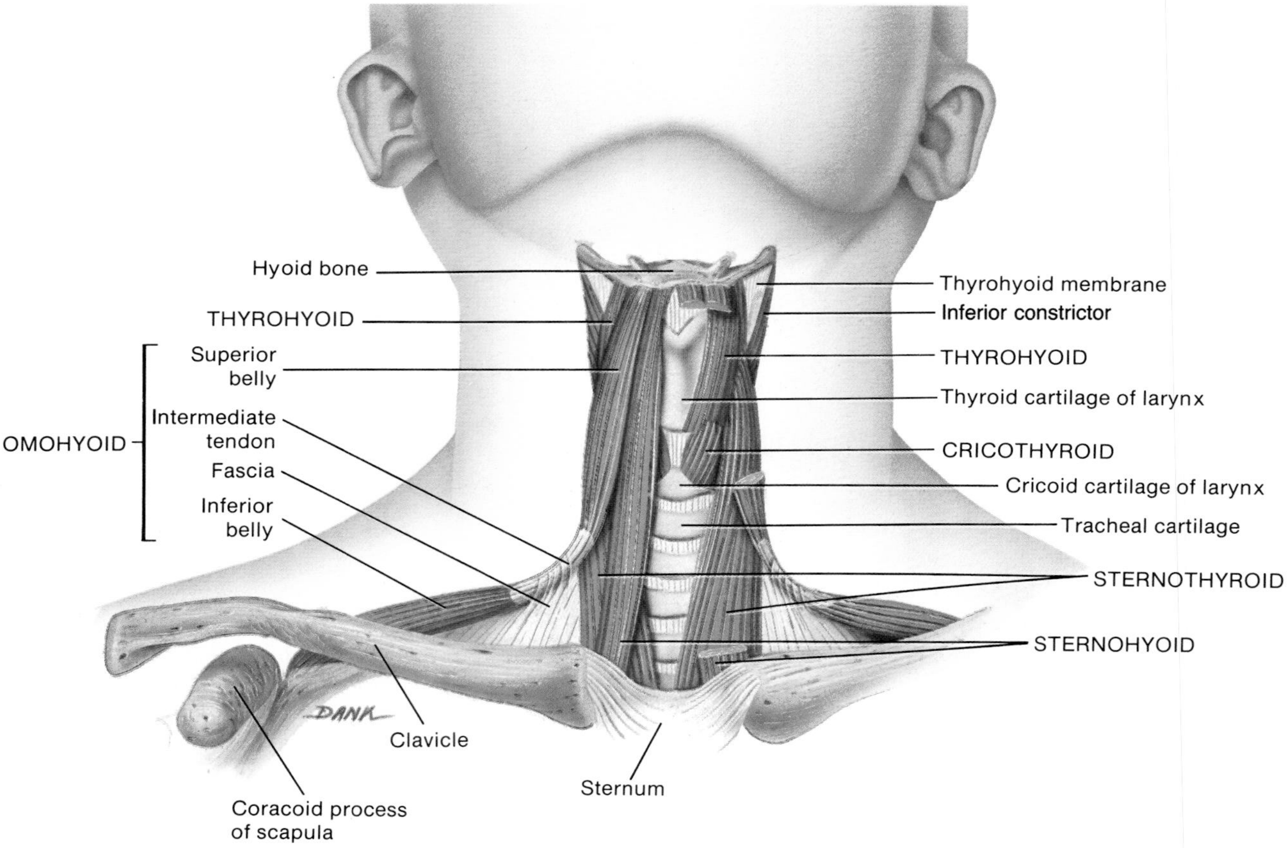

(a)

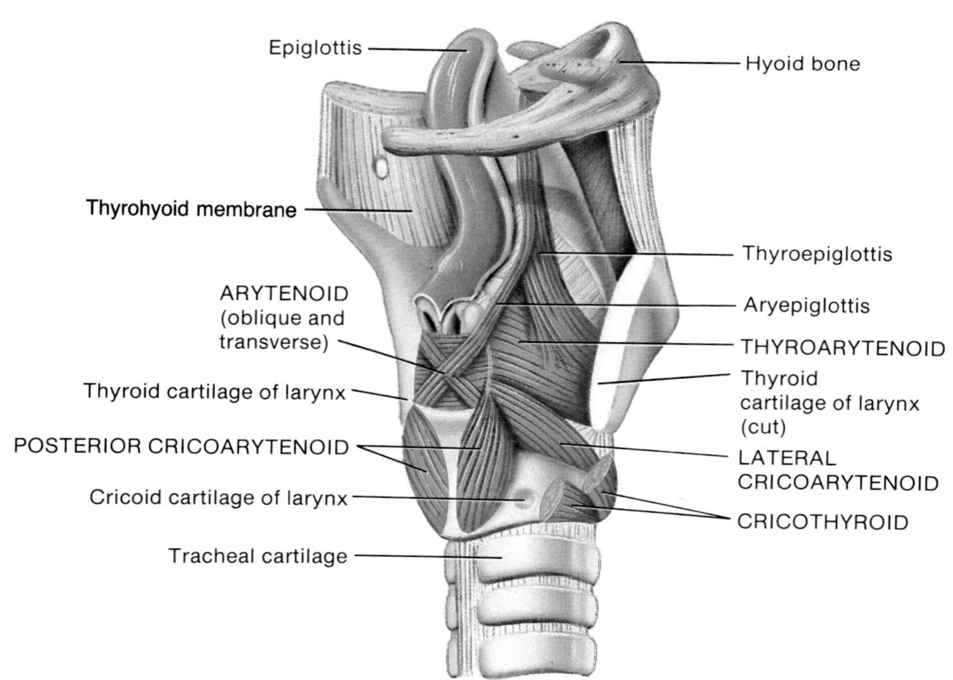

(b)

EXHIBIT 11-10 MUSCLES THAT MOVE THE HEAD

Overview: The cervical region is divided by the sternocleidomastoid muscle into two principal triangles—anterior and posterior. The ***anterior triangle*** is bordered superiorly by the mandible, inferiorly by the sternum, medially by the cervical midline, and laterally by the anterior border of the sternocleidomastoid muscle. The ***posterior triangle*** is bordered inferiorly by the clavicle, anteriorly by the posterior border of the sternocleidomastoid muscle, and posteriorly by the anterior border of the trapezius muscle. Subsidiary triangles exist within the two principal triangles.

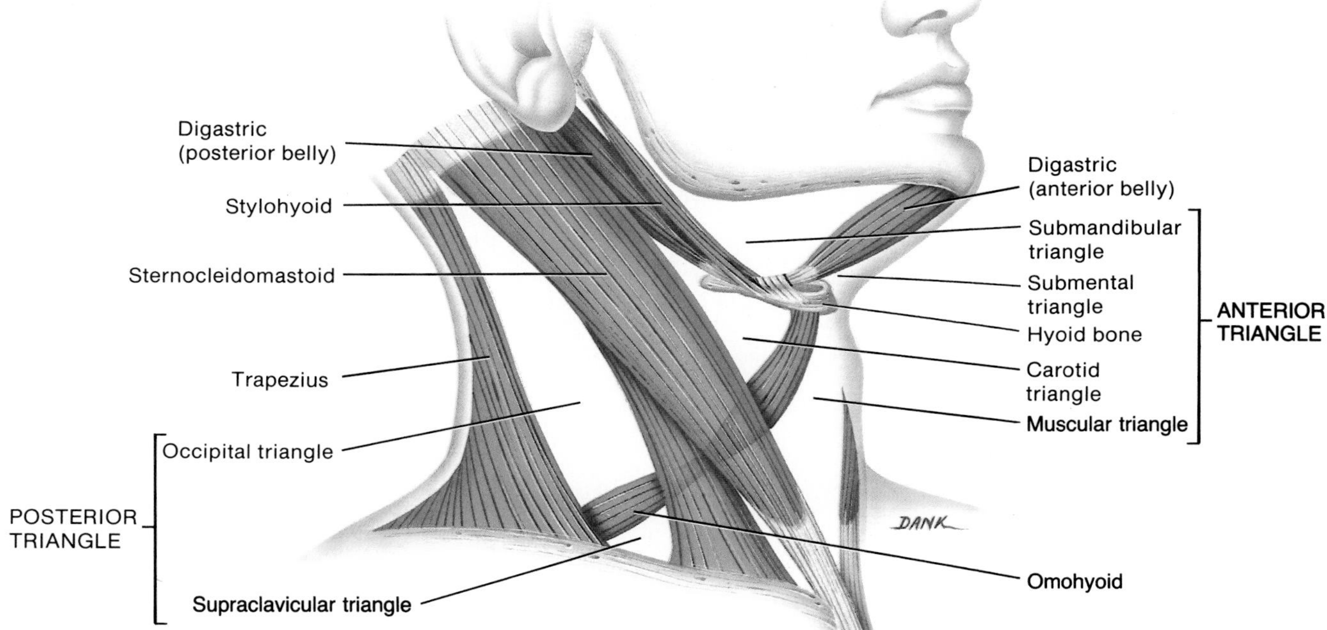

EXHIBIT 11-10 MUSCLES THAT MOVE THE HEAD (*Continued*)

Muscle	Origin	Insertion	Action	Innervation
Sternocleidomastoid (*sternum* = breast-bone; *cleido* = clavicle; *mastoid* = mastoid process of temporal bone) (see Figure 11-16)	Sternum and clavicle.	Mastoid process of temporal bone.	Contraction of both muscles flexes the cervical part of the vertebral column and draws the head forward; contraction of one muscle rotates face toward side opposite contracting muscle.	Accessory (XI) nerve; cervical nerves C2–C3.
Semispinalis capitis (*semi* = half; *spine* = spinous process; *caput* = head) (see Figure 11-21)	Articular process of seventh cervical vertebra and transverse processes of first six thoracic vertebrae.	Occipital bone between superior and inferior nuchal lines.	Both muscles extend head; contraction of one muscle rotates face toward same side as contracting muscle.	Dorsal rami of spinal nerves.
Splenius capitis (*splenion* = bandage) (see Figure 11-21)	Ligamentum nuchae and spines of seventh cervical vertebra and first four thoracic vertebrae.	Superior nuchal line of occipital bone and mastoid process of temporal bone.	Both muscles extend head; contraction of one rotates it to same side as contracting muscle.	Dorsal rami of middle and lower cervical nerves.
Longissimus capitis (*longissimus* = longest) (see Figure 11-21)	Transverse processes of last four cervical vertebrae.	Mastoid process of temporal bone.	Extends head and rotates face toward side opposite contracting muscle.	Dorsal rami of middle and lower cervical nerves.

EXHIBIT 11-11 MUSCLES THAT ACT ON THE ABDOMINAL WALL (Figure 11-12)

Overview: The anterolateral abdominal wall is composed of skin, fascia, and four pairs of flat, sheetlike muscles: rectus abdominis, external oblique, internal oblique, and transversus abdominis. The anterior surfaces of the rectus abdominis muscles are interrupted by three transverse fibrous bands of tissue called ***tendinous intersections,*** believed to be remnants of septa that separated myotomes during embryological development. The aponeuroses of the external oblique, internal oblique, and transversus abdominis muscles meet at the midline to form the ***linea alba*** (white line), a tough, fibrous band that extends from the xiphoid process of the sternum to the symphysis pubis. The inferior free border of the external oblique aponeurosis, plus some collagenous fibers, forms the ***inguinal ligament,*** which runs from the anterior superior iliac spine to the pubic tubercle (see Figure 11-22). The ligament demarcates the thigh and body wall.

Just superior to the medial end of the inguinal ligament is a triangular slit in the aponeurosis referred to as the ***superficial inguinal ring,*** the outer opening of the ***inguinal canal*** (see Figure 28-8). The canal contains the spermatic cord and inguinal nerve in males and round ligament of the uterus and ilioinguinal nerve in females.

The posterior abdominal wall is formed by the lumbar vertebrae, parts of the ilia of the hipbones, psoas major muscle (described in Exhibit 11-21), quadratus lumborum muscle, and iliacus muscle (also described in Exhibit 11-21). Whereas the anterolateral abdominal wall is contractile and distensible, the posterior abdominal wall is bulky and stable by comparison.

Muscle	Origin	Insertion	Action	Innervation
Rectus abdominis (*rectus* = fibers parallel to midline; *abdomino* = abdomen)	Pubic crest and symphysis pubis.	Cartilage of fifth to seventh ribs and xiphoid process.	Compresses abdomen to aid in defecation, urination, forced expiration, and childbirth and flexes vertebral column.	Branches of thoracic nerves T7–T12.
External oblique (*external* = closer to surface; *oblique* = fibers diagonal to midline)	Lower eight ribs.	Iliac crest and linea alba (midline aponeurosis).	Contraction of both compresses abdomen; contraction of one side alone bends vertebral column laterally.	Branches of thoracic nerves T7–T12 and iliohypogastric nerve.
Internal oblique (*internal* = farther from surface)	Iliac crest, inguinal ligament, and thoracolumbar fascia.	Cartilage of last three or four ribs.	Compresses abdomen; contraction of one side alone bends vertebral column laterally.	Branches of thoracic nerves T8–T12, iliohypogastric, and ilioinguinal nerves.
Transversus abdominis (*transverse* = fibers perpendicular to midline)	Iliac crest, inguinal ligament, lumbar fascia, and cartilages of last six ribs.	Xiphoid process, linea alba, and pubis.	Compresses abdomen.	Branches of thoracic nerves T8–T12, iliohypogastric, and ilioinguinal nerves.
Quadratus lumborum (*quad* = four; *lumbo* = lumbar region) (see Figure 11-13)	Iliac crest and iliolumbar ligament.	Lower border of twelfth rib and transverse processes of first four lumbar vertebrae.	Contraction of one side bends vertebral column laterally and pulls thoracic cage toward pelvis.	Branches of thoracic nerve T1 and lumbar nerves L1–L3 or L1–L4.

FIGURE 11-12 **Muscles of the anterior abdominal wall. (a) Superficial view. (b) Deep view.**

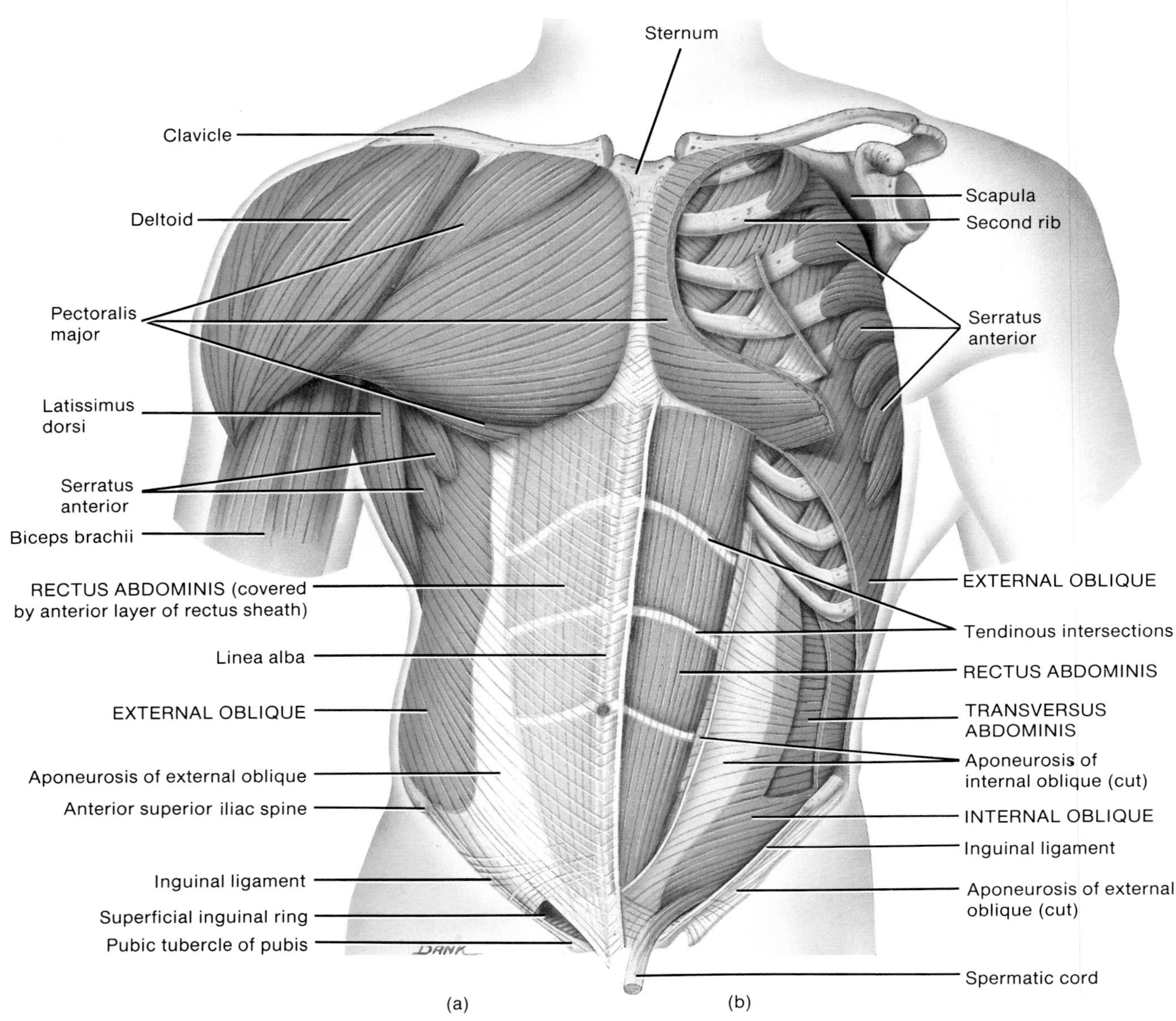

EXHIBIT 11-12 MUSCLES USED IN BREATHING (Figure 11-13)

Overview: The muscles described here are attached to the ribs and by their contraction and relaxation alter the size of the thoracic cavity during normal breathing. In forced breathing, other muscles are involved as well. Essentially, inspiration occurs when the thoracic cavity increases in size. Expiration occurs when the thoracic cavity decreases in size.

Muscle	Origin	Insertion	Action	Innervation
Diaphragm (*dia* = across; *phragma* = wall)	Xiphoid process, costal cartilages of last six ribs, and lumbar vertebrae.	Central tendon (strong aponeurosis which serves as the tendon of insertion for all muscular fibers of the diaphragm).	Forms floor of thoracic cavity; pulls central tendon downward during inspiration and thus increases vertical length of thorax.	Phrenic nerve.
External intercostals (*external* = closer to surface; *inter* = between; *costa* = rib)	Inferior border of rib above.	Superior border of rib below.	May elevate ribs during inspiration and thus increase lateral and anteroposterior dimensions of thorax.	Intercostal nerves.
Internal intercostals (*internal* = farther from surface)	Superior border of rib below.	Inferior border of rib above.	May draw adjacent ribs together during forced expiration and thus decrease lateral and anteroposterior dimensions of thorax.	Intercostal nerves.

EXHIBIT 11-13 MUSCLES OF THE PELVIC FLOOR (Figure 11-14)

Overview: The muscles of the pelvic floor, together with the fascia covering their external and internal surfaces, are referred to as the ***pelvic diaphragm.*** The diaphragm is funnel-shaped and forms the floor of the abdominopelvic cavity, where it supports the pelvic viscera. It is pierced by the anal canal and urethra in both sexes and also by the vagina in the female.

Muscle	Origin	Insertion	Action	Innervation
Levator ani (*levator* = raises; *ani* = anus)	This muscle is divisible into two parts, the pubococcygeus muscle and the iliococcygeus muscle.			
Pubococcygeus (*pubo* = pubis; *coccygeus* = coccyx)	Pubis.	Coccyx, urethra, anal canal, central tendon of perineum, and anococcygeal raphe.	Supports and slightly raises pelvic floor, resists increased intraabdominal pressure, and draws anus toward pubis and constricts it.	Sacral nerves S3–S4 or S4 and perineal branch of pudendal nerve.
Iliococcygeus (*ilio* = ilium)	Ischial spine.	Coccyx.	Supports and slightly raises pelvic floor, resists increased intraabdominal pressure, and draws anus toward pubis and constricts it.	Sacral nerves S3–S4 or S4 and perineal branch of pudendal nerve.
Coccygeus	Ischial spine.	Lower sacrum and upper coccyx.	Supports and slightly raises pelvic floor, resists intraabdominal pressure, and pulls coccyx forward following defecation or parturition.	Sacral nerve S3 or S4.

FIGURE 11-13 **Muscles used in breathing. (a) Anterior superficial view. (b) Anterior deep view.**

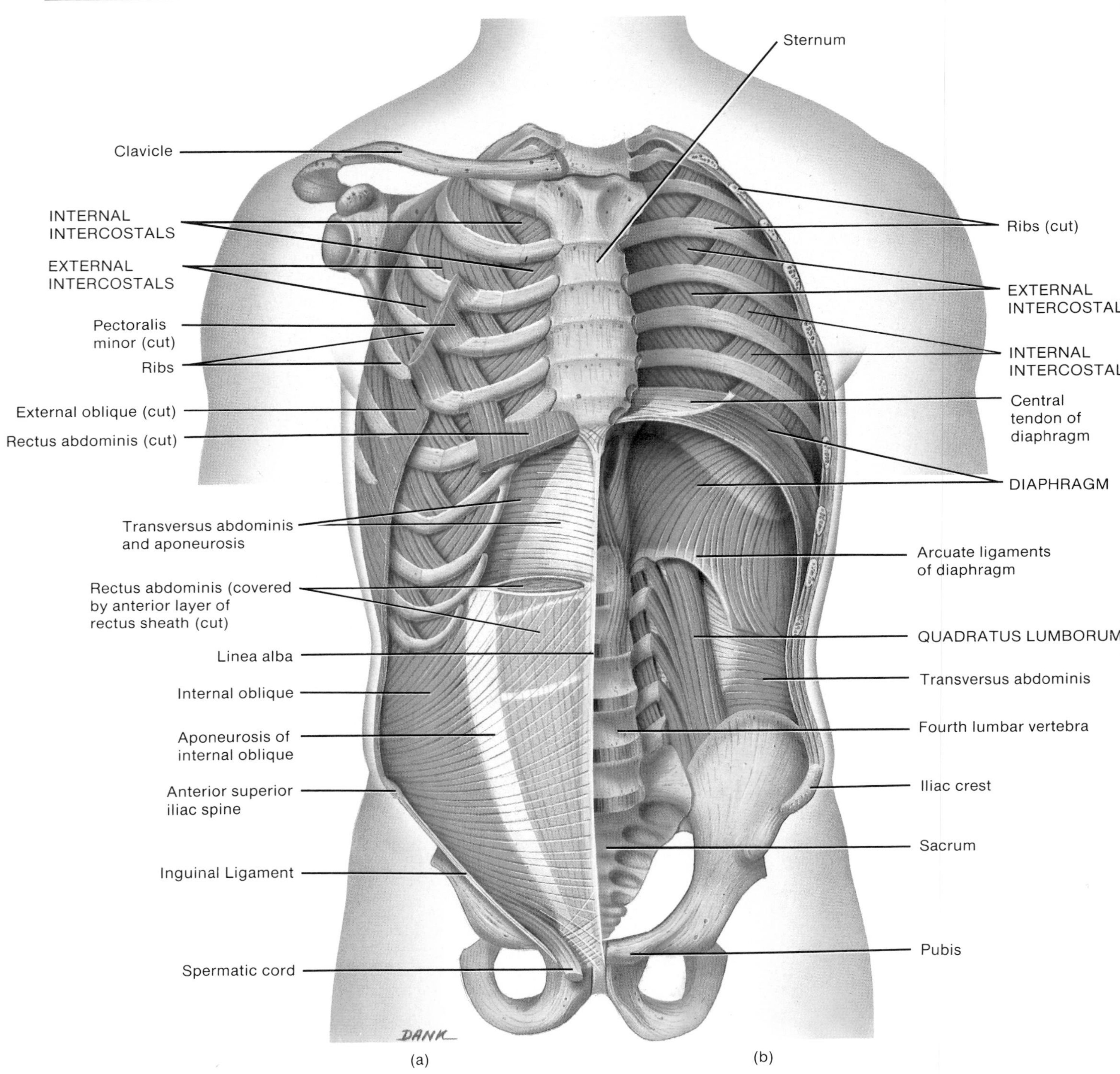

FIGURE 11-14 **Muscles of the pelvic floor seen in the female perineum.**

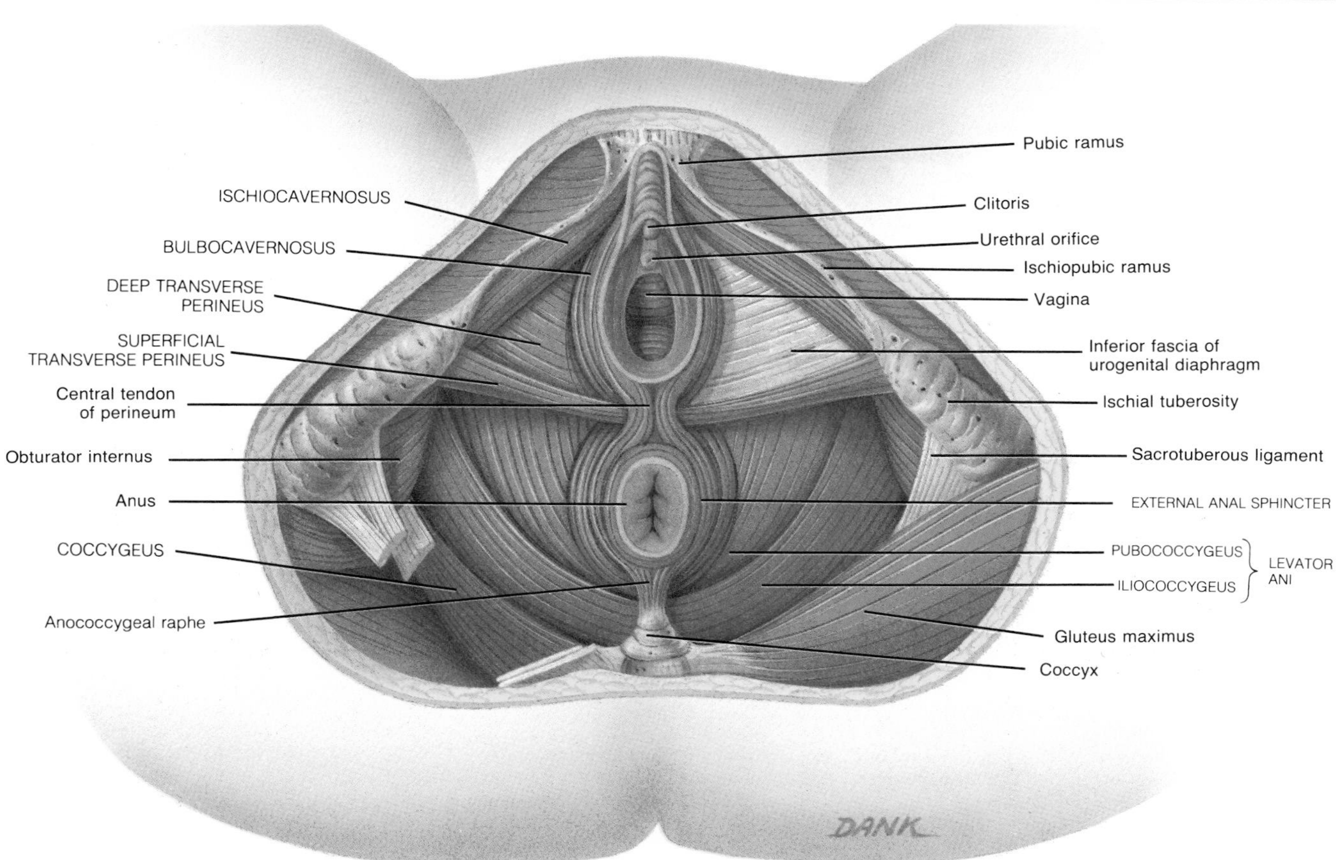

EXHIBIT 11-14 MUSCLES OF THE PERINEUM (Figure 11-15; see also Figure 11-14)

Overview: The ***perineum*** is the entire outlet of the pelvis. It is a diamond-shaped area at the lower end of the trunk between the thighs and buttocks. It is bordered anteriorly by the symphysis pubis, laterally by the ischial tuberosities, and posteriorly by the coccyx. A transverse line drawn between the ischial tuberosities divides the perineum into an anterior ***urogenital triangle*** that contains the external genitals and a posterior ***anal triangle*** that contains the anus (see Figure 28-23). The deep transverse perineus, the urethral sphincter, and a fibrous membrane constitute the ***urogenital diaphragm.*** It surrounds the urogenital ducts and helps to strengthen the pelvic floor.

Muscle	Origin	Insertion	Action	Innervation
Superficial transverse perineus (*superficial* = closer to surface; *transverse* = across; *perineus* = perineum)	Ischial tuberosity.	Central tendon of perineum.	Helps to stabilize the central tendon of the perineum.	Perineal branch of pudendal nerve.
Bulbocavernosus (*bulbus* = bulb; *caverna* = hollow place)	Central tendon of perineum.	Inferior fascia of urogenital diaphragm, corpus spongiosum of penis, and deep fascia on dorsum of penis in male; pubic arch and root and dorsum of clitoris in female.	Helps expel last drops of urine during micturition, helps propel semen along urethra, and may assist in erection of the penis in male; decreases vaginal orifice and assists in erection of clitoris in female.	Perineal branch of pudendal nerve.
Ischiocavernosus (*ischion* = hip)	Ischial tuberosity and ischial and pubic rami.	Corpus cavernosum of penis in male and clitoris in female.	May maintain erection of penis in male and clitoris in female.	Perineal branch of pudendal nerve.
Deep transverse perineus (*deep* = farther from surface)	Ischial rami.	Central tendon of perineum.	Helps expel last drops of urine and semen in male and urine in female.	Perineal branch of pudendal nerve.
Urethral sphincter (*sphincter* = circular muscle that decreases size of an opening; *urethrae* = urethra)	Ischial and pubic rami.	Median raphe in male and vaginal wall in female.	Helps expel last drops of urine and semen in male and urine in female.	Perineal branch of pudendal nerve.
External anal sphincter	Anococcygeal raphe.	Central tendon of perineum.	Keeps anal canal and orifice closed.	Sacral nerve S4 and inferior rectal branch of pudendal nerve.

FIGURE 11-15 **Muscles of the male perineum.**

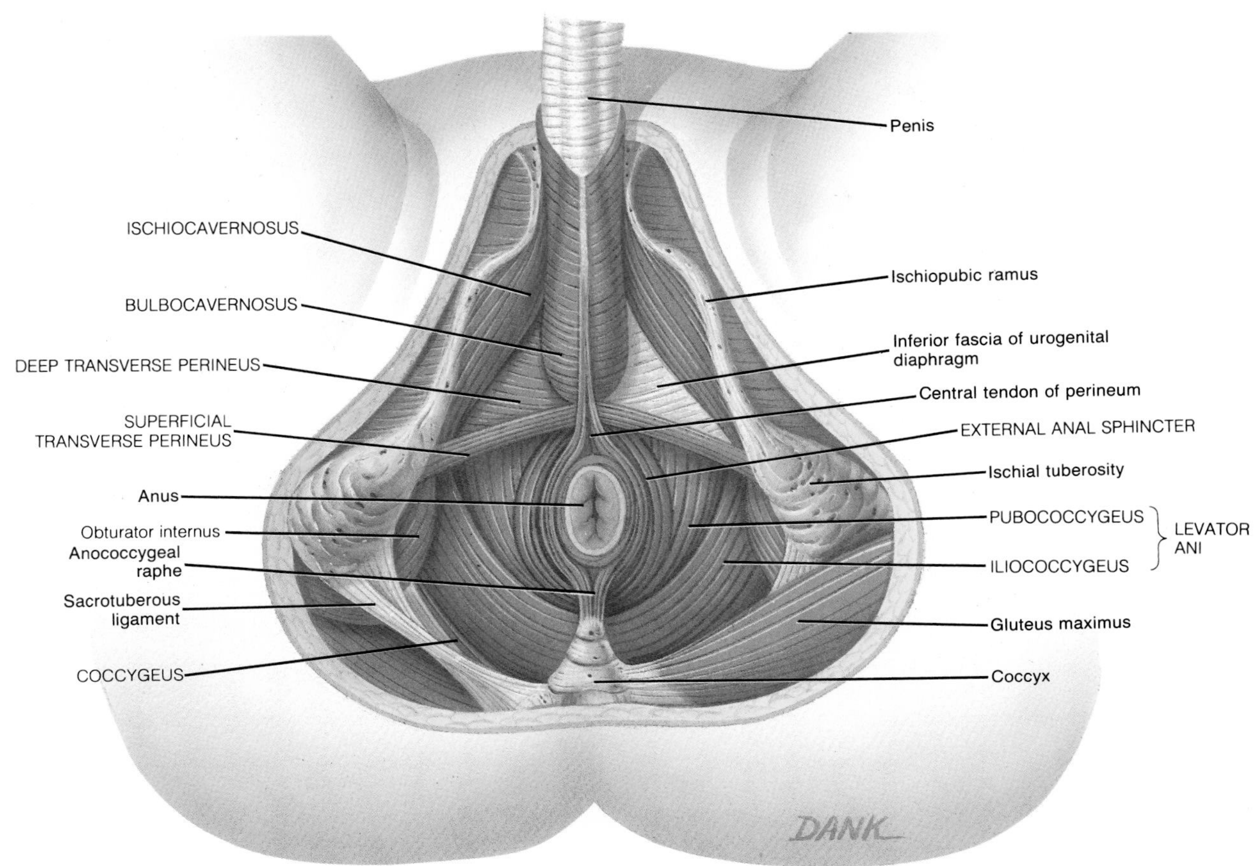
Penis
ISCHIOCAVERNOSUS
Ischiopubic ramus
BULBOCAVERNOSUS
Inferior fascia of urogenital diaphragm
DEEP TRANSVERSE PERINEUS
Central tendon of perineum
SUPERFICIAL TRANSVERSE PERINEUS
EXTERNAL ANAL SPHINCTER
Ischial tuberosity
Anus
PUBOCOCCYGEUS
Obturator internus
LEVATOR ANI
Anococcygeal raphe
ILIOCOCCYGEUS
Sacrotuberous ligament
Gluteus maximus
COCCYGEUS
Coccyx
DANK

EXHIBIT 11-15 MUSCLES THAT MOVE THE PECTORAL (SHOULDER) GIRDLE (Figure 11-16)

Overview: Muscles that move the pectoral (shoulder) girdle originate on the axial skeleton and insert on the clavicle or scapula. The muscles can be distinguished into ***anterior*** and ***posterior*** groups. The principal action of the muscles is to stabilize the scapula so that it can function as a stable point of origin for most of the muscles that move the humerus (arm).

Muscle	Origin	Insertion	Action	Innervation
ANTERIOR				
Subclavius (*sub* = under; *clavius* = clavicle)	First rib.	Clavicle.	Depresses clavicle.	Nerve to subclavius.
Pectoralis minor (*pectus* = breast, chest, thorax; *minor* = lesser)	Third through fifth ribs.	Coracoid process of scapula.	Depresses and moves scapula anteriorly and elevates third through fifth ribs during forced inspiration when scapula is fixed.	Medial pectoral nerve.
Serratus anterior (*serratus* = saw-toothed; *anterior* = front)	Upper eight or nine ribs.	Vertebral border and inferior angle of scapula.	Rotates scapula upward and laterally and elevates ribs when scapula is fixed.	Long thoracic nerve.
POSTERIOR				
Trapezius (*trapezoides* = trapezoid-shaped)	Superior nuchal line of occipital bone, ligamentum nuchae, and spines of seventh cervical and all thoracic vertebrae.	Clavicle and acromion and spine of scapula.	Elevates clavicle, adducts scapula, rotates scapula upward, elevates or depresses scapula, and extends head.	Accessory (XI) nerve and cervical nerves C3–C4.
Levator scapulae (*levator* = raises; *scapulae* = scapula)	Upper four or five cervical vertebrae.	Superior vertebral border of scapula.	Elevates scapula and slightly rotates it downward.	Dorsal scapular nerve and cervical nerves C3–C5.
Rhomboideus major (*rhomboides* = rhomboid or diamond-shaped)	Spines of second to fifth thoracic vertebrae.	Vertebral border of scapula below spine.	Adducts scapula and slightly rotates it downward.	Dorsal scapular nerve.
Rhomboideus minor	Spines of seventh cervical and first thoracic vertebrae.	Vertebral border of scapula above spine.	Adducts scapula and slightly rotates it downward.	Dorsal scapular nerve.

FIGURE 11-16 Muscles that move the pectoral (shoulder) girdle. (a) Anterior deep view. (b) Anterior deeper view.

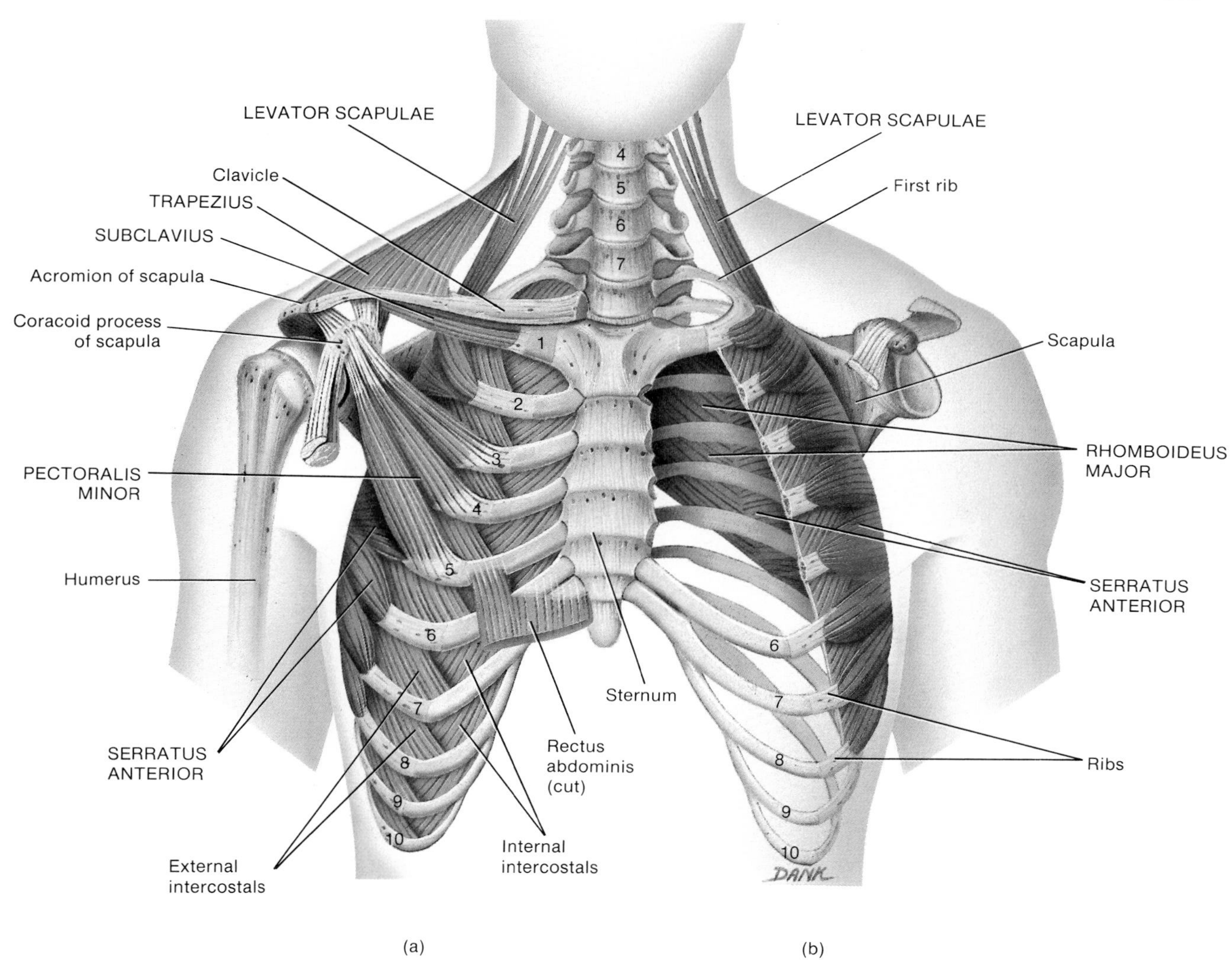

FIGURE 11-16 (*Continued*) (c) Posterior superficial view. (d) Posterior deep view.

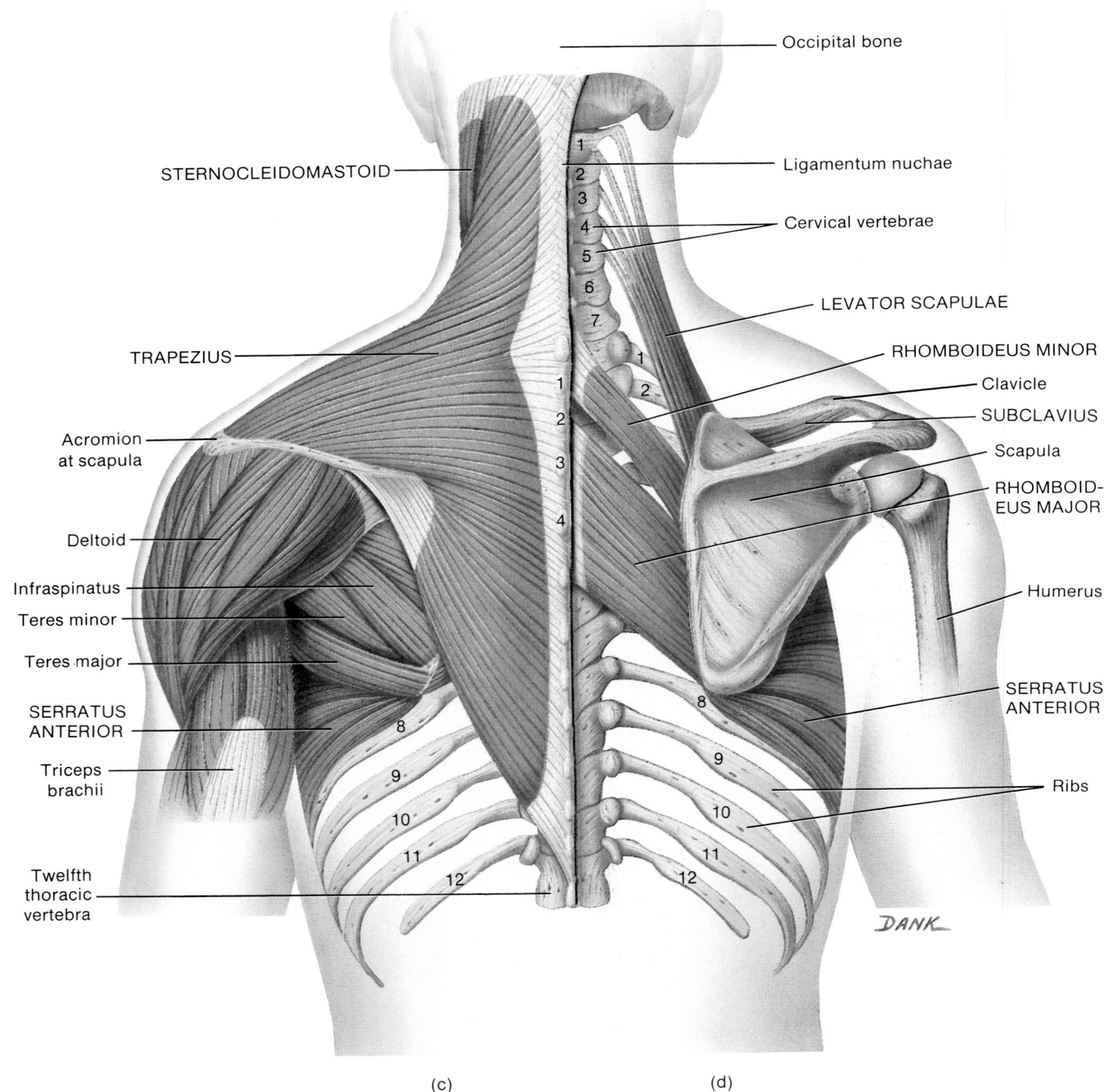

EXHIBIT 11-16 MUSCLES THAT MOVE THE ARM (HUMERUS) (Figure 11-17)

Overview: Of the nine muscles that cross the shoulder joint, only two of them (pectoralis major and latissimus dorsi) do not originate on the scapula. These two muscles are thus designated as ***axial muscles,*** since they originate on the axial skeleton. The remaining seven muscles, the ***scapular muscles,*** arise from the scapula.

The strength and stability of the shoulder joint are not provided by the shape of the articulating bones or its ligaments. Instead, four deep muscles of the shoulder and their tendons—subscapularis, supraspinatus, infraspinatus, and teres minor—strengthen and stabilize the shoulder joint. The muscles and their tendons are so arranged as to form a nearly complete circle around the joint. This arrangement is referred to as the ***rotator (musculotendinous) cuff*** and is a common site of injury to baseball pitchers, especially tearing of the supraspinatus muscle.

After you have studied the muscles in this exhibit, arrange them according to the following actions: flexion, extension, abduction, adduction, medial rotation, and lateral rotation. (The same muscle can be used more than once.)

Muscle	Origin	Insertion	Action	Innervation
AXIAL				
Pectoralis major (see also Figure 11-12a)	Clavicle, sternum, cartilages of second to sixth ribs.	Greater tubercle and intertubercular sulcus of humerus.	Flexes, adducts, and rotates arm medially.	Medial and lateral pectoral nerve.
Latissimus dorsi (*latissimus* = widest; *dorsum* = back)	Spines of lower six thoracic vertebrae, lumbar vertebrae, crests of sacrum and ilium, lower four ribs.	Intertubercular sulcus of humerus.	Extends, adducts, and rotates arm medially; draws arm downward and backward.	Thoracodorsal nerve.
SCAPULAR				
Deltoid (*delta* = triangular)	Acromial extremity of clavicle and acromion and spine of scapula.	Deltoid tuberosity of humerus.	Abducts, flexes, extends, and medially and laterally rotates arm.	Axillary nerve.
Subscapularis (*sub* = below; *scapularis* = scapula)	Subscapular fossa of scapula.	Lesser tubercle of humerus.	Rotates arm medially.	Upper and lower subscapular nerves.
Supraspinatus (*supra* = above; *spinatus* = spine of scapula)	Fossa superior to spine of scapula.	Greater tubercle of humerus.	Assists deltoid muscle in abducting arm.	Suprascapular nerve.
Infraspinatus (*infra* = below)	Fossa inferior to spine of scapula.	Greater tubercle of humerus.	Rotates arm laterally; adducts arm.	Suprascapular nerve.
Teres major (*teres* = long and round)	Inferior angle of scapula.	Intertubercular sulcus of humerus.	Extends arm; assists in adduction and medial rotation of arm.	Lower subscapular nerve.
Teres minor	Inferior lateral border of scapula.	Greater tubercle of humerus.	Rotates arm laterally; extends and adducts arm.	Axillary nerve.
Coracobrachialis (*coraco* = coracoid process)	Coracoid process of scapula.	Middle of medial surface of shaft of humerus.	Flexes and adducts arm.	Musculocutaneous nerve.

FIGURE 11-17 **Muscles that move the arm (humerus). (a) Anterior deep view.**

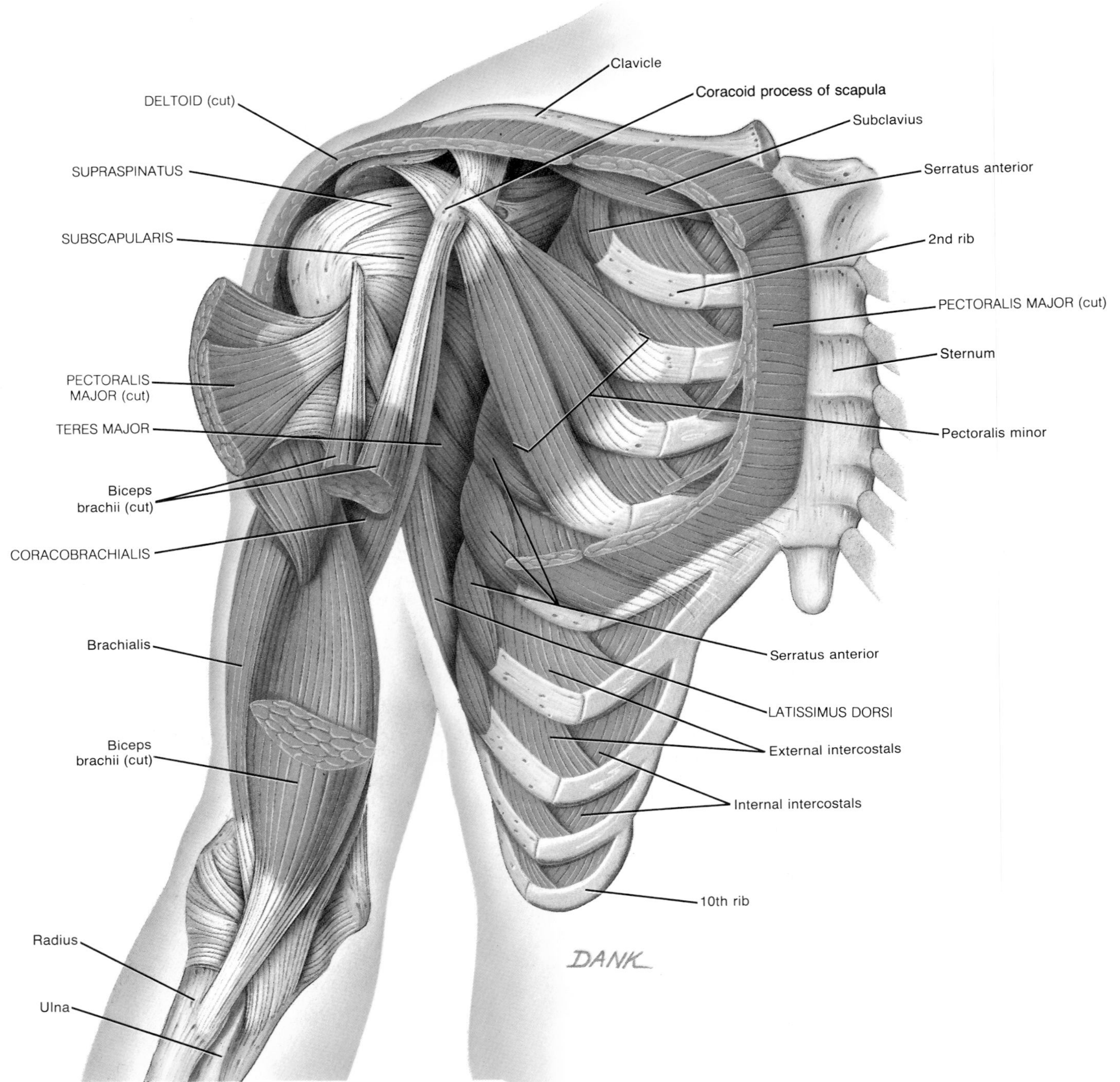

(a)

FIGURE 11-17 (*Continued*) Muscles that move the arm (humerus). (b) Posterior superficial view. (c) Posterior deep view.

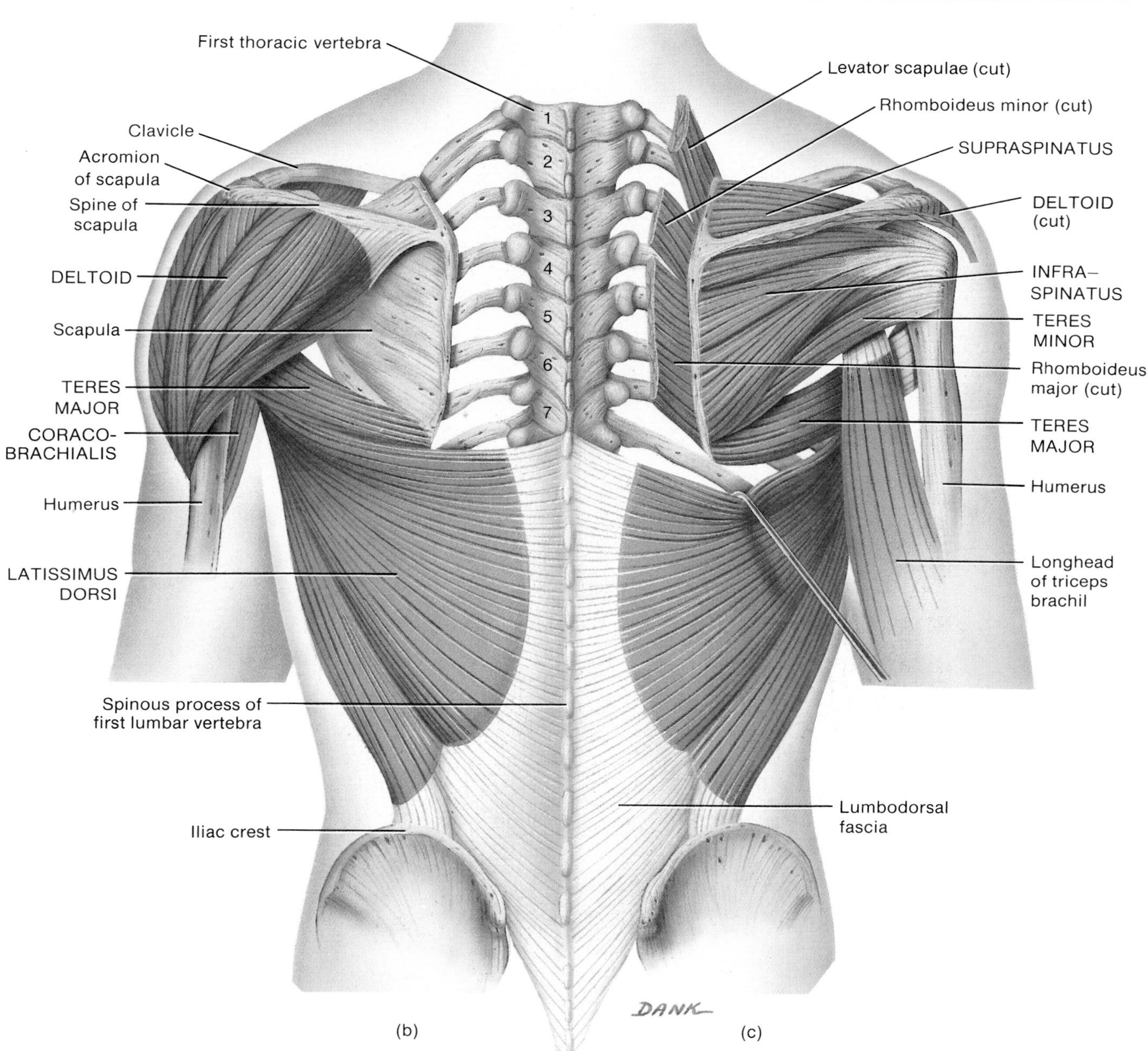

EXHIBIT 11-17 MUSCLES THAT MOVE THE FOREARM (RADIUS AND ULNA) (Figure 11-18)

Overview: Most of the muscles that move the forearm (radius and ulna) are divided into ***flexors*** and ***extensors.*** Recall that the elbow joint is a hinge joint, capable only of flexion and extension under normal conditions. Whereas the biceps brachii, brachialis, and brachioradialis are flexors of the elbow joint, the triceps brachii and anconeus are extensors. Other muscles that move the forearm are concerned with pronation and supination.

Muscle	Origin	Insertion	Action	Innervation
FLEXORS				
Biceps brachii (*biceps* = two heads of origin; *brachion* = arm)	Long head originates from tubercle above glenoid cavity; short head originates from coracoid process of scapula.	Radial tuberosity and bicipital aponeurosis.	Flexes and supinates forearm; flexes arm.	Musculocutaneous nerve.
Brachialis	Distal, anterior surface of humerus.	Tuberosity and coronoid process of ulna.	Flexes forearm.	Musculocutaneous and radial nerves.
Brachioradialis (*radialis* = radius) see also Figure 11-19a)	Supracondyloid ridge of humerus.	Superior to styloid process of radius.	Flexes forearm; semisupinates and semipronates forearm.	Radial nerve.
EXTENSORS				
Triceps brachii (*triceps* = three heads of origin)	Long head originates from infraglenoid tuberosity of scapula; lateral head originates from lateral and posterior surface of humerus superior to radial groove; medial head originates from posterior surface of humerus inferior to radial groove.	Olecranon of ulna.	Extends forearm; extends arm.	Radial nerve.
Anconeus (*anconeal* = pertaining to elbow) (see Figure 11-19c)	Lateral epicondyle of humerus.	Olecranon and superior portion of shaft of ulna.	Extends forearm.	Radial nerve.
PRONATORS				
Pronator teres (*pronation* = turning palm downward or posteriorly) (see Figure 11-19a)	Medial epicondyle of humerus and coronoid process of ulna.	Midlateral surface of radius.	Pronates forearm and hand and flexes forearm.	Median nerve.
Pronator quadratus (*quadratus* = squared, four-sided) (see Figure 11-19a,b)	Distal portion of shaft of ulna.	Distal portion of shaft of radius.	Pronates forearm and hand.	Median nerve.
SUPINATOR				
Supinator (*supination* = turning palm upward or anteriorly) (see Figure 11-19b)	Lateral epicondyle of humerus and ridge of ulna.	Lateral surface of proximal one-third of radius.	Supinates forearm and hand.	Deep radial nerve.

FIGURE 11-18 **Muscles that move the forearm. (a) Anterior view. (b) Posterior view.**

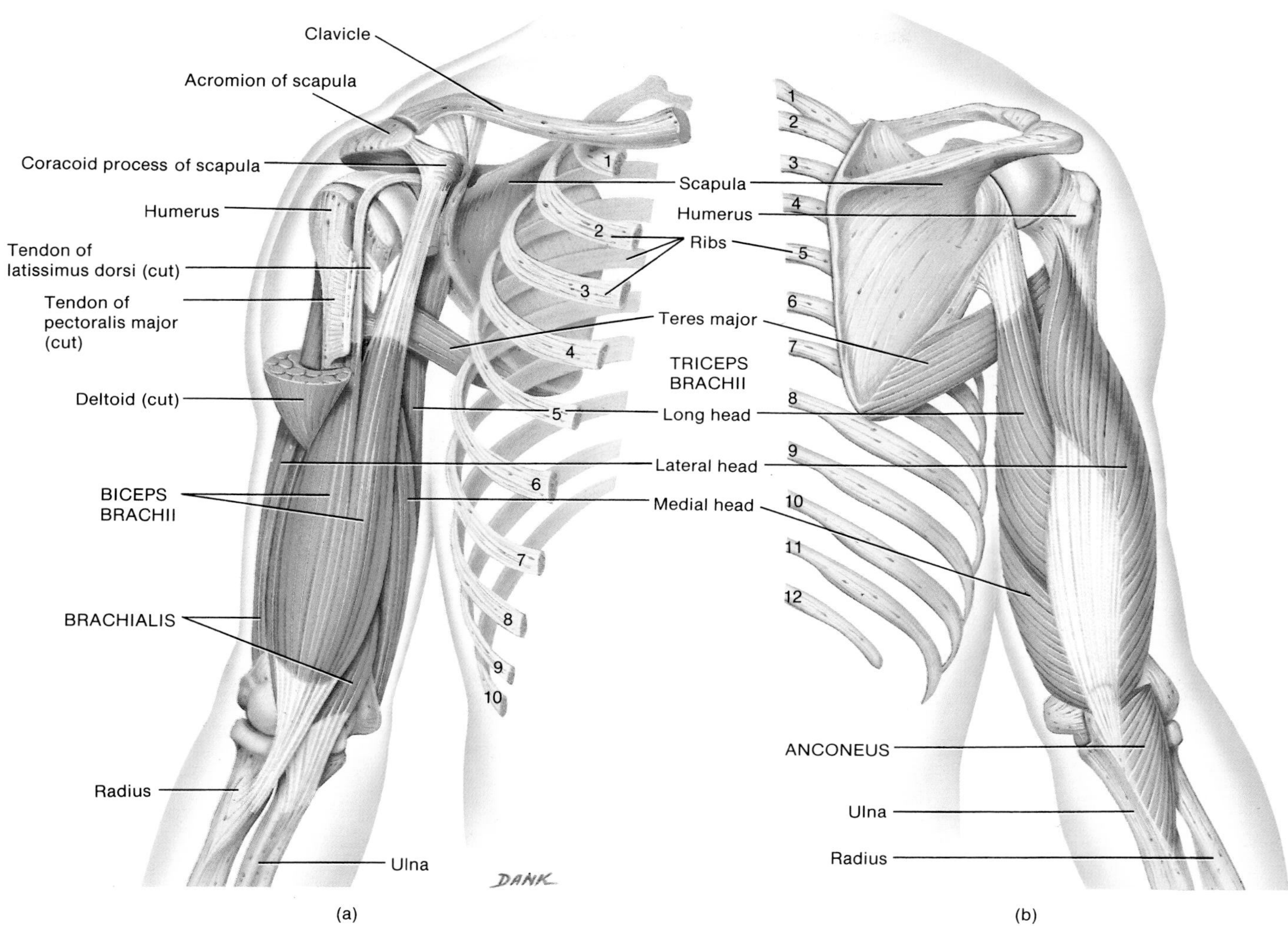

EXHIBIT 11-18 MUSCLES THAT MOVE THE WRIST, HAND, AND FINGERS (Figure 11-19)

Overview: Muscles that move the wrist, hand, and fingers are many and varied. However, as you will see, their names for the most part give some indication of their origin, insertion, or action. On the basis of location and function, the muscles are divided into two groups—anterior and posterior. The ***anterior muscles*** function as flexors. They originate on the humerus and typically insert on the carpals, metacarpals, and phalanges. The bellies of these muscles form the bulk of the proximal forearm. The ***posterior muscles*** function as extensors. These muscles arise on the humerus and insert on the metacarpals and phalanges. Each of the two principal groups is also divided into superficial and deep muscles.

The tendons of the muscles of the forearm that attach to the wrist or continue into the hand, along with blood vessels and nerves, are held close to bones by strong fascial structures. The tendons are also surrounded by tendon sheaths. At the wrist, the deep fascia is thickened into fibrous bands called retinacula (*retinere* = retain). The ***flexor retinaculum (transverse carpal ligament)*** is located over the palmar surface of the carpal bones. Through it pass the long flexor tendons of the digits and wrist and the median nerve. The ***extensor retinaculum (dorsal carpal ligament)*** is located over the dorsal surface of the carpal bones. Through it pass the extensor tendons of the wrist and digits.

After you have studied the muscles in this exhibit, arrange them according to the following actions: flexion, extension, abduction, adduction, supination, and pronation. (The same muscles can be used more than once.)

Muscle	Origin	Insertion	Action	Innervation
ANTERIOR GROUP (flexors)				
Superficial				
Flexor carpi radialis (*flexor* = decreases angle at joint; *carpus* = wrist; *radialis* = radius)	Medial epicondyle of humerus.	Second and third metacarpals.	Flexes and abducts wrist.	Median nerve.
Palmaris longus (*palma* = palm; *longus* = long)	Medial epicondyle of humerus.	Flexor retinaculum.	Flexes wrist.	Median nerve.
Flexor carpi ulnaris (*ulnaris* = ulna)	Medial epicondyle of humerus and upper posterior border of ulna.	Pisiform, hamate, and fifth metacarpal.	Flexes and adducts wrist.	Ulnar nerve.
Flexor digitorum superficialis (*digit* = finger or toe; *superficialis* = closer to surface)	Medial epicondyle of humerus, coronoid process of ulna, and oblique line of radius.	Middle phalanges.	Flexes middle phalanges of each finger.	Median nerve.
Deep				
Flexor digitorum profundus (*profundus* = deep)	Anterior medial surface of body of ulna.	Bases of distal phalanges.	Flexes distal phalanges of each finger.	Median and ulnar nerves.
Flexor pollicis longus (*pollex* = thumb)	Anterior surface of radius and interosseous membrane.	Base of distal phalanx of thumb.	Flexes thumb.	Median nerve.
POSTERIOR GROUP (extensors)				
Superficial				
Extensor carpi radialis longus (*extensor* = increases angle at joint)	Lateral epicondyle of humerus.	Second metacarpal.	Extends and abducts wrist.	Radial nerve.
Extensor carpi radialis brevis (*brevis* = short)	Lateral epicondyle of humerus.	Third metacarpal.	Extends wrist.	Radial nerve.
Extensor digitorum	Lateral epicondyle of humerus.	Second through fifth phalanges.	Extends phalanges.	Radial nerve.

FIGURE 11-19 Muscles that move the wrist, hand, and fingers. (a) Superficial anterior view. (b) Deep anterior view.

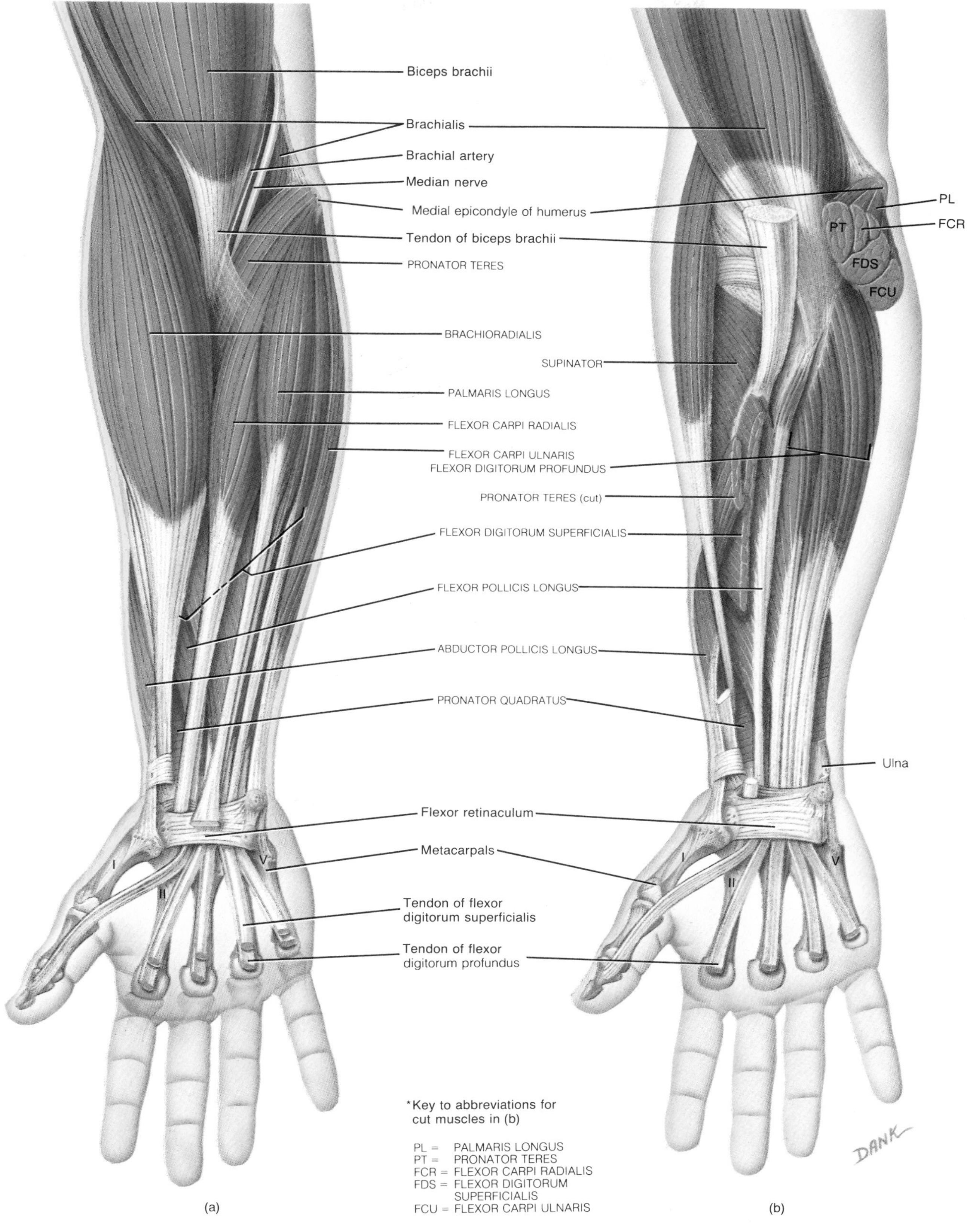

FIGURE 11-19 (*Continued*) (c) Superficial posterior view. (d) Deep posterior view.

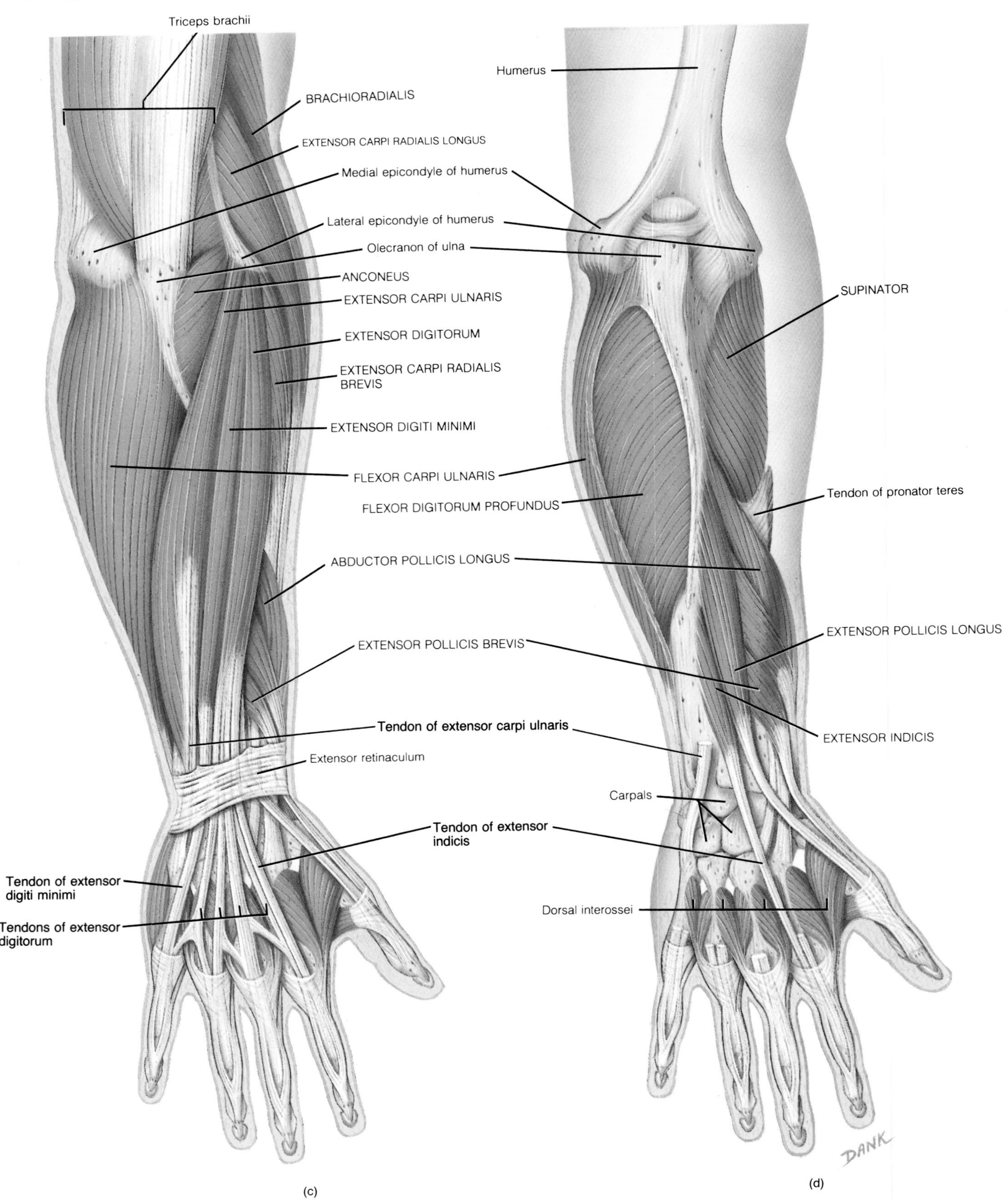

EXHIBIT 11-18 MUSCLES THAT MOVE THE WRIST, HAND, AND FINGERS (*Continued*)

Muscle	Origin	Insertion	Action	Innervation
Extensor digiti minimi (*minimi* = little finger)	Tendon of extensor digitorum.	Tendon of extensor digitorum on fifth phalanx.	Extends little finger.	Deep radial nerve.
Extensor carpi ulnaris	Lateral epicondyle of humerus and posterior border of ulna.	Fifth metacarpal.	Extends and adducts wrist.	Deep radial nerve.
Deep				
Abductor pollicis longus (*abductor* = moves part away from midline)	Posterior surface of middle of radius and ulna and interosseous membrane.	First metacarpal.	Extends thumb and abducts wrist.	Deep radial nerve.
Extensor pollicis brevis	Posterior surface of middle of radius and interosseous membrane.	Base of proximal phalanx of thumb.	Extends thumb and abducts wrist.	Deep radial nerve.
Extensor pollicis longus	Posterior surface of middle of ulna and interosseous membrane.	Base of distal phalanx of thumb.	Extends thumb and abducts wrist.	Deep radial nerve.
Extensor indicis (*indicis* = index)	Posterior surface of ulna.	Tendon of extensor digitorum of index finger.	Extends index finger.	Deep radial nerve.

FIGURE 11-19 (*Continued*) (e) Position of the flexor retinaculum seen in an anterior view of the right wrist and hand. (f) Position of the extensor retinaculum seen in a posterior view of the right wrist and hand.

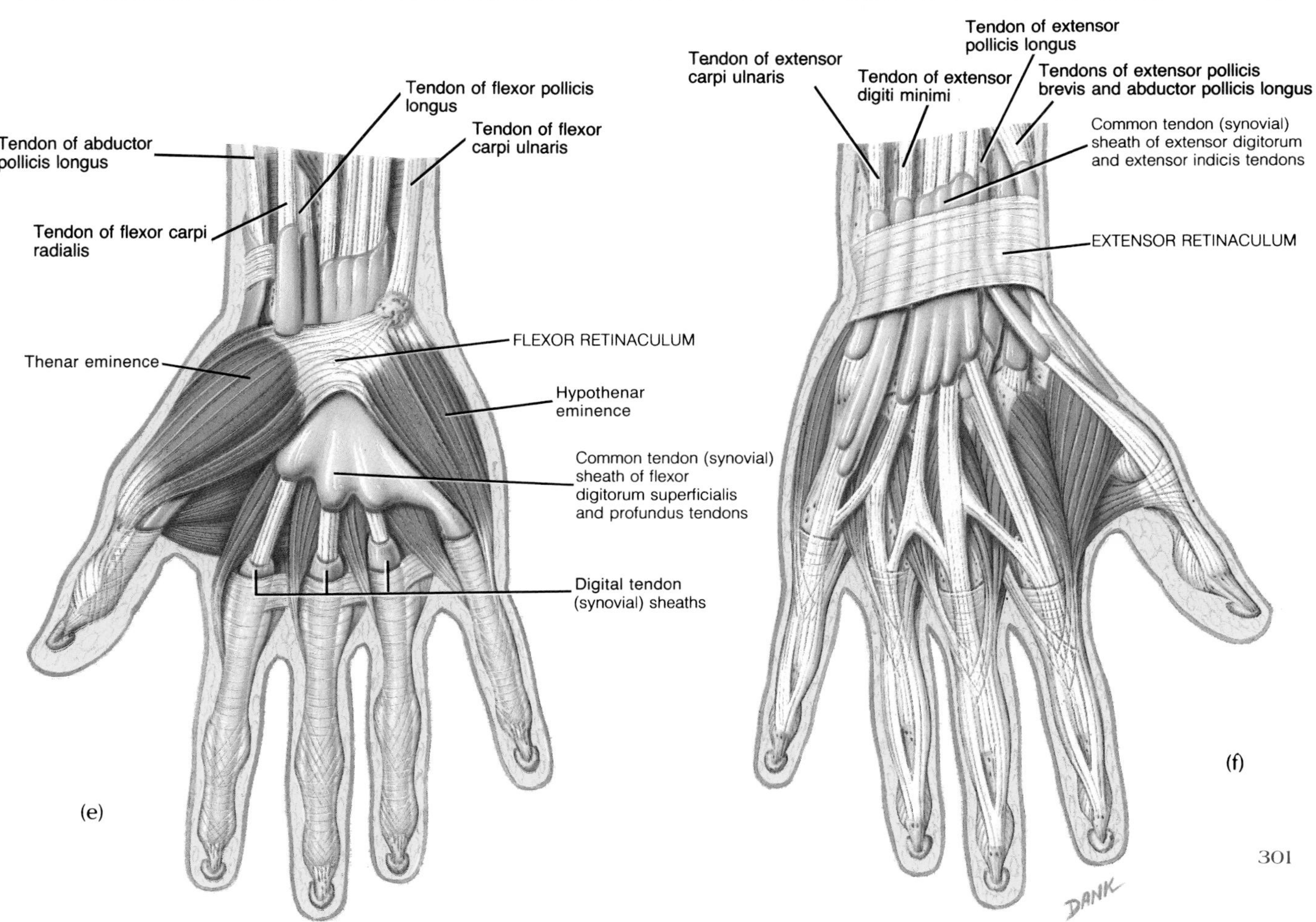

EXHIBIT 11-19 INTRINSIC MUSCLES OF THE HAND (Figure 11-20)

Overview: Several of the muscles discussed in Exhibit 11-18 help to move the digits in various ways. In addition, there are muscles in the palmar surface of the hand called ***intrinsic muscles*** that also help to move the digits. Such muscles are so named because their origins and insertions are both within the hands. These muscles assist in the intricate and precise movements that are characteristic of the human hand.

The intrinsic muscles of the hand are divided into three principal groups—thenar, hypothenar, and intermediate. The 4 ***thenar*** (THĒ-nar) ***muscles*** act on the thumb and form the ***thenar eminence.*** The 4 ***hypothenar*** (HĪ-pō-thē′-nar) ***muscles*** act on the little finger and form the ***hypothenar eminence.*** The 11 ***intermediate (midpalmar) muscles*** act on all the digits, except the thumb.

The functional importance of the hand is readily apparent when one considers that certain hand injuries can result in permanent disability. In fact, most of the efficiency of the hand depends on the movements of the thumb. The general activities of the hand are free motion, power grip (forcible movement of the fingers and thumb against the palm, as in squeezing), precision handling (a change in position of a handled object that requires exact control of finger and thumb positions, as in winding a watch or threading a needle), and pinch (compression between the thumb and index finger or between the thumb and first two fingers).

Movement of the thumb is very important in the precise activities of the hand. The five principal movements of the thumb, illustrated below, are flexion (movement of the thumb at right angles to the fingers), extension (movement in the opposite direction), abduction (movement of the thumb anteriorly and laterally), adduction (movement of the thumb toward the index finger), and opposition (movement of the thumb across the palm so that the tip of the thumb meets the tips of the fingers). Opposition is the single most distinctive digital movement that gives humans their characteristic tool-making and tool-using capacity.

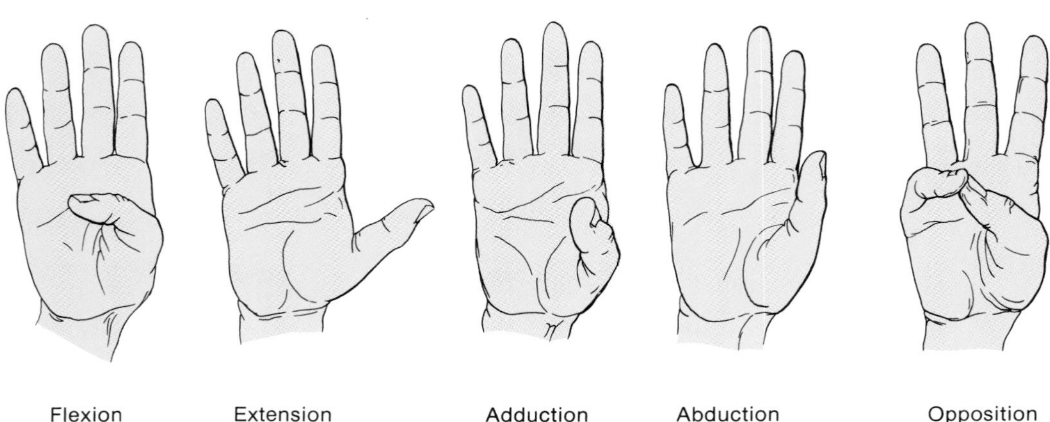

Flexion Extension Adduction Abduction Opposition

EXHIBIT 11-19 INTRINSIC MUSCLES OF THE HAND (*Continued*)

Muscle	Origin	Insertion	Action	Innervation
THENAR				
Abductor pollicis brevis (*abductor* = moves part away from middle; *pollex* = thumb; *brevis* = short)	Flexor retinaculum, scaphoid, and trapezium.	Proximal phalanx of thumb.	Abducts thumb.	Median nerve.
Opponens pollicis (*opponens* = opposes)	Flexor retinaculum and trapezium.	Metacarpal of thumb.	Draws thumb across palm to meet little finger (opposition).	Median nerve.
Flexor pollicis brevis (*flexor* = decreases angle at joint)	Flexor retinaculum, trapezium, and first metacarpal.	Proximal phalanx of thumb.	Flexes and adducts thumb.	Median and ulnar nerve.
Adductor pollicis (*adductor* = moves part toward midline)	Capitate and second and third metacarpals.	Proximal phalanx of thumb.	Adducts thumb.	Ulnar nerve.
HYPOTHENAR				
Palmaris brevis* (*palma* = palm)	Flexor retinaculum and palmar aponeurosis.	Skin on ulnar border of palm of hand.	Draws skin toward middle of palm as in clenching fist.	Ulnar nerve.
Abductor digiti minimi (*digit* = finger or toe; *minimi* = little finger)	Pisiform and tendon of flexor carpi ulnaris.	Proximal phalanx of little finger.	Abducts little finger.	Ulnar nerve.
Flexor digiti minimi brevis	Flexor retinaculum and hamate.	Proximal phalanx of little finger.	Flexes little finger.	Ulnar nerve.
Opponens digiti minimi	Flexor retinaculum and hamate.	Metacarpal of little finger.	Draws little finger across palm to meet thumb.	Ulnar nerve.
INTERMEDIATE (MIDPALMAR)				
Lumbricals (four muscles)	Tendons of flexor digitorum profundus.	Tendons of extensor digitorum.	Extend interphalangeal joints.	Median and ulnar nerve.
Dorsal interossei (four muscles; *dorsal* = back surface; *inter* = between; *ossei* = bones)	Adjacent sides of metacarpals.	Proximal phalanx of second, third, and fourth fingers.	Abduct fingers from middle finger and flex fingers at metacarpophalangeal joints.	Ulnar nerve.
Palmar interossei (three muscles)	Medial side of second metacarpal and lateral sides of fourth and fifth metacarpals.	Proximal phalanx of same finger.	Adduct fingers toward middle finger.	Ulnar nerve.

* Not illustrated.

FIGURE 11-20 Palmar view of the intrinsic muscles of the right hand. (a) Superficial view. (b) Deep view.

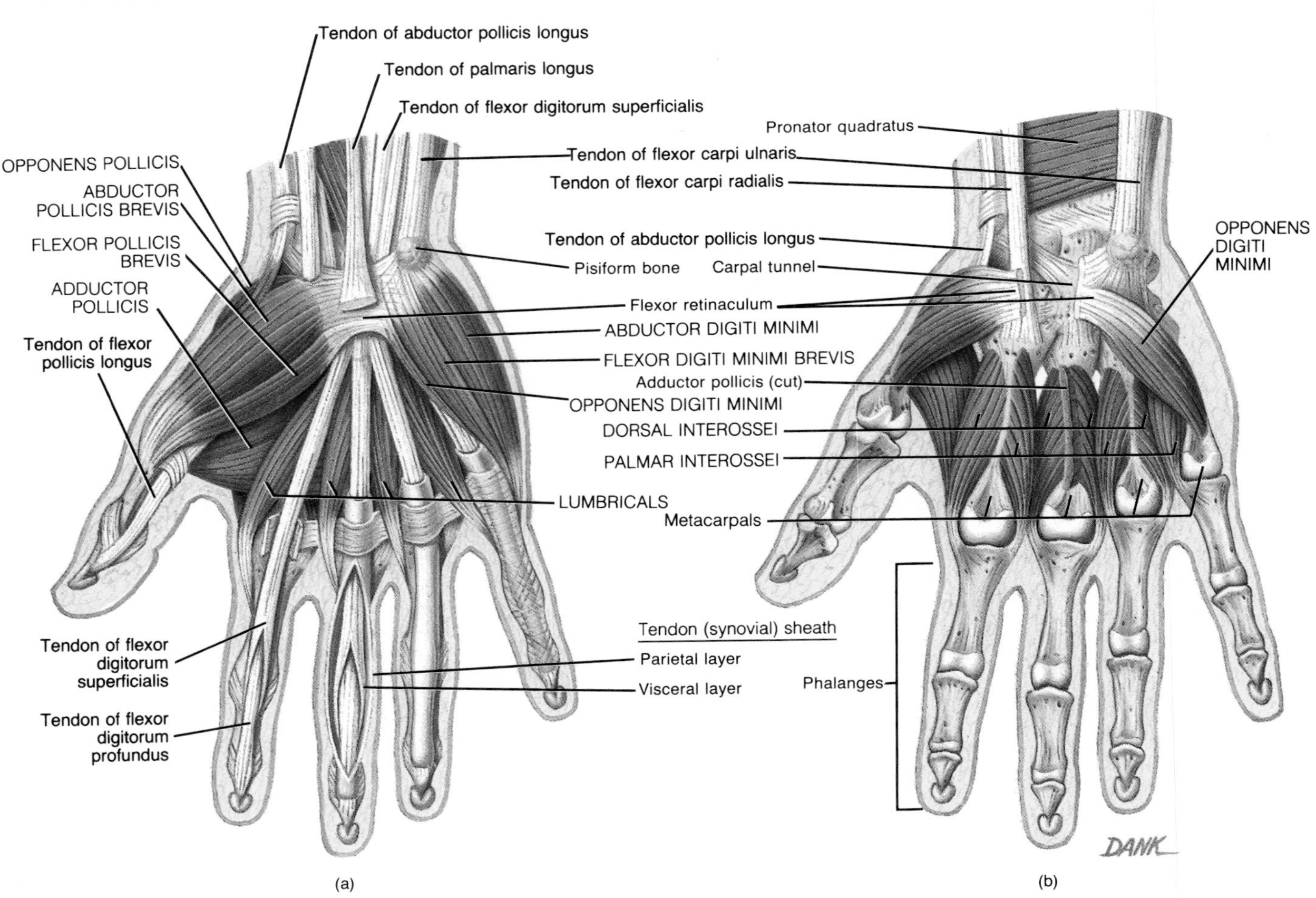

EXHIBIT 11-20 MUSCLES THAT MOVE THE VERTEBRAL COLUMN (Figure 11-21)

Overview: The muscles that move the vertebral column are quite complex because they have multiple origins and insertions and there is considerable overlapping among them. One way to group the muscles is on the basis of the general direction of the muscle bundles and their approximate lengths. For example, the ***splenius muscles*** arise from the midline and run laterally and superiorly to their insertions. The ***erector spinae (sacrospinalis) muscle*** arises from either the midline or more laterally but usually runs almost longitudinally, with neither a marked outward nor inward direction as it is traced superiorly. The ***transversospinalis muscles*** arise laterally but run toward the midline as they are traced superiorly. Deep to these three muscle groups are small ***segmental muscles*** that run between spinous processes or transverse processes of vertebrae. Since the scalene muscles also assist in moving the vertebral column, they are included in this exhibit.

Note in Exhibit 11-11 that the rectus abdominis and quadratus lumborum muscles also assume a role in moving the vertebral column.

Muscle	Origin	Insertion	Action	Innervation
SPLENIUS				
Splenius capitis (*splenium* = bandage; *caput* = head)	Ligamentum nuchae and spinous processes of seventh cervical vertebra and first three or four thoracic vertebrae.	Occipital bone and mastoid process of temporal bone.	Acting together, they extend the head and neck; acting singly, each laterally flexes and rotates head to same side.	Dorsal rami of middle cervical nerves.
Splenius cervicis (*cervix* = neck)	Spinous processes of third through sixth thoracic vertebrae.	Transverse processes of first two or four cervical vertebrae.	Acting together, they extend the head and neck; acting singly, each laterally flexes and rotates head to same side.	Dorsal rami of lower cervical nerves.
ERECTOR SPINAE (SACROSPINALIS)	This is the largest muscular mass of the back and consists of three groupings—iliocostalis, longissimus, and spinalis. These groups, in turn, consist of a series of overlapping muscles. The iliocostalis group is laterally placed, the longissimus group is intermediate in placement, and the spinalis group is medially placed.			
Iliocostalis (lateral) group				
Iliocostalis lumborum (*ilium* = flank; *costa* = rib)	Iliac crest.	Lower six ribs.	Extends lumbar region of vertebral column.	Dorsal rami of lumbar nerves.
Iliocostalis thoracis (*thorax* = chest)	Lower six ribs.	Upper six ribs.	Maintains erect position of spine.	Dorsal rami of thoracic (intercostal) nerves.
Iliocostalis cervicis	First six ribs.	Transverse processes of fourth to sixth cervical vertebrae.	Extends cervical region of vertebral column.	Dorsal rami of cervical nerves.
Longissimus (intermediate) group				
Longissimus thoracis (*longissimus* = longest)	Transverse processes of lumbar vertebrae.	Transverse processes of all thoracic and upper lumbar vertebrae and ninth and tenth ribs.	Extends thoracic region of vertebral column.	Dorsal rami of spinal nerves.
Longissimus cervicis	Transverse processes of fourth and fifth thoracic vertebrae.	Transverse processes of second to sixth cervical vertebrae.	Extends cervical region of vertebral column.	Dorsal rami of spinal nerves.
Longissimus capitis	Transverse processes of upper four thoracic vertebrae.	Mastoid process of temporal bone.	Extends head and rotates it to opposite side.	Dorsal rami of middle and lower cervical nerves.

EXHIBIT 11-20 MUSCLES THAT MOVE THE VERTEBRAL COLUMN (*Continued*)

Muscle	Origin	Insertion	Action	Innervation
Spinalis (medial) group				
Spinalis thoracis (*spinalis* = vertebral column)	Spinous processes of upper lumbar and lower thoracic vertebrae.	Spinous processes of upper thoracic vertebrae.	Extends vertebral column.	Dorsal rami of spinal nerves.
Spinalis cervicis	Ligamentum nuchae and spinous process of seventh cervical vertebra.	Spinous process of axis.	Extends vertebral column.	Dorsal rami of spinal nerves.
Spinalis capitis	Arises with semispinalis thoracis.	Inserts with spinalis thoracis.	Extends vertebral column.	Dorsal rami of spinal nerves.
TRANSVERSOSPINALIS				
Semispinalis thoracis (*semi* = partially or one-half)	Transverse processes of sixth to tenth thoracic vertebrae.	Spinous processes of first four thoracic and last two cervical vertebrae.	Extends vertebral column and rotates it to opposite side.	Dorsal rami of thoracic and cervical spinal nerves.
Semispinalis cervicis	Transverse processes of first five or six thoracic vertebrae.	Spinous processes of first to fifth cervical vertebrae.	Extends vertebral column and rotates it to opposite side.	Dorsal rami of thoracic and cervical spinal nerves.
Semispinalis capitis	Transverse processes of first six or seven thoracic vertebrae and seventh cervical vertebra and articular processes of fourth, fifth, and sixth cervical vertebrae.	Occipital bone.	Extends vertebral column and rotates it to opposite side.	Dorsal rami of cervical nerves.
Multifidus (*multi* = many; *findere* = to split)	Sacrum, ilium, transverse processes of lumbar, thoracic, and lower four cervical vertebrae.	Spinous process of a higher vertebra.	Extends vertebral column and rotates it to opposite side.	Dorsal rami of spinal nerves.
Rotatores (*rotate* = turn on an axis)	Transverse processes of all vertebrae.	Spinous process of vertebra above the one of origin.	Extends vertebral column and rotates it to opposite side.	Dorsal rami of spinal nerves.
SEGMENTAL				
Interspinales (*inter* = between)	Superior surface of all spinous processes.	Inferior surface of spinous process of vertebra above the one of origin.	Extends vertebral column.	Dorsal rami of spinal nerves.
Intertransversarii (*inter* = between)	Transverse processes of all vertebrae.	Transverse process of vertebra above the one of origin.	Laterally flexes vertebral column.	Dorsal and ventral rami of spinal nerves.
SCALENE				
Anterior scalene (*anterior* = front; *skalenos* = uneven)	Transverse processes of third through sixth cervical vertebrae.	First rib.	Flexes and rotates neck and assists in inspiration.	Ventral rami of fifth and sixth cervical nerves.
Middle scalene	Transverse processes of last six cervical vertebrae.	First rib.	Flexes and rotates neck and assists in inspiration.	Ventral rami of third through eighth cervical nerves.
Posterior scalene	Transverse processes of fourth through sixth cervical vertebrae.	Second or third rib.	Flexes and rotates neck and assists in inspiration.	Ventral rami of last three cervical nerves.

FIGURE 11-21 Muscles that move the vertebral column. (a) Posterior view of deep muscles of the neck and back.

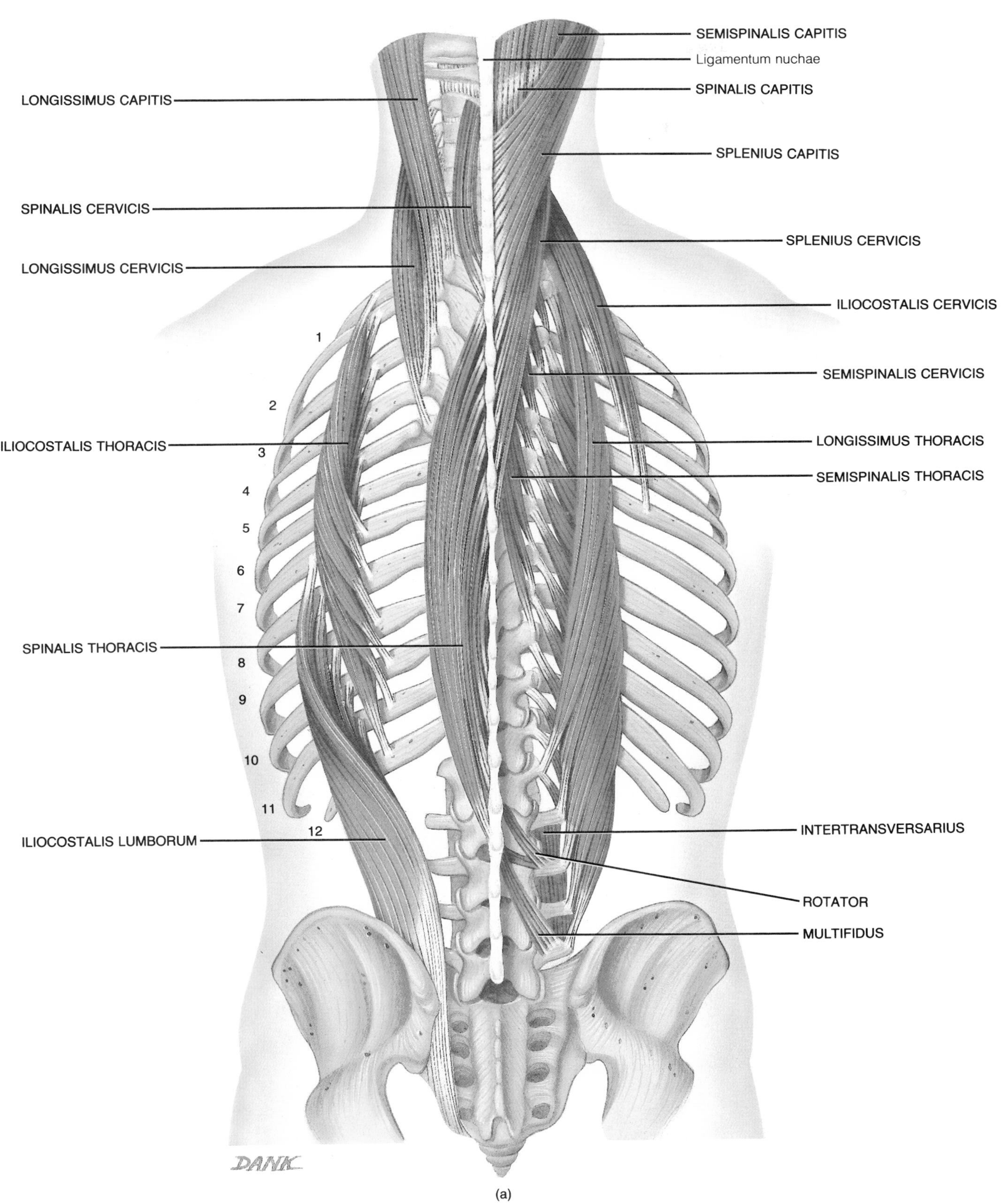

(a)

FIGURE 11-21 (*Continued*) (b) Posterolateral view of several intervertebral muscles. (c) Anterior view of scalene muscles.

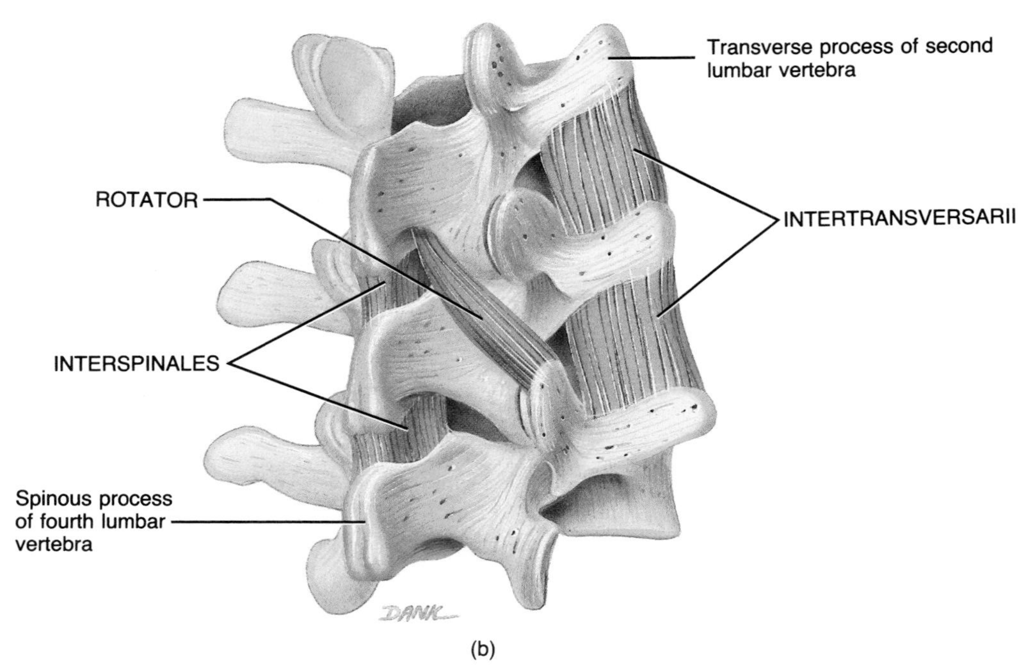

(b)

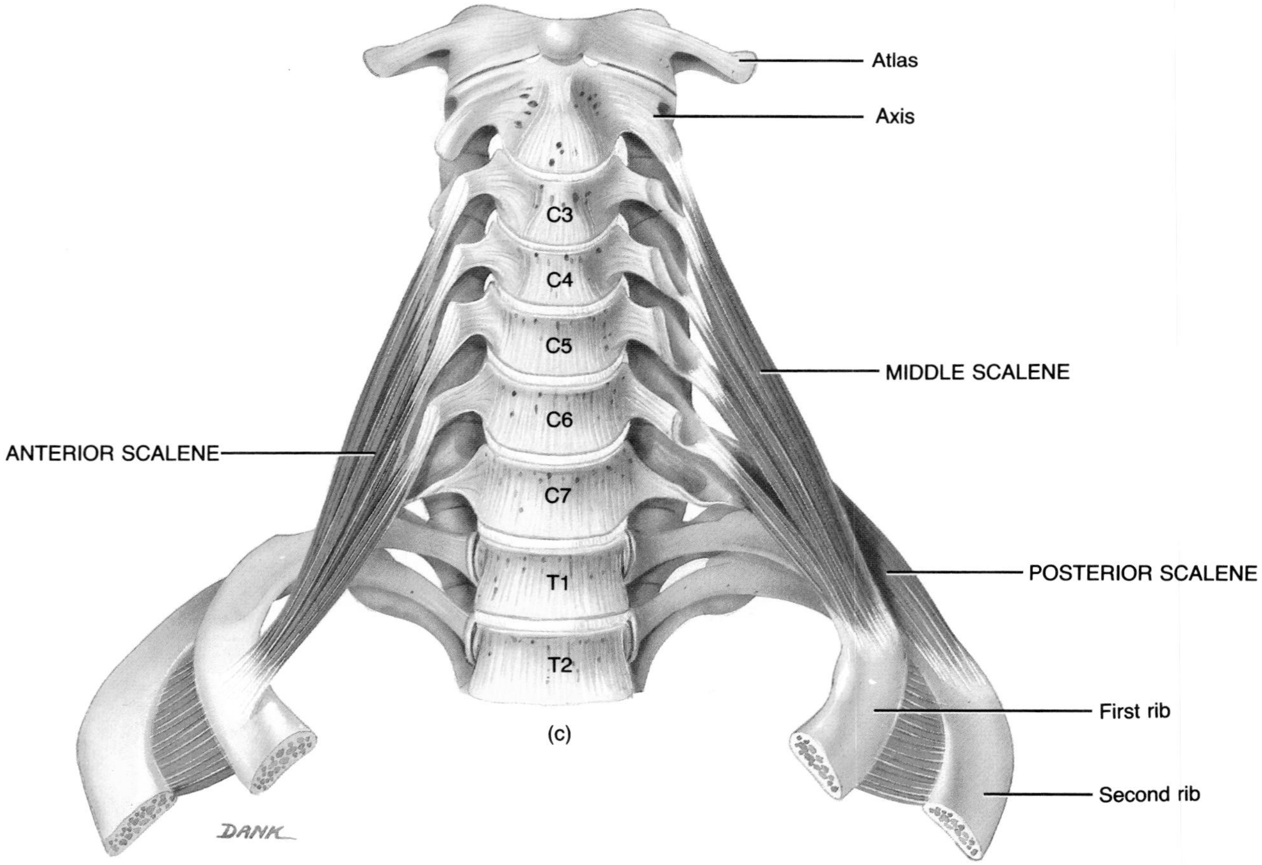

Atlas
Axis
C3
C4
C5
C6
C7
T1
T2
MIDDLE SCALENE
ANTERIOR SCALENE
POSTERIOR SCALENE
First rib
Second rib
DANK

(c)

EXHIBIT 11-21 MUSCLES THAT MOVE THE THIGH (FEMUR) (Figure 11-22)

Overview: As you will see, muscles of the lower extremities are larger and more powerful than those of the upper extremities since lower extremity muscles function in stability, locomotion, and maintenance of posture. Upper extremity muscles are characterized by versatility of movement. In addition, muscles of the lower extremities frequently cross two joints and act equally on both.

The majority of muscles that act on the thigh (femur) originate on the pelvic (hip) girdle and insert on the femur. The anterior muscles are the psoas major and iliacus, together referred to as the iliopsoas muscle. The remaining muscles (except for the pectineus, adductors, and tensor fasciae latae) are posterior muscles. Technically, the pectineus and adductors are components of the medial compartment of the thigh, but they are included in this exhibit because they act on the thigh. The tensor fasciae latae muscle is laterally placed. The ***fascia lata*** is a deep fascia of the thigh that encircles the entire thigh. It is well developed laterally, where together with the tendons of the gluteus maximus and tensor fasciae latae muscles it forms a structure called the ***iliotibial tract.*** The tract inserts into the lateral condyle of the tibia.

After you have studied the muscles in this exhibit, arrange them according to the following actions: flexion, extension, abduction, adduction, medial rotation, and lateral rotation. (The same muscles can be used more than once.)

Muscle	Origin	Insertion	Action	Innervation
Psoas major (*psoa* = muscle of loin)	Transverse processes and bodies of lumbar vertebrae.	Lesser trochanter of femur.	Flexes and rotates thigh laterally; flexes vertebral column.	Lumbar nerves L2–L3.
Iliacus (*iliac* = ilium)	Iliac fossa.	Tendon of psoas major.	Flexes and rotates thigh laterally.	Femoral nerve.
Gluteus maximus (*glutos* = buttock; *maximus* = largest)	Iliac crest, sacrum, coccyx, and aponeurosis of sacrospinalis.	Iliotibial tract of fascia lata and gluteal tuberosity of femur.	Extends and rotates thigh laterally.	Inferior gluteal nerve.
Gluteus medius (*media* = middle)	Ilium.	Greater trochanter of femur.	Abducts and rotates thigh medially.	Superior gluteal nerve.
Gluteus minimus (*minimus* = smallest)	Ilium.	Greater trochanter of femur.	Abducts and rotates thigh medially.	Superior gluteal nerve.
Tensor fasciae latae (*tensor* = makes tense; *fascia* = band; *latus* = wide)	Iliac crest.	Tibia by way of the iliotibial tract.	Flexes and abducts thigh.	Superior gluteal nerve.
Piriformis (*pirum* = pear; *forma* = shape)	Sacrum anteriorly.	Superior border of greater trochanter of femur.	Rotates thigh laterally and abducts it.	Sacral nerves S2 or S1–S2.
Obturator internus (*obturator* = obturator foramen; *internus* = inside)	Inner surface of obturator foramen, pubis, and ischium.	Greater trochanter of femur.	Rotates thigh laterally and abducts it.	Nerve to obturator internus.
Obturator externus (*externus* = outside)	Outer surface of obturator membrane.	Trochanteric fossa of femur.	Rotates thigh laterally.	Obturator nerve.
Superior gemellus (*superior* = above; *gemellus* = twins)	Ischial spine.	Greater trochanter of femur.	Rotates thigh laterally and abducts it.	Nerve to obturator internus.
Inferior gemellus (*inferior* = below)	Ischial tuberosity.	Greater trochanter of femur.	Rotates thigh laterally and abducts it.	Nerve to quadratus femoris.
Quadratus femoris (*quad* = four; *femoris* = femur)	Ischial tuberosity.	Small tubercle on posterior femur.	Laterally rotates and adducts thigh.	Nerve to quadratus femoris.
Adductor longus (*adductor* = moves part closer to midline; *longus* = long)	Pubic crest and symphysis pubis.	Linea aspera of femur.	Adducts, laterally rotates, and flexes thigh.	Obturator nerve.

FIGURE 11-22 **Muscles that move the thigh (femur). (a) Anterior superficial view.**

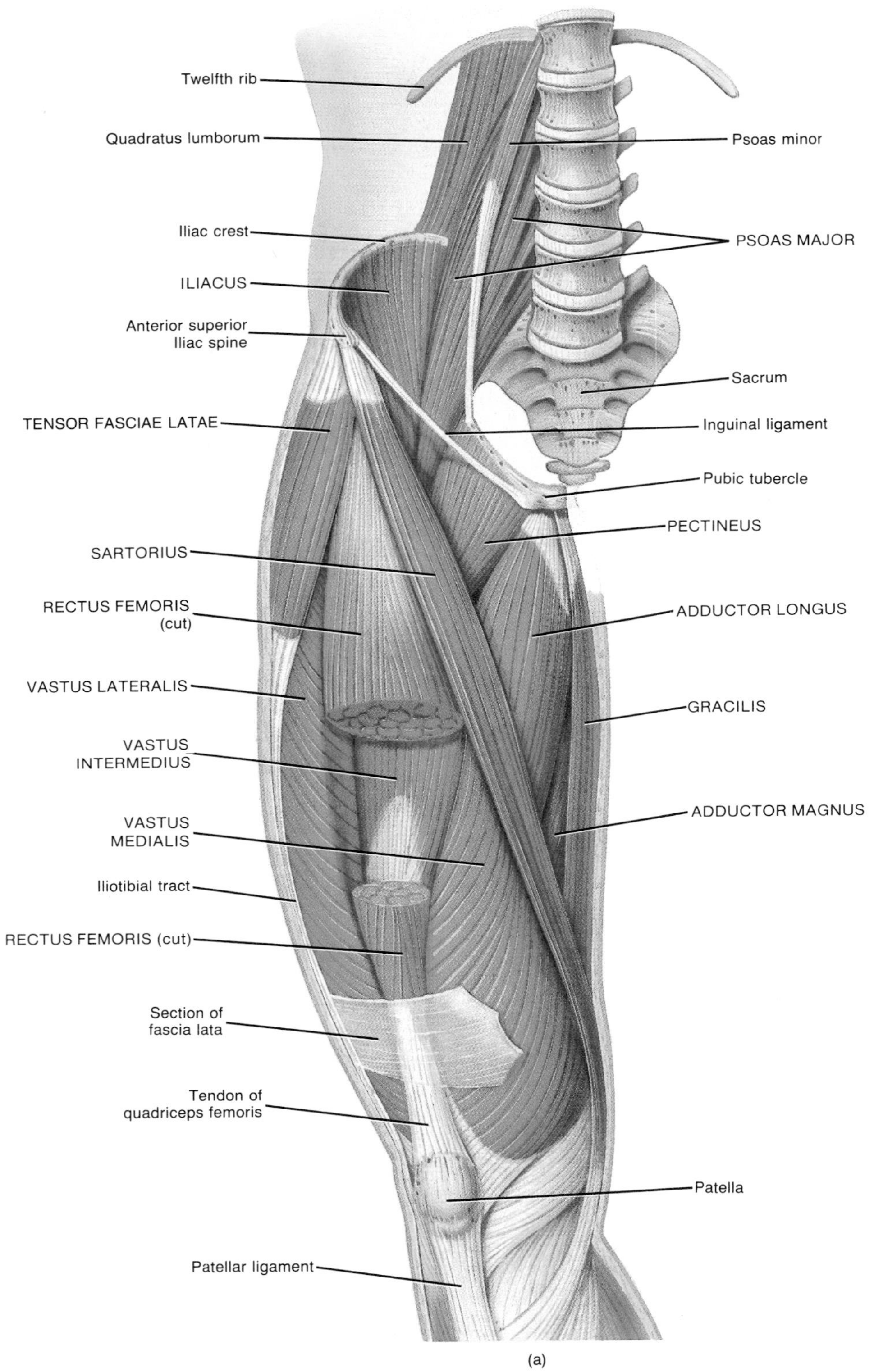

EXHIBIT 11-21 MUSCLES THAT MOVE THE THIGH (FEMUR) (*Continued*)

Muscle	Origin	Insertion	Action	Innervation
Adductor brevis (*brevis* = short)	Inferior ramus of pubis.	Upper half of linea aspera of femur.	Adducts, laterally rotates, and flexes thigh.	Obturator nerve.
Adductor magnus (*magnus* = large)	Inferior ramus of pubis and ischium to ischial tuberosity.	Linea aspera of femur.	Adducts, flexes, laterally rotates, and extends thigh (anterior part flexes; posterior part extends).	Obturator and sciatic nerves.
Pectineus (*pecten* = comb-shaped)	Superior ramus of pubis.	Pectineal line of femur, between lesser trochanter and linea aspera.	Flexes and adducts thigh.	Femoral nerve.

FIGURE 11-22 (*Continued*) (b) Anterior deep view.

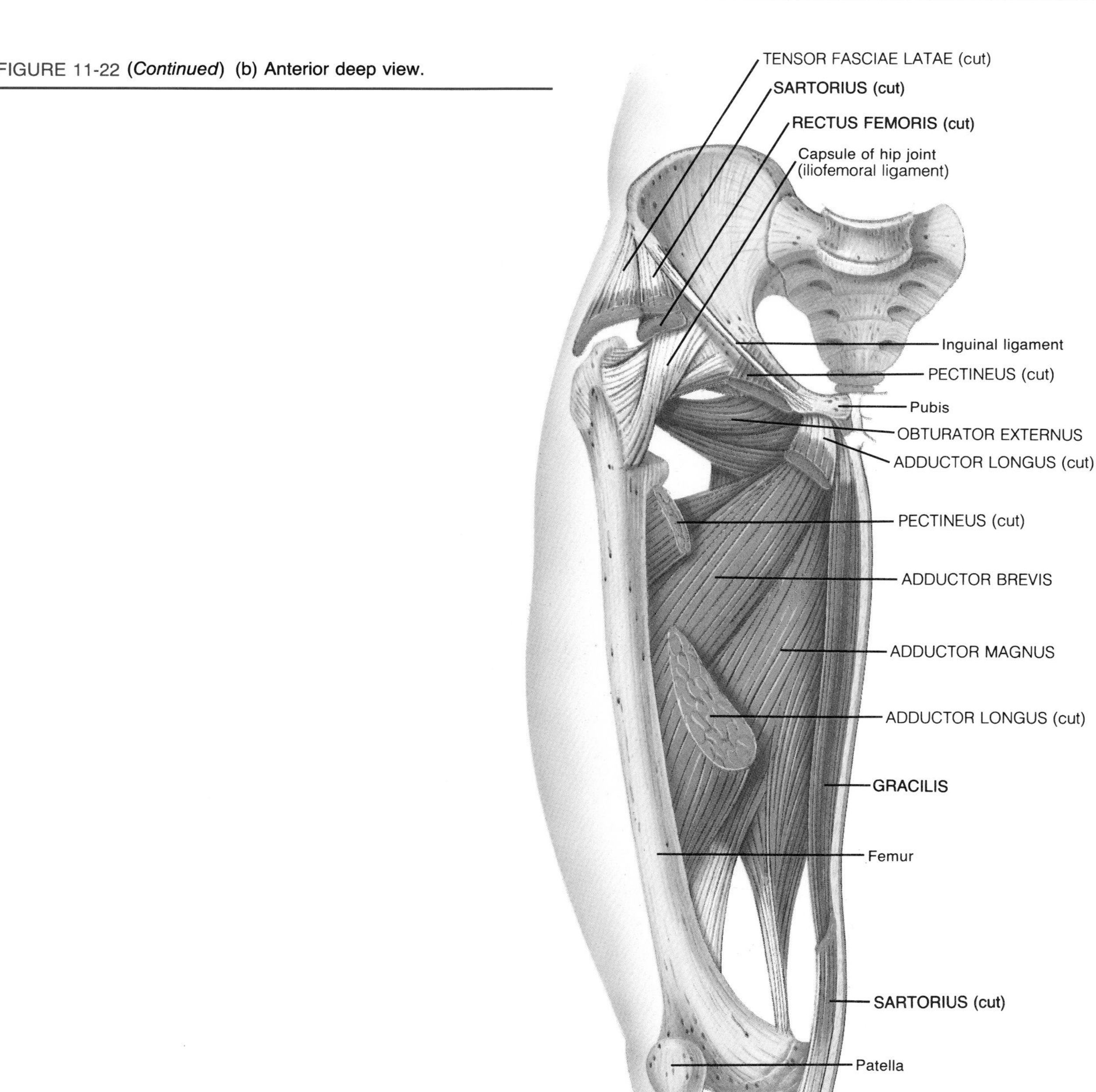

FIGURE 11-22 (*Continued*) (c) Posterior deep view.

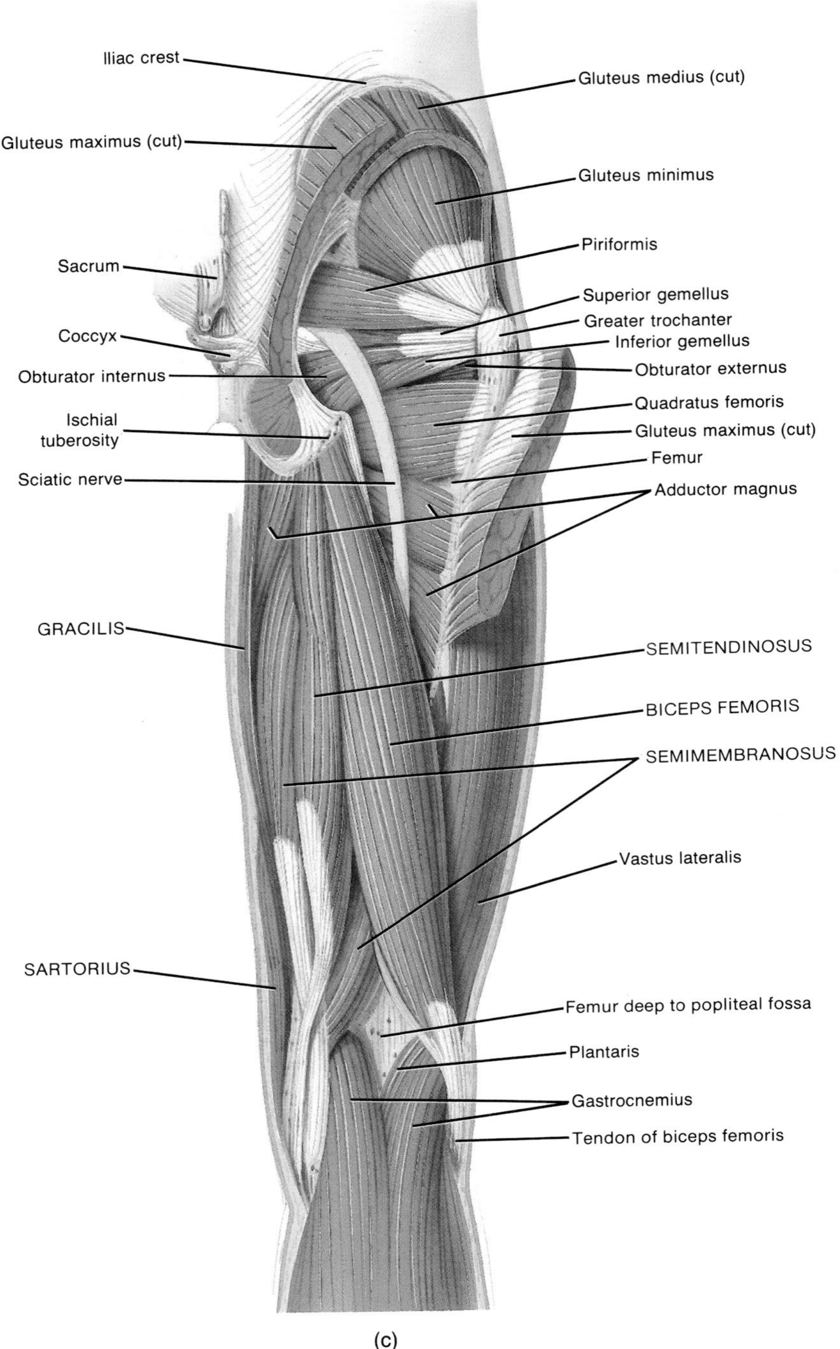
Iliac crest
Gluteus medius (cut)
Gluteus maximus (cut)
Gluteus minimus
Piriformis
Sacrum
Superior gemellus
Greater trochanter
Coccyx
Inferior gemellus
Obturator internus
Obturator externus
Quadratus femoris
Ischial tuberosity
Gluteus maximus (cut)
Femur
Sciatic nerve
Adductor magnus
GRACILIS
SEMITENDINOSUS
BICEPS FEMORIS
SEMIMEMBRANOSUS
Vastus lateralis
SARTORIUS
Femur deep to popliteal fossa
Plantaris
Gastrocnemius
Tendon of biceps femoris

(c)

FIGURE 11-22 (*Continued*) (d) Posterior deep view.

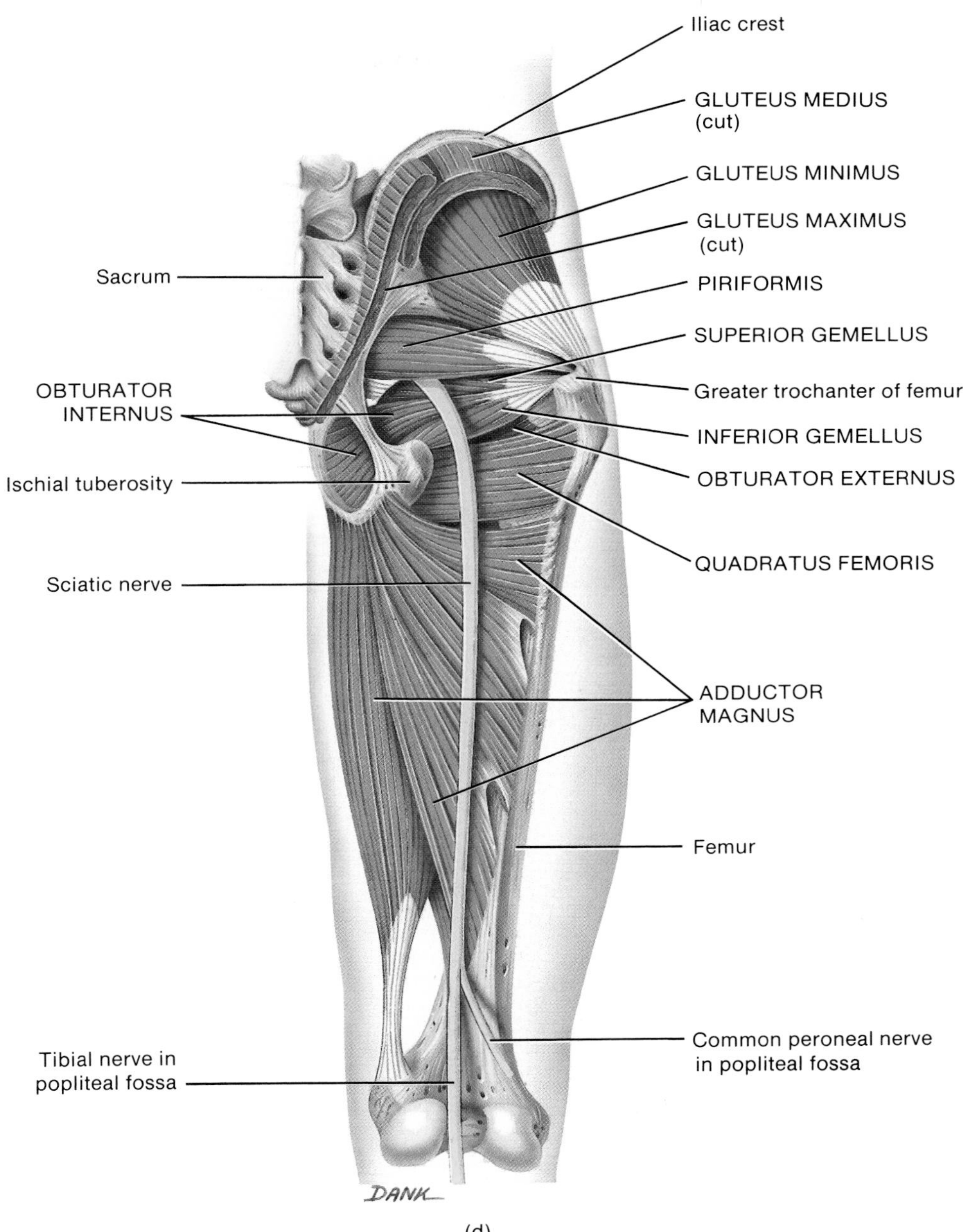

(d)

EXHIBIT 11-22 MUSCLES THAT ACT ON THE LEG (TIBIA AND FIBULA) (Figures 11-22 and 11-23)

Overview: The muscles that act on the leg (tibia and fibula) originate in the hip and thigh and are separated into compartments by deep fascia. The ***medial (adductor) compartment*** is so named because its muscles adduct the thigh. It is innervated by the obturator nerve. As noted earlier, the adductor magnus, adductor longus, adductor brevis, and pectineus muscles, components of the medial compartment, are included in Exhibit 11-21 because they act on the femur. The gracilis, the other muscle in the medial compartment, not only adducts the thigh but also flexes the leg. For this reason, it is included in this exhibit.

The ***anterior (extensor) compartment*** is so designated because its muscles act to extend the leg, and some also flex the thigh. It is composed of the quadriceps femoris and sartorius muscles and is innervated by the femoral nerve. The quadriceps femoris muscle is a composite muscle that includes four distinct parts, usually described as four separate muscles (rectus femoris, vastus lateralis, vastus intermedius, and vastus medialis). The common tendon for the four muscles is known as the ***patellar ligament*** and attaches to the tibial tuberosity. The rectus femoris and sartorius muscles are also flexors of the thigh.

The ***posterior (flexor) compartment*** is so named because its muscles flex the leg (but also extend the thigh). It is innervated by branches of the sciatic nerve. Included are the hamstrings (biceps femoris, semitendinosus, and semimembranosus). The hamstrings are so named because their tendons are long and stringlike in the popliteal area. The ***popliteal fossa*** is a diamond-shaped space on the posterior aspect of the knee bordered laterally by the tendons of the biceps femoris and medially by the semitendinosus and semimembranosus muscles.

Muscle	Origin	Insertion	Action	Innervation
MEDIAL (ADDUCTOR) COMPARTMENT				
Adductor magnus	See Exhibit 11-21			
Adductor longus	See Exhibit 11-21			
Adductor brevis	See Exhibit 11-21			
Pectineus	See Exhibit 11-21			
Gracilis (*gracilis* = slender)	Symphysis pubis and pubic arch.	Medial surface of body of tibia.	Adducts thigh and flexes leg.	Obturator nerve.
ANTERIOR (EXTENSOR) COMPARTMENT				
Quadriceps femoris (*quadriceps* = four heads of origin; *femoris* = femur)				
Rectus femoris (*rectus* = fibers parallel to midline)	Anterior inferior iliac spine.	Upper border of patella.	All four heads extend leg; rectus portion alone also flexes thigh.	Femoral nerve.
Vastus lateralis (*vastus* = large; *lateralis* = lateral)	Greater trochanter and linea aspera of femur.	Tibial tuberosity through patellar ligament (tendon of quadriceps).	All four heads extend leg; rectus portion alone also flexes thigh.	Femoral nerve.
Vastus medialis (*medialis* = medial)	Linea aspera of femur.	Tibial tuberosity through patellar ligament (tendon of quadriceps).	All four heads extend leg; rectus portion alone also flexes thigh.	Femoral nerve.
Vastus intermedius (*intermedius* = middle)	Anterior and lateral surfaces of body of femur.	Tibial tuberosity through patellar ligament (tendon of quadriceps).	All four heads extend leg; rectus portion alone also flexes thigh.	Femoral nerve.
Sartorius (*sartor* = tailor; refers to cross-legged position of tailors)	Anterior superior spine of ilium.	Medial surface of body of tibia.	Flexes leg; flexes thigh and rotates it laterally, thus crossing leg.	Femoral nerve.

EXHIBIT 11-22 MUSCLES THAT ACT ON THE LEG (TIBIA AND FIBULA) (*Continued*)

Muscle	Origin	Insertion	Action	Innervation
POSTERIOR (FLEXOR) COMPARTMENT				
Hamstrings	A collective designation for three separate muscles.			
Biceps femoris (*biceps* = two heads of origin)	Long head arises from ischial tuberosity; short head arises from linea aspera of femur.	Head of fibula and lateral condyle of tibia.	Flexes leg and extends thigh.	Tibial nerve from sciatic nerve.
Semitendinosus (*semi* = half; *tendo* = tendon)	Ischial tuberosity.	Proximal part of medial surface of body of tibia.	Flexes leg and extends thigh.	Tibial nerve from sciatic nerve.
Semimembranosus (*membran* = membrane)	Ischial tuberosity.	Medial condyle of tibia.	Flexes leg and extends thigh.	Tibial nerve from sciatic nerve.

FIGURE 11-23 Muscles that act on the leg. Cross section of the thigh, showing its musculature and related structures.

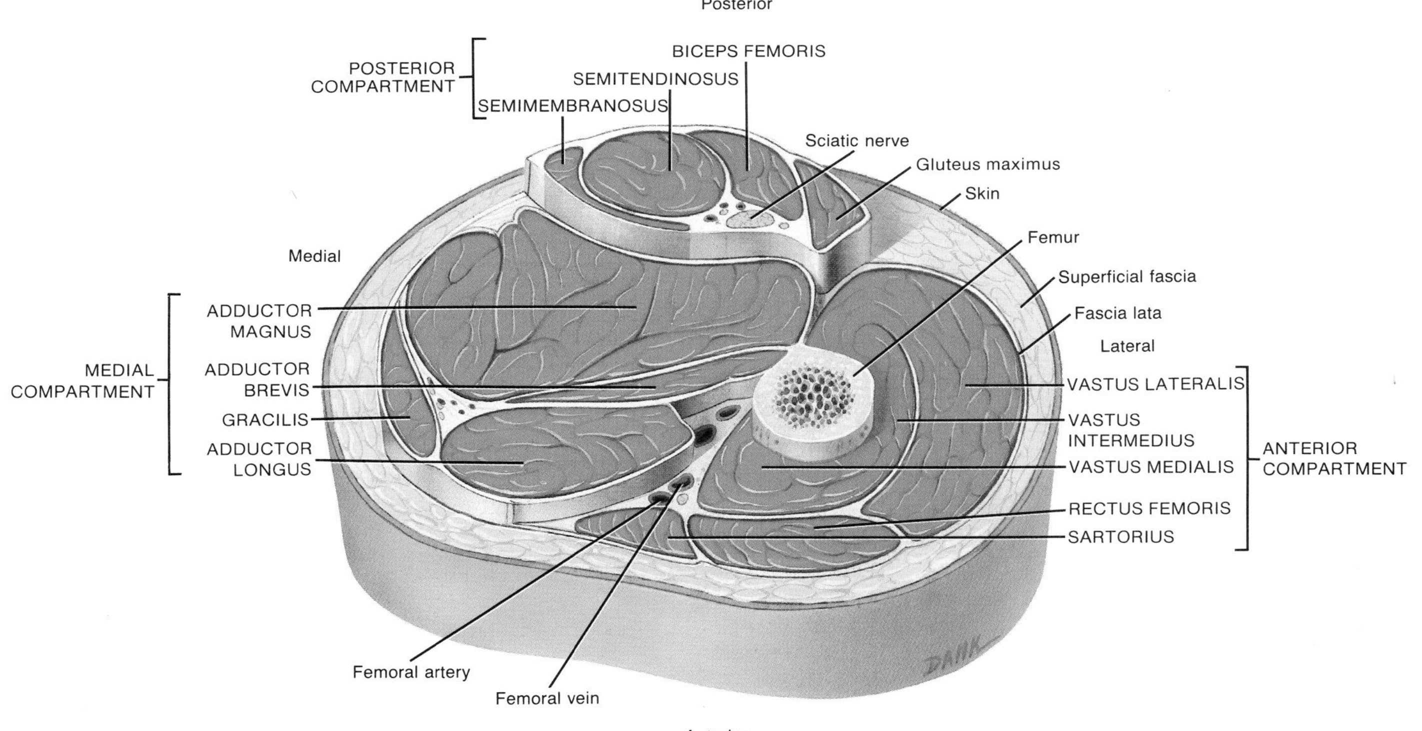

EXHIBIT 11-23 MUSCLES THAT MOVE THE FOOT AND TOES (Figure 11-24)

Overview: The musculature of the leg, like that of the thigh, can be distinguished into three compartments by deep fascia. In addition, all the muscles in the respective compartments are innervated by the same nerve. The ***anterior compartment*** consists of muscles that dorsiflex the foot and are innervated by the deep peroneal nerve. In a situation analogous to the wrist, the tendons of the muscles of the anterior compartment are held firmly to the ankle by thickenings of deep fascia called the ***superior extensor retinaculum (transverse ligament of the ankle)*** and ***inferior extensor retinaculum (cruciate ligament of the ankle).***

The ***lateral (peroneal) compartment*** contains two muscles that plantar flex and evert the foot. They are supplied by the superficial peroneal nerve.

The ***posterior compartment*** consists of muscles that are divisible into superficial and deep groups. All are innervated by the tibial nerve. All three superficial muscles share a common tendon of insertion, the calcaneal (Achilles) tendon that inserts into the calcaneus bone of the ankle. The superficial muscles are plantar flexors of the foot. Of the four deep muscles, three plantar flex the foot.

Muscle	Origin	Insertion	Action	Innervation
ANTERIOR COMPARTMENT				
Tibialis anterior (*tibialis* = tibia; *anterior* = front)	Lateral condyle and body of tibia and interosseous membrane.	First metatarsal and first (medial) cuneiform.	Dorsiflexes and inverts foot.	Deep peroneal nerve.
Extensor hallucis longus (*extensor* = increases angle at joint; *hallucis* = hallux or great toe; *longus* = long)	Anterior surface of fibula and interosseous membrane.	Distal phalanx of great toe.	Dorsiflexes and inverts foot and extends great toe.	Deep peroneal nerve.
Extensor digitorum longus	Lateral condyle of tibia, anterior surface of fibula, and interosseous membrane.	Middle and distal phalanges of four outer toes.	Dorsiflexes and everts foot and extends toes.	Deep peroneal nerve.
Peroneus tertius (*perone* = fibula; *tertius* = third)	Distal third of fibula and interosseous membrane.	Fifth metatarsal.	Dorsiflexes and everts foot.	Deep peroneal nerve.
LATERAL (PERONEAL) COMPARTMENT				
Peroneus longus	Head and body of fibula and lateral condyle of tibia.	First metatarsal and first cuneiform.	Plantar flexes and everts foot.	Superficial peroneal nerve.
Peroneus brevis (*brevis* = short)	Body of fibula.	Fifth metatarsal.	Plantar flexes and everts foot.	Superficial peroneal nerve.
POSTERIOR COMPARTMENT				
Superficial				
Gastrocnemius (*gaster* = belly; *kneme* = leg)	Lateral and medial condyles of femur and capsule of knee.	Calcaneus by way of calcaneal (Achilles) tendon.*	Plantar flexes foot and flexes leg.	Tibial nerve.
Soleus (*soleus* = sole of foot)	Head of fibula and medial border of tibia.	Calcaneus by way of calcaneal (Achilles) tendon.	Plantar flexes foot.	Tibial nerve.
Plantaris (*plantar* = sole of foot)	Femur above lateral condyle.	Calcaneus by way of calcaneal (Achilles) tendon.	Plantar flexes foot.	Tibial nerve.

* The calcaneal (Achilles) tendon, the ***strongest tendon of the body,*** is able to withstand a 1000-pound force without tearing. Despite this, however, the calcaneal tendon ruptures more frequently than any other tendon because of the tremendous pressures placed on it during competitive sports.

EXHIBIT 11-23 MUSCLES THAT MOVE THE FOOT AND TOES (*Continued*)

Muscle	Origin	Insertion	Action	Innervation
Deep				
Popliteus (*poples* = posterior surface of knee)	Lateral condyle of femur.	Proximal tibia.	Flexes and medially rotates leg.	Tibial nerve.
Flexor hallucis longus (*flexor* = decreases angle at joint)	Lower two-thirds of fibula.	Distal phalanx of great toe.	Plantar flexes and everts foot and flexes great toe.	Tibial nerve.
Flexor digitorum longus (*digitorum* = finger or toe)	Posterior surface of tibia.	Distal phalanges of four outer toes.	Plantar flexes and inverts foot and flexes toes.	Tibial nerve.
Tibialis posterior (*posterior* = back)	Tibia, fibula, and interosseous membrane.	Second, third, and fourth metatarsals; navicular; all three cuneiforms; and cuboid.	Plantar flexes and inverts foot.	Tibial nerve.

FIGURE 11-24 Muscles that move the foot and toes. (a) Superficial posterior view. (b) Deep posterior view.

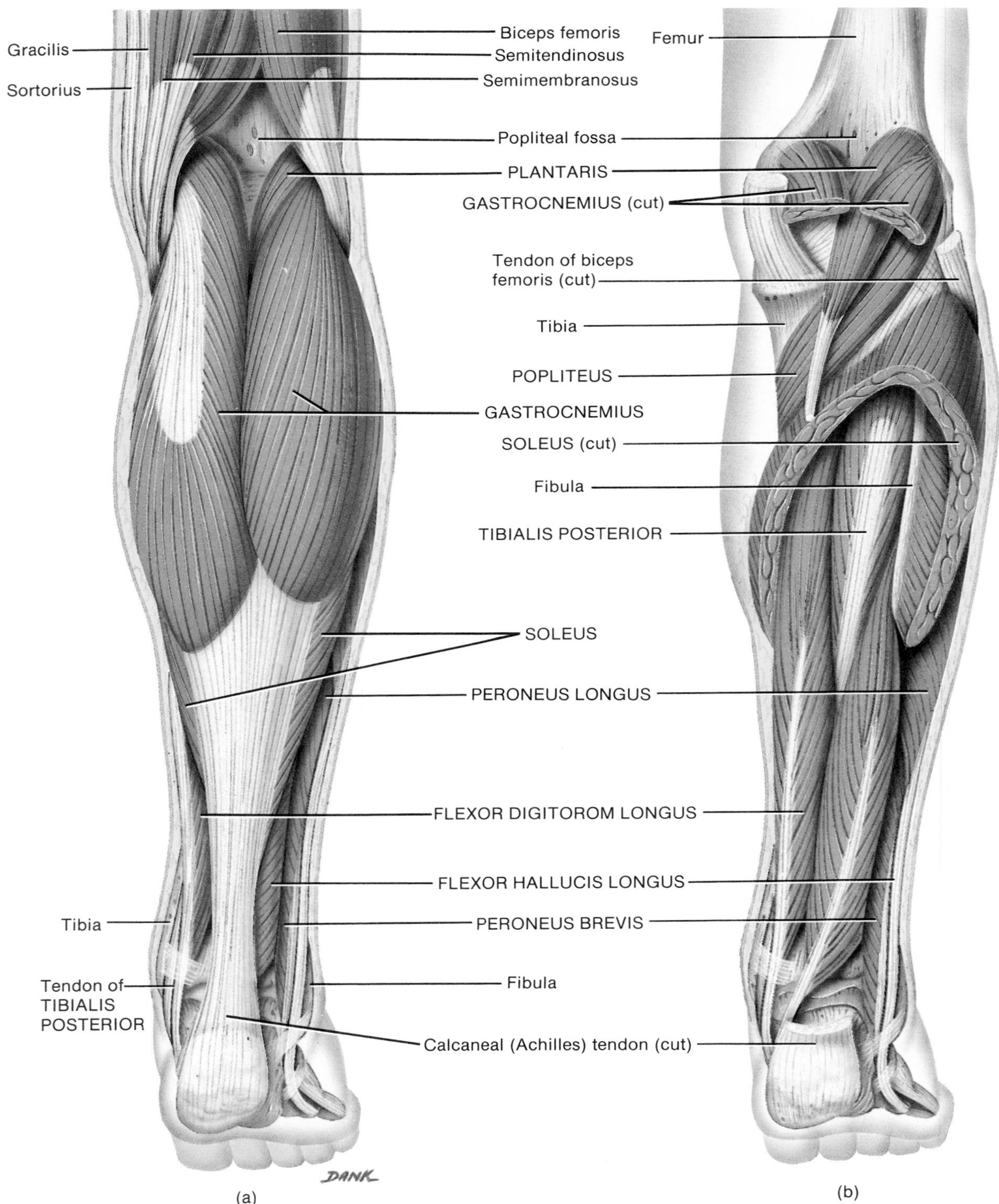

FIGURE 11-24 (*Continued*) (c) Superficial anterior view. (d) Superficial right lateral view.

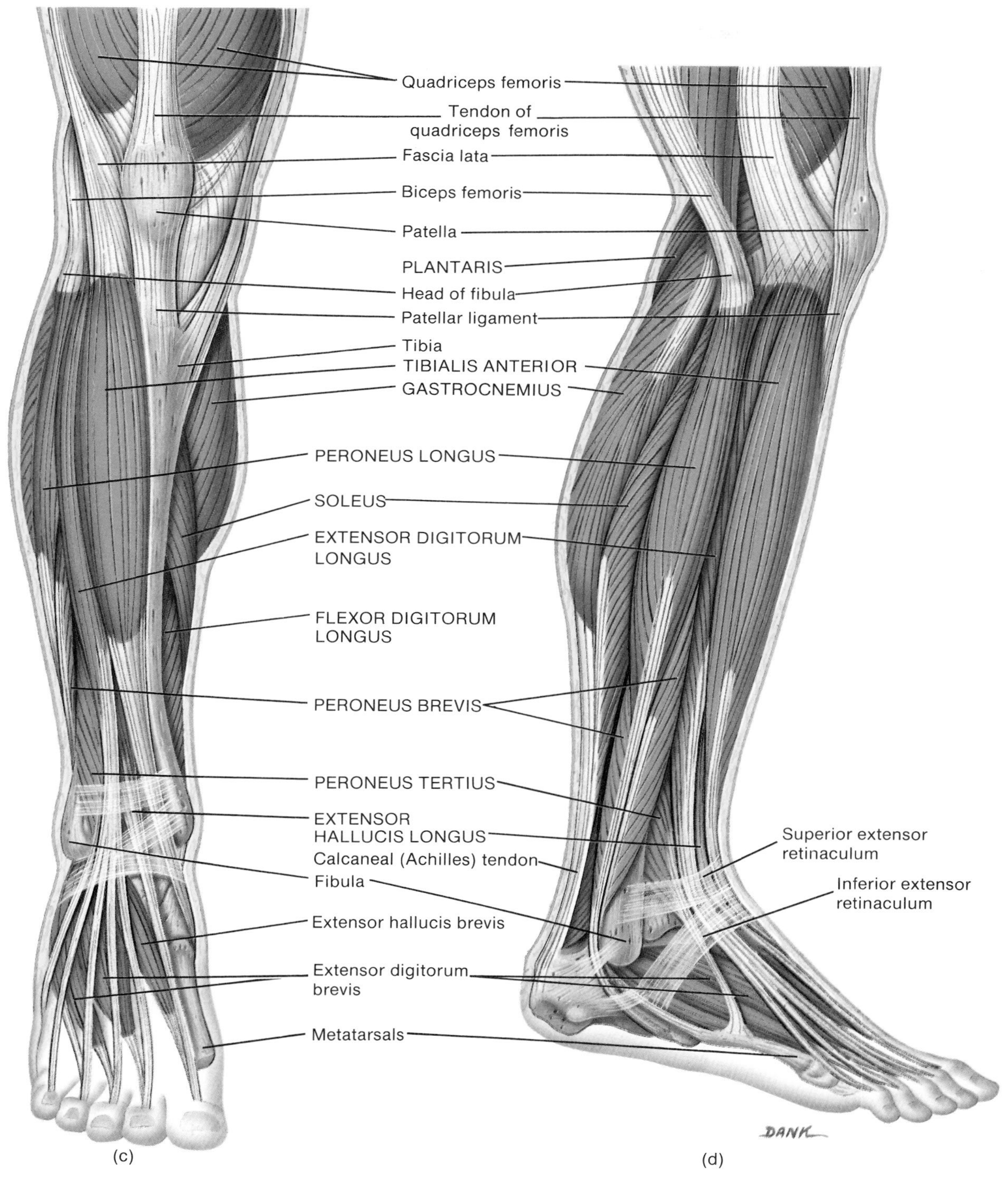

EXHIBIT 11-24 INTRINSIC MUSCLES OF THE FOOT (Figure 11-25)

Overview: The intrinsic muscles of the foot are, for the most part, comparable to those in the hand. Where the muscles of the hand are specialized for precise and intricate movements, those of the foot are limited to support and locomotion. The deep fascia of the foot forms the ***plantar aponeurosis (fascia)*** that extends from the calcaneus to the phalanges. The aponeurosis supports the longitudinal arch of the foot and transmits the flexor tendons of the foot.

The intrinsic musculature of the foot is divided into two groups—dorsal and plantar. There is only one ***dorsal muscle.*** The ***plantar muscles*** are arranged in four layers, the most superficial layer being referred to as the first layer.

Muscle	Origin	Insertion	Action	Innervation
DORSAL				
Extensor digitorum brevis* (*extensor* = increases angle at joint; *digit* = finger or toe; *brevis* = short)	Calcaneus.	Tendon of extensor digitorum longus and proximal phalanx of great toe.	Extends first through fourth toes.	Deep peroneal nerve.
PLANTAR				
First (superficial) Layer				
Abductor hallucis (*abductor* = moves part away from midline; *hallucis* = hallux or great toe)	Calcaneus and plantar aponeurosis.	Proximal phalanx of great toe.	Abducts great toe and flexes metatarsophalangeal joint.	Medial plantar nerve.
Flexor digitorum brevis (*flexor* = decreases angle at joint)	Calcaneus and plantar aponeurosis.	Middle phalanx of second through fifth toes.	Flexes second through fifth toes.	Medial plantar nerve.
Abductor digiti minimi (*minimi* = small toe)	Calcaneus and plantar aponeurosis.	Proximal phalanx of small toe.	Abducts and flexes small toe.	Lateral plantar nerve.
Second layer				
Quadratus plantae (*quad* = four; *planta* = sole of foot)	Calcaneus.	Tendons of flexor digitorum longus.	Flexes second through fifth toes.	Lateral plantar nerve.
Lumbricals	Tendons of flexor digitorum longus.	Tendons of extensor digitorum longus.	Extend second through fifth toes.	Medial and lateral plantar nerves.
Third layer				
Flexor hallucis brevis	Cuboid and third (lateral) cuneiform.	Proximal phalanx of great toe.	Flexes great toe.	Medial plantar nerve.
Adductor hallucis	Second through fourth metatarsals and ligaments of metatarsophalangeal joints.	Proximal phalanx of great toe.	Adducts and flexes great toe.	Lateral plantar nerve.

* Not illustrated.

EXHIBIT 11-24 INTRINSIC MUSCLES OF THE FOOT (*Continued*)

Muscle	Origin	Insertion	Action	Innervation
Flexor digiti minimi brevis	Fifth metatarsal.	Proximal phalanx of small toe.	Flexes small toe.	Lateral plantar nerve.
Fourth (deep) layer				
Dorsal interossei	Adjacent side of metatarsals.	Proximal phalanges, both sides of second toe, lateral side of third and fourth toes.	Abduct toes and flex proximal phalanges.	Lateral plantar nerve.
Plantar interossei	Third, fourth, and fifth metatarsals.	Proximal phalanges of same toes.	Adduct third, fourth, and fifth toes and flex proximal phalanges.	Lateral plantar nerve.

FIGURE 11-25 Intrinsic muscles of the foot. (a) Plantar view showing some superficial and deeper muscles. (b) Plantar view showing deeper muscles.

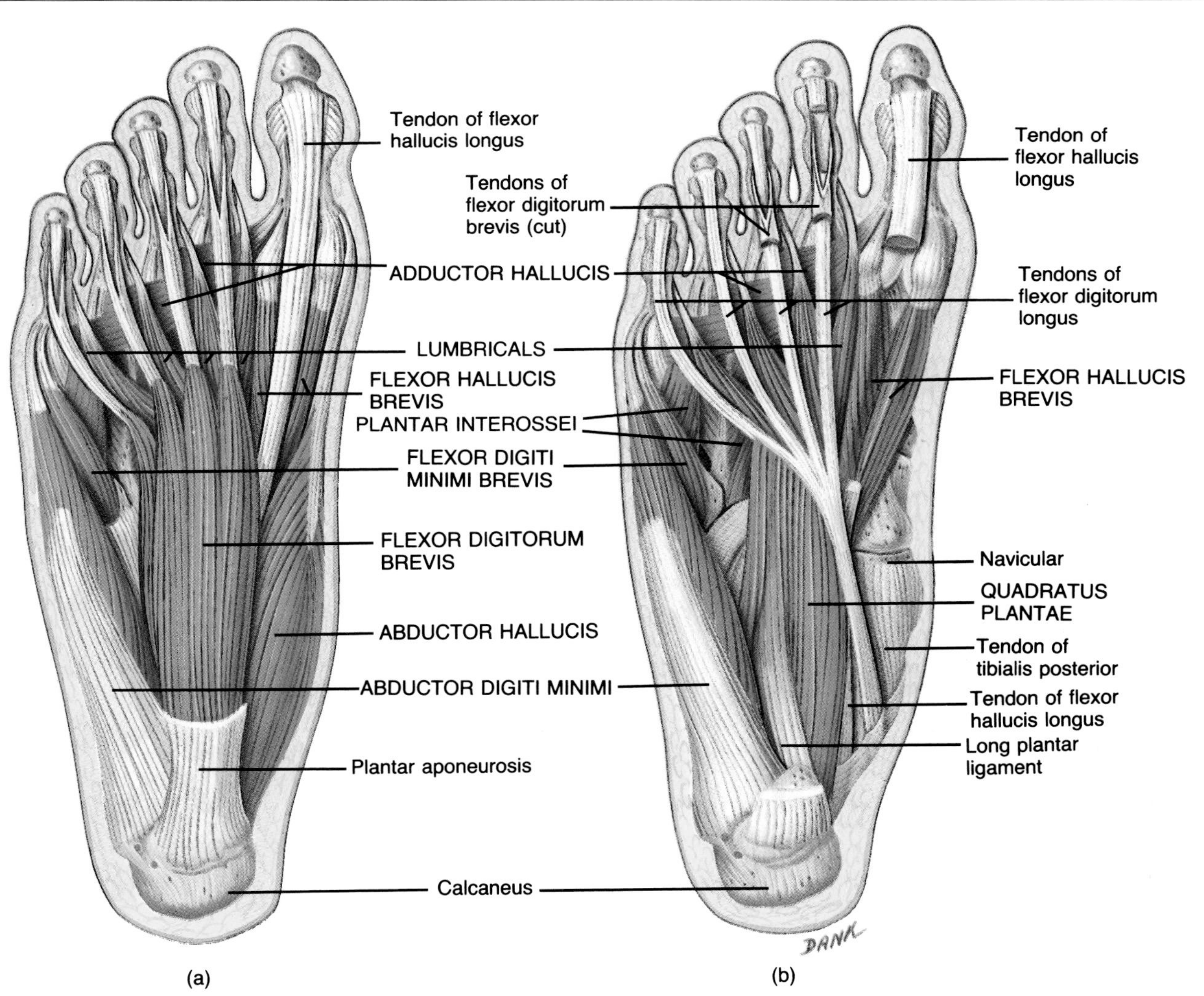

FIGURE 11-25 (*Continued*) (c) Plantar view showing plantar interossei. (d) Plantar view showing dorsal interossei.

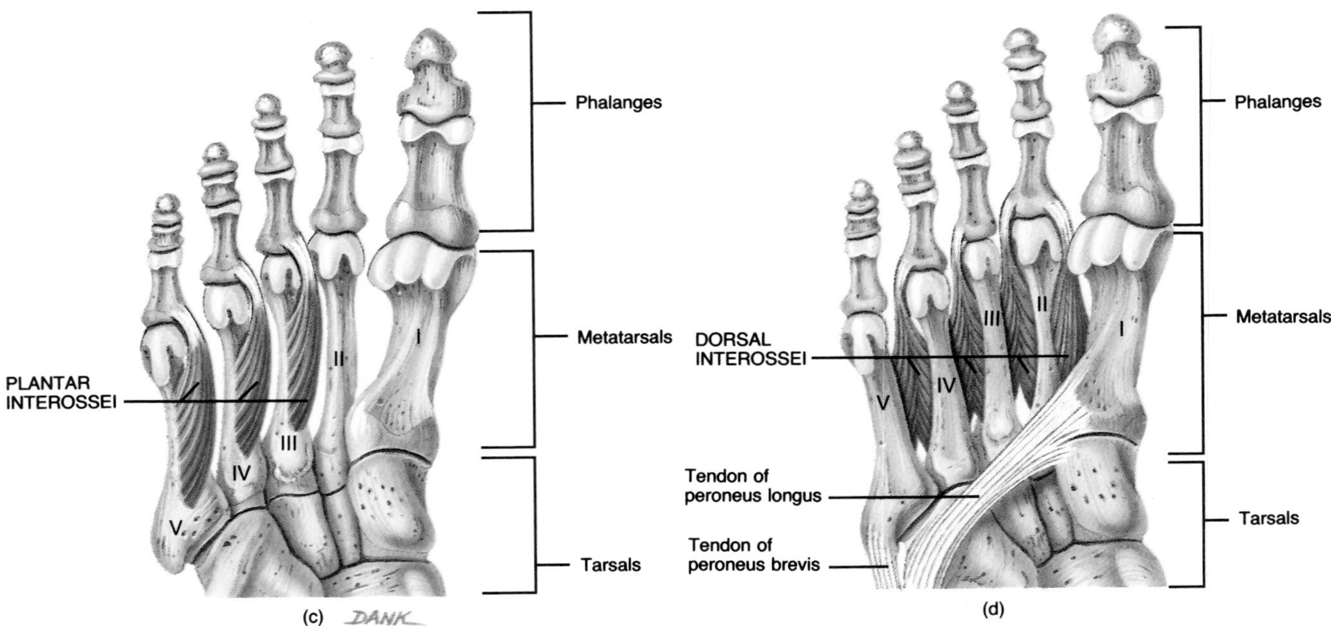

INTRAMUSCULAR (IM) INJECTIONS

An ***intramuscular (IM) injection*** penetrates the skin and subcutaneous tissue to enter the muscle itself. Intramuscular injections are preferred when prompt absorption is desired, when larger doses than can be given cutaneously are indicated, or when the drug is too irritating to give subcutaneously. The common sites for intramuscular injections include the buttock, lateral side of the thigh, and the deltoid region of the arm. Muscles in these areas, especially the gluteal muscles in the buttock, are fairly thick. Because of the large number of muscle fibers and extensive fascia, the drug has a large surface area for absorption. Absorption is further promoted by the extensive blood supply to muscles. Ideally, intramuscular injections should be given deep within the muscle and away from major nerves and blood vessels.

For many intramuscular injections, the preferred site is the ***gluteus medius muscle*** of the buttock (Figure 11-26a). The buttock is divided into quadrants, and the upper outer quadrant is used as the injection site. The iliac crest serves as a landmark for this quadrant. The spot for injection in an adult is usually about 5 to $7\frac{1}{2}$ cm (2 to 3 inches) below the iliac crest. The upper outer quadrant is chosen because the muscle in this area is quite thick and has few nerves. Injection in this area thus reduces the chance of injury to the sciatic nerve, which can cause paralysis of the lower extremity. The probability of injecting the drug into a blood vessel is also remote in this area. After the needle is inserted into the gluteus medius muscle, the plunger is pulled up for a few seconds. If the syringe fills with blood, the needle is in a blood vessel, and a different injection site on the opposite buttock is chosen. An IM antibiotic, such as penicillin, is always injected into the gluteal region.

Injections for small infants and toddlers may be given in the lateral side of the thigh in the midportion of the ***vastus lateralis muscle*** (Figure 11-26b). This site is determined by using the knee and greater trochanter of the femur as landmarks. The midportion of the muscle is located by measuring a handbreadth above the knee and a handbreadth below the greater trochanter.

The ***deltoid*** injection is given in the midportion of the muscle about two to three fingerbreadths below the acromion of the scapula and lateral to the axilla (Figure 11-26c). Deltoid IM injections in the adult are usually for immunization (tetanus toxoid booster, pneumococcal vaccine), whereas subcutaneous injections in the deltoid region include typhoid immunization and inactivated poliovirus vaccine.

RUNNING INJURIES

It is estimated that nearly 70 percent of individuals who jog or run will sustain some type of running-related injury. Even though most such injuries are minor, such as sprains

FIGURE 11-26 **Intramuscular injections. Shown are the three common sites for intramuscular injections. (a) Buttock. (b) Lateral surface of the thigh. (c) Deltoid region of the arm.**

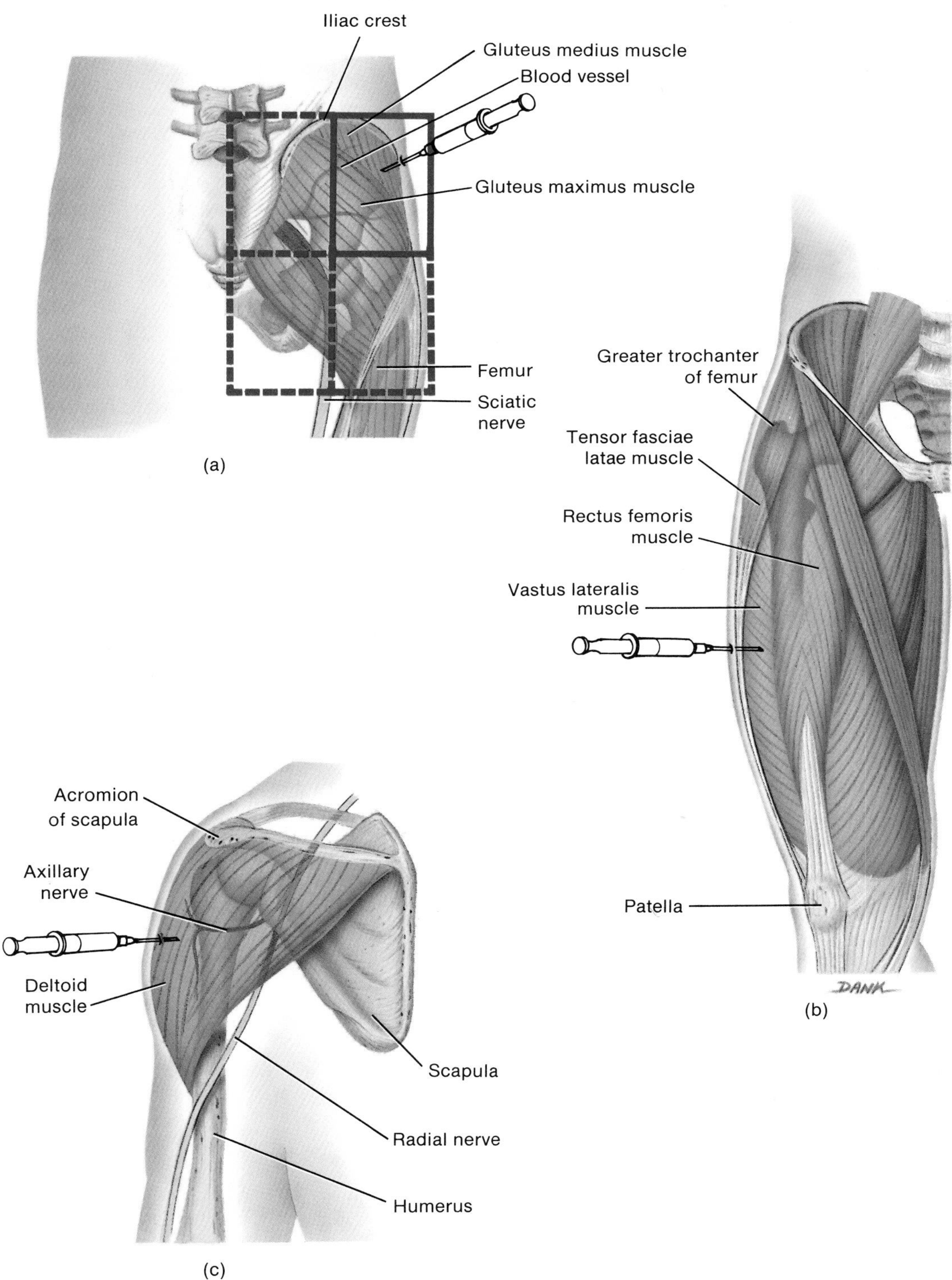

and strains, some are quite serious; moreover, untreated or inappropriately treated minor injuries may become chronic. Among runners, the knee is the most common site of injury, accounting for about 40 percent of injuries. Other common sites of injury, each accounting for about 15 percent of all injuries, include the calcaneal (Achilles) tendon, medial aspect of the tibia, hip area, and groin area. Foot and ankle injuries account for about 10 percent of running-related injuries, and about 5 percent involve the back.

Running injuries are frequently related to faulty training techniques. This may involve improper or lack of warm-up, running too much, or running too soon. Or it might involve running on hard and/or uneven surfaces. Poorly constructed or worn-out running shoes can also contribute to injury. Any biomechanical problems aggravated by running can also cause injuries. Most running injuries can be treated early with rest, ice, moist heat, and nonsteroidal antiinflammatory drugs (NSAIDs) together with an alternate fitness program.

Hip, Buttock, and Back Injuries

When back pain occurs in runners, it is frequently due to a preexisting degenerative condition that is aggravated by an increase in mileage or hill running. Such pain is usually due to an injury in the buttocks, pelvis, or lumbar spine. For example, the pain may be caused by strain of the distal attachments of the abductors (especially the gluteus medius) or the proximal attachment of the abductors to the iliac crest, both of which contribute to a ***"pulled groin";*** inflammation of bursae deep to the gluteus medius or gluteus maximus muscles ***(trochanteric bursitis);*** ischial fracture caused by severe contraction of the hamstrings; strain or partial tear of the proximal hamstrings (***hamstring strain*** or ***"pulled hamstrings"***); nerve root compression in the lumbar spine; or compression of the sciatic nerve by the piriformis muscle. Hamstring strains ("pulled hamstrings") are common sports injuries in individuals who run very hard. Sometimes the violent muscular exertion required to perform a feat tears off part of the tendinous origins of the hamstrings, especially the biceps femoris, from the ischial tuberosity. This is usually accompanied by a contusion (bruising) and tearing of some of the muscle fibers and rupture of blood vessels, producing a hematoma (collection of blood), and pain. Adequate training with good balance between the quadriceps femoris and hamstrings and stretching exercises before running or competing are important in preventing this injury.

Treatment consists of rest, ice, moist heat, NSAIDs, and sometimes local injections of corticosteroids.

Knee Injuries

Patellofemoral Stress Syndrome ("Runner's Knee")

Patellofemoral stress syndrome ("runner's knee") is the single most common problem in runners. During normal flexion and extension of the knee, the patella tracks (glides) up and down in the groove between the femoral condyles. In patellofemoral stress syndrome, normal tracking does not occur; instead, the patella tracks laterally, and the increased pressure of abnormal tracking causes the associated pain. The pain is usually described as an aching or tenderness around or under the patella. The pain typically occurs after a person has been sitting for a while, especially after exercise. The syndrome may be due to excessive pronation of the foot or tight hamstring muscles. A common cause of runner's knee is constantly walking, running, or jogging on the same side of the road. Since roads are high in the middle and slope down on the sides, the slope stresses the knee that is closer to the center of the road.

Treatment of patellofemoral syndrome consists of the application of first ice, then moist heat; cessation of running; avoidance of kneeling, stair climbing, and prolonged sitting; hamstring muscle stretching exercises; short-arc quadriceps strengthening exercises; and use of an orthotic device for the foot, knee wraps, and knee braces.

Iliotibial Tract Friction Syndrome

Another knee injury is ***iliotibial tract friction syndrome.*** As you may recall, the iliotibial tract is the thickened portion of the tensor fasciae latae muscle that passes down the lateral surface of the thigh and inserts on the lateral tibial condyle (see Figure 11-22). During running, the friction caused by the iliotibial tract rubbing against the lateral femoral condyle may cause inflammation and pain that result in the syndrome. This disorder usually occurs in runners who are bowlegged with pronated feet and in individuals who wear shoes with worn lateral soles. Treatment consists of conservative measures (ice, moist heat, NSAIDs), replacement of worn shoes, and exercises to stretch the iliotibial tract. Corticosteroid injections also help.

Leg and Foot Injuries

Shinsplint Syndrome

Shinsplint syndrome refers to pain or soreness along the tibia, specifically the medial, distal two-thirds. It may be caused by tendinitis of the posterior tibialis muscle or toe flexors, inflammation of the periosteum (periostitis) around the tibia, or stress fractures of the tibia. The tendinitis usually occurs when poorly conditioned runners run on hard or banked surfaces with poorly supportive running shoes. The condition may also occur as a result of vigorous activity of the legs following a period of relative inactivity. Rest, ice, and moist heat usually alleviate the pain. In addition, the muscles in the anterior compartment (mainly the tibialis anterior) can be strengthened to balance the stronger posterior compartment muscles. Patients who do not respond may be given local injections of antiinflammatory steroid drugs or may have to undergo minor surgery to release pressure in the soft tissues around the bone.

Anterior Compartment Syndrome

Anterior compartment syndrome refers to overload of the anterior compartment muscles of the leg (see Figure 11-23) that results in pain along the muscles or along the extensor tendons of the ankle and foot. The condition usually occurs when a runner changes running style from flat-footed to fore-footed, begins interval training on a track or hill, or wears shoes with an overly flexible sole. Treatment consists of stretching exercises, proper training techniques, and strengthening exercises for the anterior compartment muscles. In some cases, minor surgery is performed to permit expansion of the muscles in the anterior compartment.

Achilles Tendinitis

Achilles tendinitis refers to pain, with or without swelling, in the area of the calcaneal (Achilles) tendon. The condition may be caused by downhill running, toeing-off during uphill running, improperly fitted running shoes, or various biomechanical problems. Rest, ice massage, exercises that stretch the calcaneal (Achilles) tendon and anterior compartment muscles, ultrasound, and orthotic devices are used in treatment.

Plantar Fasciitis (Painful Heel Syndrome)

Plantar fasciitis (painful heel syndrome) is an inflammatory reaction due to chronic irritation of the plantar aponeurosis (fascia) at its origin on the calcaneus (heel bone) (see Figure 11-25). The condition is the most common cause of heel pain in runners and arises in response to the repeated impact of walking or running. Treatment consists of rest, heel pads, ice massage, exercises that gently stretch calf muscles, NSAIDs, and in some cases, corticosteroid injections.

Stress Fractures

Stress fractures are partial fractures that result from inability to withstand repeated stress owing to a change in training, harder surfaces, longer distances, greater speed, or an existing pathology. Such fractures can occur in the bodies of lumbar vertebrae, sacroiliac joint, symphysis pubis, iliac crest, femoral neck and body, fibula, lateral malleolus, and metatarsals. About 25 percent of all stress fractures involve the fibula, specifically the distal third. With all stress fractures, running must stop temporarily and immobilization may be needed.

STUDY OUTLINE

How Skeletal Muscles Produce Movement (p. 260)

1. Skeletal muscles produce movement by pulling on bones.
2. The attachment to the stationary bone is the origin. The attachment to the movable bone is the insertion.
3. Bones serve as levers and joints serve as fulcrums. The lever is acted on by two different forces: resistance and effort.
4. Levers are categorized into three types—first-class, second-class, and third-class—according to the position of the fulcrum, effort, and resistance on the lever.
5. Fascicular arrangements include parallel, convergent, pennate, and circular. Fascicular arrangement is correlated with the power of a muscle and the range of movement.
6. The prime mover produces the desired action. The antagonist produces an opposite action. The synergist assists the prime mover by reducing unnecessary movement. The fixator stabilizes the origin of the prime mover so that it can act more efficiently.

Naming Skeletal Muscles (p. 263)

1. Skeletal muscles are named on the basis of distinctive criteria: direction of fibers, location, size, number of origins (or heads), shape, origin and insertion, and action.

Principal Skeletal Muscles (p. 263)

1. The principal skeletal muscles of the body are grouped according to region in Exhibits 11-2 through 11-24.

Intramuscular (IM) Injections (p. 322)

1. Advantages of intramuscular injections are prompt absorption, use of larger doses than can be given cutaneously, and minimal irritation.
2. Common sites for intramuscular injections are the buttock, lateral side of the thigh, and deltoid region of the arm.

Running Injuries (p. 322)

1. Most running injuries involve the knee. Other commonly injured sites are the calcaneal (Achilles) tendon, medial aspect of tibia, hip, groin, foot, ankle, and back.
2. Running injuries are frequently treated by rest, ice, moist heat, nonsteroidal antiinflammatory drugs (NSAIDs), and an alternate fitness program.

REVIEW QUESTIONS

1. What is meant by the muscular system? (p. 260)
2. Using the terms origin, insertion, and belly in your discussion, describe how skeletal muscles produce body movements by pulling on bones. (p. 260)
3. What is a lever? Fulcrum? Apply these terms to the body and indicate the nature of the forces that act on levers. Describe the three classes of levers and provide one example for each in the body. (p. 261)
4. Describe the various arrangements of fasciculi. How is fascicular arrangement correlated with the strength of a muscle and its range of movement? (p. 261)
5. Define the role of the prime mover (agonist), antagonist, synergist, and fixator in producing body movements. (p. 263)
6. Select at random several muscles presented in Exhibits 11-2 through 11-24 and see if you can determine the criterion or criteria employed for naming each. In addition, refer to the prefixes, suffixes, roots, and definitions in each exhibit as a guide. Select as many muscles as you wish, as long as you feel you understand the concept involved.
7. What muscles would you use to do the following: (a) frown, (b) pout, (c) show surprise, (d) show your upper teeth, (e) pucker your lips, (f) squint, (g) blow up a balloon, (h) smile? (p. 267)
8. What are the principal muscles that move the mandible? Give the function of each. (p. 270)
9. What would happen if you lost tone in the masseter and temporalis muscles? (p. 270)
10. What extrinsic muscles move the eyeball? In which direction does each muscle move the eyeball? (p. 272)
11. Describe the action of each of the muscles acting on the tongue. (p. 274)
12. What tongue, facial, and mandibular muscles would you use when chewing food? (p. 267, 270, 274)
13. What muscles tense and elevate the soft palate? (p. 275)
14. What muscles constrict the pharynx? What muscle dilates the pharynx? Distinguish the circular and longitudinal layers. (p. 276)
15. Describe the muscles involved, and their actions, in moving the hyoid bone. (p. 278)
16. Describe the actions of the extrinsic and intrinsic muscles of the larynx. (p. 279)
17. What muscles are responsible for moving the head? And how do they move the head? (p. 281)
18. What muscles would you use to signify "yes" and "no" by moving your head? (p. 282)
19. Describe the composition of the anterolateral and posterior abdominal wall. (p. 283)
20. What muscles accomplish compression of the abdominal wall? (p. 283)
21. What are the principal muscles involved in breathing? What are their actions? (p. 285)
22. Describe the actions of the muscles of the pelvic floor. What is the pelvic diaphragm? (p. 285)
23. Describe the actions of the muscles of the perineum. What is the urogenital diaphragm? (p. 288)
24. In what directions is the pectoral (shoulder) girdle drawn? What muscles accomplish these movements? (p. 290)
25. What muscles are used to raise your shoulders, lower your shoulders, join your hands behind your back, and join your hands in front of your chest? (p. 290)
26. What movements are possible at the shoulder joint? What muscles accomplish these movements? (p. 290)
27. Distinguish axial and scapular muscles involved in moving the arm. (p. 293)
28. What muscles move the arm? In which directions do these movements occur? (p. 293)
29. Organize the muscles that move the forearm into flexors and extensors. What muscles move the forearm and what actions are used when striking a match? (p. 296)
30. Discuss the various movements possible at the wrist, hand, and fingers. What muscles accomplish these movements? (p. 298)
31. How many muscles and actions of the wrist, hand, and fingers that are used when writing can you list? (p. 298, 303)
32. What is the flexor retinaculum? Extensor retinaculum? (p. 298)
33. What muscles form the thenar eminence? What are their functions? (p. 302)
34. What muscles form the hypothenar eminence? What are their functions? (p. 302)
35. Discuss the various muscles and movements of the vertebral column. How are the muscles grouped? (p. 305)
36. What muscles accomplish movements of the thigh? What actions are produced by these muscles? What is the iliotibial tract? (p. 309)
37. Organize the mucles that act on the leg into medial, anterior, and posterior compartments. What is the popliteal fossa? (p. 314)
38. What muscles act at the knee joint? What kinds of movements do these muscles perform? (p. 314)
39. Determine the muscles and their actions listed in Exhibit 11-22 that you would use to climb a ladder to a diving board, dive into the water, swim the length of a pool, and then sit at pool side. (p. 314)
40. Name the muscles that plantar flex, evert, pronate, and dorsiflex the foot. What is the superior extensor retinaculum? Inferior extensor retinaculum? (p. 316)
41. How are the intrinsic muscles of the foot organized? (p. 320)
42. In which directions are the toes moved? What muscles bring about these movements? (p. 320)
43. What are the advantages of intramuscular (IM) injections? Describe how you would locate the sites for an intramuscular injection in the buttock, lateral side of the thigh, and deltoid region of the arm. (p. 322)
44. List the common sites for running injuries. How are most running injuries treated? (p. 322)
45. Define the following: "pulled hamstrings (p. 324)," "pulled groin (p. 324)," patellofemoral stress syndrome (p. 324), iliotibial tract friction syndrome, shinsplint syndrome (p. 324), anterior compartment syndrome (p. 325), Achilles tendonitis (p. 325), plantar fasciitis (p. 325), and stress fracture. (p. 325)

SELECTED READINGS

Brody, D. M. "Running Injuries: Prevention and Management," *CIBA-GEIGY Clinical Symposia,* 39 (3), 1987.

Clemente, C. D. *Anatomy: A Regional Atlas of the Human Body,* 3rd ed. Baltimore: Urban & Schwarzenberg, 1987.

Ferner, H., and J. Staubesand (eds). *Sobotta Atlas of Human Anatomy.* Vols 1–2. Baltimore: Urban & Schwarzenberg, 1983.

Gosling, J. A., P. F. Harris, J. R. Humpherson, I. Whitmore, and P. L. T. Willan. *Atlas of Human Anatomy.* Philadelphia: Lippincott, 1985.

Hinson, M. M. *Kinesiology.* Dubuque: Wm. C. Brown, 1977.

Lavin, R. J. "The High Pressure Demands of Compartment Syndrome," *RN,* February 1989.

Moore, M. P. "Shinsplints," *Postgraduate Medicine,* January 1988.

Netter, F. H. *Musculoskeletal System: Anatomy, Physiology, and Metabolic Disorders.* Summit: CIBA-GEIGY Corporation, 1987.

Rohen, J. W., and C. Yokochi. *Color Atlas of Anatomy.* New York, Tokyo: Igaku-Shoin, Ltd., 1983.

UNIT 3

CONTROL SYSTEMS OF THE HUMAN BODY

This unit will show you the significance of the nerve impulse in making rapid adjustments for maintaining homeostasis. You will learn how the nervous system detects changes in the environment, selects a course of action, and responds to the changes. We will also investigate the role of hormones in maintaining long-term homeostasis.

Chapter 12

Nervous Tissue

Chapter Contents at a Glance

Student Objectives

1. Identify the three basic functions of the nervous system in maintaining homeostasis.
2. Classify the organs of the nervous system into central and peripheral divisions.
3. Contrast the histological characteristics and functions of neuroglia and neurons.
4. List the sequence of events involved in the generation and conduction of a nerve impulse.
5. Define the all-or-none principle of nerve impulse transmission.
6. Discuss the factors that determine the speed of nerve impulse conduction.
7. Define a synapse and list the factors involved in the conduction of a nerve impulse across a synapse.
8. Compare the functions of excitatory transmitter-receptor interactions and inhibitory transmitter-receptor interactions in helping to maintain homeostasis.
9. List the necessary conditions for the regeneration of nervous tissue.
10. Explain the organization of neurons in the nervous system.

The ***nervous system*** is one of the body's principal control and integrating centers. The other is the endocrine system. In humans, the nervous system serves three broad functions: sensory, integrative, and motor. First, it senses certain changes within the body and in the outside environment; this is its sensory function. Second, it interprets the changes; this is its integrative function. Third, it responds to the interpretation by initiating action in the form of muscular contractions or glandular secretions; this is its motor function.

Through sensation, integration, and response, the nervous system represents the body's most rapid means of maintaining homeostasis. Its split-second reactions, carried out by nerve impulses, can normally make the adjustments necessary to keep the body functioning efficiently. As you will see later, the nervous system shares the maintenance of homeostasis with the endocrine system. Although the adjustments made by hormones secreted by endocrine glands are slower than those made by nerve impulses, they are no less effective.

The branch of medical science that deals with the normal functioning and disorders of the nervous system is called ***neurology*** (noo-ROL-ō-jē; *neuro* = nerve or nervous system; *logos* = study of).

The developmental anatomy of the nervous system is considered in Chapter 14.

FIGURE 12-1 Organization of the nervous system.

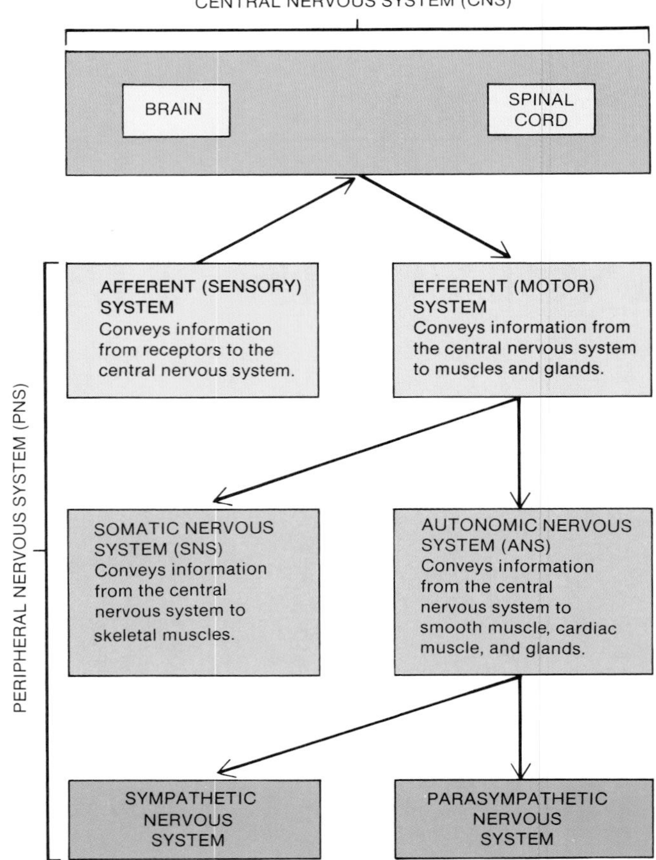

ORGANIZATION

The nervous system may be divided into two principal divisions, the central nervous system and the peripheral nervous system, and several subdivisions (Figure 12-1).

The ***central nervous system (CNS)*** is the control center for the entire nervous system and consists of the brain and spinal cord. All body sensations must be relayed from receptors to the central nervous system if they are to be interpreted and acted upon. The majority of nerve impulses that stimulate muscles to contract and glands to secrete must also originate in the central nervous system.

The various nerve processes, in the form of nerves, that connect the brain and spinal cord with receptors, muscles, and glands constitute the ***peripheral*** (pe-RIF-er-al) ***nervous system (PNS)***. The peripheral nervous system may be divided into an afferent system and an efferent system. The ***afferent*** (AF-er-ent; *ad* = toward; *fero* = to carry) ***system*** consists of nerve cells that convey information from receptors in the periphery of the body to the central nervous system. These nerve cells, called ***afferent (sensory) neurons,*** are the first cells to pick up incoming information. The ***efferent*** (EF-er-ent; *effero* = to bring out) ***system*** consists of nerve cells that convey information from the central nervous system to muscles and glands. These nerve cells are called ***efferent (motor) neurons.***

The efferent system is subdivided into a somatic nervous system and an autonomic nervous system. The ***somatic*** (*soma* = body) ***nervous system,*** or ***SNS,*** consists of efferent neurons that conduct impulses from the central nervous system to skeletal muscle tissue. The somatic nervous system produces movement only in skeletal muscle tissue. It is under conscious control and therefore voluntary. The ***autonomic*** (*auto* = self; *nomos* = law) ***nervous system,*** or ***ANS,*** by contrast, contains efferent neurons that convey impulses from the central nervous system to smooth muscle tissue, cardiac muscle tissue, and glands. Since it produces responses only in involuntary muscles and glands, it is usually considered to be involuntary.

With few exceptions, the viscera receive nerve fibers from the two divisions of the autonomic nervous system: the ***sympathetic division*** and the ***parasympathetic division.*** In general, the fibers of one division stimulate or increase an organ's activity, whereas the fibers from the other inhibit or decrease its activity (see Chapter 16).

HISTOLOGY

Despite the organizational complexity of the nervous system, it consists of only two principal kinds of cells: neurons and neuroglia. Neurons are highly specialized for nerve impulse conduction and for all special functions attributed to the nervous system: thinking, controlling muscle activity, regulating glands. Neuroglia serve as a special supporting and protective component of the nervous system.

Neuroglia

The cells of the nervous system that perform the functions of support and protection are called ***neuroglia*** (noo-ROG-lē-a; *neuro* = nerve; *glia* = glue) or ***glial cells*** (Figure 12-2). Neuroglia are believed to be derived from ectoderm. They are generally smaller than neurons and outnumber them by 5 to 10 times. Many of the glial cells form a supporting network by twining around nerve cells or lining certain structures in the brain and spinal cord. Others bind nervous tissue to supporting structures and attach the neurons to their blood vessels. A few types of glial cells also serve specialized functions. For example, some produce a phospholipid covering, called a myelin sheath, around nerve fibers in the central nervous system,

FIGURE 12-2 Histology of neuroglia. (a) Fibrous astrocyte (left) and protoplasmic astrocyte (right) associated with a blood vessel. (b) Oligodendrocyte associated with an axon of a neuron of the central nervous system. (c) Microglial cell. (d) Ependyma. Note the cilia.

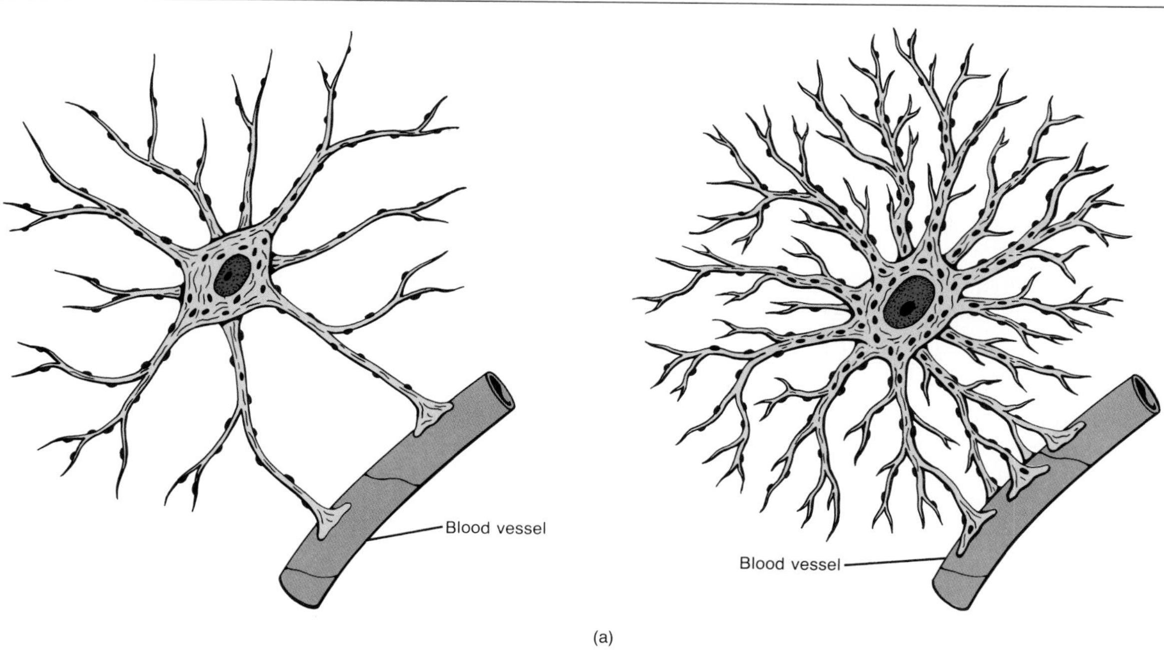

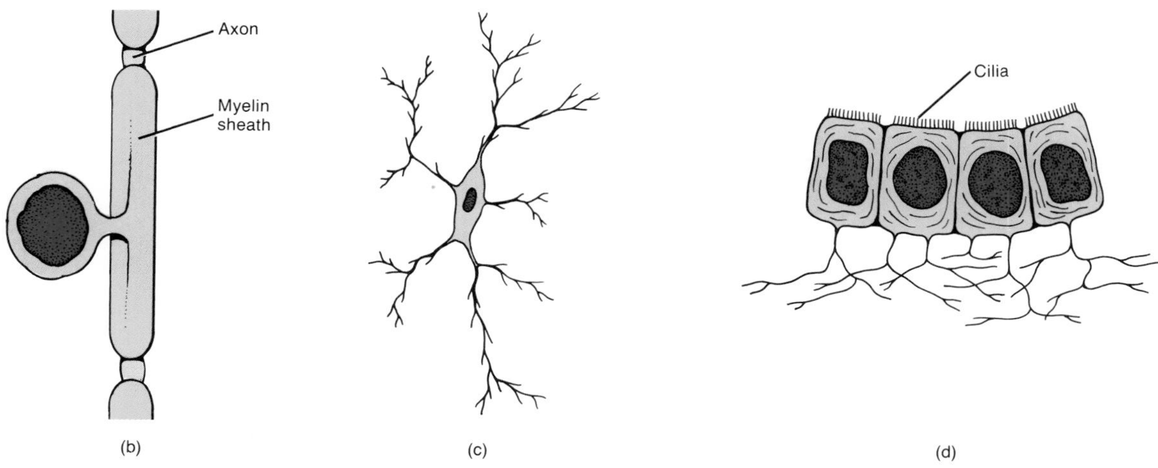

which increases the speed of nerve impulse conduction and insulates the fibers. Certain small glial cells are phagocytic; they protect the central nervous system from disease by engulfing invading microbes and clearing away debris. Neuroglia are of clinical interest because they are a common source of tumors (gliomas) of the nervous system. It is estimated that gliomas account for 40 to 45 percent of brain tumors. Unfortunately, gliomas are very invasive.

Exhibit 12-1 lists the neuroglial cells and summarizes their functions.

Neurons

Nerve cells, called ***neurons,*** are responsible for conducting nerve impulses from one part of the body to another. They are the basic information-processing units of the nervous system.

Structure

Most neurons consist of three distinct portions: (1) cell body, (2) dendrites, and (3) axon (Figure 12-3a). The ***cell body, soma,*** or ***perikaryon*** (per'-i-KAR-ē-on), contains a well-defined nucleus and nucleolus surrounded by a granular cytoplasm. Within the cytoplasm are typical organelles such as lysosomes, mitochondria, and Golgi complexes. Many neurons also contain cytoplasmic inclusions such as ***lipofuscin*** pigment that occurs as clumps of yellowish brown granules. Lipofuscin may be a by-product of lysosomal activity. Although its significance is unknown, lipofuscin is related to aging; the amount of pigment increases with age. Also located in the cytoplasm are structures characteristic of neurons: chromatophilic substance and neurofibrils. The ***chromatophilic substance (Nissl bodies)*** is an orderly arrangement of granular (rough) endoplasmic reticulum whose function is protein synthesis. Newly synthesized proteins pass from the cell body into the neuronal processes, mainly the axon, at the rate of about 1 mm (0.04 inch) per day. These proteins replace those lost during metabolism and are used for growth of neurons and regeneration of peripheral nerve fibers. ***Neurofibrils*** are long, thin fibrils composed of intermediate filaments. They may assume a function in support and the transportation of nutrients. Mature neurons do not contain a mitotic apparatus. The significance of this absence will be noted shortly. (It has recently been shown that adult neurons occasionally contain centrioles, but their function is unknown.)

Neurons have two kinds of cytoplasmic processes: dendrites and axons. ***Dendrites*** (*dendro* = tree) are usually highly branched, thick extensions of the cytoplasm of the cell body. They typically contain chromatophilic substance, mitochondria, and other cytoplasmic organelles. A neuron usually has several main dendrites. Their function is to conduct nerve impulses toward the cell body.

The second type of cytoplasmic process, called an ***axon (axis cylinder),*** is usually a single long, thin process that is highly specialized and conducts nerve impulses away from the cell body to another neuron or muscular or glandular tissue. It usually originates from the cell body as a small conical elevation called the ***axon hillock.*** An axon contains mitochondria and neurofibrils but no chro-

EXHIBIT 12-1 NEUROGLIA OF CENTRAL NERVOUS SYSTEM

Type	Description	Function
Astrocytes (*astro* = star; *cyte* = cell)	Star-shaped cells with numerous processes. ***Protoplasmic astrocytes*** are found in the gray matter of the CNS, and ***fibrous astrocytes*** are found in the white matter of the CNS.	Twine around nerve cells to form supporting network in CNS; attach neurons to their blood vessels; help form blood–brain barrier (Chapter 14).
Oligodendrocytes (*oligo* = few; *dendro* = tree)	Resemble astrocytes in some ways, but processes are fewer and shorter.	Give support by forming semirigid connective tissue rows between neurons in CNS; produce a phospholipid myelin sheath around axons of neurons of CNS.
Microglia (*micro* = small; *glia* = glue)	Small cells with few processes; derived from monocytes; normally stationary but may migrate to site of injury; also called ***brain macrophages.***	Engulf and destroy microbes and cellular debris; may migrate to area of injured nervous tissue and function as small macrophages.
Ependyma (ependymocytes) (*ependyma* = upper garment)	Epithelial cells arranged in a single layer and ranging in shape from squamous to columnar; many are ciliated.	Form a continuous epithelial lining for the ventricles of the brain (spaces that form and circulate cerebrospinal fluid) and the central canal of the spinal cord; probably assist in the circulation of cerebrospinal fluid (CSF) in these areas.

FIGURE 12-3 Structure of a typical neuron as exemplified by an efferent (motor) neuron. (a) An efferent neuron. Arrows indicate the direction in which nerve impulses travel. The break indicates that the process is actually longer than shown. (b) Sectional planes through a myelinated fiber. (c) Photomicrograph of an efferent neuron at a magnification of 640×. (Courtesy of Biophoto Associates/Photo Researchers.)

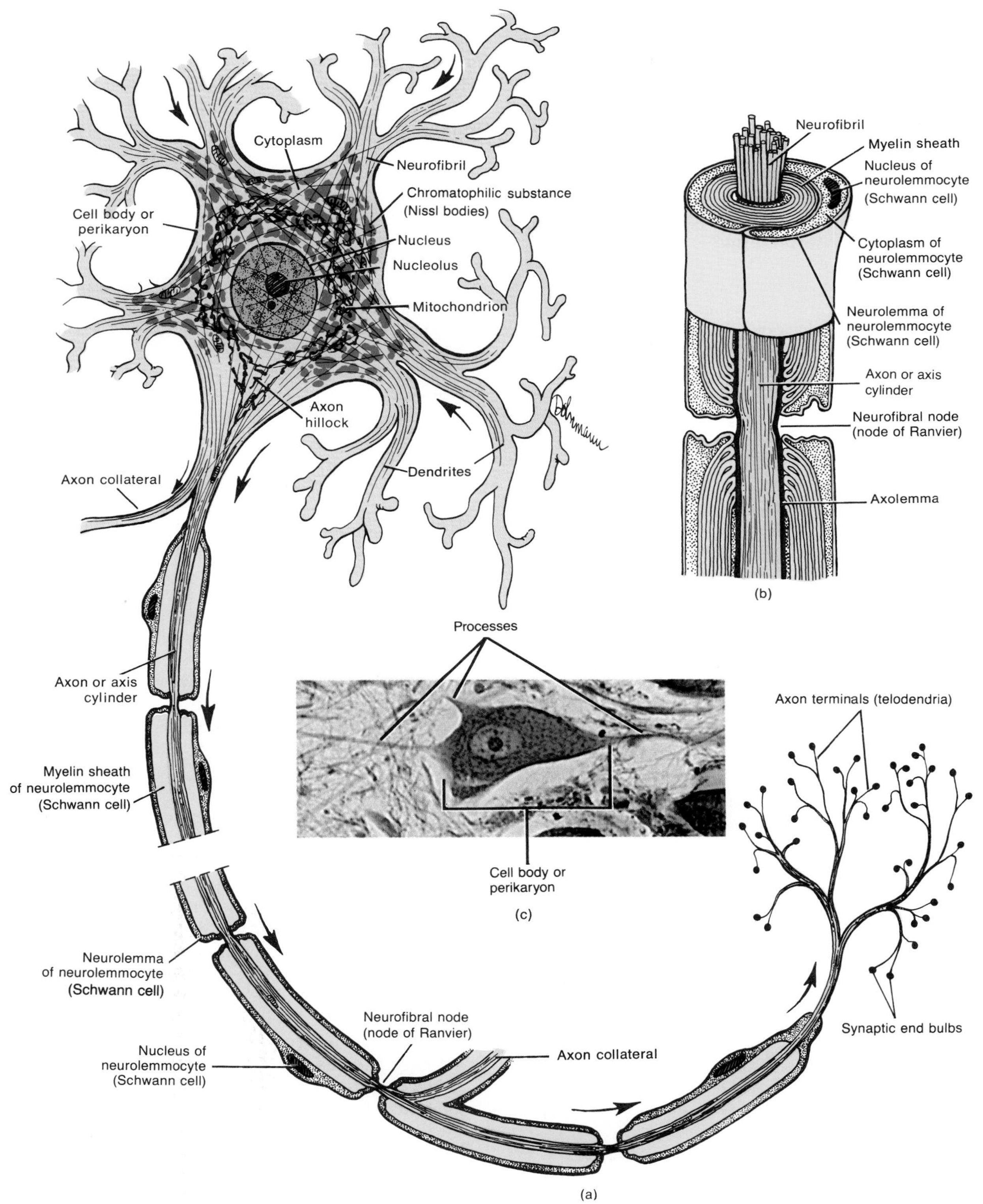

matophilic substance; thus, it does not carry on protein synthesis. Its cytoplasm, called ***axoplasm,*** is surrounded by a plasma membrane known as the ***axolemma*** (*lemma* = sheath or husk). Axons vary in length from a few millimeters (1 mm = 0.04 inch) in the brain to a meter (3.28 ft) or more between the spinal cord and toes. Along the length of an axon, there may be side branches called ***axon collaterals.*** The axon and its collaterals terminate by branching into many fine filaments called ***axon terminals (telodendria).*** The distal ends of axon terminals are expanded into bulblike structures called ***synaptic end bulbs,*** which are important in nerve impulse conduction from one neuron to another and from a neuron to muscle or glandular tissue. They contain membrane-enclosed sacs called ***synaptic vesicles*** that store chemicals called neurotransmitters that determine whether or not impulses pass from one neuron to another or from a neuron to another tissue (muscle or gland).

The cell body of a neuron is essential for the synthesis of many substances that sustain the life of the nerve cell. Neurons have two types of intracellular systems for transporting synthesized materials from the cell body. The slower one, about 1 mm per day, is called ***axoplasmic flow,*** and conveys axoplasm in one direction only—from the cell body toward axon terminals. This mechanism may occur by protoplasmic streaming and supplies new axoplasm for developing or regenerating axons and renews axoplasm in growing and mature axons. The faster type of intracellular transport, about 300 mm per day, is called ***axonal transport.*** It conveys materials in both directions—away from the cell body and toward the cell body—possibly along tracks formed by microtubules and filaments. Axonal transport moves various organelles and materials that form the membranes of the axolemma, synaptic end bulbs, and synaptic vesicles. Materials returning to the cell body are degraded or recycled.

CLINICAL APPLICATION: AXONAL TRANSPORT AND DISEASE

The route taken by materials back to the cell body by axonal transport is the route by which the ***herpes virus*** and ***rabies virus*** make their way back to nerve cell bodies, where they multiply and cause their damage. The toxin produced by the ***tetanus bacterium*** uses the same route to reach the central nervous system. In fact, the time delay between the release of the toxin and the first appearance of symptoms is in part due to the time required for movement of the toxin by axonal transport. This is why a tetanus-prone injury of the head or neck, or a bite in this region by a rabid dog, bat, or other animal, is a more serious matter than the introduction of the bacterium in the extremities. If the bacterium is closer to the brain, the incubation period is shorter, and treatment to prevent the disease requires emergency measures.

The term ***nerve fiber*** may be applied to any process projecting from the cell body. More commonly, it refers to an axon and its sheaths. Figure 12-3b shows two sectional planes of a nerve fiber of the peripheral nervous system. Many axons, especially ones outside the CNS, are surrounded by a multilayered, white, phospholipid, segmented covering called the ***myelin sheath.*** Axons containing such a covering are ***myelinated,*** whereas those without it are ***unmyelinated*** (see Figure 12-4). The function of the myelin sheath is to increase the speed of nerve impulse conduction and to insulate and maintain the axon. Myelin is responsible for the color of the white matter in the nerves, brain, and spinal cord. As you will see later, certain diseases such as multiple sclerosis and Tay-Sachs disease are related to destruction of myelin sheaths.

The myelin sheath of axons of the peripheral nervous system is produced by flattened cells, called ***neurolemmocytes (Schwann cells),*** located along the axons. In the formation of a sheath, a developing neurolemmocyte encircles the axon until its ends meet and overlap (Figure 12-4). The cell then winds around the axon many times, and as it does so, the cytoplasm and nucleus are pushed to the outside layer. The inner portion, consisting of up to 20–30 layers of neurolemmocyte membrane, is the myelin sheath. The peripheral nucleated cytoplasmic layer of the neurolemmocyte (the outer layer that encloses the sheath) is called the ***neurolemma (sheath of Schwann).***

The neurolemma is found only around fibers of the peripheral nervous system. Its function is to assist in the regeneration of injured axons and dendrites by forming a tube in which a regenerating axon or dendrite grows. Between the segments of the myelin sheath are unmyelinated gaps, called ***neurofibral nodes (nodes of Ranvier)*** (ron-VĒ-ā). Unmyelinated fibers are also enclosed by neurolemmocytes. However, they do not have multiple wrappings; they only contain a neurolemma.

Nerve fibers of the central nervous system may also be myelinated or unmyelinated. Myelination of central nervous system axons is accomplished by oligodendrocytes in somewhat the same manner that neurolemmocytes myelinate peripheral nervous system axons (see Figure 12-2b). Myelinated axons of the central nervous system also contain neurofibral nodes, but they are not so numerous. However, these axons do not have a neurolemma, and therefore they cannot regenerate.

Myelin sheaths are first laid down during the later part of fetal development and during the first year of life. The amount of myelin increases from birth to maturity, and its presence greatly increases the rate of nerve impulse conduction. Since myelination is still in progress during infancy, an infant's responses to stimuli are not as rapid or coordinated as those of an older child or an adult.

Nerve growth factor (NGF) is a protein hormone that functions as a growth factor. It is found in many different tissues, especially the submaxillary salivary glands. NGF is critical for the normal survival and de-

FIGURE 12-4 Comparison between myelinated and unmyelinated axons. (a) Stages in the formation of a myelin sheath by a neurolemmocyte (Schwann cell). (b) Unmyelinated axon.

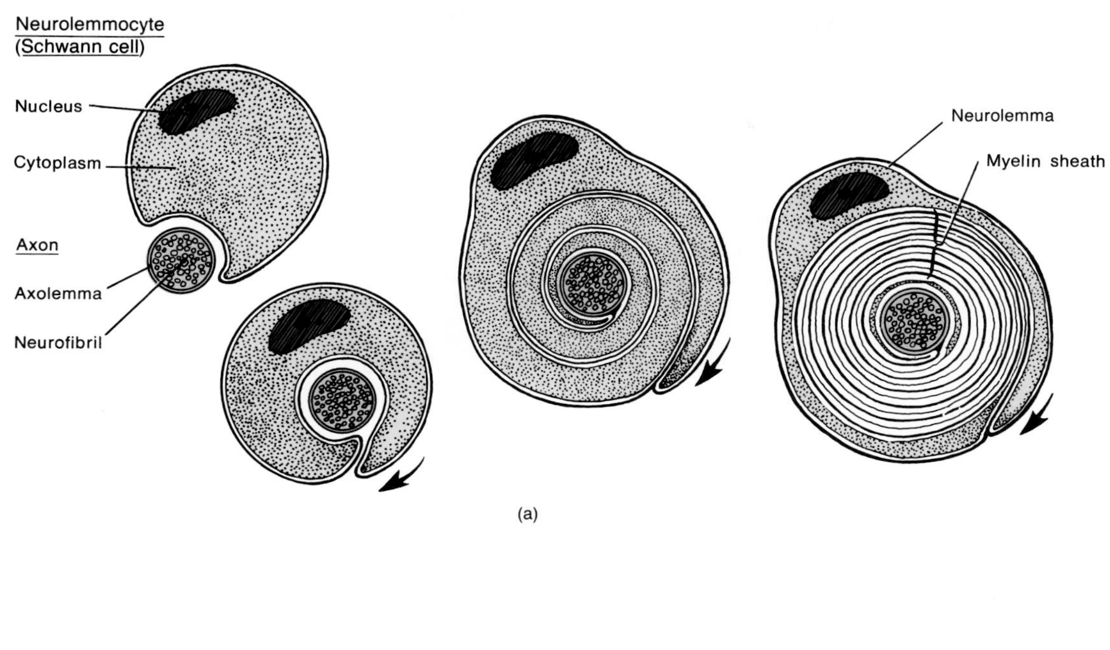

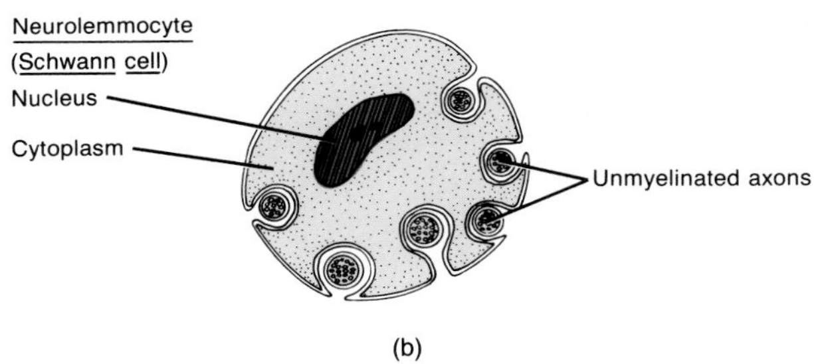

velopment of peripheral sympathetic and sensory neurons. Whereas sensory neurons are stimulated by NGF only during a short period of embryonic development, sympathetic neurons are subject to stimulation into adulthood. NGF also plays a role in the brain, where it increases synthesis of acetylcholine (ACh), stimulates neuronal hypertrophy, stimulates axon growth, and may help to maintain neuronal function.

Structural Variation

Although all neurons conform to the general plan described, there are considerable differences in structure. For example, cell bodies range in diameter from 5 μm for the smallest cells to 135 μm for large motor neurons. The pattern of dendritic branching is varied and distinctive for neurons in different parts of the body. The axons of very small neurons are only a fraction of a millimeter in length and lack a myelin sheath, whereas axons of large neurons are over a meter long and are usually enclosed in a myelin sheath.

Classification

The different neurons in the body may be classified by structure and function.

The structural classification is based on the number of processes extending from the cell body. ***Multipolar neurons*** usually have several dendrites and one axon (see Figure 12-3a). Most neurons in the brain and spinal cord are of this type. ***Bipolar neurons*** have one dendrite and one axon and are found in the retina of the eye, inner ear, and olfactory area. ***Unipolar (pseudounipolar) neurons*** have only one process extending from the cell body. The single process divides into a central branch, which functions as an axon, and a peripheral branch, which functions as a dendrite. Unipolar neurons originate in the embryo as bipolar neurons, and during development, the axon and dendrite fuse into a single process. Unipolar neurons are found in posterior (sensory) root ganglia of spinal nerves and the ganglia of cranial nerves that carry general somatic sensory impulses.

The functional classification of neurons is based on the direction in which they transmit impulses. ***Sensory (afferent) neurons*** transmit impulses from receptors in the skin, sense organs, muscles, joints, and viscera to the brain and spinal cord and from lower to higher centers of the CNS. They are usually unipolar (Figure 12-5). ***Motor (efferent) neurons*** convey impulses from the brain and spinal cord to effectors, which may be either muscles or glands (see Figure 12-3a), and from higher to lower centers of the CNS. Other neurons, called ***association (connecting*** or ***interneuron) neurons,*** carry impulses from sensory neurons to motor neurons and are located in the brain and spinal cord. Examples of association neurons are a ***stellate cell,*** a ***cell of Martinotti*** (mar'-ti-NOT-ē), a ***horizontal cell of Cajal*** (kā-HAL), and a ***pyramidal cell*** (pi-RAM-i-dal) cell. All are found in the cerebral cortex, the outer layer of the cerebrum. The ***granule cell*** and ***Purkinje*** (pur-KIN-jē) ***cell*** are association neurons in the cortex of the cerebellum. Most neurons in the body, perhaps 90 percent, are association neurons. Several association neurons are shown in Figure 12-6.

The processes of afferent and efferent neurons are arranged into bundles called ***nerves*** if outside the CNS or ***fiber tracts*** if inside the CNS. (Axons of association neurons can also form fiber tracts.) Since nerves lie outside the central nervous system, they belong to the peripheral nervous system. The functional components of nerves are the nerve fibers, which may be grouped according to the following scheme.

1. ***General somatic afferent fibers*** conduct nerve impulses from the skin, skeletal muscles, and joints to the central nervous system.
2. ***General somatic efferent fibers*** conduct nerve impulses from the central nervous system to skeletal muscles. Impulses over these fibers cause the contraction of skeletal muscles.
3. ***General visceral afferent fibers*** convey nerve impulses from the viscera and blood vessels to the central nervous system.
4. ***General visceral efferent fibers*** belong to the autonomic nervous system and are also called ***autonomic fibers.*** They convey nerve impulses from the central nervous system to help control contractions of smooth and cardiac muscle and rate of secretion by glands.

FIGURE 12-5 **Structure of a typical afferent (sensory) neuron. Arrows indicate the direction in which the nerve impulse travels. The break indicates that the process is actually longer than shown.**

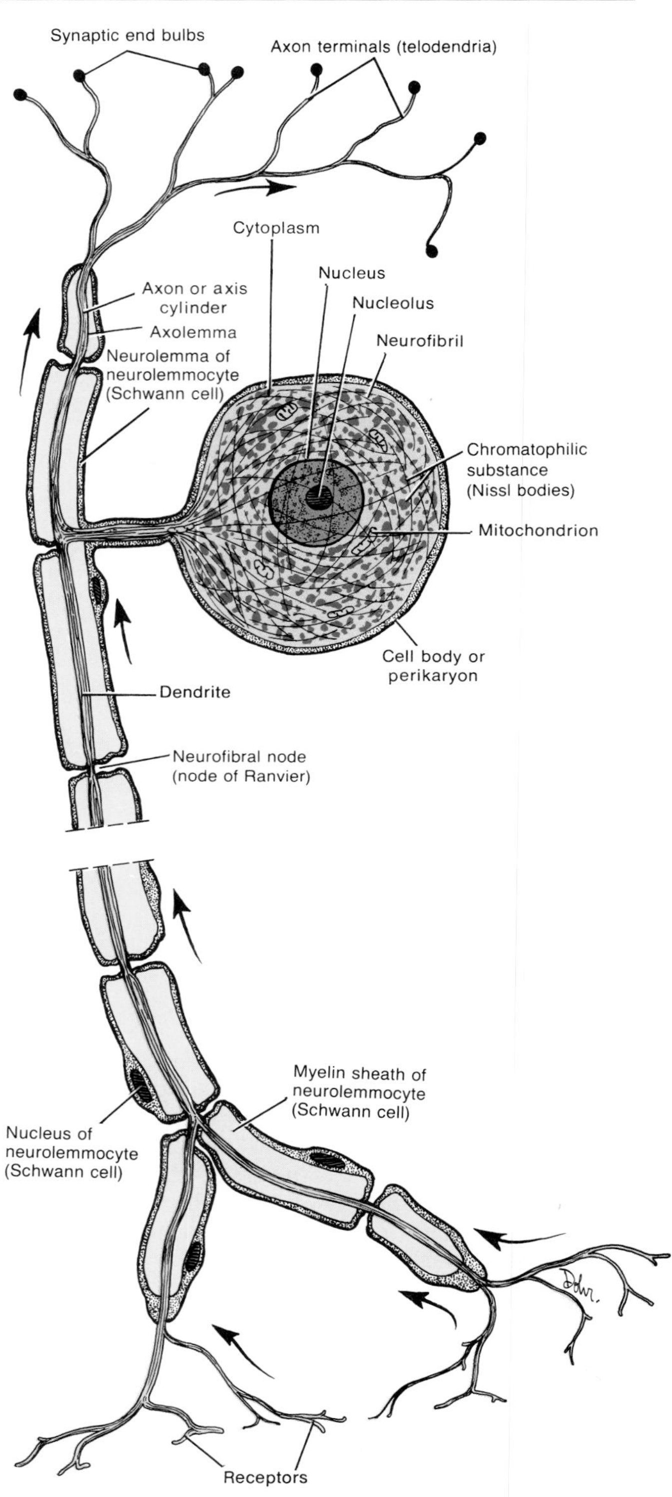

FIGURE 12-6 Representative association (connecting) neurons. Arrows indicate direction in which nerve impulse travels. (a) Pyramidal cell. (b) Granule cell. (c) Purkinje cell.

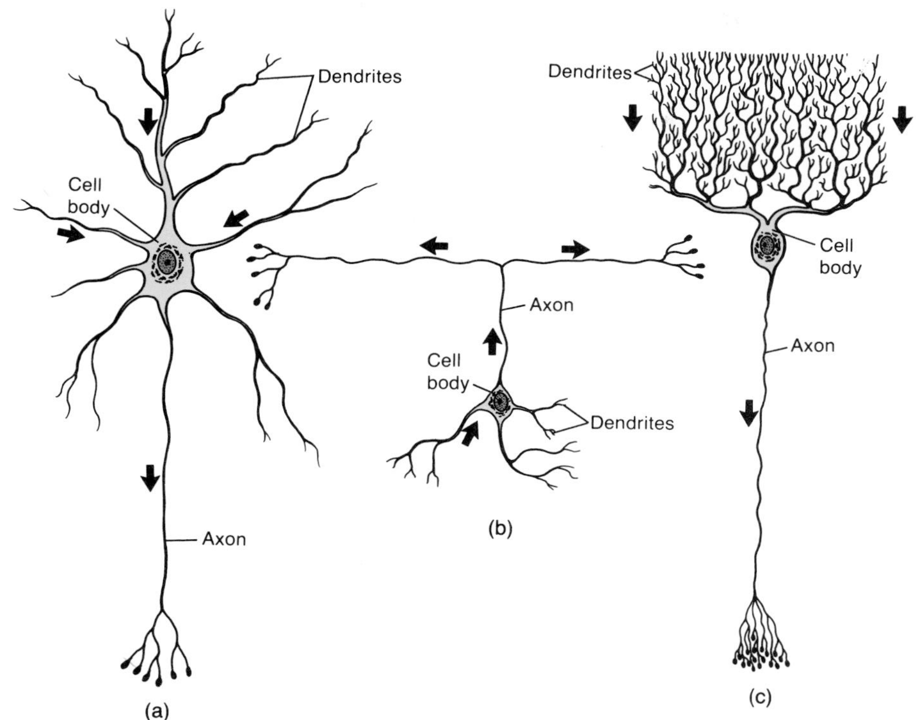

In addition to being grouped as nerves, neural tissue is also organized into other structures such as ganglia, tracts, nuclei, and horns. These are described in Chapter 13.

PHYSIOLOGY

Two striking features of neurons are (1) their highly developed ability to generate and conduct electrical messages called nerve impulses and (2) their limited ability to regenerate.

Nerve Impulse

At this point, we will consider the nerve impulse—your body's quickest way of controlling and maintaining homeostasis.

Membrane Potentials

Studies of cell membranes, especially in nerve and muscle cells, indicate that when a cell is at rest, there is a considerable difference between the ion concentration outside and inside the plasma membrane. In a resting neuron (one that is not conducting an impulse), there is a difference in electrical charges on either side of the membrane. This difference is partly the result of an unequal distribution of potassium (K^+) ions and sodium (Na^+) ions on either side of the membrane. In resting neurons, the K^+ ion concentration inside the cell is about 28 to 30 times greater than it is outside. The Na^+ ion concentration is about 14 times greater outside than inside. Another significant factor is the presence of large, nondiffusible negatively charged ions trapped in the cell. These include organic phosphate and protein anions. What causes the outside of the nerve cell membrane to differ from the inside?

Even when a nerve cell is not conducting an impulse, it is actively transporting ions across its membrane. (The mechanism of active transport may be reviewed in Figure 3-5.) Na^+ ions are actively transported out and K^+ ions are actively transported in. The membrane system by which Na^+ and K^+ ions are actively transported simultaneously is called the ***sodium-potassium pump*** (Figure 12-7a). The operation of the pump requires the expenditure of energy from ATP. The active transport of Na^+ and K^+ ions is unequal; that is, three Na^+ ions are actively transported out for every two K^+ ions actively transported in.

Neurons also contain a large number of negative ions, including organic phosphates and protein anions, on the

FIGURE 12-7 Development of the resting membrane potential. (a) Schematic representation of the sodium-potassium pump and distribution of ions. Note that the large number of Na^+ ions outside the membrane results in an external positive charge. Although there are more K^+ ions inside the cell membrane than outside, there are many more negative ions inside the membrane. This results in a net internal negative charge. (b) Simplified representation of a polarized membrane.

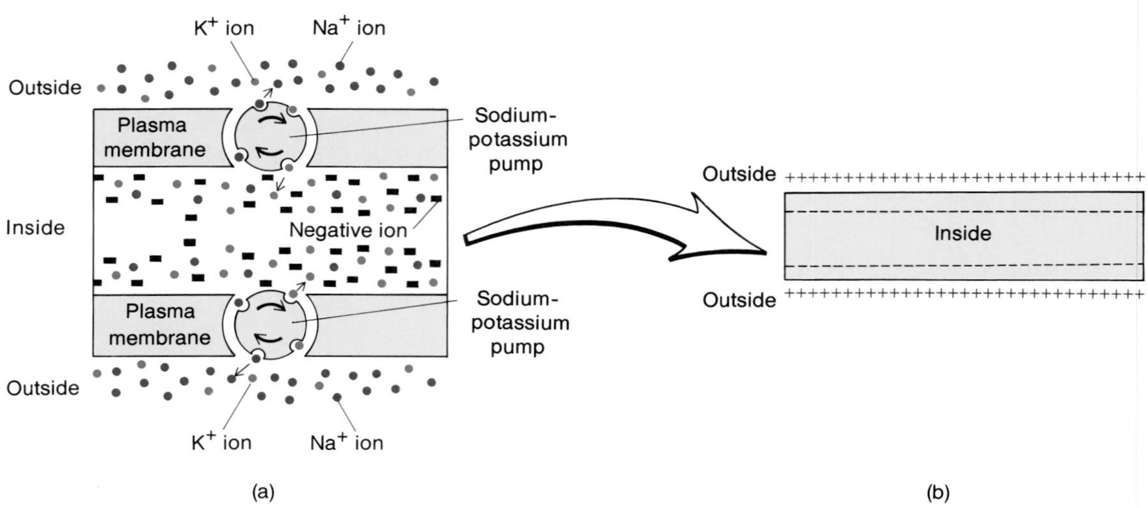

inside that cannot diffuse outside or diffuse very slowly. Since Na^+ ions are positive and are actively transported outside the cell by the sodium-potassium pump, a positive charge develops outside the membrane. Even though K^+ ions are also positive and are actively transported to the inside of the cell by the sodium-potassium pump, there are insufficient K^+ ions to equalize the even larger number of nondiffusible negative ions trapped in the cell.

In addition, as a result of the operation of the sodium-potassium pump, there is a concentration and electrical gradient for Na^+ and K^+ ions. As a result, K^+ ions tend to diffuse (leak) out of the cell, and Na^+ ions tend to diffuse into the cell. Membrane permeability to K^+ ions is 100 times greater than that to Na^+ ions and K^+ ions diffuse out of the cell along their concentration gradient more easily and quickly than Na^+ ions enter the cell along their gradient. The sodium-potassium pump not only actively transports Na^+ and K^+ ions but also establishes concentration gradients for the ions. The result is that there is a difference in charge on either side of the membrane—net positive outside and net negative inside. This difference in charge on either side of the membrane of a resting neuron is the ***resting membrane potential.*** Such a membrane is said to be ***polarized*** (Figure 12-7b). Since the sodium-potassium pump transports unequal numbers of Na^+ and K^+ ions that contribute to the resting membrane potential, the pump is referred to as ***electrogenic.***

Electrical measurements of a polarized membrane indicate a voltage of about 70 millivolts, abbreviated mV. (One mV equals one-thousandth of a volt; a 1.5-volt battery will power an ordinary flashlight.) Since positive charges predominate outside, the inside of the membrane is said to be 70 mV less positive than the outside, that is, the resting membrane potential is −70 mV. In subsequent discussions of resting membrane potentials, we will use the mV value that refers to the *inside* of the membrane, −70 mV. The mV value established is the result of the active separation of charges on either side of the membrane by the sodium-potassium pump and, when the membrane permeability toward Na^+ and/or K^+ ions is altered, the resting membrane potential will be changed. The membrane permeability, and thus the resting membrane potential, can be altered by way of neurotransmitters as well as other mechanisms.

Excitability

The events associated with the generation of nerve impulses will now be examined. The ability of nerve cells to respond to stimuli and convert them into nerve impulses is called ***excitability.*** A ***stimulus*** is any condition in the environment capable of altering the resting membrane potential.

If an excitatory stimulus of adequate strength, called a threshold stimulus, is applied to a polarized membrane, the membrane's permeability to Na^+ ions greatly increases at the point of stimulation (Figure 12-8a–d). What happens is that voltage-sensitive sodium channels open and permit the influx of Na^+ ions by diffusion. The movement of Na^+ ions inward also results from the attraction of positively charged Na^+ ions to the negative ions on

FIGURE 12-8 Initiation and propagation of a nerve impulse. (a–d) The stippled area containing the straight arrow represents the region of the membrane that is propagating the nerve impulse. The curved arrows represent local currents. (e) Record of potential changes of a nerve impulse.

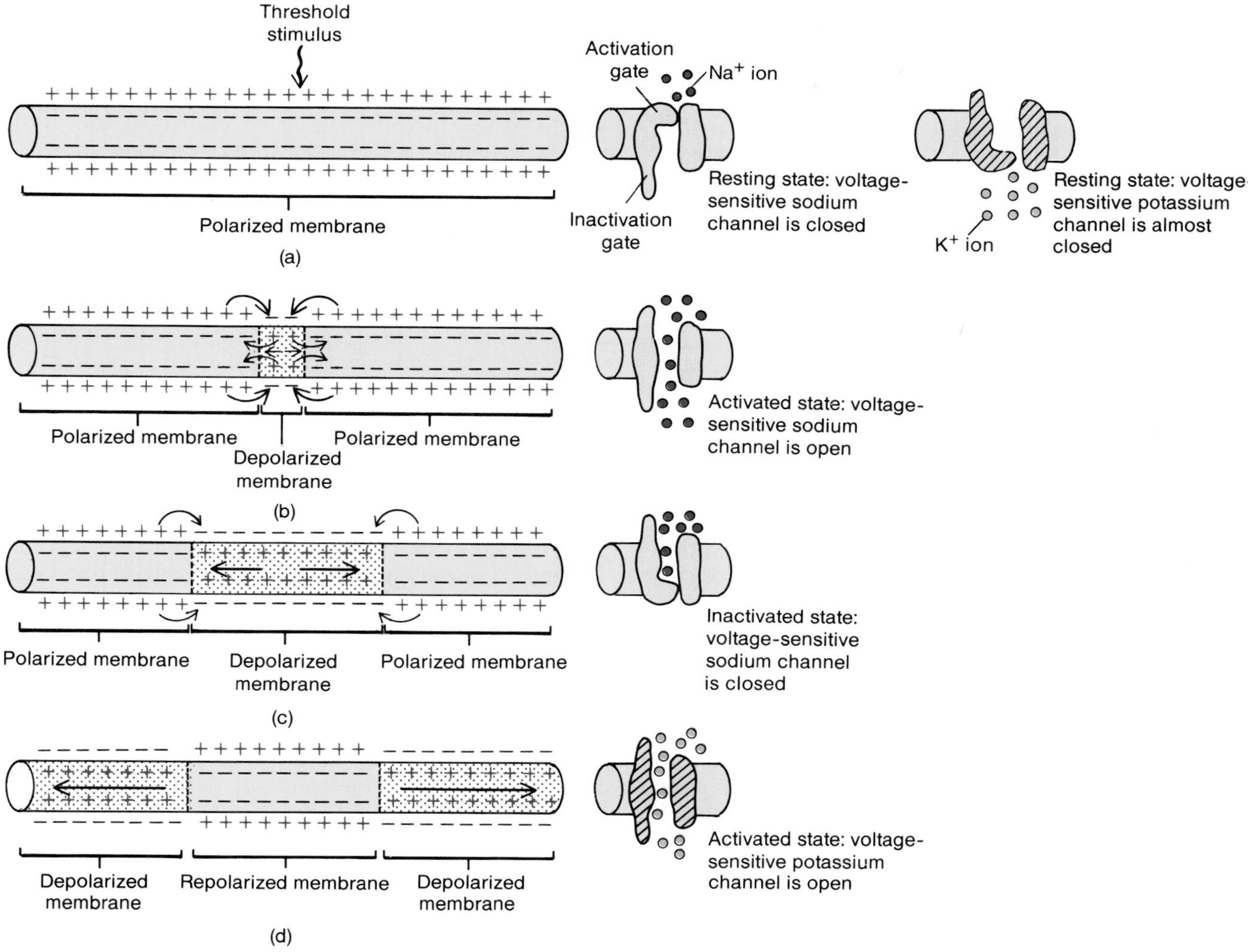

Threshold stimulus

Polarized membrane

(a)

Activation gate

Na^+ ion

Inactivation gate

Resting state: voltage-sensitive sodium channel is closed

K^+ ion

Resting state: voltage-sensitive potassium channel is almost closed

Polarized membrane

Depolarized membrane

Polarized membrane

(b)

Activated state: voltage-sensitive sodium channel is open

Polarized membrane

Depolarized membrane

Polarized membrane

(c)

Inactivated state: voltage-sensitive sodium channel is closed

Depolarized membrane

Repolarized membrane

Depolarized membrane

(d)

Activated state: voltage-sensitive potassium channel is open

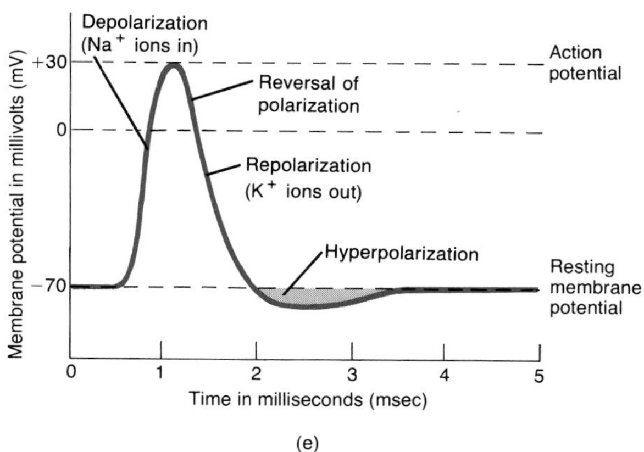

(e)

the inside of the membrane, and since there are more Na^+ ions entering than leaving, the resting membrane potential begins to change. At first, the potential inside the membrane shifts from −70 mV toward 0, and then to a positive value. This process is called ***depolarization.*** Depolarization begins at −69 mV, and from this point on, the membrane is said to be ***depolarized.*** Throughout depolarization, the Na^+ ions continue to rush inside until the resting membrane potential is *reversed*—the inside of the membrane becomes positive and the outside negative. Electrical measurements indicate that the inside of the membrane becomes +30 mV with respect to the outside. Thus, the potential inside the membrane changes from −70 mV to 0 mV to +30 mV.

Voltage-sensitive (gated) channels are ion channels composed of integral proteins in the plasma membrane. These channels contain one or more proteins that can undergo alterations in shape and thus function like gates by restricting or permitting the movement of ions in response to the membrane's voltage state (electric potential). Two types of voltage-sensitive channels involved in impulse generation are voltage-sensitive sodium channels and voltage-sensitive potassium channels.

Each ***voltage-sensitive sodium channel*** has two separate gates, an *activation gate* near the exterior of the channel and an *inactivation gate* near the interior of the channel (Figure 12-8a). In a polarized (resting) membrane, the activation gate is closed and the inactivation gate is open. As a result, the movement of Na^+ ions to the interior of the cell is inhibited. A voltage-sensitive sodium channel in this state is said to be *resting*.

When a threshold stimulus is applied to a polarized membrane, the membrane starts to lose its polarity. When it reaches a level of depolarization, usually between -70 to -50 mV, voltage-sensitive sodium channels undergo a change from a resting to an *activated* state. In this state, both the activation and inactivation gates in the channel are open and Na^+ ions move inward in sufficient numbers so that the membrane voltage changes from -70 to 0 to $+30$ mV (Figure 12-8b). The same voltage change that causes the activation gate to open also causes the inactivation gate to close (Figure 12-8c). This is called the *inactivated* state of the channel. But the closure of the inactivation gate occurs a few ten-thousandths of a second after opening of the activation gate. Thus, after a voltage-sensitive sodium channel is open for only a few ten-thousandths of a second, it immediately closes. Since Na^+ ions no longer move inward, the membrane begins to recover back to its original polarized (resting) state because voltage-sensitive potassium channels are open and K^+ rapidly diffuses *out* of the neuron to bring about repolarization of the membrane. In fact, the inactivated state will not revert to the activated state until the polarized (resting) state is restored.

The inward movement of Na^+ ions during depolarization is an example of a positive feedback system that is essential for homeostasis. As Na^+ ions continue to move inward, depolarization increases. This, in turn, causes more voltage-sensitive sodium channels to open. As more channels open, more Na^+ ions move inward, and this leads to more depolarization, and so on. The continued movement of Na^+ ions is independent of the initial stimulus; it occurs as part of a positive feedback cycle.

Once the events of depolarization have occurred, we say that a ***nerve impulse (nerve action potential)*** is initiated. (It lasts about 1 msec.) An ***action potential*** is a rapid change in membrane potential that involves a depolarization followed by a repolarization.

In order for a nerve impulse to function as a signal to communicate information from one part of the body to another, it must be propagated (transmitted) along a neuron from its site of origin. A nerve impulse that is generated at any one point on the membrane usually excites (depolarizes) adjacent portions of the membrane, causing the impulse to be propagated. This propagation relies on local electrical currents along the membrane. As Na^+ ions flow inward at the point of stimulation, local currents affect adjacent polarized areas (Figure 12-8b–d). This causes depolarization of the adjacent areas, activation of new voltage-sensitive sodium channels, inward movement of Na^+ ions, and development of new nerve impulses at successive points along the membrane. As more and more areas along the membrane depolarize in this manner, the nerve impulse spreads rapidly. In each case, the local currents cause the adjacent portions of the membrane to reverse their potential from -70 mV to 0 to $+30$ mV. The reversal repeats itself over and over until the nerve impulse is propagated along the length of the neuron. As you can see, a nerve impulse is actually a wave of depolarization (negativity) that self-propagates along the outside surface of the membrane of a neuron. Depolarization and reversal of potential require only about 0.5 msec. Of all the cells of the body, only muscle fibers and nerve cells produce action potentials. Their ability to do this is called ***excitability.*** The local currents usually move in only one direction along a neuron, since the stimulus that generates a nerve impulse normally arrives at only one end of the neuron.

By the time the nerve impulse has traveled from one point on the membrane to the next, the previous point becomes ***repolarized***—its resting potential is restored. Repolarization results from a new series of changes in membrane permeability. This involves ***voltage-sensitive potassium channels.*** When the membrane is polarized (resting), the potassium channel is in its *resting* state; that is, the gate is open to some degree, but K^+ ions are prevented from diffusing to the exterior because of an electric gradient that opposes their concentration gradient (Figure 12-8a). However, when the membrane becomes depolarized, the voltage change causes a slow alteration that results in opening of the gate and rapid passage of K^+ ions to the exterior by diffusion down both a concentration and electric gradient. In this state, the voltage-sensitive potassium channels are said to be *activated.* Owing to the slowness of the opening of the voltage-sensitive potassium channels, they open at the same time that the voltage-sensitive sodium channels are becoming inactivated (closing). The net effect is inhibition of Na^+ ions from passing inward and acceleration of K^+ ions passing outward by diffusion down both a concentration and electric gradient. As the K^+ ions move through the open voltage-sensitive potassium channels, the outer surface of the membrane becomes electrically positive. The loss of positive ions leaves the inner surface of the membrane negative again, that is, the membrane is repolarized by this outward diffusion of K^+. Eventually, any ions that have moved into or out of the nerve cell are restored to their original sites by the sodium-potassium pump.

The repolarization period returns the cell to its resting membrane potential, from $+30$ mV to -70 mV. The neuron is now prepared to receive another stimulus and conduct it in the same manner. In fact, until repolarization occurs, the neuron cannot conduct another nerve impulse. The period of time during which the neuron cannot generate another nerve action potential is called the ***refractory period.*** The ***absolute refractory period*** refers to the period of time during which a second action potential cannot be initiated, even with a very strong stimulus. It corresponds roughly with the period of sodium permeability changes. Large fibers repolarize in about 1/2500 second. Their absolute refractory period is 0.4 msec. Thus, a second nerve impulse can be transmitted 1/2500 second after the first—a total of up to 2500 impulses per second. Small fibers, on the other hand, require as much as 1/250 second to repolarize. Their absolute refractory period is 4 msec. Thus, they can transmit only 250 impulses per second. Under normal body conditions, the frequency of impulses conducted over nerve fibers may range between 10 and 500 impulses per second. The ***relative refractory period*** refers to the period of time during which a second action potential can be initiated, but only by a stronger-than-normal stimulus. It corresponds roughly to the period of increased potassium permeability, which results in a brief hyperpolarization, that is, the inside of the cell is more negative (farther from zero) than the resting level.

K^+ ions that have leaked to the outside of the membrane and Na^+ ions that have leaked to the inside of the membrane must be restored to their original sides of the membrane. Ultimately, this is accomplished by the sodium-potassium pump, by which extra K^+ ions on the outside are pumped back in and extra Na^+ ions on the inside are pumped back out. However, the pump is not required for the initial repolarization of the membrane after *each* nerve impulse. This is accomplished by the outward diffusion of K^+ ions through the voltage-sensitive potassium channels. In fact, each time a single nerve impulse is conducted, only a minute amount of Na^+ ions enters the neuron and an equal amount of K^+ ions leaves the neuron shortly after Na^+ ions diffuse inward. It literally takes thousands and thousands of nerve impulses to produce a massive shift in Na^+ and K^+ ions that results in cessation of nerve impulse conduction.

A record of the electrical changes associated with a nerve impulse is illustrated in Figure 12-8e.

A sensory neuron is generally stimulated at its dendrite's distal end by a receptor, a structure sensitive to changes in the environment. Association and motor neurons are usually stimulated at their dendrites or cell bodies by another neuron. However, if a neuron's plasma membrane is depolarized at some point other than the usual one, the impulse will travel in both directions over the cell membrane of the entire neuron.

The initiation and conduction of a ***muscle action potential*** are basically similar to a nerve action potential (nerve impulse), although there are some notable differences. Whereas the resting membrane potential of a neuron is −70 mV, it is −90 mV in skeletal muscle fibers (cells). In addition, the duration of a nerve action potential is about 0.5 msec, but a muscle action potential is considerably longer—about 1.0 to 5.0 msec. Finally, the velocity of conduction of a nerve action potential can be about 18 times faster than a muscle action potential.

All-or-None Principle

Any stimulus strong enough to initiate a nerve impulse is referred to as a ***threshold (liminal) stimulus.*** When a stimulus is of threshold strength, we say the neuron has reached its threshold of stimulation. A single nerve cell, just like a single muscle fiber, transmits an action potential according to the ***all-or-none principle:*** If a stimulus is strong enough to generate a nerve action potential, the impulse is conducted along the entire neuron at a constant and maximum strength for the existing conditions. The conduction is independent of any further intensity of the stimulus. However, conduction may be altered by conditions such as toxic materials in cells, fatigue, and malaise. An analogy may help in understanding this principle. If a long trail of gunpowder were spilled along the ground, it could be ignited at one end to send a blazing signal down the entire length of the trail. It would not matter how tiny or great the triggering spark or flame or explosion was; the blazing signal moving along the trail of gunpowder would be just the same, a maximal one (or none at all if it never started).

Any stimulus weaker than a threshold stimulus is termed a ***subthreshold (subliminal) stimulus.*** Such a stimulus, if occurring only once, is incapable of initiating a nerve impulse. If, however, a second stimulus or a series of subthreshold stimuli is quickly applied to the neuron, the cumulative effect may be sufficient to initiate an impulse. This phenomenon is called summation and is discussed shortly.

Saltatory Conduction

Thus far we have considered nerve impulse conduction for unmyelinated fibers. The step-by-step depolarization of each adjacent area of the axon or dendrite plasma membrane just described is called ***continuous conduction.*** In myelinated fibers, conduction is somewhat different. The myelin sheath surrounding a fiber contains myelin, a phospholipid that does not conduct an electric current. It thus forms an insulating layer around the fiber and virtually inhibits the movement of ions. The myelin sheath is interrupted at various intervals called neurofibral nodes (nodes of Ranvier). At the nodes, membrane depolarization can occur and nerve action potentials can be generated and conducted. But beneath the myelin sheath, depolarization is impossible since ionic movement is inhibited. When a nerve impulse is conducted along a myelinated fiber, it moves from one node to another by ionic current flow through the surrounding extracellular fluids and through the axoplasm. Thus, the impulse jumps from node to node as each node area depolarizes to threshold and so conducts the impulse as it is regenerated at each node. This type of impulse conduction, characteristic of myelinated fibers (axons and dendrites), is called ***saltatory conduction*** (*saltare* = leaping) (Figure 12-9).

Saltatory conduction is a valuable asset to your homeostasis. Since the impulse jumps long intervals as it moves from one node to the next, the speed of conduction is greatly increased. The nerve impulse travels much faster than in the step-by-step depolarization process involved in an unmyelinated fiber of equal diameter. This is especially important in situations where split-second responses are necessary. Another advantage of saltatory conduction is greater efficiency as a result of energy expenditure. Saltatory conduction prevents depolarization of large areas of the plasma membrane of the fiber, thus preventing leakage of large amounts of Na^+ ions to the inside of the membrane and K^+ ions to the outside each time a nerve impulse is transmitted. This results in a lower expenditure of energy by the sodium-potassium pump.

Speed of Nerve Impulses

The speed of a nerve impulse is independent of stimulus strength. Once a neuron reaches its threshold of stimulation, the speed of the nerve impulse is normally determined by temperature, the diameter of the fiber, and the presence or absence of myelin.

When warmed, nerve fibers conduct impulses at higher speeds; when cooled, they conduct nerve impulses at lower speeds. Localized cooling of a nerve can block nerve conduction. Pain resulting from injured tissue can be reduced by the application of cold because the nerve fibers carrying the pain sensations are partially blocked.

Fibers with large diameters conduct impulses faster than those with small ones. Fibers with the largest diameter are called ***A fibers*** and are all myelinated. The A fibers have a brief absolute refractory period and are capable of saltatory conduction. They transmit impulses at speeds up to 130 m/sec. The A fibers are located in the axons of large sensory nerves that relay impulses associated with touch, pressure, position of joints, heat, and cold. They are also found in motor nerves that convey impulses to the skeletal muscles. Sensory A fibers generally connect the brain and spinal cord with sensors that detect danger in the outside environment. Motor A fibers innervate the muscles that can do something about the situation. If you touch a hot object, information about the heat passes over sensory A fibers to the spinal cord. There it is relayed to motor A fibers that stimulate the muscles of the hand to withdraw instantaneously. The A fibers are located where split-second reaction may mean survival.

Other fibers, called B and C fibers, conduct impulses more slowly and are generally found where instantaneous response is not a life-and-death matter. ***B fibers*** have a middle-sized diameter and a somewhat longer absolute refractory period than A fibers. They are also myelinated and therefore capable of saltatory conduction. They conduct impulses at speeds of about 10 m/sec. B fibers are found in nerves that transmit impulses from the skin and viscera to the brain and spinal cord. They also constitute

FIGURE 12-9 Saltatory conduction. (a) The nerve impulse at the first node generates a local current that passes to the second node. (b) At the second node, the local current generates a nerve impulse. Then, the nerve impulse from the second node generates a local current that passes to the third node, and so on. After the nerve impulse jumps from node to node, each node becomes repolarized.

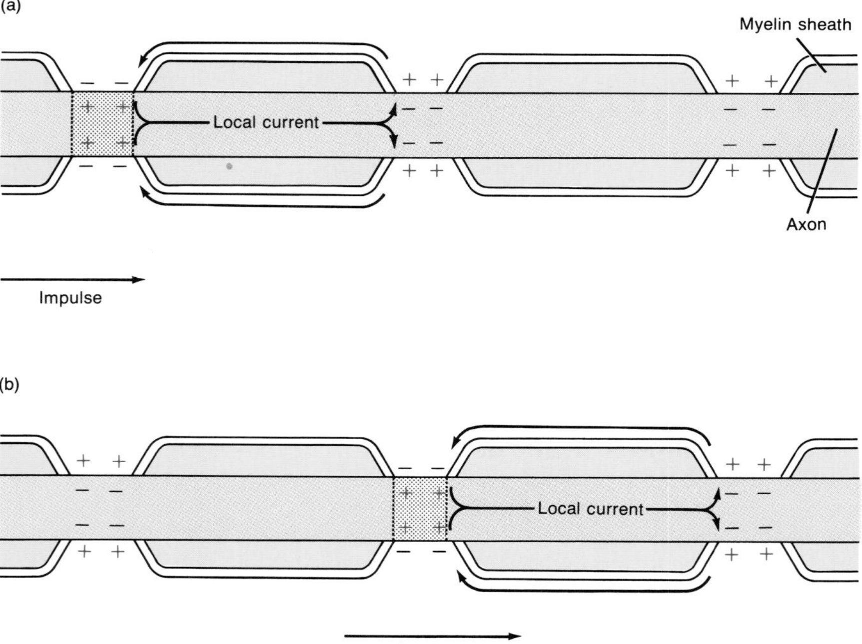

all the axons of the visceral efferent neurons located in the motor nerves that leave the lower part of the brain and spinal cord and terminate in relay stations called ***ganglia*** (GANG-lē-a). The ganglia ultimately link with other fibers that stimulate the smooth muscle and glands of the viscera.

C fibers have the smallest diameter and the longest absolute refractory periods. They conduct nerve impulses at the rate of about 0.5 m/sec. C fibers are unmyelinated and incapable of saltatory conduction. They are located in nerves that conduct impulses from the skin and in visceral nerves. These fibers conduct impulses for pain and perhaps impulses for touch, pressure, heat, and cold from the skin and pain receptors from the viscera. C fibers are located in all motor nerves that lead from the ganglia and stimulate the smooth muscle and glands of the viscera. Examples of the motor functions of B and C fibers are constricting and dilating the pupils, increasing and decreasing the heart rate, and contracting and relaxing the urinary bladder—functions of the autonomic nervous system.

Conduction Across Synapses

A nerve impulse is conducted not only along the length of a neuron but also from one neuron to another or to an effector such as a muscle or gland. Impulses are conducted from a neuron to a muscle fiber (cell) across an area of contact called a ***neuromuscular (myoneural) junction,*** which was discussed in detail in Chapter 10 (Figure 10-6). The area of contact between a neuron and glandular cells is known as a ***neuroglandular junction.*** Together, neuromuscular and neuroglandular junctions are known as ***neuroeffector junctions.***

Impulses are conducted from one neuron to another across a ***synapse***—a junction between two neurons. The term *synapsis* means "connection." The synapse is essential for homeostasis because of its ability to transmit certain impulses and inhibit others. Much of an organism's ability to learn will probably be explained in terms of synapses. Moreover, most diseases of the brain and many psychiatric disorders result from a disruption of synaptic communication. Synapses are also the sites of action for most drugs that affect the brain, including therapeutic and addictive substances. Figure 12-10 shows that within a synapse is a minute space, filled with extracellular fluid, about 20 nm across called the ***synaptic cleft.*** A ***presynaptic neuron*** is a neuron located before a synapse. A ***postsynaptic neuron*** is located after a synapse.

Axon terminals of neurons end in bulblike structures referred to as ***synaptic end bulbs.*** The synaptic end bulbs of a presynaptic neuron may synapse with the dendrites, cell body, or axon hillock of a postsynaptic neuron. Accordingly, synapses may be classified as ***axodendritic, axosomatic,*** and ***axoaxonic.*** The synaptic end bulbs

FIGURE 12-10 Impulse conduction at synapses. Shown is nerve impulse conduction from a presynaptic end bulb across a synapse to a postsynaptic dendrite in which synaptic vesicles fuse with the presynaptic membrane and discharge neurotransmitter into the synaptic cleft.

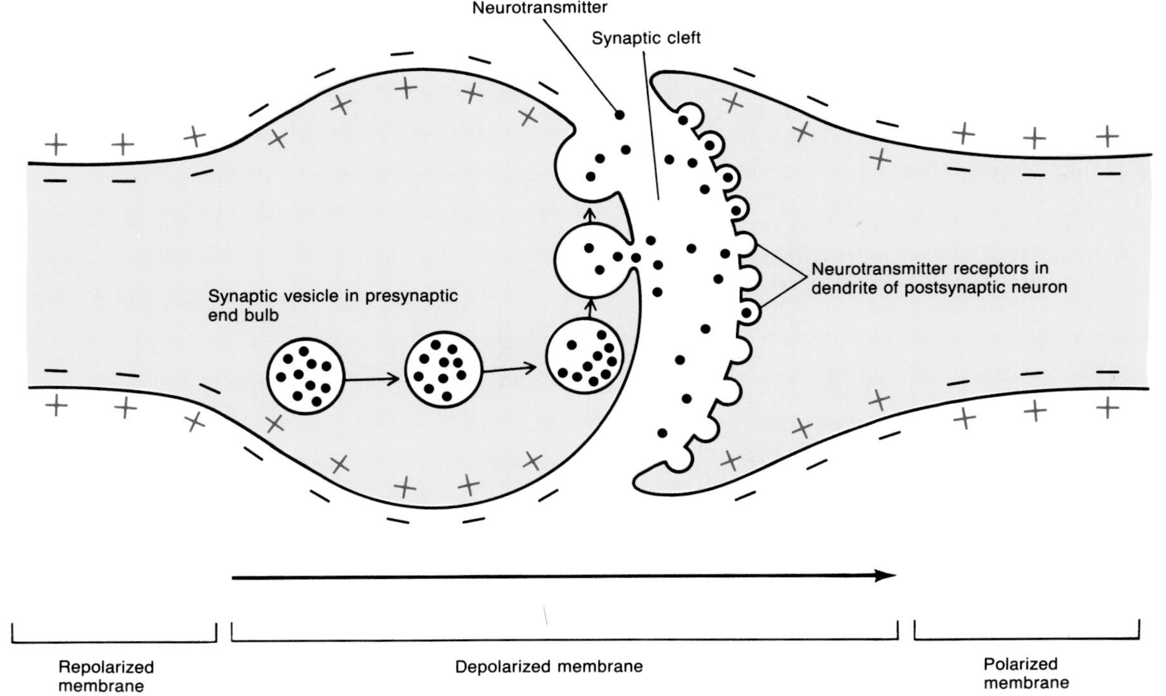

from a single presynaptic neuron may synapse with several postsynaptic neurons. Such an arrangement, called ***divergence,*** permits a single presynaptic neuron to influence several postsynaptic neurons or several muscle fibers or gland cells at the same time (see Figure 12-15a). In another arrangement, called ***convergence,*** the synaptic end bulbs of several presynaptic neurons synapse with a single postsynaptic neuron (see Figure 12-15b). This arrangement permits more effective stimulation or inhibition of the postsynaptic neuron.

Nerve impulses are conducted across one of two types of synapses: electrical and chemical. At ***electrical synapses*** a nerve impulse passes from one cell to another through small, protein, tubular structures called ***gap junctions.*** You may recall that visceral smooth muscle and cardiac muscle tissue fibers (cells) both contain gap junctions for conduction of muscle action potentials between muscle fibers (cells). Some gap junctions have been observed in different parts of the central nervous system, but their significance is not known.

In a ***chemical synapse,*** a neuron secretes a chemical substance called a ***neurotransmitter (transmitter substance)*** that acts on receptors of the next neuron at a neuronal synapse, a muscle fiber at a neuromuscular junction, or a glandular cell at a neuroglandular junction. Almost all synapses in the CNS are chemical synapses. Neurotransmitters are made by the neuron, usually from amino acids. Following its production and transportation to the synaptic end bulbs, the neurotransmitter is stored in the end bulbs in small membrane-enclosed sacs called ***synaptic vesicles*** (Figure 12-10). Each of the thousands of synaptic vesicles present may contain between 10,000 and 100,000 neurotransmitter molecules.

When a nerve impulse arrives at a synaptic end bulb of a presynaptic neuron, it is believed that some calcium ions enter the end bulb from interstitial fluid, attract synaptic vesicles to the plasma membrane, and help to liberate the neurotransmitter molecules from the vesicles into the cleft at the synapse. In general, each neuron liberates only one type of neurotransmitter at all of the synaptic end bulbs. Some scientists believe that the synaptic vesicles fuse with the plasma membrane of the presynaptic neuron, form openings, and release the neurotransmitter through the openings into the synaptic cleft, a type of exocytosis. Others think that neurotransmitter stored in cytoplasm (rather than in synaptic vesicles) leaves through small protein channels formed under the influence of calcium ions and enters the synaptic cleft. In either case, the neurotransmitter enters the synaptic cleft, and depending on the chemical nature of the neurotransmitter and the interaction of the neurotransmitter with receptors of the postsynaptic plasma membrane, several things can happen at the postsynaptic cells (described shortly).

At a chemical synapse, there is only ***one-way impulse conduction***—from a presynaptic axon to a postsynaptic dendrite, cell body, or axon hillock. This is because only synaptic end bulbs of presynaptic neurons can release neurotransmitter. As a result, nerve impulses must move forward over their pathways. They cannot back up into another presynaptic neuron, a situation that would seriously disrupt homeostasis. Such a mechanism is crucial in preventing nerve impulse conduction along improper pathways.

Excitatory Transmission

An ***excitatory transmitter-receptor interaction*** is one that can lower (make less negative) the postsynaptic neuron's membrane potential so that a new nerve impulse can be generated across the synapse. If the potential is lowered enough, that is, reaches threshold for the postsynaptic neuron, the membrane becomes depolarized, the potential inside the membrane becomes positive and the potential outside becomes negative, and a nerve impulse is initiated. Generally, the release of a neurotransmitter by a single presynaptic end bulb is not sufficient to develop an action potential in a postsynaptic neuron. However, its release does bring the resting membrane potential closer to threshold level as Na^+ ions move into the cell through sodium channels. This change from the resting membrane potential level in the direction of the threshold level is called the ***excitatory postsynaptic potential (EPSP)*** (Figure 12-11a). The EPSP is always less negative than the

FIGURE 12-11 **Comparison between (a) EPSP and (b) IPSP.**

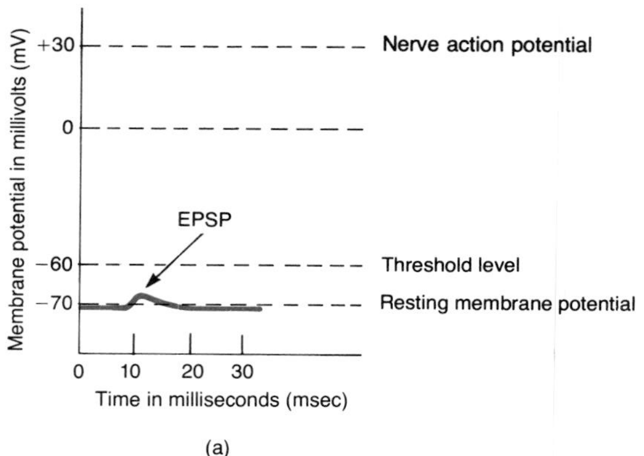

(a)

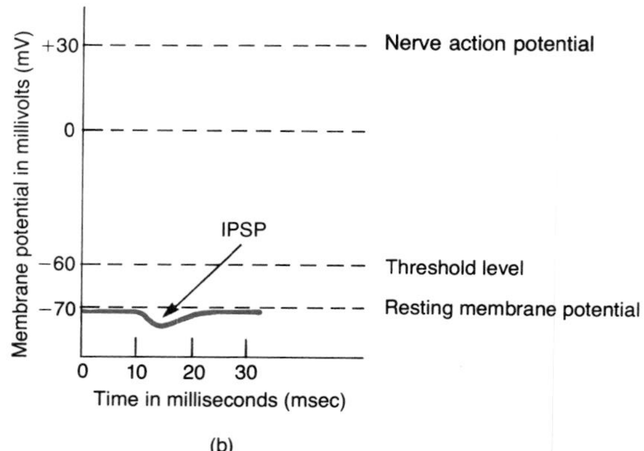

(b)

resting membrane potential of the neuron but more negative than its threshold level, that is, it is a depolarization that does not reach or exceed threshold. Even though the release of a neurotransmitter by a single impulse in a presynaptic end bulb is not sufficient to initiate an action potential in a postsynaptic neuron, the postsynaptic neuron does become more excitable to impulses from presynaptic neurons, so that the possibility for generating an impulse is greater. This depolarizing effect is called ***facilitation,*** that is, near excitation so that subsequent stimuli can more easily generate an impulse. In other words, the postsynaptic neuron is prepared for subsequent stimuli that can trigger an impulse because it is already partially depolarized and thus more likely to reach threshold when a subsequent excitatory stimulus occurs.

The EPSP lasts only a few milliseconds. If several presynaptic end bulbs release their neurotransmitter at about the same time, however, the combined effect may initiate a nerve impulse—an effect known as ***summation.*** If hundreds of presynaptic end bulbs release their neurotransmitter simultaneously, they increase the chance of initiating a nerve impulse in the postsynaptic neuron. The greater the summation, and therefore the depolarization, the greater the probability an impulse will be initiated. When the summation is the result of the accumulation of neurotransmitter from several presynaptic end bulbs, it is called ***spatial summation*** (Figure 12-12a). When summation is the result of the accumulation of neurotransmitter from a single presynaptic end bulb firing two or more times in rapid succession, it is called ***temporal summation*** (Figure 12-12b). Since the EPSP lasts only about 15 msec, the second firing must follow quickly if temporal summation is to occur. The time required for the impulse to actually cross a synaptic cleft—the ***synaptic delay***—is about 0.5 msec. This delay is caused by the liberation of neurotransmitter, its passage across the synapse, stimulation of the postsynaptic neuron to become more permeable to Na^+ ions, and the inward movement of Na^+ ions through voltage-sensitive sodium channels that initiates the nerve impulse in the postsynaptic neuron.

Excitatory transmitter-receptor interactions are thought to occur by two different mechanisms. In the first mechanism, neurotransmitters, such as acetylcholine, bind to proteins on the postsynaptic plasma membrane, called ***neurotransmitter receptors*** (see Figures 12-10 and 12-13). This binding causes sodium channels in the plasma membrane to become more permeable to Na^+ ions. The inward movement of Na^+ ions through sodium channels results in depolarization and initiation of a nerve impulse if the depolarization reaches threshold. Other neurotransmitters, such as catecholamines (norepinephrine, epinephrine, and dopamine), use a different mechanism. They bind to neurotransmitter receptors and activate an enzyme in the postsynaptic membrane called ***adenylate cyclase*** (ad-EN-il-āt SĪ-klās). Once activated, the enzyme converts ATP into ***cyclic AMP.*** Cyclic AMP then activates enzymes that increase the permeability of the postsynaptic membrane to Na^+ ions by opening sodium channels through which the Na^+ ions pass. This initiates the nerve impulse in the postsynaptic neuron if the depolarization reaches threshold. In this mechanism, the neurotransmitter is considered to be the first messenger and cyclic AMP the second messenger. The second-messenger role of cyclic AMP in mediating certain hormonal responses is considered in detail in Chapter 18.

As you will see shortly, once an excitatory neurotransmitter attaches to a receptor site, it must be rapidly inactivated or it will stimulate (depolarize) the postsynaptic neuron, muscle, or gland indefinitely. This would prevent repolarization of the membrane, which must occur in order for normal neuron function.

FIGURE 12-12 Summation. (a) Spatial summation. When two axons (a and b) are stimulated separately (arrows), threshold level is not reached. When axons a and b, plus many others, act simultaneously on the postsynaptic cell, their depolarizations summate to reach the threshold level and trigger a nerve action potential. (b) Temporal summation. Stimuli applied to the same axon (arrows) sufficiently close together in time cause the membrane to depolarize. Depolarization may reach the threshold level, triggering a nerve action potential.

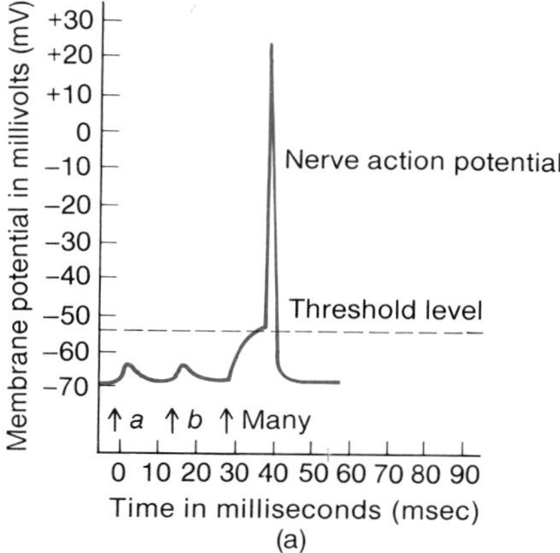

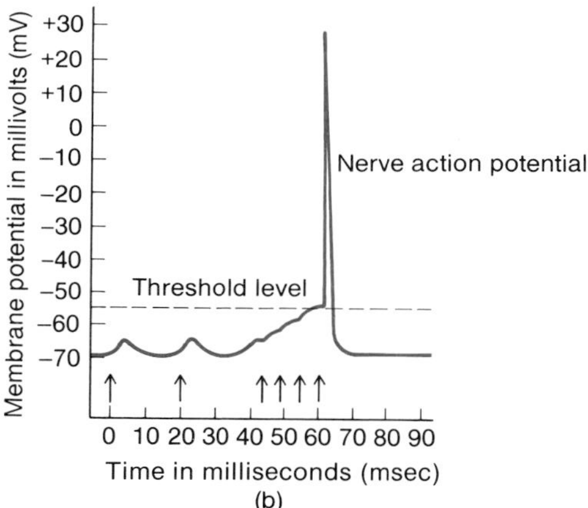

Inhibitory Transmission

An ***inhibitory transmitter-receptor interaction*** is one that can inhibit nerve impulse generation at a synapse. Whereas excitatory transmitter-receptor interactions make the postsynaptic neuron's resting membrane potential *less* negative (cause depolarization), inhibitory transmitter-receptor interactions make the postsynaptic neuron's resting membrane potential *more* negative. This is referred to as ***hyperpolarization.*** When it is at rest, the cell interior becomes even more negative in comparison to the outside, making it even more difficult for the neuron to generate an impulse because its membrane potential would be even farther from threshold than it is in a resting state. The alteration of the postsynaptic membrane in which the resting membrane potential is made more negative is called the ***inhibitory postsynaptic potential (IPSP)*** (Figure 12-11b). When the neurotransmitter attaches to the receptor site, the membrane becomes more permeable to K^+ and/or Cl^- ions. Permeability to Na^+ ions is not affected. When potassium channels are open, K^+ ions move to the exterior of the membrane. When chloride channels are open, Cl^- ions move to the interior of the membrane. As a result, there is an increase in internal negativity, or hyperpolarization. In contrast, the EPSP is less negative than the resting membrane potential of a neuron, and the IPSP is more negative than the resting membrane potential.

In addition to IPSP as a means of inhibiting nerve impulses in postsynaptic neuronal membranes (postsynaptic inhibition), there is another type of inhibition that occurs *before* a nerve impulse reaches a synapse. This is called ***presynaptic inhibition.*** In this process, a synaptic end bulb of an inhibitory neuron synapses with the synaptic end bulb of a presynaptic neuron at an excitatory synapse. When the inhibitory neuron releases neurotransmitter, it depresses the release of excitatory transmitter by the presynaptic neuron. This, in turn, decreases stimulation of the postsynaptic neuron.

Presynaptic inhibition occurs in many sensory pathways of the nervous system and provides a means by which responses of postsynaptic neurons can be regulated. Whereas presynaptic inhibition may last for several minutes or hours, postsynaptic inhibition lasts only milliseconds.

Integration at Synapses

A single postsynaptic neuron receives synapses from many presynaptic neurons. Some presynaptic end bulbs produce excitation and some produce inhibition. The sum of all the effects, excitatory and inhibitory, determines the effect on the postsynaptic neuron. Thus, the postsynaptic neuron is an ***integrator.*** It receives signals, integrates them, and then responds accordingly. The postsynaptic neuron may respond in the following ways:

1. If the excitatory effect is greater than the inhibitory effect, but less than the threshold level of stimulation, the result is ***facilitation,*** that is, near threshold excitation so that subsequent stimuli can more easily generate a nerve impulse.
2. If the excitatory effect is greater than the inhibitory effect, but equal to or higher than the threshold level of stimulation, the result is ***generation of one or many impulses,*** one after another.
3. If the inhibitory effect is greater than the excitatory effect, the membrane hyperpolarizes, and the result is ***inhibition of the postsynaptic neuron*** and thus an inability to generate a nerve impulse.

Neurotransmitters

Perhaps the best-studied neurotransmitter is ***acetylcholine*** (as'-ē-til-KŌ-lēn), or ***ACh.*** It is a neurotransmitter released by many neurons outside the brain and spinal cord and by some neurons inside the brain and cord. This neurotransmitter was discussed briefly in Chapter 10 in relation to nerve impulse conduction from a motor neuron to a muscle fiber (cell) across a neuromuscular junction. Following the arrival of an impulse at the synaptic end bulb of an axon terminal, calcium (Ca^{2+}) ions enter the end bulb and cause the release of ACh from synaptic vesicles or the cytoplasm of the end bulb (Figure 12-13). ACh is released into the synaptic cleft in multiple units called *quanta.* One quantum contains between 1,000 and 10,000 molecules. At neuromuscular junctions of skeletal muscle fibers, ACh binds to neurotransmitter receptors on the muscle fiber membrane and increases its permeability to Na^+ and K^+ ions through their respective channels. The neurotransmitter (ACh) receptor is an integral protein in the plasma membrane of the muscle fiber. When two ACh molecules bind to the receptor, a change occurs in the receptor, causing its channel to open. As a result, there is an inward movement of Na^+ ions. In the absence of ACh, the receptor channel remains closed. The inward movement of Na^+ ions depolarizes the membrane to threshold, and ultimately a muscle action potential is generated, causing the muscle fiber to contract.

As long as ACh is present in the neuromuscular junction, it can stimulate a muscle fiber almost indefinitely. The transmission of a continuous succession of action potentials by ACh is normally prevented by an enzyme called ***acetylcholinesterase (AChE),*** or simply ***cholinesterase*** (kō'-lin-ES-ter-ās). AChE is found on the outside surface of the subneural clefts of the membrane of muscle fibers. Within 1/500 sec, AChE inactivates ACh by breaking ACh down into its components, acetate and choline. This action permits the membrane of the muscle fiber to repolarize almost immediately so that another muscle action potential may be generated. When the next nerve impulse comes through, the synaptic vesicles release more ACh, another muscle action potential is generated, and AChE again inactivates ACh. This cycle is repeated over and over

FIGURE 12-13 Release, action, breakdown, and synthesis of acetylcholine (ACh). (1) A nerve action potential causes (2) calcium (Ca^{2+}) ions to (3) enter a synaptic end bulb and (4) induce the release of ACh from a synaptic vesicle. (5) ACh diffuses across the synaptic cleft and two molecules of ACh bind to ACh receptors on the muscle fiber membrane. In the absence of ACh, the channel remains closed. When two molecules of ACh bind to the receptor, (6) a change occurs, permitting positive ions, such as Na^+ ions, to pass through. The ACh–receptor interaction ultimately depolarizes the membrane and a (7) muscle action potential is generated, thus causing a muscle fiber to contract. Once it has accomplished its function, ACh is (8) broken down by AChE into acetate and choline to prevent continuous generation of action potentials. Within the synaptic end bulb, (9) acetate and choline combine to form (10) ACh.

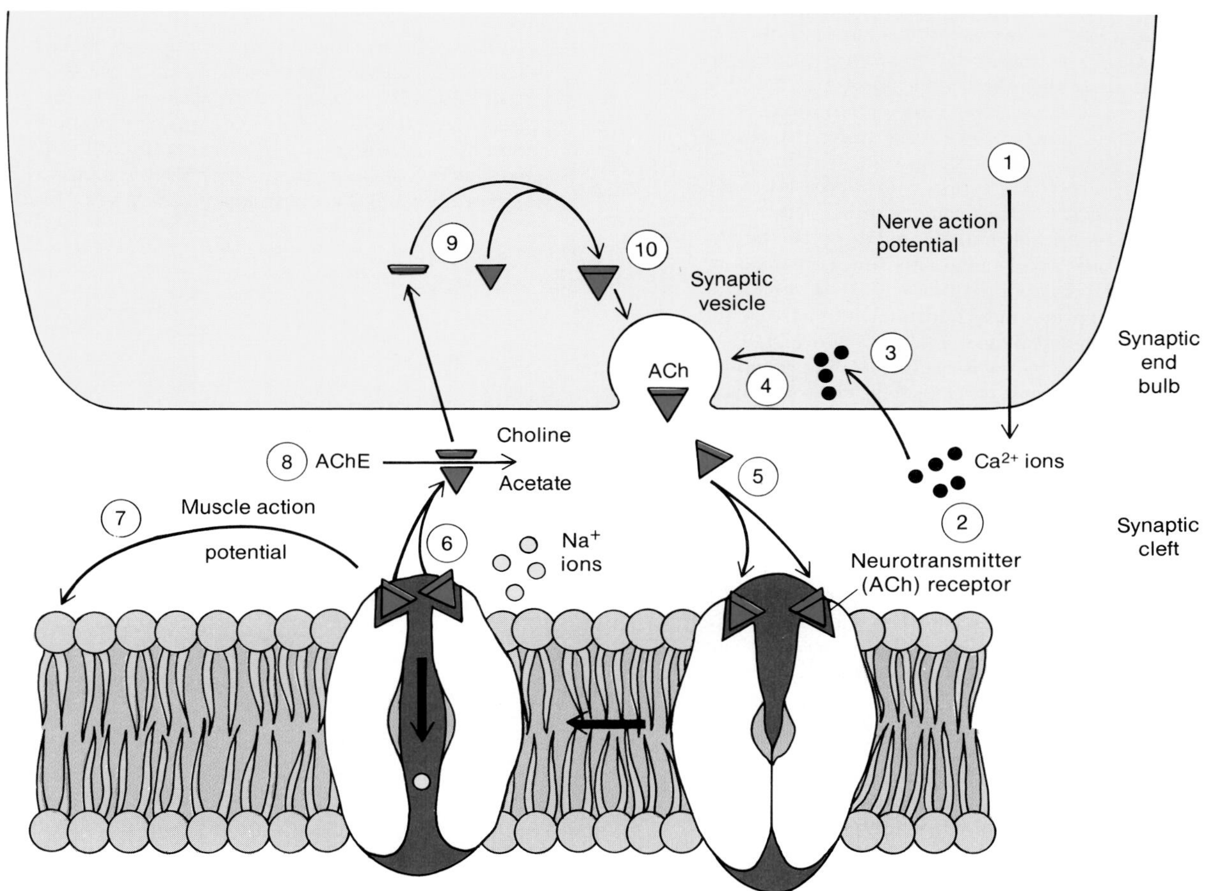

again. Eventually, choline reenters the synaptic end bulb, where it combines with acetate produced in the end bulb to form more ACh. The synthesis of ACh occurs in the presence of the enzyme choline acetyltransferase. Then, new synaptic vesicles form and ACh is distributed into vesicles and cytoplasm of the axon terminal.

ACh is released at some neuromuscular junctions that contain cardiac and smooth muscle, as well as those that contain skeletal muscle. It is also released at some neuroglandular junctions. In the brain and spinal cord, it appears that the effects of ACh are mediated through cyclic AMP. Although ACh leads to excitation in many parts of the body, it is inhibitory with respect to the heart [vagus (X) nerve].

There is considerable evidence that a substance called ***gamma aminobutyric*** (GAM-ma am-i-nō-byoo-TĒR-ik) ***acid (GABA)*** leads to inhibition of certain neurons in the central nervous system. It probably exerts its effect by hyperpolarizing (making more negative) the postsynaptic membrane according to the mechanism previously described. The amino acid ***glycine*** leads to inhibition of certain neurons in the spinal cord.

Neurotransmitters in the central nervous system will be discussed in more detail in Chapter 14. At that time, you will also learn that certain disorders such as Parkinson's disease, Alzheimer's disease, depression, anxiety, and schizophrenia involve improperly functioning neurotransmitters.

CLINICAL APPLICATION: INTERFERENCE WITH SYNAPTIC CONDUCTION

There are many ways that ***synaptic conduction can be altered*** by disease, drugs, and pressure. In Chapter 10 it was noted that *myasthenia gravis* results from antibodies directed against acetylcholine receptors on skeletal muscle fiber membranes at neuromuscular junctions, causing dysfunctions in skeletal muscular contractions. *Alkalosis,* an increase in pH above 7.45, results in increased excitability of neurons that can cause lightheadedness, numbness around the mouth, tingling in the fingertips, nervousness, muscle spasms, and convulsions. *Acidosis,* a decrease in pH below 7.35, results in a progressive depression of neuronal activity that can produce apathy, weakness, and a comatose state.

Curare also competes for acetylcholine receptor sites and can thus prevent muscular contractions. Curare-like drugs are often used in surgery to increase muscle relaxation during certain phases of an operation. *Neostigmine* and *physostigmine* are anticholinesterase agents that combine with acetylcholinesterase to inactivate it for several hours with the result being a reduction in the removal of acetylcholine from its receptors. Neostigmine is the antidote to curare that can be used to terminate curare effects during surgery or in case of accidental curare poisoning. Both neostigmine and physostigmine can be used to treat myasthenia gravis and glaucoma (physostigmine drops constrict the pupil). *Diisopropyl fluorophosphate* is a very powerful nerve gas that is found in many insecticides. It inactivates acetylcholinesterase for up to several weeks, making it a particularly lethal drug. It may cause nausea, diarrhea, sweating, bronchial constriction, excess respiratory mucus, generalized weakness, and fasciculation of skeletal muscles. The *botulism toxin* inhibits the release of acetylcholine, thus inhibiting muscle contraction. The toxin is exceedingly potent in even small amounts (less than 0.0001 mg) and is the substance involved in one type of food poisoning. *Hypnotics, tranquilizers,* and *anesthetics* depress synaptic conduction by increasing the threshold for excitation of neurons, whereas *caffeine, benzedrine,* and *nicotine* reduce the threshold for excitation of neurons and result in facilitation.

Crack, a potent form of cocaine that is smoked rather than sniffed, interferes with the normal functioning of neurotransmitters—dopamine (DA), norepinephrine (NE), and serotonin (5-HT)—that are involved in the regulation of mood and motor functions. Once a neurotransmitter has accomplished its function, it is inactivated and returned to the presynaptic neuron for resynthesis. Initially, crack inhibits the inactivation of the neurotransmitters. The resultant buildup of dopamine, in particular, has been linked to feelings of euphoria. Inhibition or inactivation of neurotransmitters also can cause convulsions, accelerated and abnormal heart rate, vasoconstriction and high blood pressure, weight loss, insomnia, and susceptibility to disease. Repeated use of crack may produce a temporary shortage of neurotransmitters, resulting in depression, anxiety, and craving for more crack. Heavy, prolonged use of crack may eventually deplete neurotransmitters to the point where euphoria no longer occurs and depression is persistent.

Pressure has an effect on nerve impulse conduction also. If excessive or prolonged pressure is applied to a nerve, as when crossing one's legs, impulse conduction is interrupted, and part of the body may "go to sleep," producing a tingling sensation. This sensation is caused by an accumulation of waste products and a depressed circulation of blood.

Regeneration

Unlike the cells of epithelial tissue, neurons have only limited powers for ***regeneration,*** that is, a natural ability to renew themselves. Around 6 months of age, the cell bodies of most developing nerve cells lose their mitotic apparatus (centrioles and mitotic spindles) and their ability to reproduce. Thus, when a neuron is damaged or destroyed, it cannot be replaced by the daughter cells of other neurons. A neuron destroyed is permanently lost, and only some types of damage may be repaired.

Damage to some types of peripheral myelinated axons or some dendrites often can be repaired if the cell body remains intact, if the cell that performs the myelination remains active, and if a completely severed nerve is surgically reattached. Some axons and dendrites in the peripheral nervous system are myelinated by neurolemmocytes (Schwann cells). They proliferate following axonal damage, and their neurolemmas form a tube that assists in regeneration (Chapter 13). Axons in the brain and spinal cord (central nervous system) are myelinated by oligodendrocytes. These cells do not form neurolemmas to assist in regeneration and do not survive following axonal damage. An added complication in the central nervous system is that following axonal damage astrocytes appear to stop axons from regenerating by activating a physiological pathway that inhibits axonal regeneration. This is the same pathway that stops axonal growth during development once a target cell has been reached. In addition, following axonal damage, the affected region is rapidly converted into a special form of scar tissue by astroglial proliferation. The scar tissue forms a physical barrier to regeneration. Thus, an injury to the brain or spinal cord is also permanent because axonal regeneration is blocked by a physiological stop-pathway and rapid scar tissue formation. An injury to a nerve in the arm (peripheral nervous system) may repair itself before scar tissue forms, and so some nerve function may be restored.

ORGANIZATION OF NEURONS

In subsequent chapters, we will be concerned with the structure and physiology of the nervous system. To give you a better understanding of how your nervous system helps maintain homeostasis, we now examine how synapses are organized into functional units.

The central nervous system contains billions of neurons. Their arrangement is not haphazard. They are organized into definite patterns called ***neuronal pools.*** Each pool differs from all others and has its own role in regulating homeostasis.

A neuronal pool may contain thousands or even millions of neurons. To illustrate the composition of a neuronal pool, a simplified version is given in Figure 12-14. This example contains only five postsynaptic neurons and two incoming presynaptic neurons. The postsynaptic neurons are subject to stimulation by the incoming presynaptic neurons. The postsynaptic neurons in the pool may be stimulated by one or several presynaptic end bulbs. In addition, varying synaptic delays may cause impulses to arrive at postsynaptic neurons at different times. Moreover, the incoming presynaptic end bulbs may produce facilitation, excitation, or inhibition. A principal feature of a neuronal pool is the location of the presynaptic neurons in relation to the postsynaptic neurons. Compare, for example, the location of presynaptic axon 1 with that of postsynaptic neuron B. Since they are aligned, more presynaptic end bulbs of axon 1 synapse with postsynaptic neuron B than with postsynaptic neurons A or C. We say

FIGURE 12-14 Relative positions of discharge and facilitated zones in a very simplified version of a neuronal pool.

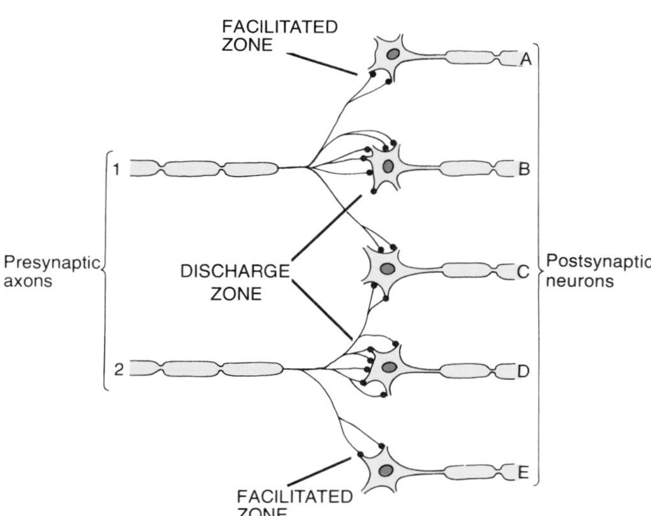

that postsynaptic neuron B is in the center of the field of presynaptic axon 1. Consequently, postsynaptic neuron B usually receives sufficient presynaptic end bulbs from axon 1 to generate nerve impulses. This region where the neuron in the pool fires is called the ***discharge zone.***

Now look outside the field of presynaptic axon 1. Note its relation to postsynaptic neuron A. Here, postsynaptic neuron A in the pool is receiving few presynaptic end bulbs from the axon supplying postsynaptic neuron B. Thus, there are insufficient presynaptic end bulbs to fire postsynaptic neuron A but enough to cause facilitation. We therefore call this area the ***facilitated zone.*** Presynaptic axon 1, when stimulated, will cause excitation of postsynaptic neuron B and facilitation of postsynaptic neuron A. In the example shown, postsynaptic neuron C can generate impulses if presynaptic neurons 1 and 2 send action potentials simultaneously.

Neuronal pools in the central nervous system are arranged in patterns over which the impulses are conducted. These are termed ***circuits. Simple series circuits*** are arranged so that a presynaptic neuron stimulates a single neuron in a pool. The single neuron then stimulates another, and so on. In other words, the impulse is relayed from one neuron to another in succession as new impulses are generated at each synapse.

Most circuits, however, are more complex. In a ***diverging circuit,*** the nerve impulse from a single presynaptic neuron causes the stimulation of increasing numbers of cells along the circuit (Figure 12-15a). An example of such a circuit is a single motor neuron in the brain stimulating numerous other motor neurons in the spinal cord that, in turn, leave the spinal cord where each stimulates many skeletal muscle fibers. Thus, a single impulse may result in the contraction of several skeletal muscle fibers. In another kind of diverging circuit, impulses from one pathway are relayed to other pathways so the same information travels in various directions at the same time. This circuit is common along sensory pathways of the nervous system.

Another kind of circuit is called a ***converging circuit*** (Figure 12-15b). In one pattern of convergence, the postsynaptic neuron receives impulses from several fibers of the same source. Here, there is the possibility of strong excitation or inhibition. In a second pattern, the postsynaptic neuron receives impulses from several different sources. Here, there is a possibility of reacting the same way to different stimuli. Suppose your reaction to vomit is distinctly unpleasant. The smell of vomit (one kind of stimulus), the sight of vomit (another kind), or just reading about vomit (still another kind) might all have the same effect on you—an unpleasant one.

Some circuits in your body are constructed so that once the presynaptic cell is stimulated, it will cause the postsynaptic cell to transmit a series of nerve impulses. One such circuit is called a ***reverberating (oscillatory) circuit*** (Figure 12-15c). In this pattern, the incoming impulse stimulates the first neuron, which stimulates the second, which stimulates the third, and so on. Branches from the second and third neurons synapse with the first,

FIGURE 12-15 **Circuits of neuronal pools. (a) Diverging. (b) Converging. (c) Reverberating. (d) Parallel after-discharge.**

Divergence in a single pathway

Divergence in several pathways

(a)

Presynaptic neuron

Postsynaptic neuron

Postsynaptic neuron

Convergence involving one presynaptic neuron

(b)

Convergence involving several presynaptic neurons

(c)

(d)

however, sending the impulse back through the circuit again and again. A central feature of the reverberating circuit is that once fired, the output signal may last from a few seconds to many hours. The duration depends on the number and arrangement of neurons in the circuit. Among the body responses thought to be the result of output signals from reverberating circuits are the rate of breathing, coordinated muscular activities, waking up, and sleeping (when reverberation stops). Some scientists think reverberating circuits are related to short-term memory. One form of epilepsy (grand mal) is probably caused by abnormal reverberating circuits.

A final circuit worth consideration is the ***parallel after-discharge circuit*** (Figure 12-15d). Like a reverberating circuit, a parallel after-discharge circuit is constructed so the postsynaptic cell transmits a series of nerve impulses. In a parallel after-discharge circuit, a single presynaptic cell stimulates a group of neurons, each of which synapses with a common postsynaptic cell. The advantage of this circuit is that the postsynaptic neuron can send out a stream of impulses in succession as they are received. The impulses leave the postsynaptic neuron once every 1/2000 second. This circuit has no feedback system. Once all the neurons in the circuit have transmitted their impulses to the postsynaptic neuron, the circuit is broken. It is thought that the parallel after-discharge circuit is employed for precise activities such as mathematical calculations.

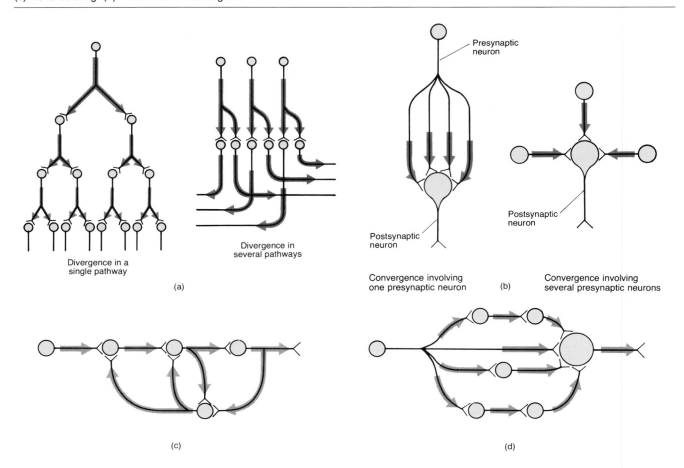

STUDY OUTLINE

Organization (p. 332)

1. The nervous system helps control and integrate all body activities by sensing changes (sensory), interpreting them (integrative), and reacting to them (motor).
2. The central nervous system (CNS) consists of the brain and spinal cord.
3. The peripheral nervous system (PNS) is classified into an afferent system and an efferent system.

4. The efferent system is subdivided into a somatic nervous system and an autonomic nervous system.
5. The somatic nervous system (SNS) consists of efferent neurons that conduct impulses from the central nervous system to skeletal muscle tissue.
6. The autonomic nervous system (ANS) contains efferent neurons that convey impulses from the central nervous system to smooth muscle tissue, cardiac muscle tissue, and glands.

Histology (p. 332)

Neuroglia (p. 333)

1. Neuroglia are specialized tissue cells that support neurons, attach neurons to blood vessels, produce the myelin sheath around axons of the CNS, and carry out phagocytosis.
2. Neuroglial cells include astrocytes, oligodendrocytes, microglia, and ependyma.

Neurons (p. 334)

1. Most neurons, or nerve cells, consist of a cell body (perikaryon), dendrites that pick up stimuli and convey nerve impulses to the cell body, and usually a single axon. The axon conducts nerve impulses from the neuron to the dendrites or cell body of another neuron or to an effector organ of the body (muscle or gland).
2. On the basis of structure, neurons are multipolar, bipolar, and unipolar.
3. On the basis of function, sensory (afferent) neurons conduct impulses from receptors to the central nervous system; association (connecting) neurons conduct impulses to other neurons, including motor neurons; and motor (efferent) neurons conduct impulses to effectors.

Physiology (p. 339)

Nerve Impulse (p. 339)

1. The nerve impulse (nerve action potential) is the body's quickest way of controlling and maintaining homeostasis.
2. The membrane of a nonconducting neuron is positive outside and negative inside owing to the permeability property of the membrane toward Na^+ ions, K^+ ions, charged proteins, and the operation of the sodium-potassium pump, and the membrane is said to be polarized.
3. When a stimulus causes the inside of the cell membrane to become positive and the outside negative, the membrane is said to have an action potential, which travels from point to point along the membrane. The traveling action potential is a nerve impulse. The ability of a neuron to respond to a stimulus and convert it into a nerve impulse is called excitability.
4. Restoration of the resting membrane potential is called repolarization. The period of time during which the membrane recovers and cannot initiate another action potential is called the absolute refractory period.
5. According to the all-or-none principle, if a stimulus is strong enough to generate an action potential, the impulse travels at a constant and maximum strength for the existing conditions. A stronger stimulus will not cause a larger impulse.
6. Nerve impulse conduction in which the impulse jumps from neurofibral node to node is called saltatory conduction.
7. Fibers with larger diameters conduct impulses faster than those with smaller diameters; myelinated fibers conduct impulses faster than unmyelinated fibers.

Conduction Across Synapses (p. 345)

1. Nerve impulse conduction can occur from one neuron to another or from a neuron to an effector.
2. The junction between neurons is called a synapse.
3. At a chemical synapse, there is only one-way nerve impulse conduction from a presynaptic axon to a postsynaptic dendrite, cell body, or axon hillock.
4. An excitatory transmitter-receptor interaction is one that can depolarize or lower (make less negative) the postsynaptic neuron's membrane potential, so that a new impulse can be generated across the synapse.
5. Facilitation refers to a state of near threshold excitation, so that subsequent stimuli can generate an impulse more easily.
6. If several presynaptic end bulbs release their neurotransmitter at about the same time, the combined effect may generate a nerve impulse, resulting in the phenomenon referred to as summation. Summation may be spatial or temporal.
7. The time required for neurotransmitter to diffuse across a synaptic cleft, about 0.5 msec, is called synaptic delay.
8. An inhibitory transmitter-receptor interaction is one that can raise (make more negative or hyperpolarize) the postsynaptic neuron's membrane potential and thus inhibit an impulse at a synapse.
9. The postsynaptic neuron is an integrator. It receives signals, integrates them, and then responds accordingly.

Neurotransmitters (p. 348)

1. It is thought that the neurotransmitter that causes excitation in a major portion of the central nervous system is acetylcholine (ACh). An enzyme called acetylcholinesterase (AChE) inactivates ACh.
2. Neurotransmitters that are probably inhibitory are gamma aminobutyric acid (GABA) and glycine.

Regeneration (p. 350)

1. Around 6 months of age, the neuronal cell body loses its mitotic apparatus and is no longer able to divide.
2. Nerve fibers (axis cylinders) that have a neurolemma are capable of regeneration.

Organization of Neurons (p. 351)

1. Neurons in the central nervous system are organized into definite patterns called neuronal pools. Each pool differs from all others and has its own role in regulating homeostasis.
2. Neuronal pools are organized into circuits. These include simple series, diverging, converging, reverberating (oscillatory), and parallel after-discharge circuits.

REVIEW QUESTIONS

1. Describe the three basic functions of the nervous system that are necessary to maintain homeostasis. (p. 332)
2. Distinguish between the central and peripheral nervous systems and describe the functions of each subdivision. (p. 332)

3. Relate the terms *voluntary* and *involuntary* to the nervous system. (p. 332)
4. What are neuroglia? List the principal types and their functions. Why are they important clinically? (p. 333)
5. Define a neuron. Diagram and label a neuron. Next to each part, list its function. (p. 335)
6. Distinguish between axoplasmic flow and axonal transport. Why is axonal transport important clinically? (p. 336)
7. What is a myelin sheath? How is it formed? (p. 336)
8. Define the neurolemma. Why is it important? (p. 336)
9. Discuss the structural classification of neurons. Give an example of each. (p. 337)
10. What are the structural differences between a typical afferent and efferent neuron? Give several examples of association neurons. (p. 338)
11. Describe the functional classification of neurons. (p. 338)
12. Distinguish among the following kinds of fibers: general somatic afferent, general somatic efferent, general visceral afferent, and general visceral efferent. (p. 338)
13. Define excitability. (p. 340)
14. Outline the principal steps in the origin and conduction of a nerve impulse. (p. 339)
15. Define the following: resting membrane potential (p. 340), polarized membrane (p. 340), nerve impulse (nerve action potential) (p. 342), depolarized membrane (p. 341), repolarized membrane (p. 342), and refractory period. (p. 343)
16. What is the all-or-none principle? Relate it to threshold stimulus, subthreshold stimulus, and summation. (p. 343)
17. What is saltatory conduction? Why is it important? (p. 343)
18. What factors determine the speed of nerve impulses? (p. 344)
19. What events are involved in the conduction of a nerve impulse across a synapse? (p. 345)
20. Distinguish between excitatory and inhibitory transmission. (p. 346, 348)
21. Why does one-way nerve impulse conduction occur at chemical synapses? (p. 346)
22. How are nerve impulses inhibited? Of what advantage is this to the body? (p. 348)
23. Why is the postsynaptic neuron called an integrator? (p. 348)
24. What is a neurotransmitter? Describe, in detail, the release, action, and inactivation of acetylcholine (ACh) at a neuromuscular junction. (p. 348)
25. List several probable neurotransmitters, indicate their locations in the nervous system and whether or not they may lead to excitation or inhibition. (p. 348)
26. How do disease, drugs, crack, and pressure affect synaptic transmission? (p. 350)
27. What factors determine neuron regeneration? (p. 350)
28. Distinguish between the discharge zone and facilitated zone and neuronal pool. (p. 351)
29. What is a neuron circuit? Distinguish among simple series, diverging, converging, reverberating (oscillatory), and parallel after-discharge circuits. (p. 351)

SELECTED READINGS

Barr, M. L., and J. A. Kiernan. *The Human Nervous System,* 5th ed. Philadelphia: Lippincott, 1988.

Bloom, F. E. "Neuropeptides," *Scientific American,* October 1981.

Byrne, J. H., and S. G. Schultz. *An Introduction to Membrane Transport and Bioelectricity.* New York: Raven Press, 1988.

Dunant, Y., and M. Israël. "The Release of Acetylcholine," *Scientific American,* March 1985.

Llinas, R. R. "Calcium in Synaptic Transmission," *Scientific American,* October 1982.

Miller, J. A. "Cell Communication Equipment: Do-It-Yourself," *Science News,* 14 April 1984.

Morell, P., and W. T. Norton. "Myelin," *Scientific American,* May 1980.

Patton, H. D., A. F. Fuchs, B. Hille, A. M. Scher, and R. Steiner. *Textbook of Physiology: Excitable Cells and Neurophysiology,* 21st ed. Philadelphia: Saunders, 1989.

Netter, F. H. *Nervous System: Anatomy and Physiology.* Summit: CIBA-GEIGY Corporation, 1983.

Stevens, C. F. "The Neuron," *Scientific American,* September 1979.

Chapter 13

The Spinal Cord and the Spinal Nerves

Chapter Contents at a Glance

Student Objectives

1. Describe how neural tissue is grouped.
2. Describe the protection, gross anatomical features, and cross sectional structure of the spinal cord.
3. List the location, origin, termination, and function of the principal ascending and descending tracts of the spinal cord.
4. Describe the components of a reflex arc and its relationship to homeostasis.
5. List and describe several clinically important reflexes.
6. Describe the composition and coverings of a spinal nerve.
7. Define a plexus and describe the composition and distribution of nerves of the cervical, brachial, lumbar, and sacral plexuses.
8. Describe spinal cord injury and list the immediate and long-range effects.
9. Identify the effects of peripheral nerve damage and conditions necessary for its regeneration.
10. Explain the causes and symptoms of neuritis, sciatica, and shingles.

In this chapter, our main concern will be to study the structure and function of the spinal cord and the nerves that originate from it. Keep in mind, however, that the spinal cord is continuous with the brain and that together they constitute the central nervous system.

GROUPING OF NEURAL TISSUE

The term ***white matter*** refers to aggregations of myelinated processes from many neurons supported by neuroglia or neurolemmocytes (Schwann cells). The lipid substance myelin has a whitish color that gives white matter its name. The ***gray matter*** of the nervous system contains either nerve cell bodies and dendrites or bundles of unmyelinated axons and neuroglia. The absence of myelin in these areas accounts for their gray color.

A ***nerve*** is a bundle of fibers (axons and/or dendrites) located outside the central nervous system. Since the dendrites of somatic afferent neurons and axons of somatic efferent neurons of the peripheral nervous system are myelinated, most nerves are white matter. Nerve cell bodies that lie outside the central nervous system are generally grouped with other nerve cell bodies to form clumps termed ***ganglia*** (GANG-lē-a; *ganglion* = knot). Ganglia, since they are made up principally of nerve cell bodies, are masses of gray matter.

A ***tract*** is a bundle of fibers in the central nervous system. Tracts may run long distances up or down the spinal cord. Tracts also exist in the brain and connect parts of the brain with each other and with the spinal cord. The chief spinal tracts that conduct impulses up the cord to tracts in the brain are concerned with sensory impulses and are called ***ascending tracts.*** Spinal tracts that carry impulses down the cord are motor tracts and are called ***descending tracts*** and are continuous with motor tracts of the brain. The major tracts consist of myelinated fibers and are therefore white matter. A ***nucleus*** is a mass of unmyelinated nerve cell bodies and dendrites in the central nervous system. Nuclei form gray matter. ***Horns (columns)*** are the chief areas of gray matter in the spinal cord. The term *horn* describes the two-dimensional appearance of the organization of gray matter in the spinal cord as seen in cross section. The term *column* describes the three-dimensional appearance of the gray matter in longitudinal columns. Since the white matter of the spinal cord is also arranged in columns, we will refer to the gray matter as being arranged in horns (see Figure 13-2a).

SPINAL CORD

Protection and Coverings

Vertebral Canal

The spinal cord is located in the vertebral (spinal) canal of the vertebral column. The canal is formed by the vertebral foramina of all the vertebrae arranged on top of each other. Since the wall of the vertebral canal is essentially a ring of bone surrounding the spinal cord, the cord is well protected. A certain degree of protection is also provided by the meninges, cerebrospinal fluid, and the vertebral ligaments.

Meninges

The ***meninges*** (me-NIN-jēz) are coverings that run continuously around the spinal cord and brain (***meninx*** is singular). Those associated specifically with the cord are known as ***spinal meninges*** (see Figure 14-2a). The outer spinal meninx is called the ***dura mater*** (DYOO-ra MĀ-ter), meaning "tough mother." It forms a tube from the level of the second sacral vertebra, where it is fused with the filum terminale, to the foramen magnum, where it is continuous with the dura mater of the brain. It is composed of dense, fibrous connective tissue. Between the dura mater and the wall of the vertebral canal is the ***epidural space,*** which is filled with fat, connective tissue, and blood vessels. It serves as padding around the cord. The epidural space inferior to the second lumbar vertebra is the site for the injection of anesthetics, such as a saddleblock for childbirth.

The middle spinal meninx is called the ***arachnoid*** (a-RAK-noyd; *arachne* = spiderlike). It is a delicate connective tissue membrane that forms a tube inside the dura mater. It is also continuous with the arachnoid of the brain. Between the dura mater and the arachnoid is the ***subdural space,*** which contains serous fluid.

The inner meninx is known as the ***pia mater*** (PĪ-a MĀ-ter), or delicate mother. It is a transparent fibrous membrane that adheres to the surface of the spinal cord and brain. It contains numerous blood vessels. Between the arachnoid and the pia mater is the ***subarachnoid space,*** where the cerebrospinal fluid circulates.

Inflammation of the meninges is known as ***meningitis.*** If only the dura mater becomes inflamed, the condition is ***pachymeningitis.*** Inflammation of the arachnoid and pia mater is ***leptomeningitis,*** the most common form of meningitis. All three spinal meninges cover the spinal nerves up to the point of exit from the spinal column through the intervertebral foramina. The spinal cord is suspended in the middle of its dural sheath by membranous extensions of the pia mater. These extensions, called the ***denticulate*** (den-TIK-yoo-lāt) ***ligaments,*** are attached laterally to the dura mater along the length of the cord between the ventral and dorsal nerve roots on either side. The ligaments protect the spinal cord against shock and sudden displacement. Essentially, the spinal cord is fixed in its position in the vertebral canal, since it is anchored to the coccyx inferiorly by the filum terminale, laterally to the dura mater by the denticulate ligaments, and superiorly to the brain.

MEDICAL TEST

Spinal (lumbar) tap (puncture)

Diagnostic Value: To withdraw cerebrospinal fluid (CSF) for diagnostic purposes; introduce antibiotics, contrast media for other procedures, and anesthetics; administer chemotherapy; measure CSF pressure; and evaluate the effects of treatment.

Procedure: A spinal tap is normally performed between the third and fourth or fourth and fifth lumbar vertebrae. (A line drawn across the highest points of the iliac crests passes through the spinous process of the fourth lumbar vertebra.) Because the spinal cord ends at the upper border of the second lumbar vertebra, a spinal tap poses no danger to the cord. In one commonly used position, the patient lies on one side, drawing the knees and chest together to separate the vertebrae slightly. A local anesthetic is given, and a long needle is inserted into the subarachnoid space. Once the needle is in place, indicated by obtaining some cerebrospinal fluid, a pressure-measuring device (manometer) is attached to measure the pressure of the CSF. Next, a sample of CSF, usually between 5 to 10 ml in 3 or more separate tubes, is carefully withdrawn for analysis. (In the event that increased intracranial pressure is suspected, the physician may order a CT scan to rule out this possibility prior to performing the tap. The removal of any CSF may be dangerous in the presence of elevated intracranial pressure. The rapid decrease in pressure can result in herniation of the brain stem.) Before the needle is withdrawn, the pressure of the CSF is again recorded. Leakage of fluid from the site of the spinal tap can lead to severe headache, and for this reason, the patient is asked to remain recumbent for 8–24 hours after the procedure, the purpose being to minimize the leakage of CSF from the puncture site. Fluids are also encouraged.

General Features

The ***spinal cord*** is a cylindrical structure that is slightly flattened anteriorly and posteriorly. It begins as a continuation of the medulla oblongata, the inferior part of the brain stem, and extends from the foramen magnum of the occipital bone to the level of the upper border of the second lumbar vertebra (Figure 13-1). The length of the adult spinal cord ranges from 42 to 45 cm (16 to 18 inches). The circumference of the cord is about 2.54 cm (1 inch) in the midthoracic region but is somewhat larger in the lower cervical and midlumbar regions.

When the cord is viewed externally, two conspicuous enlargements can be seen. The superior enlargement, the ***cervical enlargement,*** extends from the fourth cervical to the first thoracic vertebra. Nerves that supply the upper extremities arise from the cervical enlargement. The inferior enlargement, called the ***lumbar enlargement,*** extends from the ninth to the twelfth thoracic vertebra. Nerves that supply the lower extremities arise from the lumbar enlargement.

Below the lumbar enlargement, the spinal cord tapers to a conical portion known as the ***conus medullaris*** (KŌ-nus med-yoo-LAR-is). The conus medullaris ends at the level of the intervertebral disc between the first and second lumbar vertebrae. Arising from the conus medullaris is the ***filum terminale*** (FĪ-lum ter-mi-NAL-ē), a nonnervous fibrous tissue of the spinal cord that extends inferiorly to attach to the coccyx. The filum terminale consists mostly of pia mater, the innermost of three meninges that cover and protect the spinal cord and brain. Some nerves that arise from the lower portion of the cord do not leave the vertebral column immediately. They angle inferiorly in the vertebral canal like wisps of coarse hair flowing from the end of the cord. They are appropriately named the ***cauda equina*** (KAW-da ē-KWĪ-na), meaning "horse's tail."

The spinal cord is a series of 31 sections called segments, each giving rise to a pair of spinal nerves. ***Spinal segment*** refers to a region of the spinal cord from which a pair of spinal nerves arises. The cord is divided into right and left sides by two grooves (see Figure 13-3). The ***anterior median fissure*** is a deep, wide groove on the anterior (ventral) surface, and the ***posterior median sulcus*** is a shallower, narrow groove on the posterior (dorsal) surface.

Structure in Cross Section

The spinal cord consists of both gray and white matter. Figure 13-2 shows that the gray matter forms an H-shaped area within the white matter. The gray matter consists primarily of nerve cell bodies and unmyelinated axons and dendrites of association and motor neurons. The white matter consists of bundles of myelinated axons of motor and sensory neurons and forms the sensory and motor tracts of the spinal cord.

The cross bar of the H is formed by the ***gray commissure*** (KOM-mi-shur). In the center of the gray commissure is a small space called the ***central canal.*** This canal runs the length of the spinal cord and is continuous with the fourth ventricle of the medulla. It contains cerebrospinal fluid. Anterior to the gray commissure is the ***anterior (ventral) white commissure,*** which connects the white matter of the right and left sides of the spinal cord.

The upright portions of the H are further subdivided into regions. Those closer to the front of the cord are called ***anterior (ventral) gray horns.*** The regions closer to the back of the cord are referred to as ***posterior (dorsal) gray horns.*** The regions between the anterior and posterior gray horns are intermediate ***lateral gray***

FIGURE 13-1 Spinal cord and spinal nerves. (a) Diagram of posterior view.

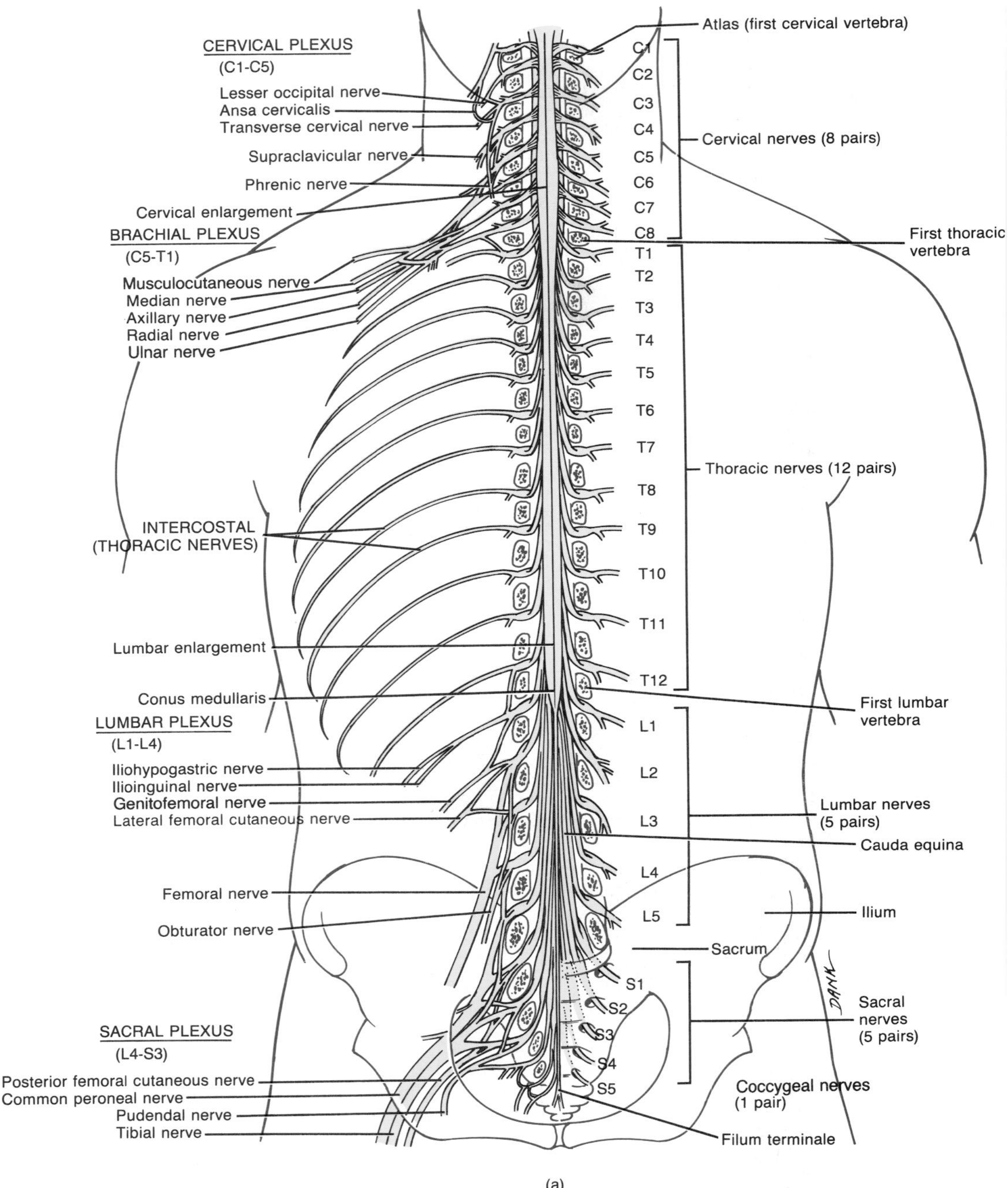

CERVICAL PLEXUS
(C1-C5)
Lesser occipital nerve
Ansa cervicalis
Transverse cervical nerve
Supraclavicular nerve
Phrenic nerve
Cervical enlargement
BRACHIAL PLEXUS
(C5-T1)
Musculocutaneous nerve
Median nerve
Axillary nerve
Radial nerve
Ulnar nerve
INTERCOSTAL
(THORACIC NERVES)
Lumbar enlargement
Conus medullaris
LUMBAR PLEXUS
(L1-L4)
Iliohypogastric nerve
Ilioinguinal nerve
Genitofemoral nerve
Lateral femoral cutaneous nerve
Femoral nerve
Obturator nerve
SACRAL PLEXUS
(L4-S3)
Posterior femoral cutaneous nerve
Common peroneal nerve
Pudendal nerve
Tibial nerve
Atlas (first cervical vertebra)
C1
C2
C3
C4
C5
C6
C7
C8
Cervical nerves (8 pairs)
First thoracic vertebra
T1
T2
T3
T4
T5
T6
T7
T8
T9
T10
T11
T12
Thoracic nerves (12 pairs)
First lumbar vertebra
L1
L2
L3
L4
L5
Lumbar nerves (5 pairs)
Cauda equina
Ilium
Sacrum
S1
S2
S3
S4
S5
Sacral nerves (5 pairs)
Coccygeal nerves (1 pair)
Filum terminale
DANK

(a)

FIGURE 13-1 (*Continued*) (b) Photograph of the posterior aspect of the conus medullaris and cauda equina. (Photograph courtesy of N. Gluhbegovic and T. H. Williams, *The Human Brain: A Photographic Guide,* Harper & Row, Publishers, Inc., Hagerstown, MD, 1980.)

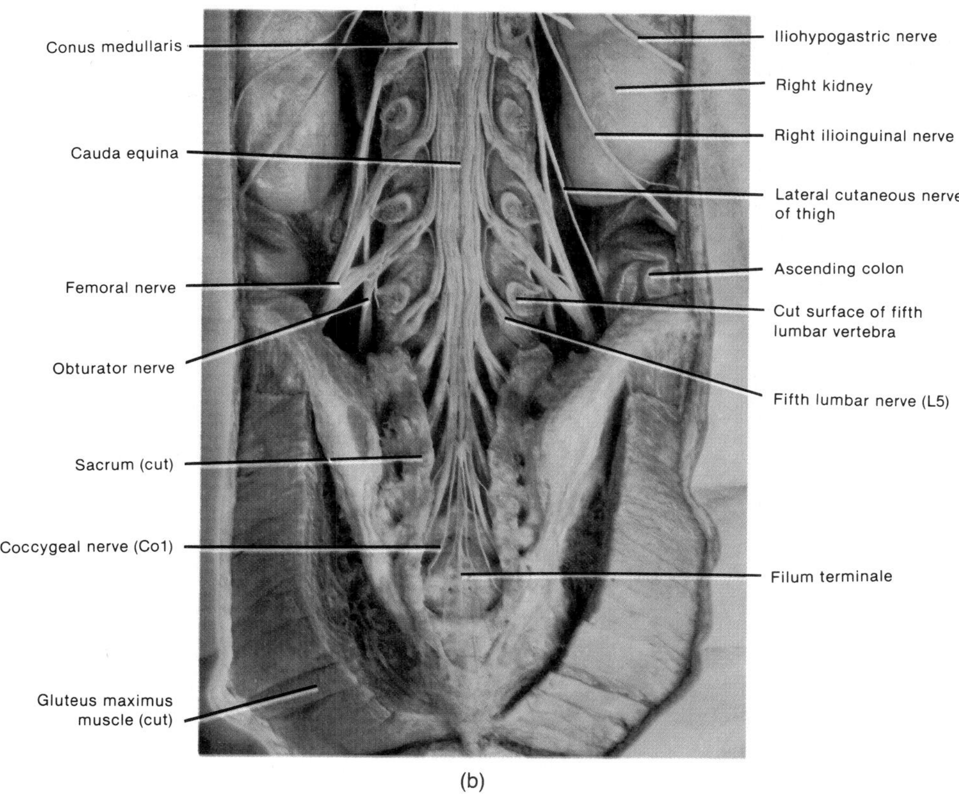

(b)

horns. The lateral gray horns are present in the thoracic, upper lumbar, and sacral segments of the cord.

The gray matter of the cord also contains several nuclei that serve as relay stations for nerve impulses and origins for certain nerves. Nuclei are clusters of nerve cell bodies and dendrites in the spinal cord and brain.

The white matter, like the gray matter, is also organized into regions. The anterior and posterior gray horns divide the white matter on each side into three broad areas: ***anterior (ventral) white columns, posterior (dorsal) white columns,*** and ***lateral white columns.*** Each column in turn consists of distinct bundles of myelinated fibers that run within the cord. These bundles are called ***tracts*** or ***fasciculi*** (fa-SIK-yoo-lī). The long ***ascending tracts*** consist of sensory axons that conduct impulses that enter the spinal cord upward to the brain. The long ***descending tracts*** consist of motor axons that conduct impulses from the brain downward into the spinal cord, where they synapse with other neurons whose axons pass out to muscles and glands. Thus, the ascending tracts are sensory tracts and the descending tracts are motor tracts.

MEDICAL TEST

Myelography (mī-e-LOG-ra-fē; *myelo* = spinal cord; *graphein* = to write)

Diagnostic Value: To demonstrate a lesion, such as a tumor or herniated (slipped) disc, within or adjacent to the spinal cord.

Procedure: The patient is placed on her or his side at the edge of a table with the knees drawn toward the chest to separate the vertebrae. A spinal tap is performed, and the patient is returned to the prone position and secured with straps. The head is hyperextended to prevent the flow of contrast medium into the cranium. With the spinal needle in place, a contrast medium is injected and the table is tilted so that the medium flows through the subarachnoid space of the spinal cord. The flow of the contrast medium is monitored fluoroscopically, and radiographs are taken for interpretation and to provide a permanent record of the findings.

FIGURE 13-2 Spinal cord. (a) The organization of gray and white matter in the spinal cord as seen in cross section. The left half of the spinal cord has been sectioned at a lower level than the right so that you can see what is inside the posterior root ganglion, posterior root of the spinal nerve, anterior root of the spinal nerve, and the spinal nerve. In this and other illustrations of cross sections of the spinal cord, dendrites are not shown in relation to cell bodies of motor or association neurons for purposes of simplicity. (b) Photograph of the spinal cord at the seventh cervical segment. (Courtesy of Victor B. Eichler, Ph.D., Wichita, Kansas.)

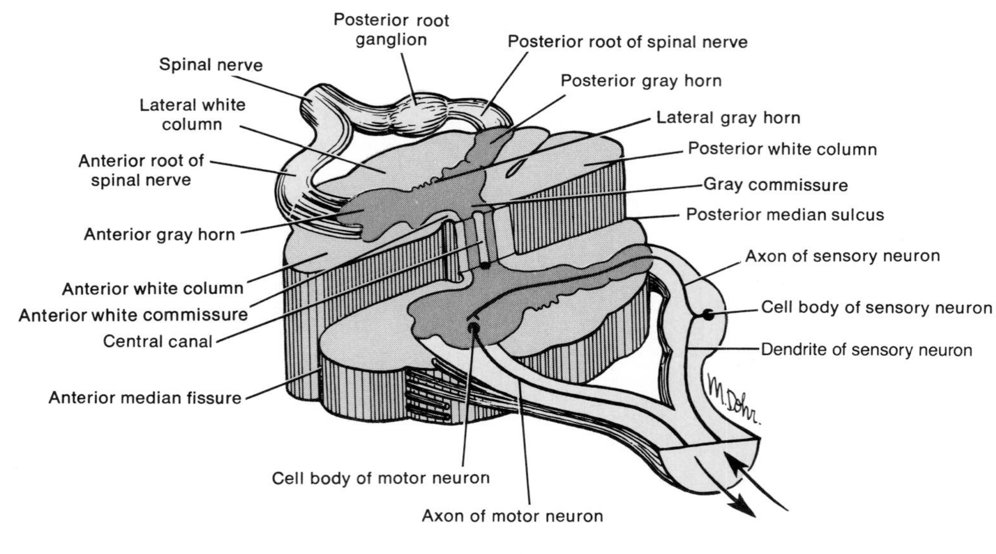

(a)

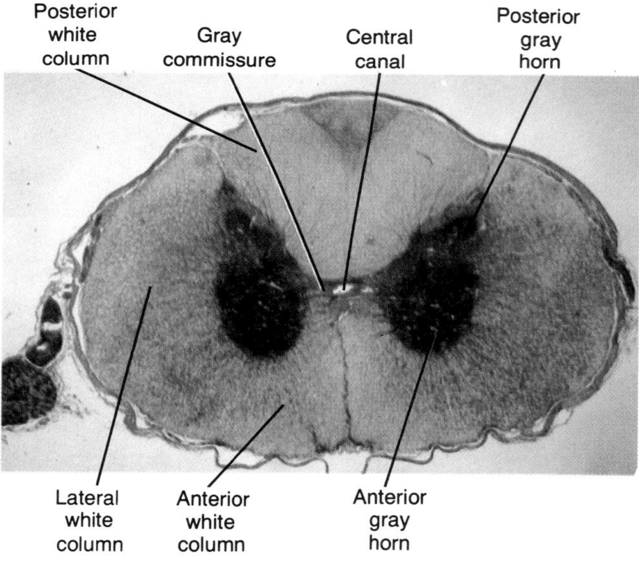

(b)

Functions

A major function of the spinal tracts in the spinal cord is to convey sensory impulses from the periphery to the brain and to conduct motor impulses from the brain to the periphery. A second principal function is to provide a means of integrating reflexes. Both functions are essential to maintaining homeostasis.

Impulse Conduction

The vital function of conveying sensory and motor information to and from the brain is carried out by the ascending and descending tracts of the cord. The names of the tracts typically indicate the white column in which the tract travels, where the cell bodies of the tract originate, and where the axons of the tract terminate. Since the origin and termination are specified, the direction of impulse conduction is also indicated by the name. For example, the anterior spinothalamic tract is located in the *anterior* white column, it originates in the *spinal cord,* and it terminates in the *thalamus* (a region of the brain). It is an ascending (sensory) tract since it conveys nerve impulses from the cord upward to the brain.

The principal ascending and descending tracts are listed in Exhibit 13-1 and shown in Figure 13-3.

EXHIBIT 13-1 SELECTED ASCENDING AND DESCENDING TRACTS OF SPINAL CORD

Tract	Location (white column)	Origin	Termination	Function
ASCENDING				
Anterior (ventral) spinothalamic	Anterior (ventral) column.	Posterior (dorsal) gray horn on one side of cord but crosses to opposite side of brain.	Thalamus; impulses eventually conveyed to cerebral cortex.	Conveys sensory impulses for touch and pressure from one side of body to opposite side of thalamus. Eventually, sensations reach cerebral cortex.
Lateral spinothalamic	Lateral column.	Posterior (dorsal) gray horn on one side of cord but crosses to opposite side of brain.	Thalamus; impulses eventually conveyed to cerebral cortex.	Conveys sensory impulses for pain and temperature from one side of body to opposite side of thalamus. Eventually, sensations reach cerebral cortex.
Fasciculus gracilis and **fasciculus cuneatus**	Posterior (dorsal) column.	Axons of afferent neurons from periphery that enter posterior (dorsal) column on one side of cord and rise to same side of brain.	Nucleus gracilis and nucleus cuneatus of medulla; impulses eventually conveyed to opposite side of cerebral cortex.	Conveys sensory impulses from one side of body to opposite side of medulla for touch; two-point discrimination (ability to distinguish that two points on skin are touched even though close together); conscious proprioception (conscious awareness of precise position of body parts and their direction of movement); stereognosis (ability to recognize size, shape, and texture of object); weight discrimination (ability to assess weight of an object); and vibration. Eventually, impulses over these tracts reach cerebral cortex which allows the sensations to be consciously perceived.
Posterior (dorsal) spinocerebellar	Posterior (dorsal) portion of lateral column.	Posterior (dorsal) gray horn on one side of cord and rises to same side of brain.	Cerebellum.	Conveys sensory impulses from one side of body to same side of cerebellum for subconscious proprioception.
Anterior (ventral) spinocerebellar	Anterior (ventral) portion of lateral column.	Posterior (dorsal) gray horn on one side of cord; contains both crossed and uncrossed fibers.	Cerebellum.	Conveys sensory impulses from both sides of body to cerebellum for subconscious proprioception.
DESCENDING				
Lateral corticospinal	Lateral column.	Cerebral cortex on one side of brain but crosses in base of medulla to opposite side of cord.	Anterior (ventral) gray horn.	Conveys motor impulses from one side of cortex to anterior gray horn of opposite side. Eventually, impulses reach skeletal muscles on opposite side of body that coordinate precise, discrete movements.
Anterior (ventral) corticospinal	Anterior (ventral) column.	Cerebral cortex on one side of brain, uncrossed in medulla, but crosses to opposite side of cord.	Anterior (ventral) gray horn.	Conveys motor impulses from one side of cortex to anterior gray horn of same side. Impulses cross to opposite side in spinal cord and reach skeletal muscles that coordinate movements of the axial skeleton.

EXHIBIT 13-1 SELECTED ASCENDING AND DESCENDING TRACTS OF SPINAL CORD (*Continued*)

Tract	Location (white column)	Origin	Termination	Function
Rubrospinal	Lateral column.	Midbrain (red nucleus) on one side of brain but crosses to opposite side of cord.	Anterior (ventral) gray horn.	Conveys motor impulses from one side of midbrain to skeletal muscles on opposite side of body that are concerned with precise, discrete movements.
Tectospinal	Anterior (ventral) column.	Midbrain on one side of brain but crosses to opposite side of cord.	Anterior (ventral) gray horn.	Conveys motor impulses from one side of midbrain to skeletal muscles on opposite side of body that control movements of head in response to auditory, visual, and cutaneous stimuli.
Vestibulospinal	Anterior (ventral) column.	Medulla on one side of brain and descends to same side of cord.	Anterior (ventral) gray horn.	Conveys motor impulses from one side of medulla to skeletal muscles on same side of body that regulate body tone in response to movements of head (equilibrium).
Lateral reticulospinal	Lateral column.	Medulla on one side of brain and descends mainly to same side of cord.	Anterior (ventral) gray horn.	Conveys motor impulses from one side of medulla to axial skeleton muscles and proximal limb muscles that inhibit extensor reflexes and muscle tone.
Anterior (ventral) or **medial reticulospinal**	Anterior (ventral) column.	Pons on one side of brain and descends mainly to same side of cord.	Anterior (ventral) gray horn.	Conveys motor impulses from one side of pons to axial skeleton muscles and proximal limb muscles that facilitate extensor reflexes and muscle tone.

FIGURE 13-3 Selected tracts of the spinal cord. Ascending (sensory) tracts are indicated in pink; descending (motor) tracts are shown in blue.

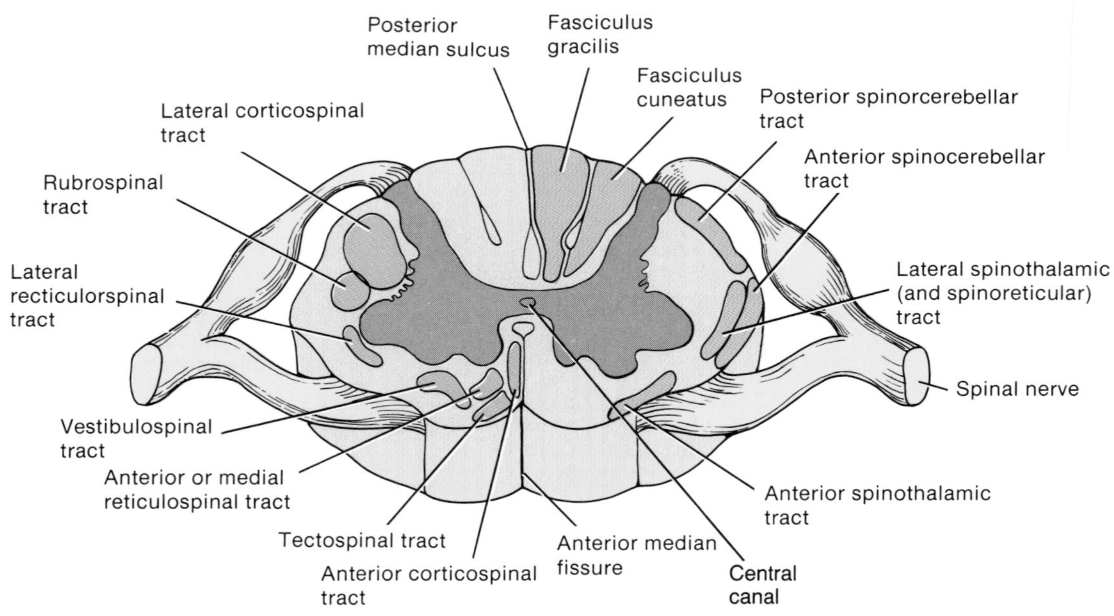

Reflex Center

The second principal function of the spinal cord is to serve as a center for some reflex actions. Spinal nerves are the paths of communication between the spinal cord tracts and the periphery. Figure 13-3 reveals that each pair of spinal nerves is connected to a segment of the cord by two points of attachment called roots. The ***posterior*** or ***dorsal (sensory) root*** contains sensory nerve fibers only and conducts nerve impulses from the periphery to the spinal cord. These fibers extend into the posterior (dorsal) gray horn. Each dorsal root also has a swelling, the ***posterior*** or ***dorsal (sensory) root ganglion,*** which contains the cell bodies of the sensory neurons from the periphery. The other point of attachment of a spinal nerve to the cord is the ***anterior*** or ***ventral (motor) root.*** It contains motor neuron axons only and conducts impulses from the spinal cord to the periphery.

The cell bodies of the motor neurons are located in the gray matter of the cord. If the motor impulse supplies a skeletal muscle, the cell bodies are located in the anterior (ventral) gray horn. If, however, the motor nerve impulse supplies smooth muscle, cardiac muscle, or a gland through the autonomic nervous system, the cell bodies are located in the lateral gray horn.

Reflex Arc and Homeostasis

The path a nerve impulse follows from its origin in the dendrites or cell body of a neuron in one part of the body to its termination elsewhere in the body is called a ***conduction pathway.*** All conduction pathways consist of circuits of neurons. One pathway is known as a ***reflex arc,*** the functional unit of the nervous system. A reflex arc contains two or more types of neurons over which nerve impulses are conducted from a receptor to the brain or spinal cord by way of sensory neurons and then to an effector by way of motor neurons. The basic components of a reflex arc are as follows (Figure 13-4).

1. **Receptor.** The distal end of a dendrite or a sensory structure associated with the distal end of a dendrite. Its role in the reflex arc is to respond to a specific change in the internal or external environment by initiating a nerve impulse in a sensory neuron by way of a receptor potential (a local depolarization of the receptor cell membrane).

2. **Sensory neuron.** Passes the nerve impulse from the receptor to its axonal termination in the central nervous system.

3. **Center.** A region in the central nervous system where an incoming sensory impulse generates an outgoing motor impulse. In the center, the impulse may be inhibited, transmitted, or rerouted. In the center of some reflex arcs, the sensory neuron directly generates the impulse in the motor neuron. The center usually contains one or more association neurons between the sensory neuron and the motor neuron leading to a muscle or a gland.

4. **Motor neuron.** Transmits the impulse generated by the sensory or association neuron in the center to the effector organ of the body that will respond, such as a muscle or gland.

5. **Effector.** The organ of the body that responds to the motor nerve impulse. This response is called a reflex and involves either an increase or decrease in muscle contraction or an increase or decrease in secretion by glands.

Reflexes are fast responses to certain changes (stimuli) in the internal or external environment that allow the body to maintain homeostasis. Reflexes are associated not only with skeletal muscle contraction but also with body functions such as heart rate, respiration, digestion, urination, and defecation, which involve cardiac and smooth muscle and glands. Reflexes carried out by the spinal cord

FIGURE 13-4 Components of a generalized reflex arc. The arrows show the direction of nerve impulse conduction.

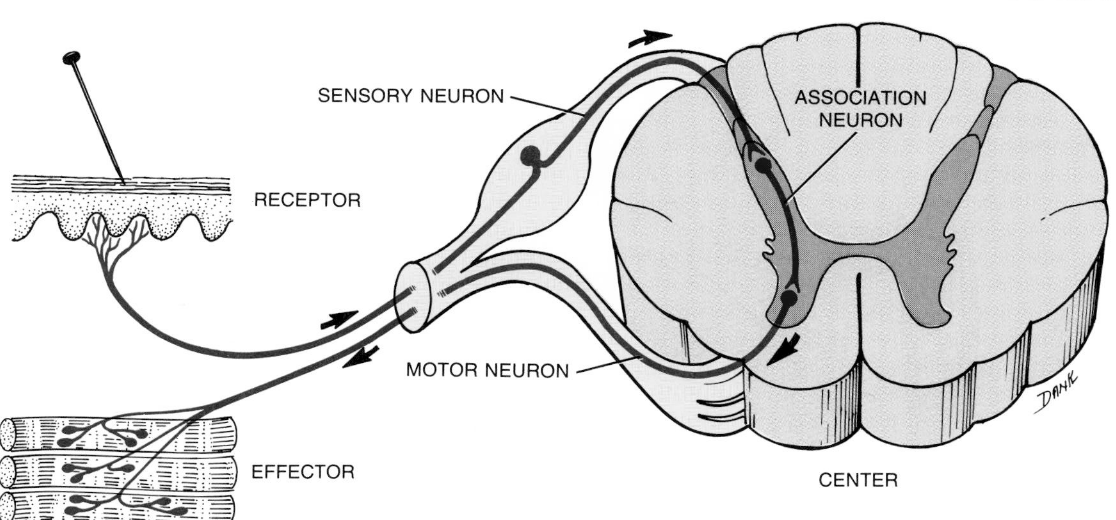

alone are called ***spinal reflexes.*** Reflexes that result in the contraction of skeletal muscles are known as ***somatic reflexes.*** Reflexes that involve brain centers and cranial nerves are called ***cranial reflexes.*** Those that cause the contraction of smooth or cardiac muscle or secretion by glands are ***visceral (autonomic) reflexes.*** Our concern at this point is to examine a few somatic spinal reflexes: the stretch reflex, tendon reflex, flexor reflex, and crossed extensor reflex.

▪ **Physiology of the Stretch Reflex** The ***stretch (tendon jerk) reflex*** is based on a two-neuron or ***monosynaptic reflex arc.*** Only two types of neurons (one sensory and one motor) are involved, and there is only one synapse in the pathway (Figure 13-5). This reflex results in the contraction of a muscle when it is stretched suddenly. Slight stretching of a muscle stimulates receptors in the muscle called ***muscle spindles*** (see Figure 15-3a). The spindles monitor changes in the length of the muscle by responding to the rate and degree of change in length. Once the spindle is stimulated (depolarized to threshold), a nerve impulse is sent along a sensory neuron through the posterior root of the spinal nerves to the spinal cord. The sensory neuron synapses with a motor neuron in the anterior gray horn. The sensory neuron generates an impulse at the synapse that is transmitted along the motor neuron. The motor neuron in the anterior root of the spinal nerve terminates in a skeletal muscle. Once the impulse reaches the stretched muscle, a muscle action potential is generated, and the muscle contracts. Thus, the stretch is counteracted by contraction which shortens the muscle that had been stretched.

Since the sensory nerve impulse enters the spinal cord on the same side that the motor nerve impulse leaves the spinal cord, the reflex arc is called an ***ipsilateral*** (ip'-si-LAT-er-al) ***reflex arc.*** All monosynaptic reflex arcs are ipsilateral as are deep tendon reflexes (discussed shortly).

The stretch reflex is essential in maintaining muscle tone and is important for muscle functions during exercise. It also helps prevent injury from overstretching of muscles. Moreover, it is the basis for several tests used in neurological examinations. One such reflex is the ***patellar reflex (knee jerk).*** This reflex involves extension of the leg by contraction of the quadriceps femoris muscle in response to tapping the patellar ligament. When the patellar ligament below the patella is tapped (stimulus), receptors (muscle spindles) in the quadriceps femoris muscle sensitive to changes in muscle length are stimulated because tapping the tendon of the muscle causes a rapid lengthening of the muscle. Sensory impulses from the receptors are sent to the spinal cord, and the returning motor nerve impulse results in generation of a muscle action potential that causes contraction of the muscle and extension of the leg at the knee, or a knee jerk as the muscle shortens during contraction.

Although the stretch reflex pathway contains only two types of neurons, other neurons are involved in the pathway in various ways. For example, the sensory neuron from the muscle spindle that enters the spinal cord also synapses with an inhibitory association neuron that, in turn, synapses with a motor neuron that controls antag-

FIGURE 13-5 Stretch reflex. The neurons of the stretch reflex pathway are indicated in color. A stretch reflex is monosynaptic since only one synapse and two different neurons—sensory neuron and motor neuron—are involved in the pathway. The synapse is between the sensory neuron from the receptor and motor neuron to the effector. This reflex is also ipsilateral. Why?

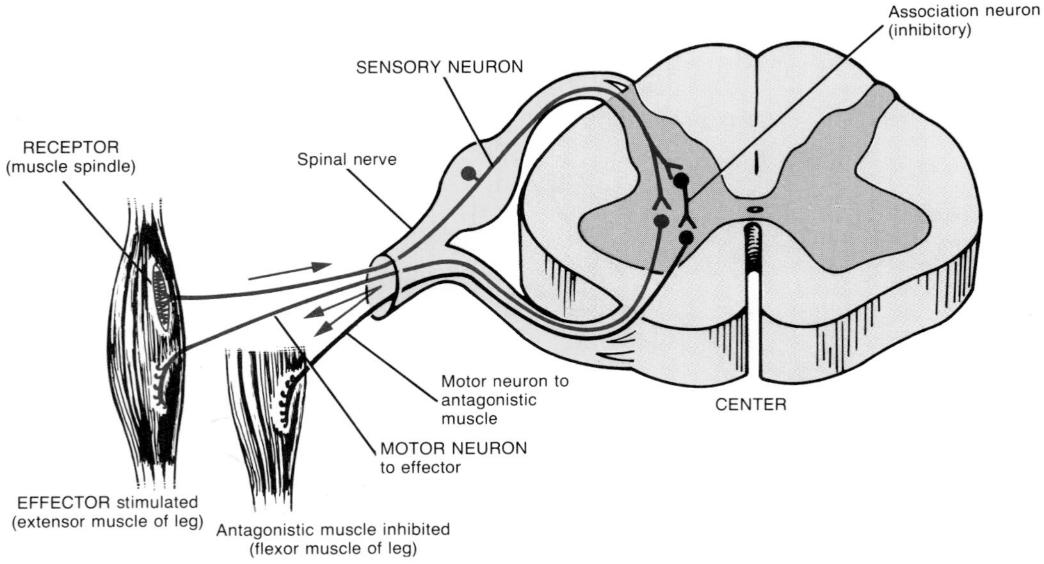

onistic muscles. The inhibitory association neuron inhibits the motor neurons that normally send excitatory impulses to antagonistic muscles. Thus, during the stretch reflex when the stretched muscle is counteracted by contraction, antagonistic muscles that oppose the contraction are inhibited. This phenomenon, by which action potentials stimulate contraction of one muscle and simultaneously inhibit contraction of antagonistic muscles, is called ***reciprocal innervation.*** It avoids conflict between prime movers and antagonists and is vital in coordinating body movements. You may recall from Chapter 11 that skeletal muscles act in groups rather than alone and that each muscle in the group (agonist, antagonist, synergist, fixator) has a specific role in bringing about the movement.

The sensory neuron from the muscle spindle also synapses with neurons that relay impulses to the brain by way of sensory tracts. In this way, the brain is provided with information about the state of stretch or contraction of skeletal muscles that enables it to coordinate muscular movements and posture.

▪ **Physiology of the Tendon Reflex** Reflexes other than the stretch reflex involve association neurons in addition to sensory and motor neurons. Since more than two types of neurons are involved, there is more than one type of synapse, and thus these reflexes are ***polysynaptic reflex arcs.*** One example of a reflex based on a polysynaptic reflex arc is the ***tendon reflex.*** The tendon reflex, like the stretch reflex, is also ipsilateral.

Just as the stretch reflex operates as a feedback mechanism to control muscle *length,* the tendon reflex operates as a feedback mechanism to control muscle *tension* by protecting tendons and their associated muscles from excessive tension. The receptors for this reflex are called ***tendon organs*** or ***Golgi tendon organs*** (see Figure 15-3b). Tendon organs lie within muscle tendons near the junction of a tendon and a muscle. Whereas muscle spindles are sensitive to changes in muscle length, tendon organs detect and respond to changes in muscle tension caused by passive stretch or muscular contraction.

When an increase in tension is applied to a tendon, the tendon organ is stimulated (depolarized to threshold) and nerve impulses are generated and transmitted to the spinal cord via a sensory neuron (Figure 13-6). Within the spinal cord, the sensory neuron synapses with an inhibitory association neuron, which, in turn, synapses with and hyperpolarizes (inhibits) a motor neuron that innervates the muscle associated with the tendon organ. Thus, as tension on the tendon organ increases, the frequency of inhibitory impulses increases, and the inhibition of the motor neurons to the muscle developing excess tension causes relaxation of the muscle. In this way, the tendon reflex protects the tendon and muscle from damage from excessive tension. Thus, the tendon reflex is protective.

The sensory neuron from the tendon organ also synapses with a stimulatory association neuron in the spinal cord. The stimulatory association neuron, in turn, synapses with motor neurons controlling antagonistic muscles. Thus, whereas the tendon reflex brings about relaxation of the muscle containing the tendon organ, it also brings about contraction of the antagonistic muscles. This is an-

FIGURE 13-6 Tendon reflex. The neurons of the tendon reflex pathway are indicated in color. This reflex arc is polysynaptic since there is more than one synapse and more than two different neurons involved in the pathway. There is one synapse between the sensory neuron from the receptor and association neuron (inhibitory) and another synapse between the association neuron (inhibitory) and motor neuron to the effector.

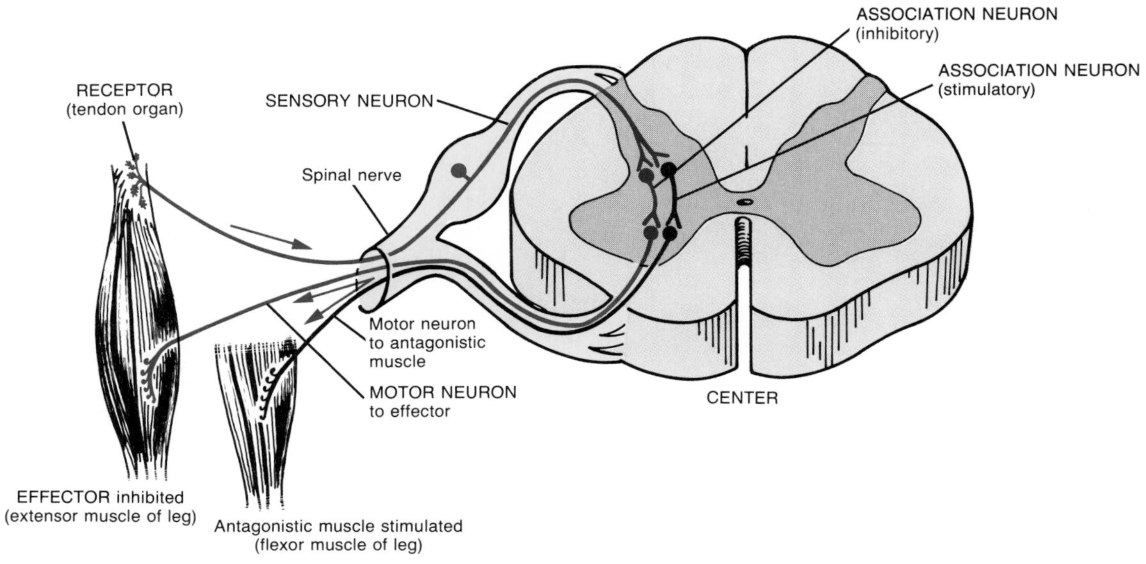

other example of reciprocal innervation. The sensory neuron also synapses with neurons that relay nerve impulses to the brain by way of sensory tracts, thus informing the brain about the state of muscle tension throughout the body.

▪ **Physiology of the Flexor Reflex and Crossed Extensor Reflex** Another example of a reflex based on a polysynaptic reflex arc is the ***flexor reflex,*** or ***withdrawal reflex*** (Figure 13-7). Suppose you step on a tack. As a result of the painful (noxious) stimulus, you immediately withdraw your foot. What has happened? A sensory neuron transmits a nerve impulse from the stimulated pain receptor to the spinal cord. A second impulse is generated in an association neuron, which generates a third impulse in a motor neuron. A motor neuron stimulates the flexor muscles and inhibits the extensor muscles of your foot, and you withdraw it. Thus, a flexor reflex is protective in that it results in the movement of an extremity to avoid pain by stimulating its flexor muscles and inhibiting its extensor muscles. The inhibition of the extensor muscles occurs as inhibitory association neurons act to hyperpolarize (inhibit) the motor neurons that innervate the extensor muscles.

The flexor reflex, like the stretch reflex, is also ipsilateral. The incoming and outgoing impulses are on the same side of the spinal cord. The flexor reflex also illustrates another feature of polysynaptic reflex arcs. In the monosynaptic stretch reflex, the returning motor impulse affects only the quadriceps muscle of the thigh. When you withdraw your entire lower or upper extremity from a painful stimulus, more than one muscle is involved, and

FIGURE 13-7 Flexor (withdrawal) reflex. This reflex arc is polysynaptic and ipsilateral. Why is the reflex arc shown also an intersegmental reflex arc?

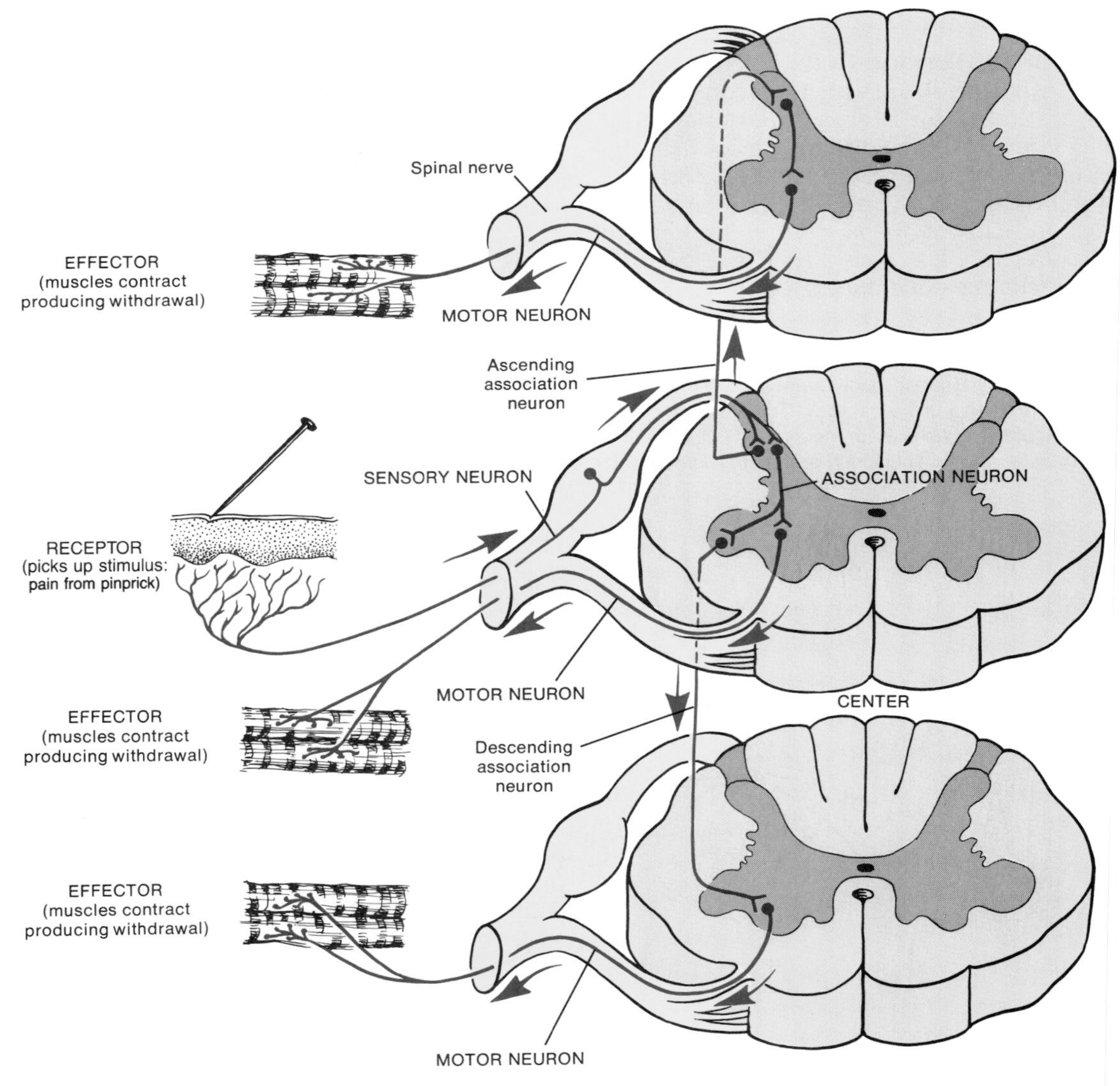

several motor neurons are simultaneously returning impulses to several upper and lower extremity muscles at the same time. Thus, a single sensory impulse generates several motor responses. This kind of reflex arc, in which a single sensory neuron splits into ascending and descending branches, each forming a synapse with association neurons at different segments of the cord, is called an ***intersegmental reflex arc.*** Because of intersegmental reflex arcs, a single sensory neuron can activate several motor neurons and thereby cause stimulation of more than one effector.

Something else may happen when you step on a tack. You may lose your balance as your body weight shifts to the other foot. Then you do whatever you can to regain your balance so you do not fall. This means motor impulses are also sent to your unstimulated limb and both upper extremities. The motor impulses cause extension at the knee, hip, and ankle so you can place your entire body weight on the foot of the limb that must now support the entire body. These impulses cross the spinal cord as shown in Figure 13-8. The incoming sensory impulse not only initiates the flexor reflex that causes you to withdraw, but it also initiates a crossed extensor reflex. The incoming sensory nerve impulse crosses to the opposite side of the

FIGURE 13-8 Crossed extensor reflex. Although the flexor reflex is shown on the left of the diagram so that you can correlate it with the crossed extensor reflex on the right, concentrate your attention on the crossed extensor reflex. Why is the crossed extensor reflex classified as a contralateral reflex arc?

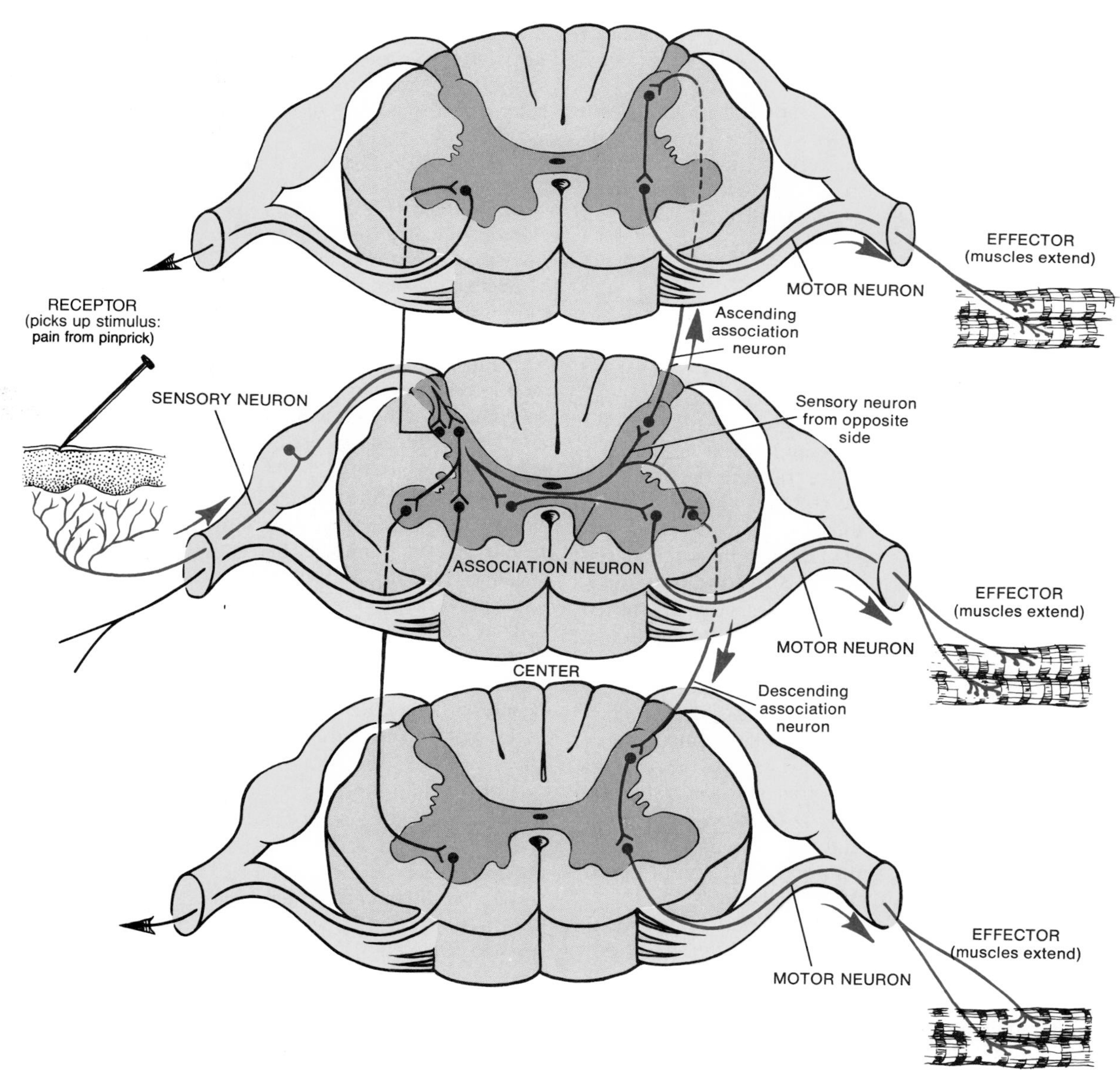

spinal cord through association neurons at that level and several levels above and below the point of sensory stimulation. From these levels, the motor neurons cause extension of the knee, hip, and ankle, thus maintaining balance. Unlike the flexor reflex, which passes over an ipsilateral reflex arc, the crossed extensor reflex passes over a ***contralateral*** (kon'-tra-LAT-er-al) ***reflex arc***—the impulse enters one side of the spinal cord and exits on the opposite side. The reflex just described, in which extension of the joints in one limb occurs in conjunction with contraction of the flexor muscles of the opposite limb, is called a ***crossed extensor reflex.***

In the flexor reflex, when the flexor muscles of your painfully stimulated lower extremity are contracting, the extensor muscles of the same extremity are being inhibited to some degree. If both sets of muscles contracted at the same time, you would not be able to flex your limb because both sets of muscles would pull on the limb bones. But because of reciprocal innervation, one set of muscles contracts while the other is being inhibited from contracting.

In the crossed extensor reflex, reciprocal innervation also occurs. While you are contracting the flexor muscles of the limb that has been stimulated by the tack, the stimulated extensor muscles of your other lower limb are producing extension to help maintain balance.

Reflexes and Diagnosis

Reflexes are often used for diagnosing disorders of the nervous system and locating injured tissue. If a reflex ceases to function or functions abnormally, the physician may suspect that the damage lies somewhere along a particular conduction pathway. Visceral reflexes, however, are usually not practical tools for diagnosis. It is difficult to stimulate visceral receptors, since they are deep in the body. In contrast, many somatic reflexes can be tested simply by tapping or stroking the body.

Superficial reflexes are withdrawal reflexes that are elicited by tactile or noxious stimuli. The stimuli are applied to the skin, mucous membranes, or cornea of the eye. Examples include the corneal reflex, abdominal reflex, and pharyngeal reflex.

Any skeletal muscle can normally be stimulated to contract by a slight sudden stretch of its tendon, which can be created by administering a light tap. Many muscle tendons are deeply buried, however, and cannot be readily tapped through the skin. Reflexes that involve a stretch stimulus to a tendon are called ***deep tendon reflexes.*** Deep tendon reflexes provide information about the integrity and function of the reflex arcs and spinal cord segments without involving the higher centers.

To obtain a substantial response when testing deep tendon reflexes, the muscle must be slightly stretched before the tap is administered. If it is stretched an appropriate amount, tapping the tendon elicits a muscle contraction.

If reflexes are weak or absent, *reinforcement* can be used. In this method, muscle groups other than those being tested are tensed voluntarily with isometric contractions to increase reflex activity in other parts of the body. For example, the person can be asked to hook the fingers together and then try to pull them apart. This action may increase the strength of reflexes involving other muscles. If the reflex can be demonstrated, it is certain that the sensory and motor nerve connections are intact between muscle and spinal cord.

Muscle reflexes can help determine the spinal cord's excitability. When a large number of facilitatory impulses are transmitted from the brain to the spinal cord, the muscle reflexes become so sensitive that simply tapping the knee tendon with the tip of one's finger may cause the leg to jump a considerable distance. On the other hand, the cord may be so intensely inhibited by other impulses from the brain that almost no degree of pounding on the muscles or tendons can elicit a response.

Neurological impairment can be evaluated by using a stopwatch to time the reflex response. Sensitivity of sensory end organs in a muscle is demonstrated by stretching it by as little as 0.05 mm and for as short a duration as $\frac{1}{20}$ sec.

Among the reflexes of clinical significance are:

1. **Patellar reflex** (knee jerk). This reflex involves extension of the leg by contraction of the quadriceps femoris muscle in response to tapping the patellar ligament (see Figure 13-5). The reflex is blocked by damaged afferent or efferent nerves to the muscle or reflex centers in the second, third, or fourth lumbar segments of the spinal cord. This reflex is also absent in people with chronic diabetes and neurosyphilis. The reflex is exaggerated in disease or injury involving the corticospinal tracts descending from the cortex to the spinal cord. This reflex may also be exaggerated by applying a second stimulus (a sudden loud noise) while tapping the patellar tendon.

2. **Achilles reflex** (ankle jerk). This reflex involves extension (plantar flexion) of the foot by contraction of the gastrocnemius and soleus muscles in response to tapping the calcaneal (Achilles) tendon. Blockage of the ankle jerk indicates damage to the nerves supplying the posterior leg muscles or to the nerve cells in the lumbosacral region of the spinal cord. This reflex is also absent in people with chronic diabetes, neurosyphilis, alcoholism, and subarachnoid hemorrhages. An exaggerated Achilles reflex indicates cervical cord compression or a lesion of the motor tracts of the first or second sacral segments of the cord.

3. **Babinski sign.** This reflex results from gentle stimulation to the outer margin of the sole of the foot. The great toe is extended, with or without fanning of the other toes. This phenomenon occurs in normal children under $1\frac{1}{2}$ years of age and is due to incomplete development of the nervous system. The myelination of fibers in the corticospinal tract has not reached completion. A positive Babinski sign after age $1\frac{1}{2}$ is considered abnormal and indicates an interruption of the corticospinal tract as the result of a lesion of the tract, usually in the upper portion. The normal response after $1\frac{1}{2}$ years of age is the ***plantar reflex,*** or negative Babinski—a curling under

of all the toes, accompanied by a slight turning in and flexion of the anterior part of the foot.

4. Abdominal reflex. This reflex involves contraction of the muscles of the abdominal wall which compress the wall in response to stroking the side of the abdomen. Two separate reflexes, the upper abdominal reflex and the lower abdominal reflex, are involved. The patient should be lying down and relaxed, with arms at the sides and knees slightly flexed. The response is an abdominal muscle contraction that results in a lateral deviation of the umbilicus to the side opposite the stimulus. Absence of this reflex is associated with lesions of the corticospinal system. It may also be absent because of lesions of the peripheral nerves, lesions of reflex centers in the thoracic part of the cord, and multiple sclerosis.

SPINAL NERVES

Names

The 31 pairs of spinal nerves are named and numbered according to the region and level of the spinal cord from which they emerge (see Figure 13-1). The first cervical pair emerges between the atlas and the occipital bone. All other spinal nerves leave the vertebral column from the intervertebral foramina between adjoining vertebrae. There are 8 pairs of cervical nerves, 12 pairs of thoracic nerves, 5 pairs of lumbar nerves, 5 pairs of sacral nerves, and 1 pair of coccygeal nerves.

During fetal life, the spinal cord and vertebral column grow at different rates, the cord growing more slowly. Thus, not all the spinal cord segments are in line with their corresponding vertebrae. Remember that the spinal cord terminates near the level of the upper border of the second lumbar vertebra. Thus, the lower lumbar, sacral, and coccygeal nerves must descend more and more to reach their foramina before emerging from the vertebral column. This arrangement constitutes the cauda equina.

Composition and Coverings

A ***spinal nerve*** has two points of attachment to the cord: a posterior root and an anterior root. The posterior and anterior roots unite to form a spinal nerve at the intervertebral foramen (see Figure 13-2a). Since the posterior root contains sensory fibers and the anterior root contains motor fibers, a spinal nerve is a ***mixed nerve,*** at least at its origin. The posterior (dorsal) root ganglion contains cell bodies of sensory neurons.

In Figure 13-9, you can see that a spinal nerve contains many fibers surrounded by different coverings. The individual fibers, whether myelinated or unmyelinated, are wrapped in a connective tissue called the **endoneurium** (en'-dō-NYOO-rē-um). Groups of fibers with their endoneurium are arranged in bundles called ***fascicles,*** and each bundle is wrapped in connective tissue called the ***perineurium*** (per'-i-NYOO-rē-um). The outermost covering around the entire nerve is the ***epineurium*** (ep'-i-NYOO-rē-um). The spinal meninges fuse with the epineurium as the nerve exits from the vertebral canal.

FIGURE 13-9 Coverings of a spinal nerve.

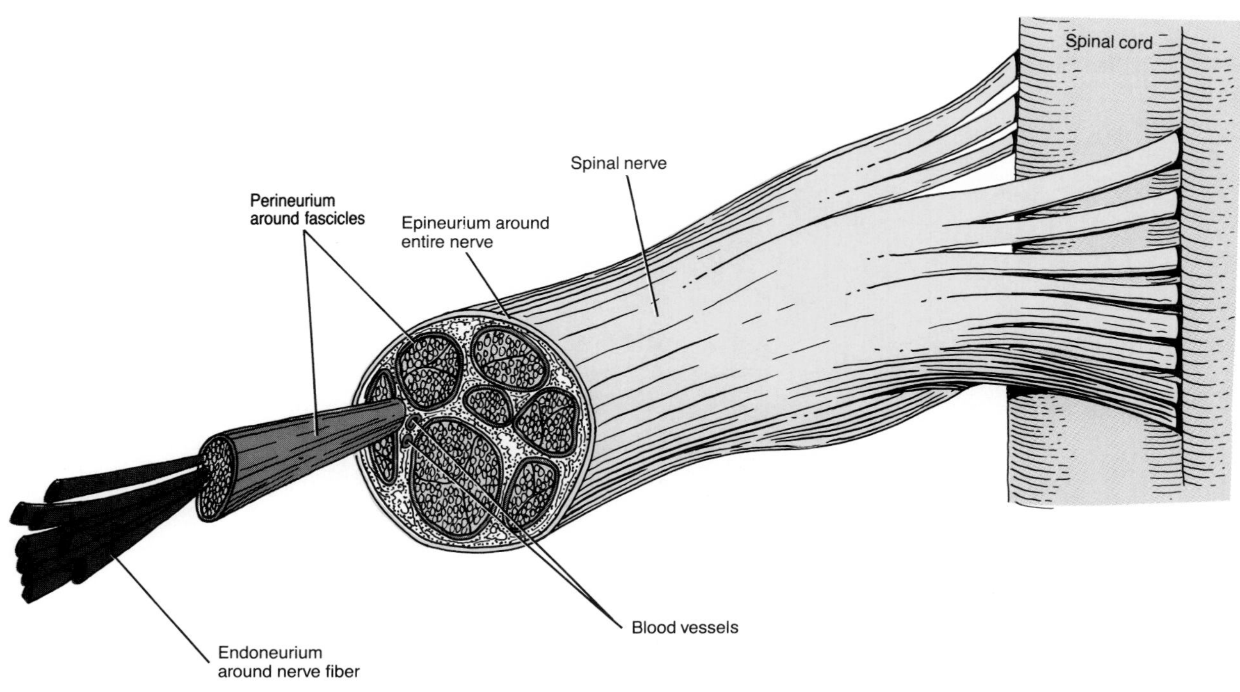

Distribution

Branches

Shortly after a spinal nerve leaves its intervertebral foramen, it divides into several branches (Figure 13-10). These branches are known as ***rami*** (RĀ-mī). The ***dorsal ramus*** (RĀ-mus) innervates the deep muscles and skin of the dorsal surface of the back. The ***ventral ramus*** of a spinal nerve innervates the muscles and structures of the extremities and the lateral and ventral trunk. In addition to dorsal and ventral rami, spinal nerves also give off a ***meningeal branch.*** This branch reenters the spinal canal through the intervertebral foramen and supplies the vertebrae, vertebral ligaments, blood vessels of the spinal cord, and the meninges. Other branches of a spinal nerve are the ***rami communicantes*** (kō-myoo-ni-KAN-tēz), components of the autonomic nervous system whose structure and function are discussed in Chapter 16.

Plexuses

The ventral rami of spinal nerves, except for thoracic nerves T2–T11, do not go directly to the structures of the body they supply. Instead, they form networks on either side of the body by joining with adjacent nerves. Such a network is called a ***plexus*** (*plexus* = braid). The principal plexuses are the cervical plexus, brachial plexus, lumbar plexus, and sacral plexus (see Figure 13-1). Emerging from the plexuses are nerves bearing names that are often descriptive of the general regions they supply or the course they take. Each of the nerves, in turn, may have several branches named for the specific structures they innervate.

▪ **Cervical Plexus** The ***cervical plexus*** is formed by the ventral rami of the first four cervical nerves (C1–C4) with contributions from C5. There is one on each side of the neck alongside the first four cervical vertebrae (Figure 13-11). The ***roots*** of the plexus indicated in the diagram are the ventral rami. The cervical plexus supplies the skin and muscles of the head, neck, and upper part of the shoulders. Branches of the cervical plexus also connect with the accessory (XI) and hypoglossal (XII) cranial nerves. The phrenic nerves are a major pair of nerves arising from the cervical plexuses that supply the motor fibers to the diaphragm. Damage to the spinal cord above the origin of the phrenic nerves results in paralysis of the diaphragm, since the phrenic nerves no longer send nerve impulses to the diaphragm. Contractions of the diaphragm are essential for normal breathing.

Exhibit 13-2 summarizes the nerves and distributions of the cervical plexus. The relationship of the cervical plexus to the other plexuses is shown in Figure 13-1a.

▪ **Brachial Plexus** The ***brachial plexus*** is formed by the ventral rami of spinal nerves C5–C8 and T1. On either side of the last four cervical and first thoracic vertebrae, the brachial plexus extends downward and laterally, passes over the first rib behind the clavicle, and then enters the axilla (Figure 13-12). The brachial plexus constitutes the entire nerve supply for the upper extremities and shoulder region.

The ***roots*** of the brachial plexus, like those of the cervical plexus, are the ventral rami of the spinal nerves. The roots of C5 and C6 unite to form the ***superior trunk,*** C7 becomes the ***middle trunk,*** and C8 and T1 form the ***inferior trunk.*** Each trunk, in turn, divides into an ***anterior division*** and a ***posterior division.*** The divisions then unite to form cords. The ***posterior cord*** is formed by the union of the posterior divisions of the superior, middle, and inferior trunks. The ***medial cord***

FIGURE 13-10 Branches of a typical spinal nerve.

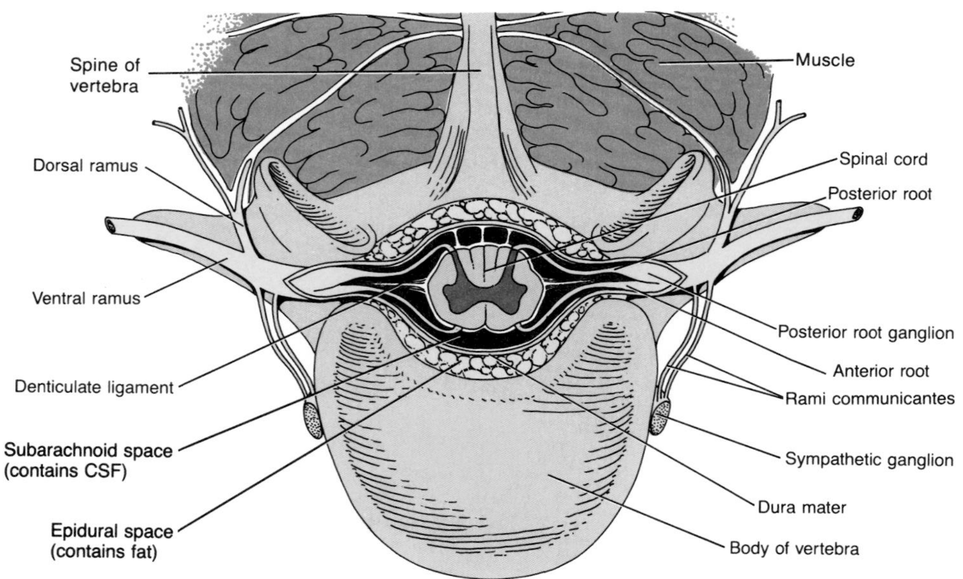

FIGURE 13-11 Cervical plexus. Consult Exhibit 13-2 so that you can determine the distribution of each of the nerves of the cervical plexus.

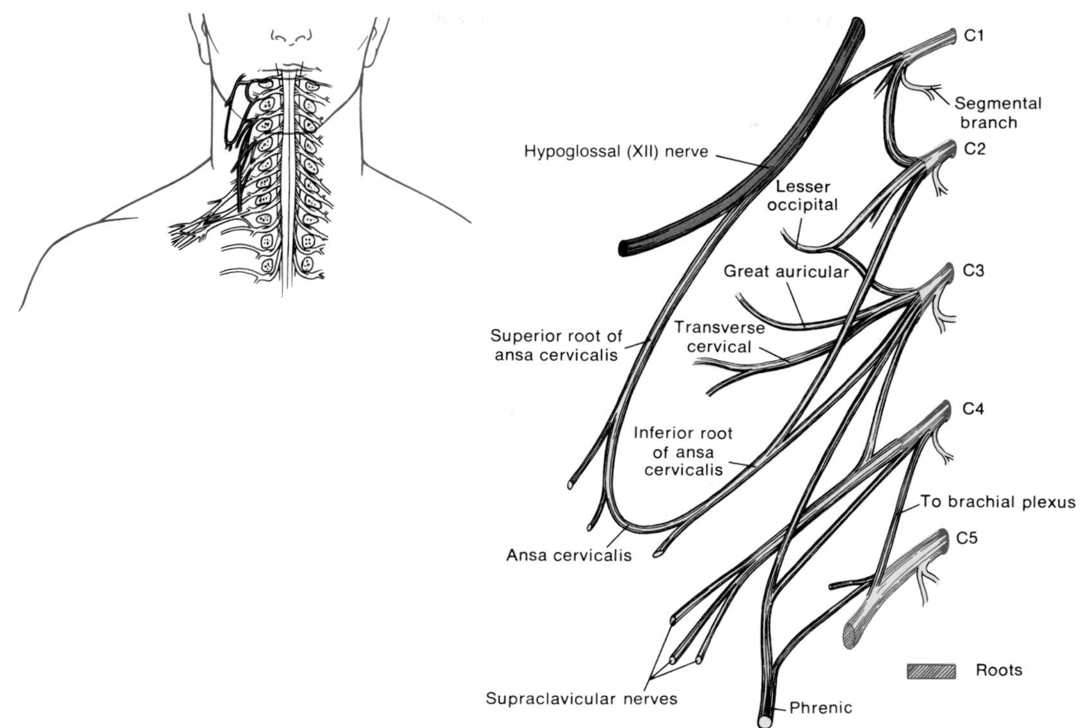

is formed as a continuation of the anterior division of the inferior trunk. The ***lateral cord*** is formed by the union of the anterior divisions of the superior and middle trunk. The peripheral nerves arise from the cords. Thus, the brachial plexus begins as roots that unite to form trunks, the trunks branch into divisions, the divisions form cords, and the cords give rise to the peripheral nerves.

Five important nerves arising from the brachial plexus are the axillary, musculocutaneous, radial, median, and ulnar. The ***axillary nerve*** supplies the deltoid and teres

EXHIBIT 13-2 CERVICAL PLEXUS

Nerve	Origin	Distribution
SUPERFICIAL OR CUTANEOUS BRANCHES		
Lesser occipital	C2.	Skin of scalp behind and above ear.
Greater auricular	C2–C3.	Skin in front, below, and over ear and over parotid glands.
Transverse cervical	C2–C3.	Skin over anterior aspect of neck.
Supraclavicular	C3–C4.	Skin over upper portion of chest and shoulder.
DEEP OR LARGELY MOTOR BRANCHES		
Ansa cervicalis	This nerve is divided into a superior root and an inferior root.	
Superior root	C1.	Infrahyoid, thyrohyoid, and geniohyoid muscles of neck.
Inferior root	C2–C3.	Omohyoid, sternohyoid, and sternothyroid muscles of neck.
Phrenic	C3–C5.	Diaphragm between thorax and abdomen.
Segmental branches	C1–C5.	Prevertebral (deep) muscles of neck, levator scapulae, and middle scalene muscles.

FIGURE 13-12 Brachial plexus. (a) Origin.

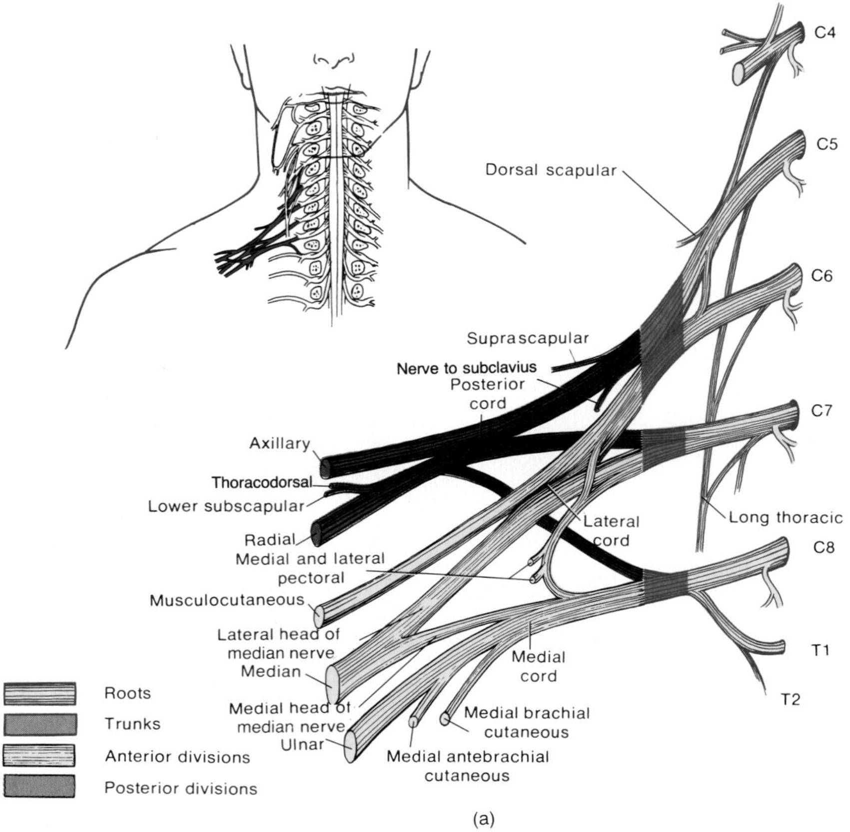

minor muscles. The ***musculocutaneous nerve*** supplies the flexors of the arm and forearm. The ***radial nerve*** supplies the muscles on the posterior aspect of the arm and forearm. The ***median nerve*** supplies most of the muscles of the anterior forearm and some of the muscles in the palm. The ***ulnar nerve*** supplies the anteromedial muscles of the forearm and most of the muscles of the palm.

CLINICAL APPLICATION: INJURIES TO BRACHIAL PLEXUS

Prolonged use of a crutch that presses into the axilla may result in injury to a portion of the brachial plexus. The usual ***crutch palsy*** involves the posterior cord of the brachial plexus or, more frequently, just the radial nerve, which, in general, supplies extensors.

Radial nerve damage is indicated by wrist drop: inability to extend the hand at the wrist. Care must be taken not to injure the radial and axillary nerves when deltoid intramuscular injections are given. The radial nerve may also be injured when a cast is applied too tightly around the midhumerus. ***Median nerve damage*** is indicated by numbness, tingling, and pain in the palm and fingers; weak thumb movements; and inability to pronate the forearm and difficulty in flexing the wrist properly. Compression of the median nerve inside the carpal tunnel, formed anteriorly by the flexor retinaculum (transverse carpal ligament) and posteriorly by the carpal bones (see Figure 11-19), is known as ***carpal tunnel syndrome.*** It may be caused by any condition that aggravates compression of the contents of the carpal tunnel, such as trauma, edema, and flexion of the wrist as a result of activities such as needlepoint, driving a car, cutting hair, and playing a piano. ***Ulnar nerve damage*** is indicated by an inability to adduct or abduct the medial four fingers (not the thumb), weakness in flexing and adducting the wrist, and loss of sensation over the little finger.

FIGURE 13-12 (*Continued*) (b) Distribution. Consult Exhibit 13-3 so that you can determine the distribution of each of the nerves of the brachial plexus.

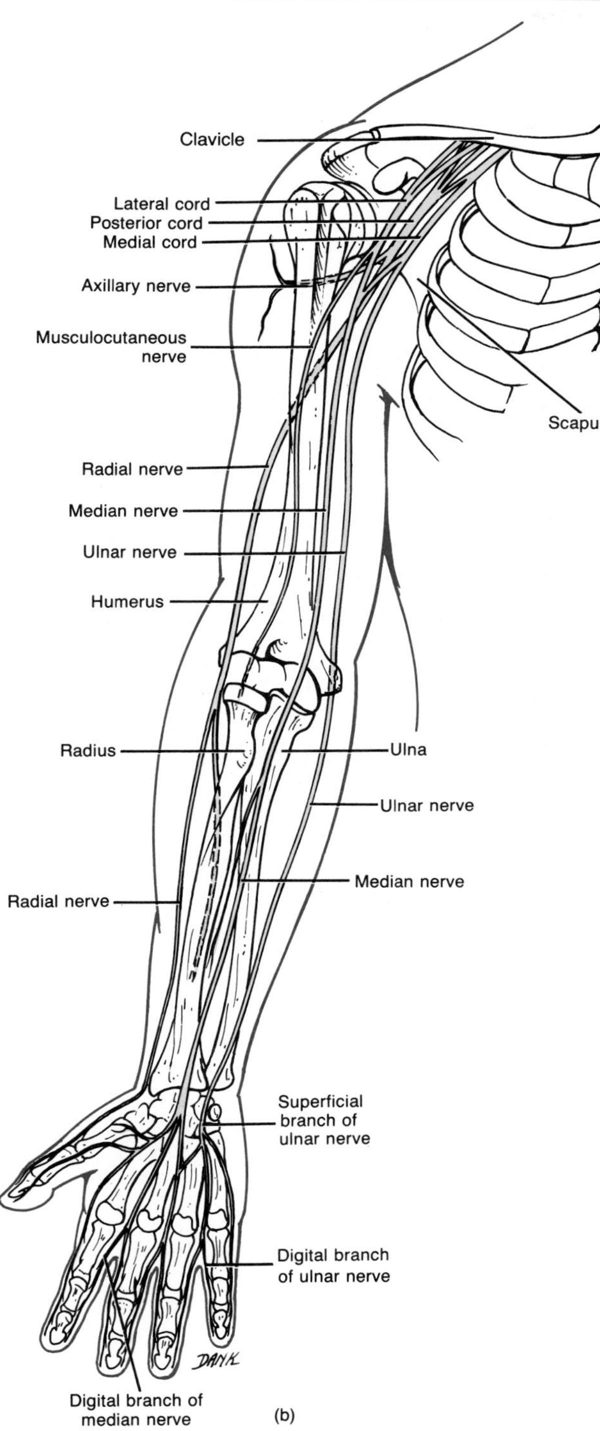

A summary of the nerves and distributions of the brachial plexus is given in Exhibit 13-3. The relationship of the brachial plexus to the other plexuses is shown in Figure 13-1a.

▪ Lumbar Plexus The ***lumbar plexus*** is formed by the ventral rami of spinal nerves L1–L4. It differs from the brachial plexus in that there is no intricate interlacing of fibers. It also consists of ***roots*** and an ***anterior*** and ***posterior division.*** On either side of the first four lumbar vertebrae, the lumbar plexus passes obliquely outward behind the psoas major muscle (posterior division) and anterior to the quadratus lumborum muscle (anterior division), and then gives rise to its peripheral nerves (Figure 13-13). The lumbar plexus supplies the anterolateral abdominal wall, external genitals, and part of the lower extremity. The largest nerve arising from the lumbar plexus is the femoral nerve.

CLINICAL APPLICATION: FEMORAL NERVE INJURY

Injury to the femoral nerve is indicated by an inability to extend the leg and by loss of sensation in the skin over the anteromedial aspect of the thigh.

A summary of the nerves and distributions of the lumbar plexus is presented in Exhibit 13-4. The relationship of the lumbar plexus to the other plexuses is shown in Figure 13-1a.

▪ Sacral Plexus The ***sacral plexus*** is formed by the ventral rami of spinal nerves L4–L5 and S1–S4. It is situated largely in front of the sacrum (Figure 13-14). Like the lumbar plexus, it contains ***roots*** and an ***anterior*** and ***posterior division.*** The sacral plexus supplies the buttocks, perineum, and lower extremities. The largest nerve arising from the sacral plexus—and in fact the largest nerve in the body—is the sciatic nerve. The sciatic nerve supplies the entire musculature of the leg and foot.

CLINICAL APPLICATION: SCIATIC NERVE INJURY

Injury to the sciatic nerve (common peroneal portion) and its branches results in pain that may extend from the buttock down the back of the leg, foot drop, an inability to dorsiflex the foot, and loss of sensation over the leg and foot. This nerve may be injured because of a slipped disc, dislocated hip, osteoarthritis of the lumbosacral spine, pressure from the uterus during pregnancy, or an improperly administered gluteal intramuscular injection.

EXHIBIT 13-3 BRACHIAL PLEXUS

Nerve	Origin	Distribution
ROOT		
Dorsal scapular	C5.	Levator scapulae, rhomboideus major, and rhomboideus minor muscles.
Long thoracic	C5–C7.	Serratus anterior muscle.
TRUNK		
Nerve to subclavius	C5–C6.	Subclavius muscle.
Suprascapular	C5–C6.	Supraspinatus and infraspinatus muscles.
LATERAL CORD		
Musculocutaneous	C5–C7.	Coracobrachialis, biceps brachii, and brachialis muscles.
Median (lateral head)	C5–C7.	See distribution for ***Median (medial head)*** in this exhibit.
Lateral pectoral	C5–C7.	Pectoralis major muscle.
POSTERIOR CORD		
Upper subscapular	C5–C6.	Subscapularis muscle.
Thoracodorsal	C6–C8.	Latissimus dorsi muscle.
Lower subscapular	C5–C6.	Subscapularis and teres major muscles.
Axillary (circumflex)	C5–C6.	Deltoid and teres minor muscles; skin over deltoid and upper posterior aspect of arm.
Radial	C5–C8 and T1.	Extensor muscles of arm and forearm (triceps brachii, brachioradialis, extensor carpi radialis longus, extensor digitorum, extensor carpi ulnaris, extensor carpi radialis brevis, extensor indicis); skin of posterior arm and forearm, lateral two-thirds of dorsum of hand, and fingers over proximal and middle phalanges.
MEDIAL CORD		
Medial pectoral	C8–T1.	Pectoralis major and pectoralis minor muscles.
Medial brachial cutaneous	C8–T1.	Skin of medial and posterior aspects of lower third of arm.
Medial antebrachial cutaneous	C8–T1.	Skin of medial and posterior aspects of forearm.
Median (medial head)	C5–C8 and T1.	Medial and lateral heads of median nerve form median nerve. Distributed to flexors of forearm (pronator teres, flexor carpi radialis, flexor digitorum superficialis, lateral half of flexor digitorum profundus) except flexor carpi ulnaris; skin of lateral two-thirds of palm of hand and fingers.
Ulnar	C8–T1.	Flexor carpi ulnaris and flexor digitorum profundus muscles; skin of medial side of hand, little finger, and medial half of ring finger.
OTHER CUTANEOUS DISTRIBUTIONS		
Intercostobrachial	Second intercostal nerve.	Skin over medial side of arm.
Upper lateral brachial cutaneous	Axillary.	Skin over deltoid muscle and down to elbow.
Posterior brachial cutaneous	Radial.	Skin over posterior aspect of arm.
Lower lateral brachial cutaneous	Radial.	Skin over lateral aspect of elbow.
Lateral antebrachial cutaneous	Musculocutaneous.	Skin over lateral aspect of forearm.
Posterior antebrachial cutaneous	Radial.	Skin over posterior aspect of forearm.

FIGURE 13-13 Lumbar plexus. (a) Origin. (b) Distribution of nerves of the lumbar and sacral plexuses in anterior view (left) and posterior view (right). Consult Exhibit 13-4 so that you can determine the distribution of the nerves of the lumbar plexus.

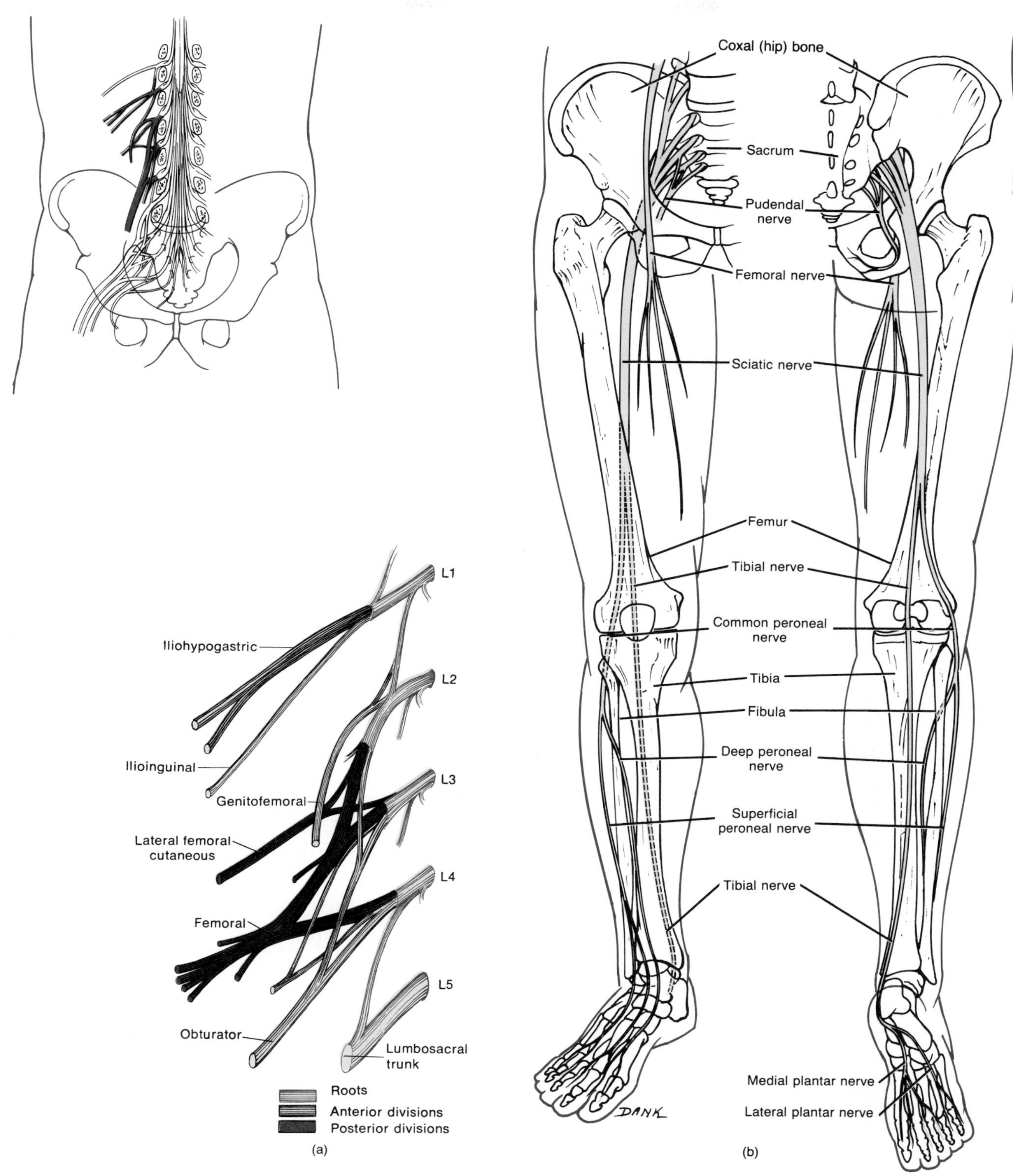

EXHIBIT 13-4 LUMBAR PLEXUS

Nerve	Origin	Distribution
Iliohypogastric	T12–L1.	Muscles of anterolateral abdominal wall (external oblique, internal oblique, transversus abdominis); skin of lower abdomen and buttock.
Ilioinguinal	L1.	Muscles of anterolateral abdominal wall as indicated above; skin of upper medial aspect of thigh, root of penis and scrotum in male, and labia majora and mons pubis in female.
Genitofemoral	L1–L2.	Cremaster muscle; skin over middle anterior surface of thigh, scrotum in male, and labia majora in female.
Lateral femoral cutaneous	L2–L3.	Skin over lateral, anterior, and posterior aspects of thigh.
Femoral	L2–L4.	Flexor muscles of thigh (iliacus, psoas major, pectineus, rectus femoris, sartorius); extensor muscles of leg (rectus femoris, vastus lateralis, vastus medialis, vastus intermedius); skin on front and over medial aspect of thigh and medial side of leg and foot.
Obturator	L2–L4.	Adductor muscles of leg (obturator externus, pectineus, adductor longus, adductor brevis, adductor magnus, gracilis); skin over medial aspect of thigh.

FIGURE 13-14 Sacral plexus. Refer to Figure 13-13b for the distribution of nerves of the sacral plexus. Consult Exhibit 13-5 so that you can determine the distribution of each of the nerves of the sacral plexus.

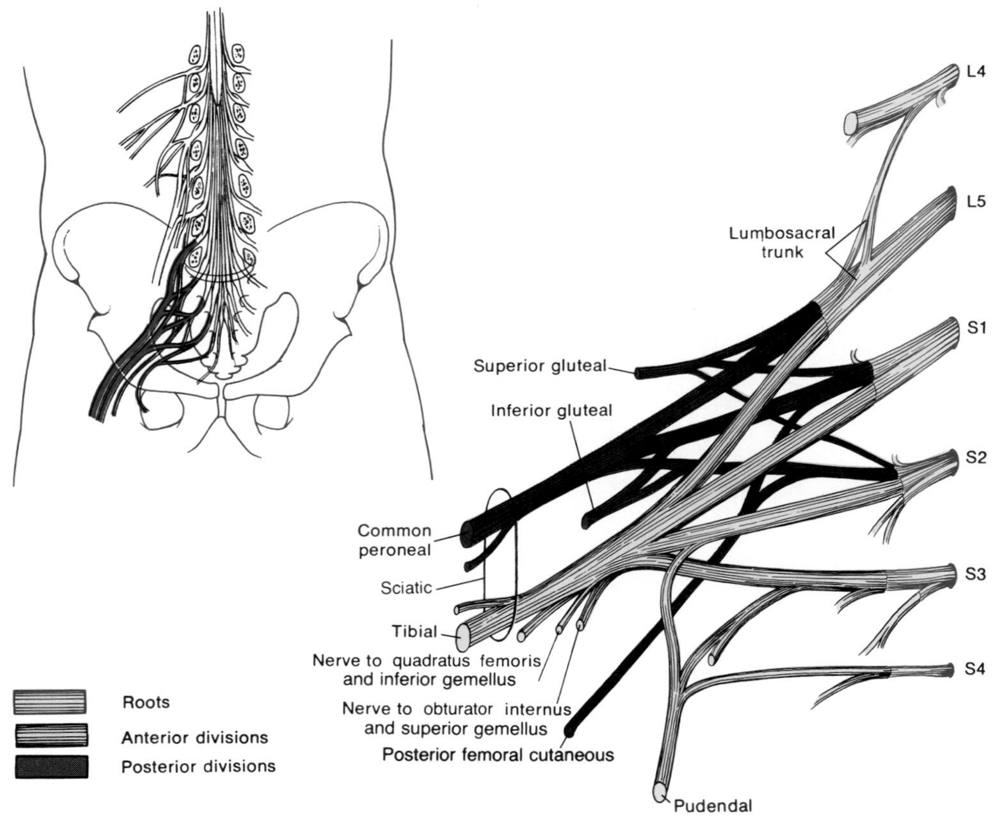

A summary of the nerves and distributions of the sacral plexus is given in Exhibit 13-5. The relationship of the sacral plexus to the other plexuses is shown in Figure 13-1a.

Intercostal (Thoracic) Nerves

The ventral rami of spinal nerves T2–T11 do not enter into the formation of plexuses and are known as ***intercostal (thoracic) nerves.*** These nerves are distributed directly to the structures they supply in intercostal spaces (see Figure 13-1). After leaving its intervertebral foramen, the ventral ramus of nerve T2 supplies the intercostal muscles of the second intercostal space and the skin of the axilla and posteromedial aspect of the arm. Nerves T3 and T6 pass in the costal grooves of the ribs and are distributed to the intercostal muscles and skin of the anterior and lateral chest wall. Nerves T7–T11 supply the intercostal muscles and the abdominal muscles and overlying skin. The dorsal rami of the intercostal nerves supply the deep back muscles and skin of the dorsal aspect of the thorax.

Dermatomes

The skin over the entire body is supplied segmentally by spinal nerves. This means that the spinal nerves innervate specific, constant segments of the skin. All spinal nerves except C1 supply branches to the skin. The skin segment supplied by the dorsal root of a spinal nerve is a ***dermatome.***

EXHIBIT 13-5 SACRAL PLEXUS

Nerve	Origin	Distribution
Superior gluteal	L4–L5 and S1.	Gluteus minimus and gluteus medius muscles and tensor fasciae latae.
Inferior gluteal	L5–S2.	Gluteus maximus muscle.
Nerve to piriformis	S1–S2.	Piriformis muscle.
Nerve to quadratus femoris	L4–L5 and S1.	Quadratus femoris and inferior gemellus muscles.
Nerve to obturator internus	L5–S2.	Obturator internus and superior gemellus muscles.
Perforating cutaneous	S2–S3.	Skin over lower medial aspect of buttock.
Posterior femoral cutaneous	S1–S3.	Skin over anal region, lower lateral aspect of buttock, upper posterior aspect of thigh, upper part of calf, scrotum in male, and labia majora in female.
Sciatic	L4–S3.	Actually two nerves: tibial and common peroneal, bound together by common sheath of connective tissue. It splits into its two divisions, usually at knee. (See below for distributions.) As sciatic nerve descends through thigh, it sends branches to hamstring muscles (biceps femoris, semitendinosus, semimembranosus) and adductor magnus.
Tibial (medial popliteal)	L4–S3.	Gastrocnemius, plantaris, soleus, popliteus, tibialis posterior, flexor digitorum longus, and flexor hallucis longus muscles. Branches of tibial nerve in foot are medial plantar nerve and lateral plantar nerve.
Medial plantar		Abductor hallucis, flexor digitorum brevis, and flexor hallucis brevis muscles; skin over medial two-thirds of plantar surface of foot.
Lateral plantar		Remaining muscles of foot not supplied by medial plantar nerve; skin over lateral third of plantar surface of foot.
Common peroneal (lateral popliteal)	L4–S2.	Divides into a superficial peroneal and a deep peroneal branch.
Superficial peroneal		Peroneus longus and peroneus brevis muscles; skin over distal third of anterior aspect of leg and dorsum of foot.
Deep peroneal		Tibialis anterior, extensor hallucis longus, peroneus tertius, and extensor digitorum longus and brevis muscles; skin over great and second toes.
Pudendal	S2–S4.	Muscles of perineum; skin of penis and scrotum in male and clitoris, labia majora, labia minora, and lower vagina in female.

In the neck and trunk, the dermatomes form consecutive bands of skin (Figure 13-15). In the trunk, there is an overlap of adjacent dermatome nerve supply. Thus, there is little loss of sensation if only a single nerve supply to a dermatome is interrupted. Most of the skin of the face and scalp is supplied by the trigeminal (V) cranial nerve.

Since physicians know which spinal nerves are associated with each dermatome, it is possible to determine which segment of the spinal cord or spinal nerve is malfunctioning. If a dermatome is stimulated and the sensation is not perceived, it can be assumed that the nerves supplying the dermatome are involved.

FIGURE 13-15 Distribution of spinal nerves to dermatomes seen in anterior view. The lines are not perfectly aligned so that there is often considerable overlap.

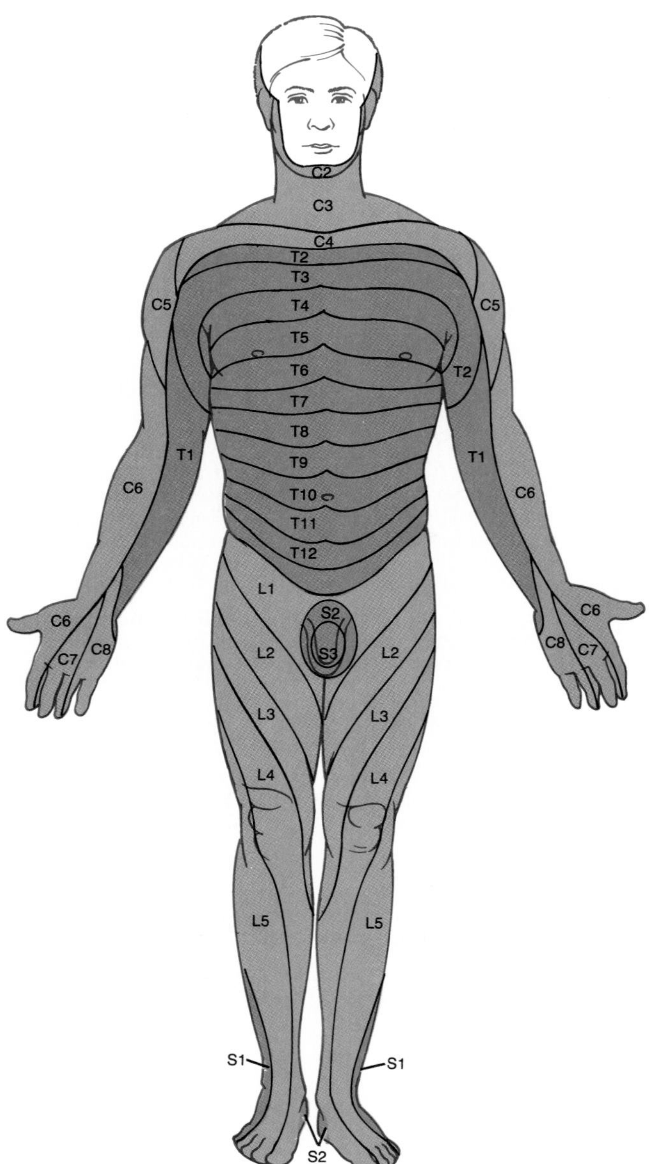

DISORDERS: HOMEOSTATIC IMBALANCES

Spinal Cord Injury

The spinal cord may be damaged by compression from a tumor within or adjacent to the spinal cord, herniated intervertebral discs, blood clots, degenerative and demyelinating disorders, fracture or dislocation of the vertebrae enclosing it, penetrating wounds caused by projectile metal fragments, or other traumatic events such as automobile accidents. Depending on the location and extent of the injury, paralysis may occur. The various types of paralysis may be classified as follows: ***monoplegia*** (*mono* = one; *plege* = stroke), paralysis of one extremity only; ***diplegia*** (*di* = two), paralysis of both upper extremities or both lower extremities; ***paraplegia*** (*para* = beyond), paralysis of both lower extremities; ***hemiplegia*** (*hemi* = half), paralysis of the upper extremity, trunk, and lower extremity on one side of the body; and ***quadriplegia*** (*quad* = four), paralysis of the two upper and two lower extremities.

Complete transection of the spinal cord means that the cord is cut transversely and severed from one side to the other, thus cutting all ascending and descending tracts. It results in a loss of all sensations and voluntary movement below the level of the transection. If the upper cervical cord is transected, quadriplegia results; if the transection is between the cervical and lumbar enlargements, paraplegia results. ***Hemisection*** of the spinal cord refers to a partial transection. It is characterized, below the hemisection, by a loss of proprioception, tactile discrimination, and feeling of vibration on the same side as the injury; paralysis on the same side; and loss of feelings of pain and temperature on the opposite side. If the hemisection is of the upper cervical cord, hemiplegia results; if the hemisection is of the thoracic cord, paralysis of one lower extremity results (monoplegia).

Following transection, there is an initial period of ***spinal shock*** that lasts from a few days to several weeks. During this period, all reflex activity is abolished, a condition called ***areflexia*** (a'-rē-FLEK-sē-a). In time, however, there is a return of reflex activity. The first reflex to return is a stretch reflex (knee jerk). Its reappearance may take several days. Next, the flexion reflexes return, over a period of up to several months. Then the crossed extensor reflexes return. Visceral reflexes such as erection and ejaculation are also affected by transection. Moreover, urinary bladder and bowel functions are no longer under voluntary control. An experimental procedure called *electroejaculation* has been used with some success in helping males with spinal cord injuries to ejaculate. In the procedure, a probe is inserted into the rectum and attached to a device that delivers an electric current in gradually increasing increments until ejaculation occurs. The husband's sperm is then used to inseminate the wife.

Until recently, severe damage resulting from transection was thought to be irreversible. However, a team of researchers has developed a technique for regenerating severed spinal cords in animals. The technique, called ***delayed nerve grafting,*** involves cutting away the crushed or injured section of the spinal cord and bridging the gap with nerve segments from the arm or leg. The original severed axons in the cord can then grow through the bridge. Delayed nerve grafting offers hope for paraplegics who have lost voluntary movements of the lower extremities, as well as urinary, bowel, and sexual functions.

Peripheral Nerve Damage and Repair

As we have seen, axons and dendrites that have a neurolemma can be repaired as long as the cell body is intact, fibers are in association with neurolemmocytes (Schwann cells), and scar tissue formation does not occur too rapidly. Most nerves that lie outside the brain and spinal cord consist of processes (axons and dendrites) that are covered with a neurolemma. A person who injures a nerve in the upper extremity, for example, has a good chance of regaining nerve function. Processes in the brain and spinal cord do not have a neurolemma. Injury there is permanent.

When there is damage to an axon (or to dendrites of somatic afferent neurons), there are usually changes that occur in the cell body of the affected neuron called chromatolysis. In addition, there are always changes that occur in the portion of the axon distal to the site of injury, called Wallerian degeneration, and in the portion of the axon proximal to the site of injury, called retrograde degeneration. Chromatolysis occurs in essentially the same way, whether the damaged fiber is in the central or peripheral nervous system. The Wallerian degeneration reaction, however, depends on whether the fiber is central or peripheral.

Chromatolysis

About 24 to 48 hours after injury to a process of a central or peripheral neuron, the chromatophilic substance (Nissl bodies), normally arranged in an orderly fashion in an uninjured cell body, breaks down into finely granular masses. This alteration is called ***chromatolysis*** (krō'-ma-TOL-i-sis; *chromo* = color; *lysis* = dissolution). It begins between the axon hillock and nucleus but spreads throughout the cell body. As a result of chromatolysis, the cell body swells, and the swelling reaches its maximum between 10 and 20 days after injury (Figure 13-16b). Chromatolysis results in a loss of ribosomes by the rough endoplasmic reticulum and an increase in the number of free ribosomes. Another sign of the chromatolysis is the off-center position of the nucleus in the cell body. This change makes it possible to iden-

FIGURE 13-16 Peripheral nerve damage and repair. (a) Normal neuron. (b) Chromatolysis. (c) Wallerian degeneration. (d) Regeneration.

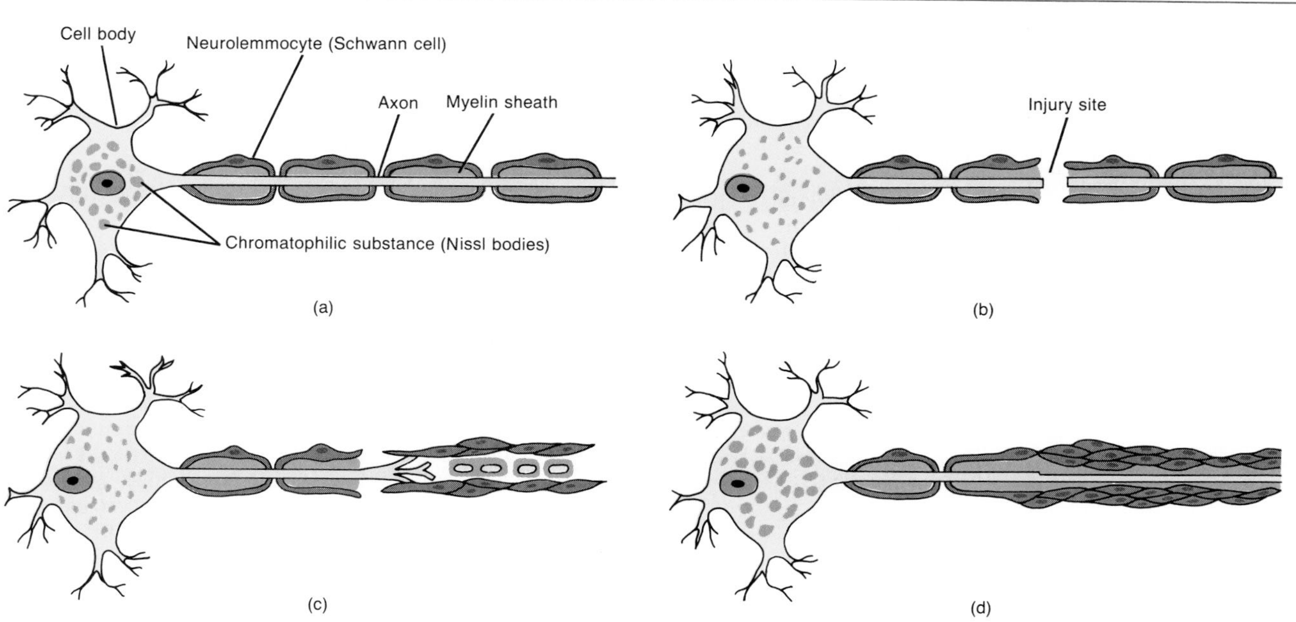

tify the cell bodies of damaged fibers through a microscope.

Wallerian Degeneration

The part of the process distal to the damage becomes slightly swollen and then breaks up into fragments by the third to fifth day. The myelin sheath around the axon or dendrite also undergoes degeneration (Figure 13-16c). Degeneration of the distal portion of the axon or dendrite and myelin sheath is called ***Wallerian degeneration.*** Following degeneration, there is phagocytosis of the remains by macrophages.

Even though there is degeneration of the axon or dendrite and myelin sheath, the neurolemma of the neurolemmocytes remains. The neurolemmocytes on either side of the site of injury multiply by mitosis and grow toward each other and attempt to form a tube across the injured area. The tube provides a means for new axons or dendrites to grow from the proximal area across the injured area into the distal area previously occupied by the original nerve fiber (Figure 13-16d). The growth of new axons or dendrites will not occur if the gap at the site of injury is too large or if the gap becomes filled with dense collagenous fibers.

Retrograde Degeneration

The changes in the proximal portion of the axon, called ***retrograde degeneration,*** are similar to those that occur during Wallerian degeneration. The main difference in retrograde degeneration is that the changes occur only as far as the first neurofibral node (node of Ranvier).

Regeneration

Following chromatolysis, there are signs of recovery in the cell body. There is an acceleration of RNA and protein syntheses, which favors ***regeneration*** of the axon. Recovery often takes several months and involves the restoration of normal levels of RNA, proteins, and the chromatophilic substance to their usual, uninjured patterns.

Accelerated protein synthesis is required for repair of the damaged axon. The proteins synthesized in the cell body pass into the empty lumen of the tube formed by neurolemmocytes, by axoplasmic flow at the rate of about 1 mm (0.04 inch) per day. The proteins assist in regenerating the damaged axon. During the first few days following damage, buds of regenerating axons or dendrites begin to invade the tube formed by the neurolemmocytes. Axons or dendrites from the proximal area grow at the rate of about 1.5 mm (0.06 inch) per day across the area of damage, find their way into the distal neurolemmal tubes, and grow toward the distally located receptors and effectors. Thus, sensory and motor connections are reestablished. In time, a new myelin sheath is also produced by the neurolemmocytes. However, function is never completely restored after a nerve is severed.

Neuritis

Neuritis is inflammation of a single nerve (***mononeuropathy***), such as would lead to sciatica; two or more nerves in separate areas (***multiple mononeuropathy***), as a result of systemic lupus erythematosus; or many nerves simultaneously (***polyneuropathy***), as typified by Guillain-Barré syndrome. It may result from irritation to the nerve produced by direct blows, bone fractures, contusions, or penetrating injuries. Additional causes include vitamin deficiency (usually thiamine) and poisons such as carbon monoxide, carbon tetrachloride, heavy metals, and some drugs.

Sciatica

Sciatica (sī-AT-i-ka) is a type of neuritis characterized by severe pain along the path of the sciatic nerve or its branches. The term is commonly applied to a number of disorders affecting this nerve. Because of its length and size, the sciatic nerve is exposed to many kinds of injury. Inflammation of or injury to the nerve causes pain that passes from the back or thigh down its length into the leg, foot, and toes.

Probably the most common cause of sciatica is a herniated (slipped) intervertebral disc. Other causes include irritation from osteoarthritis, back injuries, or pressure on the nerve from certain types of exertion. Sciatica may be associated with diabetes mellitus, gout, or vitamin deficiencies. Other cases are idiopathic (unknown).

Shingles

Shingles is an acute infection of the peripheral nervous system and is frequently a relapse of the chickenpox acquired during childhood. Shingles is caused by a virus called herpes zoster (HER-pēz ZOS-ter), the chickenpox virus. Following recovery from chickenpox, the virus retreats to posterior root ganglia, where it resides. If the virus becomes activated, the immune system usually prevents it from spreading. However, from time to time, the activated virus overcomes a weakened immune system. In this case, the virus leaves the ganglion and travels down sensory neurons, causing the pain, and invades the skin where the neurons end, causing a characteristic line of skin blisters and discoloration of the skin. The line of blisters has a shape corresponding to the distribution of a particular nerve. The intercostal nerves and thoracic spinal nerves in the waist area are most commonly affected. Almost always, the skin rash is limited to one side of the body.

STUDY OUTLINE

Grouping of Neural Tissue (p. 356)

1. White matter is an aggregation of myelinated axons and associated neuroglia.
2. Gray matter is a collection of unmyelinated nerve cell bodies and dendrites or unmyelinated axons along with associated neuroglia.
3. A nerve is a bundle of nerve axons and/or dendrites outside the central nervous system.
4. A ganglion is a collection of unmyelinated cell bodies outside the central nervous system.
5. A tract is a bundle of myelinated fibers of similar function in the central nervous system.
6. A nucleus is a mass of unmyelinated nerve cell bodies and dendrites present as gray matter in the brain and spinal cord.
7. A horn (column) is an area of gray matter in the spinal cord.

Spinal Cord (p. 356)

Protection and Coverings (p. 356)

1. The spinal cord is protected by the vertebral canal, meninges, cerebrospinal fluid, and vertebral ligaments.
2. The meninges are three coverings that run continuously around the spinal cord and brain: dura mater, arachnoid, and pia mater.
3. Removal of cerebrospinal fluid from the subarachnoid space is called a spinal (lumbar) tap (puncture). The procedure is used to diagnose pathologies and to introduce antibiotics or contrast media.

General Features (p. 357)

1. The spinal cord begins as a continuation of the medulla oblongata and terminates at about the second lumbar vertebra.
2. It contains cervical and lumbar enlargements that serve as points of origin for nerves to the extremities.
3. The tapered portion of the spinal cord is the conus medullaris, from which arise the filum terminale and cauda equina.
4. The spinal cord is partially divided into right and left sides by the anterior median fissure and posterior median sulcus.
5. The gray matter in the spinal cord is divided into horns and the white matter into funiculi (columns).
6. In the center of the spinal cord is the central canal, which runs the length of the spinal cord and contains cerebrospinal fluid.
7. There are ascending (sensory) tracts and descending (motor) tracts.

Structure in Cross Section (p. 357)

1. Parts of the spinal cord observed in cross section are the gray commissure; central canal; anterior, posterior, and lateral gray horns; anterior, posterior, and lateral white columns; and ascending and descending tracts.
2. The spinal cord conveys sensory and motor information by way of the ascending and descending tracts, respectively.

Functions (p. 360)

1. A major function of the spinal cord is to convey sensory nerve impulses from the periphery to the brain and to conduct motor impulses from the brain to the periphery.
2. Another function is to serve as a reflex center. The posterior root, posterior root ganglion, and anterior root are involved in conveying an impulse.
3. A reflex arc is the shortest route that can be taken by an impulse from a receptor to an effector. Its basic components are a receptor, a sensory neuron, a center, a motor neuron, and an effector.
4. A reflex is a quick, involuntary response to a stimulus that passes along a reflex arc. Reflexes represent the body's principal mechanisms for responding to certain changes (stimuli) in the internal and external environment.
5. Somatic spinal reflexes include the stretch reflex, tendon reflex, flexor reflex, and crossed extensor reflex; all exhibit reciprocal innervation.
6. A two-neuron or monosynaptic reflex arc contains one sensory and one motor neuron. A stretch reflex, such as the patellar reflex, is an example.
7. The stretch reflex is ipsilateral and is important in maintaining muscle tone and muscle coordination during exercise.
8. A polysynaptic reflex arc contains a sensory, association, and motor neuron. The tendon reflex, flexor reflex, and crossed extensor reflexes are examples.
9. The tendon reflex is ipsilateral and prevents damage to muscles and tendons as a result of stretching; the flexor reflex is ipsilateral and is a withdrawal reflex; the crossed extensor reflex is contralateral.
10. Among clinically important somatic reflexes are the patellar reflex, the Achilles reflex, the Babinski sign, and the abdominal reflex.

Spinal Nerves (p. 369)

Names (p. 369)

1. The 31 pairs of spinal nerves are named and numbered according to the region and level of the spinal cord from which they emerge.
2. There are 8 pairs of cervical, 12 pairs of thoracic, 5 pairs of lumbar, 5 pairs of sacral, and 1 pair of coccygeal nerves.

Composition and Coverings (p. 369)

1. Spinal nerves are attached to the spinal cord by means of a posterior root and an anterior root. All spinal nerves are mixed.
2. Spinal nerves are covered by endoneurium, perineurium, and epineurium.

Distribution (p. 370)

1. Branches of a spinal nerve include the dorsal ramus, ventral ramus, meningeal branch, and rami communicantes.
2. The ventral rami of spinal nerves, except for T2–T11, form networks of nerves called plexuses.
3. Emerging from the plexuses are nerves bearing names that are often descriptive of the general regions they supply or the course they take.
4. The cervical plexus supplies the skin and muscles of the head, neck, and upper part of the shoulders; connects with some cranial nerves; and supplies the diaphragm.

5. The brachial plexus constitutes the nerve supply for the upper extremities and a number of neck and shoulder muscles.
6. The lumbar plexus supplies the anterolateral abdominal wall, external genitals, and part of the lower extremities.
7. The sacral plexus supplies the buttocks, perineum, and lower extremities.
8. Ventral rami of nerves T2–T11 do not form plexuses and are called intercostal (thoracic) nerves. They are distributed directly to the structures they supply in intercostal spaces.

Dermatomes (p. 377)

1. All spinal nerves except C1 innervate specific, constant segments of the skin. The skin segments are called dermatomes.
2. Knowledge of dermatomes helps a physician to determine which segment of the spinal cord or a spinal nerve is malfunctioning.

Disorders: Homeostatic Imbalances (p. 378)

1. Spinal cord injury may result in paralysis, which may be classified as monoplegia, diplegia, paraplegia, hemiplegia, or quadriplegia.
2. Transection is followed by a period of loss of reflex activity called areflexia.
3. Following peripheral nerve damage, repair is accompanied by chromatolysis, Wallerian degeneration, retrograde degeneration, and regeneration.
4. Inflammation of nerves is known as neuritis.
5. Neuritis of the sciatic nerve and its branches is called sciatica.
6. Shingles is an acute infection of peripheral nerves.

REVIEW QUESTIONS

1. Define the following groupings of neural tissue: white matter, gray matter, nerve, ganglion, tract, nucleus, and horn (column). (p. 356)
2. Describe the bony covering of the spinal cord. (p. 356)
3. Explain the location and composition of the spinal meninges. Describe the location of the epidural, subdural, and subarachnoid spaces. Define meningitis. (p. 356)
4. Describe the location of the spinal cord. What are the cervical and lumbar enlargements? (p. 357)
5. Define conus medullaris, filum terminale, and cauda equina. What is a spinal segment? How is the spinal cord partially divided into a right and left side? (p. 357)
6. Based upon your knowledge of the structure of the spinal cord in cross section, define the following: gray commissure, central canal, anterior gray horn, lateral gray horn, posterior gray horn, anterior white column, lateral white column, posterior white column, ascending tract, and descending tract. (p. 357)
7. Describe the function of the spinal cord as a conduction pathway. Using Exhibit 13-1 as a guide, be sure that you can list the location, origin, termination, and function of the principal ascending and descending tracts. (p. 360, 361)
8. Describe how the spinal cord serves as a reflex center. (p. 363)
9. What is a reflex arc? List and define the components of a reflex arc. (p. 363)
10. Define a reflex. How are reflexes related to the maintenance of homeostasis? (p. 363)
11. Describe the mechanism and function of a stretch reflex, tendon reflex, and crossed extensor reflex. (p. 364)
12. Define the following terms relating to reflex arcs: monosynaptic, ipsilateral, polysynaptic, intersegmental, contralateral, and reciprocal innervation. (p. 364–368)
13. Why are reflexes important in diagnosis? Indicate the clinical importance of the following reflexes: patellar, Achilles, Babinski sign, and abdominal. (p. 368)
14. Define a spinal nerve. Why are all spinal nerves classified as mixed nerves? (p. 369)
15. Describe how a spinal nerve is attached to the spinal cord. (p. 369)
16. Explain how a spinal nerve is enveloped by its connective tissue coverings. (p. 369)
17. How are spinal nerves named and numbered? (p. 369)
18. Describe the branches and innervations of a typical spinal nerve. (p. 370)
19. What is a plexus? Describe the principal plexuses and the regions they supply. (p. 370)
20. What are intercostal (thoracic) nerves? (p. 377)
21. Define a dermatome. Why is a knowledge of dermatomes important? (p. 377)
22. Distinguish the following types of paralysis: monoplegia, diplegia, paraplegia, hemiplegia, and quadriplegia. (p. 378)
23. Define complete transection and hemisection. What are the consequences of each? What is spinal shock? (p. 378)
24. Outline the principal events that occur as part of chromatolysis, Wallerian degeneration, and retrograde degeneration, and regeneration following peripheral nerve damage. (p. 379)
25. Distinguish between sciatica and neuritis. (p. 380)
26. What is shingles? (p. 380)
27. Describe nerve injuries to the brachial plexus, femoral nerve, and sciatic nerve. (p. 372, 373)
28. Why is a spinal (lumbar) tap (puncture) performed? Describe the procedure. (p. 357)
29. What is myelography? Why is it performed? (p. 359)

SELECTED READINGS

Bruno, M., and S. Katz. "New Hope for Hurt Nerves," *Newsweek*, 7 October 1985.

Goldfinger, S. E. (ed). "Parkinson's Disease: Less Mysterious, More Manageable," *Harvard Medical School Health Letter*, November 1984.

———. "Shingles," *Harvard Medical School Health Letter*, June 1984.

Gottlieb, D. I. "GABAergic Neurons," *Scientific American,* February 1988.

Greenspan, J. "Carpal Tunnel Syndrome," *Postgraduate Medicine,* November 1988.

Johnson, G. T. (ed). "Is Spinal Anesthesia Best for You? *Mayo Clinic Health Letter,* October 1984.

Mattewson, J. "Alzheimer's Disease: Source Searching," *Science News,* 13 July 1985.

Miller, J. A. "Grow Nerves Grow," *Science News,* 29 March 1986.

Rappaport, S. "Common Peripheral Nerve Injuries," *Hospital Medicine,* June 1984.

Romeo, J. H. "Spinal Cord Injury," *RN,* May 1988.

Wurtman, R. J. "Alzheimer's Disease," *Scientific American,* January 1985.

Chapter 14

The Brain and the Cranial Nerves

Chapter Contents at a Glance

Student Objectives

1. Identify the principal parts of the brain and describe how the brain is protected.
2. Explain the formation and circulation of cerebrospinal fluid (CSF).
3. Describe the blood supply to the brain and the concept of the blood–brain barrier (BBB).
4. Compare the structure and functions of the brain stem, diencephalon, cerebrum, and cerebellum.
5. Discuss the various neurotransmitters found in the brain, as well as the different types of neuropeptides and their functions.
6. Define a cranial nerve and identify the 12 pairs of cranial nerves by name, number, type, location, and function.
7. Describe the effects of aging on the nervous system.
8. Describe the development of the nervous system.
9. List the clinical symptoms of these disorders of the nervous system: cerebrovascular accident (CVA), epilepsy, transient ischemic attack (TIA), brain tumors, poliomyelitis, cerebral palsy (CP), Parkinson's disease (PD), multiple sclerosis (MS), dyslexia, Tay-Sachs disease, headache, trigeminal neuralgia (tic douloureux), Reye's syndrome (RS), Alzheimer's disease (AD), and delirium.
10. Define medical terminology associated with the central nervous system.

Now we will consider the principal parts of the brain, how the brain is protected, and how it is related to the spinal cord and to the 12 pairs of cranial nerves.

The developmental anatomy of the brain is explained in detail at the end of the chapter.

BRAIN

Principal Parts

The ***brain*** of an average adult is made up of about 1000 billion neurons and is one of the largest organs of the body, weighing about 1300 g (3 lb). Figure 14-1 shows that the brain is mushroom-shaped and divided into four principal parts: brain stem, diencephalon, cerebrum, and cerebellum. In some cases, embryological names are retained when distinguishing the various parts of the brain. The ***brain stem,*** the stalk of the mushroom, consists of the medulla oblongata, pons, and midbrain or mesencephalon (mes-en-SEF-a-lon). The lower end of the brain stem is a continuation of the spinal cord. Above the brain stem is the ***diencephalon*** (dī-en-SEF-a-lon), consisting primarily of the thalamus and hypothalamus. The ***cerebrum*** spreads over the diencephalon. The cerebrum constitutes about seven-eighths of the total weight of the brain and occupies most of the cranium. Inferior to the cerebrum and posterior to the brain stem is the ***cerebellum.***

The brain develops very rapidly during the first few years of life. Growth is due mainly to an increase in the size of cells already present, proliferation and growth of neuroglia, development of synaptic contacts and dendritic branching, and myelination of the various fiber tracts.

Protection and Coverings

The brain is protected by the cranial bones (see Figure 7-2). Like the spinal cord, the brain is also protected by meninges. The ***cranial meninges*** surround the brain,

FIGURE 14-1 Brain. (a) Diagram of the principal parts of the medial aspect of the brain seen in sagittal section. The infundibulum and pituitary gland are discussed in conjunction with the endocrine system in Chapter 18.

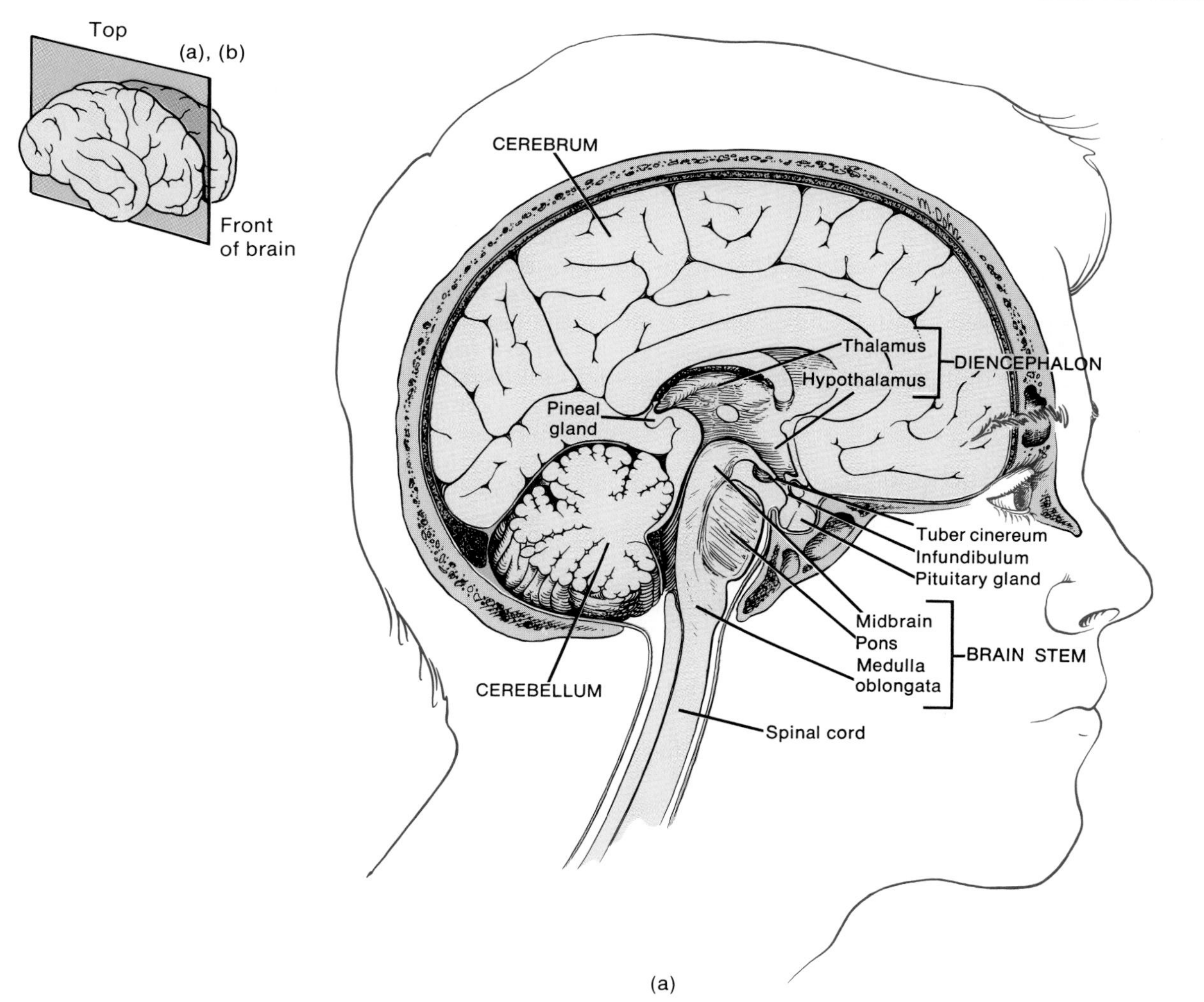

FIGURE 14-1 (*Continued*) (b) Photograph of the medial aspect of the brain seen in sagittal section. (Copyright © Kage, Peter Arnold.)

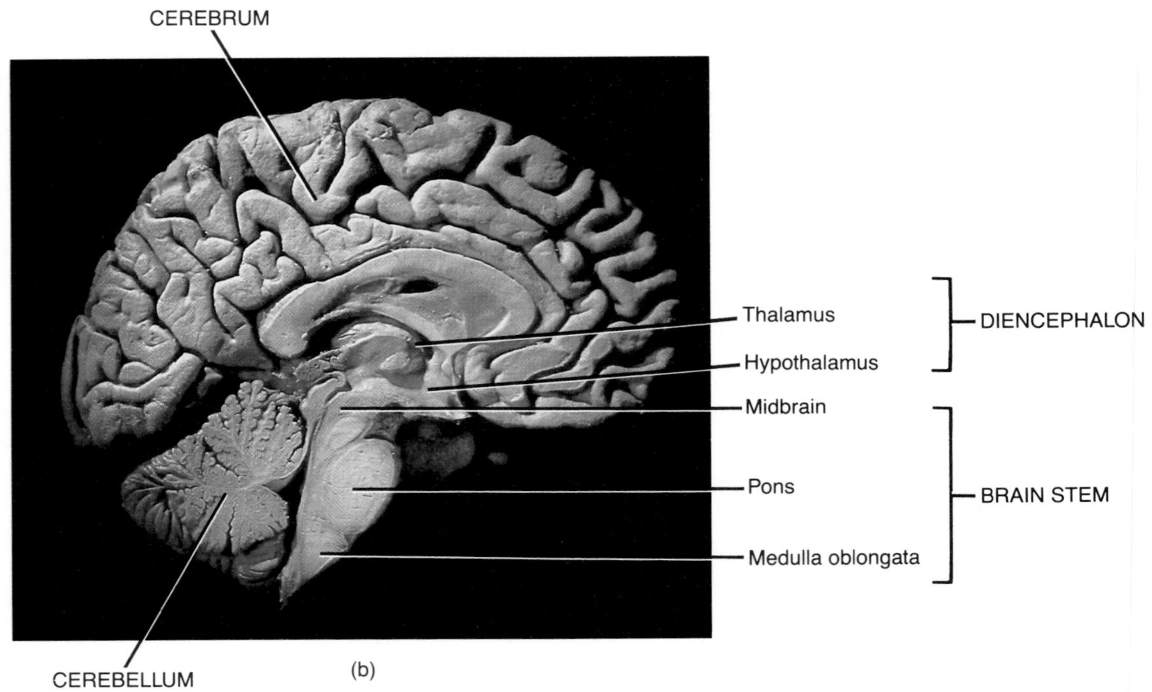

are continuous with the spinal meninges, and have the same basic structure and bear the same names as the spinal meninges: the outermost ***dura mater,*** middle ***arachnoid,*** and innermost ***pia mater*** (Figure 14-2).

The cranial dura mater consists of two layers. The thicker, outer layer (endosteal layer) tightly adheres to the cranial bones and serves as periosteum. The thinner, inner layer (meningeal layer) includes a mesothelial layer on its smooth surface. The spinal dura mater corresponds to the meningeal layer of the cranial dura mater.

Cerebrospinal Fluid (CSF)

The brain, as well as the rest of the central nervous system, is further protected against injury by ***cerebrospinal fluid (CSF).*** This fluid circulates through the subarachnoid space around the brain and spinal cord and through the ventricles of the brain. The subarachnoid space is the area between the arachnoid and pia mater. The CSF is an excellent shock absorber and thus helps protect the delicate brain and spinal cord from trauma.

The ***ventricles*** (VEN-tri-kuls) are cavities in the brain that communicate with each other, with the central canal of the spinal cord, and with the subarachnoid space (Figure 14-2a, b). Each of the two ***lateral ventricles*** is located in a hemisphere (side) of the cerebrum under the corpus callosum. The ***third ventricle*** is a vertical slit between and inferior to the right and left halves of the thalamus and between the lateral ventricles. Each lateral ventricle communicates with the third ventricle by a narrow, oval opening, the ***interventricular foramen (foramen of Monro).*** The ***fourth ventricle*** lies between the inferior brain stem and the cerebellum. It communicates with the third ventricle via the ***cerebral aqueduct,*** which passes through the midbrain. The roof of the fourth ventricle has three openings: a ***median aperture (of Magendie)*** and two ***lateral apertures (of Luschka).*** Through these openings, the fourth ventricle also communicates with the subarachnoid space of the brain and cord.

The entire central nervous system contains between 80 and 150 ml (3 to 5 oz) of cerebrospinal fluid. It is a clear, colorless liquid of watery consistency. Chemically, it contains proteins, glucose, urea, and salts. It also contains some lymphocytes. Cerebrospinal fluid has two principal functions related to homeostasis: protection and circulation. The fluid serves as a shock-absorbing medium to protect the brain and spinal cord from jolts that would otherwise cause them to crash against the bony walls of the cranial and vertebral cavities. The fluid also buoys the brain so that it "floats" in the cranial cavity. With regard to its circulatory function, cerebrospinal fluid delivers nutritive substances filtered from blood to the brain and spinal cord and removes wastes and toxic substances produced by brain and spinal cord cells.

Cerebrospinal fluid is formed primarily by filtration and secretion from networks of capillaries and ependymal

FIGURE 14-2 Meninges and ventricles of the brain. (a) Brain, ventricles, spinal cord, and meninges seen in sagittal section. Arrows indicate the direction of flow of cerebrospinal fluid.

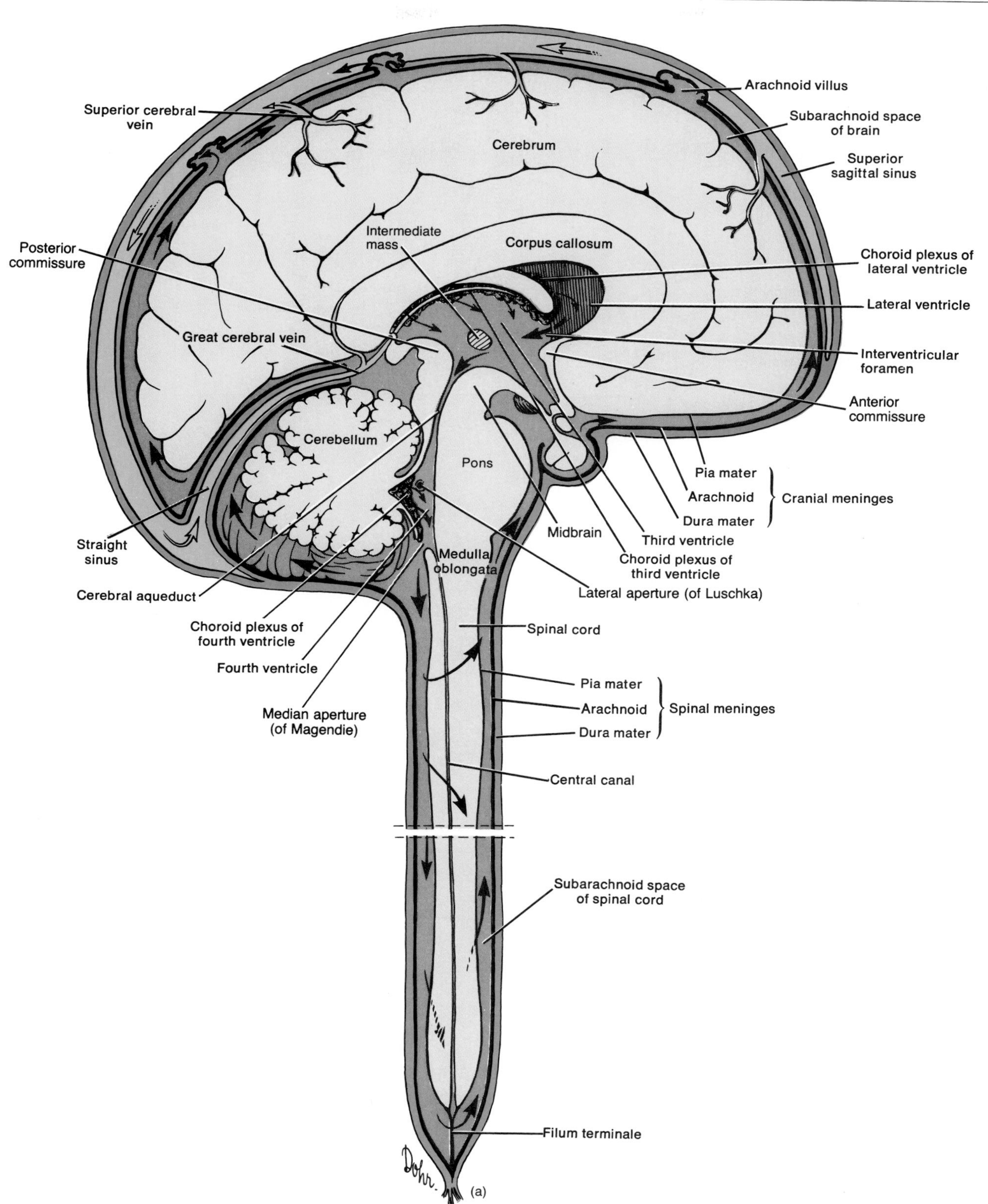

FIGURE 14-2 (*Continued*) Meninges and ventricles of the brain. (b) Diagrammatic lateral projection of the ventricles. (c) Frontal section through the superior portion of the brain, showing the relationship of the superior sagittal sinus to the arachnoid villi.

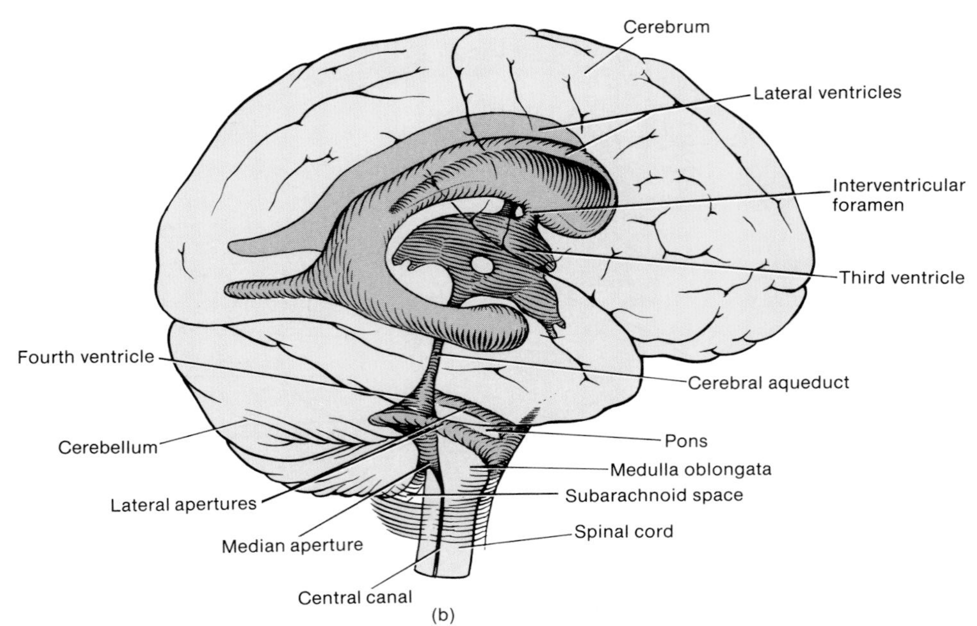

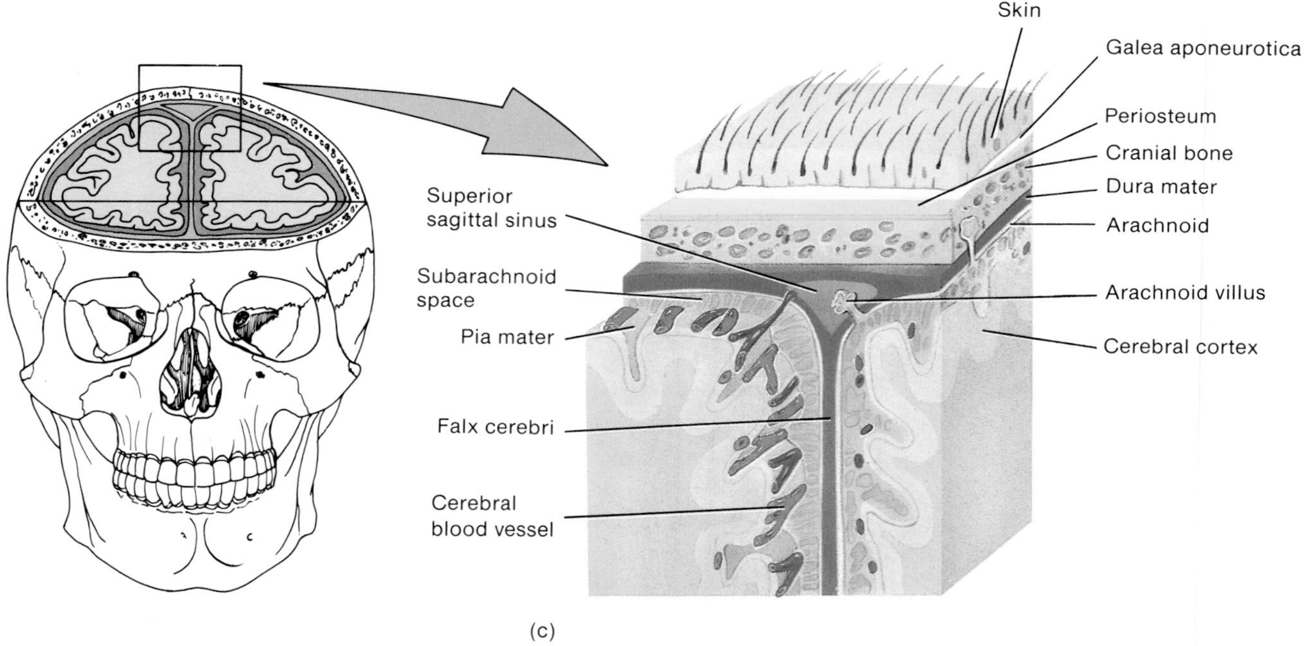

cells in the ventricles called ***choroid*** (KŌ-royd; *chorion* = delicate) ***plexuses*** (Figure 14-2a). Various components of the choroid plexuses form a ***blood–cerebrospinal fluid barrier*** that permits certain substances to enter the fluid but prohibits others. Such a barrier protects the brain and spinal cord from harmful substances. The fluid formed in the choroid plexuses of the lateral ventricles circulates through the interventricular foramina to the third ventricle, where more fluid is added by the choroid plexus of the third ventricle. It then flows through the cerebral aqueduct into the fourth ventricle. Here, there are contributions from the choroid plexus of the fourth ventricle. The fluid then circulates through the apertures of the fourth ventricle into the subarachnoid space around the back of the brain. It also passes downward to the subarachnoid space around the posterior surface of the spinal cord, up the anterior surface of the spinal cord, and around the anterior part of the brain. From there it is gradually reabsorbed into veins. Some cerebrospinal fluid may be formed by ependymal (neuroglial) cells lining the central canal of the spinal cord. This small quantity of fluid ascends to reach the fourth ventricle. Most of the fluid is absorbed into a blood vascular sinus called the superior sagittal sinus. The absorption actually occurs through ***arachnoid villi***—fingerlike projections of the arachnoid that push into the dural venous sinuses, especially the superior sagittal sinus (Figure 14-2c). Normally, cerebrospinal fluid is absorbed as rapidly as it is formed.

CLINICAL APPLICATION: HYDROCEPHALUS

If an obstruction, such as a tumor or a congenital blockage, or an inflammation arises in the brain and interferes with the drainage of cerebrospinal fluid from the ventricles into the subarachnoid space, large amounts of fluid accumulate in the ventricles. Fluid pressure inside the brain increases, and if the fontanels have not yet closed, the head bulges to relieve the pressure. This condition is called ***internal (noncommunicating) hydrocephalus*** (*hydro* = water; *enkephalos* = brain). If an obstruction interferes with drainage somewhere in the subarachnoid space and cerebrospinal fluid accumulates inside the space, the condition is termed ***external (communicating) hydrocephalus.*** Hydrocephalus responds dramatically to ventricular drainage and diversion of CSF (ventriculo-peritoneal or ventriculo-atrial shunt).

Blood Supply

The brain is well supplied with oxygen and nutrients by blood vessels that form the cerebral arterial circle (circle of Willis). Cerebral circulation is outlined in Exhibits 21-3 and 21-8 and Figures 21-14c and 21-18. Blood vessels that enter brain tissue pass along the surface of the brain, and as they penetrate inward, they are surrounded by a loose-fitting layer of pia mater. The space between the penetrating blood vessel and pia mater is called a ***perivascular space.***

Although the brain composes only about 2 percent of total body weight, it utilizes about 20 percent of the oxygen used by the entire body. The brain is one of the most metabolically active organs of the body, and the amount of oxygen it uses varies with the degree of mental activity. If the blood flow to the brain is interrupted even briefly, unconsciousness may result. A one- or two-minute interruption may weaken the brain cells by starving them of oxygen, and if the cells are totally deprived of oxygen for about four minutes, many are permanently injured. Lysosomes of brain cells are sensitive to decreased oxygen concentration. If the condition persists long enough, lysosomes break open and release enzymes that bring about self-destruction of brain cells. Occasionally during childbirth, the oxygen supply from the mother's blood is interrupted before the baby leaves the birth canal and can breathe. Often such babies are stillborn or suffer permanent brain damage that may result in mental retardation, epilepsy, or paralysis.

Blood supplying the brain also contains glucose, the principal source of energy for brain cells. Because carbohydrate storage in the brain is limited, the supply of glucose must be continuous. If blood entering the brain has a low glucose level, mental confusion, dizziness, convulsions, and loss of consciousness may occur.

Both carbon dioxide (CO_2) and oxygen (O_2) have potent effects on cerebral blood flow. Carbon dioxide increases cerebral blood flow by combining with water to form carbonic acid (H_2CO_3), which breaks down into hydrogen ions (H^+) and bicarbonate ions (HCO_3^-). The H^+ ions then cause vasodilation of cerebral vessels and increased blood flow. Dilation is almost directly proportional to an increase in H^+ ion concentration. A decrease in oxygen in the blood also causes vasodilation and increased cerebral blood flow.

Glucose, oxygen, and certain ions pass rapidly from the circulating blood into brain cells. Other substances, such as creatinine, urea, chloride, insulin, and sucrose, enter quite slowly. Still other substances—proteins and most antibiotics—do not pass at all from the blood into brain cells. The differential rates of passage of certain materials from the blood into most parts of the brain are based upon a concept called the ***blood–brain barrier (BBB).*** The barrier is either absent or less selective in the hypothalamus and roof of the fourth ventricle. Electron micrograph studies of the capillaries of the brain reveal that they differ structurally from other capillaries. Brain capillaries are constructed of more densely packed cells and are surrounded by terminations of processes of large numbers of astrocytes (one of the types of neuroglia) and a continuous basement membrane. Current evidence indicates that astrocytes produce a substance that influences the capillaries and confers on them the ability to selectively pass various substances but inhibit others. Substances that cross the barrier are soluble in lipids or water-soluble substances that require the assistance of a carrier molecule to cross by active transport. Some examples of lipid-soluble substances are nicotine, alcohol, and heroin. Water-soluble substances include glucose, certain amino acids, and sodium. The blood–brain barrier functions as a selective barrier to protect brain cells from harmful substances. An injury to the brain due to trauma, inflammation, or toxins causes a breakdown of the blood–brain barrier, permitting the passage of normally restricted substances into brain tissue. Recent evidence suggests that the AIDS virus may penetrate the blood–brain barrier, producing dementia (irreversible deterioration of mental state), acute meningitis, back spasm, and other neurologic disorders before other symptoms of AIDS become apparent.

Brain Stem: Structure and Physiology

Medulla Oblongata

The ***medulla oblongata*** (me-DULL-la ob'-long-GA-ta), or simply ***medulla,*** is a continuation of the upper portion of the spinal cord and forms the inferior part of the brain stem (Figure 14-3). Its position in relation to the other parts of the brain may be noted in Figure 14-1. It lies just superior to the level of the foramen magnum and extends upward to the inferior portion of the pons. The medulla measures 3 cm (about 1 inch) in length.

The medulla contains all ascending and descending tracts that communicate between the spinal cord and various parts of the brain. These tracts constitute the white matter of the medulla. Some tracts cross as they pass through the medulla. Let us see how this crossing occurs and what it means.

On the ventral side of the medulla are two roughly triangular structures called ***pyramids*** (Figures 14-3 and 14-4). The pyramids are composed of the largest motor tracts that pass from the outer region of the cerebrum (cerebral cortex) to the spinal cord. Just above the junction of the medulla with the spinal cord, most of the fibers in the left pyramid cross to the right side, and most of the fibers in the right pyramid cross to the left. This crossing is called the ***decussation*** (dē'-ku-SĀ-shun) ***of pyramids.*** The adaptive value, if any, of this phenomenon is unknown. Decussation explains why motor areas of one side of the cerebral cortex control muscular movements on the opposite side of the body. The principal motor fibers that undergo decussation belong to the lateral corticospinal tracts (see Figure 15-5). These tracts originate in the cerebral cortex and pass inferiorly to the medulla.

FIGURE 14-3 Brain stem. Diagram of the ventral surface of the brain, showing the structure of the brain stem in relation to the cranial nerves and associated structures.

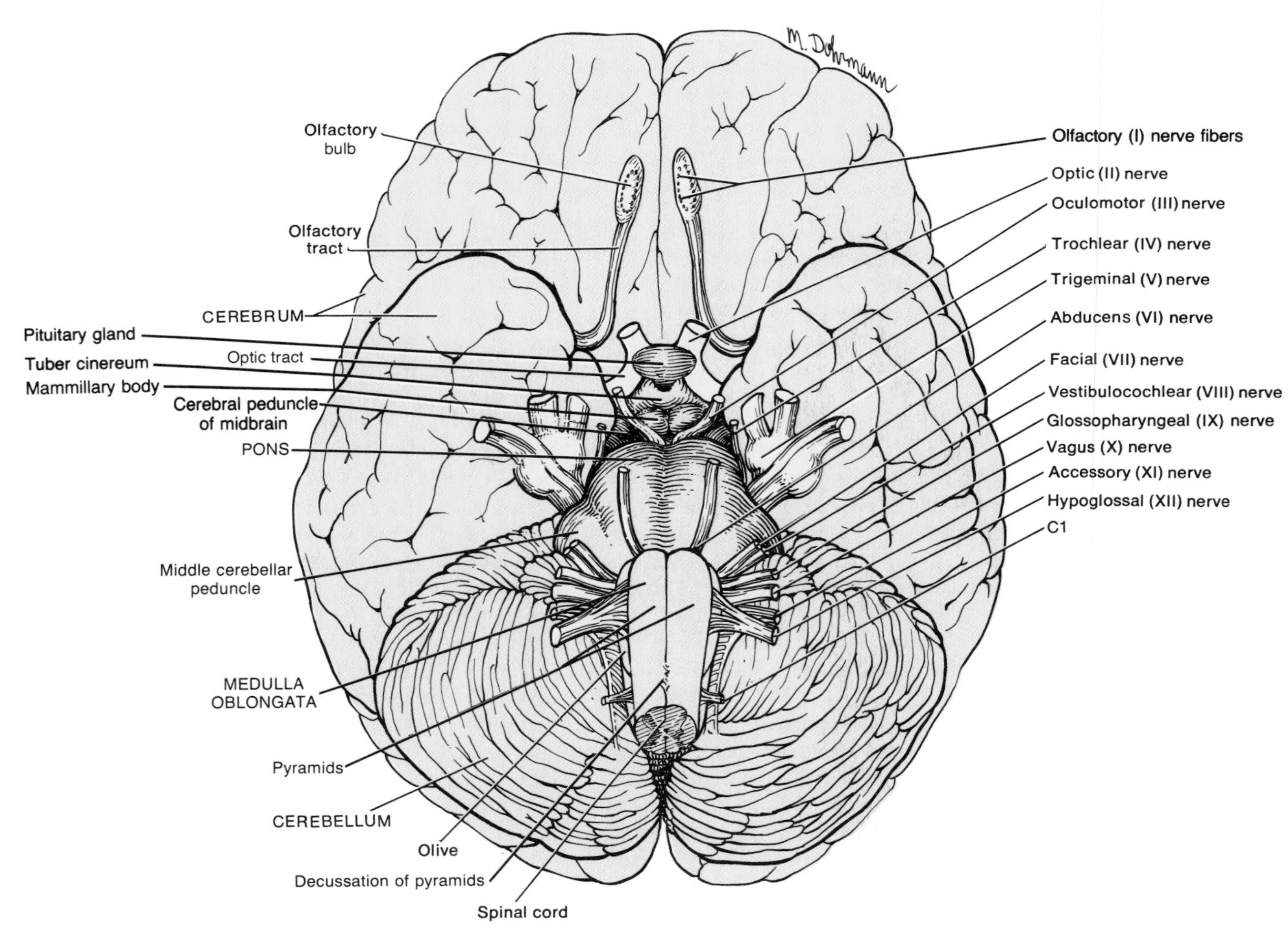

FIGURE 14-4 Medulla. (a) Dorsal aspect of the brain stem, showing the relationship of the medulla to the other components of the brain stem following removal of the cerebellum. (b) Details of the ventral surface of the medulla showing the decussation of pyramids.

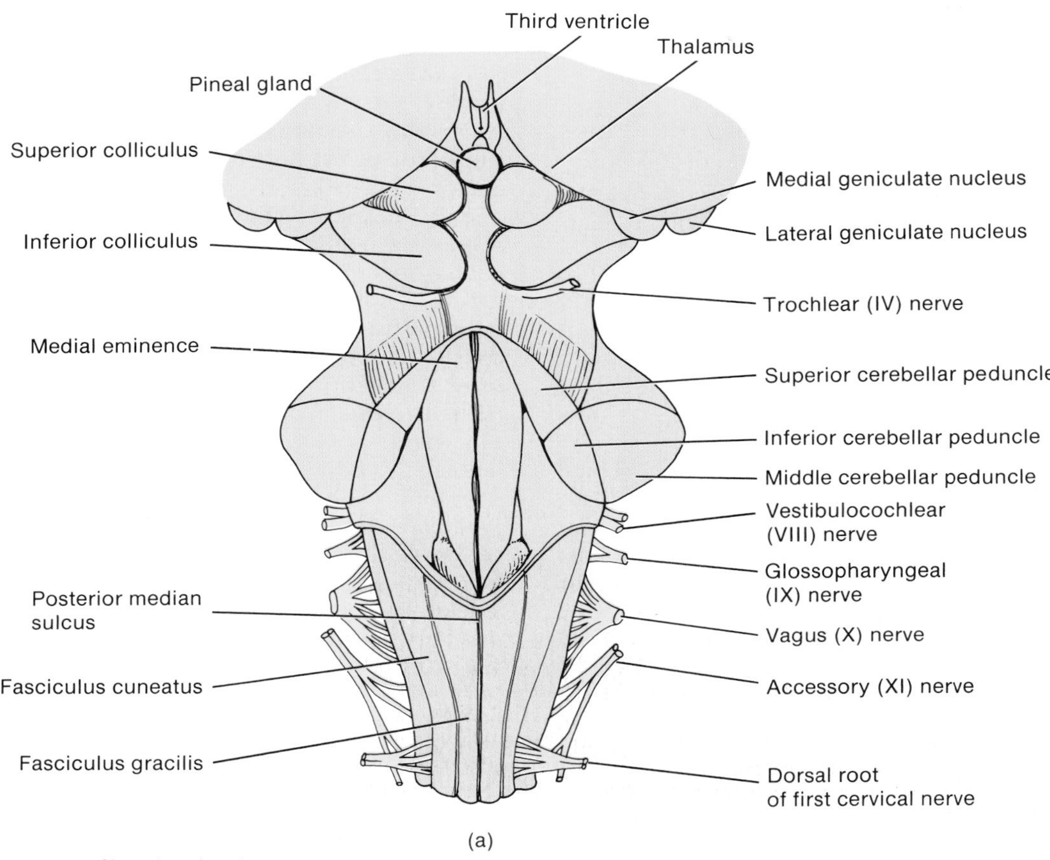

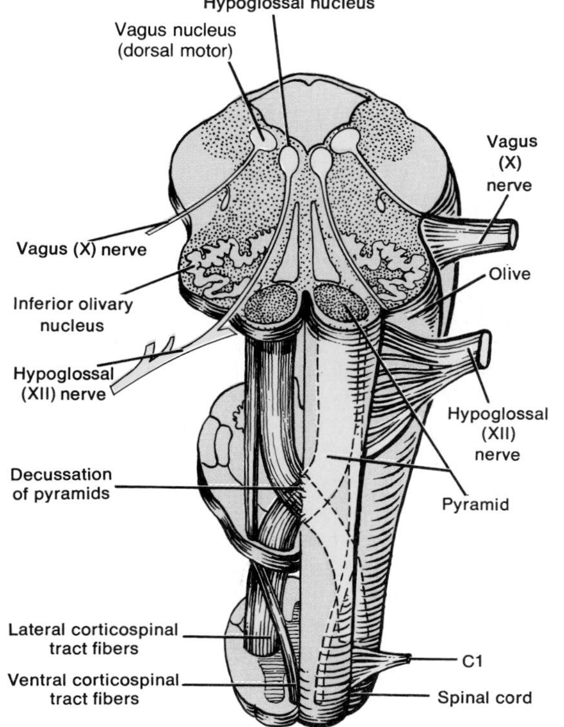

The fibers cross in the pyramids and descend in the lateral columns of the spinal cord, terminating in the anterior gray horns. Here synapses occur with motor neurons that terminate in skeletal muscles. As a result of the crossing, fibers that originate in the left cerebral cortex activate muscles on the right side of the body, and fibers that originate in the right cerebral cortex activate muscles on the left side.

The dorsal side of the medulla contains two pairs of prominent nuclei: the right and left ***nucleus gracilis*** (gras-I-lis; *gracilis* = slender) and ***nucleus cuneatus*** (kyoo-nē-Ā-tus; *cuneus* = wedge). These nuclei receive sensory fibers from ascending tracts (right and left fasciculus gracilis and fasciculus cuneatus) of the spinal cord and relay the sensory information to the opposite side of the medulla (see Figure 15-7). The information is conveyed to the thalamus and then to the sensory areas of the cerebral cortex. Nearly all sensory impulses received on one side of the body cross in the medulla or spinal cord and are perceived in the opposite side of the cerebral cortex.

In addition to its function as a conduction pathway for motor and sensory impulses between the brain and spinal cord, the medulla also contains an area of dispersed gray matter containing some white fibers. This region is called the ***reticular formation*** (see Figure 15-10). Actually, portions of the reticular formation are also located in the spinal cord, pons, midbrain, and diencephalon. The reticular formation functions in consciousness and arousal from sleep.

CLINICAL APPLICATION: UNCONSCIOUSNESS

The most common knockout blow is one that makes contact with the mandible. Such a blow twists and distorts the brain stem and overwhelms the reticular activating system (RAS) of the reticular formation by sending a sudden volley of nerve impulses to the brain, resulting in *unconsciousness.*

Within the medulla are also three vital reflex centers of the reticular system. The ***cardiac center*** regulates the rate of heartbeat and force of contraction (see Figure 20-9), the ***medullary rhythmicity area*** adjusts the basic rhythm of breathing (see Figure 23-23), and the ***vasomotor (vasoconstrictor) center*** regulates the diameter of blood vessels. Other centers in the medulla are considered nonvital and coordinate swallowing, vomiting, coughing, sneezing, and hiccuping.

CLINICAL APPLICATION: CAVITRON ULTRASONIC SURGICAL ASPIRATOR

Neurosurgeons are now using an ultrasonic device called a *cavitron ultrasonic surgical aspirator (CUSA)* that can shatter certain kinds of brain tumors. High-frequency sound waves stimulate the slender tip of the instrument to vibrate 23,000 times per second. This vibratory action disintegrates the tumors. The instrument also delivers an irrigating saline solution that aspirates the fragmented particles. One major advantage of CUSA is that it decreases the probability of damaging either adjacent normal tissues such as the brain stem, which could cause abnormal heart rhythms, or large blood vessels, which could cause hemorrhage.

The medulla also contains the nuclei of origin for several pairs of cranial nerves (Figures 14-3, 14-4, and 14-14). These are the cochlear and vestibular branches of the vestibulocochlear (VIII) nerves, which are concerned with hearing and equilibrium (there is also a nucleus for the vestibular branches in the pons); the glossopharyngeal (IX) nerves, which relay nerve impulses related to swallowing, salivation, and taste; the vagus (X) nerves, which relay nerve impulses to and from many thoracic and abdominal viscera; the cranial portion of the accessory (XI) nerves (a part of this nerve, the spinal portion, originates in the upper five cervical segments of the spinal cord), which convey nerve impulses related to head and shoulder movements; and the hypoglossal (XII) nerves, which convey nerve impulses that involve tongue movements.

On each lateral surface of the medulla is an oval projection called the ***olive*** (see Figure 14-3) that contains an ***inferior olivary nucleus*** and two ***accessory olivary nuclei.*** The nuclei are connected to the cerebellum by fibers. The inferior olivary nucleus projects fibers to the part of the cerebellum that ensures the efficiency of voluntary movements, especially precision ones. The accessory olivary nuclei project nerve fibers to regions of the cerebellum concerned with the maintenance of equilibrium, postural changes, and locomotion.

Also associated with the medulla is the greater part of the ***vestibular nuclear complex.*** This nuclear group consists of the ***lateral, medial,*** and ***inferior vestibular nuclei*** in the medulla and the ***superior vestibular nucleus*** in the pons. As you will see later (Chapter 17), the vestibular nuclei assume an important role in helping the body maintain its sense of equilibrium.

In view of the many vital activities controlled by the medulla, it is not surprising that a hard blow to the base of the skull can be fatal. Nonfatal medullary injury may be indicated by cranial nerve malfunctions on the same side of the body as the area of medullary injury, paralysis and loss of sensation on the opposite side of the body, and irregularities in respiratory control.

Pons

The relationship of the ***pons*** to other parts of the brain can be seen in Figures 14-1 and 14-3. The pons, which means "bridge," lies directly above the medulla and anterior to the cerebellum. It measures about 2.5 cm (1 inch) in length. Like the medulla, the pons consists of nuclei and white fibers present as tracts that are scattered throughout. As the name implies, the pons is a bridge connecting the spinal cord with the brain and parts of the brain with each other. These connections are provided by fibers that run in two principal directions. The transverse fibers connect with the cerebellum through the ***middle cerebellar peduncles.*** The longitudinal fibers of the pons belong to the motor and sensory tracts that connect the spinal cord or medulla with the upper parts of the brain stem.

The nuclei for certain paired cranial nerves are also contained in the pons (see Figures 14-3 and 14-14). These include the trigeminal (V) nerves, which relay nerve impulses for chewing and for sensations of the head and face; the abducens (VI) nerves, which regulate certain eyeball movements; the facial (VII) nerves, which conduct impulses related to taste, salivation, and facial expression; and the vestibular branches of the vestibulocochlear (VIII) nerves, which are concerned with equilibrium.

Other important nuclei in the reticular formation of the pons are the ***pneumotaxic*** (noo-mō-TAK-sik) ***area*** and the ***apneustic*** (ap-NOO-stik) ***area.*** (See Figure 23-23.) Together with the medullary rhythmicity area in the medulla, they help control respiration (breathing movements).

Midbrain

The ***midbrain,*** or ***mesencephalon*** (*meso* = middle; *enkephalos* = brain), extends from the pons to the lower portion of the diencephalon (Figures 14-1 and 14-3). It is about 2.5 cm (1 inch) in length. The cerebral aqueduct passes through the midbrain and connects the third ventricle above with the fourth ventricle below.

The ventral portion of the midbrain contains a pair of fiber bundles referred to as ***cerebral peduncles*** (pe-DUNG-kulz). The cerebral peduncles contain some motor fibers that convey nerve impulses from the cerebral cortex to the pons, medulla, and spinal cord. They also contain sensory fibers that pass from the spinal cord to the medulla and then from the pons to the thalamus. The cerebral peduncles constitute the main connection for tracts between upper parts of the brain and lower parts of the brain and the spinal cord.

The dorsal portion of the midbrain is called the ***tectum*** (*tectum* = roof) and contains four rounded eminences: the ***corpora quadrigemina*** (KOR-po-ra kwad-ri-JEM-in-a). Two of the eminences are known as the ***superior colliculi*** (ko-LIK-yoo-lī). These serve as reflex centers for movements of the eyeballs and head and neck in response to visual and other stimuli. The other two eminences, the ***inferior colliculi,*** serve as reflex centers for movements of the head and trunk in response to auditory stimuli. The midbrain also contains the ***substantia nigra*** (sub-STAN-shē-a NĪ-gra), a large, heavily pigmented nucleus near the cerebral peduncles.

A major nucleus in the reticular formation of the midbrain is the ***red nucleus.*** Fibers from the cerebellum and cerebral cortex terminate in the red nucleus. The red nucleus is also the origin of cell bodies of the descending rubrospinal tract. Other nuclei in the midbrain are associated with cranial nerves (see Figures 14-3 and 14-14). These include the oculomotor (III) nerves, which mediate some movements of the eyeballs and changes in pupil size and lens shape, and the trochlear (IV) nerves, which conduct impulses that move the eyeballs.

A structure called the ***medial lemniscus*** (*lemniskos* = ribbon or band) is common to the medulla, pons, and midbrain. The medial lemniscus is a band of white fibers containing axons that convey impulses for fine touch, proprioception, pressure, and vibrations from the medulla to the thalamus.

Diencephalon

The ***diencephalon*** (*dia* = through; *enkephalos* = brain) consists principally of the thalamus and hypothalamus. The relationship of these structures to the rest of the brain is shown in Figure 14-1.

Thalamus

The ***thalamus*** (THAL-a-mus; *thalamos* = inner chamber) is an oval structure above the midbrain that measures about 3 cm (1 inch) in length and constitutes four-fifths of the diencephalon. It consists of two oval masses of mostly gray matter organized into nuclei that form the lateral walls of the third ventricle (Figure 14-5). The masses are joined by a bridge of gray matter that crosses the third ventricle called the ***intermediate mass.*** Each mass is

FIGURE 14-5 Thalamus. (a) Diagram of a frontal section showing the thalamus and associated structures.

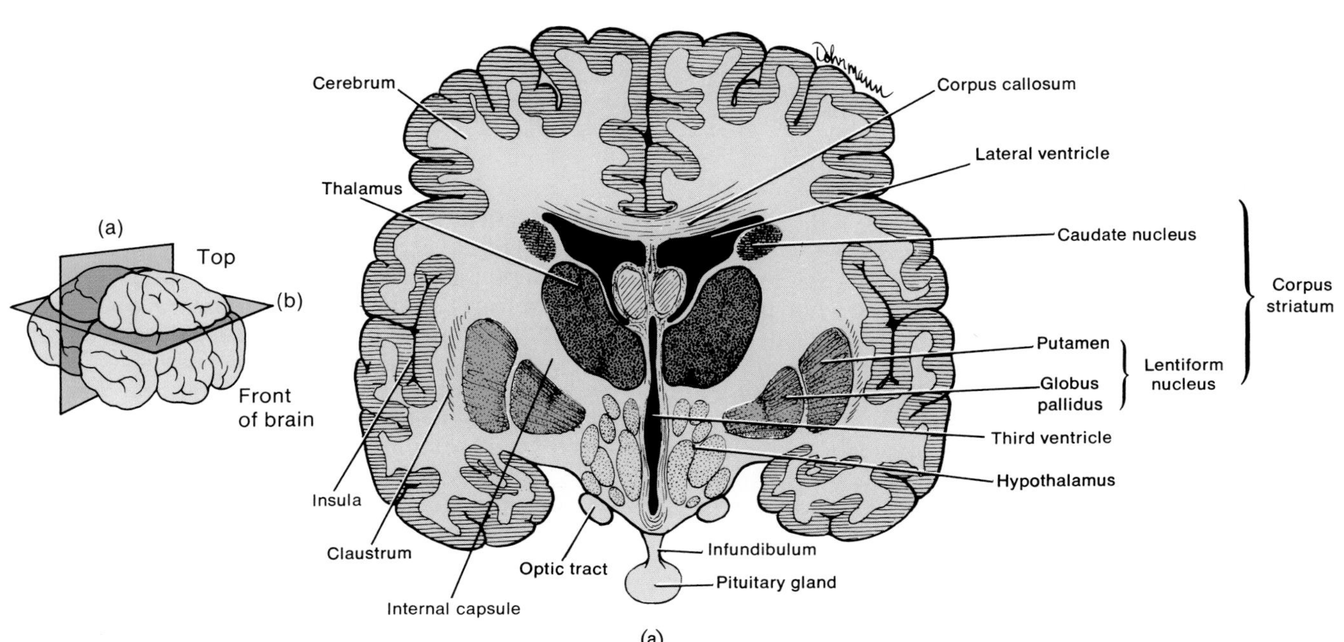

FIGURE 14-5 (*Continued*) (b) Photograph of a horizontal section of the cerebrum showing the thalamus and associated structures. (Courtesy of Stephen A. Kieffer and E. Robert Heitzman, *An Atlas of Cross-Sectional Anatomy,* Harper & Row, Publishers, Inc., Hagerstown, MD, 1979). (c) Diagram of right lateral view of the thalamic nuclei. The arrows indicate some of the connections between the thalamus and cerebral cortex.

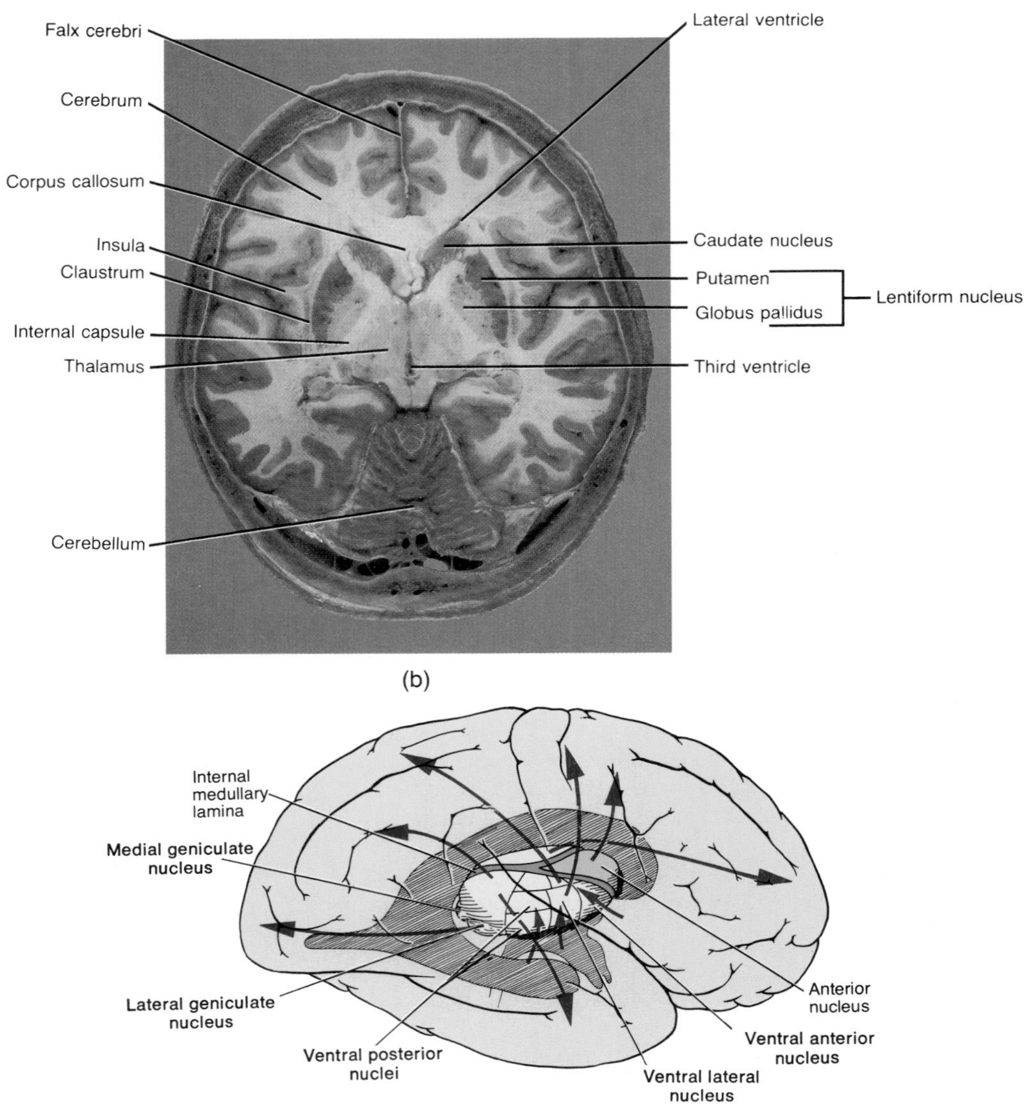

deeply embedded in a cerebral hemisphere and is bounded laterally by the ***internal capsule.***

Although the thalamic masses are primarily gray matter, some portions are white matter. Among the white matter portions are the ***stratum zonale,*** which covers the dorsal surface; the ***external medullary lamina,*** covering the lateral surface; and the ***internal medullary lamina,*** which divides the gray matter masses into an anterior nuclear group, a medial nuclear group, and a lateral nuclear group.

Within each group are nuclei that assume various roles. Some nuclei in the thalamus serve as relay stations for all sensory impulses, except smell, to the cerebral cortex. These include the ***medial geniculate*** (je-NIK-yoo-lāt) ***nuclei*** (hearing), the ***lateral geniculate nuclei*** (vision), and the ***ventral posterior nuclei*** (general sensations and taste). Other nuclei are centers for synapses in the somatic motor system. These include the ***ventral lateral nuclei*** (voluntary motor actions) and ***ventral anterior nuclei*** (voluntary motor actions and arousal). (See Figure 14-5c.) The thalamus is the principal relay station for sensory impulses that reach the cerebral cortex from the spinal cord, brain stem, cerebellum, and parts of the cerebrum.

The thalamus also functions as an interpretation center for some sensory impulses, such as pain, temperature, light touch, and pressure. The thalamus also contains a ***reticular nucleus*** in its reticular formation, which in some way seems to modify neuronal activity in the thalamus, and an ***anterior nucleus*** in the floor of the lateral ventricle, which is concerned with certain emotions and memory.

Hypothalamus

The ***hypothalamus*** (*hypo* = under) is a small portion of the diencephalon located below the thalamus. Its relationship to other parts of the brain is shown in Figures 14-1 and 14-5a. The hypothalamus forms the floor and part of the lateral walls of the third ventricle. It is partially protected by the sella turcica of the sphenoid bone.

Information from the external environment comes to the hypothalamus via afferent pathways originating in the peripheral sense organs. Impulses from sound, taste, smell, and somatic receptors all come to the hypothalamus. Afferent impulses, monitoring the internal environment, arise from the internal viscera and reach the hypothalamus. Other receptors in parts of the hypothalamus itself continually monitor water concentration, certain hormone concentrations, and temperature of blood. And, as you will see shortly, the hypothalamus has several very important connections with the pituitary gland and is itself able to produce a variety of hormones.

The hypothalamus is divided into nuclei and several major regions (Figure 14-6):

1. **Supraoptic region.** This anterior region lies above the optic chiasma and contains the paraventricular nucleus, supraoptic nucleus, anterior hypothalamic nucleus, and suprachiasmatic nucleus.

2. **Tuberal region.** This middle region is the widest portion of the hypothalamus. On its ventral surface are the ***tuber cinereum*** (si-NE-rē-um) and ***infundibulum.*** The tuber cinereum is an elevated mass of gray matter that contains neurons that transport regulating hormones (or factors) from the hypothalamus to the infundibulum, a stalklike structure that attaches the pituitary gland to the hypothalamus. Within the infundibulum, axons of neurons in the tuber cinereum form the tubero-hypophyseal tract that transports regulating hormones (or factors) into blood vessels (see also Figure 18-8). From here, the regulating hormones (or factors) are transported via blood vessels to the anterior lobe of the pituitary gland. Also found in

FIGURE 14-6 Hypothalamus. Selected areas of the hypothalamus and a three-dimensional representation of hypothalamic nuclei seen in sagittal section.

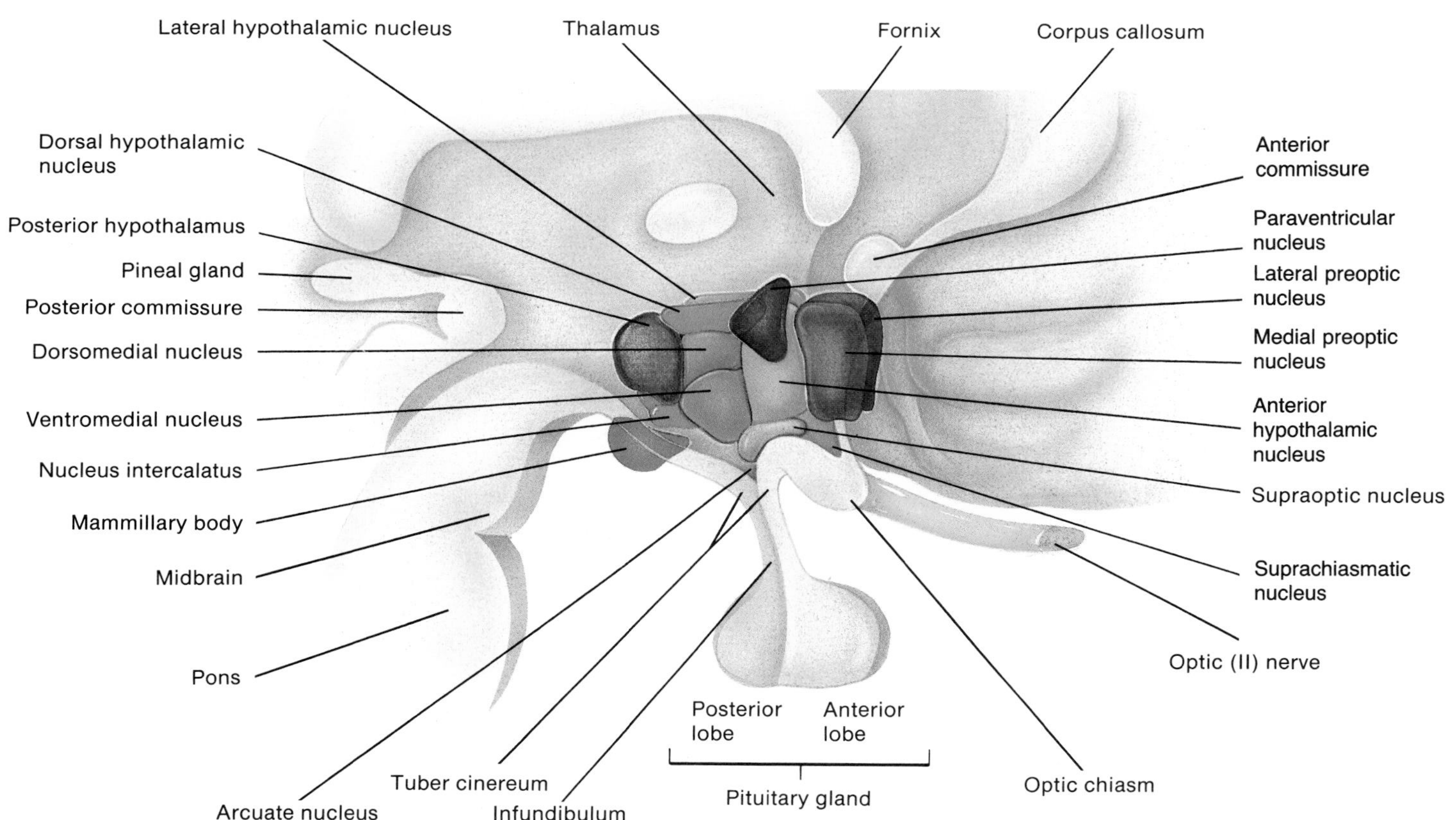

the infundibulum are nerve fibers from some hypothalamic nuclei that form the hypothalamic-hypophyseal tract. This tract transports hormones made in the nuclei (oxytocin and antidiuretic hormone) to the posterior pituitary gland where they are stored and released. Nuclei associated with the tuberal region include the ventromedial, dorsomedial, and arcuate nuclei.

3. **Mammillary region.** This region is behind the tuberal region and contains the mammillary bodies and posterior hypothalamic nucleus. The ***mammillary bodies*** are two, small, rounded bodies placed side by side that are posterior to the tuber cinereum. The bodies serve as relay stations for olfactory neurons that are involved in reflexes related to the sense of smell.

The ***preoptic region*** in front of the supraoptic region is sometimes considered part of the hypothalamus because it functions in regulating certain autonomic activities in conjunction with the hypothalamus. The preoptic region contains the preoptic periventricular nucleus, medial preoptic nucleus, and lateral preoptic nucleus.

Despite its small size, nuclei in the hypothalamus control many body activities, most of them related to homeostasis. Although differentiation of the hypothalamic nuclei is far from precise, it is possible to identify certain nuclei. Some of these nuclei are more readily identified in lower animals and are more distinct in fetuses than adults. Also, within a given nucleus there may be several kinds of cells that can be differentiated histologically. The localization of function, with a few exceptions, is not specific to the individual nuclei; certain functions tend to overlap nuclear boundaries. For this reason, functions are attributed to regions rather than specific nuclei.

The chief functions of the hypothalamus are as follows.

1. It controls and integrates the autonomic nervous system, which regulates contraction of smooth muscle and cardiac muscle and secretions of many glands. This is accomplished by axons of neurons whose dendrites and cell bodies are in hypothalamic nuclei. The axons form tracts from the hypothalamus to sympathetic and parasympathetic nuclei in the brain stem and spinal cord. Through the autonomic nervous system, the hypothalamus is the main regulator of visceral activities. It regulates heart rate, movement of food through the gastrointestinal tract, and contraction of the urinary bladder.

2. It is involved in the reception and integration of sensory impulses from the viscera.

3. It is the principal intermediary between the nervous system and the endocrine system—the two major control systems of the body. The hypothalamus lies just above the pituitary. When the hypothalamus detects certain changes in the body, it releases chemicals called regulating hormones (or factors) that stimulate or inhibit specific cells in the anterior pituitary gland. The anterior pituitary then releases or holds back hormones that regulate various physiological activities of the body. The hypothalamus also produces two hormones, antidiuretic hormone (ADH) and oxytocin (OT), which are transported to and stored in the posterior pituitary gland. ADH decreases urine volume and OT brings about uterine contractions during labor and assists in milk ejection by the mammary glands. The hormones are released from storage when needed by the body.

4. It is the center for the mind-over-body phenomenon. When the cerebral cortex interprets strong emotions, it often sends nerve impulses along the tracts that connect the cortex with the hypothalamus. The hypothalamus then directs impulses via the autonomic nervous system and also releases chemicals that stimulate the anterior pituitary gland. The result can be a wide range of changes in body activities. For instance, when you are under stress, impulses leave the hypothalamus to stimulate your heart to beat faster. Likewise, continued psychological stress can produce long-term abnormalities in body function that result in serious illness. These so-called psychosomatic disorders are definitely real.

5. It is associated with feelings of rage and aggression.

6. It controls normal body temperature. Certain cells of the hypothalamus serve as a thermostat. If blood flowing through the hypothalamus is above normal temperature, the hypothalamus directs nerve impulses along the autonomic nervous system to stimulate activities that promote heat loss. Heat can be lost through relaxation of the smooth muscle in the blood vessels, causing vasodilation of cutaneous vessels and increased heat loss from the skin. Heat loss also is enhanced by sweating. Conversely, if the temperature of the blood is below normal, the hypothalamus generates impulses that promote heat retention. Heat can be retained through the constriction of cutaneous blood vessels, cessation of sweating, and by shivering.

7. It regulates food intake through two centers. The ***feeding (hunger) center*** is responsible for hunger sensations. When sufficient food has been ingested, the ***satiety*** (sa-TĪ-e-tē) ***center*** is stimulated and sends out nerve impulses that inhibit the feeding center.

8. It contains a ***thirst center.*** Certain cells in the hypothalamus are stimulated when the extracellular fluid volume is reduced. The stimulated cells produce the sensation of thirst.

9. It is one of the centers that maintains the waking state and sleep patterns.

10. It exhibits properties of a self-sustained oscillator and, as such, acts as a pacemaker to drive many biological rhythms.

Cerebrum

Supported on the brain stem and forming the bulk of the brain is the ***cerebrum*** (see Figure 14-1). The surface of the cerebrum is composed of gray matter 2 to 4 mm (0.08 to 0.16 inch) thick and is referred to as the ***cerebral cortex*** (*cortex* = rind or bark). The cortex, containing billions of cells, consists of six layers of nerve cell bodies in most areas. Beneath the cortex lies the cerebral white matter.

During embryonic development, when there is a rapid increase in brain size, the gray matter of the cortex enlarges out of proportion to the underlying white matter. As a result, the cortical region rolls and folds upon itself. The folds are called ***gyri*** (JĪ-rī) or ***convolutions*** (Figure 14-7a, c). The deep grooves between folds are referred

FIGURE 14-7 Lobes and fissures of the cerebrum. (a) Diagram of right lateral view. Since the insula cannot be seen externally, it has been projected to the surface. It can be seen in Figure 14-5a. (b) Photograph of right lateral view showing the insula after removal of a portion of the cerebrum. (Courtesy of N. Gluhbegovic and T. H. Williams, *The Human Brain: A Photographic Guide,* Harper & Row, Publishers, Inc., Hagerstown, MD, 1980.) (c) Diagram of superior view. The insert to the left indicates the relative differences among a gyrus, sulcus, and fissure.

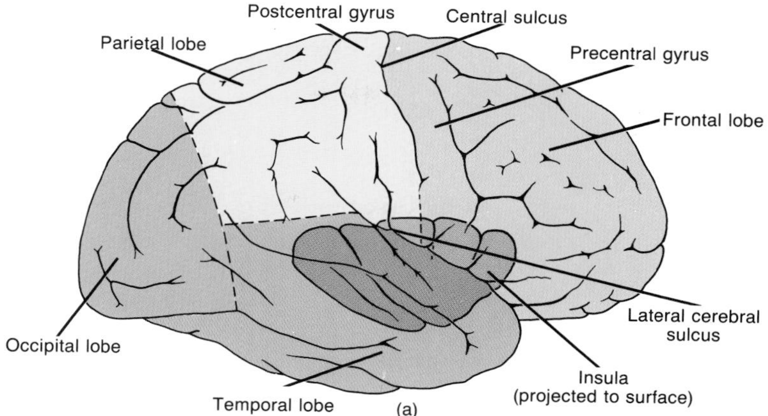

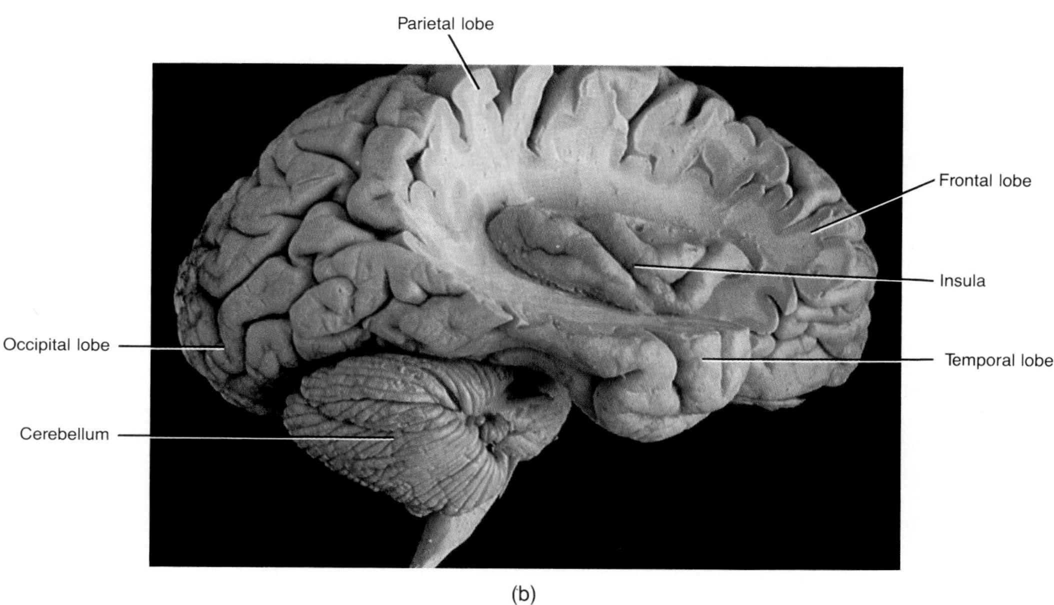

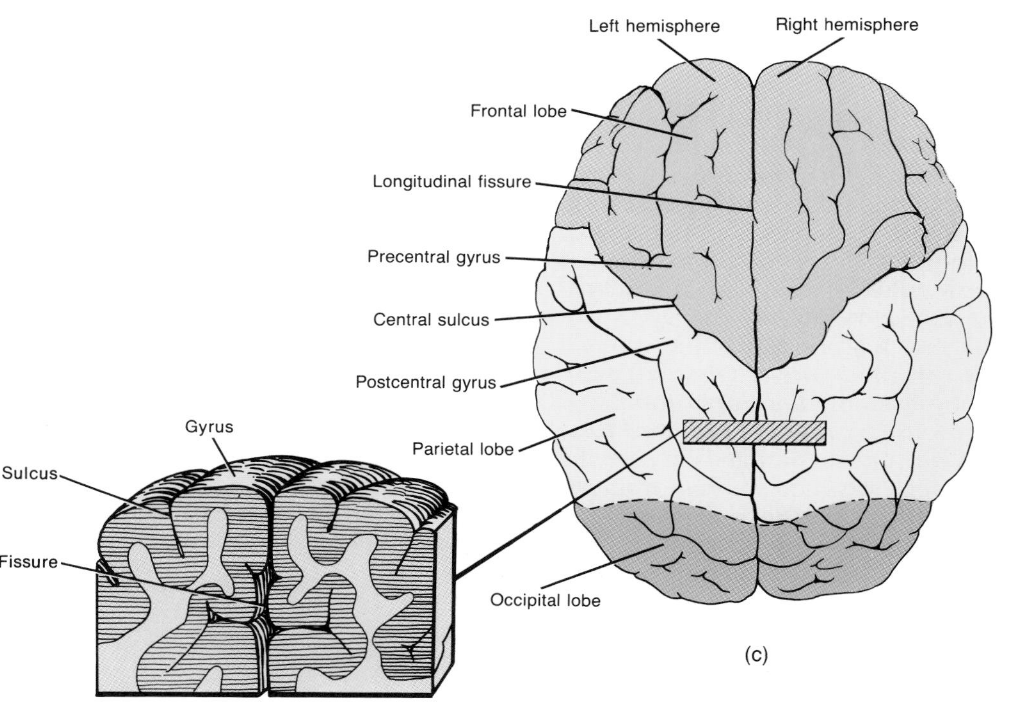

FIGURE 14-7 (*Continued*) (d) Photograph of a horizontal section through the cerebrum. (Courtesy of Stephen A. Kieffer and E. Robert Heitzman, *An Atlas of Cross-Sectional Anatomy,* Harper & Row, Publishers, Inc., Hagerstown, MD, 1979.)

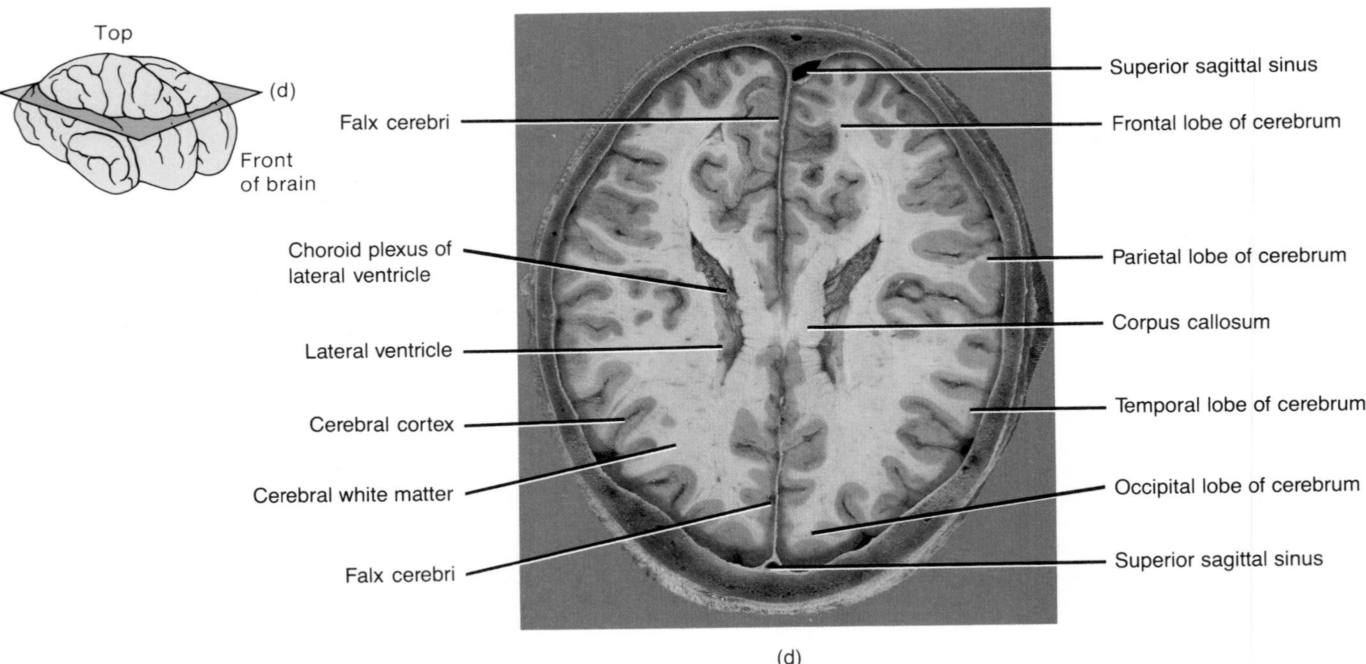

to as ***fissures;*** the shallow grooves between folds are ***sulci*** (SUL-sī). The most prominent fissure, the ***longitudinal fissure,*** nearly separates the cerebrum into right and left halves, or ***hemispheres.*** The hemispheres, however, are connected internally by a large bundle of transverse fibers composed of white matter called the ***corpus callosum*** (kal-LŌ-sum; *corpus* = body; *callosus* = hard). Between the hemispheres is an extension of the cranial dura mater called the ***falx*** (FALKS) ***cerebri (cerebral fold).*** It encloses the superior and inferior sagittal sinuses.

Lobes

Each cerebral hemisphere is further subdivided into four lobes by sulci or fissures. The ***central sulcus*** separates the ***frontal lobe*** from the ***parietal lobe.*** A major gyrus, the ***precentral gyrus,*** is located immediately anterior to the central sulcus. The gyrus is a landmark for the primary motor area of the cerebral cortex. Another major gyrus, the ***postcentral gyrus,*** is located immediately posterior to the central sulcus. This gyrus is a landmark for the general sensory area of the cerebral cortex. The ***lateral cerebral sulcus (fissure)*** separates the ***frontal lobe*** from the ***temporal lobe.*** The ***parietooccipital sulcus*** separates the ***parietal lobe*** from the ***occipital lobe.*** Another prominent fissure, the ***transverse fissure,*** separates the cerebrum from the cerebellum. The frontal lobe, parietal lobe, temporal lobe, and occipital lobe are named after the bones that cover them. A fifth part of the cerebrum, the ***insula,*** lies deep within the lateral cerebral fissure, under the parietal, frontal, and temporal lobes. It cannot be seen in an external view of the brain (Figure 14-7a, b).

As you will see later, the olfactory (I) and optic (II) nerves are associated with specific lobes of the cerebrum.

White Matter

The white matter underlying the cortex consists of myelinated axons running in three principal directions (Figure 14-8).

1. **Association fibers** connect and transmit nerve impulses between gyri in the same hemisphere.
2. **Commissural fibers** transmit impulses from the gyri in one cerebral hemisphere to the corresponding gyri in the opposite cerebral hemisphere. Three important groups of commissural fibers are the ***corpus callosum, anterior commissure,*** and ***posterior commissure.***
3. **Projection fibers** form ascending and descending tracts that transmit impulses from the cerebrum to other parts of the brain and spinal cord. The internal capsule is an example.

Basal Ganglia (Cerebral Nuclei)

The ***basal ganglia (cerebral nuclei)*** are paired masses of gray matter in each cerebral hemisphere (Figures 14-5 and 14-9). The largest of the basal ganglia of each hemisphere is the ***corpus striatum*** (strī-Ā-tum;

FIGURE 14-8 White matter tracts of the left cerebral hemisphere seen in sagittal section. (Courtesy of N. Gluhbegovic and T. H. Williams, *The Human Brain: A Photographic Guide,* Harper & Row, Publishers, Inc., Hagerstown, MD, 1980.)

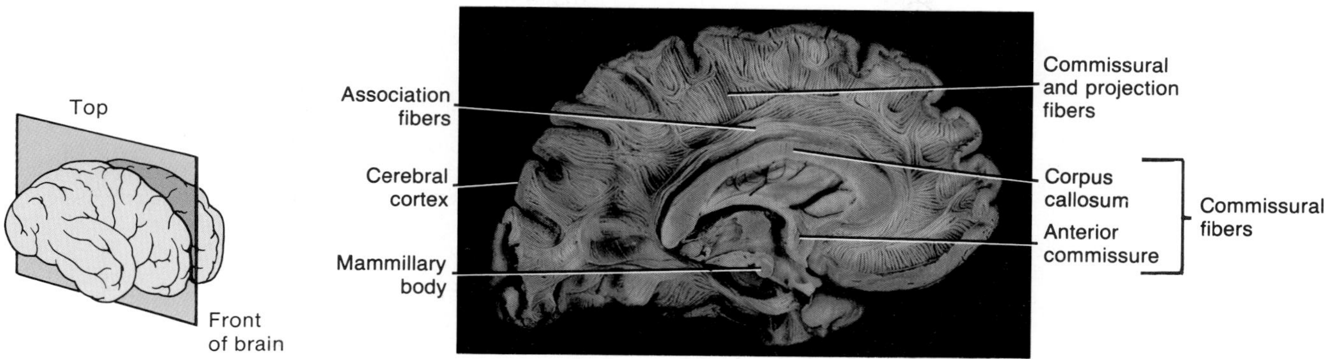

corpus = body; *striatus* = striped). It consists of the ***caudate*** (*cauda* = tail) ***nucleus*** and the ***lentiform*** (*lenticula* = shaped like a lentil or lens) ***nucleus.*** The lentiform nucleus, in turn, is subdivided into a lateral portion called the ***putamen*** (pu-TĀ-men; *putamen* = shell) and a medial portion called the ***globus pallidus*** (*globus* = ball; *pallid* = pale).

The portion of the ***internal capsule*** passing between the lentiform nucleus and the caudate nucleus and between the lentiform nucleus and thalamus is sometimes

FIGURE 14-9 Basal ganglia. (a) In this diagram of the right lateral view of the cerebrum, the basal ganglia have been projected to the surface. Refer to Figure 14-5a for the positions of the basal ganglia in the frontal section of the cerebrum. (b) Photograph of the medial surface of the left cerebral hemisphere showing portions of the basal ganglia. (Courtesy of N. Gluhbegovic and T. H. Williams, *The Human Brain: A Photographic Guide,* Harper & Row, Publishers, Inc., Hagerstown, MD, 1980.)

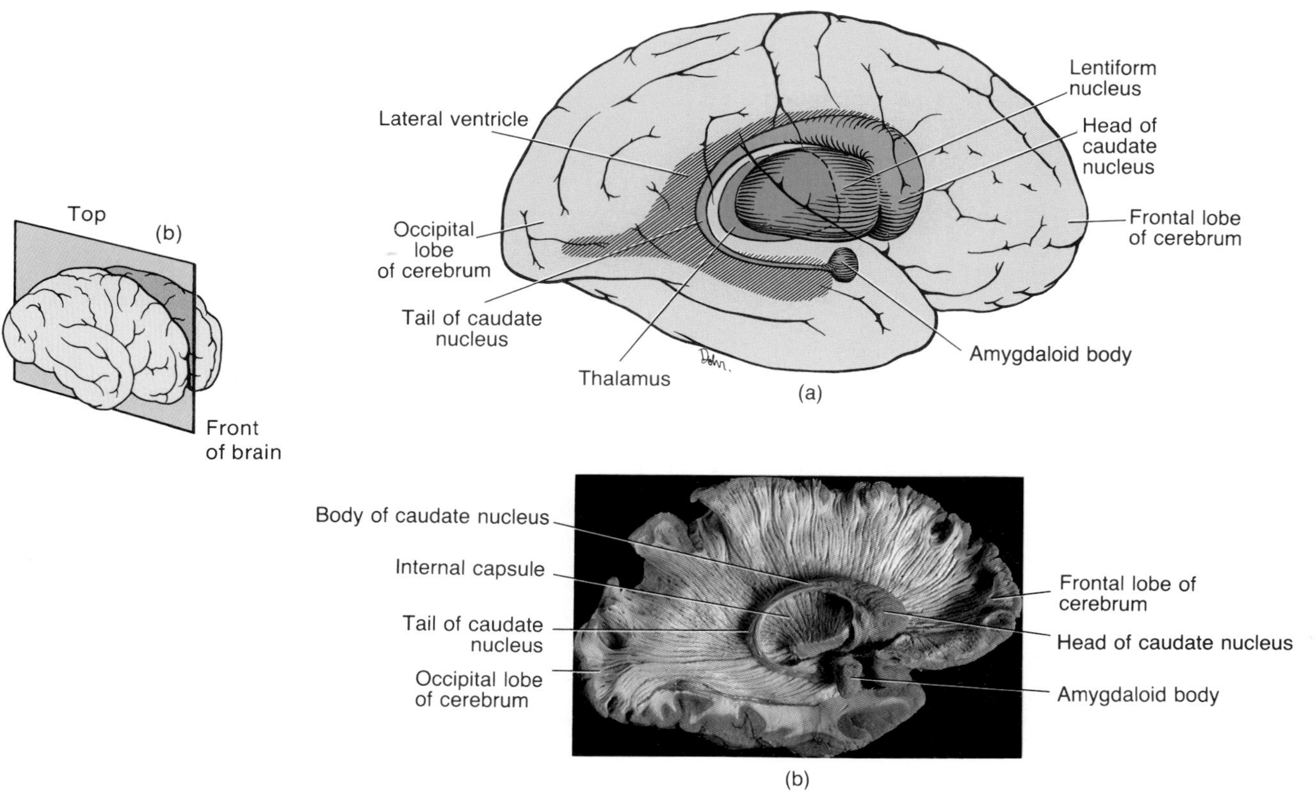

considered part of the corpus striatum. The internal capsule is made up of a group of sensory and motor white matter tracts that connect the cerebral cortex with the brain stem and spinal cord.

Other structures frequently considered part of the basal ganglia are the ***substantia nigra, subthalamic nucleus,*** and ***red nucleus.*** The substantia nigra is a large nucleus in the midbrain whose axons terminate in the caudate nucleus and putamen. The subthalamic nucleus lies against the internal capsule. Its major connection is with the globus pallidus.

The basal ganglia are interconnected by many fibers. They are also connected to the cerebral cortex, thalamus, and hypothalamus. The caudate nucleus and the putamen control large subconscious movements of skeletal muscles, such as swinging the arms while walking. Such gross movements are also consciously controlled by the cerebral cortex. The globus pallidus is concerned with the regulation of muscle tone required for specific body movements.

CLINICAL APPLICATION: DAMAGE TO BASAL GANGLIA

Damage to the basal ganglia results in abnormal body movements, such as uncontrollable shaking, called ***tremor,*** and ***involuntary movements of skeletal muscles.*** Moreover, destruction of a substantial portion of the caudate nucleus results in almost total ***paralysis*** of the side of the body opposite to the damage. The caudate nucleus is an area often affected by a stroke.

A lesion in the subthalamic nucleus results in a motor disturbance on the opposite side of the body called ***hemiballismus*** (*hemi* = half; *ballismos* = jumping), which is characterized by involuntary movements occurring suddenly with great force and rapidity. The movements are purposeless and generally of the withdrawal type, although they may be jerky. The spontaneous movements affect the proximal portions of the extremities most severely, especially the arms.

Limbic System

Certain components of the cerebral hemispheres and diencephalon constitute the ***limbic*** (*limbus* = border) ***system.*** Among its components are the following regions of gray matter (Figure 14-10).

1. Limbic lobe (cortex). Largest components are the ***parahippocampal*** and ***cingulate gyri,*** both gyri of the cerebral hemispheres, and ***hippocampus,*** an extension of the parahippocampal gyrus that extends into the floor of the lateral ventricle.

FIGURE 14-10 Selected components of the limbic system and surrounding structures.

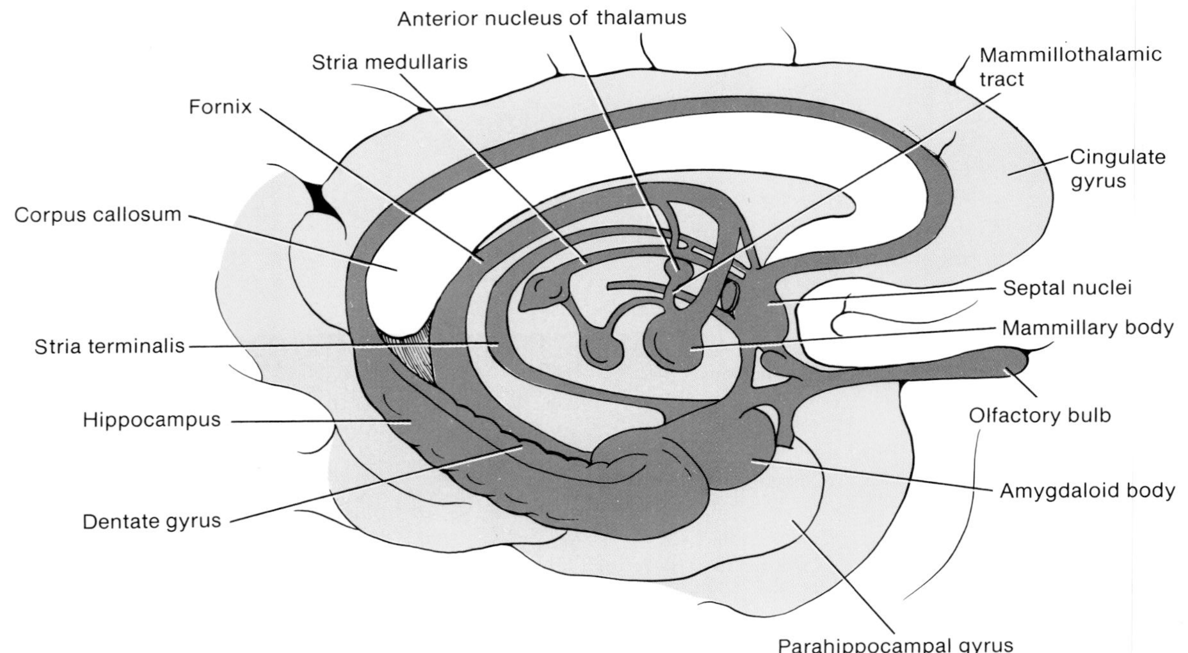

2. **Dentate gyrus.** A cerebral gyrus between the hippocampus and parahippocampal gyrus.

3. **Amygdaloid body (amygdala).** Several groups of neurons located at the tail end of the caudate nucleus.

4. **Septal nuclei.** Nuclei within the septal area, a region formed by the regions under the corpus callosum and a cerebral gyrus (paraterminal).

5. **Mammillary bodies of the hypothalamus.** Two round masses close to the midline near the cerebral peduncles.

6. **Anterior nucleus of the thalamus.** Located in the floor of the lateral ventricle.

7. **Olfactory bulbs.** Flattened bodies of the olfactory (I) nerves that rest on the cribriform plate.

8. **Bundles of interconnecting myelinated axons.** Bundles that interconnect various components of the limbic system and include the fornix, stria terminalis, stria medullaris, and mammillothalamic tract.

The limbic system is a wishbone-shaped group of structures that encircles the brain stem and functions in the emotional aspects of behavior related to survival. The hippocampus, together with portions of the cerebrum, also functions in memory. Memory impairment results from lesions in the limbic system. People with such damage forget recent events and cannot commit anything to memory. How the limbic system functions in memory is not clear. Although behavior is a function of the entire nervous system, the limbic system controls most of its involuntary aspects. Experiments on the limbic system of monkeys and other animals indicate that the amygdaloid nucleus assumes a major role in controlling the overall pattern of behavior.

Other experiments have shown that the limbic system is associated with pleasure and pain. When certain areas of the limbic system of the hypothalamus, thalamus, and midbrain are stimulated in animals, their reactions indicate they are experiencing intense punishment. When other areas are stimulated, the animals' reactions indicate they are experiencing extreme pleasure. In still other studies, stimulation of the perifornical nuclei of the hypothalamus results in a behavioral pattern called rage. The animal assumes a defensive posture—extending its claws, raising its tail, hissing, spitting, growling, and opening its eyes wide. Stimulating other areas of the limbic system results in an opposite behavioral pattern: docility, tameness, and affection. Because the limbic system assumes a primary function in emotions such as pain, pleasure, anger, rage, fear, sorrow, sexual feelings, docility, and affection, it is sometimes called the "visceral" or "emotional" brain.

CLINICAL APPLICATION: BRAIN INJURIES

Brain injuries **are commonly associated with head injuries and result from displacement and distortion of neuronal tissue at the moment of impact. The various degrees of brain injury are described by the following terms.**

1. Concussion. An abrupt but temporary loss of consciousness following a blow to the head or a sudden stopping of a moving head. A concussion produces no visible bruising of the brain, but post-traumatic amnesia may occur.

2. Contusion. A visible bruising of the brain due to trauma and blood leaking from microscopic vessels. The pia mater is stripped from the brain over the injured area and may be torn, allowing blood to enter the subarachnoid space. A contusion usually results in an extended loss of consciousness, ranging from several minutes to many hours.

3. Laceration. Tearing of the brain, usually from a skull fracture or gunshot wound. A laceration results in rupture of large blood vessels with bleeding into the brain and subarachnoid space. Consequences include cerebral hematoma, edema, and increased intracranial pressure.

Although trauma to the head can lead to brain injuries, not all of the damage is due to the impact alone. Most of the damage is probably due to the release of large numbers of free radicals, that is, charged oxygen molecules from damaged cells. (Brain cells recovering from the effects of a stroke or cardiac arrest also release excess free radicals.) Free radicals cause damage by disrupting cellular DNA and enzymes and altering plasma membrane permeability. Investigations are under way to develop compounds to counter the effects of free radicals.

Functional Areas of Cerebral Cortex

The functions of the cerebrum are numerous and complex. In a general way, the cerebral cortex is divided into sensory, motor, and association areas. The ***sensory areas*** interpret sensory impulses, the ***motor areas*** control muscular movement, and the ***association areas*** are concerned with emotional and intellectual processes.

▪ **Sensory Areas** The ***primary somesthetic*** (sō-mes-THET-ik; *soma* = body; *aisthesis* = perception) ***area*** or ***general sensory area*** is located directly posterior to the central sulcus of the cerebrum in the postcentral gyrus of the parietal lobe. It extends from the longitudinal fissure on the top of the cerebrum to the lateral cerebral sulcus. In Figure 14-11, the general sensory area is designated by the areas numbered 1, 2, and 3.*

The primary somesthetic area receives sensations from cutaneous, muscular, and visceral receptors in various

* These numbers, as well as most of the others shown, are based on K. Brodmann's cytoarchitectural map of the cerebral cortex. His map, first published in 1909, attempts to correlate structure and function.

FIGURE 14-11 Functional areas of the cerebrum. This right lateral view indicates the sensory and motor areas of the right hemisphere. Although Broca's area is in the left hemisphere of most people, it is shown here to indicate its location.

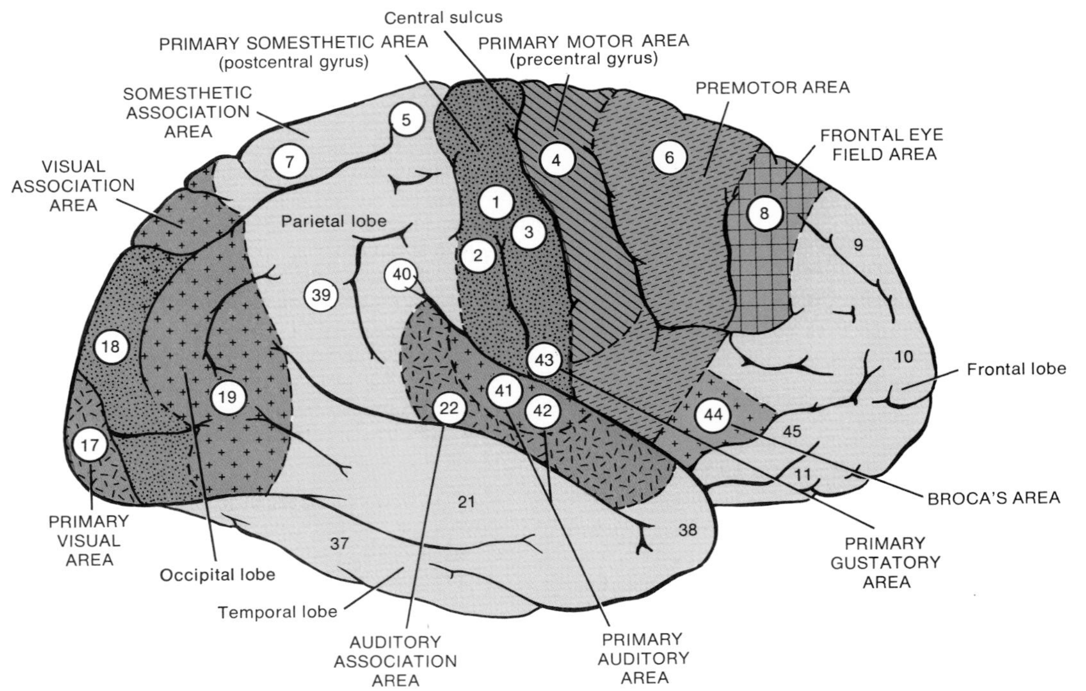

parts of the body. Each point of the area receives sensations from specific parts of the body, and essentially the entire body is spatially represented in it. The size of the portion of the sensory area receiving stimuli from body parts is not dependent on the size of the part but on the number of receptors the part contains. For example, a larger portion of the sensory area receives impulses from the lips than from the thorax (see Figure 15-4). The major function of the primary somesthetic area is to localize exactly the points of the body where the sensations originate. The thalamus is capable of localizing sensations in a general way; that is, it receives sensations from large areas of the body but cannot distinguish precisely between specific areas of stimulation. This ability is reserved for the primary somesthetic area of the cortex.

The ***secondary somesthetic area*** is a small region in the posterior wall of the lateral sulcus in line with the postcentral gyrus. It is involved mainly in less discriminative aspects of sensation.

Posterior to the primary somesthetic area is the ***somesthetic association area.*** It corresponds to the areas numbered 5 and 7 in Figure 14-11. The somesthetic association area receives input from the thalamus, other lower portions of the brain, and the primary somesthetic area. Its role is to integrate and interpret sensations. This area permits you to determine the exact shape and texture of an object without looking at it, to determine the orientation of one object to another as they are felt, and to sense the relationship of one body part to another. Another role of the somesthetic association area is the storage of memories of past sensory experiences. Thus, you can compare sensations with previous experiences.

Other sensory areas of the cortex include:

1. **Primary visual area** (area 17). Located on the medial surface of the occipital lobe and occasionally extends around to the lateral surface. It receives sensory impulses from the eyes and interprets shape, color, and movement.

2. **Visual association area** (areas 18 and 19). Located in the occipital lobe. It receives sensory impulses from the primary visual area and the thalamus. It relates present to past visual experiences with recognition and evaluation of what is seen.

3. **Primary auditory area** (areas 41 and 42). Located in the superior part of the temporal lobe near the lateral cerebral sulcus. It interprets the basic characteristics of sound such as pitch and rhythm. Whereas the anterolateral portion of the auditory area responds to low pitches, the posterolateral portion responds to high pitches.

4. **Auditory association (Wernicke's) area** (area 22). Located inferior to the primary auditory area in the temporal cortex. It determines if a sound is speech, music, or noise. It also interprets the meaning of speech by translating words into thoughts.

5. **Primary gustatory area** (area 43). Located at the base of the postcentral gyrus above the lateral cerebral sulcus in the parietal cortex. It interprets sensations related to taste.

6. Primary olfactory area. Located in the temporal lobe on the medial aspect. It interprets sensations related to smell.

7. Gnostic (NOS-tik; *gnosis* = knowledge) **area** (areas 5, 7, 39, and 40). This ***common integrative area*** is located among the somesthetic, visual, and auditory association areas. The gnostic area receives nerve impulses from these areas, as well as from the taste and smell areas, the thalamus, and lower portions of the brain stem. It integrates sensory interpretations from the association areas and impulses from other areas so that a common thought can be formed from the various sensory inputs. It then transmits signals to other parts of the brain to cause the appropriate response to the sensory signal.

In Chapter 1, we discussed the principle and clinical applications of a type of radioisotope scanning called ***positron emission tomography*** **(*PET*).** At that point it was indicated that PET is being used in the diagnosis of an assortment of diseases. It is also being used by scientists to probe the healthy brain (see Figure 1-14). By detecting and recording changes in glucose metabolism (consumption), it is possible to identify which specific areas of the brain are involved in specific sensory and motor activities.

▪ **Motor Areas** The ***primary motor area*** (area 4) is located in the precentral gyrus of the frontal lobe (Figure 14-11). Like the primary somesthetic area, the primary motor area consists of regions that control specific muscles or groups of muscles (see Figure 15-8). Stimulation of a specific point of the primary motor area results in a muscular contraction, usually on the opposite side of the body.

The ***premotor area*** (area 6) is anterior to the primary motor area. It is concerned with learned motor activities of a complex and sequential nature. It generates nerve impulses that cause a specific group of muscles to contract in a specific sequence, for example, writing. Thus, the premotor area controls skilled movements.

The ***frontal eye field area*** (area 8) in the frontal cortex is sometimes included in the premotor area. This area controls voluntary scanning movements of the eyes—searching for a word in a dictionary, for instance.

The ***language areas*** are also significant parts of the motor cortex. The translation of speech or written words into thought involves sensory areas—primary auditory, auditory association, primary visual, visual association, and gnostic—as we just described. The translation of thoughts into speech involves the ***motor speech area*** (area 44) or ***Broca's*** (BRŌ-kaz) ***area,*** located in the frontal lobe just superior to the lateral cerebral sulcus. From this area, a sequence of nerve impulses is sent to the premotor regions that control the muscles of the larynx, pharynx, and mouth. The impulses from the premotor area to the muscles result in specific, coordinated contractions that enable you to speak. Simultaneously, impulses are sent from Broca's area to the primary motor area. From here, impulses reach your breathing muscles to regulate the proper flow of air past the vocal cords. The coordinated contractions of your speech and breathing muscles enable you to translate your thoughts into speech.

CLINICAL APPLICATION: SPEECH AREA INJURIES

Broca's area and other language areas are located in the left cerebral hemisphere of most individuals, regardless of whether they are left-handed or right-handed. Injury to the sensory or motor speech areas results in *aphasia* (a-FĀ-zē-a; *a* = without; *phasis* = speech), an inability to speak; *agraphia* (*a* = without; *graph* = write), an inability to write; *word deafness,* an inability to understand spoken words; or *word blindness,* an inability to understand written words.

▪ **Association Areas** The ***association areas*** of the cerebrum are made up of association tracts that connect motor and sensory areas (see Figure 14-8). The association region of the cortex occupies the greater portion of the lateral surfaces of the occipital, parietal, and temporal lobes, and the frontal lobes anterior to the motor areas. The association areas are concerned with memory, emotions, reasoning, will, judgment, personality traits, and intelligence.

Electroencephalogram (EEG)

Brain cells can generate electrical activity as a result of literally millions of nerve impulses (nerve action potentials) and other changes in membrane potentials of individual neurons. These electrical potentials are called ***brain waves*** and indicate activity of the cerebral cortex. Brain waves pass easily through the skull and can be detected by sensors called electrodes. A record of such waves is called an ***electroencephalogram*** **(*EEG*).**

MEDICAL TEST

Electroencephalogram (e-lek'-trō-en-SEF-a-lō-gram') **(EEG** or **brain wave test)**

Diagnostic Value: To diagnose epilepsy and other seizure disorders, infectious diseases, tumors, trauma, hematomas, metabolic abnormalities, degenerative diseases, and periods of unconsciousness and confusion. In some cases, an EEG can be used to furnish information regarding sleep and wakefulness. An EEG may also be used as one criterion to confirm brain death in which two flat EEGs (complete absence of brain waves) are taken 24 hours apart. Other criteria include unconsciousness, no spontaneous breathing, absence of a light response and dilation of the pupils, absence of reflexes (although local spinal reflexes may be present), unresponsiveness, and lack of normal muscle tone or strength.

Procedure: A technician applies 16 to 30 electrodes to the scalp. They are held in place by a paste or a head cap. The electrodes are connected by wires to an amplifier and recording machine (electroencephalograph) that converts electrical signals from the brain into a series of wavy lines on a moving sheet of graph paper. During the procedure, the person must remain quiet and still, and mild sedation may be employed. Brain waves may also be recorded as the brain responds to certain stimuli. For example, the person may be asked to breathe deeply and rapidly (hyperventilate) or look at a light while the eyes are open or closed. A typical EEG takes 1½ to 2 hours. If a sleep recording is performed, 3 hours may be required. The tracing must be interpreted by a psychiatrist or neurologist.

As indicated in Figure 14-12 four kinds of waves are produced by normal individuals.

1. **Alpha waves.** These rhythmic waves occur at a frequency of about 8 to 13 cycles per second. (The unit commonly used to express frequency is the Hertz (Hz); 1 Hz = 1 cycle per second.) Alpha waves are found in the EEGs of nearly all normal individuals when awake and in the resting state with their eyes closed. These waves disappear entirely during sleep.

FIGURE 14-12 Types of brain waves recorded in an electroencephalogram (EEG).

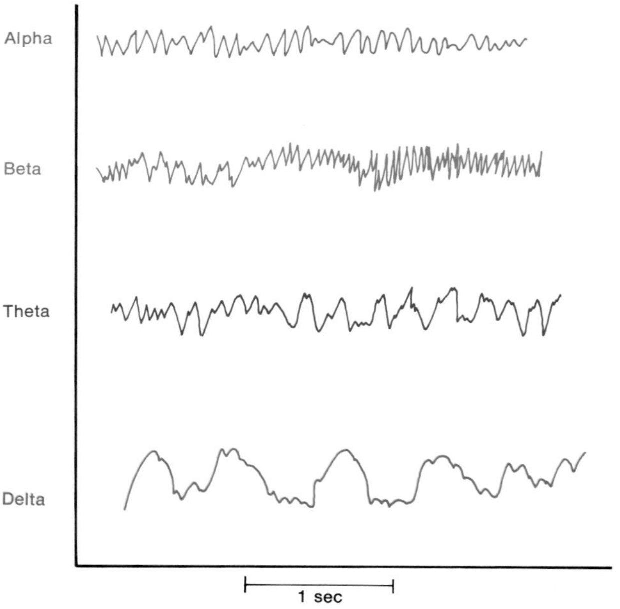

2. **Beta waves.** The frequency of these waves is between 14 to 30 Hz. Beta waves generally appear when the nervous system is active, that is, during periods of sensory input and mental activity.

3. **Theta waves.** These waves have frequencies of 4 to 7 Hz. Theta waves normally occur in children and in adults experiencing emotional stress. They also occur in many disorders of the brain.

4. **Delta waves.** The frequency of these waves is 1 to 5 Hz. Delta waves occur during deep sleep. They are normal in an awake infant. When produced by an awake adult, they indicate brain damage.

Although the EEG is still the principal procedure for recording brain electrical activity, a new procedure called ***magnetoencephalography* (*MEG*)** is being used to localize regions of the brain involved in epileptic seizures. MEG is the detection and measurement of magnetic fields outside the brain, produced by neuronal discharges (depolarizations and hyperpolarizations) within the brain.

Brain Lateralization (Split-Brain Concept)

Gross examination of the brain would suggest that it is bilaterally symmetric. However, detailed examination by the use of CT scans reveals certain anatomical differences between the two hemispheres. For example, in left-handed people the parietal and occipital lobes of the right hemisphere are usually narrower than the corresponding lobes of the left hemisphere. In addition, the frontal lobe of the left hemisphere of such individuals is typically narrower than that of the right hemisphere.

In addition to the structural differences between both sides of the brain, there are also several important functional differences. It has been shown that the left hemisphere is more important for right-hand control, spoken and written language, numerical and scientific skills, ability to use and understand sign language, and reasoning in most people. Conversely, it has been shown that the right hemisphere is more important for left-hand control; musical and artistic awareness; space and pattern perception; insight; imagination; and generating mental images of sight, sound, touch, taste, and smell in order to compare relationships.

Cerebellum

The ***cerebellum*** is the second-largest portion of the brain (almost one-eighth of the brain's mass) and occupies the inferior and posterior aspects of the cranial cavity. Specifically, it is posterior to the medulla and pons and below the occipital lobes of the cerebrum (see Figure 14-1). It is separated from the cerebrum by the ***transverse fissure*** and by an extension of the cranial dura mater called the ***tentorium*** (*tentorium* = tent) ***cerebelli.*** The tentorium cerebelli partially encloses the transverse sinuses and supports the occipital lobes of the cerebral hemispheres.

Structure

The cerebellum is shaped somewhat like a butterfly. The central constricted area is the ***vermis,*** which means "worm-shaped," and the lateral "wings" or lobes are referred to as ***hemispheres*** (Figure 14-13). Each hemisphere consists of lobes that are separated by deep and distinct fissures. The ***anterior lobe*** and ***posterior lobe*** are concerned with subconscious movements of skeletal muscles. The ***flocculonodular lobe*** is concerned with the sense of equilibrium (see Chapter 17). Between the hemispheres is another extension of the cranial dura mater: the ***falx cerebelli.*** It passes only a short distance between the cerebellar hemispheres and contains the occipital sinus.

The surface of the cerebellum, called the ***cortex,*** consists of gray matter in a series of slender, parallel ridges called ***folia.*** They are less prominent than the convolutions of the cerebral cortex. Beneath the gray matter are ***white matter tracts (arbor vitae)*** that resemble branches of a tree. Deep within the white matter are masses of gray matter, the ***cerebellar nuclei.*** The nuclei give rise to nerve fibers that convey information out of the cerebellum to other parts of the nervous system (both other brain centers and the spinal cord).

The cerebellum is attached to the brain stem by three paired bundles of fibers (tracts) called ***cerebellar peduncles*** (see Figures 14-3 and 14-4). ***Inferior cerebellar peduncles*** connect the cerebellum with the medulla at the base of the brain stem and with the spinal cord. These peduncles contain both afferent and efferent fibers and thus bring information into and out of the cerebellum. ***Middle cerebellar peduncles*** connect the cerebellum with the pons. These peduncles contain

FIGURE 14-13 Cerebellum. (a) Superior view. (b) Viewed in sagittal section. (Courtesy of Lester V. Bergman & Associates, Inc.)

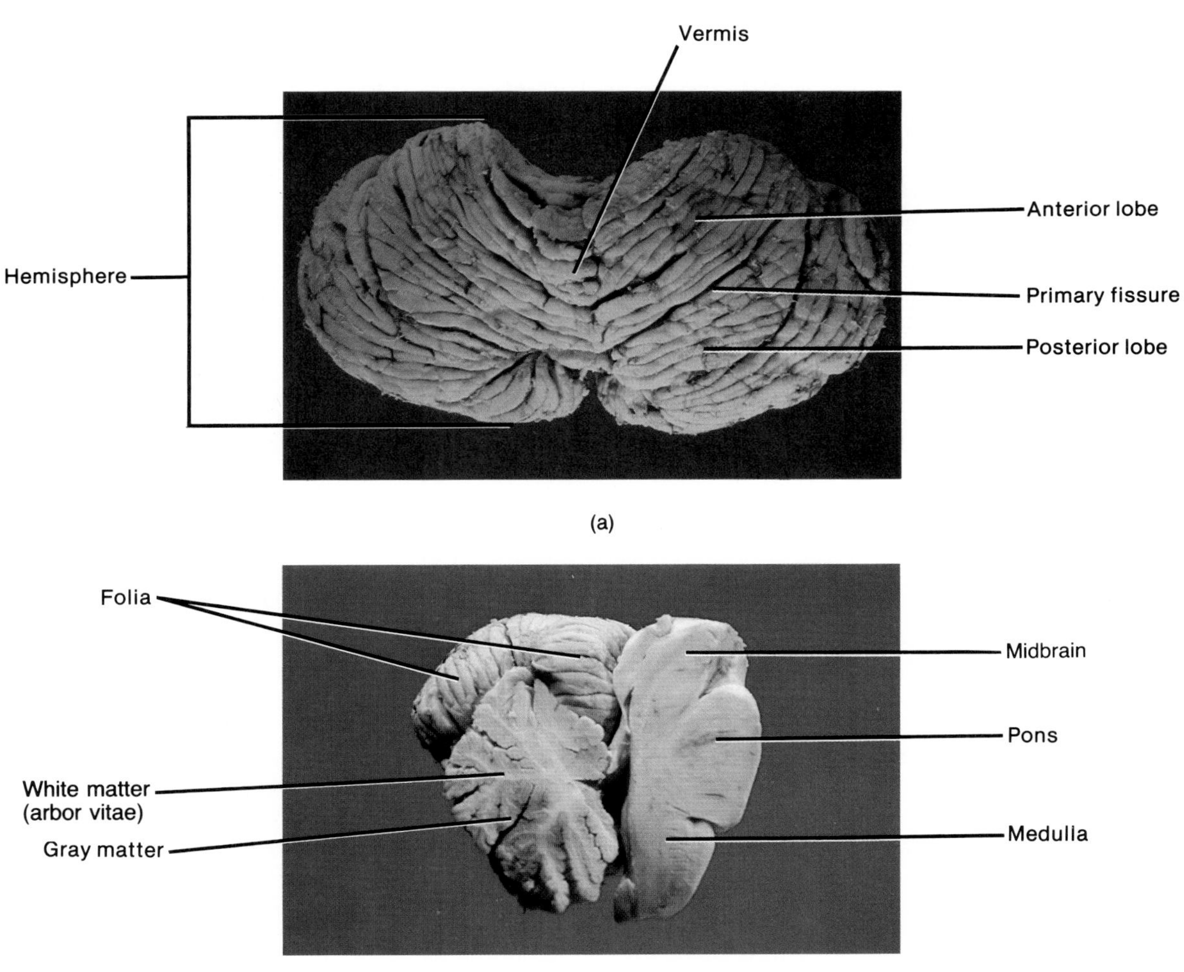

only afferent fibers and thus bring information into the cerebellum. ***Superior cerebellar peduncles*** connect the cerebellum with the midbrain. These peduncles contain mostly efferent fibers and thus mostly bring information out of the cerebellum.

Functions

Functionally, the cerebellum is a motor area of the brain concerned with coordinating subconscious movements of skeletal muscles. The cerebellar peduncles are the fiber tracts that permit information to pass into and out of the cerebellum. The cerebellum constantly receives input signals from proprioceptors in muscles, tendons, and joints, receptors for equilibrium, and visual receptors of the eyes. Such input permits the cerebellum to collect information on the physical status of the body with regard to posture, equilibrium, and all movements at joints. In addition, when other motor areas of the brain, such as the motor cortex of the cerebrum and basal ganglia, send signals to skeletal muscles, they also send a duplicate set of signals to the cerebellum. The cerebellum compares this input information regarding the actual status of the body with the intended movement determined by the other motor areas of the brain (cerebrum and basal ganglia). If the intent of these motor areas is not being attained by the skeletal muscles, the cerebellum detects the variation and sends feedback signals to the motor areas to either stimulate or inhibit the activity of skeletal muscles. This interaction produces smooth, coordinated movements by way of our body's skeletal muscles.

The cerebellum also functions in maintaining equilibrium and controlling posture. For example, receptors for equilibrium send nerve impulses to the cerebellum, informing it of body position. When the direction of movement changes, the cerebellum sends corrective signals to the motor cortex of the cerebrum. The motor cortex then sends signals over motor tracts to somatic motor neurons to skeletal muscles to reposition the body.

Another function of the cerebellum is related to predicting the future position of a body part during a particular movement. Just before a moving part of the body reaches its intended position, the cerebellum sends signals over motor tracts to somatic motor neurons to skeletal muscles to slow the moving part and stop it at a specific point. This function of the cerebellum is used in actions such as walking.

There is some evidence that the cerebellum may play a role in a person's emotional development, modulating sensations of anger and pleasure, allowing normal emotional expression and interpretation.

CLINICAL APPLICATION: DAMAGE TO CEREBELLUM

Damage to the cerebellum through trauma or disease is characterized by certain symptoms involving skeletal muscles on the same side of the body as the damage. The effects are on the same side of the body as the damaged side of the cerebellum because of a double crossing of tracts within the cerebellum. There may be lack of muscle coordination, called _ataxia_ (_a_ = without; _taxis_ = order). Blindfolded people with ataxia cannot touch the tip of their nose with a finger because they cannot coordinate movement with their sense of where a body part is located. Another sign of ataxia is a change in the speech pattern due to a lack of coordination of speech muscles. Cerebellar damage may also result in _disturbances of gait,_ in which the subject staggers or cannot coordinate normal walking movements, and _severe dizziness._

A summary of the functions of the various parts of the brain is presented in Exhibit 14-1.

NEUROTRANSMITTERS IN THE BRAIN

There are numerous substances that are either known or suspected neurotransmitters in the brain. These substances can facilitate, excite, or inhibit postsynaptic neurons. They establish the lines of communication between brain cells and are localized in certain parts of the brain. In Chapter 12, we noted that ***acetylcholine* (*ACh*)** is released at some neuromuscular junctions, at some neuroglandular junctions, and at synapses between certain brain and spinal cord cells. In most parts of the body, ACh leads to excitation. ACh is inactivated to acetate and choline by acetylcholinesterase (AChE). The amino acids ***glutamic acid*** and ***aspartic acid*** are also believed to lead to excitation in the brain.

Norepinephrine* (*NE*)** is a neurotransmitter that is released at some neuromuscular junctions and some neuroglandular junctions. NE probably causes excitation in most instances, but it may be involved in inhibition. It is concentrated in a group of neurons in the brain stem near the fourth ventricle called the ***locus coeruleus (LŌ-kus sē-ROO-lē-us). This area, which literally means "blue place," projects axons into the hypothalamus, cerebellum, cerebral cortex, and spinal cord. In these places, NE has been implicated in maintaining arousal (awakening from deep sleep), dreaming, and the regulation of mood. Inactivation of NE is different from that of ACh. After NE is released from its synaptic vesicles, it is rapidly pumped back into the synaptic end bulbs. Here it is either destroyed by the enzymes ***catechol-O-methyltransferase*** (kat'-e-kōl-ō-meth-il-TRANS-fer-ās), or ***COMT,*** and ***monoamine oxidase*** (mon-ō-AM-ēn OK-si-dās), or ***MAO,*** or recycled back into the synaptic vesicles.

EXHIBIT 14-1 SUMMARY OF FUNCTIONS OF PRINCIPAL PARTS OF BRAIN

Part	Function
BRAIN STEM	
Medulla	Relays motor and sensory impulses between other parts of the brain and the spinal cord. Reticular formation (also in pons, midbrain, and diencephalon) functions in consciousness and arousal. Vital reflex centers regulate heartbeat, breathing (together with pons), and blood vessel diameter. Nonvital reflex centers coordinate swallowing, vomiting, coughing, sneezing, and hiccuping. Contains nuclei of origin for cranial nerves VIII, IX, X, XI, and XII. Vestibular nuclear complex helps maintain equilibrium.
Pons	Relays impulses within the brain and between parts of the brain and spinal cord. Contains nuclei of origin for cranial nerves V, VI, VII, and VIII. Pneumotaxic area and apneustic area, together with the medulla, help control breathing.
Midbrain	Relays motor impulses from the cerebral cortex to the pons and spinal cord and relays sensory impulses from the spinal cord to the thalamus. Superior colliculi coordinate movements of the eyeballs in response to visual and other stimuli, and the inferior colliculi coordinate movements of the head and trunk in response to auditory stimuli. Contains nuclei of origin for cranial nerves III and IV.
DIENCEPHALON	
Thalamus	Several nuclei serve as relay stations for all sensory impulses, except smell, to the cerebral cortex. Relays motor impulses from the cerebral cortex to the spinal cord. Interprets pain, temperature, light touch, and pressure sensations. Anterior nucleus functions in emotions and memory.
Hypothalamus	Controls and integrates the autonomic nervous system. Receives sensory impulses from viscera. Regulates and controls the pituitary gland. Center for mind-over-body phenomena. Secretes regulating hormones (or factors). Functions in rage and aggression. Controls normal body temperature, food intake, and thirst. Helps maintain the waking state and sleep. Functions as a self-sustained oscillator that drives many biological rhythms.
CEREBRUM	Sensory areas interpret sensory impulses, motor areas control muscular movement, and association areas function in emotional and intellectual processes. Basal ganglia control gross muscle movements and regulate muscle tone. Limbic system functions in emotional aspects of behavior related to survival.
CEREBELLUM	Controls subconscious skeletal muscle contractions required for coordination, posture, and balance. Assumes a role in emotional development, modulating sensations of anger and pleasure.

Neurons containing the neurotransmitter ***dopamine* (*DA*)** are clustered in the midbrain in the substantia nigra. DA usually leads to inhibition. Some axons projecting from the substantia nigra terminate in the cerebral cortex, where DA is thought to be involved in emotional responses. Other axons project to the corpus striatum (basal ganglia), where the DA is involved in gross subconscious movements of skeletal muscles. The degeneration of axons in the basal ganglia results in Parkinson's disease, which is discussed at the end of this chapter.

Serotonin* (*5-HT*)** is a neurotransmitter that leads to excitation and is concentrated in the neurons in a part of the brain stem called the ***raphe nucleus. Axons projecting from the nucleus terminate in the hypothalamus, thalamus, other parts of the brain, and spinal cord. Serotonin is thought to be involved in inducing sleep, sensory perception, temperature regulation, and control of mood.

The most common neurotransmitter in the brain leading to inhibition is ***gamma aminobutyric acid* (*GABA*).** It is most highly concentrated in the superior and inferior colliculi, thalamus, hypothalamus, and occipital lobes of the cerebrum. It has been implicated as a likely target for antianxiety drugs such as diazepam (Valium), which enhance the action of GABA. A neurotransmitter that leads to inhibition in the spinal cord is ***glycine.***

In recent years, another group of chemical messengers in the brain has been identified. These are known as ***neuropeptides,*** which consist of chains of 2 to about 40 amino acids that occur naturally in the brain. Although some neuropeptides function as true neurotransmitters, most act primarily to modulate the response of or the response to a neurotransmitter. In 1975 the first neuropeptides, referred to as ***enkephalins*** (en-KEF-a-lins; *en* = without; *keph* = head), were discovered. These chemicals are several times more potent than morphine, the painkiller derived from the opium poppy. Enkephalins are concentrated in the thalamus, in the hypothalamus, in parts of the limbic system, and in those spinal cord pathways that relay impulses for pain. It has been suggested that enkephalins are the body's natural painkillers. They do this by inhibiting impulses in the pain pathway and by binding to the same receptors in the brain as morphine.

Other naturally occurring neuropeptides, called ***endorphins*** (en-DOR-fins), have subsequently been isolated from the pituitary gland. Like the enkephalins, they have morphinelike properties that suppress pain. Endorphins have also been linked to memory and learning; sexual activity; control of body temperature; regulation of hormones that affect the onset of puberty, sexual drive, and reproduction; and mental illnesses such as depression and schizophrenia. One of the better studied endorphins is ***beta-endorphin (β-endorphin).*** A neuropeptide that is believed to work with endorphins is known as ***substance P.*** It is found in sensory nerves, spinal cord pathways, and parts of the brain associated with pain transmission. When substance P is released by neurons, it conducts pain-related nerve impulses from peripheral pain receptors into the central nervous system. It is now suspected that endorphins may exert their analgesic effects by suppressing the release of substance P. Substance P has also been shown to counter the effects of certain nerve-damaging chemicals, prompting speculation that it might prove useful as a treatment for nerve degeneration.

A neuropeptide called ***dynorphin*** (*dynamis* = power) is 200 times more powerful in action than morphine and 50 times more powerful than beta-endorphin. It is found in the posterior pituitary gland, hypothalamus, and small intestine. Its exact functions have yet to be determined. However, the chemical might be a factor in controlling pain and registering emotions.

As research into neuropeptides continues, a long list of substances is being drawn up. Interestingly enough, many of these neuropeptides are also found in other parts of the body, where they serve as hormones or other regulators of physiological responses. Although we will be discussing many of them in subsequent chapters, a few will be mentioned here to give you some idea of their diversity.

1. Angiotensin II. This is a substance that can raise blood pressure and is found in the bloodstream as well. It is activated by a kidney enzyme called renin. Some evidence suggests that angiotensin II in the brain may be part of the brain's mechanism for regulating its own blood pressure.

2. Cholecystokinin (ko'-lē-sis'-tō-KĪN-in), or *CCK*. This substance (also called pancreozymin) is produced by the lining of the small intestine and causes the pancreas to release digestive juice and the gallbladder to release bile. Although the significance of cholecystokinin in the brain is unclear, there are some suggestions that it may be related to the control of feeding. Other neuropeptides associated with the gastrointestinal tract include neurotensin, vasoactive intestinal peptide (VIP), gastrin, and secretin.

3. Regulating hormones (or factors). These are chemicals produced by the hypothalamus that regulate the release of hormones by the pituitary gland. Examples include thyrotropin releasing hormone (TRH), gonadotropin releasing hormone (GnRH), and growth hormone releasing hormone (GHRH).

Exhibit 14-2 presents a summary of representative neurotransmitters and neuropeptides.

CRANIAL NERVES

Of the 12 pairs of ***cranial nerves,*** 10 originate from the brain stem, but all leave the skull through foramina of the skull. The cranial nerves are designated with Roman numerals and with names (see Figure 14-3). The Roman numerals indicate the order in which the nerves arise from the brain (front to back). The names indicate the distribution or function.

Some cranial nerves contain only sensory fibers and thus are called ***sensory nerves.*** The remainder contain both sensory and motor fibers and are referred to as ***mixed nerves.*** At one time, it was believed that some cranial nerves (oculomotor, trochlear, abducens, accessory, and hypoglossal) were entirely motor. However, it is now known that these cranial nerves also contain some sensory fibers from proprioceptors in muscles they innervate. Although these nerves are mixed, they are primarily motor in function, serving to stimulate skeletal muscle contraction. The cell bodies of sensory fibers are found outside the brain, whereas the cell bodies of motor fibers lie in nuclei within the brain (Figure 14-14).

Some motor fibers control subconscious movements, yet the somatic nervous system has been defined as a *conscious* system. The reason for this apparent contradiction is that some fibers of the autonomic nervous system leave the brain, bundled together with somatic fibers of the cranial nerves, as is the case for spinal nerves. Therefore, subconscious functions transmitted by the autonomic fibers are described along with the conscious functions of the somatic fibers of the cranial nerves.

A summary of cranial nerves and clinical applications related to dysfunction is presented in Exhibit 14-3.

AGING AND THE NERVOUS SYSTEM

One of the effects of aging on the nervous system is that neurons are lost. Associated with this decline, there is a decreased capacity for sending nerve impulses to and from

EXHIBIT 14-2 SUMMARY OF REPRESENTATIVE NEUROTRANSMITTERS AND NEUROPEPTIDES

Substance	Comment
NEUROTRANSMITTERS	
Acetylcholine (ACh)	Released by some neuromuscular and neuroglandular junctions and at synapses between certain brain and spinal cord cells; in most parts of the body, ACh leads to excitation; ACh is inactivated by acetylcholinesterase (AChE); blockage of ACh receptors in skeletal muscles leads to myasthenia gravis.
Serotonin (5-HT)	Concentrated in neurons in the raphe nucleus in the brain stem; leads to excitation and may be involved in inducing sleep, sensory perception, temperature regulation, and control of mood.
Norepinephrine (NE)	Released at some neuromuscular and neuroglandular junctions; concentrated in the locus coeruleus of the brain stem; NE probably causes excitation in most cases. NE may be related to arousal, dreaming, and regulation of mood; NE is inactivated by catechol-O-methyltransferase monoamine oxidase (MAO).
Glutamic acid and aspartic acid	Cause excitation in brain.
Gamma aminobutyric acid (GABA)	Concentrated in superior and inferior colliculi, thalamus, hypothalamus, and occipital lobes of cerebrum; leads to inhibition in brain; probably a target for antianxiety drugs.
Glycine	Leads to inhibition in spinal cord.
Dopamine (DA)	Concentrated in substantia nigra; leads to inhibition; involved in emotional responses and subconscious movements of skeletal muscles; decreased levels associated with Parkinson's disease; an excess of DA might be involved in schizophrenia.
NEUROPEPTIDES	
Enkephalins	Concentrated in thalamus, hypothalamus, parts of limbic system, and spinal cord pathways that relay pain impulses; they inhibit pain impulses by suppressing substance P.
Endorphins	Concentrated in pituitary gland; function in inhibiting pain by inhibiting substance P and may have a role in memory and learning, sexual activity, control of body temperature; have been linked to depression and schizophrenia.
Substance P	Found in sensory nerves, spinal cord pathways, and parts of brain associated with pain; stimulates perception of pain; endorphins may exert their pain-inhibiting properties by suppressing release of substance P.
Dynorphin	Found in posterior pituitary gland, hypothalamus, and small intestine; 50 times more powerful than beta-endorphin; may be related to controlling pain and registering emotions.
Angiotensin II	Produced from renin released by kidneys; may regulate blood pressure in brain.
Cholecystokinin (CCK)	Produced by small intestine; may be related to the regulation of feeding.
Regulating hormones (or factors)	Chemicals produced by hypothalamus that regulate the release of hormones by the pituitary gland.

the brain. Conduction velocity decreases, voluntary motor movements slow down, and the reflex time for skeletal muscles increases. Deep reflexes may diminish and superficial reflexes may be lost. Parkinson's disease is the most common movement disorder involving the central nervous system. Degenerative changes and disease states involving the sense organs can alter vision, hearing, taste, smell, and touch. The disorders that represent the most common visual problems and may be responsible for serious loss of vision are presbyopia (inability to focus on nearby objects), cataracts (cloudiness of the lens), and glaucoma (excessive fluid pressure in the eyeball). Impaired hearing associated with aging, known as presbycusis, is usually the result of changes in important structures of the inner ear.

DEVELOPMENTAL ANATOMY OF THE NERVOUS SYSTEM

The development of the nervous system begins early in the third week of development with a thickening of the ***ectoderm*** called the ***neural plate*** (Figure 14-15a–c). The plate folds inward and forms a longitudinal groove, the ***neural groove.*** The raised edges of the neural plate are called ***neural folds.*** As development continues, the

FIGURE 14-14 Nuclei of cranial nerves.

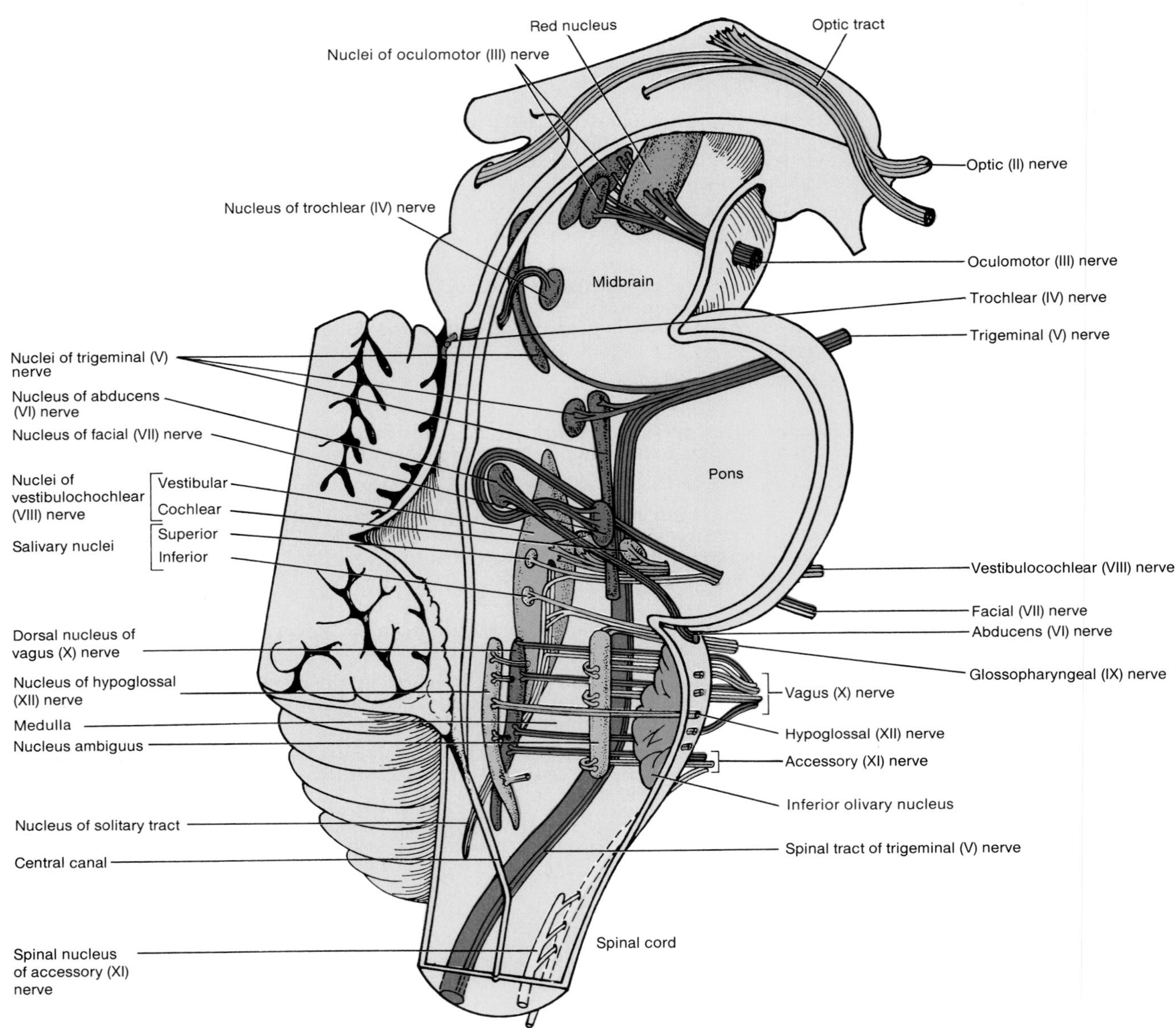

neural folds increase in height, meet, and form a tube, the ***neural tube.***

The cells of the wall that encloses the neural tube differentiate into three kinds. The outer or ***marginal layer*** develops into the *white matter* of the nervous system; the middle or ***mantle layer*** develops into the *gray matter* of the system; and the inner or ***ependymal layer*** eventually forms the *lining of the ventricles* of the central nervous system.

The ***neural crest*** is a mass of tissue between the neural tube and the ectoderm (Figure 14-15c). It becomes differentiated and eventually forms the *posterior* (*dorsal*) *root ganglia of spinal nerves, spinal nerves, ganglia of cranial nerves, cranial nerves, ganglia of the autonomic nervous system,* and the *adrenal medulla.*

When the neural tube is formed from the neural plate, the anterior portion of the neural tube develops into three enlarged areas called vesicles: (1) ***forebrain vesicle***

EXHIBIT 14-3 SUMMARY OF CRANIAL NERVES*

Nerve (Type)	Location	Function and Clinical Application
Olfactory (I) (sensory)	Arises in olfactory mucosa, passes through cribriform plate of ethmoid bone, through olfactory bulb and olfactory tract, and terminates in primary olfactory areas of cerebral cortex.	Function: smell. Clinical application: Loss of the sense of smell, called *anosmia*, may result from head injuries in which the cribriform plate of the ethmoid bone is fractured and from lesions along the olfactory pathway.
Optic (II) (sensory)	Arises in retina of the eye, passes through optic foramen, forms optic chiasma, passes through optic tracts, lateral geniculate nucleus in thalamus, and terminates in visual areas of cerebral cortex.	Function: vision. Clinical application: Fractures in the orbit, lesions along the visual pathway, and diseases of the nervous system may result in visual field defects and loss of visual acuity. A defect of vision is called *anopsia*.
Oculomotor (III) (mixed, primarily motor)	Motor portion: originates in midbrain, passes through superior orbital fissure, and is distributed to levator palpebrae superioris of upper eyelid and four extrinsic eyeball muscles (superior rectus, medial rectus, inferior rectus, and inferior oblique); parasympathetic innervation to ciliary muscle of eyeball and sphincter muscle of iris. Sensory portion: consists of afferent fibers from proprioceptors in eyeball muscles that passes through superior orbital fissure and terminates in midbrain.	Motor function: movement of eyelid and eyeball, accommodation of lens for near vision, and constriction of pupil. Sensory function: muscle sense (proprioception). Clinical application: A lesion in the nerve causes *strabismus* (squinting), *ptosis* (drooping) of the upper eyelid, pupil dilation, the movement of the eyeball downward and outward on the damaged side, a loss of accommodation for near vision, and double vision (*diplopia*).
Trochlear (IV) (mixed, primarily motor)	Motor portion: originates in midbrain, passes through superior orbital fissure, and is distributed to superior oblique muscle, an extrinsic eyeball muscle. Sensory portion: consists of afferent fibers from proprioceptors in superior oblique muscles that pass through superior orbital fissure and terminates in midbrain.	Motor function: movement of eyeball. Sensory function: muscle sense (proprioception). Clinical application: In trochlear nerve paralysis, the head is tilted to the affected side and diplopia and strabismus occur.
Trigeminal (V) (mixed)	Motor portion: originates in pons, passes through foramen ovale, and terminates in muscle of mastication, anterior belly of digastric and mylohyoid muscles. Sensory portion: consists of three branches: *ophthalmic*—contains sensory fibers from skin over upper eyelid, eyeball, lacrimal glands, nasal cavity, side of nose, forehead, and anterior half of scalp and passes through superior orbital fissure; *maxillary*—contains sensory fibers from mucosa of nose, palate, parts of pharynx, upper teeth, upper lip, and lower eyelid and passes through foramen rotundum; *mandibular*—contains sensory fibers from anterior two-thirds of tongue, lower teeth, skin over mandible, cheek and mucosa deep to it, and side of head in front of ear and passes through foramen ovale. The three branches terminate in pons. Sensory portion also consists of afferent fibers from proprioceptors in muscles of mastication.	Motor function: chewing. Sensory function: conveys sensations for touch, pain, and temperature from structures supplied; muscle sense (proprioception). Clinical application: Injury results in paralysis of the muscles of mastication and a loss of sensation of touch and temperature. *Neuralgia* (pain) of one or more branches of trigeminal nerve is called *trigeminal neuralgia* (*tic douloureux*).

* A mnemonic device used to remember the names of the nerves is: "Oh, oh, oh, to touch and feel very green vegetables—AH!" The initial letter of each word corresponds to the initial letter of each pair of cranial nerves.

EXHIBIT 14-3 SUMMARY OF CRANIAL NERVES (*Continued*)

Nerve (Type)	Location	Function and Clinical Application
Abducens (VI) (mixed, primarily motor)	Motor portion: originates in pons, passes through superior orbital fissure, and is distributed to lateral rectus muscle, and extrinsic eyeball muscle. Sensory portion: consists of afferent fibers from proprioceptors in lateral rectus muscle that pass through superior orbital fissure and terminates in pons.	Motor function: movement of eyeball. Sensory function: muscle sense (proprioception). Clinical application: With damage to this nerve, the affected eyeball cannot move laterally beyond the midpoint and the eye is usually directed medially.
Facial (VII) (mixed)	Motor portion: originates in pons, passes through stylomastoid foramen, and is distributed to facial, scalp, and neck muscles, parasympathetic distribution to lacrimal sublingual, submandibular, nasal, and palatine glands. Sensory portion: arises from taste buds on anterior two-thirds of tongue, passes through stylomastoid foramen, passes through geniculate ganglion, a nucleus in pons that sends fibers to thalamus for relay to gustatory areas of cerebral cortex. Also consists of afferent fibers from proprioceptors in muscles of face and scalp.	Motor function: facial expression and secretion of saliva and tears. Sensory function: muscle sense (proprioception). Clinical application: Injury produces paralysis of the facial muscles, called *Bell's palsy,* loss of taste, and the eyes remain open, even during sleep.
Vestibulocochlear (VIII) (sensory)	Cochlear branch: arises in spiral organ (organ of Corti), forms spiral ganglion, passes through nuclei in the medulla, and terminates in thalamus. Fibers synapse with neurons that relay impulses to auditory areas of cerebral cortex. Vestibular branch: arises in semicircular canals, saccule, and utricle and forms vestibular ganglion; fibers terminate in nuclei in thalamus.	Cochlear branch function: conveys impulses associated with hearing. Vestibular branch function: conveys impulses associated with equilibrium. Clinical application: Injury to the cochlear branch may cause *tinnitus* (ringing) or deafness. Injury to the vestibular branch may cause *vertigo* (a subjective feeling of rotation), *ataxia,* and *nystagmus* (involuntary rapid movement of the eyeball).
Glossopharyngeal (IX) (mixed)	Motor portion: originates in medulla, passes through jugular foramen, and is distributed to stylopharyngeus muscle; parasympathetic distribution to parotid gland. Sensory portion: arises from taste buds on posterior one-third of tongue and from carotid sinus, passes through jugular foramen, and terminates in thalamus. Also consists of afferent fibers from proprioceptors in swallowing muscles supplied.	Motor function: secretion of saliva. Sensory function: taste and regulation of blood pressure; muscle sense (proprioception). Clinical application: Injury results in pain during swallowing, reduced secretion of saliva, loss of sensation in the throat, and loss of taste.
Vagus (X) (mixed)	Motor portion: originates in medulla, passes through jugular foramen, and terminates in muscles of respiratory passageways, lungs, esophagus, heart, stomach, small intestine, most of large intestine, and gallbladder; parasympathetic fibers innervate involuntary muscles and glands of the gastrointestinal (GI) tract. Sensory portion: arises from essentially same structures supplied by motor fibers, passes through jugular foramen, and terminates in medulla and pons. Also consists of afferent fibers from proprioceptors in muscles supplied.	Motor function: visceral muscle movement. Sensory function: sensations from organs supplied; muscle sense (proprioception). Clinical application: Severing of both nerves in the upper body interferes with swallowing, paralyzes vocal cords, and interrupts sensations from many organs. Injury to both nerves in the abdominal area has little effect, since the abdominal organs are also supplied by autonomic fibers from the spinal cord.
Accessory (XI) (mixed, primarily motor)	Motor portion: consists of a cranial portion and a spinal portion. Cranial portion originates from medulla, passes through jugular foramen, and supplies voluntary muscles of pharynx, larynx, and soft palate. Spinal por-	Motor function: cranial portion mediates swallowing movements; spinal portion mediates movement of head. Sensory function: muscle sense (proprioception).

EXHIBIT 14-3 SUMMARY OF CRANIAL NERVES (*Continued*)

Nerve (Type)	Location	Function and Clinical Application
	tion originates from anterior gray horn of first five cervical segments of spinal cord, passes through jugular foramen, and supplies sternocleidomastoid and trapezius muscles. Sensory portion: consists of afferent fibers from proprioceptors in muscles supplied and passes through jugular foramen.	Clinical application: If damaged, the sternocleidomastoid and trapezius muscles become paralyzed, with resulting inability to turn the head or raise the shoulders.
Hypoglossal (XII) (mixed, primarily motor)	Motor portion: originates in medulla, passes through hypoglossal canal, and supplies muscles of tongue. Sensory portion: consists of fibers from proprioceptors in tongue muscles that pass through hypoglossal canal and terminate in medulla.	Motor function: movement of tongue during speech and swallowing. Sensory function: muscle sense (proprioception). Clinical application: Injury results in difficulty in chewing, speaking, and swallowing. The tongue, when protruded, curls toward the affected side and the affected side becomes atrophied, shrunken, and deeply furrowed.

FIGURE 14-15 Origin of the nervous system. (a) Dorsal view of an embryo with three pairs of somites showing the neural plate. (b) Dorsal view of an embryo with seven pairs of somites in which the neural folds have united just medial to the somites, forming the early neural tube. (c) Cross section through the embryo showing the formation of the neural tube.

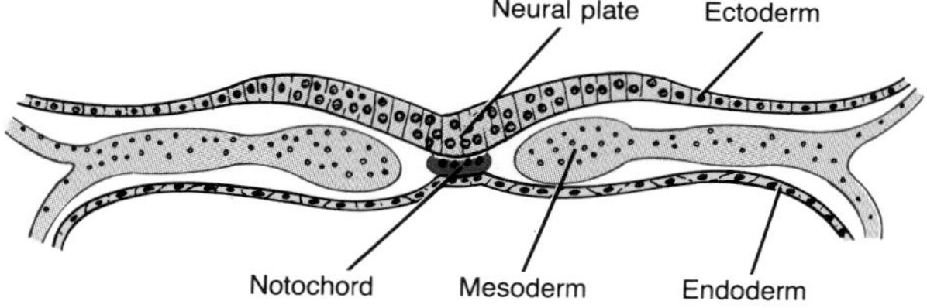

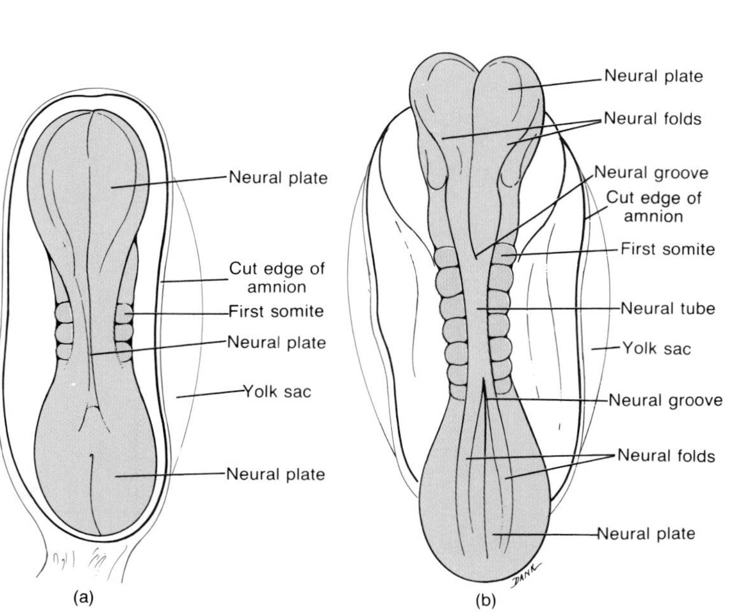

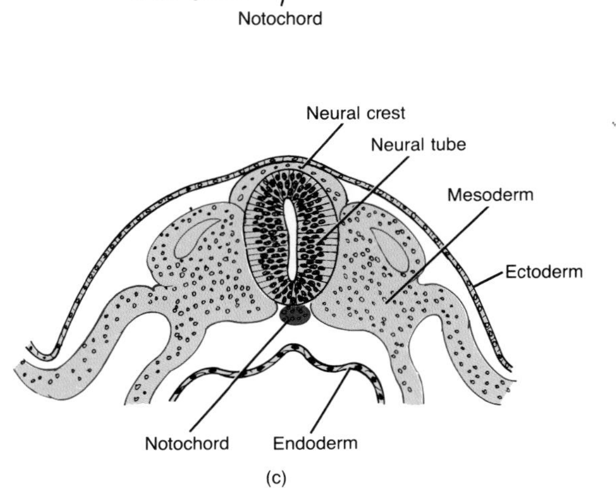

(prosencephalon), (2) ***midbrain vesicle (mesencephalon)***, and (3) ***hindbrain vesicle (rhombencephalon)*** (Figure 14-16). The vesicles are fluid-filled enlargements that develop by the fourth week of gestation. Since they are the first vesicles to form, they are called ***primary vesicles.*** As development progresses, the vesicular region undergoes several flexures (bends), resulting in subdivision of the three primary vesicles, so that by the fifth week of development the embryonic brain consists of five ***secondary vesicles.*** The forebrain vesicle (prosencephalon) divides into an anterior ***telencephalon*** and a posterior ***diencephalon;*** the midbrain vesicle (mesencephalon) remains unchanged; the hindbrain vesicle (rhombencephalon) divides into an anterior ***metencephalon*** and a posterior ***myelencephalon.***

Ultimately, the telencephalon develops into the *cerebral hemispheres* and *basal ganglia;* the diencephalon develops into the *thalamus, hypothalamus,* and *pineal gland;* the midbrain vesicle (mesencephalon) develops into the *midbrain;* the metencephalon develops into the *pons* and *cerebellum;* and the myelencephalon develops into the *medulla oblongata.* The cavities within the vesicles develop into the *ventricles* of the brain, whereas the fluid within them is *cerebrospinal fluid.* The area of the neural tube posterior to the myelencephalon gives rise to the *spinal cord.*

FIGURE 14-16 Development of the brain and spinal cord. (a) Primary vesicles of the neural tube seen in frontal section (left) and right lateral view (right) at about 3 to 4 weeks. (b) Secondary vesicles seen in frontal section (left) and right lateral view (right) at about 5 weeks.

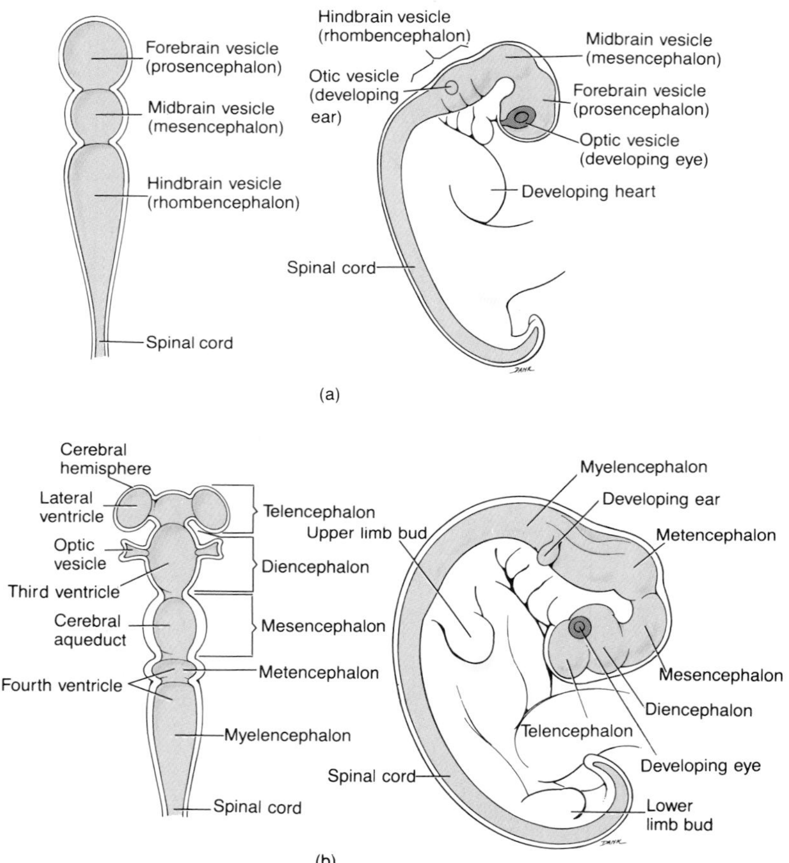

DISORDERS: HOMEOSTATIC IMBALANCES

Many disorders can affect the central nervous system. Some are caused by viruses or bacteria. Others are caused by damage to the nervous system during birth. The origins of many conditions, however, are unknown. Here we discuss the origins and symptoms of some common central nervous system disorders.

DISORDERS: HOMEOSTATIC IMBALANCES (*continued*)

Cerebrovascular Accident (CVA)

The most common brain disorder is a ***cerebrovascular accident* (*CVA*)**, also called a ***stroke*** or ***cerebral apoplexy.*** A CVA is characterized by a relatively abrupt onset of persisting neurological symptoms due to the destruction of brain tissue (infarction) resulting from disorders in the blood vessels that supply the brain. CVAs may be classified into two principal types: (1) *ischemic,* the most common type, due to a decreased blood supply; and (2) *hemorrhagic,* due to a blood vessel in the brain that bursts. Common causes of CVAs are intracerebral hemorrhage (rupture of a blood vessel in the pia mater or brain), emboli (blood clots), and atherosclerosis (formation of plaques) of the cerebral arteries.

Among the risk factors implicated in CVAs are high blood pressure, heart disease, narrowed carotid arteries, transient ischemic attacks (TIAs), diabetes, smoking, obesity, high plasma fibrinogen level, maternal history of CVA (for males), and excessive alcohol intake.

Epilepsy

Epilepsy is the second most common neurological disorder after stroke. It is characterized by short, recurrent, periodic attacks of motor, sensory, or psychological malfunction. The attacks, called ***epileptic seizures,*** are initiated by abnormal and irregular discharges of electricity from millions of neurons in the brain, probably resulting from abnormal reverberating circuits. The discharges stimulate many of the neurons to send nerve impulses over their conduction pathways. As a result, a person undergoing an attack may contract skeletal muscles involuntarily. Lights, noise, or smells may be sensed when the eyes, ears, and nose actually have not been stimulated. The electrical discharges may also inhibit certain brain centers. For instance, the waking center in the brain may be depressed so that the person loses consciousness.

Epilepsy has many causes; they are classified as symptomatic or idiopathic. Symptomatic causes include brain damage at birth, the most common cause; metabolic disturbances (hypoglycemia, hypocalcemia, uremia, hypoxia); infections (encephalitis or meningitis); toxins (alcohol, tranquilizers, hallucinogens); vascular disturbances (hemorrhage, hypotension); head injuries; and tumors and abscesses of the brain. Most epileptic seizures, however, are idiopathic; that is, they have no demonstrable cause. It should be noted that epilepsy almost never affects intelligence.

Epileptic seizures can be eliminated or alleviated by drugs that make neurons more difficult to stimulate. Many of these drugs change the permeability of the neuron cell membrane so that it does not depolarize as easily. One such drug is valproic acid, which increases the quantity of the inhibitory neurotransmitter gamma aminobutyric acid (GABA).

Transient Ischemic Attack (TIA)

A ***transient ischemic attack* (*TIA*)** is an episode of temporary, focal, nonconvulsive, cerebral dysfunction caused by an interference of the blood supply to the brain. Symptoms include dizziness, weakness, numbness, or paralysis in a limb or in one-half of the body; drooping of one side of the face; headache; slurred speech or difficulty understanding speech; or a partial loss of vision or double vision. Sometimes nausea or vomiting also occur. The onset of symptoms is sudden and reaches maximum intensity almost immediately. A TIA usually persists for 2 to 15 minutes and only rarely as long as 24 hours. Each TIA leaves no persistent neurologic deficits. The causes of the ischemia that lead to TIAs are emboli, atherosclerosis, impaired blood flow due to hemodynamic disruptions, and hematologic disorders (polycythemia, thrombocytosis, and sickle-cell anemia).

It is estimated that about one-third of patients with a TIA will have a CVA within five years. Therapy for TIAs includes antiplatelet-aggregating agents such as aspirin, anticoagulants, cerebral artery bypass grafting, and carotid endarterectomy (excision of the atheromatous tunica intima of an artery).

Brain Tumors

A ***brain tumor*** refers to any benign or malignant growth within the cranium. Tumors may arise from neuroglial cells in the cerebrum, brain stem, and cerebellum or from supporting or neighboring structures such as cranial nerve coverings, meninges, and the pituitary gland. Intracranial pressure from a growing tumor or edema associated with the tumor produce the characteristic signs and symptoms of a brain tumor. Among these are headache, altered consciousness, and vomiting. Brain tumors can also result in seizures, visual problems, cranial nerve abnormalities, hormonal syndromes, personality changes, dementia, and sensory or motor deficits. Treatment of brain tumors involves surgery, radiation therapy, and chemotherapy.

Poliomyelitis

Poliomyelitis* (*infantile paralysis*)**, or simply ***polio, is most common during childhood and is caused by a virus called poliovirus. The onset of the disease is marked by fever, severe headache, a stiff neck and back, deep muscle pain and weakness, and loss of certain somatic reflexes. In its most serious form, called ***bulbar polio,*** the virus spreads via blood to the central nervous system, where it destroys the motor nerve cell bodies, specifically those in the anterior horns of the spinal cord and in the nuclei of the cranial nerves. Injury to the spinal gray matter is the basis for the name of this disease (*polio* = gray matter; *myel* = spinal cord). Destruction of the anterior horns produces paralysis. The first sign of

DISORDERS: HOMEOSTATIC IMBALANCES (*continued*)

bulbar polio is difficulty in swallowing, breathing, and speaking. Poliomyelitis can cause death from respiratory or heart failure if the virus invades the brain cells of the vital medullary centers. The incidence of polio in the United States has decreased markedly since the availability of polio vaccines (Salk vaccine and more recently Sabin vaccine).

Cerebral Palsy (CP)

The term ***cerebral palsy* (*CP*)** refers to a group of motor disorders resulting in muscular uncoordination and loss of muscle control. It is caused by damage to the motor areas of the brain during fetal life, birth, or infancy. One cause is infection of the mother with German measles during the first three months of pregnancy when certain cells in the fetus are dividing and differentiating in order to lay down the basic structures of the brain. These cells can be abnormally changed by toxin from the measles virus. Radiation during fetal life, temporary oxygen starvation during birth, and hydrocephalus during infancy may also damage brain cells. Cases of cerebral palsy are categorized into three groups, depending on whether the cortex, the basal ganglia of the cerebrum, or the cerebellum is affected most severely. About 70 percent of cerebral palsy victims appear to be mentally retarded. The apparent mental slowness, however, is often due to the person's inability to speak or hear well. Such individuals are often more mentally acute than they appear. Cerebral palsy is not a progressive disease; it does not worsen as time elapses. Once the damage is done, however, it is irreversible. Recently, a surgical procedure called *selective posterior rhizotomy* has been used on children to reduce muscle spasticity. In the procedure, selected rootlets in posterior roots are severed, allowing the other rootlets to achieve better muscle tone.

Parkinson's Disease (PD)

Parkinson's disease* (*PD*)** or ***parkinsonism is a progressive disorder of the central nervous system that begins inconspicuously and typically affects its victims around age 60. The cause is unknown, but there are indications that an environmental agent might be involved. The disease is related to pathological changes in the substantia nigra and basal ganglia. The substantia nigra contains cell bodies of neurons that produce the neurotransmitter dopamine (DA) in their axon terminals. The axon terminals release DA in the basal ganglia of the cerebrum. Recall that basal ganglia regulate subconscious contractions of skeletal muscles that aid activities also consciously controlled by the motor areas of the cerebral cortex—swinging the arms when walking, for example. In Parkinson's disease there is a degeneration of DA-producing neurons in the substantia nigra, and the severe reduction of DA in the basal ganglia brings about most of the symptoms of Parkinson's disease.

Diminished levels of DA cause unnecessary skeletal muscle movements that often interfere with voluntary movement. For instance, the muscles of the upper extremity may alternately contract and relax, causing the hand to shake. This shaking is called ***tremor,*** the most common symptom of Parkinson's disease. The tremor may spread to the ipsilateral lower extremity and then to the contralateral extremities. Motor performance is also impaired by ***bradykinesia*** (*brady* = slow; *kinesis* = motion), in which activities such as shaving, cutting food, and buttoning a shirt take longer and become increasingly more difficult. Muscular movements are performed not only slowly but with decreasing range of motion (***hypokinesia***). For example, as handwriting continues, letters get smaller, become poorly formed, and eventually become illegible. Some muscles may contract continuously, causing ***rigidity*** of the involved body part. Rigidity of the facial muscles gives the face a masklike appearance. The expression is characterized by a wide-eyed, unblinking stare and a slightly open mouth with uncontrolled drooling. Decreased DA production also results in impaired walking, in which steps become shorter and shuffling and arm swing diminishes. There are also changes that lead to stooped posture and loss of postural reflexes, autonomic dysfunction (constipation, retention), sensory complaints (pain, numbness, tingling), and sustained muscle spasms. Vision, hearing, and intelligence are unaffected by the disorder, indicating that Parkinson's disease does not attack the cerebral cortex.

Treatment of the symptoms of Parkinson's disease is directed toward increasing levels of DA. Although people with Parkinson's disease do not manufacture enough DA, injections of it are useless; the blood–brain barrier stops it. However, symptoms are somewhat relieved by a drug developed in the 1960s called levodopa, a precursor of DA. Administered by itself, levodopa may elevate brain levels of DA, causing undesirable side effects such as low blood pressure, nausea, mental changes, and liver dysfunction. Levodopa has been combined with carbidopa, which inhibits the formation of DA outside the brain. The combined drugs diminish the undesirable side effects. The drug amantadine may relieve the symptoms of mild parkinsonism. Anticholinergic drugs such as cycrimine, benztropin mesylate, and ethopropazine have been in use for years to control symptoms of the disease. Bromocriptine stimulates DA receptors and has also been used. In addition, two experimental drugs are being tested; selegiline, which may promote the destruction of DA-producing cells, and tocopherol, which may delay the need for levodopa. Currently, methods are being devised to help drugs pass through the blood–brain barrier by combining them with fat-soluble chemicals. Once the complex passes through the barrier, the fat-soluble chemical breaks down and is excreted, leaving the drug in the brain where it exerts its effect.

In early 1987, physicians in Mexico City transplanted tissue from the adrenal glands into the brain in order to alleviate the symptoms of Parkinson's disease. The adrenal medulla produces epinephrine, a chemical similar to DA. In the procedure, adrenal medullary tissue (chro-

DISORDERS: HOMEOSTATIC IMBALANCES (*continued*)

maffin cells) was transplanted to the caudate nucleus of the brain. It appears that either the transplanted tissue produces DA or stimulates the brain to produce DA. Although the procedure seems to benefit some patients by decreasing the length of time they are incapacitated, decreasing off-time (periods when drugs are not working), and decreasing amounts of medication required, the results among patients in the United States are less dramatic than Mexican patients. Moreover, many patients who have undergone the procedure experience unusual and unexpected behavioral changes immediately after surgery that may persist for several months. These include reduced need for analgesics, altered sleep patterns, delusions, and personality changes. The first transplants of human fetal cells into brains of adults with Parkinson's disease were also performed in 1987. Results are still being evaluated.

Multiple Sclerosis (MS)

Multiple sclerosis (MS) is often called the "great imitator." It is the progressive destruction of the myelin sheaths of neurons in the central nervous system accompanied by disappearance of oligodendrocytes and the proliferation of astrocytes. The sheaths deteriorate to ***scleroses,*** which are hardened scars or plaques, in multiple regions. The destruction of myelin sheaths interferes with the transmission of nerve impulses from one neuron to another, literally short-circuiting conduction pathways. Usually, the first symptoms occur in early adult life. The average age of onset is 33. The frequency of flare-ups is greatest during the first 3 to 4 years of the disease, but a first attack, which may have been so mild as to escape medical attention, may not be followed by another attack for 10 to 20 years.

Among the first symptoms of MS are muscular weakness of one or more extremities; abnormal sensations, such as burning or pins and needles; visual impairment that includes blurring, double vision, and problems with color and light perception; uncoordination; vertigo; and sphincter impairment that results in urinary problems, such as urinary urgency. Following a period of remission during which the symptoms temporarily disappear, a new series of plaques develop, and the victim suffers a second attack. One attack follows another over the years, usually every year or two. Each time the plaques form, some neurons are damaged by the hardening of their sheaths, whereas others are uninjured by their plaques. The result is a progressive loss of function interspersed with remission periods during which the undamaged neurons regain their ability to transmit nerve impulses.

The symptoms of MS depend on the areas of the central nervous system most heavily laden with plaques. Sclerosis of the white matter of the spinal cord is common. As the sheaths of the neurons in the corticospinal tract deteriorate, the patient loses the ability to contract skeletal muscles. Damage to the ascending tracts produces numbness and short-circuits impulses related to position of body parts and flexion of joints. Damage to either set of tracts also destroys spinal cord reflexes.

The clinical course of MS is unpredictable. During typical episodes, symptoms worsen over a period of a few days to two to three weeks and then remit. Relapses occur at an average rate of 0.5 per year during the initial five years, although this rate is highly variable. Some patients experience complete remissions following relapses, whereas others gradually accumulate neurologic problems. Many patients suffer multiple attacks but are never disabled. About 20 percent of MS patients have symptoms and signs that appear slowly and steadily, without a clear relapsing-remitting pattern. Such a course often occurs in late-onset patients (over 40) and is frequently associated with severe disability. About two-thirds of MS patients are ambulatory 25 years after the onset of their disease, one-half will be working 10 years from the onset, and one-third will have unrestricted functions. MS does not predictably shorten life except in a minority of patients who are bedridden or succumb to a urinary tract infection or pneumonia.

Although the etiology of MS is unclear, there is some evidence that it might result from a viral infection that precipitates an autoimmune response. Viruses may trigger the destruction of oligodendrocytes by the antibodies and cytotoxic (killer) cells of the body's immune system. Like other demyelinating diseases, MS is incurable. However, in view of the evidence that it might be an autoimmune disease, immunosuppressive therapy is widely used (glucocorticoids such as prednisone). A recent treatment consists of administering cyclophosphamide (a powerful anticancer drug that suppresses the immune system) in combination with adrenocorticotropic hormone or ACTH (a pituitary gland hormone that stimulates secretion of hormones containing cortisone) to patients with active MS. An experimental therapy for MS is the use of a synthetic protein called copolymer (Cop 1), a decoy that spares destruction of a similar protein in oligodendrocytes. Another treatment is the use of colchicine, a drug used to treat gout. Electrical stimulation of the spinal cord can also improve function in certain patients. Improvement has also been shown in patients who are administered pure oxygen while in a pressure chamber (hyperbaric oxygen). This supports the idea that the destruction of myelin occurs preferentially in parts of the brain that are relatively low in oxygen. Treatment is also directed at management of complications such as spasticity, facial neuralgia and twitching, urinary bladder problems, and constipation.

Dyslexia

Dyslexia (dis-LEK-sē-a; *dys* = difficulty; *lexis* = words) is an impairment of the brain's ability to translate images received from the eyes or ears into understandable language. The condition is unrelated to basic intellectual capacity, but it causes a mysterious difficulty in handling words and symbols. Apparently, some peculiarity in the brain's organizational pattern distorts the ability to read,

DISORDERS: HOMEOSTATIC IMBALANCES (*continued*)

write, and count. Letters in words seem transposed, reversed, or upside down—*dog* becomes *god; b* changes identity with *d;* a sign saying "OIL" inverts into "710." Frequently, dyslectics reread portions of paragraphs, skip words, and change the order of letters in a word. Many dyslectics cannot orient themselves in the three dimensions of space and may show bodily awkwardness.

The exact cause of dyslexia is unknown, since it is unaccompanied by outward scars of detectable neurological damage and its symptoms vary from victim to victim. It occurs almost four times as often among boys as among girls. It has been variously attributed to defective vision, brain damage, abnormal brain development, lead in the air, physical trauma, or oxygen deprivation during birth. A recent theory holds that it might be related to a complex language deficiency involving inability to represent and access the sound of a word in order to help remember it, inability to break down words into components, poor vocabulary development, and difficulty discriminating grammatical differences among words and phrases.

A technique used primarily to diagnose dyslexia is called ***brain electrical activity mapping (BEAM)***. It is a noninvasive procedure that measures and displays the electrical activity of the brain on a color television screen and compares the image produced with a normal image. BEAM may have other diagnostic applications for learning disabilities, schizophrenia, depression, dementia, epilepsy, and early tumor occurrence and recurrence.

Positron emission tomography (PET) scans (see Figure 1-12) are also providing information about dyslexia. For example, PET scans have demonstrated that in persons with dyslexia the left side of the brain is more active than the right, the language region is less active than normal, and the visual discrimination region is more active.

Tay-Sachs Disease

Tay-Sachs disease is a central nervous system affliction that brings death before age 5. The Tay-Sachs gene is carried mostly by individuals descended from the Ashkenazi Jews of Eastern Europe. Approximately 1 in 30 of their offspring will carry the trait (if both parents carry the gene, 1 in 4 children will inherit the disease). The disease involves the neuronal degeneration of the central nervous system because of excessive amounts of a lipid called ganglioside in the neurons of the brain. The substance accumulates because of a deficient lysosomal enzyme. The afflicted child develops normally until the age of 4 to 8 months. Then the symptoms follow a course of progressive degeneration: paralysis, blindness, inability to eat, decubitus ulcers, and death from infection. Amniocentesis can detect the trait prenatally and a blood test can be done to detect the trait postnatally. There is no known cure.

Headache

One of the most common human afflictions is ***headache***, or ***cephalgia*** (*enkephalos* = brain; *algia* = painful condition). Based on origin, two general types are distinguished: intracranial and extracranial. Serious headaches of intracranial origin are caused by brain tumors, blood vessel abnormalities, inflammation of the brain or meninges, decrease in oxygen supply to the brain, and damage to brain cells. Extracranial headaches are related to infections of the eyes, ears, nose, and sinuses and are commonly felt as headaches because of the location of these structures.

Most headaches require no special treatment. Analgesic and tranquilizing compounds are generally effective for tension headaches but not for migraine headaches. Drugs that constrict the blood vessels can be helpful for migraine. Biofeedback training and dietary changes may also be of some help. Tension headaches are the most common form of extracranial headache; they are associated with stress, fatigue, and anxiety; and classically they occur in the occipital and temporal muscles.

Trigeminal Neuralgia (Tic Douloureux)

As noted earlier, pain arising from irritation of the trigeminal (V) nerve is known as ***trigeminal neuralgia***, or ***tic douloureux*** (doo-loo-ROO). The disorder is characterized by brief but extreme pain in the face and forehead on the affected side. The characteristic pain consists of red-hot, needlelike jabs lasting a few seconds and building to a searing pain like a hot poker being dragged or stuck into the face, lasting 10 to 15 seconds. Many patients describe sensitive regions around the mouth and nose that can cause an attack when touched. Eating, drinking, washing the face, and exposure to cold may also bring on an attack.

Treatment may be palliative (relieving symptoms without curing the disease) or surgical. One technique involves alcohol injections directly into the semilunar (Gasserian) ganglion, which controls the trigeminal (V) nerve. This method is superior to open surgery because it is safer. Moreover, it can bring lasting pain relief with preservation of touch sensation in the face and release of the patient 24 hours after treatment. Acupuncture can also provide relief in some patients.

Reye's Syndrome (RS)

Reye's syndrome (RS), first described in 1963 by the Australian pathologist R. Douglas Reye, seems to occur following a viral infection, particularly chickenpox or influenza. Aspirin at normal doses is believed to be a risk factor in the development of RS. The majority of persons affected are children or teenagers. The disease is characterized by vomiting and brain dysfunction (disorientation, lethargy, and personality changes) and may progress to coma. Also, the liver becomes infiltrated with small lipid droplets and loses some of its ability to detoxify ammonia.

Brain dysfunction and death are typically caused by swelling of brain cells. The pressure not only kills the cells directly but also results in hypoxia that kills them indirectly. The survival rate is about 70 percent. Swell-

DISORDERS: HOMEOSTATIC IMBALANCES (*continued*)

ing may result in irreversible brain damage, including mental retardation, in children who survive. Therapy is directed at controlling the swelling.

Alzheimer's Disease (AD)

Alzheimer's (ALTZ-hī-merz) ***disease,*** or ***AD,*** is a disabling neurological disorder that afflicts about 5 percent (about 1 million people) of the population over age 65. Its causes are unknown, its effects are irreversible, and it has no cure. The disease claims over 120,000 lives a year, making it the fourth leading cause of death among the elderly following heart disease, cancer, and stroke.

Victims of AD initially have trouble remembering recent events. Next they become more confused and forgetful, often repeating questions or getting lost while traveling to previously familiar places. Disorientation grows, memories of past events disappear, and there may be episodes of paranoia, hallucination, or violent changes in mood. As their minds continue to deteriorate, they lose their ability to read, write, talk, eat, walk, or take care of themselves. Finally, the disease culminates in dementia, the loss of reason. A person with AD usually dies of some complication that affects bedridden patients, such as pneumonia.

At present, there is no diagnostic test for AD. Autopsy findings, however, clearly show several characteristic pathologies in the brains of Alzheimer's victims. Although there is normally gradual loss of neurons in the brain associated with aging, the rate of loss in persons with AD is greater than normal, especially in regions of the brain that are important for memory and other intellectual processes, such as the cerebral cortex and hippocampus. Another pathological finding is the presence of ***neurofibrillary tangles,*** bundles of fibrous proteins, in the cell bodies of neurons in the cerebral cortex, hippocampus, and brain stem. Also present is ***amyloid,*** pathological protein-rich accumulations. In some cases, amyloid surrounds and invades cerebral blood vessels. This form is called ***cerebrovascular amyloid,*** which collects in the middle muscular layer of the blood vessels and eventually results in hemorrhage. Another form of amyloid is a component of ***neuritic (senile) plaques,*** which consist of abnormal axons and axon terminals that surround amyloid. Neuritic plaques are abundant in the cerebral cortex, hippocampus, and amygdala.

Amyloid forms a protein called A68 which is found in neurofibrillary tangles. Both neurofibrillary tangles and A68 are found in only two groups of adults: persons with AD and persons with Down's syndrome (DS). DS is a genetic defect that is the leading cause of mental retardation (Chapter 29). Of the two proteins, A68 is emerging as a key factor in the onset of AD. In normal infants, A68 is believed to cause programmed death of surplus neurons, a process that is completed by age 2, after the A68 disappears from the brain. However, it reappears in the brains of AD patients. It is suspected that A68 may cause AD by triggering the growth of neuro-fibrillary tangles or mediate programmed cell death as it does in normal infants.

It has recently been learned that a gene on chromosome 21 codes for amyloid, and the extra gene, as found in DS, could lead to excessive amyloid production. Such information suggests that a common genetic defect might be responsible for both AD and DS.

One hypothesis concerning the possible cause of AD centers on the observation that the enzyme choline acetyltransferase is found in very low concentrations in the brains of AD victims. The enzyme is needed for the synthesis of the neurotransmitter acetylcholine (ACh) in axon terminals. Levels of the enzyme are quite low in the cerebral cortex and hippocampus of AD patients owing to the loss of axon terminals in these regions, and thus acetylcholine levels are also low. Decreased levels of acetylcholine are corrected with reduced transmission of nerve impulses, impulses required for memory.

Other hypotheses suggest that AD may be caused by the loss of neurons due to the inheritance of faulty genes, an abnormal accumulation of proteins in the brain, a slow-acting virus, environmental toxins such as aluminum, and decreased blood flow to the brain resulting in inadequate amounts of oxygen and glucose.

Delirium

Delirium (de-LIR-ē-um; *deliria* = off the tract), also called ***acute confusional state (ACS),*** is a transient disorder of abnormal cognition (perception, thinking, and memory) and disordered attention that is accompanied by disturbances of the sleep-wake cycle and psychomotor behavior (hyperactivity or hypoactivity of movements and speech). Delirium is also accompanied by emotions that range from apathy or depression to fear or rage and signs of sympathetic nervous system arousal (tachycardia, sweating, dilated pupils, elevated blood pressure, and pallor or flushing of the face). Among the conditions that cause delirium are brain disorders (infections, tumors, trauma, epilepsy, and stroke); systemic diseases that affect the brain (infections, metabolic disturbances, and cardiovascular diseases); intoxication with medical and recreational drugs and environmental poisons; and withdrawal from substances of abuse such as alcohol and sedative-hypnotic drugs.

Treatment consists of reduction or elimination of drugs, especially anticholinergic drugs; psychiatric evaluation; maintaining proper fluid and electrolyte balance and nutrition; and sedation.

MEDICAL TERMINOLOGY ASSOCIATED WITH THE CENTRAL NERVOUS SYSTEM

Agnosia (ag-NŌ-zē-a; *a* = without; *gnosis* = knowledge) Inability to recognize the significance of sensory stimuli such as auditory, visual, olfactory, gustatory, and tactile.

Analgesia (an-al-JĒ-zē-a; *an* = without; *algia* = painful condition) Pain relief.

Anesthesia (an'-es-THĒ-zē-a; *esthesia* = feeling) Loss of feeling.

Apraxia (a-PRAK-sē-a; *pratto* = to do) Inability to carry out purposeful movements in the absence of paralysis.

Dementia (de-MEN-shē-a; *de* = away from; *mens* = mind) An organic mental disorder that results in permanent or progressive general loss of intellectual abilities such as impairment of memory, judgment, and abstract thinking and changes in personality. The most common cause is Alzheimer's disease. Others are cerebrovascular disease, central nervous system infection, brain tumors or trauma, pernicious anemia, external hydrocephalus, and neurological diseases (Parkinson's disease, multiple sclerosis, and Huntington's chorea).

Electroconvulsive therapy (ECT) (e-lek'-trō-con-VUL-siv THER-a-pē) A form of shock therapy in which convulsions are induced by the passage of a brief electric current through the brain. A patient undergoing ECT is properly anesthetized and given a muscle relaxant to minimize the convulsions. ECT is regarded as an important therapeutic option in the treatment of severe depression and acute mania. Side effects include acute confusional states and memory deficits.

Huntington's chorea (HUNT-ing-tunz kō-RĒ-a; *choreia* = dance) A rare hereditary disease characterized by involuntary jerky movements and mental deterioration that terminates in dementia.

Lethargy (LETH-ar-jē) A condition of functional torpor or sluggishness.

Nerve block Loss of sensation in a region, such as in local dental anesthesia, due to injection of a local anesthetic.

Neuralgia (noo-RAL-jē-a; *neur* = nerve) Attacks of pain along the entire course or branch of a peripheral sensory nerve.

Paralysis (pa-RAL-a-sis) Diminished or total loss of motor function resulting from damage to nervous tissue or a muscle.

Spastic (SPAS-tik; *spas* = draw or pull) An increase in muscle tone (stiffness) associated with an increase in tendon reflexes and abnormal reflexes (Babinski sign).

Stupor (STOO-por) Unresponsiveness from which a patient can be aroused only briefly and by vigorous and repeated stimulation.

Torpor (TOR-por) State of lethargy and sluggishness that precedes stupor, which precedes semicoma, which precedes coma.

Viral encephalitis (VĪ-ral en'-sef-a-LĪ-tis) An acute inflammation of the brain caused by a direct attack by various viruses or by an allergic reaction to any of the many viruses that are normally harmless to the central nervous system. If the virus affects the spinal cord as well, it is called ***encephalomyelitis.***

STUDY OUTLINE

Brain (p. 385)

Principal Parts (p. 385)

1. During embryological development, brain vesicles are formed and serve as forerunners of various parts of the brain.
2. The diencephalon develops into the thalamus and hypothalamus, the telencephalon forms the cerebrum, the mesencephalon develops into the midbrain, the myelencephalon forms the medulla, and the metencephalon develops into the pons and cerebellum.
3. The principal parts of the brain are the brain stem, diencephalon, cerebrum, and cerebellum.

Protection and Coverings (p. 385)

1. The brain is protected by cranial bones, meninges, and cerebrospinal fluid.
2. The cranial meninges are continuous with the spinal meninges and are named dura mater, arachnoid, and pia mater.

Cerebrospinal Fluid (CSF) (p. 386)

1. Cerebrospinal fluid is formed in the choroid plexuses and circulates through the subarachnoid space, ventricles, and central canal. Most of the fluid is absorbed by the arachnoid villi of the superior sagittal blood sinus.
2. Cerebrospinal fluid protects by serving as a shock absorber. It also delivers nutritive substances from the blood and removes wastes.
3. If cerebrospinal fluid accumulates in the ventricles, it is called internal hydrocephalus. If it accumulates in the subarachnoid space, it is called external hydrocephalus.

Blood Supply (p. 389)

1. The blood supply to the brain is via the cerebral arterial circle (circle of Willis).
2. Any interruption of the oxygen supply to the brain can result in weakening, permanent damage, or death of brain cells. Interruption of the mother's blood supply to a child during childbirth before it can breathe may result in paralysis, mental retardation, epilepsy, or death.
3. Glucose deficiency may produce dizziness, convulsions, and unconsciousness.
4. The blood–brain barrier (BBB) is a concept that explains the differential rates of passage of certain material from the blood into the brain.

Brain Stem: Structure and Physiology (p. 390)

1. The medulla oblongata is continuous with the upper part of the spinal cord and contains portions of both motor and sensory tracts. It contains nuclei that are reflex centers for regulation of heart rate, respiratory rate, vasoconstriction, swallowing, coughing, vomiting, sneezing, and hiccuping. It also contains the nuclei of origin for cranial nerves VIII (cochlear and vestibular branches) through XII.
2. The pons is superior to the medulla. It connects the spinal cord with the brain and links parts of the brain with one another by way of tracts. It relays nerve impulses related to voluntary skeletal movements from the cerebral cortex to the cerebellum. It contains the nuclei for cranial nerves V through VII and the vestibular branch of VIII. The reticular formation of the pons contains the pneumotaxic and apneustic centers, which help control respiration.
3. The midbrain connects the pons and diencephalon. It conveys motor impulses from the cerebrum to the cerebellum and spinal cord, sends sensory impulses from cord to thalamus, and regulates auditory and visual reflexes. It also contains the nuclei of origin for cranial nerves III and IV.

Diencephalon (p. 393)

1. The diencephalon consists primarily of the thalamus and hypothalamus.
2. The thalamus is superior to the midbrain and contains nuclei that serve as relay stations for all sensory impulses, except smell, to the cerebral cortex. It also registers conscious recognition of pain and temperature and some awareness of light touch and pressure.
3. The hypothalamus is inferior to the thalamus. It controls and integrates the autonomic nervous system, receives sensory impulses from viscera, connects the nervous and endocrine systems, secretes a variety of regulating hormones (or factors), coordinates mind-over-body phenomena, functions in rage and aggression, controls body temperature, regulates food and fluid intake, maintains the waking state and sleep patterns, and acts as a self-sustained oscillator that drives biological rhythms.

Cerebrum (p. 396)

1. The cerebrum is the largest part of the brain. Its cortex contains convolutions, fissures, and sulci.
2. The cerebral lobes are named the frontal, parietal, temporal, and occipital.
3. The white matter is under the cortex and consists of myelinated axons running in three principal directions.
4. The basal ganglia (cerebral nuclei) are paired masses of gray matter in the cerebral hemispheres. They help to control muscular movements.
5. The limbic system is found in the cerebral hemispheres and diencephalon. It functions in emotional aspects of behavior and memory.
6. The sensory areas of the cerebral cortex are concerned with the interpretation of sensory impulses. The motor areas are the regions that govern muscular movement. The association areas are concerned with emotional and intellectual processes.
7. Positron emission tomography (PET) helps identify which parts of the brain are involved in specific sensory and motor activities.
8. Brain waves generated by the cerebral cortex are recorded as an electroencephalogram (EEG). It may be used to diagnose epilepsy, infections, and tumors.

Brain Lateralization (Split-Brain Concept (p. 404)

1. Recent research indicates that the two hemispheres of the brain are not bilaterally symmetrical, either anatomically or functionally.
2. The left hemisphere is more important for right-handed control, spoken and written language, numerical and scientific skills, and reasoning.
3. The right hemisphere is more important for left-handed control, musical and artistic awareness, space and pattern perception, insight, imagination, and generating mental images of sight, sound, touch, taste, and smell.

Cerebellum (p. 404)

1. The cerebellum occupies the inferior and posterior aspects of the cranial cavity. It consists of two hemispheres and a central, constricted vermis.
2. It is attached to the brain stem by three pairs of cerebellar peduncles.
3. The cerebellum functions in the coordination of skeletal muscles and the maintenance of normal muscle tone and body equilibrium.

Neurotransmitters in the Brain (p. 406)

1. There are numerous substances that are either known or suspected neurotransmitters that can facilitate, excite, or inhibit postsynaptic neurons.
2. Examples of neurotransmitters include acetylcholine (ACh), glutamic acid, aspartic acid, norepinephrine (NE), dopamine (DA), serotonin (5-HT), gamma aminobutyric acid (GABA), and glycine.
3. Neuropeptides that act as natural painkillers in the body are enkephalins, endorphins, and dynorphin.
4. Other neuropeptides serve as hormones or other regulators of physiological responses. Examples include angiotensin, cholecystokinin, and regulating hormones (or factors) produced by the hypothalamus.

Cranial Nerves (p. 408)

1. Twelve pairs of cranial nerves originate from the brain.
2. The pairs are named primarily on the basis of distribution and numbered by order of attachment to the brain. (See Exhibit 14-3 for summary of cranial nerves.)

Aging and the Nervous System (p. 408)

1. Age-related effects involve loss of neurons and decreased capacity for sending nerve impulses.
2. Degenerative changes also affect the sense organs.

Developmental Anatomy of the Nervous System (p. 409)

1. The development of the nervous system begins with a thickening of ectoderm called the neural plate.
2. The parts of the brain develop from primary and secondary vesicles.

Disorders: Homeostatic Imbalances (p. 414)

1. A cerebrovascular accident (CVA), also called a stroke, involves brain tissue destruction due to hemorrhage, thrombosis, or atherosclerosis.
2. Epilepsy results from irregular electrical discharges of brain cells and may be diagnosed by an EEG. Depending on the

form of the disease, the victim experiences degrees of motor, sensory, or psychological malfunction.
3. A transient ischemic attack (TIA) is an episode of temporary, focal, nonconvulsive cerebral dysfunction due to interference of the blood supply to the brain.
4. Brain tumors are neoplasms within the cranium.
5. Poliomyelitis is a viral infection that results in paralysis.
6. Cerebral palsy (CP) refers to a group of motor disorders caused by damage to motor centers of the cerebral cortex, cerebellum, or basal ganglia during fetal development, childbirth, or early infancy.
7. Parkinson's disease (PD) is a progressive degeneration of the dopamine (DA)-producing neurons in the substantia nigra resulting in insufficient DA in the basal ganglia.
8. Multiple sclerosis (MS) is the destruction of myelin sheaths of the neurons of the central nervous system. Impulse transmission is interrupted.
9. Dyslexia involves an inability of an individual to comprehend written language. It may be diagnosed by using brain electrical activity mapping (BEAM).
10. Tay-Sachs disease is an inherited disorder that involves neurological degeneration of the CNS because of excessive amounts of ganglioside.
11. Headaches are of two types: intracranial and extracranial.
12. Irritation of the trigeminal (V) nerve is known as trigeminal neuralgia (tic douloureux).
13. Reye's syndrome (RS) is characterized by vomiting, brain dysfunction, and liver damage.
14. Alzheimer's disease (AD) is a disabling neurological disorder of the elderly that involves widespread intellectual impairment, personality changes, and sometimes delirium.
15. Delirium is a transient disorder of cognition and disordered attention accompanied by disturbances of the sleep-wake cycle and psychomotor behavior.

REVIEW QUESTIONS

1. Identify the four principal parts of the brain and the components of each, where applicable. What is the origin of each of the parts? (p. 385)
2. Describe the location of the cranial meninges. What is an extradural hemorrhage? (p. 385)
3. Where is cerebrospinal fluid (CSF) formed? Describe its circulation. Where is CSF absorbed? (p. 386)
4. Distinguish between internal and external hydrocephalus. (p. 389)
5. Explain the importance of oxygen and glucose to brain cells. (p. 389)
6. What is the blood–brain barrier (BBB)? Describe the passage of several drugs with respect to the BBB. (p. 389)
7. Describe the location and structure of the medulla. Define decussation of pyramids. Why is it important? List the principal functions of the medulla. (p. 390)
8. Describe the location and structure of the pons. What are its functions? (p. 392)
9. Describe the location and structure of the midbrain. What are its functions? (p. 393)
10. Describe the location and structure of the thalamus. List its functions. (p. 393)
11. Where is the hypothalamus located? Explain some of its major functions. (p. 395)
12. Where is the cerebrum located? Describe the cortex, convolutions, fissures, and sulci of the cerebrum. (p. 396)
13. List and locate the lobes of the cerebrum. How are they separated from one another? What is the insula? (p. 398)
14. Describe the organization of cerebral white matter. Be sure to indicate the function of each group of fibers. (p. 398)
15. What are basal ganglia? Name the important basal ganglia and list the function of each. Describe the effects of damage on the basal ganglia. (p. 398)
16. Define the limbic system. Explain several of its functions. (p. 400)
17. What is meant by a sensory area of the cerebral cortex? List, locate, and give the function of each sensory area. (p. 401)
18. Describe the principle of positron emission tomography (PET). What are its clinical applications? (p. 403)
19. What is meant by a motor area of the cerebral cortex? List, locate, and give the function of each motor area. (p. 403)
20. What conditions may result from damage to sensory or motor speech areas? (p. 403)
21. What is an association area of the cerebral cortex? What are its functions? (p. 403)
22. Define an electroencephalogram (EEG). List the principal waves recorded and the importance of each. What is the diagnostic value of an EEG? (p. 403)
23. Describe brain lateralization (split-brain concept). (p. 404)
24. Describe the location of the cerebellum. List the principal parts of the cerebellum. (p. 404)
25. What are cerebellar peduncles? List and explain the function of each. (p. 405)
26. Explain the functions of the cerebellum. Describe some effects of cerebellar damage. (p. 406)
27. What are neurotransmitters? How do they affect postsynaptic neurons? (p. 406)
28. Give some examples of neurotransmitters and indicate their functions. (p. 409)
29. Describe the neuropeptides that act as natural painkillers in the body. Give their particular locations in the body and their functions. (p. 409)
30. Define a cranial nerve. How are cranial nerves named and numbered? Distinguish between a mixed and a sensory cranial nerve. (p. 408)
31. For each of the 12 pairs of cranial nerves, list (a) its name, number, and type; (b) its location; and (c) its function. In addition, list the effects of damage, where applicable. (p. 411)
32. Describe the effects of aging on the nervous system. (p. 408)
33. Describe the development of the nervous system. (p. 409)
34. Define each of the following: cerebrovascular accident (CVA), epilepsy, transient ischemic attack (TIA), brain tumors, poliomyelitis, cerebral palsy (CP), Parkinson's disease (PD), multiple sclerosis (MS), dyslexia, Tay-Sachs disease, headache, trigeminal neuralgia (tic douloureux), Reye's syndrome (RS), Alzheimer's disease (AD), and delirium. (p. 414)
35. What is brain electrical activity mapping (BEAM)? What are its clinical applications? (p. 418)
36. Refer to the glossary of medical terminology associated with the nervous system. Be sure that you can define each term. (p. 420)

SELECTED READINGS

Barr, M. L., and J. A. Kiernan. *The Human Nervous System,* 5th ed. Philadelphia: Lippincott, 1988.

Begley, S., J. Carey, and R. Sawhill. "How the Brain Works," *Newsweek,* 7 February 1983.

"The Brain," *Scientific American,* September 1979. (Entire issue devoted to the brain and the human nervous system.)

Brown, M. R., and L. A. Fisher. "Brain Peptides as Intercellular Messengers," *Journal of the American Medical Association,* 9 March 1984.

Bruno, M., and S. Katz. "New Hope for Hurt Nerves," *Newsweek,* 7 October 1985.

Burden, N. "Regional Anesthesia: What Patients and Nurses Need to Know," *RN,* May 1988.

Carpenter, M. B. *Core Text of Neuroanatomy,* 3rd ed. Baltimore: Williams & Wilkins, 1985.

Edwards, D. D. "A Common Medical Denominator," *Science News,* 25 January 1986.

Gluhbegovic, N., and T. H. Williams. *The Human Brain: A Photographic Guide.* New York: Harper & Row, 1980.

Goldfinger, S. E. (ed). "Alzheimer's Disease," *Harvard Medical School Health Letter,* April 1988.

———. "Shingles," *Harvard Medical School Health Letter,* June 1984.

———. "The Endorphins—The Body's Own Opiates," *Harvard Medical School Health Letter,* January 1983.

Gorelick, P. B. "Clues to the Mystery of Multiple Sclerosis," *Postgraduate Medicine,* March 1989.

Guyton, A. C. *Basic Neuroscience: Anatomy and Physiology.* Philadelphia: Saunders, 1987.

Kiley, M. J. (ed). "A Long Goodbye," *Mayo Clinic Health Letter,* January 1988.

———. "Parkinson's Disease," *Mayo Clinic Health Letter,* May 1989.

Koller, W. C. "Diagnosis and Treatment of Parkinson's Disease," *Modern Medicine,* May 1989.

Nauta, W. J. H., and M. Feirtag. *Fundamental Neuroanatomy.* New York: W. H. Freeman, 1985.

Thompson, R. F. *The Brain: An Introduction to Neuroscience.* New York: W. H. Freeman, 1985.

Weiss, R. "Muscular Dystrophy Protein Identified," *Science News,* 2 January 1988.

Chapter 15

The Sensory, Motor, and Integrative Systems

Chapter Contents at a Glance

Student Objectives

1. Define a sensation and list the characteristics of a sensation.
2. Classify receptors on the basis of location, stimulus detected, and simplicity or complexity.
3. List the location and function of the receptors for tactile sensations (touch, pressure, vibration), thermoreceptive sensations (heat and cold), and pain.
4. Distinguish somatic, visceral, referred, and phantom pain.
5. Identify the proprioceptive receptors and indicate their functions.
6. Discuss the origin, neuronal components, and destination of the posterior column and spinothalamic pathways.
7. Explain the neural pathways for pain and temperature; light touch and pressure; and discriminative touch, proprioception, and vibration.
8. Describe how sensory input and motor responses are linked in the central nervous system.
9. Compare the location and functions of the pyramidal and extrapyramidal motor pathways.
10. Compare integrative functions such as memory, wakefulness, and sleep.

Having examined the structure of the nervous system and its activities, we will now see how its different parts cooperate in performing its three essential functions: (1) receiving sensory information; (2) transmitting motor impulses that result in movement or secretion; and (3) integration, an activity that deals with memory, sleep, and emotions.

SENSATIONS

The central nervous system requires a continual flow of information to maintain homeostasis and initiate appropriate responses to changes in the internal and external environments. At any given time, our brains receive and respond to many varieties of information. However, we are aware only of the information that we consciously focus upon. The central nervous system selects only those bits of information that are important for the moment, and it is only those bits of information that are brought to our conscious level. There is no question that we would collapse into nervous wrecks if our consciousness was forced to deal with all the information arriving at once. The conscious mind is turned off to protect itself from overstimulation.

Your ability to sense stimuli is vital to your survival. If pain could not be sensed, burns would be common. An inflamed appendix or stomach ulcer would progress unnoticed. A lack of sight would increase the risk of injury from unseen obstacles, a loss of smell would allow a harmful gas to be inhaled, a loss of hearing would prevent recognition of automobile horns, and a lack of taste would allow toxic substances to be ingested. In short, if you could not "sense" your environment and make the necessary homeostatic adjustments, you could not survive very well on your own.

Definition

In its broadest context, ***sensation*** refers to a state of awareness of external or internal conditions of the body. ***Perception*** refers to the conscious registration of a sensory stimulus. For a sensation to occur, four prerequisites must be satisfied.

1. A ***stimulus,*** or change in the environment, capable of initiating a nerve impulse by the nervous system must be present.
2. A ***receptor*** or ***sense organ*** must pick up the stimulus and transduce (convert) it to a nerve impulse by way of a generator or receptor potential (described shortly). A receptor or sense organ may be viewed as specialized nervous tissue that is extremely sensitive to certain types of changes in internal or external conditions. Such changes are termed stimuli.
3. The impulse must be ***conducted*** along a neural pathway from the receptor or sense organ to the brain.
4. A region of the brain must ***translate*** the impulse into a sensation.

Receptors are capable of converting a specific stimulus into a nerve impulse by way of generator or receptor potentials. The stimulus may be light, heat, pressure, mechanical energy, or chemical energy.

When an adequate stimulus is applied to a receptor, that is, a specialized neuronal ending, it responds by altering its membrane's permeability to small ions. This results in a change in the resting membrane potential called a ***generator potential.***

Generator potentials differ from nerve action potentials in several ways. A generator potential is a localized response that decreases in intensity as it travels along a nerve fiber, whereas a nerve action potential is propagated at a constant and maximum strength. A generator potential is a graded response; that is, within limits, the stronger and more frequent the stimulus, the greater the magnitude of the generator potential. A nerve action potential obeys the all-or-none principle. A generator potential usually lasts longer than 1 to 2 msec; a nerve action potential does not. A generator potential does not have a refractory period, whereas a nerve action potential has one that lasts for about 1 msec. This means that if a second stimulus is applied to a receptor before a generator potential resulting from the first stimulus disappears, the second stimulus can add to the effect of the first, producing an even greater generator potential. Thus, summation in producing generator potentials is possible, but summation of nerve action potentials is not. The generator potential only travels a few millimeters before dying out.

When a generator potential reaches threshold, it initiates a nerve action potential. The function of a generator potential is to convert a stimulus into a nerve action potential.

A phenomenon that has many characteristics similar to a generator potential is called a ***receptor potential.*** When a receptor cell connected to a neuron via a synapse is adequately stimulated, the receptor responds by depolarization of its membrane. This depolarization is called a receptor potential. Once developed, a receptor potential stimulates the release of neurotransmitters from a receptor cell, which alters the permeability of the neuron's membrane. If the neuron *becomes* depolarized to threshold, a nerve action potential is triggered.

A receptor may be quite simple. It may consist of the dendrites of a single neuron in the skin that are sensitive to pain stimuli; or it may be contained in a complex organ such as the eye. Regardless of complexity, all sense receptors contain the dendrites of sensory neurons. The dendrites occur either alone or in close association with specialized cells of other tissues.

Receptors are at the same time very excitable and very specialized. Except for pain receptors, each has a low threshold of response to its specific stimulus and a high threshold to all others.

Once a stimulus is received by a receptor and converted

into a nerve impulse by way of a receptor or generator potential, the impulse is conducted along an afferent pathway that enters either the spinal cord or the brain. Many sensory impulses are conducted to the sensory areas of the cerebral cortex. It is in this region that stimuli produce conscious sensations. Sensory impulses that terminate in the spinal cord or brain stem can initiate motor activities but typically do not produce conscious sensations. The thalamus detects pain and temperature sensations but cannot distinguish the intensity or location from which they arise. This is a function of the cerebrum.

Characteristics

Most conscious sensations or perceptions occur in the cortical regions of the brain. In other words, you see, hear, and feel in the brain. You seem to see with your eyes, hear with your ears, and feel pain in an injured part of your body only because the cortex interprets the sensation as coming from the stimulated sense receptor. The term ***projection*** describes this process by which the brain refers sensations to their point of stimulation.

A second characteristic of many sensations is ***adaptation,*** that is, a decrease in sensitivity to continued stimuli. In fact, the perception of a sensation may actually disappear even though the stimulus is still being applied. For example, when you first get into a tub of hot water, you probably feel a burning sensation, but soon the sensation decreases to one of comfortable warmth even though the stimulus (hot water) is still present. In time, the sensation of warmth disappears completely. Other examples of adaptation include placing a ring on your finger, putting on your shoes or hat, sitting on a chair, and pushing your glasses up onto the top of your head. Adaptation results from a change in a receptor, a change in a structure associated with a receptor, or inhibitory feedback from the brain. Receptors vary in their ability to adapt. ***Rapidly adapting (phasic) receptors,*** such as those associated with pressure, touch, and smell, adapt very quickly. Such receptors play a major role in signaling changes in a particular sensation. ***Slowly adapting (tonic) receptors,*** such as those associated with pain, body position, and detecting chemicals in blood, adapt slowly. These receptors are important in signaling information regarding steady states of the body.

Sensations may also be characterized by ***afterimages;*** that is, some sensations persist even though the stimulus has been removed. This phenomenon is the reverse of adaptation. One common example of afterimage occurs when you look at a bright light and then look away or close your eyes. You still see the light for several seconds or minutes afterward.

Another characteristic of sensations is ***modality:*** the specific type of sensation felt. The sensation may be one of temperature change, pain, pressure, touch, body position, equilibrium, hearing, vision, smell, or taste. In other words, the distinct property by which one sensation may be distinguished from another is its modality.

Classification of Receptors

Location

One convenient method of classifying receptors is by their location. ***Exteroceptors*** (eks'-ter-ō-SEP-tors) provide information about the external environment. They are sensitive to stimuli outside the body and transmit sensations of hearing, sight, smell, taste, touch, pressure, temperature, and pain. Exteroceptors are located at or near the surface of the body.

Visceroceptors (vis'-er-ō-SEP-tors), or ***enteroceptors,*** provide information about the internal environment. These sensations arise from within the body and may be felt as pain, pressure, fatigue, hunger, thirst, and nausea. Visceroceptors are located in blood vessels and viscera.

Proprioceptors (prō'-prē-ō-SEP-tors) provide information about body position and movement. Such sensations give us information about muscle tension, the position and activity of our joints, and equilibrium. These receptors are located in muscles, tendons, joints, and the internal ear.

Stimulus Detected

Another method of classifying receptors is by the type of stimuli they detect. ***Mechanoreceptors*** detect mechanical deformation of the receptor itself or in adjacent cells. Stimuli so detected include those related to touch, pressure, vibration, proprioception, hearing, equilibrium, and blood pressure. ***Thermoreceptors*** detect changes in temperature. ***Nociceptors*** detect pain, usually as a result of physical or chemical damage to tissues. ***Photoreceptors*** detect light on the retina of the eye. ***Chemoreceptors*** detect taste in the mouth, smell in the nose, and chemicals in body fluids, such as oxygen, carbon dioxide, water, and glucose.

Simplicity or Complexity

As will be described shortly, receptors may also be classified according to the simplicity or complexity of their structure and the neural pathway involved. ***Simple receptors*** and neural pathways are associated with ***general senses.*** The receptors for general sensations are numerous and widespread. Examples include cutaneous sensations such as touch, pressure, vibration, heat, cold, and pain. ***Complex receptors*** and neural pathways are associated with ***special senses.*** The receptors for each special sense are found in only one or two specific areas of the body. Among the special senses are smell, taste, sight, equilibrium, and hearing.

GENERAL SENSES

Cutaneous Sensations

Cutaneous (*cuta* = skin) ***sensations*** include tactile sensations (touch, pressure, vibration), thermoreceptive sensations (cold and heat), and pain. The receptors for these sensations are in the skin, connective tissue under the skin, mucous membranes, and the ends of the gastrointestinal tract.

The cutaneous receptors are distributed over the body surface in such a way that certain parts of the body are densely populated with receptors and other parts contain only a few. Such an unequal distribution of receptors is called ***punctate distribution.*** Areas of the body that have few cutaneous receptors are insensitive; those containing many are very sensitive.

This type of distribution can be demonstrated in the skin by using the ***two-point discrimination test*** for touch. A compass is applied to the skin, and the distance in millimeters between the two points of the compass is varied. The subject then indicates when two points are felt and when only one is felt. The compass may be placed on the tip of the tongue, an area where receptors are very densely packed. At a distance of 1.4 mm (0.06 inch), the points are able to stimulate two different receptors, and the subject feels touched by two objects. If the distance is less than 1.4 mm, the subject feels only one point even though both points are touching the tongue, because the points are so close together that they reach only one receptor. If the compass is placed on the back of the neck, the subject feels two distinctly different points only if the distance between them is 36.2 mm (1.43 inches) or greater, because the receptors are few in number and far apart. The results of this test indicate that the more sensitive the area, the closer the compass points may be placed and still be felt separately. The following order for these receptors, from greatest sensitivity to least, has been established: tip of tongue, tip of finger, side of nose, back of hand, and back of neck.

Cutaneous receptors have simple structures. They consist of the dendrites of sensory neurons that may or may not be enclosed in a capsule of epithelial or connective tissue. Nerve impulses generated by cutaneous receptors pass along somatic afferent neurons in spinal and cranial nerves, through the thalamus, to the general sensory area of the parietal lobe of the cortex (see Figures 15-5 and 15-6).

Tactile Sensations

Even though the ***tactile*** (*tact* = touch) ***sensations*** are divided into separate sensations (touch, pressure, and vibration), they are all detected by the same types of receptors, called mechanoreceptors, which are receptors subject to deformation.

▪ Touch ***Touch sensations*** generally result from stimulation of tactile receptors in the skin or tissues immediately beneath the skin. ***Light touch*** refers to the ability to perceive that something has touched the skin, although its exact location, shape, size, or texture cannot be determined. ***Discriminative touch*** refers to the ability to recognize exactly what point of the body is touched.

Tactile receptors for touch include hair root plexuses, free nerve endings, tactile discs, corpuscles of touch, and type II cutaneous mechanoreceptors (Figure 15-1). ***Hair***

FIGURE 15-1 **Structure and location of cutaneous receptors.**

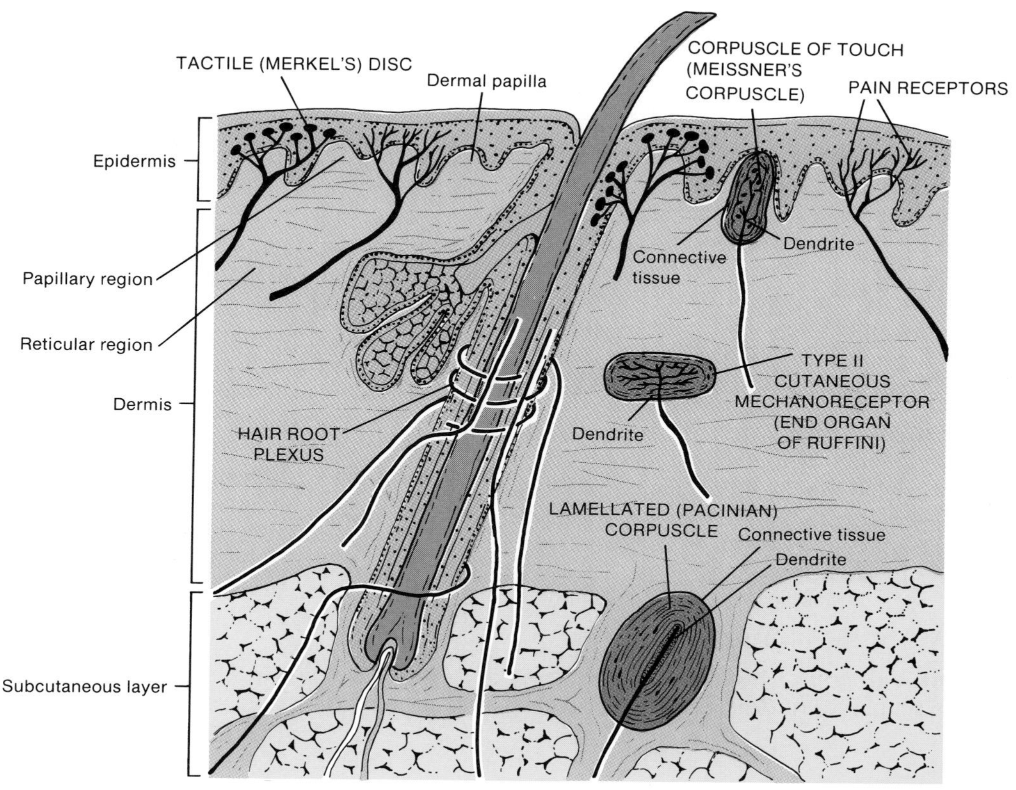

root plexuses are dendrites arranged in networks around hair follicles. They are not surrounded by supportive or protective structures. If a hair shaft is moved, the dendrites are stimulated. Hair root plexuses detect movements mainly on the surface of the body when hairs are disturbed, the hair functioning as a lever.

Other receptors that are not surrounded by supportive or protective structures are called ***free (naked) nerve endings.*** Free nerve endings are found everywhere in the skin and many other tissues. Although they are very important pain receptors, free nerve endings also respond to objects that are in continuous contact with the skin, such as clothing.

Tactile, or ***Merkel's*** (MER-kelz) ***discs,*** are modified epidermal cells in the stratum basale of hairless skin. Their basal ends are in contact with dendrites of sensory neurons. Tactile discs are distributed in many of the same locations as corpuscles of touch and also function in discriminative touch.

Corpuscles of touch, or ***Meissner's*** (MĪS-nerz) ***corpuscles,*** are egg-shaped receptors for discriminative touch containing a mass of dendrites enclosed by connective tissue. They are located in the dermal papillae of the skin and are most numerous in the fingertips, palms of the hands, and soles of the feet. They are also abundant in the eyelids, tip of the tongue, lips, nipples, clitoris, and tip of penis.

Type II cutaneous mechanoreceptors, or ***end organs of Ruffini,*** are embedded deeply in the dermis and in deeper tissues of the body. They detect heavy and continuous touch sensations.

▪ Pressure ***Pressure sensations*** generally result from stimulation of tactile receptors in deeper tissues and are longer lasting and have less variation in intensity than touch sensations. Pressure is really sustained touch. Moreover, pressure is felt over a larger area than touch.

Pressure receptors are free nerve endings, type II cutaneous mechanoreceptors, and lamellated corpuscles. ***Lamellated,*** or ***Pacinian*** (pa-SIN-ē-an), ***corpuscles*** (Figure 15-1) are oval structures composed of a capsule resembling an onion that consist of connective tissue layers enclosing dendrites. Lamellated corpuscles are located in the subcutaneous tissue under the skin, the deep submucosal tissues that lie under mucous membranes, in serous membranes, around joints and tendons, in the perimysium of muscles, in the mammary glands, in the external genitalia of both sexes, and in certain viscera.

▪ Vibration ***Vibration sensations*** result from rapidly repetitive sensory signals from tactile receptors.

The receptors for vibration sensations are corpuscles of touch and lamellated corpuscles. Whereas corpuscles of touch detect low-frequency vibration, lamellated corpuscles detect higher-frequency vibration.

Thermoreceptive Sensations

The ***thermoreceptive*** (*therm* = heat) ***sensations*** are hot and cold. The exact nature of ***thermoreceptive receptors*** is not known, but they might be free (naked) nerve endings.

Pain Sensations

Pain is indispensable for a normal life. It provides us with information about tissue-damaging (noxious) stimuli and thus often enables us to protect ourselves from greater damage. It is pain that initiates our search for medical assistance, and it is our subjective description and indication of the location of the pain that helps to pinpoint the underlying cause of disease.

The receptors for ***pain,*** called ***nociceptors*** (nō'-sē-SEP-tors; (*noci* = harmful), are simply free (naked) nerve endings, the branching ends of the dendrites of certain sensory neurons (Figure 15-1). Pain receptors are found in practically every tissue of the body. They may respond to any type of stimulus if it is strong enough to cause tissue damage. When stimuli for other sensations, such as touch, pressure, heat, and cold, reach a certain threshold, they stimulate the sensation of pain as well. Excessive stimulation of a sense organ causes pain. Additional stimuli for pain receptors include excessive distension or dilation of a structure, prolonged muscular contractions, muscle spasms, inadequate blood flow to an organ, or the presence of certain chemical substances. Pain receptors, because of their sensitivity to all stimuli, perform a protective function by identifying changes that may endanger the body. Pain receptors adapt only slightly or not at all. Adaptation is the decrease or disappearance of the perception of a sensation even though the stimulus is still present. If there were adaptation to pain, it would cease to be sensed and irreparable damage could result.

Sensory impulses for pain are conducted to the central nervous system along spinal and cranial nerves (see Figure 15-5). The lateral spinothalamic tracts of the spinal cord relay impulses to the thalamus. From here the impulses may be relayed to the postcentral gyrus of the parietal lobe. Recognition of the kind and intensity of most pain is ultimately localized in the cerebral cortex. Some awareness of pain occurs at subcortical levels such as the thalamus.

Pain may be classified on the basis of speed of onset and duration: acute and chronic. ***Acute pain*** occurs very rapidly, usually within 0.1 second after a stimulus is applied, and is not felt in deeper tissues of the body. This type of pain is also known as sharp, fast, and pricking pain. The pain felt from a needle puncture or knife cut to the skin are examples of acute pain. Impulses for acute pain are carried along A fibers. ***Chronic pain,*** by contrast, begins after a second or more and then gradually increases over a period of several seconds or minutes. This type of pain may be excruciating and is also referred to as burning, aching, throbbing, and slow pain. Chronic pain can occur both in the skin and deeper tissues or internal organs. An example is the pain associated with a toothache. Impulses for chronic pain are carried along C fibers.

Pain may be divided into two types on the basis of the location of stimulated receptors: somatic and visceral. ***So-***

matic pain arises from stimulation of receptors in the skin, in which it is called ***superficial somatic pain,*** or from stimulation of receptors in skeletal muscles, joints, tendons, and fascia, then called ***deep somatic pain. Visceral pain*** results from stimulation of receptors in the viscera.

Although receptors for somatic and visceral pain are similar, viscera do not evoke the same pain response as somatic tissue. For example, highly *localized* damage to certain viscera, such as cutting the intestine in two in a patient who is awake, causes very little, if any, pain. But, if stimulation is *diffuse,* involving large areas, visceral pain can be severe. Such stimulation might include distension, spasms, or ischemia. There are even some viscera that are almost entirely insensitive to pain of any type. Examples are the parenchyma of the liver and alveoli of the lungs.

The ability of the cerebral cortex to locate the origin of pain is related to past experience. In most instances of somatic pain and in some instances of visceral pain, the cortex accurately projects the pain back to the stimulated area. If you burn your finger, you feel the pain in your finger. If the pleural membranes are inflamed, you experience pain in the chest. In most instances of visceral pain, however, the sensation is not projected back to the point of stimulation. Rather, the pain may be felt in or just under the skin that overlies the stimulated organ. The pain may also be felt in a surface area far from the stimulated organ. This phenomenon is called ***referred pain.*** In general, the area to which the pain is referred and the visceral organ involved receive their innervation from the same segment of the spinal cord. Consider the following example. Afferent fibers from the heart as well as from the skin over the heart and along the medial aspect of the left upper extremity enter spinal cord segments T1–T4. Thus, the pain of a heart attack is typically felt in the skin over the heart and along the left arm. Figure 15-2 illustrates cutaneous regions to which visceral pain may be referred.

CLINICAL APPLICATION: PHANTOM PAIN

A kind of pain frequently experienced by patients who have had a limb amputated is called *phantom pain (phantom limb sensation).* They still experience sensations such as itching, pressure, tingling, or pain in the extremity as if the limb were still there. This probably occurs because the remain-

FIGURE 15-2 Referred pain. The colored parts of the diagrams indicate cutaneous areas to which visceral pain is referred. (a) Anterior view. (b) Posterior view.

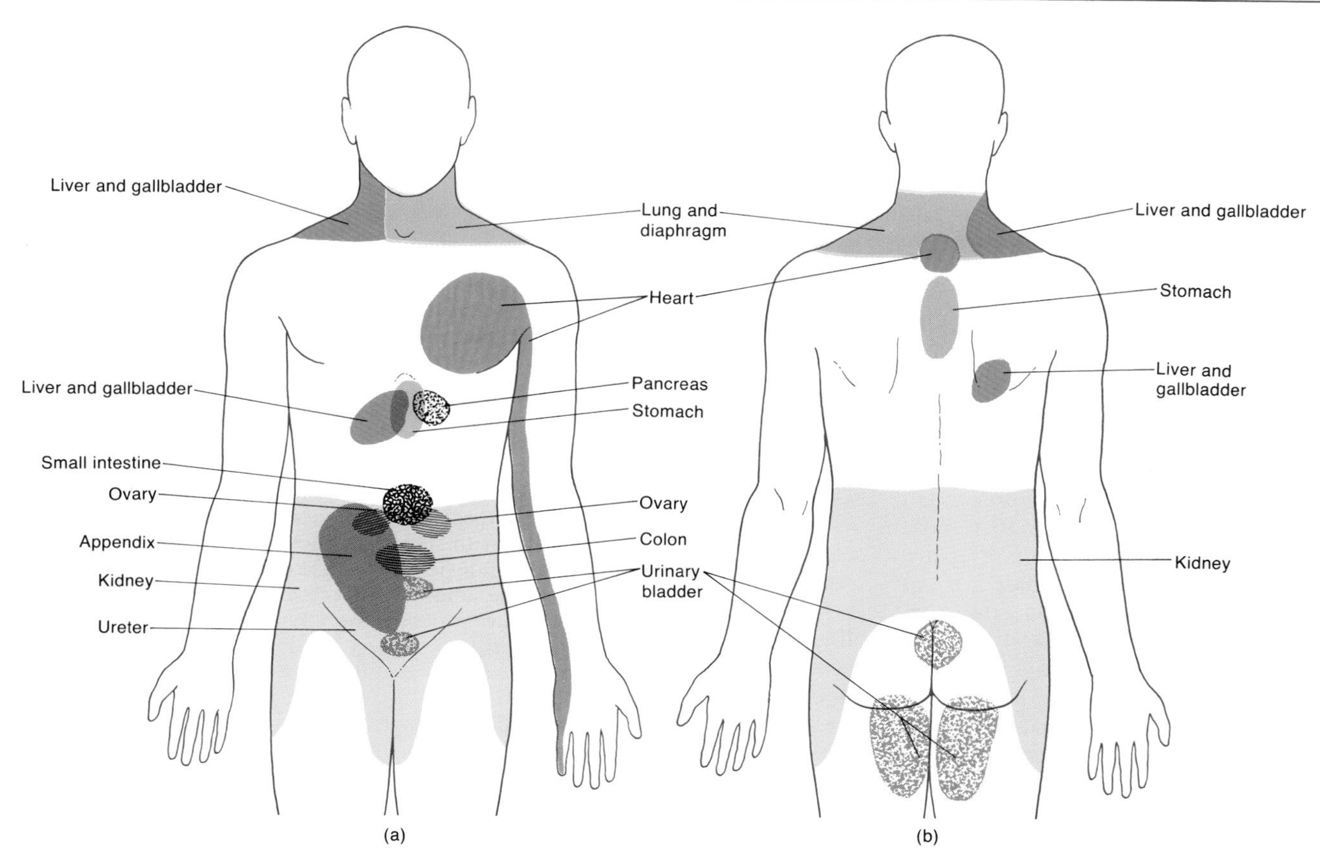

ing proximal portions of the sensory nerves that previously received nerve impulses from the limb are being stimulated by the trauma of the amputation. Stimuli from these nerves are interpreted by the brain as coming from the nonexistent (phantom) limb.

Pain sensations may be controlled by interrupting the pain impulse between the receptors and the interpretation centers of the brain. This may be done chemically, surgically, or by other means. Most pain sensations respond to pain-reducing drugs, which, in general, act to inhibit nerve impulse conduction at synapses.

Occasionally, however, pain may be controlled only by surgery. The purpose of surgical treatment is to interrupt the pain impulse somewhere between the receptors and the interpretation centers of the brain by severing the sensory nerve, its spinal root, or certain tracts in the spinal cord or brain. ***Sympathectomy*** is excision of portions of the neural tissue from the autonomic nervous system; ***cordotomy*** is severing a spinal cord tract, usually the lateral spinothalamic; ***rhizotomy*** is the cutting of sensory nerve roots; ***prefrontal lobotomy*** is the destruction of the tracts that connect the thalamus with the frontal lobe of the cerebral cortex. In each instance, the pathway for pain is severed so that pain impulses are no longer conducted to the cortex.

Another method of inhibiting pain impulses is ***acupuncture*** (*acus* = needle; *pungere* = sting). Needles are inserted through selected areas of the skin and then twirled by the acupuncturist or by a mechanical device. After 20 to 30 minutes, pain is deadened for 6 to 8 hours. The location of needle insertion depends on the part of the body the acupuncturist wishes to anesthetize. To pull a tooth, a needle is inserted in the web between thumb and index finger. For a tonsillectomy, a needle is inserted approximately 5 cm (2 inches) above the wrist. For removal of a lung, a needle is placed in the forearm midway between wrist and elbow.

Acupuncture probably works by taking advantage of the body's natural inhibitory influences that can normally block pain pathways. For example, it has been established that sensory pain fibers in posterior root ganglia release a neurotransmitter called substance P. This substance is required by neurons to produce sensations of pain. Some neurons near those in the pain pathway release enkephalin. The enkephalin released by these small neurons blocks the release of substance P from the nerve terminals of the sensory pain fibers and therefore inhibits pain transmission to the brain. Acupuncture enhances the release of these inhibitory substances, such as enkephalin, from various locations, and these substances can then be carried by the cardiovascular system to the pain fibers. The onset of the effect of acupuncture is delayed until the enkephalin level rises to inhibitory levels. Similarly, the effect of acupuncture lingers after the twirling or vibration of the needles stops.

Currently, acupuncture is used in the United States mostly for childbirth, trigeminal neuralgia (tic douloureux), arthritis, and other nonsurgical conditions. It is also being used in spinal cord injury rehabilitation to activate useful motion in limbs, provided that there is still some intact neural tissue.

CLINICAL APPLICATION: ANESTHESIA

During certain surgical or diagnostic procedures, *anesthesia* (an'-es-THĒ-zē-a; *an* = without; *algia* = painful condition) is required to remove the sensation of pain while still maintaining the stability of the patient's organ systems. Two commonly used forms of anesthesia are general and spinal. *General anesthesia* not only removes the sensation of pain but also produces unconsciousness and sometimes muscular relaxation. Usually, general anesthesia involves use of more than one drug. First, an injection of a drug is given to induce unconsciousness. Then a drug to alleviate pain is given, either by injection or by inhalation. For certain surgical procedures, a third drug is given to relax muscles in order to prevent spasms during surgery. In this case, a respirator is used to take the place of the patient's own breathing. Continuous monitoring of a patient's blood pressure, heart rate, respirations, and eye reflexes during surgery are used as indicators of the level of unconsciousness and muscular relaxation.

Spinal anesthesia, a form of local anesthesia, involves injection of a drug via a spinal puncture into the subarachnoid space in order to render a person insensitive to pain. The procedure is widely used for surgery performed below the diaphragm such as hernia repair, procedures on the hips and lower extremities, and operations involving the rectum, urinary bladder, prostate gland, and other pelvic structures. During the procedure, a person may be wide awake or, more commonly, sedated. Among the advantages of spinal anesthesia are the following: (1) there is almost no risk of toxic reactions that sometimes occur with general anesthesia, (2) patients with diseased kidneys or livers who could have a problem disposing of general anesthetic drugs are at less risk with spinal anesthesia, (3) no artificial breathing devices are required, (4) normal functioning of the lungs is unaffected so that people with respiratory problems have fewer complications, and (5) emergency surgery may be performed even if a person has not been fasting. Some disadvantages of spinal anesthesia are brief periods of headache, back pain, difficulty in urinating, and lowered blood pressure.

Proprioceptive Sensations

An awareness of the activities of muscles, tendons, and joints and of equilibrium is provided by the ***proprioceptive*** (*proprio* = one's own), or ***kinesthetic*** (kin'-es-THET-ik), ***sense.*** It informs us of the degree to which mucles are contracted, the amount of tension created in the tendons, the change of position of a joint, and the orientation of the head relative to the ground and in response to movements (equilibrium). The proprioceptive sense enables us to recognize the location and rate of movement of one body part in relation to others. It also allows us to estimate weight and determine the muscular work necessary to perform a task. With the proprioceptive sense, we can judge the position and movements of our limbs without using our eyes when we walk, type, or dress in the dark.

Proprioceptors adapt only slightly. This feature is advantageous since the brain must be apprised of the status of different parts of the body at all times so that adjustments can be made to ensure coordination.

Receptors

Proprioceptive receptors are located in skeletal muscles, tendons in and around synovial joints, and the internal ear.

Muscle Spindles Muscle spindles are delicate proprioceptive receptors interspersed among skeletal muscle fibers (cells) and oriented parallel to the fibers (Figure 15-3a). The ends of the spindles are anchored to the endomysium and perimysium. Muscle spindles consist of 3 to 10 specialized muscle fibers called ***intrafusal fibers,*** which are partially enclosed in a connective tissue capsule that is filled with lymph. The spindles are surrounded by skeletal muscle fibers of the muscle called ***extrafusal fibers.*** The central region of each intrafusal fiber has few or no actin and myosin myofilaments and an accumulation of nuclei. In some intrafusal fibers the nuclei bunch at the center **(*nuclear bag fibers*)**; in others, the nuclei form a chain at the center **(*nuclear chain fibers*).** The central region of the intrafusal fibers cannot contract and represents the sensory receptor area for a spindle.

Although the central receptor area cannot contract because it lacks myofilaments, it does contain two types of afferent (sensory) fibers. A large sensory fiber, called a ***type Ia fiber,*** innervates the exact center of the intrafusal fibers. The branches of the Ia fiber, called ***primary (annulospiral) endings,*** wrap around the center of the intrafusal fibers. When the central part of the spindle is stretched, the primary endings are stimulated and send nerve impulses to the spinal cord at exceedingly great velocities. The central receptor area of most muscle spindles is also innervated by two sensory fibers called ***type***

FIGURE 15-3 **Proprioceptive receptors. (a) Muscle spindle. (b) Tendon organ.**

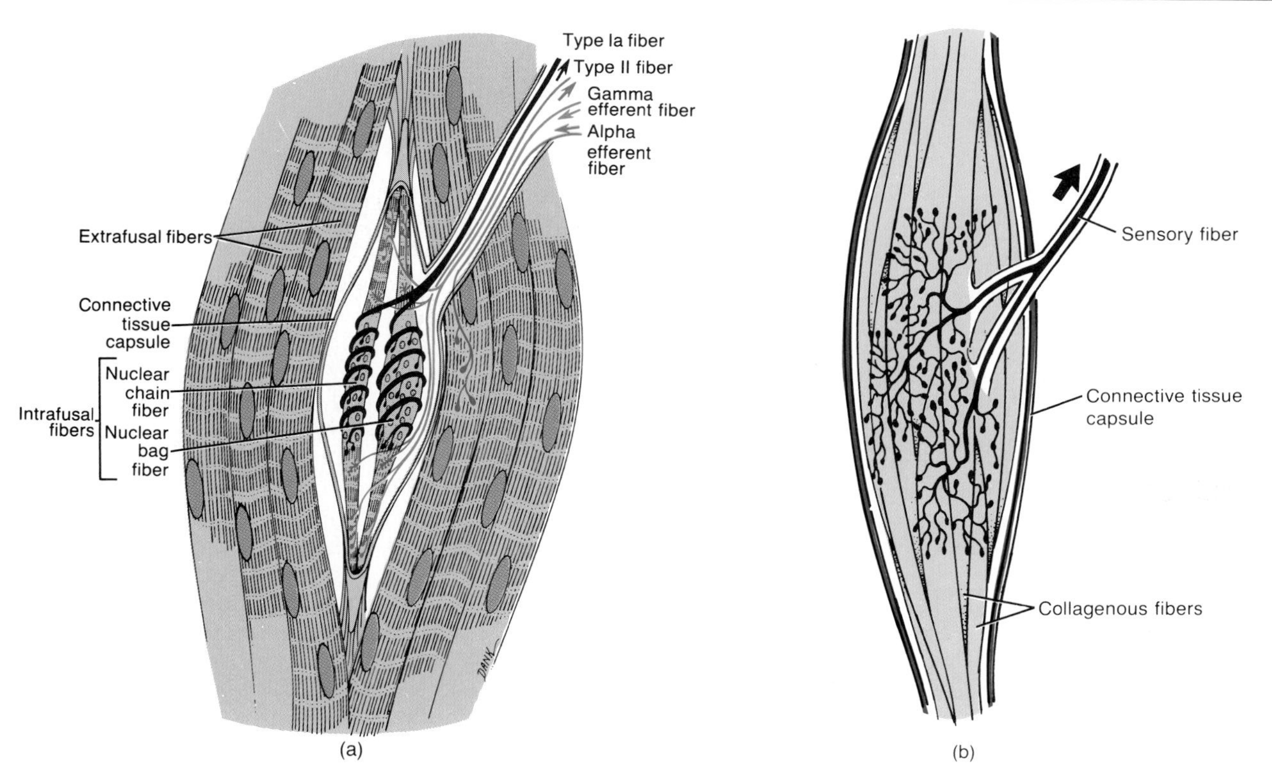

II fibers. Their branches, known as ***secondary (flower spray) endings,*** are located on either side of the primary ending. Secondary endings are also stimulated when the central part of the spindle is stretched, and they, too, send impulses to the spinal cord.

The ends of the intrafusal fibers contain actin and myosin myofilaments and represent the contractile portions of the fibers. The ends of the fibers contract when stimulated by ***gamma efferent (motor) neurons.*** These neurons are small motor neurons located in the anterior gray horn of the spinal cord. The fibers of some gamma efferent neurons terminate as motor end plates on the ends of the intrafusal fibers. Extrafusal fibers are innervated by large motor neurons called ***alpha efferent neurons.*** These neurons are also located in the anterior gray horn of the spinal cord near gamma efferent neurons.

Muscle spindles are stimulated in response to both sudden and prolonged stretch on the central areas of the intrafusal fibers. The muscle spindles monitor changes in the length of a skeletal muscle by responding to the rate and degree of change in length. This information is relayed to the central nervous system to assist in the coordination and efficiency of muscle contraction (see Figure 15-7).

▪ Tendon Organs ***Tendon organs (Golgi tendon organs)*** are proprioceptive receptors found at the junction of a tendon with a muscle. They help protect tendons and their associated muscles from damage resulting from excessive tension and also function as contraction receptors, that is, they monitor the force of contraction of each muscle. Each consists of a thin capsule of connective tissue that encloses a few collagenous fibers (Figure 15-3b). The capsule is penetrated by one or more sensory neurons whose terminal branches entwine among and around the collagenous fibers. When tension is applied to a tendon, tendon organs are stimulated, and the information is relayed to the central nervous system (see Figure 15-7).

▪ Joint Kinesthetic Receptors There are several types of ***joint kinesthetic receptors*** within and around the articular capsules of synovial joints. Encapsulated receptors, similar to type II cutaneous mechanoreceptors (end organs of Ruffini), are present in the capsules of joints and respond to pressure. Small lamellated (Pacinian) corpuscles in the connective tissue outside articular capsules are receptors that respond to acceleration and deceleration. Articular ligaments contain receptors similar to tendon organs that mediate reflex inhibition of the adjacent muscles when excessive strain is placed on the joint.

▪ Maculae and Cristae The proprioceptors in the internal ear are the macula of the saccule and the utricle and cristae in the semicircular ducts. Their function in equilibrium is discussed in Chapter 17.

The afferent pathway for muscle sense consists of impulses generated by proprioceptors via cranial and spinal nerves to the central nervous system (see Figure 15-7). Impulses for conscious proprioception pass along ascending tracts in the cord, where they are relayed to the thalamus and cerebral cortex. The sensation is registered in the general sensory area in the parietal lobe of the cerebral cortex posterior to the central sulcus. Proprioceptive impulses that have resulted in reflex action pass to the cerebellum along spinocerebellar tracts and contribute to subconscious proprioception.

Levels of Sensation

As we have said, a receptor converts a stimulus into a nerve impulse by way of a generator or receptor potential, and only after that impulse has been conducted to a region of the spinal cord or brain can it be translated into a sensation. The nature of the sensation and the type of reaction generated vary with the level of the central nervous system at which the sensation is translated.

Sensory fibers terminating in the spinal cord can generate spinal reflexes without immediate action by the brain. Sensory fibers terminating in the lower brain stem bring about far more complex motor reactions than simple spinal reflexes. When sensory impulses reach the lower brain stem, they cause subconscious motor reactions. Sensory impulses that reach the thalamus can be localized crudely in the body. At the thalamic levels, sensations are sorted by modality, that is, identified as the *specific* sensation of touch, pressure, pain, position, hearing, or taste. When sensory information reaches the cerebral cortex, we experience precise localization. It is at this level that memories of previous sensory information are stored and the perception of sensation occurs on the basis of past experience.

PHYSIOLOGY OF SENSORY PATHWAYS

Somatosensory Cortex

Most sensory information from receptors on one side of the body crosses over to the opposite side in the spinal cord or brain stem and then to the ***somatosensory cortex (primary somesthetic*** or ***general sensory area)*** of the cerebral cortex where conscious sensations result (see Figure 14-11). Areas of the somatosensory cortex have been mapped out that represent the termination of sensory information from all parts of the body. Figure 15-4 shows the location and areas of representation of the somatosensory cortex of the right cerebral hemisphere. The left cerebral hemisphere has a duplicate somatosensory cortex.

Note that some parts of the body are represented by large areas in the somatosensory cortex. These include the lips, face, tongue, and thumb. Other parts of the body, such as the trunk and lower extremities, are represented by relatively small areas. The relative sizes of the so-

FIGURE 15-4 Somatosensory cortex of the right cerebral hemisphere. (After Penfield and Rasmussen.)

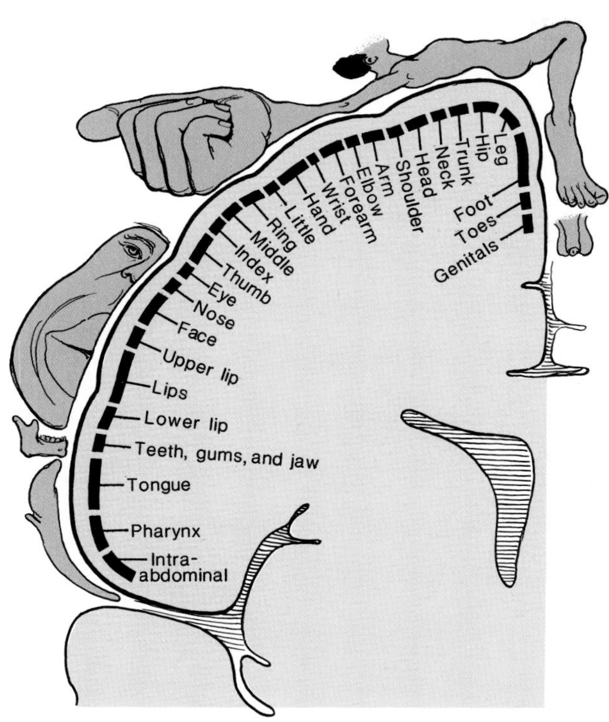

matosensory cortex are directly proportional to the number of specialized sensory receptors in each respective part of the body. Thus, there are numerous receptors in the skin of the lips but relatively few in the skin of the trunk. Essentially, the size of the area for a particular part of the body is determined by the functional importance of the part and its need for sensitivity.

Let us now examine how sensory information is transmitted from receptors to the central nervous system. You will find it helpful to review the principal ascending and descending tracts of the spinal cord (see Exhibit 13-1 and Figure 13-3). Sensory information transmitted from the spinal cord to the brain is conducted along two general pathways: the posterior column pathway and the spinothalamic pathways.

Posterior Column Pathway

In the ***posterior column pathway*** (***fasciculus gracilis*** and ***fasciculus cuneatus***) to the cerebral cortex, there are three separate sensory neurons. The ***first-order neuron*** connects the receptor with the spinal cord and medulla on the same side of the body. The cell body of the first-order neuron is in the posterior root ganglion of a spinal nerve. The axon of the first-order neuron synapses with the cell body of the ***second-order neuron,*** the axon of which passes from the medulla upward to the thalamus. The cell body of the second-order neuron is located in the nuclei cuneatus or gracilis of the medulla. Before passing into the thalamus, the axon of the second-order neuron crosses to the opposite side of the medulla and enters the medial lemniscus, a projection tract that terminates at the thalamus. In the thalamus, the axon of the second-order neuron synapses with the cell body of the ***third-order neuron,*** the axon of which terminates in the somesthetic sensory area of the cerebral cortex.

The posterior column pathway conducts impulses related to proprioception, discriminative touch, two-point discrimination, and vibrations.

Spinothalamic Pathways

The ***spinothalamic pathways*** are also composed of three orders of sensory neurons. The first-order neuron connects a receptor of the neck, trunk, and extremities with the spinal cord. The cell body of the first-order neuron is in the posterior root ganglion also. The axon of the first-order neuron synapses with the cell body of the second-order neuron, which is located in the posterior gray horn of the spinal cord. The axon of the second-order neuron crosses to the opposite side of the spinal cord and passes upward to the brain stem in the lateral spinothalamic tract or anterior spinothalamic tract. The axon from the second-order neuron terminates in the thalamus. There, the axon of the second-order neuron synapses with the cell body of the third-order neuron. The axon of the third-order neuron terminates in the somesthetic sensory area of the cerebral cortex. The spinothalamic pathways convey sensory impulses for pain and temperature (lateral spinothalamic) as well as light touch and pressure (anterior spinothalamic).

Now we can examine the anatomy of specific sensory pathways—for pain and temperature; for light touch and pressure; and for discriminative touch, proprioception, and vibration.

Pain and Temperature

The sensory pathway for pain and temperature is called the ***lateral spinothalamic pathway*** (Figure 15-5). The first-order neuron conveys the nerve impulse for pain or temperature from the appropriate receptor to the posterior gray horn on the same side of the spinal cord. In the horn, the axon of the first-order neuron synapses with the cell body of the second-order neuron. The axon of the second-order neuron crosses to the opposite side of the cord. Here it becomes a component of the ***lateral spinothalamic tract*** in the lateral white column. The axon of the second-order neuron passes upward in the tract through the brain stem to a nucleus in the thalamus called the ventral posterolateral nucleus. In the thalamus, conscious recognition of pain and temperature occurs, but not localization of these stimuli. The sensory impulse is then conveyed from the thalamus through the internal

FIGURE 15-5 Sensory pathway for pain and temperature—the lateral spinothalamic pathway.

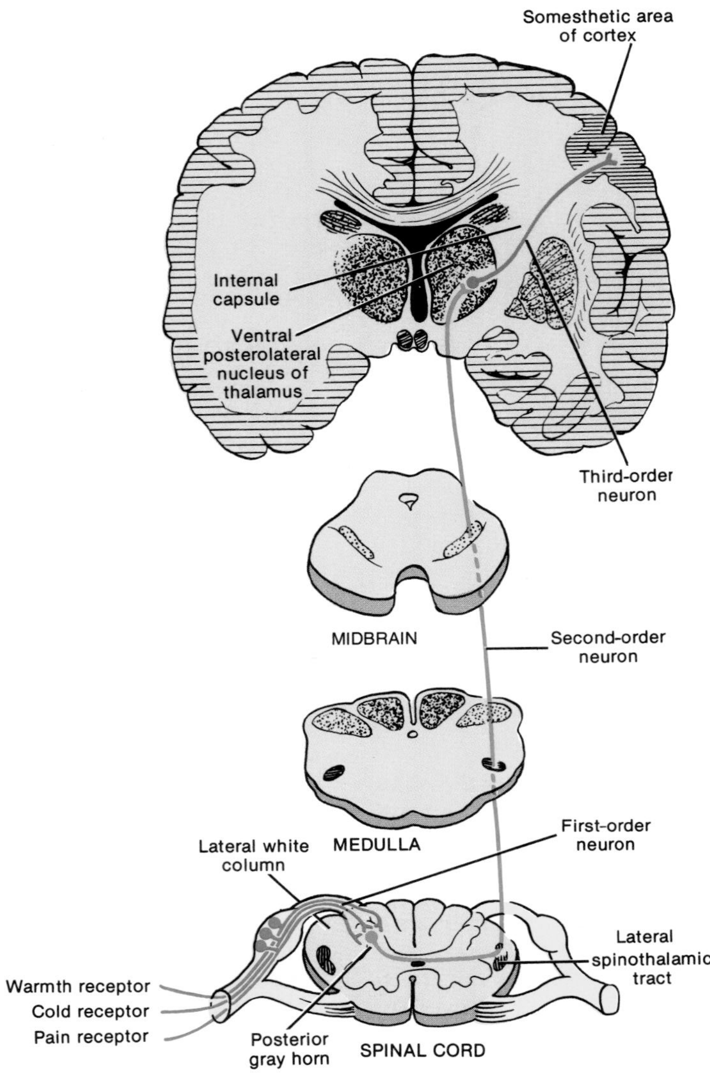

capsule to the somesthetic area of the cerebral cortex by the axon of the third-order neuron. The cortex analyzes the sensory information for the precise source, severity, and quality of the pain and temperature stimuli.

Light Touch and Pressure

The neural pathway that conducts nerve impulses for light touch and pressure is the ***anterior (ventral) spinothalamic pathway*** (Figure 15-6). The first-order neuron conveys the impulse from a light touch or pressure receptor to the posterior gray horn on the same side of the spinal cord. In the horn, the axon of the first-order neuron synapses with the cell body of the second-order neuron. The axon of the second-order neuron crosses to the opposite side of the cord and becomes a component of the ***anterior spinothalamic tract*** in the anterior white column. The axon of the second-order neuron passes upward in the tract through the brain stem to the ventral posterolateral nucleus of the thalamus. The sensory impulse is then relayed from the thalamus through the internal capsule to the somesthetic area of the cerebral cortex by the axon of the third-order neuron. Although there is some awareness of light touch and pressure at the thalamic level, it is not fully perceived nor localized until the impulses reach the cerebral cortex.

Discriminative Touch, Proprioception, and Vibration

The neural pathway for discriminative touch, conscious proprioception, and vibration is called the ***posterior col-***

umn pathway (Figure 15-7). This pathway conducts impulses that give rise to several discriminating senses.

1. **Discriminative touch,** the ability to recognize the exact location of stimulation and to make two-point discriminations.

2. **Stereognosis,** the ability to recognize by "feel" the size, shape, and texture of an object.

3. **Proprioception,** the awareness of the precise position of body parts and directions of movement.

4. **Weight discrimination,** the ability to assess the weight of an object.

5. The ability to sense ***vibrations.***

First-order neurons for the discriminating senses just noted follow a pathway different from those for pain and temperature and light touch and pressure. Instead of terminating in the posterior gray horn, the axon of the first-order neurons from an appropriate receptor passes upward in the fasciculus gracilis or fasciculus cuneatus in the posterior white column of the cord. From here, the axon of the first-order neuron enters either the nucleus gracilis or nucleus cuneatus in the medulla, where it synapses with the cell body of the second-order neurons. The axon of the second-order neuron crosses to the opposite side of the medulla and ascends to the thalamus through the medial lemniscus, a projection tract of white fibers passing through the medulla, pons, and midbrain. The axon of the second-order neuron synapses with the cell body of the third-order neuron in the ventral posterior nucleus in the thalamus. In the thalamus, there is no conscious awareness of the discriminating senses, except

FIGURE 15-6 **Sensory pathway for light touch and pressure—the anterior spinothalamic pathway.**

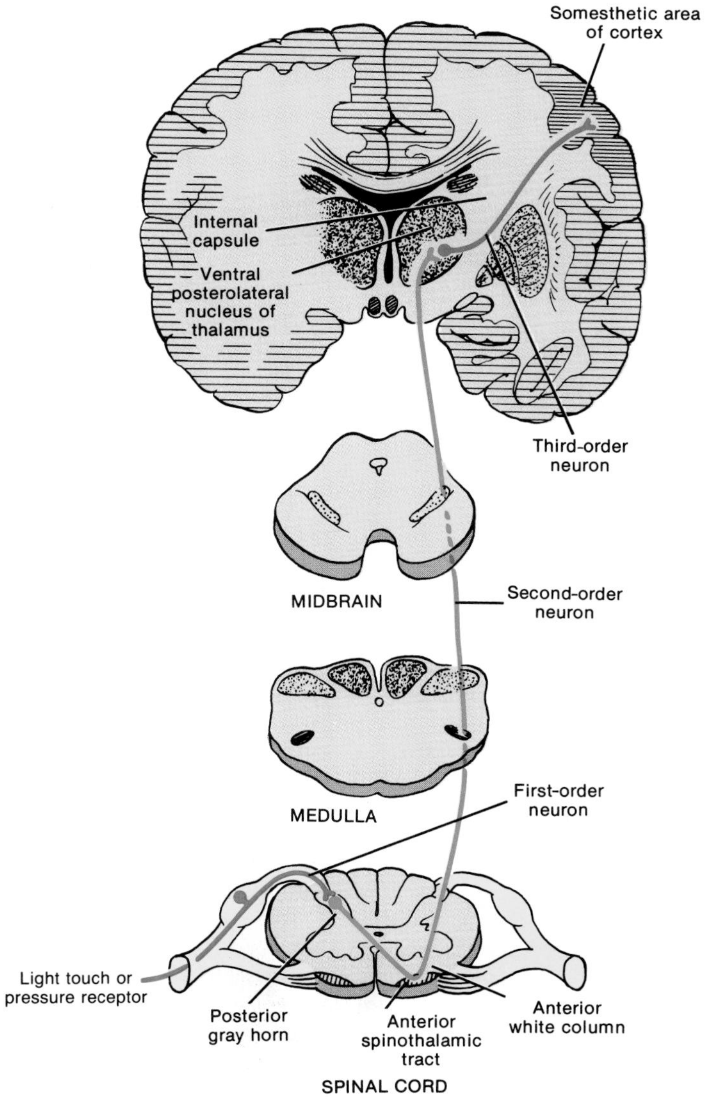

FIGURE 15-7 **Sensory pathway for discriminative touch, proprioception, and vibration—the posterior column pathway.**

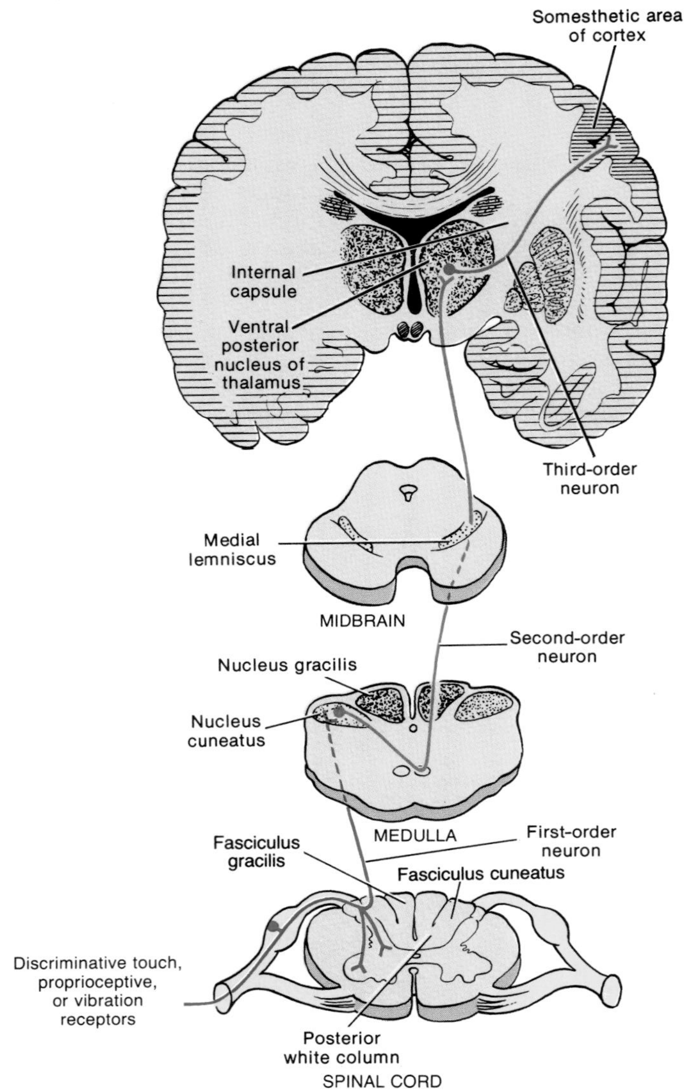

for a possible crude awareness of vibrations. The axon of the third-order neurons conveys the sensory impulses to the somesthetic area of the cerebral cortex. It is here that you perceive your conscious sense of position and movement and light touch.

Cerebellar Tracts

The ***posterior spinocerebellar tract*** is an uncrossed tract that conveys nerve impulses concerned with subconscious muscle and joint sense and thus assumes a role in reflex adjustments for posture and muscle tone. The impulses originate in neurons that run between proprioceptors in muscles, tendons, and joints and the posterior gray horn of the spinal cord. Here the axons of the neurons synapse with the cell bodies of association neurons that pass to the ipsilateral lateral white column of the cord to enter the posterior cerebellar tract. The tract enters the inferior cerebellar peduncles from the medulla and ends at the cerebellar cortex. In the cerebellum, synapses are made that ultimately result in the transmission of impulses back to the spinal cord to the anterior gray horn to synapse with the lower motor neurons leading to skeletal muscles.

The ***anterior spinocerebellar tract*** also conveys impulses for subconscious muscle sense. It, however, is made up of both crossed and uncrossed nerve fibers. Sensory neurons deliver impulses from proprioceptors to the posterior gray horn of the spinal cord. Here, a synapse occurs with neurons that make up the anterior spinocerebellar tracts. Some axons cross to the opposite side of the spinal cord in the anterior white commissure. Others pass laterally to the ipsilateral anterior spinocerebellar tract and move upward, through the brain stem, to the pons to enter the cerebellum through the superior cerebellar peduncles. Here again, the impulses for subconscious muscle and joint sense are registered.

PHYSIOLOGY OF MOTOR PATHWAYS

After receiving and interpreting sensations, the central nervous system generates nerve impulses to direct responses to that sensory input. We have already considered somatic reflex arcs and visceral efferent pathways. Now we will look at the transmission of motor impulses that result in the movement of skeletal muscles.

Linkage of Sensory Input and Motor Responses

Sensory systems function to keep the central nervous system aware of certain changes in the external and internal environment. Responses to this information are brought about by the motor systems, which enable us to move about and alter the secretory rate of certain glands and change our relationship to the world around us. As sensory information is conveyed into the central nervous system, it becomes part of a large pool of sensory input. We do not actively respond to every bit of input the central nervous system receives. Rather, the incoming information is integrated with other information arriving from all other operating sensory receptors. The integration process occurs not just once but at many stations along the pathways of the central nervous system and at both a conscious and subconscious level. It occurs within the spinal cord, brain stem, cerebellum, and cerebral motor cortex. As a result, a motor response to make a muscle contract or a gland secrete can be initiated at any of these levels. The cerebral motor cortex assumes the major role for controlling precise, discrete muscular movements and for conscious perception of sensations. The basal ganglia largely integrate semivoluntary movements like walking, swimming, and laughing. The cerebellum, although not a control center, assists the motor cortex and basal ganglia by making body movements smooth and coordinated and by contributing significantly to the maintenance of normal posture (equilibrium).

The design of the motor response system is very precise. Each skeletal muscle is connected to its own group of nerve fibers, and these fibers innervate only fibers of that muscle. This arrangement ensures that when the central nervous system initiates a response, only the desired muscle contracts. You may wish to review the description of the motor unit in Chapter 10.

Muscle actions can vary greatly in complexity, and the pattern of activation of motor units therefore also varies in complexity. Motor units can respond simultaneously to produce a sudden, forceful movement, or they can be orchestrated to respond asynchronously in a time sequence so that the force of contraction builds up more gradually. When more than one joint is involved, the sequential patterns of motor unit activation involving all the joints can be very complicated indeed. Yet when complicated movements are performed by an Olympic gymnast, for example, the beauty of the act belies the underlying complexity of the myriad of neuromuscular interactions.

Some muscular responses are simple and occur subconsciously. The quick withdrawal of a hand when it touches a hot object or the knee jerk that results when a percussion hammer taps the patellar tendon are simple responses. Playing a piano is a much more complex response. Simple or very complex, all muscle responses involve both contraction and relaxation of muscle groups. The innervation of antagonistic (opposing) muscles always acts in a reciprocal fashion. One nerve fiber set causes the contraction of the prime movers (agonists), and the other set of fibers permits relaxation of the antagonists. The complexity of the muscular movements involved in piano playing can be astonishing. Sensory input is rapidly and continually changing as the eyes scan the music score. Each bit of information initiates and modifies the reciprocal pattern of muscle innervation to the fingers, within milliseconds. Additional information from sensory receptors relating to touch, position, pressure, speed, and sound also affect the reciprocal innervation program. This almost instantaneous analysis of the total sensory input generates the incredible array of motor responses to produce beautiful music. The linking of the sensory input and motor response occurs at several major levels in the central nervous system, and each level can add richness and variety to the repertoire of movements.

As the location of the sensory–motor linkage climbs to higher levels in the central nervous system—from spinal cord to brain stem to cerebellum to basal ganglia to cerebral motor cortex—additional contributions are introduced that enrich the growing inventory of motor responses. All information about each sensory modality arises from the peripheral nervous system and is directed into the spinal cord or brain. At the level of entry, a synapse is made that directs the input of information to higher centers in the central nervous system. When information reaches the highest level, the cerebral cortex, conscious sensation results. Most conscious sensations originate from fibers connected to receptors on the exterior of the body.

Other inputs to the central nervous system never give rise to conscious sensation. These operate to coordinate and modulate the contractions of muscle groups. Nerve fibers that connect to proprioceptive receptors within muscles, tendons, and joints constantly deliver input to the central nervous system about the present state of the muscle length and its speed and strength of contraction and activity of joints. These fibers enter the spinal cord, synapse, and project their input up the cord to the cerebellum.

The number of interconnections between the regions of the central nervous system is so great that it is impossible to define the final site of sensory input. Some sensory input arrives first at the cerebral cortex and some is delivered to the cerebellum. In other cases, the sensory input is delivered to the cerebellum, which then relays some of the input to the cerebral cortex.

When the input reaches the highest center, the process of sensory–motor integration occurs. This involves not only the utilization of information contained within that center but also the information impinging on that center that is delivered from other centers within the central nervous system. After integration occurs, the output of the center is sent down the spinal cord in two major descending motor pathways: the pyramidal pathways and the extrapyramidal pathways.

Motor Cortex

Just as the somatosensory cortex has been mapped to indicate the termination of sensory information from many parts of the body, the ***motor cortex*** has been mapped to indicate which groups of muscles are controlled by its specific areas (Figure 15-8). The right cerebral hemisphere has a duplicate of the motor cortex in the left cerebral hemisphere. Note that the different muscle groups are not represented equally in the motor cortex. In general, the degree of representation is proportional to the preciseness of movement required of a particular part of the body. For example, the thumb, fingers, lips, tongue, and vocal cords have large representations. The trunk has a relatively small representation. By comparing Figures 15-4 and 15-8, you will see that somatosensory and motor representations are not identical for the same part of the body.

FIGURE 15-8 **Motor cortex of the right cerebral hemisphere. (After Penfield and Rasmussen.)**

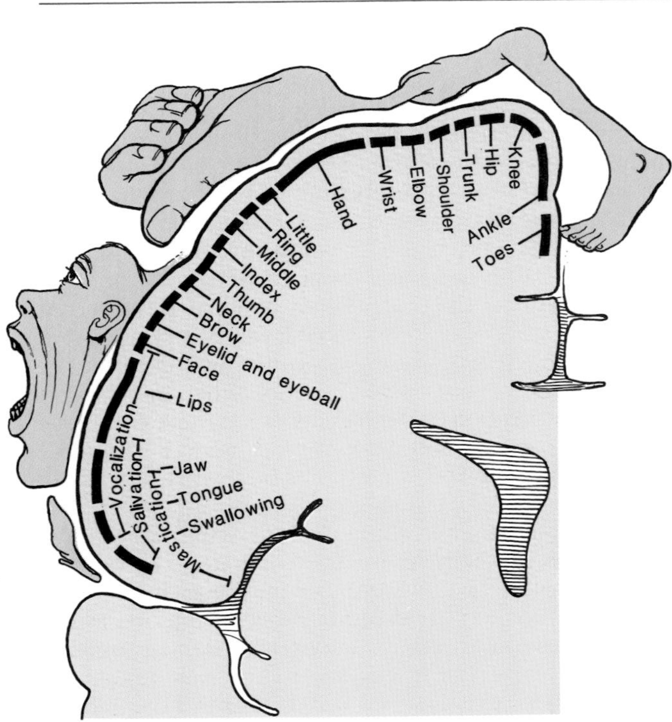

Pyramidal Pathways

Voluntary motor impulses are conveyed from the motor areas of the brain to somatic efferent neurons (voluntary motor neurons) leading to skeletal muscles via the ***pyramidal*** (pi-RAM-i-dal) ***pathways.*** Most pyramidal axons originate from cell bodies in the precentral gyrus. They descend through the internal capsule of the cerebrum, and most cross to the opposite side of the brain. They terminate in nuclei of cranial nerves that innervate voluntary muscles or in the anterior gray horn of the spinal cord. A short association neuron probably completes the connection of the pyramidal axons with the cell bodies of the somatic (voluntary) motor neurons that activate voluntary muscles.

The pathways over which the nerve impulses travel from the motor cortex to skeletal muscles have two components: ***upper motor neurons* (*pyramidal fibers*)** in the brain and spinal cord and ***lower motor neurons* (*peripheral fibers*)** which extend from the spinal cord to the muscle fibers. (Lower motor neurons also originate in motor nuclei of cranial nerves within the brain and extend out to muscle fibers.) Here we consider three tracts of the pyramidal system:

1. **Lateral corticospinal tract (pyramidal tract proper).** This tract begins in the motor cortex and descends through the internal capsule of the cerebrum, the cerebral peduncle of the midbrain, and then to the pons on the same side as the point of origin (Figure 15-9). About 85 percent of the axons of the upper motor neurons from the motor cortex cross in the medulla. After decussation (crossing), these axons descend through the spinal cord in the lateral white column in the lateral corticospinal tract. Thus, the motor cortex of the right side of the brain controls muscles on the left side of the body, and vice versa. Most upper motor neuron axons of the lateral corticospinal tract synapse with cell bodies of short association neurons in the anterior gray horn of the cord. These axons then synapse in the anterior gray horn with cell bodies of lower motor neurons, the axons of which exit all levels of the cord via the ventral roots of spinal nerves. The lower motor neurons (somatic motor neurons) terminate in skeletal muscles.

2. **Anterior corticospinal tract.** About 15 percent of the axons of upper motor neurons from the motor cortex do not cross in the medulla. These pass through the medulla and continue to descend on the same side to the anterior white column to become part of the anterior (straight or uncrossed) corti-

FIGURE 15-9 **Pyramidal pathways.**

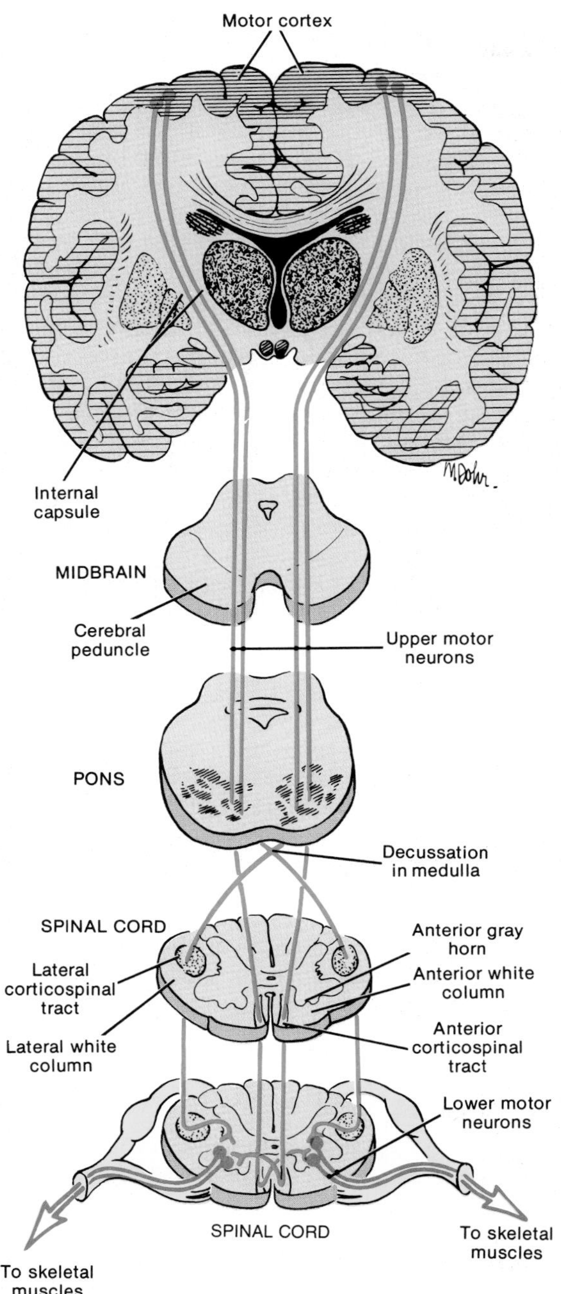

cospinal tract. The axons of these upper motor neurons decussate and synapse with cell bodies of association neurons in the anterior gray horn of the spinal cord on the side opposite the origin of the anterior corticospinal tract. The axons of association neurons in the horn synapse with cell bodies of lower motor neurons, the axons of which exit the cervical and upper thoracic segments of the cord via the ventral roots of spinal nerves. The axons of lower motor neurons terminate in skeletal muscles that control muscles of the neck and part of the trunk.

The axons of the various tracts of the central nervous system develop myelin sheaths at different times. For example, the afferent and efferent axons of spinal nerves develop myelin after the fifth fetal month, but the axons of the corticospinal tracts do not become fully myelinated until the second year of life. This, in part, explains the fact that infants and young children have a definite pattern of developmental activity. It is for this reason that they cannot be expected to walk until the nervous system is mature enough.

3. **Corticobulbar tract.** The axons of this tract arise from cell bodies of upper motor neurons in the motor cortex. The axons accompany the corticospinal tracts through the internal capsule to the brain stem, where they decussate and terminate in the nuclei of cranial nerves in the pons and medulla. These cranial nerves include the oculomotor (III), trochlear (IV), trigeminal (V), abducens (VI), facial (VII), glossopharyngeal (IX), vagus (X), accessory (XI), and hypoglossal (XII). The corticobulbar tract conveys impulses that largely control voluntary movements of the head and neck.

The various tracts of the pyramidal system convey impulses from the cortex that result in precise muscular movements.

Extrapyramidal Pathways

The ***extrapyramidal pathways*** include all descending (motor) tracts other than the pyramidal tracts. Generally, these include tracts that begin in the basal ganglia and reticular formation of the brain stem. The main extrapyramidal tracts are as follows.

1. **Rubrospinal tract.** This tract originates in the red nucleus of the midbrain (after receiving axons from the cerebellum), crosses over to descend in the lateral white column of the opposite side, and extends through the entire length of the spinal cord. The tract transmits impulses to skeletal muscles concerned with muscle tone and posture.

2. **Tectospinal tract.** This tract originates in the superior colliculus of the midbrain, crosses to the opposite side, descends in the anterior white column, and enters the anterior gray horns in the cervical segments of the cord. Its function is to transmit impulses that control movements of the head in response to visual stimuli.

3. **Vestibulospinal tract.** This tract originates in the vestibular nucleus of the medulla, descends on the same side of the cord in the anterior white column, and terminates in the anterior gray horns, mostly in the cervical and lumbosacral segments of a cord. It conveys impulses that regulate muscle tone in response to movements of the head. This tract, therefore, plays a major role in equilibrium.

4. **Lateral reticulospinal tract.** This tract originates in the medulla, descends mainly on the same side of the cord in the lateral white column, and terminates in the anterior gray horn. Its function is to inhibit extensor reflexes and muscle tone in muscles of the axial skeleton and proximal limb.

5. Anterior (*ventral*) or *medial reticulospinal tract.* This tract originates in the pons, descends mainly on the same side of the cord, and terminates in the anterior gray horn. Its function is to facilitate extensor reflexes and muscle tone in muscles of the axial skeleton and proximal limb.

Only motor neurons carry impulses from the cerebral cortex to the cranial nerve nuclei or spinal cord and are known as ***upper motor neurons.*** Also, only motor neurons in the pathway actually terminate in skeletal muscles and are referred to as ***lower motor neurons.*** These lower motor neurons, somatic (voluntary) efferent neurons, always extend from the central nervous system to skeletal muscles. Since they are the final transmitting neurons in the pathway, they are also called ***final common pathways.***

CLINICAL APPLICATION: FLACCID AND SPASTIC PARALYSIS

The lower motor and upper motor neurons are important clinically. If the lower motor neuron is damaged or diseased, there is neither voluntary nor reflex action of the muscle it innervates, and the muscle remains flaccid (decreased or lost muscle tone), a condition called *flaccid paralysis.* Injury or disease of upper motor neurons in a motor pathway is characterized by varying degrees of spasticity (increased muscle tone), exaggerated reflexes, and pathological reflexes, such as the Babinski sign. This condition is called *spastic paralysis.* The Babinski sign is dorsiflexion of the great toe accompanied by fanning of the lateral toes in response to stroking the plantar surface along the outer border of the foot. As we have seen, this response is normal only in infants; the normal adult response is plantar flexion of the toes.

Lower (voluntary) motor neurons are subjected to stimulation by many other presynaptic neurons. Some nerve impulses are excitatory; others are inhibitory. The algebraic sum of the opposing signals determines the final response of the lower motor neuron. It is not just a simple matter of the brain sending an impulse and the muscle always contracting.

Association neurons are of considerable importance in the motor pathways. Most impulses from the brain are conveyed to association neurons before being received by lower motor neurons. These association neurons integrate the pattern of muscle contraction.

The basal ganglia have many connections with other parts of the brain. Through these connections, they help to control subconscious movements. The caudate nucleus controls gross intentional movements. The caudate nucleus and putamen, together with the cerebral cortex, control patterns of movement. The globus pallidus controls positioning of the body for performing a complex movement. The subthalamic nucleus is thought to control walking and possibly rhythmic movements. Many potential functions of the basal ganglia are held in check by the cerebrum. Thus, if the cerebral cortex is damaged early in life, a person can still perform many gross muscular movements.

The role of the cerebellum is significant also. The cerebellum is connected to other parts of the brain that are concerned with movement. The vestibulocerebellar tract transmits impulses from the equilibrium apparatus in the ear to the cerebellum. The olivocerebellar tract transmits nerve impulses from the basal ganglia to the cerebellum. The corticopontocerebellar tract conveys impulses from the cerebrum to the cerebellum. The spinocerebellar tracts relay proprioceptive information to the cerebellum. Thus, the cerebellum receives considerable information regarding the overall physical status of the body. Using this information, the cerebellum generates impulses that integrate body responses so that our movements are smooth and well coordinated and we are able to maintain our equilibrium and normal posture.

Take tennis, for example. To make a good serve, you must bring your racket forward just far enough to make solid contact. How do you stop at the exact point without swinging too far? This is where the cerebellum comes in. It receives information about your body status while you are serving. Before you even hit the ball, the cerebellum had already sent information to the cerebral cortex and basal ganglia informing them that your swing must stop at an exact point. In response to cerebellar stimulation, the cortex and basal ganglia transmit motor impulses to your opposing body muscles to stop the swing. The cerebellar function of stopping overshoot when you want to zero in on a target is called its ***damping function.*** The cerebellum also helps you to coordinate different body parts while walking, running, swimming, and speaking. Finally, the cerebellum helps you maintain equilibrium and proper posture.

INTEGRATIVE FUNCTIONS

We turn now to a fascinating, though poorly understood, function of the cerebrum: integration. The ***integrative functions*** include cerebral activities such as memory, sleep and wakefulness, and emotional response. The role of the limbic system in emotional behavior was discussed in Chapter 14.

Memory

Without a memory, we would repeat mistakes and be unable to learn. Similarly, we would not be able to repeat our successes or accomplishments, except by chance. Although both memory and learning have been studied by scientists for many years, there is still no satisfactory explanation for how we recall information or how we remember. Some things, however, are known about how information is acquired and stored.

Learning is the ability to acquire knowledge or a skill through instruction or experience. Learning is very closely associated with rewards and punishments. ***Memory*** is the ability to recall thoughts. For an experience to become part of memory, it must produce changes in the central nervous system that represent the experience. Such a memory trace in the brain is called an ***engram.*** The portions of the brain thought to be associated with memory include the association cortex of the frontal, parietal, occipital, and temporal lobes; parts of the limbic system, especially the hippocampus and amygdaloid nucleus; and the diencephalon.

Memory may generally be classified into two kinds based on how long the memory lasts: short-term and long-term memory. ***Short-term (activated) memory*** lasts only seconds or hours and is the ability to recall bits of information. One example is finding an unfamiliar telephone number in a telephone book and then dialing it. If the number has no special significance, it is usually forgotten within a few seconds. ***Long-term memory,*** on the other hand, lasts from days to years. For example, if you frequently use a telephone number for any period of time, such as your home telephone number, it becomes part of long-term memory. When the number is in long-term memory, it can be retrieved for use whenever needed for quite a long period of time. Such reinforcement due to the frequent use of the telephone number is called ***memory consolidation.***

As stated earlier in the chapter, our brain receives many stimuli, but we are conscious of only a few of them. It has been estimated that of all the information that comes to our consciousness, only about 1 percent goes into long-term memory, and much of what goes into long-term memory is forgotten. It is fortunate that our brains select only a small percentage of our thoughts and lose much of what we stored; otherwise, our brain would be overwhelmed with information.

It is a feature of memory that we can remember short lists easier than long ones. This is to say the obvious, but human memory does not record everything like an endless magnetic tape. It cannot remember or record long lists of details. Another feature of memory is that even when details are lost, the concept or main idea is retained. Then, interestingly, we can often explain the idea or concept—not like replaying a tape—but with our own selection of words and ways of explanation.

Despite several decades of research, it is still impossible to explain the mechanisms of memory. However, by integrating several clinical and experimental observations, scientists have developed several theories to help us understand the mechanisms.

One theory of short-term memory states that memories may be caused by reverberating neuronal circuits—an incoming nerve impulse stimulates the first neuron, which stimulates the second, which stimulates the third, and so on (see Figure 12-15c). Branches from the second and third neurons synapse with the first, sending the impulse back through the circuit again and again. Thus, the output neuron generates continuous impulses. Once fired, the output signal may last from a few seconds to many hours, depending on the arrangement of neurons in the circuit. If this pattern is applied to short-term memory, an incoming thought—the phone number—continues in the brain even after the initial stimulus is gone. Thus, you can recall the thought only for as long as the reverberation continues.

There is some evidence to support the concept that short-term memory is related to electrical and chemical events rather than structural changes in the brain. For example, several conditions may inhibit the electrical activity of the brain. These include anesthesia, coma, electroconvulsive shock, and ischemia (reduced blood supply) of the brain. Although such states do interfere with the retention of recently acquired information, they do not usually interfere with long-term memory laid down prior to the interference with electrical activity. It is a common experience that individuals who suffer retrograde amnesia cannot remember any events that have occurred for about 30 minutes before the amnesia developed. But as the person recovers from the amnesia, the most recent memories return last; loss of consciousness in no way affects memories learned before the amnesia.

Most research on long-term memory focuses on anatomical or biochemical changes at synapses that might enhance facilitation at synapses. Anatomical changes occur in neurons when they are either stimulated or made inactive. For example, electron micrographic studies of presynaptic neurons that have been subjected to prolonged, intense activity exhibit several anatomical changes. These include an increase in the number of presynaptic terminals, enlargement of synaptic end bulbs, and an increase in the branching patterns and conductance of dendrites. There is also an increase in the number of neuroglia. Moreover, neurons grow new synaptic end bulbs with increasing age, presumably as a result of increased utilization. These changes, which are correlated with faster learning, suggest enhancement of facilitation at synapses. Such changes do not occur when neurons are inactive. In fact, in animals that have lost their eyesight, there is thinning of the cerebral cortex in the visual area.

Any of the events in the synaptic transmission could be responsible for enhanced communication between neurons. There is experimental evidence that repetitive use of a synapse causes the presynaptic membrane to hyperpolarize so that when the nerve impulse passes over the synaptic end bulb, its amplitude is much greater. The greater the amplitude, the greater the amount of neurotransmitter released. To enhance this possibility further, the neurotransmitter in the presynaptic cell increases greatly and therefore can be released with each nerve impulse. Other suggestions include increase in the number of receptor molecules in the postsynaptic cell membrane or a decrease in the rate of removal of neurotransmitter substance. Acetylcholine (ACh) seems to play a vital role in memory.

There is also interest in the possible involvement of nucleic acids in long-term memory. The molecules DNA and RNA store information, and these molecules, especially DNA, tend to persist for the lifetime of the cell. Studies have shown an increase in the RNA content of activated neurons. Conversely, there is some evidence that shows that long-term memory will not occur to any significant extent when RNA formation is inhibited.

A recent hypothesis concerning memory involves a series of chemical reactions in the brain that permanently alter connections between neurons in the brain. This alteration may be the basis for memory. Simply stated, when a nerve impulse reaches a synapse, a neurotransmitter is released which can fit into a receptor on the postsynaptic neuron. This interaction opens calcium channels in the membrane of the postsynaptic neuron, causing the influx of the ions. Inside the postsynaptic neuron, calcium ions activate an enzyme called calpain that breaks down the cytoskeleton, changing the shape of the postsynaptic neuron and somehow making it more sensitive to subsequent transmission of information and thus contributing to the formation of a memory. Such synapses might form the first link of a memory trace.

Another recent hypothesis involves neurons that have two different ways of transmitting information. Such neurons contain receptors called ***NMDA receptors,*** named after the chemical N-methyl D-aspartate that is used to detect them. These receptors permit calcium influx into neurons. All other receptors involved in neuronal firing respond to *either* neurotransmitters or a change in voltage. NMDA receptors are unique in that they must *first* be stimulated by a voltage change and *then* by a neurotransmitter (glutamic acid). Such neurons have a special mode of signal transmission that is activated only if the cell receives the signals in a row—the first signal cocks the gun; the second signal fires it. This unique type of signal transmission provides the neurons with a different way to process information related to memory formation. It is not known whether calcium ions strengthen neural connections by causing more neurotransmitters to be released or by making the receptor neuron more responsive to the neurotransmitter. There is some evidence that calcium ions lead not only to strengthening of preexisting synapses but also to the formation of new ones.

Wakefulness and Sleep

Humans sleep and awaken in a fairly constant 24-hour rhythm called a ***circadian*** (ser-KĀ-dē-an) ***rhythm.*** When the brain is aroused or awake, it is in a state of readiness and able to react consciously to various stimuli. Since neuronal fatigue precedes sleep and the signs of fatigue disappear after sleep, fatigue is apparently one cause of sleep. Moreover, EEG recordings indicate that during wakefulness the cerebral cortex is very active, sending impulses continuously through the body. During sleep, however, fewer impulses are transmitted by the cerebral cortex. The activity of the cerebral cortex is thought to be related to the reticular formation.

The reticular formation has numerous connections with the cerebral cortex (Figure 15-10). Stimulation of portions of the reticular formation results in increased cortical activity. Thus, a portion of the reticular formation is known as the ***reticular activating system (RAS).*** One part of the system, the mesencephalic part, is composed of areas of gray matter of the pons and midbrain. When this area is stimulated, many nerve impulses pass upward into the thalamus and disperse to widespread areas of the cerebral cortex. The effect is a generalized increase in cortical activity. The other part of the RAS, the thalamic part, consists of gray matter in the thalamus. When the thalamic part is stimulated, signals from specific parts of the thalamus cause activity in specific parts of the cerebral cortex. Apparently, the mesencephalic part of the RAS causes general wakefulness (consciousness), and the thalamic part causes ***arousal,*** that is, awakening from deep sleep.

For arousal to occur, the RAS must be stimulated by input signals. Almost any sensory input can activate the RAS: pain stimuli, proprioceptive signals, bright light, or an alarm clock. Once the RAS is activated, the cerebral cortex is also activated and you experience arousal. Nerve impulses from the cerebral cortex can also stimulate the RAS. Such impulses may originate in the somesthetic cortex, the motor cortex, or the limbic system. When the impulses activate the RAS, the RAS activates the cerebral cortex and arousal occurs.

Following arousal, the RAS and cerebral cortex continue to activate each other through a feedback system consisting of many circuits. The RAS also has a feedback system

FIGURE 15-10 Reticular formation.

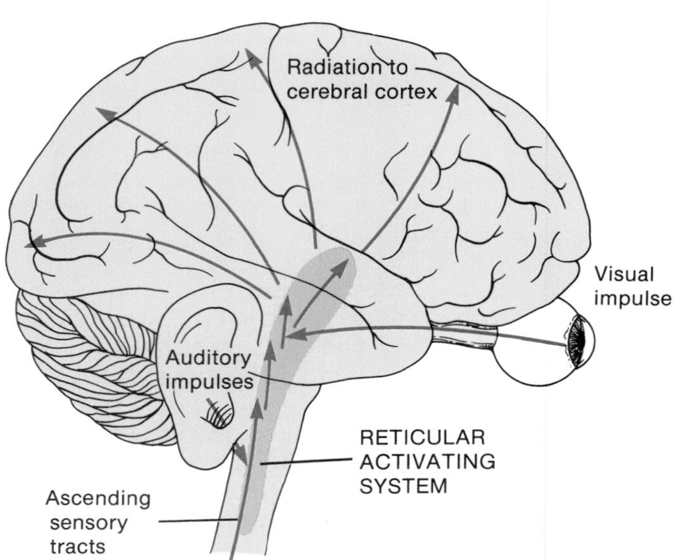

with the spinal cord that is composed of many circuits. Impulses from the activated RAS are transmitted down the spinal cord and then to skeletal muscles. Muscle activation causes proprioceptors to return impulses that activate the RAS. The two feedback systems maintain activation of the RAS, which in turn maintains activation of the cerebral cortex. The result is a state of wakefulness called ***consciousness.*** The RAS is the physical basis of consciousness, the brain's chief watchguard. It continuously sifts and selects, forwarding only the essential, unusual, or dangerous to the conscious mind. Since humans experience different levels of consciousness (alertness, attentiveness, relaxation, inattentiveness), it is assumed that the level of consciousness depends on the number of feedback currents operating at the time. During resting wakefulness alpha waves appear on an EEG.

CLINICAL APPLICATION: ALTERED CONSCIOUSNESS

Consciousness may be altered by various factors. Amphetamines probably activate the RAS to produce a state of wakefulness and alertness. Meditation produces a relaxed, focused consciousness. Anesthetics produce a state of consciousness called anesthesia. Damage to the nervous system, as well as disease, can produce coma. Drugs such as LSD and alcohol can also alter consciousness.

Coma is the final stage of brain failure that is characterized by total unresponsiveness to all external stimuli. A comatose patient lies in a sleeplike state with the eyes closed. In the lightest stages of coma, primitive brain stem and spinal cord reflexes persist, but in the deepest stages, these reflexes are lost as well as corneal, papillary, tendon, and plantar reflexes. Eventually, respiratory and cardiovascular controls disappear and the patient usually dies. Coma may be induced by lesions on various parts of the brain, poisoning, hypoxia, hypoglycemia, ischemia, infection, acid–base and electrolyte disorders, and trauma.

If the theory of wakefulness just described is accepted, how then does sleep occur if the activating feedback systems are in continual operation? One explanation is that the feedback system slows tremendously or is inhibited. Inactivation of the RAS produces a state known as ***sleep,*** a state of altered consciousness or partial unconsciousness from which an individual can be aroused by different stimuli.

Just as there are different levels of awareness when awake, there are different levels of sleep. Normal sleep consists of two types: non–rapid eye movement sleep (NREM) and rapid eye movement sleep (REM).

NREM sleep, or ***slow wave sleep,*** consists of four stages, each of which gradually merges into the next. Each stage has been identified by EEG recordings.

Stage 1. This is a transition stage between waking and sleep that normally lasts from one to seven minutes. The person is relaxing with eyes closed. During this time, respirations are regular, pulse is even, and the person has fleeting thoughts. If awakened, the person will frequently say he has not been sleeping. Alpha waves diminish and theta waves appear on the EEG.

Stage 2. This is the first stage of true sleep, even though the person experiences only light sleep. It is a little harder to awaken the person. Fragments of dreams may be experienced, and the eyes may slowly roll from side to side. The EEG shows *sleep spindles*—sudden, short bursts of sharply pointed waves that occur at 12 to 14 Hz (cycles) per second.

Stage 3. This is a period of moderately deep sleep. The person is very relaxed. Body temperature begins to fall and blood pressure decreases. It is difficult to awaken the person, and the EEG shows a mixture of sleep spindles and delta waves. This stage occurs about 20 minutes after falling asleep.

Stage 4. Deep sleep occurs. The person is very relaxed and responds slowly if awakened. When bed-wetting and sleepwalking occur, they do so during this stage. The EEG is dominated by delta waves.

In a typical seven- or eight-hour sleep period, a person goes from stage 1 to 4 of NREM sleep. Then the person ascends to stage 3 and 2 and then to REM sleep within 50 to 90 minutes. The cycle normally continues throughout the sleep period.

In ***REM sleep,*** also called ***paradoxical sleep,*** the EEG readings are similar to those of stage 1 of NREM sleep. There are significant physiological differences, however. During REM sleep, muscle tone is depressed (except for rapid movements of the eyes), and respirations and pulse rate increase and are irregular. Blood pressure also fluctuates considerably. It is during REM sleep that most dreaming occurs. Following REM sleep, the person descends again to stage 3 and 4 of NREM sleep.

REM and NREM sleep alternate throughout the night with approximately 90-minute intervals between REM periods. This cycle repeats itself from three to five times during the entire sleep period. The REM periods start out lasting from 5 to 10 minutes and gradually lengthen until the final one lasts about 50 minutes. In a normal sleep period, REM totals 90 to 120 minutes. As much as 50 percent of an infant's sleep is REM, as contrasted with 20 percent for adults. Most sedatives significantly decrease REM sleep.

As a person ages, the average time spent sleeping decreases. In addition, the percentage of REM sleep decreases. It has been suggested that the high percentage of REM sleep in infants and children reflects increased neuronal activity, which is important for the maturation of the brain. Infants apparently need this internal stimulation since the available external stimuli are restricted. Support for this idea comes from the fact that dreams, a particular kind of conscious activity in the brain, are most frequent during REM-type sleep.

Recent studies with animals suggest that two specific neural centers in the brain stem determine the occurrence of NREM and REM sleep. The NREM sleep center is found in the raphe nuclei. Its neurons contain large amounts of the neurotransmitter serotonin (5-HT). When the supply of serotonin is exhausted, the result is severe insomnia and a reduction in both NREM and REM sleep. The insomnia can be alleviated by the administration of a precursor of serotonin (5-hydroxytryptophan). Serotonin itself cannot cross the blood–brain barrier. The REM sleep center is found in the locus coeruleus. Its neurons contain large amounts of the neurotransmitter norepinephrine (NE). Destruction of the loci coerulei results in a complete disappearance of REM sleep but has no influence on NREM sleep. The administration of reserpine, a drug that exhausts the supply of both serotonin and norepinephrine, results in the elimination of both NREM and REM sleep. All these observations suggest that serotonin is important for NREM sleep, that norepinephrine is important for REM sleep, and that normally REM sleep is possible only if preceded by NREM sleep.

Natural body rhythms, especially body temperature, determine the length of sleep. The higher the body temperature, the longer a person will sleep.

A ***polysomograph*** (*poly* = many; *somnus* = sleep; *graph* = to write) is an instrument that uses electrodes to record several physiological variables during sleep. Among these variables are brain electrical activity recorded as an electroencephalogram (EEG), eye movements recorded as an electrooculogram (EOG), and muscle electrical activity recorded as an electromyogram (EMG). These recordings indicate precisely when patients fall asleep, how many wake periods they experience, and the quality and duration of sleep.

CLINICAL APPLICATION: SLEEP DISORDERS

At least three sleep disorders are clinically significant. *Narcolepsy* (NAR-kō-lep-sē; *narke* = numbness; *lepsis* = seizure) is a condition of involuntary attacks of sleep that last about 15 minutes and may occur at almost any time of the day. It is an inability, in the waking state, to inhibit REM sleep. *Insomnia* (in-SOM-nē-a; *in* = not; *somnus* = sleep) consists of difficulty in falling asleep and, usually, frequent awakening. It may be related to specific disorders or secondary factors, both medical and psychiatric. One reason sleeping pills such as barbiturates do not work to control an insomnia problem is that these drugs interfere with the normal sleeping patterns and, especially, reduce the relative amount of REM sleep. *Hypersomnia* refers to an excessively long or deep sleep from which a person can be awakened only by vigorous stimulation. It may be associated with a toxic state.

STUDY OUTLINE

Sensations (p. 425)

Definition (p. 425)

1. Sensation is a state of awareness of external and internal conditions of the body.
2. The prerequisites for a sensation to occur are reception of a stimulus, conversion of the stimulus into a nerve impulse through the generation of a receptor or generator potential by a receptor, conduction of the impulse to the brain, and translation of the impulse into a sensation by a region of the brain.
3. Each stimulus is capable of causing the membrane of a receptor to depolarize. This is called the generator potential or receptor potential.

Characteristics (p. 426)

1. Projection occurs when the brain refers a sensation to the point of stimulation.
2. Adaptation is the loss of sensation even though the stimulus is still applied.
3. An afterimage is the persistence of a sensation even though the stimulus is removed.
4. Modality is the property by which one sensation is distinguished from another.

Classification of Receptors (p. 426)

1. According to location, receptors are classified as exteroceptors, visceroceptors, and proprioceptors.
2. On the basis of type of stimulus detected, receptors are classified as mechanoreceptors, thermoreceptors, nociceptors, electromagnetic receptors, and chemoreceptors.
3. In terms of simplicity or complexity, simple receptors are associated with general senses, and complex receptors are associated with special senses.

General Senses (p. 426)

Cutaneous Sensations (p. 426)

1. Cutaneous sensations include tactile sensations (touch, pressure, vibration), thermoreceptive sensations (heat and cold), and pain. Receptors for these sensations are located in the skin, connective tissue under the skin, mucous membranes, and the ends of the gastrointestinal tract.
2. Receptors for touch are hair root plexuses, free nerve endings, tactile (Merkel's) discs, corpuscles of touch (Meissner's corpuscles), and type II cutaneous mechanoreceptors (end organs of Ruffini). Receptors for pressure are free nerve endings, type II cutaneous mechanoreceptors, and lamellated (Pacinian) corpuscles. Receptors for vibration are corpuscles of touch and lamellated corpuscles.
3. Pain receptors (nociceptors) are located in nearly every body tissue. Pain may be acute or chronic.
4. Two kinds of pain recognized in the parietal lobe of the cortex are somatic and visceral.

5. Referred pain is felt in the skin near or away from the organ sending pain impulses.
6. Phantom pain is the sensation of pain in a limb that has been amputated.
7. Pain impulses may be inhibited by drugs, surgery, or acupuncture.
8. In acupuncture, enkephalins block the release of substance P and inhibit pain transmission in the brain.

Proprioceptive Sensations (p. 431)

1. Receptors located in skeletal muscles, tendons, in and around joints, and the internal ear convey nerve impulses related to muscle tone, movement of body parts, and body position.
2. The receptors include muscle spindles, tendon organs (Golgi tendon organs), joint kinesthetic receptors, and the maculae and cristae.

Levels of Sensation (p. 432)

1. Sensory fibers terminating in the lower brain stem bring about far more complex motor reactions than simple spinal reflexes.
2. When sensory impulses reach the lower brain stem, they cause subconscious motor reactions.
3. Sensory impulses that reach the thalamus can be localized crudely in the body.
4. When sensory impulses reach the cerebral cortex, we experience precise localization.

Physiology of Sensory Pathways (p. 432)

1. Sensory information from all parts of the body terminates in a specific area of the somatosensory cortex.
2. In the posterior column pathway and the spinothalamic pathways there are first-order, second-order, and third-order neurons.
3. The neural pathway for pain and temperature is the lateral spinothalamic pathway.
4. The neural pathway for light touch and pressure is the anterior spinothalamic pathway.
5. The neural pathway for discriminative touch, proprioception, and vibration is the posterior column pathway.
6. The pathways to the cerebellum are the anterior and posterior spinocerebellar tracts.

Physiology of Motor Pathways (p. 437)

1. Incoming sensory information is added to, subtracted from, or integrated with other information arriving from all other operating sensory receptors.
2. The integration process occurs at many stations along the pathways of the central nervous system such as within the spinal cord, brain stem, cerebellum, and cerebral cortex.
3. A motor response to make a muscle contract or a gland secrete can be initiated at any of these stations or levels.
4. As the location of the sensory–motor linkage climbs to higher levels in the central nervous system, additional contributions are introduced that enrich the growing inventory of motor responses.
5. When the input reaches the highest center, the process of sensory–motor integration occurs. This involves not only the utilization of information contained within that center but also the information impinging on that center that is delivered from other centers within the central nervous system.
6. The muscles of all parts of the body are controlled by a specific area of the motor cortex.
7. Voluntary motor impulses are conveyed from the brain through the spinal cord along the pyramidal pathways and the extrapyramidal pathways.
8. Pyramidal pathways include the lateral corticospinal, anterior corticospinal, and corticobulbar tracts.
9. Major extrapyramidal tracts are the rubrospinal, tectospinal, vestibulospinal, and reticulospinal tracts.

Integrative Functions (p. 440)

1. Memory is the ability to recall thoughts and is generally classified into two kinds: short-term and long-term memory.
2. A memory trace in the brain is called an engram.
3. Short-term memory is related to electrical and chemical events; long-term memory is related to anatomical and biochemical changes at synapses.
4. Sleep and wakefulness are integrative functions that are controlled by the reticular activating system (RAS).
5. Non–rapid eye movement (NREM) sleep consists of four stages identified by EEG recordings.
6. Most dreaming occurs during rapid eye movement (REM) sleep.
7. The neurotransmitters that affect sleep are serotonin (5-HT) and norepinephrine (NE).
8. A polysomograph is an instrument that measures several physiological variables during sleep.

REVIEW QUESTIONS

1. Define a sensation and a sense receptor. What prerequisites are necessary for the perception of a sensation? (p. 425)
2. Describe the following characteristics of a sensation: projection, adaptation, afterimage, modality. (p. 426)
3. Name some examples of adaptation not discussed in the text.
4. Classify receptors on the basis of location, stimulus detected, and simplicity or complexity. (p. 426)
5. Distinguish between a general sense and a special sense. (p. 426)
6. What is a cutaneous sensation? Distinguish tactile, thermoreceptive, and pain sensations. (p. 426)
7. How are cutaneous receptors distributed over the body? Relate your response to the two-point discrimination test. (p. 427)
8. For each of the following cutaneous sensations, describe the receptor involved in terms of structure, function, and location: touch, pressure, vibration, and pain. (p. 427)
9. How do cutaneous sensations help maintain homeostasis?

10. Why are pain receptors important? Differentiate somatic pain, visceral pain, referred pain, and phantom pain. (p. 428)
11. Why is the concept of referred pain useful to the physician in diagnosing internal disorders? (p. 429)
12. What is acupuncture? Describe how acupuncture is believed to relieve pain. (p. 430)
13. How is acupuncture currently being used in the United States? (p. 430)
14. What is the proprioceptive sense? Where are the receptors for this sense located? (p. 431)
15. Describe the structure of muscle spindles, tendon organs (Golgi tendon organs), and joint kinesthetic receptors. (p. 431)
16. Relate proprioception to the maintenance of homeostasis.
17. Describe the various levels of sensation in the central nervous system. (p. 432)
18. Describe how various parts of the body are represented in the somatosensory cortex. (p. 432)
19. Distinguish between the posterior column and spinothalamic pathways. (p. 433)
20. What is the sensory pathway for pain and temperature? How does it function? (p. 433)
21. Which pathway controls light touch and pressure? How does it function? (p. 434)
22. Which pathway is responsible for discriminative touch, proprioception, and vibration? How does it function? (p. 434)
23. Describe how sensory input and motor responses are linked in the central nervous system. (p. 437)
24. Describe how various parts of the body are represented in the motor cortex. (p. 438)
25. Which pathways control voluntary motor impulses from the brain through the spinal cord? How are they distinguished? (p. 438)
26. Define memory. What are the two kinds of memory? (p. 440)
27. Define an engram and memory consolidation. (p. 441)
28. Describe the proposed mechanism of memory. (p. 441)
29. Describe how sleep and wakefulness are related to the reticular activating system (RAS). (p. 442)
30. What are the four stages of non–rapid eye movement (NREM) sleep? How is NREM sleep distinguished from rapid eye movement (REM) sleep? (p. 443)
31. Explain the roles of serotonin (5-HT) and norepinephrine (NE) in sleep. (p. 444)
32. Define narcolepsy, insomnia, and hypersomnia. (p. 444)

SELECTED READINGS

Begley, S., K. Springen, S. Katz, M. Hager, and E. Jones. "Memory," *Newsweek,* 29 September 1986.

Belgrade, M. J. "Control of Pain in Cancer Patients," *Postgraduate Medicine,* March 1989.

Bower, B. "Million-Cell Memories," *Science News,* 15 November 1986.

Guyton, A. C. *Basic Neuroscience: Anatomy and Physiology.* Philadelphia: Saunders, 1987.

Herbert, W. "Remembrance of Things Partly," *Science News,* 10 December 1983.

Mishkin, M., and T. Appenzeller. "The Anatomy of Memory," *Scientific American,* June 1987.

Morrison, A. R. "A Window on the Sleeping Brain," *Scientific American,* April 1983.

Schmidt, R. F. *Fundamentals of Sensory Physiology,* 3rd ed. New York: Springer-Verlag. 1986.

Wener, S. W. "The Pain of Cancer," *Postgraduate Medicine,* October 1988.

Chapter 16

The Autonomic Nervous System

Chapter Contents at a Glance

Student Objectives

1. Compare the structural and functional differences between the somatic efferent and autonomic portions of the nervous system.
2. Identify the principal structural features of the autonomic nervous system.
3. Compare the sympathetic and parasympathetic divisions of the autonomic nervous system in terms of structure, physiology, and neurotransmitters released.
4. Describe the various postsynaptic receptors involved in autonomic responses.
5. Describe a visceral autonomic reflex and its components.
6. Explain the role of the hypothalamus and its relationship to the sympathetic and parasympathetic divisions.
7. Explain the relationship between biofeedback and meditation and the autonomic nervous system.

The portion of the nervous system that regulates the activities of smooth muscle, cardiac muscle, and certain glands is the ***autonomic nervous system* (*ANS*).** Structurally, the system consists of visceral efferent neurons organized into nerves, ganglia, and plexuses. (There are also visceral receptors and afferent neurons involved in the ANS). Functionally, it usually operates without conscious control. The system was originally named *autonomic* because physiologists thought it functioned with no control from the central nervous system, that it was autonomous or self-governing. It is now known that the autonomic system is neither structurally nor functionally independent of the central nervous system. It is regulated by centers in the brain, in particular by the cerebral cortex, hypothalamus, and medulla oblongata. However, the old terminology has been retained, and since the autonomic nervous system does differ from the somatic nervous system in some ways, the two are separated for convenience of study.

SOMATIC EFFERENT AND AUTONOMIC NERVOUS SYSTEMS

Whereas the somatic efferent nervous system produces conscious movement in skeletal muscles, the autonomic nervous system (visceral efferent nervous system) regulates visceral activities, and it generally does so involuntarily and automatically. Examples of visceral activities regulated by the autonomic nervous system are changes in the size of the pupil, accommodation for near vision, dilation of blood vessels, adjustment of the rate and force of the heartbeat, movements of the gastrointestinal tract, and secretion by most glands. These activities usually lie beyond conscious control. They are automatic.

The autonomic nervous system is generally considered to be entirely motor. All its axons are efferent fibers, which transmit nerve impulses from the central nervous system to visceral effectors. Autonomic fibers are called ***visceral efferent fibers. Visceral effectors*** include cardiac muscle, smooth muscle, and glandular epithelium. This does not mean that there are no afferent (sensory) impulses from visceral effectors, however. Impulses that give rise to visceral sensations pass over visceral afferent neurons that have cell bodies located in the posterior (dorsal) root ganglia of spinal nerves and ganglia of some cranial nerves. Some functions of these afferent neurons were described in discussing the cranial and spinal nerves. The hypothalamus, which largely controls the autonomic nervous system, also receives impulses from the visceral sensory fibers, as well as from some somatic sensory fibers.

In the motor portion of the neural pathway of the somatic efferent system, the axon of an efferent neuron runs from the central nervous system and synapses directly on skeletal muscle fibers. In the neural pathway of the autonomic nervous system, there are two types of efferent neurons and a ganglion between them. The axon of the first neuron runs from the central nervous system to a ganglion, where it synapses with the cell body of the second efferent neuron. It is the axon of this second neuron that ultimately synapses on a visceral effector. Also, whereas fibers of somatic efferent neurons release acetylcholine (ACh) as their neurotransmitter, fibers of autonomic efferent neurons release either ACh or norepinephrine (NE).

The autonomic nervous system consists of two principal divisions: the ***sympathetic*** and the ***parasympathetic.*** Many organs innervated by the autonomic nervous system receive visceral efferent neurons from both components of the autonomic system—one set from the sympathetic division; another from the parasympathetic division. In general, impulses transmitted by the fibers of one division stimulate the organ to start or increase activity, whereas impulses from the other division decrease the organ's activity. Organs that receive impulses from both sympathetic and parasympathetic fibers are said to have ***dual innervation.*** Thus, autonomic innervation may be excitatory or inhibitory. In the somatic efferent nervous system, only one kind of motor neuron innervates a skeletal muscle and, moreover, innervation is always excitatory. When a somatic neuron stimulates a skeletal muscle, the muscle becomes active. When the neuron ceases to stimulate the muscle, contraction stops altogether.

A summary of the principal differences between the somatic efferent and autonomic nervous systems is presented in Exhibit 16-1.

STRUCTURE OF THE AUTONOMIC NERVOUS SYSTEM

Visceral Efferent Pathways

Autonomic visceral efferent pathways almost always consist of two efferent (motor) neurons. One extends from the central nervous system to a ganglion. The other extends directly from the ganglion to the effector (muscle or gland).

The first of the visceral efferent neurons in an autonomic pathway is called a ***preganglionic neuron*** (Figure 16-1). Its cell body is in the brain or spinal cord. Its myelinated axon, called a ***preganglionic fiber,*** passes out of the central nervous system as part of a cranial or spinal nerve. At some point, the fiber separates from the nerve and travels to an autonomic ganglion, where it synapses with the dendrites or cell body of the postganglionic neuron, the second neuron in the visceral efferent pathway.

The ***postganglionic neuron*** lies entirely outside the central nervous system. Its cell body and dendrites (if it has dendrites) are located in the autonomic ganglion, where the synapse with one or more preganglionic fibers occurs. The axon of a postganglionic neuron, called a ***postganglionic fiber,*** is unmyelinated and terminates in a visceral effector.

Thus, preganglionic neurons convey efferent impulses

EXHIBIT 16-1 COMPARISON OF SOMATIC EFFERENT AND AUTONOMIC NERVOUS SYSTEMS

	Somatic Efferent	Autonomic
Effectors	Skeletal muscles.	Cardiac muscle, smooth muscle, glandular epithelium.
Type of control	Voluntary.	Involuntary.
Neural pathway	The axon of one efferent neuron extends from CNS and synapses directly on skeletal muscle fibers.	The axon of one efferent neuron extends from the CNS and synapses with another efferent neuron in a ganglion; the axon of the second neuron synapses on a visceral effector.
Action on effector	Always excitatory.	May be excitatory or inhibitory, depending on whether stimulation is sympathetic or parasympathetic and the effector innervated.
Neurotransmitters	Acetylcholine (ACh).	Acetylcholine (ACh) or norepinephrine (NE).

from the central nervous system to autonomic ganglia. Postganglionic neurons relay the impulses from autonomic ganglia to visceral effectors. This is termed a ***two motor neuron pathway.***

Preganglionic Neurons

In the sympathetic division, the preganglionic neurons have their cell bodies in the lateral gray horns of the 12 thoracic segments and first 2 or 3 lumbar segments of the spinal cord (Figure 16-2). It is for this reason that the sympathetic division is also called the ***thoracolumbar*** (thō'-ra-kō-LUM-bar) ***division*** and the fibers of the sympathetic preganglionic neurons are known as the ***thoracolumbar outflow.***

The cell bodies of the preganglionic neurons of the parasympathetic division are located in the nuclei of cranial nerves III, VII, IX, and X in the brain stem and in the lateral gray horns of the second through fourth sacral segments of the spinal cord. Hence, the parasympathetic division is also known as the ***craniosacral division,*** and the fibers of the parasympathetic preganglionic neurons are referred to as the ***craniosacral outflow.***

Autonomic Ganglia

Autonomic pathways almost always include ***autonomic ganglia,*** where synapses between preganglionic and postganglionic visceral efferent neurons occur. Autonomic ganglia differ from posterior root ganglia. The latter

FIGURE 16-1 **Relationship between preganglionic and postganglionic (sympathetic) neurons.**

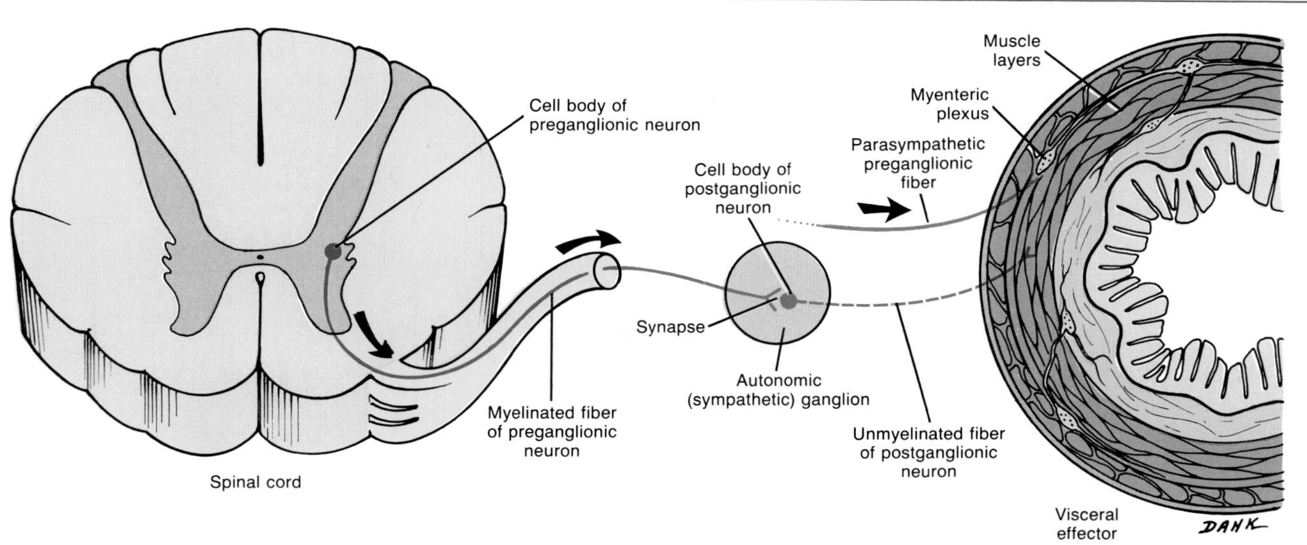

FIGURE 16-2 Structure of the autonomic nervous system. Although the parasympathetic division is shown only on the left side of the figure and the sympathetic division is shown only on the right side, keep in mind that each division is actually on both sides of the body (bilateral symmetry).

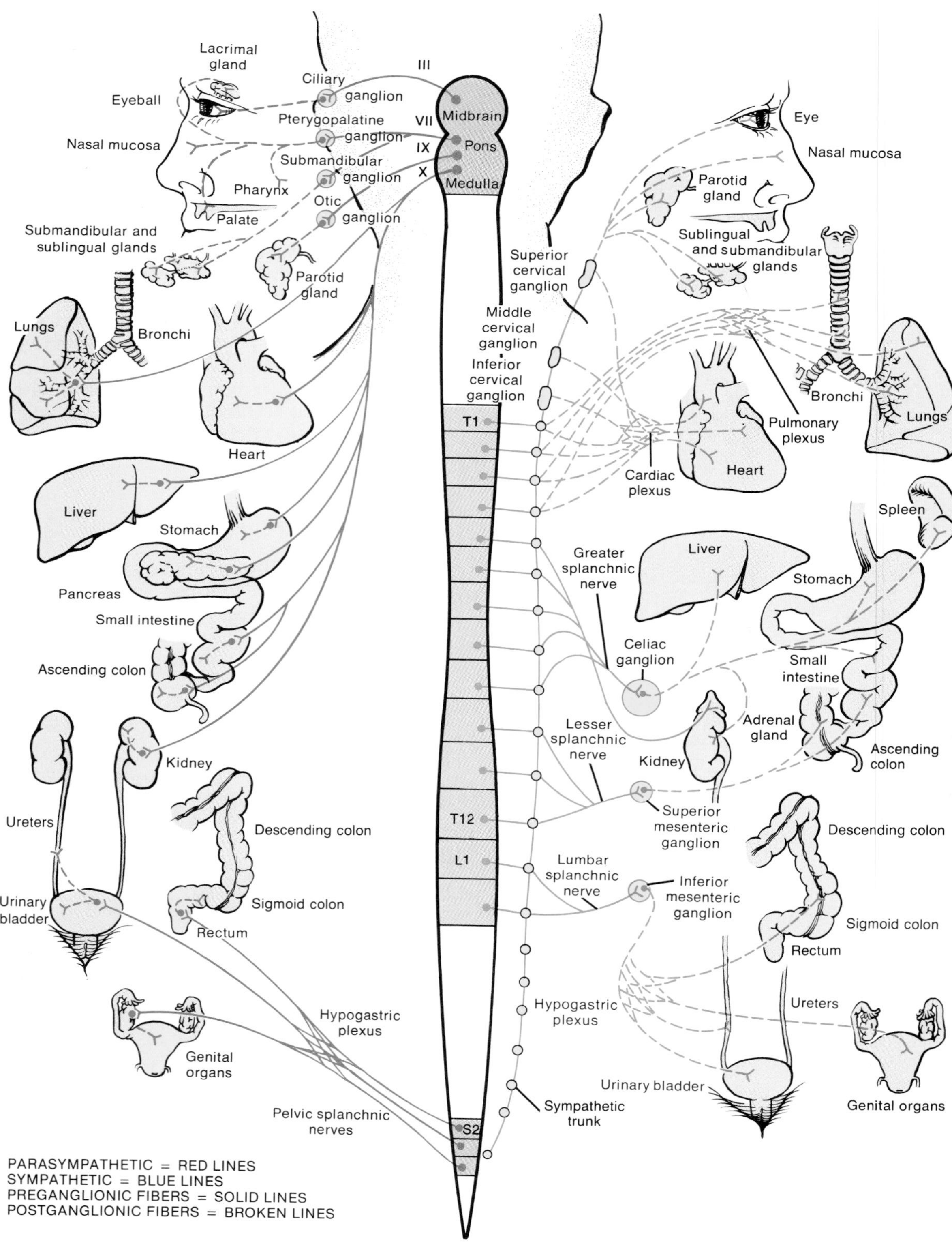

contain cell bodies of sensory neurons, and no synapses occur in them.

The autonomic ganglia may be divided into three general groups. The ***sympathetic trunk (vertebral chain) ganglia*** are a series of ganglia that lie in a vertical row on either side of the vertebral column, extending from the base of the skull to the coccyx (Figure 16-3). They are also known as ***paravertebral (lateral) ganglia.*** They receive preganglionic fibers only from the sympathetic division (Figure 16-2). Because of this, sympathetic preganglionic fibers tend to be short.

The second kind of autonomic ganglion also belongs to the sympathetic division. It is called a ***prevertebral (collateral) ganglion*** (Figure 16-3). The ganglia of this group lie anterior to the spinal column and close to the large abdominal arteries from which their names are derived. Examples of prevertebral ganglia so named are the celiac ganglion, on either side of the celiac artery just

FIGURE 16-3 **Ganglia and rami communicantes of the sympathetic division of the autonomic nervous system.**

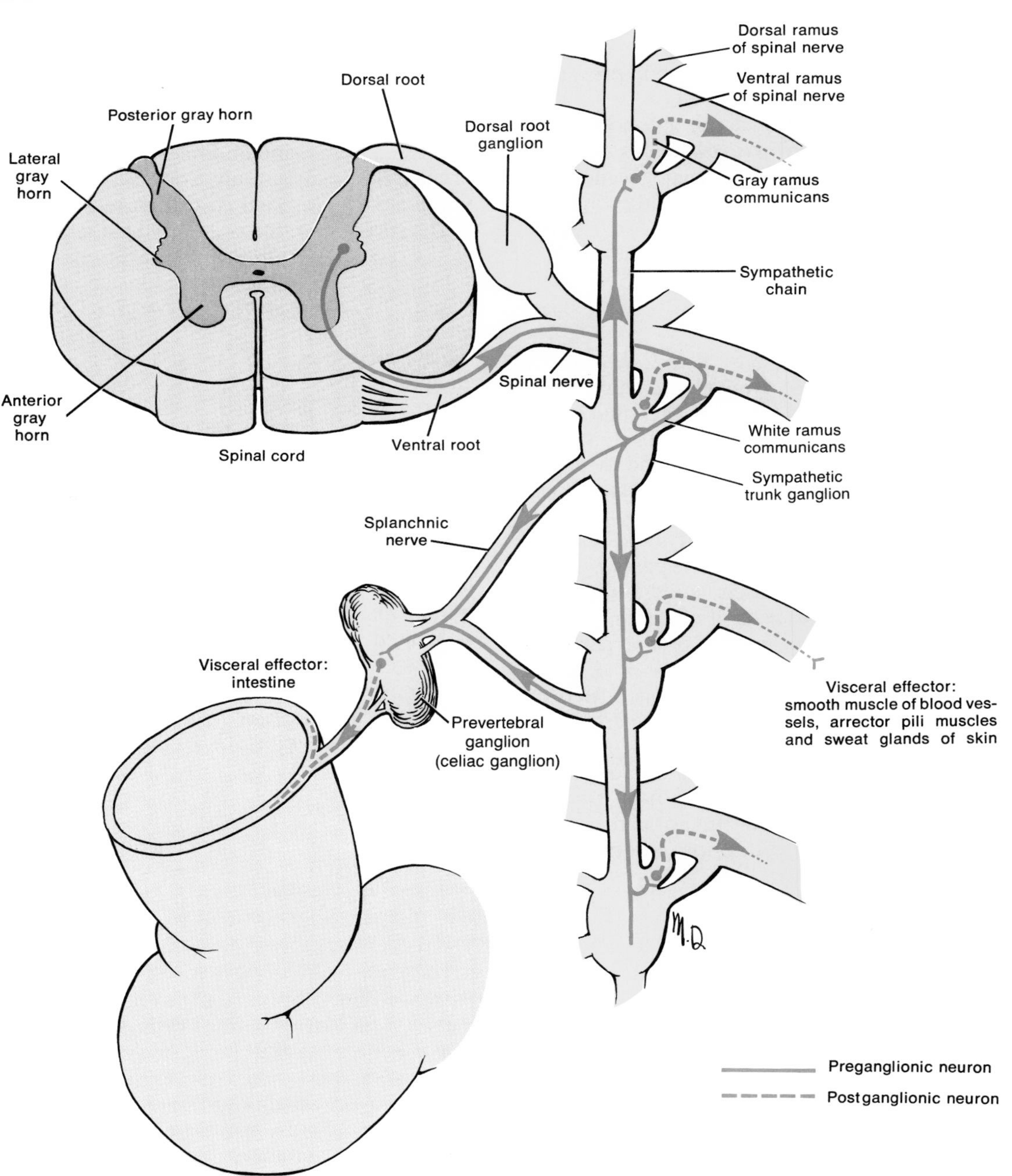

below the diaphragm; the superior mesenteric ganglion, near the beginning of the superior mesenteric artery in the upper abdomen; and the inferior mesenteric ganglion, located near the beginning of the inferior mesenteric artery in the middle of the abdomen (Figure 16-2). Prevertebral ganglia receive preganglionic fibers from the sympathetic division.

The third kind of autonomic ganglion belongs to the parasympathetic division and is called a ***terminal (intramural) ganglion.*** The ganglia of this group are located at the end of a visceral efferent pathway very close to visceral effectors or actually within the walls of visceral effectors. Terminal ganglia receive preganglionic fibers from the parasympathetic division. Since preganglionic fibers travel from the CNS to terminal ganglia close to or within the organ innervated, parasympathetic preganglionic fibers tend to be long (Figure 16-2).

In addition to autonomic ganglia, the autonomic nervous system contains ***autonomic plexuses.*** Slender nerve fibers from ganglia containing postganglionic nerve cell bodies arranged in a branching network constitute an autonomic plexus.

Postganglionic Neurons

Axons from preganglionic neurons of the sympathetic division pass to ganglia of the sympathetic trunk. They can either synapse in the sympathetic chain ganglia with postganglionic neurons, or they can continue, without synapsing, through the chain ganglia to end at a prevertebral ganglion where synapses with the postganglionic neurons can take place. Each sympathetic preganglionic fiber synapses with 20 or more postganglionic fibers in the ganglion, and the postganglionic fibers pass to several visceral effectors. After exiting their ganglia, the postsynaptic fibers innervate their visceral effectors.

Axons from preganglionic neurons of the parasympathetic division pass to terminal ganglia near or within a visceral effector. In the ganglion, the presynaptic neuron usually synapses with only four or five postsynaptic neurons to a single visceral effector. After exiting their ganglia, the postsynaptic fibers supply their visceral effectors.

With this background in mind, we can now examine some specific structural features of the sympathetic and parasympathetic divisions of the autonomic nervous system.

Sympathetic Division

The preganglionic fibers of the sympathetic division have their cell bodies located in the lateral gray horns of all the thoracic segments and first two or three lumbar segments of the spinal cord (Figure 16-2). The preganglionic fibers are myelinated and leave the spinal cord through the ventral root of a spinal nerve along with the somatic efferent fibers at the same segmental levels. After exiting through the intervertebral foramina, the preganglionic sympathetic fibers enter a white ramus to pass to the nearest sympathetic trunk ganglion on the same side. Collectively, the white rami are called the ***white rami communicantes*** (kō-myoo-ni-KAN-tēz). Their name indicates that they contain myelinated fibers. Only thoracic and upper lumbar nerves have white rami communicantes. The white rami communicantes connect the ventral ramus of the spinal nerve with the ganglia of the sympathetic trunk.

The paired sympathetic trunks are situated anterolaterally to the spinal cord, one on either side. Each consists of a series of ganglia arranged more or less segmentally. The divisions of the sympathetic trunk are named on the basis of location. Typically, there are 22 ganglia in each chain: 3 cervical, 11 thoracic, 4 lumbar, and 4 sacral. Although the trunk extends downward from the neck, thorax, and abdomen to the coccyx, it receives preganglionic fibers only from the thoracic and lumbar segments of the spinal cord (Figure 16-2).

The cervical portion of each sympathetic trunk is located in the neck anterior to the prevertebral muscles. It is subdivided into a superior, middle, and inferior ganglion (Figure 16-2). The ***superior cervical ganglion*** is posterior to the internal carotid artery and anterior to the transverse processes of the second cervical vertebra. Postganglionic fibers leaving the ganglion serve the head, where they are distributed to the sweat glands, the smooth muscle of the eye and blood vessels of the face, the nasal mucosa, and the submandibular, sublingual, and parotid salivary glands. Gray rami communicantes (described shortly) from the ganglion also pass to the upper two to four cervical spinal nerves. The ***middle cervical ganglion*** is situated near the sixth cervical vertebra at the level of the cricoid cartilage. Postganglionic fibers from it innervate the heart. The ***inferior cervical ganglion*** is located near the first rib, anterior to the transverse processes of the seventh cervical vertebra. Its postganglionic fibers also supply the heart.

The thoracic portion of each sympathetic trunk usually consists of 11 segmentally arranged ganglia, lying ventral to the necks of the corresponding ribs. This portion of the sympathetic trunk receives most of the sympathetic preganglionic fibers. Postganglionic fibers from the thoracic sympathetic trunk innervate the heart, lungs, bronchi, and other thoracic viscera.

The lumbar portion of each sympathetic trunk is found on either side of the corresponding lumbar vertebrae. The sacral portion of the sympathetic trunk lies in the pelvic cavity on the medial side of the sacral foramina. Postganglionic fibers from the lumbar and sacral sympathetic chain ganglia are distributed with the respective spinal nerves via gray rami, or they may join the hypogastric plexus via direct visceral branches.

When a preganglionic fiber of a white ramus communicans enters the sympathetic trunk, it may terminate (synapse) in several ways. Some fibers synapse in the first ganglion at the level of entry. Others pass up or down the sympathetic trunk for a variable distance to form the fibers on which the ganglia are strung. These fibers, known as ***sympathetic chains*** (Figure 16-3), may not synapse until they reach a ganglion in the cervical or sacral area. Most rejoin the spinal nerves before supplying peripheral visceral effectors such as sweat glands and the smooth muscle in blood vessels and around hair follicles in the extremities. The ***gray ramus communicans*** (kō-MYOO-ni-kanz) is the structure containing the postganglionic fibers that connect the ganglion of the sympathetic trunk to the spinal nerve (Figure 16-3). The fibers are unmyelinated. All spinal nerves have gray rami communicantes. Gray rami communicantes outnumber the white rami, since there is a gray ramus leading to each of the 31 pairs of spinal nerves.

In most cases, a sympathetic preganglionic fiber terminates by synapsing with a large number, usually 20 or more, of postganglionic cell bodies in a ganglion. Often the postganglionic fibers then terminate in widely separated organs of the body. Thus, an impulse that starts in a single preganglionic neuron may reach several visceral effectors. For this reason, most sympathetic responses have widespread effects on the body.

Some preganglionic fibers pass through the sympathetic trunk without terminating in the trunk. Beyond the trunk, they form nerves known as ***splanchnic*** (SPLANK-nik) ***nerves*** (Figure 16-2). After passing through the trunk of ganglia, the splanchnic nerves from the thoracic area terminate in the ***celiac ganglion*** (SĒ-lē-ak) or ***solar plexus.*** In the plexus, the preganglionic fibers synapse in ganglia with postganglionic cell bodies. These ganglia are prevertebral ganglia. The greater splanchnic nerve passes to the celiac ganglion of the celiac plexus. From here, postganglionic fibers are distributed to the stomach, spleen, liver, kidney, and small intestine. The lesser splanchnic nerve passes through the celiac plexus to the superior mesenteric ganglion of the superior mesenteric plexus. Postganglionic fibers from this ganglion innervate the small intestine and colon. The lowest splanchnic nerve, not always present, enters the renal plexus. Postganglionics supply the renal arterioles and ureter. The lumbar splanchnic nerve enters the inferior mesenteric plexus. In the plexus, the preganglionic fibers synapse with postganglionic fibers in the inferior mesenteric ganglion. These fibers pass through the hypogastric plexus and supply the distal colon and rectum, urinary bladder, and genital organs. As noted earlier, the postganglionic fibers leaving the prevertebral ganglia follow the course of various arteries to abdominal and pelvic visceral effectors.

Sympathetic preganglionic fibers that innervate the medulla of each of the adrenal glands travel in splanchnic nerves but do not synapse before terminating in the gland. This means that there are no sympathetic postganglionic fibers to the adrenal medulla. This is the only exception to the usual pattern of two efferent neurons in a chain in an autonomic efferent pathway.

Parasympathetic Division

The preganglionic cell bodies of the parasympathetic division are found in nuclei in the brain stem and the lateral gray horn of the second through fourth sacral segments of the spinal cord (Figure 16-2). Their fibers emerge as part of a cranial nerve or as part of the ventral root of a spinal nerve. The ***cranial parasympathetic outflow*** consists of preganglionic fibers that leave the brain stem by way of the oculomotor (III) nerves, facial (VII) nerves, glossopharyngeal (IX) nerves, and vagus (X) nerves. The ***sacral parasympathetic outflow*** consists of preganglionic fibers that leave the ventral roots of the second through fourth sacral nerves. The preganglionic fibers of both the cranial and sacral outflows end in terminal ganglia, where they synapse with postganglionic neurons. We will first look at the cranial outflow.

The cranial outflow has five components: four pairs of ganglia and the plexuses associated with the vagus nerve. The four pairs of cranial parasympathetic ganglia innervate structures in the head and are located close to the organs they innervate. The ***ciliary ganglion*** is near the back of an orbit lateral to each optic (II) nerve. Preganglionic fibers pass with the oculomotor (III) nerve to the ciliary ganglion. Postganglionic fibers from the ganglion innervate smooth muscle cells in the eyeball. Each ***pterygopalatine*** (ter'-i-gō-PAL-a-tin) ***ganglion*** is situated lateral to a sphenopalatine foramen. It receives preganglionic fibers from the facial (VII) nerve and transmits postganglionic fibers to the nasal mucosa, palate, pharynx, and lacrimal gland. Each ***submandibular ganglion*** is found near the duct of a submandibular salivary gland. It receives preganglionic fibers from the facial (VII) nerve and transmits postganglionic fibers that innervate the submandibular and sublingual salivary glands. The ***otic ganglia*** are situated just below each foramen ovale. The otic ganglion receives preganglionic fibers from the glossopharyngeal (IX) nerve and transmits postganglionic fibers that innervate the parotid salivary gland. Ganglia associated with the cranial outflow are classified as terminal ganglia. Since the terminal ganglia are close to their visceral effectors, postganglionic parasympathetic fibers are short. Postganglionic sympathetic fibers are relatively long.

The last component of the cranial outflow consists of the preganglionic fibers that leave the brain via the vagus (X) nerves. This component has the most extensive distribution of the parasympathetic fibers, providing about 80 percent of the craniosacral outflow. Each vagus (X) nerve enters into the formation of several plexuses in the thorax and abdomen. As it passes through the thorax, it sends fibers to the ***superficial cardiac plexus*** in the arch of the aorta and the ***deep cardiac plexus*** anterior to the branching of the trachea. These plexuses contain terminal ganglia, and the postganglionic parasympathetic fibers emerging from them supply the heart. Also in the thorax is the ***pulmonary plexus,*** anterior and posterior to roots of the lungs and within the lungs themselves. It receives preganglionic fibers from the vagus and transmits postganglionic parasympathetic fibers to the lungs and bronchi. Other plexuses associated with the vagus (X) nerve are described in later chapters in conjunction with the appropriate thoracic, abdominal, and pelvic viscera. Postganglionic fibers from these plexuses innervate viscera such as the liver, pancreas, stomach, kidneys, small intestine, and part of the colon.

The sacral parasympathetic outflow consists of preganglionic fibers from the ventral roots of the second through fourth sacral nerves. Collectively, they form the ***pelvic splanchnic nerves.*** They pass into the hypogastric plexus. From ganglia in the plexus, parasympathetic postganglionic fibers are distributed to the colon, ureters, urinary bladder, and reproductive organs.

The salient structural features of the sympathetic and parasympathetic divisions are compared in Exhibit 16-2.

EXHIBIT 16-2 STRUCTURAL FEATURES OF SYMPATHETIC AND PARASYMPATHETIC DIVISIONS

Sympathetic	Parasympathetic
Forms thoracolumbar outflow.	Forms craniosacral outflow.
Contains sympathetic trunk and prevertebral ganglia.	Contains terminal ganglia.
Ganglia are close to the CNS and distant from visceral effectors.	Ganglia are near or within visceral effectors.
Each preganglionic fiber synapses with many postganglionic neurons that pass to many visceral effectors.	Each preganglionic fiber usually synapses with four or five postganglionic neurons that pass to a single visceral effector.
Distributed throughout the body, including the skin.	Distribution limited primarily to head and viscera of thorax, abdomen, and pelvis.

FIGURE 16-4 **Neurotransmitters and receptors associated with the autonomic nervous system.**

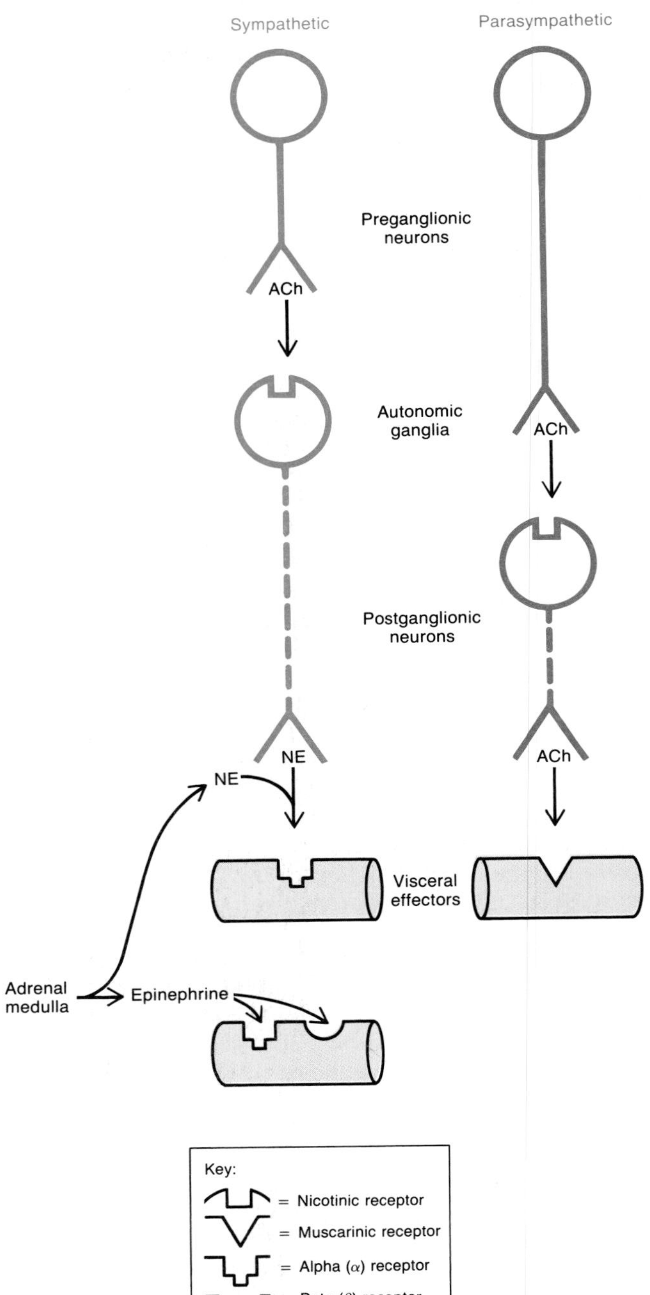

PHYSIOLOGY OF THE AUTONOMIC NERVOUS SYSTEM

Neurotransmitters

Autonomic fibers, like other axons of the nervous system, release neurotransmitters at synapses as well as at points of contact with visceral effectors (smooth and cardiac muscle and glands). These latter points are called ***neuroeffector junctions.*** Neuroeffector junctions may be either neuromuscular or neuroglandular junctions. On the basis of the neurotransmitter produced, autonomic fibers may be classified as either cholinergic or adrenergic (Figure 16-4).

Cholinergic (kō'-lin-ER-jik) ***fibers*** release ***acetylcholine (ACh)*** and include the following: (1) all sympathetic and parasympathetic preganglionic axons, (2) all parasympathetic postganglionic axons, and (3) a few sympathetic postganglionic axons. The cholinergic sympathetic postganglionic axons include those to sweat glands, arrector pili muscles, and blood vessels in skeletal muscles. Since acetylcholine is quickly inactivated by the enzyme ***acetylcholinesterase (AChE),*** the effects of cholinergic fibers are short-lived and local.

Adrenergic (ad'-ren-ER-jik) ***fibers*** produce ***norepinephrine (NE).*** Most sympathetic postganglionic axons are adrenergic. Since norepinephrine is inactivated much more slowly by ***catechol-O-methyltransferase (COMT)*** or ***monoamine oxidase (MAO)*** than acetylcholine is inactivated by acetylcholinesterase, and since norepinephrine may enter the bloodstream, the effects of sympathetic stimulation are longer lasting and more widespread than parasympathetic stimulation. The action of NE produced by sympathetic postganglionic axons is augmented by both NE and epinephrine secreted by the adrenal medulla of each of the two adrenal glands. Since both substances are secreted into the blood, their effects are more sustained than NE released by axons. NE and epinephrine released by the adrenal medullae are ultimately destroyed by enzymes in the liver once they have acted on their visceral effectors.

Receptors

ACh is synthesized and stored in an inactive form in synaptic vesicles in the axon terminals of the cholinergic fibers. The release of ACh from the axon terminals requires calcium ions (Ca^{2+}) from the interstitial fluid and comes in bursts of thousands of molecules. ACh then diffuses the short distance across the synaptic cleft to bind with receptors on the postsynaptic membrane. The membrane, if of a neuron or muscle cell, is either hyperpolarized or depolarized quickly and the ACh is rapidly inactivated by acetylcholinesterase (AChE), an enzyme that can be inhibited by a variety of drugs such as physostigmine and neostigmine.

The actual effect produced by ACh is determined by the type of postsynaptic receptor with which it interacts (Figure 16-4). The two types of ACh postsynaptic receptors are known as nicotinic receptors and muscarinic receptors. ***Nicotinic receptors*** are found on both sympathetic and parasympathetic postganglionic neurons. These receptors are so named because the actions of ACh on such receptors are similar to those produced by nicotine. ***Muscarinic receptors*** are found on all effectors innervated by parasympathetic postganglionic axons and some effectors innervated by sympathetic postganglionic axons. These postsynaptic receptors are so named because the actions of ACh on such receptors are similar to those produced by muscarine, a toxin produced by a mushroom.

NE is released from most postganglionic sympathetic fibers. It is synthesized and stored in synaptic vesicles located in the axon terminals of adrenergic fibers. When an action potential reaches the axon terminal, NE is rapidly released into the synaptic cleft. The molecules diffuse across the cleft and combine with specific receptors on the postsynaptic membrane to elicit a specific effector response.

The effects of NE and epinephrine, like those of ACh, are also determined by the type of postsynaptic receptor with which they interact. Such receptors are found on visceral effectors innervated by most sympathetic postganglionic axons and are referred to as ***alpha (α) receptors*** and ***beta (β) receptors*** (Figure 16-4). Both alpha and beta receptors are, in turn, classified into subtypes—α-1, α-2, β-1, and β-2. The receptors are distinguished by the specific responses they elicit and by their selective combination with drugs that excite or inhibit them. In general, stimulation of an alpha receptor results in an increase in permeability of a postsynaptic membrane to sodium ions (Na^+), followed by depolarization. For this reason, alpha receptors are generally excitatory. Conversely, stimulation of beta receptors generally results in an increase in permeability of a postsynaptic membrane to the outward movement of potassium ions (K^+) and the inward movement of chloride ions (Cl^-). This brings about hyperpolarization of a postsynaptic membrane and inhibition of the postsynaptic cell. Thus, beta receptors are generally inhibitory.

Although cells of most effectors contain either alpha or beta receptors, some visceral effector cells contain both. NE, in general, stimulates alpha receptors to a greater extent than beta receptors, and epinephrine, in general, stimulates both alpha and beta receptors.

Exhibit 16-3 shows the types of receptors present on the cells of visceral effectors and their response to autonomic stimulation. Note that there are some very important (and perplexing) exceptions to the general rule describing alpha and beta receptors. For example, heart muscle has beta receptors that operate as excitatory receptors and cause the heart muscle to contract more forcibly. Other visceral effector cells possess both types of receptors. In these situations, the majority rules. The beta receptors greatly outnumber the alpha receptors in the smooth muscle fibers (cells) of the blood vessels that flow through skeletal muscles. Therefore, an injection of NE, which binds only to alpha receptors, attaches onto the surface of such smooth muscle fibers and causes a vasoconstriction. An injection of epinephrine, however, would result in a vasodilation because epinephrine acts equally on both types of receptors. But since there is a preponderance of beta receptors, the effect is that of the beta receptor, that is, vasodilation due to relaxation of smooth muscle fibers. Moreover, adding to the complexity of this situation is the fact that the effects on skeletal muscle blood vessels are primarily due to responses to local metabolites.

Activities

Most visceral effectors have ***dual innervation;*** that is, they receive fibers from both the sympathetic and the parasympathetic divisions. In these cases, impulses from one division stimulate the organ's activities, whereas impulses from the other division inhibit the organ's activities.

EXHIBIT 16-3 ACTIVITIES OF AUTONOMIC NERVOUS SYSTEM

Visceral Effector	Sympathetic Receptor	Effect of Sympathetic Stimulation	Effect of Parasympathetic Stimulation
Eye			
Radial muscle of iris	α	Contraction that results in dilation of pupil (mydriasis).	No known functional innervation.
Sphincter muscle of iris	α	No known functional innervation.	Contraction that results in constriction of pupil (miosis).
Ciliary muscle	β	Relaxation that results in far vision.	Contraction that results in near vision.
Glands			
Sweat	α	Stimulates local secretion.	Stimulates generalized secretion.
Lacrimal (tear)	—	No known functional innervation.	Stimulates secretion.
Salivary	α	Vasoconstriction, which decreases secretion.	Stimulates secretion and vasodilation.
Gastric	—	Vasoconstriction, which inhibits secretion.	Stimulates secretion.
Intestinal	—	Vasoconstriction, which inhibits secretion.	Stimulates secretion.
Adrenal medulla	—	Promotes epinephrine and norepinephrine secretion.	No known functional innervation.
Fat cells	β	Promotes lipolysis.	No known functional innervation.
Lungs (bronchial muscle)	β	Dilation.	Constriction.
Heart	β	Increases rate and strength of contraction; dilates coronary vessels that supply blood to heart muscle fibers (cells).	Decreases rate and strength of contraction; constricts coronary vessels.
Arterioles			
Skin and mucosa	α	Constriction.	No known functional innervation for most.
Skeletal muscle	α,β	Constriction or dilation.	No known functional innervation.
Abdominal viscera	α,β	Constriction.	No known functional innervation for most.
Cerebral	α	Slight constriction.	No known functional innervation.
Systemic veins	α,β	Constriction and dilation.	No known functional innervation.
Liver	β	Promotes glycogenolysis and gluconeogenesis; decreases bile secretion.	Promotes glycogen synthesis; increases bile secretion.
Gallbladder and ducts	—	Relaxation.	Contraction.
Stomach	α,β	Decreases motility and tone; contracts sphincters.	Increases motility and tone; relaxes sphincters.
Intestines	α,β	Decreases motility and tone; contracts sphincters.	Increases motility and tone; relaxes sphincters.
Kidney	β	Constriction of blood vessels that results in decreased urine volume; secretion of renin.	No effect.
Ureter	α	Increases motility.	Decreases motility.

EXHIBIT 16-3 ACTIVITIES OF AUTONOMIC NERVOUS SYSTEM (*Continued*)

Visceral Effector	Sympathetic Receptor	Effect of Sympathetic Stimulation	Effect of Parasympathetic Stimulation
Pancreas	α,β	Inhibits secretion of enzymes and insulin; promotes secretion of glucagon.	Promotes secretion of enzymes and insulin.
Spleen	—	Contraction and discharge of stored blood into general circulation.	No known functional innervation.
Urinary bladder	α,β	Relaxation of muscular wall; contraction of internal sphincter.	Contraction of muscular wall; relaxation of internal sphincter.
Arrector pili of hair follicles	α	Contraction that results in erection of hairs.	No known functional innervation.
Uterus	α,β	Inhibits contraction if nonpregnant; stimulates contraction if pregnant.	Minimal effect.
Sex organs	α	In male, contraction of smooth muscle of ductus (vas) deferens, seminal vesicle, prostate; results in ejaculation. In female, reverse uterine peristalsis.	Vasodilation and erection in both sexes; secretion in females.

The stimulating division may be either the sympathetic or the parasympathetic, depending on the organ. For example, sympathetic impulses increase heart activity, whereas parasympathetic impulses decrease it. On the other hand, parasympathetic impulses increase digestive activities, whereas sympathetic impulses inhibit them. The actions of the two systems are carefully integrated to help maintain homeostasis. A summary of the activities of the autonomic nervous system is presented in Exhibit 16-3.

The parasympathetic division is primarily concerned with activities that conserve and restore body energy during times of rest or recovery of the body. It is an ***energy conservation-restorative system.*** Under normal body conditions, for instance, parasympathetic impulses to the digestive glands and the smooth muscle of the gastrointestinal tract dominate over sympathetic impulses. Thus, energy-supplying food can be digested and absorbed by the body.

The sympathetic division, in contrast, is primarily concerned with processes involving the expenditure of energy as when one is in a condition of stress due to either physical or emotional activities. When the body is in homeostasis, the main function of the sympathetic division is to counteract the parasympathetic effects just enough to carry out normal processes requiring energy. During extreme stress, however, the sympathetic dominates the parasympathetic. When people are confronted with a stress condition, for example, their bodies become alert and they sometimes perform feats of unusual strength. Fear stimulates the sympathetic division as do a variety of other emotions and physical activities.

Activation of the sympathetic division sets into operation a series of physiological responses collectively called the ***fight-or-flight response.*** It produces the following effects.

1. The pupils of the eyes dilate.
2. The heart rate and force of contraction increase and blood pressure increases.
3. The blood vessels of the skin and viscera constrict.
4. The remainder of the blood vessels dilate. This causes a faster flow of blood into the dilated blood vessels of skeletal muscles, cardiac muscle, and lungs—organs involved in fighting off danger.
5. Rapid and deeper breathing occurs and the bronchioles dilate to allow faster movement of air in and out of the lungs.
6. Blood sugar level rises as liver glycogen is converted to glucose to supply the body's additional energy needs.
7. The medullae of the adrenal glands are stimulated to produce epinephrine and norepinephrine, hormones that intensify and prolong the sympathetic effects noted previously.
8. Processes that are not essential for meeting the stress situation are inhibited. For example, muscular movements of the gastrointestinal tract and digestive secretions are slowed down or even stopped.

CLINICAL APPLICATION: HORNER'S SYNDROME

If the sympathetic trunk is cut or damaged on one side, the sympathetic supply to that side of the head is removed, resulting in *Horner's syndrome,* in which the patient exhibits (on the affected side): ptosis (drooping of the upper eyelid), slight elevation of the lower eyelid, narrowing of the palpebral fissure, enophthalmos (the eye appears sunken), miosis (constricted pupil), and anhidrosis (lack of sweating).

VISCERAL AUTONOMIC REFLEXES

A ***visceral autonomic reflex*** adjusts the activity of a visceral effector. In other words, it results in the contraction or relaxation of smooth or cardiac muscle or change in the rate of secretion by a gland. Such reflexes assume a key role in activities involved in homeostasis such as regulating heart action, blood pressure, respiration, digestion, defecation, and urinary bladder functions.

A visceral autonomic reflex arc consists of the following components.

1. **Receptor.** The receptor is the distal end of an afferent neuron in an exteroceptor or enteroceptor.
2. **Afferent neuron.** This neuron, either a somatic afferent or visceral afferent neuron, conducts the sensory impulse to the spinal cord or brain.
3. **Association neurons.** These neurons are found in the central nervous system.
4. **Visceral efferent preganglionic neuron.** In the thoracic and abdominal regions, this neuron is in the lateral gray horn of the spinal cord. The axon passes through the ventral root of the spinal nerve, the spinal nerve itself, and the white ramus communicans. It then enters a sympathetic trunk or prevertebral ganglion, where it synapses with a postganglionic neuron. In the cranial and sacral regions, the visceral efferent preganglionic axon leaves the central nervous system and passes to a terminal ganglion, where it synapses with a postganglionic neuron. The role of the visceral efferent preganglionic neuron is to convey a motor impulse from the brain or spinal cord to an autonomic ganglion.
5. **Visceral efferent postganglionic neuron.** This neuron conducts a motor impulse from a visceral efferent preganglionic neuron to the visceral effector.
6. **Visceral effector.** A visceral effector is smooth muscle, cardiac muscle, or a gland. Alteration of the rate of activity in the effector is the response.

The basic difference between a somatic reflex arc and a visceral autonomic reflex arc is that in a somatic reflex arc only one efferent neuron is involved. In a visceral autonomic reflex arc, two efferent neurons are involved.

Visceral sensations do not always reach the cerebral cortex. Most remain at subconscious levels. Under normal conditions, you are not aware of muscular contractions of the digestive organs, heartbeat, changes in the diameter of blood vessels, and pupil dilation and constriction. Your body adjusts such visceral activities by visceral reflex arcs whose centers are in the spinal cord or lower regions of the brain. Among such centers are the cardiac, respiratory, vasomotor, swallowing, and vomiting centers in the medulla and the temperature control center in the hypothalamus. Stimuli delivered by somatic or visceral afferent neurons synapse in these centers, and the returning motor impulses conducted by visceral efferent neurons bring about an adjustment in the visceral effector usually without conscious recognition. The impulses are interpreted and acted on subconsciously. Some visceral sensations do give rise to conscious recognition: hunger, nausea, fullness of the urinary bladder and rectum, and pain from damaged viscera.

CONTROL BY HIGHER CENTERS

The autonomic nervous system is not a separate nervous system. Although little is known about the specific centers in the brain that regulate specific autonomic functions, it is known that axons from many parts of the central nervous system are connected to both the sympathetic and the parasympathetic divisions of the autonomic nervous system and thus exert considerable control over it. Autonomic centers in the cerebral cortex are connected to autonomic centers of the thalamus, for example. These, in turn, are connected to the hypothalamus. In this hierarchy of command, the thalamus sorts incoming impulses before they reach the cerebral cortex. The cerebral cortex then turns over control and integration of visceral activities to the hypothalamus. It is at the level of the hypothalamus that the major control and integration of the autonomic nervous system is exerted and much of this control is by way of hypothalamic influence of centers in the medulla and spinal cord.

The hypothalamus receives input from areas of the nervous system concerned with emotions, visceral functions, olfaction, gustation, as well as changes in temperature, osmolarity, and levels of various substances in blood. Anatomically, the hypothalamus is connected to both the sympathetic and the parasympathetic divisions of the autonomic nervous system by axons of neurons whose dendrites and cell bodies are in various hypothalamic nuclei. The axons form tracts from the hypothalamus to sympathetic and parasympathetic nuclei in the brain stem and spinal cord through relays in the reticular formation. The posterior and lateral portions of the hypothalamus appear to control the sympathetic division. When these areas are stimulated, there is an increase in visceral activities—an increase in heart rate and force of beat, a rise in blood pressure due to vasoconstriction of blood vessels, an increase in the rate and depth of respiration, dilation of the pupils, and inhibition of the gastrointestinal tract. On the other hand, the anterior and medial portions of the hypothalamus seem to control the parasympathetic division. Stimulation of these areas results in a decrease in heart rate, lowering of blood pressure, constriction of the pupils, and increased secretion and motility of the gastrointestinal tract.

Control of the autonomic nervous system by the cerebral cortex occurs primarily during emotional stress. In extreme anxiety, which can result from either conscious or subconscious stimulation in the cerebral cortex, the cortex can stimulate the hypothalamus as part of the limbic system. This, in turn, stimulates the cardiac and vasomotor centers of the medulla, which increases heart rate and force of beat, and blood pressure. If the cortex is stimulated by hearing bad news or experiencing an extremely unpleasant sight, the stimulation causes vasodilation of blood vessels, a lowering of blood pressure, and fainting.

Evidence of even more direct control of visceral responses is provided by data gathered from studies of biofeedback and meditation.

Biofeedback

In the simplest terms, ***biofeedback*** is a process in which people get constant signals, or feedback, about various visceral biological functions such as blood pressure, heart rate, and muscle tension. By using special monitoring devices, they can control these visceral functions consciously to a limited degree.

In a study conducted at the Menninger Foundation,* subjects suffering from migraine headaches received instructions in the use of a monitor that registers the skin temperature of the right index finger. Subjects were also given a typewritten sheet containing two sets of phrases. The first set was designed to help them relax the entire body. The second set was designed to bring about an increased flow of blood in the hands. The subjects practiced raising their skin temperature at home for 5 to 15 minutes a day. When skin temperature increased, the monitor emitted a high-pitched sound. In time, the monitor was abandoned.

Once the subjects learned how to vasodilate blood vessels of their extremities, the migraine headaches lessened. Since migraine headaches are believed to involve a distention of blood vessels in the head, the shunting of blood from head to hands relieved the distension and thus the pain.

* Much of the following discussion of the use of biofeedback for the treatment of migraine headaches is based on information provided by Dr. Joseph D. Sargent of the Menninger Foundation, Topeka, Kansas.

Other experiments have shown that biofeedback can be applied to childbirth. Women were given monitors hooked up to their fingers and arms to measure electrical conductivity of the skin and skeletal muscle tension. Both conductivity and tension increased with nervousness and made labor difficult. Muscle tension was recorded as a sirenlike sound that became louder with nervousness. Skin conductivity was recorded as a crackling noise that also increased in intensity with nervousness. The monitors kept the women informed of their nervousness. This was the biofeedback. Having pleasant thoughts reduced the sound levels. The reward was less nervousness. The results of the study indicate that the women in a state of reduced nervousness needed less medication during labor, and labor time was shortened.

There is no way to determine where biofeedback will lead. Perhaps the outstanding contribution of biofeedback research has been to demonstrate that the autonomic nervous system is not autonomous. Visceral responses can be controlled consciously to some degree. Current therapeutic applications of biofeedback include treatment of asthma, Raynaud's disease, hypertension, gastrointestinal disorders, fecal incontinence, anxiety, pain, and neuromuscular rehabilitation following cerebrovascular accidents (CVAs).

Meditation

Yoga, which literally means "union," is defined as a higher state of consciousness achieved through a fully rested and relaxed body and a fully awake and relaxed mind. One widely practiced technique for achieving higher consciousness is called ***transcendental meditation (TM).*** One sits in a comfortable position with the eyes closed and concentrates on a suitable sound or thought.

Research indicates that transcendental meditation can alter physiological responses. Oxygen consumption decreases drastically along with carbon dioxide elimination. Subjects have experienced a reduction in metabolic rate and blood pressure. Researchers have also observed a decrease in heart rate, an increase in the intensity of alpha brain waves, a sharp decrease in the amount of lactic acid in the blood, and an increase in the skin's electrical resistance. These last four responses are characteristic of a highly relaxed state of mind. Alpha waves are found in the EEGs of almost all individuals in a resting, but awake, state; they disappear during sleep.

These responses have been called an ***integrated response***—essentially, a hypometabolic state due to inactivation of the sympathetic division of the autonomic nervous system. The response is the exact opposite of the fight-or-flight response, which is a hyperactive state of the sympathetic division. The existence of the integrated response suggests that the central nervous system does exert some control over the autonomic nervous system.

STUDY OUTLINE

Somatic Efferent and Autonomic Nervous Systems (p. 448)

1. The somatic efferent nervous system allows conscious control of movement in skeletal muscles.
2. The autonomic nervous system, or visceral efferent nervous system, regulates visceral activities, that is, activities of smooth muscle, cardiac muscle, and certain glands, and it usually operates without conscious control.
3. It is regulated by centers in the brain and spinal cord, in particular by the cerebral cortex, the hypothalamus, and the medulla oblongata.
4. A single somatic efferent neuron synapses on skeletal muscles; in the autonomic nervous system, there are two efferent neurons—one from the CNS to a ganglion and one from a ganglion to a visceral effector.
5. Somatic efferent neurons release acetylcholine (ACh), and autonomic efferent neurons release either acetylcholine (ACh) or norepinephrine (NE).

Structure of the Autonomic Nervous System (p. 448)

1. The autonomic nervous system consists of visceral efferent neurons organized into nerves, ganglia, and plexuses.
2. It is considered entirely motor but also involves visceral receptors and visceral afferent neurons. All autonomic axons are efferent fibers.
3. Efferent neurons are preganglionic (with myelinated axons) and postganglionic (with unmyelinated axons).
4. The autonomic system consists of two principal divisions: sympathetic (thoracolumbar) and parasympathetic (craniosacral).
5. Autonomic ganglia are classified as sympathetic trunk ganglia (on both sides of spinal column), prevertebral ganglia (anterior to spinal column), and terminal ganglia (near or inside visceral effectors).

Physiology of the Autonomic Nervous System (p. 454)

1. Autonomic fibers release neurotransmitters at synapses. On the basis of the neurotransmitter produced, these fibers may be classified as cholinergic or adrenergic.
2. Cholinergic fibers release acetylcholine (ACh). Adrenergic fibers produce norepinephrine (NE).
3. Acetylcholine (ACh) interacts with nicotinic receptors on postganglionic neurons and muscarinic receptors on certain visceral effectors.
4. Norepinephrine (NE) generally interacts with alpha receptors on visceral effectors, and epinephrine generally interacts with alpha and beta receptors on visceral effectors.
5. Sympathetic responses are widespread and, in general, concerned with energy expenditure. Parasympathetic responses are restricted and are typically concerned with energy restoration and conservation.

Visceral Autonomic Reflexes (p. 458)

1. A visceral autonomic reflex adjusts the activity of a visceral effector.
2. A visceral autonomic reflex arc consists of a receptor, afferent neuron, association neuron, visceral efferent preganglionic neuron, visceral efferent postganglionic neuron, and visceral effector.

Control by Higher Centers (p. 458)

1. The hypothalamus controls and integrates the autonomic nervous system. It is connected to both the sympathetic and the parasympathetic divisions.
2. Biofeedback is a process in which people learn to monitor visceral functions and to control them consciously. It has been used to control heart rate, alleviate migraine headaches, and make childbirth easier.
3. Yoga is a higher consciousness achieved through a fully rested and relaxed body and a fully awake and relaxed mind.
4. Transcendental meditation (TM) produces the following physiological responses: decreased oxygen consumption and carbon dioxide elimination, reduced metabolic rate, decreased heart rate, increased intensity of alpha brain waves, a sharp decreased amount of lactic acid in the blood, and an increased electrical resistance of the skin.

REVIEW QUESTIONS

1. What are the principal components of the autonomic nervous system? What is its general function? Why is it called involuntary? (p. 448)
2. What are the principal differences between the voluntary nervous system and the autonomic nervous system? (p. 448)
3. Relate the role of visceral efferent fibers and visceral effectors to the autonomic nervous system. (p. 448)
4. Distinguish between preganglionic neurons and postganglionic neurons with respect to location and function. (p. 449)
5. What is an autonomic ganglion? Describe the location and function of the three types of autonomic ganglia. Define white and gray rami communicantes. (p. 449)
6. On what basis are the sympathetic and parasympathetic divisions of the autonomic nervous system differentiated anatomically and functionally? (p. 452)
7. Discuss the distinction between cholinergic and adrenergic fibers of the autonomic nervous system. (p. 454)
8. How is acetylcholine (ACh) related to nicotinic and muscarinic receptors? (p. 455)
9. How are alpha and beta receptors related to norepinephrine (NE) and epinephrine? (p. 455)
10. Give examples of the antagonistic effects of the sympathetic and parasympathetic divisions of the autonomic nervous system. (p. 456)
11. Summarize the principal functional differences between the

voluntary nervous system and the autonomic nervous system. (p. 448)

12. Give the *sympathetic response* in a fear situation for each of the following body parts: hair follicles, iris of eye, lungs, spleen, adrenal medulla, kidneys, urinary bladder, stomach, intestines, gallbladder, liver, heart, arterioles of the abdominal viscera, skeletal muscles, and skin and mucosa. (p. 456)
13. Define a visceral autonomic reflex and give three examples. (p. 458)
14. Describe a complete visceral autonomic reflex in proper sequence. (p. 458)
15. Describe how the hypothalamus controls and integrates the autonomic nervous system. (p. 458)
16. Define biofeedback. Explain how it could be useful. (p. 459)
17. What is transcendental meditation (TM)? How is the integrated response related to the autonomic nervous system? (p. 459)
18. What is Horner's syndrome? (p. 458)

SELECTED READINGS

Agras, W. S. "Relaxation Therapy in Hypertension," *Hospital Practice,* May 1983.

Carney, R. M. "Clinical Applications of Relaxation Training," *Hospital Practice,* July 1983.

Carpenter, M. B. *Core Text of Neuroanatomy,* 3rd ed. Baltimore: Williams & Wilkins, 1985.

Nobach, C. R., and R. J. Demarest. *The Nervous System: Introduction and Review,* 3rd ed. New York: McGraw-Hill, 1986.

Chapter 17

The Special Senses

Chapter Contents at a Glance

Student Objectives

1. Locate the receptors for olfaction and describe the neural pathway for smell.
2. Identify the gustatory receptors and describe the neural pathway for taste.
3. List and describe the structural divisions of the eye.
4. Discuss retinal image formation by describing refraction, accommodation, constriction of the pupil, convergence, and inverted image formation.
5. Identify the afferent pathway of light impulses to the brain.
6. Describe the anatomical subdivisions of the ear.
7. List the principal events in the physiology of hearing.
8. Identify the receptor organs for static and dynamic equilibrium.
9. Contrast the causes and symptoms of cataracts, glaucoma, conjunctivitis, trachoma, deafness, labyrinthine disease, Ménière's syndrome, vertigo, otitis media, and motion sickness.
10. Define medical terminology associated with the sense organs.

The special senses—smell, taste, sight, hearing, and equilibrium—have receptor organs that are structurally more complex than receptors for general sensations. The sense of smell is the least specialized, as opposed to the sense of sight, which is the most specialized. Like the general senses, however, the special senses allow us to detect changes in our environment.

OLFACTORY SENSATIONS

Structure of Receptors

The receptors for the ***olfactory*** (ol-FAK-tō-rē; *olfact* = smell) ***sense,*** or sense of smell, are located in the nasal epithelium in the superior portion of the nasal cavity on either side of the nasal septum (Figure 17-1). The nasal epithelium consists of three principal kinds of cells: supporting, olfactory, and basal. The ***supporting (sustentacular) cells*** are columnar epithelial cells of the mucous membrane lining the nose. The ***olfactory cells*** are bipolar neurons whose cell bodies lie between the supporting cells. The distal (free) end of each olfactory cell contains a dendrite that terminates in a swelling ***(olfactory vesicle)*** from which six to eight cilia, called ***olfactory hairs,*** radiate. The hairs are believed to react to odors in the air and then to stimulate the olfactory cells, thus initiating the olfactory response. The proximal (basal) part of each olfactory cell contains a single process that represents its axon. ***Basal cells*** lie between the bases of the supporting cells and are believed to produce new supporting cells. Within the connective tissue beneath the olfactory epithelium are ***olfactory (Bowman's) glands*** that produce mucus, which is carried to the surface of the epithelium by ducts. The secretion moistens the surface of the olfactory epithelium and serves as a solvent for odoriferous substances. The continuous secretion of mucus also serves to freshen the surface film of fluid and prevents continuous stimulation of olfactory hairs by the same odor. Continuous stimulation is also prevented by rapid adaptation of olfactory receptors.

Physiology of Olfaction

In order for a substance to be smelled, it must be volatile, that is, capable of entering into a gaseous state so that the gaseous particles can enter the nostrils. You have probably noticed that smell occurs in cycles each time inspiration occurs and that sensitivity to smell is greatly increased by sniffing. The substance to be smelled must be water-soluble so that it can dissolve in the mucus to make contact with olfactory cells. The substance must also be lipid-soluble. Since the plasma membranes of olfactory hairs

FIGURE 17–1 Olfactory receptors. (a) Location of receptors in nasal cavity. (b) Enlarged aspect of olfactory receptors.

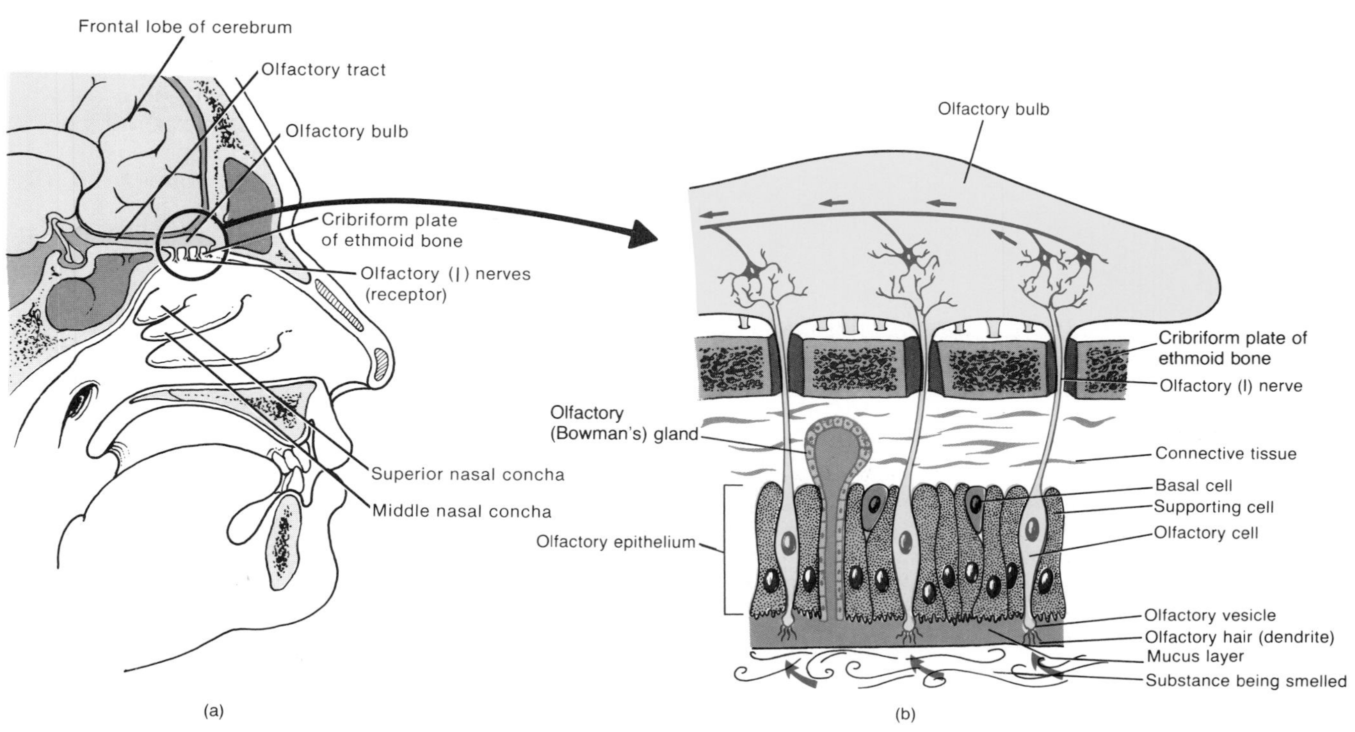

FIGURE 17-1 (*Continued*) (c) Photomicrograph of the olfactory mucosa at a magnification of 300×. (Copyright © 1985 by Michael H. Ross. Used by permission.)

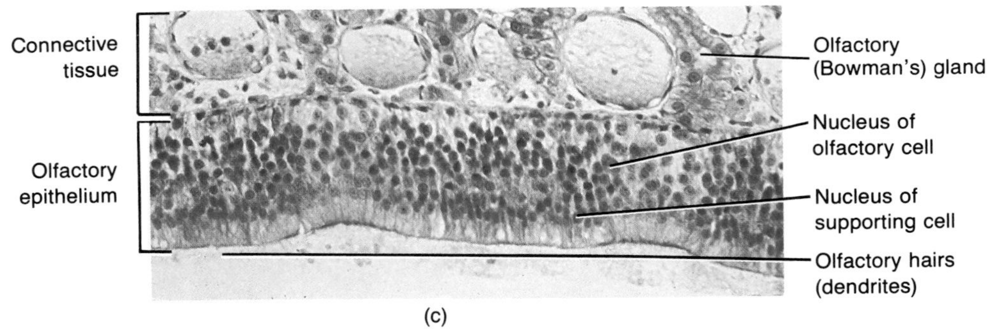

are largely lipid, the substance to be smelled must be dissolved in the lipid covering to make contact with olfactory hairs in order to initiate a nerve impulse by way of a generator potential which is a result of the odor stimulus.

Many attempts have been made to distinguish and classify the primary sensations of smell. One classification includes seven classes of primary sensations: camphoraceous, musky, floral, pepperminty, ethereal, pungent, and putrid. Recent data suggest there may be as many as 50 or more primary sensations of smell.

It is believed that olfactory cells react to olfactory stimuli in the same way that most sensory receptors react to their specific stimuli: first, a generator potential (depolarization) is developed, followed by initiation of a nerve impulse. The ***chemical theory*** assumes that there are different receptor chemicals in the membranes of olfactory hairs, each capable of reacting with a particular olfactory substance (stimulus). The interaction between chemical receptor and substance alters the permeability of the plasma membrane so that a generator potential is developed, followed by initiation of a nerve impulse.

Adaptation and Odor Thresholds

The sensation of smell happens quickly. Olfactory cells respond in milliseconds (msec). Adaptation to odors also occurs rapidly and appears to involve the central nervous system. Olfactory receptors themselves adapt about 50 percent in the first second or so after stimulation but adapt very slowly thereafter. Yet we know from common experience that complete adaptation to odors occurs in about a minute after exposure to a strong atmosphere, a rate far faster than can be explained on the basis of receptor adaptation alone. It has been suggested that most adaptation to odor involves a psychological component in the central nervous system. The mechanism and site of psychological adaptation are unknown.

One of the major characteristics of smell is its low threshold. Only a minute quantity of a substance need be present in air for it to be smelled (minimal identifiable odor, or MIO). A good example is the substance methyl mercaptan, which can be detected in concentrations as low as 1/25,000,000,000 mg per milliliter of air.

Olfactory Pathway

The unmyelinated axons of the olfactory receptor cells unite to form the ***olfactory (I) nerves,*** which pass through foramina in the cribriform plate of the ethmoid bone (Figure 17-1a,b). The olfactory (I) nerves terminate in paired masses of gray matter in the brain called the ***olfactory bulbs.*** The olfactory bulbs lie beneath the frontal lobes of the cerebrum on either side of the crista galli of the ethmoid bone. The first synapse of the olfactory neural pathway occurs in the olfactory bulbs between the axons of the olfactory (I) nerves and the dendrites of neurons inside the olfactory bulbs. Axons of these neurons run posteriorly to form the ***olfactory tract.*** From here, nerve impulses are conveyed to the primary olfactory area of the cerebral cortex. In the cerebral cortex, the impulses are interpreted as odor and give rise to the sensation of smell. Impulses related to olfaction do not pass through the thalamus.

Both the supporting cells of the nasal epithelium and olfactory glands are innervated by branches of the facial (VII) nerve. The trigeminal (V) nerve receives stimuli of pain, cold, heat, tickling, and pressure. Olfactory stimuli such as pepper, ammonia, and chloroform are irritating and may cause tearing because they stimulate the lacrimal and nasal mucosal receptors of the trigeminal (V) nerve as well as the olfactory neurons.

GUSTATORY SENSATIONS

Structure of Receptors

The receptors for ***gustatory*** (GUS-ta-tō'-rē; *gust* = taste) ***sensations,*** or sensations of taste, are located in the taste buds (Figure 17-2). The nearly 2000 taste buds are most

FIGURE 17-2 Gustatory receptors. (a) Photomicrograph of a circumvallate papilla containing taste buds at a magnification of 50×. (Copyright © 1983 by Michael H. Ross. Used by permission.) (b) Diagram of the structure of a taste bud. (c) Locations of papillae and four taste zones.

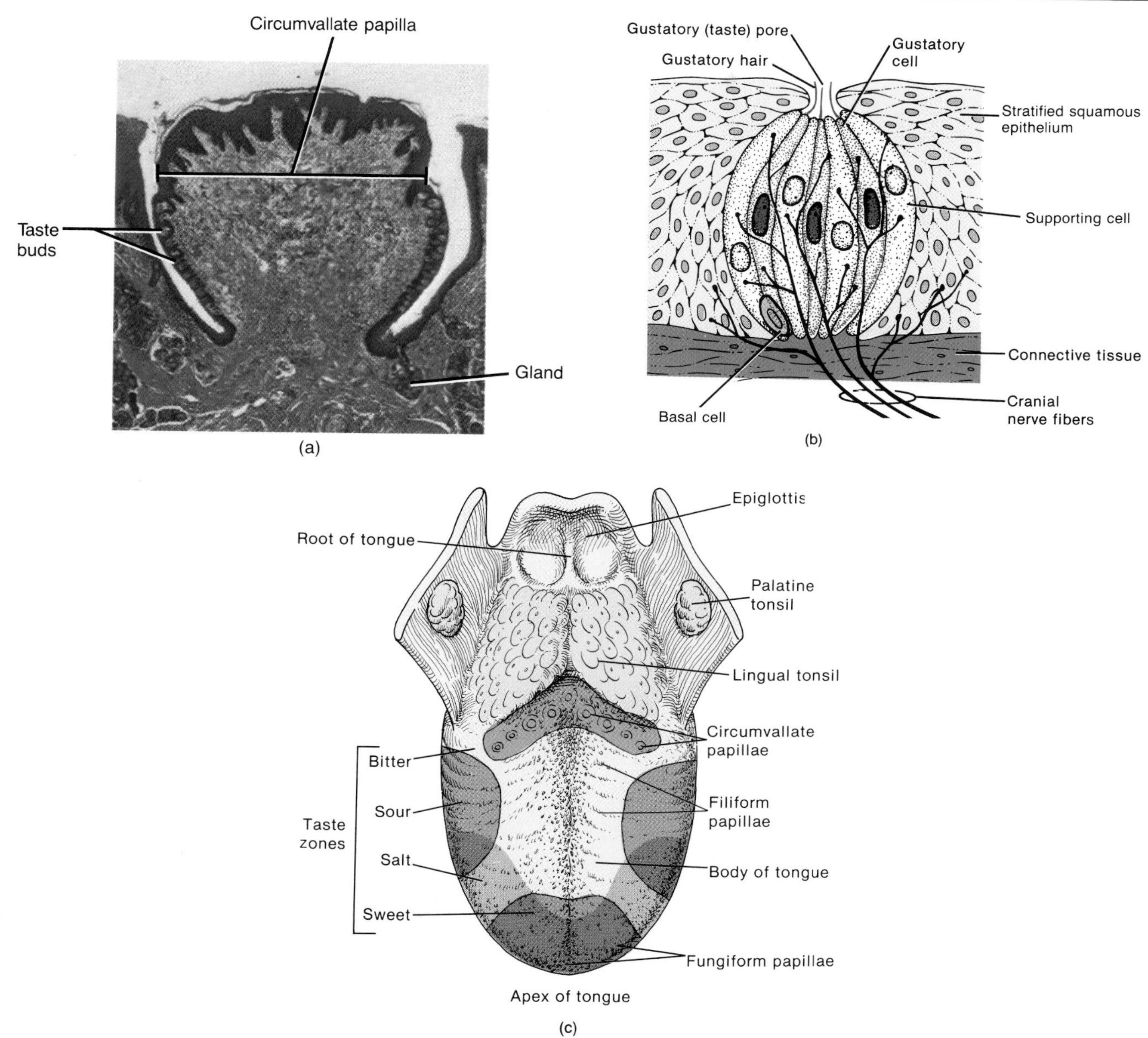

numerous on the tongue, but they are also found on the soft palate and in the throat. The ***taste buds*** are oval bodies consisting of three kinds of cells: supporting, gustatory, and basal. The ***supporting (sustentacular) cells*** are a specialized epithelium that forms a capsule, inside each of which are 4 to 20 ***gustatory cells.*** Each gustatory cell contains a hairline process ***(gustatory hair)*** that projects to the external surface through an opening in the taste bud called the ***taste pore.*** Gustatory cells make contact with taste stimuli through the taste pore. ***Basal cells*** are found at the periphery of the taste bud near the basal lamina. These cells produce supporting and gustatory cells whose average life span is about 10 days.

Taste buds are found in some elevations on the tongue called ***papillae*** (pa-PIL-ē). The papillae give the upper surface of the tongue its rough appearance. ***Circumvallate*** (ser-kum-VAL-āt) or ***vallate papillae,*** the largest type, are circular and form an inverted V-shaped row at the posterior portion of the tongue. ***Fungiform*** (FUN-ji-form; meaning mushroom-shaped) ***papillae*** are knob-like elevations found primarily on the tip and sides of the tongue. All circumvallate and most fungiform papillae contain taste buds. ***Filiform*** (FIL-i-form) ***papillae*** are pointed threadlike structures that cover the anterior two-thirds of the tongue. They rarely contain taste buds.

Physiology of Gustation

For gustatory receptor cells to be stimulated, the substances we taste must be in solution in saliva so they can enter taste pores. Once the taste substance makes contact with plasma membranes of the gustatory hairs, the generator potential is developed. This is presumed to occur as a result of a membrane receptor–taste substance interaction. Then the receptor potential initiates a nerve impulse.

Despite the many substances we seem to taste, there are basically only four primary taste sensations: sour, salt, bitter, and sweet. All other "tastes," such as chocolate, pepper, and coffee, are combinations of these four that are modified by accompanying olfactory sensations.

Persons with colds or allergies sometimes complain that they cannot taste their food. Although their taste sensations may be operating normally, their olfactory sensations are not. This illustrates that much of what we think of as taste is actually smell. Odors from foods pass upward into the nasopharynx and stimulate the olfactory system. In fact, a given concentration of a substance will stimulate the olfactory system thousands of times more than it stimulates the gustatory system.

Each of the four primary tastes is caused by a different response to different chemicals. Certain regions of the tongue react more strongly than others to certain taste sensations. Although the tip of the tongue reacts to all four primary taste sensations, it is highly sensitive to sweet and salty substances. The posterior portion of the tongue is highly sensitive to bitter substances. The lateral edges of the tongue are more sensitive to sour substances (Figure 17-2c).

Adaptation and Taste Thresholds

Adaptation to taste occurs rapidly. Complete adaptation can occur in one to five minutes of continuous stimulation. As with odors, receptor adaptation alone cannot account for the speed of complete adaptation. Although taste receptors have a period of rapid adaptation during the first two to three seconds after contact with food, adaptation is slow thereafter. Adaptation of receptors to smell contributes to taste adaptation but still does not account for its speed. As with adaptation to odors, adaptation to taste involves a psychological adaptation in the central nervous system.

The threshold for taste varies for each of the primary tastes. The threshold for bitter substances, as measured by quinine, is lowest. This may have a protective function. The threshold for sour substances, as measured by hydrochloric acid, is somewhat higher. The thresholds for salt substances, as measured by sodium chloride, and sweet substances, as measured by sucrose, are about the same and are higher than both bitter and sour substances.

Gustatory Pathway

The cranial nerves that supply afferent fibers to taste buds are the facial (VII), which supplies the anterior two-thirds of the tongue; the glossopharyngeal (IX), which supplies the posterior one-third of the tongue; and the vagus (X), which supplies the throat and epiglottis. Taste impulses are conveyed from the gustatory cells in taste buds along the nerves to the medulla and then to the thalamus. They terminate in the primary gustatory area in the parietal lobe of the cerebral cortex.

VISUAL SENSATIONS

The study of the structure, function, and diseases of the eye is known as ***ophthalmology*** (of′-thal-MOL-ō-jē; *ophthalmo* = eye; *logos* = study of). A physician who specializes in the diagnosis and treatment of eye disorders with drugs, surgery, and corrective lenses is known as an ***ophthalmologist,*** whereas an ***optometrist*** is a specialist with a doctorate in optometry who is licensed to examine and test the eyes and treat visual defects by prescribing corrective lenses. An ***optician*** is a technician who fits, adjusts, and dispenses corrective lenses on prescription of an ophthalmologist or optometrist. The structures related to vision are the eyeball, the optic (II) nerve, the brain, and a number of accessory structures.

Accessory Structures of Eye

Among the ***accessory structures*** are the eyebrows, eyelids, eyelashes, and the lacrimal (tearing) apparatus (Figure 17-3). The ***eyebrows*** form a transverse arch at the junction of the upper eyelid and forehead. Structurally, they resemble the hairy scalp. The skin of the eyebrows is richly supplied with sebaceous (oil) glands. The hairs are generally coarse and directed laterally. Deep to the skin of the eyebrows are the fibers of the orbicularis oculi muscles. The eyebrows help to protect the eyeballs from foreign objects, perspiration, and the direct rays of the sun.

The upper and lower ***eyelids,*** or ***palpebrae*** (PAL-pe-brē), have several important roles. They shade the eyes during sleep, protect the eyes from excessive light and foreign objects, and spread lubricating secretions over the eyeballs. The upper eyelid is more movable than the lower and contains in its superior region a special levator muscle known as the ***levator palpebrae superioris.*** The

FIGURE 17-3 Accessory structures of the eye. (a) Anterior view. (b) Sagittal section of the eyelids and anterior portion of the eyeball. (c) Surface anatomy of the eye. (Copyright © 1982 by Gerard J. Tortora. Courtesy of Lynne Borghesi.)

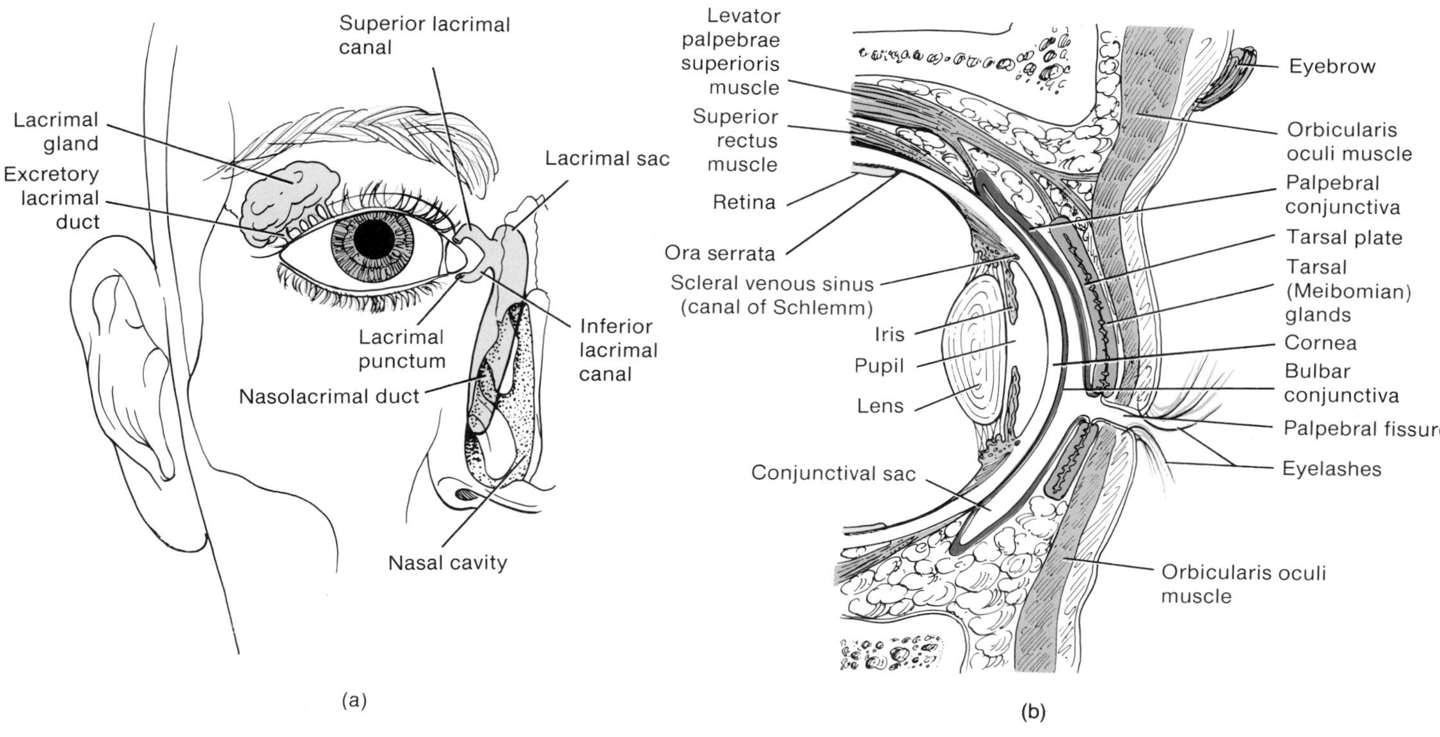

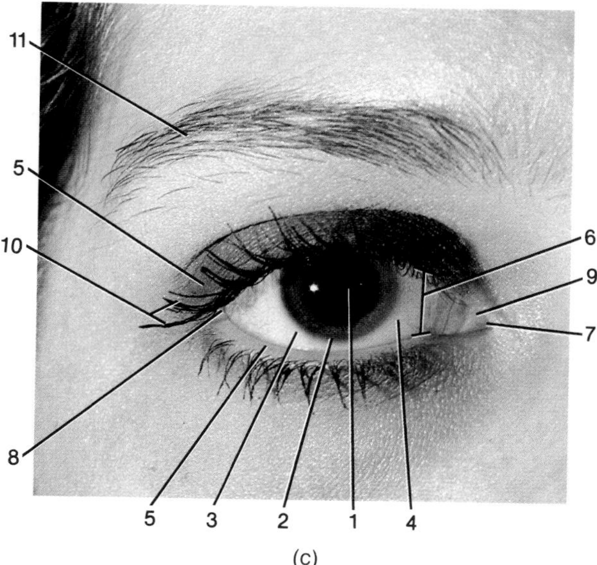

1. **Pupil.** Opening of center of iris of eyeball for light transmission.
2. **Iris.** Circular, pigmented muscular membrane behind cornea.
3. **Sclera.** "White" of eye, a coat of fibrous tissue that covers entire eyeball except for cornea.
4. **Conjunctiva.** Membrane that covers the sclera and lines eyelids.
5. **Palpebrae (eyelids).** Folds of skin and muscle lined by conjunctiva.
6. **Palpebral fissure.** Space between eyelids when they are open.
7. **Medial commissure.** Site of union of upper and lower eyelids near nose.
8. **Lateral commissure.** Site of union of upper and lower eyelids away from nose.
9. **Lacrimal caruncle.** Fleshy, yellowish projection of medial commissure that contains modified sweat and sebaceous glands.
10. **Eyelashes.** Hairs on margins of eyelids, usually arranged in two or three rows.
11. **Eyebrows.** Several rows of hairs superior to upper eyelids.

space between the upper and lower eyelids that exposes the eyeball is called the ***palpebral fissure.*** Its angles are known as the ***lateral commissure*** (KOM-i-shūr), which is narrower and closer to the temporal bone, and the ***medial commissure,*** which is broader and nearer the nasal bone. In the medial commissure, there is a small, reddish elevation, the ***lacrimal caruncle*** (KAR-ung-kul), containing sebaceous (oil) and sudoriferous (sweat) glands. A whitish material secreted by the caruncle collects in the medial commissure.

From superficial to deep, each eyelid consists of epidermis, dermis, subcutaneous areolar connective tissue, fibers of the orbicularis oculi muscle, a tarsal plate, tarsal glands, and a conjunctiva. The ***tarsal plate*** is a thick fold of connective tissue that forms much of the inner wall of each eyelid and gives form and support to the eyelids. Embedded in grooves on the deep surface of each tarsal plate is a row of elongated tarsal glands known as ***tarsal*** or ***Meibomian*** (mī-BŌ-mē-an) ***glands.*** These are modified sebaceous glands, and their oily secretion helps keep the eyelids from adhering to each other. Infection of the tarsal glands produces a tumor or cyst on the eyelid called a ***chalazion*** (ka-LĀ-zē-on). The ***conjunctiva*** (kon'-junk-TĪ-va) is a thin mucous membrane. It is called the ***palpebral conjunctiva*** when it lines the inner aspect of the eyelids. It is called the ***bulbar (ocular) conjunctiva*** when it is reflected from the eyelids onto the anterior surface of the eyeball. When the blood vessels of the bulbar conjunctiva are dilated and congested due to local irritation or infection, the person has bloodshot eyes.

Projecting from the border of each eyelid, anterior to the tarsal glands, is a row of short, thick hairs, the ***eyelashes.*** In the upper lid, they are long and turn upward; in the lower lid, they are short and turn downward. Sebaceous glands at the base of the hair follicles of the eyelashes, called ***sebaceous ciliary glands*** or ***glands of Zeis*** (ZĪS), pour a lubricating fluid into the follicles. Infection of these glands is called a ***sty.***

The ***lacrimal*** (*lacrima* = tear) ***apparatus*** is a term used for a group of structures that manufactures and drains tears. These structures are the lacrimal glands, excretory lacrimal ducts, lacrimal canals, lacrimal sacs, and nasolacrimal ducts. A ***lacrimal gland*** is a compound tubuloacinar gland located at the superior anterolateral portion of each orbit. Each is about the size and shape of an almond. Leading from the lacrimal glands are 6 to 12 ***excretory lacrimal ducts*** that empty lacrimal fluid, or tears, onto the surface of the conjunctiva of the upper lid. From here the lacrimal fluid passes medially and enters two small openings called ***lacrimal puncta*** that appear as two small pores, one in each papilla of the eyelid, at the medial commissure of the eye. The lacrimal secretion then passes into two ducts, the ***lacrimal canals,*** and is next conveyed into the lacrimal sac. The lacrimal canals are located in the lacrimal grooves of the lacrimal bones. The ***lacrimal sac*** is the superior expanded portion of the ***nasolacrimal duct,*** a canal that transports the lacrimal secretion into the inferior meatus of the nose.

The ***lacrimal secretion*** is a watery solution containing salts, some mucus, and a bactericidal enzyme called ***lysozyme.*** It cleans, lubricates, and moistens the eyeball. After being secreted by the lacrimal glands, it is spread medially over the surface of the eyeball by the blinking of the eyelids. Usually, 1 ml per day is produced by each gland.

CLINICAL APPLICATION: "WATERY" EYES

Normally, the lacrimal secretion is carried away by evaporation or by passing into the lacrimal canals and then into the nasal cavities as fast as it is produced. If, however, an irritating substance makes contact with the conjunctiva, the lacrimal glands are stimulated to oversecrete. Tears then accumulate more rapidly than they can be carried away, a condition commonly referred to as *"watery" eyes.* This is a protective mechanism, since the tears dilute and wash away the irritating substance. "Watery" eyes also occur when an inflammation of the nasal mucosa, such as a cold, obstructs the nasolacrimal ducts so that drainage of tears is blocked. Humans are unique in that they have the ability to cry to express certain emotions. In response to parasympathetic stimulation, the lacrimal glands produce excessive tears that may spill over the edges of the eyelids and even fill the nasal cavity with fluid. Crying may indicate happiness or sadness.

Structure of Eyeball

The adult ***eyeball*** measures about 2.5 cm (1 inch) in diameter. Of its total surface area, only the anterior one-sixth is exposed. The remainder is recessed and protected by the orbit into which it fits. Anatomically, the eyeball can be divided into three layers: fibrous tunic, vascular tunic, and retina or nervous tunic (Figure 17-4a).

Fibrous Tunic

The ***fibrous tunic*** is the outer coat of the eyeball. It can be divided into two regions: the posterior portion is the sclera, and the anterior portion is the cornea. The ***sclera*** (SKLE-ra; *skleros* = hard), the "white of the eye," is a white coat of dense fibrous tissue that covers all the eyeball except the most anterior portion, the iris. The sclera gives shape to the eyeball and protects its inner parts. Its posterior surface is pierced by the optic (II) nerve. The ***cornea*** (KOR-nē-a) is a nonvascular, transparent, fibrous coat through which the iris can be seen. The cornea's outer surface is covered by an epithelial layer continuous with the epithelium of the bulbar conjunctiva. At the junction of the sclera and cornea is a venous sinus known as the ***scleral venous sinus,*** or ***canal of Schlemm.*** (Surface features of the eye are shown in Figure 17-3c.)

FIGURE 17-4 **Structure of the eyeball. (a) Diagram of gross structure in transverse section.**

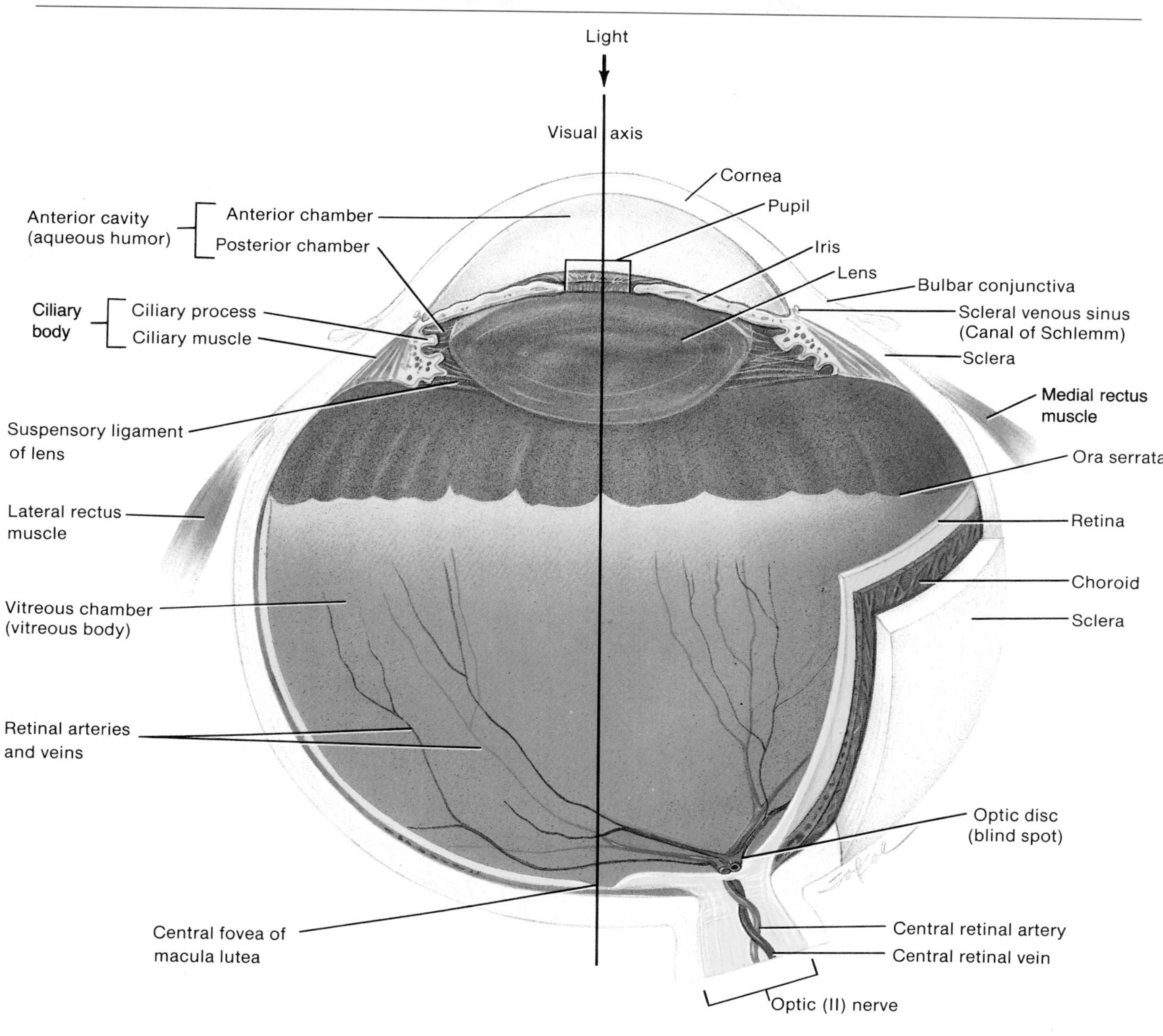

(a)

FIGURE 17-4 (*Continued*) (b) Diagram of the microscopic structure of the retina exaggerated for emphasis. The arrows pointing upward indicate the direction of the signals passing through the nervous layer of the retina that ultimately result in a nerve impulse that passes into the optic (II) nerve.

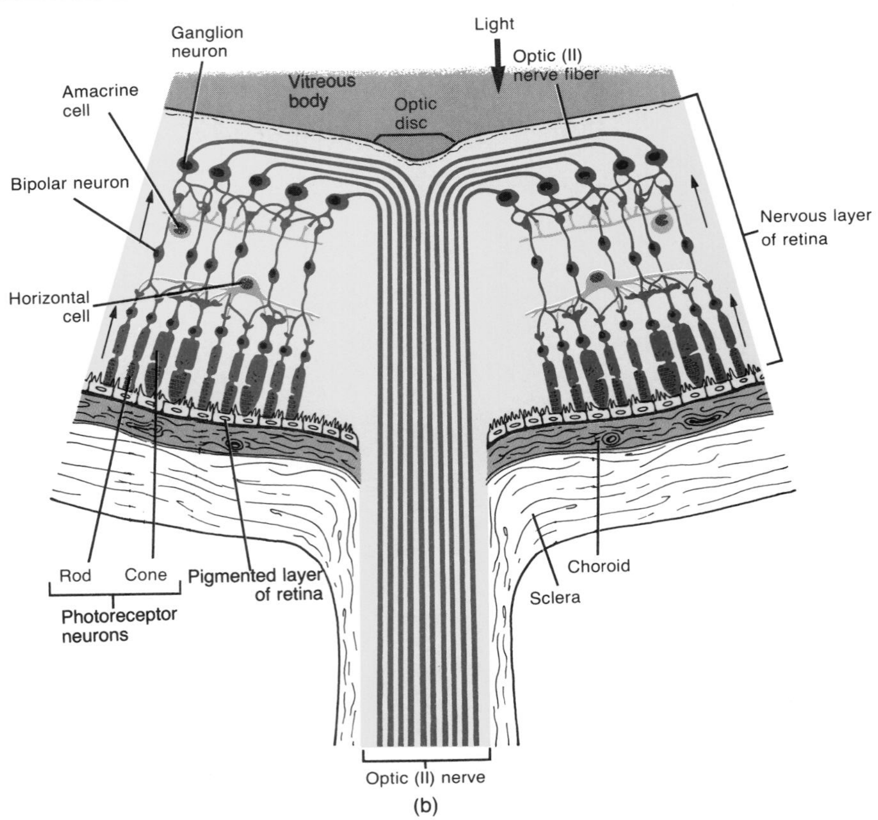

CLINICAL APPLICATION: CORNEAL SURGERY

Corneal transplants are the most common organ transplant operation, and they are considered to be the most successful type of transplant since they are rarely rejected. This is because the cornea is avascular, and antibodies that might cause rejection do not circulate there. (The cornea receives nourishment from tears and aqueous humor.) The surgical procedure is performed under an operating microscope at magnifications of between 6× and 40×. A general or local anesthesia is used. The defective cornea, usually 7 to 8 mm (about 0.3 inch) in diameter, is excised and replaced with a donor cornea of similar diameter. The transplanted cornea is sewn into place with nylon sutures which stay in place for about one year because of slow corneal healing. Patients are later fitted with hard or soft contact lenses or eyeglasses. (The shortage of donated corneas has been partially overcome by the development of artificial corneas made of plastic.)

In the process of normal vision, light rays pass through the cornea, as well as other structures, to form a clear image on the retina (nervous tunic). In myopia (nearsightedness), the cornea is too curved, and images of distant objects fall short of the retina, causing blurred vision (see Figure 17-5c). Many people with myopia may be helped by a surgical procedure called ***radial keratotomy* (*RK*).** In the procedure, several microscopic incisions, like the spokes of a wheel, are made in the cornea. This flattens the cornea and improves vision without using corrective lenses. Because of various problems associated with the procedure (undercorrected vision, overcorrected vision, and extra sensitivity to glare at night or changes in visual acuity during the day), it appears that it is best suited to people with severe myopia.

An alternative procedure to radial keratotomy for treating myopia is known as ***epikeratoplasty*** (ep'-ē-KER-a-tō-plas'-tē). In one modification of this procedure, a small circular area of defective cornea is removed and a commercially prepared piece of donor cornea, shaped to provide proper curvature, is sewn into its place.

FIGURE 17-5 Normal and abnormal refraction in the eyeball. (a) Refraction of light rays passing from air into water. (b) In the normal or emmetropic eye, light rays from an object are bent sufficiently by the four refracting media and converge on the central fovea. A clear image is formed. (c) In the nearsighted or myopic eye, the image is focused in front of the retina. The condition may result from an elongated eyeball or thickened lens. Correction is by use of a concave lens that diverges entering light rays so that they have to travel further through the eyeball and are focused directly on the retina. (d) In the farsighted or hypermetropic eye, the image is focused behind the retina. The condition results from a shortened eyeball or a thin lens. Correction is by a convex lens that converges entering light rays so that they focus directly on the retina.

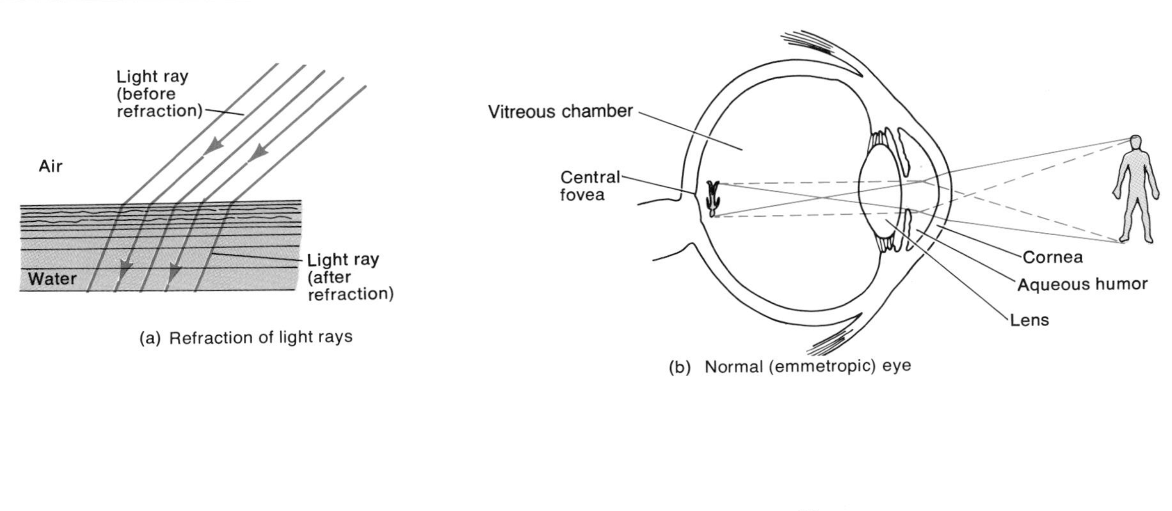

(a) Refraction of light rays

(b) Normal (emmetropic) eye

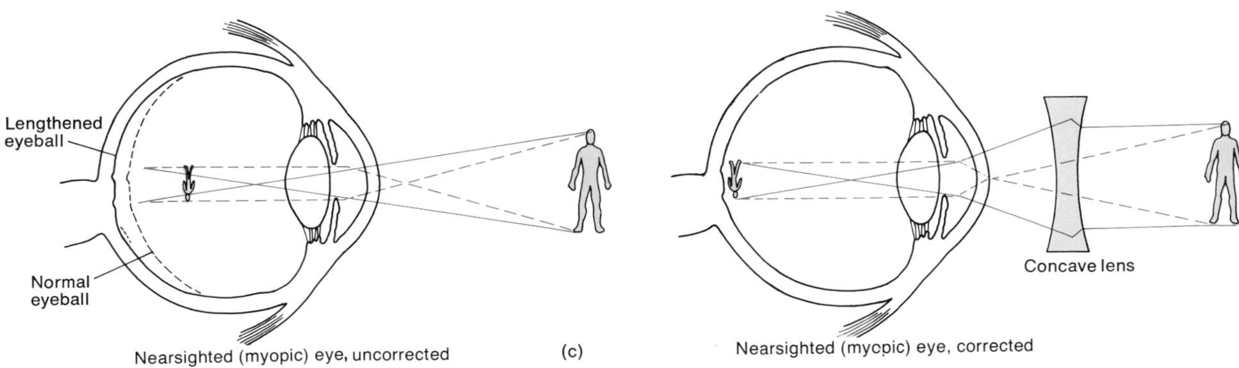

Nearsighted (myopic) eye, uncorrected (c) Nearsighted (myopic) eye, corrected

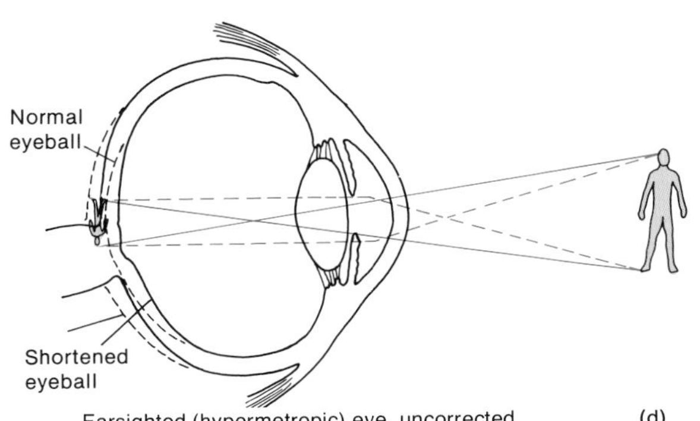

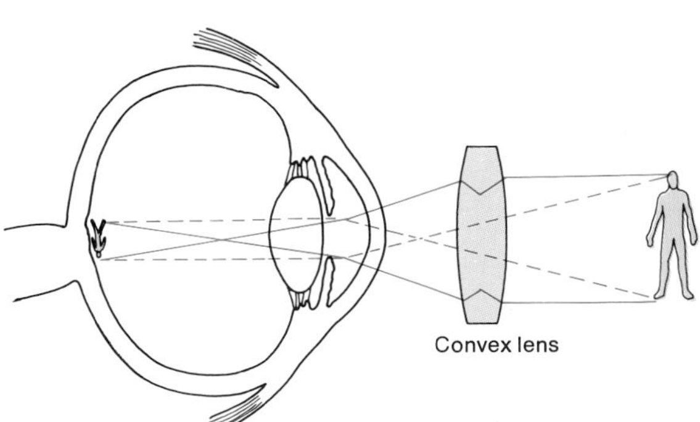

Farsighted (hypermetropic) eye, uncorrected (d) Farsighted (hypermetropic) eye, corrected

FIGURE 17-5 *(Continued)* (e) An astigmatism is a disorder of the cornea (shown on left) or lens (shown on right) characterized by irregular curvatures which prevent images from being focused on the retina. This results in blurred or distorted vision. Suitable glasses or contact lenses correct the refraction of an astigmatic eye.

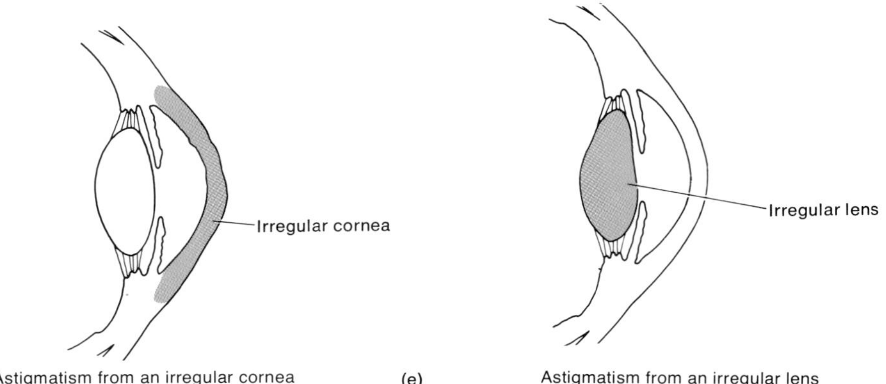

Vascular Tunic

The ***vascular tunic*** or ***uvea*** (YOO-vē-a) is the middle layer of the eyeball and is composed of three portions: choroid, ciliary body, and iris. The ***choroid*** (KŌ-royd), the posterior portion of the vascular tunic, is a thin, dark brown membrane that lines most of the internal surface of the sclera. It contains numerous blood vessels and a large amount of pigment. The choroid absorbs light rays so that they are not reflected within the eyeball. Through its blood supply, it nourishes the retina. Like the sclera, it is pierced by the optic (II) nerve at the back of the eyeball.

In the anterior portion of the vascular tunic, the choroid becomes the ***ciliary*** (SIL-ē-ar'-ē) ***body.*** It is the thickest portion of the vascular tunic. It extends from the ***ora serrata*** (Ō-ra ser-RĀ-ta) of the retina (nervous tunic) to a point just behind the sclerocorneal junction. The ora serrata is simply the jagged margin of the retina. The ciliary body consists of the ciliary processes and ciliary muscle. The ***ciliary processes*** consist of protrusions or folds on the internal surface of the ciliary body that secrete aqueous humor. The ***ciliary muscle*** is a smooth muscle that alters the shape of the lens for near or far vision.

The ***iris*** (*irid* = colored circle) is the third portion of the vascular tunic. It is the colored portion seen through the cornea and consists of circular iris and radial iris smooth muscle fibers (cells) arranged to form a doughnut-shaped structure. The black hole in the center of the iris is the ***pupil,*** the area through which light enters the eyeball. The iris is suspended between the cornea and the lens and is attached at its outer margin to the ciliary process. A principal function of the iris is to regulate the amount of light entering the posterior cavity of the eyeball. When the eye is stimulated by bright light, the circular iris (constrictor pupillae) muscles contract and decrease the size of the pupil (constriction). When the eye must adjust to dim light, the radial iris (dilator pupillae) muscles contract and increase the pupil's size (dilation).

Retina (Nervous Tunic)

The third and inner coat of the eye, the ***retina (nervous tunic),*** lies only in the posterior portion of the eye. Its primary function is image formation. The retina is one of the few places in the body where blood vessels can be seen directly. Visualization is achieved by use of a special reflecting light called an ophthalmoscope, which is described shortly. The retina consists of an inner nervous tissue layer (visual portion) and an outer pigmented layer (nonvisual portion). The retina covers the choroid. At the edge of the ciliary body, it terminates in a scalloped border called the ora serrata. This is where the nervous layer or visual portion of the retina ends. The pigmented layer extends anteriorly over the back of the ciliary body and the iris as the nonvisual portion of the retina.

CLINICAL APPLICATION: DETACHED RETINA

Detachment of the retina may occur in trauma, such as a blow to the head, or may be secondary to various intraocular disorders. The actual detachment occurs between the inner nervous tissue layer and outer pigmented layer. Fluid accumulates between these layers, forcing the thin, pliable retina to billow out toward the vitreous body, resulting in distorted vision and blindness in the corresponding field of vision. The retina may be reattached by photocoagulation by laser beam, cryosurgery, or scleral resection.

The nervous layer of the retina contains three zones of neurons. These three zones, named in the order in which they conduct nerve impulses, are ***photoreceptor neurons, bipolar neurons,*** and ***ganglion neurons.*** The dendrites of the photoreceptor neurons are called rods or cones because of their shapes. They are visual receptors highly specialized for stimulation by light rays. Functionally, rods and cones develop generator potentials. ***Rods*** are specialized for black-and-white vision in dim light. They also allow us to discriminate between different shades of dark and light and permit us to see shapes and movement. ***Cones*** are specialized for color vision and sharpness of vision ***(visual acuity)*** in bright light. This is why we cannot see color by moonlight. It is estimated that there are 3 million cones and 100 million rods. Cones are most densely concentrated in the ***central fovea,*** a small depression in the center of the macula lutea. The ***macula lutea*** (MAK-yoo-la LOO-tē-a; *macula* = spot; *lutea* = yellow) is in the exact center of the posterior portion of the retina, corresponding to the visual axis of the eye. The fovea is the area of sharpest vision because of the high concentration of cones. Rods are absent from the fovea and macula and increase in density toward the periphery of the retina. It is for this reason that you can see better at night while not looking directly at an object.

The principal blood supply of the retina is from the ***central retinal artery,*** a branch of the ophthalmic artery. The central retinal artery enters the retina at about the middle of the optic disc, an area where the optic (II) nerve is attached and that contains no photoreceptor neurons (Figure 17-4a). After emerging through the disc, the central retinal artery divides into superior and inferior branches, each of which subdivides into nasal and temporal branches. The ***central retinal vein*** drains blood from the retina through the optic disc.

CLINICAL APPLICATION: SENILE MACULAR DEGENERATION

In a disease called *senile macular degeneration (SMD)*, new blood vessels grow over the macula lutea. The effect ranges from distorted vision to blindness. SMD accounts for nearly all new cases of blindness in people over 65. Its cause is unknown. Laser beam treatment has been used effectively in arresting blood vessel proliferation and restoring normal vision in some cases. In order to determine if SMD exists, a simple test can be performed without any assistance or special instruments. Stare, one eye at a time (covering the other with your hand), at any long straight line, such as a door frame. If either eye perceives the line as bent or twisted, or if a black spot appears, a physician should be informed immediately.

When information has passed through the photoreceptor neurons, it is conducted across synapses to the bipolar neurons in the intermediate zone of the nervous layer of the retina. From here it is passed to the ganglion neurons. These cells transmit their signals through optic (II) nerve fibers to the brain in the form of nerve impulses. Many rods connect with one bipolar neuron, and many of these bipolar neurons transmit impulses to one ganglion cell. This greatly lowers visual acuity, but it permits summation effects to occur so that low levels of light can stimulate a ganglion cell that would not respond had it been connected directly to a cone. The synaptic connections thus contribute much to the difference in visual acuity and light sensitivity.

The axons of the ganglion neurons extend posteriorly to a small area of the retina called the ***optic disc (blind spot).*** This region contains openings through which the axons of the ganglion neurons exit as the optic (II) nerve. Since it contains no rods or cones, an image striking it cannot be perceived. Thus, it is called the blind spot.

Lens

In addition to the fibrous tunic, vascular tunic, and retina, the eyeball itself contains the lens, just behind the pupil and iris. The ***lens*** is constructed of proteins called ***crystallins*** arranged like the layers of an onion. These proteins not only provide structure but also function as enzymes that convert sugars into energy for the lens. Normally, the lens is perfectly transparent. It is enclosed by a clear connective tissue capsule and held in position by ***suspensory ligaments.*** The lens helps focus light rays for clear vision. A loss of transparency of the lens is known as a ***cataract.***

Interior

The interior of the eyeball is a large space divided into two cavities by the lens: anterior cavity and vitreous chamber. The ***anterior cavity,*** the division anterior to the lens, is further divided into the ***anterior chamber,*** which lies behind the cornea and in front of the iris, and the ***posterior chamber,*** which lies behind the iris and in front of the suspensory ligaments and lens. The anterior cavity is filled with a watery fluid, similar to cerebrospinal fluid, called the ***aqueous*** (*aqua* = water) ***humor.*** The fluid is believed to be secreted into the posterior chamber by choroid plexuses of the ciliary processes of the ciliary bodies behind the iris. Once the fluid is formed, it passes into the posterior chamber and then passes forward between the iris and the lens, through the pupil, into the anterior chamber. From the anterior chamber, the aqueous humor, which is continually produced, is drained off into the scleral venous sinus (canal of Schlemm) and then into the blood. The anterior chamber thus serves a function similar to the subarachnoid space around the brain and spinal cord. The scleral venous sinus is analogous to a venous sinus of the dura mater. The pressure in the eye, called ***intraocular pressure (IOP),*** is produced mainly by the aqueous humor. The intraocular pressure, along with the vitreous body, maintains the shape of the eyeball and keeps the retina smoothly applied to the choroid so the retina will form clear images. Normal intraocular pressure (about 16 mm Hg) is maintained by drainage of the aqueous humor through the scleral venous sinus. Excessive intraocular pressure, called ***glaucoma*** (glaw-KŌ-ma), results in degeneration of the retina and blindness. Besides maintaining intraocular pressure, the aqueous humor is also the principal link between the cardiovascular system and the lens and cornea. Neither the lens nor the cornea has blood vessels.

MEDICAL TEST

Tonometry (tō-NOM-e-trē; *ton* = tension; *metron* = to measure)

Diagnostic Value: To screen for glaucoma by measuring intraocular pressure and to permit follow-up evaluation of glaucoma.

Procedure: Most optometrists now employ pneumatic tonometry, the electronic measurement of pressure after a short burst of gas strikes the cornea. In applanation tonometry, the subject lies down on her or his back, and several drops of topical anesthetic are instilled in each eye. After the lights are lowered, the subject stares at a spot on the ceiling. Then the tonometer, an instrument that measures the resistance of the eyeball to slight pressure, is lowered onto the cornea. First one eye is tested and then the other. Regular and routine screening for glaucoma should be done in adults, especially those over age 40.

The second, and larger, cavity of the eyeball is the ***vitreous chamber (posterior cavity).*** It lies between the lens and the retina and contains a jellylike substance called the ***vitreous body.*** This substance contributes to intraocular pressure, helps to prevent the eyeball from collapsing, and holds the retina flush against the internal portions of the eyeball. The vitreous body, unlike the aqueous humor, does not undergo constant replacement. It is formed during embryonic life and is not replaced thereafter.

MEDICAL TEST

Ophthalmoscopy (of'-thal-MOS-kō-pē; *ophthalmo* = eye; *skopein* = to view)

Diagnostic Value: To evaluate the interior (fundus) of the eye and especially to detect retinal changes associated with hypertension, diabetes mellitus, atherosclerosis, and increased intracranial pressure resulting from head injury, brain tumor, meningitis, or encephalitis.

Procedure: An ophthalmoscope is used to examine the interior of the eyeball, with or without the use of dilating drops. The instrument contains a light source and a set of mirrors or prism to reflect light so that the interior of the eyeball can be examined. The room lights are turned down or off and the patient is instructed to look at a specific object straight ahead while the examiner directs light to each eye. Such a test permits direct examination of the retina (nervous tunic), optic disc, optic nerve (II), macula lutea, and retinal blood vessels.

A summary of structures associated with the eyeball is presented in Exhibit 17-1.

EXHIBIT 17-1 SUMMARY OF STRUCTURES ASSOCIATED WITH THE EYEBALL

Structure	Function
Fibrous tunic	
Sclera	Provides shape and protects inner parts.
Cornea	Admits and refracts light.
Vascular tunic	
Choroid	Provides blood supply and absorbs light.
Ciliary body	Secretes aqueous humor and alters shape of lens for near or far vision (accommodation).
Iris	Regulates amount of light that enters eyeball.
Retina (nervous tunic)	Receives light, converts light into generator potentials and nerve impulses, and transmits impulses to the optic (II) nerve.
Lens	Refracts light.
Anterior cavity	Contains aqueous humor that helps maintain shape of eyeball, and refracts light.
Vitreous chamber	Contains vitreous body that helps maintain shape of eyeball, keeps retina applied to choroid, and refracts light.

Physiology of Vision

Before light can reach the rods and cones of the retina to result in image formation, it must pass through the cornea, aqueous humor, pupil, lens, and vitreous body.

For vision to occur, light reaching the rods and cones must form an image on the retina. The resulting nerve impulses must then be conducted to the visual area of the cerebral cortex. In discussing the physiology of vision, let us first consider retinal image formation.

Retinal Image Formation

The formation of an image on the retina requires four basic processes, all concerned with focusing light rays: (1) refraction of light rays, (2) accommodation of the lens, (3) constriction of the pupul, and (4) convergence of the eyes. Accommodation and pupil size are functions of the smooth muscle cells of the ciliary muscle and the dilator and constrictor muscles of the iris. They are termed *intrinsic eye muscles* since they are inside the eyeball. Convergence is a function of the voluntary muscles attached to the outside of the eyeball called the *extrinsic eye muscles* (see Figure 11-6).

▪ **Refraction of Light Rays** When light rays traveling through a transparent medium (such as air) pass into a second transparent medium with a different density (such as water), they bend at the surface of the two media. This is ***refraction*** (Figure 17-5a). The eye has four such media of refraction: cornea, aqueous humor, lens, and vitreous body. Light rays entering the eye from the air are refracted at the following points: (1) the anterior surface of the cornea as they pass from the lighter air into the denser cornea, (2) the posterior surface of the cornea as they pass into the less dense aqueous humor, (3) the anterior surface of the lens as they pass from the aqueous humor into the denser lens, and (4) the posterior surface of the lens as they pass from the lens into the less dense vitreous body.

The degree of refraction that takes place at each surface in the eye is very precise. When an object is 6 m (20 ft) or more away from the viewer, the light rays reflected from the object are nearly parallel to one another. The parallel rays must be bent sufficiently to fall exactly on the central fovea, where vision is sharpest. Light rays that are reflected from objects closer than 6 m (20 ft) are divergent rather than parallel. As a result, they must be refracted toward each other to a greater extent. This change in refraction is brought about by the lens of the eye by a process called accommodation (Figures 17-5b and 17-6).

▪ **Accommodation of the Lens** If the surface of a lens curves outward, as in a convex lens, the lens will refract incoming rays toward each other so they eventually intersect. The greater the curve, the more acutely it bends the rays toward each other. Conversely, when the surface of a lens curves inward, as in a concave lens, the rays bend away from each other. The lens of the eye is biconvex. Furthermore, it has the unique ability to change the focusing power of the eye by becoming moderately curved at one moment and greatly curved the next. When the eye is focusing on a close object, the lens curves greatly in order to bend the rays toward the central fovea. This increase in the curvature of the lens is called ***accommodation*** (Figure 17-6). In near vision, the ciliary muscle contracts, pulling the ciliary process and choroid forward toward the lens. This action releases the tension on the lens and suspensory ligament. Because of its elasticity, the lens shortens, thickens, and bulges. In far vision, the ciliary muscle is relaxed and the lens is flatter. With aging, the lens loses elasticity and therefore its ability to accommodate.

CLINICAL APPLICATION: ABNORMALITIES OF REFRACTION

The normal eye, known as an *emmetropic* (em'-e-TROP-ik) *eye,* can sufficiently refract light rays from an object 6 m (20 ft) away to focus a clear image on the retina. Many individuals, however, do not have this ability because of abnormalities related to improper refraction. Among these abnormalities are *myopia* (mī-Ō-pē-a), or nearsightedness; *hyperme-*

FIGURE 17-6 Accommodation. (a) For objects 6 m (20 ft) or more away. (b) For objects closer than 6 m.

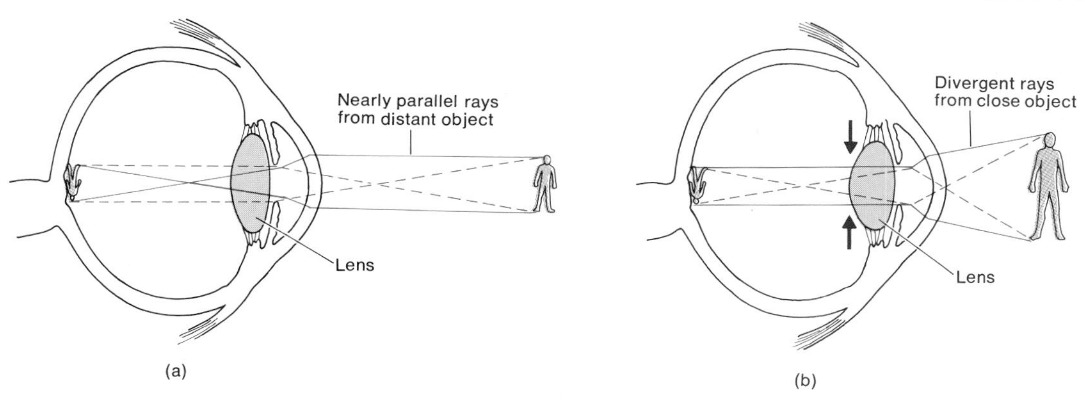

tropia (hī'-per-mē-TRŌ-pē-a), or farsightedness, also known as ***hyperopia;*** and ***astigmatism*** (a-STIG-ma-tizm), or irregularities in the surface of the lens or cornea. The conditions are illustrated and explained in Figure 17-5c–e. The inability to focus on nearby objects due to loss of elasticity of the lens with age is called ***presbyopia*** (prez-bē-Ō-pē-a). This condition usually begins in the midforties and is responsible for the fact that those already wearing glasses must then go to bifocals, and those who have not previously needed glasses may require reading glasses.

Almost 100 million Americans have various kinds of errors of vision. Although most people still wear eyeglasses, contact lenses are becoming increasingly popular. A ***contact lens*** is a lens that is worn in contact with the cornea, beneath the eyelids in front of the pupil. The outer surface of the lens is spherical and is designed to correct the visual defect, while the inner surface is precisely ground to fit the curvature of the cornea. The lens adheres to the cornea by capillary action on the normal layer of tear fluid.

Contact lenses may be worn to improve vision for several reasons: convenience during strenuous physical activity, improvement in peripheral vision, and cosmetic advantages. *Hard lenses* are semiflexible and consist of plastic and silicone that permit oxygen to pass through to the eye. Among their advantages are that their contour effectively corrects for astigmatism; vision is usually clear and stable; and they are sturdy, durable, and generally easy to care for. In some cases, they produce distortion and can scratch the corneal surface. *Soft lenses* are constructed of polymers that bind water. They have a soft consistency and conform to the curvature of the cornea. Among the disadvantages of soft lenses are their inability to correct for astigmatism, minor fluctuations in vision, less durability, and more complex care.

▪ **Constriction of Pupil** The circular muscle fibers of the iris also assume a function in the formation of clear retinal images. Part of the accommodation mechanism consists of the contraction of the sphincter muscles of the iris to constrict the pupil. ***Constriction of the pupil*** means narrowing the diameter of the hole through which light enters the eye. This action occurs simultaneously with accommodation of the lens and prevents light rays from entering the eye through the periphery of the lens. Light rays entering at the periphery would not be brought to focus on the retina and would result in blurred vision. The pupil, as noted earlier, also constricts in bright light to protect the retina from sudden or intense stimulation.

▪ **Convergence** Because of the alignment of the orbits in the skull, many animals see a set of objects off to the left through one eye and an entirely different set off to the right through the other. This characteristic doubles their field of vision and allows them to detect predators behind them. In humans, both eyes focus on only one set of objects—a characteristic called ***single binocular vision.***

Single binocular vision occurs when light rays from an object are directed toward corresponding points on the two retinas. When we stare straight ahead at a distant object, the incoming light rays are aimed directly at both pupils and are refracted to identical spots on the retinas of both eyes. But as we move closer to the object, our eyes must rotate medially for the light rays from the object to hit the same points on both retinas. The term ***convergence*** refers to this medial movement of the two eyeballs so they are both directed toward the object being viewed. The nearer the object, the greater the degree of convergence necessary to maintain single binocular vision. Convergence is brought about by the coordinated action of the extrinsic eye muscles.

▪ **Inverted Image** Images are focused upside down on the retina. They also undergo mirror reversal; that is, light reflected from the right side of an object hits the left side of the retina, and vice versa. Note in Figure 17-5b that reflected light from the top of the object crosses light from the bottom of the object and strikes the retina below the central fovea. Reflected light from the bottom of the object crosses light from the top of the object and strikes the retina above the central fovea. The reason we do not see an inverted world is that the brain learns early in life to coordinate visual images with the exact locations of objects. The brain stores memories of reaching and touching objects and automatically turns visual images right-side-up and right-side-around.

Stimulation of Photoreceptors

▪ **Excitation of Rods** After an image is formed on the retina by refraction, accommodation, constriction of the pupil, and convergence, light energy must be converted into nerve impulses. The initial step is the development of receptor potentials by rods and cones. To see how this occurs, we will first examine the role of photopigments.

A ***photopigment*** is a substance that can absorb light and undergo structural changes that lead to the development of a receptor potential. Recall from Chapter 15 that when an adequate stimulus is applied to a receptor, it responds by altering its membrane permeability to small ions by opening voltage-sensitive channels. This results in depolarization of the membrane—the receptor potential. The photopigment in rods is called ***rhodopsin (visual purple),*** which is sensitive to low levels of illumination. Rhodopsin is composed of ***scotopsin*** (a protein) and ***retinal*** (a derivative of vitamin A) (Figure 17-7). A very important characteristic of retinal is that it can exist in two forms, depending on whether or not light is present. In the absence of light, retinal occurs as *cis*-retinal, in which form it structurally fits the scotopsin portion of rhodopsin. In the presence of light, *cis*-retinal undergoes an immediate change in structure from a

FIGURE 17-7 **Rhodopsin cycle.**

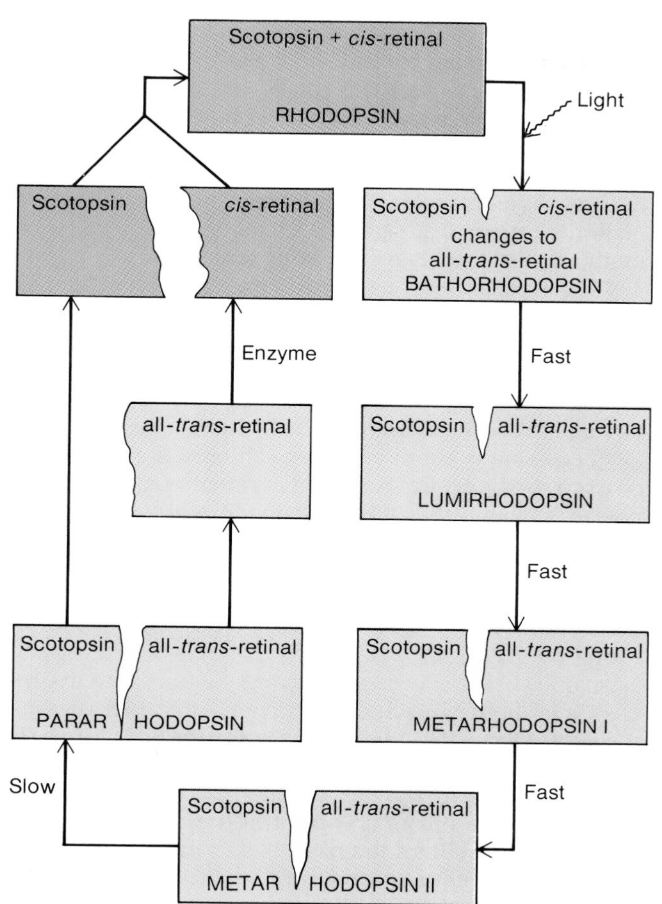

curved molecule to a straight molecule called all-*trans*-retinal. Since all-*trans*-retinal has an altered physical structure, it no longer fits the scotopsin portion of rhodopsin, and it actually pulls away from scotopsin, forming a very unstable substance called ***bathorhodopsin* (*prelumirhodopsin*)** that rapidly decomposes sequentially to ***lumirhodopsin, metarhodopsin I, metarhodopsin II,*** and ***pararhodopsin.*** All these substances are loose combinations of scotopsin and all-*trans*-retinal. Ultimately, pararhodopsin decomposes completely into scotopsin and all-*trans*-retinal. At the point where metarhodopsin I is converted to metarhodopsin II, the receptor potential develops in rods.

Following decomposition of rhodopsin into scotopsin and all-*trans*-retinal in the presence of light, rhodopsin is once again re-formed. In this process, which occurs in the absence of light, the all-*trans*-retinal is converted back to *cis*-retinal in the presence of an enzyme. Then *cis*-retinal combines with scotopsin to re-form rhodopsin. This compound is stable until its decomposition is again triggered by light energy.

The receptor potential that develops from rhodopsin breakdown in rods is different from receptor potentials that occur in other types of sensory receptors because stimulation results in hyperpolarization rather than depolarization. In rods, there is a decrease in the leakage of sodium ions (Na^+) to the interior, even though Na^+ ions are continually pumped out. The net loss of Na^+ ions from rods results in increased negativity inside the membrane of the rod, resulting in hyperpolarization. The closure of Na^+ ion channels that results in hyperpolarization involves an enzyme, ***transducin,*** that breaks down a nucleotide called ***cyclic GMP* (*guanosine monophosphate*).** This breakdown ultimately leads to closure of the Na^+ ion channels.

Rhodopsin is highly light sensitive—even the light rays from the moon or a candle will break down some of it and thereby allow us to see. The rods, then, are specialized for night vision. However, they are of only limited help for daylight vision. In bright light, the rhodopsin is broken down faster than it can be manufactured. In dim light, production is able to keep pace with a slower rate of breakdown. These characteristics of rhodopsin are responsible for the experience of having to adjust to a dark room after walking in from the sunshine. The normal period of adjustment is the time it takes for the completely dissociated rhodopsin to re-form. ***Night blindness,*** which is also referred to as ***nyctalopia*** (nik'-ta-LŌ-pē-a), is the lack of normal night vision following the adjustment period. It is most often caused by vitamin A deficiency.

▪ **Excitation of Cones** Cones are the receptors for bright light, color, and visual acuity (sharpness). As in rods, photopigment decomposition produces receptor potentials as a result of hyperpolarization. The photopigments in cones are almost the same as those in rods. Both contain retinal, but the protein portions in cones are called ***photopsins,*** which are slightly different from the scotopsin in rods. Unlike rhodopsin, the photopigments of the cones require bright light for their breakdown and they re-form quickly. It is believed that there are three types of cones, each containing a different combination of retinal and photopsin. Each of the three different types of cones has a different maximum absorption of light of different wavelengths, and thus each responds best to light of a given color. One type of cone responds best to red light, the second to green light, and the third to blue light. Just as an artist can obtain almost any color by mixing colors, it is believed that cones can perceive any color by differential stimulation.

CLINICAL APPLICATION: RED-GREEN COLOR BLINDNESS

If a single group of color receptive cones is missing from the retina, an individual cannot distinguish some colors from others and is said to be *color-*

blind. **The most common type is *red-green color blindness,* an X-linked recessive trait, in which usually one photopigment receptive to red or green light is missing. As a result, the person cannot distinguish between red and green. Color blindness is an inherited condition that affects males far more frequently than females. The inheritance of the condition is discussed in Chapter 29 and illustrated in Figure 29-20.**

▪ **Light and Dark Adaptation** If an individual is exposed to bright light for an extended period of time, a considerable amount of the photopigment in both rods and cones is broken down into retinal and opsin (scotopsin or photopsin). In addition, most of the retinal in both rods and cones is converted to vitamin A. As a result, the concentration of the photopigment in both is reduced considerably, and the sensitivity of the eye to light is greatly reduced. This is called ***light adaptation.***

Conversely, if an individual is exposed to darkness for an extended period of time, a considerable amount of retinal and opsins in the rods and cones is converted into the photopigments. Also, large amounts of vitamin A are converted into retinal, which is then converted into photopigments upon combination with opsins. As a result, the concentration of photopigments in rods and cones is increased considerably, and the sensitivity of the eye to light is greatly increased. This is called ***dark adaptation.***

Visual Pathway

Examination of Figure 17-4b reveals that the retina (nervous tunic) is composed of three principal zones of neurons: photoreceptor (rods and cones), bipolar, and ganglion. For light to stimulate rods and cones to produce receptor potentials, light must pass through ganglion and bipolar cells before reaching rods and cones. The three zones of neurons establish a direct pathway from the source of light to the optic (II) nerve. Note also that there are two special types of cells present in the retina (nervous tunic) called ***horizontal cells*** and ***amacrine cells.*** These cells form laterally directed pathways that modify and control the message that is transmitted along the direct pathway.

Once receptor potentials are developed by rods and cones, the potentials are transmitted unchanged through the bodies of the rods and cones. These receptor potentials induce signals in both bipolar neurons and horizontal cells, probably by means of neurotransmitters. In response to the neurotransmitters, bipolar neurons become excited and horizontal cells become inhibited. Functionally, bipolar neurons transmit the excitatory visual signal from rods and cones to ganglion cells (see Figure 17-4b). Horizontal cells transmit inhibitory signals to bipolar neurons in the areas lateral to excited rods and cones. This lateral inhibition enhances contrasts in the visual scene between areas of the retina (nervous tunic) that are strongly stimulated and adjacent areas that are weakly stimulated. Horizontal cells also assist in the differentiation of various colors. Amacrine cells, which are also excited by bipolar neurons, synapse with ganglion cells and transmit information to them that signals a change in the level of illumination of the retina (nervous tunic).

When bipolar neurons transmit excitatory visual signals to ganglion cells, the ganglion cells become depolarized and initiate nerve impulses. The cell bodies of the ganglion cells lie in the retina, and their axons leave the eyeball via the ***optic (II) nerve*** (Figure 17-8). The axons pass through the ***optic chiasma*** (kī-AZ-ma; *chiasma* = letter X), a crossing point of the optic (II) nerves. Some fibers cross to the opposite side. Others remain uncrossed. On passing through the optic chiasma, the fibers, now part of the ***optic tract,*** enter the brain and terminate in the lateral geniculate nucleus of the thalamus. Here the fibers synapse with third-order neurons whose axons pass to the visual areas located in the occipital lobes of the cerebral cortex.

Analysis of the afferent pathway to the brain reveals that the visual field of each eye is divided into two regions: the ***nasal (medial) half*** and the ***temporal (lateral) half.*** For each eye, light rays from an object in the nasal half of the visual field fall on the temporal half of the retina. Light rays from an object in the temporal half of the visual field fall on the nasal half of the retina (Figure 17-8a). Also, light rays from objects at the top of the visual field of each eye fall on the inferior portion of the retina, and light rays from objects at the bottom of the visual field fall on the superior portion of the retina. In the optic chiasma, nerve fibers from the nasal halves of the retinas cross and continue on to the lateral geniculate nuclei of the thalamus; nerve fibers from the temporal halves of the retinas do not cross but continue directly on to the lateral geniculate nuclei. As a result, the primary visual area of the cerebral cortex of the right occipital lobe interprets visual sensations from the left side of an object via nerve impulses from the temporal half of the retina of the right eye and the nasal half of the retina of the left eye. The primary visual area of the cerebral cortex of the left occipital lobe interprets visual sensations from the right side of an object via impulses from the nasal half of the right eye and the temporal half of the left eye.

Although we have just described the visual pathway as a more or less single pathway, it should be noted that recent evidence suggests that visual signals are actually processed by at least three separate systems, each with its own function. One system is involved in processing information related to the shape of objects, another system forms a pathway regarding color of objects, and a third system processes information about movement, location, and spatial organization.

FIGURE 17-8 **Afferent pathway for visual impulses. (a) Diagram. The dark circle in the center of the visual fields is the macula lutea. The center of the macula lutea is the central fovea, the area of sharpest vision.**

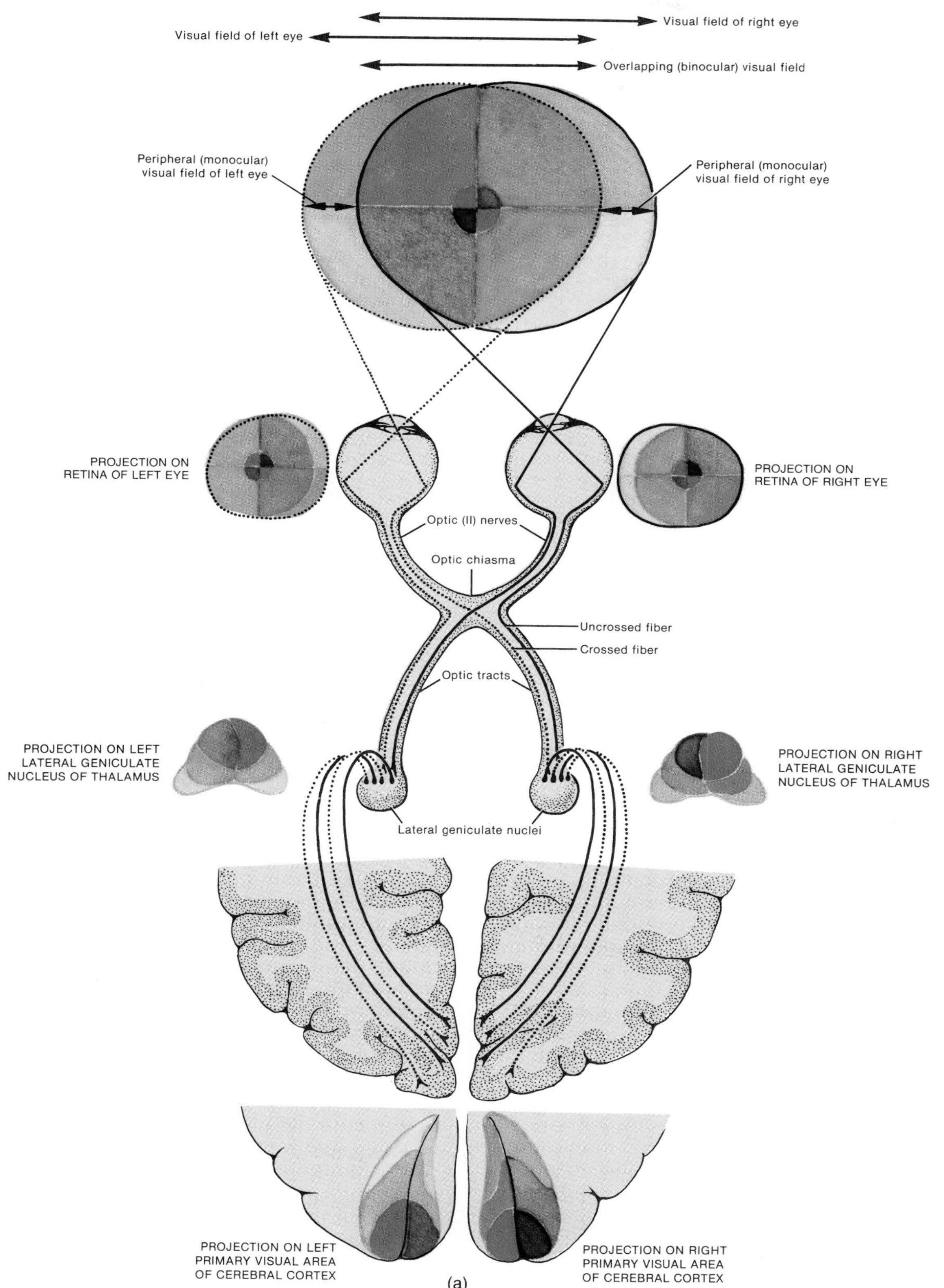

(a)

FIGURE 17-8 (*Continued*) (b) Photograph as seen from the interior aspect of the brain. (Courtesy of N. Gluhbegovic and T. H. Williams, *The Human Brain: A Photographic Guide,* Harper & Row, Publishers, Inc., Hagerstown, MD, 1980.)

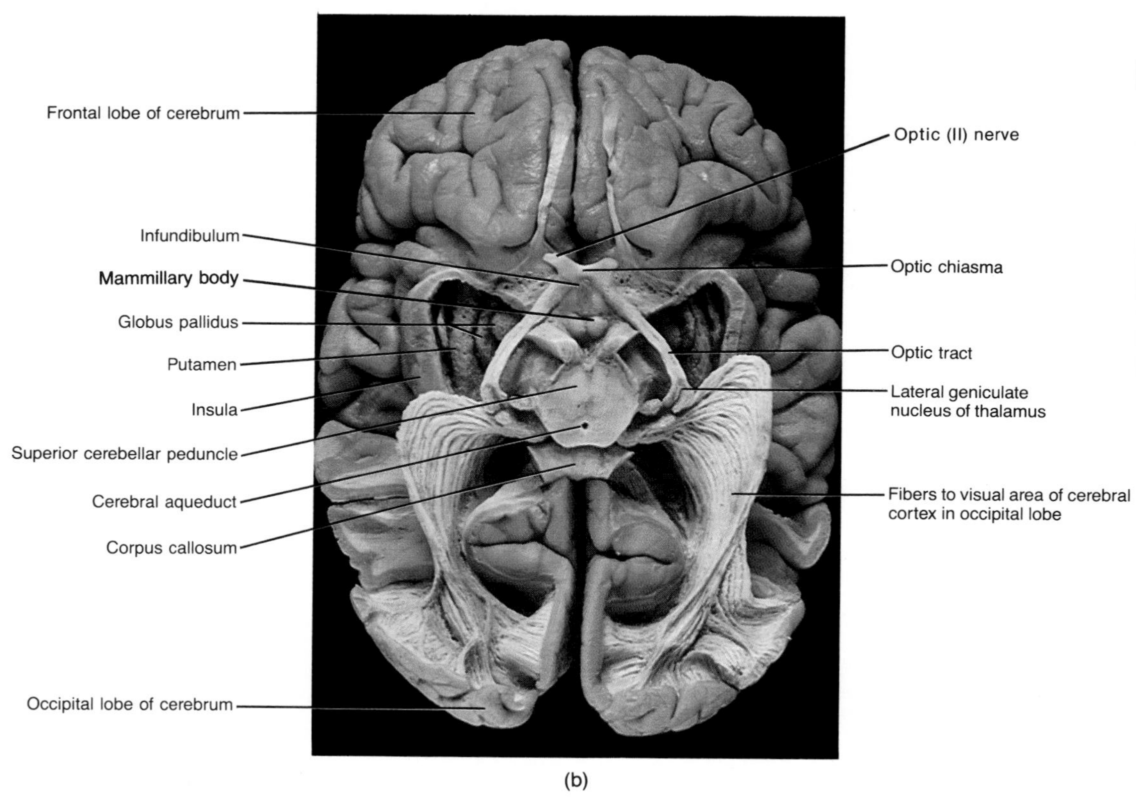

(b)

MEDICAL TEST

Snellen test of visual acuity (sharpness)

Diagnostic Value: To evaluate any problems or changes in vision.

Procedure: The subject stands 6 m (20 ft) away from a Snellen chart that contains several rows of letters in gradually decreasing size. (For children who cannot read, the letter *E* or pictures of small objects in various positions and sizes are used instead.) To begin the test, the subject removes her or his glasses or contact lenses, covers one eye, and reads the smallest line possible on the chart. The same procedure is repeated for the other eye. Finally, the test is repeated with glasses or contact lenses in place, to determine how well they correct vision.

Visual acuity is expressed as a fraction such as 20/15, 20/20, or 20/40. The top number is the distance the subject stood from the chart. The bottom number is the distance at which a normal eye can read the same line of letters. (The results may be different for each eye.) If your vision is 20/20, which is considered normal, it means that a person with normal vision could read the smallest line that you could read while standing at the same distance from the chart as you. If, however, your vision is 20/40, it means that a person with normal vision could read the smallest line that you were able to read while standing 40 feet from the chart—twice the distance. The larger the bottom number, the less accurate the vision. It is possible to have vision better than 20/20. Someone with a vision of 20/15 can see from 20 feet what the average normal person would only be able to see from 15 feet away from the chart.

AUDITORY SENSATIONS AND EQUILIBRIUM

In addition to containing receptors for sound waves, the ear also contains receptors for equilibrium. Anatomically, the ear is divided into three principal regions: the external (outer) ear, the middle ear, and the internal (inner) ear.

External (Outer) Ear

The ***external (outer) ear*** is structurally designed to collect sound waves and direct them inward (Figure 17-9a,b). It consists of the pinna, external auditory canal, and tympanic membrane, also called the eardrum.

The ***pinna (auricle)*** is a flap of elastic cartilage shaped like the flared end of a trumpet and covered by

FIGURE 17-9 Structure of the auditory apparatus. (a) Surface anatomy of the right ear seen in lateral view. (Copyright © 1982 by Gerard J. Tortora. Courtesy of James Borghesi.) (b) Divisions of the right ear into external, middle, and internal portions seen in a frontal section through the right side of the skull.

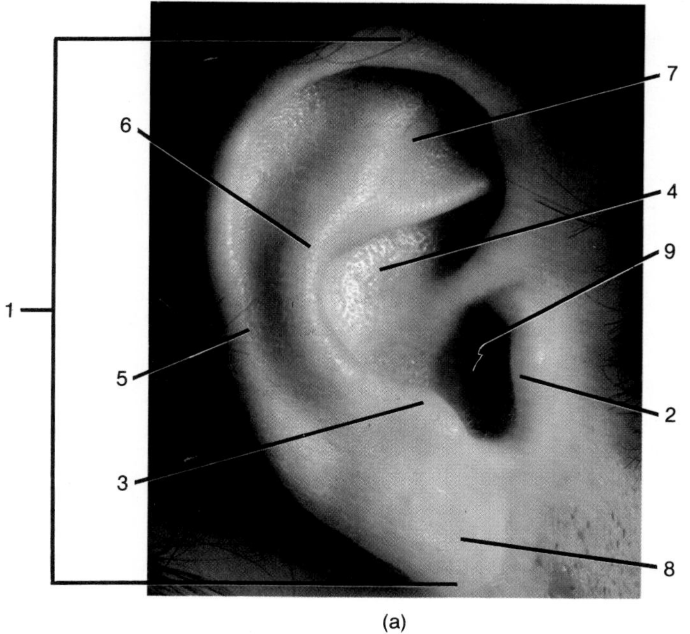

(a)

1. **Pinna.** Portion of external ear not contained in head, also called auricle or trumpet.
2. **Tragus.** Cartilaginous projection anterior to external opening to ear.
3. **Antitragus.** Cartilaginous projection opposite tragus.
4. **Concha.** Hollow of auricle.
5. **Helix.** Superior and posterior free margin of auricle.
6. **Antihelix.** Semicircular ridge posterior and superior to concha.
7. **Triangular fossa.** Depression in superior portion of antihelix.
8. **Lobule.** Inferior portion of pinna that consists of fibrous and fatty tissue, but no cartilage.
9. **External auditory canal (meatus).** Canal extending from external ear to tympanic membrane (eardrum).

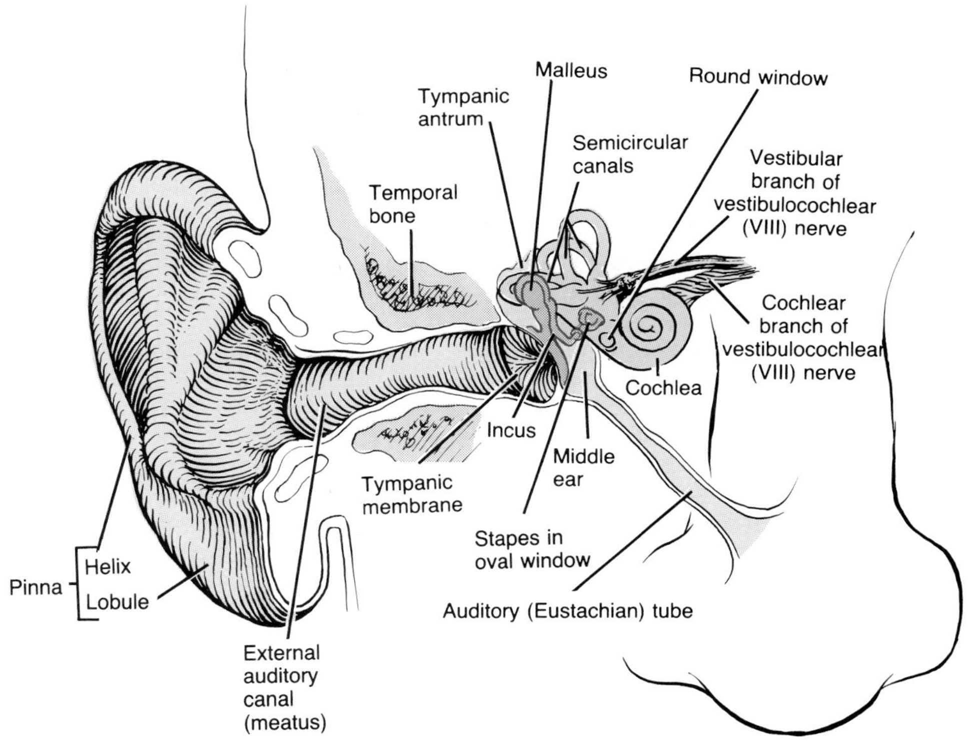

(b)

FIGURE 17-9 (*Continued*) (c) Details of the middle ear and bony labyrinth of the internal ear. (Illustration after Biagio John Melloni, Ph.D.)

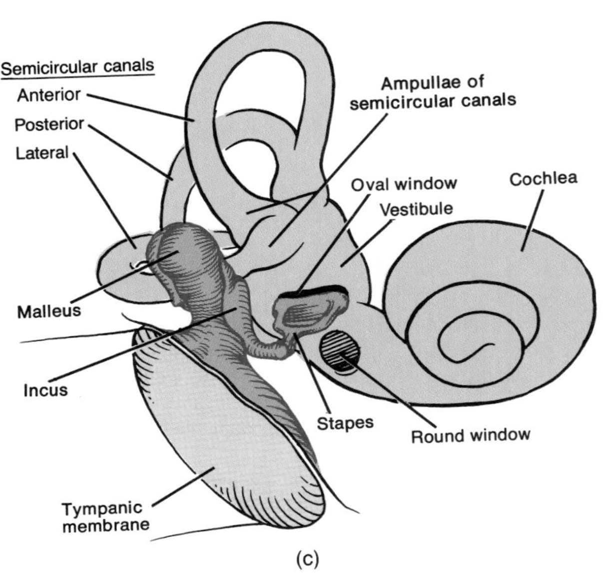

CLINICAL APPLICATION: PERFORATED EARDRUM

A ***perforated eardrum*** is a hole in the tympanic membrane that reduces sound transmission. The condition is characterized by acute pain initially, ringing or roaring in the affected ear, hearing impairment, and sometimes dizziness. Causes of perforated eardrums include compressed air (explosion, scuba diving, trauma (skull fracture or from objects such as ear swabs), or acute middle ear infections. If the condition does not resolve by itself, a surgical procedure called myringoplasty can be performed, in which a graft of skin, a vein, or fascia is used to repair the perforation. Or, the wound edges may be cauterized and a piece of blood-soaked Gelfoam is placed over the defect. New tissues will grow over the Gelfoam.

thick skin. The rim of the pinna is called the ***helix;*** the inferior portion is the ***lobule.*** The pinna is attached to the head by ligaments and muscles. Additional surface anatomy features of the external ear are shown in Figure 17-9a.

The ***external auditory*** (*audire* = hearing) ***canal (meatus)*** is a curved tube about 2.5 cm (1 inch) in length that lies in the temporal bone. It leads from the pinna to the eardrum. The wall of the canal consists of bone lined with cartilage that is continuous with the cartilage of the pinna. The cartilage in the external auditory canal is covered with thin, highly sensitive skin. Near the exterior opening, the canal contains a few hairs and specialized sebaceous glands called ***ceruminous*** (se-ROO-mi-nus) ***glands*** that secrete ***cerumen*** (earwax). The combination of hairs and cerumen (se-ROO-min) helps to prevent dust and foreign objects from entering the ear. Usually, cerumen dries up and falls out of the ear canal. Sometimes, however, it builds up in the canal and impairs hearing (impacted cerumen).

The ***tympanic*** (tim-PAN-ik; *tympano* = drum) ***membrane (eardrum)*** is a thin, semitransparent partition of fibrous connective tissue between the external auditory canal and middle ear. Its external surface is concave and covered with skin. Its internal surface is convex and covered with a mucous membrane.

Middle Ear

The ***middle ear (tympanic cavity)*** is a small, epithelial-lined, air-filled cavity hollowed out of the temporal bone (Figure 17-9b–c). It is separated from the external ear by the eardrum and from the internal ear by a thin bony partition that contains two small openings: the oval window and the round window.

The posterior wall of the middle ear communicates with the mastoid air cells of the temporal bone through a chamber called the ***tympanic antrum.*** This anatomical fact explains why a middle ear infection may spread to the temporal bone, causing mastoiditis, or even to the brain.

The anterior wall of the middle ear contains an opening that leads directly into the ***auditory (Eustachian) tube.*** The auditory tube connects the middle ear with the nasopharynx of the throat. Through this passageway, infections may travel from the throat and nose to the ear. The function of the tube is to equalize air pressure on both sides of the tympanic membrane. Abrupt changes in external or internal air pressure might otherwise cause the eardrum to rupture. During swallowing and yawning, the tube opens to allow atmospheric air to enter or leave the middle ear until the internal pressure equals the external pressure. If the pressure is not relieved, intense pain, hearing impairment, tinnitus, and vertigo could develop. Any sudden pressure changes against the eardrum may be equalized by deliberately swallowing or pinching the nose closed, closing the mouth, and gently forcing air from the lungs into the nasopharynx.

Extending across the middle ear are three exceedingly small bones called ***auditory ossicles*** (OS-si-kuls). The bones, named for their shape, are the malleus, incus, and stapes, commonly called the hammer, anvil, and stirrup, respectively. They are connected by synovial joints. The "handle" of the ***malleus*** is attached to the internal surface of the tympanic membrane. Its head articulates with the body of the incus. The ***incus*** is the intermediate bone in the series and articulates with the head of the stapes. The base or footplate of the ***stapes*** fits into a small opening in the thin bony partition between the middle and inner ear. The opening is called the ***oval window*** or ***fenestra vestibuli*** (fe-NES-tra ves-TIB-yoo-lī). Directly below the oval window is another opening, the ***round window*** or ***fenestra cochlea*** (fe-NES-tra KŌk-lē-a). This opening is enclosed by a membrane called the ***secondary tympanic membrane.*** Auditory ossicles are attached to the middle ear by means of ligaments.

In addition to the ligaments, there are also two skeletal muscles attached to the ossicles. The ***tensor tympani muscle*** draws the malleus medially, thus limiting movement and increasing tension on the tympanic membrane to prevent damage to the inner ear when exposed to loud sounds. Since this response is slow, it only protects the inner ear from prolonged loud noises, not brief ones, such as a gunshot.

The ***stapedius muscle*** is the smallest of all skeletal muscles. Its action is to draw the stapes posteriorly, thus preventing it from moving too much. Like the tensor tympani muscle, the stapedius muscle has a protective function in that it dampens (checks) large vibrations that result from loud noises. It is for this reason that paralysis of the stapedius muscle is associated with ***hyperacusia*** (acuteness of hearing).

Internal (Inner) Ear

The ***internal (inner) ear*** is also called the ***labyrinth*** (LAB-i-rinth) because of its complicated series of canals (Figure 17-10). Structurally, it consists of two main divisions: an outer bony labyrinth and an inner membranous labyrinth that fits in the bony labyrinth. The ***bony labyrinth*** is a series of cavities in the petrous portion of the temporal bone. It can be divided into three areas named on the basis of shape: the vestibule, cochlea, and semicircular canals. The bony labyrinth is lined with periosteum and contains a fluid called ***perilymph.*** This fluid,

FIGURE 17-10 Details of the internal ear. (a) Relative position of the bony labyrinth projected to the inner surface of the floor of the skull. (b) The outer, blue area belongs to the bony labyrinth. The inner, brown-colored area belongs to the membranous labyrinth. (Illustration after Biagio John Melloni, Ph.D.)

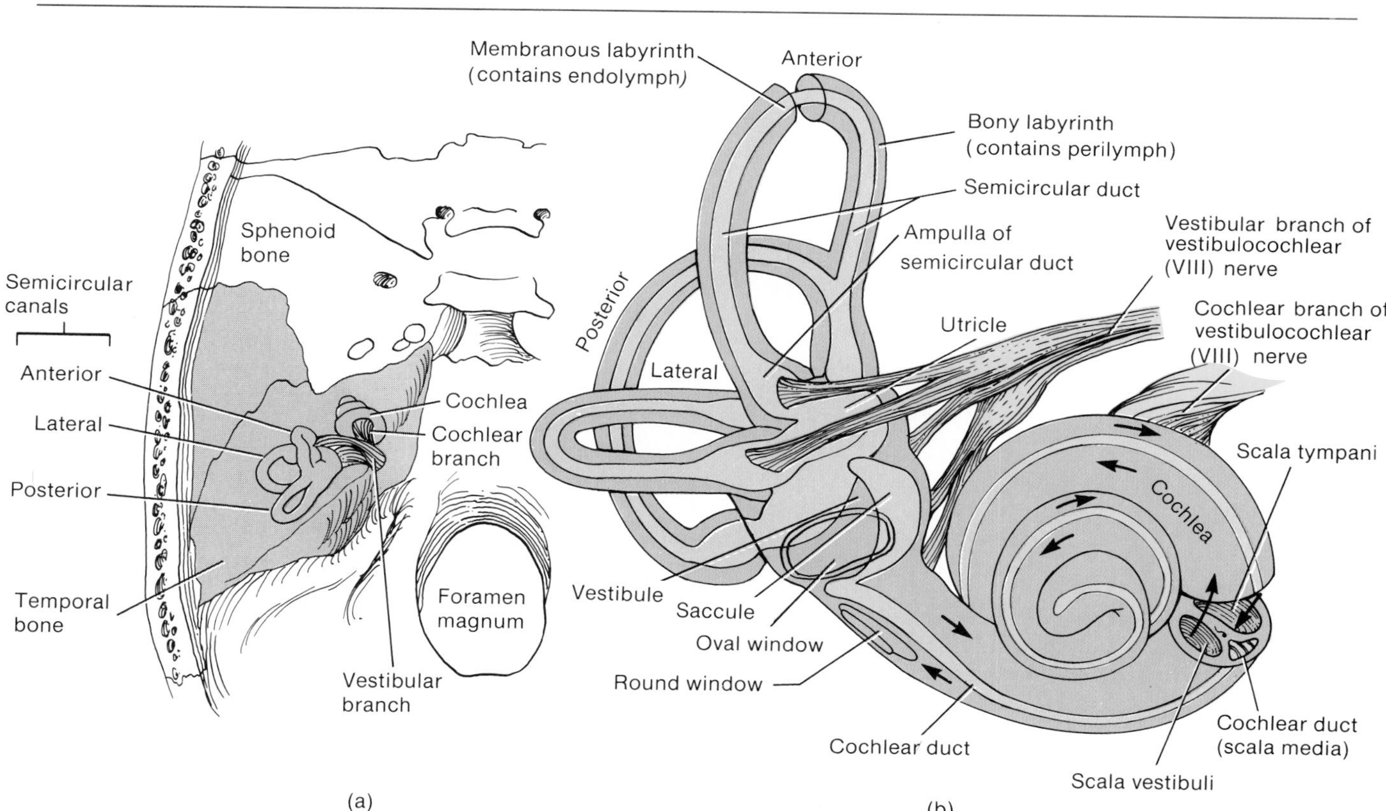

FIGURE 17-10 (*Continued*) (c) Section through the cochlea. (d) Enlargement of a section through one turn of the cochlea.

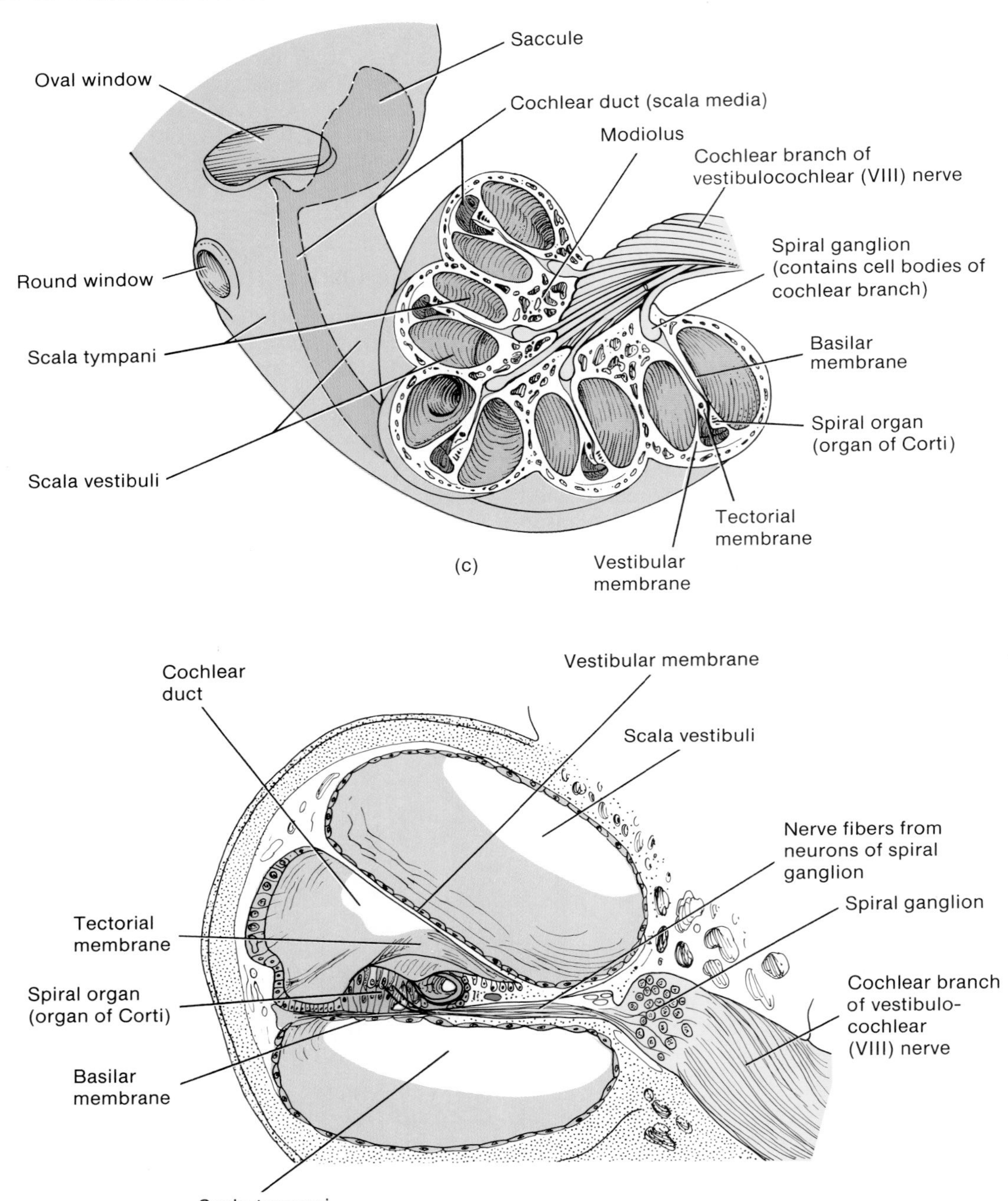

which is chemically similar to cerebrospinal fluid, surrounds the ***membranous labyrinth,*** a series of sacs and tubes lying inside and having the same general form as the bony labyrinth. The membranous labyrinth is lined with epithelium and contains a fluid called ***endolymph,*** which is chemically similar to intracellular fluid.

The ***vestibule*** constitutes the oval central portion of the bony labyrinth. The membranous labyrinth in the vestibule consists of two sacs called the ***utricle*** (YOO-tri-kul = little bag) and ***saccule*** (SAK-yool). These sacs are connected to each other by a small duct.

Projecting upward and posteriorly from the vestibule

FIGURE 17-10 (*Continued*) (e) Enlargement of the spiral organ (organ of Corti).

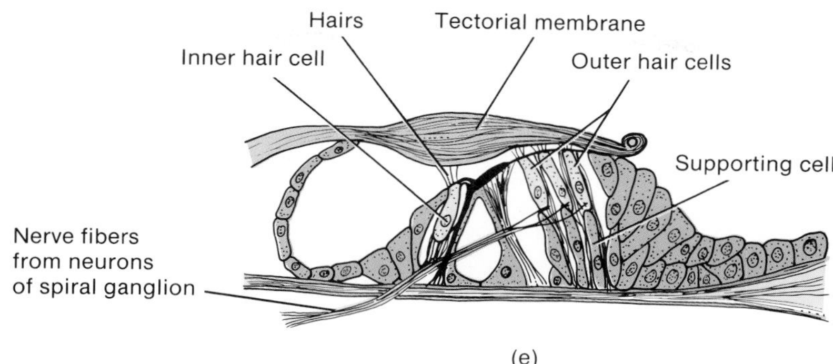

are the three bony ***semicircular canals.*** Each is arranged at approximately right angles to the other two. On the basis of their positions, they are called the anterior, posterior, and lateral canals. The anterior and posterior semicircular canals are oriented vertically; the lateral one is oriented horizontally. One end of each canal enlarges into a swelling called the ***ampulla*** (am-POOL-la; *ampulla* = little jar). The portions of the membranous labyrinth that lie inside the bony semicircular canals are called the ***semicircular ducts (membranous semicircular canals).*** These structures are almost identical in shape to the semicircular canals and communicate with the utricle of the vestibule.

Lying in front of the vestibule is the ***cochlea*** (KŌK-lē-a), so designated because of its resemblance to a snail's shell. The cochlea consists of a bony spiral canal that makes about 2¾ turns around a central bony core called the ***modiolus.*** A cross section through the cochlea shows the canal is divided into three separate channels by partitions that together have the shape of the letter Y. The stem of the Y is a bony shelf that protrudes into the canal; the wings of the Y are composed mainly of membranous labyrinth. The channel above the bony partition is the ***scala vestibuli*** which ends at the oval window; the channel below is the ***scala tympani*** which terminates at the round window. The scala vestibuli and scala tympani both contain perilymph and are completely separated except at an opening at the apex of the cochlea called the ***helicotrema.*** The cochlea adjoins the wall of the vestibule, into which the scala vestibuli opens. The perilymph of the vestibule is continuous with that of the scala vestibuli. The third channel (between the wings of the Y) is the membranous labyrinth: the ***cochlear duct (scala media).*** The cochlear duct is separated from the scala vestibuli by the ***vestibular membrane.*** It is separated from the scala tympani by the ***basilar membrane.***

Resting on the basilar membrane is the ***spiral organ (organ of Corti),*** the organ of hearing. The spiral organ is a series of epithelial cells on the inner surface of the basilar membrane. It consists of a number of supporting cells and hair cells, which are the receptors for auditory sensations. The inner hair cells are medially placed in a single row and extend the entire length of the cochlea. The outer hair cells are arranged in several rows throughout the cochlea. The hair cells have long hairlike processes at their free ends that extend into the endolymph of the cochlear duct. The basal ends of the hair cells are in contact with fibers of the cochlear branch of the vestibulocochlear (VIII) nerve. Projecting over and in contact with the hair cells of the spiral organ is the ***tectorial*** (*tectum* = cover) ***membrane,*** a delicate and flexible gelatinous membrane.

Hair cells of the spiral organ are easily damaged by exposure to high-intensity noises such as those produced by engines of jet planes, revved-up motorcycles, and loud music. As a result of the noises, hair cells become arranged in disorganized patterns, or they and their supporting cells may degenerate. High-intensity noises may produce permanent hearing damage that typically begins in the high-frequency tone range and therefore is not noticed until, in some cases, it is fairly advanced. Only a part of the hearing loss is reversible; this is why ear protection should be worn by those exposed to high-intensity noises.

Sound Waves

Sound waves result from the alternate compression and decompression of air molecules. They originate from a vibrating object, much the same way that waves travel over the surface of water. The sounds heard most acutely by human ears are those from sources that vibrate at frequencies between 1000 and 4000 cycles per second (Hz). The entire range extends from 20 to 20,000 Hz.

The frequency of vibration of the source of the sound is directly related to pitch. The greater the vibration, the higher the pitch. Also, the greater the force of the vibration, the louder the sound. Relative sound intensities (loudness) are usually measured in units called ***decibels (dB).*** In the sound intensity range for normal communication, the human ear can detect approximately a 1-dB change in sound intensity. The hearing threshold—that is, the point at which an average young adult can just detect sound from silence—is 0 dB. Between 115 and 120 dB is the threshold of pain. Just to give you some idea of how decibels relate to common sounds in the range between 0 and 120 dB, rustling leaves have a decibel rating of 15; normal conversation, 45; crowd noise, 60; a vacuum cleaner, 75; and a pneumatic drill, 90.

Physiology of Hearing

The events involved in the physiology of hearing and the conduction of auditory impulses to the brain are as follows (Figure 17-11):

1. Sound waves that reach the ear are directed by the pinna into the external auditory canal.

2. When the waves strike the tympanic membrane, the alternate compression and decompression of the air causes the membrane to vibrate (move forward and backward). The distance the membrane moves is always very small and is relative to the force and velocity of the sound waves. It vibrates slowly in response to low-frequency sounds and rapidly in response to high-frequency sounds.

3. The central area of the tympanic membrane is connected to the malleus, which also starts to vibrate. The vibration is then picked up by the incus, which transmits the vibration to the stapes.

4. As the stapes moves back and forth, it pushes the oval window in and out.

If sound waves passed directly to the oval window without passing through the tympanic membrane and auditory ossicles, hearing would be inadequate. A minimal amount of sound energy is required to transmit sound waves through the perilymph of the cochlea. Since the tympanic membrane has a surface area about 22 times larger than that of the oval window, it can collect about 22 times more sound energy. This energy is sufficient to transmit sound waves through the perilymph.

5. The movement of the oval window sets up waves in the perilymph.

6. As the oval window bulges inward, it pushes the perilymph of the scala vestibuli and pressure waves are propagated through the scala vestibuli. If the inward movement of the stapes at the oval window is slow, the pressure in the perilymph pushes the perilymph from the scala vestibuli to the scala tympani and eventually to the round window, causing it to bulge outward into the middle ear. See number 9 in the illustration.

FIGURE 17-11 **Physiology of hearing. The numbers correspond to the events listed in the text. The cochlea has been uncoiled in order to illustrate the transmission of sound waves and then subsequent distortion of the vestibular and basilar membranes of the cochlear duct.**

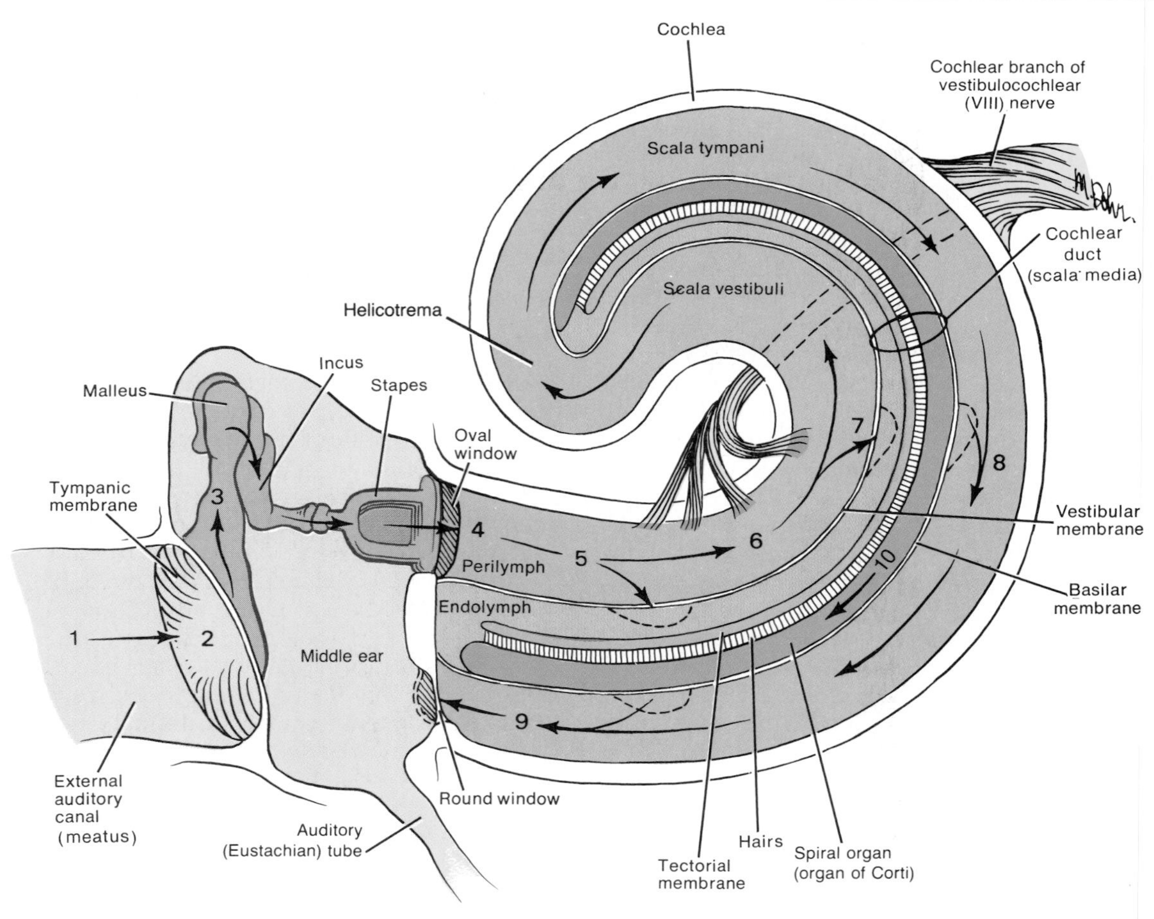

7. As the pressure moves through the perilymph of the scala vestibuli, it pushes the vestibular membrane inward and increases the pressure of the endolymph inside the cochlear duct.

8. As a result, the basilar membrane moves slightly and bulges into the scala tympani. Slow movement of the oval window has only a *very slight* effect on the basilar membrane.

9. As the pressure moves through the scala tympani, perilymph moves toward the round window, causing it to bulge outward into the middle ear. If the stapes vibrates rapidly, pressure in the perilymph does not have time to pass from the scala vestibuli to the scala tympani to the round window. Instead, pressure in the perilymph in the scala vestibuli is transmitted through the basilar membrane and eventually to the round window. As a result, the area of the basilar membrane near the oval and round windows vibrates.

10. When the basilar membrane vibrates, the hair cells of the spiral organ move against the tectorial membrane. The movement of the hairs develops receptor potentials that ultimately lead to the generation of nerve impulses.

The function of hair cells is to convert a mechanical force (stimulus) into an electrical signal (nerve impulse) by way of the development of a receptor potential. It is believed to occur as follows. When the hairs at the top of the cell are moved, ion channels in the hair cell membranes open, permitting a rapid influx of potassium (K^+) ions. This depolarizes the hair cell membrane, producing the receptor potential. Depolarization spreads through the cell and causes calcium (Ca^{2+}) ion channels in the base of the hair cell to open, resulting in an influx of Ca^{2+} ions. At the base of the hair cell are vesicles that store neurotransmitter. The Ca^{2+} ions cause the vesicles to fuse with the basal portion of the hair cell's membrane, thus releasing neurotransmitter from the hair cell, which excites a sensory nerve fiber at the base of the hair cell. From here, the nerve impulse is carried to the brain via the cochlear branch of the vestibulocochlear (VIII) nerve. Although the exact nature of the neurotransmitter is not known, there is some evidence that it might be glutamate or gamma aminobutyric acid (GABA).

Nerve impulses from the cochlear branch of the vestibulocochlear (VIII) nerve (see Figure 17-10b) pass to the cochlear nuclei in the medulla. Here, most impulses cross to the opposite side and then travel to the midbrain, to the thalamus, and finally to the auditory area of the temporal lobe of the cerebral cortex.

Differences in pitch are related to differences in width of the basilar membrane and sound waves of various frequencies that cause specific regions of the basilar membrane to vibrate more intensely than others. The membrane is less flexible at the base of the cochlea (portion closer to the oval window) and more flexible near the apex of the cochlea. High-frequency or high-pitched sounds cause the basilar membrane to vibrate near the base of the cochlea, and low-frequency or low-pitched sounds cause the basilar membrane to vibrate near the apex of the cochlea. Loudness is determined by the intensity of sound waves. High-intensity sound waves cause greater vibration of the basilar membrane. Thus, more hair cells are stimulated and more impulses reach the brain.

MEDICAL TEST

Audiometry (aw'-dē-OM-e-trē; ***audire*** = to hear; ***metron*** = to measure)

Diagnostic Value: To evaluate an individual's hearing acuity. It may be done as a screening examination prior to entry into the military or performed to evaluate a hearing problem in one or both ears. Other uses are to screen young children for hearing problems, to evaluate individuals exposed to loud noises, and to establish a baseline hearing level in a person taking an antibiotic that could cause hearing damage.

Procedure: The person is placed in a soundproof room and listens through earphones to sounds produced by an instrument called an ***audiometer*** (aw'-dē-OM-e-ter) while a technician (audiologist) notes whether or not the sounds can be heard. The sounds produced are varied in terms of pitch and intensity, and a person with a hearing problem will be unable to hear these at the normal threshold, so that higher decibel (dB) levels are required.

Normal Values: 0–25 dB (26–40 dB = mild hearing loss; 41–55 dB = moderate hearing loss; 56–70 dB = moderately severe hearing loss; 71–90 dB = severe hearing loss; 91 or more dB = profound hearing loss).

CLINICAL APPLICATION: ARTIFICIAL EARS

Artificial ears (***cochlear implants***) are devices that translate sounds into electronic signals that can be interpreted by the brain. They take the place of hair cells of the spiral organ (organ of Corti), which normally convert sound waves into electrical signals carried to the brain. The implants are used for individuals with sensorineural deafness, that is, deafness due to disease or injury that has destroyed hair cells of the spiral organ. Such people make up the majority of those who have no hearing in either ear.

The device consists of electrodes implanted in the cochlea that are connected to a plug attached to the outside of the head. A microprocessor pack, about the size of a small transistor radio, is worn on the person's belt or in a shirt pocket and connected by a cord to the plug. Sound waves enter a tiny microphone in the ear, similar to that of a usual hearing aid, and travel down the cord into the pack where they are converted into electrical signals. The signals

then travel to the electrodes in the cochlea, where they stimulate nerve endings and transmit the signals to the brain via the cochlear branch of the vestibulocochlear (VIII) nerve. The sounds heard are crude compared to normal hearing. However, the sounds provide a sense of rhythm and loudness and information about noises such as telephones, automobiles, and the pitch and cadence of speech.

Physiology of Equilibrium

There arr two kinds of ***equilibrium.*** One, called ***static equilibrium,*** refers to the orientation of the body (mainly the head) relative to the ground (gravity). The second kind, ***dynamic equilibrium,*** is the maintenance of body position (mainly the head) in response to sudden movements such as rotation, acceleration, and deceleration. The receptor organs for equilibrium are the maculae of the saccule and the utricle and cristae in the semicircular ducts.

Static Equilibrium

The walls of both the utricle and saccule contain a small, thickened region called a ***macula*** (Figure 17-12). The maculae are the receptors that are concerned with static equilibrium. They provide sensory information regarding the orientation of the head in space and are essential for maintaining posture as we stand or sit.

Microscopically, the two maculae resemble the spiral organ (organ of Corti). They consist of differentiated neuroepithelial cells that are innervated by the vestibular branch of the vestibulocochlear (VIII) nerve. The maculae are anatomically located in planes perpendicular to one another and possess two kinds of cells: ***hair (receptor) cells*** and ***supporting cells.*** Two shapes of hair cells have been identified: one is more flask-shaped and the other is more cylindrical. Both show long extensions of the cell membrane consisting of many ***stereocilia*** (they are actually microvilli) and one ***kinocilium*** (a conventional cilium) anchored firmly to its basal body and extending beyond the longest microvilli. Some of the microvilli reach lengths of over 100 μm, and some receptor cells have over 80 such projections.

The columnar supporting cells of the maculae are scattered between the hair cells. Floating directly over the hair cells is a thick, gelatinous, glycoprotein layer, probably secreted by the supporting cells, called the ***otolithic membrane.*** A layer of calcium carbonate crystals, called ***otoliths*** (*oto* = ear; *lithos* = stone), extends over the entire surface of the otolithic membrane. The specific gravity of these otoliths is about 3, which makes them much denser than the endolymph fluid that fills the rest of the utricle.

The otolithic membrane sits on top of the macula like a discus on a greased cookie sheet. If you tilt your head forward, the otolithic membrane (the discus in our analogy) slides downhill over the hair cells in the direction determined by the tilt of your head. In the same manner, if you are sitting upright in a car that suddenly jerks forward, the otolithic membrane, owing to its inertia, slides

FIGURE 17-12 **Structure of the macula. (a) Overall structure. (b) Details of a hair cell.**

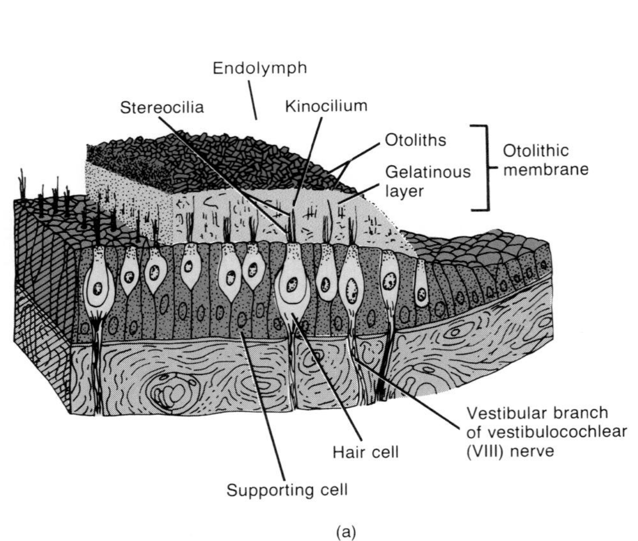

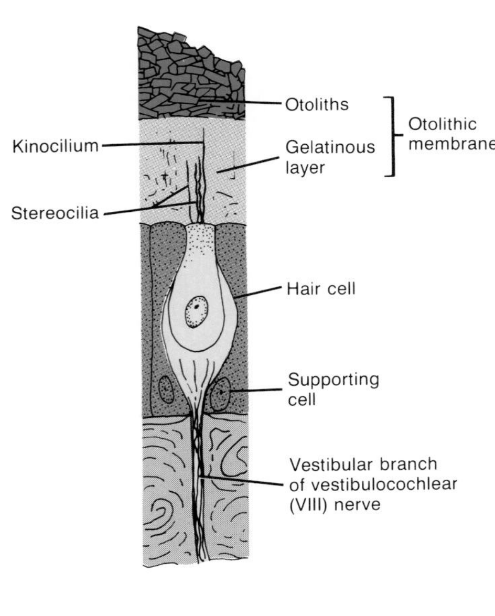

backward and stimulates the hair cells. As the otoliths move, they pull on the gelatinous layer, which pulls on the stereocilia and makes them bend. The movement of the stereocilia initiates nerve impulses by way of receptor potentials that are then transmitted to the vestibular branch of the vestibulocochlear (VIII) nerve (see Figure 17-10b).

Most of the vestibular branch fibers enter the brain stem and terminate in the vestibular nuclear complex in the medulla. The remaining fibers enter the flocculonodular lobe of the cerebellum through the inferior cerebellar peduncle. Bidirectional pathways connect the vestibular nuclei and cerebellum. Fibers from all the vestibular nuclei form the medial longitudinal tracts that extend from the brain stem into the cervical portion of the spinal cord. The tracts send nerve impulses to the nuclei of cranial nerves that control eye movements [oculomotor (III), trochlear (IV), and abducens (VI)] and to the accessory (XI) nerve nucleus that helps control movements of the head and neck. In addition, fibers from the lateral vestibular nucleus form the vestibulospinal tract which conveys impulses to skeletal muscles that regulate body tone in response to head movements. Various pathways between the vestibular nuclei, cerebellum, and cerebrum enable the cerebellum to assume a key role in helping the body maintain static equilibrium. The cerebellum continuously receives updated sensory information from the utricle and saccule concerning static equilibrium. Using this information, the cerebellum monitors and makes corrective adjustments in the motor activities that originate in the cerebral cortex. Essentially, the cerebellum sends continuous nerve impulses to the motor areas of the cerebrum, in response to input from the utricle and saccule, causing the motor system to increase or decrease its impulses to specific skeletal muscles in order to maintain static equilibrium.

Dynamic Equilibrium

The cristae in the three semicircular ducts maintain dynamic equilibrium (Figure 17-13). The ducts are positioned at right angles to one another in three planes: the anterior and posterior semicircular ducts are oriented vertically, and the lateral semicircular duct is oriented horizontally. This positioning permits detection of an imbalance in three planes. In the ampulla, the dilated portion of each duct, there is a small elevation called the ***crista.*** Each crista is composed of a group of ***hair (receptor)***

FIGURE 17-13 Semicircular ducts and dynamic equilibrium. (a) Position of the cristae relative to the membranous ampullae. (Illustration after Biagio John Melloni, Ph.D.) (b) Enlarged aspect of a crista. The ampullar nerves are branches of the vestibular division of the vestibulor cochlear (VIII) nerve.

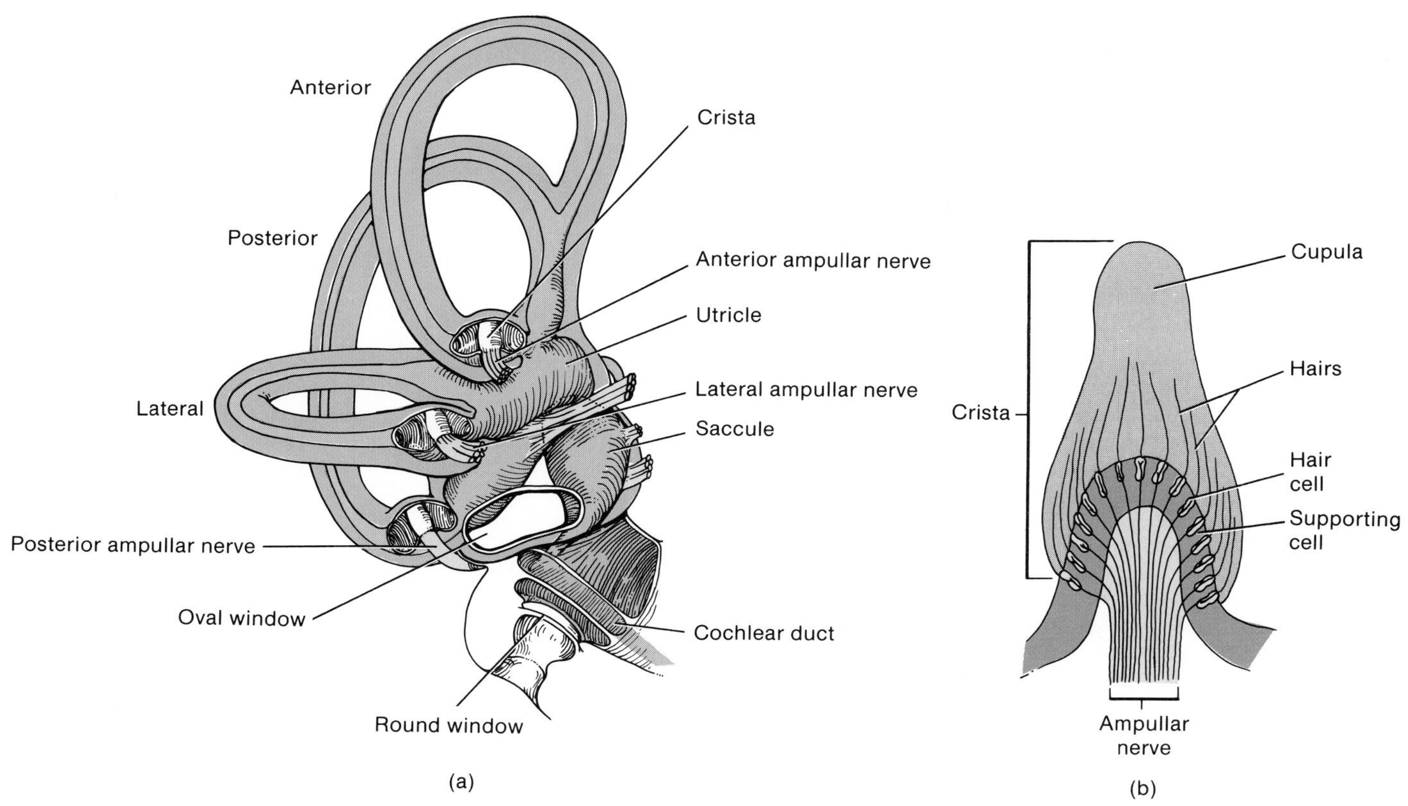

cells and ***supporting cells*** covered by a mass of gelatinous material called the ***cupula.*** When the head moves due to rotation of the body, the endolymph in the semicircular ducts flows over the hairs and bends them. The movement of the hairs stimulates sensory neurons, and the nerve impulses, produced by the resulting receptor potentials, pass over the vestibular branch of the vestibulocochlear (VIII) nerve. The impulses follow the same pathways as those involved in static equilibrium and are eventually sent to the muscles that must contract to maintain body balance in the new position.

A summary of the structures of the ear related to hearing and equilibrium is presented in Exhibit 17-2.

EXHIBIT 17-2 SUMMARY OF STRUCTURES OF THE EAR RELATED TO HEARING AND EQUILIBRIUM

Structure	Function
External (outer) ear	
Pinna (auricle)	Collects sound waves.
External auditory canal (meatus)	Directs sound waves to tympanic membrane.
Tympanic membrane	Sound waves cause it to vibrate, which, in turn, causes the malleus to vibrate.
Middle ear	
Auditory (Eustachian) tube	Equalizes pressure on both sides of the tympanic membrane.
Auditory ossicles	Transmit sound waves from tympanic membrane to oval window.
Internal (inner) ear	
Utricle	Contains macula, receptor for static equilibrium.
Saccule	Contains macula, receptor for static equilibrium.
Semicircular ducts	Contain cristae, receptors for dynamic equilibrium.
Cochlea	Contains a series of fluids, channels, and membranes that transmit sound waves to the spiral organ (organ of Corti), the organ of hearing; the spiral organ generates nerve impulses and transmits them to the cochlear branch of the vestibulocochlear (VIII) nerve.

DISORDERS: HOMEOSTATIC IMBALANCES

The special sense organs can be altered or damaged as a result of numerous disorders. The causes of disorder range from congenital origins to the effects of old age. Here we discuss a few common disorders of the eyes and ears.

Cataract

The most prevalent disorder resulting in blindness is cataract formation. A ***cataract,*** meaning "waterfall," is a clouding of the lens or its capsule so that it becomes opaque or milk white. Two basic processes involved in cataract formation are the breakdown of the normal lens protein and an influx of water into the lens. As a result, the light from an object, which normally passes directly through the lens to produce a sharp image, produces only a degraded image. If the cataract is severe enough, no image at all is produced. Cataracts are associated with aging but may also be caused by injury, exposure to radiation (ultraviolet B), certain medications (long-term use of steroids), or complications of other diseases.

Glaucoma

The second most common cause of blindness, especially in the elderly, is ***glaucoma.*** It is a group of disorders characterized by an abnormally high intraocular pressure (IOP) owing to a buildup of aqueous humor inside the eyeball. The aqueous humor does not return into the bloodstream through the scleral venous sinus (canal of Schlemm) as quickly as it is formed. The fluid accumulates and, by compressing the lens into the vitreous body, puts pressure on the neurons of the retina. If the pressure continues over a long period of time, glaucoma can progress from mild visual impairment to a point where neurons of the retina are destroyed, resulting in degeneration of the optic disc, visual field defects, and blindness. Treatment is through drugs such as timolol (Timoptic) and acetazolamide (Diamox), or laser surgery.

Conjunctivitis (Pinkeye)

Many different eye inflammations exist, but the most common type is ***conjunctivitis*** **(*pinkeye*)**, an inflam-

DISORDERS: HOMEOSTATIC IMBALANCES (*continued*)

mation of the conjunctiva. Conjunctivitis can be caused by bacteria such as pneumococci, staphylococci, or *Hemophilis influenzae.* In such cases, the inflammation is very contagious. This epidemic type is common in children. Conjunctivitis may also be caused by a number of irritants, in which case the inflammation is not contagious. Irritants include dust, smoke, wind, pollutants in the air, and excessive glare. The condition may be acute or chronic.

Trachoma

A serious form of chronic contagious conjunctivitis is known as ***trachoma*** (tra-KŌ-ma), the greatest single cause of blindness in the world. It is caused by a bacterium called *Chlamydia trachomatis,* which has characteristics of both viruses and bacteria. Trachoma is characterized by many granulations or fleshy projections on the eyelids. If untreated, these projections can irritate and inflame the cornea and reduce vision. The disease produces an excessive growth of subconjunctival tissue and the invasion of blood vessels into the upper half of the front of the cornea. The disease progresses until it covers the entire cornea, bringing about a loss of vision because of corneal opacity.

Antibiotics, such as tetracycline and the sulfa drugs, kill the organisms that cause trachoma and have reduced the seriousness of this infection.

Deafness

Deafness is the lack of the sense of hearing or significant hearing loss. It is generally divided into two principal types: sensorineural deafness and conduction deafness. ***Sensorineural deafness*** is caused by impairment of the cochlea or cochlear branch of the vestibulocochlear (VIII) nerve. ***Conduction deafness*** is caused by impairment of the external and middle ear mechanisms for transmitting sounds into the cochlea. Among the factors that contribute to deafness are genetic factors; otosclerosis, the deposition of new bone around the oval window; repeated exposure to loud noise, which destroys hair cells of the spiral organ (organ of Corti); certain drugs such as streptomycin; swimmer's ear and other infections; impacted cerumen; injury to the tympanic membrane; and aging, which results in thickening of the tympanic membrane, stiffening of the joints of the auditory ossicles, and decreased numbers of hair cells owing to diminished cell division.

Labyrinthine Disease

Labyrinthine (lab'-i-RIN-thēn) ***disease*** refers to a malfunction of the inner ear that is characterized by attacks of hearing loss, tinnitus (ringing in the ears), vertigo (sensation of spinning or movement), nausea, and vomiting. There may also be a blurring of vision, nystagmus (rapid, involuntary movement of the eyeballs), and a tendency to fall in a certain direction.

Among the causes of labyrinthine disease are: (1) infection of the inner ear, (2) trauma from brain concussion producing hemorrhage or splitting of the labyrinth, (3) cardiovascular diseases such as atherosclerosis and blood vessel disturbances, (4) congenital malformation of the labyrinth, (5) excessive formation of endolymph, (6) allergy, (7) blood abnormalities, and (8) aging.

Ménière's Syndrome

Ménière's (men-YAIRZ) ***syndrome*** is one type of labyrinthine disorder that is characterized by an increased amount of endolymph that enlarges the labyrinth. Among the symptoms are fluctuating hearing loss, attacks of vertigo, and roaring tinnitus. Etiology of Ménière's syndrome is unknown. It is now thought that there is either an overproduction or underabsorption of endolymph in the cochlear duct. The hearing loss is caused by distortions in the basilar membrane of the cochlea. The classic type of Ménière's syndrome involves both the semicircular canals and the cochlea. The natural course of Ménière's syndrome may stretch out over a period of years, with the end result being almost total destruction of hearing. Treatment consists of drugs to relieve the vertigo or surgery that creates an opening in the labyrinth, severing the vestibular branch of the vestibulocochlear (VIII) nerve, or removal of the labyrinth.

Vertigo

Vertigo (*vertex* = whorl) is a sensation of spinning or movement in which the world is revolving or the person is revolving in space. Depending on its cause, vertigo may be classified as: (1) ***peripheral,*** which originates in the ear, has a sudden onset, and lasts from minutes to hours; (2) ***central,*** which is caused by an abnormality of the central nervous system (CNS) and may persist for more than three weeks; and (3) ***psychogenic,*** in which the cause is of psychological origin.

Otitis Media

Otitis media is an acute infection of the middle ear, caused primarily by bacteria such as *Streptococcus pneumoniae* and *Hemophilus influenzae.* It is characterized by pain, malaise, fever, and a reddening and outward bulging of the eardrum, which may rupture unless prompt treatment is given (this may involve draining pus from the middle ear). Abnormal function of the auditory (Eustachian) tube appears to be the most important mechanism in the pathogenesis of middle ear disease. This allows bacteria from the nasopharynx, which are a primary cause of middle ear infection, to enter the middle ear.

Motion Sickness

Motion sickness is a functional disorder brought on by repetitive angular, linear, or vertical motion and characterized by various symptoms, primarily nausea and vomiting. Warning symptoms include yawning, salivation, pallor, hyperventilation, profuse cold sweating, and prolonged drowsiness. Specific kinds of motion sickness include seasickness, airsickness, carsickness, and swing sickness.

DISORDERS: HOMEOSTATIC IMBALANCES (*continued*)

Etiology is excessive stimulation of the vestibular apparatus by motion. Nerve impulses pass from the internal ear to the vomiting center in the medulla. Visual stimuli and emotional factors like fear and anxiety can also contribute to motion sickness. Ideally, treatment with drugs of susceptible individuals should be instituted prior to their entering into conditions that would produce motion sickness, since prevention is more successful than treatment of symptoms once they have developed. Among the drugs used to treat motion sickness are dimenhydrinate (Dramamine), meclizine (Antivert), and promethazine (Phenegran).

MEDICAL TERMINOLOGY ASSOCIATED WITH SENSORY STRUCTURES

Achromatopsia (a-krō'-ma-TOP-sē-a; *a* = without; *chrom* = color) Complete color blindness.

Ametropia (am'-e-TRŌ-pē-a; *ametro* = disproportionate; *ops* = eye) Refractive defect of the eye resulting in an inability to focus images properly on the retina.

Anopsia (an-OP-sē-a; *opsia* = vision) A defect of vision.

Astereognosis (a-ster-ē-og-NŌ-sis; *stereos* = solid; *gnosis* = knowledge) Loss of ability to recognize objects or to appreciate their form by touching them.

Blepharitis (blef-a-RĪ-tis; *blepharo* = eyelid; *itis* = inflammation of) An inflammation of the eyelid.

Dyskinesia (dis'-ki-NĒ-zē-a; *dys* = difficult; *kinesis* = movement) Abnormality of motor function characterized by involuntary, purposeless movements.

Epiphora (e-PIF-ō-ra; *epi* = above) Abnormal overflow of tears.

Eustachitis (yoo'-stā-KĪ-tis) An inflammation or infection of the auditory (Eustachian) tube.

Exotropia (ek'-sō-TRŌ-pē-a; *ex* = out; *tropia* = turning) Turning outward of the eyes.

Keratitis (ker'-a-TĪ-tis; *kerato* = cornea) An inflammation or infection of the cones.

Kinesthesis (kin'-es-THĒ-sis; *aisthesis* = sensation) The sense of perception of movement.

Labyrinthitis (lab'-i-rin-THĪ-tis) An inflammation of the labyrinth (inner ear).

Mydriasis (mi-DRĒ-a-sis) Dilated pupil.

Myringitis (mir'-in-JĪ-tis; *myringa* = eardrum) An inflammation of the eardrum; also called **tympanitis.**

Nystagmus (nis-TAG-mus; *nystazein* = to nod) A rapid involuntary movement of the eyeballs, possibly caused by a disease of the central nervous system. It is associated with conditions that cause vertigo.

Otalgia (o-TAL-jē-a; *oto* = ear; *algia* = pain) Earache.

Otosclerosis (ō'-tō-skle-RŌ-sis; *oto* = ear; *sclerosis* = hardening) Pathological process that may be hereditary in which new bone is deposited around the oval window. The result may be immobilization of the stapes, leading to deafness.

Photophobia (fō'-tō-FŌ-bē-a; *photo* = light; *phobia* = fear) Abnormal visual intolerance to light.

Presbyopia (pres'-bē-Ō-pē-a; *presby* = old) Inability to focus on nearby objects owing to loss of elasticity of the crystalline lens. The loss is usually caused by aging.

Ptosis (TŌ-sis; *ptosis* = fall) Falling or drooping of the eyelid. (This term is also used for the slipping of any organ below its normal position.)

Retinoblastoma (ret'-i-nō-blas-TŌ-ma; *blast* = bud; *oma* = tumor) A tumor arising from immature retinal cells and accounting for 2 percent of childhood malignancies.

Scotoma (skō-TŌ-ma; *scotoma* = darkness) An area of reduced or lost vision in the visual field. Also called a **blind spot** (other than the normal blind spot or optic disc).

Strabismus (stra-BIZ-mus) An imbalance in the extrinsic eye muscles that produces a squint (formerly referred to as "cross-eyes"). **Amblyopia** is the term used to describe the loss of vision in an otherwise normal eye that, because of muscle imbalance, cannot focus in sync with the other eye.

Tinnitus (ti-NĪ-tus) A ringing, roaring, or clicking in the ears.

STUDY OUTLINE

Olfactory Sensations (p. 463)

1. The receptors for olfaction, the olfactory cells, are in the nasal epithelium.
2. Substances to be smelled must be volatile, water-soluble, and lipid-soluble.
3. Adaptation to odors occurs quickly, and the threshold of smell is low.
4. Olfactory cells convey nerve impulses to olfactory (I) nerves, olfactory bulbs, olfactory tracts, and cerebral cortex.

Gustatory Sensations (p. 464)

1. The receptors for gustation, the gustatory cells, are located in taste buds.

2. Substances to be tasted must be in solution in saliva.
3. The four primary tastes are salt, sweet, sour, and bitter.
4. Adaptation to taste occurs quickly, and the threshold varies with the taste involved.
5. Gustatory cells convey nerve impulses to cranial nerves V, VII, IX, and X, medulla, thalamus, and cerebral cortex.

Visual Sensations (p. 466)

1. Accessory structures of the eyes include the eyebrows, eyelids, eyelashes, and the lacrimal apparatus.
2. The eye is constructed of three coats: (a) fibrous tunic (sclera and cornea), (b) vascular tunic (choroid, ciliary body, and iris), and (c) retina (nervous tunic), which contains rods and cones.
3. The anterior cavity contains aqueous humor; the vitreous chamber contains the vitreous body.
4. The refractive media of the eye are the cornea, aqueous humor, lens, and vitreous body.
5. Retinal image formation involves refraction of light, accommodation of the lens, constriction of the pupil, convergence, and inverted image formation.
6. Improper refraction may result from myopia (nearsightedness), hypermetropia (farsightedness), and astigmatism (corneal or lens abnormalities).
7. Rods and cones develop receptor potentials and ganglion cells initiate nerve impulses.
8. Impulses from ganglion cells are conveyed through the retina to the optic (II) nerve, the optic chiasma, the optic tract, the thalamus, and the cortex.

Auditory Sensations and Equilibrium (p. 480)

1. The ear consists of three anatomical subdivisions: (a) the external or outer ear (pinna, external auditory canal, and tympanic membrane), (b) the middle ear (auditory or Eustachian tube, ossicles, oval window, and round window), and (c) the internal or inner ear (bony labyrinth and membranous labyrinth). The internal ear contains the spiral organ (organ of Corti), the organ of hearing.
2. Sound waves enter the external auditory canal, strike the tympanic membrane, pass through the ossicles, strike the oval window, set up waves in the perilymph, strike the vestibular membrane and scala tympani, increase pressure in the endolymph, strike the basilar membrane, and stimulate hairs on the spiral organ (organ of Corti). A sound impulse is then initiated.
3. Static equilibrium is the orientation of the body relative to the pull of gravity. The maculae of the utricle and saccule are the sense organs of static equilibrium.
4. Dynamic equilibrium is the maintenance of body position in response to movement. The cristae in the semicircular ducts are the sense organs of dynamic equilibrium.

Disorders: Homeostatic Imbalances (p. 490)

1. Cataract is the loss of transparency of the lens or capsule.
2. Glaucoma is abnormally high intraocular pressure (IOP), which destroys neurons of the retina.
3. Conjunctivitis (pinkeye) is an inflammation of the conjunctiva.
4. Trachoma is a chronic, contagious inflammation of the conjunctiva.
5. Deafness is the lack of the sense of hearing or significant hearing loss. It is classified as sensorineural and conduction.
6. Labyrinthine disease is basically a malfunction of the inner ear that has a variety of causes.
7. Ménière's syndrome is the malfunction of the inner ear that may cause deafness and loss of equilibrium.
8. Vertigo is a sensation of motion that is classified as peripheral, central, or psychogenic.
9. Otitis media is an acute infection of the middle ear.
10. Motion sickness is a functional disorder precipitated by repetitive angular, linear, or vertical motion.

REVIEW QUESTIONS

1. Describe the structure of olfactory receptors. (p. 463)
2. What are the necessary conditions for substances to be smelled? (p. 463)
3. Discuss the origin and path of a nerve impulse that results in olfaction. (p. 464)
4. Describe the structure of gustatory receptors. (p. 464)
5. How are gustatory receptors stimulated? (p. 466)
6. Discuss how a nerve impulse for gustation travels from a taste bud to the brain. (p. 466)
7. Describe the structure and importance of the following accessory structures of the eye: eyelids, eyelashes, and eyebrows. (p. 466)
8. What is the function of the lacrimal apparatus? Explain how it operates. (p. 468)
9. By means of a labeled diagram, indicate the principal anatomical structures of the eye. (p. 469)
10. Describe the location and contents of the chambers of the eye. What is intraocular pressure (IOP)? How is the scleral venous sinus (canal of Schlemm) related to this pressure? (p. 473)
11. Explain how each of the following events is related to the physiology of vision: (a) refraction of light, (b) accommodation of the lens, (c) constriction of the pupil, (d) convergence, and (e) inverted image formation. (p. 475)
12. Distinguish emmetropia, myopia, hypermetropia, and astigmatism by means of a diagram. (p. 475)
13. How are rods and cones excited? Relate your discussion to the rhodopsin cycle by means of a diagram. (p. 476)
14. What is night blindness? What causes it? (p. 477)
15. Distinguish between light adaptation and dark adaptation. (p. 477)
16. Describe the path of a visual impulse from the optic (II) nerve to the brain. (p. 478)
17. Define visual field. Relate the visual field to image formation on the retina. (p. 478)
18. Diagram the principal parts of the external, middle, and internal ear. Describe the function of each part labeled. (p. 480)
19. What are sound waves? How are sound intensities measured? (p. 485)
20. Explain the events involved in the transmission of sound from the pinna to the spiral organ (organ of Corti). (p. 486)

21. What is the afferent pathway for sound impulses from the cochlear branch of the vestibulocochlear (VIII) nerve to the brain? (p. 486)
22. Compare the function of the maculae in the saccule and utricle in maintaining static equilibrium with the role of the cristae in the semicircular ducts in maintaining dynamic equilibrium. (p. 488)
23. Describe the path of a nerve impulse that results in static and dynamic equilibrium. (p. 489)
24. Define each of the following: cataract, glaucoma, conjunctivitis, trachoma, deafness, labyrinthine disease, Ménière's syndrome, vertigo, otitis media, and motion sickness. (p. 490)
25. Define the following: "watery" eyes (p. 468), corneal transplant (p. 470), radial keratotomy (p. 470), epikeratoplasty (p. 470), detached retina (p. 472), senile macular degeneration (SMD) (p. 473), contact lens (p. 476), red-green color blindness (p. 477), perforated eardrum (p. 482), and artificial ear. (p. 487)
26. Briefly describe the diagnostic value and procedure for the following: tonometry (p. 474), ophthalmoscopy (p. 474), visual acuity (p. 480), and audiometry. (p. 487)
27. Refer to the medical terminology associated with the sense organs. Be sure that you can define each term listed. (p. 492)

SELECTED READINGS

Franklin, D. "Crafting Sound from Silence," *Science News,* 20 October 1984.

———. "Hearing Loss and Hearing Aids," *Harvard Medical School Health Letter,* April 1989.

Hogan, M. J. (ed). "Color Blindness," *Mayo Clinic Health Letter,* March 1989.

Hudspeth, A. J. "The Hair Cells of the Inner Ear," *Scientific American,* January 1983.

Kiely, J. M. (ed). "Cataracts," *Mayo Clinic Health Letter,* February 1989.

———. "Retinal Detachment," *Mayo Clinic Health Letter,* January 1988.

———. "Your Eyes Can Be Windows to Your Health," *Mayo Clinic Health Letter,* March 1985.

Koretz, J. F., and G. H. Handelman. "How the Human Eye Focuses," *Scientific American,* July 1988.

Loeb, G. E. "The Functional Replacement of the Ear," *Scientific American,* February 1985.

Morrison, A. R. "A Window on the Sleeping Brain," *Scientific American,* April 1983.

Von Bekesy, G. "The Ear," *Scientific American,* August 1975.

Wiet, R. J. "Help for the Hearing-Impaired," *Postgraduate Medicine,* November 1988.

Chapter 18

The Endocrine System

Chapter Contents at a Glance

Student Objectives

1. Discuss the functions of the endocrine system in maintaining homeostasis.
2. Explain the mechanisms of hormonal action by interaction with plasma membrane receptors and intracellular receptors.
3. Describe the control of hormonal secretions via feedback cycles and give several examples.
4. Describe the release of hormones stored in the neurohypophysis.
5. Describe the location, histology, hormones, and functions of the following endocrine glands: pituitary, thyroid, parathyroids, adrenals (suprarenals), pancreas, ovaries, testes, pineal, and thymus.
6. Discuss the symptoms of pituitary dwarfism, giantism, acromegaly, diabetes insipidus, cretinism, myxedema, exophthalmic goiter, simple goiter, tetany, osteitis fibrosa cystica, aldosteronism, Addison's disease, Cushing's syndrome, adrenogenital syndrome, pheochromocytomas, diabetes mellitus, and hyperinsulinism.
7. Describe the development of the endocrine system.
8. Describe the effects of aging on the endocrine system.
9. Define the general adaptation syndrome (GAS) and compare homeostatic responses and stress responses.
10. Define medical terminology associated with the endocrine system.

Two regulatory systems are involved in transmitting messages and correlating various body functions: the nervous system and endocrine system. The nervous system controls homeostasis through impulses delivered via neurons. The endocrine system affects bodily activities by releasing chemical messengers, called hormones, into the bloodstream. Whereas the nervous system sends messages to a specific set of cells (muscle fibers, gland cells, or other neurons), the endocrine system as a whole sends messages to cells in virtually any part of the body. The nervous system causes muscles to contract or relax and glands to secrete more or less of their product; the endocrine system brings about changes in the metabolic activities of almost all body tissues. Thus the endocrine system not only helps to regulate activity of smooth and cardiac muscle and some glands, it significantly affects virtually all other tissues as well. Neurons tend to act within a few milliseconds; hormones can take up to several hours or more to bring about their responses. Also, the effects of nervous system stimulation are generally brief compared with the effects of endocrine stimulation.

Obviously, the body could not function if the two great control systems were to pull in opposite directions. The nervous and endocrine systems coordinate their activities like an interlocking supersystem. Certain parts of the nervous system stimulate or inhibit the release of hormones, and certain hormones, in turn, are quite capable of stimulating or inhibiting the flow of nerve impulses.

Although the effects of hormones are many and varied, their actions can be categorized into five broad areas:

1. They help to control the internal environment by regulating its chemical composition and volume.
2. They respond to marked changes in environmental conditions to help the body cope with emergency demands such as infection, trauma, emotional stress, dehydration, starvation, hemorrhage, and temperature extremes.
3. They assume a role in the smooth, sequential integration of growth and development.
4. They contribute to the basic processes of reproduction, including gamete (egg and sperm) production, fertilization, nourishment of the embryo and fetus, delivery, and nourishment of the newborn.
5. They help regulate organic metabolism and energy balance.

The science concerned with the structure and functions of the endocrine glands and the diagnosis and treatment of disorders of the endocrine system is called ***endocrinology*** (en'-dō-kri-NOL-ō-jē; *endo* = within; *crin* = to secrete; *logos* = study of).

The developmental anatomy of the endocrine system is considered later in the chapter.

ENDOCRINE GLANDS

The endocrine glands make up the ***endocrine system.*** The body contains two kinds of glands: exocrine and endocrine. ***Exocrine glands*** secrete their products into ducts, and the ducts carry the secretions into body cavities, into the lumens of various organs, or to the body's surface. Exocrine glands include sudoriferous (sweat), sebaceous (oil), mucous, and digestive glands. ***Endocrine glands,*** by contrast, secrete their products (hormones) into the extracellular space around the secretory cells, rather than into ducts. The secretion then passes into capillaries to be transported in the blood. The endocrine glands of the body include the pituitary (hypophysis), thyroid, parathyroids, adrenals (suprarenals), pineal (epiphysis cerebri), and thymus gland. In addition, there are several organs of the body that contain endocrine tissue but are not exclusively endocrine glands. These include the pancreas, ovaries, testes, kidneys, stomach, small intestine, skin, heart, and placenta. The location of many organs of the endocrine system and endocrine-containing organs is illustrated in Figure 18-1.

CHEMISTRY OF HORMONES

The secretions of endocrine glands are called ***hormones*** (*hormone* = set in motion). Although hormones are chemically diverse, they can be grouped into three principal classes: (1) amines, (2) proteins and peptides, and (3) steroids.

1. **Amines.** Structurally, these are the simplest hormone molecules. They are modified from the amino acid tyrosine. Examples of amines are thyroid hormones (triiodothyronine and tetraiodothyronine), produced by the thyroid gland, and epinephrine and norepinephrine (NE), together called catecholamines, secreted by the medulla (inner zone) of the adrenal glands (Figure 18-2a).

2. **Proteins and peptides.** These hormones consist of chains of amino acids, which may vary in size from relatively small molecules, as in the case of oxytocin (OT) secreted by the hypothalamus, to fairly large molecules, as in the case of insulin secreted by the pancreas (Figure 18-2b). Protein and peptide hormones are synthesized in the manner described for protein synthesis in Chapter 3.

Other endocrine glands that synthesize protein or peptide hormones include the anterior pituitary, thyroid (a hormone called calcitonin), and parathyroids. Protein, peptide, and amine hormones are water-soluble.

3. **Steroids.** These hormones are derived from cholesterol. Examples include aldosterone, cortisol, and androgen secreted by the cortex (outer zone) of the adrenal glands, testosterone secreted by the testes, and estrogens and progesterone secreted by the ovaries (Figure 18-2c). Steroid hormones are lipid-soluble and are produced in both mitochondria and agranular endoplasmic reticulum.

FIGURE 18-1 **Location of many endocrine glands, organs containing endocrine tissue, and associated structures.**

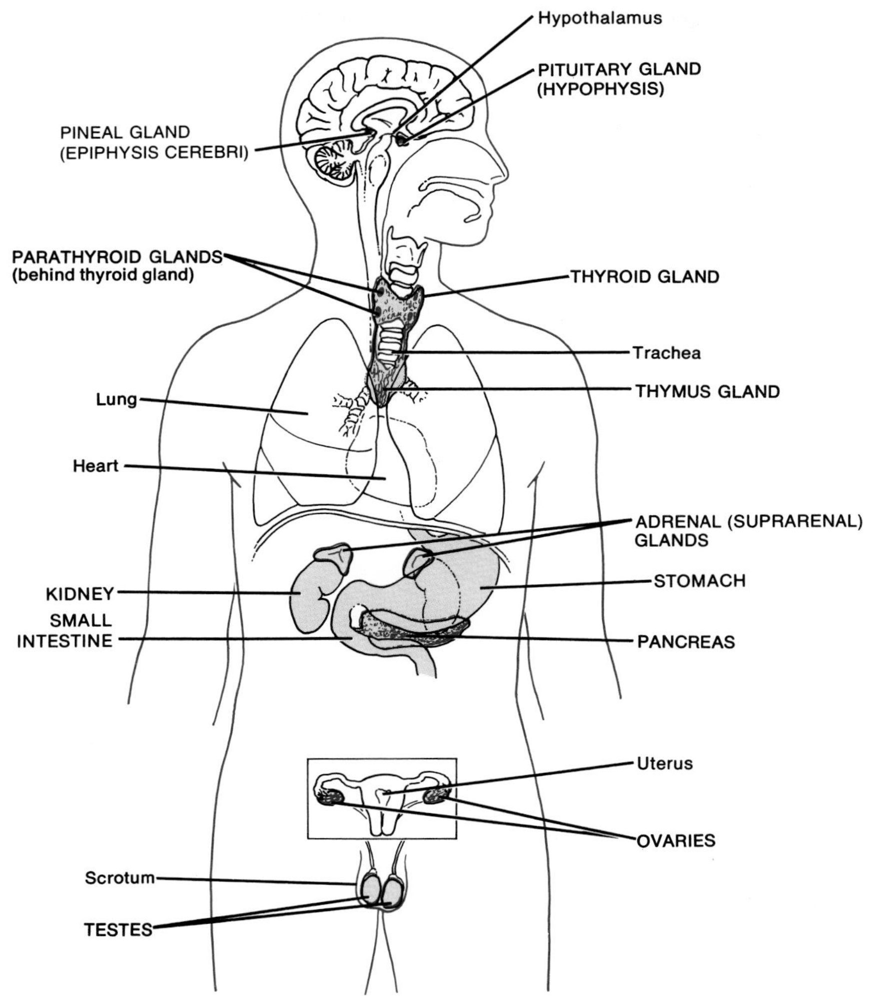

FIGURE 18-2 **Representative hormones based on chemical structure. (a) Amines.**

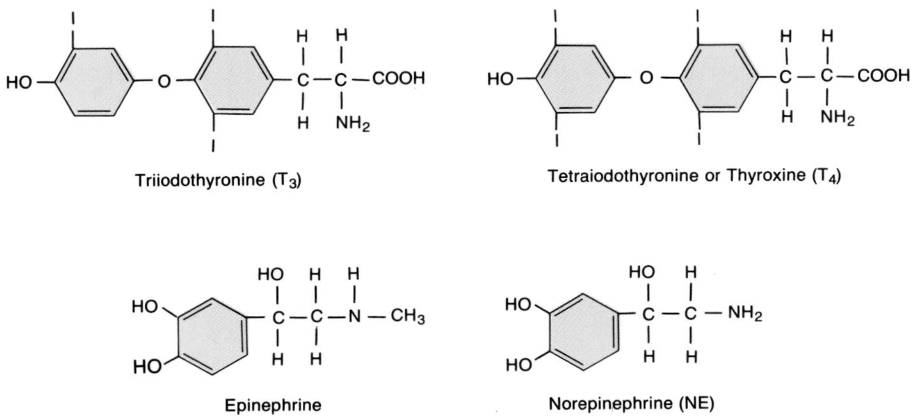

Triiodothyronine (T_3)

Tetraiodothyronine or Thyroxine (T_4)

Epinephrine

Norepinephrine (NE)

(a) AMINES

FIGURE 18-2 (*Continued*) Representative hormones based on chemical structure.
(b) Proteins and peptides. (c) Steroids.

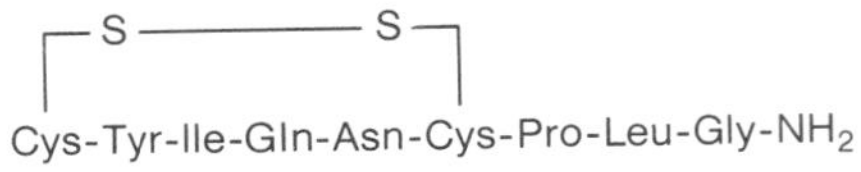

S — S

Cys-Tyr-Ile-Gln-Asn-Cys-Pro-Leu-Gly-NH_2

Oxytocin (OT)

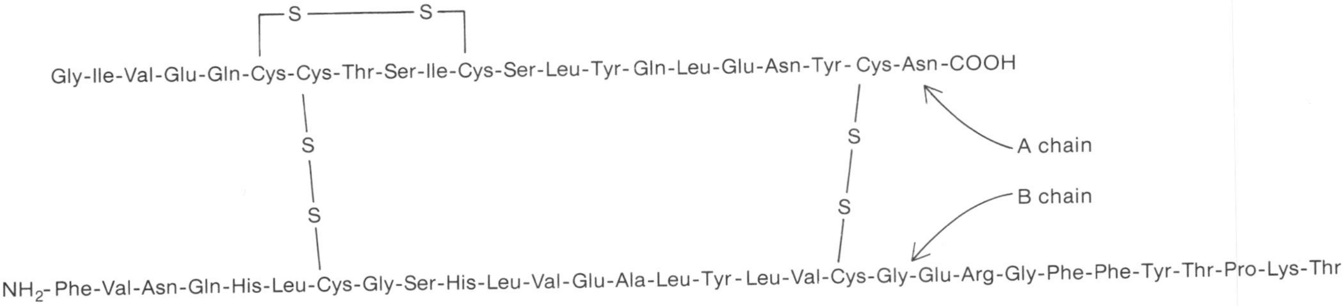

S — S

Gly-Ile-Val-Glu-Gln-Cys-Cys-Thr-Ser-Ile-Cys-Ser-Leu-Tyr-Gln-Leu-Glu-Asn-Tyr-Cys-Asn-COOH

S S

S S

A chain

B chain

NH_2-Phe-Val-Asn-Gln-His-Leu-Cys-Gly-Ser-His-Leu-Val-Glu-Ala-Leu-Tyr-Leu-Val-Cys-Gly-Glu-Arg-Gly-Phe-Phe-Tyr-Thr-Pro-Lys-Thr

Insulin

Key to abbrevations for amino acids:

Alanine (Ala)
Arginine (Arg)
Asparagine (Asn)
Aspartic acid (Asp)
Cystine (Cys)
Glutamic acid (Glu)
Glutamine (Gln)
Glycine (Gly)
Histidine (His)
Isoleucine (Ile)
Leucine (Leu)
Lysine (Lys)
Methionine (Met)
Phenylalanine (Phe)
Proline (Pro)
Serine (Ser)
Theronine (Thr)
Tryptophan (Try)
Tyrosine (Tyr)
Valine (Val)

(b) PROTEINS AND PEPTIDES

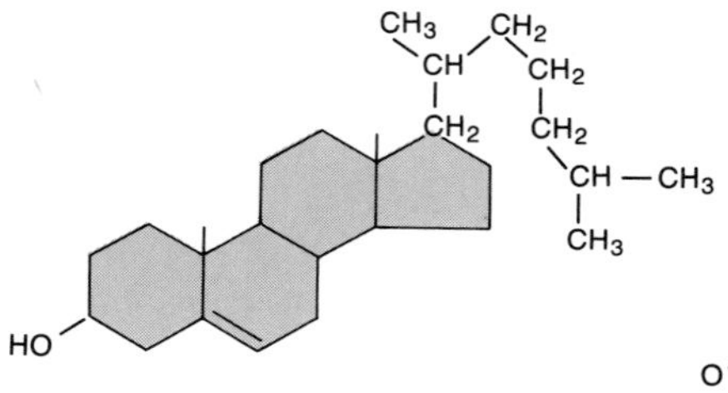

CH_3 CH_2 CH CH_2 CH_2 CH_2 CH — CH_3 CH_3 HO

Cholesterol

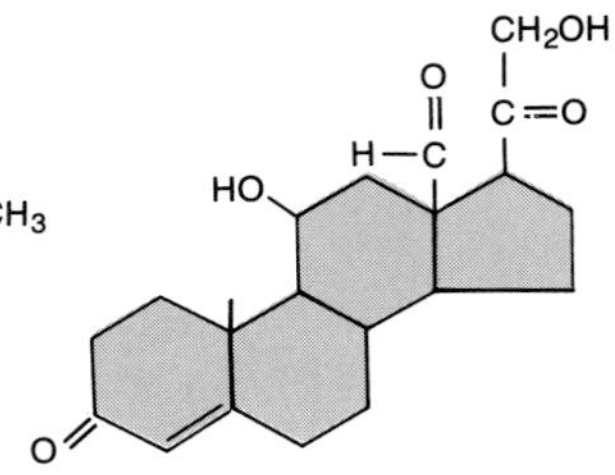

CH_2OH O C=O H—C HO O

Aldosterone

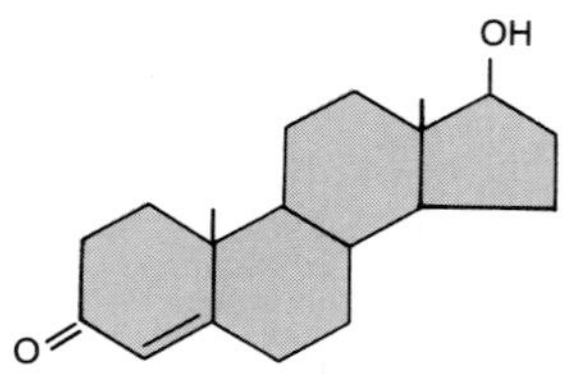

OH O

Testosterone

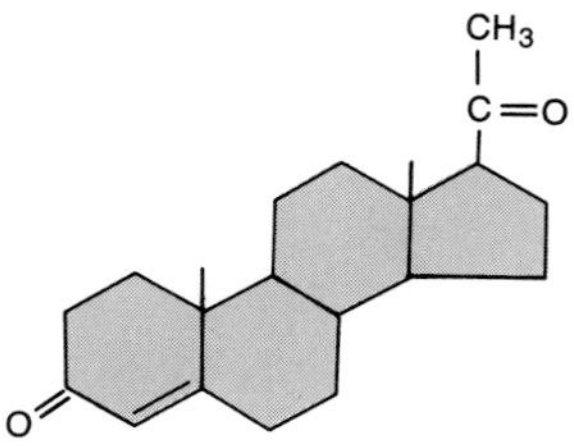

CH_3 C=O O

Progesterone

(c) STEROIDS

As you will learn in Chapter 29, all tissues and organs of the body are derived from three basic tissues of an embryo called the endoderm, mesoderm, and ectoderm. Endocrine glands that originate from the endoderm include the parathyroids and pancreas. Endocrine glands of ectodermal origin include the adrenal medulla and pituitary gland. Endocrine glands of mesodermal origin include the adrenal cortex, ovaries, and testes.

The one thing all hormones have in common is the function of maintaining homeostasis by changing the rate of the physiological activities of cells.

PHYSIOLOGY OF HORMONAL ACTION

Overview

The amount of hormone released by an endocrine gland or tissue is determined by the body's *need* for the hormone at any given time (Figure 18-3). This is the basis on which the endocrine system operates. Hormone-producing cells have available to them information from sensing and signaling systems that permit the hormone-producing cells to regulate the amount and duration of hormone release. Secretion is normally regulated so that there is no overproduction or underproduction of a particular hormone.

FIGURE 18-3 Overview of physiology of the endocrine system.

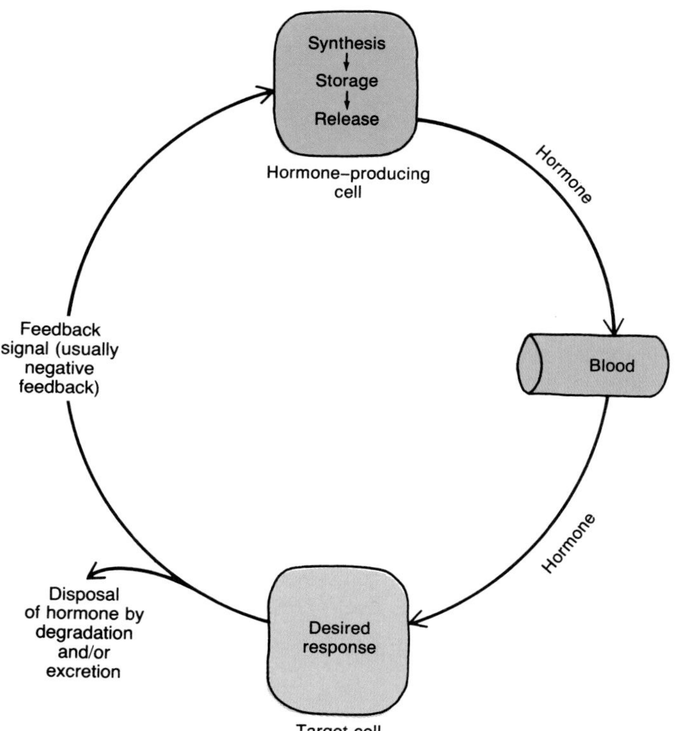

Once a hormone is released by a secretory cell, it is carried by the blood to ***target cells,*** cells that respond to the hormone. All cells are target cells for one or more hormones, but not all cells respond to a particular hormone. Whereas catecholamines, protein, and peptide hormones are generally carried in free form in blood, steroid and thyroid hormones are carried by one or more plasma proteins synthesized in the liver. Target cells contain receptors (described shortly) that bind the hormone so that it can produce the effect. Once the target cells respond to the hormone, the response must be recognized by the secretory cell by some type of feedback signal. Hormones that have accomplished their goals are degraded by target cells or the liver or removed by the kidneys in urine.

Receptors and Hormone Specificity

Cells that produce hormones represent only a limited number of cell types. By contrast, practically all cells of the body are target cells for one hormone or another. As a rule, most of the over 50 or so hormones affect a wide range of target cells. Since all body cells are exposed to equal concentrations of hormones, why is it that some target cells respond to particular hormones and others do not? The answer is ***receptors,*** large protein molecules that may be found in the plasma membrane (integral proteins), cytoplasm, and nucleus of target cells. Only cells that have receptors to certain hormones can respond. Receptors are quite specific in that they recognize only certain hormones. Because of the specific complementary structure of hormone and receptor, only certain hormones bind to certain receptors (see Figure 18-4). It is for this very reason that a hormone influences its target cells but not other cells in the body. For example, whereas thyroid-stimulating hormone (TSH) interacts with receptors on the surface of cells of the thyroid gland, it does not bind to cells of the ovaries because they do not have TSH receptors. In addition, different types of cells can possess the same receptors for the same hormone, but the responses of the cells are quite different from each other. For example, the hormone insulin acts on fat cells to stimulate glucose transport and lipid synthesis; it acts on liver cells to stimulate amino acid transport and glycogen synthesis; and it acts on certain pancreatic cells to inhibit their secretion of a hormone called glucagon. An increase in the concentration of one hormone can also produce an increase in the number of receptors for a different hormone. For example, an increase in estrogens can bring about an increase in the number of progesterone receptors. Both hormones act to promote uterine growth for the possible implantation of a fertilized egg. During the menstrual cycle, secretion of estrogens precedes that of progesterone, thus allowing progesterone to have a greater effect. There are generally 2,000 to 100,000 receptors per target cell.

Once a hormone is bound to a specific receptor, the combination activates a chain of biochemical events within the target cell in which the physiological effects of the hormone are expressed. Receptors, like other cell proteins, are constantly synthesized and degraded and change in concentration and affinity in response to changes within the body. For example, when a hormone (or neurotransmitter) is present in excess, usually there is a decrease in the number of receptors, a phenomenon referred to as ***down-regulation.*** This occurs in testicular cells exposed to high concentrations of luteinizing hormone (LH). Such a mechanism decreases the responsiveness of target cells to the hormone over a period of time. Conversely, when a hormone (or neurotransmitter) is deficient, usually there is an increase in the number of receptors, a process known as ***up-regulation.*** When an insulin deficiency occurs, more receptors are produced for the hormone. This mechanism makes a target tissue more sensitive to the stimulating effect of a hormone. Up-regulation is far less common than down-regulation.

The binding of many hormones to specific receptors can occur in two quite different ways. One involves interaction of hormones with plasma membrane receptors; the other involves interaction of hormones with intracellular receptors.

Interaction with Plasma Membrane Receptors

One mechanism of hormone action involves an interaction at the cell surface between the hormone and a receptor site on the plasma membrane. A hormone released from an endocrine gland circulates in the blood, reaches a target cell, and brings a specific message to that cell. The hormone is called the ***first messenger.*** To give the cell its message, the hormone must attach to a specific receptor site (integral protein) on the plasma membrane.

Once a hormone attaches to a specific receptor, there is an increase in the synthesis of ***cyclic AMP*** (see Figure 2-14). Cyclic AMP is synthesized from ATP, the main energy-storing chemical in cells, in a process that requires the presence of an enzyme, ***adenylate cyclase,*** on the inner surface of the plasma membrane. When the first messenger (hormone) attaches to its receptor, there is an activation of adenylate cyclase in the plasma membrane. Then adenylate cyclase converts ATP into cyclic AMP in the cytoplasm of the cell (Figure 18-4a). Cyclic AMP then acts as a ***second messenger,*** altering cell function according to the message indicated by the hormone. Cyclic AMP has also been shown to serve as a second messenger for some neurotransmitters and other substances, as well as for hormones. Cyclic AMP is the best understood of the second messengers.

Cyclic AMP does not directly produce a given physiological response, as indicated by the hormone. Instead, cyclic AMP activates one or more enzymes collectively known as ***protein kinases,*** which may be free in the cytoplasm or membrane-bound. Protein kinases have the ability to add a phosphate group (phosphorylation) from

FIGURE 18-4 **Proposed mechanisms of hormonal action. (a) Interaction with plasma membrane receptors in which there is an increase in the synthesis of cyclic AMP.**

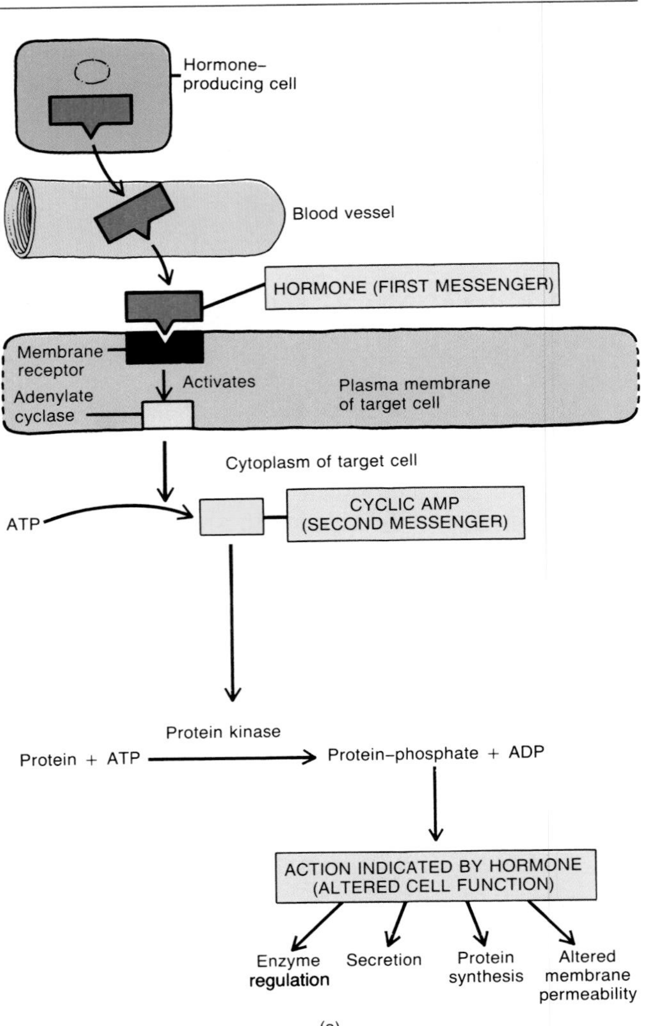

ATP to a protein. As a result, the protein, usually an enzyme itself, is altered, thus catalyzing a physiological response indicated by the hormone. Such responses include regulating enzymes, inducing secretion, activating protein synthesis, and altering plasma membrane permeability. The reason that hormones are so effective in low concentrations is that the original activation of adenylate cyclase initiates a cascade effect in which there is an amplification of products. That is, each hormone molecule activates the formation of several protein kinases. These, in turn, induce the formation of many molecules of another enzyme, and so on. The net effect is that a small amount of hormone produces a response of great magnitude.

A unique property of protein kinases is their ability to catalyze a wide variety of responses. Each of the different protein kinases typically has a different protein as a sub-

FIGURE 18-4 (*Continued*) (b) Activation of genes by a steroid hormone.

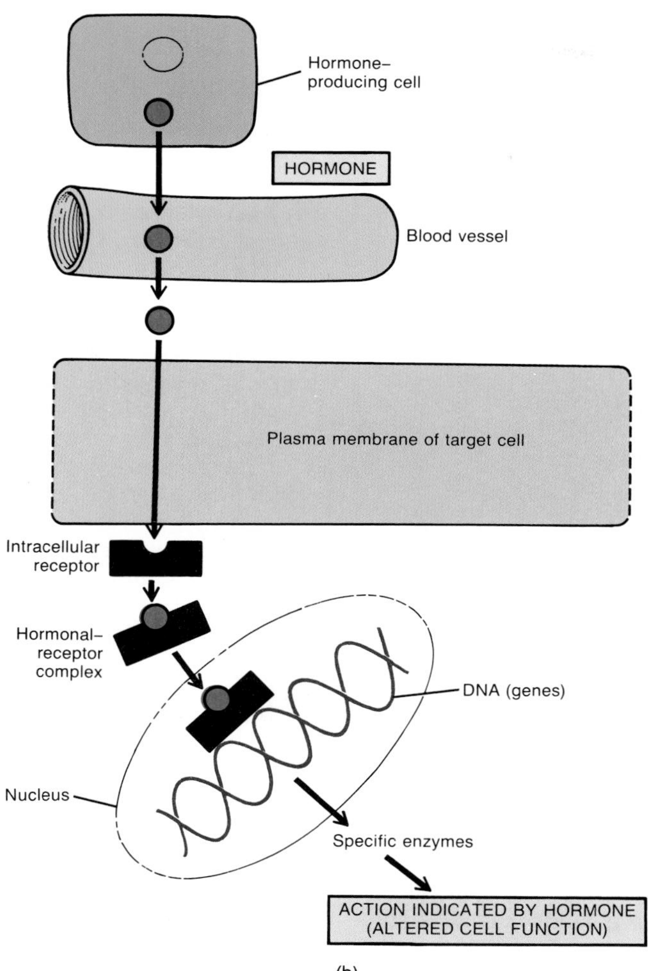

strate, and each exists within different target cells and within different organelles of the same target cell. As a result, one protein kinase might be involved with glycogen synthesis, another with lipid breakdown, another with protein synthesis, and so on. Moreover, protein kinases can inhibit as well as activate enzymes.

High levels of cyclic AMP persist only briefly because they are rapidly degraded by ***phosphodiesterase.*** At least some physiological responses of many hormones exert their effects through the increased synthesis of cyclic AMP. These include antidiuretic hormone (ADH), oxytocin (OT), follicle-stimulating hormone (FSH), luteinizing hormone (LH), thyroid-stimulating hormone (TSH), adrenocorticotropic hormone (ACTH), calcitonin (CT), parathyroid hormone (PTH), glucagon, epinephrine, norepinephrine (NE), and hypothalamic releasing hormones (or factors).

Along with cyclic AMP, calcium ions (Ca^{2+}) may sometimes be involved as second messengers. The binding of some hormones to their receptors causes the entry of calcium ions into cells via open calcium channels in the plasma membrane. Once the level of calcium ions is increased, they bind to an intracellular protein called ***calmodulin,*** a substance closely related to troponin, the calcium-binding protein involved in muscle contraction. Once activated, calmodulin can, in turn, activate or inhibit certain enzymes, many of which are protein kinases. The binding of calcium ions activates calmodulin and troponin; they have no intrinsic activity of their own.

Other substances that serve as second messengers are *cyclic guanosine monophosphate* (*cyclic GMP*), a nucleotide that functions like cyclic AMP but activates different enzymes; *inositol triphosphate,* which activates still other enzymes; and probably *prostaglandins.*

Interaction with Intracellular Receptors

Steroid hormones and thyroid hormones both alter cell function by activation of genes. Since these hormones are lipid-soluble, they easily pass through the plasma membrane of the target cell. Upon entering the cell, the hormones bind to intracellular protein receptor sites, localized primarily within the nucleus. After the binding, the receptor undergoes a transformation and interacts with specific genes of the nuclear DNA and activates them to form the proteins, usually enzymes, necessary to produce the effect that is characteristic of the hormone (Figure 18-4b).

Hormonal Interactions

The degree to which a target cell responds to a hormone is determined by the number of receptors and concentration of a hormone (down-regulation and up-regulation). The manner in which hormones interact with other hormones is also important. One type of interaction is referred to as a ***permissive effect.*** In this interaction, the effect of one hormone on a target cell requires a previous or simultaneous exposure to another hormone(s). Such previous exposure enhances the response of a target cell or increases the activity of another hormone. An example of a permissive effect, as noted earlier, is the exposure of the uterus first to estrogens and then to progesterone in order to prepare the organ for implantation.

Another type of hormonal interaction is known as a ***synergistic effect.*** In this situation, the effects of two or more hormones complement each other in such a way that the target cell responds effectively to the sum of the hormones involved. For example, the production and secretion of milk by the mammary glands requires, among others, the synergistic effects of estrogens, progesterone, prolactin (PRL), and oxytocin (OT).

A final example of a hormonal interaction is an ***antagonistic effect.*** Here, the effect of one hormone on a target cell is opposed by another hormone. An example is calcitonin (CT), which lowers blood calcium level, and parathyroid hormone (PTH), which raises blood calcium level (see Figure 18-14). Another antagonistic situation involves insulin, which lowers blood sugar level, and glucagon, which raises it (see Figure 18-23).

Prostaglandins (PGs) and Hormones

Prostaglandins (pros'-ta-GLAN-dins), or ***PGs,*** are membrane-associated, biologically active lipids that are secreted into blood in minute quantities and are potent in their action. Prostaglandins are also called ***local*** or ***tissue hormones*** because their site of action is the immediate area in which they are produced. This differentiates them from ***circulating hormones,*** which act on distant targets. In addition, prostaglandins are synthesized not by specialized endocrine tissues, as are circulating hormones, but by nearly every mammalian cell and tissue. Chemical and mechanical stimuli, as well as anaphylaxis, lead to prostaglandin release.

Prostaglandins are synthesized from ***arachidonic acid*** which can be released from phospholipids in the plasma membrane. In an alternate pathway, arachidonic acid may be converted to leukotrines, substances related to prostaglandins. As you will see in Chapter 22, both prostaglandins and leukotrines participate in the inflammatory response. Steroidal antiinflammatory drugs, such as cortisone, exert their action by inhibiting the release of arachidonic acid from phospholipids. Nonsteroidal antiinflammatory drugs, such as aspirin, inhibit prostaglandin synthesis but not the formation of leukotrines.

Chemically, prostaglandins are composed of 20-carbon fatty acids containing 5 carbon atoms joined to form a cyclopentane ring. They are classified into several groups designated by letters *A* through *I.* Thus, prostaglandins are designated as *PGA* through *PGI,* respectively. In addition, within each group there are further subdivisions based on the number of double bonds in the fatty acids. Thus, PGE_2 has two double bonds, and PGE_1 has one. Prostaglandins are believed to be regulators or modulators of cell metabolism. Varying with the tissue and species, prostaglandins either increase or decrease cyclic AMP formation. In this way, prostaglandins can alter the responses of cells to a hormone whose action involves cyclic AMP. In this respect, prostaglandins might be modulators of cyclic AMP–induced responses. Prostaglandins are rapidly inactivated, especially in the lungs, liver, and kidneys. The broad range of prostaglandin biological activity in relation to smooth muscle, secretion, blood flow, reproduction, platelet function, respiration, nerve impulse transmission, fat metabolism, immune responses, and other life processes, as well as their involvement in inflammation, neoplasia, promoting fever, intensifying pain, and other conditions, indicates their importance in both normal physiology and pathology. Certain drugs such as aspirin and acetaminophen (Tylenol) inhibit prostaglandin synthesis and thus reduce fever and decrease pain.

What has intrigued investigators even more than the physiological role of prostaglandins has been their pharmacological effects and their implications for potential therapy. These effects include lowering or raising blood pressure; reducing gastric secretion; bronchodilation or bronchoconstriction; stimulating or inhibiting platelet aggregation; contracting or relaxing intestinal and uterine smooth muscle; shrinking nasal passages; blocking or augmenting norepinephrine release; mediating inflammation; increasing intraocular pressure; causing sedation, stupor, and fever; inducing labor in pregnancy; stimulating steroid production; promoting natriuresis (excretion of sodium in the urine); promoting diuresis; potentiating the pain-producing effect of kinins; and many others.

CONTROL OF HORMONAL SECRETIONS: FEEDBACK CONTROL

As indicated earlier, the amount of hormone released by an endocrine gland or tissue is determined by the body's need for the hormone at any given time. Most hormones are released in short bursts, with little or no release between bursts. When properly stimulated, an endocrine gland will release hormone in more frequent bursts, and thus blood levels of the hormone increase. Conversely, in the absence of stimulation, bursts are minimal or inhibited, and thus blood levels of the hormone decrease. Secretion is normally regulated so that there is no overproduction or underproduction of a particular hormone. This regulation is one of the very important ways that the body attempts to maintain homeostasis. Unfortunately, there are times when the regulating mechanism does not operate properly and hormonal levels are excessive or deficient. When this happens, disorders result, several of which are discussed later.

The typical way in which hormonal secretions are regulated is by ***negative feedback control*** (Chapter 1). As applied to hormones, information regarding the hormone level or its effect is fed back to the gland, which then responds accordingly. Here we will describe three negative feedback systems in which the input that stimulates or inhibits hormonal secretion varies in each situation.

In one type of negative feedback system, the regulation of hormonal secretion does not involve direct participation by the nervous system. For example, blood calcium level is controlled in part by parathyroid hormone (PTH), produced by the parathyroid glands. If, for some reason, blood calcium level is low, this serves as a stimulus for the parathyroids to release more PTH (see Figure 18-14). PTH then exerts its effects in various parts of the body until the blood calcium level is raised to normal. A high blood calcium level serves as a stimulus for the parathyroids to cease their production of PTH. In the absence of the hormone, other mechanisms take over until blood calcium level is lowered to normal. Note that in negative feedback control the body's response (increased or decreased calcium level) is opposite (negative) to the stimulus (low or high calcium level). Other hormones that are regulated without direct involvement of the nervous system include calcitonin (CT), produced by the thyroid gland, which is controlled by blood levels of calcium; insulin, produced by the pancreas, which is controlled principally by blood levels of glucose; and aldosterone, produced by the adrenal cortex, which is controlled by blood volume and blood levels of potassium.

In other negative feedback systems, the hormone is released as a direct result of nerve impulses that stimulate the endocrine gland. Epinephrine and norepinephrine (NE) are released from the adrenal medulla in response to sympathetic nerve impulses. Antidiuretic hormone (ADH) is released from the posterior pituitary in response to nerve impulses from the hypothalamus (see Figure 26-10).

There are also negative feedback systems that involve the nervous system through chemical secretions from the hypothalamus, called ***regulating hormones*** (or ***factors***). If the structure of the secretion is known, it is called a ***regulating hormone;*** if its structure is unknown, it is referred to as a ***regulating factor*** (see Figure 18-7). Some hypothalamic secretions, called ***releasing hormones*** (or ***factors***), stimulate the release of the hormone into the blood, so that it can exert its influence. Others, called ***inhibiting hormones*** (or ***factors***), prevent the release of the hormone.

One of the few exceptions to the rule of negative feedback control is oxytocin (OT). The regulating system for the release of OT from the pituitary gland in response to nerve impulses from the hypothalamus is a positive feedback cycle; that is, the output intensifies the input (see Figure 18-9). Another exception is the luteinizing hormone (LH) surge that results in ovulation (Chapter 28).

As we discuss the effects of various hormones in this chapter, we will also describe how the secretions of the hormones are controlled. At that time, you will be able to see which type of negative feedback system is operating.

PITUITARY (HYPOPHYSIS)

The hormones of the ***pituitary gland,*** also called the ***hypophysis*** (hī-POF-i-sis), regulate so many body activities that the pituitary gland has been nicknamed the "master gland." It is a round structure and surprisingly small, measuring about 1.3 cm (0.5 inch) in diameter. The pituitary gland lies in the sella turcica of the sphenoid bone and is attached to the hypothalamus of the brain via a stalklike structure, the ***infundibulum*** (see Figure 18-5).

The pituitary gland is divided structurally and functionally into an anterior lobe and a posterior lobe. Both are closely associated with the hypothalamus, but only the posterior lobe is neurally connected to the hypothalamus. The ***anterior lobe*** constitutes about 75 percent of the total weight of the gland. It is derived from an outgrowth of ectoderm in the embryonic pharyngeal region called the hypophyseal (Rathke's) pouch (see Figure 18-24b). Accordingly, the anterior lobe contains many glandular epithelial cells and forms the glandular part of the pituitary. A system of blood vessels connects the anterior lobe with the hypothalamus.

The ***posterior lobe*** is also derived from the ectoderm, but from an outgrowth called the neurohypophyseal bud (see Figure 18-24b). Accordingly, the posterior lobe contains axon terminations of neurons whose cell bodies are located in the hypothalamus. The nerve fibers that terminate in the posterior lobe are supported by cells called pituicytes. Other nerve fibers connect the posterior lobe directly with the hypothalamus.

Between the lobes is a small, relatively avascular zone, the ***pars intermedia.*** Although much larger and more clearly defined in structure and function in some lower animals, its role in humans is obscure.

Adenohypophysis

The anterior lobe of the pituitary is also called the ***adenohypophysis*** (ad'-e-nō-hī-POF-i-sis; *adeno* = glandular). It releases hormones that regulate a whole range of bodily activities from growth to reproduction. The release of these hormones is either stimulated or inhibited by chemical secretions from the hypothalamus called ***regulating hormones*** (or ***factors***). Regulating hormones (or factors), which will be considered along with each of the anterior pituitary hormones, constitute an important link between the nervous and endocrine systems.

The hypothalamic regulating hormones (or factors) are delivered to the adenohypophysis through a series of blood vessels. The blood supply to the adenohypophysis and infundibulum is derived principally from several ***superior hypophyseal*** (hī'-po-FIZ-ē-al) ***arteries.*** These arteries are branches of the internal carotid and posterior communicating arteries (Figure 18-5). The superior hypophyseal arteries form a network or plexus of capillaries, the ***primary plexus,*** in the infundibulum near the inferior portion of the hypothalamus. Regulating hormones (or factors) from the hypothalamus diffuse into this plexus. This plexus drains into the ***hypophyseal portal veins*** that pass down the infundibulum. At the inferior portion of the infundibulum, the veins form a ***secondary plexus*** of capillaries in the adenohypophysis. From this plexus, hormones of the adenohypophysis pass into the anterior hypophyseal veins for distribution to tissue cells. Such a delivery system permits regulating hormones (or factors) to act quickly on the adenohypophysis without first circulating through the heart. The short route prevents dilution or destruction of the regulating hormones (or factors).

When the adenohypophysis receives proper stimulation from the hypothalamus via regulating hormones (or factors), its glandular cells secrete any one of seven hormones.* Special staining techniques have established the division of glandular cells into five principal types (Figure 18-6):

1. ***Somatotroph cells*** produce ***human growth hormone (hGH),*** which controls general body growth.
2. ***Lactotroph cells*** synthesize ***prolactin (PRL),*** which initiates milk production by the mammary glands.
3. ***Corticolipotroph cells*** synthesize ***adrenocorticotropic hormone (ACTH),*** which stimulates the adrenal cortex to secrete its hormones, and ***melanocyte-stimulating hormone (MSH),*** which is believed to be related to skin pigmentation.
4. ***Thyrotroph cells*** manufacture ***thyroid-stimulating hormone (TSH),*** which controls the thyroid gland.
5. ***Gonadotroph cells*** produce ***follicle-stimulating hormone (FSH),*** which stimulates the production of eggs and sperm in the ovaries and testes, respectively, and ***luteinizing hormone (LH),*** which stimulates other sexual and reproductive activities.

* Human chorionic gonadotropin (hCG) is a hormone that until recently was believed to be produced only by the placenta to help maintain pregnancy. Recent evidence suggests that it is also produced by the pituitary gland and that its production is stimulated by a regulating hormone produced by the hypothalamus called gonadotropin releasing hormone (GnRH).

FIGURE 18-5 **Blood supply of the pituitary gland (hypophysis).**

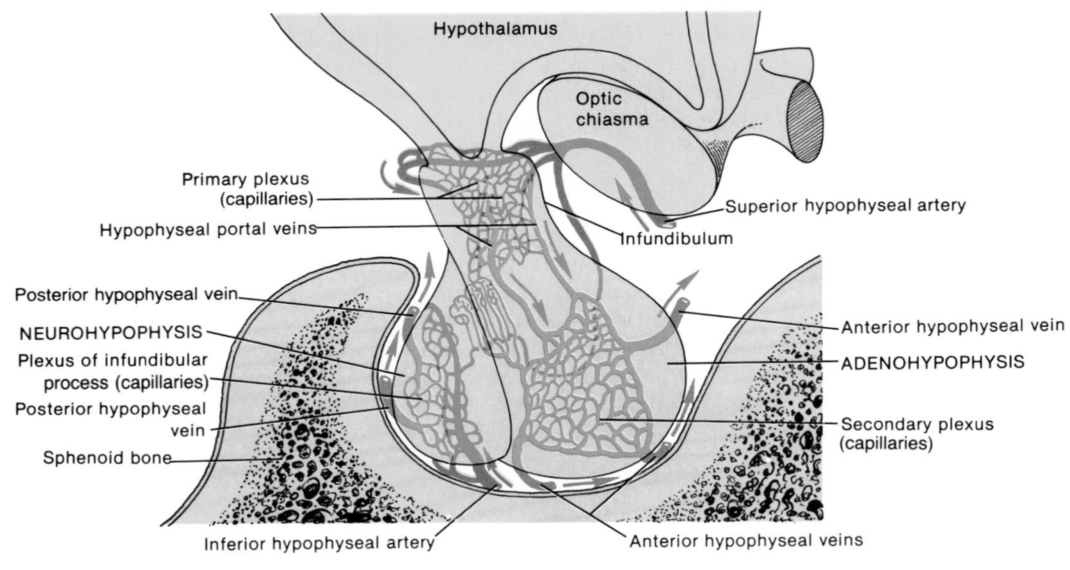

FIGURE 18-6 Cells of the adenohypophysis as revealed by special stains. Most cells that produce human growth hormone (somatotrophs) and prolactin (lactotrophs) are separate cells. However, some normal and tumor cells are single cells that produce both human growth hormone and prolactin. The corticolipotroph cell produces adrenocorticotropic hormone and melanocyte-stimulating hormone. Thyrotroph cells synthesize thyroid-stimulating hormone. Most gonadotroph cells produce both follicle-stimulating hormone and luteinizing hormone. However, a few separate cells may exist, some producing follicle-stimulating hormone and some producing luteinizing hormone. (Adapted from a slide provided by Calvin Ezrin, M.D., Clinical Professor of Medicine, U.C.L.A., and Adjunct Professor of Pathology, University of Toronto.)

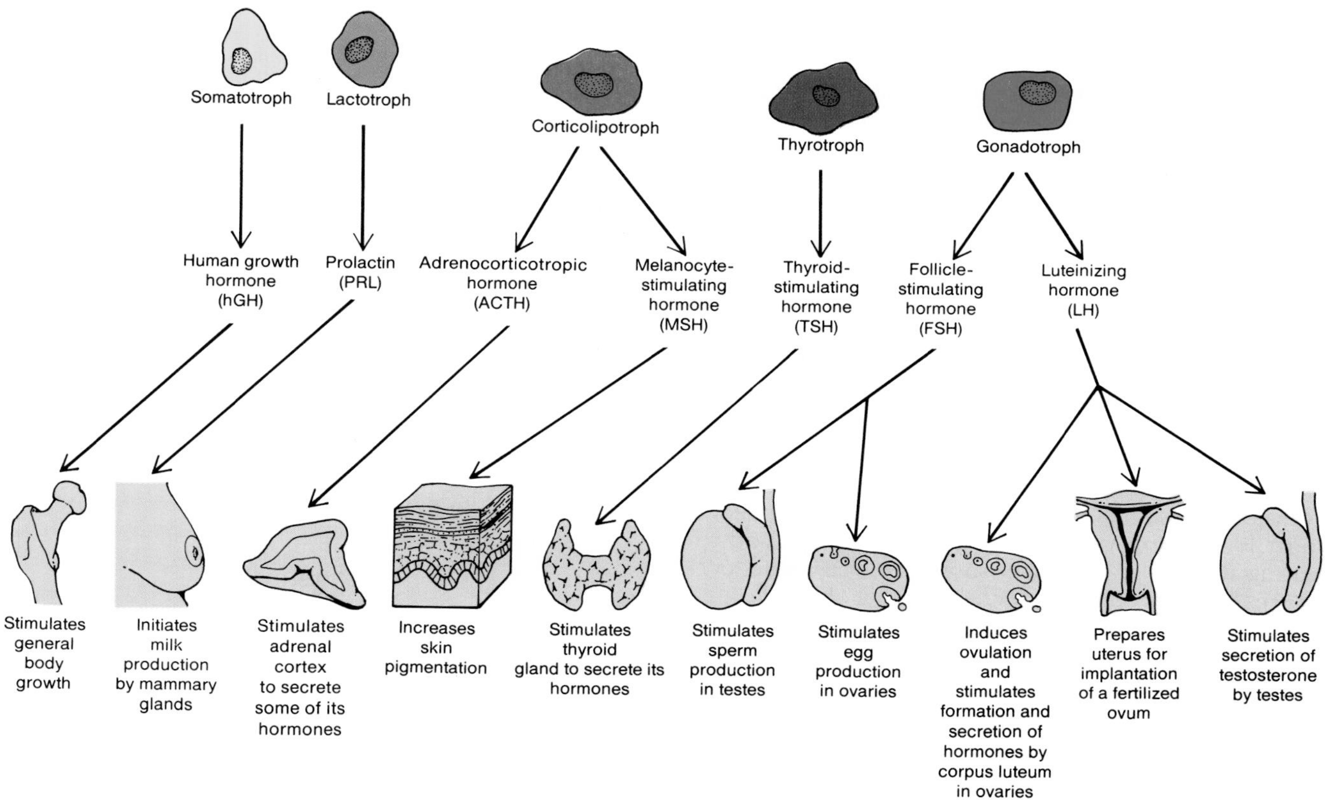

Except for human growth hormone (hGH), melanocyte-stimulating hormone (MSH), and prolactin (PRL), all the secretions are referred to as ***tropic hormones*** (*trop* = turn on), which means that they stimulate other endocrine glands. Follicle-stimulating hormone (FSH) and luteinizing hormone (LH) are also called ***gonadotropic*** (gō-nad-ō-TRŌ-pik) ***hormones*** because they regulate the functions of the gonads (ovaries and testes). The gonads are the endocrine glands that produce sex hormones that are steroids.

Human Growth Hormone (hGH)

Human growth hormone (hGH or ***GH),*** also known as ***somatotropin*** (sō'-ma-tō-TRŌ-pin) and ***somatotropic hormone (STH),*** causes body cells to grow. Its principal function is to act on the skeleton and skeletal muscles, in particular, to increase their rate of growth and maintain their size once growth is attained. hGH causes cells to grow and multiply by directly increasing the rate at which amino acids enter cells and are built up into proteins. hGH is considered to be a hormone of protein anabolism, since it increases the rate of protein synthesis. At the same time, hGH decreases the breakdown of proteins and the utilization of proteins and amino acids for energy. hGH also promotes fat catabolism; that is, it causes cells to switch from burning carbohydrates and proteins to burning fats for energy. For example, it stimulates adipose tissue to release fat, and it stimulates other cells to break down the released fat molecules. hGH also decreases glucose utilization by cells, decreases glucose uptake by cells, accelerates the rate at which glycogen stored in the liver is converted into glucose, and therefore increases blood glucose concentration, a condition called ***hyperglycemia*** (hī'-per-glī-SĒ-mē-a). The high blood sugar level results because, in the presence of hGH, cells use fats for energy and not glucose. This phenomenon is called the ***diabetogenic*** (dī'-a-bet'-ō-JEN-ik) ***effect*** because it mimics the elevated blood glucose level of diabetes mellitus.

The effects of hGH are not due to the direct action of the hormone itself. Instead, hGH stimulates the liver to synthesize and secrete small proteins called ***somatomedins*** (sō'-ma-tō-MĒ-dins). These substances mediate most of the effects of hGH and are structurally and functionally similar to insulin. However, their growth-promoting effects are much more potent than those of insulin.

The control of hGH secretion is believed to occur as follows. Its release from the anterior pituitary is apparently controlled by at least two regulating hormones from the hypothalamus: ***growth hormone releasing hormone (GHRH)***, or ***somatocrinin;*** and ***growth hormone inhibiting hormone (GHIH)***, or ***somatostatin.*** When GHRH is released by the hypothalamus into the bloodstream, it circulates to the anterior pituitary and stimulates the release of hGH. On the other hand, GHIH inhibits the release of hGH.

Among the stimuli that promote hGH secretion is ***hypoglycemia,*** that is, low blood sugar level. Other promoting stimuli are listed in Exhibit 18-1. When blood sugar level is low, the hypothalamus is stimulated to secrete GHRH (Figure 18-7a). Upon reaching the anterior pituitary, GHRH causes the anterior pituitary to release hGH. Somatomedins, under the influence of hGH, raise blood sugar level by converting glycogen into glucose and releasing it into the blood. As soon as blood sugar level returns to normal, GHRH secretion shuts off.

One stimulus that inhibits hGH secretion is ***hyperglycemia,*** high blood sugar level (Figure 18-7a). Other inhibiting stimuli are listed in Exhibit 18-1. An abnormally high blood sugar level stimulates the hypothalamus to secrete the regulating hormone GHIH. GHIH inhibits the release of hGH. As a result, blood sugar level decreases. You will see later that certain endocrine cells of the pancreas, called delta cells, also secrete GHIH, and it can inhibit the secretion of pancreatic hormones (insulin and glucagon) just as it inhibits secretion of hGH.

EXHIBIT 18-1 REPRESENTATIVE STIMULI THAT INFLUENCE HUMAN GROWTH HORMONE (hGH) SECRETION

Promotes	Inhibits
Hypoglycemia	Hyperglycemia
Decreased fatty acids	Increased fatty acids
Increased amino acids	Decreased amino acids
Low levels of hGH	High levels of hGH
Somatomedins	Growth hormone inhibiting hormone (GHIH)
Growth hormone releasing hormone (GHRH)	REM sleep
States 3 and 4 of NREM sleep	Emotional deprivation
Stress	Obesity
Vigorous physical exercise	Hypothyroidism
Estrogens	
Glucagon	
Insulin	
Glucocorticoids	
Dopamine (DA)	
Acetylcholine (ACh)	

The regulation of secretion of hGH illustrates two phenomena that are typical of secretions of the adenohypophysis. First, each hormone is believed to be controlled by its own regulating hormone (or factor) from the hypothalamus. Some regulating hormones (or factors) stimulate hormone secretion; other hormones (or factors) inhibit secretion. Second, secretion is generally regulated through negative feedback systems. Since hormones are chemical regulators of homeostasis, these feedback systems are hardly surprising. Continued heavy secretion of a hormone would overshoot the goal and send the body out of balance in the opposite direction.

CLINICAL APPLICATION: PITUITARY DWARFISM, GIANTISM, AND ACROMEGALY

Disorders of the endocrine system, in general, involve ***hyposecretion*** (underproduction) of hormones or ***hypersecretion*** (overproduction).

Among the clinically interesting disorders related to the adenohypophysis are those involving hGH. If hGH is hyposecreted during the growth years, bone growth is slow, and the epiphyseal plates close before normal height is reached. This condition is called ***pituitary dwarfism.*** Other organs of the body also fail to grow, and the pituitary dwarf is childlike in many physical respects. Treatment requires administration of hGH during childhood before the epiphyseal plates close. Dwarfism is also caused by other conditions, in which administration of hGH is not corrective.

Hypersecretion of hGH during childhood results in ***giantism (gigantism)***, an abnormal increase in the length of long bones. As a result, the person grows to be very large, but body proportions are about normal. Hypersecretion during adulthood is called ***acromegaly*** (ak'-rō-MEG-a-lē), which is shown in Figure 18-7b. Acromegaly cannot produce further lengthening of the long bones because the epiphyseal plates are already closed. Instead, the bones of the hands, feet, cheeks, and jaws thicken.

FIGURE 18-7 Human growth hormone (hGH). (a) Regulation of the secretion of human growth hormone. Like other hormones of the adenohypophysis, the secretion of hGH is controlled by regulating hormones (or factors). Like most hormones of the body, hGH secretion and inhibition involve negative feedback systems. (b) Photograph of an individual with acromegaly. (Courtesy of Lester V. Bergman & Associates, Inc.)

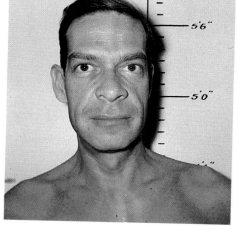

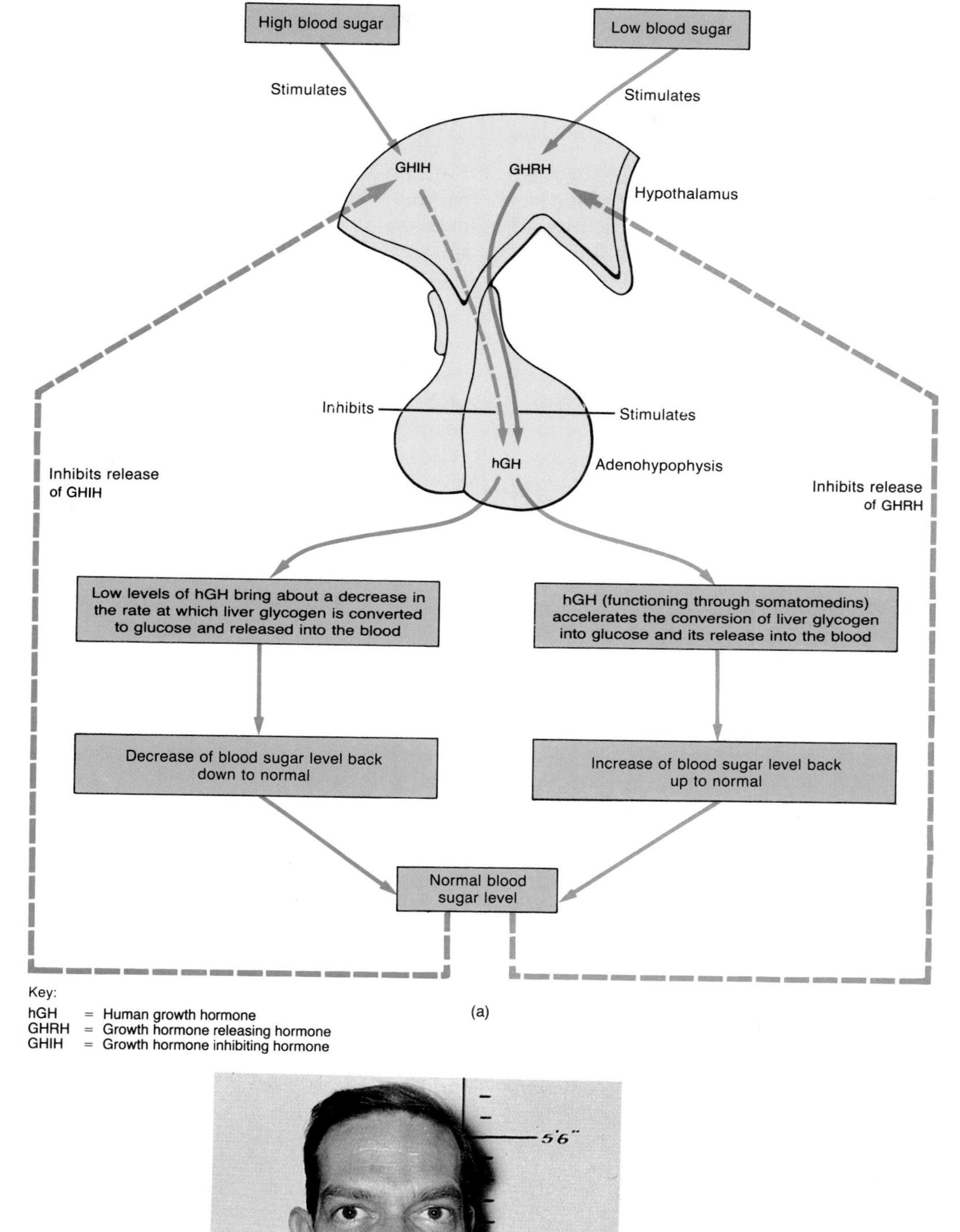

(b)

Other tissues also grow. The eyelids, lips, tongue, and nose enlarge, and the skin thickens and furrows, especially on the forehead and soles of the feet. Diabetes mellitus is also a complication.

Thyroid-Stimulating Hormone (TSH)

Thyroid-stimulating hormone (TSH), also called ***thyrotropin*** (thī-rō-TRŌ-pin), stimulates the synthesis and secretion of the hormones produced by the thyroid gland. Secretion is controlled by a regulating hormone produced by the hypothalamus called ***thyrotropin releasing hormone (TRH)***. Release of TRH depends on blood levels of thyroid hormones, blood glucose level, and the body's metabolic rate, among other factors, and operates according to a negative feedback system.

Adrenocorticotropic Hormone (ACTH)

Adrenocorticotropic hormone (ACTH) is also called ***adrenocorticotropin*** (ad-rē'-nō-kor'-ti-kō-TRŌ-pin). Its tropic function is to control the production and secretion of certain adrenal cortex hormones. Secretion of ACTH is governed by a regulating hormone produced by the hypothalamus called ***corticotropin releasing hormone (CRH)***. Release of CRH depends on a number of stimuli that cause stress, such as low blood glucose or physical trauma, and operates as a negative feedback system. As will be explained later, ACTH secretions can also be initiated by a substance produced by macrophages called interleukin-1 (IL-1).

Follicle-Stimulating Hormone (FSH)

In the female, ***follicle-stimulating hormone (FSH)*** is transported from the adenohypophysis by the blood to the ovaries, where it initiates the development of ova each month. FSH also stimulates cells in the ovaries to secrete estrogens, or female sex hormones. In the male, FSH stimulates the testes to initiate sperm production. Secretion of FSH is subject to a regulating hormone produced by the hypothalamus called ***gonadotropin releasing hormone (GnRH)***. GnRH is released in response to estrogens and possibly progesterone in the female, and to testosterone in the male, and involves a negative feedback system.

Luteinizing Hormone (LH)

In the female, ***luteinizing*** (LOO-tē-in'-īz-ing) ***hormone (LH)***, together with FSH, stimulates the ovary to release a secondary oocyte (ovulation). LH also stimulates formation of the corpus luteum in the ovary, which secretes progesterone (another female sex hormone). Estrogens and progesterone prepare the uterus for implantation of a fertilized ovum and prepare the mammary glands for milk secretion. In the male, LH stimulates the interstitial endocrinocytes in the testes to develop and secrete large amounts of testosterone. Because of this function, LH is also referred to as ***interstitial cell-stimulating hormone (ICSH)*** in males. Secretion of LH, like that of FSH, is controlled by GnRH. Release of GnRH is governed by a negative feedback system involving estrogens, progesterone, and testosterone.

GnRH analogues (modified compounds with similar structures) are used therapeutically in a number of clinical situations. Some analogues have been used with varying degrees of success to treat precocious puberty, endometriosis, polycystic ovaries, endometrial hyperplasia, breast and prostate cancer, and premenstrual syndrome (PMS). One GnRH analogue also is used in contraception.

Prolactin (PRL)

Prolactin (PRL), or the ***lactogenic hormone***, together with other hormones, initiates and maintains milk secretion by the mammary glands. The actual ejection of milk by the mammary glands is controlled by a hormone stored in the posterior lobe called oxytocin. Together, milk secretion and ejection are referred to as lactation. PRL acts directly on tissues. By itself, it has little effect; it requires preparation by estrogens, progesterone, corticosteroids, human growth hormone (hGH), thyroxine, and insulin. When the mammary glands have been primed by these hormones, PRL brings about milk secretion.

PRL has both an inhibitory and an excitatory negative control system. During menstrual cycles, ***prolactin inhibiting factor (PIF)***, a regulating factor from the hypothalamus, inhibits the release of PRL from the anterior pituitary. As the levels of estrogens and progesterone fall during the late secretory phase of the menstrual cycle, the secretion of PIF diminishes and the blood level of PRL rises. However, its rising level does not last long enough to have much effect on the breasts, which may be tender because of the presence of PRL just before menstruation. As the menstrual cycle starts up again and the level of estrogens rises again, PIF is again secreted and the PRL level drops.

PRL levels rise during pregnancy. Apparently, a regulating factor from the hypothalamus, called ***prolactin releasing factor (PRF)***, stimulates PRL secretion after long periods of inhibition. PRL levels fall after delivery and rise again during breast feeding. A nursing infant causes a reduction in hypothalamic secretion of PIF. Mechanical stimulation of the nonlactating female breast brings about increased secretion of PRL; stimulation of the male breast does not.

In males, PRL enhances the effect of luteinizing hormone in promoting the production of testosterone.

Melanocyte-Stimulating Hormone (MSH)

Melanocyte-stimulating hormone (MSH) increases skin pigmentation by stimulating the dispersion of melanin granules in melanocytes in amphibians. Its exact role in humans is unknown, but continued administration of

MSH for several days does produce a darkening of the skin. In the absence of the hormone, the skin may be pallid. Secretion of MSH is stimulated by a hypothalamic regulating factor called ***melanocyte-stimulating hormone releasing factor (MRF).*** It is inhibited by a ***melanocyte-stimulating hormone inhibiting factor (MIF).***

Before leaving our discussion of the adenohypophysis, it should be noted that several hormones, as well as neuropeptides, are produced by cells of the adenohypophysis from a large protein precursor molecule called ***pro-opiomelanocortin.*** When this molecule is split, it can give rise to adrenocorticotropic hormone (ACTH), melanocyte-stimulating hormone (MSH), enkephalin, endorphin, and beta-lipotropin (β-LPH). This last substance is believed to have a role in lipid breakdown in cells.

A summary of anterior pituitary hormones, their principal actions, associated hypothalamic regulating hormones (or factors), and selected related disorders is presented in Exhibit 18-2.

EXHIBIT 18-2 ANTERIOR PITUITARY HORMONES, PRINCIPAL ACTIONS, ASSOCIATED HYPOTHALAMIC REGULATING HORMONES (OR FACTORS), AND SELECTED DISORDERS

Hormone	Principal Actions	Associated Hypothalamic Regulating Hormones (or Factors)	Selected Disorders
Human growth hormone (hGH)	Growth of body cells; protein anabolism; elevation of blood glucose concentration.	Growth hormone releasing hormone (GHRH); growth hormone inhibiting hormone (GHIH).	Hyposecretion of hGH during the growth years results in pituitary dwarfism; hypersecretion of hGH during the growth years results in giantism; hypersecretion of hGH during adulthood results in acromegaly.
Thyroid-stimulating hormone (TSH)	Controls secretion of thyroid hormones by thyroid gland.	Thyrotropin releasing hormone (TRH).	Hypersecretion of thyroid hormones through the action of TSH causes exophthalmic goiter (discussed later).
Adrenocorticotropic hormone (ACTH)	Controls secretion of some hormones by adrenal cortex (mainly cortisol).	Corticotropin releasing hormone (CRH).	Hyposecretion of glucocorticoids through the action of ACTH results in Addison's disease (discussed later).
Follicle-stimulating hormone (FSH)	In female, initiates development of ova and induces ovarian secretion of estrogens. In male, stimulates testes to produce sperm.	Gonadotropin releasing hormone (GnRH).	
Luteinizing hormone (LH) [Also called interstitial cell-stimulating hormone (ICSH) in male.]	In female, together with FSH, stimulates ovulation and formation of progesterone-producing corpus luteum. In male, stimulates interstitial cells in testes to develop and produce testosterone.	Gonadotropin releasing hormone (GnRH).	
Prolactin (PRL)	In females, together with other hormones, initiates and maintains effect of luteinizing hormone in promoting milk secretion by the mammary glands; in males, it enhances the production of testosterone.	Prolactin inhibiting factor (PIF); prolactin releasing factor (PRF).	
Melanocyte-stimulating hormone (MSH)	Stimulates dispersion of melanin granules in melanocytes.	Melanocyte-stimulating hormone releasing factor (MRF); melanocyte-stimulating hormone inhibiting factor (MIF).	

Neurohypophysis

In a strict sense, the posterior lobe, or ***neurohypophysis,*** is not an endocrine gland, since it does not synthesize hormones. Instead, it stores and releases two hormones. The posterior lobe consists of cells called ***pituicytes*** (pi-TOO-i-sītz), which are similar in appearance to the neuroglia of the nervous system. It also contains axon terminals of secretory neurons of the hypothalamus (Figure 18-8). Such neurons are called ***neurosecretory cells.*** The cell bodies of the neurons originate in nuclei (paraventricular and supraoptic) in the hypothalamus. The fibers project from the hypothalamus, form the ***hypothalamic-hypophyseal tract,*** and terminate on blood capillaries in the neurohypophysis. The cell bodies of the neurosecretory cells produce two hormones: ***oxytocin (OT)*** and ***antidiuretic hormone (ADH).***

Following their production in the cell bodies of neurosecretory cells, the hormones are transported by small proteins called ***neurophysins*** (noo-rō-FĪ-sins) through the neuron fibers to the neurohypophysis and then stored in the axon terminals. The transport from the cell bodies to the axon terminations takes about 10 hours. Neurophysins aid in storing the hormones and are important in the release mechanism. The bound hormone is released from the nerve endings in response to nerve impulses reaching the axon terminals. Thus, these fibers perform two tasks. First, they act as conduits for the transport of the hormone molecules from their site of production in the hypothalamic nucleus to their site of secretion in the neurohypophysis. Second, the fibers carry the releasing nerve impulses down to the axon terminals where the hormone is stored. The mechanism of hormone release is a familiar one. The nerve impulses depolarize the axon terminal membrane, and calcium ions (Ca^{2+}) enter the neuronal endings. When released into the blood, the hormone–neurophysin bonds are broken, probably owing to a pH change.

FIGURE 18-8 Hypothalamic–hypophyseal tract.

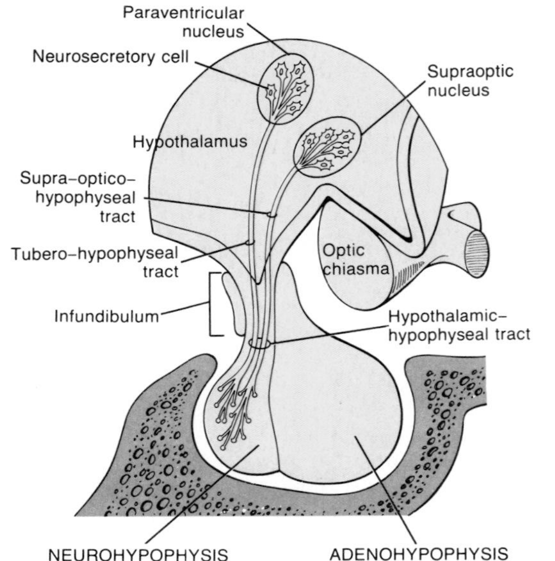

The blood supply to the neurohypophysis is from the ***inferior hypophyseal arteries,*** derived from the internal carotid arteries (see Figure 18-5). In the neurohypophysis, the inferior hypophyseal arteries form a plexus of capillaries called the ***plexus of the infundibular process.*** From this plexus, hormones stored in the neurohypophysis pass into the ***posterior hypophyseal veins*** for distribution to tissue cells.

Oxytocin (OT)

Oxytocin (ok'-sē-TŌ-sin), or ***OT,*** stimulates the contraction of the smooth muscle cells in the pregnant uterus and the contractile cells around the ducts and glandular cells (alveoli) of the mammary glands. It is released in large quantities just prior to giving birth (Figure 18-9). When labor begins, the cervix of the uterus is distended. This distension initiates afferent impulses to the neurosecretory cells of the hypothalamic nuclei (paraventricular and supraoptic) that stimulate the synthesis of OT. The OT is transported by neurophysins to the neurohypophysis. The nerve impulses also cause the neurohypophysis to release OT into the blood. It is then carried by the blood to the uterus to reinforce uterine contractions. As the contractions become more forceful, the resulting afferent impulses stimulate the synthesis of more OT. In addition, as the baby's head passes through the uterine cervix, further distension of the cervix triggers the release of more OT. Thus, a positive feedback cycle is established. The cycle is broken by the birth of the infant. Note also that the afferent part of the cycle is neural, whereas the efferent part is hormonal.

OT is used clinically to induce labor (the trade name for such preparations is Pitocin). It is also used in the immediate postpartum period to increase uterine tone and control hemorrhage.

OT affects milk ejection. Milk formed by the glandular cells of the breasts is stored until the baby begins active sucking. This initiates afferent impulses from the nipple to the hypothalamus. According to a mechanism similar to that involved in forming and releasing OT for uterine muscle contractions, OT is transported from the neurohypophysis via the blood to the mammary glands, where it stimulates smooth muscle cells around the glandular cells and ducts to contract and eject milk. This response is called the ***milk let-down reflex.*** Ejection of milk by the reflex does not occur for about 30 seconds to 1 minute (latent period) after nursing begins. The time is required for the responses that occur between initiation of the afferent impulses and actual milk ejection. However, since some milk is stored in lactiferous sinuses near the nipple, milk is available to the infant during the latent period. Even in cases of let-down failure, infants can still obtain

FIGURE 18-9 Regulation of the secretion of oxytocin (OT) during labor. This is a positive feedback cycle.

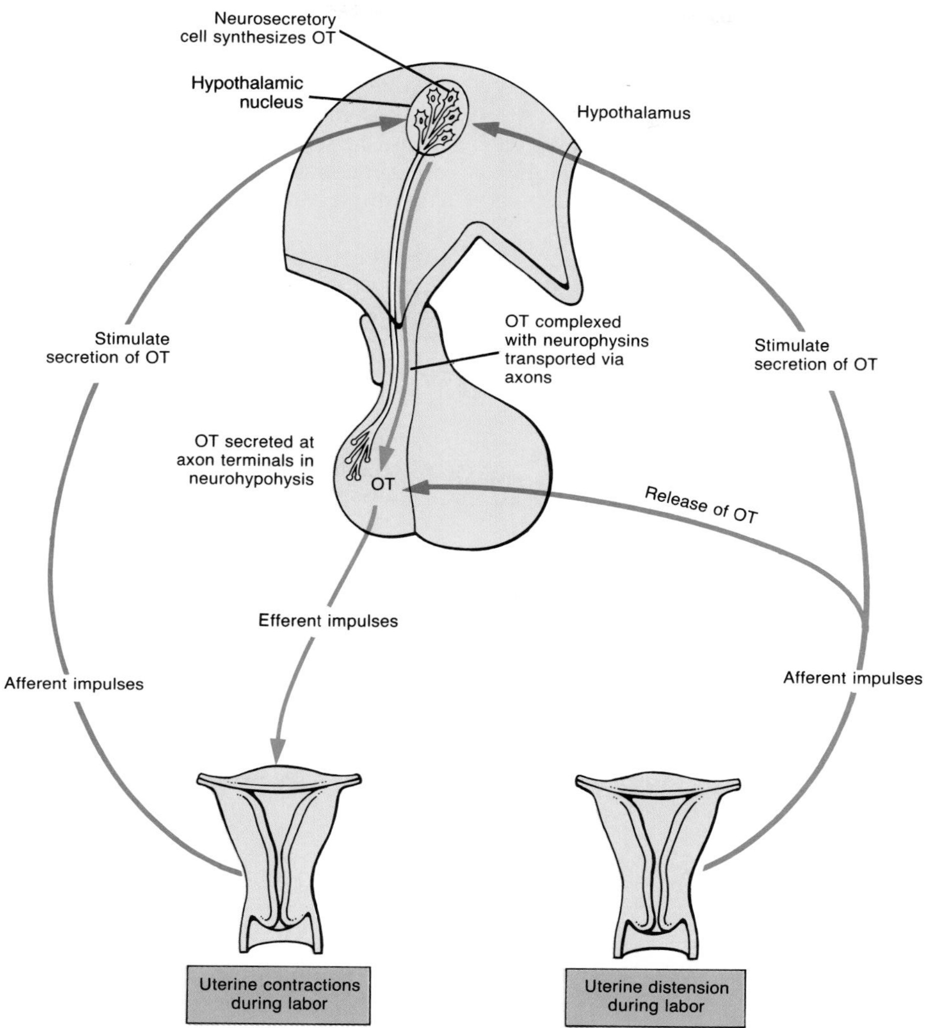

one-third of the breast's milk. But pre–let-down milk has a different protein-to-fat ratio.

OT is inhibited by progesterone but works together with estrogens. Estrogens and progesterone inhibit, via PIF, the release of PRL. The sucking stimulation that produces the release of OT also inhibits the release of PIF. This results in an increased secretion of PRL, which maintains lactation.

Antidiuretic Hormone (ADH)

An ***antidiuretic*** is any chemical substance that prevents excessive urine production. The principal physiological activity of ***antidiuretic hormone (ADH)*** is its effect on urine volume. ADH causes the kidneys to remove water from fluid that will become urine and return it to the bloodstream, thus decreasing urine volume ***(antidiuresis)*** and conserving body water. This involves an increase in the permeability of plasma membranes of kidney water-reabsorbing cells so that more water passes from newly forming urine and is returned into kidney cells. In the absence of ADH, urine output may be increased more than 10-fold from normal to 25 liters a day.

ADH can also raise blood pressure by bringing about constriction of arterioles. For this reason, ADH is also referred to as ***vasopressin.*** This effect is noted if there is a severe loss of blood volume due to hemorrhage.

The amount of ADH normally secreted varies with the body's needs (Figure 18-10). When the body is dehydrated, the concentration of water in the blood is below normal limits and the change in the salt-to-water ratio results in an increase in osmotic pressure. Receptors in the hypothalamus called ***osmoreceptors*** detect the low water

FIGURE 18-10 Regulation of the secretion of antidiuretic hormone (ADH).

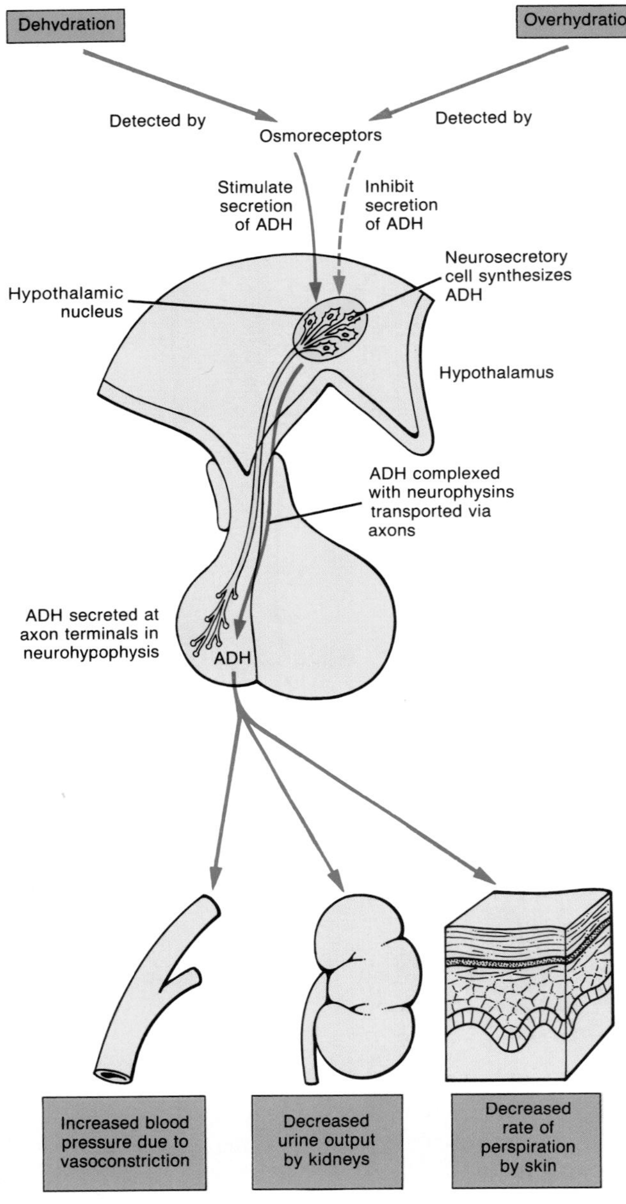

concentration (high osmotic pressure) in the blood and stimulate the neurosecretory cells of the hypothalamic paraventricular and supraoptic nuclei to synthesize ADH. The hormone is transported by neurophysins to the neurohypophysis. It is then released into the bloodstream and transported to the kidneys. The kidneys respond by decreasing urine output, and more water is retained by the body. ADH also decreases the rate at which perspiration is produced during dehydration. By contrast, if the blood contains a higher-than-normal water concentration, the receptors detect the increase and hormone secretion is reduced or stopped. The kidneys can then release large quantities of urine, and the volume of body fluid is brought down to normal.

Secretion of ADH can also be altered by a number of other conditions. Changes in blood volume, pain, stress, trauma, anxiety, acetylcholine (ACh), nicotine, and drugs, such as morphine, tranquilizers, and some anesthetics, stimulate secretion of the hormone. Alcohol inhibits secretion and thereby increases urine output. This may be why thirst is one symptom of a hangover.

CLINICAL APPLICATION: DIABETES INSIPIDUS

The principal abnormality associated with dysfunction of the neurohypophysis is *diabetes insipidus* (in-SIP-i-dus). *Diabetes* means "overflow" and *insipidus* means "tasteless." This disorder should not be confused with diabetes mellitus (*meli* = honey), a disorder of the pancreas characterized by glucose in the urine. Diabetes insipidus is the result of a hyposecretion of ADH, usually caused by damage to the neurohypophysis or the hypothalamic paraventricular and supraoptic nuclei. Symptoms include excretion of large amounts of urine and subsequent dehydration and thirst. A major problem associated with this condition is inability to concentrate urine. Because of this, a person with severe diabetes insipidus may die of dehydration if deprived of water for only a day or so. Diabetes insipidus is treated by administering ADH.

A summary of posterior pituitary hormones, their principal actions, control of secretion, and a selected disorder is presented in Exhibit 18-3.

THYROID

The ***thyroid gland*** is located just below the larynx. The right and left ***lateral lobes*** lie one on either side of the trachea. The lobes are connected by a mass of tissue called an ***isthmus*** (IS-mus) that lies in front of the trachea, just below the cricoid cartilage (Figure 18-11). The ***pyramidal lobe,*** when present, extends upward from the isthmus. The gland has a rich blood supply, receiving about 80 to 120 ml of blood per minute. Thus, the thyroid gland can deliver high levels of hormones in a short period of time, if necessary.

Histologically, the thyroid gland is composed of spherical sacs called ***thyroid follicles*** (Figure 18-12). The wall of each follicle consists of two types of cells. Those that reach the surface of the lumen of the follicle are called ***follicular cells,*** and those that do not reach the lumen are called ***parafollicular cells,*** or ***C cells.*** When the cells are inactive, they tend to be low cuboidal to squamous, but when actively secreting hormones, they become more columnar. The follicular cells manufacture ***thyrox-***

EXHIBIT 18-3 SUMMARY OF POSTERIOR PITUITARY HORMONES, PRINCIPAL ACTIONS, CONTROL OF SECRETION, AND SELECTED DISORDER

Hormone	Principal Actions	Control of Secretion	Selected Disorder
Oxytocin (OT)	Stimulates contraction of smooth muscle cells of pregnant uterus during labor and stimulates contraction of contractile cells of mammary glands for milk ejection.	Neurosecretory cells of hypothalamus secrete OT in response to uterine distension and stimulation of nipples.	
Antidiuretic hormone (ADH)	Principal effect is to decrease urine volume; also raises blood pressure by constricting arterioles during severe hemorrhage.	Neurosecretory cells of hypothalamus secrete ADH in response to low water concentration of the blood, pain, stress, trauma, anxiety, acetylcholine (ACh), nicotine, morphine, and tranquilizers; alcohol inhibits its secretion.	Hyposecretion of ADH results in diabetes insipidus.

ine (thī-ROK-sēn), or ***T***$_4$, since it contains four atoms of iodine, and ***triiodothyronine*** (trī-ī'-ōd-ō-THĪ-rō-nēn), or ***T***$_3$, since it contains three atoms of iodine. Together these hormones are referred to as the ***thyroid hormones.*** Thyroxine is normally secreted in greater quantity than triiodothyronine, but triiodothyronine is three to four times more potent. Moreover, in peripheral tissues, especially the liver and lungs, much of the thyroxine is converted into triiodothyronine. Both hormones are functionally similar. The parafollicular cells produce ***calcitonin*** (kal-si-TŌ-nin), or ***CT.***

Formation, Storage, and Release of Thyroid Hormones

One of the thyroid gland's unique features is its ability to store hormones and release them in a steady flow over a long period of time. The first stage in the synthesis of thyroid hormones is the active transport of iodine ions (I^-) from the blood into the follicular cells. Iodide is the ionized form of iodine as it appears in the blood. Under normal conditions, the iodide concentration in the cells is about 40 times that of the blood; during periods of

FIGURE 18-11 Location and blood supply of the thyroid gland in anterior view. The pyramidal lobe of the thyroid, when present, may be attached to the hyoid bone by a fibrous or fibromuscular band.

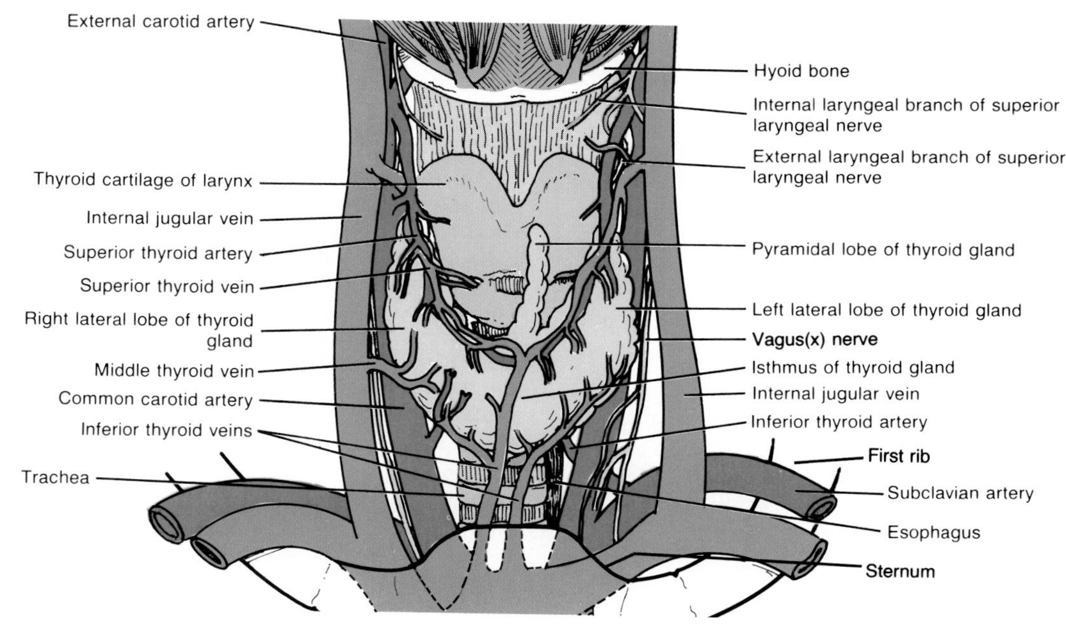

FIGURE 18-12 Histology of the thyroid gland. (a) Photomicrograph at a magnification of 230×. (Copyright © 1983 by Michael H. Ross. Used by permission.) (b) Diagram showing details of a single thyroid follicle.

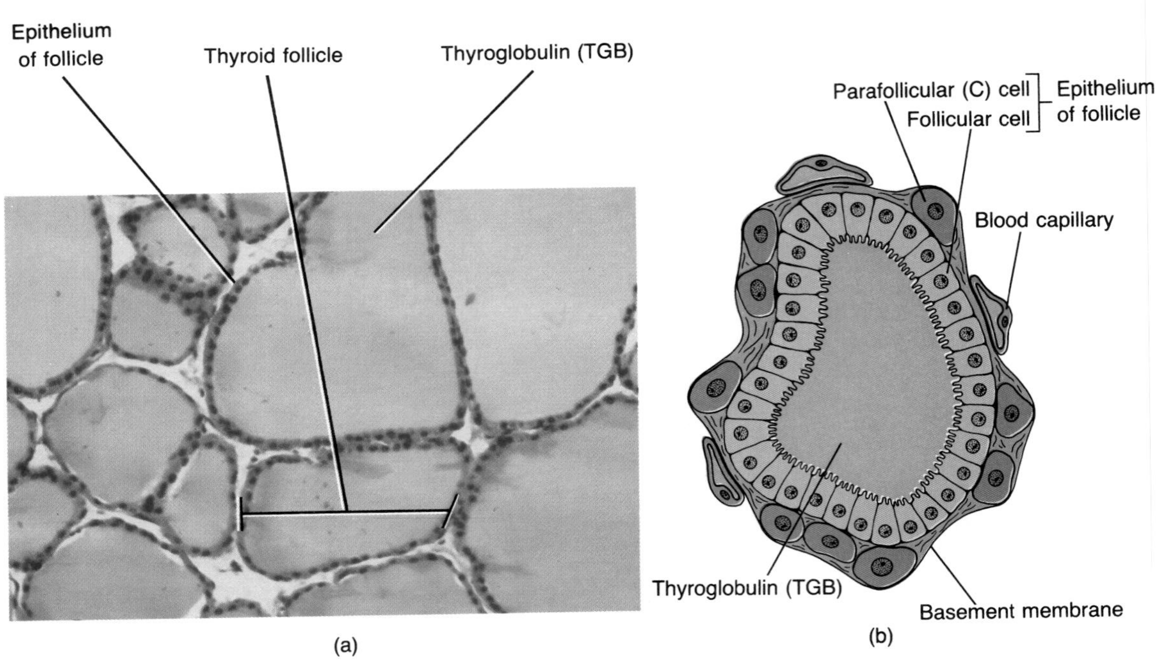

maximal activity, the concentration can increase to over 300 times that of the blood. In the follicular cells, the iodide is oxidized to iodine. Through a series of enzymatically controlled reactions, the iodine combines with the amino acid tyrosine to form the thyroid hormones. This combination occurs within a large glycoprotein molecule, called ***thyroglobulin (TGB),*** which is secreted by the follicular cells into the follicle.

The thyroid hormones are an integral part of TGB and, as such, are stored (sometimes for months) until needed. The entire complex in the follicle, consisting of TGB and the stored hormones, constitutes the ***thyroid colloid.*** Before the thyroid hormones can be released into the blood, droplets of colloid are taken into the follicular cells by endocytosis (pinocytosis), and the thyroid hormones are split from TGB. Following their release into the blood,

MEDICAL TEST

Thyroid function tests (Radioiodine uptake, serum T_4 concentration, and serum T_3 concentration)

Diagnostic Value: To evaluate a swelling or lump in the thyroid gland, to ascertain symptoms of excessive or deficient thyroid hormone levels, to monitor the response of thyroid disease to therapy, and to screen newborns for cretinism.

The radioiodine uptake (RAIU) test evaluates thyroid function by measuring the amount of orally ingested radioactive isotopes of iodine (^{123}I or ^{131}I) that accumulates in the thyroid gland. The test indicates the ability of the thyroid gland to trap and retain iodine.

The serum T_4 concentration test measures the total circulating level of thyroxine in the blood (free and bound to TBG).

The serum T_3 concentration test measures the total circulating level of triiodothyronine in the blood (free and bound to TBG). Other blood tests measure the levels and percentage of free thyroxine and TBG.

If a blood level of a thyroid hormone is abnormal, other tests may be ordered to determine the cause. These tests may include thyroid ultrasound to evaluate the size, shape, and overall structure of the thyroid gland; thyroid scan to evaluate the size and structure of the thyroid gland; and thyroid biopsy to test for thyroid cancer.

Procedure: The serum T_4 concentration and serum T_3 concentration tests are done on a blood sample.

Normal Values: *RAIU: 5–30 percent*
Serum T_4: 5–11 μg/dl
Serum T_3: 80–180 ng/dl

most of the thyroid hormones combine with plasma proteins, mainly ***thyroxine-binding globulin (TBG).*** Thyroid hormones combined with plasma proteins are referred to as ***protein-bound iodine (PBI).***

Function and Control of Thyroid Hormones

The thyroid hormones have three principal effects on the body: (1) regulation of organic metabolism and energy balance, (2) regulation of growth and development, and (3) regulation of the activity of the nervous system. With respect to the regulation of organic metabolism, the thyroid hormones stimulate virtually all aspects of carbohydrate and lipid catabolism in most cells of the body. They also increase the rate of protein synthesis. Since their overall effect is to increase catabolism, they increase the basal metabolic rate. The energy produced raises body temperature as heat is given off, and this phenomenon is called the ***calorigenic effect.***

The thyroid hormones help to regulate tissue growth and development, especially in children. They work with hGH to accelerate body growth, particularly the growth of nervous tissue. Deficiency of the hormones during fetal development can result in fewer and smaller neurons, defective myelination of axons, and mental retardation. During the early years of life, deficiency of the hormones results in small stature and poor development of certain organs such as the brain and reproductive structures.

Finally, the thyroid hormones increase the reactivity of the nervous system. This results in increased blood flow, increased and more forceful heartbeats, increased blood pressure, increased motility of the gastrointestinal tract, and increased nervousness.

The secretion of thyroid hormones is stimulated by several factors (Figure 18-13a). If thyroid hormone levels in the blood fall below normal, chemical sensors in the hypothalamus detect the change in blood chemistry and stimulate the hypothalamus to secrete a regulating factor called thyrotropin releasing hormone (TRH). TRH stimulates the adenohypophysis to secrete thyroid-stimulating

FIGURE 18-13 **Thyroid hormones. (a) Regulation of the secretion of thyroid hormones.**

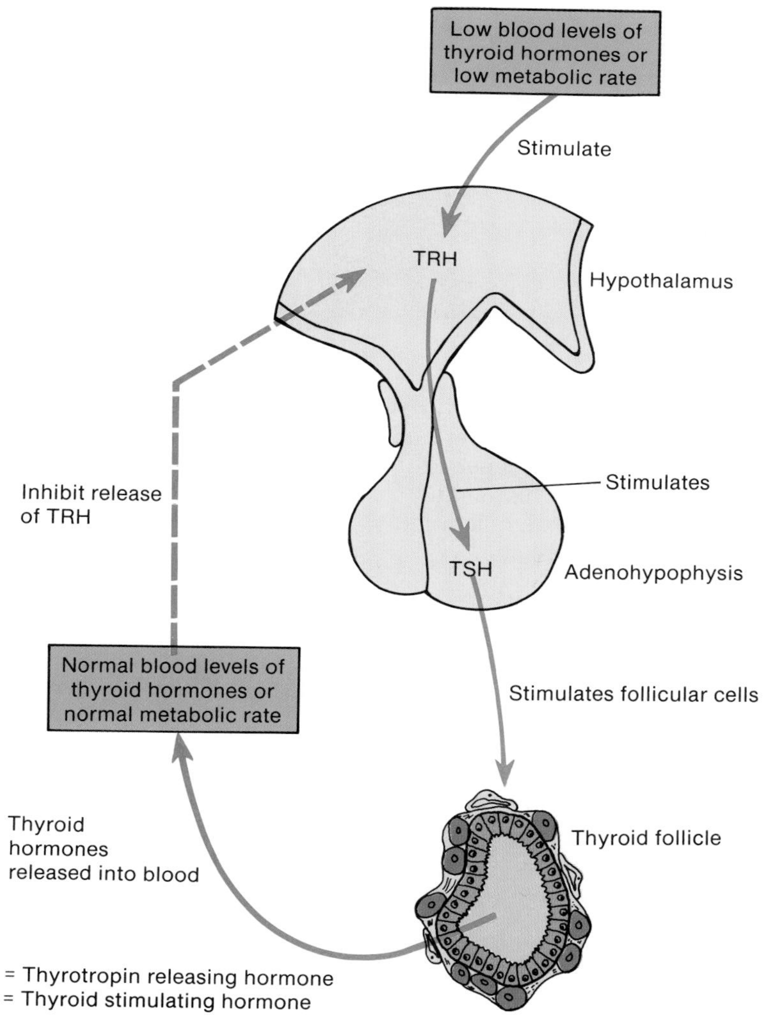

FIGURE 18-13 (*Continued*) (b) Photograph of an individual with cretinism. (c) Photograph of an individual with exophthalmos. (d) Photograph of an individual with simple goiter.

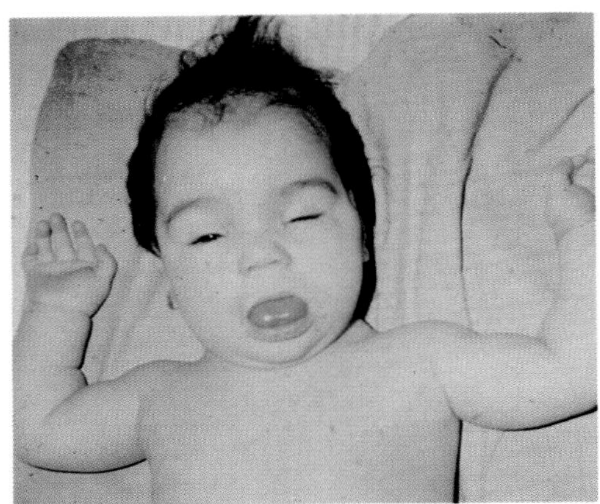

(b)

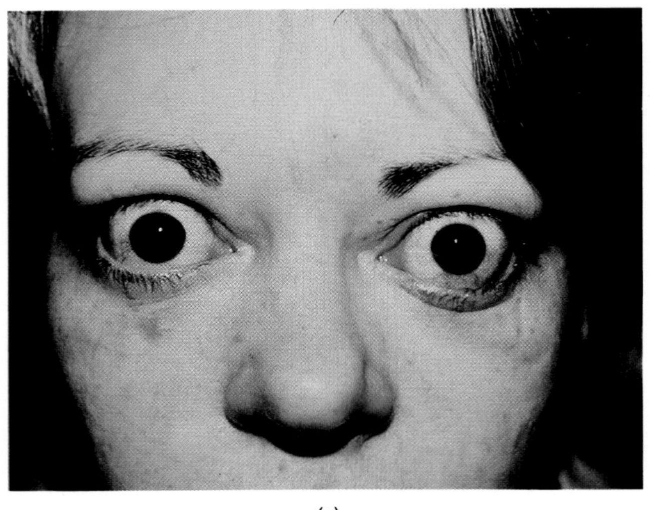

(c)

(d)

hormone (TSH). Then TSH stimulates the thyroid to release thyroid hormones until the metabolic rate returns to normal. Conditions that increase the body's need for energy—a cold environment, hypoglycemia, high altitude, pregnancy—also trigger this negative feedback system and increase the secretion of thyroid hormones.

Thyroid activity can be inhibited by a number of other factors. When large amounts of certain sex hormones (estrogens and androgens) are circulating in the blood, for example, TSH secretion diminishes. Aging slows down the activities of most glands, and thyroid production may decrease.

CLINICAL APPLICATION: CRETINISM, MYXEDEMA, AND GOITER

Hyposecretion of thyroid hormones during fetal life or infancy results in *cretinism* (KRĒ-tin-izm), which is shown in Figure 18-13b. Two outstanding clinical symptoms of the cretin are dwarfism and mental retardation. The first is caused by failure of the skeleton to grow and mature; the second is caused by failure of the brain to develop fully. Recall that one function of thyroid hormones is to control tissue growth and development. Cretins also exhibit retarded sexual development and a yellowish skin color. Flat pads of fat develop, giving the cretin a characteristic round face and thick nose; a large, thick, protruding tongue; and protruding abdomen. Be-

cause the energy-producing metabolic reactions are slow, the cretin has a low body temperature and suffers from general lethargy. Carbohydrates are stored rather than utilized, and heart rate is also slow. If the condition is diagnosed early, the symptoms can be eliminated by administering thyroid hormones.

Hypothyroidism during the adult years produces ***myxedema*** (mix-e-DĒ-ma). A hallmark of this disorder is an edema that causes the facial tissues to swell and look puffy. Like the cretin, the person with myxedema suffers from slow heart rate, low body temperature, sensitivity to cold, hypersensitivity to certain drugs (narcotics, barbiturates, and anesthetics), dry hair and skin, muscular weakness, general lethargy, and a tendency to gain weight easily. The long-term effect of a slow heart rate may overwork the heart muscle, causing the heart to enlarge. Because the brain has already reached maturity, the person with myxedema does not experience mental retardation. However, in moderately severe cases, nerve reactivity may be dulled so that the person lacks mental alertness. Myxedema occurs about five times more frequently in females than in males. Its symptoms are alleviated by the administration of thyroid hormones.

Hypersecretion of thyroid hormones may be due to an autoimmune disease called ***exophthalmic*** (ek'-sof-THAL-mik) ***goiter*** (GOY-ter), also called ***Grave's disease.*** This disease, like myxedema, is also more frequent in females. One of its primary symptoms is an enlarged thyroid, called a ***goiter,*** which may be two to three times its original size. Two other symptoms are an edema behind the eye, which causes the eye to protrude (***exophthalmos***), which is shown in Figure 18-13c, and an abnormally high metabolic rate. The high metabolic rate produces a range of effects that are generally opposite to those of myxedema—increased pulse, high body temperature, heat intolerance, and moist, flushed skin. The person loses weight and is usually full of "nervous" energy. The thyroid hormones also increase the responsiveness of the nervous system, causing the person to become irritable and exhibit tremors of the extended fingers. Hyperthyroidism is usually treated by administration of antithyroid drugs that suppress thyroid hormone synthesis, by treatment with radioactive iodine that selectively destroys thyroid cells, or by surgical removal of part of the gland.

Goiter is a symptom of many thyroid disorders. In North America, it is typically caused by ingestion of excess iodine (owing to its presence in bread, iodized salt, and certain vegetables, such as turnips and rutabagas). It may also be genetically caused. This condition is called ***simple goiter*** (Figure 18-13d). Simple goiter may also be caused by a lower-than-average amount of iodine in the diet. This cause is more prevalent in Third World countries. It may also develop if iodine intake is not increased during certain conditions that put a high demand on the body for thyroxine, such as pregnancy, frequent exposure to cold, and high-fat and -protein diets.

Calcitonin (CT)

The hormone produced by the parafollicular cells of the thyroid gland is ***calcitonin*** (kal-si-TŌ-nin), or ***CT.*** It is involved in the homeostasis of blood calcium (Ca^{2+}) and phosphate (HPO_4^{2-}) levels. CT lowers the amount of calcium and phosphate in the blood by inhibiting bone breakdown and accelerating the uptake of calcium and phosphate by bones. It appears to exert its effect in lowering calcium and phosphate blood levels by inhibiting the action of osteoclasts (bone-destroying cells) and the parathyroid hormone (PTH). If CT is administered to a person with a normal level of blood calcium, it causes ***hypocalcemia*** (low blood calcium level). Hypocalcemia is also a complication of magnesium deficiency. If CT is given to a person with ***hypercalcemia*** (high blood calcium level), the level returns to normal. CT is used to treat postmenopausal osteoporosis, usually in conjunction with adequate calcium and vitamin dietary intake. It is suspected that the blood calcium level directly controls the secretion of CT according to a negative feedback system that does not involve the pituitary gland (Figure 18-14).

A summary of hormones produced by the thyroid gland, their principal actions, control of secretion, and selected disorders is presented in Exhibit 18-4.

PARATHYROIDS

Typically embedded on the posterior surfaces of the lateral lobes of the thyroid gland are small, round masses of tissue called the ***parathyroid glands.*** Usually, two parathyroids, superior and inferior, are attached to each lateral thyroid lobe (Figure 18-15).

Histologically, the parathyroids contain two kinds of epithelial cells (Figure 18-16). The more numerous cells, called ***principal (chief) cells,*** are believed to be the major synthesizer of ***parathyroid hormone (PTH).*** Some researchers believe that the other kind of cell, called an ***oxyphil cell,*** synthesizes a reserve capacity of hormone.

Parathyroid Hormone (PTH)

Parathyroid hormone (PTH), or ***parathormone,*** controls the homeostasis of ions in the blood, especially calcium (Ca^{2+}) and phosphate (HPO_4^{2-}) ions. If adequate amounts of vitamin D are present, PTH increases the rate of calcium, phosphate, and some magnesium (Mg^{2+}) absorption from the gastrointestinal tract into the blood. It also leads to the activation of vitamin D. Moreover, PTH increases the number and activity of osteoclasts, or bone-

FIGURE 18-14 Regulation of the secretion of the parathyroid hormone (PTH) and calcitonin (CT).

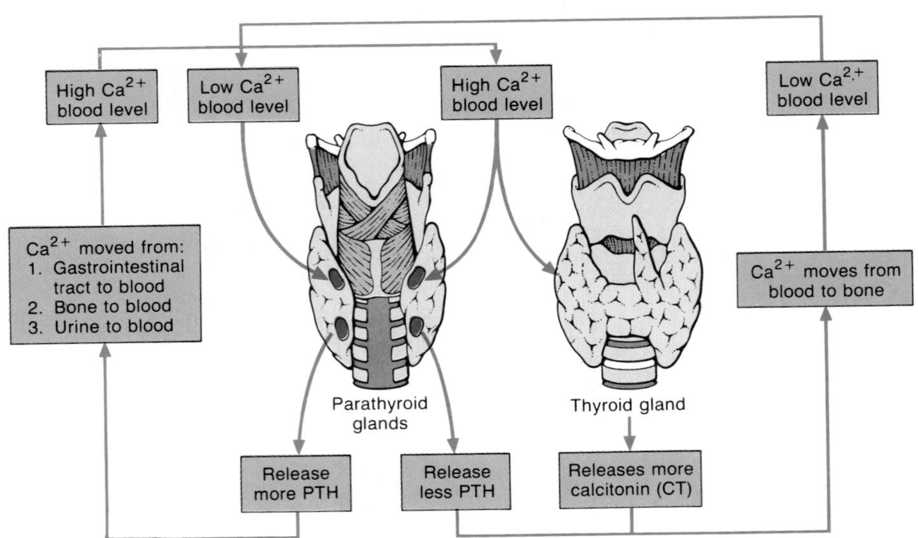

destroying cells. As a result, bone tissue is broken down, and calcium and phosphate are released into the blood. PTH produces two changes in the kidneys. It increases the rate at which the kidneys remove calcium and magnesium from urine that is being formed and returns them to the blood. It inhibits the transport of phosphate from urine into blood so that phosphate is excreted in urine. More phosphate is lost through the urine than is gained from the bones.

The overall effect of PTH with respect to ions, then, is to decrease blood phosphate level and increase blood calcium and magnesium levels. As far as blood calcium level is concerned, PTH and CT are antagonists; that is, they have opposite actions.

EXHIBIT 18-4 SUMMARY OF THYROID GLAND HORMONES, PRINCIPAL ACTIONS, CONTROL OF SECRETION, AND SELECTED DISORDERS

Hormone	Principal Actions	Control of Secretion	Selected Disorders
Thyroid hormones			
Thyroxine (T_4)	Regulates organic metabolism, growth and development, and activity of nervous system.	Thyrotropin releasing hormone (TRH) is released from hypothalamus in response to low thyroid hormone levels, low metabolic rate, cold, pregnancy, and high altitudes; TRH secretion is inhibited in response to high thyroid hormone levels, high metabolic rate, high levels of estrogens and androgens, and aging.	Hyposecretion of thyroid hormones during the growth years results in cretinism; hypothyroidism during adult years results in myxedema; hypersecretion of thyroid hormones results in exophthalmic goiter; iodine excess or deficiency produces simple goiter.
Triiodothyronine (T_3)	Same as above.	Same as above.	
Calcitonin (CT)	Lowers blood levels of calcium by accelerating calcium absorption by bones.	High blood calcium levels stimulate secretion; low levels inhibit secretion.	

FIGURE 18-15 **Location and blood supply of the parathyroid glands in posterior view.**

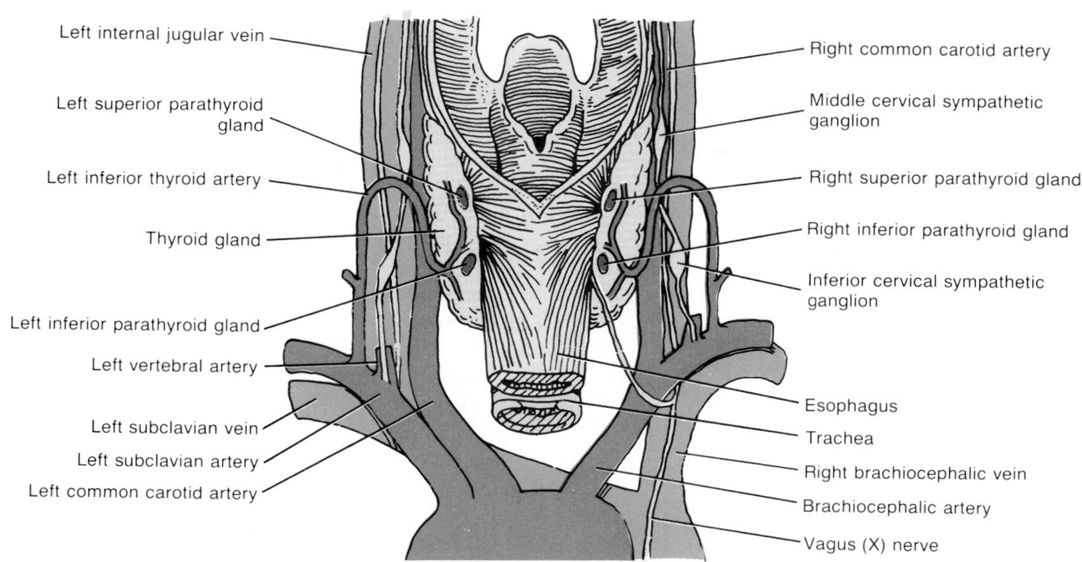

PTH secretion is not controlled by the pituitary gland. When the calcium level of the blood falls, more PTH is released (see Figure 18-14). Conversely, when the calcium level of the blood rises, less PTH (and more CT) is secreted. This is another example of a negative feedback control system that does not involve the pituitary gland.

CLINICAL APPLICATION: TETANY AND OSTEITIS FIBROSA CYSTICA

A normal amount of calcium in the extracellular fluid is necessary to maintain the resting state of

FIGURE 18-16 **Histology of the parathyroid glands. (a) Photomicrograph at a magnification of 180×. (Copyright © 1983 by Michael H. Ross. Used by permission.) (b) Diagram showing details of a portion of a parathyroid gland in relation to a portion of the thyroid gland.**

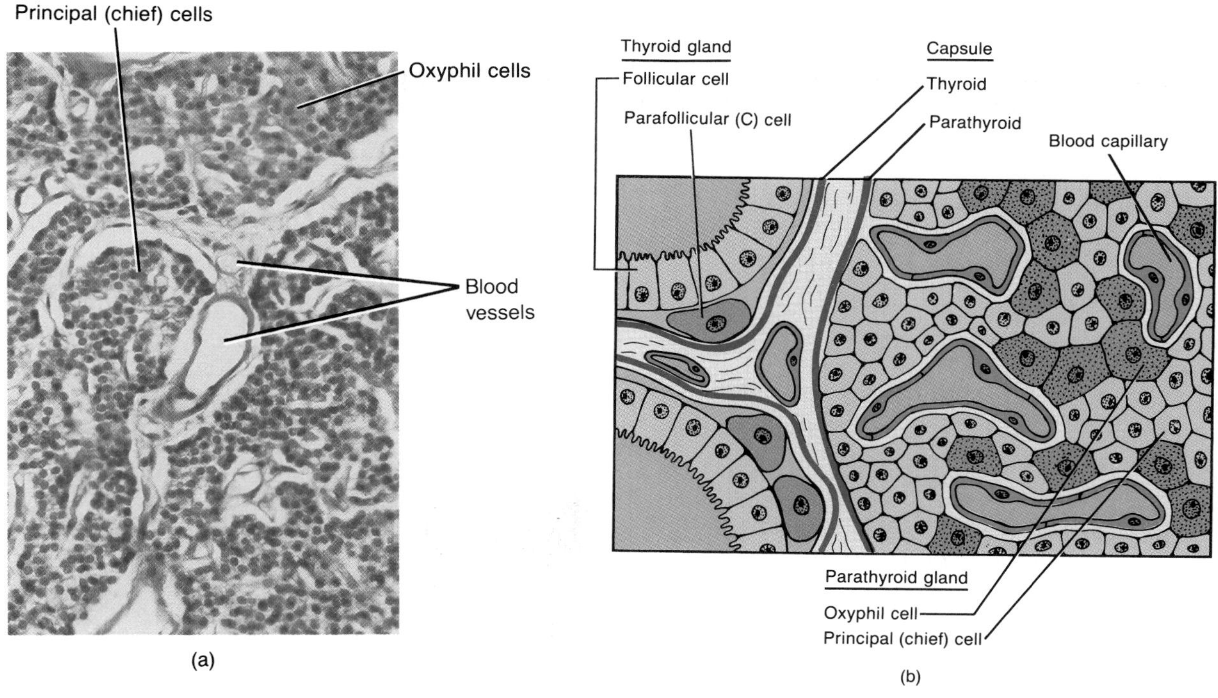

neurons. A deficiency of calcium caused by ***hypoparathyroidism*** causes neurons to depolarize without the usual stimulus. As a result, nervous impulses increase and result in muscle twitches, spasms, and convulsions. This condition is called ***tetany.*** Hypoparathyroidism results from surgical removal of the parathyroids or from damage caused by parathyroid disease, infection, hemorrhage, or mechanical injury.

Hyperparathyroidism causes demineralization of bone. If uncorrected, this condition may lead to ***osteitis fibrosa cystica,*** so named because the areas of destroyed bone tissue are replaced by cavities that fill with fibrous tissue. The bones thus become deformed and are highly susceptible to fracture. Hyperparathyroidism is usually caused by a tumor in the parathyroids.

A summary of the principal actions, control of secretion, and selected disorders related to parathyroid hormone (PTH) is presented in Exhibit 18-5.

ADRENALS (SUPRARENALS)

The body has two ***adrenal (suprarenal) glands,*** one of which is located superior to each kidney (Figure 18-17). Each adrenal gland is structurally and functionally differentiated into two regions: the outer ***adrenal cortex,*** which makes up the bulk of the gland, and the inner ***adrenal medulla*** (Figure 18-18). Whereas the adrenal cortex is derived from mesoderm of a developing embryo, the adrenal medulla is derived from the ectoderm. Since their origins are different, they also produce different hormones. Covering the gland is a connective tissue capsule. The adrenals, like the thyroid, are among the more vascular organs of the body.

Adrenal Cortex

Histologically, the adrenal cortex is subdivided into three zones (Figure 18-18). Each zone has a different cellular arrangement and secretes different groups of steroid hormones. The outer zone, directly underneath the connective tissue capsule, is referred to as the ***zona glomerulosa.*** It composes about 15 percent of the total cortical volume. Its cells are arranged in arched loops or round balls. Its primary secretions are a group of hormones called mineralocorticoids (min'-er-al-ō-KOR-ti-koyds).

The middle zone, or ***zona fasciculata,*** is the widest of the three zones and consists of cells arranged in long, straight cords. The zona fasciculata secretes mainly glucocorticoids (gloo'-kō-KOR-ti-koyds).

The inner zone, the ***zona reticularis,*** contains cords of cells that branch freely. This zone synthesizes minute amounts of hormones, mainly the sex hormones called gonadocorticoids (gō-na-dō-KOR-ti-koyds), and of these, primarily male hormones called androgens.

Mineralocorticoids

Mineralocorticoids help control water and electrolyte homeostasis, particularly the concentrations of sodium (Na^+) and potassium (K^+) ions. Although the adrenal cortex secretes at least three different hormones classified as mineralocorticoids, one of these hormones is responsible for about 95 percent of the mineralocorticoid activity—***aldosterone*** (al-do-STĒR-ōn). Aldosterone acts on the tubule cells in the kidneys and causes them to increase their reabsorption of sodium. As a result, sodium ions are removed from the fluid that will become urine and returned to the blood. In this manner, aldosterone prevents rapid depletion of sodium from the body. On the other hand, aldosterone increases excretion of potassium ions so that large amounts of potassium are lost in the urine.

EXHIBIT 18-5 SUMMARY OF PARATHYROID GLAND HORMONE, PRINCIPAL ACTIONS, CONTROL OF SECRETION, AND SELECTED DISORDERS

Hormone	Principal Actions	Control of Secretion	Selected Disorders
Parathyroid hormone (PTH)	Increases blood calcium and magnesium levels and decreases blood phosphate level by increasing rate of calcium and magnesium absorption from gastrointestinal (GI) tract into blood; increases number and activity of osteoclasts; increases calcium absorption by kidneys; increases phosphate excretion by kidneys; and activates vitamin D.	Low blood calcium levels stimulate secretion; high levels inhibit secretion.	Hypoparathyroidism results in tetany. Hyperparathyroidism produces osteitis fibrosa cystica.

FIGURE 18-17 Location and blood supply of the adrenal (suprarenal) glands.

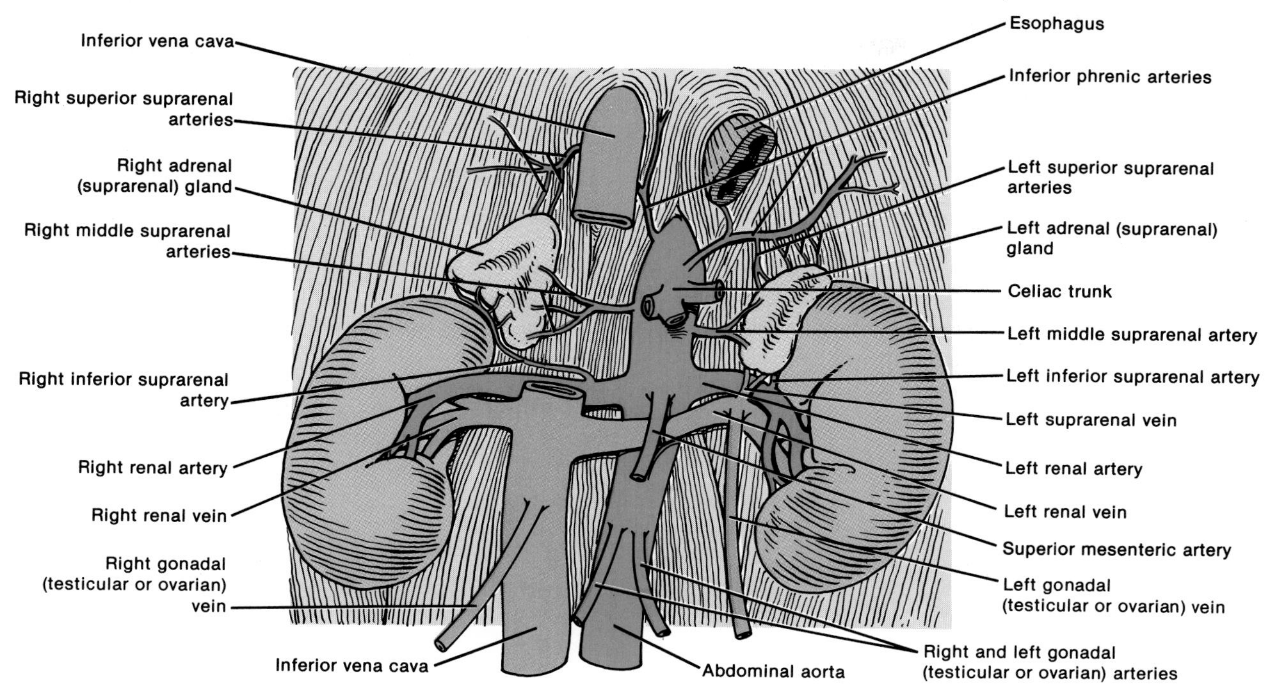

FIGURE 18-18 Histology of the adrenal (suprarenal) glands. (a) Photomicrograph of a section of the adrenal gland showing its subdivisions and zones at a magnification of 45×. (Courtesy of Andrew Kuntzman.) (b) Hormones produced by the cortical zones and medulla.

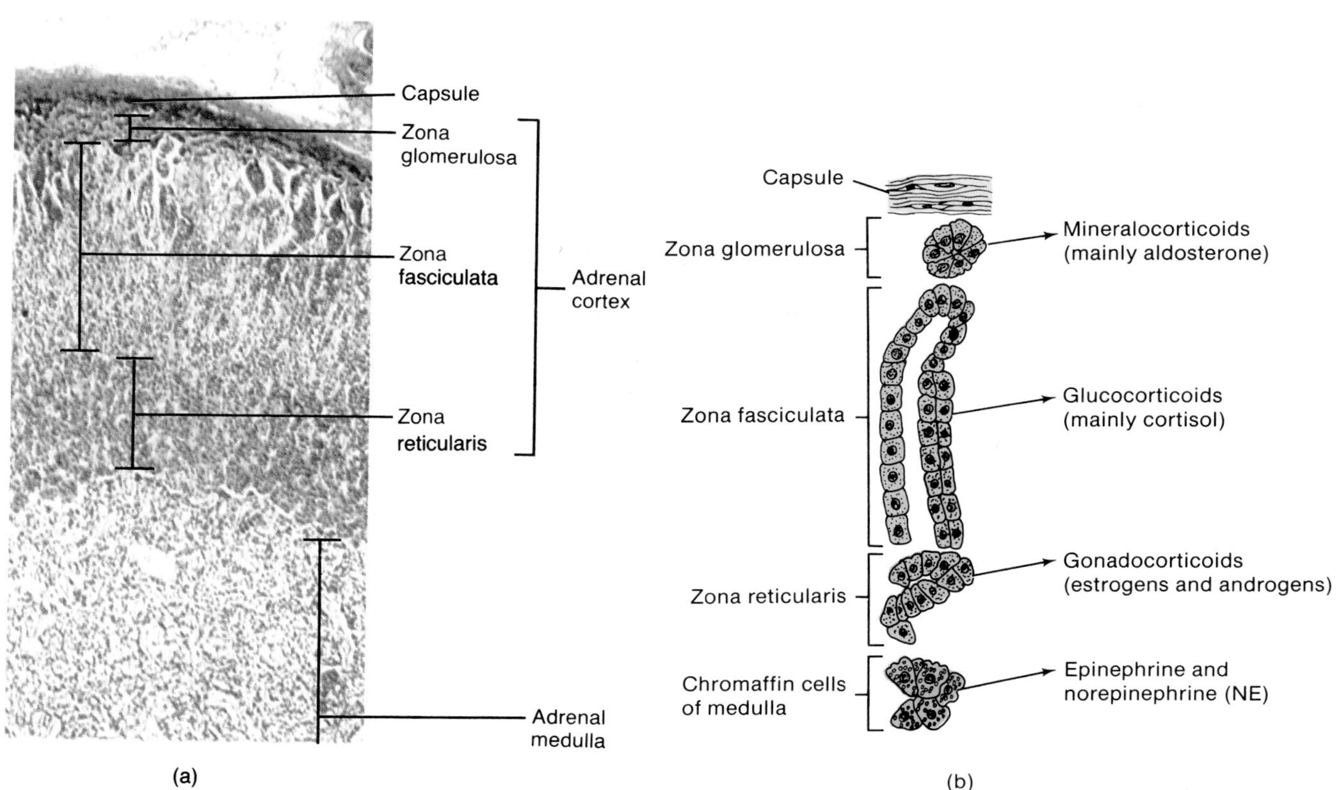

These two basic functions—conservation of sodium and elimination of potassium—cause a number of secondary effects. For example, a large proportion of the sodium reabsorption occurs through an exchange reaction whereby positive hydrogen (H^+) ions pass into the urine to replace the positive sodium ions. Since this mechanism removes hydrogen ions, which are acid ions, it makes the blood less acidic and prevents acidosis. The movement of sodium ions also sets up a positively charged field in the blood vessels around the kidney tubules. As a result, negatively charged chloride (Cl^-) and bicarbonate (HCO_3^-) ions are drawn out of the newly forming fluid that will become urine and back into the blood. Finally, the increase in sodium ion concentration in the blood vessels causes water to move by osmosis from the fluid that eventually becomes urine into the blood. When ADH is present, even more water is reabsorbed. In summary, aldosterone causes potassium excretion and sodium reabsorption. The sodium reabsorption leads to the elimination of H^+ ions; the retention of Na^+, Cl^-, and HCO_3^- ions; and the retention of water.

The control of aldosterone secretion is complex. Apparently, several mechanisms operate. One of these is the ***renin–angiotensin*** (an'-jē-ō-TEN-sin) ***pathway*** (Figure 18-19). A decrease in blood volume from dehydration, sodium ion (Na^+) deficiency, or hemorrhage brings about a drop in blood pressure. The low blood pressure stimulates certain kidney cells, called juxtaglomerular cells, to secrete into the blood an enzyme called ***renin*** (RĒ-nin) (see Figure 26-9). In this pathway, renin converts ***angiotensinogen,*** a plasma protein produced by the liver, into ***angiotensin I,*** which is then converted into ***angiotensin II*** by a plasma enzyme in the lungs. Angiotensin II stimulates the adrenal cortex to produce more aldosterone. In the kidneys, aldosterone brings about increased Na^+ reabsorption and water follows. Some of this

FIGURE 18-19 **Proposed mechanism for the regulation of the secretion of aldosterone by the renin–angiotensin pathway.**

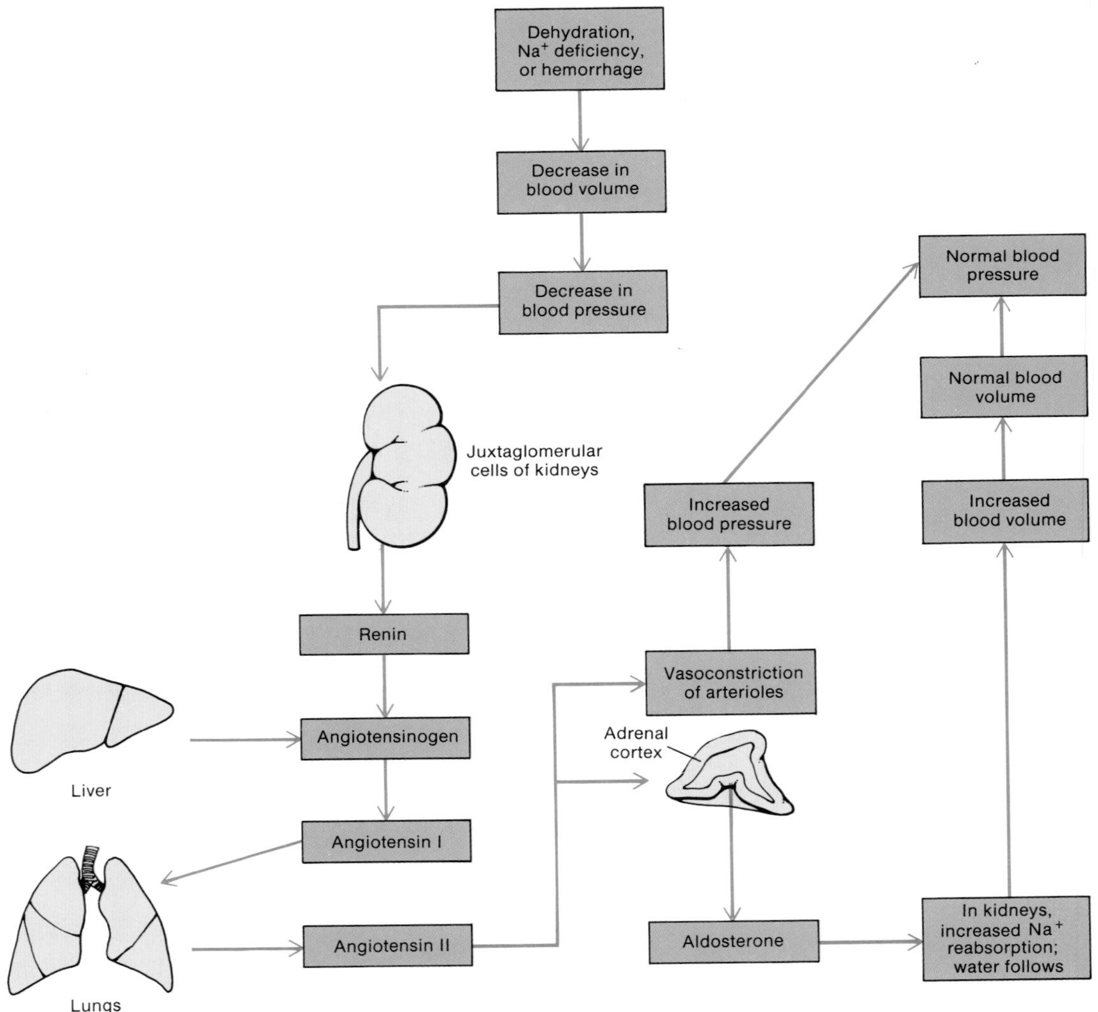

reabsorption occurs in parts of the kidney called distal convoluted tubules and collecting tubules (Chapter 26) and leads to an increase in extracellular fluid volume and a restoration of blood pressure to normal. Since angiotensin II is a powerful vasoconstrictor, this action also helps to elevate blood pressure.

A second mechanism for the control of aldosterone involves potassium (K^+) ion concentration. An increased K^+ concentration in extracellular fluid directly stimulates aldosterone secretion by the adrenal cortex and causes the elimination of excess K^+ by the kidneys. A decreased K^+ concentration in extracellular fluid decreases aldosterone production, and thus less K^+ than usual is eliminated by the kidneys.

Adrenocorticotropic hormone (ACTH) from the anterior pituitary has a minor effect on the secretion of aldosterone. Only in the complete absence of ACTH is there a mild to moderate deficiency of aldosterone. The major controlling effect of ACTH is on glucocorticoids, as will be discussed shortly.

CLINICAL APPLICATION: ALDOSTERONISM

Hypersecretion of the mineralocorticoid aldosterone results in *aldosteronism,* characterized by an increase in sodium and decrease in potassium concentration in the blood and a usually correctable form of hypertension (high blood pressure). If potassium depletion is great, neurons cannot depolarize, and muscular paralysis results. Hypersecretion also brings about excessive retention of sodium and water. The water increases the volume of the blood and causes the hypertension.

Glucocorticoids

The ***glucocorticoids*** are a group of hormones concerned with normal organic metabolism and resistance to stress. Three glucocorticoids are ***cortisol (hydrocortisone), corticosterone,*** and ***cortisone.*** Of the three, cortisol is the most abundant and is responsible for about 95 percent of glucocorticoid activity. The glucocorticoids have the following effects on the body.

1. Glucocorticoids work with other hormones in promoting normal organic metabolism. Their role is to make sure enough energy is available. They increase the rate at which proteins are catabolized and amino acids are removed from cells, primarily muscle fibers (cells), and transported to the liver. The amino acids may be synthesized into new proteins, such as the enzymes needed for metabolic reactions. If the body's reserves of glycogen and fat are low, the liver may convert certain amino acids to glucose. This conversion of a substance other than carbohydrate into glucose is called ***gluconeogenesis*** (gloo'-kō-nē'-ō-JEN-e-sis). Glucocorticoids also release fatty acids from adipose tissue.

2. Glucocorticoids work in many ways to provide resistance to stress. A sudden increase in available glucose by way of gluconeogenesis from amino acids makes the body more alert. Additional glucose gives the body energy for combating a range of stresses: fright, temperature extremes, high altitude, bleeding, infection, surgery, trauma, and almost any debilitating disease, as well as fasting. Glucocorticoids also make the blood vessels more sensitive to vessel-constricting chemicals. They thereby raise blood pressure. This effect is advantageous if the stress happens to be blood loss, which causes a drop in blood pressure.

3. Glucocorticoids are antiinflammatory compounds that inhibit the cells and secretions that participate in inflammation. Specifically, glucocorticoids reduce the number of mast cells to inhibit the release of histamines, stabilize lysosomal membranes to reduce release of enzymes, decrease blood capillary permeability, and depress phagocytosis by monocytes. Unfortunately, they also retard connective tissue regeneration and are thereby responsible for slow wound healing. High doses of glucocorticoids cause atrophy of the thymus gland, spleen, and lymph nodes, thus depressing immune responses. High doses can also result in severe mental disturbances. However, high doses may be useful in the treatment of rheumatism.

The control of glucocorticoid secretion is a typical negative feedback mechanism (Figure 18-20a). The two principal stimuli are stress and low blood level of glucocorticoids. Either condition stimulates the hypothalamus to secrete a regulating hormone called ***corticotropin releasing hormone (CRH).*** This secretion initiates the release of ACTH from the anterior lobe of the pituitary. ACTH is carried through the blood to the adrenal cortex, where it then stimulates glucocorticoid secretion.

As part of the discussion of stress at the end of the chapter, you will see that ACTH secretion is also controlled by a substance called interleukin-1 (IL-1) that is secreted by macrophages of the immune system in response to infection or inflammation.

CLINICAL APPLICATION: ADDISON'S DISEASE AND CUSHING'S SYNDROME

Hyposecretion of glucocorticoids (and aldosterone) results in the condition called *Addison's disease (primary adrenocortical insufficiency).* Clinical symptoms include mental lethargy, anorexia, nausea and vomiting, weight loss, and hypoglycemia, which leads to muscular weakness. Increased potassium and decreased sodium lead to low blood pressure, dehydration, decreased cardiac output, arrhythmias, and potential cardiac arrest. Excessive skin pigmentation, especially in sun-exposed areas, and in mucous membranes, also occurs.

Cushing's syndrome **is a hypersecretion of glucocorticoids, especially cortisol and cortisone (Figure 18-20b). The condition is characterized by the redistribution of fat. The result is spindly legs accompanied by a "moon face," "buffalo hump" on the back, and pendulous abdomen. Facial skin is flushed, and the skin covering the abdomen develops stretch marks (striae). The individual also bruises easily, and wound healing is poor. Other symptoms include hyperglycemia, osteoporosis, weakness, hypertension,**

FIGURE 18-20 **Glucocorticoids. (a) Regulation of the secretion of glucocorticoids. (b) Photograph of an individual with Cushing's syndrome. (c) Photograph of an individual with adrenogenital syndrome. (Courtesy of Lester V. Bergman & Associates, Inc.)**

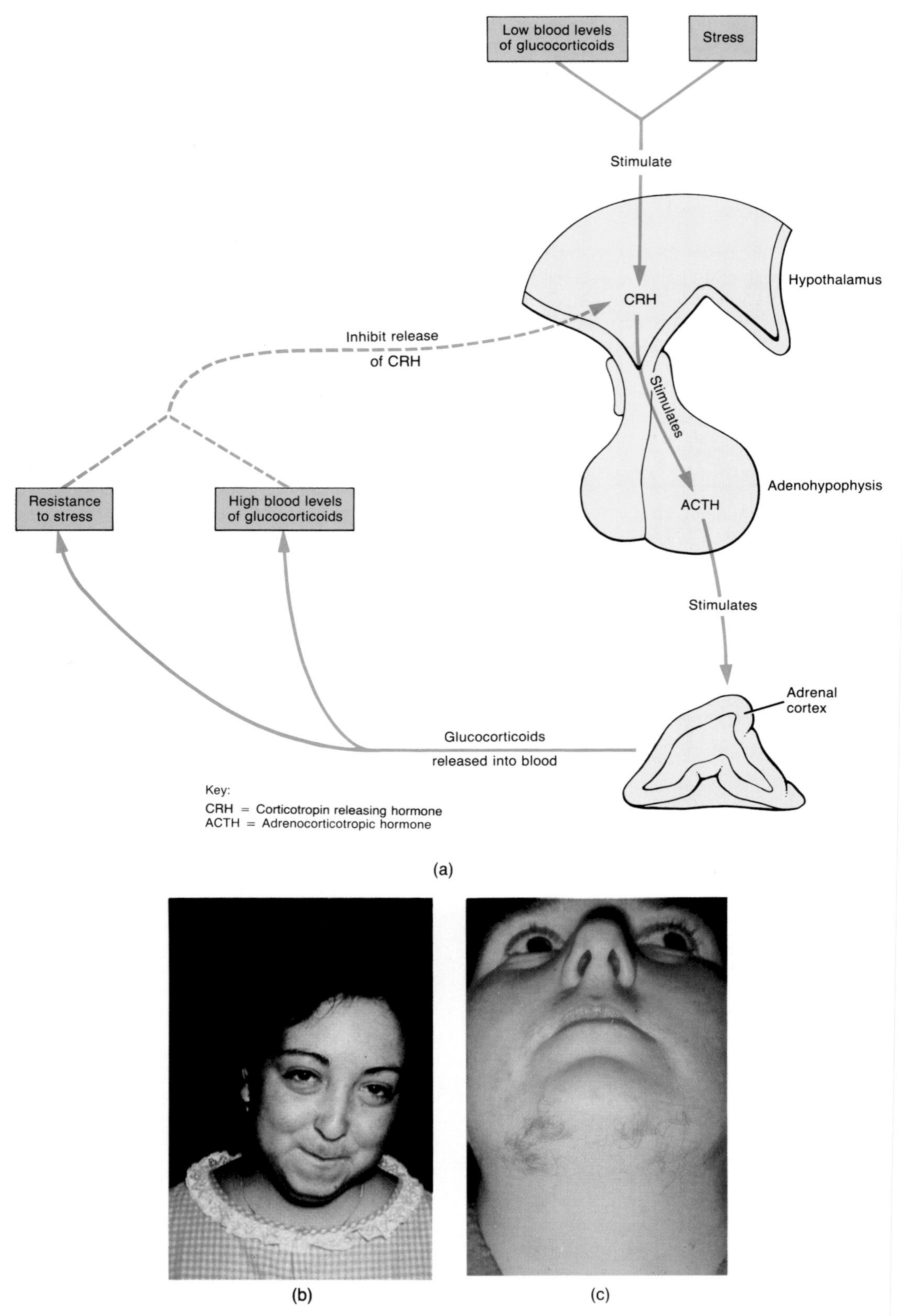

increased susceptibility to infection, decreased resistance to stress, and mood swings. The most common cause of a cushinoid appearance is the administration of a steroid hormone such as prednisone to a transplant recipient or in the treatment of asthma, arthritis, or a blood disorder.

Gonadocorticoids

The adrenal cortex secretes both male and female ***gonadocorticoids (sex hormones).*** These are estrogens and androgens. Estrogens are several closely related female sex hormones that are also produced by the ovaries and placenta. Androgens are hormones that exert masculinizing effects. An important androgen, called testosterone, is produced by the testes. The concentration of sex hormones secreted by normal adult male adrenals is usually so low that their effects are insignificant. In females, androgens contribute to sex drive (libido).

CLINICAL APPLICATION: ADRENOGENITAL SYNDROME AND GYNECOMASTIA

The *adrenogenital syndrome* usually refers to a group of enzyme deficiencies that block the synthesis of glucocorticoids. In an attempt to compensate, the anterior pituitary secretes more ACTH. As a result, excess androgens are produced, causing *virilism,* or masculinization. For instance, the female develops extremely virile characteristics such as growth of a beard (Figure 18-20c), development of a much deeper voice, occasionally the development of baldness, development of a masculine distribution of hair on the body and on the pubis, growth of the clitoris that resembles a penis, atrophy of the breasts, infrequent or absent menstruation, and deposition of proteins in the skin and muscles, producing typical masculine characteristics. Such virilism may also result from tumors of the adrenal gland called *virilizing adenomas* (*aden* = gland; *oma* = tumor).

In the prepubertal male, the syndrome causes the same characteristics as in the female, plus rapid development of the male sexual organs and creation of male sexual desires. In the adult male, the virilizing characteristics of the adrenogenital syndrome are usually completely obscured by the normal virilizing characteristics of the testosterone secreted by the testes. As a result, it is often difficult to make a diagnosis of adrenogenital syndrome in the male adult. However, an occasional adrenal tumor secretes sufficient quantities of feminizing hormones (estrogens) that the male patient develops *gynecomastia* (*gyneca* = woman; *mast* = breast), which means excessive growth (benign) of the male mammary glands. Such a tumor is called a *feminizing adenoma.* Gynecomastia is also associated with androgen-deficiency states, pulmonary diseases, chest wall trauma, psychological stress, and certain drugs (such as alcohol, cimetidine, and digitalis derivatives). As a rule, specific treatment of gynecomastia is not indicated; however, medical treatment (antiestrogen or synthetic androgen) may be given or surgery may be undertaken for cosmetic reasons or a chronic condition.

Adrenal Medulla

The adrenal medulla consists of hormone-producing cells, called ***chromaffin*** (krō-MAF-in) ***cells,*** which surround large blood-containing sinuses (see Figure 18-18). Chromaffin cells develop from the same source as the postganglionic cells of the sympathetic division of the autonomic nervous system. They are directly innervated by preganglionic cells of the sympathetic division of the autonomic nervous system and may be regarded as postganglionic cells that are specialized to secrete. In all other visceral effectors, preganglionic sympathetic fibers first synapse with postganglionic neurons before innervating the effector. In the adrenal medulla, however, the preganglionic fibers pass directly into the chromaffin cells of the gland. The secretion of hormones from the chromaffin cells is directly controlled by the autonomic nervous system, and innervation by the preganglionic fibers allows the gland to respond rapidly to a stimulus.

Epinephrine and Norepinephrine (NE)

The two principal hormones synthesized by the adrenal medulla are ***epinephrine*** and ***norepinephrine (NE),*** also called adrenaline and noradrenaline, respectively. Epinephrine constitutes about 80 percent of the total secretion of the gland. Both hormones are ***sympathomimetic*** (sim'-pa-thō-mi-MET-ik); that is, they produce effects that mimic those brought about by the sympathetic division of the autonomic nervous system. To a large extent, they are responsible for the fight-or-flight response. Like the glucocorticoids of the adrenal cortices, these hormones help the body resist stress. However, unlike the cortical hormones, the medullary hormones are not essential for life.

Under stress, impulses received by the hypothalamus are conveyed to sympathetic preganglionic neurons, which cause the chromaffin cells to increase their output of epinephrine and norepinephrine. Epinephrine and norepinephrine increase blood pressure by increasing heart rate and force of contraction and constricting blood vessels. They accelerate the rate of respiration, dilate respiratory passageways, decrease the rate of digestion, increase the efficiency of muscular contractions, increase blood sugar level, and stimulate cellular metabolism. Hypoglycemia (low blood sugar level) may also stimulate medullary secretion of epinephrine and norepinephrine (NE).

CLINICAL APPLICATION: PHEOCHROMOCYTOMAS

Tumors of the chromaffin cells of the adrenal medulla, called ***pheochromocytomas*** (fē-ō-krō'-mō-sī-TŌ-mas), cause hypersecretion of the medullary hormones. Such tumors are usually benign. The oversecretion causes rapid heart rate, headache, high blood pressure, high levels of sugar in the blood and urine, an elevated basal metabolic rate (BMR), flushing of the face, nervousness, sweating, decreased gastrointestinal motility, and vertigo. Since the medullary hormones create the same effects as sympathetic nervous stimulation, hypersecretion puts the individual into a prolonged version of the fight-or-flight response. This condition ultimately wears out the body, and the individual eventually suffers from general weakness. The only definitive treatment of pheochromocytomas is surgical removal of the tumor(s). Pheochromocytoma is one of the curable causes of hypertension that is tested for when a person has episodes of hypertension.

A summary of the hormones produced by the adrenal glands, their principal actions, control of secretion, and selected disorders is presented in Exhibit 18-6.

EXHIBIT 18-6 SUMMARY OF HORMONES PRODUCED BY THE ADRENAL GLANDS, PRINCIPAL ACTIONS, CONTROL OF SECRETION, AND SELECTED DISORDERS

Hormone	Principal Actions	Control of Secretion	Selected Disorders
Adrenal Cortical Hormones			
Mineralocorticoids (mainly aldosterone)	Increase blood levels of sodium and water and decrease blood levels of potassium.	Decreased blood volume or sodium levels initiate renin–angiotensin pathway to stimulate aldosterone secretion; increased blood levels of potassium stimulate aldosterone secretion; ACTH has only a minor effect in promoting aldosterone secretion.	Hypersecretion of aldosterone results in aldosteronism.
Glucocorticoids (mainly cortisol)	Help promote normal organic metabolism, resistance to stress, and counter inflammatory response.	ACTH release is stimulated by corticotropin releasing hormone (CRH) in response to stress and low blood levels of glucocorticoids.	Hyposecretion of glucocorticoids produces Addison's disease; hypersecretion results in Cushing's syndrome.
Gonadocorticoids	Concentrations secreted by adults are so low that their effects are usually insignificant.	Discussed in detail in Chapter 28.	Adrenogenital syndrome inhibits synthesis of glucocorticoids that results in excess production of ACTH and androgens, causing virilism. The release of sufficient feminizing hormones in males causes gynecomastia.
Adrenal Medullary Hormones			
Epinephrine	Sympathomimetic, that is, produces effects that mimic those of the sympathetic division of the autonomic nervous system (ANS) during stress.	Sympathetic preganglionic neurons stimulate secretion by chromaffin cells.	Hypersecretion of medullary hormones results in a prolonged fight-or-flight response.
Norepinephrine (NE)	Same as above.		

PANCREAS

The ***pancreas*** can be classified as both an endocrine and an exocrine gland. Thus, it is referred to as a ***heterocrine gland.*** We shall treat its endocrine functions at this point; its exocrine functions are discussed in the chapter on the digestive system (Chapter 24). The pancreas is a flattened organ located posterior and slightly inferior to the stomach (Figure 18-21). The adult pancreas consists of a head, body, and tail.

The endocrine portion of the pancreas consists of about one million clusters of cells called ***pancreatic islets*** or ***islets of Langerhans*** (LAHNG-er-hanz) (Figure 18-22). Three kinds of cells are found in these clusters: (1) ***alpha cells,*** which secrete the hormone glucagon that raises blood sugar level; (2) ***beta cells,*** which secrete the hormone insulin that lowers blood sugar level; and (3) ***delta cells,*** which secrete growth hormone inhibiting hormone (GHIH) or somatostatin, which inhibits the secretion of insulin and glucagon. The islets are infiltrated by blood capillaries and surrounded by clusters of cells (acini) that form the exocrine part of the gland. Glucagon and insulin are the endocrine secretions of the pancreas and are concerned with regulation of blood sugar level.

Glucagon

The product of the alpha cells is ***glucagon*** (GLOO-ka-gon), a hormone whose principal physiological activity is to increase the blood glucose level (Figure 18-23). Glucagon does this by accelerating the conversion of glycogen in the liver into glucose (glycogenolysis) and the conversion in the liver of other nutrients, such as amino acids, glycerol, and lactic acid, into glucose (gluconeogenesis). The liver then releases the glucose into the blood, and the blood sugar level rises. Secretion of glucagon is directly controlled by the level of blood sugar via a negative feedback system. When the blood sugar level falls below normal, chemical sensors in the alpha cells of the islets stimulate the cells to secrete glucagon. When blood sugar rises, the cells are no longer stimulated and production slackens. If for some reason the self-regulating device fails and the alpha cells secrete glucagon continuously, hyperglycemia (high blood sugar level) may result. Exercise and largely or entirely protein meals that raise the amino acid level of the blood also cause an increase in glucagon secretion. Glucagon secretion is inhibited by GHIH (somatostatin).

FIGURE 18-21 Location and blood supply of the pancreas.

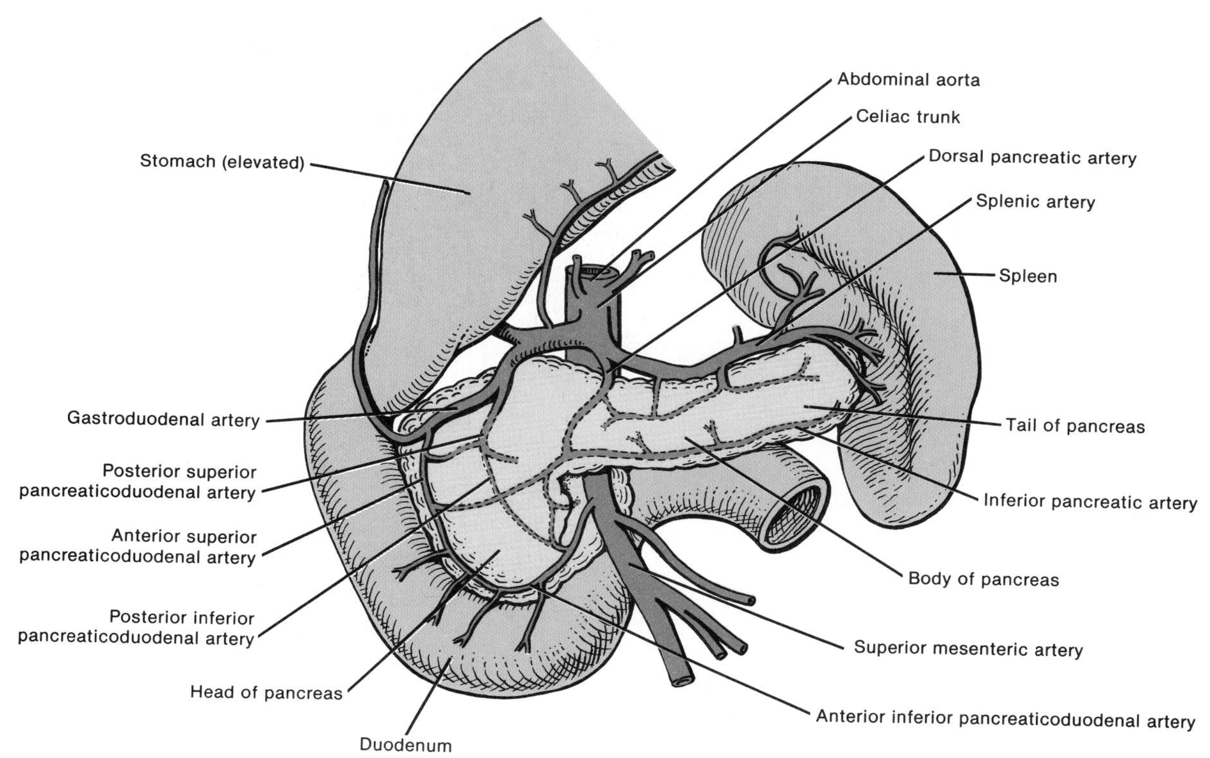

FIGURE 18-22 Histology of the pancreas. (a) Photomicrograph at a magnification of 600×. (Copyright © 1983 by Michael H. Ross. Used by permission.) (b) Diagram. Of the nearly almost one million islets in the pancreas, most are located in the tail of the pancreas.

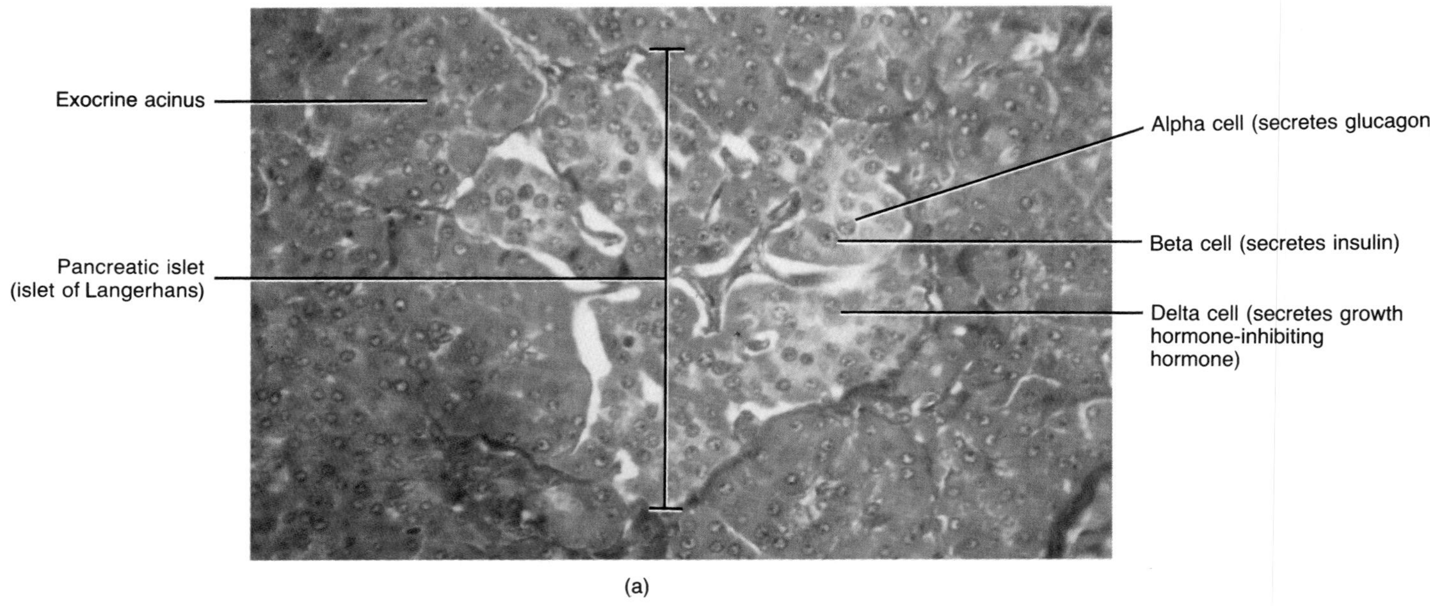

(a)

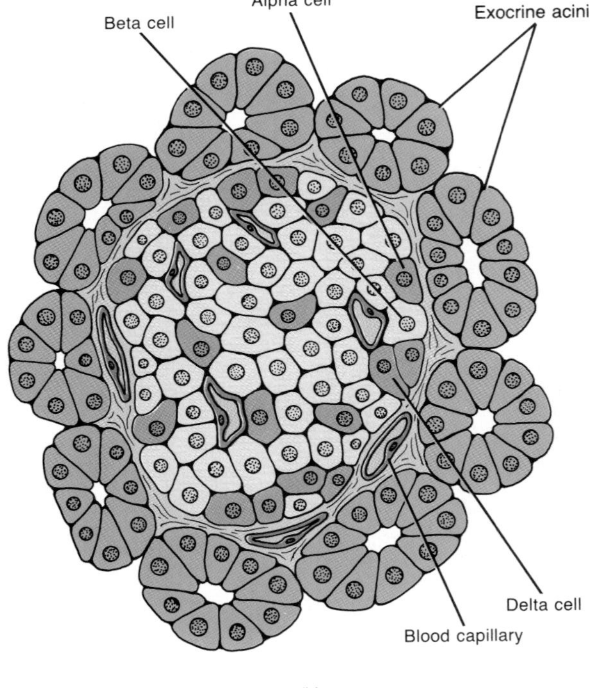

(b)

FIGURE 18-23 Regulation of the secretion of glucagon and insulin.

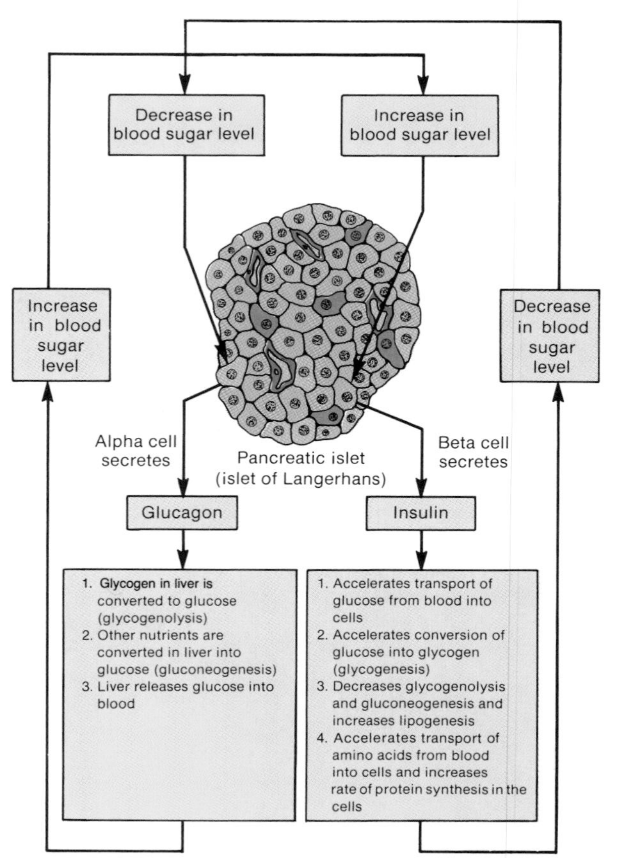

Insulin

The beta cells of the islets produce the hormone ***insulin,*** which acts to lower blood glucose levels. Its chief physiological action is opposite that of glucagon. Insulin decreases blood sugar level in several ways (Figure 18-23). It accelerates the transport of glucose from the blood into cells, especially skeletal muscle fibers (cells). Glucose entry into cells depends on the presence of insulin receptors on the surfaces of target cells. It also accelerates the conversion of glucose into glycogen (glycogenesis). Insulin also decreases glycogenolysis and gluconeogenesis, stimulates the conversion of glucose or other nutrients into fatty acids (lipogenesis), and helps stimulate protein synthesis.

The regulation of insulin secretion, like that of glucagon secretion, is directly determined by the level of sugar in the blood and is based on a negative feedback system. Other hormones can indirectly affect insulin production, however. For instance, human growth hormone (hGH) raises blood glucose level, and the rise in glucose level triggers insulin secretion. Adrenocorticotropic hormone (ACTH), by stimulating the secretion of glucocorticoids, brings about hyperglycemia and also indirectly stimulates the release of insulin. Increased blood levels of amino acids stimulate the release of insulin. Gastrointestinal hormones (discussed in Chapter 24) like stomach and intestinal gastrin, secretin, cholecystokinin (CCK), and gastric inhibitory peptide (GIP) also stimulate insulin secretion. GHIH (somatostatin) inhibits the secretion of insulin.

CLINICAL APPLICATION: DIABETES MELLITUS AND HYPERINSULINISM

Diabetes mellitus (MEL-i-tus) is not a single hereditary disease but a heterogeneous group of diseases, all of which ultimately lead to an elevation of glucose in the blood (hyperglycemia) and loss of glucose in the urine as hyperglycemia increases. Diabetes mellitus is also characterized by the three "polys": an inability to reabsorb water, resulting in increased urine production (***polyuria***); excessive thirst (***polydipsia***); and excessive eating (***polyphagia***).

Two major types of diabetes mellitus are distinguished: type I and type II. ***Type I diabetes,*** which occurs abruptly, is characterized by an absolute deficiency of insulin due to a marked decline in the number of insulin-producing beta cells (perhaps caused by the autoimmune destruction of beta cells), even though target cells contain insulin receptors. Type I diabetes is called ***insulin-dependent diabetes*** because periodic administration of insulin is required to treat it. It is also known as ***juvenile-onset diabetes*** because it most commonly develops in people younger than age 20, though it persists throughout life. Although people who develop type I diabetes appear to have certain genes that make them more susceptible, some triggering factor is required. Viral infection seems to be such a factor. The deficiency of insulin accelerates the breakdown of the body's reserve of fat, resulting in the production of organic acids called ketones. This causes a form of acidosis called ***ketosis,*** which lowers the pH of the blood and can result in death. The catabolism of stored fats and proteins also causes weight loss. As lipids are transported by the blood from storage depots to cells, lipid particles are deposited on the walls of blood vessels. The deposition leads to atherosclerosis and a multitude of cardiovascular problems including cerebrovascular insufficiency, ischemic heart disease, peripheral vascular disease, and gangrene. One of the major complications of diabetes is loss of vision due to cataracts (excessive blood sugar chemically attaches to lens proteins, causing cloudiness) or damage to blood vessels of the retina. Severe kidney problems also may result from damage to renal blood vessels.

Based on studies with the drug cyclosporine, which is used to counter the rejection of an organ following a transplant, there is strong evidence that type I diabetes is probably an autoimmune disease. Several studies have shown that cyclosporine can interrupt the destruction of beta cells.

Research is now under way to prevent or improve type I diabetes by using transplants of islet cells. In some cases, patients are given clusters of islet cells that have been treated to render them incapable of inducing rejection by the recipient. In other cases, clusters of islet cells are encapsulated so that the insulin can get out, but elements responsible for rejection cannot get in. In still another procedure, patients have been injected with fetal islet cells. Transplantation of the entire pancreas in conjunction with immunosuppressive drugs is also being tried. Before any of these procedures can be used in larger numbers of patients, considerably more research is needed.

Type II diabetes is much more common than type I, representing more than 90 percent of all cases. Type II diabetes most often occurs in people who are over 40 and overweight. Since type II diabetes usually occurs later in life, it is called ***maturity-onset diabetes.*** Clinical symptoms are mild, and the high glucose levels in the blood can usually be controlled by diet, exercise, and/or with antidiabetic drugs such as *glyburide* (DiaBeta). Many type II diabetics have a sufficient amount or even a surplus of insulin in the blood. For these individuals, diabetes arises not from a shortage of insulin but probably from defects in the molecular machinery that mediates the action of insulin on its target cells. Cells in many parts of the body, especially skeletal muscles and the liver, become less sensitive to insulin because they have fewer insulin receptors. Type II diabetes is therefore called ***non–insulin-dependent diabetes.*** However, some elderly persons with this type of diabetes are required to administer insulin to themselves daily since the cells may be less sensitive to insulin.

Hyperinsulinism (***hypoglycemia***) is much

rarer than hyposecretion and is generally the result of a malignant tumor in an islet. It can also be caused by hyperplasia of islets. The principal symptom is a decreased blood glucose level, which stimulates the secretion of epinephrine, glucagon, and hGH. As a consequence, anxiety, sweating, tremor, increased heart rate, hunger, and weakness occur. Moreover, brain cells do not have enough glucose to function efficiently. This condition leads to mental disorientation, convulsions, unconsciousness, and shock. Surgery may be required to remove the tumor or to resect the hyperplastic pancreatic tissue.

A summary of the hormones produced by the pancreas, their principal actions, control of secretion, and selected disorders is presented in Exhibit 18-7.

OVARIES AND TESTES

The female gonads, called the ***ovaries,*** are paired oval bodies located in the pelvic cavity. The ovaries produce female sex hormones called ***estrogens*** and ***progesterone (PROG).*** These hormones are responsible for the development and maintenance of female sexual characteristics. Along with the gonadotropic hormones of the pituitary gland, the sex hormones also regulate the menstrual cycle, maintain pregnancy, and prepare the mammary glands for lactation. The ovaries (and placenta) also produce a hormone called ***relaxin (RLX),*** which relaxes the symphysis pubis and helps dilate the uterine cervix toward the end of pregnancy. The ovaries also produce ***inhibin,*** a hormone that inhibits secretion of FSH (and to a lesser extent, LH) and that might be important in decreasing secretion of FSH and LH toward the end of the menstrual cycle.

The male has two oval glands, called ***testes,*** that lie in the scrotum. The testes produce ***testosterone,*** the primary male sex hormone, which stimulates the development and maintenance of male sexual characteristics. The testes also produce the hormone inhibin that, like in the female, inhibits secretion of FSH. The detailed structure of the ovaries and testes and the specific roles of gonadotropic hormones and sex hormones will be discussed in Chapter 28.

A summary of hormones produced by the ovaries and testes and their principal actions is presented in Exhibit 18-8.

PINEAL (EPIPHYSIS CEREBRI)

The endocrine gland attached to the roof of the third ventricle is known as the ***pineal*** (PĪN-ē-al) ***gland*** (because of its resemblance to a pine cone), or ***epiphysis cerebri*** (see Figure 18-1). The gland is covered by a capsule formed by the pia mater and consists of masses of ***neuroglial cells*** and parenchymal secretory cells called ***pinealocytes*** (pin-ē-AL-ō-sīts). Around the cells are scattered postganglionic sympathetic fibers. The pineal gland starts to accumulate calcium at about the time of

EXHIBIT 18-7 SUMMARY OF HORMONES PRODUCED BY THE PANCREAS, PRINCIPAL ACTIONS, CONTROL OF SECRETION, AND SELECTED DISORDERS

Hormone	Principal Actions	Control of Secretion	Selected Disorders
Glucagon	Raises blood sugar level by accelerating breakdown of glycogen into glucose in liver (glycogenolysis) and conversion of other nutrients into glucose in liver (gluconeogenesis) and releasing glucose into blood.	Decreased blood level of glucose, exercise, and largely protein meals stimulate glucagon secretion; GHIH (somatostatin) inhibits glucagon secretion.	
Insulin	Lowers blood sugar level by accelerating transport of glucose into cells, converting glucose into glycogen (glycogenesis), and decreasing glycogenolysis and gluconeogenesis; also increases lipogenesis and stimulates protein synthesis.	Increased blood level of glucose, hGH, ACTH, and gastrointestinal hormones stimulate insulin secretion; GHIH (somatostatin) inhibits insulin secretion.	An absolute deficiency of insulin or a defect in the molecular machinery that mediates the action of insulin on target cells produces diabetes mellitus. Hypersecretion of insulin results in hyperinsulinism.
Growth hormone inhibiting hormone (GHIH) or **somatostatin**	Inhibits secretion of insulin and glucagon.		

EXHIBIT 18-8 SUMMARY OF HORMONES OF THE OVARIES AND TESTES AND THEIR PRINCIPAL ACTIONS

Hormone	Principal Actions
Ovarian hormones	
Estrogens and progesterone (PROG)	Development and maintenance of female sexual characteristics. Together with gonadotropic hormones of the adenohypophysis, they also regulate the menstrual cycle, maintain pregnancy, prepare the mammary glands for lactation, and regulate oogenesis.
Relaxin (RLX)	Relaxes symphysis pubis and helps dilate uterine cervix near the end of pregnancy.
Inhibin	Inhibits secretion of FSH toward the end of the menstrual cycle.
Testicular hormones	
Testosterone	Development and maintenance of male sexual characteristics, regulation of spermatogenesis, and stimulation of descent of testes before birth.
Inhibin	Inhibits secretion of FSH to control sperm production.

puberty. Such calcium deposits are referred to as ***brain sand.*** Contrary to a once widely held belief, there is no evidence that the pineal atrophies with age and that the presence of brain sand is an indication of atrophy. In fact, the presence of brain sand may even indicate increased secretory activity.

Although many anatomical facts concerning the pineal gland have been known for years, its physiology is still somewhat obscure. One hormone secreted by the pineal gland is ***melatonin,*** which appears to inhibit reproductive activities by inhibiting gonadotropic hormones. Melatonin is produced during darkness; its formation is interrupted when light reaches the eyes. Light entering the eyes stimulates retinal neurons that transmit impulses to the pineal gland that, in turn, inhibit melatonin secretion. The release of melatonin is based on a cycle of active and nonactive periods determined by internal mechanisms and reoccurring about every 24 hours. Some evidence also exists that the pineal secretes a second hormone called ***adrenoglomerulotropin*** (a-drē′-nō-glō-mer′-yoo-lō-TRŌ-pin). This hormone may stimulate the adrenal cortex to secrete aldosterone. Other substances found in the pineal gland include norepinephrine (NE), serotonin, histamine, gonadotropin releasing hormone (GnRH), and gamma aminobutyric acid (GABA).

A summary of hormones produced by the pineal gland and their principal actions is presented in Exhibit 18-9.

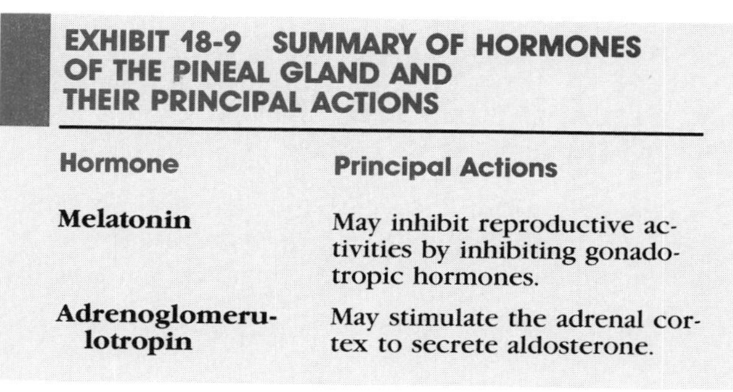

EXHIBIT 18-9 SUMMARY OF HORMONES OF THE PINEAL GLAND AND THEIR PRINCIPAL ACTIONS

Hormone	Principal Actions
Melatonin	May inhibit reproductive activities by inhibiting gonadotropic hormones.
Adrenoglomerulotropin	May stimulate the adrenal cortex to secrete aldosterone.

THYMUS

Because of its role in immunity, the details of the structure and functions of the ***thymus gland*** are discussed in Chapter 22, which deals with the lymphatic system and immunity. At this point, only its hormonal role in immunity will be discussed.

One of the types of white blood cells is referred to as lymphocytes. These, in turn, are divided into two types, called T cells and B cells, on the basis of their specific roles in immunity. Hormones produced by the thymus gland, called ***thymosin, thymic humoral factor (THF), thymic factor (TF),*** and ***thymopoietin,*** promote the proliferation and maturation of T cells. There is also some evidence that thymic hormones may retard the aging process.

A summary of hormones produced by the thymus gland and their principal actions is presented in Exhibit 18-10.

EXHIBIT 18-10 SUMMARY OF HORMONES PRODUCED BY THE THYMUS GLAND AND THEIR PRINCIPAL ACTIONS

Hormones	Principal Actions
Thymosin, thymic humoral factor (THF), thymic factor (TF), and thymopoietin	Promote proliferation and maturation of T cells.

AGING AND THE ENDOCRINE SYSTEM

The endocrine system exhibits a variety of changes, and many researchers look to this system with the hope of finding the key to the aging process. Disorders of the endocrine system are not frequent, but when they do occur, most often they are related to pathologic changes rather than age. Diabetes mellitus and thyroid disorders, especially hypothyroidism, are relatively common endocrine problems that have a significant effect on health and function. Male hormone levels decrease with age, but elderly males can still produce active sperm in normal numbers.

DEVELOPMENTAL ANATOMY OF THE ENDOCRINE SYSTEM

The development of the endocrine system is not as localized as the development of other systems. The endocrine organs develop in widely separated parts of the embryo.

The *pituitary gland* (*hypophysis*) originates from two different regions of the ***ectoderm.*** The *neurohypophysis* (posterior lobe) is derived from an outgrowth of ectoderm called the ***neurohypophyseal bud,*** located on the floor of the hypothalamus (Figure 18-24a,b). The *infundibulum,* also an outgrowth of the neurohypophyseal bud, connects the neurohypophysis to the hypothalamus. The *adenohypophysis* (anterior lobe) is derived from an outgrowth of ***ectoderm*** from the roof of the stomodeum (mouth) called the ***hypophyseal (Rathke's) pouch.*** The pouch grows toward the neurohypophyseal bud, and the pouch loses its connection with the roof of the mouth.

The *thyroid gland* develops as a midventral outgrowth of ***endoderm,*** called the ***thyroid diverticulum,*** from the floor of the pharynx at the level of the second pair

FIGURE 18-24 Development of the endocrine system. (a) Position of the neurohypophyseal bud, hypophyseal (Rathke's) pouch, and thyroid diverticulum. (b) Development of the pituitary gland.

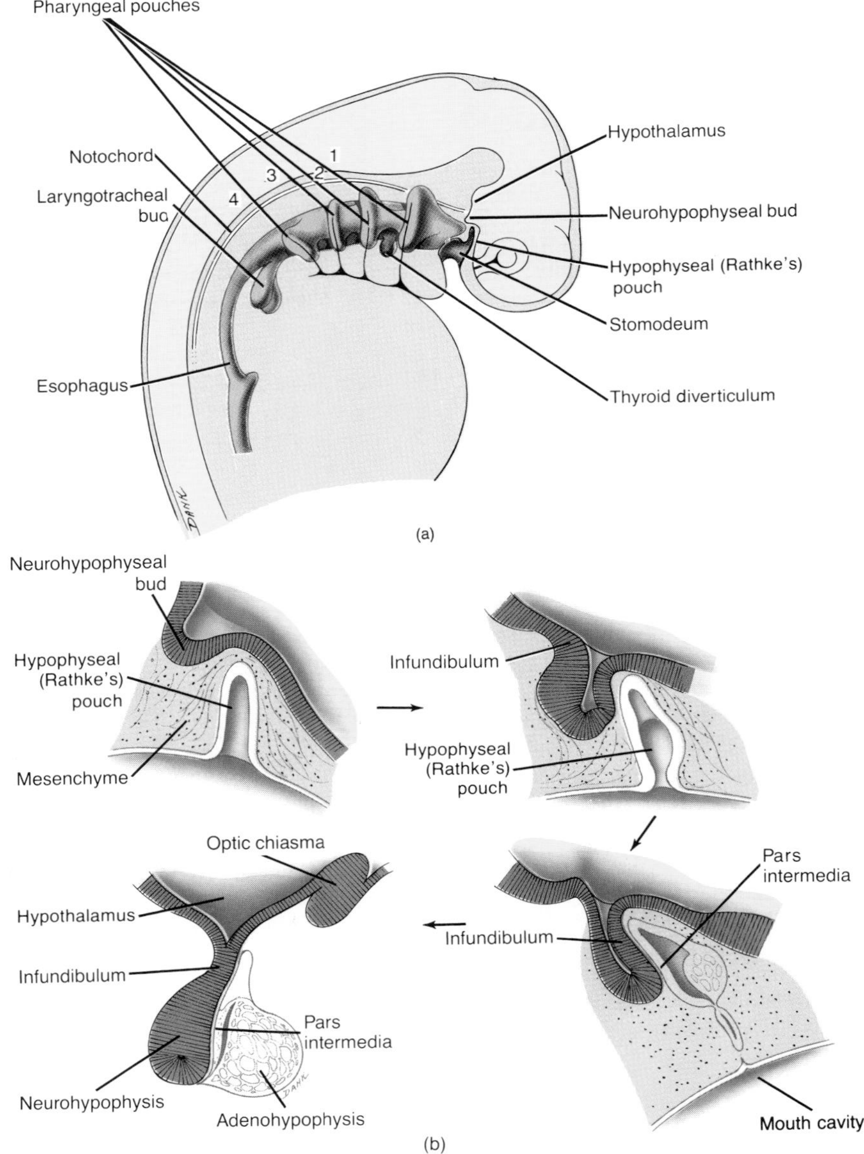

of pharyngeal pouches (Figure 18-24b). The outgrowth projects inferiorly and differentiates into the right and left lateral lobes and the isthmus of the gland.

The *parathyroid glands* develop from ***endoderm*** as outgrowths from the third and fourth ***pharyngeal pouches*** (Figure 18-24a).

The adrenal cortex and adrenal medulla have completely different embryological origins. The *adrenal cortex* is derived from intermediate ***mesoderm*** from the same region that produces the gonads (see Figure 28-27). The *adrenal medulla* is ***ectodermal*** in origin and is derived from the ***neural crest,*** which also produces sympathetic ganglia and other structures of the nervous system (see Figure 14-15c).

The *pancreas* develops from ***endoderm*** from dorsal and ventral outgrowths of the part of the ***foregut*** that later becomes the duodenum (see Figure 24-24). The two outgrowths eventually fuse to form the pancreas. The origin of the ovaries and testes is discussed in the section on the reproductive system.

The *pineal gland* arises from ***ectoderm*** of the ***diencephalon*** (see Figure 14-16b), as an outgrowth between the thalamus and colliculi.

The *thymus gland* arises from ***endoderm*** from the third ***pharyngeal pouches*** (Figure 18-24b).

OTHER ENDOCRINE TISSUES

Before leaving our discussion of hormones, it should be noted that body tissues other than those usually classified as endocrine glands also contain endocrine tissue and thus secrete hormones. The gastrointestinal tract synthesizes several hormones that regulate digestion in the stomach and small intestine. Among these hormones are ***stomach gastrin, enteric gastrin, secretin, cholecystokinin (CCK), enterocrinin,*** and ***gastric inhibitory peptide (GIP).*** The actions of these hormones and their control of secretion are summarized in Exhibit 24-4.

The placenta produces ***human chorionic gonadotropin (hCG), estrogens, progesterone (PROG), relaxin (RLX),*** and ***human chorionic somatomammotropin (hCS),*** all of which are related to pregnancy (Chapter 29).

When the kidneys (and liver, to a lesser extent) become hypoxic (subject to below-normal levels of oxygen), it is believed that they release an enzyme called ***renal erythropoietic factor.*** It is secreted into the blood, where it acts on a plasma protein to convert it to a hormone called ***erythropoietin*** (ē-rith'-rō-POY-ē-tin), which stimulates red blood cell production (Chapter 19). The kidneys also help bring about the activation of the hormone ***vitamin D.***

The skin produces vitamin D from inactive precursor molecules in the presence of sunlight (Chapter 5).

Cardiac muscle fibers (cells) of the atria (upper chambers) of the heart produce a peptide hormone called ***atrial natriuretic factor (ANF)*** that is secreted when they are stretched, as might occur in response to increased blood volume, a factor that increases blood pressure. ANF seems to act as a calcium-channel blocker and may also be secreted by the brain. The general effect of ANF is to modify the renin–angiotensin pathway by somehow inhibiting the secretion of renin by the juxtaglomerular cells of the kidneys and thus the secretion of aldosterone by the adrenal cortex. Because of this inhibition, there is an increase in sodium and water excretion and blood vessel dilation; the result is a decrease in blood pressure. ANF also acts directly on the kidneys to bring about increases in sodium and water excretion and on the hypothalamus to inhibit the secretion of ADH. Again, these conditions lead to a decrease in blood pressure. The overall function of ANF is therefore to lower blood pressure.

GROWTH FACTORS

In addition to some of the well-defined hormones that stimulate cell growth and division (somatomedins, thymosin, insulin, human growth hormone, prolactin, and erythropoietin), a number of recently discovered hormones called ***growth factors*** also assume a similar role. A summary of these growth factors is presented in Exhibit 18-11.

STRESS AND THE GENERAL ADAPTATION SYNDROME (GAS)

Homeostatic mechanisms are geared toward counteracting the everyday stresses of living. If they are successful, the internal environment maintains normal physiological limits of chemistry, temperature, and pressure. If a stress is extreme, unusual, or long-lasting, however, the normal mechanisms may not be sufficient. In this case, the stress triggers a wide-ranging set of bodily changes called the ***general adaptation syndrome (GAS).*** Hans Selye, a world authority on stress, was responsible for the concept of the general adaptation syndrome. Unlike the homeostatic mechanisms, the general adaptation syndrome does not maintain a normal internal environment. In fact, it does just the opposite. For instance, blood pressure and blood sugar level are raised above normal. The purpose of these changes in the internal environment is to gear up the body to meet an emergency.

It should be pointed out that it is impossible to remove all stress from our everyday lives. In fact, some stress prepares us to meet certain challenges. Thus, some stress is productive, whereas other stress is harmful. In this regard, the term ***eustress*** is used to distinguish productive stress, and the term ***distress*** is used to refer to harmful stress. One recent finding suggests that distress lowers resistance to infection by temporarily inhibiting certain components of the immune system (described shortly).

EXHIBIT 18-11 SUMMARY OF GROWTH FACTORS

Growth Factor	Comment
Epidermal growth factor (EGF)	Produced in submaxillary (salivary) glands; stimulates proliferation of epithelial cells and fibroblasts; suppresses some cancer cells and secretion of gastric juice by the stomach.
Platelet-derived growth factor (PDGF)	Produced in blood platelets; stimulates proliferation of several cell types, including glial cells, smooth muscle fibers, and fibroblasts; appears to have a role in wound healing; may contribute to the development of atherosclerosis.
Fibroblast growth factor (FGF)	Found in pituitary gland and brain; stimulates proliferation of many cells derived from embryonic mesoderm (fibroblasts, adrenocortical cells, smooth muscle fibers, chondrocytes, and endothelial cells); also stimulates cell migration and growth and production of the adhesion protein, fibronectin.
Nerve growth factor (NGF)	Produced in submaxillary (salivary) glands and parts of brain (hippocampus); stimulates the growth of ganglia in embryonic life, maintains sympathetic nervous system; stimulates hypertrophy and differentiation of mature nerve cells.
Tumor angiogenesis factors (TAFs)	Produced by normal and tumor cells; stimulates growth of new capillaries, organ regeneration, and wound healing.
Insulinlike growth factor (IGF)	A somatomedin produced in the liver; stimulates growth in chondrocytes, fibroblasts, and other cells.
Lymphokines (Cytokines)	Produced by activated lymphocytes (white blood cells) and other cells; the various lymphokines and their roles in immunity are discussed in Chapter 22.

Stressors

The hypothalamus can be called the body's watchdog. It has sensors that detect changes in the chemistry, temperature, and pressure of the blood. It is informed of emotions through tracts that connect it with the emotional centers of the cerebral cortex. When the hypothalamus senses stress, it initiates a chain of reactions that produce the general adaptation syndrome. The stimuli that produce the syndrome are called ***stressors.*** A stressor may be almost any disturbance—heat or cold, environmental poisons, poisons given off by bacteria during a raging infection, heavy bleeding from a wound or surgery, or a strong emotional reaction. Stressors differ among different people and even in the same person at different times.

When a stressor appears, it stimulates the hypothalamus to initiate the syndrome through two pathways. The first pathway is stimulation of the sympathetic division of the autonomic nervous system and adrenal medulla. This stimulation produces an immediate set of responses called the alarm reaction. The second pathway, called the resistance reaction, involves the anterior pituitary gland and adrenal cortex. The resistance reaction is slower to start, but its effects last longer.

Alarm Reaction

The ***alarm reaction,*** or ***fight-or-flight response,*** is the body's initial reaction to a stressor (Figure 18-25a). It is actually a complex of reactions initiated by the hypothalamic stimulation of the sympathetic division of the autonomic nervous system and the adrenal medulla. The responses of the visceral effectors are immediate and short-lived. They are designed to counteract a danger by mobilizing the body's resources for immediate physical activity. In essence, the alarm reaction brings tremendous amounts of glucose and oxygen to the organs that are most active in warding off danger. These are the brain, which must become highly alert; the skeletal muscles, which may have to fight off an attacker; and the heart, which must work furiously to pump enough materials to the brain and muscles. The hyperglycemia associated with sympathetic activity is produced by liver glycogenolysis, stimulated by epinephrine and norepinephrine (NE) from the adrenal medulla, and liver gluconeogenesis, stimulated by fat mobilization and glucose-sparing by hGH and protein mobilization by glucocorticoids.

Among the stress responses that characterize the alarm stage are the following:

1. The heart rate and the strength of cardiac muscle contraction increase. This response circulates substances in the blood very quickly to areas where they are needed to combat the stress.

2. Blood vessels supplying the skin and viscera, except the heart and lungs, undergo constriction. At the same time, blood vessels supplying the skeletal muscles and brain undergo dilation. These responses route more blood to organs active in the stress responses while decreasing blood supply to organs that do not assume an immediate, active role.

FIGURE 18-25 **Responses to stressors during the general adaptation syndrome (GAS). (a) During the alarm stage. (b) During the resistance stage. Colored arrows indicate immediate reactions. Black arrows indicate long-term reactions.**

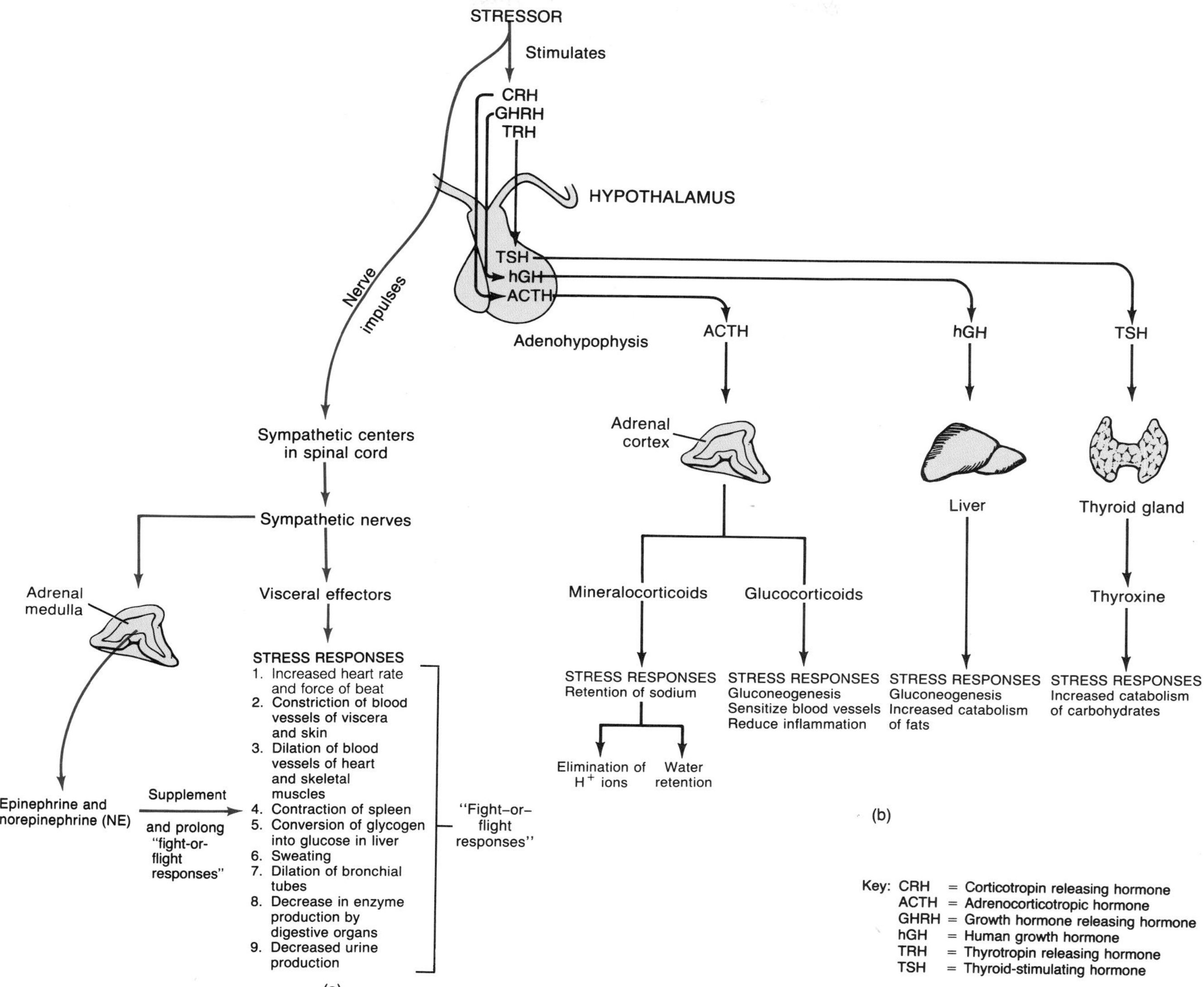

3. The spleen contracts and discharges its stored blood into the general circulation to provide additional blood. Moreover, red blood cell production is accelerated, and the ability of the blood to clot is increased. These preparations are made to combat bleeding.

4. The liver transforms large amounts of stored glycogen into glucose and releases it into the bloodstream. The glucose is broken down by the active cells to provide the energy needed to meet the stressor.

5. The rate of breathing increases, and the respiratory passageways widen to accommodate more air. This response enables the body to acquire more oxygen, which is needed in the decomposition reactions of catabolism. It also allows the body to eliminate more carbon dioxide, which is produced as a side product during catabolism.

6. Production of saliva, stomach enzymes, and intestinal enzymes decreases. This reaction takes place since digestive activity is not essential for counteracting the stress.

7. Sympathetic impulses to the adrenal medulla increase its secretion of epinephrine and norepinephrine (NE). These hormones supplement and prolong many sympathetic responses—increasing heart rate and strength, increasing blood pressure, constricting blood vessels, accelerating the rate of breathing, widening respiratory passageways, increasing the rate of catabolism, decreasing the rate of digestion, and increasing blood sugar level.

If you group the stress responses of the alarm stage by function, you will note that they are designed to increase circulation rapidly, promote catabolism for energy production, and decrease nonessential activities. If the stress is great enough, the body mechanisms may not be able to cope and death can result. Note that during the alarm reaction digestive, urinary, and reproductive activities are inhibited.

Resistance Reaction

The second stage in the stress response is the ***resistance reaction*** (Figure 18-25b). Unlike the short-lived alarm reaction that is initiated by nerve impulses from the hypothalamus, the resistance reaction is initiated by regulating hormones secreted by the hypothalamus and is a long-term reaction. The regulating hormones are corticotropin releasing hormone (CRH), growth hormone releasing hormone (GHRH), and thyrotropin releasing hormone (TRH).

CRH stimulates the adenohypophysis to increase its secretion of ACTH. ACTH stimulates the adrenal cortex to secrete more of its hormones. The adrenal cortex is also indirectly stimulated by the alarm reaction. During the alarm reaction, kidney activity is cut back because of decreased blood circulation to the kidneys, since it is not essential for meeting sudden danger.

The mineralocorticoids secreted by the adrenal cortex bring about the conservation of sodium ions by the body. A secondary effect of sodium conservation is the elimination of hydrogen ions which tend to build up in high concentrations as a result of increased catabolism and tend to make the blood more acidic. Thus, during stress, a lowering of body pH is prevented. Sodium retention also leads to water retention, thus maintaining the high blood pressure that is typical of the alarm reaction. It also helps make up for fluid lost through severe bleeding.

The glucocorticoids, which are produced in high concentrations during stress, bring about the following reactions:

1. The glucocorticoids accelerate protein catabolism and the conversion of amino acids into glucose so that the body has a large supply of energy long after the immediate stores of glucose have been used up. The glucocorticoids also stimulate the removal of proteins from cell structures and stimulate the liver to break them down into amino acids. The amino acids can then be rebuilt into enzymes that are needed to catalyze the increased chemical activities of the cells or be converted to glucose.

2. The glucocorticoids make blood vessels more sensitive to stimuli that bring about their constriction. This response counteracts a drop in blood pressure caused by bleeding.

3. The glucocorticoids also inhibit the production of fibroblasts, which develop into connective tissue cells. Injured fibroblasts release chemicals that play a role in stimulating the inflammatory response. Thus, the glucocorticoids reduce inflammation and prevent it from becoming disruptive rather than protective. Unfortunately, through their effect on fibroblasts, the glucocorticoids also discourage connective tissue formation. Wound healing is therefore slow during a prolonged resistance stage.

Two other regulating hormones are secreted by the hypothalamus in response to a stressor: TRH and GHRH. TRH causes the adenohypophysis to secrete thyroid-stimulating hormone (TSH); GHRH causes it to secrete human growth hormone (hGH). TSH stimulates the thyroid to secrete thyroxine and triiodothyronine which decrease the catabolism of carbohydrates. hGH stimulates the catabolism of fats and the conversion of glycogen to glucose. The combined actions of TSH and hGH increase catabolism and thereby supply additional energy for the body.

The resistance stage of the general adaptation syndrome allows the body to continue fighting a stressor long after the effects of the alarm reaction have dissipated. It increases the rate at which life processes occur. It also provides the energy, functional proteins, and circulatory changes required for meeting emotional crises, performing strenuous tasks, fighting infection, or resisting the threat of bleeding to death. During the resistance stage, blood chemistry returns to nearly normal. The cells use glucose at the same rate it enters the bloodstream. Thus, blood sugar level returns to normal. Blood pH is brought under control by the kidneys as they excrete more hydrogen ions. However, blood pressure remains abnormally high because the retention of water increases the volume of blood.

All of us are confronted by stressors from time to time, and we have all experienced the resistance stage. Generally, this stage is successful in seeing us through a stressful situation, and our bodies then return to normal. Occasionally, the resistance stage fails to combat the stressor, however. In this case, the general adaptation syndrome moves into the stage of exhaustion.

Exhaustion

A major cause of exhaustion is loss of potassium ions. When the mineralocorticoids stimulate the kidneys to retain sodium ions, potassium and hydrogen ions are traded off for sodium ions and secreted in the fluid that will become urine. As the chief positive ion in cells, potassium is partly responsible for controlling the water concentration of the cytoplasm. As the cells lose more and more potassium, they function less and less effectively. Finally, they start to die. This condition is called the ***stage of exhaustion.*** Unless it is rapidly reversed, vital organs cease functioning and the person dies. Another cause of exhaustion is depletion of the adrenal glucocorticoids. In this case, blood glucose level suddenly falls, and the cells do not receive enough nutrients. A final cause of exhaustion is weakening organs. A long-term or strong resistance reaction puts heavy demands on the body, particularly on the heart, blood vessels, and adrenal cortex.

They may not be up to handling the demands, or they may suddenly fail under the strain. In this respect, ability to handle stressors is determined to a large degree by general health.

Stress and Disease

Scientists have long known about the effects of environmental stress on laboratory animals. Although the exact role of stress in human diseases is not known, it is becoming quite clear that stress can lead to certain diseases. Among the conditions that are related to stress are gastritis, ulcerative colitis, irritable bowel syndrome, peptic ulcers, hypertension, asthma, rheumatoid arthritis (RA), migraine headaches, anxiety, and depression. It has also been shown that individuals under stress are at a greater risk of developing chronic disease or dying prematurely.

As noted earlier, stress is believed to increase susceptibility to infection by temporarily inhibiting certain components of the immune system. A very important link between stress and immunity has recently been discovered. It is interleukin-1 (IL-1), a lymphokine secreted by macrophages of the immune system (Chapter 22). In response to infection, inflammation, and other stressors, IL-1 stimulates production of immune substances by the liver, increases the number of circulating neutrophils (white blood cells that are phagocytes), activates cells that participate in immunity, and induces fever. All these responses result in a powerful immune response. However, in order to control the response, it is believed that IL-1 also stimulates the secretion of ACTH. ACTH, in turn, stimulates the production of glucocorticoids (cortisol or corticosterone), hormones that not only provide resistance to stress and inflammation, but also suppress further production of IL-1. Such a negative feedback system would keep the immune response in check once it has accomplished its goal. Since glucocorticoids, such as cortisol, do suppress certain aspects of the immune response, they are used as immunosuppressants following organ transplantation. Thus, the immune system turns on the stress response.

It is not known whether IL-1 acts directly on the anterior pituitary gland or indirectly through the hypothalamus to stimulate the release of corticotropin releasing hormone (CRH), which in turn stimulates secretion of ACTH.

MEDICAL TERMINOLOGY ASSOCIATED WITH THE ENDOCRINE SYSTEM

Hyperplasia (hī'-per-PLĀ-zē-a; *hyper* = over; *plas* = grow) Increase in the number of cells owing to an increase in the frequency of cell division.

Hypertrophy (hī-PER-trō-fē) Increase in the size of cells without cell division.

Hypoplasia (hi'-pō-PLĀ-zē-a; *hypo* = under) Defective development of tissue.

Neuroblastoma (noo'-rō-blas-TŌ-ma; *neuro* = nerve) Malignant tumor arising from the adrenal medulla associated with metastases to bones.

Thyroid (THĪ-roid) **storm** An aggravation of all symptoms of hyperthyroidism characterized by unregulated hypermetabolism with fever and rapid heart rate; results from trauma, surgery, and unusual emotional stress or labor.

STUDY OUTLINE

Endocrine Glands (p. 496)

1. Both the endocrine and nervous systems assume a role in maintaining homeostasis.
2. Hormones help regulate the internal environment, respond to stress, help regulate growth and development, and contribute to reproductive processes.
3. Exocrine glands (sudoriferous, sebaceous, digestive) secrete their products through ducts into body cavities or onto body surfaces.
4. Endocrine glands secrete hormones into the blood.

Chemistry of Hormones (p. 497)

1. Chemically, hormones are classified as amines, proteins and peptides, and steroids.
2. Amines are modified from the amino acid tyrosine (e.g., thyroid hormones); protein and peptide hormones consist of chains of amino acids (e.g., oxytocin); and steroid hormones are derived from cholesterol (e.g., aldosterone).

Physiology of Hormonal Action (p. 499)

1. The amount of hormone released is determined by the body's need for the hormone.
2. Cells that respond to the effects of hormones are called target cells.
3. Specific receptors are found in the plasma membrane, cytoplasm, and nucleus of target cells.
4. The combination of hormone and receptor activates a chain of events in a target cell in which the physiological effects of the hormone are expressed.
5. One mechanism of hormonal action involves interaction of hormone with plasma membrane receptors; the second messenger may be cyclic AMP or calcium ions.

6. Another mechanism of hormonal action involves interaction of a hormone with intracellular receptors.
7. Three hormonal interactions are permissive, synergistic, and antagonistic.
8. Prostaglandins (PGs) are membrane-associated lipids that can increase or decrease cyclic AMP formation and thus modulate hormone responses that use cyclic AMP.

Control of Hormonal Secretions: Feedback Control (p. 503)

1. A negative feedback control mechanism prevents overproduction or underproduction of a hormone.
2. Hormone secretions are controlled by levels of circulating hormone itself, nerve impulses, and regulating hormones (or factors).

Pituitary (Hypophysis) (p. 503)

1. The pituitary gland is located in the sella turcica and is differentiated into the adenohypophysis (the anterior lobe and glandular portion), the neurohypophysis (the posterior lobe and nervous portion), and pars intermedia (avascular zone between lobes).
2. Hormones of the adenohypophysis are released or inhibited by regulating hormones (or factors) produced by the hypothalamus.
3. The blood supply to the adenohypophysis is from the superior hypophyseal arteries. It transports hypothalamic regulating factors.
4. Histologically, the adenohypophysis consists of somatotroph cells that produce human growth hormone (hGH); lactotroph cells that produce prolactin (PRL); thyrotroph cells that secrete thyroid-stimulating hormone (TSH); gonadotroph cells that synthesize follicle-stimulating hormone (FSH) and luteinizing hormone (LH); and corticolipotroph cells that secrete adrenocorticotropin hormone (ACTH) and melanocyte-stimulating hormone (MSH).
5. hGH stimulates body growth through somatomedins and is controlled by GHIH (growth hormone inhibiting hormone, or somatostatin) and GHRH (growth hormone releasing hormone, or somatocrinin).
6. Disorders associated with improper levels of hGH are pituitary dwarfism, giantism, and acromegaly.
7. TSH regulates thyroid gland activities and is controlled by TRH (thyrotropin releasing hormone).
8. ACTH regulates the activities of the adrenal cortex and is controlled by CRH (corticotropin releasing hormone).
9. FSH regulates the activities of the ovaries and testes and is controlled by GnRH (gonadotropin releasing hormone).
10. LH regulates the activities of the ovaries and testes and is controlled by GnRH.
11. PRL helps initiate milk secretion and is controlled by PIF (prolactin inhibiting factor) and PRF (prolactin releasing factor).
12. MSH increases skin pigmentation and is controlled by MRF (melanocyte-stimulating hormone releasing factor) and MIF (melanocyte-stimulating hormone inhibiting factor).
13. The neural connection between the hypothalamus and neurohypophysis is via the hypothalamic–hypophyseal tract.
14. Hormones made by the hypothalamus and stored in the neurohypophysis are oxytocin (OT) (stimulates contraction of uterus and ejection of milk) and antidiuretic hormone (ADH) (stimulates water reabsorption by the kidneys and arteriole constriction).
15. OT secretion is controlled by uterine distension and nursing; ADH is controlled primarily by water concentration of the blood.
16. A disorder associated with dysfunction of the neurohypophysis is diabetes insipidus.

Thyroid (p. 512)

1. The thyroid gland is located below the larynx.
2. Histologically, the thyroid consists of thyroid follicles composed of follicular cells, which secrete the thyroid hormones thyroxine (T_4) and triiodothyronine (T_3), and parafollicular cells, which secrete calcitonin (CT).
3. Thyroid hormones are synthesized from iodine and tyrosine within thyroglobulin (TGB) and carried in the blood with plasma proteins, mostly thyroxine-binding globulin (TBG).
4. Thyroid hormones regulate the rate of metabolism, growth and development, and the reactivity of the nervous system. Secretion is controlled by TRH.
5. Cretinism, myxedema, exophthalmic goiter, and simple goiter are disorders associated with the thyroid gland.
6. Calcitonin (CT) lowers the blood level of calcium. Secretion is controlled by calcium level in blood.

Parathyroids (p. 517)

1. The parathyroids are embedded on the posterior surfaces of the lateral lobes of the thyroid.
2. The parathyroids consist of principal and oxyphil cells.
3. Parathyroid hormone (PTH) regulates the homeostasis of calcium and phosphate by increasing blood calcium level and decreasing blood phosphate level. Secretion is controlled by calcium level in blood.
4. Tetany and osteitis fibrosa cystica are disorders associated with the parathyroid glands.

Adrenals (Suprarenals) (p. 520)

1. The adrenal glands are located superior to the kidneys. They consist of an outer cortex and inner medulla.
2. Histologically, the cortex is divided into a zona glomerulosa, zona fasciculata, and zona reticularis; the medulla consists of chromaffin cells.
3. Cortical secretions are mineralocorticoids, glucocorticoids, and gonadocorticoids.
4. Mineralocorticoids (e.g., aldosterone) increase sodium and water reabsorption and decrease potassium reabsorption. Secretion is controlled by the renin–angiotensin pathway and blood level of potassium.
5. A dysfunction related to aldosterone secretion is aldosteronism.
6. Glucocorticoids (e.g., cortisol) promote normal organic metabolism, help resist stress, and serve as antiinflammatories. Secretion is controlled by CRH.
7. Disorders associated with glucocorticoid secretion are Addison's disease and Cushing's syndrome.
8. Gonadocorticoids (estrogens and androgens) secreted by the adrenal cortex usually have minimal effects. Excessive production results in adrenogenital syndrome.
9. Medullary secretions are epinephrine and norepinephrine

(NE), which produce effects similar to sympathetic responses. They are released under stress.
10. Tumors of medullary chromaffin cells are called pheochromocytomas.

Pancreas (p. 527)

1. The pancreas is posterior and slightly inferior to the stomach.
2. Histologically, it consists of pancreatic islets or islets of Langerhans (endocrine cells) and acini (enzyme-producing cells). Three types of cells in the endocrine portion are alpha cells, beta cells, and delta cells.
3. Alpha cells secrete glucagon, beta cells secrete insulin, and delta cells secrete growth hormone inhibiting hormone (GHIH), or somatostatin.
4. Glucagon increases blood sugar level. Secretion is controlled by glucose level in the blood.
5. Insulin decreases blood sugar level. Secretion is controlled by glucose level in the blood.
6. Disorders associated with insulin production are diabetes mellitus and hyperinsulinism.

Ovaries and Testes (p. 530)

1. Ovaries are located in the pelvic cavity and produce sex hormones related to development and maintenance of female sexual characteristics, menstrual cycle, pregnancy, lactation, and normal reproductive functions.
2. Testes lie inside the scrotum and produce sex hormones related to the development and maintenance of male sexual characteristics and normal reproductive functions.

Pineal (Epiphysis Cerebri) (p. 530)

1. The pineal is attached to the roof of the third ventricle.
2. Histologically, it consists of secretory parenchymal cells called pinealocytes, neuroglial cells, and scattered postganglionic sympathetic fibers. Calcium-containing deposits are referred to as brain sand.
3. The pineal secretes melatonin (possibly regulates reproductive activities by inhibiting gonadotropic hormones) and adrenoglomerulotropin (may stimulate adrenal cortex to secrete aldosterone).

Thymus (p. 531)

1. The thymus gland secretes several hormones related to immunity.
2. Thymosin, thymic humoral factor (THF), thymic factor (TF), and thymopoietin promote the maturation of T cells.

Aging and the Endocrine System (p. 531)

1. Most endocrine disorders are related to pathologies rather than age.
2. Diabetes mellitus and thyroid disorders are common endocrine disorders.

Developmental Anatomy of the Endocrine System (p.532)

1. The development of the endocrine system is not as localized as other systems.
2. The pituitary, adrenal medulla, and pineal gland develop from ectoderm; the adrenal cortex develops from mesoderm; and the thyroid, parathyroids, pancreas, and thymus develop from endoderm.

Other Endocrine Tissues (p. 533)

1. The gastrointestinal (GI) tract synthesizes stomach and intestinal gastrin, secretin, cholecystokinin (CCK), enterocrinin, and gastric inhibitory peptide (GIP).
2. The placenta produces human chorionic gonadotropin (hCG), estrogens, progesterone (PROG), relaxin (RLX), and human chorionic somatomammotropin (hCS).
3. The kidneys release an enzyme that causes the production of erythropoietin.
4. The skin synthesizes vitamin D, and the kidneys activate it.
5. The atria of the heart produces atrial natriuretic factor (ANF).

Growth Factors (p. 533)

1. Growth factors are hormones that stimulate cell growth and division.
2. Examples include epidermal growth factor (EGF), fibroblast growth factor (FGF), nerve growth factor (NGF), lymphokines, and tumor necrosis factor (TNF).

Stress and the General Adaptation Syndrome (GAS) (p. 533)

1. If the stress is extreme or unusual, it triggers a wide-ranging set of bodily changes called the general adaptation syndrome (GAS).
2. Unlike the homeostatic mechanisms, this syndrome does not maintain a constant internal environment.

Stressors (p. 534)

1. The stimuli that produce the general adaptation syndrome are called stressors.
2. Stressors include surgical operations, poisons, infections, fever, and strong emotional responses.

Alarm Reaction (p. 534)

1. The alarm reaction is initiated by nerve impulses from the hypothalamus to the sympathetic division of the autonomic nervous system and adrenal medulla.
2. Responses are the immediate and short-lived fight-or-flight responses that increase circulation, promote catabolism for energy production, and decrease nonessential activities.

Resistance Reaction (p. 536)

1. The resistance reaction is initiated by regulating hormones secreted by the hypothalamus.
2. The regulating hormones are CRH, GHRH, and TRH.
3. CRH stimulates the adenohypophysis to increase its secretion of ACTH, which in turn stimulates the adrenal cortex to secrete hormones.
4. Resistance reactions are long term and accelerate catabolism to provide energy to counteract stress.
5. Glucocorticoids are produced in high concentrations during stress. They create many distinct physiological effects.

Exhaustion (p. 536)

1. The stage of exhaustion results from dramatic changes during alarm and resistance reactions.

2. Exhaustion is caused mainly by loss of potassium, depletion of adrenal glucocorticoids, and weakened organs, and if stress is too great, it may lead to death.

Stress and Disease (p. 537)

1. It appears that stress can lead to certain diseases.
2. Among stress-related conditions are gastritis, ulcerative colitis, irritable bowel syndrome, asthma, anxiety, depression, peptic ulcers, hypertension, and migraine headaches.
3. A very important link between stress and immunity is interleukin-1 (IL-1) produced by macrophages; it stimulates secretion of ACTH.

REVIEW QUESTIONS

1. Contrast endocrine and nervous system control of homeostasis. (p. 496)
2. Distinguish between an endocrine gland and an exocrine gland. What are the four principal actions of hormones? (p. 496)
3. What is a hormone? Distinguish between tropic and gonadotropic hormones. (p. 496)
4. Describe the chemical classification of hormones. Give an example of each. (p. 496)
5. Explain how receptors are related to hormones and describe three types of hormonal interactions. (p. 499)
6. Describe the mechanism of hormonal action involving (a) interaction with plasma membrane receptors and (b) interaction with intracellular receptors. (p. 500)
7. Define a prostaglandin (PG). Give several functions of prostaglandins. How are prostaglandins related to hormonal action? (p. 502)
8. How are negative feedback systems related to hormonal control? Discuss three models of operation. (p. 503)
9. In what respect is the pituitary gland actually two glands? (p. 503)
10. Describe the histology of the adenohypophysis. Why does the anterior lobe of the gland have such an abundant blood supply? (p. 504)
11. What hormones are produced by the adenohypophysis? What are their functions? How are they controlled? (p. 504)
12. Relate the importance of regulating hormones (or factors) to secretions of the adenohypophysis. (p. 504)
13. Describe the clinical symptoms of pituitary dwarfism, gigantism, and acromegaly. (p. 506)
14. Discuss the histology of the neurohypophysis and the function and regulation of its hormones. Describe the structure and importance of the hypothalamic–hypophyseal tract. (p. 510)
15. What are the clinical symptoms of diabetes insipidus? (p. 512)
16. Describe the location and histology of the thyroid gland. (p. 512)
17. How are the thyroid hormones made, stored, and secreted? (p. 513)
18. Discuss the physiological effects of the thyroid hormones. How are these hormones regulated? (p. 515)
19. Discuss the clinical symptoms of cretinism, myxedema, exophthalmic goiter, and simple goiter. (p. 516)
20. Describe the function and control of calcitonin (CT). (p. 517)
21. Where are the parathyroids located? What is their histology? (p. 517)
22. What are the functions of parathyroid hormone (PTH)? (p. 517)
23. Discuss the clinical symptoms of tetany and osteitis fibrosa cystica. (p. 520)
24. Compare the adrenal cortex and adrenal medulla with regard to location and histology. (p. 520)
25. Describe the hormones produced by the adrenal cortex in terms of type, normal function, and control. (p. 520)
26. Describe the clinical symptoms of aldosteronism, Addison's disease, Cushing's syndrome, and adrenogenital syndrome. (p. 523, 525)
27. What relationship does the adrenal medulla have to the autonomic nervous system? What is the action of adrenal medullary hormones? (p. 525)
28. What is a pheochromocytoma? (p. 526)
29. Describe the location of the pancreas and the histology of the pancreatic islets (islets of Langerhans). (p. 527)
30. What are the actions of glucagon and insulin? How are the hormones controlled? (p. 527)
31. Describe the clinical symptoms of diabetes mellitus and hyperinsulinism. Distinguish the two types of diabetes mellitus. (p. 529)
32. Why are the ovaries and testes considered to be endocrine glands? (p. 530)
33. Where is the pineal gland located? What are its assumed functions? (p. 530)
34. How are hormones of the thymus gland related to immunity? (p. 531)
35. Describe the effects of aging on the endocrine system. (p. 531)
36. Describe the development of the endocrine system. (p. 532)
37. List the hormones secreted by the gastrointestinal tract, placenta, kidneys, skin, and heart. (p. 533)
38. What are growth factors? List several and explain their actions. (p. 533)
39. Define the general adaptation syndrome (GAS). What is a stressor? (p. 533)
40. How do homeostatic responses differ from stress responses? (p. 533)
41. Outline the reactions of the body during the alarm stage, resistance stage, and stage of exhaustion when placed under stress. What is the central role of the hypothalamus during stress? (p. 534)
42. Explain how stress and immunity are related. (p. 537)
43. Why are thyroid function tests performed? (p. 514)
44. Refer to the glossary of medical terminology associated with the endocrine system. Be sure that you can define each term. (p. 537)

SELECTED READINGS

Clark, M. "The Power of Hormones," *Newsweek on Health,* Spring 1987.

Crapo, L. *Hormones: Messengers of Life.* New York: W. H. Freeman, 1985.

de los Santos, E. T., and E. L. Mazzaferri. "Thyroid Function Tests," *Postgraduate Medicine,* April 1989.

Fellman, B. "A Clockwork Gland," *Science 85,* May 1985.

Fry, W. F. "Benefits of Stress," *Healthline,* January 1984.

Goldfinger, S. E. (ed). "Diabetes: The Long Run," *Harvard Medical School Health Letter,* April 1985.

———. "Insulin and the Diabetic," *Harvard Medical School Health Letter,* March 1985.

———. "The Diseases Called Sugar Diabetes," (Part 1), *Harvard Medical School Health Letter,* February 1985.

Goodman, H. M. *Basic Medical Endocrinology.* New York: Raven Press. 1980

Griffin, J. E., and S. R. Ojeda. *Textbook of Endocrine Physiology.* New York: Oxford University Press. 1988.

Miller, J. "Eye to (Third) Eye," *Science News,* 9 November 1985.

Molitch, M. E. "Diabetes Mellitus," *Postgraduate Medicine,* March 1989.

Muldoon, T. G., and A. C. Evans. "Hormones and Their Receptors," *Archives of Internal Medicine,* April 1988.

Orci, L., J. D. Vassalli, and A. Perrelet. "The Insulin Factory," *Scientific American,* September 1988.

Seyle, H. *Stress in Health and Disease.* London: Butterworths, 1976.

Williams, R. *Textbook of Endocrinology.* 7th ed. Philadelphia: Saunders, 1985.

UNIT 4

MAINTENANCE OF THE HUMAN BODY

This unit explains how the body maintains homeostasis on a day-to-day basis. In these chapters, you will be studying the interrelations among the cardiovascular, lymphatic, respiratory, digestive, and urinary systems. You will also learn about metabolism, fluid and electrolyte balance, and acid–base dynamics.

Chapter 19

The Cardiovascular System: The Blood

Chapter Contents at a Glance

Student Objectives

1. Contrast the general roles of blood, lymph, and interstitial fluid in maintaining homeostasis.
2. Define the principal physical characteristics of blood and the functions of its various components.
3. Compare the origins, histology, and functions of the formed elements in blood.
4. Explain the significance of a differential white blood cell count.
5. List the components of plasma and explain their importance.
6. Identify the stages involved in blood clotting and explain the various factors that promote and inhibit blood clotting.
7. Explain ABO and Rh blood grouping.
8. Contrast the causes of nutritional, pernicious, hemorrhagic, hemolytic, aplastic, and sickle-cell anemia (SCA).
9. Identify the clinical symptoms of infectious mononucleosis (IM), chronic fatigue syndrome, and leukemia.
10. Define medical terminology associated with blood.

The more differentiated cells become, the less capable they are of carrying on an independent existence. For instance, specialized cells are less able to protect themselves from extreme temperatures, toxic chemicals, and changes in pH. They cannot seek food or devour whole bits of food. And, if they are firmly implanted in a tissue, they cannot move away from their own wastes. The substance that bathes cells and carries out these vital functions for them is called ***interstitial fluid*** (also known as ***intercellular*** or ***tissue fluid***). You may recall that fluids outside of cells such as interstitial fluid, blood, plasma, and other fluids are together referred to as extracellular fluid (see Figure 1-14).

The interstitial fluid, in turn, is serviced by blood and lymph. The blood picks up oxygen from the lungs, nutrients from the gastrointestinal tract, and hormones from endocrine glands. It transports these substances to all the tissues, where they diffuse from capillaries (microscopic blood vessels) into interstitial fluid. In the interstitial fluid, the substances are passed on to the cells and exchanged for wastes.

Since the blood services all the tissues of the body, it can be an important medium for the transport of disease-causing organisms. To protect itself from the spread of disease, the body has a lymphatic system—a collection of lymphatic tissue and lymph vessels that contain a fluid called lymph. Lymph picks up materials, including wastes, from interstitial fluid, cleanses them of bacteria, and returns them to the blood. The blood then carries the wastes to the lungs, kidneys, and sweat glands, where they are eliminated from the body. The blood also takes wastes to the liver, where they are detoxified.

Blood inside blood vessels, interstitial fluid around body cells, and lymph inside lymph vessels constitute the ***internal environment*** of the human organism. Because the body cells are too specialized to adjust to more than very limited changes in their environment, the internal environment must be kept within normal physiological limits. This condition we have called homeostasis. In preceding chapters, we have discussed how the internal environment is kept in homeostasis. Now we will look at that environment itself.

The blood, heart, and blood vessels constitute the ***cardiovascular system.*** The lymph, lymph vessels, and structures and organs containing lymphatic tissue (large numbers of white blood cells called lymphocytes) make up the ***lymphatic system*** (Chapter 22). We will first take a look at the substance known as blood. The branch of science concerned with the study of blood, blood-forming tissues, and the disorders associated with them is called ***hematology*** (hēm-a-TOL-ō-jē; *hem* = blood; *logos* = study of).

The developmental anatomy of blood and blood vessels is considered in Chapter 21.

PHYSICAL CHARACTERISTICS OF BLOOD

The red body fluid that flows through all the vessels except the lymph vessels is called ***blood.*** Blood is a viscous fluid—it is thicker than water. Water is considered to have a viscosity of 1.0. The viscosity of blood, by comparison, ranges from 4.5 to 5.5. It flows more slowly than water, at least in part because of its viscosity. The adhesive quality of blood, or its stickiness, may be demonstrated by touching it. Blood is also slightly heavier than water.

Other physical characteristics of blood include a temperature of about 38°C (100.4°F), a pH range of 7.35 to 7.45 (slightly alkaline), and a salt (NaCl) concentration of 0.90 percent.

Blood constitutes about 8 percent of the total body weight. The blood volume of an average-sized male is 5 to 6 liters (5 to 6 qt). The average-sized female has 4 to 5 liters.

The physical characteristics of blood are summarized in Exhibit 19-1.

EXHIBIT 19-1 SUMMARY OF PHYSICAL CHARACTERISTICS OF BLOOD

Viscosity	4.5–5.5
Temperature	38°C (100.4°F)
pH	7.35–7.45
Salinity	0.90 percent
Total body weight	8 percent
Volume	5–6 liters for average male; 4–5 liters for average female

CLINICAL APPLICATION: WITHDRAWING BLOOD

Blood samples for laboratory testing may be obtained in several ways. ***Venipuncture,*** the most frequently used procedure, involves withdrawal of blood from a vein. Veins are used instead of arteries because they are closer to the surface and more readily accessible and they contain blood at a much lower pressure. A commonly used vein is the median cubital vein in front of the elbow (see Figure 21-19). A tourniquet is wrapped around the arm to stop blood flow through the veins. This makes the veins below the tourniquet stand out. Opening and closing the fist has the same effect.

Another procedure used to withdraw blood is the ***fingerstick.*** Using a sterile needle or lancet, a drop or two of capillary blood is taken from a finger, earlobe, or heel of the foot for evaluation.

Finally, an ***arterial stick*** may be used to withdraw blood. The sample is most frequently taken from the radial artery in the wrist or the femoral artery in the thigh.

FUNCTIONS OF BLOOD

Despite its simple appearance, blood is a complex liquid that performs a number of critical functions.

1. ***It transports:*** oxygen from the lungs to the cells of the body; carbon dioxide from the cells to the lungs; nutrients from the gastrointestinal tract to the cells; waste products from cells; hormones from endocrine glands to the cells; heat from various cells.

2. ***It regulates:*** pH through buffers; normal body temperature through the heat-absorbing and coolant properties of its water content; the water content of cells, principally through dissolved sodium ions (Na^+) and proteins.

3. ***It protects against:*** blood loss through the clotting mechanism; foreign microbes and toxins through certain white blood cells that are phagocytic or specialized proteins such as antibodies, interferon, and complement.

COMPONENTS OF BLOOD

Blood is composed of two portions: formed elements (cells and cell-like structures) and plasma (liquid containing dissolved substances). The formed elements compose about 45 percent of the volume of blood; plasma constitutes about 55 percent (Figure 19-1).

Formed Elements

In clinical practice, the most common classification of the ***formed elements*** of the blood is the following (Figure 19-2).

Erythrocytes (red blood cells)
Leucocytes (white blood cells)

FIGURE 19-1 Components of blood in a normal adult.

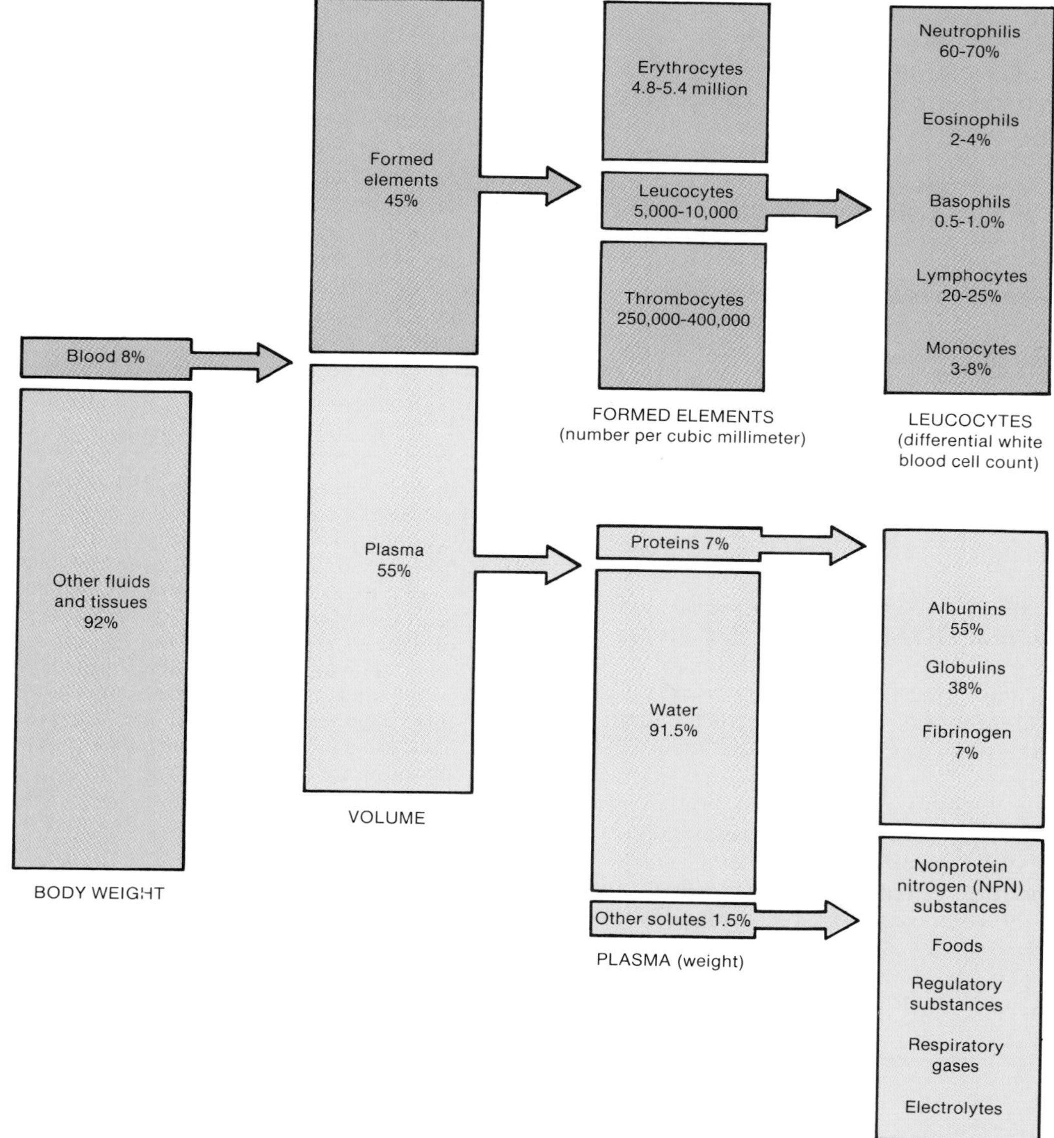

Granular leucocytes (granulocytes)
- Neutrophils
- Eosinophils
- Basophils

Agranular leucocytes (agranulocytes)
- Lymphocytes
- Monocytes

Thrombocytes (platelets)

Origin

The process by which blood cells are formed is called ***hemopoiesis*** (hē-mō-poy-Ē-sis), or ***hematopoiesis.*** During embryonic and fetal life, there are several centers for blood cell production. The yolk sac, liver, spleen, thymus gland, lymph nodes, and bone marrow all participate at various times in producing the formed elements. In the adult, however, we can pinpoint the production process to the red bone marrow (myeloid tissue) in the proximal epiphyses of the humerus and femur; flat bones of the cranium, sternum, ribs, vertebrae, and pelvis; and lymphoid tissue. Red blood cells, granular leucocytes, and platelets are produced in red bone marrow. Agranular leucocytes arise from both myeloid tissue and from lymphoid tissue—spleen, tonsils, thymus gland, lymph nodes. Undifferentiated mesenchymal cells in red bone marrow are transformed into ***hemocytoblasts*** (hē'-mō-SĪ-tō-blasts), immature cells that are eventually capable of developing into mature blood cells (Figure 19-2a). The hemocytoblasts undergo differentiation into five types of cells from which the major types of blood cells develop.

1. ***Proerythroblasts (rubriblasts)*** form mature erythrocytes.
2. ***Myelobasts*** form mature neutrophils, eosinophils, and basophils.
3. ***Megakaryoblasts*** form mature thrombocytes (platelets).
4. ***Lymphoblasts*** form mature lymphocytes.
5. ***Monoblasts*** form mature monocytes.

Erythrocytes

Structure Microscopically, ***erythrocytes*** (e-RITH-rō-sīts), or ***red blood cells (RBCs)***, or ***red blood corpuscles,*** appear as biconcave discs averaging about 8 μm in diameter (Figure 19-2b). Mature red blood cells are quite simple in structure. They lack a nucleus and can neither reproduce nor carry on extensive metabolic activities. The plasma membrane is selectively permeable and consists of protein (stromatin) and lipids (lecithin and cholesterol). The membrane encloses cytoplasm and a red pigment called ***hemoglobin.*** Hemoglobin, which constitutes about 33 percent of the cell weight, is responsible for the red color of blood. Normal values for hemoglobin are 14 to 20 g/100 ml of blood in infants, 12 to 15 g/100 ml in adult females, and 14 to 16.5 g/100 ml in adult males. As you will see later, certain proteins (antigens) on the surfaces of red blood cells are responsible for the various blood groups. The ABO and Rh groups are examples.

Functions The hemoglobin in erythrocytes combines with oxygen to form oxyhemoglobin, and with carbon dioxide to form carbaminohemoglobin, and then transports them through blood vessels. A hemoglobin molecule consists of a protein called ***globin,*** composed of four polypeptide chains, and four nonprotein pigments called ***hemes,*** each of which contains iron (as Fe^{2+}) that can combine reversibly with an oxygen molecule. As erythrocytes pass through the lungs, each of the four iron atoms in the hemoglobin molecules combines with a molecule of oxygen. The oxygen is transported in this state to other tissues of the body. In the tissues, the iron–oxygen reaction reverses, and the oxygen is released to diffuse into the interstitial fluid and from there into cells. On the return trip, the globin portion combines with a molecule of carbon dioxide from the interstitial fluid to form carbaminohemoglobin. This complex is transported to the lungs, where the carbon dioxide is released and then exhaled. Although about 23 percent of carbon dioxide is transported by hemoglobin in this manner, the greater portion, about 70 percent, is transported in blood plasma as the bicarbonate ion, HCO_3^- (Chapter 23).

It is interesting to note that the substitution of the amino acid valine for the amino acid glutamic acid in two of the polypeptide chains of globin is due to a genetic defect and results in sickle-cell anemia (SCA). When this abnormal hemoglobin is exposed to low oxygen, it forms crystals inside the red blood cells that deform them into their characteristic sickle shape (see Figure 19-8). Sickle-cell anemia is discussed at the end of the chapter.

CLINICAL APPLICATION: INDUCED ERYTHROCYTEMIA (BLOOD DOPING)

In recent years, some athletes have become involved in a practice called *induced erythrocytemia,* commonly known as *blood doping.* It involves removing red blood cells from the body and then reinjecting them a few days before an athletic event. Since red blood cells carry oxygen, it is presumed that the increased oxygen-carrying capacity of the blood increases muscular performance. In one major study, it has been shown that induced erythrocytemia does improve athletic performance in endurance events. However, the procedure is considered dishonest by the International Olympics Committee, and there is still the possibility that it can lead to some negative effects.

Red blood cells are highly specialized for their transport function. They contain a large number of hemoglobin molecules, which increases their oxygen-carrying capacity. One estimate is that each erythrocyte can carry 280

FIGURE 19-2 **Blood cells. (a) Diagram of origin, development, and structure.**

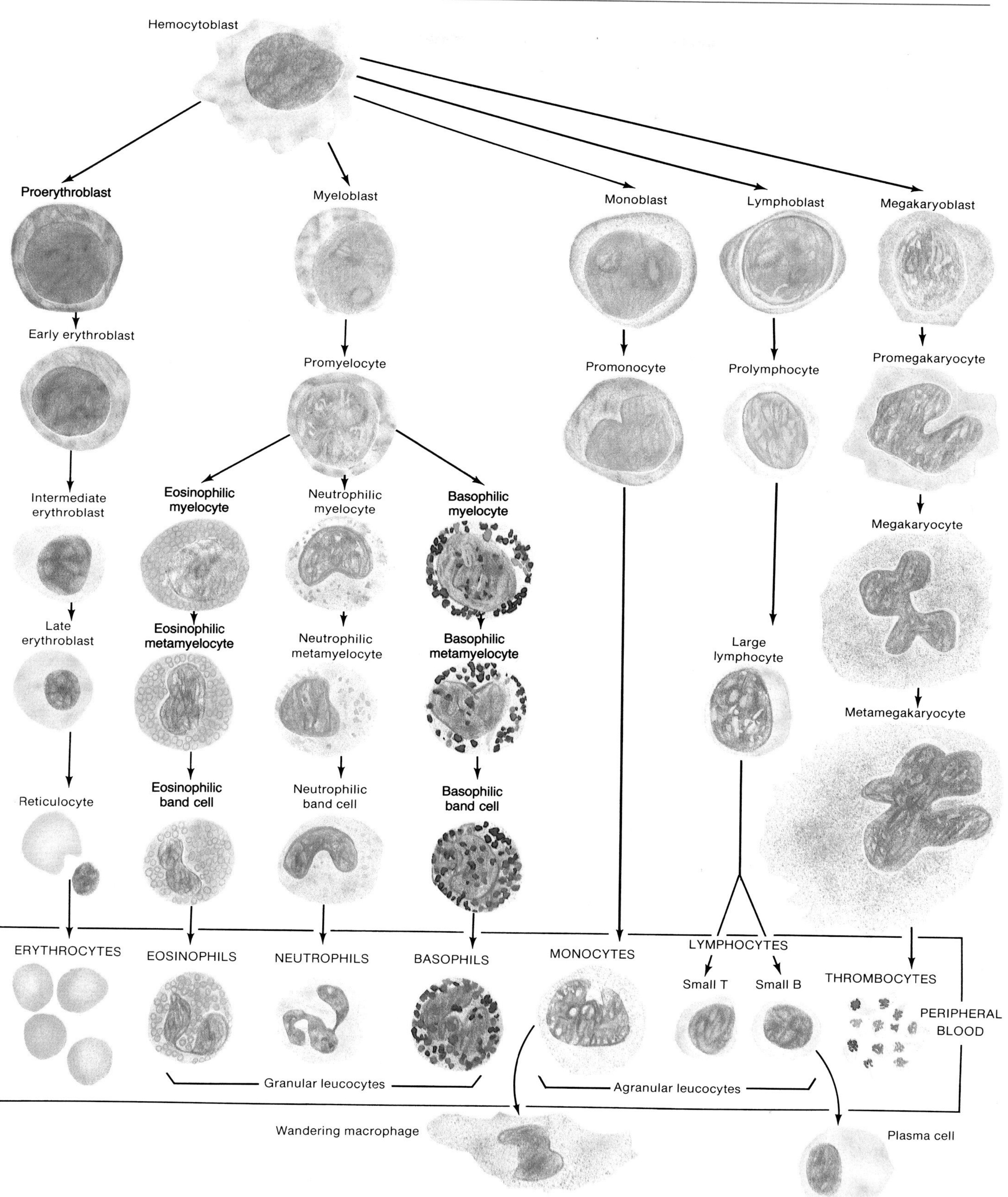

(a)

FIGURE 19-2 (*Continued*) (b) Photomicrograph of blood cells. Shown in the center is a blood smear at a magnification of 400×. The two larger cells with darker-staining nuclei are white blood cells; the more numerous cells are red blood cells, which lack nuclei; and the specklike objects are platelets. The principal types of white blood cells are shown in the large circles at a magnification of 1800×. These white blood cells are surrounded by red blood cells. Note the platelets, indicated by the arrows. (Courtesy of Michael H. Ross and Edward J. Reith, *Histology: A Text and Atlas,* Copyright © 1985, by Michael H. Ross and Edward J. Reith, Harper & Row, Publishers, Inc., New York.)

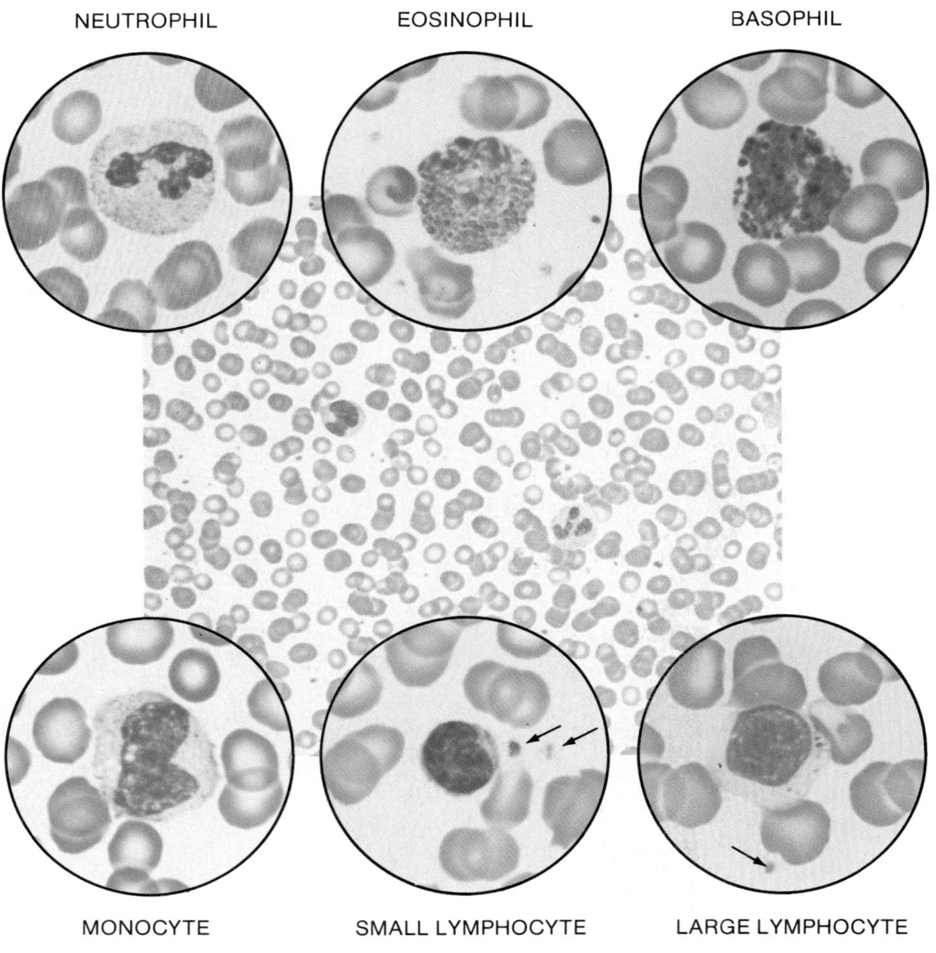

million molecules of hemoglobin. The biconcave shape of a red blood cell has a much greater surface area than, say, a sphere or a cube. The erythrocyte thus presents a large surface area for the diffusion of gas molecules that pass through the membrane to combine with hemoglobin.

Red blood cells, as well as stomach cells and kidney cells, contain an abundant supply of an enzyme called carbonic anhydrase. As you will see in Chapter 23, the enzyme speeds up a reaction in which carbon dioxide and water combine to form carbonic acid (in body tissues) or in which carbonic acid breaks down into carbon dioxide and water (in the lungs). This enzyme is essential for the normal transport of carbon dioxide by the blood.

CLINICAL APPLICATION: BLOOD SUBSTITUTE

A ***blood substitute*** called *Fluosol-DA* is a slippery, white liquid that has a very high capacity for oxygen transport. Since its primary function is to transport oxygen, it is actually a hemoglobin substitute. It may eventually turn out to be useful in providing oxygen to tissues that are oxygen-deficient as a result of various disorders. In clinical trials, Fluosol-DA is being used to oxygenate the heart during procedures in which blood flow to the heart is blocked, to oxygenate tumors to make them more responsive to chemotherapy or radiation therapy, to reperfuse tissues while not causing further damage following clot removal, and to protect heart muscle during a heart attack. Some researchers believe that Fluosol-DA may be useful in keeping donor organs oxygenated while they are readied for transplant and treating cerebral ischemia. Since Fluosol-DA is also a plasma expander, it is useful for emergency situations in which hypovolemia (below normal plasma volume) is a factor. Use of Fluosol-DA is valuable since patients would be free of the risks of transfusion reactions and posttransfusion AIDS and hepatitis infections.

▪ **Life Span and Number** The plasma membrane of a red blood cell becomes fragile and the cell is nonfunctional in about 120 days. The plasma membranes and other parts of wornout red blood cells are removed from circulation by macrophages (phagocytes) in the spleen, liver, and bone marrow. The hemoglobin is broken down into iron which associates with a protein to form hemosiderin, an iron-containing pigment; bilirubin, a non–iron-containing pigment; and globin, a protein. The hemosiderin is stored or used in bone marrow to produce new hemoglobin for new red blood cells. Bilirubin is secreted by the liver into bile, and globin is metabolized by the liver releasing its amino acids for cell use.

A healthy male has about 5.4 million red blood cells per cubic millimeter (mm^3) of blood, and a healthy female has about 4.8 million. The higher value in the male is caused by higher levels of testosterone which stimulates the production of red blood cells. To maintain normal quantities of erythrocytes, the body must produce new mature cells at the astonishing rate of at least 2 million per second. In the adult, production takes place in the red bone marrow in the spongy bone of the cranium, ribs, sternum, bodies of vertebrae, and proximal epiphyses of the humerus and femur.

MEDICAL TEST

Erythrocyte sedimentation rate (ESR), or **sed rate**

Diagnostic Value: Used as a screening test for a wide variety of infections, inflammations, and cancers. The test does not distinguish between specific diseases but is used to monitor the status of persons with rheumatoid arthritis and certain other rheumatic conditions.

Procedure: A sample of blood is placed vertically in a tube to allow the red blood cells, by gravity, to form a sediment at the bottom of the tube. The distance, in millimeters (mm), that the cells fall in one hour can be measured and is called the erythrocyte sedimentation rate. With certain diseases, the rate of sedimentation is increased.

Normal Values: Males: Under 50 years, less than 15 mm in 1 hr
Over 50 years, less than 20 mm in 1 hr
Females: Under 50 years, less than 20 mm in 1 hr
Over 50 years, less than 30 mm in 1 hr
See Appendix B also.

▪ **Production** The process by which erythrocytes are formed is called ***erythropoiesis*** (e-rith'-rō-poy-Ē-sis). Erythropoiesis starts in red bone marrow with the transformation of a hemocytoblast into a proerythroblast (Figure 19-2a). The ***proerythroblast (rubriblast)*** gives rise to an ***early erythroblast (prorubricyte),*** which then develops into an ***intermediate erythroblast (rubricyte),*** the first cell in the sequence that begins to synthesize hemoglobin. The intermediate erythroblast next develops into a ***late erythroblast (metarubricyte),*** in which hemoglobin synthesis is at a maximum. In the next stage, the late erythroblast develops into a ***reticulocyte,*** a cell in the developmental sequence that contains about 34 percent hemoglobin, retains some mitochondria and ribosomes, and loses its nucleus. After the nucleus is lost, the cell is known as a reticulocyte. Reticulocytes pass from bone marrow into the bloodstream by squeezing between the endothelial cells of blood capillaries. Reticulocytes generally become ***erythrocytes,*** or mature red blood cells, within one to two days after their release from bone marrow. (A nucleated red blood cell found in red marrow, but rarely found in blood is called a ***normoblast.***) The normal proportion of reticulocytes in blood is between 0.5 and 1.5 percent. Aged erythrocytes are destroyed by fixed phagocytic macrophages in the liver and spleen. The hemoglobin molecules are split apart, the iron is reused, and the rest of the molecule is converted into other substances for reuse (globin) or elimination (heme).

Normally erythropoiesis and red blood cell destruction proceed at the same pace. But if the body suddenly needs more erythrocytes or if erythropoiesis is not keeping up with red blood cell destruction, a homeostatic mechanism steps up erythrocyte production. The mechanism is triggered by the reduced supply of oxygen for body cells that results from a reduced number of erythrocytes or other dysfunctions that reduce oxygen delivery to tissues (hypoxia). If certain kidney cells become oxygen-deficient, they release an enzyme called ***renal erythropoietic factor*** that converts a plasma protein (possibly produced by the liver) into the hormone ***erythropoietin*** (*poiem* = to make). This hormone circulates through the blood to the red bone marrow, where it stimulates more hemocytoblasts to develop into red blood cells. Erythropoietin is also produced by other body tissues, particularly the liver. As you will see, erythropoietin is one of a group of hormones, called ***hemopoietins,*** involved in the production of blood cells.

MEDICAL TESTS

Reticulocyte (re-TIK-yoo-lō-sīt) **count**

Diagnostic Value: To measure the rate of erythropoiesis and thus evaluate the bone marrow's response to anemia (low hemoglobin) or to monitor treatment for anemia.

Procedure: A blood sample is stained and examined to determine the percentage of reticulocytes in the total number of red blood cells.

Normal Values: 0.5–1.5 percent
A high reticulocyte count might indicate the response to bleeding, hemolysis (rapid breakdown of erythrocytes), or the response to iron therapy in someone who is iron deficient. Low reticulocyte count in the presence of anemia might indicate inability of the bone marrow to respond, owing to a nutritional deficiency, pernicious anemia, or leukemia.

Hematocrit (he-MAT-ō-krit), or **Hct**

Diagnostic Value: The hematocrit is the percentage of red blood cells in blood. A hematocrit of 40 means that 40 percent of the volume of blood is composed of red blood cells. The test is used to diagnose anemia and polycythemia (an increase in the percentage of the red blood cells) and abnormal states of hydration.

Procedure: A sample of venous blood is centrifuged (spun at high speed, causing erythrocytes and plasma to separate), and the ratio of red blood cells to whole blood is measured.

Normal Values: Females: 38–46 percent (average 42)
Males: 40–54 percent (average 47)
See Appendix B also.
A significant drop in hematocrit constitutes anemia, which may vary from mild (hematocrit of 35) to severe (hematocrit of less than 15). Polycythemic blood may have a hematocrit of 65 percent or higher. Athletes not uncommonly have a higher-than-average hematocrit, and the average hematocrit of persons living in mountainous terrain is greater than that of persons living at sea level.

Cellular oxygen deficiency, called ***hypoxia*** (hī-POKS-ē-a), may occur if you do not breathe in enough oxygen. This deficiency commonly occurs at high altitudes, where the air contains less oxygen. Oxygen deficiency may also occur because of anemia. Consequently, the erythrocytes are unable to transport enough oxygen from the lungs to the cells. Anemia has many causes; lack of iron, lack of certain amino acids, and lack of vitamin B_{12} are but a few. Iron is needed for the oxygen-carrying part of the hemoglobin molecule. The amino acids are needed for the protein, or globin, part. Vitamin B_{12} helps the red bone marrow to produce erythrocytes. This vitamin is obtained from meat, especially liver, but it cannot be absorbed by the lining of the small intestine without the help of another substance—***intrinsic factor*** **(*IF*)** produced by the parietal cells of the mucosa of the stomach. Intrinsic factor facilitates the absorption of vitamin B_{12} (extrinsic factor). Once absorbed, it is stored in the liver. (Several types of anemia are described later in the chapter.)

Leucocytes

Structure and Types Unlike red blood cells, ***leucocytes,*** also spelled ***leukocytes*** (LOO-kō-sīts), or ***white blood cells* (*WBCs*),** have nuclei and do not contain hemoglobin (Figure 19-2).

Leucocytes fall into two major groups. The first group is the ***granular leucocytes.*** They develop from red bone marrow, have conspicuous granules in the cytoplasm, and possess lobed nuclei. The three kinds of granular leucocytes (***polymorphs, polymorphonuclear leucocytes,*** or ***PMNS***) are ***neutrophils*** (10 to 12 μm in diameter), ***eosinophils*** (10 to 12 μm in diameter), and ***basophils*** (8 to 10 μm in diameter). The nuclei of neutrophils have two or six lobes, connected by very thin strands. As the cells age, the extent of lobulation increases. Fine, evenly distributed pale lilac colored granules are found in the cytoplasm when a commonly used stain is applied to the cells. Eosinophils contain nuclei that are usually bilobed, with the lobes connected by a thin strand or thick isthmus. The cytoplasm is packed with large, uniform-sized granules that do not cover or obscure the nucleus. The granules stain red-orange with a commonly used stain. The nuclei of basophils are bilobed or irregular in shape, often in the form of a letter S. The cytoplasmic granules are round, variable in size, stain blue-black, and commonly obscure the nucleus.

The second principal group of leucocytes is the ***agranular leucocytes.*** They develop from lymphoid and myeloid tissue (red bone marrow), and no cytoplasmic granules can be seen under a light microscope, owing to their small size. The two kinds of agranular leucocytes are ***lymphocytes*** (7 to 15 μm in diameter) and ***monocytes*** (14 to 19 μm in diameter). The nuclei of lymphocytes are darkly stained, round, or slightly indented. The cytoplasm stains sky blue, and forms a rim around the nucleus. The nuclei of monocytes are usually indented or kidney-shaped, and the cytoplasm has a foamy appearance.

Just as red blood cells have surface proteins, so do white blood cells and all other nucleated cells in the body. These proteins, called ***HLA (human leucocyte associated) antigens,*** are unique for each person (except for identical twins) and can be used to identify a tissue. HLA antigens are determined by a group of genes on a single chromosome (chromosome b). This group of genes is called the ***major histocompatibility complex (MHC).*** Since HLA antigens are determined by the MHC, they are also referred to as ***histocompatibility antigens.***

MEDICAL TEST

Histocompatibility (his'-tō-kom-pat-i-BIL-i-tē) testing (HLA antigen typing or tissue typing)

Diagnostic Value: The success of a proposed organ or tissue transplant depends on the tissue compatibility between the donor and the recipient (histocompatibility). The more similar the HLA antigens, the higher the degree of histocompatibility, and the greater the probability the transplant will not be rejected. Histocompatibility testing is done prior to any organ transplant. And, in fact, because of a computerized matching system, tissue typing can be used to determine the match between a donor kidney that becomes available in Seattle and a potential recipient in Chicago. Also, in cases of disputed parentage, tissue typing can be used to establish paternity.

Procedure: Following removal of a sample of venous blood, white blood cell antigens are detected using specific antisera.

Functions The skin and mucous membranes of the body are continuously exposed to microbes and their toxins. Some of these microbes are capable of invading deeper tissues to cause disease, and once they enter the body, the general function of leucocytes is to combat them by phagocytosis or antibody production. Neutrophils and monocytes are actively ***phagocytotic***—they can ingest bacteria and dispose of dead matter (see Figure 3-6a,b). Neutrophils (NOO-trō-fils) are the most active leucocytes in responding to tissue destruction by bacteria. In addition to carrying on phagocytosis, they release the enzyme lysozyme, which destroys certain bacteria. Apparently, monocytes (MON-ō-sīts) take longer to reach the site of infection than do neutrophils, but once they arrive, they do so in larger numbers and destroy more microbes. Monocytes that have migrated to infected tissues and differentiated into phagocytes are called ***wandering macrophages.*** They clean up cellular debris and microbes following an infection.

A number of different chemicals in inflamed tissue attract phagocytes toward the tissue. This phenomenon is called ***chemotaxis.*** Among the substances that provide stimuli for chemotaxis are toxins produced by microbes and degenerative products of damaged tissues and specialized products called kinins.

Most leucocytes possess, to some degree, the ability to migrate through the minute spaces between the cells that form the walls of capillaries and into connective and epithelial tissue. This movement through capillary walls is called ***diapedesis*** (di'-a-pe-DĒ-sis). First, part of the cell membrane projects outward. Then the cytoplasm and nucleus flow into the projection. Finally, the rest of the membrane snaps up into place. Another projection is made, and so on, until the cell has migrated to its destination (see Figure 22-8).

Neutrophils also contain ***defensins,*** amino acids that exhibit a broad range of antibiotic activity against bacteria, fungi, and viruses.

Eosinophils (ē'-ō-SIN-ō-fils) are believed to release substances that combat the effects of histamine and other mediators of inflammation in allergic reactions. Eosinophils leave the capillaries, enter the tissue fluid, and phagocytize antigen–antibody complexes. Eosinophils are also effective against certain parasitic worms. Thus, a high eosinophil count frequently indicates an allergic condition or a parasitic infection.

Basophils (BĀ-sō-fils) are also believed to be involved in allergic reactions. Basophils leave the capillaries, enter the tissues, and liberate heparin, histamine, and serotonin. These substances intensify the overall inflammatory reaction and are involved in hypersensitivity (allergic) reactions (Chapter 22).

Lymphocytes (LIM-fō-sīts) are involved in the production of antibodies. ***Antibodies*** (AN-ti-bod'-ēz) are special proteins that inactivate antigens. ***Antigens*** (AN-ti-jens) are substances that will stimulate the production of antibodies and are capable of reacting specifically with the antibody. Most antigens are proteins, and most are not synthesized by the body. Many of the proteins that make up the cell structures and enzymes of bacteria are antigens. The toxins released by bacteria are also antigens. When antigens enter the body, they react chemically with substances in the lymphocytes and stimulate some lymphocytes, called B cells, to become ***plasma cells*** (Figure 19-3). The plasma cells then produce antibodies, globulin-type proteins that attach to antigens in a very precise way. A specific antibody will generally attach only to a certain antigen. However, unlike enzymes, which enhance the reactivity of the substrate, antibodies "cover" their antigens so the antigens cannot come in contact with other chemicals in the body. In this way, bacterial poisons can be sealed up and rendered harmless. The bacteria themselves are destroyed by the antibodies. This process is called the ***antigen–antibody response.*** Phagocytes in tissues destroy the antigen–antibody complexes. The antigen–antibody response is discussed in more detail in Chapter 22.

Other lymphocytes are called ***T cells.*** One group of T cells, the ***cytotoxic (killer) T cells,*** are activated by

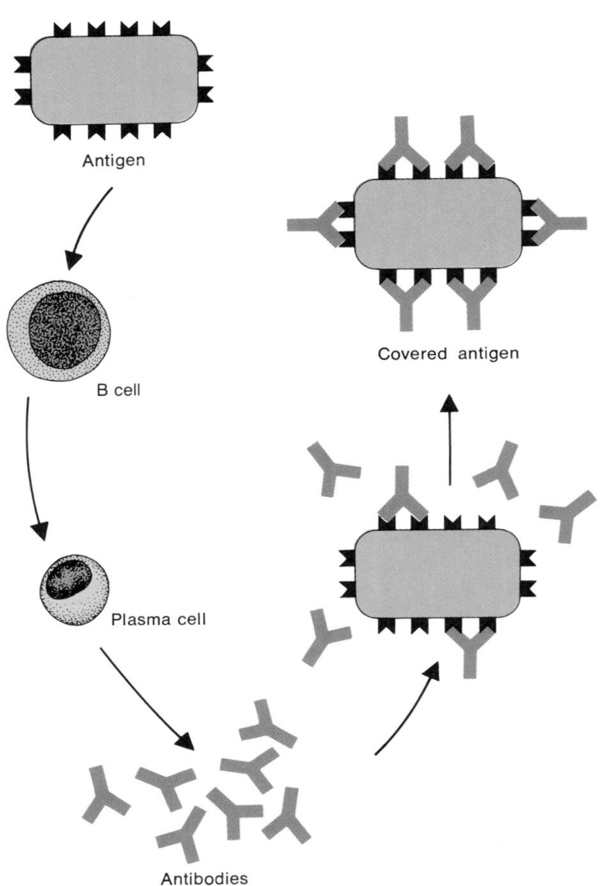

FIGURE 19-3 Antigen–antibody response. An antigen entering the body stimulates a B cell to develop into an antibody-producing plasma cell. The antibodies attach to the antigen, cover it, and render it harmless.

certain antigens and react by destroying them directly or indirectly by recruiting other lymphocytes and macrophages. T cells are especially effective against bacteria, viruses, fungi, transplanted cells, and cancer cells. Other types of T cells and their functions are discussed in Chapter 22.

The antigen–antibody response helps us combat infection and gives us immunity to some diseases. It is also responsible for transfusion reactions, allergies, and the body's rejection of organs transplanted from an individual with a different genetic makeup.

An increase in the number of white cells present in the blood typically indicates a state of inflammation or infection. Because each type of white cell plays a different role, determining the *percentage* of each type in the blood assists in diagnosing the condition. This test is called a differential white blood cell count and should not be confused with a total count which determines the *number* of white blood cells per cubic millimeter (mm^3) of blood.

MEDICAL TEST

Differential white blood cell count

Diagnostic Value: A routine part of the complete blood count (CBC) that may be helpful in evaluating infection or inflammation, determining the effects of possible poisoning by chemicals or drugs, monitoring blood disorders (e.g., leukemia) and effects of chemotherapy, or detecting allergic reactions and parasitic infections.

Procedure: A sample of blood is spread on a glass slide and stained. Then the percentage of each type of white blood cell is determined.

Normal Values:

	Percent
Neutrophils	60–70
Eosinophils	2–4
Basophils	0.5–1
Lymphocytes	20–25
Monocytes	3–8
	100

In interpreting the results of a differential white blood cell count, particular attention is paid to neutrophils. An increased neutrophil count might result from infections (bacterial), burns, stress, or inflammation; a decreased count might be caused by radiation, certain drugs, vitamin B_{12} deficiency, or systemic lupus erythematosus (SLE). An increase in the number of eosinophils could indicate allergic reactions, parasitic infections, autoimmune disease, or adrenal insufficiency; a decrease could be caused by certain drugs, stress, or Cushing's syndrome. Basophils could be elevated in some types of allergic responses, leukemias, cancers, and hypothyroidism; decreases could occur during pregnancy, ovulation, stress, and hyperthyroidism. High lymphocyte counts could indicate viral infections, immune diseases, and some leukemias; low counts might occur as a result of prolonged severe illness, high steroid levels, and immunosuppression. Finally, a high monocyte count could result from certain viral or fungal infections, tuberculosis (TB), some leukemias, and chronic diseases; below normal monocyte levels rarely occur.

▪ **Life Span and Number** Bacteria exist everywhere in the environment and have continuous access to the body through the mouth, nose, and pores of the skin. Furthermore, many cells, especially those of epithelial tissue, age and die, and their remains must be disposed of daily. Even when the body is healthy, the leucocytes actively ingest bacteria and debris. However, a leucocyte can phagocytize only a certain number of substances before they interfere with the leucocyte's normal metabolic activities and bring on its death. Consequently, the life span of most leucocytes is very short. In a healthy body, some white blood cells can live up to several months, but most live only a few days. During a period of infection, they may live only a few hours.

Leucocytes are far less numerous than red blood cells, averaging from 5,000 to 10,000 cells per cubic millimeter (mm^3) of blood. Red blood cells, therefore, outnumber white blood cells about 700:1. The term ***leucocytosis*** (loo'-kō-sī-TŌ-sis) refers to an increase in the number of white blood cells. If the increase exceeds 10,000/mm^3, a pathological condition is usually indicated. An abnormally low level of white blood cells (below 5000/mm^3) is termed ***leucopenia*** (loo-kō-PĒ-nē-a).

▪ **Production** Granular leucocytes are produced in red bone marrow (myeloid tissue); agranular leucocytes are produced in both myeloid and lymphoid tissue. The developmental sequences for the five types of leucocytes are shown in Figure 19-2a.

Recall that red blood cells develop under the influence of the hormone erythropoietin, one of the hematopoietins. White blood cells also develop under the influence of hematopoietins, but these are called ***colony-stimulating factors (CSFs).*** The classes of CSFs and their functions are as follows:

1. **Macrophage CSF (M-CSF).** Stimulates development of monocytes and macrophages.
2. **Granulocyte, Monocyte CSF (GM-CSF).** Stimulates development of granulocytes and monocytes.
3. **Granulocyte CSF (G-CSF).** Stimulates development of only granulocytes.
4. **Multi-CSF or interleukin-3.** Stimulates development of several types of white blood cells.

In addition to stimulating the development of various white blood cells, CSFs also enhance their function. Because of their roles in immunity (Chapter 22), CSFs are also being tested to determine their effects on bolstering body defenses. Trials are in progress to evaluate the effectiveness of CSFs in treating AIDS, cancer, and bone marrow suppression.

CLINICAL APPLICATION: BONE MARROW TRANSPLANT

Bone marrow, as indicated earlier, contains hemocytoblasts that differentiate into erythrocytes, leucocytes, and platelets. Among the leucocytes are cells that combat infections and cause tissue rejection. Until recently, the use of *bone marrow transplant,* the transfer of bone marrow from a donor to a recipient, required that the marrow of the donor be obtained from a relative and closely matched to that of the recipient. In cases where the match is not identical, T cells in donor marrow, which are part of the immune system, recognize the recipient's tissues as foreign and attack them. The principal tissues rejected in this way are in the skin, liver, and gastrointestinal (GI) tract. Several recent advances now permit bone marrow transplants in which the tissue match between donor and recipient is not even close.

Donor marrow is aspirated from the hip bones, mixed with heparin (an anticoagulant), and then passed through screens. The suspension of bone marrow cells is treated to remove T cells and given to the recipient just like a blood transfusion. The cells in the suspension pass through the lungs, enter general circulation, and reseed and grow in the marrow cavities of the recipient's bones.

Bone marrow transplants have been used to treat aplastic anemia, certain types of leukemia, and severe combined immunodeficiency disease (SCID), an inherited deficiency of infection-fighting blood cells. The technique is also being expanded to treat other kinds of leukemia, non-Hodgkin's lymphoma, thalassemia, multiple myeloma, sickle-cell anemia (SCA), and hemolytic anemia.

Thrombocytes

▪ **Structure** In addition to the immature cell types that develop into erythrocytes and leucocytes, hemocytoblasts differentiate into still another kind of cell, called a megakaryoblast (see Figure 19-2a). Megakaryoblasts are ultimately transformed into megakaryocytes, large cells that shed fragments of cytoplasm. Each fragment becomes enclosed by a piece of the cell membrane and is called a ***thrombocyte*** (THROM-bō-sīt), or ***platelet.*** Thrombocytes break off from the megakaryocytes in bone marrow and then enter the blood circulation, whereas the megakaryocytes stay in the marrow. Platelets are round or oval discs without a nucleus. They average from 2 to 4 μm in diameter.

▪ **Function** Platelets repair slightly damaged blood vessels and initiate a chain of reactions that results in blood clotting. This mechanism is described shortly.

▪ **Life Span and Number** Platelets have a short life span, normally just five to nine days. Between 250,000 and 400,000 platelets are present in each cubic millimeter (mm^3) of blood.

▪ **Production** Platelets are produced in red bone marrow according to the development sequence shown in Figure 19-2a.

A summary of the formed elements in blood is presented in Exhibit 19-2.

EXHIBIT 19-2 SUMMARY OF THE FORMED ELEMENTS IN BLOOD

Formed Element	Number	Diameter (in μm)	Life Span	Function
Erythrocyte (red blood cell)	4.8 million/mm^3 in females; 5.4 million/mm^3 in males.	8	120 days.	Transports oxygen and carbon dioxide.
Leucocyte (white blood cell)	5,000–10,000/mm^3.		Few hours to a few days.*	
Granular				
Neutrophil	60–70% of total.	8–10		Phagocytosis.
Eosinophil	2–4% of total.	10–12		Combats the effects of histamine in allergic reactions, phagocytizes antigen–antibody complexes, and destroys certain parasitic worms.
Basophil	0.5–1% of total.	8–10		Liberates heparin, histamine, and serotonin in allergic reactions that intensify the overall inflammatory response.
Agranular				
Lymphocyte	20–25% of total.	7–15		Immunity (antigen–antibody reactions).
Monocyte	3–8% of total.	14–19		Phagocytosis (after transforming into wandering macrophages).
Thrombocyte (platelet)	250,000–400,000/ mm^3.	2–4	5–9 days.	Blood clotting.

* Some lymphocytes, called T and B memory cells, can live throughout life once they are established. Most white blood cells, however, have life spans ranging from a few hours to a few days.

MEDICAL TEST

Complete blood count (CBC)

Diagnostic Value: A valuable test that screens for anemia and various infections by assessing the white blood cell count, red blood cell count, and platelet count.

Procedure: The test is performed on a blood sample and consists of cell counting by a Coulter counter and blood smears for evaluating the white blood cells and red blood cells. Usually included are a determination of red blood cell count, hemoglobin, hematocrit, white blood cell count, differential white blood cell count, and platelet count.

Plasma

When the formed elements are removed from blood, a straw-colored liquid called ***plasma*** is left. Exhibit 19-3 outlines the chemical composition of plasma. Some of the proteins in plasma are also found elsewhere in the body, but in blood they are called ***plasma proteins. Albumins,*** which constitute 55 percent of plasma proteins, along with red blood cells, are largely responsible for blood's viscosity. The concentration of albumins is about four times higher in plasma than in interstitial fluid and they are the principal reason that the osmotic pressure of the blood is greater than that of tissue fluid. Albumins can therefore help regulate blood volume by preventing the water in the blood from moving into the interstitial fluid and can draw water from the tissue fluid. Recall that water moves by osmosis from an area of high water (low solute) concentration to an area of low water (high solute) concentration. ***Globulins,*** which comprise 38 percent of plasma proteins, include the antibody proteins released by plasma cells. Gamma globulin is especially well known because it is able to form an antigen–antibody complex with the protein of the hepatitis and measles viruses and the tetanus bacterium, among others. ***Fibrinogen*** makes up about 7 percent of plasma proteins and takes part in the blood-clotting mechanism along with the platelets.

EXHIBIT 19-3 CHEMICAL COMPOSITION AND DESCRIPTION OF SUBSTANCES IN PLASMA

Constituent	Description
WATER	Liquid portion of blood; consti-

EXHIBIT 19-3 CHEMICAL COMPOSITION AND DESCRIPTION OF SUBSTANCES IN PLASMA (*Continued*)

Constituent	Description
	tutes about 91.5 percent of plasma. Ninety percent of water derived from absorption from gastrointestinal (GI) tract; 10 percent from cellular respiration. Acts as solvent and suspending medium for solid components of blood and absorbs, transports, and releases heat.
SOLUTES	Constitute about 8.5 percent of plasma.
Proteins	
Albumins	Smallest plasma proteins. Produced by liver and provide blood with viscosity, a factor related to maintenance and regulation of blood pressure. Also exert considerable osmotic pressure to maintain water balance between blood and tissues and regulate blood volume.
Globulins	Protein group to which antibodies belong. Gamma globulins attack measles, hepatitis, and polio viruses, and tetanus bacterium.
Fibrinogen	Produced by liver. Plays essential role in clotting.
Nonprotein nitrogen (NPN) substances	Contain nitrogen but are not proteins. Represent breakdown products of protein metabolism and are carried by blood to organs of excretion. Include urea, uric acid, creatine, creatinine, and ammonium salts.
Food substances	Products of digestion passed into blood for distribution to all body cells. Include amino acids (from proteins), glucose (from carbohydrates), and fatty acids and glycerol (from fats).
Regulatory substances	Enzymes, produced by body cells, to catalyze chemical reactions. Hormones, produced by endocrine glands, to regulate growth and development in body.
Respiratory gases	Oxygen (O_2) and carbon dioxide (CO_2). Whereas O_2 is more closely associated with hemoglobin or red blood cells, CO_2 is more closely associated with plasma.
Electrolytes	Inorganic salts of plasma. Cations include Na^+, K^+, Ca^{2+}, Mg^{2+}; anions include Cl^-, HPO_4^{2-}, SO_4^{2-}, HCO_3^-. Help maintain osmotic pressure, normal pH, physiological balance between tissues and blood.

CLINICAL APPLICATIONS: PLASMAPHERESIS

Plasmapheresis (plaz'-ma-fe-RĒ-sis; *aphairesis* = removal), also called ***therapeutic plasma exchange (TPE)***, refers to a procedure in which blood is withdrawn from the body, its components are selectively separated, the undesirable component causing disease is removed, and the remainder is returned to the body. Among the substances removed are toxins, metabolic substances, and antibodies. Blood is withdrawn by a needle or catheter, mixed with an anticoagulant, and pumped through a separator where red blood cells, white blood cells, platelets, and plasma are separated by spinning the sample. The process usually takes three to five hours. No more than 15 percent of the patient's total blood volume is allowed outside the body at any one time.

The ability of plasmapheresis to remove antibodies and other immunologically active substances has made the procedure useful for neurological conditions in which autoimmunity is believed to play a role. About half of plasmapheresis procedures are done on patients with neurological disorders. Among the neurological disorders for which plasmapheresis benefits patients are myasthenia gravis, Eaton-Lambert syndrome, chronic inflammatory demyelinating polyneuropathy, and Guillain-Barré syndrome. Unfortunately, the procedure shows no benefit in amyotrophic lateral sclerosis (ALS). Plasmapheresis may also be used in the treatment of aplastic anemia, hemolytic anemia, multiple myeloma, thrombocytemia, posttransfusion Rh incompatibility, life-threatening sickle-cell crisis, kidney diseases, systemic lupus erythematosus (SLE), rheumatoid arthritis (RA), and certain drug overdoses. Because of some potentially serious complications of plasmapheresis (hypovolemia, shock, congestive heart failure, pulmonary edema, muscle twitching, thrombosis), the procedure is used discriminately and for short periods of time when more conservative therapy is unsuccessful.

PHYSIOLOGY OF HEMOSTASIS

Hemostasis (hē'-mō-STĀ-sis) refers to the stoppage of bleeding. When blood vessels are damaged or ruptured, three basic mechanisms operate to prevent blood loss: (1) vascular spasm, (2) platelet plug formation, and (3) blood coagulation (clotting). Although these mechanisms are useful for preventing hemorrhage in smaller (microcirculation) blood vessels, extensive hemorrhage from larger vessels usually requires voluntary intervention of some type.

Vascular Spasm

When a blood vessel other than a capillary is damaged, the circularly arranged smooth muscle in its wall contracts immediately. Such a ***vascular spasm*** reduces blood loss for several minutes to several hours, during which time the other hemostatic mechanisms can go into operation. The spasm is probably caused by damage to the smooth muscle and from reflexes initiated by pain receptors.

Platelet Plug Formation

In their unstimulated state, platelets are disc-shaped. Structurally, the platelet surface membrane is composed of carbohydrates, lipids, and proteins. Just under the membrane are tubelike structures composed of actin and myosin. Among the contents present in the cytoplasm are (1) *alpha granules,* which contain clotting factors and platelet-derived growth factor (PDGF), which can cause vascular endothelial cells, vascular smooth muscle cells, and fibroblasts to proliferate in order to help repair damaged blood vessel walls; (2) *dense granules,* which contain ADP, ATP, and serotonin; enzyme systems that produce prostaglandins; *fibrin-stabilizing factor,* which helps to strengthen a blood clot; lysosomes; some mitochondria; membrane systems that take up and store calcium and provide channels for release of the contents of granules; and glycogen.

In the first phase of platelet plug formation, platelets come into contact with parts of a damaged blood vessel, such as collagen under the endothelium. This process is called ***platelet adhesion.*** As a result of adhesion, the characteristics of platelets change drastically. They begin to enlarge and become irregularly shaped with numerous projections, which enable the platelets to come into contact with each other. In addition, the platelets secrete ADP, prostaglandins, serotonin, enzymes, calcium ions, and several substances involved in blood coagulation. This phase of secretion is referred to as the ***platelet release reaction.*** ADP plays a major role in that it acts on nearby platelets to activate them as well. One prostaglandin (thromboxane A2) is also a platelet activator. Serotonin and thromboxane A2 function as vasoconstrictors, which help decrease blood flow to the injured area. The release of ADP also makes other platelets in the area sticky, and the stickiness of the newly recruited and activated platelets causes them to adhere to the originally activated platelets. Eventually, the accumulation and attachment of large numbers of platelets form a mass called a ***platelet plug.*** The plug is very effective in preventing blood loss in a small vessel. Although the platelet plug is initially loose, it becomes quite tight when reinforced by fibrin threads formed during coagulation. A platelet plug can stop blood loss completely if the hole in a blood vessel is small.

Physiology of Coagulation

Normally, blood maintains its liquid state as long as it remains in the vessels. If it is drawn from the body, however, it thickens and forms a gel. Eventually, the gel separates from the liquid. The straw-colored liquid, called ***serum,*** is simply plasma minus its clotting proteins. The gel is called a ***clot*** and consists of a network of insoluble protein fibers called fibrin in which the cellular components of blood are trapped.

The process of clotting is called ***coagulation.*** If the blood clots too easily, the result can be ***thrombosis***—clotting in an unbroken blood vessel. If the blood takes too long to clot, a hemorrhage can result.

Clotting involves various chemicals known as ***coagulation factors.*** In plasma, these factors are called ***plasma coagulation factors.*** A few ***platelet coagulation factors*** are released by platelets. One coagulation factor (tissue factor; also called thromboplastin) is found on the surfaces of various body cells. The various plasma coagulation factors and their synonyms are listed in Exhibit 19-4.

EXHIBIT 19-4 PLASMA COAGULATION FACTORS AND THEIR SYNONYMS

Coagulation Factor	Synonym
I	Fibrinogen.
II	Prothrombin.
III	Tissue factor (thromboplastin).
IV	Calcium ions.
V	Proaccelerin, labile factor, or accelerator globulin.
VII	Serum prothrombin conversion accelerator (SPCA), stable factor, or proconvertin.
VIII	Antihemophilic factor (AHF), antihemophilic factor A, or antihemophilic globulin (AHG).
IX	Christmas factor, plasma thromboplastin component (PTC), or antihemophilic factor B.
X	Stuart factor, Power factor, thrombokinase.
XI	Plasma thromboplastin antecedent (PTA) or antihemophilic factor C.
XII	Hageman factor, glass factor, or contact factor.
XIII	Fibrin stabilizing factor (FSF) or fibrinase.

Clotting is a complex process in which the activated form of one coagulation factor catalyzes the activation of the next factor in the clotting sequence. Once the process is initiated, there is a cascade of events that results in the formation of large quantities of product. For purposes of discussion, we will describe the clotting mechanism as a sequence of three basic stages:

Stage 1. Formation of ***prothrombin activator.***

Stage 2. ***Conversion of prothrombin*** (a plasma protein formed by the liver) into the enzyme ***thrombin*** by prothrombin activator.

Stage 3. ***Conversion of soluble fibrinogen*** (another plasma protein formed by the liver) ***into insoluble fibrin*** by thrombin. Fibrin forms the threads of the clot. (Cigarette smoke contains at least two substances that interfere with fibrin formation.)

The formation of prothrombin activator is initiated by the interplay of two mechanisms: the extrinsic and intrinsic pathways of blood clotting.

Extrinsic Pathway

The ***extrinsic pathway*** of blood clotting has fewer steps than the intrinsic pathway and occurs rapidly, within a matter of seconds if trauma is severe. It is so named because the formation of prothrombin activator is initiated by a protein called ***tissue factor (TF),*** also known as ***thromboplastin,*** found on the surfaces of cells throughout the body, especially brain, lung, and intestinal cells—all cells *outside* (*extrinsic*) the blood itself. In damaged tissues, tissue factor combines with coagulation factor VII, thus activating it (Figure 19-4a). Activated coagulation factor VII next combines with coagulation factor X, thus activating it. Once factor X is activated, it reacts with factor V and calcium ions (Ca^{2+}) to form prothrombin activator. This completes the extrinsic pathway and stage 1 of clotting. Next, the prothrombin activator, together with Ca^{2+} ions, brings about the conversion of prothrombin into thrombin in stage 2. In stage 3, thrombin, in the presence of Ca^{2+} ions, converts fibrinogen, which is soluble, to fibrin, which is insoluble. Thrombin also activates coagulation factor XIII, which strengthens and stabilizes the fibrin clot. Factor XIII occurs in plasma, and it is also released by platelets trapped in the clot.

Thrombin has a positive feedback effect, through factor V, in accelerating the formation of prothrombin activator. Once clotting begins and thrombin is formed, a positive feedback cycle is set up in which thrombin accelerates production of prothrombin activator, which, in turn, accelerates the production of more thrombin, and so on. If unchecked, a clot would continue to get larger and larger, as a result of the positive feedback cycle. One condition that prevents this is blood flow. The movement of blood within vessels is usually sufficient to carry away factors necessary for clotting, so that their concentrations will not rise enough to promote further clotting. In addition, as you will see shortly, there are certain substances in blood that inhibit or destroy coagulating factors.

Intrinsic Pathway

The ***intrinsic pathway*** of blood clotting is more complex than the extrinsic pathway, and it occurs more slowly, usually requiring several minutes. The intrinsic pathway is so named because the formation of prothrombin activator is initiated by tissue factor (TF) found on the surfaces of endothelial cells, cells that line blood vessels and monocytes—cells in direct contact with blood or contained *within* (*intrinsic*) the blood itself. The intrinsic pathway is triggered when blood comes into contact with tissue factor of damaged endothelial cells or monocytes (Figure19-4b). The tissue factor of these cells initiates clotting in the same way as tissue factor in the extrinsic pathway. In addition, trauma to endothelial cells causes damage to blood platelets, resulting in the release of phospholipids by the platelets. Once factor VII is activated, it acts together with Ca^{2+} ions and platelet phospholipids to activate factor X. Once factor X is activated, it reacts with platelet phospholipids, factor V, and Ca^{2+} ions to form prothrombin activator. This completes the intrinsic pathway and stage 1 of clotting. From this point on, the reactions for stages 2 and 3 are similar to those of the extrinsic pathway. Since stages 2 and 3 are similar for the extrinsic and intrinsic pathways, these stages together are referred to as the common pathway. Once thrombin is formed, it causes more platelets to adhere to each other, resulting in the release of more platelet phospholipids. This is another example of a positive feedback cycle.

Once the clot is formed, it plugs the ruptured area of the blood vessel and thus prevents hemorrhage (bleeding). Permanent repair of the blood vessel can then take place. In time, fibroblasts form connective tissue in the ruptured area, and new endothelial cells repair the lining.

CLINICAL APPLICATION: HEMOPHILIA

Hemophilia (*hemo* = blood; *philein* = to love) refers to several different hereditary deficiencies of coagulation in which bleeding may occur spontaneously or after only minor trauma. The effects of all the forms of the disorder are so similar that they are hardly distinguishable from one another, but each is a deficiency of a different blood clotting factor. For example, persons with the most common type of hemophilia, hemophilia A (classic hemophilia), lack factor VIII, the antihemophilic factor. Persons with hemophilia B lack factor IX. Those with hemophilia C lack factor XI. Hemophilia A and B occur primarily among males; hemophilia C is a mild form affecting both males and females.

Hemophilia is characterized by spontaneous or traumatic subcutaneous and intramuscular hemorrhaging, nosebleeds, blood in the urine, and joint pain and damage due to joint hemorrhaging. Treatment involves the application of pressure to accessible bleeding sites and transfusions of fresh plasma or

FIGURE 19-4 Blood clotting. (a) Extrinsic pathway. (b) Intrinsic pathway. (c) Scanning electron micrograph of fibrin threads and red blood cells at a magnification of about 1500 to 2000×. (Courtesy of Fisher Scientific Company and S.T.E.M. Laboratories, Inc., Copyright 1975.)

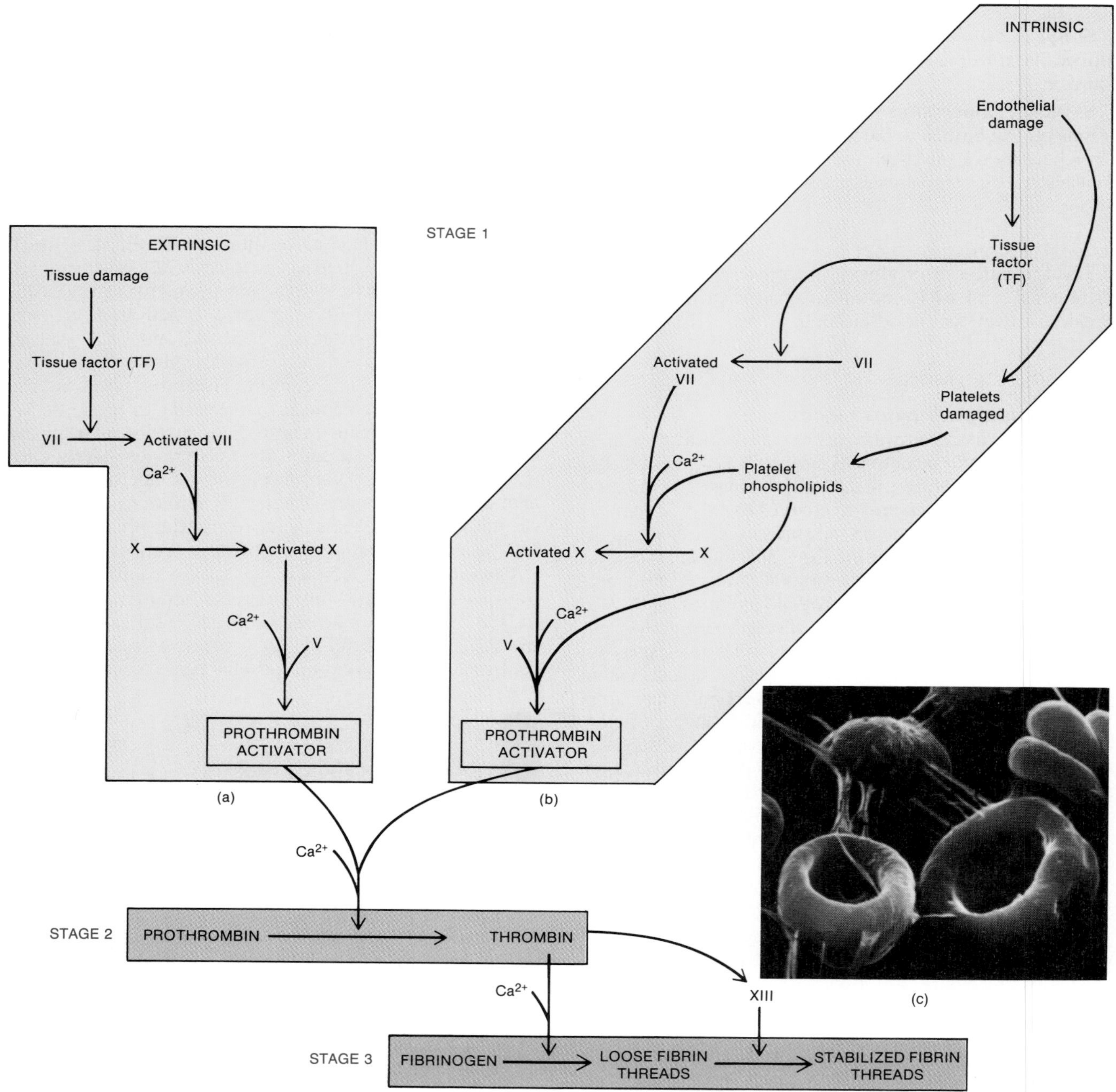

the appropriate deficient clotting factor to relieve the bleeding tendency. The inheritance of hemophilia is discussed in Chapter 29.

Retraction and Fibrinolysis

Normal coagulation involves two additional events after clot formation: clot retraction and fibrinolysis. ***Clot retraction*** or ***syneresis*** (si-NER-e-sis), is the consolidation or tightening of the fibrin clot. The fibrin threads attached to the damaged surfaces of the blood vessel gradually contract owing to platelets pulling on them. As the clot retracts, it pulls the edges of the damaged vessel closer together. Thus, the risk of hemorrhage is further decreased. During retraction, some serum escapes between the fibrin threads, but the formed elements in blood remain trapped in the fibrin threads. Normal clot retraction depends on an adequate number of platelets. Platelets in the clot bind various fibrin threads together; pull on them; and release factor XIII, which strengthens and stabilizes the clot, and factors that help compress the clot.

The second event following clot formation—***fibrinolysis*** (fī-brin-OL-i-sis)—involves dissolution of the blood clot. When a clot is formed, an inactive plasma enzyme called ***plasminogen*** is incorporated into the clot. Both body tissues and blood contain substances that can activate plasminogen to ***plasmin,*** an active plasma enzyme. Among these substances are thrombin, activated factor XII, lysosomal enzymes from damaged tissues, and substances from the endothelial lining of blood vessels. Once plasmin is formed, it can dissolve the clot by digesting fibrin threads and inactivating substances such as fibrinogen, prothrombin, and factors V, VIII, and XII. In addition to dissolving large clots in tissues, plasmin is an important enzyme in removing very small clots in small intact blood vessels before the clot can grow to a size that seriously impairs blood flow to the tissue supplied by the blood vessel.

CLINICAL APPLICATION: THROMBOLYTIC (CLOT-DISSOLVING) AGENTS

Thrombolytic (clot-dissolving) agents **are chemical substances injected into the body that dissolve blood clots to restore circulation. They are used to dissolve blood clots associated with pulmonary embolism, deep-vein thrombosis, peripheral arterial occlusion, and heart attacks. All such agents in current use either directly or indirectly activate plasminogen, an inactive plasma enzyme. The first thrombolytic agent, approved for use in 1982, is** ***streptokinase*** **(Kabikinase, Streptase). It has been shown that its effectiveness is enhanced by use with aspirin. A recently developed thrombolytic agent in use since 1987 is a normal enzyme referred to as** ***tissue plasminogen activator (t-PA).*** **This substance, marketed under the brand name Activase, is a genetically engineered, artificial version of an enzyme normally found in very small amounts in the body.**

Clot formation is a vital mechanism that prevents excessive loss of blood from the body. To form clots, the body needs calcium and vitamin K. Vitamin K is not involved in actual clot formation, but it is required for the synthesis of prothrombin (factor II) and factors VII, IX, and X. The vitamin is normally produced by bacteria that live in the large intestine. Because it is fat-soluble, it can be absorbed through the mucosa of the intestine and into the blood only if it is attached to fat. People suffering from disorders that prevent absorption of fat often experience uncontrolled bleeding. Clotting may be encouraged by applying a thrombin or fibrin spray, a rough surface such as gauze, or heat.

Hemostatic Control Mechanisms

It was noted earlier that even though thrombin has a positive feedback effect on producing a blood clot, clot formation normally occurs locally at the site of damage; it does not extend beyond the wound site into general circulation. One reason for this is that some of the coagulation factors are carried away by the blood so that their concentrations are not high enough to bring about widespread clotting. In addition, fibrin itself has the ability to absorb and inactivate up to nearly 90 percent of thrombin formed from prothrombin. This helps stop the spread of thrombin into the blood and thus inhibits spread of the clot beyond the site of damage.

Several other mechanisms that control blood clotting also operate. For example, both endothelial cells and white blood cells produce a derivative of prostaglandins called ***prostacyclin.*** This substance is a very powerful inhibitor of platelet adhesion and release. Also, a number of substances that inhibit coagulation are present in blood. Such substances are called ***anticoagulants.*** These include ***antithrombin III,*** which inhibits formation of thrombin and coagulation factor X; ***alpha-2-macroglobulin,*** which inactivates thrombin and plasmin; ***alpha-1-antitrypsin,*** which inhibits coagulation factor XI; and ***protein C,*** which inactivates coagulation factors V and VIII and stimulates plasminogen activators.

Heparin is another anticoagulant. It is produced by mast cells and basophils and is located in endothelial cells that line blood vessels. Its anticoagulant activity is to combine with antithrombin III to increase the effectiveness of antithrombin III in removing thrombin. Heparin is also a pharmacologic anticoagulant extracted from donated animal lung tissue and intestinal mucosa. It is frequently used in open heart surgery and during hemodialysis. The pharmaceutical preparation ***warfarin (Coumadin)*** may be given to patients who are thrombosis-prone. It acts as an antagonist to vitamin K and thus lowers the level of prothrombin. Warfarin is slower acting than heparin and is used primarily as a preventative. ***CPD (citrate phosphate dextrose)***, ***ACD (acid citrate dextrose)***, and ***EDTA (ethylenediamine tetracetic acid)*** are used by laboratories and blood banks to prevent blood samples from clotting. These substances react with Ca^{2+} to form insoluble compounds. In this way, the blood Ca^{2+} is tied up and is no longer free to catalyze the conversion of prothrombin to thrombin.

Intravascular Clotting

Even though the human body has certain anticoagulating mechanisms, blood clots sometimes form within the cardiovascular system. Such clots may be initiated by roughened endothelial surfaces of a blood vessel as a result of atherosclerosis, trauma, or infection. These conditions induce adhesion of platelets. Intravascular clots may also form when blood flows too slowly (stasis), allowing coagulation factors in local areas to increase in high enough concentrations to initiate coagulation. Clotting in an unbroken blood vessel (usually a vein) is called ***thrombosis.*** The clot itself is a ***thrombus*** (*thrombo* = clot). A thrombus may dissolve spontaneously, but if it remains intact, there is the possibility that the thrombus will become dislodged and be carried with the blood to the lungs. If it occurs in an artery, the clot may block the circulation to a vital organ. A blood clot, bubble of air, fat from broken bones, or a piece of debris transported by the bloodstream is called an ***embolus*** (*em* = in; *bolus* = a mass). When an embolus becomes lodged in the lungs, the condition is called ***pulmonary embolism.***

MEDICAL TESTS

Prothrombin time (PT or Pro Time)

Diagnostic Value: Detects abnormalities of the extrinsic and common pathways of blood clotting by determining the amount of prothrombin in blood. The test is used to evaluate a bleeding disorder, to monitor the effects of oral anticoagulant therapy, and to evaluate for liver disorders.

Procedure: A blood sample is treated with CPD or similar compound to tie up Ca^{2+} so that the blood will not coagulate. This, Ca^{2+}, thromboplastin, and plasma coagulation factors V and VII are mixed with the blood sample. The length of time required for the blood to clot is the prothrombin time; as a control, a normal blood sample is tested simultaneously.

Normal Values: 11–15 seconds, depending on the concentration of the thrombin reagent used. See Appendix B also.

Partial thromboplastin time (PTT)

Diagnostic Value: Same as for prothrombin time (PT) test, except for monitoring effects of oral anticoagulant therapy. The test detects abnormalities of the intrinsic pathway of blood clotting. In addition, the partial thromboplastin time test may be used for screening before surgery and to monitor heparin therapy.

Procedure: A sample of blood plasma is combined with partial thromboplastin and calcium chloride ($CaCl_2$) in a test tube. After 30 seconds, the tube is observed for clotting.

Normal Values: 30–45 seconds.

Bleeding time

Diagnostic Value: Measures the time required for bleeding to stop from a small skin puncture. The test identifies platelet function defects as well as the integrity of small vessels.

Procedure: In one method (Ivy method), a blood pressure cuff is placed around the arm and inflated to 40 mm Hg to produce a constant blood pressure in capillaries. The skin of the forearm is then punctured. The edge of a piece of filter paper is used to collect blood by touching the drop every 30 seconds until the bleeding stops. Then the time is recorded.

Normal Values: 2–8 minutes. See Appendix B also.

GROUPING (TYPING) OF BLOOD

The surfaces of erythrocytes contain genetically determined antigens called ***agglutinogens*** (ag'-loo-TIN-ō-jens), or ***isoantigens.*** There are at least 300 blood group systems that can be detected on the surface of red blood cells. The two major blood group classifications—ABO and Rh—are the ones we will discuss in detail. Among the others are the Lewis, Kell, Kidd, and Duffy systems.

ABO

The ***ABO blood grouping*** is based on two agglutinogens symbolized as *A* and *B* (Figure 19-5). Individuals whose erythrocytes manufacture only agglutinogen *A* are said to have blood type A. Those who manufacture only agglutinogen *B* are type B. Individuals who manufacture both *A* and *B* are type AB. Those who manufacture neither are type O.

People inherit *ABO* agglutinogens according to simple Mendelian principles (Chapter 29). Every person inherits two genes, one from each parent, that are responsible for production of these agglutinogens. The six possible genotypes (genetic makeups) are: *OO, AO, AA, BO, BB,* and *AB.* Both *A* and *B* are inherited as dominant traits; *O* is inherited as a recessive trait. The six genotypes may be expressed (phenotypes) as follows:

1. Genotype *OO* produces phenotype O.
2. Genotypes *AO* and *AA* produce phenotype A.
3. Genotypes *BO* and *BB* produce phenotype B.
4. Genotype *AB* produces phenotype AB.

FIGURE 19-5 Agglutinogens (antigens) and agglutinins (antibodies) involved in the ABO blood grouping system.

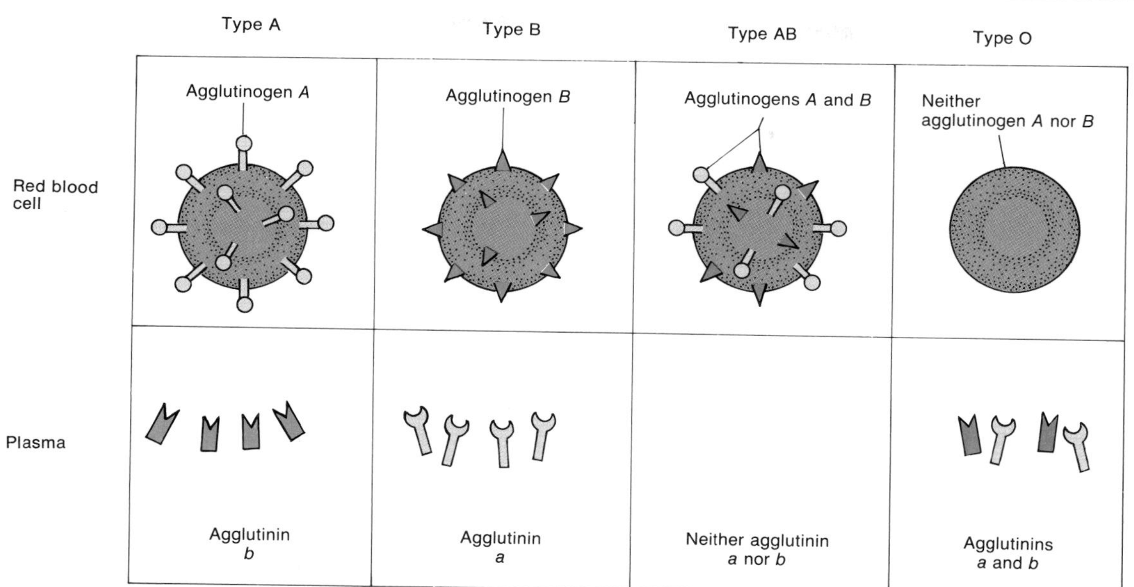

These four blood types are not equally distributed. The incidence in the white population in the United States is: type A, 41 percent; type B, 10 percent; type AB, 4 percent; type O, 45 percent. Among blacks, the frequencies are: type A, 27 percent; type B, 20 percent; type AB, 7 percent; type O, 46 percent.

The blood plasma of many people contains genetically determined antibodies referred to as ***agglutinins*** (a-GLOO-ti-nins), or ***isoantibodies.*** These are agglutinin *a* (anti-A), which attacks agglutinogen *A*, and agglutinin *b* (anti-B), which attacks agglutinogen *B*. The agglutinins formed by each of the four blood types are shown in Figure 19-5. You do not have agglutinins that attack the agglutinogens of your own erythrocytes, but you do have an agglutinin against any agglutinogen you do not synthesize. In an incompatible blood transfusion, the donated erythrocytes are attacked by the recipient agglutinins, causing the blood cells to ***agglutinate*** (clump). Agglutinated cells become lodged in small capillaries throughout the body, and over a period of hours, the cells swell, rupture, and release hemoglobin into the blood. Such a reaction as it relates to erythrocytes is called ***hemolysis*** (*lysis* = dissolve). The degree of agglutination depends on the titer (strength or amount) of agglutinin in the blood. This reaction is another example of an antigen–antibody response (Figure 19-6).

Blood transfusions are most frequently given when blood volume is low, for example, during circulatory shock. Other indications for blood transfusions include anemia, hemophilia, and hemolytic disease of the newborn (erythroblastosis fetalis). When blood is transfused, care must be taken to avoid incompatibilities that can result in agglutination, for agglutinated cells can block blood vessels and may lead to kidney or brain damage and death, and the liberated hemoglobin may also cause kidney damage.

FIGURE 19-6 Compatibility of blood types as viewed through a microscope. (a) Since the blood type shown has no agglutinins to attack the agglutinogens, no agglutination (clumping) occurs. (b) But if a blood type has incompatible agglutinins, they will attack the agglutinogens and agglutination (clumping) occurs. (Photomicrographs Copyright © 1987 by Michael H. Ross. Used by permission.)

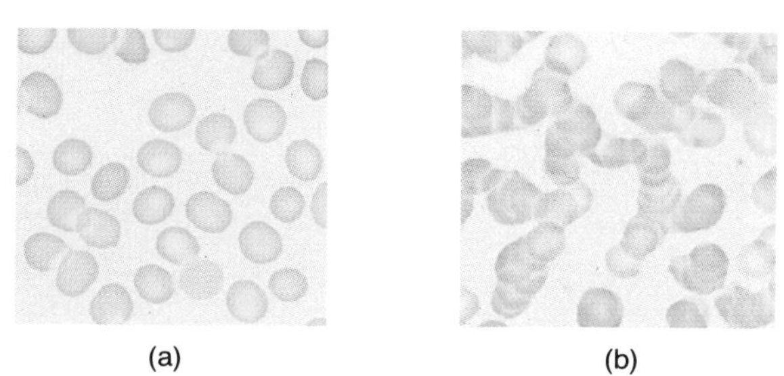

As an example of an incompatible blood transfusion, consider what happens if an individual with type A blood receives a transfusion of type B blood. The recipient's blood (type A) contains *A* agglutinogens and *b* agglutinins. The donor's blood (type B) contains *B* agglutinogens and *a* agglutinins. Given this situation, two things can happen. First, the *b* agglutinins in the recipient's plasma will attack the *B* agglutinogens on the donor's erythrocytes, causing agglutination and hemolysis of the red blood cells. Second, the *a* agglutinins in the donor's plasma will attack the *A* agglutinogens on the recipient's erythrocytes, causing agglutination and hemolysis. However, the latter reaction is usually not serious because the donor's *a* agglutinins become so diluted in the recipient's plasma that they do not cause any significant hemolysis of the recipient's erythrocytes. Thus, a person with type A blood may not receive type B or AB blood. But a person with A blood may receive type A or type O blood. The interactions of the four blood types of the ABO system are summarized in Exhibit 19-5.

Individuals with type AB blood do not have any *a* or *b* agglutinins in their plasma, and such people are sometimes called universal recipients because they can *theoretically* receive blood from donors of all four blood types; there are no agglutinins to attack donated erythrocytes. Individuals with type O blood have no *A* or *B* agglutinogens on their erythrocytes and are sometimes referred to as universal donors because they can *theoretically* donate blood to recipients of all four blood types. Type O persons requiring blood may receive only type O blood. In practice, use of the terms *universal recipient* and *universal donor* is misleading and dangerous because there are other agglutinogens and agglutinins in blood, besides those associated with the ABO system, that can cause transfusion problems. Thus, blood should be carefully cross-matched before transfusion, except in extreme emergency.

Knowledge of blood types is also used in disproving paternity, linking suspects to crimes, and as part of anthropology studies to establish a relationship among races.

In about 80 percent of the population (called ***secretors***), soluble antigens of the ABO type appear in saliva and other bodily fluids. In criminal investigations it has been possible to type such fluids from saliva residues on a cigarette or from semen in cases of rape.

Rh

The ***Rh system*** of blood classification is so named because it was first worked out in the blood of the *Rhesus* monkey. Like the ABO grouping, the Rh system is based on agglutinogens that lie on the surfaces of erythrocytes. Individuals whose erythrocytes have the Rh agglutinogens (D antigens) are designated *Rh*$^+$. Those who lack Rh agglutinogens are designated *Rh*$^-$. It is estimated that 85 percent of whites and 88 percent of blacks in the United States are Rh$^+$, whereas 15 percent of whites and 12 percent of blacks are Rh$^-$.

MEDICAL TEST

Blood typing

Diagnostic Value: To determine blood groups before a person receives a transfusion, before an individual donates blood, during evaluation of a new pregnancy, in establishing paternity, and in criminal investigations.

Procedure: In the procedure for determining ABO blood grouping, single drops of blood are mixed with different antisera, solutions that contain agglutinins (antibodies). One drop of blood is mixed with anti-A serum, which contains agglutinins (*a*) that will agglutinate red blood cells that contain *A* agglutinogens (antigens). The other drop is mixed with anti-B serum, which contains agglutinins (*b*) that will agglutinate red blood cells that contain *B* agglutinogens. If only the red blood cells mixed with anti-A serum agglutinate, the blood is type A. If only the red blood cells mixed with anti-B serum agglutinate, the blood is type B. If both drops agglutinate, the sample is type AB, and if neither sample agglutinates, the sample is type O (see Figure 19-6).

In the procedure for determining Rh factor, a drop of blood is mixed with anti-D serum, a solution that contains agglutinins that will agglutinate red blood cells that contain Rh agglutinogens. If the sample agglutinates, it is Rh$^+$; no agglutination indicates Rh$^-$.

Under normal circumstances, human plasma does not contain anti-Rh agglutinins. However, if an Rh$^-$ person receives Rh$^+$ blood, the body starts to make anti-Rh agglutinins that will remain in the blood. If a second transfusion of Rh$^+$ blood is given later, the previously formed anti-Rh agglutinins will react against the donated blood, and a severe reaction may occur.

CLINICAL APPLICATION: HEMOLYTIC DISEASE OF THE NEWBORN

One of the most common problems with Rh incompatibility arises from pregnancy (Figure 19-7). Dur-

EXHIBIT 19-5 SUMMARY OF ABO SYSTEM INTERACTIONS

Blood Type	Agglutinogen (antigen)	Agglutinin (antibody)	Compatible Donor Blood Types	Incompatible Donor Blood Types	Genotype (genetic makeup)	Phenotype (expressed genotype)
A	*A*	*b*	A, O	B, AB	*AO* and *AA*	A
B	*B*	*a*	B, O	A, AB	*BO* and *BB*	B
AB	*A, B*	Neither *a* nor *b*	A, B, AB, O	—	*AB*	AB
O	Neither *A* nor *B*	*a, b*	O	A, B, AB	*OO*	O

FIGURE 19-7 Development of hemolytic disease of the newborn (erythroblastosis fetalis). (a) Rh^+ father. (b) Rh^- mother and Rh^+ fetus. If an Rh^- female is impregnated by an Rh^+ male and the fetus is Rh^+, fetal Rh^+ agglutinogens may enter the maternal blood via the placenta during delivery. (c) Upon exposure to the fetal Rh^+ agglutinogens, the mother will make anti-Rh agglutinins. (d) If the female becomes pregnant again, her anti-Rh agglutinins will cross the placenta into the fetal blood. If the fetus is Rh^+, hemolytic disease of the newborn will result.

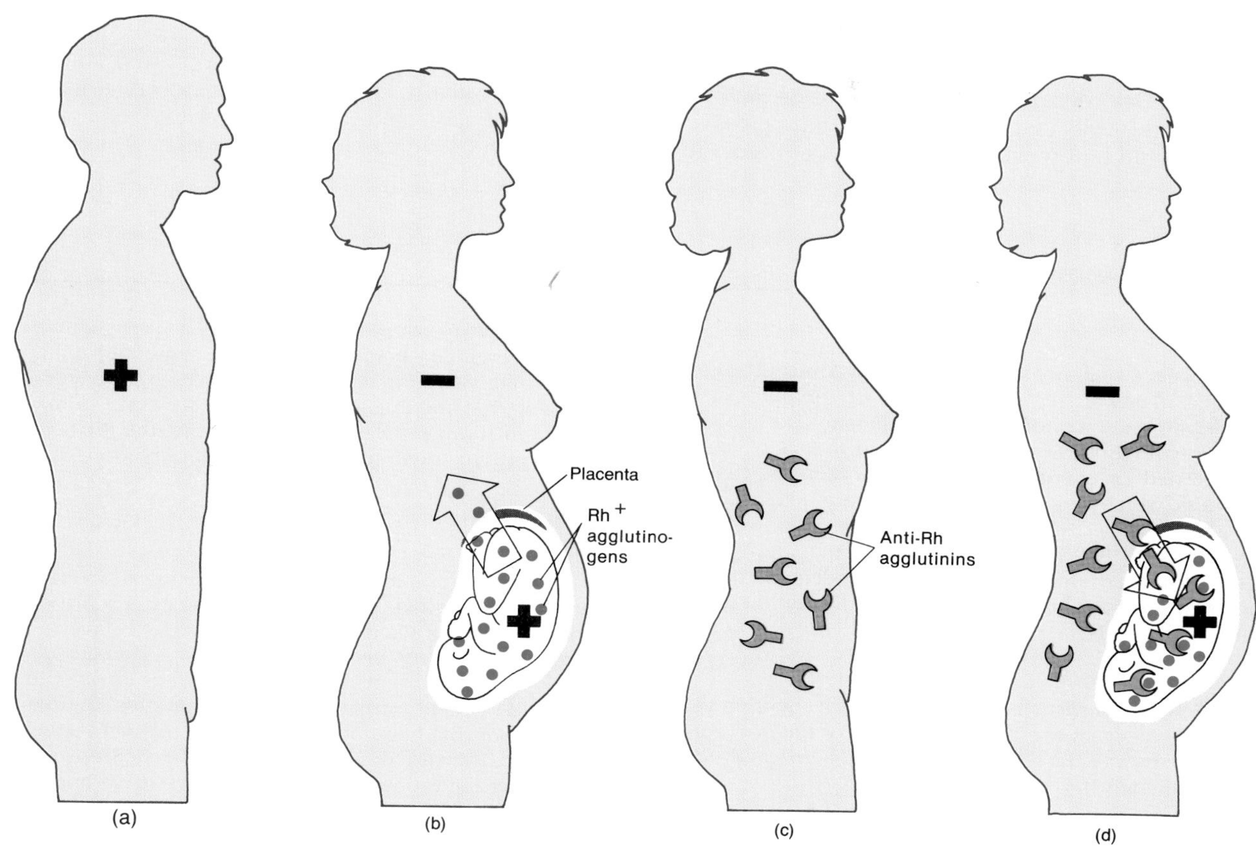

ing pregnancy, a small amount of the fetus's blood may leak from the placenta into the mother's bloodstream, with the greatest possibility of transfer occurring at delivery. If the fetus is Rh^+ and the mother is Rh^-, she, upon exposure to the Rh^+ fetal cells or cellular fragments will make anti-Rh agglutinins. If she becomes pregnant again, her anti-Rh agglutinins will cross the placenta and make their way into the bloodstream of the baby. If the fetus is Rh^-, no problem will occur, since Rh^- blood does not have the Rh agglutinogen. If the fetus is Rh^+, hemolysis may occur in the fetal blood. The hemolysis brought on by fetal-maternal incompatibility is called ***hemolytic disease of the newborn (HDN)***, also called ***erythroblastosis fetalis.*** When a baby is born with this condition, blood is slowly removed and replaced a little at a time, with Rh^- blood. It is even possible to transfuse blood into the unborn child if the disease is diagnosed before birth. More important, though, is the fact that the disorder can be prevented with an injection of an anti-Rh gamma$_2$-globulin preparation, administered to Rh^- mothers right after delivery, miscarriage, or abortion. These agglutinins tie up the fetal agglutinogens if present so the mother cannot respond to the foreign agglutinogens by producing agglutinins. Thus, the fetus of the next pregnancy is protected. In the case of an Rh^+ mother, there are no complications, since she cannot make anti-Rh agglutinins. In cases where Rh incompatibility is not implicated in hemolytic disease of the newborn, incompatibility of other blood groups such as the Kell, Kidd, and Duffy groups is frequently involved.

INTERSTITIAL FLUID AND LYMPH

Whole blood does not flow into the tissue spaces; it remains in closed vessels. However, certain constituents of the plasma do move through the capillary walls, and once they move out of the blood, they are called interstitial fluid. Interstitial fluid and lymph are basically the same. The major difference between the two is location. When the fluid bathes the cells, it is called ***interstitial fluid,***

intercellular fluid, or ***tissue fluid.*** When it flows through the lymphatic vessels, it is called ***lymph.***

Both fluids are similar in composition to plasma. The principal chemical difference is that they contain less protein than plasma because the larger protein molecules are not easily filtered through the cells that form the walls of the capillaries. The transfer of materials between blood and interstitial fluid occurs by osmosis, diffusion, and filtration across the cells that make up the capillary walls.

Interstitial fluid and lymph also differ from plasma in that they contain variable numbers of leucocytes. Leucocytes can enter the tissue fluid by diapedesis, and the lymphoid tissue itself is one of the sites of agranular leucocyte production. Like plasma, interstitial fluid and lymph lack erythrocytes and platelets.

Other substances, especially organic molecules, in interstitial fluid and lymph vary in relation to the location of the sample analyzed. The lymph vessels that drain the organs of the gastrointestinal tract, for example, contain a great deal of lipid absorbed from food.

DISORDERS: HOMEOSTATIC IMBALANCES

Anemia

Anemia is a sign, not a diagnosis. Many kinds of anemia exist, all characterized by insufficient erythrocytes or hemoglobin. These conditions lead to fatigue and intolerance to cold, both of which are related to lack of oxygen needed for energy and heat production, and to paleness, which is due to low hemoglobin content.

Nutritional Anemia

Nutritional anemia arises from an inadequate diet, one that provides insufficient amounts of iron, the necessary amino acids, or vitamin B_{12}.

Pernicious Anemia

Pernicious anemia is the insufficient production of erythrocytes resulting from an inability of the body to produce intrinsic factor. As a result, the person cannot absorb vitamin B_{12}.

Hemorrhagic Anemia

An excessive loss of erythrocytes through bleeding is called ***hemorrhagic anemia.*** Common causes are large wounds, stomach ulcers, and heavy menstrual bleeding. If bleeding is extraordinarily heavy, the anemia is termed *acute.* Excessive blood loss can be fatal. Slow, prolonged bleeding is apt to produce a *chronic* anemia; the chief symptom is fatigue.

Hemolytic Anemia

If erythrocyte cell membranes rupture prematurely, the cells remain as "ghosts," and their hemoglobin pours out into the plasma. A characteristic sign of this condition, called ***hemolytic anemia,*** is distortion in the shape of erythrocytes that are progressing toward hemolysis. There may also be a sharp increase in the number of reticulocytes, since the destruction of red blood cells stimulates erythropoiesis.

The premature destruction of red blood cells may result from inherent defects, such as hemoglobin defects, abnormal red blood cell enzymes, or defects of the red blood cell membrane. Agents that may cause hemolytic anemia are parasites, toxins, and antibodies from incompatible blood (Rh^- mother and Rh^+ fetus, for instance). Hemolytic disease of the newborn (erythroblastosis fetalis) is an example of a hemolytic anemia.

The term ***thalassemia*** (thal'-a-SĒ-mē-a) represents a group of hereditary hemolytic anemias, resulting from a defect in the synthesis of hemoglobin, which produces extremely thin and fragile erythrocytes. It occurs primarily in populations from countries bordering the Mediterranean Sea. Treatment generally consists of blood transfusions.

Aplastic Anemia

Destruction or inhibition of the red bone marrow results in ***aplastic anemia.*** Typically, the marrow is replaced by fatty tissue, fibrous tissue, or tumor cells. Toxins, gamma radiation, and certain medications are causes. Many of the medications inhibit the enzymes involved in hemopoiesis. Bone marrow transplants can now be done with a reasonable hope of success in patients with aplastic anemia. They are done early, before the victim is sensitized by transfusions. Immunosuppressive drugs are given for several days before the transplant and with decreasing frequency afterward.

Sickle-Cell Anemia (SCA)

The erythrocytes of a person with ***sickle-cell anemia (SCA)*** manufacture an abnormal kind of hemoglobin. When such an erythrocyte gives up its oxygen to the interstitial fluid, the abnormal hemoglobin tends to lose its integrity in places of low oxygen tension and forms long, stiff, rodlike structures that bend the erythrocyte into a sickle shape (Figure 19-8). The sickled cells rupture easily. Even though erythropoiesis is stimulated by the loss of the cells, it cannot keep pace with the hemolysis. The individual consequently suffers from a hemolytic anemia that reduces the amount of oxygen that can be supplied to the tissues. Prolonged oxygen reduction may eventually cause extensive tissue damage. Furthermore, because of the shape of the sickled cells, they tend to get stuck in blood vessels and can cut off blood supply to an organ altogether.

Sickle-cell anemia is characterized by several symptoms. In young children, hand–foot syndrome is present, in which there is swelling and pain in the wrists and feet. Older patients experience pain in the back and extremities without swelling and abdominal pain. Complications include neurological disorders (meningitis, sei-

DISORDERS: HOMEOSTATIC IMBALANCES (*continued*)

FIGURE 19-8 Scanning electron micrograph of erythrocytes in sickle-cell anemia at a magnification of about 2000×. (Courtesy of Fisher Scientific Company and S.T.E.M. Laboratories, Inc., Copyright © 1975.)

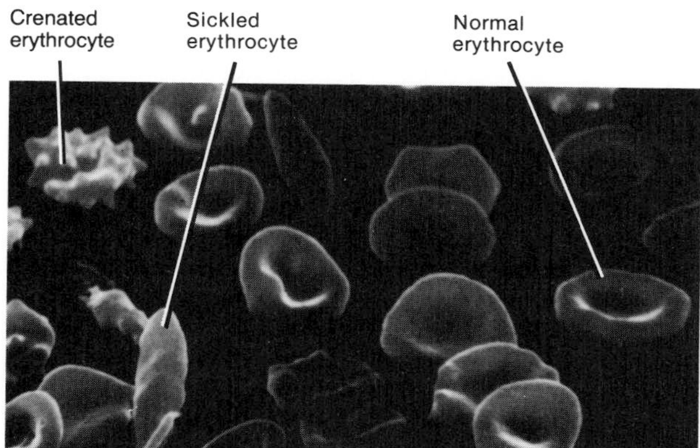

zures, stroke), impaired pulmonary function, orthopedic abnormalities (femoral head necrosis, osteomyelitis), genitourinary tract disorders (involuntary urination, blood in urine, kidney failure), ocular disturbances (hemorrhage, detached retina, blindness), and obstetric complications (convulsions, coma, infection).

Sickle-cell anemia is inherited. The gene responsible for the tendency of the erythrocytes to sickle during hypoxia also seems to prevent erythrocytes from rupturing during a malarial crisis. The gene also alters the permeability of the plasma membranes of sickled cells, causing potassium to leak out. Low levels of potassium kill the malarial parasites that infect sickled cells. Sickle-cell genes are found primarily among populations, or descendants of populations, that live in the malaria belt around the world, including parts of Mediterranean Europe and subtropical Africa and Asia. A person with only one of the sickling genes is said to have ***sickle-cell trait.*** Such an individual has a high resistance to malaria—a factor that may have tremendous survival value—but does not develop the anemia. Only people who inherit a sickling gene from both parents get sickle-cell anemia.

There are several blood tests designed to determine sickle-cell anemia and sickle-cell trait. In the most common screening test, a small amount of blood is taken from a finger, oxygen is removed, and the sample is examined for sickled cells. The distinction between sickle-cell anemia and sickle-cell trait is made by hemoglobin electrophoresis in which the different types of hemoglobin are separated.

Treatment consists of administration of analgesics to relieve pain, antibiotics to counter infections, and transfusion therapy.

Polycythemia

The term ***polycythemia*** (pol'-ē-sī-THĒ-mē-a) refers to a disorder characterized by a hematocrit (Hct) that is elevated significantly above the normal upper limit of about 55. There is an increased blood viscosity associated with the elevated hematocrit. The increased viscosity causes a rise in blood pressure and contributes to thrombosis and hemorrhage. The thrombosis results from too many red blood cells piling up as they try to enter smaller vessels. The hemorrhage is due to widespread hyperemia (unusually large amount of blood in an organ).

Infectious Mononucleosis (IM)

Infectious mononucleosis (IM) is a contagious disease primarily affecting lymphoid tissue throughout the body. It is caused by the *Epstein-Barr virus (EBV)*, the same agent that has been linked to Burkitt's lymphoma, nasopharyngeal carcinoma, and Hodgkin's disease. It occurs mainly in children and young adults, with the peak incidence at 15 to 20 years of age. The ratio of females to males affected by IM is 3:1. The virus most commonly enters the body through intimate oral contact, multiplies in lymphoid tissues, and spreads into the blood where it infects and multiplies in B lymphocytes, the primary host cells. As a result of this infection, the B lymphocytes become enlarged and abnormal in appearance and resemble monocytes, the primary reason for which the disease receives its name, mononucleosis. Infectious mononucleosis is characterized by an elevated white blood cell count with an abnormally high percentage of lymphocytes. Symptoms include fatigue, headache, dizziness, sore throat, lymphadenopathy (enlarged and tender lymph nodes), fever, brilliant red throat and soft palate, stiff neck, cough, and malaise. The spleen may also enlarge. Secondary complications may be hematologic (hemolytic anemia, neutropenia), respiratory (pneumonia), neurologic (meningitis, encephalitis, seizures, Guillain-Barré syndrome), cardiovascular (myocarditis), gastrointestinal (hepatitis, splenic rupture), and renal (nephritis, nephrosis). There is no cure for infectious mononucleosis, and treatment consists of watching for and treating complications. Usually the disease runs its course in a few weeks, and the individual generally suffers no permanent ill effects.

Chronic Fatigue Syndrome

Several years ago, a new disease referred to as chronic Epstein-Barr virus syndrome was reported. Recent findings indicate that the disease is neither new nor primarily caused by the Epstein-Barr virus. The revised name of the disease is ***chronic fatigue syndrome,*** which typically occurs among young adults, primarily females. Diagnosis of the disease must include the following: (1) new, extreme fatigue that impairs normal activities for at least six months; (2) the absence of known diseases (cancer, infections, drug abuse, toxicity, or psychiatric

DISORDERS: HOMEOSTATIC IMBALANCES (*continued*)

disorders) that might produce similar symptoms; and (3) at least 8 of these 11 indications that persist or recur over six months: mild fever or chills, sore throat, painful lymph nodes, general muscle weakness, muscle pain, fatigue for more than 24 hours after mild exercise, headache that differs in type and severity from past ones, joint pain without swelling, neuropsychological complaints (irritability, confusion, depression), sleep disturbances, and development of the initial symptoms over a few hours to a few days. The diagnosis can also be made if the patient reports 6 of the 11 symptoms just indicated plus observation by a physician of two of these three physical signs: low-grade fever, inflamed throat, and enlarged lymph nodes in the neck or axilla.

Among the possible causes of chronic fatigue syndrome are emotional factors such as depression and excess stress and physical factors such as a viral or bacterial infection or reactivation of an existing virus. There is no effective treatment for the disease.

Leukemia

Clinically, ***leukemia*** is classified on the basis of the duration and character of the disease, that is, acute or chronic. In its simplest terms, acute leukemia refers to a malignant disease of blood-forming tissues characterized by uncontrolled production and accumulation of immature leucocytes and many of the cells fail to reach maturity. In chronic leukemia, there is an accumulation of mature leucocytes in the bloodstream because they do not die at the end of their normal life span. The *human T cell leukemia-lymphoma virus-1* (*HTLV-1*) is strongly associated with leukemia. Leukemia is also classified according to the identity and site of origins of the predominant cell involved, such as myelocytic (myelogenous, myeloblastic, granulocytic), lymphocytic (lymphogenous, lymphatic), and monocytic.

In acute leukemia, the anemia and bleeding problems commonly seen result from the crowding out of normal bone marrow cells by the overproduction of immature cells, preventing normal production of red blood cells and platelets. One cause of death from acute leukemia is internal hemorrhaging, especially cerebral hemorrhage that destroys the vital centers in the brain. Perhaps the most frequent cause of death is uncontrolled infection due to lack of mature or normal white blood cells. The abnormal accumulation of immature leucocytes may be reduced by using X rays and antileukemic drugs. Partial or complete remissions may be induced, with some lasting as long as 15 years.

MEDICAL TERMINOLOGY ASSOCIATED WITH BLOOD

Autologous (aw-TOL-o-gus; *auto* = self) **transfusion** (trans-FYOO-zhun) Donating one's own blood for up to six weeks before elective surgery to ensure an abundant supply and reduce transfusion complications such as those that may be associated with AIDS and hepatitis. It is not indicated for individuals who cannot donate 500 ml of blood in a short period of time and who have preexisting anemia, bleeding disorders, infection, unstable blood pressure, and malignancy involving bone marrow. Also called **predonation.**

Autologous intraoperative transfusion (AIT) Procedure in which blood lost during surgery is suctioned from the patient, treated with an anticoagulant, filtered of debris, and centrifuged to recover the red blood cells. Then the red blood cells are washed in saline solution and reinfused into the patient.

Blood bank A stored supply of blood for future use by the donor or other individuals. Since blood banks have now assumed additional and diverse functions (immunohematology reference work, continuing medical education, bone and tissue storage, and clinical consultation), they are more appropriately referred to as **centers of transfusion medicine.**

Citrated (SIT-rāt-ed) **whole blood** Whole blood protected from coagulation by CPD (citrate phosphate dextrose) or a similar compound.

Cyanosis (sī'-a-NŌ-sis; *cyano* = blue) Slightly bluish, dark purple skin coloration due to oxygen deficiency in systemic blood.

Direct (immediate) transfusion (*trans* = through) Transfer of blood directly from one person to another without exposing the blood to a storage container.

Exchange transfusion Removing blood from the recipient while alternately replacing it with donor blood. This method is used for treating hemolytic disease of the newborn (HDN) and poisoning. This may be done intrauterine.

Gamma globulin (GLOB-yoo-lin) Solution of globulins from nonhuman blood consisting of antibodies that react with specific pathogens, such as measles, epidemic hepatitis, tetanus, and possibly poliomyelitis viruses. It is prepared by injecting the specific virus into animals, removing blood from the animals after antibodies have accumulated, isolating antibodies, and injecting them into a human for short-term immunity.

Hemochromatosis (hē-mō-krō'-ma-TŌ-sis; *heme* = iron; *chroma* = color) Disorder of iron metabolism characterized by excess deposits of iron in tissues, especially the liver and pancreas, that result in bronze coloration of the skin, cirrhosis, diabetes mellitus, and bone and joint abnormalities.

Hemorrhage (HEM-or-ij; *rrhage* = bursting forth)

MEDICAL TERMINOLOGY ASSOCIATED WITH BLOOD (*continued*)

Bleeding, either internal (from blood vessels into tissues) or external (from blood vessels directly to the surface of the body).

Indirect (mediate) transfusion Transfer of blood from a donor to a container and then to the recipient, permitting blood to be stored for an emergency. The blood may be separated into its components so that a patient will receive only a needed part.

Multiple myeloma (mī'-e-LŌ-ma) Malignant disorder of plasma cells in bone marrow; symptoms (pain, osteoporosis, hypercalcemia, thrombocytopenia, kidney damage) are caused by the growing tumor cell mass or antibodies produced by malignant cells.

Platelet (PLĀT-let) **concentrates** A preparation of platelets obtained from freshly drawn whole blood and used for transfusions in platelet-deficiency disorders such as hemophilia.

Reciprocal (re-CIP-rō-cal) **transfusion** Transfer of blood from a person who has recovered from a contagious infection into the vessels of a patient suffering with the same infection. An equal amount of blood is returned from the patient to the well person.

Septicemia (sep'-ti-SĒ-mē-a; *sep* = decay; *emia* = condition of blood) Toxins or disease-causing bacteria in the blood. Also called "blood poisoning."

Thrombocytopenia (throm'-bō-sī'-tō-PĒ-nē-a; *thrombo* = clot; *penia* = poverty) Very low platelet count that results in a tendency to bleed from capillaries.

Transfusion (trans-FYOO-zhun) Transfer of whole blood, blood components (red blood cells only or plasma only), or bone marrow directly into the blood stream.

Venesection (vēn'-e-SEK-shun; *veno* = vein) Opening of a vein for withdrawal of blood. Although **phlebotomy** (fle-BŌT-ō-mē; *phlebo* = vein; *tome* = to cut) is a synonym for venesection, in clinical practice, phlebotomy refers to therapeutic bloodletting, such as might be done to remove a pint of blood to lower the viscosity of blood of a patient with polycythemia.

Whole blood Blood containing all formed elements, plasma, and plasma solutes in natural concentration.

STUDY OUTLINE

Physical Characteristics of Blood (p. 546)

1. The cardiovascular system consists of blood, the heart, and blood vessels. The lymphatic system consists of lymph, lymph vessels, and lymph glands.
2. Physical characteristics of blood include viscosity, 4.5 to 5.5; temperature, 38°C (100.4°F); pH, 7.35 to 7.45; and a salt (NaCl) concentration of 0.90 percent. Blood constitutes about 8 percent of body weight.
3. Blood samples may be obtained by venepuncture, fingerstick, or arterial stick.

Functions of Blood (p. 547)

1. Blood transports oxygen, carbon dioxide, nutrients, wastes, and hormones.
2. It helps to regulate pH, body temperature, and water content of cells.
3. It prevents blood loss through clotting and combats toxins and microbes through special combat-unit cells.

Components of Blood (p. 547)

1. The formed elements in blood include erythrocytes (red blood cells), leucocytes (white blood cells), and thrombocytes (platelets).
2. Blood cells are formed by a process called hemopoiesis.
3. Red bone marrow (myeloid tissue) is responsible for producing red blood cells, granular leucocytes, and platelets; lymphoid tissue and myeloid tissue produce agranular leucocytes.

Erythrocytes (p. 548)

1. Erythrocytes are biconcave discs without nuclei and containing hemoglobin.
2. The function of red blood cells is to transport oxygen and carbon dioxide.
3. Red blood cells live about 120 days. A healthy male has about 5.4 million/mm^3 of blood; a healthy female, about 4.8 million/mm^3.
4. An erythrocyte sedimentation rate (ESR) is a screening test for a wide variety of disorders but does not distinguish specific diseases.
5. Erythrocyte formation, called erythropoiesis, occurs in adult red marrow of certain bones.
6. A reticulocyte count is a diagnostic test that indicates the rate of erythropoiesis.
7. A hematocrit (Hct) measures the percentage of red blood cells in whole blood.

Leucocytes (p. 552)

1. Leucocytes are nucleated cells. Two principal types are granular (neutrophils, eosinophils, basophils) and agranular (lymphocytes and monocytes).
2. Histocompatibility testing is performed to determine compatibility between a donor and recipient.
3. The general function of leucocytes is to combat inflammation

and infection. Neutrophils and monocytes (wandering macrophages) do so through phagocytosis.
4. Eosinophils combat the effects of histamine in allergic reactions, phagocytize antigen–antibody complexes, and combat parasitic worms; basophils liberate heparin, histamine, and serotonin in allergic reactions that intensify the inflammatory response.
5. Lymphocytes, in response to the presence of foreign substances called antigens, differentiate into tissue plasma cells that produce antibodies. Antibodies attach to the antigens and render them harmless. This antigen–antibody response combats infection and provides immunity.
6. A differential white blood cell count is a diagnostic test in which specific white blood cells are enumerated.
7. White blood cells usually live for only a few hours or a few days. Normal blood contains 5,000 to 10,000/mm^3.

Thrombocytes (p. 555)

1. Thrombocytes are disc-shaped structures without nuclei.
2. They are formed from megakaryocytes and are involved in clotting.
3. Normal blood contains 250,000 to 400,000/mm^3.
4. A complete blood count (CBC) is used to determine blood cell counts, hemoglobin, and hematocrit and comments about blood cell morphology.

Plasma (p. 556)

1. The liquid portion of blood, called plasma, consists of 91.5 percent water and 8.5 percent solutes.
2. Principal solutes include proteins (albumins, globulins, fibrinogen), nonprotein nitrogen (NPN) substances, foods, hormones, respiratory gases, and electrolytes.
3. Plasmapheresis involves withdrawing blood from the body, selectively separating its components, removing undesirable components, and returning it to the body.

Physiology of Hemostasis (p. 557)

1. Hemostasis refers to the prevention of blood loss.
2. It involves vascular spasm, platelet plug formation, and blood coagulation.
3. In vascular spasm, the smooth muscle of a blood vessel wall contracts to stop bleeding.
4. Platelet plug formation involves the clumping of platelets to stop bleeding.
5. A clot is a network of insoluble protein (fibrin) in which formed elements of blood are trapped.
6. The chemicals involved in clotting are known as coagulation factors. There are two kinds: plasma and platelet coagulation factors.
7. Blood clotting involves two pathways: the intrinsic and the extrinsic.
8. Hemophilia refers to several different hereditary deficiencies of coagulation in which bleeding may occur spontaneously or only after minor trauma.
9. Normal coagulation also involves clot retraction (tightening of the clot) and fibrinolysis (dissolution of the clot). Clots may be dissolved by injection of streptokinase or tissue plasminogen activator (t-PA).
10. Clotting in an unbroken blood vessel is called thrombosis. A thrombus that moves from its site of origin is called an embolus.
11. Anticoagulants (e.g., heparin) prevent clotting.
12. Clinically important clotting tests are prothrombin time, partial thromboplastin time, and bleeding time.

Grouping (Typing) of Blood (p. 562)

1. ABO and Rh systems are based on antigen–antibody responses.
2. In the ABO system, agglutinogens (antigens) *A* and *B* determine blood type. Plasma contains agglutinins (antibodies), designated as *a* and *b,* that clump agglutinogens that are foreign to the individual.
3. In the Rh system, individuals whose erythrocytes have Rh agglutinogens are classified as Rh^+. Those who lack the antigen are Rh^-.
4. A disorder due to Rh incompatibility between mother and fetus is called hemolytic disease of the newborn (erythroblastosis fetalis).

Interstitial Fluid and Lymph (p. 565)

1. Interstitial fluid bathes body cells, whereas lymph is found in lymphatic vessels.
2. These fluids are similar in chemical composition. They differ chemically from plasma in that both contain less protein and a variable number of leucocytes. Like plasma, they contain no platelets or erythrocytes.

Disorders: Homeostatic Imbalances (p. 566)

1. Anemia is a decreased erythrocyte count or hemoglobin deficiency. Kinds of anemia include nutritional, pernicious, hemorrhagic, hemolytic, aplastic, and sickle-cell anemia (SCA).
2. Polycythemia is an abnormal increase in the number of erythrocytes.
3. Infectious mononucleosis (IM) is a contagious disease that primarily affects lymphoid tissue. It is characterized by an elevated white blood cell count, with an abnormally high percentage of lymphocytes. The cause is the Epstein-Barr virus (EBV).
4. Chronic fatigue syndrome is characterized by extreme fatigue for at least six months and the absence of known diseases that might produce similar symptoms. Among the symptoms are sore throat, headache, muscular aches, fever and chills, fatigue, joint pain, and neurological defects.
5. Leukemia is a malignant disease of blood-forming tissues characterized by the uncontrolled production of white blood cells that interferes with normal clotting and vital body activities.

REVIEW QUESTIONS

1. How are blood, interstitial fluid, and lymph related to the maintenance of homeostasis? (p. 546)
2. Distinguish between the cardiovascular system and lymphatic system. (p. 546)

3. List the principal physical characteristics of blood. (p. 546)
4. List the functions of blood and their relationship to other systems of the body. (p. 547)
5. How are blood samples obtained for laboratory testing? (p. 546)
6. Distinguish between plasma and formed elements. (p. 547)
7. Describe the origin of blood cells. (p. 548)
8. Describe the microscopic appearance of erythrocytes. What is the function of erythrocytes? What is induced erythrocytemia (blood doping)? (p. 548)
9. Define erythropoiesis. Relate erythropoiesis to red blood cell count. What factors accelerate and decelerate erythropoiesis? (p. 551)
10. Describe the classification of leucocytes and describe the microscopic appearance. What are their functions? (p. 552)
11. What is the importance of diapedesis, chemotaxis, and phagocytosis in fighting bacterial invasion? (p. 553)
12. Describe a bone marrow transplant and its use in the treatment of diseases such as multiple myeloma. (p. 555)
13. Distinguish between leucocytosis and leucopenia. (p. 555)
14. Describe the antigen–antibody response. How is it protective? (p. 553)
15. Describe the structure and function of thrombocytes. (p. 555)
16. Compare erythrocytes, leucocytes, and thrombocytes with respect to size, number per mm^3, and life span. (p. 556)
17. What are the advantages of the blood substitute Fluosol-DA? (p. 550)
18. What are the major constituents in plasma? What do they do? What is the difference between plasma and serum? (p. 557)
19. What is plasmapheresis? What are its clinical applications? (p. 557)
20. Define hemostasis. Explain the mechanism involved in vascular spasm and platelet plug formation. (p. 557)
21. Briefly describe the process of clot formation. What is fibrinolysis? Why does blood usually not remain clotted in vessels? (p. 558)
22. How do the extrinsic and intrinsic pathways of blood clotting differ? (p. 559)
23. What is hemophilia? Describe its signs and symptoms. (p. 559)
24. Define the following: thrombus, embolus, anticoagulant. (p. 562)
25. What is the basis for ABO blood grouping? What are agglutinogens and agglutinins? (p. 562)
26. What is the basis for the Rh system? How does hemolytic disease of the newborn (erythroblastosis fetalis) occur? How may it be prevented? (p. 564)
27. Compare interstitial fluid and lymph with regard to location, chemical composition, and function. (p. 565)
28. Define anemia. Contrast the causes of nutritional, pernicious, hemorrhagic, hemolytic, aplastic, and sickle-cell anemia (SCA). (p. 566)
29. What is infectious mononucleosis (IM)? (p. 567)
30. Describe the symptoms of chronic fatigue syndrome. (p. 567)
31. What is leukemia, and what are the causes of some of its symptoms? (p. 568)
32. Briefly describe the diagnostic value and procedure for the following blood tests: erythrocyte sedimentation rate (ESR) (p. 551), reticulocyte count (p. 552), hematocrit (Hct) (p. 552), histocompatibility testing (p. 553), differential white blood cell count (p. 554), complete blood count (CBC) (p. 556), prothrombin time (PT) (p. 562), partial thromboplastin time (PTT) (p. 562), bleeding time (p. 562), and blood typing. (p. 564)
33. Refer to the glossary of medical terminology associated with blood. Be sure that you can define each term. (p. 568)

SELECTED READINGS

Aledort, L. M. "Current Concepts in Diagnosis and Management of Hemophilia," *Hospital Practice,* October 1982.

Bloom, M. (ed). "New Jobs for Man-Made Blood," *Physician's Weekly,* 17 February 1986.

Bock, H. (ed). "New Method for Saving Your Own Blood," *Medical Update,* October 1988.

Dixon, B. "Of Different Bloods," *Science 84,* November 1984.

Doolittle, R. F. "Fibrinogen and Fibrin," *Scientific American,* December 1981.

England, J. A. "The Many Faces of Epstein-Barr Virus," *Postgraduate Medicine,* February 1988.

Froberg, J. H. "The Anemias: Causes and Course of Action," *RN,* January, March, and May 1989.

Golde, D. W., and J. C. Ganon. "Hormones that Stimulate the Growth of Blood Cells," *Scientific American,* July 1988.

Goldfinger, S. E. (ed). "Chronic Fatigue Syndrome," *Harvard Medical School Health Letter,* July 1988.

Kiley, J. M. (ed). "Chronic Mononucleosis," *Mayo Clinic Health Letter,* June 1988.

Lefant, C. "Dissolving Blood Clots," *Medical Update,* March 1988.

Sabetta, J. R. "Diagnosis: Infectious Mononucleosis," *Hospital Medicine,* March 1984.

Silberner, J. "Clot Dissolver May Save Heart Tissue," *Science News,* 26 November 1983.

Spivak, J. *Fundamentals of Clinical Hematology,* 2nd ed. New York: Harper & Row, 1984.

Thorup, O. A., et al. *Leavell and Thorup's Fundamentals of Clinical Hematology,* 5th ed. Philadelphia: Saunders, 1987.

Vichinsky, E. P., D. Hurst, and B. Lubin. "Sickle Cell Disease: Basic Concepts," *Hospital Medicine,* September 1983.

Zucker, M. D. "The Functioning of Blood Platelets," *Scientific American,* June 1980.

Chapter 20

The Cardiovascular System: The Heart

Chapter Contents at a Glance

Student Objectives

1. Describe the structure and functions of the chambers, great vessels, and valves of the heart.
2. Explain the structural and functional features of the conduction system of the heart.
3. Describe the principal events of a cardiac cycle.
4. Contrast the sounds of the heart and their clinical significance.
5. Explain the various factors that affect heart rate and cardiac output (CO).
6. List the risk factors involved in heart disease.
7. Describe how atherosclerosis and coronary artery spasm contribute to coronary artery disease (CAD).
8. Describe coarctation of the aorta, patent ductus arteriosus, septal defects, valvular stenosis, and tetralogy of Fallot as congenital heart defects and atrioventricular (AV) block, atrial flutter, atrial fibrillation, and ventricular fibrillation as abnormalities of the conduction system of the heart (arrhythmias).
9. Define congestive heart failure (CHF) and cor pulmonale (CP).
10. Define medical terminology associated with the heart.

The ***heart*** is the center of the cardiovascular system. Whereas the term *cardio* refers to the heart, the term *vascular* refers to blood vessels (or an abundant blood supply). The heart is a hollow, muscular organ that weighs between 250 and 350 grams (9–12 oz) and beats over 100,000 times a day to pump 7000 liters (1835 gallons) of blood per day through over 60,000 miles of blood vessels. The blood vessels form a network of tubes that carry blood from the heart to the tissues of the body and then return it to the heart. The specific aspects of the heart that we will consider are its location, covering, wall and chambers, great vessels, valves, surface anatomy, conduction system, electrocardiogram (ECG), cardiac cycle (heartbeat), cardiac output (CO), and autonomic control. Several disorders related to the heart will also be considered.

The study of the normal heart and diseases associated with it is known as ***cardiology*** (kar-dē-OL-ō-jē; *cardio* = heart).

The developmental anatomy of the heart is considered at the end of the chapter.

LOCATION OF HEART

The heart is situated between the lungs and is a component of the mediastinum, the mass of tissue between the lungs that extends from the sternum to the vertebral column. About two-thirds of the mass of the heart lies to the left of the body's midline (Figure 20-1). The heart is shaped like a blunt cone about the size of your closed fist—12 cm (5 inches) long, 9 cm ($3\frac{1}{2}$ inches) wide at its broadest point, and 6 cm ($2\frac{1}{2}$ inches) thick.

Its pointed end, the ***apex,*** is formed by the tip of the left ventricle and projects inferiorly, anteriorly, and to the left. The ***left border*** is formed almost entirely by the left ventricle, although the left atrium forms part of the upper end of the border. The ***superior border,*** where the great vessels enter and leave the heart, is formed by both atria. The ***base*** of the heart projects superiorly, posteriorly, and to the right. It is formed by the atria, mostly the left atrium. The ***right border*** is formed by the right atrium. The ***inferior border*** is formed by the right ventricle and slightly by the left ventricle. The ***sternocostal***

FIGURE 20-1 Position of the heart and associated blood vessels in the thoracic cavity. In this and subsequent illustrations, vessels that carry oxygenated blood are shown in red; vessels that carry deoxygenated blood are shown in blue.

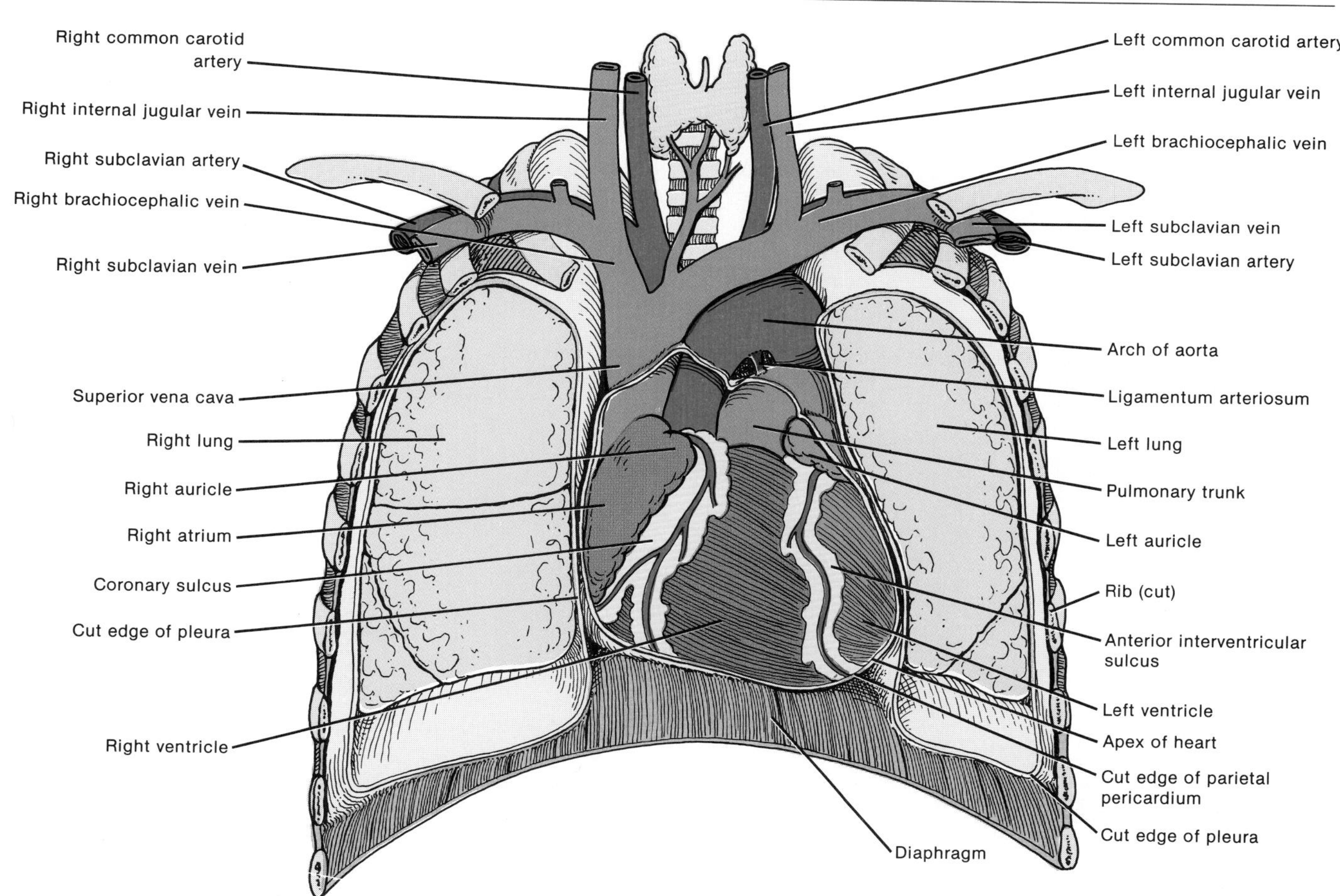

(anterior) surface is formed mainly by the right ventricle, left ventricle, and right atrium, whereas the ***diaphragmatic (inferior) surface*** is formed by the left and right ventricles, mostly the left.

PERICARDIUM

The heart is enclosed and held in place by the ***pericardium,*** an ingenious structure designed to confine the heart to its position in the mediastinum, yet allow it sufficient freedom of movement so that it can contract vigorously and rapidly when the need arises.

The pericardium consists of two portions referred to as the fibrous pericardium and the serous pericardium (Figure 20-2a). The outer ***fibrous pericardium*** consists of very heavy fibrous connective tissue. The fibrous pericardium resembles a bag that rests on and is attached to the diaphragm with its open end fused to the connective tissues of the great vessels entering and leaving the heart. The fibrous pericardium prevents overdistension of the heart, provides a tough protective membrane around the heart, and anchors the heart in the mediastinum. The inner ***serous pericardium*** is a thinner, more delicate membrane that forms a double layer around the heart (Figure 20-2b). The outer ***parietal layer*** of the serous pericardium is directly beneath and fused to the fibrous pericardium. The inner ***visceral layer*** of the serous pericardium, also called the ***epicardium,*** is beneath the parietal layer, attached to the myocardium (muscle) of the heart. Between the parietal and visceral layers of the serous pericardium is a thin film of serous fluid that holds the two layers together much like a thin film of water does to two microscope slides. The serous fluid, known as ***pericardial fluid,*** is an ultrafiltrate of plasma and prevents friction between the membranes as the heart moves. There are up to 50 ml of pericardial fluid. The space occupied by the pericardial fluid is a potential space (not an actual space) called the ***pericardial cavity.***

An inflammation of the pericardium is known as ***pericarditis.*** Pericarditis with a buildup of pericardial fluid or extensive bleeding into the pericardium, if untreated, is a life-threatening condition. Since the pericardium cannot stretch to accommodate the excessive fluid or blood buildup, the heart is subjected to compression. This compression is known as ***cardiac tamponade*** (tam'-pon-ĀD) and can result in cardiac failure.

HEART WALL

The wall of the heart (Figure 20-2a) is divided into three layers: the epicardium (external layer), myocardium (middle layer), and endocardium (inner layer). The ***epicardium*** (also called the ***visceral layer of the serous pericardium***) is the thin, transparent outer layer of the wall. It is composed of serous tissue and mesothelium.

The ***myocardium,*** which is cardiac muscle tissue, constitutes the bulk of the heart. Cardiac muscle fibers (cells) are involuntary, striated, and branched, and the tissue is arranged in interlacing bundles of fibers. The myocardium is responsible for the contraction of the heart.

The ***endocardium*** is a thin layer of endothelium overlying a thin layer of connective tissue. It lines the inside of the myocardium and covers the valves of the heart and the tendons that attach to some of them. It is continuous with the endothelial lining of the large blood vessels of the heart and the remainder of the cardiovascular system.

Inflammation of the epicardium, myocardium, and endocardium is referred to as ***epicarditis, myocarditis,*** and ***endocarditis,*** respectively.

FIGURE 20-2 Pericardium and heart wall. (a) Structure of the pericardium and heart wall. (b) Relationship of the serous pericardium to the heart.

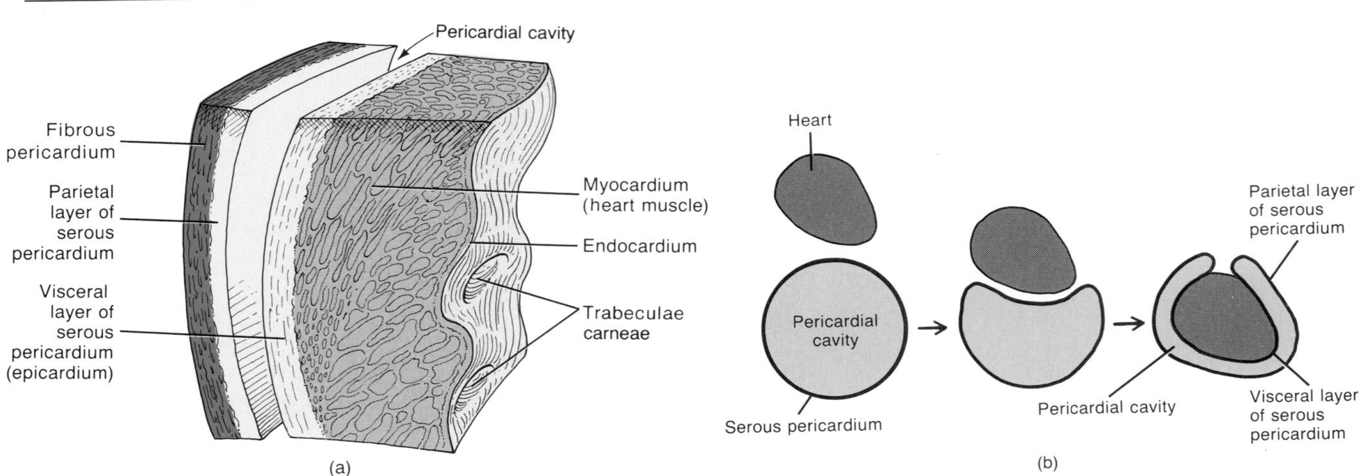

CHAMBERS OF THE HEART

The interior of the heart is divided into four cavities called ***chambers*** that receive the circulating blood (Figure 20-3). The two superior chambers are called the right and left ***atria*** (*atrium* = court or hall). Each atrium has an appendage called an ***auricle*** (OR-i-kul; *auris* = ear), so named because its shape resembles a dog's ear. The auricle increases the atrium's surface area. The lining of the atria is smooth, except for their anterior walls and the lining of the auricles, which contain projecting muscle bundles that are parallel to one another and resemble the teeth of a comb: they are called ***pectinate*** (PEK-ti-nāt) ***muscles.*** These bundles give the lining of the auricles a ridged appearance.

The atria are separated by a partition called the ***interatrial septum.*** A prominent feature of this septum is an oval depression, the ***fossa ovalis,*** which corresponds to the site of the foramen ovale, an opening in the interatrial septum of the fetal heart. The fossa ovalis faces the opening of the inferior vena cava and is located in the septal wall of the right atrium (see Figure 21-24).

The two inferior chambers are the right and left ***ventricles.*** They are separated from each other by an ***interventricular septum.***

The muscle tissue of the atria and ventricles is separated by connective tissue that also forms the valves. This "cardiac skeleton" effectively divides the myocardium into two separate muscle masses. Externally, a groove known as the ***coronary sulcus*** (SUL-kus) separates the atria from the ventricles. It encircles the heart and houses the coronary sinus and circumflex branch of the left coronary artery. The ***anterior interventricular sulcus*** and ***posterior interventricular sulcus*** separate the right and left ventricles externally. The sulci contain coronary blood vessels and a variable amount of fat (Figure 20-3a, c).

FIGURE 20-3 Structure of the heart. (a) Diagram of anterior external view.

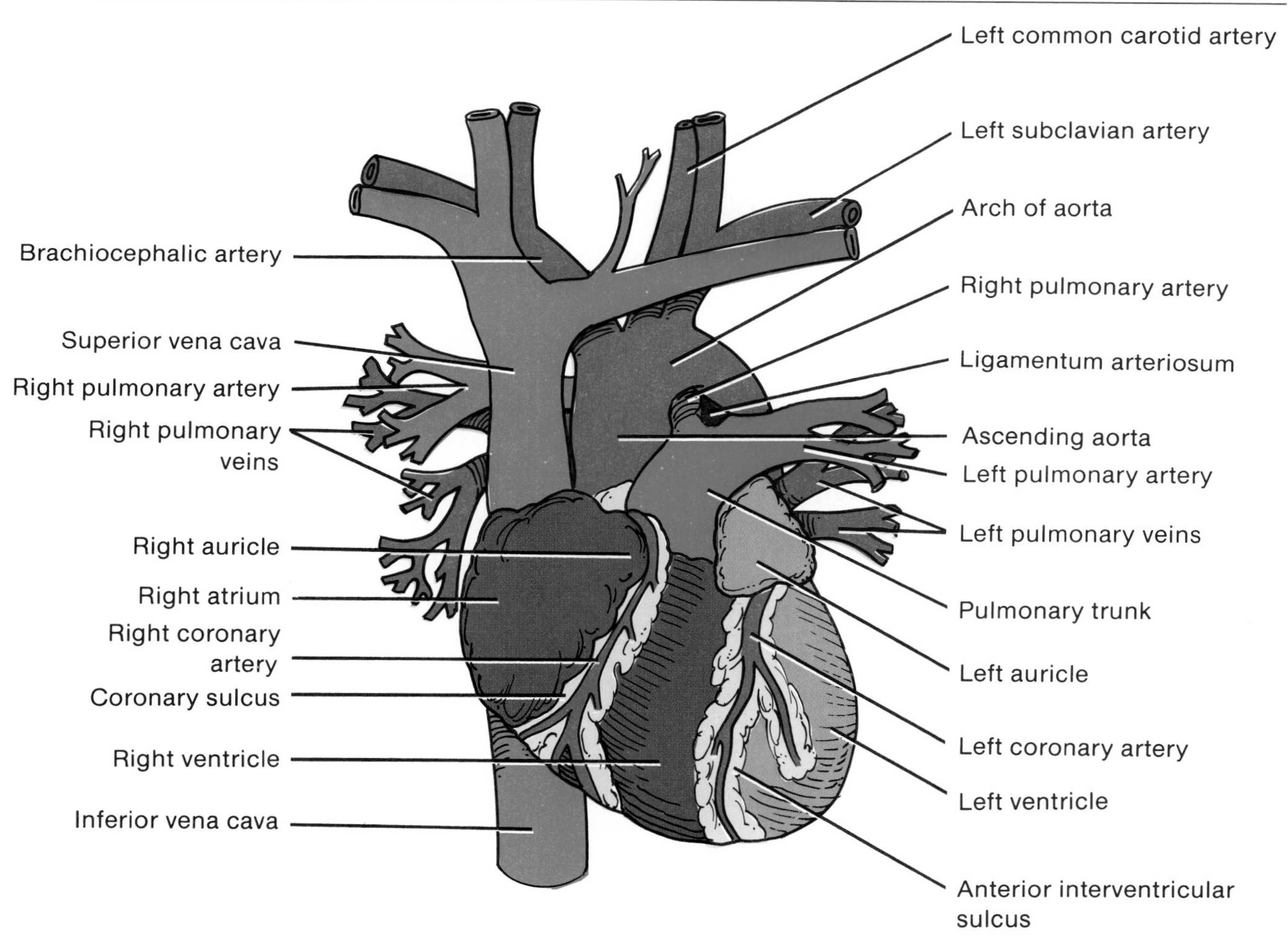

FIGURE 20-3 (*Continued*) (b) Photograph of anterior external view. (Courtesy of J. A. Gosling, P. F. Harris, et al., *Atlas of Human Anatomy,* Gower Medical Publishing Ltd., 1985.)

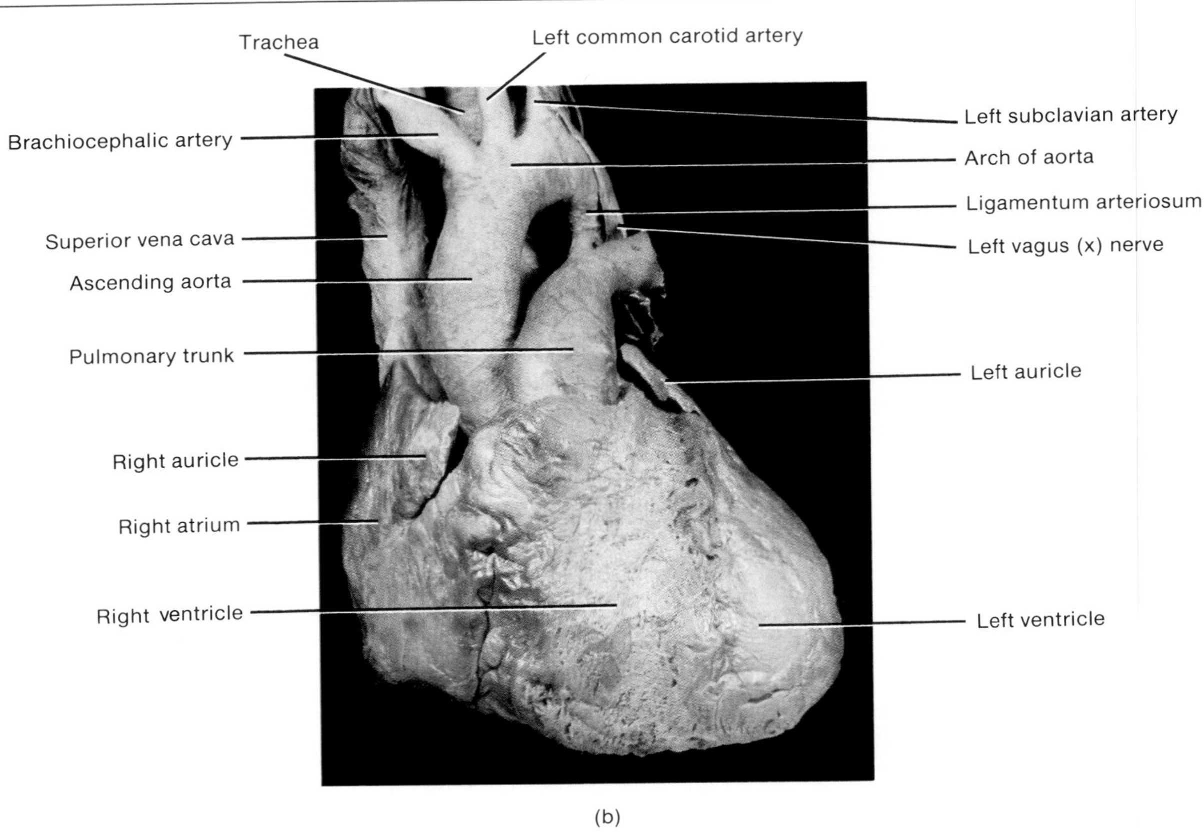

(b)

GREAT VESSELS OF THE HEART

The right atrium receives blood from all parts of the body except the lungs. It receives the blood through three veins. In general, the ***superior vena cava (SVC)*** brings blood from parts of the body superior to the heart; in general, the ***inferior vena cava (IVC)*** brings blood from parts of the body inferior to the heart; and the ***coronary sinus*** drains blood from most of the vessels supplying the wall of the heart (Figure 20-3c). The right atrium then delivers the blood into the right ventricle, which pumps it into the ***pulmonary trunk.*** The pulmonary trunk divides into a ***right*** and ***left pulmonary artery,*** each of which carries blood to the lungs. In the lungs, the blood releases its carbon dioxide and takes on oxygen. Blood returns to the heart via four ***pulmonary veins*** that empty into the left atrium. The blood then passes into the left ventricle, which pumps the blood into the ***ascending aorta*** (*aorte* = to suspend, because the aorta was once believed to suspend the heart). From here the blood is passed into the ***coronary arteries, arch of the aorta, thoracic aorta,*** and ***abdominal aorta.*** These blood vessels and their branches transport the blood to the heart and all other body parts, except the lungs.

During fetal life, there is a temporary blood vessel, called the ductus arteriosus, that connects the pulmonary trunk with the aorta (see Figure 21-24). Its purpose is to redirect blood so that only a small volume enters the nonfunctioning fetal lungs. The ductus arteriosus normally closes shortly after birth, leaving a remnant known as the ***ligamentum arteriosum.***

The thickness of the four chambers varies according to function (Figure 20-3d). The atria are thin-walled because they need only enough cardiac muscle tissue to deliver the blood into the ventricles with the aid of gravity and a reduced pressure created by the expanding ventricles. The right ventricle has a thicker layer of myocardium than the atria, since it must send blood to the lungs and back around to the left atrium. The left ventricle has the thickest wall, since it must pump blood at high pressure through literally thousands of miles of vessels in the head, trunk, and extremities.

VALVES OF THE HEART

As each chamber of the heart contracts, it pushes a portion of blood into a ventricle or out of the heart through an artery. In order to keep the blood from flowing backward,

FIGURE 20-3 (*Continued*) (c) Diagram of posterior external view.

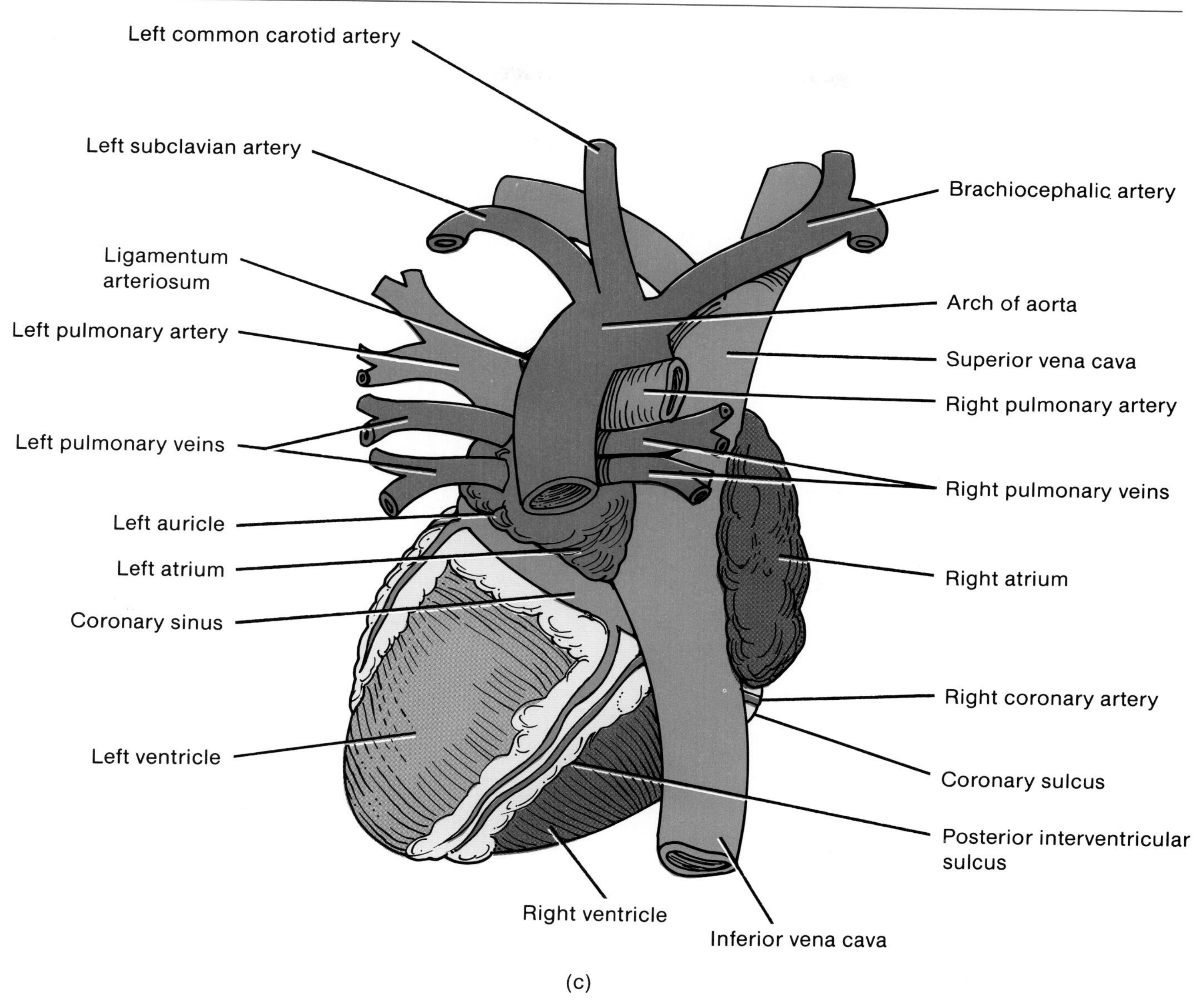

(c)

the heart has structures composed of dense connective tissue covered by endothelium called ***valves.***

Atrioventricular (AV) Valves

Atrioventricular (AV) valves lie between the atria and ventricles (Figure 20-3d). The right atrioventricular valve between the right atrium and right ventricle is also called the ***tricuspid valve*** because it consists of three cusps (flaps). These cusps are fibrous tissues that grow out of the walls of the heart and are covered with endocardium. The pointed ends of the cusps project into the ventricle. Tendonlike fibrous cords called ***chordae tendineae*** (KOR-dē ten-DIN-ē-ē) connect the pointed ends and undersurfaces to small conical projections—the ***papillary muscles*** (muscular columns)—located on the inner surface of the ventricles. The irregular surface of ridges and folds of the myocardium in the ventricles is known as the ***trabeculae carneae*** (tra-BEK-yoo-lē KAR-nē-ē). The chordae tendineae and their papillary muscles keep the cusps of the atrioventricular valves from being inverted into the atria when the ventricles contract. The left atrioventricular valve between the left atrium and left ventricle is called the ***bicuspid (mitral) valve.*** It has two cusps that work in the same way as the cusps of the tricuspid valve. Its cusps are also attached by way of the chordae tendineae to papillary muscles.

In order for blood to pass from an atrium to a ventricle, an atrioventricular (AV) valve must open. The opening and closing of the valves is due to pressure differences across the valves. When blood moves from an atrium to

FIGURE 20-3 (*Continued*) (d) Diagram of anterior internal view. (e) Path of blood through the heart.

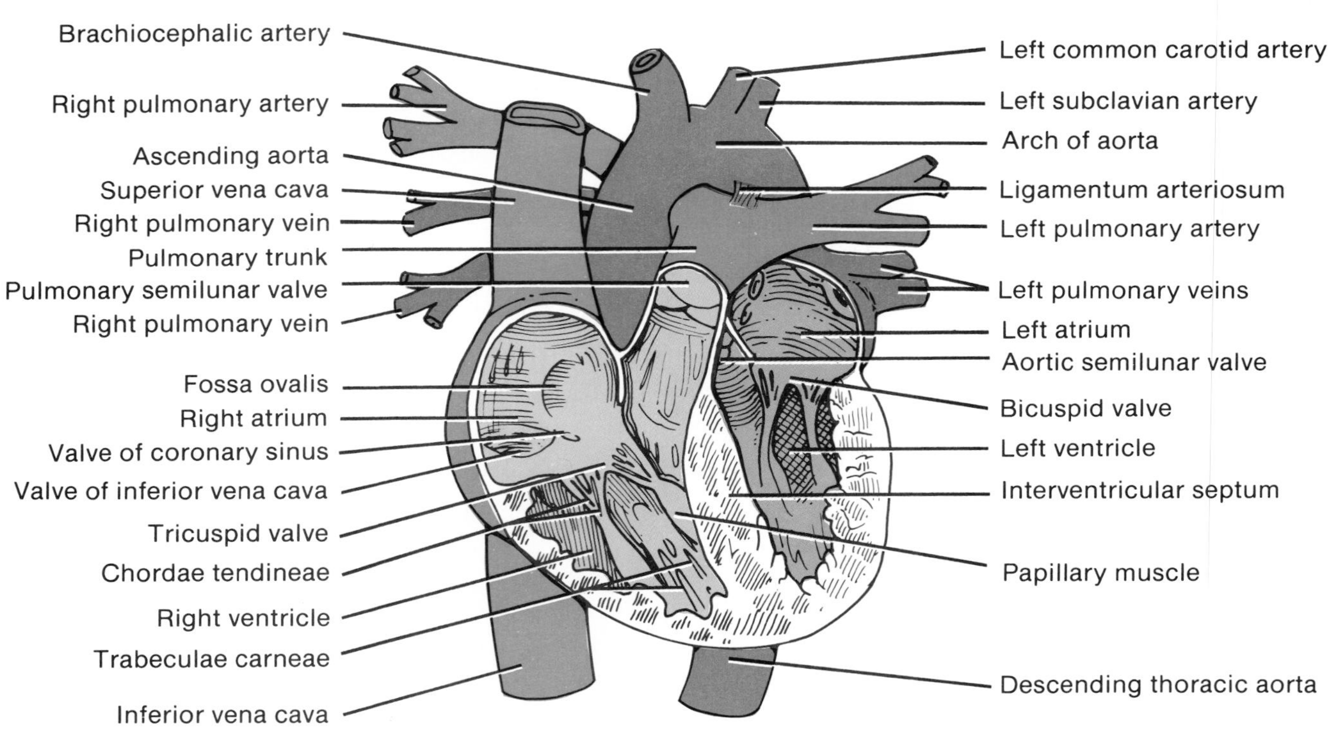

(d)

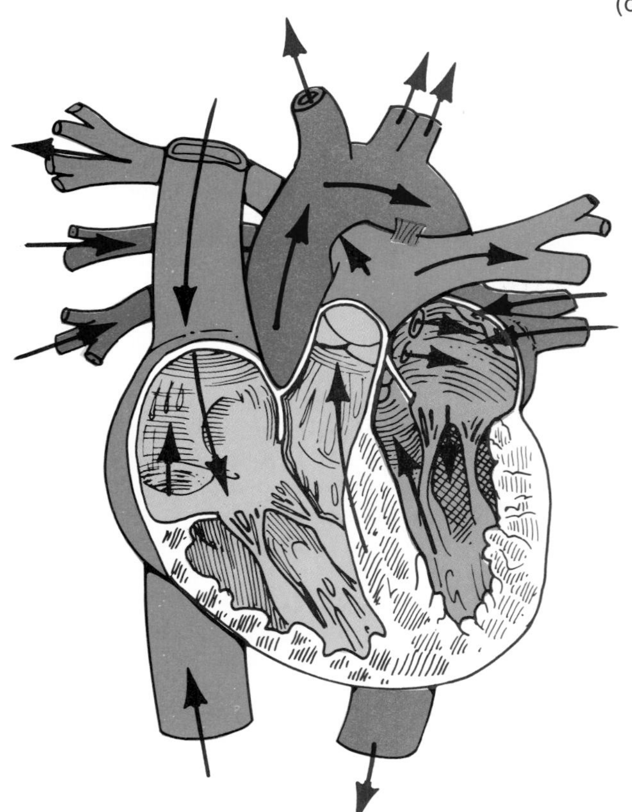

(e)

a ventricle, the valve is pushed open, the papillary muscles relax, and the chordae tendineae slacken (Figure 20-4a). When a ventricle contracts, the pressure of the blood drives the cusps upward until their edges meet and close the opening (Figure 20-4b). At the same time, contraction of the papillary muscles and tightening of the chordae tendineae help prevent the valve from swinging upward into the atrium.

Semilunar Valves

Both arteries that leave the heart have a valve that prevents blood from flowing back into the heart. These are the ***semilunar valves*** (see Figure 20-3d). The ***pulmonary semilunar valve*** lies in the opening where the pulmonary trunk leaves the right ventricle. The ***aortic semilunar valve*** is situated at the opening between the left ventricle and the aorta.

Both valves consist of three semilunar (half-moon or crescent-shaped) cusps. Each cusp is attached by its convex margin to the artery wall. The free borders of the cusps curve outward and project into the opening inside the blood vessel. Like the atrioventricular valves, the semilunar valves permit blood to flow in one direction only—in this case, the flow is from the ventricles into the arteries.

FIGURE 20-4 Atrioventricular (AV) valves. (a) Bicuspid valve open. (b) Bicuspid valve closed. The tricuspid valve operates in a similar manner. (c) Photograph of the tricuspid and bicuspid valves. (d) Photograph of the valves of the heart in superior view in which the atria have been removed. (Photographs courtesy of J. Willis Hurst et al., *Atlas of the Heart,* Gower Medical Publishing Ltd., 1988.)

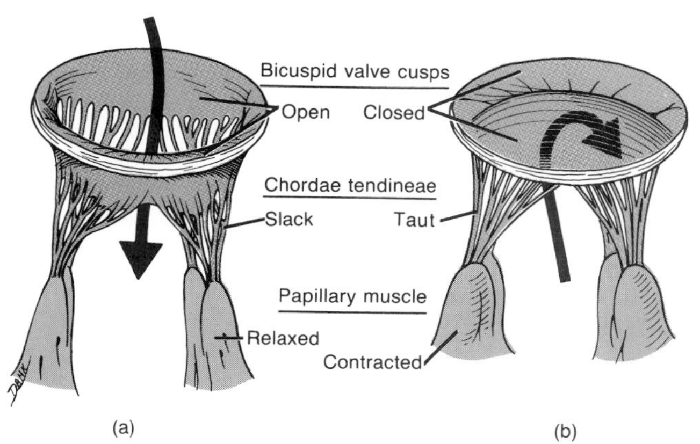

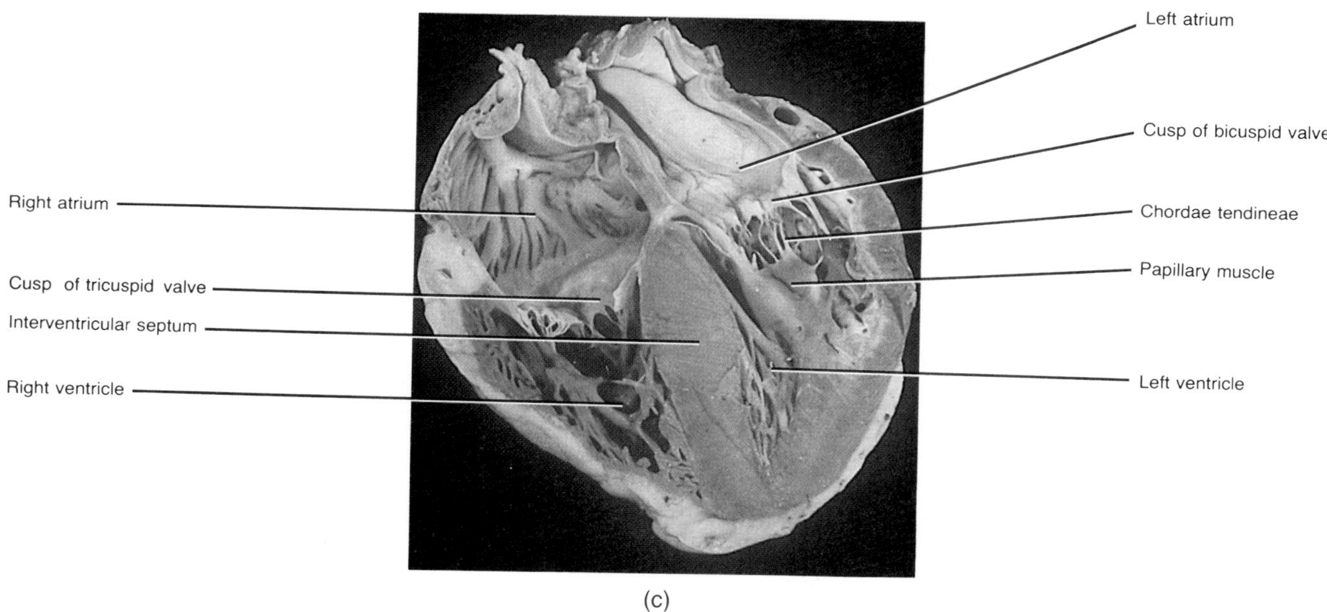

(c)

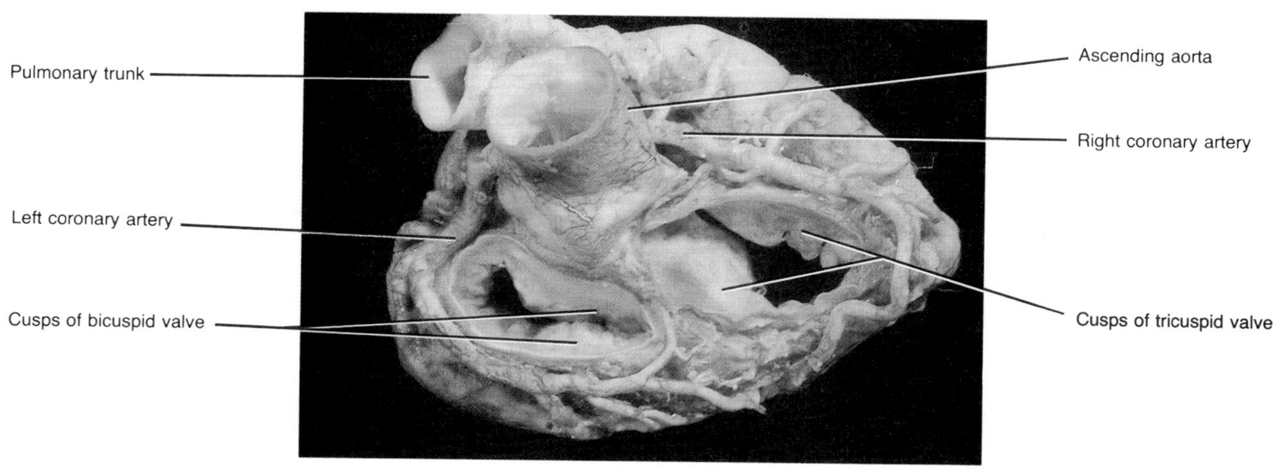

(d)

CLINICAL APPLICATION: RHEUMATIC FEVER

An infection with group A, β-hemolytic *Streptococcus pyogenes* bacteria may result in an inflammation of the heart that damages the valves, particularly the bicuspid and aortic semilunar valves. This condition is called ***rheumatic fever.*** It is usually precipitated by a streptococcal sore throat. The symptoms of rheumatic fever include arthritis, carditis, fever, subcutaneous nodules, skin rash, chorea (involuntary, abrupt, nonrepetitive limb movements and characteristic grimaces), and malaise. Although the exact mechanism by which streptococci produce rheumatic fever is still obscure, it is an immunologic reaction that is localized in the joints and heart. Antibodies that recognize molecules carried by streptococci attack similar molecules in joints, heart valves, and other tissues.

Surface Projection

The location of the valves of the heart may be identified by surface projection (Figure 20-5). The pulmonary and aortic semilunar valves are represented on the surface by a line about 2.5 cm (1 inch) in length. The pulmonary semilunar valve lies horizontally behind the inner end of the left third costal cartilage and the adjoining part of the sternum. The aortic semilunar valve is placed obliquely behind the left side of the sternum, at the level of the third intercostal space. The tricuspid valve lies behind the sternum, extending from the midline at the level of the fourth costal cartilage down toward the right sixth chondrosternal junction. The bicuspid valve lies behind the left side of the sternum obliquely at the level of the fourth costal cartilage. It is represented by a line about 3 cm in length.

Although heart sounds are produced in part by the closure of valves, they are not necessarily heard best over these valves. Each sound tends to be clearest in a slightly different location closest to the surface of the body (Figure 20-5).

BLOOD SUPPLY

The wall of the heart, like any other tissue, has its own blood vessels. Nutrients could not possibly diffuse through all the layers of cells that make up the heart tissue. The flow of blood through the numerous vessels that pierce the myocardium is called ***coronary (cardiac) circulation*** (Figure 20-6). The term *coronary* refers to the fact that the blood vessels of the heart resemble a crown.

The vessels that serve the myocardium include the ***left coronary artery,*** which originates as a branch of the ascending aorta. This artery runs under the left atrium and divides into the anterior interventricular and circumflex branches. The ***anterior interventricular branch*** follows the anterior interventricular sulcus and supplies oxygenated blood to the walls of both ventricles. The ***circumflex branch*** distributes oxygenated blood to the walls of the left ventricle and left atrium.

In January 1988, Pete Maravich, the leading scorer in National Collegiate Athletic Association history and a member of the Basketball Hall of Fame, collapsed during a pickup basketball game and died a short time later. He had a very rare defect, a missing left coronary artery. As a result, his heart was enlarged and weakened owing to a continuous decreased oxygen supply. Most people who have this condition usually die before age 20.

FIGURE 20-5 Surface projection of the heart. The red circles indicate where heart sounds caused by the respective valves are best heard.

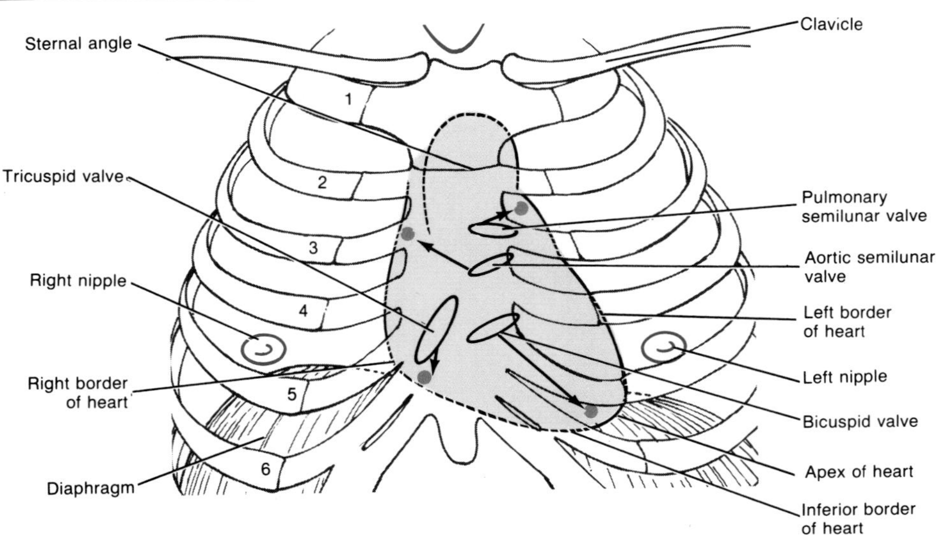

FIGURE 20-6 Coronary (cardiac) circulation. (a) Anterior view of arterial distribution. (b) Anterior view of venous drainage.

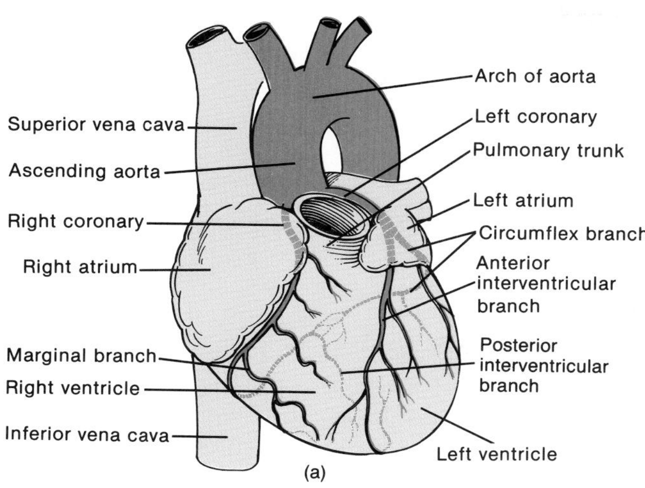

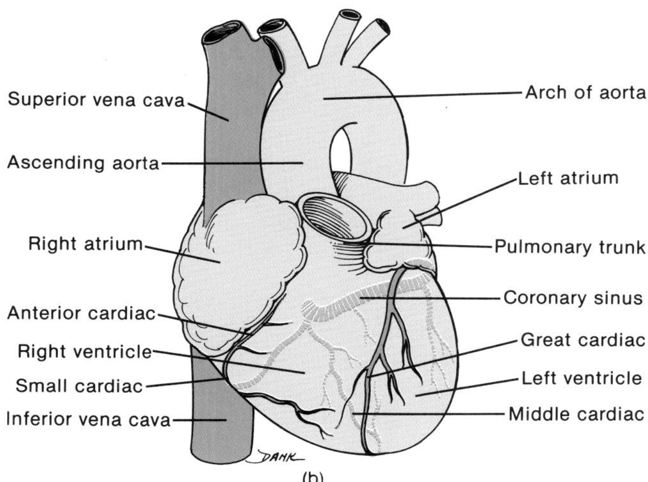

The ***right coronary artery*** also originates as a branch of the ascending aorta. In its course, it supplies small branches to the right atrium. It runs under the right atrium and divides into the posterior interventricular and marginal branches. The ***posterior interventricular branch*** follows the posterior interventricular sulcus and supplies the walls of the two ventricles with oxygenated blood. The ***marginal branch*** transports oxygenated blood to the myocardium of the right ventricle. The left ventricle receives the most abundant blood supply because of the enormous work it must do.

As blood passes through the coronary system of the heart, it delivers oxygen and nutrients and collects carbon dioxide and wastes. Most of the deoxygenated blood, which carries the carbon dioxide and wastes, is collected by a large vein, the ***coronary sinus,*** which empties into the right atrium. A vascular sinus is a vein with a thin wall that has no smooth muscle to alter its diameter. The principal tributaries of the coronary sinus are the ***great cardiac vein,*** which drains the anterior aspect of the heart, and the ***middle cardiac vein,*** which drains the posterior aspect of the heart.

Most parts of the body receive branches from more than one artery, and where two or more arteries supply the same region, they usually connect with each other. This is referred to as *anastomosis.* Anastomoses between arteries provide collateral circulation (alternate routes) for blood to reach a particular organ or tissue. The myocardium contains numerous anastomoses either connecting branches of one coronary artery or between branches of different coronary arteries. When a major coronary blood vessel is about 90 percent obstructed, blood will flow through the collateral vessels. Although most collaterals in the heart are quite small, heart muscle can remain alive as long as it receives as little as 10 to 15 percent of its normal supply.

CLINICAL APPLICATION: ANGINA PECTORIS AND MYOCARDIAL INFARCTION

Most heart problems result from faulty coronary circulation. If a reduced oxygen supply weakens cells but does not actually kill them, the condition is called ***ischemia*** (is-KĒ-mē-a). ***Angina pectoris*** (an-JĪ-na, or AN-ji-na, PEK-to-ris), meaning "chest pain," results from ischemia of the myocardium. Angina is typically described as a tightness or choking sensation or a squeezing, pressure-type discomfort and is usually of short duration. It occurs during exertion but is relieved by rest. (Pain impulses originating from most visceral muscles are referred to an area on the surface of the body. The pain associated with angina pectoris is referred to the neck, chin, or down the left arm to the elbow.) Common causes of angina include stress, strenuous exertion after a heavy meal, atherosclerosis, coronary artery spasm, hypertension, fever, anemia, hyperthyroidism, and aortic stenosis. The symptoms of angina pectoris include chest pain, accompanied by tightness or pressure, labored breathing, and a sensation of foreboding. Sometimes weakness, dizziness, and perspiration occur. Some individuals have angina pectoris but are unaware of it. This is known as ***silent myocardial ischemia.***

A much more serious problem is ***myocardial infarction*** (in-FARK-shun), or ***MI,*** commonly called a heart attack. ***Infarction*** means the death of an area of tissue because of an interrupted blood supply. Myocardial infarction may result from a thrombus or embolus in one of the coronary arteries. The tissue distal to the obstruction dies and is replaced by noncontractile scar tissue. Thus, the heart muscle loses at least some of its strength. The aftereffects depend partly on the size and location of the infarcted, or dead, area. In addition to killing normal heart tissue, the infarction may disturb the conducting system of the heart. This is why heart attacks may cause sudden death (ventricular fibrillation), which may, nev-

ertheless, be reversed by the timely and effective use of cardiopulmonary resuscitation (CPR). The most modern treatment for a myocardial infarction is to perform coronary angiography and inject a thrombolytic agent such as t-PA or to do coronary angioplasty (discussed at the end of the chapter).

Whenever a disease or injury deprives a tissue of oxygen, reestablishing the blood flow (reperfusion), and therefore reintroducing oxygen, may damage the tissue further. Some researchers believe that oxygen ***free radicals*** may play a major role in tissue destruction. Free radicals are electrically charged molecules in which the number of electrons is not equal to the number of protons. Such molecules are unstable and highly reactive because they must balance the charges. When an oxygen free radical takes an electron from one molecule, that molecule becomes unstable and borrows an electron from another molecule, which, in turn, becomes unstable. Such a chain reaction leads to cellular damage and death. Among the molecules attacked by oxygen free radicals are proteins (such as enzymes), neurotransmitters, nucleic acids, and phospholipids of plasma membranes. Free radicals have been implicated in diseases such as heart disease, cancer, Alzheimer's disease, Parkinson's disease, cataracts, and rheumatoid arthritis. They may also contribute to aging. In order to counter the effects of oxygen free radicals, the body produces enzymes—such as *superoxide dismutase, catalase,* and *glutathione peroxidase*—that convert free radicals to less toxic substances. Among the nutrients that counter oxygen free radicals are vitamins E and C and beta-carotene. Many studies are now under way to develop antiradical therapy for a number of disorders.

There has been considerable publicity concerning the use of low doses of aspirin to reduce the rate of acute myocardial infarction. However, the benefits of aspirin may not be the same for everyone, and aspirin may not be indicated for people at low risk for MI because of the risk of cerebral bleeding (aspirin interferes with blood clotting). Moreover, some people develop gastrointestinal problems associated with taking aspirin. Most experts agree that routine aspirin therapy should be discouraged. The drug should be considered on a case-by-case basis.

MEDICAL TEST

Serum enzyme studies

Diagnostic Value: Probably the most common test, after an ECG, to diagnose and monitor heart attacks. In response to muscle damage, such as might occur following a heart attack, certain enzymes are released into the blood: creatine phosphokinase (CPK), serum glutamic oxaloacetic transaminase (SGOT), and lactic dehydrogenase (LDH), among others. These enzyme levels also go up after injury or inflammation of skeletal muscles and may be elevated in infectious hepatitis (SGOT, LDH) or in muscular dystrophy (CPK).

Procedure: Test is performed on a blood sample.

Normal Values: Creatine phosphokinase (CPK): 55–170 U/l in males and 30–135 U/l in females
Serum glutamic oxaloacetic transaminase (SGOT): 5–40 IU/l
Lactic dehydrogenase (LDH): 71–207 IU/l
See also Appendix B.

CONDUCTION SYSTEM

Recall from Chapter 10 that cardiac muscle fibers form two separate networks—one atrial and one ventricular. Each fiber is in physical contact with other fibers in the networks by transverse thickenings of the sarcolemma called ***intercalated discs*** (see Figure 10-15b). Within the discs are gap junctions that aid in the conduction of muscle action potentials between cardiac muscle fibers. The gap junctions provide low-resistance bridges for the spread of excitation (muscle action potentials) from one fiber to another. Thus, the spread of excitation between fibers is exceedingly rapid, and the atria contract as one unit and the ventricles as another. The intercalated discs also function in adhesion of cardiac muscle fibers so that they do not pull apart. Each network contracts as a functional unit.

The heart is innervated by the autonomic nervous system (Chapter 16), but the autonomic neurons only increase or decrease the time it takes to complete a cardiac cycle (heartbeat); that is, they do not initiate contraction. The chamber walls can go on contracting and relaxing, contracting and relaxing, without any direct stimulus from the nervous system. This action is possible because the heart has an intrinsic regulating system called the ***conduction system.*** The conduction system is composed of specialized muscle tissue that generates and distributes the electrical impulses that stimulate the cardiac muscle fibers (cells) to contract. These tissues are the sinoatrial (sinuatrial) or SA node, the atrioventricular (AV) node, the atrioventricular (AV) bundle (of His), the bundle branches, and the conduction myofibers (Purkinje fibers). The cells of the conduction system develop during embryological life from certain cardiac muscle fibers and provide autorhythmicity for the heart.

Before discussing the spread of action potentials through the conduction system that results in contraction of the heart, we will first examine a characteristic of nodal cells called ***self-excitability,*** their ability to spontaneously and rhythmically generate action potentials. Most smooth muscle fibers and neurons of the central nervous system are also self-excitable. The membranes of cells of the sinoatrial node (described shortly) are periodically very permeable to sodium (Na^+) ions, even in the resting state. As a result, Na^+ ions diffuse through sodium ion channels into the cell, causing the membrane potential to move toward a more positive value (depolarization). Once the membrane potential reaches its threshold level, an action potential is generated. After the action potential begins, the membrane becomes less permeable to Na^+ ions but very permeable to potassium (K^+) ions. As K^+ ions diffuse out of the cell through potassium ion channels, the inside of the cell becomes more negative (repolarization). This reversal of charges stops the action potential. Through the operation of the sodium and potassium pumps, Na^+ ions are actively transported out of the cells and K^+ ions are actively transported into the cells. Then, the influx of Na^+ ions initiates another action potential, and the process of self-excitation repeats itself over and over. In humans, the normal resting rate of self-excitation of the sinoatrial node is about 75 times per minute.

The ***sinoatrial (sinuatrial) node,*** also known as the ***SA node,*** or ***pacemaker,*** is a compact mass of cells located in the right atrial wall inferior to the opening of the superior vena cava (Figure 20-7a). The SA node initiates each cardiac cycle and thereby sets the basic pace for the heart rate. The SA node spontaneously depolarizes and generates action potentials faster than other components of the conduction system and myocardium. As a result, action potentials from the SA node spread to other areas of the conduction system and myocardium and stimulate them so frequently that they are not able to generate action potentials at their own inherent rates. Thus, the faster rate of discharge of the SA node sets the rhythm for the rest of the heart—hence, its common name, pacemaker. The rate set by the SA node may be altered by nerve impulses from the autonomic nervous system or by certain blood-borne chemicals such as thyroid hormones and epinephrine.

Once a muscle action potential is initiated by the SA node, the impulse spreads out over both atria, causing them to contract, and at the same time depolarizes the

FIGURE 20-7 Conduction system of the heart. (a) Location of the nodes and bundles of the conduction system.

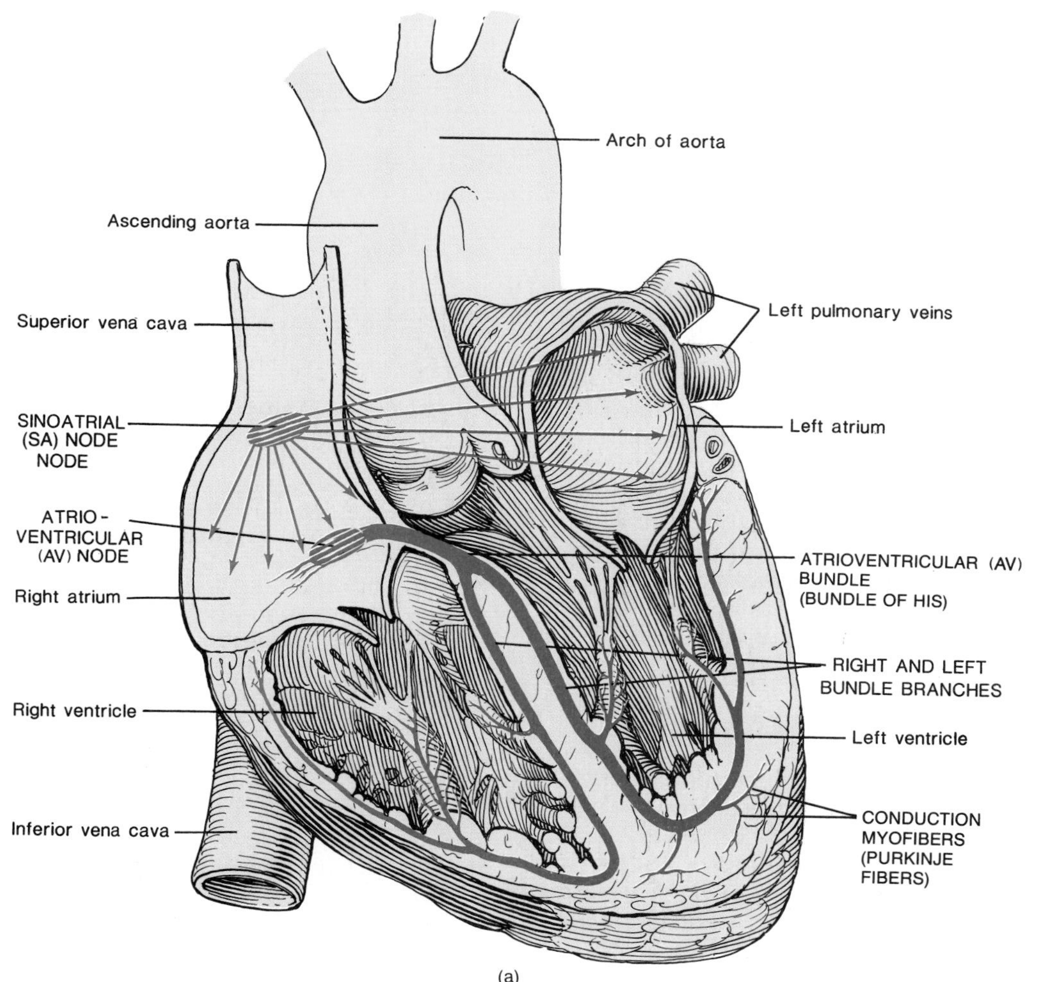

FIGURE 20-7 (*Continued*) The arrows indicate the flow of electricity (impulses) through the atria. (b) Normal electrocardiogram of a single heartbeat, enlarged for emphasis.

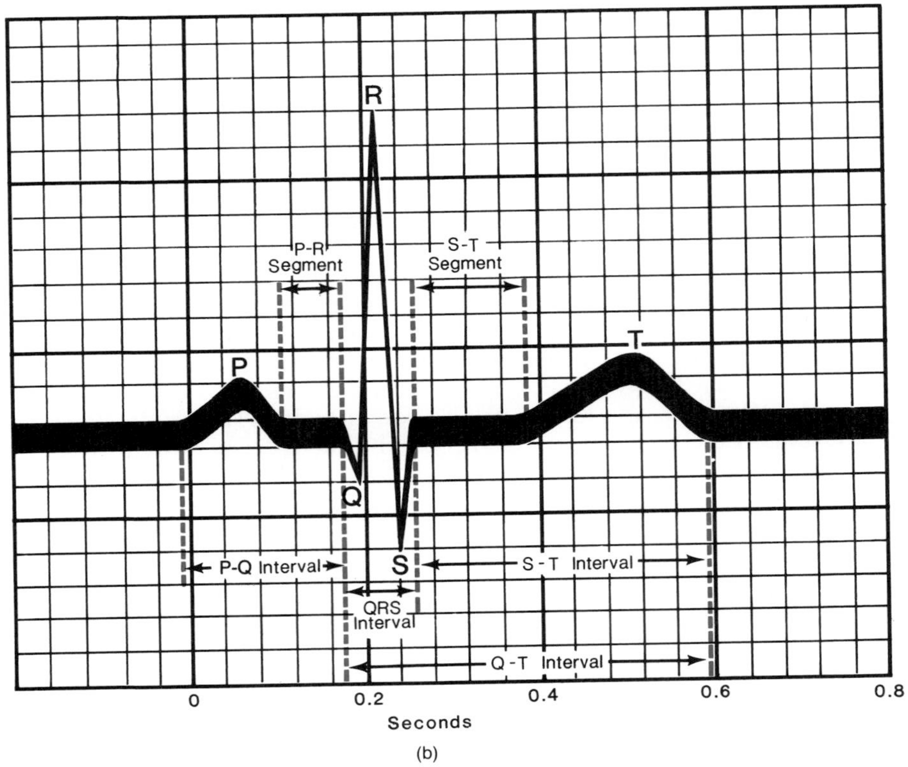

atrioventricular* (AV) *node. Because of its location near the inferior portion of the interatrial septum, the AV node is one of the last portions of the atria to be depolarized.

From the AV node, a tract of conducting fibers called the ***atrioventricular* (AV) *bundle* (*bundle of His*)** runs to the top of the interventricular septum. It then continues down both sides of the septum as the ***right*** and ***left bundle branches.*** The atrioventricular bundle distributes the action potential over the medial surfaces of the ventricles. Actual contraction of the ventricles is stimulated by ***conduction myofibers* (*Purkinje fibers*)** that emerge from the bundle branches and pass into the fibers of the myocardium of the ventricles.

ELECTROCARDIOGRAM (ECG)

Impulse transmission through the conduction system generates electrical currents that can be detected on the body's surface. A recording of the electrical changes that accompany the cardiac cycle is called an ***electrocardiogram*** (e-lek'-trō-KAR-dē-ō-gram) **(*ECG* or *EKG*).** The instrument used to record the changes is an ***electrocardiograph.*** There are three types of electrocardiograms: resting, stress, and ambulatory. We will first examine a resting ECG.

MEDICAL TEST

Resting electrocardiogram (ECG, EKG)

Diagnostic Value: Performed to evaluate possible symptoms of heart disease (chest pain, palpitations, dizziness, faintness), to detect abnormal cardiac rhythms and conduction patterns, to follow the course of recovery after a heart attack, to monitor the effectiveness and side effects of certain drugs that might affect the heart, to check the functioning of artificial pacemakers, and to evaluate the status of the heart prior to surgery. Some physicians include an ECG as part of a routine physical examination for individuals over 40. In addition to its screening value, this baseline ECG might be compared with a subsequent ECG done after onset of symptoms. Although an ECG monitors cardiac status, it does not detect all instances of heart disease, and it cannot predict what cardiac event might occur in the future.

Procedure: The wrists, ankles, and chest are cleansed with alcohol to remove skin oils and perspiration. Then the cleansed areas are dabbed with a special jelly or paste to improve electrical conduction. Next, metal electrodes, or leads, are attached to each arm and leg with rubber straps. By small rubber suction cups, other electrodes are placed on the chest. Once the electrodes are in place, the person is asked to remain quiet and still. The electrocardiograph is started, and the electrical impulses generated by the heart and detected by the electrodes are amplified by the machine and converted to a recording pen that graphs the impulses as a series of up-and-down wavy lines called ***deflection waves.*** Each portion of the cardiac cycle produces a different electrical impulse. The test requires only a few minutes.

The standard ECG is referred to as a 12-lead ECG, since 12 different tracings (electrical "views" of the heart) are recorded from the limb and chest leads. The 12 tracings provide a comprehensive view of the electrical activity of the various regions of the heart.

In a typical record (Figure 20-7b), three clearly recognizable waves accompany each cardiac cycle. The first, called the ***P wave,*** is a small upward wave. It indicates atrial depolarization—the spread of an impulse from the SA node through the two atria. A fraction of a second after the P wave begins, the atria contract. The second wave, called the ***QRS wave (complex),*** begins as a downward deflection, continues as a large, upright, triangular wave, and ends as a downward wave at its base. This deflection represents ventricular depolarization, that is, the spread of the electrical impulse through the ventricles. Very shortly after the QRS wave begins, the ventricles undergo contraction. The third recognizable deflection is a dome-shaped, upward ***T wave.*** This wave indicates ventricular repolarization. There is no deflection to show atrial repolarization because the stronger QRS wave masks this event.

In reading an electrocardiogram, it is important to note the size of the deflection waves at certain time intervals. Enlargement of the P wave, for example, indicates enlargement of the atrium, as in mitral stenosis. In this condition, the mitral valve narrows, blood backs up into the left atrium, and there is expansion of the atrial wall.

The ***P-Q (PR) interval*** is measured from the beginning of the P wave to the beginning of the QRS wave. It represents the conduction time from the beginning of atrial excitation to the beginning of ventricular excitation. The P-Q interval is the time required for an impulse to travel through the atria and atrioventricular node to the remaining conducting tissues. The lengthening of this interval, as in atherosclerotic heart disease and rheumatic fever, occurs because the heart tissue covered by the P-Q interval, namely the atria and atrioventricular node, is scarred or inflamed. Thus, the impulse must travel at a slower rate, and the interval is lengthened. The normal P-Q interval covers no more than 0.2 sec.

An enlarged Q wave may indicate a myocardial infarction (heart attack). An enlarged R wave generally indicates enlarged ventricles.

The ***S-T segment*** begins at the end of the S wave and terminates at the beginning of the T wave. It represents the time between the end of the spread of the impulse through the ventricles and repolarization of the ventricles. The S-T segment is elevated in acute myocardial infarction and depressed when the heart muscle receives insufficient oxygen.

The T wave represents ventricular repolarization. It is flat when the heart muscle is receiving insufficient oxygen, as in atherosclerotic heart disease. It may be elevated when the body's potassium level is increased.

For other individuals, it is necessary to evaluate the heart's response to the stress of physical exercise. Such a test is called a stress electrocardiogram, or stress test. It is based on the principle that blocked or narrowed coronary arteries may function adequately while a person is at rest, but during exercise will be unable to meet the heart's increased need for oxygen, creating deflection wave changes that can be noted on an electrocardiogram.

MEDICAL TEST

Stress electrocardiogram (ECG), or stress test

Diagnostic Value: To determine the cause of unexplained chest pain or to evaluate the severity of heart disease, as after recovery from myocardial infarction. Monitoring of the capacity of the heart to withstand rigorous exercise is one of the most frequent indications for stress testing. In this regard, stress testing is used to identify arrhythmias that occur during exercise and evaluate the effectiveness of antianginal or antiarrhythmic therapy.

Procedure: Several ECG electrodes are placed on the chest, and a sphygmomanometer is wrapped around the arm. Then the subject begins to exercise on a moving treadmill or stationary bicycle. As the test progresses, the speed and/or incline of the treadmill is increased, or the speed and resistance of pedaling is increased. During the test, which usually lasts from 15 to 30 minutes, ECG, heart rate, and blood pressure are monitored. After the test, monitoring is continued for at least 5 to 10 minutes.

For some people, an ***ambulatory ECG*** is indicated. In this procedure a ***Holter monitor*** is used. It is a portable, light-weight recorder that is connected to several electrodes attached to the person's chest. These detect electrical impulses from the heart that are recorded on magnetic tape in the monitor and later analyzed by computer. While wearing a Holter monitor for 24 hours or longer, an individual goes about normal daily activities, thus, the term *ambulatory* (walking) ECG. An ambulatory ECG is especially important (1) in diagnosing rhythm disorders in the conduction system that might not show up during the brief period of time used for recording a resting ECG and (2) in evaluating the effectiveness of heart surgery, drugs, or an artificial pacemaker.

CLINICAL APPLICATION: ARTIFICIAL PACEMAKER

When major elements of the conduction system are disrupted, an irregular heart rhythm may occur. In one type of rhythm disturbance, the ventricles fail to receive atrial impulses, causing the ventricles and atria to beat independently of each other. In patients with such a condition, normal heart rhythm can be restored and maintained with an ***artificial pacemaker,*** a device that sends out small electrical charges that stimulate the heart. It consists of three basic parts: a *pulse generator,* which contains the battery cells and produces the impulse; a *lead,* which is a flexible wire connected to the pulse generator that delivers the impulse to the electrode; and an *electrode,* which makes contact with a portion of the heart and delivers the charge to the heart. Many of the newer pacemakers, called activity-adjusted pacemakers, automatically speed up the heartbeat during exercise.

BLOOD FLOW THROUGH THE HEART

Two phenomena control the movement of blood through the heart: the opening and closing of the valves and the contraction and relaxation of the myocardium. Both these activities can occur without direct stimulation from the nervous system. The valves are controlled by blood pressure changes in each heart chamber. The contraction of cardiac muscle is stimulated by its conduction system.

Blood flows from an area of higher blood pressure to an area of lower blood pressure. The blood pressure developed in a heart chamber is related primarily to its force of contraction as a result of changes in the chamber's size. For example, if the chamber size decreases, the pressure increases. The pressure in the atria is called ***atrial pressure,*** that in the ventricles is called ***ventricular pressure,*** and pressure in the aorta and pulmonary trunk is referred to as ***arterial pressure.*** The relationship of these pressures to a heartbeat in the left side of the heart is shown in Figure 20-8b. Although the pressures in the right side of the heart are considerably lower, about one-fifth as great, each ventricle pumps the same amount of blood per beat and the same pattern exists for both chambers.

PHYSIOLOGY OF CARDIAC CYCLE

In a normal heartbeat, the two atria contract while the two ventricles relax. Then, when the two ventricles contract, the two atria relax. The term ***systole*** (SIS-tō-lē; *systellein* = draw together) refers to the phase of contractions; ***diastole*** (dī-AS-tō-lē; = pause) is the phase of relaxation. A ***cardiac cycle,*** or complete heartbeat, consists of a systole and diastole of both atria plus the systole and diastole of both ventricles.

For purposes of our discussion, we will divide the cardiac cycle into the following phases: As you read the description, refer to Figure 20-8.

1. **Atrial systole (contraction).** Under normal conditions, blood flows continuously from the superior vena cava, inferior vena cava, and coronary sinus into the right atrium and from the pulmonary veins into the left atrium. The bulk of blood, about 70 percent, flows passively from the atria into the ventricles, even before atrial contraction occurs. When the SA node fires, the atria depolarize. This is followed by atrial contraction. Atrial depolarization produces the P wave on the ECG. Contraction of the atria forces its remaining blood into the ventricles. This final push accounts for only about 30 percent of the blood passed into the ventricles. Thus, atrial contraction is not really necessary for adequate ventricular filling at normal heart rates. During atrial systole, deoxygenated blood passes through the open tricuspid valve from the right atrium into the right ventricle, and oxygenated blood passes through the open bicuspid valve from the left atrium to the left ventricle.

If you look at the atrial pressure curve in Figure 20-8b, you will notice that there are several pressure elevations, the a, c, and v waves. The *a wave* is produced by atrial contractions. The *c wave* is caused by ventricular contraction, which makes the atrioventricular valves bulge into the atria. The *v wave* is due to atrial filling while the atrioventricular valves are closed during ventricular contraction.

2. **Ventricular filling.** When the ventricles are contracting, the atrioventricular valves are closed and atrial pressure increases as blood fills the relaxed atria. During atrial diastole (relaxation), deoxygenated blood from various parts of the body enters the right atrium and oxygenated blood from the lungs enters the left atrium. But once ventricular contraction is over, ventricular pressure falls. The higher atrial pressure pushes the atrioventricular valves open and blood fills the ventricles.

The major part of ventricular filling occurs immediately upon opening of the atrioventricular valves. The first third of ventricular filling is the ***period of rapid ventricular filling.*** During the middle third, called ***diastasis,*** only a small amount of blood flows into the ventricles. This is blood that continuously empties into the right atrium from the superior vena cava, inferior vena cava, and coronary sinus, and into the left atrium from the pulmonary veins, and passes through the atria directly into the ventricles. During the last third of ventricular filling, called the ***period of slow ventricular filling,*** the atria contract. As noted earlier, this accounts for about 30 percent of the blood that fills the ventricles.

3. **Ventricular systole (contraction).** Near the end of atrial systole, the action potential from the SA node passes to the AV node and through the ventricles, causing them to depolarize and contract. This is represented as the QRS complex in the ECG. The beginning of ventricular contraction coincides with the first heart sound. At the onset of ventricular contraction, there is an abrupt rise in ventricular pressure that causes the atrioventricular valves to close. The first 0.05 seconds of ventricular systole is known as ***isovolumetric contraction*** be-

FIGURE 20-8 Cardiac cycle. (a) ECG related to the cardiac cycle. (b) Left atrial, left ventricular, and arterial (aortic) pressure changes along with the opening and closing of valves during the cardiac cycle. The letters a, c, and v represent the elevation waves of the left atrial pressure curve. (c) Left ventricular volume during the cardiac cycle. (d) Heart sounds related to the cardiac cycle.

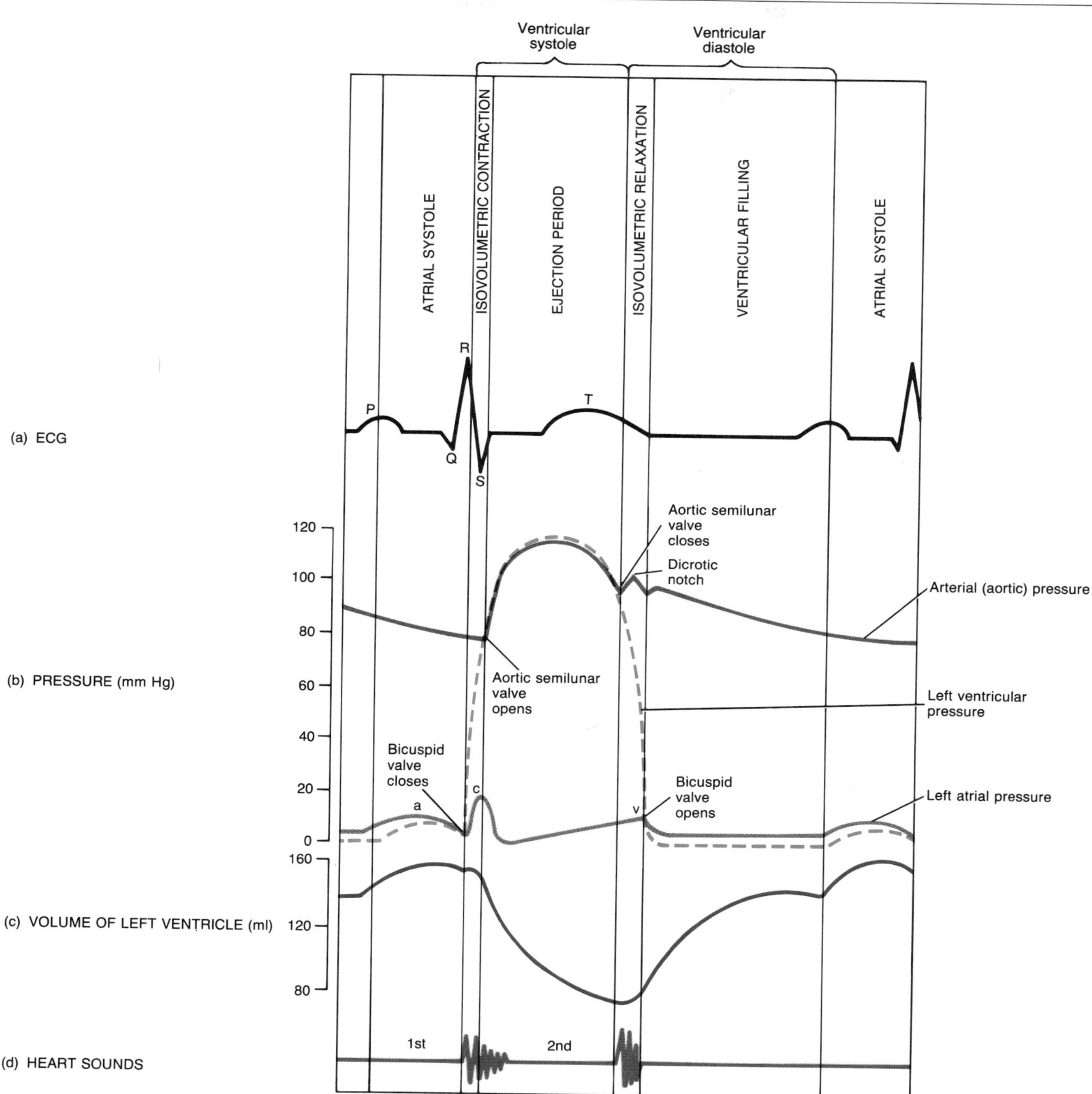

cause ventricular volume is constant (Figure 20-8c). It is the interval of time between the start of ventricular systole and the opening of the semilunar valves. During this time, there is contraction of the ventricles but no emptying; there is a rapid rise in ventricular pressure.

Once ventricular pressure exceeds arterial pressure, the semilunar valves open, and blood is forced from the ventricles into their respective arteries. This is called the ***ejection period*** and lasts for about 0.25 seconds, until the semilunar valves close. The blood pumped by each ventricle during ejection is about half the content of the ventricle and is known as stroke volume (discussed shortly).

4. Ventricular diastole (relaxation). At the end of ventricular contraction, the ventricles suddenly begin their relaxation. The period of time, about 0.05 seconds, between the closing of the semilunar valves and the opening of the atrioventricular valves is termed ***isovolumetric relaxation.*** It is characterized by a drastic decrease in ventricular pressure without a change in ventricular volume (Figure 20-8c). The higher arterial pressure causes blood to flow back toward the ventricles. This results in closure of the semilunar valves. Closure of the aortic semilunar valve produces a brief rise in arterial (aortic) pressure (dicrotic notch), shown in Figure 20-8b, and the second heart sound.

Firing of the SA node results in atrial depolarization, followed by atrial contraction, and the beginning of another cardiac cycle.

Timing

If we assume that the average heart rate (HR) is about 75 times per minute, then each cardiac cycle requires about 0.8 sec. During the first 0.1 sec, the atria contract and the ventricles relax. The atrioventricular valves are open, and the semilunar valves are closed. For the next 0.3 sec, the atria are relaxing and the ventricles are contracting. During the first part of this period, all valves are closed; during the second part, the semilunar valves are open. The last 0.4 sec of the cycle is the ***relaxation (quiescent) period,*** and all chambers are in diastole. In a complete cycle, then, the atria are in systole 0.1 sec and in diastole 0.7 sec; the ventricles are in systole 0.3 sec and in diastole 0.5 sec. For the first part of the relaxation period, all valves are closed; during the latter part, the atrioventricular valves open and blood starts draining into the ventricles. When the heart beats faster than normal, the relaxation period is shortened accordingly.

Sounds

The act of listening to sounds within the body is called ***auscultation*** (aws-kul-TĀ-shun; *auscultare* = to listen), and it is usually done with a stethoscope. The sound of the heartbeat comes primarily from turbulence in blood flow created by the closure of the valves, not from the contraction of the heart muscle (Figure 20-8b,d). The first sound, which can be described as a ***lubb*** ($\overline{\text{oo}}$) sound, is louder and a bit longer than the second sound. The lubb is the sound created by the closure of the atrioventricular valves soon after ventricular systole begins. The second sound, which is not as loud and shorter than the first, can be described as a ***dupp*** (ŭ) sound. Dupp is the sound created as the semilunar valves close toward the end of ventricular systole. A pause between the second sound and the first sound of the next cycle is about two times longer than the pause between the first and second sound of each cycle. Thus, the cardiac cycle can be heard as a lubb, dupp, pause; lubb, dupp, pause; lubb, dupp, pause. When a sound is heard before or after the normal heart tones, this may represent a heart murmur or some other abnormality.

CLINICAL APPLICATION: HEART MURMUR

Heart sounds provide valuable information about the valves. Besides lubb-dupp, some persons also have a third or fourth heart sound. Because this sound may suggest the cantering of a horse, the term "gallop rhythm" is used to describe these extra heart sounds, which are often associated with heart disease. Other abnormal sounds include "snaps," "knocks," "rubs," and "clicks." A *heart murmur* is an abnormal sound that consists of a flow noise that is heard before or after the lubb-dupp, or that may mask the normal heart sounds. Some murmurs are caused by turbulent blood flow around valves due to abnormal anatomy or increased volume of flow. Among the valvular abnormalities that may contribute to murmurs are *mitral stenosis* (narrowing of the mitral valve by scar formation or a congenital defect), *mitral insufficiency* (backflow of blood from the left ventricle into the left atrium due to a damaged mitral valve or ruptured chordae tendineae), *aortic stenosis* (narrowing of the aortic semilunar valve), and *aortic insufficiency* (backflow of blood from the atria into the left ventricle. Another cause of a heart murmur is due to *mitral valve prolapse (MVP)*, an inherited disorder in which a portion of a mitral valve is pushed back too far (prolapsed) during contraction owing to expansion of the cusps and elongation of the chordae tendineae. This condition is usually asymptomatic and may be found in up to 10 percent of otherwise healthy young men and women. Nevertheless, in some patients the condition is associated with various problems, and the individual who has it must be examined periodically by a cardiologist. Not all murmurs are abnormal.

PHYSIOLOGY OF CARDIAC OUTPUT (CO)

Although the heart possesses properties that enable it to beat independently, its operation is regulated by events occurring in the rest of the body. All body cells must receive a certain amount of oxygenated blood each minute to maintain health and life. When cells are very active, as during exercise, they need even more blood. During rest periods, cellular need is reduced, and the heart cuts back on its output.

The amount of blood ejected from the left ventricle (or right ventricle) into the aorta per minute is called the ***cardiac output (CO),*** or ***cardiac minute output.*** Cardiac output is determined by (1) the amount of blood pumped by the left ventricle (or right ventricle) during each beat and (2) the number of heartbeats per minute. The amount of blood ejected by a ventricle during each systole is called the ***stroke volume (SV).*** In a resting adult, stroke volume averages 70 ml and heart rate is about 75 beats per minute. The average cardiac output, then, in a resting adult is

Cardiac output = stroke volume × beats per minute
= 70 ml × 75/min
= 5250 ml/min or 5.25 liters/min

Factors that increase stroke volume or heart rate tend to increase cardiac output. For example, during mild exercise (stroke volume at 109 ml and heart rate at 100 beats/minute), CO is 10.90 l/min. During more vigorous exercise (stroke volume at 112 ml and heart rate at 120 beats/min), CO is 13.44 l/min. Factors that decrease stroke volume or heart rate tend to decrease cardiac output.

CLINICAL APPLICATION: SYNCOPE

Syncope **(SIN-kō-pē), or faint, refers to a sudden, temporary loss of consciousness associated with loss of postural tone and followed by spontaneous recovery. Syncope is most commonly due to cerebral ischemia (lack of sufficient blood). Clinically, it may be preceded by uneasiness, malaise, lightheadedness, nausea, vertigo, confusion, disturbances in vision, weakness, sweating, or tinnitus. Among the common causes of syncope are sudden emotional stress or real, threatened, or fantasized injury (vasodepressor syncope); micturition, defecation, or severe coughing (situational syncope); drugs such as antihypertensives, diuretics, vasodilators, and tranquilizers (drug-induced syncope); an excessive decrease in blood pressure or assuming an upright position (orthostatic hypotension); maneuvers that stretch the carotid sinus, such as hyperextension of the head, tight collars, or carrying shoulder loads (carotid sinus syncope); and reduced cardiac output.**

Stroke Volume (SV)

The actual amount of blood ejected by a ventricle with each heartbeat depends on how much blood enters the ventricle during diastole and how much blood is left in the ventricle following its systole.

End-Diastolic Volume (EDV)

The term ***end-diastolic volume (EDV)*** refers to the volume of blood that enters a ventricle during diastole (relaxation). As noted earlier, about 70 percent of the blood flows from the atria into the ventricles before atrial systole (contraction); atrial systole causes the additional 30 percent to enter the ventricles. During diastole, the ventricles normally increase their volume to 120 to 130 ml, the end-diastolic volume. This volume is determined principally by the length of ventricular diastole and venous pressure. When heart rate increases, the duration of diastole is shorter, the ventricles may contract before they are adequately filled, and the EDV is reduced. When venous pressure increases, a greater volume of blood is forced into the ventricles, and the EDV is increased. As you will see shortly, this results in a more forceful ventricular contraction.

End-Systolic Volume (ESV)

End-systolic volume (ESV) refers to the volume of blood still left in a ventricle following its systole (contraction). Since end-diastolic volume is about 120 to 130 ml and stroke volume is about 70 ml, end-systolic volume is about 50 to 60 ml (end-diastolic volume minus stroke volume). Stroke volume is equal to end-diastolic volume minus end-systolic volume.

End-systolic volume is determined principally by arterial pressure and the force of ventricular contraction. Arterial pressure in the aorta and pulmonary trunk just prior to ventricular systole is the pressure that must be overcome for the semilunar valves to open. If arterial pressure is significantly elevated, the resistance prevents the ventricles from pumping out as much as they should, and stroke volume decreases.

One factor that determines the force of ventricular contraction is the length of cardiac muscle fibers. Within limits, the greater the length of stretched fibers, the stronger the contraction. This relationship is referred to as ***Starling's law of the heart.*** The situation is somewhat like stretching a rubber band; the more you stretch it, the harder it contracts. The operation of Starling's law of the heart is important in maintaining equal blood output from both ventricles. For example, if for some reason the right ventricle begins to pump more blood than the left, the increased blood volume to the left ventricle stretches it, resulting in an increased cardiac output equal to that of the right ventricle. If the right ventricle pumps more blood than the left, without compensation by the left, blood would accumulate in the lungs. The operation of Starling's law of the heart is also important in individuals who have constricted arteries or improperly functioning heart valves. Arterial constriction creates a resistance to blood flow in the aorta (or pulmonary trunk). This prevents normal emptying of the ventricle, causing it to overfill with blood and stretch beyond normal. As a result, the ventricle contracts more forcefully, thus maintaining normal cardiac output. This is what happens to individuals who have hypertension (high blood pressure). A similar situation occurs with improperly functioning heart valves. If blood backflows into an atrium or ventricle, the chamber overfills with blood and contracts harder. Since the increased blood leaks backward during diastole (relaxation), net cardiac output remains normal.

The force of ventricular contraction is also determined by factors such as nervous control, chemicals (such as hormones), temperature, emotions, sex, and age. These factors are considered as part of the discussion dealing with the regulation of heart rate later in the chapter.

The maximum percentage that the cardiac output can increase above normal is referred to as ***cardiac reserve.*** For example, during strenuous exercise the cardiac output of a normal adult may increase to about four times normal. Since this is an increase of 400 percent above normal, we say that the cardiac reserve is 400 percent. Well-trained athletes may have a cardiac reserve as high as 600 percent, an increase in cardiac output of about six times normal. Although cardiac reserve depends on many factors, it is markedly influenced by heart conditions such as ischemic heart disease, valvular disorders, and myocardial damage.

Heart Rate

Cardiac output depends on heart rate as well as stroke volume. In fact, changing heart rate is the body's principal mechanism of short-term control over cardiac output and blood pressure. The sinoatrial (SA) node initiates contraction and left to itself would set an unvarying heart rate. However, the body's need for blood supply varies under different conditions. There are several regulatory mechanisms stimulated by factors such as chemicals present in the body, temperature, emotional state, and age.

During certain pathological conditions, stroke volume may fall dangerously low. If the ventricular myocardium is weak or damaged by an infarction, it cannot contract strongly. Or blood volume may be reduced by excessive bleeding. Stroke volume then falls because the cardiac fibers are not sufficiently stretched. In these cases, the body attempts to maintain a safe cardiac output by increasing the rate and strength of contraction.

The heart rate is regulated by several factors. The most important control of heart rate and strength of contraction is the autonomic nervous system.

Autonomic Control

Within the medulla of the brain is a group of neurons called the ***cardioacceleratory center (CAC).*** Arising from this center are sympathetic fibers that travel down a tract in the spinal cord and then pass outward in the ***cardiac (accelerator) nerves*** and innervate the SA node, AV node, and portions of the myocardium (Figure 20-9). When the cardioacceleratory center is stimulated, nerve impulses travel along the sympathetic fibers. This causes them to release norepinephrine (NE), which in-

FIGURE 20-9 Innervation of the heart by the autonomic nervous system.

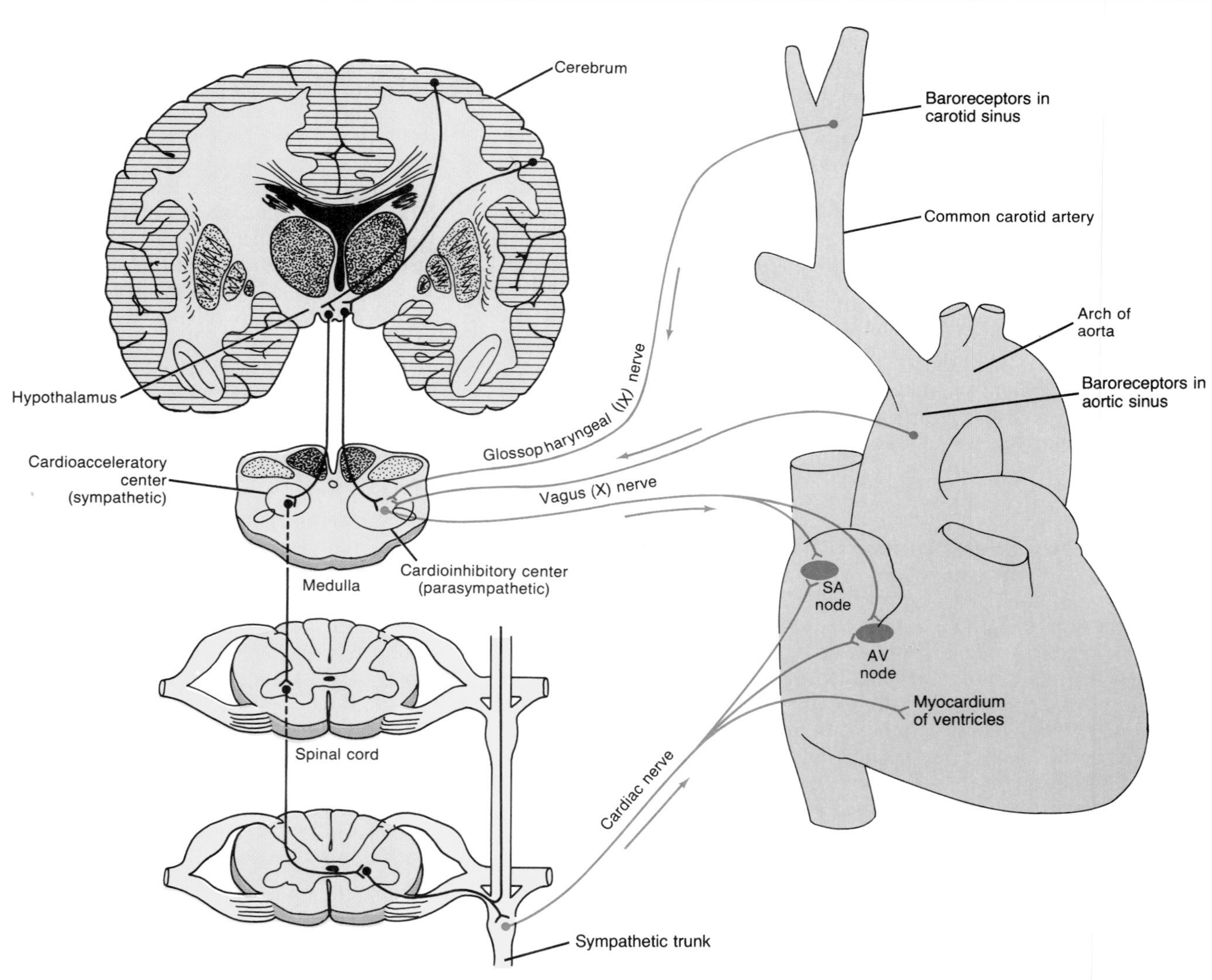

creases the rate of heartbeat and the strength of contraction.

The medulla also contains a group of neurons that form the ***cardioinhibitory center (CIC).*** Arising from this center are parasympathetic fibers that reach the heart via the ***vagus (X) nerve.*** These fibers innervate the SA node and AV node. When this center is stimulated, nerve impulses transmitted along the parasympathetic fibers cause the release of acetylcholine (ACh). This decreases the rate of heartbeat and force of contraction.

The autonomic control of the heart is therefore the result of opposing sympathetic (stimulatory) and parasympathetic (inhibitory) influences. Sensory impulses from receptors in different parts of the cardiovascular system act upon the centers so that a balance between stimulation and inhibition is maintained. Nerve cells capable of responding to changes in blood pressure are called ***baroreceptors (pressoreceptors).*** These receptors affect the rate of heartbeat and are involved in three reflex pathways: the carotid sinus reflex, the aortic reflex, and the right heart (atrial) reflex.

▪ Carotid Sinus Reflex The ***carotid sinus reflex*** is concerned with maintaining normal blood pressure in the brain. The ***carotid sinus*** is a small widening of the internal carotid artery just above the point where it branches off from the common carotid artery (Figure 20-9). In the wall of the carotid sinus or attached to it are baroreceptors. Any increase in blood pressure stretches the wall of the sinus, and the stretching stimulates the baroreceptors. The impulses then travel from the baroreceptors over sensory neurons in the glossopharyngeal (IX) nerves. Within the medulla, the impulses stimulate the cardioinhibitory center and inhibit the cardioacceleratory center. Consequently, more parasympathetic impulses pass from the cardioinhibitory center via the vagus (X) nerves to the heart and fewer sympathetic impulses pass from the cardioacceleratory center via cardiac (accelerator) nerves to the heart. The result is a decrease in heart rate and force of contraction. There is a subsequent decrease in cardiac output and a decrease in arterial blood pressure to normal.

If blood pressure falls, reflex acceleration of the heart rate and increased force of contraction take place. The baroreceptors in the carotid sinus do not stimulate the cardioinhibitory center, and the cardioacceleratory center is free to dominate. The heart then beats faster and more forcefully to restore normal blood pressure. This inverse relationship between blood pressure and heart rate is referred to as ***Marey's law of the heart*** (Figure 20-10).

The ability of the carotid sinus reflex (and the aortic reflex, to be described shortly) to maintain a relatively constant blood pressure is very important when a person sits or stands from a lying position. Immediately upon moving from a prone to erect position, blood pressure in the head and upper part of the body falls. The decrease in pressure, however, is counteracted by the reflexes. If the pressure were to fall markedly, unconsciousness could occur.

FIGURE 20-10 Control of heart rate and force of contraction by baroreceptors of the carotid sinus reflex and aortic reflex.

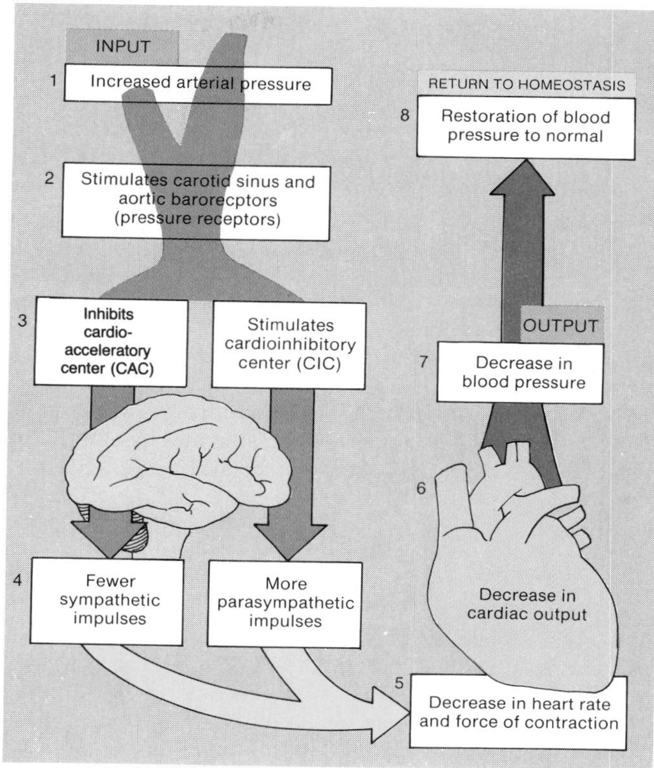

▪ Aortic Reflex The ***aortic reflex*** is concerned with general systemic blood pressure. It is initiated by baroreceptors in the wall of the arch of the aorta or attached to the arch (see Figure 20-9), and it operates like the carotid sinus reflex.

▪ Right Heart (Atrial) Reflex The ***right heart (atrial) reflex*** responds to increases in venous blood pressure. It is initiated by baroreceptors in the atria. When venous pressure increases, the baroreceptors send impulses through the vagus (X) nerves that stimulate the cardioaccelerator center. Returning impulses via vagal and sympathetic nerves increase heart rate and force of contraction. This mechanism is called the ***Bainbridge reflex.***

Chemicals

Certain chemicals present in the body also have an effect on heart rate and force of contraction. For example, epinephrine, produced by the adrenal medulla in response to sympathetic stimulation, increases the excitability of the SA node and affects the myocardium. This, in turn, increases the rate and strength of contraction. Elevated levels of potassium or sodium decrease the heart rate and strength of contraction. It appears that excess potassium interferes with the generation of nerve impulses, and excess sodium interferes with calcium participation in muscular contraction. An excess of calcium increases heart rate and strength of contraction.

Temperature

Increased body temperature, such as occurs during fever or strenuous exercise, causes the SA and AV nodes to discharge impulses faster and thereby increases heart rate. Decreased body temperature resulting from exposure to cold or deliberately cooling the body prior to surgery decreases heart rate and strength of contraction.

Emotions

Strong emotions such as fear, anger, and anxiety, along with a multitude of physiological stressors, increase heart rate through the general adaptation syndrome (see Figure 18-25). Mental states such as depression and grief tend to stimulate the cardioinhibitory center and decrease heart rate.

Sex and Age

Sex is another factor—the heartbeat is somewhat faster in normal females than normal males. Age is yet another factor among normal individuals—the heartbeat is fastest at birth, moderately fast in youth, average in adulthood, and below average in old age.

A summary of the factors that influence cardiac output is presented in Figure 20-11.

ARTIFICIAL HEART

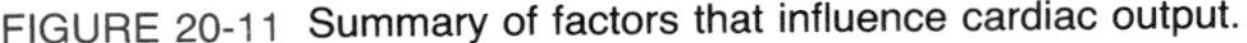

On December 2, 1982, Dr. Barney B. Clark, a 61-year-old retired dentist, made medical history by becoming the first human to receive an ***artificial heart.*** The $7\frac{1}{2}$-hour operation was performed by Dr. William DeVries, who was a surgeon at the University of Utah Medical Center at the time.

The artificial heart, known as Jarvik-7 after its inventor, Dr. Robert Jarvik, consists of an aluminum base and a pair of rigid plastic chambers that serve as ventricles. Each chamber contains a flexible diaphragm driven by compressed air that forces blood past mechanical valves into the major blood vessels. The surgery consisted of removing the ventricles from Dr. Clark's heart but leaving the atria intact. Then Dacron connectors were sutured onto the atria, aorta, and pulmonary trunk. The artificial heart was then snapped into position via the connectors. Essentially, the artificial heart is designed to pump sufficient blood to sustain only moderate activity.

The artificial heart has come under some criticism because of complications experienced by recipients. Complications include acute renal failure, postoperative bleeding, hemolytic anemia, thrombi, infections, seizures, and strokes. Research is under way to develop an artificial heart that uses an electrically powered implant to replace the current model, which uses a bulky external power source to drive a mechanical pump inside the body by compressed air. Also under development is a ***left ventricular assist device (LVAD),*** which boosts the action of a failing heart (left ventricle) rather than replacing both ventricles. The LVAD takes blood from a patient's weakened left ventricle and pumps it into the aorta. It is powered by a battery pack strapped outside the body. Both artificial hearts and LVADs are used primarily as ***bridges to transplants;*** that is, they are used as temporary mechanical support devices until a donor heart transplantation can be performed. A more recent development is the ***Hemopump.*** It consists of a propellerlike pump, about $\frac{1}{4}$ inch wide and $\frac{1}{2}$ inch long, that is threaded through an artery in the groin into the left ventricle. There the blades of the pump whirl at about 25,000 revolutions per minute, pulling blood out of the left ventricle into the aorta.

FIGURE 20-11 Summary of factors that influence cardiac output.

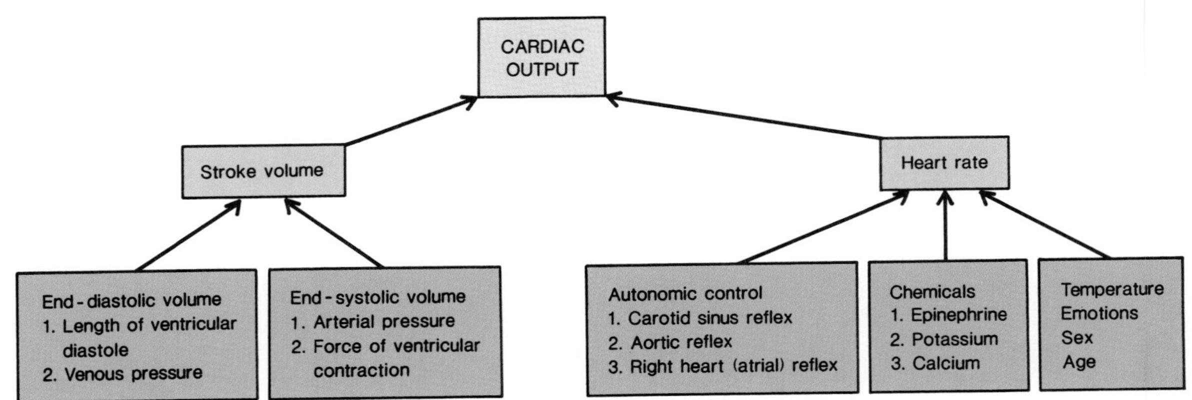

CLINICAL APPLICATION: THREE-WAY TRANSPLANT

Teams of surgeons at two hospitals in Baltimore performed a ***three-way transplant*** in May 1987. At the University of Maryland Hospital, physicians removed the heart and lungs from an accident victim who had been declared brain dead. The organs were then transported to Johns Hopkins Medical Center and transplanted into a person whose lungs were destroyed by cystic fibrosis (CF) but whose heart was healthy. The healthy heart was removed and transplanted into another person who was in need of a heart transplant. The procedure, referred to as a "domino donor" organ exchange, marked the first time in U.S. medical history that a healthy human heart was taken from a living person and transplanted into another human.

DIAGNOSING HEART DISORDERS

The various diagnostic tests for cardiovascular disorders may be grouped into several categories: (1) serum enzyme studies (discussed earlier in the chapter) that are useful in detecting myocardial infarction (MI); (2) radiography, including X ray films of the heart, to assess cardiac or vascular abnormalities; (3) graphic recording, such as electrocardiography (discussed earlier in the chapter) and phonocardiography (graphic recording of heart sounds), to evaluate electrical and acoustical properties of the heart; (4) ultrasonography, such as echocardiography, to evaluate cardiac shape, size, and structure; (5) nuclear medicine, such as thallium imaging, to evaluate cardiac blood flow; (6) coronary angiography to detect blocked arteries; and (7) cardiac catheterization to diagnose cardiac and vascular dysfunction.

In this section, we will discuss echocardiography, thallium imaging, coronary angiography, and cardiac catheterization.

MEDICAL TESTS

Echocardiogram (ek-ō-KAR-dē-ō-gram; *echo* = returned sound; *cardio* = heart; *gram* = record of) **(cardiac echo** or **ultrasound cardiogram)**

Diagnostic Value: To evaluate congenital heart disorders, tumors, or valvular disorders; detect fluid around the heart; assess the effects of coronary artery disease on the heart; and evaluate the size, shape, and motion of the heart.

Procedure: High-frequency (ultrasonic) sound waves are directed at the heart from a hand-held transducer, an instrument resembling a microphone. As the sound waves bounce off the heart, their echoes are picked up by the transducer and converted to a computerized image that appears on a screen. There are two types of echocardiogram. The time-motion (M-mode) echocardiogram produces a series of waves on graph paper that illustrate the linings, valves, and muscular motion of the heart. The real-time (B-mode) echocardiogram provides a moving image of various regions of the beating heart. Usually, both types of echo study are performed.

Thallium (THAL-ē-um) **imaging** or **scan (cold spot myocardial imaging)**

Diagnostic Value: To evaluate blood flow through coronary arteries; to demonstrate the location and extent of a myocardial infarction (MI); and to evaluate (1) the patency (open state) of blood vessels following coronary artery bypass grafting (CABG) and (2) the effectiveness of drug therapy and percutaneous transluminal coronary angioplasty (PCTA).

Procedure: A small amount of radioisotopic thallium is injected into a vein in the arm. Thallium concentrates in healthy myocardial tissue only. A scintillation camera above the chest measures the radioactivity emitted by thallium. The signal is relayed to a computer for translation, and an image of the heart is displayed. Areas of heart muscle that have a poor blood supply or damaged muscle cells appear as "cold spots" on the image (scan). Thallium imaging may be done at rest or after vigorous exercise, usually by comparing initial and four-hour delayed images.

Coronary (KOR-ō-na-rē) **angiography** (an'-jē-OG-ra-fē) or **coronary arteriography**

Diagnostic Value: To determine the severity and location of blocked coronary arteries by injecting contrast dyes into the coronary arteries and to inject clot-dissolving chemicals, such as streptokinase or tissue plasminogen activator (t-PA), into a coronary artery to dissolve an obstructing thrombus.

Procedure: With the patient on a tilt-top table, the heart is monitored through ECG leads, and an IV line is begun. After injection of a local anesthetic at the catheterization site, the tip of a long plastic tube (catheter) is introduced into the femoral artery and passed to the left atrium, ventricle, and base of the aorta, where the coronary artery openings are located. The catheter may carry a radiopaque contrast material for visualization. At the end of the procedure, the catheter is removed and a pressure dressing is applied at the site.

Cardiac catheterization (kath'-e-ter-i-ZĀ-shun)

Diagnostic Value: To measure pressure in the heart and blood vessels; to assess left ventricular function, cardiac output (CO), and diastolic properties of the left ventricle; to measure the flow of blood through the heart and blood vessels, the oxygen content of blood, status of heart valves and conduction system; and to identify the exact location of septal and valvular defects.

Procedure: The heart is monitored through ECG leads and an IV line is begun. After injection of a local anesthetic at the catheterization site, the tip of a long plastic tube (catheter) may be introduced into a vein in the arm or groin. The catheter is radiopaque so that its insertion can be guided with a fluoroscope. It is threaded into the right atrium, right ventricle, and pulmonary artery (right heart catheterization). The catheter can also be inserted into an artery in the arm or groin and passed through the aorta into the coronary artery orifices and/or left ventricle (left heart catheterization). A radiopaque contrast medium is introduced for visualization. Sometimes vasodilator or vasoconstrictor drugs are given to monitor their effects on coronary arteries. At the end of the procedure, the catheter is removed and a pressure dressing is applied to the site.

HEART–LUNG MACHINE

Some surgical procedures, such as those that involve heart transplantation, open-heart surgery (OHS), and coronary artery bypass surgery, make it necessary to use a ***heart–lung machine.*** This device performs two complex roles simultaneously: (1) it pumps blood (functioning as a heart) and (2) removes carbon dioxide (CO_2) from blood and adds oxygen (O_2) to it (functioning as the lungs). The principle of the heart–lung machine is fairly simple. Deoxygenated blood from the venae cavae is removed from the body and passed through an oxygenator, where carbon dioxide is removed and oxygen is added. Then the oxygenated blood is passed through a temperature controller, where it may be rewarmed to body temperature or cooled. Next, the blood passes through a filter to remove emboli and is finally returned to the body into the arterial system at the proper pressure. If necessary, drugs, anesthetics, and transfusions may be added to the circuit.

To surgically repair certain heart abnormalities, it is necessary to slow down a patient's heart. One such method is called ***hypothermia*** (hī-pō-THER-mē-a), which refers to a low body temperature. In various surgical procedures, it refers to a deliberate cooling of the body to slow metabolism and reduce the oxygen needs of the tissues. Thus, the heart and brain can withstand short periods of interrupted or reduced blood flow. Lost blood is then replaced by transfusion during and after the operation.

RISK FACTORS IN HEART DISEASE

It is estimated that one in every five persons who reaches age 60 will have a myocardial infarction (heart attack). One in every four persons between 30 and 60 has the potential to be stricken. Heart disease is epidemic in the United States, despite the fact that some of the causes can be foreseen and prevented. The results of research indicate that people who develop combinations of certain risk factors eventually have heart attacks. ***Risk factors*** are characteristics, symptoms, or signs present in a person free of disease that are statistically associated with an excessive rate of development of a disease. Among the major risk factors in heart disease are:

1. High blood cholesterol level.
2. High blood pressure.
3. Cigarette smoking.
4. Obesity.
5. Lack of regular exercise.
6. Diabetes mellitus.
7. Genetic predisposition (family history of heart disease at an early age).
8. Sex.
9. Age.
10. Fibrinogen level.
11. Left ventricular hypertrophy (enlarged left ventricle).

The first five risk factors all contribute to increasing the heart's work load. High blood cholesterol is discussed shortly, and hypertension is discussed in the next chapter. Cigarette smoking, through the effects of nicotine, stimulates the adrenal gland to oversecrete aldosterone, epinephrine, and norepinephrine (NE)—the latter two being powerful vasoconstrictors. Overweight people develop miles of extra capillaries to nourish fat tissue. The heart has to work harder to pump the blood through more vessels. Without exercise, venous return gets less help from contracting skeletal muscles. In addition, regular exercise strengthens the smooth muscle of blood vessels and enables them to assist general circulation. Exercise also increases cardiac efficiency and output. In diabetes mellitus, fat metabolism dominates glucose metabolism. As a result, cholesterol levels get progressively higher and result in plaque formation, a situation that may lead to high blood pressure. High blood pressure drives fat into the vessel wall, encouraging atherosclerosis. Up to age 50, there is a 10- to 15-year lag in the extent of heart disease in females compared with males; after that, the rates of disease in both sexes are similar. Regardless of sex, the incidence of heart disease increases with age. Regarding fibrinogen, the higher the level, the higher the risk of heart disease (more so in males than females). Fibrinogen enhances blood clot formation. Both hypertension and obesity contribute to left ventricular hypertrophy.

DEVELOPMENTAL ANATOMY OF THE HEART

The *heart,* a derivative of ***mesoderm,*** begins to develop before the end of the third week of gestation. It begins its development in the ventral region of the embryo beneath the foregut (see Figure 24-24). The first step is the formation of a pair of tubes, the ***endothelial (endocardial) tubes,*** from mesodermal cells (Figure 20-12). These tubes then unite to form a common tube, the ***primitive heart tube.*** Next, the primitive heart tube develops into five regions: ***ventricle, bulbus cordis, atrium, sinus venosus,*** and ***truncus arteriosus.*** Since the bulbus cordis and ventricle subsequently grow more rapidly than the others, and the heart grows more rapidly than its superior and inferior attachments, the heart assumes a U-shape and later an S-shape. The flexures of the heart reorient the regions so that the atrium and sinus venosus eventually come to lie superior to the bulbus cordis, ventricle, and truncus arteriosus.

At about the seventh week, a partition, the ***interatrial septum,*** forms in the atrial region, dividing it into a *right* and *left atrium.* The opening in the partition is the ***foramen ovale,*** which normally closes at birth and later forms a depression called the *fossa ovalis.* An ***interventricular septum*** also develops and partitions the ventricular region into a *right* and *left ventricle.* The bulbus cordis and truncus arteriosus divide into two vessels, the *aorta* (arising from the left ventricle) and the *pulmonary trunk* (arising from the right ventricle). Recall that the ductus arteriosus is a temporary vessel between the aorta and pulmonary trunk until birth. The great veins of the heart, *superior vena cava* and *inferior vena cava,* develop from the venous end of the primitive heart tube.

FIGURE 20-12 **Development of the heart. The arrows show the direction of the flow of blood from the venous to the arterial end.**

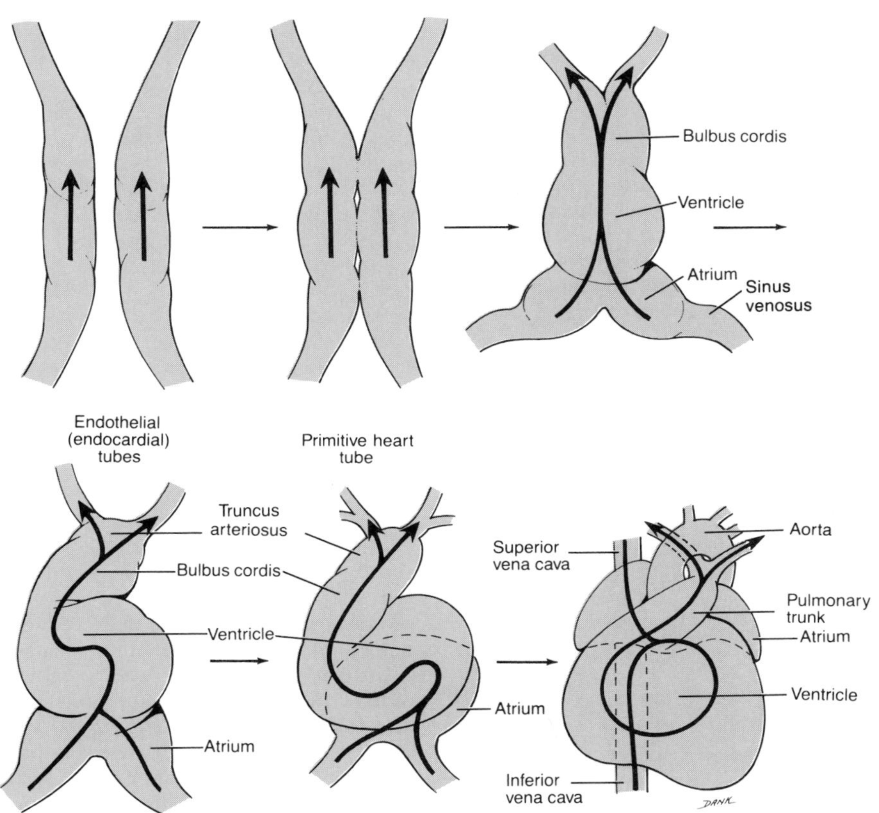

DISORDERS: HOMEOSTATIC IMBALANCES

Coronary Artery Disease (CAD)

Coronary artery disease (CAD) is a condition in which the heart muscle receives an inadequate amount of blood because of an interruption of its blood supply. It is presently the leading cause of death in the United States. Depending on the degree of interruption, symptoms can range from a mild chest pain to a full-scale heart attack. Generally, symptoms manifest themselves when there is about a 75 percent narrowing of coronary artery lumina. The underlying causes of CAD are many

DISORDERS: HOMEOSTATIC IMBALANCES (*continued*)

and varied. Two of the principal ones are atherosclerosis and coronary artery spasm. Another is a thrombus or an embolus in a coronary artery.

Atherosclerosis

Atherosclerosis (ath'-er-ō-skle-RŌ-sis) is a process in which fatty substances, especially cholesterol (CH) and triglycerides (ingested fats), are deposited in the walls of medium-sized and large arteries in response to certain stimuli. It is believed that the first event in atherosclerosis is damage to the endothelial lining of the artery. Contributing factors include high blood pressure (hypertension), carbon monoxide in cigarettes, diabetes mellitus, and a high cholesterol level. Following endothelial damage, monocytes begin to stick to the smooth inner endothelial lining of the tunica interna (inner coat) of the artery. Then, they slip between endothelial cells, penetrate the endothelium, and develop into macrophages. Here, the macrophages begin to take up cholesterol from low-density lipoproteins (described shortly). This cholesterol, it has recently been learned, must first undergo oxidation before it can accumulate in arterial walls. At the same time, smooth muscle fibers (cells) in the tunica media (middle coat) of the artery also accumulate cholesterol and proliferate in response to growth factors secreted by macrophages. The more cholesterol there is in the blood, the faster the process occurs. The accumulated cholesterol and cells form a lesion called an ***atherosclerotic plaque*** (Figure 20-13). The plaque looks like a pearly gray or yellow mound of tissue. As it grows, it may obstruct blood flow in the affected artery and damage the tissues the artery supplies. An additional danger is that the plaque provides a roughened surface that causes blood platelets to release *platelet-derived growth factor* (*PDGF*), a hormone that promotes the proliferation of smooth muscle fibers (cells). In fact, macrophages and endothelial cells also produce PDGF. This further complicates the process since PDGF causes the lesion to grow larger. Platelets in the area of the atherosclerotic plaque also release clot-forming chemicals. Thus, a thrombus may form. If the clot breaks off and forms an embolus, it may obstruct small arteries and capillaries quite a distance from the site of formation.

FIGURE 20-13 Photomicrograph of a cross section of an artery partially obstructed by an atherosclerotic plaque. (Courtesy of Sklar/Photo Researchers.)

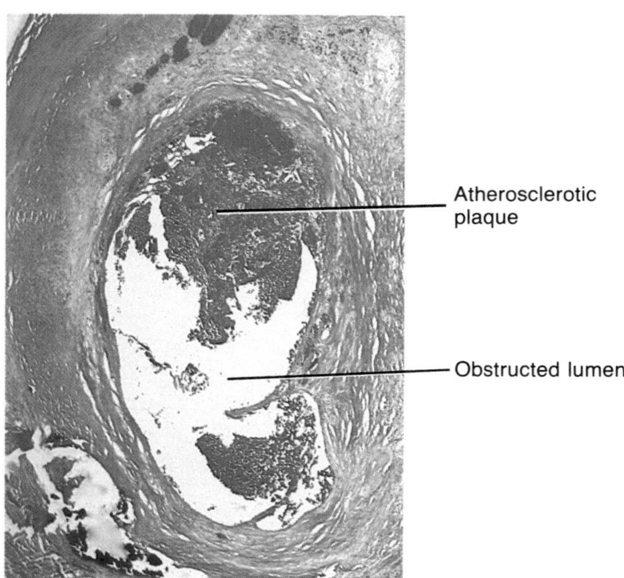

Scientists have identified a mechanism by which atherosclerosis may trigger hypertension, stroke, and heart disease. Secretions of blood platelets induce endothelial cells that line blood vessels to produce a substance called *endothelium-derived relaxing factor* (*EDRF*). EDRF relaxes smooth muscle fibers beneath the endothelium. However, atherosclerosis interferes with the production of EDRF, and this imbalance causes smooth muscle fibers to contract, that is, to go into vascular spasm. These spasms not only cause pain but can also lead to transient ischemic attacks (TIAs), stroke, high blood pressure, and heart disease.

Cholesterol is a major building block of all plasma membranes and a key compound for the synthesis of steroid hormones and bile salts. Some cholesterol is present in foods (eggs, dairy products, organ meats, beef, pork, and processed luncheon meats); most is synthesized by the liver. As it turns out, foods containing cholesterol are not the main source of cholesterol in the blood. It is the type of fat eaten. When saturated fats are broken down in the body, the liver uses some of the breakdown products to produce cholesterol. Saturated fats (see Chapter 2) can increase blood cholesterol level as much as 25 percent.

Cholesterol and triglycerides cannot dissolve in water and thus cannot travel in the blood in their unaltered forms. They are made water-soluble by combining with proteins produced by the liver and intestine (apoproteins), and the complexes thus formed are called ***lipoproteins,*** which vary in size, weight, and density. Two of the major classes are called ***low-density lipoproteins (LDLs)*** and ***high-density lipoproteins (HDLs).*** There is an important difference between the two: LDL contains 60 to 70 percent of total serum cholesterol and seems to pick up cholesterol and deposit it in body cells including, under abnormal conditions, smooth muscle fibers (cells) in arteries. As blood levels of LDL *increase,* the risk of coronary artery disease *increases.* LDL may be calculated by using the following formula: LDL = total cholesterol − HDL − (triglycerides/5). HDL, which contains 20 to 30 percent of total serum cholesterol, seems to gather cholesterol from body cells and transport it to the liver for elimination. As blood levels of HDL *increase,* the risk of coronary artery disease *decreases.* Most tissues of the body contain LDL receptors. The liver, suprarenal (adrenal) glands, and ovaries, because of their large requirement for cholesterol, have

DISORDERS: HOMEOSTATIC IMBALANCES (*continued*)

the most LDL receptors. Once LDL attaches to its receptor, it is taken into the cell by receptor-mediated endocytosis (Chapter 3). Within the cell, the LDL is broken down, and the cholesterol is released to serve the cell's needs. In this way, LDL is removed from the blood. Once a cell has sufficient cholesterol for its activities, a negative feedback system inhibits the cell from synthesizing new LDL receptors. It is believed that some people have too few LDL receptors, owing to various environmental and genetic factors, and thus are more susceptible to atherosclerosis. A third type of lipoprotein, known as a ***very low-density lipoprotein (VLDL),*** contains about 10 to 15 percent of total serum cholesterol and a large amount of triglyceride (fat). It also plays a role in the development of atherosclerotic plaques, but its role is not clear.

The amount of total cholesterol (HDL, LDL, and VLDL) in blood is one commonly used indicator of risk for CAD. In general, as the total cholesterol level increases above 150 mg/dl (a deciliter is $\frac{1}{10}$ of a liter), the risk of CAD slowly begins to rise. Above 200 mg/dl, the risk increases even more. The chance of a heart attack has been found to double with every 50 mg/dl increase in total cholesterol once the level goes over 200 mg/dl. The risk of developing CAD may be predicted by determining the ratio of HDL to total cholesterol. The risk ratio is determined by dividing total cholesterol by HDL. For example, a person with a total cholesterol of 200 and an HDL of 65 has a risk ratio of about 3. Ratios above 4 are considered undesirable; the higher the ratio, the greater the risk of developing CAD. Several ways to improve the HDL:LDL ratio are regular sustained exercise, diet (reducing animal fats, especially), and giving up cigarettes.

Unsaturated oils from fish, seal, and whale contain fatty acids referred to as ***omega-3 fatty acids.*** These fatty acids appear to have a role in protecting against heart disease. Specifically, they are believed to decrease serum cholesterol and low-density lipoprotein (LDL) levels; decrease the tendency of blood platelets to clump together, thus decreasing the likelihood of blood clot formation; decrease blood viscosity; suppress production of clotting factors; enhance the production of an anticlotting agent; and discourage renarrowing of arteries previously opened by insertion of a tiny balloon. Oily fish, such as salmon, bluefish, herring, sardines, anchovies, and mackerel, are good sources of omega-3 fatty acids. Shellfish also contain omega-3 fatty acids. Among the potential side effects of taking high concentrations of fish oil supplements are weight gain, vitamin A and D toxicity, vitamin E deficiency, and possibly increased LDL levels. At the present time, most researchers recommend increasing the intake of seafood rather than taking supplements.

MEDICAL TESTS

Lipid profile

Diagnostic Value: Since high blood cholesterol levels are associated with atherosclerosis, the test is used to assess the risk for cardiovascular disease including coronary artery disease (CAD). An abnormal cholesterol test may also occur in liver disorders and in hypothyroidism.

Procedure: Test is performed on a venous blood sample. As part of a lipid profile, usually done when the screening cholesterol is elevated, the total cholesterol, HDL, LDL, and triglycerides are measured.

Normal Values*: Total cholesterol under 200 mg/dl: desirable blood cholesterol
Total cholesterol 200–239 mg/dl: borderline-high blood cholesterol
Total cholesterol above 240 mg/dl: high blood cholesterol
HDL: greater than 40 mg/dl
LDL: less than 160 mg/dl
Triglycerides: 10–190 mg/dl

* Vary according to sex and age.

Among the therapies used to reduce cholesterol levels are exercise, diet, and drugs. Regular physical activity at aerobic and nearly aerobic levels tends to raise HDL. Dietary changes are aimed at reducing the intake of total fat, saturated fats, and cholesterol. This means restricting intake of foods containing butterfat (butter, cheese, ice cream, cream, and whole milk), fatty meats, egg yolks, organ meats, and certain oils (coconut and palm). Oat bran and rice bran are also believed to have a role in reducing total cholesterol. Among the drugs used to treat high blood cholesterol levels are cholestyramine (Questran) and colestipol (Colestid), both bile acid sequestrants; nicotinic acid (Lipo-nicin); lovastatin (Mevacor), an HMG CoA reductase inhibitor; gemifbrozil (Lopid); and probucol (Lorelco).

Treatment of CAD varies with the nature and urgency of symptoms. Among the treatment options are drug therapy (nitroglycerine, beta blockers, and thrombolytic agents) and various surgical and nonsurgical procedures.

Coronary artery bypass grafting (CABG) is one way of increasing the blood supply to the heart. It is a surgical procedure in which a portion of a blood vessel is removed from another part of the body. (At one time, the saphenous vein from the leg was frequently used.

DISORDERS: HOMEOSTATIC IMBALANCES *(continued)*

Now, however, the internal mammary artery from the chest is used more frequently. Research has shown that obstructions reform in grafted veins more readily than arteries.) Once the heart is exposed, circulation is maintained by a heart–lung machine. Next, a clamp is placed across the aorta above the openings of the coronary arteries. This stops the blood supply to the heart. A segment of the grafted blood vessel is then sutured between the aorta and the unblocked portion of the coronary artery, distal to the obstruction (Figure 20-14a). If more than one artery is clogged, additional bypasses may be made. Once the repair is completed, the aortic clamp is removed.

An innovative procedure called ***cardiomyoplasty*** (*cardio* = heart; *myo* = muscle; *plasty* = to mold or shape) is providing hope for patients with end-stage heart failure who cannot withstand the rigors of coronary artery bypass grafting or a transplant. In cardiomyoplasty, surgeons remove a rib, free one of the latissimus dorsi muscles from its connective tissue tendons, and wrap it over the heart. The remainder of the muscle, which is not wrapped around the heart, is left attached to the posterior body wall along with its blood and nerve supply. A pacemaker is placed between the heart and latissimus dorsi muscle to stimulate the muscle to contract in synchrony with the heart. The pacemaker is needed to coordinate the type II (fast-twitch) fibers of the latissimus dorsi muscle with the type I (slow-twitch) fibers of the heart muscle.

A nonsurgical procedure used to treat CAD is referred to as ***percutaneous transluminal coronary angioplasty (PTCA)*** (*percutaneous* = through the skin; *trans* = across; *lumen* = channel in a tube; *angio* = blood vessel; *plasty* = to mold or shape). Like coronary artery bypass grafting, it is an attempt to increase the blood supply to the heart muscle. In this procedure, a balloon catheter is inserted into an artery of an arm or leg and gently guided through the arterial system under X ray observation (Figure 20-14b). Once in place in an obstructed coronary artery, the balloon is inflated for a few seconds, thus compressing the obstructing material against the vessel wall and permitting increased blood flow. PTCA is most frequently used to relieve angina pectoris. Since about 30 percent of balloon-opened arteries restenose (narrow from reformed obstructions), special devices called ***stents*** are being used to keep the arteries patent (open). A stent is a stainless steel device, resembling a spring coil, that is permanently placed in an artery to maintain patency, permitting blood to circulate (Figure 20-14c). A stent is inserted in a blood vessel via a balloon catheter. In addition, it has been found that megadoses of omega-3 fatty acids discourage restenosis. A modification of PTCA, called ***percutaneous balloon valvuloplasty,*** is used to treat faulty heart valves.

Another nonsurgical technique for opening clogged arteries is a procedure called ***laser angioplasty.*** In one variation of this procedure, a laser vaporizes the atherosclerotic plaque and makes a channel through the blood vessel obstruction. Then a balloon catheter is inserted, and the balloon is inflated to widen the vessel.

Two of the latest nonsurgical techniques for clearing arteries are balloon–laser welding and catheter artherectomy. In ***balloon–laser welding,*** an artery is first widened by PTCA. On the last balloon inflation, a laser heats the surrounding tissue sufficiently to stretch and weld the arterial wall into a smooth surface. In ***catheter artherectomy,*** a rotating drill shaves off plaque. Shavings are trapped for removal.

Coronary Artery Spasm

Atherosclerosis results in a fixed obstruction to blood flow. Obstruction can also be caused by ***coronary artery spasm,*** a condition in which the smooth muscle of a coronary artery undergoes a sudden contraction, resulting in vasoconstriction. Coronary artery spasm typically occurs in individuals with atherosclerosis and may result in chest pain during rest (variant angina), chest pain during exertion (typical angina), heart attacks, and sudden death. Although the causes of coronary artery spasm are unknown, several factors are receiving attention. These include smoking, stress, and a vasoconstrictor chemical released by platelets. There is considerable interest in aspirin and other drugs that might inhibit the vasoconstrictor chemical. Coronary artery spasm is treated with long-acting nitrates (nitroglycerin) and calcium-blocking agents.

Congenital Defects

A defect that exists at birth, and usually before, is called a ***congenital defect.*** Many defects are not serious and may go unnoticed for a lifetime. Others heal themselves. But some are life-threatening and must be mended by surgical techniques ranging from simple suturing to replacement of malfunctioning parts with synthetic materials.

One congenital defect that can occur is ***coarctation*** (kō'-ark-TĀ-shun) ***of the aorta.*** In this condition, a segment of the aorta is too narrow. As a result, the flow of oxygenated blood to the body is reduced, the left ventricle is forced to pump harder, and high blood pressure develops. The condition may be corrected by insertion of a synthetic graft.

Another common congenital defect is ***patent ductus arteriosus.*** The ductus arteriosus between the aorta and the pulmonary trunk normally closes shortly after birth in response to increased oxygenation of blood flowing through the ductus arteriosus. Only a few hours after birth, the muscular wall of the ductus arteriosus constricts, and within a week, constriction is sufficient to stop all blood flow. During the second month, the ductus arteriosus becomes occluded by the formation of fibrous tissue in its lumen. In some babies, the ductus arteriosus remains open. As a result, aortic blood flows into the lower-pressure pulmonary trunk, thus increasing the pulmonary trunk blood pressure and overworking both ventricles and the heart. Treatment consists of a

DISORDERS: HOMEOSTATIC IMBALANCES *(continued)*

FIGURE 20-14 Procedures for reestablishing blood flow in occluded coronary arteries. (a) Coronary artery bypass grafting (CABG). The internal mammary artery is removed from the patient's chest. At a point distal to the obstruction, the grafted artery is sutured to the coronary artery. (b) Percutaneous transluminal coronary angioplasty (PTCA). (c) Stent in an artery.

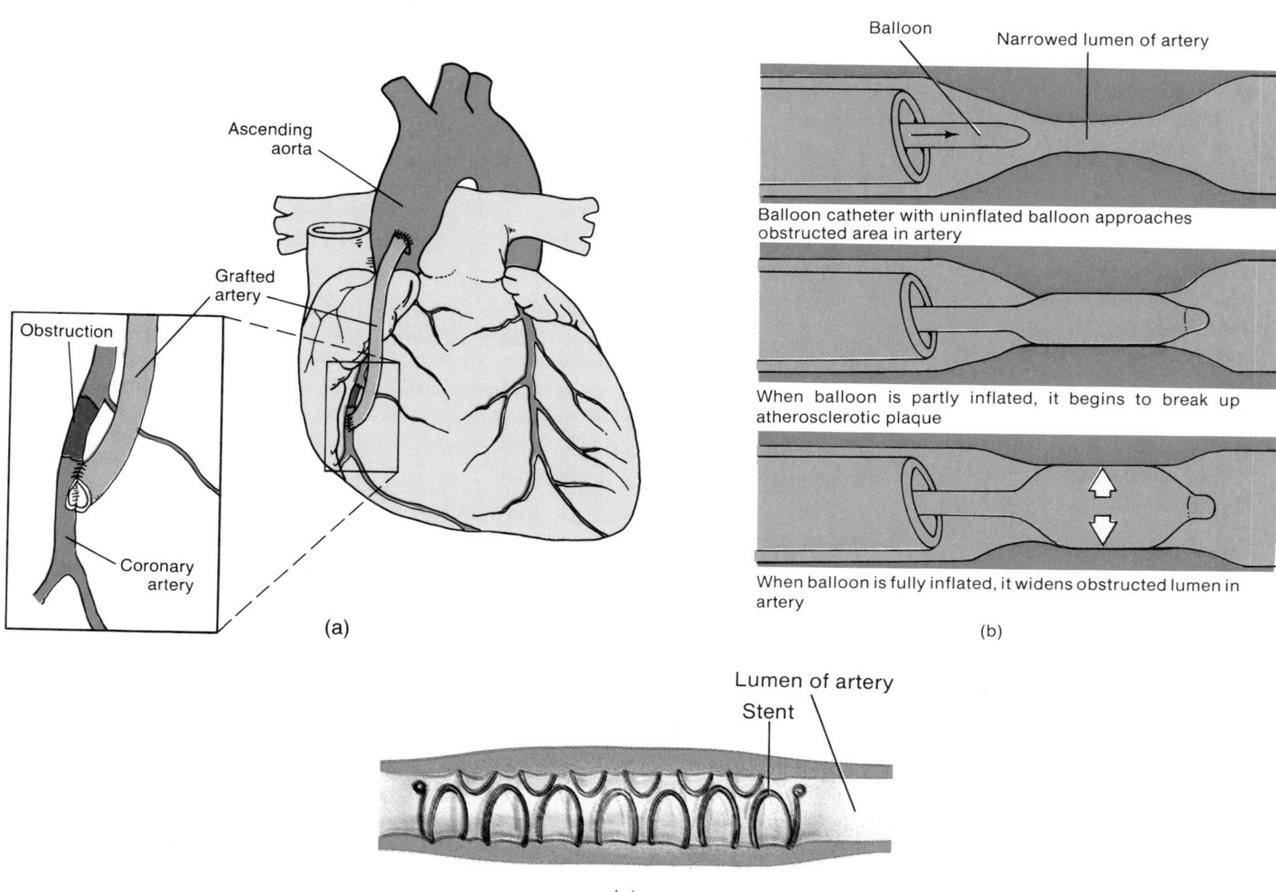

surgical procedure in which the ductus arteriosus is ligated or cut so that the cut ends are tied off.

A ***septal defect*** is an opening in the septum that separates the interior of the heart into a left and right side. In one type of ***interatrial septal defect*** there is a failure of the fetal foramen ovale between the two atria to close after birth. Because pressure in the right atrium is low, interatrial septal defect generally allows a good deal of blood to flow from the left atrium to the right without going through systemic circulation. This defect, an example of left-to-right shunt, overloads the pulmonary circulation, produces fatigue, and increases respiratory infections. If it occurs early in life, it inhibits growth because the systemic circulation may be deprived of a considerable portion of the blood destined for the organs and tissues of the body. ***Interventricular septal defect*** is caused by an incomplete closure of the interventricular septum. Owing to the higher pressure in the left ventricle, blood is shunted directly from the left ventricle into the right ventricle. This is another example of a left-to-right shunt. Septal openings can now be sewn shut or covered with synthetic patches inserted via a catheter.

Valvular stenosis is a narrowing of one of the valves regulating blood flow in the heart. Stenosis may occur in the valve itself, most commonly in the mitral valve from rheumatic heart disease or in the aortic valve from sclerosis or rheumatic fever. Or it may occur near a valve. All stenoses are serious because they place a severe work load on the heart by making it work harder to push the blood through the abnormally narrow valve openings. As a result of mitral stenosis, blood pressure is

DISORDERS: HOMEOSTATIC IMBALANCES (*continued*)

increased. Angina pectoris and heart failure may accompany this disorder. Most stenosed valves are totally replaced with artificial valves.

Tetralogy of Fallot (tet-RAL-ō-jē of fal-Ō) is a combination of four defects: an interventricular septal defect, an aorta that emerges from both ventricles instead of from the left ventricle only, a stenosed pulmonary semilunar valve, and an enlarged right ventricle. The condition is an example of a right-to-left shunt, in which blood is shunted from the right ventricle to the left ventricle without going through pulmonary circulation. Because there is stenosis of the pulmonary semilunar valve, the increased right ventricular pressure forces deoxygenated blood from the right ventricle to enter the left ventricle through the interventricular septum. As a result, deoxygenated blood gets mixed with the oxygenated blood that is pumped into systemic circulation. Also, because the aorta emerges from the right ventricle and the pulmonary artery is stenosed, very little blood gets to the lungs, and pulmonary circulation is bypassed almost completely. ***Cyanosis*** (sī-a-NŌ-sis) refers to a reduced hemoglobin (unoxygenated) concentration in blood of more than 5 g/dl. It causes a blue or dark purple discoloration that is most easily seen in nail beds and mucous membranes. For this reason, tetralogy of Fallot is one of the conditions that causes a "blue baby." Chronic lung disorders and suffocation also result in cyanosis.

It is possible to correct cases of tetralogy of Fallot when the patient is of proper age and condition. Open-heart operations are performed in which the narrowed pulmonary valve is cut open and the interventricular septal defect is sealed with a Dacron patch.

Arrhythmias

Arrhythmia (a-RITH-mē-a) is a general term that refers to an abnormality or irregularity in the heart rhythm. More physicians are using the term ***dysrhythmia*** since it implies an abnormal rhythm, whereas arrhythmia implies no rhythm. An arrhythmia results when there is a disturbance in the conduction system of the heart, either due to faulty production of electrical impulses or faulty conduction of impulses as they pass through the system.

There are many different types of arrhythmias that can occur, some normal and some quite serious. Arrhythmias may be caused by factors such as caffeine, nicotine, alcohol, anxiety, certain drugs, hyperthyroidism, potassium deficiency, and certain heart diseases. Serious arrhythmias can result in cardiac arrest if the heart cannot supply its own oxygen demands, as well as those of the rest of the body. They can be controlled, and the normal heart rhythm can be reestablished, if they are detected and treated early enough.

Heart Block

One serious arrhythmia is called a ***heart block.*** Perhaps the most common blockage is in the atrioventricular (AV) node, which conducts impulses from the atria to the ventricles. This disturbance is called ***atrioventricular (AV) block.*** It usually indicates a myocardial infarction, atherosclerosis, rheumatic heart disease, diphtheria, or syphilis. With complete AV block, patients may have vertigo, unconsciousness, or convulsions. These symptoms result from a decreased cardiac output with diminished cerebral blood flow and cerebral hypoxia or lack of sufficient oxygen.

Among the causes of AV block are excessive stimulation by the vagus (X) nerves (which depresses conductivity of the junctional fibers), destruction of the AV bundle as a result of coronary infarct, atherosclerosis, myocarditis, or depression caused by various drugs. Heart block is most often corrected by an artificial pacemaker.

Flutter and Fibrillation

Two other abnormal rhythms are ***flutter*** and ***fibrillation.*** In ***atrial flutter*** the atrial rhythm averages between 240 and 360 beats per minute. The condition is essentially rapid atrial contractions accompanied by a second-degree AV block. Flutter may occur from rheumatic heart disease, atherosclerotic heart disease, or certain congenital heart diseases. ***Atrial fibrillation*** is asynchronous contraction of the atrial muscles that causes the atria to contract irregularly and still faster. The atria send impulses to the AV node so rapidly that not all can pass through and cause ventricular contraction, and the heart rhythm is irregular. Atrial fibrillation may occur in myocardial infarction, acute and chronic rheumatic heart disease, and hyperthyroidism. Atrial fibrillation results in complete uncoordination of atrial contraction so that atrial pumping ceases altogether. When the muscle fibrillates, the muscle fibers of the atrium quiver individually instead of contracting together. The quivering cancels out the pumping of the atrium. In a strong heart, atrial fibrillation reduces the pumping effectiveness of the heart by only 25 to 30 percent. Treatment choices are to give an electrical shock to convert the rhythm to sinus (*cardioversion*) or to put the patient on digitalis, which will increase tone in the AV node and slow the heart to an acceptable rate (70 to 80 beats per minute at rest).

Ventricular fibrillation (VF) is another abnormality that almost always indicates imminent cardiac arrest and death unless corrected quickly. It is characterized by asynchronous, haphazard, ventricular muscle contractions. The rate may be rapid or slow. The impulse travels to the different parts of the ventricles at different rates. Thus, part of the ventricle may be contracting while other parts are still unstimulated. Ventricular contraction becomes ineffective, and circulatory failure and death occur. Ventricular fibrillation may be caused by coronary occlusion. It sometimes occurs during surgical procedures on the heart or pericardium. It may be the cause of death in electrocution. It is possible to pass a very strong electrical current through the ventricles for a short interval of time in order to stop ventricular fibril-

DISORDERS: HOMEOSTATIC IMBALANCES (*continued*)

lation. This is accomplished by placing electrodes on two sides of the heart and is called ***defibrillation.***

Ventricular Premature Contraction (VPC)

Another form of arrhythmia arises when a small region of the heart outside the pacemaker becomes more excitable than normal, causing an occasional abnormal impulse to be generated between normal impulses. The region from which the abnormal impulse is generated is called an ***ectopic focus.*** As a wave of depolarization spreads outward from the ectopic focus, it causes a ***ventricular premature contraction (VPC),*** also called a ***premature ventricular contraction (PVC).*** The contraction occurs early in diastole before the SA node is normally scheduled to discharge its impulses. A person with such a ventricular premature contraction might feel a thump in the chest. The contractions may be relatively benign and may be caused by emotional stress, excessive intake of stimulants such as caffeine or nicotine, and lack of sleep. In other cases, the contractions may indicate an underlying pathology.

Congestive Heart Failure (CHF)

Congestive heart failure (CHF) may be defined as a chronic or acute state that results when the heart is not capable of supplying the oxygen demands of the body. Symptoms and signs of CHF include fatigue, peripheral and pulmonary edema, and visceral congestion. These symptoms are produced by diminished blood flow to the various tissues of the body and by accumulation of excess blood in the various organs because the heart is unable to pump out the blood returned to it by the great veins. Both diminished blood flow and accumulation of excess blood in the organs occur together, but certain symptoms result from congestion, whereas others are produced by poor tissue nutrition. Among the causes of CHF are long-standing hypertension and myocardial infarction.

Cor Pulmonale (CP)

Cor pulmonale (kor pul-mōn-ALE; *cor* = heart; *pulmon* = lung), or ***CP,*** refers to right ventricular hypertrophy from disorders that bring about hypertension (high blood pressure) in the pulmonary circulation. Certain disorders such as those involving the respiratory center in the brain, lung diseases, diseases involving airways in the lungs, deformities of the thoracic cage, and neuromuscular disorders can result in hypoxia in the lungs. As a result, vasoconstriction of pulmonary arteries and arterioles occurs, and blood flow to the lungs is decreased. Also, any conditions such as tumors, aneurysms, and diseased blood vessels in the lungs can result in decreased blood flow. The reduction in blood flow and increased resistance in the lungs bring about pulmonary hypertension. Ultimately, this causes increased pressure in the pulmonary arteries and pulmonary trunk, resulting in right ventricular hypertrophy.

Individuals with CP may exhibit fatigue, weakness, dyspnea, right upper quadrant (RUQ) pain due to an enlarged liver, warm and cyanotic extremities, cyanosis of the face, distended neck veins, chest pain, and edema.

MEDICAL TERMINOLOGY ASSOCIATED WITH THE HEART

Angiocardiography (an'-jē-ō-kar'-dē-OG-ra-fē; *angio* = vessel; *cardio* = heart; *graph* = writing) X ray examination of the heart and great blood vessels after injection of a radiopaque dye into the bloodstream.

Cardiac arrest (KAR-dē-ak a-REST) A clinical term meaning cessation of an effective heartbeat. The heart may be completely stopped (cardiac standstill) or quivering ineffectively (ventricular fibrillation).

Cardiomegaly (kar'-dē-ō-MEG-a-lē; *mega* = large) Heart enlargement. Long-distance runners and weight lifters frequently develop heart enlargement as a natural adaptation to increased work load produced by regular exercise (physiologic cardiomegaly). The enlargement may also be due to disease (pathologic cardiomegaly).

Commissurotomy (kom'-i-shur-OT-ō-mē) An operation that is performed to widen the opening in a heart valve that has become narrowed by scar tissue.

Compliance (kom-PLĪ-ans) The passive or diastolic stiffness properties of the left ventricle. A hypertrophied or fibrosed heart with a stiff wall, for example, has decreased compliance. Also, the stiffness properties of the lung.

Constrictive pericarditis (kon-STRIK-tiv per'-i-kar-DĪ-tis) A shrinking and thickening of the pericardium that prevents heart muscle from expanding and contracting normally.

Incompetent valve (in-KOM-pe-tent VALV) Any valve that does not close properly, thus permitting a backflow of blood; also called **valvular insufficiency.**

Palpitation (pal'-pi-TĀ-shun) A fluttering of the heart or abnormal rate or rhythm of the heart.

Pancarditis (pan'-kar-DĪ-tis; *pan* = all) Inflammation of the whole heart including the inner layer (endocardium), heart muscle (myocardium), and outer sac (pericardium).

Paroxysmal tachycardia (par'-ok-SIZ-mal tak'-e-KAR-dē-ā) A period of rapid heartbeats that begins and ends suddenly.

Stokes–Adams syndrome Sudden attacks of unconsciousness, sometimes with convulsions, that may accompany heart block.

Sudden cardiac death The unexpected cessation of circulation and breathing due to an underlying heart disease such as ischemia, myocardial infarction, or a disturbance in cardiac rhythm.

STUDY OUTLINE

Location of Heart (p. 573)

1. The heart is situated between the lungs in the mediastinum.
2. About two-thirds of its mass is to the left of the midline.

Pericardium (p. 574)

1. The pericardium consists of an outer fibrous layer and an inner serous pericardium.
2. The serous pericardium is composed of a parietal and visceral layer.
3. Between the parietal and visceral layers of the serous pericardium is the pericardial cavity, a potential space filled with pericardial fluid that prevents friction between the two membranes.

Wall; Chambers; Vessels; and Valves (pp. 574–580)

1. The wall of the heart has three layers: epicardium, myocardium, and endocardium.
2. The chambers include two upper atria and two lower ventricles.
3. The blood flows through the heart from the superior and inferior venae cavae and the coronary sinus to the right atrium, through the tricuspid valve to the right ventricle, through the pulmonary trunk to the lungs, through the pulmonary veins into the left atrium, through the bicuspid valve to the left ventricle, and out through the aorta.
4. Valves prevent backflow of blood in the heart.
5. Atrioventricular (AV) valves, between the atria and their ventricles, are the tricuspid valve on the right side of the heart and the bicuspid (mitral) valve on the left.
6. The chordae tendineae and their muscles keep the flaps of the valves pointing in the direction of the blood flow and stop blood from backing into the atria.
7. The two arteries that leave the heart both have a semilunar valve.
8. The heart valves may be identified by surface projection.

Blood Supply (p. 580)

1. The coronary (cardiac) circulation delivers oxygenated blood to the myocardium and removes carbon dioxide from it.
2. Deoxygenated blood returns to the right atrium via the coronary sinus.
3. Complications of this system are angina pectoris and myocardial infarction (MI).

Conduction System (p. 582)

1. The conduction system consists of tissue specialized for generation and conduction of action potentials.
2. Components of this system are the sinoatrial (SA) node (pacemaker), atrioventricular (AV) node, atrioventricular (AV) bundle (bundle of His), bundle branches, and conduction myofibers (Purkinje fibers).

Electrocardiogram (ECG) (p. 584)

1. The record of electrical changes during each cardiac cycle is referred to as an electrocardiogram (ECG).
2. A normal ECG consists of a P wave (spread of impulse from SA node over atria), QRS wave (spread of impulse through ventricles), and T wave (ventricular repolarization).
3. The P-Q interval represents the conduction time from the beginning of atrial excitation to the beginning of ventricular excitation. The S-T segment represents the time between the end of the spread of the impulse through the ventricles and repolarization of the ventricles.
4. The ECG is invaluable in diagnosing abnormal cardiac rhythms and conduction patterns, detecting the presence of fetal life, determining the presence of several fetuses, and following the course of recovery from a heart attack.
5. An artificial pacemaker may be used to restore an abnormal cardiac rhythm.

Blood Flow through the Heart (p. 586)

1. Blood flows through the heart from an area of higher to lower pressure.
2. The pressure developed is related to the size and volume of a chamber.
3. The movement of blood through the heart is controlled by the opening and closing of the valves and the contraction and relaxation of the myocardium.

Physiology of Cardiac Cycle (p. 586)

1. A cardiac cycle consists of the systole (contraction) and diastole (relaxation) of both atria, plus the systole and diastole of both ventricles followed by a short pause.
2. The phases of the cardiac cycle are: (1) atrial systole, (2) ventricular filling, (3) ventricular systole, and (4) ventricular diastole.
3. With an average heartbeat of 75 beats/min, a complete cardiac cycle requires 0.8 sec.
4. The first heart sound (lubb) represents the closing of the atrioventricular valves. The second sound (dupp) represents the closing of semilunar valves.
5. An abnormal heart sound is called a murmur.

Physiology of Cardiac Output (CO) (p. 588)

1. Cardiac output (CO) is the amount of blood ejected by the left ventricle into the aorta per minute. It is calculated as follows: CO = stroke volume × beats per minute.
2. Stroke volume (SV) is the amount of blood ejected by a ventricle during each systole.
3. Stroke volume (SV) depends on how much blood enters a ventricle during diastole (end-diastolic volume) and how much blood is left in a ventricle following its systole (end-systolic volume).
4. According to Starling's law of the heart, the greater the length of stretched cardiac fibers, within limits, the stronger the contraction, and thus the greater the stroke volume.
5. The maximum percentage that cardiac output can be increased above normal is cardiac reserve.
6. Heart rate and strength of contraction may be increased by sympathetic stimulation from the cardioacceleratory center (CAC) in the medulla and decreased by parasympathetic stimulation from the cardioinhibitory center (CIC) in the medulla.

7. Baroreceptors (pressure receptors) are nerve cells that respond to changes in blood pressure. They act on the cardiac centers in the medulla through three reflex pathways: carotid sinus reflex, aortic reflex, and right heart (atrial) reflex.
8. Other influences on heart rate include chemicals (epinephrine, sodium, potassium), temperature, emotion, sex, and age.

Artificial Heart (p. 592)

1. The Jarvik artificial heart consists of an aluminum base and rigid plastic chambers that serve as ventricles.
2. When it is implanted, the recipient's atria are left intact.
3. Two recently developed bridges to transplant devices are the left ventricular assist device (LVAD) and the Hemopump.

Diagnosing Heart Disorders (p. 593)

1. Diagnosis of heart disorders may involve several procedures.
2. Among these are cardiac enzyme studies, radiography, graphic recording, ultrasonography, nuclear medicine, coronary angiography, and cardiac catheterization.

Heart–Lung Machine (p. 594)

1. Hypothermia (deliberate body cooling) and the heart–lung bypass permit open-heart surgery.
2. The heart–lung machine pumps blood (functioning as a heart) and oxygenates blood and removes carbon dioxide (functioning as lungs).

Risk Factors in Heart Disease (p. 594)

1. Risk factors in heart disease include high blood cholesterol, high blood pressure, cigarette smoking, obesity, lack of regular exercise, diabetes mellitus, genetic disposition, sex, age, fibrinogen level, and left ventricular hypertrophy.

Developmental Anatomy of the Heart (p. 594)

1. The heart develops from mesoderm.
2. The endothelial tubes develop into the four-chambered heart and great vessels of the heart.

Disorders: Homeostatic Imbalances (p. 595)

1. Coronary artery disease (CAD) refers to a condition in which the myocardium receives inadequate blood due to atherosclerosis, coronary artery spasm, or thrombi or emboli; it may be treated with drugs or surgically.
2. Congenital heart defects include coarctation of the aorta, patent ductus arteriosus, septal defects, valvular stenosis, and tetralogy of Fallot.
3. Arrhythmias result in heart block, flutter, and fibrillation.
4. Congestive heart failure (CHF) results when the heart cannot supply the oxygen demands of the body.
5. Cor pulmonale (CP) refers to right ventricular hypertrophy secondary to pulmonary hypertension.

REVIEW QUESTIONS

1. Describe the location of the heart in the mediastinum and identify the various borders and surfaces of the heart. (p. 573)
2. Distinguish the subdivisions of the pericardium. What is the purpose of this structure? (p. 574)
3. Compare the three layers of the heart wall according to composition, location, and function. (p. 574)
4. Define atria and ventricles. What vessels enter or exit the atria and ventricles? (p. 575)
5. Describe the principal valves in the heart and how they operate. (p. 576)
6. Describe how the valves of the heart may be identified by surface projection. (p. 580)
7. Describe the route of blood in coronary (cardiac) circulation. (p. 580)
8. Describe the structure and function of the heart's conduction system. (p. 582)
9. Define and label the deflection waves of a normal electrocardiogram (ECG). (p. 584)
10. Discuss the pressure changes associated with the flow of blood through the heart. (p. 586)
11. Define cardiac cycle and list the principal events of atrial systole, ventricular filling, ventricular systole, and ventricular diastole. (p. 586)
12. By means of a labeled diagram, relate the events of the cardiac cycle to time. What is the relaxation period? (p. 587)
13. Describe the significance of the heart sounds. (p. 588)
14. What is cardiac output (CO)? How is it calculated? (p. 588)
15. Define stroke volume (SV). Explain how end-diastolic volume (EDV) and end-systolic volume (ESV) are related to stroke volume. (p. 589)
16. What is Starling's law of the heart? What is its significance? (p. 589)
17. Define cardiac reserve. Why is it important? (p. 589)
18. Distinguish between the cardioacceleratory center (CAC) and cardioinhibitory center (CIC) with respect to the regulation of heart rate. (p. 590)
19. What is a baroreceptor? Outline the operation of the carotid sinus reflex, aortic reflex, and right heart (atrial) reflex. p. 591)
20. Explain how each of the following affects heart rate: chemicals, temperature, emotions, sex, and age. (p. 591)
21. Describe the Jarvik artificial heart. What is the left ventricular assist device (LVAD)? How does the Hemopump operate? (p. 592)
22. How are heart disorders diagnosed? (p. 593)
23. Describe the operation of the heart–lung machine. (p. 594)
24. Describe the risk factors involved in heart disease. (p. 594)
25. Describe how the heart develops. (p. 594)
26. What is coronary artery disease (CAD)? What factors contribute to it? Explain. (p. 595)
27. Describe how CAD is treated. (p. 597)
28. Define each of the following congenital heart defects: co-

arctation of the aorta, patent ductus arteriosus, septal defects, valvular stenosis, and tetralogy of Fallot. (p. 598)
29. Define an arrhythmia. Give several examples. (p. 600)
30. Describe the symptoms of congestive heart failure (CHF) and cor pulmonale (CP). (p. 601)
31. Define the following: rheumatic fever (p. 580), ischemia (p. 581), angina pectoris (p. 581), myocardial infarction (MI) (p. 581), free radical (p. 582), artificial pacemaker (p. 586), heart murmur (p. 588), three-way transplant (p. 593), syncope (p. 589), percutaneous transluminal coronary angioplasty (PTCA) (p. 598), and laser angioplasty. (p. 598)
32. Explain the diagnostic value of the following: serum enzyme studies (p. 582), electrocardiogram (ECG)—resting, stress, and ambulatory (pp. 584–585), echocardiogram (p. 593); thallium imaging (p. 593); coronary angiography (p. 593); cardiac catheterization (p. 593); and lipid profile. (p. 597)
33. Refer to the glossary of medical terminology associated with the heart. Be sure that you can define each term. (p. 601)

SELECTED READINGS

Alpert, J. S. "Management of Acute and Chronic Myocardial Ischemia," *Modern Medicine,* April 1989.

Brown, M. S., and J. L. Goldstein. "How LDL Receptors Influence Cholesterol and Atherosclerosis," *Scientific American,* November 1984.

Cantin, M., and J. Genest. "The Heart as an Endocrine Gland," *Scientific American,* February 1986.

Comroe, J. H. "Doctor, You Have Six Minutes," *Science 84,* January/February 1984.

DeVries, W. C. "The Permanent Artificial Heart," *Journal of the American Medical Association,* 12 February 1988.

Dolan, B., and J. M. Nash. "Searching For Life's Elixir," *Newsweek,* 12 December 1988.

Dusheck, J. "Fish, Fatty Acids, and Physiology," *Science News,* 19 October 1985.

Eisenberg, M. S., L. Bergner, A. P. Hallstrom, and R. O. Cummins. "Sudden Cardiac Death," *Scientific American,* May 1986.

Escher, D. J. W. "Use of Cardiac Pacemakers," *Hospital Practice,* September 1981.

Falk, J. S., B. Kaufman, and M. H. Weil. "Cardiopulmonary Resuscitation: An Update," *Hospital Medicine,* January 1984.

Fowler, R. E. "Acute Myocardial Infarction," *Postgraduate Medicine,* November 1988.

Goerke, J. and A. H. Mines. *Cardiovascular Physiology.* New York: Raven Press, 1988.

Goldfinger, S. E. (ed). "Balloons: Expanding," *Harvard Medical School Health Letter,* May 1987.

———. "Keeping Up with Cholesterol," *Harvard Medical School Health Letter,* June 1985.

Gwynn, J. T., and M. K. Lawrence. "Current Concepts in the Evaluation and Treatment of Hypercholesterolemia," *Modern Medicine,* March 1989.

Hogan, M. J. (ed). "Cholesterol," *Mayo Clinic Health Letter,* March 1988.

———. "Coronary Atherectomy," *Mayo Clinic Health Letter,* April 1989.

Honig, C. R. *Modern Cardiovascular Physiology.* Boston: Little, Brown, 1988.

Jarvik, R. K. "The Total Artificial Heart," *Scientific American,* January 1981.

Kiley, J. M. (ed). "Echocardiography," *Mayo Clinic Health Letter,* May 1985.

McIntyre, K. "Cardiac Arrest," *Hospital Medicine,* November 1984.

Netter, F. *The Heart,* CIBA Collection of Medical Illustrations. Summit: CIBA Pharmaceutical Company, 1971.

O'Keefe, J. H., C. J. Lavie, and J. O. O'Keefe. "Dietary Prevention of Coronary Artery Disease," *Postgraduate Medicine,* May 1989.

O'Toole, M. T., and A. R. Waldman. "Chest Pain," *RN,* April 1989.

Rios, J. C. (ed). "Breakthrough Discoveries on How to Avoid a Fatal Heart Attack," *Cardiac Alert,* 1988.

Rogers, W. J. "Diagnosis: Coronary Artery Disease," *Hospital Medicine,* August 1984.

Shabeti, R. "Answers to Questions on Cardiac Catheterization," *Hospital Medicine,* August 1982.

Silberner, J. "Lasers Powering through Coronary Arteries," *Science News,* 23 November 1985.

———. "Anatomy of Atherosclerosis," *Science News,* 16 March 1985.

Skluth, H., K. Grauer, and J. Gums. "Ventricular Arrhythmias," *Postgraduate Medicine,* May 1989.

Toufexis, A. "The Latest Word on What to Eat," *Time,* 13 March 1989.

Wehrmacher, W. H. "Acute Myocardial Infarction," *Postgraduate Medicine,* February 1989.

Chapter 21

The Cardiovascular System: Vessels and Routes

Chapter Contents at a Glance

Student Objectives

1. Contrast the structure and function of the various types of blood vessels.
2. Explain how blood pressure is developed and controlled.
3. Discuss the various pressures involved in the movement of fluids between capillaries and interstitial spaces.
4. Explain how the return of venous blood to the heart is accomplished.
5. Define pulse and blood pressure (BP) and contrast the clinical significance of systolic, diastolic, and pulse pressures.
6. Identify the principal arteries and veins of systemic, pulmonary, hepatic portal, and fetal circulation.
7. Explain the effects of exercise and aging on the cardiovascular system.
8. Describe the development of blood vessels and blood.
9. List the causes and symptoms of shock, hypertension, aneurysms, coronary artery disease (CAD), and deep-venous thrombosis (DVT).
10. Define medical terminology associated with the cardiovascular system.

Blood vessels form a network of tubes that carry blood away from the heart, transport it to the tissues of the body, and then return it to the heart. ***Arteries*** are vessels that carry blood from the heart to the tissues. Large, elastic arteries leave the heart and divide into medium-sized, muscular arteries that branch out into the various regions of the body. Medium-sized arteries then divide into small arteries, which, in turn, divide into still smaller arteries called ***arterioles.*** As the arterioles enter a tissue, they branch into countless microscopic vessels called ***capillaries.*** Through the walls of capillaries, substances are exchanged between the blood and body tissues. Before leaving the tissue, groups of capillaries reunite to form small veins called ***venules.*** These, in turn, merge to form progressively larger tubes called veins. ***Veins,*** then, are blood vessels that convey blood from the tissues back to the heart. Since blood vessels require oxygen and nutrients just like other tissues of the body, they also have blood vessels in their own walls called ***vasa vasorum.***

The developmental anatomy of blood vessels and blood is considered later in the chapter.

ARTERIES

Arteries (*aer* = air; *tereo* = to carry; arteries found empty at death were once thought to contain only air) have walls constructed of three coats or tunics and a hollow core, called a ***lumen,*** through which blood flows (Figure 21-1). The inner coat of an arterial wall, the ***tunica interna (intima),*** is composed of a lining of *endothelium* (simple squamous epithelium) that is in contact with the blood, a *basement membrane,* and a layer of elastic tissue called the *internal elastic lamina.* The middle coat, or ***tunica media,*** is usually the thickest layer. It consists of elastic fibers and smooth muscle. The outer coat, the ***tunica externa (adventitia),*** is composed principally of elastic and collagenous fibers. An *external elastic lamina* may separate the tunica externa from the tunica media.

As a result of the structure of the middle coat especially, arteries have two major properties: elasticity and contractility. When the ventricles of the heart contract and eject blood into the large arteries, they expand to accommodate the extra blood. Then, as the ventricles relax, the elastic recoil of the arteries forces the blood onward. The contractility of an artery comes from its smooth muscle, which is arranged longitudinally and in rings around the lumen somewhat like a doughnut and is innervated by sympathetic branches of the autonomic nervous system. When there is sympathetic stimulation, the smooth muscle contracts, squeezes the wall around the lumen, and narrows the vessel. Such a decrease in the size of the lumen is called ***vasoconstriction.*** Conversely, when sympathetic stimulation is removed, the smooth muscle fibers relax and the size of the arterial lumen increases. This increase is called ***vasodilation*** and is usually due to the inhibition of vasoconstriction.

The contractility of arteries also serves a function in stopping bleeding. This is called vascular spasm, one of the three mechanisms involved in hemostasis (Chapter 19). The blood flowing through an artery is under a great deal of pressure. Thus, great quantities of blood can be quickly lost from a broken artery. When an artery is cut, its wall constricts due to contraction of the smooth muscle so that blood does not escape quite so rapidly. However, there is a limit as to how much vasoconstriction can help.

Elastic (Conducting) Arteries

Large arteries are referred to as ***elastic (conducting) arteries.*** They include the aorta and brachiocephalic, common carotid, subclavian, vertebral, and common iliac arteries. The wall of elastic arteries is relatively thin in proportion to their diameter, and their tunica media contains more elastic fibers and less smooth muscle. As the heart alternately contracts and relaxes, the rate of blood flow tends to be intermittent. When the heart contracts and forces blood into the aorta, the wall of the elastic arteries stretches to accommodate the surge of blood and store the pressure energy. During relaxation of the heart, the wall of the elastic arteries recoils to create pressure, moving the blood forward in a more continuous flow. Elastic arteries are called conducting arteries because they *conduct* blood from the heart to medium-sized muscular arteries.

Muscular (Distributing) Arteries

Medium-sized arteries are called ***muscular (distributing) arteries.*** They include the axillary, brachial, radial, intercostal, splenic, mesenteric, femoral, popliteal, and tibial arteries. Their tunica media contains more smooth muscle than elastic fibers, and they are capable of greater vasoconstriction and vasodilation to adjust the volume of blood to suit the needs of the structure supplied. The wall of muscular arteries is relatively thick, mainly due to the large amounts of smooth muscle. Muscular arteries are called distributing arteries because they *distribute* blood to various parts of the body.

Anastomoses

Most parts of the body receive branches from more than one artery. In such areas the distal ends of the vessels unite. The junction of two or more vessels supplying the same body region is called an ***anastomosis*** (a-nas-tō-MŌ-sis). Anastomoses may also occur between the origins of veins and between arterioles and venules. Anastomoses between arteries provide alternate routes by which blood can reach a tissue or organ. Thus, if a vessel is occluded by disease, injury, or surgery, circulation to a part of the body is not necessarily stopped. The alternate route of blood to a body part through an anastomosis is known as

FIGURE 21-1 Comparative structure of (a) an artery, (b) a vein, and (c) a capillary. The relative size of the capillary is enlarged. Histology of blood vessels. (d) Photomicrograph of a portion of the wall of an artery (right) and its accompanying vein (left) at a magnification of 250×. Note that the lumen of a vein is larger than that of an artery, but the wall of the vein is thinner and the vein frequently appears collapsed (flattened). (Courtesy of Andrew Kuntzman.)

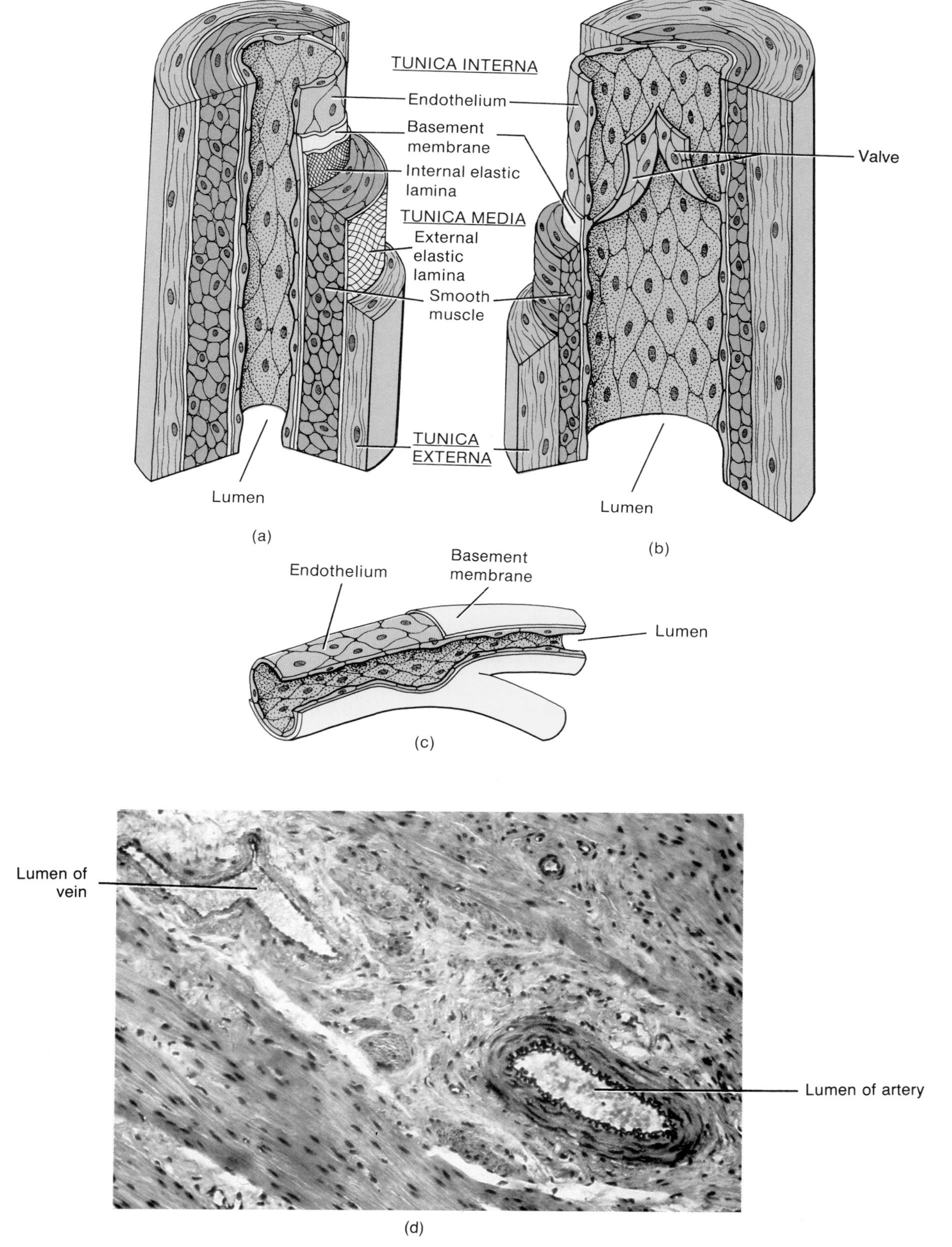

collateral circulation. An alternate blood route may also be from nonanastomosing vessels that supply the same region of the body.

Arteries that do not anastomose are known as ***end arteries.*** Occlusion of an end artery interrupts the blood supply to a whole segment of an organ, producing necrosis (death) of that segment.

ARTERIOLES

An ***arteriole*** is a very small, almost microscopic artery that delivers blood to capillaries. Arterioles closer to the arteries from which they branch have a tunica interna like that of arteries, a tunica media composed of smooth muscle and very few elastic fibers, and a tunica externa composed mostly of elastic and collagenous fibers. As arterioles get smaller in size, the tunics change character so that arterioles closest to capillaries consist of little more than a layer of endothelium surrounded by a few scattered smooth muscle fibers (cells) (Figure 21-2).

Arterioles play a key role in regulating blood flow from arteries into capillaries. The smooth muscle of arterioles, like that of arteries, is subject to vasoconstriction and vasodilation. During vasoconstriction, blood flow into capillaries is restricted; during vasodilation, the flow is significantly increased. A change in diameter of arterioles can also significantly affect blood pressure. The relationship of arterioles to blood flow will be considered in detail later in the chapter.

Two medical tests that are extremely valuable in diagnosing abnormalities of blood vessels are digital subtraction angiography (DSA) and Doppler ultrasound, both of which have been described briefly in Chapter 1.

CAPILLARIES

Capillaries are microscopic vessels that usually connect arterioles and venules. They are found near almost every cell in the body. The distribution of capillaries in the body varies with the activity of the tissue. For example, in places where activity is higher, such as muscles, the liver, kidneys, lungs, and nervous system, there are rich capillary supplies. In areas where activity is lower, such as tendons and ligaments, the capillary supply is not as extensive. The epidermis, cornea and lens of the eye, and cartilage are devoid of capillaries.

The primary function of capillaries is to permit the exchange of nutrients and wastes between the blood and tissue cells. The structure of the capillaries is admirably suited to this purpose. Capillary walls are composed of only a single layer of cells (endothelium) and a basement membrane (see Figure 21-1c). They have no tunica media or tunica externa. Thus, a substance in the blood must pass through the plasma membrane of just one cell to reach tissue cells. This vital exchange of materials occurs only through capillary walls—the thick walls of arteries and veins present too great a barrier.

Although capillaries pass directly from arterioles to venules in some places in the body, in other places they form extensive branching networks. These networks increase the surface area for diffusion and filtration and thereby allow a rapid exchange of large quantities of materials. In most tissues, blood normally flows through only a small portion of the capillary network when metabolic needs are low. But when a tissue becomes active, the entire capillary network fills with blood.

The flow of blood through capillaries is regulated by vessels with smooth muscle in their walls. A ***metarteriole*** (*met* = beyond) is a vessel that emerges from an arteriole, passes through the capillary network, and empties into a venule (Figure 21-3). The proximal portions of metarterioles are surrounded by scattered smooth muscle fibers (cells) whose contraction and relaxation help to regulate the amount and force of blood. The distal portion of a metarteriole has no smooth muscle fibers and is called a ***thoroughfare channel.*** It serves as a low-resistance channel that increases blood flow. ***True capillaries*** emerge from arterioles or metarterioles and are not on the direct flow route from arteriole to venule. At their sites of origin, there is a ring of smooth muscle fibers called a ***precapillary sphincter*** that controls the flow of blood entering a true capillary. The various factors that regulate the contraction of smooth muscle fibers of arterioles and precapillary sphincters are discussed later.

Some capillaries of the body, such as those found in muscle tissue and other locations, are referred to as ***continuous capillaries,*** so named because the cytoplasm of the endothelial cells is continuous when viewed in cross section through a microscope; the cytoplasm appears as an uninterrupted ring, except for the endothelial junction. Other capillaries of the body are referred to as ***fenestrated capillaries.*** They differ from continuous capillaries in that their endothelial cells have numerous fenestrae (pores) where the cytoplasm is very thin or absent. The fenestrae range from 70 to 100 nm in diameter and are closed by a thin diaphragm, except in the capillaries in the kidneys where they are assumed to be open. Fenestrated capillaries are also found in the villi of the

FIGURE 21-2 Structure of an arteriole.

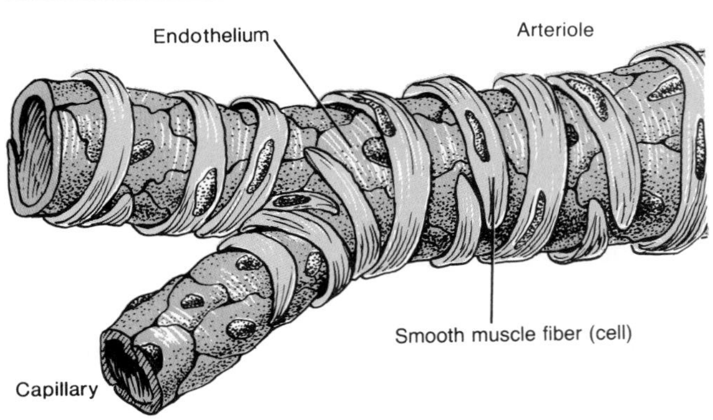

FIGURE 21-3 **Capillaries (a) Details of a capillary network. (b) Photograph of a blood capillary. Note that some of the red blood cells pass through sideways. (Courtesy of Lennart Nilsson, *Behold Man,* Delacorte Press, Dell Publishing.)**

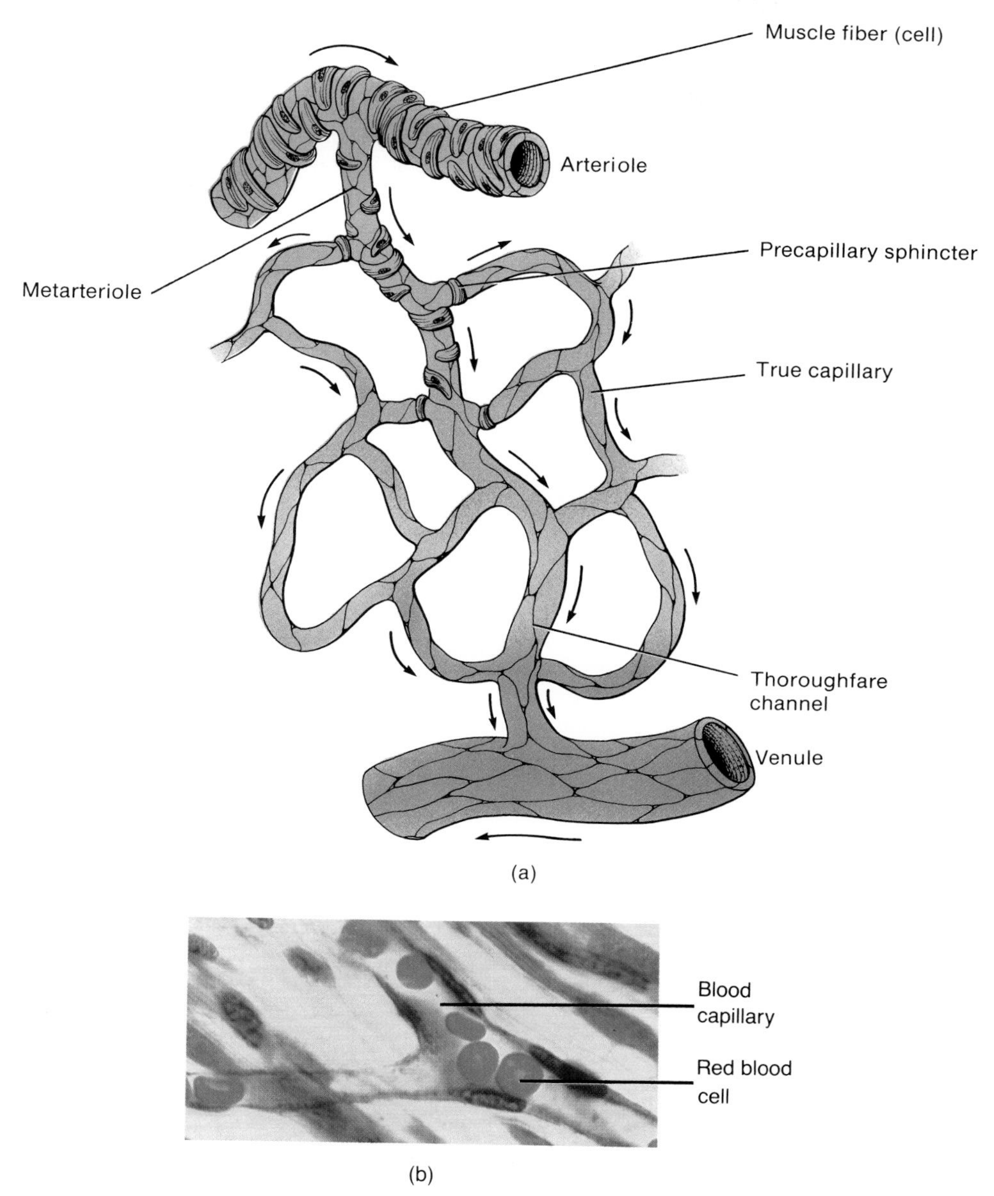

small intestine, choroid plexuses of the ventricles in the brain, ciliary processes of the eyes, and endocrine glands.

Microscopic blood vessels in certain parts of the body, such as the liver, are termed ***sinusoids.*** They are wider than capillaries and more tortuous. Also, instead of the usual endothelial lining, sinusoids contain spaces between endothelial cells, and the basal lamina is incomplete or absent. In addition, sinusoids contain specialized lining cells that are adapted to the function of the tissue. For example, in the liver, sinusoids contain phagocytic cells called ***stellate reticuloendothelial (Kupffer's) cells.*** Like capillaries, sinusoids convey blood from arterioles to venules. Other regions containing sinusoids include the spleen, adenohypophysis, parathyroid glands, adrenal cortex, and bone marrow.

In performing their function of permitting the exchange of substances between blood and body cells, substances enter and leave capillaries through four basic routes: through junctions that anchor endothelial cells, via pinocytic vesicles, directly across capillary membranes, and through fenestrations.

VENULES

When several capillaries unite, they form small veins called ***venules.*** Venules collect blood from capillaries and drain it into veins. The venules closest to the capillaries consist of a tunica interna of endothelium and a tunica externa of connective tissue. As the venules approach the veins, they also contain the tunica media characteristic of veins.

VEINS

Veins are composed of essentially the same three coats as arteries, but there are variations in their relative thickness. The tunica interna of veins is extremely thin compared with that of their accompanying arteries. In addition, the tunica media of veins is much thinner than that of accompanying arteries, and the tunica externa is thicker in veins (see Figure 21-1b). Despite these differences, veins are still distensible enough to adapt to variations in the volume and pressure of blood passing through them.

By the time the blood leaves the capillaries and moves into the veins, it has lost a great deal of pressure. The difference in pressure can be observed in the blood flow from a cut vessel; blood leaves a cut vein in an even, slow flow rather than in the rapid spurts characteristic of arteries. Most of the structural differences between arteries and veins reflect this pressure difference. For example, the walls of veins are not as strong as those of arteries. The low pressure in veins, however, has its disadvantages. When you stand, the pressure pushing blood up the veins in your lower extremities is barely enough to balance the force of gravity pushing it back down. For this reason, many veins, especially those in the limbs, contain valves that prevent backflow (see Figure 21-9). Normal valves ensure the flow of blood toward the heart.

CLINICAL APPLICATION: VARICOSE VEINS

In people with weak venous valves, gravity forces large quantities of blood back down into distal parts of the vein. This pressure overloads the vein and pushes its wall outward. After repeated overloading, the walls lose their elasticity and become stretched and flabby. Such dilated and tortuous veins caused by incompetent valves are called *varicose veins* (*VVs*). They may be due to heredity, mechanical factors (prolonged standing and pregnancy), or aging. Because a varicosed wall is not able to exert a firm resistance against the blood, blood tends to accumulate in the pouched-out area of the vein, causing it to swell and forcing fluid into the surrounding tissue. Veins close to the surface of the legs are highly susceptible to varicosities. Veins that lie deeper are not as vulnerable because surrounding skeletal muscles prevent their walls from overstretching.

Varicose veins may be treated by several methods, depending on the severity of the condition. These include: (1) frequent periods of rest with elevation of the lower extremities; (2) external pressure with elastic stockings or bandages; (3) a procedure known as sclerotherapy, an intravenous injection of sclerosing chemicals that collapses the veins and prevents blood from flowing into them, thus eliminating the purple-blue discoloration; and (4) surgery, in which the saphenous veins with their incompetent valves are ligated and removed ("stripping").

A mild form of varicose veins in the legs is referred to as *spider burst veins,* a condition that is most commonly treated by sclerotherapy.

A ***vascular (venous) sinus*** is a vein with a thin endothelial wall that has no smooth muscle to alter its diameter. Surrounding dense connective tissue replaces the tunica media and tunica externa to provide support. Intracranial vascular sinuses, which are supported by the dura mater, return cerebrospinal fluid and deoxygenated blood from the brain to the heart. Another example of a vascular sinus is the coronary sinus of the heart.

BLOOD RESERVOIRS

The volume of blood in various parts of the cardiovascular system varies considerably. Veins, venules, and venous sinuses contain about 59 percent of the blood in the system; arteries, about 13 percent; pulmonary vessels, about 12 percent; the heart, about 9 percent; and arterioles and capillaries, about 7 percent. Since systemic veins contain so much of the blood, they are referred to as ***blood reservoirs.*** They serve as storage depots for blood, which can be moved quickly to other parts of the body if the need arises. When there is increased muscular activity, the vasomotor center sends increasing sympathetic impulses to veins that serve as blood reservoirs. The result is vasoconstriction, which permits the distribution of blood from venous reservoirs to skeletal muscles, where it is needed most. A similar mechanism operates in cases of hemorrhage, when blood volume and pressure decrease. Vasoconstriction of veins in venous reservoirs helps to compensate for the blood loss. Among the principal blood reservoirs are the veins of the abdominal organs (especially the liver and spleen) and the veins of the skin.

PHYSIOLOGY OF CIRCULATION

Blood Flow

Blood flow refers to the amount of blood that passes through a blood vessel in a given period of time. Blood flow is determined by two principal factors: (1) blood pressure and (2) resistance (opposition), the force of friction as blood travels through blood vessels.

Blood Pressure

***Blood pressure* (*BP*)** is the pressure exerted by blood on the wall of a blood vessel. In clinical use, the term refers to pressure in arteries. BP is generated by cardiac output, which is determined by the rate and force of heartbeat and the resistance to the flow of blood through vessels. An increase in heart rate or force of contraction and an increase in resistance would increase blood pressure. Blood flows through its system of closed vessels because of different blood pressures in various parts of the cardiovascular system. Blood flow is directly proportional to blood pressure. Blood always flows from regions of higher blood pressure to regions of lower blood pressure and pressure differences are related to cardiac output and resistance. The mean (average) pressure in the aorta is about 100 millimeters of mercury (mm Hg). Since the heart pumps in a pulsating manner, the systemic arterial pressure in a resting young adult fluctuates between 120 mm Hg (systolic) and 80 mm Hg (diastolic). As blood leaves the left ventricle and flows through systemic circulation, its pressure falls progressively to 0 mm Hg by the time it reaches the right atrium (Figure 21-4). Resistance in the aorta is nearly 0, but it is close to the powerfully contracting left ventricle and thus the pressure is high (100 mm Hg). Likewise, resistance in large arteries is also quite low, but since these are close to the aorta, the mean arterial pressure is still high (about 95 to 97 mm Hg). In small arteries, resistance increases rapidly and pressure within the arteries decreases rapidly to about 85 mm Hg at the beginning of arterioles. Resistance in arterioles is the highest in the cardiovascular system, accounting for about one-half the total resistance to blood flow. Pressure decreases in arterioles from 85 to about 30 mm Hg. As blood next passes into the arteriole end of a capillary, its pressure is about 30 mm Hg, and at the venous end of a capillary, the pressure is about 10 mm Hg.

The decrease in pressure through the venous system is not as abrupt as in the arterial system. The pressure at the beginning of the venous system, that is, in the venules, is about 10 mm Hg. The pressure continues to decrease in veins and the venae cavae to about 0 mm Hg in the right atrium. A great deal of the resistance in veins is caused by their compression from surrounding tissues and pressure in the abdomen.

FIGURE 21-4 **Blood pressures in various portions of systemic circulation of the cardiovascular system. The dashed line represents mean (arterial) pressure.**

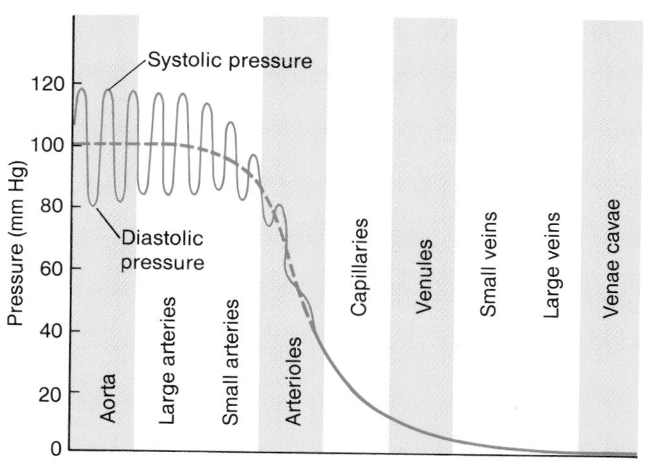

Resistance

As noted earlier, ***resistance*** refers to the opposition or impedance to blood flow principally as a result of the force of friction between blood and the walls of blood vessels and viscosity created by red blood cells and plasma proteins. Resistance is related to: (1) blood viscosity, (2) blood vessel length, and (3) blood vessel radius.

- **Blood Viscosity.** The viscosity ("thickness") of blood is a function of the ratio of red blood cells and solutes to fluid. Resistance to blood flow is directly proportional to the viscosity of blood. Any condition that increases the viscosity of blood, such as dehydration, an unusually high number of red blood cells (polycythemia), or severe burns, increases blood pressure. A depletion of plasma proteins or red blood cells, as a result of anemia or hemorrhage, decreases blood viscosity and blood pressure.

- **Blood Vessel Length.** Resistance to blood flow through a vessel is directly proportional to the length of the blood vessel. The longer a blood vessel, the greater the resistance as blood flows through it.

- **Blood Vessel Radius.** Resistance is inversely proportional to the fourth power of the radius of the blood vessel. The smaller the radius of the blood vessel, the greater resistance it offers to blood flow. As an example, if the radius of a blood vessel is decreased by one-half, its resistance to blood flow would increase sixteen times and this increased resistance would significantly increase if the smaller radius involved many arterioles.

Factors That Affect Arterial Blood Pressure

Here we will identify several factors that influence arterial blood pressure.

Cardiac Output (CO)

***Cardiac output* (*CO*),** the amount of blood ejected by the left ventricle into the aorta each minute, is the principal determinant of blood pressure. As noted in Chapter 20, CO is calculated by multiplying stroke volume by heart rate. In a normal, resting adult, it is about 5.25 liters/min (70 ml × 75 beats/min). Blood pressure varies directly with CO. If CO is increased by any increase in stroke volume or heart rate, then blood pressure increases. A decrease in CO causes a decrease in blood pressure.

Blood Volume

Blood pressure is directly proportional to the ***volume of blood*** in the cardiovascular system. The normal volume of blood in a human body is about 5 liters (5 qt). Any decrease in this volume, as from hemorrhage, decreases the amount of blood that is circulated through the arteries each minute. As a result, blood pressure drops. Conversely, anything that increases blood volume, such as high salt intake and therefore water retention, increases blood pressure.

Peripheral Resistance

Peripheral resistance refers to all the vascular resistances offered by the cardiovascular system, that is, all the factors that oppose blood flow. A major function of arterioles is to control peripheral resistance—and, therefore, blood pressure and flow—by changing their diameters. The principal center for this regulation is the vasomotor center in the medulla, which will be described shortly.

The factors that affect arterial blood pressure are summarized in Figure 21-5.

FIGURE 21-5 **Summary of factors that affect arterial blood pressure.**

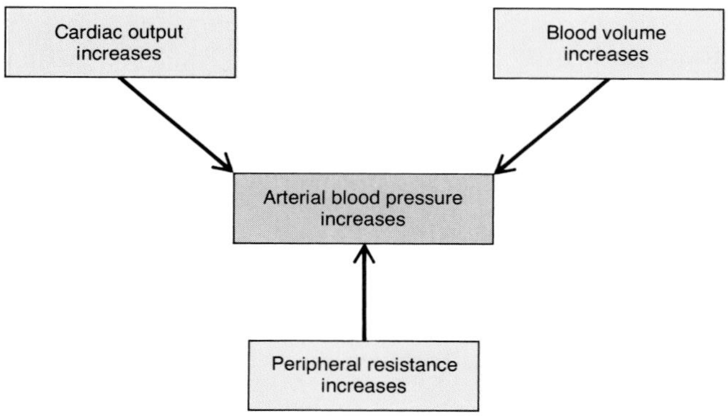

Control of Blood Pressure

In Chapter 20 we considered the increase and decrease of heart rate and force of contraction which is partially regulated by the cardioacceleratory center (CAC) and the cardioinhibitory center (CIC). We also took a look at how certain chemicals (epinephrine, potassium, sodium, and calcium), temperature, emotions, sex, and age affect heart rate. By their effect on heart rate and force of contraction, these factors also control blood pressure. Any increase in heart rate and force of contraction increases blood pressure. Conversely, any decrease will lower blood pressure, assuming that peripheral resistance remains constant.

At this point we will examine those factors that help to regulate blood pressure by acting on the blood vessels themselves.

Vasomotor Center

Within the medulla is a cluster of neurons referred to as the ***vasomotor*** (*vas* = vessel; *motor* = movement) ***center.*** The function of this center is to control the diameter of blood vessels, especially arterioles of the skin and abdominal viscera. It continually sends impulses to the smooth muscle in arteriole walls that result in a moderate state of vasoconstriction at all times. This state of tonic contraction, called ***vasomotor tone,*** assumes a role in maintaining peripheral resistance and blood pressure. The vasomotor center brings about vasoconstriction by increasing the number of sympathetic impulses above normal. It causes vasodilation by decreasing the number of sympathetic impulses below normal. In other words, in this case, the sympathetic division of the autonomic nervous system (ANS) can bring about both vasoconstriction and vasodilation depending on the frequency of impulses which travel over sympathetic vasoconstrictor nerves which innervate the smooth muscle of blood vessels.

The vasomotor center also brings about constriction of veins when arterial blood pressure decreases below normal. This action forces blood out of reservoirs (spleen, liver, large abdominal veins, and veins of the skin) into other portions of the cardiovascular system. This directly raises peripheral venous pressure which leads to increased venous return, cardiac output, and arterial blood pressure. Dilation of veins causes an increase of blood in reservoirs, decreased peripheral venous pressure, decreased venous return, decreased cardiac output, and decreased arterial blood pressure.

Sympathetic vasomotor nerve fibers from the vasomotor center leave the spinal cord through all thoracic and the first one or two lumbar spinal nerves. The fibers pass into the sympathetic trunk ganglia. From here, they pass through sympathetic nerves that innervate blood vessels in viscera and spinal nerves that innervate blood vessels in peripheral areas. Sympathetic stimulation of small arteries and arterioles causes vasoconstriction and thus raises blood pressure. Sympathetic stimulation of veins results in constriction that moves blood from reservoirs to increase blood pressure.

Recall that sympathetic stimulation of the heart increases heart rate and force of contraction, whereas parasympathetic stimulation decreases heart rate.

The vasomotor center is affected by several factors, all of which influence blood pressure.

Baroreceptors

It will be recalled that stimulated baroreceptors (pressure receptors) in the wall of the carotid sinus and aorta or attached to them send impulses to the cardiac center that result in increased or decreased cardiac output to help regulate blood pressure (see Figure 20-9). This reflex acts not only on the heart but also on the arterioles. For example, if there is an increase in blood pressure, the baroreceptors stimulate the cardioinhibitory center (CIC) and inhibit the cardioacceleratory center (CAC). The result is a decrease in cardiac output and a decrease in blood pressure. The baroreceptors also send impulses to the vasomotor center. In response, the vasomotor center decreases sympathetic stimulation to arterioles and veins. The result is vasodilation and a decrease in blood pressure (Figure 21-6). If there is a decrease in blood pressure, the baroreceptors inhibit the cardioinhibitory center and stimulate the cardioacceleratory center. The result is an increase in cardiac output and an increase in blood pressure. Also, the baroreceptors send impulses to the vasomotor center that increase sympathetic stimulation to arterioles and veins. The result is vasoconstriction and an increase in blood pressure.

FIGURE 21-6 Vasomotor center control of blood pressure by stimulation of baroreceptors.

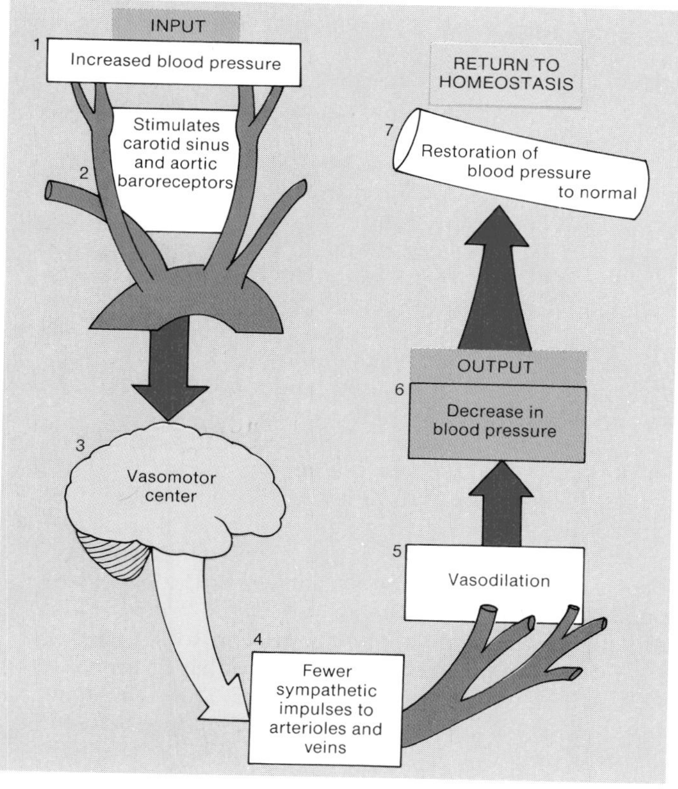

Chemoreceptors

Receptors sensitive to chemicals in the blood are called ***chemoreceptors.*** Chemoreceptors near or on the carotid sinus and aorta are called ***carotid*** and ***aortic bodies,*** respectively. They are sensitive to arterial blood levels of oxygen, carbon dioxide, and hydrogen ions. Given a severe deficiency of oxygen (hypoxia), an increase in hydrogen ion concentration, or an excess of carbon dioxide (hypercapnia), the chemoreceptors are stimulated and send impulses to the vasomotor center. In response, the vasomotor center increases sympathetic stimulation to arterioles and veins. This brings about vasoconstriction and an increase in blood pressure. As you will see in Chapter 23, chemoreceptors also stimulate the medullary rhythmicity area in response to these chemical stimuli to adjust the rate of breathing.

Control by Higher Brain Centers

In response to strong emotions, ***higher brain centers,*** such as the cerebral cortex, can have a significant influence on blood pressure. For example, during periods of intense anger, the cerebral cortex stimulates the vasomotor center to fire sympathetic impulses to arterioles and veins. This causes vasoconstriction and an increase in blood pressure. When a person is depressed or grieving, impulses from higher brain centers cause a decrease in sympathetic stimulation by the vasomotor center. This produces vasodilation and a decrease in blood pressure. A frequent result is fainting because blood flow to the brain is diminished.

Chemicals

Several ***chemicals*** affect blood pressure by causing vasoconstriction. Epinephrine and norepinephrine (NE), produced by the adrenal medulla, increase the rate and force of heart contractions and bring about vasoconstriction of abdominal and cutaneous arterioles and veins. They also bring about dilation of cardiac and skeletal muscle arterioles. Antidiuretic hormone (ADH), produced by the hypothalamus and released from the neurohypophysis, causes vasoconstriction if there is a severe loss of blood due to hemorrhage. Angiotensin II helps to raise blood pressure by stimulating secretion of aldosterone (increases sodium ion concentration and water reabsorption) and causing vasoconstriction. Histamine produced by mast cells and kinins found in plasma are vasodilators that assume key functions during the inflammatory response.

Autoregulation (Local Control)

Autoregulation refers to a local, automatic adjustment of blood flow in a given region of the body in response to the particular needs of the tissue. In most body tissues, oxygen is the principal, though not direct, stimulus for autoregulation. A suggested mechanism for autoregulation is as follows. In response to low oxygen supplies, the cells in the immediate area produce and release ***vasodilator substances.*** Such substances are thought to include potassium ions, hydrogen ions, carbon dioxide, lactic acid, and adenosine. Once released, the vasodilator substances produce a local dilation of arterioles and relaxation of precapillary sphincters. The result is an increased flow of blood into the tissue, which restores oxygen levels to normal. The autoregulation mechanism is important in meeting the nutritional demands of active tissues, such as muscle tissue, where the demand might increase as much as 10-fold.

Capillary Exchange

For reasons to be discussed shortly, you will see that although blood pressure decreases consistently from the aorta to the venae cavae, the velocity of blood decreases as it flows through the aorta, arterioles, and capillaries and then increases as it passes into venules and veins. The velocity of blood flow in capillaries is the slowest in the cardiovascular system, and this is important because it allows adequate time for the exchange of materials between blood and body tissues.

Blood usually does not flow in a continuous manner through capillary networks. Rather, it flows intermittently, because of contraction and relaxation of the smooth muscle fibers (cells) of metarterioles and the precapillary sphincters of true capillaries. The intermittent contraction and relaxation, which may occur 5 to 10 times per minute, is called ***vasomotion.*** The most important factor in controlling vasomotion is the oxygen concentration in tissues, a mechanism similar to autoregulation in arterioles.

The movement of water and dissolved substances, except proteins, through capillary walls is mostly by diffusion (but also involves filtration) and is dependent on several opposing forces, or pressures. Some forces push fluid out of capillaries into the surrounding interstitial (tissue) spaces, resulting in filtration of fluid. To prevent fluid from moving in one direction only and accumulating in interstitial spaces, opposing forces push fluid from interstitial spaces into blood capillaries, resulting in reabsorption of fluid. Let us now see how these forces operate.

First we will consider the hydrostatic pressures involved. These pressures are due to the pressure of water in the fluids. Blood pressure in capillaries, called ***blood hydrostatic pressure* (*BHP*)**, tends to move fluids out of capillaries into interstitial fluid. BHP averages about 30 mm Hg at the arterial end of a capillary and about 15 mm Hg at the venous end, although the pressures may vary considerably, depending on the activity of the arterioles and venules (Figure 21-7a). Opposing BHP is the pressure of the interstitial fluid, called ***interstitial fluid hydrostatic pressure* (*IFHP*)**, which tends to move fluid out of interstitial spaces into capillaries. Because of difficulties in measuring IFHP, its reported values are variable, ranging from small positive to small negative values. For purposes of our discussion, we will assume, however, that the value of IFHP is 0 mm Hg at both ends of capillaries. Regardless of its exact value, the basic principles of fluid movement still apply.

Now let us consider the osmotic pressures involved in fluid movement. Osmotic pressures are due to the presence of nondiffusible proteins in blood and interstitial fluid. ***Blood osmotic pressure* (*BOP*)**, which tends to move fluid from interstitial spaces into capillaries, averages about 28 mm Hg at both ends of capillaries. Opposing BOP is ***interstitial fluid osmotic pressure* (*IFOP*)**, which tends to move fluid out of capillaries into interstitial fluid and averages about 6 mm Hg at both ends of capillaries.

Whether fluids leave or enter capillaries depends on how the pressures relate to each other. If the forces that tend to move fluid out of capillaries are greater than the forces that tend to move fluid into capillaries, the fluid will move from capillaries into interstitial spaces (filtration). If, on the other hand, the forces that tend to move fluid out of interstitial spaces into capillaries are greater than the forces that tend to move fluid out of capillaries, then fluid will move from interstitial spaces into capillaries (reabsorption). The term ***effective filtration pressure* (*Peff*)** is used to show the direction of fluid movement. It is calculated as follows:

$$Peff = (\text{BHP} + \text{IFOP}) - (\text{IFHP} + \text{BOP})$$

Substituting values at the arterial end of a capillary,

$$\begin{aligned} Peff &= (30 + 6) - (0 + 28) \\ &= (36) - (28) \\ &= 8 \text{ mm Hg} \end{aligned}$$

Substituting measured values at the venous end of a capillary,

$$\begin{aligned} Peff &= (15 + 6) - (0 + 28) \\ &= (21) - (28) \\ &= -7 \text{ mm Hg} \end{aligned}$$

Thus, at the arterial end of a capillary, there is a *net outward force* (8 mm Hg), and fluid moves out of the capillary (filtered) into interstitial spaces. At the venous end of a capillary, the negative value (−7 mm Hg) represents a *net inward force,* and fluid moves into the capillary (reabsorbed) from tissue spaces.

Not all the fluid filtered at one end of the capillary is reabsorbed at the other end. When we discuss fluid dynamics in more detail in Chapter 27, you will see that some of the filtered fluid and any proteins that escape from blood into interstitial fluid are returned by the lymphatic system to the cardiovascular system. Under normal conditions, there is a state of near equilibrium at the arterial and venous ends of a capillary in which filtered fluid and absorbed fluid, in addition to that returned to the lymphatic system, are nearly equal. This near equilibrium is known as ***Starling's law of the capillaries.***

CLINICAL APPLICATION: EDEMA

Occasionally, the balance of filtration and reabsorption between interstitial fluid and plasma is disrupted, allowing an abnormal increase in interstitial fluid called *edema* (*oidema* = swelling) (Figure 21-7b). In general, edema is not detectable in tissues until interstitial fluid volume has increased by about 30 percent above normal.

FIGURE 21-7 Dynamics of capillary exchange. (a) Hydrostatic and osmotic pressure forces involved in moving fluid out of capillaries (filtration) and into capillaries (reabsorption). (b) Factors that contribute to edema.

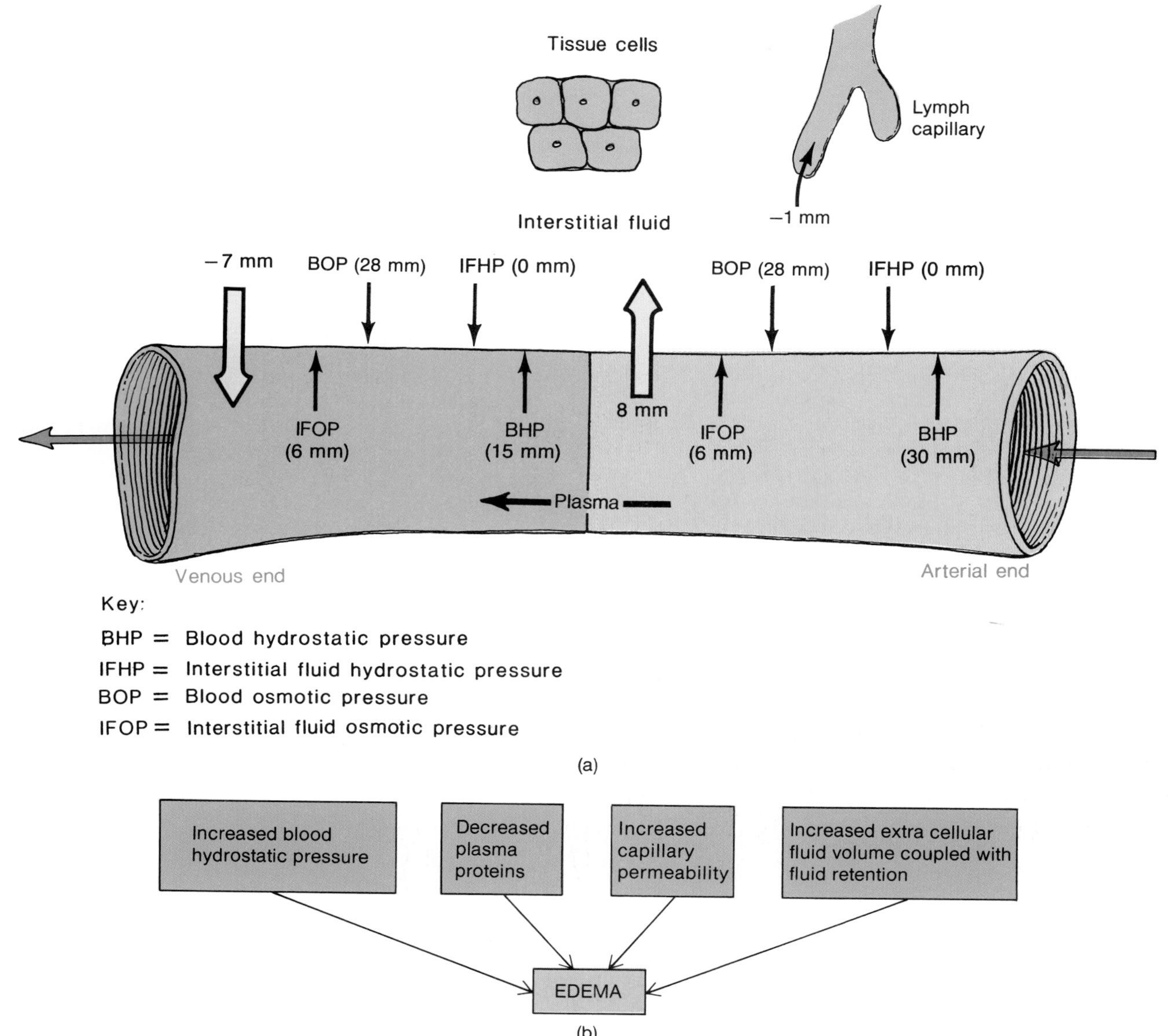

Edema may result from several main causes:

1. Increased blood hydrostatic pressure in capillaries from an increase in venous pressure. This may result from obstruction of venous return due to cardiac failure or blood clots.

2. Decreased plasma proteins that lower blood osmotic pressure. Protein loss may result from burns, malnutrition, liver disease, and kidney disease.

3. Increased permeability of capillaries that raises interstitial fluid osmotic pressure by allowing significant amounts of plasma proteins to leave the blood and enter tissue fluid. This may be caused by chemical, bacterial, thermal, or mechanical agents.

4. Increased extracellular fluid volume as a result of conditions associated with fluid retention. When a person has difficulty voiding large amounts of water, but continues to intake normal amounts of water, extracellular fluid in the body increases. Some of the fluid enters blood and increases blood hydrostatic pressure.

Factors That Aid Venous Return

The establishment of a pressure gradient for blood is the primary reason why blood flows. A number of factors help the blood to return through the veins by increasing the magnitude of this pressure gradient between the veins and the right atrium: contractions of the skeletal muscles, valves, and breathing.

Velocity of Blood Flow

The velocity of blood flow is inversely related to the cross-sectional area of the blood vessels. Blood flows most rapidly where the cross-sectional area is least (Figure 21-8). Each time an artery branches, the total cross-sectional area of all the branches is greater than that of the original vessel. When branches combine, the resulting cross-sectional area is less than that of the original branches. The cross-sectional area of the aorta is 2.5 cm^2, and the velocity of the blood there is 40 cm/sec. Capillaries have a cross-sectional area of 2500 cm^2, and the velocity is less than 0.1 cm/sec. In the venae cavae, the cross-sectional area is 8 cm^2, and the velocity is 5 to 20 cm/sec.

Thus, the velocity of blood decreases as it flows from the aorta to arteries to arterioles to capillaries. The velocity of blood in the capillaries is the slowest in the cardiovascular system; so there is adequate exchange (diffusion) time between the capillaries and adjacent tissues. As blood vessels leave capillaries and approach the heart, their cross-sectional area decreases. Therefore, the velocity of blood increases as it flows from capillaries to venules to veins to the heart.

Blood flow can be measured to diagnose certain circulatory disorders. ***Circulation time*** is the time required for blood to pass from the right atrium, through pulmonary circulation, back to the left ventricle, through systemic circulation down to the foot, and back again to the right atrium. Such a trip usually takes about 1 min.

Skeletal Muscle Contractions and Valves in Veins

The combination of skeletal muscle contractions and valves in veins is important in returning venous blood to the heart. Many veins, especially those in the extremities, contain valves. When skeletal muscles contract, they tighten around the veins running through them which increases the venous blood pressure, and the valves open. This pressure drives the blood toward the heart—the action is called ***milking*** (Figure 21-9). When the muscles relax, the valves close to prevent the backflow of blood away from the heart. Individuals who are immobilized through injury or disease cannot take advantage of these contractions. As a result, the return of venous blood to the heart is slower, and the heart has to work harder. For this reason, periodic massage is helpful.

Breathing

Another factor that is important in maintaining venous circulation is breathing. During inspiration, the diaphragm moves downward. This causes a decrease in pressure in

FIGURE 21-8 Relationship between velocity of blood flow and total cross-sectional area in various blood vessels of the cardiovascular system. (a) Graph. (b) Exhibit.

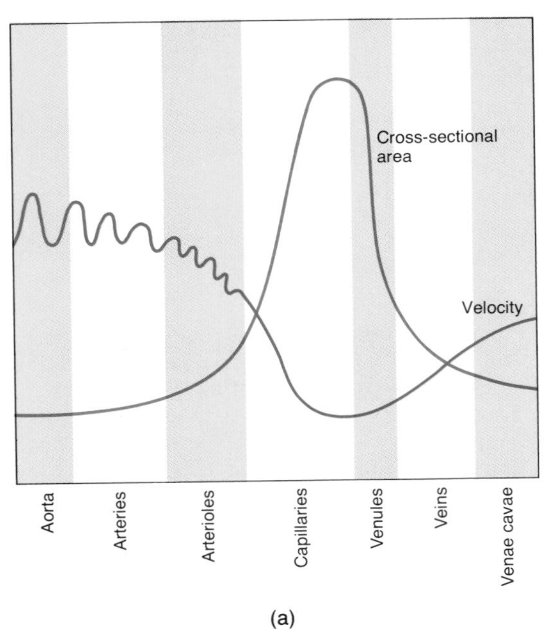

(a)

RELATIONSHIP BETWEEN BLOOD VELOCITY AND CROSS-SECTIONAL AREA

Vessel	Cross-Sectional Area (cm^2)	Velocity (cm/sec)
Aorta	2.5	40
Arteries	20	10–40
Arterioles	40	0.1
Capillaries	2500	less than 0.1
Venules	250	0.3
Veins	80	0.3–5.0
Venae cavae	8	5–20

(b)

FIGURE 21-9 Role of skeletal muscle contractions and venous valves in returning blood to the heart. (a) When skeletal muscles contract, the valves open, and blood is forced toward the heart. (b) When skeletal muscles relax, the valves close to prevent the backflowing of blood from the heart. (c) Photograph of a one-way valve in a vein (cross section above, longitudinal section below). (Courtesy of J. A. Gosling, P. F. Harris, et al., *Atlas of Human Anatomy,* Gower Medical Publishing Ltd., 1985.)

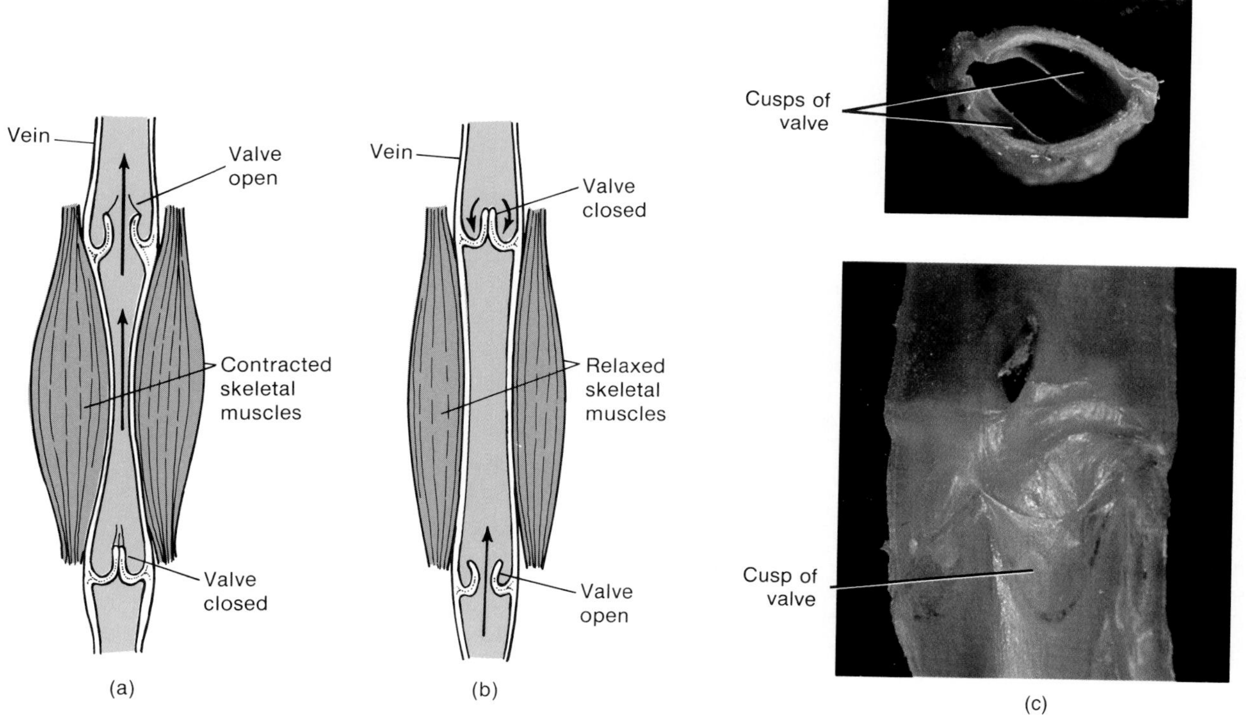

the thoracic (chest) cavity and an increase in pressure in the abdominal cavity. As a result, a greater volume of blood moves from the abdominal veins into the thoracic veins. When the pressures reverse during expiration, blood in the veins is prevented from backflowing by the valves.

SHOCK AND HOMEOSTASIS

Shock is a failure of the cardiovascular system to deliver adequate amounts of oxygen and nutrients to meet the metabolic needs of body cells due to an inadequate cardiac output. As a result, cellular membranes dysfunction, cellular metabolism is abnormal, and cellular death may eventually occur without proper treatment.

Signs and Symptoms

The signs and symptoms of shock vary with the severity of the condition. Among the characteristic ones are the following:

1. Hypotension in which the systolic blood pressure is lower than 90 mm Hg as a result of generalized vasodilation and decreased cardiac output.
2. Clammy, cool, pale skin due to vasoconstriction of skin blood vessels.
3. Sweating due to sympathetic stimulation and increased levels of epinephrine.
4. Reduced urine formation due to hypotension and increased levels of aldosterone and antidiuretic hormone (ADH).
5. Altered mental status due to cerebral ischemia.
6. Acidosis due to buildup of lactic acid.
7. Tachycardia due to sympathetic stimulation and increased levels of epinephrine.
8. Weak, rapid pulse due to generalized vasodilation and reduced cardiac output.
9. Thirst due to loss of extracellular fluid.

Classification

The causes of shock are many and varied, but all are characterized by inadequate tissue perfusion. One type of shock, called ***hypovolemic*** (hī-pō-vō-LĒ-mik) ***shock*** refers to decreased intravascular volume resulting from loss of blood or plasma. Factors that lead to hypovolemic shock are *acute hemorrhage* due to conditions such as trauma, gastrointestinal bleeding, and hematomas; and *excessive fluid loss* as a result of excess vomiting, diarrhea, sweating, dehydration, excessive urine production, and burns.

Another type of shock called ***cardiogenic shock*** refers to inadequate cardiac function. Causes include myocardial infarction (MI), inflammation of the heart, dysfunctioning heart valves, dysrhythmias, cardiac rupture, and congestive heart failure.

Obstructive shock refers to mechanical disruption of blood flow at any site in the cardiovascular system. Causes include pulmonary embolism, cardiac tamponade, cardiac tumors, and certain aneurysms.

Other types of shock result from changes in vascular resistance or permeability. Loss of vasomotor tone, also called ***neurogenic shock,*** results in widespread vasodilation and may be caused by general or spinal anesthesia, spinal cord or brain damage, certain drugs, and anaphylaxis. ***Septic shock*** results from tissue injury as a result of toxins produced by certain bacteria, especially gram-negative ones. The toxins interact with cell membranes and components of the coagulation system and thus activate blood clotting, injure cells, and alter blood flow. Alteration in vasomotor tone and permeability of blood capillaries are responsible for inadequate tissue perfusion. Septic shock brings about a number of complications including coagulation defects and respiratory, renal, and cardiac failure.

Stages

Since hypovolemic shock has been studied so extensively, we will describe the stages of shock as they apply to this type of shock. The development of shock occurs in three principal stages, which merge with one another.

Stage I. Compensated (Nonprogressive) Shock

During Stage I, when symptoms and signs are minimal, certain homeostatic mechanisms of the cardiovascular system compensate for the shock so that no serious damage results. If the initiating cause does not get any worse, a full recovery follows. The major mechanisms of compensation are negative feedback systems that attempt to return cardiac output (CO) and arterial blood pressure to normal. These compensatory adjustments are mediated through the sympathetic nervous system and the release of various substances. In an otherwise healthy individual, acute blood loss of as much as 10 percent of the total volume occurs by compensatory mechanisms. Among the adjustments are the following (Figure 21-10a):

1. A decrease in general blood pressure, detected by baroreceptors, initiates powerful sympathetic responses throughout most of the body that results in vasoconstriction of arterioles and veins of the skin and abdominal viscera (vasoconstriction does not occur in the brain or heart). This increases peripheral resistance, helps maintain adequate venous return, and increases heart rate and force of contraction. Vasoconstriction is most marked in the skin, where it accounts for coolness and paleness; kidneys; and other abdominal viscera.

2. Decreased blood flow to the kidneys causes the kidneys to secrete renin and initiates the renin-angiotensin pathway (see Figure 18-19). Recall that angiotensin II stimulates the adrenal cortex to secrete aldosterone—a hormone that increases sodium and indirectly water reabsorption to help raise blood pressure and angiotensin II also serves as a vasoconstrictor which helps to raise blood pressure by increasing peripheral resistance.

3. Intense sympathetic stimulation leads to increased blood levels of epinephrine and norepinephrine (NE) by the adrenal medulla. These hormones intensify vasoconstriction and increase heart rate and force of beat, all of which helps to raise blood pressure.

4. In response to decreased blood pressure, the posterior pituitary releases more antidiuretic hormone (ADH). This hormone brings about water conservation by the kidneys and vasoconstriction.

5. In response to hypoxia (low-level oxygen availability), affected cells produce vasodilator substances (potassium ions, hydrogen ions, carbon dioxide, lactic acid, and adenosine) that dilate arterioles and relax precapillary sphincters. This increases regional blood flow and restores oxygen levels to normal. (Such vasodilation has the potentially harmful effect of decreasing peripheral resistance and thus lowering blood pressure.)

The various compensatory mechanisms may require from 30 seconds to 48 hours. Eventually, recovery occurs, provided the shock does not intensify and enter the second stage.

Stage II. Decompensated (Progressive) Shock

In stage II, there is a reduction in blood volume of 15 to 25 percent, and the shock becomes steadily worse as compensatory mechanisms are no longer sufficient to maintain adequate perfusion. As the cardiovascular system progressively deteriorates, cardiac output falls dramatically. This stage is characterized by a number of positive feedback systems that lead to further reductions in blood pressure and cardiac output and even more cardiovascular deterioration.

Among the positive feedback cycles that contribute to decreased cardiac output are the following (Figure 21-10b):

1. Depression of cardiac activity. When blood pressure falls below 60 mm Hg, the pressure is no longer adequate to force blood through coronary arteries and the myocardium becomes ischemic. This weakens the heart muscle, decreases cardiac out-

FIGURE 21-10 **Shock. (a) Responses of the body during the compensated (nonprogressive) stage of shock. The cycles shown are negative feedback cycles.**

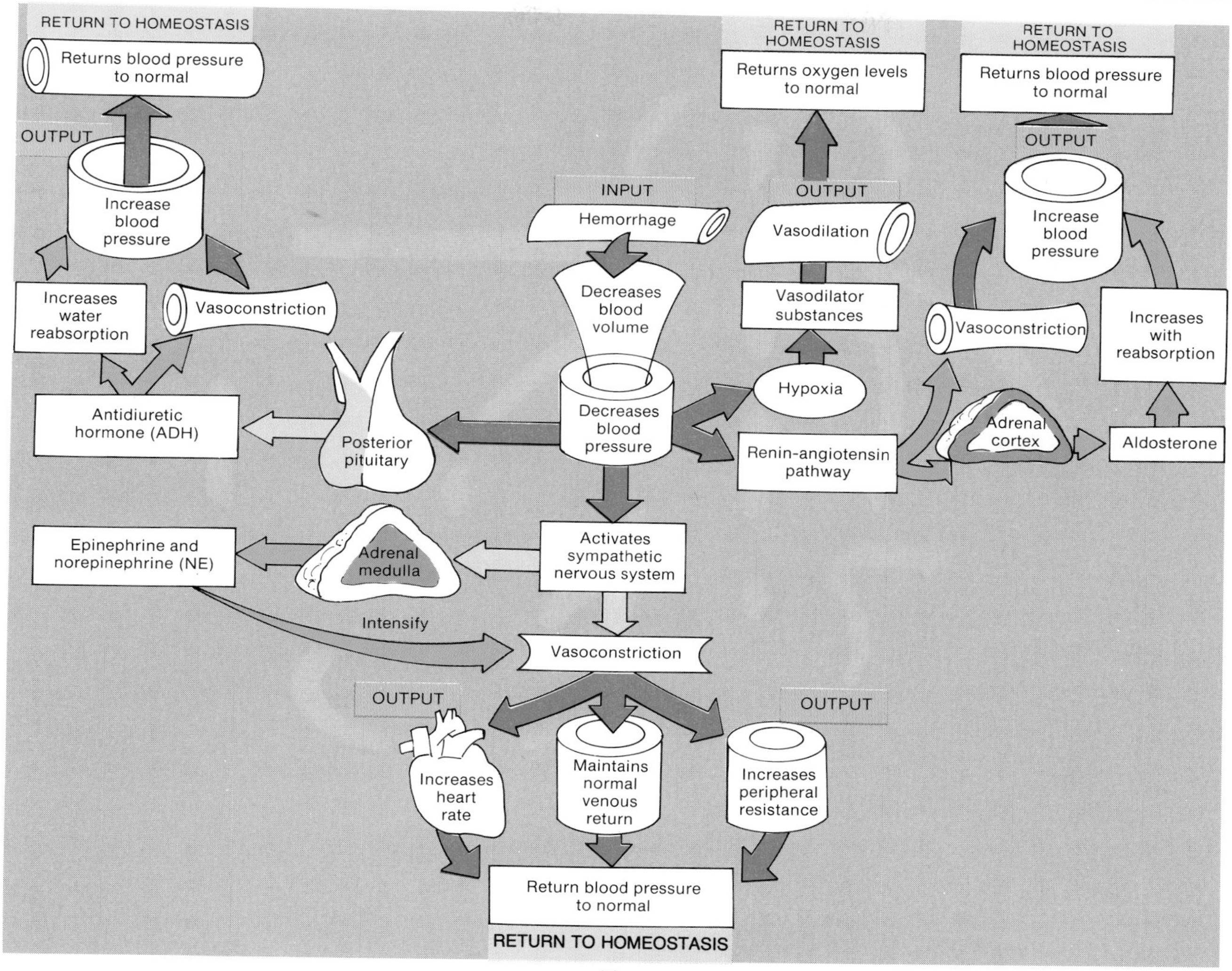

(a)

put further, and depresses blood pressure even more. The decreased cardiac output and blood pressure produce additional ischemia and even more severe depression of cardiac output and blood pressure.

2. Depression of vasoconstriction. Decreased blood pressure in the vasomotor center depresses the center, and it becomes progressively less active. Decreased activity of the center results in lower blood pressure due to generalized vasodilation and even more depression of the center. This positive feedback cycle becomes operative when blood pressure falls below 40 to 50 mm Hg.

3. Increased permeability of capillaries. In very late stages of prolonged shock, hypoxia causes an increase in blood capillary permeability. As more blood plasma constituents move from the capillaries into tissue spaces, blood volume decreases. This decreases cardiac output, the reduced cardiac output intensifies the hypoxia, and the cycle leads to progressively worse shock.

4. Intravascular clotting. Decreased cardiac output results in the formation of blood clots and platelet aggregation. The resulting obstructions further reduce cardiac output.

5. Cellular destruction. In response to shock, cellular destruction occurs throughout the body, especially the liver. The cellular changes include lysosomal rupture, depressed mitochondrial activity, diminished active transport, and decreased metabolism.

6. Acidosis. As a result of metabolic dysfunction, cells produce

FIGURE 21-10 (*Continued*) (b) Responses of the body during the decompensated (progressive) stage of shock. The cycles shown are positive feedback cycles. Without immediate, medical intervention, irreversible shock and death occur.

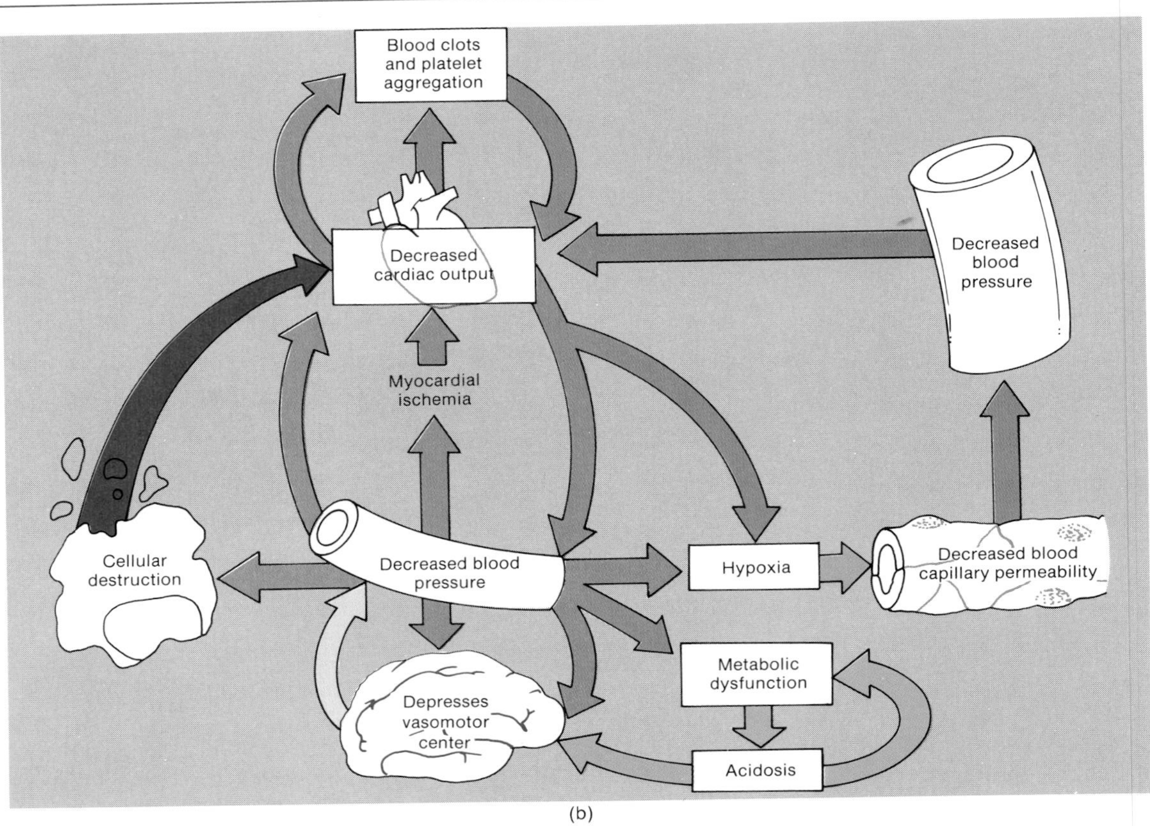

(b)

excess lactic acid which results in acidosis, a condition in which the pH of blood ranges from 7.35 to 6.80 or lower. The principal physiological effect of acidosis is depression of the central nervous system.

To attempt to reverse the changes that occur during the decompensated stage, immediate medical intervention is required. If this fails, shock progresses to a third stage in which even medical intervention is insufficient to maintain cardiac output.

Stage III. Irreversible Shock

In stage III, there is rapid deterioration of the cardiovascular system that cannot be helped by compensatory mechanisms or medical intervention. As the shock cycle perpetuates itself, there are life-threatening reductions in cardiac output, blood pressure, and tissue perfusion. Ultimately, high-energy phosphate reserves (ATP) are depleted, especially in the heart and liver, and the heart deteriorates so much that it can no longer pump blood.

CHECKING CIRCULATION

Pulse

The alternate expansion and elastic recoil of the wall of an artery with each systole and diastole of the left ventricle is called the ***pulse.*** Pulse is strongest in the arteries closest to the heart. It becomes weaker as it passes over the arterial system, and it disappears altogether in the capillaries. The pulse may be felt in any artery that lies near the surface of the body and over a bone or other firm tissue. The radial artery at the wrist is most commonly used. Other arteries that may be used for determining pulse are the:

1. Temporal artery, lateral to the orbit of the eye.
2. Facial artery, at the lower jawbone on a line with the corners of the mouth.
3. Common carotid artery, lateral to the larynx (voice box).
4. Brachial artery, along the medial side of the biceps brachii muscle.

5. Femoral artery, inferior to the inguinal ligament.
6. Popliteal artery, behind the knee.
7. Posterior tibial artery, posterior to the medial malleolus of the tibia.
8. Dorsalis pedis artery, superior to the instep of the foot.

The pulse rate is the same as the heart rate and averages between 70 and 90 beats per minute in the resting state. The term ***tachycardia*** (tak'-i-KAR-dē-a; *tachy* = fast) is applied to a rapid heart or pulse rate (over 100/min). The term ***bradycardia*** (brād'-i-KAR-dē-a; *brady* = slow) indicates a slow heart or pulse rate (under 50/min).

Other characteristics of the pulse may give additional information about circulation. For example, the intervals between beats should be equal in length. If a pulse is missed at intervals, the pulse is said to be irregular. Also, each pulse beat should be of equal strength. Irregularities in strength may indicate a lack of muscle tone in the heart or arteries.

Measurement of Blood Pressure (BP)

In clinical use, the term ***blood pressure (BP)*** refers to the pressure in arteries exerted by the left ventricle when it undergoes systole and the pressure remaining in the arteries when the ventricle is in diastole. Blood pressure is usually taken in the left brachial artery, and it is measured by a ***sphygmomanometer*** (sfig'-mō-ma-NOM-e-ter; *sphygmo* = pulse). A commonly used sphygmomanometer consists of a rubber cuff attached by a rubber tube to a compressible hand pump or bulb. Another tube attaches to the cuff and to a column of mercury or pressure dial marked off in millimeters. This column measures the pressure. The cuff is wrapped around the arm over the brachial artery and inflated by squeezing the bulb. The inflation creates a pressure on the artery. The bulb is squeezed until the pressure in the cuff exceeds the pressure in the artery. At this point, the walls of the brachial artery are compressed tightly against each other, and no blood can flow through. Compression of the artery may be evidenced in two ways. First, if a stethoscope is placed over the artery below the cuff, no sound can be heard. Second, no pulse can be felt by placing the fingers over the radial artery at the wrist.

Next the cuff is deflated gradually until the pressure in the cuff is slightly less than the maximal pressure in the brachial artery. At this point, the artery opens, a spurt of blood passes through, and a sound may be heard through the stethoscope due to the turbulence of the blood flow. As cuff pressure is further reduced, the sound suddenly becomes faint as the blood turbulence reduces significantly. Finally, the sound disappears altogether. When the first sound is heard, a reading on the mercury column is made. This sound corresponds to ***systolic blood pressure (SBP)***—the force with which blood is pushing against arterial walls during ventricular contraction. The pressure recorded on the mercury column when the sounds suddenly become faint is called ***diastolic blood pressure (DBP).*** It measures the force of blood in arteries during ventricular relaxation. Whereas systolic pressure indicates the force of the left ventricular contraction, diastolic pressure provides information about the resistance of blood vessels. The various sounds that are heard while taking blood pressure are called ***Korotkoff*** (kō-ROT-kof) ***sounds.***

CLINICAL APPLICATION: PULSE PRESSURE

The average blood pressure of a young adult male is about 120 mm Hg systolic and 80 mm Hg diastolic, expressed as 120/80. In young adult females, the pressures are 8 to 10 mm Hg less. The difference between systolic and diastolic pressure is called *pulse pressure.* This pressure, which averages 40 mm Hg, provides information about the condition of the arteries. For example, conditions such as atherosclerosis and patent ductus arteriosus greatly increase pulse pressure. The normal ratio of systolic pressure to diastolic pressure to pulse pressure is about 3:2:1.

CIRCULATORY ROUTES

Arteries, arterioles, capillaries, venules, and veins are organized into definite routes to circulate blood throughout the body. We can now look at the basic routes the blood takes as it is transported through its vessels.

Figure 21-11 shows a number of basic ***circulatory routes*** through which the blood travels. ***Systemic circulation*** includes all the oxygenated blood that leaves the left ventricle through the aorta and reaches all systemic capillaries and the deoxygenated blood that returns to the right atrium after traveling to all the organs including the nutrient arteries to the lungs. Two of the many subdivisions of the systemic circulation are the ***coronary (cardiac) circulation*** (see Figure 20-6), which supplies the myocardium of the heart, and the ***hepatic portal circulation,*** which runs from the gastrointestinal tract to the liver (see Figure 21-23). Blood leaving the aorta and traveling through the systemic arteries is a bright red color. As it moves through capillaries, it loses some of its oxygen and takes on carbon dioxide, so that blood in systemic veins is a dark red color.

When blood returns to the heart from the systemic route, it is eventually pumped out of the right ventricle through the ***pulmonary circulation*** to the lungs (see Figure 21-22). In pulmonary capillaries of the air sacs of the lungs, it loses some of its carbon dioxide and takes on oxygen. It is now bright red again. It returns to the left atrium of the heart and reenters the systemic circulation as it is pumped out by the left ventricle.

FIGURE 21-11 Circulatory routes. Systemic circulation is indicated by heavy black arrows; pulmonary circulation, by thin black arrows in the pulmonary blood vessels; and hepatic portal circulation, by thin red arrows. Refer to Figure 20-6 for the details of coronary (cardiac) circulation and to Figure 21-23 for the details of fetal circulation.

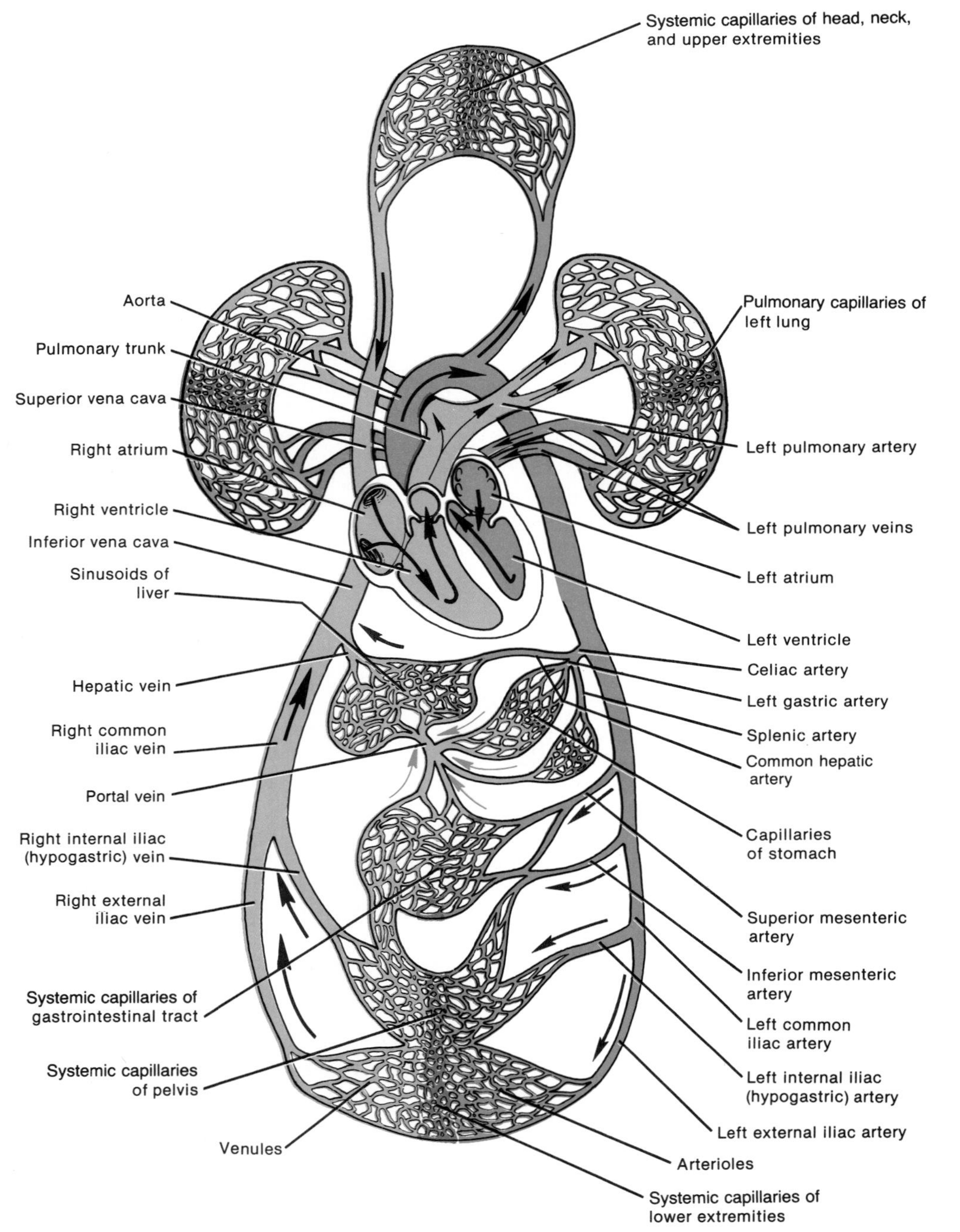

Another major route—***fetal circulation***—exists only in the fetus and contains special structures that allow the developing fetus to exchange materials with its mother (see Figure 21-24).

Cerebral circulation (cerebral arterial circle, or circle of Willis) is discussed in Exhibit 21-3.

Systemic Circulation

The flow of blood from the left ventricle to all parts of the body (except the air sacs of the lungs) and back to the right atrium is called the ***systemic circulation.*** The purpose of systemic circulation is to carry oxygen and nutrients to body tissues and to remove carbon dioxide and other wastes and heat from the tissues. All systemic arteries branch from the ***aorta,*** which arises from the left ventricle of the heart.

As the aorta emerges from the left ventricle, it passes upward and deep to the pulmonary trunk. At this point, it is called the ***ascending aorta.*** The ascending aorta gives off two coronary branches to the heart muscle. Then it turns to the left, forming the ***arch of the aorta,*** before descending to the level of the fourth thoracic vertebra as the ***descending aorta.*** The descending aorta lies close to the vertebral bodies, passes through the diaphragm, and divides at the level of the fourth lumbar vertebra into two ***common iliac arteries,*** which carry blood to the lower extremities. The section of the descending aorta between the arch of aorta and the diaphragm is referred to as the ***thoracic aorta.*** The section between the diaphragm and the common iliac arteries is termed the ***abdominal aorta.*** Each section of the aorta gives off arteries that continue to branch into distributing arteries leading to organs and finally into the arterioles and capillaries that service the systemic tissues (except the air sacs of lungs).

Blood is returned to the heart through the systemic veins. All the veins of the systemic circulation flow into either the ***superior*** or ***inferior venae cavae*** or the ***coronary sinus.*** They in turn empty into the right atrium. The principal arteries and veins of systemic circulation are described and illustrated in Exhibits 21-1 through 21-12 and Figures 21-12 through 21-21.

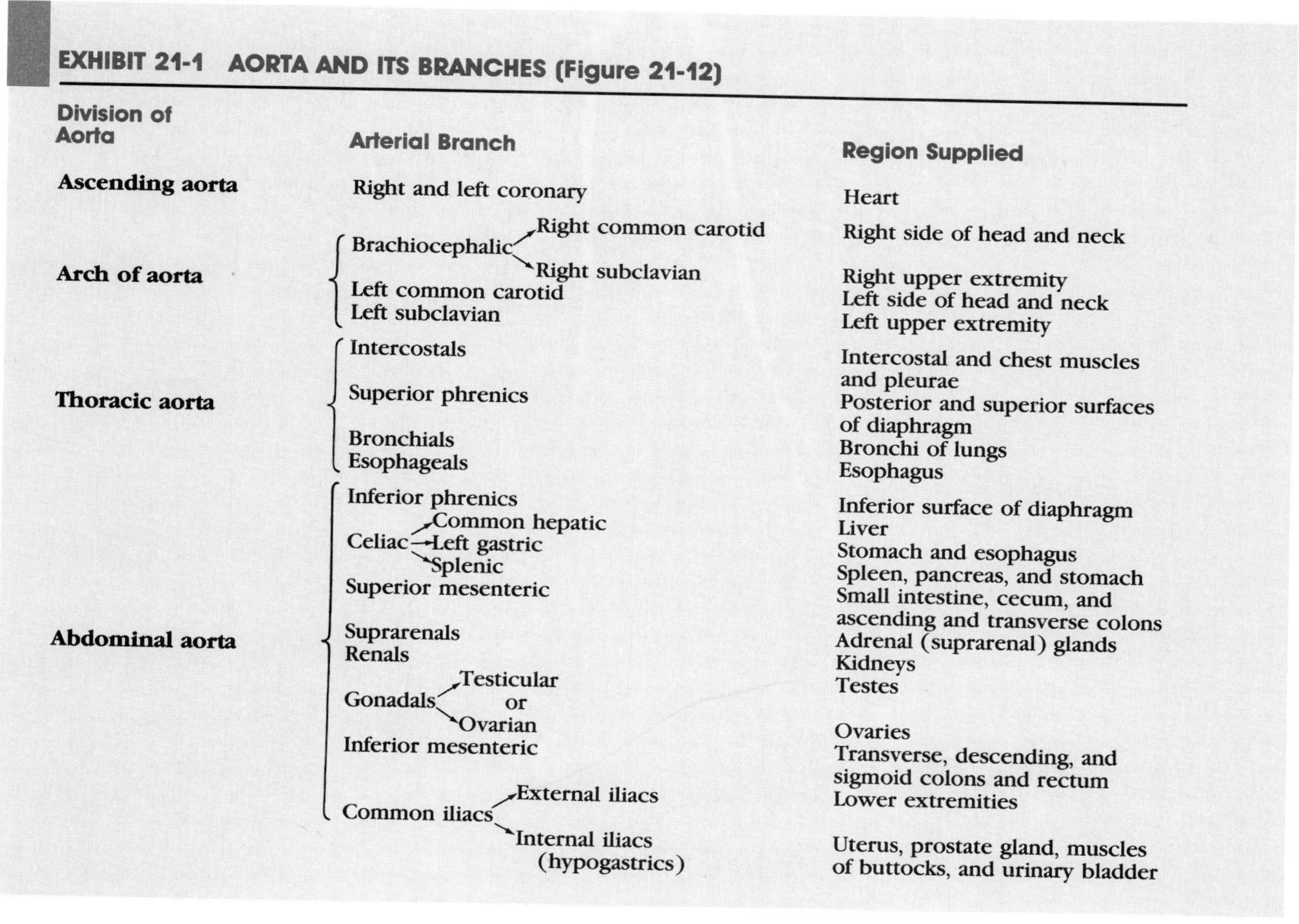

EXHIBIT 21-1 AORTA AND ITS BRANCHES (Figure 21-12)

Division of Aorta	Arterial Branch	Region Supplied
Ascending aorta	Right and left coronary	Heart
Arch of aorta	Brachiocephalic → Right common carotid	Right side of head and neck
	Brachiocephalic → Right subclavian	Right upper extremity
	Left common carotid	Left side of head and neck
	Left subclavian	Left upper extremity
Thoracic aorta	Intercostals	Intercostal and chest muscles and pleurae
	Superior phrenics	Posterior and superior surfaces of diaphragm
	Bronchials	Bronchi of lungs
	Esophageals	Esophagus
Abdominal aorta	Inferior phrenics	Inferior surface of diaphragm
	Celiac → Common hepatic	Liver
	Celiac → Left gastric	Stomach and esophagus
	Celiac → Splenic	Spleen, pancreas, and stomach
	Superior mesenteric	Small intestine, cecum, and ascending and transverse colons
	Suprarenals	Adrenal (suprarenal) glands
	Renals	Kidneys
	Gonadals → Testicular or	Testes
	Gonadals → Ovarian	Ovaries
	Inferior mesenteric	Transverse, descending, and sigmoid colons and rectum
	Common iliacs → External iliacs	Lower extremities
	Common iliacs → Internal iliacs (hypogastrics)	Uterus, prostate gland, muscles of buttocks, and urinary bladder

FIGURE 21-12 Aorta and its principal branches in anterior view.

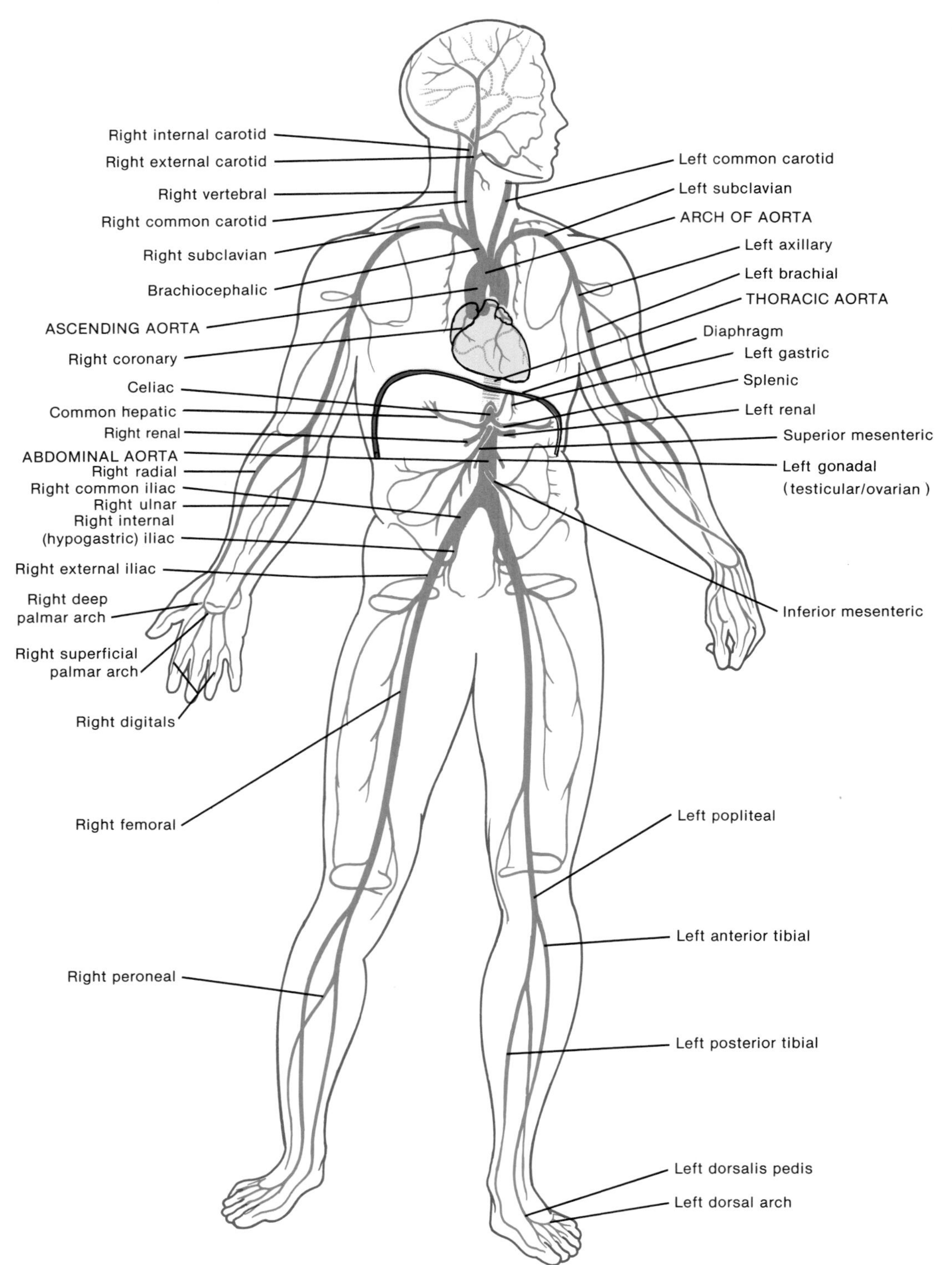
Right internal carotid
Right external carotid
Right vertebral
Right common carotid
Right subclavian
Brachiocephalic
ASCENDING AORTA
Right coronary
Celiac
Common hepatic
Right renal
ABDOMINAL AORTA
Right radial
Right common iliac
Right ulnar
Right internal (hypogastric) iliac
Right external iliac
Right deep palmar arch
Right superficial palmar arch
Right digitals
Right femoral
Right peroneal
Left common carotid
Left subclavian
ARCH OF AORTA
Left axillary
Left brachial
THORACIC AORTA
Diaphragm
Left gastric
Splenic
Left renal
Superior mesenteric
Left gonadal (testicular/ovarian)
Inferior mesenteric
Left popliteal
Left anterior tibial
Left posterior tibial
Left dorsalis pedis
Left dorsal arch

EXHIBIT 21-2 ASCENDING AORTA (Figure 21-13)

Branch	Description and Region Supplied
Coronary arteries	Right and left branches arise from ascending aorta just superior to aortic semilunar valve. They form crown around heart, giving off branches to atrial and ventricular myocardium.

SCHEME OF DISTRIBUTION

Ascending aorta

Right coronary artery ← → Left coronary artery

FIGURE 21-13 **Ascending aorta and its branches in anterior view.**

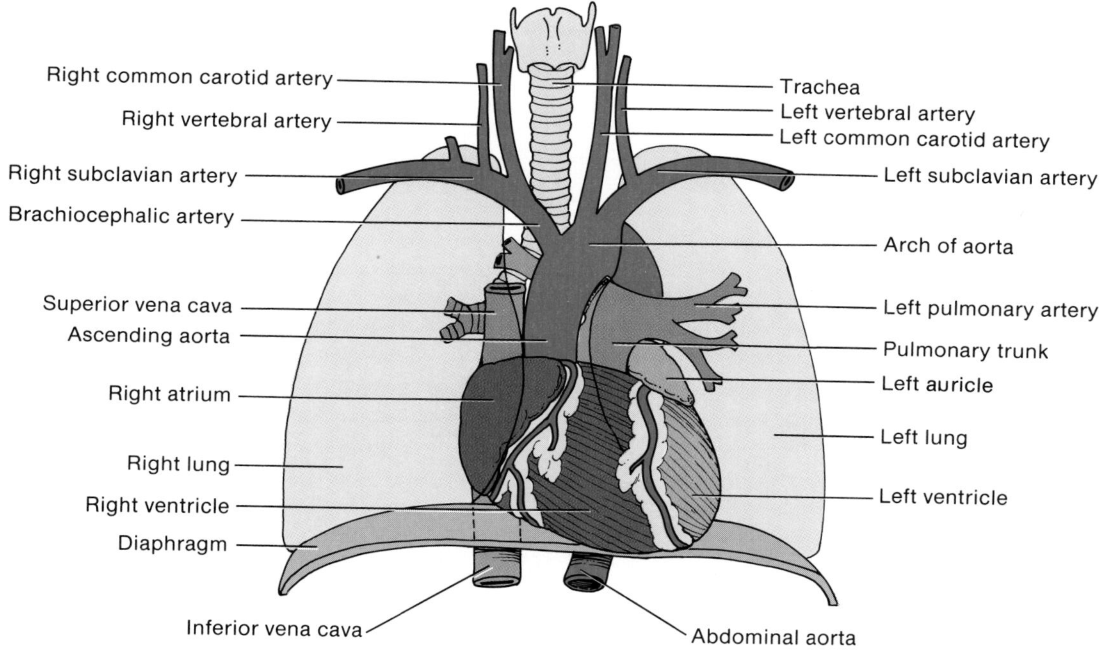

EXHIBIT 21-3 ARCH OF AORTA (Figure 21-14)

Branch	Description and Region Supplied
Brachiocephalic	***Brachiocephalic artery*** is first branch off arch of aorta. It divides to form right subclavian artery and right common carotid artery. ***Right subclavian artery*** extends from brachiocephalic to first rib and then passes into armpit (axilla) and supplies arm, forearm, and hand. Continuation of right subclavian into axilla is called ***axillary artery.**** From here, it continues into arm as ***brachial artery.*** At bend of elbow, brachial artery divides into medial ***ulnar*** and lateral ***radial arteries.*** These vessels pass down to palm, one on each side of forearm. In palm, branches of two arteries anastomose to form two palmar arches—***superficial palmar arch*** and ***deep palmar arch.*** From these arches arise ***digital arteries,*** which supply fingers and thumb. Before passing into axilla, right subclavian gives off major branch to brain called ***vertebral artery.*** Right vertebral artery passes through foramina of transverse processes of cervical vertebrae and enters skull through foramen magnum to reach undersurface of brain. Here it unites with left vertebral artery to form ***basilar artery.***

EXHIBIT 21-3 ARCH OF AORTA (*Continued*)

Right common carotid artery passes upward in neck. At upper level of larynx, it divides into ***right external*** and ***right internal carotid arteries.*** External carotid supplies right side of thyroid gland, tongue, throat, face, ear, scalp, and dura mater. Internal carotid supplies brain, right eye, and right sides of forehead and nose.

Inside the cranium, anastomoses of left and right internal carotids along with basilar artery form a somewhat hexagonal arrangement of blood vessels at base of brain near sella turcica called ***cerebral arterial circle*** (***circle of Willis***). From this circle arise arteries supplying most of the brain. Essentially, cerebral arterial circle is formed by union of ***anterior cerebral arteries*** (branches of internal carotids) and ***posterior cerebral arteries*** (branches of basilar artery). Posterior cerebral arteries are connected with internal carotids by ***posterior communicating arteries.*** Anterior cerebral arteries are connected by ***anterior communicating arteries.*** The ***internal carotid arteries*** are also considered part of cerebral arterial circle. The function of the cerebral arterial circle is to equalize blood pressure to brain and provide alternate routes for blood to brain, should arteries become damaged.

Left common carotid — ***Left common carotid*** is second branch off arch of aorta (see Figure 21-13). Corresponding to right common carotid, it divides into basically same branches with same names, except that arteries are now labeled "left" instead of "right."

Left subclavian — ***Left subclavian artery*** is third branch off arch of aorta (see Figure 21-13). It distributes blood to left vertebral artery and vessels of left upper extremity. Arteries branching from left subclavian are named like those of right subclavian.

SCHEME OF DISTRIBUTION

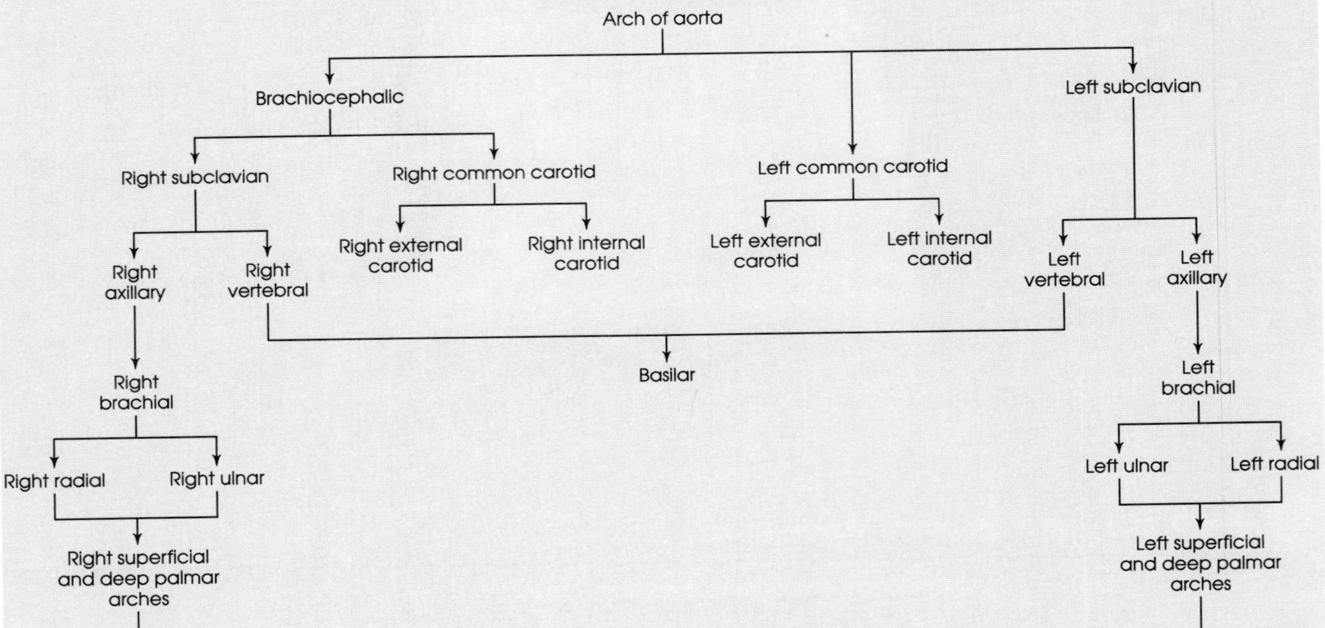

* The right subclavian artery is a good example of the practice of giving the same vessel different names as it passes through different regions.

FIGURE 21-14 Arch of the aorta and its branches. (a) Anterior view of the principal arteries of the right upper extremity. (b) Right lateral view of the principal arteries of the neck and head. (c) Principal arteries of the base of the brain. Note the arteries that comprise the cerebral arterial circle (circle of Willis).

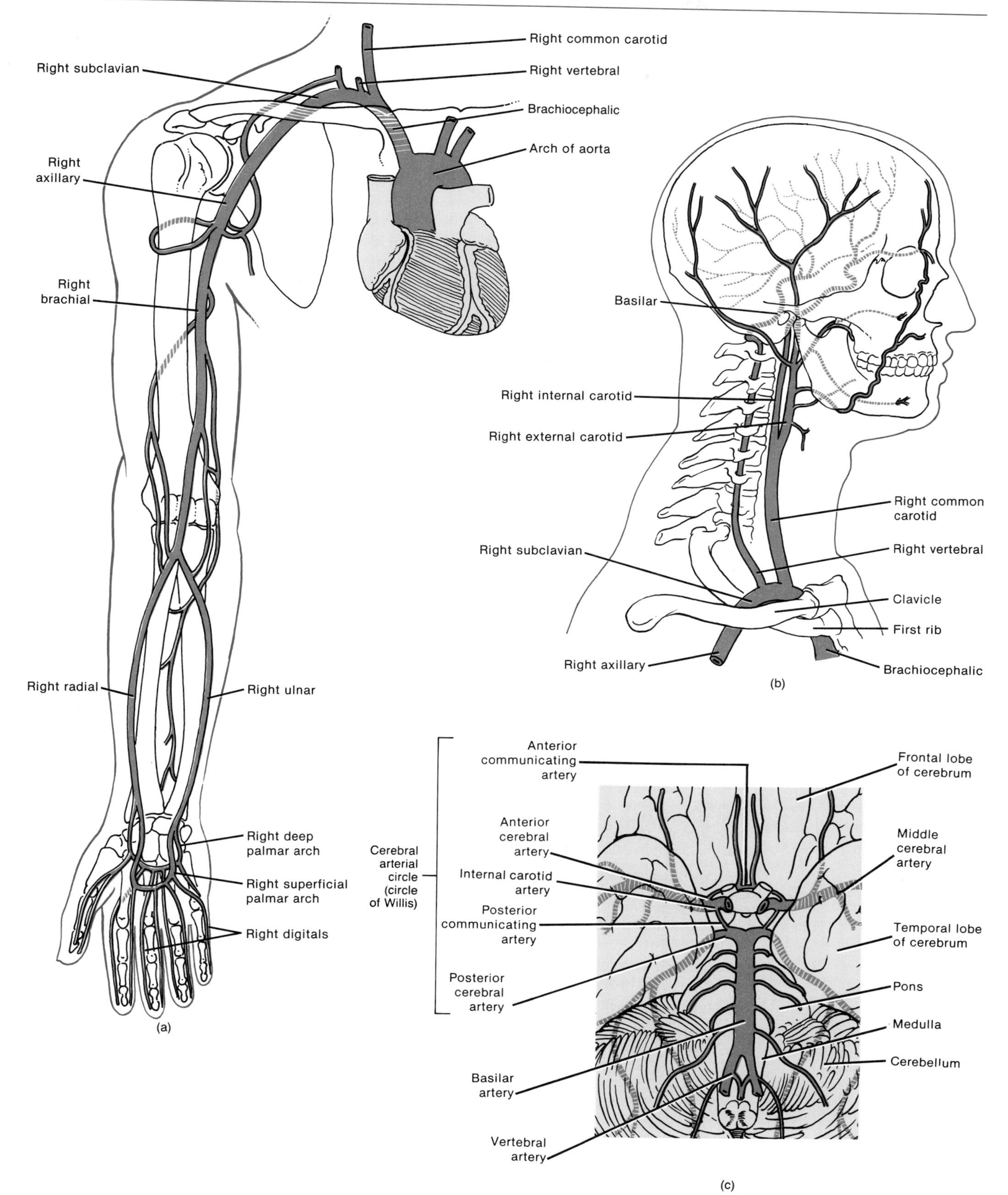

EXHIBIT 21-4 THORACIC AORTA (Figure 21-15)

Branch	Description and Region Supplied
	Thoracic aorta runs from fourth to twelfth thoracic vertebra. Along its course, it sends off numerous small arteries to viscera and skeletal muscles of the chest. Branches of an artery that supply viscera are called ***visceral branches.*** Those that supply body wall structures are called ***parietal branches.***
VISCERAL	
Pericardial	Several minute ***pericardial arteries*** supply blood to dorsal aspect of pericardium.
Bronchial	One right and two left ***bronchial arteries*** supply the bronchial tubes, visceral pleurae, bronchial lymph nodes, and esophagus.
Esophageal	Four or five ***esophageal arteries*** supply the esophagus.
Mediastinal	Numerous small ***mediastinal arteries*** supply blood to structures in the posterior mediastinum.
PARIETAL	
Posterior intercostal	Nine pairs of ***posterior intercostal arteries*** supply the intercostal, pectoral, and abdominal muscles; overlying subcutaneous tissue and skin; mammary glands; and vertebral canal and its contents.
Subcostal	The left and right ***subcostal arteries*** have a distribution similar to that of the posterior intercostals.
Superior phrenic	Small ***superior phrenic arteries*** supply the posterior surface of the diaphragm.

EXHIBIT 21-5 ABDOMINAL AORTA (Figure 21-15)

Branch	Description and Region Supplied
VISCERAL	
Celiac	***Celiac artery (trunk)*** is first visceral aortic branch below diaphragm. It has three branches: (1) ***common hepatic artery,*** (2) ***left gastric artery,*** and (3) ***splenic artery.*** The common hepatic artery has three main branches: (1) ***hepatic artery proper,*** a continuation of the common hepatic artery, which supplies the liver and gallbladder; (2) ***right gastric artery,*** which supplies the stomach and duodenum; and (3) ***gastroduodenal artery,*** which supplies the stomach, duodenum, and pancreas. The left gastric artery supplies the stomach, and its ***esophageal branch*** supplies the esophagus. The splenic artery supplies the spleen and has three main branches: (1) ***pancreatic arteries,*** which supply the pancreas; (2) ***left gastroepiploic artery,*** which supplies the stomach and greater omentum; and (3) ***short gastric arteries,*** which supply the stomach.
Superior mesenteric	The ***superior mesenteric artery*** has several principal branches: (1) ***inferior pancreaticoduodenal artery,*** which supplies the pancreas and duodenum; (2) ***jejunal*** and ***ileal arteries,*** which supply the jejunum and ileum, respectively; (3) ***ileocolic artery,*** which supplies the ileum and ascending colon; (4) ***right colic artery,*** which supplies the ascending colon; and (5) ***middle colic artery,*** which supplies the transverse colon.
Suprarenals	Right and left ***suprarenal arteries*** supply blood to adrenal (suprarenal) glands. The glands are also supplied by branches of the renal and inferior phrenic arteries.
Renals	Right and left ***renal arteries*** carry blood to kidneys and adrenal (suprarenal) glands.
Gonadals (testiculars or ovarians)	Right and left ***testicular arteries*** extend into scrotum and terminate in testes; right and left ***ovarian arteries*** are distributed to ovaries.

FIGURE 21-15 **Thoracic and abdominal aorta. (a) Thoracic and abdominal aorta and their principal branches in anterior view.**

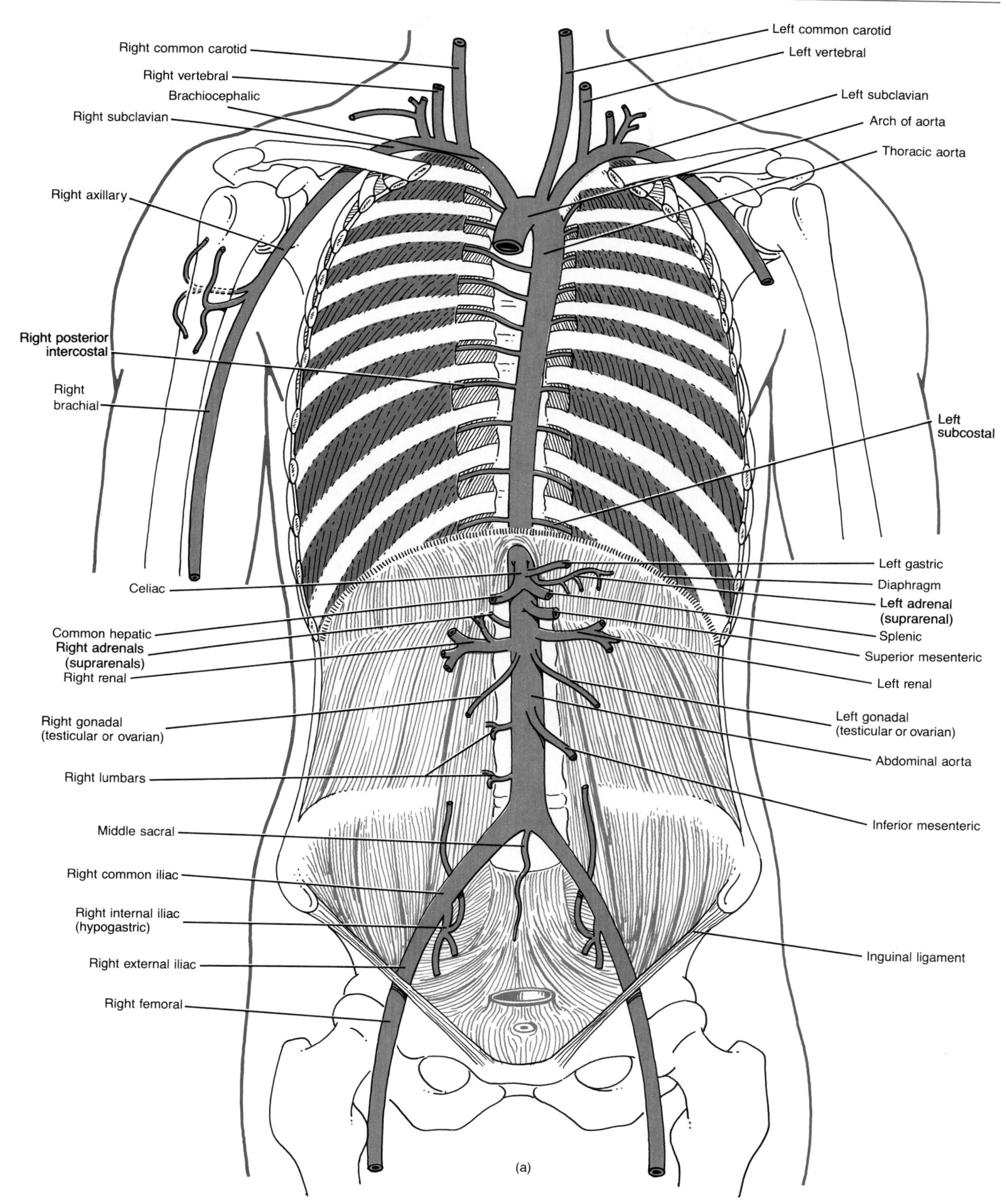

EXHIBIT 21-5 ABDOMINAL AORTA (*Continued*)

Inferior mesenteric	The principal branches of the ***inferior mesenteric artery*** are the (1) ***left colic artery,*** which supplies the transverse and descending colons; (2) ***sigmoid arteries,*** which supply the descending and sigmoid colons; and (3) ***superior rectal artery,*** which supplies the rectum.
PARIETAL	
Inferior phrenics	***Inferior phrenic arteries*** are distributed to undersurface of diaphragm and adrenal (suprarenal) glands.
Lumbars	***Lumbar arteries*** supply spinal cord and its meninges and muscles and skin of lumbar region of back.
Middle sacral	***Middle sacral artery*** supplies sacrum, coccyx, and rectum.

FIGURE 21-15 (*Continued*) (b) Diagram of branches of common hepatic, left gastric, and splenic arteries in anterior view.

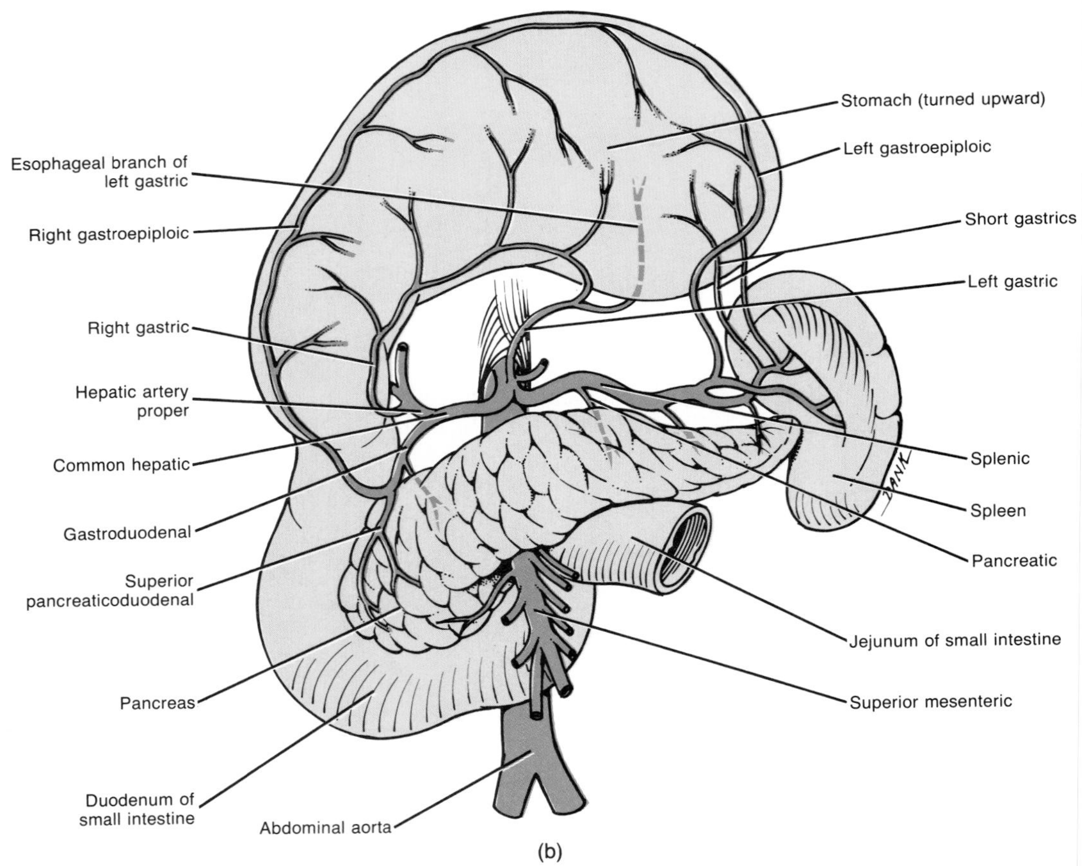

FIGURE 21-15 (*Continued*) (c) Diagram of branches of superior mesenteric artery. (d) Diagram of branches of inferior mesenteric artery.

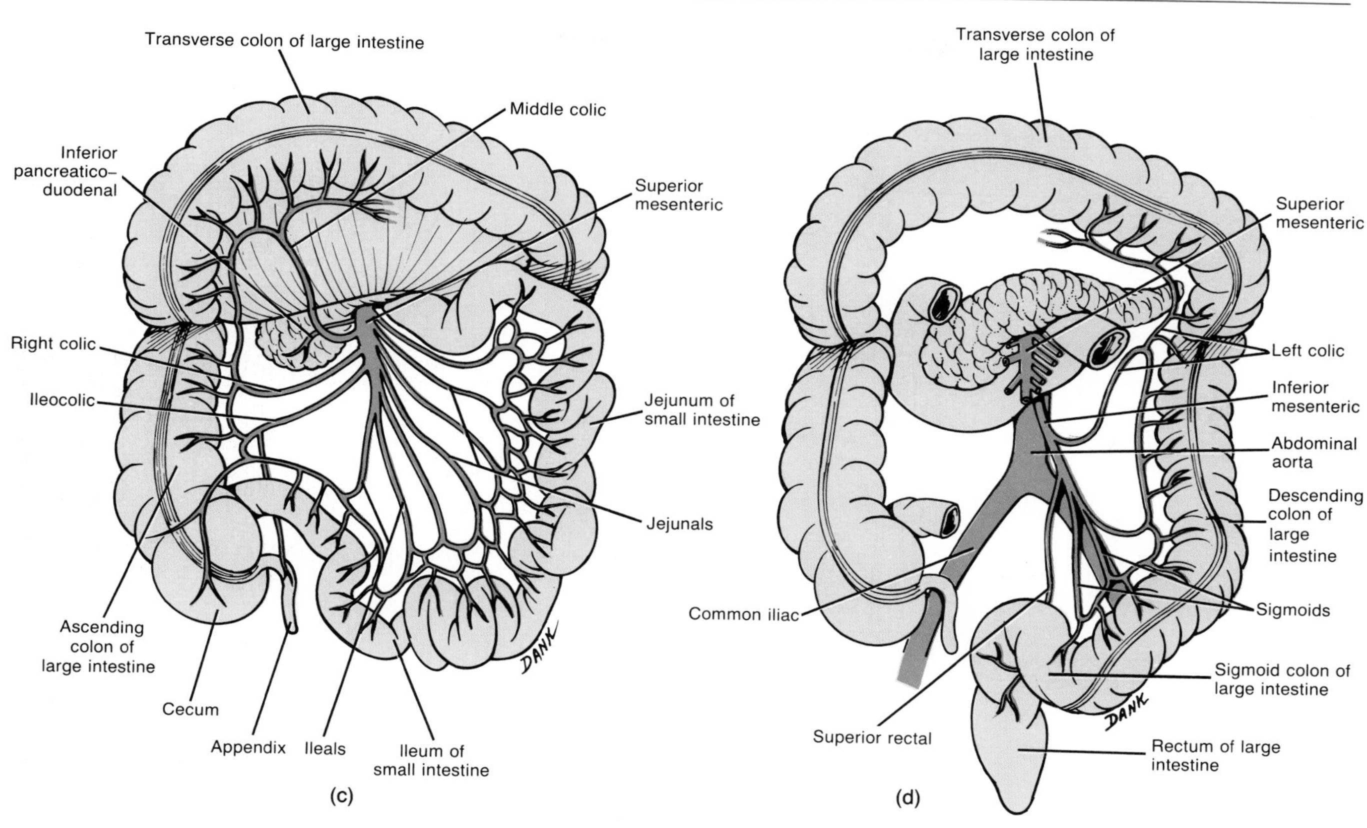

EXHIBIT 21-6 ARTERIES OF PELVIS AND LOWER EXTREMITIES (Figure 21-16)

Branch	Description and Region Supplied
Common iliacs	At about level of fourth lumbar vertebra, abdominal aorta divides into right and left ***common iliac arteries.*** Each passes downward about 5 cm (2 inches) and gives rise to two branches: internal iliac and external iliac.
Internal iliacs	***Internal iliac (hypogastric) arteries*** form branches that supply psoas major, quadratus lumborum, and medial side of each thigh, urinary bladder, rectum, prostate gland, ductus (vas) deferens, uterus, and vagina.
External iliacs	***External iliac arteries*** diverge through greater (false) pelvis, enter thighs, and here become right and left ***femoral arteries.*** Both femorals send branches back up to genitals and wall of abdomen. Other branches run to muscles of thigh. Femoral continues down medial and posterior side of thigh at back of knee joint, where it becomes ***popliteal artery.*** Between knee and ankle, popliteal runs down back of leg and is called ***posterior tibial artery.*** Below knee, ***peroneal artery*** branches off posterior tibial to supply structures on medial side of fibula and calcaneus. In calf, ***anterior tibial artery*** branches off popliteal and runs along front of leg. At ankle, it becomes ***dorsalis pedis artery.*** At ankle, posterior tibial divides into ***medial*** and ***lateral plantar arteries.***

FIGURE 21-16 Principal arteries of the pelvis and right lower extremity. (a) Anterior view. (b) Posterior view.

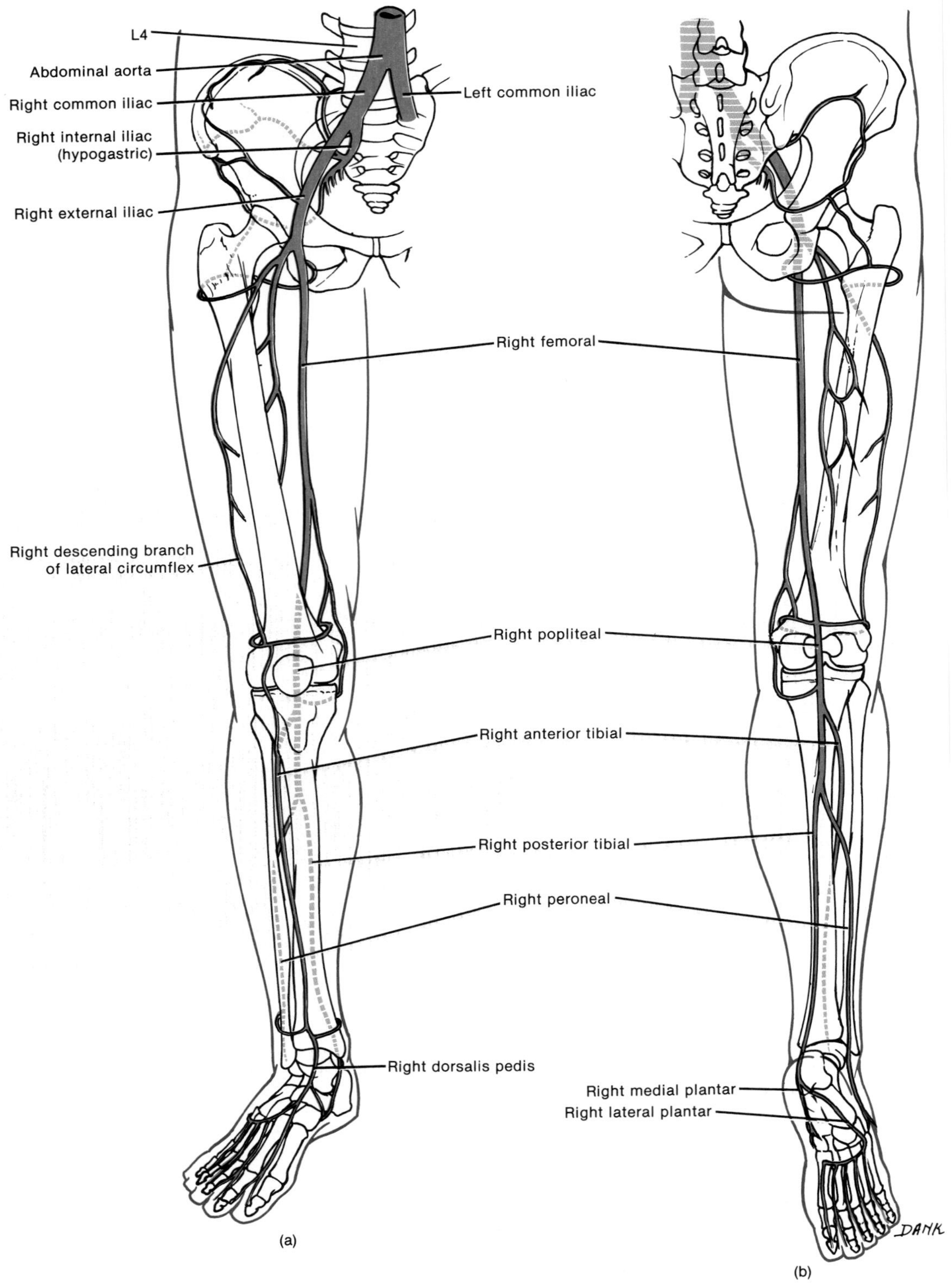

EXHIBIT 21-6 ARTERIES OF PELVIS AND LOWER EXTREMITIES (*Continued*)

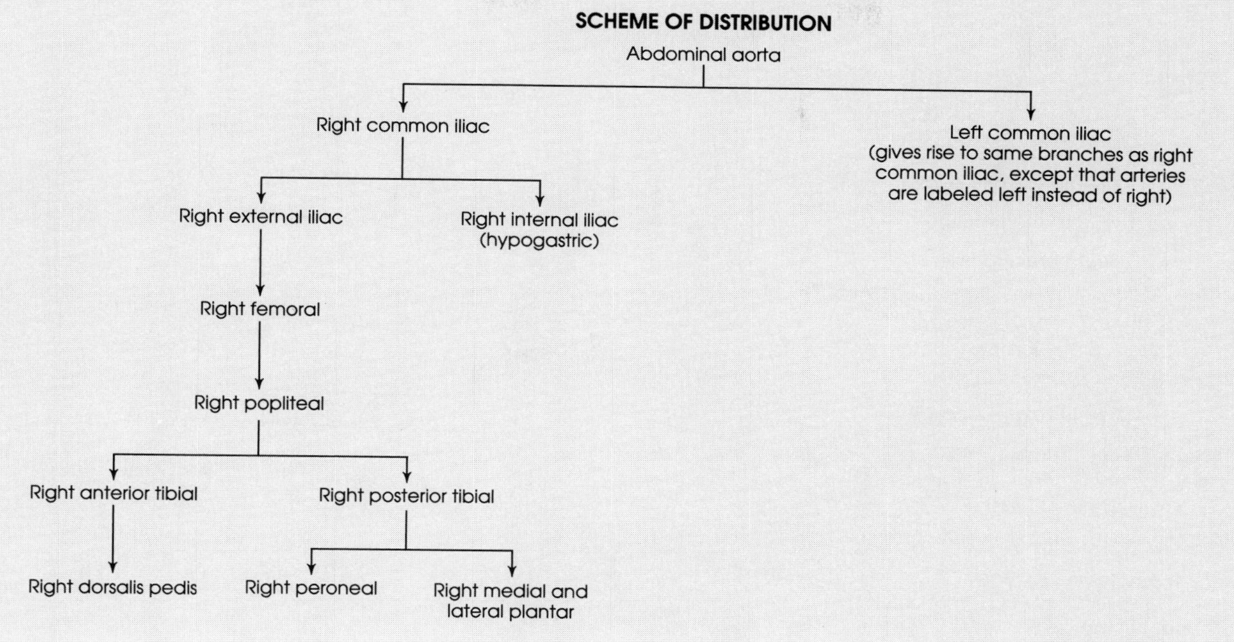

EXHIBIT 21-7 VEINS OF SYSTEMIC CIRCULATION (Figure 21-17)

Vein	Description and Region Drained
Coronary sinus	All systemic and coronary (cardiac) veins return blood to the right atrium of the heart through one of three large vessels. Return flow in coronary circulation is taken up by ***cardiac veins,*** which empty into the large vein of the heart, the ***coronary sinus.*** From here, the blood empties into the right atrium of the heart (see Figure 20-6). Return flow in systemic circulation empties into the superior vena cava or inferior vena cava.
Superior vena cava (SVC)	Veins that empty into the ***superior vena cava*** are veins of the head and neck, upper extremities, and some from the thorax.
Inferior vena cava (IVC)	Veins that empty into the ***inferior vena cava*** are some from the thorax and veins of the abdomen, pelvis, and lower extremities.

The inferior vena cava is commonly compressed during the later stages of pregnancy owing to the enlargement of the uterus. This produces edema of the ankles and feet and temporary varicose veins.

FIGURE 21-17 **Principal veins in anterior view.**

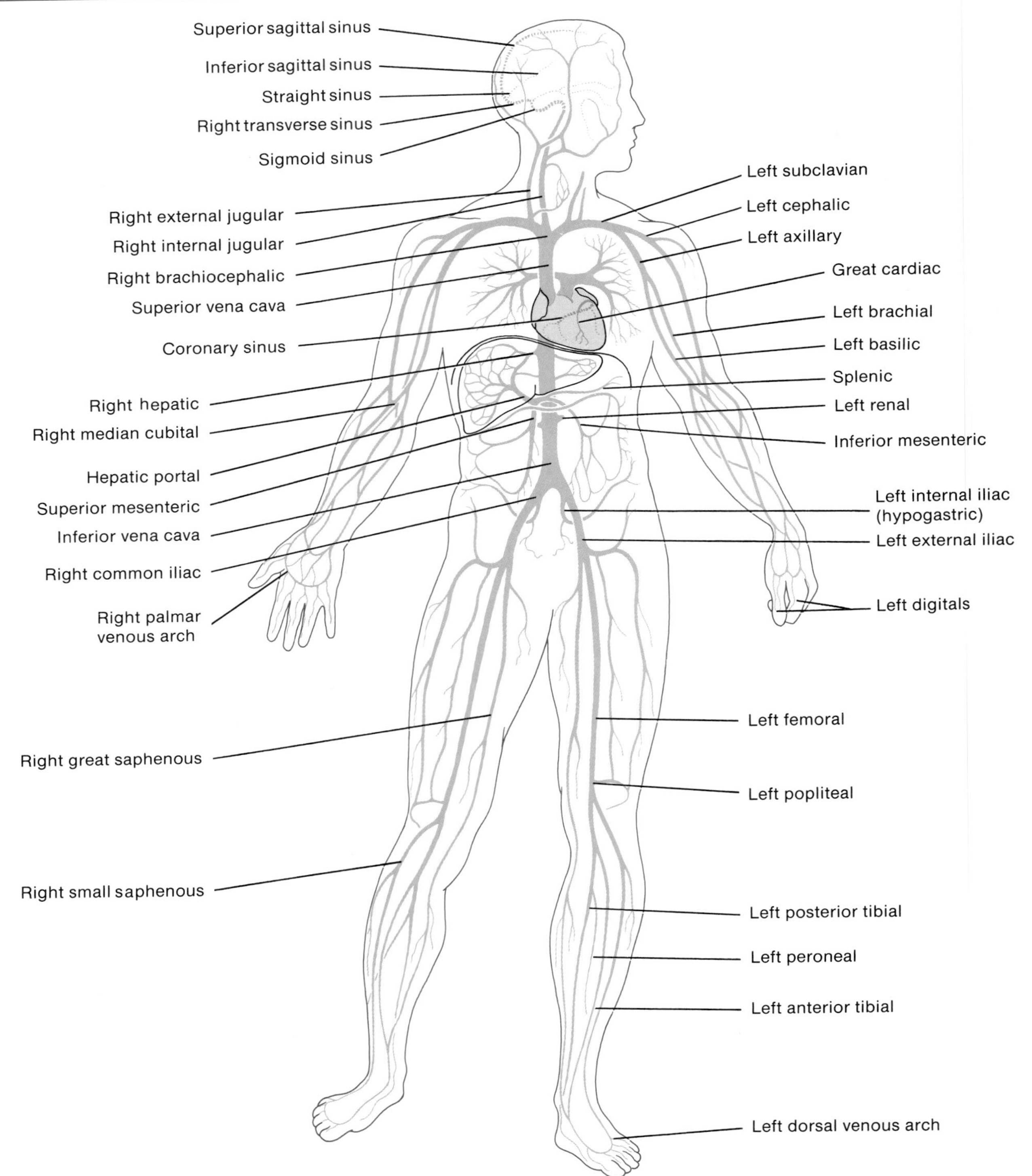
Superior sagittal sinus
Inferior sagittal sinus
Straight sinus
Right transverse sinus
Sigmoid sinus
Right external jugular
Right internal jugular
Right brachiocephalic
Superior vena cava
Coronary sinus
Right hepatic
Right median cubital
Hepatic portal
Superior mesenteric
Inferior vena cava
Right common iliac
Right palmar
venous arch
Right great saphenous
Right small saphenous
Left subclavian
Left cephalic
Left axillary
Great cardiac
Left brachial
Left basilic
Splenic
Left renal
Inferior mesenteric
Left internal iliac
(hypogastric)
Left external iliac
Left digitals
Left femoral
Left popliteal
Left posterior tibial
Left peroneal
Left anterior tibial
Left dorsal venous arch

EXHIBIT 21-8 VEINS OF HEAD AND NECK (Figure 21-18)

Vein	Description and Region Drained
Internal jugulars	Right and left ***internal jugular veins*** receive blood from face and neck. They arise as continuation of ***sigmoid sinuses*** at base of skull. Intracranial vascular sinuses are located between layers of dura mater and receive blood from brain. Other sinuses that drain into internal jugular include ***superior sagittal sinus, inferior sagittal sinus, straight sinus,*** and ***transverse (lateral) sinuses.*** Internal jugulars descend on either side of neck and pass behind clavicles, where they join with right and left subclavian veins. Unions of internal jugulars and subclavians form right and left brachiocephalic veins. From here blood flows into superior vena cava.
External jugulars	Right and left ***external jugular veins*** run down neck along outside of internal jugulars. They drain blood from parotid (salivary) glands, facial muscles, scalp, and other superficial structures into subclavian veins.

In cases of heart failure, the venous pressure in the right atrium may rise. In such patients the pressure in the column of blood in the external jugular vein rises so that, even with the patient at rest and sitting in a chair, the external jugular vein will be visibly distended. Temporary distention of the vein is often seen in healthy adults when the intrathoracic pressure is raised in coughing and physical exertion.

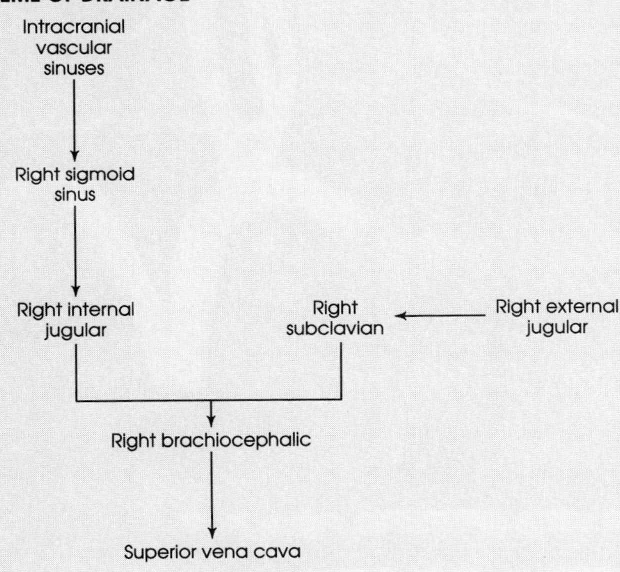

FIGURE 21-18 **Principal veins of the head and neck.**

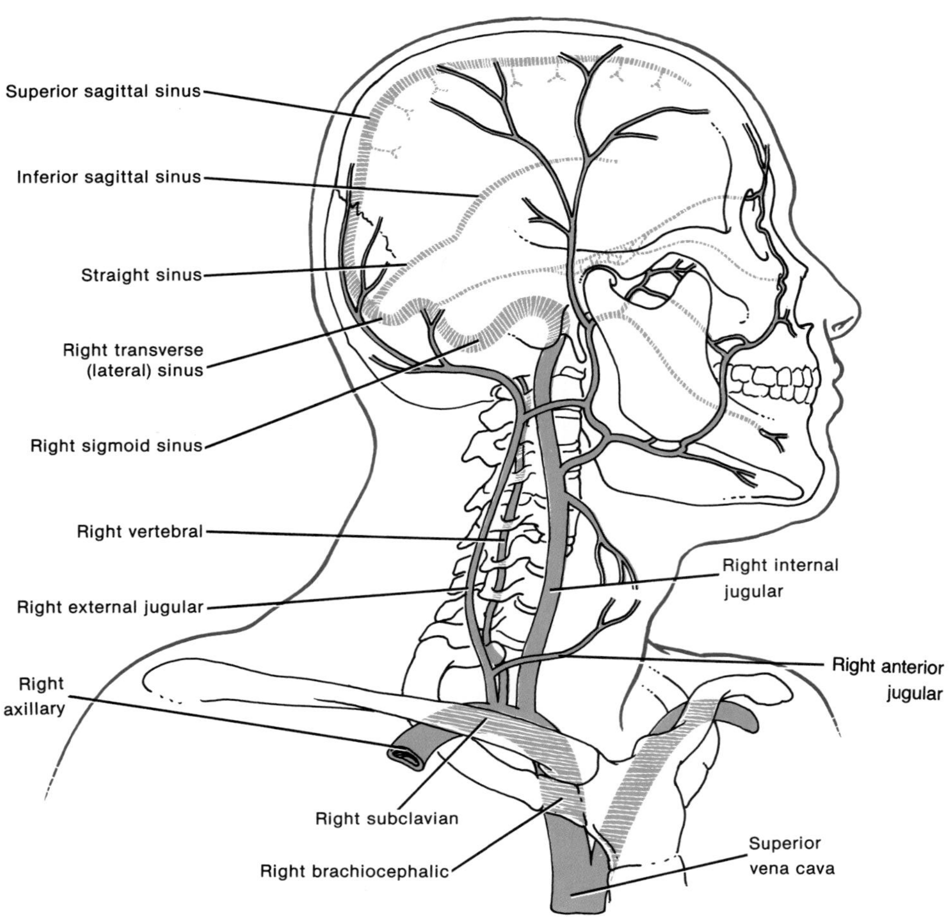

EXHIBIT 21-9 VEINS OF UPPER EXTREMITIES (Figure 21-19)

Vein	Description and Region Drained
	Blood from each upper extremity is returned to the heart by superficial and deep veins. Both sets of veins contain valves. ***Superficial veins*** are located just below the skin and are often visible. They anastomose extensively with each other and deep veins. ***Deep veins*** are located deep in the body. They usually accompany arteries, and many have the same names as corresponding arteries.
SUPERFICIAL	
Cephalics	***Cephalic vein*** of each upper extremity begins in the medial part of ***dorsal venous arch*** and winds upward around radial border of forearm. In front of elbow, it is connected to basilic vein by the ***median cubital vein.*** Just below elbow, cephalic vein unites with ***accessory cephalic vein*** to form cephalic vein of upper extremity. Ultimately, cephalic vein empties into axillary vein.
Basilics	***Basilic vein*** of each upper extremity originates in the ulnar part of ***dorsal venous arch.*** It extends along posterior surface of ulna to point below elbow where it receives ***median cubital vein.*** If a vein must be punctured for an injection, transfusion, or removal of a blood sample, median cubitals are preferred. The median cubital vein joins the basilic vein to form the axillary vein.

EXHIBIT 21-9 VEINS OF UPPER EXTREMITIES (*Continued*)

Median antebrachials — ***Median antebrachial veins*** drain ***palmar venous arch,*** ascend on ulnar side of anterior forearm, and end in median cubital veins.

DEEP

Radials — ***Radial veins*** receive ***dorsal metacarpal veins.***

Ulnars — ***Ulnar veins*** receive tributaries from ***palmar venous arch.*** Radial and ulnar veins unite in bend of elbow to form brachial veins.

Brachials — Located on either side of brachial artery, ***brachial veins*** join into axillary veins.

Axillaries — ***Axillary veins*** are a continuation of brachials and basilics. Axillaries end at first rib, where they become subclavians.

Subclavians — Right and left ***subclavian veins*** unite with internal jugulars to form brachiocephalic veins. Thoracic duct of lymphatic system delivers lymph into left subclavian vein at junction with internal jugular. Right lymphatic duct delivers lymph into right subclavian vein at corresponding junction.

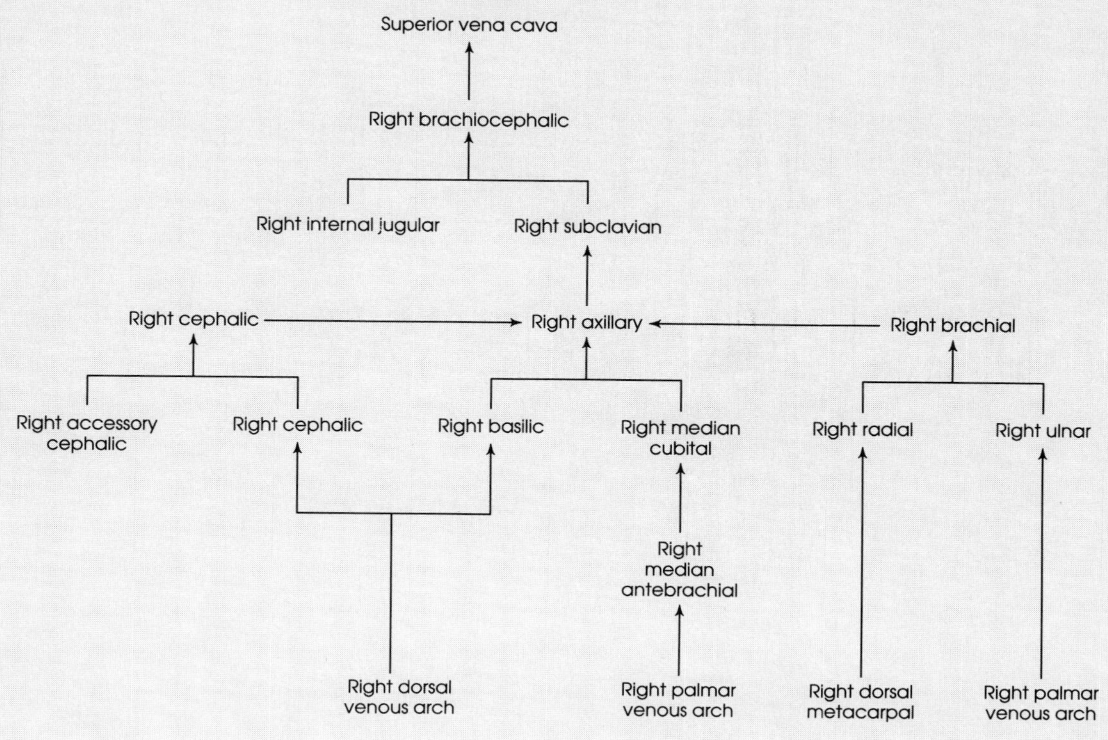

FIGURE 21-19 **Principal veins of the right upper extremity in anterior view.**

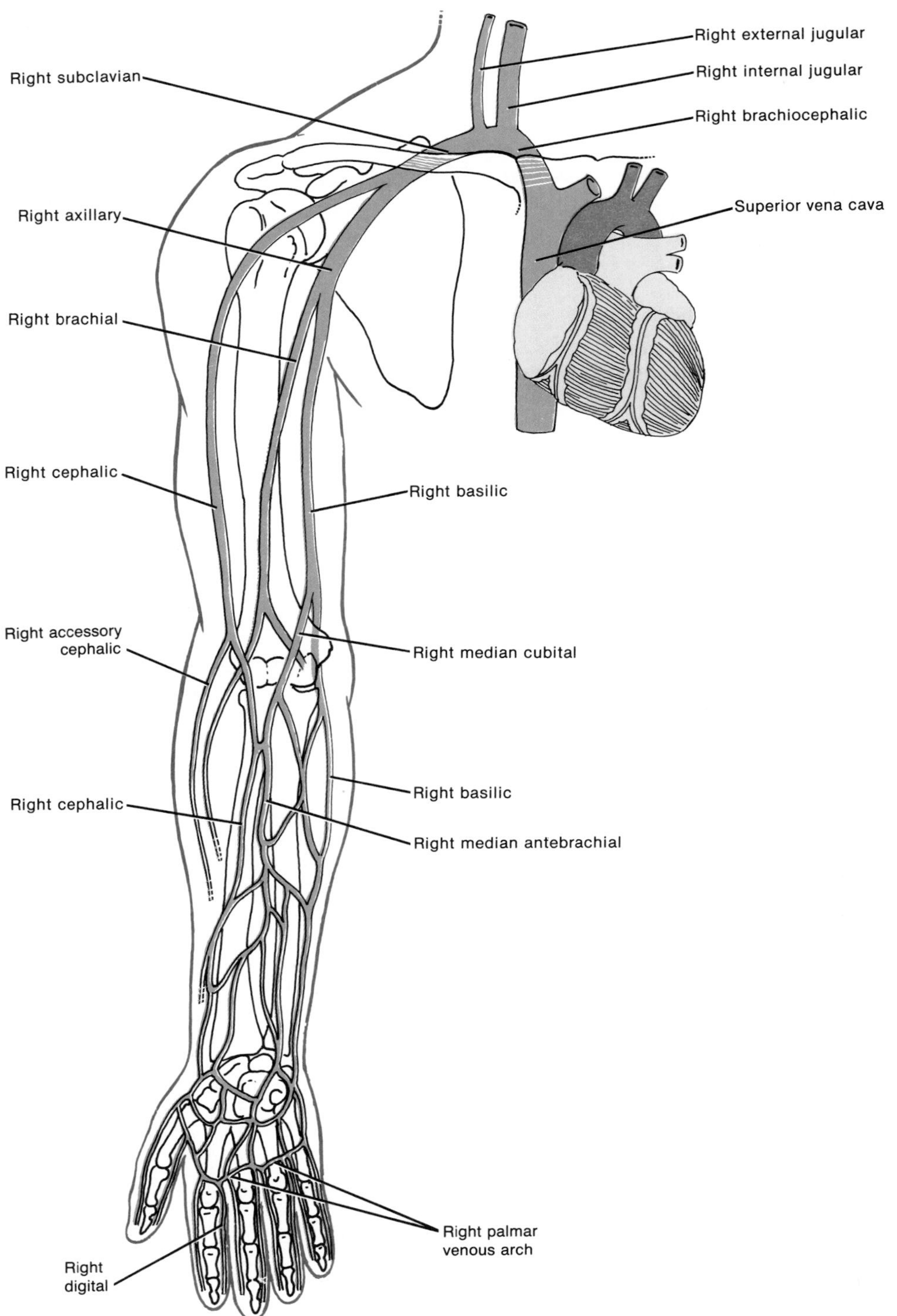

EXHIBIT 21-10 VEINS OF THORAX (Figure 21-20)

Vein	Description and Region Drained
Brachiocephalic	Right and left ***brachiocephalic veins,*** formed by union of subclavians and internal jugulars, drain blood from head, neck, upper extremities, mammary glands, and upper thorax. Brachiocephalics unite to form superior vena cava.
Azygos veins	***Azygos veins,*** besides collecting blood from thorax, may serve as bypass for inferior vena cava that drains blood from lower body. Several small veins directly link azygos veins with inferior vena cava. Large veins that drain lower extremities and abdomen dump blood into azygos. If inferior vena cava or hepatic portal vein becomes obstructed, azygos veins can return blood from lower body to superior vena cava.
Azygos	***Azygos vein*** lies in front of vertebral column, slightly right of midline. It begins as continuation of right ascending lumbar vein. It connects with inferior vena cava, right common iliac, and lumbar veins. Azygos receives blood from ***right intercostal veins*** that drain chest muscles; from hemiazygos and accessory hemiazygos veins; from several ***esophageal, mediastinal,*** and ***pericardial veins;*** and from right ***bronchial vein.*** Vein ascends to fourth thoracic vertebra, arches over right lung, and empties into superior vena cava.
Hemiazygos	***Hemiazygos vein*** is in front of vertebral column and slightly left of midline. It begins as continuation of left ascending lumbar vein. It receives blood from lower four or five ***intercostal veins*** and some ***esophageal*** and ***mediastinal veins.*** At level of ninth thoracic vertebra, it joins azygos vein.
Accessory hemiazygos	***Accessory hemiazygos vein*** is also in front and to left of vertebral column. It receives blood from three or four ***intercostal veins*** and left ***bronchial vein.*** It joins azygos at level of eighth thoracic vertebra.

EXHIBIT 21-11 VEINS OF ABDOMEN AND PELVIS (Figure 21-20)

Vein	Description and Region Drained
Inferior vena cava	***Inferior vena cava*** is the largest vein of the body. It is formed by union of two common iliac veins that drain lower extremities and abdomen. Inferior vena cava extends upward through abdomen and thorax to right atrium. Numerous small veins enter the inferior vena cava. Most carry return flow from branches of abdominal aorta, and names correspond to names of arteries.
Common iliacs	***Common iliac veins*** are formed by union of internal (hypogastric) and external iliac veins and represent distal continuation of inferior vena cava at its bifurcation.
Internal iliacs	Tributaries of ***internal iliac (hypogastric) veins*** basically correspond to branches of internal iliac arteries. Internal iliacs drain gluteal muscles, medial side of thigh, urinary bladder, rectum, prostate gland, ductus (vas) deferens, uterus, and vagina.
External iliacs	***External iliac veins*** are continuation of femoral veins and receive blood from lower extremities and inferior part of anterior abdominal wall.
Renals	***Renal veins*** drain kidneys.
Gonadals (testicular or ovarian)	***Testicular veins*** drain testes (left testicular vein empties into left renal vein); ***ovarian veins*** drain ovaries (left ovarian vein empties into left renal vein).
Suprarenals	***Suprarenal veins*** drain adrenal (suprarenal) glands (left suprarenal vein empties into left renal vein).
Inferior phrenics	***Inferior phrenic veins*** drain diaphragm (left inferior phrenic vein sends tributary to left renal vein).
Hepatics	***Hepatic veins*** drain liver.
Lumbars	A series of parallel ***lumbar veins*** drain blood from both sides of posterior abdominal wall. Lumbars connect at right angles with right and left ***ascending lumbar veins,*** which form origin of corresponding azygos or hemiazygos vein. Lumbars drain blood into ascending lumbars and then run to inferior vena cava, where they release remainder of flow.

FIGURE 21-20 **Principal veins of the thorax, abdomen, and pelvis in anterior view.**

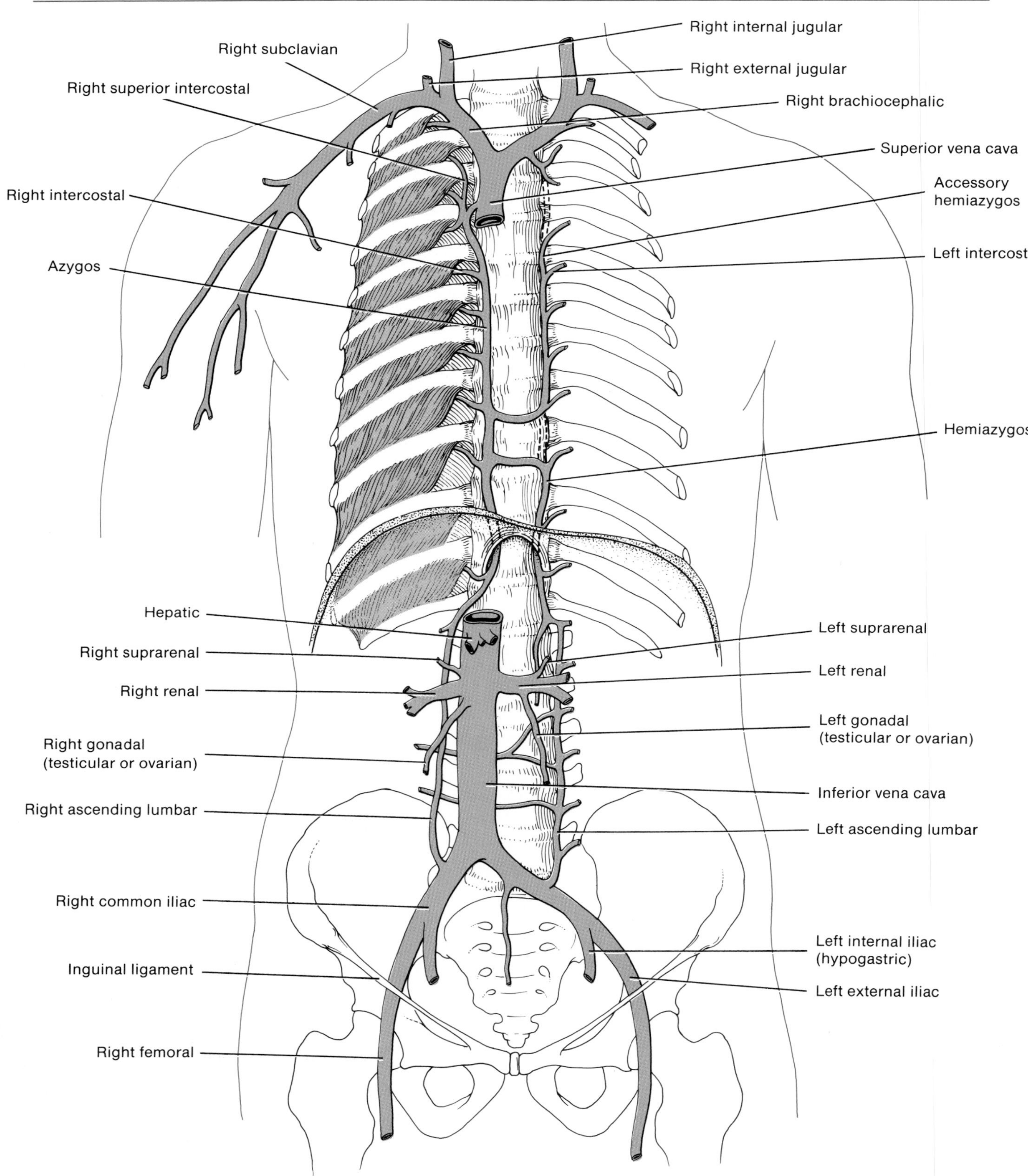

EXHIBIT 21-12 VEINS OF LOWER EXTREMITIES (Figure 21-21)

Vein	Description and Region Drained
	Blood from each lower extremity is returned by superficial set and deep set of veins.
SUPERFICIAL VEINS	
Great saphenous	***Great saphenous vein,*** longest vein in body, begins at medial end of ***dorsal venous arch*** of foot. It passes in front of medial malleolus and then upward along medial aspect of leg and thigh. It receives tributaries from superficial tissues and connects with deep veins as well. It empties into femoral vein in groin.
Small saphenous	***Small saphenous vein*** begins at lateral end of dorsal venous arch of foot. It passes behind lateral malleolus and ascends under skin of back of leg. It receives blood from foot and posterior portion of leg. It empties into popliteal vein behind knee.
DEEP VEINS	
Posterior tibial	***Posterior tibial vein*** is formed by union of ***medial*** and ***lateral plantar veins*** behind medial malleolus. It ascends deep in muscle at back of leg, receives blood from ***peroneal vein,*** and unites with anterior tibial vein just below knee.
Anterior tibial	***Anterior tibial vein*** is upward continuation of ***dorsalis pedis veins*** in foot. It runs between tibia and fibula and unites with posterior tibial to form popliteal vein.
Popliteal	***Popliteal vein,*** just behind knee, receives blood from anterior and posterior tibials and small saphenous vein.
Femoral	***Femoral vein*** is upward continuation of popliteal just above knee. Femorals run up posterior of thighs and drain deep structures of thighs. After receiving great saphenous veins in groin, they continue as right and left external iliac veins.

The great saphenous vein is very constant in its position anterior to the medial malleolus. It is frequently used for prolonged administration of intravenous fluids. This is particularly important in very young babies and in patients of any age who are in shock and whose veins are collapsed. It and the small saphenous vein are subject to varicosity.

SCHEME OF DRAINAGE

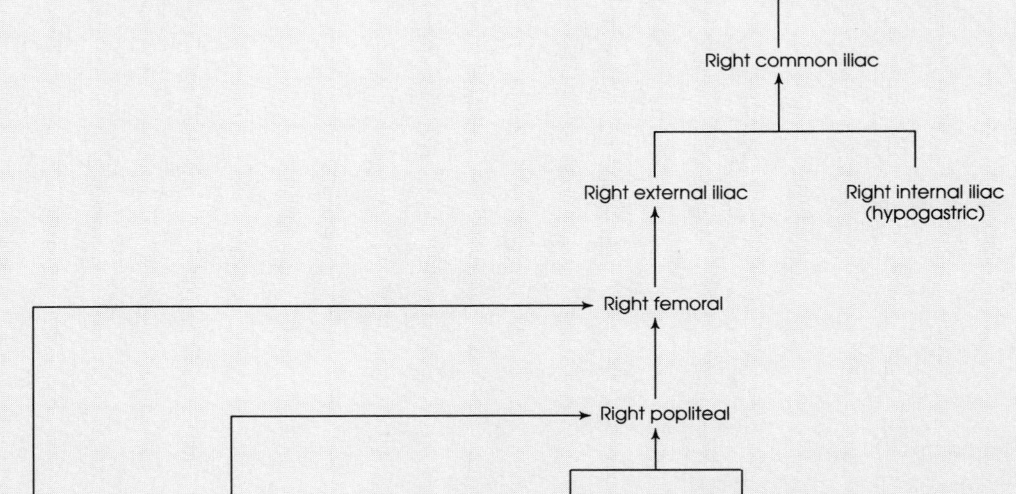

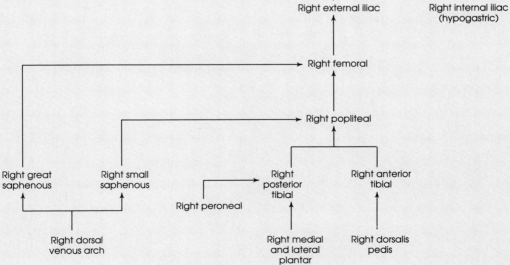

FIGURE 21-21 Principal veins of the pelvis and right lower extremity. (a) Anterior view. (b) Posterior view.

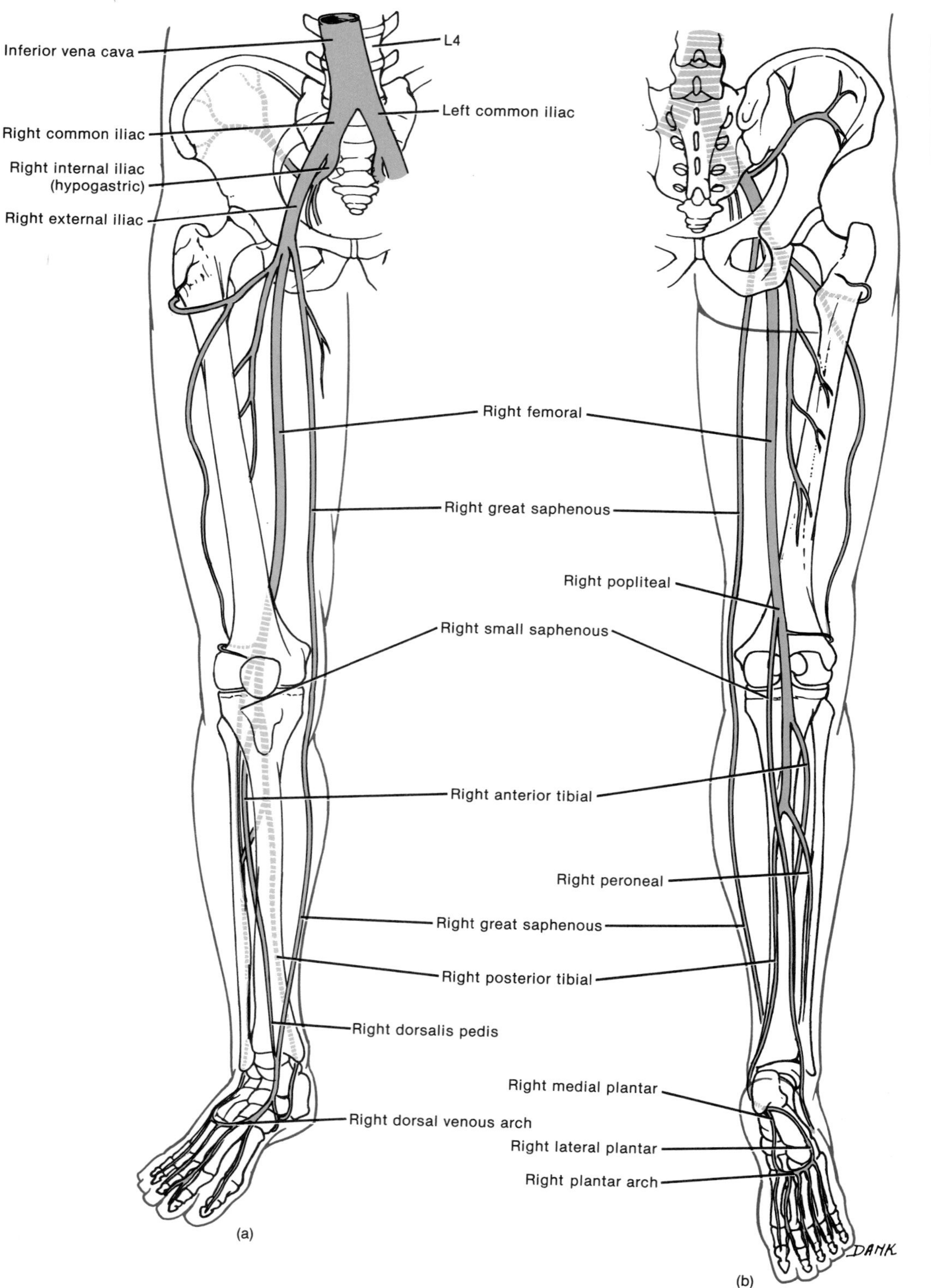

The flow of deoxygenated blood from the right ventricle to the air sacs of the lungs and the return of oxygenated blood from the air sacs of the lungs to the left atrium is called ***pulmonary circulation*** (Figure 21-22). The ***pulmonary trunk*** emerges from the right ventricle and passes upward, backward, and to the left. It then divides into two branches: the ***right pulmonary artery*** runs to the right lung; the ***left pulmonary artery*** goes to the left lung. On entering the lungs, the branches divide and subdivide until ultimately they form capillaries around the alveoli (air sacs) in the lungs. Carbon dioxide is passed from the blood into the alveoli to be breathed out of the lungs. Oxygen breathed in by the lungs is passed from the alveoli into the blood. The capillaries unite, venules and veins are formed, and eventually two ***pulmonary veins*** exit from each lung and transport the oxygenated blood to the left atrium. The pulmonary veins are the only postnatal (after birth) veins that carry oxygenated blood. Contractions of the left ventricle then send the blood into the systemic circulation.

Blood flow through the lungs differs in several ways from systemic circulation. The pulmonary arteries have larger diameters than their counterpart systemic arteries and they have thinner walls and less elastic tissue than

FIGURE 21-22 Pulmonary circulation. (a) Diagram. (b) Scheme of circulation.

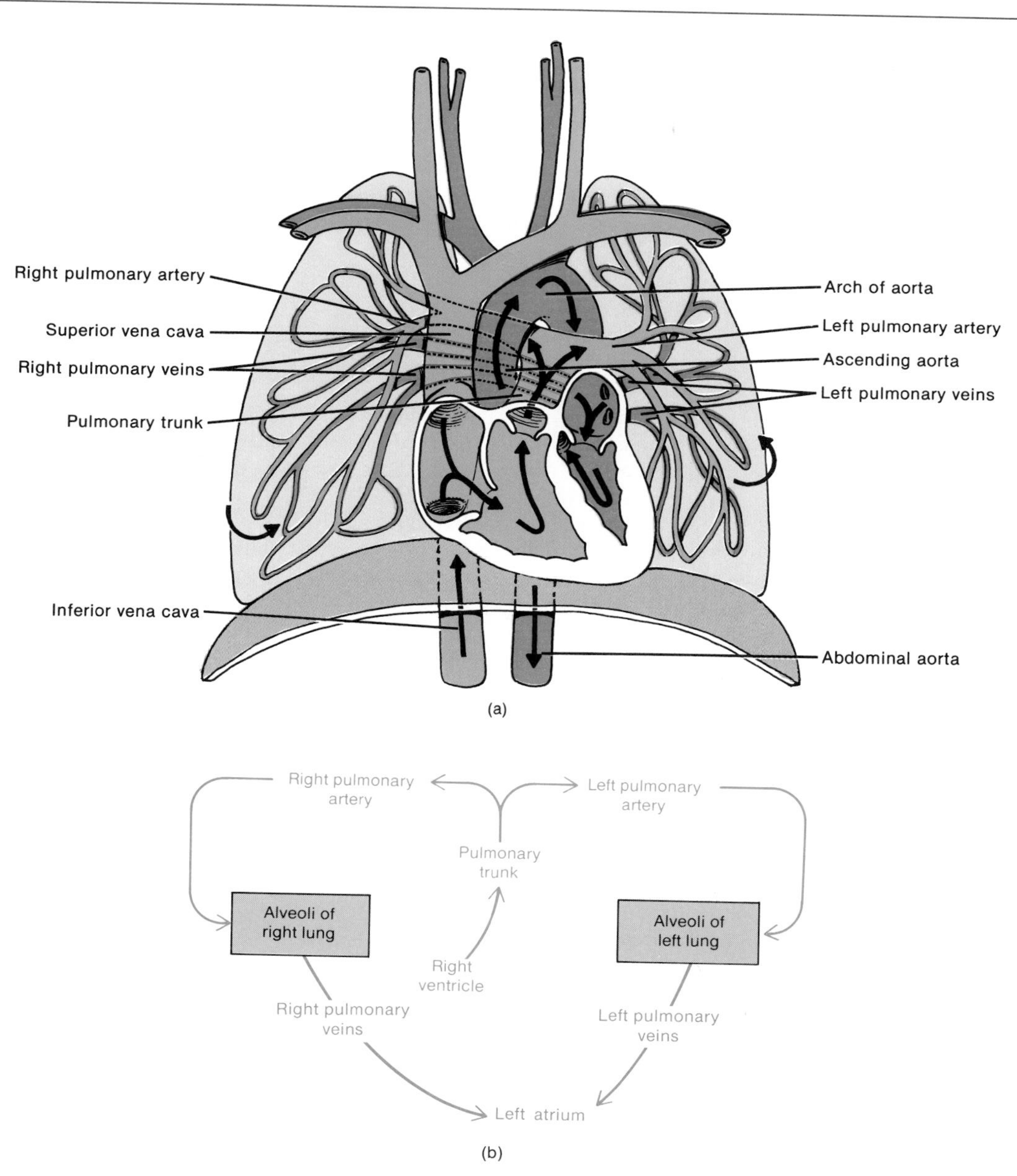

arteries. As a result, there is little resistance to blood flow. This means that less pressure is required to move blood through the lungs. In fact, as noted earlier, the systolic pressure of the right ventricle is only about one-fifth that of the left ventricle.

Another important difference between pulmonary and systemic circulation relates to the autoregulatory mechanism in response to levels of oxygen. In systemic circulation, blood vessels dilate in response to low oxygen concentration. In pulmonary circulation, blood vessels constrict in response to low levels of oxygen. This mechanism is very important in the distribution of blood in the lungs where it is needed most. For example, if some alveoli (air sacs) receive less blood (oxygen) than normal, the blood vessels in the affected area constrict and blood will largely bypass the poorly functioning areas. As a result, most of the blood flows to other areas of the lung that are better oxygenated.

One more difference between pulmonary and systemic circulation relates to capillary hydrostatic pressure. Normal pulmonary capillary hydrostatic pressure, the principal force that moves fluid out of capillaries into interstitial fluid, is only 15 mm Hg compared to 25 mm Hg, the average systemic capillary pressure. Despite this low value of 15 mm Hg, there is still normally a net fluid flow from pulmonary capillaries because the interstitial fluid protein concentration in the lungs is fairly high compared to that in pulmonary capillaries. The relatively lower protein concentration in pulmonary capillaries also favors net fluid flow from capillaries into interstitial fluid. The small continuous leak of fluid out of pulmonary capillaries is drained off by lymph capillaries so that pulmonary edema does not develop. However, pulmonary edema may develop from an increase in capillary blood pressure (due to increased left atrial pressure as may occur in mitral valve stenosis) or increased capillary permeability (as may occur from bacterial toxins). Pulmonary edema results in a reduced rate of diffusion of oxygen and carbon dioxide and thus inhibits the exchange of respiratory gases in the lungs.

Hepatic Portal Circulation

Blood enters the liver from two sources. The hepatic artery delivers oxygenated blood from the systemic circulation; the hepatic portal vein delivers deoxygenated blood from the gastrointestional tract, spleen, pancreas, and gallbladder. The term ***hepatic portal circulation*** refers to the flow of venous blood from the gastrointestinal organs and spleen to the liver before returning to the heart (Figure 21-23). During the absorptive state, hepatic portal blood is rich with substances absorbed from the gastrointestinal tract. The liver monitors these substances before they pass into the general circulation. For example, the liver stores nutrients such as glucose. It also modifies other digested substances so they may be used by cells, detoxifies harmful substances that have been absorbed by the gastrointestinal tract, and destroys bacteria by phagocytosis.

The hepatic portal system includes veins that drain blood from the pancreas, spleen, stomach, intestines, and gallbladder and transport it to the hepatic portal vein of the liver. The ***hepatic portal vein*** is formed by the union of the superior mesenteric and splenic veins. The ***superior mesenteric vein*** drains blood from the small intestine and portions of the large intestine and stomach. The ***splenic vein*** drains the spleen and receives tributaries from the stomach, pancreas, and portions of the colon. The tributaries from the stomach are the ***gastric, pyloric,*** and ***gastroepiploic veins.*** The ***pancreatic veins*** come from the pancreas, and the ***inferior mesenteric veins*** come from the portions of the colon. Before the hepatic portal vein enters the liver, it receives the ***cystic vein*** from the gallbladder and other veins. Ultimately, deoxygenated blood leaves the liver through the ***hepatic veins,*** which enter the inferior vena cava.

As a result of various conditions, specifically, those that increase resistance to venous flow (cirrhosis; occlusion of hepatic portal, splenic, or hepatic veins; cardiac disease), the pressure within the hepatic portal vein or its tributaries may increase above normal. This is called ***portal hypertension.*** The disorder is characterized by bleeding of varicosed veins of the esophagus, splenomegaly (enlarged spleen), and ascites (accumulation of fluid in the peritoneal cavity).

Fetal Circulation

The circulatory system of a fetus, called ***fetal circulation,*** differs from an adult's because the lungs, kidneys, and gastrointestinal tract of a fetus are not functioning. The fetus derives its oxygen and nutrients from the maternal blood and eliminates its carbon dioxide and wastes into the maternal blood (Figure 21-24).

The exchange of materials between fetal and maternal circulation occurs through a structure called the ***placenta*** (pla-SEN-ta). It is attached to the umbilicus (navel) of the fetus by the umbilical (um-BIL-i-kal) cord, and it communicates with the mother through countless small blood vessels that emerge from the uterine wall. The umbilical cord contains blood vessels that branch into capillaries in the placenta. Wastes from the fetal blood diffuse out of the capillaries, into spaces containing maternal blood (intervillous spaces) in the placenta, and finally into the mother's uterine blood vessels (see Figure 29-9a). Nutrients travel the opposite route—from the maternal blood vessels to the intervillous spaces to the fetal capillaries. Normally, there is no mixing of maternal and fetal blood since all exchanges occur through capillaries.

Blood passes from the fetus to the placenta via two ***umbilical arteries.*** These branches of the internal iliac (hypogastric) arteries are included in the umbilical cord. At the placenta, the blood picks up oxygen and nutrients and eliminates carbon dioxide and wastes. The oxygenated blood returns from the placenta via a single ***um-***

FIGURE 21-23 Hepatic portal circulation. (a) Diagram. (b) Scheme of blood flow through the liver, including arterial circulation. Deoxygenated blood is indicated in blue; oxygenated blood in red.

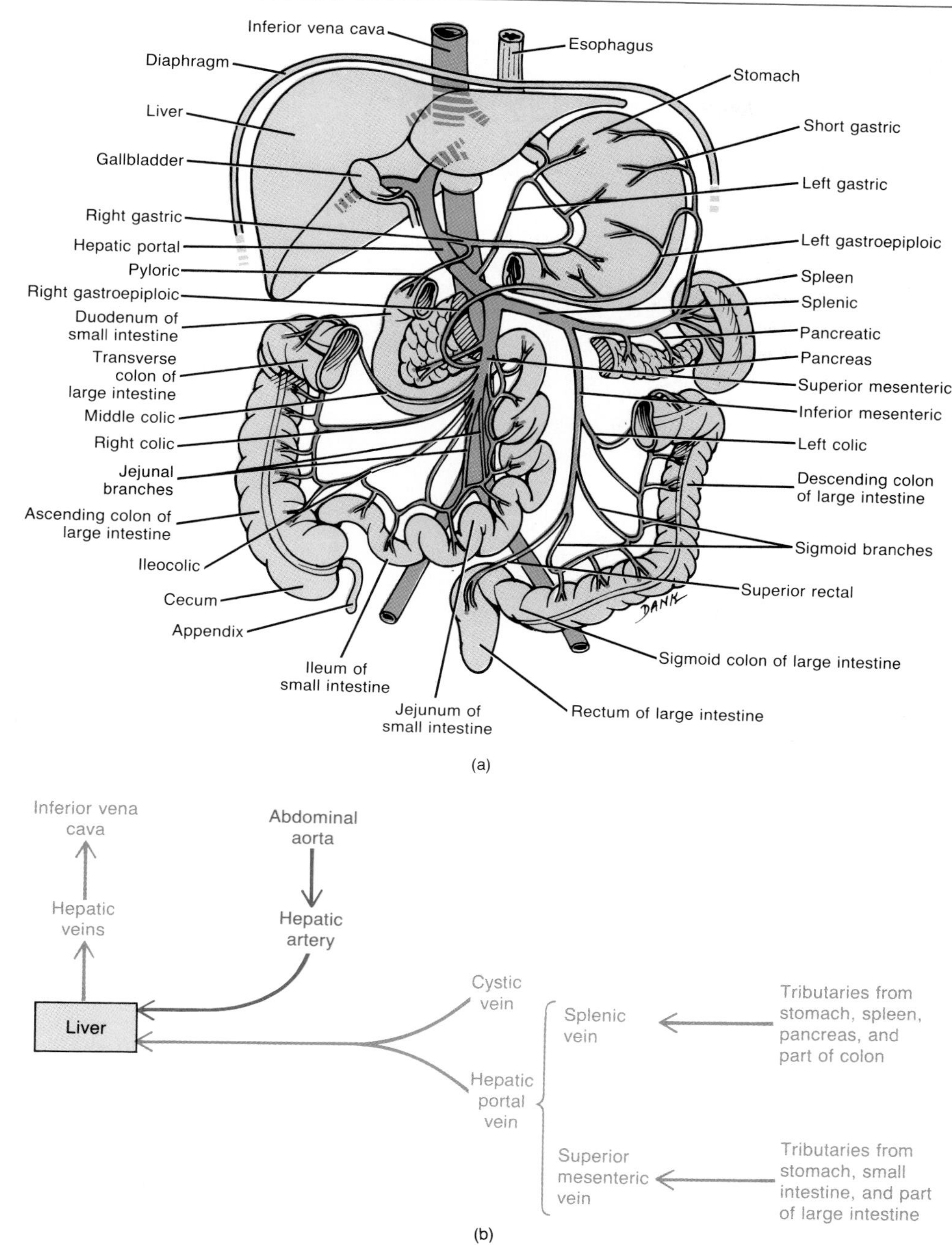

bilical vein. This vein ascends to the liver of the fetus, where it divides into two branches. Some blood flows through the branch that joins the hepatic portal vein and enters the liver. Although the fetal liver manufactures red blood cells, it does not function in digestion. Therefore, most of the blood flows into the second branch, the ***ductus venosus*** (DUK-tus ve-NŌ-sus). The ductus venosus eventually passes its blood to the inferior vena cava, bypassing the liver.

In general, circulation through other portions of the fetus is not unlike postnatal circulation. Deoxygenated blood returning from the lower regions is mingled with oxygenated blood from the ductus venosus in the inferior vena cava. This mixed blood then enters the right atrium. The circulation of blood through the upper portion of the fetus is also similar to postnatal flow. Deoxygenated blood returning from the upper regions of the fetus is collected by the superior vena cava, and it also passes into the right atrium.

Most of the blood does not pass through the right ventricle to the lungs, as it does in postnatal circulation, since the fetal lungs do not operate. In the fetus, an opening

FIGURE 21-24 Fetal circulation. The various colors represent different degrees of oxygenation of blood ranging from greatest (red) to intermediate (two shades of purple) to least (blue).

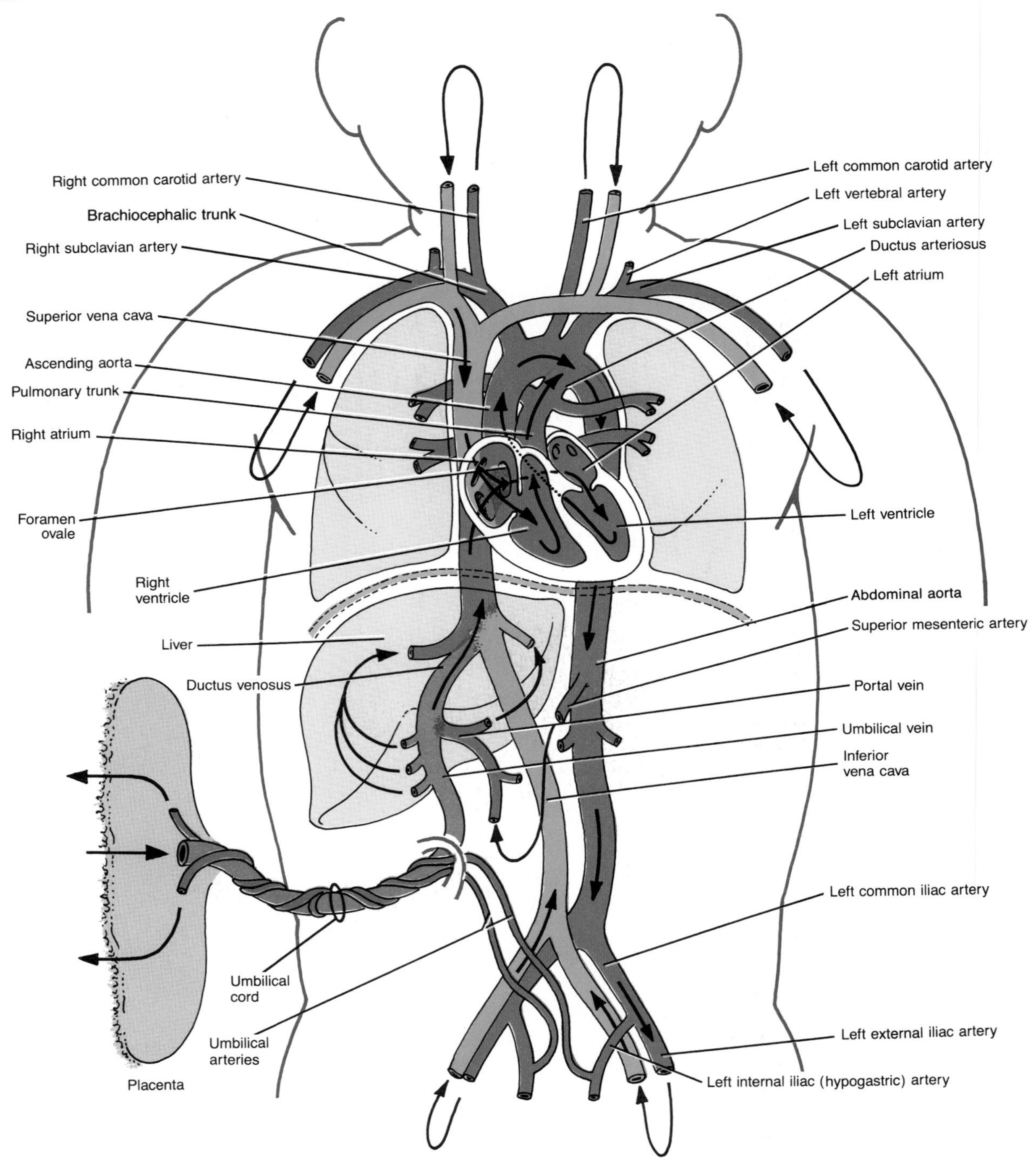

called the ***foramen ovale*** (fō-RĀ-men ō-VAL-ē) exists in the septum between the right and left atria. A valve in the inferior vena cava directs about a third of the blood through the foramen ovale so that it may be sent directly into the systemic circulation. The blood that does descend into the right ventricle is pumped into the pulmonary trunk, but little of this blood actually reaches the lungs. Most blood in the pulmonary trunk is sent through the ***ductus arteriosus*** (ar-tē-rē-Ō-sus). This small vessel connecting the pulmonary trunk with the aorta enables most blood to bypass the fetal lungs. The blood in the aorta is carried to all parts of the fetus through its systemic branches. When the common iliac arteries branch into the external and internal iliacs, part of the blood flows into the internal iliacs. It then goes to the umbilical arteries and back to the placenta for another exchange of materials. The only vessel that carries fully oxygenated blood is the umbilical vein.

At birth, when pulmonary (lung), renal, digestive, and liver functions are established, the special structures of fetal circulation are no longer needed and the following changes occur.

1. The umbilical arteries vasoconstrict shut and atrophy to become the ***medial umbilical ligaments.***
2. The umbilical vein vasoconstricts shut and becomes the ***round ligament*** of the liver.
3. The placenta is delivered by the mother as the ***"afterbirth."***
4. The ductus venosus vasoconstricts shut and becomes the ***ligamentum venosum,*** a fibrous cord in the liver.
5. The foramen ovale normally closes shortly after birth to become the ***fossa ovalis,*** a depression in the interatrial septum.
6. The ductus arteriosus closes by vasoconstriction and atrophies and becomes the ***ligamentum arteriosum.***

Anatomical defects resulting from failure of these changes to occur are described in Chapter 20.

EXERCISE AND THE CARDIOVASCULAR SYSTEM

No matter what a person's level of fitness, it can be improved at any age with regular participation in exercise. Of the various types of exercise, some are more effective than others for improving the health of the cardiovascular system because they involve movements of large body muscles. One example is ***aerobics,*** any activity that works large muscles, increases the flow of blood to the heart, and accelerates metabolic rate for a prolonged period of time (at least 20 minutes). Three to five sessions a week are usually recommended. Brisk walking, running, bicycling, cross-country skiing, and swimming are examples of aerobic exercises.

Sustained exercise increases the oxygen demand of the muscles, and whether the demand is met depends primarily on the adequacy of cardiac output and proper function of the respiratory system. After several weeks of training, the healthy individual increases cardiac output and thereby increases the rate of oxygen delivery to the tissues.

Physical conditioning also causes an interesting effect upon systemic blood pressure. After a conditioning period, hypertensive individuals show a reduction in systolic pressure amounting to an average of 13 mm Hg. Blood pressure response to a given work load also is less after training, whereas lower pressure itself helps reduce myocardial oxygen requirements. So conditioning also may be useful in the management of systemic hypertension.

Additional benefits to be gained from physical conditioning are an increase in high-density lipoprotein (HDL), a substance that seems to counter the impact of cholesterol in heart disease, a decrease in triglyceride levels, and improved lung function. Exercise also helps to reduce anxiety and depression, control weight, and increase the body's ability to dissolve blood clots by increasing fibrinolytic activity. Intense exercise increases levels of endorphins, the body's natural painkillers. This may explain the psychological "high" that runners experience with strenuous training and the "low" they feel when they miss regular workouts. Exercise also helps make bones stronger and thus may be a factor in inhibiting and treating osteoporosis. Some research indicates that exercise may even offer some protection against cancer and diabetes.

A well-trained athlete can achieve a cardiac output up to about 6 times that of an untrained individual during activity. This is because of hypertrophy (enlargement) of the heart as a result of training. Even though the heart of a well-trained athlete is larger, *resting* cardiac output is about the same as in a healthy untrained person. This occurs because stroke volume is increased while heart rate is decreased. The heart rate of a trained athlete is about 40 to 60 beats per minute.

AGING AND THE CARDIOVASCULAR SYSTEM

General changes associated with aging and the cardiovascular system include loss of extensibility of the aorta, reduction in cardiac muscle fiber (cell) size, progressive loss of cardiac muscular strength, a reduced output of blood by the heart, a decline in maximum heart rate, and an increase in systolic blood pressure. Total blood cholesterol tends to increase with age, as does low-density lipoprotein (LDL); high-density lipoprotein (HDL) tends to decrease. There is an increase in the incidence of coronary artery disease (CAD), the major cause of heart disease and death in older Americans. Congestive heart failure (CHF), a set of symptoms associated with impaired pumping performance of the heart, also occurs. Changes in blood vessels, such as hardening of the arteries and cholesterol deposits in arteries that serve brain tissue, reduce nourishment to the brain and result in the malfunction or death of brain cells. By age 80, cerebral blood flow is 20 percent less and renal blood flow is 50 percent less than in the same person at age 30.

DEVELOPMENTAL ANATOMY OF BLOOD AND BLOOD VESSELS

Since the human egg and yolk sac have little yolk to nourish the developing embryo, blood and blood vessel formation starts as early as 15 to 16 days. The development begins in the ***mesoderm*** of the yolk sac, chorion, and body stalk.

Blood vessels develop from isolated masses and cords of mesenchyme in the mesoderm called ***blood islands*** (Figure 21-25). Spaces soon appear in the islands and become the lumens of the blood vessels. Some of the mesenchymal cells immediately around the spaces give rise to the *endothelial lining of the blood vessels.* Mesenchyme around the endothelium forms the *tunics* (intima, media, externa) of the larger blood vessels. Growth and fusion of blood islands form an extensive network of blood vessels throughout the embryo.

Blood plasma and *blood cells* are produced by the endothelial cells and appear in the blood vessels of the yolk sac and allantois quite early. Blood formation in the embryo itself begins at about the second month in the liver and spleen, a little later in bone marrow, and much later in lymph nodes.

FIGURE 21-25 **Development of blood vessels and blood cells from blood islands.**

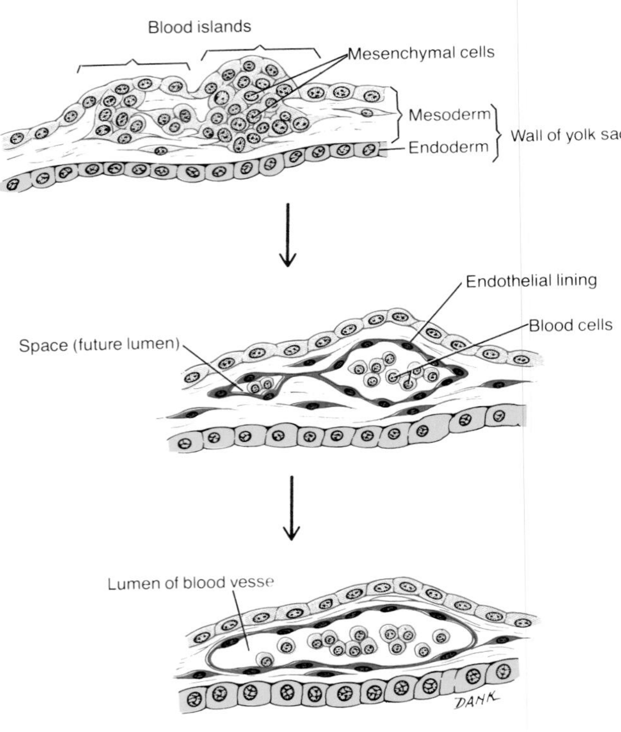

DISORDERS: HOMEOSTATIC IMBALANCES

Hypertension

Hypertension, or high blood pressure, is the most common disease affecting the heart and blood vessels. Statistics indicate that hypertension afflicts one out of every five American adults. Although there is some disagreement as to what defines hypertension, a strong consensus has emerged suggesting that a blood pressure of 120/80 is normal and desirable in a healthy adult. Borderline high blood pressure is defined as diastolic pressure between 85 and 89. Mild high blood pressure is diastolic pressure between 90 and 104. Moderate high blood pressure is diastolic pressure between 105 and 114. Severe high blood pressure is diastolic pressure of 115 or higher. Isolated systolic hypertension is systolic pressure greater than 160 in those whose diastolic pressure is less than 90. As noted in Chapter 20, the lower the blood pressure, the less the risk of coronary artery disease (CAD). The lowest functioning pressure for an individual is the optimum pressure for that person.

Primary hypertension (essential hypertension) is a persistently elevated blood pressure that cannot be attributed to any particular organic cause. Approximately 90 to 95 percent of all hypertension cases fit this definition. The remaining percent are ***secondary hypertension.*** Secondary hypertension has an identifiable underlying cause such as kidney disease and adrenal hypersecretion. Kidney disease and obstruction of blood flow may cause the kidneys to release renin into the blood. This enzyme catalyzes the formation of angiotensin II from a plasma protein. Angiotensin II is a powerful blood vessel constrictor—and the most potent agent known for raising blood pressure. It also stimulates aldosterone release. Aldosteronism, the hypersecretion of aldosterone, may also cause an increase in blood pressure. Aldosterone is the adrenal cortex hormone that promotes the retention of salt and water by the kidneys. It thus tends to increase plasma volume. ***Pheochromocytoma*** (fē-ō-krō'-mō-sī-TŌ-ma) is a tumor of the adrenal

DISORDERS: HOMEOSTATIC IMBALANCES (*continued*)

medulla. It produces and releases into the blood large quantities of norepinephrine and epinephrine. These hormones also raise blood pressure. Epinephrine causes an increase in heart rate and force of contraction and norepinephrine causes vasoconstriction.

High blood pressure is of considerable concern because of the harm it can do to the heart, brain, and kidneys if it remains uncontrolled. The heart is most commonly affected by high blood pressure. When pressure is high, the heart uses more energy in pumping against the increased resistance caused by the elevated arterial blood pressure. Because of the increased effort, the heart muscle thickens and the heart becomes enlarged. The heart also needs more oxygen. If it cannot meet the demands put on it, angina pectoris or even myocardial infarction may develop. Hypertension is also a factor in the development of atherosclerosis. Continued high blood pressure may produce a cerebral vascular accident (CVA), or stroke. In this case, severe strain has been imposed on the cerebral arteries that supply the brain. These arteries are usually less protected by surrounding tissues than are the major arteries in other parts of the body. These weakened cerebral arteries may finally rupture, and a brain hemorrhage follows.

The kidneys are also prime targets of hypertension. The principal site of damage is in the arterioles that supply them. The continual high blood pressure pushing against the walls of the arterioles causes them to thicken, thus narrowing the lumen. The blood supply to the kidneys is thereby gradually reduced. In response, the kidneys may secrete renin, which raises the blood pressure even higher and complicates the problem. The reduced blood flow to the kidney cells may eventually lead to the death of the cells.

Although medical science cannot cure essential hypertension, certain forms of secondary hypertension, if diagnosed early, can be cured by removing the underlying cause. And even in some cases of essential hypertension, measures such as weight loss in an obese person or salt reduction in one with a history of high salt (NaCl) intake may result in a dramatic reduction in blood pressure. Evidence suggests that less fat and more potassium and calcium may also lead to a reduction in blood pressure, whereas magnesium deficiency may cause hypertension. Since nicotine is a vasoconstrictor, it elevates blood pressure. Stopping smoking may help to decrease blood pressure. As indicated earlier, regular exercise can help to reduce hypertension. Since alcohol and caffeine raise blood pressure, lowering intake can help to reduce it. In recent years, some physicians have advocated relaxation techniques (yoga, meditation, and biofeedback) to treat hypertension. For those individuals who require medication, a number of drugs are available. Many people can be treated with *diuretics,* which eliminate large amounts of water and sodium, thus decreasing blood volume and reducing blood pressure. *Vasodilators* are often used in combination with diuretics. They relax the smooth muscle in arterial walls, causing vasodilation and thus lowering blood pressure by lowering peripheral resistance. *Beta blockers* reduce blood pressure by inhibiting secretion of renin and decreasing rate and force of heartbeat. Beta blockers are often used in combination with diuretics. *Calcium channel blockers* prevent movement of Ca^{2+} ions into blood vessels and heart cells, causing relaxation and lowering of blood pressure.

Aneurysm

An ***aneurysm*** (AN-yoo-rizm) is a thin, weakened section of the wall of an artery or a vein that bulges outward, forming a balloonlike sac of the blood vessel. Common causes of aneurysms include atherosclerosis, syphilis, congenital blood vessel defects, and trauma. If an aneurysm goes untreated, it grows larger and larger until the blood vessel wall becomes so thin that it may burst, causing massive hemorrhage with shock, severe pain, stroke, or death, depending on which vessel is involved. Even an unruptured aneurysm can lead to damage by interrupting blood flow or putting pressure on adjacent blood vessels, organs, or bones.

Surgical repair of an aneurysm consists of temporarily clamping the damaged artery above and below the aneurysm and then excising (surgically removing) the aneurysm. A graft, usually of Dacron, is then sutured to healthy segments of the artery to reestablish normal blood flow.

Coronary Artery Disease (CAD)

In Chapter 20 it was indicated that the common causes of heart disease are related to inadequate coronary blood supply, anatomical disorders, and arrhythmias. ***Coronary artery disease (CAD)*** is a condition in which the heart muscle receives inadequate blood because of an interruption of blood supply. Depending on the degree of interruption, symptoms can range from a mild chest pain to a full-scale heart attack. The underlying causes of CAD are many and varied. Two of the principal ones are atherosclerosis and coronary artery spasm, both of which were discussed in detail in Chapter 20.

Deep-Venous Thrombosis (DVT)

Venous thrombosis, the presence of a thrombus in a vein, typically occurs in deep veins of the lower extremities. In such cases, the condition is referred to as ***deep-venous thrombosis (DVT).*** The two most serious complications of DVT are *pulmonary embolism,* in which the thrombus dislodges and finds its way into the pulmonary arterial blood flow, and *postphlebitic syndrome,* which consists of edema, pain, and skin changes due to destruction of venous valves. Treatment consists of anticoagulant therapy, elevation of the extremity, fibrinolytic therapy (streptokinase or urokinase), and in rare cases, thrombectomy.

MEDICAL TERMINOLOGY ASSOCIATED WITH BLOOD VESSELS

Aortography (ā'-or-TOG-ra-fē) X-ray examination of the aorta and its main branches after injection of radiopaque dye.

Arteritis (ar'-te-RĪ-tis; *itis* = inflammation of) Inflammation of an artery, probably due to an autoimmune response.

Carotid endarterectomy (ka-ROT-id end'-ar-ter-EK-tō-mē) The removal of atherosclerotic plaque from the carotid artery to restore greater blood flow to the brain.

Claudication (klaw'-di-KĀ-shun) Pain and lameness or limping caused by defective circulation of the blood in the vessels of the limbs.

Hypercholesterolemia (hī-per-kō-les'-ter-ol-Ē-mē-a; *hyper* = over; *heme* = blood) An excess of cholesterol in the blood.

Hypotension (hī'-pō-TEN-shun; *hypo* = below; *tension* = pressure) Low blood pressure; most commonly used to describe an acute drop in blood pressure, as occurs in hypovolemic shock.

Normotensive (nor'-mō-TEN-siv) Characterized by normal blood pressure.

Occlusion (o-KLOO-shun) The closure or obstruction of the lumen of a structure such as a blood vessel.

Orthostatic (or'-tho-STAT-ik) **hypotension** (*ortho* = straight; *statikos* = causing to stand) An excessive lowering of systemic blood pressure with the assumption of an erect or semierect posture; it is usually a sign of a disease. May be caused by excessive fluid loss, certain drugs (antihypertensives), and cardiovascular or neurogenic factors. Also called **postural hypotension.**

Phlebitis (fle-BĪ-tis; *phleb* = vein) Inflammation of a vein, often in a leg.

Raynaud's (rā-NOZ) **disease** A vascular disorder, primarily of females, characterized by bilateral attacks of ischemia, usually of the fingers and toes, in which the skin becomes pale and exhibits burning and pain; it is brought on by cold or emotional stimuli. If the condition is secondary to another disorder, it is called **Raynaud's phenomenon.**

Shunt A passage between two blood vessels or between the two sides of the heart.

Thrombectomy (throm-BEK-tō-mē; *thrombo* = clot) An operation to remove a blood clot from a blood vessel.

Thrombophlebitis (throm'-bō-fle-BĪ-tis) Inflammation of a vein with clot formation. Superficial thrombophlebitis occurs in veins under the skin, especially the calf.

White coat (office) hypertension A syndrome found in patients who have elevated blood pressures while being examined by health-care personnel, but are otherwise normotensive.

STUDY OUTLINE

Arteries (p. 606)

1. Arteries carry blood away from the heart. The wall of an artery consists of a tunica interna, tunica media (which maintains elasticity and contractility), and tunica externa.
2. Large arteries are referred to as elastic (conducting) arteries, and medium-sized arteries are called muscular (distributing) arteries.
3. Many arteries anastomose—the distal ends of two or more vessels unite. An alternate blood route from an anastomosis is called collateral circulation. Arteries that do not anastomose are called end arteries.

Arterioles (p. 608)

1. Arterioles are small arteries that deliver blood to capillaries.
2. Through constriction and dilation, they assume a key role in regulating blood flow from arteries into capillaries and in altering peripheral resistance and thus arterial blood pressure.

Capillaries (p. 608)

1. Capillaries are microscopic blood vessels through which materials are exchanged between blood and tissue cells; some capillaries are continuous, whereas others are fenestrated.
2. Capillaries branch to form an extensive capillary network throughout the tissue. This network increases the surface area, allowing a rapid exchange of large quantities of materials.
3. Precapillary sphincters regulate blood flow through capillaries.
4. Microscopic blood vessels in the liver are called sinusoids.

Venules (p. 610)

1. Venules are small vessels that continue from capillaries and merge to form veins.
2. They drain blood from capillaries into veins.

Veins (p. 610)

1. Veins consist of the same three tunics as arteries but have less elastic tissue and smooth muscle.
2. They contain valves to prevent backflow of blood.
3. Weak valves can lead to varicose veins or hemorrhoids.
4. Vascular (venous) sinuses are veins with very thin walls.

Blood Reservoirs (p. 610)

1. Systemic veins are collectively called blood reservoirs.
2. They store blood, which through vasoconstriction can move to other parts of the body if the need arises.
3. The principal reservoirs are the veins of the abdominal organs (liver and spleen) and skin.

Physiology of Circulation (p. 610)

1. Blood flow is determined by blood pressure and resistance.
2. Blood flows from regions of higher to lower pressure. The established pressure gradient for the systemic circulation is from aorta to large arteries to small arteries to arterioles to capillaries to venules to veins to venae cavae to right atrium.
3. Resistance refers to the opposition to blood as a result of friction between blood and the walls of blood vessels.
4. Resistance is determined by blood viscosity, blood vessel length, and blood vessel radius.
5. Any factor that increases cardiac output (CO) increases blood pressure.
6. As blood volume increases, blood pressure increases.
7. Peripheral resistance refers to all the vascular resistance factors of the cardiovascular system.
8. Factors that determine heart rate and force of contraction, and therefore blood pressure, are the autonomic nervous system through the cardiac center, chemicals, temperature, emotions, sex, and age.
9. Factors that regulate blood pressure by acting on blood vessels include the vasomotor center in the medulla together with baroreceptors, chemoreceptors, and higher brain centers; chemicals; and autoregulation.
10. The movement of water and dissolved substances (except proteins) through capillaries is dependent on hydrostatic and osmotic pressures.
11. The near equilibrium at the arterial and venous ends of a capillary by which fluids exit and enter is called Starling's law of the capillaries.
12. Edema is an abnormal increase in interstitial fluid.
13. Blood return to the heart is maintained by several factors including skeletal muscular contractions, valves in veins (especially in the extremities), and breathing.

Shock and Homeostasis (p. 617)

1. Shock is a failure of the cardiovascular system to deliver adequate amounts of oxygen and nutrients to meet the metabolic needs of cells.
2. Signs and symptoms include hypotension; clammy, cool, pale skin; sweating; decreased urinary output; altered mental state; acidosis; tachycardia; weak, rapid pulse; and thirst.
3. Types of shock include hypovolemic, cardiogenic, obstructive, neurogenic, and septic.
4. Stages of shock are: (1) compensated, in which negative feedback cycles restore homeostasis; (2) decompensated, in which positive feedback cycles intensify the shock and immediate medical intervention is required; and (3) irreversible, in which the cardiovascular system collapses and death results.

Checking Circulation (p. 620)

Pulse (p. 620)

1. Pulse is the alternate expansion and elastic recoil of an artery wall with each heartbeat. It may be felt in any artery that lies near the surface or over a hard tissue.
2. A normal pulse rate is between 70 and 80 beats per minute.

Measurement of Blood Pressure (BP) (p. 621)

1. Blood pressure is the pressure exerted by blood on the wall of an artery when the left ventricle undergoes systole and then diastole. It is measured by the use of a sphygmomanometer.
2. Systolic blood pressure (SBP) is the force of blood recorded during ventricular contraction. Diastolic blood pressure (DBP) is the force of blood recorded during ventricular relaxation. The average blood pressure is 120/80 mm Hg.
3. Pulse pressure is the difference between systolic and diastolic pressure. It averages 40 mm Hg and provides information about the condition of arteries.

Circulatory Routes (p. 621)

1. The largest circulatory route is the systemic circulation.
2. Two of the several subdivisions of the systemic circulation are coronary (cardiac) circulation and hepatic portal circulation.
3. Other routes include the cerebral, pulmonary, and fetal circulation.

Systemic Circulation (p. 623)

1. The systemic circulation takes oxygenated blood from the left ventricle through the aorta to all parts of the body, including lung tissue (but does *not* supply the air sacs of the lungs) and returns the deoxygenated blood to the right atrium.
2. The aorta is divided into the ascending aorta, the arch of the aorta, and the descending aorta. Each section gives off arteries that branch to supply the whole body.
3. Blood is returned to the heart through the systemic veins. All the veins of the systemic circulation flow into either the superior or inferior venae cavae or the coronary sinus. They in turn empty into the right atrium.

Pulmonary Circulation (p. 643)

1. The pulmonary circulation takes deoxygenated blood from the right ventricle to the air sacs of the lungs and returns oxygenated blood from the lungs to the left atrium.
2. It allows blood to be oxygenated for systemic circulation.

Hepatic Portal Circulation (p. 644)

1. The hepatic portal circulation collects blood from the veins of the pancreas, spleen, stomach, intestines, and gallbladder and directs it into the hepatic portal vein of the liver.
2. This circulation enables the liver to utilize nutrients and detoxify harmful substances in the blood.

Fetal Circulation (p. 644)

1. The fetal circulation involves the exchange of materials between fetus and mother.
2. The fetus derives its oxygen and nutrients and eliminates its carbon dioxide and wastes through the maternal blood supply by means of a structure called the placenta.
3. At birth, when pulmonary (lung), digestive, and liver functions are established, the special structures of fetal circulation are no longer needed.

Exercise and the Cardiovascular System (p. 647)

1. Aerobic exercises provide useful benefits for the cardiovascular system.
2. Among the benefits are increased cardiac output, increased delivery of oxygen to tissues, reduced systolic blood pres-

sure, increased high-density lipoprotein (HDL), weight control, and increased ability to dissolve blood clots.

Aging and the Cardiovascular System (p. 647)

1. General changes include loss of elasticity of blood vessels, reduction in cardiac muscle size, and reduced cardiac output.
2. The incidence of coronary artery disease (CAD), congestive heart failure (CHF), and atherosclerosis increases with age.

Developmental Anatomy of Blood and Blood Vessels (p. 648)

1. Blood vessels develop from isolated masses of mesenchyme in mesoderm called blood islands.
2. Blood is produced by the endothelium of blood vessels.

Disorders: Homeostatic Imbalances (p. 648)

1. Hypertension, or high blood pressure, is classifed as primary and secondary.
2. An aneurysm is a thin, weakened section of the wall of an artery or vein that bulges outward, forming a balloonlike sac.
3. Deep-venous thrombosis (DVT) refers to a blood clot in a deep vein, especially in the lower extremities.

REVIEW QUESTIONS

1. Describe the structural and functional differences among arteries, arterioles, capillaries, venules, and veins. (p. 606)
2. Discuss the importance of the elasticity and contractility of arteries. (p. 606)
3. Distinguish between elastic (conducting) and muscular (distributing) arteries in terms of location, histology, and function. What is an anastomosis? What is collateral circulation? (p. 606)
4. Describe how capillaries are structurally adapted for exchanging materials between blood and body cells. (p. 608)
5. Define varicose veins and describe how they are treated. (p. 610)
6. What are blood reservoirs? Why are they important? (p. 610)
7. What is blood flow? How is it determined? (p. 610)
8. Why does blood flow faster in arteries and veins than in capillaries? (p. 611)
9. Define resistance and explain the factors that contribute to it. (p. 611)
10. Describe how each of the following affects blood pressure: cardiac output (CO), blood volume, and peripheral resistance. (p. 611)
11. What is the vasomotor center? What are its functions? (p. 612)
12. Discuss how the vasomotor center operates with baroreceptors, chemoreceptors, and higher brain centers to control blood pressure. (p. 612)
13. Explain the effects of different chemicals such as epinephrine, norepinephrine (NE), antidiuretic hormone (ADH), angiotensin II, histamine, and kinins on blood pressure. (p. 613)
14. What is meant by autoregulation? Describe its probable mechanism. (p. 613)
15. Describe how hydrostatic and osmotic pressures determine fluid movement through capillaries. Set up the equation for effective filtration pressure (*Peff*) to substantiate your response. How is edema produced? (p. 614)
16. What is Starling's law of the capillaries? (p. 614)
17. Describe the factors that assist the return of venous blood to the heart. (p. 616)
18. Define shock. What are its signs and symptoms? How is shock classified. (p. 617)
19. Describe the three stages of shock. (p. 618)
20. Define pulse. Where may pulse be felt? Contrast tachycardia and bradycardia. (p. 620)
21. What is blood pressure (BP)? Describe how systolic and diastolic blood pressure are recorded by means of a sphygmomanometer. (p. 621)
22. Compare the clinical significance of systolic and diastolic pressure. How are these pressures written? (p. 621)
23. What is meant by a circulatory route? Define systemic circulation. (p. 621)
24. Diagram the major divisions of the aorta, their principal arterial branches, and the regions supplied. (p. 623)
25. Trace a drop of blood from the arch of the aorta through its systemic circulatory route to the tip of the big toe on your left foot and back to the heart again. Remember that the major branches of the arch are the brachiocephalic artery, left common carotid artery, and left subclavian artery. Be sure to also indicate which veins return the blood to the heart. (p. 625)
26. What is the cerebral arterial circle (circle of Willis)? Why is it important? (p. 626)
27. What are visceral branches of an artery? Parietal branches? What major organs are supplied by branches of the thoracic aorta? How is blood returned from these organs to the heart? (p. 628)
28. What organs are supplied by the celiac, superior mesenteric, renal, inferior mesenteric, inferior phrenic, and middle sacral arteries? How is blood returned to the heart? (p. 628)
29. Trace a drop of blood from the brachiocephalic artery into the digits of the right upper extremity and back again to the right atrium. (p. 625)
30. What are the three major groups of systemic veins? (p. 633)
31. Define pulmonary circulation. Prepare a diagram to indicate the route. What is the purpose of the route? (p. 643)
32. What is hepatic portal circulation? Describe the route by means of a diagram. Why is this route significant? (p. 644)
33. Discuss in detail the anatomy and physiology of fetal cir-

culation. Be sure to indicate the function of the umbilical arteries, umbilical vein, ductus venosus, foramen ovale, and ductus arteriosus. (p. 644)
34. What are the effects of exercise on the cardiovascular system? (p. 647)
35. Describe the effects of aging on the cardiovascular system. (p. 647)
36. Describe the development of blood vessels and blood. (p. 648)
37. How does hypertension affect the body? How is hypertension treated? (p. 648)
38. What is an aneurysm? Why is an aneurysm a serious problem? (p. 649)
39. What is deep-venous thrombosis (DVT)? (p. 649)
40. Refer to the glossary of medical terminology associated with blood vessels. Be sure that you can define each term. (p. 650)

SELECTED READINGS

Blake, P. "Precision Moves That Counter Cardiogenic Shock," *RN,* May 1989.

Bock, H. (ed). "New Ways to Control Blood Pressure," *Medical Update,* July 1988.

Garvas, I., M. Bursztyn, and H. Garvas. "Changing Trends in Hypertension Therapy," *Modern Medicine,* October 1988.

Goldfinger, S. E. (ed). "Exercise and Well-Being," *Harvard Medical School Health Letter,* February 1985.

———. "High Blood Pressure: Newer Treatments," *Harvard Medical School Health Letter,* January 1989.

———. "High Blood Pressure: A New Look," *Harvard Medical School Health Letter,* December 1988.

Johnson, G. T. (ed). "The Ups and Downs of Blood Pressure Numbers," *Harvard Medical School Health Letter,* November 1987.

Rios, J. C. (ed). "Exercise and the Heart," *Cardiac Alert,* February 1984.

Rippe, J. M., A. Ward, J. P. Pocari, and P. S. Freedson. "Walking for Health and Fitness," *Journal of the American Medical Association,* 13 May 1988.

Chapter 22

The Lymphatic System and Immunity

Chapter Contents at a Glance

Student Objectives

1. Describe the components of the lymphatic system and list their functions.
2. Discuss how edema develops.
3. Discuss the roles of the skin and mucous membranes, phagocytosis, inflammation, fever, and antimicrobial substances as components of nonspecific resistance.
4. Define immunity and explain the relationship between an antigen (Ag) and an antibody (Ab).
5. Contrast the role of T cells in cellular immunity and the role of B cells in humoral immunity.
6. Discuss the relationship of immunology to cancer.
7. Describe the effects of aging on the immune system.
8. Describe the development of the lymphatic system.
9. Describe the clinical symptoms of the following disorders: acquired immune deficiency syndrome (AIDS), autoimmune diseases, severe combined immunodeficiency (SCID), hypersensitivity (allergy), tissue rejection, and Hodgkin's disease (HD).
10. Define medical terminology associated with the lymphatic system.

The ***lymphatic*** (lim-FAT-ik) ***system*** consists of a fluid called lymph, vessels that transport lymph called lymphatic vessels (lymphatics), and a number of structures and organs that contain lymphatic (lymphoid) tissue (see Figure 22-2b). Essentially, lymphatic tissue is a specialized form of reticular connective tissue that contains large numbers of lymphocytes. The stroma (framework) of lymphatic tissue is a meshwork of reticular fibers and reticular cells (fibroblasts and fixed macrophages). One exception to this is the thymus gland, which has a stroma that is somewhat different, composed of epithelioreticular tissue (discussed later in the chapter).

Lymphatic tissue occurs in the body in various ways. Lymphatic tissue not enclosed by a capsule is referred to as ***diffuse lymphatic tissue.*** This is the simplest form of lymphatic tissue and is found in the lamina propria (connective tissue) of mucous membranes of the gastrointestinal tract, respiratory passageways, urinary tract, and reproductive tract. It is also normally found in small amounts in the stroma of almost every organ of the body.

Lymphatic nodules also do not have capsules and are oval-shaped concentrations of lymphatic tissue that usually consist of a central, lighter-staining region consisting of large lymphocytes **(*germinal center*)** and a peripheral, darker-staining region of small lymphocytes **(*cortex*).** Most lymphatic nodules are solitary, small, and discrete. Such nodules are found randomly in the lamina propria of mucous membranes of the gastrointestinal tract, respiratory passageways, urinary tract, and reproductive tract. Some lymphatic nodules occur in multiple, large aggregations in specific parts of the body. Among these are the tonsils in the pharyngeal region and aggregated lymphatic follicles (Peyer's patches) in the ileum of the small intestine (Chapter 24). Aggregations of lymphatic nodules also occur in the appendix.

Lymphatic organs of the body—the lymph nodes, spleen, and thymus gland—all contain lymphatic tissue enclosed by a connective tissue capsule. Since bone marrow produces lymphocytes, it may also be considered a component of the lymphatic system.

The lymphatic system has several functions. Lymphatic vessels drain tissue spaces that have protein-containing fluid (interstitial fluid) that escapes from blood capillaries. The proteins, which cannot be directly reabsorbed by blood vessels, are returned to the cardiovascular system by lymphatic vessels. Lymphatic vessels also transport fats from the gastrointestinal tract to the blood. Lymphatic tissue also functions in surveillance and defense; that is, lymphocytes, with the aid of macrophages, protect the body from foreign cells, microbes, and cancer cells. Lymphocytes recognize foreign cells and substances, microbes, and cancer cells and respond to them in two general ways. Some lymphocytes (T cells) destroy them directly or indirectly by releasing various substances. Other lymphocytes (B cells) differentiate into plasma cells that secrete antibodies against foreign substances to help eliminate them. Overall, the lymphatic system concentrates foreign substances in certain lymphatic organs, circulates lymphocytes through the organs to make contact with the foreign substances, and destroys the foreign substances and eliminates them from the body.

The developmental anatomy of the lymphatic system is considered later in the chapter.

LYMPHATIC VESSELS

Lymphatic vessels originate as microscopic vessels in spaces between cells called ***lymph capillaries*** (Figure 22-1a). Lymph capillaries may occur singly or in extensive plexuses. They originate throughout the body, but not in avascular tissue, the central nervous system, splenic pulp, and bone marrow. They are slightly larger and more permeable in one direction only than blood capillaries.

Lymph capillaries also differ from blood capillaries in that they end blindly; blood capillaries have an arterial and a venous end. In addition, lymph capillaries are structurally adapted to ensure the return of proteins to the cardiovascular system when they leak out of blood capillaries. Close examination of lymph capillaries reveals that there are minute openings between endothelial cells making up the capillary wall that permit fluid to flow easily into the capillary but prevent the flow of fluid out of the capillary, much like a one-way valve would operate (Figure 22-1b). Note also that the outer surfaces of the endothelial cells of the capillary wall are attached to the surrounding tissue by structures called ***anchoring filaments.*** During edema, there is an excessive accumulation of fluid in the tissue, causing tissue swelling. This swelling produces a pull on the anchoring filaments, making the openings between cells even larger so that more fluid can flow into the lymph capillary.

FIGURE 22-1 Lymph capillaries. (a) Relationship of lymph capillaries to tissue cells and blood capillaries.

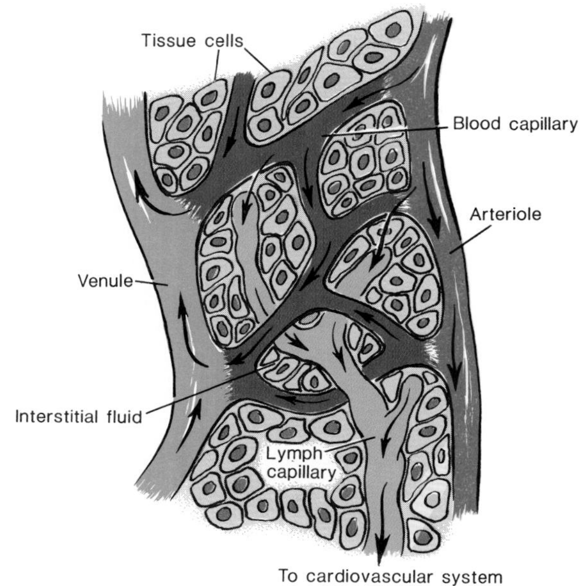

FIGURE 22-1 (*Continued*) (b) Details of a lymph capillary.

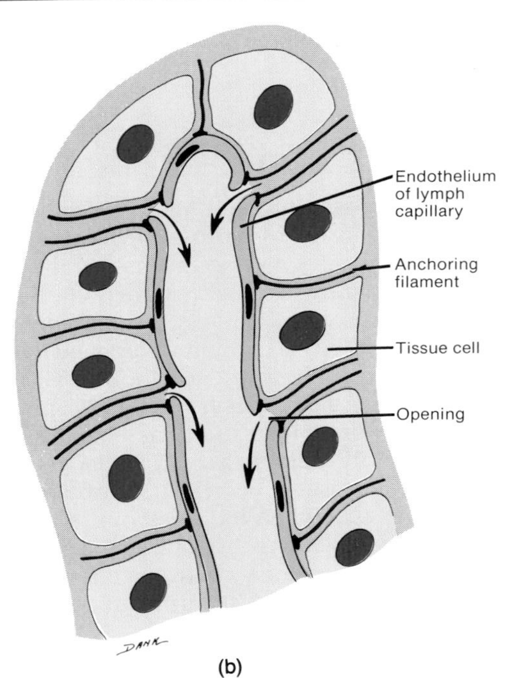

MEDICAL TEST

Lymphangiography (lim-fan'-jē-OG-ra-fē)

Diagnostic Value: To detect and stage lymphomas (lymphatic tissue tumors), to identify cancerous involvement of lymph nodes, to help determine the cause of certain forms of edema, to locate enlarged lymph nodes for surgical or radiotherapeutic treatment, and to evaluate the effectiveness of chemotherapy and radiation therapy in treating cancer.

Procedure: A radiopaque substance is injected into lymphatic vessels, usually in the foot, at a rate of 0.1 to 0.2 ml/minute for about 1½ hours to avoid injuring delicate lymphatic vessels. Fluoroscopy may be used to monitor filling of the lymphatic circulation. X ray films are then taken of the lymphatic system. Such x ray films are called ***lymphangiograms*** (lim-FAN-jē-ō-grams). X rays are taken immediately after injection and 24 hours later. The contrast substance remains in lymph nodes for up to six months so that additional films could be taken during that time to monitor progress.

Just as blood capillaries converge to form venules and veins, lymph capillaries unite to form larger and larger lymph vessels called ***lymphatic vessels*** (Figure 22-2a, b). Lymphatic vessels resemble veins in structure but have thinner walls and more valves and contain lymph nodes at various intervals along their length. Lymphatic vessels of the skin travel in loose subcutaneous tissue and generally follow veins. Lymphatic vessels of the viscera gen-

FIGURE 22-2 Lymphatic system. (a) Schematic representation of the relationship of the lymphatic system to the cardiovascular system.

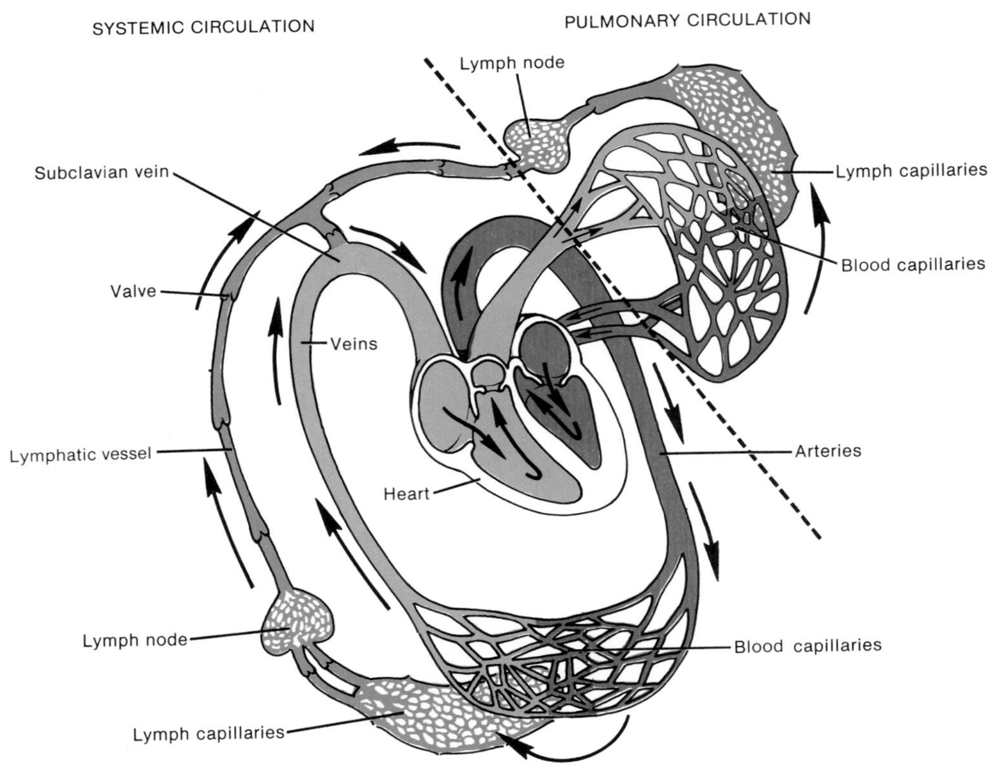

FIGURE 22-2 (*Continued*) (b) Location of the principal components of the lymphatic system. (c) The light gold area indicates those portions of the body drained by the right lymphatic duct. All other areas of the body are drained by the thoracic duct.

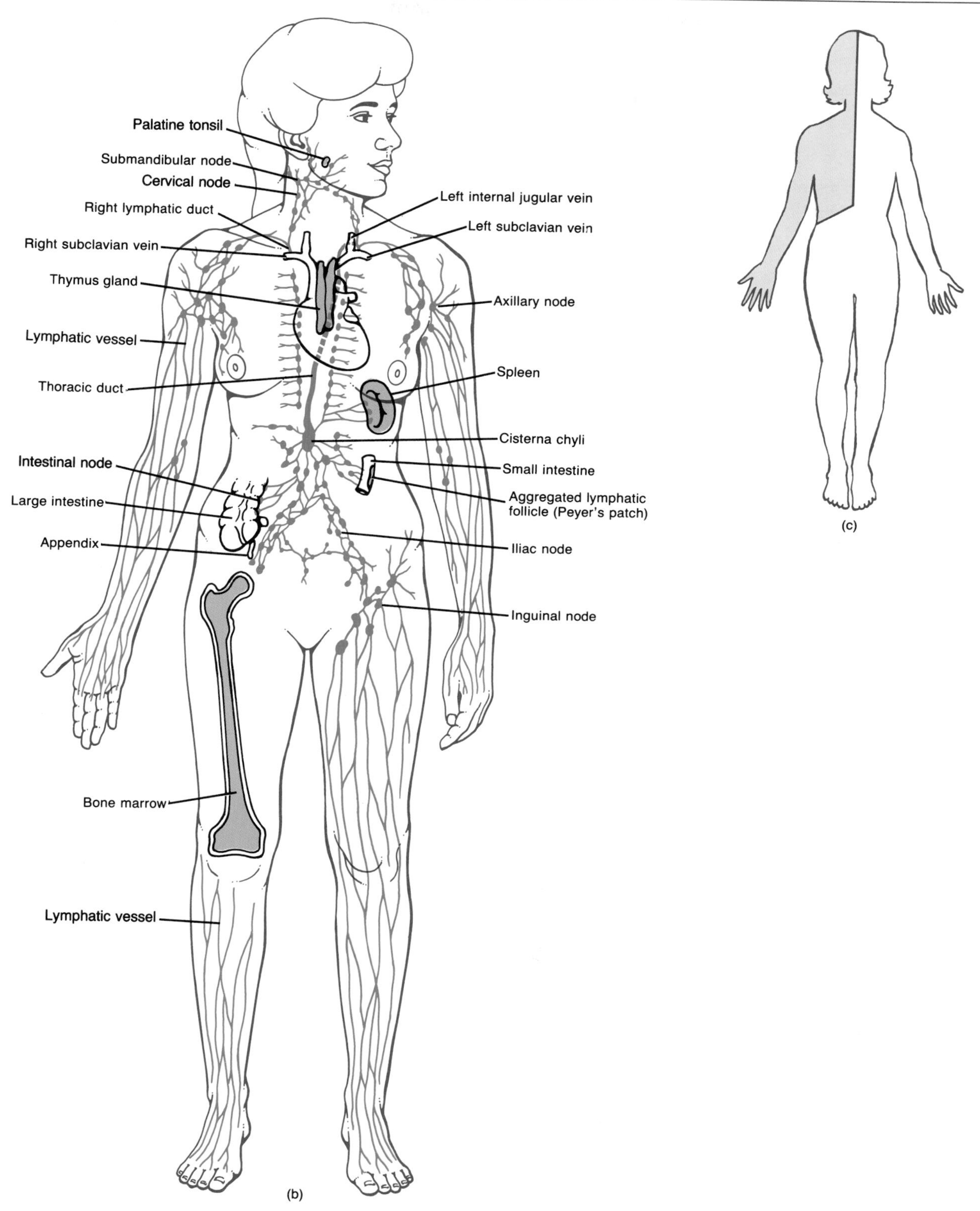

erally follow arteries, forming plexuses around them. Ultimately, lymphatic vessels deliver their lymph into two main channels—the thoracic duct and the right lymphatic duct. These will be described shortly.

LYMPHATIC TISSUE

Lymph Nodes

The oval or bean-shaped structures located along the length of lymphatic vessels are called ***lymph nodes.*** They range from 1 to 25 mm (0.04 to 1 inch) in length. A lymph node contains a slight depression on one side called a ***hilus*** (HĪ-lus), or ***hilum,*** where blood vessels and efferent lymphatic vessels leave the node (Figure 22-3a). Each node is covered by a ***capsule*** of dense connective tissue that extends into the node. The capsular extensions are called ***trabeculae*** (tra-BEK-yoo-lē). Internal to the capsule is a supporting network of reticular fibers and reticular cells (fibroblasts and macrophages). The capsule, trabeculae, and reticular fibers and cells constitute the stroma (framework) of a lymph node. The parenchyma of a lymph node is specialized into two regions: cortex and medulla. The outer ***cortex*** contains densely packed lymphocytes arranged in masses called ***lymphatic nodules.*** The nodules often contain lighter-staining central areas, the ***germinal centers,*** where lymphocytes are produced. The inner region of a lymph node is called the ***medulla.*** In the medulla, the lymphocytes are arranged in strands called ***medullary cords.*** These cords also contain macrophages and plasma cells.

The circulation of lymph through a node involves afferent (to convey toward a center) lymphatic vessels, sinuses in the node, and efferent (to convey away from a center) lymphatic vessels. ***Afferent lymphatic vessels*** enter the convex surface of the node at several points. They contain valves that open toward the node so that the lymph is directed *inward.* Once inside the node, the lymph enters the sinuses, which are a series of irregular channels. Lymph from the afferent lymphatic vessels enters the ***cortical sinuses*** just inside the capsule. From here it circulates to the ***medullary sinuses*** between the medullary cords. From these sinuses the lymph usually circulates into one or two ***efferent lymphatic vessels.*** Efferent lymphatic vessels are wider than the afferent vessels and contain valves that open away from the node to convey lymph *out* of the node.

Lymph nodes are scattered throughout the body, usually in groups (see Figure 22-2b). Typically, these groups are arranged in two sets: ***superficial*** and ***deep.***

Lymph passing from tissue spaces through lymphatic vessels on its way back to the cardiovascular system is filtered through lymph nodes. As lymph passes through the nodes, it is filtered of foreign substances. These substances are trapped by the reticular fibers within the node. Then macrophages destroy the foreign substances by phagocytosis, T cells may destroy them by releasing various products, and/or B cells develop into plasma cells that produce antibodies that destroy them. Lymph nodes also produce lymphocytes, some of which can circulate to other parts of the body.

Histological features of a lymph node are shown in Figure 22-3b.

CLINICAL APPLICATION: METASTASIS THROUGH LYMPHATIC SYSTEM

Knowledge of the location of the lymph nodes and the direction of lymph flow is important in the diagnosis and prognosis of the spread of cancer by *metastasis.* Cancer cells may be spread by way of the lymphatic system and produce aggregates of tumor cells where they lodge. Such secondary tumor sites are predictable by the direction of lymph flow from the organ primarily involved. (Cancer may also spread via the cardiovascular system and local extension.) The hallmark of a lymph node that is cancerous is that it feels enlarged, firm, and nontender. Most lymph nodes that undergo infectious enlargement, by contrast, are not firm and are very tender.

Tonsils

Tonsils are multiple aggregations of large lymphatic nodules embedded in a mucous membrane. The tonsils are arranged in a ring at the junction of the oral cavity and pharynx. The single ***pharyngeal*** (fa-RIN-jē-al) ***tonsil*** or ***adenoid*** is embedded in the posterior wall of the nasopharynx (see Figure 23-2b). The paired ***palatine*** (PAL-a-tīn) ***tonsils*** are situated in the tonsillar fossae between the pharyngopalatine and glossopalatine arches (see Figure 17-2c). These are the ones commonly removed by a tonsillectomy. The paired ***lingual*** (LIN-gwal) ***tonsils*** are located at the base of the tongue and may also have to be removed by a tonsillectomy (see Figure 17-2c).

The tonsils are situated strategically to protect against invasion of foreign substances. Functionally, the tonsils produce lymphocytes and antibodies.

Spleen

The oval ***spleen*** is the largest mass of lymphatic tissue in the body, measuring about 12 cm (5 inches) in length. It is situated in the left hypochondriac region between the fundus of the stomach and diaphragm (see Figure 1-8d). Its ***visceral surface*** contains the contours of the organs adjacent to it—the gastric impression (stomach), renal impression (left kidney), and colic impression (left flexure of colon). The ***diaphragmatic*** (dī-a-fra-MAT-ik)

FIGURE 22-3 Structure of a lymph node. (a) Diagram showing the path taken by circulating lymph. (b) Photomicrograph of a lymph node at a magnification of 25×. (Photomicrograph Copyright © 1983 by Michael H. Ross. Used by permission.)

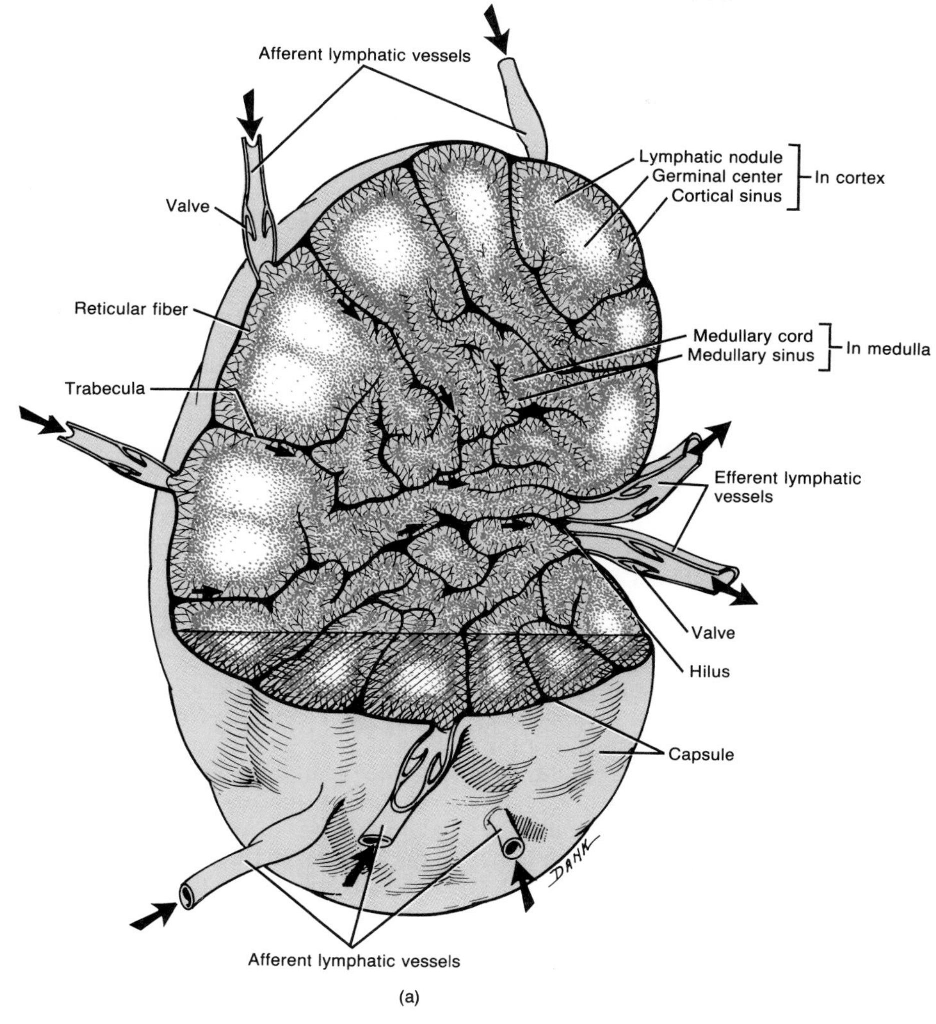

(a)

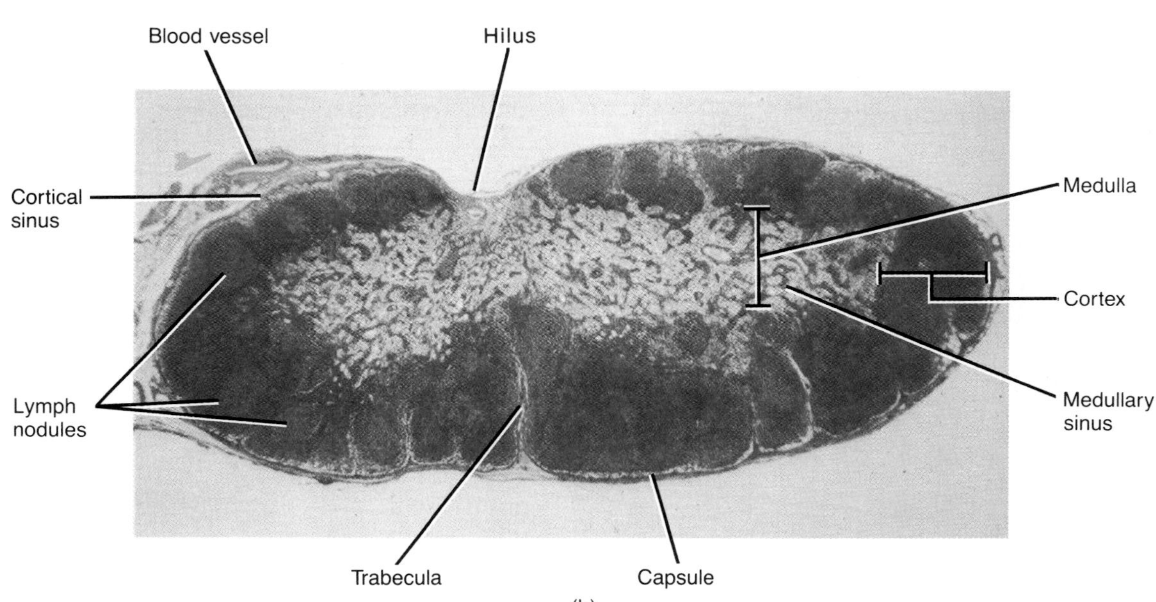

(b)

surface is smooth and convex and conforms to the concave surface of the diaphragm to which it is adjacent.

The spleen is surrounded by a capsule of dense connective tissue and scattered smooth muscle fibers (cells). The capsule, in turn, is covered by a serous membrane, the peritoneum. Like lymph nodes, the spleen contains a hilus, trabeculae, reticular fibers, and reticular cells. The capsule, trabeculae, reticular fibers, and reticular cells constitute the stroma of the spleen.

The parenchyma of the spleen consists of two different kinds of tissue called white pulp and red pulp (Figure 22-4). ***White pulp*** is essentially lymphatic tissue, mostly lymphocytes, arranged around arteries called central arteries. In various areas, the lymphocytes are thickened into lymphatic nodules referred to as ***splenic nodules (Malpighian corpuscles).*** The ***red pulp*** consists of ***venous sinuses*** filled with blood and cords of splenic tissue called ***splenic (Billroth's) cords.*** Veins are closely associated with the red pulp. Splenic cords consist of erythrocytes, macrophages, lymphocytes, plasma cells, and granulocytes.

The splenic artery and vein and the efferent lymphatic vessels pass through the hilus. Since the spleen has no afferent lymphatic vessels or lymph sinuses, it does not filter lymph. One key splenic function related to immunity is the production of B cells, which develop into antibody-producing plasma cells. The spleen also phagocytizes bacteria and worn-out and damaged red blood cells and platelets. In addition, the spleen stores and releases blood in case of demand, such as during hemorrhage. Sympathetic impulses cause the smooth muscle of the capsule of the spleen to contract. During early fetal development, the spleen participates in blood cell formation.

About 10 percent of the population has ***accessory spleens.*** They are most commonly found near the hilus of the primary spleen or embedded in the tail of the pancreas. In general, accessory spleens are about 1 cm (0.5 inch) or less in diameter.

FIGURE 22-4 **Histology of the spleen. Photomicrograph of a portion of the spleen at a magnification of 60×. (Copyright © 1983 by Michael H. Ross. Used by permission.)**

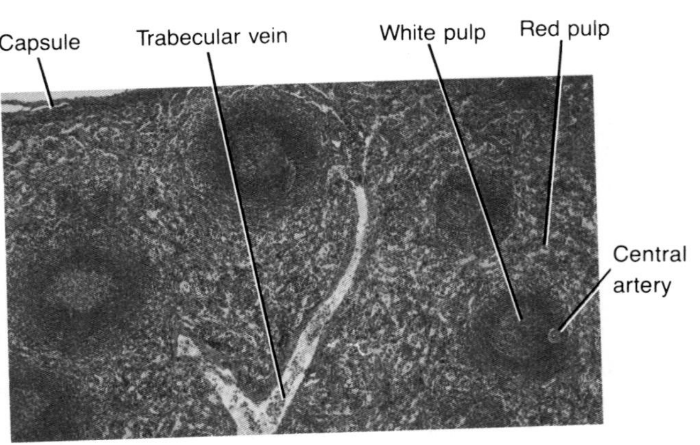

CLINICAL APPLICATION: RUPTURED SPLEEN AND SPLENECTOMY

The spleen is the most frequently damaged organ in cases of abdominal trauma, particularly those involving severe blows over the lower left chest or upper abdomen that fracture the protecting ribs. Such a crushing injury may *rupture the spleen,* which causes severe intraperitoneal hemorrhage and shock. Prompt removal of the spleen, called a *splenectomy,* is needed to prevent the patient from bleeding to death. The functions of the spleen are then assumed by other structures, particularly bone marrow and the liver.

Thymus Gland

Usually a bilobed lymphatic organ, the ***thymus gland*** is located in the superior mediastinum, posterior to the sternum and between the lungs (Figure 22-5a). The two ***thymic lobes*** are held in close proximity by an enveloping layer of connective tissue. Each lobe is enclosed by a connective tissue ***capsule.*** The capsule gives off extensions into the lobes called ***trabeculae,*** which divide the lobes into ***lobules*** (Figure 22-5b). Each lobule consists of a deeply staining peripheral ***cortex*** and a lighter-staining central ***medulla.*** The cortex is composed almost entirely of small, medium, and 'arge tightly packed lymphocytes held in place by reticular tissue fibers. Since the reticular tissue of the thymus gland differs in origin and structure from that usually found in other lymphatic organs, it is referred to as ***epithelioreticular*** supporting tissue. The medulla consists mostly of epithelial cells and more widely scattered lymphocytes, and its reticulum is more cellular than fibrous. In addition, the medulla contains characteristic ***thymic (Hassall's) corpuscles,*** concentric layers of epithelial cells. Their significance is unknown.

The thymus gland is conspicuous in the infant, and it reaches its maximum size of about 40 grams during puberty. After puberty, much of the thymic tissue is replaced by fat and connective tissue. By the time the person reaches maturity, the gland has atrophied but still continues to be functional.

Its role in immunity is to help produce and distribute to other lymphoid organs T cells that destroy invading microbes directly or indirectly by producing various substances. Recall from Chapter 18 that hormones produced by the thymus gland (thymosin, thymic humoral factor, thymic factor, and thymopoietin) promote the proliferation and maturation of T cells. The role of the thymus gland in immunity will be discussed shortly. A summary

FIGURE 22-5 Thymus gland. (a) Diagram showing location of the thymus gland in a young child. (b) Photomicrograph of several lobules at a magnification of 40×. (Photomicrograph courtesy of Andrew Kuntzman.)

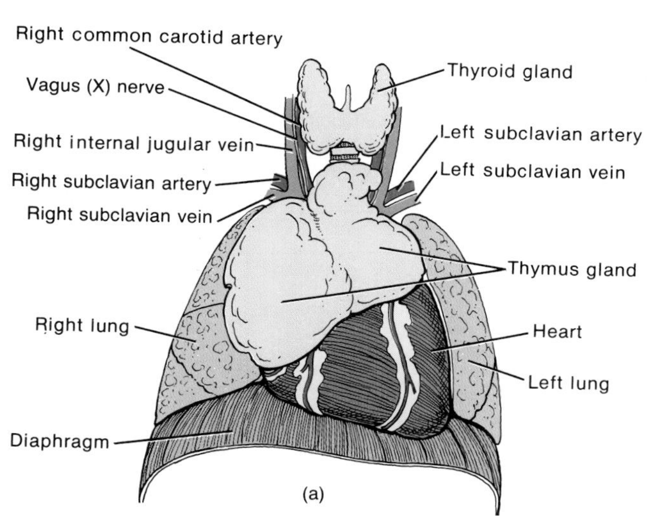

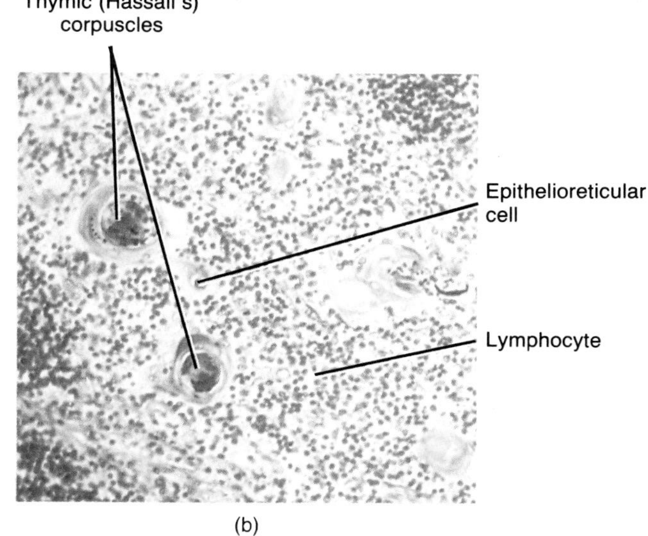

of the functions of the lymphatic system is presented in Exhibit 22-1.

EXHIBIT 22-1 SUMMARY OF LYMPHATIC SYSTEM

Component	Functions
Lymphatic vessels	Drain, from tissue spaces, protein-containing fluid that escapes from blood capillaries.
Lymph nodes	Filter lymph of foreign substances through phagocytosis by macrophages, T cells that destroy directly or indirectly by secreting destructive products, and B cells producing plasma cells that secrete antibodies.
Tonsils	Produce lymphocytes and antibodies.
Spleen	Produces antibody-producing plasma cells, phagocytizes bacteria and worn-out and damaged blood cells; stores and releases blood.
Thymus gland	Produces T cells that destroy microbes directly or indirectly by producing various substances.

LYMPH CIRCULATION

Route

When plasma is filtered by blood capillaries, it passes into the interstitial spaces; it is then known as interstitial fluid. Recall from Chapter 21 that fluid movement between blood capillaries and body cells depends on hydrostatic and osmotic pressures. When this fluid passes from interstitial spaces into lymph capillaries, it is called ***lymph*** (*lympha* = clear water). Lymph from lymph capillaries is then passed to lymphatic vessels that run toward lymph nodes. At the nodes, afferent vessels penetrate the capsules at numerous points, and the lymph passes through the sinuses of the nodes. Efferent vessels from the nodes either run with afferent vessels into another node of the same group or pass on to another group of nodes. From the most proximal group of each chain of nodes, the efferent vessels unite to form ***lymph trunks.*** The principal trunks are the ***lumbar, intestinal, bronchomediastinal, subclavian,*** and ***jugular trunks*** (Figure 22-6).

Thoracic (Left Lymphatic) Duct

The principal trunks pass their lymph into two main channels, the thoracic duct and the right lymphatic duct. The ***thoracic (left lymphatic) duct*** is about 38 to 45 cm (15 to 18 inches) in length and begins as a dilation in front of the second lumbar vertebra called the ***cisterna chyli*** (sis-TER-na KĪ-lē). (See Figure 22-2b also.) The tho-

FIGURE 22-6 **Relationship of lymph trunks to the thoracic duct and right lymphatic duct. Also see Figure 22-2b.**

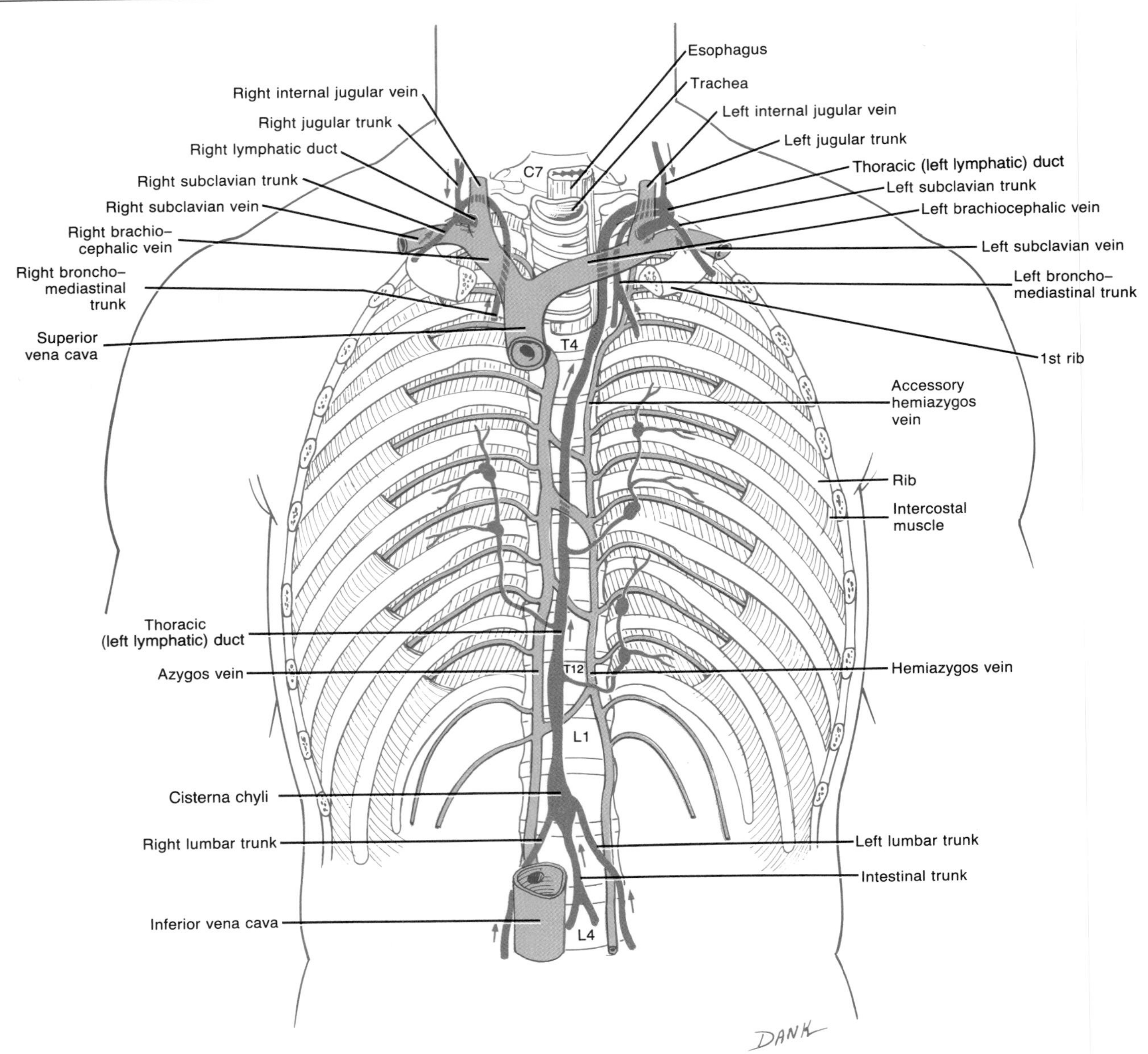

racic duct is the main collecting duct of the lymphatic system and receives lymph from the left side of the head, neck, and chest, the left upper extremity, and the entire body below the ribs.

The cisterna chyli receives lymph from the right and left lumbar trunks and from the intestinal trunk. The lumbar trunks drain lymph from the lower extremities, wall and viscera of the pelvis, kidneys, adrenals (suprarenals), and the deep lymphatics from most of the abdominal wall. The intestinal trunk drains lymph from the stomach, intestines, pancreas, spleen, and visceral surface of the liver. In the neck, the thoracic duct also receives lymph from the left jugular, left subclavian, and left bronchomediastinal trunks. The left jugular trunk drains lymph from the left side of the head and neck; the left subclavian trunk drains lymph from the left upper extremity; and the left bronchomediastinal trunk drains lymph from the left side of the head and neck; and the left bronchomediastinal trunk drains lymph from the left side of the deeper parts of the anterior thoracic wall, upper part of the anterior abdominal wall, anterior part of the diaphragm, left lung, and left side of the heart.

Right Lymphatic Duct

The ***right lymphatic duct*** is about 1.25 cm (0.5 inch) long and drains lymph from the upper right side of the body (see Figure 22-2b). The right lymphatic duct collects lymph from its trunks as follows (Figure 22-6). It receives lymph from the right jugular trunk, which drains the right side of the head and neck, from the right subclavian trunk, which drains the right upper extremity, and from the right bronchomediastinal trunk, which drains the right side of the thorax, right lung, right side of the heart, and part of the convex surface of the liver.

Ultimately, the thoracic duct empties all its lymph into the junction of the left internal jugular vein and left subclavian vein, and the right lymphatic duct empties all its lymph into the junction of the right internal jugular vein and right subclavian vein. Thus, lymph is drained back into the blood, and the cycle repeats itself continuously.

Maintenance

The flow of lymph from tissue spaces to the large lymphatic ducts to the subclavian veins is maintained primarily by the milking action of skeletal muscles. Skeletal muscle contractions compress lymphatic vessels and force lymph toward the subclavian veins. Lymphatic vessels, like veins, contain valves, and the valves ensure the movement of lymph toward the subclavian veins.

Another factor that maintains lymph flow is respiratory movements. These movements create a pressure gradient between the two ends of the lymphatic system. Lymph flows from the abdominal region, where the pressure is higher, toward the thoracic region, where it is lower as each inhalation occurs.

NONSPECIFIC RESISTANCE TO DISEASE

The human body continually attempts to maintain homeostasis by counteracting the activities of disease-producing organisms, called ***pathogens*** (PATH-ō-jens), or their toxins. The ability to ward off disease through our defenses is called ***resistance.*** Vulnerability or lack of resistance is called ***susceptibility.*** Defenses against disease may be grouped into two broad areas: nonspecific resistance and specific resistance. ***Nonspecific resistance*** is inherited and represents a wide variety of body reactions that provide a general response against invasion by a wide range of pathogens. ***Specific resistance,*** or ***immunity,*** involves the production of a specific antibody or specific cells against a specific pathogen or its toxin. Immunity is developed during a person's life; it is not inherited. We will first consider mechanisms of nonspecific resistance that include the skin and mucous membranes, phagocytosis, inflammation, fever, and production of antimicrobial substances (other than antibodies).

Skin and Mucous Membranes

The skin and mucous membranes of the body are the first line of defense against pathogens. They possess certain mechanical and chemical factors that are involved in combating the initial attempt of a microbe or foreign substance to cause disease.

Mechanical Factors

The ***intact skin,*** as noted in Chapter 5, consists of two distinct portions called the epidermis (outer epithelial portion) and dermis (inner connective tissue portion). If you consider the closely packed cells of the epidermis, its many layers, and the presence of keratin, you can see that the intact skin provides a formidable physical barrier to the entrance of microbes. In addition, periodic shedding of epidermal cells helps to remove microbes on the skin. The intact surface of healthy epidermis seems to be rarely penetrated by bacteria. But when the epithelial surface is broken, a subcutaneous infection often develops. The bacteria most likely to cause such an infection are staphylococci, which normally inhabit the hair follicles and sudoriferous (sweat) glands of the skin. Infections of the skin and underlying structures frequently occur as a result of burns, cuts, stab wounds, or other conditions that break the skin. Moreover, when the skin is moist, as in hot, humid climates, dermal infections are quite common, especially fungus infections such as athlete's foot. (The role of the skin in immunity is considered later in the chapter.)

Mucous membranes also consist of an epithelial layer and an underlying connective tissue layer. But unlike the skin, a mucous membrane lines a body cavity that opens to the exterior. Examples include the membranes lining the entire gastrointestinal, respiratory, urinary, and reproductive tracts. The epithelial layer of a mucous membrane secretes ***mucus,*** which prevents the cavities from drying out. Since mucus is slightly viscous, it traps many microbes and other foreign substances. The mucous membrane of the nose has mucus-coated ***hairs*** that trap and filter air containing microbes, dust, and pollutants. The mucous membrane of the upper respiratory tract contains ***cilia,*** microscopic hairlike projections of the epithelial cells (see Figure 23-5). These cilia move in such a manner that they pass inhaled dust and microbes that have become trapped in mucus toward the throat. This so-called ciliary escalator keeps the "mucus blanket" moving toward the throat at a rate of 1 to 3 cm per hour. Coughing and sneezing speed up the escalator. Microbes are additionally prevented from entering the lower respiratory tract by a small lid of cartilage called the ***epiglottis*** that covers the voice box during swallowing (see Figure 23-3b).

Since some pathogens can thrive on the moist secretions of a mucous membrane, the microbes are able to penetrate the membrane if present in sufficient numbers. This penetration may be related to toxic products produced by the microbes, prior injury by viral infections, or mucosal irritations. Although mucous membranes do inhibit the entrance of many microbes, they are less effective than the skin.

There are several other mechanical factors that help to protect epithelial surfaces of the skin and mucous membranes. One that protects the eyes is the ***lacrimal*** (LAK-ri-mal) ***apparatus*** (see Figure 17-3a), a group of structures that manufactures and drains away tears. Normally, tears are carried away by evaporation or pass into the nose as fast as they are produced. Tears are spread over the surface of the eyeball by blinking. The continual washing action of tears helps to keep microbes from settling on the surface of the eye. If an irritating substance or large number of microbes make contact with the eye, the lacrimal glands start to secrete heavily. Tears then accumulate more rapidly than they can be carried away. This is a protective mechanism to dilute and wash away the irritating substance or microbes.

Saliva, produced by the salivary glands, washes microbes from the surfaces of the teeth and the mucous membrane of the mouth, much as tears wash the eyes. This helps to prevent colonization by microbes.

The cleansing of the urethra by the ***flow of urine*** represents another mechanical factor that prevents microbial colonization in the urinary system. Vaginal secretions likewise move microbes out of the female body.

Chemical Factors

Mechanical factors alone do not account for the high degree of resistance of skin and mucous membranes to microbial invasion. Certain chemical factors also play important roles.

Sebaceous (oil) glands of the skin secrete an oily substance called ***sebum*** that prevents hair from drying and becoming brittle and also forms a protective film over the surface of the skin. One of the components of sebum is unsaturated fatty acids, which inhibit the growth of certain pathogenic bacteria and fungi. The low pH of the skin, between pH 3 and 5, is caused in part by the secretion of fatty acids and lactic acid. The skin's acidity probably discourages the growth of many other microorganisms. Certain bacteria commonly found in the skin metabolize sebum, and this metabolism forms fatty acids that cause the inflammatory response associated with acne. A treatment for a very severe type of acne, called cystic acne, is isotretinoin (Accutane), a derivative of vitamin A that prevents sebum formation.

The sudoriferous (sweat) glands of the skin produce ***perspiration,*** which helps to maintain body temperature, eliminate certain wastes, and flush microorganisms from the surface of the skin. Perspiration also contains ***lysozyme,*** an enzyme capable of breaking down cell walls of certain bacteria under certain conditions. Lysozyme is also found in tears, saliva, nasal secretions, and tissue fluids, where it exhibits its antimicrobial activity. ***Hyaluronic acid,*** located in loose connective tissue, assumes a role in preventing the spread of noxious agents in localized infections.

Gastric juice is produced by the glands of the stomach. It is a mixture of hydrochloric acid, enzymes, and mucus. The very high acidity of gastric juice (pH 1.2 to 3.0) is sufficient to preserve the usual sterility of the stomach. This acidity destroys bacteria and most bacterial toxins.

Phagocytosis

When microbes penetrate the skin and mucous membranes, or bypass the antimicrobial substances in blood, there is another nonspecific resistance of the body called phagocytosis. Very simply, ***phagocytosis*** (*phagein* = to eat; *cyto* = cell) means that ingestion and destruction of microbes or any foreign particulate matter by cells called phagocytes, which are certain types of white blood cells and cells derived from them.

Kinds of Phagocytes

The kinds of phagocytes that participate in phagocytosis fall into two broad categories: granulocytes (microphages) and macrophages. The ***granulocytes*** of blood vary with regard to the degree to which they function as phagocytes. Neutrophils have the most prominent phagocytic activity. Eosinophils are believed to have some phagocytic capability, and the role of basophils in phagocytosis is debatable.

When an infection occurs, both granulocytes (especially neutrophils) and monocytes migrate to the infected area. During this migration, the monocytes enlarge and develop into actively phagocytic cells called ***macrophages*** (MAK-rō-fā-jez). Since these cells leave the blood and migrate to infected areas, they are called ***wandering macrophages.*** Some of the macrophages, called ***fixed macrophages,*** remain in certain tissues and organs of the body. Fixed macrophages are found in the skin and subcutaneous layer (histiocytes), liver (stellate reticuloendothelial cells), lungs (alveolar macrophages), brain (microglia), spleen, lymph nodes, and bone marrow (tissue macrophages). The combination of wandering and fixed macrophages is referred to as the ***tissue macrophage (mononuclear phagocytic) system,*** formerly called the reticuloendothelial system.

Mechanism

For convenience of study, phagocytosis will be divided into four phases: chemotaxis, adherence, ingestion, and digestion.

▪ Chemotaxis ***Chemotaxis*** (kē'mō-TAK-sis; *chemo* = chemicals; *taxis* = arrangement) is the chemical attraction of phagocytes to microorganisms. Among the chemotactic chemicals that attract phagocytes are microbial products, components of white blood cells and damaged tissue cells, and chemicals derived from complement.

▪ **Adherence** ***Adherence*** is the *attachment* of the cell membrane of a phagocyte to the surface of a microorganism or other foreign material (Figure 22-7a). In some instances, adherence occurs easily, and the microorganism is readily phagocytized (see Figure 3-6a, b). In other cases, adherence is more difficult, but the particle can be phagocytized if the phagocyte traps the particle to be ingested against a rough surface, like a blood vessel, blood clot, or connective tissue fibers, where it cannot slide away. This is sometimes called ***nonimmune (surface) phagocytosis.***

Microorganisms can be more readily phagocytized if they are first coated with certain plasma proteins (complement) that promote the attachment of the microorganism to the phagocyte. This coating process is ***opsonization.***

▪ **Ingestion** Following adherence, ingestion occurs. During the process of ***ingestion,*** the cell membrane of the phagocyte extends projections, called pseudopods, that engulf the microorganism (Figure 22-7). Once the microorganism is surrounded, the pseudopods meet and fuse, surrounding the microorganism with a sac called a ***phagocytic vesicle.***

FIGURE 22-7 Phagocytosis. (a) Diagram showing phases. (b) Scanning electron micrograph of two macrophages in the human lung; the lower macrophage is ingesting a foreign body. (Courtesy of Dr. A. Brody, Science Photo Library, Photo Researchers, Inc.)

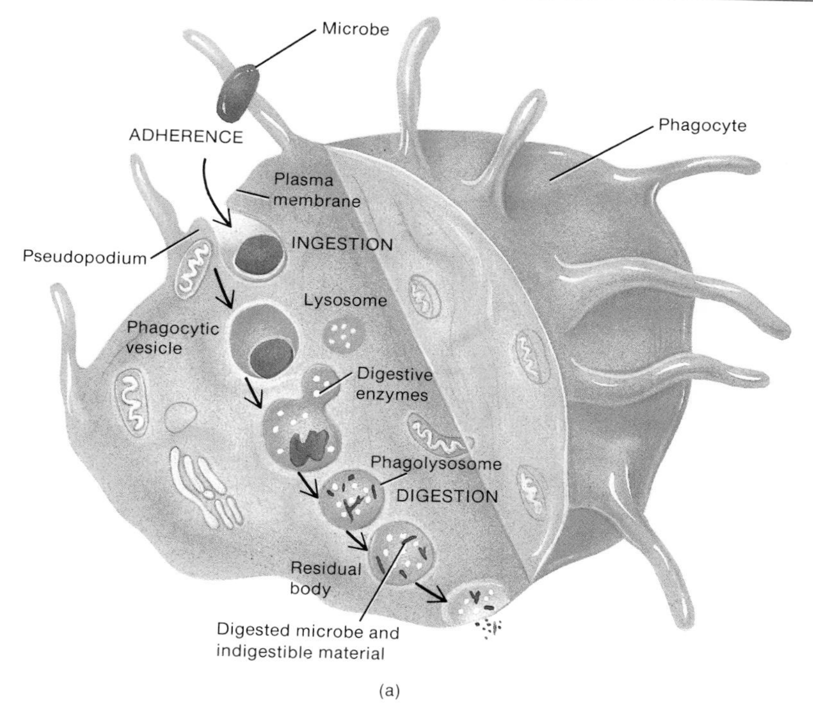

(a)

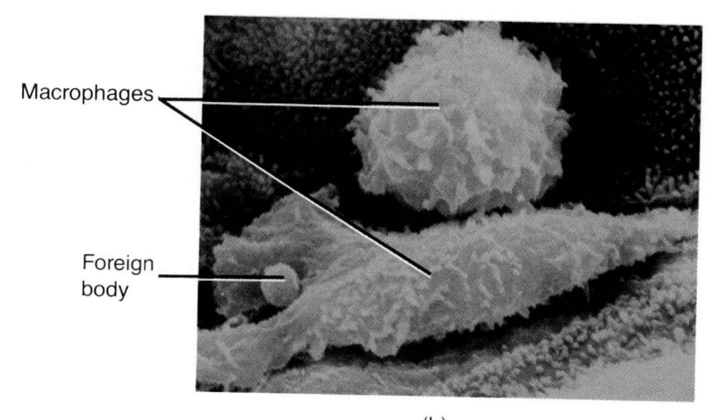

(b)

▪ **Digestion** In this phase of phagocytosis, the phagocytic vesicle pinches off from the membrane and enters the cytoplasm (Figure 22-7a). Within the cytoplasm, it contacts lysosomes that contain digestive enzymes and bactericidal substances. Upon contact, the phagocytic vesicle and lysosome membranes fuse to form a single, larger structure called a ***phagolysosome,*** or ***digestive vesicle.*** The contents of the phagolysosome kill many types of bacteria in only 10 to 30 minutes. It is assumed that microbial destruction occurs because of the contents of the lysosomes: presence of lactic acid (which lowers the pH in the phagolysosome), the production of hydrogen peroxide, presence of lysozyme, and the destructive capabilities of enzymes that degrade carbohydrates, proteins, lipids, and nucleic acids. Any indigestible materials that cannot be degraded further are found in structures called ***residual bodies.*** The cell disposes of materials in residual bodies by exocytosis, a process in which the residual body migrates to the plasma membrane, fuses to it, ruptures, and releases its contents.

Some microbes, such as toxin-producing staphylococci, may be ingested but are not necessarily killed. In fact, their toxins can actually kill the phagocytes. Other microbes, such as the tubercle bacillus, may multiply within the phagolysosome and eventually destroy the phagocyte. Still other microbes, such as the causative agents of tularemia and brucellosis, may remain dormant in phagocytes for months or years at a time.

Inflammation

When cells are damaged by microbes, physical agents, or chemical agents, the injury sets off an inflammation (inflammatory response). The injury may be viewed as a form of stress.

Symptoms

Inflammation is a defensive response of the body to stress due to tissue damage that is usually characterized by four fundamental symptoms: ***redness, pain, heat,*** and ***swelling.*** A fifth symptom can be the ***loss of function*** in the injured area. Whether loss of function occurs depends on the site and extent of the injury. The inflammatory response serves a protective and defensive role. It is an attempt to dispose of microbes, toxins, or foreign material at the site of injury, to prevent their spread to other organs, and to prepare the site for tissue repair. Thus, the inflammatory response is an attempt to restore tissue homeostasis.

Stages

The inflammatory response is one of the body's nonspecific internal systems of defense. The response of a tissue to a rusty nail wound is similar to other inflammatory responses, such as those that result from burns, radiation, or bacterial or viral invasion. These are the basic stages of the inflammatory response.

▪ **Vasodilation and Increased Permeability of Blood Vessels** Immediately following tissue damage, there is vasodilation and increased permeability of blood vessels in the area of the injury. ***Vasodilation*** is an increase in diameter in the blood vessels. ***Increased permeability*** means that substances normally retained in blood are permitted to pass from the blood vessels. Vasodilation allows more blood to go to the damaged area, and increased permeability permits defensive substances in the blood to enter the injured area. Such defensive substances include antibodies, phagocytes, and clot-forming chemicals. The increased blood supply also removes toxic products and dead cells, preventing them from complicating the injury. These toxic substances include waste products released by invading microorganisms.

Among these substances that contribute to vasodilation, increased permeability, and other aspects of the inflammatory response are the following (Figure 22-8):

Histamine. This substance is found in many body cells, especially mast cells in connective tissue, basophils, and blood platelets. Histamine is released in response to any injured cells that contain it. Phagocytes (neutrophils and macrophages) attracted to the site of injury also stimulate release of histamine. Histamine causes vasodilation and increased permeability.

Kinins. These polypeptides are present in blood and induce vasodilation and increased permeability. They also serve as chemotactic agents for phagocytes.

Prostaglandins (PGs). These substances, especially of the E series, are released by damaged cells and intensify the effects of histamine and kinins. PGs also may stimulate the migration of phagocytes through capillary walls.

Leukotrienes. These substances are produced by basophils and mast cells. They cause increased permeability and also function in adherence and as chemotactic agents for phagocytes.

Complement. As will be described shortly, complement is a group of plasma proteins. These proteins stimulate histamine release by mast cells, basophils, and platelets; attract neutrophils by chemotaxis; and promote phagocytosis. They can also destroy bacteria.

Within minutes after an injury, dilation of arterioles and increased permeability of capillaries produce heat, redness, and swelling. The heat results from the large amount of warm blood that accumulates in the area and, to a certain extent, from the heat energy produced by the metabolic reactions. The large amounts of blood in the area are also responsible for the redness (erythema).

Pain, whether immediate or delayed, is a cardinal symptom of inflammation. It can result from an injury to nerve fibers or from an irritation caused by the release of toxic chemicals from microorganisms. Kinins affect some nerve endings, causing much of the pain associated with in-

FIGURE 22-8 Inflammatory response. Histamine kinins, prostaglandins, leukotrienes, and complement together stimulate vasodilation, increased permeability of blood vessels, chemotaxis, diapedesis, and phagocytosis.

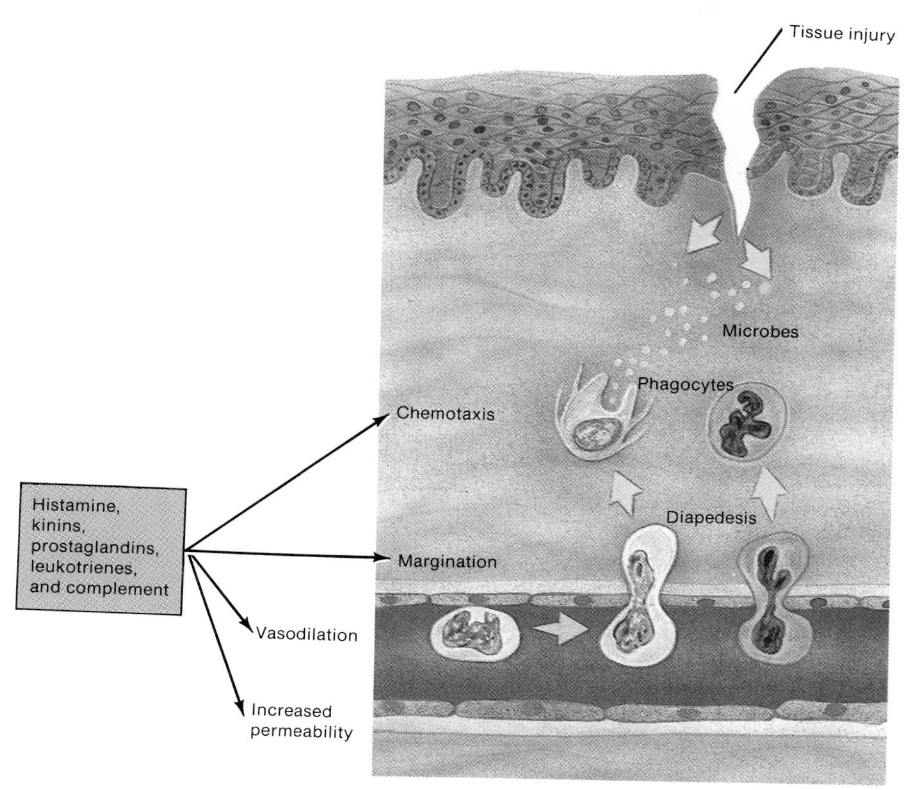

flammation. Prostaglandins intensify and prolong the pain associated with inflammation. Pain may also be due to increased pressure from edema.

▪ **Fibrin Formation** Blood contains a soluble protein called ***fibrinogen*** (fī-BRIN-ō-jen). The increased permeability of capillaries causes leakage of fibrinogen into tissues. Fibrinogen is then converted to an insoluble, thick network called ***fibrin*** (FĪ-brin), which localizes and traps the invading organisms, preventing their spread. This network eventually forms a clot that isolates the invading microbes or their toxins.

▪ **Phagocyte Migration** Generally, within an hour after the inflammatory process is initiated, phagocytes appear on the scene. As the flow of blood decreases, neutrophils begin to stick to the inner surface of the endothelium (lining) of blood vessels. This is called ***margination*** (Figure 22-8). Then the neutrophils begin to squeeze through the wall of the blood vessel to reach the damaged area. This migration, which resembles amoeboid movement, is called ***diapedesis*** (dī'-a-pe-DĒ-sis). It can take as little as two minutes. The movement of neutrophils depends on chemotaxis. Neutrophils are attracted by microbes, kinins, complement, and other neutrophils. A steady stream of neutrophils is ensured by the production and release of additional cells from bone marrow. This is brought about by a substance called ***leucocytosis-promoting factor,*** which is released from inflamed tissues. Neutrophils attempt to destroy the invading microbes by phagocytosis. Neutrophils also contain chemicals with antibiotic activity called ***defensins,*** so named because of their apparent role in preventing and overcoming infections. Defensins are active against bacteria, fungi, and viruses, unlike other antibiotics that are directed against specific microbes.

As the inflammatory response continues, monocytes follow the neutrophils into the infected area. Once in the tissue, monocytes become transformed into wandering macrophages that augment the phagocytic activity of fixed macrophages. In addition, fixed macrophages mobilize under the stimulus of inflammation and, as wandering macrophages, also migrate to the infected area. The neutrophils predominate in the early stages of infection but tend to die off rapidly. Macrophages enter the picture during a later stage of the infection, once neutrophils have accomplished their function. Macrophages are several times more phagocytic than neutrophils and large enough to engulf tissue that has been destroyed, neutrophils that have been destroyed, and invading microbes.

▪ **Pus Formation** In all but very mild inflammations, ***pyogenesis*** (*pyo* = pus; *genesis* = to produce) occurs. ***Pus*** is a thick fluid that contains living, as well as nonliving, white blood cells and debris from other dead tissue. Pus formation usually continues until the infection subsides. At times, the pus pushes to the surface of the body or into an internal cavity for dispersal. On other occasions, the pus remains even after the infection is terminated. In this case, the pus is gradually destroyed over a period of days and is absorbed by the body.

CLINICAL APPLICATION: ABSCESS AND ULCERS

If the pus cannot drain out of the body, an abscess develops. An ***abscess*** is simply an excessive accumulation of pus in a confined space. Common examples are pimples and boils. When inflamed tissue is shed many times, it produces an open sore, called an ***ulcer,*** on the surface of an organ or tissue. Ulcers may result from a prolonged inflammatory response to a continuously injured tissue. For instance, overproduction of hydrochloric acid in the stomach may cause a steady erosion of the epithelial tissue lining the stomach. People with poor circulation are susceptible to ulcers in the tissues of their legs. The ulcers, called stasis ulcers, develop because of poor oxygen and nutrient supply to tissues that then become very susceptible to even a very mild injury or infectious process.

Fever

The most frequent cause of ***fever,*** an abnormally high body temperature, is infection from bacteria (and their toxins) and viruses. The high body temperature inhibits some microbial growth and speeds up body reactions that aid repair. (Fever is discussed in more detail in Chapter 25.)

Antimicrobial Substances

In addition to the mechanical and chemical barriers of the skin and mucous membranes, the body also produces certain antimicrobial substances. Among these are interferon, complement, and properdin.

Interferon (IFN)

Body cells infected with viruses produce a protein called **interferon** (in'-ter-FĒR-on), or ***IFN.*** Actually, there are three principal types of interferon called alpha, beta, and gamma. In humans, IFN has been shown to be produced by lymphocytes and other leucocytes and fibroblasts, and each can have a slightly different effect on the body. Once produced by and released from virus-infected cells, IFN diffuses to uninfected neighboring cells and binds to surface receptors. This somehow induces uninfected cells to synthesize antiviral proteins (AVPs) that inhibit viral replication within body cells. Viruses can cause disease only if they can replicate within cells. Interferon appears to be the body's first line of defense against infection by many different viruses.

The importance of interferon in protecting the body against viruses, as well as its potential as an anticancer agent, has made its production in large quantities a top public health priority (discussed later in the chapter).

A form of alpha interferon, called alpha-2 interferon, has been incorporated into a nasal spray that may prevent the spread of colds among family members.

Complement

Another antimicrobial substance that is very important to nonspecific resistance (and immunity) is complement. ***Complement*** is actually a group of at least 20 proteins found in normal blood serum. The system is called complement because it "complements" certain immune reactions involving antibodies. These will be discussed shortly. The function of an antibody is to recognize the microbe as a foreign organism, form an antigen–antibody complex, and activate complement for attack. The antigen–antibody complex also fixes (attaches) the complement to the surface of the invading microbe. Once complement is activated, it attacks and destroys microbes as follows:

1. Some complement proteins initiate a series of reactions that form holes in the plasma membrane of the microbe. This causes the contents of the microbes to leak out, a process called ***cytolysis.***

2. Some complement proteins contribute to the development of ***inflammation*** by causing the release of ***histamine*** from mast cells, basophils, and platelets. Histamine increases the permeability of blood capillaries, a process that enables leucocytes to penetrate tissues in order to combat infection or allergy. Other complement proteins serve as ***chemotactic agents*** that attract phagocytes to the site of inflammation and increase blood flow to the area by causing vasodilation of arterioles.

3. Some complement proteins bind to the surface of a microbe and interact with receptors on phagocytes, promoting phagocytosis. This is called ***opsonization*** (op-sō-ni-ZĀ-shun), or ***immune adherence.***

Properdin

Properdin (prō-PER-din; *pro* = in behalf of; *perdere* = to destroy) is a complex of three proteins and, like complement, is also found in serum. Properdin, acting together with complement, leads to the destruction of several types of bacteria by cytolysis, the enhancement of phagocytosis, and triggering of inflammatory responses.

A summary of some of the various components of nonspecific resistance is presented in Exhibit 22-2.

IMMUNITY (SPECIFIC RESISTANCE TO DISEASE)

Despite the variety of mechanisms of nonspecific resistance, they all have one thing in common. They are designed to protect the body from any kind of pathogen or other foreign substance. They are not specifically directed against a particular microbe. Specific resistance to disease, called ***immunity,*** involves the production of a specific type of cell or specific molecule (antibody) to destroy a particular antigen. If antigen 1 invades the body, antibody 1 is produced against it. If antigen 2 invades the body, antibody 2 is produced against it, and so on. The branch of science that deals with the responses of the body when challenged by antigens is referred to as ***immunology*** (im'-yoo-NOL-ō-jē; *immunis* = free). Immunology is a rapidly advancing science as a result of the explosion of research in the field. Every attempt has been made to present the very latest information.

We will examine the components of specific resistance by first examining the nature of antigens and antibodies.

Antigens (Ags)

Definition

An ***antigen (Ag),*** or ***immunogen,*** is any chemical substance that, when introduced into the body, causes the body to produce specific antibodies and/or specific cells called T cells, which can react with the antigen. Antigens thus have two important characteristics. The first is ***immunogenicity*** (im'-yoo-nō-jen-IS-it-ē), the ability to stimulate the formation of specific antibodies. The second is ***reactivity,*** the ability of the antigen to react specifically with the produced antibodies. An antigen with both of these characteristics is called a ***complete antigen.***

Characteristics

Chemically, the vast majority of antigens are proteins, nucleoproteins (nucleic acid + protein), lipoproteins

EXHIBIT 22-2 SUMMARY OF NONSPECIFIC RESISTANCE

Component	Functions
Skin and mucous membranes	
Mechanical factors	
Intact skin	Forms a physical barrier to the entrance of microbes.
Mucous membranes	Inhibit the entrance of many microbes, but not as effective as intact skin.
Mucus	Traps microbes in respiratory and gastrointestinal tracts.
Hairs	Filter microbes and dust in nose.
Cilia	Together with mucus, trap and remove microbes and dust from upper respiratory tract.
Epiglottis	Prevents microbes and dust from entering trachea.
Lacrimal apparatus	Tears dilute and wash away irritating substances and microbes.
Saliva	Washes microbes from surfaces of teeth and mucous membranes of mouth.
Urine	Washes microbes from urethra.
Chemical factors	
Acid pH of skin	Discourages growth of many microbes.
Unsaturated fatty acids	Antibacterial substance in sebum.
Lysozyme	Antimicrobial substance in perspiration, tears, saliva, nasal secretions, and tissue fluids.
Hyaluronic acid	Prevents spread of noxious agents in localized infection.
Gastric juice	Destroys bacteria and most toxins in stomach.
Phagocytosis	Ingestion and destruction of foreign particulate matter by granulocytes and macrophages.
Inflammation	Confines and destroys microbes and repairs tissues.
Fever	Inhibits microbial growth and speeds up body reactions that aid repair.
Antimicrobial substances	
Interferon (IFN)	Protects uninfected host cells from viral infection.
Complement	Causes cytolysis of microbes, promotes phagocytosis, and contributes to inflammation.
Properdin	Works with complement to bring about same responses as complement.

(lipid + protein), glycoproteins (carbohydrate + protein), and certain large polysaccharides. In general, they have molecular weights of 10,000 or greater.

The entire microbe, such as a bacterium or virus, or components of microbes may act as an antigen. For example, bacterial structures such as flagella, capsules, and cell walls are also antigenic; bacterial toxins are highly antigenic. Nonmicrobial examples of antigens include pollen, egg white, incompatible blood cells, and transplanted tissues and organs. The myriad of antigens in the environment provides many opportunities for the production of antibodies by the body.

Antibodies do not form against the whole antigen. At specific regions on the surface of the antigen, called ***antigenic determinant sites*** (Figure 22-9), specific chemical groups of the antigen combine with the antibody. This combination depends upon the size and shape of the determinant site and the manner in which it corresponds to the chemical structure of the antibody. The combination is very much like the lock-and-key analogy used to describe the combination of enzyme and substrate molecule.

The number of antigenic determinant sites on the surface of an antigen is called ***valence.*** Most antigens are ***multivalent;*** that is, they have several antigenic determinant sites. It is believed that in order to induce antibody formation, an antigen must have at least two determinant sites ***(bivalent).*** If an antigen is chemically broken down, it is possible to separate the determinant site from the rest of the antigen. Its molecular weight, compared with that of the antigen as a whole, is small. It might be only 200 to 1000. An isolated determinant site still has the ability to react with an antibody in response to the original antigen (reactivity) but does not have the ability to stimulate the production of antibodies (immunogenicity) when injected into an animal. Its lower molecular weight prevents it from serving as a complete antigen. A determinant site that has reactivity but not immunogenicity is called a ***partial antigen*** or ***hapten*** (*haptein* = to grasp). A hapten can stimulate an immune response if it is attached to a larger carrier molecule so that the combined molecule has two determinant sites. Some drugs of low molecular weight, like penicillin, may combine with proteins of high molecular weight in the body. This combination forms two determinant sites, and the complex becomes antigenic. Antibodies formed against the complex are responsible for the allergic reactions to drugs and other chemicals.

FIGURE 22-9 Relationship of an antigen to an antibody. Since most antigens contain more than one antigenic determinant site, they are referred to as multivalent. Most human antibodies are bivalent; they have two reaction sites that are complementary to the antigenic determinant sites for which they are specific. Each antibody has reaction sites for specific antigenic determinant sites only.

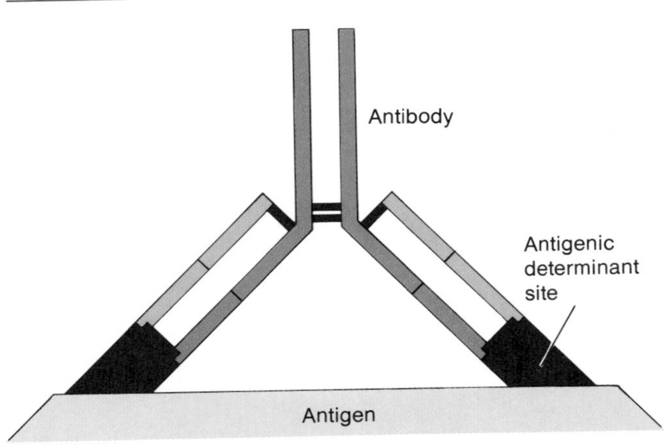

As a rule, antigens are foreign substances. They are not usually part of the chemistry of the body. The body's own substances, recognized as "self," do not act as antigens; substances identified as "nonself," however, stimulate antibody production. However, there are certain conditions in which the distinction between self and nonself breaks down, and antibodies that attack the body are produced. These conditions (to be described later) result in autoimmune disease.

Now let us look at some of the characteristics of antibodies.

Antibodies (Abs)

Definition

An ***antibody (Ab)*** is a protein produced by the body in response to the presence of an antigen and is capable of combining specifically with the antigen. This is essentially the complementary definition of an antigen.

The specific fit of the antibody with antigen depends not only on the size and shape of the antigenic determinant site but also on the corresponding antibody site, much like the lock-and-key analogy (Figure 22-9). An antibody, like an antigen, also has a valence. Whereas most antigens are multivalent, antibodies are bivalent or multivalent. The majority of human antibodies are bivalent.

Antibodies belong to a group of proteins called globulins, and for this reason they are also known as ***immunoglobulins*** (im'-yoo-nō-GLOB-yoo-lins), or ***Ig.*** Five different classes of immunoglobulins are known to exist in humans. These are designated as IgG, IgA, IgM, IgD, and IgE. Each has a distinct chemical structure and a specific biological role. IgG antibodies are the most abundant antibodies. They are found in blood, lymph, and the intestines. Functionally, they protect against bacteria and viruses by enhancing phagocytosis, neutralizing toxins, and triggering the complement system. IgA antibodies are found in tears, saliva, mucus, milk, gastrointestinal secretions, blood, and lymph and provide localized protection on mucous membranes. IgM antibodies are the first antibodies to appear after an initial exposure to any antigen. They are found in blood, in lymph, and on the surfaces of B cells and are especially effective against microbes by causing agglutination and lysis. They are the predominant antibodies involved in ABO blood groups. IgD antibodies are found in blood, in lymph, and on the surfaces of B cells and may be involved in stimulating antibody-producing cells to manufacture antibodies. IgE antibodies are located on mast and basophil cells and are involved in allergic reactions. It is interesting to note that during stress IgA levels decrease. Depending on an individual's personality, this decrease can lower resistance to infection.

Structure

Since antibodies are proteins, they consist of polypeptide chains. Most antibodies contain two pairs of polypeptide chains (Figure 22-10). Two of the chains are identical to each other and are called ***heavy (H) chains.*** Each consists of more than 400 amino acids. The other two chains, also identical to each other, are called ***light (L) chains,*** and each consists of 200 amino acids. The antibody consists of identical halves held together by disulfide bonds (S—S). Each half consists of a heavy and a light chain, also held together by a disulfide bond (S—S). Overall, the antibody molecule sometimes assumes the shape of a letter *Y*. At other times, it resembles a letter *T*. The reasons for these differences in shape will be explained shortly.

Within each H and L chain, there are two distinct regions. The tops of the H and L chains, called the ***variable portions,*** contain the antigen-binding site. The variable portion is different for each kind of antibody, and it is this portion that allows the antibody to recognize and attach specifically to a particular type of antigen. Since most antibodies have two variable portions for attachment of antigens, they are bivalent. The remainder of each polypeptide chain is called the ***constant portion.*** The constant portion is the same in all antibodies of the same class and is responsible for the type of antigen–antibody reaction that occurs. However, the constant portion differs from one class of antibody to another, and its structure serves as a basis for distinguishing the classes.

In performing its role in immunity, the antibody is believed to behave as a switch. When antibodies are viewed in the electron microscope before combination with antigen, they resemble the letter **T**. But after combination with antigen, they appear to be smaller and resemble the letter **Y**. It is possible that the binding of antigen causes a rearrangement of the structure of antibody. If you examine Figure 22-10 again, you will note that there is a hinge area along the H chains; this hinge provides the antibody molecule with flexibility. When antigen and antibody are combined, the rearrangement of antibody might consist of a pivoting movement in which the H chain moves from a **T** shape into a **Y** shape. This pivoting could cause exposure of the constant portion, which then functions in a particular antigen–antibody reaction, such as fixing complement.

Cellular and Humoral Immunity

The ability of the body to defend itself against invading agents such as bacteria, toxins, viruses, and foreign tissues consists of two closely allied immune responses. One response consists of the formation of specially sensitized lymphocytes that have the capacity to attach to the foreign agent and destroy it. This is called ***cellular (cell-mediated) immunity*** and is particularly effective against

FIGURE 22-10 General structure of the portions of an antibody molecule. (a) Diagram. (b) Computer-generated model of an IgG antibody (Courtesy of Science VU-NIH, R. Feldman/Visuals Unlimited.)

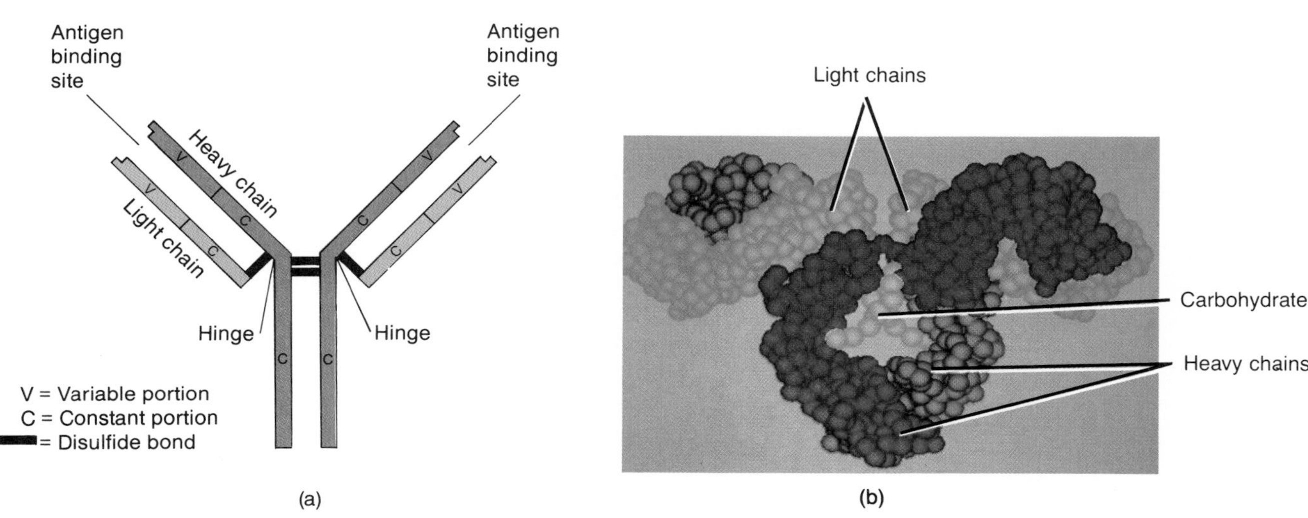

fungi, parasites, intracellular viral infections, cancer cells, and foreign tissue transplants. In the other response, the body produces circulating antibodies that are capable of attacking an invading agent. This is called ***humoral (antibody-mediated) immunity*** and is particularly effective against bacterial and viral infections.

Cellular immunity and humoral immunity are the product of the body's lymphoid tissue. The bulk of lymphoid tissue is located in the lymph nodes, but it is also found in the spleen and gastrointestinal tract and, to a lesser extent, in bone marrow. The placement of lymphoid tissue is strategically designed to intercept an invading agent before it can spread too extensively into the general circulation.

Formation of T Cells and B Cells

Lymphoid tissue consists primarily of lymphocytes that may be distinguished into two kinds. ***T cells*** are responsible for cellular immunity. ***B cells*** develop into specialized cells (plasma cells) that produce antibodies and provide humoral immunity.

Both types of lymphocytes are derived originally in the embryo from lymphocytic stem cells in bone marrow (Figure 22-11). Before migrating to their positions in lymphoid tissues, the descendants of the stem cells follow two distinct pathways. About half of them first migrate to the thymus gland, where they are processed to become T cells. The name T cell is derived from the processing that occurs in the thymus gland. The thymus gland in some way confers what is called ***immunologic competence*** on the T cells. This means that they develop the ability to differentiate into cells that perform specific immune reactions. They then leave the thymus gland and become embedded in lymphoid tissue. Immunologic competence is conferred by the thymus gland shortly before birth and for a few months after birth. Removal of the thymus gland from an animal before it has processed the T cells results in a failure to develop cellular immunity. Even after T cells leave the thymus gland to take up residence in other tissues, thymic hormones continue to stimulate them and their descendants.

The remaining stem cells are processed in some unknown area of the body, possibly bone marrow, fetal liver and spleen, or gut-associated lymphoid tissue. They become the B cells. They are given this name because in birds they are processed in the bursa of Fabricius, a small pouch of lymphoid tissue attached to the intestine. Once processed, B cells migrate to lymphoid tissue and take up their positions there. Although both T cells and B cells occupy the same lymphoid tissues, they localize in separate areas of the tissues.

FIGURE 22-11 Origin and differentiation of T cells and B cells.

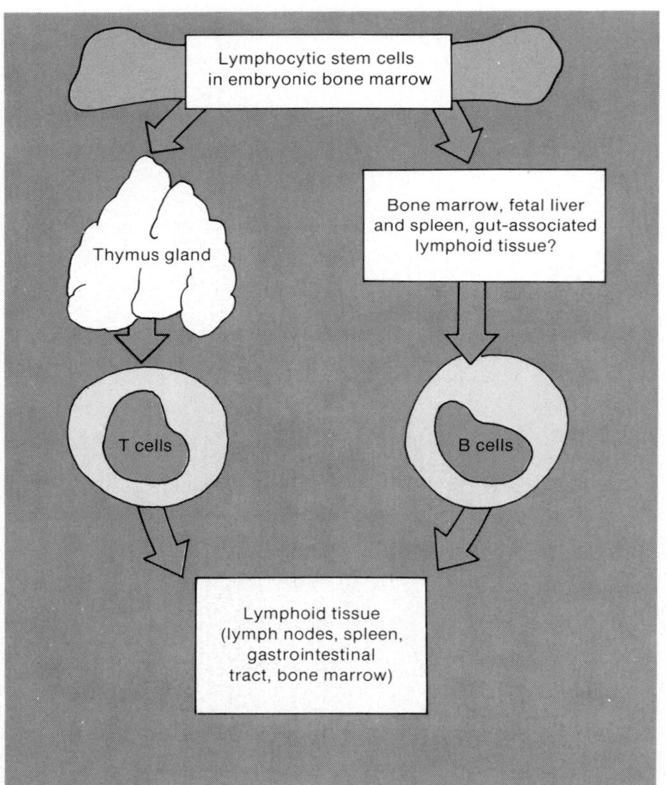

T Cells and Cellular Immunity

Unlike the various nonspecific resistance factors previously discussed, the responses of T cells very much depend on the ability of the cells to recognize specific antigens. This specific recognition is based in part on the specific fit between antigenic determinant sites of antigens and variable portions of antibodies. However, for T cells to perform their vital functions in immunity, another cell must participate. This cell, which has already been discussed in previous chapters, is called a macrophage, a phagocyte derived from a monocyte. Macrophages serve as antigen-processing cells in that they process and present antigens to T cells (Figure 22-12). The processing occurs when the macrophages phagocytize the antigens; as part of phagocytosis, the antigens are digested in phagolysosomes. In the presentation phase, the partially digested antigen is displayed on the surface of the macrophage and presented to the T cell for recognition. However, before recognition can occur, the antigen must be presented together with human leucocyte associated (HLA) antigens produced by the major histocompatibility complex (MHC). Recall from Chapter 19 that the MHC is a group of genes that codes for HLA antigens, also called histocompatibility antigens. Once the partially digested antigen and HLA antigen are displayed, receptors on T cells interact with them. Thus, in order for a T cell to recognize an antigen, it must be presented along with some component of the HLA system. During the processing and presentation of antigen, macrophages secrete lymphokines (cytokines), growth factors that participate

FIGURE 22-12 Role of T cells in cellular immunity.

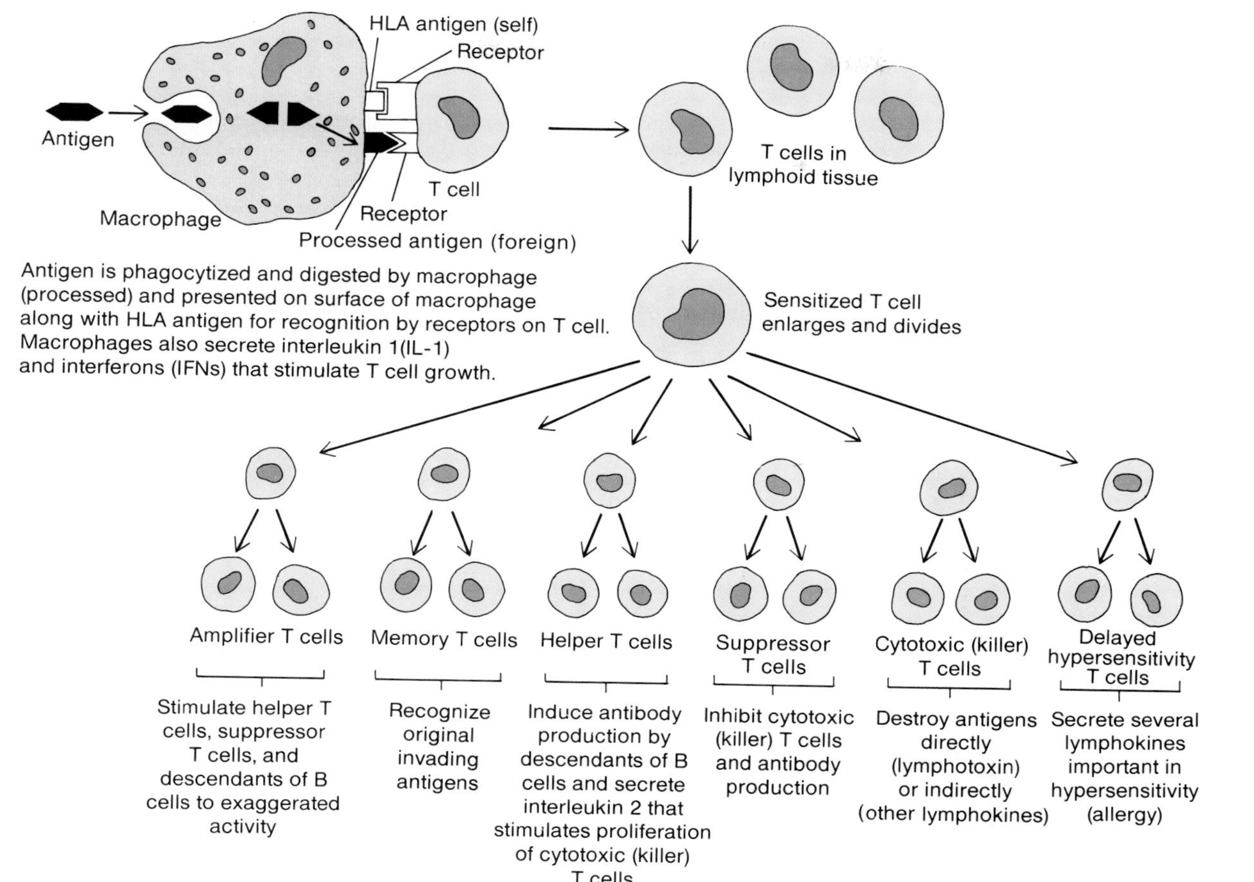

HLA antigen (self)
Receptor
Antigen
Macrophage
T cell
Receptor
Processed antigen (foreign)
T cells in lymphoid tissue
Antigen is phagocytized and digested by macrophage (processed) and presented on surface of macrophage along with HLA antigen for recognition by receptors on T cell. Macrophages also secrete interleukin 1(IL-1) and interferons (IFNs) that stimulate T cell growth.
Sensitized T cell enlarges and divides
Amplifier T cells
Memory T cells
Helper T cells
Suppressor T cells
Cytotoxic (killer) T cells
Delayed hypersensitivity T cells
Stimulate helper T cells, suppressor T cells, and descendants of B cells to exaggerated activity
Recognize original invading antigens
Induce antibody production by descendants of B cells and secrete interleukin 2 that stimulates proliferation of cytotoxic (killer) T cells
Inhibit cytotoxic (killer) T cells and antibody production
Destroy antigens directly (lymphotoxin) or indirectly (other lymphokines)
Secrete several lymphokines important in hypersensitivity (allergy)

in the immune response. These lymphokines are interleukin-1 (IL-1) and interferons (IFNs) that stimulate T cell growth.

There are literally thousands of different types of T cells, and each is capable of responding to a specific antigen or set of antigens. This forms the basis for cellular immunity. At any given time, most T cells are inactive. When an antigen enters the body, only the particular T cell specifically programmed to react with the antigen becomes activated. Such an activated T cell is said to be ***sensitized.*** Activation occurs when macrophages phagocytize the antigen and process and present it along with HLA antigens to receptors on T cells. Sensitized T cells then increase in size, differentiate, and divide, each differentiated cell giving rise to a ***clone,*** or population of cells identical to itself (Figure 22-12). Several subpopulations of cells within the clone can be recognized: cytotoxic (killer) T cells, helper T cells, suppressor T cells, delayed hypersensitivity T cells, amplifier T cells, and memory T cells.

Cytotoxic (killer) T cells leave the lymphoid tissue and migrate to the site of invasion, where several things happen. They attach to the invading cell and secrete a lymphokine (cytokine) called ***lymphotoxin (LT)*** that destroys the antigen directly (Figure 22-13). One lymphotoxin that is released by cytotoxic T cells is called ***perforin*** because it perforates membranes. Next, the cytotoxic T cell attaches to its target cell and releases perforin, which produces holes in the plasma membrane of the target cell, resulting in the lysis of the cell. Most of the protection provided by T cells is the result of secretion of other lymphokines. For example, the cells can release a lymphokine called ***transfer factor (TF),*** a substance that reacts with nonsensitized lymphocytes at the site of invasion, causing them to take on the same characteristics as the sensitized cytotoxic T cells. In this way, additional lymphocytes are recruited, and the effect of the sensitized cytotoxic T cells is intensified. Cytotoxic T cells also secrete a lymphokine called ***macrophage chemotactic factor,*** a substance that attracts macrophages to the site of invasion where they destroy the antigens by phagocytosis. Another lymphokine released by cytotoxic T cells, called ***macrophage activating factor (MAF),*** greatly increases the phagocytic activity of macrophages. A lym-

FIGURE 22-13 Cytotoxic (killer) T cell. (a) Scanning electron micrograph of a cytotoxic T cell making contact with a tumor cell at a magnification of 2500×. (b) Scanning electron micrograph of a cytotoxic T cell destroying a tumor cell at a magnification of 2500×. (Courtesy of Dr. A. Liepins, Science Photo Library, Photo Researchers, Inc.)

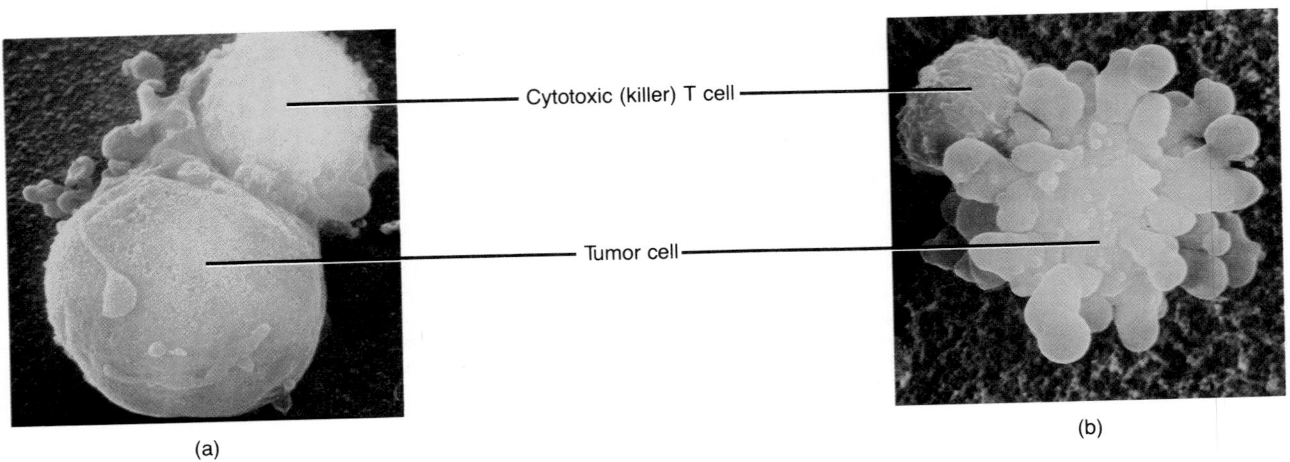

phokine called ***macrophage migration inhibition factor*** prevents macrophages from migrating away from the site of infection, and a lymphokine called ***mitogenic factor*** causes uncommitted or nonsensitized lymphocytes to divide more rapidly. In summary, then, cytotoxic T cells can destroy antigens directly by releasing lymphotoxin and indirectly by secreting other lymphokines that recruit additional lymphocytes, attract macrophages, intensify phagocytosis by macrophages, prevent macrophages from migrating away from the site of infection, and induce rapid division of uncommitted or nonsensitized lymphocytes. Cytotoxic T cells also secrete ***interferons,*** which exert inhibition of viral replication and enhance the killing action of cytotoxic T cells themselves. Cytotoxic T cells are especially effective against cancer cells, slowly developing bacterial diseases (such as tuberculosis and brucellosis), some viruses, fungi, and transplanted cells.

Helper T cells cooperate with B cells to help amplify antibody production. Although helper T cells do not themselves secrete antibodies, antigens first interact with them before inducing antibody production by the descendants of B cells. Helper T cells also secrete ***interleukin-2 (IL-2),*** a lymphokine that stimulates proliferation of cytotoxic T cells. The secretion of interleukin-2 is stimulated by interleukin-1 that is released by macrophages. Helper T cells also secrete proteins that amplify the inflammatory response and killing by macrophages.

Suppressor T cells dampen parts of the immune response by shutting down certain activities several weeks after infection activates them. They can inhibit secretion of injurious substances by cytotoxic (killer) T cells and inhibit production of antibodies by plasma cells. The normal ratio of helper T cells to suppressor T cells is 2:1.

Delayed hypersensitivity T cells are the source of several lymphokines, such as macrophage activating factor and macrophage migration inhibition factor, that are key factors in the hypersensitivity (allergy) response and rejection of transplanted tissue.

Amplifier T cells somehow stimulate helper T cells, suppressor T cells, and descendants of B cells to higher levels of activity.

Memory T cells are programmed to recognize the original invading antigen. Should the pathogen invade the body at a later date, the memory cells initiate a far swifter reaction than during the first invasion. In fact, the second response is so swift that the pathogens are usually destroyed before any signs or symptoms of the disease occur.

Natural Killer (NK) Cells

There is another population of lymphocytes in the body that have certain characteristics similar to killer T cells. These lymphocytes, referred to as ***natural killer (NK) cells,*** have the ability to kill certain cells spontaneously, without interacting with lymphocytes or antigens. Like cytotoxic (killer) T cells, NK cells lyse target cells by releasing the lymphotoxin perforin. Although NK cells may be a separate line of T cells, they are not cytotoxic T cells. NK cells provide an immediate first line of defense against cancer cells and perhaps other cells infected by agents other than viruses. NK cells can produce interferons that inhibit viral replication and enhance their own killing action by lysis. It has been observed that the number of NK cells is decreased in cancer patients and that the decrease is related to the severity of the disease.

B Cells and Humoral Immunity

The body contains not only thousands of different T cells but also thousands of different B cells, each capable of responding to a specific antigen. Whereas killer T cells leave their reservoirs of lymphoid tissue to meet a foreign antigen, B cells respond differently. They differentiate into cells called plasma cells that produce specific antibodies that then circulate in the lymph and blood to reach the site of invasion. In the presence of a foreign antigen, B cells specific for the antigen in lymph nodes, the spleen, or lymphoid tissue in the gastrointestinal tract become activated and stimulated to differentiate into antibody-producing plasma cells. In this process, an antigen binds to antibodies on B cells. The antigen is then processed and presented on the surface of B cells along with human leucocyte associated (HLA) antigens. The presented antigen and HLA antigens can then be recognized by receptors on a helper T cell, and the T cell produces substances that promote B cell division and differentiation. Some of the B cells enlarge and divide and differentiate into a clone of ***plasma cells*** under the influence of thymic hormones (Figure 22-14). Plasma cells secrete the antibody. B cell proliferation and differentiation into plasma cells are also influenced by interleukin-1, secreted by macrophages. The phenomenal rate of antibody secretion is about 2000 molecules per second for each cell, and it occurs for several days until the plasma cell dies (in four to five days). The activated B cells that do not differentiate into plasma cells remain as ***memory B cells,*** ready to respond more rapidly and forcefully should the same antigen appear at a future time.

Different antigens stimulate different B cells to develop into plasma cells and their accompanying memory B cells because the B cells of a particular clone are capable of secreting only one kind of antibody. Each specific antigen activates only those B cells already predetermined to secrete antibody specific to that antigen. The specific antigen selects a specific B cell because that particular B cell has on its surface the antibody molecules that it is capable of producing. The surface antibodies serve as receptor sites with which the antigen can combine.

Antibodies produced by B cells enter circulation and form antigen–antibody complexes with foreign antigens. These antibodies activate complement enzymes for attack and fix the complement to the surface of the antigens.

FIGURE 22-14 **Role of B cells in humoral immunity. Although**

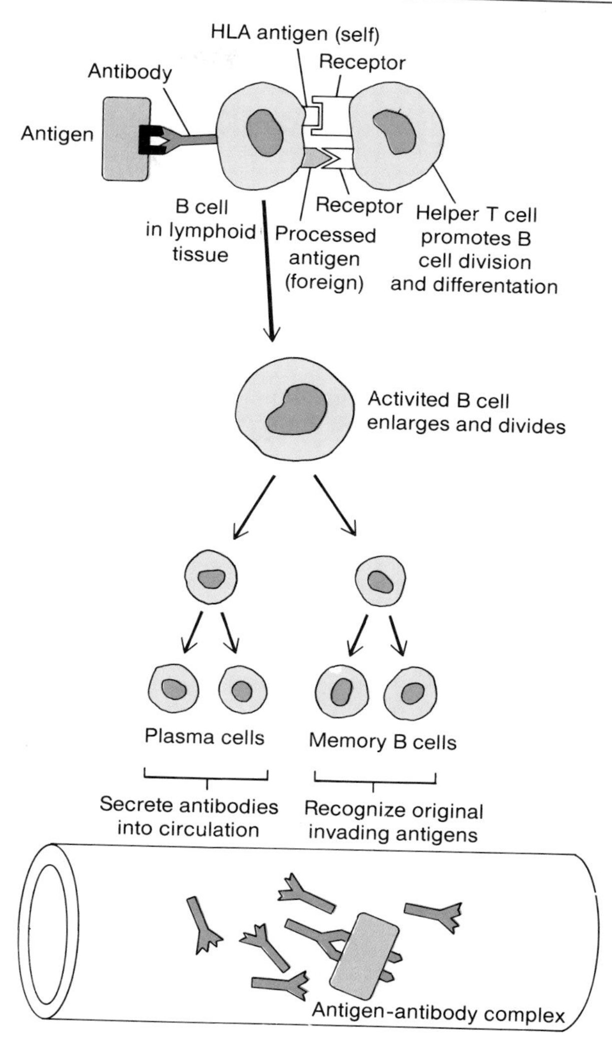

CLINICAL APPLICATION: ANAMNESTIC (SECONDARY) RESPONSE

The immune response of the body, whether cellular or humoral, is much more intense after a second or subsequent exposure to an antigen than after the initial exposure. This can be demonstrated by measuring the amount of antibody in serum, called the *antibody titer* (TĪ-ter). After an initial contact with an antigen, there is a period of several days during which no antibody is present. Then there is a slow rise in the antibody titer, followed by a gradual decline. Such a response of the body to the first contact is called the ***primary response.*** During the primary response, in which the body is said to be primed or sensitized, there is proliferation of immunocompetent lymphocytes. When the antigen is contacted again, whether for the second time or the two hundredth, there is an immediate proliferation of immunocompetent lymphocytes, and the antibody titer is far greater than that for the primary response. This accelerated, more intense response is called the ***anamnestic*** (an'-am-NES-tik; *anamne* = recall), or ***secondary, response.*** The reason for the anamnestic response is that some of the immunocompetent lymphocytes formed during the primary response remain as memory cells. These memory cells not only add to the pool of cells that can respond to the antigen—their response is more intense as well.

Primary and anamnestic responses occur during microbial infection. When you recover from an infection without taking antibiotics, it is usually because of the primary response. If, at a later time, you contact the same microbe, the anamnestic response could be so swift that the microbes are quickly destroyed and you do not exhibit any signs or symptoms.

You have probably guessed by now that the anamnestic response provides the basis for immunization against certain diseases. When you receive the initial immunization, your body is sensitized. Should you encounter the pathogen again as an infecting microbe or "booster dose," your body experiences the anamnestic response. The booster dose literally boosts the antibody titer to a higher level. Since their effects are not permanent, boosters must be given periodically, and immunization schedules are designed to maintain high antibody titers.

A summary of the various cells involved in immune responses is presented in Exhibit 22-3, and a summary of the actions of various lymphokines is presented in Exhibit 22-4.

The Skin and Immunity

The skin not only provides nonspecific resistance to disease because of its mechanical and chemical properties; it also is an active component of the immune system. Recall from Chapter 5 that the epidermis of the skin contains four principal kinds of cells: melanocytes, keratinocytes, and nonpigmented granular dendrocytes (Langerhans cells and Granstein cells). Keratinocytes and nonpigmented granular dendrocytes assume a role in immunity, presumably according to the following mechanism.

If an antigen penetrates the upper layers of keratinized cells of the epidermis, it binds to either Langerhans or

EXHIBIT 22-3 SUMMARY OF CELLS THAT ARE IMPORTANT IN IMMUNE RESPONSES

Cell	Function
Macrophage	Phagocytosis; processing and presentation of foreign antigens to T cells; secretion of interleukin-1 that stimulates secretion of interleukin-2 by helper T cells (stimulates proliferation of cytotoxic T cells) and induces proliferation of B cells; secretion of interferons that stimulate T cell growth.
Cytotoxic (killer) T cell	Lysis of foreign cells by lymphotoxin and release of various other lymphokines that recruit and intensify cytotoxic T cell actions (transfer factor), attract macrophages (macrophage chemotactic factor), increase phagocytic activity of macrophages (macrophage activating factor), prevent macrophage migration from site of action (macrophage migration inhibition factor), and proliferation of uncommitted or nonsensitized lymphocytes (mitogenic factor). Cytotoxic T cells also secrete interferons.
Helper T cell	Cooperates with B cells to amplify antibody production by plasma cells and secretes interleukin-2, which stimulates proliferation of cytotoxic T cells.
Suppressor T cell	Inhibits secretion of injurious substances by cytotoxic T cells and inhibits antibody production by plasma cells.
Delayed hypersensitivity T cell	Secretes macrophage activating factor and macrophage migration inhibition factor, key substances related to hypersensitivity (allergy).
Amplifier T cell	Stimulates helper T cells, suppressor T cells, and B cells to exaggerated levels of activity.
Memory T cell	Remains in lymphoid tissue and recognizes original invading antigens, even years after infection.
Natural killer (NK) cell	Lymphocyte that destroys foreign cells by lysis and produces interferon.
B cell	Differentiates into antibody-producing plasma cell.
Plasma cell	Descendant of B cell that produces antibodies.
Memory B cell	Ready to respond more rapidly and forcefully than initially should the same antigen challenge the body in the future.

EXHIBIT 22-4 SUMMARY OF LYMPHOKINES (CYTOKINES)

Lymphokine	Function
Lymphotoxin (LT)	Destroys antigens directly by lysis.
Perforin	Perforates cell membranes of target cells.
Transfer factor (TF)	Recruits additional lymphocytes and converts them into sensitized cytotoxic cells.
Macrophage chemotactic factor	Attracts macrophages to site of invasion to destroy antigens by phagocytosis.
Macrophage activating factor (MAF)	Increases phagocytic activity of macrophages.
Macrophage migration inhibiting factor	Prevents macrophages from migrating away from site of infection.
Mitogenic factor	Induces uncommitted or nonsensitized lymphocytes to divide more rapidly.
Interleukin-1 (IL-1) (lymphocyte-activating factor)	Produced by antigen-stimulated macrophages and stimulates T cell and B cell growth; stimulates secretion of interleukin-2 by helper T cells.
Interleukin-2 (IL-2) (T cell growth factor)	Produced by helper T cells to stimulate the proliferation of cytotoxic (killer) T cells and natural killer cells.
Interleukin-3 (IL-3) (multipotential CSF)	Produced by activated T cells; supports the growth of bone marrow stem cells and is a growth factor for mast cells.
Interleukin-4 (IL-4) (B cell stimulating factor 1)	Produced by activated T cells; growth factor for activated B cells and resting T cells.
Colony stimulating factors (CSFs)	Produced by white blood cells and function in their proliferation (see Chapter 19).
Interferons (IFNs)	Produced by virus-infected cells to inhibit viral replication in uninfected cells; produced by antigen-stimulated macrophages to stimulate T cell growth; produced by cytotoxic T cells to augment killing action of cytotoxic T cells; produced by natural killer cells to inhibit viral replication.
Tumor necrosis factor (TNF) (cachectin)	Produced by macrophages in the presence of bacterial endotoxins; kills some tumor cells, stimulates synthesis of lymphokines, activates macrophages, and mediates inflammation.

Granstein cells. Langerhans cells present the antigen to helper T cells, which are also present in epidermis, thus activating the T cells. In response to interleukin-1 secreted by keratinocytes, T cells are stimulated to produce interleukin-2. This substance causes T cells to proliferate so that the number of T cells responding to antigen is greatly increased. The T cells then enter the lymphatic system and spread throughout the body.

Just as Langerhans cells interact with helper T cells, Granstein cells interact with suppressor T cells in the epidermis. Normally, however, the helper T cell response predominates to destroy the antigens. In those instances where Langerhans cells might be destroyed by ultraviolet radiation or bypassed by antigens that react directly with suppressor T cells, the suppressor T cell response predominates.

Monoclonal Antibodies (MAbs)

Scientists have known for many years how to induce laboratory animals (or humans) to produce antibodies. By injecting a particular antigen, antibodies are produced against the antigen by plasma cells (descendants of B cells). However, since the antibodies are produced by many different plasma cells, they vary physically and chemically; they are not pure. What scientists have learned to do is isolate a single B cell and fuse it with a tumor cell that is capable of proliferating endlessly. The resulting hybrid cell is called a ***hybridoma.*** Hybridoma cells are a long-term source of substantial quantities of pure antibodies called ***monoclonal antibodies* (*MAbs*).** One of a group of genetically identical descendants is called a clone; so antibodies produced by a single hybridoma clone are called monoclonal antibodies. Such antibodies will combine with one and only one antigen.

One clinical use for monoclonal antibodies is to measure levels of a drug in a patient's blood. Other uses include the diagnosis of pregnancy, allergies, and diseases such as hepatitis, rabies, and certain sexually transmitted diseases. Compared with conventional diagnostic tests, monoclonal antibodies offer greater sensitivity, specificity, and speed. They have also been used to detect cancer at an early stage and to determine the extent of metastasis. One of the most exciting applications is in the area of treating diseases. Monoclonal antibodies, either alone or in combination with radioactivity, toxins, or drugs, are being used in trial studies to treat cancer. This approach has the advantage of destroying diseased tissues only, while sparing healthy tissues, thus overcoming some of the major adverse effects of chemotherapy and radiation treatment. Monoclonal antibodies may also be useful in preparing vaccines to counteract the rejection associated with transplants and to treat autoimmune diseases. One such disease that has been treated with some success is severe combined immunodeficiency (SCID), whose victims are born lacking all immune defenses for fighting disease. This is the rare disease that affected the boy who was internationally known as the "boy in a bubble."

Immunology and Cancer

When a normal cell becomes transformed into a cancer cell, the tumor cell assumes cell surface components called ***tumor-specific antigens.*** It is believed that the immune system usually recognizes tumor-specific antigens as "nonself" and destroys the cancer cells carrying them. Such an immune response is called ***immunologic surveillance.*** Most investigators believe that cellular immunity is the basic mechanism involved in tumor destruction. In this regard, it is presumed that sensitized cytotoxic T cells react with tumor-specific antigens, initiating lysis of tumor cells. Sensitized macrophages may also be involved.

Despite the mechanism of immunologic surveillance, some cancer cells escape destruction, a phenomenon called ***immunologic escape.*** One possible explanation is that tumor cells may shed their tumor-specific antigens, thus evading recognition. Another theory is that antibodies produced by plasma cells bind to tumor-specific antigens, preventing recognition by cytotoxic T cells. It is reasonable to assume that immunologic techniques using monoclonal antibodies and toxins or radioactive compounds may be used in the future to prevent cancer. Hopefully, the antibody would selectively locate the cancer cell, and the attached toxic agent or radioactive substance would destroy the cell but cause little or no damage to healthy tissue. There is great expectation that this approach will become analogous to the use of antibiotics in treating infectious diseases.

CLINICAL APPLICATION: IMMUNOTHERAPY

In addition to using monoclonal antibodies to treat disease, scientists also use various lymphokines (cytokines). For many years, researchers have been trying to induce the immune system to mount an attack against cancer, a technique called *immunotherapy.* Although such methods have been generally successful in laboratory animals, results in humans have not been as impressive. However, advances in the past few years may translate into effective therapies against at least some types of human cancer.

The first lymphokines that were shown to be effective against any human cancer were interferons (IFNs), especially alpha IFN. It showed remarkable improvement in patients with a rare cancer called hairy cell leukemia and some beneficial effects against certain other types of tumors, such as renal cell carcinoma and Kaposi's sarcoma. Disappointingly, it produces little or no improvement in lung, breast, or colon cancers. Some patients taking IFNs have side effects such as fever, chills, fatigue, depression, weight loss, cough, body aches, bone marrow suppression, and liver damage.

Tumor necrosis factor (TNF), also called cachectin, is another lymphokine that has proved effective in animal tumor experiments. Although TNF does kill cancer cells, it also releases stored fat from adipose tissue and inhibits enzymes needed to store fat. This results in a continuous severe weight loss called cachexia that is experienced by cancer patients. Other problems with TNF include fever, anemia, and shock. If TNF is to be an effective anticancer drug, it must be given in small enough doses to prevent cachexia but large enough doses to kill tumors.

Of the interleukins, the one most widely employed to fight cancer is interleukin-2 (IL-2). After inactive cancer-fighting cytotoxic T cells and natural killer (NK) cells are removed from a cancer patient, they are combined with IL-2 to stimulate them. Such cells are called lymphokine-activated killer (LAK) cells. The LAK cells are then transfused back into the patient's blood in an attempt to locate and destroy cancer cells. Although some tumor regression does occur with this procedure, severe complications affect almost all patients. Among the adverse effects are high fever, severe weakness, vomiting, difficult breathing, anemia, thrombocytopenia, hypotension, gastritis, jaundice, paranoid delusions, and hallucinations. Until these side effects can be reduced or eliminated, immunotherapy using LAK cells will remain experimental.

Colony stimulating factors (CSFs) are also being used in clinical trials in immunotherapy. They are given to patients with cancer, AIDS, and bone marrow disorders. The primary therapeutic effect of CSFs is to bolster the activities of neutrophils, eosinophils, and macrophages. Side effects also occur with use of CSFs.

AGING AND THE IMMUNE SYSTEM

As the immune system ages, elderly individuals become more susceptible to all types of infections and malignancies. Their response to vaccines is decreased and they tend to produce more autoantibodies. In addition, the cellular and humoral immune systems exhibit lowered levels of function.

DEVELOPMENTAL ANATOMY OF THE LYMPHATIC SYSTEM

The lymphatic system begins its development by the end of the fifth week. *Lymphatic vessels* develop from ***lymph sacs*** that arise from developing veins. Thus, the lymphatic system is also derived from ***mesoderm.***

The first lymph sacs to appear are the paired ***jugular lymph sacs*** at the junction of the internal jugular and subclavian veins (Figure 22-15). From the jugular lymph sacs, capillary plexuses spread to the *thorax, upper extremities, neck,* and *head.* Some of the plexuses enlarge and form lymphatic vessels in their respective regions. Each jugular lymph sac retains at least one connection with its jugular vein, the left one developing into the superior portion of the thoracic duct (left lymphatic duct).

The next lymph sac to appear is the unpaired ***retroperitoneal lymph sac*** at the root of the mesentery of the intestine. It develops from the primitive vena cava and mesonephric (primitive kidney) veins. Capillary plexuses and lymphatic vessels spread from the retroperitoneal lymph sac to the *abdominal viscera* and *diaphragm.* The sac establishes connections with the cisterna chyli but loses its connections with neighboring veins.

At about the time the retroperitoneal lymph sac is developing, another lymph sac, the ***cisterna chyli,*** develops below the diaphragm on the posterior abdominal wall. It gives rise to the inferior portion of the *thoracic duct* and the *cisterna chyli* of the thoracic duct. Like the retroperitoneal lymph sac, the cisterna chyli also loses its connections with surrounding veins.

The last of the lymph sacs, the paired ***posterior lymph sacs,*** develop from the iliac veins at their union with the posterior cardinal veins. The posterior lymph sacs produce capillary plexuses and lymphatic vessels of the *abdominal wall, pelvic region,* and *lower extremity.* The posterior lymph sacs join the cisterna chyli and lose their connections with adjacent veins.

FIGURE 22-15 **Development of the lymphatic system.**

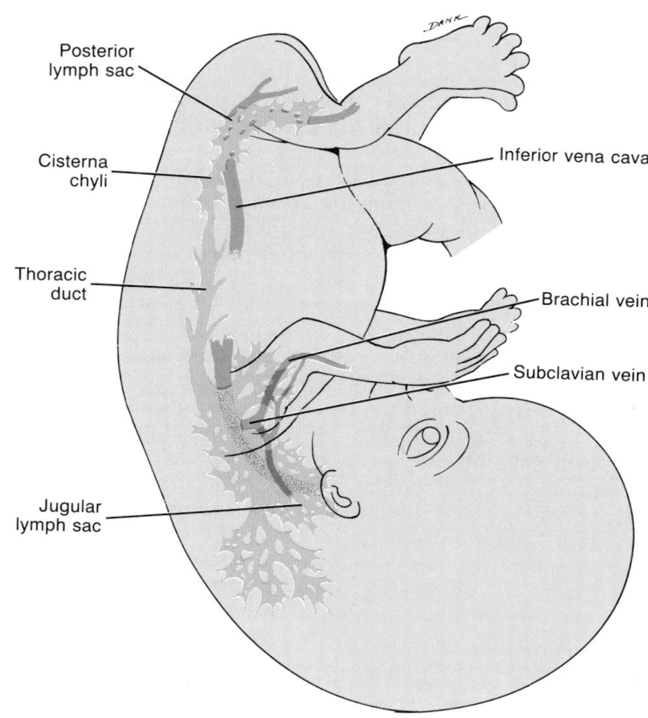

With the exception of the anterior part of the sac from which the cisterna chyli develops, all lymph sacs become invaded by ***mesenchymal cells*** and are converted into groups of *lymph nodes.*

DISORDERS: HOMEOSTATIC IMBALANCES

Acquired Immune Deficiency Syndrome (AIDS)

Never before has science been confronted with an epidemic in which the primary disease only lowers the victim's immunity, and then a second unrelated disease produces the symptoms that may result in death. The primary disease is called ***acquired immune deficiency syndrome (AIDS).*** It is an illness characterized by a positive antibody test that indicates the presence of the AIDS virus and by the presence of ***indicator diseases.*** Indicator diseases are diseases that are found in AIDS sufferers much more frequently than in others and to which the AIDS sufferer is more than usually susceptible. One example is a cancer called Kaposi's sarcoma (KS), a deadly form of skin cancer prevalent in equatorial Africa but previously almost unknown in the United States. It arises from endothelial cells of blood vessels and produces painless purple or brownish lesions that resemble bruises on the skin or on the side of the mouth, nose, or rectum. There is some evidence that the AIDS virus may, by itself, be carcinogenic.

AIDS first surfaced in June 1981 as a result of reports from the Los Angeles area to the Centers for Disease Control (CDC) of several cases of a very rare type of pneumonia caused by a fungus, not a protozoan, as originally assumed. The pneumonia, called *Pneumocystis carinii* pneumonia (PCP), occurred among homosexual males. It results in shortness of breath, persistent dry cough, sharp chest pains, and difficulty in breathing. At about the same time, the CDC also received reports from New York and Los Angeles concerning an increase in the incidence of Kaposi's sarcoma among homosexual males. A group of scientists in France and the United

DISORDERS: HOMEOSTATIC IMBALANCES (*continued*)

States first isolated the virus that causes the disease in 1983. The virus is called ***human immunodeficiency virus (HIV).***

The primary victims are homosexual men, intravenous drug users, and hemophilia patients. Other high-risk groups are the heterosexual partners of AIDS sufferers and transfusion patients. Ninety percent of AIDS victims in the United States are between the ages of 20 and 49 years, and 94 percent are males. AIDS has occurred in all 50 states. The average incubation period (time interval from infection with HIV to date of diagnosis) is nearly eight years. Forty-eight percent of known AIDS cases have died within two years of diagnosis.

It is estimated that by the end of 1992 there will be over 365,000 cases of AIDS in the United States. The number of infected Americans is estimated to be between 1 and 1.5 million, and most do not even know they have the disease. All carriers of the virus, whether they have the disease or not, are assumed to be infected for life and are capable of transmitting the virus to others. The risk of developing AIDS increases yearly after infection with the virus; the longer a person is infected, the greater the chances of developing AIDS.

HIV: Structure and Pathogenesis

Viruses consist of a core of DNA or RNA surrounded by a protein coat (capsid). Some viruses also contain an envelope (outer layer) composed of a double layer of lipid penetrated by proteins (Figure 22-16). Outside a living host cell, a virus has no metabolic functions and is unable to replicate. However, once a virus makes contact with a host cell, the viral nucleic acid enters the host cell. Once inside, the viral nucleic acid uses the host cell's enzymes, ribosomes, nutrients, and other resources to make copies of itself and new protein coats and envelopes. As these components accumulate, they are assembled into a large number of viruses that can then leave the host cell to infect other cells. As viruses go through their cycle of replication, they can damage or kill host cells in various ways. These include shutting down the synthesis of host cell proteins, RNA, and DNA; inhibiting cell division; damaging DNA; rupturing lysosomes that can lead to autolysis; and inducing toxic effects (viral proteins). Moreover, the body's own defenses, attacking the infected cells, can kill the cell as well as the viruses it harbors. Not only can the AIDS virus damage and kill host cells, it can also lie dormant for years before causing its effects, attacking more different types of host cells than imagined, and it has more complex means of destroying the immune system than suspected.

The AIDS virus belongs to a group of viruses called ***retroviruses,*** which are distinguished by an enzyme called ***reverse transcriptase.*** Retroviruses carry their genetic material in RNA and copy this genetic coding into DNA by reverse transcriptase (RNA → DNA). In other words, retroviruses reverse the ordinary flow of genetic information (DNA → RNA → proteins). Once the AIDS virus produces its DNA from RNA, the DNA is integrated into the host cell's DNA. There it can remain dormant, giving no sign of its presence, or it can take

FIGURE 22-16 Human immunodeficiency virus (HIV), the causative agent of AIDS. (a) Diagram. (b) Electron micrograph of several HIVs targeting a white blood cell at a magnification of 93,500 ×. (Courtesy of Wagner, Herbert Stock/Phototake.)

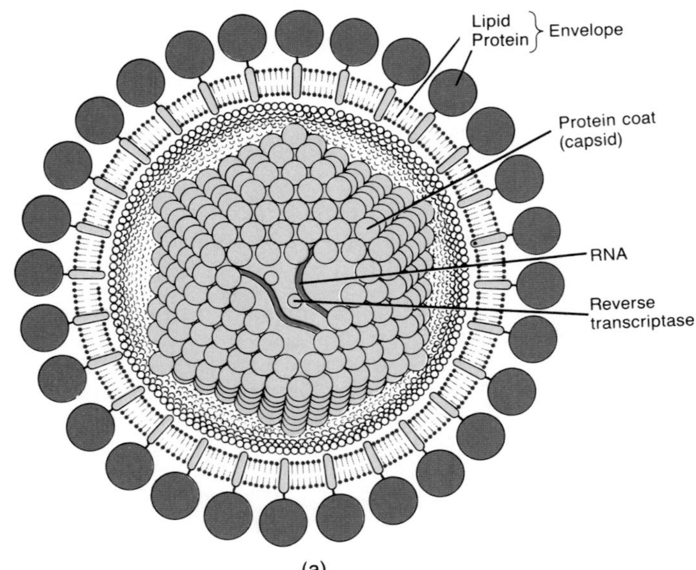

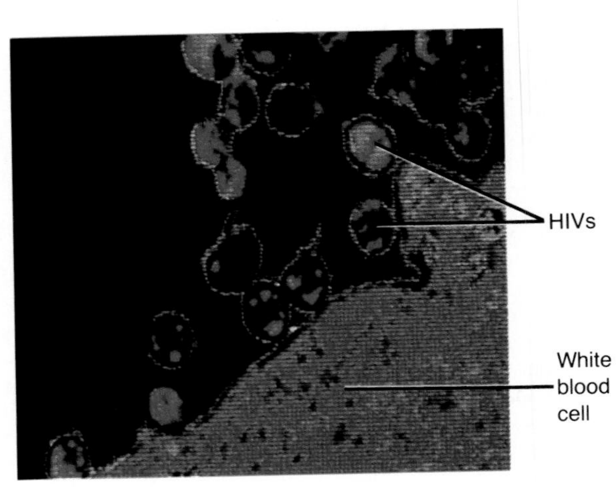

DISORDERS: HOMEOSTATIC IMBALANCES (*continued*)

over the host cell's genetic machinery to produce RNA, some of which is translated into viral proteins to make new AIDS viruses. Other retroviruses include Epstein-Barr virus (EBV), which is associated with infectious mononucleosis and Burkitt's lymphoma; cytomegalovirus (CMV), which is associated with a mononucleosislike syndrome and some cases of hepatitis; hepatitis B virus (HBV), which causes hepatitis B; human T cell lymphotropic virus type I (HTLV I), which is associated with adult T cell leukemia; and human T cell lymphotropic virus type II (HTLV II), which is associated with hairy cell leukemia.

Infection with HIV begins when the virus binds to a protein receptor, called the CD4 receptor, on the surface of helper T (T4) cells. Through very complicated mechanisms, helper T cells are killed, and as the process progresses, there is a decline in immune functioning. Recall that helper T cells cooperate with B cells to amplify antibody production. The reduction in the number of T cells inhibits antibody production by descendants of B cells (plasma cells) against HIV. Helper T cells also stimulate the destruction of infected cells by cytotoxic (killer) T cells and natural killer (NK) cells. Helper T cells also influence the activity of monocytes and macrophages, which engulf infected cells and foreign particles. Since helper T cells orchestrate a large portion of our immune defense, their destruction leads to collapse of the immune system and susceptibility to opportunistic infections such as Kaposi's sarcoma (KS) and *Pneumocystis carinii* pneumonia (PCP). AIDS victims are also subject to tuberculosis, thrush, oral hairy leukoplakia, cytomegalovirus, a form of herpes that attacks the central nervous system, cryptococcal meningitis, central nervous system toxoplasmosis, oral and esophageal candidiasis, cryptosporidiosis, a severe form of arthritis called Reiter's syndrome, salmonellosis, and psoriasis. The AIDS virus also attacks macrophages; brain cells (where it causes AIDS dementia complex); endothelial cells that line various organs, body cavities, and blood vessels; bone marrow cells; and possibly colon cells. Heart and liver cells are also possible target cells. The virus might kill these cells or multiply in them, thus spreading the virus.

Presently, macrophages are receiving a great deal of attention as pivotal cells in the development of AIDS. Much research suggests that macrophages may actually be the first and possibly the only cells invaded by the AIDS virus. It seems that once the virus invades macrophages, it then spreads to helper T cells. As a result, helper T cells are killed and macrophages do not function properly in the immune response. Macrophages serve as reservoirs for the virus, and even though the viruses multiply within the macrophages, they are not themselves killed. A very important consequence of this is that commonly used screening tests that detect antibodies to the AIDS virus in the blood may be negative even though the person has the AIDS virus hidden in macrophages. While stored in macrophages, the virus does not trigger the production of antibodies that would appear in blood. A new test, called the Cetus test, is used to detect the presence of AIDS viruses in macrophages, rather than AIDS antibodies in blood.

Soon after infection with the AIDS virus, the host develops antibodies against several proteins in the virus. The presence of these antibodies in blood is used as a basis for diagnosing AIDS. Normally, antibodies so developed are protective, but in the case of the AIDS virus, this is not necessarily so. In fact, one of the most remarkable aspects of the AIDS virus is its ability to produce a carrier state in a high proportion of infected people even though antibodies are present.

Symptoms

The various stages through which an individual passes are correlated with decreasing helper T cell counts. Following exposure to HIV, there are usually no symptoms at first, and it may take from six weeks to one year before HIV is detected by standard antibody tests. Although most people have no symptoms when AIDS is first diagnosed, some develop mononucleosislike symptoms such as fatigue, fever, swollen glands, headache, and even encephalitis. Within a few weeks, these symptoms disappear. In the next stage of the disease, which lasts from three to five years, chronically swollen lymph nodes develop in the neck, armpits, and groin. Following this, helper T cell counts decline even more, and the patients fail to respond to most skin tests that measure delayed hypersensitivity, a measure of the individual's ability to produce a cellular immune response against specific proteins injected under the skin. Progression to the next stage is evidenced by a total absence of delayed hypersensitivity. It is during this stage that opportunistic infections develop (thrush, hairy leukoplakia, Kaposi's sarcoma, *Pneumocystis carinii* pneumonia, toxoplasmosis, candidiasis, and so on). In the terminal stages of HIV infection, many patients suffer from AIDS dementia complex, which is characterized by a progressive loss of function in motor activities, cognition, and behavior.

HIV Outside the Body

Outside the body, the AIDS virus is fragile and can be eliminated in a number of ways. Dishwashing and clotheswashing by exposing the virus to 135°F (56°C) for 10 minutes will kill HIV. Chemicals such as hydrogen peroxide (H_2O_2), rubbing alcohol, Lysol, household bleach, and germicidal skin cleaner (such as Betadine and Hibiclens) are also very effective against the virus. Standard chlorination is also sufficient to kill HIV in swimming pools and hot tubs.

Transmission

Although HIV has been isolated from a number of body fluids, such as blood, semen, vaginal fluid, cerebrospinal fluid, tears, and saliva, it appears that the fluids that provide sufficient virus for transmission seem to be limited to blood, semen, and vaginal secretions. The virus is found in macrophages in these fluids. The macrophages also serve as a reservoir of the AIDS virus. It is possible

DISORDERS: HOMEOSTATIC IMBALANCES (*continued*)

that macrophages pass the virus to helper T cells, resulting in their destruction. The sites best suited for establishment of infection after exposure appear to be the cardiovascular system, open wounds of the skin, the penis, vagina, and rectum.

HIV is effectively transmitted by sexual contact between males, from males to females, and from females to males through vaginal or anal intercourse with infected persons. HIV is also effectively transmitted through exchanges of blood, such as by contaminated hypodermic needles and needle stick, open wound, or mucous membrane exposure in health-care workers. The virus is also transmitted from infected mothers to their infants before or during birth or through breast-feeding. It does not appear that individuals become infected as a result of routine, nonintimate contacts. Thousands of family members, coworkers, and friends of AIDS patients do not have AIDS.

No evidence exists that AIDS can be spread through *normal* kissing (some experts feel that prolonged, vigorous, wet deep kissing that results in breaks or tears in the lining of the mouth or if preexisting sores are present could theoretically transmit the virus). Also, although mosquitoes can carry the AIDS virus for several days, there is no evidence that the virus can multiply inside mosquitoes or that they are capable of transmitting the disease. It also appears that health-care personnel who take proper routine barrier precautions (gloves, masks, safety glasses) are not at risk.

Drugs and Vaccines against HIV

Medical scientists are mounting what is probably the greatest concentrated effort ever to find a cure for a single virus disease. One of the problems with finding a drug to treat AIDS is that HIV can lie dormant in body cells. In addition, HIV can infect a variety of cells, including those in the central nervous system that are protected by the blood–brain barrier. Added to this is the problem of opportunistic infections, which may be very difficult to treat. Antiviral therapy is aimed at disrupting various points in the viral life cycle. Some research centers on preventing the attachment of the virus to the host cell plasma membrane. Other research is designed to counterattack the virus once it gets inside a host cell and takes over the machinery of the host cell. Still other research is aimed at inhibiting assembly of viruses within the host cell and their subsequent release. Most drugs now used against the AIDS virus are directed against reverse transcriptase.

The drug currently most widely tried against the AIDS virus and given approval by the Food and Drug Administration (FDA) is zidovudine (Retrovir), formerly known as azidothymidine (AZT). The action of this drug is to inhibit the action of reverse transcriptase, thus preventing the virus from making DNA from RNA. Clinical improvements among patients taking zidovudine are weight gain, increased energy, and neurological improvements (reversal of loss of mental function and dementia). Side effects include severe bone marrow damage, anemia, and suppression of the immune system. Moreover, patients must take the medication every four hours and are required to receive weekly or bimonthly transfusions. Dideoxycytidine (ddC), a drug similar in action to zidovudine but with fewer toxic effects, is now in early human trials, as is phosphonoformate. Other drugs that also inhibit reverse transcriptase are dideoxyadenosine (ddA), dideoxyinosine (ddI), and rifabutin. A drug called ribavirin (Virazole) appears to prevent the synthesis of viral proteins. It is not yet approved for AIDS patients in the United States. Alpha interferon is believed to attach at the final stage of virus production. It is being tried in trials both alone and in combination with other drugs. Other experimental drugs being considered are granulocyte-macrophage colony-stimulating factor (GM-MCS), which boosts the number of white blood cells; AL-721, peptide T, soluble CD4, and dextran sulfate, which hamper viral attachment to cells; IM reg-1, which may decrease the likelihood of developing full-fledged AIDS; and fusidic acid, whose mechanism of action is unknown.

In addition to drug development and testing, considerable emphasis is also being given to development of a vaccine against AIDS. The purpose of a vaccine is to stimulate the production of antibodies against the virus that will kill the virus directly and to bolster immune defenses after the virus has invaded the body. The principal experimental vaccines under study make use of various subunits of the AIDS virus. In a process that is largely trial and error, researchers have selected different elements of the virus (proteins from the envelope or coat) that are believed to be most likely to produce the broadest range of antibodies. Although some scientists propose using the entire killed AIDS virus in a vaccine to stimulate antibody production, there is some concern that some of the viruses might remain alive and thus cause disease.

The development of an effective AIDS vaccine has been impeded because of lack of understanding of protective immunity against HIV, the ability of HIV to mutate so quickly, and inability to use most animals for experimentation since the virus will not grow within them. In addition, since the AIDS virus can remain protected in certain cells, antibodies cannot attach it. Moreover, clinical trials have raised safety concerns, and there may be a shortage of volunteers in the United States.

The objective of developing an AIDS vaccine is to determine which part of HIV elicits the most powerful immune response to inhibit the virus. At present, most studies are concentrated on proteins in the envelope of the virus. Others focus on internal components of HIV rather than envelope antigens.

Prevention of Transmission

At the present time, there are no drugs or vaccines to prevent AIDS. The only means of prevention is to stop transmission of the virus. Sexual transmission of the AIDS virus can be prevented if infected persons do not have vaginal, oral, or anal intercourse with susceptible persons, or, if during intercourse, effective barrier tech-

DISORDERS: HOMEOSTATIC IMBALANCES (*continued*)

niques (condoms and spermicides such as nonoxynol 9) are used. Infection from donated blood and blood products can be prevented by testing blood for evidence of the AIDS virus. AIDS transmitted by needles and syringes could be avoided if the use of intravenous drugs is stopped or if unsterilized infection paraphernalia are not used. Mother-to-infant infections could be avoided if infected females would not become pregnant. If these measures are to be effective, they must be part of an overall program involving education, counseling, screening individuals at high risk, tracing contacts, and modifying behavior.

Until there is effective drug therapy or an effective vaccine, preventing the spread of AIDS must rely on education and safer sexual practices.

Autoimmune Diseases

Under normal conditions, the body's immune mechanism is able to recognize its own tissues and chemicals. It normally does not produce T cells or B cells against its own substances. Such recognition of self is called ***immunologic tolerance.*** Although the mechanism of tolerance is not completely understood, it is believed that suppressor T cells may inhibit the differentiation of B cells into antibody-producing plasma cells or inhibit helper T cells that cooperate with B cells to amplify antibody production.

At times, however, immunologic tolerance breaks down, and this leads to an ***autoimmune disease (autoimmunity).*** For reasons still not understood, certain tissues undergo changes causing the immune system to recognize them as foreign antigens and produce an immune attack against them. Autoimmune diseases are immunologic responses mediated by antibodies against a person's own tissue antigens. Among human autoimmune diseases are rheumatoid arthritis (RA), systemic lupus erythematosus (SLE), thyroiditis, rheumatic fever, glomerulonephritis, encephalomyelitis, hemolytic and pernicious anemias, Addison's disease, Grave's disease, type I diabetes mellitus, myasthenia gravis, and multiple sclerosis (MS).

Severe Combined Immunodeficiency (SCID)

Severe combined immunodeficiency (SCID) is a rare immunodeficiency disease in which both B cells and T cells are missing or inactive in providing immunity. Perhaps the most famous patient with SCID was David, the "bubble boy," who lived in a sterile plastic chamber for all but 15 days of his life; he died on February 22, 1984, at age 12. He was the oldest untreated survivor of SCID.

David was placed in the sterile chamber shortly after birth to protect him from microbes that his body could not fight. In an effort to correct his disorder, David underwent a bone marrow transplant from his older sister, who was the most closely but still not perfectly matched donor available. Eighty days after the transplant and still in a germ-free environment, David developed some of the symptoms of infectious mononucleosis (IM), a condition caused by the Epstein-Barr virus (EBV). David was brought out of isolation for easier treatment, with the hope that the transplant would provide the same protection as his sterile plastic chamber. Unfortunately, the transplant was not successful, and David died about four months after it was performed. It was discovered that what killed David was not the immediate failure of the transplant but cancer. His B cells had proliferated as a result of the EBV. This is a very clear demonstration of a virus causing a cancer in humans.

Hypersensitivity (Allergy)

A person who is overly reactive to an antigen is said to be ***hypersensitive (allergic).*** Whenever an allergic reaction occurs, there is tissue injury. The antigens that induce an allergic reaction are called ***allergens.*** Almost any substance can be an allergen for some individual. Common allergens include certain foods (milk, peanuts, shellfish, eggs), antibiotics (penicillin, tetracycline), vitamins (thiamine, folic acid), protein drugs (insulin, ACTH, estradiol), vaccines (pertussis, typhoid), venoms (honeybee, wasp, snake), cosmetics, chemicals in plants such as poison ivy, pollens, dust, molds, iodine-containing dyes used in certain x-ray procedures, and even microbes.

There are four basic types of hypersensitivity reactions: type I (anaphylaxis), type II (cytotoxic), type III (immune complex), and type IV (cell-mediated). The first three involve antibodies; the last involves T cells.

Type I (anaphylaxis) reactions are the most common and occur within a few minutes after a person sensitized to an allergen is reexposed to it. ***Anaphylaxis*** (an'-a-fi-LAK-sis) literally means "against protection" and results from the interaction of antibodies (IgE) with mast cells and basophils. In response to certain allergens, some people produce IgE antibodies that bind to the surfaces of mast cells and basophils. This binding is what causes a person to be allergic to the allergen. Whereas basophils circulate in blood, mast cells are especially numerous in connective tissue of the skin, respiratory system, and endothelium of blood vessels. In response to the attachment of IgE antibodies to basophils and mast cells, the cells release chemicals called ***mediators of anaphylaxis,*** among which are histamine, prostaglandins (PGs), leukotrines, and kinin. Collectively, the mediators increase blood capillary permeability, increase smooth muscle contraction in the lungs, and increase mucus secretion. As a result, a person may experience inflammatory responses, difficulty in breathing from constricted bronchial tubes, and a "runny" nose from excess mucus secretion.

Some anaphylactic reactions, such as hives, eczema, swelling of the lips or tongue, abdominal cramps, and diarrhea are referred to as ***localized*** (affecting one part or a limited area). Other anaphylactic reactions are considered ***systemic*** (affecting several parts or the entire body). An example is acute anaphylaxis (anaphylactic shock), which may occur in a susceptible individual who has just received a triggering drug or been stung by

DISORDERS: HOMEOSTATIC IMBALANCES (*continued*)

a wasp or (more rarely) ingested a certain causative food such as seafood in persons with a marked sensitivity to iodine. The person develops respiratory symptoms (wheezing and shortness of breath) as bronchioles constrict, usually accompanied by cardiovascular failure and collapse due to vasodilation and fluid loss from blood. The treatment includes epinephrine and usually injectable corticosteroids which reverse the respiratory and cardiovascular effects, as well as supportive measures. It also includes the prevention of future such events by education and by having the person purchase and wear a Medic-Alert bracelet or necklace.

Type II (cytotoxic) reactions are caused by antibodies (IgG or IgM) directed against antigens on a person's blood cells (red blood cells, lymphocytes, or platelets) or tissue cells. The reaction of antibodies and antigens usually leads to activation of complement. Type II reactions, which may occur in incompatible transfusion reactions, damage cells by lysis.

Type III (immune complex) reactions involve antigens (not part of a host tissue cell), antibodies (IgA or IgM), and complement. When certain ratios of antigen to antibody occur, the complexes are small and escape phagocytosis. The complexes become trapped in the basement membrane under the endothelium of blood vessels, activate complement, and cause an inflammation. Conditions that so arise include glomerulonephritis, systemic lupus erythematosus (SLE), and rheumatoid arthritis (RA).

Type IV (cell-mediated) reactions involve T cells and often are not apparent for a day or more. Type IV reactions occur when allergens that bind to tissue cells are phagocytized by macrophages and presented to receptors on the surfaces of T cells. This results in the proliferation of T cells, which respond by destroying the allergens. An example of a type IV reaction that involves the skin is the familiar skin test for tuberculosis. Tissue rejection is another example.

Tissue Rejection

Transplantation involves the replacement of an injured or diseased tissue or organ. Usually, the body recognizes the proteins in the transplanted tissue or organ as foreign and produces antibodies against them. This phenomenon is known as ***tissue rejection.*** Rejection can be somewhat reduced by matching donor and recipient HLA antigens and by administering drugs that inhibit the body's ability to form antibodies. Recall from Chapter 19 that white blood cells and other nucleated cells have surface antigens called human leucocyte associated (HLA) antigens. These are unique for each person, except identical twins. The more closely matched the HLA antigens between donor and recipient, the less the likelihood of tissue rejection.

Types of Transplants

The most successful transplants are ***autografts,*** transplants in which one's own tissue is grafted to another part of the body (such as skin grafts for burn treatment or plastic surgery), and ***isografts,*** transplants in which the donor and recipient have identical genetic backgrounds.

An ***allograft*** is a transplant between individuals of the same species but with different genetic backgrounds. The success of this type of transplant has been moderate. Frequently, it is used as a temporary measure until the damaged or diseased tissue is able to repair itself. Skin transplants from other individuals and blood transfusions might properly be considered allografts. The one organ allograft that has been quite successful is the thymus. Children born without a thymus can now receive the gland from an aborted fetus. The thymus-deficient child cannot produce antibodies and thus cannot reject the transplant. Rejection later on indicates that the child is manufacturing antibodies and no longer needs the organ.

A ***xenograft*** is a transplant between animals of different species. This type of transplantation is used primarily as a physiological dressing over severe burns.

Immunosuppressive Therapy

Until recently, ***immunosuppressive drugs*** suppressed not only the recipient's immune rejection of the donor organ but also the immune response to all antigens as well. This causes patients to become very susceptible to infectious diseases. A drug called *cyclosporine,* derived from a fungus, has largely overcome this problem with regard to kidney, heart, and liver transplants. A selective immunosuppressive drug, it inhibits T cells that are responsible for tissue rejection but has only a minimal effect on B cells. Thus, rejection is avoided and resistance against disease is still maintained.

As an alternative to immunosuppressive drugs, some scientists are experimenting with total irradiation of lymph nodes, a procedure called ***total lymphoid irradiation (TLI).*** The radiation kills cytotoxic (killer) T cells, helper T cells, and suppressor T cells which prevents rejection. It is believed that the cells reappear from 1 to 12 months after irradiation and that the suppressor T cells become sensitive to the transplant and inhibit cytotoxic and helper T cells that would attack the transplant.

It is recommended that testing for AIDS antibodies or virus should be done for all donors of tissues or organs intended for transplantation, as well as donors of blood, semen, or ova.

Hodgkin's Disease (HD)

Hodgkin's disease (HD) is a form of cancer, usually arising in lymph nodes, the cause of which is unknown. It may, however, arise from a combination of genetic predisposition, disturbance of the immune system, and an infectious agent (Epstein-Barr virus). The histological diagnosis of the disease is made by the presence of large, malignant, multinucleate cells in the affected lymph node called Reed-Sternberg cells. The disease is initially

DISORDERS: HOMEOSTATIC IMBALANCES (*continued*)

characterized by a painless, nontender enlargement of one or more lymph nodes, most commonly in the neck but occasionally in the axilla, inguinal, or femoral region. About one-quarter to one-third of patients also have an unexplained and persistent fever and/or night sweats. Fatigue and weight loss are also associated complaints, as is pruritus (itching). Treatment consists of radiation therapy, chemotherapy, and combinations of the two. Hodgkin's disease is considered to be a curable malignancy.

MEDICAL TERMINOLOGY ASSOCIATED WITH THE LYMPHATIC SYSTEM

Adenitis (ad'-e-NĪ-tis; *adeno* = gland; *itis* = inflammation of) Enlarged, tender, and inflamed lymph nodes resulting from an infection.

Elephantiasis (el'-e-fan-TĪ-a-sis) Long-standing edema of one or both lower extremities, and sometimes of the arms or other body parts, that is due to lymphatic obstruction. The involved part is tremendously swollen and hardened, and the skin surface folds and produces fissures, causing it to resemble the leg of an elephant. Elephantiasis may be due to filariasis, an infestation by a parasitic worm. Other causes include congestive heart failure and chronic obstruction of the lymphatic vessels.

Hypersplenism (hī'-per-SPLĒN-izm; *hyper* = over) Abnormal splenic activity due to splenic enlargement and associated with an increased rate of destruction of normal blood cells.

Lymphadenectomy (lim-fad'-e-NEK-tō-mē; *ectomy* = removal) Removal of a lymph node.

Lymphadenopathy (lim-fad'-e-NOP-a-thē; *patho* = disease) Enlarged, sometimes tender lymph glands.

Lymphangioma (lim-fan'-jē-Ō-ma; *angio* = vessel; *oma* = tumor). A benign tumor of the lymphatic vessels.

Lymphangitis (lim'-fan-JĪ-tis) Inflammation of the lymphatic vessels.

Lymphedema (lim'-fe-DĒ-ma; *edema* = swelling) Accumulation of lymph fluid producing subcutaneous tissue swelling.

Lymphoma (lim'-FŌ-ma) Any tumor composed of lymph tissue.

Lymphostasis (lim-FŌS-tā-sis; *stasis* = halt) A lymph flow stoppage.

Splenomegaly (splē'-nō-MEG-a-lē; *mega* = large) Enlarged spleen.

STUDY OUTLINE

Lymphatic Vessels (p. 655)

1. The lymphatic system consists of lymph, lymphatic vessels, and structures and organs that contain lymphatic tissue (specialized reticular tissue containing large numbers of lymphocytes).
2. Among the lymphatic tissue-containing components of the lymphatic system are diffuse lymphatic tissue, lymphatic nodules, and lymphatic organs (lymph nodes, spleen, and thymus gland).
3. Lymphatic vessels begin as blind-ended lymph capillaries in tissue spaces between cells.
4. Lymph capillaries merge to form larger vessels, called lymphatic vessels, which ultimately converge into the thoracic duct or right lymphatic duct.
5. Lymphatic vessels have thinner walls and more valves than veins.

Lymphatic Tissue (p. 658)

1. Lymph nodes are oval structures located along lymphatic vessels.
2. Lymph enters nodes through afferent lymphatic vessels and exits through efferent lymphatic vessels.
3. Lymph passing through the nodes is filtered. Lymph nodes also produce lymphocytes.
4. Tonsils are multiple aggregations of large lymphatic nodules embedded in mucous membranes. They include the pharyngeal, palatine, and lingual tonsils.
5. The spleen is the largest mass of lymphatic tissue in the body and functions in production of lymphocytes and antibodies, phagocytosis of bacteria and worn-out red blood cells, and storage of blood.
6. The thymus gland functions in immunity by producing T cells.

Lymph Circulation (p. 661)

1. The passage of lymph is from interstitial fluid to lymph capillaries to lymphatic vessels to lymph trunks to the thoracic duct or right lymphatic duct to the subclavian veins.
2. Lymph flows as a result of skeletal muscle contractions and respiratory movements. It is also aided by valves in the lymphatics.

Nonspecific Resistance to Disease (p. 663)

1. The ability to ward off disease using a number of defenses is called resistance. Lack of resistance is called susceptibility.
2. Nonspecific resistance refers to a wide variety of body responses against a wide range of pathogens.
3. Nonspecific resistance includes mechanical factors (skin, mucous membranes, lacrimal apparatus, saliva, mucus, cilia, epiglottis, and flow of urine), chemical factors (gastric juice, acid pH of skin, unsaturated fatty acids, and lysozyme), phagocytosis, inflammation, fever, and antimicrobial substances (interferon, complement, and properdin).

Immunity (Specific Resistance to Disease) (p. 669)

1. Specific resistance to disease involves the production of a specific lymphocytes or antibody against a specific antigen and is called immunity.
2. Antigens (Ags) are chemical substances that, when introduced into the body, stimulate the production of antibodies that react with the antigen.
3. Examples of antigens are microbes, microbial structures, pollen, incompatible blood cells, and transplants.
4. Antigens are characterized by immunogenicity, reactivity, and multivalence.
5. Antibodies (Abs) are proteins produced in response to antigens.
6. Based on chemistry and structure, antibodies are distinguished into five principal classes, each with specific biological roles (IgG, IgA, ·IgM, IgD, and IgE).
7. Antibodies consist of heavy and light chains and variable and constant portions.
8. Cellular immunity refers to destruction of antigens by T cells, and humoral immunity refers to destruction of antigens by antibodies.
9. T cells are processed in the thymus gland; B cells may be processed in bone marrow, fetal liver and spleen, or gut-associated lymphoid tissue.
10. Macrophages process and present antigens to T cells and B cells and secrete interleukin-1, which induces proliferation of T cells and B cells.
11. T cells consist of several subpopulations; cytotoxic (killer) T cells secrete lymphotoxin that destroys, by lysis, antigens directly and other lymphokines (transfer factor, macrophage chemotactic factor, macrophage activating factor, macrophage inhibiting factor, and mitogenic factor) that destroy antigens indirectly; helper T cells cooperate with B cells to help amplify antibody production and secrete interleukin-2, which stimulates proliferation of cytotoxic T cells; suppressor T cells inhibit secretion of injurious substances by cytotoxic T cells and antibody production by plasma cells; delayed hypersensitivity T cells produce lymphokines and are important in hypersensitivity responses; amplifier T cells stimulate helper and suppressor T cells and plasma cells to greater levels of activity; and memory T cells recognize antigens at a later date.
12. Natural killer (NK) cells are lymphocytes with certain characteristics similar to those of cytotoxic T cells.
13. B cells develop into antibody-producing plasma cells under the influence of thymic hormones and interleukin-1; memory B cells recognize the original, invading antigen.
14. The anamnestic response provides the basis for immunization against certain diseases.
15. The immunologic properties of the skin are due to keratinocytes and nonpigmented granular dendrocytes (Langerhans cells and Granstein cells).
16. Monoclonal antibodies (MAbs) are pure antibodies produced by fusing a B cell with a tumor cell; they are important in diagnosis, detection of disease, treatment, preparing vaccines, and countering rejection by transplants and autoimmune diseases.
17. Cancer cells contain tumor-specific antigens and are frequently destroyed by the body's immune system (immunologic surveillance); some cancer cells escape detection and destruction, a phenomenon called immunologic escape.
18. Induction of the immune system against cancer is called immunotherapy.

Aging and the Immune System (p. 678)

1. With advancing age, individuals become more susceptible to infections and malignancies, response to vaccines is decreased, and more autoantibodies are produced.
2. Cellular and humoral responses also diminish.

Developmental Anatomy of the Lymphatic System (p. 679)

1. Lymphatic vessels develop from lymph sacs, which develop from veins. Thus, they are derived from mesoderm.
2. Lymph nodes develop from lymph sacs that become invaded by mesenchymal cells.

Disorders: Homeostatic Imbalances (p. 679)

1. Acquired immune deficiency syndrome (AIDS) lowers the body's immunity by decreasing the number of helper T cells and reversing the ratio of helper T cells to suppressor T cells. AIDS victims frequently develop Kaposi's sarcoma (KS) and *Pneumocystis carinii* pneumonia (PCP).
2. Autoimmune diseases result when the body does not recognize "self" antigens and produces antibodies against them. Several human autoimmune diseases are rheumatoid arthritis (RA), systemic lupus erythematosus (SLE), rheumatic fever, hemolytic and pernicious anemias, myasthenia gravis, and multiple sclerosis (MS).
3. Severe combined immunodeficiency (SCID) is an immunodeficiency disease in which both B cells and T cells are missing or inactive in providing immunity.
4. Hypersensitivity (allergy) is overreactivity to an antigen. Localized anaphylactic reactions include hay fever, asthma, eczema, and hives; acute anaphylaxis is a severe reaction with systemic effects.
5. Tissue rejection of a transplanted tissue or organ involves antibody production against the proteins (antigens) in the transplant. It may be overcome with immunosuppressive drugs.
6. Hodgkin's disease (HD) is a malignant disorder, usually arising in lymph nodes.

REVIEW QUESTIONS

1. Identify the components and functions of the lymphatic system. (p. 655)
2. How do lymphatic vessels originate? Compare veins and lymphatic vessels with regard to structure. (p. 655)
3. Describe the structure of a lymph node. What functions do lymph nodes serve? (p. 658)
4. Identify the tonsils by location. (p. 658)
5. Describe the location, gross anatomy, histology, and functions of the spleen. What is a splenectomy? (p. 658)
6. Describe the role of the thymus gland in immunity. (p. 660)
7. Construct a diagram to indicate the route of lymph circulation. (p. 662)
8. List and explain the various factors involved in the maintenance of lymph circulation. (p. 663)
9. Describe the various mechanical and chemical factors involved in nonspecific resistance. (p. 663)
10. What is phagocytosis? Describe the four phases involved in phagocytosis. (p. 664)
11. Define an inflammation. Describe the principal symptoms associated with inflammation and outline its stages. (p. 666)
12. Outline the role of the following antimicrobial substances: interferon (IFN), complement, and properdin. (p. 668)
13. Define immunity. Distinguish between an antigen (Ag) and an antibody (Ab). (p. 669)
14. List the various characteristics of antigens and antibodies. (pp. 669, 670)
15. How are T cells and B cells processed? How do they differ in function? (p. 672)
16. What is the function of macrophages in immunity? (p. 672)
17. Describe the role of T cells in cellular immunity. Be sure to cite the roles of each type of T cell. (p. 672)
18. What are natural killer (NK) cells? How do they differ from cytotoxic (killer) T cells? (p. 674)
19. Describe the role of B cells in humoral immunity. (p. 674)
20. Discuss the importance of the secondary (anamnestic) response of the body to an antigen. (p. 675)
21. Describe the role of the skin in immunity. (p. 676)
22. What are monoclonal antibodies (MAbs)? How are they produced? Why are they clinically important? (p. 677)
23. Explain how immunology is related to cancer. (p. 678)
24. Describe the effects of aging on the immune system. (p. 678)
25. Describe how the lymphatic system develops. (p. 679)
26. Describe the cause and symptoms of acquired immune deficiency syndrome (AIDS). What are the complications of AIDS? (p. 679)
27. Define an autoimmune disease. Give several examples. (p. 683)
28. What is severe combined immunodeficiency (SCID)? (p. 683)
29. Define hypersensitivity (allergy) and distinguish among the four types of hypersensitivity reactions. (p. 683)
30. Why does tissue rejection occur? How is this problem overcome? (p. 684)
31. How are transplants classified? (p. 684)
32. Define Hodgkin's disease (HD). (p. 684)
33. Define the following: metastasis (p. 658), splenectomy (p. 660), abscess (p. 668), ulcer (p. 668), primary response (p. 675), secondary response (p. 675), and immunotherapy. (p. 678)
34. What is lymphangiography? What is its diagnostic value? (p. 656)
35. Refer to the glossary of medical terminology associated with the lymphatic system. Be sure that you can define each term. (p. 685)

SELECTED READINGS

Ada, G. L., and G. Nossal. "The Clonal Selection Theory," *Scientific American,* August 1987.

Barnes, D. M. "Obstacles to an AIDS Vaccine," *Science,* 6 May 1988.

Bolognesi, D. P. "Prospects for Prevention of and Early Intervention Against HIV," *Journal of the American Medical Association,* 26 May 1989.

Carpenter, C. B., and T. B. Strom. "Transplantation: Immunogenetic and Clinical Aspects," *Hospital Practice,* December 1982.

Cohen, I. R. "The Self, the World and Immunity," *Scientific American,* April 1988.

Ding, E. Y., and Z. A. Cohn. "How Killer Cells Kill," *Scientific American,* January 1988.

Frolich, E. D. "Research into AIDS: Status, Prospects, Byproducts," *Hospital Medicine,* August 1988.

Gallo, R. C., and L. Montagnier. "AIDS in 1988," *Scientific American,* October 1988.

Haseltine, W. A., and F. Wong-Staal. "The Molecular Biology of the AIDS Virus," *Scientific American,* October 1988.

Hayward, W. L., and J. W. Curran. "The Epidemiology of AIDS in the US," *Scientific American,* October 1988.

Henry, K., J. Thurn, and D. Anderson. "Testing for Human Immunodeficiency Virus," *Postgraduate Medicine,* January 1989.

Johnson, R. B. "Current Concepts: Immunology," *New England Journal of Medicine,* 24 March 1988.

Koop, C. E. "Surgeon General's Report on Acquired Immune Deficiency Syndrome," *U.S. Public Health Service Office,* 22 October 1986.

Marx, J. L. "AIDS Drugs—Coming but Not Here," *Science,* 21 April 1989.

Matthews, T. J., and D. P. Bolognese. "AIDS Vaccines," *Scientific American,* October 1988.

Redfield, R. R., and D. S. Burke. "HIV Infections: The Clinical Picture," *Scientific American,* October 1988.

Robertson, S. "Drugs That Keep AIDS Patients Alive," *RN,* February 1989.

Scutchfield, F. D., and A. S. Benenson. "AIDS Update," *Postgraduate Medicine,* March 1989.

Silberner, J. "AIDS: Disease, Research Efforts Advance," *Science News,* 27 April 1985.

Soto-Aguilar, M. C., R. D. de Shazo, and N. P. Waring. "Anaphylaxis," *Postgraduate Medicine,* October 1987.

Springate, C. F., M. A. Flaum, and E. S. Cooper. "The Therapeutic Potential of Interleukins," *Modern Medicine,* October 1987.

Tonegawa, S. "The Molecules of the Immune System," *Scientific American,* October 1985.

Weber, J. N., and R. A. Weiss. "HIV Infection: The Cellular Picture," *Scientific American,* October 1988.

Weiss, R. "HIV Can Linger Years with No Antibodies," *Science News,* 3 June 1989.

Yarchoan, R., and S. Broder. "AIDS Therapies," *Scientific American,* October 1988.

Chapter 23

The Respiratory System

Chapter Contents at a Glance

Student Objectives

1. Identify the organs of the respiratory system and describe their functions.
2. Explain the structure of the alveolar-capillary (respiratory) membrane and describe its function in the diffusion of respiratory gases.
3. Describe the events involved in inspiration and expiration.
4. Explain how respiratory gases are carried by blood.
5. Describe the various factors that control the rate of respiration.
6. Describe the effects of aging on the respiratory system.
7. Describe the development of the respiratory system.
8. Define bronchogenic carcinoma (lung cancer), bronchial asthma, bronchitis, emphysema, pneumonia, tuberculosis (TB), respiratory distress syndrome (RDS) of the newborn, respiratory failure, sudden infant death syndrome (SIDS), coryza (common cold), influenza (flu), pulmonary embolism, pulmonary edema, and smoke inhalation injury as disorders of the respiratory system.
9. Define medical terminology associated with the respiratory system.

Cells need a continuous supply of oxygen (O_2) for various metabolic reactions that release energy from nutrient molecules, some of which is stored in ATP for cellular use. As a result of these reactions, cells also release quantities of carbon dioxide (CO_2). (As you will see in Chapter 25, oxygen consumption and carbon dioxide production occur in mitochondria due to the cellular respiration which occurs there.) Since an excessive amount of carbon dioxide produces acid conditions that are poisonous to cells, the excess gas must be eliminated quickly and efficiently. The two systems that supply oxygen and eliminate carbon dioxide are the cardiovascular system and the respiratory system. The ***respiratory system*** consists of the nose, pharynx, larynx, trachea, bronchi, and lungs (Figure 23-1). The cardiovascular system transports the gases in the blood between the lungs and the cells. The term ***upper respiratory system*** refers to the nose, throat, and associated structures. The ***lower respiratory system*** refers to the larynx, trachea, bronchi, and lungs.

The overall exchange of gases between the atmosphere, blood, and cells is called ***respiration.*** Three basic processes are involved. The first process, ***pulmonary*** (*pulmo* = lung) ***ventilation,*** or breathing, is the inspiration (inflow) and expiration (outflow) of air between the atmosphere and the lungs. The second and third processes involve the exchange of gases within the body. ***External respiration*** is the exchange of gases between the lungs and blood. ***Internal respiration*** is the exchange of gases between the blood and the cells.

The respiratory and cardiovascular systems participate equally in respiration. Failure of either system has the same effect on the body: disruption of homeostasis and rapid death of cells from oxygen starvation and buildup of waste products.

The developmental anatomy of the respiratory system is considered later in the chapter.

ORGANS

Nose

Anatomy

The ***nose*** has an external portion and an internal portion inside the skull (Figure 23-2a–c). The external portion consists of a supporting framework of bone and cartilage covered with skin and lined with mucous membrane. The bridge of the nose is formed by the nasal bones, which hold it in a fixed position. Because it has a framework of pliable cartilage, the rest of the external nose is somewhat flexible. On the undersurface of the external nose are two openings called the ***external nares*** (NA-rēz; *sing.,* naris), or ***nostrils.*** The surface anatomy of the nose is shown in Figure 23-2a.

The internal portion of the nose is a large cavity in the skull that lies inferior to the cranium and superior to the mouth. Anteriorly, the internal nose merges with the external nose, and posteriorly it communicates with the throat (pharynx) through two openings called the ***internal nares (choanae).*** Four paranasal sinuses (frontal, sphenoidal, maxillary, and ethmoidal) and the nasolacrimal ducts also open into the internal nose. The lateral walls of the internal nose are formed by the ethmoid, maxillae, lacrimal, palatine, and inferior nasal conchae bones. The ethmoid also forms the roof. The floor is formed by the soft palate and palatine bones and palatine process of the maxilla, which together comprise the hard palate.

The inside of both the external and internal nose consists of a ***nasal cavity,*** divided into right and left sides by a vertical partition called the ***nasal septum.*** The anterior portion of the septum is made primarily of cartilage. The remainder is formed by the vomer, perpendicular plate of the ethmoid, maxillae, and palatine bones (see Figure 7-7a). The anterior portion of the nasal cavity, just inside the nostrils, is called the ***vestibule.*** It is surrounded by cartilage. The upper nasal cavity is surrounded by bone.

CLINICAL APPLICATION: RHINOPLASTY

Rhinoplasty **(RĪ-nō-plas'-tē; *rhino* = nose; *plassein* = to form), also called a "nose job," is a surgical procedure in which the structure of the external nose is altered. Although it is frequently done for cosmetic reasons, it is sometimes performed to repair a fractured nose or deviated nasal septum (DNS). In the procedure, a local anesthetic is given, and with instruments inserted through the nostrils, the nasal bones are fractured and repositioned to achieve the desired shape. Any extra bone fragments or cartilage are shaved down and removed through the nostrils. Both an internal packing and an external splint keep the nose in the desired position while it heals. In some cases, only the cartilage has to be reshaped to achieve the desired appearance.**

Physiology

The interior structures of the nose are specialized for three functions: incoming air is warmed, moistened, and filtered; olfactory stimuli are received; and large, hollow resonating chambers are provided for speech sounds.

When air enters the nostrils, it passes first through the vestibule. The vestibule is lined by skin containing coarse hairs that filter out large dust particles. The air then passes into the upper nasal cavity. Three shelves formed by projections of the superior, middle, and inferior nasal conchae extend out of the lateral wall of the cavity. The conchae, almost reaching the septum, subdivide each side of the nasal cavity into a series of groovelike passageways—the ***superior, middle,*** and ***inferior meatuses*** (singular is ***meatus***). Mucous membrane lines the cavity and

FIGURE 23-1 Organs of the respiratory system in relation to surrounding structures. (a) Diagram. (b) Photograph of lungs after removal of anterolateral thoracic wall and parietal pleura. In this specimen, the lungs occupy more of the mediastinum than normal. (Courtesy of J. A. Gosling, P. F. Harris, et al., *Atlas of Human Anatomy,* Gower Medical Publishing Ltd., 1985.)

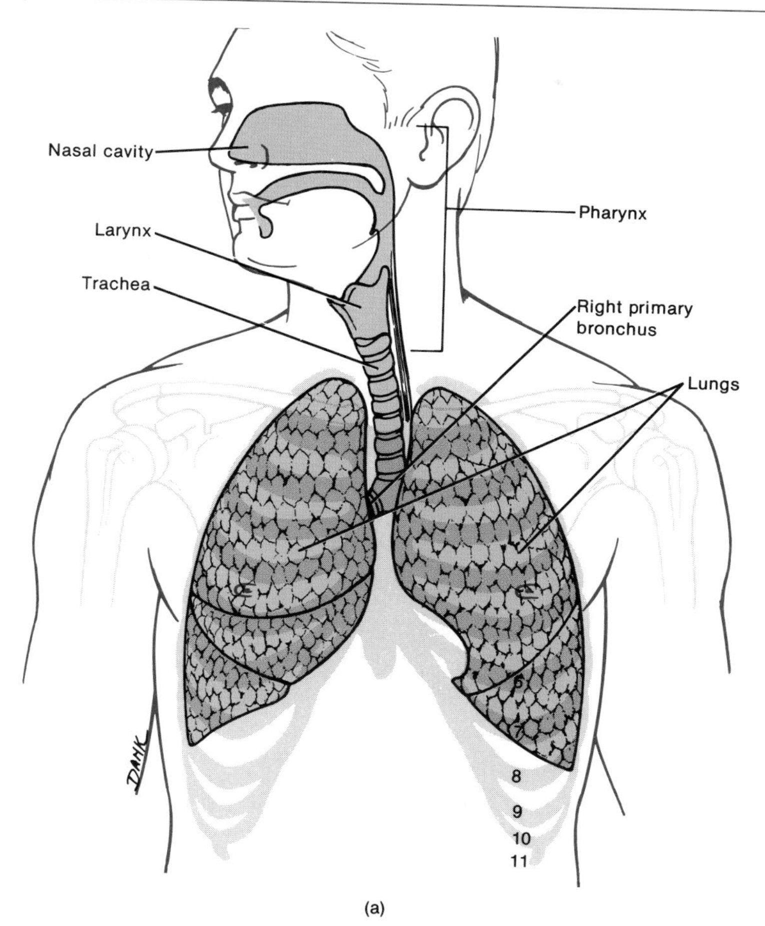

(a)

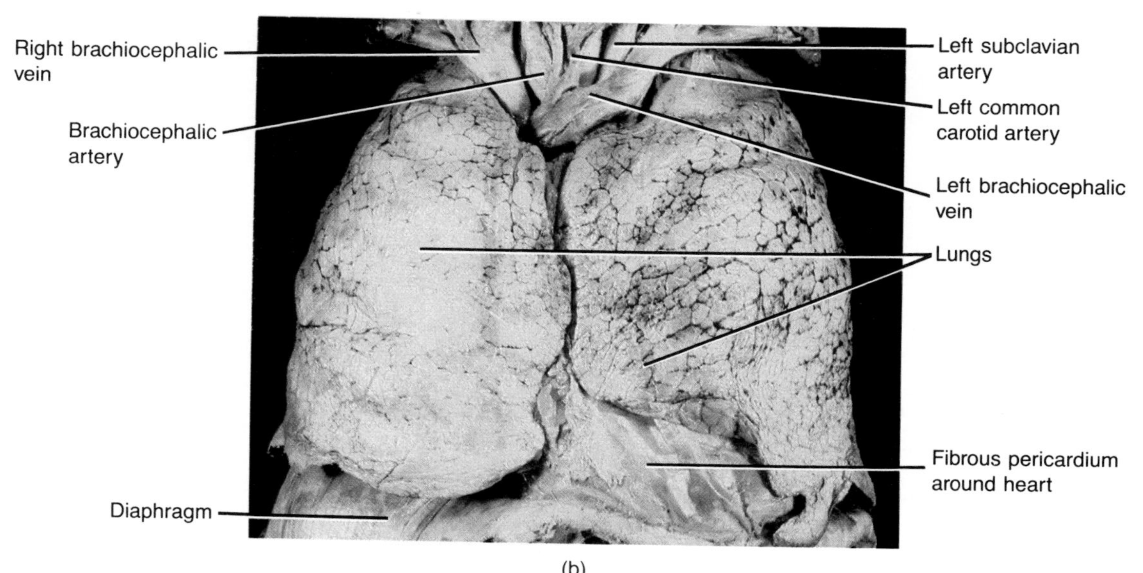

(b)

FIGURE 23-2 Respiratory organs in the head and neck. (a) Surface anatomy of the nose in anterior view. (Copyright © 1982 by Gerard J. Tortora. Courtesy of Lynne Tortora.) (b) Diagram of the left side of the head and neck seen in sagittal section with the nasal septum removed.

1. **Root.** Superior attachment of nose at forehead located between eyes.
2. **Apex.** Tip of nose.
3. **Dorsum nasi.** Rounded anterior border connecting root and apex; in profile, may be straight, convex, concave, or wavy.
4. **Nasofacial angle.** Point at which side of nose blends with tissues of face.
5. **Ala.** Convex flared portion of inferior lateral surface; unites with upper lip.
6. **External naris.** External opening into nose.
7. **Bridge.** Superior portion of dorsum nasi, superficial to nasal bones.
8. **Philtrum.** Vertical groove in medial portion of upper lip.
9. **Lips.** Upper and lower fleshy borders of oral cavity.

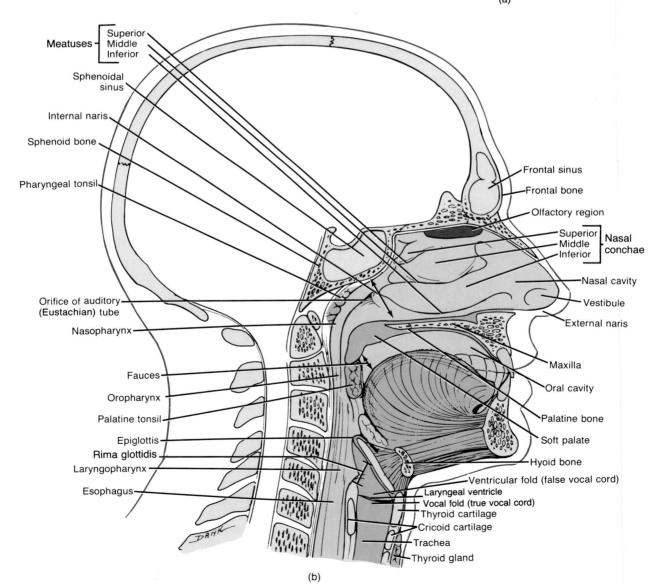

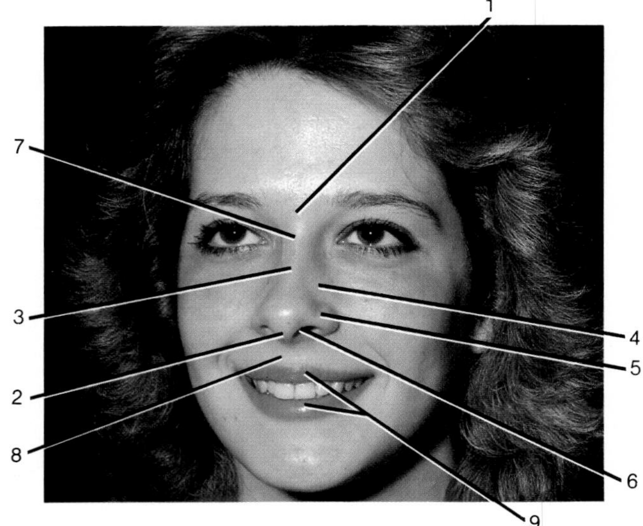

(a)

Meatuses
Superior
Middle
Inferior
Sphenoidal sinus
Internal naris
Sphenoid bone
Pharyngeal tonsil
Orifice of auditory (Eustachian) tube
Nasopharynx
Fauces
Oropharynx
Palatine tonsil
Epiglottis
Rima glottidis
Laryngopharynx
Esophagus
Frontal sinus
Frontal bone
Olfactory region
Nasal conchae
Superior
Middle
Inferior
Nasal cavity
Vestibule
External naris
Maxilla
Oral cavity
Palatine bone
Soft palate
Hyoid bone
Ventricular fold (false vocal cord)
Laryngeal ventricle
Vocal fold (true vocal cord)
Thyroid cartilage
Cricoid cartilage
Trachea
Thyroid gland
DANK

(b)

FIGURE 23-2 (*Continued*) (c) Photograph of the left side of the nasal cavity seen in sagittal section with the nasal septum removed. (Courtesy of J. A. Gosling, P. F. Harris, et al., *Atlas of Human Anatomy,* Gower Medical Publishing Ltd., 1985.)

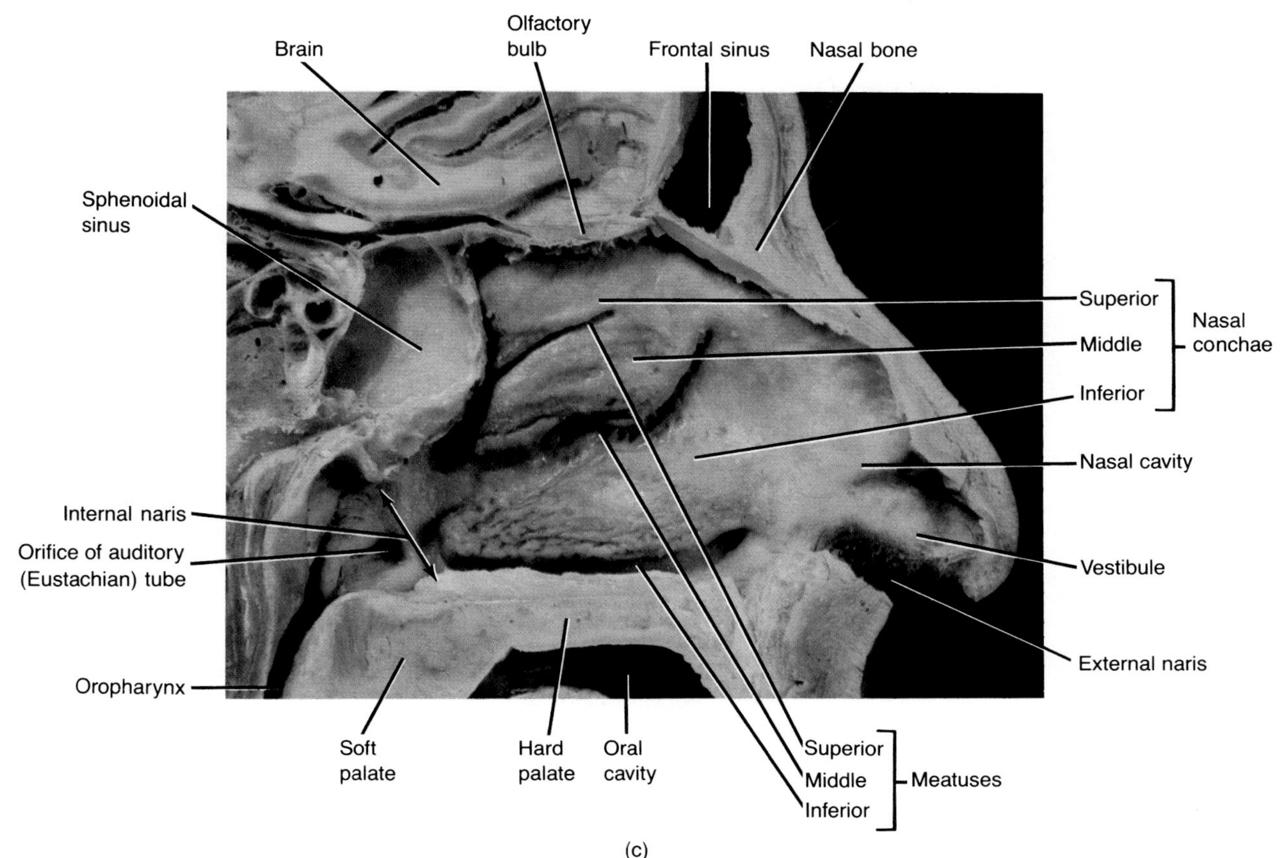

its shelves. The olfactory receptors lie in the membrane lining the area superior to the superior nasal conchae, called the ***olfactory region.*** Below the olfactory region, the membrane contains capillaries and pseudostratified ciliated columnar cells with many goblet cells. As the air whirls around the conchae and meatuses, it is warmed by the capillaries. Mucus secreted by the goblet cells moistens the air and traps dust particles. Drainage from the lacrimal ducts and perhaps secretions from the paranasal sinuses also help moisten the air. The cilia move the mucus–dust packages along the pharynx so they can be eliminated from the body.

Pharynx

The ***pharynx*** (FAR-inks), or throat, is a somewhat funnel-shaped tube about 13 cm (5 inches) long that starts at the internal nares and extends to the level of the cricoid cartilage (Figure 23-3). It lies just posterior to the nasal cavity, oral cavity, and larynx and just anterior to the cervical vertebrae. Its wall is composed of skeletal muscles and lined with mucous membrane. The functions of the pharynx are to serve as a passageway for air and food and to provide a resonating chamber for speech sounds.

The branch of medicine that deals with the diagnosis and treatment of diseases of the ears, nose, and throat is called ***otorhinolaryngology*** (ō'-tō-rī'-nō-lar'-in-GOL-ō-jē; *oto* = ear; *rhino* = nose).

The uppermost portion of the pharynx, called the ***nasopharynx,*** lies posterior to the internal nasal cavity and extends to the plane of the soft palate. There are four openings in its wall: two internal nares and two openings that lead into the auditory (Eustachian) tubes. The posterior wall also contains the pharyngeal tonsil, or adenoid. Through the internal nares, the nasopharynx receives air from the nasal cavities and receives the packages of dust-laden mucus. It is lined with pseudostratified ciliated epithelium, and the cilia move the mucus down toward the pharynx. The nasopharynx also exchanges small amounts of air with the auditory (Eustachian) tubes so that the air pressure inside the middle ear equals the pressure of the atmospheric air flowing through the nose and pharynx.

The middle portion of the pharynx, the ***oropharynx,*** lies posterior to the oral cavity and extends from the soft palate inferiorly to the level of the hyoid bone. It has

FIGURE 23-3 **Photograph of the head and neck seen in sagittal section. (Courtesy of J. A. Gosling, P. F. Harris, et al., *Atlas of Human Anatomy,* Gower Medical Publishing Ltd., 1985.)**

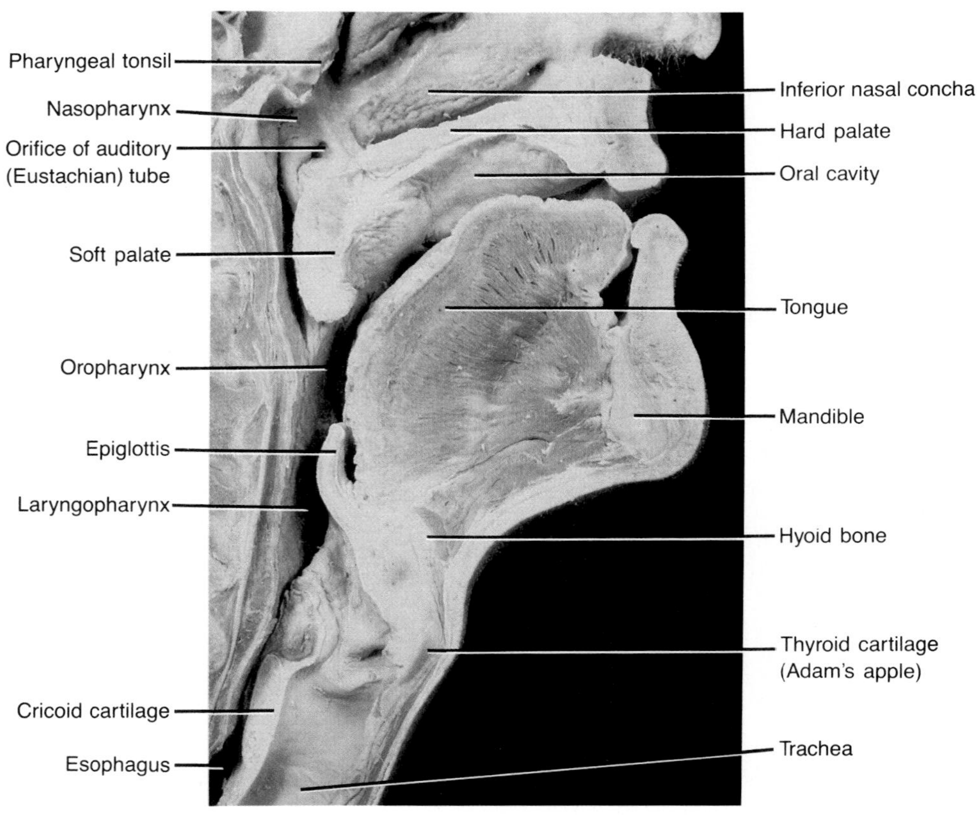

only one opening, the ***fauces*** (FAW-sēz), the opening from the mouth. It is lined by stratified squamous epithelium. This portion of the pharynx is both respiratory and digestive in function, since it is a common passageway for air, food, and drink. Two pairs of tonsils, the palatine and lingual tonsils, are found in the oropharynx. The lingual tonsil lies at the base of the tongue (see also Figure 17-2c).

The lowest portion of the pharynx, the ***laryngopharynx*** (la-rin'-gō-FAR-inks), extends downward from the hyoid bone and becomes continuous with the esophagus (food tube) posteriorly and the larynx (voice box) anteriorly. Like the oropharynx, the laryngopharynx is a respiratory and a digestive pathway and is lined by stratified squamous epithelium.

Larynx

The ***larynx,*** or voice box, is a short passageway that connects the pharynx with the trachea. It lies in the midline of the neck anterior to the fourth through sixth cervical vertebrae (C4–C6).

Anatomy

The wall of the larynx is composed of nine pieces of cartilage (Figure 23-4). Three are single and three are paired. The three single pieces are the thyroid cartilage, epiglottic cartilage (epiglottis), and cricoid cartilage. Of the paired cartilages, the arytenoid cartilages are the most important. The paired corniculate and cuneiform cartilages are of lesser significance.

The ***thyroid cartilage (Adam's apple)*** consists of two fused plates that form the anterior wall of the larynx and give it its triangular shape. It is larger in males than in females.

The ***epiglottis*** (*epi* = above; *glotta* = tongue) is a large, leaf-shaped piece of cartilage lying on top of the larynx (see also Figure 23-3). The "stem" of the epiglottis is attached to the thyroid cartilage, but the "leaf" portion is unattached and free to move up and down like a trap door. During swallowing, there is elevation of the larynx. This causes the free edge of the epiglottis to form a lid over the glottis, closing it off. The ***glottis*** consists of the vocal folds (true vocal cords) in the larynx and the space between them ***(rima glottidis).*** In this way, the larynx

FIGURE 23-4 Larynx. (a) Anterior view. (b) Posterior view.

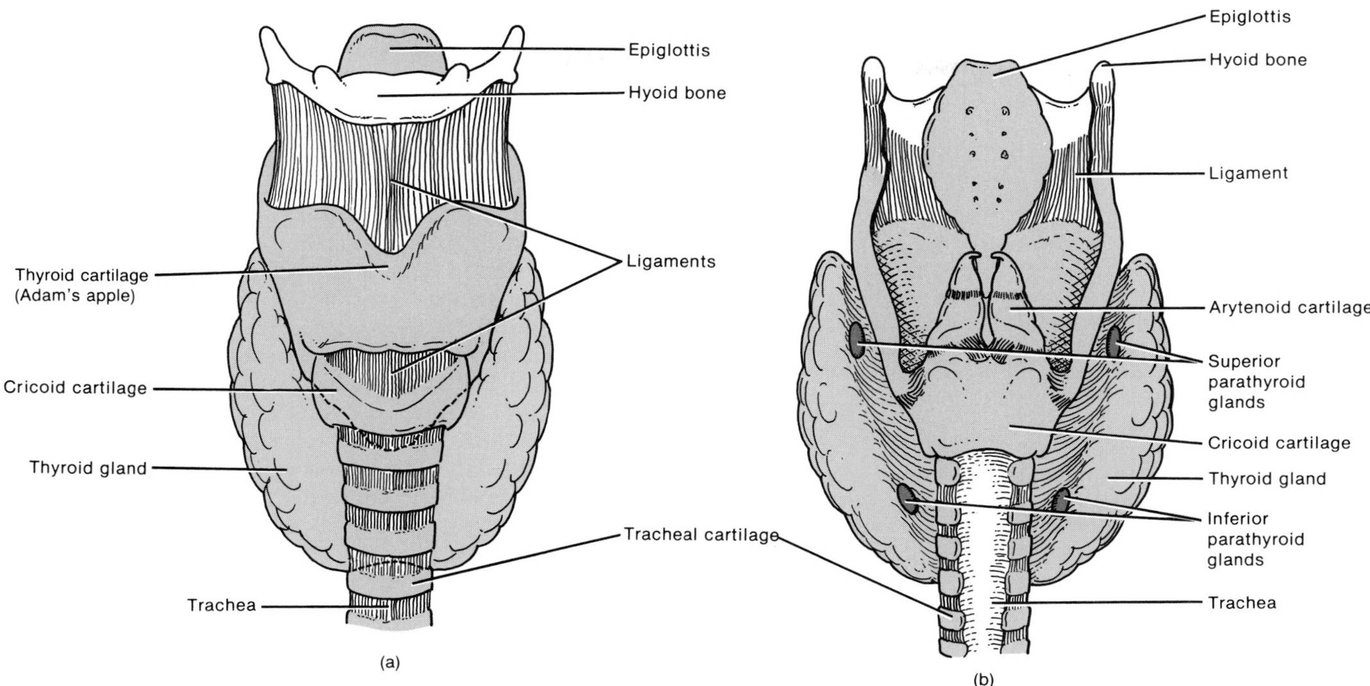

is closed off, and liquids and foods are routed into the esophagus and kept out of the larynx and air passageways below it. When anything but air passes into the larynx, a cough reflex attempts to expel the material.

The ***cricoid*** (KRĪ-koyd) ***cartilage*** is a ring of cartilage forming the inferior wall of the larynx. It is attached to the first ring of cartilage of the trachea.

The paired ***arytenoid*** (ar'-i-TĒ-noyd) ***cartilages*** are pyramidal in shape and located at the posterior, superior border of the cricoid cartilage. They attach to the vocal folds and intrinsic pharyngeal muscles and by their action can move the vocal folds.

The paired ***corniculate*** (kor-NIK-yoo-lāt) ***cartilages*** are horn-shaped. One is located at the apex of each arytenoid cartilage. The paired ***cuneiform*** (kyoo-NĒ-i-form) ***cartilages*** are club-shaped cartilages anterior to the corniculate cartilages and located in the aryepiglottic fold.

The epithelium lining the larynx below the vocal folds is pseudostratified. It consists of ciliated columnar cells, goblet cells, and basal cells, and it helps trap dust not removed in the upper passages.

Voice Production

The mucous membrane of the larynx is arranged into two pairs of folds—an upper pair called the ***ventricular folds (false vocal cords)*** and a lower pair called simply the ***vocal folds (true vocal cords).*** The space between the ventricular folds is known as the ***rima vestibuli.*** The ***laryngeal ventricle*** is a lateral expansion of the middle portion of the laryngeal cavity between the vestibular folds above and the vocal folds below. When the ventricular folds are brought together, they function in holding the breath against pressure in the thoracic cavity, such as might occur when a person exerts a strain while lifting a heavy weight. The mucous membrane of the vocal folds is lined by nonkeratinized stratified squamous epithelium. Under the membrane lie bands of elastic ligaments stretched between pieces of rigid cartilage like the strings on a guitar. Skeletal muscles of the larynx, called intrinsic muscles, are attached internally to the pieces of rigid cartilage and to the vocal folds themselves. When the muscles contract, they pull the strings of elastic ligaments tight and stretch the vocal folds out into the air passageways so that the glottis is narrowed. If air is directed against the vocal folds, they vibrate and set up sound waves in the column of air in the pharynx, nose, and mouth. The greater the pressure of air, the louder the sound.

Pitch is controlled by the tension on the vocal folds. If they are pulled taut by the muscles, they vibrate more rapidly, and a higher pitch results. Lower sounds are produced by decreasing the muscular tension on the vocal folds. Vocal folds are usually thicker and longer in males than in females, and therefore they vibrate more slowly. Thus, men generally have a lower range of pitch than women.

Sound originates from the vibration of the vocal folds, but other structures are necessary for converting the sound into recognizable speech. The pharynx, mouth, nasal cavity, and paranasal sinuses all act as resonating chambers that give the voice its human and individual quality. By constricting and relaxing the muscles in the wall of the pharynx, we produce the vowel sounds. Muscles of the face, tongue, and lips help us enunciate words.

CLINICAL APPLICATION: LARYNGITIS AND CANCER OF LARYNX

Laryngitis is an inflammation of the larynx that is most often caused by a respiratory infection or irritants such as cigaratte smoke. Inflammation of the vocal folds causes hoarseness or loss of voice by interfering with the contraction of the folds or by causing them to swell to the point where they cannot vibrate freely. Many long-term smokers acquire a permanent hoarseness from the damage done by chronic inflammation.

Cancer of the larynx is found almost exclusively in individuals who smoke. The condition is characterized by hoarseness, pain on swallowing, or pain radiating to an ear. Treatment consists of radiation therapy and/or surgery.

Trachea

The ***trachea*** (TRĀ-kē-a), or windpipe, is a tubular passageway for air about 12 cm (4.5 inch) in length and 2.5 cm (1 inch) in diameter. It is located anterior to the esophagus and extends from the larynx to the fifth thoracic vertebra (T5), where it divides into right and left primary bronchi (see Figure 23-6).

The wall of the trachea consists of a mucosa, submucosa, cartilaginous layer, and adventitia (outer layer of loose connective tissue). The tracheal epithelium of the mucosa is pseudostratified. It consists of ciliated columnar cells that reach the luminal surface, goblet cells, and basal cells that do not reach the luminal surface (Figure 23-5). The epithelium provides the same protection against dust as the membrane lining the larynx. Seromucous glands and their ducts are present in the submucosa. The cartilaginous layer consists of 16 to 20 horizontal incomplete rings of hyaline cartilage that look like a series of letter *C*s stacked one on top of another. The open parts of the *C*s face the esophagus and permit it to expand slightly into the trachea during swallowing. Transverse smooth muscle fibers, called the ***trachealis muscle,*** and elastic connective tissue, attach the open ends of the cartilage rings. The solid parts of the *C*s provide a rigid support so the tracheal wall does not collapse inward and obstruct the air passageway. In some situations, however, such as crushing injuries to the chest, the rings of cartilage may not be strong enough to overcome collapse and obstruction of the trachea.

At the point where the trachea bifurcates into right and left primary bronchi, there is an internal ridge called the ***carina*** (ka-RĪ-na). It is formed by a posterior and somewhat inferior projection of the last tracheal cartilage. The mucous membrane of the carina is one of the most sensitive areas of the respiratory system and is associated with the cough reflex. Widening and distortion of the carina, which can be seen in an examination by bronchoscopy, is a serious prognostic sign, since it usually indicates a carcinoma of the lymph nodes around the bifurcation of the trachea.

FIGURE 23-5 Photomicrograph of enlarged aspect of the tracheal epithelium at a magnification of 600×. (Courtesy of Andrew Kuntzman.)

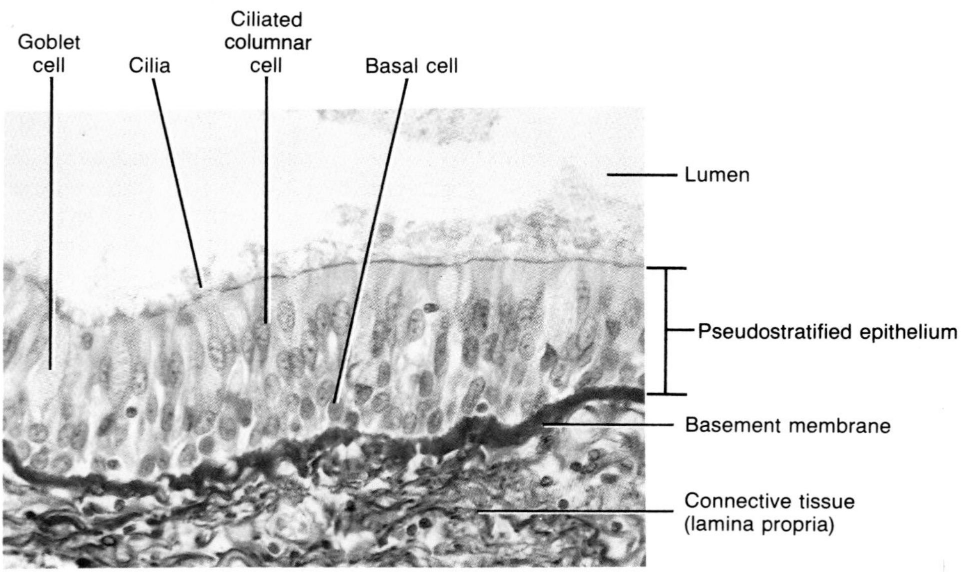

CLINICAL APPLICATION: TRACHEOSTOMY AND INTUBATION

Occasionally, the respiratory passageways are unable to protect themselves from obstruction. The rings of cartilage may accidentally be crushed; the mucous membrane may become inflamed and swell so much that it closes off the air passageways; inflamed membranes secrete a great deal of mucus that may clog the lower respiratory passageways; a large object may be breathed in (aspirated) while the glottis is open; or an aspirated foreign object may cause spasm of the laryngeal muscles. The passageways must be cleared quickly. If the obstruction is above the level of the larynx, a ***tracheostomy*** (trā-kē-OS-tō-mē) may be performed. A skin incision is made, followed by a short longitudinal incision into the trachea inferior to the cricoid cartilage. The patient breathes through a metal or plastic tracheal tube inserted through the incision. Another method is ***intubation.*** A tube is inserted into the mouth or nose and passed down through the larynx and trachea. The firm wall of the tube pushes back any flexible obstruction, and the inside of the tube provides a passageway for air. If mucus is clogging the trachea, it can be suctioned out through the tube.

Bronchi

The trachea terminates in the chest by dividing at the sternal angle into a ***right primary bronchus*** (BRON-kus), which goes to the right lung, and a ***left primary bronchus,*** which goes to the left lung (Figure 23-6). The right primary bronchus is more vertical, shorter, and wider than the left. As a result, foreign objects in the air passageways are more likely to enter it than the left and frequently lodge in it. Like the trachea, the primary bronchi (BRON-kē) contain incomplete rings of cartilage and are lined by pseudostratified ciliated epithelium.

On entering the lungs, the primary bronchi divide to form smaller bronchi—the ***secondary (lobar) bronchi,*** one for each lobe of the lung (the right lung has three lobes; the left lung has two). The secondary bronchi continue to branch, forming still smaller bronchi, called ***tertiary (segmental) bronchi,*** that divide into ***bronchioles.*** Bronchioles, in turn, branch into even smaller tubes called ***terminal bronchioles.*** This continuous branching from the trachea resembles a tree trunk with its branches and is commonly referred to as the ***bronchial tree. Bronchography*** (bron-KOG-ra-fē) is a technique for examining the bronchial tree. An intratracheal catheter is passed into the mouth or nose, through the

FIGURE 23-6 Bronchial tree.

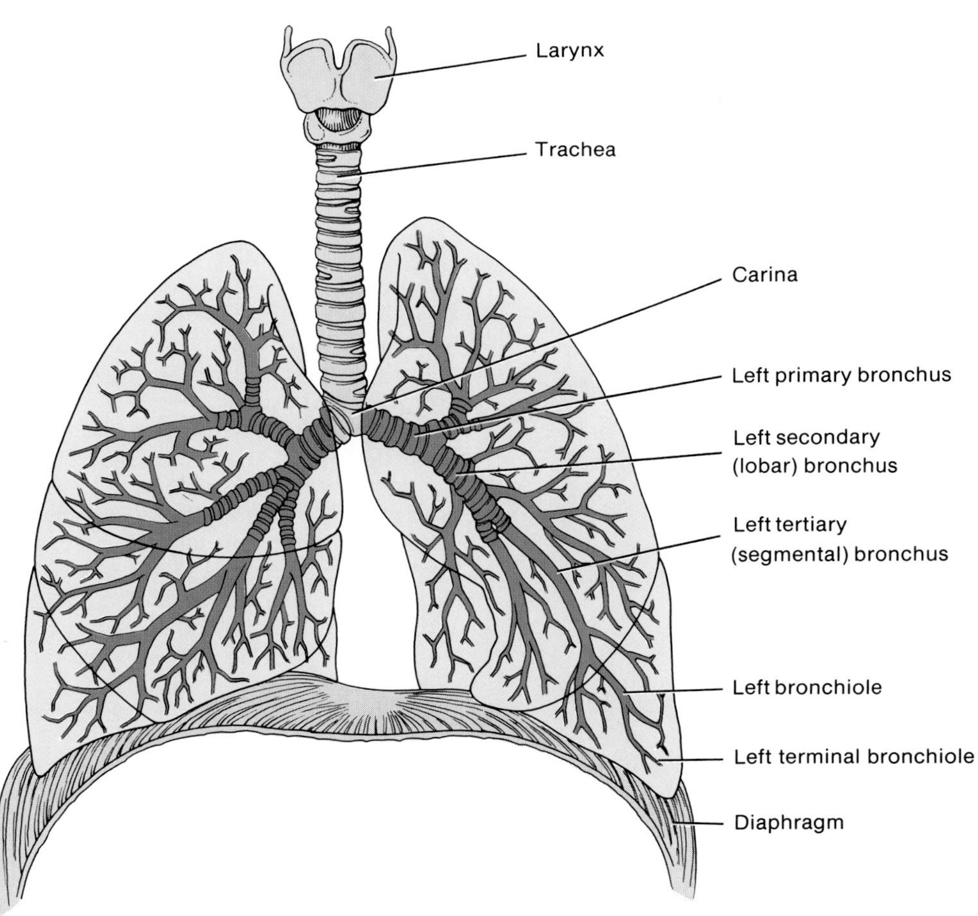

glottis, and into the trachea. Then an opaque contrast medium, usually containing iodine, is introduced by means of gravity into the trachea and distributed through the bronchial branches. Roentgenograms of the chest in various positions are taken, and the developed film, a ***bronchogram*** (BRON-kō-gram), provides a picture of the tree.

As the branching becomes more extensive in the bronchial tree, several structural changes may be noted. First, rings of cartilage are replaced by plates of cartilage that finally disappear in the bronchioles. Second, as the cartilage decreases, the amount of smooth muscle increases. Third, the epithelium changes from pseudostratified ciliated to simple cuboidal in the terminal bronchioles.

CLINICAL APPLICATION: ASTHMA ATTACK

The fact that the walls of the bronchioles contain a great deal of smooth muscle but no cartilage is clinically significant. During an *asthma attack* these smooth muscles contract, reducing the diameter of the airways. Because there is no supporting cartilage, the spasms can even close off the air passageways. Movement of air through constricted tubes results in shortness of breath (dyspnea) and may or may not cause wheezing.

MEDICAL TEST

Bronchoscopy (bron-KOS-kō-pē; *bronchos* = windpipe; *skopein* = to examine)

Diagnostic Value: Using a lighted instrument, called a **bronchoscope,** the examiner can view the interior of the trachea and bronchi to biopsy a tumor, clear an obstructing object or secretions from an airway, take cultures or smears for microscopic examination, stop bleeding, or deliver drugs.

Procedure: In flexible bronchoscopy, an anesthetic solution is sprayed into the back of the throat following an intravenous sedation. A flexible (fiberoptic) bronchoscope is introduced into the nose or mouth and threaded through the larynx and trachea to the bronchial tree. Each step of the way, anesthesia is introduced so that the area in front of the bronchoscope is anesthetized. In rigid bronchoscopy, rarely done anymore, the patient may require a general anesthesia.

Lungs

The ***lungs*** (*lunge* = light, since the lungs float) are paired cone-shaped organs lying in the thoracic cavity. They are separated from each other by the heart and other structures in the mediastinum (see Figure 20-1). Two layers of serous membrane, collectively called the ***pleural membrane,*** enclose and protect each lung. The outer layer is attached to the wall of the thoracic cavity and is called the ***parietal pleura.*** The inner layer, the ***visceral pleura,*** covers the lungs themselves. Between the visceral and parietal pleura is a small potential space, the ***pleural cavity,*** which contains a lubricating fluid secreted by the membranes (see Figure 1-7d). This fluid prevents friction between the membranes and allows them to move easily on one another during breathing.

CLINICAL APPLICATION: PNEUMOTHORAX, HEMOTHORAX, AND PLEURISY

In certain conditions, the pleural cavity may fill with air (*pneumothorax; pneumo* = air or breath), blood (*hemothorax*), or pus. Air in the pleural cavity, most commonly introduced in a surgical opening of the chest or as a result of a stab or gunshot wound, may cause the lung to collapse (atelectasis). Fluid can be drained from the pleural cavity by inserting a needle, usually posteriorly through the seventh intercostal space. The needle is passed along the superior border of the lower rib to avoid damage to the intercostal nerves and blood vessels. Below the seventh intercostal space there is danger of penetrating the diaphragm.

Inflammation of the pleural membrane, or *pleurisy,* may in its early stages cause pain due to friction between the parietal and visceral layers of the pleura. If the infection persists, fluid accumulates in the pleural space, a condition known as *pleural effusion.* One cause of pleural effusion is cancer that may have arisen in the lung or reached the pleura from some other cancer site.

Gross Anatomy

The lungs extend from the diaphragm to a point about 1.5 to 2.5 cm (0.75–1 inches) superior to the clavicles and lie against the ribs anteriorly and posteriorly. The broad inferior portion of the lung, the ***base,*** is concave and fits over the convex area of the diaphragm (Figure 23-7). The narrow superior portion of the lung is termed the ***apex (cupula).*** The surface of the lung lying against the ribs, the ***costal surface,*** is rounded to match the curvature of the ribs. The ***mediastinal (medial) surface*** of each lung contains a region, the ***hilus,*** through which bronchi, pulmonary vessels, lymphatic vessels, and nerves enter and exit. These structures are held together by the pleura and connective tissue and constitute the ***root*** of the lung. Medially, the left lung also contains a concavity, the ***cardiac notch,*** in which the heart lies.

The right lung is thicker and broader than the left. It is also somewhat shorter than the left because the diaphragm is higher on the right side to accommodate the liver that lies below it.

FIGURE 23-7 Lungs. (a) Right lung, lateral view. (b) Left lung, lateral view. (c) Right lung, medial view. (d) Left lung, medial view.

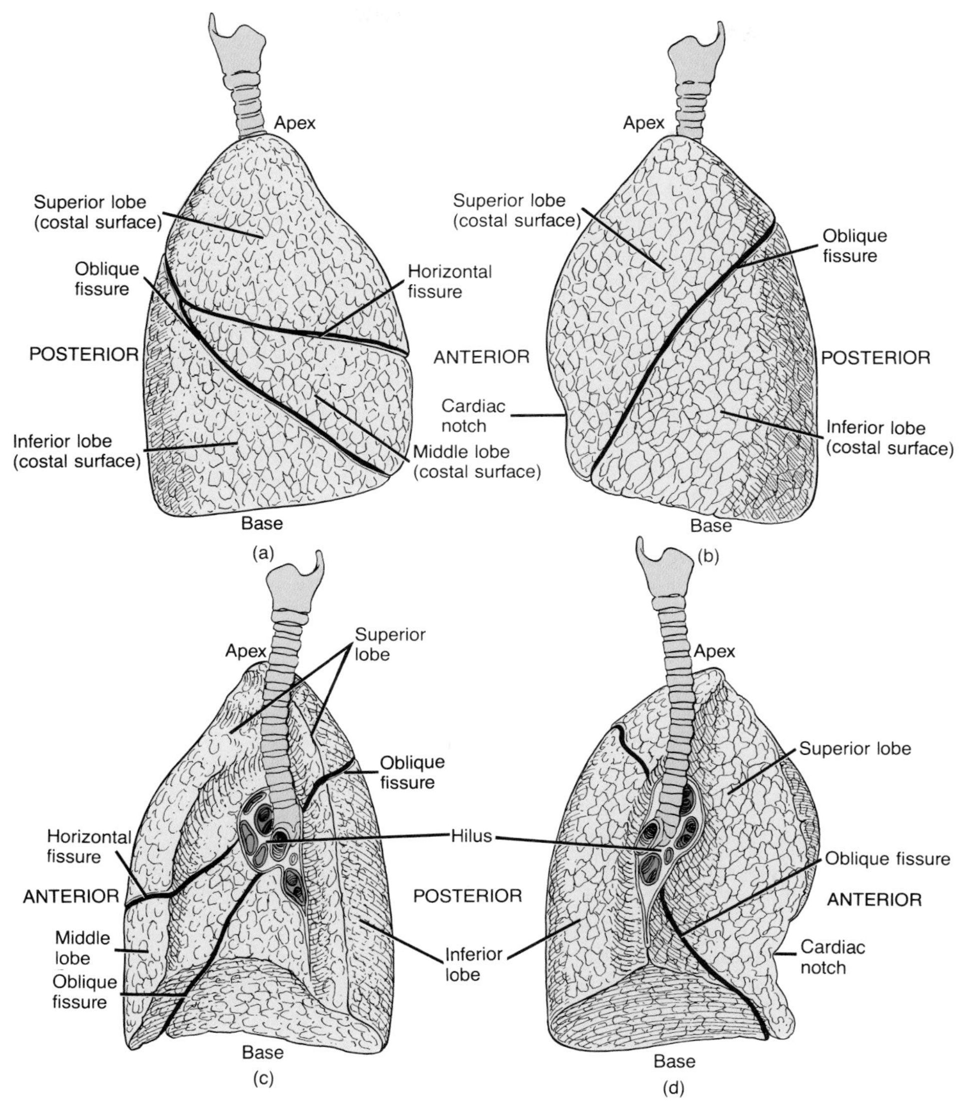

Lobes and Fissures

Each lung is divided into lobes by one or more fissures (Figure 23-7). Both lungs have an ***oblique fissure,*** which extends downward and forward. The right lung also has a ***horizontal fissure.*** The oblique fissure in the left lung separates the ***superior lobe*** from the ***inferior lobe.*** The upper part of the oblique fissure of the right lung separates the superior lobe from the inferior lobe, whereas the lower part of the oblique fissure separates the inferior lobe from the ***middle lobe.*** The horizontal fissure of the right lung subdivides the superior lobe, thus forming a middle lobe.

Each lobe receives its own secondary (lobar) bronchus. Thus, the right primary bronchus gives rise to three secondary (lobar) bronchi called the ***superior, middle,*** and ***inferior secondary (lobar) bronchi.*** The left primary bronchus gives rise to a ***superior*** and an ***inferior secondary (lobar) bronchus.*** Within the substance of the lung, the secondary bronchi give rise to the ***tertiary (segmental) bronchi,*** which are constant in both origin and distribution. The segment of lung tissue that each supplies is called a ***bronchopulmonary segment.*** Bronchial and pulmonary disorders, such as tumors or abscesses, may be localized in a bronchopulmonary segment and may be surgically removed without seriously disrupting surrounding lung tissue.

FIGURE 23-8 Histology of the lungs. (a) Diagram of a portion of a lobule of the lung. (b) Photomicrograph of alveolar ducts, alveolar sacs, and alveoli at a magnification of 55×. (Courtesy of Michael H. Ross and Edward J. Reith, *Histology: A Text and Atlas,* Copyright © 1985, by Michael H. Ross and Edward J. Reith, Harper & Row, Publishers, Inc., New York.)

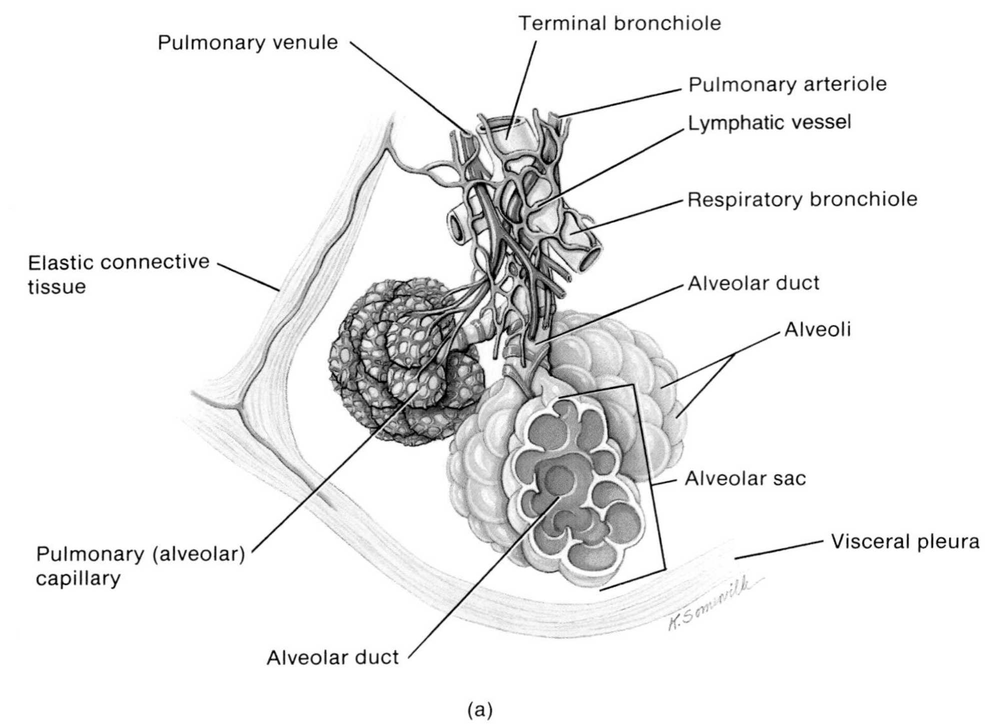

(a)

Lobules

Each bronchopulmonary segment of the lungs is broken up into many small compartments called ***lobules*** (Figure 23-8a). Each lobule is wrapped in elastic connective tissue and contains a lymphatic vessel, an arteriole, a venule, and a branch from a terminal bronchiole. Terminal bronchioles subdivide into microscopic branches called ***respiratory bronchioles.*** As the respiratory bronchioles penetrate more deeply into the lungs, the epithelial lining changes from cuboidal to squamous. Respiratory bronchioles, in turn, subdivide into several (2–11) ***alveolar ducts (atria).***

Around the circumference of the alveolar ducts are numerous alveoli and alveolar sacs. An ***alveolus*** (al-VĒ-ō-lus) is a cup-shaped outpouching lined by epithelium and supported by a thin elastic basement membrane. ***Alveolar sacs*** are two or more alveoli that share a common opening (Figure 23-8a,b). The alveolar walls consist of several types of epithelial cells (Figure 23-9). The ***squamous pulmonary epithelial cells*** are large cells that form a continuous lining of the alveolar wall, except for occasional septal cells. ***Septal cells*** are much smaller, somewhat cuboidal in shape, and are dispersed among the squamous pulmonary epithelial cells. Septal cells pro-

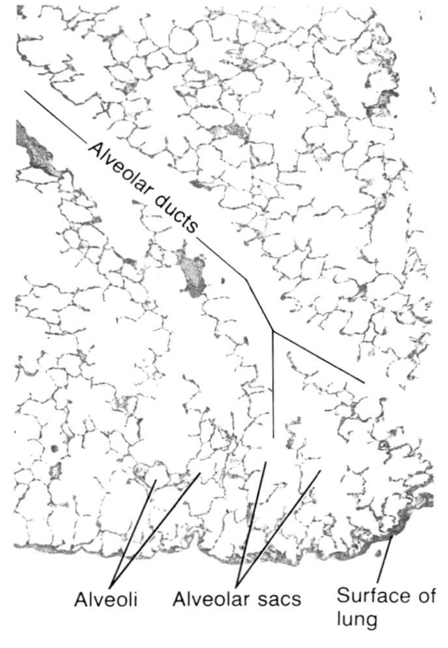

(b)

FIGURE 23-9 Structure of an alveolus. (a) Details of an alveolus. (b) Details of alveolar-capillary (respiratory) membrane.

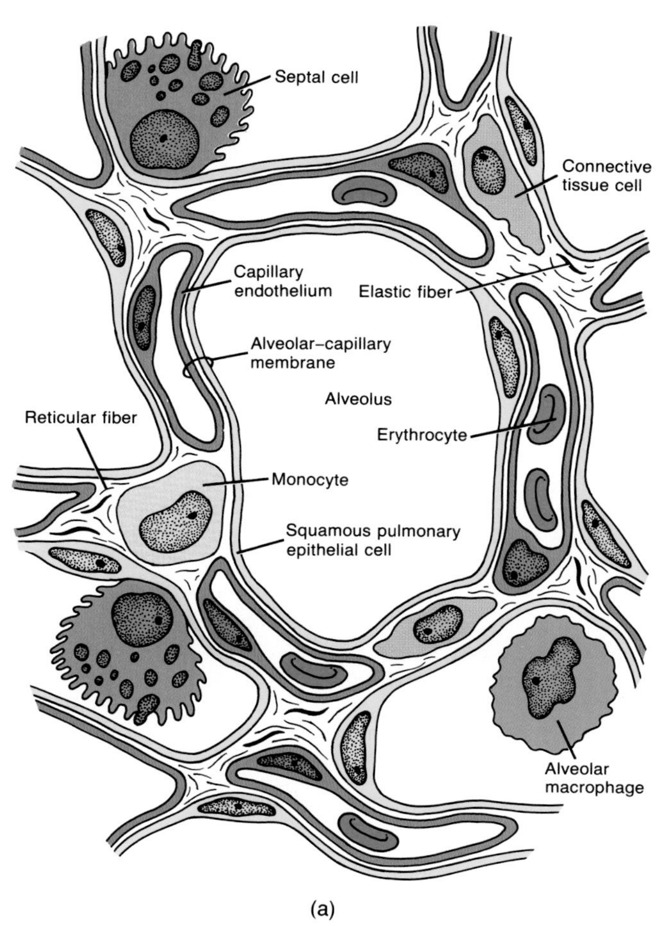

(a)

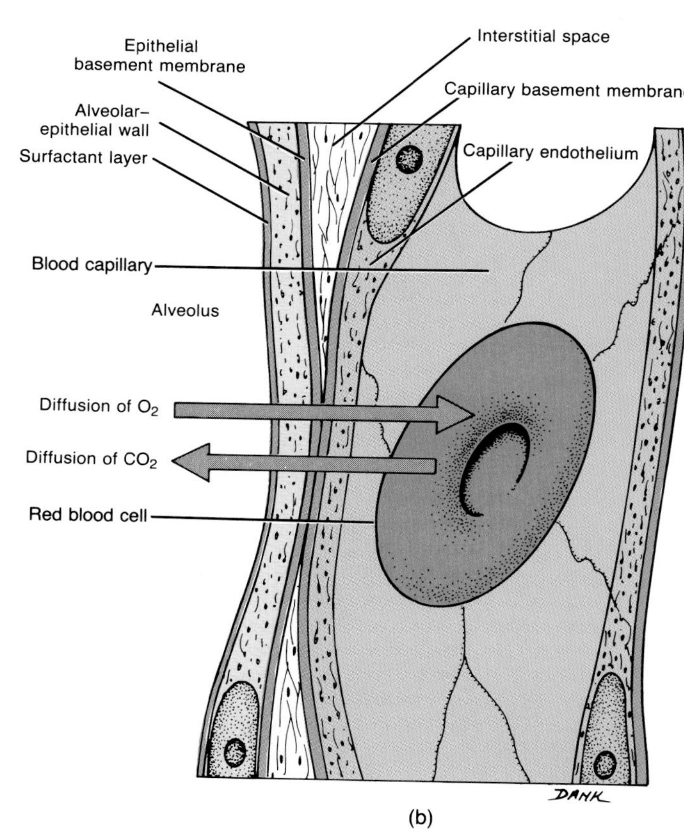

(b)

duce a phospholipid substance called ***surfactant*** (sur-FAK-tant), which lowers surface tension (described shortly). Free ***alveolar macrophages (dust cells)*** are highly phagocytic cells that remove dust particles or other debris from alveolar spaces. Also found are monocytes, white blood cells that become transformed into alveolar macrophages, and fibroblasts. Also present between lining cells of the alveolar wall are reticular and elastic fibers. Deep to the layer of squamous pulmonary epithelial cells is an elastic basement membrane. Over the alveoli, the arteriole and venule disperse into a capillary network. The blood capillaries consist of a single layer of endothelial cells and basement membrane.

CLINICAL APPLICATION: NEBULIZATION

Many respiratory disorders are treated by means of nebulization (neb-yoo-li-ZĀ-shun). This procedure consists of administering medication in the form of droplets that are suspended in air to select areas of the respiratory tract. The patient inhales the medication as a fine mist. Nebulization therapy can be used with many different types of drugs, such as chemicals that relax the smooth muscle of the respiratory passageways, chemicals that reduce the thickness of mucus, and antibiotics.

Alveolar-Capillary (Respiratory) Membrane

The exchange of respiratory gases between the lungs and blood takes place by diffusion across alevolar and capillary walls. This membrane, through which the respiratory gases move, is collectively known as the ***alveolar-capillary (respiratory) membrane*** (Figure 23-9b). It consists of:

1. A layer of squamous pulmonary epithelial cells with septal cells and free alveolar macrophages that constitute the alveolar (epithelial) wall.

2. An epithelial basement membrane underneath the alveolar wall.

3. A capillary basement membrane that is often fused to the epithelial basement membrane.

4. The endothelial cells of the capillary.

Despite the several layers, the alveolar-capillary membrane averages only 0.5 μm in thickness, about $\frac{1}{16}$ the diameter of a red blood cell. This is of considerable importance to the rapid diffusion of respiratory gases. Moreover, it has been estimated that the lungs contain 30 million alveoli, providing an immense surface area of 70 m^2 (753 ft^2) for the exchange of gases.

MEDICAL TEST

Lung scan

Diagnostic Value: To evaluate for pulmonary embolism (PE), pneumonia, or cancer and to assess arterial perfusion to the lungs.

Procedure: A ***perfusion scan*** detects blockages in blood flow from the heart to the lungs. In this procedure, a radioactive substance is injected into veins and detected in the lungs by a scanning camera. Cold spots, that is, areas of low radioactive uptake, indicate poor blood flow (perfusion); hot spots, or areas of high radioactive uptake, indicate normal blood flow. The scanner translates the information into an image. A ***ventilation scan*** measures the flow of air into and out of the lungs. In this procedure, a radioactive gas is inhaled, and once it is in the lungs, a scanning camera is used to produce an image. Normal findings include an equal distribution of radioactive gas in both lungs.

Blood Supply

There is a double blood supply to the lungs. Deoxygenated blood passes through the pulmonary trunk, which divides into a left pulmonary artery that enters the left lung and a right pulmonary artery that enters the right lung. The venous return of the oxygenated blood is by way of the pulmonary veins, typically two in number on each side—the right and left superior and inferior pulmonary veins. All four veins drain into the left atrium (see Figure 21-22).

Oxygenated blood is delivered through bronchial arteries, direct branches of the aorta. There are communications between the two systems, and most blood returns via pulmonary veins. Some blood, however, drains into bronchial veins, branches of the azygos system.

Recall from Chapter 21 that pulmonary circulation differs from systemic circulation in that pulmonary blood vessels provide less resistance to blood flow and less pressure is required to move blood through pulmonary circulation. In addition, pulmonary blood vessels constrict in response to low oxygen levels so that pulmonary blood can bypass poorly aerated areas. Moreover, the small continuous leak of fluid out of pulmonary capillaries is drained off by the lymphatic system to prevent pulmonary edema.

PHYSIOLOGY OF RESPIRATION

The principal purpose of ***respiration*** is to supply the cells of the body with oxygen and remove the carbon dioxide produced by cellular activities. The three basic processes of respiration are pulmonary ventilation, external respiration, and internal respiration.

Pulmonary Ventilation

Pulmonary ventilation (breathing) is the process by which gases are exchanged between the atmosphere and lung alveoli. Air flows between the atmosphere and lungs for the same reason that blood flows through the body—a pressure gradient exists. Air moves into the lungs when the pressure inside the lungs is less than the air pressure in the atmosphere. Air moves out of the lungs when the pressure inside the lungs is greater than the pressure in the atmosphere. Let us examine the mechanics of pulmonary ventilation by first looking at inspiration.

Inspiration

Breathing in is called ***inspiration (inhalation).*** Just before each inspiration, the air pressure inside the lungs equals the pressure of the atmosphere, which is about 760 mm Hg, or 1 atmosphere (atm), at sea level. For air to flow into the lungs, the pressure inside the lungs must become lower than the pressure in the atmosphere. This condition is achieved by increasing the volume (size) of the lungs.

The pressure of a gas in a closed container is inversely proportional to the volume of the container. If the size of a closed container is increased, the pressure of the air inside the container decreases. If the size of the container is decreased, then the pressure inside it increases. This is referred to as ***Boyle's law*** and may be demonstrated as follows. Suppose we place a gas in a cylinder that has a movable piston and a pressure gauge, and the initial pressure is 1 atm (Figure 23-10). This pressure is created by the gas molecules striking the wall of the container. If the piston is pushed down, the gas is concentrated in a smaller volume. This means that the same number of gas molecules are striking less wall space. The gauge shows that the pressure doubles as the gas is compressed to half its volume. In other words, the same number of molecules in half the space produces twice the pressure. Conversely, if the piston is raised to increase the volume, the pressure decreases. Thus, the volume of a gas varies inversely with

FIGURE 23-10 Boyle's law. The volume of a gas varies inversely with the pressure. If the volume is decreased to one-fourth, what happens to the pressure?

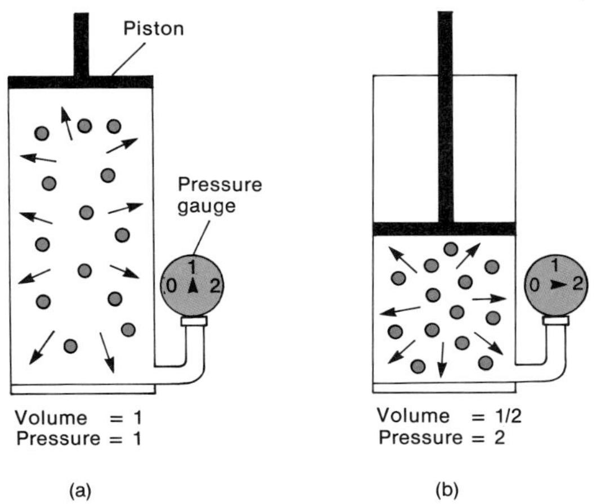

pressure (assuming that temperature is constant). Boyle's law applies to the operation of a bicycle pump and the blowing up of a balloon. Differences in pressure force air into our lungs when we inhale and force the air out when we exhale.

In order for inspiration to occur, the lungs must be expanded. This increases lung volume and thus decreases the pressure in the lungs. The first step toward increasing lung volume involves contraction of the principal inspiratory muscles—the diaphragm and external intercostals (Figure 23-11; see also Figure 11-12). The diaphragm, the most important muscle of inspiration, is a dome-shaped skeletal muscle that forms the floor of the thoracic cavity and is innervated by the phrenic nerve. Contraction of the diaphragm causes it to flatten, lowering its dome. This increases the vertical dimension of the thoracic cavity and accounts for the movement of about 75 percent of the air that enters the lungs during inspiration. The distance the diaphragm moves during inspiration ranges from 1 cm (0.4 inch) during normal quiet breathing up to about 10 cm (4 inches) during high levels of breathing. Advanced pregnancy, excessive obesity, or confining abdominal clothing can prevent a complete descent of the diaphragm. At the same time the diaphragm contracts, the external intercostals contract. These muscles run obliquely downward and forward between adjacent ribs, and when these muscles contract, the ribs are pulled upward and the sternum is pushed forward. This increases the anterior-posterior diameter of the thoracic cavity.

The term applied to normal quiet breathing is ***eupnea*** (yoop-NĒ-a; *eu* = normal). Eupnea involves shallow, deep, or combined shallow and deep breathing. Shallow (chest) breathing is called ***costal breathing.*** It consists of an upward and outward movement of the chest as a result of contraction of the external intercostal muscles. Deep (abdominal) breathing is called ***diaphragmatic breathing.*** It consists of the outward movement of the abdomen as a result of the contraction and descent of the diaphragm. During deep, labored inspiration, accessory muscles of inspiration also participate in increasing the size of the thoracic cavity. These include the sternocleidomastoid, which elevates the sternum, the scalenes which elevate the superior two ribs, and the pectoralis minor which elevates the third through fifth ribs. Inspiration is referred to as an active process because it is initiated by muscle contraction.

During normal breathing, the pressure between the two pleural layers, called ***intrapleural (intrathoracic) pressure,*** is always subatmospheric. (It may become temporarily positive only during modified respiratory movement such as coughing or straining during defecation.) Just before inspiration, it is about 756 mm Hg (Figure 23-12). The overall increase in the size of the thoracic cavity causes intrapleural pressure to fall to about 754 mm Hg. Consequently, the walls of the lungs are sucked outward. Expansion of the lungs is further aided by movement of the pleura. The parietal and visceral pleurae are normally strongly attached to each other owing to subatmospheric pressure and to surface tension created by their moist adjoining surfaces. As the thoracic cavity expands, the parietal pleura lining the cavity is pulled in all directions, and the visceral pleura and lungs are pulled along with it.

When the volume of the lungs increases, the pressure inside the lungs, called the ***intrapulmonic (intraalveolar) pressure,*** drops from 760 to 758 mm Hg. A pressure gradient is thus established between the atmosphere and the alveoli. Air rushes from the atmosphere into the lungs due to a gas pressure difference, and inspiration takes place. Air continues to move into the lungs as long as the pressure difference exists.

A summary of inspiration is presented in Figure 23-13a.

Expiration

Breathing out, called ***expiration (exhalation),*** is also achieved by a pressure gradient, but in this case the gradient is reversed so that the pressure in the lungs is greater than the pressure of the atmosphere. Normal expiration during quiet breathing, unlike inspiration, is a passive process since no muscular contractions are involved. It depends on the elasticity of the lungs. Expiration starts when the inspiratory muscles relax. As the external intercostals relax, the ribs move downward, and as the diaphragm relaxes, its dome moves upward owing to its elasticity. These movements decrease the vertical and anterior-posterior dimensions of the thoracic cavity, and it returns to its resting size (see Figure 23-11).

Expiration becomes active during higher levels of ventilation and when air movement out of the lungs is impeded. During these times, muscles of expiration—abdominal and internal intercostals—contract. Contraction of the abdominal muscles moves the inferior ribs downward and compresses the abdominal viscera, thus forcing

FIGURE 23-11 Pulmonary ventilation: muscles of inspiration and expiration. (a) Inspiratory muscles and their actions (left), and expiratory muscles and their actions (right). The pectoralis minor muscle, an accessory inspiratory muscle, is not shown here, but is illustrated in Figure 11-16a. (b) Changes in size of thoracic cavity during inspiration (blue) and expiration (black).

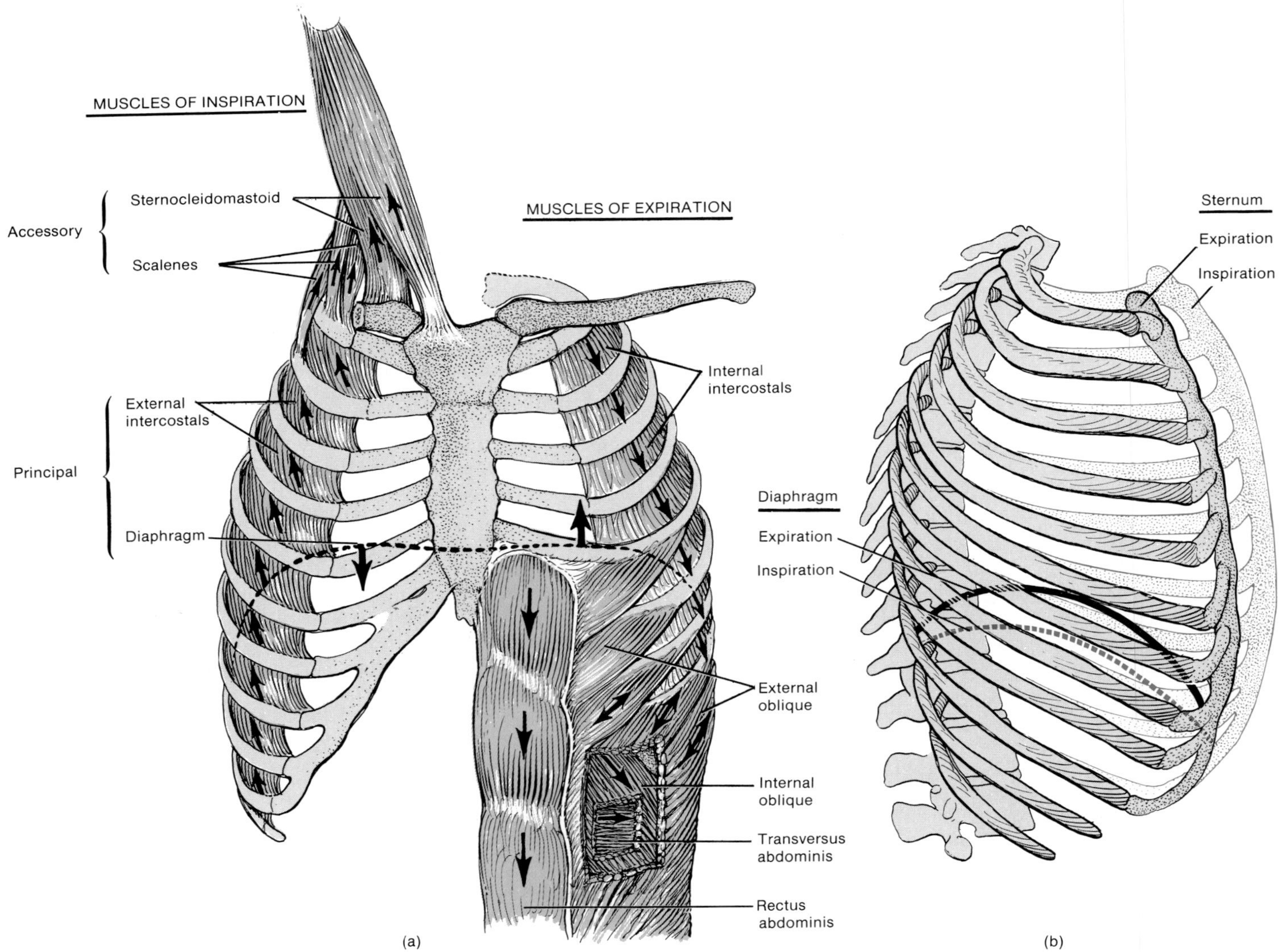

the diaphragm upward. Contraction of the internal intercostals, which run downward and backward between adjacent ribs, moves the ribs downward.

As intrapleural pressure returns to its preinspiration value (756 mm Hg), the walls of the lungs are no longer sucked out. The elastic basement membranes of the alveoli and elastic fibers in bronchioles and alveolar ducts recoil, and lung volume decreases. Intrapulmonic pressure increases to 763 mm Hg, and air moves from the area of higher pressure in the alveoli to the area of lower pressure in the atmosphere (see Figure 23-12c).

A summary of expiration is presented in Figure 23-13b.

Atelectasis (Collapsed Lung)

It was noted earlier that intrapleural pressure is normally subatmospheric. The pleural cavities are sealed off from the outside environment and cannot equalize their pressure with that of the atmosphere. Nor can the diaphragm and rib cage move inward enough to bring the intrapleural pressure up to atmospheric pressure. Maintenance of a low intrapleural pressure is vital to the functioning of the lungs. The alveoli are so elastic that at the end of an expiration they attempt to recoil inward and collapse on themselves like the walls of a deflated balloon. A collapsed

FIGURE 23-12 Pulmonary ventilation: pressure changes. (a) Lungs and pleural cavity just before inspiration. (b) Chest expanded and intrapleural pressure decreased; lungs pulled outward and intrapulmonic pressure decreased. (c) Chest relaxes, intrapleural pressure rises, and lungs snap inward. Intrapulmonic pressure raised, forcing air out until intrapulmonic pressure equals atmospheric pressure (a).

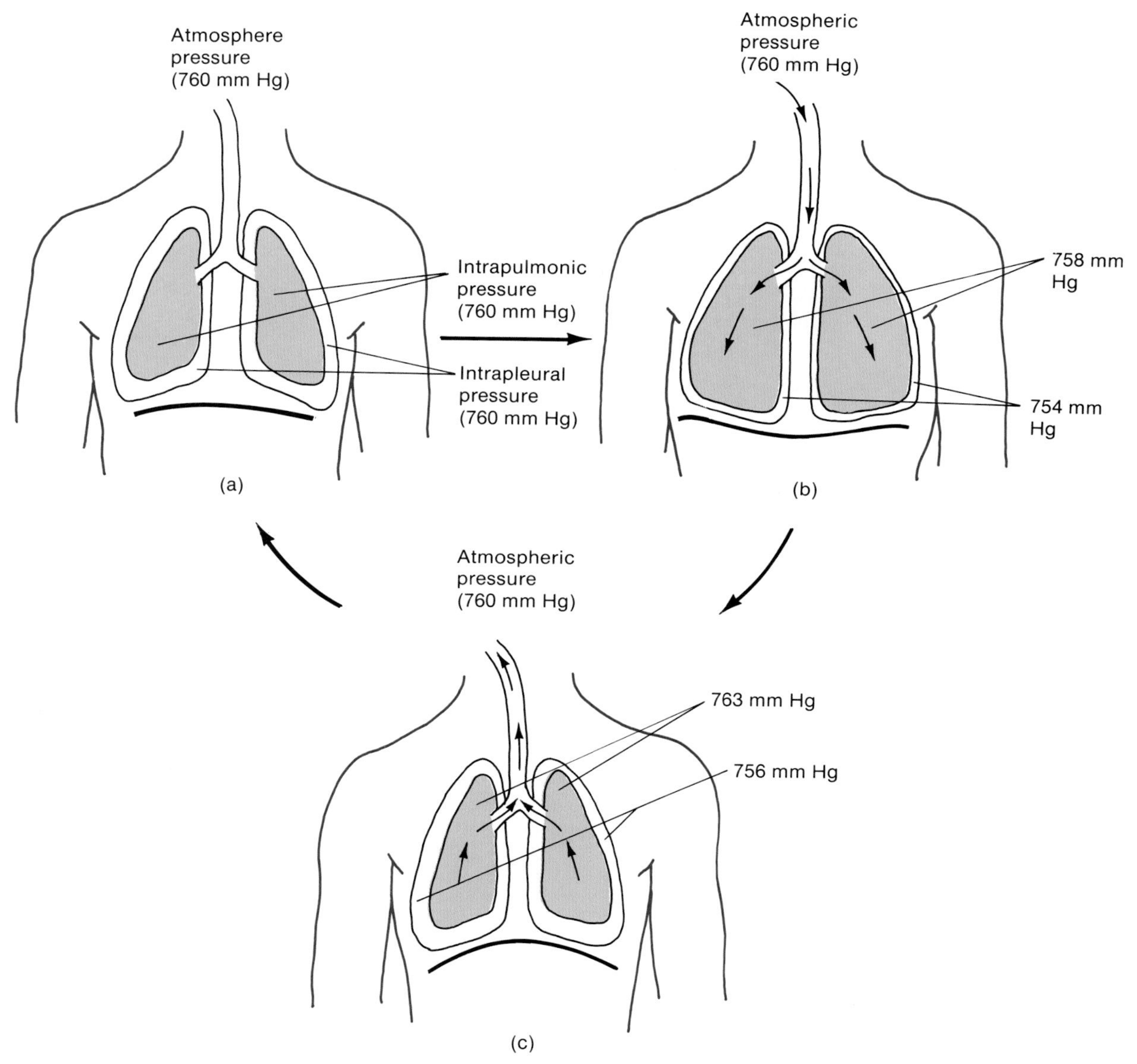

lung or portion of a lung is called ***atelectasis*** (at'-ē-LEK-ta-sis; *ateles* = incomplete; *ektasis* = dilation).

One factor that prevents the collapse of alveoli is the presence of ***surfactant,*** the phospholipid produced by the septal cells of the alveolar walls. Surfactant alters surface tension in the lungs. That is, it forms a thin lining on the alveoli and prevents them from collapsing and sticking together following expiration. Thus, as alveoli become smaller—for example, following expiration—the tendency of alveoli to collapse is minimized because the surface tension does not increase. As you will see later, a deficiency of surfactant in infants results in a disorder called respiratory distress syndrome (RDS) of the newborn.

Compliance

Compliance refers to the ease with which the lungs and thoracic wall can be expanded. High compliance means that the lungs and thoracic wall expand easily; low com-

FIGURE 23-13 Summary of inspiration and expiration.

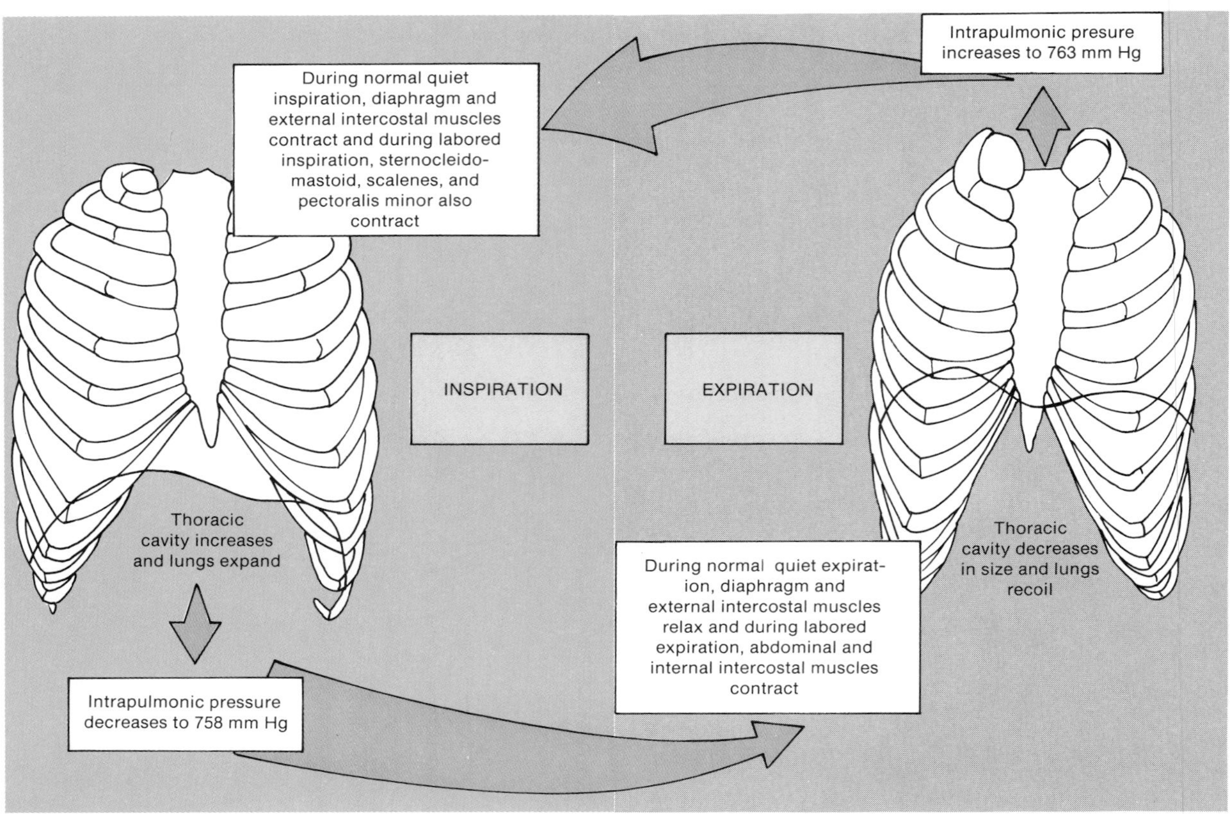
Intrapulmonic presure increases to 763 mm Hg
During normal quiet inspiration, diaphragm and external intercostal muscles contract and during labored inspiration, sternocleidomastoid, scalenes, and pectoralis minor also contract
INSPIRATION
EXPIRATION
Thoracic cavity increases and lungs expand
Thoracic cavity decreases in size and lungs recoil
During normal quiet expiration, diaphragm and external intercostal muscles relax and during labored expiration, abdominal and internal intercostal muscles contract
Intrapulmonic pressure decreases to 758 mm Hg

pliance means that they resist expansion. Compliance is related to two principal factors: elasticity and surface tension. The presence of elastic fibers in lung tissue results in high compliance. If surface tension within lung tissue were high, the tissues would resist expansion, but surfactant lowers surface tension and thus increases compliance. Any condition that destroys lung tissue, causes it to become filled with fluid, produces a deficiency in surfactant, or in any way impedes lung expansion or contraction decreases compliance.

Airway Resistance

The walls of the respiratory passageways, especially the bronchi and bronchioles, offer some resistance to the normal flow of air into the lungs. The muscular contraction of normal inspiration not only expands the thoracic cavity but also helps to overcome resistance to airflow by increasing the diameter of bronchi and bronchioles. Any condition that obstructs the air passageways increases resistance, and more pressure to force air through is required. During a forced expiration, as in coughing, straining, or playing a wind instrument, intrapleural pressure may increase from its normally subatmospheric (negative) value to a positive one. This greatly increases airway resistance because it results in compression of the airways. This is an important consideration for people with chronic obstructive pulmonary disease (COPD), in which there is some degree of obstruction of air passageways and airway resistance.

Modified Respiratory Movements

Respirations also provide humans with methods for expressing emotions such as laughing, yawning, sighing, and sobbing. Moreover, respiratory air can be used to expel foreign matter from the upper air passages through actions such as sneezing and coughing. Some of the modified respiratory movements that express emotion or clear the air passageways are listed in Exhibit 23-1. All these movements are reflexes, but some of them also can be initiated voluntarily.

Pulmonary Air Volumes and Capacities

In clinical practice, the word ***respiration (ventilation)*** means one inspiration plus one expiration. The healthy adult averages about 12 respirations a minute while at

EXHIBIT 23-1 MODIFIED RESPIRATORY MOVEMENTS

Movement	Comment
Coughing	A long-drawn and deep inspiration followed by a complete closure of the glottis, which results in a strong expiration that suddenly pushes the glottis open and sends a blast of air through the upper respiratory passages. Stimulus for this reflex act may be a foreign body lodged in the larynx, trachea, or epiglottis.
Sneezing	Spasmodic contraction of muscles of expiration that forcefully expels air through the nose and mouth. Stimulus may be an irritation of the nasal mucosa.
Sighing	A long-drawn and deep inspiration immediately followed by a shorter but forceful expiration.
Yawning	A deep inspiration through the widely opened mouth producing an exaggerated depression of the lower jaw. It may be stimulated by drowsiness, fatigue, or someone else's yawning, but precise stimulus–receptor cause is unknown.
Sobbing	A series of convulsive inspirations followed by a single prolonged expiration. The glottis closes earlier than normal after each inspiration so only a little air enters the lungs with each inspiration.
Crying	An inspiration followed by many short convulsive expirations, during which the glottis remains open and the vocal cords vibrate; accompanied by characteristic facial expressions and tears.
Laughing	The same basic movements as crying, but the rhythm of the movements and the facial expressions usually differ from those of crying. Laughing and crying are sometimes indistinguishable.
Hiccuping	Spasmodic contraction of the diaphragm followed by a spasmodic closure of the glottis to produce a sharp inspiratory sound. Stimulus is usually irritation of the sensory nerve endings of the gastrointestinal tract.

rest. During each respiration, the lungs exchange various amounts of air with the atmosphere. A lower-than-normal amount of exchange is usually a sign of pulmonary malfunction.

Spirometry

The apparatus commonly used to measure the amount of air exchanged during breathing and the rate of ventilation is a ***spirometer*** (*spiro* = breathe) or ***respirometer.*** A spirometer consists of a weighted drum inverted over a chamber of water. The drum usually contains oxygen or air. A tube connects the air-filled chamber with the subject's mouth. During inspiration, air is removed from the chamber, the drum sinks, and an upward deflection is recorded by a stylus on graph paper on the rotating drum. During expiration, air is added, the drum rises, and a downward deflection is recorded. The record is called a ***spirogram*** (Figure 23-14).

Pulmonary Volumes

During the process of normal quiet breathing, about 500 ml of air moves into the respiratory passageways with each inspiration. The same amount moves out with each expiration. This volume of air inspired (or expired) is called ***tidal volume*** (Figure 23-14). Only about 350 ml of the tidal volume actually reaches the alveoli. The other 150 ml remains in air spaces of the nose, pharynx, larynx, trachea, bronchi, and bronchioles and is known as ***anatomic dead space.*** The total air taken in during one minute is called the ***minute volume of respiration (MVR).*** It is calculated by multiplying the tidal volume by the normal breathing rate per minute. An average volume would be 500 ml times 12 respirations per minute, or 6000 ml/min.

By taking a very deep breath, we can inspire a good deal more than 500 ml. This excess inhaled air, called the ***inspiratory reserve volume,*** averages 3100 ml above the 500 ml of tidal volume. Thus, the respiratory system can pull in 3600 ml of air. In fact, even more air can be pulled in if inspiration follows forced expiration.

If we inhale normally and then exhale as forcibly as possible, we should be able to push out 1200 ml of air in addition to the 500 ml tidal volume. This extra 1200 ml is called the ***expiratory reserve volume.***

Even after the expiratory reserve volume is expelled, a good deal of air remains in the lungs because the lower intrapleural pressure keeps the alveoli slightly inflated, and some air also remains in the noncollapsible air passageways. This air, the ***residual volume,*** amounts to about 1200 ml.

Opening the thoracic cavity allows the intrapleural pressure to equal the atmospheric pressure, forcing out some

FIGURE 23-14 Spirogram of pulmonary volumes and capacities.

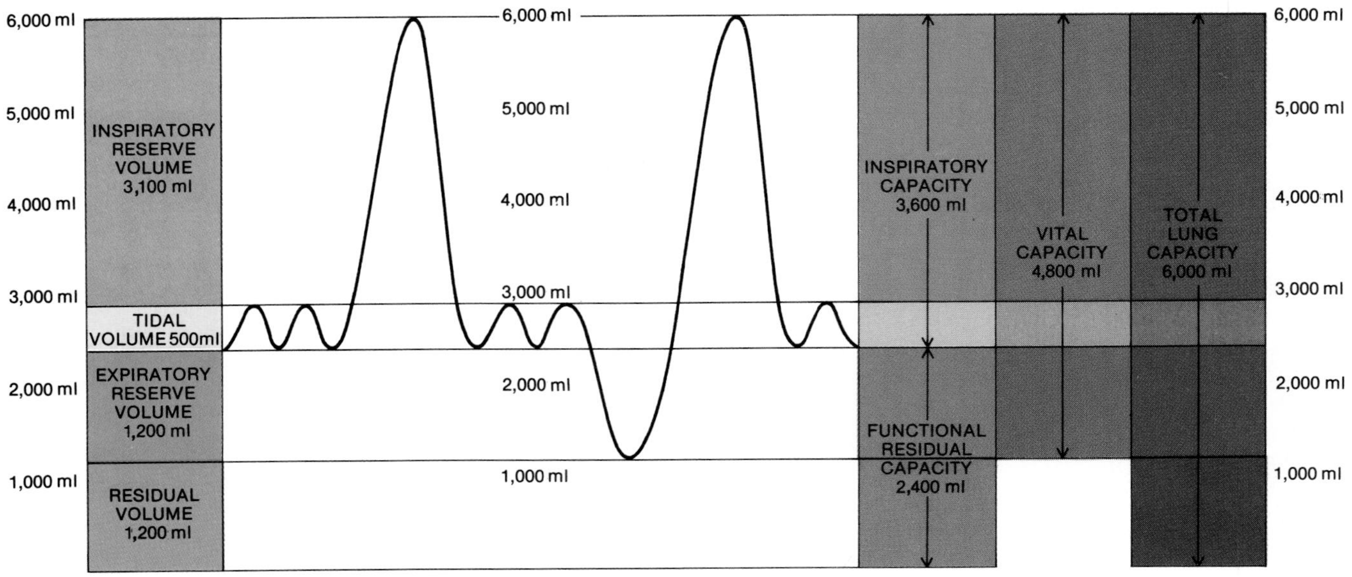

Spirogram of pulmonary volumes and capacities.

of the residual volume. The air remaining is called the ***minimal volume.*** Minimal volume provides a medical and legal tool for determining whether a baby was born dead or died after birth. The presence of minimal volume can be demonstrated by placing a piece of lung in water and watching it float. Fetal lungs contain no air, and so the lung of a stillborn will not float in water.

Pulmonary Capacities

Lung capacity can be calculated by combining specific lung volumes (Figure 23-14). ***Inspiratory capacity,*** the total inspiratory ability of the lungs, is the sum of tidal volume plus inspiratory reserve volume (3600 ml). ***Functional residual capacity*** is the sum of residual volume plus expiratory reserve volume (2400 ml). ***Vital capacity*** is the sum of inspiratory reserve volume, tidal volume, and expiratory reserve volume (4800 ml). Finally, ***total lung capacity*** is the sum of all volumes (6000 ml).

MEDICAL TEST

Pulmonary (PUL-mo-ner-ē) **function tests (PFTs)**

Diagnostic Value: To evaluate for the presence or severity of lung disease and to measure the extent of pulmonary impairment so that the effectiveness of therapy can be monitored.

Procedure: All pulmonary function tests involve a spirometer, and measurements are recorded on a spirogram. The tests may be performed in a physician's office, at bedside, or in a laboratory. After a noseclip is applied to stop air movement through the nose, the subject breathes into a mouthpiece that is connected to the testing apparatus. The more frequently ordered tests are:

1. **Forced vital capacity (FVC).** Measures the maximum volume of air a person can exhale after inhaling as deeply as possible.
2. **Forced expiratory volume in one second (FEV_1).** Measures the volume of air exhaled during the first second of a forced exhalation after a maximum inhalation. (A useful test for diagnosing obstructive lung diseases* such as emphysema and asthma, in which the FEV_1 [expressed as a percentage of the FVC] is significantly reduced.
3. **Maximum midexpiratory flow (MMEF).** Measures the maximum rate at which air flows during the middle of forced exhalation. A reduced MMEF is specific for obstructive airway disease.
4. **Maximum voluntary ventilation (MVV).** Measures the maximum volume of air that can be inhaled or exhaled in one minute.

* An *obstructive lung disease* results from a limitation of expiratory airflow, regardless of cause. A *restrictive lung disease* refers to a restriction of or limitation to the amount of gas in the lungs, characterized by reduction in lung volumes or capacities.

Exchange of Respiratory Gases

As soon as the lungs fill with air, oxygen diffuses from the alveoli into the blood, through the interstitial fluid, and finally into the cells. Carbon dioxide diffuses in the opposite direction—from the cells, through interstitial fluid to the blood, and to the alveoli. To understand how respiratory gases are exchanged in the body, you need to know a few gas laws.

Charles' Law

According to ***Charles' law,*** the volume of a gas is directly proportional to its absolute temperature, assuming that the pressure remains constant. Recall the cylinder we used to demonstrate Boyle's law. Suppose now that the gas in the cylinder exerts an initial pressure of 1 atm when the piston is halfway down (Figure 23-15). When the gas is heated, the gas molecules move faster and the number of collisions within the cylinder increases. Assuming that the piston moves freely (no external pressure is applied), the force of the molecules hitting it moves it upward. As the gas expands, the movement of the piston provides a measure of the increase in volume. As the space in the cylinder increases, the molecules have farther to travel, so the number of collisions decreases as the space increases. The pressure of 1 atm is maintained, and the volume increases in direct proportion to the temperature increase. As gases enter the warmer lungs, the gases expand, thus increasing lung volume.

FIGURE 23-15 **Charles' law. The volume of a gas is directly proportional to its absolute temperature, assuming that the pressure is constant.**

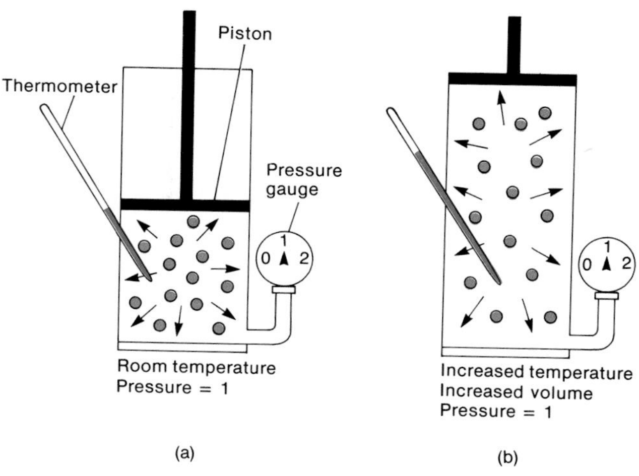

Dalton's Law

According to ***Dalton's law,*** each gas in a mixture of gases exerts its own pressure as if all the other gases were not present. This *partial pressure* is denoted as *P*. The total pressure of the mixture is calculated by simply adding all the partial pressures. Atmospheric air is a mixture of several gases—oxygen, carbon dioxide, nitrogen, water vapor, and a number of other gases that appear in such small quantities that we will ignore them. Atmospheric pressure is the sum of the pressures of all these gases:

$$\text{Atmospheric pressure} = PO_2 + PCO_2 + PN_2 + PH_2O$$
(760 mm Hg)

We can determine the partial pressure exerted by each component in the mixture by multiplying the percentage of the gas in the mixture by the total pressure of the mixture. For example, to find the partial pressure of oxygen in the atmosphere, multiply the percentage of atmospheric air composed of oxygen (21 percent) by the total atmospheric pressure (760 mm Hg):

$$\text{Atmospheric } PO_2 = 21\% \times 760 \text{ mm Hg}$$
$$= 159.60 \text{ or } 160 \text{ mm Hg}$$

Since the percentage of CO_2 in the atmosphere is 0.04,

$$\text{Atmospheric } PCO_2 = 0.04\% \times 760 \text{ mm Hg}$$
$$= 0.3040 \text{ or } 0.3 \text{ mm Hg}$$

The partial pressures of the respiratory gases and nitrogen in the atmosphere, alveoli, blood, and tissue cells are shown in Exhibit 23-2. These partial pressures are important in determining the movement of oxygen and carbon dioxide between the atmosphere and lungs, the lungs and blood, and the blood and body cells. When a mixture of gases diffuses across a permeable membrane, each gas diffuses from the area where its partial pressure is greater to the area where its partial pressure is less. Every gas is on its own and behaves as if the other gases in the mixture did not exist.

EXHIBIT 23-2 PARTIAL PRESSURES (mm Hg) OF RESPIRATORY GASES AND NITROGEN IN ATMOSPHERIC AIR, ALVEOLAR AIR, BLOOD, AND TISSUE CELLS

	Atmospheric Air (sea level)	Alveolar Air	Deoxygenated Blood	Oxygenated Blood	Tissue Cells
PO_2	160	105	40	105	40
PCO_2	0.3	40	45	40	45
PN_2	597	569	569	569	569

The amounts of respiratory gases vary in inspired (atmospheric), alveolar, and expired air (Exhibit 23-3). Inspired (atmospheric) air contains about 21 percent oxygen and 0.04 percent carbon dioxide. Expired air contains less oxygen (about 16 percent) and more carbon dioxide (about 4.5 percent) than inspired air. Compared with alveolar air, expired air contains more oxygen (about 16 percent versus 14 percent) and less carbon dioxide (about 4.5 percent versus 5.5 percent) because some of the expired air is in the anatomic dead space and has not participated in gaseous exchange. Expired air is actually a mixture of inspired and alveolar air.

EXHIBIT 23-3 APPROXIMATE PERCENTAGE OF OXYGEN AND CARBON DIOXIDE IN INSPIRED AIR, ALVEOLAR AIR, AND EXPIRED AIR

	Inspired Air	Alveolar Air	Expired Air
Oxygen	21	14	16
Carbon dioxide	0.04	5.50	4.5

Henry's Law

You have probably noticed that a bottle of soda makes a hissing sound when the top is removed, and bubbles rise to the surface for some time afterward. The gas dissolved in carbonated beverages is carbon dioxide. The ability of a gas to stay in solution depends on its partial pressure and solubility coefficient, that is, its physical or chemical attraction for water. The solubility coefficient of carbon dioxide is high (0.57), that of oxygen is lower (0.024), and that of nitrogen is still lower (0.012). The higher the partial pressure of a gas over a liquid and the higher the solubility coefficient, the more gas will stay in solution. Since the soda is bottled under pressure and capped, the CO_2 remains dissolved as long as the bottle is unopened. Once you remove the cap, the pressure is released and the gas begins to bubble out. This phenomenon is explained by ***Henry's law:*** The quantity of a gas that will dissolve in a liquid is proportional to the partial pressure of the gas and its solubility coefficient, when the temperature remains constant.

Henry's law explains two conditions resulting from changes in the solubility of nitrogen in body fluids. Even though the air we breathe contains about 79 percent nitrogen, this gas has no known effect on bodily functions since very little of it dissolves in blood plasma because of its low solubility coefficient at sea level pressure. But, when a deep-sea diver, ***scuba (self-contained underwater breathing apparatus)*** diver, or caisson worker (person who builds tunnels under water) breathes air under high pressure, the nitrogen in the mixture can affect the body. Partial pressure is a function of total pressure, and therefore the partial pressure of all the components of a mixture increases as the total increases. Since the partial pressure of nitrogen is higher in a mixture of compressed air than in air at sea level pressure, a considerable amount of nitrogen goes into solution in plasma and interstitial fluid. Excessive amounts of dissolved nitrogen may produce giddiness and other symptoms similar to alcohol intoxication. The condition is called ***nitrogen narcosis*** or "rapture of the depths." The greater the depth, the more severe the condition.

If a diver is brought to the surface slowly, the dissolved nitrogen can be eliminated through the lungs. However, if a diver ascends too rapidly, the nitrogen comes out of solution too quickly to be eliminated by respiration. Instead, it forms gas bubbles in the tissues that result in ***decompression sickness*** (***caisson disease,*** or ***bends***). The effects of decompression sickness typically result from bubbles in nervous tissue and can be mild or severe, depending on the amount of bubbles formed. Symptoms include joint pain, especially in the arms and legs, dizziness, shortness of breath, extreme fatigue, paralysis, and unconsciousness. Decompression sickness can be prevented by a slow ascent or by the use of a special decompression tank within five minutes after arriving at the surface. The use of helium-oxygen mixtures instead of air containing nitrogen may reduce the dangers of decompression sickness since helium is only about 40 percent as soluble as nitrogen in blood.

CLINICAL APPLICATION: HYPERBARIC OXYGENATION (HBO)

A major clinical application of Henry's law is ***hyperbaric*** (*hyper* = over; *baros* = pressure) ***oxygenation (HBO).*** Using pressure to cause more oxygen to dissolve in the blood is an effective technique in treating patients infected by anaerobic bacteria, such as those that cause tetanus and gangrene. (Anaerobic bacteria cannot live in the presence of free oxygen.) A person undergoing hyperbaric oxygenation is placed in a hyperbaric chamber, which contains oxygen at a pressure of 3 to 4 atm (2280 to 3040 mm Hg). The body tissues pick up the oxygen, and the bacteria are killed. Hyperbaric chambers may also be used for treating certain heart disorders, carbon monoxide poisoning, gas embolisms, crush injuries, cerebral edema, certain hard-to-treat bone infections, smoke inhalation, near-drowning, asphyxia, vascular insufficiencies, and burns.

Physiology of External Respiration

External respiration is the exchange of oxygen and carbon dioxide between the alveoli of the lungs and pulmonary blood capillaries (Figure 23-16a). It results in the conversion of ***deoxygenated blood*** (more CO_2 than O_2) coming from the heart to ***oxygenated blood*** (more O_2 than CO_2) returning to the heart. During inspiration, atmospheric air containing oxygen enters the alveoli. De-

FIGURE 23-16 Partial pressures involved in respiration. (a) External. (b) Internal. All pressures are in mm Hg.

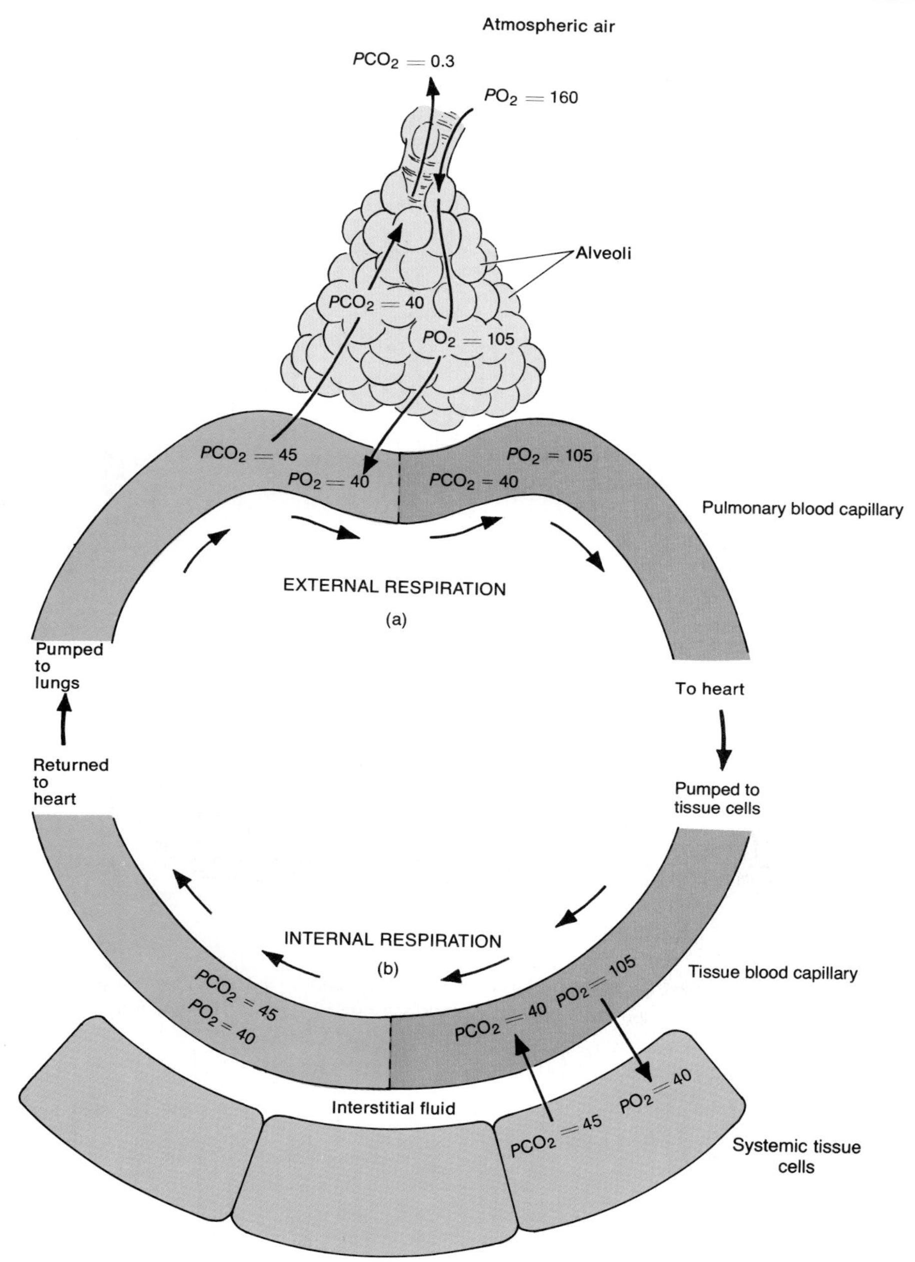

oxygenated blood is pumped from the right ventricle through the pulmonary arteries into the pulmonary capillaries overlying the alveoli. The PO_2 of alveolar air is 105 mm Hg. The PO_2 of the deoxygenated blood entering the pulmonary capillaries is only 40 mm Hg. As a result of this difference in PO_2, oxygen diffuses from the alveoli into the deoxygenated blood until equilibrium is reached, and the PO_2 of the now oxygenated blood is 105 mm Hg. While oxygen diffuses from the alveoli into deoxygenated blood, carbon dioxide diffuses in the opposite direction. On arriving at the lungs, the PCO_2 of pulmonary deoxygenated blood is 45 mm Hg, whereas that of the alveoli is 40 mm Hg. Because of this difference in PCO_2, carbon dioxide diffuses from pulmonary deoxygenated blood into the alveoli until the PCO_2 of the blood decreases to 40 mm Hg, the PCO_2 of pulmonary oxygenated blood. Thus, the PO_2 and PCO_2 of oxygenated blood leaving the lungs are the same as in alveolar air. The carbon dioxide that diffuses into the alveoli is eliminated from the lungs during expiration.

External respiration is aided by several anatomic adaptations. The total thickness of the alveolar-capillary (respiratory) membranes is only 0.5 μm. Thicker membranes would restrict diffusion. The surface area over which diffusion may occur is large. The total surface area of the alveoli is about 70 m^2 (753 ft^2), many more times the total surface area of the skin. Lying over the alveoli are countless capillaries—so many that 100 ml of blood is able to participate in gas exchange at any time. Finally, the capillaries are so narrow that the red blood cells must flow through them in single file. This feature gives each red blood cell maximum exposure to the available oxygen.

The efficiency of external respiration depends on several factors. One of the most important is altitude. As long as alveolar PO_2 is higher than venous blood PO_2, oxygen diffuses from the alveoli into the blood. As a person ascends in altitude, the atmospheric PO_2 decreases, the alveolar PO_2 decreases correspondingly, and less oxygen diffuses into the blood. For example, at sea level, PO_2 is 160 mm Hg. At 10,000 ft, it decreases to 110 mm Hg; at 20,000 ft, to 73 mm Hg; and at 50,000 ft, to 18 mm Hg. The common symptoms of ***high altitude sickness (acute mountain sickness)***—shortness of breath, headache, fatigue, insomnia, nausea, dizziness—are attributable to the low concentrations of oxygen in the blood.

Another factor that affects external respiration is the total surface area available for O_2–CO_2 exchange. Any pulmonary disorder that decreases the functional surface area formed by the alveolar-capillary membranes decreases the efficiency of external respiration.

A third factor that influences external respiration is the minute volume of respiration. Certain drugs, such as morphine, slow down the respiration rate, thereby decreasing the amount of oxygen and carbon dioxide that can be exchanged between the alveoli and the blood.

A fourth factor that affects external respiration is diffusion distance which is greater in pulmonary edema.

Physiology of Internal Respiration

As soon as external respiration is completed, oxygenated blood leaves the lungs through the pulmonary veins and returns to the heart. From here, it is pumped from the left ventricle into the aorta and through the systemic arteries to capillaries to tissue cells. The exchange of oxygen and carbon dioxide between tissue blood capillaries and tissue cells is called ***internal respiration*** (Figure 23-16b). It results in the conversion of oxygenated blood into deoxygenated blood. Oxygenated blood entering tissue capillaries has a PO_2 of 105 mm Hg, whereas tissue cells have an average PO_2 of 40 mm Hg. Because of this difference in PO_2, oxygen diffuses from the oxygenated blood through interstitial fluid and into tissue cells until the PO_2 in the blood decreases to 40 mm Hg, the PO_2 of tissue capillary deoxygenated blood. At rest, only about 25 percent of the available oxygen in oxygenated blood actually enters tissue cells. This amount is sufficient to support the needs of resting cells. During exercise, more oxygen is released. While oxygen diffuses from the tissue blood capillaries into tissue cells, carbon dioxide diffuses in the opposite direction. The PCO_2 of tissue cells is 45 mm Hg, whereas that of tissue capillary oxygenated blood is 40 mm Hg. As a result, carbon dioxide diffuses from tissue cells through interstitial fluid into the oxygenated blood until the PCO_2 in the blood increases to 45 mm Hg, the PCO_2 of tissue capillary deoxygenated blood. The deoxygenated blood now returns to the heart. From here it is pumped to the lungs for another cycle of external respiration.

MEDICAL TEST

Arterial blood gas (ABG)

Diagnostic Value: To measure the partial pressures of oxygen and carbon dioxide and pH of blood, to monitor respiratory therapy, and to evaluate acid-base status.

Procedure: A blood sample is taken from an artery, since blood coming directly from the heart is oxygen rich. The radial, femoral, or brachial artery are preferred sites for obtaining the blood sample. The various values of the ABG analysis are measured by electrodes.

Normal Values: Oxygen (PO_2): arterial blood—80–100 mm Hg
Carbon dioxide (PCO_2): arterial blood—35–45 mm Hg
pH: arterial blood—7.35–7.45
Oxygen saturation: 94–100%
Bicarbonate (HCO_3^-): 22–26 mEq/l

Transportation of Respiratory Gases

The transportation of respiratory gases between the lungs and body tissues is a function of the blood. When oxygen and carbon dioxide enter the blood, certain physical and chemical changes occur that aid in gas transport and exchange.

Oxygen

Oxygen does not dissolve easily in water, and, therefore, very little oxygen is carried in the dissolved state in the water in blood plasma. In fact, 100 ml of oxygenated blood contains only about 3 percent oxygen dissolved in plasma. The remainder of the oxygen, about 97 percent, is carried in chemical combination with hemoglobin in red blood cells (Figure 23-17).

FIGURE 23-17 Transportation of respiratory gases in respiration. (a) External. (b) Internal. The role of red blood cells in the transportation of carbon dioxide is detailed in Figure 23-22.

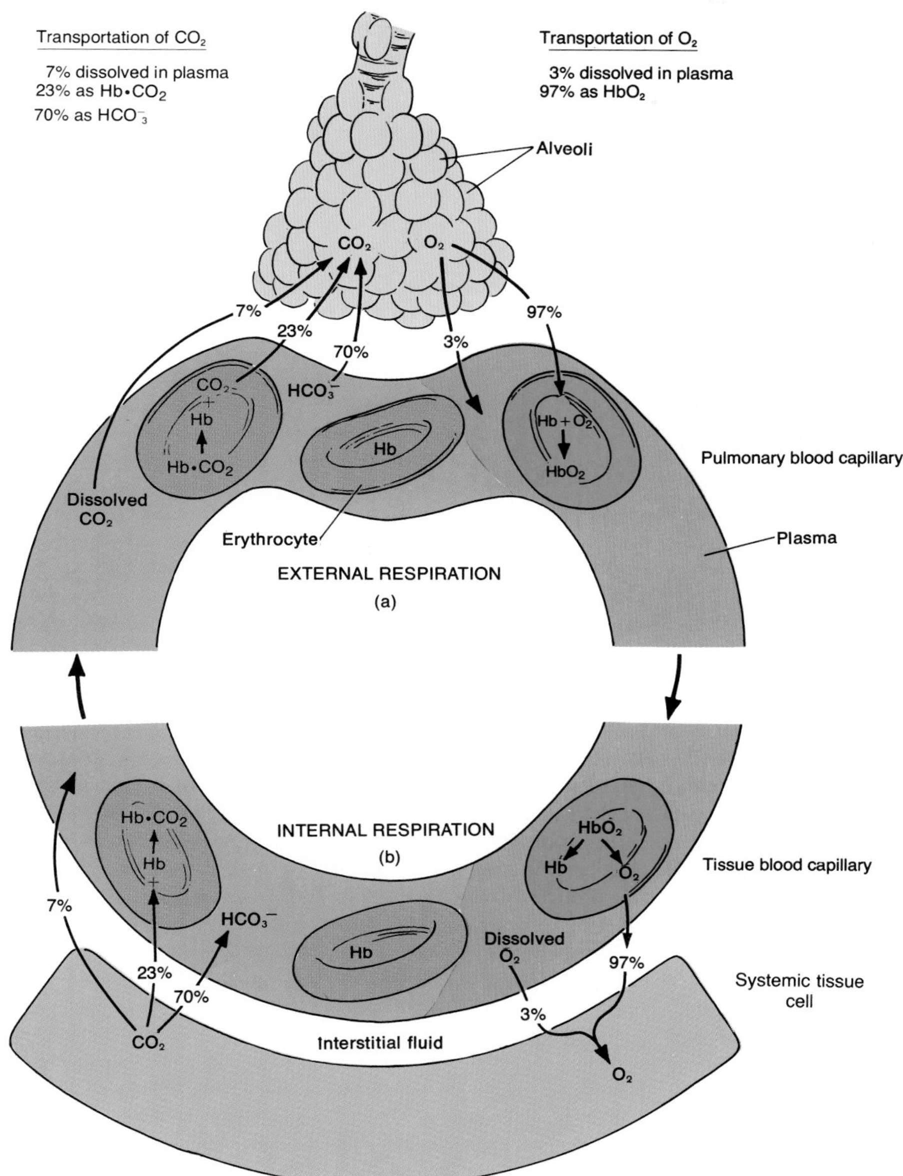

Hemoglobin consists of a protein portion called globin and a pigment portion called heme. The heme portion contains four atoms of iron, each capable of combining with a molecule of oxygen. Oxygen and hemoglobin combine in an easily reversible reaction to form ***oxyhemoglobin*** as follows:

Hb	+	O_2	⇌	HbO_2
Reduced (deoxygenated) hemoglobin (uncombined hemoglobin)		Oxygen		Oxyhemoglobin (combined hemoglobin)

Hemoglobin and PO_2 The most important factor that determines how much oxygen combines with hemoglobin is PO_2. When Hb (reduced or deoxygenated hemoglobin) is completely converted to HbO_2, it is ***fully saturated.*** When hemoglobin consists of a mixture of Hb and HbO_2, it is ***partially saturated.*** The ***percent saturation of hemoglobin*** is the percent of HbO_2 in total hemoglobin. The relationship between the percent saturation of hemoglobin and PO_2 is illustrated in Figure 23-18, the oxygen–hemoglobin dissociation curve. Note that when the PO_2 is high, hemoglobin binds with large

FIGURE 23-18 **Oxygen–hemoglobin dissociation curve at normal body temperature showing the relationship between hemoglobin saturation and *P*O₂. As *P*O₂ increases, more oxygen combines with hemoglobin. The percentages of saturation of Hb shown in the table are approximate.**

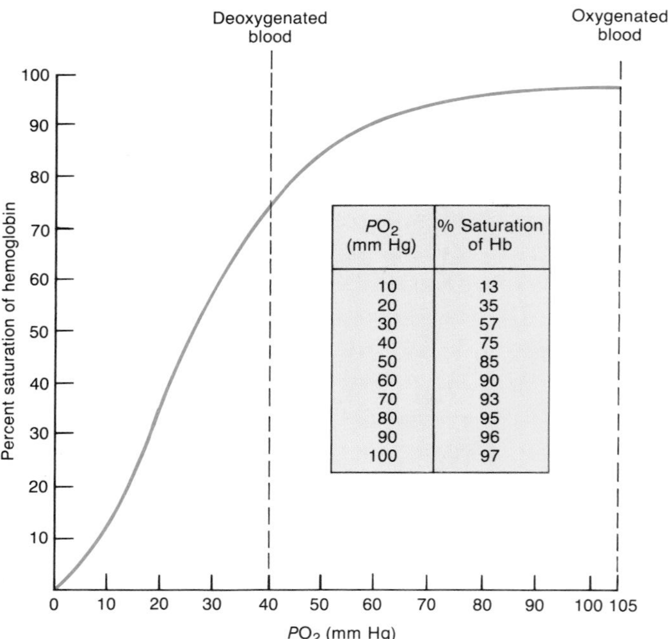

PO_2 (mm Hg)	% Saturation of Hb
10	13
20	35
30	57
40	75
50	85
60	90
70	93
80	95
90	96
100	97

amounts of oxygen and is almost fully saturated. When PO_2 is low, hemoglobin is only partially saturated and oxygen is released from hemoglobin. In other words, the greater the PO_2, the more oxygen will combine with hemoglobin, until the available hemoglobin molecules are saturated. Therefore, in pulmonary capillaries, where PO_2 is high, a lot of oxygen binds with hemoglobin, but in tissue capillaries, where the PO_2 is lower, hemoglobin does not hold as much oxygen, and the oxygen is released for diffusion into the tissue cells. Recall that only about 25 percent of the available oxygen splits from hemoglobin under resting conditions.

Note in the oxygen–hemoglobin dissociation curve that when the PO_2 is between 60 and 100 mm Hg, hemoglobin is 90 percent or more saturated with oxygen. Thus, blood picks up a nearly full load of oxygen from the lungs even when the PO_2 there is as low as 60 mm Hg. This explains why people can still perform well at high altitudes or when they have certain cardiac and pulmonary diseases, even though PO_2 may drop to as low as 60 mm Hg. Hemoglobin is 100 percent saturated with oxygen at a PO_2 of about 250 mm Hg. Note also in the curve that at a considerably lower PO_2 of 40 mm Hg, hemoglobin is 75 percent saturated with oxygen but drops to only 13 percent saturated at 10 mm Hg. This means that between 10 and 40 mm Hg large amounts of oxygen are released in response to only small changes in PO_2. In active tissues such as contracting muscles, PO_2 may decrease to well below 40 mm Hg. At such pressures, nearly all the oxygen is released from hemoglobin, oxygen required for contracting muscles.

▪ **Hemoglobin and pH** The amount of oxygen released from hemoglobin is determined by several factors in addition to PO_2. For example, in an acid environment, oxygen splits more readily from hemoglobin (Figure 23-19). This is referred to as the ***Bohr effect*** and is based on the belief that when hydrogen ions (H^+) bind to hemoglobin, they alter the structure of hemoglobin and thereby decrease its oxygen-carrying capacity.

Low blood pH (acidic condition) results from high PCO_2. As carbon dioxide is taken up by the blood, much of it is temporarily converted to carbonic acid. This conversion is catalyzed by an enzyme in red blood cells called *carbonic anhydrase*.

$$\underset{\text{Carbon dioxide}}{CO_2} + \underset{\text{Water}}{H_2O} \xrightleftharpoons{\text{carbonic anhydrase}} \underset{\text{Carbonic acid}}{H_2CO_3} \rightleftharpoons \underset{\text{Hydrogen ion}}{H^+} + \underset{\text{Bicarbonate ion}}{HCO_3^-}$$

The carbonic acid thus formed in red blood cells dissociates into hydrogen ions and bicarbonate ions. As the hydrogen ion concentration increases, pH decreases.

FIGURE 23-19 **Oxygen–hemoglobin dissociation curve at normal body temperature showing the relationship between hemoglobin saturation and pH (*P*CO₂). As pH increases (*P*CO₂ decreases), more oxygen combines with hemoglobin and less is available to tissues. Or, stated another way, as pH decreases (*P*CO₂ increases), less oxygen combines with hemoglobin and more is available to tissues. These relationships are emphasized by the broken lines.**

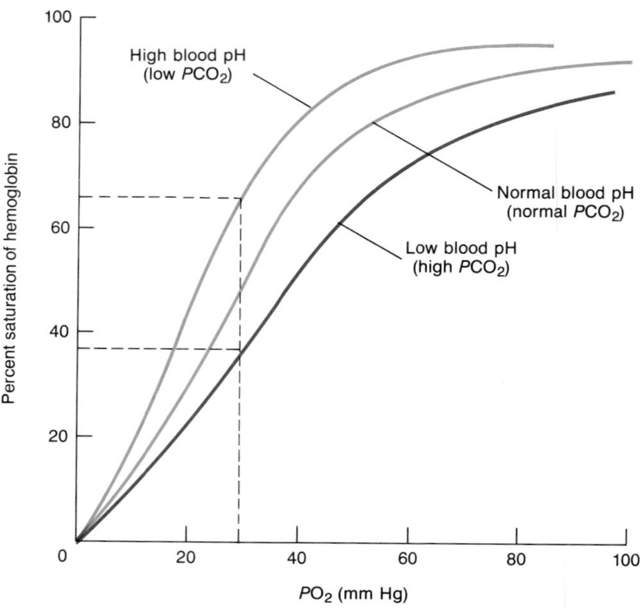

Thus, an increased PCO_2 produces a more acid environment that helps to split oxygen from hemoglobin. Low blood pH can also result from lactic acid, a by-product of anaerobic muscle contraction.

Hemoglobin and Temperature Within limits, as temperature increases, so does the amount of oxygen released from hemoglobin (Figure 23-20). Heat energy is a byproduct of the metabolic reactions of all cells, and contracting muscle fibers (cells) release an especially large amount of heat. Splitting the oxyhemoglobin molecule is another example of how homeostatic mechanisms adjust body activities to cellular needs. Active cells require more oxygen, and active cells liberate more acid and heat. The acid and heat, in turn, stimulate the oxyhemoglobin to release its oxygen.

Hemoglobin and DPG A final factor that helps to release oxygen from hemoglobin is a substance found in red blood cells called ***2,3-diphosphoglycerate,*** or simply ***DPG***, a substance formed in red blood cells during glycolysis (Chapter 25). It has the ability to combine reversibly with hemoglobin and thus alter its structure to release oxygen. The greater the level of DPG, the more oxygen is released from hemoglobin. Certain hormones, such as thyroxine, human growth hormone, epinephrine, norepinephrine, and testosterone increase the formation of DPG.

Fetal Hemoglobin Fetal hemoglobin differs from adult hemoglobin in structure and in its affinity for oxygen. Fetal hemoglobin has a higher affinity for oxygen because it cannot bind to DPG and thus can carry as much as 20 to 30 percent more oxygen than maternal hemoglobin (Figure 23-21). Thus, as the maternal blood enters the placenta, oxygen is readily transferred to fetal blood. This is very important since the O_2 saturation in maternal blood in the placenta is quite low and the fetus might suffer hypoxia were it not for the affinity of fetal hemoglobin for oxygen.

FIGURE 23-20 Oxygen–hemoglobin dissociation curve showing the relationship between hemoglobin saturation and temperature. As temperature increases, less oxygen combines with hemoglobin.

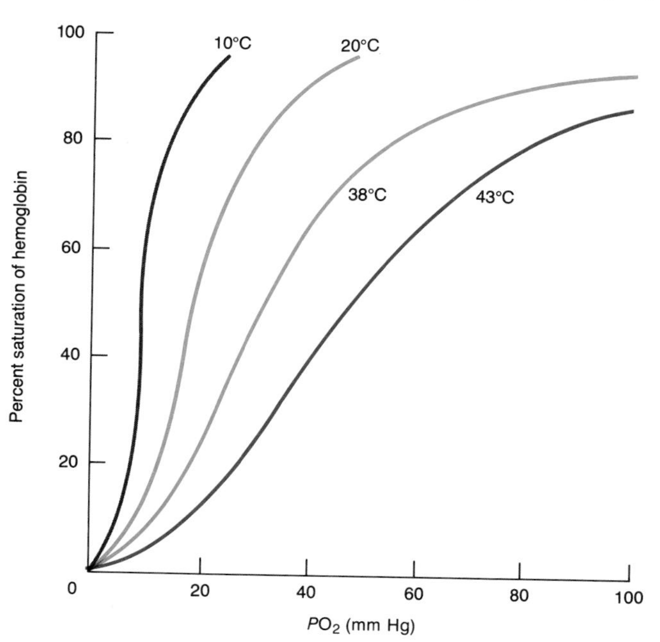

FIGURE 23-21 Oxygen–hemoglobin dissociation curves comparing fetal and maternal blood. Note that fetal blood can carry a greater quantity of oxygen than maternal blood.

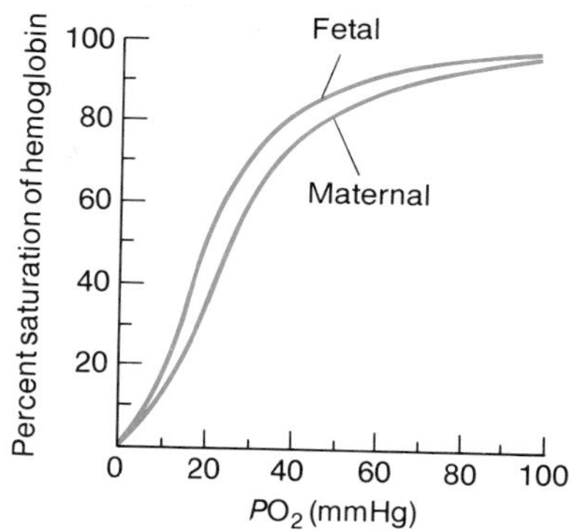

CLINICAL APPLICATION: CARBON MONOXIDE (CO) POISONING

Carbon monoxide (CO) is a colorless and odorless gas found in exhaust fumes from automobiles and in tobacco smoke. It is a by-product of burning carbon-containing materials such as coal and wood. One of its interesting features is that it combines with hemoglobin very much as oxygen does, except that the combination of carbon monoxide and hemoglobin is over 200 times as tenacious as the combination of oxygen and hemoglobin. In addition, in concentrations as small as 0.1 percent (PCO = 0.5 mm Hg), carbon monoxide will combine with half the hemoglobin molecules. Thus, the oxygen-carrying capacity of the blood is reduced by one-half. Increased levels of carbon monoxide lead to hypoxia, and the result is ***carbon monoxide (CO) poisoning.*** The condition may be treated by administering pure oxygen (PO_2 = 600 mm Hg), which slowly replaces the carbon monoxide combined with the hemoglobin.

Oxygen Toxicity Breathing pure oxygen can be hazardous. Animal experiments have shown that when guinea pigs breathe 100 percent oxygen at atmospheric pressure, they develop pulmonary edema. In patients breathing 100 percent oxygen, there has been evidence of impaired gas exchange. One reported hazard of breathing 100 percent oxygen has been observed in premature infants. When baby incubators were provided with 100 percent oxygen, blindness occurred in some of the infants. The formation of fibrous tissue behind the lens is triggered by a local vasoconstriction caused by the high PO_2. The hazard is removed simply by reducing the percentage of oxygen in the incubator to below 40 percent.

Hypoxia

Hypoxia (hī-POK-sē-a; *hypo* = below or under) refers to a general low level of oxygen availability; physiologically, it refers to a deficiency of oxygen at the tissue level. Based on the cause, we can classify hypoxias as follows:

1. **Hypoxic hypoxia.** This is caused by a low PO_2 in arterial blood. The condition may be the result of high altitude, obstructions in air passageways, or fluid in the lungs.
2. **Anemic hypoxia.** In this case, there is too little functioning hemoglobin in the blood. Among the causes are hemorrhage, anemia, or failure of hemoglobin to carry its normal complement of oxygen, as in carbon monoxide (CO) poisoning.
3. **Stagnant hypoxia.** This condition results from the inability of blood to carry oxygen to tissues fast enough to sustain their needs. It may be caused by heart failure or circulatory shock, both of which diminish delivery of oxygen to tissues.
4. **Histotoxic hypoxia.** In this condition, the blood delivers adequate oxygen to tissues, but the tissues are unable to use it properly. A commonly cited cause is cyanide poisoning, in which cyanide blocks the metabolic machinery of cells related to oxygen utilization.

Carbon Dioxide

Under normal resting conditions, each 100 ml of deoxygenated blood contains 4 ml of carbon dioxide (CO_2), which is carried by the blood in several forms (see Figure 23-16). The smallest percentage, about 7 percent, is dissolved in plasma. Upon reaching the lungs, it diffuses into the alveoli. A somewhat higher percentage, about 23 percent, combines with the globin portion of hemoglobin to form ***carbaminohemoglobin*** **(*Hb* · CO_2).**

$$\underset{\text{Hemoglobin}}{\text{Hb}} + \underset{\text{Carbon dioxide}}{CO_2} \rightleftharpoons \underset{\text{Carbaminohemoglobin}}{\text{Hb}\cdot CO_2}$$

The formation of carbaminohemoglobin is greatly influenced by PCO_2. For example, in tissue capillaries PCO_2 is relatively high, and this encourages the formation of carbaminohemoglobin. But in pulmonary capillaries, PCO_2 is relatively low, and the CO_2 readily splits apart from globin and enters the alveoli by diffusion.

The greatest percentage of CO_2, about 70 percent, is transported in plasma as bicarbonate ions. The reaction that brings about this method of transportation of CO_2 is the same one noted earlier.

$$\underset{\text{Carbon dioxide}}{CO_2} + \underset{\text{Water}}{H_2O} \xrightleftharpoons{\text{carbonic anhydrase}} \underset{\text{Carbonic acid}}{H_2CO_3} \rightleftharpoons \underset{\text{Hydrogen ion}}{H^+} + \underset{\text{Bicarbonate ion}}{HCO_3^-}$$

As CO_2 diffuses into tissue capillaries and enters the red blood cells, it reacts with water, in the presence of carbonic anhydrase, to form carbonic acid. The carbonic acid dissociates into H^+ ions and HCO_3^- ions. The H^+ ions combine mainly with hemoglobin. The HCO_3^- ions leave the red blood cells and enter the plasma. In exchange, chloride ions (Cl^-) diffuse from plasma into the red blood cells. This exchange of negative ions maintains the ionic balance between plasma and red blood cells and is known as the ***chloride shift*** (Figure 23-22a). The Cl^- ions that enter red blood cells combine with potassium ions (K^+) to form the salt potassium chloride (KCl). The HCO_3^- ions that enter plasma from the red blood cells combine with sodium (Na^+), the principal positive ion in extracellular fluid, to form sodium bicarbonate ($NaHCO_3$). The net effect of all these reactions is that CO_2 is carried from tissue cells as bicarbonate ions in plasma.

Deoxygenated blood returning to the lungs contains CO_2 dissolved in plasma, CO_2 combined with globin as carbaminohemoglobin, and CO_2 incorporated in bicarbonate ions. In the pulmonary capillaries, the events are reversed. The CO_2 dissolved in plasma diffuses into the alveoli. The CO_2 combined with globin splits from the globin and diffuses into the alveoli. The CO_2 carried as bicarbonate is released as follows (Figure 23-22b). As the hemoglobin in pulmonary blood picks up oxygen, the H^+ ions are released from hemoglobin. The Cl^- ions simultaneously split from K^+ ions, and HCO_3^- ions reenter the red blood cells after splitting from Na^+ ions. The H^+ and HCO_3^- ions recombine to form H_2CO_3, which splits into CO_2 and H_2O. The CO_2 leaves the red blood cells and diffuses into the alveoli. The direction of the carbonic acid reaction depends mostly on PCO_2. In tissue capillaries, where PCO_2 is high, bicarbonate is formed. In pulmonary capillaries, where PCO_2 is low, CO_2 and H_2O are formed.

Just as an increase in CO_2 in blood causes oxygen to split from hemoglobin, the binding of O_2 to hemoglobin causes the release of CO_2 from blood. In the presence of O_2, less CO_2 binds in the blood. This reaction, called the ***Haldane effect,*** occurs because when O_2 combines with hemoglobin, the hemoglobin becomes a stronger acid. In this state, hemoglobin combines with less CO_2. Also, the more acidic hemoglobin releases more H^+ ions that bind to HCO_3^- ions to form H_2CO_3; the H_2CO_3 breaks down into $H_2O + CO_2$, and the CO_2 is released from blood into alveoli. In tissue capillaries, blood picks up more CO_2 as O_2 is removed from hemoglobin. In pulmonary capillaries, blood releases more CO_2 as O_2 is picked up by hemoglobin.

FIGURE 23-22 Transportation of carbon dioxide. (a) Between tissue cells and tissue capillaries. (b) Between pulmonary capillaries and alveoli. The amount of carbon dioxide transported in blood is significantly influenced by PO_2. As carbon dioxide (CO_2) leaves tissue cells and enters blood cells, it causes more oxygen (O_2) to dissociate from hemoglobin, the Bohr effect, and thus more CO_2 combines with hemoglobin and more bicarbonate (HCO_3^-) ions are produced. As O_2 passes from alveoli into blood cells, hemoglobin becomes saturated with O_2 and becomes a stronger acid. The more acidic hemoglobin releases more hydrogen (H^+) ions, which bind to HCO_3^- ions to form carbonic acid (H_2CO_3). The H_2CO_3 dissociates into $H_2O + CO_2$ and the CO_2 is released from blood into alveoli.

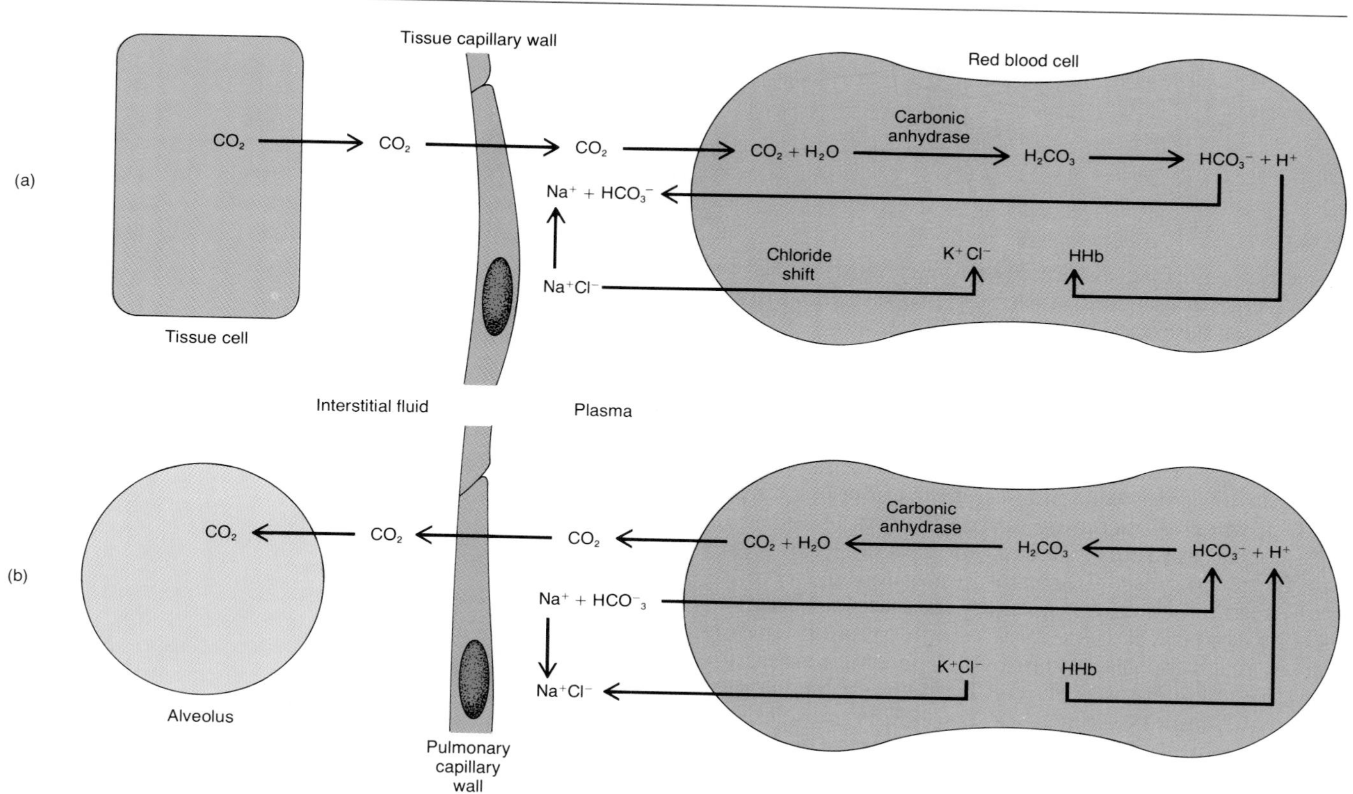

CONTROL OF RESPIRATION

The basic rhythm of respiration is controlled by portions of the nervous system in the medulla and pons. In response to the demands of the body, this rhythm can be modified. We will first examine the principal mechanisms involved in the nervous control of the rhythm of respiration.

Nervous Control

The size of the thorax is affected by the action of the respiratory muscles. These muscles contract and relax as a result of nerve impulses transmitted to them from centers in the brain. The area from which nerve impulses are sent to respiratory muscles is located bilaterally in the reticular formation of the brain stem; it is referred to as the ***respiratory center.*** The respiratory center consists of a widely dispersed group of neurons that is functionally divided into three areas: (1) the medullary rhythmicity area, in the medulla; (2) the pneumotaxic (noo-mō-TAK-sik) area, in the pons; and (3) the apneustic (ap-NOO-stik) area, also in the pons (Figure 23-23).

Medullary Rhythmicity Area

The function of the ***medullary rhythmicity area*** is to control the basic rhythm of respiration. In the normal resting state, inspiration usually lasts for about two seconds and expiration for about three seconds. This is the basic rhythm of respiration. Within the medullary rhythmicity area are found both inspiratory and expiratory neu-

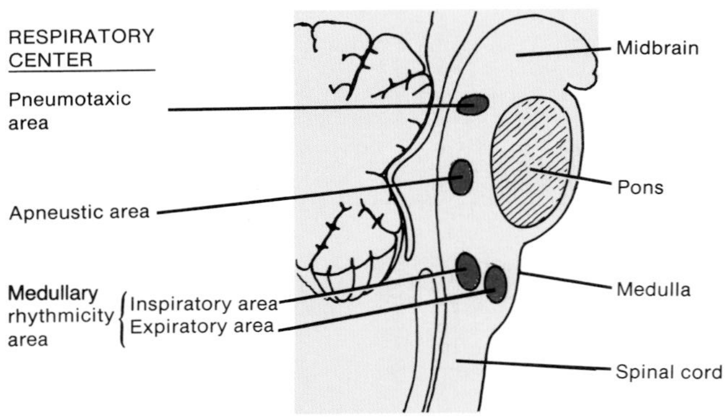

FIGURE 23-23 **Approximate location of areas of the respiratory center.**

rons that comprise inspiratory and expiratory areas, respectively. Let us first consider the proposed role of the inspiratory neurons in respiration.

The basic rhythm of respiration is determined by nerve impulses generated in the inspiratory area (Figure 23-24a). At the beginning of expiration, the inspiratory area is inactive, but after three seconds it suddenly and automatically becomes active. This activity seems to result from an intrinsic excitability of the inspiratory neurons. In fact, when all incoming nerve connections to the inspiratory area are cut or blocked, the area still rhythmically discharges impulses that result in inspiration. Nerve impulses from the active inspiratory area last for about two seconds and travel to the muscles of inspiration. The impulses reach the diaphragm by the phrenic nerves, and the external intercostal muscles by the intercostal nerves. When the impulses reach the inspiratory muscles, the muscles contract and inspiration occurs. At the end of two seconds, the inspiratory muscles become inactive again, and the cycle repeats itself over and over.

It is believed that the expiratory neurons remain inactive during most normal quiet respiration. During quiet respiration, inspiration is accomplished by active contraction of the inspiratory muscles, and expiration results from passive elastic recoil of the lungs and thoracic wall as the inspiratory muscles relax. However, during high levels of ventilation, it is believed that impulses from the inspiratory area activate the expiratory area (Figure 23-24b). Impulses discharged from the expiratory area cause contraction of the internal intercostals and abdominal muscles that decrease the size of the thoracic cavity to cause forced (labored) expiration.

Pneumotaxic Area

Although the medullary rhythmicity area controls the basic rhythm of respiration, other parts of the nervous system help coordinate the transition between inspiration and expiration. One of these is the ***pneumotaxic area*** in the upper pons (see Figure 23-23). It continuously transmits inhibitory impulses to the inspiratory area. The major effect of these impulses is to help turn off the inspiratory area before the lungs become too full of air. In other words, the impulses limit inspiration and thus facilitate expiration.

Apneustic Area

Another part of the nervous system that coordinates the transition between inspiration and expiration is the ***apneustic area*** in the lower pons (see Figure 23-23). It sends stimulatory impulses to the inspiratory area that activate it and prolong inspiration, thus inhibiting expiration. This occurs when the pneumotaxic area is inactive.

FIGURE 23-24 **Proposed role of the medullary rhythmicity area in controlling the basic rhythm of respiration. (a) During normal quiet breathing. (b) During high levels of respiration.**

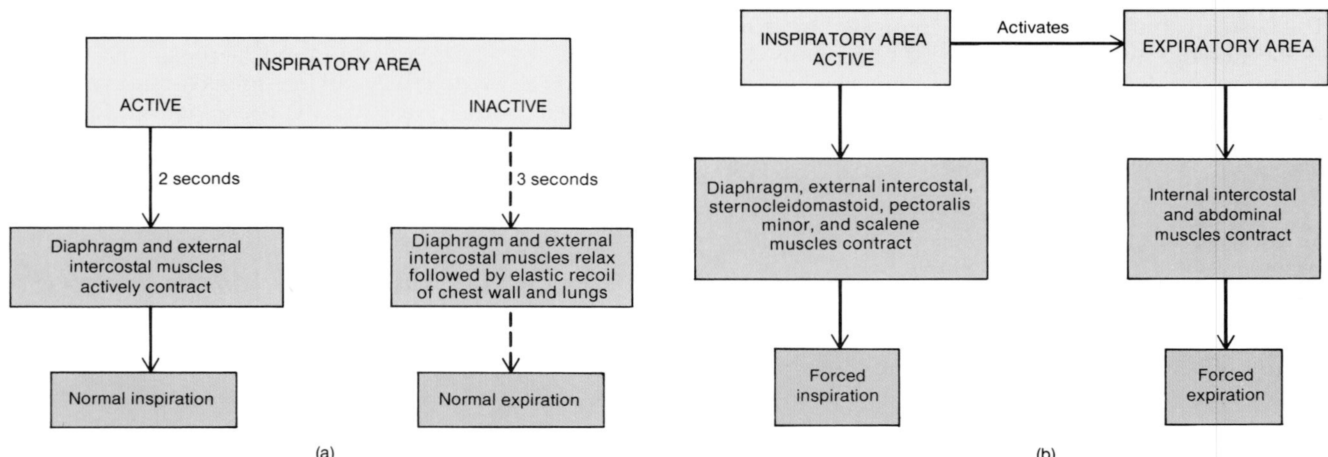

When the pneumotaxic area is active, it overrides the apneustic area.

A summary of the nervous control of respiration is presented in Figure 23-25.

Regulation of Respiratory Center Activity

Although the basic rhythm of respiration is set and coordinated by the respiratory center, the rhythm can be modified in response to the demands of the body by nerve impulses to the center.

Cortical Influences

The respiratory center has connections from the cerebral cortex, which means we can voluntarily alter our pattern of breathing. We can even refuse to breathe at all for a short time. Voluntary control is protective because it enables us to prevent water or irritating gases from entering the lungs. The ability to stop breathing is limited by the buildup of CO_2 and H^+ in the blood, however. When PCO_2 and H^+ increase to a certain level, the inspiratory area is stimulated, nerve impulses are sent to inspiratory muscles, and breathing resumes whether or not the person wishes. It is impossible for people to kill themselves by holding their breath. Even if a person faints, breathing resumes when consciousness is lost.

Inflation Reflex

Located in the walls of bronchi and bronchioles throughout the lungs are ***stretch receptors.*** When the receptors become overstretched, nerve impulses are sent along the vagus (X) nerves to the inspiratory area and apneustic area. In response, the inspiratory area is inhibited and the apneustic area is inhibited from activating the inspiratory area. The result is that expiration follows. As air leaves the lungs during expiration, the lungs deflate and the stretch receptors are no longer stimulated. Thus, the inspiratory and apneustic areas are no longer inhibited, and a new inspiration begins. This reflex is referred to as the ***inflation (Hering–Breuer) reflex.*** Some evidence suggests that the reflex is mainly a protective mechanism for preventing overinflation of the lungs rather than a key component in the regulation of respiration.

Chemical Stimuli

Certain chemical stimuli determine how fast and deeply we breathe. The ultimate goal of the respiratory system is to maintain proper levels of carbon dioxide (CO_2) and

FIGURE 23-25 **Summary of the nervous control of respiration.**

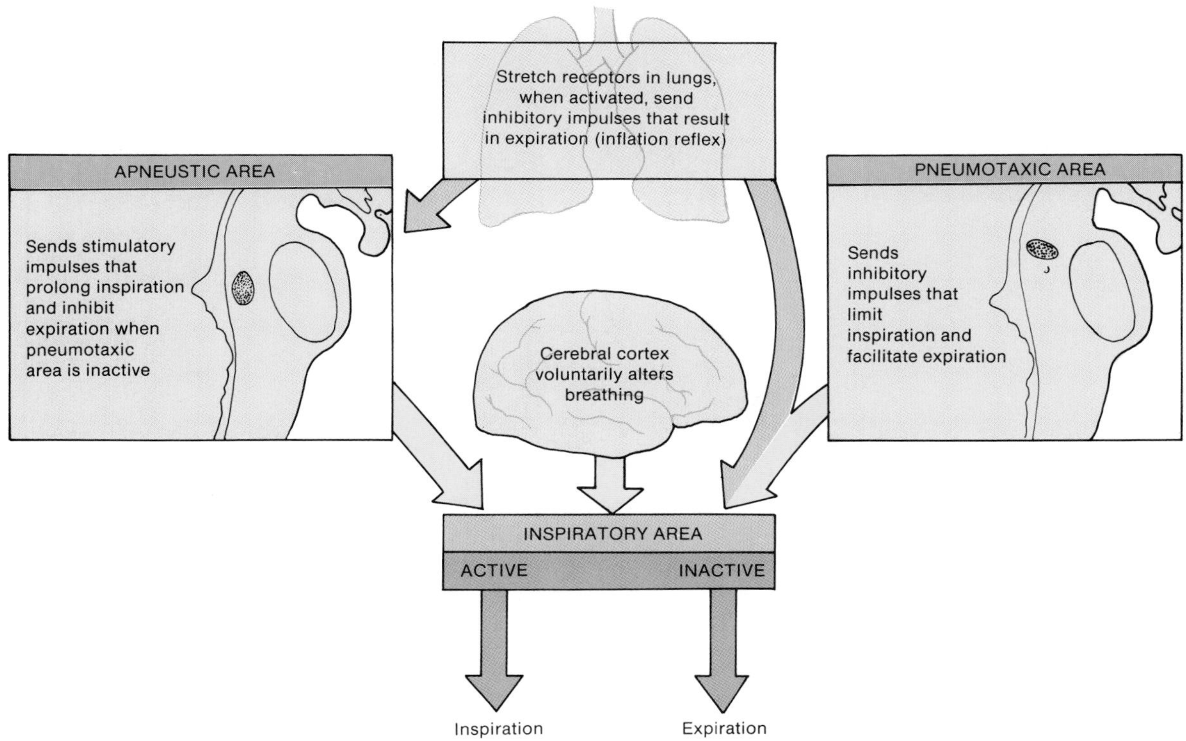

oxygen (O_2), and the system is highly responsive to changes in the blood levels of either. Although it is convenient to speak of carbon dioxide as the most important chemical stimulus for regulating the rate of respiration, it is most likely hydrogen ions (H^+) that assume this role. For example, CO_2 in the blood combines with water (H_2O) to form carbonic acid (H_2CO_3). But the carbonic acid quickly breaks down into H^+ ions and bicarbonate (HCO_3^-) ions. Any increase in CO_2 will cause an increase in H^+ ions, and any decrease in CO_2 will cause a decrease in H^+ ions. In effect, it is primarily the H^+ ions that alter the rate of respiration rather than the CO_2 molecules. Although the following discussion refers to the levels of CO_2 and their effect on respiration, keep in mind that it is really the H^+ ions that cause the effects.

Within the medulla is an area called the ***chemosensitive area*** that is highly sensitive to blood concentrations of CO_2. Outside the central nervous system are ***chemoreceptors*** that are sensitive to changes in CO_2 and O_2 levels in the blood. They are located within the carotid bodies near the bifurcation (branching) of the common carotid arteries and in the aortic bodies near the arch of the aorta. The ***carotid bodies*** are small, oval nodules about 4 to 5 mm long, located in the space between the internal and external carotid arteries. The ***aortic bodies*** are clustered in the region between the arch of the aorta and the dorsal surface of the pulmonary artery. The afferent nerve fibers from the carotid bodies join with those from the carotid sinus to form the carotid sinus nerve, which joins the glossopharyngeal (IX) nerve. The afferent fibers from aortic bodies run into the vagus (X) nerve.

Under normal circumstances, arterial blood PCO_2 is 40 mm Hg. If there is even a slight increase in PCO_2—a condition called ***hypercapnia***—the chemosensitive area in the medulla and chemoreceptors in the carotid and aortic bodies are stimulated (Figure 23-26). Stimulation of the chemosensitive area and chemoreceptors causes the inspiratory area to become highly active, and the rate and depth of respiration increases. This increased rate, ***hyperventilation,*** allows the body to expel more CO_2 until PCO_2 and H^+ are lowered to normal. If arterial PCO_2 is lower than 40 mm Hg, a condition called ***hypocapnia,*** the chemosensitive area and chemoreceptors are not stimulated and stimulatory impulses are not sent to the inspiratory area. Consequently, the area sets its own moderate pace until CO_2 accumulates and the PCO_2 rises to 40 mm Hg. A slow rate of respiration is called ***hypoventilation.***

FIGURE 23-26 **Effects of increased blood levels of PCO_2 (blue arrows) and decreased blood levels of PO_2 down to 50 mm Hg (red arrows).**

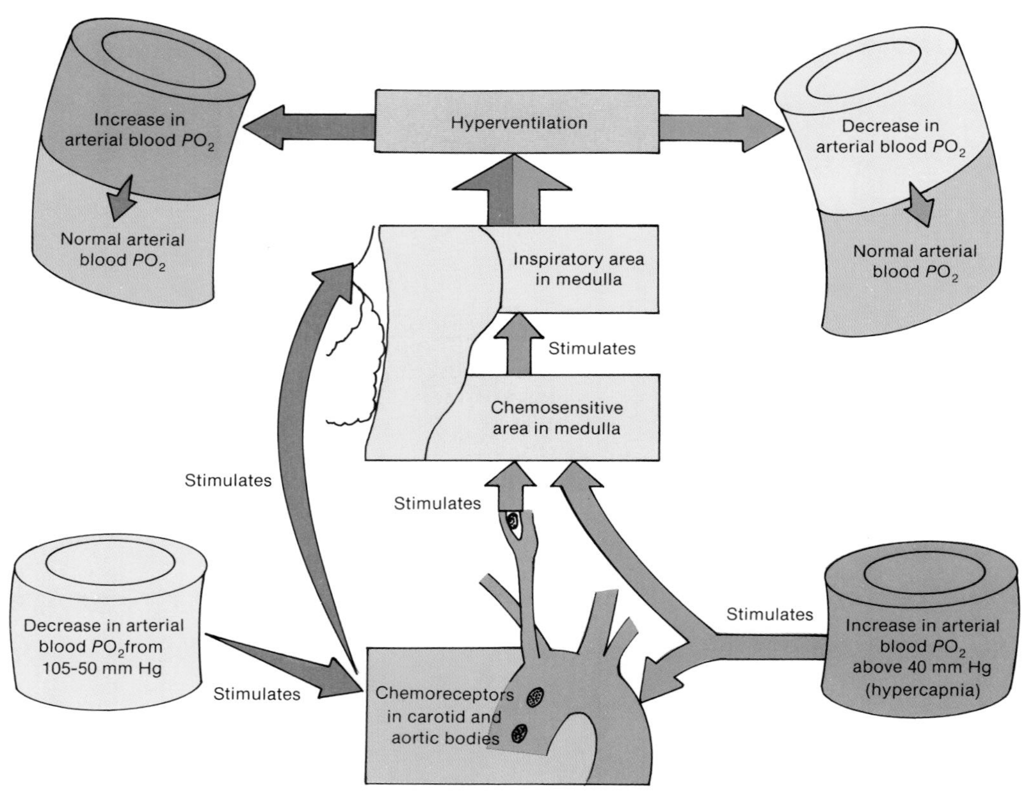

The oxygen chemoreceptors are sensitive only to large decreases in the PO_2 since hemoglobin remains about 85 percent or more saturated at PO_2 values all the way down to 50 mm Hg (see Figure 23-18). If arterial PO_2 falls from a normal of 105 mm Hg to about 50 mm Hg, the oxygen chemoreceptors become stimulated and send impulses to the inspiratory area and respiration increases (Figure 23-26). But if the PO_2 falls much below 50 mm Hg, the cells of the inspiratory area suffer oxygen starvation and do not respond well to any chemical receptors. They send fewer impulses to the inspiratory muscles, and the respiration rate decreases or breathing ceases altogether.

Other Influences

The carotid and aortic sinuses also contain baroreceptors (pressure receptors) that detect changes in blood pressure. Although these baroreceptors are concerned mainly with the control of circulation (see Figure 20-10), they affect respiration. For example, a sudden rise in blood pressure decreases the rate of respiration, and a drop in blood pressure increases the respiratory rate.

These are other factors that control respiration:

1. An increase in body temperature, as during a fever or vigorous muscular exercise, increases the rate of respiration. A decrease in body temperature decreases respiratory rate. A sudden cold stimulus such as plunging into cold water causes ***apnea*** (ap-NĒ-a), a temporary cessation of breathing.
2. A sudden, severe pain brings about apnea, but a prolonged pain triggers the general adaptation syndrome and increases respiratory rate.
3. Stretching the anal sphincter muscle increases the respiratory rate. This technique is sometimes employed to stimulate respiration during emergencies.
4. Irritation of the pharynx or larynx by touch or chemicals brings about an immediate cessation of breathing followed by coughing.

A summary of stimuli that affect the rate of respiration is presented in Exhibit 23-4.

AGING AND THE RESPIRATORY SYSTEM

The cardiovascular and respiratory systems operate as a unit, and damage or disease in one of these organ systems is often secondarily reflected in the other. In pulmonary heart disease, the right side of the heart enlarges in response to certain lung diseases. This condition is known as right ventricular hypertrophy, or cor pulmonale. With advancing age, the airways and tissues of the respiratory tract, including the air sacs, become less elastic and more rigid. In addition to the lungs becoming less elastic, the chest wall also becomes more rigid. As a result, there is a decrease in pulmonary lung capacity. In fact, vital capacity (the amount of air moved by maximal inspiration followed by maximal expiration) can decrease as much as 35 percent by age 70. Also, there is a decrease in blood levels of oxygen, decreased activity of alveolar macrophages, and diminished ciliary action of the epithelium lining the respiratory tract. Owing to all these age-related factors, elderly people are more susceptible to pneumonia, bronchitis, emphysema, and other pulmonary disorders.

DEVELOPMENTAL ANATOMY OF THE RESPIRATORY SYSTEM

The development of the mouth and pharynx is considered in Chapter 24 on the digestive system. Here we will consider the remainder of the respiratory system.

At about four weeks of fetal development, the respiratory system begins as an outgrowth of the ***endoderm*** of the foregut (precursor of some digestive organs) just

EXHIBIT 23-4 SUMMARY OF STIMULI THAT AFFECT THE RATE OF RESPIRATION

Stimuli That Increase Rate of Respiration	Stimuli That Decrease (or Inhibit) Rate of Respiration
Increase in arterial blood PCO_2 above 40 mm Hg (actually H^+ level).	Decrease in arterial blood PCO_2 below 40 mm Hg (actually H^+ level).
Decrease in arterial blood PO_2 from 105 mm Hg to 50 mm Hg.	Decrease in arterial blood PO_2 below 50 mm Hg.
Decrease in blood pressure.	Increase in blood pressure.
Increase in body temperature.	Decrease in body temperature decreases rate of respiration, and sudden cold stimulus causes apnea.
Prolonged pain.	Severe pain causes apnea.
Stretching anal sphincter.	Irritation of pharynx or larynx by touch or chemicals causes apnea.

FIGURE 23-27 Development of the bronchial tubes and lungs.

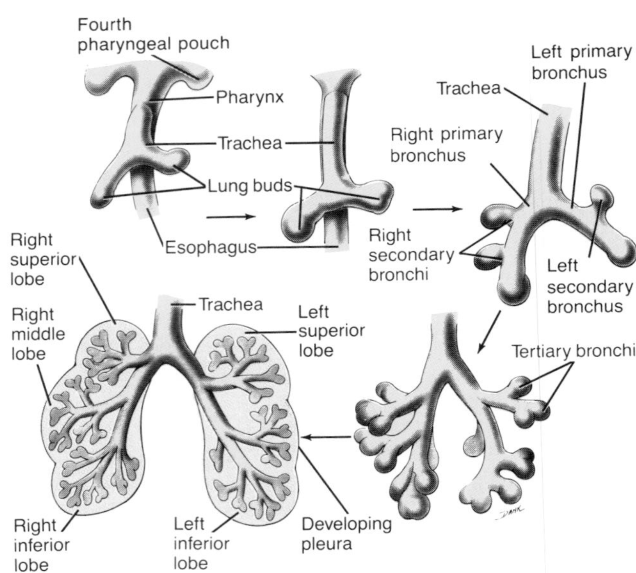

behind the pharynx. This outgrowth is called the ***laryngotracheal bud*** (see Figure 18-24). As the bud grows, it elongates and differentiates into the future *larynx* and other structures as well. Its proximal end maintains a slitlike opening into the pharynx called the *glottis*. The middle portion of the bud gives rise to the *trachea*. The distal portion divides into two ***lung buds,*** which grow into the *bronchi* and *lungs* (Figure 23-27).

As the lung buds develop, they branch and rebranch and give rise to all the *bronchial tubes*. After the sixth month, the closed terminal portions of the tubes dilate and become the *alveoli* of the lungs. The smooth muscle, cartilage, and connective tissues of the bronchial tubes and the pleural sacs of the lungs are contributed by ***mesenchymal (mesodermal) cells.***

DISORDERS: HOMEOSTATIC IMBALANCES

Bronchogenic Carcinoma (Lung Cancer)

As part of ordinary breathing, many irritating substances are inhaled. Almost all pollutants, including inhaled smoke, have an irritating effect on the bronchial tubes and lungs and may be regarded as stresses or irritating stimuli. The effects of irritation on the bronchial epithelium are important clinically because a common lung cancer, ***bronchogenic carcinoma,*** starts in the walls of the bronchi.

Microscopic examination of the pseudostratified epithelium of a bronchial tube reveals three kinds of cells, all of which are in contact with the basement membrane. The columnar cells contain cilia and extend upward from the basement membrane and reach the luminal surface. At intervals between the ciliated columnar cells are the mucus-secreting goblet cells. The basal cells do not reach the luminal surface. The basal cells divide continuously, replacing the columnar cells as they wear down and are sloughed off.

The constant irritation by inhaled smoke and pollutants causes an enlargement of the goblet cells of the bronchial epithelium. They respond by secreting excessive mucus. The basal cells also respond to the stress by undergoing cell division so fast that they push into the area occupied by the goblet and columnar cells. As many as 20 rows of basal cells may be produced. Many researchers believe that if the stress is removed at this point, the epithelium can return to normal.

If the stress persists, more and more mucus is secreted, and the cilia become less effective. As a result, mucus is not carried toward the throat but remains trapped in the bronchial tubes. The individual then develops a "smoker's cough." Moreover, the constant irritation from the pollutant slowly destroys the alveoli, which are replaced with thick, inelastic connective tissue. Alveoli are destroyed by protein, digesting enzymes produced by leucocytes, and macrophages in response to the stress. Mucus that has accumulated becomes trapped in the air sacs. Millions of sacs rupture, reducing the diffusion surface for the exchange of oxygen and carbon dioxide. The individual has now developed emphysema. If the stress is removed at this point, there is little chance for improvement. Alveolar tissue that has been destroyed cannot be repaired. But removal of the stress can stop further destruction of lung tissue.

If the stress still persists, the emphysema gets progressively worse, and the basal cells of the bronchial tubes continue to divide and break through the basement membrane. At this point the stage is set for bronchogenic carcinoma. Columnar and goblet cells disappear and may be replaced with squamous cancer cells. If this happens, the malignant growth spreads throughout the lung and may block a bronchial tube. If the obstruction occurs in a large bronchial tube, very little oxygen enters the lung, and disease-producing bacteria thrive on the mucoid secretions. In the end, the patient may develop emphysema, carcinoma, and a host of infectious diseases. Treatment involves surgical removal of the diseased lung. However, metastasis (spreading) of the growth through the lymphatics or blood vessels may result in new growths in other parts of the body such as the brain and liver.

Other factors may be associated with lung cancer. For instance, breast, stomach, and prostate malignancies can metastasize to the lungs. People who apparently have not been exposed to pollutants do occasionally develop bronchogenic carcinoma. However, the occurrence of

DISORDERS: HOMEOSTATIC IMBALANCES (*continued*)

bronchogenic carcinoma is probably over 20 times higher in heavy cigarette smokers than it is in nonsmokers. (In 1988 a report of the Surgeon General warned that nicotine in cigarettes and other tobacco products is as addictive as cocaine or heroine.)

Bronchial Asthma

Bronchial asthma is a reversible obstructive airway disease characterized by periods of coughing, difficult breathing, and wheezing that reverse either spontaneously or with treatment. Attacks are brought on by spasms of the smooth muscles that lie in the walls of the smaller bronchi and bronchioles, causing the passageways to close partially. The patient has trouble exhaling, and the alveoli may remain inflated during expiration. Usually, the mucous membranes that line the respiratory passageways become irritated and secrete excessive amounts of mucus that may clog the bronchi and bronchioles and worsen the attack. About three out of four asthma victims are allergic to edible or airborne substances as common as wheat or dust. Others are sensitive to the proteins of harmless bacteria that inhabit the paranasal sinuses, nose, and throat. Asthma might also have a psychosomatic origin. The two major varieties of asthma are *extrinsic asthma,* occurring in childhood or early adulthood, and due to respiratory allergies, and *intrinsic asthma,* which has its onset after age 45, and the cause is other than just allergies.

Bronchitis

Bronchitis is inflammation of the bronchi characterized by hypertrophy and hyperplasia of seromucous glands and goblet cells lining the bronchial airways. The typical symptom is a productive cough in which a thick greenish-yellow sputum is raised. This secretion signifies the presence of the underlying infection that is causing excessive secretion of mucus. Cigarette smoking remains the most important cause of chronic bronchitis, that is, bronchitis that lasts for at least three months of the year for two successive years. Other factors that influence development of the disease are family history of bronchitis or other lung disease, air pollution, carbon monoxide, respiratory infections, and deficient antibodies, especially IgA antibodies.

Emphysema

In ***emphysema*** (em'-fi-SĒ-ma), the alveolar walls lose their elasticity and remain filled with air during expiration. The name means "blown up" or "full of air." Reduced forced expiratory volume is the first symptom. Later, alveoli in other areas of the lungs are damaged. Many alveoli may merge to form larger air sacs with a reduced overall volume. The lungs become permanently inflated because they have lost elasticity. To adjust to the increased lung size, the size of the chest cage increases, resulting in a "barrel chest." The patient has to work voluntarily to exhale. Oxygen diffusion does not occur as easily across the damaged alveolar-capillary membrane, blood O_2 is somewhat lowered, and any mild exercise that raises the oxygen requirements of the cells leaves the patient breathless. As the disease progresses, the alveoli are replaced with thick fibrous connective tissue. Even carbon dioxide does not diffuse easily through this fibrous tissue. If the blood cannot buffer all the hydrogen ions that accumulate, the blood pH drops or unusually high amounts of carbon dioxide may dissolve in the plasma. High carbon dioxide levels produce acid conditions that are toxic to brain cells. Consequently, the inspiratory area becomes less active and the respiration rate slows down, further aggravating the problem. The compressed and damaged capillaries around the deteriorating alveoli may no longer be able to receive blood. As a result, resistance to blood flow increases in the pulmonary trunk, and the right ventricle overworks as it attempts to force blood through the remaining capillaries.

Emphysema is generally caused by a long-term irritation. Air pollution, occupational exposure to industrial dust, and cigarette smoke are the most common irritants. Some evidence suggests that destruction of alveolar sacs may be caused by an imbalance between enzymes called proteases, one of the most important ones called elastase, and a molecule (alpha$_1$-antitrypsin) that inhibits proteases. It seems that when there is a decreased production by the liver of the plasma protein alpha$_1$-antitrypsin, elastase is not inhibited and is free to attack the connective tissue in the walls of alveolar sacs. Cigarette smoke not only deactivates a protein that is seemingly crucial in preventing emphysema but also prevents the repair of affected lung tissue. Chronic bronchial asthma also may produce alveolar damage. Cases of emphysema are becoming more frequent in the United States. The irony is that the disease can be prevented and the progressive deterioration can be stopped by eliminating the harmful stimuli.

Diseases such as bronchial asthma, bronchitis, and emphysema have in common some degree of obstruction of the air passageways. The term ***chronic obstructive pulmonary disease* (*COPD*)** is used to refer to these disorders. Among the symptoms that might indicate significant airflow obstruction are cough, wheezing, and dyspnea.

Pneumonia

The term ***pneumonia*** means an acute infection or inflammation of the alveoli. It is the most common infectious cause of death in the United States. In this disease, the alveolar sacs fill up with fluid and dead white blood cells, reducing the amount of air space in the lungs. (Remember that one of the cardinal signs of inflammation is edema.) Oxygen has difficulty diffusing through the inflamed alveoli, and the blood oxygen may be drastically reduced. Blood carbon dioxide usually remains normal because carbon dioxide diffuses through the alveoli more easily than oxygen does.

The most common cause of pneumonia is the pneu-

DISORDERS: HOMEOSTATIC IMBALANCES (*continued*)

mococcus bacterium (*Streptococcus pneumoniae*), but other bacteria or fungi, protozoans, or viruses may be the source of trouble. Among those who are susceptible to pneumonia are the elderly, infants, immunocompromised individuals (persons with AIDS, malignancy, or those taking immunosuppressive drugs), cigarette smokers, and persons with obstructive lung disease.

Tuberculosis (TB)

The bacterium *Mycobacterium tuberculosis* produces an infectious, communicable disease called ***tuberculosis (TB)***. The disease most often affects the lungs and the pleurae. The bacteria destroy parts of the lung tissue, and the tissue is replaced by fibrous connective tissue. Because the connective tissue is inelastic and thick, the affected areas of the lungs do not snap back during expiration and larger amounts of air are retained. Gases no longer diffuse easily through the fibrous tissue. Patients with AIDS have a higher-than-normal incidence of tuberculosis, and this is one factor in recent increases in the number of reported cases of tuberculosis.

Tuberculosis bacteria are spread by inhalation and exhalation. Although they can withstand exposure to many disinfectants, they die quickly in sunlight. Thus, tuberculosis is sometimes associated with crowded, poorly lighted housing. Many drugs, such as isoniazid and rifampin, are successful in treating tuberculosis. Rest, sunlight, and good diet are vital parts of treatment.

Respiratory Distress Syndrome (RDS) of the Newborn

Respiratory distress syndrome (RDS) of the newborn, also called ***glassy-lung disease*** or ***hyaline membrane disease (HMD)***, is responsible for approximately 20,000 newborn infant deaths per year. Before birth, the respiratory passages are filled with fluid. Part of this fluid is amniotic fluid inhaled during respiratory movements in utero. The remainder is produced by the submucosal glands and the goblet cells of the respiratory epithelium.

At birth, this fluid-filled passageway must become an air-filled passageway, and the collapsed primitive alveoli (terminal sacs) must expand and function in gaseous exchange. The success of this transition depends largely on surfactant, a phospholipid produced by alveolar cells that lowers surface tension in the fluid layer lining the primitive alveoli once air enters the lungs. Surfactant is present in the fetus's lungs as early as 23 weeks, and by 28 to 32 weeks the amount present is sufficient to prevent alveolar collapse during breathing. Surfactant is produced continuously by septal alveolar cells. The presence of surfactant can be detected by amniocentesis.

Although in a normal, full-term infant the second and subsequent breaths require less respiratory effort than the first, breathing is not completely normal until about 40 minutes after birth. The entire lung is not inflated fully with the first one or two breaths. In fact, for the first 7 to 10 days, small areas of the lungs may remain uninflated.

In the newborn whose lungs are deficient in surfactant, the effort required for the first breath is essentially the same as that required in normal newborns. However, the surface tension of the alveolar fluid is 7 to 14 times higher than the surface tension of alveolar fluid with a monomolecular layer of surfactant. Consequently, during expiration after the first inspiration, the surface tension of the alveoli increases as the alveoli deflate. The alveoli collapse almost to their original uninflated state.

Idiopathic RDS of the newborn usually appears within a few hours after birth. (An idiopathic condition is one that occurs spontaneously in an individual from an unknown or obscure cause.) Affected infants show difficult and labored breathing with retraction of the intercostal and subcostal spaces. Death may occur soon after onset of respiratory difficulty or may be delayed for a few days, although many infants survive. At autopsy, the lungs are underinflated and areas of atelectasis (collapse of lung) are prominent. If the infant survives for at least a few hours after developing respiratory distress, the alveoli are often filled with a fluid of high-protein content that resembles a hyaline (or glassy) membrane. RDS of the newborn occurs frequently in premature infants and also in infants of diabetic mothers, particularly if the diabetes is untreated or poorly controlled.

A treatment called PEEP (positive end expiratory pressure) is used to treat RDS of the newborn. It consists of passing a tube through the air passages to the top of the lungs to provide needed oxygen-rich air at continuous pressures of up to 14 millimeters of mercury (mm Hg). Continuous pressure keeps the baby's alveoli open and available for gas exchange. The addition of human surfactant, obtained from the amniotic fluid of infants born by caesarian section, to the oxygen-rich air counters the effects of RDS of the newborn more effectively and lessens lung damage.

Respiratory Failure

Respiratory failure refers to a condition in which the respiratory system cannot supply sufficient oxygen to maintain metabolism or cannot eliminate enough carbon dioxide to prevent respiratory acidosis, a lower-than-normal pH. Specifically, respiratory failure occurs when the oxygen level of arterial blood is too low or when the carbon dioxide level of arterial blood is too high. Respiratory failure always causes dysfunction in other organs as well.

Among the causes of respiratory failure are lung disorders (chronic obstructive pulmonary disease, pneumonia, and adult respiratory distress syndrome); mechanical disorders that disturb the chest wall or neuromusculature (anesthetic blocking agents, cervical spinal cord injuries, myasthenia gravis, and amyotrophic lateral sclerosis); depression of the respiratory center by drugs, strokes, or trauma; and carbon monoxide poisoning.

Symptoms of respiratory failure include disorientation,

DISORDERS: HOMEOSTATIC IMBALANCES (*continued*)

malaise, headache, muscular weakness, difficulty with sleep, palpitations, breathlessness, coughing, tachycardia, dysrhythmia, cyanosis, hypertension, edema, stupor, and coma. Treatment consists of providing adequate oxygenation and reversing the respiratory acidosis.

Sudden Infant Death Syndrome (SIDS)

***Sudden infant death syndrome* (*SIDS*),** also called crib death or cot death, kills approximately 10,000 American babies every year. It kills more infants between the ages of 1 week and 12 months than any other disease. SIDS occurs without warning. Although about half of its victims had an upper respiratory infection within two weeks of death, the babies tended to be otherwise remarkably healthy. Two other findings are important: the apparently *silent* nature of death and the disarray frequently found at the scene.

Given this death scene, it is not surprising that for centuries "crib death" was mistakenly thought to be due to suffocation. Most medical people, however, believe that a healthy child cannot smother in its bedclothes. The diaper of a SIDS victim is usually full of stool and urine, and the urinary bladder is almost always empty. External examination reveals only a blood-tinged froth that often exudes from the nostrils. These signs all point to a sudden lethal episode associated with tonic motor activity, not to a gradually deepening coma leading to death.

The similarity of findings in SIDS cases strengthens the theory that there is one underlying cause. The seasonal distribution of SIDS incidences, being lowest during the summer months and highest in the late fall and winter, strongly suggests an infectious agent, viruses being the most likely possibility. Most cases occur at a time in life when gamma globulin levels are low and the infant is at a critical period of susceptibility.

Many hypotheses have been proposed as explanations for SIDS. Although the precise pathogenic mechanisms have yet to be clarified, several valuable observations have emerged as a result of investigations during the past decade. Most, if not all, such cases appear to share a common mechanism ("final common pathway"). Thus, SIDS is probably a single disease entity. According to one study, a likely cause of death is laryngospasm, possibly triggered by a previous mild, nonspecific virus infection of the upper respiratory tract. Another possible factor includes periods of prolonged apnea as a result of malfunction of the respiratory center. Hypoxia may also be involved. It has recently been learned that the blood of infants whose cause of death was listed as SIDS have elevated levels of hemoglobin F (fetal hemoglobin). Normally, hemoglobin F is replaced by adult hemoglobin during the first six months after birth. The delayed switch from hemoglobin F to adult hemoglobin could affect the delivery of oxygen to sensitive tissue sites. Other suspected causes are an allergic reaction, bacterial poisoning, excess dopamine (DA), overheating, and altered cardiac repolarization as reflected by a prolonged QT interval.

Coryza (Common Cold) and Influenza (Flu)

Several viruses are responsible for ***coryza* (*common cold*).** A group of viruses, called rhinoviruses, is responsible for about 40 percent of all colds in adults. Typical symptoms include sneezing, excessive nasal secretion, dry cough, and congestion. The uncomplicated common cold is not usually accompanied by a fever. Complications include sinusitis, asthma, bronchitis, ear infections, and laryngitis. Direct transfer of nasal secretions on hands accounts for nearly all transmission of viruses that cause coryza, not sneezing. Since common colds are caused by viruses, antibiotics are of no use in treatment. Although over-the-counter drugs (aspirin, antihistamines, and decongestants) may lessen the severity of certain symptoms of the common cold, they do not lessen recovery time. An experimental antiviral nasal spray containing alpha-2 interferon may prevent the spread of colds among family members. The interferon is effective mainly against rhinoviruses. Several studies show that neither cold temperature nor dampness increases susceptibility to common colds.

***Influenza* (*flu*)** is also caused by a virus. Its symptoms include chills, fever (usually higher than 101°F), headache, and muscular aches. Coldlike symptoms appear as the fever subsides.

Pulmonary Embolism

Pulmonary embolism refers to the presence of a blood clot or other foreign substance in a pulmonary arterial vessel that obstructs circulation to lung tissue. Most pulmonary emboli originate from thrombi in the proximal deep veins of the lower extremities. From there, they pass through the right side of the heart to a lung via a pulmonary artery. The immediate effect of pulmonary embolism is complete or partial obstruction of the pulmonary arterial blood flow to the lung, resulting in dysfunction of the affected lung tissue. A large embolus can produce death in a few minutes. Pulmonary embolism usually develops in patients with one or more risk factors. Among these are immobilization in bed for more than three days; heart disease (especially with congestive heart failure); trauma, such as fractures; malignant disease; obesity; pregnancy; polycythemia; and the use of oral contraceptives.

Among the clinical findings associated with pulmonary embolism are sudden and unexplained dyspnea (shortness of breath), chest pain, apprehension, rapid breathing, rales, cough, deep-venous thrombosis, pulmonary infarction as evidenced by hemoptysis (expectoration of blood or blood-stained sputum), and syncope.

Treatment of pulmonary embolism consists of intravenous (IV) injections of heparin (an anticoagulant); administration of oral anticoagulants, such as warfarin; administration of clot-dissolving enzymes, such as streptokinase, tissue plasminogen activator, or urokinase;

DISORDERS: HOMEOSTATIC IMBALANCES *(continued)*

bed rest; analgesia; and oxygen administration. In some cases, surgery (embolectomy) is indicated to reverse the lethal effect of massive pulmonary embolism.

Pulmonary Edema

Pulmonary edema refers to an abnormal accumulation of interstitial fluid in the interstitial spaces and alveoli of the lungs. The edema may arise from increased pulmonary capillary permeability (pulmonary origin) or increased pulmonary capillary pressure (cardiac origin). This latter cause may coincide with congestive heart failure. The most common symptom is dyspnea. Others include wheezing, tachypnea (rapid respirations), restlessness, feeling of suffocation, cyanosis, pallor (paleness), and diaphoresis (excessive perspiration). Treatment consists of administering oxygen, drugs that dilate the bronchioles and lower blood pressure, and drugs that correct acid–base imbalance; suctioning of airways; and mechanical ventilation.

Smoke Inhalation Injury

When smoke is inhaled, the lungs are injured directly by heat from flames and substances in fumes. ***Smoke inhalation injury*** has three components that occur in sequence: (1) inhibition of oxygen delivery and utilization, (2) upper airway injury from heat, and (3) lung damage from acids and aldehydes in smoke.

Inhibition of oxygen delivery and utilization is due to inhalation of mainly carbon monoxide (CO) and, to a lesser extent, cyanide. CO binds to hemoglobin and components of the cytochrome system and thus impairs oxygen delivery and utilization and also causes metabolic acidosis. The onset of symptoms (confusion and disorientation with moderate toxicity, and coma and cardiac arrest with severe toxicity) is immediate.

Upper airway injury from heat may cause edema of the laryngotracheal mucosa and obstruction of upper airways with atelectasis (collapsed lung). Symptoms, such as labored breathing and hypoxemia, occur after 18 to 24 hours.

The acids and aldehydes in inhaled smoke result in bronchoconstriction, atelectasis, bronchiolar edema, impairment of mucociliary action, and destruction of surfactant. The onset of symptoms (labored breathing, hypovolemia, hypoxemia, and bronchopneumonia) is both early and delayed.

Treatment consists of administering 90 to 100 percent oxygen, inserting an endotracheal tube, performing pulmonary suction, administering bronchodilators, and restoring and maintaining fluid balance.

MEDICAL TERMINOLOGY ASSOCIATED WITH THE RESPIRATORY SYSTEM

Asphyxia (as-FIK-sē-a; *sphyxis* = pulse) Oxygen starvation due to low atmospheric oxygen or interference with ventilation, external respiration, or internal respiration.

Aspiration (as'-pi-RĀ-shun; *spirare* = breathe) Inhalation of a foreign substance such as water, food, or foreign body into the bronchial tree; drawing of a substance in or out by suction.

Bronchiectasis (bron'-kē-EK-ta-sis; *ektasis* = dilation) A chronic dilation of the bronchi or bronchioles.

Cardiopulmonary resuscitation (kar'-dē-ō-PUL-mo-ner-ē re-sus'-i-TA-shun) **(CPR)** The artificial establishment of normal or near-normal respiration and circulation. The **A, B, C's** of cardiopulmonary resuscitation are **Airway, Breathing,** and **Circulation,** meaning the rescuer must establish an airway, provide artificial ventilation if breathing has stopped, and reestablish circulation if there is inadequate cardiac action. The procedure should be performed in that order (A, B, C).

Cheyne-Stokes respiration (CHĀN STŌKS res'-pi-RĀ-shun) A repeated cycle of irregular breathing beginning with shallow breaths that increase in depth and rapidity, then decrease and cease altogether for 15 to 20 seconds. Cheyne-Stokes is normal in infants. It is also often seen just before death from pulmonary, cerebral, cardiac, and kidney disease.

Diphtheria (dif-THĒ-rē-a; *diphthera* = membrane) An acute bacterial infection that causes the mucous membranes of the oropharynx, nasopharynx, and larynx to enlarge and become leathery. Enlarged membranes may obstruct airways and cause death from asphyxiation.

Dyspnea (DISP-nē-a; *dys* = painful, difficult; *pnoia* = breath) Painful or labored breathing.

Epistaxis (ep'-i-STAK-sis) Loss of blood from the nose due to trauma, infection, allergy, neoplasms, and bleeding disorders. It can be arrested by cautery with silver nitrate, electrocautery, and firm packing. Also called **nosebleed.**

Heimlich (abdominal thrust) maneuver (HĪM-lik ma-NOO-ver) First-aid procedure designed to clear the air passageways of obstructing objects. It is performed by applying a quick upward thrust that causes sudden elevation of the diaphragm and forceful, rapid expulsion of air in the lungs; this action forces air out

MEDICAL TERMINOLOGY ASSOCIATED WITH THE RESPIRATORY SYSTEM *(continued)*

the trachea to eject the obstructing object. The Heimlich maneuver is also used to expel water from the lungs of near-drowning victims before resuscitation is begun.

Hemoptysis (hē-MOP-ti-sis; *hemo* = blood; *ptein* = to spit) Spitting of blood from the respiratory tract.

Orthopnea (or'-thop-NĒ-a; *ortho* = straight) Dyspnea that occurs in the horizontal position. It is an abnormal finding because a normal individual can tolerate the reduction in vital capacity that accompanies the recumbent position.

Pneumonectomy (noo'-mō-NEK-tō-mē; *pneumo* = lung; *tome* = cutting) Surgical removal of a lung.

Rales (RĀLS) Sounds sometimes heard in the lungs that resemble bubbling or rattling. Rales are to the lungs what murmurs are to the heart. There are many types of rales, but each is due to the presence of an abnormal amount or type of fluid or mucus inside the bronchi or alveoli, or to bronchoconstriction so that air cannot enter or leave the lungs normally.

Respirator (RES-pi-rā'-tor) An apparatus fitted to a mask over the nose and mouth, or hooked directly to an endotracheal or tracheotomy tube, that is used to assist or support ventilation or to provide nebulized medication to the air passages under pressure.

Rhinitis (rī-NĪ-tis; *rhino* = nose) Chronic or acute inflammation of the mucous membrane of the nose.

Tachypnea (tak'-ip-NĒ-a; *tachy* = rapid; *pnoia* = breath) Rapid breathing.

STUDY OUTLINE

Organs (p. 690)

1. Respiratory organs include the nose, pharynx, larynx, trachea, bronchi, and lungs.
2. They act with the cardiovascular system to supply oxygen and remove carbon dioxide from the blood.

Nose (p. 690)

1. The external portion is made of cartilage and skin and is lined with mucous membrane. Openings to the exterior are the external nares.
2. The internal portion communicates with the paranasal sinuses and nasopharynx through the internal nares.
3. The nasal cavity is divided by a septum. The anterior portion of the cavity is called the vestibule.
4. The nose is adapted for warming, moistening, and filtering air, for olfaction, and for speech.

Pharynx (p. 693)

1. The pharynx (throat) is a muscular tube lined by a mucous membrane.
2. The anatomic regions are nasopharynx, oropharynx, and laryngopharynx.
3. The nasopharynx functions in respiration. The oropharynx and laryngopharynx function both in digestion and in respiration.

Larynx (p. 694)

1. The larynx (voice box) is a passageway that connects the pharynx with the trachea.
2. It contains the thyroid cartilage (Adam's apple); the epiglottis, which prevents food from entering the larynx; the cricoid cartilage, which connects the larynx and trachea; and the paired arytenoid, corniculate, and cuneiform cartilages.
3. The larynx contains vocal folds, which produce sound. Taut folds produce high pitches, and relaxed ones produce low pitches.

Trachea (p. 696)

1. The trachea (windpipe) extends from the larynx to the primary bronchi.
2. It is composed of smooth muscle and C-shaped rings of cartilage and is lined with pseudostratified epithelium.
3. Two methods of bypassing obstructions from the respiratory passageways are tracheostomy and intubation.

Bronchi (p. 697)

1. The bronchial tree consists of the trachea, primary bronchi, secondary bronchi, tertiary bronchi, bronchioles, and terminal bronchioles. Walls of bronchi contain rings of cartilage; walls of bronchioles contain smooth muscle.
2. A bronchogram is a roentgenogram of the tree after introduction of an opaque contrast medium usually containing iodine.

Lungs (p. 698)

1. Lungs are paired organs in the thoracic cavity. They are enclosed by the pleural membrane. The parietal pleura is the outer layer; the visceral pleura is the inner layer.
2. The right lung has three lobes separated by two fissures; the left lung has two lobes separated by one fissure and a depression, the cardiac notch.
3. The secondary bronchi give rise to branches called segmental bronchi, which supply segments of lung tissue called bronchopulmonary segments.
4. Each bronchopulmonary segment consists of lobules, which contain lymphatics, arterioles, venules, terminal bronchioles, respiratory bronchioles, alveolar ducts, alveolar sacs, and alveoli.
5. Gas exchange occurs across the alveolar-capillary (respiratory) membranes.

Physiology of Respiration (p. 702)

Pulmonary Ventilation (p. 702)

1. Pulmonary ventilation, or breathing, consists of inspiration and expiration.
2. The movement of air into and out of the lungs depends on pressure changes governed in part by Boyle's law, which states that the volume of a gas varies inversely with pressure, assuming that temperature is constant.
3. Inspiration occurs when intrapulmonic pressure falls below atmospheric pressure. Contraction of the diaphragm and external intercostal muscles increases the size of the thorax, thus decreasing the intrapleural pressure so that the lungs expand. Expansion of the lungs decreases intrapulmonic pressure so that air moves along the pressure gradient from the atmosphere into the lungs.
4. During forced inspiration, accessory muscles of inspiration (sternocleidomastoids, scalenes, and pectoralis minor) are also used.
5. Expiration occurs when intrapulmonic pressure is higher than atmospheric pressure. Relaxation of the diaphragm and external intercostal muscles results in elastic recoil of the chest wall and lungs which increases intrapleural pressure, lung volume decreases, and intrapulmonic pressure increases so that air moves from the lungs to the atmosphere.
6. Forced expiration employs contraction of the internal intercostals and abdominal muscles.
7. Compliance is the ease with which the lungs and thoracic wall expand.
8. The walls of the respiratory passageways offer some resistance to breathing.

Modified Respiratory Movements (p. 706)

1. Modified respiratory movements are used to express emotions and to clear air passageways.
2. Coughing, sneezing, sighing, yawning, sobbing, crying, laughing, and hiccuping are types of modified respiratory movements.

Pulmonary Air Volumes and Capacities (p. 706)

1. Air volumes exchanged during breathing and rate of respiration are measured with a spirometer.
2. Among the pulmonary air volumes exchanged in ventilation are tidal, inspiratory reserve, expiratory reserve, residual, and minimal volumes.
3. Pulmonary lung capacities, the sum of two or more volumes, include inspiratory, functional residual, vital, and total.
4. The minute volume of respiration is the total air taken in during one minute (tidal volume times 12 respirations per minute).

Exchange of Respiratory Gases (p. 708)

1. According to Charles' law, the volume of a gas is directly proportional to the absolute temperature, assuming the pressure remains constant.
2. The partial pressure of a gas is the pressure exerted by that gas in a mixture of gases. It is symbolized by *P*.
3. According to Dalton's law, each gas in a mixture of gases exerts its own pressure as if all the other gases were not present.
4. Henry's law states that the quantity of a gas that will dissolve in a liquid is proportional to the partial pressure of the gas and its solubility coefficient, when the temperature remains constant.
5. A major clinical application of Henry's law is hyperbaric oxygenation (HBO).

External Respiration; Internal Respiration (p. 710; p. 712)

1. In internal and external respiration, O_2 and CO_2 move from areas of their higher partial pressure to areas of their lower partial pressures.
2. External respiration is the exchange of gases between alveoli and pulmonary blood capillaries. It is aided by a thin alveolar-capillary (respiratory) membrane, a large alveolar surface area, and a rich blood supply.
3. Internal respiration is the exchange of gases between tissue blood capillaries and tissue cells.

Transportation of Respiratory Gases (p. 712)

1. In each 100 ml of oxygenated blood, 3 percent of the O_2 is dissolved in plasma and 97 percent is carried with hemoglobin as oxyhemoglobin (HbO_2).
2. The association of oxygen and hemoglobin is affected by PO_2, PCO_2, temperature, and DPG.
3. Hypoxia refers to oxygen deficiency at the tissue level and is classified as hypoxic, anemic, stagnant, and histotoxic.
4. In each 100 ml of deoxygenated blood, 7 percent of CO_2 is dissolved in plasma, 23 percent combines with hemoglobin as carbaminohemoglobin ($Hb{\cdot}CO_2$), and 70 percent is converted to the bicarbonate ion (HCO_3^-).
5. Carbon monoxide poisoning occurs when CO combines with hemoglobin instead of oxygen. The result is hypoxia.

Control of Respiration (p. 717)

Nervous Control (p. 717)

1. The respiratory center consists of a medullary rhythmicity area (inspiratory and expiratory area), pneumotaxic area, and apneustic area.
2. The inspiratory area has an intrinsic excitability that sets the basic rhythm of respiration.
3. The pneumotaxic and apneustic areas coordinate the transition between inspiration and expiration.

Regulation of Respiratory Center Activity (p. 719)

1. Respirations may be modified by a number of factors, both in the brain and outside.
2. Among the modifying factors are cortical influences, the inflation reflex, chemical stimuli, such as O_2 and CO_2 (actually H^+) levels, blood pressure, temperature, pain, and irritation to the respiratory mucosa.

Aging and the Respiratory System (p. 721)

1. Aging results in decreased vital capacity, decreased blood level of oxygen, and diminished alveolar macrophage activity.
2. Elderly people are more susceptible to pneumonia, emphysema, bronchitis, and other pulmonary disorders.

Developmental Anatomy of the Respiratory System (p. 721)

1. The respiratory system begins as an outgrowth of endoderm called the laryngotracheal bud.
2. Smooth muscle, cartilage, and connective tissue of the bronchial tubes and pleural sacs develop from mesoderm.

Disorders: Homeostatic Imbalances (p. 722)

1. In bronchogenic carcinoma, bronchial epithelial cells are replaced by cancer cells after constant irritation has disrupted the normal growth, division, and function of the epithelial cells.
2. Bronchial asthma is characterized by spasms of smooth muscle in bronchial tubes that result in partial closure of air passageways; inflammation; inflated alveoli; and excess mucus production.
3. Bronchitis is an inflammation of the bronchial tubes.
4. Emphysema is characterized by deterioration of alveoli, leading to loss of their elasticity. Symptoms are reduced expiratory volume, inflated lungs at the end of expiration, and enlarged chest.
5. Pneumonia is an acute inflammation or infection of alveoli.
6. Tuberculosis (TB) is an inflammation of pleurae and lungs produced by the organism *Mycobacterium tuberculosis.*
7. Respiratory distress syndrome (RDS) of the newborn is an infant disorder in which surfactant is lacking and alveolar ducts and alveoli have a glassy appearance.
8. Respiratory failure means the respiratory system cannot supply sufficient oxygen or eliminate sufficient carbon dioxide.
9. Sudden infant death syndrome (SIDS) has recently been linked to laryngospasm, possibly triggered by a viral infection of the upper respiratory tract.
10. Coryza (common cold) is caused by viruses and is usually not accompanied by a fever, whereas influenza (flu) is usually accompanied by a fever greater than 101°F.
11. Pulmonary embolism refers to a blood clot or other foreign substance in a pulmonary arterial vessel that obstructs blood flow to lung tissue.
12. Pulmonary edema is an accumulation of interstitial fluid in interstitial spaces and alveoli of the lungs.
13. Smoke inhalation injury has three sequential components: (a) inhibition of oxygen delivery and utilization, (b) upper airway injury from heat, and (c) lung damage from acids and aldehydes in smoke.

REVIEW QUESTIONS

1. What organs make up the respiratory system? What functions do the respiratory and cardiovascular systems have in common? (p. 690)
2. Describe the structure of the external and internal nose, and describe their functions in filtering, warming, and moistening air. (p. 690)
3. What is the pharynx? Differentiate the three anatomical regions of the pharynx and indicate their roles in respiration. (p. 693)
4. Describe the structure of the larynx and explain how it functions in respiration and voice production. (p. 694)
5. Describe the location and structure of the trachea. (p. 696)
6. What is the bronchial tree? Describe its structure. What is a bronchogram? (p. 697)
7. Where are the lungs located? Distinguish the parietal pleura from the visceral pleura. (p. 698)
8. Define each of the following parts of a lung: base, apex, costal surface, medial surface, hilus, root, cardiac notch, and lobe. (p. 698)
9. What is a lobule of the lung? Describe its composition and function in respiration. (p. 700)
10. What is a bronchopulmonary segment? (p. 699)
11. Describe the histology and function of the alveolar-capillary (respiratory) membrane. (p. 701)
12. What are the basic differences among pulmonary ventilation, external respiration, and internal respiration? (p. 702)
13. Discuss the basic steps involved in inspiration and expiration. Be sure to include values for all pressures involved. (p. 702)
14. Distinguish between quiet and forced inspiration and expiration. (p. 703)
15. Describe how compliance and airway resistance relate to pulmonary ventilation. (p. 705)
16. Define the various kinds of modified respiratory movements. (p. 707)
17. What is a spirometer? Define the various lung volumes and capacities. How is the minute volume of respiration (MVR) calculated? (p. 707)
18. Define the partial pressure of a gas. How is it calculated? (p. 709)
19. Define Boyle's law (p. 702), Dalton's law (p. 709), Charles' law (p. 709), and Henry's law. (p. 710)
20. How does decompression sickness occur? (p. 710)
21. Construct a diagram to illustrate how and why the respiratory gases move during external and internal respiration. (p. 711)
22. What factors affect external respiration? (p. 712)
23. Describe the relationship between hemoglobin and PO_2, PCO_2, temperature, and DPG. (p. 713)
24. Define hypoxia and distinguish the principal types. (p. 716)
25. Explain how CO_2 is picked up by tissue capillary blood and then discharged into the alveoli. (p. 716)
26. How does the medullary rhythmicity area function in controlling respiration? How are the apneustic area and pneumotaxic area related to the control of respiration? (p. 717)
27. Explain how each of the following modifies respiration: cerebral cortex, inflation reflex, CO_2, O_2, blood pressure, temperature, pain, and irritations of the respiratory mucosa. (p. 719)
28. How does the control of respiration demonstrate the principle of homeostatsis?
29. Describe the effects of aging on the respiratory system. (p. 721)
30. Describe the development of the respiratory system. (p. 721)
31. For each of the following disorders, list the principal clinical symptoms: bronchogenic carcinoma (lung cancer), bronchial asthma, bronchitis, emphysema, pneumonia, tuberculosis (TB), respiratory distress syndrome (RDS) of the newborn, respiratory failure, sudden infant death syndrome (SIDS), coryza (common cold), influenza (flu), pulmonary embolism, pulmonary edema, and smoke inhalation injury. (p. 722)
32. Define the following terms: rhinoplasty (p. 690), laryngitis

(p. 696), cancer of the larynx (p. 696), tracheostomy (p. 697), intubation (p. 697), asthma attack (p. 698), pneumothorax (p. 698), hemothorax (p. 698), pleurisy (p. 698), nebulization (p. 701), hyperbaric oxygenation (HBO) (p. 710), high altitude sickness (p. 712), and carbon monoxide (CO) poisoning. (p. 715)

33. Describe the diagnostic value and procedure for the following tests: bronchoscopy (p. 698), lung scan (p. 702), pulmonary function tests (PFTs) (p. 708), and arterial blood gas (ABG). (p. 712)
34. Refer to the glossary of medical terminology associated with the respiratory system. Be sure that you can define each term. (p. 726)

SELECTED READINGS

Amin, N. M. "Evaluation and Treatment of Pneumonia in Adult Patients," *Modern Medicine,* January 1989.

Carr, D. T. "Malignant Lung Disease," *Hospital Practice,* January 1981.

Eberhart, J. "Common Cold Preventive," *Science News,* 11 January 1986.

Edwards, D. D. "Nicotine: A Drug of Choice? *Science News,* 18 January 1986.

Heimlich, H. J. "A Life-Saving Maneuver to Prevent Food Choking," *Journal of the American Medical Association,* 27 October 1975.

Kelly, D. H., and D. C. Shannon. "Sudden Infant Death Syndrome—Parts I and II," *Hospital Medicine,* October and November 1982.

Kiley, J. (ed). "Sudden Infant Death Syndrome," *Mayo Clinic Health Letter,* April 1989.

Mines, A. H. *Respiratory Physiology,* 2nd ed. New York: Raven Press, 1986.

Niewuehner, D. E. "Diagnosis Management, and Monitoring of Chronic Pulmonary Obstructive Disease," *Modern Medicine,* February 1989.

Weiss, R. "TB Troubles," *Science News,* 6 February 1988.

Wenger, N. K. "Pulmonary Embolism," *Postgraduate Medicine,* August 1988.

Chapter 24

The Digestive System

Chapter Contents at a Glance

Chapter Contents at a Glance

Student Objectives

1. Describe the mechanism that regulates food intake.
2. Identify the organs of the gastrointestinal (GI) tract and the accessory organs of digestion and their functions in digestion.
3. Describe the mechanical movements of the gastrointestinal tract.
4. Explain how salivary secretion, gastric secretion, gastric emptying, pancreatic secretion, bile secretion, and small intestinal secretion are regulated.
5. Define absorption and explain how the end products of digestion are absorbed.
6. Define the processes involved in the formation of feces and defecation.
7. Describe the effects of aging on the digestive system.
8. Describe the development of the digestive system.
9. Describe the clinical symptoms of the following disorders: dental caries, periodontal disease, peritonitis, peptic ulcers, appendicitis, gastrointestinal tumors, diverticulitis, cirrhosis, hepatitis, gallstones, anorexia nervosa, and bulimia.
10. Define medical terminology associated with the digestive system.

Food is vital for life because it is the source of energy that drives the chemical reactions occurring in every cell and provides matter that is used to form new tissue or to repair damaged tissue. Energy is needed for muscle contraction, conduction of nerve impulses, and secretory and absorptive activities of many cells. Food as it is consumed, however, is not in a state suitable for use as an energy source by any cell. The food must be broken down into molecules small enough to be transported through the plasma (cell) membranes. The breaking down of larger food molecules into molecules small enough for use by body cells is called ***digestion,*** and the organs that collectively perform this function compose the ***digestive system.***

The medical specialty that deals with the structure, function, diagnosis, and treatment of diseases of the stomach and intestines is called ***gastroenterology*** (gas′-trō-en′-ter-OL-ō-jē; *gastro* = stomach; *enteron* = intestines).

The developmental anatomy of the digestive system is considered later in the chapter.

DIGESTIVE PROCESSES

The digestive system prepares food for consumption by the cells through five basic activities.

1. **Ingestion.** Taking food into the body (eating).
2. **Movement of food.** Passage of food along the gastrointestinal tract.
3. **Digestion.** The breakdown of food by both chemical and mechanical processes.
4. **Absorption.** The passage of digested food from the gastrointestinal tract into the cardiovascular and lymphatic systems for distribution to cells.
5. **Defecation.** The elimination of indigestible substances from the gastrointestinal tract.

Chemical digestion is a series of catabolic (hydrolysis) reactions that break down the large carbohydrate, lipid, and protein molecules that we eat into smaller molecules that are absorbable and usable by body cells. These products of digestion are small enough to pass through the epithelial cells of the wall of the gastrointestinal tract, into the blood and lymph capillaries, and eventually into the body's cells. ***Mechanical digestion*** consists of various movements of the gastrointestinal tract that aid chemical digestion. Food is prepared by the teeth before it can be swallowed. Then the smooth muscles of the stomach and small intestine churn the food so it is thoroughly mixed with enzymes that digest foods.

ORGANIZATION

The organs of digestion are traditionally divided into two main groups. First is the ***gastrointestinal (GI) tract,*** or ***alimentary canal,*** a continuous tube running through the ventral body cavity and extending from the mouth to the anus (Figure 24-1). The relationship of the digestive organs to the nine regions of the abdominopelvic cavity may be reviewed in Figure 1-8b. The length of a tract taken from a cadaver is about 9 m (30 ft). In a living person it is somewhat shorter because the muscles in its wall are in a state of tone. Organs composing the gastrointestinal tract include the mouth, pharynx, esophagus, stomach, small intestine, and large intestine. The GI tract contains the food from the time it is eaten until it is digested and prepared for elimination. Muscular contractions in the wall of the GI tract break down the food physically by churning it. Secretions produced by cells along the tract break down the food chemically.

The second group of organs composing the digestive system consists of the ***accessory structures***—the teeth, tongue, salivary glands, liver, gallbladder, and pancreas. Teeth protrude into the GI tract and aid in the physical breakdown of food. The tongue assists in mastication (chewing) and deglutition (swallowing). The other accessory structures, except for the tongue, lie totally outside the tract and produce or store secretions that aid in the chemical breakdown of food. These secretions are released into the tract through ducts.

General Histology

The wall of the GI tract, especially from the esophagus to the anal canal, has the same basic arrangement of tissues. The four layers or tunics of the tract from the inside out are the mucosa, submucosa, muscularis, and serosa or adventitia (Figure 24-2).

Mucosa

The ***mucosa,*** or inner lining of the tract, is a mucous membrane. Three layers compose the membrane in the GI tract: (1) a lining layer of ***epithelium*** in direct contact with the contents of the GI tract, (2) an underlying layer of loose connective tissue called the ***lamina propria,*** and (3) a thin layer of smooth muscle called the ***muscularis mucosae.***

The epithelial layer is composed of nonkeratinized cells that are stratified in the mouth and esophagus but are simple throughout the rest of the tract. The functions of the stratified epithelium are protection and secretion. The functions of the simple epithelium are secretion and absorption.

The lamina propria is made of loose connective tissue containing many blood and lymph vessels and scattered lymphatic nodules, masses of lymphatic tissue that are not encapsulated. This layer supports the epithelium, binds it to the muscularis mucosae, and provides it with a blood and lymph supply. The blood and lymph vessels are the avenues by which nutrients in the tract reach the other tissues of the body. The lymphatic tissue also protects against disease. Remember that the GI tract is in contact

FIGURE 24-1 Organs of the digestive system and related structures.

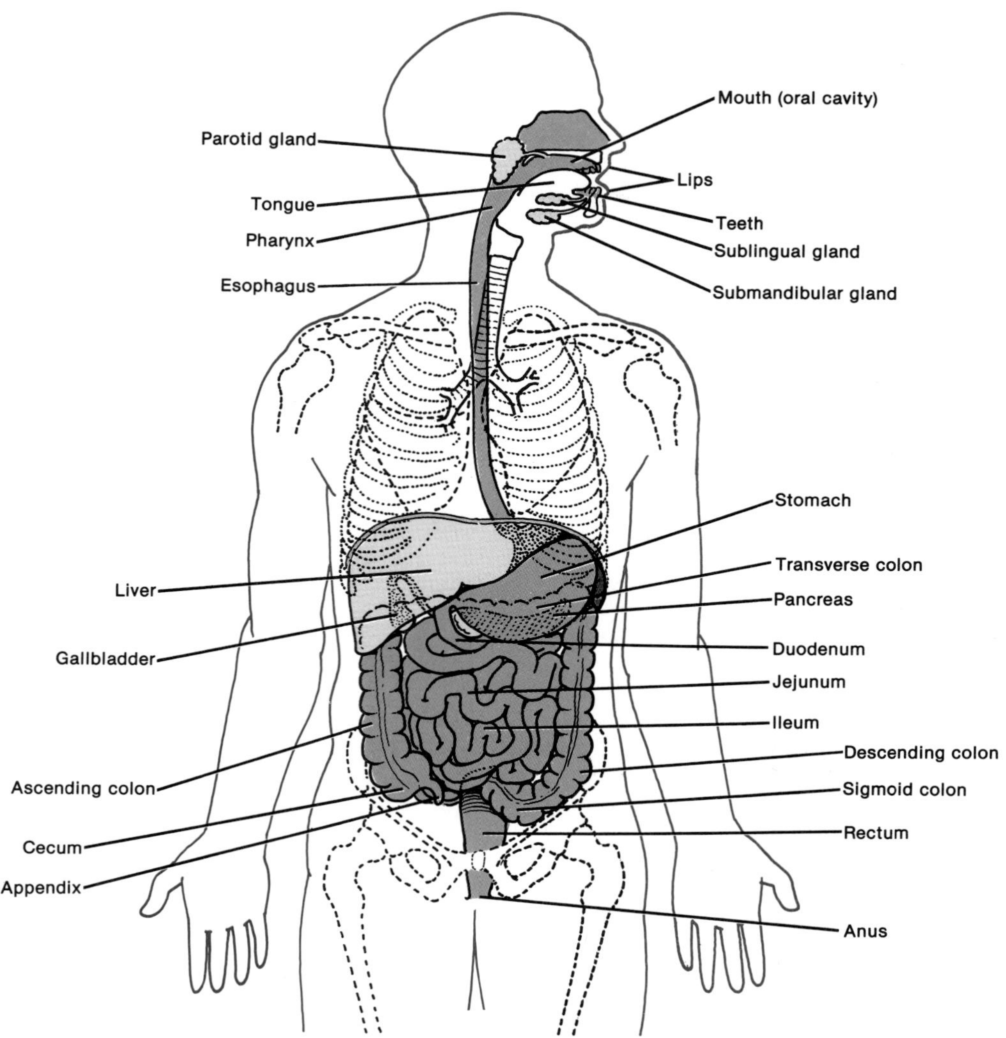

with the outside environment and contains food that often carries harmful bacteria. Unlike the skin, the mucous membrane of the tract is not protected from bacterial entry by keratin.

The muscularis mucosae contains smooth muscle fibers (cells) that throw the mucous membrane of the intestine into small folds that increase the digestive and absorptive area. With one exception, which will be described later, the other three coats of the intestine contain no glandular epithelium.

Submucosa

The ***submucosa*** consists of dense connective tissue that binds the mucosa to the third tunic, the muscularis. It is highly vascular and contains a portion of the ***submucosal plexus (plexus of Meissner),*** which is part of the autonomic nerve supply to the muscularis mucosae. This plexus is also important in controlling secretions by the GI tract.

Muscularis

The ***muscularis*** of the mouth, pharynx, and upper esophagus consists in part of skeletal muscle that produces voluntary swallowing. Throughout the rest of the tract, the muscularis consists of smooth muscle that is generally found in two sheets: an inner sheet of circular fibers and an outer sheet of longitudinal fibers. Contractions of the smooth muscles help break down food physically, mix it with digestive secretions, and propel it through the tract. The muscularis also contains the major nerve supply to the gastrointestinal tract—the ***myenteric plexus (plexus of Auerbach),*** which consists of fibers from

FIGURE 24-2 Composite of various sections of the gastrointestinal tract seen in a three-dimensional drawing depicting the various layers and related structures.

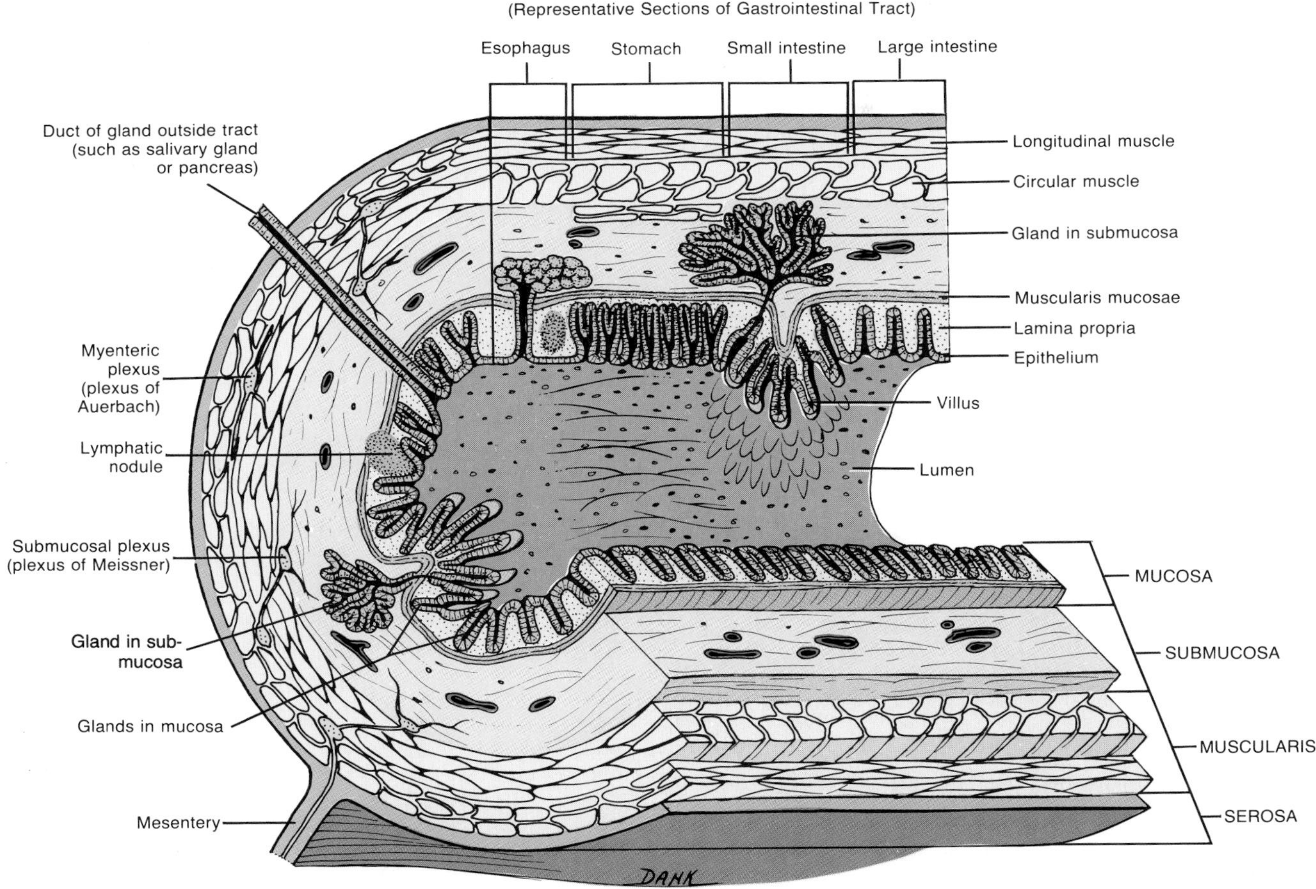

both autonomic divisions. This plexus mostly controls GI motility.

Serosa

The ***serosa*** is the outermost layer of most portions of the GI tract. It is a serous membrane composed of connective tissue and epithelium. This layer is also called the ***visceral peritoneum*** and forms a portion of the peritoneum, which we shall now describe in detail.

Peritoneum

The ***peritoneum*** (per'-i-tō-NĒ-um; *peri* = around; *tonos* = tension) is the largest serous membrane of the body. Serous membranes are also associated with the heart (pericardium) and lungs (pleurae). Serous membranes consist of a layer of simple squamous epithelium (called mesothelium) and an underlying supporting layer of connective tissue. The ***parietal peritoneum*** lines the wall of the abdominal cavity. The ***visceral peritoneum*** covers some of the organs and constitutes their serosa. The potential space between the parietal and visceral portions of the peritoneum is called the ***peritoneal cavity*** and contains serous fluid. In certain diseases, the peritoneal cavity may become distended by several liters of fluid so that it forms an actual space. Such an accumulation of serous fluid is called ***ascites*** (a-SĪ-tēz). As you will see later, some organs lie on the posterior abdominal wall and are covered by peritoneum on their anterior surfaces only. Such organs, including the kidneys and pancreas, are said to be ***retroperitoneal.***

Unlike the pericardium and pleurae, the peritoneum contains large folds that weave between the viscera. The folds bind the organs to each other and to the walls of the cavity and contain the blood and lymph vessels and the nerves that supply the abdominal organs. One extension of the peritoneum is called the ***mesentery*** (MEZ-en-ter'-ē; *meso* = middle; *enteron* = intestine). It is an outward fold of the serous coat of the small intestine (Figure 24-3). The tip of the fold is attached to the posterior

FIGURE 24-3 **Sagittal section through the abdomen and pelvis indicating the relationship of the peritoneal extensions to each other.**

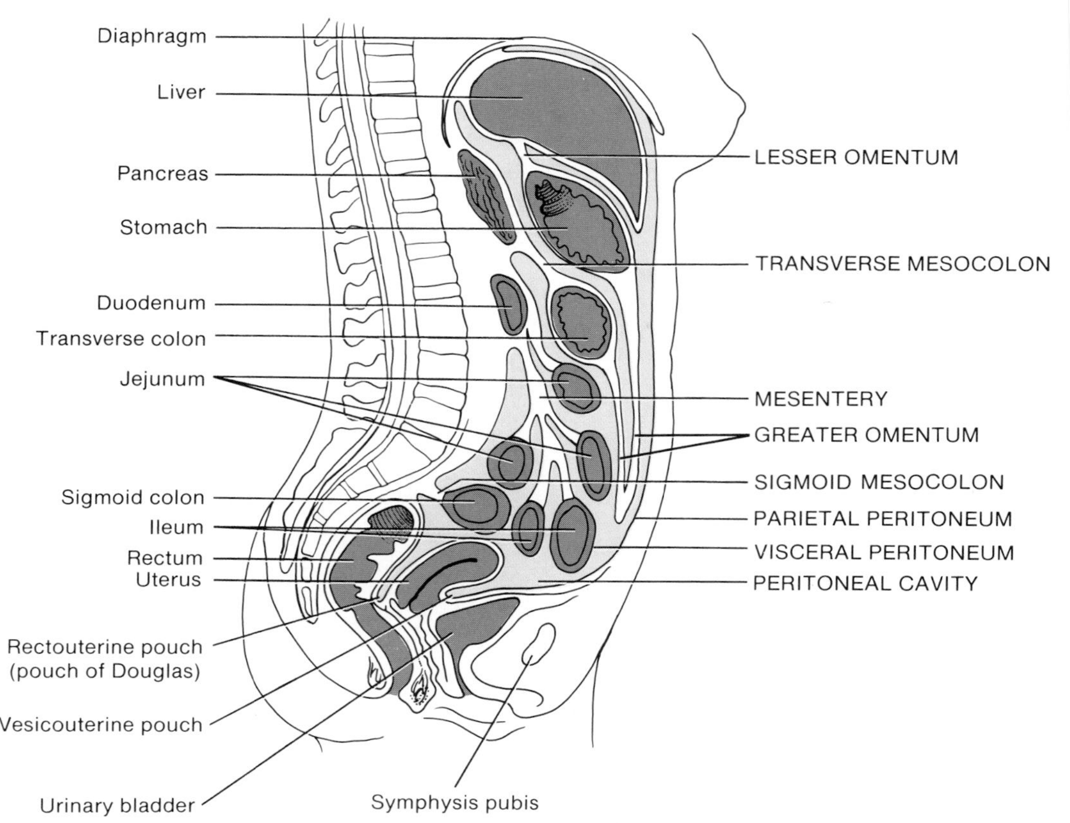

abdominal wall. The mesentry binds the small intestine to the wall. A similar fold of parietal peritoneum, called the ***mesocolon*** (mez'-ō-KŌ-lon), binds the large intestine to the posterior body wall. It also carries blood vessels and lymphatics to the intestines.

Other important peritoneal folds are the falciform ligament, lesser omentum, and greater omentum. The ***falciform*** (FAL-si-form) ***ligament*** attaches the liver to the anterior abdominal wall and diaphragm. The ***lesser omentum*** (ō-MENT-um) arises as two folds in the serosa of the stomach and duodenum suspending the stomach and duodenum from the liver. The ***greater omentum*** is the largest peritoneal fold that drapes over the transverse colon and coils of the small intestine. It is a double sheet that folds upon itself and thus is a four-layered structure. It is attached along the stomach and duodenum, passes downward over the small intestine for a variable distance, and then turns upward to the transverse colon where it is attached to it. Because the greater omentum contains large quantities of adipose tissue, it commonly is called the "fatty apron." The greater omentum contains numerous lymph nodes. If an infection occurs in the intestine, plasma cells formed in the lymph nodes combat the infection and help prevent it from spreading to the peritoneum.

MOUTH (ORAL CAVITY)

The ***mouth,*** also referred to as the ***oral*** or ***buccal*** (BUK-al) ***cavity,*** is formed by the cheeks, hard and soft palates, and tongue (Figure 24-4). Forming the lateral walls of the oral cavity are the ***cheeks***—muscular structures covered on the outside by skin and lined by nonkeratinized stratified squamous epithelium. The anterior portions of the cheeks terminate in the superior and inferior lips.

The ***lips (labia)*** are fleshy folds surrounding the orifice of the mouth. They are covered on the outside by skin and on the inside by a mucous membrane. The transition zone where the two kinds of covering tissue meet is called the ***vermilion*** (ver-MIL-yon). This portion of the lips is nonkeratinized, and the color of the blood in the underlying blood vessels is visible through the transparent surface layer of the vermilion. The inner surface of each lip is attached to its corresponding gum by a midline fold of mucous membrane called the ***labial frenulum*** (LĀ-bē-al FREN-yoo-lum).

The orbicularis oris muscle and connective tissue lie between the external integumentary covering and the internal mucosal lining. During chewing, the cheeks and lips help keep food between the upper and lower teeth. They also assist in speech.

FIGURE 24-4 Structures of the mouth (oral cavity).

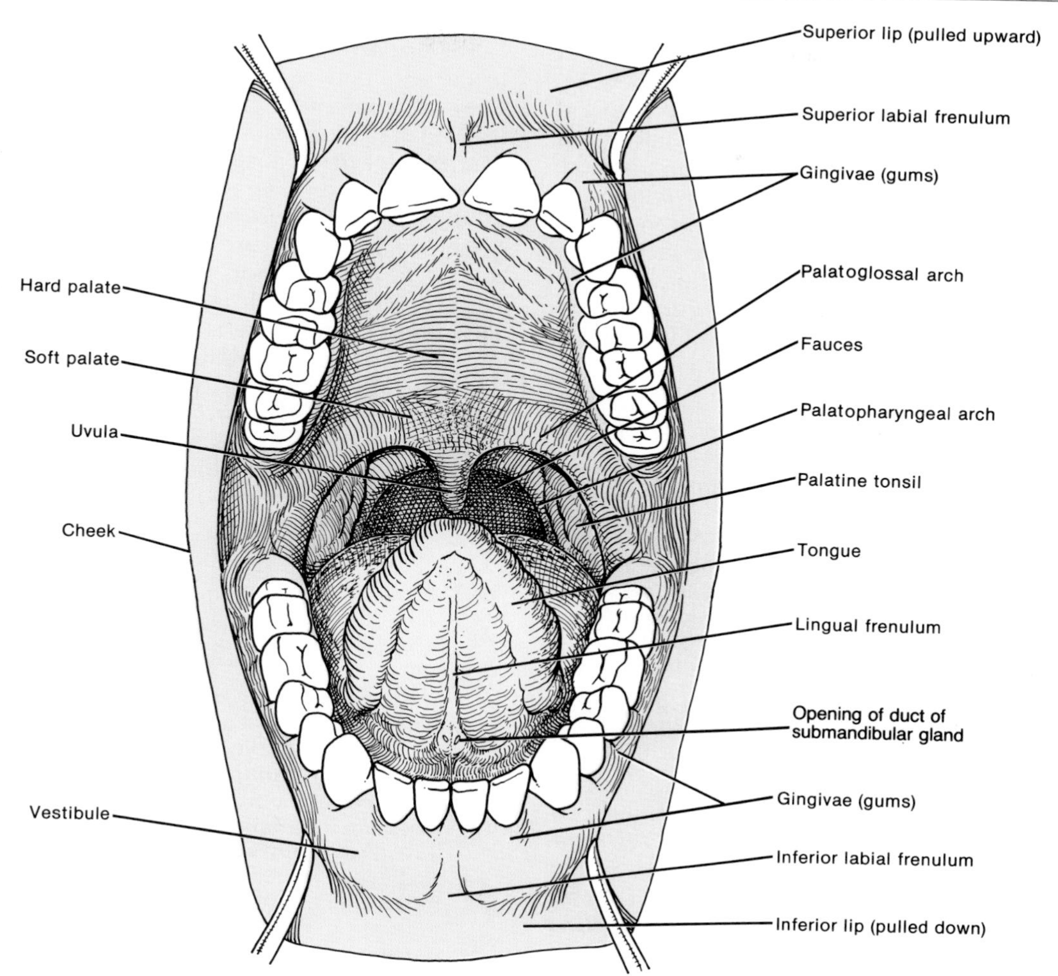

The ***vestibule*** of the oral cavity is bounded externally by the cheeks and lips and internally by the gums and teeth. The ***oral cavity proper*** extends from the vestibule to the ***fauces*** (FAW-sēs; *fauces* = passages), the opening between the oral cavity and the pharynx or throat.

The ***hard palate,*** the anterior portion of the roof of the mouth, is formed by the maxillae and palatine bones, is covered by mucous membrane, and forms a bony partition between the oral and nasal cavities. The ***soft palate*** forms the posterior portion of the roof of the mouth. It is an arch-shaped muscular partition between the oropharynx and nasopharynx and is lined by mucous membrane.

Hanging from the free border of the soft palate is a conical muscular process called the ***uvula*** (YOU-vyoo-la). On either side of the base of the uvula are two muscular folds that run down the lateral side of the soft palate. Anteriorly, the ***palatoglossal arch (anterior pillar)*** extends inferiorly, laterally, and anteriorly to the side of the base of the tongue. Posteriorly, the ***palatopharyngeal*** (PAL-a-tō-fa-rin'-jē-al) ***arch (posterior pillar)*** projects inferiorly, laterally, and posteriorly to the side of the pharynx. The palatine tonsils are situated between the arches, and the lingual tonsil is situated at the base of the tongue. At the posterior border of the soft palate, the mouth opens into the oropharynx through the fauces.

Tongue

The ***tongue,*** together with its associated muscles, forms the floor of the oral cavity. It is an accessory structure of the digestive system composed of skeletal muscle covered with mucous membrane (see Figure 17-2). The tongue is divided into symmetrical lateral halves by a median septum that extends throughout its entire length and is attached inferiorly to the hyoid bone. Each half of the tongue consists of an identical complement of extrinsic and intrinsic muscles.

The ***extrinsic muscles*** of the tongue originate outside the tongue and insert into it. They include the hyoglossus, genioglossus, and styloglossus (see Figure 11-7). The extrinsic muscles move the tongue from side to side and in and out. These movements maneuver food for chewing, shape the food into a rounded mass, and force the food to the back of the mouth for swallowing. They also form the floor of the mouth and hold the tongue in position. The ***intrinsic muscles*** originate and insert within the tongue and alter the shape and size of the tongue for speech and swallowing. The intrinsic muscles include the longitudinalis superior, longitudinalis inferior, transversus linguae, and verticalis linguae. The ***lingual frenulum,*** a fold of mucous membrane in the midline of the undersurface of the tongue, aids in limiting the movement of the tongue posteriorly.

CLINICAL APPLICATION: ANKYLOGLOSSIA

If the lingual frenulum is too short, tongue movements are restricted, speech is faulty, and the person is said to be "tongue-tied." This congenital problem is referred to as *ankyloglossia* (ang'-ki-lō-GLOSS-ē-a). It can be corrected by cutting the lingual frenulum.

The upper surface and sides of the tongue are covered with ***papillae*** (pa-PIL-ē), projections of the lamina propria covered with epithelium (Figure 17-2c). ***Filiform papillae*** are conical projections distributed in parallel rows over the anterior two-thirds of the tongue. They are whitish and contain no taste buds. ***Fungiform papillae*** are mushroomlike elevations distributed among the filiform papillae and are more numerous near the tip of the tongue. They appear as red dots on the surface of the tongue, and most of them contain taste buds. ***Circumvallate papillae,*** 10 to 12 in number, are arranged in the form of an inverted V on the posterior surface of the tongue, and all of them contain taste buds. Note the taste zones of the tongue in Figure 17-2c.

Salivary Glands

Saliva is a fluid that is continuously secreted by glands in or near the mouth. Ordinarily, just enough saliva is secreted to keep the mucous membranes of the mouth and pharynx moist, but when food enters the mouth, secretion increases so the saliva can lubricate, dissolve, and begin the chemical breakdown of the food. The mucous membrane lining the mouth contains many small glands, the ***buccal glands,*** that secrete small amounts of saliva. However, the major portion of saliva is secreted by the ***salivary glands,*** accessory structures that lie outside the mouth and pour their contents into ducts that empty into the oral cavity. There are three pairs of salivary glands: parotid, submandibular (submaxillary), and sublingual glands (Figure 24-5a).

The ***parotid glands*** are located inferior and anterior to the ears between the skin and the masseter muscle. They are compound tubuloacinar glands. Each secretes into the oral cavity vestibule via a duct, called the ***parotid (Stensen's) duct,*** that pierces the buccinator muscle to open into the vestibule opposite the upper second molar tooth. The ***submandibular glands,*** which are compound acinar glands, are found beneath the base of the tongue in the posterior part of the floor of the mouth (Figure 24-5b). Their ducts, the ***submandibular (Wharton's) ducts,*** run superficially under the mucosa on either side of the midline of the floor of the mouth and enter the oral cavity proper on either side of the lingual frenulum. The ***sublingual glands,*** also compound acinar glands, are anterior to the submandibular glands, and their ducts, the ***lesser sublingual (Rivinus's) ducts,*** open into the floor of the mouth in the oral cavity proper.

CLINICAL APPLICATION: MUMPS

Although any of the salivary glands may become infected as a result of a nasopharyngeal infection, the parotids are typically the target of the mumps virus (myxovirus). *Mumps* is an inflammation and enlargement of the parotid glands accompanied by moderate fever, malaise, and extreme pain in the throat, especially when swallowing sour foods or acid juices. Swelling occurs on one or both sides of the face, just anterior to the ramus of the mandible. In about 20 to 35 percent of males past puberty, the testes may also become inflamed, and although it rarely occurs, sterility is a possible consequence. In some people, aseptic meningitis, pancreatitis, and hearing loss may also occur as complications.

Composition of Saliva

The fluids secreted by the buccal glands and the three pairs of salivary glands constitute ***saliva.*** Amounts of saliva secreted daily vary considerably but range from 1000 to 1500 ml. Chemically, saliva is 99.5 percent water and 0.5 percent solutes. Among the solutes are salts—chlorides, bicarbonates, and phosphates of sodium and potassium. Some dissolved gases and various organic substances including urea and uric acid, serum albumin and globulin, mucin, the bacteriolytic enzyme lysozyme, and the digestive enzyme salivary amylase are also present.

Each saliva-producing gland supplies different proportions of ingredients to saliva. The parotids contain cells that secrete a watery serous liquid containing the enzyme salivary amylase. The submandibular glands contain cells

FIGURE 24-5 Salivary glands. (a) Location of the salivary glands. (b) Photomicrograph of the submandibular gland showing mostly serous acini and a few mucous acini at a magnification of 120×. The parotids consist of all serous acini, and the sublinguals consist of mostly mucous acini and a few serous acini. (Courtesy of Andrew Kuntzman.)

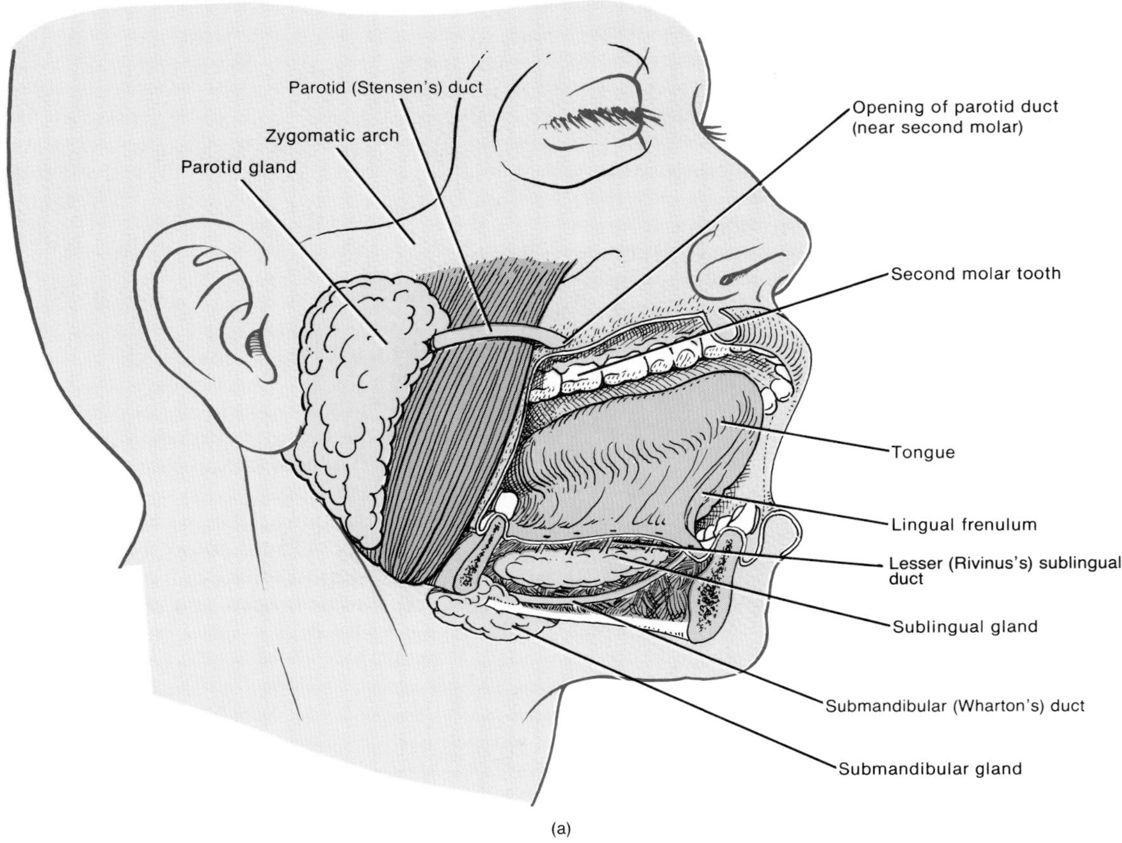

(a)

similar to those found in the parotids and some mucous cells. Therefore, they secrete a fluid that is thickened with mucus but still contains quite a bit of enzyme. The sublingual glands contain mostly mucous cells, so they secrete a much thicker fluid that contributes only a small amount of enzyme to the saliva.

The water in saliva provides a medium for dissolving foods so they can be tasted and for initiating digestive reactions. The chlorides in the saliva activate the salivary amylase. The bicarbonates and phosphates buffer chemicals that enter the mouth and keep the saliva at a slightly acidic pH of 6.35 to 6.85. Urea and uric acid are found in saliva because the saliva-producing glands (like the sweat glands of the skin) help the body to get rid of wastes. Mucin is a protein that forms mucus when dissolved in water. Mucus lubricates the food so it can be easily moved about in the mouth, formed into a ball, and swallowed. The enzyme lysozyme destroys bacteria, thereby protecting the mucous membrane from infection and thus the teeth from decay.

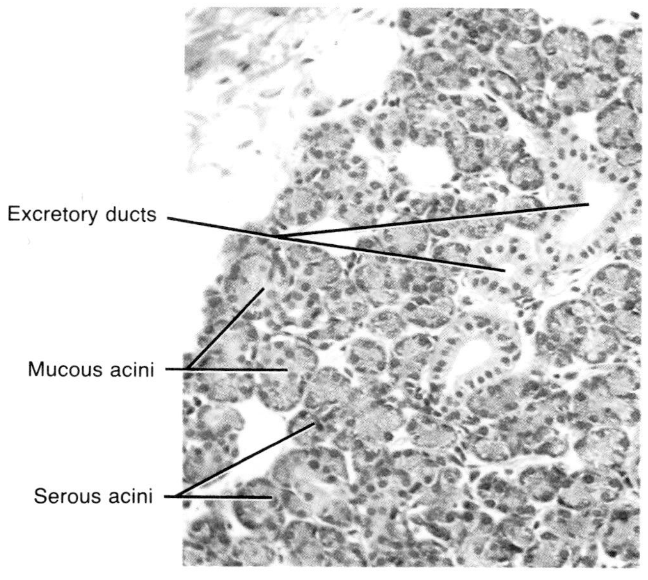

(b)

Secretion of Saliva

Salivation is entirely under nervous control. Normally, moderate amounts of saliva are continuously secreted in response to parasympathetic stimulation to keep the mucous membranes moist and to lubricate the movements of the tongue and lips during speech. The saliva is then swallowed and reabsorbed to prevent fluid loss. Dehydration causes the salivary and buccal glands to cease secreting saliva to conserve water. The subsequent feeling of dryness in the mouth promotes sensations of thirst. This phenomenon is also noted during fear or anxiety, when sympathetic stimulation dominates.

Food stimulates the glands to secrete heavily. When food is taken into the mouth, chemicals in the food stimulate receptors in taste buds on the tongue. Rolling a dry, indigestible object over the tongue produces friction, which also may stimulate the receptors. Impulses are conveyed from the receptors to two salivary nuclei in the brain stem called the ***superior*** and ***inferior salivatory nuclei.*** The nuclei are located at about the junction of the medulla and pons. Returning parasympathetic autonomic impulses from the nuclei activate the secretion of saliva.

The smell, sight, touch, or sound of food preparation also stimulates increased saliva secretion. These stimuli constitute psychological activation and involve learned behavior. Memories stored in the cerebral cortex that associate the stimuli with food are activated. The cortex sends impulses to the nuclei in the brain stem via extrapyramidal pathways, and the salivary glands are activated. Psychological activation of the glands has some benefit to the body because it allows the mouth to start chemical digestion as soon as the food is ingested.

Salivation also occurs in response to swallowing irritating foods or during nausea. Reflexes originating in the stomach and upper small intestine stimulate salivation. This mechanism presumably helps to dilute or neutralize the irritating substance.

Saliva continues to be secreted heavily some time after food is swallowed. This flow of saliva washes out the mouth and dilutes and buffers the chemical remnants of irritating substances.

Teeth

The ***teeth (dentes)*** are accessory structures of the digestive system located in sockets of the alveolar processes of the mandible and maxillae. The alveolar processes are covered by the ***gingivae*** (jin-JI-vē), or gums, which extend slightly into each socket forming the gingival sulcus (Figure 24-6). The sockets are lined by the ***periodontal ligament,*** which consists of dense fibrous connective tissue and is attached to the socket walls and the cemental surface of the roots. Thus, it anchors the teeth in position and also acts as a shock absorber to dissipate the forces of chewing.

A typical tooth consists of three principal portions. The ***crown*** is the exposed portion above the level of the gums. The ***root*** consists of one to three projections embedded in the socket. The ***neck*** is the constricted junction line of the crown and the root near the gumline.

Teeth are composed primarily of ***dentin,*** a calcified connective tissue that gives the tooth its basic shape and rigidity. The dentin encloses a cavity. The enlarged part of the cavity, the ***pulp cavity,*** lies in the crown and is filled with ***pulp,*** a connective tissue containing blood vessels, nerves, and lymphatic vessels. Narrow extensions of the pulp cavity run through the root of the tooth and are called ***root canals.*** Each root canal has an opening at its base, the ***apical foramen.*** Through the foramen enter blood vessels bearing nourishment, lymphatics affording protection, and nerves providing sensation. The dentin of the crown is covered by ***enamel*** that consists primarily of calcium phosphate and calcium carbonate. Enamel is the hardest substance in the body and protects the tooth from the wear of chewing. It is also a barrier against acids that easily dissolve the dentin. The dentin of the root is covered by ***cementum,*** another bonelike substance, which attaches the root to the periodontal ligament.

The branch of dentistry that is concerned with the prevention, diagnosis, and treatment of diseases that affect the pulp, root, periodontal ligament, and alveolar bone is known as ***endodontics*** (en′-dō-DON-tiks; *endo* = within; *odous* = tooth).

CLINICAL APPLICATION: ROOT CANAL THERAPY

Root canal therapy refers to a procedure, accomplished in several phases, in which all traces of pulp tissue are removed from the pulp cavity and root canal of a badly diseased tooth. After a hole is made in the tooth, the root canal is filed out and irrigated to remove bacteria. Then the canal is treated with medication and sealed tightly. The damaged crown is then repaired.

Dentitions

Everyone has two ***dentitions,*** or sets of teeth. The first of these—the ***deciduous (primary) teeth, milk teeth,*** or ***baby teeth***—begin to erupt at about 6 months of age, and one pair appears at about each month thereafter until all 20 are present. Figure 24-7a illustrates the deciduous teeth. The incisors, which are closest to the midline, are chisel-shaped and adapted for cutting into food. They are referred to as either ***central*** or ***lateral incisors*** on the basis of their position. Next to the incisors, moving posteriorly, are the ***cuspids (canines),*** which have a pointed surface called a cusp. Cuspids are used to tear and shred food. The incisors and cuspids have only one root apiece. Behind them lie the ***first*** and ***second molars,*** which have four cusps. Upper molars

FIGURE 24-6 Parts of a typical tooth as seen in a section of a mandibular molar.

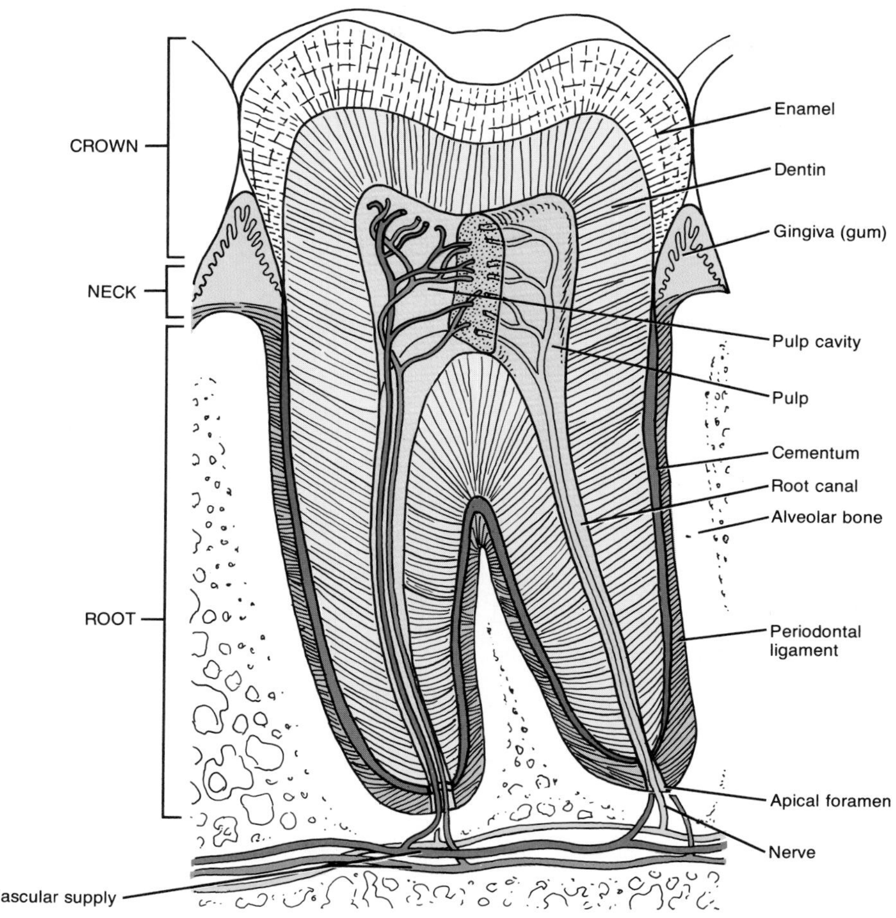

have three roots; lower molars have two roots. The molars crush and grind food.

All the deciduous teeth are lost—generally between 6 and 12 years of age—and are replaced by the ***permanent (secondary) dentition*** (Figure 24-7b). The permanent dentition contains 32 teeth that appear between age 6 and adulthood. It resembles the deciduous dentition with the following exceptions. The deciduous molars are replaced with the ***first*** and ***second premolars (bicuspids),*** which have two cusps and one root (upper first bicuspids have two roots) and are used for crushing and grinding. The permanent molars erupt into the mouth behind the bicuspids. They do not replace any deciduous teeth and erupt as the jaw grows to accommodate them—the ***first molars*** at age 6, the ***second molars*** at age 12, the ***third molars (wisdom teeth)*** after age 18. The human jaw has become smaller through time and often does not afford enough room behind the second molars for the eruption of the third molars. In this case, the third molars remain embedded in the alveolar bone and are said to be "impacted." Most often they cause pressure and pain and must be surgically removed. In some individuals, third molars may be dwarfed in size or may not develop at all.

Physiology of Digestion in the Mouth

Mechanical

Through chewing, or ***mastication,*** the tongue manipulates food, the teeth grind it, and the food is mixed with saliva. As a result, the food is reduced to a soft, flexible mass called a ***bolus*** that is easily swallowed.

Chemical

The enzyme ***salivary amylase*** (AM-i-lās), formerly known as ptyalin, initiates the breakdown of starch. This is the only chemical digestion that occurs in the mouth. Carbohydrates are either monosaccharide and disaccharide sugars or polysaccharide starches (see Chapter 2). Most of the carbohydrates we eat are polysaccharides.

FIGURE 24-7 Dentitions and times of eruptions. The times of eruptions are indicated in parentheses. (a) Deciduous (primary) dentition. (b) Permanent (secondary) dentition.

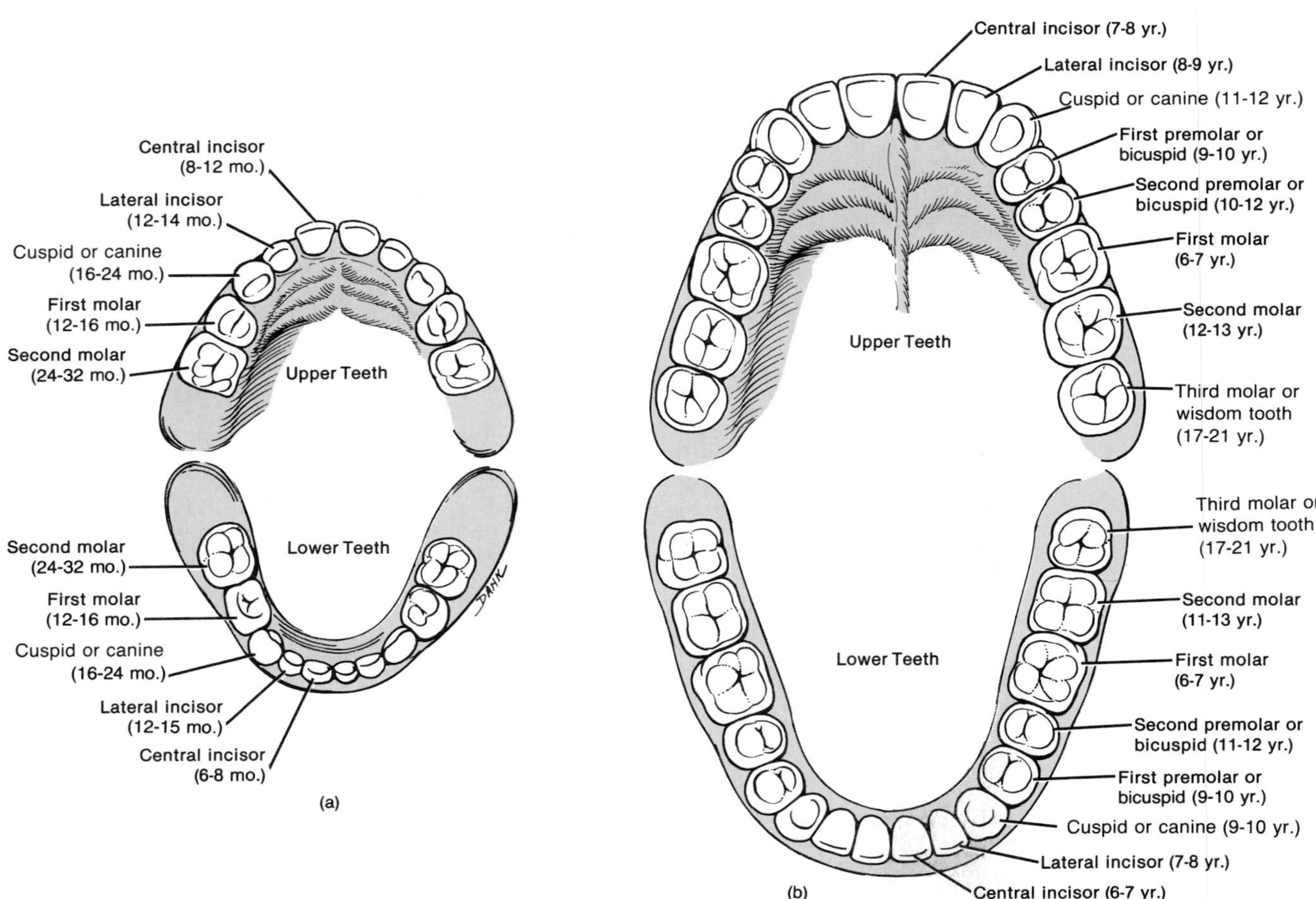

Since only monosaccharides can be absorbed into the bloodstream, ingested disaccharides and polysaccharides must be broken down. The function of salivary amylase is to break the chemical bonds between some of the monosaccharides in the starches to reduce the long-chain polysaccharides to the disaccharide maltose. Food usually is swallowed too quickly for all the starches to be reduced to disaccharides in the mouth. However, salivary amylase in the swallowed food continues to act on starches for another 15 to 30 minutes in the stomach before the stomach acids eventually inactivate it.

Exhibit 24-1 summarizes digestion in the mouth.

Physiology of Deglutition

Swallowing, or ***deglutition*** (dē-gloo-TISH-un), is a mechanism that moves food from the mouth to the stomach. It is facilitated by saliva and mucus and involves the mouth, pharynx, and esophagus. Swallowing is conveniently divided into three stages: (1) the voluntary stage, in which the bolus is moved into the oropharynx; (2) the pharyngeal stage, the involuntary passage of the bolus through the pharynx into the esophagus; and (3) the esophageal stage, the involuntary passage of the bolus through the esophagus into the stomach (described in conjunction with the esophagus).

Swallowing starts when the bolus is forced to the back of the mouth cavity and into the oropharynx by the movement of the tongue upward and backward against the palate. This represents the ***voluntary stage*** of swallowing.

With the passage of bolus into the oropharynx, the involuntary ***pharyngeal stage*** of swallowing begins (Figure 24-8b). The respiratory passageways close, and breathing is temporarily interrupted. The bolus stimulates receptors in the oropharynx, which send impulses to the ***deglutition center*** in the medulla and lower pons of the brain stem. The returning impulses cause the soft palate and uvula to move upward to close off the nasopharynx, and the larynx is pulled forward and upward under the tongue. As the larynx rises, the epiglottis moves

EXHIBIT 24-1 DIGESTION IN THE MOUTH

Structure	Activity	Result
Cheeks	Keep food between teeth during mastication.	Foods uniformly chewed.
Lips	Keep food between teeth during mastication.	Foods uniformly chewed.
Tongue		
Extrinsic muscles	Move tongue from side to side and in and out.	Food maneuvered for mastication, shaped into bolus, and maneuvered for deglutition.
Intrinsic muscles	Alter shape of tongue.	Deglutition and speech.
Taste buds	Serve as receptors for food stimulus.	Secretion of saliva stimulated by nerve impulses from taste buds to salivatory nuclei in brain stem to salivary glands.
Buccal glands	Secrete saliva.	Lining of mouth and pharynx moistened and lubricated.
Salivary glands	Secrete saliva.	Lining of mouth and pharynx moistened and lubricated. Saliva softens, moistens, and dissolves food, coats food with mucin, cleanses mouth and teeth. Salivary amylase reduces polysaccharides to the disaccharide maltose.
Teeth	Cut, tear, and pulverize food.	Solid foods reduced to smaller particles for swallowing.

FIGURE 24-8 Deglutition. (a) Position of structures prior to deglutition. (b) During the pharyngeal stage of deglutition, the tongue rises against the palate, the nasopharynx is closed off, the larynx rises, the epiglottis seals off the larynx, and the bolus is passed into the esophagus.

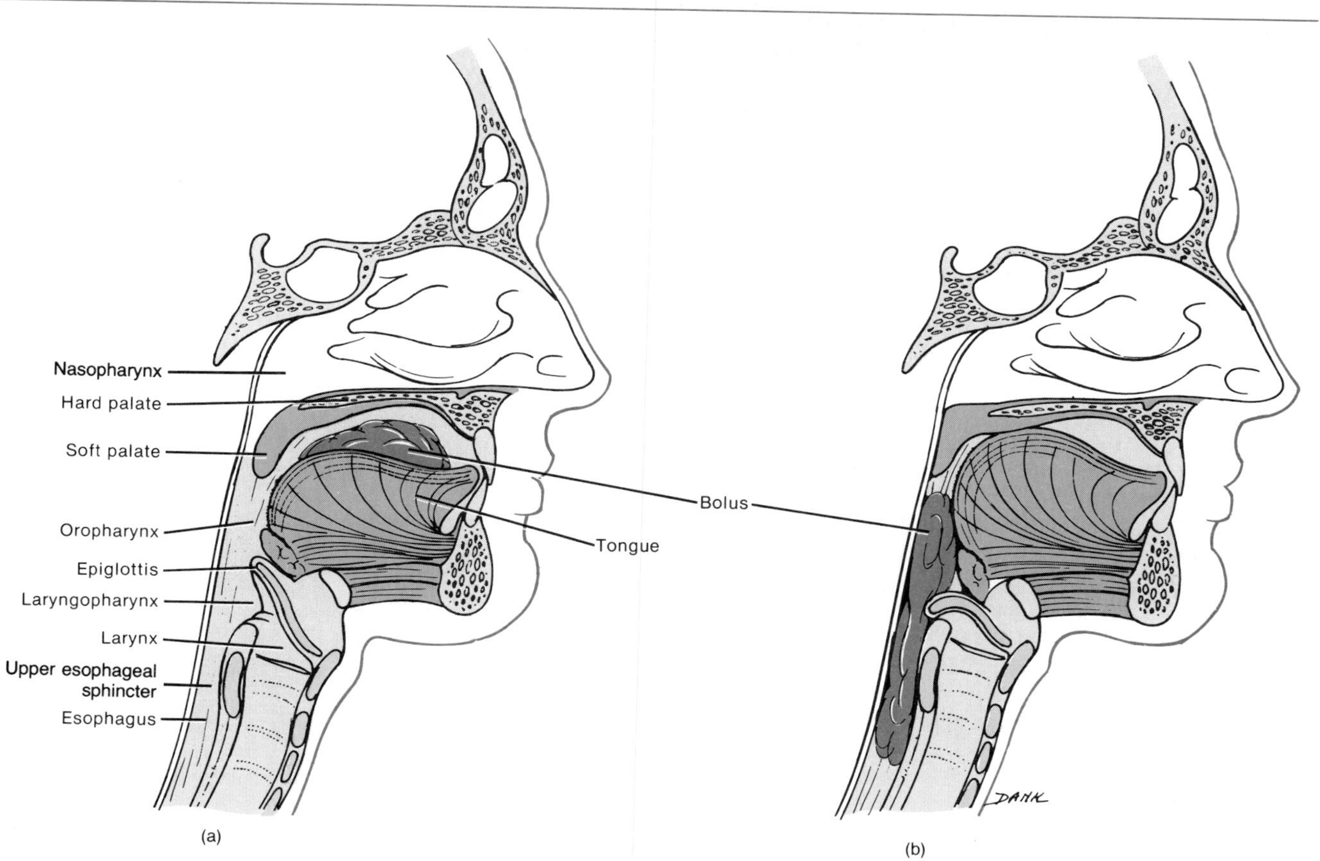

FIGURE 24-8 (*Continued*) (c) Esophageal stage of deglutition.

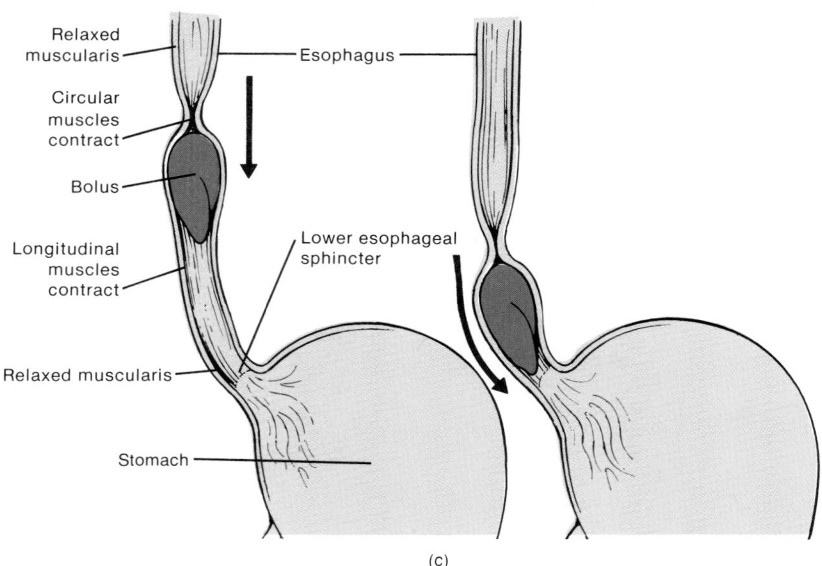

backward and downward and seals off the glottis. The movement of the larynx also pulls the vocal cords together, further sealing off the respiratory tract, and widens the opening between the laryngopharynx and esophagus. The bolus passes through the laryngopharynx and enters the esophagus in one to two seconds. The respiratory passageways then reopen and breathing resumes.

ESOPHAGUS

The ***esophagus*** (e-SOF-a-gus), the third organ involved in deglutition, is a muscular, collapsible tube that lies behind the trachea. It is about 23 to 25 cm (10 inches) long and begins at the end of the laryngopharynx, passes through the mediastinum anterior to the vertebral column, pierces the diaphragm through an opening called ***esophageal hiatus,*** and terminates in the superior portion of the stomach (see Figure 24-1).

Histology

The ***mucosa*** of the esophagus consists of nonkeratinized stratified squamous epithelium, lamina propria, and a muscularis mucosae (Figure 24-9). Near the stomach, the mucosa of the esophagus also contains mucous glands. The ***submucosa*** contains connective tissue, blood vessels, and mucous glands. The ***muscularis*** of the upper third is striated, the middle third is striated and smooth, and the lower third is smooth. The outer layer is known as the ***adventitia*** (ad-ven-TISH-ya) rather than the serosa because the loose connective tissue of the layer is not covered by epithelium (mesothelium) and because the

FIGURE 24-9 Histology of the esophagus. Photomicrograph of a portion of the wall of the esophagus at a magnification of 60×. An enlarged aspect of the mucosa of the esophagus is shown in Exhibit 4-1, stratified squamous. (Courtesy of Andrew Kuntzman.)

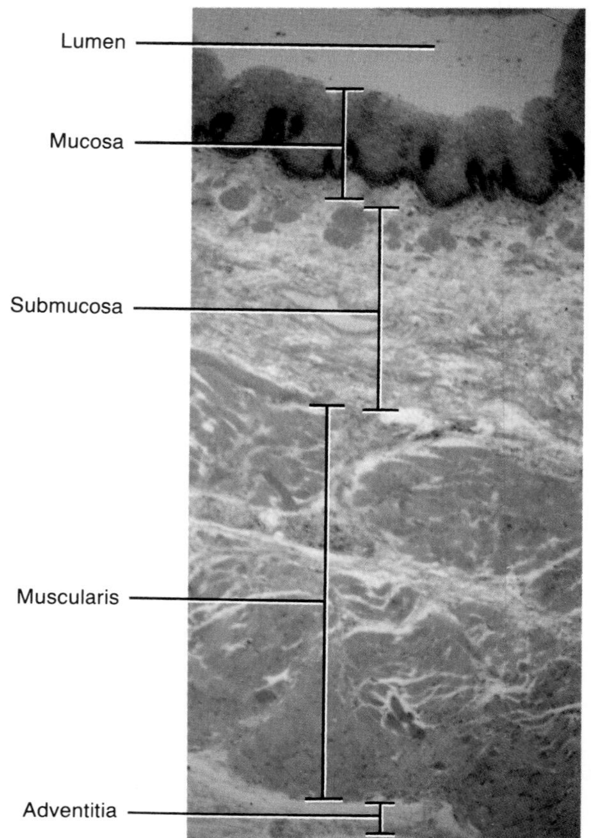

connective tissue merges with the connective tissue of surrounding structures.

Physiology

The esophagus does not produce digestive enzymes and does not carry on absorption. It secretes mucus and transports food to the stomach. The passage of food from the laryngopharynx into the esophagus is regulated by a sphincter (thick circle of muscle around an opening) at the entrance to the esophagus called the ***upper esophageal*** (e-sof'-a-JĒ-al) ***sphincter.*** It consists of the cricopharyngeus muscle attached to the cricoid cartilage. The elevation of the larynx during the pharyngeal stage of swallowing causes the sphincter to relax, and the bolus enters the esophagus. The sphincter also relaxes during expiration.

During the ***esophageal stage*** of swallowing (deglutition), food is pushed through the esophagus by involuntary muscular movements called ***peristalsis*** (per'-is-STAL-sis) (see Figure 24-8c). Peristalsis is a function of the muscularis and is controlled by the medulla. In the section of the esophagus lying just above and around the top of the bolus, the circular muscle fibers (cells) contract. The contraction constricts the esophageal wall and squeezes the bolus downward. Meanwhile, longitudinal fibers lying around the bottom of and just below the bolus also contract. Contraction of the longitudinal fibers shortens this lower section, pushing its walls outward so it can receive the bolus. The contractions are repeated in a wave that moves down the esophagus, pushing the food toward the stomach. Passage of the bolus is further facilitated by glands that secrete mucus. The passage of solid or semisolid food from the mouth to the stomach takes four to eight seconds. Very soft foods and liquids pass through in about one second.

Just above the level of the diaphragm, the esophagus is slightly narrowed. This narrowing has been attributed to a physiological sphincter in the inferior part of the esophagus known as the ***lower esophageal (gastroesophageal) sphincter.*** The lower esophageal sphincter relaxes during swallowing and thus aids the passage of the bolus from the esophagus into the stomach.

CLINICAL APPLICATION: ACHALASIA AND HEARTBURN

If the lower esophageal sphincter fails to relax normally as food approaches, the condition is called *achalasia* (ak'-a-LĀ-zē-a; *a* = without; *chalasis* = relaxation). As a result, food passage from the esophagus into the stomach is greatly impeded. A whole meal may become lodged in the esophagus, entering the stomach very slowly. Distension of the esophagus results in chest pain that is often confused with pain originating from the heart. The condition is caused by malfunction of the myenteric plexus (plexus of Auerbach).

If, on the other hand, the lower esophageal sphincter fails to close adequately after food has entered the stomach, the stomach contents can enter the lower esophagus. Hydrochloric acid (HCl) from the stomach contents can irritate the esophageal wall, resulting in a burning sensation. The sensation is known as *heartburn* because it is experienced in the region very near the heart, although it is not related to any cardiac problem. Heartburn can be treated by taking antacids (Tums, Gelusil, Rolaids, Maalox) that neutralize the hydrochloric acid and lessen the severity of the burning sensation and discomfort. In addition to the use of antacids for heartburn, other measures are important. Symptoms are less likely to occur if food is eaten in smaller amounts at a meal (no gorging), and heartburn is less of a problem if the person does not lie down right after a meal. Some people with persistent heartburn have a hiatal hernia (defined later).

Exhibit 24-2 summarizes the digestion-related activities of the pharynx and esophagus.

EXHIBIT 24-2 DIGESTIVE ACTIVITIES OF THE PHARYNX AND ESOPHAGUS

Structure	Activity	Result
Pharynx	Pharyngeal stage of deglutition.	Moves bolus from oropharynx to laryngopharynx and into esophagus; closes air passageways.
	Relaxation of upper esophageal sphincter.	Moves bolus from laryngopharynx into esophagus.
Esophagus	Esophageal stage of deglutition (peristalsis).	Forces bolus down esophagus.
	Relaxation of lower esophageal sphincter.	Moves bolus into stomach.
	Secretion of mucus.	Lubricates esophagus for smooth passage of bolus.

STOMACH

The ***stomach*** is a J-shaped enlargement of the GI tract directly under the diaphragm in the epigastric, umbilical, and left hypochondriac regions of the abdomen (see Figure 1-8b). The superior portion of the stomach is a continuation of the esophagus. The inferior portion empties into the duodenum, the first part of the small intestine. Within each individual, the position and size of the stomach vary continually. For instance, the diaphragm pushes the stomach downward with each inspiration and pulls it upward with each expiration. Empty, it is about the size of a large sausage, but it can stretch to accommodate large amounts of food.

Anatomy

The stomach is divided by gross anatomists into four areas: cardia, fundus, body, and pylorus (Figure 24-10). The ***cardia*** surrounds the lower esophageal sphincter. The rounded portion above and to the left of the cardia is the ***fundus*** (*fundus* = bottom). Below the fundus is the large central portion of the stomach, called the ***body.*** The narrow, inferior region is the ***pylorus.*** The concave medial border of the stomach is called the ***lesser curvature,*** and the convex lateral border is the ***greater curvature.*** The pylorus communicates with the duodenum of the small intestine via a sphincter called the ***pyloric*** (*pyle* = gate; *ouros* = guard) ***sphincter (valve).***

CLINICAL APPLICATION: PYLOROSPASM AND PYLORIC STENOSIS

Two abnormalities of the pyloric sphincter can occur in infants. *Pylorospasm* is characterized by failure of the muscle fibers encircling the opening to relax normally. It can be caused by hypertrophy or continuous spasm of the sphincter and usually occurs between the second and twelfth weeks after birth. Ingested food does not pass easily from the stomach to the small intestine, the stomach becomes overly full, and the infant vomits frequently to relieve the pressure. Pylorospasm is treated by drugs that relax the muscle fibers of the sphincter. *Pyloric stenosis* is a narrowing of the pyloric sphincter caused by a tumorlike mass that apparently is formed by enlargement of the circular muscle fibers. The hallmark symptom is projectile vomiting—the spraying of liquid vomitus some feet from the infant. Pyloric stenosis must be surgically corrected.

Histology

The stomach wall is composed of the same four basic layers as the rest of the GI tract, with certain modifications. When the stomach is empty, the ***mucosa*** lies in large folds, called ***rugae*** (ROO-jē), that can be seen with the naked eye (Figure 24-10b). Microscopic inspection of the mucosa reveals a layer of simple columnar epithelium (surface mucous cells) containing many narrow openings that extend down into the lamina propria called ***gastric pits*** (Figure 24-11). At the bottoms of the pits are the orifices of ***gastric glands.*** Each gland consists of four types of secreting cells: zymogenic, parietal, mucous, and enteroendocrine. The ***zymogenic (peptic)*** or ***chief cells*** secrete the principal gastric enzyme precursor, pep-

FIGURE 24-10 External and internal anatomy of the stomach. (a) Diagram in anterior view. (b) Photograph of the internal surface in anterior view showing rugae. (Courtesy of J. A. Gosling, P. F. Harris, et al., *Atlas of Human Anatomy,* Gower Medical Publishing Ltd., 1985.)

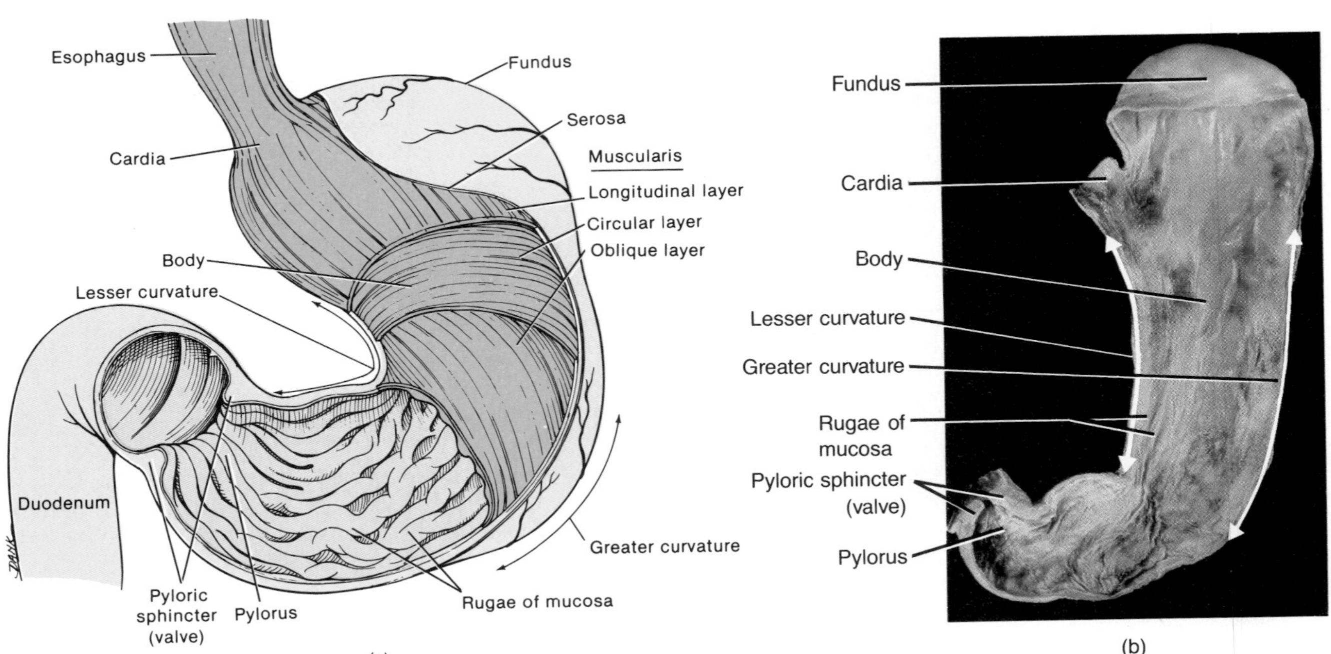

FIGURE 24-11 Histology of the stomach. (a) Diagram showing four principal layers. (b) Diagram of details of gastric mucosa.

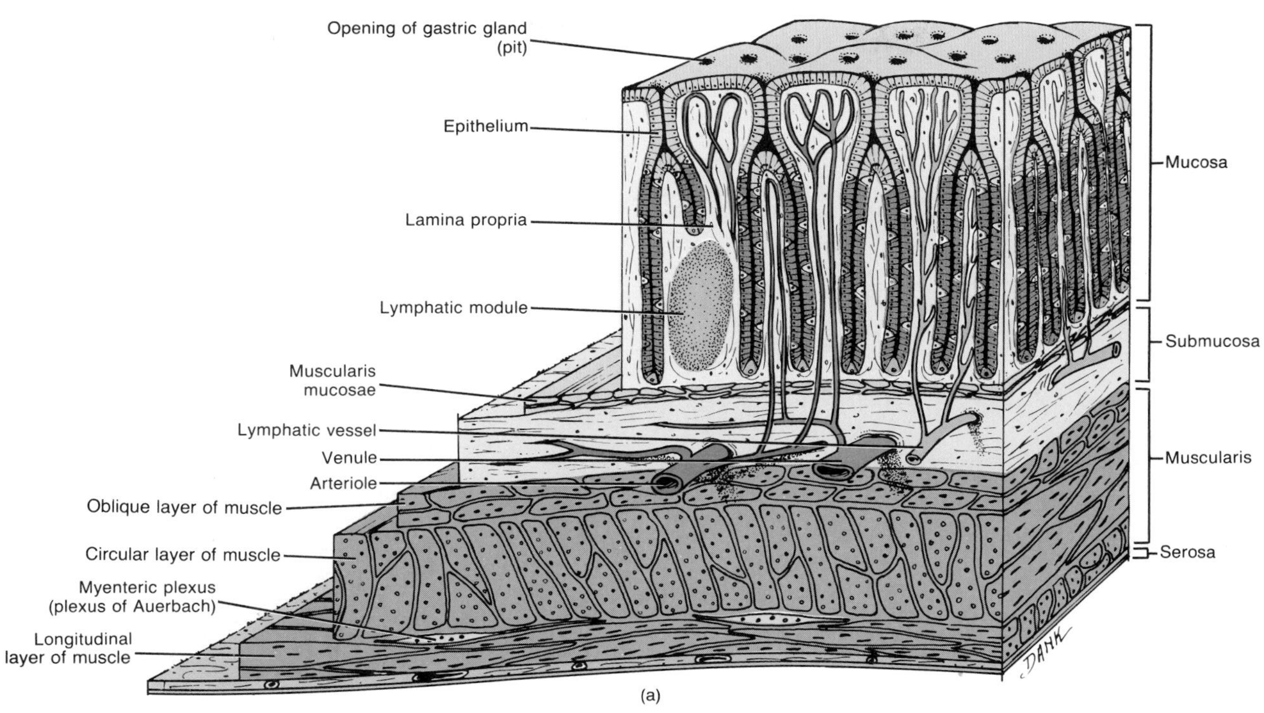

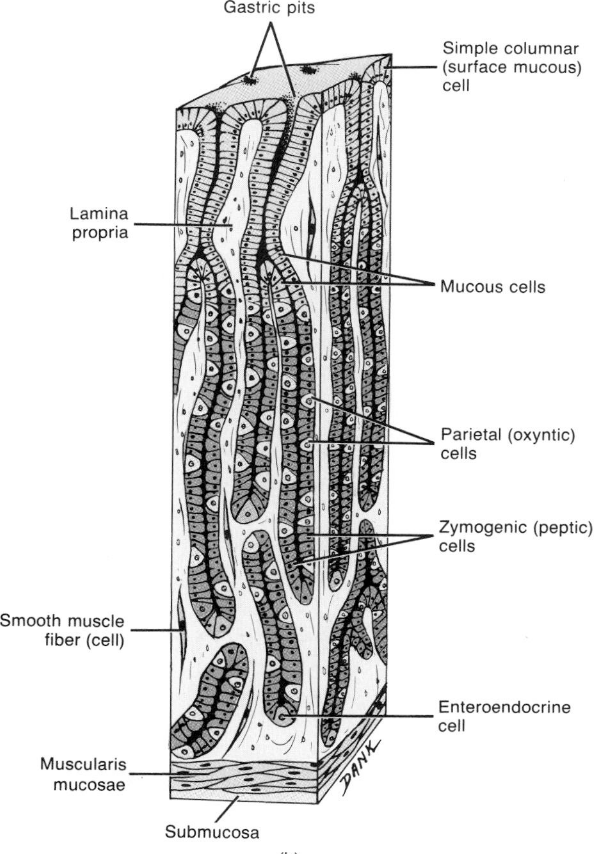

sinogen. Hydrochloric acid, involved in the conversion of pepsinogen to the active enzyme pepsin, and intrinsic factor, involved in the absorption of vitamin B_{12} for red blood cell production, are produced by the ***parietal (oxyntic) cells.*** You may recall from Chapter 19 that inability to produce intrinsic factor can result in pernicious anemia. The ***mucous cells*** secrete mucus. Secretions of the zymogenic, parietal, and mucous cells are collectively called ***gastric juice.*** The ***enteroendocrine cells*** secrete stomach gastrin, a hormone that stimulates secretion of hydrochloric acid and pepsinogen, contracts the lower esophageal sphincter, mildly increases motility of the GI tract, and relaxes the pyloric sphincter.

MEDICAL TEST

Gastroscopy* (gas-TROS-kō-pē; *gastro* = stomach; *skopein* = to examine)

Diagnostic Value: Using a lighted viewing instrument called a gastroscope, the examiner can view the interior of the stomach directly to evaluate an ulcer, tumor, inflammation, or bleeding source. It is also possible to biopsy lesions, stop bleeding, and remove foreign objects with the gastroscope.

Procedure: The patient is sedated, and a local anesthetic is applied to the back of the throat. The gas-

FIGURE 24-11 (*Continued*) (c) Photomicrograph of the gastric mucosa showing surface mucous cells at a magnification of 740×. (d) Photomicrograph of a portion of the fundic wall of the stomach at a magnification of 18×. (Photomicrographs courtesy of Andrew Kuntzman.)

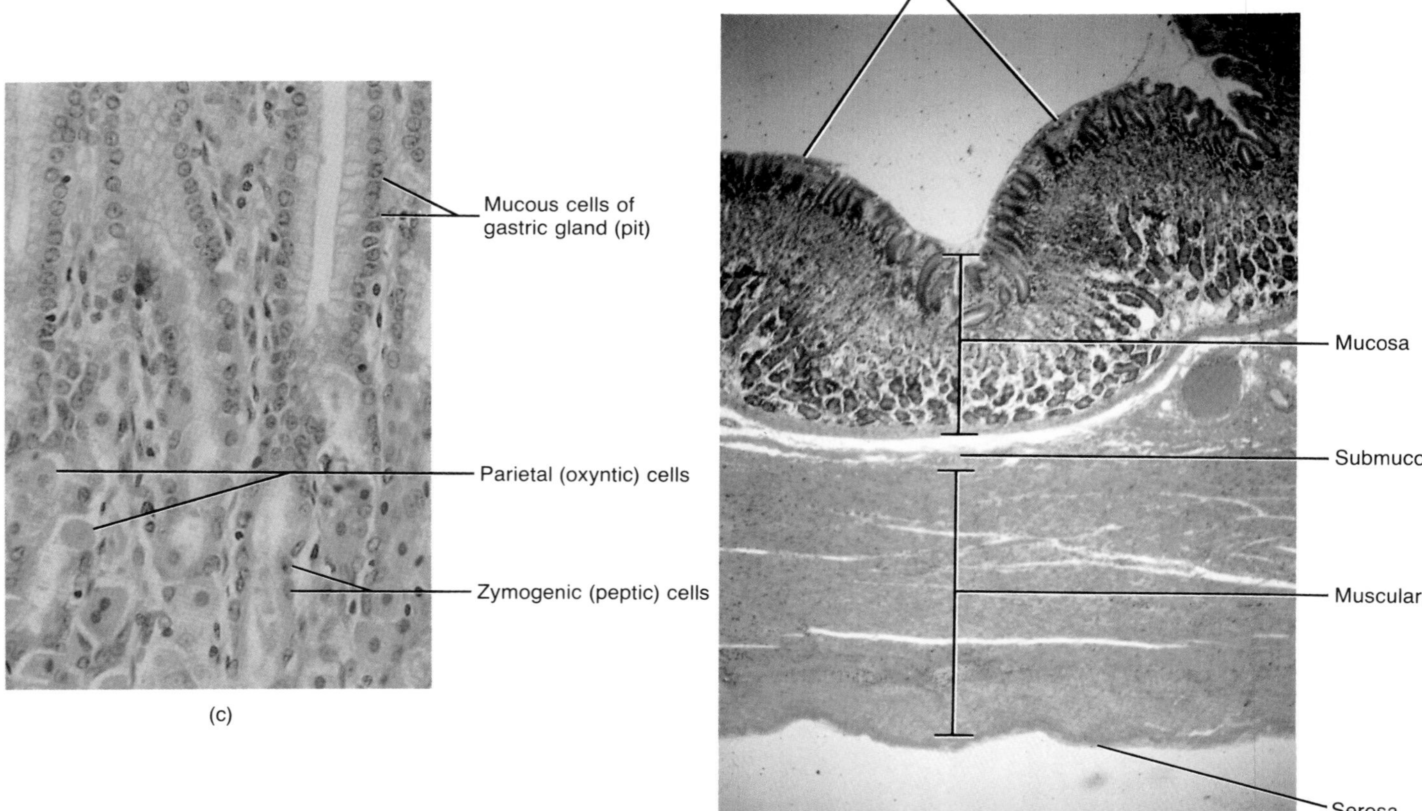

troscope is guided through the mouth, throat, and esophagus into the stomach.

* The general term ***endoscopy*** refers to visual inspection of any cavity of the body using an endoscope, an illuminated tube with lenses. Endoscopes can be used to visualize the entire gastrointestinal tract, as well as other systems of the body. *Fiberoptic endoscopy* is the preferred method, since fiberoptic technology makes possible the use of a flexible endoscope. This type of endoscope can be adapted to the many curvatures in the mouth, throat, stomach, and other organs and has improved all aspects of endoscopy.

The ***submucosa*** of the stomach is composed of loose (areolar) connective tissue, which connects the mucosa to the muscularis.

The ***muscularis,*** unlike that in other areas of the gastrointestinal tract, has three layers of smooth muscle: an outer longitudinal layer, a middle circular layer, and an inner oblique layer. The oblique layer is limited mostly to the body of the stomach. This arrangement of fibers allows the stomach to contract in a variety of ways to churn food, break it into small particles, mix it with gastric juice, and pass it to the duodenum.

The ***serosa*** covering the stomach is part of the visceral peritoneum. At the lesser curvature, the two layers of the visceral peritoneum come together and extend upward to the liver as the lesser omentum. At the greater curvature, the visceral peritoneum continues downward as the greater omentum hanging over the intestines.

Physiology of Digestion in the Stomach

Mechanical

Several minutes after food enters the stomach, gentle, rippling, peristaltic movements called ***mixing waves*** pass over the stomach every 15 to 25 seconds. These waves macerate food, mix it with the secretions of the gastric glands, and reduce it to a thin liquid called ***chyme*** (kīm). Few mixing waves are observed in the fundus, which is primarily a storage area. Foods may remain in the fundus for an hour or more without becoming mixed with gastric juice. During this time, salivary digestion continues.

As digestion proceeds in the stomach, more vigorous mixing waves begin at the body of the stomach and intensify as they reach the pylorus. The pyloric sphincter normally remains almost, but not completely, closed. As food reaches the pylorus, each mixing wave forces a small amount of the gastric contents into the duodenum through the pyloric sphincter. Most of the food is forced back into the body of the stomach, where it is subjected to further mixing. The next wave pushes it forward again and forces a little more into the duodenum. The forward and backward movement of the gastric contents are responsible for almost all the mixing in the stomach.

Chemical

The principal chemical activity of the stomach is to begin the digestion of proteins. In the adult, digestion is achieved primarily through the enzyme ***pepsin.*** Pepsin breaks certain peptide bonds between the amino acids making up proteins. Thus, a protein chain of many amino acids is broken down into smaller fragments called ***peptides.*** Pepsin is most effective in the very acidic environment of the stomach (pH 2). It becomes inactive in an alkaline environment.

What keeps pepsin from digesting the protein in stomach cells along with the food? First, pepsin is secreted in an inactive form called ***pepsinogen,*** so it cannot digest the proteins in the zymogenic cells that produce it. It is not converted into active pepsin until it comes in contact with the hydrochloric acid secreted by the parietal cells. Second, the stomach cells are protected by an alkaline mucus, especially after pepsin has been activated. The mucus coats the mucosa to form a barrier between it and the gastric juices.

Another enzyme of the stomach is ***gastric lipase.*** Gastric lipase splits the short chain triglycerides in butterfat molecules found in milk. This enzyme operates best at a pH of 5 to 6 and has a limited role in the adult stomach. Adults rely almost exclusively on an enzyme secreted by the pancreas ***(pancreatic lipase)*** into the small intestine to digest fats.

The infant stomach also secretes ***rennin,*** which is important in the digestion of milk. Rennin and calcium act on the casein of milk to produce a curd. The coagulation prevents too rapid a passage of milk from the stomach into the duodenum (first part of the small intestine). Rennin is absent in the gastric secretions of adults.

The principal activities of gastric digestion are summarized in Exhibit 24-3.

Regulation of Gastric Secretion

Stimulation

The secretion of gastric juice is regulated by both nervous and hormonal mechanisms (Figure 24-12). Parasympathetic impulses from nuclei in the medulla are transmitted via the vagus (X) nerves and stimulate the gastic glands to secrete pepsinogen, hydrochloric acid, mucus, and stomach gastrin. Stomach gastrin is also secreted by gastric glands in response to certain foods (partially digested proteins) that enter the stomach.

EXHIBIT 24-3 SUMMARY OF GASTRIC DIGESTION

Structure	Activity	Result
Mucosa		
Zymogenic (peptic) cells	Secrete pepsinogen.	Precursor of pepsin is produced.
Parietal (oxyntic) cells	Secrete hydrochloric acid.	Converts pepsinogen into pepsin, which digests proteins into peptides.
	Secrete intrinsic factor.	Required for absorption of vitamin B_{12}, which is required for normal erythrocyte formation.
Mucous cells	Secrete mucus.	Prevents digestion of stomach wall.
Enteroendocrine cells	Secrete stomach gastrin.	Stimulates gastric secretion, contracts lower esophageal sphincter, increases motility of the stomach, and relaxes pyloric sphincter.
Muscularis	Mixing waves.	Macerate food, mix it with gastric juice, reduce food to chyme, and force chyme through pyloric sphincter.
Pyloric sphincter (valve)	Opens to permit passage of chyme into duodenum.	Prevents backflow of food from duodenum to stomach.

FIGURE 24-12 **Summary of factors that stimulate gastric secretion.**

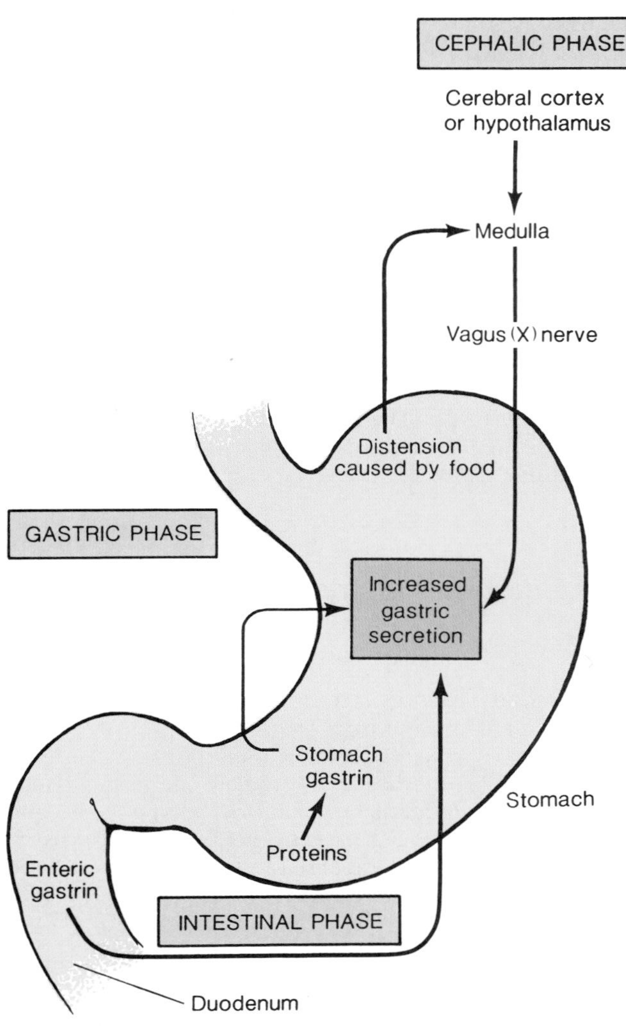

▪ **Cephalic (Reflex) Phase** The ***cephalic (reflex) phase*** of gastric secretions occurs before food enters the stomach and prepares the stomach for digestion. The sight, smell, taste, or thought of food initiates this reflex. Nerve impulses from the cerebral cortex or feeding center in the hypothalamus send impulses to the medulla. The medulla relays impulses over the parasympathetic fibers in the vagus (X) nerve to stimulate the gastric glands to secrete.

▪ **Gastric Phase** Once the food reaches the stomach, both nervous and hormonal mechanisms ensure that gastric secretion continues. This is the ***gastric phase*** of secretion. Food of any kind causes distention and stimulates receptors in the wall of the stomach. These receptors send impulses to the medulla and back to the gastric glands, and they may send messages to the glands as well. The impulses stimulate the flow of gastric juice. Emotions such as anger, fear, and anxiety may slow down digestion in the stomach because they stimulate the sympathetic nervous system, which inhibits gastric activity.

Partially digested proteins and caffeine stimulate the pyloric mucosa to secrete the hormone ***stomach gastrin.*** It is absorbed into the bloodstream, circulated through the body, and finally reaches its target cells, the gastric glands, where it stimulates secretion of large amounts of gastric juice. It also contracts the lower esophageal sphincter, increases motility of the GI tract, and relaxes the pyloric sphincter and ileocecal sphincter. Stomach gastrin secretion is inhibited when the pH of gastric juice (HCl) reaches 2.0. This negative feedback mechanism helps to provide the optimal pH for the functioning of enzymes in the stomach.

▪ **Intestinal Phase** The ***intestinal phase*** occurs when foods leave the stomach and enter the small intestine. When partially digested proteins leave the stomach and enter the duodenum, they stimulate the duodenal mucosa to release ***enteric gastrin,*** a hormone that stimulates the gastric glands to continue their secretion. However, this mechanism produces relatively small amounts of gastric juice. Carbohydrates and lipids in chyme, however, inhibit gastric secretion (described next). The *net* effect of food in the small intestine is the inhibition of gastric secretion.

Inhibition

Even though chyme stimulates gastric secretion during the gastric phase, it usually inhibits secretion during the intestinal phase. For example, the presence of food in the small intestine during the intestinal phase initiates an ***enterogastric reflex*** in which nerve impulses carried to the medulla from the duodenum return to the stomach and inhibit gastric secretion. These impulses ultimately inhibit parasympathetic stimulation and stimulate sympathetic activity. Stimuli that initiate this reflex are distension of the duodenum, the presence of acid or partially digested proteins in food in the duodenum, or irritation of the duodenal mucosa.

Several intestinal hormones also inhibit gastric secretion. In the presence of acid, partially digested carbohydrates, proteins, and fats, hypertonic or hypotonic fluids, or irritating substances in chyme, the intestinal mucosa releases ***secretin*** (se-KRĒ-tin), ***cholecystokinin*** (kō'-lē-sis'-tō-KĪN-in) or ***CCK,*** and ***gastric inhibiting peptide (GIP).*** All three hormones inhibit gastric secretion and decrease motility of the GI tract. GIP also stimulates release of insulin. Secretin and cholecystokinin are also important in the control of pancreatic and intestinal secretion, and cholecystokinin also helps regulate secretion of bile from the gallbladder (Exhibit 24-4).

EXHIBIT 24-4 HORMONAL CONTROL OF GASTRIC SECRETION, PANCREATIC SECRETION, AND SECRETION AND RELEASE OF BILE

Hormone	Where Produced	Stimulant	Action
Stomach gastrin	Pyloric mucosa.	Partially digested proteins and caffeine in stomach.	Stimulates secretion of gastric juice, constricts lower esophageal sphincter, increases motility of GI tract, and relaxes pyloric sphincter and ileocecal sphincter.
Enteric gastrin	Intestinal mucosa.	Partially digested proteins in chyme in small intestine.	Same as above.
Secretin	Intestinal mucosa.	Acid, partially digested carbohydrates, proteins, and fats, and hypertonic or hypotonic fluids that enter the small intestine.	Inhibits secretion of gastric juice, decreases motility of the GI tract, stimulates secretion of pancreatic juice rich in sodium bicarbonate ions, stimulates secretion of bile by hepatic cells of liver, and stimulates secretion of intestinal juice.
Cholecystokinin (CCK)	Intestinal mucosa.	Fats and proteins that enter the small intestine.	Inhibits secretion of gastric juice, decreases motility of the GI tract, stimulates secretion of pancreatic juice rich in digestive enzymes, causes ejection of bile from the gallbladder and opening of the sphincter of the hepatopancreatic ampulla (sphincter of Oddi), stimulates secretion of intestinal juice, and induces satiety (feeling full to satisfaction).
Gastric inhibiting peptide (GIP)	Intestinal mucosa.	Fats in the small intestine.	Inhibits secretion of gastric juice and decreases motility of the GI tract.

Regulation of Gastric Emptying

Gastric emptying is stimulated by two principal factors: (1) nerve impulses in response to distension and (2) stomach gastrin released in the presence of certain types of foods. As noted earlier, during the gastric phase of secretion, distension and the presence of partially digested proteins and caffeine stimulate secretion of gastric juice and stomach gastrin. In the presence of stomach gastrin, the lower esophageal sphincter contracts, the motility of the stomach increases, and the pyloric sphincter relaxes. The net effect of these actions is stomach emptying (Figure 24-13a).

The stomach empties all its contents into the duodenum within two to six hours after ingestion. Foods rich in carbohydrate spend the least time in the stomach. Protein foods are somewhat slower, and emptying is slowest after a meal containing large amounts of fat.

Stomach emptying is inhibited by the enterogastric reflex and hormones released in response to certain constituents in chyme. The enterogastric reflex not only inhibits gastric secretion but also inhibits gastric motility. The hormones secretin, cholecystokinin (CCK), and gastric inhibiting peptide (GIP) also inhibit gastric secretion and inhibit gastric motility (Figure 24-13b). The rate of stomach emptying is limited to the amount of chyme that the small intestine can process.

CLINICAL APPLICATION: VOMITING

Excessive gastric emptying in the wrong direction sometimes occurs. ***Vomiting*** is the forcible expulsion of the contents of the upper GI tract (stomach and sometimes duodenum) through the mouth. The strongest stimuli for vomiting are irritation and distension of the stomach. Other stimuli include unpleasant sights, dizziness, and certain drugs such as morphine and derivatives of digitalis. Nerve impulses are transmitted to the vomiting center in the medulla, and returning impulses to the upper GI tract organs, diaphragm, and abdominal muscles bring about the vomiting act. Basically, vomiting involves squeezing the stomach between the diaphragm and abdominal muscles and expelling of the contents through the open esophageal sphincters. Prolonged vomiting, especially in infants and elderly people, can be serious because the loss of gastric juice and fluids can lead to disturbances in fluid and acid–base balance.

Absorption

The stomach wall is impermeable to the passage of most materials into the blood, so most substances are not absorbed until they reach the small intestine. However, the

FIGURE 24-13 **Factors that (a) stimulate gastric emptying and (b) inhibit gastric emptying.**

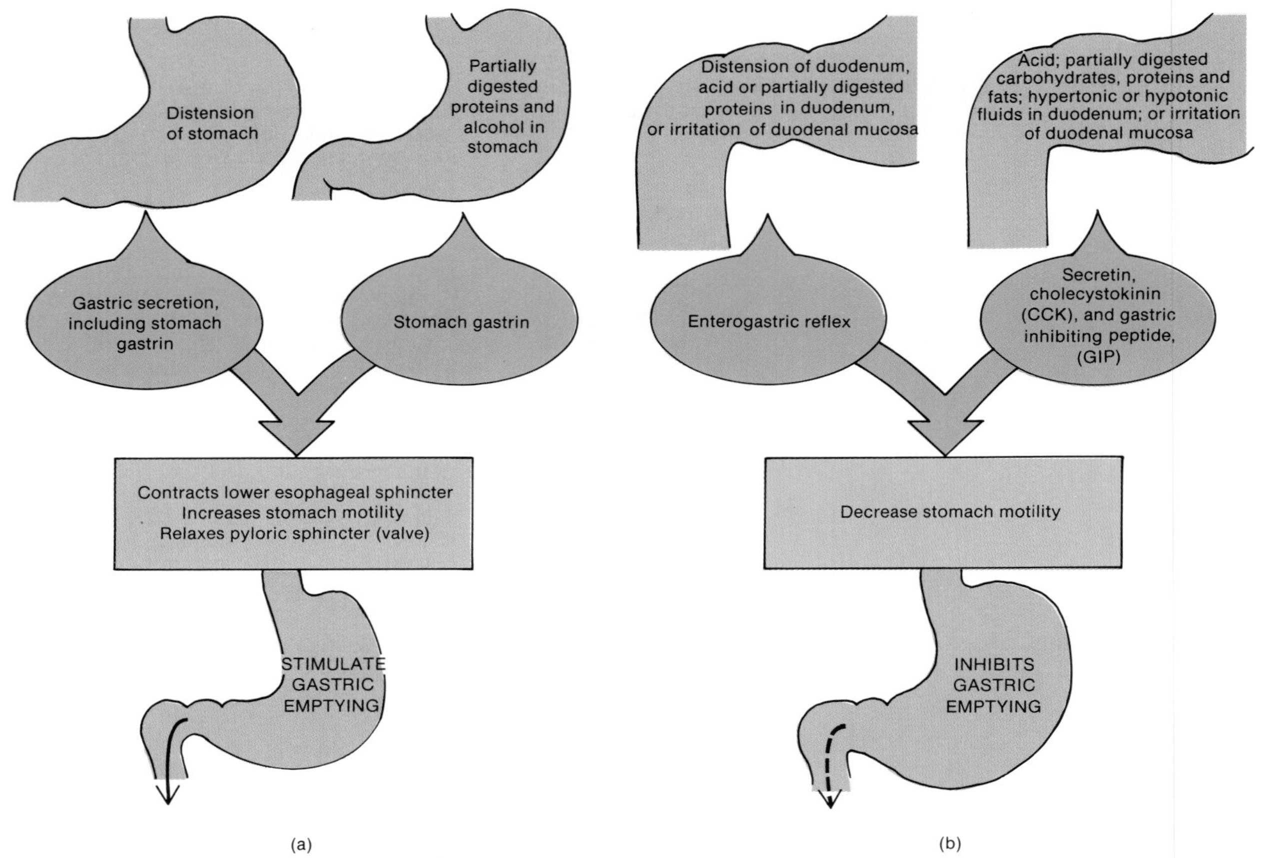

stomach does participate in the absorption of some water, electrolytes, certain drugs (especially aspirin), and alcohol.

PANCREAS

The next organ of the GI tract involved in the breakdown of food is the small intestine. Chemical digestion in the small intestine depends not only on its own secretions but also on activities of three accessory structures of digestion outside the gastrointestinal tract: the pancreas, liver, and gallbladder. We will first consider the activities of the accessory structures and then examine their contributions to digestion in the small intestine.

Anatomy

The ***pancreas*** is an oblong tubuloacinar gland about 12.5 cm (5 inches) long and 2.5 cm (1 inch) thick. It lies posterior to the greater curvature of the stomach and is connected, usually by two ducts, to the duodenum. The pancreas is divided into a head, body, and tail. The ***head*** is the expanded portion near the C-shaped curve of the duodenum. Moving superiorly and to the left of the head are the centrally located ***body*** and the terminal tapering ***tail*** (Figure 24-14).

Pancreatic secretions pass from the secreting cells in the pancreas to small ducts that unite to form the two ducts that convey the secretions into the small intestine. The larger of the two ducts is called the ***pancreatic duct (duct of Wirsung).*** In most people, the pancreatic duct unites with the common bile duct from the liver and gallbladder and enters the duodenum in a common duct called the ***hepatopancreatic ampulla (ampulla of Vater).*** The ampulla opens on an elevation of the duodenal mucosa known as the ***duodenal papilla,*** about 10 cm (4 inches) below the pylorus of the stomach. The smaller of the two ducts is the ***accessory duct (duct of Santorini),*** which leads from the pancreas and empties into the duodenum about 2.5 cm (1 inch) above the hepatopancreatic ampulla.

FIGURE 24-14 **Pancreas. Relation of pancreas to liver, gallbladder, and duodenum. The insert shows details of the common bile duct and pancreatic duct forming the hepatopancreatic ampulla (of Vater) and emptying into the duodenum.**

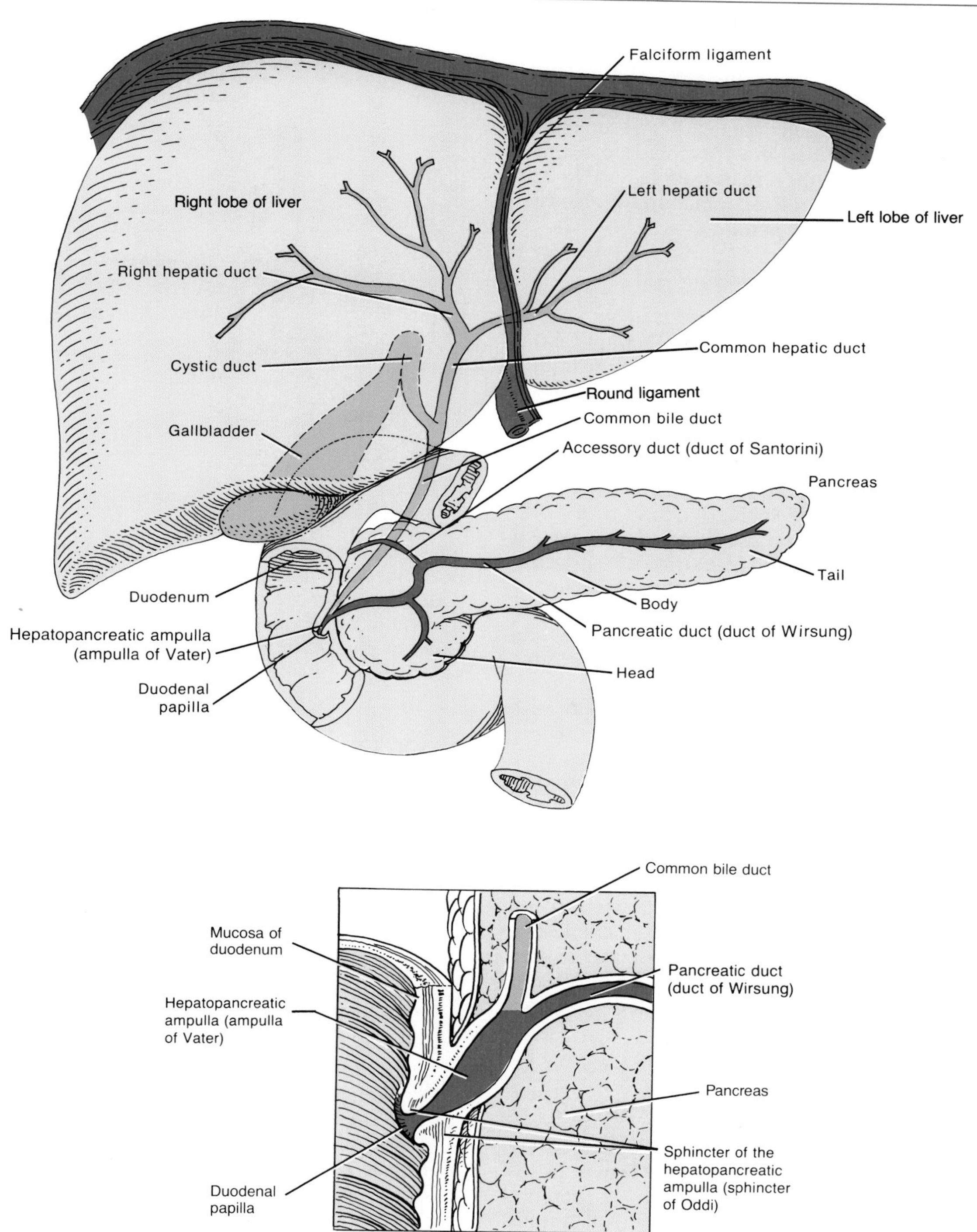

Histology

The pancreas is made up of small clusters of glandular epithelial cells. About 1 percent of the cells, the ***pancreatic islets (islets of Langerhans),*** form the endocrine portion of the pancreas and consist of alpha, beta, and delta cells that secrete hormones (glucagon, insulin, and somatostatin, respectively). The functions of these hormones may be reviewed in Chapter 18. The remaining 99 percent of the cells, called ***acini*** (AS-i-nī), constitute the exocrine portions of the organ (see Figure 18-22). Secreting cells of the acini release a mixture of digestive enzymes called ***pancreatic juice.***

Pancreatic Juice

Each day the pancreas produces 1200 to 1500 ml (about 1.2 to 1.5 qt) of pancreatic juice, a clear, colorless liquid. It consists mostly of water, some salts, sodium bicarbonate, and enzymes. The sodium bicarbonate gives pancreatic juice a slightly alkaline pH (7.1–8.2) that stops the action of pepsin from the stomach and creates the proper environment for the enzymes in the small intestine. The enzymes in pancreatic juice include a carbohydrate-digesting enzyme called ***pancreatic amylase;*** several protein-digesting enzymes called ***trypsin*** (TRIP-sin), ***chymotrypsin*** (kī'-mō-TRIP-sin), and ***carboxypeptidase*** (kar-bok'-sē-PEP-ti-dās); the principal fat-digesting enzyme in the adult body called ***pancreatic lipase;*** and nucleic acid–digesting enzymes called ***ribonuclease*** and ***deoxyribonuclease.***

Just as pepsin is produced in the stomach in an inactive form (pepsinogen), so too are the protein-digesting enzymes of the pancreas. This prevents the enzymes from digesting cells of the pancreas. The active enzyme trypsin is secreted in an inactive form called ***trypsinogen*** (trip-SIN-ō-jen). Its activation to trypsin is accomplished in the small intestine by an enzyme secreted by the intestinal mucosa when chyme comes in contact with the mucosa. The activating enzyme is called ***enterokinase*** (en'-ter-ō-KĪ-nās). Chymotrypsin is activated in the small intestine by trypsin from its inactive form, ***chymotrypsinogen.*** Carboxypeptidase is also activated in the small intestine by trypsin. Its inactive form is called ***procarboxypeptidase.***

Regulation of Pancreatic Secretions

Pancreatic secretion, like gastric secretion, is regulated by both nervous and hormonal mechanisms (Figure 24-15). When the cephalic and gastric phases of gastric secretion occur, parasympathetic impulses are simultaneously transmitted along the vagus (X) nerves to the pancreas that result in the secretion of pancreatic enzymes.

In response to acid chyme that enters the small intestine, the small intestinal mucosa secretes secretin, and in response to fats and proteins in the small intestine, the intestinal mucosa secretes cholecystokinin (CCK), two hormones that affect pancreatic secretion. Secretin stimulates the pancreas to secrete pancreatic juice that is rich in sodium bicarbonate ions. Cholecystokinin (CCK) stimulates a pancreatic secretion rich in digestive enzymes (see Exhibit 24-4).

FIGURE 24-15 Control of pancreatic secretion.

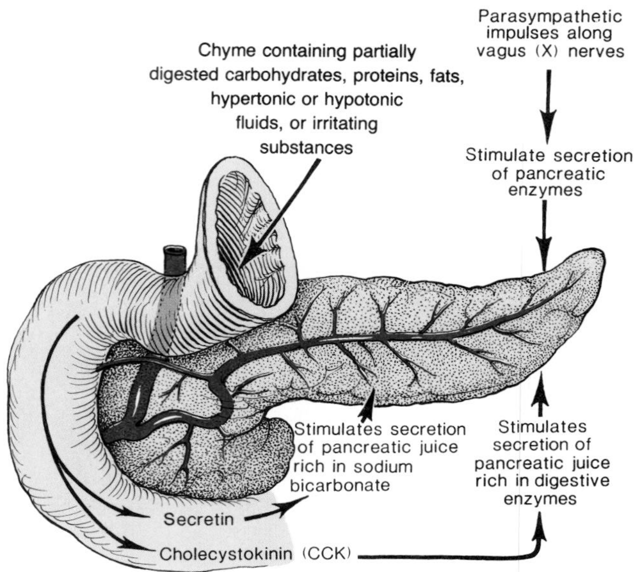

LIVER

The ***liver*** weighs about 1.4 kg (about 3 lb) in the average adult. It is located under the diaphragm and occupies most of the right hypochondrium and part of the epigastrium of the abdomen (see Figure 1-8c).

Anatomy

The liver is almost completely covered by peritoneum and completely covered by a dense connective tissue layer that lies beneath the peritoneum. It is divided into two principal lobes—a large ***right lobe*** and a smaller ***left lobe***—separated by the ***falciform ligament*** (Figure 24-16). The right lobe is considered by many anatomists to consist of an inferior ***quadrate lobe*** and a posterior ***caudate lobe.*** However, on the basis of internal morphology, primarily the distribution of blood, the quadrate and caudate lobes more appropriately belong to the left lobe. The falciform ligament is a reflection of the parietal peritoneum, which extends from the undersurface of the diaphragm to the superior surface of the liver, between the two principal lobes of the liver. In the free border of the falciform ligament is the ***ligamentum teres (round***

FIGURE 24-16 External anatomy of the liver. (a) Diagram of inferior (visceral) surface. (b) Photograph of inferior (visceral) surface. The anterior view is illustrated in Figure 24-14. (Courtesy of J. A. Gosling, P. F. Harris, et al., *Atlas of Human Anatomy,* Gower Medical Publishing Ltd., 1985.)

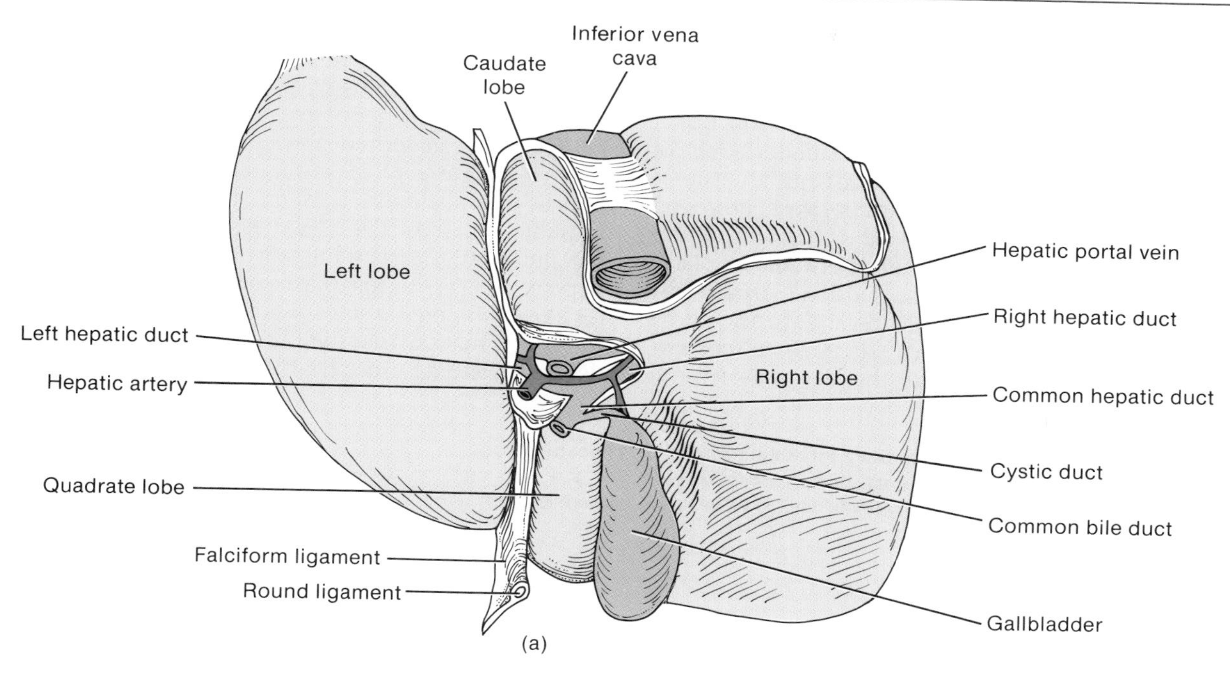

(a)

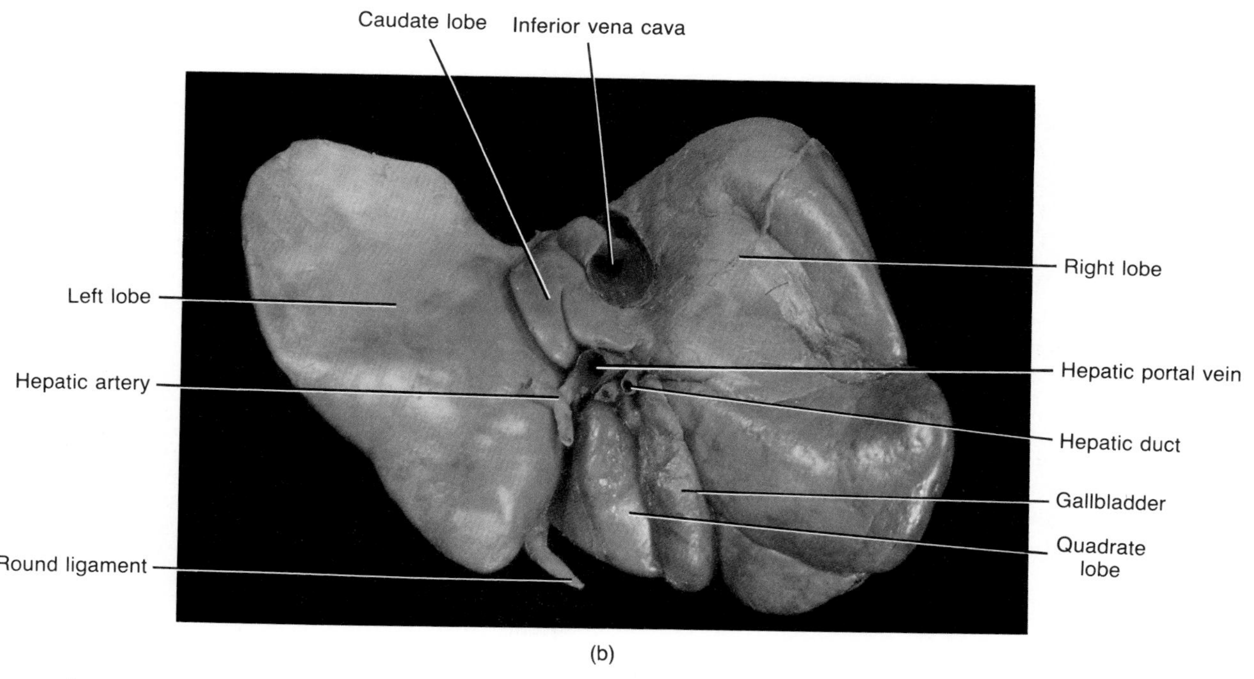

(b)

***ligament*).** It extends from the liver to the umbilicus. The ligamentum teres is a fibrous cord derived from the umbilical vein of the fetus.

Bile, one of the liver's products, enters ***bile capillaries*** or ***canaliculi*** (kan'-a-LIK-yoo-lī) that empty into small ducts. These small ducts eventually merge to form the larger ***right*** and ***left hepatic ducts,*** which unite to leave the liver as the ***common hepatic duct*** (Figures 24-14 and 24-16a). Further on, the common hepatic duct joins the ***cystic duct*** from the gallbladder. The two tubes become the ***common bile duct.*** The common bile duct and pancreatic duct enter the duodenum in a common duct called the ***hepatopancreatic ampulla (ampulla of Vater).***

Histology

The lobes of the liver are made up of numerous functional units called ***lobules,*** which may be seen under a microscope (Figure 24-17). A lobule consists of epithelial cells, called ***hepatic (liver) cells*** or ***hepatocytes,*** arranged in irregular, branching, interconnected plates around a ***central vein.*** These cells secrete bile. Between the plates of cells are endothelial-lined spaces called ***sinusoids,*** through which blood passes. The sinusoids are also partly lined with phagocytic cells, termed ***stellate reticuloendothelial (Kupffer's) cells,*** that destroy worn-out white and red blood cells, bacteria, and toxic substances. The liver contains sinusoids instead of typical capillaries.

Blood Supply

The liver receives a double supply of blood. From the hepatic artery it obtains oxygenated blood, and from the hepatic portal vein it receives deoxygenated blood containing newly absorbed nutrients during the absorptive state (see Figures 21-23 and 24-17). Branches of both the hepatic artery and the hepatic portal vein carry the blood into the sinusoids of the lobules, where oxygen, most of the nutrients, and certain poisons are extracted by the hepatic cells. The phagocytic reticuloendothelial (Kupffer's) cells lining the sinusoids remove microbes and bits of foreign matter from the blood. Nutrients are stored or used to make new materials. The poisons are stored or

FIGURE 24-17 Histology of the liver. (a) Diagram of the microscopic appearance of a portion of a liver lobule. (b) Photomicrograph of a portion of a liver lobule at a magnification of 65×. (Copyright © 1983 by Michael H. Ross. Used by permission.)

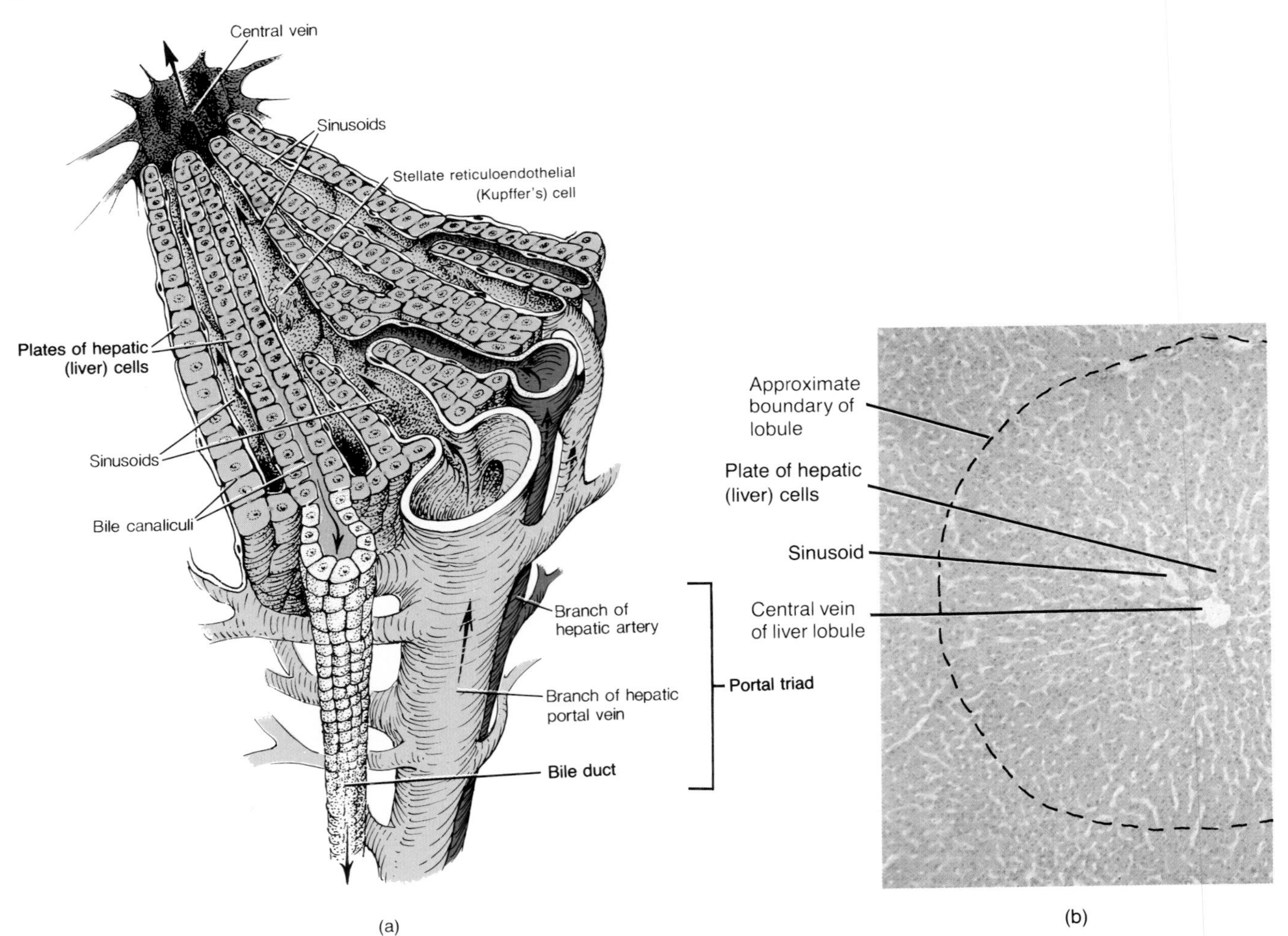

detoxified. Products manufactured by the hepatic cells and nutrients needed by other cells are secreted back into the blood. The blood then drains into the central vein and eventually passes into a hepatic vein. Unlike the other products of the liver, bile normally is not secreted into the bloodstream.

Branches of the hepatic portal vein, hepatic artery, and bile duct typically accompany each other in their distribution through the liver. Collectively, these three structures are referred to as a ***portal triad*** (Figure 24-17a).

Bile

Each day, the hepatic cells secrete 800 to 1000 ml (about 1 qt) of ***bile,*** a yellow, brownish, or olive-green liquid. It has a pH of 7.6 to 8.6. Bile consists mostly of water and bile salts, cholesterol, a phospholipid called lecithin, bile pigments, and several ions.

Bile is partially an excretory product and partially a digestive secretion. Bile salts assume a role in ***emulsification,*** the breakdown of large fat globules into a suspension of fat droplets about 1 μm in diameter, and absorption of fats following their digestion. The tiny fat droplets present a very large surface area for the action of the pancreatic lipase necessary for rapid fat digestion. Cholesterol is made soluble in bile by bile salts and lecithin. The principal bile pigment is ***bilirubin.*** When worn-out red blood cells are broken down, iron, globin, and bilirubin (derived from heme) are released. The iron and globin are recycled, but some of the bilirubin is excreted into the bile ducts. Bilirubin eventually is broken down in the intestine, and one of its breakdown products (urobilinogen) gives feces their color.

CLINICAL APPLICATION: JAUNDICE

Jaundice is a yellowish coloration of the sclerae, skin, and mucous membranes due to a buildup of bilirubin in the body. After bilirubin is formed from the breakdown of the heme pigment in senescent red blood cells, it is transported to the liver where it is processed and eventually excreted into bile. The three main categories of jaundice are:

1. *Prehepatic (hemolytic) jaundice.* This is due to excess production of bilirubin. In hemolytic anemia, for example, bilirubin may be produced so rapidly and in such quantities that the liver cannot excrete it fast enough to avoid the occurrence of jaundice. Since the liver of a newborn functions poorly for the first week or so, many babies experience a mild form of jaundice called *neonatal (physiological) jaundice* that disappears as the liver matures. Neonatal jaundice is usually treated by exposing the infant to blue light, which converts bilirubin into substances the liver can excrete. A new treatment consists of administering a new drug (Sn-protoporphyrin) that combines with the enzyme that catalyzes heme breakdown, thus blocking production of bilirubin.

2. *Hepatic (medical) jaundice.* This is due to dysfunction of liver cells. Certain congenital liver diseases result in jaundice because of a derangement of hepatic enzyme systems. Jaundice accompanying cirrhosis of the liver results from a loss of functional hepatic lobules. Hepatitis, by producing swelling within the liver lobules, blocks the bile canaliculi and produces jaundice.

3. *Extrahepatic (surgical) jaundice.* This is due to interference with the removal of bile from the hepatobiliary system. It is also known as *obstructive jaundice.* Causes include gallstones and cancer of the bowel or the head of the pancreas. Surgery must usually be performed to relieve the obstruction.

Regulation of Bile Secretion

The rate at which bile is secreted is determined by several factors. Vagal stimulation can increase the production of bile to more than twice the normal rate. Secretin, the hormone that stimulates the synthesis of pancreatic juice rich in sodium bicarbonate, also stimulates the secretion of bile (see Exhibit 24-4). Within limits, as blood flow through the liver increases, so does the secretion of bile. Finally, the presence of large amounts of bile salts in the blood also increases the rate of bile production.

Physiology of the Liver

The liver performs many vital functions, many of which are related to metabolism and are discussed in Chapter 25. Among the functions of the liver are the following:

1. ***Carbohydrate metabolism.*** In carbohydrate metabolism, the liver is especially important in maintaining a normal blood glucose level. For example, the liver can convert glucose to glycogen (glycogenesis) when blood sugar level is high and convert glycogen to glucose (glycogenolysis) when blood sugar level is low. The liver can also convert amino acids to glucose (gluconeogenesis) when blood sugar level is low: convert other sugars, such as fructose and galactose into glucose; and convert glucose to fats.

2. ***Fat metabolism.*** With respect to fat metabolism, the liver breaks down fatty acids from acetyl coenzyme A, a process called beta oxidation; converts excess acetyl coenzyme A into ketone bodies (ketogenesis); synthesizes lipoproteins which transport fatty acids, fats, and cholesterol to and from body cells; synthesizes cholesterol and phospholipids and breaks down cholesterol to bile salts; and stores fats.

3. ***Protein metabolism.*** Without the role of the liver in protein metabolism, death would occur in a few days. Among the functions of the liver related to protein metabolism are:

deamination (removal of the amino group, NH_2) of amino acids so that they can be used for energy or converted to carbohydrates or fats; conversion of ammonia (NH_3), a toxic substance, into the much less toxic urea for excretion in urine (ammonia is produced from deamination and by bacteria in the gastrointestinal tract); synthesis of most plasma proteins, such as alpha and beta globulins, albumin, prothrombin, and fibrinogen (together with mast cells, the liver also produces the anticoagulant, heparin); and transamination, the transfer of an amino group from an amino acid to another substance (α-keto acid) in order to convert one amino acid into another.

4. ***Removal of drugs and hormones.*** The liver can detoxify or excrete into bile drugs such as penicillin, ampicillin, erythromycin, and sulfonamides. It can also chemically alter or excrete steroid hormones, such as estrogens and aldosterone, and thyroxine.

5. ***Excretion of bile.*** As noted earlier, bilirubin, derived from the heme of worn-out red blood cells, is absorbed by the liver from the blood and secreted into bile. Most of the bilirubin in bile is metabolized in the intestine by bacteria and eliminated in feces.

6. ***Synthesis of bile salts.*** Bile salts are used in the small intestine for the emulsification and absorption of fats, cholesterol, phospholipids, and lipoproteins.

7. ***Storage.*** In addition to glucogen, the liver stores vitamins (A, B_{12}, D, E, and K) and minerals (iron and copper). Liver cells contain a protein called ***apoferretin*** which combines with iron to form ***ferretin,*** the form in which iron is stored in the liver. The iron is released from the liver when needed elsewhere in the body.

8. ***Phagocytosis.*** The stellate reticuloendothelial (Kupffer's) cells of the liver phagocytize worn-out red and white blood cells and some bacteria.

9. ***Activation of vitamin D.*** The liver and kidneys participate in the activation of vitamin D.

A battery of blood tests (liver battery) is performed to evaluate the status of the liver. Such ***liver function tests (LFTs)*** include serum glutamic pyruvic transaminase (SGPT), alphafetoprotein (AFP), alkaline phosphatase, bilirubin, serum glutamic oxalacetic transaminase (SGOT), and lactic dehydrogenase (LDH). Since the SGOT and LDH tests have already been discussed in Chapter 20 as part of serum enzyme studies, only SGPT, alkaline phosphatase, and bilirubin will be described here.

MEDICAL TESTS

Serum glutamic pyruvic transaminase (SGPT) This enzyme is found primarily in the liver and is released into blood from a damaged or necrotic liver.

Diagnostic Value: To screen for liver disease, to follow the response of the patient to therapy, and to monitor liver function in individuals receiving medications that may be toxic to the liver.

Procedure: The test is performed on a venous blood sample.

Normal Values: 10–30 IU/l
Values increase in liver cell damage due to hepatitis, drug or chemical toxicity, infectious mononucleosis (IM), obstructive jaundice, cholecystitis, cirrhosis, liver cancer, congestive heart failure (CHF), or following alcohol ingestion.

Alkaline phosphatase This enzyme is released into blood by the liver in response to certain diseases, by the bone in response to disease, or during growth or fracture healing.

Diagnostic Value: To evaluate certain liver disorders (obstructive jaundice, tumor, abscess, hepatitis, and cirrhosis) and bone diseases (Paget's disease, metastatic bone disease, osteomalacia, and rickets) and to evaluate fracture repair.

Procedure: The test is performed on a venous blood sample.

Normal Values: Adults, 4–13 U/dl (King-Armstrong)
See also Appendix B.

Bilirubin

Diagnostic Value: To evaluate clinical jaundice or to screen for its presence and to determine if newborns need exchange transfusions or bili light therapy. The fraction(s) that is elevated helps to distinguish between prehepatic, hepatic, and extrahepatic causes.

Procedure: Bilirubin levels may be determined in a venous blood sample, and the urine may also be checked for the presence of bilirubin. (In newborns, blood for bilirubin is obtained by heelstick.)

Normal Values: Blood: Unconjugated, 0.2–0.7 mg/dl (indirect)
Conjugated, 0.1–0.4 mg/dl (direct)
Total, 0.2–1.2 mg/dl (1.0–12.0 mg/dl in newborns)
Urine: Negative
See also Appendix B.

GALLBLADDER (GB)

The ***gallbladder (GB)*** is a pear-shaped sac about 7 to 10 cm (3 to 4 inches) long. It is located in a fossa of the visceral surface of the liver (see Figures 24-14 and 24-16a).

Histology

The mucosa of the gallbladder consists of simple columnar epithelium arranged in rugae resembling those of the stomach. The gallbladder lacks a submucosa. The middle, muscular coat of the wall consists of smooth muscle fibers (cells). Contraction of these fibers by hormonal stimulation ejects the contents of the gallbladder into the ***cystic duct.*** The outer coat is the visceral peritoneum.

Physiology

The functions of the gallbladder are to store and concentrate bile (up to 10-fold) until it is needed in the small intestine. In the concentration process, water and many ions are absorbed by the mucosa of the gallbladder. Bile from the liver enters the small intestine through the common bile duct. When the small intestine is empty, a valve around the hepatopancreatic ampulla (ampulla of Vater) called the ***sphincter of the hepatopancreatic ampulla (sphincter of Oddi)*** closes, and the backed-up bile flows into the cystic duct to the gallbladder for storage (see Figure 24-14).

Emptying of the Gallbladder

In order for the gallbladder to eject bile into the small intestine to participate in the digestive process, the muscularis must contract to force bile into the common bile duct, and the sphincter of the hepatopancreatic ampulla must relax. Chyme entering the duodenum that contains particularly high concentrations of fats or partially digested proteins stimulates the intestinal mucosa to secrete cholecystokinin (CCK). This hormone brings about contraction of the muscularis coupled with relaxation of the sphincter of the hepatopancreatic ampulla, resulting in emptying of the gallbladder (see Exhibit 24-4).

MEDICAL TEST

Oral cholecystogram (kō-lē-SIS-tō-gram; *chol* = bile; *kystis* = bladder; *gram* = record of)

Diagnostic Value: To evaluate for the presence of gallstones and to diagnose inflammations and tumors of the gallbladder.

Procedure: At noon on the day before the test is done, a fatty meal is eaten to empty the gallbladder of bile. In the late afternoon or early evening, a fat-free meal is eaten, enabling the gallbladder to fill with bile. Then, at bedtime, tablets containing an x ray contrast medium are swallowed. The iodinated contrast medium is absorbed into the bloodstream, excreted by the liver, and concentrated by the gallbladder, which can be visualized the following day by a series of x rays. Sometimes a fatty meal or agent is given during the course of the x rays to observe results. If the gallbladder does not contract following this, the patient may have cholecystitis or obstruction of the common bile duct. Nonvisualization or poor visualization may indicate acute or chronic gallbladder disease, liver disease, or obstruction of the biliary tree.

SMALL INTESTINE

The major portions of digestion and absorption occur in a long tube called the ***small intestine.*** The small intestine begins at the pyloric sphincter of the stomach, coils through the central and lower part of the abdominal cavity, and eventually opens into the large intestine. It averages 2.5 cm (1 inch) in diameter and about 6.35 m (21 ft) in length in a cadaver.

Anatomy

The small intestine is divided into three segments (see Figure 24-1). The ***duodenum*** (doo'-ō-DĒ-num), the shortest part, originates at the pyloric sphincter of the stomach and extends about 25 cm (10 inches) until it merges with the jejunum. *Duodenum* means "12"; the structure is 12 fingers' breadth in length. The ***jejunum*** (jē-JOO-num) is about 2.5 m (8 ft) long and extends to the ileum. *Jejunum* means "empty," since at death it is found empty. The final portion of the small intestine, the ***ileum*** (IL-ē-um; *eileos* = twisted) measures about 3.6 m (12 ft) and joins the large intestine at the ***ileocecal*** (il'-ē-ō-SĒ-kal) ***sphincter (valve).***

Histology

The wall of the small intestine is composed of the same four coats that make up most of the GI tract. However, both the mucosa and the submucosa are modified to allow the small intestine to complete the processes of digestion and absorption (Figure 24-18).

The ***mucosa*** contains many pits lined with glandular epithelium. These pits—the ***intestinal glands (crypts of Lieberkühn)***—secrete intestinal juice. The submucosa of the duodenum contains ***duodenal (Brunner's) glands*** that secrete an alkaline mucus to protect the wall of the small intestine from the action of the enzymes and to aid in neutralizing acid in the chyme. Some of the epithelial cells in the mucosa and submucosa have been transformed to goblet cells, which secrete additional mucus.

Since almost all the digestion and absorption of nutrients occurs in the small intestine, its structure is specially adapted for this function. Its length alone provides a large surface area for digestion and absorption, and that area

FIGURE 24-18 Small intestine. Shown are various structures that adapt the small intestine for digestion and absorption. (a) Photograph of a section of the jejunum cut open to expose the circular folds (plicae circulares). (Courtesy of J. A. Gosling, P. F. Harris, et al., *Atlas of Human Anatomy,* Gower Medical Publishing Ltd., 1985.) (b) Diagram of villi in relation to the four principal layers.

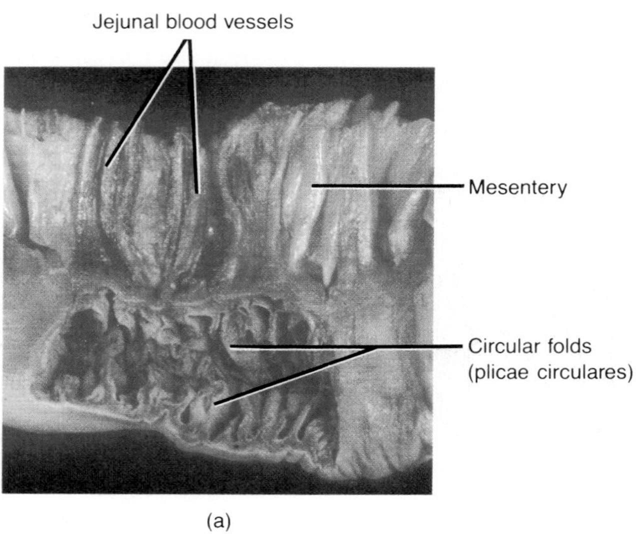

(a)

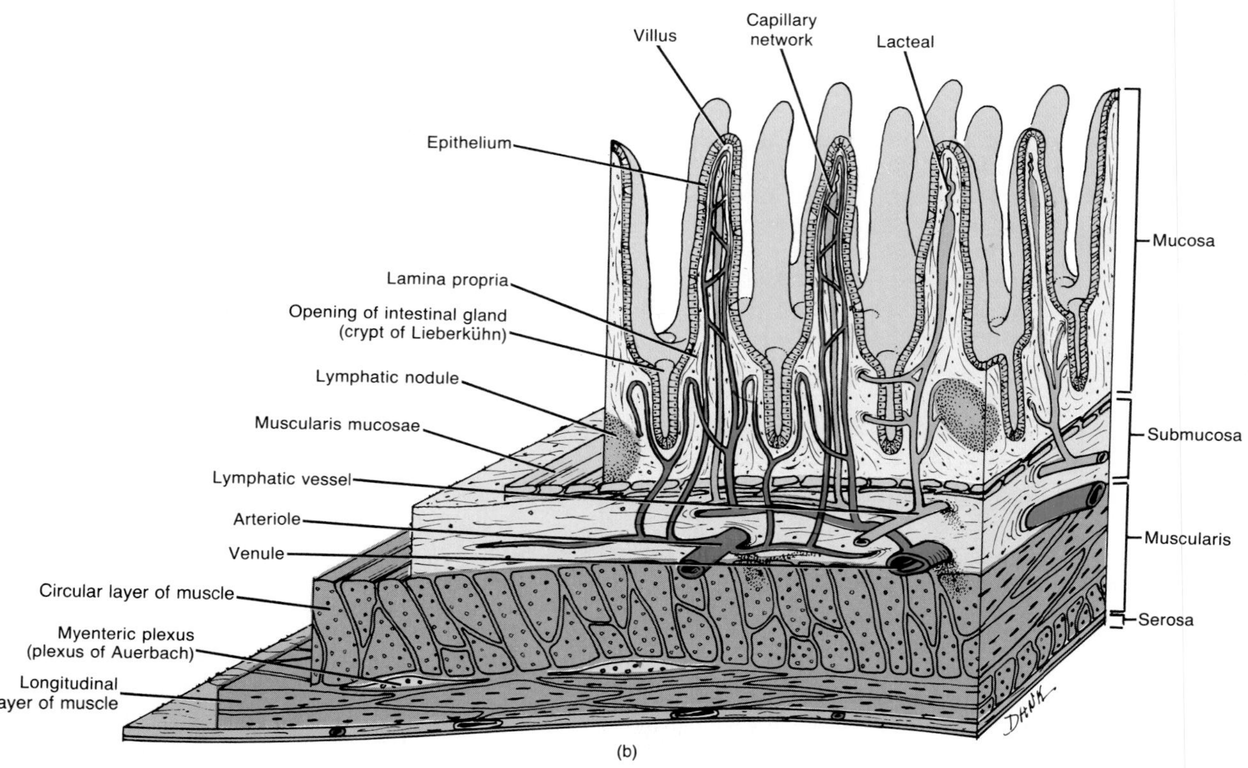

(b)

FIGURE 24-18 (*Continued*) (c) Diagram of enlarged aspect of a single villus.

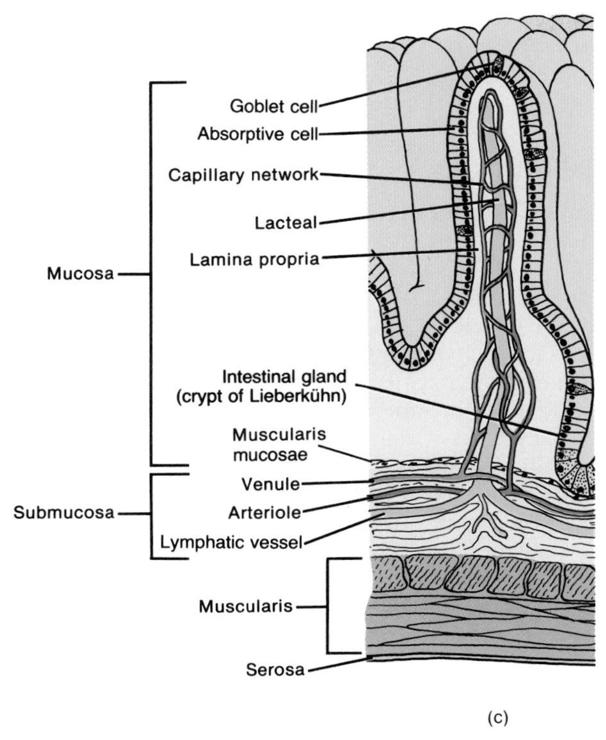

is further increased by modifications in the structure of its wall. The epithelium covering and lining the mucosa consists of simple columnar epithelium and contains goblet cells and absorptive cells. The absorptive cells possess ***microvilli,*** fingerlike projections of the plasma membrane. Larger amounts of digested nutrients diffuse into the absorptive cells of the intestinal wall because the microvilli increase the surface area of the plasma membrane. They also increase the surface area for digestion.

The mucosa lies in a series of ***villi,*** projections 0.5 to 1 mm high, giving the intestinal mucosa its velvety appearance. The enormous number of villi (10 to 40 per square millimeter) vastly increases the surface area of the epithelium available for absorption and digestion. Each villus has a core of lamina propria, the connective tissue layer of the mucosa. Embedded in this connective tissue are an arteriole, a venule, a capillary network, and a ***lacteal*** (LAK-tē-al) or lymphatic vessel. Nutrients that diffuse through the epithelial cells that cover the villus are able to pass through the capillary walls and the lacteal and enter the cardiovascular and lymphatic systems, respectively.

In addition to the microvilli and villi, a third set of projections called ***circular folds,*** or ***plicae circulares*** (PLĪ-kē SER-kyoo-lar-es), further increases the surface area for absorption and digestion. The folds are permanent ridges, about 10 mm (0.4 inch) high, in the mucosa. Some of the folds extend all the way around the intestine, and others extend only partway around. The folds begin near the proximal portion of the duodenum and terminate at about the midportion of the ileum. The circular folds enhance absorption by causing the chyme to spiral, rather than to move in a straight line, as it passes through the small intestine. Since the folds and villi decrease in size in the distal ileum, most absorption occurs in the duodenum and jejunum.

The ***muscularis*** of the small intestine consists of two layers of smooth muscle. The outer, thinner layer contains longitudinally arranged fibers (cells). The inner, thicker layer contains circularly arranged fibers. Except for a major portion of the duodenum, the serosa (or visceral peritoneum) completely covers the small intestine. Additional histological aspects of the small intestine are shown in Figure 24-19.

There is an abundance of lymphatic tissue in the form

FIGURE 24-19 Histology of the small intestine. (a) Photomicrograph of a portion of the wall of the duodenum at a magnification of 40×.

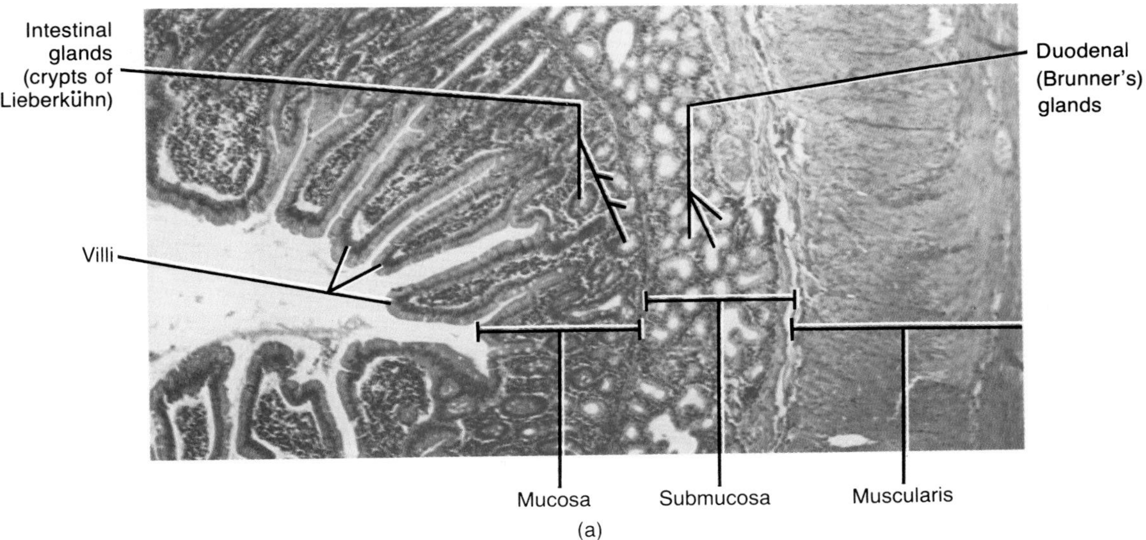

FIGURE 24-19 (*Continued*) (b) Photomicrograph of an enlarged aspect of two villi from the ileum at a magnification of 120×. (Photomicrographs copyright © 1983 by Michael H. Ross. Used by permission.) (c) Transmission electron micrograph of several microvilli at a magnification of 130,000×. (Courtesy of Visuals Unlimited.)

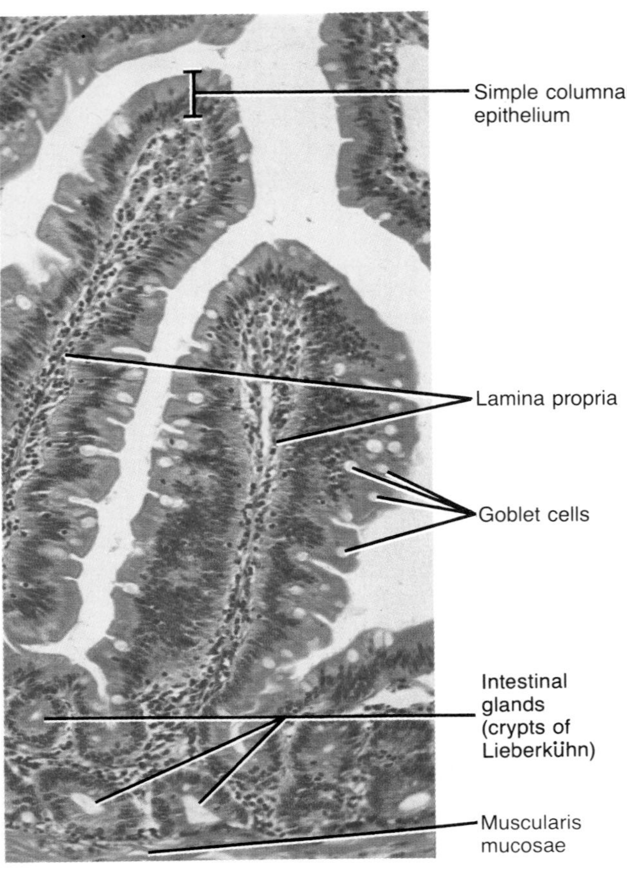

(b)

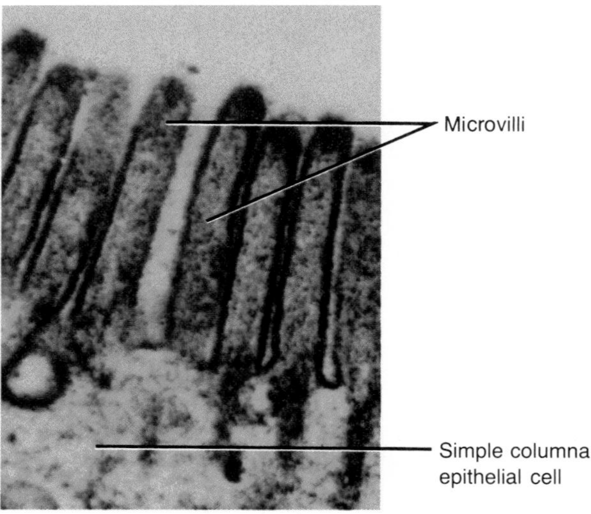

(c)

of lymphatic nodules, masses of lymphatic tissue not covered by a capsule wall. ***Solitary lymphatic nodules*** are most numerous in the lower part of the ileum. Groups of lymphatic nodules, referred to as ***aggregated lymphatic follicles (Peyer's patches),*** are numerous in the ileum.

Intestinal Juice

Intestinal juice is a clear yellow fluid secreted in amounts of about 2 to 3 liters (about 2 to 3 qt) a day. It has a pH of 7.6, which is slightly alkaline, and contains water and mucus. The juice is rapidly reabsorbed by the villi and provides a vehicle for the absorption of substances from chyme as they come in contact with the villi. The intestinal enzymes are formed in the epithelial cells that line the villi. Some digestion by enzymes of the small intestine occurs within the cells on the surfaces of their microvilli. As small intestinal cells containing enzymes slough off into the lumen of the intestine, they break apart and release small quantities of enzymes that digest food in the chyme. Thus, some digestion by enzymes of the small intestine occurs in or on the epithelial cells that line the villi, rather than in the lumen exclusively, as in other parts of the gastrointestinal tract. Among the enzymes produced by small intestinal cells are three carbohydrate-digesting enzymes called ***maltase, sucrase,*** and ***lactase;*** protein-digesting enzymes called ***peptidases (aminopeptidase*** and ***dipeptidase);*** and two nucleic acid–digesting enzymes, ***ribonuclease*** and ***deoxyribonuclease.***

MEDICAL TEST

Barium swallow

Diagnostic Value: To evaluate the upper gastrointestinal (UGI) tract (throat, esophagus, stomach, and duodenum) for a condition such as difficulty in swallowing, ulcer, pain after eating, heartburn, bleeding, or motility disorders.

Procedure: Barium, a chalky liquid, is swallowed. As the barium moves through the upper gastrointestinal tract, it is monitored by a fluoroscope, and several x rays are taken of the parts being examined.

A procedure called a ***barium enema (BE)*** or ***lower GI series*** is used to examine the colon for cancer, polyps, or inflammation. In this test, barium is inserted into the colon via an enema, under fluoroscopic guidance, and x rays are taken to provide a permanent record of the study.

Following the procedure, it is important to make sure that the barium is expelled rectally within two or three days. The barium can harden and result in obstruction of the intestine.

When both a barium swallow and barium en-

ema are ordered, the barium enema should be done first, because the swallowed barium may interfere with proper visualization on the radiographs.

Physiology of Digestion in the Small Intestine

Mechanical

The movements of the small intestine are divided into two types: segmentation and peristalsis. ***Segmentation*** is the major movement of the small intestine. It is strictly a localized contraction in areas containing food. It mixes chyme with the digestive juices and brings the particles of food into contact with the mucosa for absorption. It does not push the intestinal contents along the tract. Segmentation starts with the contractions of circular muscle fibers in a portion of the small intestine, an action that constricts the intestine into segments. Next, muscle fibers that encircle the middle of each segment also contract, dividing each segment again. Finally, the fibers that contracted first relax, and each small segment unites with an adjoining small segment so that large segments are formed. This sequence of events is repeated 12 to 16 times a minute, sloshing the chyme back and forth (Figure 24-20). Segmentation depends mainly on intestinal distension, which initiates nerve impulses to the central nervous system. Returning parasympathetic impulses increase motility. Sympathetic impulses decrease intestinal motility.

Peristalsis propels the chyme onward through the intestinal tract. Peristaltic contractions in the small intestine are normally very weak compared with those in the esophagus or stomach. Chyme moves through the small intestine at a rate of about 1 cm/min. Thus, chyme remains in the small intestine for three to five hours. Peristalsis, like segmentation, is initiated by distension and controlled by the autonomic nervous system.

Chemical

In the mouth, salivary amylase converts starch (polysaccharide) to maltose (disaccharide). In the stomach, pepsin converts proteins to peptides (small fragments of proteins). Thus, chyme entering the small intestine contains partially digested carbohydrates, partially digested proteins, and essentially undigested lipids. The completion of the digestion of carbohydrates, proteins, and lipids is a collective effort of pancreatic juice, bile, and intestinal juice in the small intestine.

FIGURE 24-20 Segmentation. Localized contractions thoroughly mix the contents of the small intestine.

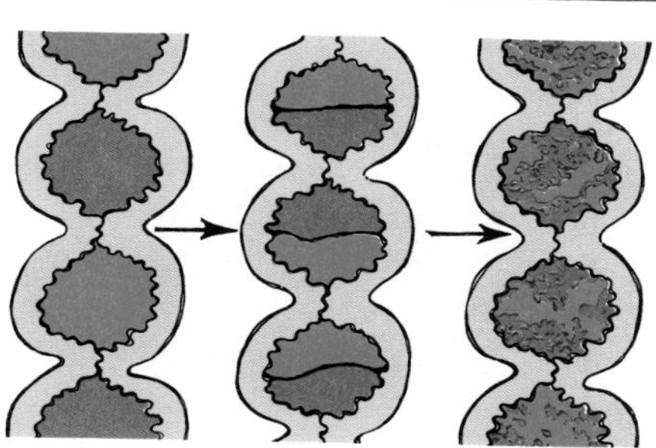

▪ Carbohydrates Even though the action of ***salivary amylase*** may continue in the stomach for some time, its activity is blocked by the acid pH of the stomach which denatures salivary amylase. Thus, few starches are reduced to maltose by the time chyme leaves the stomach. Any starches not already broken down into the disaccharide are converted by ***pancreatic amylase,*** an enzyme in pancreatic juice that acts in the small intestine. Although amylase acts on both glycogen and starches, it does not act on another polysaccharide called cellulose, an indigestible plant fiber.

Sucrose and lactose, two disaccharides, are ingested as such and are not acted upon until they reach the small intestine. Three enzymes in the intestinal juice digest the disaccharides into monosaccharides. ***Maltase*** splits maltose into two molecules of glucose. ***Sucrase*** breaks sucrose into a molecule of glucose and a molecule of fructose. ***Lactase*** digests lactose into a molecule of glucose and a molecule of galactose. This completes the digestion of carbohydrates since these monosaccharides are small enough to be absorbed.

CLINICAL APPLICATION: LACTOSE INTOLERANCE

In some individuals the mucosal cells of the small intestine fail to produce lactase, which is essential for the digestion of lactose. This condition is called *lactose intolerance.* Undigested lactose retains fluid and bacterial fermentation of lactose results in the production of gases, producing the symptoms of lactose intolerance. Its symptoms include diarrhea, gas, bloating, and abdominal cramps following consumption of milk and other dairy products. The severity of symptoms varies, perhaps owing to self-adjustment of diet, from relatively minor to sufficiently serious to require medical attention. Lactose intolerance sometimes occurs following gastric surgery.

▪ Proteins ***Protein*** digestion starts in the stomach, where proteins are fragmented by the action of ***pepsin*** into peptides. Enzymes found in pancreatic juice (trypsin, chymotrypsin, and carboxypeptidase) continue the digestion. ***Trypsin*** and ***chymotrypsin*** continue to break down proteins into peptides. Although pepsin, trypsin, and chymotrypsin all convert whole proteins into peptides, their actions differ somewhat since each splits peptide bonds between different amino acids. ***Carboxypeptidase*** acts on peptides and breaks the peptide bond that attaches the terminal amino acid to the carboxyl (acid) end of the peptide. Protein digestion is completed by the ***peptidases*** produced by small intestinal cells: aminopeptidase and dipeptidase. ***Aminopeptidase*** acts on peptides and breaks the peptide bonds that attach amino acids to the amino end of the peptide. ***Dipeptidase*** splits dipeptides (two amino acids joined by a peptide bond) into amino acids that can be absorbed.

Since trypsin and chymotrypsin act on interior peptide bonds of a protein molecule, they are referred to as ***endopeptidases.*** Carboxypeptidases and aminopeptidases which hydrolize amino acids at the carboxyl end and amino end of proteins, respectively, are known as ***exopeptidases.***

Lipids In an adult, almost all lipid digestion occurs in the small intestine. The first step in the process involves the preparation of neutral fats (triglycerides) by bile salts. Neutral fats, or just simply fats, are the most abundant lipids in the diet. They are called triglycerides because they consist of a molecule of glycerol bonded to three molecules of fatty acid (see Figure 2-10). Bile salts break the globules of fat into droplets about 1 μm in diameter. This process is called ***emulsification.*** It is necessary to increase surface area so that the fat-splitting enzyme can get at more of the lipid molecules. In the second step, ***pancreatic lipase,*** an enzyme found in pancreatic juice, hydrolyzes each fat molecule into fatty acids and monoglycerides, end products of fat digestion. Lipase removes two of the three fatty acids from glycerol; the third remains attached to the glycerol, thus forming monoglycerides.

Nucleic Acids Both intestinal juice and pancreatic juice contain ***nucleases*** that digest nucleotides into their constituent pentoses and nitrogenous bases. ***Ribonuclease*** acts on ribonucleic acid nucleotides, and ***deoxyribonuclease*** acts on deoxyribonucleic acid nucleotides.

A summary of digestive enzymes in terms of source, substrate acted on, and product is presented in Exhibit 24-5.

EXHIBIT 24-5 SUMMARY OF DIGESTIVE ENZYMES

Enzyme	Source	Substrate	Product
Salivary amylase	Salivary glands.	Starches (polysaccharides).	Maltose (disaccharide).
Pepsin (activated from pepsinogen by hydrochloric acid)	Stomach (zymogenic cells).	Proteins.	Peptides.
Pancreatic amylase	Pancreas.	Starches (polysaccharides).	Maltose (disaccharide).
Trypsin (activated from trypsinogen by enterokinase)	Pancreas.	Proteins.	Peptides.
Chymotrypsin (activated from chymotrypsinogen by trypsin)	Pancreas.	Proteins.	Peptides.
Carboxypeptidase (activated from procarboxypeptidase by trypsin)	Pancreas.	Terminal amino acid at carboxyl (acid) end of peptides.	Peptides and amino acids.
Pancreatic lipase	Pancreas.	Neutral fats (triglycerides) that have been emulsified by bile salts.	Fatty acids and monoglycerides.
Maltase	Small intestine.	Maltose.	Glucose.
Sucrase	Small intestine.	Sucrose.	Glucose and fructose.
Lactase	Small intestine.	Lactose.	Glucose and galactose.
Peptidases			
Aminopeptidase	Small intestine.	Terminal amino acids at amino end of peptides.	Amino acids.
Dipeptidase	Small intestine.	Dipeptides.	Amino acids.
Nucleases			
Ribonuclease	Pancreas and small intestine.	Ribonucleic acid nucleotides.	Pentoses and nitrogenous bases.
Deoxyribonuclease	Pancreas and small intestine.	Deoxyribonucleic acid nucleotides.	Pentoses and nitrogenous bases.

Regulation of Intestinal Secretion

The most important means for regulating small intestinal secretion is local reflexes in response to the presence of chyme. Also, secretin and cholecystokinin (CCK) stimulate the production of intestinal juice.

Physiology of Absorption

All the chemical and mechanical phases of digestion from the mouth down through the small intestine are directed toward changing food into forms that can pass through the epithelial cells lining the mucosa into the underlying blood and lymphatic vessels. These forms are monosaccharides (glucose, fructose, and galactose), amino acids, fatty acids, glycerol, and glycerides. Passage of these digested nutrients from the gastrointestinal tract into the blood or lymph is called ***absorption.***

About 90 percent of all absorption of nutrients takes place throughout the length of the small intestine. The other 10 percent occurs in the stomach and large intestine. Any undigested or unabsorbed material left in the small intestine is passed on to the large intestine. Absorption of materials in the small intestine occurs specifically through the villi and depends on diffusion, facilitated diffusion, osmosis, and active transport.

Carbohydrates

Essentially all carbohydrates are absorbed as monosaccharides. Glucose and galactose are transported into epithelial cells of the villi by an active process that is coupled with the active transport of sodium. It appears that the carrier protein has receptor sites for glucose, galactose, and sodium, and unless all three sites are simultaneously filled, none of the substances is transported. Fructose is transported by facilitated diffusion. Transported monosaccharides then move out of the epithelial cells by facilitated diffusion and enter the capillaries of the villi. From here they are transported in the bloodstream to the liver via the hepatic portal system. After their passage through the liver, they move through the heart and then enter general circulation (Figure 24-21).

Proteins

Most proteins are absorbed as amino acids, and the process occurs mostly in the duodenum and jejunum. Amino acid transport into epithelial cells of the villi is an active transport process also coupled with active sodium transport. There appear to be several (perhaps four) different transport systems for amino acids, one each for the various chemical groups of amino acids. Amino acids move out of the epithelial cells by diffusion to enter the bloodstream. They follow the same route as that taken by monosaccharides. At times, dipeptides and tripeptides can be taken in by epithelial cells by active transport. Most of them are hydrolyzed to amino acids in the cells and then passed into the capillaries of the villi (Figure 24-21). From here, the amino acids are transported in the blood to the liver by way of the hepatic portal system. Amino acids not removed by liver cells will then reach the heart and be pumped out through the general circulation.

Lipids

As a result of emulsification and fat digestion, neutral fats (triglycerides) are broken down into monoglycerides and fatty acids. Recall that pancreatic lipase removes two of the three fatty acids from glycerol during fat digestion; the other fatty acid remains attached to glycerol, thus forming monoglycerides. Short-chain fatty acids (those with fewer than 10 to 12 carbon atoms) pass into the epithelial cells by diffusion and follow the same route taken by monosaccharides and amino acids (Figure 24-21).

Most fatty acids are long-chain fatty acids. They and the monoglycerides are transported differently. Bile salts form spherical aggregates called ***micelles*** (mī-SELZ). They are about 2.5 nm in diameter and consist of 20 to 50 molecules of bile salt. Despite their large size, micelles have the ability to dissolve in water in the intestinal fluid. During fat digestion, fatty acids and monoglycerides dissolve in the center of the micelles, and it is in this form that they reach the epithelial cells of the villi. On coming into contact with the surfaces of the epithelial cells, fatty acids and monoglycerides diffuse into the cells, leaving the micelles behind in chyme. The micelles continually repeat this ferrying function. The majority of bile salts in the small intestine are ultimately reabsorbed in the ileum to be returned by the blood to the liver for resecretion. This cycle is called ***enterohepatic circulation.*** Insufficient bile salts, due to obstruction of the biliary ducts or removal of the gallbladder, can result in the loss of up to 40 percent of lipids in feces due to improper lipid absorption. Also, when lipids are not absorbed properly, the fat-soluble vitamins (A, D, E, K) are not adequately absorbed.

Within the epithelial cells, many monoglycerides are further digested by lipase in the cells to glycerol and fatty acids. Then, the fatty acids and glycerol are recombined to form triglycerides in the smooth endoplasmic reticulum of the epithelial cell. The triglycerides, while still in the endoplasmic reticulum, aggregate into globules along with phospholipids and cholesterol and become coated with proteins. These masses are called ***chylomicrons.*** The protein coat keeps the chylomicrons suspended and from sticking to each other. The chylomicrons leave the epithelial cell and enter the lacteal of a villus. From here, they are transported by way of lymphatic vessels to the thoracic duct and enter the cardiovascular system at the left subclavian vein. Finally, they arrive at the liver through the hepatic artery (Figure 24-21).

Within a few hours after consuming a meal that contains a large amount of fat, most chylomicrons are removed from blood as they pass through blood capillaries in the liver and adipose tissue. This is accomplished by an en-

FIGURE 24-21 Absorption. (a) Movement of digested nutrients through epithelial cells of the villi. For simplicity, all digested foods are shown in the lumen of the small intestine, even though some nutrients are digested within or on the surfaces of small intestinal epithelial cells.

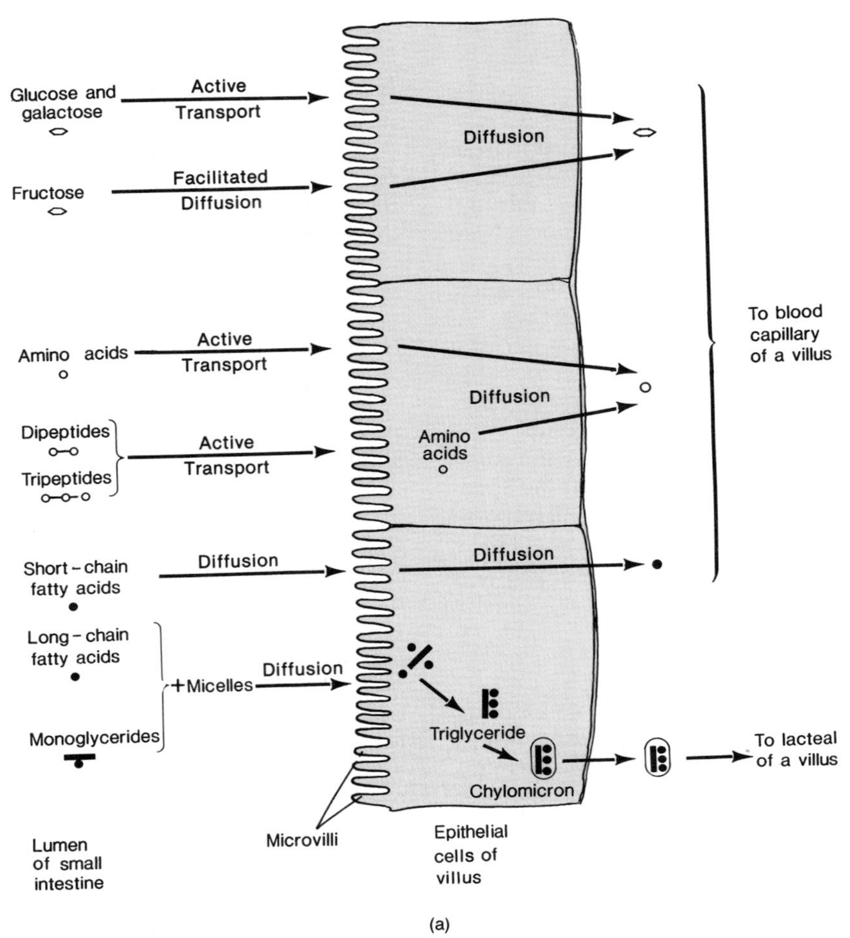

zyme called ***lipoprotein lipase*** found in capillary endothelial cells. The enzyme breaks down triglycerides in chylomicrons into fatty acids and glycerol. The fatty acids diffuse into liver and fat cells and recombine with glycerol produced by the cells to re-form triglycerides.

The plasma lipids—fatty acids, triglycerides, cholesterol—are insoluble in water and body fluids. In order to be transported in blood and utilized by body cells, the lipids must be combined with protein transporters, called ***apoproteins*** to make them soluble, and the combination of lipid and protein is referred to as a ***lipoprotein.*** Most lipoproteins are synthesized in the liver and all contain triglycerides, phospholipids, cholesterol, and protein in varying combinations. In addition to chylomicrons, there are several other types of lipoproteins, known as ***high-density lipoproteins (HDLs), low-density lipoproteins (LDLs),*** and ***very low-density lipoproteins (VLDLs).*** The function of VLDLs is to transport triglycerides synthesized in the liver to adipose tissue and muscle tissue where the VLDLs are catabolized and the triglycerides are deposited. The products of VLDL catabolism are LDLs which are rich in cholesterol with some phospholipids. LDLs contain 60 to 70 percent of the total serum cholesterol. The function of LDLs is to transport cholesterol to peripheral tissues, where it is used for activities such as steroid hormone synthesis, bile acid production, and manufacture of cell membranes. High LDL levels are associated with the development of atherosclerosis since some of the cholesterol may be deposited in arteries ("bad" cholesterol). HDLs are rich in phospholipids and cholesterol and transport 20 to 30 percent of the total serum cholesterol. Their function is to transport cholesterol from peripheral tissues to the liver where some of it is catabolized to become a component of bile (bile salts). High HDL levels are associated with a decreased risk of cardiovascular disease since they may remove cholesterol from arterial walls ("good" cholesterol) and do not deposit cholesterol in arterial walls.

FIGURE 24-21 (*Continued*) (b) Movement of digested nutrients into the cardiovascular and lymphatic systems.

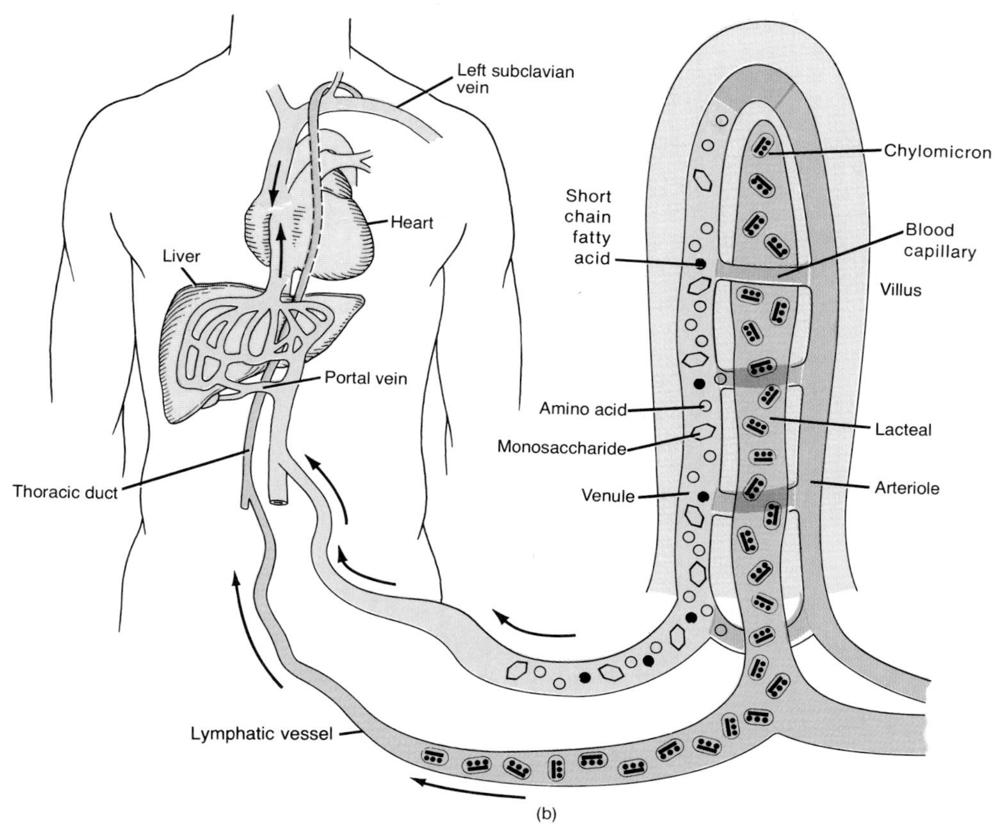

Water

The total volume of fluid that enters the small intestine each day is about 9 liters (about 9 qt). This fluid is derived from ingestion of liquids (about 1.5 liters) and from various gastrointestinal secretions (about 7.5 liters). Roughly 8 to 8.5 liters of the fluid in the small intestine is absorbed; the remainder, about 0.5 to 1.0 liter, passes into the large intestine. There, most of the rest of it is also absorbed.

The absorption of water by the small intestine occurs by osmosis from the lumen of the small intestine through epithelial cells and into the blood capillaries in the villi. The normal rate of absorption is about 200 to 400 ml/hour. Water can move across the intestinal mucosa in both directions. The absorption of water from the small intestine is associated with the absorption of electrolytes and digested foods in order to maintain an osmotic balance with the blood. The absorbed solutes establish a concentration gradient for water that allows the water to be absorbed by osmosis.

Electrolytes

The electrolytes absorbed by the small intestine are mostly constituents of gastrointestinal secretions. Some are also components of ingested foods and liquids. Sodium is able to move in and out of epithelial cells by diffusion. It can also move into mucosal cells by active transport for removal from the small intestine. Chloride, iodide, and nitrate ions can passively follow sodium ions or be actively transported. Calcium ions are also actively transported, and their movement depends on parathyroid hormone (PTH) and vitamin D. Other electrolytes such as iron, potassium, magnesium, and phosphate also move by active transport.

Vitamins

Fat-soluble vitamins (A, D, E, and K) are absorbed along with ingested dietary fats in micelles. In fact, they cannot be absorbed unless they are ingested with some fat. Most water-soluble vitamins, such as the B vitamins and C, are absorbed by diffusion. Vitamin B_{12}, you may recall, requires combination with intrinsic factor produced by the stomach for its absorption.

A summary of the digestive and absorptive activities of the small intestine is presented in Exhibit 24-6.

EXHIBIT 24-6 SUMMARY OF DIGESTION AND ABSORPTION IN THE SMALL INTESTINE

Structure	Activity
Pancreas	Delivers pancreatic juice into the duodenum via the pancreatic duct (see Exhibit 24-5) for pancreatic enzymes and their functions.
Liver	Produces bile (bile salts), necessary for emulsification of fats.
Gallbladder	Stores, concentrates, and delivers bile into the duodenum via the common bile duct.
Small intestine	
Mucosa and submucosa	
Intestinal glands	Secrete intestinal juice (see Exhibit 24-5 for intestinal enzymes and their functions).
Duodenal (Brunner's) glands	Secrete mucus for protection and lubrication.
Microvilli	Fingerlike projections of epithelial cells that increase surface area for absorption and digestion.
Villi	Projections of mucosa that are the sites of absorption of digested food and also increase the surface area for absorption and digestion.
Circular folds	Folds of mucosa and submucosa that increase surface area for absorption and digestion.
Muscularis	
Segmentation	Consists of alternating contractions of circular fibers that produce segmentation and resegmentation of portions of the small intestine; mixes chyme with digestive juices and brings food into contact with the mucosa for absorption.
Peristalsis	Consists of mild waves of contraction and relaxation of circular and longitudinal muscle passing the length of the small intestine; moves chyme toward ileocecal sphincter.

LARGE INTESTINE

The overall functions of the large intestine are the completion of absorption, the manufacture of certain vitamins, the formation of feces, and the expulsion of feces from the body.

Anatomy

The ***large intestine*** is about 1.5 m (5 ft) in length and averages 6.5 cm (2.5 inches) in diameter. It extends from the ileum to the anus and is attached to the posterior abdominal wall by its ***mesocolon*** of visceral peritoneum. Structurally, the large intestine is divided into four principal regions: cecum, colon, rectum, and anal canal (Figure 24-22a).

The opening from the ileum into the large intestine is guarded by a fold of mucous membrane called the ***ileocecal sphincter (valve).*** This structure allows materials from the small intestine to pass into the large intestine. Hanging below the ileocecal valve is the ***cecum,*** a blind pouch about 6 cm (2.5 inches) long. Attached to the cecum is a twisted, coiled tube, measuring about 8 cm (3 inches) in length, called the ***vermiform appendix*** (*vermis* = worm; *appendix* = appendage). The visceral peritoneum of the appendix, called the ***mesoappendix,*** attaches the appendix to the inferior part of the ileum and adjacent part of the posterior abdominal wall.

The open end of the cecum merges with a long tube called the ***colon*** (*kolon* = food passage). The colon is divided into ascending, transverse, descending, and sigmoid portions. The ***ascending colon*** ascends on the right side of the abdomen, reaches the undersurface of the liver, and turns abruptly to the left. Here it forms the ***right colic (hepatic) flexure.*** The colon continues across the abdomen to the left side as the ***transverse colon.*** It curves beneath the lower end of the spleen on the left side as the ***left colic (splenic) flexure*** and passes downward to the level of the iliac crest as the ***descending colon.*** The ***sigmoid colon*** begins near the left iliac crest, projects inward to the midline, and terminates as the rectum at about the level of the third sacral vertebra.

The ***rectum,*** the last 20 cm (8 inches) of the GI tract, lies anterior to the sacrum and coccyx. The terminal 2 to 3 cm (1 inch) of the rectum is called the ***anal canal*** (Figure 24-22b). The mucous membrane of the anal canal is arranged in longitudinal folds called ***anal columns*** that contain a network of arteries and veins. The opening of the anal canal to the exterior is called the ***anus.*** It is guarded by an internal sphincter of smooth muscle (in-

FIGURE 24-22 Large intestine. (a) Anatomy of the large intestine. (b) Anal canal seen in longitudinal section.

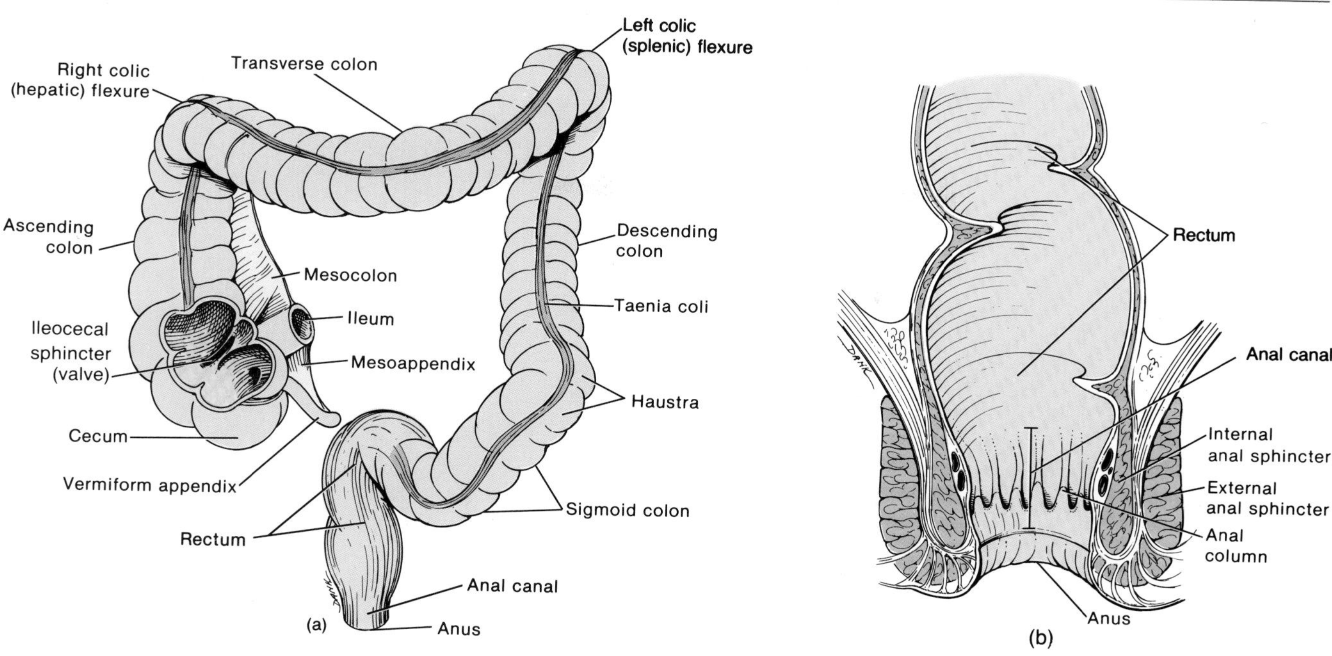

voluntary) and an external sphincter of skeletal muscle (voluntary). Normally the anus is closed except during the elimination of the wastes of digestion.

The medical specialty that deals with the diagnosis and treatment of disorders of the rectum and anus is called ***proctology*** (prok-TOL-ō-jē; *proct* = rectum; *logos* = study of).

CLINICAL APPLICATION: HEMORRHOIDS

As noted in Chapter 21, varicosities in any veins involve inflammation and enlargement. Varicosities of the rectal veins are known as ***hemorrhoids* (*piles*).** Hemorrhoids develop when the veins are put under pressure and become engorged with blood. If the pressure continues, the wall of the vein stretches. Such a distended vessel oozes blood, and bleeding or itching are usually the first signs that a hemorrhoid has developed. Stretching of a vein also favors clot formation, further aggravating swelling and pain. Initially contained within the anus (first degree), they gradually enlarge until they prolapse or extend outward on defecation (second degree) and finally remain prolapsed through the anal orifice (third degree). Hemorrhoids may be caused by constipation. Repeated straining during defecation forces blood down into the superior hemorrhoidal plexus, increasing pressure in these veins. Constipation is related to low-fiber diets, especially in North America. One hypothesis links increased intraabdominal pressure, caused by straining during evacuation of firm feces, directly to hemorrhoids or varicose veins. The most widely used treatment for hemorrhoids that bleed consistently and cause severe pain and discomfort is ***rubber band ligation,*** in which a rubber band is tied around the hemorrhoid, cutting off its blood supply. In a few days, the hemorrhoid dries up and falls off. An even newer treatment is called ***infrared photocoagulation,*** in which a very high-energy light beam coagulates the hemorrhoid. (A laser can be used for the same purpose.) Other treatments include sclerotherapy (injection of a substance that causes atrophy of the hemorrhoid) and surgery.

Histology

The wall of the large intestine differs from that of the small intestine in several respects. No villi or permanent circular folds are found in the mucosa, which does, however, contain simple columnar epithelium with numerous goblet cells (Figure 24-23). The columnar cells function primarily in water absorption. The goblet cells secrete mucus that lubricates the colonic contents as they pass through the colon. Both columnar and mucous cells are located in long, straight, tubular intestinal glands that extend the full thickness of the mucosa. Solitary lymphatic nodules are also found in the mucosa. The submucosa of the large intestine is similar to that found in the rest of

FIGURE 24-23 **Histology of the large intestine. (a) Diagram showing four principal layers. (b) Diagram of enlarged aspect of mucosa.**

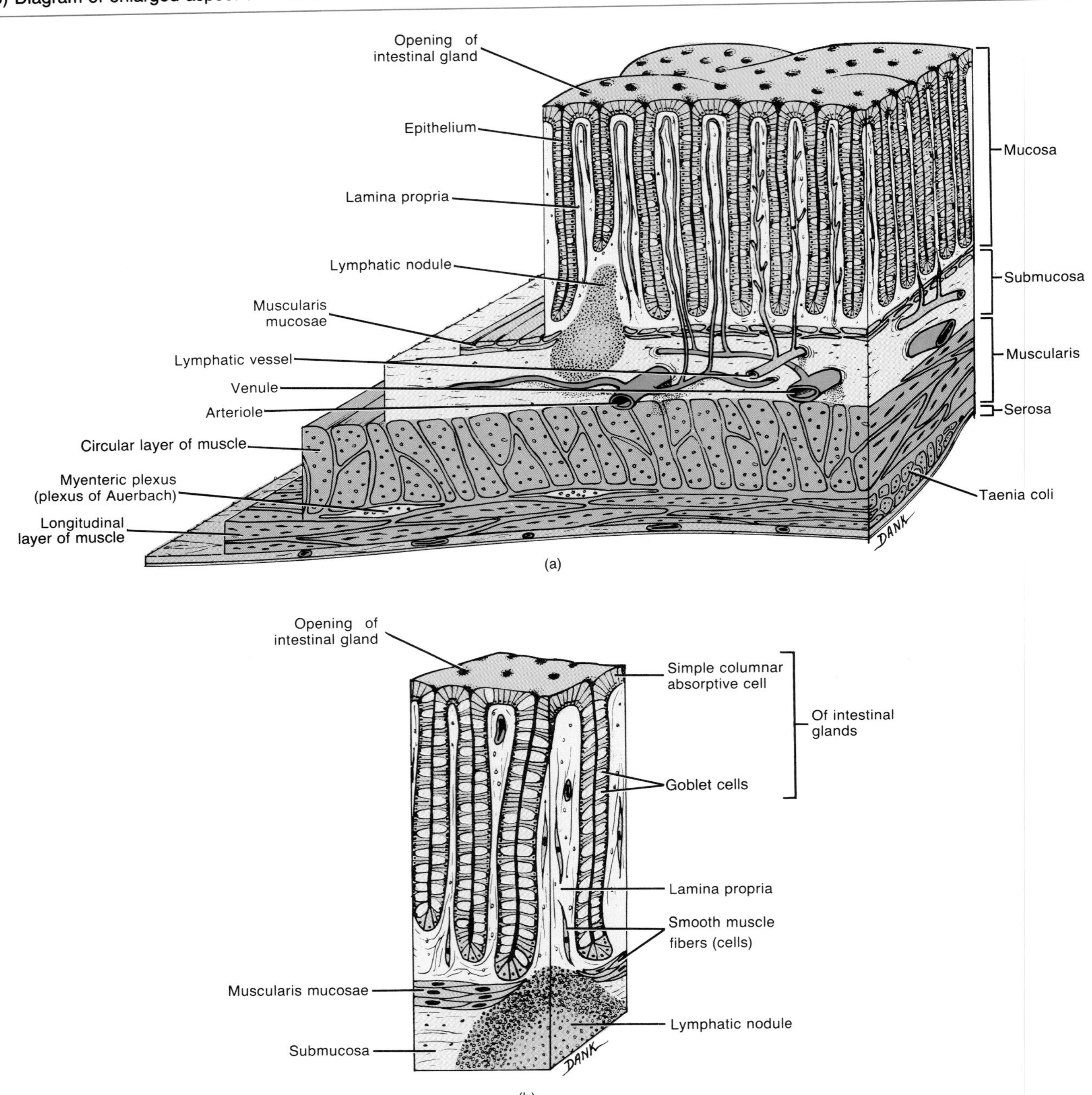

the gastrointestinal tract. The muscularis consists of an external layer of longitudinal muscles and an internal layer of circular muscles. Unlike other parts of the gastrointestinal tract, portions of the longitudinal muscles are thickened, forming three conspicuous longitudinal bands referred to as ***taeniae coli*** (TĒ-nē-a KŌ-lī). Each band runs the length of most of the large intestine (see Figure 24-22a). Tonic contractions of the bands gather the colon into a series of pouches called ***haustra*** (HAWS-tra; *haustrum* = shaped like a pouch), which give the colon its puckered appearance. The serosa of the large intestine is part of the visceral peritoneum. Small pouches of visceral peritoneum filled with fat are attached to taeniae coli and are called ***epiploic appendages.***

FIGURE 24-23 (*Continued*) (c) Photomicrograph of a portion of the wall of the large intestine at a magnification of 2000 ×. (Kuntzman.) (d) Scanning electron micrograph of an enlarged aspect of the mucosa of the large intestine at a magnification of 2000×. (Copyright © 1983 by Michael H. Ross. Used by permission.)

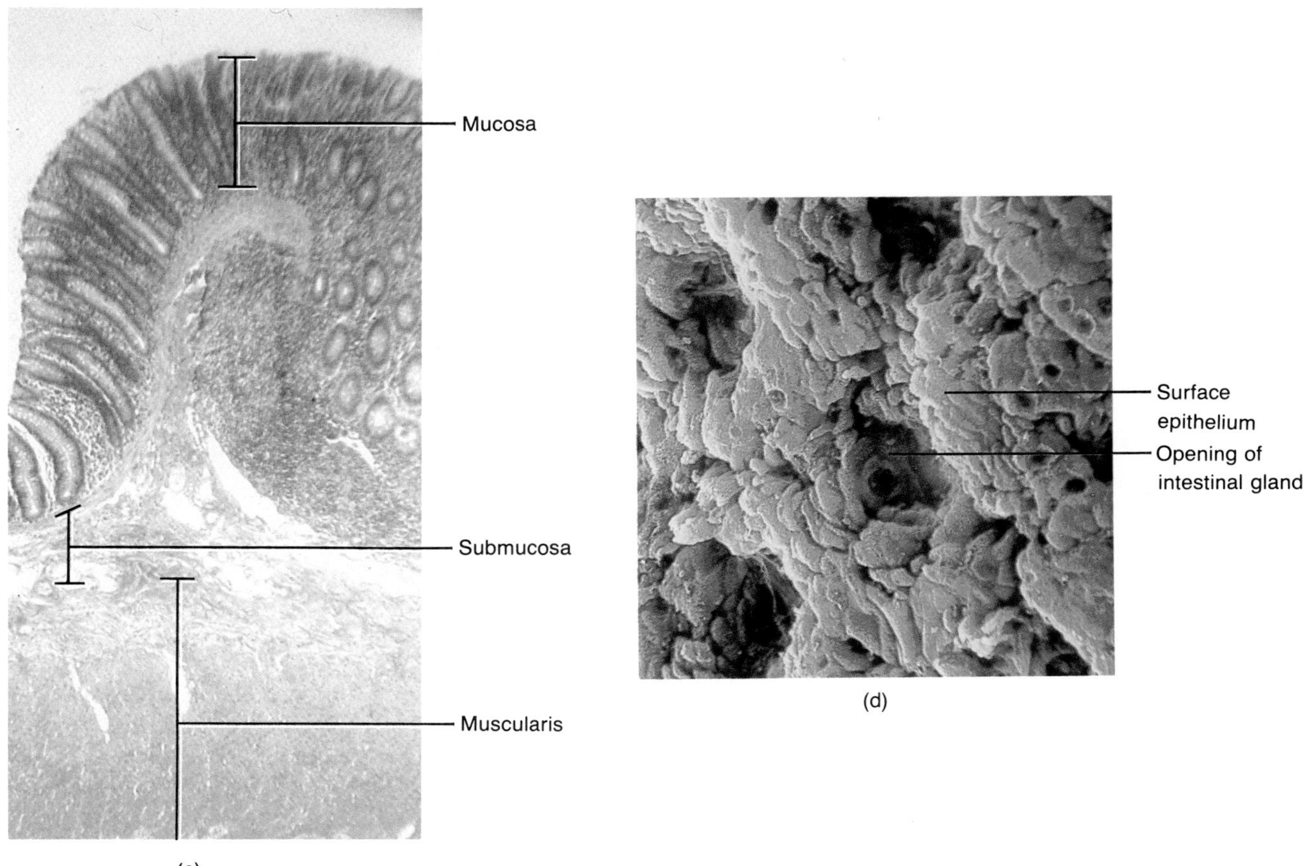

Physiology of Digestion in the Large Intestine

Mechanical

The passage of chyme from the ileum into the cecum is regulated by the action of the ileocecal sphincter. The valve normally remains mildly contracted so that the passage of chyme into the cecum is usually a slow process. Immediately following a meal, there is a ***gastroileal reflex*** in which ileal peristalsis is intensified and any chyme in the ileum is forced into the cecum. The hormone stomach gastrin also relaxes the sphincter. Whenever the cecum is distended, the degree of contraction of the ileocecal sphincter is intensified.

Movements of the colon begin when substances enter through the ileocecal sphincter. Since chyme moves through the small intestine at a fairly constant rate, the time required for a meal to pass into the colon is determined by gastric evacuation time. As food passes through the ileocecal sphincter, it fills the cecum and accumulates in the ascending colon.

One movement characteristic of the large intestine is ***haustral churning.*** In this process, the haustra remain relaxed and distended while they fill up. When the distension reaches a certain point, the walls contract and squeeze the contents into the next haustrum. ***Peristalsis*** also occurs, although at a slower rate than in other portions of the tract (3 to 12 contractions per minute). A final type of movement is ***mass peristalsis,*** a strong peristaltic wave that begins at about the middle of the transverse colon and drives the colonic contents into the rectum. Food in the stomach initiates this reflex action in the colon. Thus, mass peristalsis usually takes place three or four times a day, during a meal or immediately after.

Chemical

The last stage of digestion occurs through bacterial action, not enzymes secreted by the colon. Mucus is secreted by the glands of the large intestine, but no enzymes are secreted. Chyme is prepared for elimination by the action of bacteria. These bacteria ferment any remaining carbohydrates and release hydrogen, carbon dioxide, and methane gas. These gases contribute to flatus (gas) in the colon. They also convert remaining proteins to amino acids and break down the amino acids into simpler substances: indole, skatole, hydrogen sulfide, and fatty acids. Some of the indole and skatole is carried off in the feces and contributes to their odor. The rest are absorbed and transported to the liver, where they are converted to less toxic compounds and excreted in the urine. Bacteria also decompose bilirubin to simpler pigments (urobilinogen), which give feces their brown color. Several vitamins needed for normal metabolism, including some B vitamins and vitamin K, are synthesized by bacterial action and absorbed.

Absorption and Feces Formation

By the time the chyme has remained in the large intestine 3 to 10 hours, it has become solid or semisolid as a result of absorption principally of water and is now known as ***feces.*** Chemically, feces consist of water, inorganic salts, sloughed-off epithelial cells from the mucosa of the gastrointestinal tract, bacteria, products of bacterial decomposition, and undigested parts of food.

Although most water absorption occurs in the small intestine, the large intestine absorbs enough to make it an important organ in maintaining the body's water balance. Of the 0.5 to 1.0 liter that enters the large intestine, all but about 100 ml is absorbed. The absorption is greatest in the cecum and ascending colon. The large intestine also absorbs electrolytes, including sodium and chloride, and some vitamins.

Physiology of Defecation

Mass peristaltic movements push fecal material from the sigmoid colon into the rectum. The resulting distension of the rectal wall stimulates pressure-sensitive receptors, initiating a reflex for ***defecation,*** the emptying of the rectum. The defecation reflex occurs as follows. In response to distension of the rectal wall, the receptors send nerve impulses to the sacral spinal cord. Motor impulses from the cord travel along parasympathetic nerves back to the descending colon, sigmoid colon, rectum, and anus. Contraction of the longitudinal rectal muscles shortens the rectum, thereby increasing the pressure inside it. The pressure along with voluntary contractions of the diaphragm and abdominal muscles forces the internal sphincter open, and the feces are expelled through the anus. The external sphincter is voluntarily controlled. If it is voluntarily relaxed, defecation occurs; if it is voluntarily constricted, defecation can be postponed. Voluntary contractions of the diaphragm and abdominal muscles aid defecation by increasing the pressure inside the abdomen, which pushes the walls of the sigmoid colon and rectum inward. If defecation does not occur, the feces back into the sigmoid colon until the next wave of mass peristalsis again stimulates the pressure-sensitive receptors, creating the desire to defecate.

In infants, the defecation reflex causes automatic emptying of the rectum without the voluntary control of the external anal sphincter. In certain instances of spinal cord injury, the reflex is abolished and defecation requires supportive measures, such as cathartics (laxatives).

Diarrhea refers to frequent defecation of liquid feces caused by increased motility of the intestines. Since chyme passes too quickly through the small intestine and feces pass too quickly through the large intestine, there is not enough time for absorption. Like vomiting, diarrhea can result in dehydration and electrolyte imbalances. Diarrhea may be caused by stress and microbes that irritate the gastrointestinal mucosa.

Constipation refers to infrequent or difficult defecation. It is caused by decreased motility of the intestines in which feces remain in the colon for prolonged periods of time. As it does so, there is considerable water absorption, and feces become dry and hard. Constipation may be caused by improper bowel habits, spasms of the colon, insufficient bulk in the diet, inadequate fluid intake, lack of exercise, and emotions. Usual treatment for constipation is a mild cathartic, such as milk of magnesia, that induces defecation. However, many physicians maintain that laxatives are habit-forming, and that adding bulk to the diet, increasing one's amount of exercise, and improving fluid intake are safer ways of controlling this common problem.

Activities of the large intestine are summarized in Exhibit 24-7.

Screening for Colorectal Cancer

Colorectal cancer, described later, may be diagnosed by several methods. Among these are fecal occult blood testing, digital rectal examination, sigmoidoscopy, colonoscopy, and barium enema. At this point, three of the screening tests will be described: fecal occult blood testing, sigmoidoscopy, and colonoscopy.

MEDICAL TESTS

Fecal occult blood testing (occult blood is hidden; not detectable by the human eye)

Diagnostic Value: To screen for the slow bleeding that may be a sign of colorectal cancer and to detect blood from other areas of the gastrointestinal tract.

Procedure: Several tests are available for at-home use to detect small amounts of blood in the feces.

EXHIBIT 24-7 DIGESTIVE ACTIVITIES OF LARGE INTESTINE

Structure	Action	Function
Mucosa	Secretes mucus.	Lubricates colon and protects mucosa.
	Absorbs water and other soluble compounds.	Maintains water balance; solidifies feces. Vitamins and electrolytes absorbed and toxic substances sent to liver to be detoxified.
	Bacterial activity.	Breaks down undigested carbohydrates, proteins, and amino acids into products that can be expelled in feces or absorbed and detoxified by liver. Certain B vitamins and vitamin K synthesized.
Muscularis	Haustral churning.	Contents moved from haustrum to haustrum by muscular contractions.
	Peristalsis.	Contents moved along length of colon by contractions of circular and longitudinal muscles.
	Mass peristalsis.	Contents forced into sigmoid colon and rectum by strong peristaltic wave.
	Defecation.	Feces eliminated by contractions in sigmoid colon and rectum.

The tests, which usually require three stool specimens, are based on color changes when reagents are added to feces. Certain colors indicate the presence of blood; other colors indicate its absence. (Urine may also be tested at home for occult blood by using dip-and-read reagent strips.)

Sigmoidoscopy (sig'-moy-DOS-kō-pē; *skopein* = to examine) and **colonoscopy** (kō'-lon-OS-kō-pē)

Diagnostic Value: To screen for colon cancer based on symptoms such as unexplained rectal bleeding, change in bowel habits, lower abdominal pain, and abnormalities found on a barium enema, or to follow the progress of polyps or inflammations.

Procedure: After the colon has been evacuated of its contents by taking a laxative and enema, the patient is usually sedated. A lighted viewing instrument is inserted into the rectum and slowly advanced into the colon. (An anoscope is a short tube used to examine the anal canal. A proctoscope is slightly longer and is used to examine the anal canal and rectum. A rigid sigmoidoscope is about 30 cm [12 inches] long and is used to examine the rectum and a portion of the lower colon. A flexible fiberoptic sigmoidoscope allows a more complete view of the colon. A colonoscope permits visual inspection of the entire colon.) Sometimes air is introduced via the scope to clear a path, permit better viewing, and to help the physician to locate the lumen of the colon. Also, a suction device may be used to remove feces, mucus, or blood to improve the visual field. It is also possible during a sigmoidoscopy or colonoscopy to collect biopsy samples, to remove polyps, to gather specimens for culture, and to photograph the intestinal mucosa.

Carcinoembryonic (car'-sin-ō-em-brē-ON-ik) ***antigen (CEA)*** is a glycoprotein that is secreted by normally developing fetal tissue during the first or second trimester. Its secretion stops before birth. Secretion of CEA may begin if certain malignant conditions (neoplasm of the colon, pancreas, stomach, lung, and breast) and benign conditions (chronic pulmonary disease, alcoholic cirrhosis, hepatitis, and irritable bowel disease) develop. Although CEA is not specifically diagnostic for cancer, it is useful for monitoring treatment and recurrence of colorectal cancer.

AGING AND THE DIGESTIVE SYSTEM

Overall general changes associated with aging of the digestive system include decreasing secretory mechanisms, decreasing motility (muscular movement) of the digestive organs, loss of strength and tone of the muscular tissue and its supporting structures, changes in neurosensory feedback regarding enzyme and hormone release, and diminished response to pain and internal sensations. Specific changes include reduced sensitivity to mouth irritations and sores, loss of taste, pyorrhea, difficulty in swallowing, hiatal hernia, cancer of the esophagus, gastritis, peptic ulcer, and gastric cancer. Changes in the small intestine include duodenal ulcers, appendicitis, malabsorption, and maldigestion. Other pathologies that increase in incidence are gallbladder problems, jaundice, cirrhosis, and acute pancreatitis. Large intestinal changes such as constipation, cancer of the colon or rectum, hemorrhoids, and diverticular disease of the colon also occur.

DEVELOPMENTAL ANATOMY OF THE DIGESTIVE SYSTEM

About the fourteenth day after fertilization, the cells of the endoderm form a cavity referred to as the ***primitive gut*** (Figure 24-24). Soon after the mesoderm forms and splits into two layers (somatic and splanchnic), the splanchnic mesoderm associates with the endoderm of

FIGURE 24-24 **Development of the digestive system.**

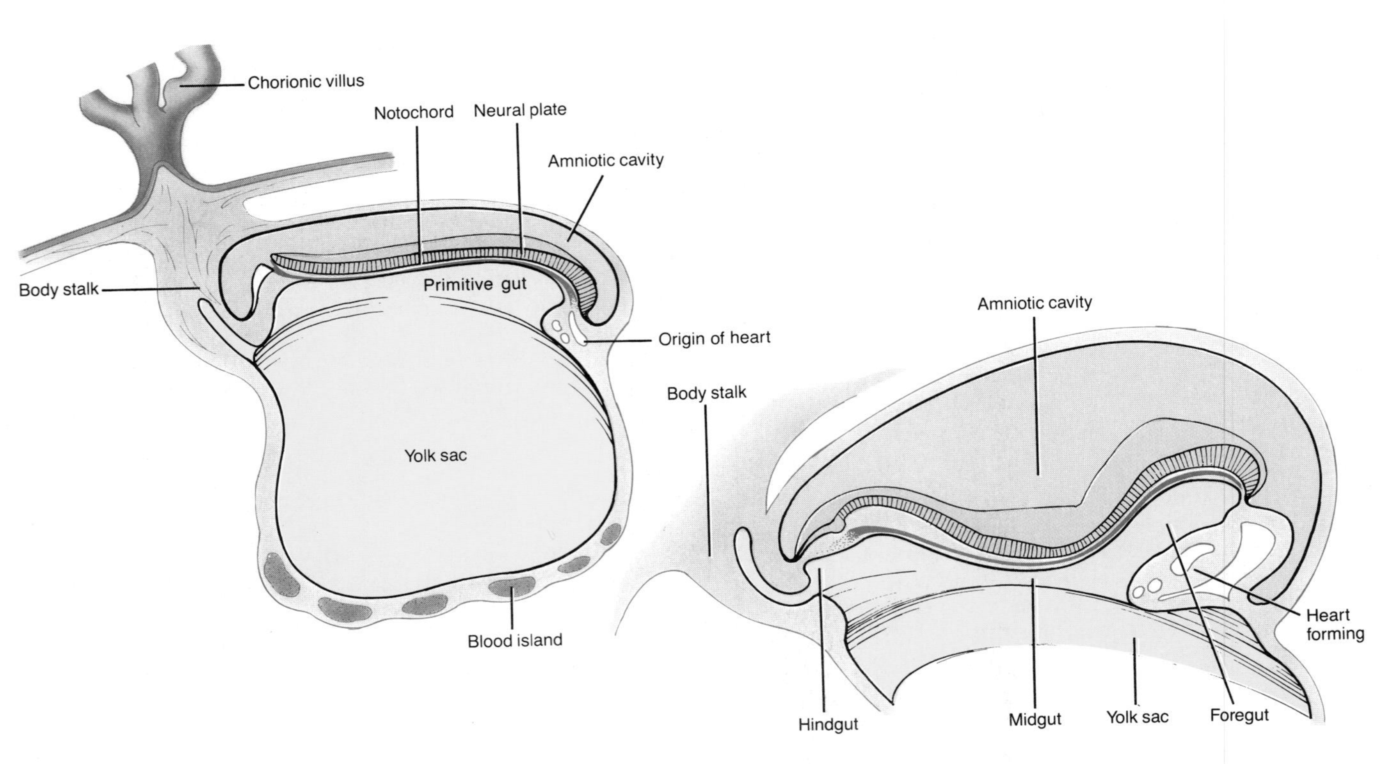

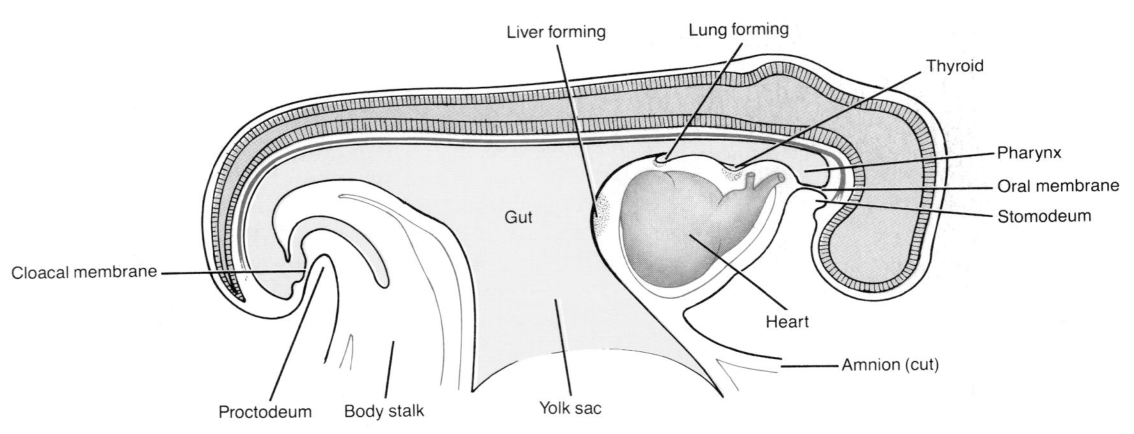

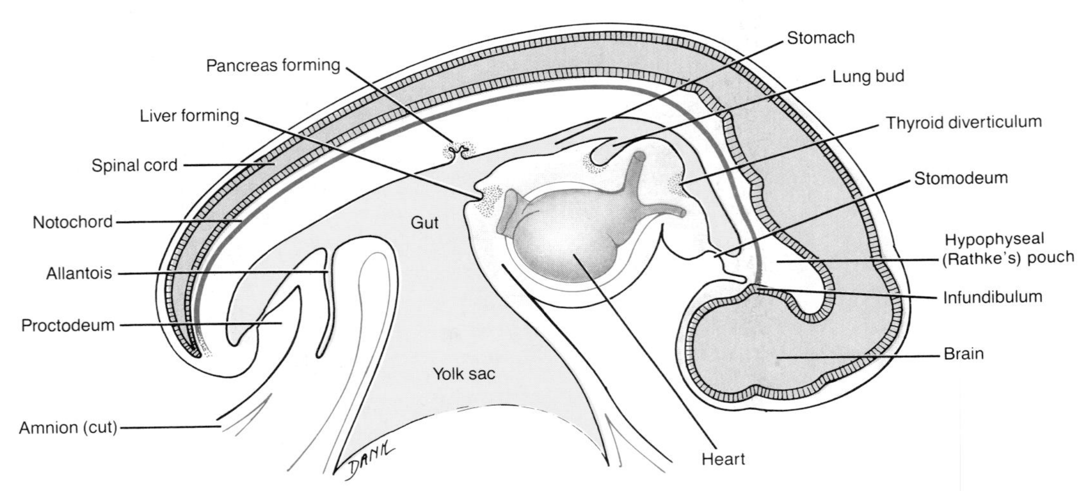

the primitive gut. Thus, the primitive gut has a double-layered wall. The ***endodermal layer*** gives rise to the *epithelial lining* and *glands* of most of the gastrointestinal tract, and the ***mesodermal layer*** produces its *smooth muscle* and *connective tissue*.

The primitive gut elongates, and about the latter part of the third week, it differentiates into an anterior ***foregut,*** a central ***midgut,*** and a posterior ***hindgut.*** Until the fifth week of development, the midgut opens into the yolk sac. After that time, the yolk sac constricts, detaches from the midgut, and the midgut seals. In the region of the foregut, a depression consisting of ***ectoderm,*** the ***stomodeum,*** appears. This develops into the *oral cavity*. The ***oral membrane*** that separates the foregut from the stomodeum ruptures during the fourth week of development, so that the foregut is continuous with the outside of the embryo through the oral cavity. Another depression consisting of ***ectoderm,*** the ***proctodeum,*** forms in the hindgut and goes on to develop into the *anus*. The ***cloacal membrane,*** which separates the hindgut from the proctodeum, ruptures, so that the hindgut is continuous with the outside of the embryo through the anus. Thus, the gastrointestinal tract forms a continuous tube from the mouth to the anus.

The foregut develops into the *pharynx, esophagus, stomach,* and a *portion of the duodenum*. The midgut is transformed into the *remainder of the duodenum,* the *jejunum,* the *ileum,* and *portions of the large intestine* (cecum, appendix, ascending colon, and most of the transverse colon). The hindgut develops into the *remainder of the large intestine,* except for a portion of the anal canal that is derived from the proctodeum.

As development progresses, the endoderm at various places along the foregut develops into hollow buds that grow into the mesoderm. These buds will develop into the *salivary glands, liver, gallbladder,* and *pancreas*. Each of the glands retains a connection with the gastrointestinal tract through ducts.

DISORDERS: HOMEOSTATIC IMBALANCES

Dental Caries

Dental caries, or tooth decay, involve a gradual demineralization (softening) of the enamel and dentin. If untreated, various microorganisms may invade the pulp, causing inflammation and infection with subsequent death (necrosis) of the pulp and abscess of the alveolar bone surrounding the root's apex. Such teeth are treated by root canal therapy.

The process of dental caries is initiated when bacteria act on sugars, giving off acids that demineralize the enamel. Microbes that digest sugar into lactic acid are common in the mouth cavity. One that seems to be cariogenic (caries causing) is the bacterium *Streptococcus mutans.* ***Dextran,*** a sticky polysaccharide produced from sucrose, forms a capsule around the bacteria, causing them to stick to the teeth. Masses of bacterial cells, dextran, and other debris adhering to teeth are collectively called ***dental plaque.*** Saliva cannot reach the tooth surface to buffer the acid because the plaque covers the teeth. Brushing the teeth immediately after eating removes the plaque from flat surfaces before the bacteria have a chance to go to work. Dentists also suggest that the plaque between the teeth be removed every 24 hours with dental floss or by flushing with a water irrigation device or by using an antiplaque dental rinse.

Preventive measures other than brushing, flossing, and irrigation include prenatal diet supplements (chiefly vitamin D, calcium, and phosphorus), fluoride treatments to protect against acids during the period when teeth are being calcified, and dental sealing. In this last procedure, pits and fissures that serve as reservoirs for dental plaque are sealed by the application of a permanent, durable plastic sealant. The sealant is applied to the prepared biting surfaces of the molar teeth. Dental sealants are used primarily for children and adolescents, although some adults could benefit from them.

Periodontal Disease

Periodontal disease is a collective term for a variety of conditions characterized by inflammation and degeneration of the gingivae, alveolar bone, periodontal ligament, and cementum. One such condition is called ***pyorrhea.*** The initial symptoms are enlargement and inflammation of the soft tissue and bleeding gums. Without treatment, the soft tissue may deteriorate and the alveolar bone may be resorbed, causing loosening of the teeth and recession of the gums.

Periodontal diseases are frequently caused by poor oral hygiene; by local irritants, such as bacteria, impacted food, and cigarette smoke; or by a poor "bite." Of the several types of bacteria implicated in periodontal disease, *Bacteroides gingivalis* is receiving a great deal of attention. Periodontal diseases may also be caused by allergies, vitamin deficiencies (especially vitamin C), and a number of systemic disorders, especially those that affect bone, connective tissue, or circulation.

Several new technologies and procedures are being used to treat periodontal disease. New synthetic materials (calcium phosphate salts and beta tricalcium phosphate), similar to mineral crystals in natural bone, are used to rebuild parts of destroyed alveolar bone. An experimental technique called guided tissue regeneration is being employed to build back connective tissues that attach the tooth to underlying bone.

Peritonitis

Peritonitis is an acute inflammation of the serous membrane lining the abdominal cavity and covering the ab-

DISORDERS: HOMEOSTATIC IMBALANCES (*continued*)

dominal viscera. One possible cause is contamination of the peritoneum by pathogenic bacteria from the external environment. This contamination could result from accidental or surgical wounds in the abdominal wall or from perforation or rupture of organs with consequent exposure to the outside environment. Another possible cause is perforation of the walls of organs that contain bacteria or chemicals beneficial to the organ but toxic to the peritoneum. For example, the large intestine contains colonies of bacteria that live on undigested nutrients and break them down so they can be eliminated. But if the bacteria enter the peritoneal cavity, they attack the cells of the peritoneum for food and produce acute infection. Moreover, the peritoneum has no natural barriers that keep it from being irritated or digested by chemical substances such as bile and digestive enzymes.

Although it contains a great deal of lymphatic tissue and can combat infection fairly well, the peritoneum is in contact with most of the abdominal organs. If infection gets out of hand, it may destroy vital organs and bring on death. For these reasons, perforation of the gastrointestinal tract from an ulcer is considered serious. A surgeon planning to do extensive surgery on the colon may give the patient high doses of antibiotics for several days prior to surgery to kill intestinal bacteria and reduce the risk of peritoneal contamination.

Peptic Ulcers

An ***ulcer*** is a craterlike lesion in a membrane. Ulcers that develop in areas of the gastrointestinal tract exposed to acid gastric juice are called ***peptic ulcers.*** Peptic ulcers occasionally develop in the lower end of the esophagus, but most occur on the lesser curvature of the stomach, where they are called ***gastric ulcers,*** or in the first part of the duodenum, where they are called ***duodenal ulcers.*** Most peptic ulcers are duodenal.

Hypersecretion of acid gastric juice seems to be the immediate cause of duodenal ulcers. In gastric ulcer patients, because the stomach wall is highly adapted to resist gastric juice through the secretion of mucus, the cause may be hyposecretion of mucus. Hypersecretion of pepsin also may contribute to ulcer formation.

Among the factors believed to stimulate an increase in acid secretion are emotions, cigarette smoking, certain foods or medications (alcohol, coffee, aspirin), and overstimulation of the vagus (X) nerve. Normally, the mucous membrane lining the stomach and duodenal walls resists the secretions of hydrochloric acid and pepsin. In some people, however, this resistance breaks down and an ulcer develops. Some evidence suggests that peptic ulcers may be caused by the bacterium *Campylobacter pyloridis.*

The most common complication of peptic ulcers is bleeding. Another is perforation, erosion of the ulcer all the way through the wall of the stomach or duodenum. Perforation allows bacteria and partially digested food to pass into the peritoneal cavity, producing peritonitis. A third complication is obstruction.

Appendicitis

Appendicitis is an inflammation of the vermiform appendix. It is preceded by obstruction of the lumen of the appendix by fecal material, inflammation, a foreign body, carcinoma of the cecum, stenosis, or kinking of the organ. The infection that follows may result in edema, ischemia, gangrene, and perforation. Rupture of the appendix develops into peritonitis. Loops of the intestines, the omentum, and the parietal peritoneum may become adherent and form an abscess, either at the site of the appendix or elsewhere in the abdominal cavity.

Typically, appendicitis begins with referred pain in the umbilical region of the abdomen, followed by anorexia (lack or loss of appetite for food), nausea, and vomiting. After several hours, the pain localizes in the right lower quadrant (RLQ) and is continuous, dull or severe, and intensified by coughing, sneezing, or body movements.

Early appendectomy or APPY (removal of the appendix) is recommended in all suspected cases because it is safer to operate than to risk gangrene, rupture, and peritonitis. Appendectomy may be performed through a muscle-splitting incision in the right lower quadrant (RLQ) in which the cecum is brought into the incision. The base of the appendix is tied, the appendix is excised, and the stump is usually cauterized and then invaginated into the cecum.

Tumors

Both benign and malignant ***tumors*** can occur in all parts of the gastrointestinal tract. Although the benign growths are much more common than malignant ones, cancers of the gastrointestinal tract are responsible for 30 percent of all deaths from cancer in the United States.

Colorectal cancer is one of the most common malignant diseases, ranking second to cancer of the lungs in males and breasts in females. The overall mortality rate is nearly 60 percent. Over 50 percent of colorectal cancers occur in the sigmoid colon and rectum. Both environmental and genetic factors affect the development of colorectal cancer. Dietary fiber, retinoids, calcium, and selenium may be protective, whereas intake of animal fat and protein may cause an increase in the disease. Genetics plays a very important role in that an inherited predisposition contributes to more than half of all cases of colorectal cancer. There is considerable evidence that most colorectal cancers develop from benign polyps. Signs and symptoms of colorectal cancer include changes in the normal pattern of bowel habits (diarrhea, constipation), cramping, abdominal pain, and rectal bleeding, either visible or occult. Screening for colorectal cancer includes fecal occult blood testing, digital rectal examination, sigmoidoscopy, colonoscopy, and barium enema. The only definitive treatment for

DISORDERS: HOMEOSTATIC IMBALANCES (*continued*)

gastrointestinal tumors, if they cannot be removed endoscopically, is surgery.

Diverticulitis

Diverticula are saclike outpouchings of the wall of the colon in places where the muscularis has become weak. The development of diverticula is called ***diverticulosis.*** Many people who develop diverticulosis are asymptomatic and experience no complications. About 15 percent of people with diverticulosis will eventually develop an inflammation within diverticula, a condition known as ***diverticulitis.***

Research indicates that diverticula form because of lack of sufficient bulk in the colon during segmentation. The powerful contractions, working against insufficient bulk, create a pressure so high that it causes the colonic walls to bulge.

The increase in diverticular disease has been attributed to a shift to low-fiber diets. Patients treated for diverticular disease with high-fiber diets show marked relief of symptoms.

Treatment consists of bed rest, cleansing enemas, and drugs to reduce infection. In severe cases, portions of the affected colon may require surgical removal and temporary colostomy.

Cirrhosis

Cirrhosis refers to a distorted or scarred liver as a result of chronic inflammation. The parenchymal (functional) liver cells are replaced by fibrous or adipose connective tissue, a process called stromal repair. The liver has a high capacity for parenchymal regeneration, and stromal repair occurs whenever a parenchymal cell is killed or cells are damaged continuously for a long time. The symptoms of cirrhosis include jaundice, edema in the legs, uncontrolled bleeding, and increased sensitivity to drugs. Cirrhosis may be caused by hepatitis (inflammation of the liver), certain chemicals that destroy liver cells, parasites that infect the liver, and alcoholism.

Hepatitis

Hepatitis refers to inflammation of the liver and can be caused by viruses, drugs, and chemicals, including alcohol. Clinically, several types are recognized.

Hepatitis A (infectious hepatitis) is caused by hepatitis A virus and is spread by fecal contamination of food, clothing, toys, eating utensils, and so forth (fecal–oral route). It is generally a mild disease of children and young adults characterized by anorexia, malaise, nausea, diarrhea, fever, and chills. Eventually, jaundice appears. It does not cause lasting liver damage. Most people recover in four to six weeks.

Hepatitis B (serum hepatitis) is caused by hepatitis B virus and is spread primarily by sexual contact and contaminated syringes and transfusion equipment. It can also be spread by saliva and tears. Hepatitis B virus can be present for years or even a lifetime and can produce cirrhosis and possibly cancer of the liver. Persons who harbor the active hepatitis B virus are at risk for cirrhosis and also become carriers. A vaccine produced through recombinant DNA technology (Recombivax HB) is available for hepatitis B.

Non-A, non-B (NANB) hepatitis is a form of hepatitis that cannot be traced to either hepatitis A or hepatitis B viruses. It is clinically similar to hepatitis B and is often spread by blood transfusions. It is believed to account for considerably more posttransfusion hepatitis than that related to hepatitis B. The NANB hepatitis virus can cause cirrhosis and possibly liver cancer.

Gallstones

The cholesterol in bile may crystallize at any point between bile canaliculi, where it is first apparent, and the hepatopancreatic ampulla, where the bile enters the duodenum. The fusion of single crystals is the beginning of 95 percent of all ***gallstones (biliary calculi).*** Following their formation, gallstones gradually grow in size and number and may cause minimal, intermittent, or complete obstruction to the flow of bile from the gallbladder into the duct system. If obstruction of the outlet occurs and the gallbladder cannot empty as it normally does after eating, the pressure within it increases, and the individual may have intense pain or discomfort ***(biliary colic).*** Jaundice, due to the inability to secrete bilirubin into the intestine, will accompany complete biliary obstruction.

Treatment of gallstones consists of using gallstone-dissolving drugs, lithotripsy, or surgery. One such drug is methyl tert-butyl ether (MTBE), a cholesterol solvent that under local anesthesia is pumped in and out of the gallbladder through a catheter to dissolve the gallstones. Other drugs such as ursodeoxycholic acid and chenodeoxycholic acid are taken orally. Lithotripsy (shock-wave therapy) is used successfully in selected patients to fragment gallstones in conjunction with oral gallstone-dissolving drugs. For individuals in whom drugs or lithotripsy are not indicated, removal of the gallbladder and its contents (cholecystectomy) is necessary.

Anorexia Nervosa

Anorexia nervosa is a chronic disorder characterized by self-induced weight loss, body-image and other perceptual disturbances, and physiologic changes that result from nutritional depletion. The subconsciously self-imposed starvation appears to be a response to emotional conflicts about self-identification and acceptance of a normal adult sex role. Patients with anorexia nervosa have a fixation on weight control and, often, the insistence of having a bowel movement every day despite a lack of adequate food intake. They abuse cathartics, which worsens the fluid/electrolyte/nutrient deficiencies. The disorder is found predominantly in young, single females and may be inherited. Abnormal patterns of menstruation, amenorrhea (absence of menstruation), and a lowered basal metabolic rate reflect the depressant effects of the starvation. Individuals may become emaci-

DISORDERS: HOMEOSTATIC IMBALANCES (*continued*)

ated and may ultimately die of starvation or one of its complications. Also associated with the disorder are osteoporosis, depression, and brain abnormalities coupled with impaired mental performance. Treatment consists of psychotherapy and dietary regulation.

Bulimia

A disorder that typically affects single, middle-class, young, white females is known as ***bulimia*** (*bous* = ox; *limos* = hunger), or ***binge–purge syndrome.*** It is characterized by overeating at least twice a week followed by purging by self-induced vomiting, strict dieting or fasting, vigorous exercise, or use of laxatives or diuretics. This binge–purge cycle occurs in response to fears of being overweight, stress, depression, and physiological disorders such as hypothalamic tumors.

Bulimia can upset the body's electrolyte balance and increase susceptibility to flu, salivary gland infections that result in bilateral parotid gland enlargement, pharyngeal scratches from self-induced gagging, dry skin, acne, muscle spasms, loss of hair, kidney and liver diseases, erosion of dental enamel from stomach acids, ulcers, hernias, constipation, and hormone imbalances. Although the cause of bulimia is unknown, some evidence suggests that it may be related to impaired release of cholecystokinin (CCK), a hormone that induces satiety, the sensation of being full to satisfaction. Treatment of bulimia combines nutrition counseling, psychotherapy, and medical treatment.

Dietary Fiber (Roughage) and GI Disorders

A deficiency in our diet that has received recent attention is lack of dietary fiber or roughage. Dietary fiber consists of indigestible plant substances, such as cellulose, lignin, and pectin, found in fruits, vegetables, grains, and beans. Fiber may be classified as ***insoluble,*** which does not dissolve in water, and ***soluble,*** which does dissolve in water. Insoluble fiber includes the woody or structural parts of plants such as fruit and vegetable skins and the bran coating around wheat and corn kernels. Insoluble fiber passes through the GI tract largely unchanged and speeds up the passage of material through the tract. Soluble fiber is found in abundance in beans, oats, barley, broccoli, prunes, apples, and citrus fruits. It has the consistency of a gel and tends to slow the passage of material through the tract. Both insoluble and soluble fiber are important for digestive functioning. People who choose a fiber-rich, unrefined diet will greatly reduce their chances of developing disease of overnutrition (such as obesity, diabetes, gallstones, and coronary heart disease), diseases of the underworked mouth (caries and periodontal disease), and diseases of an underfed large bowel (constipation, varicose veins, hemorrhoids, colon spasm, diverticulitis, appendicitis, and large intestinal cancer). Each of these conditions is directly related to the digestion and metabolism of food and the operation of the digestive system. There is also evidence that insoluble fiber may help protect against colon cancer and that soluble fiber may help lower cholesterol (LDL).

MEDICAL TERMINOLOGY ASSOCIATED WITH THE DIGESTIVE SYSTEM

Borborygmus (bor'-bō-RIG-mus) A rumbling noise caused by the propulsion of gas through the intestines.

Botulism (BOCH-yoo-lism; *botulus* = sausage) A type of food poisoning caused by a toxin produced by *Clostridium botulinum.* The bacterium is ingested when improperly cooked or preserved foods are eaten. The toxin inhibits nerve impulse conduction at synapses by inhibiting the release of acetylcholine. Symptoms include paralysis, nausea, vomiting, blurred or double vision, difficulty in speech, difficulty in swallowing, dryness of mouth, and general weakness.

Canker (KANG-ker) **sore** Painful ulcer on the mucous membrane of the mouth that affects females more frequently than males and usually occurs between ages 10 to 40; may be an autoimmune reaction. Also known as **aphthous stomatitis.**

Cholecystitis (kō'-lē-sis-TĪ-tis; *chole* = bile; *kystis* = bladder; *itis* = inflammation of) Inflammation of the gallbladder that often leads to infection. Some cases are caused by obstruction of the cystic duct with bile stones. Stagnating bile salts irritate the mucosa. Dead mucosal cells provide a medium for the growth of bacteria.

Cholelithiasis (kō'-lē-li-THĪ-a-sis; *lithos* = stone) The presence of gallstones.

Colitis (ko-LĪ-tis) Inflammation of the mucosa of the colon and rectum in which absorption of water and salts is reduced, producing watery, bloody feces and, in severe cases, dehydration and salt depletion. Spasms of the irritated muscularis produce cramps.

Colostomy (ko-LOS-tō-mē; *stomoun* = provide an opening) The diversion of the fecal stream through an opening in the colon, creating a surgical "stoma" (artificial opening) that is affixed to the exterior of the abdominal wall. This opening serves as a substitute anus through which feces are eliminated. A temporary colostomy may be done to allow a badly inflamed colon

MEDICAL TERMINOLOGY ASSOCIATED WITH THE DIGESTIVE SYSTEM (continued)

to rest and heal. If the rectum is removed for malignancy, the colostomy provides a permanent outlet for feces.

Dysphagia (dis-FĀ-jē-a; *dys* = abnormal; *phagein* = to eat) Difficulty in swallowing that may be caused by inflammation, paralysis, obstruction, or trauma.

Enteritis (en'-ter-Ī-tis; *enteron* = intestine) An inflammation of the intestine, particularly the small intestine.

Flatus (FLĀ-tus) Air (gas) in the stomach or intestine, usually expelled through the anus. If the gas is expelled through the mouth, it is called **eructation** or **belching** (burping). Flatus may result from gas released during the breakdown of foods in the stomach or from swallowing air or gas-containing substances such as carbonated drinks.

Gastrectomy (gas-TREK-tō-mē; *gastro* = stomach; *tome* = excision) Removal of a portion of or the entire stomach.

Hernia (HER-nē-a) Protrusion of an organ or part of an organ through a membrane or cavity wall, usually the abdominal cavity. Diaphragmatic (hiatal) hernia is the protrusion of the lower esophagus, stomach, or intestine into the thoracic cavity through the opening in the diaphragm (esophageal hiatus) that allows passage of the esophagus. Umbilical hernia is usually a mild defect that contains the protrusion of a portion of peritoneum through the navel area of the abdominal wall. Inguinal hernia is the protrusion of the hernial sac into the inguinal opening. It may contain a portion of the bowel in an advanced stage and may extend into the scrotal compartment in males, causing strangulation of the herniated part.

Inflammatory bowel (in-FLAM-a-tō'-rē BOW-el) **disease** Disorder that exists in two forms: (1) Crohn's disease (inflammation of the gastrointestinal tract, especially the distal ileum and proximal colon, in which the inflammation may extend from the mucosa through the serosa) and (2) ulcerative colitis (inflammation of the mucosa of the gastrointestinal tract, usually limited to the large intestine and usually accompanied by rectal bleeding).

Irritable bowel (IR-i-ta-bul BOW-el) **syndrome (IBS)** Disease of the entire gastrointestinal tract in which persons with this condition may react to stress by developing symptoms such as cramping and abdominal pain associated with alternating patterns of diarrhea and constipation. Excessive amounts of mucus may appear in the stools, and other symptoms include flatulence, nausea, and loss of appetite. The condition is also known as **irritable colon** or **spastic colitis.**

Malocclusion (mal'-ō-KLOO-zhun; *mal* = disease; *occlusio* = to fit together) Condition in which the upper and lower teeth do not close together.

Nausea (NAW-sē-a; *nausia* = seasickness) Discomfort characterized by a loss of appetite and the sensation of impending vomiting. Its causes include local irritation of the gastrointestinal tract, a systemic disease, brain disease or injury, overexertion, or the effects of medication or drug overdosage.

Pancreatitis (pan'-krē-a-TĪ-tis) Inflammation of the pancreas, as may occur in association with mumps. In a more severe condition, known as **acute pancreatitis,** which is associated with heavy alcohol intake or biliary tract obstruction, the pancreatic cells may release trypsin instead of trypsinogen, and the trypsin begins to digest the pancreatic cells. The patient with acute pancreatitis usually responds to treatment, but recurrent attacks are the rule.

Traveler's diarrhea Infectious disease of the gastrointestinal tract that results in loose, urgent bowel movements, cramping, abdominal pain, malaise, nausea, and occasionally fever and dehydration. It is acquired through ingestion of food or water that has become contaminated with fecal material containing mostly bacteria (especially *Escherichia coli*). Viruses or protozoan parasites are less frequently involved. Commonly referred to as **Montezuma's revenge, turista,** and **Tut's tummy.**

STUDY OUTLINE

Digestive Processes (p. 733)

1. Food is prepared for use by cells by five basic activities: ingestion, movement, mechanical and chemical digestion, absorption, and defecation.
2. Chemical digestion is a series of catabolic (hydrolysis) reactions that break down large carbohydrate, lipid, and protein food molecules into smaller molecules that are usable by body cells.
3. Mechanical digestion consists of movements that aid chemical digestion.
4. Absorption is the passage of end products of digestion from the gastrointestinal tract into blood or lymph for distribution to cells.
5. Defecation is emptying of the rectum.

Organization (p. 733)

1. The organs of digestion are usually divided into two main groups: those composing the gastrointestinal (GI) tract, or alimentary canal, and accessory structures.
2. The GI tract is a continuous tube running through the ventral body cavity from the mouth to the anus.

3. The accessory structures include the teeth, tongue, salivary glands, liver, gallbladder, and pancreas.
4. The basic arrangement of layers in the alimentary canal from the inside outward is the mucosa, submucosa, muscularis, and serosa (visceral peritoneum).
5. Extensions of the peritoneum include the mesentery, mesocolon, falciform ligament, lesser omentum, and greater omentum.

Mouth (Oral Cavity) (p. 736)

1. The mouth is formed by the cheeks, hard and soft palates, lips, and tongue, which aid mechanical digestion.
2. The vestibule is the space between the cheeks and lips and teeth and gums.
3. The oral cavity proper extends from the vestibule to the fauces.

Tongue (p. 737)

1. The tongue, together with its associated muscles, forms the floor of the oral cavity. It is composed of skeletal muscle covered with mucous membrane.
2. The upper surface and sides of the tongue are covered with papillae. Some papillae contain taste buds.

Salivary Glands (p. 738)

1. The major portion of saliva is secreted by the salivary glands, which lie outside the mouth and pour their contents into ducts that empty into the oral cavity.
2. There are three pairs of salivary glands: parotid, submandibular (submaxillary), and sublingual glands.
3. Saliva lubricates food and starts the chemical digestion of carbohydrates.
4. Salivation is entirely under nervous control.

Teeth (p. 740)

1. The teeth, or dentes, project into the mouth and are adapted for mechanical digestion.
2. A typical tooth consists of three principal portions: crown, root, and neck.
3. Teeth are composed primarily of dentin and are covered by enamel, the hardest substance in the body.
4. There are two dentitions—deciduous and permanent.

Physiology of Digestion in the Mouth (p. 741)

1. Through mastication, food is mixed with saliva and shaped into a bolus.
2. Salivary amylase converts polysaccharides (starches) to disaccharides (maltose).

Physiology of Deglutition (p. 742)

1. Deglutition, or swallowing, moves a bolus from the mouth to the stomach.
2. It consists of a voluntary stage, pharyngeal stage (involuntary), and esophageal stage (involuntary).

Esophagus (p. 744)

1. The esophagus is a collapsible, muscular tube that connects the pharynx to the stomach.
2. It passes a bolus into the stomach by peristalsis.
3. It contains an upper and lower esophageal sphincter.

Stomach (p. 745)

Anatomy; Histology (p. 746)

1. The stomach begins at the bottom of the esophagus and ends at the pyloric sphincter.
2. The gross anatomic subdivisions of the stomach include the cardia, fundus, body, and pylorus.
3. Adaptations of the stomach for digestion include rugae; glands that produce mucus, hydrochloric acid, a protein-digesting enzyme, intrinsic factor, and stomach gastrin; and a three-layered muscularis for efficient mechanical movement.

Physiology of Digestion in the Stomach (p. 748)

1. Mechanical digestion consists of mixing waves.
2. Chemical digestion consists of the conversion of proteins into peptides by pepsin.

Regulation of Gastric Secretion (p. 749)

1. Gastric secretion is regulated by nervous and hormonal mechanisms.
2. Stimulation occurs in three phases: cephalic (reflex), gastric, and intestinal.

Regulation of Gastric Emptying (p. 751)

1. Gastric emptying is stimulated in response to distension, and stomach gastrin is released in response to the presence of certain types of foods.
2. Gastric emptying is inhibited by the enterogastric reflex and hormones (secretin, CCK, and GIP).

Absorption (p. 751)

1. The stomach wall is impermeable to most substances.
2. Among the substances absorbed are some water, certain electrolytes and drugs, and alcohol.

Pancreas (p. 752)

1. The pancreas is divisible into a head, body, and tail and is connected to the duodenum via the pancreatic duct (duct of Wirsung) and accessory duct (duct of Santorini).
2. Pancreatic islets (islets of Langerhans) secrete hormones, and acini secrete pancreatic juice.
3. Pancreatic juice contains enzymes that digest starch (pancreatic amylase), proteins (trypsin, chymotrypsin, and carboxypeptidase), fats (pancreatic lipase), and nucleotides (nucleases).
4. Pancreatic secretion is regulated by nervous and hormonal mechanisms.

Liver (p. 754)

1. The liver is divisible into left and right lobes; associated with the right lobe are the caudate and quadrate lobes.
2. The lobes of the liver are made up of lobules that contain hepatocytes (liver cells), sinusoids, stellate reticuloendothelial (Kupffer's) cells, and a central vein.
3. Hepatic cells (hepatocytes) of the liver produce bile that is transported by a duct system to the gallbladder for concentration and temporary storage.
4. Bile's contribution to digestion is the emulsification of neutral fats.
5. The liver also functions in carbohydrate, fat, and protein metabolism; excretion of bile; synthesis of bile salts; removal

of drugs and hormones; storage of vitamins and minerals; phagocytosis; and activation of vitamin D.
6. Bile secretion is regulated by nervous and hormonal mechanisms.

Gallbladder (GB) (p. 758)

1. The gallbladder (GB) is a sac located in a fossa on the visceral surface of the liver.
2. The gallbladder stores and concentrates bile.
3. Bile is ejected into the common bile duct under the influence of cholecystokinin (CCK).

Small Intestine (p. 759)

Anatomy; Histology (p. 759)

1. The small intestine extends from the pyloric sphincter to the ileocecal sphincter.
2. It is divided into duodenum, jejunum, and ileum.
3. It is highly adapted for digestion and absorption. Its glands produce enzymes and mucus, and the microvilli, villi, and circular folds of its wall provide a large surface area for digestion and absorption.
4. Some intestinal enzymes break down foods inside epithelial cells of the mucosa.

Physiology of Digestion in the Small Intestine (p. 763)

1. Intestinal enzymes break down maltose to glucose (maltase), sucrose to glucose and fructose (sucrase), lactose to glucose and galactose (lactase), terminal amino acids at the amino ends of peptides (aminopeptidase), dipeptides to amino acids (dipeptidase), and nucleotides to pentoses and nitrogenous bases (nucleases).
2. Mechanical digestion in the small intestine involves segmentation and peristalsis.

Regulation of Intestinal Secretion (p. 765)

1. The most important mechanism is local reflexes.
2. Hormones also assume a role.

Physiology of Absorption (p. 765)

1. Absorption is the passage of the end products of digestion from the gastrointestinal tract into the blood or lymph.
2. Monosaccharides, amino acids, and short-chain fatty acids pass into the blood capillaries.
3. Long-chain fatty acids and monoglycerides are absorbed as part of micelles, resynthesized to triglycerides, and transported as chylomicrons.
4. Chylomicrons are taken up by the lacteal of a villus.
5. The small intestine also absorbs water, electrolytes, and vitamins.

Large Intestine (p. 768)

Anatomy; Histology (p. 768, 769)

1. The large intestine extends from the ileocecal sphincter to the anus.
2. Its subdivisions include the cecum, colon, rectum, and anal canal.
3. The mucosa contains numerous goblet cells, and the muscularis consists of taeniae coli.

Physiology of Digestion in the Large Intestine (p. 771)

1. Mechanical movements of the large intestine include haustral churning, peristalsis, and mass peristalsis.
2. The last stages of chemical digestion occur in the large intestine through bacterial, rather than enzymatic, action. Substances are further broken down and some vitamins are synthesized.

Absorption and Feces Formation (p. 772)

1. The large intestine absorbs water, electrolytes, and vitamins.
2. Feces consists of water, inorganic salts, epithelial cells, bacteria, and undigested foods.

Physiology of Defecation (p. 772)

1. The elimination of feces from the rectum is called defecation.
2. Defecation is a reflex action aided by voluntary contractions of the diaphragm and abdominal muscles.

Screening for Colorectal Cancer (p. 772)

1. Over one-half of colorectal cancers occur in the sigmoid colon and rectum.
2. Screening tests for colorectal cancer include fecal occult blood testing, digital rectal examination, sigmoidoscopy, colonoscopy, and barium enema.

Aging and the Digestive System (p. 773)

1. General changes include decreased secretory mechanisms, decreased motility, and loss of tone.
2. Specific changes include loss of taste, pyorrhea, hernias, ulcers, constipation, hemorrhoids, and diverticular diseases.

Developmental Anatomy of the Digestive System (p. 773)

1. The endoderm of the primitive gut forms the epithelium and glands of most of the gastrointestinal tract.
2. The mesoderm of the primitive gut forms the smooth muscle and connective tissue of the gastrointestinal tract.

Disorders: Homeostatic Imbalances (p. 775)

1. Dental caries are started by acid-producing bacteria that reside in dental plaque.
2. Periodontal diseases are characterized by inflammation and degeneration of gingivae, alveolar bone, periodontal membrane, and cementum.
3. Peritonitis is inflammation of the peritoneum.
4. Peptic ulcers are craterlike lesions that develop in the mucous membrane of the gastrointestinal tract in areas exposed to gastric juice.
5. Appendicitis is an inflammation of the vermiform appendix resulting from obstruction of the lumen of the appendix by inflammation, a foreign body, carcinoma of the cecum, stenosis, or kinking of the organ.
6. Tumors may occur in any portion of the gastrointestinal tract. One of the most common malignancies is colorectal cancer.
7. Diverticulitis is the inflammation of diverticula in the colon.
8. Cirrhosis is a condition in which parenchymal cells of the liver damaged by chronic inflammation are replaced by fibrous or adipose connective tissue.

9. Hepatitis is an inflammation of the liver. Types include hepatitis A; hepatitis B; and non-A, non-B (NANB) hepatitis.
10. The fusion of individual crystals of cholesterol is the beginning of 95 percent of all gallstones. Gallstones can cause obstruction to the outflow of bile in any portion of the duct system.
11. Anorexia nervosa is a disorder characterized by bizzare eating patterns.
12. Bulimia is a binge–purge syndrome of behavior in which uncontrollable overeating is followed by forced vomiting or overdoses of laxatives.

REVIEW QUESTIONS

1. Define digestion and describe the five phases involved. Distinguish between chemical and mechanical digestion. (p. 733)
2. Identify, in sequence, the organs of the gastrointestinal (GI) tract. How does the gastrointestinal tract differ from the accessory structures of digestion? (p. 733)
3. Describe the structure of each of the four layers of the gastrointestinal tract. (p. 733)
4. What is the peritoneum? Describe the location and function of the mesentery, mesocolon, falciform ligament. lesser omentum, and greater omentum. (p. 735)
5. What structures form the mouth or oral cavity? (p. 736)
6. Make a simple diagram of the tongue. Indicate the location of the papillae and the four taste zones. What is ankyloglossia? (p. 465, 738)
7. Describe the location of the salivary glands and their ducts. What is mumps? (p. 738)
8. How are salivary glands distinguished histologically? (p. 738)
9. Describe the composition of saliva and the role of each of its components in digestion. What is the pH of saliva? (p. 738)
10. How is salivary secretion regulated? (p. 740)
11. What are the principal parts of a typical tooth? What are the functions of each part? What is root canal therapy? (p. 740)
12. Compare deciduous and permanent dentitions with regard to number of teeth and time of eruption. (p. 740)
13. Contrast the functions of incisors, cuspids, premolars, and molars. (p. 740)
14. What is a bolus? How is it formed? (p. 741)
15. Define deglutition. List the sequence of events involved in passing a bolus from the mouth to the stomach. Be sure to discuss the voluntary, pharyngeal, and esophageal stages of swallowing. (p. 742)
16. Describe the location and histology of the esophagus. What is its role in digestion? (p. 744)
17. Explain the operation of the upper and lower esophageal sphincters. How is achalasia related to heartburn? (p. 745)
18. Describe the location of the stomach. List and briefly explain the anatomical features of the stomach. (p. 745)
19. Distinguish between pylorospasm and pyloric stenosis. (p. 746)
20. What is the importance of rugae, zymogenic cells, parietal cells, mucous cells, and enteroendocrine cells in the stomach? (p. 746)
21. Describe mechanical digestion in the stomach. (p. 748)
22. What is the role of pepsin? Why is it secreted in an inactive form? (p. 749)
23. Outline the factors that stimulate and inhibit gastric secretion. Be sure to discuss the cephalic, gastric, and intestinal phases. (p. 749)
24. How is gastric emptying stimulated and inhibited? (p. 749)
25. Explain how vomiting occurs. (p. 751)
26. Describe the role of the stomach in absorption. (p. 751)
27. Where is the pancreas located? Describe the duct system connecting the pancreas to the duodenum. (p. 752)
28. What are pancreatic acini? Contrast their functions with those of the pancreatic islets (islets of Langerhans). (p. 754)
29. Describe the composition of pancreatic juice and the digestive functions of each component. (p. 754)
30. How is pancreatic juice secretion regulated? (p. 754)
31. Where is the liver located? What are its principal functions? (p. 754)
32. Describe the anatomy of the liver. Draw a labeled diagram of a liver lobule. (p. 754)
33. How is blood carried to and from the liver? (p. 756)
34. Once bile has been formed by the liver, how is it collected and transported to the gallbladder for storage? (p. 757)
35. What is the function of bile? What causes gallstones? (p. 757)
36. How is bile secretion regulated? Define jaundice and distinguish the principal types. (p. 757)
37. Where is the gallbladder (GB) located? How is it connected to the duodenum? (p. 758)
38. Describe the function of the gallbladder. How is emptying of the gallbladder regulated? (p. 759)
39. What are the subdivisions of the small intestine? How are the mucosa and submucosa of the small intestine adapted for digestion and absorption? (p. 759)
40. Describe the movements in the small intestine. (p. 763)
41. Explain the function of each enzyme in intestinal juice. (p. 763)
42. What is lactose intolerance? (p. 763)
43. How is small intestinal secretion regulated? (p. 765)
44. Define absorption. How are the end products of carbohydrate and protein digestion absorbed? How are the end products of fat digestion absorbed? (p. 765)
45. What routes are taken by absorbed nutrients to reach the liver? (p. 765)
46. Describe the absorption of water, electrolytes, and vitamins by the small intestine. (p. 767)
47. What are the principal subdivisions of the large intestine? How does the muscularis of the large intestine differ from that of the rest of the gastrointestinal tract? What are haustra? (p. 768)
48. What are hemorrhoids? How are they treated? (p. 769)
49. Describe the mechanical movements that occur in the large intestine. (p. 771)
50. Explain the activities of the large intestine that change its contents into feces. (p. 772)
51. Define defecation. How does it occur? (p. 772)
52. Distinguish between diarrhea and constipation. (p. 772)
53. Describe the effect of aging on the digestive system. (p. 773)
54. Describe the development of the digestive system. (p. 773)

55. Describe the causes (where known) and clinical symptoms for: dental caries, periodontal disease, peritonitis, peptic ulcers, appendicitis, tumors, diverticulitis, cirrhosis, hepatitis, gallstones, anorexia nervosa, and bulimia. (p. 775)
56. Explain the diagnostic value of the following: gastroscopy (p. 747), liver function tests (LFTs) (p. 758), oral cholecystogram (p. 759), barium swallow and enema (p. 762), fecal occult blood testing (p. 772), sigmoidoscopy (p. 773), and colonoscopy. (p. 773)
57. Refer to the glossary of medical terminology associated with the digestive system. Be sure that you can define each term. (p. 778)

SELECTED READINGS

Bruckstein, A. H. "Chronic Hepatitis." *Postgraduate Medicine,* May 1989.

Cunha, B. A. "The Yellow Patient," *Postgraduate Medicine,* October 1988.

Davenport, H. W. *Physiology of the Digestive System,* 5th ed. Chicago: Year Book Medical Publishers, 1982.

Farley, D. "Today's Dentistry: A Mouthful of Marvels," *Healthline,* October 1985.

Friedman, G. "Peptic Ulcer Disease," *Clinical Symposia,* Vol. 40, No. 5, 1988.

Goldfinger, S. E. (ed). "Shock Waves for Gallstones," *Harvard Medical School Health Letter,* July 1988.

———. "Screening for Bowel Cancer," *Harvard Medical School Health Letter,* April 1986.

———. "Sealing Out Tooth Decay," *Harvard Medical School Health Letter,* February 1984.

Jenkins, E. "Large Bowel Exam: When to Use What Test," *Modern Medicine,* March 1984.

Johnson, D. A. "Fecal Occult Blood Testing," *Postgraduate Medicine,* April 1989.

Johnson, G. T. (ed). "The Tragedy of Anorexia Nervosa," *Harvard Medical School Health Letter,* December 1981.

Leff, E. "Hemorrhoids," *Postgraduate Medicine,* November 1987.

Levine, M. P. "Eating Disorders," *Postgraduate Medicine,* November 1988.

Steinberg, S. "Cancer and Cuisine," *Science News,* 1 October 1983.

Chapter 25

Metabolism

Chapter Contents at a Glance

Student Objectives

1. Define a nutrient and list the functions of the six principal classes of nutrients.
2. Define metabolism and explain the role of ATP in anabolism and catabolism.
3. Describe the characteristics and importance of enzymes in metabolism.
4. Describe the metabolism of carbohydrates, lipids, and proteins.
5. Distinguish between the absorptive and postabsorptive (fasting) states.
6. Compare the sources, functions, and importance of minerals and vitamins in metabolism.
7. Describe the various mechanisms involved in heat production and heat loss.
8. Define basal metabolic rate (BMR) and explain several factors that affect it.
9. Explain how normal body temperature is maintained and describe fever, heat cramp, heatstroke, heat exhaustion, and hypothermia as abnormalities of temperature regulation.
10. Define obesity, malnutrition, phenylketonuria (PKU), cystic fibrosis (CF), and celiac disease.

In this chapter, we will discuss the various types of nutrients and how they are metabolized by the body. First, however, we will look at how food intake is regulated.

REGULATION OF FOOD INTAKE

Within the hypothalamus are two centers related to food intake. One is a cluster of nerve cells in the lateral nuclei called the ***feeding (hunger) center.*** When this area is stimulated in animals, they begin to eat heartily, even if they are already full. The second center is a cluster of neurons in the ventromedial nuclei of the hypothalamus referred to as the ***satiety center.*** When this center is stimulated in animals, it causes them to stop eating, even if they have been starved for days. Apparently the feeding center is constantly active, but it is inhibited by the satiety center. Other parts of the brain that assume a function in feeding and satiety are the brain stem, amygdala, and limbic system.

One substance that is related to food intake is glucose. According to the ***glucostatic theory,*** when blood glucose levels are low, feeding increases. It is believed that in response to low levels of blood glucose the rate of glucose utilization by neurons of the satiety center becomes low. As a result, the activity of the neurons decreases to the point where the satiety center can no longer inhibit the feeding center, and the individual eats. Conversely, when blood glucose levels are high, the activity of the neurons of the satiety center is sufficiently high to inhibit the feeding center, and feeding is depressed. Low levels of amino acids in the blood also enhance feeding, whereas high levels depress eating. This mechanism, however, is not as powerful as that for glucose.

Lipids are also believed to be related to the regulation of food intake. It has been observed that as the amount of adipose tissue increases in the body, the rate of feeding decreases. It is theorized that some substance or substances, perhaps fatty acids, are released from stored fats in proportion to the total fat content of the body. The released substances then activate neurons of the satiety center, which, in turn, inhibit the feeding center.

Another factor that affects food intake is body temperature. Whereas a cold environment enhances eating, a warm environment depresses it. Food intake is also regulated by distension of the gastrointestinal tract, particularly the stomach and duodenum. When these organs are stretched, a reflex is initiated that activates the satiety center and depresses the feeding center. It has also been shown that the hormone cholecystokinin (CCK), secreted when fat enters the small intestine, inhibits eating.

NUTRIENTS

Nutrients are chemical substances in food that provide energy, form new body components, or assist in the functioning of various body processes. There are six principal classes of nutrients: carbohydrates, lipids, proteins, minerals, vitamins, and water. Carbohydrates, proteins, and lipids are digested by enzymes in the gastrointestinal tract. The end products of digestion that ultimately reach body cells are monosaccharides, amino acids, fatty acids, glycerol, and monoglycerides. Some are used to synthesize new structural molecules in cells or to synthesize new regulatory molecules, such as hormones and enzymes. Most are used to produce energy to sustain life processes. This energy is used for processes such as active transport, DNA replication, synthesis of proteins and other molecules, muscle contraction, and nerve impulse conduction. Until the energy is needed, it is stored in ATP (adenosine triphosphate).

Some minerals and many vitamins are part of enzyme systems that catalyze the reactions undergone by carbohydrates, proteins, and lipids.

Water has five major functions. It is an excellent solvent and suspending medium, it participates in hydrolysis reactions, it acts as a coolant, it lubricates, and it helps to maintain a constant body temperature owing to its ability to release and absorb heat slowly.

METABOLISM

Metabolism (me-TAB-ō-lizm; *metabole* = change) refers to all the chemical reactions of the body. Because chemical reactions either require or release energy, the body's metabolism may be thought of as an energy-balancing act between anabolic (synthesis) and catabolic (degradative) reactions.

Anabolism

In living cells, those chemical reactions that combine simple substances into more complex molecules are collectively known as ***anabolism*** (a-NAB-ō-lizm; *ana* = upward). Overall, anabolic processes often involve dehydration synthesis reactions (reactions that release water) and require energy to form new chemical bonds. One example of an anabolic process is the formation of peptide bonds between amino acids, thereby building up the amino acids into the protein portions of cytoplasm, enzymes, hormones, and antibodies. Fats also participate in the body's anabolism. For instance, fats can be built into the phospholipids that form the plasma membrane. They are also part of the steroid hormones. Through anabolism, simple sugars are also built up into polysaccharides.

Catabolism

The chemical reactions that break down complex organic compounds into simple ones are collectively known as ***catabolism*** (ka-TAB-ō-lizm; *cata* = downward). Catabolic reactions are generally hydrolysis reactions (reactions that use water to break chemical bonds) that release the available chemical energy in organic molecules. An example of a catabolic reaction is chemical digestion in which the breaking of bonds of food molecules releases energy. Another example is a process called oxidation (cellular respiration). It involves the breakdown of absorbed nutrients, resulting in a release of energy, some of which is used to synthesize most of a cell's ATP.

Whereas anabolic reactions are mostly energy-requiring reactions, catabolic reactions provide the energy needed to drive anabolic reactions. This coupling of energy-requiring and energy-releasing reactions is achieved through adenosine triphosphate (ATP). ATP stores energy derived from catabolic reactions and releases it later to drive anabolic reactions. Recall from Chapter 2 that a molecule of ATP consists of an adenine molecule, a ribose molecule, and three phosphate groups (see Figure 2-13). When the terminal phosphate group, P, is split from ATP, adenosine diphosphate (ADP) is formed, and the energy released is used to drive anabolic reactions ($ATP \rightleftharpoons ADP + P + energy$). Then the energy from catabolic reactions is used to combine ADP and the terminal phosphate group (P) to resynthesize ATP ($ADP + P + energy \rightleftharpoons ATP$). The importance of the concept of coupled reactions will be discussed shortly.

The role of ATP in linking anabolic and catabolic reactions is shown in Figure 25-1. Actually, only a small part of the energy released in catabolism is available for cellular functions. Part of the energy is lost to the environment as heat. Thus, there is a continuous need for new external sources of energy for the cell in order for the cell to synthesize enough ATP to sustain life.

FIGURE 25-1 **Catabolism, anabolism, and ATP. When simple compounds are combined to form complex compounds (anabolism), ATP provides the energy for synthesis. When large compounds are split apart (catabolism), much of the energy is given off as heat. Some is transferred to and trapped in ATP and then utilized to drive anabolic reactions.**

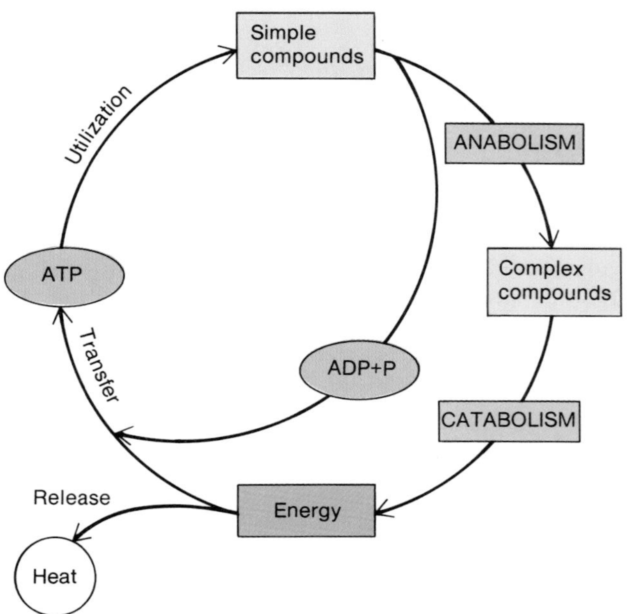

Before we discuss the cell's energy production methods, let us first consider the principal properties of a group of proteins, called enzymes, that are involved in increasing the rate of almost all biologically important chemical reactions. A cell's metabolic pathways are determined by its enzymes, which are in turn determined by the cell's genetic makeup.

Metabolism and Enzymes

It was noted in Chapter 2 that chemical reactions occur when chemical bonds are made or broken. In order for reactions to take place, atoms, ions, or molecules must collide with each other. The effectiveness of the collision depends on the velocity of the particles, the amount of energy required for the reaction to occur ***(activation energy)***, and the specific configuration (shape) of the particles. Normal body temperature and pressure are too low for chemical reactions to occur at a rate rapid enough to maintain life. Although raising the temperature, pressure, and the amount of the reacting molecules can increase the frequency of collisions and also increase the rate of chemical reactions, such changes could damage or kill cells. The living cell's solution to this problem is enzymes. Enzymes speed up chemical reactions by increasing the frequency of collisions, lowering the activation energy, and properly orienting the colliding molecules. And they do this without increasing the temperature or pressure—in other words, without disrupting or killing the cell. Substances that can speed up a chemical reaction by increasing the frequency of collisions or lowering the activation energy requirement, *without themselves being altered,* are called ***catalysts.*** In living cells, ***enzymes*** function as biological catalysts.

Characteristics of Enzymes

As catalysts, enzymes are specific. Each particular enzyme will affect only specific ***substrates***—molecules with which enzymes react and which undergo reactions to produce products. The specificity of enzymes is made possible by their structures which allow them to bond to only certain substrates. Enzymes are generally large proteins that range in molecular weight from about 10,000 to somewhere in the millions. Of the thousand or more known enzymes, each has a characteristic three-dimensional shape with a specific surface configuration, due to its primary, secondary, and tertiary structures, and this determines the specific substrate with which it can chemically bond.

Enzymes are extremely efficient. Under optimum conditions, they can catalyze reactions at rates that are from

10^8 to 10^{10} times (up to 10 billion times) more rapid than those of comparable reactions occurring without enzymes. The ***turnover number*** (number of substrate molecules converted to product per enzyme molecule per second) is generally between 1 and 10,000 and can be as high as 500,000.

Yet, as previously mentioned, enzymes are specific in the reactions they catalyze as well as in the substrates with which they react. From the large number of diverse molecules in a cell, an enzyme must "find" the correct substrate. Moreover, the enzyme-catalyzed reactions tend to take place in aqueous solutions and at relatively low temperatures—conditions that otherwise would not favor rapid movement of molecules or rapid chemical reactions.

Enzymes are also subject to various cellular controls. Their rate of synthesis and their concentration at any given time are under the control of a cell's genes and are influenced by various other molecules in the cell. Many enzymes occur in a cell in both active and inactive forms. The rate at which the inactive form becomes active or the active form becomes inactive is determined by the cellular environment.

The names of enzymes usually end in the suffix ***-ase.*** All enzymes can be grouped according to the types of chemical reactions they catalyze. For example, ***oxidoreductases*** are involved with oxidation–reduction reactions. (Enzymes that remove hydrogen are called ***dehydrogenases;*** those that add oxygen are called ***oxidases.***) ***Transferases*** transfer groups of atoms, ***hydrolases*** add water, ***isomerases*** rearrange atoms within a molecule, and ***ligases*** join two molecules in a reaction in which ATP is broken down.

Components of an Enzyme

Some enzymes consist entirely of proteins. Most enzymes, however, contain a protein called an ***apoenzyme*** that is inactive without a nonprotein component called the ***cofactor.*** Together, the apoenzyme and cofactor are an activated ***holoenzyme,*** or whole enzyme (Figure 25-2). If the cofactor is removed, the apoenzyme will not function. The cofactor can be a metal ion or a complex organic molecule called a ***coenzyme.***

Examples of metal ions that serve as cofactors are iron, magnesium, zinc, and calcium. Their function is to serve as a bridge between apoenzyme and substrate so that the holoenzyme can bring about transformation of the substrate. Coenzymes assist in the enzyme's activity by accepting atoms removed from the substrate or donating atoms needed by the substrate. Many coenzymes are derived from vitamins. Three coenzymes commonly used by living cells to carry hydrogen atoms are nicotinamide adenine dinucleotide (NAD) and nicotinamide adenine dinucleotide phosphate (NADP), both of which are derivatives of the B vitamin niacin, and flavin adenine dinucleotide (FAD), a derivative of vitamin B_2 (riboflavin). Another important coenzyme is coenzyme A (CoA) derived from pantothenic acid, another B vitamin. CoA functions in decarboxylation (removal of carbon dioxide) and is associated with an acetyl group, a useful substance in catabolic reactions.

FIGURE 25-2 **Components of a holoenzyme. Many enzymes require both an apoenzyme (protein portion) and a cofactor (nonprotein portion) to become active. The cofactor can be a metal ion or an organic molecule, called a coenzyme (as shown here). The apoenzyme and cofactor together make up the holoenzyme, or complete enzyme. The substrate is the substance with which the enzyme reacts.**

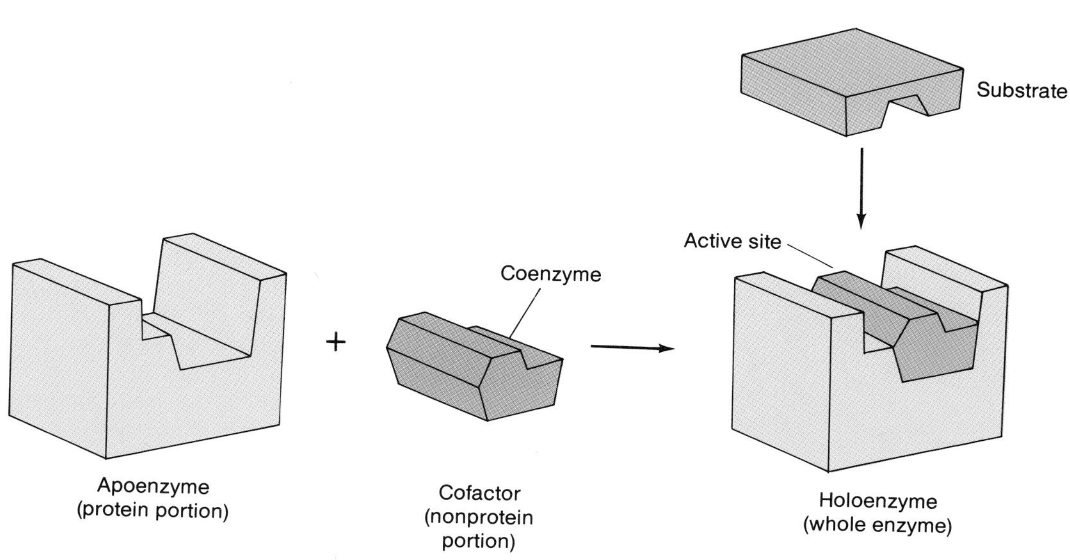

Mechanism of Enzymatic Action

Exactly how enzymes lower activation energy is not completely understood. However, the following general sequence is believed to occur.

1. The surface of the substrate makes contact with a specific region on the surface of the enzyme molecule known as the ***active site.***

2. A temporary intermediate compound called an ***enzyme-substrate complex*** forms.

3. The substrate molecule is transformed by rearrangement of existing atoms, breakdown of the substrate molecule, or combination of several substrate molecules.

4. The transformed substrate molecules, now called the ***products*** of the reaction, move away from the enzyme molecule.

5. After the reaction is completed, the products of the reaction move away from the unchanged enzyme, and the enzyme is free to attach to another substrate molecule.

PHYSIOLOGY OF ENERGY PRODUCTION

Oxidation–Reduction Reactions

Nutrient molecules, like all molecules, have energy stored in the bonds between their atoms. When this energy is spread throughout the molecule, it is difficult for the cell to use the energy. Various reactions in catabolic pathways, however, concentrate the energy as it is released into the high-energy bonds of ATP, which serves as a convenient energy carrier. Before discussing metabolic pathways, we will first consider two important aspects of energy production—the concept of oxidation–reduction and mechanisms of ATP generation.

Oxidation is the removal of electrons or hydrogen (H^+) ions from a molecule and results in a decrease in the energy content of the molecule. In many cellular oxidations, two electrons and two hydrogen ions (H^+) are removed at the same time; this is equivalent to the removal of two hydrogen atoms ($2e^- + 2H^+ \rightarrow 2H$). Because most biological oxidations involve the loss of hydrogen atoms, they are called dehydrogenation reactions. An example of an oxidation is the conversion of lactic acid into pyruvic acid:

$$\underset{\text{Lactic acid}}{\begin{array}{c} COOH \\ | \\ CHOH \\ | \\ CH_3 \end{array}} \xrightarrow{-2H\text{(oxidation)}} \underset{\text{Pyruvic acid}}{\begin{array}{c} COOH \\ | \\ C{=}O \\ | \\ CH_3 \end{array}}$$

Reduction is the addition of elecrons or hydrogen ions (hydrogen atoms) to a molecule and results in an increase in the energy content of the molecule. Reduction is the opposite of oxidation. An example of reduction is the conversion of pyruvic acid into lactic acid:

$$\underset{\text{Pyruvic acid}}{\begin{array}{c} COOH \\ | \\ C{=}O \\ | \\ CH_3 \end{array}} \xrightarrow{+2H\text{(oxidation)}} \underset{\text{Lactic acid}}{\begin{array}{c} COOH \\ | \\ CHOH \\ | \\ CH_3 \end{array}}$$

Within a cell, oxidation and reduction reactions are always coupled; that is, whenever a substance is oxidized, another is almost simultaneously reduced. This coupling of reactions is simply referred to as ***oxidation–reduction.***

When a substance is oxidized, the liberated hydrogen atoms do not remain free in the cell but are transferred immediately by coenzymes to another compound. Two coenzymes commonly used by living cells to carry hydrogen atoms are nicotinamide adenine dinucleotide (NAD^+) and nicotinamide adenine dinucleotide phosphate ($NADP^+$). The oxidation and reduction states of NAD^+ and $NADP^+$ can be represented as follows:

$$\underset{\text{Oxidized}}{NAD^+} \underset{-2H(2H^+ + 2e^-)}{\overset{+2H(2H^+ + 2e^-)}{\rightleftharpoons}} \underset{\text{Reduced}}{NADH + H^+}$$

$$\underset{\text{Oxidized}}{NADP^+} \underset{-2H(2H^+ + 2e^-)}{\overset{+2H(2H^+ + 2e^-)}{\rightleftharpoons}} \underset{\text{Reduced}}{NADPH + H^+}$$

In the preceding equations, 2H ($2H^+ + 2e^-$) means that two neutral hydrogen atoms (2H) are equivalent to two hydrogen ions ($2H^+$) and two electrons ($2e^-$). When NAD^+ is reduced to $NADH + H^+$, two hydrogen ions ($2H^+$) and two electrons ($2e^-$) are gained by NAD^+. (Actually, one H^+ ion and two electrons are delivered to NAD^+, and the other H^+ ion is released into the surrounding solution.) The addition of a H^+ ion and $2e^-$ to NAD^+ neutralizes the charge on NAD^+ and adds a hydrogen atom so that the reduced form is $NADH + H^+$. The same holds true for the reduction of $NADP^+$ to $NADPH + H^+$.

Conversely, when $NADH + H^+$ is oxidized to NAD^+, two hydrogen ions ($2H^+$) and two electrons ($2e^-$) are lost by $NADH + H^+$. (Actually, one H^+ ion and two electrons are lost from $NADH + H^+$ and the other H^+ ion is picked up from the surrounding medium.) Then, the loss of a H^+ ion and $2e^-$ by $NADH + H^+$ results in one less hydrogen atom and an additional positive charge, so that the oxidized form is NAD^+. $NADPH + H^+$ is oxidized to $NADP^+$ by the same mechanism.

Thus, when lactic acid is *oxidized* to form pyruvic acid, the two hydrogen atoms removed in the reaction are used to *reduce* NAD. This coupled oxidation–reduction reaction may be written as follows:

$$\underset{\text{Oxidized}}{NAD^+} \underset{-2H(2H^+ + 2e^-)}{\overset{+2H(2H^+ + 2e^-)}{\rightleftharpoons}} \underset{\text{Reduced}}{NADH + H^+}$$

$$\underset{\text{Oxidized}}{NADP^+} \underset{-2H(2H^+ + 2e^-)}{\overset{+2H(2H^+ + 2e^-)}{\rightleftharpoons}} \underset{\text{Reduced}}{NADPH + H^+}$$

An important point to remember about oxidation–reduction reactions is that oxidation is usually an energy-producing reaction. Cells take nutrients, some of which serve as energy sources, and degrade them from energy rich, highly reduced compounds (with many hydrogen atoms) to lower energy, highly oxidized compounds (with many oxygen atoms or multiple bonds). For example, when a cell oxidizes a molecule of glucose ($C_6H_{12}O_6$), the energy in the glucose molecule is removed in a stepwise manner and ultimately some of the energy is trapped by using it in the form of ATP, which then serves as an energy source for energy-requiring reactions. Compounds such as glucose that have many hydrogen atoms are highly reduced compounds, containing more potential energy than oxidized compounds. For this reason, glucose is a valuable nutrient for organisms.

Generation of ATP

Some of the energy released during oxidation reactions is trapped within a cell as ATP is formed. To summarize what was discussed at the beginning of this chapter, a phosphate group (symbolized P), is added to ADP, with an input of energy, to form ATP.

$$\underset{\text{ADP}}{\text{Adenosine—P} \sim \text{P}} + \text{Energy} + \text{P} \rightarrow$$

$$\underset{\text{ATP}}{\text{Adenosine—P} \sim \text{P} \sim \text{P}}$$

The symbol ~ designates a high-energy bond, one that can readily be broken to release usable energy. The high-energy bond that attaches the third phosphate group contains the energy stored in this reaction. This addition of a phosphate group to a chemical compound is called ***phosphorylation*** (fos'-for-i-LĀ-shun) and this raises the energy level of the molecule. Organisms use three mechanisms of phosphorylation to generate ATP.

1. In ***substrate-level phosphorylation,*** ATP is generated when a high-energy phosphate group is transferred directly from an intermediate phosphorylated metabolic compound to ADP. In cells of the human body, this occurs in the cytoplasm. The following example shows only the carbon skeleton and the phosphate group of the metabolic compound:

$$\text{C—C—C} \sim \text{P} + \text{ADP} \rightarrow \text{C—C—C} + \text{ATP}$$

2. In ***oxidative phosphorylation,*** electrons are removed from organic compounds (usually by NAD^+) and passed through a series of electron acceptors to molecules of oxygen (O_2) or other inorganic molecules. This process occurs in the inner mitochondrial membrane of human cells. The series of electron acceptors used in oxidative phosphorylation is called the ***electron transport chain*** (see Figure 25-7). The transfer of electrons from one electron acceptor to the next releases energy, which is used to generate ATP from ADP and P through a process called chemiosmosis, to be described shortly.

3. The third mechanism of phosphorylation, ***photophosphorylation,*** which will not be discussed here, occurs only in photosynthetic cells, which contain light-trapping pigments such as the chlorophylls.

Chemiosmotic Mechanism of ATP Generation

Substances diffuse passively across membranes from regions of their higher concentration to regions of their lower concentration; this diffusion yields energy. The movement of substances against such a concentration gradient *requires* energy; in such an active transport of molecules or ions across biological membranes, the required energy is usually provided by ATP. ***Chemiosmosis*** (kem'-ē-oz-MŌ-sis) is a process in which the energy released when a substance moves along a gradient is used to *synthesize* ATP. The substance in this case refers to protons (H^+). The basic mechanism of chemiosmosis is as follows (Figure 25-3):

1. Energy from NADH + H^+ (in oxidative phosphorylation) or sunlight powers some of the carriers of the electron transport chain to pump (actively transport) hydrogen ions from one side of the membrane to the other. Because hydrogen ions are in fact protons, such carrier molecules are called ***proton pumps.***

FIGURE 25-3 **Chemiosmosis. In the human body, the process occurs on the inner mitochondrial membrane.**

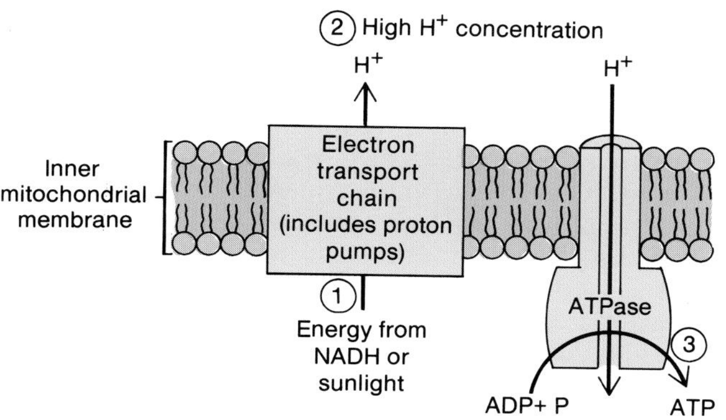

2. The phospholipid membrane is normally impermeable to protons, so this one-directional pumping establishes a proton gradient (a difference in concentration of protons on either side of the membrane). Thus, the fluid on one side of the membrane acquires a positive charge, whereas the fluid on the other side is left with a negative charge. The proton gradient has potential energy and is called the ***proton motive force.***

3. The protons on the side of the membrane with the higher proton concentration can diffuse across the membrane with the help of an enzyme in the membrane called ***adenosine triphosphatase (ATPase).*** When this movement occurs, energy is released and is used by the enzyme to synthesize ATP from ADP and P.

We will now consider the metabolism of carbohydrates, lipids, and proteins in body cells.

PHYSIOLOGY OF CARBOHYDRATE METABOLISM

During the process of digestion, polysaccharides and disaccharides are hydrolyzed into the monosaccharides glucose, fructose, and galactose. Glucose represents about 80 percent of the monosaccharides. Some fructose is converted into glucose as it is absorbed through the intestinal epithelial cells. The three monosaccharides are absorbed into the capillaries of the villi of the small intestine. They are then carried through the hepatic portal vein to the liver, where much of the remaining fructose and practically all the galactose are converted to glucose. Thus, the story of carbohydrate metabolism is really the story of glucose metabolism.

Fate of Carbohydrates

Since glucose is the body's preferred source of energy, the fate of absorbed glucose depends on the body cells' energy needs. If the cells require immediate energy, the glucose is oxidized by the cells. Each gram of carbohydrate produces about 4.0 kilocalories (kcal). (The kilocalorie content of a food is a measure of the heat it releases upon oxidation. Determination of caloric value is described later in the chapter.) The glucose not needed for immediate use is handled in several ways. First, the liver can convert excess glucose to glycogen (glycogenesis) and then store it. Skeletal muscle fibers (cells) can also store glycogen. Later, when the cells need more energy, the glycogen can be converted back to glucose, which is released into the bloodstream so it can be transported to cells for oxidation. Second, if the glycogen storage areas are filled up, liver cells and fat cells can transform the glucose to fat that can be stored in adipose tissue. Third, excess glucose can be excreted in the urine. Normally, this happens only when a meal containing mostly carbohydrates and no fats is eaten. Without the inhibiting effect of fats, the stomach empties its contents quickly, and the carbohydrates are all digested at the same time. As a result, large numbers of monosaccharides suddenly flood into the bloodstream. Since the liver is unable to process all of them simultaneously, the blood glucose level rises.

Glucose Movement into Cells

Before glucose can be used by body cells, it must first pass through the plasma membrane and then enter the cytoplasm. Where glucose absorption in the gastrointestinal tract (and kidney tubules) is accomplished by active transport, glucose movement from blood into body cells occurs by facilitated diffusion (Chapter 3), and the rate of glucose transport into cells is greatly increased by insulin (Chapter 18). Immediately upon entry into cells, glucose combines with a phosphate group, produced by the breakdown of ATP. This addition of a phosphate group to a molecule as noted earlier is called phosphorylation. The product of this enzymatically catalyzed reaction is glucose-6-phosphate (see Figure 25-9). In most cells of the body, phosphorylation serves to capture glucose in the cell so that it cannot move back out. Liver cells, kidney tubule cells, and intestinal epithelial cells have the necessary enzyme (phosphatase) to remove the phosphate group and transport glucose out of the cell (see Figure 25-9).

Glucose Catabolism

The ***oxidation*** of glucose is also known as ***cellular respiration.*** It occurs in every cell in the body (except red blood cells which lack mitochondria) and provides the cell's chief source of energy. The complete oxidation of glucose to carbon dioxide and water produces large amounts of energy. It occurs in three successive stages: glycolysis, the Krebs cycle, and the electron transport chain (see Figure 25-8).

Glycolysis

The term ***glycolysis*** (glī-KOL-i-sis; *glyco* = sugar; *lysis* = breakdown) refers to a series of chemical reactions in the cytoplasm of a cell that converts a six-carbon molecule of glucose into two three-carbon molecules of pyruvic acid. The simplified overall reaction for glycolysis is:

$$\underset{\text{Glucose}}{C{-}C{-}C{-}C{-}C{-}C} \longrightarrow 2\ \underset{\text{Pyruvic acid}}{C{-}C{-}C} + 2\ \text{ATP}$$

The details of glycolysis are shown in Figure 25-4a. The essential features of the process are:

1. Each of the 10 reactions is catalyzed by a specific enzyme. (The names of the enzymes are indicated in the rust color.)

2. All the reactions are reversible.

FIGURE 25-4 Glycolysis. As a result of glycolysis, there is a net gain of two molecules of ATP. (a) Detailed version (continues on page 792).

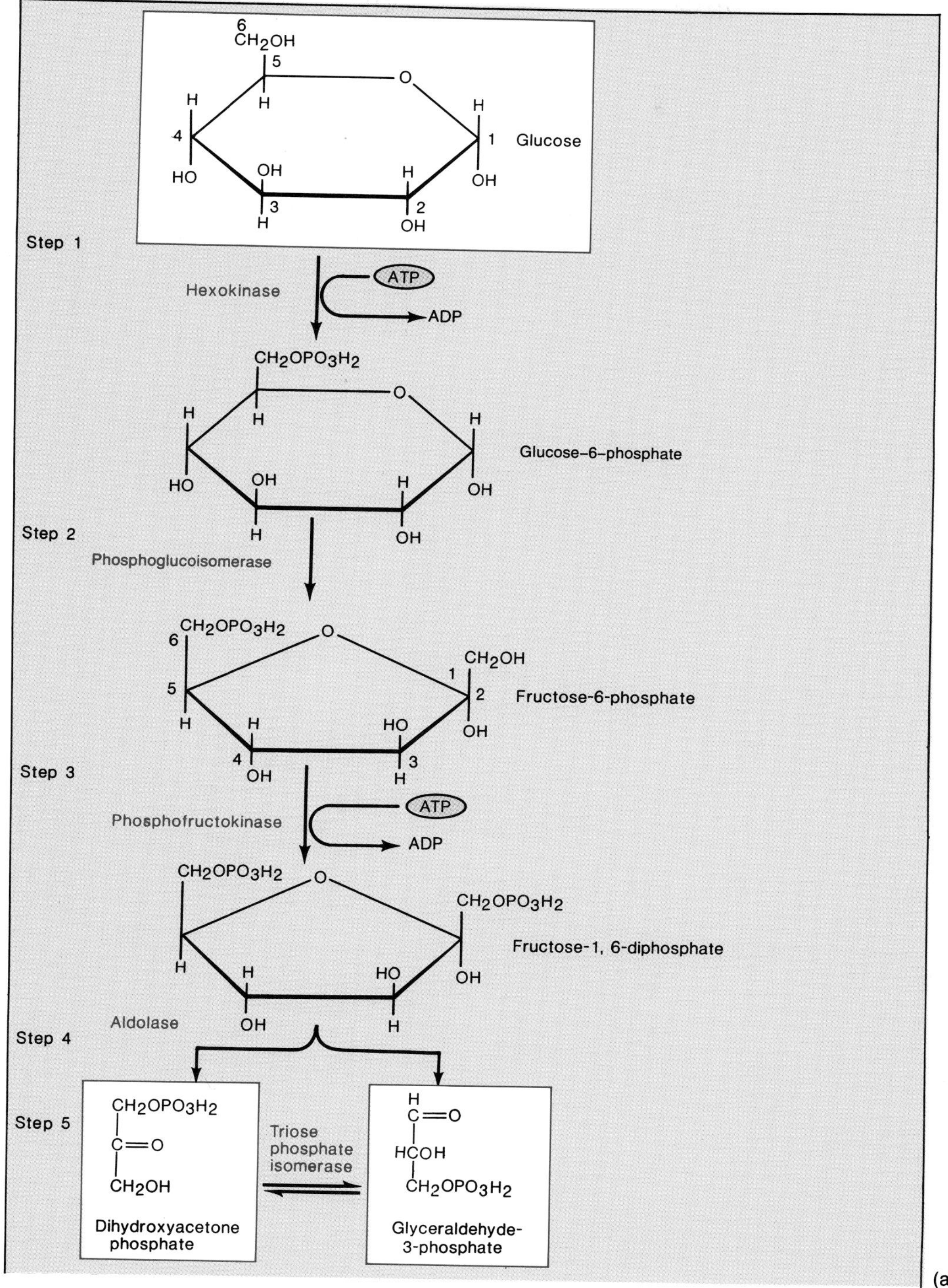

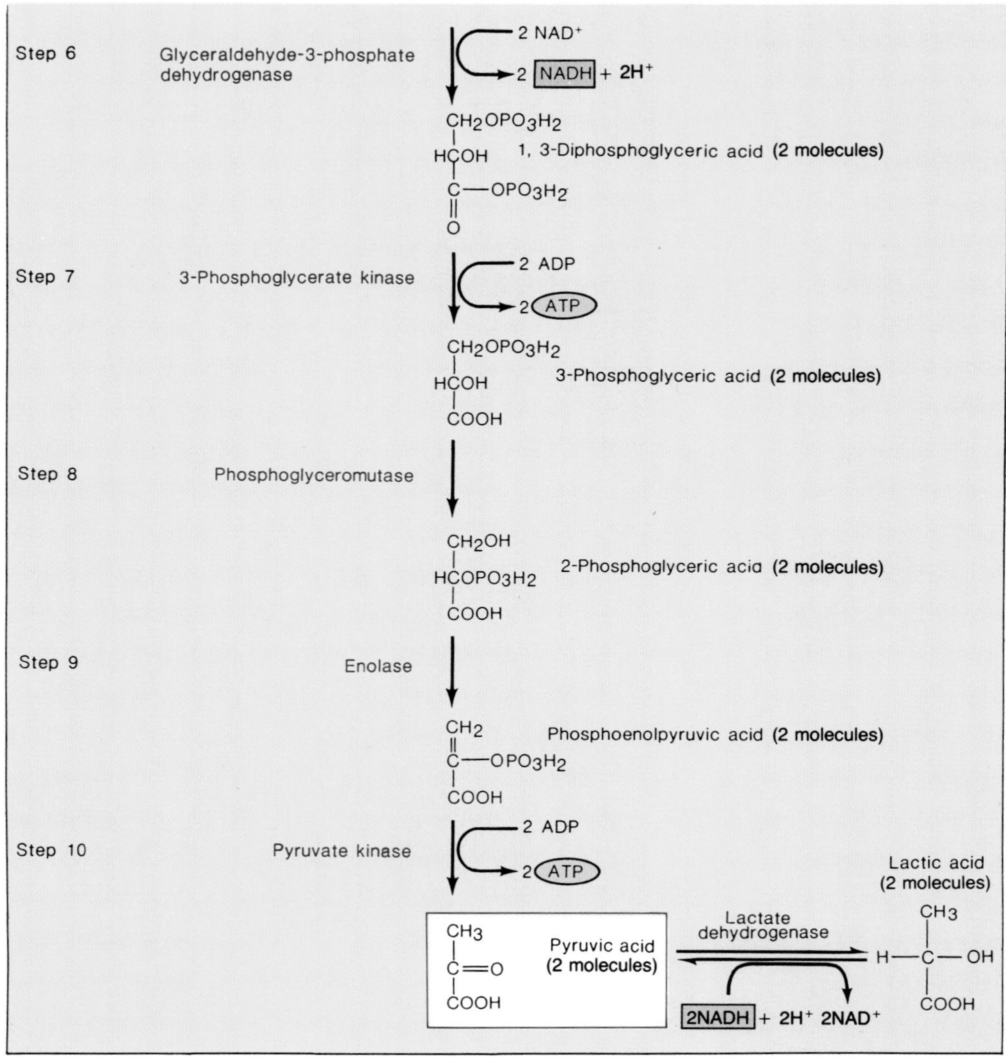

(a)

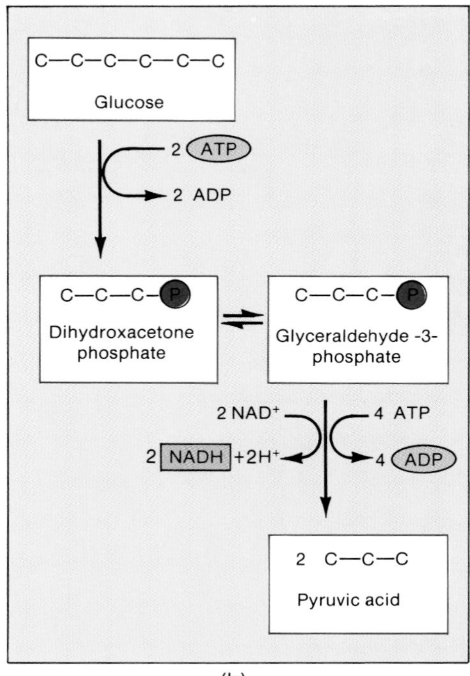

(b)

FIGURE 25-4 (*Continued*) (a) Detailed version (continued). (b) Simplified version.

3. The first three reactions involve the addition of a phosphate group (phosphorylation) to glucose, its conversion to fructose, and the addition of another phosphate group to fructose to prime the pathway. This involves an input of energy in which two molecules of ATP are converted to ADP.

4. The double phosphorylated molecule of fructose splits into two three-carbon compounds, glyceraldehyde-3-phosphate and dihydroxyacetone phosphate. These two compounds are interconvertible.

5. Each of the two three-carbon compounds is degraded to form a molecule of pyruvic acid. In the process, each three-carbon compound generates two molecules of ATP.

6. Since two molecules of ATP are needed to get glycolysis started, and four molecules of ATP are generated to complete the process, there is a *net gain of two molecules of ATP for each molecule of glucose that is oxidized* (Figure 25-4b).

7. Most of the energy produced during glycolysis is used to generate ATP. The remainder is expended as heat energy, some of which helps to maintain body temperature.

8. There is also the production of two molecules of NADH + H^+.

The fate of pyruvic acid depends on the availability of oxygen. If anaerobic (without oxygen) conditions exist in certain cells (as during strenuous exercise), pyruvic acid is reduced by the addition of two hydrogen atoms to form lactic acid (Figure 25-4a). The lactic acid may be transported to the liver, where it is converted back to pyruvic acid. Or, it may remain in the cells until aerobic (with oxygen) conditions are restored (recall the mechanisms for the repayment of the oxygen debt described in Chapter 10), and it is then converted to pyruvic acid in the cells. (Neurons do not produce lactic acid.)

Under aerobic conditions, the process of the complete oxidation of glucose continues in mitochondria, and pyruvic acid is oxidized to form carbon dioxide and water in two sets of reactions: the Krebs cycle and the electron transport chain. This is known as ***aerobic (cellular) respiration.*** However, before pyruvic acid can enter the Krebs cycle, it must be prepared. This process occurs in the matrix of mitochondria. Although the outer mitochondrial membrane is permeable to most molecules, the inner mitochondrial membrane is impermeable to most. In general, the only molecules that cross the inner membrane are those for which specific transport proteins are present. It is through this mechanism that pyruvic acid enters mitochondria.

Formation of Acetyl Coenzyme A

Each step in the oxidation of glucose requires a different enzyme and often a coenzyme as well. We are interested in only one coenzyme at this point: ***coenzyme A (CoA).***

During the transitional step between glycolysis and the Krebs cycle, pyruvic acid is prepared for entrance into the cycle. Essentially, pyruvic acid is converted to a two-carbon compound by the loss of carbon dioxide. The loss of a molecule of CO_2 by a substance is called ***decarboxylation*** (dē-kar-bok'-si-LĀ-shun). (It is indicated in the diagram by the arrow labeled CO_2.) This two-carbon fragment, called an ***acetyl group,*** attaches itself to coenzyme A, and the whole complex is called ***acetyl coenzyme A*** (Figure 25-5). During this reaction, NAD^+ is reduced to NADH + H^+. Remember that the oxidation of one glucose molecule produces two molecules of pyruvic acid, so for each molecule of glucose two molecules of carbon dioxide are lost, and two NADH + H^+ are produced. Each molecule of NADH will later yield three molecules of ATP in the electron transport chain. Once the pyruvic acid has undergone decarboxylation and its derivative (the acetyl group) has attached to CoA, the resulting compound (acetyl CoA) is ready to enter the Krebs cycle.

Krebs Cycle

The ***Krebs cycle*** is also called the ***citric acid cycle,*** or the ***tricarboxylic acid (TCA) cycle.*** It is a series of biochemical reactions that occur in the matrix of mitochondria of cells in which the large amount of potential energy stored in intermediate substances ultimately derived from pyruvic acid is released step by step (Figure 25-6a,b). In this cycle, a series of oxidations and reductions transfers the potential energy, in the form of electrons, to a number of coenzymes. The pyruvic acid derivatives are oxidized, whereas the coenzymes are reduced.

FIGURE 25-5 Formation of acetyl coenzyme A from pyruvic acid for entrance into the Krebs cycle.

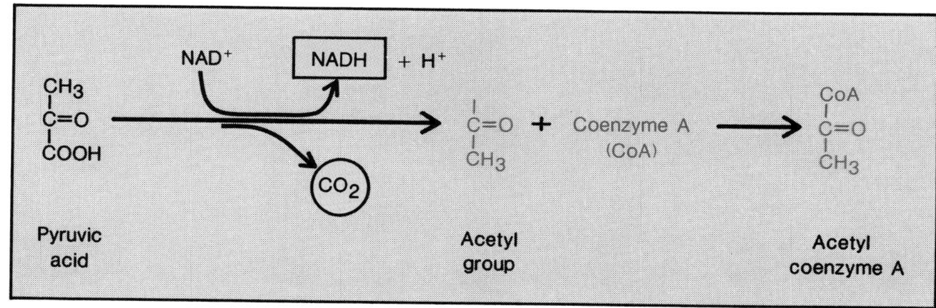

FIGURE 25-6 **Krebs cycle. The net results of the Krebs cycle are the production of reduced coenzymes (NADH + H^+ and $FADH_2$), which contain stored energy; the generation of GTP, a high-energy compound that is used to produce ATP; and the formation of CO_2, which is transported to the lungs for expiration. (a) Detailed version.**

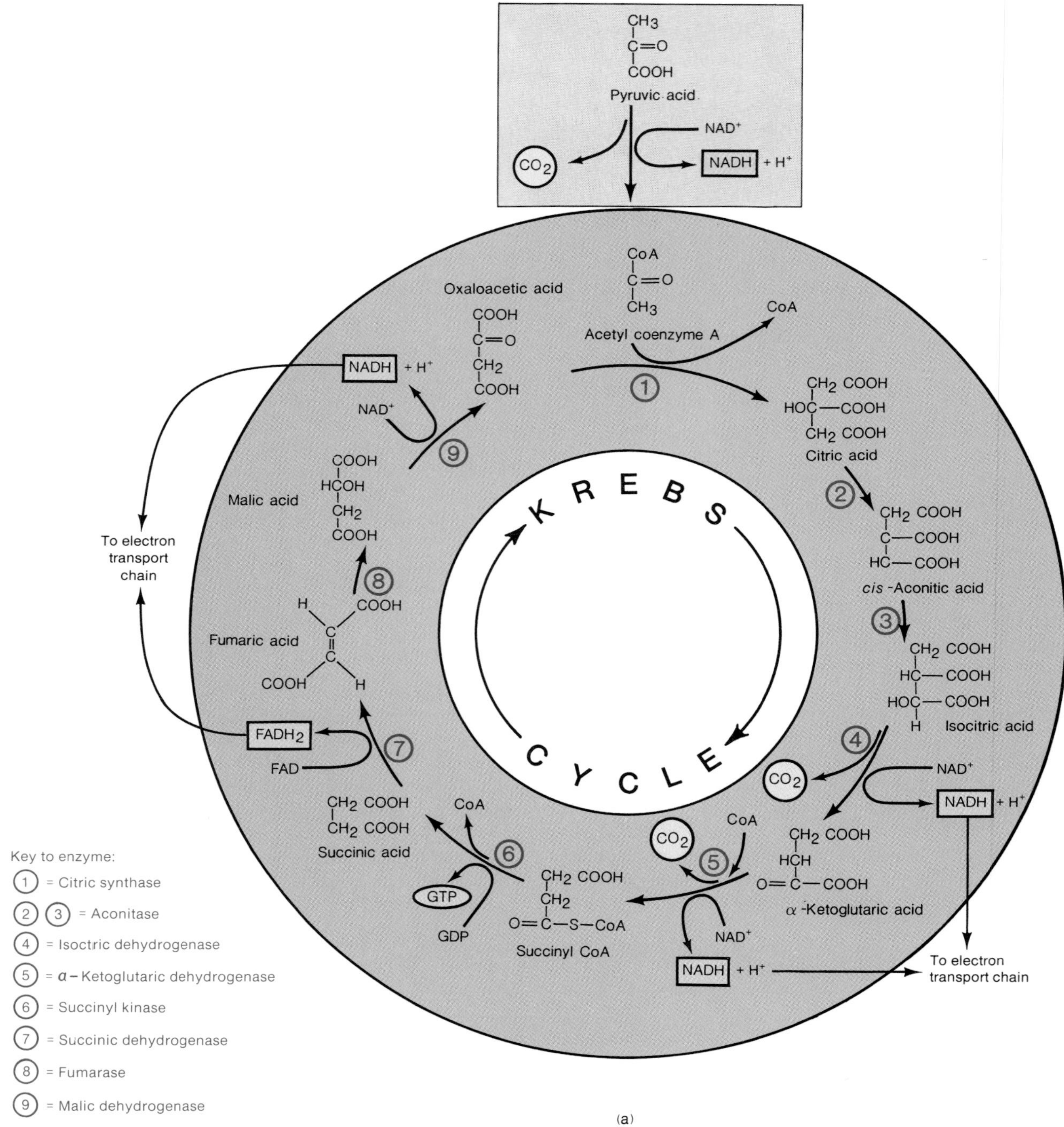

(a)

FIGURE 25-6 (*Continued*) (b) Simplified version.

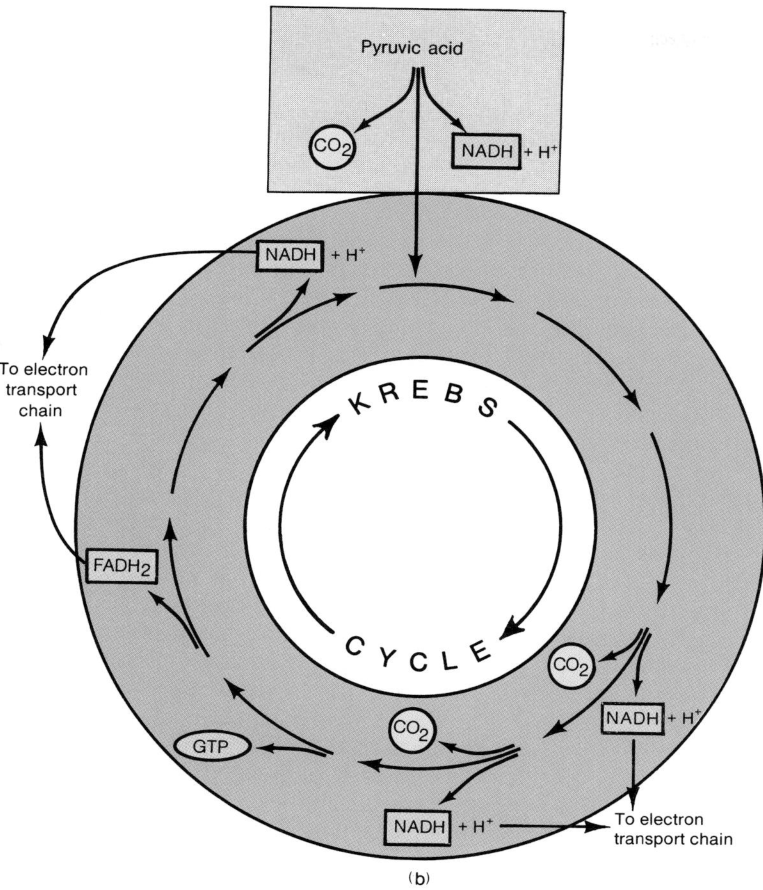

Coenzyme A carries an acetyl group to a mitochondrion, detaches itself, and goes back into the cytoplasm to pick up another fragment. The acetyl group combines with oxaloacetic acid to form citric acid. From this point, the Krebs cycle consists mainly of a series of decarboxylation and oxidation–reduction reactions, each controlled by a different enzyme.

Let us look at the decarboxylation reactions first. As we have seen (Figure 25-5), in preparation for entering the cycle, pyruvic acid is decarboxylated to form an acetyl group. In the cycle, isocitric acid, a six-carbon compound, loses a molecule of CO_2 to form a five-carbon compound called α-ketoglutaric acid. In the next step, α-ketoglutaric acid is decarboxylated (and picks up a molecule of CoA) to form succinyl CoA, a four-carbon compound. Thus, each time pyruvic acid enters the Krebs cycle, three molecules of CO_2 are liberated by decarboxylation. The molecules of CO_2 leave the mitochondria, diffuse through the cytoplasm of the cell to the plasma membrane, and then diffuse into the blood by internal respiration. Eventually, the CO_2 is transported by the blood to the lungs by external respiration. Once it reaches the lungs, it is exhaled.

Now let us look at the oxidation–reduction reactions (Figure 25-6a). Remember, when a molecule is oxidized, it loses hydrogen atoms (electrons and hydrogen ions). When a molecule is reduced, it gains hydrogen atoms (electrons and hydrogen atoms). Every oxidation is coupled with a reduction.

1. In the conversion of pyruvic acid to acetyl CoA (see Figure 25-5), pyruvic acid loses two hydrogen atoms; that is, it is oxidized. The hydrogen atoms are picked up by a coenzyme called ***nicotinamide adenine dinucleotide*** **(NAD^+).** NAD^+ contains two adenine nucleotides and is derived from a B vitamin called niacin. Since NAD^+ picks up the hydrogen atoms, it is reduced and represented as NADH + H^+ (indicated in the diagram by the curved arrow entering and then leaving the reaction).

2. Isocitric acid is oxidized to form α-ketoglutaric acid. NAD^+ is reduced to NADH + H^+.

3. α-Ketoglutaric acid is oxidized in the conversion to succinyl CoA. Again, NAD^+ is reduced to NADH + H^+.

4. Succinic acid is oxidized to fumaric acid. In this case, the coenzyme that picks up the two hydrogen atoms is ***flavin adenine dinucleotide*** **(*FAD*).** FAD, like NAD^+, contains two adenine nucleotides, but it is derived from vitamin B_2 (riboflavin). It belongs to a class of compounds called ***flavoproteins.*** FAD is reduced to $FADH_2$.

5. Malic acid is oxidized to oxalocetic acid. NAD^+ is reduced to NADH + H^+.

If we look at the Krebs cycle as a whole, we see that for every two molecules of acetyl CoA that enter the cycle four molecules of carbon dioxide are liberated by decarboxylation, 6 NADH + $6H^+$ and 2 $FADH_2$ are produced by oxidation–reduction reactions, and two molecules of guanosine triphosphate (GTP, the equivalent of ATP) are generated by substrate-level phosphorylation. Many of the intermediates in the Krebs cycle also play a role in other pathways, especially in amino acid biosynthesis (discussed later in this chapter). In the electron transport chain, the 6 NADH + $6H^+$ will later yield 18 ATP molecules, and the 2 $FADH_2$ will yield 4 ATP molecules. The reduced coenzymes (NADH + H^+ and $FADH_2$) are the most important outcome of the Krebs cycle because they contain the energy originally stored in glucose and then stored in pyruvic acid. During the next phase of aerobic respiration, a series of reductions transfers the energy stored in the coenzymes to ADP + P to form ATP. These reactions are collectively called the electron transport chain and occur on the cristae of the inner mitochondrial membrane.

Electron Transport Chain

The ***electron transport chain*** consists of a sequence of carrier molecules on the inner mitochondrial membrane that are capable of oxidation and reduction. As electrons are passed through the chain, there is a stepwise release of energy from the electrons for the generation of ATP. In aerobic respiration, the terminal electron acceptor of the chain is molecular oxygen (O_2) and the final oxidation is irreversible.

The respiration of organic compounds in electron transport chains involves three classes of carrier molecules. The first are called ***flavoproteins.*** These proteins contain flavin, a coenzyme derived from riboflavin (vitamin B_2), and are capable of performing alternating oxidations and reductions. One important flavin coenzyme is called ***flavin mononucleotide (FMN).*** The second class of carrier molecules is referred to as ***cytochromes*** (SĪ-tō-krōms), proteins with an iron-containing group (heme) capable of existing alternately as a reduced form (Fe^{2+}) and as an oxidized form (Fe^{3+}). The several cytochromes involved in the electron transport chain described shortly are cytochrome *b* (cyt *b*), cytochrome c_1 (cyt c_1), cytochrome *c* (cyt *c*), cytochrome *a* (cyt *a*), and cytochrome a_3 (cyt a_3). The third class of carriers is known as ***ubiquinones (coenzyme Q),*** symbolized ***Q,*** nonprotein carriers of low molecular weight.

The first step in the electron transport chain involves the transfer of high-energy electrons from NADH + H^+ to FMN, the first carrier in the chain (Figure 25-7a). Two NADH + $2H^+$ are generated from glycolysis, two from the formation of acetyl CoA, and six from the Krebs cycle. This transfer actually involves the passage of a hydrogen atom with two electrons to FMN, which then picks up an additional H^+ from the surrounding aqueous medium. As a result of the first transfer, NADH + H^+ is oxidized to NAD^+, and FMN is reduced to $FMNH_2$.

In the second step in the electron transport chain, $FMNH_2$ passes electrons (hydrogen atoms each with two electrons) to Q, which picks up an additional H^+ from the surrounding aqueous medium. As a result, $FMNH_2$ is oxidized to FMN.

The next sequence in the electron transport chain involves mostly cytochromes located between Q and molecular oxygen. Electrons are passed successively from Q to cyt *b,* cyt c_1, cyt *c,* cyt *a,* and cyt a_3. Each cytochrome in the chain is reduced as it picks up electrons and is oxidized as it gives up electrons. The last cytochrome, cyt a_3, passes its electrons to one-half of a molecule of oxygen (O_2), which becomes negatively charged and then picks up $2H^+$ from the surrounding medium to form H_2O.

Note in Figure 25-7a that $FADH_2$, derived from the Krebs cycle, is another source of electrons. However, $FADH_2$ adds its electrons to the electron transport chain at a lower level than NADH + H^+. Because of this, the electron transport chain produces about one-third less energy for ATP generation when $FADH_2$ donates electrons as compared with NADH + H^+.

An important feature of the electron transport chain is the presence of some carriers, such as FMN and Q, that accept H^+ as well as electrons, and other carriers, such as cytochromes, that transfer electrons only. As a result, electron flow down the chain is accompanied at several points by the active transport (pumping) of protons (H^+) from the matrix side of the inner mitochondrial membrane to the opposite side of the membrane. The result is the proton gradient described earlier, which is responsible for the chemiosmotic mechanism for the generation of ATP (see Figure 25-3).

Within the mitochrondrial membrane, the carriers of the electron transport chain are organized into three complexes: (1) ***NADH dehydrogenase complex,*** which contains FMN; (2) ***cytochrome b–c_1 complex,*** which contains cytochromes *b* and c_1; and (3) ***cytochrome oxidase complex,*** which contains cytochromes *a* and a_3 (Figure 25-7b). Q transfers electrons between the first and second complexes, and cytochrome *c* transfers electrons between the second and third complexes. In summary, there are three components of the electron transport chain that pump protons (H^+): the NADH dehydrogenase complex, the carrier Q, and the cytochrome *b*–c_1 complex.

Recall that the potential energy stored in the proton gradient is the energy source for the conversion of ADP and P to ATP by the adenosine triphosphatase (ATPase) complexes in the membrane. By coupling electron transport to the generation of ATP, the proton gradient functions as the main mechanism of oxidative phosphorylation. The chemiosmotic generation of ATP is responsible for most of the ATP produced during cellular respiration.

The various electron transfers in the electron transport chain generate 34 ATP molecules from each molecule of glucose oxidized: 3 from each of the 10 molecules of NADH + H^+ (30) and 2 from each of the 2 molecules of $FADH_2$ (4). During aerobic respiration, a total of 38 ATPs can be generated from one molecule of glucose. Note that four of those ATPs come from substrate-level phosphorylation in glycolysis and the Krebs cycle. Exhibit 25-1 provides a detailed accounting of the ATP yield during aerobic respiration.

The overall reaction for aerobic respiration is as follows:

$$\underset{\text{Glucose}}{C_6H_{12}O_6} + \underset{\text{Oxygen}}{6O_2} + 38\text{ ADP} + 38\text{ P} \longrightarrow \underset{\text{Carbon dioxide}}{6CO_2} + \underset{\text{Water}}{6H_2O} + 38\text{ ATP}$$

About 43 percent of the energy originally in glucose is captured by ATP; the remainder is given off as heat. Compared with man-made machines, this energy yield is quite efficient.

A summary of the various stages of aerobic respiration is presented in Figure 25-8.

The ATP generated in mitochondria must be transported out of these organelles into the cytoplasm for use elsewhere in a cell. This is accomplished by transport proteins in the inner mitochondrial membrane that couple the outward movement of ATP with the inner movement of ADP, that is formed from metabolic reactions in cytoplasm.

FIGURE 25-7 **Electron transport chain. (a) The energy drop for electrons passing through the chain is gradual and occurs in a stepwise fashion. (b) Organization of electron carries into three complexes. Note the three points at which protons (H^+) are pumped from the matrix side of the mitochondrial membrane to the opposite side.**

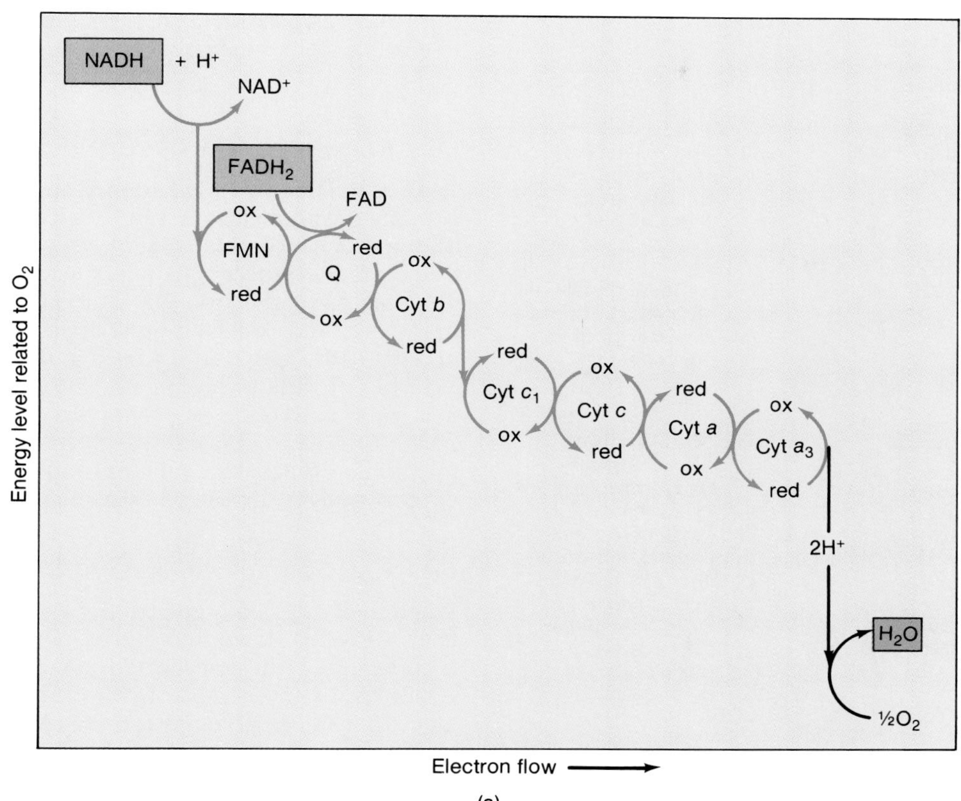

(a)

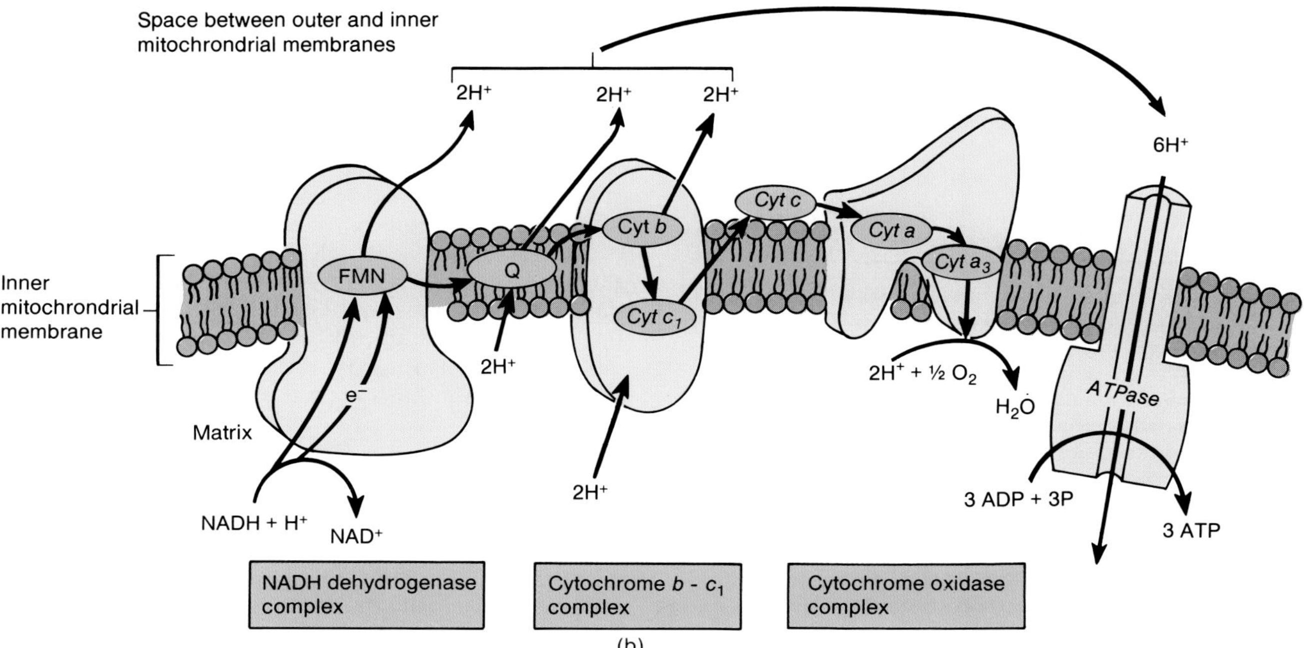

(b)

EXHIBIT 25-1 SUMMARY OF ATP PRODUCED DURING AEROBIC RESPIRATION OF ONE GLUCOSE MOLECULE

Source	ATP Yield (Method)
GLYCOLYSIS	
1. Oxidation of glucose to pyruvic acid	2 ATP (substrate-level phosphorylation)
2. Production of 2 NADH + 2H$^+$	6 ATP (oxidative phosphorylation in electron transport chain)
FORMATION OF ACETYL CoA	
2 NADH + 2H$^+$	6 ATP (oxidative phosphorylation in electron transport chain)
KREBS CYCLE	
1. Oxidation of succinyl CoA to succinic acid	2 GTP (equivalent of ATP; substrate-level phosphorylation)
2. Production of 6 NADH + 6H$^+$	18 ATP (oxidative phosphorylation in electron transport chain)
3. Production of 2 FADH$_2$	4 ATP (oxidative phosphorylation in electron transport chain)
Total:	38 ATP

Glycolysis, the Krebs cycle, and especially the electron transport chain provide all the ATP for cellular activities. And because the Krebs cycle and electron transport chain are aerobic processes, the cells cannot carry on their activities for long without sufficient oxygen.

CLINICAL APPLICATION: CARBOHYDRATE LOADING

During exercise, the body preferentially uses carbohydrates as a source of energy. Carbohydrates are stored as glycogen in the liver and skeletal muscles. Since glycogen stores are limited and can be depleted after running more than 32 km (20 mi), many marathon runners consume large amounts of carbohydrates prior to the event. This practice, called *carbohydrate loading,* is designed to increase body stores of carbohydrates to provide additional energy for an athletic event. It is frequently done for three days prior to the run. Some potential side effects of carbohydrate loading are muscle stiffness, diarrhea, chest pain, depression, and lethargy.

Glucose Anabolism

Most of the glucose in the body is catabolized to supply energy. However, some glucose participates in a number of anabolic reactions. One is the synthesis of glycogen from many glucose molecules. Another is the manufacture of glucose from the breakdown products of proteins and lipids.

Glucose Storage: Glycogenesis

If glucose is not needed immediately for energy, it is combined with many other molecules of glucose to form a long-chain molecule called glycogen. This process is called ***glycogenesis*** (*glyco* = sugar; *genesis* = origin). The body can store about 500 g (about 1.1 lb) of glycogen in the liver and skeletal muscle fibers (cells). Roughly 80 percent of the glycogen is stored in the skeletal muscles.

In the process of glycogenesis (Figure 25-9), glucose that enters cells is first phosphorylated to glucose-6-phosphate. This is then converted to glucose-1-phosphate, then to uridine diphosphate glucose, and finally to glycogen. Glycogenesis is stimulated by insulin from the pancreas.

Glucose Release: Glycogenolysis

When the body needs energy, the glycogen stored in the liver is broken down into glucose and released into the bloodstream to be transported to cells, where it will be catabolized. The process of converting glycogen back to glucose is called ***glycogenolysis*** (*lysis* = breakdown). Glycogenolysis usually occurs between meals.

Glycogenolysis does not occur exactly by the reverse process involved in glycogenesis (Figure 25-9). The first step in glycogenolysis involves the splitting off of glucose molecules from the branched glycogen molecule by phosphorylation to form glucose-1-phosphate. Phosphorylase, the enzyme that catalyzes this reaction, is activated by the hormones glucagon from the pancreas and epinephrine from the adrenal medulla. Glucose-1-phosphate is then converted to glucose-6-phosphate and finally to glucose. Phosphatase, the enzyme that converts glucose-6-phosphate into glucose and releases glucose into the blood, is present in liver cells but absent in skeletal muscle cells. In skeletal muscle cells, glycogen is broken down into glucose-1-phosphate, which is then catabolized for energy via glycolysis and the Krebs cycle. Moreover, pyruvic acid and lactic acid, produced by glycolysis in muscle cells, can be converted to glucose in the liver. In this way, muscle glycogen is an indirect source of blood glucose.

Formation of Glucose from Proteins and Fats: Gluconeogenesis

When your liver runs low on glycogen, it is time to eat. If you do not eat, your body starts catabolizing more fats and proteins. Actually, the body normally catabolizes some of its fats and a few of its proteins. But large-scale fat and

FIGURE 25-8 **Summary of the complete oxidation of glucose. (a) Overall process. (b) Sites of principal events.**

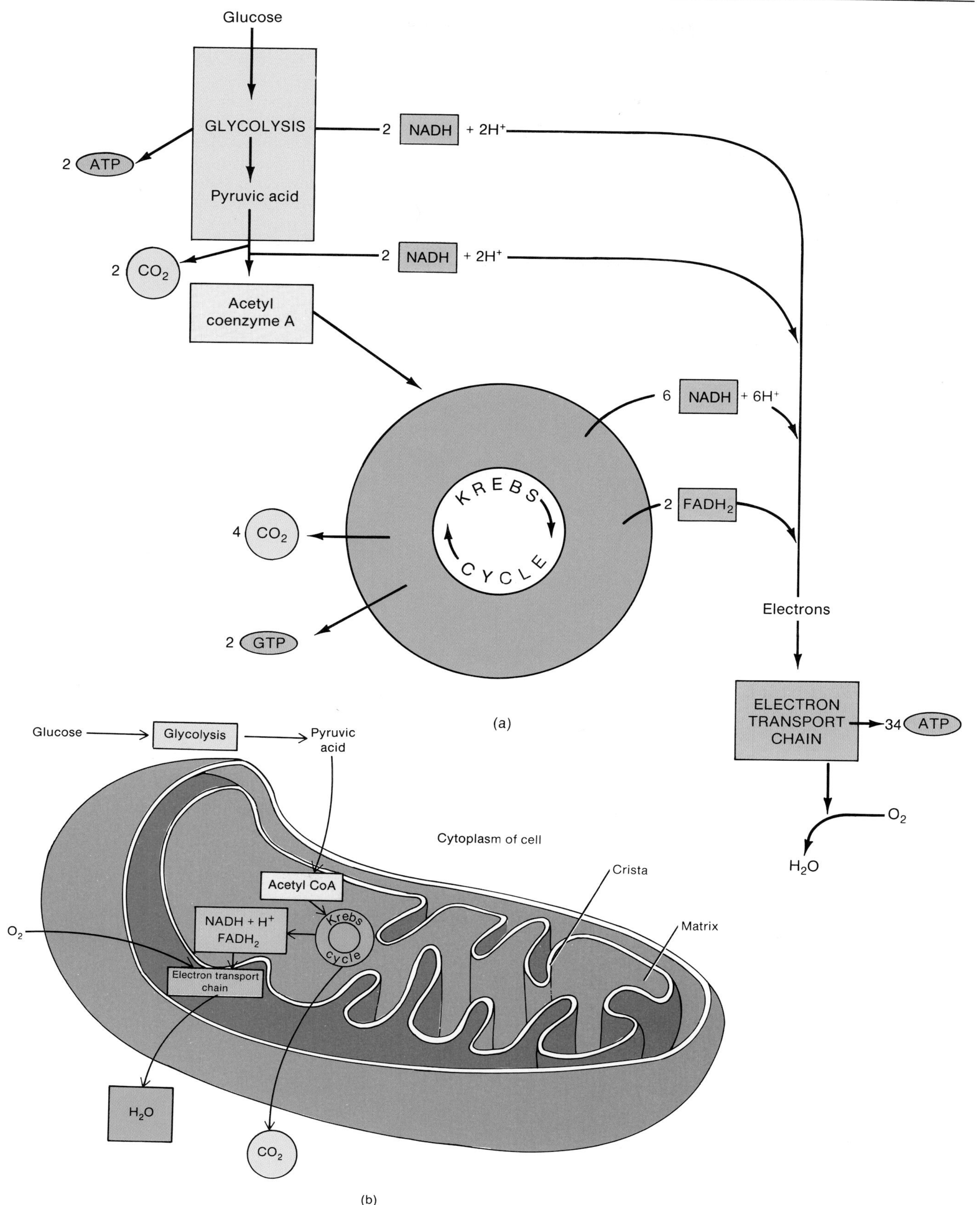

protein catabolism does not happen unless you are starving, eating meals that contain very few carbohydrates, or suffering from an endocrine disorder.

Both fat molecules and protein molecules may be converted in the liver to glucose. The process by which glucose is formed from noncarbohydrate sources is called ***gluconeogenesis*** (gloo'-kō-nē'-ō-JEN-e-sis).

In the process of gluconeogenesis, moderate quantities of glucose can be formed from certain amino acids and the glycerol portion of fat molecules (Figure 25-10). About 60 percent of the amino acids in the body can undergo this conversion. Amino acids such as alanine, cysteine, glycine, serine, and threonine are converted to pyruvic acid. The pyruvic acid may be resynthesized into glucose or enter the Krebs cycle. Glycerol may be converted into glyceraldehyde-3-phosphate, which may also be resynthesized into glucose or form pyruvic acid. Figure 25-10 also shows how gluconeogenesis is related to other metabolic reactions.

Gluconeogenesis is stimulated by cortisol, one of the glucocorticoid hormones of the adrenal cortex, and thyroxine from the thyroid gland. Cortisol mobilizes proteins from body cells, making them available in the form of

FIGURE 25-9 Glycogenesis and glycogenolysis. The glycogenesis pathway, the conversion of glucose into glycogen, is indicated by red arrows. The glycogenolysis pathway, the conversion of glycogen into glucose, is indicated by blue arrows.

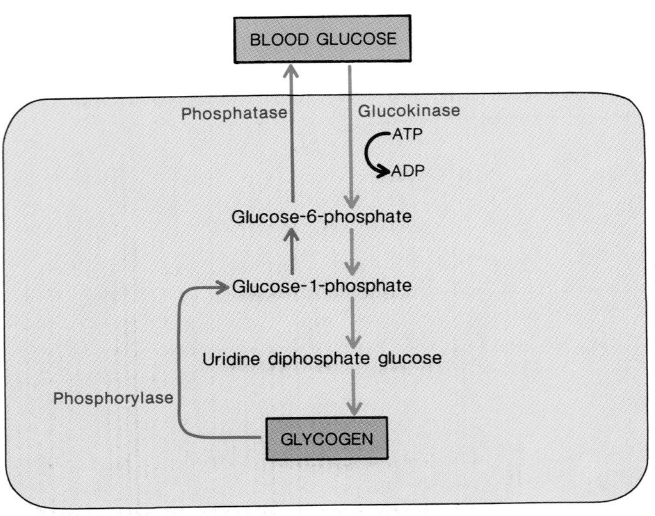

FIGURE 25-10 Gluconeogenesis. This process, which occurs in the liver, involves the conversion of noncarbohydrate sources (amino acids and glycerol) into glucose. Note also where glycogenesis and glycogenolysis occur.

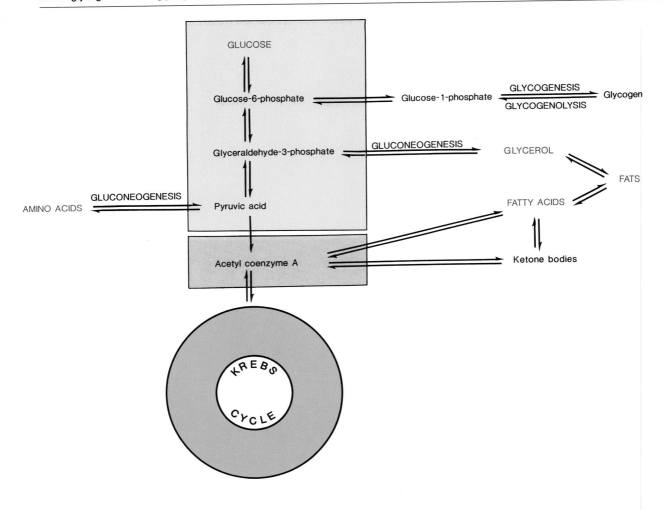

GLUCOSE
Glucose-6-phosphate
Glucose-1-phosphate
GLYCOGENESIS
GLYCOGENOLYSIS
Glycogen
Glyceraldehyde-3-phosphate
GLUCONEOGENESIS
GLYCEROL
FATS
AMINO ACIDS
GLUCONEOGENESIS
Pyruvic acid
FATTY ACIDS
Acetyl coenzyme A
Ketone bodies
KREBS CYCLE

amino acids, thus supplying a pool of amino acids for gluconeogenesis. Thyroxine also mobilizes proteins and may mobilize fats from fat depots by making glycerol available for gluconeogenesis. Gluconeogenesis is also stimulated by epinephrine, glucagon, and human growth hormone (hGH).

PHYSIOLOGY OF LIPID METABOLISM

Lipids are second to carbohydrates as a source of energy. More frequently they are used as building blocks to form essential structures. When neutral fats (triglycerides) are eaten, they are ultimately digested into fatty acids and monoglycerides. Short-chain fatty acids enter the blood capillaries in villi, whereas long-chain fatty acids and monoglycerides are carried in micelles to epithelial cells of the villi for entrance. Once inside, they are further digested to glycerol and fatty acids, recombined to form triglycerides, and transported as chylomicrons through the lymph of the lacteals of villi into the thoracic duct, subclavian vein, and to the liver.

Fate of Lipids

Lipids, such as fatty acids and glycerol, like carbohydrates, may be oxidized to produce ATP. Each gram of fat produces about 9.0 kilocalories. If the body has no immediate need to utilize fats this way, they are stored in adipose tissue (fat depots) throughout the body and in the liver. Other lipids are used as structural molecules or to synthesize other essential substances. For example, phospholipids are constituents of plasma membranes, lipoproteins are used to transport cholesterol throughout the body, thromboplastin is needed for blood clotting, and myelin sheaths speed up nerve impulse conduction. Cholesterol, another lipid, is used in the synthesis of bile salts and steroid hormones (adrenocortical hormones and sex hormones). The various functions of lipids in the body may be reviewed in Exhibit 2-4.

Fat Storage

The major function of adipose tissue is to store fats until they are needed for energy in other parts of the body. It also insulates and protects. About 50 percent of stored fat is deposited in subcutaneous tissue—approximately 12 percent around the kidneys, 10 to 15 percent in the omenta, 20 percent in genital areas, and 5 to 8 percent between muscles. Fat is also stored behind the eyes, in the sulci of the heart, and in the folds of the large intestine.

Adipose cells contain lipases that catalyze the deposition of fats from chylomicrons and hydrolyze fats into fatty acids and glycerol. Because of the rapid exchange of fatty acids, fats are renewed about once every two to three weeks. Thus, the fat stored in your adipose tissue today is not the same fat that was present last month. Moreover, fat is continually released from storage, transported through the blood, and redeposited in other adipose tissue cells.

Lipid Catabolism

Fats stored in adipose tissue constitute the largest reserve of energy. The body can store much more fat than it can store glycogen. Moreover, the energy yield of fats is more than twice that of carbohydrates. Nevertheless, fats are for the most part only the body's second-favorite source of energy because they are more difficult to catabolize than carbohydrates. (The heart, however, oxidizes fatty acids preferentially over glucose.)

Glycerol

Before fat molecules can be metabolized as an energy source, they must be split into glycerol and fatty acids and released from fat depots. Several hormones have significant effects on fat breakdown into fatty acids and glycerol. These include epinephrine, norepinephrine, glucocorticoids, thyroxine, and to a lesser extent, human growth hormone. Glycerol and fatty acids are then catabolized separately (Figure 25-11). The energy yield for fatty acids is considerably greater than that for glycerol.

Glycerol is converted easily by many cells of the body to glyceraldehyde-3-phosphate, one of the compounds also formed during the catabolism of glucose. The cells then transform glyceraldehyde-3-phosphate into glucose or continue the catabolic sequence to pyruvic acid. This is one example of gluconeogenesis. Since glyceraldehyde-3-phosphate is an intermediate product in the conversion of glycerol to glucose, glycerol thus enters the glycolytic pathway to be used for energy.

Fatty Acids

Fatty acids are catabolized differently, and the process occurs in the matrix of mitochondria. The first step in fatty acid catabolism involves a series of reactions called ***beta oxidation.*** Through a series of complex reactions involving dehydration, hydration, and cleavage, enzymes remove pairs of carbon atoms at a time from the long chain of carbon atoms composing a fatty acid. The result of beta oxidation is a series of two-carbon fragments, ***acetyl coenzyme A (CoA).*** Since the majority of fatty acids have even numbers of carbon atoms, the number of acetyl CoA molecules produced is easily calculated by dividing the number of carbon atoms in the fatty acid by two. Thus, palmitic acid, a fatty acid with 16 carbon atoms, will produce eight acetyl CoA molecules upon beta oxidation.

In the second step of fatty acid catabolism, the acetyl CoA formed as a result of beta oxidation enters the Krebs

FIGURE 25-11 Metabolism of lipids. Glycerol may be converted to glyceraldehyde-3-phosphate, which can then be converted to glucose or enter the Krebs cycle for oxidation. Fatty acids undergo beta oxidation and enter the Krebs cycle via acetyl coenzyme A. Fatty acids also can be converted into ketone bodies (ketogenesis). Lipogenesis is the synthesis of lipids from glucose or amino acids.

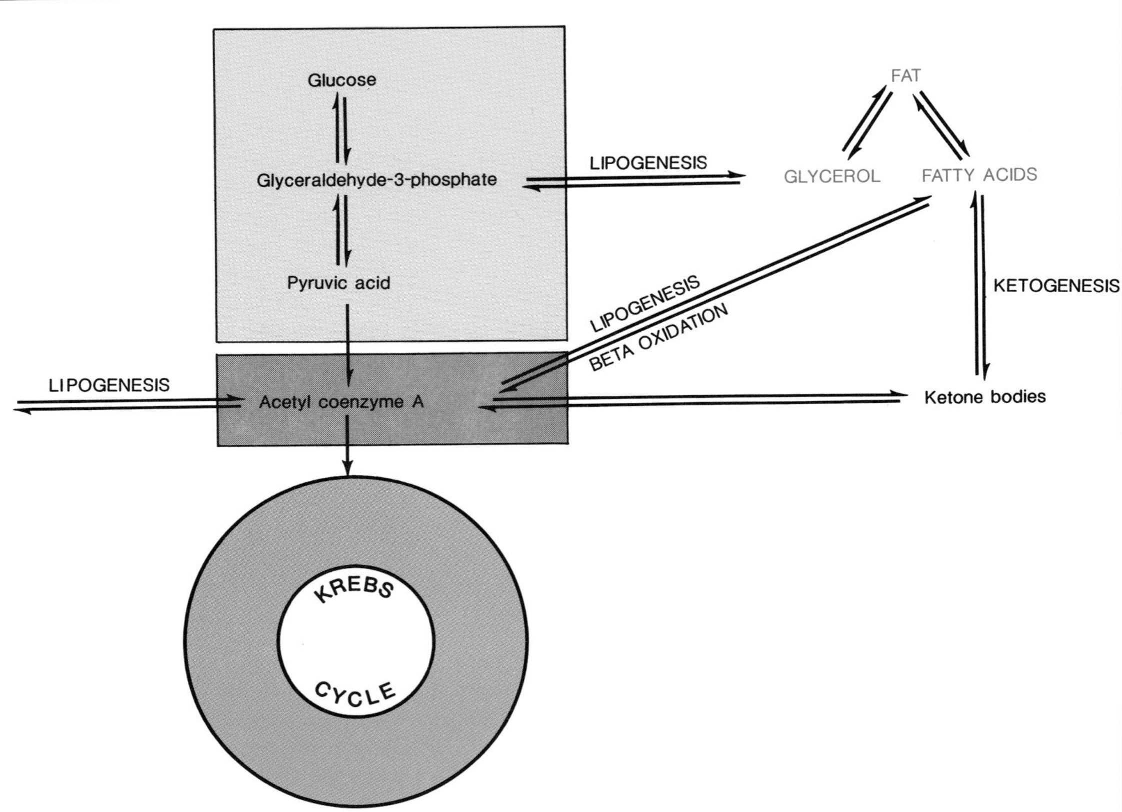

cycle (Figure 25-11). In terms of energy yield, a 16-carbon fatty acid, such as palmitic acid, can yield a net of 131 ATPs upon its complete oxidation via the Krebs cycle and electron transport chain.

As part of normal fatty acid catabolism, the liver can take two acetyl CoA molecules at a time and condense them to form a substance called ***acetoacetic acid,*** which is converted mostly into β-hydroxybutyric acid and partially into ***acetone,*** substances collectively known as ***ketone bodies.*** The formation of ketone bodies is called ***ketogenesis*** (Figure 25-11). They then leave the liver, enter the bloodstream, and diffuse into other body cells where they are broken down into acetyl CoA, which enters the Krebs cycle for oxidation. During periods of excessive beta oxidation, large amounts of acetyl CoA are produced. This might occur following a meal rich in fats, or during fasting or starvation, because essentially no carbohydrates are available for catabolism. It may also occur in diabetes mellitus because insulin is not available to stimulate glycogenesis and glucose transport into cells. When a diabetic becomes seriously insulin deficient, one of the telltale signs is a sweet smell of acetone on the breath.

CLINICAL APPLICATION: KETOSIS

Since the body for the most part prefers glucose as a source of energy, ketone bodies are generally produced in very small quantities. When the number of ketone bodies in the blood rises above normal—a condition called *ketosis*—the ketone bodies, most of which are acids, must be buffered by the body. If too many accumulate, they use up the body's buffers and the blood pH falls. Thus, extreme or prolonged ketosis can lead to *acidosis,* or abnormally low blood pH.

Lipid Anabolism: Lipogenesis

Liver cells and fat cells can synthesize lipids from glucose or amino acids through a process called ***lipogenesis*** (Figure 25-11). Lipogenesis occurs when a greater quantity of carbohydrate enters the body than can be used for energy or stored as glycogen. The excess carbohydrates are synthesized into fats. The steps in the conversion of glucose to lipids are complex and involve the formation of glyceraldehyde-3-phosphate, which can be converted to glycerol, and acetyl CoA, which can be converted to fatty acids (see Figure 25-10). The process is enhanced by insulin. The resulting glycerol and fatty acids (1) can undergo anabolic reactions to become fat that can be stored or (2) can go through a series of anabolic reactions that produce other lipids such as lipoproteins, phospholipids, and cholesterol.

Many amino acids can be converted into acetyl CoA, which can then be converted into fats (Figure 25-11). When people have more proteins in their diets than can be utilized as such, much of the excess protein is converted to and stored as fats.

PHYSIOLOGY OF PROTEIN METABOLISM

During the process of digestion, proteins are broken down into their constituent amino acids. The amino acids are then absorbed by the blood capillaries in villi and transported to the liver via the hepatic portal vein.

Fate of Proteins

Amino acids enter body cells by active transport. This process is stimulated by human growth hormone (hGH) and insulin. Almost immediately after entrance, they are synthesized into proteins. Other amino acids are stored as fat or glycogen or used for energy. Each gram of protein produces about 4.0 kcal. Many proteins function as enzymes. Other proteins are involved in transportation (hemoglobin); others serve as antibodies, clotting chemicals (fibrinogen), hormones (insulin), and contractile elements in muscle fibers (cells) (actin and myosin). Several proteins serve as structural components of the body (collagen, elastin, keratin, and nucleoproteins) The various functions of proteins in the body may be reviewed in Exhibit 2-5.

Protein Catabolism

A certain amount of protein catabolism occurs in the body each day, although much of this is only partial catabolism. Proteins are extracted from worn-out cells, such as red blood cells, and broken down into free amino acids. Some amino acids are converted into other amino acids, peptide bonds are reformed, and new proteins are made as part of the constant state of turnover in all cells.

If other energy sources are used up or if other sources are inadequate and protein intake is high, the liver can convert protein to fat or glucose or oxidize it to carbon dioxide and water. However, before amino acids can be catabolized, they must first be converted to various substances that can enter the Krebs cycle. One such conversion consists of removing the amino group (NH_2) from the amino acid, a process called ***deamination.*** The liver cells then convert the NH_2 to ammonia (NH_3) and finally to urea, which is excreted in the urine. Other conversions are decarboxylation and dehydrogenation. The fate of the remaining part of the amino acid depends on what kind of amino acid it is. Figure 25-12 shows that amino acids enter the Krebs cycle at different points. They may be converted to pyruvic acid, acetyl CoA, α-ketoglutaric acid, succinyl CoA, fumaric acid, oxaloacetic acid, or acetoacetyl CoA. The point is that amino acids can be altered in various ways to enter the Krebs cycle at various locations.

The gluconeogenesis of amino acids into glucose may be reviewed in Figure 25-10. The conversion of amino acids into fatty acids or ketone bodies is shown in Figure 25-11.

Protein Anabolism

Protein anabolism involves the formation of peptide bonds between amino acids to produce new proteins. Protein anabolism, or synthesis, is carried out on the ribosomes of almost every cell in the body, directed by the cells' DNA and RNA (see Figure 3-15). Protein synthesis is stimulated by human growth hormone (hGH), thyroxine, and insulin. The synthesized proteins are the primary constituents of enzymes, antibodies, clotting chemicals, hormones, structural components of cells, and so forth. Because proteins are a primary ingredient of most cell structures, high-protein diets are essential during the growth years, during pregnancy, and when tissue has been damaged by disease or injury.

Of the naturally occurring amino acids, 10 are referred to as ***essential amino acids.*** These amino acids cannot be synthesized by the human body from molecules present in the body. They are synthesized by plants or bacteria, and so foods containing these amino acids are "essential" for human growth and must be part of the diet. ***Nonessential amino acids*** can be synthesized by body cells by a process called ***transamination,*** the transfer of an amino group from an amino acid to a substance such as pyruvic acid or an acid of the Krebs cycle. Once the appropriate essential and nonessential amino acids are present in cells, protein synthesis occurs rapidly.

A summary of carbohydrate, lipid, and protein metabolism is presented in Exhibit 25-2.

ABSORPTIVE AND POSTABSORPTIVE STATES

The body actually alternates between two metabolic states, referred to as absorptive and postabsorptive. During the ***absorptive state,*** ingested nutrients are entering the cardiovascular and lymphatic systems from the gastro-

FIGURE 25-12 **Various points at which amino acids, shown in gold boxes, enter the Krebs cycle for oxidation.**

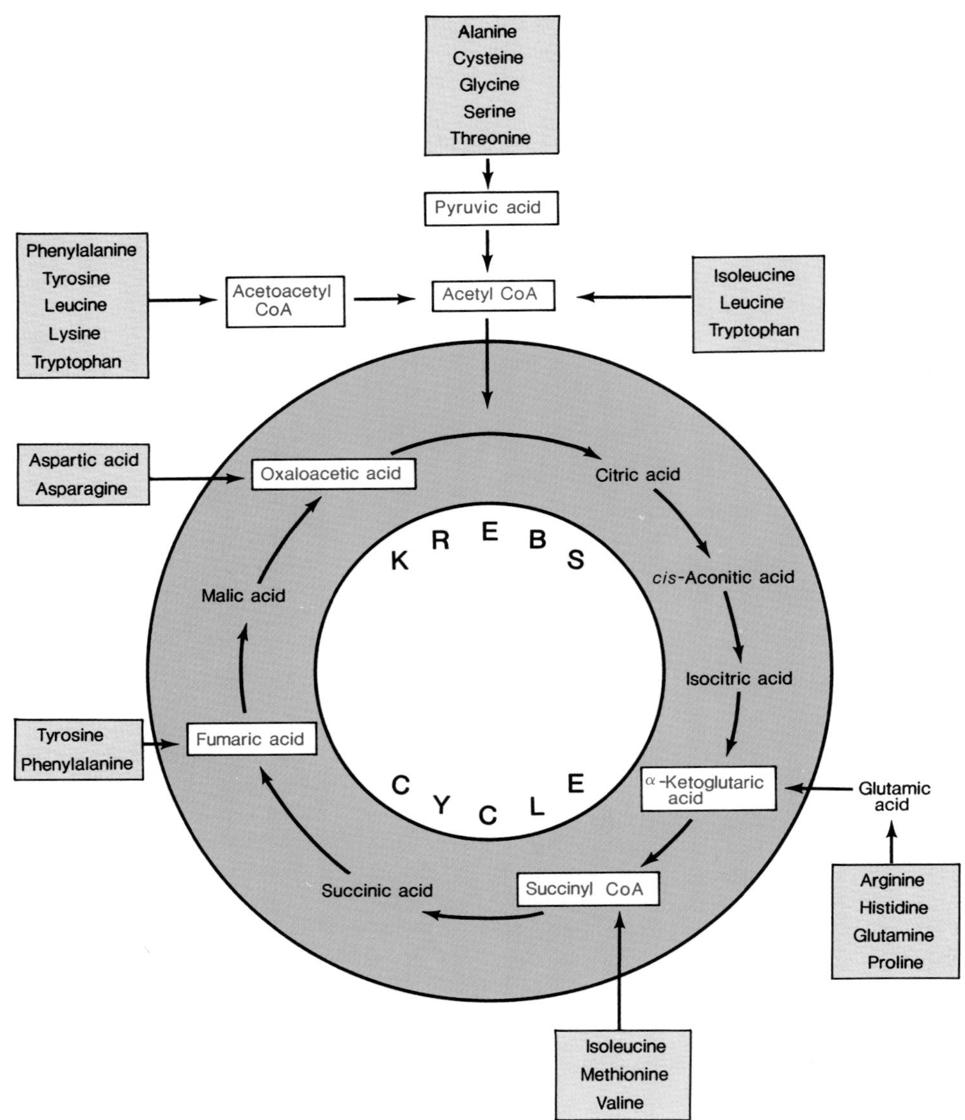

EXHIBIT 25-2 SUMMARY OF METABOLISM

Process	**Comment**
CARBOHYDRATE METABOLISM	
Glucose catabolism	Complete oxidation of glucose, also referred to as cellular respiration, is chief source of energy in cells. The process requires glycolysis, Krebs cycle, and electron transport chain. The complete oxidation of 1 molecule of glucose results in the net production of 38 molecules of ATP.
Glycolysis	Conversion of glucose into pyruvic acid results in the production of some ATP. Reactions do not require oxygen.
Krebs cycle	Cycle includes series of oxidation–reduction reactions in which coenzymes (NAD^+ and FAD) pick up hydrogen atoms from oxidized organic

EXHIBIT 25-2 SUMMARY OF METABOLISM (*Continued*)

Process	Comment
	acids, and some ATP is produced. CO_2 and H_2O are by-products. Reactions are aerobic.
Electron transport chain	Third set of reactions in glucose catabolism is another series of oxidation–reduction reactions, in which electrons are passed between FAD, coenzyme Q, and cytochromes, and most of the ATP is produced. Reactions are aerobic.
Glucose anabolism	Some glucose is converted into glycogen (glycogenesis) for storage if not needed immediately for energy. Glycogen can be converted to glucose (glycogenolysis) if needed for energy. The conversion of fats and proteins into glucose is called gluconeogenesis.
LIPID METABOLISM	
Catabolism	Glycerol may be converted into glucose (gluconeogenesis) or catabolized via glycolysis. Fatty acids are catabolized via beta oxidation into acetyl CoA that is catabolized in the Krebs cycle. Acetyl CoA can also be converted into ketone bodies.
Anabolism	The synthesis of lipids from glucose and amino acids is lipogenesis. Fats are stored in adipose tissue.
PROTEIN METABOLISM	
Catabolism	Amino acids are oxidized via the Krebs cycle after conversion by processes such as deamination. Ammonia resulting from the conversions is converted into urea in the liver and excreted in urine. Amino acids may be converted into glucose (glyconeogenesis) and fatty acids or ketone bodies.
Anabolism	Protein synthesis is directed by DNA and utilizes the cells' RNA and ribosomes.

intestinal tract. During the ***postabsorptive (fasting) state,*** absorption is complete and the energy needs of the body must be satisfied by nutrients already in the body. By analyzing the major events of both states, we can also gain a better understanding of how the metabolic pathways are interrelated.

Absorptive State

An average meal usually requires about 4 hours for complete absorption, and given three meals a day, the body spends about 12 hours each day in the absorptive state. The other 12 hours, during late morning, late afternoon, and most of the evening, are spent in the postabsorptive state.

During the absorptive state, glucose transported to the liver is mostly converted to fats or glycogen; little is oxidized for energy in the liver. Some fat synthesized in the liver remains there, but most of it is released into the blood for storage in adipose tissue. Adipose tissue cells also take up glucose not picked up by the liver and convert it into fat for storage. Some blood glucose is stored as glycogen in skeletal muscles. The majority of blood glucose is used by body cells for oxidation to carbon dioxide and water with the release of energy to form ATP (Figure 25-13).

During the absorptive state, most fat is stored in adipose tissue; only a small portion is used for synthesis. Adipose cells derive the fat from chylomicrons, from synthetic activities of the liver, and from their own synthetic reactions (Figure 25-13).

Many absorbed amino acids that enter liver cells are converted to carbohydrates called keto acids. These are catabolized to carbon dioxide and water to provide energy. Some keto acids are also synthesized into fatty acids in liver cells. These fatty acids may be synthesized into fats and stored in adipose tissue. Some amino acids that enter liver cells are used to synthesize proteins, such as plasma proteins. Amino acids not taken up by liver cells enter other cells of the body, such as muscle cells, for synthesis into proteins or regulatory chemicals such as hormones or enzymes.

Postabsorptive (Fasting) State

The principal concern of the body during the postabsorptive state is to maintain normal blood glucose level (70 to 110 mg/100 ml of blood), even though no absorption is occurring. The maintenance of this level is especially important to the functioning of the nervous system, which cannot use any other nutrient for energy.

In general, blood glucose concentrations are main-

FIGURE 25-13 **Principal pathways during the absorptive state.**

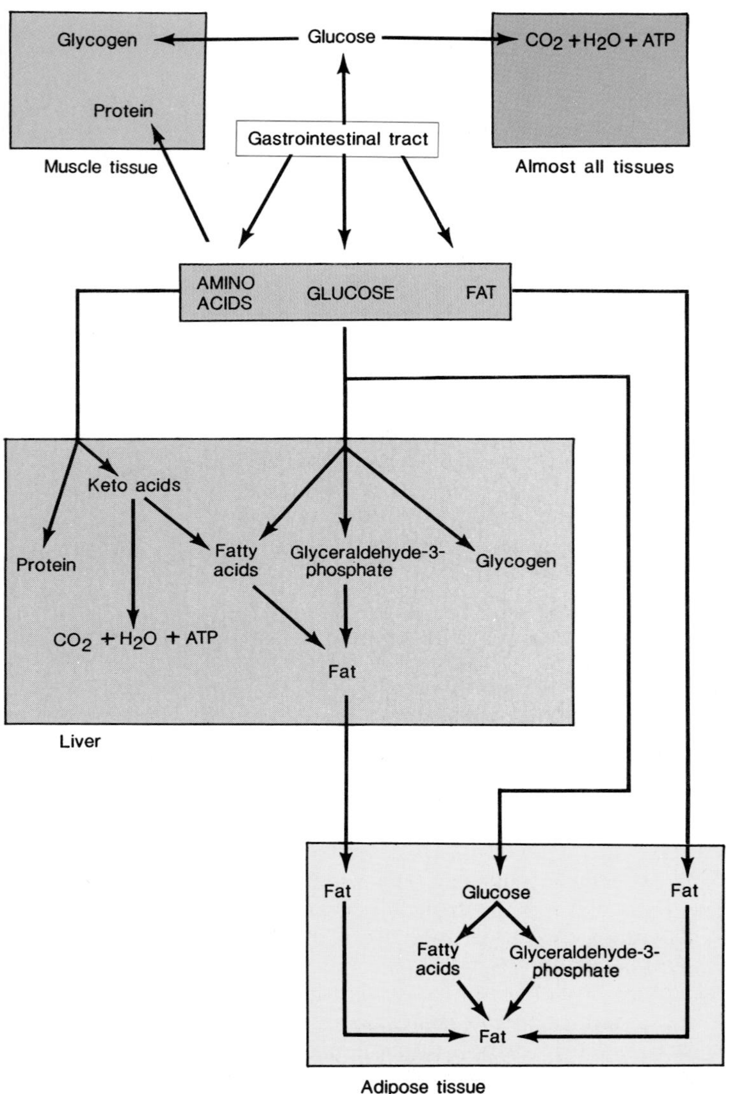

tained by utilization of various glucose sources and by glucose-sparing reactions such as the utilization of fat or possibly protein for energy. A major source of blood glucose during fasting is liver glycogen, which can provide only about a four-hour supply of glucose. Liver glycogen is continually being formed and broken down as needed in response to the concentration of glucose in the blood (Figure 25-14). Another source of blood glucose is glycerol, produced by the hydrolysis of triglycerides, primarily in adipose tissue. However, only the glycerol produced from fat breakdown can form glucose; the fatty acids cannot be utilized because acetyl CoA cannot be readily converted to pyruvic acid. But the fatty acids are oxidized directly by entering the Krebs cycle as acetyl CoA, thereby sparing the use of glucose for energy. A third source of glucose may be muscle glycogen during periods of vigorous exercise when significant amounts of lactic acid are produced by anaerobic glycolysis in muscle. Muscle lactic acid is released into the blood and carried to the liver, where it is converted back to glucose and released into the blood. Finally, tissue protein, primarily from skeletal muscle, can contribute to blood glucose. During fasting, large amounts of amino acids from protein breakdown are released from muscle to be converted to glucose in the liver by gluconeogenesis. Or the amino acids may be oxidized directly, thus sparing blood glucose. During fasting, amino acids from muscle contribute to blood glucose after liver glycogen and fat stores are depleted.

Despite all these mechanisms for supplying blood glucose, they cannot maintain blood glucose level for very long. Thus, a major body adjustment must be made during the postabsorptive state. Although the nervous system continues to utilize blood glucose normally, all other body tissues reduce their oxidation of glucose and switch over to fat as their energy source. Accordingly, fats in adipose tissue are broken down and fatty acids are released into the blood. The fatty acids are picked up by body cells, except nervous tissue, and oxidized to carbon dioxide and water, releasing energy used to form ATP. The liver converts fatty acids to ketone bodies, which enter most body cells and are then oxidized to carbon dioxide and water, releasing energy used to form ATP. As a result of this glucose sparing and fat utilization, an individual can fast for several weeks, provided water is consumed, and the glucose level will not drop more than 25% from its normal range.

REGULATION OF METABOLISM

Absorbed nutrients have several alternatives, based upon the needs of the body. They may be oxidized for energy, stored, or converted into other molecules. The pathway taken by a particular nutrient is enzymatically controlled and is regulated by hormones. In the discussion of the metabolism of carbohydrates, lipids, and proteins, reference has been made to the hormonal regulation involved. The actions of the principal hormones related to metabolism are summarized in Exhibit 25-3.

Hormones are the primary regulators of metabolism. However, hormonal control is ineffective without the proper minerals and vitamins. Some minerals and many vitamins are components of the enzyme systems that catalyze the metabolic reactions.

MINERALS

Minerals are inorganic substances. They may appear in combination with each other or in combination with organic compounds or as ions in solution. Minerals constitute about 4 percent of the total body weight, and they are concentrated most heavily in the skeleton. Minerals

FIGURE 25-14 **Principal pathways during the postabsorptive (fasting) state.**

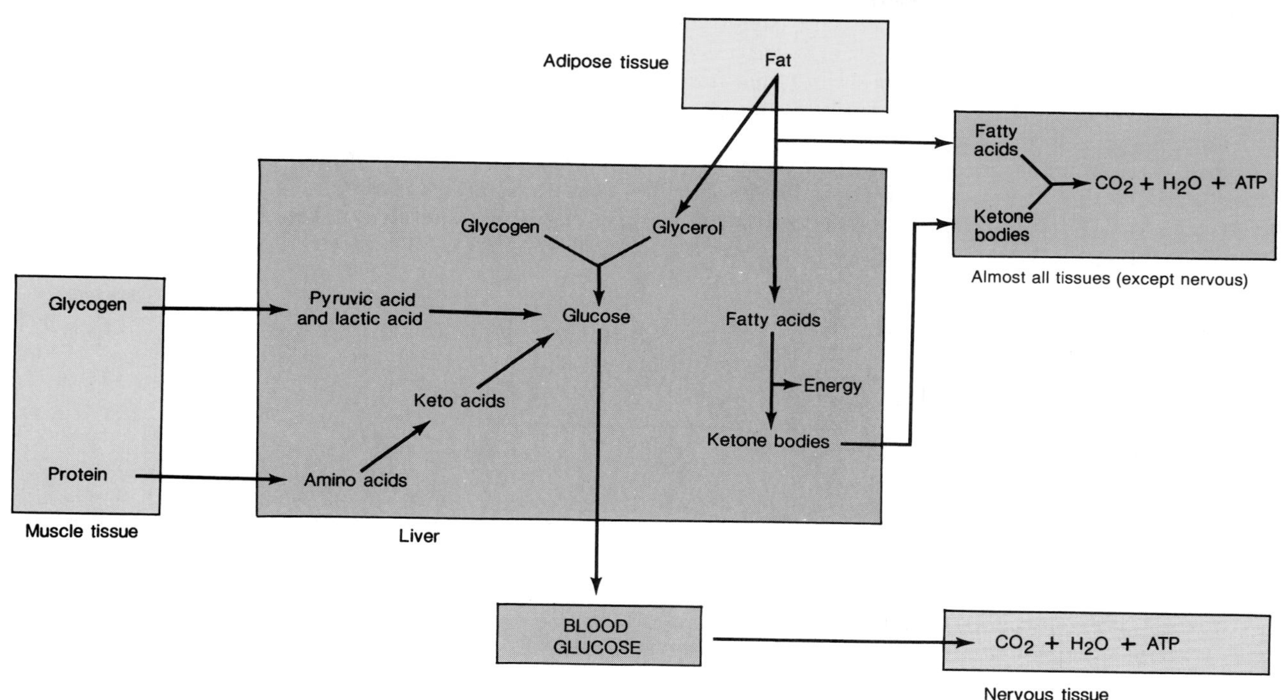

EXHIBIT 25-3 SUMMARY OF HORMONAL REGULATION OF THE METABOLISM OF CARBOHYDRATES, LIPIDS AND PROTEINS

Hormone	Actions
Insulin	Increases glucose uptake by cells, especially muscle and adipose cells. Stimulates conversion of glucose into glycogen (glycogenesis). Stimulates fat synthesis and inhibits fat breakdown. Stimulates active transport of amino acids into cells, especially muscle cells. Stimulates protein synthesis.
Glucagon	Stimulates conversion of glycogen into glucose (glycogenolysis). Stimulates conversion of noncarbohydrates into glucose (gluconeogenesis). Stimulates breakdown of fats (fat mobilization).
Epinephrine	Stimulates conversion of glycogen into glucose (glycogenolysis). Stimulates conversion of noncarbohydrates into glucose (gluconeogenesis). Stimulates breakdown of fats (fat mobilization).
Human growth hormone (hGH)	Stimulates active transport of amino acids into body cells, especially muscle cells. Stimulates protein synthesis. Stimulates conversion of noncarbohydrates into glucose (gluconeogenesis). Stimulates breakdown of fats (fat mobilization).
Thyroxine	Stimulates protein synthesis. Stimulates conversion of noncarbohydrates into glucose (gluconeogenesis). Stimulates breakdown of fats (fat mobilization).
Cortisol	Stimulates conversion of noncarbohydrates into glucose (gluconeogenesis).
Testosterone	Increases protein deposition in body cells, especially muscle cells.

known to perform functions essential to life include potassium, magnesium, chloride, iodine, calcium, phosphorus, sodium, manganese, cobalt, copper, zinc, selenium, and chromium. Other minerals—aluminum, silicon, arsenic, and nickel—are present in the body, but their functions have not yet been determined.

Calcium and phosphorus form part of the structure of bone. But since minerals do not form long-chain compounds, they are otherwise poor building materials. Their chief role is to help regulate body processes. Calcium, iron, magnesium, and manganese are constituents of some coenzymes. Magnesium also serves as a catalyst for the conversion of ADP to ATP. Without these minerals, metabolism would stop and the body would die. Minerals such as sodium and phosphorus work in buffer systems. Sodium helps regulate the osmosis of water and, along with other ions, is involved in the generation of nerve impulses. Exhibit 25-4 describes the functions of some minerals vital to the body. Note that the body generally uses the ions of the minerals rather than the nonionized form. Some minerals, such as chlorine, are toxic or even fatal to the body if ingested in the nonionized form.

EXHIBIT 25-4 MINERALS VITAL TO THE BODY

Mineral	Comments	Importance
Calcium	Most abundant cation in body. Appears in combination with phosphorus in ratio of 2:1.5. About 99 percent is stored in bone and teeth. Remainder stored in muscle, other soft tissues, and blood plasma. Blood calcium level controlled by calcitonin (CT) and parathyroid hormone (PTH). Absorption occurs only in the presence of vitamin D. Most is excreted in feces and small amount in urine. Sources are milk, egg yolk, shellfish, green leafy vegetables.	Formation of bones and teeth, blood clotting, normal muscle and nerve activity, endocytosis and exocytosis, cellular motility, chromosome movement prior to cell division, glycogen metabolism, and synthesis and release of neurotransmitters.
Phosphorus	About 80 percent found in bones and teeth. Remainder distributed in muscle, brain cells, blood. More functions than any other mineral. Blood phosphorous level controlled by calcitonin (CT) and parathyroid hormone (PTH). Most excreted in urine; small amount eliminated in feces. Sources are dairy products, meat, fish, poultry, nuts.	Formation of bones and teeth. Constitutes a major buffer system of blood. Plays important role in muscle contraction and nerve activity. Component of many enzymes. Involved in transfer and storage of energy (ATP). Component of DNA and RNA.
Iron	About 66 percent found in hemoglobin of blood. Remainder distributed in skeletal muscles, liver, spleen, enzymes. Normal losses of iron occur by shedding of hair, epithelial cells, and mucosal cells, and in sweat, urine, feces, and bile. Sources are meat, liver, shellfish, egg yolk, beans, legumes, dried fruits, nuts, cereals.	As component of hemoglobin, carries O_2 to body cells. Component of cytochromes involved in formation of ATP from catabolism. Large amounts of stored iron are associated with an increased risk of cancer. This may be related to the ability of iron to catalyze the production of oxygen radicals and serve as a nutrient for cancer cells.
Iodine	Essential component of thyroid hormones. Excreted in urine. Sources are seafood, cod-liver oil, iodized salt, and vegetables grown in iodine-rich soils.	Required by thyroid gland to synthesize thyroid hormones, hormones that regulate metabolic rate.
Copper	Some stored in liver and spleen. Most excreted in feces. Sources include eggs, whole-wheat flour, beans, beets, liver, fish, spinach, asparagus.	Required with iron for synthesis of hemoglobin. Component of enzyme necessary for melanin pigment formation.
Sodium	Most found in extracellular fluids, some in bones. Excreted in urine and perspiration. Normal intake of NaCl (table salt) supplies required amounts.	As most abundant cation in extracellular fluid, strongly affects distribution of water through osmosis. Part of bicarbonate buffer system. Functions in nerve impulse conduction.
Potassium	Principal cation in intracellular fluid. Most is excreted in urine. Normal food intake supplies required amounts.	Functions in transmission of nerve impulses and muscle contraction.
Chlorine	Found in extracellular and intracellular fluids. Principal anion of extracellular fluid. Most excreted in urine. Normal intake of NaCl supplies required amounts.	Assumes role in acid–base balance of blood, water balance, and formation of HCl in stomach.

EXHIBIT 25-4 MINERALS VITAL TO THE BODY (*Continued*)

Mineral	Comments	Importance
Magnesium	Component of soft tissues and bone. Excreted in urine and feces. Widespread in various foods, such as green leafy vegetables, seafood, and whole-grain cereals.	Required for normal functioning of muscle and nervous tissue. Participates in bone formation. Constituent of many coenzymes. Deficiency linked to diabetes, hypertension, high blood levels of cholesterol, pregnancy problems, and spasms in blood vessels, based on data largely from animal studies.
Sulfur	Constituent of many proteins (such as insulin) and some vitamins (thiamine and biotin). Excreted in urine. Sources include beef, liver, lamb, fish, poultry, eggs, cheese, beans.	As component of hormones and vitamins, regulates various body activities.
Zinc	Important component of certain enzymes. Widespread in many foods, especially meats.	As a component of carbonic anhydrase, important in carbon dioxide metabolism. Necessary for normal growth and wound healing, proper functioning of prostate gland, normal taste sensations and appetite, and normal sperm counts in males. As a component of peptidases, it is involved in protein digestion. Deficiency may be involved in impaired immunity and slow learning. Excess may raise cholesterol level.
Fluorine	Component of bones, teeth, other tissues.	Appears to improve tooth structure and inhibit tooth decay. In clinical trials, sodium fluoride and calcium have altered the progression of spinal osteoporosis.
Manganese	Some stored in liver and spleen. Most excreted in feces.	Activates several enzymes. Needed for hemoglobin synthesis, urea formation, growth, reproduction, lactation, bone formation, and possibly production and release of insulin, and inhibiting cell damage.
Cobalt	Constituent of vitamin B_{12}.	As part of B_{12}, required for erythropoeisis.
Chromium	Found in high concentrations in brewer's yeast. Also found in wine and some brands of beer.	Necessary for the proper utilization of dietary sugars and other carbohydrates by optimizing the production and effects of insulin. Helps increase blood levels of HDL, while decreasing levels of LDL.
Selenium	Found in seafood, meat, chicken, grain cereals, egg yolk, milk, mushrooms, and garlic.	An antioxidant. Prevents chromosome breakage and may assume a role in preventing certain birth defects and certain types of cancer (esophageal).

VITAMINS

Organic nutrients required in minute amounts to maintain growth and normal metabolism are called ***vitamins.*** Unlike carbohydrates, fats, or proteins, vitamins do not provide energy or serve as building materials. The essential function of vitamins is the regulation of physiological processes. Of the vitamins whose functions are known, most serve as coenzymes.

Most vitamins cannot be synthesized by the body. They may be ingested in food or pills. Other vitamins, such as vitamin K, are produced by bacteria in the gastrointestinal tract. The body can assemble some vitamins if the raw materials called ***provitamins*** are provided. Vitamin A is produced by the body from the provitamin carotene, a chemical present in spinach, carrots, liver, and milk. No single food contains all the required vitamins—one of the best reasons for eating a balanced diet.

The term ***avitaminosis*** refers to a deficiency of any vitamin in the diet. ***Hypervitaminosis*** refers to an excess of one or more vitamins. Conditions related to megadoses of certain vitamins and minerals are discussed shortly.

On the basis of solubility, vitamins are divided into two principal groups: fat-soluble and water-soluble. ***Fat-soluble*** vitamins are absorbed along with ingested dietary fats by the small intestine as micelles. In fact, they cannot be absorbed unless they are ingested with some fat. Fat-soluble vitamins are generally stored in cells, particularly liver cells, so reserves can be built up. Examples of fat-soluble vitamins are vitamins A, D, E, and K. ***Water-soluble*** vitamins, by contrast, are absorbed along with water in the gastrointestinal tract and dissolve in the body fluids. Excess quantities of these vitamins are excreted in the urine. Thus, the body does not store water-soluble vitamins well. Examples of water-soluble vitamins are the B vitamins and vitamin C.

You have probably noticed the initials ***RDA*** on the label of vitamin and/or mineral supplements. These letters stand for ***Recommended Dietary Allowance.*** They are the levels of intake of essential nutrients considered, in the judgment of the Committee on Dietary Allowances of the Food and Nutrition Body on the basis of available scientific knowledge, to be adequate to meet the known nutritional needs of practically all healthy persons.

Exhibit 25-5 lists the principal vitamins, their sources, functions, and related disorders.

EXHIBIT 25-5 THE PRINCIPAL VITAMINS

Vitamin	Comment and Source	Function	Deficiency Symptoms and Disorders
FAT-SOLUBLE			
A	Formed from provitamin carotene (and other provitamins) in GI tract. Requires bile salts and fat for absorption. Stored in liver. Sources of carotene and other provitamins include yellow and green vegetables; sources of vitamin A include fish-liver oils, milk, butter.	Maintains general health and vigor of epithelial cells. Its potential role in cancer prevention is currently under investigation.	Deficiency results in atrophy and keratinization of epithelium, leading to dry skin and hair, increased incidence of ear, sinus, respiratory, urinary, and digestive system infections, inability to gain weight, drying of cornea and ulceration (***xerophthalmia***), nervous disorders, and skin sores.
		Essential for formation of rhodopsin, light-sensitive chemical in rods of retina.	***Night blindness*** or decreased ability for dark adaptation.
		Aids in growth of bones and teeth by apparently helping to regulate activity of osteoblasts and osteoclasts.	Slow and faulty development of bones and teeth.
D	In presence of sunlight, 7-dehydrocholesterol in the skin is converted to cholecalciferol (vitamin D_3). In the liver, cholecalciferol is converted to 25-hydroxycholecalciferol. In the kidneys, 25-hydroxycholecalciferol is converted to 1,25-dihydroxycalciferol, the active form of vitamin D. Dietary vitamin D requires moderate amounts of bile salts and fat for absorption. Stored in tissues to slight extent. Most excreted via bile. Sources include fish-liver oils, egg yolk, fortified milk.	Essential for absorption and utilization of calcium and phosphorus from gastrointestinal tract. May work with parathyroid hormone (PTH) that controls calcium metabolism.	Defective utilization of calcium by bones leads to ***rickets*** in children and ***osteomalacia*** in adults. Possible loss of muscle tone.
E (tocopherols)	Stored in liver, adipose tissue, and muscles. Requires bile salts and fat for absorption. Sources include fresh nuts and wheat germ, seed oils, green leafy vegetables.	Probably functions as an antioxidant. Believed to inhibit catabolism of certain fatty acids that help form cell structures, especially membranes. Involved in formation of DNA, RNA, and red blood cells. May	May cause the oxidation of monosaturated fats, resulting in abnormal structure and function of mitochondria, lysosomes, and plasma membranes, a possible consequence being hemolytic anemia. Deficiency

EXHIBIT 25-5 THE PRINCIPAL VITAMINS (*Continued*)

Vitamin	Comment and Source	Function	Deficiency Symptoms and Disorders
		promote wound healing, contribute to the normal structure and functioning of the nervous system, prevent scarring, and reduce the severity of visual loss associated with retrolental fibroplasia (an eye disease in premature infants caused by too much oxygen in incubators) by functioning as an antioxidant. Believed to help protect liver from toxic chemicals like carbon tetrachloride.	also causes muscular dystrophy in monkeys and sterility in rats.
K	Produced in considerable quantities by intestinal bacteria. Requires bile salts and fat for absorption. Stored in liver and spleen. Other sources include spinach, cauliflower, cabbage, liver.	Coenzyme believed essential for synthesis of prothrombin by liver and several clotting factors. Also known as antihemorrhagic vitamin.	Delayed clotting time results in excessive bleeding.
WATER-SOLUBLE			
B_1 (thiamine)	Rapidly destroyed by heat. Not stored in body. Excessive intake eliminated in urine. Sources include whole-grain products, eggs, pork, nuts, liver, yeast.	Acts as coenzyme for many different enzymes that break carbon-to-carbon bonds and are involved in carbohydrate metabolism of pyruvic acid to CO_2 and H_2O. Essential for synthesis of acetylcholine.	Improper carbohydrate metabolism leads to buildup of pyruvic and lactic acids and insufficient energy for muscle and nerve cells. Deficiency leads to two syndromes: (1) ***beri-beri***—partial paralysis of smooth muscle of GI tract, causing digestive disturbances, skeletal muscle paralysis, atrophy of limbs; (2) ***polyneuritis***—due to degeneration of myelin sheaths; reflexes related to kinesthesia are impaired, impairment of sense of touch, decreased intestinal motility, stunted growth in children, and poor appetite.
B_2 (riboflavin)	Not stored in large amounts in tissues. Most is excreted in urine. Small amounts supplied by bacteria of GI tract. Other sources include yeast, liver, beef, veal, lamb, eggs, whole-grain products, asparagus, peas, beets, peanuts.	Component of certain coenzymes (e.g., FAD and FMN) concerned with carbohydrate and protein metabolism, especially in cells of eye, integument, mucosa of intestine, blood.	Deficiency may lead to improper utilization of oxygen resulting in blurred vision, cataracts, and corneal ulcerations. Also dermatitis and cracking of skin, lesions of intestinal mucosa, and development of one type of anemia.
Niacin (nicotinamide)	Derived from amino acid tryptophan. Sources include yeast, meats, liver, fish, whole-grain products, peas, beans, nuts.	Essential component of NAD and NADP, coenzymes concerned with energy-releasing (oxidation–reduction) reactions. In	Principal deficiency is ***pellagra,*** characterized by dermatitis, diarrhea, and psychological disturbances.

EXHIBIT 25-5 THE PRINCIPAL VITAMINS (*Continued*)

Vitamin	Comment and Source	Function	Deficiency Symptoms and Disorders
		lipid metabolism, inhibits production of cholesterol and assists in fat breakdown.	
B_6 (pyridoxine)	Synthesized by bacteria of GI tract. Stored in liver, muscle, brain. Other sources include salmon, yeast, tomatoes, yellow corn, spinach, whole-grain products, liver, yogurt.	Essential coenzyme for normal amino acid metabolism. Assists production of circulating antibodies. May function as coenzyme in fat metabolism and neuronal excitability.	Most common deficiency symptom is dermatitis of eyes, nose, and mouth. Other symptoms are retarded growth and nausea.
B_{12} (cyanocobalamin)	Only B vitamin not found in vegetables; only vitamin containing cobalt. Absorption from GI tract dependent on HCl and intrinsic factor secreted by gastric mucosa. Sources include liver, kidney, milk, eggs, cheese, meat.	Coenzyme necessary for red blood cell formation, formation of amino acid methionine, entrance of some amino acids into Krebs cycle, and manufacture of choline (chemical similar in function to acetylcholine).	Pernicious anemia, neuropsychiatric abnormalities (ataxia, memory loss, weakness, personality and mood changes, and abnormal sensations), and impaired osteoblast activity.
Pantothenic acid	Stored primarily in liver and kidneys. Some produced by bacteria of GI tract. Other sources include kidney, liver, yeast, green vegetables, cereal.	Constituent of coenzyme A essential for transfer of pyruvic acid into Krebs cycle, conversion of lipids and amino acids into glucose, and synthesis of cholesterol and steroid hormones.	Experimental deficiency tests indicate fatigue, muscle spasms, neuromuscular degeneration, insufficient production of adrenal steroid hormones.
Folic acid	Synthesized by bacteria of GI tract. Other sources include green leafy vegetables and liver.	Component of enzyme systems synthesizing purines and pyrimidines built into DNA and RNA. Essential for normal production of red and white blood cells.	Production of abnormally large red blood cells (macrocytic anemia).
Biotin	Synthesized by bacteria of GI tract. Other sources include yeast, liver, egg yolk, kidneys.	Essential coenzyme for conversion of pyruvic acid to oxaloacetic acid and synthesis of fatty acids and purines.	Mental depression, muscular pain, dermatitis, fatigue, nausea.
C (ascorbic acid)	Rapidly destroyed by heat. Some stored in glandular tissue and plasma. Sources include citrus fruits, tomatoes, green vegetables.	Exact role not understood. Promotes many metabolic reactions, particularly protein metabolism, including laying down of collagen in formation of connective tissue. As coenzyme may combine with poisons, rendering them harmless until excreted. Works with antibodies. Promotes wound healing. Role in cancer prevention is under investigation.	Scurvy; anemia; many symptoms related to poor connective tissue growth and repair including tender swollen gums, loosening of teeth (alveolar processes also deteriorate), poor wound healing, bleeding (vessel walls fragile because of connective tissue degeneration), and retardation of growth.

Abnormal conditions related to several minerals and vitamins are presented in Exhibit 25-6.

METABOLISM AND BODY HEAT

We will now consider the relationship of foods to body heat, mechanisms of heat gain and loss, and the regulation of body temperature.

Measuring Heat

Heat is a form of kinetic energy that can be measured as ***temperature*** and expressed in units called calories. A ***calorie (cal)*** is the amount of heat energy required to raise the temperature of 1 g of water 1°C, from 14° to 15°C. Since the calorie is a small unit and large amounts of energy are stored in foods, the ***kilocalorie (kcal)*** is used instead. A kilocalorie is equal to 1000 cal and is defined as the amount of heat required to raise the temperature of 1000 g of water 1°C. The kilocalorie is the unit we use to express the heating value of foods and to measure the body's metabolic rate.

The apparatus used to determine the caloric value of foods is called a ***calorimeter.*** A weighed sample of a dehydrated food is burned completely in an insulated metal container. The energy released by the burning food is absorbed by the container and transferred to a known volume of water that surrounds the container. The change in the water's temperature is directly related to the number of kilocalories released by the food. Knowing the caloric value of foods is important; if we know the amount of energy the body uses for various activities, we can adjust our food intake. In this way we can control body weight by taking in only enough kilocalories to sustain our activities.

Production of Body Heat

Most of the heat produced by the body comes from oxidation of the food we eat. The rate at which this heat is produced—the ***metabolic rate***—is also measured in

EXHIBIT 25-6 ABNORMAL CONDITIONS ASSOCIATED WITH MEGADOSES OF SELECTED MINERALS AND VITAMINS*

Substance	Abnormality
MINERALS	
Calcium	Depresses nerve function, causes drowsiness, extreme lethargy, calcium deposits, kidney stones.
Iron	Damage to liver, heart, and pancreas.
Zinc	Masklike fixed expression, difficulty in walking, slurred speech, hand tremor, involuntary laughter.
Cobalt	Goiter, polycythemia, and heart damage.
Selenium	Nausea, vomiting, fatigue, irritability, and loss of fingernails and toenails.
VITAMINS	
A	Blurred vision, dizziness, ringing in the ears, headache, insomnia, irritability, apathy, ataxia, stupor, skin rash, nausea, vomiting, diarrhea, hair loss, joint pain, menstrual irregularities, fatigue, liver damage, abnormal bone growth, and damage to nervous system.
D	Calcium deposits, deafness, nausea, fatigue, headache, loss of appetite, kidney stones, weak bones, hypertension, and high cholesterol.
E	Thrombophlebitis, pulmonary embolism, hypertension, muscular weakness, severe fatigue, breast tenderness, and slow wound healing.
Niacin	Acute flushing, peptic ulcers, liver dysfunction, gout, faintness, dizziness, tingling of fingertips, arrhythmias, and hyperglycemia.
B_6	Impaired sense of position and vibration, diminished tendon reflexes, numbness and loss of sensations in hands and feet, difficulty in walking, impaired memory, depression, headache, and fatigue.
C	Dependence on megadoses may lead to scurvy when withdrawn, kidney stones, diarrhea, hemolysis, hot flashes, headache, fatigue, and insomnia.
E	Headache, muscular weakness, diarrhea, dizziness, low-grade hepatitis, visual complaints, fatigue, and faintness.

* A ***megadose*** is defined as taking 10 or more times the adult recommended dietary allowances (RDAs) set by the National Research Council.

kilocalories. Among the factors that affect metabolic rate are the following:

1. **Exercise.** During strenuous exercise, the metabolic rate may increase to as much as 15 times the normal rate. In well-trained athletes, the rate may increase up to 20 times.

2. **Nervous system.** In a stress situation, the sympathetic nervous system is stimulated and the nerves release norepinephrine (NE), which increases the metabolic rate of body cells.

3. **Hormones.** In addition to norepinephrine (NE), several other hormones affect metabolic rate. Epinephrine is secreted in stress situations. Increased secretions of thyroid hormones increase the metabolic rate. Testosterone and human growth hormone (hGH) also increase metabolic rate.

4. **Body temperature.** The higher the body temperature, the higher the metabolic rate. Each 1°C rise in temperature increases the rate of biochemical reactions by about 10 percent. The metabolic rate may be substantially increased during fever.

5. **Ingestion of food.** The ingestion of food can raise metabolic rate by as much as 10 to 20 percent. This effect is called ***specific dynamic action (SDA)*** and is greatest with proteins and less with carbohydrates and lipids.

6. **Age.** The metabolic rate of a child, in relation to its size, is about double that of an elderly person, as the high rates of reactions related to growth metabolic rate decrease with age.

7. **Others.** Other factors that affect metabolic rate are sex (lower in females, except during pregnancy and lactation), climate (lower in tropical regions), sleep (lower), and malnutrition (lower).

Basal Metabolic Rate (BMR)

Since many factors affect metabolic rate, it is measured under standard conditions designed to reduce or eliminate those factors as much as possible. These conditions of the body are called the ***basal state,*** and the measurement obtained is the ***basal metabolic rate (BMR).*** The person should not exercise for 30 to 60 minutes before the measurement is taken. The individual must be completely at rest but awake. Air temperature should be comfortable. The person should fast for at least 12 hours. Body temperature should be normal. Basal metabolic rate is a measure of the rate at which the quiet, resting, fasting body breaks down foods (and therefore releases heat). BMR is also a measure of how much thyroxine the thyroid gland is producing, since thyroxine regulates the rate of food breakdown and is not a controllable factor under basal conditions.

Basal metabolic rate is most often measured indirectly by measuring oxygen consumption using a respirometer. If a given amount of food releases a given amount of heat energy when it is oxidized, it must combine with a given amount of oxygen. Thus, by measuring the amount of oxygen needed for the metabolism of foods, we can determine how many kilocalories are produced. The amount of heat energy released when 1 liter of oxygen combines with carbohydrates is 5.05 kcal; with fats, the heat released is 4.70 kcal; with proteins, the heat released is 4.60 kcal. The average of the three values is 4.783 kcal. Therefore, every time a liter of oxygen is consumed, 4.783 kcal are produced.

Basal metabolic rate is usually expressed in kilocalories per square meter of body surface area per hour (kcal/m^2/hr). Suppose you use 1.8 liters of oxygen in 6 min as recorded on a respirometer. This means that your oxygen consumption in an hour would be 18 liters (1.8×10). Your basal metabolic rate would be 18×4.783 kcal or 86.85 kcal/hr. To express the kilocalories per square meter of body surface, a standardized chart is used. Such a chart shows square meters of body surface relative to height in centimeters and weight in kilograms. If you weigh 75 kg (165 lb) and are 190 cm (75 inches) tall, your body surface area is 2 m^2. Your basal metabolic rate is equal to 86.85 kcal divided by 2, or about 43.43 kcal/m^2/hr.

The normal basal metabolic rates for various age groups by sex are also listed in standardized charts. Suppose a 27-year-old male has a recorded basal metabolic rate of 45.7 kcal/m^2/hr. The chart would show that his basal metabolic rate should be about 40.3. What does this mean? The recorded basal metabolic rate is 5.4 kcal/m^2/hr, or about 14 percent, above normal. Since a basal metabolic rate between $+15$ and -15 percent is considered normal, this male is judged to be normal. Values above or below 15 percent may indicate an excess or deficiency of thyroid hormones. When the thyroid is secreting extreme quantities of thyroid hormones, the basal metabolic rate can go as high as $+100$ percent. If, on the other hand, the thyroid is secreting very little of its thyroid hormones, the basal metabolic rate may be as low as -50 percent.

Loss of Body Heat

Most body heat is produced by the oxidation of foods we eat. This heat must be removed continuously or body temperature would rise steadily. The principal routes of heat loss include radiation, conduction, convection, and evaporation.

Radiation

Radiation is the transfer of heat as infrared heat rays from a warmer object to a cooler one without physical contact. Your body loses heat by the radiation of heat waves to cooler objects nearby such as ceilings, floors, and walls. If these objects are at a higher temperature, you absorb heat by radiation. Incidentally, the air temperature has no relationship to the radiation of heat to and from objects. Skiers can remove their shirts in bright sunshine even though the air temperature is very low because the radiant heat from the sun is adequate to warm them. In a room at 21°C (70°F), about 60 percent of heat loss is by radiation.

Conduction

Another method of heat transfer is ***conduction.*** In this process, body heat is transferred to a substance or object in contact with the body, such as chairs, clothing, jewelry, air, or water. About 3 percent of body heat is lost via conduction.

Convection

Convection is the transfer of heat by the movement of a liquid or gas between areas of different temperature. When cool air makes contact with the body, it becomes warmed and therefore less dense and is carried away by convection currents as the less dense air rises. Then more cool air makes contact with the body and is carried away as it warms by conduction and becomes less dense. The faster the air moves, the faster the rate of convection. About 15 percent of body heat is lost to the air by convection.

Evaporation

Evaporation is the conversion of a liquid to a vapor. Water has a high heat of evaporation. The *latent heat of evaporation* is the amount of heat necessary to evaporate 1 g of water at 30°C (86°F). Because of water's high latent heat of evaporation, every gram of water evaporating from the skin takes with it a great deal of heat—about 0.58 kcal per gram of water. Under normal conditions, about 22 percent of heat loss occurs through evaporation. The evaporation of only 150 ml of water per hour is enough to remove all the heat produced by the body under basal conditions. Under extreme conditions, about 4 liters (1 gal) of perspiration is produced each hour, and this volume can remove 2000 kcal of heat from the body. This is approximately 32 times the basal level of heat production. The rate of evaporation is inversely related to ***relative humidity,*** the ratio of the actual amount of moisture in the air to the greatest amount it can hold at a given temperature. The higher the relative humidity, the lower the rate of evaporation due to a reduced diffusion rate for water leaving the surface of the body.

A summary of heat-producing and heat-losing mechanisms is presented in Exhibit 25-7.

EXHIBIT 25-7 SUMMARY OF HEAT-PRODUCING AND HEAT-LOSING MECHANISMS

Mechanism	Comment
Heat production	Most body heat is produced by the oxidation of foods, and the rate at which it is produced is called the metabolic rate. The rate is affected by exercise, strong sympathetic stimulation, hormones, and body temperature.

EXHIBIT 25-7 SUMMARY OF HEAT-PRODUCING AND HEAT-LOSING MECHANISMS (*Continued*)

Mechanism	Comment
Heat loss	
Radiation	Transfer of heat from the body to an object without physical contact. Example is losing heat to a cool object such as a floor.
Conduction	Transfer of heat from the body to any object in physical contact with the body such as clothing.
Convection	When cool air makes contact with the body, it is warmed and carried away by convection currents.
Evaporation	The conversion of a liquid to a vapor in which the evaporating substance (e.g., perspiration) removes heat from the body.

Homeostasis of Body Temperature Regulation

As noted in Chapter 5, humans are ***homeotherms;*** that is, a proper body temperature is maintained not only by voluntary acts but also by a group of reflex responses that are integrated in the hypothalamus. Even though there are wide fluctuations in environmental temperature, these mechanisms can maintain a normal range for the internal body temperature. The homeothermic response is an excellent example of homeostasis. If the amount of heat production equals the amount of heat loss, you maintain a constant core temperature near 37°C (98.6°F). ***Core temperature*** refers to the body's temperature in body structures below the skin and subcutaneous tissue. ***Shell temperature*** refers to the body's temperature at the surface, that is, the skin and subcutaneous tissue. Core temperature is usually higher than shell temperature. If your heat-producing mechanisms generate more heat than is lost by your heat-losing mechanisms, your core temperature rises. If your heat-losing mechanisms give off more heat than is generated by heat-producing mechanisms, your core temperature falls. Too high a core temperature kills by denaturing body proteins, while too low a core temperature causes cardiac arrhythmias that result in death.

Hypothalamic Thermostat

Body temperature is regulated by mechanisms that attempt to keep heat production and heat loss in balance. A center of control for those mechanisms that are reflex in nature is found in the hypothalamus in a group of neurons in the anterior portion referred to as the ***preoptic area.*** This area receives input from temperature receptors in the skin and mucous membranes (peripheral thermoreceptors) and in internal structures (central thermoreceptors), including the hypothalamus. If blood temperature increases, the neurons of the preoptic area fire nerve impulses more rapidly. If something causes the blood's temperature to decrease, these neurons fire nerve impulses less frequently. The preoptic area is adjusted to maintain normal body temperature and thus serves as your thermostat.

Nerve impulses from the preoptic area are sent to other portions of the hypothalamus known as the heat-losing center and the heat-promoting center. The ***heat-losing center,*** when stimulated by the preoptic area, sets into operation a series of responses that lower body temperature. The ***heat-promoting center,*** when stimulated by the preoptic area, sets into operation a series of responses that raise body temperature. The heat-losing center is mainly parasympathetic in function; the heat-promoting center is primarily sympathetic.

Mechanisms of Heat Production

Suppose the environmental temperature is low or blood temperature falls below normal. Both stresses stimulate thermoreceptors to the preoptic area. The preoptic area, in turn, activates the heat-promoting center. In reflex response, the heat-promoting center discharges nerve impulses that automatically set into operation a number of responses designed to increase and retain body heat and bring body temperature back up to normal. This cycle is a negative feedback system that attempts to raise body temperature to normal (Figure 25-15a).

Vasoconstriction Nerve impulses from the heat-promoting center stimulate sympathetic nerves that cause blood vessels of the skin to constrict. The net effect of vasoconstriction is to decrease the flow of warm blood from the internal organs to the skin, thus decreasing the transfer of heat from the internal organs to the skin. This reduction in heat loss helps raise the internal body temperature.

Sympathetic Stimulation Another response triggered by the heat-promoting center is the sympathetic stimulation of metabolism. The heat-promoting center stimulates sympathetic nerves leading to the adrenal medulla. This stimulation causes the medulla to secrete epinephrine and norepinephrine (NE) into the blood. The hormones, in turn, bring about an increase in cellular metabolism, a reaction that also increases heat production. This effect is called ***chemical thermogenesis.***

FIGURE 25-15 Regulation of body temperature by the hypothalamus. (a) Responses to low body temperature. (b) Responses to high body temperature.

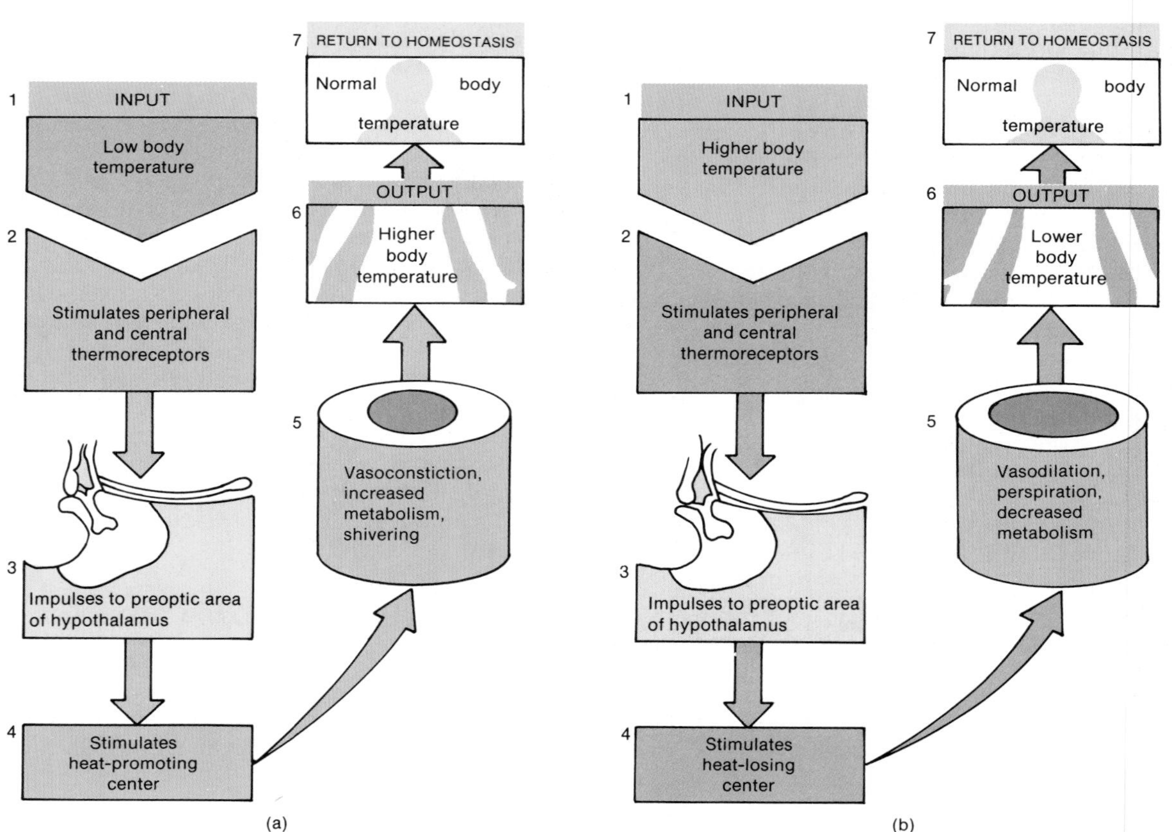

▪ **Skeletal Muscles** Heat production is also increased by responses of skeletal muscles. For example, stimulation of the heat-promoting center causes stimulation of parts of the brain that increase muscle tone and hence heat production. As the muscle tone increases, the stretching of the agonist muscle initiates the stretch reflex and the muscle contracts. This contraction causes the antagonist muscle to stretch, and it too develops a stretch reflex. The repetitive cycle—called ***shivering***—increases the rate of heat production. During maximal shivering, body heat production can rise to about four times the normal rate in just a few minutes.

▪ **Thyroxine** Another body response that increases heat production is increased production of thyroxine. A cold environmental temperature causes the secretion of thyrotropin regulating hormone (TRH) produced by the preoptic area of the hypothalamus. The TRH in turn stimulates the anterior pituitary to secrete thyroid stimulating hormone (TSH), which causes the thyroid to release thyroxine into the blood. Since increased levels of thyroxine increase the metabolic rate, body temperature is raised.

Mechanisms of Heat Loss

Now suppose some stress raises body temperature above normal. The stress or higher temperature of the blood stimulates the preoptic area, which in turn stimulates the heat-losing center and inhibits the heat-promoting center. Instead of blood vessels in the skin constricting, they dilate. The skin becomes warm, and the excess heat is lost to the environment as an increased volume of blood flows from the core of the body into the skin. At the same time, sweat glands produce perspiration, which evaporates and cools the skin, and the metabolic rate and shivering are decreased. The high temperature of the blood (by way of hypothalamic activation of sympathetic nerves) activates sweat glands of the skin to produce perspiration. As the water of the perspiration evaporates from the surface of the skin, the skin is cooled. All these responses reverse the heat-promoting effects and bring body temperature down to normal. This negative feedback cycle attempts to lower body temperature to normal (Figure 25-15b).

Voluntary activities are used, along with all the reflex responses discussed above, to help achieve homeostasis of core body temperature.

Body Temperature Abnormalities

Fever

A ***fever*** is an abnormally high body temperature. The most frequent cause of fever is infection from bacteria (and their toxins) and viruses. Other causes are heart attacks, tumors, tissue destruction by x-rays, surgery, or trauma, and reactions to vaccines. The mechanism of fever production is believed to occur as follows. When phagocytes, namely monocytes and macrophages, ingest certain bacteria, a portion of the cell wall of the bacteria is released, causing the phagocytes to secrete interleukin-1. Interleukin-1 circulates to the anterior hypothalamus and induces neurons of the preoptic area to secrete prostaglandins, particularly of the E series. Prostaglandins reset the hypothalamic thermostat at a higher temperature, and temperature regulating reflex mechanisms will then act to bring the core body temperature up to this new setting. Aspirin, acetaminophen (Tylenol), and ibuprofen (Medipren) reduce fever by inhibiting synthesis of prostaglandins. (Fever may also be reduced by peripheral cooling in which a cooling blanket is used.)

Suppose that as a result of exogenous pyrogens the thermostat is set at 39.4°C (103°F). Now the heat-promoting mechanisms (vasoconstriction, increased metabolism, shivering) are operating at full force. Thus, even though body temperature is climbing higher than normal—say, 38.3°C (101°F)—the skin remains cold, and shivering occurs. This condition, called a ***chill,*** is a definite sign that body temperature is rising. After several hours, body temperature reaches the setting of the thermostat and the chills disappear. But the body will continue to regulate temperature at 39.4°C (103°F) until the stress is removed. When the stress is removed, the thermostat is reset at normal—37°C (98.6°F). Since body temperature remains high in the beginning, the heat-losing mechanisms (vasodilation and sweating) go into operation to decrease body temperature. The skin becomes warm and the person begins to sweat. This phase of the fever is called the ***crisis*** and indicates that body temperature is falling.

Up to a point, fever is beneficial. Interleukin-1 helps step up production of T cells. High body temperature intensifies the effect of interferon. The high body temperature is believed to inhibit the growth of some bacteria and viruses. Fever also increases heart rate so that white blood cells are delivered to sites of infection more rapidly and their secretions are increased. In addition, antibody production and T cell proliferation are increased. Moreover, heat speeds up the rate of chemical reactions. This increase may help body cells to repair themselves more quickly during a disease. Among the complications of fever are dehydration, acidosis, and permanent brain damage. As a rule, death results if body temperature rises to 44.4 to 45.5°C (112° to 114°F). On the other end of the scale, death usually results when body temperature falls to 21.2 to 23.9°C (70° to 75°F).

Heat Cramps

Heat cramps occur as a result of profuse sweating that removes water and salt (NaCl) from the body. The salt loss causes painful contractions of muscles called heat cramps. The cramps tend to occur in muscles used while working but do not appear until the person relaxes after work. Salted liquids usually lead to rapid improvement.

Heat Exhaustion

In ***heat exhaustion (heat prostration)***, the body temperature is generally normal, or a little below, and the skin is cool and clammy (moist) due to profuse perspiration. Heat exhaustion is normally characterized by fluid and electrolyte loss, especially salt. The salt loss results in muscle cramps, dizziness, vomiting, and fainting. Fluid loss may cause low blood pressure. Complete rest and salt tablets are recommended.

Heatstroke

Heatstroke (sunstroke) is brought about when the temperature and relative humidity are high, making it difficult for the body to lose heat by radiation or evaporation. There is a decreased flow of blood to the skin, perspiration is greatly reduced, and body temperature rises sharply. The skin is thus dry and hot—the temperature may reach 43.7°C (110°F). Brain cells are quickly affected and may be destroyed permanently and as a result, the body temperature regulating reflexes fail to operate. Treatment must be undertaken immediately and consists of cooling the body by immersing the victim in cool water and by administering fluids and electrolytes.

Hypothermia

Hypothermia refers to a lowering of body temperature to 35°C (95°F) or below. It may be caused by an overwhelming cold stress (immersion in icy water), metabolic diseases (hypoglycemia, adrenal insufficiency, hypothyroidism), drugs (alcohol, antidepressants, sedatives, tranquilizers), burns, malnutrition, transection of the cervical spinal cord, and lowering of body temperature for surgery. Hypothermia is characterized by the following as body temperature falls: sensation of cold, shivering, confusion, vasoconstriction, muscle rigidity, bradycardia, acidosis, hypoventilation, hypotension, ventricular fibrillation, no reflexes, and loss of spontaneous movement, coma, and possibly death usually caused by cardiac arrhythmias. The elderly have lessened metabolic protection against a cold environment coupled with reduced perception of cold. As a result, they are at greater risk for hypothermia.

In 1986, a $2\frac{1}{2}$-year-old girl was submerged in icy water for 66 minutes, the longest documented submersion in which the individual had no neurological damage. Part of the successful outcome was due to use of a heart–lung bypass machine that rewarmed the blood of the child and brought body temperature back to normal, a new and novel use for this machine.

DISORDERS: HOMEOSTATIC IMBALANCES

Obesity

Obesity is defined as a body weight 10 to 20 percent above a desirable standard as the result of an excessive accumulation of fat. There is little doubt that even moderate obesity is hazardous to health. It is implicated as a risk factor in cardiovascular disease, hypertension, pulmonary disease, diabetes mellitus (type II), arthritis, certain cancers (uterus and colon), varicose veins, and gallbladder disease. Also, loss of body fat in obese individuals has been shown to elevate HDL cholesterol, the type associated with prevention of cardiovascular disease.

Classification

Clinically, obesity is classified as hypertrophic and hyperplastic. In ***hypertrophic (adult-onset) obesity***, there is an increase in the amount of fat in adipocytes but no increase in the number of fat cells. Such individuals tend to be thin or average in weight until about age 20 to 40, at which time weight gain begins. The gain may be associated with an imbalance between caloric intake and utilization. People with hypertrophic obesity tend to have fat distributed in central locations ("middle-age spread"), and their problem is easier to treat. The second category of obesity is ***hyperplastic (lifelong) obesity***, in which there is an increase in the number of adipocytes, as well as an increase in the amount of fat within them. These individuals tend to be obese as children and have a large spurt in weight gain during adolescence. After adolescence, the number of adipocytes remains about the same throughout life. Such people are generally grossly obese, and the distribution of fat is both central and peripheral. Treatment is considerably more difficult.

Causes

In a relatively few cases, obesity may result from trauma or tumors in the food-regulating centers in the hypothalamus. In most cases of obesity, no specific cause can be identified. Contributing factors include eating habits taught early in life, overeating to relieve tension, and social customs. Recently, obesity has been strongly linked to genetic factors. Studies indicate that some people inherit a low metabolic rate and that they become obese not because they eat too much but because they burn calories too slowly. In addition, it has been learned that abnormally low levels of a protein, called adipsin, produced by adipocytes, may be linked to obesity. A possible explanation is that adequate amounts of adipsin may signal the satiety center of the hypothalamus to diminish appetite.

Treatment

Reduction of body weight involves keeping caloric intake well below energy expenditure. The goals during weight decrease are:

1. Loss of body fat with a minimal accompanying breakdown of lean tissue.

DISORDERS: HOMEOSTATIC IMBALANCES (*continued*)

2. Maintenance of physical and emotional fitness during the reducing period.
3. Establishment of eating and exercise habits to maintain weight at the recommended level.

Morbid obesity refers to obese individuals who are over 100 percent of their ideal weight. The condition is so named because it is associated with serious and life-threatening conditions such as hypertension, diabetes mellitus, and atherosclerosis. For individuals with morbid obesity, surgery may be an alternative to control weight. The suitable patients for surgical procedures are individuals whose weight is more than 45.5 kg (100 pounds) over the ideal body weight for height or 100 percent over the ideal and who have tried and failed to permanently lose substantial weight on medically supervised reduction regimens.

Most operations for morbid obesity either produce a certain amount of malabsorption, such as the intestinal bypass, or reduce the size of the stomach so that it will hold less food, such as the gastric bypass or gastroplasty. The ***intestinal bypass*** is the oldest operation in which the upper jejunum is cut, the distal end is closed, and the proximal end is anastomosed to the distal ileum. The procedure has been largely abandoned because of its serious side effects, such as kidney stones, arthritis, and uncontrollable electrolyte and mineral imbalances. Both ***gastric bypass*** and ***gastroplasty*** are gastric-reduction operations in which a new, small stomach is surgically constructed from the upper part of the patient's stomach. A carefully measured opening connects the upper and lower portions of the stomach. Both procedures are generally successful.

In recent years, ***gastric balloons*** have been placed in the stomach to treat morbid obesity. The balloon is inserted with an endoscope and inflated, and the endoscope is withdrawn. The sensation of fullness created by the balloon is supposed to curb appetite. At present, the effectiveness of this procedure is under investigation.

Malnutrition

Causes

Whereas ***nutrition*** refers to the intake of adequate essential nutrients and calories to maintain health, ***malnutrition*** refers to a state of bad or poor nutrition. One cause of malnutrition is undernutrition. This may be due to inadequate food intake as a result of conditions such as fasting, anorexia nervosa, deprivation, diabetes mellitus, cancer, gastrointestinal obstructions, inability to swallow, renal disease, and poor dentition. Other causes of malnutrition are imbalance of nutrients, malabsorption of nutrients, improper distribution of nutrients, increased nutrient requirements (due to fever, infections, burns, fractures, stress, and exposure to heat or cold), increased nutrient losses (diarrhea, bleeding, glycosuria, and fistula drainage), and overnutrition (excess vitamins, minerals, and calories).

Stages

One of the major types of undernutrition is known as protein–calorie undernutrition, which occurs when there is inadequate intake of protein and/or calories to meet a person's nutritional requirements. Using protein–calorie nutrition as the example, we will describe the stages of ***starvation,*** that is, loss of energy stores in the form of glycogen, fats, and proteins.

In the first stage of starvation, carbohydrate stores in the body are depleted first. During approximately the first day of fasting, low glucose levels stimulate glucagon secretion by the pancreas. As a result, glycogen is converted to glucose (glycogenolysis) and released from the liver. This restores blood glucose levels to normal and makes the glucose available for use by body cells, including brain cells.

Once glycogen stores are depleted, the next stage of starvation occurs. During this time, the primary energy source for most body cells is fatty acids from lipid stores. As the liver metabolizes the fatty acids, ketone bodies are produced in large quantities and transported to body cells. Even brain cells use ketone bodies as a source of energy. However, since body cells are limited in the amount of ketone bodies they can metabolize, excess ketone bodies appear in the blood, resulting in a condition called ketosis. This, in turn, leads to metabolic acidosis, a decrease in the pH of blood below normal, since the body cannot neutralize the excess ketone bodies. As you will see in Chapter 27, metabolic acidosis results in depression of the central nervous system that may lead to coma. Ketosis and metabolic acidosis are frequently associated with crash dieting, low-carbohydrate diets, high-protein diets, and other fad diets. During the early stages of starvation, large quantities of muscle protein not essential to cellular functioning are broken down to amino acids. These, in turn, are converted by the liver into glucose (gluconeogenesis). Although this glucose is used to maintain a fairly normal blood sugar level, muscle and other tissues use ketone bodies as an energy source. The length of the second stage of starvation is primarily determined by the amount of stored fat in the body.

When fat reserves are depleted, the third stage of starvation occurs. During this time, even the proteins needed to maintain cellular functions are broken down as a source of energy. It is estimated that once protein stores are depleted to about one-half of their normal level, death results. The third stage of starvation occurs very rapidly once it has begun, frequently leading to death in 24 hours.

Physiological Changes

Virtually every organ of the body can undergo structural and functional changes in response to undernutrition. The most obvious change is a loss in body weight. Both adipose tissue and lean body mass are depleted. Among the other important changes associated with undernutrition are:

DISORDERS: HOMEOSTATIC IMBALANCES (*continued*)

1. Decreased heart mass and cardiac output.
2. Reduced blood volume, hematocrit, albumin, and lymphocytes.
3. Pulmonary dysfunction (decreased vital capacity, tidal volume, and minute volume of respiration) due to loss of mass and strength of respiratory muscles.
4. Decreased kidney mass and filtration of blood.
5. Decreased motility and secretion of the gastrointestinal tract and decreased absorption of nutrients.
6. Decreased liver mass and secretion, especially serum proteins.
7. Decreased function of virtually all aspects of the immune system.
8. Decreased wound healing.
9. Reduced metabolic rate.
10. Diminished reproductive activities.
11. Dysfunctional endocrine glands, as indicated by high levels of human growth hormone (hGH), low levels of thyroxine, and low levels of testosterone.

Clinical Types

Protein–calorie undernutrition may be classified into two types based on which factor is lacking in the diet. In one type, called ***kwashiorkor*** (kwash-ē-OR-kor), protein intake is deficient despite normal or nearly normal calorie intake. Some dietary proteins are referred to as complete proteins; that is, they contain adequate amounts of essential amino acids. Sources of complete proteins are primarily animal products such as milk, meat, fish, poultry, and eggs. Incomplete proteins lack essential amino acids. An example is zein, the protein in corn, which lacks the essential amino acids tryptophan and lysine. The diet of many African natives consists largely of cornmeal. As a result, many African children especially develop kwashiorkor. It is characterized by edema of the abdomen, enlarged liver, decreased blood pressure, bradycardia, hypothermia, anorexia, lethargy, dry and hyperpigmented skin, easily pluckable hair, and sometimes mental retardation.

Another type of protein–calorie undernutrition is called ***marasmus*** (mar-AZ-mus). It results from inadequate intake of both protein and calories. Its characteristics are less dramatic than those of kwashiorkor and include retarded growth, low weight, muscle wasting, emaciation, dry skin, and thin, dry, dull hair. There is less lethargy and no edema, enlarged abdomen, or dermatitis.

Phenylketonuria (PKU)

By definition, ***phenylketonuria*** (fen'-il-kē'-tō-NOO-rē-a) or ***PKU*** is a genetic error of metabolism characterized by an elevation of the amino acid phenylalanine in the blood. It is frequently associated with mental retardation. The DNA of children with phenylketonuria lacks the gene that normally programs the manufacture of the enzyme phenylalanine hydroxylase. This enzyme is necessary for converting phenylalanine into the amino acid tyrosine, an amino acid that enters the Krebs cycle. As a result, phenylalanine cannot be metabolized, and what is not used in protein synthesis builds up in the blood. High levels of phenylalanine are toxic to the brain during the early years of life when the brain is developing. Mental retardation can be prevented, when the condition is detected early, by restricting the child to a diet that supplies only the amount of phenylalanine necessary for growth. As noted earlier, in Chapter 2, the artificial sweetener aspartame contains phenylalanine, and its consumption should be restricted in children with PKU since they cannot metabolize it and it may produce neuronal damage.

PKU may be detected by a test that measures the level of phenylalanine in the blood of a newborn. A normal sample contains less than 4 mg/dl of phenylalanine. PKU can also be detected by a urine test that measures the level of phenylpyruvic acid.

Cystic Fibrosis (CF)

Cystic fibrosis (***CF***) is an inherited disease of the exocrine glands that affects the pancreas, respiratory passageways, and salivary and sweat glands. It is the most common lethal genetic disease of Caucasians—5 percent of the population are thought to be genetic carriers. The cause of cystic fibrosis is linked to an inability of chloride (Cl^-) ions to cross epithelial cells in affected regions of the body.

Among the most common signs and symptoms are pancreatic insufficiency, pulmonary involvement, and cirrhosis of the liver. It is characterized by the production of thick exocrine secretions that do not drain easily from the respiratory passageways. The buildup of the secretions leads to inflammation and replacement of injured cells with connective tissue that blocks these passageways. One of the prominent features is blockage of the pancreatic ducts so that the digestive enzymes cannot reach the intestine. Since pancreatic juice contains the only fat-digesting enzyme, the person fails to absorb fats or fat-soluble vitamins and thus suffers from vitamin A, D, and K deficiency diseases. Calcium needs fat to be absorbed, so tetany also may result.

A child suffering from cystic fibrosis is given pancreatic extract and large doses of vitamins A, D, and K. The therapeutic diet is low, but not lacking, in fats and high in carbohydrates and proteins that can be used for energy and can also be converted by the liver into the lipids essential for life processes. Currently, the median survival for patients with cystic fibrosis is about 20 years, and many survive to the third and fourth decades. Some survive to age 50 and beyond.

DISORDERS: HOMEOSTATIC IMBALANCES (continued)

Celiac Disease

Celiac disease results in malabsorption by the intestinal mucosa due to the ingestion of gluten. *Gluten* is the water-insoluble protein fraction of wheat, rye, barley, and oats. In susceptible persons, ingestion of gluten induces destruction of villi and inhibition of enzyme secretion accompanied by a variable amount of malabsorption. The condition is easily remedied by administering a diet that excludes all cereal grains except rice and corn.

STUDY OUTLINE

Regulation of Food Intake (p. 785)

1. Two centers in the hypothalamus related to regulation of food intake are the feeding center and satiety center; the feeding center is constantly active but may be inhibited by the satiety center.
2. Among the stimuli that affect the feeding and satiety centers are glucose, amino acids, lipids, body temperature, distension, and cholecystokinin (CCK).

Nutrients (p. 785)

1. Nutrients are chemical substances in food that provide energy, act as building blocks in forming new body components, or assist in the functioning of various body processes.
2. There are six major classes of nutrients: carbohydrates, lipids, proteins, minerals, vitamins, and water.

Metabolism (p. 785).

1. Metabolism refers to all chemical reactions of the body and has two phases: catabolism and anabolism.
2. Anabolism consists of a series of synthesis reactions whereby small molecules are built up into larger ones that form the body's structural and functional components. Anabolic reactions use energy.
3. Catabolism is the term for decomposition reactions that provide energy.
4. The coupling of anabolism and catabolism is via ATP.
5. Metabolic reactions are catalyzed by enzymes, proteins that are very efficient, specific for their substrates, and subject to cellular controls. Names of enzymes typically end in *-ase.*
6. A complete enzyme, consisting of an apoenzyme and cofactor, is known as a holoenzyme.
7. The essential function of an enzyme is to catalyze chemical reactions.

Physiology of Energy Production (p. 788)

1. Oxidation is the removal of hydrogen atoms from a substance; reduction is the addition of hydrogen atoms to a substance; oxidation–reduction reactions are coupled.
2. ATP is generated by substrate-level phosphorylation, oxidative phosphorylation, and photophosphorylation.
3. In chemiosmosis, the energy released when a substance moves along a gradient is usually provided by ATP.

Physiology of Carbohydrate Metabolism (p. 790)

1. During digestion, polysaccharides and disaccharides are converted to monosaccharides, which are absorbed through capillaries in villi and transported to the liver via the hepatic portal vein.
2. Carbohydrate metabolism is primarily concerned with glucose metabolism.

Fate of Carbohydrates (p. 790)

1. Some glucose is oxidized by cells to provide energy; it moves into cells by facilitated diffusion and becomes phosphorylated to glucose-6-phosphate; insulin stimulates glucose movement into cells.
2. Excess glucose can be stored by the liver and skeletal muscles as glycogen or converted to fat.

Glucose Catabolism (p. 790)

1. Glucose oxidation is also called cellular respiration.
2. The complete oxidation of glucose to CO_2 and H_2O involves glycolysis, the Krebs cycle, and the electron transport chain.

Glycolysis (p. 790)

1. Glycolysis refers to the breakdown of glucose into two molecules of pyruvic acid.
2. When oxygen is in short supply, pyruvic acid is converted to lactic acid; under aerobic conditions, pyruvic acid enters the Krebs cycle.
3. As a result of glycolysis, there is a net production of two molecules of ATP.

Krebs Cycle (p. 793)

1. Pyruvic acid is prepared for entrance into the Krebs cycle by conversion to a two-carbon compound (acetyl group) followed by the addition of coenzyme A to form acetyl coenzyme A.
2. The Krebs cycle involves decarboxylations and oxidations and reductions of various organic acids.
3. Each molecule of pyruvic acid that enters the Krebs cycle produces three molecules of CO_2, four molecules of NADH + H^+, one molecule of $FADH_2$, and one molecule of GTP.
4. The energy originally in glucose and then pyruvic acid is primarily in the reduced coenzymes NADH + H^+ and $FADH_2$.

Electron Transport Chain (p. 795)

1. The electron transport chain is a series of oxidation–reduction reactions in which the energy in NADH + H^+ and $FADH_2$ is liberated and transferred to ATP for storage.
2. The carrier molecules involved include FMN, coenzyme Q, and cytochromes.
3. The electron transport chain yields 34 molecules of ATP and H_2O.
4. The complete oxidation of glucose can be represented as follows:

$$C_6H_{12}O_6 + 6O_2 \rightarrow 38ATP + 6CO_2 + 6H_2O$$

Glucose Anabolism (p. 798)

1. The conversion of glucose to glycogen for storage in the liver and skeletal muscle is called glycogenesis. The process occurs in the liver and is stimulated by insulin.
2. The body can store about 500 g of glycogen.
3. The conversion of glycogen back to glucose is called glycogenolysis.
4. It occurs between meals and is stimulated by glucagon and epinephrine.
5. Gluconeogenesis is the conversion of protein molecules into glucose. It is stimulated by cortisol, thyroxine, epinephrine, glucagon, and human growth hormone (hGH).
6. Glycerol may be converted to glyceraldehyde-3-phosphate, and some amino acids may be converted to pyruvic acid.

Physiology of Lipid Metabolism (p. 801)

1. During digestion, fats are ultimately broken down into fatty acids and monoglycerides.
2. Long-chain fatty acids and monoglycerides are carried in micelles for entrance into villi, digested to glycerol and fatty acids in epithelial cells, recombined to form triglycerides, and transported by chylomicrons through the lacteals of villi into the thoracic duct.

Fate of Lipids (p. 801)

1. Some fats may be oxidized to produce ATP.
2. Some fats are stored in adipose tissue.
3. Other lipids are used as structural molecules or to synthesize essential molecules. Examples include phospholipids of plasma membranes, lipoproteins that transport cholesterol, thromboplastin for blood clotting, and cholesterol used to synthesize bile salts and steroid hormones.

Fat Storage (p. 801)

1. Fats are stored in adipose tissue, mostly in the subcutaneous layer.
2. Adipose cells contain lipases that catalyze the deposition of fats from chylomicrons and hydrolyze fats into fatty acids and glycerol.

Lipid Catabolism (p. 801)

1. Fat is split into fatty acids and glycerol under the influence of hormones such as epinephrine, norepinephrine, and glucocorticoids and released from fat depots.
2. Glycerol can be converted into glucose by conversion into glyceraldehyde-3-phosphate.
3. In beta oxidation, carbon atoms are removed in pairs from fatty acid chains; the resulting molecules of acetyl coenzyme A enter the Krebs cycle.
4. The formation of ketone bodies by the liver is a normal phase of fatty acid catabolism, but an excess of ketone bodies, called ketosis, may cause acidosis.

Lipid Anabolism: Lipogenesis (p. 802)

1. The conversion of glucose or amino acids into lipids is called lipogenesis. The process is stimulated by insulin.
2. The intermediary links in lipogenesis are glyceraldehyde-3-phosphate and acetyl coenzyme A.

Physiology of Protein Metabolism (p. 803)

1. During digestion, proteins are hydrolyzed into amino acids.
2. Amino acids are absorbed by the capillaries of villi and enter the liver via the hepatic portal vein.

Fate of Proteins (p. 803)

1. Amino acids, under the influence of human growth hormone (hGH) and insulin, enter body cells by active transport.
2. Inside cells, amino acids are synthesized into proteins that function as enzymes, hormones, structural elements, and so forth; stored as fat or glycogen; or used for energy.

Protein Catabolism (p. 803)

1. Before amino acids can be catabolized, they must be converted to substances that can enter the Krebs cycle; these conversions involve deamination, decarboxylation, and hydrogenation.
2. Amino acids may also be converted into glucose, fatty acids, and ketone bodies.

Protein Anabolism (p. 803)

1. Protein synthesis is stimulated by human growth hormone (hGH), thyroxine, and insulin.
2. The process is directed by DNA and RNA and carried out on the ribosomes of cells.

Absorptive and Postabsorptive States (p. 803)

1. During the absorptive state, ingested nutrients enter the blood and lymph from the GI tract.
2. During the absorptive state, most blood glucose is used by body cells for oxidation. Glucose transported to the liver is converted to glycogen or fat. Most fat is stored in adipose tissue. Amino acids in liver cells are converted to carbohydrates, fats, and proteins.
3. During the postabsorptive (fasting) state, absorption is complete and the energy needs of the body are satisfied by nutrients already present in the body.
4. The major concern of the body during the postabsorptive state is to maintain normal blood glucose level. This involves conversion of liver and skeletal muscle glycogen into glucose, conversion of glycerol into glucose, and conversion of amino acids into glucose. The body also switches from glucose oxidation to fatty acid oxidation.

Regulation of Metabolism (p. 806)

1. Absorbed nutrients may be oxidized, stored, or converted, based on the needs of the body.
2. The pathway taken by a particular nutrient is enzymatically controlled and is regulated by hormones (see Exhibit 25-3).

Minerals (p. 806)

1. Minerals are inorganic substances that help regulate body processes.
2. Minerals known to perform essential functions are calcium, phosphorus, sodium, chlorine, potassium, magnesium, iron, sulfur, iodine, manganese, cobalt, copper, zinc, selenium, and chromium. Their functions are summarized in Exhibit 25-4.

Vitamins (p. 809)

1. Vitamins are organic nutrients that maintain growth and normal metabolism. Many function in enzyme systems.
2. Fat-soluble vitamins are absorbed with fats and include A, D, E, and K.
3. Water-soluble vitamins are absorbed with water and include the B vitamins and vitamin C.
4. The functions and deficiency disorders of the principal vitamins are summarized in Exhibit 25-5.
5. Abnormalities associated with megadoses of selected minerals and vitamins are summarized in Exhibit 25-6.

Metabolism and Body Heat (p. 813)

1. A kilocalorie (kcal) is the amount of energy required to raise the temperature of 1000 g of water 1°C from 14° to 15°C.
2. The kilocalorie is the unit of heat used to express the caloric value of foods and to measure the body's metabolic rate.
3. The apparatus used to determine the caloric value of foods is called a calorimeter.

Production of Body Heat (p. 813)

1. Most body heat is a result of oxidation of the food we eat. The rate at which this heat is produced is known as the metabolic rate.
2. Metabolic rate is affected by exercise, the nervous system, hormones, body temperature, ingestion of food, age, sex, climate, sleep, and malnutrition.

Basal Metabolic Rate (BMR) (p. 814)

1. Measurement of the metabolic rate under basal conditions is called the basal metabolic rate (BMR).
2. BMR is expressed in kilocalories per square meter of body area per hour ($kcal/m^2/hr$).

Loss of Body Heat (p. 814)

1. Radiation is the transfer of heat as infrared heat rays from one object to another without physical contact.
2. Conduction is the transfer of body heat to a substance or object in contact with the body.
3. Convection is the transfer of body heat by the movement of air that has been warmed by the body.
4. Evaporation is the conversion of a liquid to a vapor.

Homeostasis of Body Temperature Regulation (p. 815)

1. A normal body temperature is maintained by a delicate balance between heat-producing and heat-losing mechanisms.
2. The hypothalamic thermostat is the preoptic area.
3. Mechanisms that produce or retain heat are vasoconstriction, sympathetic stimulation, skeletal muscle contraction, and thyroxine production.
4. Mechanisms of heat loss include vasodilation, decreased metabolic rate, decreased skeletal muscle contraction, and perspiration.

Body Temperature Abnormalities (p. 817)

1. Fever is an abnormally high body temperature most commonly caused by prostaglandins, with interleukin-1 and mediators.
2. Heat cramps are painful skeletal muscle contractions due to loss of salt and water.
3. Heat exhaustion results in a normal or below normal body temperature, profuse perspiration, nausea, cramps, and dizziness. Rest and salt tablets are indicated.
4. Heatstroke results in decreased blood flow to skin, reduced perspiration, and high body temperature. Fluid therapy and body cooling are indicated.
5. Hypothermia refers to a lowering of body temperature.

Disorders: Homeostatic Imbalances (p. 818)

1. Obesity is defined as a body weight 10 to 20 percent above desirable standard as the result of excessive accumulation of fat. It is classified as hypertrophic and hyperplastic.
2. Malnutrition refers to a state of bad or poor nutrition.
3. Phenylketonuria (PKU) is a genetic error of metabolism characterized by an elevation of phenylalanine in the blood.
4. Cystic fibrosis (CF) is a metabolic disease of the exocrine glands in which absorption of vitamins A, D, and K and calcium is inadequate.
5. Celiac disease is a condition in which the ingestion of gluten causes morphological changes in the small intestinal mucosa resulting in malabsorption.

REVIEW QUESTIONS

1. Discuss how food intake is regulated. (p. 785)
2. Define a nutrient. List the six classes of nutrients and indicate the function of each. (p. 785)
3. What is metabolism? Distinguish between anabolism and catabolism and give examples of each. (p. 785)
4. How does ATP couple anabolism and catabolism? (p. 786)
5. List and explain the characteristics of enzymes. Describe the parts of an enzyme. (p. 786)
6. Briefly describe the mechanism of enzyme action. (p. 788)
7. Define oxidation and reduction and give an example of each. (p. 788)
8. Describe how ATP is generated. What is chemiosmosis? (p. 789)
9. How are carbohydrates absorbed and what are their fates in the body? How does glucose move into body cells? (p. 790)

10. Define glycolysis. Describe its principal events and outcome. (p. 790)
11. Describe how acetyl coenzyme A is formed. (p. 793)
12. Outline the principal events and outcomes of the Krebs cycle. (p. 793)
13. Explain what happens in the electron transport chain. (p. 795)
14. Summarize the outcomes of the complete oxidation of a molecule of glucose. (p. 798)
15. Define glycogenesis and glycogenolysis. Under what circumstances does each occur? (p. 798)
16. Why is gluconeogenesis important? Give specific examples to substantiate your answer. (p. 798)
17. How are fats absorbed and what are their fates in the body? Where are fats stored in the body? (p. 801)
18. Explain the principal events of the catabolism of glycerol and fatty acids. (p. 801)
19. What are ketone bodies? What is ketosis? (p. 802)
20. Define lipogenesis and explain its importance. (p. 802)
21. How are proteins absorbed and what are their fates in the body? (p. 803)
22. Relate deamination to amino acid catabolism. (p. 803)
23. Summarize the major steps involved in protein synthesis. (p. 76)
24. Distinguish between essential and nonessential amino acids. (p. 803)
25. What is the absorptive state? Outline its principal events. (p. 803)
26. What is the postabsorptive (fasting) state? Outline its principal events. (p. 805)
27. Indicate the roles of the following hormones in the regulation of metabolism: insulin, glucagon, epinephrine, human growth hormone (hGH), thyroxine, cortisol, and testosterone. (p. 807)
28. What is a mineral? Briefly describe the functions of the following minerals: calcium, phosphorus, iron, iodine, copper, sodium, potassium, chlorine, magnesium, sulfur, zinc, fluorine, manganese, cobalt, chromium, and selenium. (p. 808)
29. Define a vitamin. Explain how we obtain vitamins. Distinguish between a fat-soluble and a water-soluble vitamin. (p. 809)
30. Compare avitaminosis and hypervitaminosis. (p. 809)
31. For each of the following vitamins, indicate its principal function and effect of deficiency: A, D, E, K, B_1, B_2, niacin, B_6, B_{12}, pantothenic acid, folic acid, biotin, and C. (p. 810)
32. Describe abnormalities associated with megadoses of several minerals and vitamins. (p. 813)
33. Define a kilocalorie (kcal). How is the unit used? (p. 813)
34. How is the caloric value of foods determined? (p. 813)
35. What is metabolic rate? What factors affect it? (p. 813)
36. Define basal metabolic rate (BMR). How is it measured? (p. 814)
37. Define each of the following mechanisms of heat loss: radiation, conduction, convection, and evaporation. (p. 814)
38. Explain how body temperature is regulated by describing the mechanisms of heat production and heat loss. (p. 815)
39. Contrast fever, heat cramps, heat exhaustion, heatstroke, and hypothermia as body temperature abnormalities. (p. 817)
40. What causes fever? What are its benefits? (p. 817)
41. What is obesity? How is it classified? How is it treated? (p. 818)
42. What is malnutrition? Describe the stages of starvation. (p. 819)
43. Define phenylketonuria (PKU), cystic fibrosis (CF), and celiac disease. (p. 820)
44. What is carbohydrate loading? (p. 798)

SELECTED READINGS

Ackerman, S. "The Management of Obesity," *Hospital Practice,* March 1983.

Cunha, B. A. "The Patient with Fever," *Postgraduate Medicine,* April 1989.

Friedman, R. B. "Very Low-Calorie Diets: How Successful?" *Postgraduate Medicine,* May 1988.

Goldfinger, S. E. "The Overweight Problem," *Harvard Medical School Health Letter,* September 1986.

Hogan, M. J. (ed.). "Liquid Diet Supplements," *Mayo Clinic Health Letter,* January 1989.

Kiester, E. "A Little Fever Is Good for You," *Science 84,* November 1984.

Miller, J. A. "Obesity: If the Gene Fits . . . ," *Science News,* 25 January 1986.

Rios, J. C. (ed). "Reviewing the Most Popular Diets," *Cardiac Alert,* February 1984.

Rosenblatt, E. "Weight-Loss Programs," *Postgraduate Medicine,* May 1988.

Rosenthal, T. C., and D. A. Silverstein. "Fever," *Postgraduate Medicine,* June 1988.

Sachs, A. "Drinking Yourself Skinny," *Time,* 19 December 1988.

Schimke, R. N. "Metabolic Diseases," *Hospital Practice,* January 1983.

Stevens, R. G., D. Y. Jones, M. S. Micozzi, and P. R. Taylor. "Body Iron Stores and the Risk of Cancer," *New England Journal of Medicine,* 20 October, 1988.

Tepperman, J. *Metabolic and Endocrine Physiology,* 4th ed. Chicago: Year Book Medical Publishers, 1980.

Chapter 26

The Urinary System

Chapter Contents at a Glance

Student Objectives

1. Identify the external and internal gross anatomical features of the kidneys.
2. Define the structural adaptations of a nephron for urine formation.
3. Discuss the process of urine formation through glomerular filtration, tubular reabsorption, and tubular secretion.
4. Discuss the structure and physiology of the ureters, urinary bladder, and urethra.
5. List and describe the physical characteristics and normal chemical constituents of urine.
6. Define albuminuria, glycosuria, hematuria, pyuria, ketosis, bilirubinuria, urobilinogenuria, casts, and renal calculi.
7. Describe the effects of aging on the urinary system.
8. Describe the development of the urinary system.
9. Discuss the causes of gout, glomerulonephritis, pyelitis, pyelonephritis, cystitis, nephrosis, polycystic disease, renal failure, and urinary tract infections (UTIs).
10. Define medical terminology associated with the urinary system.

The metabolism of nutrients results in the production of wastes by body cells, including carbon dioxide and excess water and heat. Protein catabolism produces toxic nitrogenous wastes such as ammonia and much less toxic urea. In addition, many of the essential ions such as sodium, chloride, sulfate, phosphate, and hydrogen tend to accumulate in excess of the body's needs. All the toxic materials and the excess essential materials must be eliminated.

The primary function of the ***urinary system*** is to help keep the body in homeostasis by controlling the composition and volume of blood. It does so by removing and restoring selected amounts of water and solutes. Two kidneys, two ureters, one urinary bladder, and a single urethra make up the system (Figure 26-1). The kidneys regulate the composition and volume of the blood and remove wastes from the blood in the form of urine. They excrete selected amounts of various wastes, assume a role in erythropoiesis by forming renal erythropoietic factor, help control blood pH, help regulate blood pressure by secreting renin (which activates the renin–angiotensin pathway), and participate in the activation of vitamin D. Urine is excreted from each kidney through its ureter and is stored in the urinary bladder until it is expelled from the body through the urethra. Other systems that aid in waste elimination are the respiratory, integumentary, and digestive systems (see Exhibit 26-3).

The specialized branch of medicine that deals with structure, function, and diseases of the male and female urinary systems and the male reproductive system is known as ***nephrology*** (nef-ROL-ō-jē; *neph* = kidney; *logos* = study of). The branch of surgery related to male and female urinary systems and male reproductive system is referred to as ***urology*** (yoo-ROL-ō-jē; *uro* = urine or urinary tract).

The developmental anatomy of the urinary system is considered at the end of the chapter.

FIGURE 26-1 Organs of the male urinary system in relation to surrounding structures.

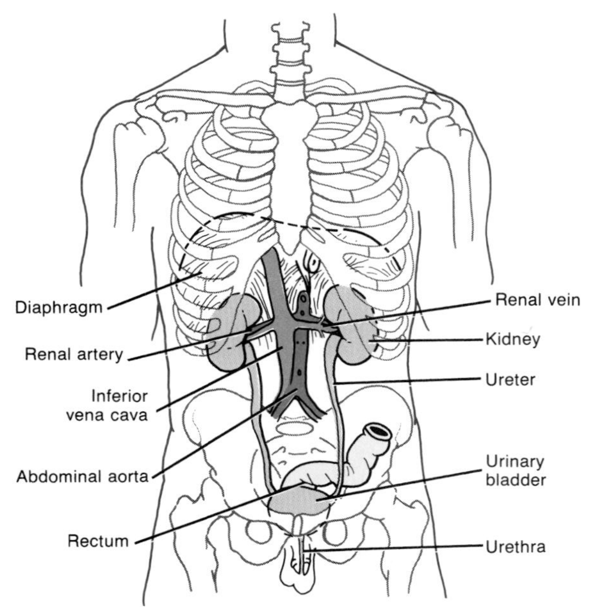

KIDNEYS

The paired ***kidneys*** are reddish organs that resemble kidney beans in shape. They are found just above the waist between the parietal peritoneum and the posterior wall of the abdomen. Since they are external to the peritoneal lining of the abdominal cavity, their placement is described as ***retroperitoneal*** (re'-trō-per-i-tō-NĒ-al). Other retroperitoneal structures include the ureters and adrenal (suprarenal) glands. Relative to the vertebral column, the kidneys are located between the levels of the last thoracic and third lumbar vertebrae and are partially protected by the eleventh and twelfth pairs of ribs. The right kidney is slightly lower than the left because of the large area occupied by the liver.

External Anatomy

The average adult kidney measures about 10 to 12 cm (4 to 5 inches) long, 5.0 to 7.5 cm (2 to 3 inches) wide, and 2.5 cm (1 inch) thick. Its concave medial border faces the vertebral column. Near the center of the concave border is a notch called the ***hilus,*** through which the ureter leaves the kidney. Blood and lymphatic vessels and nerves also enter and exit the kidney through the hilus (Figure 26-2). The hilus is the entrance to a cavity in the kidney called the ***renal sinus.***

Three layers of tissue surround each kidney. The innermost layer, the ***renal capsule,*** is a smooth, transparent, fibrous membrane that can easily be stripped off the kidney and is continuous with the outer coat of the ureter at the hilus. It serves as a barrier against trauma and the spread of infection to the kidney. The second layer, the ***adipose capsule,*** is a mass of fatty tissue surrounding the renal capsule. It also protects the kidney from trauma and holds it firmly in place within the abdominal cavity. The outermost layer, the ***renal fascia,*** is a thin layer of fibrous connective tissue that anchors the kidney to its surrounding structures and to the abdominal wall.

CLINICAL APPLICATION: FLOATING KIDNEY (NEPHROTOSIS)

Floating kidney, or ***nephrotosis*** (nef'-rō-TŌ-sis; *ptosis* = falling), is a downward displacement or dropping of the kidney. It occurs when the kidney is no longer held in place securely by the adjacent organs or its covering of fat and slips from its normal position. Individuals, especially thin people, in

FIGURE 26-2 Kidney. Coronal section of the right kidney illustrating the internal anatomy. (a) Diagram. (b) Photograph. (Courtesy of J. A. Gosling, P. F. Harris, et al., *Atlas of Human Anatomy,* Gower Medical Publishing Ltd., 1985.)

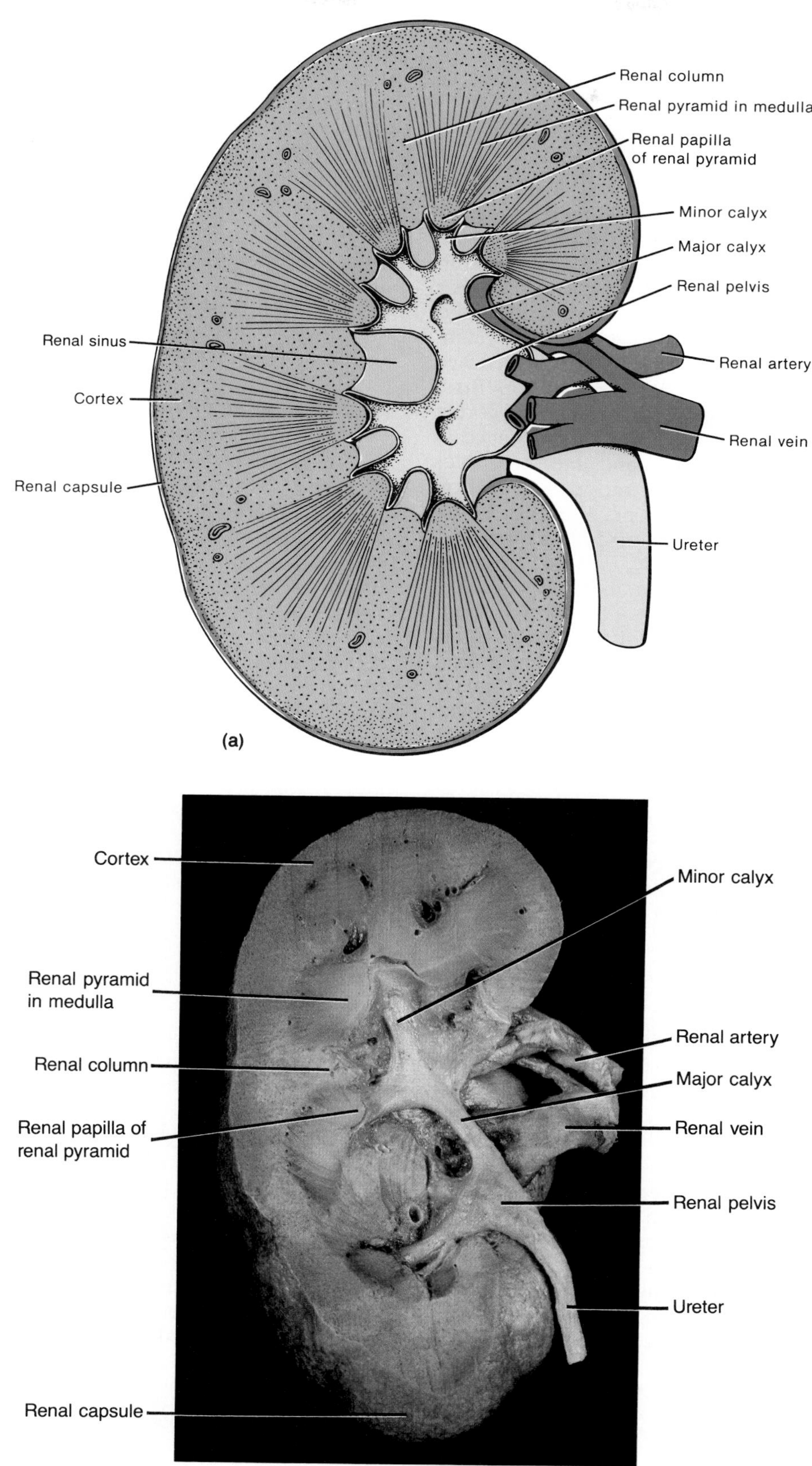

whom either the adipose capsule or renal fascia is deficient, may develop nephrotosis. It is dangerous because it may cause kinking of the ureter with reflux of urine and retrograde pressure. Pain occurs if the ureter is twisted. Also, if the kidneys drop below the rib cage, they become susceptible to blows and penetrating injuries.

Internal Anatomy

A coronal (frontal) section through a kidney reveals an outer reddish area called the ***cortex*** and an inner reddish-brown region called the ***medulla*** (Figure 26-2). Within the medulla are 8 to 18 striated triangular structures termed ***renal (medullary) pyramids.*** The striated (striped) appearance is due to the presence of straight tubules and blood vessels. The bases of the pyramids face the cortical area, and their apices, called ***renal papillae,*** are directed toward the center of the kidney. The cortex is the smooth-textured area extending from the renal capsule to the bases of the pyramids and into the spaces between them. The cortex is divided into an outer cortical zone and an inner juxtamedullary zone. The cortical substance between the renal pyramids forms the ***renal columns.***

Together the cortex and renal pyramids constitute the ***parenchyma*** (functional portion) of the kidney. Structurally, the parenchyma of each kidney consists of approximately 1 million microscopic units called nephrons, the functional units of the kidney. They help regulate blood composition and form urine. Associated with nephrons are collecting tubules and a vascular supply.

In the renal sinus of the kidney is a large cavity called the ***renal pelvis.*** The edge of the pelvis contains cuplike extensions called ***major*** or ***minor calyces*** (KĀ-li-sēz; *calyx* = cup). There are 2 or 3 major calyces and 8 to 18 minor calyces. Each minor calyx collects urine from collecting tubules of the pyramids. From the major calyces, the urine drains into the pelvis and out through the ureter.

Nephron

The functional unit of the kidney is the ***nephron*** (NEF-ron) (Figure 26-3). Nephrons have several functions related to homeostasis. They filter blood; that is, they permit some substances to pass into the kidneys, while keeping others out. As the filtered liquid (filtrate) moves through the nephrons, it is further processed by nephrons by the addition of some substances (wastes and excess substances) and the removal of others (useful materials). As a result of the activities of nephrons, urine is formed.

Essentially, a nephron consists of two portions: a ***renal tubule*** and a tuft (knot) of capillaries, called the glomerulus. The renal tubule begins as a double-walled epithelial cup, called the ***glomerular (Bowman's) capsule,*** lying in the cortex of the kidney. The outer wall, or ***parietal layer,*** is composed of simple squamous epithelium (Figure 26-4). It is separated from the inner wall, known as the ***visceral layer,*** by the ***capsular space.*** The visceral layer consists of epithelial cells called podocytes. The capsule surrounds a capillary network called the ***glomerulus*** (glō-MER-yoo-lus; *glomus* = ball; *ulus* = small). Collectively, the glomerular capsule and its enclosed glomerulus constitute a ***renal corpuscle*** (KŌR-pus-sul; *corpus* = body; *cle* = tiny).

The visceral layer of the glomerular capsule and the endothelium of the glomerulus form an ***endothelial-capsular membrane.*** This membrane consists of the following parts, listed here in the order in which substances filtered by the kidney must pass through them.

1. *Endothelium of the glomerulus.* This single layer of endothelial cells has completely opened pores (fenestrated) averaging 50–100 mm in diameter. It restricts the passage of blood cells.
2. *Basement membrane of the glomerulus.* This extracellular membrane lies beneath the endothelium and contains no pores. It consists of fibrils in a glycoprotein matrix. It restricts the passage of large-sized proteins.
3. *Epithelium of visceral layer of the glomerular (Bowman's) capsule.* These epithelial cells, because of their peculiar shape, are called ***podocytes.*** The podocytes contain footlike structures called ***pedicels*** (PED-i-sels). The pedicels are arranged parallel to the circumference of the glomerulus and cover the basement membrane, except for spaces between them called ***filtration slits (slit pores).*** Pedicels contain numerous thin contractile filaments that are believed to regulate the passage of substances through the filtration slits. In addition, another factor that helps regulate the passage of substances through the filtration slits is a thin membrane, the ***slit membrane,*** that extends between filtration slits. This membrane restricts the passage of intermediate-sized proteins.

The endothelial-capsular membrane filters water and small solutes from blood plasma. Large molecules, such as proteins, and the formed elements in blood do not normally pass through it. The water and solutes that are filtered out of the blood pass into the capsular space between the visceral and parietal layers of the glomerular (Bowman's) capsule and then into the renal tubule.

The glomerular capsule opens into the first section of the renal tubule, called the ***proximal convoluted tubule,*** which also lies in the cortex. *Convoluted* means the tubule is coiled rather than straight; *proximal* signifies that the glomerular (Bowman's) capsule is the origin of the tubule. The wall of the proximal convoluted tubule consists of cuboidal epithelium with microvilli. These surface specializations, like those of the small intestine, increase the surface area for reabsorption and secretion.

Nephrons are frequently classified into two kinds. A ***cortical nephron*** usually has its glomerulus in the outer cortical zone, and the remainder of the nephron rarely penetrates the medulla. A ***juxtamedullary nephron*** usually has its glomerulus close to the corticomedullary

FIGURE 26-3 Nephrons. (a) Juxtamedullary nephron. (b) Cortical nephron. The nephrons are colored pink.

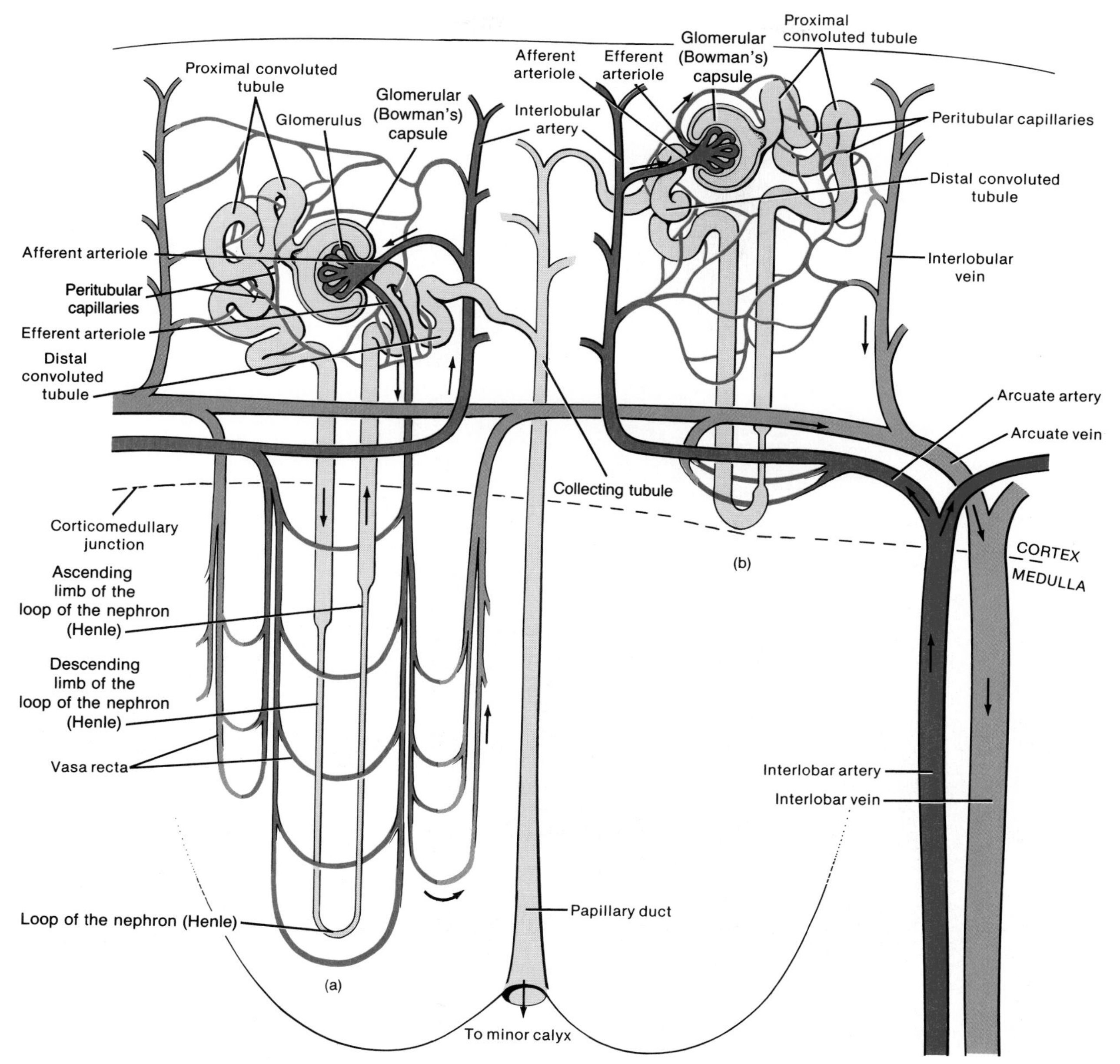

junction, and other parts of the nephron penetrate deeply into the medulla (see Figure 26-3).

In a juxtamedullary nephron, the proximal convoluted tubule straightens, becomes thinner, and dips into the medulla, where it is called the ***descending (thin) limb of the loop of the nephron.*** This section consists of squamous epithelium. The tubule then increases in diameter after it bends into a U-shaped structure called the ***loop of the nephron (loop of Henle).*** It then ascends toward the cortex as the ***ascending (thick) limb of the loop of the nephron,*** which consists of cuboidal and low columnar epithelium.

In the cortex, the tubule again becomes convoluted. Because of its distance from the point of origin at the glomerular (Bowman's) capsule, this section is referred to as the ***distal convoluted tubule.*** The cells of the distal tubule, like those of the proximal tubule, are cuboidal. Unlike the cells of the proximal tubule, however,

FIGURE 26-4 Endothelial-capsular membrane. (a) Parts of a renal corpuscle. (b) Enlarged aspect of a portion of the endothelial-capsular membrane. The size of the filtration slits has been exaggerated for emphasis.

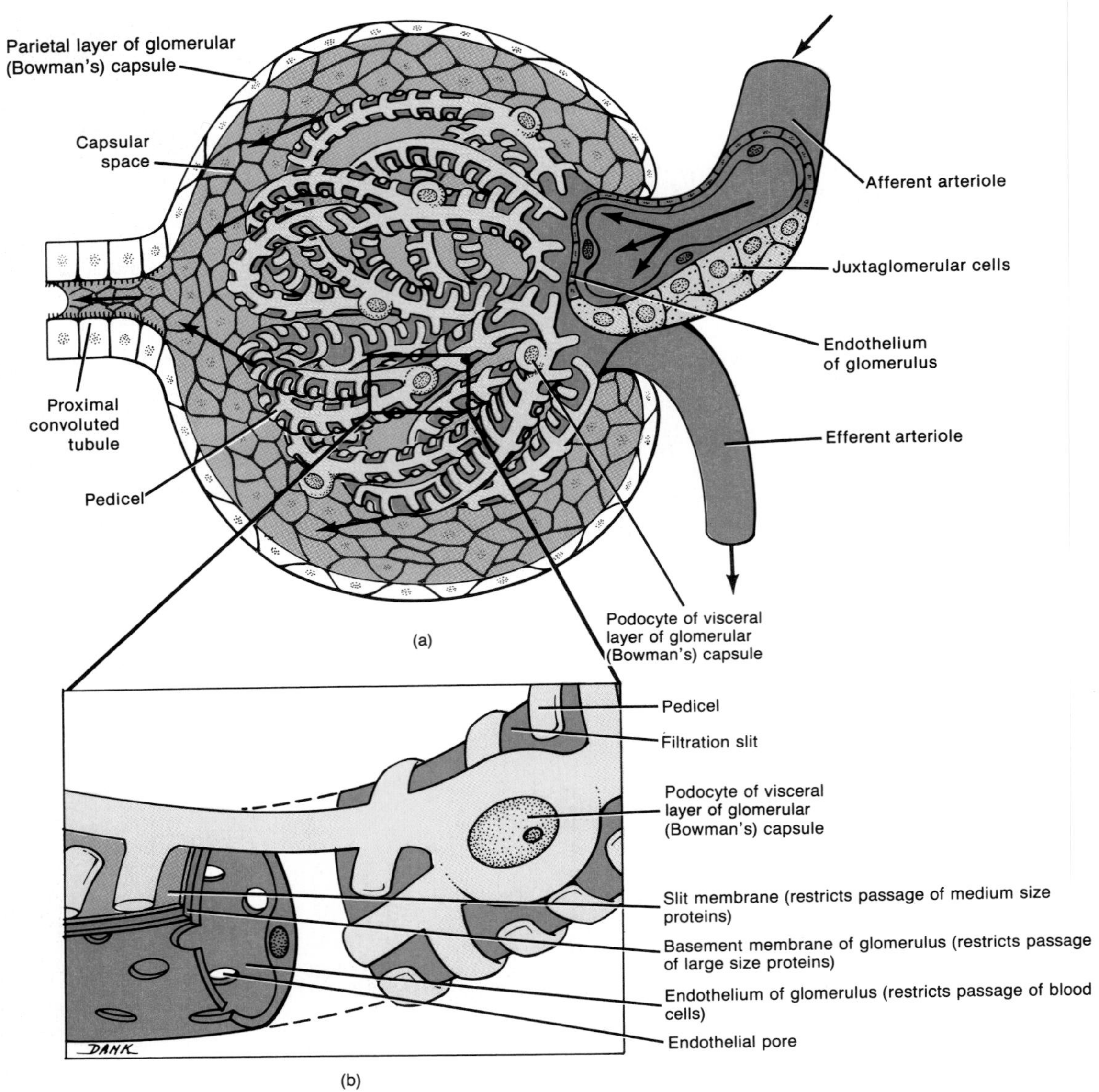

the cells of the distal tubule have few microvilli. In a cortical nephron, the proximal section runs into the distal tubule without the connecting limbs and loop of the nephron. In the cortex of both types of nephrons, the distal tubules empty into ***collecting tubules.***

In the medulla, the collecting tubules combine to form ***papillary ducts*** which open at the renal papillae into the minor calyces. On the average, there are 30 papillary ducts per renal papilla. Cells of the collecting tubules are cuboidal; those of the papillary ducts are columnar.

The histology of a nephron and glomerulus is shown in Figure 26-5.

Blood and Nerve Supply

Nephrons are largely responsible for removing wastes from the blood and regulating its fluid and electrolyte content. Thus, they are abundantly supplied with blood vessels. The right and left ***renal arteries*** transport about one-fourth the total cardiac output to the kidneys (Figure 26-6). Approximately 1200 ml passes through the kidneys every minute.

Before or immediately after entering the hilus, the renal artery typically divides into a larger anterior branch and a smaller posterior branch. From these branches, five

FIGURE 26-5 **Histology of a nephron. Photomicrograph of the cortex of the kidney showing a renal corpuscle and surrounding renal tubules at a magnification of 400×. (Courtesy of Andrew Kuntzman.)**

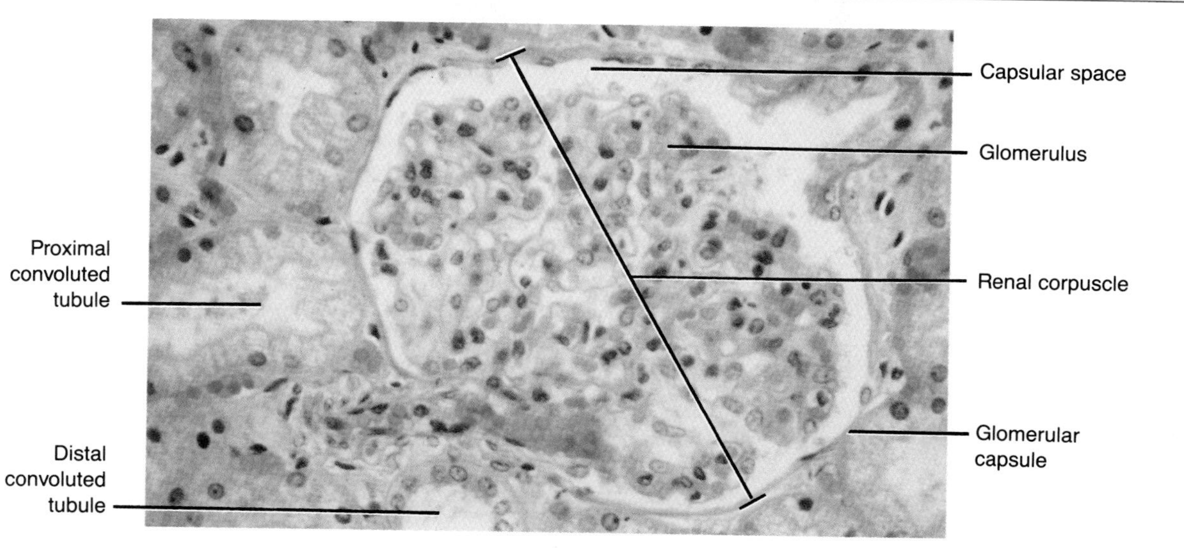

FIGURE 26-6 **Blood supply of the right kidney. (a) Diagram in coronal section.**

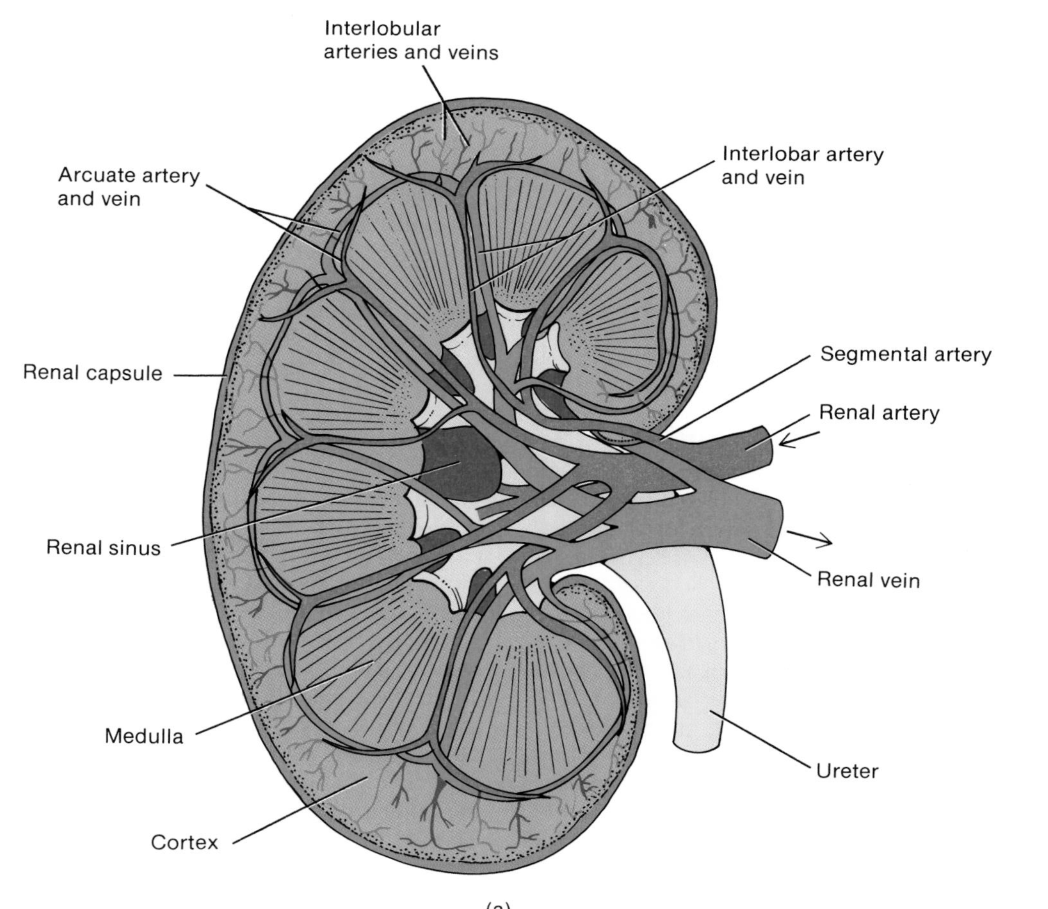

FIGURE 26-6 (*Continued*) (b) Scheme of circulation. This view is designed to show the *sequence* of blood flow, not the anatomical location of blood vessels, which is shown in (a).

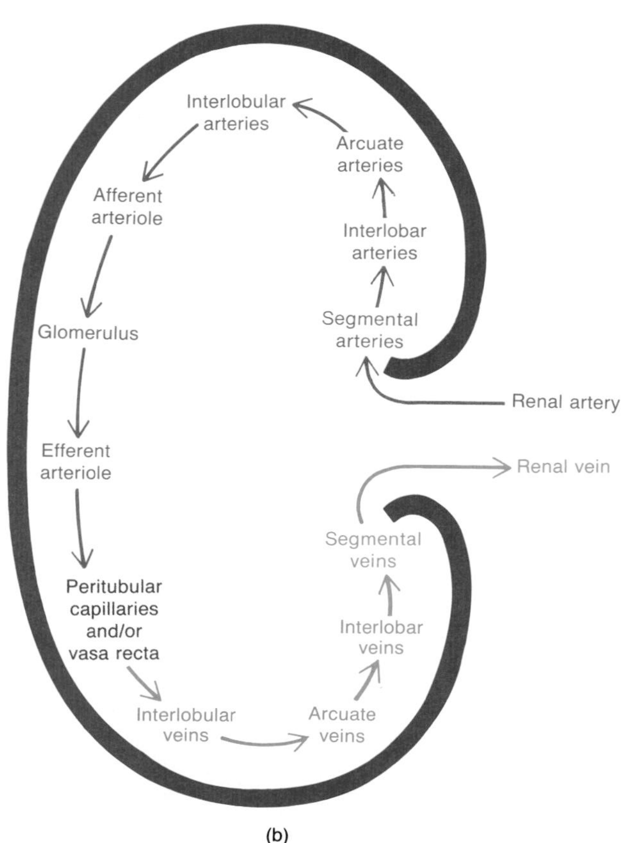

segmental arteries originate, each supplying a particular segment of the kidneys. Each segmental artery gives off several branches that enter the parenchyma and pass as the ***interlobar arteries*** between the renal pyramids in the renal columns. At the bases of the pyramids, the interlobar arteries arch between the medulla and cortex; here they are known as the ***arcuate arteries.*** Divisions of the arcuate arteries produce a series of ***interlobular arteries,*** which enter the cortex and divide into ***afferent arterioles*** (see Figure 26-3).

One afferent arteriole is distributed to each glomerular (Bowman's) capsule, where the arteriole divides into the tangled capillary network called the ***glomerulus.*** The glomerular capillaries then reunite to form an ***efferent arteriole,*** which leads away from the capsule and is smaller in diameter than the afferent arteriole. This variation in diameter helps raise the glomerular blood pressure to a level significantly higher than the pressure in capillaries elsewhere in the body. The afferent-efferent arteriole situation is unique because blood usually flows out of capillaries into venules and not into other arterioles.

Each efferent arteriole of a cortical nephron divides to form a network of capillaries, called the ***peritubular capillaries,*** around the convoluted tubules. The efferent arteriole of a juxtamedullary nephron also forms peritubular capillaries. In addition, it forms long loops of thin-walled vessels called ***vasa recta*** that dip down alongside the loop of the nephron into the medullary region of the papilla.

The peritubular capillaries eventually reunite to form ***peritubular venules*** and then ***interlobular veins.*** The blood then drains through the ***arcuate veins*** to the ***interlobar veins*** running between the pyramids, then the ***segmental veins,*** and leaves the kidney through a single ***renal vein*** that exits at the hilus. The vasa recta pass blood into the interlobular veins. From here, it goes to the arcuate veins, the interlobar veins, and then into the renal vein.

The nerve supply to the kidneys is derived from the ***renal plexus*** of the sympathetic division of the autonomic nervous system. Nerves from the plexus accompany the renal arteries and their branches and are distributed to the vessels. Because the nerves are vasomotor, they regulate the circulation of blood in the kidney by regulating the diameters of the arterioles.

Juxtaglomerular Apparatus (JGA)

The smooth muscle fibers (cells) of the tunica media adjacent to the afferent arteriole (and sometimes efferent arteriole) are modified in several ways. Their nuclei are round (instead of long), and their cytoplasm contains granules (instead of myofibrils). Such modified muscle fibers are called ***juxtaglomerular cells.*** The cells of the distal convoluted tubule adjacent to the afferent and efferent arterioles are considerably narrower and taller than the other cells. Collectively, these cells are known as the ***macula densa.*** Together with the modified cells of the afferent arteriole they constitute the ***juxtaglomerular apparatus,*** or ***JGA*** (Figure 26-7), which helps regulate renal blood pressure.

Physiology

The major work of the urinary system is done by the nephrons. The other parts of the system are primarily passageways and storage areas. Nephrons carry out three important functions: (1) they control blood concentration and volume by removing selected amounts of water and solutes; (2) they help regulate blood pH; and (3) they also remove toxic wastes from the blood. As the nephrons go about these activities, they remove many materials from the blood, return the ones that the body requires, and eliminate the remainder. The eliminated materials are collectively called ***urine.*** The entire volume of blood in the body is filtered by the kidneys approximately 60 times a day. Urine formation requires three principal processes: glomerular filtration, tubular reabsorption, and tubular secretion.

FIGURE 26-7 Juxtaglomerular apparatus (JGA). (a) External view. (b) Cells of the juxtaglomerular apparatus seen in cross section. The macula densa adjacent to the efferent arteriole is not illustrated.

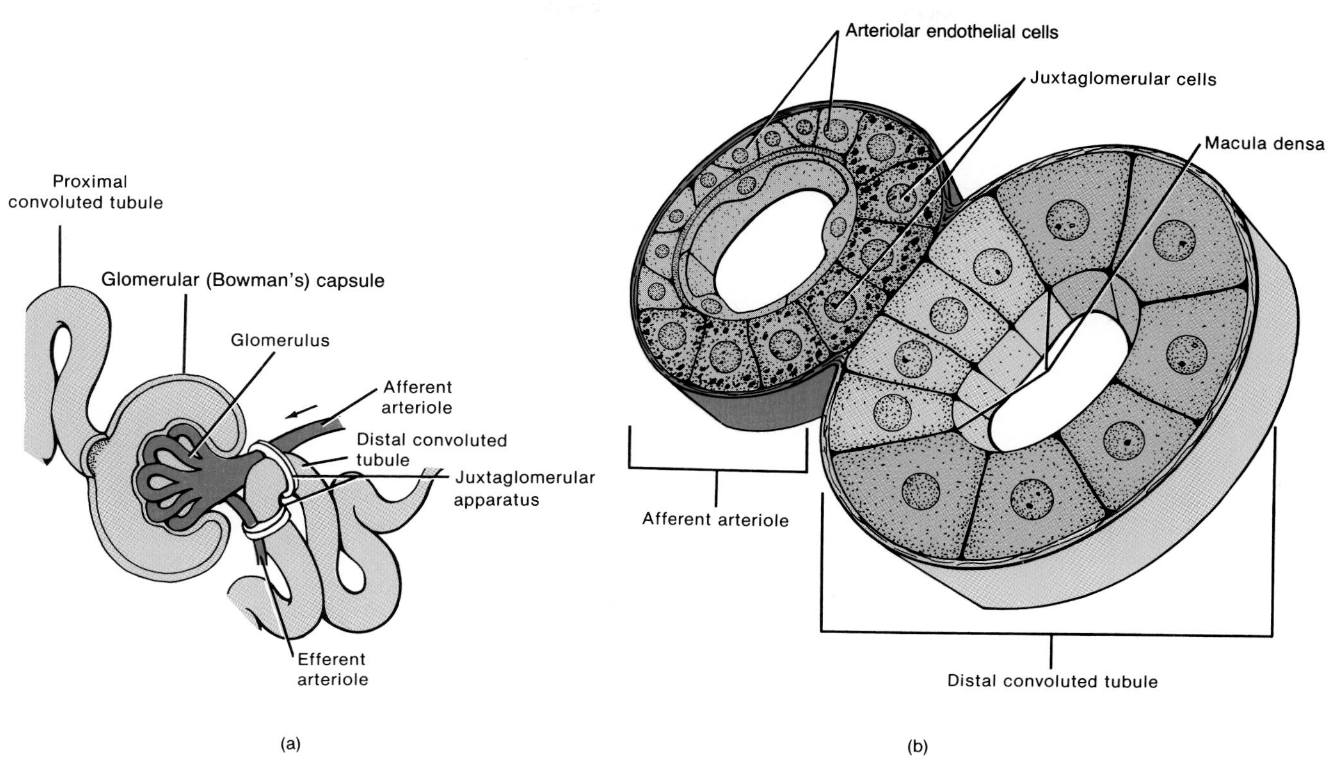

Glomerular Filtration

The first step in the production of urine is called ***glomerular filtration.*** Filtration—the forcing of fluids and dissolved substances through a membrane by pressure—occurs in the renal corpuscle of the kidneys across the endothelial-capsular membrane. When blood enters the glomerulus, the blood pressure forces water and dissolved components through the endothelial pores of the capillaries, basement membrane, and on through the filtration slits of the adjoining visceral wall of the glomerular (Bowman's) capsule (see Figure 26-4). The resulting fluid is called the ***filtrate.*** In a healthy person, the filtrate consists of all the materials present in the blood except for the formed elements and most proteins, which are too large to pass through the endothelial-capsular barrier. Exhibit 26-1 compares the constituents of plasma, glomerular filtrate, and urine during a 24-hour period. Although the values shown are typical, they vary considerably according to diet.

The amount of filtrate that flows out of all the renal corpuscles of both kidneys every minute is called the ***glomerular filtration rate (GFR).*** In the normal adult, this rate is about 125 ml/min—about 180 liters (48 gal) a day.

Renal corpuscles are especially structured for filtering blood. First, each capsule contains a tremendous length of highly coiled glomerular capillaries presenting a vast surface area for filtration. Second, the endothelial-capsular membrane is structurally adapted for filtration. Although the endothelial pores generally do not restrict the passage of substances, the basement membrane permits the passage of only smaller molecules. Thus, water, glucose, vitamins, amino acids, small proteins, nitrogenous wastes, and ions pass into the glomerular capsule. Large proteins and the formed elements in blood do not normally pass through the basement membrane. The filtration slits permit only the limited passage of very small plasma proteins such as albumins. Third, the efferent arteriole is smaller in diameter than the afferent arteriole, so there is usually high resistance to the outflow of blood from the glomerulus. Consequently, blood pressure is higher in the glomerular capillaries than in other capillaries. Glomerular blood pressure averages about 60 mm Hg, whereas the blood pressure of other capillaries averages only 30 mm Hg. The higher pressure, within limits, promotes filtration. Fourth, the endothelial-capsular membrane separating the blood from the space in the glomerular capsule is very thin (0.1 μm).

The filtering of the blood depends on a number of

EXHIBIT 26-1 CHEMICALS IN PLASMA, FILTRATE, AND URINE DURING 24-HOUR PERIOD[a]

Chemical	Plasma[b]	Filtrate Immediately after Passing into Glomerular Capsule[c]	Reabsorbed from Filtrate[d]	Urine
Water	180,000 ml	180,000 ml	178,500 ml	1,500 ml
Proteins	7,000–9,000	10–20	10–20	0[e]
Chloride (Cl^-)	630	630	625	5
Sodium (Na^+)	540	540	537	3
Bicarbonate (HCO_3^-)	300	300	299.7	0.3
Glucose	180	180	180	0
Urea	53	53	28	25
Potassium (K^+)	28	28	24	4
Uric acid	8.5	8.5	7.7	0.8
Creatinine	1.5	1.5	0	1.5

[a] All values, except for water, are expressed in grams. The chemicals are arranged in sequence from highest to lowest concentration in plasma.
[b] These substances are present in glomerular blood plasma before filtration.
[c] These substances pass from glomerular blood plasma through endothelial-capsular membrane before reabsorption.
[d] These substances have been filtered.
[e] Although trace amounts of protein (170 to 250 mg) normally appear in urine, we will assume for purposes of discussion that all of it is reabsorbed from filtrate.

pressures. The chief one is the ***glomerular blood hydrostatic pressure.*** *Hydrostatic* (*hydro* = water) *pressure* is the force that a fluid under pressure exerts against the walls of its container. Glomerular blood hydrostatic pressure means the blood pressure in the glomerulus (Figure 26-8). This pressure tends to move fluid out of the glomeruli at a force averaging about 60 mm Hg.

However, glomerular blood hydrostatic pressure is opposed by two other forces. The first of these, ***capsular hydrostatic pressure,*** develops in the following way. When the filtrate is forced into the capsular space between the walls of the glomerular (Bowman's) capsule, it meets with two forms of resistance: the walls of the capsule and the fluid that has already filled the renal tubule. As a result, some filtrate is pushed back into the capillary. The amount of "push" is the capsular hydrostatic pressure. It usually measures about 20 mm Hg.

The second force opposing filtration into the glomerular capsule is the ***blood colloidal osmotic pressure.*** *Osmotic pressure* is the pressure required to prevent the movement of pure water into a solution containing solutes when the solutions are separated by a selectively permeable membrane (Chapter 3). The greater the solute concentration of the solution, the greater its osmotic pressure. Hydrostatic pressure develops because of a force outside a solution. Osmotic pressure develops because of the concentration of the solution itself. Since the blood contains a much higher concentration of proteins than the filtrate does, water would move out of the filtrate and back into the blood vessel if it were not for the fact that blood pressure in the glomerulus normally is greater than blood colloidal osmotic pressure. This blood colloidal osmotic pressure is normally about 30 mm Hg.

To determine how much filtration pressure exists, we have to subtract the forces that oppose filtration from the glomerular blood hydrostatic pressure. The net result is called the ***effective filtration pressure,*** which is abbreviated ***Peff.***

$$Peff = \begin{pmatrix} \text{glomerular} \\ \text{blood} \\ \text{hydrostatic} \\ \text{pressure} \end{pmatrix} - \begin{pmatrix} \text{capsular} \\ \text{hydrostatic} \\ \text{pressure} \end{pmatrix} + \begin{matrix} \text{blood} \\ \text{colloidal} \\ \text{osmotic} \\ \text{pressure} \end{matrix}$$

By substituting the values just discussed, a normal *Peff* may be calculated as follows:

$$\begin{aligned} Peff &= (60 \text{ mm Hg}) - (20 \text{ mm Hg} + 30 \text{ mm Hg}) \\ &= (60 \text{ mm Hg}) - (50 \text{ mm Hg}) \\ &= 10 \text{ mm Hg} \end{aligned}$$

This means that a pressure of about 10 mm Hg causes a normal amount of plasma to filter from the glomerulus into the glomerular (Bowman's) capsule. This produces

FIGURE 26-8 **Forces involved in effective filtration pressure (*P*eff).**

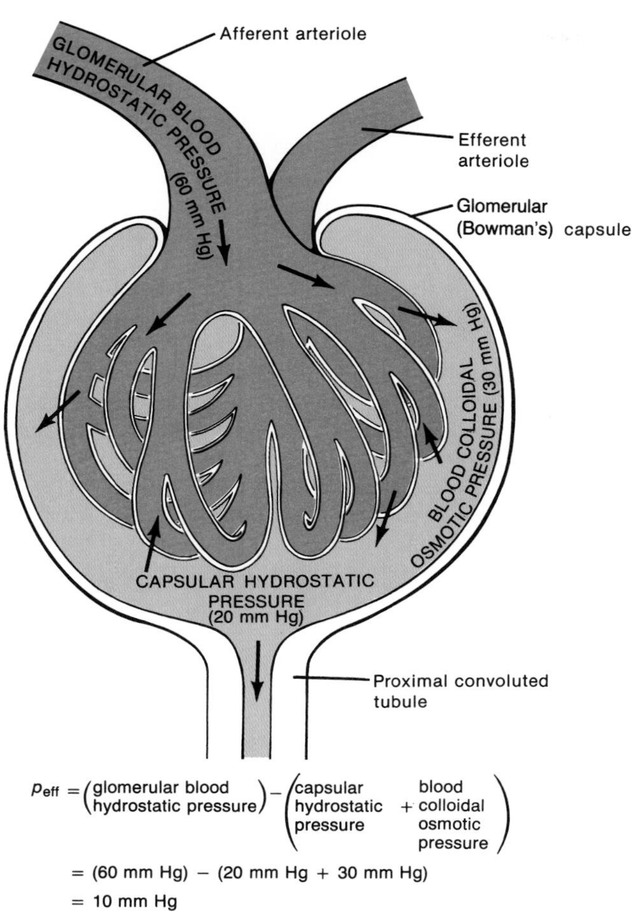

about 125 ml of filtrate per minute in both kidneys. The term ***filtration fraction*** refers to the percentage of plasma entering the nephrons that actually becomes glomerular filtrate. Although filtration fraction averages between 16 and 20 percent, the value does vary considerably in both health and disease.

Certain conditions may alter these pressures and thus the *P*eff. In some forms of kidney disease, such as glomerulonephritis, glomerular capillaries become so permeable that the plasma proteins are able to pass from the blood into the filtrate. As a result, the capsular filtrate exerts an osmotic pressure that draws water out of the blood. Thus, if a capsular osmotic pressure develops, the *P*eff will increase. At the same time, blood colloid osmotic pressure decreases, further increasing the *P*eff.

The *P*eff is also affected by changes in general arterial blood pressure. Severe hemorrhaging produces a drop in general blood pressure, which also decreases the glomerular blood hydrostatic pressure. If the blood pressure falls to the point where the hydrostatic pressure in the glomeruli reaches 50 mm Hg, no filtration occurs, because the glomerular blood hydrostatic pressure equals the opposing forces. Such a condition is called ***anuria*** (a-NOO-rē-a), a daily urine output of less than 50 ml. It may be caused by insufficient pressure to accomplish filtration or by inflammation of the glomeruli so that plasma is prevented from entering the glomerulus.

A final factor that may affect the *P*eff is the regulation of the size of the afferent and efferent arterioles. In this case, glomerular blood hydrostatic pressure is regulated separately from the general blood pressure. Sympathetic impulses and small doses of epinephrine cause constriction of both afferent and efferent arterioles. However, intense sympathetic impulses and large doses of epinephrine cause greater constriction of afferent than efferent arterioles. This intense stimulation results in a decrease in glomerular blood hydrostatic pressure even though blood pressure in other parts of the body may be normal or even higher than normal. Intense sympathetic stimulation is most likely to occur during the alarm reaction of the general adaptation syndrome. Blood may also be shunted away from the kidneys during hemorrhage.

Tubular Reabsorption

As the filtrate passes through the renal tubules, about 99 percent of it is reabsorbed into the blood. Thus, only about 1 percent of the filtrate actually leaves the body (about 1.5 liters a day). The movement of certain components of the filtrate back into the blood of the peritubular capillaries or vasa recta is called ***tubular reabsorption.***

Tubular reabsorption, carried out by epithelial cells throughout the renal tubule, is a very discriminating process. Only specific amounts of certain substances are reabsorbed, depending on the body's needs at the time. The maximum amount of a substance that can be reabsorbed under any condition is called the substance's ***tubular transport maximum (Tm).*** Materials that are reabsorbed include water, glucose, amino acids, urea, and ions such as Na^+, K^+, Ca^{2+}, Cl^-, HCO_3^-, and HPO_4^{2-}. Tubular reabsorption allows the body to retain most of its nutrients. Wastes such as urea are only partially reabsorbed. Exhibit 26-1 compares the values for the chemicals in the filtrate immediately after passing into the glomerular (Bowman's) capsule with those reabsorbed from the filtrate. It will give you an idea of how much of the various substances the kidneys reabsorb.

Reabsorption is carried out through both passive and active transport mechanisms. It is believed that glucose and amino acids are reabsorbed by an active process involving a carrier system. The carrier, which is a protein, exists in the membranes of the tubular epithelial cells, especially proximal convoluted tubule cells, in a fixed and limited amount. Glucose or amino acids bind with the same carrier that transports sodium (Na^+) ions. Normally, all the glucose filtered by the glomeruli (125 mg/100 ml of filtrate/min) is reabsorbed by the tubules.

CLINICAL APPLICATION: GLYCOSURIA

The capacity of the glucose carrier system is limited. If the blood concentration of glucose is above normal, the glucose transport mechanism cannot reabsorb it all, and the excess remains in the urine. Also, if there is a malfunction in the tubular carrier mechanism, glucose appears in the urine even though the blood sugar level is normal. This condition is called ***glycosuria*** (glī'-kō-SOO-rē-a) or ***glucosuria.*** The presence of unresorbed glucose in the distal tubules raises the osmotic pressure in the tubules and reduces water reabsorption. As a result, urine output increases. This explains two of the cardinal symptoms of diabetes mellitus, polyuria and polydipsia (see Chapter 18).

Most sodium ions (and potassium ions) are actively transported (reabsorbed) from the proximal convoluted tubules and ascending limb of the loop of the nephron. The reabsorption of the remaining sodium ions (Na^+) and potassium ions (K^+) occurs in the distal convoluted tubules and collecting tubules. Na^+ reabsorption varies with blood pressure. When the Na^+ concentration of the blood is low, there is a drop in blood pressure, and the ***renin–angiotensin pathway*** goes into operation. In response to low blood pressure, the juxtaglomerular cells of the kidneys secrete an enzyme called renin, which converts angiotensinogen (synthesized by the liver) into angiotensin I. As it passes through the lungs, angiotensin I is converted into angiotensin II, which is an active hormone. Angiotensin II causes constriction of efferent arterioles that lead away from the glomerulus, which causes glomerular blood pressure to increase. This ensures normal filtration pressure in the glomerulus. Angiotensin II also stimulates the zona glomerulosa of the adrenal cortex to produce and secrete aldosterone. Aldosterone brings about increased Na^+ and water reabsorption by the distal convoluted tubules and collecting tubules (Figure 26-9b). Extracellular fluid volume increases and blood pressure is restored to normal. In the absence of aldosterone, the Na^+ in the distal convoluted tubules and collecting tubules is not reabsorbed. It thus passes into urine for excretion.

FIGURE 26-9 **Role of the juxtaglomerular apparatus (JGA) in maintaining normal blood volume. (a) Afferent arteriolar vasodilation feedback cycle. (b) Efferent arteriolar vasoconstriction feedback cycle.**

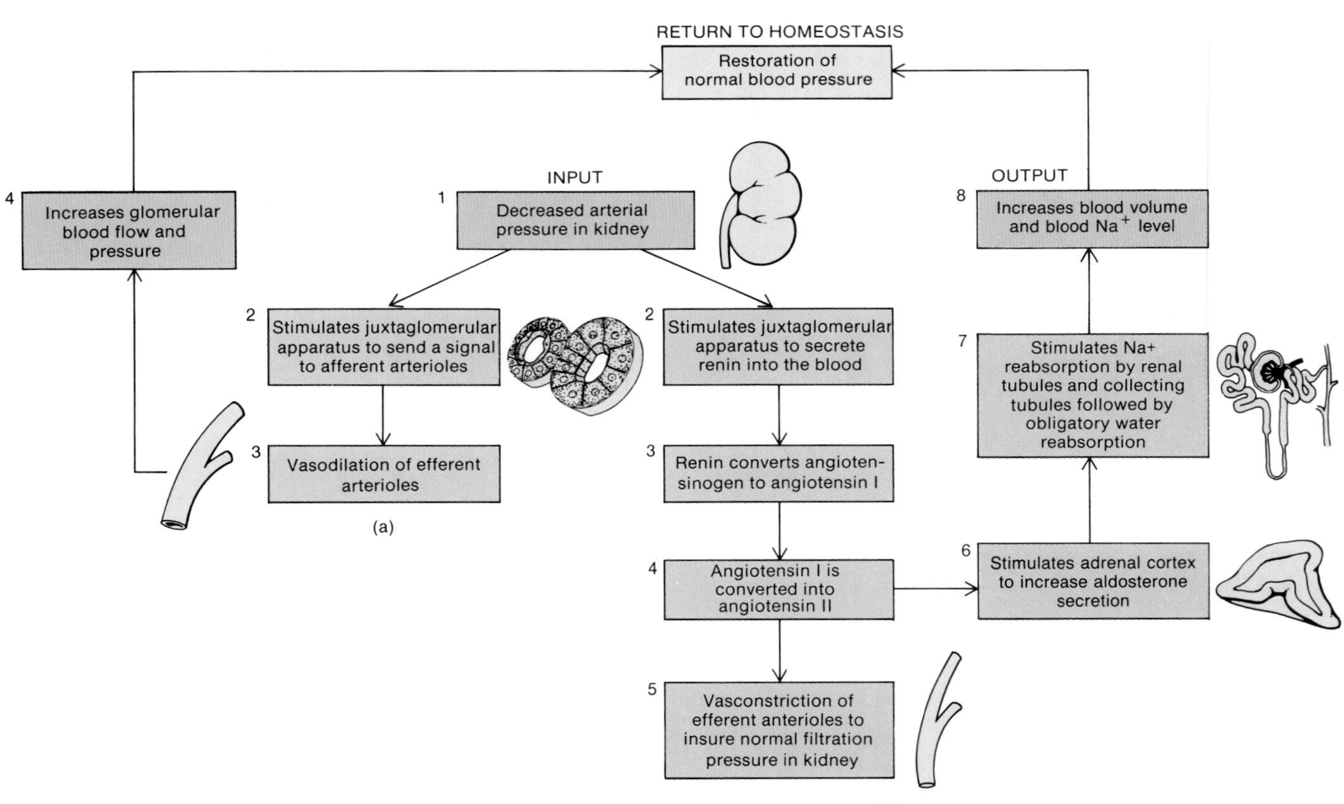

Increased amounts of water are also secreted due to the osmotic pressure created by the nonreabsorbed Na^+.

When the Na^+ concentration of blood is low, another negative feedback mechanism also goes into operation. The decreased Na^+ concentration initiates a signal from the juxtaglomerular apparatus that causes dilation of the afferent arterioles that lead into glomeruli. This increases glomerular blood flow, which increases glomerular pressure. The increased pressure returns filtration pressure back to normal (Figure 26-9a).

When Na^+ ions move out of the tubules into the peritubular blood, the blood momentarily becomes more electropositive than the filtrate. Chloride (Cl^-), a negatively charged ion, follows the positive sodium ion out of the tubule by electrostatic attraction. Thus, the movement of Na^+ ions influences the movement of Cl^- ions and other anions, such as bicarbonate (HCO_3^-) ions, into the blood.

Cl^- ions, and not Na^+ ions, are actively transported from the ascending limb of the loop of the nephron. In this portion of the nephron, then, Cl^- ions move actively and Na^+ ions follow passively. Whether Na^+ ions move actively and Cl^- ions follow passively, or Cl^- ions move actively and Na^+ ions follow passively, the end result is the same.

Water reabsorption is driven by sodium transport. As Na^+ ions are transported from the proximal convoluted tubules into the blood, the osmotic pressure of the blood becomes higher than that of the filtrate. Water follows the Na^+ ions into the blood in order to reestablish the osmotic equilibrium. About 80 percent of the water is reabsorbed by this method from the proximal convoluted tubule. This portion of reabsorbed water occurs as a function of osmosis, so it is referred to as ***obligatory water reabsorption.*** The proximal convoluted tubule has no control over osmosis because it is always permeable to water (Figure 26-10). The descending limb of the loop of the nephron is passively permeable to the passage of water.

Passage of most of the remaining water in the filtrate can be regulated. The permeability of the cells of the distal and collecting tubules is controlled by the antidiuretic hormone (ADH), produced by the hypothalamus and released into the blood by the pituitary gland. When the blood–water concentration is low—that is, osmotic pressure is increased—osmoreceptors in the hypothalamus detect the stimulus and secrete more ADH. The hormone then passes into the posterior pituitary gland where it is released into the blood. ADH increases the permeability of the plasma membranes of the distal tubule and collecting tubule cells. As the membranes become more permeable, more water molecules pass into the cells and then into blood. This type of absorption is called *facultative* (FAK-ul-tā'-tiv), meaning that it occurs under some conditions but not others. ***Facultative water reabsorption*** is responsible for about 20 percent of the water reabsorbed from the filtrate, and, as noted above, is regulated by ADH. It is a major mechanism for controlling water content of the blood (Figure 26-10).

FIGURE 26-10 **Factors that control water reabsorption.**

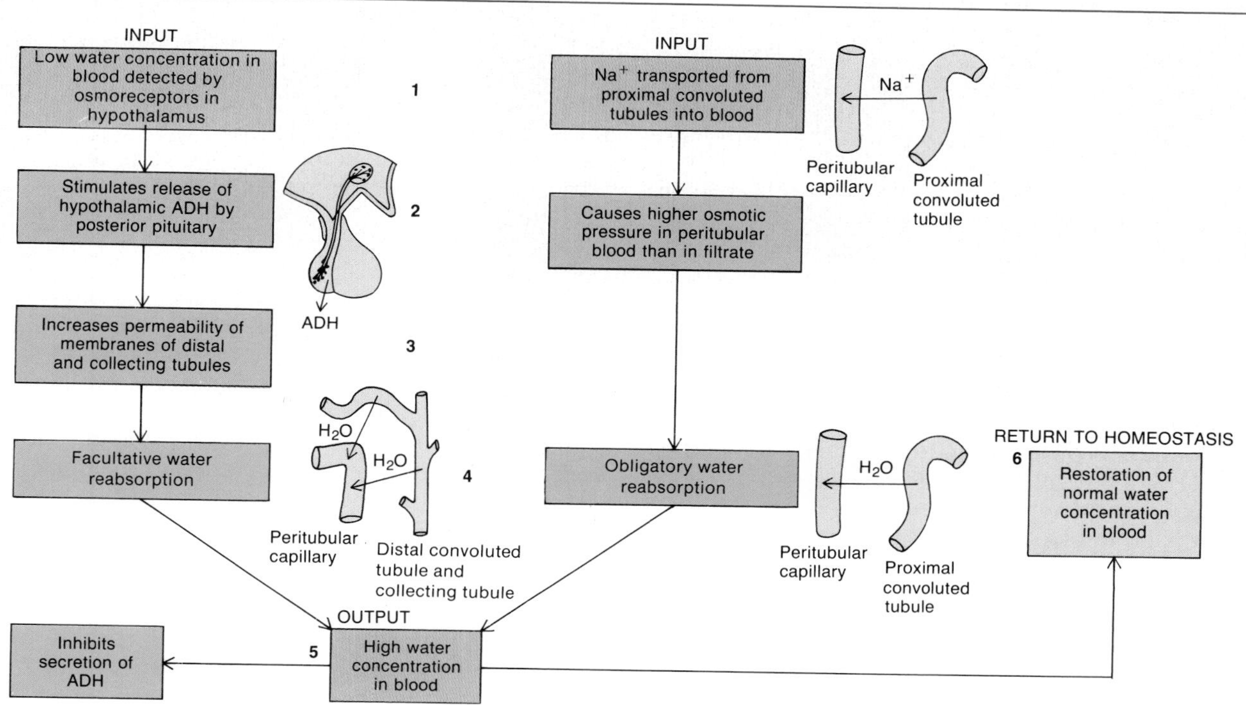

Within limits, glomerular filtration rate (GFR) remains fairly constant because of the operation of two negative feedback systems that originate in the juxtaglomerular apparatus. When there is a low rate of glomerular filtration, there is an excess reabsorption of chloride (Cl^-) ions from the filtrate. In response to diminished levels of Cl^- ions, cells of the macula densa cause vasodilation of afferent arterioles. This increases the rate of blood flow into the glomerulus, thus raising glomerular filtration rate back to normal. Diminished levels of Cl^- ions also cause the juxtaglomerular cells to release renin. This initiates a response (renin–angiotensin pathway) in which vasoconstriction of the efferent arterioles causes glomerular pressure to rise, thus also raising glomerular filtration rate back to normal.

Tubular Secretion

The third process involved in urine formation is ***tubular secretion.*** Whereas tubular reabsorption removes substances from the filtrate into the blood, tubular secretion adds materials to the filtrate from the blood. These secreted substances include potassium and hydrogen ions, ammonia, creatinine, and the drugs penicillin and para-aminohippuric acid. Tubular secretion has two principal effects. It rids the body of certain materials, and it helps control blood pH.

Normally, most potassium ions (K^+) are actively secreted into the distal convoluted tubules; a small amount is secreted into collecting tubules. As Na^+ is reabsorbed, K^+ is secreted, although this is not a one-to-one exchange. The exchange probably occurs because of a negative potential that is created when Na^+ is reabsorbed, and this attracts K^+ ions. Secretion of K^+ is controlled by (1) aldosterone (in the presence of more aldosterone, more K^+ is secreted); (2) K^+ concentration in plasma (when plasma K^+ concentration is high, K^+ secretion increases); and (3) Na^+ concentration in distal convoluted tubules (high levels of Na^+ increase the rate of Na^+ absorption and K^+ secretion). Secretion of K^+ is very important because if the K^+ concentration in plasma nearly doubles, cardiac arrhythmias may develop. At even higher concentrations, cardiac arrest may occur.

The body has to maintain normal blood pH (7.35 to 7.45) despite the fact that a normal diet provides more acid-producing foods than alkali-producing foods. To raise blood pH, the renal tubules secrete hydrogen ions (H^+) into the filtrate, which makes the urine acidic.

The kidneys also participate in the regulation of blood pH—that is, the regulation of H^+ concentration—by increasing or decreasing the base bicarbonate (HCO_3^-) concentration in blood. In order to accomplish this, the kidneys regulate the amount of HCO_3^- reabsorbed from the glomerular filtrate and also generate HCO_3^- to replace that lost in buffering strong acids formed in the body. Both reabsorption of filtered HCO_3^- and generation of new HCO_3^- are accomplished by one process—the secretion of H^+ by tubular epithelial cells. Let us see how this occurs.

Secretion of H^+ takes place in epithelial cells of proximal convoluted tubules and collecting tubules and begins when carbon dioxide (CO_2) diffuses from peritubular blood or tubular fluid into epithelial cells or is produced from metabolic reactions within epithelial cells (Figure 26-11a). Here, in the presence of the enzyme carbonic anhydrase, the CO_2 combines with water (H_2O) to form carbonic acid (H_2CO_3). The H_2CO_3 then dissociates into H^+ and HCO_3^- ions, and the H^+ is secreted into the tubular fluid (glomerular filtrate). In the proximal tubule, the secretion of H^+ is primarily coupled to the passive movement of sodium (Na^+) ions from tubular fluid into epithelial cells, a mechanism known as ***Na^+–H^+ countertransport.*** In the collecting tubules, H^+ secretion is primarily an active process. An important feature of H^+ secretion is that HCO_3^-, formed during the dissociation of H_2CO_3, can diffuse from the epithelial cells into peritubular blood. For every H^+ secreted, one HCO_3^- is returned to blood.

In order to understand the relationship between H^+ secretion and HCO_3^- reabsorption, we will have to follow the fate of the H^+ ions. Most of the H^+ ions secreted into tubular fluid react with HCO_3^- to form H_2CO_3, which, in turn, dissociates into CO_2 and H_2O (Figure 26-11b). The CO_2 diffuses into the epithelial cells to generate additional H^+ for secretion, or it diffuses through the epithelial cells into peritubular blood for eventual elimination by the lungs. In the proximal tubular fluid, the dissociation of H_2CO_3 into CO_2 and H_2O is catalyzed by carbonic anhydrase. Each time a secreted H^+ ion reacts with HCO_3^- in proximal tubular fluid, an HCO_3^- is lost from the fluid. However, the secretion of H^+ also results in the addition of an HCO_3^- into peritubular blood. Thus, when H^+ reacts with HCO_3^- in tubular fluid, the *net* result of H^+ secretion is reabsorption of HCO_3^-. The basic mechanism whereby the kidneys reabsorb filtered HCO_3^- is based on the reaction of secreted H^+ with HCO_3^- in tubular fluid.

Like the reabsorption of filterd HCO_3^-, the generation of new HCO_3^- is accomplished by H^+ secretion in which an HCO_3^- is returned to peritubular blood for each H^+ secreted (Figure 26-11c). If the secreted H^+ reacts with HCO_3^- in tubular fluid, the overall result is HCO_3^- reabsorption, since the HCO_3^- returned to peritubular blood simply replaces a HCO_3^- lost from tubular fluid in the reaction with H^+. If instead, however, *excess* H^+ is secreted, the HCO_3^- returned to peritubular blood represents new HCO_3^-. Much of the excess H^+ thus generated cannot be excreted from the body as free H^+ ions; it must be excreted in combination with buffers, the most important of which are the ammonia-ammonium ion buffer system and the phosphate buffer system.

Ammonia (NH_3), in certain concentrations, is a poisonous waste product derived from the deamination of amino acids. The liver converts much of the ammonia to a less toxic compound called urea. Urea and ammonia both become part of the tubular fluid and are subsequently expelled from the body. Any ammonia produced by the deamination of amino acids in the tubule is secreted into the tubular fluid. When ammonia (NH_3) forms in the distal

FIGURE 26-11 Role of the kidneys in maintaining blood pH. (a) Mechanism for H^+ secretion in a proximal (or collecting) tubule cell. The green circle represents the $Na^+ - H^+$ countertransport mechanism. (b) Reabsorption of HCO_3^- by H^+ secretion. (c) Generation of new HCO_3^- by H^+ secretion. The upper part of the figure also illustrates excretion of NH_4^+ as NH_4Cl; the lower part of the figure also illustrates excretion of $H_2PO_4^-$ as NaH_2PO_4.

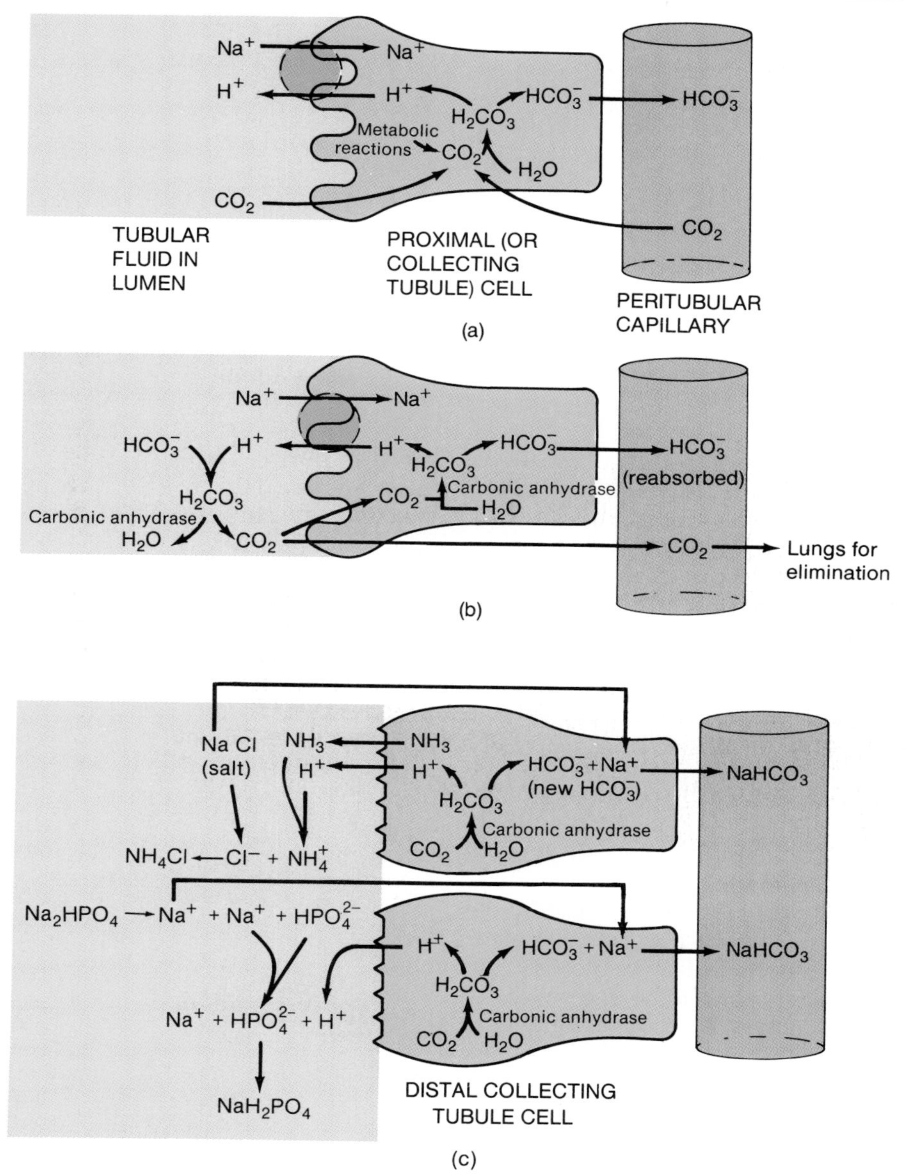

and collecting tubule epithelial cells, it diffuses out of the cells into tubular fluid and combines with H^+ to form the ammonium ion (NH_4^+). NH_4^+ is then excreted from the body (Figure 26-11c), in combination with chloride (Cl^-) ions as ammonium chloride (NH_4Cl). The formation of NH_3 in epithelial cells is dependent on tubular fluid pH; formation increases in acidosis and decreases in alkalosis. The HCO_3^- that is produced during the dissociation of H_2CO_3 combines with Na^+ to form sodium bicarbonate ($NaHCO_3$).

The phosphate buffer system is the second important buffer for secreted H^+ ions. The H^+ ions can react with the sodium salt of monohydrogen phosphate (HPO_4^{2-}) ions in tubular fluid (Na_2HPO_4) to form the sodium salt of dihydrogen phosphate ($H_2PO_4^-$) ions, which are then excreted from the body (NaH_2PO_4) (Figure 26-11c). The HCO_3^- produced during the dissociation of H_2CO_3 combines with Na^+ to form sodium bicarbonate ($NaHCO_3$). Phosphate buffers are more efficient in collecting tubules rather than proximal tubules, but are less effective than ammonia buffers

As a result of H^+ ion secretion, urine normally has an acidic pH of 6. The relationship of renal tubule ion excretion to blood pH level is summarized in Figure 26-12.

Exhibit 26-2 and Figure 26-13 summarize filtration, reabsorption, and secretion in the nephrons.

FIGURE 26-12 **Summary of kidney mechanisms that maintain the homeostasis of blood pH.**

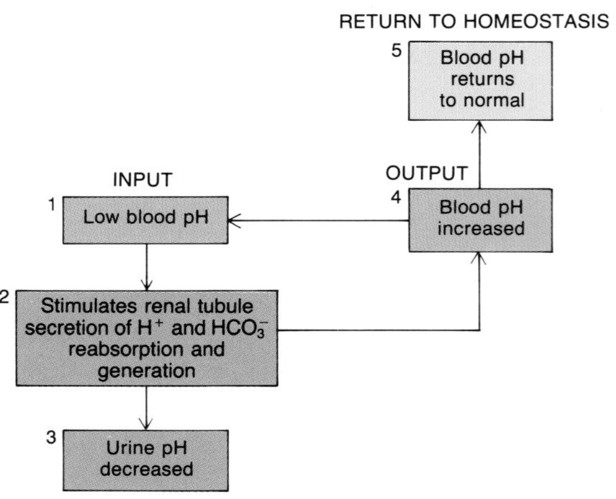

EXHIBIT 26-2 SUMMARY OF FILTRATION, REABSORPTION, AND SECRETION

Region of Nephron	Activity
Renal corpuscle (endothelial-capsular membrane)	Filtration of glomerular blood under hydrostatic pressure results in the formation of filtrate that contains water, glucose, some amino acids, Na^+, Cl^-, HCO_3^-, K^+, urea, uric acid, and creatinine. Plasma proteins and cellular elements of blood normally do not pass through the endothelial-capsular membrane and are not found in filtrate.
Proximal convoluted tubule	Reabsorption of physiologically important solutes such as glucose, amino acids, Na^+, Cl^-, HCO_3^-, K^+. Reabsorption of urea. Obligatory water reabsorption by osmosis, secretion of H^+.
Descending limb of the loop of the nephron	Passive reabsorption of water.
Ascending limb of the loop of the nephron	Reabsorption of Na^+, Cl^-, and urea.
Distal convoluted tubule	Reabsorption of Na^+ under influence of aldosterone, Cl^-, HCO_3^-, and urea. Facultative water reabsorption under influence of ADH. Secretion of H^+ and K^+, NH_3, creatinine, and certain drugs. Generation of new HCO_3^-.
Collecting tubules	Reabsorption of Na^+ under influence of aldosterone, reabsorption of Cl^- and urea. Facultative water reabsorption under influence of ADH. Secretion of H^+ and K^+.

Renal Clearance

Measurement of glomerular filtration rate is accomplished by renal clearance tests. ***Renal clearance*** refers to the ability of the kidneys to clear (remove) a specific substance from the blood in a given period of time. Clearance is usually expressed in milliliters per minute. A renal clearance test is performed as follows. A known amount of a substance is given intravenously in a steady concentration over a period of time. As the substance is filtered by the kidneys, urine is collected and analyzed to determine its concentration. The blood concentration of the substance is also determined. Renal clearance is calculated from the following equation:

$$\text{Renal clearance} = \frac{UV}{P}$$

U represents the concentration of the substance in urine, expressed in mg/ml. P represents the concentration of the same substance in plasma, V represents the volume of urine excreted and is expressed in ml/min.

A substance frequently used as a standard to determine renal clearance is a polysaccharide called ***inulin.*** This substance is especially useful because it is small enough to pass through the endothelial-capsular membrane and is neither reabsorbed (passed from filtrate back into blood) nor secreted (passed from blood into filtrate). The renal clearance of inulin is 125 ml/min. If a substance has

FIGURE 26-13 Summary of functions of a nephron. ① As blood flows from the glomerulus into the glomerular (Bowman's) capsule, it is filtered by the endothelial-capsular membrane. As the filtrate passes through the nephron ②, certain components are selectively reabsorbed into blood and ③ other substances are secreted into the filtrate for elimination in urine.

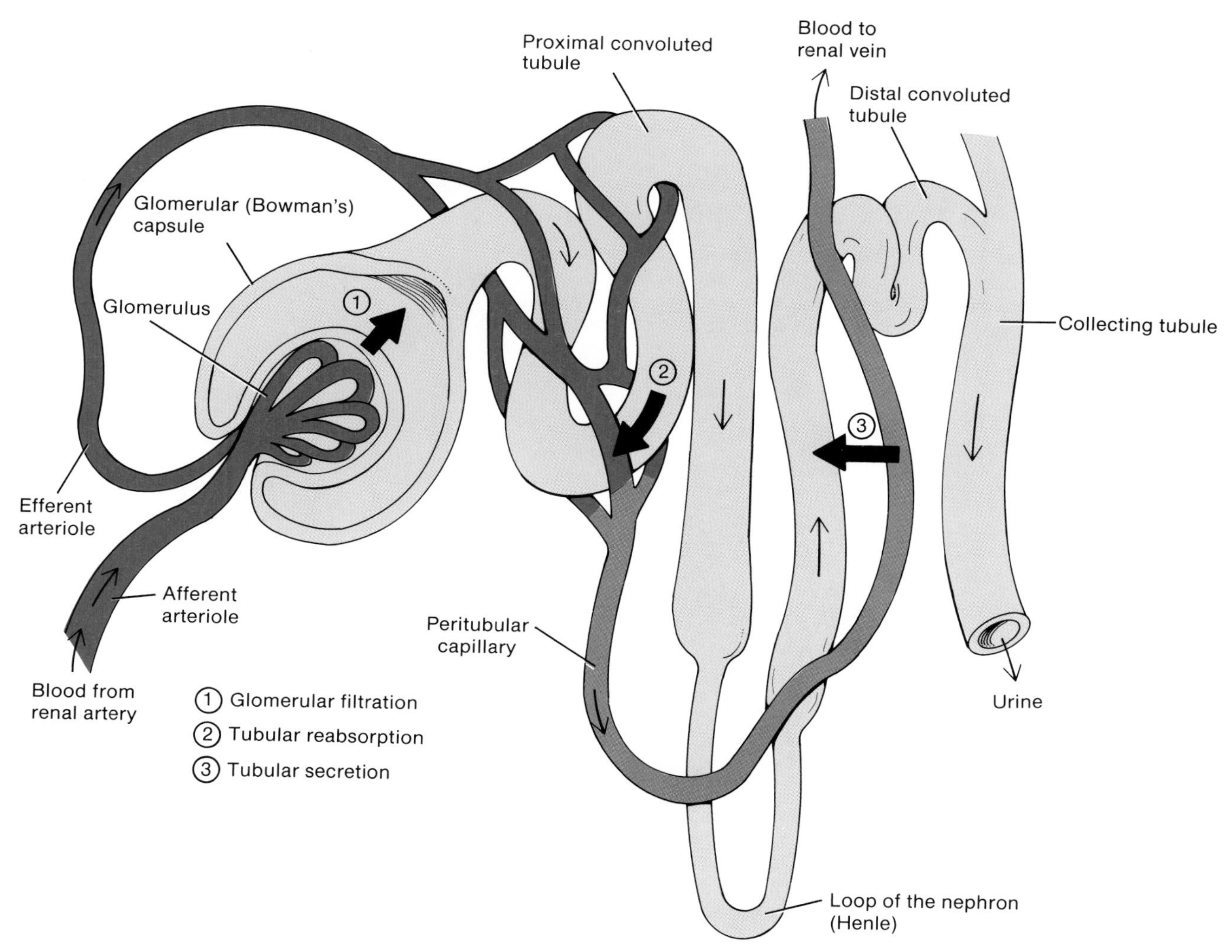

a value less than that of inulin, it is partially reabsorbed. For example, uric acid has a clearance of 14.0 ml/min. A substance with a renal clearance value of 0, such as glucose, is completely reabsorbed. If a substance has a value greater than that of inulin, then the substance is secreted into the filtrate. An example of such a substance is creatinine (140.0 ml/min).

Mechanism of Urine Dilution

The rate at which water is lost from the body depends principally on antidiuretic hormone (ADH). As noted earlier, ADH controls the permeability of distal convoluted tubules and collecting tubules to water. In the absence of ADH, the tubules are virtually impermeable to water, and much more water is expelled into urine. However, in the presence of ADH, the collecting tubules become quite permeable to water, and more water is reabsorbed back into blood. Thus, less water is expelled into urine.

For the kidneys to produce a dilute urine, they must form a urine that contains more water than normal for the amount of solutes. This is accomplished when the renal tubules allow an increased amount of water to be eliminated. The kidneys produce a dilute urine as follows. The normal concentration of glomerular filtrate as it enters the proximal convoluted tubule in the cortex of the kidney is about 300 milliosmols (mOsm) per liter (Figure

26-14a).* A milliosmol represents the concentration of substances in the filtrate, mainly NaCl. It is a measure of the osmotic properties of a solution. The concentration of glomerular filtrate is isotonic to plasma. (See Chapter 3 for a discussion of isotonic solutions.) The ascending limb of the loop of the nephron (loop of Henle) is quite impermeable to water but actively reabsorbs chloride (Cl^-) ions. The ions pass into interstitial fluid (fluid between tubules, ducts, and capillaries) and then into peritubular capillaries. This causes peritubular blood to become more electronegative than the filtrate, passively drawing out positive ions, mostly sodium (Na^+) ions. As a result, the concentration of these ions in the filtrate is reduced from its original concentration of 300 mOsm to about 100 mOsm, but the water remains. The filtrate leaving the ascending limb is thus diluted. As the filtrate continues through the distal convoluted tubule and collecting tubules, some additional ions are reabsorbed, producing an even more dilute urine. In these regions, Na^+ ions are actively reabsorbed and negative ions, mainly Cl^-, are passively reabsorbed. In the absence of ADH, urine may be only one-fourth as concentrated as blood plasma and

* A ***milliosmol*** is a unit of osmotic pressure equal to one-thousandth gram molecular weight of a substance divided by the number of particles (or ions) into which a substance dissociates in one liter of solution.

FIGURE 26-14 **Mechanism of urine dilution and urine concentration: countercurrent multiplier. All concentrations are in milliosmols (mOsm). (a) Urine dilution. The portions of the renal tubules that are indicated by heavy lines represent areas that are virtually impermeable to water in the absence of ADH and where various ions are reabsorbed.**

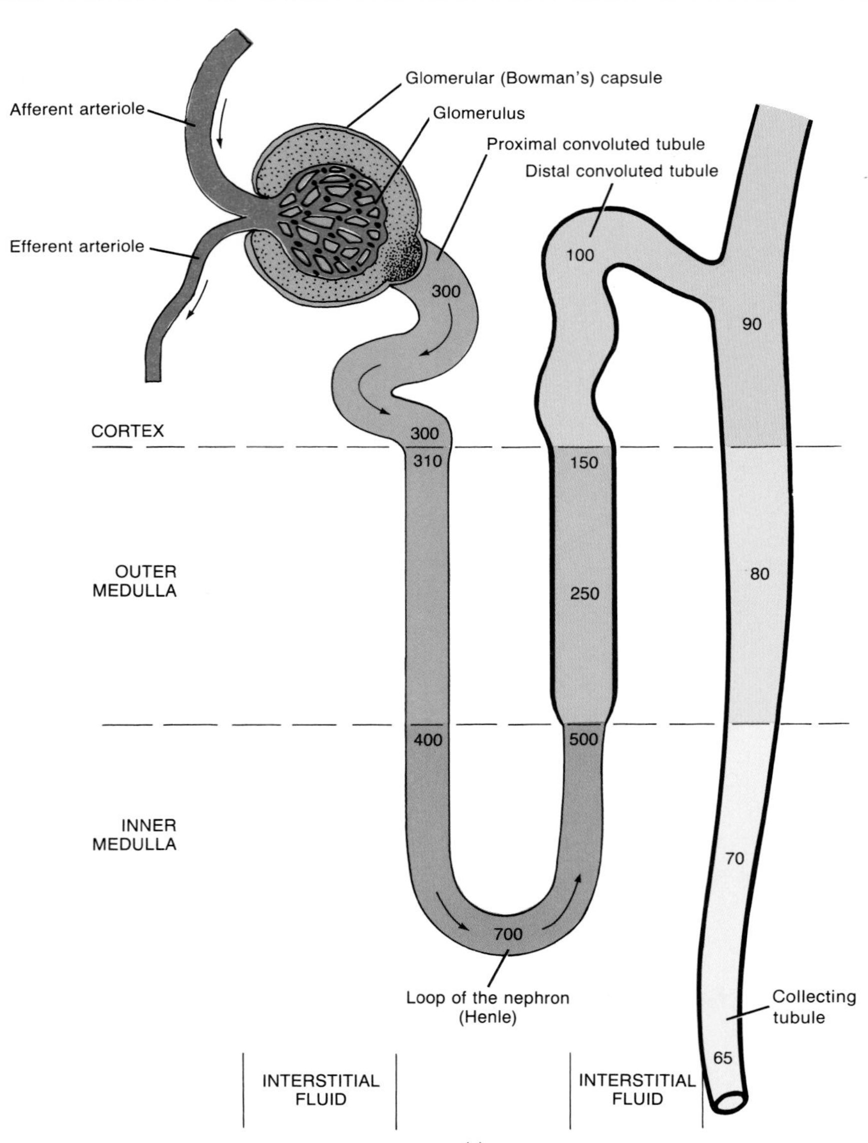

FIGURE 26-14 (*Continued*) (b) Urine concentration. The illustration on the left represents the cardiovascular component (vasa recta), whereas the illustration on the right represents the nephron component.

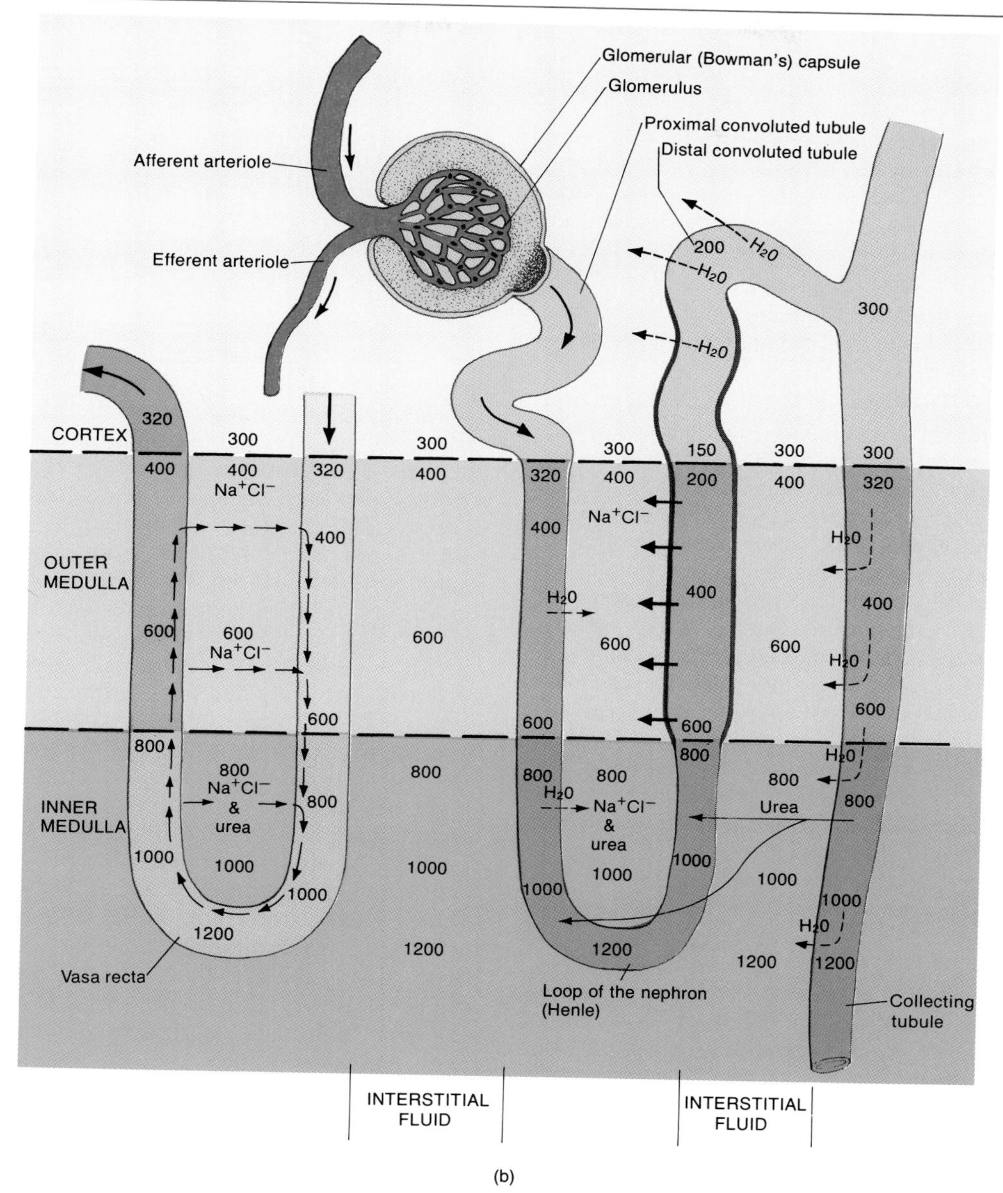

glomerular filtrate. By the time the dilute filtrate passes into the terminal collecting tubules, its concentration can be as low as 65 to 70 milliosmols. Such a dilute urine is said to be ***hyposmotic (hypotonic)*** to blood plasma.

Mechanism of Urine Concentration

During times of low water intake, the kidneys still must eliminate wastes and excess ions while conserving water. Basically, this is accomplished by increasing the volume of water that is reabsorbed into blood and thus producing the excretion of a more concentrated urine. Such urine is said to be ***hyperosmotic (hypertonic)*** to blood plasma. Whereas diluting urine simply involves the reabsorption of more solutes than water, concentrating urine is somewhat more complex. The ability of the kidneys to produce hyperosmotic urine depends in part on the countercurrent mechanism.

The excretion of concentrated urine begins with a high concentration of solutes in the interstitial fluid in the medulla of the kidney. If you examine Figure 26-14b, you will note that the solute concentration of the interstitial fluid in the kidney increases from about 300 mOsm in the cortex to about 1200 mOsm in the inner medulla. The high medullary concentration is maintained by two principal factors. The first is solute reabsorption from various parts of the renal tubules:

1. In the ascending limb of the loop of the nephron, Cl^- ions are actively transported from the filtrate into the interstitial fluid of the outer medulla. Na^+ ions, as well as other cations, follow passively. All the ions become concentrated in the fluid and are carried downward into the inner medulla by the blood flowing in the vasa recta (Figure 26-14b).

2. In the collecting tubule, Na^+ ions are actively transported from the filtrate into the interstitial fluid of the inner medulla. Cl^- ions follow passively.

3. In the presence of high levels of ADH, the collecting tubule becomes very permeable to water. Water quickly moves out of the ducts by osmosis into the interstitial fluid of the inner medulla, greatly increasing the concentration of urea in the collecting tubule. As a result, urea moves passively by diffusion into the interstitial fluid of the medulla. This movement of urea increases the concentration of urea in the interstitial fluid above that in the filtrate in the loop of the nephron. Now, because of this difference in concentration, some urea diffuses into the filtrate in the loop of the nephron, thus raising its concentration of urea (Figure 26-14b). As the filtrate moves into the ascending limb and distal convoluted tubule, both of which are virtually impermeable to urea, and then into the collecting tubule, more water moves out of the collecting tubule by osmosis in the presence of ADH. This *further* increases the concentration of urea in the collecting tubule, and *more* urea diffuses into the interstitial fluid of the inner medulla and the cycle repeats itself. Thus, in the presence of ADH, water movement from the collecting tubule can concentrate urea to very high levels as the cycle repeats itself.

The second factor that maintains the high medullary concentration of solutes (sodium, chloride, and urea) is the ***countercurrent mechanism,*** which is based on the anatomical arrangement of the juxtamedullary nephrons and vasa recta.

If you examine Figure 26-14b, you will note that the descending limb of the loop of the nephron carries filtrate downward from the cortex deep into the medulla. The ascending limb of the loop of the nephron carries filtrate upward from deep in the medulla toward the cortex. Thus, we have a situation in which fluid flowing in one tube runs parallel and counter (opposite) to fluid flowing in another tube. This is a countercurrent flow. The descending limb is relatively permeable to water and relatively impermeable to solutes. Since the interstitial fluid outside the descending limb is more concentrated than the filtrate within it, water moves out of the descending limb by osmosis. This water movement causes the concentration of the filtrate to increase. As the filtrate continues down the descending limb, more water leaves it by osmosis, and the concentration of the filtrate increases even more. In effect, a positive feedback mechanism is established. In fact, as the filtrate passes through the loop of the nephron (loop of Henle), its concentration is 1200 mOsm per liter. As noted earlier, the ascending limb is relatively impermeable to water, but Na^+ and Cl^- ions move from the limb into the interstitial fluid of the medulla. As the filtrate moves upward through the ascending limb and the ions move out, the concentration of filtrate in the ascending limb progressively decreases. Near the cortex, the concentration has decreased to 200 mOsm per liter, a concentration below that of plasma. The overall effect of the countercurrent flow is that filtrate is progressively concentrated as it flows down the descending limb and progressively diluted as it moves up the ascending limb, and NaCl is highly concentrated in the medulla. Remember that urea is also concentrated there as well.

If you again examine Figure 26-14b, you can see that the vasa recta also consists of a descending portion and an ascending portion that are parallel to each other. The various solutes remain concentrated within the medulla because of the arrangement of the ascending and descending portions of the vasa recta. Just as filtrate flows in opposite directions in the limbs of the nephron, blood flows in opposite directions in the ascending and descending portions of the vasa recta. This again is a countercurrent flow. Blood entering the vasa recta has a solute concentration of about 300 mOsm per liter. As it flows down the descending portion into the medulla, where the interstitial fluid becomes more and more concentrated, NaCl and urea diffuse into blood. As the blood becomes more concentrated, it flows into the ascending portion of the vasa recta, where the interstitial fluid becomes less and less concentrated. As a result, NaCl and urea diffuse from the blood into the interstitial fluid so that blood leaving the vasa recta has a concentration only slightly higher than blood entering. The essential feature of the countercurrent flow in the vasa recta is that high medullary concentrations of solutes are maintained, since blood passing through the vasa recta removes very few solutes. Moreover, the flow of blood in the vasa recta is very slow, and this sluggish flow also helps prevent the rapid removal of solutes from the medulla.

The actual excretion of concentrated urine depends on the presence of ADH. When ADH is present, there is a rapid movement of water from the collecting tubules into interstitial fluid. This causes a progressive increase in the solute concentration in the collecting tubules. Thus, a concentrated urine is excreted, that is, one with a high solute concentration (up to 1200 mOsm) and little water. Urine can be four times more concentrated than blood plasma and glomerular filtrate.

HEMODIALYSIS THERAPY

If the kidneys are so impaired by disease or injury that they are unable to excrete nitrogenous wastes and regulate pH and electrolyte and water concentration of the plasma, the blood must be filtered by an artificial device. Such filtering of the blood is called ***hemodialysis.*** *Dialysis* means the separation of large nondiffusible particles from smaller diffusible ones through a selectively permeable membrane. One of the best-known devices for accomplishing dialysis is the artificial kidney machine (Figure 26-15). A tube connects it with the patient's radial artery. The blood is pumped from the artery through the tubes to one side of a selectively permeable dialyzing membrane made of cellulose acetate. The other side of the membrane is continually washed with an artificial solution called the dialysate. The blood that passes through the artificial kidney is treated with an anticoagulant (heparin). Only about 500 ml of the patient's blood is in the machine at a time. This volume is easily compensated for by vasoconstriction and increased cardiac output.

All substances (including wastes) in the blood except protein molecules and blood cells can diffuse back and forth across the selectively permeable membrane. The electrolyte level of the plasma is controlled by keeping the dialysate electrolytes at the same concentration found in normal plasma. Any excess plasma electrolytes move down the concentration gradient and into the dialysate. If the plasma electrolyte level is normal, it is in equilibrium with the dialysate, and there is no net gain or loss of electrolytes. Since the dialysate contains no wastes, substances such as urea move down the concentration gradient and into the dialysate. Thus, wastes are removed and normal electrolyte balance is maintained. Hemodialysis typically is performed three times a week, each session lasting for 4 to 6 hours.

A great advantage of the kidney machine is that nutrition can be bolstered by placing large quantities of glucose in the dialysate. While the blood gives up its wastes, the glucose diffuses into the blood. Thus, the kidney machine beautifully accomplishes the principal function of the fundamental unit of the kidney—the nephron.

FIGURE 26-15 **Operation of an artificial kidney. The blood route is indicated in red and blue. The route of the dialysate is indicated in gold.**

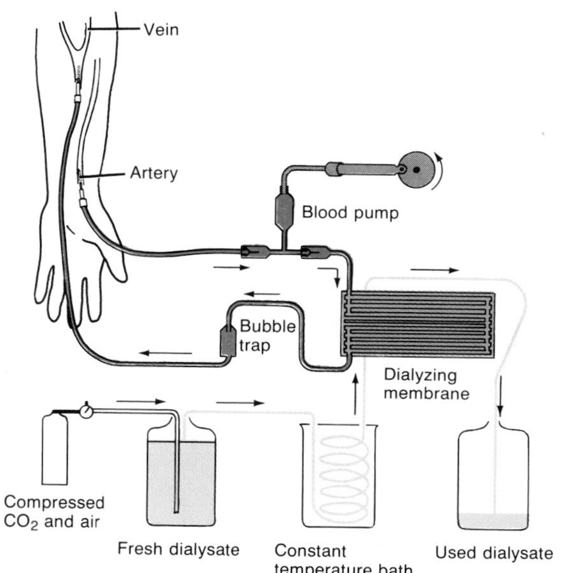

There are obvious drawbacks to the artificial kidney, however. Anticoagulants must be added to the blood during dialysis. A large amount of the patient's blood must flow through this apparatus to make the treatment effective, and so the slow rate at which the blood can be processed makes the treatment time-consuming and blood cells can be damaged in the process. To date, no artificial kidney has been implanted permanently.

Continuous ambulatory peritoneal dialysis (CAPD) is more convenient and less time-consuming for many patients. CAPD uses the peritoneum instead of cellulose acetate as the dialyzing membrane. Since the peritoneum is a selectively permeable membrane, it permits rapid bidirectional transfer of substances. A catheter is placed in the patient's peritoneal cavity and connected to a supply of dialysate. Gravity feeds the solution into the abdominal cavity from its plastic container. When the process is complete, the dialysate is returned from the abdominal cavity to the plastic container and then discarded. Danger of infection is always present when this process is used.

HOMEOSTASIS

Excretion is one of the primary ways in which the volume, pH, and chemistry of the body fluids are kept in homeostasis. The kidneys assume a good deal of the burden for excretion, but they share this responsibility with several other organ systems.

The lungs, integument, and gastrointestinal tract all perform special excretory functions (Exhibit 26-3). A major responsibility for regulating body temperature through the excretion of water is assumed by sudoriferous (sweat) glands of the skin. The lungs maintain blood–gas homeostasis through the elimination of carbon dioxide. One way in which the kidneys maintain homeostasis is by coordinating their activities with other excretory organs. When the integument increases its excretion of water, the renal tubules increase their reabsorption of water, and blood volume is maintained. When the lungs fail to eliminate enough carbon dioxide, the kidneys attempt to compensate. They change some of the carbon dioxide into sodium bicarbonate, which becomes part of the blood buffer systems, and they secrete more H^+.

EXHIBIT 26-3 EXCRETORY ORGANS AND PRODUCTS ELIMINATED

Excretory Organs	Products Eliminated	
	Primary	**Secondary**
Kidneys	Water, nitrogenous wastes from protein catabolism, and inorganic salts.	Heat and carbon dioxide.

EXHIBIT 26-3 EXCRETORY ORGANS AND PRODUCTS ELIMINATED (*Continued*)

Excretory Organs	Products Eliminated: Primary	Products Eliminated: Secondary
Lungs	Carbon dioxide.	Heat and water.
Skin (sudoriferous glands)	Heat.	Carbon dioxide, water, salts, and urea.
Gastrointestinal (GI) tract	Solid wastes and secretions.	Carbon dioxide, water, salts, and heat.

URETERS

Once urine is formed by the nephrons and passed into collecting tubules, it drains through papillary ducts into the calyces surrounding the renal papillae. The minor calyces join to become the major calyces that unite to become the renal pelvis. From the pelvis, the urine drains into the ureters and is carried by peristalsis to the urinary bladder. From the urinary bladder, the urine is discharged from the body through the single urethra. From the minor calyces on, the urine is in no way modified in either volume or composition.

Structure

The body has two ***ureters*** (YOO-re-ters)—one for each kidney. Each ureter is an extension of the pelvis of the kidney and extends 25 to 30 cm (10 to 12 inches) to the urinary bladder (see Figure 26-1). As the ureters descend, their thick walls increase in diameter, but at their widest point they measure less than 1.7 cm (0.7 inch) in diameter. Like the kidneys, the ureters are retroperitoneal in placement. The ureters enter the urinary bladder at the superior lateral angle of its base.

Although there are no anatomical valves at the openings of the ureters into the urinary bladder, there is a functional one that is quite effective. Since the ureters pass obliquely through the wall of the urinary bladder, pressure in the urinary bladder compresses the ureters and prevents backflow of urine when pressure builds up in the urinary bladder as it fills during urination. When this physiological valve is not operating, it is possible for cystitis (urinary bladder inflammation) to develop into kidney infection.

Histology

Three coats of tissue form the wall of the ureters. The inner coat, or mucosa, is a mucous membrane with transitional epithelium (see Exhibit 4-1, transitional epithelium). The solute concentration and pH of urine differs drastically from the internal environment of cells that form the wall of the ureters. Mucus secreted by the mucosa prevents the cells from coming in contact with urine. Throughout most of the length of the ureters, the second or middle coat, the muscularis, is composed of inner longitudinal and outer circular layers of smooth muscle. The muscularis of the proximal third of the ureters also contains a layer of outer longitudinal muscle. Peristalsis is the major function of the muscularis. The third, or external, coat of the ureters is a fibrous coat. Extensions of the fibrous coat anchor the ureters in place.

Physiology

The principal function of the ureters is to transport urine from the renal pelvis into the urinary bladder. Urine is carried through the ureters primarily by peristaltic contractions of the muscular walls of the ureters, but hydrostatic pressure and gravity also contribute. Peristalic waves pass from the kidney to the urinary bladder, varying in rate from one to five per minute, depending on the amount of urine formation.

URINARY BLADDER

The ***urinary bladder*** is a hollow muscular organ situated in the pelvic cavity posterior to the symphysis pubis. In the male, it is directly anterior to the rectum. In the female, it is anterior to the vagina and inferior to the uterus. It is a freely movable organ held in position by folds of the peritoneum. The shape of the urinary bladder depends on how much urine it contains. When empty, the size of the lumen is decreased, and the wall appears thicker. It becomes spherical when slightly distended. As urine volume increases, it becomes pear-shaped and rises into the abdominal cavity.

Structure

At the base of the urinary bladder is a small triangular area, the ***trigone*** (TRĪ-gōn), that points anteriorly (Figure 26-16). The opening to the urethra is found in the apex of this triangle. At the two points of the base, the ureters drain into the urinary bladder. It is easily identified because the mucosa is firmly bound to the muscularis so that the trigone is typically smooth.

Histology

Four coats make up the wall of the urinary bladder (Figure 26-17). The mucosa, the innermost coat, is a mucous membrane containing transitional epithelium. Transitional epithelium is able to stretch—a marked advantage for an organ that must continually inflate and deflate. Rugae

FIGURE 26-16 Urinary bladder and female urethra. See also Figure 28-12c.

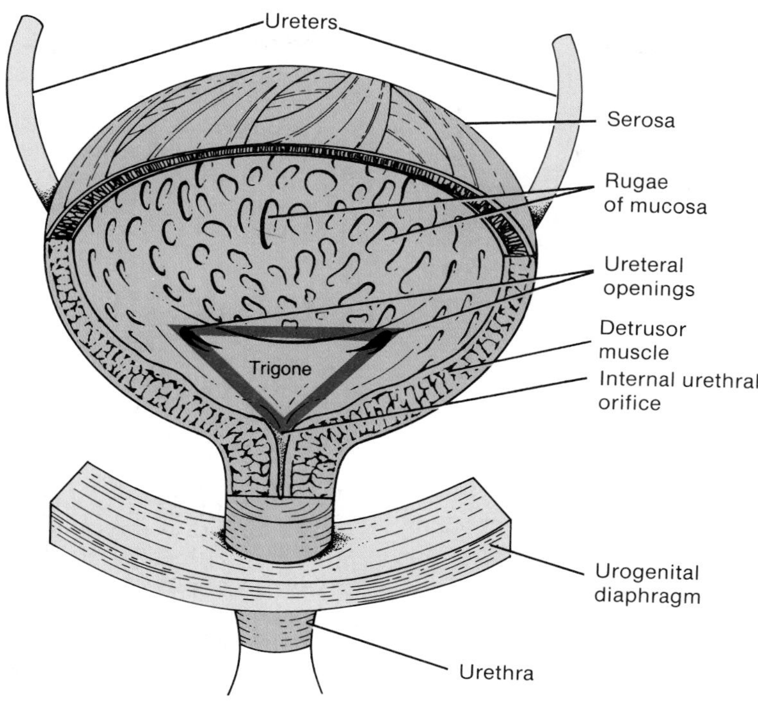

FIGURE 26-17 Histology of the urinary bladder. Photomicrograph of a portion of the wall of the urinary bladder at a magnification of 50×. (Copyright © 1983 by Michael H. Ross. Used by permission.) The details of the mucosa of the urinary bladder are shown in Exhibit 4-1, stratified transitional.

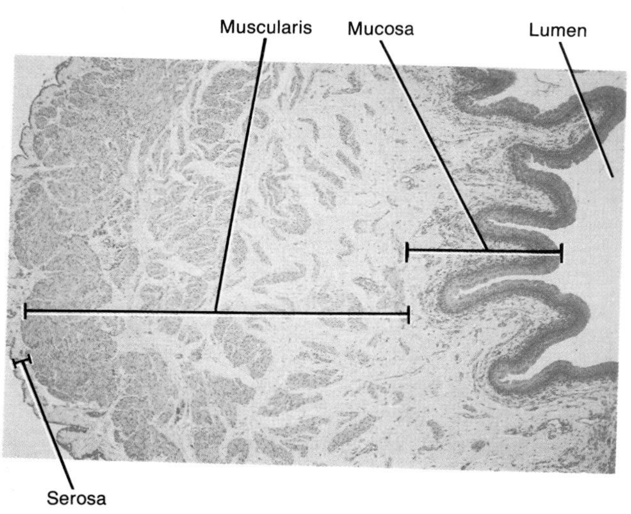

(folds in the mucosa) are also present. The second coat, the submucosa, is a layer of connective tissue that connects the mucosa and muscular coats. The third coat—a muscular one called the ***detrusor*** (de-TROO-ser) ***muscle***—consists of three layers of smooth muscle: inner longitudinal, middle circular, and outer longitudinal muscles. In the area around the opening to the urethra, the circular fibers form an ***internal sphincter*** muscle. Below the internal sphincter is the ***external sphincter,*** which is composed of skeletal muscle and is a modification of the urogenital diaphragm muscle. The outermost coat, the serous coat, is formed by the peritoneum and covers only the superior surface of the organ.

MEDICAL TEST

Cystoscopy (sis-TOS-kō-pē; *cysto* = urinary bladder; *skopein* = to examine)

Diagnostic Value: This procedure permits direct visual examination of the lower urinary tract and, in males, access to the prostate gland as well. It is performed with a cystoscope, a metallic tube with a telescopic lens and fiberoptic lighting. The procedure is valuable in evaluating strictures (narrowed areas) in the urethra, problems associated with urination, the presence of blood in the urine, and the presence in the bladder of stones, a tumor, or other filling defects as noted on an x-ray examination of the bladder. In some cases, cystoscopy can be used to remove tissue for biopsy, kidney stones, urinary bladder tumors, and urine samples. In other instances, it is used for evaluation prior to open surgery on the urinary bladder.

Procedure: A sedative is given an hour before the procedure (some patients require general anesthesia). After the pubic area is cleansed with an antiseptic, a local anesthetic is instilled in the urethra. Then the cystoscope is gently introduced into the urethral orifice and slowly guided into the urinary bladder under endoscopic visualization.

Physiology

Urine is expelled from the urinary bladder by an act called ***micturition*** (mik'-too-RISH-un), commonly known as urination or voiding. This response is brought about by a combination of involuntary and voluntary nerve impulses. The average capacity of the urinary bladder is 700 to 800 ml. When the amount of urine in the urinary bladder exceeds 200 to 400 ml, stretch receptors in the urinary bladder wall transmit impulses to the lower portion of the spinal cord. These impulses, by way of sensory tracts to the cortex, initiate a conscious desire to expel urine and, by way of a center in the sacral cord, a subconscious reflex referred to as the ***micturition reflex.*** Parasympathetic impulses transmitted from the micturition reflex center of the sacral area of the spinal cord reach the urinary bladder wall and internal urethral sphincter, bringing about contraction of the detrusor muscle of the urinary bladder and relaxation of the internal sphincter. Then the conscious portion of the brain sends impulses to the external sphincter, the sphincter relaxes, and urination takes place. Although emptying the urinary bladder is controlled by reflex, it may be initiated voluntarily and stopped at will because of cerebral control of the external sphincter and certain muscles of the urogenital (pelvic) diaphragm.

CLINICAL APPLICATION: INCONTINENCE AND RETENTION

A lack of voluntary control over micturition is referred to as ***incontinence.*** In infants about 2 years old and under, incontinence is normal because neurons to the external sphincter muscle are not completely developed. Infants void whenever the urinary bladder is sufficiently distended to arouse a reflex stimulus. Proper training overcomes incontinence if the latter is not caused by emotional stress or irritation of the urinary bladder.

Involuntary micturition in the adult may occur as a result of unconsciousness, injury to the spinal nerves controlling the urinary bladder, irritation due to abnormal constituents in urine, disease of the urinary bladder, damage to the external sphincter, and inability of the detrusor muscle to relax due to emotional stress.

Retention, a failure to completely or normally void urine, may be due to an obstruction in the urethra or neck of the urinary bladder, nervous contraction of the urethra, or lack of sensation to urinate.

URETHRA

The ***urethra*** is a small tube leading from the floor of the urinary bladder to the exterior of the body (see Figure 26-16). In females, it lies directly posterior to the symphysis pubis and is in front of the anterior wall of the vagina. Its undilated diameter is about 6 mm (0.25 inch), and its length is approximately 3.8 cm (1.5 inches). The female urethra is directed obliquely, inferiorly, and anteriorly. The opening of the urethra to the exterior, the ***urethral orifice,*** is located between the clitoris and vaginal opening.

In males, the urethra is about 20 cm (8 inches) long. Immediately below the urinary bladder it passes vertically through the prostate gland (prostatic urethra), then pierces the urogenital diaphragm (membranous urethra), and finally pierces the penis (spongy urethra) and takes a curved course through its body (see Figures 28-1 and 28-10).

Histology

The wall of the female urethra consists of three coats: an inner mucous coat, an intermediate thin layer of spongy tissue containing a plexus of veins, and an outer muscular coat that is continuous with that of the urinary bladder and consists of circularly arranged fibers (cells) of smooth muscle. The mucosa is usually lined with transitional epithelium near the urinary bladder. The remainder consists of stratified squamous epithelium with areas of stratified columnar or pseudostratified epithelium.

The male urethra is composed of two coats: an inner mucous membrane and an outer submucous tissue that connects the urethra with the structures through which it passes. The mucosa varies in different regions. The mucosa of the ***prostatic urethra*** is continuous with that of the urinary bladder and is lined by transitional epithelium. The mucosa of the ***membranous urethra*** is lined by pseudostratified epithelium. The spongy urethra is lined mostly by pseudostratified epithelium. Near its opening to the exterior it is lined by stratified squamous epithelium. In the spongy urethra, especially, there are glands, called ***urethral (Littré) glands,*** that produce mucus for lubrication during sexual intercouse.

Physiology

The urethra is the terminal portion of the urinary system. It serves as the passageway for discharging urine from the body. The male urethra also serves as the duct through which reproductive fluid (semen) is discharged from the body.

URINE

The by-product of the kidneys' activities is ***urine.*** In a healthy person its volume, pH, and solute concentration vary with the needs of the internal environment. During certain pathological conditions, the characterististics of urine may change drastically. An analysis of the volume and physical, chemical, and microscopic properties of urine tells us much about the state of the body. Such an analysis is called a ***urinalysis (UA).***

MEDICAL TEST

Urinalysis (UA)

Diagnostic Value: Performed as part of a medical examination and to evaluate urinary symptoms such as painful or frequent urination or other symptoms such as abdominal pain and fever. Urine testing can also monitor disease such as diabetes mellitus, urinary tract infections (UTIs), kidney stones, and chronic renal disease. Presence in the urine of bile pigments causes a dark discoloration that is useful in the evaluation of jaundice.

Procedure: Once the urethral meatus has been cleansed, the stream is started and, after a few seconds, collected in a sterile container ("clean-catch urine"). If the specimen is not analyzed within an hour, it should be refrigerated. A complete urinalysis includes visual examination of the urine for color and clarity; measurement of concentration (specific gravity) and pH; various chemical tests; and microscopic examination for the presence of blood cells, epithelial cells, casts, crystals, and bacteria. The chemical tests are performed with a dipstick, a plastic strip containing a series of chemically treated pads. Upon exposure to urine, the pads change color if glucose, protein, or certain chemical substances are present in the urine. The dipstick is compared with standard charts to determine if the urine is abnormal and, if so, to what degree.

Normal Values: Refer to Exhibit B-2, Urine Tests, in Appendix B, for a listing of components of urine that may be tested as part of a urinalysis, their normal values, and clinical implications when values increase or decrease.

Volume

The volume of urine eliminated per day in the normal adult varies between 1000 and 2000 ml (1 to 2 qt). Urine volume is influenced by a number of factors: blood pressure, blood concentration (blood osmotic pressure), diet, temperature, diuretics, mental state, and general health.

Blood Pressure

The cells of the juxtaglomerular apparatus are particularly sensitive to changes in blood pressure. As noted earlier, when renal blood pressure falls below normal, the juxtaglomerular apparatus secretes renin, and the renin-angiotensin pathway is activated (see Figure 26-9). The two principal effects of the pathway are: (1) constriction of efferent arterioles and (2) water reabsorption, both of which raise blood pressure. By raising the blood pressure, the juxtaglomerular apparatus ensures that the kidney cells receive enough oxygen and that the glomerular hydrostatic pressure is high enough to maintain a normal *P*eff. The juxtaglomerular apparatus also regulates blood pressure throughout the body.

Blood Concentration

The concentrations of water and solutes in the blood also affect urine volume. If you have gone without water all day and water concentration of your blood becomes low, that is, your plasma is hyperosmotic, osmotic receptors in the hypothalamus stimulate the posterior pituitary to increase release of antidiuretic hormone (ADH). The hormone stimulates the cells of the distal convoluted tubule and collecting tubule to allow water out of the filtrate and into the blood by facultative water reabsorption. Thus, water is conserved and urine volume decreases.

If you have just drunk an excessive amount of liquid, urine volume may be increased through two mechanisms. First, the blood–water concentration increases above normal, that is, your plasma is hypo-osmotic. This means that the osmoreceptors in the hypothalamus are no longer stimulated to secrete as much ADH, and facultative water reabsorption slows or stops. Second, the excess water causes the blood pressure to rise. In response, more blood is brought to the glomeruli and the filtration rate increases.

The concentration of sodium ions in the blood also influences urine volume. Sodium concentration affects aldosterone secretion, which in turn affects both sodium reabsorption and the obligatory reabsorption of water.

Temperature

When the internal or external temperature rises above normal, the rate of perspiration increases and cutaneous vessels dilate and fluid diffuses from the capillaries to the surface of the skin. As water volume decreases, ADH is secreted and facultative water reabsorption increases. In addition, the increase in temperature stimulates the abdominal vessels to constrict, so the blood flow in the glomeruli and filtration decrease. Both mechanisms reduce the volume of urine.

If the body is exposed to low temperatures, the cutaneous vessels constrict and the abdominal vessels dilate. More blood is shunted to the glomeruli, glomerular blood hydrostatic pressure increases, and urine volume increases.

Diuretics

Certain chemicals can inhibit sodium ion reabsorption. The resulting reduction in plasma sodium ion concentration reduces ADH concentration and thus inhibits facultative water reabsorption. Such chemicals are called ***diuretics,*** and the abnormal increase in urine flow is called *diuresis.* Coffee, tea, and alcoholic beverages are diuretics.

Emotions

Some emotional states can affect urine volume. Nervousness, for example, can cause an enormous discharge of urine because impulses from the brain cause an increase in blood pressure, resulting in an increased glomerular filtration rate.

Physical Characteristics

Color

Normal urine is usually a yellow or amber-colored, transparent liquid with a characteristic odor. The color is caused by urochrome, a pigment derived from the metabolism of bile. The color of urine varies considerably with the ratio of solutes to water in the urine. The less water there is, the darker the urine. Fever decreases urine volume in the same way that high environmental temperatures do, sometimes making the urine quite concentrated. It is not uncommon for a feverish person to have dark yellow or brown urine. The color of urine may also be affected by diet, such as a reddish color from beets, and by the presence of abnormal constituents, such as certain drugs. A red or brown to black color may indicate the presence of red blood cells or hemoglobin from bleeding in the urinary system.

Turbidity (Cloudiness)

Fresh urine is usually transparent. Turbid (cloudy) urine does not necessarily indicate a pathological condition, since turbidity may result from mucin secreted by the lining of the urinary tract. The presence of mucin above a critical level usually denotes an abnormality, however.

Odor

The odor of urine may vary. For example, some individuals have inherited the ability to form a substance called methyl mercaptan from the digestion of asparagus. This substance gives urine a characteristic odor. In cases of diabetes, urine has a "sweetish" odor because of the presence of acetone. Stale urine develops the odor of ammonia due to ammonium carbonate formation as a result of urea decomposition.

pH

Normal urine is slightly acid. Its pH ranges between 4.6 and 8.0. Variations in urine pH are closely related to diet. These variations are due to differences in the end products of metabolism. Whereas a high-protein diet increases acidity, a diet composed largely of vegetables increases alkalinity. High altitude, fasting, and exercise also cause variations in urinary pH. Ammonium carbonate forms in standing urine. Since it can dissociate into ammonium ions and form a strong base, the presence of ammonium carbonate tends to make urine more alkaline.

Specific Gravity

Specific gravity is the ratio of the weight of a volume of a substance to the weight of an equal volume of distilled water. Water has a specific gravity of 1.000. The specific gravity of urine depends on the amount of solid materials in solution and ranges from 1.001 to 1.035 in normal urine. The greater the concentration of solutes, the higher the specific gravity. Above-normal specific gravity readings may be due to the presence of blood cells, casts, or bacteria in urine (described later).

The physical characteristics of urine are summarized in Exhibit 26-4.

EXHIBIT 26-4 PHYSICAL CHARACTERISTICS OF NORMAL URINE

Characteristic	Description
Volume	1–2 liters in 24 hours but varies considerably.
Color	Yellow or amber but varies with concentration and diet.
Turbidity	Transparent when freshly voided but becomes turbid upon standing.
Odor	Aromatic but becomes ammonialike upon standing.
pH	4.6–8.0; average 6.0; varies considerably with diet.
Specific gravity	1.001–1.035.

Chemical Composition

Water accounts for about 95 percent of the total volume of urine. The remaining 5 percent consists of solutes derived from cellular metabolism and outside sources such as drugs. The solutes are described in Exhibit 26-5.

Several screening tests for renal functions are available. Among these are the blood urea nitrogen (BUN) test and creatinine test.

MEDICAL TESTS

Blood urea nitrogen (BUN)

Diagnostic Value: This is the most commonly ordered test for renal function. It is used to evaluate for the presence of kidney disease, dehydration, or urinary tract obstruction.

EXHIBIT 26-5 PRINCIPAL SOLUTES IN URINE OF ADULT MALE ON MIXED DIET

Constituent	Amount (g)*	Comments
ORGANIC		
Urea	25.0–35.0	Composes 60 to 90 percent of all nitrogenous material. Derived primarily from metabolism (deamination) of amino acids into ammonia (ammonia combines with CO_2 to form urea).
Creatinine	1.5	Normal alkaline constituent of blood. Derived primarily from creatine (nitrogenous substance in muscle tissue).
Uric acid	0.4–1.0	Product of catabolism of nucleic acids derived from food or cellular destruction. Because of insolubility, tends to crystallize and is common component of kidney stones.
Hippuric acid	0.7	Form in which benzoic acid (toxic substance in fruits and vegetables) is believed to be eliminated from body. High-vegetable diets increase quantity of hippuric acid excreted.
Indican	0.01	Potassium salt of indole. Indole results from putrefaction of protein in large intestine and is carried by blood to liver, where it is probably changed to indican (less poisonous substance).
Ketone bodies	0.04	Also called acetone bodies. Normally found in small amounts. In cases of diabetes mellitus and acute starvation, ketone bodies appear in high concentrations.
Other substances	2.9	May be present in minute quantities, depending on diet and general health. Include carbohydrates, pigments, fatty acids, mucin, enzymes, and hormones.
INORGANIC		
NaCl	15.0	Principal inorganic salt. Amount excreted varies with intake.
K^+	3.3	Occurs as chloride, sulfate, and phosphate salts.
SO_4^{2-}	2.5	Derived from amino acids.
PO_4^{3-}	2.5	Occurs as sodium compounds (monosodium and disodium phosphate) that serve as buffers in blood.
NH_4^+	0.7	Occurs as ammonium salts. Derived from protein catabolism and from glutamine in kidneys. Amount produced by kidney may vary with need of body for conserving Na^+ ions to offset acidity of blood and tissue fluids.
Mg^{2+}	0.1	Occurs as chloride, sulfate, and phosphate salts.
Ca^{2+}	0.3	Occurs as chloride, sulfate, and phosphate salts.

* Values are for a urine sample collected over 24 hours.

Procedure: Test is performed on a venous blood sample.

Normal Values: 8–26 mg/dl
See also Appendix B.

Creatinine

Diagnostic Value: Creatinine is a by-product of the breakdown of phosphocreatine in muscle tissue and is removed by the kidneys. Above-normal levels of creatinine in blood indicate a kidney disorder or urinary tract obstruction.

Procedure: Test is performed on a venous blood sample.

Normal Values: Females: 0.5–1.0 mg/dl
Males: 0.6–1.2 mg/dl
See also Appendix B.

Abnormal Constituents

If the body's chemical processes are not operating efficiently, traces of substances not normally present may appear in the urine, or normal constituents may appear in abnormal amounts. Following are representative abnormal constituents in urine that may be detected as part of a urinalysis.

Albumin

Albumin is a normal constituent of plasma, but it usually appears in only very small amounts in urine because the particles are too large to pass through the pores in the capillary walls and the small quantity that does pass through is largely reabsorbed by pinocytosis. The presence of excessive albumin in the urine—***albuminuria***—indicates an increase in the permeability of the endothelial-capsular membrane. Conditions that lead to albuminuria include injury to the endothelial-capsular membrane as a result of disease, increased blood pressure, and irritation of kidney cells by substances such as bacterial toxins, ether, or heavy metals. Other proteins, such as globulin and fibrinogen, may also appear in the urine under certain conditions.

Glucose

The presence of sugar in the urine is termed ***glycosuria*** or ***glucosuria.*** Normal urine contains such small amounts of glucose that clinically it may be considered absent. The most common cause of glycosuria is a high blood sugar level. Remember that glucose is filtered into the glomerular (Bowman's) capsule. Later, in the proximal convoluted tubules, the tubule cells actively transport the glucose back into the blood. However, the number of glucose carrier molecules is limited. If more carbohydrates are ingested than can be used or stored as glycogen or fat, more sugar is filtered into the glomerular (Bowman's) capsule than can be removed by the carriers. This type of glycosuria is not considered pathological. Another nonpathological cause is emotional stress. Stress can cause excessive amounts of epinephrine to be secreted. Epinephrine stimulates the breakdown of glycogen and the liberation of glucose from the liver. A more serious type of glycosuria results from diabetes mellitus. In this case, there is a frequent or continuous elimination of glucose because the pancreas fails to produce sufficient insulin.

Erythrocytes

The appearance of red blood cells in the urine is called ***hematuria.*** Hematuria generally indicates a pathological condition. One cause is acute inflammation of the urinary organs as a result of disease or irritation from kidney stones. Other causes include tumors, trauma, and kidney disease. Whenever blood is found in the urine, additional tests are performed to ascertain the part of the urinary tract that is bleeding. One should also make sure the sample was not contaminated with menstrual blood from the vagina.

Some long-distance runners may develop ***exercise-induced hematuria*** from jarring of the kidneys or bladder, and the symptom disappears after the athlete rests or reduces strenuous activity for several days. Another cause of urine discoloration following running is ***march hemoglobinuria,*** related to an exercise-induced breakdown of red blood cells and originally noted in military recruits following a long march carrying heavy packs.

Leucocytes

The presence of leucocytes and other components of pus in the urine, referred to as ***pyuria*** (pī-YOU-rē-a), indicates infection in the kidney or other urinary organs. Again, care should be taken that the urine is not contaminated.

Ketone Bodies

Ketone (acetone) bodies appear in normal urine in small amounts. Their appearance in high quantities, a condition called ***ketosis (acetonuria),*** may indicate abnormalities. It may be caused by diabetes mellitus, starvation, or simply too little carbohydrate in the diet. Whatever the cause, excessive quantities of fatty acids are oxidized in the liver, and the ketone bodies are filtered from the plasma into the glomerular capsule.

Bilirubin

As noted in Chapter 24, when red blood cells are destroyed by reticuloendothelial cells, the globin portion of hemoglobin is split off and the heme is converted to biliverdin. In humans, most of the biliverdin is converted to bilirubin which gives bile its major pigmentation. Bilirubin is carried to liver cells by an albumin molecule and is referred to as ***unconjugated (free) bilirubin.*** When it reaches the liver, the albumin is released. The bilirubin is then combined with certain substances in liver cells (glucuronic acid or sulfate) and is referred to as ***conjugated bilirubin.*** In this form, it is secreted by the liver into the biliary system and then into the small intestine. Intestinal bacteria convert conjugated bilirubin into urobilinogen, which gives feces its characteristic color. Above-normal levels of bilirubin in urine are referred to as ***bilirubinuria.***

Urobilinogen

Part of the urobilinogen formed in the intestine is excreted with feces. Another portion is absorbed and returned to the liver where it is metabolized and excreted in bile. A small percentage of the urobilinogen that is absorbed escapes the liver and enters general circulation. The kidneys handle this amount of urobilinogen as though it were a foreign substance; that is, the tubular cells do not actively reabsorb the filtered urobilinogen. The presence of urobilinogen in urine is called ***urobilinogenuria.*** Thus, traces of urobilinogen in urine are normal. When urobilinogenuria is above normal, it indicates an increase in the production of bilirubin and inability of the liver to remove reabsorbed urobilinogen from the blood. Conditions that contribute to increased urobilinogen are hemolytic and pernicious anemia, infectious hepatitis, biliary obstruction, jaundice, cirrhosis, congestive heart failure, and infectious mononucleosis.

Casts

Microscopic examination of urine may reveal ***casts***—tiny masses of material that have hardened and assumed the shape of the lumens of the tubules and were then flushed out of the tubules by a buildup of filtrate behind them. Casts are named after the substances that compose them or for their appearance. There are white blood cell casts, red blood cell casts, epithelial casts that contain cells from the walls of the tubules, granular casts that contain decomposed cells that form granules, and fatty casts from cells that have become fatty.

Renal Calculi

Occasionally, the crystals of salts found in urine may solidify into insoluble stones called ***renal calculi (kidney stones)*** (Figure 26-18). They may be formed in any portion of the urinary tract from the kidney tubules to the external opening. Conditions leading to stone formation include the ingestion of excessive mineral salts, a decrease in the amount of water intake, abnormally alkaline or acidic urine, and overactivity of the parathyroid glands. Common constituents of stones are calcium oxalate, uric acid, and calcium phosphate crystals. Calcium oxalate crystals are the most common type. A protein isolated from

FIGURE 26-18 **Photograph of a renal calculus adjacent to a dime for size comparison. (Courtesy of Matt Iacobino.)**

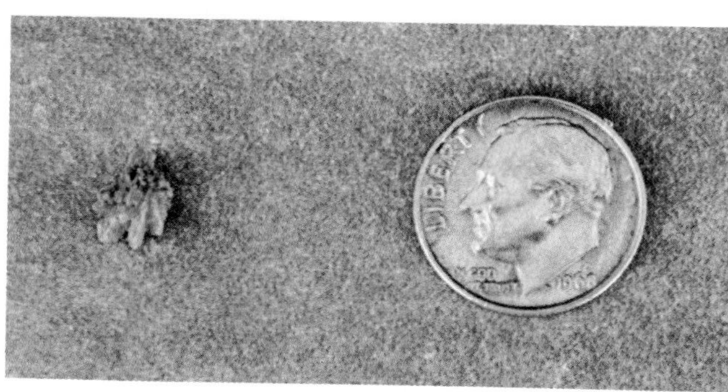

urine called glycoprotein crystal-growth inhibitor (GCI) inhibits formation of calcium oxalate stones. Individuals who do not synthesize GCI may form the calcium oxalate stones. Kidney stones usually form in the pelvis of the kidney, where they cause pain, hematuria, and pyuria. The pain of stone passage is called ***renal colic.***

For kidney stones that become painful or obstructive, there are alternatives to conventional surgery. One of the newer ones is called ***extracorporeal*** (eks'-tra-kor-PŌ-rē-al) ***shock wave lithotripsy*** (LITH-ō-trip'-sē; *litho* = stone; *tribein* = to rub), or ***ESWL.*** This is a noninvasive procedure, in clinical use since 1980, in which shock waves generated outside the body (extracorporeal) by an instrument called a ***lithotriptor*** are used to pulverize kidney stones. The patient is given a general, spinal, or epidural anesthetic and placed in a gantry and then lowered into a tank of warm water. (A recent model uses a fluid bag between the patient and lithotriptor instead of a tank of water and does not require anesthesia.) Once the location of the stones is visualized on x-ray film, the patient is positioned for maximum effect of the shock waves on the stones. Shock waves are given in cycles of 50, with a one-minute interruption between cycles to permit a urologist to monitor the progress of the stone's disintegration. The procedure normally takes 30 to 45 minutes for a stone to crumble to sandlike particles, which are then passed in urine. Recent reports indicate that as many as 25 percent of the patients receiving this treatment have some degree of renal impairment caused by the shock waves.

Microbes

In a properly collected and processed specimen, the finding of bacteria may be of considerable importance. The various bacteria in urine are identified by different microbiological tests. The number and type of bacteria vary with specific infections in the urinary tract. The most common fungus to appear in urine is *Candida albicans,* a common cause of vaginitis. The most frequent protozoan seen in urine is *Trichomonas vaginalis,* a cause of vaginitis in females and urethritis in males.

AGING AND THE URINARY SYSTEM

The effectiveness of kidney function decreases with aging, and by age 70 the filtering mechanism is only about one-half what it was at age 40. Urinary incontinence and urinary tract infections are two common problems associated with aging of the urinary system. Other pathologies include polyuria (excessive urine production), nocturia (excessive urination at night), increased frequency of urination, dysuria (painful urination), retention (failure to release urine from the urinary bladder), and hematuria (blood in the urine). Changes and diseases in the kidney include acute and chronic kidney inflammations and renal calculi (kidney stones). Since water balance and thirst are altered, elderly persons are susceptible to dehydration. The prostate gland is often implicated in various disorders of the urinary tract, and cancer of the prostate is the most frequent malignancy in elderly males.

DEVELOPMENTAL ANATOMY OF THE URINARY SYSTEM

As early as the third week of fetal development, a portion of the mesoderm along the posterior half of the dorsal side of the embryo, the ***intermediate mesoderm,*** differentiates into the kidneys. Three pairs of kidneys form within the intermediate mesoderm in successive time periods: pronephros, mesonephros, and metanephros (Figure 26-19). Only the last one remains as the functional kidneys of the adult.

The first kidney to form, the ***pronephros,*** is the superior of the three. Associated with its formation is a tube, the ***pronephric duct.*** This duct empties into the ***cloaca,*** which is the dilated caudal end of the gut derived from ***endoderm.*** The pronephros begins to degenerate during the fourth week and is completely gone by the sixth week. The pronephric ducts, however, remain.

The pronephros is replaced by the second kidney, the ***mesonephros.*** With its appearance, the retained portion of the pronephric duct, which connects to the mesonephros, becomes known as the ***mesonephric duct.*** The mesonephros begins to degenerate by the sixth week and is just about completely gone by the eighth week.

At about the fifth week, an outgrowth, called a ***ureteric bud,*** develops from the distal end of the mesonephric duct near the cloaca. This bud is the developing ***metanephros.*** As it grows toward the head of the embryo, its end widens to form the *pelvis* of the kidney with its *calyces* and associated *collecting tubules.* The unexpanded portion of the bud, the ***metanephric duct,*** becomes the *ureter.* The *nephrons,* the functional units of the kidney, arise from the intermediate mesoderm around each ureteric bud.

During development, the cloaca divides into a ***urogenital sinus,*** into which urinary and genital ducts empty, and a *rectum* that discharges into the anal canal. The *urinary bladder* develops from the urogenital sinus. In the female, the *urethra* develops from lengthening of the short duct that extends from the urinary bladder to the urogenital sinus. The *vestibule,* into which the urinary and genital ducts empty, is also derived from the urogenital sinus. In the male, the urethra is considerably longer and more complicated but is also derived from the urogenital sinus.

DISORDERS: HOMEOSTATIC IMBALANCES

Gout

Gout is a hereditary condition associated with an excessively high level of uric acid in the blood. When nucleic acids are catabolized, a certain amount of uric acid is produced as a waste. Uric acid accumulates in the body and tends to solidify into crystals that are deposited in joints, kidney tissue, and soft tissues. When the crystals are deposited in the joints, the condition is called gouty arthritis (Chapter 9). Gout is aggravated by excessive use of diuretics, dehydration, and starvation.

Glomerulonephritis (Bright's Disease)

Glomerulonephritis (***Bright's disease***) is an inflammation of the kidney that involves the glomeruli. One of the most common causes is an allergic reaction to the toxins given off by streptococci bacteria that have recently infected another part of the body, especially the throat. The glomeruli become so inflamed, swollen, and engorged with blood that the endothelial-capsular membranes become highly permeable and allow blood cells and proteins to enter the filtrate. Thus, the urine contains many erythrocytes and much protein. The glomeruli may be permanently changed, leading to acute or chronic renal failure.

Pyelitis and Pyelonephritis

Pyelitis is an inflammation of the renal pelvis and its calyces. ***Pyelonephritis,*** an inflammation of one or both kidneys, involves the nephrons and the renal pelvis. The disease is generally a complication of infection elsewhere in the body. In females, it is often a complication of lower urinary tract infections. The causative agent in about 75 percent of the cases is the bacterium *Escherichia coli.* Should pyelonephritis become chronic, the infection kills nephrons, and this severely impairs their function.

FIGURE 26-19 **Development of the urinary system.**

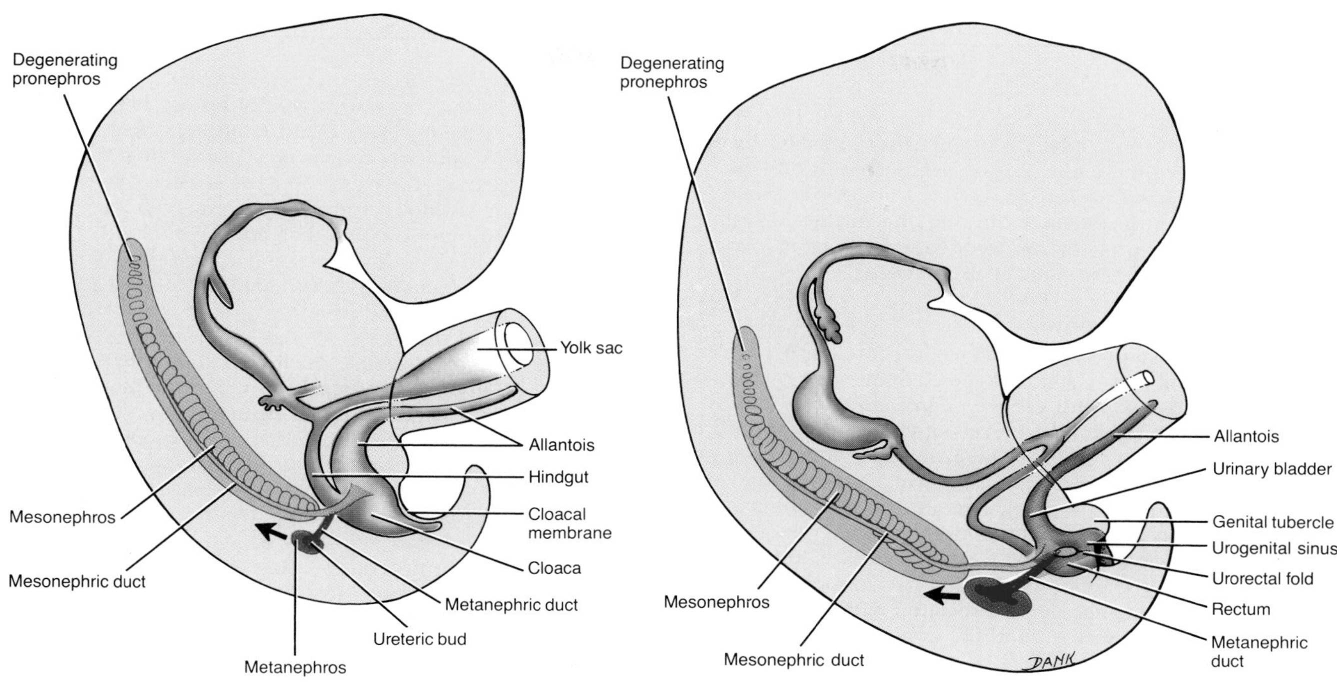

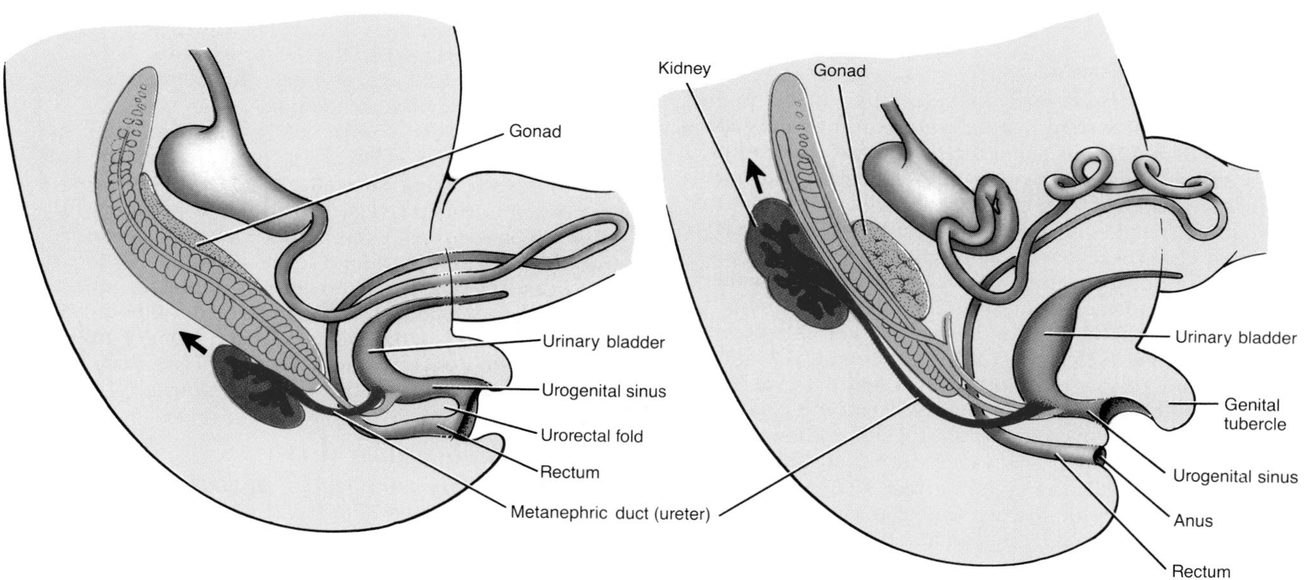

DISORDERS: HOMEOSTATIC IMBALANCES (*continued*)

Cystitis

Cystitis is an inflammation of the urinary bladder involving principally the mucosa and submucosa. It may be caused by bacterial infection, chemicals, or mechanical injury. Symptoms include burning on urination or painful urination, urgency and frequent urination, and low back pain. Bed-wetting may also occur.

Nephrotic Syndrome

Nephrotic syndrome is defined as proteinuria, primarily albumin, that results in hypoalbuminemia (low blood level of albumin) and varying degrees of edema and hyperlipidemia (high blood levels of cholesterol, phospholipids, and triglycerides). The proteinuria is due to an increased permeability of the glomerular-capsular membrane, permitting large amounts of protein (mostly albumin) to escape from blood into urine. Loss of albumin results in a decline in the blood level of albumin since liver production of albumin fails to meet increased urinary losses. The edema associated with nephrotic syndrome, usually seen around the eyes, ankles, feet, and abdomen, probably occurs because loss of albumin from the blood causes a decrease in blood colloidal osmotic pressure. This leads to fluid movement from blood into interstitial spaces. Among the causes of nephrotic syndrome are diabetes mellitus, rheumatoid arthritis, systemic lupus erythematosus, lymphoma, leukemia, bacterial and viral infections, certain drugs (nonsteroidal antiinflammatories, street heroin), heavy metals (gold, mercury), hypertension, and sickle-cell anemia. Treatment is aimed at alleviating the proteinuria.

Polycystic Disease

Polycystic disease may be caused by a defect in the renal tubular system that deforms nephrons and results in cystlike dilations along their course. It is the most common inherited disorder of the kidneys. The kidney tissue is riddled with cysts, small holes, and fluid-filled bubbles ranging in size from a pinhead to the diameter of an egg. These cysts gradually increase until they squeeze out the normal tissue, interfering with kidney function and causing uremia. The chief symptom is weight gain. The kidneys themselves may enlarge from their normal 0.25 kg (0.5 lb) to as much as 14 kg (30 lb). Many people lead normal lives without ever knowing they have the disease, however, and the condition is sometimes discovered only after death. Although the disease is progressive, its advance can be slowed by diet, drugs, and regulation of fluid intake. Kidney failure as a result of polycystic disease seldom occurs before the midforties and frequently can be controlled until the sixties.

Renal Failure

Renal failure, a decrease or cessation of glomerular filtration, is classified into two types: acute and chronic. ***Acute renal failure (ARF)*** is the sudden worsening of renal function that most commonly follows an episode of hypovolemic shock as after severe bleeding. The person with "shock kidney" stops putting out a normal volume of urine, a condition known as *oliguria* (*olig* = scanty), in which the daily urine output is usually less than 250 ml. In addition, *anuria* may occur with a reduction in urine output to less than about 50 ml a day. Acute obstruction of both ureters, and a reaction to nephrotoxic drugs—including certain antibiotics—are additional causes of acute renal failure.

Chronic renal failure (CRF) refers to the progressive and generally irreversible decline in glomerular filtration rate (GFR) that may result from chronic glomerulonephritis, pyelonephritis, congenital polycystic disease, and traumatic loss of kidney tissue, among others. CRF develops in three stages. In the first stage, *diminished renal reserve,* nephrons are destroyed until about 75 percent of the functioning nephrons are lost. At this stage, the person may show no symptoms since the remaining nephrons enlarge and take over the function of those that have been lost. When more nephrons are lost, the balance between glomerular filtration and tubular reabsorption is destroyed, and any change in diet or fluid intake brings on the symptoms. Once 75 percent of the nephrons are lost, the person enters the second stage, called *renal insufficiency.* In this stage, there is a decrease in GFR and increased blood levels of nitrogenous wastes and creatinine. Also, the kidneys cannot effectively concentrate or dilute urine. The final stage, called *end stage renal failure* (*uremia*), occurs when about 90 percent of the nephrons have been lost. At this stage, GFR diminishes to about 10 percent normal, and blood levels of nitrogenous wastes and creatinine increase further. The low GFR results in oliguria. Individuals with CRF are candidates for hemodialysis therapy and kidney transplantation.

Among the effects of renal failure are edema from salt and water retention; acidosis due to inability of the kidneys to excrete acidic substances; increased levels of nonprotein nitrogen (NPN) substances, especially urea, due to impaired renal excretion of metabolic waste products; elevated potassium levels that can lead to cardiac arrest; anemia, since the kidneys no longer produce renal erythropoietic factor required for red blood cell production; and osteomalacia, since the kidneys are no longer able to convert vitamin D to its active form for calcium absorption from the small intestine.

Urinary Tract Infections (UTIs)

The term ***urinary tract infection (UTI)*** is used to describe either an infection of a part of the urinary system or the presence of large numbers of microbes in urine. Included are ***significant bacteriuria*** (the presence of bacteria in urine in sufficient numbers to indicate active infection), ***asymptomatic bacteriuria*** (the multiplication of large numbers of bacteria in urine without producing symptoms), ***urethritis*** (inflammation of the urethra), ***cystitis*** (inflammation of the urinary bladder), and ***pyelonephritis*** (inflammation of the kidneys).

DISORDERS: HOMEOSTATIC IMBALANCES (*continued*)

Acute UTI is much more common in females than males and is most often caused by bacteria (*Escherichia coli*). Among the individuals at increased risk are pregnant women and those with renal disease, hypertension, and diabetes. Symptoms associated with UTI include burning on urination or painfulurination, urinary urgency and frequency, pubic and back pain, passage of cloudy or blood-tinged urine, chills, fever, nausea, vomiting, and urethral discharge, usually in males.

MEDICAL TERMINOLOGY ASSOCIATED WITH THE URINARY SYSTEM

Azotemia (az-ō-TĒ-mē-a; *azo* = nitrogen-containing; *emia* = condition of blood) Presence of urea or other nitrogenous elements in the blood.

Cystocele (SIS-tō-sēl; *cyst* = bladder; *cele* = cyst) Hernia of the urinary bladder.

Dysuria (dis-YOO-rē-a; *dys* = painful; *uria* = urine) Painful urination.

Enuresis (en'-yoo-RĒ-sis; *enourein* = to void urine) Bed-wetting; may be due to faulty toilet training, to some psychological or emotional disturbance, or rarely to some physical disorder. Also referred to as **nocturia.**

Intravenous pyelogram (in'-tra-VĒ-nus PĪ-e-lō-gram'), or **IVP** (*intra* = within; *veno* = vein; *pyelo* = pelvis of kidney; *gram* = written or recorded) X-ray film of the kidneys after venous injection of a dye.

Nephroblastoma (nef'-rō-blas-TŌ-ma; *neph* = kidney; *blastos* = germ or forming; *oma* = tumor) Embryonal carcinosarcoma; a malignant tumor of the kidneys arising from epithelial and connective tissue.

Polyuria (pol'-ē-YOO-rē-a; *poly* = much) Excessive urine formation.

Stricture (STRIK-chur) Narrowing of the lumen of a canal or hollow organ, as may occur in the ureter, urethra, or any other tubular structure in the body.

Uremia (yoo-RĒ-mē-a; *emia* = condition of blood) Toxic levels of urea in the blood resulting from severe malfunction of the kidneys.

Urethritis (yoo'-rē-THRĪ-tis) Inflammation of the urethra, caused by highly acidic urine, the presence of bacteria, or constriction of the urethral passage.

STUDY OUTLINE

1. The primary function of the urinary system is to regulate the concentration and volume of blood by removing and restoring selected amounts of water and solutes. It also excretes wastes.
2. The organs of the urinary system are the kidneys, ureters, urinary bladder, and urethra.

Kidneys (p. 826)

External Anatomy; Internal Anatomy (p. 826)

1. The kidneys are retroperitoneal organs attached to the posterior abdominal wall.
2. Three layers of tissue surround the kidneys: renal capsule, adipose capsule, and renal fascia.
3. Internally, the kidneys consist of a cortex, medulla, pyramids, papillae, columns, calyces, and a pelvis.
4. The nephron is the functional unit of the kidneys.
5. Each juxtamedullary nephron consists of a glomerular (Bowman's) capsule, glomerulus, proximal convoluted tubule, descending limb of the loop of the nephron, loop of the nephron (loop of Henle), ascending limb of the loop of the nephron, distal convoluted tubule, and collecting tubule.
6. The filtering unit of a nephron is the endothelial-capsular membrane. It consists of the glomerular endothelium, glomerular basement membrane, and epithelium (podocytes) of the visceral layer of the glomerular (Bowman's) capsule.
7. The extensive flow of blood through the kidney begins in the renal artery and terminates in the renal vein.
8. The nerve supply to the kidney is derived from the renal plexus.
9. The juxtaglomerular apparatus (JGA) consists of the juxtaglomerular cells and the macula densa of the distal convoluted tubule.

Physiology (p. 832)

1. Nephrons are the functional units of the kidneys. They help form urine and regulate blood composition.
2. The nephrons form urine by glomerular filtration, tubular reabsorption, and tubular secretion.
3. The primary force behind glomerular filtration is hydrostatic pressure.
4. Filtration of blood depends on the force of glomerular blood hydrostatic pressure in relation to two opposing forces: capsular hydrostatic pressure and blood colloidal osmotic pressure. This relationship is called effective filtration pressure (*Peff*).

5. If glomerular blood hydrostatic pressure falls to 50 mm Hg, renal suppression occurs because the glomerular blood hydrostatic pressure exactly equals the opposing pressures.
6. Most substances in plasma are filtered by the glomerular (Bowman's) capsule. Normally, blood cells and most proteins are not filtered.
7. Tubular reabsorption retains substances needed by the body, including water, glucose, amino acids, and ions. The maximum amount of a substance that can be absorbed is called tubular transport maximum (Tm).
8. About 80 percent of the reabsorbed water is returned by obligatory water reabsorption, the rest by facultative water reabsorption.
9. When glomerular filtration rate is low, vasodilation of afferent arterioles and vasoconstriction of efferent arterioles bring the rate back to normal.
10. Chemicals not needed by the body are discharged into the urine by tubular secretion. Included are ions, nitrogenous wastes, and certain drugs.
11. The kidneys help maintain blood pH by secreting H^+ ions and increasing or decreasing HCO_3^- concentration.
12. The kidneys produce dilute urine in the absence of ADH; renal tubules absorb more solutes than water.
13. The kidneys secrete concentrated urine in the presence of ADH; large amounts of water are reabsorbed from the filtrate into interstitial fluid, increasing solute concentration. The countercurrent mechanism also contributes to the secretion of concentrated urine.
14. Renal clearance refers to the ability of the kidneys to clear (remove) a specific substance from blood.

Hemodialysis Therapy (p. 844)

1. Filtering blood through an artificial device is called hemodialysis.
2. The kidney machine filters the blood of wastes and adds nutrients; a recent variation is called continuous ambulatory peritoneal dialysis (CAPD).

Homeostasis (p. 845)

1. The kidneys assume the major role of excretion.
2. Besides the kidneys, the lungs, integument, and gastrointestinal tract assume excretory functions.

Ureters (p. 846)

1. The ureters are retroperitoneal and consist of a mucosa, muscularis, and fibrous coat.
2. The ureters transport urine from the renal pelvis to the urinary bladder, primarily by peristalsis.

Urinary Bladder (p. 846)

1. The urinary bladder is posterior to the symphysis pubis. Its function is to store urine prior to micturition.
2. Histologically, the urinary bladder consists of a mucosa (with rugae), a muscularis (detrusor muscle), and a serous coat.
3. A lack of control over micturition is called incontinence; failure to void urine completely or normally is referred to as retention.

Urethra (p. 848)

1. The urethra is a tube leading from the floor of the urinary bladder to the exterior.
2. Its function is to discharge urine from the body.

Urine (p. 849)

1. Urine volume is influenced by blood pressure, blood concentration (blood osmotic pressure), temperature, diuretics, and emotions.
2. The physical characteristics of urine evaluated in a urinalysis (UA) are color, odor, turbidity, pH, and specific gravity.
3. Chemically, normal urine contains about 95 percent water and 5 percent solutes. The solutes include urea, creatinine, uric acid, hippuric acid, indican, ketone bodies, salts, and ions.
4. Abnormal constituents diagnosed through urinalysis include albumin, glucose, erythrocytes, leucocytes, ketone bodies, bilirubin, urobilinogen, casts, renal calculi, and microbes.

Aging and the Urinary System (p. 854)

1. After age 40, kidney function decreases.
2. Common problems related to aging include incontinence, urinary tract infections, prostate disorders, and renal calculi.

Developmental Anatomy of the Urinary System (p. 854)

1. The kidneys develop from intermediate mesoderm.
2. They develop in the following sequence: pronephros, mesonephros, metanephros.

Disorders: Homeostatic Imbalances (p. 854)

1. Gout is a high level of uric acid in the blood.
2. Glomerulonephritis is an inflammation of the glomeruli of the kidney.
3. Pyelitis is an inflammation of the kidney pelvis and calyces; pyelonephritis is an interstitial inflammation of one or both kidneys.
4. Cystitis is an inflammation of the urinary bladder.
5. Nephrotic syndrome is characterized by protein in the urine due to increased endothelial-capsular membrane permeability.
6. Polycystic disease is an inherited kidney disease in which nephrons are deformed.
7. Renal failure is classified as acute and chronic.
8. The term urinary tract infection (UTI) refers to either an infection of a part of the urinary system or the presence of large numbers of microbes in urine.

REVIEW QUESTIONS

1. What are the functions of the urinary system? What organs compose the system? (p. 826)
2. Describe the location of the kidneys. Why are they said to be retroperitoneal? (p. 826)

3. Prepare a labeled diagram that illustrates the principal external and internal features of the kidney. What is a floating kidney? (p. 827)
4. What is a nephron? List and describe the parts of a nephron from the glomerular (Bowman's) capsule to the collecting tubule. (p. 828)
5. Describe the structure of the endothelial-capsular membrane. How is the membrane adapted for filtration? (p. 828)
6. Distinguish between cortical and juxtaglomerular nephrons. How are nephrons supplied with blood? (p. 830)
7. Describe the structure and importance of the juxtaglomerular apparatus (JGA). (p. 832)
8. What is glomerular filtration? Define filtrate. (p. 833)
9. Set up an equation to indicate how effective filtration pressure (*Peff*) is calculated. (p. 834)
10. What are the major chemical differences among plasma, filtrate, and urine? (p. 834)
11. Define tubular reabsorption. Why is the process physiologically important? What is glomerular filtration rate (GFR)? (p. 835)
12. What chemical substances are normally reabsorbed by the kidneys? What is tubular transport maximum (Tm)? (p. 835)
13. Describe how glucose and sodium are reabsorbed by the kidneys. Where does the process occur? (p. 835)
14. How is chloride reabsorption related to sodium reabsorption? (p. 837)
15. Distinguish obligatory from facultative water reabsorption. How is facultative reabsorption controlled? (p. 837)
16. Define tubular secretion. Why is it important? List some substances that are secreted. (p. 838)
17. Explain the mechanisms by which the kidneys help to control body pH. (p. 838)
18. Define renal clearance. Why is it important? (p. 840)
19. Describe how the kidneys produce a dilute or concentrated urine. (p. 841)
20. Define the countercurrent multiplier mechanism. Why is it important? (p. 844)
21. What is hemodialysis? Briefly describe the operation of an artificial kidney. What is continuous ambulatory peritoneal dialysis (CAPD)? (p. 844)
22. Contrast the functions of the lungs, integument, and gastrointestinal tract as excretory organs. (p. 845)
23. Describe the structure, histology, and function of the ureters. (p. 846)
24. How is the urinary bladder adapted to its storage function? (p. 846)
25. What is micturition? Describe the micturition reflex. (p. 848)
26. Contrast the causes of incontinence and retention. (p. 848)
27. Compare the location of the urethra in the male and female. (p. 848)
28. What is urine? Describe the effects of blood pressure, blood concentration, temperature, diuretics, and emotions on the volume of urine formed. (p. 849)
29. Describe the following physical characteristics of normal urine: color, turbidity, odor, pH, and specific gravity. (p. 850)
30. Describe the chemical composition of normal urine. (p. 851)
31. Define each of the following: albuminuria, glycosuria, hematuria, pyuria, ketosis, bilirubinuria, urobilinogenuria, casts, and renal calculi. (p. 852)
32. Describe the effects of aging on the urinary system. (p. 854)
33. Describe the development of the urinary system. (p. 854)
34. Define each of the following: gout, glomerulonephritis, pyelitis, pyelonephritis, cystitis, nephrotic syndrome, polycystic disease, renal failure, and urinary tract infections (UTIs). (p. 854)
35. Explain the diagnostic value of the following: cystoscopy (p. 847), urinalysis (UA) (p. 849), blood urea nitrogen (BUN) test (p. 851), and creatinine test. (p. 852)
36. Refer to the glossary of medical terminology associated with the urinary system. Be sure that you can define each term. (p. 857)

SELECTED READINGS

Cochran, J. S., S. N. Robinson, V. S. Crane, and D. G. Jones. "Extracorporeal Shock Wave Lithotripsy," *Postgraduate Medicine,* May 1988.

Dretler, S. P. "Stone Crushers," *Harvard Medical School Health Letter,* December 1985.

Nolph, K. D., A. S. Lindblad, and J. W. Novak. "Continuous Ambulatory Peritoneal Dialysis," *New England Journal of Medicine,* 16 June 1988.

Reilly, N. J., and L. C. Torosian. "The New Wave in Lithotripsy," *RN,* March 1988.

Smith, D. A. "Incontinence," *RN,* March 1989.

Spiegel, D. M., M. Burnier, and R. W. Schrier. "Acute Renal Failure," *Postgraduate Medicine,* September 1987.

Twardowski, Z. J. "Peritoneal Dialysis," *Postgraduate Medicine,* April 1989.

Vander, A. J. *Renal Physiology,* 2nd ed. New York: McGraw-Hill, 1980.

Chapter 27

Fluid, Electrolyte, and Acid–Base Dynamics

Chapter Contents at a Glance

Student Objectives

1. Define the processes available for fluid intake and fluid output.
2. Compare the mechanisms regulating fluid intake and fluid output.
3. Contrast the electrolytic concentration of the three major fluid compartments.
4. Explain the functions and regulation of sodium, chloride, potassium, calcium, phosphate, and magnesium.
5. Describe the factors involved in the movement of fluid between plasma and interstitial fluid, and between interstitial fluid and intracellular fluid.
6. Explain the relationship between electrolyte imbalance and fluid imbalance.
7. Compare the role of buffers, respirations, and kidney excretion in maintaining body pH.
8. Define acid–base imbalances and their effects on the body.
9. Explain the appropriate treatments for acidosis and alkalosis.

The term ***body fluid*** refers to the body water and its dissolved substances. Fluid composes 45 to 75 percent of the body weight.

FLUID COMPARTMENTS AND FLUID BALANCE

About two-thirds of the fluid is located in cells and is termed ***intracellular fluid (ICF).*** The other third, called ***extracellular fluid (ECF),*** includes all the rest of the body fluids (see Figure 1-14). Examples of ECF are interstitial fluid; plasma; lymph; cerebrospinal fluid; gastrointestinal tract fluids; synovial fluid; the fluids of the eyes (aqueous humor and vitreous body) and ears (endolymph and perilymph); pleural, pericardial, and peritoneal fluids; and glomerular filtrate in the kidneys.

Body fluids are separated into distinct compartments by selectively permeable membranes. A compartment may be as small as the interior of a single cell or as large as the combined interiors of the heart and vessels. Fluids exist in compartments, but keep in mind that they are in constant motion from one compartment to another. In the healthy individual, the volume of fluid in each compartment remains relatively stable—another example of homeostasis.

Water constitutes the bulk of all the body fluids. When we say that the body is in ***fluid balance,*** we mean it contains the required amount of water and that amount is distributed to the various compartments according to their needs.

Osmosis is the primary way in which water moves in and out of body compartments. The concentration of solutes in the fluids is therefore a major determinant of fluid balance. Most solutes in body fluids are electrolytes—compounds that dissociate into ions. Fluid balance, then, means water balance, but it also implies electrolyte balance. The two are inseparable.

WATER

Water is by far the largest single constituent of the body, making up from 45 to 75 percent of the total body weight. (The functions of water may be reviewed in Chapter 2.) The percentage varies from person to person and depends primarily on age and the amount of fat present. Water proportion also decreases with age. An infant has the highest amount of water per body weight. In a normal adult male, water averages about 65 percent of the body weight. Females have more subcutaneous fat than males, and in a normal female, water averages about 55 percent. Since fat is basically water-free, lean people have a greater proportion of water to total body weight than fat people.

Fluid Intake and Output

The primary source of body fluid is water derived from ingested liquids (1600 ml) and foods (700 ml) that has been absorbed from the gastrointestinal tract. This water, called ***preformed water,*** amounts to about 2300 ml/day. Another source of fluid is ***metabolic water,*** the water produced through catabolism. This amounts to about 200 ml/day. Thus, total fluid input averages about 2500 ml/day.

There are several avenues of fluid output. The kidneys on the average lose about 1500 ml/day, the skin about 500 ml/day (400 ml/day through evaporation and 100 ml/day through perspiration, the lungs about 300 ml/day and the gastrointestinal tract about 200 ml/day. Fluid output thus totals 2500 ml/day. Under normal circumstances, fluid intake equals fluid output, so the body maintains a constant volume (Exhibit 27-1).

Regulation of Intake

Fluid intake is partially regulated by thirst. According to one theory, when water loss is greater than water intake, the resulting ***dehydration*** stimulates thirst through both

EXHIBIT 27-1 SUMMARY OF FLUID INTAKE AND OUTPUT PER DAY UNDER NORMAL CONDITIONS

Intake		Output	
Ingested liquids	1600 ml	Kidneys	1500 ml
Ingested foods	700 ml	Skin	
Metabolic water	200 ml	Evaporation	400 ml
	Total 2500 ml	Perspiration	100 ml
		Lungs	300 ml
		GI tract	200 ml
			Total 2500 ml

local and general responses. Locally, it leads to a decrease in the flow of saliva that produces a dryness of the mucosa of the mouth and pharynx (Figure 27-1). Dryness is interpreted by the brain as a sensation of thirst. In addition, dehydration raises blood osmotic pressure. Receptors (osmoreceptors) in the thirst center of the hypothalamus are stimulated by the increase in osmotic pressure and initiate impulses that are also interpreted as a sensation of thirst. The response, a desire to drink fluids, thus balances the fluid loss.

The initial quenching of thirst results from wetting the mucosa of the mouth and pharynx, but the major inhibition of thirst is believed to occur as a result of distension of the stomach or intestine and a decrease in osmotic pressure in fluids of the hypothalamus. Apparently, stimulated stretch receptors in the wall of the intestine send impulses that inhibit the thirst center in the hypothalamus.

FIGURE 27-1 **Regulation of fluid volume by the adjustment of intake to output.**

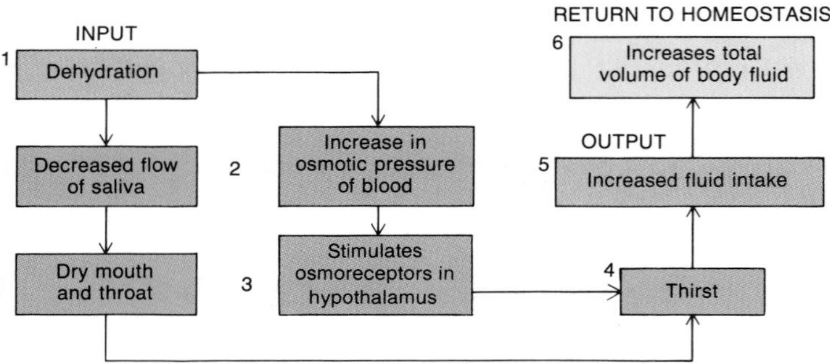

Regulation of Output

Under normal circumstances, fluid output is adjusted by antidiuretic hormone (ADH) and aldosterone, both of which regulate urine production (Chapter 26). Under abnormal conditions, other factors may influence output heavily. If the body is dehydrated, blood pressure falls, glomerular filtration rate decreases accordingly, and water is conserved. Conversely, excessive fluid in blood results in an increase of blood pressure, glomerular filtration rate, and fluid output. Hypertension produces the same effect. Hyperventilation leads to increased fluid output through the loss of water vapor by the lungs. Vomiting and diarrhea result in fluid loss from the gastrointestinal tract. Finally, fever, heavy perspiration, and destruction of extensive areas of the skin from burns bring about excessive water loss through the skin.

ELECTROLYTES

The body fluids contain a variety of dissolved chemicals. Some of these chemicals are compounds with covalent bonds; that is, the atoms that compose the molecule share electrons and do not form ions. Such compounds are called ***nonelectrolytes.*** Nonelectrolytes include most organic compounds, such as glucose, urea, and creatine. Other compounds, called ***electrolytes,*** have at least one ionic bond. When they dissolve in a body fluid, they dissociate into positive and negative ions. The positive ions are called ***cations;*** the negative ions are ***anions.*** Acids, bases, and salts are electrolytes. Most electrolytes are inorganic compounds, but a few are organic. For example, some proteins form ionic bonds. When the protein is put in solution, the ion detaches, and the rest of the protein molecule carries the opposite charge.

Electrolytes serve three general functions in the body. First, many are essential minerals. Second, they control the osmosis of water between body compartments. Third, they help maintain the acid–base balance required for normal cellular activities.

Concentration

During osmosis, water moves to the area with the greater number of particles in solution. A particle may be a whole molecule or an ion. An electrolyte exerts a far greater effect on osmosis than a nonelectrolyte because an electrolyte molecule dissociates into at least two particles, both of them charged. Suppose the nonelectrolyte glucose and two different electrolytes are placed in solution:

$$\underset{\text{Glucose}}{C_6H_{12}O_6} \xrightarrow{H_2O} C_6H_{12}O_6$$

$$\underset{\text{Sodium chloride}}{NaCl} \xrightarrow{H_2O} Na^+ + Cl^-$$

$$\underset{\text{Calcium chloride}}{CaCl_2} \xrightarrow{H_2O} Ca^{2+} + Cl^- + Cl^-$$

Because glucose does not break apart when dissolved in water, a molecule of glucose contributes only one particle to the solution. Sodium chloride, on the other hand, contributes two ions, or particles, and calcium chloride contributes three. Thus, calcium chloride has three times as great an effect on solute concentration as glucose.

Just as important, once the electrolyte dissociates, its ions can attract other ions of the opposite charge. If equal amounts of Ca^{2+} and Na^+ are placed in solution, the calcium ions will attract about twice as many chloride ions to their area as the sodium ions.

To determine how much effect an electrolyte has on concentration, we must look at the concentrations of its individual ions. The concentration of an ion is commonly expressed in ***milliequivalents per liter (mEq/l)***—the number of electrical charges in each liter of solution. The mEq/l equals the number of ions in solution times the number of charges the ions carry. Ion concentration can be calculated as shown in Exhibit 27-2.

EXHIBIT 27-2 CALCULATING CONCENTRATION OF IONS IN SOLUTION

The number of milliequivalents of an ion in each liter of solution is expressed by the following equation:

$$\text{mEq/l} = \frac{\text{milligrams of ion per liter of solution}}{\text{atomic weight}} \times \text{number of charges on one ion}$$

The atomic weight of an element is the number of protons and neutrons in an atom. Dividing the total weight of a solute by its atomic weight tells us how many ions there are in solution. The atomic weight of calcium is 40, whereas that of sodium is 23. Calcium is therefore a heavier element, and 100 g of calcium contains fewer atoms than does 100 g of sodium. The atomic weights of the elements can be found in a periodic table.

Using the preceding formula, we can calculate the milliequivalents per liter for calcium. In 1 liter of plasma there are normally 100 mg of calcium. Thus, by substituting this value in the formula, we arrive at:

$$\text{mEq/l} = \frac{100}{\text{atomic weight}} \times \text{number of charges}$$

The atomic weight of calcium is 40, and its number of charges is 2. By substituting these values, we arrive at:

$$\text{mEq/l} = \frac{100}{40} \times 2 = 5$$

Let us now find the milliequivalents per liter of plasma for sodium.

Milligrams of ions per liter = 3300
Number of charges = 1
Atomic weight = 23

$$\text{mEq/l} = \frac{3300}{23} \times 1 = 143.0$$

Even though calcium has a greater number of charges than sodium, the body retains many more sodium ions than calcium ions. Therefore, the milliequivalent for sodium in plasma is higher.

Distribution

Figure 27-2 compares the principal chemical constituents of plasma, interstitial fluid, and intracellular fluid. The chief difference between plasma and interstitial fluid is that plasma contains quite a few protein anions, whereas interstitial fluid has hardly any. Since normal capillary membranes are practically impermeable to protein, the protein stays in the plasma and does not move out of the blood into the interstitial fluid. Plasma also contains more Na^+ ions but fewer Cl^- ions than the interstitial fluid. In most other respects the two fluids are similar.

Intracellular fluid varies considerably from extracellular fluid, however. In extracellular fluid, the most abundant cation is Na^+, and the most abundant anion is Cl^-. In intracellular fluid, the most abundant cation is K^+, and the most abundant anion is HPO_4^{2-} (dibasic phosphate). Also, there are more protein anions in intracellular fluid than in extracellular fluid.

Functions and Regulation

Sodium

Sodium (Na^+), the most abundant extracellular ion, represents about 90 percent of extracellular cations. Normal blood (serum) sodium is 136–142 mEq/l. Sodium is necessary for the transmission of impulses in nervous and muscle tissue. Its location and concentration also play a significant role in fluid and electrolyte balance by creating most of the osmotic pressure of extracellular fluid (ECF). The average daily intake of sodium far exceeds the body's normal daily requirements. The kidneys excrete excess sodium and conserve it during periods of sodium restriction.

CLINICAL APPLICATION: HYPONATREMIA AND HYPERNATREMIA

Sodium loss from the body may occur through excessive perspiration, vomiting, or diarrhea; therapy with certain diuretics; and burns. Such a loss can result in ***hyponatremia*** (hī'-pō-na-TRĒ-mē-a; *natrium* = sodium), a lower-than-normal blood sodium level. Hyponatremia is characterized by muscular weakness, dizziness, headache, hypotension, tachycardia, and shock. Severe sodium loss can result in mental confusion, stupor, and coma. ***Hypernatremia*** is characterized by a higher-than-normal blood sodium level. It may occur with water loss, water deprivation, or sodium gain. Since sodium is the major determinant of the osmotic pressure of ECF, hypernatremia causes hypertonicity of ECF. As a result, water moves out of body cells into ECF, resulting in cellular dehydration. Symptoms of hypernatremia include intense thirst, fatigue, restlessness, agitation, and coma.

FIGURE 27-2 Comparison of electrolyte concentrations in plasma, interstitial fluid, and intracellular fluid. The height of each column represents the total electrolyte concentration.

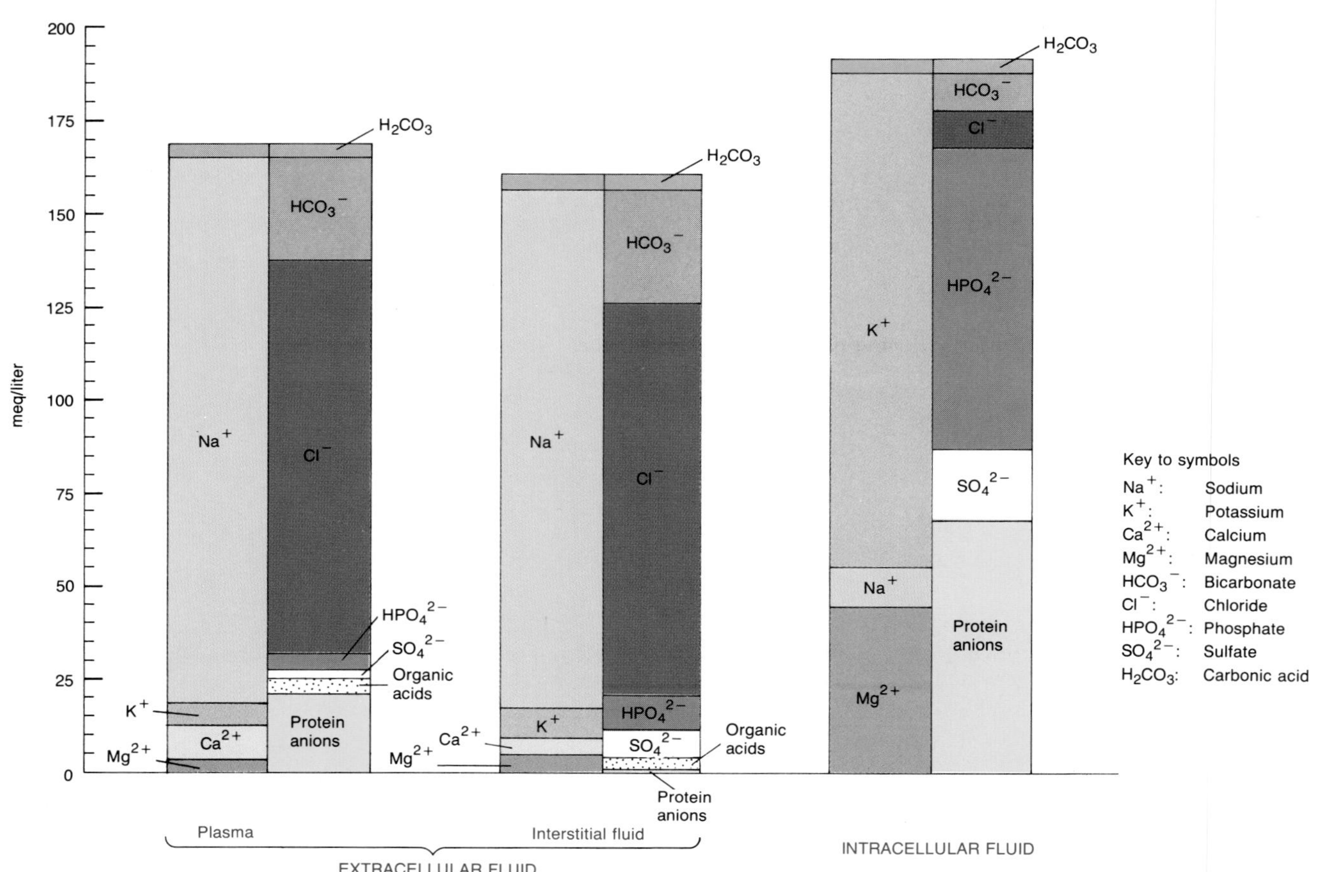

The sodium level in the blood is controlled primarily by the hormone aldosterone from the adrenal cortex (see Figure 18-19). Aldosterone acts on the distal convoluted tubules of the nephron of the kidneys and causes them to increase their reabsorption of sodium. The sodium thus moves from the filtrate back into the blood and establishes an osmotic gradient with the result being that water follows, moving from the filtrate back into the blood. Aldosterone is secreted in response to reduced blood volume or cardiac output, decreased extracellular sodium, increased extracellular potassium, and physical stress. A decrease in sodium concentration inhibits the release of antidiuretic hormone (ADH) by the posterior pituitary gland. This, in turn, permits more excretion of water in urine and restoration of sodium level to normal. Recall from Chapter 18 that the atria of the heart produce the hormone atrial natriuretic factor (ANF) that increases sodium and water excretion by the kidneys.

Chloride

Chloride (Cl^-) is the major extracellular anion. However, it can easily diffuse between the extracellular and intracellular compartments. This movement makes chloride important in regulating osmotic pressure differences between compartments. In the gastric mucosal glands, chloride combines with hydrogen to form hydrochloric acid. Normal blood (serum) chloride is 95–103 mEq/l.

CLINICAL APPLICATION: HYPOCHLOREMIA

An abnormally low level of chloride in the blood, called *hypochloremia* (hī'-pō-klō-RĒ-mē-a) may be caused by excessive vomiting, dehydration, and therapy with certain diuretics. Symptoms include muscle spasms, alkalosis, depressed respirations, and even coma.

Part of the regulation of chloride is indirectly under the control of aldosterone. Aldosterone regulates sodium reabsorption, and chloride follows sodium passively by electrical attraction since the negatively charged chloride ions will follow the positively charged sodium ions.

Potassium

Potassium (K^+) is the most abundant cation in intracellular fluid. Normal blood (serum) potassium is 3.8–5.0 mEq/l. Potassium assumes a key role in the functioning of nervous and muscle tissue and abnormal serum potassium levels adversely affect neuromuscular and cardiac function. Potassium also helps maintain fluid volume in cells and its extracellular fluid concentration influences pH in that fluid. When potassium ions move out of the cell, they are replaced by sodium ions and hydrogen ions. This shift of hydrogen ions helps regulate pH.

CLINICAL APPLICATION: HYPOKALEMIA AND HYPERKALEMIA

A lower-than-normal level of potassium, called *hypokalemia* (hī'-pō-ka-LĒ-mē-a; *kalium* = potassium), may result from vomiting, diarrhea, high sodium intake, kidney disease, and therapy with some diuretics. Symptoms include cramps and fatigue, flaccid paralysis, nausea, vomiting, mental confusion, increased urine output, shallow respirations, and changes in the electrocardiogram, including a lengthening of the Q-T interval and flattening of the T-wave.

A higher than normal blood potassium level, called *hyperkalemia,* is characterized by irritability, anxiety, abdominal cramping, diarrhea, weakness (especially of the lower extremities), and paresthesia (abnormal sensation, such as burning or prickling). Hyperkalemia can cause death by way of fibrillation of the heart.

The blood level of potassium is under the control of mineralocorticoids, mainly aldosterone. The mechanism is exactly opposite that of sodium. When sodium concentration is low, aldosterone secretion increases and more sodium is reabsorbed. But when potassium concentration is high, more aldosterone is secreted and more potassium is excreted. This process occurs in the distal convoluted tubules of the kidneys by way of tubular secretion.

Calcium

Calcium (Ca^{2+}) the most abundant ion in the body, is principally an extracellular electrolyte. It is primarily combined with phosphate (HPO_4^{2-}) as a mineral salt, mostly in bone matrix. Normal blood calcium is 4.6–5.5 mEq/l. About 98 percent of the calcium in the adult is present in the skeleton (and teeth) as the crystal lattice composed of calcium and phosphorus salts. The remaining calcium is found in extracellular fluid and within cells in a variety of tissues, especially skeletal muscle. Calcium is not only a structural component of bones and teeth, but it also is involved in blood coagulation, neurotransmitter release, neuromuscular conduction, maintenance of muscle tone, and excitability of nervous and muscle tissue.

CLINICAL APPLICATION: HYPOCALCEMIA AND HYPERCALCEMIA

An abnormally low level of calcium is called *hypocalcemia* (hī'-pō-kal-SĒ-mē-a). It may be due to increased calcium loss, reduced calcium intake, elevated levels of phosphate (as one goes up, the other goes down), or altered regulation as might occur in hypoparathyroidism. Hypocalcemia is characterized by numbness and tingling of the fingers, hyperactive reflexes, muscle cramps, tetany, and convulsions. Bone fractures may also occur. In addition, hypocalcemia may cause spasms of laryngeal muscles that can cause death by asphyxiation.

***Hypercalcemia,* an abnormally high level of calcium, is characterized by lethargy, weakness, anorexia, nausea, vomiting, polyuria, itching, bone pain, depression, confusion, paresthesia, stupor, and coma.**

Calcium blood levels are regulated principally by parathyroid hormone (PTH) and calcitonin (CT) (see Figure 18-14). More PTH is released when calcium blood level is low. PTH stimulates osteoclasts in bone tissue to release calcium (and phosphate) from mineral salts of bone matrix. Thus, PTH increases bone resorption. PTH also increases calcium ion absorption from the gastrointestinal tract by activating vitamin D and enhances reabsorption of calcium ions in glomerular filtrate through renal tubule cells and back into blood. CT, produced by the thyroid gland, is released in increased quantities when blood calcium levels are high. It decreases blood level of calcium by stimulating osteoblasts and inhibiting osteoclasts. In the presence of CT, osteoblasts remove calcium (and phosphate) from blood and deposit it in bone matrix as calcium phosphate salts.

Phosphate

Phosphate (HPO_4^{2-}) is principally an intracellular electrolyte. Normal blood (serum) phosphate is 1.7–2.6 mEq/l. About 85 percent of the phosphate in the adult is present in bone as calcium phosphate salts. The remainder is mostly combined with lipids (phospholipids), proteins, carbohydrates, and other organic molecules as components of structures such as plasma membranes and other cellular membranes, nucleic acids (DNA and RNA), and high-energy compounds such as adenosine triphosphate (ATP). Like calcium, phosphate is a structural component of bone and teeth. In addition, it is necessary for the synthesis of nucleic acids and high-energy compounds (ATP) and substances that serve as buffers (the phosphate buffer system is discussed later in the chapter).

CLINICAL APPLICATION: HYPOPHOSPHATEMIA AND HYPERPHOSPHATEMIA

An abnormally low level of phosphate, called ***hypophosphatemia*** (hī'-pō-fos'-fa-TĒ-mē-a), may occur through transient intracellular shifts, increased urinary losses, decreased intestinal absorption, increased utilization, and alcoholism. Hypophosphatemia is characterized by confusion, seizures, coma, chest and muscle pain, increased susceptibility to infection, numbness and tingling of the fingers, uncoordination, memory loss, and lethargy.

Hyperphosphatemia occurs most often in response to renal insufficiency in which the kidneys fail to excrete excess phosphate, increased intake of phosphates, extracellular shifts, and cellular destruction. The primary complication of hyperphosphatemia is metastatic calcification, the precipitation of calcium phosphate in soft tissues, joints, and arteries. Symptoms of hyperphosphatemia include anorexia, nausea, vomiting, muscular weakness, hyper reflexes, tetany, and tachycardia.

Phosphate blood levels are also regulated by PTH and CT. PTH stimulates osteoclasts to release phosphate from mineral salts of bone matrix and causes renal tubular cells to excrete phosphate. CT exerts its phosphate lowering effect by stimulating osteoblasts and inhibiting osteoclasts. In the presence of CT, osteoblasts remove phosphate from blood and deposit it in bone matrix after combining it with calcium to form the mineral salts in bone matrix.

Magnesium

Magnesium (Mg^{2+}) is primarily an intracellular electrolyte and the body's fourth most abundant cation. Normal blood (serum) magnesium is 1.3–2.1 mEq/l. In the adult, about 50 percent of the magnesium is found in bone, about 45 percent is in intracellular fluid, and about 5 percent is in extracellular fluid. Functionally, magnesium activates enzymes involved in the metabolism of carbohydrates and proteins, triggers the sodium-potassium pump, and preserves the structure of DNA, RNA, and ribosomes. Magnesium is also important in neuromuscular activity, neural transmission within the central nervous system (CNS), and myocardial functioning.

CLINICAL APPLICATION: HYPOMAGNESEMIA AND HYPERMAGNESEMIA

Magnesium deficiency, called ***hypomagnesemia*** (hī'-pō-mag'-ne-SĒ-mē-a), may be caused by malabsorption, diarrhea, nasogastric suction with administration of magnesium-free parenteral fluids, alcoholism, malnutrition, excessive lactation, diabetes mellitus, and diuretic therapy. The condition is characterized by weakness, irritability, tetany, delirium, convulsions, confusion, anorexia, nausea, vomiting, paresthesia, and cardiac arrhythmias.

Hypermagnesemia, or magnesium excess, occurs almost exclusively in persons with renal failure who have an increased intake of magnesium, as might occur with magnesium-containing medications. Other causes include Addison's disease, acute diabetic acidosis, severe dehydration, and hypothermia. The condition is characterized by flaccidity, hypotension, muscular weakness or paralysis, nausea, vomiting, and altered mental functioning.

Magnesium level is regulated by aldosterone. When magnesium concentration is low, increased aldosterone secretion acts on the kidneys so that more magnesium is reabsorbed.

MOVEMENT OF BODY FLUIDS

Between Plasma and Interstitial Compartments

The movement of fluid between plasma and interstitial compartments occurs across capillary membranes. This movement was discussed in detail in Chapter 21 (see Figure 21-7a), but we will review it here. Basically, the fluid movement is dependent on four principal pressures: (1) blood hydrostatic pressure (BHP), (2) interstitial fluid hydrostatic pressure (IFHP), (3) blood osmotic pressure (BOP), and (4) interstitial fluid osmotic pressure (IFOP).

The difference between the two forces that move fluid out of plasma and the two forces that push it into plasma is the ***effective filtration pressure* (*P*eff).** The *P*eff at the arterial end of a capillary is 8 mm Hg, whereas that at the venous end is −7 mm Hg. Thus, at the arterial end of a capillary, fluid moves out (filtered) from plasma into the interstitial compartment at a pressure of 8 mm Hg; at the venous end of a capillary, fluid moves in (reabsorbed) from the interstitial to plasma compartment at a pressure of −7 mm Hg. Thus, not all the fluid filtered at one end of the capillary is reabsorbed at the other. The fluid not reabsorbed and any proteins that escape from capillaries pass into lymph capillaries. From here, the fluid (lymph) moves through lymphatic vessels to the thoracic duct or right lymphatic duct for entrance into the cardiovascular system via the subclavian veins. Under normal conditions, there is a state of near equilibrium at the arterial and venous ends of a capillary in which filtered fluid and absorbed fluid, as well as that picked up by the lymphatic system, are nearly equal. This near equilibrium is ***Starling's law of the capillaries.***

Between Interstitial and Intracellular Compartments

Intracellular fluid and interstitial fluid have the same osmotic pressures under normal circumstances. The principal cation inside the cell is K^+, whereas the principal cation outside is Na^+ (see Figure 27-2). When a fluid imbalance between these two compartments occurs, it is usually caused by a change in the Na^+ or K^+ concentration.

Sodium balance in the body normally is controlled by aldosterone and ADH. ADH regulates extracellular fluid electrolyte concentration by adjusting the amount of water reabsorbed into the blood by the distal convoluted tubules and collecting tubules of the kidneys. Aldosterone regulates extracellular fluid volume by adjusting the amount of sodium reabsorbed by the blood from the kidneys which, in turn, directly affects the amount of water reabsorbed from the filtrate. Certain conditions, however, may result in an eventual decrease in the sodium concentration in interstitial fluid. For instance, during sweating the skin excretes sodium as well as water. Sodium also may be lost through vomiting and diarrhea. Coupled with replacement of fluid volume with plain water, these conditions can quickly produce a sodium deficit (Figure 27-3). The decrease in sodium concentration in the interstitial fluid lowers the interstitial fluid osmotic pressure and establishes an effective water concentration gradient between the interstitial fluid and the intracellular fluid. Water moves from the hypo-osmotic interstitial fluid into the cells, producing two results that can be quite serious.

The first result, an increase in intracellular water concentration, called ***overhydration,*** is particularly disruptive to nerve cell function. In fact, severe overhydration, or ***water intoxication,*** produces neurological symptoms ranging from disoriented behavior to convulsions, coma, and even death. The second result of the fluid shift is a loss of interstitial fluid volume that leads to a decrease in the interstitial fluid hydrostatic pressure. As the interstitial hydrostatic pressure drops, water moves out of the plasma, resulting in a loss of blood volume that may lead to circulatory shock.

FIGURE 27-3 **Interrelations between fluid imbalance and electrolyte imbalance.**

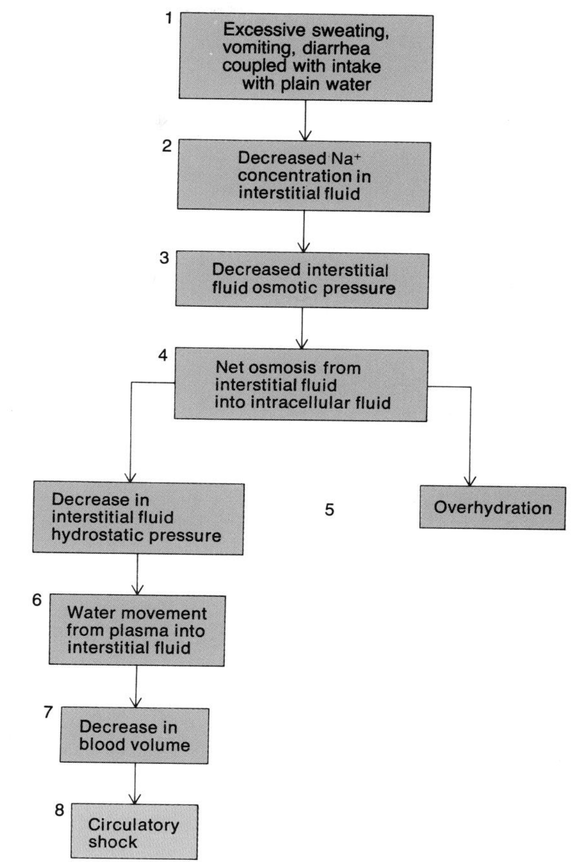

ACID–BASE BALANCE

Before reading this section, you might want to review the discussion of acids, bases, and pH in Chapter 2. In addition to controlling water movement, electrolytes also help regulate the body's acid–base balance. The overall acid–base balance is maintained by controlling the hydrogen ion (H^+) concentration of body fluids, particularly extracellular fluid. In a healthy person, the pH of the extracellular fluid is stabilized between 7.35 and 7.45. Homeostasis of this narrow range is essential to survival and depends on three major mechanisms: buffer systems, respirations, and kidney excretion.

Buffer Systems

Most ***buffer systems*** of the body consist of a weak acid and the salt of that acid, which functions as a weak base. Buffers function to prevent rapid, drastic changes in the pH of a body fluid by changing strong acids and bases into weak acids and bases. Buffers work within fractions of a second. Recall that a strong acid dissociates into H^+ ions more easily than does a weak acid. Strong acids, therefore, lower pH more than weak ones because strong acids contribute more H^+ ions. Similarly, strong bases raise pH more than weak ones because strong bases dissociate more easily into OH^- ions. The principal buffer systems of the body fluids are the carbonic acid–bicarbonate system, the phosphate system, the hemoglobin–oxyhemoglobin system, and the protein system.

Carbonic Acid–Bicarbonate

The ***carbonic acid–bicarbonate buffer system*** is actually based on the bicarbonate ion (HCO_3^-), which can act as a weak base and carbonic acid, which can act as a weak acid. Thus, the buffer system can compensate for either an excess or a shortage of H^+ ions. For example, if there is an excess of H^+ ions, an acid condition, HCO_3^- can function as a weak base and remove the excess H^+ ions as follows:

H^+	+	HCO_3^-	⇌	H_2CO_3	⇌	H_2O	+	CO_2
Hydrogen ion		Bicarbonate ion (weak base)		Carbonic acid		Water		Carbon dioxide

If, on the other hand, there is a shortage of H^+ ions, an alkaline condition, H_2CO_3 can function as a weak acid and provide H^+ ions as follows:

H_2CO_3	⇌	H^+	+	HCO_3^-
Weak acid		Hydrogen ion		Bicarbonate ion

A typical bicarbonate buffer system consists of a mixture of carbonic acid (H_2CO_3) and its salt, sodium bicarbonate ($NaHCO_3$). The carbonic acid–bicarbonate buffer system is an important regulator of blood pH. When a strong acid, such as hydrochloric acid (HCl) is added to a buffer solution containing sodium bicarbonate, which behaves like a weak base, the following reaction occurs:

HCl	+	$NaHCO_3$	⇌	$NaCl$	+	H_2CO_3
Hydrochloric acid (strong acid)		Sodium bicarbonate (weak base)		Sodium chloride (salt)		Carbonic acid (weak acid)

If a strong base, such as sodium hydroxide (NaOH) is added to a buffer solution containing a weak acid, carbonic acid, the following reaction occurs:

$NaOH$	+	H_2CO_3	⇌	H_2O	+	$NaHCO_3$
Sodium hydroxide (strong base)		Carbonic acid (weak acid)		Water		Sodium bicarbonate (weak base)

Normal body processes produce more acids than bases and thus tend to acidify the blood rather than make it more alkaline. Accordingly, the body needs more bicarbonate salt than it needs carbonic acid. In fact, when extracellular pH is normal (7.4), bicarbonate molecules outnumber carbonic acid 20 : 1.

Phosphate

The ***phosphate buffer system*** acts in essentially the same manner as the carbonic acid–bicarbonate buffer system. The components of the phosphate buffer system are the sodium salts of dihydrogen phosphate and sodium monohydrogen phosphate ions. The dihydrogen phosphate ion acts as the weak acid and is capable of buffering strong bases.

$NaOH$	+	NaH_2PO_4	⇌	H_2O	+	Na_2HPO_4
Sodium hydroxide (strong base)		Sodium dihydrogen phosphate (weak acid)		Water		Sodium monohydrogen phosphate (weak base)

The monohydrogen phosphate ion acts as the weak base and is capable of buffering strong acids.

HCl	+	Na_2HPO_4	⇌	$NaCl$	+	NaH_2PO_4
Hydrochloric acid (strong acid)		Sodium monohydrogen phosphate (weak base)		Sodium chloride (salt)		Sodium dihydrogen phosphate (weak acid)

The phosphate buffer system is an important regulator of pH both in red blood cells and in the kidney tubular fluids. NaH_2PO_4 is formed when excess H^+ ions in the kidney tubules combine with Na_2HPO_4. In this reaction, the sodium released from Na_2HPO_4 forms sodium bicarbonate ($NaHCO_3$) and is passed into the blood. The H^+ ion that replaces sodium becomes part of the NaH_2PO_4 that is passed into the urine. This reaction is one of the mechanisms by which the kidneys help maintain pH by the acidification of urine (see Figure 26-11).

Hemoglobin–Oxyhemoglobin

The ***hemoglobin–oxyhemoglobin buffer system*** is an effective method for buffering carbonic acid in red blood cells. When blood moves from the arterial end of a capillary to the venous end, the carbon dioxide given up by body cells enters the erythrocytes and combines with water to form carbonic acid (Figure 27-4). Simultaneously, oxyhemoglobin gives up its oxygen to the body cells, and some becomes reduced hemoglobin, and carries a negative charge. The hemoglobin anion attracts the hydrogen ion from the carbonic acid and becomes an acid that is even weaker than carbonic acid. When the hemoglobin–oxyhemoglobin system is active, the ex-

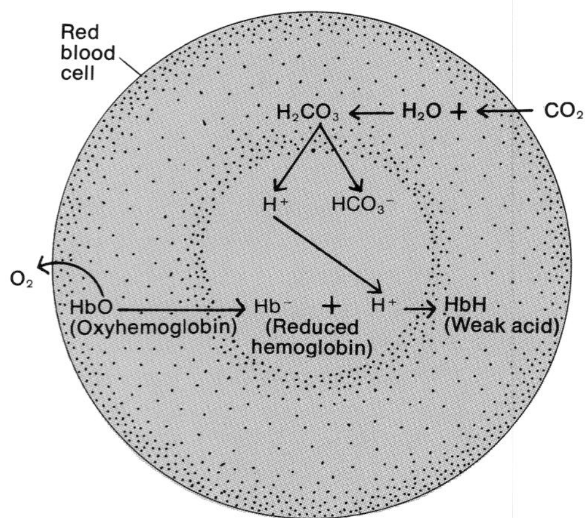

FIGURE 27-4 Hemoglobin–oxyhemoglobin buffer system. Oxyhemoglobin gives up its oxygen in an acid medium and buffers the acid. The bicarbonate (HCO_3^-) may remain in the cell and combine with potassium (K^+), or it may move out of the cell and combine with sodium (Na^+). In this way, much of the carbon dioxide (CO_2) is carried back to the lungs in the form of potassium bicarbonate ($KHCO_3$) or sodium bicarbonate ($NaHCO_3$).

change reaction that occurs shows why the erythrocyte tends to give up its oxygen when PCO_2 is high.

Protein

The ***protein buffer system*** is the most abundant buffer in body cells and plasma. Proteins are composed of amino acids. An amino acid is an organic compound that contains at least one carboxyl group (COOH) and at least one amine group (NH_2). The carboxyl group acts like an acid by releasing hydrogen (H^+) ions when pH rises and can dissociate in this way:

$$NH_2—\underset{H}{\overset{R}{C}}—COOH \rightleftharpoons NH_2—\underset{H}{\overset{R}{C}}—COO^- + H^+$$

The hydrogen ion is then able to react with any excess hydroxide (OH^-) ion in the solution to form water.

The amine group can act as a base by combining with hydrogen ions when pH falls as follows:

$$COOH—\underset{H}{\overset{R}{C}}—NH_2 + H^+ \rightleftharpoons COOH—\underset{H}{\overset{R}{C}}—NH_3^+$$

The hydroxide ion can dissociate, react with excess hydrogen ions, and also form water. Thus, proteins act as both acidic and basic buffers. Such compounds, which can act as either acid or basic components of a buffer, are said to be amphoteric (am'-fō-TER-ik).

Respirations

Respirations also assume a role in maintaining the pH of the body. An increase in the carbon dioxide concentration in body fluids as a result of cellular respiration lowers the pH (makes it more acid). This is illustrated by the following equation:

$$CO_2 + H_2O \rightleftharpoons H_2CO_3 \rightleftharpoons H^+ + HCO_3^-$$

Conversely, a decrease in the carbon dioxide concentration of body fluids raises the pH (makes it more basic).

The pH of body fluids may be adjusted by a change in the rate and depth of breathing, an adjustment that usually takes from one to three minutes. If the rate and depth of breathing is increased, more carbon dioxide is exhaled, and the blood pH rises. Slowing down the respiration rate means less carbon dioxide is exhaled, and the blood pH falls. Doubling the breathing rate increases the pH by about 0.23. Thus, it can be increased from 7.4 to 7.63. Reducing the breathing rate to one-quarter its normal rate lowers the pH by 0.4. Thus, it can be decreased from 7.4 to 7.0. If you consider that breathing rate can be altered up to eight times the normal rate, it should become obvious that alterations in the pH of body fluids may be greatly influenced by respiration.

The pH of body fluids, in turn, affects the rate of breathing (Figure 27-5). If, for example, the blood becomes more acidic, the increase in hydrogen ions stimulates the respiratory center in the medulla, and respirations increase in rate and depth. The same effect is achieved if the blood concentration of carbon dioxide increases. The increased rate and depth of respiration remove more carbon dioxide from blood to reduce the hydrogen ion concentration. On the other hand, if the pH of the blood increases, the respiratory center is inhibited and respirations decrease. A decrease in the carbon dioxide concentration of blood has the same effect. The decreased rate and depth of respirations causes carbon dioxide to accumulate in blood and the hydrogen ion concentration increases. The respiratory mechanism normally can eliminate more acid or base than can all the buffers combined.

Kidney Excretion

Since the role of the kidneys in maintaining pH has already been discussed in Chapter 26, we shall simply refer you to that section. Review Figure 26-11 very carefully.

FIGURE 27-5 **Relationship between pH and respirations.**

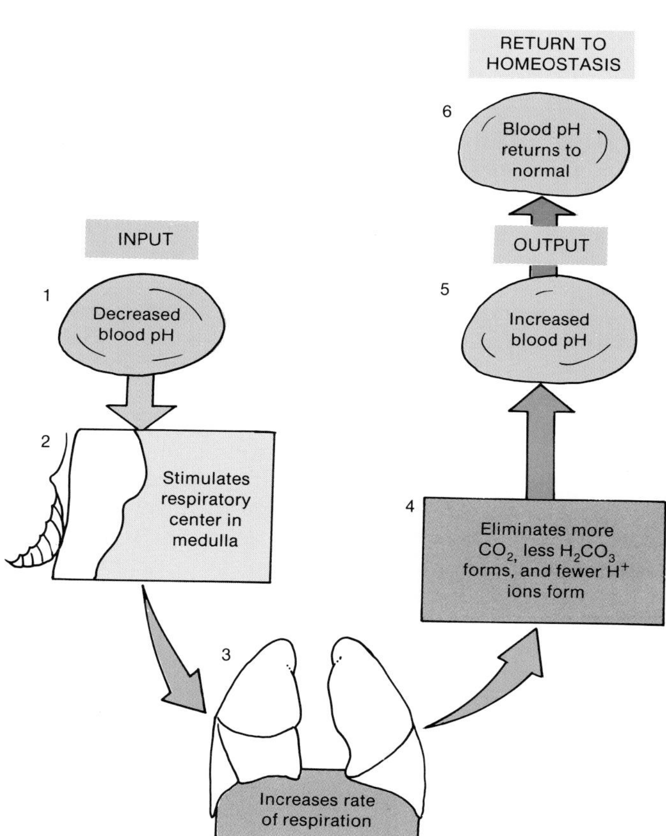

A summary of the mechanisms that maintain body pH is presented in Exhibit 27-3.

EXHIBIT 27-3 MECHANISMS THAT MAINTAIN BODY pH

Mechanism	Comments
Buffer systems	Most consist of weak acid and the salt of the acid which functions as a weak base. Prevent drastic changes in body fluid pH.
Carbonic acid–bicarbonate	Important regulator of blood pH. Most abundant buffer in extracellular fluid (ECF).
Phosphate	Important buffer in red blood cells and kidney tubule cells.
Hemoglobin–oxyhemoglobin	Buffers carbonic acid in red blood cells in blood.
Protein	Most abundant buffer in body cells and plasma.
Respirations	Regulate CO_2 level of body fluids and this indirectly helps to regulate H^+ concentration of body fluids.
Increased	Raises pH.
Decreased	Lowers pH.
Kidneys	Excrete H^+ and NH_4^+ and conserve bicarbonate.

ACID–BASE IMBALANCES

The normal blood pH range is 7.35 to 7.45. Any considerable deviation from this value falls under the category of acidosis or alkalosis. ***Acidosis*** is a condition in which blood pH ranges from 7.35 to 6.80 or lower. ***Alkalosis*** is a pH range of 7.45 to 8.00 or higher.

A change in blood pH that leads to acidosis or alkalosis can be compensated to return pH to normal. ***Compensation*** refers to the physiological response to an acid–base imbalance. If a person has an altered pH due to metabolic causes, respiratory mechanisms (hyperventilation or hypoventilation) can help compensate for the alteration. Respiratory compensation occurs within minutes and is maximized within hours. Conversely, if a person has an altered pH due to respiratory causes, metabolic mechanisms (kidney excretion) can compensate for the alteration. Metabolic compensation may begin in minutes but takes days to be maximized.

Physiological Effects

The principal physiological effect of acidosis is depression of the central nervous system through depression of synaptic transmission. If the blood pH falls below 7, depression of the nervous system is so acute that the individual becomes disoriented and comatose and dies. In fact, patients with severe acidosis usually die in a state of coma. On the other hand, the major physiological effect of alkalosis is overexcitability of the nervous system through facilitation of synaptic transmission. The overexcitability occurs both in the central nervous system and in peripheral nerves. Because of the overexcitability, nerves conduct impulses repetitively, even when not stimulated by normal stimuli, resulting in nervousness, muscle spasms, and even convulsions and death.

In the discussion that follows, note that both respiratory acidosis and alkalosis are primary disorders of blood PCO_2. On the other hand, both metabolic acidosis and alkalosis are primary disorders of bicarbonate (HCO_3^-) concentration.

Respiratory Acidosis

Respiratory acidosis is characterized by an elevated PCO_2 of arterial blood and decreased pH and is caused by hypoventilation or other causes of reduced gas exchange in the lungs. It occurs as a result of any condition that decreases the movement of carbon dioxide from the blood to the alveoli of the lungs and, therefore, causes a buildup of carbon dioxide, carbonic acid, and hydrogen ions. Such conditions include emphysema, pulmonary edema, injury to the respiratory center of the medulla, airway obstructions, or disorders of the muscles involved in breathing. Compensation is renal in which there is increased excretion of H^+ ions and increased reabsorption of HCO_3^- by the kidneys. In uncompensated respiratory acidosis, the normal bicarbonate–carbonic acid ratio shifts from 20 : 1 to 10 : 1 or 8 : 1, and blood pH decreases.

Respiratory Alkalosis

Respiratory alkalosis is characterized by a decreased arterial blood PCO_2 (lower than 35 mm Hg) and increased pH and is caused by hyperventilation. It occurs as a result of any condition that stimulates the respiratory center. Such conditions include oxygen deficiency due to high altitude or pulmonary disease, cerebrovascular accident (CVA), severe anxiety, and aspirin overdose. Compensation is renal by decreased excretion of H^+ ions and decreased reabsorption of HCO_3^- by the kidneys. In uncompensated respiratory alkalosis, the normal bicarbonate–carbonic acid ratio shifts from 20 : 1 to 20 : 0.5, and blood pH increases.

Metabolic Acidosis

Metabolic acidosis is characterized by a decrease in bicarbonate concentration (less than 22 mEq/l) and a decrease in pH. It may be caused by loss of bicarbonate, such as may occur with severe diarrhea or renal dysfunction; accumulation of an acid, other than carbonic acid, as may occur in ketosis; or failure of the kidneys to excrete H^+ ions derived from metabolism of dietary proteins. Compensation is respiratory by hyperventilation. In uncompensated metabolic acidosis, the ratio of bicarbonate to carbonic acid is 12.5 : 1.

Metabolic Alkalosis

Metabolic alkalosis is characterized by increased bicarbonate concentration (up to 40–45 mm Hg) and increased pH. It is caused by a nonrespiratory loss of acid by the body or excessive intake of alkaline drugs. Excessive vomiting of gastric contents results in a substantial loss of hydrochloric acid and is probably the most frequent cause of metabolic alkalosis. Other causes of metabolic alkalosis include gastric suctioning, use of certain diuretics, endocrine disorders, and administration of alkali. Compensation is respiratory by hypoventilation. In uncompensated metabolic alkalosis, the ratio of bicarbonate to carbonic acid is 31.6 : 1.

Treatment

The primary ***treatment of respiratory acidosis*** is aimed at increasing the exhalation of carbon dioxide. Excessive secretions may be suctioned out of the respiratory tract, and artificial respiration may be given. In addition, intravenous administration of bicarbonate and ventilation therapy to remove excessive buildup of carbon dioxide may be used. ***Treatment of metabolic acidosis*** consists of intravenous solutions of sodium bicarbonate and correcting the cause of acidosis.

Treatment of respiratory alkalosis is aimed at increasing the level of carbon dioxide in the body. One corrective measure is to have the patient breathe into a paper bag and then rebreathe the exhaled mixture of CO_2 and oxygen from the bag. ***Treatment of metabolic alkalosis*** consists of fluid therapy to replace chloride, potassium, and other electrolyte deficiencies and correcting the cause of alkalosis.

A summary of acidosis and alkalosis is presented in Exhibit 27-4.

EXHIBIT 27-4 SUMMARY OF ACIDOSIS AND ALKALOSIS

Condition	Definition	Common Causes	Compensatory Mechanism
Respiratory acidosis	Increased PCO_2 and decreased pH.	Hypoventilation due to emphysema, pulmonary edema, trauma to respiratory center, airway obstructions, dysfunction of muscles of respiration.	Renal: increased excretion of H^+ ions; increased reabsorption of HCO_3^-.
Respiratory alkalosis	Decreased PCO_2 and increased pH.	Hyperventilation due to oxygen deficiency, pulmonary disease, cerebrovascular accident (CVA), anxiety, aspirin overdose.	Renal: decreased excretion of H^+ ions; decreased reabsorption of HCO_3^-.
Metabolic acidosis	Decreased bicarbonate and decreased pH.	Loss of bicarbonate due to diarrhea, accumulation of acid (ketosis), renal dysfunction.	Respiratory: hyperventilation.
Metabolic alkalosis	Increased bicarbonate and increased pH.	Loss of acid or excessive intake of alkaline drugs due to vomiting, gastric suctioning, use of certain diuretics, and administration of alkali.	Respiratory: hypoventilation.

STUDY OUTLINE

Fluid Compartments and Fluid Balance (p. 861)

1. Body fluid is water and its dissolved substances.
2. About two-thirds of the body's fluid is located in cells and is called intracellular fluid (ICF).
3. The other third is called extracellular fluid (ECF). It includes interstitial fluid; plasma and lymph; cerebrospinal fluid; gastrointestinal tract fluids; synovial fluid; fluids of the eyes and ears; pleural, pericardial, and peritoneal fluids; and the glomerular filtrate.
4. Fluid balance means that the various body compartments contain the required amount of water.
5. Fluid balance and electrolyte balance are inseparable.

Water (p. 861)

1. Water is the largest single constituent in the body, varying from 45 to 75 percent of body weight, depending on the amount of fat present and age.
2. Primary sources of fluid intake are ingested liquids and foods and water produced by catabolism.
3. Avenues of fluid output are the kidneys, skin, lungs, and gastrointestinal tract.
4. The stimulus for fluid intake is dehydration resulting in thirst sensations. Under normal conditions, fluid output is adjusted by aldosterone and antidiuretic hormone (ADH).

Electrolytes (p. 862)

1. Electrolytes are chemicals that dissolve in body fluids and dissociate into either cations (positive ions) or anions (negative ions).
2. Electrolyte concentration is expressed in milliequivalents per liter (mEq/l).
3. Electrolytes have a greater effect on osmosis than nonelectrolytes.
4. Plasma, interstitial fluid, and intracellular fluid contain varying kinds and amounts of electrolytes.
5. Electrolytes are needed for normal metabolism, proper fluid movement between compartments, and regulation of pH.
6. Sodium (Na^+) is the most abundant extracellular ion. It is involved in impulse transmission, muscle contraction, and fluid and electrolyte balance. Its level is controlled by aldosterone.
7. Chloride (Cl^-) is the major extracellular anion. It assumes a role in regulating osmotic pressure and forming HCl. Its level is controlled indirectly by aldosterone.
8. Potassium (K^+) is the most abundant cation in intracellular fluid. It is involved in maintaining fluid volume, impulse conduction, muscle contraction, and regulating pH. Its level is controlled by aldosterone.
9. Calcium (Ca^{2+}), the most abundant ion in the body, is principally an extracellular ion that is a structural component of bones and teeth. It also functions in blood clotting, neurotransmitter release, muscle contraction, and heartbeat. Its level is controlled by parathyroid hormone (PTH) and calcitonin (CT).
10. Phosphate (HPO_4^{2-}) is principally an intracellular ion that is a structural component of bones and teeth. It is also required for the synthesis of nucleic acids and ATP and for buffer reactions. Its level is controlled by PTH and CT.
11. Magnesium (Mg^{2+}) is primarily an intracellular electrolyte that activates several enzyme systems. Its level is controlled by aldosterone.

Movement of Body Fluids (p. 866)

1. At the arterial end of a capillary, fluid moves from plasma into interstitial fluid. At the venous end, fluid moves in the opposite direction.
2. The state of near equilibrium at the arterial and venous ends of a capillary between filtered fluid and absorbed fluid, as well as that picked up by the lymphatic system, is referred to as Starling's law of the capillaries.
3. Fluid movement between interstitial and intracellular compartments depends on the movement of sodium and potassium and the secretion of aldosterone and antidiuretic hormone (ADH).
4. Fluid imbalance may lead to edema and overhydration (water intoxication).

Acid–Base Balance (p. 867)

1. The overall acid–base balance of the body is maintained by controlling the H^+ concentration of body fluids, especially extracellular fluid.
2. The normal pH of extracellular fluid is 7.35 to 7.45.
3. Homeostasis of pH is maintained by buffers, respirations, and kidney excretion.
4. The important buffer systems include: carbonic acid–bicarbonate, phosphate, hemoglobin–oxyhemoglobin, and protein.
5. An increase in rate of respirations increases pH; a decrease in rate decreases pH.
6. The kidneys excrete H^+ and NH_4^+.

Acid–Base Imbalances (p. 870)

1. Acidosis is a blood pH between 7.35 and 6.80 and lower. Its principal effect is depression of the central nervous system (CNS).
2. Alkalosis is a blood pH between 7.45 and 8.00 and higher. Its principal effect is overexcitability of the CNS.
3. Respiratory acidosis is characterized by an elevated PCO_2 and is caused by hypoventilation; metabolic acidosis is characterized by a decreased bicarbonate level and results from an abnormal increase in acid metabolic products (other than CO_2) and loss of bicarbonate.
4. Respiratory alkalosis is characterized by a decreased PCO_2 and is caused by hyperventilation; metabolic alkalosis is characterized by increased bicarbonate and results from nonrespiratory loss of acid or excess intake of alkaline drugs.
5. Metabolic acidosis or alkalosis is compensated by respiratory mechanisms; respiratory acidosis or alkalosis is compensated by renal mechanisms.

REVIEW QUESTIONS

1. Define body fluid. List several compartments and describe how they are separated. (p. 861)
2. What is meant by fluid balance? How are fluid balance and electrolyte balance related? (p. 861)
3. Describe the avenues of fluid intake and fluid output. Be sure to indicate volumes in each case. (p. 861)
4. Discuss the role of thirst in regulating fluid intake. (p. 861)
5. Explain how aldosterone and antidiuretic hormone (ADH) adjust normal fluid output. What are some abnormal routes of fluid output? (p. 862)
6. Define a nonelectrolyte and an electrolyte. Give specific examples of each. (p. 862)
7. Distinguish between a cation and an anion. Give several examples of each. (p. 862)
8. How is the ionic concentration of a fluid expressed? Calculate the ionic concentration of sodium in a body fluid. (p. 862)
9. Describe the functions of electrolytes in the body. (p. 862)
10. Describe some of the major differences in the electrolytic concentrations of the three major fluid compartments in the body. (p. 863)
11. Name three important extracellular electrolytes and three important intracellular electrolytes. (p. 864)
12. Indicate the function and regulation of each of the following electrolytes: sodium, chloride, potassium, calcium, phosphate, and magnesium. (p. 863)
13. Describe the physiological effects of hyponatremia (p. 863), hypernatremia (p. 863), hypochloremia (p. 864), hypokalemia (p. 865), hyperkalemia (p. 865), hypocalcemia (p. 865), hypercalcemia (p. 865), hypophosphatemia (p. 866), hyperphosphatemia (p. 866), hypomagnesemia (p. 866), and hypermagnesemia. (p. 866)
14. Explain the forces involved in moving fluid between plasma and interstitial fluid. Summarize these forces by setting up an equation to express effective filtration pressure (*Peff*). (p. 866)
15. What is Starling's law of the capillaries? (p. 866)
16. Explain the factors involved in fluid movement between the interstitial fluid and the intracellular fluid. (p. 867)
17. Explain how the following buffer systems help to maintain the pH of body fluids: carbonic acid–bicarbonate, phosphate, hemoglobin–oxyhemoglobin, and protein. (p. 867)
18. Describe how respirations are related to the maintenance of pH. (p. 869)
19. Briefly discuss the role of the kidneys in maintaining pH. (p. 869)
20. Define acidosis and alkalosis. Distinguish between respiratory and metabolic acidosis and alkalosis. (p. 870)
21. What are the principal physiological effects of acidosis and alkalosis? (p. 870)
22. How are acidosis and alkalosis compensated and treated? (p. 870)

SELECTED READINGS

Arieff, A., and R. A. DeFonzo. *Fluid, Electrolyte, and Acid–Base Disorders.* New York: Churchill Livingstone, Inc. 1985.

Barta, M. A. "Correcting Electrolyte Imbalances," *RN,* February 1987.

Halperin, M. L., and M. C. Goldstein. *Fluid, Electrolyte, and Acid–Base Emergencies.* Philadelphia: W. B. Saunders Co. 1988.

Hobbs, J. "Disturbances in Acid–Base Metabolism," *Postgraduate Medicine,* February 1988.

Horne, M. M., and P. L. Swearingen. *Pocket Guide to Fluids and Electrolytes.* St. Louis: C. V. Mosby Co. 1989.

Tepper, D., and R. S. Aronson. "Effects of Potassium and Calcium Abnormalities," *Hospital Medicine,* February 1984.

UNIT 5

CONTINUITY

This unit is designed to show you how the human organism is adapted for reproduction. It also traces the developmental sequence involved in pregnancy and discusses principles of inheritance.

Chapter 28

The Reproductive Systems

Chapter Contents at a Glance

Chapter Contents at a Glance

Student Objectives

1. Define reproduction and classify the organs of reproduction by function.
2. Explain the structure, histology, and functions of the ducts, accessory sex glands, and penis.
3. Describe the location, histology, and functions of the ovaries, uterine (Fallopian) tubes, uterus, vagina, vulva, and mammary glands.
4. Compare the principal events of the menstrual and ovarian cycles.
5. Explain the roles of the male and female in sexual intercourse.
6. Contrast the various kinds of birth control (BC) and their effectiveness.
7. Describe the effects of aging on the reproductive systems.
8. Describe the development of the reproductive systems.
9. Explain the symptoms and causes of sexually transmitted diseases (STDs) such as gonorrhea, syphilis, genital herpes, chlamydia, trichomoniasis, and genital warts.
10. Describe the symptoms and causes of male disorders (testicular cancer, prostate dysfunctions, impotence, and infertility) and female disorders (amenorrhea, dysmenorrhea, premenstrual syndrome [PMS], toxic shock syndrome [TSS], ovarian cysts, endometriosis, infertility, breast tumors, cervical cancer, pelvic inflammatory disease [PID], and vulvovaginal candidiasis).
11. Define medical terminology associated with the reproductive systems.

Reproduction is the mechanism by which the thread of life is sustained. In one sense, reproduction is the process by which a single cell duplicates its genetic material, allowing an organism to grow and repair itself; thus, reproduction maintains the life of the individual. But reproduction is also the process by which genetic material is passed from generation to generation. In this regard, reproduction maintains the continuation of the species.

The organs of the male and female reproductive systems may be grouped by function. The testes and ovaries, also called ***gonads*** (*gonos* = seed), function in the production of gametes—sperm cells and ova, respectively. The gonads also secrete hormones. The production of gametes and their discharge into ducts classifies the gonads as exocrine glands, whereas their production of hormones classifies them as endocrine glands. The ***ducts*** transport, receive, and store gametes. Still other reproductive organs, called ***accessory sex glands,*** produce materials that support gametes.

The developmental anatomy of the reproductive systems is considered later in the chapter.

MALE REPRODUCTIVE SYSTEM: STRUCTURE AND PHYSIOLOGY

The organs of the male reproductive system are the testes, or male gonads, that produce sperm and hormones; a number of ducts that either store or transport sperm to the exterior; accessory sex glands that add secretions constituting part of the semen; and several supporting structures, including the penis (Figure 28-1a,b).

Scrotum

The ***scrotum*** (SKRŌ-tum) is a cutaneous outpouching of the abdomen consisting of loose skin and superficial fascia (Figure 28-1). It is the supporting structure for the testes. Externally, it looks like a single pouch of skin separated into lateral portions by a median ridge called the ***raphe*** (RĀ-fē; *rafe* = seam). Internally, it is divided by a septum into two sacs, each containing a single testis. The septum consists of superficial fascia and contractile

FIGURE 28-1 Male organs of reproduction and surrounding structures seen in sagittal section. (a) Diagram.

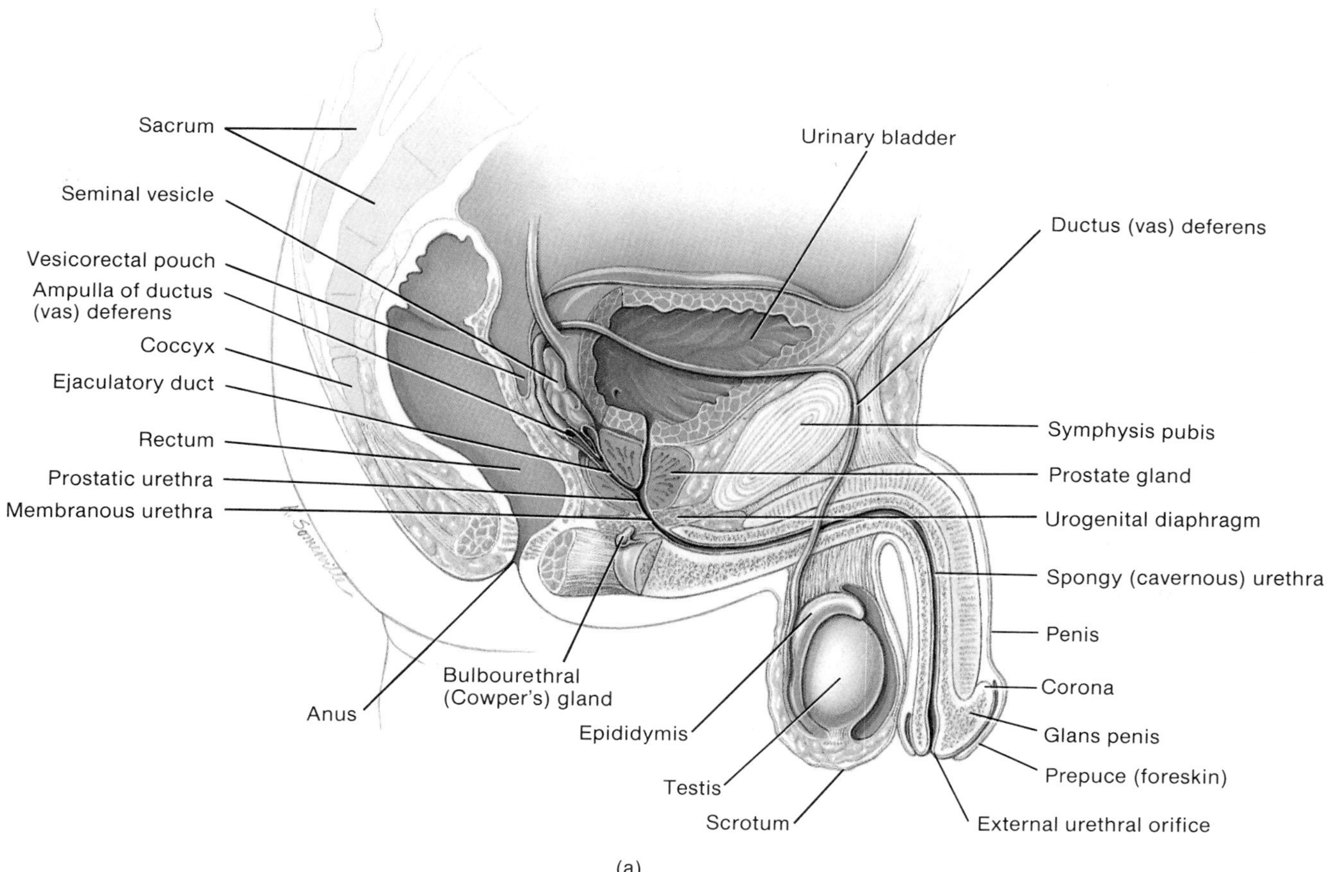

FIGURE 28-1 (*Continued*) (b) Photograph (taken with the cadaver on its back). (Courtesy of J. A. Gosling, P. F. Harris, et al., *Atlas of Human Anatomy,* Gower Medical Publishing Ltd., 1985.)

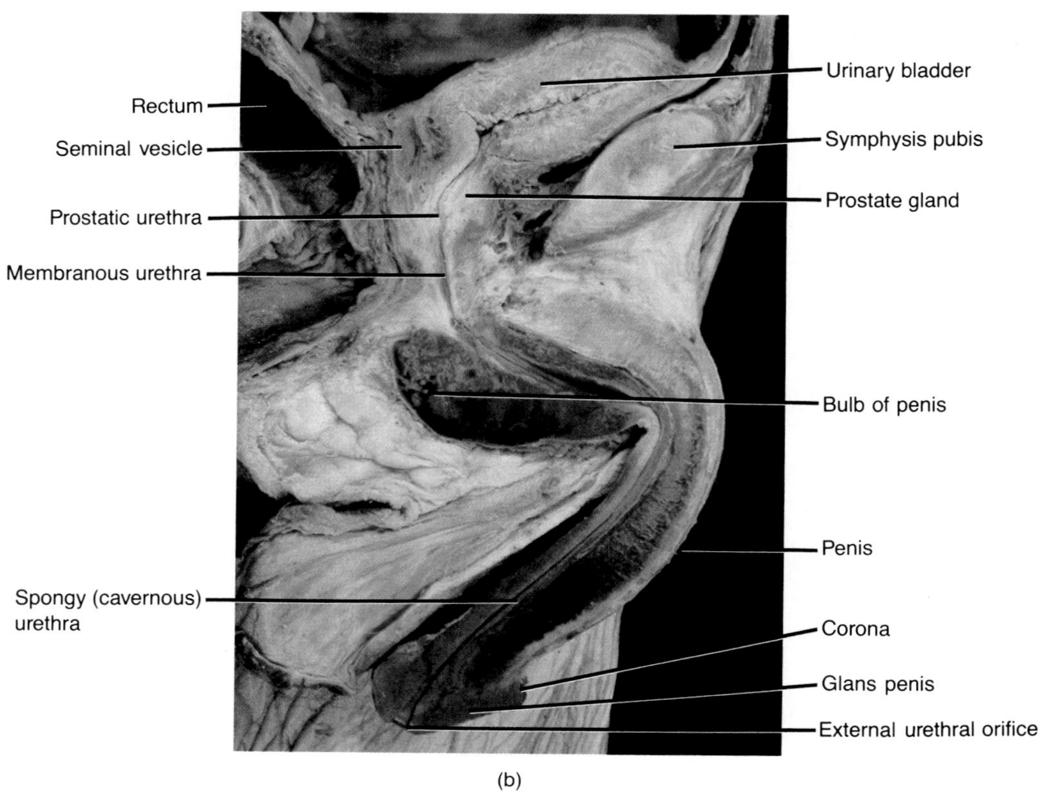

(b)

tissue called the ***dartos*** (DAR-tōs), which consists of bundles of smooth muscle fibers (cells). The dartos is also found in the subcutaneous tissue of the scrotum and is directly continuous with the subcutaneous tissue of the abdominal wall. The dartos causes wrinkling of the skin of the scrotum.

The location of the scrotum and the contraction of its muscle fibers regulate the temperature of the testes. The production and survival of sperm require a temperature that is lower than normal core body temperature. Because the scrotum is outside the body cavities, it provides an environment about 3°C below normal body temperature. The ***cremaster*** (krē-MAS-ter; *kremaster* = suspender) ***muscle*** (see Figure 28-8a), a small band of skeletal muscle, elevates the testes during sexual arousal and on exposure to cold, moving them closer to the pelvic cavity where they can absorb body heat. Exposure to warmth reverses the process. The dartos also is reflexly controlled to help assure that testis temperature is maintained about 3°C below core body temperature.

Testes

The ***testes,*** or ***testicles,*** are paired oval glands measuring about 5 cm (2 inches) in length and 2.5 cm (1 inch) in diameter (Figure 28-2). Each weighs between about 10 and 15 g. The testes develop high on the embryo's posterior abdominal wall and usually begin their descent into the scrotum through the inguinal canals during the latter half of the seventh month of fetal development (see Figure 28-8).

CLINICAL APPLICATION: CRYPTORCHIDISM

When the testes do not descend, the condition is referred to as ***cryptorchidism*** (krip-TOR-ki-dizm). The condition occurs in about 3 percent of full-term infants and about 30 percent of premature infants. Bilateral cryptorchidism results in sterility because the cells involved in the initial development of sperm cells are destroyed by the higher body temperature of the pelvic cavity. However, the testes will still secrete testosterone. Also, the probability of testicular cancer is 30 to 50 times greater in cryptorchid testes. The testes of about four-fifths of boys with cryptorchidism will descend spontaneously during the first year of life. When the testes remain undescended, injections of human chorionic gonadotropin (hCG) given at 2 to 5 years of age may stimulate descent. Surgical descent of the testes, known as *orchidopexy,* may become necessary and should be done at about age 5.

FIGURE 28-2 Anatomy of the testes. (a) Photograph of transverse section through the scrotum showing a testis. (b) Photograph of lateral aspect of a testis. (c) Diagram of a sagittal section illustrating the internal anatomy of a testis. (Courtesy of J. A. Gosling, P. F. Harris, et al., *Atlas of Human Anatomy*, Gower Medical Publishing Ltd., 1985.)

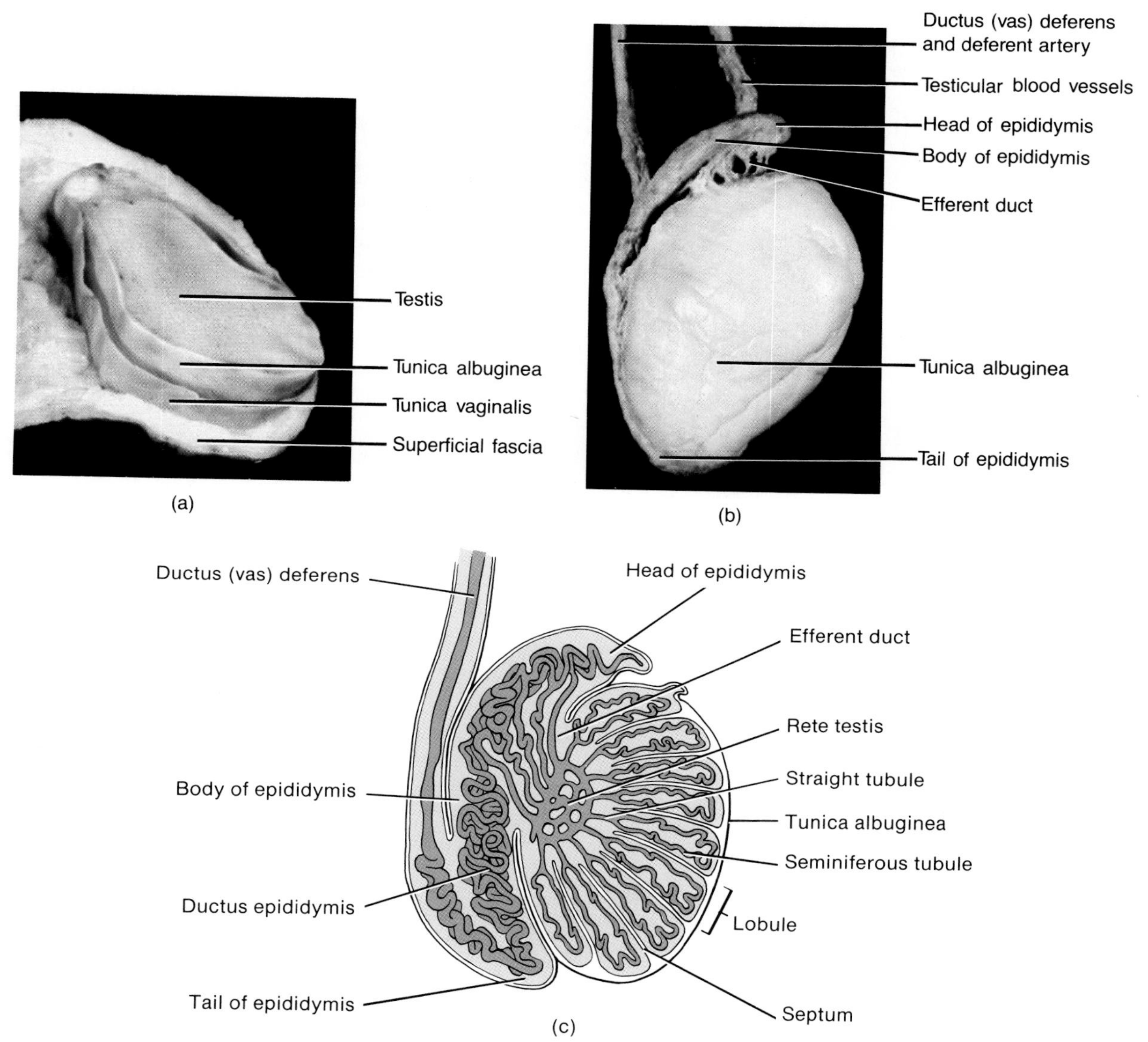

The testes are covered by a serous membrane called the ***tunica vaginalis,*** an outpocketing of the peritoneum formed during the descent of the testes. Internal to the tunica vaginalis is a dense layer of white fibrous tissue, the ***tunica albuginea*** (al'-byoo-JIN-ē-a), that extends inward and divides each testis into a series of internal compartments called ***lobules.*** Each of the 200 to 300 lobules contains one to three tightly coiled tubules, the convoluted ***seminiferous tubules,*** that produce sperm by a process called ***spermatogenesis.*** This process is considered shortly.

A cross section through a seminiferous tubule reveals that it is lined with spermatogenic cells in various stages of development (Figure 28-3a,b). Spermatogenic cells represent successive stages in a continuous process of differentiation of male germ cells. The most immature spermatogenic cells, the ***spermatogonia,*** are located against the basement membrane. Toward the lumen of the tube, one can see layers of progressively more mature cells. In order of advancing maturity, these are primary spermatocytes, secondary spermatocytes, and spermatids. By the time a ***sperm cell,*** or ***spermatozoon*** (sper'-ma-tō-ZŌ-on), has nearly reached maturity, it is in the lumen of the tubule and begins to be moved through a series of ducts. Embedded between the developing sperm cells in the tubules are ***sustentacular*** (sus'-ten-TAK-yoo-lar),

FIGURE 28-3 Histology of the testes. (a) Diagram of a cross section of a portion of a seminiferous tubule showing the stages of spermatogenesis. (b) Photomicrograph of an enlarged aspect of several seminiferous tubules at a magnification of 350×. (Copyright © 1983 by Michael H. Ross. Used by permission.)

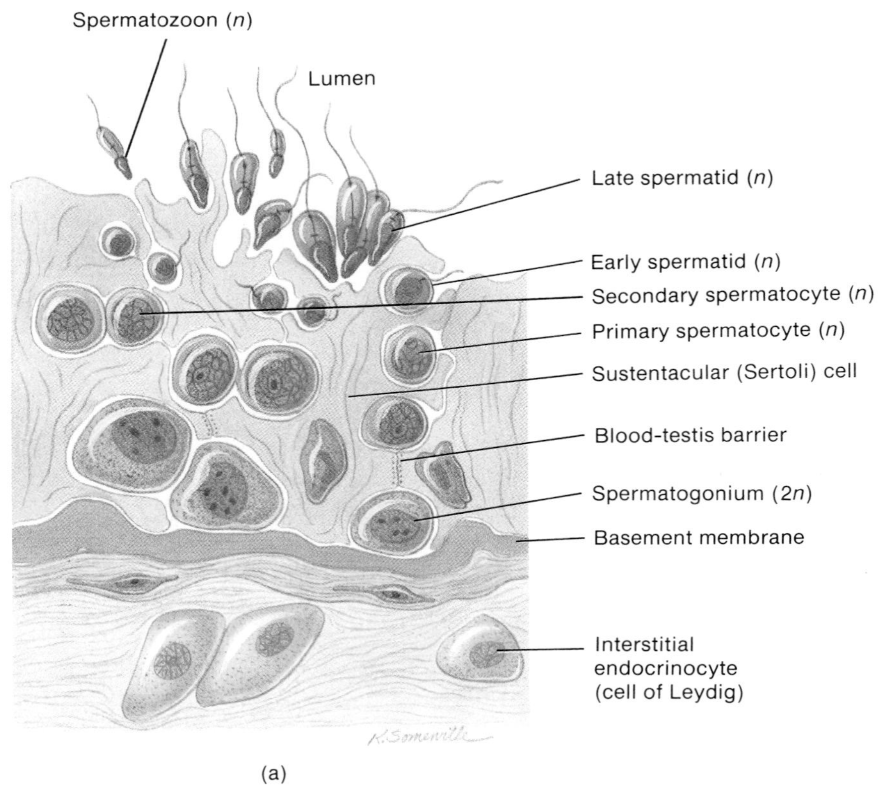

(a)

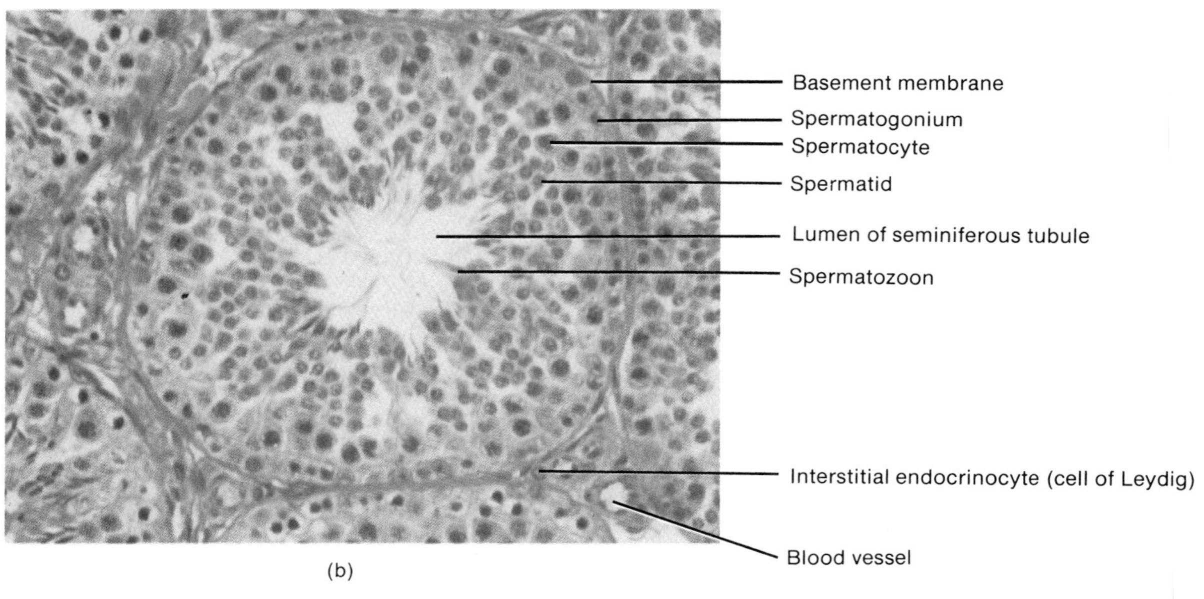

(b)

or ***Sertoli, cells.*** Just internal to the basement membrane, sustentacular cells are joined to one another by junctional points that form a ***blood–testis barrier.*** The barrier is important because spermatozoa and developing cells produce surface antigens that are recognized as foreign by the immune system. The barrier prevents an immune response against the antigens by isolating the cells from the blood. Such an immune response is seen following vasectomy (described shortly) in which sperm-specific antibodies, produced in cells of the immune system, are exposed to spermatozoa that no longer remain isolated in the reproductive tract. Sustentacular cells support and protect developing spermatogenic cells; nourish spermatocytes, spermatids, and spermatozoa; phagocytize degenerating spermatogenic cells; control movements of spermatogenic cells and the release of spermatozoa into the lumen of the seminiferous tubule; and secrete the hormone inhibin that helps regulate sperm production and androgen-binding protein, a substance required for sperm production that concentrates testosterone in the seminiferous tubule. Between the seminiferous tubules are clusters of ***interstitial endocrinocytes (interstitial cells of Leydig).*** These cells secrete the male hormone testosterone, the most important androgen.

Spermatogenesis

The process by which the seminiferous tubules of the testes produce haploid (n) spermatozoa involves several phases, including meiosis and mitosis, and is called ***spermatogenesis*** (sper'-ma-tō-JEN-e-sis). Before reading the following discussion of spermatogenesis, you should review the details of meiosis in Chapter 3 (see Figure 3-18). At this point, a few key concepts should be kept in mind.

1. In sexual reproduction, a new organism is produced by the union and fusion of sex cells called ***gametes*** (*gameto* = to marry). Male gametes, produced in the testes, are called sperm cells, and female gametes, produced in the ovaries, are called ova.

2. The cell resulting from the union and fusion of gametes, called a ***zygote*** (*zygo* = joined), contains a mixture of chromosomes (DNA) from the two parents. Through repeated mitotic cell divisions, a zygote develops into a new organism.

3. Gametes differ from all other body cells (somatic cells) in that they contain the ***haploid*** (one-half) ***chromosome number,*** symbolized as n. In humans, this number is 23, which composes a single set of chromosomes. Uninucleated somatic cells contain the ***diploid chromosome number,*** symbolized $2n$. In humans, this number is 46, which composes two sets of chromosomes.

4. In a diploid cell, two chromosomes that belong to a pair are called ***homologous*** (*homo* = same) ***chromosomes (homologues).*** In human diploid cells, 22 of the 23 pairs of chromosomes are morphologically similar and are called ***autosomes.*** The other pair comprises the ***sex chromosomes,*** designated as X and Y. In the female, the homologous pair of sex chromosomes consists of two X chromosomes; in the male, the pair consists of an X and a Y. Sex determination will be discussed in more detail in Chapter 29 as part of inheritance of sex.

5. If gametes were diploid ($2n$), like somatic cells, the zygote would contain twice the diploid number ($4n$), and with every succeeding generation, the chromosome number would continue to double and normal development could not occur.

6. This continual doubling of the chromosome number does not occur because of meiosis, a process of cell division by which gametes produced in the testes and ovaries receive the haploid chromosome number. Thus, when haploid (n) gametes fuse, the zygote contains the diploid chromosome number ($2n$) and can undergo normal development.

In humans, spermatogenesis takes about 74 days. The seminiferous tubules are lined with immature cells called ***spermatogonia*** (sper'-ma-tō-GŌ-nē-a; *sperm* = seed; *gonium* = generation or offspring), or sperm mother cells (Figures 28-3a and 28-4). Spermatogonia contain the diploid ($2n$) chromosome number and represent a heterogenous group of cells in which three subtypes can be distinguished. These are referred to as *pale type A, dark type A,* and *type B* and are distinguished by the appearance of their nuclear chromatin. Pale type A spermatogonia remain relatively undifferentiated and capable of extensive mitotic division. Following division, some of the daughter cells remain undifferentiated and serve as a reservoir of precursor cells to prevent depletion of the stem cell population. Such cells remain near the basement membrane. The remainder of the daughter cells differentiate into type B spermatogonia. These cells lose contact with the basement membrane of the seminiferous tubule, undergo certain developmental changes, and become known as ***primary spermatocytes*** (SPER-ma-tō-sītz'). Primary spermatocytes, like spermatogonia, are diploid ($2n$); that is, they have 46 chromosomes. Dark type A spermatogonia are believed to represent reserve stem cells, only becoming activated if pale type A cells become critically depleted.

Reduction Division (Meiosis I) Each primary spermatocyte enlarges before dividing. Then two nuclear divisions take place as part of meiosis. In the first, DNA is replicated and 46 chromosomes (each made up of two chromatids) form and move toward the equatorial plane of the nucleus. There they line up by homologous pairs so that there are 23 pairs of duplicated chromosomes in the center of the nucleus. This pairing of homologous chromosomes is called ***synapsis.*** The four chromatids of each homologous pair then become associated with each other to form a ***tetrad.*** In a tetrad, portions of one chromatid may be exchanged with portions of another. This process, called ***crossing-over,*** permits an exchange of genes among chromatids (see Figure 3-19) that results in the recombination of genes. Thus, the spermatozoa eventually produced are genetically unlike each other and unlike the cell that produced them—one reason for the

FIGURE 28-4 Spermatogenesis.

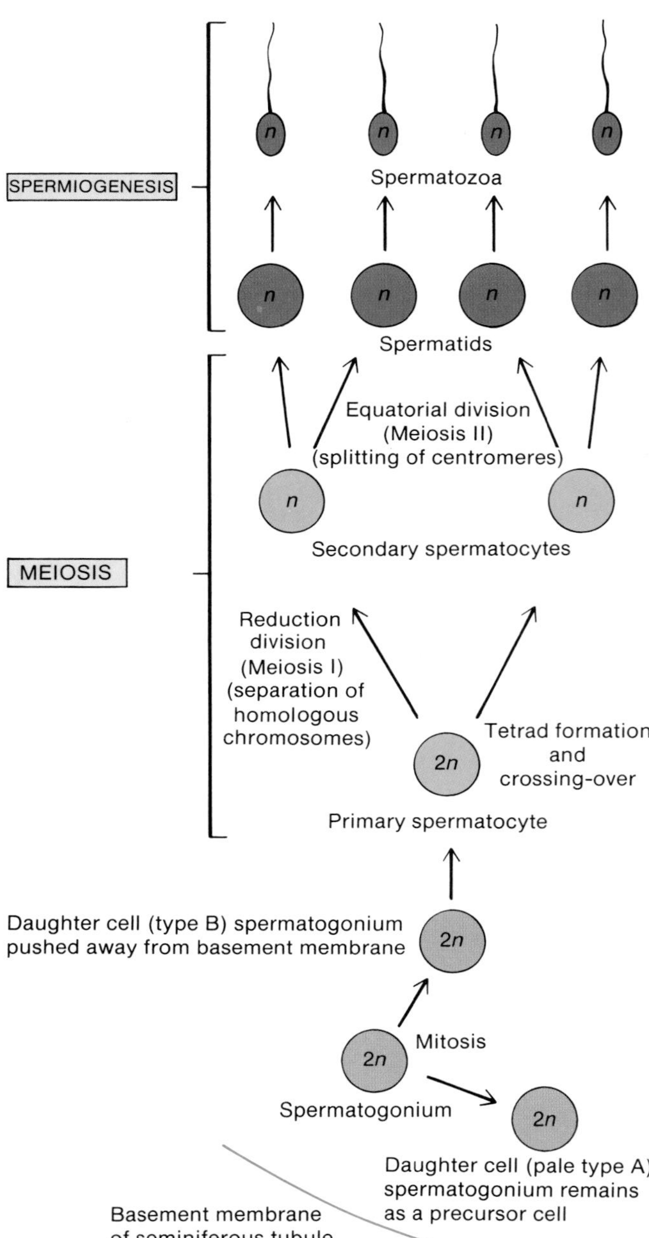

great variation among humans. Next, the meiotic spindle forms and the chromosomal microtubules produced by the centromeres of the paired chromosomes extend toward the poles of the cell. As the pairs separate, one member of each pair migrates to opposite poles of the dividing nucleus. The random arrangement of chromosome pairs on the spindle is another reason for variation among humans. The cells formed by the first nuclear division (reduction division) are called ***secondary spermatocytes.*** Each cell has 23 chromosomes—the haploid number. Each chromosome of the secondary spermatocytes, however, is made up of two chromatids. Moreover, the genes of the chromosomes of secondary spermatocytes may be rearranged as a result of crossing-over.

▪ Equatorial Division (Meiosis II) The second nuclear division of meiosis is ***equatorial division.*** There is no replication of DNA. The chromosomes (each composed of two chromatids) line up in single file around the equatorial plane, and the chromatids of each chromosome separate from each other. The cells formed from the equatorial division are called ***spermatids.*** Each contains half the original chromosome number, or 23 chromosomes, and is haploid. Each primary spermatocyte therefore produces four spermatids by meiosis (reduction division and equatorial division). Spermatids lie close to the lumen of the seminiferous tubule.

During spermatogenesis, a very interesting and unique process occurs. As the sperm cells proliferate, they fail to complete cytoplasmic separation (cytokinesis) so that all the daughter cells, except for the least-differentiated spermatogonia, remain continuous via cytoplasmic bridges. These cytoplasmic bridges persist until development of the spermatozoa is complete, at which point they float out individually into the lumen of the seminiferous tubule. Thus, the offspring of an original spermatogonium remain in cytoplasmic communication through their entire development. This pattern of development undoubtedly accounts for the synchronized production of spermatozoa in any given area of a seminiferous tubule. This pattern may have survival value in that half the spermatozoa contain an X chromosome and half a Y chromosome. The X chromosome probably carries many essential genes that are lacking on the Y chromosome, and if it were not for the cytoplasmic bridges between the developing sperm, it may be that the Y-bearing spermatozoon could not survive, with the result that no males could be produced in the next generation.

▪ Spermiogenesis The final stage of spermatogenesis, called ***spermiogenesis*** (sper'-mē-ō-JEN-e-sis), involves the maturation of spermatids into spermatozoa. Each spermatid embeds in a sustentacular (Sertoli) cell and develops a head with an acrosome (described shortly) and a flagellum (tail). Sustentacular cells extend from the basement membrane to the lumen of the seminiferous tubule where they nourish the developing spermatids. Since there is no cell division in spermiogenesis, each spermatid develops into a single ***spermatozoon (sperm cell).*** The release of a spermatozoon from a sustentacular cell is known as ***spermiation.***

Spermatozoa enter the lumen of the seminiferous tubule and migrate to the ductus epididymis, where in 10 to 14 days they complete their maturation and become capable of fertilizing an ovum. Spermatozoa are also stored in the ductus (vas) deferens. Here, they can retain their fertility for up to several weeks.

Spermatozoa

Spermatozoa are produced or matured at the rate of about 300 million per day and, once ejaculated, have a life expectancy of about 48 hours within the female reproductive tract. A spermatozoon is highly adapted for reaching and penetrating a female ovum. It is composed of a head, a midpiece, and a tail (Figure 28-5). Within the ***head*** are the nuclear material and a dense granule called the ***acrosome*** (*acro* = atop), which develops from the Golgi complex and contains enzymes (hyaluronidase and proteinases) that facilitate penetration of the sperm cell into a secondary oocyte. The acrosome is basically a specialized lysosome. Numerous mitochondria in the ***midpiece*** carry on the metabolism that provides energy for locomotion. The ***tail,*** a typical flagellum, propels the sperm along its way.

Testosterone and Inhibin

Secretions of the anterior pituitary gland assume a major role in the developmental changes associated with puberty (described shortly). At the onset of puberty, the anterior pituitary starts to secrete gonadotropic hormones called follicle-stimulating hormone (FSH) and luteinizing hormone (LH). Their release is controlled from the hypothalamus by gonadotropin releasing hormone (GnRH). Once secreted, the gonadotropic hormones have profound effects on male reproductive organs. FSH acts on the seminiferous tubules to initiate spermatogenesis and stimulate sustentacular (Sertoli) cells. LH also assists the seminiferous tubules to develop mature sperm, but its chief function is to stimulate the interstitial endocrinocytes (interstitial cells of Leydig) to secrete the hormone testosterone (tes-TOS-te-rōn).

Testosterone is synthesized from cholesterol or acetyl coenzyme A in the testes. It is the principal male hormone (androgen) and has a number of effects on the body. It controls the development, growth, and maintenance of the male sex organs. It also stimulates bone growth and epiphyseal closure, protein anabolism, sexual behavior, final maturation of sperm, and the development of male secondary sex characteristics. These characteristics, which appear at puberty, include muscular and skeletal development resulting in wide shoulders and narrow hips; body hair patterns that include pubic hair, axillary and chest hair (within hereditary limits), facial hair, and temporal hairline recession; and enlargement of the thyroid cartilage of the larynx, producing deepening of the voice. Testosterone also stimulates descent of the testes just prior to birth.

The interaction of LH with testosterone illustrates the operation of another negative feedback system (Figure 28-6). LH stimulates the production of testosterone, but once the testosterone concentration in the blood reaches a certain level, it inhibits the release of GnRH by the hypothalamus. This inhibition, in turn, inhibits the release of LH by the anterior pituitary. Thus, testosterone pro-

FIGURE 28-5 Spermatozoa. (a) Diagram of the parts of a spermatozoon. (b) Scanning electron micrograph of spermatozoon in contact with the surface of a secondary oocyte at a magnification of 1100×. (Courtesy of Fawcett/Phillips, Science Photo Library, Photo Researchers.)

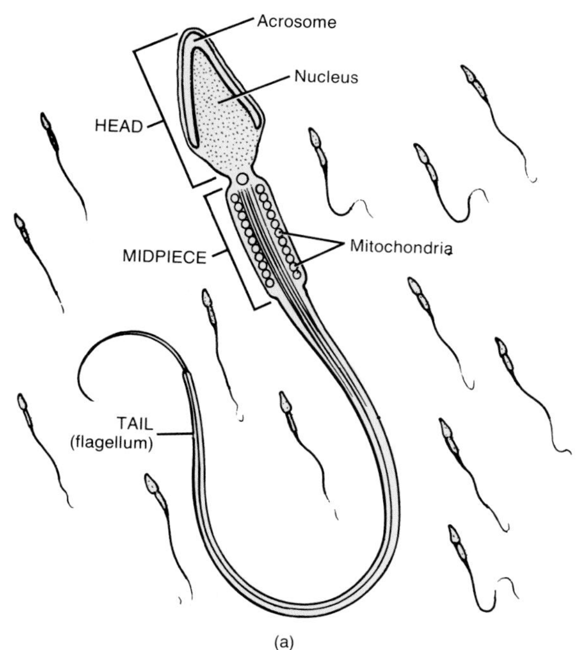

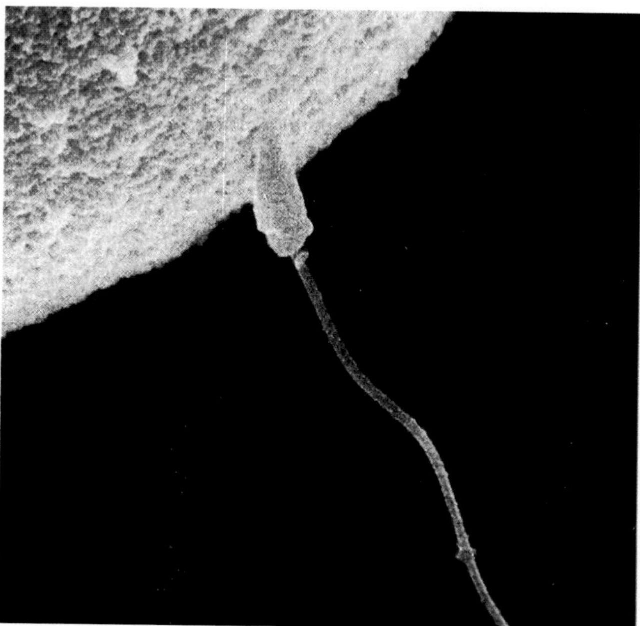

FIGURE 28-6 **Secretion, physiological effects, and control of testosterone and inhibin.**

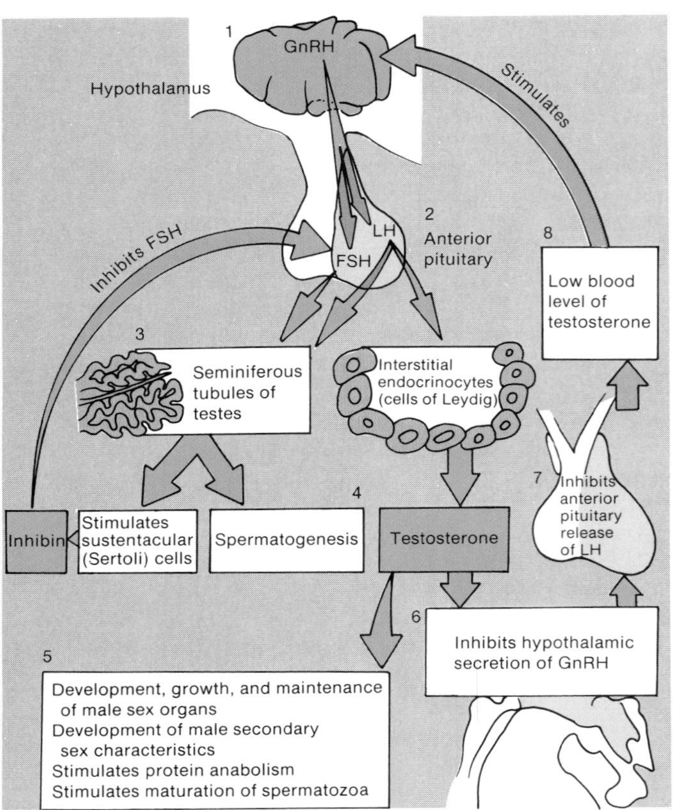

EXHIBIT 28-1 SUMMARY OF HORMONES SECRETED BY TESTES AND OVARIES

Hormone	Functions
Testosterone	Secreted by testes and controls development, growth, and maintenance of male sex organs; stimulates bone growth, protein anabolism, sexual behavior, final maturation of sperm, and development of male secondary sex characteristics; stimulates descent of testes.
Inhibin	Secreted by testes and ovaries and inhibits secretion of FSH.
Estrogens	Secreted by ovaries and control development and maintenance of female reproductive structures, especially the endometrium, secondary sex characteristics, and breasts; control fluid and electrolyte balance; increase protein anabolism.
Progesterone	Secreted by ovaries and works with estrogens to prepare endometrium for implantation of a fertilized ovum and mammary glands for milk secretion.
Relaxin	Secreted by ovaries and relaxes symphysis pubis and helps dilate uterine cervix to facilitate delivery.

duction is decreased. However, once the testosterone concentration in the blood decreases to a certain level, GnRH is released by the hypothalamus. This release of GnRH stimulates the release of LH by the anterior pituitary and stimulates testosterone production. Thus this testosterone–LH cycle is complete. Testosterone also inhibits LH release by a direct action on the anterior pituitary gland.

Inhibin is a protein hormone that has a direct effect on the anterior pituitary by inhibiting the secretion of FSH. FSH brings about spermatogenesis and stimulates sustentacular cells. Once the degree of spermatogenesis required for male reproductive functions has been achieved, sustentacular cells secrete inhibin. The hormone feeds back negatively to the anterior pituitary to inhibit FSH and thus to decrease spermatogenesis (Figure 28-6). If spermatogenesis is proceeding too slowly, lack of inhibin production permits FSH secretion and an increased rate of spermatogenesis. As you will see later, inhibin is also secreted by the ovaries during the menstrual cycle.

The functions of testosterone and inhibin are summarized in Exhibit 28-1.

Male Puberty

Puberty (PŪ-ber-tē; *puber* = marriageable age) refers to the period of time when secondary sex characteristics begin to develop and the potential for sexual reproduction is reached. Male puberty begins at an average age of 10 to 11 and ends at an average age of 15 to 17. The factors that determine the onset of puberty are poorly understood, but the sequence of events is well-established. During the prepubertal years, plasma levels of LH, FSH, and testosterone are low. At around age six or seven, boys experience an increase in secretion of adrenal androgens (adrenarche), probably under the influence of ACTH. Part of the prepubertal growth spurt and early development of axillary and pubic hair is probably related to stimulation by the adrenal androgens. Before the onset of puberty, low levels of LH are under feedback control by testosterone.

The onset of puberty is signaled by sleep-associated surges in LH and, to a lesser extent, FSH secretion. As puberty advances, elevated LH and FSH levels are present throughout the day and are accompanied by increased levels of testosterone. The rise in LH and FSH are believed to result from increased GnRH secretion and enhanced responsiveness of the anterior pituitary to GnRH. With sexual maturity, the hypothalamic-pituitary system becomes less sensitive to the feedback inhibition of testosterone on LH and FSH secretion.

The changes in the testes that occur during puberty include maturation of sustentacular (Sertoli) cells and initiation of spermatogenesis. The anatomical and functional changes associated with puberty are the result of increased testosterone secretion. Usually, the first sign is enlargement of the testes. About a year later, the penis increases in size. The prostate gland, seminal vesicles, bulbourethral (Cowper's) glands, and epididymis increase in size over a period of several years. Development of the secondary sex characteristics occurs and a growth spurt takes place as elevated testosterone levels increase both bone and muscle growth.

FIGURE 28-7 **Histology of the ductus epididymis. Photomicrograph of the ductus epididymis seen in cross section at a magnification of 160×. (Copyright © 1983 by Michael H. Ross. Used by permission.)**

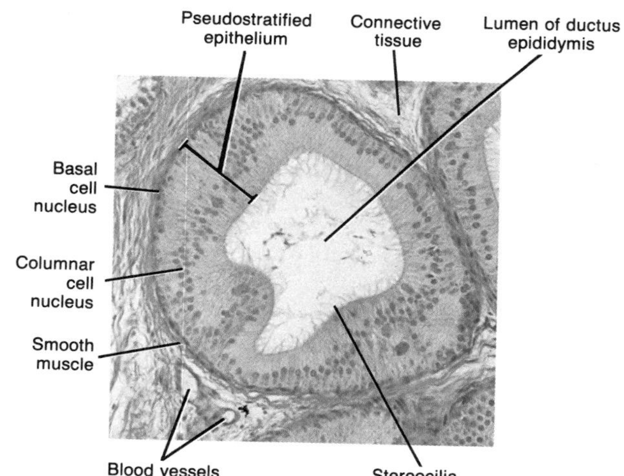

Ducts

Ducts of the Testis

Following their production, spermatozoa are moved through the convoluted seminiferous tubules to the ***straight tubules*** (see Figure 28-2b). The straight tubules lead to a network of ducts in the testis called the ***rete*** (RĒ-tē) ***testis.*** Some of the cells lining the rete testis possess cilia that probably help move the sperm along. The sperm are next transported out of the testis.

Epididymis

The sperm are transported out of the testis through a series of coiled ***efferent ducts*** in the epididymis that empty into a single tube called the ductus epididymis. Morphological changes occur in the spermatozoa during their passage through the epididymis.

The ***epididymis*** (ep'-i-DID-i-mis; *epi* = above; *didymos* = testis) is a comma-shaped organ that lies along the posterior border of the testis (see Figures 28-1 and 28-2) and consists mostly of a tightly coiled tube, the ***ductus epididymis.*** The larger, superior portion of the epididymis is known as the ***head.*** In the head, the efferent ducts join the ductus epididymis. The ***body*** is the narrow midportion of the epididymis. The ***tail*** is the smaller, inferior portion. At its distal end, the tail of the epididymis continues as the ductus (vas) deferens.

The ductus epididymis is a tightly coiled structure that would measure about 6 m (20 ft) in length and 1 mm in diameter if it were straightened out. The epididymis measures only about 3.8 cm (1.5 inches). The ductus epididymis is lined with pseudostratified columnar epithelium and encircled by layers of smooth muscle. The free surfaces of the columnar cells contain long, branching microvilli called ***stereocilia*** (Figure 28-7).

Functionally, the ductus epididymis is the site of sperm maturation. They require between 10 and 14 days to complete their maturation, that is, to become capable of fertilizing an ovum. The ductus epididymis also stores spermatozoa and propels them toward the urethra during emission by peristaltic contraction of its smooth muscle. Spermatozoa may remain in storage in the ductus epididymis for up to four weeks. After that, they are expelled from the epididymis or reabsorbed.

Ductus (Vas) Deferens

Within the tail of the epididymis, the ductus epididymis becomes less convoluted, its diameter increases, and at this point it is referred to as the ***ductus (vas) deferens*** or ***seminal duct*** (see Figure 28-2). The ductus (vas) deferens, about 45 cm (18 inches) long, ascends along the posterior border of the testis, penetrates the inguinal canal, and enters the pelvic cavity, where it loops over the side and down the posterior surface of the urinary bladder (see Figure 28-1a). The dilated terminal portion of the ductus (vas) deferens is known as the ***ampulla*** (am-POOL-la). The ductus (vas) deferens is lined with pseudostratified epithelium and contains a heavy coat of three layers of muscle. Functionally, the ductus (vas) deferens stores sperm and conveys sperm from the epididymis toward the urethra during emission by peristaltic contractions of the muscular coat.

CLINICAL APPLICATION: VASECTOMY

One method of sterilization of males is called *vasectomy,* a relatively uncomplicated procedure, typically performed under local anesthesia, in which a

portion of each ductus (vas) deferens is removed. In the procedure, an incision is made in the scrotum, the ducts are located, each is tied in two places, and the portion between the ties is excised. Although sperm production continues in the testes, the sperm cannot reach the exterior because the ducts are cut, and the sperm degenerate and are destroyed by phagocytosis. Vasectomy has no effect on sexual desire and performance, and if performed correctly, it is virtually 100 percent effective.

Traveling with the ductus (vas) deferens as it ascends in the scrotum are the testicular artery, autonomic nerves, veins that drain the testes (pampiniform plexus), lymphatic vessels, and the cremaster muscle. These structures constitute the ***spermatic cord,*** a supporting structure of the male reproductive system (Figure 28-8). The cremaster muscle, which also surrounds the testes, elevates the testes during sexual stimulation and exposure to cold. The spermatic cord and ilioinguinal nerve pass through the ***inguinal*** (IN-gwin-al) ***canal*** in the male. The canal is an oblique passageway in the anterior abdominal wall just superior and parallel to the medial half of the inguinal ligament. The canal is about 4 to 5 cm (1.6 to 2.0 inches) in length. It originates at the ***deep (abdominal) inguinal ring,*** a slitlike opening in the aponeurosis of the transversus abdominis muscle. The canal terminates at the ***superficial (subcutaneous) inguinal ring,*** a somewhat triangular opening in the aponeurosis of the external oblique muscle. In the female, the round ligament of the uterus and ilioinguinal nerve pass through the inguinal canal.

Ejaculatory Ducts

Posterior to the urinary bladder are the ***ejaculatory*** (e-JAK-yoo-la-tō'-rē) ***ducts*** (Figure 28-9). Each duct is about 2 cm (1 inch) long and is formed by the union of the duct from the seminal vesicle and ductus (vas) deferens. The ejaculatory ducts eject spermatozoa into the prostatic urethra just prior to ejaculation.

FIGURE 28-8 Spermatic cord and inguinal canal. The left spermatic cord has been opened to expose its contents.

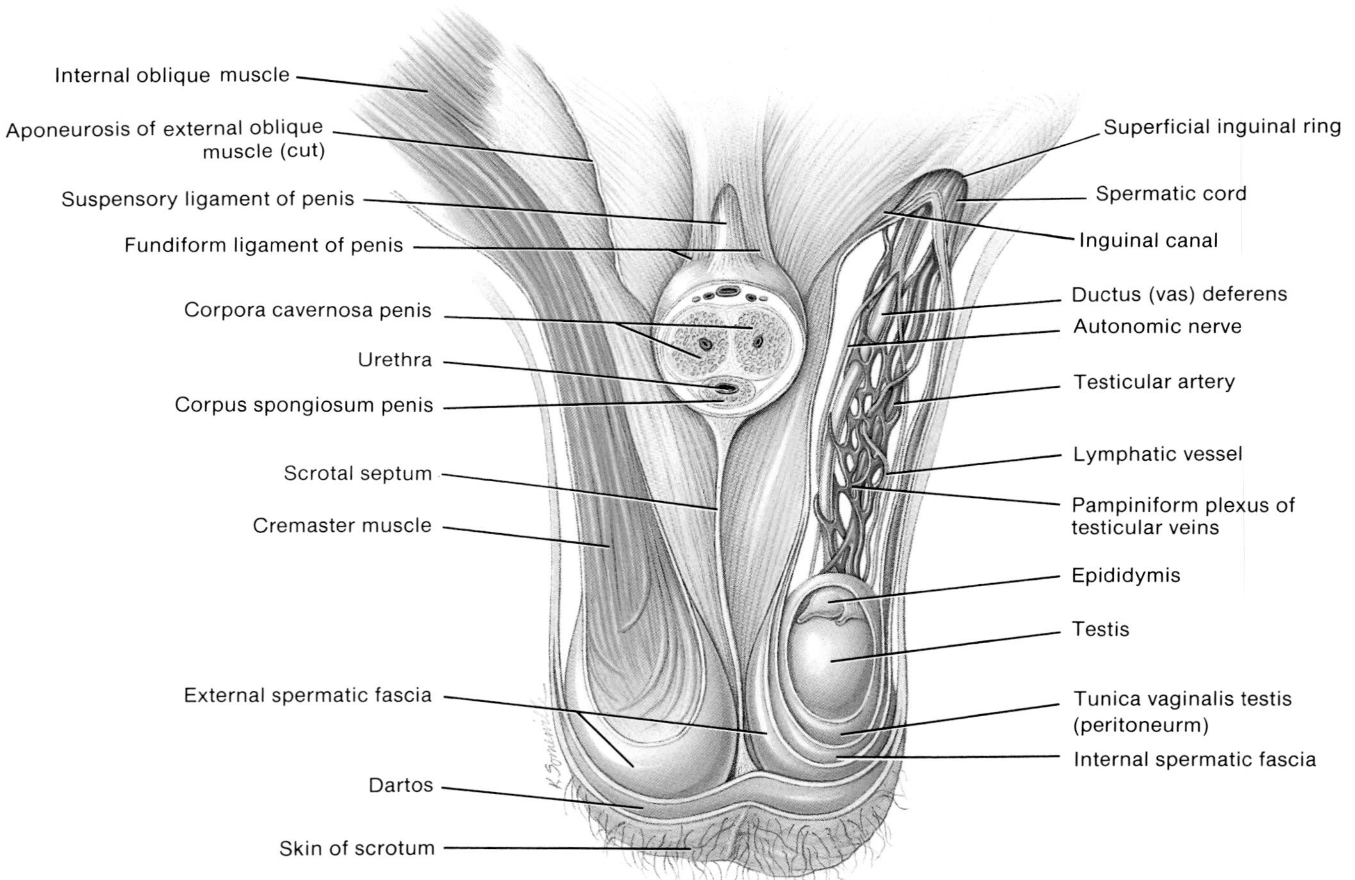

FIGURE 28-9 **Male reproductive organs in relation to surrounding structures. (a) Diagram in posterior view. (b) Photograph of parasagittal section. (Courtesy of J. A. Gosling, P. F. Harris, et al., *Atlas of Human Anatomy,* Gower Medical Publishing Ltd., 1985.)**

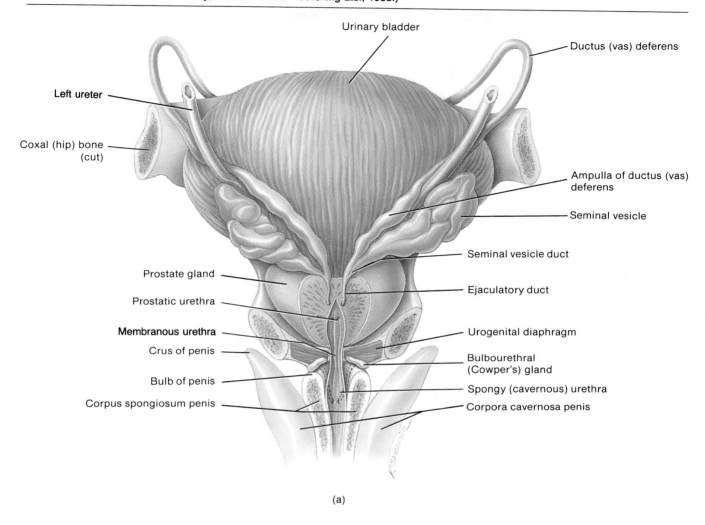

Urinary bladder
Ductus (vas) deferens
Left ureter
Coxal (hip) bone (cut)
Ampulla of ductus (vas) deferens
Seminal vesicle
Seminal vesicle duct
Prostate gland
Ejaculatory duct
Prostatic urethra
Membranous urethra
Urogenital diaphragm
Crus of penis
Bulbourethral (Cowper's) gland
Bulb of penis
Spongy (cavernous) urethra
Corpus spongiosum penis
Corpora cavernosa penis

(a)

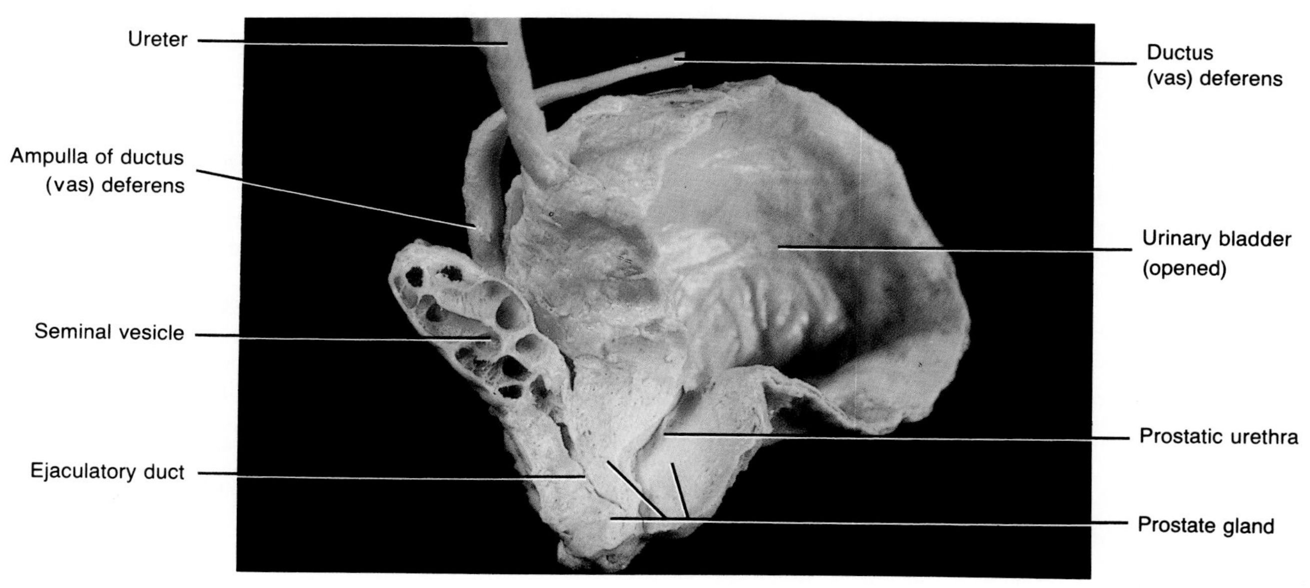

(b)

Urethra

The ***urethra*** is the terminal duct of the system, serving as a passageway for spermatozoa or urine. In the male, the urethra passes through the prostate gland, the urogenital diaphragm, and the penis. It measures about 20 cm (8 inches) in length and is subdivided into three parts (see Figures 28-1 and 28-9). The ***prostatic urethra*** is 2 to 3 cm (1 inch) long and passes through the prostate gland. It continues inferiorly, and as it passes through the urogenital diaphragm, a muscular partition between the two ischiopubic rami, it is known as the ***membranous urethra.*** The membranous portion is about 1 cm (0.5 inch) in length. As it passes through the corpus spongiosum of the penis, it is known as the ***spongy (cavernous) urethra.*** This portion is about 15 cm (6 inches) long. The spongy urethra enters the bulb of the penis and terminates at the ***external urethral orifice.*** The histology of the male urethra may be reviewed in Chapter 26.

Accessory Sex Glands

Whereas the ducts of the male reproductive system store and transport sperm cells, the ***accessory sex glands*** secrete most of the liquid portion of semen. The paired ***seminal vesicles*** (VES-i-kuls) are convoluted pouchlike structures, about 5 cm (2 inches) in length, lying posterior to and at the base of the urinary bladder in front of the rectum (Figure 28-9). They secrete an alkaline, viscous fluid, rich in the sugar fructose, and pass it into the ejaculatory duct. This secretion provides a carbohydrate (fructose) that is used as an energy source by sperm. It constitutes about 60 percent of the volume of semen. The alkaline nature of the fluid helps to neutralize acid in the female tract. This acid would inactivate and kill sperm if not neutralized.

The ***prostate*** (PROS-tāt) ***gland*** is a single, doughnut-shaped gland about the size of a chestnut (Figure 28-9). It is inferior to the urinary bladder and surrounds the superior portion of the urethra. The prostate secretes a slightly acid fluid rich in citric acid, prostatic acid phosphatase, and prostaglandins into the prostatic urethra through numerous prostatic ducts. The prostatic secretion constitutes 13 to 33 percent of the volume of semen and contributes to sperm motility and viability. The prostate gland slowly increases in size from birth to puberty, and then a rapid growth spurt occurs. The size attained by the third decade remains stable until about age 45, when enlargement may occur.

CLINICAL APPLICATION: CANCER OF THE PROSTATE GLAND

In cases of *cancer of the prostate gland, prostatic acid phosphatase* is released by the prostate gland into the blood. Elevated levels of prostatic acid phosphatase typically indicate prostate cancer and that the tumor has spread to other parts of the body, especially bone. Another test measures the level of *prostate-specific antigen* in blood. This substance is a protease produced only by prostate epithelial cells, normal and malignant. The amount of prostate-specific antigen is also elevated in cancer of the prostate gland.

The paired ***bulbourethral*** (bul'-bō-yoo-RĒ-thral), or ***Cowper's, glands*** are about the size of peas. They are located beneath the prostate on either side of the membranous urethra within the urogenital diaphragm (Figure 28-9). The bulbourethral glands secrete an alkaline substance that protects sperm by neutralizing the acid environment of the urethra and mucus that lubricates the end of the penis during sexual intercourse. Their ducts open into the spongy urethra.

Semen (Seminal Fluid)

Semen (seminal fluid) is a mixture of sperm and the secretions of the seminal vesicles, prostate gland, and bulbourethral glands. The average volume of semen for each ejaculation is 2.5 to 5 ml, and the average range of spermatozoa ejaculated is 50 to 150 million/ml. When the number of spermatozoa falls below 20 million/ml, the male is likely to be infertile. The very large number is required because only a small percentage eventually reach the ovum. And although only a single spermatozoon fertilizes an ovum, fertilization seems to require the combined action at the ovum of a larger number of them. The intercellular material of the cells covering the ovum presents a barrier to the sperm. This barrier is digested by the hyaluronidase and proteinases secreted by the acrosomes of sperm, resulting in the dispersion of the cells surrounding the ovum. A single sperm does not produce enough of these enzymes to dissolve the barrier. A passageway through which one sperm may enter can be created only by the action of many sperm cells.

Semen has a slightly alkaline pH of 7.20 to 7.60. The prostatic secretion gives semen a milky appearance, and fluids from the seminal vesicles and bulbourethral glands give it a mucoid consistency. Semen provides spermatozoa with a transportation medium and nutrients. It neutralizes the acid environment of the male urethra and the female vagina. It also contains enzymes that activate sperm after ejaculation.

Semen contains an antibiotic, ***seminalplasmin,*** that has the ability to destroy a number of bacteria. Since both semen and the lower female reproductive tract contain bacteria, the antibiotic activity of seminalplasmin may keep these bacteria under control to help ensure fertilization.

Once ejaculated into the vagina, liquid semen coagulates rapidly because of a clotting enzyme produced by the prostate gland that acts on a substance produced by the seminal vesicle. This clot liquefies in about 5 to 20 minutes because of another enzyme produced by the prostate gland. Abnormal or delayed liquefaction of coagulated semen may cause complete or partial immobilization of spermatozoa, thus inhibiting their movement through the cervix of the uterus.

MEDICAL TEST

Semen analysis (sperm count, male fertility test)

Diagnostic Value: To determine if sterility is related to sperm production and, following vasectomy, to determine if any sperm are present in semen. The test is also used for medicolegal purposes to detect semen on the body or clothing of a suspected rape victim.

Procedure: After a semen sample is collected, it is analyzed within two hours in a physician's office or laboratory. Among the criteria analyzed are the following:

1. ***Volume.*** A low volume might suggest an anatomical or functional defect or inflammation.
2. ***Motility.*** This refers to the percentage of motile spermatozoa (40 to 60 percent) and quality of movement (forward and progressive).
3. ***Count.*** Sperm counts below 20 million/ml could indicate sterility.
4. ***Liquefaction.*** Delayed liquefaction of more than two hours suggests inflammation of accessory sex glands or enzyme defects in the secretory products of the glands.
5. ***Morphology.*** No more than about 35 percent of spermatozoa should have abnormal morphology.
6. ***Autoagglutination.*** Agglutination does not occur normally.
7. ***pH.*** A rise in pH above its normally slightly alkaline state could indicate prostatitis.
8. ***Fructose.*** This sugar is present in a normal ejaculate. Its abscence indicates obstruction or congenital absence of the ejaculatory ducts or seminal vesicles.

A normal semen analysis does not guarantee fertility; the absence of spermatozoa or zero motility is the only definitive sign of sterility.

Penis

The ***penis*** is used to introduce spermatozoa into the vagina (Figure 28-10). The penis is cylindrical in shape and consists of a body, root, and glans penis. The ***body*** of the penis is composed of three cylindrical masses of tissue, each bound by fibrous tissue **(*tunica albuginea*).** The two dorsolateral masses are called the ***corpora cavernosa penis.*** The smaller midventral mass, the ***corpus spongiosum penis,*** contains the spongy urethra. All three masses are enclosed by fascia and skin and consist of erectile tissue permeated by blood sinuses. Under the influence of sexual stimulation (visual, tactile, auditory, olfactory, and imaginative), the arteries supplying the penis dilate, and large quantities of blood enter the blood sinuses. Expansion of these spaces compresses the veins draining the penis, so most entering blood is retained. These vascular changes result in an ***erection,*** a parasympathetic reflex. The penis returns to its flaccid state when the arteries constrict and pressure on the veins is relieved. Details of erection are presented later in the chapter. During ejaculation, which is a sympathetic reflex, the smooth muscle sphincter at the base of the urinary bladder is closed. Thus, urine is not expelled during ejaculation, and semen does not enter the urinary bladder.

The ***root*** of the penis is the attached portion and consists of the ***bulb of the penis,*** the expanded portion of the base of the corpus spongiosum penis, and the ***crura*** (sing., ***crus***) ***of the penis,*** the separated and tapered portion of the corpora cavernosa penis. The bulb of the penis is attached to the inferior surface of the urogenital diaphragm and enclosed by the bulbocavernosus muscle. Each crus of the penis is attached to the ischial and pubic rami and surrounded by the ischiocavernosus muscle.

The distal end of the corpus spongiosum penis is a slightly enlarged region called the ***glans penis,*** which means "shaped like an acorn." The margin of the glans penis is referred to as the ***corona.*** The distal urethra enlarges within the glans penis and forms a terminal slitlike opening, the ***external urethral orifice* (*meatus*).** Covering the glans is the loosely fitting ***prepuce*** (PRE-pyoos), or ***foreskin.***

CLINICAL APPLICATION: CIRCUMCISION

Circumcision (*circumcido* = to cut around) is a surgical procedure in which part or all of the prepuce is removed. It is usually performed in the delivery room or by the third or fourth day after birth (or on the eighth day as part of a Jewish religious rite). There is no consensus among physicians regarding circumcision. Some physicians are opposed to circumcision, except for religious reasons, and feel that there is no medical justification for performing it. Other physicians feel that circumcised boys have a far lower risk of urinary tract infections, protection against penile cancer, and possibly a lower risk for many sexually transmitted diseases.

FIGURE 28-10 Internal structure of the penis. (a) Coronal section. (b) Cross section. The insert shows details of the skin and fascia.

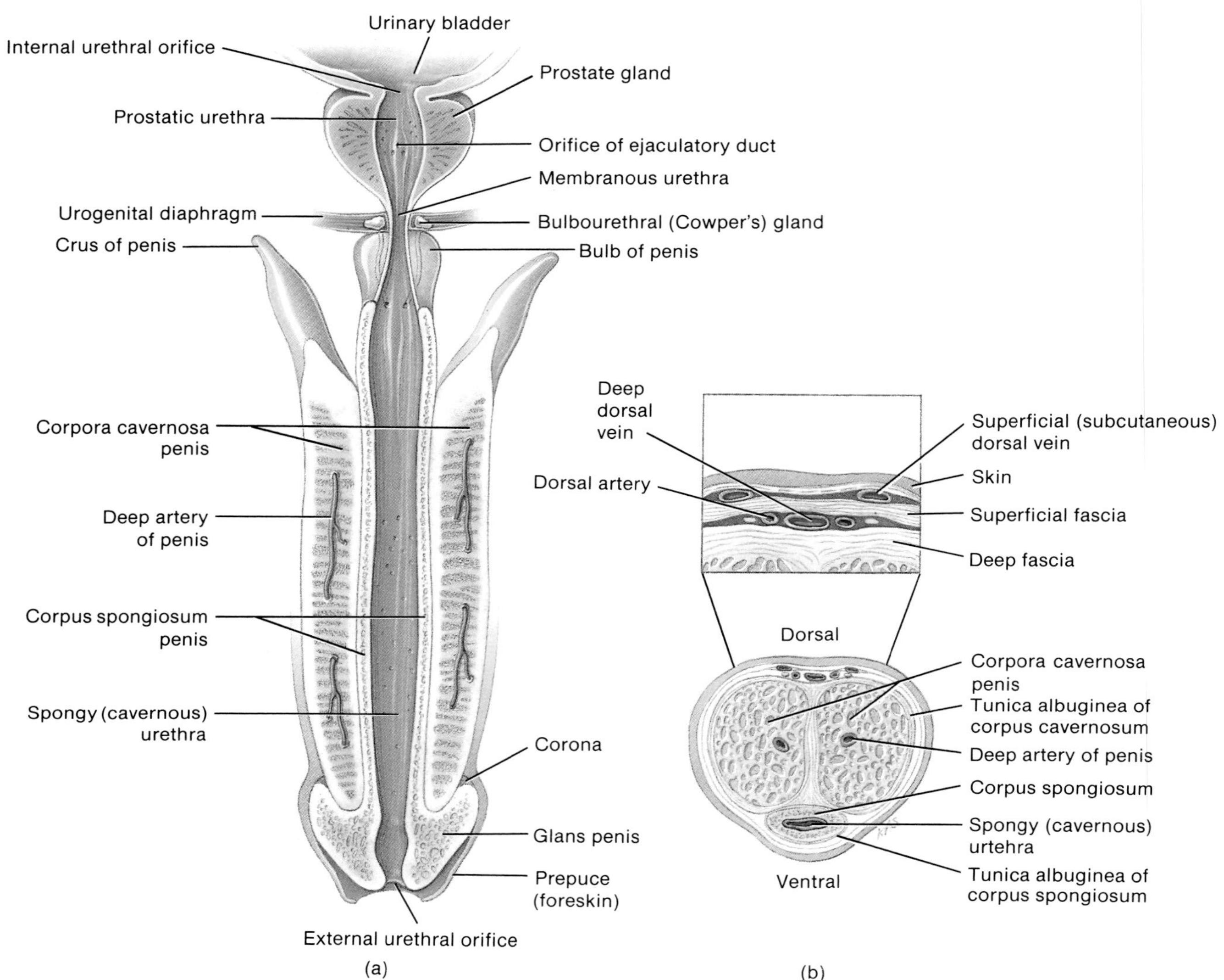

FEMALE REPRODUCTIVE SYSTEM: STRUCTURE AND PHYSIOLOGY

The female organs of reproduction include the ovaries, which produce secondary oocytes (cells that develop into mature ova or eggs following fertilization) and the female sex hormones progesterone, estrogens, and relaxin; the uterine (Fallopian) tubes, which transport ova to the uterus (womb); the vagina; and external organs that constitute the vulva, or pudendum (Figure 28-11). The mammary glands also are considered part of the female reproductive system.

The specialized branch of medicine that deals with the diagnosis and treatment of diseases of the female reproductive system is called ***gynecology*** (gī'-ne-KOL-ō-jē; *gyneco* = woman).

Ovaries

The ***ovaries*** (*ovarium* = egg receptacle), or female gonads, are paired glands resembling unshelled almonds in size and shape. They are homologous to the testes. (***Homologous*** means that two organs correspond in structure, position, and origin.) The ovaries descend to the brim of the pelvis during the third month of development. They are positioned in the upper pelvic cavity, one on

FIGURE 28-11 **Female organs of reproduction and surrounding structures seen in sagittal section.**

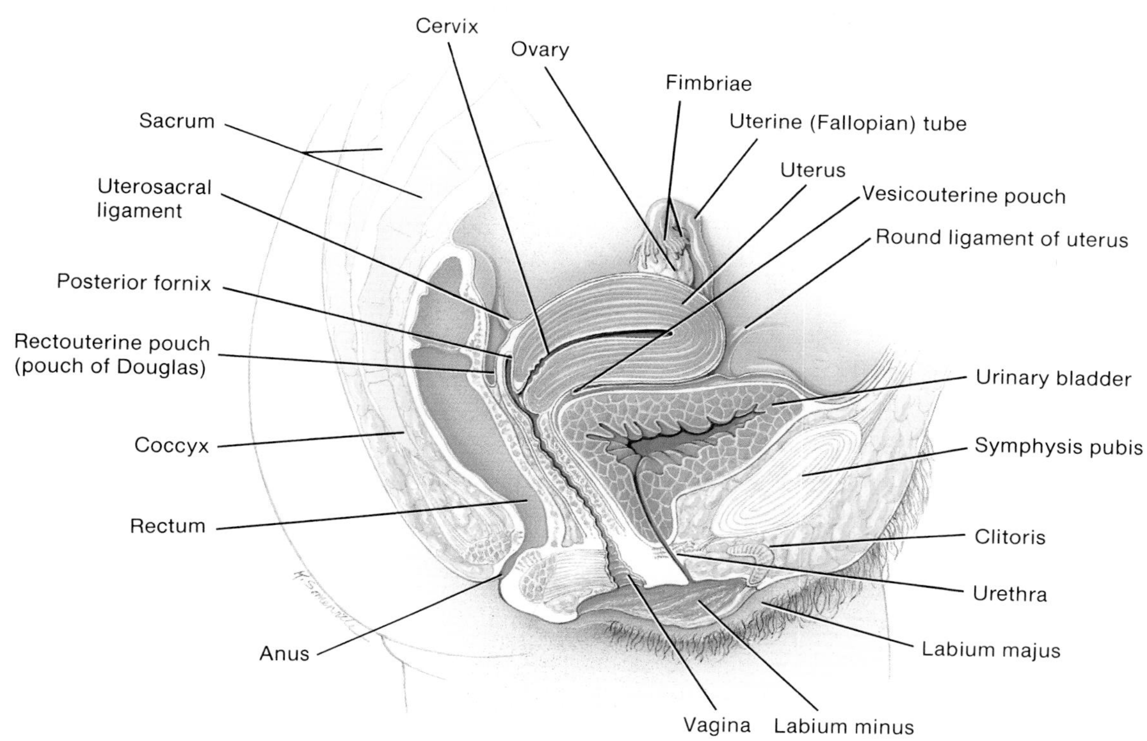

each side of the uterus. The ovaries are maintained in position by a series of ligaments (Figure 28-12). They are attached to the broad ligament of the uterus, which is itself part of the parietal peritoneum, by a double-layered fold of peritoneum called the ***mesovarium.*** The ovaries are anchored to the uterus by the ***ovarian ligament*** and are attached to the pelvic wall by the ***suspensory ligament.*** Each ovary also contains a ***hilus,*** the point of entrance for blood vessels and nerves and along which the mesovarium is attached.

The microscope reveals that each ovary consists of the following parts (Figure 28-13).

1. *Germinal epithelium.* A layer of simple epithelium (low cuboidal or squamous) that covers the free surface of the ovary and is continuous with the mesothelium that covers the mesovarium. The term *germinal epithelium* is a misnomer since it does not give rise to ova, although at one time it was believed that it did. It is now known that the cells that give rise to ova arise from the endoderm of the yolk sac and migrate to the ovaries.
2. *Tunica albuginea.* A capsule of collagenous connective tissue immediately deep to the germinal epithelium.
3. *Stroma.* A region of connective tissue deep to the tunica albuginea and composed of an outer, dense layer called the **cortex** and an inner, loose layer known as the **medulla.** The cortex contains ovarian follicles.
4. *Ovarian follicles.* Oocytes (immature ova) and their surrounding tissues in various stages of development.
5. *Vesicular ovarian (Graafian) follicle.* A relatively large, fluid-filled follicle containing an immature ovum and its surrounding tissues. The follicle secretes hormones called estrogens.
6. *Corpus luteum.* Glandular body that develops from a vesicular ovarian follicle after extrusion of a secondary oocyte (potential mature ovum), a process known as ovulation. The corpus luteum produces the hormones progesterone, estrogens, relaxin, and inhibin.

The ovaries produce secondary oocytes, discharge secondary oocytes (ovulation), and secrete the sex hormones progesterone, estrogens, relaxin, and inhibin.

Oogenesis

The formation of haploid (n) ova in the ovary involves several phases, including meiosis, and is referred to as ***oogenesis*** (ō'-ō-JEN-e-sis). With some exceptions, oogenesis occurs in essentially the same manner as spermatogenesis.

▪ Reduction Division (Meiosis I)

During early fetal development, primordial (primitive) germ cells migrate from the endoderm of the yolk sac to the ovaries. There,

FIGURE 28-12 Uterus and associated structures seen in posterior view. (a) Diagram. The left side of the figure has been sectioned to show internal structures. (b) Photograph. Part of the posterior wall of the uterus has been removed. (Courtesy of J. A. Gosling, P. F. Harris, et al., *Atlas of Human Anatomy,* Gower Medical Publishing Ltd., 1985.)

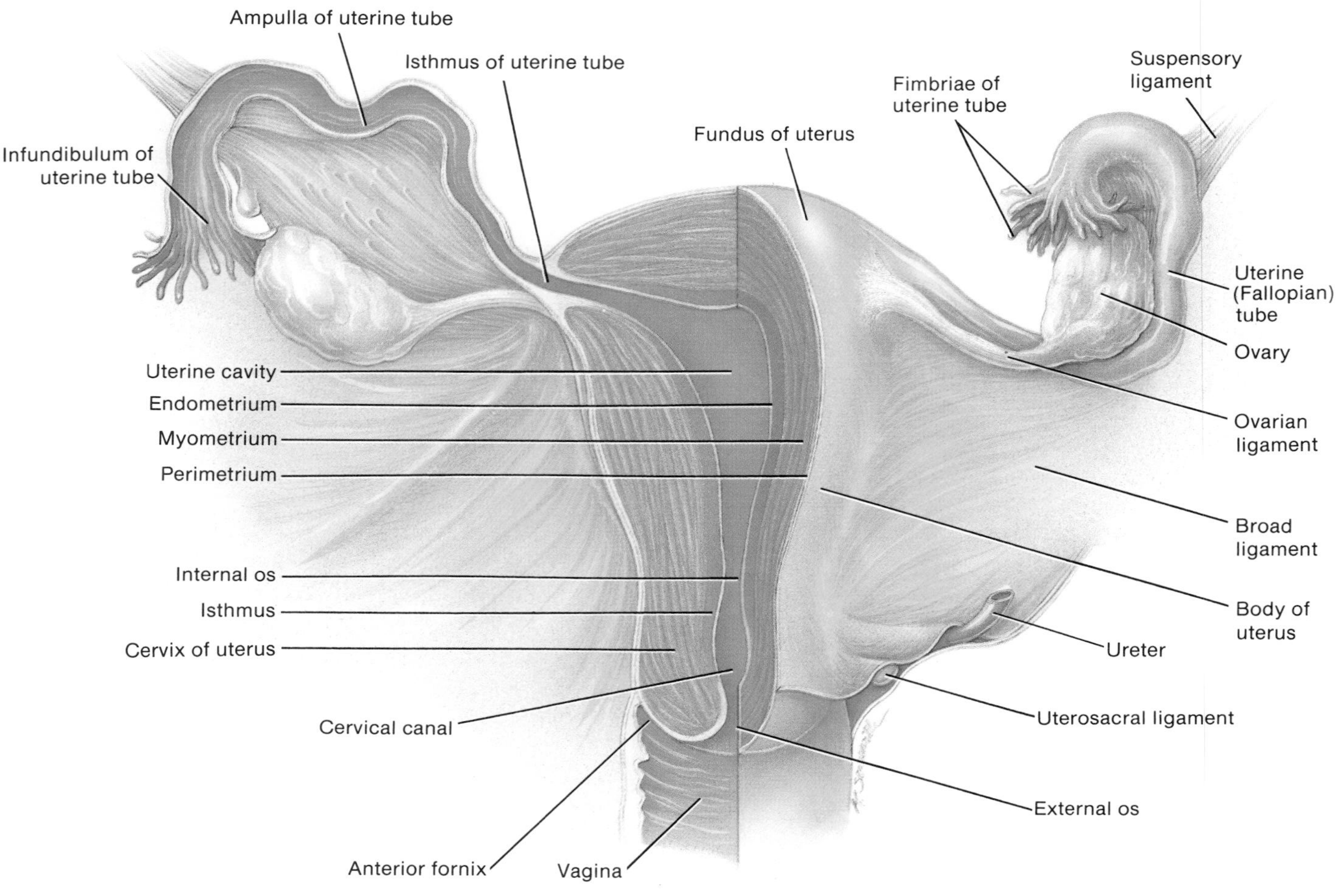

Ampulla of uterine tube
Isthmus of uterine tube
Infundibulum of uterine tube
Fundus of uterus
Fimbriae of uterine tube
Suspensory ligament
Uterine (Fallopian) tube
Ovary
Uterine cavity
Endometrium
Myometrium
Perimetrium
Ovarian ligament
Broad ligament
Internal os
Isthmus
Cervix of uterus
Body of uterus
Ureter
Uterosacral ligament
Cervical canal
External os
Anterior fornix
Vagina

(a)

Isthmus of uterine tube
Uterine (Fallopian) tube
Infundibulum
Ovary
Fundus of uterus
Ovarian ligament
Uterine cavity
Broad ligament
Body of uterus
Cervical canal
Vagina

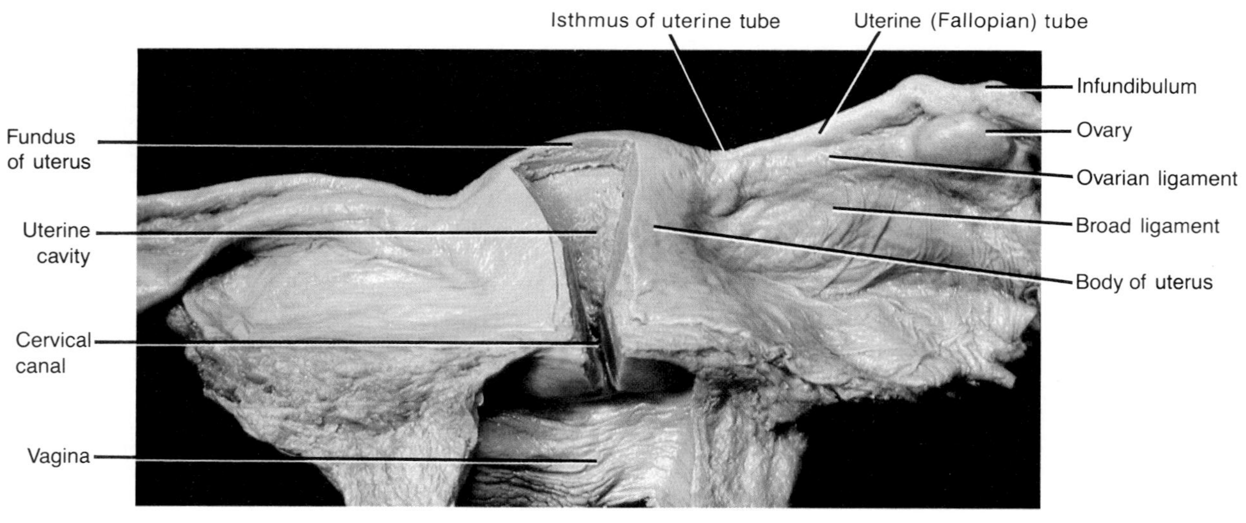

(b)

FIGURE 28-12 (*Continued*) **(c) Photograph of uterus and vagina in sagittal section. (Courtesy of J. A. Gosling, P. F. Harris, et al., *Atlas of Human Anatomy,* Gower Medical Publishing Ltd., 1985.)**

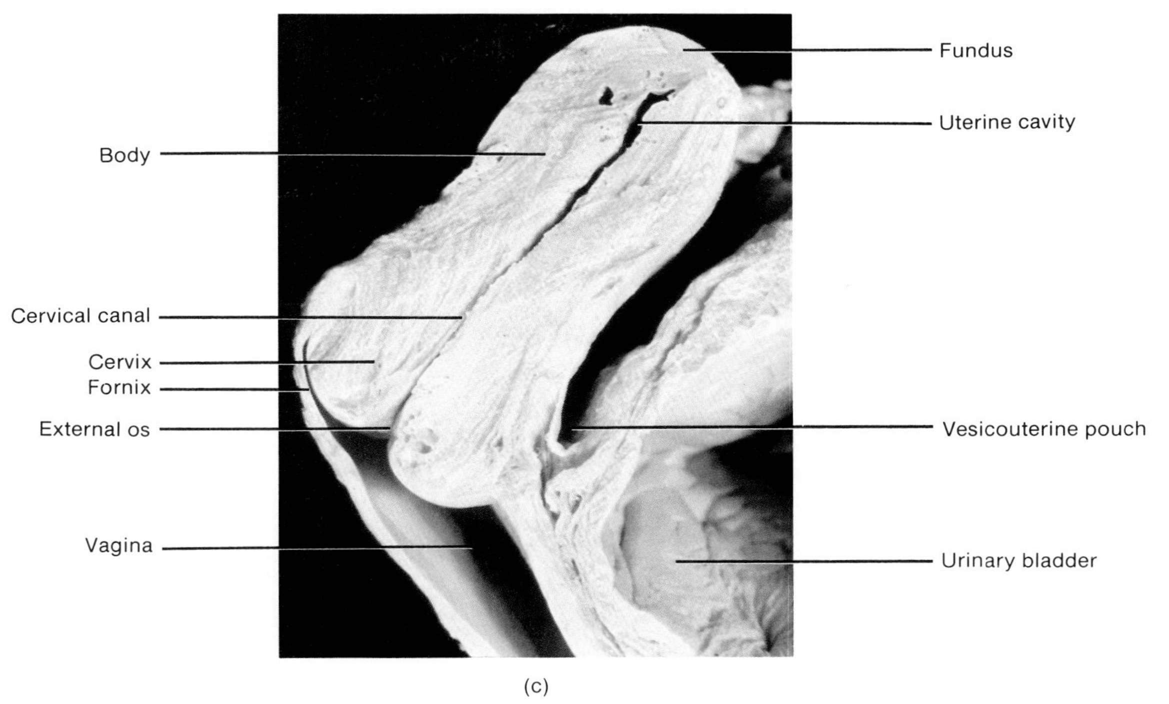

FIGURE 28-13 **Histology of the ovary. (a) Diagram of the parts of an ovary seen in sectional view. The arrows indicate the sequence of developmental stages that occurs as part of the ovarian cycle.**

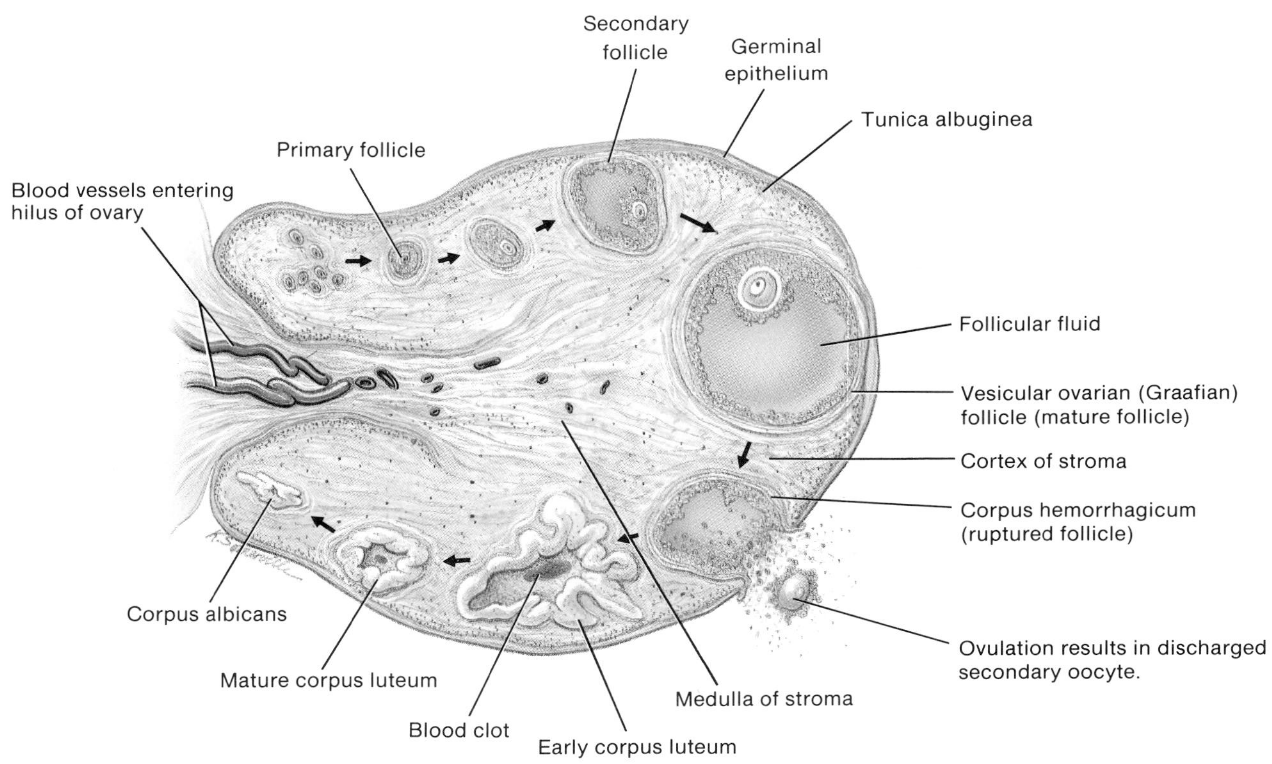

FIGURE 28-13 (*Continued*) (b) Photomicrograph of the cortex of an ovary at a magnification of 60×. (c) Photomicrograph of an enlarged aspect of a secondary follicle at a magnification of 160×. The theca interna and theca externa are connective tissue coverings around the secondary follicle. (Photomicrographs copyright © 1983 by Michael H. Ross. Used by permission.)

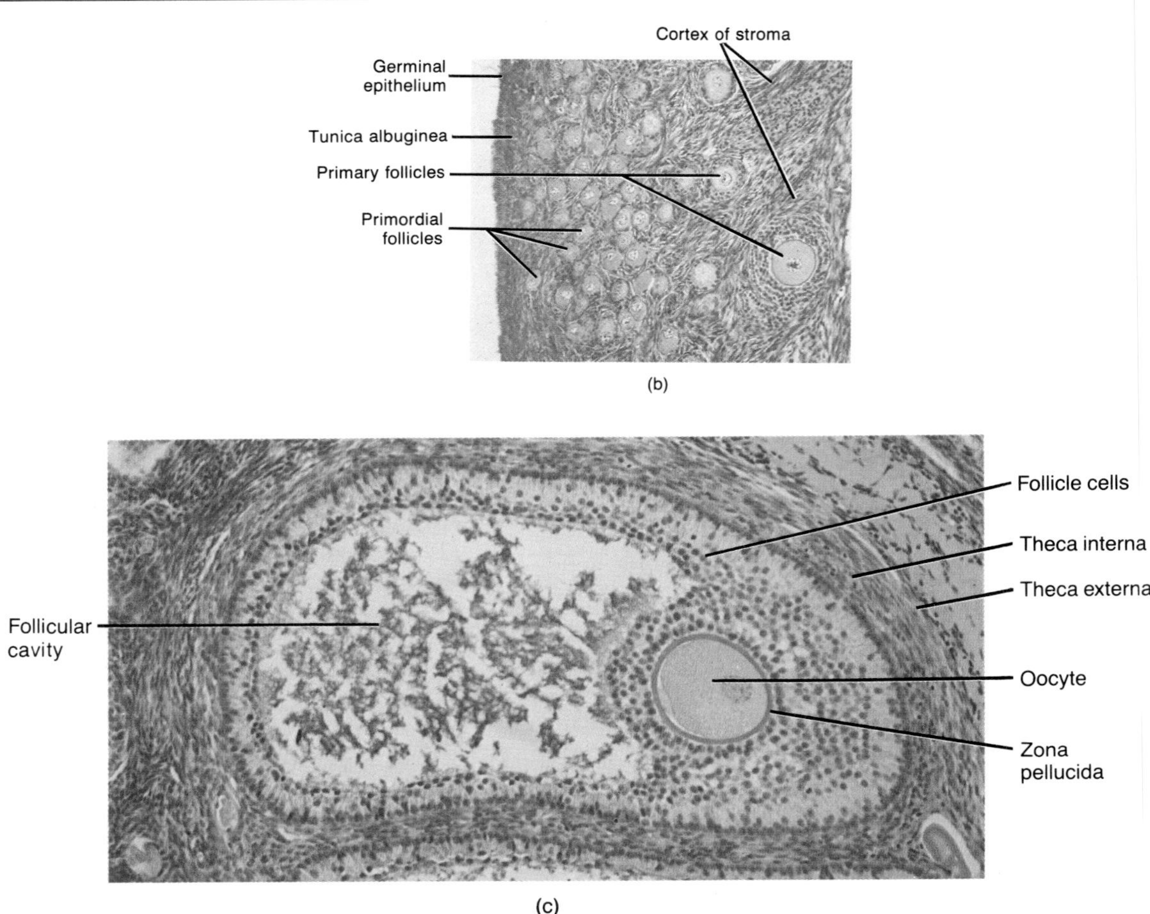

germ cells differentiate within the ovaries into ***oogonia*** (ō'-o-GŌ-nē-a; *oo* = egg), cells that can give rise to cells that develop into ova (Figure 28-14). Oogonia are diploid (2*n*) cells that divide mitotically to produce a large population of cells. At about the third month of prenatal development, oogonia divide and develop into larger diploid (2*n*) cells called ***primary oocytes*** (Ō-o-sītz). These cells enter prophase of reduction division (meiosis I) but do not complete it until after the female reaches puberty. Each primary follicle is surrounded by a single layer of flattened epithelial cells (follicular), and the entire structure is called a primary follicle. Primary follicles do not begin further development until they are stimulated by follicle-stimulating hormone (FSH) from the anterior pituitary gland, which has responded to gonadotropin releasing hormone (GnRH) from the hypothalamus.

Starting with puberty, several primary follicles respond each month to the rising level of FSH. As the preovulatory phase of the menstrual cycle proceeds and luteinizing hormone (LH) is secreted from the anterior pituitary, one of the primary follicles reaches a stage in which meiosis resumes and the diploid primary oocyte completes reduction division (meiosis I). Synapsis, tetrad formation, and crossing-over occur, and two cells of unequal size, both with 23 chromosomes (*n*) of two chromatids each, are produced. The smaller cell, called the ***first polar body,*** is essentially a packet of discarded nuclear material. The larger cell, known as the ***secondary oocyte,*** receives most of the cytoplasm. Each secondary oocyte is surrounded by several layers of first cuboidal then columnar epithelial cells, and the entire structure is called a secondary (growing) follicle. Once a secondary oocyte is formed, it proceeds to the metaphase of equatorial division (meiosis II) and then stops at this stage. The equatorial division (meiosis II) is completed following ovulation and fertilization.

FIGURE 28-14 **Oogenesis.**

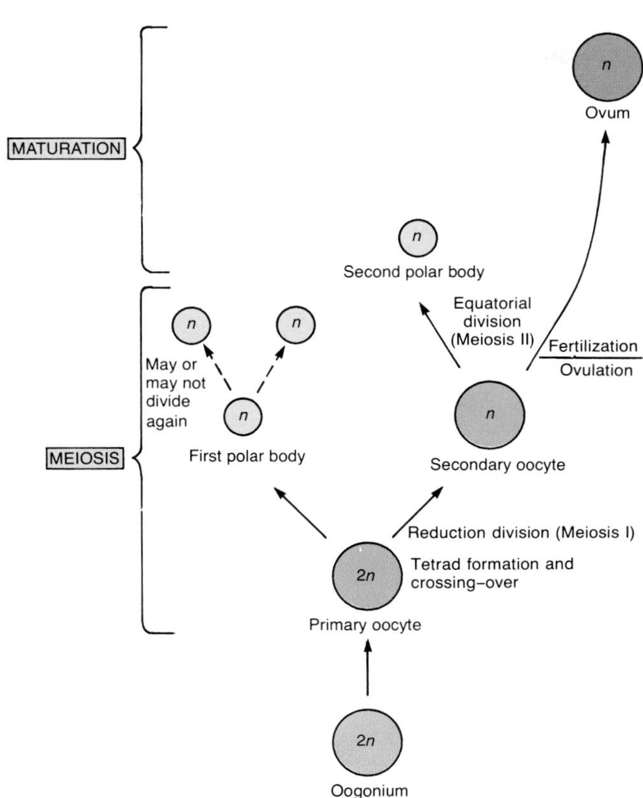

▪ **Equatorial Division (Meiosis II)** At ovulation, the secondary oocyte with its polar body and some surrounding supporting cells is discharged. The discharged secondary oocyte enters the uterine (Fallopian) tube, and if spermatozoa are present and fertilization occurs, the second division, the equatorial division (meiosis II), is completed.

▪ **Maturation** The secondary oocyte produces two cells of unequal size, both of them haploid (n). The larger cell eventually develops into an ***ovum,*** or mature egg; the smaller is the ***second polar body.***

The first polar body may undergo another division to produce two polar bodies. If it does, meiosis of the primary oocyte results in a single haploid (n), secondary oocyte and three haploid (n) polar bodies. In any event, all polar bodies disintegrate. Thus, each oogonium produces a single secondary oocyte, whereas each spermatocyte produces four spermatozoa.

Uterine (Fallopian) Tubes

The female body contains two ***uterine (Fallopian) tubes,*** also called ***oviducts,*** that extend laterally from the uterus and transport the ova from the ovaries to the uterus (see Figure 28-12). Measuring about 10 cm (4 inches) long, the tubes are positioned between the folds of the broad ligaments of the uterus. The funnel-shaped open distal end of each tube, called the ***infundibulum,*** lies close to the ovary and is surrounded by a fringe of fingerlike projections called ***fimbriae*** (FIM-brē-ē). One fimbria is attached to the lateral end of the ovary. As you will see later, fimbriae help to carry a secondary oocyte into the uterine tube following ovulation. From the infundibulum, the uterine tube extends medially and inferiorly and attaches to the superior lateral angle of the uterus. The ***ampulla*** of the uterine tube is the widest, longest portion, making up about two-thirds of its length. The ***isthmus*** of the uterine tube is the short, narrow, thick-walled portion that joins the uterus.

Histologically, the uterine tubes are composed of three layers. The internal ***mucosa*** contains ciliated columnar cells and secretory cells, which are believed to aid the movement and nutrition of the ovum (Figure 28-15). The middle layer, the ***muscularis,*** is composed of a thick, circular region of smooth muscle and an outer, thin, longitudinal region of smooth muscle. Peristaltic contractions of the muscularis and the ciliated action of the mucosa help move the ovum down into the uterus. The outer layer of the uterine tubes is a serous membrane, the ***serosa.***

About once a month a vesicular ovarian (Graafian) follicle (developed from a secondary follicle) ruptures, re-

FIGURE 28-15 **Histology of the uterine (Fallopian) tube. Scanning electron micrograph shows the ciliated epithelium at a magnification of 5.650 x. (Courtesy of Photoresearchers, Inc. © D. W. Fawcett/Gaddum-Rosse.)**

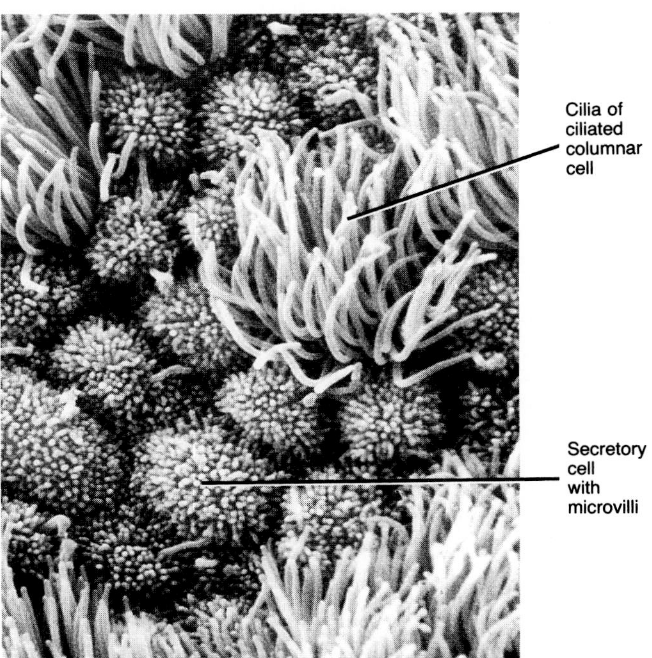

leasing a secondary oocyte, a process called ***ovulation.*** The oocyte is swept into the uterine tube by the ciliary action of the epithelium of the infundibulum, which becomes associated with the surface of the most mature vesicular ovarian follicle just before ovulation occurs. The oocyte is then moved along the tube by ciliary action that is supplemented by the peristaltic contractions of the muscularis. If the oocyte is fertilized by a sperm cell, it usually occurs in the ampulla of the uterine tube. Fertilization may occur at any time up to about 24 hours following ovulation. With fertilization, the secondary oocyte completes meiosis II in which the oocyte produces a larger cell that develops into an ovum (mature egg) and a smaller second polar body. The fertilized ovum is now referred to as a zygote. Following a number of cell divisions, it descends into the uterus within 7 days and at this point is called a blastocyst. An unfertilized secondary oocyte disintegrates.

Uterus

Part of the pathway for sperm to reach the uterine (Fallopian) tubes is the ***uterus.*** It is also the site of menstruation, implantation of a fertilized ovum, development of the fetus during pregnancy, and labor. Situated between the urinary bladder and the rectum, the uterus is shaped like an inverted pear (see Figures 28-11 and 28-12). Before the first pregnancy, the adult uterus measures approximately 7.5 cm (3 inches) long, 5 cm (2 inches) wide, and 2.5 cm (1 inch) thick.

Anatomical subdivisions of the uterus include the dome-shaped portion above the uterine tubes called the ***fundus,*** the major tapering central portion called the ***body,*** and the inferior narrow portion opening into the vagina called the ***cervix.*** The secretory cells of the mucosa of the cervix produce a secretion called ***cervical mucus,*** a mixture of water, glycoprotein, serum-type proteins, lipids, enzymes, and inorganic salts. Females of reproductive age secrete 20 to 60 ml of mucus per day. The uterine cervix and its secretions are important in reproduction. Cervical mucus is less viscous and thus more receptive to spermatozoa at or near the time of ovulation. At other times, the mucus is more viscous and forms a plug that impedes sperm penetration. The mucus also supplements the energy requirements of spermatozoa. Both the cervix and mucus serve as a sperm reservoir, protect spermatozoa from the hostile environment of the vagina, protect spermatozoa from phagocytes, and may assume a role in capacitation—a functional change that spermatozoa undergo in the female reproductive tract in order to fertilize a secondary oocyte. Between the body and the cervix is the ***isthmus*** (IS-mus), a constricted region about 1 cm (0.5 inch) long. The interior of the body of the uterus is called the ***uterine cavity,*** and the interior of the narrow cervix is called the ***cervical canal.*** The junction of the isthmus with the cervical canal is the ***internal os.*** The ***external os*** is the place where the cervix opens into the vagina.

Normally, the uterus is flexed between the uterine body and the cervix. This is called ***anteflexion.*** In this position, the body of the uterus projects anteriorly and slightly superiorly over the urinary bladder, and the cervix projects inferiorly and posteriorly and enters the anterior wall of the vagina at nearly a right angle. Several structures that are either extensions of the parietal peritoneum or fibromuscular cords, referred to as ligaments, maintain the position of the uterus. The paired ***broad ligaments*** are double folds of parietal peritoneum attaching the uterus to either side of the pelvic cavity. Uterine blood vessels and nerves pass through the broad ligaments. The paired ***uterosacral ligaments,*** also peritoneal extensions, lie on either side of the rectum and connect the uterus to the sacrum. The ***cardinal (lateral cervical) ligaments*** extend below the bases of the broad ligaments between the pelvic wall and the cervix and vagina. These ligaments contain smooth muscle, uterine blood vessels, and nerves and are the chief ligaments that maintain the position of the uterus and help keep it from dropping down into the vagina. The ***round ligaments*** are bands of fibrous connective tissue between the layers of the broad ligament. They extend from a point on the uterus just below the uterine (Fallopian) tubes to a portion of the labia majora of the external genitalia. Although the ligaments normally maintain the anteflexed position of the uterus, they also afford the uterine body some movement. As a result, the uterus may become malpositioned. A posterior tilting of the uterus is called ***retroflexion.***

Histologically, the uterus consists of three layers of tissue (Figure 28-16). The outer layer, the ***perimetrium (serosa)*** is part of the visceral peritoneum. Laterally, it becomes the broad ligament. Anteriorly, it is reflected over the urinary bladder and forms a shallow pouch, the ***vesicouterine*** (ves'-i-kō-YOO-ter-in) ***pouch*** (see Figure 28-12c). Posteriorly, it is reflected onto the rectum and forms a deep pouch, the ***rectouterine*** (rek-tō-YOO-ter-in) ***pouch (pouch of Douglas)***—the lowest point in the pelvic cavity.

The middle layer of the uterus, the ***myometrium,*** forms the bulk of the uterine wall. This layer consists of three layers of smooth muscle fibers and is thickest in the fundus and thinnest in the cervix. During childbirth, coordinated contractions of the muscles help expel the fetus from the body of the uterus.

The inner layer of the uterus, the ***endometrium,*** is very vascular and is composed of (1) a surface layer of simple columnar epithelium (ciliated and secretory cells), (2) uterine (endometrial) glands that develop as invaginations of the surface epithelium, and (3) endometrial stroma, a very thick region of lamina propria (connective tissue). The endometrium is divided into two layers. The ***stratum functionalis,*** the layer closer to the uterine cavity, is shed during menstruation. The second layer, the ***stratum basalis*** (bā-SAL-is), is permanent. Its function is to produce a new functionalis following menstruation.

Blood is supplied to the uterus by branches of the internal iliac artery called ***uterine arteries*** (Figure 28-17). Branches called ***arcuate arteries*** are arranged in

FIGURE 28-16 **Histology of the uterus. Photomicograph of a portion of the uterine wall at a magnification of 25×. (Courtesy of Andrew Kuntzman.)**

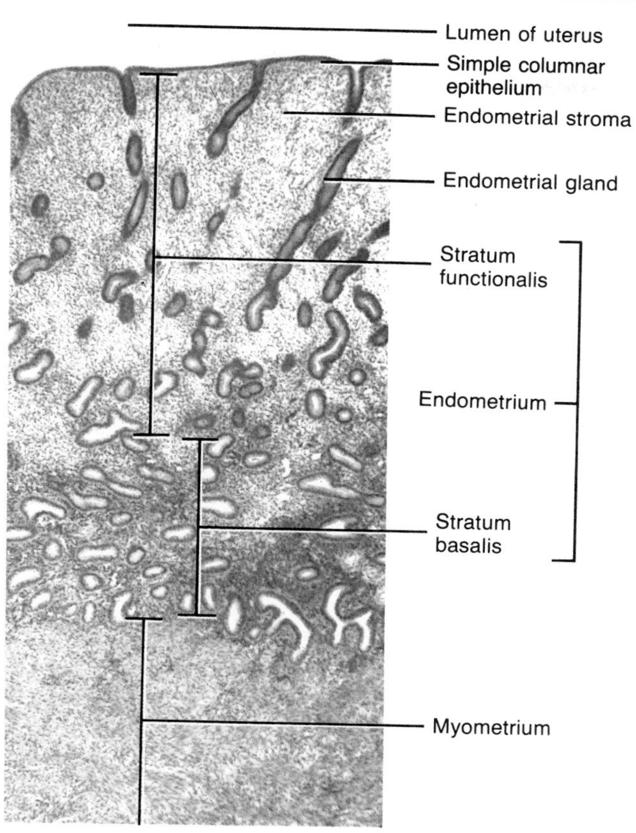

a circular fashion in the myometrium and give off ***radial arteries*** that penetrate deeply into the myometrium. Just before the branches enter the endometrium, they divide into two kinds of arterioles. The ***straight arterioles*** terminate in the basalis and supply it with the materials necessary to regenerate the functionalis. The ***spiral arterioles*** penetrate the functionalis and change markedly during the menstrual cycle. The uterus is drained by the ***uterine veins.***

Early diagnosis of cancer of the uterus is accomplished by a Papanicolaou test (Pap smear). If a Pap smear is abnormal, the next test usually performed is colposcopy. If the diagnosis is still in doubt, a cone biopsy is indicated. A dilatation and curettage (D and C) may also be performed. If cancer has spread beyond the uterine lining, treatment may involve removal of the uterus, called hysterectomy (*hyster* = uterus), or radiation treatment.

Following are descriptions of several tests used to diagnose or treat cancer of the uterus.

MEDICAL TESTS

Papanicolaou (pap'-a-NIK-ō-la-oo) **test, or Pap smear**

Diagnostic Value: To detect cancerous and precancerous cells from the cervix, uterus, or vagina

FIGURE 28-17 **Blood supply of the uterus. The insert shows details of the blood vessels of the endometrium.**

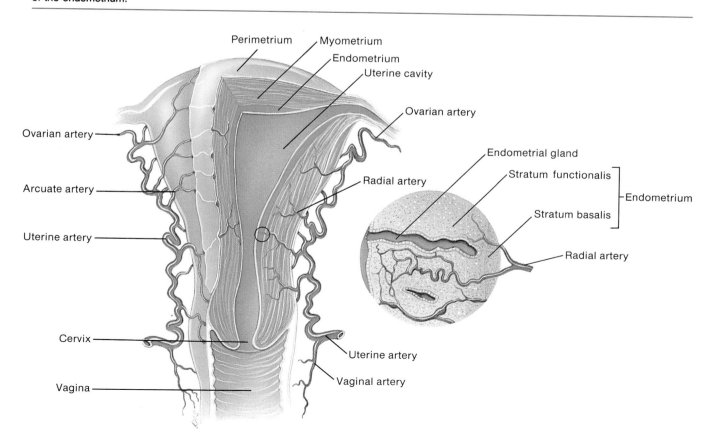

and to monitor responses to radiation and/or chemotherapy.

Procedure: During a pelvic examination, cells are obtained with a cervical spatula from the wall of the vagina below the uterine cervix, from around the opening of the cervix, and from the inside of the cervix. The secretions are spread separately on microscope slides, sprayed with a fixative, and then sent to a laboratory for microscopic analysis.

Colposcopy (kol-POS-kō-pē)

Diagnostic Value: The procedure is often the first test done if an individual has had an abnormal Pap smear. Together with biopsy, it can help determine if cancer is present.

Procedure: A ***colposcope*** is a magnifying device similar to a low-power microscope that can magnify the vaginal and cervical mucosa 10 to 40 times. After a speculum is inserted into the vagina and the cervix is swabbed with an acetic acid solution to remove mucus and enhance the appearance of the mucosa, the colposcope is positioned at the entrance of the vagina to examine the vaginal and cervical mucosa.

Cone biopsy

Diagnostic Value: To remove a tissue sample to help determine the presence and/or extent of cancer of the uterus.

Procedure: Under anesthesia, a speculum is inserted into the vagina and an inverted, cone-shaped specimen is excised from around the cervical os. Then the tissue is examined microscopically for evidence of cancer.

Dilatation and curettage (ku-re-TAZH; *curette* = scraper), or **D and C**

Diagnostic Value: To help determine the presence of cancer of the uterus, to diagnose and treat the cause of a change in menstrual bleeding pattern, and to remove any tissue following an abortion or miscarriage.

Procedure: Following general or spinal anesthesia, a series of dilators of progressive size are used to dilate the cervix. This permits the insertion of the curette into the uterine cavity, and the endometrium can then be scraped away with this spoon-shaped instrument. Samples of tissue are sent to a laboratory for examination.

Endocrine Relations: Menstrual and Ovarian Cycles

The principal events of the menstrual cycle can be correlated with those of the ovarian cycle and changes in the endometrium. All are hormonally controlled events.

The ***menstrual cycle*** is a series of changes in the endometrium of a nonpregnant female. Each month, the endometrium is prepared to receive an already fertilized ovum that eventually normally develops into an embryo and then into a fetus until delivery. If no fertilization occurs, the stratum functionalis portion of the endometrium is shed. The ***ovarian cycle*** is a monthly series of events associated with the maturation of an ovum.

Hormonal Control

The menstrual cycle, ovarian cycle, and other changes associated with puberty in the female are controlled by a regulating factor from the hypothalamus called gonadotropin releasing hormone (GnRH). Its influence is shown in Figure 28-18. GnRH stimulates the release of follicle-stimulating hormone (FSH) from the anterior pituitary. FSH stimulates the initial development of the ovarian follicles and the secretion of estrogens by the follicles. GnRH also stimulates the release of another anterior pituitary hormone—the luteinizing hormone (LH), which stimulates the further development of ovarian follicles, brings about ovulation, and stimulates the production of estrogens, progesterone, inhibin, and relaxin by ovarian cells of the corpus luteum.

At least six different estrogens have been isolated from the plasma of human females. However, only three are present in significant quantities. These are *beta* (β)*-estradiol, estrone,* and *estriol.* Of these, β-estradiol exerts the major effect. It is synthesized from cholesterol or acetyl coenzyme A in the ovaries. As reference is made to estrogens in subsequent discussions, keep in mind that β-estradiol is the principal estrogen.

Estrogens, the hormones of growth, have three main functions. First is the development and maintenance of female reproductive structures, especially the endometrial lining of the uterus, secondary sex characteristics, and the breasts. The secondary sex characteristics include fat distribution to the breasts, abdomen, mons pubis, and hips; voice pitch; broad pelvis; and hair pattern. Second, they help control fluid and electrolyte balance. Third, they increase protein anabolism. In this regard, estrogens are synergistic with human growth hormone (hGH). High levels of estrogens in the blood inhibit the release of GnRH by the hypothalamus, which in turn inhibits the secretion of FSH by the anterior pituitary gland. This inhibition provides the basis for the action of one kind of contraceptive pill.

Progesterone (PROG), the hormone of maturation, works with estrogens to prepare the endometrium for implantation of a fertilized ovum and the mammary glands for milk secretion. High levels of progesterone also inhibit GnRH and prolactin (PRL).

Inhibin is secreted by the corpus luteum and sustentacular (Sertoli) cells of the testis. Its function is to inhibit secretion of FSH and, to a lesser extent, LH. Inhibin might be important in decreasing secretion of FSH and LH toward the end of the menstrual cycle.

Relaxin is produced by the corpus luteum during pregnancy (and placenta) and goes into operation near the end of pregnancy. It relaxes the symphysis pubis and helps dilate the uterine cervix to facilitate delivery.

FIGURE 28-18 **Secretion and physiological effects of estrogens, progesterone, and relaxin.**

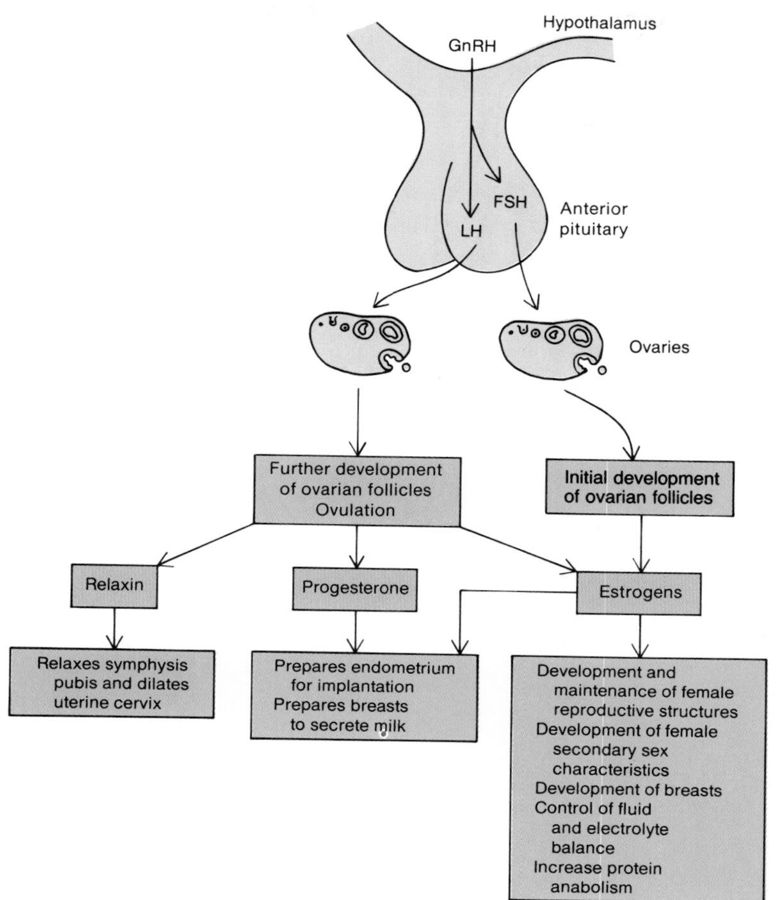

The function of estrogens, progesterone, inhibin, and relaxin are summarized in Exhibit 28-1.

Menstrual Phase (Menstruation)

The duration of the menstrual cycle ranges from 24 to 35 days. For this discussion, we shall assume an average duration of 28 days. Events occurring during the menstrual cycle may be divided into three phases: the menstrual phase, preovulatory phase, and postovulatory phase (Figure 28-19).

The ***menstrual phase,*** also called ***menstruation*** or the ***menses,*** is the periodic discharge of 25 to 65 ml of blood, tissue fluid, mucus, and epithelial cells. It is caused by a sudden reduction in estrogens and progesterone and lasts for approximately the first 5 days of the cycle. The first day of the ovarian cycle is designated as the first day of menstruation. The discharge is associated with endometrial changes in which the stratum functionalis layer degenerates and patchy areas of bleeding develop. Small areas of the stratum functionalis detach one at a time (total detachment would result in hemorrhage), the uterine glands discharge their contents and collapse, and tissue fluid is discharged. The menstrual flow passes from the uterine cavity to the cervix and through the vagina to the exterior. Generally, the flow terminates by the fifth day of the cycle. At this time the entire stratum functionalis has been shed, and the endometrium is very thin because only the stratum basalis remains.

During the menstrual phase, the ovarian cycle is also in operation. Ovarian follicles, called ***primary follicles,*** begin their development. At birth, each ovary contains about 200,000 such follicles, each consisting of a primary oocyte (potential ovum) surrounded by a single flattened layer of epithelial (follicular) cells. During the early part of each menstrual phase, 20 to 25 primary follicles start to produce very low levels of estrogens. Toward the end of the menstrual phase (days 4 to 5), about 20 of the primary follicles develop into ***secondary (growing) follicles.*** A secondary follicle consists of a secondary oocyte and several layers of cells formed by division of the single layer of epithelial cells around a primary follicle.

FIGURE 28-19 Correlation of menstrual and ovarian cycles with the hypothalamic and anterior pituitary gland hormones. In the cycle shown, fertilization and implantation have not occurred.

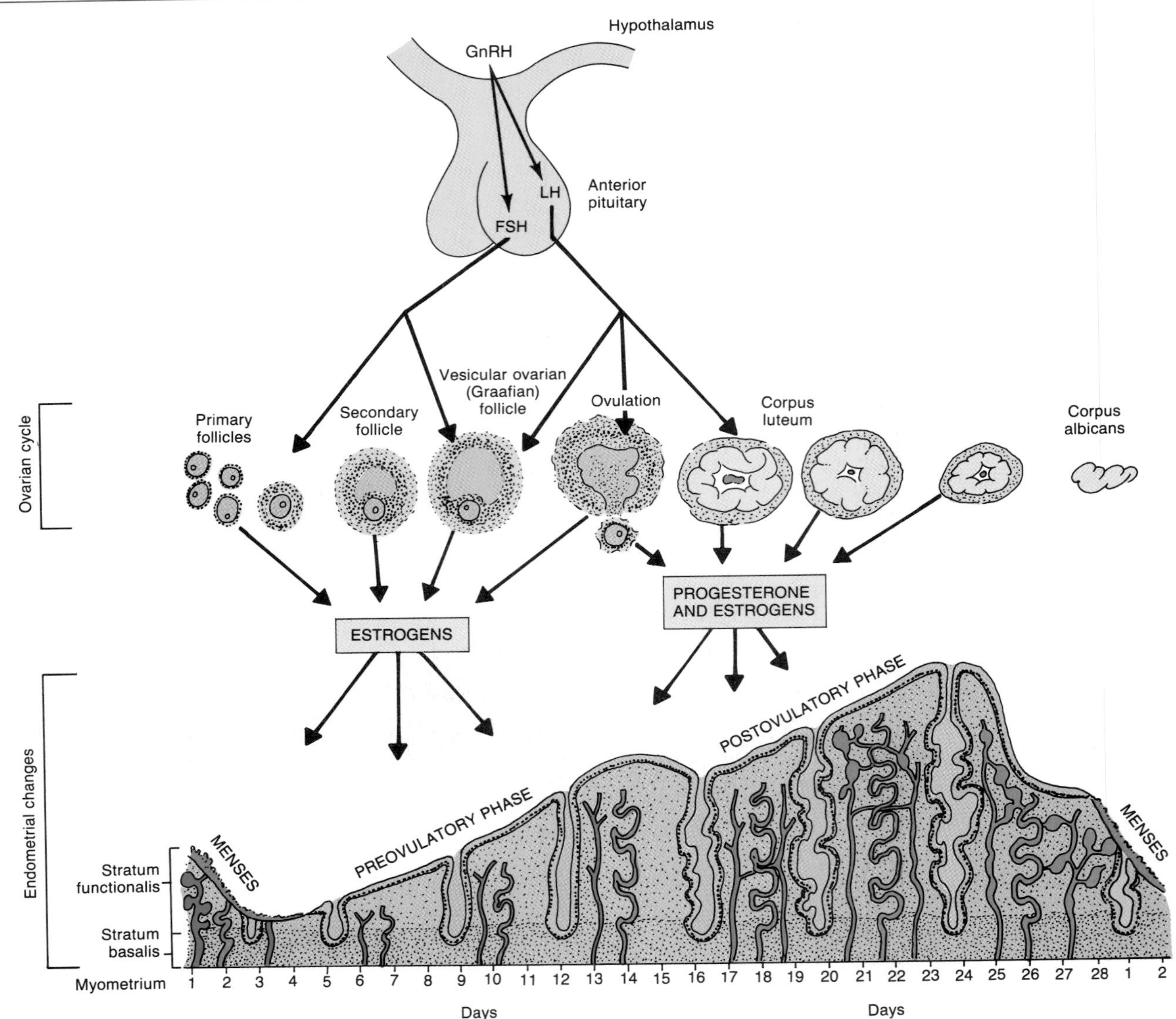

The epithelial cells of the secondary follicle are cuboidal and later columnar and are called *granular* (*granulosa*) *cells.* As a secondary follicle continues to grow, it forms a clear glycoprotein layer between the secondary oocyte and granular cells called the ***zona pellucida*** (pe-LOO-si-da). Also, the granular cells secrete follicular (fō-LIK-yoo-lar) fluid that forces the secondary oocyte to the edge of the secondary follicle and fills the follicular cavity or antrum (see Figure 28-13c). The production of estrogens by the secondary follicles elevates the level of estrogens in the blood slightly. Ovarian follicle development is the result of GnRH secretion by the hypothalamus, which in turn stimulates FSH production by the anterior pituitary. During this part of the cycle, FSH secretion is relatively high. Although about 20 follicles begin development each cycle, usually only 1 attains maturity. The others undergo atresia (death).

Preovulatory Phase

The ***preovulatory phase,*** the second phase of the menstrual cycle, is the time between menstruation and ovulation. This phase of the menstrual cycle is more variable in length than the other phases. It lasts from days 6 to 13 in a 28-day cycle.

FSH and LH stimulate the ovarian follicles to produce more estrogens, and this increase in estrogens stimulates the repair of the endometrium. Cells of the stratum basalis undergo mitosis and produce a new stratum functionalis. As the endometrium thickens, the short, straight endometrial glands develop and the arterioles coil and lengthen as they penetrate the stratum functionalis. The thickness of the endometrium approximately doubles to about 4 to 6 mm. Because of the proliferation of endometrial cells, the preovulatory phase is also termed the ***proliferative phase.*** Still another name is the ***follicular phase*** because of increasing secretion of estrogens by the developing follicle. Functionally, estrogens are the dominant ovarian hormones during this phase of the menstrual cycle.

During the preovulatory phase, one of the secondary follicles in the ovary matures into a ***vesicular ovarian (Graafian) follicle*** or ***mature follicle,*** a follicle ready for ovulation. This follicle produces a bulge on the surface of the ovary. During the maturation process, the follicle increases its estrogen production. Early in the preovulatory phase, FSH is the dominant hormone of the anterior pituitary, but close to the time of ovulation, LH is secreted in increasing quantities (Figure 28-20). Moreover, small amounts of progesterone may be produced by the vesicular ovarian (Graafian) follicle a day or two before ovulation.

FIGURE 28-20 **Relative concentrations of anterior pituitary hormones (FSH and LH) and ovarian hormones (estrogens and progesterone) during a normal menstrual cycle.**

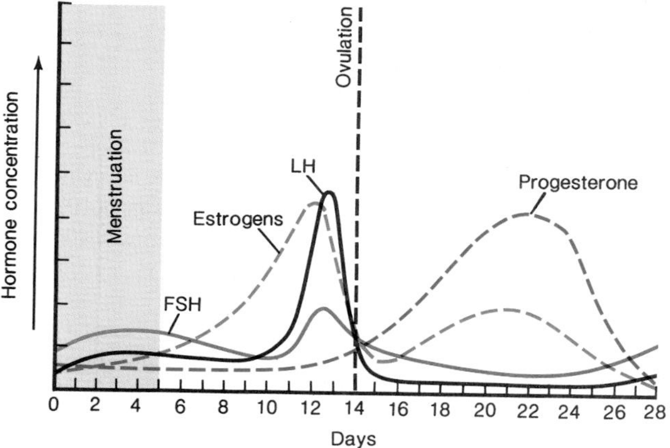

Ovulation

Ovulation, the rupture of the vesicular ovarian (Graafian) follicle with release of the secondary oocyte into the pelvic cavity, usually occurs on day 14 in a 28-day cycle. During ovulation, the secondary oocyte remains surrounded by its zona pellucida and a covering of follicle cells directly around it. These cells are referred to as the ***corona radiata.*** It generally takes 10 to 14 days for a primary follicle to develop into a vesicular ovarian (Graafian) follicle, and it is during this time that the developing ovum completes reduction division (meiosis I) and reaches metaphase of equatorial division (meiosis II). The developing ovum is in this stage when it is discharged during ovulation. The fimbriae of the uterine tubes drape over the ovaries and become active near the time of ovulation. Movements of the fimbriae and ciliary action create currents in the peritoneal serous fluid that carry the secondary oocyte into the uterine tube.

Just prior to ovulation, the high level of estrogens that developed during the preovulatory phase exert a positive feedback directly on both LH and GnRH. LH release increases sharply because of the direct effect of estrogens on the anterior pituitary and also because of the increase in secretion of GnRH by the hypothalamus. This causes the anterior pituitary to release a surge of LH. Without this surge of LH, ovulation will not occur. (An over-the-counter home test that detects the LH surge associated with ovulation is now available. The test predicts ovulation a day in advance.) FSH also increases at this time, but not as dramatically as LH because FSH is stimulated only by the increase in GnRH. Following ovulation, the vesicular ovarian (Graafian) follicle collapses, and blood within it forms a clot called the ***corpus hemorrhagicum.*** The clot is eventually absorbed by the remaining follicular cells. In time, the follicular cells enlarge, change character, and form the ***corpus luteum,*** or yellow body, under the influence of LH, which also stimulates the corpus luteum to secrete estrogens and progesterone.

CLINICAL APPLICATION: SIGNS OF OVULATION

One ***sign of ovulation*** involves ***basal temperature*** (body temperature at rest). When menstruation ceases, the temperature is taken immediately upon awakening each morning and marked on a chart. An increase in temperature, usually between 0.4 to 0.6°F, typically occurs about 14 days after the start of the last menstrual cycle and is due to increasing levels of progesterone. The 24 hours following this rise in temperature is the period immediately following ovulation and is generally considered the best time to become pregnant. The accuracy of the determination depends on many factors including individual variations, the accuracy of the temperature readings, and any factor other than the ovarian cycle that might affect body temperature.

Another sign of ovulation is the amount and consistency of ***cervical mucus.*** Secretion of cervical mucus is regulated by estrogens and progesterone. At midcycle, near the time of ovulation, increasing levels of estrogens cause secretory cells of the cervix to produce large amounts of cervical mucus. About a day or two before ovulation, the quantity of mucus frequently begins to decrease and usually disappears a few days after ovulation. More important, as ovulation approaches, the mucus becomes clear, stretchy (it may stretch from 2.54 to 15.24 cm, that is, 1 to 6 inches), and slippery and causes feelings of lubrication, slipperiness, or wetness on the outer lips (labia majora) of the external genitals. This is the more fertile type of mucus and indicates the time of greatest fertility. Around the day of ovulation, cervical mucus becomes nonstretchy, tacky, thicker, and more opaque and then disappears by a few days after ovulation. This less fertile mucus is produced in response to the influence of progesterone.

The cervix also exhibits signs of ovulation. The external os opens, the cervix rises, and the cervix becomes softer. There is also abundant cervical mucus.

Some females also experience a pain in the area of one or both ovaries around the time of ovulation. Such pain is called ***mittelschmerz*** (MIT-el-shmarts), meaning "pain in the middle," and may last for several hours to a day or two.

Postovulatory Phase

The ***postovulatory phase*** of the menstrual cycle is the most constant in duration and lasts from days 15 to 28 in a 28-day cycle. It represents the time between ovulation and the onset of the next menses. Following ovulation, LH secretion stimulates the development of the corpus luteum. The corpus luteum then secretes increasing quantities of estrogens and progesterone. Progesterone is responsible for preparing the endometrium to receive a fertilized ovum. Preparatory activities include secretory activity of the endometrial glands that causes them to appear tortuously coiled, vascularization of the superficial endometrium, thickening of the endometrium, glycogen storage, and an increase in the amount of tissue fluid. These preparatory changes are maximal about one week after ovulation, and they correspond to the anticipated arrival of the fertilized ovum. During the latter postovulatory phase, FSH secretion again gradually increases and LH secretion decreases. The functionally dominant ovarian hormone during this phase is progesterone. The relation of progesterone to prostaglandins in causing painful menstruation will be considered at the end of the chapter.

If fertilization and implantation do not occur, the rising levels of progesterone and estrogens from the corpus luteum inhibit GnRH and LH secretion. As a result, the corpus luteum degenerates and becomes the ***corpus albicans,*** or white body. The decreased secretion of progesterone and estrogens by the degenerating corpus luteum then initiates another menstrual period. In addition, the decreased levels of progesterone and estrogens in the blood bring about a new output of the anterior pituitary hormones—especially FSH in response to an increased output of GnRH by the hypothalamus. Thus, a new ovarian cycle is initiated. A summary of these hormonal interactions is presented in Figure 28-21.

If, however, fertilization and implantation do occur, the corpus luteum is maintained until the placenta takes over its hormone-producing functions. During this time, the corpus luteum secretes estrogens and progesterone. The corpus luteum is maintained by ***human chorionic*** (kō-rē-ON-ik) ***gonadotropin (hCG),*** a hormone produced by the developing placenta. As you will see later, the presence of hCG is an indication that a female is pregnant. The placenta itself secretes estrogens to support pregnancy and progesterone to support pregnancy and breast development for lactation. Once the placenta begins its secretion, the role of the corpus luteum becomes minor.

Menarche and Menopause

The menstrual cycle normally occurs once each month from ***menarche*** (me-NAR-kē), the first menses, to ***menopause*** (*mens* = monthly; *pausa* = to stop), the last menses. The advent of menopause is signaled by the ***climacteric*** (klī-MAK-ter-ik)—menstrual cycles become less frequent. The climacteric, which typically begins between ages 40 and 50, results from the failure of the ovaries to respond to the stimulation of gonadotropic hormones from the anterior pituitary. Some women experience hot flashes, copious sweating, headache, hair loss, muscular pains, vaginal dryness, insomnia, depression, weight gain, and emotional instability. In the postmenopausal woman there will be some atrophy of the ovaries, uterine (Fallopian) tubes, uterus, vagina, external genitalia, and breasts. Osteoporosis is also a possible occurrence.

The cause of menopause is related to a decreasing ability of aging ovaries to respond to FSH and LH. As a result, there is a decrease in the production of estrogens, progesterone, and ova by the ovaries. Throughout a woman's sexual life, some of the primary ovarian follicles grow into vesicular ovarian follicles with each sexual cycle, and eventually most of them degenerate. As the number of primary follicles diminishes, the production of estrogens by the ovary decreases. Alterations in GnRH release patterns and decreased responsiveness to it by cells of the anterior pituitary gland that secrete LH also contribute to the onset of menopause.

Female Puberty

As occurs with male puberty, the factors that determine the onset of female puberty are poorly understood. In addition, prepubertal levels of LH, FSH, and estrogens are low. At around age seven or eight, girls experience an increase in the secretion of adrenal androgens (adrenarche), which are responsible for the growth of pubic and axillary hair. The onset of puberty is signaled by sleep–associated surges in LH and FSH. As puberty progresses, LH and FSH levels increase throughout the day. The rising levels stimulate the ovaries to secrete estrogens. These

FIGURE 28-21 Summary of hormonal interactions of the menstrual and ovarian cycles.

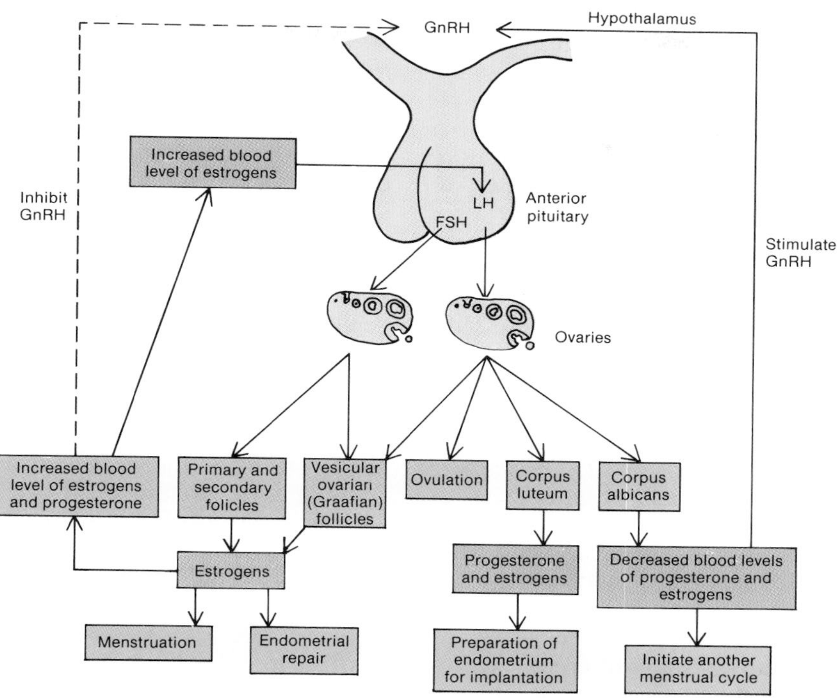

hormones are responsible for the development of the secondary sex characteristics. (Budding of the breasts is the first observable sign of puberty.) Additionally, the hormones stimulate the growth of the uterine (Fallopian) tubes, uterus, and vagina. Also associated with estrogens is menarche which occurs at an average of 12 years of age. However, the first ovulation does not take place until 6–9 months after menarche because the positive feedback of estrogens on LH and FSH is the last step in the maturation of the hypothalamic–pituitary–ovarian axis. As you will see in Chapter 29, a female must have a minimum amount of body fat in order to begin and maintain a normal menstrual cycle.

Vagina

The ***vagina*** serves as a passageway for spermatozoa and the menstrual flow. It is also the receptacle for the penis during coitus, or sexual intercourse, and the lower portion of the birth canal. It is a tubular, fibromuscular organ lined with mucous membrane and measures about 10 cm (4 inches) in length, extending from the cervix to the vestibule (see Figures 28-11 and 28-12). Situated between the urinary bladder and the rectum, it is directed superiorly and posteriorly, where it attaches to the uterus. A recess, called the ***fornix*** (*fornix* = arch or vault), surrounds the vaginal attachment to the cervix. The fornix makes it possible for a female to use contraceptive diaphragms.

Histologically, the mucosa of the vagina is continuous with that of the uterus and consists of stratified squamous epithelium and connective tissue that lies in a series of transverse folds, the ***rugae.*** The muscularis is composed of smooth muscle that can stretch considerably. This distension is important because the vagina receives the penis during sexual intercourse and serves as the lower portion of the birth canal. At the lower end of the vaginal opening, the ***vaginal orifice,*** there may be a thin fold of vascularized mucous membrane called the ***hymen*** (*hymen* = membrane), which forms a border around the orifice, partially closing it (see Figure 28-22).

CLINICAL APPLICATION: IMPERFORATE HYMEN

Sometimes the hymen completely covers the orifice, a condition called *imperforate* (im-PER-fō-rāt) *hymen.* Surgery is required to open the orifice to permit the discharge of the menstrual flow.

The mucosa of the vagina contains large amounts of glycogen, which upon decomposition produces organic acids. These acids create a low pH environment that retards microbial growth. However, the acidity is also injurious to sperm cells. Semen neutralizes the acidity of the vagina to ensure survival of the sperm.

Vulva

The term ***vulva*** (VUL-va; *volvere* = to wrap around), or ***pudendum*** (pyoo-DEN-dum), is a collective designation for the external genitalia of the female (Figure 28-22). Its components are as follows.

The ***mons pubis,*** an elevation of adipose tissue covered by skin and coarse pubic hair, is situated over the symphysis pubis. It lies anterior to the vaginal and urethral openings. From the mons pubis, two longitudinal folds of skin, the ***labia majora*** (LĀ-bē-a ma-JŌ-ra), extend inferiorly and posteriorly. The labia majora are homologous to the scrotum. The labia majora contain an abundance of adipose tissue and sebaceous (oil) and sudoriferous (sweat) glands; they are covered by pubic hair. Medial to the labia majora are two folds of skin called the ***labia minora*** (mī-NŌ-ra). Unlike the labia majora, the labia minora are devoid of pubic hair and fat and have few sudoriferous (sweat) glands. They do, however, contain numerous sebaceous (oil) glands.

The ***clitoris*** (KLI-to-ris) is a richly innervated structure that contains a small, cylindrical mass of erectile tissue and nerves. It is located at the anterior junction of the labia minora. A layer of skin called the ***prepuce*** (foreskin) is formed at the point where the labia minora unite and covers the body of the clitoris. The exposed portion of the clitoris is the ***glans.*** The clitoris is homologous to the penis of the male. Like the penis, the clitoris is capable of enlargement upon tactile stimulation and assumes a role in sexual excitement of the female.

The cleft between the labia minora is called the ***vestibule.*** Within the vestibule are the hymen (if present), vaginal orifice, urethral orifice, and the openings of several ducts. The ***vaginal orifice,*** the opening of the vagina to the exterior, occupies the greater portion of the vestibule and is bordered by the hymen. The ***bulb of the vestibule*** consists of two elongated masses of erectile tissue just deep to the labia on either side of the vaginal orifice. The bulb becomes engorged with blood during sexual arousal, narrowing the vaginal orifice and placing pressure on the penis during intercourse. The bulb is homologous to the corpus spongiosum penis and bulb of the penis. Anterior to the vaginal orifice and posterior to the clitoris is the ***external urethral orifice,*** the opening of the urethra to the exterior. On either side of the urethral orifice are the openings of the ducts of the ***paraurethral (Skene's) glands.*** These glands are embedded in the wall of the urethra and secrete mucus. The paraurethral glands are homologous to the male prostate. On either side of the vaginal orifice itself are the ***greater vestibular (Bartholin's) glands.*** These glands open by ducts into a groove between the hymen and labia minora and produce a mucoid secretion that supplements lubrication during sexual intercourse. The greater vestib-

FIGURE 28-22 **Components of the vulva.**

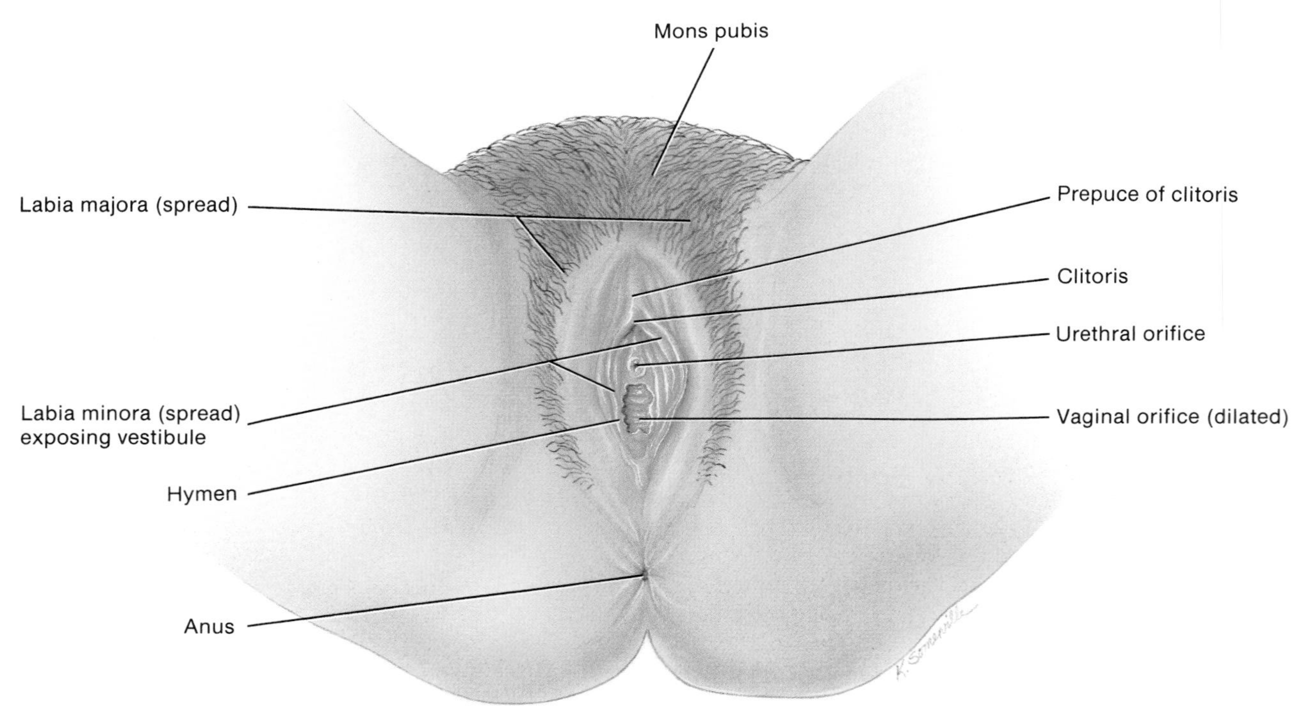

ular glands are homologous to the male bulbourethral (Cowper's) glands. A number of ***lesser vestibular glands,*** whose orifices are microscopic, open into the vestibule.

Perineum

The ***perineum*** (per'-i-NĒ-um) is the diamond-shaped area between the thighs and buttocks of both males and females. It is bounded anteriorly by the symphysis pubis, laterally by the ischial tuberosities, and posteriorly by the coccyx. A transverse line drawn between the ischial tuberosities divides the perineum into an anterior ***urogenital*** (yoo'-rō-JEN-i-tal) ***triangle*** that contains the external genitalia and a posterior ***anal triangle*** that contains the anus (Figure 28-23).

CLINICAL APPLICATION: EPISIOTOMY

The perineal region is stretched during childbirth as the fetal head stretches the vaginal epithelium, subcutaneous fat, and superficial transverse perineal muscle. *Episiotomy* (e-piz'-ē-OT-ō-mē) is an option that the mother may choose in order to prevent undue stretching and even tearing of this region. The purpose of episiotomy—a cut made with surgical scissors—is to enlarge the perineal opening to make room for the fetal head. In effect, a controlled cut is substituted for a jagged, uncontrolled laceration. The incision is closed in layers with a continuous suture that is resorbed by the body within a few weeks, so that stitches do not have to be removed. However, not all women undergoing childbirth need or elect to have an episiotomy.

Mammary Glands

Structure

The ***mammary glands*** are modified sudoriferous (sweat) glands (branched tubuloalveolar) that lie over the pectoralis major and serratus anterior muscles and are attached to them by a layer of connective tissue (Figure 28-24). Internally, each mammary gland consists of 15 to 20 ***lobes,*** or compartments, separated by adipose tissue. The amount of adipose tissue determines the size of the breasts. However, breast size has nothing to do with the amount of milk produced. In each lobe are several smaller compartments called ***lobules,*** composed of connective tissue in which milk-secreting cells referred to as ***alveoli*** are embedded (Figure 28-25). Alveoli are arranged in

FIGURE 28-23 **Perineum. Borders seen in the female.**

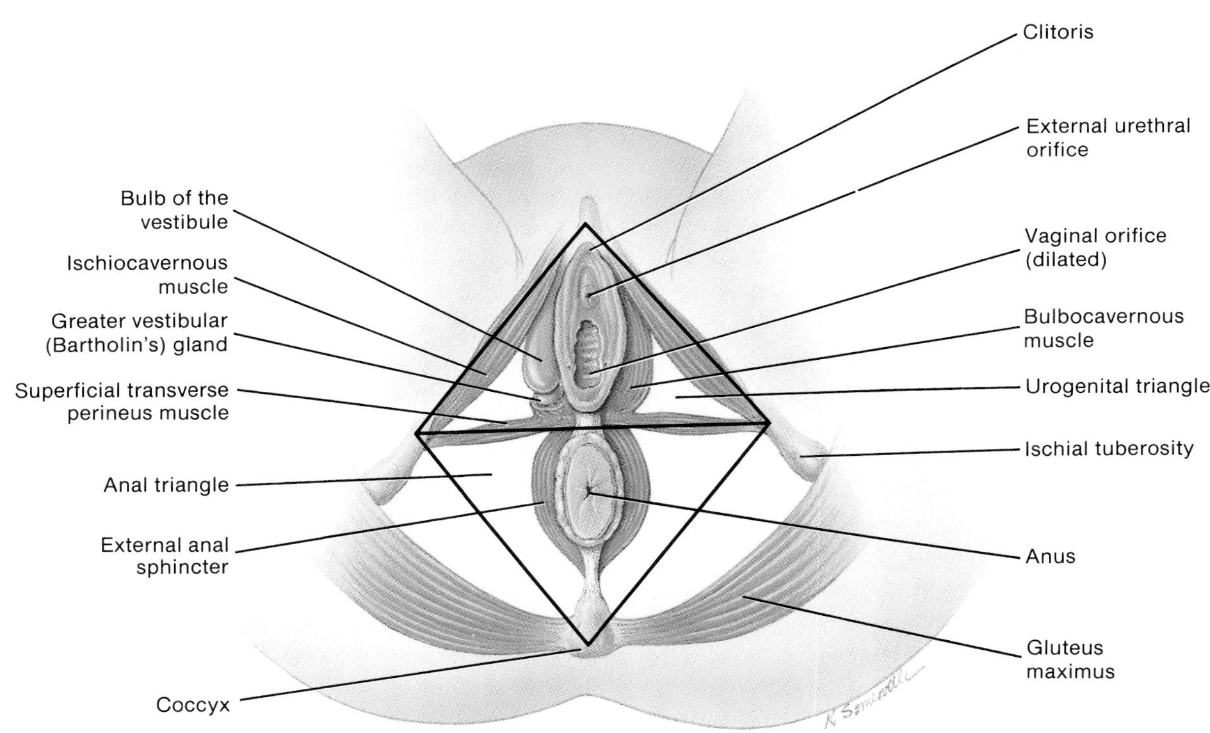

FIGURE 28-24 Mammary glands. (a) Diagram in sagittal section. (b) Diagram in anterior view, partially sectioned. (c) Photograph in sagittal section. (Courtesy of J. A. Gosling, P. F. Harris, et al., *Atlas of Human Anatomy,* Gower Medical Publishing Ltd., 1985.)

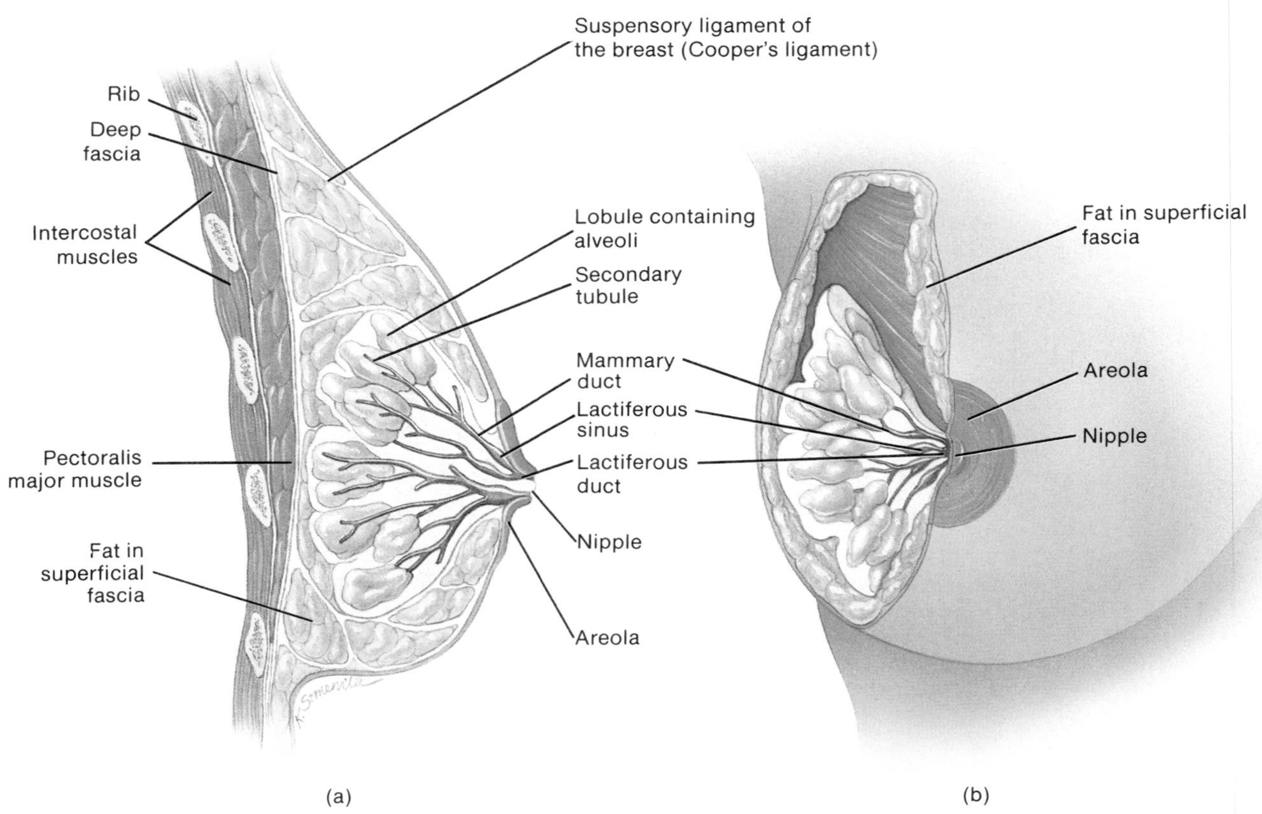

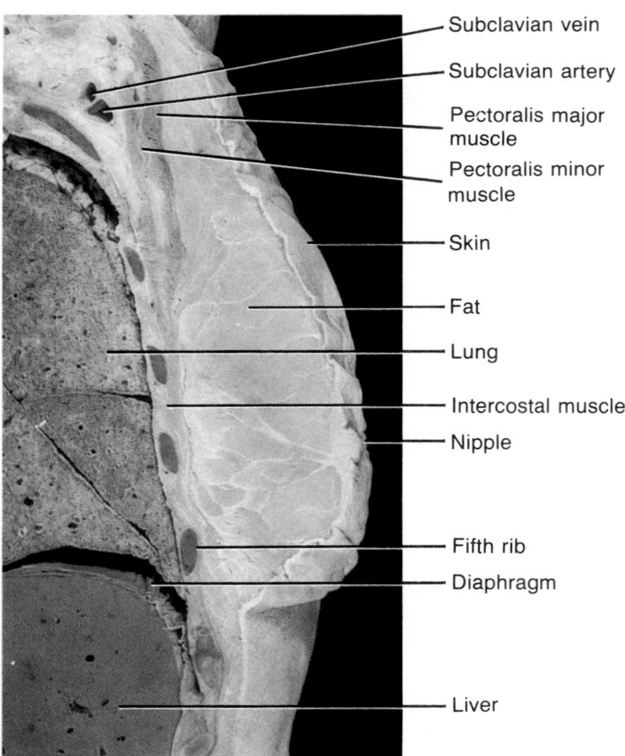

grapelike clusters. Alveoli convey the milk into a series of ***secondary tubules.*** From here the milk passes into the ***mammary ducts.*** As the mammary ducts approach the nipple, they expand to form sinuses called ***lactiferous sinuses,*** where milk may be stored. The sinuses continue as ***lactiferous ducts*** that terminate in the ***nipple.*** Each lactiferous duct conveys milk from one of the lobes to the exterior, although some may join before reaching the surface. The circular pigmented area of skin surrounding the nipple is called the ***areola*** (a-RĒ-ō-la). It appears rough because it contains modified sebaceous (oil) glands. Strands of connective tissue called the ***suspensory ligaments of the breast (Cooper's ligaments)*** run between the skin and deep fascia and support the breast.

Development

At birth, both male and female mammary glands are undeveloped and appear as slight elevations on the chest. With the onset of puberty, the female breasts begin to develop—the ductile system matures, extensive fat deposition occurs, and the areola and nipple grow and become pigmented. These changes are correlated with an

FIGURE 28-25 Histology of the mammary glands. Photomicrograph of several alveoli in a nonlactating mammary gland at a magnification of 60×. (Copyright © 1983 by Michael H. Ross. Used by permission.)

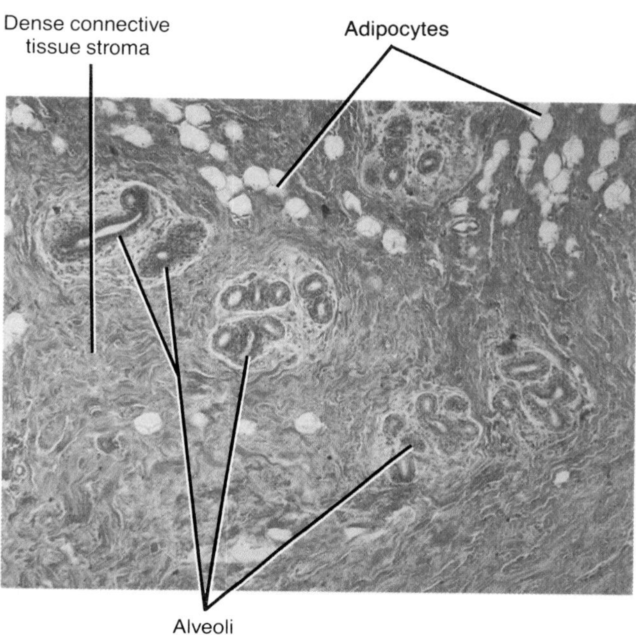

increased output of estrogens by the ovary. Further mammary development occurs at sexual maturity with the onset of ovulation and the formation of the corpus luteum. During adolescence, increased levels of progesterone cause the alveoli to proliferate, enlarge, and become secretory. Also, fat deposition continues, increasing the size of the glands. Although the changes in mammary gland development are associated with estrogens and progesterone secretion by the ovaries, ovarian secretion is ultimately controlled by FSH and LH, which are secreted in response to GnRH by the hypothalamus.

Functions

The essential function of the mammary glands is milk secretion and ejection, together called ***lactation.*** The secretion of milk is due largely to the hormone prolactin (PRL), with contributions from progesterone and estrogens. The ejection of milk occurs in the presence of oxytocin (OT), which is released from the posterior pituitary gland in response to sucking. Lactation is discussed in more detail in Chapter 29.

Breast Cancer

Early detection—especially by breast self-examination (BSE) and mammography—is still the most promising method to increase the survival rate for ***breast cancer.*** It is estimated that 95 percent of breast cancer is first detected by women themselves. Each month after the menstrual period the breasts should be thoroughly examined for lumps, puckering of the skin, or nipple retraction or discharge.

MEDICAL TEST

Mammography (mam-OG-ra-fē; *mammae* = breast; *graphein* = to record)

Diagnostic Value: To evaluate symptoms of breast disease such as lumps, nipple discharges, retraction of nipples, dimpling of skin, or persistent breast pain. The test is also used to screen asymptomatic women as well as those having an increased risk for breast cancer. The test may be performed between ages 35 to 39 to provide a baseline mammogram, an image to be used later for comparison.

Procedure: The mammographic image, called a ***mammogram*** (Figure 28-26), is obtained by placing the breasts, one at a time, on a flat surface and using a compressor to smooth the breast for better imaging. Usually, two images are taken of each breast, one from the top and one from the side. There are two basic types of mammography: xeromammography and film-screen mammography. In *xeromammography,* x-rays are beamed onto a specially coated metal plate, and the blue-on-white image produced provides the physician with detail of thicker as well as thinner portions of the breast. In *film-screen mammography,* x-rays are beamed onto a fluorescent screen, and the image is produced on x-ray film. Breast compression is the key to effective mammography. It increases image detail, decreases blurring and overlap of tissue, and reduces radiation dosage.

One of the most recent breast cancer detecting procedures is ***ultrasound (US).*** The procedure is performed while the patient lies on her stomach in a specially designed hospital bed with her breasts immersed in a tank of water. Ultrasound produces images using a device that first emits a pulse of high-frequency sound and then records the echo on a monitor. Although ultrasound can neither detect microcalcifications or tumors less than 1 cm in diameter, it can be used to determine whether a lump is a benign cyst or a malignant tumor.

Another technique combines ***computed tomography with mammography (CT/M)*** and appears to overcome some of the limitations of mammography. The procedure is based on the fact that breast carcinoma has an

FIGURE 28-26 Photographs of xeroradiographic mammograms. (a) Normal breast. (b) Malignant breast. (Courtesy of Xerox Medical Systems, Pasadena, CA.)

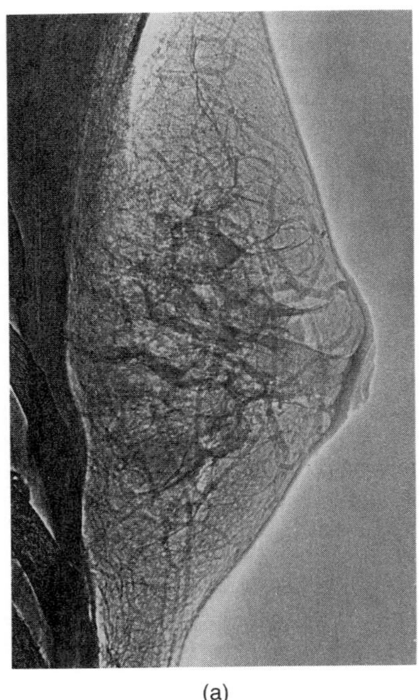

(a)

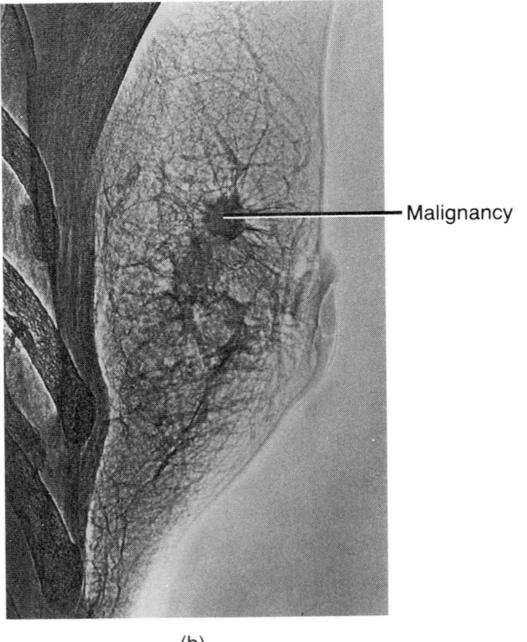

(b)

abnormal affinity for iodide. CT scans are made before and after the rapid intravenous infusion of an iodide contrast material. Comparison of the initial density of a suspected lesion with the density following infusion of the iodide gives an indication of the status of the tumor. CT/M affords definitive diagnostic help in instances where the mammographic and physical examinations are inconclusive and appears to be a significantly improved method of breast cancer diagnosis.

Long-term studies are now under way to determine the effectiveness of an experimental home device for breast cancer detection. It is called a ***breast cancer screening indicator.*** It is composed of two plastic discs that contain heat-sensitive chemicals. The discs are worn inside a female's brassiere for 15 minutes each month so that she can observe color changes that could indicate a breast abnormality. The screening device is said to detect tumors as small as 2 mm in diameter, their size about 2 to 10 years before they become palpable or can be seen by mammography.

Treatment for breast cancer may involve hormone therapy, chemotherapy, ***lumpectomy*** (removal of just the tumor and immediate surrounding tissue), a modified or radical mastectomy, or a combination of these. A ***radical mastectomy*** involves removal of the affected breast along with the underlying pectoral muscles and the axillary lymph nodes. Metastasis of cancerous cells is usually through the lymphatic vessels or blood vessels. Radiation treatment and chemotherapy may follow the surgery to ensure the destruction of any stray cancer cells.

Among the factors that clearly increase the risk of breast cancer development are (1) a family history of breast cancer, especially in a mother or sister; (2) never having a child or having a first child after age 34; (3) previous cancer in one breast; (4) exposure to ionizing radiation; and (5) excessive fat and alcohol intake. Females who take birth control pills do not have a higher risk of developing breast cancer than females who do not. But cigarette smoking may increase the incidence of breast cancer, especially in postmenopausal females.

The American Cancer Society recommends the following steps in order to help diagnose breast cancer as early as possible:

1. A mammogram should be taken between the ages of 35 and 39, to be used later for comparison (baseline mammogram).
2. A physician should examine the breasts every three years when a female is between the ages of 20 and 40, and every year after 40.
3. Females with no symptoms should have a mammogram every year or two between ages 40 and 49, and every year after 50.
4. Females of any age with a history of breast cancer, a strong family history of the disease, or other risk factors should consult a physician to determine a schedule for mammography.
5. All females over 20 should develop the habit of monthly breast self-examination (BSE).

By using silicone implants and skin, fat, and muscles from other parts of the body, breast reconstruction following a radical mastectomy can be accomplished. Using these techniques, it is possible to reconstruct a natural-looking breast.

PHYSIOLOGY OF SEXUAL INTERCOURSE

Sexual intercourse, or ***copulation*** (in humans, called ***coitus***), is the process by which spermatozoa are deposited in the vagina.

CLINICAL APPLICATION: DONOR INSEMINATION

Donor (artificial) insemination (*in* = into; *seminatus* = sown; *semen* = seed) refers to the use of seminal fluid from the husband or another male that is artificially deposited by a physician at a time during the menstrual cycle when pregnancy is most likely to occur. If the husband's seminal fluid is used, the process is known as ***homologous insemination.*** Such a procedure might be carried out if there is a developmental anomaly that prevents placement of the penis in the vagina or normal ejaculation. A small volume of seminal fluid is also an indication for homologous insemination. If seminal fluid from a donor is used, the process is called ***heterologous insemination.*** In such cases, an anonymous donor is selected on the basis of race, blood type compatibility, physical appearance, general health, and genetic background.

It has been documented that hepatitis B can be transmitted by semen donors to donor inseminated females. In order to decrease the risk of passing on sexually transmitted diseases, guidelines have been established for screening semen donors.

Male Sexual Act

Erection

The male role in the sexual act starts with ***erection,*** the enlargement and stiffening of the penis. An erection may be initiated in the cerebrum by stimuli such as anticipation, memory, and visual sensation, or it may be a reflex brought on by stimulation of the touch receptors in the penis, especially in the glans. In any case, parasympathetic impulses pass from the gray matter of the second, third, and fourth sacral segments of the spinal cord in pelvic splanchnic nerves to the penis. The impulses cause dilation of the arteries of the penis, allowing blood to fill the cavernous spaces of the spongy bodies. This raises the internal pressure of the penis, resulting in enlargement and rigidity (erection).

Lubrication

Parasympathetic impulses from the sacral cord also cause the bulbourethral (Cowper's) glands and urethral glands (glands of Littré) to secrete mucus, which affords only a small amount of ***lubrication*** for intercourse. The mucus flows through the urethra. The major portion of lubricating fluid is produced by the cervical mucosa of the female. Without satisfactory lubrication, the male sexual act is difficult since unlubricated intercourse causes pain impulses that inhibit rather than promote coitus.

Orgasm

Tactile stimulation of the penis brings about emission and ejaculation. When sexual stimulation becomes intense, rhythmic sympathetic impulses leave the spinal cord at the levels of the first and second lumbar vertebrae and pass to the genital organs. These impulses cause peristaltic contractions of the ducts in the testes, epididymides, and ductus (vas) deferens that propel spermatozoa into the urethra—a process called ***emission.*** Simultaneously, peristaltic contractions of the seminal vesicles and prostate gland expel seminal and prostatic fluid along with the spermatozoa. All these mix with the mucus of the bulbourethral glands, resulting in the fluid called semen. Other rhythmic impulses sent from the spinal cord at the levels of the first and second sacral vertebrae reach the skeletal muscles at the base of the penis, and the penis expels the semen from the urethra to the exterior. The propulsion of semen from the urethra to the exterior constitutes an ***ejaculation.*** A number of sensory and motor activities accompany ejaculation, including a rapid heart rate, an increase in blood pressure, an increase in respiration, and pleasurable sensations. These activities, together with the muscular events involved in ejaculation, are referred to as an ***orgasm.***

During sexual intercourse, ejaculation introduces millions of sperm into the vagina. Next, the sperm move into the cervix where muscular contractions aid their movement into the uterus. Inside the uterus, rhythmic contractions of the muscular wall greatly aid the sperm as they swim toward the uterine (Fallopian) tubes. There is some evidence that suggests the hormone oxytocin (OT) may be released from the posterior pituitary during orgasm and/or that prostaglandins in seminal fluid function to cause the uterine contractions. Of the total number of spermatozoa that enter the vagina, less than 1 percent come in proximity to the ovum.

Female Sexual Act

Erection

The female role in the sex act, like that of the male, also involves erection, lubrication, and orgasm. Stimulation of the female, as in the male, depends on both psychic and tactile responses. Under appropriate conditions, stimulation of the female genitalia, especially the clitoris, results in erection and widespread sexual arousal. This response is controlled by parasympathetic impulses sent from the sacral spinal cord to the external genitalia.

Lubrication

Parasympathetic impulses from the sacral spinal cord also cause the bulk of ***lubrication*** of the vagina. The impulses result in the secretion of a mucoid fluid from the epithelium of the cervical mucosa. Some mucus is also produced by the greater vestibular (Bartholin's) glands. As was noted earlier, lack of sufficient lubrication results in pain impulses that inhibit rather than promote coitus.

Orgasm (Climax)

When tactile stimulation of the genitalia reaches maximum intensity, reflexes are initiated that cause the female ***orgasm (climax).*** Female orgasm is analogous to male ejaculation. The perineal muscles contract rhythmically from spinal reflexes (sympathetic nerve impulses) similar to those that occur in the male ejaculation.

BIRTH CONTROL (BC)

Although there is no single, ideal method of ***birth control (BC),*** several types of contraceptive methods are available, each with their own advantages and disadvantages. The methods discussed here are sterilization, hormonal, intrauterine, barrier, chemical, physiologic, coitus interruptus (withdrawal), and induced abortion.

Sterilization

One means of ***sterilization*** of males is ***vasectomy*** (discussed earlier in the chapter). Sterilization in females generally is achieved by performing a ***tubal ligation*** (lī-GĀ-shun). In one procedure, an incision is made into the abdominal cavity, the uterine (Fallopian) tubes are squeezed, and a small loop called a knuckle is made. A suture is tied tightly at the base of the knuckle, and the knuckle is then cut. After four or five days the suture is digested by body fluids, and the two severed ends of the tubes separate. The ovum is thus prevented from passing to the uterus, and the sperm cannot reach the ovum. Tubal ligation may also be performed by laparoscopy.

MEDICAL TEST

Laparoscopy (lap'-a-ROS-kō-pē; *lapara* = flank; *skopein* = to examine)

Diagnostic Value: To evaluate abdominal or pelvic masses or pain, abnormal menstrual periods, female infertility, metastasis of cancer, and abdominal trauma. The procedure can also be used for sterilization by tubal ligation.

Procedure: Following local or general anesthesia, a small incision is made in the abdominal wall in or just below the umbilicus (navel). A needle is inserted through the incision, and gas is slowly injected to inflate the abdomen. (This creates a space inside the abdomen for better visualization and easier manipulation of instruments.) After the needle is withdrawn, a ***laparoscope*** (lighted tube) is inserted through the incision so that the physician can view the abdominal and pelvic viscera. Other instruments, such as forceps, probes, and small scissors, may be inserted through a second incision, usually at the pubic hairline. Thus, the technique may also be used to remove fluids and tissues for biopsy, drain ovarian cysts, cut adhesions, and perform tubal ligation.

Hormonal

The ***hormonal method,*** also called ***oral contraception (OC)*** or ***the pill,*** has found rapid and widespread use. Although several pills are available, the one most commonly used contains a high concentration of progesterone and a low concentration of estrogens (combination pill). These two hormones act on the anterior pituitary to decrease the secretion of FSH and LH by inhibiting GnRH by the hypothalamus. The low levels of FSH and LH usually prevent ovulation, and thus pregnancy cannot occur. Even if ovulation does occur, as it does in some cases, oral contraceptives also alter cervical mucus so that it is more hostile to spermatozoa and may make the endometrium less receptive to implantation.

Women for whom all oral contraceptives are contraindicated include those with a history of thromboembolic disorders (predisposition to blood clotting), cerebral blood vessel damage, hypertension, liver malfunction, heart disease, or cancer of the breast or reproductive system. Recent reports link pill users with an increased risk of infertility. About 40 percent of all pill users experience side effects—generally minor problems such as nausea, weight gain, headache, irregular menses, spotting between periods, and amenorrhea. The statistics on the life-threatening conditions associated with the pill such as blood clots, heart attacks, liver tumors, and gallbladder disease are somewhat more reassuring. For all the problems combined, fewer than 3 deaths occur per 100,000 users under age 30, 4 among those 30 to 35, 10 among those 35 to 39, and 18 among women over 40. The major exception is that women who take the pill and smoke face far higher odds of developing heart attack and stroke than do nonsmoking pill users.

Among the noncontraceptive benefits of oral contraceptives are menstrual regulation, decreased menstrual flow, and prevention of functional ovarian cysts. Some evidence also suggests a protective effect of the pill against endometrial and ovarian cancer.

In late 1986, a team of French physicians reported a new approach to birth control. It involves using a substance called ***mifepristone (RU486)*** that blocks the action of progesterone. Progesterone prepares the uterine endometrium for implantation and then maintains the uterine lining after implantation. If progesterone levels fall during pregnancy or if the hormone is inhibited from acting, menstruation occurs, and the embryo is sloughed off along with the uterine lining. Mifepristone occupies the endometrial receptor sites for progesterone and, in effect, blocks the access of progesterone to the endometrium and causes a miscarriage. Mifepristone can be taken up to five weeks after conception. One side effect of the drug is uterine bleeding, which averages 11 days and ranges from 5 to 21 days. The drug is not yet available in the United States but is available in France and China. In 1988, Dutch scientists reported another drug used to terminate early pregnancy (during the first eight weeks) called ***epostane.*** It inhibits progesterone synthesis. The most frequent side effect following epostane use is nausea. Epostane is yet unavailable in the United States. Also not available in the United States are ***Norplant,*** capsules surgically implanted under the skin that slowly release hormones that inhibit ovulation, and ***Depo-Provera,*** an injectable hormone given every three months, which is released slowly from muscle and inhibits ovulation.

Modified human chorionic gonadotropin (hCG), a hormone produced by the placenta, has shown promise in laboratory animals as a chemical method of birth control. In its modified form, hCG prevents implantation of a fertilized egg or terminates an already established pregnancy. Modified versions of gonadotropin releasing hormone (GnRH) are also being tested as contraceptives that inhibit ovulation.

The quest for an efficient male oral contraceptive has been generally disappointing. Trials are now under way to test the effectiveness of inhibin in male (and female) birth control. The hormone inhibits FSH secretion by the pituitary gland. Modified versions of GnRH have also been shown to inhibit testosterone production, with a resultant decrease in sperm count and motility. However, the development of impotence has made this approach unpopular.

Intrauterine Devices (IUDs)

An ***intrauterine device (IUD)*** is a small object made of plastic, copper, or stainless steel that is inserted into the cavity of the uterus. It is not clear how IUDs operate. Some investigators believe they cause changes in the uterine lining, which, in turn, produce a substance that destroys either the sperm or fertilized ovum. The dangers associated with the use of IUDs in some females include pelvic inflammatory disease (PID), infertility, and excess menstrual bleeding and pain. (Females in monogamous relationships are at a lesser risk of developing PID.) Because of this, IUDs are rapidly declining in popularity. Lawsuits resulting from damage claims have caused most manufacturers of copper IUDs in the United States to stop production and sales. An IUD, similar to the famous copper T, is now available in the United States. It is called Para Gard model T380A. For a female to use the IUD, she must sign or initial a consent form to indicate that she is well aware of the potential risks involved.

Barrier

Barrier methods are designed to prevent spermatozoa from gaining access to the uterine cavity and uterine tubes. Among the barrier methods are condoms, diaphragms, and cervical caps.

The ***condom*** is a nonporous, elastic (latex or similar material) covering placed over the penis that prevents deposition of sperm in the female reproductive tract. Proper use of condoms, especially when used with sperm-killing chemicals, with each act of sexual intercourse can reduce, but not eliminate, risk of STDs. Individuals likely to become infected or known to be infected with human immunodeficiency virus (HIV) should be aware that condom use cannot completely eliminate the risk of transmission to themselves or to others since condoms have a failure rate of up to 10 percent.

The ***diaphragm*** is a dome-shaped structure that fits over the cervix and is generally used in conjunction with a sperm-killing chemical. The diaphragm stops the sperm from passing into the cervix. The chemical kills the sperm cells. Toxic shock syndrome (TSS) and recurrent urinary tract infections are associated with diaphragm use in some females.

The ***cervical cap*** (Prentif cavity-Rim), is a thimble-shaped contraceptive device made of latex or plastic that measures about 3.9 cm (1.5 inches) in diameter. It fits snugly over the cervix of the uterus and is held in position by suction. Like the diaphragm, the cervical cap is used with a spermicide. Also, like the diaphragm, it must be fitted initially by a physician or other trained personnel. Among the advantages of the cervical cap over the diaphragm are the following: the cap can be worn up to 48 hours versus 24 hours for the diaphragm, and since the cap fits tightly and rarely leaks, it is not necessary to reintroduce spermicide before intercourse. The cervical cap is prescribed only to females with normal Pap smears, and users should undergo a follow-up Pap smear after three months. The cervical cap is contraindicated for females with toxic shock syndrome (TSS), known or suspected cervical or uterine malignancies, and current vaginal or cervical infections.

Chemical

Chemical methods of contraception include spermicidal agents. Various foams, creams, jellies, suppositories, and douches that contain spermicidal agents make the vagina and cervix unfavorable for sperm survival. The most widely used spermicides are nonoxynol-9 and octoxynol-9. The action is to disrupt plasma membranes of spermatozoa, thus killing them. They are most effective when used in conjunction with a diaphragm. Several studies have found no increase in the overall frequency of birth defects in association with the use of spermicides. A recent spermicidal development is a ***contraceptive sponge*** (Today). It is a nonprescription polyurethane sponge that contains nonoxynol-9. This spermicide decreases the incidence of chlamydia and gonorrhea but slightly increases the risk of developing vaginal infections caused by *Candida.* The sponge is placed in the vagina where it releases spermicide for up to 24 hours and also acts as a physical barrier to sperm. The sponge is equal to the diaphragm in effectiveness. Some cases of toxic shock syndrome (TSS) have been reported among users of the contraceptive sponge.

Physiologic

Physiologic methods are based upon knowledge of certain physiologic events that occur during the menstrual cycle. In females with normal menstrual cycles, especially, the physiologic events help to predict during which days of the cycle ovulation is likely to occur.

The first physiologic method, developed in the 1930s, is known as the ***rhythm method.*** It takes advantage of the fact that a secondary oocyte is fertilizable for only 24 hours and is available only during a period of three to five days in each menstrual cycle. During this time, the couple refrains from intercourse (three days before ovulation, the day of ovulation, and three days after ovulation). Its effectiveness is limited by the fact that few women have absolutely regular cycles.

Another natural family planning system, developed during the 1950s and 1960s, is the ***sympto-thermal method.*** According to this method, couples are instructed to know and understand certain signs of fertility and infertility. Recall that the signs of ovulation include increased basal body temperature; the production of clear, stretchy cervical mucus; opening of the external os; elevation of the cervix; softening of the cervix; abundant cervical mucus; and pain associated with ovulation (mittelschmerz). If the couple refrains from sexual intercourse when the signs of ovulation are present, the likelihood of pregnancy is significantly decreased.

Coitus Interruptus (Withdrawal)

Coitus (KŌ-i-tus; *coitio* = coming together) ***interruptus*** refers to withdrawal of the penis from the vagina just prior to ejaculation. Failures with this method are related to either failure to withdraw before ejaculation or pre-ejaculatory escape of sperm-containing fluid from the urethra.

Induced Abortion

Abortion refers to the premature expulsion from the uterus of the products of conception. An abortion may be spontaneous (naturally occurring) or induced (intentionally performed). When birth control methods are not practiced or are not successful, ***induced abortion*** may be performed. Induced abortions may involve vacuum aspiration (suction), a saline solution, surgical evacuation (scraping), or use of drugs such as mifepristone (RU486) and epostane.

A summary of methods of birth control is presented in Exhibit 28-2.

AGING AND THE REPRODUCTIVE SYSTEMS

Although there are major age-specific physical changes in structure and function, few age-specific disorders are associated with the reproductive system. In the male, the decreasing production of testosterone produces less muscle strength, fewer viable sperm, and decreased sexual desire. However, abundant spermatozoa may be found even in old age. Most of the age-dependent pathologies do not affect general health, except for prostate problems that could become serious and fatal.

The female reproductive system has a time-limited span of fertility between menarche and menopause. The system demonstrates an age-dependent decline in fertility, possibly as a result of less frequent ovulation and the declining ability of the uterine (Fallopian) tubes and uterus to support the young embryo. There is a decrease in the production of progesterone and estrogens. Menopause is only one of a series of phases that leads to reduced fertility, irregular or absent menstruation, and a variety of physical changes. Uterine cancer peaks at about 65 years of age, but cervical cancer is more common in younger women, and breast cancer is the leading cause of death among women between the ages of 40 and 60. Prolapse (falling down or sinking) of the uterus is possibly the most common complaint among female geriatric patients.

DEVELOPMENTAL ANATOMY OF THE REPRODUCTIVE SYSTEMS

The *gonads* develop from the ***intermediate mesoderm.*** By the sixth week, they appear as bulges that protrude into the ventral body cavity (Figure 28-27). The gonads develop near the mesonephric ducts. A second pair of ducts, the ***paramesonephric (Müller's) ducts,***

EXHIBIT 28-2 SUMMARY OF BIRTH CONTROL (BC) METHODS

Method	Comments
Sterilization	Procedure involving severing ductus (vas) deferens in males (vasectomy) and uterine (Fallopian) tubes in females (tubal ligation and laparoscopy). Failure rate is less than 1 percent.*
Hormonal	Except for total abstinence or surgical sterilization, oral contraception is the most effective contraceptive known. Side effects include nausea, occasional light bleeding between periods, breast tenderness or enlargement, fluid retention, and weight gain. Should not be used by women who have cardiovascular conditions (thromboembolic disorders, cerebrovascular disease, heart disease, hypertension), liver malfunction, cancer or neoplasia of breast or reproductive organs, or by women who smoke. Pill users may have an increased risk of infertility. Failure rate is about 2 percent.
Intrauterine device (IUD)	Small object (loop, coil, T, or 7) made of plastic, copper, or stainless steel and inserted into uterus by physician. May be left in place for long periods of time (some must be changed every 2–3 years). Does not require continued attention by user. Some women cannot use them because of expulsion, bleeding, or discomfort. Not recommended for women who have not had children because uterus is too small and cervical canal too narrow. Use of IUDs has diminished because of problems experienced by some females such as pelvic inflammatory disease (PID) and infertility. Para Gard model T380A is now available in the United States, but a consent form indicating the potential risks must be initialed or signed. Failure rate is about 5 percent.
Barrier	A condom is a thin, strong sheath of rubber or similar material worn by male to prevent sperm from entering vagina. Failures caused by sheath tearing or slipping off after climax or not putting the sheath on soon enough. If used correctly and consistently, especially with a spermicide, their effectiveness is similar to that of diaphragm. Failure rate is about 10 percent. A diaphragm is a flexible rubber dome inserted into vagina to cover cervix, providing barrier to sperm. Usually used with spermicidal cream or jelly. Must be left in place at least 6 hours after intercourse and may be left in place as long as 24 hours. Must be fitted by physician or other trained personnel and refitted every two years and after each pregnancy. Offers high level of protection if used with spermicide. Occasional failures caused by improper insertion or displacement during sexual intercourse. Failure rate is about 20 percent. A cervical cap is a thimble-shaped latex device that fits snugly over the cervix of the uterus. Used with a spermicide, must be fitted by a physician or other trained personnel. May be left in place for up to 48 hours, and it is not necessary to reintroduce spermicide before sexual intercourse. Prescribed for females with normal Pap smears, and users should undergo a follow-up after three months. Failure rate is about 10 to 20 percent.
Chemical	Sperm-killing chemicals inserted into vagina to coat vaginal surfaces and cervical opening. Provide protection for about one hour. Effective when used alone but significantly more effective when used with diaphragm or condom. The contraceptive sponge is made of polyurethane and releases a spermicide for up to 24 hours. A few cases of toxic shock syndrome (TSS) have been reported among users of the contraceptive sponge. Failure rate is about 20 percent.
Physiologic	In the rhythm method, sexual intercourse is avoided just before and just after ovulation (about seven days). Failure rate is about 20 percent even in females with regular menses. In the sympto-thermal method, signs of ovulation are noted (increased basal body temperature, clear and stretchy cervical mucus, opening of the external os, elevation and softening of the cervix, abundant cervical mucus, and pain associated with ovulation), and sexual intercourse is avoided. Failure rate is about 20 percent.
Coitus interruptus	Withdrawal of penis from vagina before ejaculation occurs. Failure rate is 20 to 25 percent.
Induced abortion	Involves vacuum aspiration (suction), saline solution, surgical evacuation (scraping) to remove prematurely products of conception, or drugs such as mifepristone (RU486) and epostane.

* Failure rates are based on estimated pregnancy rates (percentage) in the first year of use, assuming variations in consistency of use.

FIGURE 28-27 **Development of the internal reproductive systems.**

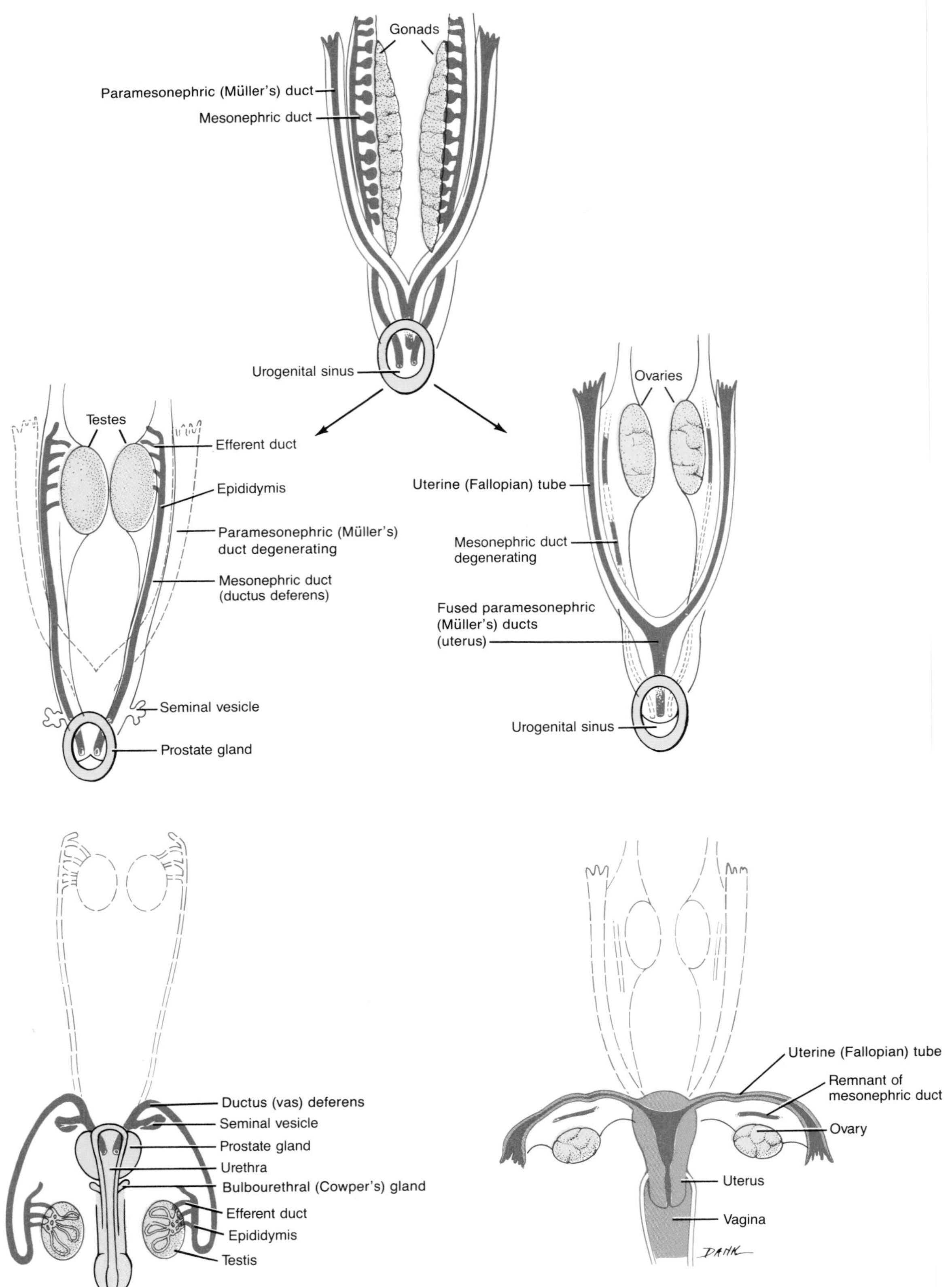

develop lateral to the mesonephric ducts. Both sets of ducts empty into the urogenital sinus. An embryo contains primitive gonads for both sexes. Differentiation into a male depends on the presence of a gene on the Y chromosome called *testis-determining factor (TDF)* and the release of testosterone, whereas differentiation into a female depends on the absence of TDF and the absence of testosterone. At about the seventh week, the gonads are clearly differentiated into ovaries or testes.

In the male embryo, the *testes* connect to the mesonephric duct through a series of tubules. These tubules become the *seminiferous tubules.* Continued development of the mesonephric ducts produces the *efferent ducts, ductus epididymis, ductus (vas) deferens, ejaculatory ducts,* and *seminal vesicle.* The *prostate* and *bulbourethral (Cowper's) glands* are ***endodermal*** outgrowths of the urethra. Shortly after the gonads differentiate into testes, the paramesonephric ducts degenerate without contributing any functional structures to the male reproductive system.

In the female embryo, the gonads develop into *ovaries.* At about the same time, the distal ends of the paramesonephric ducts fuse to form the *uterus* and *vagina.* The unfused portions become the *uterine (Fallopian) tubes.* The *greater (Bartholin's)* and *lesser vestibular glands* develop from ***endodermal*** outgrowths of the vestibule. The mesonephric ducts in the female degenerate without contributing any functional structures to the female reproductive system.

The *external genitals* of both male and female embryos also remain undifferentiated until about the eighth week. Before differentiation, all embryos have an elevated region, the ***genital tubercle,*** a point between the tail (future coccyx) and the umbilical cord, where the mesonephric and paramesonephric ducts open to the exterior (Figure 28-28). The tubercle consists of a ***urethral groove*** (opening into the urogenital sinus), paired ***urethral folds,*** and paired ***labioscrotal swellings.***

In the male embryo, the genital tubercle elongates and develops into a *penis.* Fusion of the urethral folds forms the *spongy (cavernous) urethra* and leaves an opening only at the distal end of the penis, the *urethral orifice.* The labioscrotal swellings develop into the *scrotum.* In the female, the genital tubercle gives rise to the *clitoris.* The urethral folds remain open as the *labia minora,* and the labioscrotal swellings become the *labia majora.* The urethral groove becomes the *vestibule.*

FIGURE 28-28 **Development of the external genitals.**

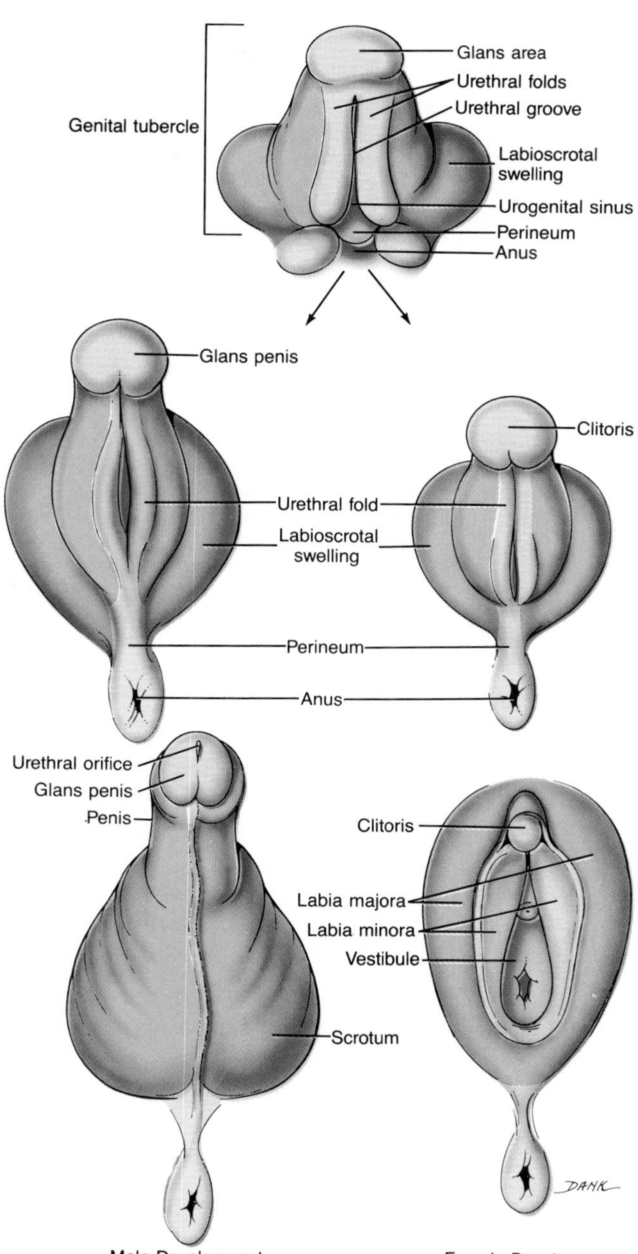

DISORDERS: HOMEOSTATIC IMBALANCES

Sexually Transmitted Diseases (STDs)

The general term *sexually transmitted disease (STD)* is applied to any of the large group of diseases that can be spread by sexual contact. The group includes conditions traditionally specified as *venereal diseases (VD)*, from Venus, goddess of love, such as gonorrhea, syphilis, and genital herpes, and several other conditions that are contracted sexually, or may be contracted otherwise, but are then transmitted to a sexual partner.

DISORDERS: HOMEOSTATIC IMBALANCES (*continued*)

Gonorrhea

Gonorrhea (or ***"clap"***) is an infectious sexually transmitted disease that affects primarily the mucous membrane of the urogenital tract, the rectum, and occasionally the eyes. The disease is caused by the bacterium *Neisseria gonorrhoeae.* In the United States, nearly 1 million new cases of gonorrhea are reported annually. However, the actual number is probably between 3 to 4 million. Most cases are in the 15- to 24-year-old age group. Discharges from the involved mucous membranes are the source of infection, and the bacteria are transmitted by direct contact, usually sexual or during passage of a newborn through the birth canal.

Males usually suffer inflammation of the urethra with pus and painful urination. Fibrosis sometimes occurs in an advanced stage, causing stricture of the urethra. There also may be involvement of the epididymis and prostate gland. In females, infection may occur in the urethra, vagina, and cervix, and there may be a discharge of pus. However, infected females often harbor the disease without any symptoms until it has progressed to a more advanced stage. If the uterine (Fallopian) tubes become involved, pelvic inflammation may follow. Peritonitis, or inflammation of the peritoneum, is a very serious disorder. The infection should be treated and controlled immediately because, if neglected, sterility or death may result. Although antibiotics have greatly reduced the mortality rate of acute peritonitis, it is estimated that between 50,000 and 80,000 women are made sterile by gonorrhea every year as a result of scar tissue formation that closes the uterine tubes. If the bacteria are transmitted to the eyes of a newborn in the birth canal, blindness can result.

Administration of a 1 percent silver nitrate solution or penicillin in the infant's eyes prevents infection. Penicillin or tetracycline are the drugs of choice for the treatment of gonorrhea in adults. However, the incidence of antibiotic-resistant gonorrhea has increased substantially since 1984.

Syphilis

Syphilis is a sexually transmitted disease caused by the bacterium *Treponema pallidum.* In the United States, it affects about 85,000 persons per year. Although the incidence among male homosexuals is 25 to 30 percent higher than in the general population, significant increases in reported cases among heterosexuals have been observed since 1985. The highest incidence is in the 20- to 39-year-old age group. It is acquired through sexual contact or transmitted through the placenta to a fetus. The disease progresses through several stages: primary, secondary, latent, and sometimes tertiary. During the ***primary stage,*** the chief symptom is an open sore, called a ***chancre*** (pronounced SHANKG-ker), at the point of contact. The chancre heals within one to five weeks. From 6 to 24 weeks later, symptoms such as a skin rash, fever, and aches in the joints and muscles usher in the ***secondary stage.*** These symptoms also eventually disappear (in about 4 to 12 weeks), and the disease ceases to be infectious, but a blood test for the presence of the bacteria generally remains positive. During this "symptomless" period, called the ***latent stage,*** the bacteria may invade body organs. When signs of organ degeneration appear, the disease is said to be in the ***tertiary stage.***

If the syphilis bacteria attack the organs of the nervous system, the tertiary stage is called ***neurosyphilis.*** Neurosyphilis may take different forms, depending on the tissue involved. For instance, about two years after the onset of the disease, the bacteria may attack the meninges, producing meningitis. The blood vessels that supply the brain may also become infected. In this case, symptoms depend on the parts of the brain destroyed by oxygen and glucose starvation. Cerebellar damage is manifested by uncoordinated movements in such activities as writing. As the motor areas become extensively damaged, victims may be unable to control urine and bowel movements. Eventually, they may become bedridden, unable even to feed themselves. Damage to the cerebral cortex produces memory loss and personality changes that range from irritability to hallucinations.

Infection of the fetus with syphilis can occur after the fifth month. Infection of the mother is not necessarily followed by fetal infection, provided that the placenta remains intact. But once the bacteria gain access to fetal circulation, there is nothing to impede their growth and multiplication. As many as 80 percent of children born to untreated syphilitic mothers will be infected in the uterus if the fetus is exposed at the onset or in the early stages of the disease. About 25 percent of the fetuses will die within the uterus. Most of the survivors will arrive prematurely, but 30 percent will die shortly after birth. Of the infected and untreated children surviving infancy, about 40 percent will develop symptomatic syphilis during their lifetimes.

Syphilis can be treated with antibiotics (penicillin) during the primary, secondary, and latent periods. Certain forms of neurosyphilis may also be successfully treated, but the prognosis for others is very poor. Noticeable symptoms do not always appear during the first two stages of the disease. Syphilis, however, is usually diagnosed through a blood test whether noticeable symptoms appear or not. The importance of these blood tests and follow-up treatments cannot be overemphasized.

Some evidence suggests that AIDS may alter the course of neurosyphilis by accelerating its progression, possibly by impairing macrophages and antibody production and facilitating penetration of the AIDS virus into the central nervous system.

Genital Herpes

Another sexually transmitted disease, ***genital herpes,*** is common in the United States. Each year, between 400,000 and 600,000 new cases are reported. The sexual transmission of the herpes simplex virus is well established. Unlike syphilis and gonorrhea, genital herpes

DISORDERS: HOMEOSTATIC IMBALANCES (*continued*)

is incurable. Type I herpes simplex virus is the virus that causes the majority of infections above the waist such as cold sores. Type II herpes simplex virus causes most infections below the waist such as painful genital blisters on the prepuce, glans penis, and penile shaft in males and on the vulva or sometimes high up in the vagina in females. The blisters disappear and reappear in most patients, but the virus itself remains in the body.

Genital herpes virus infection causes considerable discomfort, and there is an extraordinarily high rate of recurrence of the symptoms. The infection is usually characterized by fever, chills, flulike symptoms, lymphadenopathy, and numerous clusters of genital blisters. For pregnant women with genital herpes symptoms at the time of delivery, a cesarean section will usually prevent complications in the child. Complications range from a mild asymptomatic infection to central nervous system (CNS) damage to death.

Treatment of the symptoms involves pain medication, saline compresses, sexual abstinence for the duration of the eruption, and use of an oral drug called acyclovir (Zovirax). This drug interferes with viral DNA replication but not with host cell DNA replication. Acyclovir speeds the healing and sometimes reduces the pain of initial genital herpes infections and shortens the duration of lesions in patients with recurrent genital herpes. A topically applied ointment that contains Inter Vir-A (Immuvir), an antiviral substance, is another drug used to treat genital herpes. Inter Vir-A provides rapid relief for the pain, itching, and burning associated with genital herpes. An experimental genital herpes vaccine will involve human testing shortly.

Chlamydia

Chlamydia (kla-MID-ē-a) is a sexually transmitted disease caused by the bacterium *Chlamydia trachomatis* (*chlamys* = cloak; the bacterium cannot grow outside the body; it cloaks itself inside cells to divide). At present, chlamydia is the most prevalent and one of the most damaging of the sexually transmitted diseases. It affects between 3 and 5 million persons annually. In males, urethritis is the principal result. It is characterized by burning on urination, frequency of urination, painful urination, and low back pain. In females, urethritis may spread through the reproductive tract and develop into inflammation of the uterine (Fallopian) tubes, which increases the risk of ectopic pregnancy and sterility. As in gonorrhea, the organism may be passed from mother to infant during childbirth, infecting the eyes. Treatment consists of the administration of tetracycline or doxycycline.

Trichomoniasis

The microorganism *Trichomonas vaginalis,* a flagellated protozoan (one-celled animal), causes ***trichomoniasis,*** an inflammation of the mucous membrane of the vagina in females and the urethra in males. *T. vaginalis* is a common inhabitant of the vagina of females and urethra of males. If the normal acidity of the vagina is disrupted, the protozoan may overgrow the normal microbial population and cause trichomoniasis. Symptoms include a yellow vaginal discharge with a particularly offensive odor and severe vaginal itch in women. Men can have it without symptoms but can transmit it to women nonetheless. Sexual partners must be treated simultaneously. The drug of choice is metronidazole.

Genital Warts

Warts are an infectious disease caused by viruses. Sexual transmission of ***genital warts*** is common and is caused by the human papilloma virus (HPV). It is estimated that nearly 1 million persons a year develop genital warts. Patients with a history of genital warts may be at increased risk for certain types of cancer (cervical, vaginal, anal, vulval, and penile). There is no cure for genital warts. Treatment consists of cryotherapy with liquid nitrogen, electrocautery, excision, laser surgery, and topical application of podophyllin in tincture of benzoin. Alpha interferon is also used to treat genital warts.

Male Disorders

Testicular Cancer

Testicular cancer occurs most often between the ages of 15 and 34 and is one of the most common cancers seen in young males. Although the cause is unknown, the condition is associated with males with a history of undescended testes or late-descended testes. Most testicular cancers arise from the sperm-producing cells. An early sign of testicular cancer is a mass in the testis, often associated with pain or discomfort. Treatment involves removal of the diseased testis.

Prostate Disorders

The prostate gland is susceptible to infection, enlargement, and benign and malignant tumors. Because the prostate surrounds the urethra, any of these disorders can obstruct the flow of urine. Prolonged obstruction may result in serious changes in the urinary bladder, ureters, and kidneys and may perpetuate urinary tract infections. An experimental treatment that consists of widening a narrowed urethra with a balloon catheter (balloon urethroplasty) is undergoing clinical trials. If the obstruction cannot be relieved by other means, surgical removal of part of or the entire gland is indicated. The surgical procedure is called ***prostatectomy*** (pros'-ta-TEK-tō-mē).

Acute and chronic infections of the prostate gland are common in postpubescent males, often in association with inflammation of the urethra. In ***acute prostatitis,*** the prostate gland becomes swollen and tender. Appropriate antibiotic therapy, bed rest, and above-normal fluid intake are effective treatment.

Chronic prostatitis is one of the most common chronic infections in men of the middle and later years. On examination, the prostate gland feels enlarged, soft, and extremely tender, and its surface outline is irregular.

DISORDERS: HOMEOSTATIC IMBALANCES (*continued*)

This disease frequently produces no symptoms, but the prostate is believed to harbor infectious microorganisms responsible for some allergic conditions, arthritis, and inflammation of nerves (neuritis), muscles (myositis), and the iris (iritis).

An enlarged prostate gland, increasing to two to four times larger than normal, occurs in approximately one-third of all males over age 60. The condition is called ***benign prostatic hyperplasia* (*BPH*)** and is characterized by nocturia (bed-wetting), hesitancy in urination, decreased force of urinary stream, postvoiding dribbling, and a sensation of incomplete emptying. The condition may be corrected surgically by a procedure called ***transurethral resection of the prostate* (*TURP*)**, in which pieces of the gland are removed using a special cystoscope inserted into the urethra. The enlarged condition usually can be detected by rectal examination.

Prostate cancer is the second leading cause of death from cancer in men in the United States, and it is responsible for approximately 26,000 deaths annually. Its incidence is related to age, race, occupation, geography, and ethnic origin. Both benign and malignant growths are common in elderly men. Both types of tumors put pressure on the urethra, making urination painful and difficult. At times, the excessive back pressure destroys kidney tissue and gives rise to an increased susceptibility to infection. Therefore, even when the tumor is benign, surgery is indicated. Prostate cancer may be detected by digital rectal examination and fine-needle aspiration. In addition, a new procedure called ***transrectal ultrasonography*** is used. In the procedure, a rectal probe is used to bounce sound waves off the prostate gland. The returning echoes are converted into an image that can be viewed on a monitor and printed on paper. The procedure can detect tumors as small as a grain of rice.

Sexual Functional Abnormalities

Impotence (*impotenia* = lack of strength) is the inability of an adult male to attain or hold an erection long enough for sexual intercourse. Impotence could be the result of diabetes mellitus, physical abnormalities of the penis, systemic disorders such as syphilis, vascular disturbances (arterial or venous obstructions), neurological disorders, testosterone deficiency, drugs (alcohol, antidepressants, antihistamines, antihypertensives, narcotics, nicotine, and tranquilizers), or psychic factors such as fear of causing pregnancy, fear of sexually transmitted diseases, religious inhibitions, and emotional immaturity. In selected individuals, penile implants may be indicated. Penile injections of papaverine (Pavabid), a vasodilator, and phentolamine mesylate (Regitine), an alpha-adrenergic blocker, can produce excellent effects in overcoming both physical and psychological impotence.

***Male infertility* (*sterility*)** is an inability to fertilize the ovum. It does not imply impotence. Male fertility requires production of adequate amounts of viable, normal spermatozoa by the testes, unobstructed transportation of sperm through the seminal tract, and satisfactory deposition in the vagina. The tubules of the testes are sensitive to many factors—x-rays, infections, toxins, malnutrition, and significantly higher-than-normal scrotal temperatures—that may cause degenerative changes and produce male sterility. If inadequate spermatozoa production is suspected, a sperm analysis should be performed. At least one type of infertility may be improved by administration of vitamin C.

Female Disorders

Menstrual Abnormalities

Because menstruation reflects not only the health of the uterus but also the health of the endocrine glands that control it, the ovaries and the pituitary gland, disorders of the female reproductive system frequently involve menstrual disorders.

Amenorrhea (ā-men'-ō-RĒ-a; *a* = without; *men* = month; *rhein* = to flow) is the absence of menstruation. If a woman has never menstruated, the condition is called ***primary amenorrhea.*** Primary amenorrhea can be caused by endocrine disorders, most often in the pituitary gland and hypothalamus, or by a genetically caused abnormal development of the ovaries or uterus. ***Secondary amenorrhea,*** the skipping of one or more periods, is commonly experienced by women at some time during their lives. Changes in body weight, either gains or losses, often cause amenorrhea. Obesity may disturb ovarian function, and similarly, the extreme weight loss that characterizes anorexia nervosa often leads to a suspension of menstrual flow. When amenorrhea is unrelated to weight, analysis of levels of estrogens often reveals deficiencies of pituitary and ovarian hormones. Amenorrhea may also be caused by continuous involvement in rigorous athletic training (exercise-associated amenorrhea). This may be related to increased levels of glucocorticoids and decreased levels of estrogens, progesterone, and prolactin.

Dysmenorrhea (dis'-men-ō-RĒ-a; *dys* = difficult) refers to pain associated with menstruation and is usually reserved to describe an individual with menstrual symptoms that are severe enough to prevent her from functioning normally for one or more days each month. ***Primary dysmenorrhea*** is painful menstruation with no detectable organic disease. The pain of primary dysmenorrhea is thought to result from uterine contractions, probably associated with uterine muscle ischemia and prostaglandins produced by the uterus. Prostaglandins are known to stimulate uterine contractions, but they cannot do so in the presence of high levels of progesterone. As we have noted earlier, progesterone levels are high during the last half of the menstrual cycle. During this time, prostaglandins are apparently inhibited by progesterone from producing uterine contractions. However, if pregnancy does not occur, progesterone levels drop rapidly and prostaglandin production increases. This causes the uterus to contract and slough off its lin-

DISORDERS: HOMEOSTATIC IMBALANCES (*continued*)

ing and may result in dysmenorrhea. In addition to pain, other signs and symptoms may include headache, nausea, diarrhea or constipation, and urinary frequency. Primary dysmenorrhea is less of a problem after pregnancy and vaginal delivery, perhaps because of enlargement of the endocervical canal. Drugs that inhibit prostaglandin synthesis (naproxen and ibuprofen) are used to treat primary dysmenorrhea.

Secondary dysmenorrhea is painful menstruation that is frequently associated with a pelvic pathology. Some cases are caused by uterine tumors, ovarian cysts, pelvic inflammatory disease (PID), endometriosis, and intrauterine devices (IUDs). Treatment is aimed at correction of the underlying cause.

Abnormal uterine bleeding includes menstruation of excessive duration or excessive amount, diminished menstrual flow, too frequent menstruation, intermenstrual bleeding, and postmenopausal bleeding. These abnormalities may be caused by disordered hormonal regulation, emotional factors, fibroid tumors of the uterus, and systemic diseases.

Premenstrual syndrome (PMS) is a term usually reserved for severe physical and emotional distress occurring late in the postovulatory phase of the menstrual cycle and sometimes overlapping with menstruation. Signs and symptoms usually increase in severity until the onset of menstruation and then dramatically disappear. Among the signs and symptoms are edema, weight gain, breast swelling and tenderness, abdominal distension, backache, joint pain, constipation, skin eruptions, fatigue and lethargy, greater need for sleep, depression or anxiety, irritability, mood swings, headache, poor coordination and clumsiness, and cravings for sweet or salty foods. The basic cause of PMS is unknown. Although PMS is related to the cyclic production of ovarian hormones, the symptoms are not directly due to changes in the levels of these hormones. Treatment is individualized, depending on the type and severity of symptoms, and may include dietary changes, exercise, over-the-counter drugs (aspirin or acetaminophen), psychoactive drugs (sedatives, tranquilizers, and antidepressants), prostaglandins (Ponstel), diuretics, hormone therapy (progesterone), and vitamin B_6. One approach to management of PMS provides medical, psychological, and social support that may include education of the patient and her family; elimination of fears or inappropriate beliefs regarding the menstrual cycle; alteration in coping style; change in lifestyle, occupation, or family relationships; and use of appropriate medications.

Toxic Shock Syndrome (TSS)

Toxic shock syndrome (TSS), first described in 1978, is primarily a disease of previously healthy, young, menstruating females who use tampons. It is also recognized in males, children, and nonmenstruating females. Clinically, TSS is characterized by high fever up to 40.6°C (105°F), sore throat or very tender mouth, headache, fatigue, irritability, muscle soreness and tenderness, conjunctivitis, diarrhea and vomiting, abdominal pain, vaginal irritation, and erythematous rash. Other symptoms include lethargy, unresponsiveness, memory loss, hypotension, peripheral vasoconstriction, respiratory distress syndrome, intravascular coagulation, decreased platelet count, renal failure, circulatory shock, and liver involvement.

Toxin-producing strains of the bacterium *Staphylococcus aureus* are necessary for development of the disease. Actually, it appears that a virus has become incorporated into *S. aureus,* causing the bacterium to produce the toxins. Although all tampon users are at some risk for developing TSS, the risk is increased considerably by females who use highly absorbent tampons. Apparently, high-absorbency tampons provide a substrate on which the bacteria grow and produce toxins. There is some evidence that the absorption of magnesium by the fibers of the tampon is a factor in toxic shock syndrome. As the metal is absorbed, its absence slows microbial growth in the reproductive tract but stimulates the staphylococci to produce toxin. TSS can also occur as a complication of influenza and influenzalike illness and use of contraceptive sponges. Initial therapy is directed at correcting all homeostatic imbalances as quickly as possible. Antistaphylococcal antibiotics, such as penicillin or clindamycin, are also administered. In severe cases, high doses of corticosteroids are also administered.

Ovarian Cysts

Ovarian cysts are fluid-containing tumors of the ovary. Follicular cysts may occur in the ovaries of elderly women, in ovaries that have inflammatory diseases, and in menstruating females. They have thin walls and contain a serous albuminous material. Cysts may also arise from the corpus luteum or the endometrium.

Endometriosis

Endometriosis (en'-dō-mē-trē-Ō-sis; *endo* = within; *metri* = uterus; *osis* = condition) is a benign condition characterized by the growth of endometrial tissue outside the uterus. The tissue enters the pelvic cavity via the open uterine (Fallopian) tubes and may be found in any of several sites—on the ovaries, rectouterine pouch, surface of the uterus, sigmoid colon, pelvic and abdominal lymph nodes, cervix, abdominal wall, kidneys, and urinary bladder. One theory for the development of endometriosis is that there is regurgitation of menstrual flow through the uterine tubes. Another theory is that migrational events during embryonic development are somehow altered. Endometriosis is common in women 25 to 40 years of age who have not had children. Symptoms include premenstrual pain or unusual menstrual pain. The unusual pain is caused by the displaced tissue sloughing off at the same time the normal uterine endometrium is being shed during menstruation. Infertility can be a conse-

DISORDERS: HOMEOSTATIC IMBALANCES (*continued*)

quence. Treatment usually consists of hormone therapy, modified GnRH (nafarelin), videolaseroscopy (laparoscope with camera and laser), or conventional surgery. Endometriosis disappears at menopause or when the ovaries are removed.

Female Infertility

Female infertility, or the inability to conceive, occurs in about 10 percent of married females in the United States. Once it is established that ovulation occurs regularly, the reproductive tract is examined for functional and anatomical disorders to determine the possibility of union of the sperm and the ovum in the uterine tube. Female infertility may be caused by tubal obstruction, ovarian disease, and certain conditions of the uterus. An upset in hormone balance, so that the endometrium is not adequately prepared to receive the fertilized ovum, may also be the problem. Some research suggests that an autoimmune disease might underlie many cases of infertility. Infertility treatment may involve the use of fertility drugs, donor (artificial) insemination, or surgery. Gynecologists are now using a procedure called ***transcervical balloon tuboplasty*** to clear obstructions in the uterine tubes. The technique, borrowed from the cardiology procedure to unclog coronary arteries, consists of inserting a catheter through the cervix of the uterus and into the uterine tube. Then a balloon is inflated, compressing the obstruction.

Disorders Involving the Breasts

The breasts of females are highly susceptible to cysts and tumors. Men are also susceptible to breast tumors, but certain breast cancers are 100 times more common in women.

In the female, the benign ***fibroadenoma*** is a common tumor of the breast. It occurs most frequently in young women. Fibroadenomas have a firm rubbery consistency and are easily moved about within the mammary tissue. The usual treatment is excision of the growth. The breast itself is not removed.

Breast cancer has one of the highest fatality rates of all cancers affecting women, but it is rare in men. In the female, breast cancer is rarely seen before age 30, and its occurrence rises rapidly after menopause. Breast cancer is generally not painful until it becomes quite advanced, so often it is not discovered early or, if noted, is ignored. Any lump, no matter how small, should be reported to a doctor at once. New evidence links some breast cancers to loss of protective anti-oncogenes.

Cervical Cancer

Another common disorder of the female reproductive tract is ***cervical cancer,*** carcinoma of the cervix of the uterus. The condition starts with ***cervical dysplasia*** (dis-PLĀ-sē-a), a change in the shape, growth, and number of the cervical cells. If the condition is minimal, the cells may regress to normal. If it is severe, it may progress to cancer. Cervical cancer may be detected in most cases in its earliest stages by a Pap smear. There is some evidence linking cervical cancer to penile virus (papillomavirus) infections of male sexual partners. Depending on the progress of the disease, treatment may consist of excision of lesions, radiotherapy, chemotherapy, and hysterectomy.

Pelvic Inflammatory Disease (PID)

Pelvic inflammatory disease* (PID)** is a collective term for any extensive bacterial infection (primarily involving ***Chlamydia trachomatis, Neisseria gonorrhoeae, Bacteroides, Peptostreptococcus, and ***Gardnerella vaginalis***) of the pelvic organs, especially the uterus, uterine (Fallopian) tubes, or ovaries. A vaginal or uterine infection may spread into the uterine tube (***salpingitis***) or even farther into the abdominal cavity, where it infects the peritoneum (***peritonitis***). Diagnosis of PID depends on three findings: abdominal tenderness; cervical tenderness; and ovarian, uterine tube, and uterine ligament tenderness. In addition, diagnosis is based on at least one of the following: fever, leucocytosis, pelvic abscess or inflammation, purulent cervical discharge, and the presence of certain bacteria in smears. Early treatment with bed rest and antibiotics (cefoxitin, penicillin, tetracycline, doxycycline) can stop the spread of PID.

Vulvovaginal Candidiasis

Candida albicans is a yeastlike fungus that commonly grows on mucous membranes of the gastrointestinal and genitourinary tracts. The organism is responsible for ***vulvovaginal candidiasis,*** the most common form of vaginitis. It is characterized by severe itching; a thick, yellow, cheesy discharge; a yeasty odor; and pain. The disorder, experienced at least once by about 75 percent of females, is usually a result of proliferation of the fungus following antibiotic therapy for another condition. Predisposing conditions include use of oral contraceptives, cortisonelike medications, pregnancy, and diabetes. Treatment is by topical (clostrinazole) or oral (ketoconazole) antibiotics.

MEDICAL TERMINOLOGY ASSOCIATED WITH THE REPRODUCTIVE SYSTEMS

Castration (kas-TRĀ-shun; *castrare* = to prune) Removal, inactivation, or destruction of the testes.

Colpotomy (kol-POT-ō-mē; *colp* = vagina; *tome* = cutting) Incision of the vagina.

MEDICAL TERMINOLOGY ASSOCIATED WITH THE REPRODUCTIVE SYSTEMS (*continued*)

Culdoscopy (kul-DOS-kō-pē; *skopein* = to examine) A procedure in which a culdoscope (endoscope) is used to view the pelvic cavity. The approach is through the vagina.

Culdotomy (kul-DOT-ō-mē; *tome* = cutting) An incision into (or needle aspiration of) the cul-de-sac to test for the presence of pelvic bleeding.

Hermaphroditism (her-MAF-rō-dī-tizm') Presence of both male and female sex organs in one individual.

Hypospadias (hī'-pō-SPĀ-dē-as; *hypo* = below; *span* = to draw) A displaced urethral opening. In the male, the opening may be on the underside of the penis, at the penoscrotal junction, between the scrotal folds, or in the perineum. In the female, the urethra opens into the vagina.

Leukorrhea (loo'-kō-RĒ-a; *leuco* = white; *rrhea* = discharge) A nonbloody vaginal discharge that may occur at any age and affects most women at some time.

Oophorectomy (ō'-of-ō-REK-tō-mē; *oophoro* = bearing eggs) Removal of the ovaries.

Salpingectomy (sal'-pin-JEK-tō-mē; *salpingo* = tube) Excision of a uterine (Fallopian) tube.

Smegma (SMEG-ma; *smegma* = soap) The secretion, consisting principally of desquamated epithelial cells, found chiefly about the external genitalia and especially under the foreskin of the male.

Vaginitis (vaj'-i'-NĪ-tis) Inflammation of the vagina.

STUDY OUTLINE

Male Reproductive System: Structure and Physiology (p. 879)

1. Reproduction is the process by which genetic material is passed on from one generation to the next.
2. The organs of reproduction are grouped as gonads (produce gametes), ducts (transport and store gametes), and accessory sex glands (produce materials that support gametes).
3. The male structures of reproduction include the testes, ductus epididymis, ductus (vas) deferens, ejaculatory duct, urethra, seminal vesicles, prostate gland, bulbourethral (Cowper's) glands, and penis.

Scrotum (p. 879)

1. The scrotum is a cutaneous outpouching of the abdomen that supports the testes.
2. It regulates the temperature of the testes by contraction of the cremaster muscle, which elevates them and brings them closer to the pelvic cavity or relaxes causing testes to move farther from the pelvic cavity.

Testes (p. 880)

1. The testes are oval-shaped glands (gonads) in the scrotum containing seminiferous tubules, in which sperm cells are made; sustentacular (Sertoli) cells, which nourish sperm cells and secrete inhibin; and interstitial endocrinocytes (cells of Leydig), which produce the male sex hormone testosterone.
2. Failure of the testes to descend is called cryptorchidism.
3. Ova and sperm are collectively called gametes, or sex cells, and are produced in gonads.
4. Uninucleated somatic cells divide by mitosis, the process in which each daughter cell receives the full complement of 23 chromosome pairs (46 chromosomes). Somatic cells are said to be diploid ($2n$).
5. Immature gametes divide by meiosis, in which the pairs of chromosomes are split so that the mature gamete has only 23 chromosomes. It is said to be haploid (n).
6. Spermatogenesis occurs in the testes. It results in the formation of four haploid spermatozoa from each primary spermatocyte.
7. Spermatogenesis is a process in which immature spermatogonia develop into mature spermatozoa. The spermatogenesis sequence includes reduction division (meiosis I), equatorial division (meiosis II), and spermiogenesis.
8. Mature spermatozoa consist of a head, midpiece, and tail. Their function is to fertilize an ovum.
9. At puberty, gonadotropin releasing hormone (GnRH) stimulates anterior pituitary secretion of FSH and LH. FSH initiates spermatogenesis, and LH assists spermatogenesis and stimulates production of testosterone.
10. Testosterone controls the growth, development, and maintenance of sex organs; stimulates bone growth, protein anabolism, and sperm maturation; and stimulates development of male secondary sex characteristics.
11. Inhibin is produced by sustentacular (Sertoli) cells. Its inhibition of FSH helps to regulate the rate of spermatogenesis.
12. Puberty refers to the period of time when secondary sex characteristics begin to develop and the potential for sexual reproduction is reached.
13. The onset of male puberty is signaled by increased levels of LH, FSH, and testosterone.

Ducts (p. 887)

1. The duct system of the testes includes the seminiferous tubules, straight tubules, and rete testis.
2. Sperm are transported out of the testes through the efferent ducts.
3. The ductus epididymis is lined by stereocilia and is the site of sperm maturation and storage.
4. The ductus (vas) deferens stores sperm and propels them toward the urethra during ejaculation.
5. Alteration of the ductus (vas) deferens to prevent fertilization is called vasectomy.
6. The ejaculatory ducts are formed by the union of the ducts

from the seminal vesicles and ductus (vas) deferens and eject spermatozoa into the prostatic urethra.
7. The male urethra is subdivided into three portions: prostatic, membranous, and spongy (cavernous).

Accessory Sex Glands (p. 890)

1. The seminal vesicles secrete an alkaline, viscous fluid that constitutes about 60 percent of the volume of semen and contributes to sperm viability.
2. The prostate gland secretes a slightly acid fluid that constitutes about 13 to 33 percent of the volume of semen and contributes to sperm motility.
3. The bulbourethral (Cowper's) gland secretes mucus for lubrication and a substance that neutralizes acid.
4. Semen (seminal fluid) is a mixture of spermatozoa and accessory sex gland secretions that provides the fluid in which spermatozoa are transported, provides nutrients, and neutralizes the acidity of the male urethra and female vagina.

Penis (p. 891)

1. The penis is the male organ of copulation that consists of a root, body, and glans penis.
2. Expansion of its blood sinuses under the influence of sexual excitation is called erection.

Female Reproductive System: Structure and Physiology (p. 892)

1. The female organs of reproduction include the ovaries (gonads), uterine (Fallopian) tubes, uterus, vagina, and vulva.
2. The mammary glands are considered part of the reproductive system.

Ovaries (p. 892)

1. The ovaries are female gonads located in the upper pelvic cavity, on either side of the uterus.
2. They produce secondary oocytes, discharge secondary oocytes (ovulation), and secrete estrogens, progesterone, inhibin and relaxin.
3. Oogenesis occurs in the ovaries. It results in the formation of a single haploid secondary oocyte.
4. The oogenesis sequence includes reduction division (meiosis I), equatorial division (meiosis II), and maturation.

Uterine (Fallopian) Tubes (p. 897)

1. The uterine (Fallopian) tubes transport ova from the ovaries to the uterus and are the normal sites of fertilization.
2. Ciliated cells and peristaltic contractions help move a secondary oocyte toward the uterus.

Uterus (p. 898)

1. The uterus is an organ shaped like an inverted pear that functions in transporting spermatozoa, menstruation, implantation of a fertilized ovum, development of a fetus during pregnancy, and labor.
2. The uterus is normally held in position by a series of ligaments.
3. Histologically, the uterus consists of an outer perimetrium, middle myometrium, and inner endometrium.

Endocrine Relations: Menstrual and Ovarian Cycles (p. 900)

1. The function of the menstrual cycle is to prepare the endometrium each month for the reception of a fertilized egg.
2. The menstrual and ovarian cycles are controlled by GnRH, which stimulates the release of FSH and LH.
3. FSH stimulates the initial development of ovarian follicles and secretion of estrogens by the ovaries. LH stimulates further development of ovarian follicles, ovulation, and the secretion of estrogens and progesterone by the ovaries.
4. Estrogens stimulate the growth, development, and maintenance of female reproductive structures; stimulate the development of secondary sex characteristics; regulate fluid and electrolyte balance; and stimulate protein anabolism.
5. Progesterone (PROG) works with estrogens to prepare the endometrium for implantation and the mammary glands for milk secretion.
6. Relaxin relaxes the symphysis pubis and helps dilate the uterine cervix to facilitate delivery, and increases sperm motility.
7. During the menstrual phase, the stratum functionalis layer of the endometrium is shed with a discharge of blood, tissue fluid, mucus, and epithelial cells. Primary follicles develop into secondary follicles.
8. During the preovulatory phase, endometrial repair occurs. A secondary follicle develops into a vesicular (Graafian) follicle. Estrogens are the dominant ovarian hormones.
9. Ovulation is the rupture of a vesicular (Graafian) follicle and the release of a secondary oocyte into the pelvic cavity brought about by inhibition of FSH and a surge of LH. Signs of ovulation include increased basal body temperature; clear, stretchy cervical mucus; changes in the uterine cervix, and ovarian pain.
10. During the postovulatory phase, the endometrium thickens in anticipation of implantation. Progesterone is the dominant ovarian hormone.
11. If fertilization and implantation do not occur, the corpus luteum degenerates, and low levels of estrogens and progesterone initiate another menstrual and ovarian cycle.
12. If fertilization and implantation do occur, the corpus luteum is maintained by placental hCG, and the corpus luteum and placenta secrete estrogens and progesterone to support pregnancy and breast development for lactation.
13. The female climacteric is the time immediately before menopause, the cessation of the sexual cycles.
14. The onset of female puberty is signaled by increased levels of LH, FSH, and estrogens.

Vagina (p. 905)

1. The vagina is a passageway for spermatozoa and the menstrual flow, the receptacle of the penis during sexual intercourse, and the lower portion of the birth canal.
2. It is capable of considerable distension to accomplish its functions.

Vulva (p. 906)

1. The vulva is a collective term for the external genitals of the female.
2. It consists of the mons pubis, labia majora, labia minora, clitoris, vestibule, vaginal and urethral orifices, hymen, bulb of the vestibule, and the paraurethral (Skene's), greater vestibular (Bartholin's), and lesser vestibular glands.

Perineum (p. 907)

1. The perineum is a diamond-shaped area at the inferior end of the trunk between the thighs and buttocks.
2. An incision in the perineal skin prior to delivery is called an episiotomy.

Mammary Glands (p. 907)

1. The mammary glands are modified sweat glands (branched tubuloalveolar) over the pectoralis major muscles. Their function is to secrete and eject milk (lactation).
2. Mammary gland development is dependent on estrogens and progesterone.
3. Milk secretion is mainly due to prolactin (PRL), and milk ejection is stimulated by oxytocin (OT).
4. Screening for breast cancer may involve mammography, ultrasound (US), and computed tomography combined with mammography (CT/M).

Physiology of Sexual Intercourse (p. 911)

1. The role of the male in the sex act involves erection, minimal lubrication, and orgasm.
2. The female role also involves erection, lubrication, and orgasm (climax).

Birth Control (BC) (p. 912)

1. Methods include sterilization (vasectomy, tubal ligation), hormonal, intrauterine devices, barriers (condom, diaphragm, cervical cap), chemical (spermicides), physiologic (rhythm, sympto-thermal method), coitus interruptus, induced abortion).
2. Contraceptive pills of the combination type contain estrogens and progesterone in concentrations that decrease the secretion of FSH and LH and thereby inhibit ovulation.

Aging and the Reproductive Systems (p. 914)

1. In the male, decreased levels of testosterone decrease muscle strength, sexual desire, and viable sperm; prostate disorders are common.
2. In the female, levels of progesterone and estrogens decrease, resulting in changes in menstruation; uterine and breast cancer increase in incidence.

Developmental Anatomy of the Reproductive Systems (p. 914)

1. The gonads develop from intermediate mesoderm and are differentiated into ovaries or testes by about the seventh week of fetal development.
2. The external genitals develop from the genital tubercle.

Disorders: Homeostatic Imbalances (p. 917)

1. Sexually transmitted diseases (STDs) are diseases spread by sexual contact and include gonorrhea, syphilis, genital herpes, chlamydia, trichomoniasis, and genital warts.
2. Testicular cancer originates in sperm-producing cells.
3. Conditions that affect the prostate gland are prostatitis, benign prostatic hyperplasia (BPH), and cancer.
4. Impotence is the inability of the male to attain or hold an erection long enough for intercourse.
5. Male infertility is the inability of a male's sperm to fertilize an ovum.
6. Menstrual disorders include amenorrhea, dysmenorrhea, abnormal bleeding, and premenstrual syndrome (PMS).
7. Toxic shock syndrome (TSS) includes widespread homeostatic imbalances and is a reaction to toxins produced by *Staphylococcus aureus*.
8. Ovarian cysts are tumors that contain fluid.
9. Endometriosis refers to the growth of uterine tissue outside the uterus.
10. Female infertility is the inability of the female to conceive.
11. The mammary glands are susceptible to benign fibroadenomas and malignant tumors. The removal of a malignant breast, pectoral muscles, and lymph nodes is called a radical mastectomy.
12. Cervical cancer starts with dysplasia and can be diagnosed by a Pap test.
13. Pelvic inflammatory disease (PID) refers to bacterial infection of pelvic organs.
14. Vulvovaginal candidiasis is a form of vaginitis caused by the yeastlike fungus *Candida albicans*.

REVIEW QUESTIONS

1. Define reproduction. Describe how the reproductive organs are classified and list the male and female organs of reproduction. (p. 879)
2. Describe the function of the scrotum in protecting the testes from temperature fluctuations. (p. 879)
3. Describe the internal structure of a testis. Where are the sperm cells made? (p. 880)
4. Describe the principal events of spermatogenesis. Why is meiosis important? Distinguish between haploid (n) and diploid ($2n$) cells. (p. 883)
5. Identify the principal parts of a spermatozoon. List the functions of each. (p. 885)
6. Explain the effects of FSH and LH on the male reproductive system. How are these hormones controlled by GnRH? (p. 885)
7. Describe the physiological effects of testosterone and inhibin on the male reproductive system. How is testosterone level controlled? (p. 885)
8. Which ducts are involved in transporting sperm *within* the testes? (p. 887)
9. Describe the location, structure, and functions of the ductus epididymis, ductus (vas) deferens, and ejaculatory duct. (p. 887)
10. What is the spermatic cord? (p. 888)
11. Give the location of the three subdivisions of the male urethra. (p. 890)
12. Trace the course of spermatozoa through the male system of ducts from the seminiferous tubules through the urethra. (p. 881)
13. Briefly explain the locations and functions of the seminal vesicles, prostate gland, and bulbourethral (Cowper's) glands. How is cancer of the prostate gland detected by a blood test? (p. 890)
14. What is semen? What is its function? (p. 890)
15. How is the penis structurally adapted as an organ of copulation? How does an erection occur? (p. 891)
16. How are the ovaries held in position in the pelvic cavity? (p. 892)
17. Describe the microscopic structure of an ovary. What are the functions of the ovaries? (p. 893)

18. Describe the principal events of oogenesis. (p. 893)
19. Where are the uterine (Fallopian) tubes located? What is their function? (p. 897)
20. Diagram the principal parts of the uterus. (p. 898)
21. Describe the arrangement of ligaments that hold the uterus in its normal position. What is retroflexion? (p. 898)
22. Discuss the blood supply to the uterus. Why is an abundant blood supply important? (p. 898)
23. Describe the histology of the uterus. (p. 898)
24. What is the function of each of the following in the menstrual and ovarian cycles: GnRH, FSH, LH, estrogens, progesterone, and inhibin? (p. 900)
25. Briefly outline the major events of each phase of the menstrual cycle and correlate them with the events of the ovarian cycle. (p. 901)
26. Prepare a labeled diagram of the principal hormonal interactions involved in the menstrual and ovarian cycles. (p. 902)
27. What are the signs that ovulation has occurred? (p. 903)
28. What is menarche? Define female climacteric and menopause. (p. 904)
29. What is the function of the vagina? Describe its histology. (p. 905)
30. List the parts of the vulva and the functions of each part. (p. 906)
31. Describe the structure of the mammary glands. How are they supported? (p. 907)
32. Describe the passage of milk from the alveolar cells of the mammary gland to the nipple. (p. 908)
33. Explain the roles of estrogens and progesterone in the development of the mammary glands. (p. 908)
34. Define lactation. How is it controlled? (p. 909)
35. How is breast cancer detected? How is breast cancer treated? (p. 909)
36. Explain the role of the male's erection, lubrication, and orgasm in the sex act. How do the female's erection, lubrication, and orgasm (climax) contribute to the sex act? (p. 911)
37. Briefly describe the various methods of birth control (BC) and the effectiveness of each. (p. 912)
38. Explain the effects of aging on the reproductive systems. (p. 914)
39. Describe the development of the reproductive systems. (p. 914)
40. Define a sexually transmitted disease (STD). Describe the cause, clinical symptoms, and treatment of gonorrhea, syphilis, genital herpes, chlamydia, trichomoniasis, and genital warts. (p. 917)
41. What is testicular cancer? (p. 919)
42. Describe several disorders that affect the prostate gland. (p. 919)
43. What are some of the causes of amenorrhea, dysmenorrhea, and abnormal uterine bleeding? (p. 920)
44. Describe the clinical symptoms of premenstrual syndrome (PMS) and toxic shock syndrome (TSS). (p. 921)
45. What are ovarian cysts? Define endometriosis. (p. 921)
46. What is cervical cancer? Relate the condition to cervical dysplasia. (p. 922)
47. Define pelvic inflammatory disease (PID) and vulvovaginal candidiasis. (p. 922)
48. Define the following: cryptorchidism (p. 880), vasectomy (p. 887), circumcision (891), imperforate hymen (p. 905), episiotomy (907), and donor insemination. (p. 911)
49. Explain the diagnostic value of the following: semen analysis (p. 891), Papanicolaou smear (p. 899), colposcopy (p. 900), cone biopsy (p. 900), dilatation and curettage (p. 900), and mammography. (p. 909)
50. Refer to the glossary of medical terminology at the end of the chapter. Be sure that you can define each term. (p. 922)

SELECTED READINGS

Cole, H. M. "Intrauterine Devices," *Journal of the American Medical Association,* 14 April 1989.

Copeland, L. J. "Screening for Cervical Cancer: The Role of the Pap Test," *Modern Medicine,* January 1987.

Danforth, D. N., and J. R. Scott (eds). *Obstetrics and Gynecology,* 5th ed. Philadelphia: Lippincott, 1986.

Goldfinger, S. E. (ed). "Breast Cancer: Early Decisions," *Harvard Medical School Health Letter,* March 1988.

———. "Circumcision and Urinary Tract Infections," *Harvard Medical School Health Letter,* April 1989.

———. Prostate Cancer," *Harvard Medical School Health Letter,* September 1988.

Johnson, G. T. (ed). "Cesarean Section," *Harvard Medical School Health Letter,* June 1981.

———. "Sexually Transmitted Diseases," *Harvard Medical School Health Letter,* April 1981.

Kiely, J. M. (ed). "Endometriosis," *Mayo Clinic Health Letter,* March 1987.

———. "Premenstrual Syndrome," *Mayo Clinic Health Letter,* February 1987.

McElhose, P. "The Other STDs: As Dangerous As Ever," *RN,* June 1988.

McKeon, V. A. "Cruel Myths and Clinical Facts About Menopause," *RN,* June 1989.

Mishell, D. R. "Contraception," *New England Journal of Medicine,* 23 March 1989.

Nero, F. A. "When Couples Ask About Infertility," *RN,* November 1988.

Orshan, S. A. "The Pill, the Patient, and You," *RN,* July 1988.

Rivers, R. R. "Breast Cancer," *Postgraduate Medicine,* November 1988.

Sweet, R. L. "Pelvic Inflammatory Disease," *Modern Medicine,* April 1989.

Townsend, C. M. "Management of Breast Cancer," *CIBA–GEIGY Clinical Symposia,* 39(4), 1987.

Chapter 29

Development and Inheritance

Chapter Contents at a Glance

Student Objectives

1. Explain the activities associated with fertilization, morula formation, blastocyst development, and implantation.
2. Describe how in vitro fertilization (IVF) is performed.
3. Discuss the principal body changes associated with embryonic and fetal growth.
4. Compare the sources and functions of the hormones secreted during pregnancy.
5. Describe how pregnancy is diagnosed and describe some of the anatomical and physiological changes associated with gestation.
6. Explain the respiratory and cardiovascular adjustments that occur in an infant at birth.
7. Discuss potential hazards to an embryo and fetus associated with chemicals and drugs, irradiation, alcohol, and cigarette smoking.
8. Discuss the physiology and control of lactation.
9. Define inheritance and describe the inheritance of several traits.
10. Define medical terminology associated with development and inheritance.

Developmental anatomy is the study of the sequence of events from the fertilization of a secondary oocyte to the formation of an adult organism. As we look at the sequence from fertilization to birth, we will consider fertilization, implantation, placental development, embryonic development, fetal growth, gestation, parturition, and labor. The sequential development of an embryo and fetus is an exceedingly complex series of events that are precisely coordinated and controlled.

DEVELOPMENT DURING PREGNANCY

Once spermatozoa and ova are developed through meiosis and maturation, and the spermatozoa are deposited in the vagina, pregnancy can occur. ***Pregnancy*** is a sequence of events that normally includes fertilization, implantation, embryonic growth, and fetal growth that terminates in birth.

Fertilization and Implantation

Fertilization

Once a spermatozoon makes contact with a secondary oocyte, a process called ***syngamy,*** the spermatozoon penetrates a secondary oocyte, and their nuclei meet and fuse, a process called ***fertilization.*** Of the 300 to 500 million sperm cells introduced into the vagina, very few, perhaps only several hundred to several thousand, arrive in the vicinity of the oocyte. Fertilization normally occurs in the uterine (Fallopian) tube when the oocyte is about one-third of the way down the tube, usually within 24 hours after ovulation. Peristaltic contractions and the action of cilia transport the oocyte through the uterine tube. The mechanism by which sperm reach the uterine tube is apparently related to several factors. Sperm probably swim up the female tract by means of whiplike movements of their flagella. In addition, the acrosome of sperm produces an enzyme called ***acrosin*** that stimulates sperm motility and migration within the female reproductive tract. Finally, sperm are probably transported by muscular contractions of the uterus stimulated by prostaglandins in semen.

In addition to assisting in the transport of sperm, the female reproductive tract also confers on sperm the capacity to fertilize a secondary oocyte. Although sperm undergo maturation in the epididymis, they are still not able to fertilize an oocyte until they have remained in the female reproductive tract for about 10 hours. The functional changes that sperm undergo in the female reproductive tract that allow them to fertilize a secondary oocyte are referred to as ***capacitation*** (ka'-pas'-i'-TĀ-shun). During this process, the membrane around the acrosome becomes fragile so that several enzymes—hyaluronidase, acrosin, and neuraminidase—are secreted by the acrosomes. The enzymes help penetrate the ***corona radiata,*** several layers of follicular cells around the oocyte, and a gelatinous glycoprotein layer internal to the corona radiata called the ***zona pellucida*** (pe-LOO-si-da) (Figure 29-1a). Spermatozoa bind to receptors in the zona pellucida. Once this is accomplished, normally only one spermatozoon enters and fertilizes a secondary oocyte because once union is achieved, the electrical changes in

FIGURE 29-1 Fertilization and implantation. (a) Diagram of a spermatozoon moving through the corona radiata and zona pellucida on its way to reach the nucleus of a secondary oocyte.

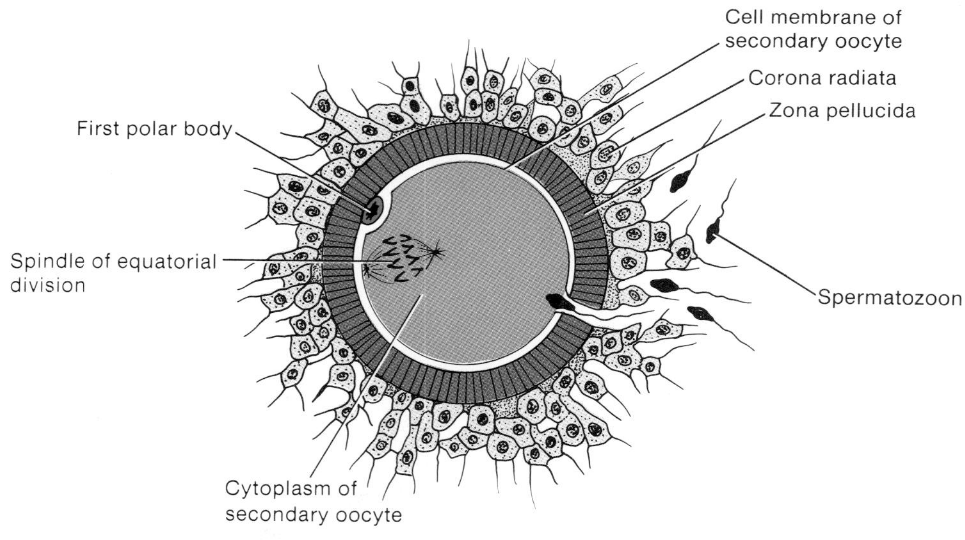

FIGURE 29-1 (*Continued*) (b) Scanning electron micrograph of a spermatozoon moving through the zona pellucida on its way to reach the nucleus of a secondary oocyte. (Courtesy of Fawcett, Photo Researchers, Inc.) (c) Photomicrograph showing female and male pronuclei. (Courtesy of Carolina Biological Supply Company.) Figure 28-5b shows a spermatozoon in contact with the surface of a secondary oocyte.

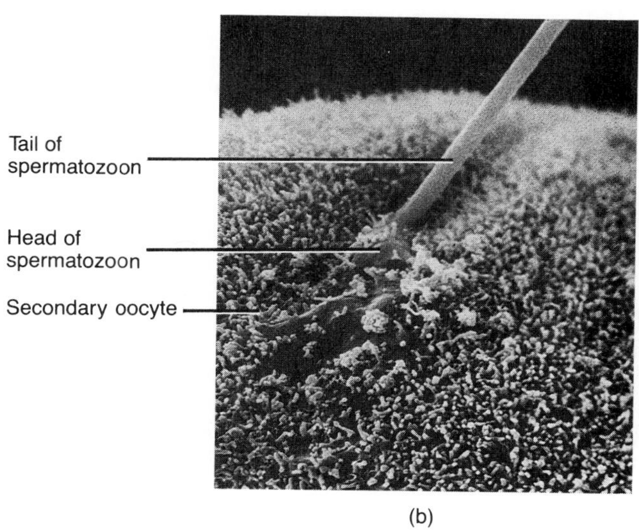

(b)

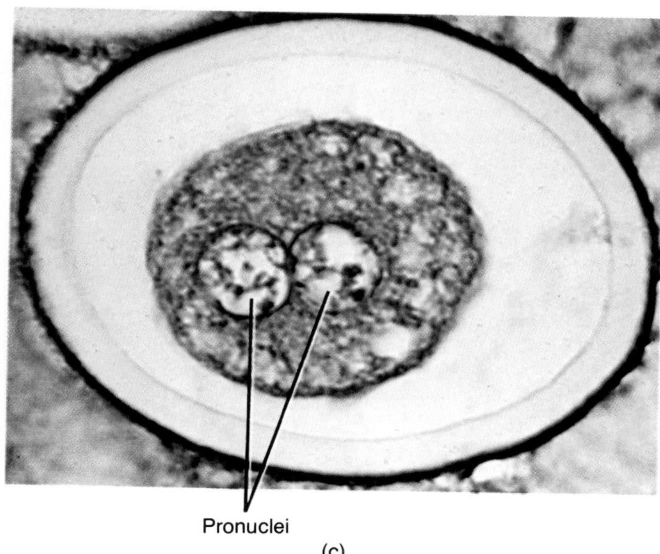

(c)

the surface of the oocyte block the entry of other sperm, and enzymes produced by the fertilized ovum (egg) alter receptor sites so that sperm already bound are detached and others are prevented from binding. In this way, polyspermy, fertilization by more than one spermatozoon, is prevented.

When a spermatozoon has entered a secondary oocyte, the tail is shed and the nucleus in the head develops into a structure called the ***male pronucleus.*** The nucleus of the oocyte develops into a ***female pronucleus*** (Figure 29-1c). After the pronuclei are formed, they fuse to produce a ***segmentation nucleus.*** The segmentation nucleus contains 23 chromosomes (n) from the male pronucleus and 23 chromosomes (n) from the female pronucleus. Thus, the fusion of the haploid (n) pronuclei restores the diploid number ($2n$). Once a spermatozoon has entered a secondary oocyte, the oocyte completes equatorial division (meiosis II). The secondary oocyte divides into a larger ovum (mature egg) and a smaller second polar body that fragments and disintegrates. The fertilized ovum, consisting of a segmentation nucleus, cytoplasm, and zona pellucida, is called a ***zygote.***

Dizygotic (fraternal) twins are produced from the independent release of two ova and the subsequent fertilization of each by different spermatozoa. They are the same age and are in the uterus at the same time, but they are genetically as dissimilar as any other siblings. They may or may not be the same sex. ***Monozygotic (identical) twins*** are derived from a single fertilized ovum that splits at an early stage in development. They contain the same genetic material and are always the same sex.

On rare occasions, monozygotic twins may be joined in varying degrees, from slight skin fusion to sharing of limbs, trunks, and viscera. Such twins are called ***conjoined (Siamese) twins.*** In September 1987, a surgical team at Johns Hopkins University performed a unique operation in which 7-month-old conjoined twins were successfully separated after being joined at the head since birth. The 22-hour procedure, performed by a 70-member team, was exceedingly complex and involved a total stoppage of circulation and hypothermia for about an hour. This was necessary to prevent hemorrhage while separating a shared sagittal sinus between the infants and to reduce brain activity to near zero to reduce brain swelling. In all known previous attempts to separate conjoined twins joined at their heads, one or both infants died or suffered serious neurological impairment. In this case, the twins suffered only slight brain damage but may have vision problems.

CLINICAL APPLICATION: ECTOPIC PREGNANCY

Ectopic (*ektos* = outside; *topos* = place) ***pregnancy (EP)*** refers to the development of an embryo or fetus outside the uterine cavity. The majority occur in the uterine (Fallopian) tube, usually in

the ampullar and infundibular portions. Some occur in the ovaries, abdomen, uterine cervix, and broad ligaments. The basic cause of a tubal pregnancy is impaired passage of the fertilized ovum through the uterine tube related to factors such as pelvic inflammatory disease (PID), previous uterine tube surgery, previous ectopic pregnancy, repeated elective abortions, pelvic tumors, and developmental abnormalities. Ectopic pregnancy may be characterized by one or two missed periods, followed by bleeding and acute abdominal and pelvic pain. Strong risk factors for EP include current use of an intrauterine device (IUD), a history of pelvic inflammatory disease (PID) or infertility, and prior surgery involving the uterine tubes.

Unless removed or discharged from a uterine tube, the developing embryo can rupture the tube, often resulting in death of the mother. Standard practice for terminating an ectopic pregnancy in the uterine tube is to remove the tube, occasionally with its associated ovary. An alternative procedure involves injecting prostaglandin F into the uterine tube directly above the improperly implanted embryo. Prostaglandin F induces contractions that expel the embryo.

Formation of the Morula

Immediately after fertilization, rapid mitotic cell division of the zygote takes place. This early division of the zygote is called ***cleavage.*** During this time, the dividing cells are contained by the zona pellucida. Although cleavage increases the number of cells, it does not result in an increase in the size of the developing organism.

The first cleavage is completed after about 36 hours, and each succeeding division takes slightly less time (Figure 29-2). By the second day after conception, the second cleavage is completed. By the end of the third day, there are 16 cells. The progressively smaller cells produced by cleavage are called ***blastomeres*** (BLAS-tō-mērz; *blast* = germ, sprout). A few days after fertilization, the successive cleavages have produced a solid mass of cells, still surrounded by zona pellucida, the ***morula*** (MOR-yoo-la) or mulberry, which is about the same size as the original zygote.

Development of the Blastocyst

As the number of cells in the morula increases, it moves from the original site of fertilization down through the ciliated uterine (Fallopian) tube toward the uterus and enters the uterine cavity. By this time, the dense cluster of cells is altered to form a hollow ball of cells. The mass is now referred to as a ***blastocyst*** (Figure 29-3).

The blastocyst is differentiated into an outer covering of cells called the ***trophoblast*** (TRŌF-ō-blast; *troph* = nourish), an ***inner cell mass (embryoblast),*** and an internal fluid-filled cavity called the ***blastocoele*** (BLAS-tō-sēl). The trophoblast ultimately forms part of the membranes composing the fetal portion of the placenta; the inner cell mass develops into the embryo.

Implantation

The blastocyst remains free within the cavity of the uterus from two to four days before it actually attaches to the uterine wall. During this time, the zona pellucida degenerates and nourishment is provided by secretions of glands of the endometrium, sometimes called uterine milk. The attachment of the blastocyst to the endometrium occurs seven to eight days after fertilization and is called ***implantation*** (Figure 29-4). At this time, the endometrium is in its postovulatory phase. As the blastocyst becomes implanted, the trophoblast separates into two layers in the region of contact between the blastocyst and endometrium. These layers are an outer ***syncytiotrophoblast*** (sin-sīt'-ē-ō-TRŌF-ō-blast) that contains no cell boundaries and an inner ***cytotrophoblast*** (sī-tō-TRŌF-ō-blast) that is composed of distinct cells. During implantation, the syncytiotrophoblast secretes enzymes that enable the blastocyst to penetrate the uterine lining. The enzymes digest and liquefy the endometrial cells. The fluid and nutrients further nourish the burrowing blastocyst for about a week after implantation. Eventually, the blastocyst becomes buried in the endometrium, usually on the posterior wall of the fundus or body of the uterus. The blastocyst becomes oriented so that the inner cell mass is toward the endometrium. Eventually, nutrients are delivered through the placenta for the subsequent growth and development of the embryo and fetus.

FIGURE 29-2 **Formation of the morula. About 72 hours after fertilization, the morula enters the uterus at about the 8- to 12-cell stage.**

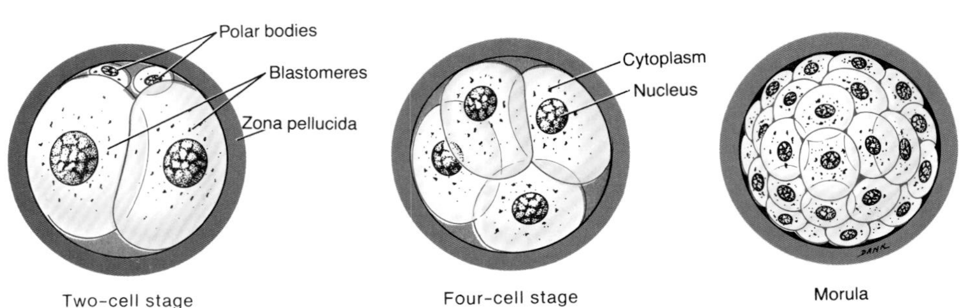

FIGURE 29-3 Blastocyst. (a) External view. (b) Internal view.

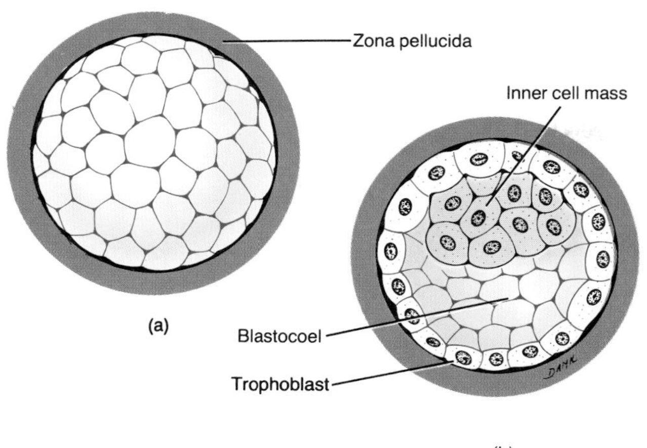

Since a developing embryo, and later a fetus, contains genes from the father as well as the mother, it is essentially a foreign graft and should be rejected as such by the mother. The trophoblast is the only tissue of the developing organism that makes contact with the mother's uterus. Even though trophoblast cells have paternal antigens that could provoke a rejection response, some mechanism in the uterus prevents this in order to permit development of the fetus to term. One suggestion for the mechanism that prevents rejection is that the mother makes antibodies that mask paternal antigens so that other components of her immune system cannot recognize and attack the antigens.

A summary of the principal events associated with fertilization and implantation is shown in Figure 29-5.

FIGURE 29-4 Implantation. (a) External view of the blastocyst in relation to the endometrium of the uterus about five days after fertilization. (b) Internal view of the blastocyst in relation to the endometrium about six days after fertilization. (c) Internal view of the blastocyst at implantation about seven days after fertilization. (d) Photomicrograph of implantation. (From *From Conception to Birth: The Drama of Life's Beginnings* by Roberts Rugh, Landrum B. Shettles with Richard Einhorn. Copyright © 1971 by Roberts Rugh and Landrum B. Shettles. By permission of Harper & Row, Publishers, Inc.)

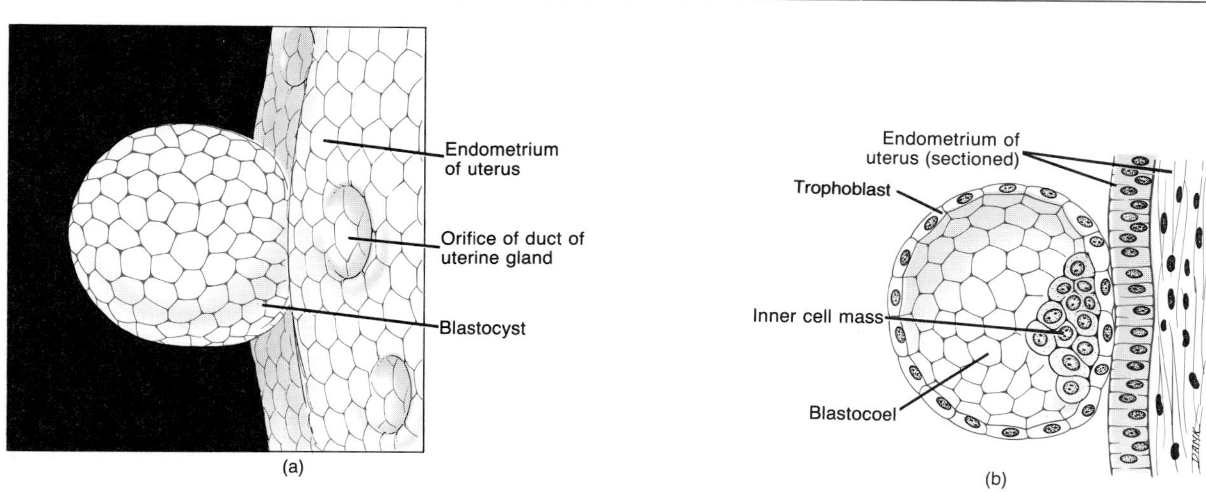

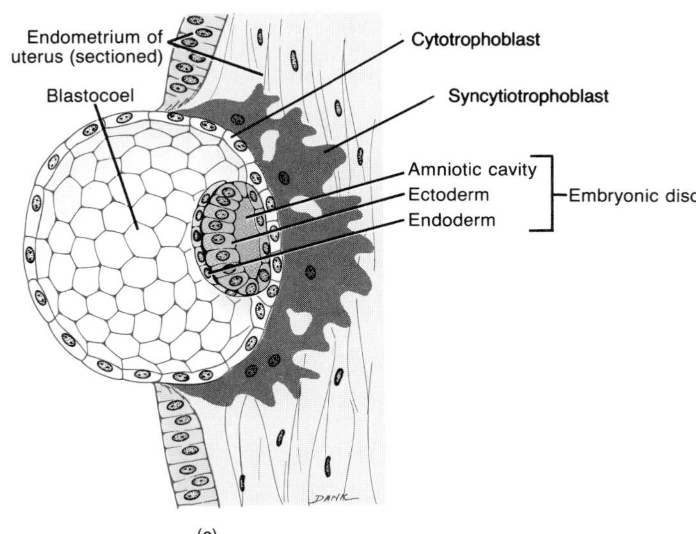

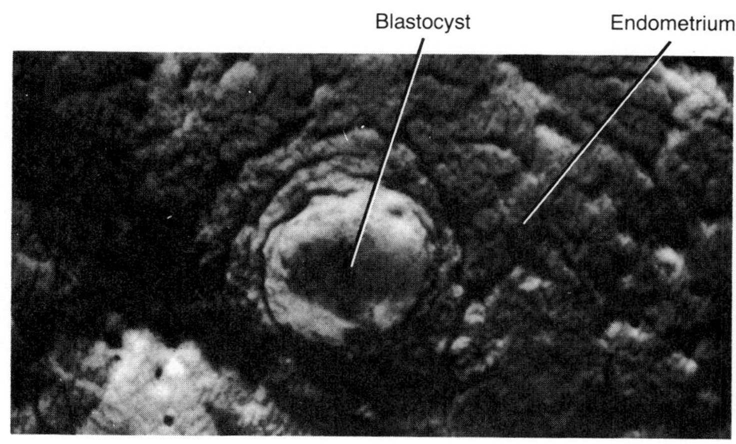

FIGURE 29-5 **Summary of events associated with fertilization and implantation.**

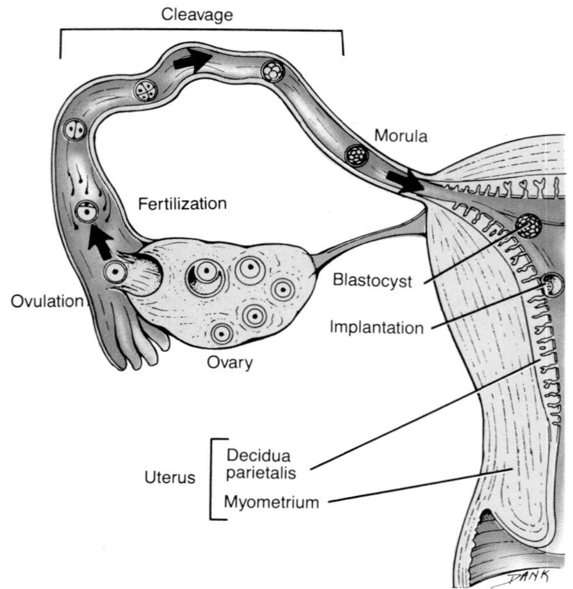

CLINICAL APPLICATION: MORNING SICKNESS

In the early months of pregnancy, *emesis gravidarum (morning sickness)* may occur, characterized by episodes of nausea and possibly vomiting that are most likely to occur in the morning. The cause is unknown, but the high levels of human chorionic gonadotropin (hCG) secreted by the placenta have been implicated. In some individuals the severity of these symptoms requires hospitalization for intravenous feeding, and the condition is then known as *hyperemesis gravidarum.*

In Vitro Fertilization (IVF)

On July 12, 1978, Louise Joy Brown was born near Manchester, England. Her birth was the first recorded case of ***in vitro fertilization (IVF)***—fertilization in a glass dish. In this procedure for IVF, the female is given follicle-stimulating hormone (FSH) soon after menstruation, so that several secondary oocytes, rather than the typical single one, will be produced (superovulation). Administration of luteinizing hormone (LH) may also ensure the maturation of the secondary oocytes. Next, a small incision is made near the umbilicus, and the secondary oocytes are aspirated from the follicles and placed in a medium that simulates the fluids in the female reproductive tract. The secondary oocytes are then transferred to a solution of the male's sperm. Once fertilization has taken place, the fertilized ovum is put in another medium and observed for cleavage. When the fertilized ovum reaches the 8-cell or 16-cell stage, it is introduced into the uterus for implantation and subsequent growth. The growth and developmental sequences that occur are similar to those in internal fertilization. It is also possible to freeze unused embryos (cryopreservation) to permit parents a successive pregnancy several years later or allow a second attempt at implantation if the first attempt is unsuccessful. The first instance in the United States of two successful pregnancies from a single IVF procedure occurred on November 19, 1988, when a woman gave birth to two daughters two years after the birth of her son.

Embryo transfer is a type of IVF in which a husband's seminal fluid is used to artificially inseminate a fertile secondary oocyte donor, and following fertilization, the morula or blastocyst is transferred from the donor to the infertile wife who carries it to term. Embryo transfer is indicated for females who are infertile; who have surgically untreatable blocked uterine (Fallopian) tubes; or who are afraid of passing on their own genes because they are carriers of a serious genetic disorder.

In the procedure, the donor is monitored to ascertain the time of ovulation by checking her blood levels of luteinizing hormone (LH) and by ultrasound. The wife is also monitored to make sure that her ovarian cycle is synchronized with that of the donor. Once ovulation occurs in the donor, she is donor (artificially) inseminated with the husband's (or another male's) seminal fluid. Four days later, a morula or blastocyst is flushed from the donor's uterus through a soft plastic catheter and transferred to the uterus of the wife, where it grows and develops until the time of birth. Embryo transfer is an office procedure that requires no anesthesia.

Gamete intrafallopian transfer (GIFT) is a type of IVF that is essentially an attempt to mimic the normal process of conception by uniting sperm and secondary oocyte in the prospective mother's uterine (Fallopian) tubes. In the procedure, the female is given FSH and LH to stimulate the production of several secondary oocytes. The secondary oocytes are aspirated with a laparoscope fitted with a suction device, mixed with a solution of the male's sperm outside the body, and then immediately inserted into the uterine (Fallopian) tubes.

In ***transvaginal oocyte retrieval,*** another type of IVF, the female is given hormones to produce several secondary oocytes, a needle is placed through the vaginal wall and guided to the ovaries by ultrasound, suction is applied to the needle, and the secondary oocytes are removed and placed in a solution outside the body. The male's sperm are added to the solution, and the fertilized ova are then implanted in the uterus.

EMBRYONIC DEVELOPMENT

The first two months of development are generally considered the ***embryonic period.*** During this period the developing human is called an ***embryo*** (*bryein* = grow).

The study of development from the fertilized egg through the eighth week in utero is referred to as ***embryology*** (em-brē-OL-ō-jē). The months of development after the second month are considered the ***fetal period,*** and during this time the developing human is called a ***fetus*** (*feo* = to bring forth). By the end of the embryonic period, the rudiments of all the principal adult organs are present, the embryonic membranes are developed, and the placenta is functioning.

Beginnings of Organ Systems

Following implantation, the inner cell mass of the blastocyst begins to differentiate into the three ***primary germ layers:*** ectoderm, endoderm, and mesoderm. They are the embryonic tissues from which all tissues and organs of the body will develop. The various movements of cell groups leading to the establishment of the primary germ layers are referred to as ***gastrulation.***

In the human, the germ layers form so quickly that it is difficult to determine the exact sequence of events. Within eight days after fertilization, the top layer of cells of the inner cell mass proliferates and forms the amnion (a fetal membrane) and a space, the ***amniotic (amnionic) cavity,*** over the inner cell mass. The upper layer of cells of the inner cell mass that is closer to the amniotic cavity develops into the ***ectoderm.*** The bottom layer of the inner cell mass that borders the blastocoel develops into the ***endoderm.***

About the twelfth day after fertilization, striking changes appear (Figure 29-6a). The cells below the amniotic cavity are called the ***embryonic disc.*** They will form the embryo. At this stage, the embryonic disc contains ectodermal and endodermal cells; the mesodermal cells are scattered external to the disc. The cells of the endodermal layer have been dividing rapidly, so that groups of them now extend downward in a circle, forming the yolk sac, another fetal membrane. The cells of the ***mesoderm,*** which develop between the ectodermal and endodermal layers, also have been dividing, and many have left the area of the embryonic disc and can be seen around the structures that are becoming fetal membranes.

About the fourteenth day, the cells of the embryonic disc differentiate into three distinct layers: the upper ectoderm, the middle mesoderm, and the lower endoderm (Figure 29-6b). The mesoderm in the disc soon splits into two layers, and the space between the layers becomes the ***extraembryonic coelom.***

As the embryo develops (Figure 29-6c), the endoderm becomes the epithelial lining of the gastrointestinal tract, respiratory tract, and a number of other organs. The mesoderm forms the peritoneum, muscle, bone, and other connective tissue. The ectoderm develops into the skin and nervous system. Exhibit 29-1 provides more details about the fates of these primary germ layers.

Embryonic Membranes

During the embryonic period, the ***embryonic membranes*** form (Figure 29-7). These membranes lie outside the embryo and protect and nourish the embryo and, later, the fetus. The membranes are the yolk sac, amnion, chorion, and allantois.

The ***yolk sac*** is an endoderm-lined membrane that, in many species, provides the primary or exclusive nutrient for the embryo (Figure 29-8; see also Figures 29-6c and 29-7). However the human embryo receives its

FIGURE 29-6 Formation of the primary germ layers and associated structures. (a) Internal view of the developing embryo about 12 days after fertilization.

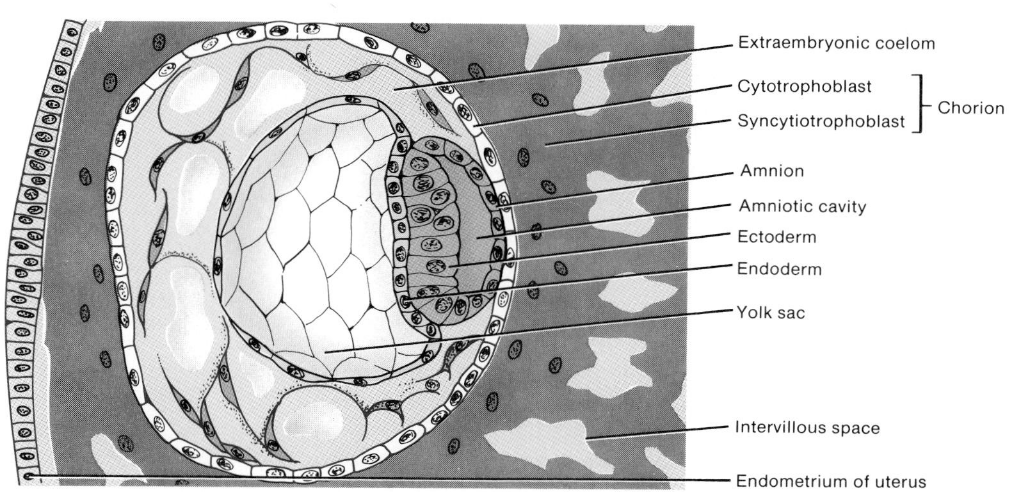

FIGURE 29-6 (*Continued*) (b) Internal view of the developing embryo about 14 days after fertilization.(c) External view of the developing embryo about 25 days after fertilization.

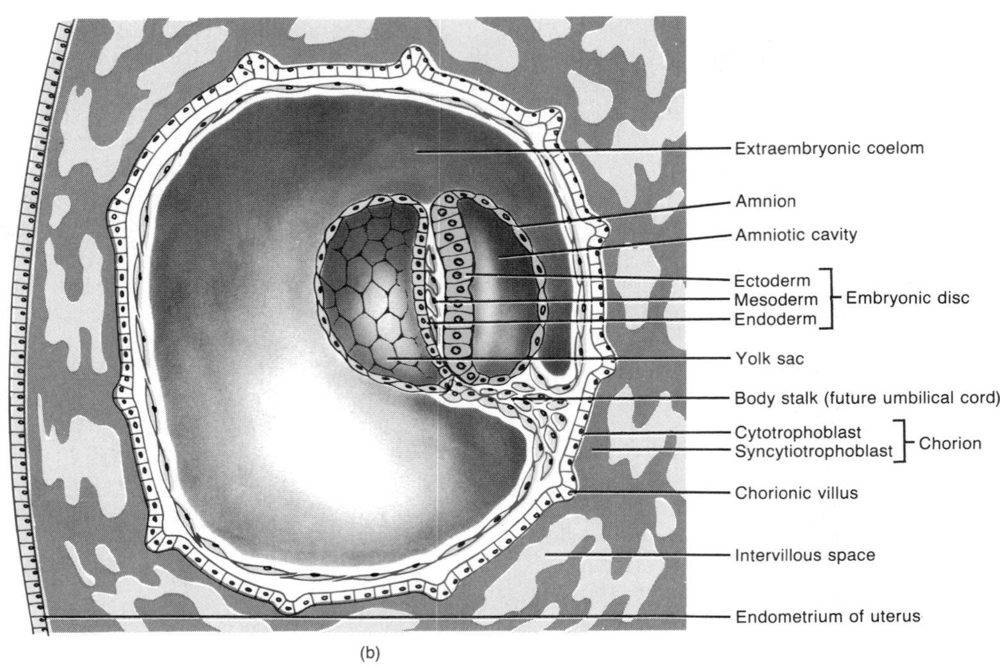

(b)

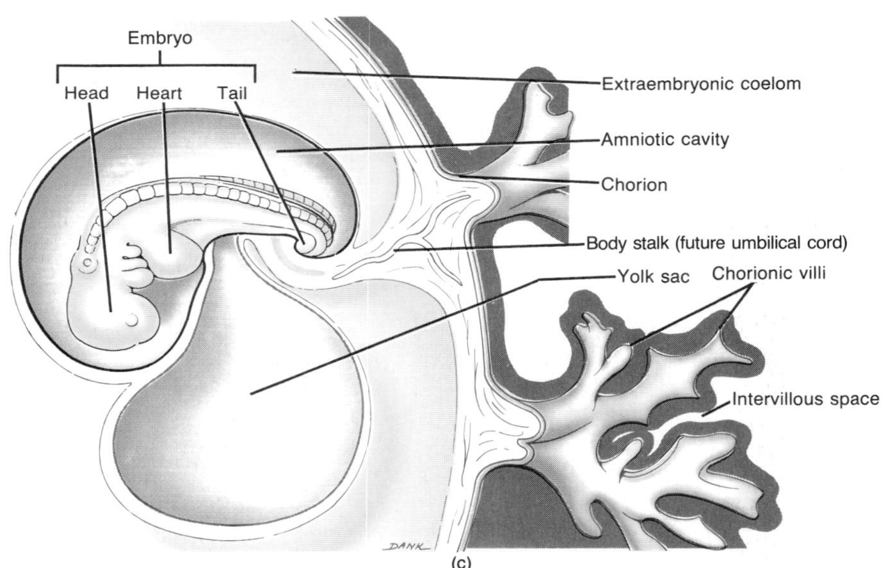

(c)

nourishment from the endometrium, and the yolk sac remains small. In humans, the yolk sac provides an early site of blood formation. The yolk sac also contains cells which differentiate into the primitive germ cells (spermatogonia and oogonia). During an early stage of development, it becomes a nonfunctional part of the umbilical cord.

The ***amnion*** is a thin, protective membrane that initially overlies the embryonic disc and is formed by the eighth day following fertilization. As the embryo grows, the amnion entirely surrounds the embryo, creating a cavity which becomes filled with ***amniotic fluid,*** or ***AF*** (Figure 29-8). Most amniotic fluid is initially derived from a filtrate of maternal blood. Later, the fetus makes daily

EXHIBIT 29-1 STRUCTURES PRODUCED BY THE THREE PRIMARY GERM LAYERS

Endoderm	Mesoderm	Ectoderm
Epithelium of gastrointestinal tract (except the oral cavity and anal canal) and the epithelium of its glands.	All skeletal, most smooth, and all cardiac muscle.	All nervous tissue.
Epithelium of urinary bladder, gallbladder, and liver.	Cartilage, bone, and other connective tissues.	Epidermis of skin.
Epithelium of pharynx, auditory (Eustachian) tube, tonsils, larynx, trachea, bronchi, and lungs.	Blood, bone marrow, and lymphoid tissue.	Hair follicles, arrector pili muscles, nails, and epithelium of skin glands (sebaceous and sudoriferous).
Epithelium of thyroid, parathyroid, pancreas, and thymus glands.	Endothelium of blood vessels and lymphatics.	Lens, cornea, and internal eye muscles.
Epithelium of prostate and bulbourethral (Cowper's) glands, vagina, vestibule, urethra, and associated glands such as the greater (Bartholin's) vestibular and lesser vestibular glands.	Dermis of skin.	Internal and external ear.
	Fibrous tunic and vascular tunic of eye.	Neuroepithelium of sense organs.
	Middle ear.	Epithelium of oral cavity, nasal cavity, paranasal sinuses, salivary glands, and anal canal.
	Mesothelium of ventral body and joint cavities.	Epithelium of pineal gland, pituitary gland (hypophysis), and adrenal medulla.
	Epithelium of kidneys and ureters.	
	Epithelium of adrenal cortex.	
	Epithelium of gonads and genital ducts.	

FIGURE 29-7 Embryonic membranes.

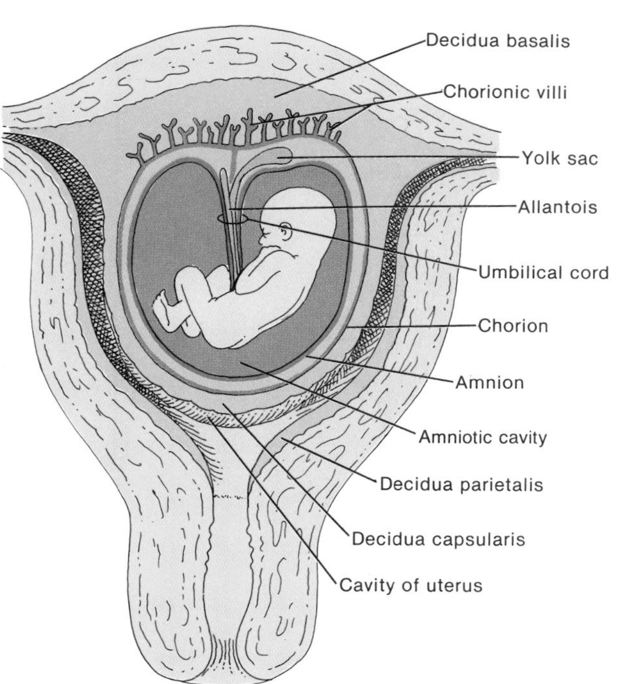

FIGURE 29-8 Ten-week fetus in which the amnion, yolk sac, umbilical cord, and placenta are clearly visible. (From *From Conception to Birth: The Drama of Life's Beginnings* by Roberts Rugh, Landrum B. Shettles with Richard Einhorn. Copyright © 1971 by Roberts Rugh and Landrum B. Shettles. By permission of Harper & Row, Publishers, Inc.)

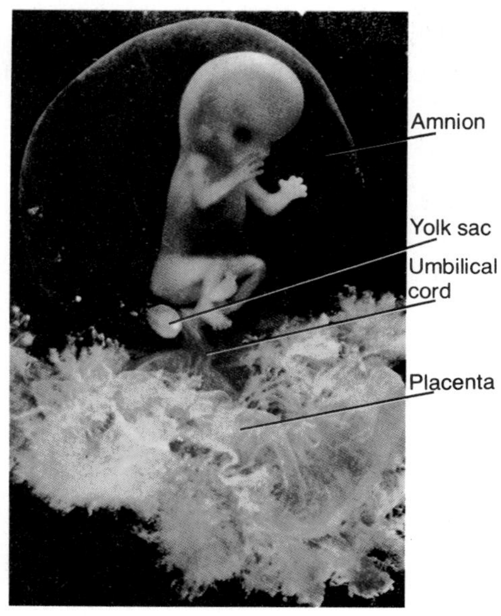

contributions to the fluid by excreting urine into the amniotic cavity. Amniotic fluid serves as a shock absorber for the fetus and assists in the regulation of fetal body temperature and prevents adhesions between the skin of the fetus and surrounding tissues. Embryonic cells are sloughed off into amniotic fluid; they can be examined in the procedure called ***amniocentesis*** (am'-nē-ō-sen-TĒ-sis), which is discussed later. The amnion usually ruptures just before birth and with its fluid constitutes the "bag of waters" (BOW).

The ***chorion*** (KŌ-rē-on) is derived from the trophectoderm of the blastocyst and the mesoderm that lines the trophoblast. It surrounds the embryo and, later, the fetus. Eventually, the chorion becomes the principal embryonic part of the placenta, the structure through which materials are exchanged between mother and fetus. The amnion also surrounds the fetus and eventually fuses to the inner layer of the chorion.

The ***allantois*** (a-LAN-tō-is) is a small vascularized membrane. It serves as an early site of blood formation. Later its blood vessels serve as connections in the placenta between mother and fetus. This connection is the umbilical cord.

Placenta and Umbilical Cord

Development of the placenta, the third major event of the embryonic period, is accomplished by the third month of pregnency. The ***placenta*** has the shape of a flat cake when fully developed and is formed by the chorion of the embryo and a portion of the endometrium (decidua basalis) of the mother (Figure 29-9). Functionally, the

FIGURE 29-9 **Placenta and umbilical cord. (a) Diagram of the structure of the placenta and umbilical cord.**

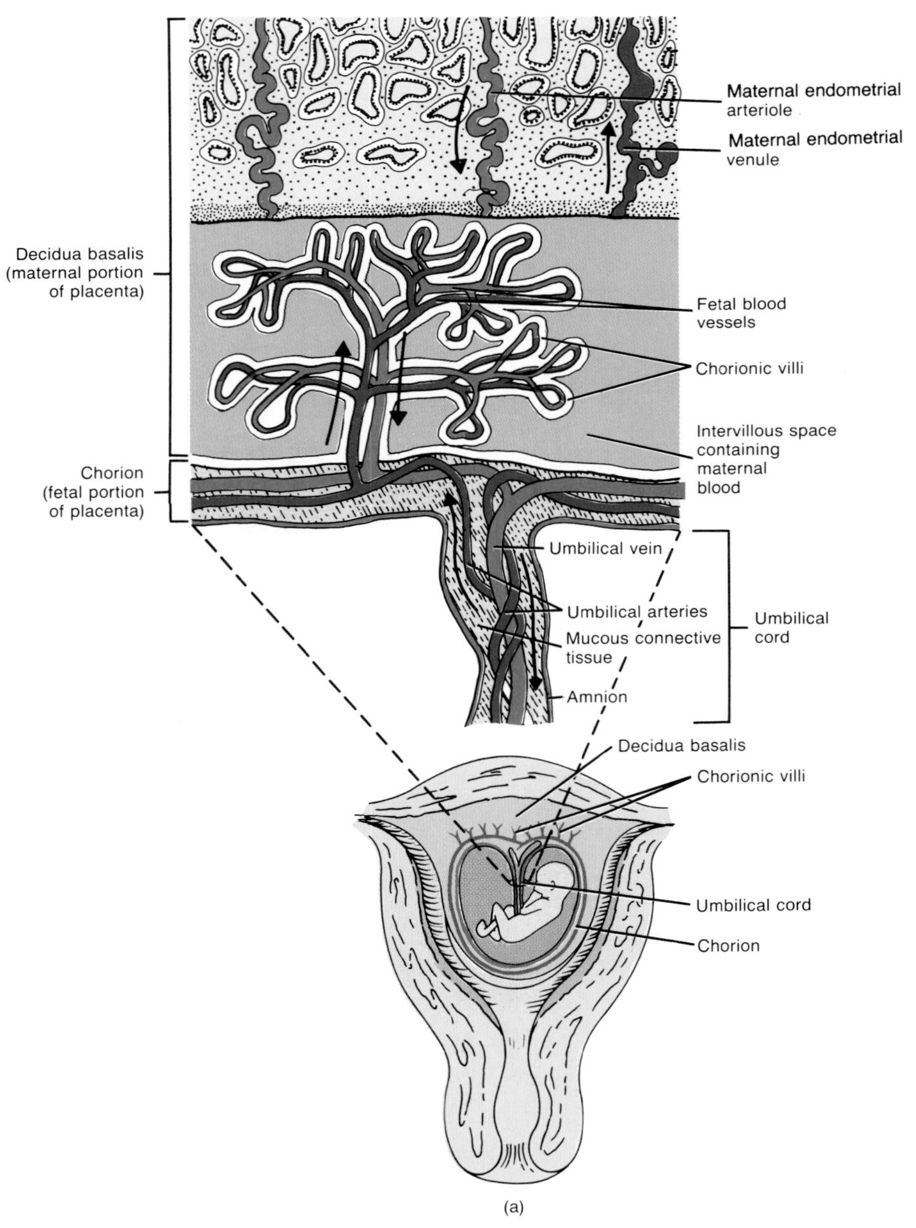

FIGURE 29-9 (*Continued*) (b) Photograph of the fetal aspect of the placenta. Whereas the umbilical arteries carry deoxygenated blood, the umbilical vein carries oxygenated blood. (c) Position of the placenta during the second stage of labor. (Courtesy of M. A. Colin England, *A Colour Atlas of Life Before Birth,* Year Book Medical Publishers, Chicago.)

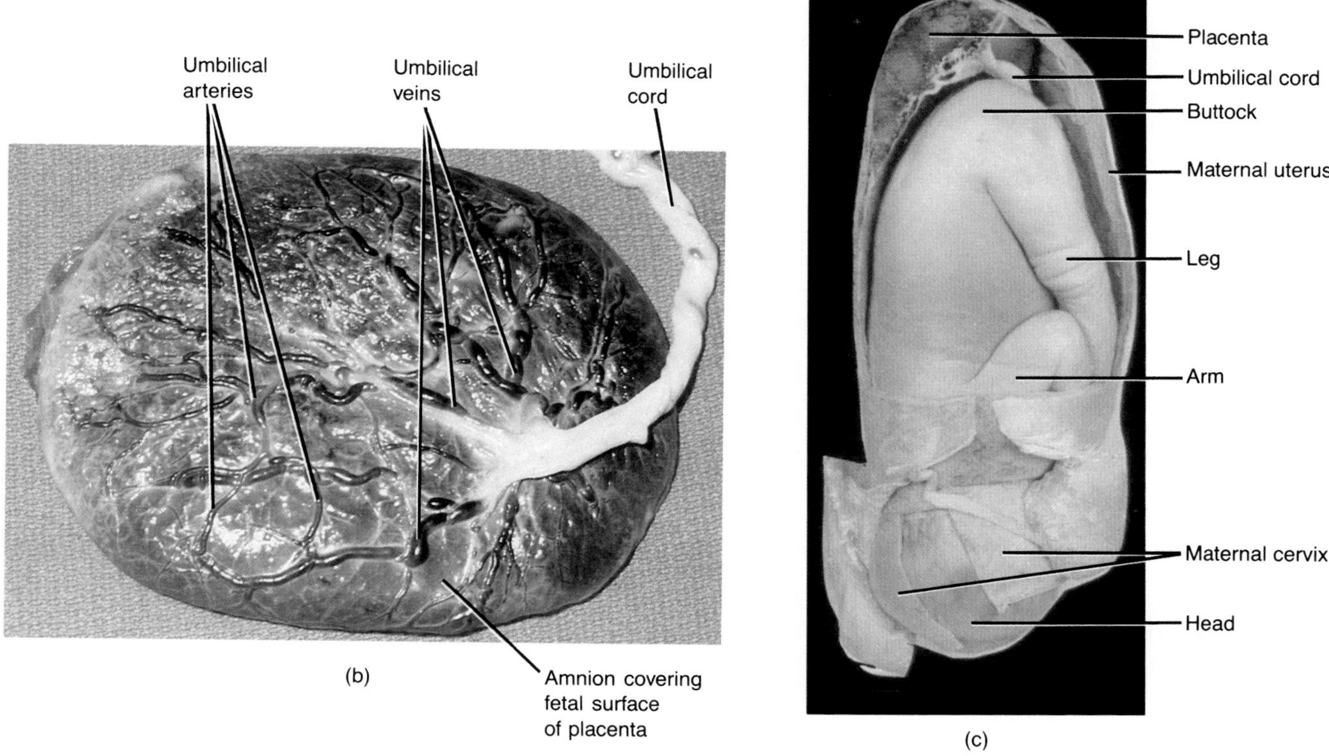

placenta allows oxygen and nutrients to diffuse into fetal blood from maternal blood; and carbon dioxide and wastes to diffuse from fetal blood into maternal blood. In addition, the placenta provides some degree of protection since most microorganisms cannot cross it. The placenta also stores nutrients such as carbohydrates, proteins, calcium, and iron, which are released into fetal circulation as required. Finally, the placenta produces several hormones that are necessary to maintain pregnancy (discussed later).

If implantation occurs, a portion of the endometrium becomes modified and is known as the ***decidua*** (dē-SID-yoo-a). The decidua includes all but the deepest layer of the endometrium and is shed when the fetus is delivered. Different regions of the decidua, all areas of the stratum functionalis, are named on the basis of their positions relative to the site of the implanted, fertilized ovum (Figure 29-10). The ***decidua parietalis*** (par-rī-e-TAL-is) is the portion of the modified endometrium that lines the entire pregnant uterus, except for the area where the placenta is forming. The ***decidua capsularis*** is the portion of the endometrium between the embryo and the uterine cavity. The ***decidua basalis*** is the portion of

FIGURE 29-10 **Regions of the decidua.**

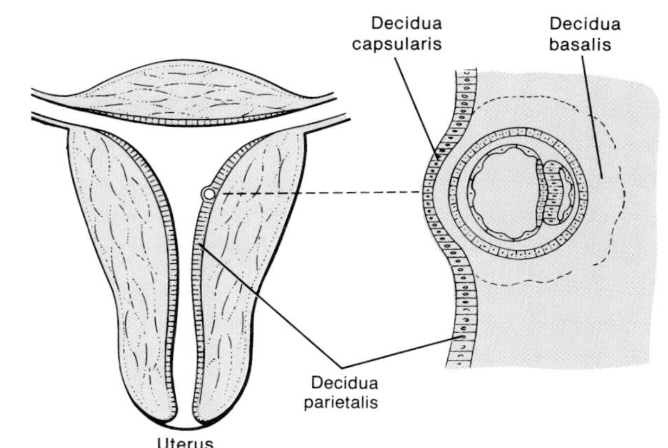

the endometrium between the chorion and the stratum basalis of the uterus. The decidua basalis becomes the maternal part of the placenta.

During embryonic life, fingerlike projections of the chorion, called ***chorionic villi*** (kō'-rē-ON-ik VIL-ī), grow into the decidua basalis of the endometrium (see also Figures 29-6 and 29-7). These will contain fetal blood vessels of the allantois. They continue growing until they are bathed in maternal blood sinuses called ***intervillous*** (in-ter-VIL-us) ***spaces.*** Thus, maternal and fetal blood vessels are brought into proximity. It should be noted, however, that maternal and fetal blood do not normally mix. Oxygen and nutrients from the mother's blood diffuse into the capillaries of the villi where all exchange occurs between fetal and maternal blood. From the capillaries the nutrients circulate into the umbilical vein. Wastes leave the fetus through the umbilical arteries, pass into the capillaries of the villi, and diffuse into the maternal blood. The ***umbilical cord*** consists of an outer layer of amnion containing the umbilical arteries and umbilical vein, supported internally by mucous connective tissue (Wharton's jelly) from the allantois.

At delivery, the placenta detaches from the uterus and is referred to as the ***afterbirth.*** At this time, the umbilical cord is severed, leaving the baby on its own. The scar that marks the site of the entry of the fetal umbilical cord into the abdomen is the ***umbilicus (navel).***

Pharmaceutical houses use human placenta as a source of hormones, drugs, and blood. Portions of placentas are also used for burn coverage. The placental and umbilical cord veins are used in blood vessel grafts.

CLINICAL APPLICATION: PLACENTA PREVIA, FETOMATERNAL HEMORRHAGE, AND UMBILICAL CORD ACCIDENTS

In some cases, part or all of the placenta becomes implanted in the lower portion of the uterus, near or over the internal os of the cervix. This condition is called ***placenta previa*** (PRĒ-vē-a; *previa* = before or in front of) and occurs in approximately 1 in 250 live births. The condition occurs 10 to 20 times more frequently in association with spontaneous abortion. It is also associated with fetal abnormalities, twin gestation, and multiple uterine curettages. The most important symptom is sudden, painless, bright red vaginal bleeding in the third trimester. Cesarean section is the preferred method of delivery in placenta previa.

Fetomaternal hemorrhage refers to the entrance of blood into maternal circulation brought on by a dysfunction in placental circulation. Although the condition occurs in at least 50 percent of all pregnancies, in most instances blood loss is so small that the pregnancy is not adversely affected. In some situations, however, the hemorrhage can compromise the fetus and lead to serious complications later in the same pregnancy or a future pregnancy. Among the causes of fetomaternal hemorrhage are trauma, rapid deceleration, placental and umbilical cord abnormalities, amniocentesis, intrauterine fetal surgery, umbilical vein thrombosis, and operative delivery (e.g., cesarean section). As a result of fetomaternal hemorrhage, certain complications may result. Examples include intrauterine fetal death, hypovolemic shock and anemia in the newborn, edema of the newborn, fetal cardiac arrhythmia, and anaphylactic shock in the mother.

Two of the most frequently encountered ***umbilical cord accidents*** are prolapse and entrapment. In ***prolapse,*** the umbilical cord descends in advance of the fetus during delivery. In ***entrapment,*** circulation through the cord is compromised because of pressure between the fetus and uterine wall. These, plus other conditions, such as knots in the cord, strictures, and thrombosis, could compromise the oxygen supply to the fetus resulting in damage to or death of the fetus.

FETAL GROWTH

During the ***fetal period,*** organs established by the primary germ layers grow rapidly. The organism takes on a human appearance. A summary of changes associated with the embryonic and fetal periods is presented in Exhibit 29-2.

CLINICAL APPLICATION: FETAL SURGERY AND FETAL-TISSUE IMPLANTATION

Fetal surgery is a new medical field that had its beginnings in 1985. In a pioneering operation, a team of surgeons removed a 23-week-old fetus from its mother's uterus, operated to correct a blocked urinary tract, and then returned the fetus to the uterus. Nine weeks later, a healthy baby was delivered. Surgeons are now experimenting on animals with fetal surgical procedures that could repair diaphragmatic hernias and spina bifida and correct hydrocephalus. It has been observed that surgery on fetuses does not leave any scars, although the reason is not known. It is hoped that surgeons can perform cranio-facial surgery before birth to correct conditions such as cleft lip without leaving scars.

Another relatively new therapy for treating certain diseases is known as ***fetal-tissue implantation.*** In the procedure, tissue is used from aborted fetuses in order to correct certain defects. For example, surgeons in China have been transplanting fetal pancreatic islet (islet of Langerhans) cells to treat type I diabetes since 1982. In 1988, surgeons in Mexico City and the United States transplanted fetal brain cells into the brains of patients with Parkinson's disease. It is hoped that fetal liver tissue may be transplanted to cure hereditary blood disorders such as thalassemia.

EXHIBIT 29-2 CHANGES ASSOCIATED WITH EMBRYONIC AND FETAL GROWTH

End of Month	Approximate Size and Weight	Representative Changes
1	0.6 cm (3/16 inch)	Eyes, nose, and ears not yet visible. Backbone and vertebral canal form. Small buds that will develop into arms and legs form. Heart forms and starts beating. Body systems begin to form.
2	3 cm (1¼ inches) 1 g (1/30 oz)	Eyes far apart, eyelids fused, nose flat. Ossification begins. Limbs become distinct as arms and legs. Digits are well formed. Major blood vessels form. Many internal organs continue to develop.
3	7.5 cm (3 inches) 28 g (1 oz)	Eyes almost fully developed but eyelids still fused, nose develops bridge, and external ears are present. Ossification continues. Appendages are fully formed and nails develop. Heartbeat can be detected. Body systems continue to develop.
4	18 cm (6½–7 inches) 113 g (4 oz)	Head large in proportion to rest of body. Face takes on human features and hair appears on head. Skin bright pink. Many bones ossified, and joints begin to form. Continued development of body systems.
5	25–30 cm (10–12 inches) 227–454 g (½–1 lb)	Head less disproportionate to rest of body. Fine hair (lanugo) covers body. Skin still bright pink. Rapid development of body systems.
6	27–35 cm (11–14 inches) 567–781 g (1¼–1½ lb)	Head becomes even less disproportionate to rest of body. Eyelids separate and eyelashes form. Skin wrinkled and pink.
7	32–42 cm (13–17 inches) 1135–1362 g (2½–3 lb)	Head and body more proportionate. Skin wrinkled and pink. Seven-month fetus (premature baby) is capable of survival.
8	41–45 cm (16½–18 inches) 2043–2270 g (4½–5 lb)	Subcutaneous fat deposited. Skin less wrinkled. Testes descend into scrotum. Bones of head are soft. Chances of survival much greater at end of eighth month.
9	50 cm (20 inches) 3178–3405 g (7–7½ lb)	Additional subcutaneous fat accumulates. Lanugo shed. Nails extend to tips of fingers and maybe even beyond.

Fetal cells have the advantage of being immunologically naive; that is, they have not yet developed all the antigens that allow a recipient's immune system to reject the cells. In addition, fetal cells are usually not mature enough to cause graft-versus-host disease, in which the tissues of a transplant recipient are attacked by implanted adult cells. Moreover, fetal nerve cells have the ability to regenerate and thus have the potential to repair damaged brain or spinal cord tissue.

HORMONES OF PREGNANCY

The corpus luteum is maintained for at least the first three or four months of pregnancy, during which time it continues to secrete ***estrogens*** and ***progesterone (PROG)***. Both these hormones maintain the lining of the uterus during pregnancy and prepare the mammary glands to secrete milk. The amount secreted by the corpus luteum, however, is only slightly more than that produced after ovulation in a normal menstrual cycle. The high levels of estrogens and progesterone needed to maintain pregnancy and develop the mammary glands for lactation are provided by the placenta from the third month through the rest of the gestation period. The chorion of the placenta secretes ***human chorionic gonadotropin (hCG)***. The primary role of hCG is to provide the stimulus for the continued production of progesterone by the corpus luteum—an activity necessary for the continued attachment of the embryo and fetus to the lining of the uterus (Figure 29-11). hCG is excreted in the urine of pregnant women from about the eighth day of pregnancy, reaching its peak of excretion about the ninth week of pregnancy. The hCG level decreases sharply during the fourth and fifth months and then levels off until childbirth. Excretion of hCG in the urine serves as the basis for most home pregnancy tests. hCG can be detected in blood by a laboratory test even before a period is missed.

The placenta begins to secrete estrogens after the first three or four weeks and progesterone by the sixth week of pregnancy. They are secreted in increasing quantities until the time of birth. By the fourth month, when the placenta is established, the secretion of hCG is greatly reduced because the secretions of the corpus luteum are

FIGURE 29-11 Hormones of pregnancy. (a) Summary of sources and functions. (b) Blood levels of hCG, estrogens, and progesterone.

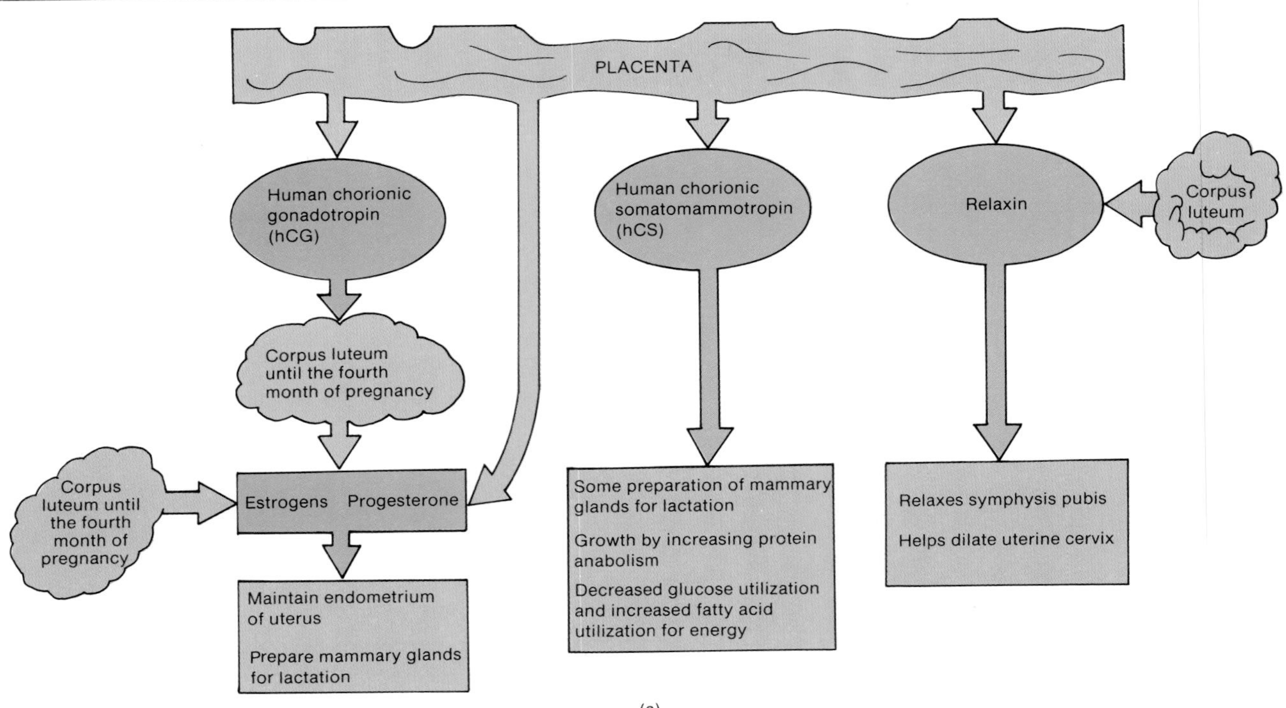

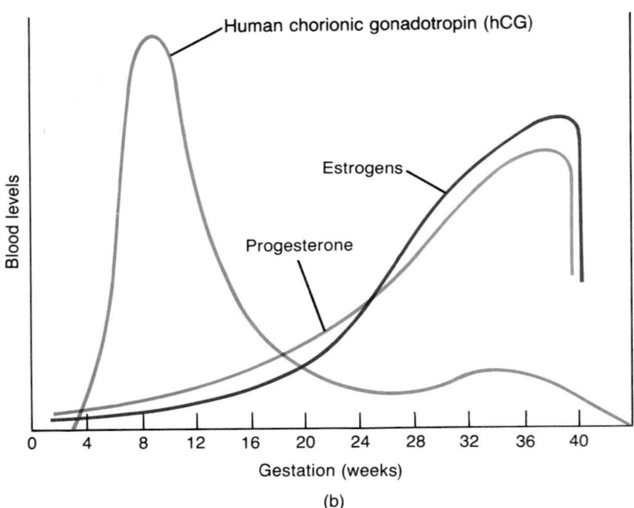

no longer essential. Thus, from the third to ninth month, the placenta supplies the levels of estrogens and progesterone needed to maintain the pregnancy. The placental hormones thus take over management of the mother's body in preparation for parturition (birth) and lactation. Following delivery, estrogens and progesterone in the blood decrease to normal levels.

Another hormone produced by the chorion of the placenta is ***human chorionic somatomammotropin (hCS)***, also known as ***human placental lactogen (hPL)***. Its secretion begins at about the same time as that of hCG, but its pattern of secretion is quite different. The rate of secretion of hCS increases in proportion to placental mass, reaching maximum levels after 32 weeks and remaining relatively constant after that. hCS is believed to stimulate some development of breast tissue for lactation, enhance growth by causing protein deposition in tissues, and regulate certain aspects of metabolism. For example, hCS causes decreased utilization of glucose by the mother, thus making more available for fetal metabolism. Also, hCS promotes the release of fatty acids from fat depots, providing an alternative source of energy for the mother's metabolism.

Relaxin is a hormone produced by the placenta and ovaries. Its physiological role is to relax the symphysis pubis and ligaments of the sacroiliac and sacrococcygeal joints and help dilate the uterine cervix toward the end of pregnancy. Both of these actions assist in delivery.

Inhibin, the hormone produced by the ovaries and testes, has also been found in the human placenta at term. Its role is to inhibit secretion of follicle-stimulating hormone (FSH) and might regulate secretion of hCG.

CLINICAL APPLICATION: FERTILITY AND BODY FAT

Recent evidence suggests that body fat has a regulatory role in reproduction. In order to begin and maintain a normal menstrual cycle, a female must have a minimum amount of body fat. A moderate loss of fat, from 10 to 15 percent below normal weight for height, may delay the onset of menstruation (menarche), inhibit ovulation during the menstrual cycle, or induce the cessation of the menstrual cycle (amenorrhea). Both dieting and intensive exercise may reduce body fat below the minimum amount and lead to infertility. The resulting infertility is reversible following weight gain or reduction of intensive exercise or both.

It appears that in underweight or very lean females, the secretion of gonadotropin releasing hormone (GnRH) by the hypothalamus is abnormal in quantity and timing. GnRH stimulates release of follicle-stimulating hormone (FSH) and luteinizing hormone (LH) from the anterior pituitary. Both hormones, in turn, control development of ovarian follicles, ovulation, and secretion of progesterone and estrogens by ovarian follicles. For some reason, the same factors that cause infertility in underweight or athletic females also provide a degree of protection against cancers that are sensitive to sex hormones, such as breast cancer.

Studies of very obese females also indicate they, like very lean ones, experience problems with amenorrhea and infertility. Males, like females, also experience problems related to reproduction in response to undernutrition and weight loss. For example, they produce less prostatic fluid, spermatozoa with decreased motility, and reduced numbers of spermatozoa.

DIAGNOSIS OF PREGNANCY

Several methods may be used to establish the diagnosis of pregnancy. The various signs and symptoms that aid in the diagnosis of pregnancy may be divided into three groups: presumptive, probable, and positive.

Presumptive Evidence

Presumptive evidence of pregnancy consists of signs and symptoms that can be recognized by the patient. These include amenorrhea (absence of menstruation), frequency of urination, breast changes, congestion and bluish-violet coloration of the vulva and vagina, increased skin pigmentation (lower abdomen, bridge of nose, under eyes), nausea, fatigue, and the perception of fetal movements.

Probable Evidence

Probable evidence of pregnancy includes enlargement of the abdomen, uterine and cervical changes, palpation of the fetus, uterine contractions, and results of endocrine tests. During early pregnancy, the uterus changes in size, shape, and consistency. At about the sixth week of gestation, there is a softening of the uterus (Hegar's sign), and by the beginning of the second month, there is a softening of the cervix (Goodell's sign).

Endocrine tests (pregnancy tests) are designed to detect the presence of human chorionic gonadotropin (hCG). Home pregnancy tests are of two types: test tube and dipstick. Test tube tests are based on agglutination (clumping). If hCG is present in urine, particles clump together when urine is mixed with the reagent provided in the test kit. Usually, the particles form a ring-shaped deposit with a hole in the center like a donut at the bottom of the test tube. If no ring forms, there is not enough hCG, and the test is negative. However, home pregnancy tests are not 100 percent accurate. A false negative result (test is negative, but the female is still pregnant) may occur from testing too soon, ectopic pregnancy, or not following instructions carefully. A false positive result (test is a positive test, but the female is not pregnant) may be due to reading the test at the wrong time; excessive heat, sunlight, or vibration; excess protein or blood in urine; or hCG production due to a rare type of uterine cancer.

With the dipstick test, a few drops of urine are mixed with the reagent solution in the test kit, and a test pad on the end of a dipstick is placed in the mixture. Then the pad is thoroughly rinsed with water and placed in a developing solution. If the pad turns blue, hCG is present and the test is positive.

Blood tests for pregnancy are more accurate and sensitive than urine tests. The tests are more expensive and require analysis in a laboratory.

Positive Evidence

The positive signs of pregnancy are demonstration of a fetal heart distinct from the mother's, detection of fetal movement by someone other than the mother, and visualization of the fetus by using a technique such as ultrasound.

GESTATION

The time a zygote, embryo, or fetus is carried in the female reproductive tract is called ***gestation*** (jes-TĀ-shun). The total human gestation period is about 280 days from the beginning of the last menstrual period. The specialized branch of medicine that deals with pregnancy, labor, and the period of time immediately following delivery is called ***obstetrics*** (ob-STET-riks; *obstetrix* = midwife).

Anatomical and Physiological Changes

By about the end of the third month of gestation, the uterus occupies most of the pelvic cavity, and as the fetus continues to grow, the uterus extends higher and higher into the abdominal cavity. In fact, toward the end of a full-term pregnancy, the uterus occupies practically all the abdominal cavity, reaching above the costal margin nearly to the xiphoid process of the sternum (Figure 29-12), causing displacement of the maternal intestines, liver, and stomach upward, elevation of the diaphragm, and widening of the thoracic cavity. In the pelvic cavity, there is compression of the ureters and urinary bladder.

In addition to the anatomical changes associated with pregnancy, there are also certain pregnancy-induced physiological changes. General changes include weight gain due to the fetus, amniotic fluid, placenta, uterine enlargement, and increased total body water; increased proteins, fat, and mineral storage; marked breast enlargement in anticipation of lactation; and lower back pain due to lordosis.

With respect to the cardiovascular system, there is an increase in stroke volume by about 30 percent; a rise in cardiac output (CO) by 20 to 30 percent by the twenty-seventh week due to increased maternal blood flow to the placenta and increased metabolism; an increase in heart rate by about 10 to 15 percent; and an increase in blood volume up to 30 to 50 percent, mostly during the latter half of pregnancy. When a pregnant female is lying on her back, the enlarged uterus may compress the aorta, resulting in diminished blood flow to the uterus. Hormonal changes associated with pregnancy and compression of the inferior vena cava also produce varicose veins.

Pulmonary function is also altered during pregnancy in that tidal volume can increase 30 to 40 percent, expiratory reserve volume can decrease up to 40 percent, functional residual capacity can decrease up to 25 percent, minute volume of respiration (MVR) can increase up to 40 percent, and airway resistance in the bronchial tree can decrease up to 36 percent. There is also an increase in total body oxygen consumption by about 10 to 20 percent. Dyspnea also occurs.

FIGURE 29-12 Normal fetal position during a full-term pregnancy. (a) Diagram.

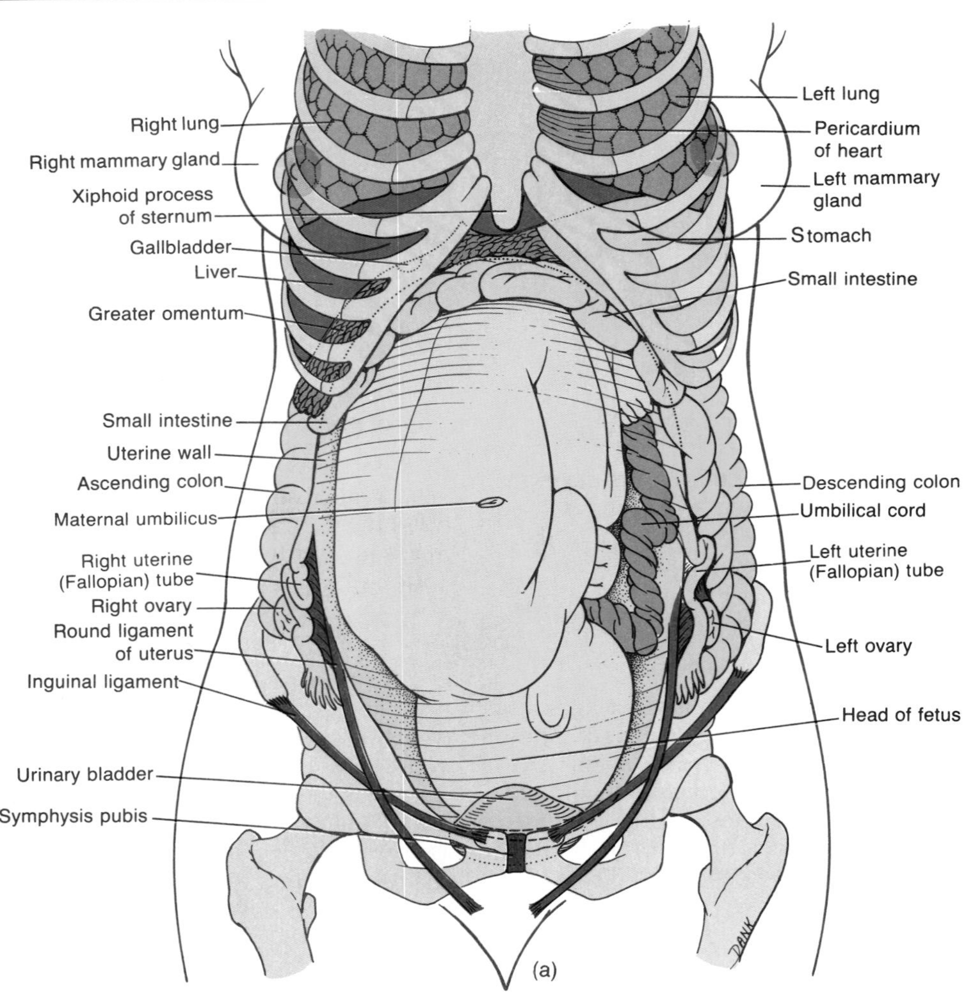

FIGURE 29-12 (*Continued*) (b) Photograph. (Courtesy of M. A. Colin England, *A Colour Atlas of Life Before Birth,* Year Book Medical Publishers, Chicago.)

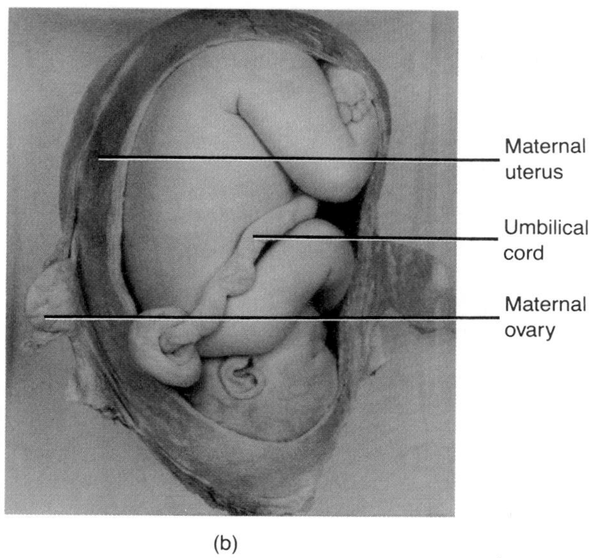

(b)

With regard to the gastrointestinal tract, there is an increase in appetite and a general decrease in motility that can result in constipation and a delay in gastric emptying time. Nausea, vomiting, and heartburn also occur.

Pressure on the urinary bladder by the enlarging uterus can produce urinary symptoms, such as frequency, urgency, and stress incontinence. Other conditions related to the urinary system include an increase in renal plasma flow up to 35 percent, an increase in glomerular filtration rate (GFR) up to 40 percent, and a decrease in ureteral muscle tone.

Changes in the skin during pregnancy are more apparent in some patients than others. Included are increased pigmentation around the eyes and cheekbones in a masklike pattern (chloasma), in the areolae of the breasts, and in the linea alba of the lower abdomen (linea nigra). Striae (stretch marks) over the abdomen occur as the uterus enlarges, and hair loss also increases.

Changes in the reproductive system include edema and increased vascularity of the vulva and increased pliability and vascularity of the vagina. The uterus increases in weight from its nonpregnant state of 60 to 80 g to 900 to 1200 g at term. This increase is due to hyperplasia of muscle fibers (cells) in the myometrium in early pregnancy and hypertrophy of muscle fibers during the second and third trimesters.

Exercise and Pregnancy

Since pregnancy results in so many major body changes, it has an impact on a female's ability to exercise. In early pregnancy, there are only a few changes that affect exercise. Accordingly, the mother may tire much earlier than usual, or she may become lethargic. Morning sickness may also curtail regular exercise. As the pregnancy develops, weight is gained and posture changes. As a result, more energy is needed to perform activities, and certain maneuvers (sudden stopping, changes in direction, rapid movements) are difficult to execute. In even later stages of pregnancy, certain joints, especially the symphysis pubis, become less stable in response to elevated levels of progesterone. As compensation, many females walk with widely spread legs and a shuffling motion.

Although during exercise blood shifts from viscera (including the uterus) to the muscles and skin, there is no evidence of placental insufficiency. The heat generated during exercise may cause dehydration and further increase body temperature. During early pregnancy, especially, excessive exercise and heat buildup should be avoided since elevated body temperature has been implicated in neural tube defects. Exercise has no known effect on lactation, provided the female remains hydrated and wears a bra with good support. Moderate physical activity does not endanger the fetuses of healthy females who have a normal pregnancy.

Among the benefits of exercise during pregnancy are improvement in oxygen capacity, greater sense of well-being, and fewer minor complaints.

PRENATAL DIAGNOSTIC TESTS

Several tests are available to detect genetic disorders and assess fetal well-being. Here we will describe amniocentesis, chorionic villi sampling (CVS), fetal ultrasonography, and the alphafetoprotein (AFP) test.

MEDICAL TESTS

Amniocentesis (am'-nē-ō-sen-TĒ-sis; ***amnio*** = amnion; ***kentesis*** = puncture)

Diagnostic Value: To test for the presence of certain genetic disorders, such as Down's syndrome (DS), spina bifida, hemophilia, Tay-Sachs disease, sickle-cell anemia, and certain muscular dystrophies or to determine fetal maturity and well-being near the time of the delivery. About 300 chromosomal disorders and over 50 biochemical defects can be detected through amniocentesis. When both parents are known or suspected to be genetic carriers of any one of these disorders, or when maternal age approaches 35, amniocentesis is advised. The procedure is also advised when there is concern of a pre-term delivery, a medical condition that necessitates an early delivery, and for patients who are Rh^- sensitized. Amniocentesis can also determine gender.

Procedure: Using ultrasound and palpation, the position of the fetus and placenta are first determined. After the skin is prepared with an antiseptic, a local anesthetic is given, a hypodermic needle is inserted into the amniotic cavity, and about 10 to 20 ml of fluid is removed (Figure 29-13). The test is usually done at 14–16 weeks of gestation. Cells and fluid are subjected to microscopic examination, biochemical testing, and chromosome studies.

Chorionic (ko-rē-ON-ik) **villus** (VIL-us) **sampling (CVS)**

Diagnostic Value: Although the procedure determines the same defects as amniocentesis, it offers several advantages. It can be performed earlier, usually at 8 to 10 weeks of gestation. Moreover, the procedure does not require penetration of the abdomen, uterus, or amniotic cavity. The safety of the procedure is believed to be comparable to that for amniocentesis, although some feel that the risk to the fetus is slightly greater than with amniocentesis. (In a recently developed variation of chorionic villus sampling, an ultrasound probe is inserted into the vagina, where it bounces sound waves off the uterus. The probe, which has a needle on the end, is guided through the cervix into the uterus, where a sample of chorionic villi is taken for analysis. This method permits prenatal diagnosis as early as the sixth week of pregnancy and is indicated for females in whom standard prenatal techniques are not possible.)

Procedure: A catheter is placed through the vagina (or through the abdominal cavity) into the uterus and then to the chorionic villi under ultrasound guidance. About 30 mg of tissue is suctioned out and prepared for chromosomal analysis. Chorion cells and fetal cells contain identical genetic information.

Fetal ultrasonography (ul'-tra-son-OG-ra-fē)

Diagnostic Value: In the United States today, most obstetricians perform ultrasound examination only when there is some clinical question about the normal progress of the pregnancy. By far the most common use of diagnostic ultrasound is to determine true fetal age when the date of conception is uncertain. It is also used to evaluate fetal viability and growth, determine fetal position, ascertain multiple pregnancies, identify fetal-maternal abnormalities, and serve as an adjunct to special procedures such as amniocentesis. Ultrasound is not used routinely to determine the sex of a fetus; it is performed only for a specific medical indication.

Procedure: An instrument (transducer) that emits high-frequency sound waves is passed back and forth over the abdomen. The reflected sound waves from the developing fetus are picked up by the transducer and converted to an image on a screen (Figure 29-14). This image is called a **sonogram.** Since the urinary bladder serves as a landmark during the procedure, the patient needs to drink liquids and not void in order to maintain a full bladder.

Alphafetoprotein (al'-fa-fē'-tō-PRŌ-tēn), or **(AFP)** This substance is produced by liver cells of fetuses and adults and by the ovaries and testes.

Diagnostic Value: In adults, the level of AFP may be elevated with cancer of the liver, testicles, and ovaries. AFP can also be used to monitor response to therapy for these malignancies. When done on pregnant women, the test can help detect a neural tube defect, a twin pregnancy, or the need for further studies such as fetal ultrasonography.

FIGURE 29-13 Amniocentesis.

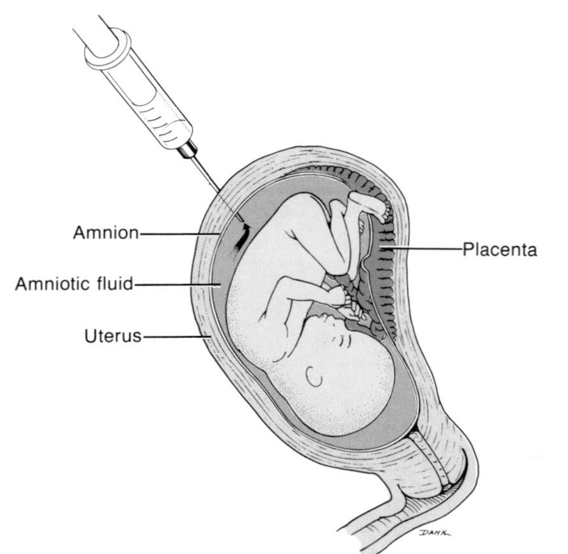

FIGURE 29-14 Sonogram of a transverse section through the fetal head at week 33. (Courtesy of Lynne, James, and James Gerard Borghesi). Labels provided by Professor Carol Schanel, Bergen Community College.

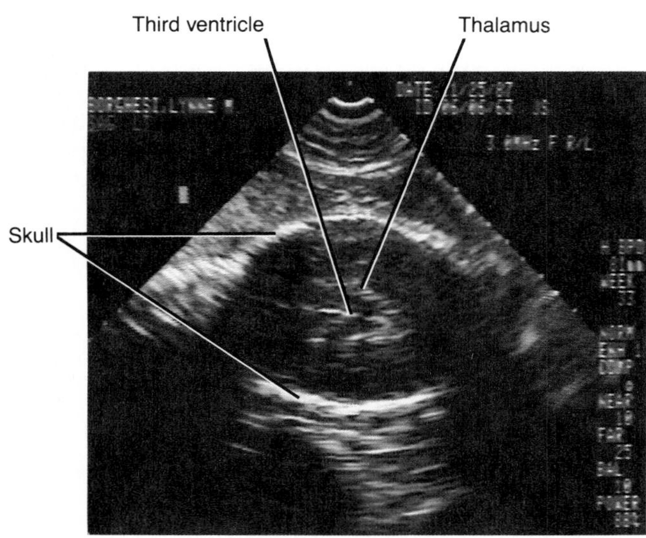

Procedure: May be determined by a blood sample or amniotic fluid (AF) sample, which should be obtained at 15 to 16 weeks of gestation.

Normal Values: In nonpregnant adults: <25 ng/ml
In pregnant females: values rise from 70 ng/ml at 8 weeks gestation to 430 ng/ml at 36 weeks, then gradually decrease soon after birth.

PARTURITION AND LABOR

The term ***parturition*** (par'-too-RISH-un) refers to birth. Parturition is accompanied by a sequence of events commonly called ***labor.*** The onset of labor is apparently related to a complex interaction of many factors. Just prior to birth, the muscles of the uterus contract rhythmically and forcefully. Both placental and ovarian hormones seem to play a role in these contractions. Since progesterone inhibits uterine contractions, labor cannot take place until its effects are diminished. At the end of gestation, the level of estrogens in the mother's blood is sufficient to overcome the inhibiting effects of progesterone, since the progesterone level falls, and labor commences. It has been suggested that some factor released by the placenta, fetus, or mother rather suddenly overcomes the inhibiting effects of progesterone so that estrogens can exert their effect. Prostaglandins may also play a role in labor. Oxytocin (OT) from the posterior pituitary gland also stimulates uterine contractions (see Figure 18-9), and relaxin assists by relaxing the symphysis pubis and helping to dilate the uterine cervix.

Uterine contractions occur in waves, quite similar to peristaltic waves, that start at the top of the uterus and move downward. These waves expel the fetus. ***True labor*** begins when pains occur at regular intervals. The pains correspond to uterine contractions. As the interval between contractions shortens, the contractions intensify. Another sign of true labor in some females is localization of pain in the back, which is intensified by walking. A reliable indication of true labor is the "show" and dilation of the cervix. The "show" is a discharge of a blood-containing mucus that accumulates in the cervical canal during labor. In ***false labor,*** pain is felt in the abdomen at irregular intervals. The pain does not intensify and is not altered significantly by walking. There is no "show" and no cervical dilation.

Labor can be divided into three stages (Figure 29-15).

1. The ***stage of dilation*** is the time from the onset of labor to the complete dilation of the cervix. During this stage, there are regular contractions of the uterus, usually a rupturing of the amniotic sac, and complete dilation (10 cm) of the cervix. If the amniotic sac does not rupture spontaneously, it is done artificially.

2. The ***stage of expulsion*** is the time from complete cervical dilation to delivery.

3. The ***placental stage*** is the time after delivery until the placenta or "afterbirth" is expelled by powerful uterine contractions. These contractions also constrict blood vessels that were torn during delivery. In this way, the possibility of hemorrhage is reduced.

During labor, the fetus may be squeezed through the birth canal for up to several hours. As a result, the fetal head is compressed, and there is some degree of intermittent hypoxia due to compression of the umbilical cord and placenta during uterine contractions. In response to this compression, the adrenal medulla of a fetus secretes very high levels of epinephrine and norepinephrine (NE), the "fight-or-flight" hormones. Much of the protection afforded against the stresses of the birth process and preparation of the infant to survive extrauterine life are provided by the adrenal medullary hormones. Among other functions, the hormones clear the lungs and alter their physiology for breathing outside the uterus, mobilize readily usable nutrients for cellular metabolism, and promote a rich vascular supply to the brain and heart.

CLINICAL APPLICATION: DYSTOCIA AND CESAREAN SECTION

Dystocia (dis-TŌ-sē-a), or difficult labor, may result from impaired uterine forces, an abnormal position (presentation) of the fetus, or a birth canal of inadequate size to permit vaginal birth. In these instances, and in certain conditions of fetal or maternal distress occurring during labor, it may be necessary to deliver the baby via a ***cesarean*** (*caedere* = to cut) ***section (C-section).*** In this procedure, a low, horizontal incision is made through the abdominal wall near the pubic hairline and lower portion of the uterus, through which the baby and placenta are removed. Even a history of multiple C-sections need not exclude a pregnant woman from attempting a vaginal delivery.

MEDICAL TEST

Electronic fetal monitoring (EFM)

Diagnostic Value: The test records fetal heart rate and maternal uterine contractions and is used to evaluate fetal well-being during labor and to detect any early signs of potential problems. EFM is also used to monitor the fetus during special tests that may be given prior to labor (nonstress test and oxytocin challenge test). Generally, EFM is used to monitor high-risk pregnancies.

Procedure: In ***external fetal monitoring,*** two rubber straps are placed around the abdomen. At-

FIGURE 29-15 Parturition. (a) Fetal position prior to birth. (b) Dilation. Protrusion of amnionic sac through partly dilated cervix (left). Amnionic sac ruptured and complete dilation of cervix (right). (c) Stage of expulsion. (d) Placental stage.

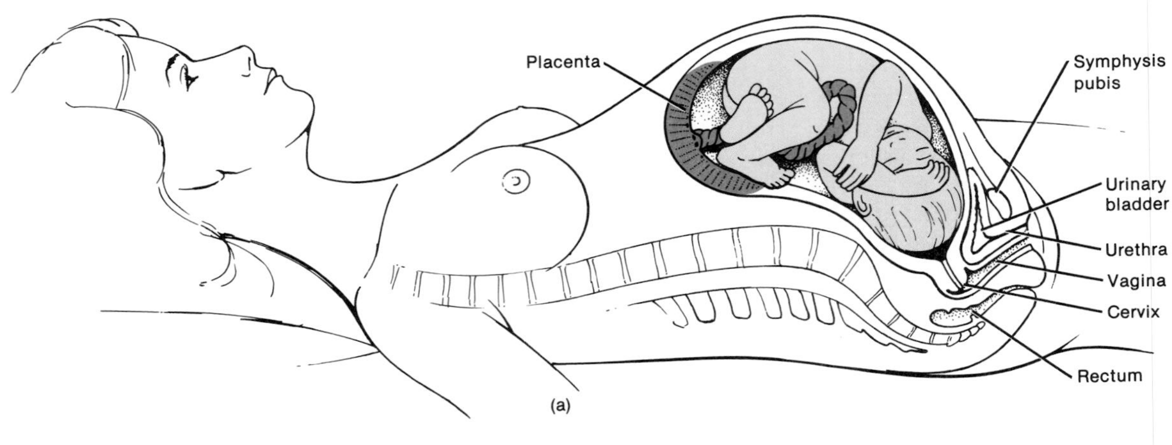

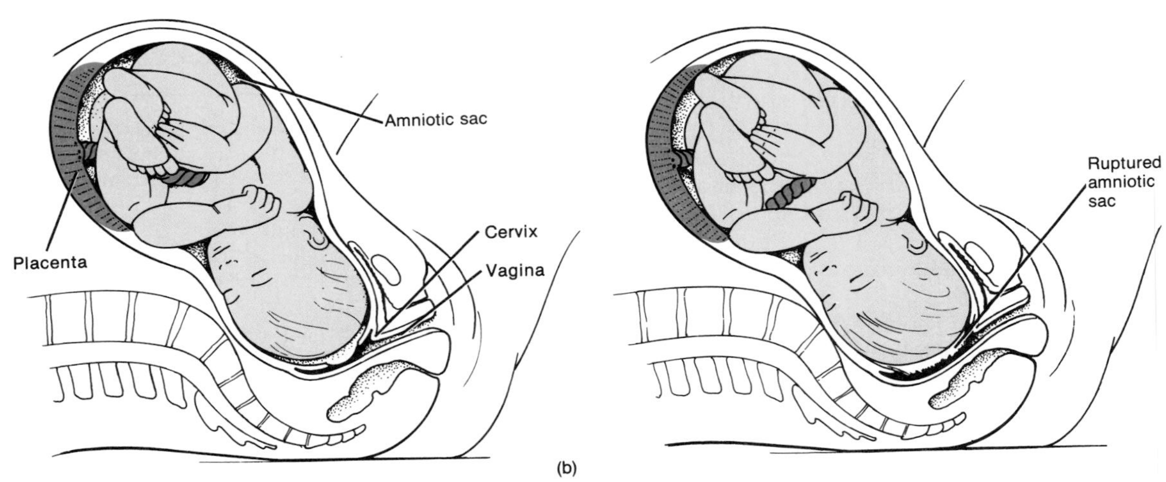

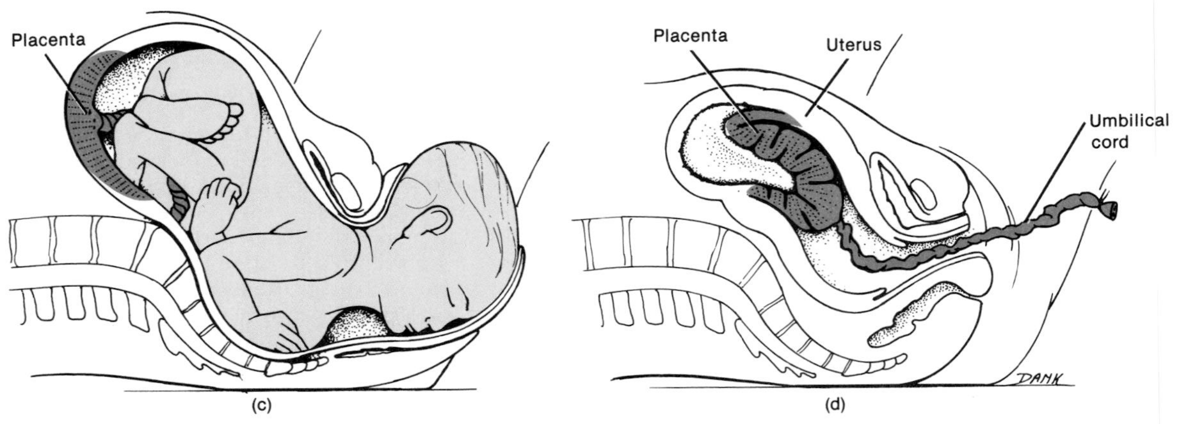

tached to the straps are sensors that detect fetal heart rate using ultrasound (US). The other measures uterine contractions or fetal movements. Measurements are recorded by a small machine that traces them on moving graph paper. External fetal monitoring can be done at any time, including early labor before the cervix dilates and the amniotic sac ruptures. In ***internal fetal monitoring,*** fetal heart rate is measured through an electrode placed through the mother's vagina and attached to the fetal scalp. A sensor attached to a catheter is also inserted through the vagina and placed in the uterus. Measurements are also made on graph paper. Internal fetal monitoring can be used only if cervical dilation has occurred and the amniotic sac has ruptured.

The ***nonstress test*** uses EFM to check a fetus's well-being before labor begins. After an external monitoring belt is applied to the abdomen, fetal movements are noted on the recording of the fetal heart rate. In this way, fetal movements and heart rate are timed simultaneously. A normal test shows that the fetus moves at least two to three times during a 20-minute period and the heart rate increases with each movement.

The ***oxytocin (OT) challenge*** or ***stress test*** uses EFM to check fetal well-being also before labor begins. After two external monitoring belts are applied to the abdomen, fetal heart rate and maternal uterine contractions are measured. A small amount of oxytocin (OT) is given intravenously to bring about uterine contractions. Fetal response to the contractions is observed. A normal fetus can adjust to the decreased amount of oxygen that accompanies a contraction, as evidenced by a heart rate that remains the same or increases. This suggests fetal well-being when natural contractions occur during labor. If, instead, the test shows a decreased heart rate during contractions, this suggests that fetal distress may occur during delivery and a cesarean section may be indicated.

Roughly 7 percent of pregnant females do not deliver two weeks after their due date, and some females do not go into labor for up to five weeks past their due date. In such cases, there is increased risk of brain damage to the fetus and even fetal death. This is due to inadequate oxygen and nutrients from an aging placenta. Postterm deliveries may be facilitated by induced labor or cesarean section.

ADJUSTMENTS OF THE INFANT AT BIRTH

During pregnancy, the embryo and later the fetus is totally dependent on the mother for its existence. The mother supplies the fetus with oxygen and nutrients, eliminates its carbon dioxide and other wastes, and protects it against shocks, temperature changes, and certain harmful microbes. However, a delivery of a physiologically immature baby carries certain risks.

A ***premature infant*** or "preemie" is generally considered to be one who weighs less than 2500 g (5 lb, 8 oz) at birth. It appears that poor prenatal care, a history of an earlier premature delivery, and mother's age below 16 or above 35 increase the risk of premature delivery. The problems related to survival of premature infants are due to the fact that they are not yet ready to take over functions the mother's body should be performing. The major problem with delivery of an infant under 36 weeks is infant respiratory distress syndrome (RDS), a condition that can be helped by a ventilator that delivers oxygen until the lungs can operate on their own. Researchers are now working on ways to inject surfactant to overcome infant RDS. Brain hemorrhages are also a major problem, and blindness is on the increase again. A procedure that freezes the surface of the eye can decrease the risk of blindness. It is also possible to supply nutrients intravenously until the digestive system matures.

At birth, a physiologically mature baby becomes self-supporting, and the newborn's body systems must make various adjustments. Following are some changes that occur in the respiratory and cardiovascular systems.

Respiratory System

The respiratory system is fairly well developed at least two months before birth as evidenced by the fact that premature babies delivered at seven months are able to breathe and cry. The fetus depends entirely on the mother for obtaining oxygen and eliminating carbon dioxide. The fetal lungs are either collapsed or partially filled with amniotic fluid, which is absorbed at birth. After delivery, the baby's supply of oxygen from the mother is stopped. Circulation in the baby continues, and as the blood level of carbon dioxide increases, the respiratory center in the medulla is stimulated. This causes the respiratory muscles to contract, and the baby draws its first breath. Since the first inspiration is unusually deep because the lungs contain no air, the baby exhales vigorously and naturally cries. A full-term baby may breathe 45 times a minute for the first two weeks after exposure to air. The rate is gradually reduced until it approaches a normal rate.

Cardiovascular System

Following the first inspiration by the baby, the cardiovascular system must make several adjustments. The foramen ovale of the fetal heart between atria closes at the moment of birth. This diverts deoxygenated blood to the lungs for the first time. The foramen ovale is closed by two flaps of heart tissue that fold together and permanently fuse. The remnant of the foramen ovale is the fossa ovalis. Once the lungs begin to function, the ductus arteriosus is shut off by contractions of the muscles in its wall. The ductus arteriosus generally does not completely and irreversibly close for about three months following birth. Incomplete closing, as you already know, results in patent ductus arteriosus.

The ductus venosus of fetal circulation connects the umbilical vein directly with the inferior vena cava. It forces any remaining umbilical blood directly into the fetal liver and from there to the heart. When the umbilical cord is severed, all visceral blood of the fetus goes directly to the fetal heart via the inferior vena cava. This shunting of blood usually occurs within minutes after birth but may take a week or two to complete. The ligamentum venosum, the remnant of the ductus venosus, is well established by the eighth postnatal week.

At birth, the infant's pulse may be from 120 to 160 per minute and may go as high as 180 following excitation. Several days after birth, there is a greater independent need for oxygen, which stimulates an increase in the rate of erythrocyte and hemoglobin production. This increase usually lasts for only a few days. Moreover, the white blood cell count at birth is very high, sometimes as much as 45,000 cells per cubic millimeter, but this decreases rapidly by the seventh day.

Finally, the infant's liver may not be adjusted at birth to control the production of bile pigment. As a result of this and other complicating factors, a temporary jaundice may result in as many as 50 percent of normal newborns by the third or fourth day after birth.

An excellent method for assessing the overall status of a newborn infant soon after birth (one minute) is the ***Apgar score.*** The score is based on heart rate, respiratory effort, muscle tone, reflex irritability, and color. The scoring system is as follows:

A score of 0 each is given for no heartbeat, no respiratory effort, no muscle tone, no reflex response to stimulation of the sole of the foot, and a pale or blue color.

A score of 1 each is given for heart rate below 100, slow or weak cry, some flexion of extremities, a grimace in response to stimulation of the sole of the foot, and a pink body and blue extremities.

A score of 2 each is given for heart rate over 100, good or strong cry, active motion, strong cry in response to stimulation of sole of foot, and complete healthy coloration.

POTENTIAL HAZARDS TO THE DEVELOPING EMBRYO AND FETUS

The developing fetus is susceptible to a number of potential hazards that can be transmitted from the mother. Such hazards include infectious microbes, chemicals and drugs, alcohol, and cigarette smoking. In addition, certain environmental conditions and pollutants can damage the fetus or even cause fetal death.

Chemicals and Drugs

Since the placenta is known to be an ineffective barrier between the maternal and fetal circulations, actually any drug or chemical dangerous to the infant may be considered potentially dangerous to the fetus when given to the mother. Many chemicals and drugs have been proven to be toxic and teratogenic to the developing embryo and fetus. A ***teratogen*** (*terato* = monster) is any agent or influence that causes physical defects in the developing embryo. Examples are pesticides, defoliants, industrial chemicals, some hormones, antibiotics, oral anticoagulants, anticonvulsants, antitumor agents, thyroid drugs, thalidomide, diethylstilbestrol (DES), LSD, marijuana, and cocaine. A pregnant female who uses cocaine subjects the fetus to several potential problems. These include retarded growth, subtle neurological abnormalities, hyperirritability, a tendency to stop breathing, higher risk of crib death, malformed or missing organs, strokes, and seizures. The risks of spontaneous abortion (miscarriage), premature birth, and stillbirth will increase from fetal exposure to cocaine.

Irradiation

Ionizing radiations are potent teratogens. Treatment of pregnant mothers with large doses of x-rays and radium during the embryo's susceptible period of development may cause microcephaly (small size of head in relation to the rest of the body), mental retardation, and skeletal malformations. Caution is advised for diagnostic x-rays during the first trimester of pregnancy.

Alcohol

Alcohol has been a suspected teratogen for centuries, but only recently has a relationship been recognized between maternal alcohol intake and the characteristic pattern of malformations in the fetus. The term applied to the effects of intrauterine exposure to alcohol is ***fetal alcohol syndrome (FAS).*** Studies to date have indicated that in the general population the incidence of FAS may exceed 1 per 1000 live births and may be by far the number one fetal teratogen. The symptoms shown by children may include slow growth before and after birth, small head, facial irregularities such as narrow eye slits and sunken nasal bridge, defective heart and other organs, malformed arms and legs, genital abnormalities, and mental retardation. There are also behavioral problems, such as hyperactivity, extreme nervousness, and a poor attention span. In humans, acetaldehyde is one of the toxic products of alcohol metabolism and may contribute to fetal damage. The human placenta may transfer acetaldehyde from maternal to fetal circulation and may oxidize ethanol to acetaldehyde.

Cigarette Smoking

The latest evidence not only indicates a causal relationship between cigarette smoking during pregnancy and low infant birth weight but also points to a strong probable association between smoking and a higher fetal and infant mortality rate. Infants nursing from smoking mothers have also been found to have an increased incidence of gastrointestinal disturbances. Other pathologies of infants of smoking mothers include an increased incidence of respiratory problems during the first year of life, including bronchitis and pneumonia. Cigarette smoking may be teratogenic and cause cardiac abnormalities and anencephaly (a developmental anomaly with absence of neural tissue in the cranium). Maternal smoking also appears to be a significant etiologic factor in the development of cleft lip and palate and has been tentatively linked with sudden infant death syndrome (SIDS).

PHYSIOLOGY OF LACTATION

The term ***lactation*** refers to the secretion and ejection of milk by the mammary glands. A principal hormone in promoting lactation is ***prolactin (PRL)*** from the anterior pituitary gland. It is released in response to the hypothalamic prolactin releasing factor (PRF). Even though PRL levels increase as the pregnancy progresses, there is no milk secretion because estrogens and progesterone inhibit the PRL from being effective. Following delivery, the levels of estrogens and progesterone in the mother's blood decrease, and the inhibition is removed.

The principal stimulus in maintaining prolactin secretion during lactation is the sucking action of the infant. Sucking initiates impulses from receptors in the nipples to the hypothalamus. The impulses inhibit prolactin inhibiting factor (PIF) production, and PRL is released by the anterior pituitary. The sucking action also initiates impulses to the posterior pituitary via the hypothalamus. These impulses stimulate the release of the hormone ***oxytocin (OT)*** by the posterior pituitary gland. Thus, a positive feedback cycle is established. OT induces smooth muscle cells surrounding the outer walls of the alveoli to contract, thereby compressing the alveoli and ejecting milk. The compression moves milk from the alveoli of the mammary gland into the ducts, where it can be sucked. This process is referred to as ***milk let-down.*** Although the actual ejection of milk does not occur from 30 seconds to 1 minute after nursing begins, some milk is stored in lactiferous sinuses near the nipple. Thus, some milk is available during the latent period. (Recall from Chapter 18 that OT, as part of a positive feedback cycle, also stimulates smooth muscle contraction of the pregnant uterus to facilitate delivery. This action of OT on uterine smooth muscle cells in nursing mothers results in a more rapid return of the uterus to its prepregnant size than occurs in nonnursing mothers and compresses torn placental vessels at delivery and thus reduces blood loss by the mother.)

During late pregnancy and the first few days after birth, the mammary glands secrete a cloudy fluid called ***colostrum.*** Although it is not as nutritious as true milk, since it contains less lactose and virtually no fat, it serves adequately until the appearance of true milk on about the fourth day. Colostrum and maternal milk are thought to contain antibodies that protect the infant during the first few months of life.

Following birth of the infant, the PRL level starts to return to the nonpregnant level, but each time the mother nurses the infant, nerve impulses from the nipples to the hypothalamus cause the release of PRF and the secretion of a 10-fold increase in PRL by the anterior pituitary that lasts about an hour. PRL acts on the mammary glands to provide milk for the next nursing period. If this surge of PRL is blocked by injury or disease, or if nursing is discontinued, the mammary glands lose their ability to secrete milk in a few days. However, milk secretion can continue for several years if the child continues to suckle. Milk secretion normally decreases considerably within seven to nine months.

Lactation often prevents the occurrence of female ovarian cycles for the first few months following delivery, if the frequency of sucking is about 8 to 10 times a day, although there is no guarantee. Ovulation will normally precede the resumption of the period, so there is always an unknown factor. The contraceptive effect of breast feeding is, therefore, not a very effective birth control measure. However, the suppression of ovulation during lactation is believed to occur as follows. During breast feeding, neural input from the nipple reaches the hypothalamus and causes it to produce beta (β)-endorphin. This, in turn, suppresses the release of gonadotropin releasing hormone (GnRH). Consequently, there is a decreased production in LH and FSH, and ovulation is inhibited.

Advocates of ***breast feeding*** feel that it offers the following advantages to the infant.

1. It establishes early and prolonged contact between mother and infant.
2. The infant is more in control of intake.
3. Fats and iron in human milk are better absorbed than those in cow's milk, and the amino acids in human milk are more readily metabolized. Also, the lower sodium content of human milk is more suited to the infant's needs.
4. Breast feeding provides important antibodies that prevent gastroenteritis. Immunity to respiratory infections and meningitis is also greater.
5. Premature infants benefit from breast feeding because the milk produced by mothers of premature infants seems to be specially adapted to the infant's needs by having a higher protein content than the milk of mothers of full-term infants.
6. There is less likelihood of an allergic reaction in the baby to the milk of the mother.
7. The act of sucking on the breast promotes development of the jaw, facial muscles, and teeth.
8. Several proteins in breast milk may stimulate the infant's immune system by increasing B lymphocyte maturation.

INHERITANCE

Inheritance is the passage of hereditary traits from one generation to another. It is the process by which you acquired your characteristics from your parents and will transmit your characteristics to your children. The branch of biology that deals with inheritance is called ***genetics*** (je-NET-iks).

Genotype and Phenotype

The nuclei of all human cells except gametes contain 23 pairs of chromosomes—the diploid number. One chromosome from each pair comes from the mother, and the other comes from the father. Homologous chromosomes, the two chromosomes in a pair, contain genes that control the same traits. If a chromosome contains a gene for height, its homologue will contain a gene for height.

The relationship of genes to heredity is illustrated admirably by the disorder called ***phenylketonuria*** or ***PKU*** (see Figure 29-16). People with PKU are unable to manufacture the enzyme phenylalanine hydroxylase (see Chapter 25). It is believed the PKU is brought about by an abnormal gene, which can be symbolized as p. The normal gene will be symbolized as P. The chromosome concerned with directions for phenylalanine hydroxylase production will have either p or P on it. Its homologue will also have p or P. Thus, every individual will have one of the following genetic makeups, or ***genotypes*** (JĒ-nō-tīps): PP, Pp, or pp. Although people with genotypes of Pp have the abnormal gene, only those with genotype pp suffer from the disorder because the normal gene, when present, dominates over and inhibits the abnormal one. A gene that dominates is called the ***dominant gene,*** and the trait expressed is said to be a dominant trait. The gene that is inhibited is called the ***recessive gene,*** and the trait it controls is called the recessive trait. Genes that control the same inherited trait—for example, height, eye color, or hair color—and that occupy the same position on homologous chromosomes are known as ***alleles.***

By tradition, we symbolize the dominant gene with a capital letter and the recessive one with a lowercase letter. An individual with the same genes on homologous chromosomes (for example, PP or pp) is said to be ***homozygous*** for the trait. An individual with different genes on homologous chromosomes (for example, Pp) is said to be ***heterozygous*** for the trait. ***Phenotype*** (FĒ-nō-tīp; *pheno* = showing) refers to how the genetic makeup is expressed in the body. A person with Pp has a different genotype from one with PP, but both have the same phenotype—which in this case is normal production of phenylalanine hydroxylase.

To determine how gametes containing haploid chromosomes unite to form diploid fertilized eggs, special charts called ***Punnett squares*** are used. Usually, the male gametes (sperm cells) are placed at the side of the chart, and the female gametes (ova) at the top (as in Figure 29-16). The four spaces on the chart represent the possible combinations (genotypes) of male and female gametes that could form fertilized eggs.

Now consider ***sickle-cell anemia (SCA).*** The gene for normal hemoglobin is designated as Hb^A; the gene for the abnormal hemoglobin associated with sickle-cell anemia is designated as Hb^S. Normal individuals have the genotype Hb^AHb^A (homozygous); individuals with sickle-cell trait have only minor problems with anemia and have the genotype Hb^AHb^S (heterozygous); individuals with the sickle-cell disease have severe anemia and have the genotype Hb^SHb^S (homozygous).

ABO blood grouping also illustrates the relationship of genes to heredity (Chapter 19). Antigens *A* and *B* are inherited as dominant traits; *O* is inherited as a recessive trait. The various possible and impossible ABO blood grouping phenotypes are illustrated in Figure 29-17.

Exhibit 29-3 lists some of the variety of simple inherited structural and functional traits in humans.

Normal traits do not always dominate over abnormal ones, but genes for severe disorders are more frequently recessive than dominant. An exception is ***Huntington's chorea***—a major disorder caused by a dominant gene and characterized by degeneration of nervous tissue, usually leading to mental disturbance and death. The first signs of Huntington's chorea do not occur until adulthood, very often after the person has already produced offspring but a genetic test is now available that can determine whether or not a child carries the gene for this disorder.

FIGURE 29-16 Inheritance of phenylketonuria (PKU).

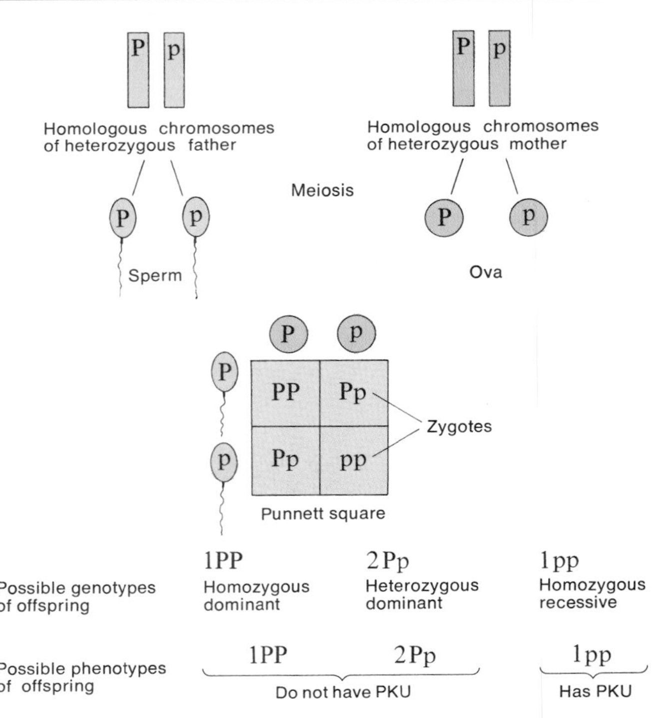

FIGURE 29-17 Inheritance of possible and impossible phenotypes of the ABO blood grouping system.

Parents	Possible phenotypes	Impossible phenotypes
A A	A,O	B,AB
A B	A,B,AB,O	None
A AB	A,B,AB	O
A O	A,O	B,AB
B B	B,O	A,AB
B AB	A,B,AB	O
B O	B,O	A,AB
AB AB	A,B,AB	O
AB O	A,B	AB,O
O O	O	A,B,AB

Key: ■ possible phenotypes
■ impossible phenotypes

EXHIBIT 29-3 SELECTED HEREDITARY TRAITS IN HUMANS

Dominant	Recessive
Curly hair	Straight hair
Dark brown hair	All other colors
Coarse body hair	Fine body hair
Pattern baldness (dominant in males)	Baldness (recessive in females)
Normal skin pigmentation	Albinism
Brown eyes	Blue or gray eyes
Near- or farsightedness	Normal vision
Normal hearing	Deafness
Normal color vision	Color blindness
Broad lips	Thin lips
Large eyes	Small eyes
Polydactylism (extra digits)	Normal digits
Brachydactylism (short digits)	Normal digits
Syndactylism (webbed digits)	Normal digits
Hypertension	Normal blood pressure
Diabetes insipidus	Normal excretion
Huntington's chorea	Normal nervous system
Normal mentality	Schizophrenia

EXHIBIT 29-3 SELECTED HEREDITARY TRAITS IN HUMANS (*Continued*)

Dominant	Recessive
Migraine headaches	Normal
Normal resistance to disease	Susceptibility to disease
Enlarged spleen	Normal spleen
Enlarged colon	Normal colon
A or B blood factor	O blood factor
Rh blood factor	No Rh blood factor

Inheritance of Sex

Microscopic examination of the chromosomes in cells reveals that one pair differs in males and in females (Figure 29-18a). In females, the pair consists of two chromosomes designated as X chromosomes. One X chromosome is

FIGURE 29-18 Inheritance of sex. (a) Normal human male chromosomes grouped according to size during the metaphase stage of cell division. Sex chromosomes are the twenty-third pair, highlighted in the blue-colored box. (b) Sex determination.

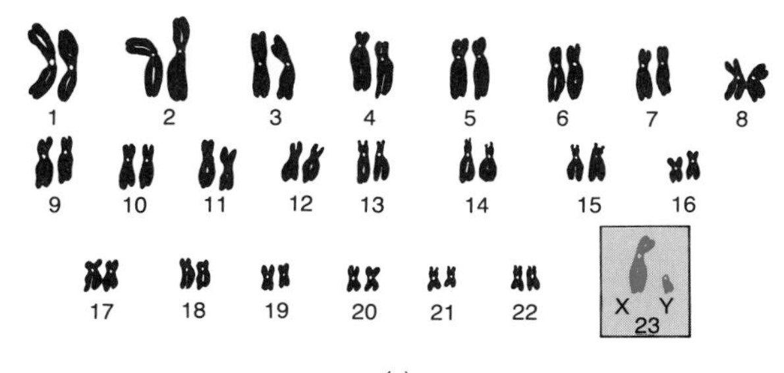

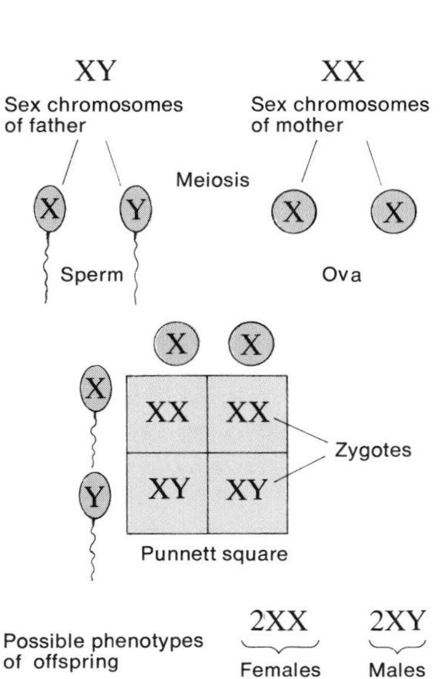

also present in males, but its mate is a chromosome called a Y chromosome. The XX pair in the female and the XY pair in the male are called the ***sex chromosomes.*** All other chromosomes are called ***autosomes.***

The sex chromosomes are responsible for the sex of the individual (Figure 29-18b). When a spermatocyte undergoes meiosis to reduce its chromosome number, one daughter cell will contain the X chromosome, and the other will contain the Y chromosome. Oocytes have no Y chromosomes and produce only X-containing ova. If the secondary oocyte is subsequently fertilized by an X-bearing sperm, the offspring normally will be female (XX). Fertilization by a Y sperm normally produces a male (XY). Thus, sex is determined at fertilization. Although X and Y sperm are produced in equal amounts, more males are born than females. It is speculated that Y sperm swim faster than X sperm because X sperm contain more genetic material.

Both female and male embryos develop identically until about seven weeks after fertilization. At that point, a gene sets into motion a cascade of events that leads to the development of a male. In the absence of the gene, the development of a female occurs. Scientists have known since the 1920s that sex chromosomes exist and since 1959 that the Y chromosome contains a gene called ***testis-determining factor (TDF).*** The TDF gene is contained in a fragment of the Y chromosome and has about 140,000 nitrogenous base pairs of DNA subunits (the entire Y chromosome has an estimated 70 million nitrogenous base pairs). The TDF gene apparently produces a protein similar to other proteins that bind to DNA to control the activity of genes. Whenever the TDF gene is present in a fertilized ovum, the fetus will develop testes and differentiate into a male; in the absence of TDF, the fetus will develop ovaries and differentiate into a female.

Down's syndrome (DS) is a disorder that results from an error in cell division called ***nondisjunction.*** In this situation, sister chromatids fail to separate properly during anaphase of mitosis (or equatorial division of meiosis), or homologous chromosomes fail to separate properly during reduction division of meiosis. As a result, the chromatids or chromosomes pass to the same daughter cell. Down's syndrome is characterized by mental retardation, retarded physical development (short stature and stubby fingers), distinctive facial structures (large tongue, flat profile, broad skull, slanting eyes, and round head), and malformation of the heart, ears, hands, and feet (Figure 29-19). Sexual maturity is rarely attained. Individuals with the disorder usually have 47 chromosomes instead of the normal 46 (an extra chromosome in the twenty-first pair). A cell that has one or more chromosomes of a set added or deleted is called an ***aneuploid*** (an'-yoo-PLOID). A monosomic cell ($2n - 1$) has a missing chromosome; a trisomic cell ($2n + 1$) has an added chromosome. Since the most common form of Down's syndrome is characterized by an extra chromosome in the twenty-first pair, it is also known as ***trisomy 21.***

In addition to autosomal chromosome aneuploids, sex chromosome aneuploids also occur in humans. One example is ***Turner's syndrome,*** due to the presence of only one X chromosome (X0). Such females are sterile with virtually no ovaries and limited development of secondary sex characteristics. Other features include short stature, webbed neck, a shieldlike chest with underdeveloped breasts, and widely spaced nipples. There is usually no mental retardation. Another sex chromosome aneuploid condition is called ***metafemale syndrome,*** characterized by at least three X chromosomes (XXX). These females have underdeveloped genital organs and limited fertility and are generally mentally retarded. ***Klinefelter's syndrome*** is usually due to trisomy XXY. Such individuals are sterile males. They have undeveloped testes, scanty body hair, and enlarged breasts and are characteristically somewhat mentally disadvantaged.

FIGURE 29-19 Photograph of an individual with Down's syndrome. (Courtesy of Michael Mohan and Ohio School Pictures.)

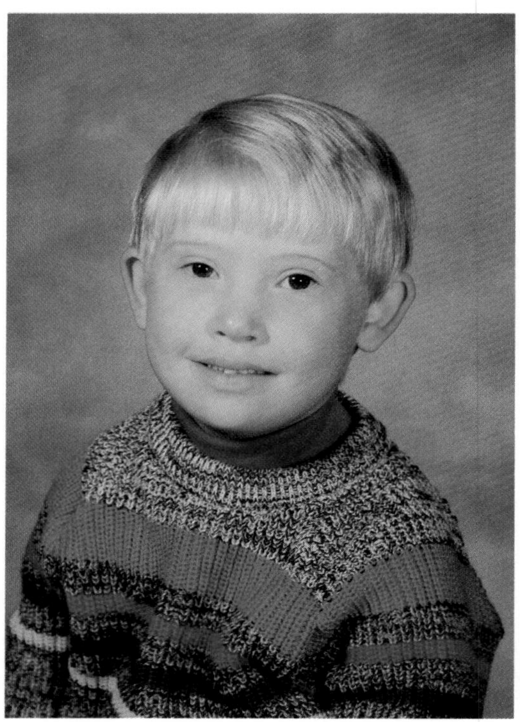

CLINICAL APPLICATION: KARYOTYPING

A *karyotype* (KAR-ē-ō-tīp; *karyon* = nucleus) is an arrangement of chromosomes from a cell based on their shape, size, and position of centromeres. A karyotype is prepared by photographing the chromosomes, usually in a white blood cell, and then cutting them out and arranging them in standard or-

der (see Figure 29-18a). A normal human karyotype contains 22 matching pairs of autosomes and one pair of sex chromosomes (two X chromosomes in females and one X and one Y chromosome in males), all of which are normal in terms of shape, size, and structure.

Karyotyping is done when a chromosomal abnormality is suspected that might be responsible for a disease or developmental problem. It is frequently done to investigate birth defects, abnormal growth, mental retardation, delayed puberty, abnormal sexual development, infertility, or certain inherited disorders. Missing, additional, or abnormal chromosomes in a karyotype are indicative of disorders such as Down's syndrome, Klinefelter's syndrome, and Turner's syndrome.

Color Blindness and X-Linked Inheritance

The sex chromosomes also are responsible for the transmission of a number of nonsexual traits. Genes for these traits appear on X chromosomes, but many of these genes are absent from Y chromosomes. This feature produces a pattern of heredity that is different from the pattern described earlier. Let us consider color blindness. The gene for ***color blindness*** is a recessive one designated c. Normal color vision, designated C, dominates. The C/c genes are located on the X chromosome. The Y chromosome does not contain the segment of DNA that programs this aspect of vision. Thus, the ability to see colors depends entirely on the X chromosomes. The genetic possibilities are:

X^CX^C	Normal female
X^CX^c	Normal female carrying the recessive gene
X^cX^c	Color-blind female
X^CY	Normal male
X^cY	Color-blind male

Only females who have two X^c chromosomes are color blind. In X^CX^c females the trait is inhibited by the normal, dominant gene. Males, on the other hand, do not have a second X chromosome that would inhibit the trait. Therefore, all males with an X^c chromosome will be color blind. The inheritance of color blindness is illustrated in Figure 29-20.

Traits inherited in the manner just described are called ***X-linked traits.*** Another X-linked trait is ***hemophilia***—a condition in which the blood fails to clot or clots very slowly after an injury (Chapter 19). Like the trait for color blindness, hemophilia is caused by a recessive gene. If H represents normal clotting and h represents abnormal clotting, then X^hX^h females will have the disorder. Males with X^HY will be normal; males with X^hY will be hemophiliac. Actually, clotting time varies somewhat among hemophiliacs, so the condition may be affected by other genes as well.

FIGURE 29-20 **Inheritance of color blindness.**

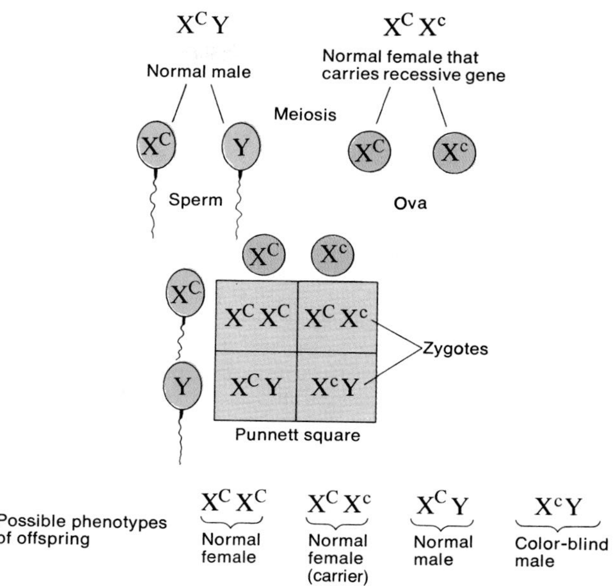

Fragile X syndrome is a recently recognized disorder due to a defective gene on the X chromosome. It is so named because a small portion of the tip of the X chromosome seems susceptible to breakage under certain conditions. It is a leading cause of mental retardation among newborns, ranking second only to Down's syndrome. Fragile X syndrome affects males more than females and results in learning difficulties, mental retardation, and physical abnormalities such as oversized ears, elongated forehead, enlarged testes, and double jointedness. The syndrome may also be involved in autism in which the individual exhibits extreme withdrawal and refusal to communicate. Fragile X syndrome can be diagnosed by amniocentesis. Although fragile X syndrome is largely X-linked, about 20 percent of males who inherit the trait are unaffected, and about one-third of females who are carriers are affected.

A few other X-linked traits in humans are nonfunctional sweat glands, certain forms of diabetes, some types of deafness, uncontrollable rolling of the eyeballs, absence of central incisors, night blindness, one form of cataract, white forelocks, juvenile glaucoma, and juvenile muscular dystrophy.

CLINICAL APPLICATION: MAPPING THE HUMAN GENOME

A *genome* (JĒ-nōm) is the complete gene complement of an organism. Of the approximately 100,000 genes in the human organism, about 4500 have been identified and only about 1500 of these have been roughly located on various chromosomes. At pres-

ent, an attempt is being made to map and sequence the entire human genome. The project, which is coordinated by the National Institutes of Health (NIH), is expected to take about 15 years and cost about 3 billion dollars. A major goal of the project is the application of molecular genetics to medical research in order to identify and hopefully cure certain diseases.

MEDICAL TERMINOLOGY ASSOCIATED WITH DEVELOPMENT AND INHERITANCE

Abortion (a-BOR-shun) Premature expulsion from the uterus of the products of conception—embryo or nonviable fetus. May be caused by abnormal development of embryo, placental abnormalities, endocrine disturbances, certain diseases, trauma, and stress. ***Induced (nonspontaneous) abortions*** are brought on intentionally by methods such as vacuum aspiration (suction curettage), up to the twelfth week of pregnancy; dilation and evacuation (D & E), commonly used between the thirteenth and fifteenth weeks of pregnancy and sometimes up to twenty weeks; and use of saline or prostaglandin preparations to induce labor and delivery, usually after the fifteenth week of pregnancy. ***Spontaneous abortions (miscarriages)*** occur without apparent cause. Recent evidence suggests that as many as one-third of all successful fertilizations end in spontaneous abortions, most of which occur even before a female or her physician is aware that she is pregnant.

Breech presentation A malpresentation in which the fetal buttocks or lower extremities present into the maternal pelvis; most common cause is prematurity.

Lethal gene (LĒ-thal jēn; *lethum* = death) A gene that, when expressed, results in death either in the embryonic state or shortly after birth.

Lochia (LŌ-kē-a) The discharge from the birth canal consisting initially of blood and later of serous fluid occurring after childbirth. The discharge is derived from the former placental site and may last up to about two to four weeks.

Mutation (myoo-TĀ-shun; *mutare* = change) A permanent heritable change in a gene that causes it to have a different effect than it had previously.

Preeclampsia (prē'-e-KLAMP-sē-a) A syndrome characterized by sudden hypertension, large amounts of protein in urine, and generalized edema; might be related to autoimmune or allergic reaction due to the presence of a fetus; when the condition is also associated with convulsions and coma, it is referred to as **eclampsia.**

Puerperal (pyoo-ER-per-al; *puer* = child; *parere* = to bring forth) **fever** Infectious disease of childbirth, also called puerperal sepsis and childbed fever. The disease results from an infection originating in the birth canal and affects the endometrium. It may spread to other pelvic structures and lead to septicemia.

STUDY OUTLINE

Development During Pregnancy (p. 928)

1. Pregnancy is a sequence of events that includes fertilization.
2. Its various events are hormonally controlled.

Fertilization and Implantation (p. 928)

1. Fertilization refers to the penetration of a secondary oocyte by a sperm cell and the subsequent union of the sperm and oocyte nuclei to form a zygote.
2. Penetration is facilitated by enzymes produced by sperm acrosomes.
3. Normally, only one sperm fertilizes a secondary oocyte.
4. Early rapid cell division of a zygote is called cleavage, and the cells produced by cleavage are called blastomeres.
5. The solid mass of cells produced by cleavage is a morula.
6. The morula develops into a blastocyst, a hollow ball of cells differentiated into a trophoblast (future embryonic membranes) and inner cell mass (future embryo).
7. The attachment of a blastocyst to the endometrium is called implantation; it occurs by enzymatic degradation of the endometrium.
8. In vitro fertilization (IVF) refers to the fertilization of a secondary oocyte outside the body and the subsequent implantation of the zygote. Variations include embryo transfer, gamete intrafallopian transfer (GIFT), and transvaginal oocyte retrieval.

Embryonic Development (p. 932)

1. During embryonic growth, the primary germ layers and embryonic membranes are formed and the placenta is functioning.
2. The primary germ layers—ectoderm, mesoderm, and endoderm—form all tissues of the developing organism.
3. Embryonic membranes include the yolk sac, amnion, chorion, and allantois.
4. Fetal and maternal materials are exchanged through the placenta.

Fetal Growth (p. 938)

1. During the fetal period, organs established by the primary germ layers grow rapidly.
2. The principal changes associated with fetal growth are summarized in Exhibit 29-2.

Hormones of Pregnancy (p. 939)

1. Pregnancy is maintained by human chorionic gonadotropin (hCG), estrogens, and progesterone (PROG).
2. Human chorionic somatomammotropin (hCS) assumes a role in breast development, protein anabolism, and glucose and fatty acid catabolism.
3. Relaxin relaxes the symphysis pubis and helps dilate the uterine cervix toward the end of pregnancy.
4. Inhibin inhibits secretion of FSH and might regulate secretion of hGH.

Diagnosis of Pregnancy (p. 941)

1. Diagnosis of pregnancy is based on presumptive, probable, and positive evidence.
2. Presumptive evidence includes amenorrhea, breast changes, nausea, and changes in skin pigmentation.
3. Probable evidence includes enlargement of the abdomen, uterine changes, and resulting endocrine tests, which are based on the presence of hCG in urine or blood.
4. Positive evidence includes detection of a fetal heartbeat and fetal movements (other than by the mother) and visualization of the fetus.

Gestation (p. 941)

1. The time an embryo or fetus is carried in the uterus is called gestation.
2. Human gestation lasts about 280 days from the beginning of the last menstrual period.
3. During gestation, several anatomical and physiological changes occur.

Prenatal Diagnostic Tests (p. 943)

1. Amniocentesis is the withdrawal of amniotic fluid. It can be used to diagnose inherited biochemical defects and chromosomal disorders, such as hemophilia, Tay-Sachs disease, sickle-cell anemia, and Down's syndrome.
2. Chorionic villi sampling (CVS) involves withdrawal of chorionic villi for chromosomal analysis.
3. CVS can be done sooner than amniocentesis, and the results are available sooner.
4. In fetal ultrasonography, an image of a fetus is displayed on a screen.

Parturition and Labor (p. 945)

1. Parturition refers to birth and is accompanied by a sequence of events called labor.
2. The birth of a baby involves dilation of the cervix, expulsion of the fetus, and delivery of the placenta.

Adjustments of the Infant at Birth (p. 947)

1. The fetus depends on the mother for oxygen and nutrients, removal of wastes, and protection.
2. Following birth the respiratory and cardiovascular systems undergo changes in adjusting to self-supporting postnatal life.
3. Overall status of a newborn soon after birth is evaluated by an Apgar score.

Potential Hazards to the Developing Embryo and Fetus (p. 948)

1. The developing embryo and fetus are susceptible to many potential hazards that can be transmitted from the mother.
2. Examples are infections, microbes, chemicals and drugs, alcohol, and smoking.

Physiology of Lactation (p. 949)

1. Lactation refers to the secretion and ejection of milk by the mammary glands.
2. Secretion is influenced by prolactin (PRL), estrogens, and progesterone.
3. Ejection is influenced by oxytocin (OT).

Inheritance (p. 950)

1. Inheritance is the passage of hereditary traits from one generation to another.
2. The genetic makeup of an organism is called its genotype. The traits expressed are called its phenotype.
3. Dominant genes control a particular trait; expression of recessive genes is inhibited by dominant genes.
4. Sex is determined by a testis-determining factor (TDF) on the Y chromosome of the male at fertilization.
5. Sex chromosome aneuploids include Turner's syndrome, metafemale syndrome, and Klinefelter's syndrome.
6. Color blindness and hemophilia primarily affect males because there are no counterbalancing dominant genes on the Y chromosomes.

REVIEW QUESTIONS

1. Define developmental anatomy. (p. 928)
2. Define fertilization. Where does it normally occur? How is a morula formed? (p. 928)
3. Explain how dizygotic (fraternal) and monozygotic (identical) twins are produced. (p. 929)
4. Describe the components of a blastocyst. (p. 930)
5. What is implantation? How does the fertilized ovum implant itself? Why is an implanted ovum usually not rejected by the mother? (p. 930)
6. Describe the various types of in vitro fertilization (IVF). (p. 932)
7. Define the embryonic period and the fetal period. (p. 932)
8. List several body structures formed by the endoderm, mesoderm, and ectoderm. (p. 935)
9. What is an embryonic membrane? Describe the functions of the four embryonic membranes. (p. 933)
10. Explain the importance of the placenta and umbilical cord to fetal growth. (p. 936)
11. Outline some of the major developmental changes during fetal growth. (p. 939)

12. List the hormones involved in pregnancy and describe the functions of each. (p. 939)
13. Describe the various types of presumptive, probable, and positive evidence used to diagnose pregnancy. What is the basis for home pregnancy tests? (p. 941)
14. Define gestation and parturition. (p. 941)
15. Describe several anatomical and physiological changes that occur during gestation. (p. 942)
16. Explain the effects of pregnancy on exercise and exercise on pregnancy. (p. 943)
17. Distinguish between false and true labor. Describe what happens during the stage of dilation, the stage of expulsion, and the placental stage of delivery. (p. 945)
18. Discuss the principal respiratory and cardiovascular adjustments made by an infant at birth. (p. 947)
19. Explain the basis of the Apgar score to assess well-being of newborns. (p. 948)
20. Explain in detail some of the potential hazards for the developing embryo and fetus. (p. 948)
21. What is lactation? Name the hormones involved and their functions. (p. 949)
22. Define inheritance. What is genetics? (p. 950)
23. Define the following terms: genotype, phenotype, dominant, recessive, homozygous, heterozygous. (p. 950)
24. What is a Punnett square? (p. 950)
25. List several dominant and recessive traits inherited in humans. (p. 951)
26. Set up Punnett squares to show the inheritance of the following traits: sex, color blindness, hemophilia. (p. 951)
27. What is Down's syndrome (DS)? What causes it? Why is it also called trisomy 21? (p. 952)
28. Describe three sex chromosome aneuploids. (p. 952)
29. What is X-linked inheritance? (p. 953)
30. Define the following: ectopic pregnancy (EP) (p. 929), emesis gravidarium (morning sickness) (p. 932), placenta previa (p. 938), fetomaternal hemorrhage (p. 938), umbilical cord accident (p. 938), fetal surgery (p. 938), fetal-tissue implantation (p. 938), dystocia (p. 945), cesarean section (p. 945), and karyotype. (p. 952)
31. Explain the procedure and diagnostic value of the following: amniocentesis (p. 943), chorionic villus sampling (CVS) (p. 944), fetal ultrasonography (p. 944), and electronic fetal monitoring (EFM). (p. 945)
32. Refer to the glossary of medical terminology associated with development and inheritance. Be sure that you can define each term. (p. 954)

SELECTED READINGS

Begley, S., and J. Carey. "How Human Life Begins," *Newsweek,* 11 January 1982.

Carlson, B. M. *Patten's Foundations of Human Embryology,* 4th ed. New York: McGraw-Hill, 1981.

England, M. A. *Color Atlas of Life Before Birth.* Chicago: YearBook Medical Publishers, 1983.

Fuchs, F. "Genetic Amniocentesis," *Sciencific American,* June 1980.

Gehring, W. J. "The Molecular Basis of Development," *Scientific American,* October 1985.

Gold, M. "The Baby Makers," *Science 85,* April 1985.

Goldfinger, S. E. (ed). "Pregnancy: Age and Outcome," *Harvard Medical School Health Letter,* October 1985.

Grimes, D. A. "Reversible Contraception for the 1980's," *Journal of the American Medical Association,* 3 January 1986.

Hacker, N. F., and J. G. Moore. *Essentials of Obstetrics and Gynecology.* Philadelphia: Saunders, 1986.

Henahan, J. F. "Fertilization, Embryo Transfer Procedures Raise Many Questions," *Journal of the American Medical Association,* 17 August 1984.

Hogan, M. J. "Cesarean Birth," *Mayo Clinic Health Letter,* February 1988.

Holliday, R. "A Different Kind of Inheritance," *Scientific American,* June 1989.

Kantrowitz, B., P. Wingert, and M. Hager. "Preemies," *Newsweek,* 16 May 1988.

Lawn, R. M., and G. A. Vehar, "The Molecular Genetics of Hemophilia," *Scientific American,* March 1986.

Leaf, D. A. "Exercise During Pregnancy," *Postgraduate Medicine,* January 1989.

McKusick, V. "Mapping and Sequencing the Human Genome," *New England Journal of Medicine,* 6 April 1989.

Miller, J. A. "Window on the Womb," *Science News,* 2 February 1985.

Moore, K. L. *Essentials of Human Embryology.* Toronto: B. C. Decker, 1988.

Seibel, M. M. "A New Era in Reproductive Technology," *New England Journal of Medicine,* 31 March 1988.

Silberner, J. "Babymaking: Expanding Horizons," *Science News,* 14 December 1985.

Singer, S. *Human Genetics.* New York: W. H. Freeman, 1985.

Wassarman, P. M. "Fertilization in Mammals," *Scientific American,* December 1988.

Zuckerman, B. "Effects of Maternal Marijuana and Cocaine Use on Fetal Growth," *New England Journal of Medicine,* 23 March 1989.

APPENDIX A: MEASUREMENTS

UNITS OF MEASUREMENT

When you measure something, you are comparing it with some standard scale to determine its *magnitude.* How long is it? How much does it weigh? How fast is it going? Some measurements are made directly by comparing the unknown quantity with the known unit of the same kind, for example, weighing a patient on a scale and taking the reading directly in pounds. Other measurements are indirect and are done by calculation, for example, counting a person's blood cells in a certain number of squares on a microscope slide and then calculating the total blood count.

Regardless of how a measurement is taken, it always requires two things: a *number* and a *unit.* When recording the weight of a patient, you would not just say 145. You have to give both the number (145) and the unit (pounds). When you count blood cells, you report the measurement as 10,000 (number) white blood cells per cubic millimeter of blood (unit).

All the units in use can be expressed in terms of one of three special units called ***fundamental units.*** These fundamental units have been established arbitrarily as length, mass, and time. Mass is perhaps an unfamiliar term to you. ***Mass*** is the amount of matter an object contains. The mass of this textbook is the same whether it is measured in a laboratory, under the sea, on top of a mountain, or even on the moon. No matter where you take it, it still has the same quantity of matter. ***Weight,*** on the other hand, is determined by the pull of gravity on an object. This textbook will not have the same weight on earth as on the moon because of the differences in gravitation. However, as long as we are dealing only with earthbound objects, weight and mass may be considered synonymous terms because the force of gravity on the surface of the earth is nearly constant. Thus, weight remains nearly the same regardless of where the measurements are taken.

All units other than the fundamental ones are ***derived units***—they can always be written as some combination of the three fundamental units. For example, units of volume are derived from units of length (the volume of a cube = length × width × height). Units of speed are combinations of distance and time (miles per hour).

Units are grouped into systems of measurement. The two principal systems of measurement commonly used in this country are the U.S. and the metric systems. The apothecary system is used by physicians and pharmacists.

U.S. SYSTEM

The ***U.S. system*** of measurement is used in everyday household work, industry, and some fields of engineering. The fundamental units in the U.S. system are the foot (length), the pound (mass), and the second (time).

The basic problem with the U.S. system is that there is no *uniform* progression from one unit to another. If you want to convert a measurement of $2\frac{1}{2}$ yd to feet, you have to multiply it by 3 because there are 3 ft in a yard. If you want to convert the same length to inches, you have to multiply by 3 and then by 12 (or by 36) because there

EXHIBIT A-1 U.S. UNITS OF MEASUREMENT

Fundamental or Derived Unit	Units and U.S. Equivalents
Length	1 inch = 0.083 foot
	1 foot (ft) = 12 in.
	= 0.333 yard
	1 yard (yd) = 3 ft = 36 in.
	1 mile (mi) = 1,760 yd = 5,280 ft
Mass	1 grain (gr) = 0.002285 ounce
	1 dram (dr) = 27.34 gr
	= 0.063 ounce
	1 ounce (oz) = 16 dr = 437.5 gr
	1 pound (lb) = 16 oz = 7,000 gr
	1 ton = 2,000 lb
Time	1 second (sec) = 1/86,400 of a day
	1 minute (min) = 60 sec
	1 hour (hr) = 60 min = 3,600 sec
	1 day = 24 hr = 1,440 min
	= 86,400 sec
Volume	1 fluidram (fl dr) = 0.125 fluidounce
	1 fluidounce (fl oz) = 8 fl dr
	= 0.0625 quart
	= 0.008 gallon
	1 pint (pt) = 16 fl oz = 128 fl dr
	1 quart (qt) = 2 pt = 32 fl oz
	= 256 fl dr
	= 0.25 gallon
	1 gallon (gal) = 4 qt = 8 pt
	= 128 fl oz
	= 1.024 fl dr

FIGURE A-1 Metric and U.S. units of length.

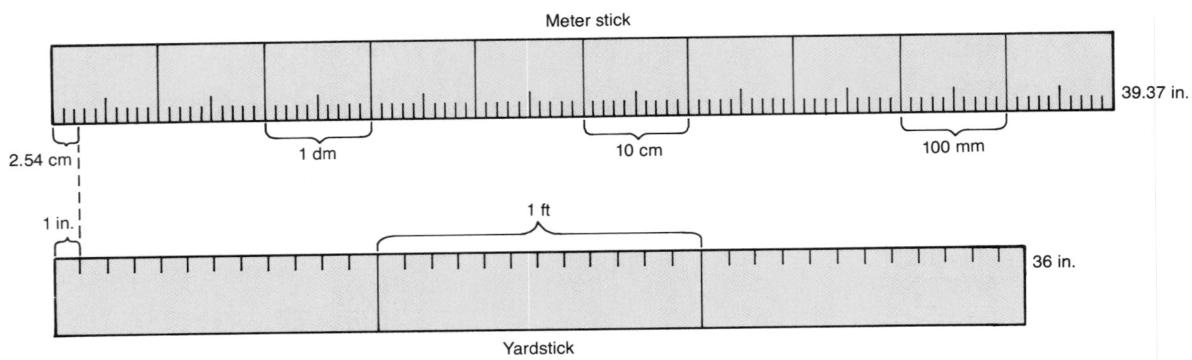

are 12 inches in a foot. In other words, to convert one unit of length to another, it is necessary to use *different* numbers each time. Conversions in the metric system are much easier since they are based on progressions of the number 10.

Exhibit A-1 lists U.S. units of measurement.

METRIC SYSTEM

The ***metric system,*** introduced in France in 1790, is now used by all major countries except the United States. Scientific observations are almost universally expressed in metric units.

LENGTH

The standard of length in the metric system is the *meter* (m). It was originally defined in 1790 as one ten-millionth of the distance from the North Pole to the Equator. In 1889 it was redefined as the distance measured at 0°C between two lines on a bar of platinum–iridium kept at the International Bureau of Weights and Measures in France. In 1960 the meter was redefined as the length of 1,650,763.73 light waves emitted by atoms of the gas krypton under strictly specified conditions. The meter is now defined as the distance light travels in 1/299,792,458 second. The meter is equal to 39.37 inches.

A major advantage of the metric system is that units are

EXHIBIT A-2 METRIC UNITS OF LENGTH AND SOME U.S. EQUIVALENTS

Metric Unit	Meaning of Prefix	Metric Equivalent	U.S. Equivalent
1 kilometer (km)	kilo = 1,000	1,000 m = 10^3 m	3,280.84 ft = 0.62 mi 1 mi = 1.61 km
1 hectometer (hm)	hecto = 100	100 m = 10^2 m	328 ft
1 dekameter (dam)	deka = 10	10 m = 10^1 m	32.8 ft
1 meter (m)		Standard unit of length	39.37 in. = 3.28 ft = 1.09 yd
1 decimeter (dm)	deci = $\frac{1}{10}$	0.1 m = 10^{-1} m	3.94 in.
1 centimeter (cm)	centi = $\frac{1}{100}$	0.01 m = 10^{-2} m	0.394 in. 1 in. = 2.54 cm
1 millimeter (mm)	milli = $\frac{1}{1,000}$	0.001 m = $\frac{1}{10}$ cm = 10^{-3} m	0.0394 in.
1 micrometer (μm) [formerly micron (μ)]	micro = $\frac{1}{1,000,000}$	0.000,001 m = $\frac{1}{10,000}$ cm = 10^{-6} m	3.94×10^{-5} in.
1 nanometer (nm) [formerly millimicron (mμ)]	nano = $\frac{1}{1,000,000,000}$	0.000,000,001 m = $\frac{1}{10,000,000}$ cm = 10^{-9} m	3.94×10^{-8} in.

related to one another by factors of 10. Thus 1 m is 10 decimeters (dm) or 100 centimeters (cm) or 1,000 millimeters (mm). Conversion from one unit to another is simple. Figure A-1 illustrates the differences between metric and U.S. conversions by comparing the meter stick and the yardstick. Exhibit A-2 lists the metric units of length with U.S. equivalents.

Since numbers with many zeros (very large numbers or very small fractions) are cumbersome to work with, they are expressed in *exponential form,* that is, as powers of 10. The form of exponential notation is

$$M \times 10^n$$

You can determine M and n in two steps. For example, how is 0.0000000001 written in exponential form? First, determine M by moving the decimal point so that only one nonzero digit is to the left of it:

0.0000000001.

The digit to the left of the decimal is 1; therefore $M = 1$. Second, determine n by counting the number of places you moved the decimal point. If you moved the point to the left, make the number positive; if you moved it to the right, it is negative. Since you moved the decimal point 10 places to the right, $n = -10$. Thus

$$0.0000000001 = 1 \times 10^{-10}$$

Now do a problem on your own. The wavelength of yellow light is about 0.000059 cm. Convert the centimeters to exponential form. If your answer is 5.9×10^{-5}, you are ready to continue. If you got the wrong answer, reread the discussion.

When we are working with a very large number, the same rules apply, but our exponential value will be positive rather than negative. Refer to Exhibit A-2. Note that 1 km equals 1,000 m. Even though 1,000 is not a cumbersome number, we can still convert it into exponential form. First, move the decimal point so there is only one nonzero digit to the left of it to determine M:

1,000.

Now, because the decimal has been moved three places to the left, n equals +3 or simply 3. Thus

$$1 \text{ km} = 1 \times 10^3 \text{ m}$$

Do another problem on your own. The speed of light is about 30,000,000,000 cm/sec. Convert the centimeters to exponential form. Your answer should be 3×10^{10} cm.

Review Exhibit A-2 and note some common metric and U.S. equivalents. Note also the exponential forms.

MASS

Now let us look at the second fundamental unit of the metric system: mass. The standard unit of mass is the *kilogram* (kg). A kilogram is defined as the mass of a platinum–iridium cylinder kept at the International Bureau of Weights and Measures in France. The standard pound is defined in terms of standard kilogram: 1 lb equals 0.4536 kg.

Exhibit A-3 lists metric units of mass and some U.S. equivalents.

EXHIBIT A-3 METRIC UNITS OF MASS AND SOME U.S. EQUIVALENTS

Metric Unit	Metric Equivalent	U.S. Equivalent
1 kilogram (kg)	1,000 g	2.205 lb 1 ton = 907 kg
1 hectogram (hg)	100 g	
1 dekagram (dag)	10 g	0.0353 oz
1 gram (g)	1 g	1 lb = 453.6 g 1 oz = 28.35 g
1 decigram (dg)	0.1 g	
1 centigram (cg)	0.01 g	
1 milligram (mg)	0.001 g	0.015 g
1 microgram (μg)	0.000,001 g	
1 nanogram (ng)	0.000,000,001 g	
1 picogram (pg)	0.000,000,000,001 g	

TIME

The third fundamental unit of both the metric and the U.S. systems is time. The standard of time is the *second.* Formerly, the second was defined as 1/86,400 of a mean solar day. (A mean solar day is the average of the lengths of all days throughout the year.) Currently, the second is defined as the time required for 9,192,631,770 vibrations of cesium atoms when they are vibrating in a specific manner. Units of time are used in measuring pulse and heart rate, metabolic rate, x-ray exposure, and intervals between medications.

Exhibit A-1 lists the units of time.

VOLUME

Units of volume, or capacity, are derived units based on length. *Volume* in the U.S. system may be expressed as cubic feet (ft^3), cubic inches ($inch^3$), and cubic yards (yd^3), or as a unit of volume such as the quart. Volume in the metric system may be expressed in cubic units of length such as cubic centimeters (cm^3) or in terms of the basic unit of volume, the *liter*. A liter is defined as the volume occupied by 1,000 g of pure water at 4°C. Since 1 cm^3 of water at this temperature weighs 1 g, then 1,000 g of water occupies a volume of 1,000 cm^3. This means that a liter is equal to 1,000 cm^3 and 1 milliliter (ml) is equal to 1 cm^3. Because of this relationship, many volume-measuring devices, such as hypodermic needles, may be graduated in either milliliters or cubic centimeters.

Exhibit A-4 lists metric units of volume and some U.S. equivalents.

EXHIBIT A-4 METRIC UNITS OF VOLUME AND SOME EQUIVALENTS

Metric Unit	Metric Equivalent	U.S. Equivalent
1 liter (l)	1,000 ml	33.81 fl oz = 1.057 qt 946 ml = 1 qt
1 milliliter (ml)	0.001 liter	0.0338 fl oz 30 ml = 1 fl oz 5 ml = 1 teaspoon
1 cubic centimeter (cm^3)	0.999972 ml	0.0338 fl oz

APOTHECARY SYSTEM

In addition to the U.S. and metric systems, there is the ***apothecary system.*** This system is commonly used by physicians prescribing medications and by pharmacists preparing them. Exhibit A-5 lists the important units and equivalents of the apothecary system. Note that the units of mass have the same names as in the U.S. system, but they are not equivalent (1 oz = 28.35 g; 1 oz ap = 30 g) and they do not have the same relationship to one another (1 oz = 16 dr; 1 oz ap = 8 dr ap). The units of volume are the same in both systems.

EXHIBIT A-5 APOTHECARY SYSTEM OF MASS AND VOLUME WITH METRIC EQUIVALENTS

Fundamental Unit	Apothecary Unit and Conversion	Metric Equivalent
Mass	1 grain (gr) = 0.002083 ounce	1 g = 15 gr
	1 dram (dr ap) = 60 gr	4 g = 1 dr
	1 ounce (oz ap) = 8 dr ap	30 g = 1 oz
	1 pound (lb ap) = 12 oz ap	1 kg = 32 oz
Volume	1 fluidram (fl dr) = 60 minims (min)	1 ml (or cm^3 = 15 min
	1 fluidounce (fl oz) = 8 fl dr	4 ml (or cm^3) = 1 fl dr
	1 pint (pt) = 16 fl oz	30 ml (or cm^3) = 1 fl oz
		500 ml (or cm^3) = 1 pt
		1,000 ml (or cm^3) = 1 qt

FREQUENTLY USED CONVERSIONS BASED ON MILLIGRAM

1,000 mg (1 g)	= 15 grains
600 mg (0.6 g)	= 10 grains
300 mg (0.3 g)	= 5 grains
60 mg (0.06 g)	= 1 grain
30 mg (0.03 g)	= 0.50 ($^1/_2$) grain
20 mg (0.02 g)	= 0.33 ($^1/_3$) grain
10 mg (0.01 g)	= 0.166 ($^1/_6$) grain
5 mg (0.005 g)	= 0.083 ($^1/_{12}$) grain
4 mg (0.004 g)	= 0.66 ($^1/_{15}$) grain
1 mg (0.001 g)	= 0.016 ($^1/_{60}$) grain
0.5 mg (0.0005 g)	= 0.0083 ($^1/_{120}$) grain
0.1 mg (0.0001 g)	= 0.0016 ($^1/_{600}$) grain

APPENDIX B: NORMAL VALUES FOR SELECTED BLOOD AND URINE TESTS

KEY TO SYMBOLS

mm^3 = cubic millimeter	mEq/1 = milliequivalent per liter
dl = deciliter	mg = milligram
g = gram	ml = milliliter
> = greater than	mm = millimeter
IU = international unit	mm Hg = millimeter of mercury
kg = kilogram	mM = millimole
< = less than	mOsm = milliosmole
l = liter	% = percent
μg = microgram	U = unit

EXHIBIT B-1 WHOLE BLOOD (WB), SERUM (S), AND PLASMA (P) TESTS

Test (Specimen)	Normal Values	Clinical Implication
Albumin (S)	3.2–5.6 g/dl (electrophoresis)	Values increase in nephritis, fever, trauma, and severe anemia and leukemia; values decrease following severe burns.
Alphafetoprotein (AFP) (WB or amniotic fluid)	Nonpregnant adult: 25 ng/ml Pregnant female: values rise in pregnancy	Values increase from 70 ng/ml at 8 weeks gestation to 430 ng/ml at 36 weeks. Values gradually decrease soon after birth.
Aminotransferases (S) **Aspartate aminotransferase**	3–21 IU/l 5–25 Reitman–Frankel units	Values increase in myocardial infarction, liver disease, trauma to skeletal muscles, and severe burns; values decrease in beriberi and uncontrolled diabetes mellitus with acidosis.
Alanine aminotransferase	5–24 IU/l 5–35 Reitman–Frankel units	Values increase in liver disease.
Ammonia (P)	20–120 μg/dl (diffusion)	Values increase in liver disease, heart failure, emphysema, pneumonia, cor pulmonale, and hemolytic disease of newborn (erythroblastosis fetalis).
Amylase (S)	60–160 Somogyi units/dl	Values increase in acute pancreatitis, mumps, and obstruction of pancreatic duct; values decrease in hepatitis, cirrhosis, burns, and toxemia of pregnancy.
Bilirubin (S)	Conjugated: 0.1–0.4 mg/dl Unconjugated: 0.2–1.0 mg/dl Total: 0.3–1.4 mg/dl Children: 0.2–0.8 mg/dl Newborn: 1.0–12.0 mg/dl	An increase in conjugated bilirubin probably results from liver dysfunction or biliary obstruction; an increase in unconjugated bilirubin probably results from excessive hemolysis of red blood cells.
Bleeding time (WB)	4–8 minutes (Simplate)	Values increase in thrombocytopenia, severe liver disease, leukemia, and aplastic anemia.

EXHIBIT B-1 WHOLE BLOOD (WB), SERUM (S), AND PLASMA (P) TESTS (*Continued*)

Test (Specimen)	Normal Values	Clinical Implication
Blood urea nitrogen (BUN) (S)	8–26 mg/dl	Values increase in kidney disease, shock, dehydration, diabetes, and acute myocardial infarction (MI); values decrease in liver failure, impaired absorption, and overhydration.
Calcium (Ca^{2+}) (S)	Adults: 4.6–5.5 mEq/l	Values increase in cancer, hyperparathyroidism, Addison's disease, hyperthyroidism, and Paget's disease; values decrease in hypoparathyroidism, chronic renal failure, osteomalacia, rickets, and diarrhea.
Carbon dioxide (CO_2), content (WB)	Arterial: 19–24 mM Venous: 22–26 mM	Values increase in severe vomiting, emphysema, and aldosteronism; values decrease in severe diarrhea, starvation, and acute renal failure.
Carbon dioxide, partial pressure (PCO_2) (WB)	Arterial: 40 mm Hg Venous: 45 mm Hg	Values increase in hypoventilation, obstructive lung disease, and emphysema; values decrease in hyperventilation, hypoxia, and pregnancy.
Carcinoembryonic antigen (CEA) (P)	0–25 mg/ml	Values increase in carcinoma of the colon, rectum, breast, ovary, liver and pancreas; inflammatory bowel disease (IBD), cirrhosis, and chronic cigarette smoking.
Carotene, beta (S)	40–200 µg/dl (varies with diet)	Values increase in myxedema, diabetes mellitus, and excessive dietary intake; values decrease in fat malabsorption, liver disease, and poor dietary intake.
Chloride (Cl^-) (S)	95–103 mEq/l	Values increase in dehydration, Cushing's syndrome, and anemia; values decrease in severe vomiting, severe burns, diabetic acidosis, and fever.
Cholesterol, total (S)	Under 200 mg/dl is desirable (varies with diet, sex, and age)	Values increase in diabetes mellitus, cardiovascular disease, nephrosis, and hypothyroidism; values decrease in liver disease, hyperthyroidism, fat malabsorption, pernicious anemia, severe infections, and terminal stages of cancer.
HDL cholesterol (P)	greater than 40 mg/dl	
LDL cholesterol (P)	less than 160 mg/dl	
VLDL cholesterol (P)	0–40 mg/dl	
Cortisol (hydrocortisone) (P)	8 A.M.–10 A.M.: 5–23 µg/dl 4 P.M.–6 P.M.: 3–13 µg/dl	Values increase in hyperthyroidism, stress, obesity, and Cushing's syndrome; values decrease in hypothyroidism, liver disease, and Addison's disease.
Creatine (S or P)	Male: 0.1–0.4 mg/dl Female: 0.2–0.7 mg/dl	Values increase in nephritis, muscular dystrophy, damage to muscle tissue, and pregnancy.
Creatine phosphokinase (CPK) (S)	Male: 55–170 U/l Female: 30–135 U/l	Values increase in myocardial infarction, progressive muscular dystrophy, myxedema, convulsions, hypothyroidism, and pulmonary edema.
Creatinine (S)	0.5–1.2 mg/dl	Values increase in impaired renal function, giantism, and acromegaly; values decrease in muscular dystrophy.
Erythrocyte sedimentation rate (ESR) (WB)	(Westergren) Female: Under 50 years: less than 20 mm/hr Over 50 years: less than 30 mm/hr Male: Under 50 years: less than 15 mm/hr Over 50 years: less than 20 mm/hr	Values increase in pregnancy, infection, carcinoma, tissue destruction, and nephritis; values decrease in sickle-cell anemia and congestive heart failure (CHF).
Fetal hemoglobin (WB)	Newborns: 60–90% Before age 2: 0–4% Adults: 0–2%	Values increase in thalassemia, sickle-cell anemia, and leakage of fetal blood into maternal bloodstream.
Gamma-glutamyl transferase (GGT) (S)	5–40 IU/l	Values increase in cirrhosis of the liver, metastatic cancer of the liver, cholelithiasis, congestive heart failure (CHF), and alcoholism.

EXHIBIT B-1 WHOLE BLOOD (WB), SERUM (S), AND PLASMA (P) TESTS (*Continued*)

Test (Specimen)	Normal Values	Clinical Implication
Globulins (S)	2.3–3.5 g/dl	Values increase in chronic infections.
Glucose (S)	70–110 mg/dl	Values increase in diabetes mellitus, acute stress, hyperthyroidism, chronic liver disease, and nephritis; values decrease in Addison's disease, hypothyroidism, and cancer of the pancreas.
Hemoglobin (S or P)	Male: 13.5–18 g/100 ml Female: 12–16 g/100 ml Newborn: 14–20 g/100 ml	Values increase in polycythemia, congestive heart failure (CHF), chronic obstructive pulmonary disease, and at high altitudes; values decrease in anemia, hyperthyroidism, cirrhosis of the liver, and severe hemorrhage.
Hematocrit (WB)	Male: 40–54% (average 47%) Female: 38–47% (average 42%)	Values increase in polycythemia, severe dehydration, and shock; values decrease in anemia, leukemia, cirrhosis, and hyperthyroidism.
Immunoglobins (S)		IgG values increase in infections of all types, liver disease, and severe malnutrition; IgA values increase in cirrhosis of the liver, chronic infections, and autoimmune disorders and decrease in immunologic deficiency states; IgM values increase in trypanosomiasis and decrease in lymphoid aplasia; IgD values increase in chronic infections and myelomas; IgE values increase in hay fever, asthma, and anaphylactic shock.
IgG	800–1,801 mg/dl	
IgA	113–563 mg/dl	
IgM	54–222 mg/dl	
IgD	0.5–3.0 mg/dl	
IgE	0.01–0.04 mg/dl	
Iron, total (S)	60–150 μg/dl	Values increase in liver disease and various anemias; values decrease in iron-deficiency anemia.
Ketone bodies (acetone) (S or P)	Negative Toxic level: 20 mg/dl	Values increase in ketoacidosis, fever, anorexia, fasting, starvation, high-fat diet, low-carbohydrate diet, and following vomiting.
Lactic acid ($CH_3 \cdot CHOH \cdot COOH$) (WB)	Arterial: 3–7 mg/dl Venous: 5–20 mg/dl	Values increase during muscular activity, congestive heart failure (CHF), shock, and severe hemorrhage.
Lactic dehydrogenase (LDH) (S)	71–207 IU/l	Values increase in myocardial infarction, liver disease, skeletal muscle necrosis, and extensive cancer.
Lipids (S)		Values increase in hyperlipidemia, diabetes mellitus, and hypothyroidism; values decrease in fat malabsorption.
Total	400–800 mg/dl	
Cholesterol	150–250 mg/dl	
Triglycerides	10–190 mg/dl	
Phospholipids	150–380 mg/dl	
Fatty acids (free)	9.0–15.0 mM/l	
Mucoprotein (S)	80–200 mg/dl	Values increase in cancer, infections, and rheumatoid arthritis; values decrease in hepatitis A (infectious) and cirrhosis.
Osmolality (S)	285–295 mOsm/kg	Values increase in cirrhosis, congestive heart failure (CHF), and high-protein diets; values decrease in aldosteronism, diabetes insipidus, and hypercalcemia.
Oxygen (O_2) content (WB)	Arterial: 15–23 volume %	Values increase in polycythemia; values decrease in chronic obstructive lung disease.
Oxygen, partial pressure (PO_2) (WB)	Arterial: 105 mm Hg	Values increase in polycythemia and hyperventilation; values decrease in anemias, insufficient atmospheric oxygen, and hypoventilation.
pH (WB)	Arterial: 7.35–7.45	Values increase in vomiting, hyperventilation, excessive bicarbonate, and lack of oxygen; values decrease in renal failure, diabetic ketoacidosis, hypoxia, airway obstruction, and shock.
Phosphatase (S)		
acid	4–13 U/dl (King–Armstrong)	Values increase in prostatic cancer, some liver disease, hyperparathyroidism, myocardial infarction (MI), and pulmonary embolism.
alkaline	4–13 U/dl (King–Armstrong)	Values increase in some liver and bone diseases, hyperparathyroidism, and pregnancy.

EXHIBIT B-1 WHOLE BLOOD (WB), SERUM (S), AND PLASMA (P) TESTS (*Continued*)

Test (Specimen)	Normal Values	Clinical Implication
Phosphorous, inorganic (S)	Adult: 1.8–4.1 mEq/l Children: 2.3–4.1 mEq/l	Values increase in renal disease, hypoparathyroidism, hypocalcemia, bone tumors, Addison's disease, and acromegaly; values decrease in hyperparathyroidism, rickets, osteomalacia, and diabetic coma.
Platelet count (WB)	250,000–400,000/mm^3	Values increase in cancer, trauma, heart disease, and cirrhosis; values decrease in anemias, allergic conditions, and during cancer chemotherapy.
Protein (S) **Total** **Albumin** **Globulin** **A/G ratio**	 6.0–7.8 g/dl 3.2–4.5 g/dl 2.3–3.5 g/dl 1.5:1 to 2.5:1	Total protein values increase in dehydration, shock, systemic lupus erythematosus, (SLE), rheumatoid arthritis (RA), chronic infections, and chronic liver disease; total protein values decrease in insufficient protein intake, hemorrhage, malabsorption, diarrhea, and severe burns.
Protein-bound iodine (PBI) (S)	4.0–8.0 µg/dl	Values increase in hyperthyroidism and acute thyroiditis; values decrease in hypothyroidism, chronic thyroiditis, myxedema, and nephrosis.
Prothrombin time (PT) (WB)	11–15 seconds	Values increase in prothrombin and vitamin K deficiency, liver disease, and hypervitaminosis A.
Red blood cell count (WB)	Male: 5.4 million/mm^3 Female: 4.8 million/mm^3	Values increase in polycythemia, dehydration, and following hemorrhaging; values decrease in systemic lupus erythematosus (SLE), anemias, and Addison's disease.
Reticulocyte count (WB)	0.5–1.5%	Values increase in hemolytic anemia, metastatic carcinoma, and leukemia; values decrease in iron-deficiency and pernicious anemia, radiation therapy, and kidney disease in which kidney cells do not make erythropoietin.
Serum glutamic oxaloacetic transaminase (SGOT) (S)	5–140 IU/l	Values increase in heart and skeletal muscle damage and in infectious hepatitis.
Serum glutamic pyruvic transaminase (SGPT) (S)	10–30 IU/l	Values increase in liver cell damage due to hepatitis, toxicity, infectious mononucleosis, jaundice, cirrhosis, liver cancer, or congestive heart failure.
Sodium (Na^+) (S)	136–142 mEq/l	Values increase in dehydration, aldosteronism, coma, Cushing's disease, and diabetes insipidus; values decrease in severe burns, vomiting, diarrhea, Addison's disease, nephritis, excessive sweating, and edema.
Thyroxine (T_4) (S)	4–11 µg/dl	Values increase in hyperthyroidism; values decrease in hypothyroidism.
Thyroxine-binding globulin (TBG) (S)	10–26 µg/dl	Values increase in hypothyroidism; values decrease in hyperthyroidism.
Uric acid (S)	Male: 4.0–8.5 mg/dl Female: 2.7–7.3 mg/dl	Values increase in impaired renal function, gout, metastatic cancer, shock, and starvation; values decrease in persons treated with uricosuric drugs.
White blood cell count, differential (WB) **Neutrophils** **Eosinophils** **Basophils** **Lymphocytes** **Monocytes**	 60–70% 2–4% 0.5–1% 20–25% 3–8%	Neutrophils increase in acute infections; eosinophils and basophils increase in allergic reactions; lymphocytes increase during antigen-antibody reactions; monocytes increase in chronic infections.
White blood cell count, total (WB)	5000–10,000/mm^3	Values increase in acute infections, trauma, malignant diseases, and cardiovascular diseases; values decrease in diabetes mellitus, anemias, and following cancer chemotherapy.

EXHIBIT B-2 URINE TESTS

Test	Sample	Normal Values	Clinical Implication
Ammonia nitrogen	24 hour	20–70 mEq/l	Values increase in diabetes mellitus; values decrease in liver diseases.
Amylase	2 hour	35–260 Somogyi units/hr	Values increase in pancreatitis and choledocholethiasis.
Bilirubin	Random	Negative	Values increase in liver disease and obstructive biliary disease.
Blood occult	Random	Negative	Values increase in renal disease, extensive burns, transfusion reactions, and hemolytic anemia.
Calcium (Ca^{2+})	Random 24 hour Average diet: Low Ca^{2+} diet: High Ca^{2+} diet:	10 mg/dl 100–240 mg/24 hr 150 mg/24 hr 240–300 mg/24 hr	Values increase in hyperparathyroidism, metastatic malignancies, and primary cancer of breasts and lungs; values decrease in hypoparathyroidism and vitamin D deficiency.
Casts			
Epithelial	24 hour	Occasional	Values increase in nephrosis and heavy metal poisoning.
Granular	24 hour	Occasional	Values increase in nephritis and pyelonephritis.
Hyaline	24 hour	Occasional	Values increase in glomerular membrane damage and fever.
Red blood cell	24 hour	Occasional	Values increase in pyelonephritis, kidney stones, and cystitis.
White blood cell	24 hour	Occasional	Values increase in kidney infections.
Chloride (Cl^-)	24 hour	140–250 mEq/24 hr	Values increase in Addison's disease, dehydration, and starvation; values decrease in pyloric obstruction, diarrhea, and emphysema.
Color	Random	Yellow, straw, amber	Varies with many disease states, hydration, and diet.
Concentration test (Fishberg)	Random and after restriction	Sp.Gr.: 1.025 Osmolality: 850 mOsm/l	Values decrease in renal disease, hyperparathyroidism, and bone diseases.
Creatinine	24 hour	Male: 1.0–2.0 g/24 hr Female: 0.8–1.8 g/24 hr	Values increase in infections; values decrease in muscular atrophy, anemia, and kidney diseases.
Glucose	Random	Negative	Values increase in diabetes mellitus, brain injury, and myocardial infarction.
Hydroxycorticosteroids	24 hour	Male: 5–15 mg/24 hr Female: 2–13 mg/24 hr	Values increase in Cushing's syndrome, burns, and infections; values decrease in Addison's disease.
Ketone bodies (acetone)	Random	Negative	Values increase in diabetic acidosis, fever, anorexia, fasting, and starvation.
17-ketosteroids (KS)	24 hour	Male: 8–25 mg/24 hr Female: 5–15 mg/24 hr	Values decrease in surgery, burns, infections, adrenogenital syndrome, and Cushing's syndrome.
Odor	Random	Aromatic	Becomes acetonelike in diabetic ketosis.
Osmolality	24 hour	500–800 mOsm/kg water	Values increase in cirrhosis, congestive heart failure (CHF), and high protein diets; values decrease in aldosteronism, diabetes insipidus, and hypokalemia.
pH	Random	4.6–8.0	Values increase in urinary tract infections and severe alkalosis; values decrease in acidosis, emphysema, starvation, and dehydration.
Phenylpyruvic acid	Random	Negative	Values increase in phenylketonuria (PKU).

EXHIBIT B-2 URINE TESTS (*Continued*)

Potassium (K^+)	24 hour	40–80 mEq/l	Values increase in chronic renal failure, dehydration, starvation, and Cushing's syndrome; values decrease in diarrhea, malabsorption syndrome, and adrenal cortical insufficiency.
Protein (albumin)	Random	Negative	Values increase in nephritis, fever, severe anemias, trauma, and hyperthyroidism.
Sodium (Na^+)	24 hour	75–200 mg/24 hr	Values increase in dehydration, starvation, and diabetic acidosis; values decrease in diarrhea, acute renal failure, emphysema, and Cushing's syndrome.
Specific gravity	Random	1.001–1.035 (normal fluid intake) 1.001–1.035 (range)	Values increase in diabetes mellitus and excessive water loss; values decrease in absence of antidiuretic hormone (ADH) and severe renal damage.
Urea	Random	25–35 g/24 hr	Values increase in response to increased protein intake; values decrease in impaired renal function.
Uric acid	24 hour	0.4–1.0 g/24 hr	Values increase in gout, leukemia, and liver disease; values decrease in kidney disease.
Urobilinogen	2 hour	0.3–1.0 Ehrlich units	Values increase in anemias, hepatitis A (infectious), biliary disease, and cirrhosis; values decrease in cholelithiasis and renal insufficiency.
Volume, total	24 hour	1000–2000 ml/24 hr	Varies with many factors.

Normal values researched and provided by Professor John Lo Russo, Bergen Community College, Division of Allied Health Sciences. The values in the exhibit are not intended to be definitive since they may vary among different laboratories.

GLOSSARY OF COMBINING FORMS, WORD ROOTS, PREFIXES, AND SUFFIXES

PRONUNCIATION KEY

1. The strongest accented syllable appears in capital letters, for example, bilateral (bī-LAT-er-al) and diagnosis (dī-ag-NŌ-sis).

2. If there is a secondary accent, it is noted by a single quote mark ('), for example, constitution (kon'-sti-TOO-shun) and physiology (fiz'-ē-OL-ō-jē). Any additional secondary accents are also noted by a single quote mark, for example, decarboxylation (dē'-kar-bok'-si-LĀ-shun).

3. Vowels marked with a line above the letter are pronounced with the long sound as in the following common words:
ā as in *māke*
ē as in *bē*
ī as in *īvy*
ō as in *pōle*

4. Vowels not so marked are pronounced with the short sound, as in the following words:
e as in *bet*
i as in *sip*
o as in *not*
u as in *bud*

5. Other phonetic symbols are used to indicate the following sounds:
a as in *above*
oo as in *sue*
yoo as in *cute*
oy as in *oil*

Many medical terms are "compound" words; that is, they are made up of one or more word roots or combining forms of word roots with prefixes or suffixes. For example, *leucocyte* (white blood cell) is a combination of *leuco,* the combining form for the word root meaning "white," and *cyt,* the word root meaning "cell." Learning the medical meanings of the fundamental word parts will enable you to analyze many long, complicated terms.

The following list includes some of the most commonly used combining forms, word roots, prefixes, and suffixes used in making medical terms and an example for each.

COMBINING FORMS AND WORD ROOTS

Acou-, Acu- ***hearing*** Acoustics (a-KOO-stiks), the science of sounds or hearing.

Acr-, Acro- ***extremity*** Acromegaly (ak'-rō-MEG-a-lē), hyperplasia of the nose, jaws, fingers, and toes.

Aden-, Adeno- ***gland*** Adenoma (ad-en'-Ō-ma), a tumor with a glandlike structure.

Alg-, Algia- ***pain*** Neuralgia (nyoo-RAL-ja), pain along the course of a nerve.

Angi- ***vessel*** Angiocardiography (an'-jē-ō-kard-ē-OG-ra-fē), roentgenography of the great blood vessels and heart after intravenous injection of radiopaque fluid.

Arthr-, Arthro- ***joint*** Arthropathy (ar-THROP-a-thē), disease of a joint.

Aut-, Auto- ***self*** Autolysis (aw-TOL-i-sis), destruction of cells of the body by their own enzymes, even after death.

Bio- ***life, living*** Bioposy (BĪ-op-sē), examination of tissue removed from a living body.

Blast- ***germ, bud*** Blastocyte (BLAS-tō-sīt), an embryonic or undifferentiated cell.

Blephar- ***eyelid*** Blepharitis (blef-a-RĪT-is), inflammation of the eyelids.

Brachi- ***arm*** Brachialis (brā-kē-AL-is), muscle that flexes the forearm.

Bronch- ***trachea, windpipe*** Bronchoscopy (bron-KOS-kō-pē), direct visual examination of the bronchi.

Bucc-, cheek Buccocervical (bū-kō-SER-vi-kal), pertaining to the cheek and neck.

Capit- ***head*** Decapitate (dē-KAP-i-tāt), to remove the head.

Carcin- ***cancer*** Carcinogenic (kar-sin-ō-JEN-ik), causing cancer.

Cardi-, Cardia-, Cardio- ***heart*** Cardiogram (KARD-ē-o-gram), a recording of the force and form of the heart's movements.

Cephal- ***head*** Hydrocephalus (hī-drō-SEF-a-lus), enlargement of the head due to an abnormal accumulation of fluid.

Cerebro- ***brain*** Cerebrospinal (se-rē'-brō-SPĪN-al) fluid, fluid contained within the cranium and spinal canal.

Cheil- ***lip*** Cheilosis (kī-LŌ-sis), dry scaling of the lips.

Chole- ***bile, gall*** Cholecystogram (kō-lē-SIS-tō-gram), roentgenogram of the gallbladder.

Chondr-, Chondri-, Chondrio- ***cartilage*** Chondrocyte (KON-drō-sīt), a cartilage cell.

Chrom-, Chromat-, Chromato- ***color*** Hyperchromic (hī-per-KRŌ-mik), highly colored.

Cili- ***eyelash*** Supercilia (soo'-per-SIL-ē-a), eyebrow (hairs above eyelash).

Colpo- ***vagina*** Colpotomy (kol-POT-ō-mē), incision into the wall of the vagina.

Cor-, coron- ***heart*** Coronary (KOR-ō-na-rē), arteries supplying blood to the heart muscle.

Cost- ***rib*** Costal (KOS-tal), pertaining to a rib.

Crani- ***skull*** Craniotomy (krā-nē-OT-ō-mē), surgical opening of the skull.

Cry-, Cryo- ***cold*** Cryosurgery (krī-ō-SERJ-e-rē), surgical procedure using a very cold liquid nitrogen probe.

Cut- ***skin*** Subcutaneous (sub-kyoo-TĀ-nē-us), under the skin.

Cysti-, Cysto- ***sac, bladder*** Cystoscope (SIS-tō-skōp), instrument for interior examination of the urinary bladder.

Cyt-, Cyto-, Cyte- ***cell*** Cytology (sī-TOL-ō-jē), the study of cells.

Dactyl-, Dactylo- ***digits*** (usually fingers, but sometimes toes) Polydactylism (pol-ē-DAK-til-ism), above normal number of fingers or toes.

Derma-, Dermato- ***skin*** Dermatosis (der-ma-TŌ-sis), any skin disease.

Dura- ***hard*** Dura mater (DYOO-ra MĀ-ter), outer membrane covering brain and spinal cord.

Entero- ***intestine*** Enteritis (ent-e-RĪT-is), inflammation of the intestine.

Erythro- ***red*** Erythrocyte (e-RITH-rō-sīt), red blood cell.

Galacto- ***milk*** Galactose (ga-LAK-tōse), a milk sugar.

Gastr- ***stomach*** Gastrointestinal (gas'-trō-in-TES-tin-al), pertaining to the stomach and intestine.

Gloss-, Glosso- ***tongue*** Hypoglossal (hī'-pō-GLOS-al), located under the tongue.

Glyco- ***sugar*** Glycosuria (glī'-kō-SUR-ē-a), sugar in the urine.

Gravid- ***pregnant*** Gravidity (gra-VID-i-tē), condition of being pregnant.

Gyn-, Gyne-, Gynec- ***female, women*** Gynecology (gīn'-e-KOL-ō-jē), the medical specialty dealing with disorders of the female reproductive system.

Hem-, Hemat- ***blood*** Hematoma (hē'-ma-TŌ-ma), a tumor or swelling filled with blood.

Hepar-, Hepato- ***liver*** Hepatitis (hep-a-TĪT-is), inflammation of the liver.

Hist-, Histio- ***tissue*** Histology (his-TOL-ō-jē), the study of tissues.

Hydr- ***water*** Hydrocele (HĪ-drō-sēl), accumulation of fluid in a saclike cavity.

Hyster- ***uterus*** Hysterectomy (his'-te-REK-tō-mē), surgical removal of the uterus.

Ileo- ***ileum*** Ileocecal (il'-ē-ō-SĒ-kal) valve, folds at the opening between ileum and cecum.

Ilio- ***ilium*** Iliosacral (il'-ē-ō-SĀ-kral), pertaining to ilium and sacrum.

Kines- ***motion*** Kinesiology (ki-nē-sē-OL-ō-jē), study of movement of body parts.

Labi- ***lip*** Labial (LĀ-bē-al), pertaining to a lip.

Lachry-, Lacri- ***tears*** Nasolacrimal (nā-zō-LAK-rim-al), pertaining to the nose and lacrimal apparatus.

Laparo- ***loin, flank, abdomen*** Laparoscopy (lap'-a-ROS-kō-pē), examination of the interior of the abdomen by means of a laparoscope.

Leuco-, Leuko- ***white*** Leucocyte (LYOO-kō-sīt), white blood cell.

Lingua- ***tongue*** Lingual (LIN-gwal), pertaining to the tongue.

Lip-, Lipo ***fat*** Lipoma (lī-PŌ-ma), a fatty tumor.

Lith- ***stone*** Lithiasis (li-THĒ-a-sis), the formation of stones.

Lumbo- ***lower back, loin*** Lumbar (LUM-bar), pertaining to the loin.

Macul- ***spot, blotch*** Macula (MAK-yoo-la), spot or blotch.

Malign- ***bad, harmful*** Malignant (ma-LIG-nant), condition that gets worse and results in death.

Mamm- ***breast*** Mammography (ma-MOG-ra-fē), roentgenography of the mammary gland.

Mast- ***breast*** Mastitis (ma-STĪT-is), inflammation of the mammary gland.

Meningo- ***membrane*** Meningitis (men-in-JĪT-is), inflammation of the membranes of spinal cord and brain.

Metro- ***uterus*** Endometrium (en'-dō-MĒ-trē-um), lining of the uterus.

Morpho- ***form, shape*** Morphology (mor-FOL-o-jē), the study of form and structure of the things.

Myelo- ***marrow, spinal cord*** Poliomyelitis (pō'-lē-ō-mī'-a-LĪT-is), inflammation of the gray matter of the spinal cord.

Myo- ***muscle*** Myocardium (mī-ō-KARD-ē-um), heart muscle.

Necro- ***corpse, dead*** Necrosis (ne-KRŌ-sis), death of areas of tissue surrounded by healthy tissue.

Nephro- ***kidney*** Nephrosis (ne-FRŌ-sis), degeneration of kidney tissue.

Neuro- ***nerve*** Neuroblastoma (nyoor'-ō-blas-TŌ-ma), malignant tumor of the nervous system composed of embryonic nerve cells.

Oculo- ***eye*** Binocular (bī-NOK-yoo-lar), pertaining to the two eyes.

Odont- ***tooth*** Orthodontic (or-thō-DONT-ik), pertaining to the proper positioning and relationship of the teeth.

Onco- ***mass, tumor*** Oncology (ong-KOL-ō-jē), study of tumors.

Oo- ***egg*** Oocyte (Ō-ō-sīt), original egg cell.

Oophor- ***ovary, egg carrier*** Oophorectomy (ō'-ōf-o-REK-tō-mē), surgical removal of ovaries.

Ophthalm- ***eye*** Ophthalmology (of'-thal-MOL-ō-jē), the study of the eye and its diseases.

Or- ***mouth*** Oral (Ō-ral), pertaining to the mouth.

Orchido- ***testicle*** Orchidectomy (or'-ki-DEK-tō-me), surgical removal of a testicle.

Osmo- ***odor, sense of smell*** Anosmia (an-OZ-mē-a), absence of sense of smell.

Oss-, Osseo-, Osteo- ***bone*** Osteoma (os-tē-Ō-ma), bone tumor.

Oto- ***ear*** Otosclerosis (ō'-tō-skle-RŌ-sis), formation of bone in the labyrinth of the ear.

Palpebr- ***eyelid*** Palpebra (PAL-pe-bra), eyelid.

Part- ***birth, delivery, labor*** Parturition (par'-too-RISH-un), act of giving birth.

Patho-* *disease Pathogenic (path'-ō-JEN-ik), causing disease.

Ped-* *children Pediatrician (pēd-ē-a-TRISH-an), medical specialist in the treatment of children.

Peps-* *digest Peptic (PEP-tik), pertaining to digestion.

Phag-, Phago-* *to eat Phagocytosis (fag'-ō-sī-TŌ-sis), the process by which cells ingest particulate matter.

Philic-, Philo-* *to like, have an affinity for Hydrophilic (hī-drō-FIL-ik), having an affinity for water.

Phleb-* *vein Phlebitis (fle-BĪT-is), inflammation of the veins.

Phon-* *voice, sound Phonogram (FŌ-nō-gram), record made of sound.

Phren-* *diaphragm Phrenic (FREN-ik), pertaining to the diaphragm.

Pilo-* *hair Depilatory (de-PIL-a-tō-re), hair remover.

Pneumo-* *lung, air Pneumothorax (nyoo-mō-THŌR-aks), air in the thoracic cavity.

Pod-* *foot Podiatry (po-DĪ-a-trē), the diagnosis and treatment of foot disorders.

Procto-* *anus, rectum Proctoscopy (prok-TOS-kō-pē), instrumental examination of the rectum.

Psycho-* *soul, mind Psychiatry (sī-KĪ-a-trē), treatment of mental disorders.

Pulmon-* *lung Pulmonary (PUL-mō-ner'-ē), pertaining to the lungs.

Pyle-, Pyloro* *opening, passage Pyloric (pī-LOR-ik), pertaining to the pylorus of the stomach.

Pyo-* *pus Pyuria (pī-YOOR-ē-a), pus in the urine.

Ren-* *kidneys Renal (RĒ-nal), pertaining to the kidney.

Rhin-* *nose Rhinitis (ri-NĪT-is), inflammation of nasal mucosa.

Salpingo-* *uterine (Fallopian) tube Salpingitis (sal'-pin-JĪ-tis), inflammation of the uterine (Fallopian) tubes.

Scler-, Sclero-* *hard Atherosclerosis (ath'-er-ō-skle-RŌ-sis), hardening of the arteries.

Sep-, Septic-* *toxic condition due to microorganisms Septicemia (sep'-ti-SĒ-mē-a), presence of bacterial toxins in the blood (blood poisoning).

Soma-, Somato-* *body Somatotropic (sō-mat-ō-TRŌ-pik), having a stimulating effect on body growth.

Somni-* *sleep Insomnia (in-SOM-nē-a), inability to sleep.

Sten-* *narrow Stenosis (ste-NŌ-sis), narrowing of a duct or canal.

Stasis-, Stat-* *stand still Homeostasis (hō'-mē-ō-STĀ-sis), achievement of a steady state.

Tegument-* *skin, covering Integumentary (in-teg-yoo-MEN-ta-rē), pertaining to the skin.

Therm-* *heat Thermometer (ther-MOM-et-er), instrument used to measure and record heat.

Thromb-* *clot, lump Thrombus (THROM-bus), clot in a blood vessel or heart.

Tox-, Toxic-* *poison Toxemia (tok-SĒ-mē-a), poisonous substances in the blood.

Trich-* *hair Trichosis (trik-Ō-sis), disease of the hair.

Tympan-* *eardrum Tympanic (tim-PAN-ik) membrane, eardrum.

Vas-* *vessel, duct Cerebrovascular (se-rē-brō-VAS-kyoo-lar), pertaining to the blood vessels of the cerebrum of the brain.

Viscer-* *organ Visceral (VIS-e-ral), pertaining to the abdominal organs.

Zoo-* *animal Zoology (zō-OL-o-jē), the study of animals.

PREFIXES

A-, An-* *without, lack of, deficient Anesthesia (an'-es-THĒ-zha), without sensation.

Ab-* *away from, from Abnormal (ab-NOR-mal), away from normal.

Ad-* *to, near, toward Adduction (a-DUK-shun), movement of an extremity toward the axis of the body.

Alb-* *white Albino (al-BĪ-no), person whose skin, hair, and eyes lack the pigment melanin.

Alveol-* *cavity, socket Alveolus (al-VĒ-ō-lus), air sac in the lung.

Ambi-* *both sides Ambidextrous (am'-bi-DEK-strus), able to use either hand.

Ambly-* *dull Amblyaphia (am-blē-A-fē-a), dull sense of touch.

Andro-* *male, masculine Androgen (AN-drō-jen), male sex hormone.

Ankyl(o)-* *bent, fusion Ankylosed (ANG-ki-lōsd), fused joint.

Ante-* *before Antepartum (ant-ē-PAR-tum), before delivery of a baby.

Anti-* *against Anticoagulant (an-tī-kō-AG-yoo-lant), a substance that prevents coagulation of blood.

Basi-* *base, foundation Basal (BĀ-sal), located near the base.

Bi-* *two, double, both Biceps (BĪ-seps), a muscle with two heads of origin.

Bili-* *bile, gall Biliary (BIL-ē-er-ē), pertaining to bile, bile ducts, or gallbladder.

Brachy-* *short Brachyesophagus (brā-kē-e-SOF-a-gus), short esophagus.

Brady-* *slow Bradycardia (brād'-ē-KARD-ē-a), abnormal slowness of the heartbeat.

Cata-* *down, lower, under, against Catabolism (ka-TAB-a-lizm), metabolic breakdown into simpler substances.

Circum-* *around Circumrenal (ser-kum-RĒN-al), around the kidney.

Cirrh-* *yellow Cirrhosis (si-RŌ-sis), liver disorder that causes yellowing of skin.

Co-, Con-, Com-* *with, together Congenital (kon-JEN-i-tal), existing at birth.

Contra-* *against, opposite Contraception (kon-tra-SEP-shun), the prevention of conception.

Crypt-* *hidden, concealed Cryptorchidism (krip-TOR-ka-dizm'), undescended or hidden testes.

Cyano-* *blue Cyanosis (sī-a-NŌ-sis), bluish discoloration due to inadequate oxygen.

De-* *down, from Decay (de-KĀ), waste away from normal.

Demi-, hemi-* *half Hemiplegia (hem'-ē-PLĒ-jē-a), paralysis on one side of the body.

Di-, Diplo-* *two Diploid (DIP-loyd), having double the haploid number of chromosomes.

Dis-* *separation, apart, away from Disarticulate (dis'-ar-TIK-yoo-lāt'), to separate at a joint.
Dys-* *painful, difficult Dyspnea (disp-NĒ-a), difficult breathing.

E-, Ec-, Ex-* *out from, out of Eccentric (ek-SEN-trik), not located at the center.
Ecto-, Exo-* *outside Ectopic (ek-TOP-ik) pregnancy, gestation outside the uterine cavity.
Em-, En-* *in, on Empyema (em'-pī-Ē-ma), pus in a body cavity.
End-, Endo-* *inside Endocardium (en'-dō-KARD-ē-um), membrane lining the inner surface of the heart.
Epi-* *upon, on, above Epidermis (ep'-i-DER-mis), outermost layer of skin.
Eu-* *well Eupnea (YOOP-nē-a), normal breathing.
Ex-, Exo-* *out, away from Exocrine (EK-sō-krin), excreting outwardly or away from.
Extra-* *outside, beyond, in addition to Extracellular (ek'-stra-SEL-yoo-lar), outside of the cell.

Fore-* *before, in front of Forehead (FOR-hed), anterior part of head.

Gen-* *originate, produce, form Pathogen (PATH-ō-jen), disease producer.
Gingiv-* *gum Gingivitis (jin'-je-VĪ-tus), inflammation of the gums.

Hemi-* *half Hemiplegia (hem-ē-PLĒ-jē-a), paralysis of only half of the body.
Heter-, Hetero-* *other, different Heterogeneous (het'-e-rō-JEN-ē-us), composed of different substances.
Homeo-, Homo-* *unchanging, the same, steady Homeostasis (hō'-mē-ō-STĀ-sis), achievement of a steady state.
Hyper-* *beyond, excessive Hyperglycemia (hī-per-glī-SĒ-mē-a), excessive amount of sugar in the blood.
Hypo-* *under, below, deficient Hypodermic (hī-pō-DER-mik), below the skin or dermis.

Idio-* *self, one's own, separate Idiopathic (id'-ē-ō-PATH-ik), a disease without recognizable cause.
In-, Im-* *in, inside, not Incontinent (in-KON-ti-nent), not able to retain urine or feces.
Infra-* *beneath Infraorbital (in'-fra-OR-bi-tal), beneath the orbit.
Inter-* *among, between Intercostal (int'-er-KOS-tal), between the ribs.
Intra* *within, inside Intracellular (in'-tra-SEL-yoo-lar), inside the cell.
Iso-* *equal, like Isogenic (ī-sō-JEN-ik), alike in morphological development.

Later-* *side Lateral (LAT-er-al), pertaining to a side or farther from the midline.
Lepto-* *small, slender, thin Leptodermic (lep'-tō-DER-mik), having thin skin.

Macro-* *large, great Macrophage (MAK-rō-fāj), large phagocytic cell.
Mal-* *bad, abnormal Malnutrition (mal'-noo-TRISH-un), lack of necessary food substances.
Medi-, Meso-* *middle Medial (MĒD-ē-al), nearer to midline.
Mega-, Megalo-* *great, large Megakaryocyte (meg'-a-KAR-ē-ō-sīt), giant cell of bone marrow.
Melan-* *black Melanin (MEL-a-nin), black or dark brown pigment found in skin and hair.
Meta-* *after, beyond Metacarpus (met'-a-KAR-pus), the part of the hand between the wrist and fingers.
Micro-* *small Microtome (MĪ-krō-tōm), instrument for preparing very thin slices of tissue for microscopic examination.
Mono-* *one Monorchid (mon-OR-kid), having one testicle.

Neo-* *new Neonatal (nē-ō-NĀT-al), pertaining to the first 4 weeks after birth.
Noct(i)-* *night Nocturia (nok-TOO-rē-a), urination occurring at night.
Null(i)-* *none Nullipara (nu-LIP-a-ra), woman with no children.
Nyct-* *night Nyctalopia (nik'-ta-LŌ-pē-a), night blindness.

Oligo-* *small, deficient Oliguria (ol-ig-YOO-rē-a), abnormally small amount of urine.
Ortho-* *straight, normal Orthopnea (or-thop-NĒ-a), inability to breathe in any position except when straight or erect.

Pan-* *all Pancarditis (pan-kar-DĪ-tis), inflammation of the entire heart.
Para-* *near, beyond, apart from, beside Paranasal (par-a-NĀ-zal), near the nose.
Per-* *through Percutaneous (per'-kyoo-TĀ-nē-us), through the skin.
Peri-* *around Pericardium (per'-i-KARD-ē-um), membrane or sac around the heart.
Poly-* *much, many Polycythemia (pol'-i-sī-THĒ-mē-a), an excess of red blood cells.
Post-* *after, beyond Postnatal (pōst-NĀT-al), after birth.
Pre-, Pro-* *before, in front of Prenatal (prē-NĀT-al), before birth.
Prim-* *first Primary (PRĪ-me-rē), first in time or order.
Proto-* *first Protocol (PRŌ-tō-kol), clinical report made from first notes taken.
Pseud-, Pseudo-* *false Pseudoangina (soo'-dō-an-JĪ-na), false angina.

Retro-* *backward, located behind Retroperitoneal (re'-trō-per'-it-on-Ē-al), located behind the peritoneum.

Schizo-* *split, divide Schizophrenia (skiz'-ō-FRE-nē-a), split personality mental disorder.
Semi-* *half Semicircular (sem'-i-SER-kyoo-lar) canals, canals in the shape of a half circle.
Sub-* *under, beneath, below Submucosa (sub'-myoo-KŌ-sa), tissue layer under a mucous membrane.
Super-* *above, beyond Superficial (soo-per-FISH-al), confined to the surface.
Supra-* *above, over Suprarenal (soo-pra-RĒN-al), adrenal gland above the kidney.
Sym-, Syn-* *with, together, joined Syndrome (SIN-drōm), all the symptoms of a disease considered as a whole.

Tachy-* *rapid Tachycardia (tak'-i-KARD-ē-a), rapid heart action.
Terat(o)-* *malformed fetus Teratogen (TER-a-tō-jen), an agent that caused development of a malformed fetus.
Tetra-, quadra-* *four Tetrad (TET-rad), group of four with something in common.
Trans-* *across, through, beyond Transudation (trans-yoo-DĀ-shun), oozing of a fluid through pores.

Tri-* *three Trigone (TRĪ-gon), a triangular space, as at the base of the bladder.

SUFFIXES

able* *capable of, having ability to Viable (VĪ-a-bal), capable of living.

-ac, -al* *pertaining to Cardiac (KARD-ē-ak), pertaining to the heart.

-agra* *severe pain Myagra (mī-AG-ra), severe muscle pain.

-an, -ian* *pertaining to Circadian (ser-KĀ-dē-an), pertaining to a cycle of active and inactive periods.

-ant* *having the characteristic of Malignant (ma-LIG-nant), having the characteristic of badness.

-ary* *connected with Ciliary (SIL-ē-ar-ē), resembling any hairlike structure.

-asis, -asia, -esis, -osis* *condition or state of Hemostasis (hē-mō-STĀ-sis), stopping of bleeding or circulation.

-asthenia* *weakness Myasthenia (mi-as-THĒ-nē-a), weakness of skeletal muscles.

-ation* *process, action, condition Inspiration (in-spi-RĀ-shun), process of drawing air into lungs.

-cel, -cele* *swelling, an enlarged space or cavity Meningocele (men-IN-gō-sēl), enlargement of the meninges.

-centesis* *puncture, usually for drainage Amniocentesis (am'-nē-ō-sen-TĒ-sis), withdrawal of amniotic fluid.

-cid, -cide, -cis* *cut, kill, destroy Germicide (jer-mi-SĪD), a substance that kills germs.

-ectasia, -ectasis* *stretching, dilation Bronchiectasis (bron-kē-EK-ta-sis), dilation of a bronchus or bronchi.

-ectomize, ectomy* *excision of, removal of Thyroidectomy (thī-royd-EK-tō-mē), surgical removal of a thyroid gland.

-ema* *swelling, distension Emphysema (em'-fi-SĒ-ma), swelling of air sacs in lungs.

-emia* *condition of blood Lipemia (lip-Ē-mē-a), abnormally high concentration of fat in the blood.

-esis* *condition, process Enuresis (en'-yoo-RĒ-sis), condition of involuntary urination.

-esthesia* *sensation, feeling Anesthesia (an'-es-THĒ-zē-a), total or partial loss of feeling.

-ferent* *carry Efferent (EF-e-rent), carrying away from a center.

-form* *shape Fusiform (FYOO-zi-form), spindle-shaped.

-gen* *agent that produces or originates Pathogen (PATH-ō-jen), microorganism or substance capable of producing a disease.

-genic* *produced from, producing Pyogenic (pī-ō-JEN-ik), producing pus.

-gram* *record, that which is recorded Electrocardiogram (e-lek'-trō-KARD-ē-ō-gram), record of heart action.

-graph* *instrument for recording Electroencephalograph (e-lek'-trō-en-SEF-a-lō-graf), instrument for recording electrical activity of the brain.

-ia* *state, condition Hypermetropia (hī'-per-me-TRŌ-pē-a), condition of farsightedness.

-iatrics, iatry* *medical practice specialities Pediatrics (pēd-ē-A-triks), medical science relating to care of children and treatment of their diseases.

-ician* *person associated with Technician (tek-NISH-an), person skilled in a technical field.

-ics* *art or science of Optics (OP-tiks), science of light and vision.

-ion* *action, condition resulting from action Incision (in-SIZH-un), act or result of cutting into flesh.

-ism,* *condition, state Rheumatism (ROO-ma-tizm), inflammation, especially of muscles and joints.

-ist* *one who practices Internist (in-TER-nist), one who practices internal medicine.

-itis* *inflammation Neuritis (nyoo-RĪT-is), inflammation of a nerve or nerves.

-ive* *relating to Sedative (SED-a-tive), relating to a pain or tension reliever.

-logy, -ology* *the study or science of Physiology (fiz-ē-OL-ō-jē), the study of function of body parts.

-lyso, -lysis* *solution, dissolve, loosening Hemolysis (hē-MOL-i-sis), dissolution of red blood cells.

-malacia* *softening Osteomalacia (os'-tē-ō-ma-LĀ-shē-a), softening of bone.

-megaly* *enlarged Cardiomegaly (kar'-dē-ō-MEG-a-lē), enlarged heart.

-oid* *resembling Lipoid (li-POYD), resembling fat.

-ologist* *specialist Dermatologist (der-ma-TOL-ō-gist), specialist in the study of the skin.

-oma* *tumor Fibroma (fi-BRŌ-ma), tumor composed mostly of fibrous tissue.

-ory* *pertaining to Sensory (SENS-o-rē), pertaining to sensation.

-ose* *full of Adipose (AD-i-pōz), characterized by presence of fat.

-osis* *condition, disease Necrosis (ne-KRŌ-sis), condition of death of cells.

-ostomy* *create an opening Colostomy (kō-LOS-tō-me), surgical creation of an opening between the colon and body surface.

-otomy* *surgical incision Tracheotomy (trā-kē-OT-ō-me), surgical incision of the trachea.

-pathy* *disease Neuropathy (nyoo-ROP-a-thē), disease of the peripheral nervous system.

-penia* *deficiency Thrombocytopenia (throm'-bō-sīt'-o-PĒ-nē-a), deficiency of thrombocytes in the blood.

-phobe, -phobia* *fear of, aversion to Hydrophobia (hī-drō-FŌ-bē-a), fear of water.

-plasia, -plasty* *development, formation Rhinoplasty (RĪ-nō-plas-tē), surgical reconstruction of the nose.

-plegia, -plexy* *stroke, paralysis Apoplexy (AP-ō-plek-sē), sudden loss of consciousness and paralysis.

-pnea* *to breathe Apnea (AP-nē-a), temporary absence of respiration, following a period of overbreathing.

-poiesis* *production Hematopoiesis (he-mat'-a-poy-Ē-sis), formation and development of red blood cells.

-ptosis* *falling, sagging Blepharoptosis (blef'-a-rō-TŌ-sis), dropping of upper eyelid.

-rrhage* *bursting forth, abnormal discharge Hemorrhage (HEM-or-rij), bursting forth of blood.

-rrhea* *flow, discharge Diarrhea (dī-a-RĒ-a), abnormal frequency of bowel evacuation, the stools with a more or less fluid consistency.

***-scope* instrument for viewing** Bronchoscope (BRON-kō-skōp), instrument used to examine the interior of a bronchus.

***-stomy* creation of a mouth or artificial opening** Tracheostomy (trā-kē-OST-ō-mē), creation of an opening in the trachea.

***-tic, -ulnar* pertaining to** Diagnostic (dī'-ag-NOS-tik), pertaining to diagnosis.

***-tomy* cutting into, incision into** Laparatomy (lap-a-ROT-ō-mē), an abdominal incision to gain access to the peritoneal cavity.

***-tripsy* crushing** Lithotripsy (LITH-ō-trip'-sē), crushing of a calculus (stone).

***-trophy* state relating to nutrition or growth** Hypertrophy (hī-PER-trō-fē), excessive growth of an organ or part.

***-tropic* turning toward, influencing, changing** Gonadotropic (gō-nad-a-TRŌ-pic), influencing the gonads.

***-uria* urine** Polyuria (pol-ē-YOOR-ē-a), excessive secretion of urine.

GLOSSARY OF TERMS

PRONUNCIATION KEY

1. The strongest accented syllable appears in capital letters, for example, bilateral (bī-LAT-er-al) and diagnosis (dī-ag-NŌ-sis).

2. If there is a secondary accent, it is noted by a single quote mark ('), for example, constitution (kon'-sti-TOO-shun) and physiology (fiz'-ē-OL-ō-jē). Any additional secondary accents are also noted by a single quote mark, for example, decarboxylation (dē'-kar-bok'-si-LĀ-shun).

3. Vowels marked with a line above the letter are pronounced with the long sound, as in the following common words:
ā as in *māke*
ē as in *bē*
ī as in *īvy*
ō as in *pōle*

4. Vowels not so marked are pronounced with the short sound, as in the following words:
e as in *bet*
i as in *sip*
o as in *not*
u as in *bud*

5. Other phonetic symbols are used to indicate the following sounds:
a as in *above*
oo as in *sue*
yoo as in *cute*
oy as in *oil*

Abatement (a-BĀT-ment) A decrease in the seriousness of a disorder or in the severity of pain or other symptoms.

Abdomen (ab-DŌ-men or AB-dō-men) The area between the diaphragm and pelvis.

Abdominal (ab-DŌM-i-nal) ***cavity*** Superior portion of the abdominopelvic cavity that contains the stomach, spleen, liver, gallbladder, pancreas, small intestine, and most of the large intestine.

Abdominal thrust maneuver A first-aid procedure for choking. Employs a quick, upward thrust against the diaphragm that forces air out of the lungs with sufficient force to eject any lodged material. Also called the ***Heimlich*** (HĪM-lik) ***maneuver.***

Abdominopelvic (ab-dom'-i-nō-PEL-vic) ***cavity*** Inferior component of the ventral body cavity that is subdivided into an upper abdominal cavity and a lower pelvic cavity.

Abduction (ab-DUK-shun) Movement away from the axis or midline of the body or one of its parts.

Abortion (a-BOR-shun) The premature loss (spontaneous) or removal (induced) of the embryo or nonviable fetus; any failure in the normal process of developing or maturing.

Abrasion (a-BRĀ-shun) A portion of skin that has been scraped away.

Abscess (AB-ses) A localized collection of pus and liquefied tissue in a cavity.

Absorption (ab-SORP-shun) The taking up of liquids by solids or of gases by solids or liquids; intake of fluids or other substances by cells of the skin or mucous membranes; the passage of digested foods from the gastrointestinal tract into blood or lymph.

Absorptive state Metabolic state during which ingested nutrients are being absorbed by the blood or lymph from the gastrointestinal tract.

Accessory duct A duct of the pancreas that empties into the duodenum about 2.5 cm (1 in.) superior to the ampulla of Vater (hepatopancreatic ampulla). Also called the ***duct of Santorini*** (san'-tō-RE-ne).

Accommodation (a-kom-ō-DĀ-shun) A change in the curvature of the eye lens to adjust for vision at various distances; focusing.

Accretion (a-KRĒ-shun) A mass of material that has accumulated in a space or cavity; the adhesion of parts.

Acetabulum (as'-e-TAB-yoo-lum) The rounded cavity on the external surface of the coxal (hip) bone that receives the head of the femur.

Acetylcholine (as'-ē-til-KŌ-lēn) ***(ACh)*** A neurotransmitter, liberated at synapses in the central nervous system, that stimulates skeletal muscle contraction.

Achille's tendon *See* **Calcaneal tendon.**

Achlorhydria (ā-klōr-HĪ-drē-a) Absence of hydrochloric acid in the gastric juice.

Acid (AS-id) A proton donor, or substance that dissociates into hydrogen ions (H^+) and anions, characterized by an excess of hydrogen ions and a pH less than 7.

Acidosis (as-i-DŌ-sis) A condition in which blood pH ranges from 7.35 to 6.80 or lower.

Acinar (AS-i-nar) Flasklike.

Acini (AS-i-nē) Masses of cells in the pancreas that secrete digestive enzymes.

Acne (AK-nē) Inflammation of sebaceous (oil) glands that usually begins at puberty; the basic acne lesions in order of increasing severity are comedones, papules, pustules, and cysts.

Acoustic (a-KOOS-tik) Pertaining to sound or the sense of hearing.

Acquired immune deficiency syndrome (AIDS) A disorder characterized by a positive HIV-antibody test and certain indicator diseases (Kaposi's sarcoma, *Pneumocystis carinii* pneumonia, tuberculosis, fungus diseases, etc.). A deficiency of helper T cells and a reversed ratio of helper T cells to suppressor T cells that results in fever or night sweats, coughing, sore throat, fatigue, body aches, weight loss, and enlarged lymph nodes. Caused by a virus called human immunodeficiency virus (HIV).

Acromegaly (ak'-rō-MEG-a-lē) Condition caused by hypersecretion of human growth hormone (hGH) during adulthood characterized by thickened bones and enlargement of other tissues.

Acrosome (AK-rō-sōm) A dense granule in the head of a spermatozoon that contains enzymes that facilitate the penetration of a spermatozoon into a secondary oocyte.

Actin (AK-tin) The contractile protein that makes up thin myofilaments in muscle fiber (cell).

Action potential A wave of negativity that self-propagates along the outside surface of the membrane of a neuron or muscle fiber (cell); a rapid change in membrane potential that involves a depolarization following a repolarization. Also called a ***nerve action potential nerve impulse*** as it relates to a neuron and a ***muscle action potential*** as it relates to a muscle fiber (cell).

Activation (ak'-ti-VĀ-shun) ***energy*** The minimum amount of energy required for a chemical reaction to occur.

Active transport The movement of substances, usually ions, across cell membranes, against a concentration gradient, requiring the expenditure of energy (ATP).

Acuity (a-KYOO-i-tē) Clearness or sharpness, usually of vision.

Acupuncture (AK-ū-punk'-chur) The insertion of a needle into a tissue for the purpose of drawing fluid or relieving pain. It is also an ancient Chinese practice employed to cure illnesses by inserting needles into specific locations of the skin.

Acute (a-KYOOT) Having rapid onset, severe symptoms, and a short course; not chronic.

Adam's apple *See* ***Thyroid cartilage.***

Adaptation (ad'-ap-TĀ-shun) The adjustment of the pupil of the eye to light variations. The property by which a neuron relays a decreased frequency of action potentials from a receptor even though the strength of the stimulus remains constant. The decrease in perception of a sensation over time while the stimulus is still present.

Addison's (AD-i-sonz) ***disease*** Disorder caused by hyposecretion of glucocorticoids (and aldosterone) characterized by muscular weakness, hypoglycemia, mental lethargy, anorexia, nausea and vomiting, weight loss, low blood pressure, dehydration, and excessive skin and mucous membrane pigmentation.

Adduction (ad-DUK-shun) Movement toward the axis or midline of the body or one of its parts.

Adenohypophysis (ad'-e-nō-hī-POF-i-sis) The anterior portion of the pituitary gland.

Adenoids (AD-e-noyds) The pharyngeal tonsils.

Adenosine triphosphate (a-DEN-ō-sēn trī-FOS-fāt) ***(ATP)*** The universal energy-carrying molecule manufactured in all living cells as a means of capturing and storing energy. It consists of the purine base *adenine* and the five-carbon sugar *ribose,* to which are added, in linear array, three *phosphate* molecules.

Adenylate cyclase (a-DEN-i-lāt SĪ-klās) An enzyme in the postsynaptic membrane that is activated when certain neurotransmitters bind to their receptors; the enzyme converts ATP into cyclic AMP.

Adherence (ad-HĒR-ens) Firm contact between the plasma membrane of a phagocyte and an antigen or other foreign substance.

Adhesion (ad-HĒ-zhun) Abnormal joining of parts to each other.

Adipocyte (AD-i-pō-sīt) Fat cell, derived from a fibroblast.

Adrenal cortex (a-DRĒ-nal KOR-teks) The outer portion of an adrenal gland, divided into three zones, each of which has a different cellular arrangement and secretes different hormones.

Adrenal (a-DRĒ-nal) ***glands*** Two glands located superior to each kidney. Also called the ***suprarenal*** (soo'-pra-RĒ-nal) ***glands.***

Adrenal medulla (me-DUL-a) The inner portion of an adrenal gland, consisting of cells that secrete epinephrine and norepinephrine (NE) in response to the stimulation of preganglionic sympathetic neurons.

Adrenergic (ad'-ren-ER-jik) ***fiber*** A nerve fiber that when stimulated releases norepinephrine (noradrenaline) at a synapse.

Adrenocorticotropic (ad-rē'-nō-kor-ti-kō-TRŌP-ik) ***hormone (ACTH)*** A hormone produced by the adenohypophysis (anterior lobe) of the pituitary gland that influences the production and secretion of certain hormones of the adrenal cortex.

Adrenoglomerulotropin (a-drē'-nō-glō-mer'-yoo-lō-TRŌ-pin) A hormone secreted by the pineal gland that may stimulate aldosterone secretion.

Adventitia (ad-ven-TISH-ya) The outermost covering of a structure or organ.

Aerobic (air-Ō-bik) Requiring molecular oxygen.

Afferent arteriole (AF-er-ent ar-TĒ-rē-ōl) A blood vessel of a kidney that breaks up into the capillary network called a glomerulus; there is one afferent arteriole for each glomerulus.

Afferent neuron (NOO-ron) A neuron that carries a nerve impulse toward the central nervous system. Also called a ***sensory neuron.***

Afterimage Persistence of a sensation even though the stimulus has been removed.

Agglutination (a-gloo'-ti-NĀ-shun) Clumping of microorganisms or blood corpuscles; typically an antigen–antibody reaction.

Agglutinin (a-GLOO-ti-nin) A specific principle or antibody in blood serum capable of causing the clumping of bacteria, blood corpuscles, or particles. Also called an ***isoantibody.***

Agglutinogen (ag'-loo-TIN-ō-gen) A genetically determined antigen located on the surface of erythrocytes; basis for the ABO grouping and Rh system of blood classification. Also called an ***isoantigen.***

Aggregated lymphatic follicles Aggregated lymph nodules that are most numerous in the ileum. Also called ***Peyer's*** (PĪ-erz) ***patches.***

Aging Normal process accompanied by a progressive alteration of the body's homeostatic adaptive responses.

Agnosia (ag-NŌ-zē-a) A loss of the ability to recognize the meaning of stimuli from the various senses (visual, auditory, touch).

Agraphia (a-GRAF-ē-a) An inability to write.

Albinism (AL-bin-izm) Abnormal, nonpathological, partial or total absence of pigment in skin, hair, and eyes.

Albumin (al-BYOO-min) The most abundant (60 percent) and smallest of the plasma proteins, which functions primarily to regulate osmotic pressure of plasma.

Albuminuria (al-byoo'-min-UR-ēa) Presence of albumin in the urine.

Aldosterone (al-do-STĒR-ōn) A mineralocorticoid produced by the adrenal cortex that brings about sodium and water reabsorption and potassium excretion.

Aldosteronism (al'-do-STER-ōn-izm') Condition caused by hypersecretion of aldosterone that results in increased sodium concentration and decreased potassium concentration in blood and characterized by muscular paralysis, high blood pressure, and edema.

Alimentary (al-i-MEN-ta-rē) Pertaining to nutrition.

Alkaline (AL-ka-līn) Containing more hydroxyl ions (OH^-) than hydrogen ions (H^+) to produce a pH of more than 7.

Alkalosis (al-ka-LŌ-sis) A condition in which blood pH ranges from 7.45 to 8.00 or higher.

Allantois (a-LAN-tō-is) A small, vascularized membrane between the chorion and amnion of the fetus that serves as an early site for blood formation.

Allele (a-LĒL) Genes that control the same inherited trait (such as height or eye color) that are located on the same position (locus) on homologous chromosomes.

Allergen (AL-er-jen) An antigen that evokes a hypersensitivity reaction.

Allergic (a-LER-jik) Pertaining to or sensitive to an allergen.

All-or-none principle In muscle physiology, muscle fibers (cells) of a motor unit contract to their fullest extent or not at all. In neuron physiology, if a stimulus is strong enough to initiate an action potential, a nerve impulse is transmitted along the entire neuron at a constant and minimum strength.

Alpha (AL-fa) ***cell*** A cell in the pancreatic islets (islets of Langerhans) in the pancreas that secretes glucagon.

Alpha receptor Receptor found on visceral effectors innervated by most sympathetic postganglionic axons; in general, stimulation of alpha receptors leads to excitation.

Alveolar-capillary (al-VĒ-ō-lar) ***membrane*** Structure in the lungs consisting of the alveolar wall and basement membrane and a capillary endothelium and basement membrane through which the diffusion of respiratory gases occurs. Also called the ***respiratory membrane.***

Alveolar duct Branch of a respiratory bronchiole around which alveoli and alveolar sacs are arranged.

Alveolar macrophage (MAK-rō-fāj) Cell found in the alveolar walls of the lungs that is highly phagocytic. Also called a ***dust cell.***

Alveolar sac A collection or cluster of alveoli that share a common opening.

Alveolus (al-VĒ-ō-lus) A small hollow or cavity; an air sac in the lungs; milk-secreting portion of a mammary gland.

Alzheimer's (ALTZ-hī-merz) ***disease (AD)*** Disabling neurological disorder characterized by dysfunction and death of specific cerebral neurons resulting in widespread intellectual impairment, personality changes, and fluctuations in alertness.

Ambulatory (AM-byoo-la-tō'-rē) Capable of walking.

Amenorrhea (ā-men-ō-RĒ-a) Absence of menstruation.

Amino acid An organic acid, containing a carboxyl group (COOH) and an amino group (NH_2), that is the building unit from which proteins are formed.

Amnesia (am-NĒ-zē-a) A lack or loss of memory.

Amniocentesis (am'-nē-ō-sen-TĒ-sis) Removal of amniotic fluid by inserting a needle transabdominally into the amniotic cavity.

Amnion (AM-nē-on) The innermost fetal membrane; a thin transparent sac that holds the fetus suspended in amniotic fluid. Also called the ***"bag of waters."***

Amniotic (am'-nē-OT-ik) ***fluid*** Fluid in the amniotic cavity, the space between the developing embryo (or fetus) and amnion; the fluid is initially produced as a filtrate from maternal blood and later from fetal urine.

Amorphous (a-MOR-fus) Without definite shape or differentiation in structure; pertains to solids without crystalline structure.

Amphiarthrosis (am'-fē-ar-THRŌ-sis) Articulation midway between diarthrosis and synarthrosis, in which the articulating bony surfaces are separated by an elastic substance to which both are attached, so that the mobility is slight.

Ampulla (am-POOL-la) A saclike dilation of a canal.

Ampulla of Vater *See* ***Hepatopancreatic ampulla.***

Amyotrophic (a-mē-ō-TROF-ik) ***lateral sclerosis (ALS)*** Progressive neuromuscular disease characterized by degeneration of motor neurons in the spinal cord that leads to muscular weakness. Also called ***Lou Gehrig's disease.***

Anabolism (a-NAB-ō-lizm) Synthetic energy-requiring reactions whereby small molecules are built up into larger ones.

Anaerobic (an-AIR-ō-bik) Not requiring molecular oxygen.

Anal (Ā-nal) ***canal*** The terminal 2 or 3 cm (1 in.) of the rectum; opens to the exterior of the anus.

Anal column A longitudinal fold in the mucous membrane of the anal canal that contains a network of arteries and veins.

Analgesia (an-al-JĒ-zē-a) Pain relief.

Anal triangle The subdivision of the female or male perineum that contains the anus.

Anamnestic (an'-am-NES-tik) ***response*** Accelerated, more intense production of antibodies upon a subsequent exposure to an antigen after the initial exposure.

Anaphase (AN-a-fāz) The third stage of mitosis in which the chromatids that have separated at the centromeres move to opposite poles of the cell.

Anaphylaxis (an'-a-fi-LAK-sis) Against protection; a hypersensitivity (allergic) reaction in which IgE antibodies attach to mast cells and basophils, causing them to produce mediators of anaphylaxis (histamine, leukotrines, kinins, and prostaglandins) that bring about increased blood permeability, increased smooth muscle contraction, and increased mucus production. Examples are hayfever, hives, and anaphylactic shock.

Anastomosis (a-nas-tō-MŌ-sis) An end-to-end union or joining together of blood vessels, lymphatics, or nerves.

Anatomic dead space The volume of air that is inhaled but remains in spaces in the upper respiratory system and does not reach the alveoli to participate in gas exchange; about 150 ml.

Anatomical (an'-a-TOM-i-kal) ***position*** A position of the body universally used in anatomical descriptions in which the body is erect, facing the observer, the upper extremities are at the sides, the palms of the hands are facing forward, and the feet are on the floor.

Anatomy (a-NAT-ō-mē) The structure or study of structure of the body and the relation of its parts to each other.

Androgen (AN-drō-jen) Substance producing or stimulating male characteristics, such as the male hormone testosterone.

Anemia (a-NĒ-mē-a) Condition of the blood in which the number of functional red blood cells or their hemoglobin content is below normal.

Anesthesia (an'-es-THĒ-zē-a) A total or partial loss of feeling or sensation, usually defined with respect to loss of pain sensation; may be general or local.

Aneuploid (an'-yoo-PLOID) A cell that has one or more chromosomes of a set added or deleted.

Aneurysm (AN-yoo-rizm) A saclike enlargement of a blood vessel caused by a weakening of its wall.

Angina pectoris (an-JĪ-na *or* AN-ji-na PEK-tō-ris) A pain in the chest related to reduced coronary circulation that may or may not involve heart or artery disease.

Angiography (an-jē-OG-ra-fē) X-ray examination of blood vessels after injection of a radiopaque substance.

Angiotensin (an-jē-ō-TEN-sin) Either of two forms of a protein associated with regulation of blood pressure. Angiotensin I is produced by the action of renin on angiotensinogen and is converted by the action of a plasma enzyme into angiotensin II, which stimulates aldosterone secretion by the adrenal cortex.

Anion (AN-ī-on) A negatively charged ion. An example is the chloride ion (Cl^-).

Ankyloglossia (ang'-ki-lō-GLOSS-ē-a) "Tongue-tied"; restriction of tongue movements by a short lingual frenulum.

Ankylosis (ang'-ki-LŌ-sus) Severe or complete loss of movement at a joint.

Anomaly (a-NOM-a-lē) An abnormality that may be a developmental (congenital) defect; a variant from the usual standard.

Anopsia (an-OP-sē-a) A defect of vision.

Anorexia nervosa (an-ō-REK-sē-a ner-VŌ-sa) A chronic disorder characterized by self-induced weight loss, body-image and other perceptual disturbances, and physiologic changes that result from nutritional depletion.

Anosmia (an-OZ-mē-a) Loss of the sense of smell.

Anoxia (an-OK-sē-a) Deficiency of oxygen.

Antagonist (an-TAG-ō-nist) A muscle that has an action opposite that of the prime mover (agonist) and yields to the movement of the prime mover.

Antagonistic (an-tag-ō-NIST-ik) ***effect*** A hormonal interaction in which the effect of one hormone on a target cell is opposed by another hormone. For example, calcitonin (CT) lowers blood calcium level, whereas parathormone (PTH) raises it.

Antepartum (an-tē-PAR-tum) Before delivery of the child; occurring (to the mother) before childbirth.

Anterior (an-TĒR-ē-or) Nearer to or at the front of the body. Also called ***ventral.***

Anterior root The structure composed of axons of motor or efferent fibers that emerges from the anterior aspect of the spinal cord and extends laterally to join a posterior root, forming a spinal nerve. Also called a ***ventral root.***

Antibiotic (an'-ti-bī-OT-ik) Literally, "antilife"; a chemical produced by a microorganism that is able to inhibit the growth of or kill other microorganisms.

Antibody (AN-ti-bod'-ē) A protein produced by certain cells in the body in the presence of a specific antigen; the antibody combines with that antigen to neutralize, inhibit, or destroy it.

Anticoagulant (an-tī-cō-AG-yoo-lant) A substance that is able to delay, suppress, or prevent the clotting of blood.

Antidiuretic (an'-ti-dī-yoo-RET-ik) Substance that inhibits urine formation.

Antidiuretic hormone (ADH) Hormone produced by neurosecretory cells in the paraventricular and supraoptic nuclei of the hypothalamus that stimulates water reabsorption from kidney cells into the blood and vasoconstriction of arterioles.

Antigen (AN-ti-jen) Any substance that when introduced into the tissues or blood induces the formation of antibodies and reacts only with its specific antibodies.

Anti-oncogene (ONG-kō-jēn) A gene that can cause cancer when inappropriately inactivated.

Antrum (AN-trum) Any nearly closed cavity or chamber, especially one within a bone, such as a sinus.

Anulus fibrosus (AN-yoo-lus fī-BRŌ-sus) A ring of fibrous tissue and fibrocartilage that encircles the pulpy substance (nucleus pulposus) of an intervertebral disc.

Anuria (a-NOO-rē-a) A daily urine output of less than 50 ml.

Anus (Ā-nus) The distal end and outlet of the rectum.

Aorta (ā-OR-ta) The main systemic trunk of the arterial system of the body; emerges from the left ventricle.

Aortic (ā-OR-tik) ***body*** Receptor on or near the arch of the aorta that responds to alterations in blood levels of oxygen, carbon dioxide, and hydrogen ions.

Aortic reflex A reflex concerned with maintaining normal general systemic blood pressure.

Aperture (AP-er-chur) An opening or orifice.

Apex (Ā-peks) The pointed end of a conical structure, such as the apex of the heart.

Apgar (AP-gar) ***score*** A method for assessing the overall status of an infant soon after birth based on evaluation of heart rate, respiratory effort, muscle tone, reflex irritability, and color.

Aphasia (a-FĀ-zē-a) Loss of ability to express oneself properly through speech or loss of verbal comprehension.

Apnea (ap-NĒ-a) Temporary cessation of breathing.

Apneustic (ap-NOO-stik) ***area*** Portion of the respiratory center in the pons that sends stimulatory nerve impulses to the inspiratory area that activate and prolong inspiration and inhibit expiration.

Apocrine (AP-ō-krin) ***gland*** A type of gland in which the secretory products gather at the free end of the secreting cell and are pinched off, along with some of the cytoplasm, to become the secretion, as in mammary glands.

Aponeurosis (ap'-ō-noo-RŌ-sis) A sheetlike tendon joining one muscle with another or with bone.

Appendage (a-PEN-dij) A structure attached to the body.

Appendicitis (a-pen-di-SĪ-tis) Inflammation of the vermiform appendix.

Appositional (ap'-ō-ZISH-o-nal) ***growth*** Growth due to surface deposition of material, as in the growth in diameter of cartilage and bone. Also called ***exogenous*** (eks-OJ-e-nus) ***growth.***

Aqueduct (AK-we-duct) A canal or passage, especially for the conduction of a liquid.

Aqueous humor (AK-wē-us HYOO-mor) The watery fluid, similar in composition to cerebrospinal fluid, that fills the anterior cavity of the eye.

Arachnoid (a-RAK-noyd) The middle of the three coverings (meninges) of the brain or spinal cord.

Arachnoid villus (VIL-us) Berrylike tuft of arachoid that protrudes into the superior sagittal sinus and through which cerebrospinal fluid is reabsorbed into the bloodstream.

Arbor vitae (AR-bōr VĒ-tē) The treelike appearance of the white matter tracts of the cerebellum when seen in midsagittal section. A series of branching ridges within the cervix of the uterus.

Arch of the aorta (ā-OR-ta) The most superior portion of the aorta, lying between the ascending and descending segments of the aorta.

Areflexia (a'-rē-FLEK-sē-a) Absence of reflexes.

Areola (a-RĒ-ō-la) Any tiny space in a tissue. The pigmented ring around the nipple of the breast.

Arm The portion of the upper extremity from the shoulder to the elbow.

Arrector pili (a-REK-tor PI-lē) Smooth muscles attached to hairs; contraction pulls the hairs into a more vertical position, resulting in "goose bumps."

Arrhythmia (a-RITH-mē-a) Irregular heart rhythm. Also called a ***dysrhythmia.***

Arteriogram (ar-TĒR-ē-ō-gram) A roentgenogram of an artery after injection of a radiopaque substance into the blood.

Arteriole (ar-TĒ-rē-ōl) A small, almost microscopic, artery that delivers blood to a capillary.

Artery (AR-ter-ē) A blood vessel that carries blood away from the heart.

Arthritis (ar-THRĪ-tis) Inflammation of a joint.

Arthrocentesis (ar-thrō-sen-TĒ-sis) Insertion of a needle into a synovial (joint) cavity to remove a sample of synovial fluid, relieve pressure, or inject anesthesia or medication.

Arthrology (ar-THROL-ō-jē) The study or description of joints.

Arthroscopy (ar-THROS-co-pē) A procedure for examining the interior of a joint, usually the knee, by inserting an arthroscope into a small incision; used to determine extent of damage, remove torn cartilage, repair cruciate ligaments, and obtain samples for analysis.

Arthrosis (ar-THRŌ-sis) A joint or articulation.

Articular (ar-TIK-yoo-lar) ***capsule*** Sleevelike structure around a synovial joint composed of a fibrous capsule and a synovial membrane.

Articular cartilage (KAR-ti-lij) Hyaline cartilage attached to articular bone surfaces.

Articular disc Fibrocartilage pad between articular surfaces of bones of some synovial joints. Also called a ***meniscus*** (men-IS-cus).

Articulate (ar-TIK-yoo-lāt) To join together as a joint to permit motion between parts.

Articulation (ar-tik'-yoo-LĀ-shun) A joint; a point of contact between bones, cartilage and bones, or teeth and bones.

Artificial pacemaker A device that generates and delivers electrical signals to the heart to maintain a regular heart rhythm.

Arytenoid (ar'-i-TĒ-noyd) Ladle-shaped.

Arytenoid (ar'-i-TĒ-noyd) ***cartilages*** A pair of small, pyramidal cartilages of the larynx that attach to the vocal folds and intrinsic pharyngeal muscles and can move the vocal folds.

Ascending colon (KŌ-lon) The portion of the large intestine that passes upward from the cecum to the lower edge of the liver where it bends at the right colic (hepatic) flexure to become the transverse colon.

Ascites (as-SĪ-tēz) Serous fluid in the peritoneal cavity.

Aseptic (ā-SEP-tik) Free from any infectious or septic material.

Asphyxia (as-FIX-ē-a) Unconsciousness due to interference with the oxygen supply of the blood.

Aspiration (as'-pi-RĀ-shun) Inhalation of a foreign substance (water, food, or foreign body) into the bronchial tree; drainage of a substance in or out by suction.

Association area A portion of the cerebral cortex connected by many motor and sensory fibers to other parts of the cortex. The association areas are concerned with motor patterns, memory, concepts of word-hearing and word-seeing, reasoning, will, judgment, and personality traits.

Association neuron (NOO-ron) A nerve cell lying completely within the central nervous system that carries nerve impulses from sensory neurons to motor neurons. Also called a ***connecting neuron.***

Astereognosis (as-ter'-ē-ōg-NŌ-sis) Inability to recognize objects or forms by touch.

Asthenia (as-THĒ-nē-a) Lack or loss of strength; debility.

Astigmatism (a-STIG-ma-tizm) An irregularity of the lens or cornea of the eye causing the image to be out of focus and producing faulty vision.

Astrocyte (AS-trō-sīt) A neuroglial cell having a star shape that supports neurons in the brain and spinal cord and attaches the neurons to blood vessels.

Ataxia (a-TAK-sē-a) A lack of muscular coordination, lack of precision.

Atelectasis (at'-ē-LEK-ta-sis) A collapsed or airless state of all or part of the lung, which may be acute or chronic.

Atherosclerosis (ath'-er-ō-skle-RŌ-sis) A process in which fatty substances (cholesterol and triglycerides) are deposited in the walls of medium and large arteries in response to certain stimuli (hypertension, carbon monoxide, dietary cholesterol). Following endothelial damage, monocytes stick to the tunica interna, develop into macrophages, and take up cholesterol and low-density lipoproteins. Smooth muscle fibers (cells) in the tunica media ingest cholesterol. This results in the formation of an atherosclerotic plaque that decreases the size of the arterial lumen.

Atherosclerotic (ath'-er-ō-skle-RO-tic) ***plaque*** (PLAK) A lesion that results from accumulated cholesterol and smooth muscle fibers (cells) of the tunica media of an artery; may become obstructive.

Atom Unit of matter that comprises a chemical element; consists of a nucleus and electrons.

Atomic number Number of protons in an atom.

Atomic weight Total number of protons and neutrons in an atom.

Atresia (a-TRĒ-zē-a) Abnormal closure of a passage, or absence of a normal body opening.

Atrial fibrillation (Ā-trē-al fib-ri-LĀ-shun) Asynchronous contraction of the atria that results in the cessation of atrial pumping.

Atrial natriuretic (na'-trē-yoo-RET-ik) ***factor (ANF)*** Peptide hormone produced by the atria of the heart in response to stretching that inhibits aldosterone production and thus lowers blood pressure.

Atrioventricular (AV) (ā'-trē-ō-ven-TRIK-yoo-lar) ***bundle*** The portion of the conduction system of the heart that begins at the atrioventricular (AV) node, passes through the cardiac skeleton separating the atria and the ventricles, then runs a short distance down the interventricular septum before splitting into right and left bundle branches. Also called the ***bundle of His (HISS).***

Atrioventricular (AV) node The portion of the conduction system of the heart made up of a compact mass of conducting cells located near the orifice of the coronary sinus in the right atrial wall.

Atrioventricular (AV) valve A structure made up of membranous flaps or cusps that allows blood to flow in one direction only, from an atrium into a ventricle.

Atrium (Ā-trē-um) A superior chamber of the heart.

Atrophy (AT-rō-fē) Wasting away or decrease in size of a part, due to a failure, abnormality of nutrition, or lack of use.

Audiometry (aw'-dē-OM-e-trē) Evaluation of an individual's hearing acuity.

Auditory ossicle (AW-di-tō-rē OS-si-kul) One of the three small bones of the middle ear called the malleus, incus, and stapes.

Auditory tube The tube that connects the middle ear with the nose and nasopharynx region of the throat. Also called the ***Eustachian*** (yoo-STĀ-kē-an) ***tube.***

Aura (OR-a) A feeling or sensation that precedes an epileptic seizure or any paroxysmal attack (like those of bronchial asthma).

Auscultation (aws-kul-TĀ-shun) Examination by listening to sounds in the body.

Autograft (AW-tō-graft) A graft of tissue from a donor site to a recipient site of the same individual.

Autoimmunity An immunologic response against a person's own tissue antigens.

Autologous transfusion (aw-TOL-ō-gus trans-FYOO-zhun) Donating one's own blood up to six weeks before elective surgery to ensure an abundant supply and reduce transfusion complications such as those that may be associated with diseases such as AIDS and hepatitis. Also called ***predonation.***

Autolysis (aw-TOL-i-sis) Spontaneous self-destruction of cells by their own digestive enzymes upon death or a pathological process or during normal embryological development.

Autonomic ganglion (aw'-tō-NOM-ik GANG-lē-on) A cluster of sympathetic or parasympathetic cell bodies located outside the central nervous system.

Autonomic nervous system (ANS) Visceral efferent neurons, both sympathetic and parasympthetic, that transmit nerve impulses from the central nervous system to smooth muscle, cardiac muscle, and glands; so named because this portion of the nervous system was thought to be self-governing or spontaneous.

Autonomic plexus (PLEK-sus) An extensive network of sympathetic and parasympathetic fibers; the cardiac, celiac, and pelvic plexuses are located in the thorax, abdomen, and pelvis, respectively.

Autophagy (aw-TOF-a-jē) Process by which worn-out organelles are digested within lysosomes.

Autopsy (AW-top-sē) The examination of the body after death.

Autoregulation (aw-tō-reg-yoo-LĀ-shun) A local, automatic adjustment of blood flow in a given region of the body in response to tissue needs.

Autosome (AW-tō-sōm) Any chromosome other than the pair of sex chromosomes.

Avitaminosis (ā-vī-ta-min-Ō-sis) A deficiency of a vitamin in the diet.

Axilla (ak-SIL-a) The small hollow beneath the arm where it joins the body at the shoulders. Also called the ***armpit.***

Axon (AK-son) The usually single, long process of a nerve cell that carries a nerve impulse away from the cell body.

Axon terminal Terminal branch of an axon and its collateral. Also called a ***telodendrium*** (tel-ō-DEN-drē-um).

Azygos (AZ-ī-gos) An anatomical structure that is not paired; occurring singly.

Babinski (ba-BIN-skē) ***sign*** Extension of the great toe, with or without fanning of the other toes, in response to stimulation of the outer margin of the sole of the foot; normal up to $1\frac{1}{2}$ years of age.

Back The posterior part of the body; the dorsum.

Bainbridge (BĀN-bridge) ***reflex*** The increased heart rate that follows increased pressure or distension of the right atrium.

Ball-and-socket joint A synovial joint in which the rounded surface of one bone moves within a cup-shaped depression or fossa of another bone, as in the shoulder or hip joint. Also called a ***spheroid*** (SFĒ-roid) ***joint.***

Barium (BA-rē-um) ***swallow*** X-ray examination of the upper gastrointestinal tract to evaluate for ulcers, tumors, and bleeding.

Baroreceptor (bar'-ō-re-SEP-tor) Nerve cell capable of responding to changes in blood pressure. Also called a ***pressoreceptor.***

Bartholin's glands *See* ***Greater vestibular glands.***

Basal ganglia (GANG-glē-a) Paired clusters of cell bodies that make up the central gray matter in each cerebral hemisphere, including the caudate nucleus, lentiform nucleus, claustrum, and amygdaloid body. Also called ***cerebral nuclei*** (SER-e-bral NOO-klē-ī).

Basal metabolic (BĀ-sal met'-a-BOL-ik) ***rate (BMR)*** The rate of metabolism measured under standard or basal conditions.

Base The broadest part of a pyramidal structure. A nonacid or a proton acceptor, characterized by excess of hydroxide ions (OH^-) and a pH greater than 7. A ring-shaped, nitrogen-containing organic molecule that is one of the components of a nucleotide, for example, adenine, guanine, cytosine, thymine, and uracil.

Basement membrane Thin, extracellular layer consisting of basal lamina secreted by epithelial cells and reticular lamina secreted by connective tissue cells.

Basilar (BAS-i-lar) ***membrane*** A membrane in the cochlea of the inner ear that separates the cochlear duct from the scala tympani and on which the spiral organ (organ of Corti) rests.

Basophil (BĀ-sō-fil) A type of white blood cell characterized by a pale nucleus and large granules that stain readily with basic dyes.

B cell A lymphocyte that develops into a plasma cell that produces antibodies or a memory cell.

Belly The abdomen. The gaster or prominent, fleshy part of a skeletal muscle.

Benign (be-NĪN) Not malignant; favorable for recovery; a mild disease.

Beta (BĀ-ta) ***cell*** A cell in the pancreatic islets (islets of Langerhans) in the pancreas that secretes insulin.

Beta receptor Receptor found on visceral effectors innervated by most sympathetic postganglionic axons; in general, stimulation of beta receptors leads to inhibition.

Bicuspid (bī-KUS-pid) ***valve*** Atrioventricular (AV) valve on the left side of the heart. Also called the ***mitral valve.***

Bifurcate (bī-FUR-kāt) Having two branches or divisions; forked.

Bilateral (bī-LAT-er-al) Pertaining to two sides of the body.

Bile (bīl) A secretion of the liver consisting of water, bile salts, bile pigments, cholesterol, lecithin, and several ions; it assumes a role in emulsification of fats prior to their digestion.

Biliary (BIL-ē-er-ē) Relating to bile, the gallbladder, or the bile ducts.

Biliary calculi (BIL-ē-er-ē CAL-kyoo-lē) Gallstones formed by the crystallization of cholesterol in bile.

Bilirubin (bil-ē-ROO-bin) A red pigment that is one of the end products of hemoglobin breakdown in the liver cells and is excreted as a waste material in the bile.

Bilirubinuria (bil-ē-roo-bi-NOO-rē-a) The presence of above-normal levels of bilirubin in urine.

Biliverdin (bil-ē-VER-din) A green pigment that is one of the first products of hemoglobin breakdown in the liver cells and is converted to bilirubin or excreted as a waste material in bile.

Biofeedback Process by which an individual gets constant signals (feedback) about varous visceral biological functions.

Biopsy (BĪ-op-sē) Removal of tissue or other material from the living body for examination, usually microscopic.

Blastocoel (BLAS-tō-sēl) The fluid-filled cavity within the blastocyst.

Blastocyst (BLAS-tō-sist) In the development of an embryo, a hollow ball of cells that consists of a blastocoel (the internal cavity), trophoblast (outer cells), and inner cell mass.

Blastomere (BLAS-tō-mēr) One of the cells resulting from the cleavage of a fertilized ovum.

Blastula (BLAS-tyoo-la) An early stage in the development of a zygote.

Bleeding time The time required for the cessation of bleeding from a small skin puncture; identifies platelet function defects and integrity of small blood vessels; ranges from 4 to 8 minutes.

Blepharism (BLEF-a-rizm) Spasm of the eyelids; continuous blinking.

Blind spot Area in the retina at the end of the optic (II) nerve in which there are no light receptor cells.

Blood The fluid that circulates through the heart, arteries, capillaries, and veins and that constitutes the chief means of transport within the body.

Blood–brain barrier (BBB) A special mechanism that prevents the passage of materials from the blood to the cerebrospinal fluid and brain.

Blood island Isolated mass and cord of mesenchyme in the mesoderm from which blood vessels develop.

Blood pressure (BP) Pressure exerted by blood as it presses against and attempts to stretch blood vessels, especially arteries; the force is generated by the rate and force of heartbeat; clinically, a measure of the pressure in arteries during ventricular systole and ventricular diastole.

Blood reservoir (REZ-er-vwar) Systemic veins that contain large amounts of blood that can be moved quickly to parts of the body requiring the blood.

Blood–testis barrier A barrier formed by sustentacular (Sertoli) cells that prevents an immune response against antigens produced by spermatozoa and developing cells by isolating the cells from the blood.

Body cavity A space within the body that contains various internal organs.

Bohr (BŌR) ***effect*** In an acid environment, oxygen splits more readily from hemoglobin because when hydrogen ions (H^+) bind to hemoglobin, they alter the structure of hemoglobin and this reduces its oxygen-carrying capacity.

Bolus (BŌ-lus) A soft, rounded mass, usually food, that is swallowed.

Bone scan Procedure in which a radioisotope is injected and the radiation emitted from bone is measured.

Bony labyrinth (LAB-i-rinth) A series of cavities within the petrous portion of the temporal bone forming the vestibule, cochlea, and semicircular canals of the inner ear.

Bowman's capsule *See* ***Glomerular capsule.***

Brachial plexus (BRĀ-kē-al PLEK-sus) A network of nerve fibers of the anterior rami of spinal nerves C5, C6, C7, C8, and T1. The nerves that emerge from the brachial plexus supply the upper extremity.

Brain A mass of nervous tissue located in the cranial cavity.

Brain electrical activity mapping (BEAM) Noninvasive procedure that measures and displays the electrical activity of the brain; used primarily to diagnose epilepsy.

Brain sand Calcium deposits in the pineal gland that are laid down starting at puberty.

Brain stem The portion of the brain immediately superior to the spinal cord, made up of the medulla oblongata, pons, and midbrain.

Brain wave Electrical activity produced as a result of action potentials of brain cells.

Bright's disease *See* ***Glomerulonephritis.***

Broad ligament A double fold of parietal peritoneum attaching the uterus to the side of the pelvic cavity.

Broca's (BRŌ-kaz) ***area*** Motor area of the brain in the frontal lobe that translates thoughts into speech. Also called the ***motor speech area.***

Bronchi (BRONG-kē) Branches of the respiratory passageway including primary bronchi (the two divisions of the trachea), secondary or lobar bronchi (divisions of the primary that are distributed to the lobes of the lung), and tertiary or segmental bronchi (divisions of the secondary that are distributed to bronchopulmonary segments of the lung).

Bronchial asthma (BRONG-kē-al AZ-ma) Usually allergic reaction characterized by smooth muscle spasms in bronchi resulting in wheezing and difficult breathing.

Bronchial tree The trachea, bronchi, and their branching structures.

Bronchiectasis (brong'-kē-EK-ta-sis) A chronic disorder in which there is a loss of the normal tissue and expansion of lung air passages; characterized by difficult breathing, coughing, expectoration of pus, and foul breath.

Bronchiole (BRONG-kē-ol) Branch of a tertiary bronchus further dividing into terminal bronchioles (distributed to lobules of the lung), which divide into respiratory bronchioles (distributed to alveolar sacs).

Bronchitis (brong-KĪ-tis) Inflammation of the bronchi characterized by hypertrophy and hyperplasia of seromucous glands and goblet cells that line the bronchi and resulting in a productive cough.

Bronchogenic carcinoma (brong'-kō-JEN-ik kar'-si-NŌ-ma) Cancer originating in the bronchi.

Bronchogram (BRONG-kō-gram) A roentgenogram of the bronchial tree.

Bronchography (bron-KOG-ra-fē) Technique for examining the bronchial tree in which an opaque contrast medium is introduced into the trachea for distribution to the bronchial branches. The roentgenogram produced is called a bronchogram.

Bronchopulmonary (brong'-kō-PUL-mō-ner-ē) ***segment*** One of the smaller divisions of a lobe of a lung supplied by its own branches of a bronchus.

Bronchoscope (BRONG-kō-skōp) An instrument used to examine the interior of the bronchi of the lungs.

Bronchoscopy (brong-KOS-kō-pē) Visual examination of the interior of the trachea and bronchi with a bronchoscope to biopsy a tumor, clear an obstruction, take cultures, stop bleeding, or deliver drugs.

Bronchus (BRONG-kus) One of the two large branches of the trachea. *Plural,* ***bronchi*** (BRONG-kē).

Brunner's gland *See* ***Duodenal gland.***

Buccal (BUK-al) Pertaining to the cheek or mouth.

Buffer (BUF-er) ***system*** A pair of chemicals, one a weak acid and one the salt of the weak acid, which functions as a weak base, that resists changes in pH.

Bulb of penis Expanded portion of the base of the corpus spongiosum penis.

Bulbourethral (bul'-bō-yoo-RĒ-thral) ***gland*** One of a pair of glands located inferior to the prostate gland on either side of the urethra that secretes an alkaline fluid into the cavernous urethra. Also called a ***Cowper's*** (KOW-perz) ***gland.***

Bulimia (boo-LIM-ē-a) A disorder characterized by overeating, at least twice a week, followed by purging by self-induced vomiting, strict dieting or fasting, vigorous exercise, or use of laxatives.

Bulk flow The movement of large numbers of ions, molecules, or particles in the same direction as a result of forces that push them (osmotic or hydrostatic pressure).

Bullae (BYOOL-ē) Blisters beneath or within the epidermis.

Bundle branch One of the two branches of the atrioventricular (AV) bundle made up of specialized muscle fibers (cells) that transmit electrical impulses to the ventricles.

Bundle of His See ***Atrioventricular (AV) bundle.***

Bunion (BUN-yun) Lateral deviation of the great toe that produces inflammation and thickening of the bursa, bone spurs, and calluses.

Burn An injury in which proteins are destroyed (denatured) as a result of heat (fire, steam), chemicals, electricity, or the ultraviolet rays of the sun.

Bursa (BUR-sa) A sac or pouch of synovial fluid located at friction points, especially about joints.

Bursitis (bur-SĪ-tis) Inflammation of a bursa.

Buttocks (BUT-oks) The two fleshy masses on the posterior aspect of the lower trunk, formed by the gluteal muscles.

Cachexia (kah-KEK-sē-ah) A state of ill health, malnutrition, and wasting.

Calcaneal tendon The tendon of the soleus, gastrocnemius, and plantaris muscles at the back of the heel. Also called the ***Achilles*** (a-KIL-ēz) ***tendon.***

Calcify (KAL-si-fī) To harden by deposits of calcium salts.

Calcitonin (kal-si-TŌ-nin) ***(CT)*** A hormone produced by the thyroid gland that lowers the calcium and phosphate levels of the blood by inhibiting bone breakdown and accelerating calcium absorption by bones.

Calculus (KAL-kyoo-lus) A stone, or insoluble mass of crystallized salts or other material, formed within the body, as in the gallbladder, kidney, or urinary bladder.

Callus (KAL-lus) A growth of new bone tissue in and around a fractured area, ultimately replaced by mature bone. An acquired, localized thickening.

Calmodulin (kal-MOD-yoo-lin) An intracellular protein that binds with calcium ions and activates or inhibits enzymes, many of which are protein kinases, to elicit physiological responses of hormones.

Calorie (KAL-ō-rē) A unit of heat. A calorie (cal) is the standard unit and is the amount of heat necessary to raise 1 g of water 1°C from 14° to 15°C. The kilocalorie (kcal), used in metabolic and nutrition studies, is the amount of heat necessary to raise 1,000 g of water 1°C and is equal to 1,000 cal.

Calyx (KĀL-iks) Any cuplike division of the kidney pelvis. *Plural,* ***calyces*** (KĀ-li-sēz).

Canal (ka-NAL) A narrow tube, channel, or passageway.

Canaliculus (kan'-a-LIK-yoo-lus) A small channel or canal, as in bones, where they connect lacunae. *Plural,* ***canaliculi*** (kan'-a-LIK-yoo-lī).

Canal of Schlemm *See* ***Scleral venous sinus.***

Cancellous (KAN-sel-us) Having a reticular or latticework structure, as in spongy tissue of bone.

Cancer (KAN-ser) A malignant tumor of epithelial origin tending to infiltrate and give rise to new growths or metastases. Also called ***carcinoma*** (kar'-si-NŌ-ma).

Canker (KANG-ker) ***sore*** Painful ulcer on the mucous membrane of the mouth that may result from an autoimmune response.

Capacitation (ka'-pas-i-TĀ-shun) The functional changes that sperm undergo in the female reproductive tract that allow them to fertilize a secondary oocyte.

Capillary (KAP-i-lar'-ē) A microscopic blood vessel located between an arteriole and venule through which materials are exchanged between blood and body cells.

Carbohydrate (kar'-bō-HĪ-drāt) An organic compound containing carbon, hydrogen, and oxygen in a particular amount and arrangement and comprised of sugar subunits; usually has the formula $(CH_2O)_n$.

Carbon monoxide (CO) poisoning Hypoxia due to increased levels of carbon monoxide as a result of its preferential and tenacious combination with hemoglobin rather than with oxygen.

Carcinoembryonic (car'-sin-ō-em-brē-ON-ik) ***antigen (CEA)*** A glycoprotein secreted by normally developing fetal tissue during the first or second trimester, after birth, and in certain malignant and benign conditions.

Carcinogen (kar-SIN-ō-jen) Any substance that causes cancer.

Carcinoma (kar'-si-NŌ-ma) A malignant tumor consisting of epithelial cells.

Cardiac (KAR-dē-ak) ***arrest*** Cessation of an effective heartbeat in which the heart is completely stopped or in ventricular fibrillation.

Cardiac catheterization (KAR-dē-ak kath'-e-ter-i-ZĀ-shun) Introduction of a catheter into the heart and/or its blood vessels to measure pressure; assess left ventricular function and cardiac output; measure blood flow, oxygen content of blood, and the status of valves and conduction system; and identify valvular and septal defects.

Cardiac (KAR-dē-ak) ***cycle*** A complete heartbeat consisting of systole (contraction) and diastole (relaxation) of both atria plus systole and diastole of both ventricles.

Cardiac muscle An organ specialized for contraction, composed of striated muscle fibers (cells), forming the wall of the heart, and stimulated by an intrinsic conduction system and visceral efferent neurons.

Cardiac notch An angular notch in the anterior border of the left lung.

Cardiac output (CO) The volume of blood pumped from one ventricle of the heart (usually measured from the left ventricle) in 1 min; about 5.2 liters/min under normal resting conditions.

Cardiac reserve The maximum percentage that cardiac output can increase above normal.

Cardiac tamponade (tam'-pon-ĀD) Compression of the heart due to excessive fluid or blood in the pericardial sac that could result in cardiac failure.

Cardinal ligament A ligament of the uterus, extending laterally from the cervix and vagina as a continuation of the broad ligament.

Cardioacceleratory (kar-dē-ō-ak-SEL-er-a-tō-rē) ***center (CAC)*** A group of neurons in the medulla from which cardiac nerves (sympathetic) arise; nerve impulses along the nerves release epinephrine that increases the rate and force of heartbeat.

Cardioinhibitory (kar-dē-ō-in-HIB-i-tō-rē) ***center (CIC)*** A group of neurons in the medulla from which parasympathetic fibers that reach the heart via the vagus (X) nerve arise; nerve impulses along the nerves release acetylcholine that decreases the rate and force of heartbeat.

Cardiology (kar-dē-OL-ō-jē) The study of the heart and diseases associated with it.

Cardiopulmonary resuscitation (rē-sus-i-TĀ-shun) ***(CPR)*** A technique employed to restore life or consciousness to a person apparently dead or dying; includes external respiration (exhaled air respiration) and external cardiac massage.

Carina (ka-RĪ-na) A ridge on the inside of the division of the right and left primary bronchi.

Carotid (ka-ROT-id) ***body*** Receptor on or near the carotid sinus that responds to alterations in blood levels of oxygen, carbon dioxide, and hydrogen ions.

Carotid sinus A dilated region of the internal carotid artery immediately above the bifurcation of the common carotid artery that contains receptors that monitor blood pressure.

Carotid sinus reflex A reflex concerned with maintaining normal blood pressure in the brain.

Carpus (KAR-pus) A collective term for the eight bones of the wrist.

Cartilage (KAR-ti-lij) A type of connective tissue consisting of chondrocytes in lacunae embedded in a dense network of collagenous and elastic fibers and a matrix of chondroitin sulfate.

Cartilaginous (kar'-ti-LAJ-i-nus) ***joint*** A joint without a synovial (joint) cavity where the articulating bones are held tightly together by cartilage, allowing little or no movement.

Caruncle (KAR-ung-kul) A small fleshy eminence, often abnormal.

Cast A small mass of hardened material formed within a cavity in the body and then discharged from the body; can originate in different areas and be composed of various materials.

Castration (kas-TRĀ-shun) The removal of the testes.

Catabolism (ka-TAB-ō-lizm) Chemical reactions that break down complex organic compounds into simple ones with the release of energy.

Cataract (KAT-a-rakt) Loss of transparency of the lens of the eye or its capsule or both.

Catheter (KATH-i-ter) A tube that can be inserted into a body cavity through a canal or into a blood vessel; used to remove fluids, such as urine and blood, and to introduce diagnostic materials or medication.

Cation A positively charged ion. An example is a sodium ion (Na^+).

Cauda equina (KAW-da ē-KWĪ-na) A taillike collection of roots of spinal nerves at the inferior end of the spinal canal.

Caudal (KAW-dal) Pertaining to any taillike structure; inferior in position.

Cecum (SĒ-kum) A blind pouch at the proximal end of the large intestine to which the ileum is attached.

Celiac (SĒ-lē-ak) Pertaining to the abdomen.

Celiac plexus (PLEK-sus) A large mass of ganglia and nerve fibers located at the level of the upper part of the first lumbar vertebra. Also called the ***solar plexus.***

Cell The basic structural and functional unit of all organisms; the smallest structure capable of performing all the activities vital to life.

Cell division Process by which a cell reproduces itself that consists of a nuclear division (mitosis) and a cytoplasmic division (cytokinesis); types include somatic and reproductive cell division.

Cell inclusion A lifeless, often temporary, constituent in the cytoplasm of a cell as opposed to an organelle.

Cellular immunity That component of immunity in which specially sensitized lymphocytes (T cells) attach to antigens to destroy them. Also called ***cell-mediated immunity.***

Cementum (se-MEN-tum) Calcified tissue covering the root of a tooth.

Center An area in the brain where a particular function is localized.

Center of ossification (os'-i-fi-KĀ-shun) An area in the cartilage model of a future bone where the cartilage cells hypertrophy and then secrete enzymes that result in the calcification of their matrix, resulting in the death of the cartilage cells, followed by the invasion of the area by osteoblasts that then lay down bone.

Central canal A circular channel running longitudinally in the center of an osteon (Haversian system) of mature compact bone, containing blood and lymphatic vessels and nerves. Also called a ***Haversian*** (ha-VĒR-shun) ***canal.*** A microscopic tube running the length of the spinal cord in the gray commissure.

Central fovea (FŌ-vē-a) A cuplike depression in the center of the macula lutea of the retina, containing cones only; the area of clearest vision.

Central nervous system (CNS) That portion of the nervous system that consists of the brain and spinal cord.

Centrioles (SEN-trē-ōlz) Paired, cylindrical structures within a centrosome, each consisting of a ring of microtubules and arranged at right angles to each other; function in cell division.

Centromere (SEN-trō-mēr) The clear, constricted portion of a chromosome where the two chromatids are joined; serves as the point of attachment for the chromosomal microtubules.

Centrosome (SEN-trō-sōm) A rather dense area of cytoplasm, near the nucleus of a cell, containing a pair of centrioles.

Cephalic (se-FAL-ik) Pertaining to the head; superior in position.

Cerebellar peduncle (ser-e-BEL-ar pe-DUNG-kul) A bundle of nerve fibers connecting the cerebellum with the brain stem.

Cerebellum (ser-e-BEL-um) The portion of the brain lying posterior to the medulla and pons, concerned with coordination of movements.

Cerebral aqueduct (SER-ē-bral AK-we-dukt) A channel through the midbrain connecting the third and fourth ventricles and containing cerebrospinal fluid.

Cerebral arterial circle A ring of arteries forming an anastomosis at the base of the brain between the internal carotid and basilar arteries and arteries supplying the brain. Also called the ***circle of Willis.***

Cerebral cortex The surface of the cerebral hemispheres, 2–4 mm thick, consisting of six layers of nerve cell bodies (gray matter) in most areas.

Cerebral palsy (PAL-zē) A group of motor disorders resulting in muscular uncoordination and loss of muscle control and caused by damage to motor areas of the brain (cerebral cortex, basal ganglia, and cerebellum) during fetal life, birth, or infancy.

Cerebral peduncle (pe-DUNG-kul) One of a pair of nerve fiber bundles located on the ventral surface of the midbrain, conducting nerve impulses between the pons and the cerebral hemispheres.

Cerebrospinal (se-rē'-brō-SPĪ-nal) ***fluid (CSF)*** A fluid produced in the choroid plexuses and ependymal cells of the ventricles of the brain that circulates in the ventricles and the subarachnoid space around the brain and spinal cord.

Cerebrovascular (se-rē'-brō-VAS-kyoo-lar) ***accident (CVA)*** Destruction of brain tissue (infarction) resulting from disorders of blood vessels that supply the brain. Also called a ***stroke.***

Cerebrum (SER-ē-brum) The two hemispheres of the forebrain, making up the largest part of the brain.

Ceruminous (se-ROO-mi-nus) ***gland*** A modified sudoriferous (sweat) gland in the external auditory meatus that secretes cerumen (ear wax).

Cervical dysplasia (dis-PLĀ-sē-a) A change in the shape, growth, and number of cervical cells of the uterus that, if severe, may progress to cancer.

Cervical ganglion (SER-vi-kul GANG-glē-on) A cluster of nerve cell bodies of postganglionic sympathetic neurons located in the neck, near the vertebral column.

Cervical mucus A mixture of water, glycoprotein, serum-type proteins, lipids, enzymes, and inorganic salts produced by secreting cells of the mucosa of the cervix.

Cervical plexus (PLEK-sus) A network of neuron fibers formed by the anterior rami of the first four cervical nerves.

Cervix (SER-viks) Neck; any constricted portion of an organ, such as the lower cylindrical part of the uterus.

Cesarean (se-SA-rē-an) ***section*** Procedure in which a low, horizontal incision is made through the abdominal wall and uterus for removal of the baby and placenta. Also called a ***C-section.***

Chalazion (ka-LĀ-zē-on) A small tumor of the eyelid.

Chemical bond Force of attraction in a molecule that holds atoms together. Examples include ionic, covalent, and hydrogen bonds.

Chemical element Unit of matter that cannot be decomposed into a simpler substance by ordinary chemical reactions. Examples include hydrogen (H), carbon (C), and oxygen (O).

Chemical reaction The combination or breaking apart of atoms in which chemical bonds are formed or broken and new products with different properties are produced.

Chemiosmosis (kem'-ē-oz-MŌ-sis) Process by which energy released is used to generate ATP when a substance moves along a gradient.

Chemonucleolysis (kē'-mō-noo'-klē-OL-i-sis) Dissolution of the nucleus pulposus of an intervertebral disc by injection of a proteolytic enzyme (chymopapain) to relieve the pressure and pain associated with a herniated (slipped) disc.

Chemoreceptor (kē'-mō-rē-SEP-tor) Receptor outside the central nervous system on or near the carotid and aortic bodies that detects the presence of chemicals.

Chemotaxis (kē-mō-TAK-sis) Attraction of phagocytes to microbes by a chemical stimulus.

Chemotherapy (kē-mō-THER-a-pē) The treatment of illness or disease by chemicals.

Chiasma (kī-AZ-ma) A crossing; especially the crossing of the optic (II) nerve fibers.

Chiropractic (kī'-rō-PRAK-tik) A system of treating disease by using one's hands to manipulate body parts, mostly the vertebral column.

Chlamydia (kla-MID-ē-a) A sexually transmitted disease characterized by burning on urination, frequent and painful urination, and low back pain; may spread to uterine (Fallopian) tubes in females.

Chloride shift Diffusion of bicarbonate ions (HCO_3^-) from the red blood cells into plasma and of chloride ions (Cl^-) from plasma into red blood cells that maintains ionic balance between red blood cells and plasma.

Choana (KŌ-a-na) A funnel-shaped structure; the posterior opening of the nasal fossa, or internal naris.

Cholecystectomy (kō'-lē-sis-TEK-tō-mē) Surgical removal of the gallbladder.

Cholesterol (kō-LES-te-rol) Classified as a lipid, the most abundant steroid in animal tissues; located in cell membranes and used for the synthesis of steroid hormones and bile salts.

Cholinergic (kō'-lin-ER-jik) ***fiber*** A nerve ending that liberates acetylcholine at a synapse.

Cholinesterase (kō'-lin-ES-ter-ās) An enzyme that hydrolyzes acetylcholine.

Chondrocyte (KON-drō-sīt) Cell of mature cartilage.

Chondroitin (kon-DROY-tin) ***sulfate*** An amorphous matrix material found outside the cell.

Chordae tendineae (KOR-dē TEN-di-nē-ē) Tendonlike, fibrous cords that connect the heart valves with the papillary muscles.

Chorion (KŌ-rē-on) The outermost fetal membrane that becomes the principal embryonic portion of the placenta; serves a protective and nutritive function.

Chorionic villus (kō'-rē-ON-ik VIL-lus) Fingerlike projection of the chorion that grows into the decidua basalis of the endometrium and contains fetal blood vessels.

Chorionic villus sampling (CVS) The removal of a sample of chorionic villus tissue by means of a catheter to analyze the tissue for prenatal genetic defects.

Choroid (KŌ-royd) One of the vascular coats of the eyeball.

Choroid plexus (PLEK-sus) A vascular structure located in the roof of each of the four ventricles of the brain; produces cerebrospinal fluid.

Chromaffin (krō-MAF-in) ***cell*** Cell that has an affinity for chrome salts, due in part to the presence of the precursors of the neurotransmitter epinephrine; found, among other places, in the adrenal medulla.

Chromatid (KRŌ-ma-tid) One of a pair of identical connected nucleoprotein strands that are joined at the centromere and separate during cell division, each becoming a chromosome of one of the two daughter cells.

Chromatin (KRŌ-ma-tin) The threadlike mass of the genetic material consisting principally of DNA, which is present in the nucleus of a nondividing or interphase cell.

Chromatolysis (krō'-ma-TOL-i-sis) The breakdown of chromatophilic substance (Nissl bodies) into finely granular masses in the cell body of a central or peripheral neuron whose process (axon or dendrite) has been damaged.

Chromatophilic substance Rough endoplasmic reticulum in the cell bodies of neurons that functions in protein synthesis. Also called ***Nissl bodies.***

Chromosomal microtubule (mī-krō-TOOB-yool) Microtubule formed during prophase of mitosis that originates from centromeres, extends from a centromere to a pole of the cell, and assists in chromosomal movement; constitutes a part of the mitotic spindle.

Chromosome (KRŌ-mō-sōm) One of the 46 small, dark-staining bodies that appear in the nucleus of a human diploid ($2n$) cell during cell division.

Chronic (KRON-ik) Long-term or frequently recurring; applied to a disease that is not acute.

Chyle (kīl) The milky fluid found in the lacteals of the small intestine after digestion.

Chyme (kīm) The semifluid mixture of partly digested food and digestive secretions found in the stomach and small intestine during digestion of a meal.

Cicatrix (SIK-a-triks) A scar left by a healed wound.

Ciliary (SIL-ē-ar'-ē) ***body*** One of the three portions of the vascular tunic of the eyeball, the others being the choroid and the iris; includes the ciliary muscle and the ciliary processes.

Ciliary ganglion (GANG-glē-on) A very small parasympathetic ganglion whose preganglionic fibers come from the oculomotor (III) nerve and whose postganglionic fibers carry nerve impulses to the ciliary muscle and the sphincter muscle of the iris.

Cilium (SIL-ē-um) A hair or hairlike process projecting from a cell that may be used to move the entire cell or to move substances along the surface of the cell.

Circadian (ser-KĀ-dē-an) ***rhythm*** A cycle of active and nonactive periods in organisms determined by internal mechanisms and repeating about every 24 hours.

Circle of Willis *See* ***Cerebral arterial circle.***

Circular folds Permanent, deep, transverse folds in the mucosa and submucosa of the small intestine that increase the surface area for absorption. Also called ***plicae circulares*** (PLĪ-kē SER-kyoo-lar-ēs).

Circulation time Time required for blood to pass from the right atrium, through pulmonary circulation, back to the left ventricle, through systemic circulation to the foot, and back again to the right atrium; normally about 1 min.

Circumcision (ser'-kum-SIZH-un) Surgical removal of the foreskin (prepuce), the fold of skin over the glans penis.

Circumduction (ser'-kum-DUK-shun) A movement at a synovial joint in which the distal end of a bone moves in a circle while the proximal end remains relatively stable.

Circumvallate papilla (ser'-kum-VAL-āt pa-PIL-a) One of the circular projections that is arranged in an inverted V-shaped row at the posterior portion of the tongue; the largest of the elevations on the upper surface of the tongue containing taste buds.

Cirrhosis (si-RŌ-sis) A liver disorder in which the parenchymal cells are destroyed and replaced by connective tissue.

Cisterna chyli (sis-TER-na KĪ-lē) The origin of the thoracic duct.

Cleavage The rapid mitotic divisions following the fertilization of a secondary oocyte, resulting in an increased number of progressively smaller cells, called blastomeres, so that the overall size of the zygote remains the same.

Cleft palate Condition in which the palatine processes of the maxillae do not unite before birth; cleft lip, a split in the upper lip, is often associated with cleft palate.

Climacteric (klī-mak-TER-ik) Cessation of the reproductive function in the female or diminution of testicular activity in the male.

Climax The peak period or moments of greatest intensity during sexual excitement.

Clitoris (KLI-to-ris) An erectile organ of the female located at the anterior junction of the labia minora that is homologous to the male penis.

Clone (KLŌN) A population of cells identical to itself.

Clot The end result of a series of biochemical reactions that changes liquid plasma into a gelatinous mass; specifically, the conversion of fibrinogen into a tangle of polymerized fibrin molecules.

Clot retraction (rē-TRAK-shun) The consolidation of a fibrin clot to pull damaged tissue together.

Coagulation (cō-ag-yoo-LĀ-shun) Process by which a blood clot is formed.

Coarctation (kō'-ark-TĀ-shun) ***of the aorta*** Congenital condition in which the aorta is too narrow and results in reduced blood supply, increased ventricular pumping, and high blood pressure.

Coccyx (KOK-six) The fused bones at the end of the vertebral column.

Cochlea (KŌK-lē-a) A winding, cone-shaped tube forming a portion of the inner ear and containing the spiral organ (organ of Corti).

Cochlear duct The membranous cochlea consisting of a spirally arranged tube enclosed in the bony cochlea and lying along its outer wall. Also called the ***scala media*** (SCA-la MĒ-dē-a).

Coenzyme A type of cofactor; a nonprotein organic molecule that is associated with and activates an enzyme; many are derived from vitamins. An example is nicotinamide adenine dinucleotide (NAD), derived from the B vitamin niacin.

Coitus (KŌ-i-tus) Sexual intercourse. Also called ***copulation*** (cop-yoo-LĀ-shun).

Colitis (ko-LĪ-tis) Inflammation of the mucosa of the colon and rectum in which absorption of water and salts is reduced, producing watery, bloody feces, and, in severe cases, dehydration and salt depletion. Spasms of the irritated muscularis produce cramps.

Collagen (KOL-a-jen) A protein that is the main organic constituent of connective tissue.

Collateral circulation The alternate route taken by blood through an anastomosis.

Colliculus (ko-LIK-yoo-lus) A small elevation.

Collision theory Theory that explains how chemical reactions occur. It states that all atoms, ions, and molecules are constantly moving and colliding, and the energy transferred during collision might disrupt their electron structures so that chemical bonds are broken or formed.

Colon The division of the large intestine consisting of ascending, transverse, descending, and sigmoid portions.

Colony-stimulating factor (CSF) One of a group of hematopoietins that stimulates development of white blood cells. Examples are macrophage CSF and granulocyte CSF.

Color blindness Any deviation in the normal perception of colors, resulting from the lack of usually one of the photopigments of the cones.

Colostomy (kō-LOS-tō-mē) The diversion of feces through an opening in the colon, creating a surgical opening at the exterior of the abdominal wall.

Colostrum (kō-LOS-trum) A thin, cloudy fluid secreted by the mammary glands a few days prior to or after delivery before true milk is secreted.

Colposcopy (kol-POS-kō-pē) Direct examination of the vaginal and cervical mucosa using a magnifying device; frequently the first procedure performed following an abnormal Pap smear.

Coma (KŌ-ma) Final stage of brain failure that is characterized by total unresponsiveness to all external stimuli.

Commissure (KOM-i-shūr) The angular junction of the eyelids at either corner of the eyes.

Common bile duct A tube formed by the union of the common hepatic duct and the cystic duct that empties bile into the duodenum at the hepatopancreatic ampulla (ampulla of Vater).

Compact (dense) bone Bone tissue with no apparent spaces in which the layers of lamellae are fitted tightly together.

Compact bone is found immediately deep to the periosteum and external to spongy bone.

Complement (KOM-ple-ment) A group of at least 20 proteins found in serum that forms a component of nonspecific resistance and immunity by bringing about cytolysis, inflammation, and opsonization.

Complete blood count (CBC) Hematology test that usually includes hemoglobin determination, hematocrit, red and white blood cell count, differential white blood cell count, and platelet count.

Compliance The ease with which the lungs and thoracic wall can be expanded.

Compound A substance that can be broken down into two or more other substances by chemical means.

Computed tomography (tō-MOG-ra-fē) ***(CT)*** X-ray technique that provides a cross-sectional image of any area of the body. Also called ***computed axial tomography (CAT).***

Concha (KONG-ka) A scroll-like bone found in the skull. *Plural,* ***conchae*** (KONG-kē).

Concussion (kon-KUSH-un) Traumatic injury to the brain that produces no visible bruising but may result in abrupt, temporary loss of consciousness.

Conduction myofiber Muscle fiber (cell) in the subendocardial tissue of the heart specialized for conducting an action potential to the myocardium; part of the conduction system of the heart. Also called a ***Purkinje*** (pur-KIN-jē) ***fiber.***

Conduction system An intrinsic regulating system composed of specialized muscle tissue that generates and distributes electrical impulses that stimulate cardiac muscle fibers (cells) to contract.

Conductivity (kon'-duk-TIV-i-tē) The ability to carry the effect of a stimulus from one part of a cell to another; highly developed in nerve and muscle fibers (cells).

Cone The light-sensitive receptor in the retina concerned with color vision.

Cone biopsy (BĪ-op-sē) Removal of a sample of tissue from the cervical os to evaluate for cancer of the uterus.

Congenital (kon-JEN-i-tal) Present at the time of birth.

Congestive heart failure (CHF) Chronic or acute state that results when the heart is not capable of supplying the oxygen demands of the body.

Conjunctiva (kon'-junk-TĪ-va) The delicate membrane covering the eyeball and lining the eyes.

Conjunctivitis (kon-junk'-ti-VĪ-tis) Inflammation of the conjunctiva, the delicate membrane covering the eyeball and lining the eyelids.

Connective tissue The most abundant of the four basis tissue types in the body, performing the functions of binding and supporting; consists of relatively few cells in a great deal of intercellular substance.

Constipation (con-sti-PĀ-shun) Infrequent or difficult defecation caused by decreased motility of the intestines.

Contact inhibition Phenomenon by which migration of a growing cell is stopped when it makes contact with another cell of its own kind.

Continuous microtubules (mī-krō-TOOB-yoolz) Microtubules formed during prophase of mitosis that originate from the vicinity of the centrioles, grow toward each other, extend from one pole of the cell to another, and assist in chromosomal movement; constitute a part of the mitotic spindle.

Contraception (kon'-tra-SEP-shun) The prevention of conception or impregnation without destroying fertility.

Contractility (kon'-trak-TIL-i-tē) The ability of cells or parts of cells actively to generate force to undergo shortening and change form for purposeful movements. Muscle fibers (cells) exhibit a high degree of contractility.

Contralateral (kon'-tra-LAT-er-al) On the opposite side; affecting the opposite side of the body.

Contusion (kon-TOO-shun) Condition in which tissue below the skin is damaged, but the skin is not broken.

Conus medullaris (KŌ-nus med-yoo-LAR-is) The tapered portion of the spinal cord below the lumbar enlargement.

Convergence (con-VER-jens) An anatomical arrangement in which the synaptic end bulbs of several presynaptic neurons terminate on one postsynaptic neuron. The medial movement of the two eyeballs so that both are directed toward a near object being viewed in order to produce a single image.

Convulsion (con-VUL-shun) Violent, involuntary, tetanic contractions of an entire group of muscles.

Cornea (KOR-nē-a) The nonvascular, transparent fibrous coat through which the iris can be seen.

Corona (kō-RŌ-na) Margin of the glans penis.

Corona radiata Several layers of follicle cells surrounding a secondary oocyte.

Coronal (kō-RŌ-nal) ***plane*** A plane that runs vertical to the ground and divides the body into anterior and posterior portions. Also called ***frontal plane.***

Coronary angiography (KOR-ō-na-rē an'-jē-OG-ra-fē) Procedure in which the severity and location of blocked coronary arteries are visualized by injection of contrast dyes or in which clot-dissolving drugs may be injected into coronary arteries.

Coronary (KOR-ō-na-rē) ***artery bypass grafting (CABG)*** Surgical procedure in which a portion of a blood vessel is removed from another part of the artery and grafted onto a coronary artery so as to bypass an obstruction in the coronary artery.

Coronary (KOR-ō-na-rē) ***artery disease (CAD)*** A condition in which the heart muscle receives inadequate blood due to an interruption of its blood supply.

Coronary artery spasm A condition in which the smooth muscle of a coronary artery undergoes a sudden contraction, resulting in vasoconstriction.

Coronary circulation The pathway followed by the blood from the ascending aorta through the blood vessels supplying the heart and returning to the right atrium. Also called ***cardiac circulation.***

Coronary sinus (SĪ-nus) A wide venous channel on the posterior surface of the heart that collects the blood from the coronary circulation and returns it to the right atrium.

Corpora quadrigemina (KOR-por-a kwad-ri-JEM-in-a) Four small elevations (superior and inferior colliculi) on the dorsal region of the midbrain concerned with visual and auditory functions.

Cor pulmonale (kor pul-mōn-ALE) ***(CP)*** Right ventricular hypertrophy from disorders that bring about hypertension in pulmonary circulation.

Corpus (KOR-pus) The principal part of any organ; any mass or body.

Corpus albicans (KOR-pus AL-bi-kanz) A white fibrous patch in the ovary that forms after the corpus luteum regresses.

Corpus callosum (kal-LŌ-sum) The great commissure of the brain between the cerebral hemispheres.

Corpuscle of touch The sensory receptor for the sensation of touch; found in the dermal papillae, especially in palms and soles. Also called a ***Meissner's*** (MĪS-nerz) ***corpuscle.***

Corpus luteum (LOO-tē-um) A yellow endocrine gland in the ovary formed when a follicle has discharged its secondary oocyte; secretes estrogens, progesterone, and relaxin.
Corpus striatum (strī-Ā-tum) An area in the interior of each cerebral hemisphere composed of the caudate and lentiform nuclei of the basal ganglia and white matter of the internal capsule, arranged in a striated manner.
Cortex (KOR-teks) An outer layer of an organ. The convoluted layer of gray matter covering each cerebral hemisphere.
Costal (KOS-tal) Pertaining to a rib.
Costal cartilage (KOS-tal KAR-ti-lij) Hyaline cartilage that attaches a rib to the sternum.
Countercurrent mechanism One mechanism involved in the ability of the kidneys to produce a hyperosmotic urine.
Cowper's gland *See* ***Bulbourethral gland.***
Cramp A spasmodic, especially a tonic, contraction of one of many muscles, usually painful.
Cranial (KRĀ-nē-al) ***cavity*** A subdivision of the dorsal body cavity formed by the cranial bones and containing the brain.
Cranial nerve One of 12 pairs of nerves that leave the brain, pass through foramina in the skull, and supply the head, neck, and part of the trunk; each is designated by a Roman numeral and a name.
Craniosacral (krā-nē-ō-SĀ-kral) ***outflow*** The fibers of parasympathetic preganglionic neurons, which have their cell bodies located in nuclei in the brain stem and in the lateral gray matter of the sacral portion of the spinal cord.
Craniotomy (krā'-nē-OT-ō-mē) Any operation on the skull, as for surgery on the brain or decompression of the fetal head in difficult labor.
Cranium (KRĀ-nē-um) The skeleton of the skull that protects the brain and the organs of sight, hearing, and balance; includes the frontal, parietal, temporal, occipital, sphenoid, and ethmoid bones.
Crenation (krē-NĀ-shun) The shrinkage of red blood cells into knobbed, starry forms when placed in a hypertonic solution.
Cretinism (KRĒ-tin-izm) Severe congenital thyroid deficiency during childhood leading to physical and mental retardation.
Crista (KRIS-ta) A crest or ridged structure. A small elevation in the ampulla of each semicircular duct that serves as a receptor for dynamic equilibrium.
Crossed extensor reflex A reflex in which extension of the joints in one limb occurs in conjunction with contraction of the flexor muscles of the opposite limb.
Crossing-over The exchange of a portion of one chromatid with another in a tetrad during meiosis. It permits an exchange of genes among chromatids and is one factor that results in genetic variation.
Crus (krus) ***of penis*** Separated, tapered portion of the corpora cavernosa penis. *Plural,* ***crura*** (KROO-ra).
Cryosurgery (KRĪ-ō-ser-jer-ē) The destruction of tissue by application of extreme cold.
Crypt of Lieberkühn *See* ***Intestinal gland.***
Cryptorchidism (krip-TOR-ki-dizm) The condition of undescended testes.
Cupula (KUP-yoo-la) A mass of gelatinous material covering the hair cells of a crista, a receptor in the ampulla of a semicircular canal stimulated when the head moves.
Curvature (KUR-va-tūr) A nonangular deviation of a straight line, as in the greater and lesser curvatures of the stomach. Abnormal curvatures of the vertebral column include kyphosis, lordosis, and scoliosis.
Cushing's syndrome Condition caused by a hypersecretion of glucocorticoids characterized by spindly legs, "moon face," "buffalo hump," pendulous abdomen, flushed facial skin, poor wound healing, hyperglycemia, osteoporosis, weakness, hypertension, and increased susceptibility to disease.
Cutaneous (kyoo-TĀ-nē-us) Pertaining to the skin.
Cyanosis (sī-a-NŌ-sis) Reduced hemoglobin (unoxygenated) concentration of blood of more than 5 g/dl that results in a blue or dark purple discoloration that is most easily seen in nail beds and mucous membranes.
Cyclic AMP (cyclic adenosine-3',5'-monophosphate) Molecule formed from ATP by the action of the enzyme adenylate cyclase; serves as an intracellular messenger (second messenger) for some hormones.
Cyst (SIST) A sac with a distinct connective tissue wall, containing a fluid or other material.
Cystic (SIS-tik) ***duct*** The duct that transports bile from the gallbladder to the common bile duct.
Cystitis (sis-TĪ-tis) Inflammation of the urinary bladder.
Cystoscope (SIS-to-skōp) An instrument used to examine the inside of the urinary bladder.
Cystoscopy (sis-TOS-kō-pē) Direct visual examination of the urinary tract (and prostate gland in males as well) using a cystoscope to evaluate urinary tract disorders and remove tissue for biopsy, kidney stones, urinary bladder tumors, and urine samples.
Cytochrome (SĪ-tō-krōm) A protein with an iron-containing group (heme) capable of alternating between a reduced form (Fe^{2+}) and an oxidized form (Fe^{3+}).
Cytokinesis (sī'-tō-ki-NĒ-sis) Division of the cytoplasm.
Cytology (sī-TOL-ō-jē) The study of cells.
Cytoplasm (SĪ-tō-plazm) Substance that surrounds organelles and located within a cell's plasma membrane and external to its nucleus. Also called ***protoplasm.***
Cytoskeleton Complex internal structure of cytoplasm consisting of microfilaments, microtubules, and intermediate filaments.

Dartos (DAR-tōs) The contractile tissue under the skin of the scrotum.
Deafness Lack of the sense of hearing or a significant hearing loss.
Debility (dē-BIL-i-tē) Weakness of tonicity in functions or organs of the body.
Decibel (DES-i-bel) **(db)** A unit that measures relative sound intensity (loudness).
Decidua (dē-SID-yoo-a) That portion of the endometrium of the uterus (all but the deepest layer) that is modified for pregnancy and shed after childbirth.
Deciduous (dē-SID-yoo-us) Falling off or being shed seasonally or at a particular stage of development. In the body, referring to the first set of teeth.
Decompression sickness A condition characterized by joint pains and neurologic symptoms; follows from a too-rapid reduction of environmental pressure or decompression, so that nitrogen that dissolved in body fluid under pressure comes out of solution as bubbles that form air emboli and occlude blood vessels. Also called ***caisson*** (KĀ-son) ***disease*** or ***bends.***
Decubitus (dē-KYOO-bi-tus) ***ulcer*** Tissue destruction due to a constant deficiency of blood to tissues overlying a bony projection that has been subjected to prolonged pressure against an object such as a bed, cast, or splint. Also called ***bedsore, pressure sore,*** or ***trophic ulcer.***

Decussation (dē'-ku-SĀ-shun) A crossing-over; usually refers to the crossing of most of the fibers in the large motor tracts to opposite sides in the medullary pyramids.

Deep Away from the surface of the body.

Deep fascia (FASH-ē-a) A sheet of connective tissue wrapped around a muscle to hold it in place.

Deep inguinal (IN-gwi-nal) ***ring*** A slitlike opening in the aponeurosis of the transversus abdominis muscle that represents the origin of the inguinal canal.

Deep-venous thrombosis (DVT) The presence of a thrombus in a vein, usually a deep vein of the lower extremities.

Defecation (def-e-KĀ-shun) The discharge of feces from the rectum.

Defibrillation (dē-fib-ri-LĀ-shun) Delivery of a very strong electrical current to the heart in an attempt to stop ventricular fibrillation.

Degeneration (dē-jen-er-Ā-shun) A change from a higher to a lower state; a breakdown in structure.

Deglutition (dē-gloo-TISH-un) The act of swallowing.

Dehydration (dē-hī-DRĀ-shun) Excessive loss of water from the body or its parts.

Delirium (de-LIR-ē-um) A transient disorder of abnormal cognition (perception, thinking, and memory) and disordered attention that is accompanied by disturbances of the sleep-wake cycle and psychomotor behavior (hyperactivity or hypoactivity of movements and speech). Also called ***acute confusional state (ACS).***

Delta cell A cell in the pancreatic islets (islets of Langerhans) in the pancreas that secretes somatostatin.

Dementia (de-MEN-shē-a) An organic mental disorder that results in permanent or progressive general loss of intellectual abilities such as impairment of memory, judgment, and abstract thinking and changes in personality; most common cause is Alzheimer's disease.

Demineralization (de-min'-er-al-i-ZĀ-shun) Loss of calcium and phosphorus from bones.

Denaturation (de-nā-chur-Ā-shun) Disruption of the tertiary structures of a protein by agents, such as heat, changes in pH, or other physical or chemical methods in which the protein loses its physical properties and biological properties.

Dendrite (DEN-drīt) A nerve cell process that carries a nerve impulse toward the cell body.

Dens (denz) Tooth.

Dental caries (KA-rēz) Gradual demineralization of the enamel and dentin of a tooth that may invade the pulp and alveolar bone. Also called ***tooth decay.***

Denticulate (den-TIK-yoo-lāt) Finely toothed or serrated; characterized by a series of small, pointed projections.

Dentin (DEN-tin) The osseous tissues of a tooth enclosing the pulp cavity.

Dentition (den-TI-shun) The eruption of teeth. The number, shape, and arrangement of teeth.

Deoxyribonucleic (dē-ok'-sē-ri'-bō-nyoo-KLĒ-ik) ***acid (DNA)*** A nucleic acid in the shape of a double helix constructed of nucleotides consisting of one of four nitrogenous bases (adenine, cytosine, guanine, or thymine), deoxyribose, and a phosphate group; encoded in the nucleotides is genetic information.

Depolarization (dē-pō-lar-i-ZĀ-shun) Used in neurophysiology to describe the reduction of voltage across a cell membrane; expressed as a movement toward less negative (more positive) voltages on the interior side of the cell membrane.

Depression (dē-PRESS-shun) Movement in which a part of the body moves downward.

Dermal papilla (pa-PILL-a) Fingerlike projection of the papillary region of the dermis that may contain blood capillaries or corpuscles of touch (Meissner's corpuscles).

Dermatology (der-ma-TOL-ō-jē) The medical specialty dealing with diseases of the skin.

Dermatome (DER-ma-tōm) An instrument for incising the skin or cutting thin transplants of skin. The cutaneous area developed from one embryonic spinal cord segment and receiving most of its innervation from one spinal nerve.

Dermis (DER-mis) A layer of dense connective tissue lying deep to the epidermis; the true skin or corium.

Descending colon (KŌ-lon) The part of the large intestine descending from the left colic (splenic) flexure to the level of the left iliac crest.

Detritus (de-TRI-tus) Particulate matter produced by or remaining after the wearing away or disintegration of a substance or tissue; scales, crusts, or loosened skin.

Detrusor (de-TROO-ser) ***muscle*** Muscle in the wall of the urinary bladder.

Developmental anatomy The study of development from the fertilized egg to the adult form. The branch of anatomy called embryology is generally restricted to the study of development from the fertilized egg through the eighth week in utero.

Diabetes insipidus (dī-a-BĒ-tēz in-SIP-i-dus) Condition caused by hyposecretion of antidiuretic hormone (ADH) and characterized by excretion of large amounts of urine and thirst.

Diabetes mellitus (MEL-i-tus) Hereditary condition caused by hyposecretion of insulin and characterized by hyperglycemia, increased urine production, excessive thirst, and excessive eating.

Diagnosis (dī-ag-NŌ-sis) Distinguishing one disease from another or determining the nature of a disease from signs and symptoms by inspection, palpation, laboratory tests, and other means.

Dialysis (dī-AL-i-sis) The process of separating small molecules from large by the difference in their rates of diffusion through a selectively permeable membrane.

Diapedesis (dī-a-pe-DĒ-sis) The passage of white blood cells through intact blood vessel walls.

Diaphragm (DĪ-a-fram) Any partition that separates one area from another, especially the dome-shaped skeletal muscle between the thoracic and abdominal cavities. Also a dome-shaped structure that fits over the cervix, usually with a spermicide, to prevent contraception.

Diaphysis (dī-AF-i-sis) The shaft of a long bone.

Diarrhea (dī-a-RĒ-a) Frequent defecation of liquid feces caused by increased motility of the intestines.

Diarthrosis (dī-'ar-THRŌ-sis) Articulation in which opposing bones move freely, as in a hinge joint.

Diastole (dī-AS-tō-lē) In the cardiac cycle, the phase of relaxation or dilation of the heart muscle, especially of the ventricles.

Diastolic (di-as-TOL-ik) ***blood pressure*** The force exerted by blood on arterial walls during ventricular relaxation; the lowest blood pressure measured in the large arteries, about 80 mm Hg under normal conditions for a young, adult male.

Diencephalon (dī'-en-SEF-a-lon) A part of the brain consisting primarily of the thalamus and the hypothalamus.

Differential (dif-fer-EN-shal) ***white blood cell count***

Determination of the number of each kind of white blood cell in a sample of 100 cells for diagnostic purposes.

Differentiation (dif'-e-ren'-shē-Ā-shun) Acquisition of specific functions different from those of the original general type.

Diffusion (dif-YOO-zhun) A passive process in which there is a net or greater movement of molecules or ions from a region of high concentration to a region of low concentration until equilibrium is reached.

Digestion (dī-JES-chun) The mechanical and chemical breakdown of food to simple molecules that can be absorbed and used by body cells.

Digital subtraction angiography (an-jē-OG-ra-fē) ***(DSA)*** A medical imaging technique that compares an x-ray image of the same artery of the body before and after a contrast substance containing iodine has been introduced intravenously.

Dilate (DĪ-lāte) To expand or swell.

Dilation (dī-LĀ-shun) ***and curettage*** (ku-re-TAZH) Following dilation of the uterine cervix, the uterine endometrium is scraped with a curette (spoon-shaped instrument). Also called a ***D and C.***

Diploid (DIP-loyd) Having the number of chromosomes characteristically found in the somatic cells of an organism. Symbolized $2n$.

Diplopia (di-PLŌ-pē-a) Double vision.

Disease Any change from a state of health.

Dislocation (dis-lō-KĀ-shun) Displacement of a bone from a joint with tearing of ligaments, tendons, and articular capsules. Also called ***luxation*** (luks-Ā-shun).

Dissect (DIS-sekt) To separate tissues and parts of a cadaver (corpse) or an organ for anatomical study.

Dissociation (dis'-sō-sē-Ā-shun) Separation of inorganic acids, bases, and salts into ions when dissolved in water. Also called ***ionization*** (ī'-on-i-ZĀ-shun).

Distal (DIS-tal) Farther from the attachment of an extremity to the trunk or a structure; farther from the point of origin.

Diuretic (dī-yoo-RET-ik) A chemcial that inhibits sodium reabsorption, reduces antidiuretic hormone (ADH) concentration, and increases urine volume by inhibiting facultative reabsorption of water.

Diurnal (dī-UR-nal) Daily.

Divergence (di-VER-jens) An anatomical arrangement in which the synaptic end bulbs of one presynaptic neuron terminate on several postsynaptic neurons.

Diverticulitis (dī-ver-tik-yoo-LĪ-tis) Inflammation of diverticula, saclike outpouchings of the colonic wall, when the muscularis becomes weak.

Diverticulum (dī-ver-TIK-yoo-lum) A sac or pouch in the wall of a canal or organ, especially in the colon.

Dominant gene A gene that is able to override the influence of the complementary gene on the homologous chromosome; the gene that is expressed.

Donor insemination (in-sem'-i-NĀ-shun) The deposition of seminal fluid within the vagina or cervix at a time during the menstrual cycle when pregnancy is most likely to occur. It may be homologous (using the husband's semen) or heterologous (using a donor's semen). Also called ***artificial insemination.***

Dorsal body cavity Cavity near the dorsal surface of the body that consists of a cranial cavity and vertebral canal.

Dorsal ramus (RĀ-mus) A branch of a spinal nerve containing motor and sensory fibers supplying the muscles, skin, and bones of the posterior part of the head, neck, and trunk.

Dorsiflexion (dor'-si-FLEK-shun) Bending the foot in the direction of the dorsum (upper surface).

Down-regulation Phenomenon in which there is a decrease in the number of receptors in response to an excess of a hormone or neurotransmitter.

Down's syndrome (DS) An inherited defect due to an extra copy of chromosome 21. Symptoms include mental retardation; a small skull, flattened from front to back; a short, flat nose; short fingers; and a widened space between the first two digits of the hand and foot. Also called ***trisomy 21.***

Dropsy (DROP-sē) A condition in which there is abnormal accumulation of water in the tissues and cavities.

Duct of Santorini *See* ***Accessory duct.***

Duct of Wirsung *See* ***Pancreatic duct.***

Ductus arteriosus (DUK-tus ar-tē-rē-Ō-sus) A small vessel connecting the pulmonary trunk with the aorta; found only in the fetus.

Ductus (vas) deferens (DEF-er-ens) The duct that conducts spermatozoa from the epididymis to the ejaculatory duct. Also called the ***seminal duct.***

Ductus epididymis (ep'-i-DID-i-mis) A tightly coiled tube inside the epididymis, distinguished into a head, body, and tail, in which spermatozoa undergo maturation.

Ductus venosus (ve-NŌ-sus) A small vessel in the fetus that helps the circulation bypass the liver.

Duodenal (doo-ō-DĒ-nal) ***gland*** Gland in the submucosa of the duodenum that secretes an alkaline mucus to protect the lining of the small intestine from the action of enzymes and to help neutralize the acid in chyme. Also called ***Brunner's*** (BRUN-erz) ***gland.***

Duodenal papilla (pa-PILL-a) An elevation on the duodenal mucosa that receives the hepatopancreatic ampulla (ampulla of Vater).

Duodenum (doo'-ō-DĒ-num) The first 25 cm (10 in.) of the small intestine.

Dura mater (DYOO-ra MĀ-ter) The outer membrane (meninx) covering the brain and spinal cord.

Dynamic equilibrium (ē-kwi-LIB-rē-um) The maintenance of body position, mainly the head, in response to sudden movements such as rotation.

Dynamic spatial reconstruction (DSR) A technique that has the ability to construct moving, three-dimensional, life-size images of all or part of an internal organ from any view desired.

Dysfunction (dis-FUNK-shun) Absence of complete normal function.

Dyslexia (dis-LEX-sē-a) Impairment of the brain's ability to translate images received from the eyes or ears into understandable language.

Dysmenorrhea (dis'-men-ō-RĒ-a) Painful menstruation.

Dysphagia (dis-FĀ-jē-a) Difficulty in swallowing.

Dysplasia (dis-PLĀ-zē-a) Change in the size, shape, and organization of cells due to chronic irritation or inflammation; may revert to normal if stress is removed or progresses to neoplasia.

Dyspnea (DISP-nē-a) Shortness of breath.

Dystocia (dis-TŌ-sē-a) Difficult labor due to factors such as pelvic deformities, malpositioned fetus, and premature rupture of fetal membranes.

Dystrophia (dis-TRŌ-fē-a) Progressive weakening of a muscle.

Dysuria (dis-SOO-rē-a) Painful urination.

Echocardiogram (ek-ō-KAR-dē-ō-gram) A procedure in which high-frequency sound waves directed at the heart are bounced back and the echoes are picked up by a transducer and converted to an image.

Ectoderm The outermost of the three primary germ layers that gives rise to the nervous system and the epidermis of skin and its derivatives.

Ectopic (ek-TOP-ik) Out of the normal location, as in ectopic pregnancy.

Eczema (EK-ze-ma) A skin rash characterized by itching, swelling, blistering, oozing, and scaling of the skin.

Edema (e-DĒ-ma) An abnormal accumulation of interstitial fluid.

Effective filtration pressure (Peff) Net pressure that expresses the relationship between the force that promotes glomerular filtration and the forces that oppose it.

Effector (e-FEK-tor) The organ of the body, either a muscle or a gland, that responds to a motor neuron impulse.

Efferent arteriole (EF-er-ent ar-TĒ-rē-ōl) A vessel of the renal vascular system that transports blood from the glomerulus to the peritubular capillary.

Efferent (EF-er-ent) ***ducts*** A series of coiled tubes that transport spermatozoa from the rete testis to the epididymis.

Efferent neuron (NOO-ron) A neuron that conveys nerve impulses from the brain and spinal cord to effectors that may be either muscles or glands. Also called a ***motor neuron.***

Effusion (e-FYOO-zhun) The escape of fluid from the lymphatics or blood vessels into a cavity or into tissues.

Ejaculation (e-jak-yoo-LĀ-shun) The reflex ejection or expulsion of semen from the penis.

Ejaculatory (e-JAK-yoo-la-tō'-rē) ***duct*** A tube that transports spermatozoa from the ductus (vas) deferens to the prostatic urethra.

Elasticity (e-las-TIS-i-tē) The ability of tissue to return to its original shape after contraction or extension.

Electrocardiogram (e-lek'-trō-KAR-dē-ō-gram) ***(ECG*** or ***EKG)*** A recording of the electrical changes that accompany the cardiac cycle and can be recorded on the surface of the body; may be resting, stress, or ambulatory.

Electroencephalogram (e-lek'-trō-en-SEF-a-lō-gram) ***(EEG)*** A recording of the electrical impulses of the brain to diagnose certain diseases (such as epilepsy), furnish information regarding sleep and wakefulness, and confirm brain death.

Electrolyte (ē-LEK-trō-līt) Any compound that separates into ions when dissolved in water and is able to conduct electricity.

Electromyography (e-lek'-trō-mī-OG-ra-fē) Evaluation of the electrical activity of resting and contracting muscle to ascertain causes of muscular weakness, paralysis, involuntary twitching, and abnormal levels of muscle enzymes; also used as part of biofeedback studies.

Electron transport chain A series of oxidation-reduction reactions in the catabolism of glucose that occur on the inner mitochondrial membrane and in which energy is released and transferred for storage to ATP.

Eleidin (el-Ē-i-din) A translucent substance found in the skin.

Elevation (el-e-VĀ-shun) Movement in which a part of the body moves upward.

Ellipsoidal (e-lip-SOY-dal) ***joint*** A synovial joint structured so that an oval-shaped condyle of one bone fits into an elliptical cavity of another bone, permitting side-to-side and back-and-forth movements, as at the joint at the wrist between the radius and carpals. Also called a ***condyloid*** (KON-di-loid) ***joint.***

Embolism (EM-bō-lizm) Obstruction or closure of a vessel by an embolus.

Embolus (EM-bō-lus) A blood clot, bubble of air, fat from broken bones, mass of bacteria, or other debris or foreign material transported by the blood.

Embryo (EM-brē-ō) The young of any organism in an early stage of development; in humans, the developing organism from fertilization to the end of the eighth week in utero.

Embryology (em'-brē-OL-ō-jē) The study of development from the fertilized egg to the end of the eighth week in utero.

Embryo transfer A type of *in vitro* fertilization in which semen is used to artificially inseminate a fertile secondary oocyte donor and the morula or blastocyst is then transferred from the donor to the infertile woman, who then carries it to term.

Emesis (EM-e-sis) Vomiting.

Emmetropia (em'-e-TRŌ-pē-a) The ideal optical condition of the eyes.

Emphysema (em'-fi-SĒ-ma) A swelling or inflation of air passages due to loss of elasticity in the alveoli.

Emulsification (ē-mul'-si-fi-KĀ-shun) The dispersion of large fat globules to smaller uniformly distributed particles in the presence of bile.

Enamel (e-NAM-el) The hard, white substance covering the crown of a tooth.

Endocardium (en-dō-KAR-dē-um) The layer of the heart wall, composed of endothelium and smooth muscle, that lines the inside of the heart and covers the valves and tendons that hold the valves open.

Endochondral ossification (en'-dō-KON-dral os'-i-fi-KĀ-shun) The replacement of cartilage by bone. Also called ***intracartilaginous*** (in'-tra-kar'-ti-LAJ-i-nus) ***ossification.***

Endocrine (EN-dō-krin) ***gland*** A gland that secretes hormones into the blood; a ductless gland.

Endocrinology (en'-dō-kri-NOL-ō-jē) The science concerned with the structure and functions of endocrine glands and the diagnosis and treatment of disorders of the endocrine system.

Endocytosis (en'-dō-sī-TŌ-sis) The uptake into a cell of large molecules and particles in which a segment of plasma membrane surrounds the substance, encloses it, and brings it in; includes phagocytosis, pinocytosis, and receptor-mediated endocytosis.

Endoderm The innermost of the three primary germ layers of the developing embryo that gives rise to the gastrointestinal tract, urinary bladder and urethra, and respiratory tract.

Endodontics (en'-dō-DON-tiks) The branch of dentistry concerned with the prevention, diagnosis, and treatment of diseases that affect the pulp, root, periodontal ligament, and alveolar bone.

Endogenous (en-DOJ-e-nus) Growing from or beginning within the organism.

Endolymph (EN-dō-lymf') The fluid within the membranous labyrinth of the inner ear.

Endometriosis (en'-dō-MĒ-trē-ō'-sis) The growth of endometrial tissue outside the uterus.

Endometrium (en'-dō-MĒ-trē-um) The mucous membrane lining the uterus.

Endomysium (em'-dō-MĪZ-ē-um) Invagination of the perimysium separating each individual muscle fiber (cell).

Endoneurium (en'-dō-NYOO-rē-um) Connective tissue wrapping around individual nerve fibers (cells).

Endoplasmic reticulum (en'-do-PLAZ-mik re-TIK-yoo-lum) ***(ER)*** A network of channels running through the cytoplasm of a cell that serves in intracellular transportation, support, storage, synthesis, and packaging of molecules. Portions of ER where ribosomes are attached to the outer surface are called ***granular*** or ***rough reticulum;*** portions that have no ribosomes are called ***agranular*** or ***smooth reticulum.***

End organ of Ruffini *See* ***Type II cutaneous mechanoreceptor.***

Endorphin (en-DOR-fin) A neuropeptide in the central nervous system that acts as a painkiller.

Endoscope (EN-dō-skōp') An illuminated tube with lenses used to look inside hollow organs such as the stomach (gastroscope) or urinary bladder (cystoscope).

Endoscopy (en-DOS-kō-pē) The visual examination of any cavity of the body using an endoscope, an illuminated tube with lenses.

Endosteum (en-DOS-tē-um) The membrane that lines the medullary cavity of bones, consisting of osteoprogenitor cells and scattered osteoclasts.

Endothelial-capsular (en-dō-THĒ-lē-al) ***membrane*** A filtration membrane in a nephron of a kidney consisting of the endothelium and basement membrane of the glomerulus and the epithelium of the visceral layer of the glomerular (Bowman's) capsule.

Endothelium (en'-dō-THĒ-lē-um) The layer of simple squamous epithelium that lines the cavities of the heart and blood and lymphatic vessels.

End-systolic (sis-TO-lik) ***volume (ESV)*** The volume of blood, about 50 to 60 ml, remaining in a ventricle following its systole (contraction).

Energy The capacity to do work.

Enkephalin (en-KEF-a-lin) A peptide found in the central nervous system that acts as a painkiller.

Enteroendocrine (en-ter-ō-EN-dō-krin) ***cell*** A stomach cell that secretes the hormone stomach gastrin.

Enterogastric (en-te-rō-GAS-trik) ***reflex*** A reflex that inhibits gastric secretion; initiated by food in the small intestine.

Enuresis (en'-yoo-RĒ-sis) Involuntary discharge of urine, complete or partial, after age 3.

Enzyme (EN-zīm) A substance that affects the speed of chemical changes; an organic catalyst, usually a protein.

Eosinophil (ē'-ō-SIN-ō-fil) A type of white blood cell characterized by granular cytoplasm readily stained by eosin.

Ependyma (e-PEN-de-ma) Neuroglial cells that line ventricles of the brain and probably assist in the circulation of cerebrospinal fluid (CSF). Also called ***ependymocytes*** (e-PEN-di-mō-sītz).

Epicardium (ep'-i-KAR-dē-um) The thin outer layer of the heart wall, composed of serous tissue and mesothelium. Also called the ***visceral pericardium.***

Epidemic (ep'-i-DEM-ik) A disease that occurs above the expected level among individuals in a population.

Epidemiology (ep'-i-dē-mē-OL-ō-jē) Medical science concerned with the occurrence and distribution of disease in human populations.

Epidermis (ep-i-DERM-is) The outermost, thinner layer of skin, composed of stratified squamous epithelium.

Epididymis (ep'-i-DID-i-mis) A comma-shaped organ that lies along the posterior border of the testis and contains the ductus epididymis, in which sperm undergo maturation. *Plural,* ***epididymides*** (ep'-i-DID-i-mi-dēz).

Epidural (ep'-i-DOO-ral) ***space*** A space between the spinal dura mater and the vertebral canal, containing loose connective tissue and a plexus of veins.

Epiglottis (ep'-i-GLOT-is) A large, leaf-shaped piece of cartilage lying on top of the larynx, with its "stem" attached to the thyroid cartilage and its "leaf" portion unattached and free to move up and down to cover the glottis (vocal folds and rima glottidis).

Epilepsy (EP-i-lep'-sē) Neurological disorder characterized by short, periodic attacks of motor, sensory, or psychological malfunction.

Epimysium (ep'-i-MĪZ-ē-um) Fibrous connective tissue around muscles.

Epinephrine (ep-ē-NEF-rin) Hormone secreted by the adrenal medulla that produces actions similar to those that result from sympathetic stimulation. Also called ***adrenaline*** (a-DREN-a-lin).

Epineurium (ep'-i-NYOO-rē-um) The outermost covering around the entire nerve.

Epiphyseal (ep'-i-FIZ-ē-al) ***line*** The remnant of the epiphyseal plate in a long bone.

Epiphyseal (ep'-i-FIZ-ē-al) ***plate*** The cartilaginous plate between the epiphysis and diaphysis that is responsible for the lengthwise growth of long bones.

Epiphysis (ē-PIF-i-sis) The end of a long bone, usually larger in diameter than the shaft (diaphysis).

Epiphysis cerebri (se-RĒ-brē) Pineal gland.

Episiotomy (e-piz'-ē-OT-ō-mē) A cut made with surgical scissors to avoid tearing of the perineum at the end of the second stage of labor.

Epistaxis (ep'-i-STAK-sis) Loss of blood from the nose due to trauma, infection, allergy, neoplasm, and bleeding disorders. Also called ***nosebleed.***

Epithelial (ep'-i-THĒ-lē-al) ***tissue*** The tissue that forms glands or the outer part of the skin and lines blood vessels, hollow organs, and passages that lead externally from the body.

Eponychium (ep'-ō-NIK-ē-um) Narrow band of stratum corneum at the proximal border of a nail that extends from the margin of the nail wall. Also called the ***cuticle.***

Erection (ē-REK-shun) The enlarged and stiff state of the penis (or clitoris) resulting from the engorgement of the spongy erectile tissue with blood.

Eructation (e-ruk'-TĀ-shun) The forceful expulsion of gas from the stomach. Also called ***belching.***

Erythema (er'-e-THĒ-ma) Skin redness usually caused by engorgement of the capillaries in the lower layers of the skin.

Erythematosus (er-i'-them-a-TŌ-sus) Pertaining to redness.

Erythrocyte (e-RITH-rō-sīt) Red blood cell.

Erythrocyte sedimentation rate (ESR) A test that measures the distances, in millimeters (mm), that red blood cells fall in one hour when a sample of blood is placed in a vertical tube; used as a screening test for infections, inflammations, and cancers.

Erythropoiesis (e-rith'-rō-poy-Ē-sis) The process by which erythrocytes (red blood cells) are formed.

Erythropoietin (e-rith'-rō-POY-ē-tin) A hormone formed from a plasma protein that stimulates erythrocyte (red blood cell) production.

Esophagus (e-SOF-a-gus) A hollow muscular tube connecting the pharynx and the stomach.

Essential amino acids Those 10 amino acids that cannot be synthesized by the human body at an adequate rate to meet its needs and therefore must be obtained from the diet.

Estrogens (ES-tro-jens) Female sex hormones produced by the ovaries concerned with the development and maintenance of female reproductive structures and secondary sex characteristics, fluid and electrolyte balance, and protein anabolism. Examples are β-estradiol, estrone, and estriol.

Etiology (ē'-tē-OL-ō-jē) The study of the causes of disease, including theories of origin and the organisms, if any, involved.

Euphoria (yoo-FŌR-ē-a) A subjectively pleasant feeling of well-being marked by confidence and assurance.

Eupnea (yoop-NĒ-a) Normal quiet breathing.

Eustachian tube *See* ***Auditory tube.***

Euthanasia (yoo'-tha-NĀ-zē-a) The practice of ending a life in case of incurable disease.

Eversion (ē-VER-zhun) The movement of the sole outward at the ankle joint.

Exacerbation (eg-zas'-er-BĀ-shun) An increase in the severity of symptoms or of disease.

Excitability (ek-sīt'-a-BIL-i-tē) The ability of muscle tissue to receive and respond to stimuli; the ability of nerve cells to respond to stimuli and convert them into nerve impulses.

Excitatory postsynaptic potential (EPSP) The slight decrease in negative voltage seen on the postsynaptic membrane when it is stimulated by a presynaptic terminal. The EPSP is a localized event that decreases in strength from the point of excitation.

Excrement (EKS-kre-ment) Material cast out from the body as waste, especially fecal matter.

Excretion (eks-KRĒ-shun) The process of eliminating waste products from a cell, tissue, or the entire body; or the products excreted.

Exocrine (EK-sō-krin) ***gland*** A gland that secretes substances into ducts that empty at covering or lining epithelium or directly onto a free surface.

Exocytosis (ex'-ō-sī-TŌ-sis) A process of discharging cellular products too big to go through the membrane. Particles for export are enclosed by Golgi membranes when they are synthesized. Vesicles pinch off from the Golgi complex and carry the enclosed particles to the interior surface of the cell membrane, where the vesicle membrane and plasma membrane fuse and the contents of the vesicle are discharged.

Exogenous (ex-SOJ-e-nus) Originating outside an organ or part.

Exon (EX-on) A region of DNA that codes for synthesis of a protein.

Exophthalmic goiter (ek'-sof-THAL-mik GOY-ter) An autoimmune disease that may result in hypersecretion of thyroid hormones characterized by protrusion of the eyeballs (exophthalmos) and an enlarged thyroid (goiter). Also called ***Graves disease.***

Exophthalmos (ek'-sof-THAL-mus) An abnormal protrusion or bulging of the eyeball.

Expiration (ek-spi-RĀ-shun) Breathing out; expelling air from the lungs into the atmosphere. Also called ***exhalation.***

Expiratory (eks-PĪ-ra-tō-rē) ***reserve volume*** The volume of air in excess of tidal volume that can be exhaled forcibly; about 1,200 ml.

Extensibility (ek-sten'-si-BIL-i-tē) The ability of muscle tissue to be stretched when pulled.

Extension (ek-STEN-shun) An increase in the angle between two bones; restoring a body part to its anatomical position after flexion.

External Located on or near the surface.

External auditory (AW-di-tōr-ē) ***canal*** or ***meatus*** (mē-Ā-tus) A curved tube in the temporal bone that leads to the middle ear.

External ear The outer ear, consisting of the pinna, external auditory canal, and tympanic membrane or eardrum.

External nares (NA-rēz) The external nostrils, or the openings into the nasal cavity on the exterior of the body.

External respiration The exchange of respiratory gases between the lungs and blood.

Exteroceptor (eks'-ter-ō-SEP-tor) A receptor adapted for the reception of stimuli from outside the body.

Extracellular fluid (ECF) Fluid outside body cells, such as interstitial fluid and plasma.

Extracorporeal (eks'-tra-kor-PŌ-rē-al) The circulation of blood outside the body.

Extravasation (eks-trav-a-SĀ-shun) The escape of fluid, especially blood, lymph, or serum, from a vessel into the tissues.

Extrinsic (ek-STRIN-sik) Of external origin.

Extrinsic clotting pathway Sequence of reactions leading to blood clotting that is initiated by the release of a substance (tissue factor or thromboplastin) *outside* blood itself, from damaged blood vessels or surrounding tissues.

Exudate (EKS-yoo-dāt) Escaping fluid or semifluid material that oozes from a space that may contain serum, pus, and cellular debris.

Eyebrow The hairy ridge above the eye.

Face The anterior aspect of the head.

Facilitated diffusion (fa-SIL-i-tā-ted dif-YOO-zhun) Diffusion in which a substance not soluble by itself in lipids is transported across a selectively permeable membrane by combining with a carrier substance.

Facilitation (fa-sil-i-TĀ-shun) The process in which a nerve cell membrane is partially depolarized by a subliminal stimulus so that a subsequent subliminal stimulus can further depolarize the membrane to reach the threshold of nerve impulse initiation.

Facultative (FAK-ul-tā-tive) ***water reabsorption*** The absorption of water from distal convoluted tubules and collecting tubules of nephrons in response to antidiuretic hormone (ADH).

Falciform ligament (FAL-si-form LIG-a-ment) A sheet of parietal peritoneum between the two principal lobes of the liver. The ligamentum teres, or remnant of the umbilical vein, lies within its fold.

Fallopian tube *See* ***Uterine tube.***

Falx cerebelli (FALKS ser'-e-BEL-lē) A small triangular process of the dura mater attached to the occipital bone in the posterior cranial fossa and projecting inward between the two cerebellar hemispheres.

Falx cerebri (SER-e-brē) A fold of the dura mater extending down into the longitudinal fissure between the two cerebral hemispheres.

Fascia (FASH-ē-a) A fibrous membrane covering, supporting, and separating muscles.

Fascicle (FAS-i-kul) A small bundle or cluster, especially of nerve or muscle fibers (cells). Also called a ***fasciculus*** (fa-SIK-yoo-lus). *Plural,* ***fasciculi*** (fa-SIK-yoo-lī).

Fasciculation (fa-sik'-yoo-LĀ-shun) Involuntary brief twitch of a muscle that is visible under the skin and is not associated with the movement of the affected muscle.

Fat A lipid compound formed from one molecule of glycerol and three molecules of fatty acids; the body's most highly concentrated source of energy. Adipose tissue, composed of adipocytes specialized for fat storage and present in the form of soft pads between various organs for support, protection, and insulation.

Fauces (FAW-sēz) The opening from the mouth into the pharynx.

Febrile (FĒ-bril) Feverish; pertaining to a fever.

Feces (FĒ-sēz) Material discharged from the rectum and made up of bacteria, excretions, and food residue. Also called ***stool.***

Feedback system A circular sequence of events in which information about the status of a situation is continually reported (fed back) to a central control region.

Feeding (hunger) center A cluster of neurons in the lateral nuclei of the hypothalamus that, when stimulated, brings about feeding.

Fenestration (fen-e-STRĀ-shun) Surgical opening made into the labyrinth of the ear for some conditions of deafness.

Fertilization (fer'-ti-li-ZĀ-shun) Penetration of a secondary oocyte by a spermatozoon and subsequent union of the nuclei of the cells.

Fetal (FĒ-tal) ***alcohol syndrome (FAS)*** Term applied to the effects of intrauterine exposure to alcohol, such as slow growth, defective organs, and mental retardation.

Fetal circulation The cardiovascular system of the fetus, including the placenta and special blood vessels involved in the exchange of materials between fetus and mother.

Fetus (FĒ-tus) The latter stages of the developing young of an animal; in humans, the developing organism in utero from the beginning of the third month to birth.

Fever An elevation in body temperature above its normal temperature of 37°C (98.6°F).

Fibrillation (fi-bri-LĀ-shun) Involuntary brief twitch of a muscle that is not visible under the skin and is not associated with movement of the affected muscle.

Fibrin (FĪ-brin) An insoluble protein that is essential to blood clotting; formed from fibrinogen by the action of thrombin.

Fibrinogen (fī-BRIN-ō-jen) A high-molecular-weight protein in the blood plasma that by the action of thrombin is converted to fibrin.

Fibrinolysis (fī-brin-OL-i-sis) Dissolution of a blood clot by the action of a proteolytic enzyme that converts insoluble fibrin into a soluble substance.

Fibroblast (FĪ-brō-blast) A large, flat cell that forms collagenous and elastic fibers and intercellular substance of loose connective tissue.

Fibrocyte (FĪ-brō-sīt) A mature fibroblast that no longer produces fibers or intercellular substance in connective tissue.

Fibromyalgia (fī-bro-mī-AL-jē-a) Groups of common nonarticular rheumatic disorders characterized by pain, tenderness, and stiffness of muscles, tendons, and surrounding tissues. Examples of fibromyalgia are lumbago and charleyhorse.

Fibroplasia (fī-brō-PLĀ-zē-a) Period of scar tissue formation.

Fibrosis (fī-BRŌ-sis) Abnormal formation of fibrous tissue.

Fibrous (FĪ-brus) ***joint*** A joint that allows little or no movement, such as a suture and syndesmosis.

Fibrous tunic (TOO-nik) The outer coat of the eyeball, made up of the posterior sclera and the anterior cornea.

Fight-or-flight response The effect of the stimulation of the sympathetic division of the autonomic nervous system.

Filiform papilla (FIL-i-form pa-PIL-a) One of the conical projections that are distributed in parallel rows over the anterior two-thirds of the tongue and contain no taste buds.

Filtrate (fil-TRĀT) The fluid produced when blood is filtered by the endothelial-capsular membrane.

Filtration (fil-TRĀ-shun) The passage of a liquid through a filter or membrane that acts like a filter.

Filtration fraction The percentage of plasma entering the nephrons that actually becomes glomerular filtrate.

Filum terminale (FĪ-lum ter-mi-NAL-ē) Nonnervous fibrous tissue of the spinal cord that extends inferiorly from the conus medullaris to the coccyx.

Fimbriae (FIM-brē-ē) Fingerlike structures, especially the lateral ends of the uterine (Fallopian) tubes.

Fissure (FISH-ur) A groove, fold, or slit that may be normal or abnormal.

Fistula (FIS-choo-la) An abnormal passage between two organs or between an organ cavity and the outside.

Fixator A muscle that stabilizes the origin of the prime mover so that the prime mover can act more efficiently.

Fixed macrophage (MAK-rō-fāj) Stationary phagocytic cell found in the liver, lungs, brain, spleen, lymph nodes, subcutaneous tissue, and bone marrow. Also called a ***histiocyte*** (HIS-tē-ō-sīt).

Flaccid (FLAS-sid) Relaxed, flabby, or soft; lacking muscle tone.

Flagellum (fla-JEL-um) A hairlike, motile process on the extremity of a bacterium or protozoan. *Plural,* ***flagella*** (fla-JEL-a).

Flatfoot A condition in which the ligaments and tendons of the arches of the foot are weakened and the height of the longitudinal arch decreases.

Flatus (FLĀ-tus) Air (gas) in the stomach or intestines, commonly used to denote passage of gas rectally.

Flexion (FLEK-shun) A folding movement in which there is a decrease in the angle between two bones.

Flexor reflex A protective reflex in which flexor muscles are stimulated while extensor muscles are inhibited.

Fluid mosaic (mō-ZĀ-ik) ***model*** Model of plasma membrane structure that depicts the membrane as a mosaic of proteins that move laterally in the phospholipid bilayer.

Fluoroscope (FLOOR-ō-skōp) An instrument for visual observation of the body by means of x-ray.

Follicle (FOL-i-kul) A small secretory sac or cavity.

Follicle-stimulating (FOL-i-kul) ***hormone (FSH)*** Hormone secreted by the adenohypophysis (anterior lobe) of the pituitary gland that initiates development of ova and stimulates the ovaries to secrete estrogens in females and initiates sperm production in males.

Fontanel (fon'-ta-NEL) A membrane-covered spot where bone formation is not yet complete, especially between the cranial bones of an infant's skull.

Foot The terminal part of the lower extremity.

Foramen (fo-RĀ-men) A passage or opening; a communication between two cavities of an organ or a hole in a bone for passage of vessels or nerves.

Foramen ovale (ō-VAL-ē) An opening in the fetal heart in the septum between the right and left atria. A hole in the greater wing of the sphenoid bone that transmits the mandibular branch of the trigeminal (V) nerve.

Forearm (FOR-arm) The part of the upper extremity between the elbow and the wrist.

Fornix (FOR-niks) An arch or fold; a tract in the brain made up of association fibers, connecting the hippocampus with the mammillary bodies; a recess around the cervix of the uterus where it protrudes into the vagina.

Fossa (FOS-a) A furrow or shallow depression.

Fourth ventricle (VEN-tri-kul) A cavity within the brain lying between the cerebellum and the medulla and pons.

Fracture (FRAK-chur) Any break in a bone.

Fragile X syndrome Inherited disorder characterized by learning difficulties, mental retardation, and physical abnormalities; due to a defective gene on the X chromosome.

Frenulum (FREN-yoo-lum) A small fold of mucous membrane that connects two parts and limits movement.

Frontal plane A plane at a right angle to a midsagittal plane that divides the body or organs into anterior and posterior portions. Also called a ***coronal*** (kō-RŌ-nal) ***plane.***

Fulminate (FUL-mi-nāt') To occur suddenly with great intensity.

Functional residual (re-ZID-yoo-al) ***volume*** The sum of residual volume plus expiratory reserve volume; about 2,400 ml.

Fundus (FUN-dus) The part of a hollow organ farthest from the opening.

Fungiform papilla (FUN-ji-form pa-PIL-a) A mushroomlike elevation on the upper surface of the tongue appearing as a red dot; papillae contain taste buds.

Furuncle (FYOOR-ung-kul) A boil; painful nodule caused by bacterial infection and inflammation of a hair follicle or sebaceous (oil) gland.

Gallbladder A small pouch that stores bile, located under the liver, which is filled with bile and emptied via the cystic duct.

Gallstone A concretion, usually consisting of cholesterol, formed anywhere between bile canaliculi in the liver and the hepatopancreatic ampulla (ampulla of Vater), where bile enters the duodenum. Also called a ***biliary calculus.***

Gamete (GAM-ēt) A male or female reproductive cell; the spermatozoon or ovum.

Gamete intrafallopian transfer (GIFT) A type of in vitro fertilization in which aspirated secondary oocytes are combined with a solution containing sperm outside the body and then the secondary oocytes are inserted into the uterine (Fallopian) tubes.

Ganglion (GANG-glē-on) A group of nerve cell bodies that lie outside the central nervous system. *Plural,* ***ganglia*** (GANG-glē-a).

Gangrene (GANG-rēn) Death and rotting of a considerable mass of tissue that usually is caused by interruption of blood supply followed by bacterial (*Clostridium*) invasion.

Gastroenterology (gas'-trō-en'-ter-OL-ō-jē) The medical specialty that deals with the structure, function, diagnosis, and treatment of diseases of the stomach and intestines.

Gastrointestinal (gas-trō-in-TES-ti-nal) ***(GI) tract*** A continuous tube running through the ventral body cavity extending from the mouth to the anus. Also called the ***alimentary*** (al'-i-MEN-tar-ē) ***canal.***

Gastroscopy (gas-TROS-kō-pē) Diagnostic procedure in which the interior of the stomach is examined with a gastroscope to detect lesions, biopsy lesions, stop bleeding, and remove foreign objects.

Gastrulation (gas'-troo-LĀ-shun) The various movements of groups of cells that lead to the establishment of the primary germ layers.

Gavage (ga-VAZH) Feeding through a tube passed through the esophagus and into the stomach.

Gene (jēn) Biological unit of heredity; an ultramicroscopic, self-reproducing DNA particle located in a definite position on a particular chromosome.

General adaptation syndrome (GAS) Wide-ranging set of bodily changes triggered by a stressor that gears the body to meet an emergency.

Generator potential The graded depolarization that results in a change in the resting membrane potential in a receptor (specialized neuronal ending).

Genetic engineering The manufacture and manipulation of genetic material.

Genetics The study of heredity.

Genital herpes (JEN-i-tal HER-pēz) A sexually transmitted disease caused by type II herpes simplex virus.

Genitalia (jen'-i-TĀL-ya) Reproductive organs.

Genome (JĒ-nōm) The complete gene complement of an organism.

Genotype (JĒ-nō-tīp) The total hereditary information carried by an individual; the genetic makeup of an organism.

Geriatrics (jer'-ē-AT-riks) The branch of medicine devoted to the medical problems and care of elderly persons.

Germinal (JER-mi-nal) ***epithelium*** A layer of epithelial cells that covers the ovaries and lines the seminiferous tubules of the testes.

Germinativum (jer'-mi-na-TĒ-vum) Skin layers where new cells are germinated.

Gestation (jes-TĀ-shun) The period of intrauterine fetal development.

Giantism (GĪ-an-tizm) Condition caused by hypersecretion of human growth hormone (hGH) during childhood characterized by excessive bone growth and body size. Also called ***gigantism.***

Gingivae (jin-JI-vē) Gums. They cover the alveolar processes of the mandible and maxilla and extend slightly into each socket.

Gingivitis (jin'-je-VĪ-tis) Inflammation of the gums.

Gland Single or group of specialized epithelial cells that secrete substances.

Glans penis (glanz PĒ-nis) The slightly enlarged region at the distal end of the penis.

Glaucoma (glaw-KŌ-ma) An eye disorder in which there is increased intraocular pressure due to an excess of aqueous humor.

Gliding joint A synovial joint having articulating surfaces that are usually flat, permitting only side-to-side and back-and-forth movements, as between carpal bones, tarsal bones, and the scapula and clavicle. Also called an ***arthrodial*** (ar-THRŌ-dē-al) ***joint.***

Glomerular (glō-MER-yoo-lar) ***capsule*** A double-walled globe at the proximal end of a nephron that encloses the glomerulus. Also called ***Bowman's*** (BŌ-manz) ***capsule.***

Glomerular filtration The first step in urine formation in which substances in blood are filtered at the endothelial-capsular membrane and the filtrate enters the proximal convoluted tubule of a nephron.

Glomerular filtration rate (GFR) The total volume of fluid that enters all the glomerular (Bowman's) capsules of the kidneys in 1 min; about 125 ml/min.

Glomerulonephritis (glō-mer-yoo-lō-nef-RĪ-tis) Inflammation of the glomeruli of the kidney that increases the permeability of the endothelial-capsular membrane and permits blood cells and proteins to enter the filtrate. Also called ***Bright's disease.***

Glomerulus (glō-MER-yoo-lus) A rounded mass of nerves or blood vessels, especially the microscopic tuft of capillaries that is surrounded by the glomerular (Bowman's) capsule of each kidney tubule.

Glottis (GLOT-is) The vocal folds (true vocal cords) in the larynx and the space between them (rima glottidis).

Glucagon (GLOO-ka-gon) A hormone produced by the alpha cells of the pancreas that increases the blood glucose level.

Glucocorticoids (gloo-kō-KOR-ti-koyds) A group of hormones of the adrenal cortex.

Gluconeogenesis (gloo'-kō-nē'-ō-JEN-e-sis) The conversion of a substance other than carbohydrate into glucose.

Glucose (GLOO-kōs) A six-carbon sugar, $C_6H_{12}O_6$; the major energy source for every cell type in the body. Its metabolism is possible by every known living cell for the production of ATP.

Glycogen (GLĪ-kō-jen) A highly branched polymer of glucose containing thousands of subunits; functions as a compact store of glucose molecules in liver and muscle fibers (cells).

Glycogenesis (glī'-kō-JEN-e-sis) The process by which many molecules of glucose combine to form a molecule called glycogen.

Glycogenolysis (glī-kō-je-NOL-i-sis) The process of converting glycogen to glucose.

Glycosuria (glī'-kō-SOO-rē-a) The presence of glucose in the urine; may be temporary or pathological. Also called ***glucosuria.***

Gnostic (NOS-tik) Pertaining to the faculties of perceiving and recognizing.

Gnostic area Sensory area of the cerebral cortex that receives and integrates sensory input from various parts of the brain so that a common thought can be formed.

Goblet cell A goblet-shaped unicellular gland that secretes mucus. Also called a ***mucus cell.***

Goiter (GOY-ter) An enlargement of the thyroid gland.

Golgi (GOL-jē) ***complex*** An organelle in the cytoplasm of cells consisting of four to eight flattened channels, stacked upon one another, with expanded areas at their ends; functions in packaging secreted proteins, lipid secretion, and carbohydrate synthesis.

Golgi tendon organ *See* ***Tendon organ.***

Gomphosis (gom-FŌ-sis) A fibrous joint in which a cone-shaped peg fits into a socket.

Gonad (GŌ-nad) A gland that produces gametes and hormones; the ovary in the female and the testis in the male.

Gonadocorticoids (gō-na-dō-KOR-ti-koydz) Sex hormones secreted by the adrenal cortex.

Gonadotropic (gō'-nad-ō-TRŌ-pik) ***hormone*** A hormone that regulates the functions of the gonads.

Gonorrhea (gon'-ō-RĒ-a) Infectious, sexually transmitted disease caused by the bacterium *Neisseria gonorrhoeae* and characterized by inflammation of the urogenital mucosa, discharge of pus, and painful urination.

Gout (gowt) Hereditary condition associated with excessive uric acid in the blood; the acid crystallizes and deposits in joints, kidneys, and soft tissue.

Graafian follicle *See* ***Vesicular ovarian follicle.***

Gray commissure (KOM-i-shur) A narrow strip of gray matter connecting the two lateral gray masses within the spinal cord.

Gray matter Area in the central nervous system and ganglia consisting of nonmyelinated nerve tissue.

Gray ramus communicans (RĀ-mus kō-MYOO-ni-kans) A short nerve containing postganglionic sympathetic fibers; the cell bodies of the fibers are in a sympathetic chain ganglion, and the nonmyelinated axons run by way of the gray ramus to a spinal nerve and then to the periphery to supply smooth muscle in blood vessels, arrector pili muscles, and sweat glands. *Plural,* ***rami communicantes*** (RĀ-mē kō-myoo-ni-KAN-tēz).

Greater omentum (ō-MEN-tum) A large fold in the serosa of the stomach that hangs down like an apron over the front of the intestines.

Greater vestibular (ves-TIB-yoo-lar) ***glands*** A pair of glands on either side of the vaginal orifice that open by a duct into the space between the hymen and the labia minora. Also called ***Bartholin's*** (BAR-to-linz) ***glands.***

Groin (groyn) The depression between the thigh and the trunk; the inguinal region.

Gross anatomy The branch of anatomy that deals with structures that can be studied without using a microscope. Also called ***macroscopic anatomy.***

Growth An increase in size due to an increase in the number of cells or an increase in the size of existing cells as internal components increase in size or an increase in the size of intercellular substances.

Gustatory (GUS-ta-tō'-rē) Pertaining to taste.

Gynecology (gī'-ne-KOL-ō-jē) The branch of medicine dealing with the study and treatment of disorders of the female reproductive system.

Gynecomastia (gīn'-e-kō-MAS-tē-a) Excessive growth (benign) of the male mammary glands due to secretion of sufficient estrogens by an adrenal gland tumor (feminizing adenoma).

Gyrus (JĪ-rus) One of the folds of the cerebral cortex of the brain. *Plural,* ***gyri*** (JĪ-rī). Also called a ***convolution.***

Hair A threadlike structure produced by hair follicles that develops in the dermis. Also called ***pilus*** (PI-lus).

Hair follicle (FOL-li-kul) Structure composed of epithelium surrounding the root of a hair from which hair develops.

Hair root plexus (PLEK-sus) A network of dendrites arranged around the root of a hair as free or naked nerve endings that are stimulated when a hair shaft is moved.

Haldane (HAWL-dān) ***effect*** In the presence of oxygen, less carbon dioxide binds in the blood because when oxygen combines with hemoglobin, the hemoglobin becomes a stronger acid, which combines with less carbon dioxide.

Hallucination (ha-loo'-si-NĀ-shun) A sensory perception of something that does not really exist in the world, that is, a sensory experience created from within the brain.

Hand The terminal portion of an upper extremity, including the carpus, metacarpus, and phalanges.

Haploid (HAP-loyd) Having half the number of chromosomes characteristically found in the somatic cells of an organism; characteristic of mature gametes. Symbolized *n.*

Hard palate (PAL-at) The anterior portion of the roof of the mouth, formed by the maxillae and palatine bones and lined by mucous membrane.

Haustra (HAWS-tra) The sacculated elevations of the colon.

Haversian canal *See* ***Central canal.***

Haversian system *See* ***Osteon.***

Head The superior part of a human, cephalic to the neck. The superior or proximal part of a structure.

Heart A hollow muscular organ lying slightly to the left of the midline of the chest that pumps the blood through the cardiovascular system.

Heart block An arrhythmia (dysrhythmia) of the heart in which the atria and ventricles contract independently because of a blocking of electrical impulses through the heart at a critical point in the conduction system.

Heartburn Burning sensation in the esophagus due to reflux of hydrochloric acid (HCl) from the stomach.

Heart-lung machine A device that pumps blood, functioning as a heart, and removes carbon dioxide from blood and oxygenates it, functioning as lungs; used during heart transplantation, open-heart surgery, and coronary artery bypass grafting.

Heart murmur (MER-mer) An abnormal sound that consists of a flow noise that is heard before the normal lubb-dupp or that may mask normal heart sounds.

Heat exhaustion Condition characterized by cool, clammy skin, profuse perspiration, and fluid and electrolyte (especially salt) loss that results in muscle cramps, dizziness, vomiting, and fainting. Also called ***heat prostration.***

Heatstroke Condition produced when the body cannot easily lose heat and characterized by reduced perspiration and elevated body temperature. Also called ***sunstroke.***

Heimlich maneuver *See* ***Abdominal thrust maneuver.***

Hematocrit (hē-MAT-ō-krit) ***(Hct)*** The percentage of blood made up of red blood cells. Usually calculated by centrifuging a blood sample in a graduated tube and then reading off the volume of red blood cells and total blood.

Hematology (hē'-ma-TOL-ō-jē) The study of blood.

Hematoma (hē'-ma-TŌ-ma) A tumor or swelling filled with blood.

Hematopoiesis (hem'-a-tō-poy-Ē-sis) Blood cell production occurring in the red marrow of bones. Also called ***hemopoiesis*** (hē-mō-poy-Ē-sis).

Hematuria (hē'-ma-TOOR-ē-a) Blood in the urine.

Hemiballismus (hem'-i-ba-LIZ-mus) Violent muscular restlessness of half of the body, especially of the upper extremity.

Hemiplegia (hem-i-PLĒ-jē-a) Paralysis of the upper extremity, trunk, and lower extremity on one side of the body.

Hemocytoblast (hē'-mō-SĪ-tō-blast) Immature stem cell in bone marrow that develops along different lines into all the different mature blood cells.

Hemodialysis (hē'-mō-dī-AL-i-sis) Filtering of the blood by means of an artificial device so that certain substances are removed from the blood as a result of the difference in rates of their diffusion through a selectively permeable membrane while the blood is being circulated outside the body.

Hemodynamics (hē-mō-dī-NA-miks) The study of factors and forces that govern the flow of blood through blood vessels.

Hemoglobin (hē'-mō-GLŌ-bin) ***(Hb)*** A substance in erythrocytes (red blood cells) consisting of the protein globin and the iron-containing red pigment heme and constituting about 33 percent of the cell volume; involved in the transport of oxygen and carbon dioxide.

Hemolysis (hē-MOL-i-sis) The escape of hemoglobin from the interior of the red blood cell into the surrounding medium; results from disruption of the integrity of the cell membrane by toxins or drugs, freezing or thawing, or hypotonic solutions.

Hemolytic disease of the newborn A hemolytic anemia of a newborn child that results from the destruction of the infant's red blood cells by antibodies produced by the mother; usually the antibodies are due to an Rh blood type incompatibility. Also called ***erythroblastosis fetalis*** (e-rith'-rō-blas-TŌ-sis fe-TAL-is).

Hemophilia (hē'-mō-FĒL-ē-a) A hereditary blood disorder where there is a deficient production of certain factors involved in blood clotting, resulting in excessive bleeding into joints, deep tissues, and elsewhere.

Hemoptysis (hē-MOP-ti-sis) Spitting of blood from the respiratory tract.

Hemorrhage (HEM-or-rij) Bleeding; the escape of blood from blood vessels, especially when it is profuse.

Hemorrhoids (HEM-ō-royds) Dilated or varicosed blood vessels (usually veins) in the anal region. Also called ***piles.***

Hemostasis (hē-MŌS-tā-sis) The stoppage of bleeding.

Hemostat (HĒ-mō-stat) An agent or instrument used to prevent the flow or escape of blood.

Hepatic (he-PAT-ik) Refers to the liver.

Hepatic duct A duct that receives bile from the bile capillaries. Small hepatic ducts merge to form the larger right and left hepatic ducts that unite to leave the liver as the common hepatic duct.

Hepatic portal circulation The flow of blood from the gastrointestinal organs to the liver before returning to the heart.

Hepatitis (hep-a-TĪ-tis) Inflammation of the liver due to a virus, drugs, and chemicals.

Hepatopancreatic (hep'-a-tō-pan'-krē-A-tik) ***ampulla*** A small, raised area in the duodenum where the combined common bile duct and main pancreatic duct empty into the duodenum. Also called the ***ampulla of Vater*** (VA-ter).

Hering–Breuer reflex *See* ***Inflation reflex.***

Hernia (HER-nē-a) The protrusion or projection of an organ or part of an organ through a membrane or cavity wall, usually the abdominal cavity.

Herniated (her'-nē-Ā-ted) ***disc*** A rupture of an intervertebral disc so that the nucleus pulposus protrudes into the vertebral cavity. Also called a ***slipped disc.***

Heterocrine (HET-er-ō-krin) ***gland*** A gland, such as the pancreas, that is both an exocrine and an endocrine gland.

Heterozygous (he-ter-ō-ZĪ-gus) Possessing a pair of different genes on homologous chromosomes for a particular hereditary characteristic.

Hiatus (hī-Ā-tus) An opening; a foramen.

High altitude sickness Disorder caused by decreased levels of alveolar PO_2 as altitude increases and characterized by headache, fatigue, insomnia, shortness of breath, nausea, and dizziness. Also called ***acute mountain sickness.***

Hilus (HĪ-lus) An area, depression, or pit where blood vessels and nerves enter or leave an organ. Also called a ***hilum.***

Hinge joint A synovial joint in which a convex surface of one bone fits into a concave surface of another bone, such as the elbow, knee, ankle, and interphalangeal joints. Also called a ***ginglymus*** (JIN-gli-mus) ***joint.***

Hirsutism (HER-soot-izm) An excessive growth of hair in females and children, with a distribution similar to that in adult males, due to the conversion of vellus hairs into large terminal hairs in response to higher-than-normal levels of androgens.

Histamine (HISS-ta-mēn) Substance found in many cells, especially mast cells, basophils, and platelets, released when the cells are injured; results in vasodilation, increased permeability of blood vessels, and bronchiole constriction.

Histocompatibility (his'-tō-kom-pat-i-BIL-i-tē) ***testing*** Comparison of human leucocyte associated (HLA) antigens between donor and recipient to determine histocompatibility, the degree of compatibility between the two. Also called ***HLA antigen typing*** or ***tissue typing.***

Histology (hiss-TOL-ō-jē) Microscopic study of the structure of tissues.

Hives (HĪVZ) Condition of the skin marked by reddened elevated patches that are often itchy; may be caused by infections, trauma, medications, emotional stress, food additives, and certain foods.

Hodgkin's disease (HD) A malignant disorder, usually arising in lymph nodes.

Holocrine (HŌL-ō-krin) ***gland*** A type of gland in which the entire secreting cell, along with its accumulated secretions, makes up the secretory product of the gland, as in the sebaceous (oil) glands.

Holter monitor Electrocardiograph worn by a person while going about everyday routines.

Homeostasis (hō'-mē-ō-STĀ-sis) The condition in which the body's internal environment remains relatively constant, within physiological limits.

Homologous (hō-MOL-ō-gus) Correspondence of two organs in structure, position, and origin.

Homologous chromosomes Two chromosomes that belong to a pair. Also called ***homologues.***

Horizontal plane A plane that runs parallel to the ground and divides the body or organs into superior and inferior portions. Also called a ***transverse plane.***

Hormone (HOR-mōn) A secretion of endocrine tissue that alters the physiological activity of target cells of the body.

Horn Principal area of gray matter in the spinal cord.

Human chorionic gonadotropin (hCG) (kō-rē-ON-ik gō-nad-ō-TRŌ-pin) A hormone produced by the developing placenta that maintains the corpus luteum.

Human chorionic somatomammotropin (sō-mat-ō-mam-ō-TRŌ-pin) ***(hCS)*** A hormone produced by the chorion of the placenta that may stimulate breast tissue for lactation, enhance body growth, and regulate metabolism.

Human growth hormone (hGH) Hormone secreted by the adenohypophysis (anterior lobe) of the pituitary that brings about growth of body tissues, especially skeletal and muscular. Also known as ***somatotropin*** and ***somatotropic hormone (STH).***

Human leucocyte associated (HLA) antigens Surface proteins on white blood cells and other nucleated cells that are unique for each person (except for identical twins) and are used to type tissues and help prevent rejection.

Humoral (YOO-mor-al) ***immunity*** That component of immunity in which lymphocytes (B cells) develop into plasma cells that produce antibodies that destroy antigens. Also called ***antibody-mediated immunity.***

Hunger center A cluster of neurons in the lateral nuclei of the hypothalamus that, when stimulated, brings about feeding.

Hyaluronic (hī'-a-loo-RON-ik) ***acid*** A viscous, amorphous extracellular material that binds cells together, lubricates joints, and maintains the shape of the eyeballs.

Hyaluronidase (hī'-a-loo-RON-i-dās) An enzyme that breaks down hyaluronic acid, increasing the permeability of connective tissues by dissolving the substances that hold body cells together.

Hydrocele (HĪ-drō-sēl) A fluid-containing sac or tumor. Specifically, a collection of fluid formed in the space along the spermatic cord and in the scrotum.

Hydrocephalus (hī-drō-SEF-a-lus) Abnormal accumulation of cerebrospinal fluid on the brain.

Hydrophobia (hī'-drō-FŌ-bē-a) Rabies; a condition characterized by severe muscle spasms when attempting to drink water. Also, an abnormal fear of water.

Hymen (HĪ-men) A thin fold of vascularized mucous membrane at the vaginal orifice.

Hyperbaric oxygenation (hī'-per-BA-rik ok'-sē-je-NĀ-shun) ***(HBO)*** Use of pressure supplied by a hyperbaric chamber to cause more oxygen to dissolve in blood to treat patients infected with anaerobic bacteria (tetanus and gangrene bacteria). Also used to treat carbon monoxide poisoning, asphyxia, smoke inhalation, and certain heart disorders.

Hypercalcemia (hī'-per-kal-SĒ-mē-a) An excess of calcium in the blood.

Hypercapnia (hī'-per-KAP-nē-a) An abnormal increase in the amount of carbon dioxide in the blood.

Hyperemia (hī'-per-Ē-mē-a) An excess of blood in an area or part of the body.

Hyperextension (hī'-per-ek-STEN-shun) Continuation of extension beyond the anatomical position, as in bending the head backward.

Hyperglycemia (hī'-per-glī-SĒ-mē-a) An elevated blood sugar level.

Hypermetropia (hī'-per-mē-TRŌ-pē-a) A condition in which visual images are focused behind the retina with resulting defective vision of near objects; farsightedness.

Hyperphosphatemia (hī-per-fos'-fa-TĒ-mē-a) An abnormally high level of phosphate in the blood.

Hyperplasia (hī'-per-PLĀ-zē-a) An abnormal increase in the number of normal cells in a tissue or organ, increasing its size.

Hyperpolarization (hī'-per-PŌL-a-ri-zā'-shun) Increase in the internal negativity across a cell membrane, thus increasing the voltage and moving it farther away from the threshold value.

Hypersecretion (hī'-per-se-KRĒ-shun) Overactivity of glands resulting in excessive secretion.

Hypersensitivity (hī'-per-sen-si-TI-vi-tē) Overreaction to an allergen that results in pathological changes in tissues. Also called ***allergy.***

Hypertension (hī'-per-TEN-shun) High blood pressure.

Hyperthermia (hī'-per-THERM-ē-a) An elevated body temperature.

Hypertonia (hī-per-TŌ-nē-a) Increased muscle tone that is expressed as spasticity or rigidity.

Hypertonic (hī'-per-TON-ik) Having an osmotic pressure greater than that of a solution with which it is compared.

Hypertrophy (hī-PER-trō-fē) An excessive enlargement or overgrowth of tissue without cell division.

Hyperventilation (hī'-per-ven-ti-LĀ-shun) A rate of respiration higher than that required to maintain a normal level of plasma PCO_2.

Hypervitaminosis (hī'-per-vī'-ta-min-Ō-sis) An excess of one or more vitamins.

Hypocalcemia (hī'-pō-kal-SĒ-mē-a) A below normal level of calcium in the blood.

Hypochloremia (hī'-pō-klō-RĒ-mē-a) Deficiency of chloride in the blood.

Hypoglycemia (hī'-pō-glī-SĒ-mē-a) An abnormally low concentration of glucose in the blood; can result from excess insulin (injected or secreted).

Hypokalemia (hī'-pō-kā-LĒ-mē-a) Deficiency of potassium in the blood.

Hypomagnesemia (hī'-pō-mag'-ne-SĒ-mē-a) Deficiency of magnesium in the blood.

Hyponatremia (hī'-pō-na-TRĒ-mē-a) Deficiency of sodium in the blood.

Hyponychium (hī'-pō-NIK-ē-um) Free edge of the fingernail.

Hypophosphatemia (hī-pō-fos'-fa-TĒ-mē-a) An abnormally low level of phosphate in the blood.

Hypophyseal (hī'-pō-FIZ-ē-al) ***pouch*** An outgrowth of ectoderm from the roof of the stomodeum (mouth) from which the adenohypophysis (anterior lobe) of the pituitary gland develops.

Hypophysis (hī-POF-i-sis) Pituitary gland.

Hypoplasia (hī-pō-PLĀ-zē-a) Defective development of tissue.

Hyposecretion (hī'-pō-se-KRĒ-shun) Underactivity of glands resulting in diminished secretion.

Hypospadias (hī'-pō-SPĀ-dē-as) A displaced urethral opening. In the male, the opening may be on the underside of the penis, at the penoscrotal junction, between the scrotal folds, or in the perineum. In the female, the urethra opens into the vagina.

Hypothalamic-hypophyseal (hī'-pō-thal-AM-ik hī'-po-FIZ-ē-al) ***tract*** A bundle of nerve processes made up of fibers that have their cell bodies in the hypothalamus but release their neurosecretions in the posterior pituitary gland or neurohypophysis.

Hypothalamus (hī'-pō-THAL-a-mus) A portion of the diencephalon, lying beneath the thalamus and forming the floor and part of the wall of the third ventricle.

Hypothermia (hī-pō-THER-mē-a) Lowering of body temperature below 35°C (95°F); in surgical procedures, it refers to deliberate cooling of the body to slow down metabolism and reduce oxygen needs of tissues.

Hypotonia (hī'-pō-TŌ-nē-a) Decreased or lost muscle tone in which muscles appear flaccid.

Hypotonic (hī'-pō-TON-ik) Having an osmotic pressure lower than that of a solution with which it is compared.

Hypoventilation (hī-pō-ven-ti-LĀ-shun) A rate of respiration lower than that required to maintain a normal level of plasma PCO_2.

Hypovolemic (hī-pō-vō-LĒ-mik) ***shock*** A type of shock characterized by decreased intravascular volume resulting from blood loss; may be caused by acute hemorrhage or excessive fluid loss.

Hypoxia (hī-POKS-ē-a) Lack of adequate oxygen at the tissue level.

Hysterectomy (his-te-REK-tō-mē) The surgical removal of the uterus.

Ileocecal (il'-ē-ō-SĒ-kal) ***sphincter*** A fold of mucous membrane that guards the opening from the ileum into the large intestine. Also called the ***ileocecal valve.***

Ileum (IL-ē-um) The terminal portion of the small intestine.

Immunity (i-MYOON-i-tē) The state of being resistant to injury, particularly by poisons, foreign proteins, and invading parasites, due to the presence of antibodies.

Immunogenicity (im-yoo-nō-jen-IS-it-ē) Ability of an antigen to stimulate antibody production.

Immunoglobulin (im-yoo-nō-GLOB-yoo-lin) ***(Ig)*** An antibody synthesized by plasma cells derived from B lymphocytes in response to the introduction of antigen. Immunoglobulins are divided into five kinds (IgG, IgM, IgA, IgD, IgE) based primarily on the larger protein component present in the immunoglobulin.

Immunology (im'-yoo-NOL-ō-jē) The branch of science that deals with the responses of the body when challenged by antigens.

Immunosuppression (im'-yoo-nō-su-PRESH-un) Inhibition of the immune response.

Immunotherapy (im-yoo-nō-THER-a-pē) Attempt to induce the immune system to mount an attack against cancer cells.

Imperforate (im-PER-fō-rāt) Abnormally closed.

Impetigo (im'-pe-TĪ-go) A contagious skin disorder characterized by pustular eruptions.

Implantation (im-plan-TĀ-shun) The insertion of a tissue or a part into the body. The attachment of the blastocyst to the lining of the uterus 7–8 days after fertilization.

Impotence (IM-pō-tens) Weakness; inability to copulate; failure to maintain an erection long enough for sexual intercourse.

Incontinence (in-KON-ti-nens) Inability to retain urine, semen, or feces, through loss of sphincter control.

Infant respiratory distress syndrome (RDS) A disease of newborn infants, especially premature ones, in which insufficient amounts of surfactant are produced and breathing is labored. Also called ***hyaline*** (HĪ-a-lin) ***membrane disease (HMD).***

Infarction (in-FARK-shun) The presence of a localized area of necrotic tissue, produced by inadequate oxygenation of the tissue.

Infection (in-FEK-shun) Invasion and multiplication of microorganisms in body tissues, which may be inapparent or characterized by cellular injury.

Infectious mononucleosis (mon-ō-nook'-lē-Ō-sis) ***(IM)*** Contagious disease caused by the Epstein-Barr virus (EBV) and characterized by an elevated mononucleocyte and lymphocyte count, fever, sore throat, stiff neck, cough, and malaise.

Inferior (in-FĒR-ē-or) Away from the head or toward the lower part of a structure. Also called ***caudad*** (KAW-dad).

Inferior vena cava (VĒ-na CĀ-va) ***(IVC)*** Large vein that collects blood from parts of the body inferior to the heart and returns it to the right atrium.

Infertility Inability to conceive or to cause conception. Also called ***sterility.***

Inflammation (in'-fla-MĀ-shun) Localized, protective response to tissue injury designed to destroy, dilute, or wall off the infecting agent or injured tissue; characterized by redness, pain, heat, swelling, and sometimes loss of function.

Inflammatory bowel (in-FLAM-a-tō'-rē BOW-el) ***disease*** Disorder that exists in two forms: (1) Crohn's disease (inflammation of the gastrointestinal tract, especially the distal ileum and proximal colon, in which the inflammation may extend from the mucosa through the serosa); and (2) ulcerative colitis (inflammation of the mucosa of the gastrointestinal tract, usually limited to the large intestine and usually accompanied by rectal bleeding).

Inflation reflex Reflex that prevents overinflation of the lungs. Also called ***Hering–Breuer reflex.***

Infraspinatous (in'-fra-SPĪ-na-tus) Bony process found below the spine of the scapula used for muscle attachment.

Infundibulum (in'-fun-DIB-yoo-lum) The stalklike structure that attaches the pituitary gland (hypophysis) to the hypothalamus of the brain. The funnel-shaped, open, distal end of the uterine (Fallopian) tube.

Ingestion (in-JES-chun) The taking in of food, liquids, or drugs, by mouth.

Inguinal (IN-gwi-nal) Pertaining to the groin.

Inguinal canal An oblique passageway in the anterior abdominal wall just superior and parallel to the medial half of the inguinal ligament that transmits the spermatic cord and ilioinguinal nerve in the male and round ligament of the uterus and ilioinguinal nerve in the female.

Inheritance The acquisition of body characteristics and qualities by transmission of genetic information from parents to offspring.

Inhibin A male sex hormone secreted by sustentacular (Sertoli) cells that inhibits FSH release by the adenohypophysis (anterior pituitary) and thus spermatogenesis.

Inhibitory postsynaptic potential (IPSP) The increase in the internal negativity of the membrane potential so that the voltage moves further from the threshold value.

Inner cell mass A region of cells of a blastocyst that differentiates into the three primary germ layers—ectoderm, mesoderm, and endoderm—from which all tissues and organs develop; also called an ***embryoblast.***

Inorganic (in'-or-GAN-ik) ***compound*** Compound that usually lacks carbon, usually small, and contains ionic bonds. Examples include water and many acids, bases, and salts.

Insertion (in-SER-shun) The manner or place of attachment of a muscle to the bone that it moves.

Insomnia (in-SOM-nē-a) Difficulty in falling asleep and, usually, frequent awakening.

Inspiration (in-spi-RĀ-shun) The act of drawing air into the lungs.

Insula (IN-su-la) A triangular area of cerebral cortex that lies deep within the lateral cerebral fissue, under the parietal, frontal, and temporal lobes, and cannot be seen in an external view of the brain. Also called the ***island*** or ***isle of Reil*** (RĪL).

Insulin (IN-su-lin) A hormone produced by the beta cells of the pancreas that decreases the blood glucose level.

Integrin (IN-te-grin) Receptor on a plasma membrane that interacts with an adhesion protein found in intercellular material and blood.

Integumentary (in-teg'-yoo-MEN-tar-ē) Relating to the skin.

Intercalated (in-TER-ka-lāt-ed) ***disc*** An irregular transverse thickening of sarcolemma that contains desmosomes that hold cardiac muscle fibers (cells) together and gap junctions that aid in conduction of muscle action potentials.

Intercostal (in'-ter-KOS-tal) ***nerve*** A nerve supplying a muscle located between the ribs.

Interferon (in'-ter-FĒR-on) ***(IFN)*** Three principal types of protein (alpha, beta, gamma) naturally produced by virus-infected host cells that induce uninfected cells to synthesize antiviral proteins (AVPs) that inhibit intracellular viral replication in uninfected host cells; artificially synthesized through recombinant DNA techniques.

Intermediate Between two structures, one of which is medial and one of which is lateral.

Intermediate filament Cytoplasmic structure, ranging from 8 to 12 nm in diameter, that may provide structural reinforcement and assist in contraction.

Internal Away from the surface of the body.

Internal capsule A tract of projection fibers connecting various parts of the cerebral cortex and lying between the thalamus and the caudate and lentiform nuclei of the basal ganglia.

Internal ear The inner ear or labyrinth, lying inside the temporal bone, containing the organs of hearing and balance.

Internal nares (NA-rēz) The two openings posterior to the nasal cavities opening into the nasopharynx. Also called the ***choanae*** (kō-A-nē).

Internal respiration The exchange of respiratory gases between blood and body cells.

Interphase (IN-ter-fāz) The period during its life cycle when a cell is carrying on every life process except division; the stage between two mitotic divisions. Also called ***metabolic phase.***

Interstitial cell of Leydig *See* ***Interstitial endocrinocyte.***

Interstitial (in'-ter-STISH-al) ***endocrinocyte*** A cell located in the connective tissue between seminiferous tubules in a mature testis that secretes testosterone. Also called an ***interstitial cell of Leydig*** (LĪ-dig).

Interstitial (in'-ter-STISH-al) ***fluid*** The portion of extracellular fluid that fills the microscopic spaces between the cells of tissues; the internal environment of the body. Also called ***intercellular*** or ***tissue fluid.***

Interstitial growth Growth from within, as in the growth of cartilage. Also called ***endogenous*** (en-DOJ-e-nus) ***growth.***

Interventricular foramen (in'-ter-ven-TRIK-yoo-lar) A narrow, oval opening through which the lateral ventricles of the brain communicate with the third ventricle. Also called the ***foramen of Monro.***

Intervertebral (in'-ter-VER-te-bral) ***disc*** A pad of fibrocartilage located between the bodies of two vertebrae.

Intestinal gland Simple tubular gland that opens onto the surface of the intestinal mucosa and secretes digestive enzymes. Also called a ***crypt of Lieberkühn*** (LĒ-ber-kyoon).

Intracellular (in'-tra-SEL-yoo-lar) ***fluid (ICF)*** Fluid located within cells.

Intrafusal (in'-tra-FYOO-zal) ***fibers*** Three to ten specialized muscle fibers (cells), partially enclosed in a connective tissue capsule that is filled with lymph; the fibers compose muscle spindles.

Intramembranous ossification (in'-tra-MEM-bra-nus os'-i'-fi-KĀ-shun) The method of bone formation in which the bone is formed directly in membranous tissue.

Intraocular (in-tra-OC-yoo-lar) ***pressure (IOP)*** Pressure in the eyeball, produced mainly by aqueous humor.

Intrapleural pressure Air pressure between the two pleural layers of the lungs, usually subatmospheric. Also called ***intrathoracic pressure.***

Intrapulmonic pressure Air pressure within the lungs. Also called ***intraalveolar pressure.***

Intrauterine device (IUD) A small metal or plastic object inserted into the uterus for the purpose of preventing pregnancy.

Intrinsic (in-TRIN-sik) Of internal origin; for example, the intrinsic factor, a mucoprotein formed by the gastric mucosa that is necessary for the absorption of vitamin B_{12}.

Intrinsic clotting pathway Sequence of reactions leading to blood clotting that is initiated by the release of a substance (tissue factor or thromboplastin) contained *within* blood itself or cells in direct contact with blood.

Intrinsic factor (IF) A glycoprotein synthesized and secreted by the parietal cells of the gastric mucosa that facilitates vitamin B_{12} absorption.

Intron (IN-tron) A region of DNA that does not code for the synthesis of a protein.

Intubation (in'-too-BĀ-shun) Insertion of a tube through the nose or mouth into the larynx and trachea for entrance of air or to dilate a stricture.

Intussusception (in'-ta-sa-SEP-shun) The infolding (invagination) of one part of the intestine within another segment.

In utero (YOO-ter-ō) Within the uterus.

Invagination (in-vaj'-i-NĀ-shun) The pushing of the wall of a cavity into the cavity itself.

Inversion (in-VER-zhun) The movement of the sole inward at the ankle joint.

In vitro (VĒ-trō) Literally, in glass; outside the living body and in artificial environment such as a laboratory test tube.

In vivo (VĒ-vō) In the living body.

Ion (Ī-on) Any charged particle or group of particles; usually formed when a substance, such as a salt, dissolves and dissociates.

Ipsilateral (ip'-si-LAT-er-al) On the same side, affecting the same side of the body.

Iris The colored portion of the eyeball seen through the cornea that consists of circular and radial smooth muscle; the black hole in the center of the iris is the pupil.

Irritable bowel (IR-i-ta-bul BOW-el) ***syndrome (IBS)*** Disease of the entire gastrointestinal tract in which persons with the condition may react to stress by developing symptoms such as cramping and abdominal pain associated with alternating patterns of diarrhea and constipation. Excessive amounts of mucus may appear in the stools, and other symptoms include flatulence, nausea, and loss of appetite. The condition is also known as ***irritable colon*** or ***spastic colitis.***

Ischemia (is-KĒ-mē-a) A lack of sufficient blood to a part due to obstruction of circulation.

Island of Reil *See* ***Insula.***

Islet of Langerhans *See* ***Pancretic islet.***

Isometric contraction A muscle contraction in which tension on the muscle increases, but there is only minimal muscle shortening so that no movement is produced.

Isotonic (ī'-sō-TON-ik) Having equal tension or tone. Having equal osmotic pressure between two different solutions or between two elements in a solution.

Isotope (Ī-sō-tōpe') A chemical element that has the same atomic number as another but a different atomic weight. Radioactive isotopes change into other elements with the emission of certain radiations.

Isovolumetric (ī-sō-vol-yoo'-MET-rik) ***contraction*** The period of time, about 0.05 sec, between the start of ventricular systole and the opening of the semilunar valves; there is contraction of the ventricles, but no emptying, and there is a rapid rise in ventricular pressure.

Isovolumetric relaxation The period of time, about 0.05 sec, between the opening of the atrioventricular (AV) valves and the closing of the semilunar valves; there is a drastic decrease in ventricular pressure without a change in ventricular volume.

Isthmus (IS-mus) A narrow strip of tissue or narrow passage connecting two larger parts.

Jaundice (JAWN-dis) A condition characterized by yellowness of skin, white of eyes, mucous membranes, and body fluids because of a buildup of bilirubin.

Jejunum (jē-JOO-num) The middle portion of the small intestine.

Joint kinesthetic (kin'-es-THET-ik) ***receptor*** A proprioceptive receptor located in a joint, stimulated by joint movement.

Juxtaglomerular (juks-ta-glō-MER-yoo-lar) ***apparatus (JGA)*** Consists of the macula densa (cells of the distal convoluted tubule adjacent to the afferent and efferent arteriole) and juxtaglomerular cells (modified cells of the afferent and sometimes efferent arteriole); secretes renin when blood pressure starts to fall.

Karyotype (KAR-ē-ō-tīp) An arrangement of chromosomes based on shape, size, and position of centromeres.

Keratin (KER-a-tin) An insoluble protein found in the hair, nails, and other keratinized tissues of the epidermis.

Keratinocyte (ker-A-tin'-ō-sīt) The most numerous of the epidermal cells that function in the production of keratin.

Keratohyalin (ker'-a-tō-HĪ-a-lin) A compound involved in the formation of keratin.

Keratosis (ker'-a-TŌ-sis) Formation of a hardened growth of tissue.

Ketone (KĒ-ton) ***bodies*** Substances produced primarily during excessive fat metabolism, such as acetone, acetoacetic acid, and β-hydroxybutyric acid.

Ketosis (kē-TŌ-sis) Abnormal condition marked by excessive production of ketone bodies.

Kidney (KID-nē) One of the paired reddish organs located in the lumbar region that regulates the composition and volume of blood and produces urine.

Kidney stone A concretion, usually consisting of calcium oxalate, uric acid, and calcium phosphate crystals, that may form in any portion of the urinary tract. Also called a ***renal calculus.***

Kilocalorie (KIL-ō-kal'-ō-rē) ***(kcal)*** The amount of heat required to raise the temperature of 1,000 g of water 1°C; the unit used to express the heating value of foods and to measure metabolic rate.

Kinesiology (ki-nē'-sē-OL-ō-jē) The study of the movement of body parts.

Kinesthesia (kin-is-THĒ-szē-a) Ability to perceive extent, direction, or weight of movement; muscle sense.

Korotkoff (kō-ROT-kof) ***sounds*** The various sounds that are heard while taking blood pressure.

Krebs cycle A series of energy-yielding chemical reactions that occur in the matrix of mitochondria in which energy is transferred to carrier molecules for subsequent liberation and carbon dioxide is formed. Also called the ***citric acid cycle*** and ***tricarboxylic acid (TCA) cycle.***

Kupffer's cell *See* ***Stellate reticuloendothelial cell.***

Kyphosis (kī-FŌ-sis) An exaggeration of the thoracic curve of the vertebral column, resulting in a "round-shouldered" or hunchback appearance.

Labial frenulum (LĀ-bē-al FREN-yoo-lum) A medial fold of mucous membrane between the inner surface of the lip and the gums.

Labia majora (LĀ-bē-a ma-JO-ra) Two longitudinal folds of skin extending downward and backward from the mons pubis of the female.

Labia minora (min-OR-a) Two small folds of mucous membrane lying medial to the labia majora of the female.

Labium (LĀ-bē-um) A lip. A liplike structure. *Plural,* ***labia*** (LĀ-bē-a).

Labor The process by which the product of conception is expelled from the uterus through the vagina.

Labyrinth (LAB-i-rinth) Intricate communicating passageway, especially in the internal ear.

Labyrinthine (lab-i-RIN-thēn) ***disease*** Malfunction of the internal ear characterized by deafness, tinnitus, vertigo, nausea, and vomiting.

Laceration (las'-er-Ā-shun) Wound or irregular area of the skin.

Lacrimal (LAK-ri-mal) Pertaining to tears.

Lacrimal (LAK-ri-mal) ***canal*** A duct, one on each eyelid, commencing at the punctum at the medial margin of an eyelid and conveying tears medially into the nasolacrimal sac.

Lacrimal gland Secretory cells located at the superior anterolateral portion of each orbit that secrete tears into excretory ducts that open onto the surface of the conjunctiva.

Lacrimal sac The superior expanded portion of the nasolacrimal duct that receives the tears from a lacrimal canal.

Lactation (lak-TĀ-shun) The secretion and ejection of milk by the mammary glands.

Lacteal (LAK-tē-al) One of many intestinal lymphatic vessels in villi that absorb fat from digested food.

Lactose intolerance Inability to digest lactose because of failure of small intestinal mucosal cells to produce lactase.

Lacuna (la-KOO-na) A small, hollow space, such as that found in bones in which the osteoblasts lie. *Plural,* ***lacunae*** (la-KOO-nē).

Lambdoidal (lam-DOY-dal) ***suture*** The line of union in the skull between the parietal bones and the occipital bone; sometimes contains sutural (Wormian) bones.

Lamellae (la-MEL-ē) Concentric rings found in compact bone.

Lamellated corpuscle Oval pressure receptor located in subcutaneous tissue and consisting of concentric layers of connective tissue wrapped around an afferent nerve fiber. Also called a ***Pacinian*** (pa-SIN-ē-an) ***corpuscle.***

Lamina (LAM-i-na) A thin, flat layer or membrane, as the flattened part of either side of the arch of a vertebra. *Plural,* ***laminae*** (LAM-i-nē).

Lamina propria (PRO-prē-a) The connective tissue layer of a mucous membrane.

Lanugo (lan-YOO-gō) Fine downy hairs that cover the fetus.

Laparoscopy (lap'-a-ROS-kō-pē) A procedure in which a laparoscope is inserted through an incision in the abdominal wall to view abdominal and pelvic viscera, remove fluids and tissues for biopsy, drain ovarian cysts, cut adhesions, stop bleeding, and perform tubal ligation.

Large intestine The portion of the gastrointestinal tract extending from the ileum of the small intestine to the anus, divided structurally into the cecum, colon, rectum, and anal canal.

Laryngitis (la-rin-JĪ-tis) Inflammation of the mucous membrane lining the larynx.

Laryngopharynx (la-rin'-gō-FAR-inks) The inferior portion of the pharynx, extending downward from the level of the hyoid bone to divide posteriorly into the esophagus and anteriorly into the larynx.

Laryngoscope (la-RIN-gō-skōp) An instrument for examining the larynx.

Laryngotracheal (la-rin'-gō-TRA-kē-al) ***bud*** An outgrowth of endoderm of the foregut from which the respiratory system develops.

Larynx (LAR-inks) The voice box, a short passageway that connects the pharynx with the trachea.

Lateral (LAT-er-al) Farther from the midline of the body or a structure.

Lateral ventricle (VEN-tri-kul) A cavity within a cerebral hemisphere that communicates with the lateral ventricle in the other cerebral hemisphere and with the third ventricle by way of the interventricular foramen.

Learning The ability to acquire knowledge or a skill through instruction or experience.

Leg The part of the lower extremity between the knee and the ankle.

Lens A transparent organ constructed of proteins (crystallins) lying posterior to the pupil and iris of the eyeball and anterior to the vitreous body.

Lesion (LĒ-zhun) Any localized, abnormal change in tissue formation.

Lesser omentum (ō-MEN-tum) A fold of the peritoneum that extends from the liver to the lesser curvature of the stomach and the commencement of the duodenum.

Lesser vestibular (ves-TIB-yoo-lar) ***gland*** One of the paired mucus-secreting glands that have ducts that open on either side of the urethral orifice in the vestibule of the female.

Lethargy (LETH-ar-jē) A condition of drowsiness or indifference.

Leucocyte (LOO-kō-sīt) A white blood cell.

Leucocytosis (loo'-kō-sī-TŌ-sis) An increase in the number of white blood cells, characteristic of many infections and other disorders.

Leucopenia (loo-kō-PĒ-nē-a) A decrease of the number of white blood cells below 5,000/mm^3.

Leukemia (loo-KĒ-mē-a) A malignant disease of the blood-forming tissues characterized by either uncontrolled production and accumulation of immature leucocytes in which many cells fail to reach maturity (acute) or an accumulation of mature leucocytes in the blood because they do not die at the end of their normal life span (chronic).

Leukoplakia (loo-kō-PLĀ-kē-a) A disorder in which there are white patches in the mucous membranes of the tongue, gums, and cheeks.

Libido (li-BĒ-dō) The sexual drive, conscious or unconscious.

Ligament (LIG-a-ment) Dense, regularly arranged connective tissue that attaches bone to bone.

Ligand (LĪ-gand) Chemical in interstitial fluid, usually in a concentration lower than in cells.

Limbic system A portion of the forebrain, sometimes termed the visceral brain, concerned with various aspects of emotion and behavior, that includes the limbic lobe, dentate gyrus, amygdaloid body, septal nuclei, mammillary bodies, anterior thalamic nucleus, olfactory bulbs, and bundles of myelinated axons.

Lingual frenulum (LIN-gwal FREN-yoo-lum) A fold of mucous membrane that connects the tongue to the floor of the mouth.

Lipase (LĪ-pās) A fat-splitting enzyme.

Lipid An organic compound composed of carbon, hydrogen, and oxygen that is usually insoluble in water, but soluble in alcohol, ether, and chloroform; examples include fats, phospholipids, steroids, and prostaglandins.

Lipid profile Blood test that measures total cholesterol, high-density lipoprotein, low-density lipoprotein, and triglycerides, to assess risk for cardiovascular disease.

Lipogenesis (li-pō-GEN-e-sis) The synthesis of lipids from glucose or amino acids by liver cells.

Lipoma (li-PŌ-ma) A fatty tissue tumor, usually benign.

Lipoprotein (lip'-ō-PRŌ-tēn) Protein containing lipid that is produced by the liver and combines with cholesterol and triglycerides to make it water-soluble for transportation by the cardiovascular system; high levels of low-density lipoproteins (LDL) are associated with increased risk of atherosclerosis, while high levels of high-density lipoproteins (HDL) are associated with decreased risk of atherosclerosis.

Lithotripsy (LITH-ō-trip'-sē) A noninvasive procedure in which shock waves generated by a lithotriptor are used to pulverize kidney stones or gallstones.

Liver Large gland under the diaphragm that occupies most of the right hypochondriac region and part of the epigastric region; functionally, it produces bile salts, heparin, and plasma proteins; converts one nutrient into another; detoxifies substances; stores glycogen, minerals, and vitamins; carries on phagocytosis of blood cells and bacteria; and helps activate vitamin D.

Lobe (lōb) A curved or rounded projection.

Locus coeruleus (LŌ-kus sē-ROO-lē-us) A group of neurons in the brain stem where norepinephrine (NE) is concentrated.

Lordosis (lor-DŌ-sis) An exaggeration of the lumbar curve of the vertebral column.

Lou Gehrig's disease *See* ***Amyotrophic lateral sclerosis.***

Lower extremity The appendage attached at the pelvic (hip) girdle, consisting of the thigh, knee, leg, ankle, foot, and toes.

Lumbar (LUM-bar) Region of the back and side between the ribs and pelvis; loin.

Lumbar plexus (PLEK-sus) A network formed by the anterior branches of spinal nerves L1 through L4.

Lumen (LOO-men) The space within an artery, vein, intestine, or a tube.

Lung One of the two main organs of respiration, lying on either side of the heart in the thoracic cavity.

Lung scan A diagnostic test in which a radioactive substance is detected in the lungs by a scanning camera; used to evaluate for pulmonary embolism, pneumonia, or cancer.

Lunula (LOO-nyoo-la) The moon-shaped white area at the base of a nail.

Luteinizing (LOO-tē-in'-īz-ing) ***hormone (LH)*** A hormone secreted by the adenohypophysis (anterior lobe) of the pituitary gland that stimulates ovulation, progesterone secretion by the corpus luteum, and readies the mammary glands for milk secretion in females and stimulates testosterone secretion by the testes in males.

Lymph (limf) Fluid confined in lymphatic vessels and flowing through the lymphatic system to be returned to the blood.

Lymphangiography (lim-fan'-jē-OG-ra-fē) A procedure by which lymphatic vessels and lymph organs are filled with a radiopaque substance in order to be x-rayed.

Lymphatic (lim-FAT-ik) ***vessel*** A large vessel that collects lymph from lymph capillaries and converges with other lymphatic vessels to form the thoracic and right lymphatic ducts.

Lymphatic tissue A specialized form of reticular tissue that contains large numbers of lymphocytes.

Lymph capillary Blind-ended microscopic lymph vessel that begins in spaces between cells and converges with other lymph capillaries to form lymphatic vessels.

Lymph node An oval or bean-shaped structure located along lymphatic vessels.

Lymphocyte (LIM-fō-sīt) A type of white blood cell, found in lymph nodes, associated with the immune system.

Lymphokines (LIM-fō-kīns) Powerful proteins secreted by T cells that endow T cells with their ability to assist in immunity.

Lysosome (LĪ-sō-sōm) An organelle in the cytoplasm of a cell, enclosed by a single membrane and containing powerful digestive enzymes.

Lysozyme (LĪ-sō-zīm) A bactericidal enzyme found in tears, saliva, and perspiration.

Macrophage (MAK-rō-fāj) Phagocytic cell derived from a monocyte. May be fixed or wandering.

Macula (MAK-yoo-la) A discolored spot or a colored area. A small, thickened region on the wall of the utricle and saccule that serves as a receptor for static equilibrium.

Macula lutea (LOO-tē-a) The yellow spot in the center of the retina.

Magnetic resonance imaging (MRI) A diagnostic procedure that focuses on the nuclei of atoms of a single element in a tissue, usually hydrogen, to determine if they behave normally in the presence of an external magnetic force; used to indicate the biochemical activity of a tissue. Formerly called ***nuclear magnetic resonance (NMR).***

Malaise (ma-LĀYZ) Discomfort, uneasiness, and indisposition, often indicative of infection.

Malignant (ma-LIG-nant) Referring to diseases that tend to become worse and cause death; especially the invasion and spreading of cancer.

Malignant melanoma (mel'-a-NŌ-ma) A usually dark, malignant tumor of the skin containing melanin.

Malnutrition (mal'-nu-TRISH-un) State of bad or poor nutrition that may be due to inadequate food intake, imbalance of nutrients, malabsorption of nutrients, improper distribution of nutrients, increased nutrient requirements, increased nutrient losses, or overnutrition.

Mammary (MAM-ar-ē) ***gland*** Modified sudoriferous (sweat) gland of the female that secretes milk for the nourishment of the young.

Mammillary (MAM-i-ler-ē) ***bodies*** Two small rounded bodies posterior to the tuber cinereum that are involved in reflexes related to the sense of smell.

Mammography (mam-OG-ra-fē) Procedure for imaging the breasts (xeromammography or film-screen mammography) to evaluate for breast disease or screen for breast cancer.

Marfan (MAR-fan) ***syndrome*** Inherited disorder that results in abnormalities of connective tissue, especially in the skeleton, eyes, and cardiovascular system.

Marrow (MAR-ō) Soft, spongelike material in the cavities of bone. Red marrow produces blood cells; yellow marrow, formed mainly of fatty tissue, has no blood-producing function.

Mast cell A cell found in loose connective tissue along blood vessels that produces heparin, an anticoagulant. The name given to a basophil after it has left the bloodstream and entered the tissues.

Mastectomy (mas-TEK-tō-mē) Surgical removal of breast tissue.

Mastication (mas'-ti-KĀ-shun) Chewing.

Maximal oxygen uptake Maximum rate of oxygen consumption during aerobic catabolism of pyruvic acid that is determined by age, sex, and body size.

Meatus (mē-Ā-tus) A passage or opening, especially the external portion of a canal.

Mechanoreceptor (me-KAN-ō-rē'-sep-tor) Receptor that detects mechanical deformation of the receptor itself or adjacent cells; stimuli so detected include those related to touch, pressure, vibration, proprioception, hearing, equilibrium, and blood pressure.

Medial (MĒ-dē-al) Nearer the midline of the body or a structure.

Medial lemniscus (lem-NIS-kus) A flat band of myelinated nerve fibers extending through the medulla, pons, and midbrain and terminating in the thalamus on the same side. Sensory neurons in this tract transmit impulses for proprioception, fine touch, pressure, and vibration sensations.

Median aperture (AP-er-choor) One of the three openings in the roof of the fourth ventricle through which cerebrospinal fluid enters the subarachnoid space of the brain and cord. Also called the ***foramen of Magendie.***

Mediastinum (mē'-dē-as-TĪ-num) A broad, median partition, actually a mass of tissue found between the pleurae of the lungs that extends from the sternum to the vertebral column.

Medulla (me-DULL-la) An inner layer of an organ, such as the medulla of the kidneys.

Medulla oblongata (ob'-long-GA-ta) The most inferior part of the brain stem.

Medullary (MED-yoo-lar'-ē) ***cavity*** The space within the diaphysis of a bone that contains yellow marrow. Also called the ***marrow cavity.***

Medullary rhythmicity (rith-MIS-i-tē) ***area*** Portion of the respiratory center in the medulla that controls the basic rhythm of respiration.

Meibomian gland *See* ***Tarsal gland.***

Meiosis (mē-Ō-sis) A type of cell division restricted to sex-cell production involving two successive nuclear divisions that result in daughter cells with the haploid (*n*) number of chromosomes.

Meissner's corpuscle *See* ***Corpuscle of touch.***

Melanin (MEL-a-nin) A dark black, brown, or yellow pigment found in some parts of the body such as the skin.

Melanoblast (MEL-a-nō-blast) Precursor cell in the epidermis that gives rise to melanocytes, cells that produce melanin.

Melanocyte (MEL-a-nō-sīt') A pigmented cell located between or beneath cells of the deepest layer of the epidermis that synthesizes melanin.

Melanocyte-stimulating hormone (MSH) A hormone secreted by the adenohypophysis (anterior lobe) of the pituitary gland that stimulates the dispersion of melanin granules in melanocytes in amphibians; continued administration produces darkening of skin in humans.

Melatonin (mel-a-TŌN-in) A hormone secreted by the pineal gland that may inhibit reproductive activities.

Membrane A thin, flexible sheet of tissue composed of an epithelial layer and an underlying connective tissue layer, as in an epithelial membrane, or of loose connective tissue only, as in a synovial membrane.

Membranous labyrinth (mem-BRA-nus LAB-i-rinth) The portion of the labyrinth of the inner ear that is located inside the bony labyrinth and separated from it by the perilymph; made up of the membranous semicircular canals, the saccule and utricle, and the cochlear duct.

Memory The ability to recall thoughts; commonly classified as short-term (activated) and long-term.

Menarche (me-NAR-kē) Beginning of the menstrual function.

Ménière's (men-YAIRZ) ***syndrome*** A type of labyrinthine disease characterized by fluctuating loss of hearing, vertigo, and tinnitus due to an increased amount of endolymph that enlarges the labyrinth.

Meninges (me-NIN-jēz) Three membranes covering the brain and spinal cord, called the dura mater, arachnoid, and pia mater. *Singular,* ***meninx*** (MEN-inks).

Meningitis (men-in-JĪ-tis) Inflammation of the meninges, most commonly the pia mater and arachnoid.

Menopause (MEN-ō-pawz) The termination of the menstrual cycles.

Menstrual (MEN-stroo-al) ***cycle*** A series of changes in the endometrium of a nonpregnant female that prepares the lining of the uterus to receive a fertilized ovum.

Menstruation (men'-stroo-Ā-shun) Periodic discharge of blood, tissue fluid, mucus, and epithelial cells that usually lasts for 5 days; caused by a sudden reduction in estrogens and progesterone. Also called the ***menstrual phase*** or ***menses.***

Merocrine (MER-ō-krin) ***gland*** A secretory cell that remains intact throughout the process of formation and discharge of the secretory product, as in the salivary and pancreatic glands.

Mesenchyme (MEZ-en-kīm) An embryonic connective tissue from which all other connective tissues arise.

Mesentery (MEZ-en-ter'-ē) A fold of peritoneum attaching the small intestine to the posterior abdominal wall.

Mesocolon (mez'-ō-KŌ-lon) A fold of peritoneum attaching the colon to the posterior abdominal wall.

Mesoderm The middle of the three primary germ layers that gives rise to connective tissues, blood and blood vessels, and muscles.

Mesothelium (mez'-ō-THĒ-lē-um) The layer of simple squamous epithelium that lines serous cavities.

Mesovarium (mez'-ō-VAR-ē-um) A short fold of peritoneum that attaches an ovary to the broad ligament of the uterus.

Metabolism (me-TAB-ō-lizm) The sum of all the biochemical reactions that occur within an organism, including the synthetic (anabolic) reactions and decomposition (catabolic) reactions.

Metacarpus (met'-a-KAR-pus) A collective term for the five bones that make up the palm of the hand.

Metaphase (MET-a-phāz) The second stage of mitosis in which chromatid pairs line up on the equatorial plane of the cell.

Metaphysis (me-TAF-i-sis) Growing portion of a bone.

Metaplasia (met'-a-PLĀ-zē-a) The transformation of one cell into another.

Metarteriole (met'-ar-TĒ-rē-ōl) A blood vessel that emerges from an arteriole, traverses a capillary network, and empties into a venule.

Metastasis (me-TAS-ta-sis) The spread of cancer to surrounding tissues (local) or to other body sites (distant).

Metatarsus (met'-a-TAR-sus) A collective term for the five bones located in the foot between the tarsals and the phalanges.

Micelle (mī-SEL) A spherical aggregate of bile salts that dissolves fatty acids and monoglycerides so that they can be transported into small intestinal epithelial cells.

Microcephalus (mi-krō-SEF-a-lus) An abnormally small head; premature closing of the anterior fontanel so that the brain has insufficient room for growth, resulting in mental retardation.

Microfilament (mī-krō-FIL-a-ment) Rodlike cytoplasmic structure about 6 nm in diameter; comprises contractile units in muscle fibers (cells) and provides support, shape, and movement in nonmuscle cells.

Microglia (mī-krō-GLĒ-a) Neuroglial cells that carry on phagocytosis. Also called ***brain macrophages*** (MAK-rō-fāj-ez).

Microphage (MĪK-rō-fāj) Granular leucocyte that carries on phagocytosis, especially neutrophils and eosinophils.

Microtomography (mī-krō-tō-MOG-ra-fē) A procedure that combines the principles of electron microscopy and computed tomography to produce highly magnified, three-dimensional images of living cells.

Microtrabeculae (mī-krō-tra-BEK-yoo-lē) Three-dimensional meshwork of fine filaments, about 10–15 nm in diameter, that hold together microfilaments, microtubules, and intermediate filaments that together constitute the microtrabecular lattice.

Microtrabecular (mī-krō-tra-BEK-yoo-lar) ***lattice*** (LAT-is) Collective term for microfilaments, microtubules, and intermediate filaments held together by microtrabeculae in cytoplasm.

Microtubule (mī-krō-TOOB-yool') Cylindrical cytoplasmic structure, ranging in diameter from 18 to 30 nm, consisting of the protein tubulin; provides support, structure, and transportation.

Microvilli (mī'-krō-VIL-ē) Microscopic, fingerlike projections of the cell membranes of small intestinal cells that increase surface area for absorption.

Micturition (mik'-too-RISH-un) The act of expelling urine from the urinary bladder. Also called ***urination*** (yoo-ri-NĀ-shun).

Midbrain The part of the brain between the pons and the diencephalon. Also called the ***mesencephalon*** (mes'-en-SEF-a-lon).

Middle ear A small, epithelial-lined cavity hollowed out of the temporal bone, separated from the external ear by the eardrum and from the internal ear by a thin bony partition containing the oval and round windows; extending across the middle ear are the three auditory ossicles. Also called the ***tympanic*** (tim-PAN-ik) ***cavity.***

Midline An imaginary vertical line that divides the body into equal left and right sides.

Midsagittal plane A vertical plane through the midline of the body that divides the body or organs into *equal* right and left sides. Also called a ***median plane.***

Milk let-down reflex Contraction of alveolar cells to force milk into ducts of mammary glands, stimulated by oxytocin (OT), which is released from the posterior pituitary in response to suckling action.

Mineral Inorganic, homogeneous solid substance that may perform a function vital to life; examples include calcium, sodium, potassium, iron, phosphorus, and chlorine.

Mineralocorticoids (min'-er-al-ō-KOR-ti-koyds) A group of hormones of the adrenal cortex.

Minimal volume The volume of air in the lungs even after the thoracic cavity has been opened forcing out some of the residual volume.

Minute volume of respiration (MVR) Total volume of air taken into the lungs per minute; about 6,000/ml.

Mitochondrion (mī'-tō-KON-drē-on) A double-membraned organelle that plays a central role in the production of ATP; known as the "powerhouse" of the cell.

Mitosis (mī-TŌ-sis) The orderly division of the nucleus of a cell that ensures that each new daughter nucleus has the same number and kind of chromosomes as the original parent nucleus. The process includes the replication of chromosomes and the distribution of the two sets of chromosomes into two separate and equal nuclei.

Mitotic apparatus Collective term for continuous and chromosomal microtubules and centrioles; involved in cell division.

Mitotic spindle The combination of continuous and chromosomal microtubules, involved in chromosomal movement during mitosis.

Mitral (MĪ-tral) ***insufficiency*** Backflow of blood from the left ventricle into the left atrium due to a damaged mitral valve or ruptured chordae tendineae.

Mitral (MĪ-tral) ***stenosis*** (ste-NŌ-sis) Narrowing of the mitral valve by scar formation or a congenital defect.

Mitral (MĪ-tral) ***valve prolapse*** (PRŌ-laps) or ***MVP*** An inherited disorder in which a portion of a mitral valve is pushed back too far (prolapsed) during contraction due to expansion of the cusps and elongation of the chordae tendineae.

Mittelschmerz (MIT-el-shmerz) Abdominopelvic pain that supposedly indicates the release of a secondary oocyte from the ovary.

Modality (mō-DAL-i-tē) Any of the specific sensory entities, such as vision, smell, or taste.

Modiolus (mō-DĪ-ō'-lus) The central pillar or column of the cochlea.

Mole The weight, in grams, of the combined atomic weights of the atoms that comprise a molecule of a substance.

Molecule (MOL-e-kyool) The chemical combination of two or more atoms.

Monoclonal antibody (MAb) Antibody produced by in vitro clones of B cells hybridized with cancerous cells.

Monocyte (MON-ō-sīt') A type of white blood cell characterized by agranular cytoplasm; the largest of the leucocytes.

Monosaturated fat A fat that contains one double covalent bond between its carbon atoms; it is not completely saturated with hydrogen atoms. Examples are olive and peanut oil.

Mons pubis (monz PYOO-bis) The rounded, fatty prominence over the symphysis pubis, covered by coarse pubic hair.

Morbid (MOR-bid) Diseased; pertaining to disease.

Morula (MOR-yoo-la) A solid mass of cells produced by successive cleavages of a fertilized ovum a few days after fertilization.

Motor area The region of the cerebral cortex that governs muscular movement, particularly the precentral gyrus of the frontal lobe.

Motor end plate Portion of the sarcolemma of a muscle fiber (cell) in close approximation with an axon terminal.

Motor unit A motor neuron together with the muscle fibers (cells) it stimulates.

Mucin (MYOO-sin) A protein found in mucus.

Mucous (MYOO-kus) ***cell*** A unicellular gland that secretes mucus. Also called a ***goblet cell.***

Mucous membrane A membrane that lines a body cavity that opens to the exterior. Also called the ***mucosa*** (myoo-KŌ-sa).

Mucus The thick fluid secretion of mucous glands and mucous membranes.

Multiple motor unit summation Type of summation in which stimuli occur at the same time but at different locations (different motor units).

Multiple sclerosis (skler-Ō-sis) Progressive destruction of myelin sheaths of neurons in the central nervous system, short-circuiting conduction pathways.

Mumps Inflammation and enlargement of the parotid glands accompanied by fever and extreme pain during swallowing.

Muscarinic (mus'-ka-RIN-ik) ***receptor*** Receptor found on all effectors innervated by parasympathetic postganglionic axons and some effectors innervated by sympathetic postganglionic axons; so named because the actions of acetylcholine (ACh) on such receptors are similar to those produced by muscarine.

Muscle An organ composed of one of three types of muscle tissue (skeletal, cardiac, or visceral), specialized for contraction to produce voluntary or involuntary movement of parts of the body.

Muscle action potential A stimulating impulse that travels along a sarcolemma and then into transverse tubules; it is generated by acetylcholine from synaptic vesicles which alters permeability of the sarcolemma to sodium (Na^+) ions.

Muscle fatigue (fa-TĒG) Inability of a muscle to maintain its strength of contraction or tension; may be related to insufficient oxygen, depletion of glycogen, and/or lactic acid buildup.

Muscle spindle An encapsulated receptor in a skeletal muscle, consisting of specialized muscle fiber (cell) and nerve endings, stimulated by changes in length or tension of muscle fibers; a proprioceptor. Also called a ***neuromuscular*** (noo-rō-MUS-kyoo-lar) ***spindle.***

Muscle tissue A tissue specialized to produce motion in response to muscle action potentials by its qualities of contractility, extensibility, elasticity, and excitability.

Muscle tone A sustained, partial contraction of portions of a skeletal muscle in response to activation of stretch receptors.

Muscular dystrophies (DIS-trō-fēz') Inherited muscle-destroying diseases, characterized by degeneration of the individual muscle fibers (cells), which leads to progressive atrophy of the skeletal muscle.

Muscularis (MUS-kyoo-la'-ris) A muscular layer (coat or tunic) of an organ.

Muscularis mucosae (myoo-KŌ-sē) A thin layer of smooth muscle fibers (cells) located in the outermost layer of the mucosa of the gastrointestinal tract, underlying the lamina propria of the mucosa.

Mutation (myoo-TĀ-shun) Any change in the sequence of bases in the DNA molecule resulting in a permanent alteration in some inheritable characteristic.

Myasthenia (mī-as-THĒ-nē-a) ***gravis*** Weakness of skeletal muscles caused by antibodies directed against acetylcholine receptors that inhibit muscle contraction.

Myelin (MĪ-e-lin) ***sheath*** A white, phospholipid, segmented covering, formed by neurolemmocytes (Schwann cells), around the axons and dendrites of many peripheral neurons.

Myelography (mī-e-LOG-ra-fē) Introduction of a contrast medium into the subarachnoid space of the spinal cord to demonstrate tumors or herniated (slipped) discs within or near the spinal cord.

Myenteric plexus A network of nerve fibers from both autonomic divisions located in the muscularis coat of the small intestine. Also called the ***plexus of Auerbach*** (OW-er-bak).

Myocardial infarction (mī'-ō-KAR-dē-al in-FARK-shun) ***(MI)*** Gross necrosis of myocardial tissue due to interrupted blood supply. Also called a ***heart attack.***

Myocardium (mī'-ō-KAR-dē-um) The middle layer of the heart wall, made up of cardiac muscle, comprising the bulk of the heart, and lying between the epicardium and the endocardium.

Myofibril (mī'-ō-FĪ-bril) A threadlike structure, running longitudinally through a muscle fiber (cell) consisting mainly of thick myofilaments (myosin) and thin myofilaments (actin).

Myoglobin (mī-ō-GLŌ-bin) The oxygen-binding, iron-containing conjugated protein complex present in the sarcoplasm of muscle fibers (cells); contributes the red color to muscle.

Myogram (MĪ-ō-gram) The record or tracing produced by the myograph, the apparatus that measures and records the effects of muscular contractions.

Myology (mī-OL-ō-jē) The study of the muscles.

Myometrium (mī'-ō-MĒ-trē-um) The smooth muscle layer of the uterus.

Myopia (mī-Ō-pē-a) Defect in vision so that objects can be seen distinctly only when very close to the eyes; nearsightedness.

Myosin (MĪ-ō-sin) The contractile protein that makes up the thick myofilaments of muscle fibers (cells).

Myotonia (mī-ō-TŌ-nē-a) A continuous spasm of muscle; increased muscular irritability and tendency to contract, and less ability to relax.

Myxedema (mix-e-DĒ-ma) Condition caused by hypothyroidism during the adult years characterized by swelling of facial tissues.

Nail A hard plate, composed largely of keratin, that develops from the epidermis of the skin to form a protective covering on the dorsal surface of the distal phalanges of the fingers and toes.

Nail matrix (MĀ-triks) The part of the nail beneath the body and root from which the nail is produced.

Narcosis (nar-KŌ-sis) Unconscious state due to narcotics.

Nasal (NĀ-zal) ***cavity*** A mucosa-lined cavity on either side of the nasal septum that opens onto the face at an external naris and into the nasopharynx at an internal naris.

Nasal septum (SEP-tum) A vertical partition composed of bone (perpendicular plate of ethmoid and vomer) and cartilage, covered with a mucous membrane, separating the nasal cavity into left and right sides.

Nasolacrimal (nā'-zō-LAK-ri-mal) ***duct*** A canal that transports the lacrimal secretion (tears) from the nasolacrimal sac into the nose.

Nasopharynx (nā'-zō-FAR-inks) The uppermost portion of the pharynx, lying posterior to the nose and extending down to the soft palate.

Nausea (NAW-sē-a) Discomfort characterized by loss of appetite and sensation of impending vomiting.

Nebulization (neb'-yoo-li-ZĀ-shun) Administration of medication to selected portions of the respiratory tract by droplets suspended in air.

Neck The part of the body connecting the head and the trunk. A constricted portion of an organ such as the neck of the femur or uterus.

Necrosis (ne-KRŌ-sis) Death of a cell or group of cells as a result of disease or injury.

Negative feedback The principle governing most control systems; a mechanism of response in which a stimulus initiates actions that reverse or reduce the stimulus.

Neonatal (nē'-ō-NĀ-tal) Pertaining to the first 4 weeks after birth.

Neoplasm (NĒ-ō-plazm) A new growth that may be benign or malignant.

Nephritis (ne-FRĪT-is) Inflammation of the kidney.

Nephron (NEF-ron) The functional unit of the kidney.

Nephrotic (ne-FROT-ik) ***syndrome*** A condition in which the endothelial-capsular membrane leaks, allowing large amounts of protein to escape into urine.

Nerve A cordlike bundle of nerve fibers (axons and/or dendrites) and their associated connective tissue coursing together outside the central nervous system.

Nerve impulse A wave of negativity (depolarization) that self-propagates along the outside surface of the plasma membrane of a neuron; also called a ***nerve action potential.***

Nervous tissue Tissue that initiates and transmits nerve impulses to coordinate homeostasis.

Neuralgia (noo-RAL-jē-a) Attacks of pain along the entire course or branch of a peripheral sensory nerve.

Neural plate A thickening of ectoderm that forms early in the third week of development and represents the beginning of the development of the nervous system.

Neuritis (noo-RĪ-tis) Inflammation of a single nerve, two or more nerves in separate areas, or many nerves simultaneously.

Neuroeffector (noo-rō-e-FEK-tor) ***junction*** Collective term for neuromuscular and neuroglandular junctions.

Neurofibral (noo-rō-FĪ-bral) ***node*** A space, along a myelinated nerve fiber, between the individual neurolemmocytes (Schwann cells) that form the myelin sheath and the neurolemma. Also called ***node of Ranvier*** (ron-VĒ-ā).

Neurofibril (noo-rō-FĪ-bril) One of the delicate threads that forms a complicated network in the cytoplasm of the cell body and processes of a neuron.

Neuroglandular (noo-rō-GLAND-yoo-lar) ***junction*** Area of contact between a motor neuron and a gland.

Neuroglia (noo-RŌG-lē-a) Cells of the nervous system that are specialized to perform the functions of connective tissue. The neuroglia of the central nervous system are the astrocytes, oligodendrocytes, microglia, and ependyma; neuroglia of the peripheral nervous system include the neurolemmocytes (Schwann cells) and the ganglion satellite cells. Also called ***glial*** (GLĒ-al) ***cells.***

Neurohypophyseal (noo'-rō-hī'-po-FIZ-ē-al) ***bud*** An outgrowth of ectoderm located on the floor of the hypothalamus that gives rise to the neurohypophysis (posterior lobe) of the pituitary gland.

Neurohypophysis (noo-rō-hī-POF-i-sis) The posterior lobe of the pituitary gland.

Neurolemma (noo-rō-LEM-ma) The peripheral, nucleated cytoplasmic layer of the neurolemmocyte (Schwann cell). Also called ***sheath of Schwann*** (SCHVON).

Neurolemmocyte A neuroglial cell of the peripheral nervous system that forms the myelin sheath and neurolemma of a nerve fiber by wrapping around a nerve fiber in a jelly-roll fashion. Also called a ***Schwann*** (SCHVON) ***cell.***

Neurology (noo-ROL-ō-jē) The branch of science that deals with the normal functioning and disorders of the nervous system.

Neuromuscular (noo-rō-MUS-kyoo-lar) ***junction*** The area of contact between the axon terminal of a motor neuron and a portion of the sarcolemma of a muscle fiber (cell). Also called a ***myoneural*** (mi-o-NOO-ral) ***junction.***

Neuron (NOO-ron) A nerve cell, consisting of a cell body, dendrites, and an axon.

Neuropeptide (noo-rō-PEP-tīd) Chain of 2 to about 40 amino acids that occurs naturally in the brain that acts primarily to modulate the response of or to a neurotransmitter. Examples are enkephalins and endorphins.

Neurophysin (noo-rō-FĪ-sin) Small protein that aids in the transportation and storage of oxytocin (OT) and antidiuretic hormone (ADH) in the neurohypophysis and their subsequent release.

Neurosecretory (noo-rō-SĒC-re-tō-rē) ***cell*** A cell in a nucleus (paraventricular and supraoptic) in the hypothalamus that produces oxytocin (OT) or antidiuretic hormone (ADH), hormones stored in the neurohypophysis of the pituitary gland.

Neurosyphilis (noo-rō-SIF-i-lis) A form of the tertiary stage of syphilis in which various types of nervous tissue are attacked by bacteria and degenerate.

Neurotransmitter One of a variety of molecules synthesized within the nerve axon terminals, released into the synaptic cleft in response to a nerve impulse, and affecting the membrane potential of the postsynaptic neuron. Also called a ***transmitter substance.***

Neutrophil (NOO-trō-fil) A type of white blood cell characterized by granular cytoplasm that stains as readily with acid or basic dyes.

Nicotinic (nik'-ō-TIN-ik) ***receptor*** Receptor found on both sympathetic and parasympathetic postganglionic neurons so named because the actions of acetylcholine (ACh) in such receptors are similar to those produced by nicotine.

Night blindness Poor or no vision in dim light or at night, although good vision is present during bright illumination; frequently caused by a deficiency of vitamin A. Also referred to as ***nyctalopia*** (nik'-ta-LŌ-pē-a).

Nipple A pigmented, wrinkled projection on the surface of the mammary gland that is the location of the openings of the lactiferous ducts for milk release.

Nissl bodies *See* ***Chromatophilic substance.***

Nociceptor (nō'-sē-SEP-tor) A free (naked) nerve ending that detects pain.

Node of Ranvier *See* ***Neurofibral node.***

Nondisjunction (non'-dis-JUNGK-shun) Failure of sister chromatids to separate properly during anaphase of mitosis (or equatorial division of meiosis) or failure of homologous chromosomes to separate properly during reduction division of meiosis in which chromatids or chromosomes pass into the same daughter cell.

Nonessential amino acid An amino acid that can be synthesized by body cells through transamination, the transfer of an amino group from an amino acid to another substance.

Nonpigmented granular dendrocytes Two distinct cell types found in the epidermis, formerly known as ***Langerhans cells*** and ***Granstein cells,*** that differ in their sensitivity to damage by ultraviolet (UV) radiation and their functions in immunity.

Norepinephrine (nor'-ep-ē-NEF-rin) ***(NE)*** A hormone secreted by the adrenal medulla that produces actions similar to those that result from sympathetic stimulation. Also called ***noradrenaline*** (nor-a-DREN-a-lin).

Notochord (NŌ-tō-cord) A flexible rod of embryonic tissue that lies where the future vertebral column will develop.

Nuclear medicine The branch of medicine concerned with the use of radioisotopes in the diagnosis of disease and therapy.

Nuclease (NOO-klē-ās) An enzyme that breaks nucleotides into pentoses and nitrogenous bases; examples are ribonuclease and deoxyribonuclease.

Nucleic (noo-KLĒ-ic) ***acid*** An organic compound that is a long polymer of nucleotides, with each nucleotide containing a pentose sugar, a phosphate group, and one of four possible nitrogenous bases (adenine, cytosine, guanine, and thymine or uracil).

Nucleolus (noo-KLĒ-ō-lus) Nonmembranous spherical body within the nucleus composed of protein, DNA, and RNA that functions in the synthesis and storage of ribosomal RNA.

Nucleosome (NOO-klē-ō-sōm) Elementary structural subunit of a chromosome consisting of histones and DNA.

Nucleus (NOO-klē-us) A spherical or oval organelle of a cell that contains the hereditary factors of the cell, called genes. A cluster of unmyelinated nerve cell bodies in the central nervous system. The central portion of an atom made up of protons and neutrons.

Nucleus cuneatus (kyoo-nē-Ā-tus) A group of nerve cells in the inferior portion of the medulla in which fibers of the fasciculus cuneatus terminate.

Nucleus gracilis (gras-I-lis) A group of nerve cells in the inferior portion of the medulla in which fibers of the fasciculus gracilis terminate.

Nucleus pulposus (pul-PŌ-sus) A soft, pulpy, highly elastic substance in the center of an intervertebral disc, a remnant of the notochord.

Nutrient A chemical substance in food that provides energy, forms new body components, or assists in the functioning of various body processes.

Nystagmus (nis-TAG-mus) Rapid, involuntary, rhythmic movement of the eyeballs; horizontal, rotary, or vertical.

Obesity (ō-BĒS-i-tē) Body weight 10–20 percent over a desirable standard as a result of excessive accumulation of fat. Types of obesity are hypertrophic (adult-onset) and hyperplastic (lifelong).

Obligatory water reabsorption The absorption of water from proximal convoluted tubules of nephrons as a function of osmosis.

Obstetrics (ob-STET-riks) The specialized branch of medicine that deals with pregnancy, labor, and the period of time immediately following delivery.

Obturator (OB-tyoo-rā'-ter) Anything that obstructs or closes a cavity or opening.

Occlusion (ō-KLOO-zhun) The act of closure or state of being closed.

Occult (o-KULT) Obscure or hidden from view, as for example, occult blood in stools or urine.

Olfactory (ōl-FAK-tō-rē) Pertaining to smell.

Olfactory bulb A mass of gray matter at the termination of an olfactory (I) nerve, lying beneath the frontal lobe of the cerebrum on either side of the crista galli of the ethmoid bone.

Olfactory cell A bipolar neuron with its cell body lying between supporting cells located in the mucous membrane lining the upper portion of each nasal cavity.

Olfactory tract A bundle of axons that extends from the olfactory bulb posteriorly to the olfactory portion of the cortex.

Oligodendrocyte (o-lig-ō-DEN-drō-sīt) A neuroglial cell that supports neurons and produces a phospholipid myelin sheath around axons of neurons of the central nervous system.

Oligospermia (ol'-i-gō-SPER-mē-a) A deficiency of spermatozoa in the semen.

Oliguria (ol'-i-GYOO-rē-a) Daily urinary output usually less than 250 ml.

Olive A prominent oval mass on each lateral surface of the superior part of the medulla.

Oncogene (ONG-kō-jēn) Gene that has the ability to transform a normal cell into a cancerous cell.

Oncology (ong-KOL-ō-jē) The study of tumors.

Oogenesis (ō'-ō-JEN-e-sis) Formation and development of the ovum.

Oophorectomy (ō'-of-ō-REK-tō-mē) The surgical removal of the ovaries.

Ophthalmic (of-THAL-mik) Pertaining to the eye.

Ophthalmologist (of'-thal-MOL-ō-jist) A physician who specializes in the diagnosis and treatment of eye disorders with drugs, surgery, and corrective lenses.

Ophthalmology (of'-thal-MOL-ō-jē) The study of the structure, function, and diseases of the eye.

Ophthalmoscopy (of'-thal-MOS-kō-pē) Examination of the interior fundus of the eyeball to detect retinal changes associated with hypertension, diabetes mellitus, atherosclerosis, and increased intracranial pressure.

Opsonization (op-sō-ni-ZĀ-shun) The action of some antibodies that renders bacteria and other foreign cells more susceptible to phagocytosis. Also called ***immune adherence.***

Optic (OP-tik) Refers to the eye, vision, or properties of light.

Optic chiasma (kī-AZ-ma) A crossing point of the optic (II) nerves, anterior to the pituitary gland.

Optic disc A small area of the retina containing openings through which the fibers of the ganglion neurons emerge as the optic (II) nerve. Also called the ***blind spot.***

Optician (op-TISH-an) A technician who fits, adjusts, and dispenses corrective lenses on prescription of an ophthalmologist or optometrist.

Optic tract A bundle of axons that transmits nerve impulses from the retina of the eye between the optic chiasma and the thalamus.

Optometrist (op-TOM-e-trist) Specialist with a doctorate degree in optometry who is licensed to examine and test the eyes and treat visual defects by prescribing corrective lenses.

Oral cholecystogram (kō-lē-SIS-to-gram) X-ray examination of the gallbladder to evaluate for the presence of gallstones, inflammations, and tumors.

Oral contraceptive (OC) A hormone compound, usually a high concentration of progesterone and a low concentration of estrogens, that is swallowed and prevents ovulation, and thus pregnancy. Also called ***"the pill."***

Ora serrata (Ō-ra ser-RĀ-ta) The irregular margin of the retina lying internal and slightly posterior to the junction of the choroid and ciliary body.

Orbit (OR-bit) The bony, pyramid-shaped cavity of the skull that holds the eyeball.

Organ A structure composed of two or more different kinds of tissues with a specific function and usually a recognizable shape.

Organelle (or-gan-EL) A permanent structure within a cell with characteristic morphology that is specialized to serve a specific function in cellular activities.

Organic (or-GAN-ik) ***compound*** Compound that always contains carbon and hydrogen and the atoms are held together by covalent bonds. Examples include carbohydrates, lipids, protein, and nucleic acids (DNA and RNA).

Organism (OR-ga-nizm) A total living form; one individual.

Orgasm (OR-gazm) Sensory and motor events involved in ejaculation for the male and involuntary contraction of the perineal muscles in the female at the climax of sexual intercourse.

Orifice (OR-i-fis) Any aperture or opening.

Origin (OR-i-jin) The place of attachment of a muscle to the more stationary bone, or the end opposite the insertion.

Oropharynx (or'-ō-FAR-inks) The second portion of the pharynx, lying posterior to the mouth and extending from the soft palate down to the hyoid bone.

Orthopedics (or'-thō-PĒ-diks) The branch of medicine that deals with the preservation and restoration of the skeletal system, articulations, and associated structures.

Orthopnea (or'-thop-NĒ-a) Dyspnea that occurs in the horizontal position.

Osmoreceptor (oz'-mō-re-CEP-tor) Receptor in the hypothalamus that is sensitive to changes in blood osmotic pressure and, in response to high osmotic pressure (low water concentration), causes synthesis and release of antidiuretic hormone (ADH).

Osmosis (os-MŌ-sis) The net movement of water molecules through a selectively permeable membrane from an area of high water concentration to an area of lower water concentration until an equilibrium is reached.

Osmotic pressure The pressure required to prevent the movement of pure water into a solution containing solutes when the solutions are separated by a selectively permeable membrane.

Osseous (OS-ē-us) Bony.

Ossicle (OS-si-kul) Small bone, as in the middle ear (malleus, incus, stapes).

Ossification (os'-i-fi-KĀ-shun) Formation of bone. Also called ***osteogenesis.***

Osteoblast (OS-tē-ō-blast') Cell formed from an osteoprogenitor cell that participates in bone formation by secreting some organic components and inorganic salts.

Osteoclast (OS-tē-ō-clast') A large multinuclear cell that develops from a monocyte and destroys or resorbs bone tissue.

Osteocyte (OS-tē-ō-sīt') A mature bone cell that maintains the daily activities of bone tissue.

Osteogenic (os'-tē-ō-JEN-ik) ***layer*** The inner layer of the periosteum that contains cells responsible for forming new bone during growth and repair.

Osteology (os'-tē-OL-ō-jē) The study of bones.

Osteomalacia (os'-tē-ō-ma-LĀ-shē-a) A deficiency of vitamin D in adults causing demineralization and softening of bone.

Osteomyelitis (os'-tē-ō-mī-i-LĪ-tis) Inflammation of bone marrow or of the bone and marrow.

Osteon The basic unit of structure in adult compact bone, consisting of a central (Haversian) canal with its concentrically arranged lamellae, lacunae, osteocytes, and canaliculi. Also called a ***Haversian*** (ha-VĒR-shun) ***system.***

Osteoporosis (os'-tē-ō-pō-RŌ-sis) Age-related disorder characterized by decreased bone mass and increased susceptibility to fractures as a result of decreased levels of estrogens.

Osteoprogenitor (os'-tē-ō-prō-JEN-i-tor) ***cell*** Stem cell derived from mesenchyme that has mitotic potential and the ability to differentiate into an osteoblast.

Otalgia (ō-TAL-jē-a) Pain in the ear; earache.

Otic (Ō-tik) Pertaining to the ear.

Otitis media (ō-TĪ-tus MĒ-dē-a) Acute infection of the middle ear characterized by pain, malaise, fever, and an inflamed tympanic membrane, subject to rupture.

Otolith (Ō-tō-lith) A particle of calcium carbonate embedded in the otolithic membrane that functions in maintaining static equilibrium.

Otolithic (ō-tō-LITH-ik) ***membrane*** Thick, gelatinous, glycoprotein layer located directly over hair cells of the macula in the saccule and utricle of the inner ear.

Otorhinolaryngology (ō'-tō-rī-nō-lar'-in-GOL-ō-jē) The branch of medicine that deals with the diagnosis and treatment of diseases of the ears, nose, and throat.

Oval window A small opening between the middle ear and inner ear into which the footplate of the stapes fit. Also called the ***fenestra vestibuli*** (fe-NES-tra ves-TIB-yoo-lē).

Ovarian (ō-VAR-ē-an) ***cycle*** A monthly series of events in the ovary associated with the maturation of an ovum.

Ovarian follicle (FOL-i-kul) A general name for oocytes (immature ova) in any stage of development, along with their surrounding epithelial cells.

Ovarian ligament (LIG-a-ment) A rounded cord of connective tissue that attaches the ovary to the uterus.

Ovary (Ō-var-ē) Female gonad that produces ova and the hormones estrogens, progesterone, and relaxin.

Ovulation (ō-vyoo-LĀ-shun) The rupture of a vesicular ovarian (Graafian) follicle with discharge of a secondary oocyte into the pelvic cavity.

Ovum (Ō-vum) The female reproductive or germ cell; an egg cell.

Oxidation (ok-si-DĀ-shun) The removal of electrons and hydrogen ions (hydrogen atoms) from a molecule or, less commonly, the addition of oxygen to a molecule that results in a decrease in the energy content of the molecule. The oxidation of glucose in the body is also called ***cellular respiration.***

Oxygen debt The volume of oxygen required to oxidize the lactic acid produced by muscular exercise.

Oxyhemoglobin (ok'-sē-HĒ-mō-glō-bin) ***(HbO_2)*** Hemoglobin combined with oxygen.

Oxyphil cell A cell found in the parathyroid gland that secretes parathyroid hormone (PTH).

Oxytocin (ok'-sē-TŌ-sin) ***(OT)*** A hormone secreted by neurosecretory cells in the paraventricular and supraoptic nuclei of the hypothalamus that stimulates contraction of the smooth muscle fibers (cells) in the pregnant uterus and contractile cells around the ducts of mammary glands.

Pacinian corpuscle *See* ***Lamellated corpuscle.***

Paget's (PAJ-ets) ***disease*** A disorder characterized by a greatly accelerated remodeling process in which osteoclastic resorption is massive and new bone formation by osteoblasts is extensive. As a result, there is an irregular thickening and softening of the bones.

Palate (PAL-at) The horizontal structure separating the oral and the nasal cavities; the roof of the mouth.

Palliative (PAL-ē-a-tiv) Serving to relieve or alleviate without curing.

Palpate (PAL-pāt) To examine by touch; to feel.

Palpitation (pal'-pi-TĀ-shun) A fluttering of the heart or abnormal rate or rhythm of the heart.

Pancreas (PAN-krē-as) A soft, oblong organ lying along the greater curvature of the stomach and connected by a duct to the duodenum. It is both exocrine (secreting pancreatic juice) and endocrine (secreting insulin, glucagon, and somatostatin).

Pancreatic (pan'-krē-AT-ik) ***duct*** A single, large tube that unites with the common bile duct from the liver and gallbladder and drains pancreatic juice into the duodenum at the hepatopancreatic ampulla (ampulla of Vater). Also called the ***duct of Wirsung.***

Pancreatic islet A cluster of endocrine gland cells in the pancreas that secretes insulin, glucagon, and somatostatin. Also called an ***islet of Langerhans*** (LANG-er-hanz).

Papanicolaou (pap'-a-NIK-ō-la-oo) ***test*** A cytological staining test for the detection and diagnosis of premalignant and malignant conditions of the female genital tract. Cells scraped from the genital epithelium are smeared, fixed, stained, and examined microscopically. Also called a ***Pap smear.***

Papilla (pa-PIL-a) A small nipple-shaped projection or elevation.

Paralysis (pa-RAL-a-sis) Loss or impairment of motor function due to a lesion of nervous or muscular origin.

Paranasal sinus (par'-a-NĀ-zal SĪ-nus) A mucus-lined air cavity in a skull bone that communicates with the nasal cavity. Paranasal sinuses are located in the frontal, maxillary, ethmoid, and sphenoid bones.

Paraplegia (par-a-PLĒ-jē-a) Paralysis of both lower extremities.

Parasagittal plane A vertical plane that does not pass through the midline and that divides the body or organs into *unequal* left and right portions.

Parasympathetic (par'-a-sim-pa-THET-ik) ***division*** One of the two subdivisions of the autonomic nervous system, having cell bodies of preganglionic neurons in nuclei in the brain stem and in the lateral gray matter of the sacral portion of the spinal cord; primarily concerned with activities that conserve and restore body energy. Also called the ***craniosacral*** (krā-nē-ō-SĀ-kral) ***division.***

Parathyroid (par'-a-THĪ-royd) ***gland*** One of four small endocrine glands embedded on the posterior surfaces of the lateral lobes of the thyroid gland.

Parathyroid hormone (PTH) A hormone secreted by the parathyroid glands that decreases blood phosphate level and increases blood calcium level.

Paraurethral (par'-a-yoo-RĒ-thral) ***gland*** Gland embedded in the wall of the urethra whose duct opens on either side of the urethral orifice and secretes mucus. Also called ***Skene's*** (SKĒNZ) ***gland.***

Parenchyma (par-EN-ki-ma) The functional parts of any organ, as opposed to tissue that forms its stroma or framework.

Parenteral (par-EN-ter-al) Situated or occurring outside the intestines; referring to introduction of substances into the body other than by way of the intestines such as intradermal, subcutaneous, intramuscular, intravenous, or intraspinal.

Parietal (pa-RĪ-e-tal) Pertaining to or forming the outer wall of a body cavity.

Parietal cell The secreting cell of a gastric gland that produces hydrochloric acid and intrinsic factor. Also called an ***oxyntic cell.***

Parietal pleura (PLOO-ra) The outer layer of the serous pleural membrane that encloses and protects the lungs; the layer that is attached to the wall of the pleural cavity.

Parkinson's disease Progressive degeneration of the basal ganglia and substantia nigra of the cerebrum resulting in decreased production of dopamine (DA) that leads to tremor, slowing of voluntary movements, and muscle weakness. Also called ***Parkinsonism.***

Parotid (pa-ROT-id) ***gland*** One of the paired salivary glands located inferior and anterior to the ears connected to the oral cavity via a duct (Stensen's) that opens into the inside of the cheek opposite the upper second molar tooth.

Paroxysm (PAR-ok-sizm) A sudden periodic attack or recurrence of symptoms of a disease.

Pars intermedia A small avascular zone between the adenohypophysis and neurohypophysis of the pituitary gland.

Parturition (par'-too-RISH-un) Act of giving birth to young; childbirth, delivery.

Patellar (pa-TELL-ar) ***reflex*** Extension of the leg by contraction of the quadriceps femoris muscle in response to tapping the patellar ligament. Also called the ***knee jerk.***

Patent ductus arteriosus Congenital anatomical heart defect in which the fetal connection between the aorta and pulmonary trunk remains open instead of closing completely after birth.

Pathogen (PATH-ō-jen) A disease-producing organism.

Pathogenesis (path'-ō-JEN-e-sis) The development of disease or a morbid or pathological state.

Pathological (path'-ō-LOJ-i-kal) Pertaining to or caused by disease.

Pathological (path'-ō-LOJ-i-kal) ***anatomy*** The study of structural changes caused by disease.

Pectinate (PEK-ti-nāt) ***muscles*** Projecting muscle bundles of the anterior atrial walls and the lining of the auricles.

Pectoral (PEK-tō-ral) Pertaining to the chest or breast.

Pediatrician (pē'-dē-a-TRISH-un) A physician who specializes in the care and treatment of children and their illnesses.

Pedicel (PED-i-sel) Footlike structure, as on podocytes of a glomerulus.

Pedicle (PED-i-kul) A short, thick process found on vertebrae.

Pelvic (PEL-vik) ***cavity*** Inferior portion of the abdominopelvic cavity that contains the urinary bladder, sigmoid colon, rectum, and internal female and male reproductive structures.

Pelvic inflammatory disease (PID) Collective term for any extensive bacterial infection of the pelvic organs, especially the uterus, uterine (Fallopian) tubes, and ovaries.

Pelvic splanchnic (PEL-vik SPLANGK-nik) ***nerves*** Preganglionic parasympathetic fibers from the levels of S2, S3, and S4 that supply the urinary bladder, reproductive organs, and the descending and sigmoid colon and rectum.

Pelvimetry (pel-VIM-e-trē) Measurement of the size of the inlet and outlet of the birth canal.

Pelvis The basinlike structure formed by the two pelvic (hip) bones, the sacrum, and the coccyx. The expanded, proximal portion of the ureter, lying within the kidney and into which the major calyces open.

Penis (PĒ-nis) The male copulatory organ, used to introduce spermatozoa into the female vagina.

Pepsin Protein-digesting enzyme secreted by zymogenic (chief) cells of the stomach as the inactive form pepsinogen, which is converted to active pepsin by hydrochloric acid.

Peptic ulcer An ulcer that develops in areas of the gastrointestinal tract exposed to hydrochloric acid; classifed as a gastric ulcer if in the lesser curvature of the stomach and as a duodenal ulcer if in the first part of the duodenum.

Percussion (per-KUSH-un) The act of striking (percussing) an underlying part of the body with short, sharp blows as an aid in diagnosing the part by the quality of the sound produced.

Perforating canal A minute passageway by means of which blood vessels and nerves from the periosteum penetrate into compact bone. Also called ***Volkmann's*** (FŌLK-manz) ***canal.***

Pericardial (per'-i-KAR-dē-al) ***cavity*** Small potential space between the visceral and parietal layers of the serous pericardium that contains pericardial fluid.

Pericardium (per'-i-KAR-dē-um) A loose-fitting membrane that encloses the heart, consisting of an outer fibrous layer and an inner serous layer.

Perichondrium (per'-i-KON-drē-um) The membrane that covers cartilage.

Perikaryon (per'-i-KAR-ē-on) The nerve cell body that contains the nucleus and other organelles.

Perilymph (PER-i-lymf) The fluid contained between the bony and membranous labyrinths of the inner ear.

Perimetrium (per-i-MĒ-trē-um) The serosa of the uterus.

Perimysium (per'-i-MĪZ-ē-um) Invagination of the epimysium that divides muscles into bundles.

Perineum (per'-i-NĒ-um) The pelvic floor; the space between the anus and the scrotum in the male and between the anus and the vulva in the female.

Perineurium (per'-i-NYOO-rē-um) Connective tissue wrapping around fascicles in a nerve.

Periodontal (per-ē-ō-DON-tal) ***disease*** A collective term for conditions characterized by degeneration of gingivae, alveolar bone, periodontal ligament, and cementum.

Periodontal membrane The periosteum lining the alveoli (sockets) for the teeth in the alveolar processes of the mandible and maxillae.

Periosteum (per'-ē-OS-tē-um) The membrane that covers bone and consists of connective tissue, osteoprogenitor cells, and osteoblasts and is essential for bone growth, repair, and nutrition.

Peripheral (pe-RIF-er-al) Located on the outer part or a surface of the body.

Peripheral nervous system (PNS) The part of the nervous system that lies outside the central nervous system—nerves and ganglia.

Peripheral resistance (pe-RIF-er-al re-ZIS-tans) Resistance (impedance) to blood flow as a result of the force of friction between blood and the walls of blood vessels that is related to viscosity of blood and blood vessel length and diameter.

Periphery (pe-RIF-er-ē) Outer part or a surface of the body; part away from the center.

Peristalsis (per'-i-STAL-sis) Successive muscular contractions along the wall of a hollow muscular structure.

Peritoneum (per'-i-tō-NĒ-um) The largest serous membrane of the body that lines the abdominal cavity and covers the viscera.

Peritonitis (per'-i-tō-NĪ-tis) Inflammation of the peritoneum.

Permissive (per-MIS-sive) ***effect*** A hormonal interaction in which the effect of one hormone on a target cell requires previous or simultaneous exposure to another hormone(s) to enhance the response of a target cell or increase the activity of another hormone. Exposure of the uterus first to estrogens and then progesterone (PROG) in preparation for implantation is an example.

Pernicious (per-NISH-us) Fatal.

Peroxisome (pe-ROKS-ī-sōm) Organelle similar in structure to a lysosome that contains enzymes related to hydrogen peroxide metabolism; abundant in liver cells.

Perspiration Substance produced by sudoriferous (sweat) glands containing water, salts, urea, uric acid, amino acids, ammonia, sugar, lactic acid, and ascorbic acid; helps maintain body temperature and eliminate wastes.

Peyer's patches *See* ***Aggregated lymphatic follicles.***

pH A symbol of the measure of the concentration of hydrogen ions in a solution. The pH scale extends from 0 to 14, with a value of 7 expressing neutrality, values lower than 7 expressing increasing acidity, and values higher than 7 expressing increasing alkalinity.

Phagocytosis (fag'-ō-sī-TŌ-sis) The process by which cells (phagocytes) ingest particulate matter; especially the ingestion and destruction of microbes, cell debris, and other foreign matter.

Phalanx (FĀ-lanks) The bone of a finger or toe. *Plural,* ***phalanges*** (fa-LAN-jēz).

Phantom pain A sensation of pain as originating in a limb that has been amputated.

Pharmacology (far'-ma-KOL-ō-jē) The science that deals with the effects and uses of drugs in the treatment of disease.

Pharynx (FAR-inks) The throat; a tube that starts at the internal nares and runs partway down the neck where it opens into the esophagus posteriorly and the larynx anteriorly.

Phenotype (FĒ-nō-tīp) The observable expression of genotype; physical characteristics of an organism determined by genetic makeup and influenced by interaction between genes and internal and external environmental factors.

Phenylketonuria (fen'-il-kē'-tō-NOO-rē-a) ***(PKU)*** A disorder characterized by an elevation of the amino acid phenylalanine in the blood.

Pheochromocytoma (fē-ō-krō'-mō-sī-TŌ-ma) Tumor of the chromaffin cells of the adrenal medulla that results in hypersecretion of medullary hormones.

Phlebitis (fle-BĪ-tis) Inflammation of a vein, usually in the lower extremities.

Phlebotomy (fle-BOT-ō-me) The cutting of a vein to allow the escape of blood.

Phosphocreatine (fos'-fō-KRĒ-a-tin) High-energy molecule in skeletal muscle fibers (cells) that is used to generate ATP rapidly; upon decomposition, phosphocreatine breaks down into creatine, phosphate, and energy—the energy is used to generate ATP from ADP.

Phospholipid (fos'-fō-LIP-id) ***bilayer*** Arrangement of phospholipid molecules in two parallel rows in which the hydrophilic "heads" face outward and the hydrophobic "tails" face inward.

Phosphorylation (fos'-for-i-LĀ-shun) The addition of a phosphate group to a chemical compound; types include substrate-level, oxidative, and photophosphorylation.

Photopigment A substance that can absorb light and undergo structural changes that can lead to the development of a receptor potential. An example is rhodopsin.

Photoreceptor Receptor that detects light on the retina of the eye.

Physiology (fiz'-ē-OL-ō-jē) Science that deals with the functions of an organism or its parts.

Pia mater (PĪ-a MĀ-ter) The inner membrane (meninx) covering the brain and spinal cord.

Piezoelectric (pē-e-zō-e-LEK-trik) ***effect*** Response of bone, mainly collagen, to stress in which very minute currents of electricity are produced; believed to stimulate osteoblasts to make new bone cells.

Pilonidal (pī-lō-NĪ-dal) Containing hairs resembling a tuft inside a cyst or sinus.

Pineal (PĪN-ē-al) ***gland*** The cone-shaped gland located in the roof of the third ventricle. Also called the ***epiphysis cerebri*** (ē-PIF-i-sis se-RĒ-brē).

Pinealocyte (pin-ē-AL-ō-sīt) Secretory cell of the pineal gland that produces hormones.

Pinna (PIN-na) The projecting part of the external ear composed of elastic cartilage and covered by skin and shaped like the flared end of a trumpet. Also called the ***auricle*** (OR-i-kul).

Pinocytosis (pi'-nō-sī-TŌ-sis) The process by which cells ingest liquid.

Pituicyte (pi-TOO-i-sīt) Supporting cell of the posterior lobe of the pituitary gland.

Pituitary (pi-TOO-i-tar'-ē) ***dwarfism*** Condition caused by hyposecretion of human growth hormone (hGH) during the growth years and characterized by childlike physical traits in an adult.

Pituitary gland A small endocrine gland lying in the sella turcica of the sphenoid bone and attached to the hypothalamus by the infundibulum; nicknamed the "master gland." Also called the ***hypophysis*** (hī-POF-i-sis).

Pivot joint A synovial joint in which a rounded, pointed, or conical surface of one bone articulates with a ring formed partly by another bone and partly by a ligament, as in the joint between the atlas and axis and between the proximal ends of the radius and ulna. Also called a ***trochoid*** (TRŌ-koid) ***joint.***

Placenta (pla-SEN-ta) The special structure through which the exchange of materials between fetal and maternal circulations occurs. Also called the ***afterbirth.***

Plantar flexion (PLAN-tar FLEK-shun) Bending the foot in the direction of the plantar surface (sole).

Plaque (plak) A cholesterol-containing mass in the tunica media of arteries. A mass of bacterial cells, dextran (polysaccharide), and other debris that adheres to teeth.

Plasma (PLAZ-ma) The extracellular fluid found in blood vessels; blood minus the formed elements.

Plasma cell Cell that produces antibodies and develops from a B cell (lymphocyte).

Plasma (cell) membrane Outer, limiting membrane that

separates the cell's internal parts from extracellular fluid and the external environment.

Plasmapheresis (plaz'-ma-fe-RĒ-sis) A procedure in which blood is withdrawn from the body, its components are selectively separated, the undesirable component causing disease is removed, and the remainder is returned to the body. Among the substances removed are toxins, metabolic substances, and antibodies. Also called ***therapeutic plasma exchange (TPE).***

Platelet plug Aggregation of thrombocytes at a damaged blood vessel to prevent blood loss.

Pleura (PLOOR-a) The serous membrane that covers the lungs and lines the walls of the chest and diaphragm.

Pleural cavity Small potential space between the visceral and parietal pleurae.

Plexus (PLEK-sus) A network of nerves, veins, or lymphatic vessels.

Plexus of Auerbach See ***Myenteric plexus.***

Plexus of Meissner See ***Submucosal plexus.***

Pneumonia (noo-MŌ-nē-a) Acute infection or inflammation of the alveoli of the lungs.

Pneumotaxic (noo-mō-TAK-sik) ***area*** Portion of the respiratory center in the pons that continually sends inhibitory nerve impulses to the inspiratory area that limit inspiration and facilitate expiration.

Podiatry (pō-DĪ-a-trē) The diagnosis and treatment of foot disorders.

Polar body The smaller cell resulting from the unequal division of cytoplasm during the meiotic divisions of an oocyte. The polar body has no function and is resorbed.

Polarized A condition in which opposite effects or states exist at the same time. In electrical contexts, having one portion negative and another positive; for example, a polarized nerve cell membrane has the outer surface positively charged and the inner surface negatively charged.

Poliomyelitis (pō'-lē-ō-mī-e-LĪ-tis) Viral infection marked by fever, headache, stiff neck and back, deep muscle pain and weakness, and loss of certain somatic reflexes; a serious form of the disease, ***bulbar polio,*** results in destruction of motor neurons in anterior horns of spinal nerves that leads to paralysis.

Polycythemia (pol'-ē-sī-THĒ-mē-a) Disorder characterized by a hematocrit above the normal level of 55 in which hypertension, thrombosis, and hemorrhage occur.

Polyp (POL-ip) A tumor on a stem found especially on a mucous membrane.

Polysaccharides (pol'-ē-SAK-a-rīds) A carbohydrate in which three or more monosaccharides are joined chemically.

Polyunsaturated fat A fat that contains more than one double covalent bond between its carbon atoms; examples are corn oil, safflower oil, and cottonseed oil.

Polyuria (pol'-ē-YOO-rē-a) An excessive production of urine.

Pons (ponz) The portion of the brain stem that forms a "bridge" between the medulla and the midbrain, anterior to the cerebellum.

Positron emission tomography (PET) A type of radioactive scanning based on the release of gamma rays when positrons collide with negatively charged electrons in body tissues; it indicates where radioisotopes are used in the body.

Postabsorptive state Metabolic state during which absorption is complete and energy needs of the body must be satisfied.

Postcentral gyrus A gyrus immediately posterior to the central sulcus that contains the general sensory area of the cerebral cortex.

Posterior (pos-TĒR-ē-or) Nearer to or at the back of the body. Also called ***dorsal.***

Posterior root The structure composed of afferent (sensory) fibers lying between a spinal nerve and the dorsolateral aspect of the spinal cord. Also called the ***dorsal (sensory) root.***

Posterior root ganglion A group of cell bodies of sensory (afferent) neurons and their supporting cells located along the posterior root of a spinal nerve. Also called a ***dorsal (sensory) root ganglion*** (GANG-glē-on).

Postganglionic neuron (pōst'-gang-lē-ON-ik NOO-ron) The second visceral efferent neuron in an autonomic pathway, having its cell body and dendrites located in an autonomic ganglion and its unmyelinated axon ending at cardiac muscle, smooth muscle, or a gland.

Postpartum (pōst-PAR-tum) After parturition; occurring after the delivery of a baby.

Postsynaptic (pōst-sin-AP-tik) ***neuron*** The nerve cell that is activated by the release of a neurotransmitter substance from another neuron and carries nerve impulses away from the synapse.

Pouch of Douglas See ***Rectouterine pouch.***

Precapillary sphincter (SFINGK-ter) A ring of smooth muscle fibers (cells) at the site of origin of true capillaries that regulate blood flow into true capillaries.

Precentral gyrus A gyrus immediately anterior to the central sulcus that contains the primary motor area of the cerebral cortex.

Preeclampsia (pre'-e-KLAMP-sē-a) A syndrome characterized by sudden hypertension, large amounts of protein in urine, and generalized edema; it might be related to an autoimmune or allergic reaction due to the presence of a fetus.

Preganglionic (prē'-gang-lē-ON-ik) ***neuron*** The first visceral efferent neuron in an autonomic pathway, with its cell body and dendrites in the brain or spinal cord and its myelinated axon ending at an autonomic ganglion, where it synapses with a postganglionic neuron.

Pregnancy Sequence of events that normally includes fertilization, implantation, embryonic growth, and fetal growth that terminates in birth.

Premenstrual syndrome (PMS) Severe physical and emotional stress occurring late in the postovulatory phase of the menstrual cycle and sometimes overlapping with menstruation.

Premonitory (prē-MON-i-tō-rē) Giving previous warning; as premonitory symptoms.

Prepuce (PRĒ-pyoos) The loose-fitting skin covering the glans of the penis and clitoris. Also called the ***foreskin.***

Presbyopia (prez-bē-Ō-pē-a) A loss of elasticity of the lens of the eye due to advancing age with resulting inability to focus clearly on near objects.

Presynaptic (prē-sin-AP-tik) ***inhibition*** Inhibition of a nerve impulse before it reaches a synapse in which neurotransmitter released by an inhibitory neuron depresses the release of excitatory transmitter at an excitatory neuron.

Presynaptic (prē-sin-AP-tik) ***neuron*** A nerve cell that carries nerve impulses toward a synapse.

Prevertebral ganglion (prē-VERT-e-bral GANG-lē-on) A cluster of cell bodies of postganglionic sympathetic neurons anterior to the spinal column and close to large abdominal arteries. Also called a ***collateral ganglion.***

Primary germ layer One of three layers of embryonic tissue, called ectoderm, mesoderm, and endoderm, that give rise to all tissues and organs of the organism.

Primary motor area A region of the cerebral cortex in the precentral gyrus of the frontal lobe of the cerebrum that controls specific muscles or groups of muscles.

Primary somesthetic (sō-mes-THET-ik) ***area*** A region of the cerebral cortex posterior to the central sulcus in the postcentral gyrus of the parietal lobe of the cerebrum that localizes exactly the points of the body where sensations originate.

Prime mover The muscle directly responsible for producing the desired motion. Also called an ***agonist*** (AG-ō-nist).

Primigravida (prī-mi-GRAV-i-da) A woman pregnant for the first time.

Primitive gut Embryonic structure composed of endoderm and mesoderm that gives rise to most of the gastrointestinal tract.

Primordial (prī-MŌR-dē-al) Existing first; especially primordial egg cells in the ovary.

Principal cell Cell found in the parathyroid glands that secretes parathyroid hormone (PTH). Also called a ***chief cell.***

Proctology (prok-TOL-ō-jē) The branch of medicine that treats the rectum and its disorders.

Progeny (PROJ-e-nē) Refers to offspring or descendants.

Progesterone (prō-JES-te-rōn) ***(PROG)*** A female sex hormone produced by the ovaries that helps prepare the endometrium for implantation of a fertilized ovum and the mammary glands for milk secretion.

Prognosis (prog-NŌ-sis) A forecast of the probable results of a disorder; the outlook for recovery.

Projection (prō-JEK-shun) The process by which the brain refers sensations to their point of stimulation.

Prolactin (prō-LAK-tin) ***(PRL)*** A hormone secreted by the adenohypophysis (anterior lobe) of the pituitary gland that initiates and maintains milk secretion by the mammary glands.

Prolapse (PRŌ-laps) A dropping or falling down of an organ, especially the uterus or rectum.

Proliferation (pro-lif'-er-Ā-shun) Rapid and repeated reproduction of new parts, especially cells.

Pronation (prō-NĀ-shun) A movement of the forearm in which the palm of the hand is turned posteriorly or inferiorly.

Properdin (prō-PER-din) A protein found in serum capable of destroying bacteria and viruses.

Prophase (PRŌ-fāz) The first stage of mitosis during which chromatid pairs are formed and aggregate around the equatorial plane region of the cell.

Proprioception (prō-prē-ō-SEP-shun) The receipt of information from muscles, tendons, and the labyrinth that enables the brain to determine movements and position of the body and its parts. Also called ***kinesthesia*** (kin'-es-THĒ-zē-a).

Proprioceptor (prō-'prē-ō-SEP-tor) A receptor located in muscles, tendons, or joints that provides information about body position and movements.

Prostaglandin (pros'-ta-GLAN-din) ***(PG)*** A membrane-associated lipid composed of 20-carbon fatty acids with 5 carbon atoms joined to form a cyclopentane ring; synthesized in small quantities and basically mimics the activities of hormones.

Prostatectomy (pros'-ta-TEK-tō-mē) The surgical removal of part of or the entire prostrate gland.

Prostate (PROS-tāt) ***gland*** A doughnut-shaped gland inferior to the urinary bladder that surrounds the superior portion of the male urethra and secretes a slightly acid solution that contributes to sperm motility and viability.

Prosthesis (pros-THĒ-sis) An artificial device to replace a missing body part.

Protein An organic compound consisting of carbon, hydrogen, oxygen, nitrogen, and sometimes sulfur and phosphorus, and made up of amino acids linked by peptide bonds.

Prothrombin (prō-THROM-bin) An inactive protein synthesized by the liver, released into the blood, and converted to active thrombin in the process of blood clotting.

Proto-oncogene (prō'-tō-ONG-kō-jēn) Gene responsible for some aspect of normal growth and development; it may transform into an oncogene, a gene capable of causing cancer.

Protraction (prō-TRAK-shun) The movement of the mandible or shoulder girdle forward on a plane parallel with the ground.

Proximal (PROK-si-mal) Nearer the attachment of an extremity to the trunk or a structure; nearer to the point of origin.

Pruritus (proo'-RĪ-tus) Itching.

Pseudopodia (soo'-dō-PŌ-dē-a) Temporary, protruding projections of cytoplasm.

Psoriasis (sō-RĪ-a-sis) Chronic skin disease characterized by reddish plaques or papules covered with scales.

Psychosomatic (sī'-kō-sō-MAT-ik) Pertaining to the relation between mind and body. Commonly used to refer to those physiological disorders thought to be caused entirely or partly by emotional disturbances.

Pterygopalatine ganglion (ter'-i-gō-PAL-a-tīn GANG-glē-on) A cluster of cell bodies of parasympathetic postganglionic neurons ending at the lacrimal and nasal glands.

Ptosis (TŌ-sis) Drooping, as of the eyelid or the kidney (nephrotosis).

Puberty (PYOO-ber-tē) The time of life during which the secondary sex characteristics begin to appear and the capability for sexual reproduction is possible; usually between the ages of 10 and 17.

Puerperium (pyoo'-er-PER-ē-um) The state immediately after childbirth, usually 4–6 weeks.

Pulmonary (PUL-mo-ner'-ē) Concerning or affected by the lungs.

Pulmonary circulation The flow of deoxygenated blood from the right ventricle to the lungs and the return of oxygenated blood from the lungs to the left atrium.

Pulmonary edema (e-DĒ-ma) An abnormal accumulation of interstitial fluid in the tissue spaces and alveoli of the lungs due to increased pulmonary capillary permeability or increased pulmonary capillary pressure.

Pulmonary embolism (EM-bō-lizm) ***(PE)*** The presence of a blood clot or other foreign substance in a pulmonary arterial blood vessel that obstructs circulation to lung tissue.

Pulmonary (PUL-mo-ner-ē) ***function tests*** Any number of tests (forced vital capacity, forced expiratory volume in one second, maximum midrespiratory flow, maximum voluntary ventilation) designed to evaluate lung disease and measure pulmonary impairment.

Pulmonary ventilation The inflow (inspiration) and outflow (expiration) of air between the atmosphere and the lungs. Also called ***breathing.***

Pulp cavity A cavity within the crown and neck of a tooth, filled with pulp, a connective tissue containing blood vessels, nerves, and lymphatics.

Pulsating electromagnetic fields (PEMFs) A procedure that uses electrotherapy to treat improperly healing fractures.

Punctate (PUNK-tāt) ***distribution*** Unequal distribution of cutaneous receptors.

Pulse pressure The difference between the maximum (systolic) and minimum (diastolic) pressures; normally a value of about 40 mm Hg.

Pupil The hole in the center of the iris, the area through which light enters the posterior cavity of the eyeball.

Purkinje fiber *See* ***Conduction myofiber.***

Pus The liquid product of inflammation containing leucocytes or their remains and debris of dead cells.

P wave The deflection wave of an electrocardiogram that records atrial depolarization (contraction).

Pyelitis (pī'-e-LĪ-tis) Inflammation of the kidney pelvis and its calyces.

Pyemia (pī-Ē-mē-a) Infection of the blood, with multiple abscesses, caused by pus-forming microorganisms.

Pyloric (pī-LOR-ik) ***sphincter*** A thickened ring of smooth muscle through which the pylorus of the stomach communicates with the duodenum. Also called the ***pyloric valve.***

Pyogenesis (pī'-ō-JEN-e-sis) Formation of pus.

Pyorrhea (pī-ō-RĒ-a) A discharge or flow of pus, especially in the alveoli (sockets) and the tissues of the gums.

Pyramid (PIR-a-mid) A pointed or cone-shaped structure; one of two roughly triangular structures on the ventral side of the medulla composed of the largest motor tracts that run from the cerebral cortex to the spinal cord; a triangular-shaped structure in the renal medulla composed of the straight segments of renal tubules.

Pyramidal (pi-RAM-i-dal) ***pathways*** Collections of motor nerve fibers arising in the brain and passing down through the spinal cord to motor cells in the anterior horns.

Pyrexia (pī-REK-sē-a) A condition in which the temperature is above normal.

Pyuria (pī-YOO-rē-a) The presence of leucocytes and other components of pus in urine.

QRS wave The deflection wave of an electrocardiogram that records ventricular depolarization (contraction) and atrial relaxation.

Quadrant (KWOD-rant) One of four parts.

Quadriplegia (kwod'-ri-PLĒ-jē-a) Paralysis of the two upper and two lower extremities.

Radiographic (rā'-dē-ō-GRAF-ic) ***anatomy*** Diagnostic branch of anatomy that includes the use of x rays.

Rales (RALS) Sounds sometimes heard in the lungs that resemble bubbling or rattling due to the presence of an abnormal amount or type of fluid or mucus inside the bronchi or alveoli, or to bronchoconstriction so that air cannot enter or leave the lungs normally.

Rami communicantes (RĀ-mē ko-myoo-ni-KAN-tēz) Branches of a spinal nerve. *Singular,* ***ramus communicans*** (RĀ-mus ko-MYOO-ni-kans).

Rapid eye movement (REM) sleep A level of sleep characterized by symmetrical flutter of the eyes and eyelids and brain wave patterns similar to those of an awake person.

Rathke's pouch *See* ***Hypophyseal pouch.***

Raynaud's (rā-NOZ) ***disease*** A vascular disorder, primarily of females, characterized by bilateral attacks of ischemia, usually of the fingers and toes, in which the skin becomes pale and exhibits burning and pain; it is brought on by cold or emotional stimuli.

Reactivity (rē-ak-TI-vi-tē) Ability of an antigen to react specifically with the antibody whose formation it induced.

Receptor A specialized cell or a nerve cell terminal modified to respond to some specific sensory modality, such as touch, pressure, cold, light, or sound, and convert it to a nerve impulse by way of a generator or receptor potential. A specific molecule or arrangement of molecules organized to accept only molecules with a complementary shape.

Receptor-mediated endocytosis A highly selective process in which cells take up large molecules or particles (ligands). In the process, successive compartments called vesicles, endosomes, and CURLs form. Ligands are eventually broken down by enzymes in lysosomes.

Receptor potential Depolarization of the plasma membrane of a receptor cell which stimulates release of neurotransmitter from the cell; if the neuron connected to the receptor cell becomes depolarized to threshold, a nerve action potential (nerve impulse) is triggered.

Recessive gene A gene that is not expressed in the presence of a dominant gene on the homologous chromosome.

Reciprocal innervation (re-SIP-rō-kal in-ner-VĀ-shun) The phenomenon by which action potentials stimulate contraction of one muscle and simultaneously inhibit contraction of antagonistic muscles.

Recombinant DNA Synthetic DNA, formed by joining a fragment of DNA from one source to a portion of DNA from another.

Recruitment (rē-KROOT-ment) The process of increasing the number of active motor units. Also called ***motor unit summation.***

Rectouterine pouch A pocket formed by the parietal peritoneum as it moves posteriorly from the surface of the uterus and is reflected onto the rectum; the lowest point in the pelvic cavity. Also called the ***pouch*** or ***cul de sac of Douglas.***

Rectum (REK-tum) The last 20 cm (7 in.) of the gastrointestinal tract, from the sigmoid colon to the anus.

Recumbent (re-KUM-bent) Lying down.

Red nucleus A cluster of cell bodies in the midbrain, occupying a large portion of the tectum and sending fibers into the rubroreticular and rubrospinal tracts.

Red pulp That portion of the spleen that consists of venous sinuses filled with blood and cords of splenic tissue called splenic (Billroth's) cords.

Reduction The addition of electrons and hydrogen ions (hydrogen atoms) to a molecule or, less commonly, the removal of oxygen from a molecule that results in an increase in the energy content of the molecule.

Referred pain Pain that is felt at a site remote from the place of origin.

Reflex Fast response to a change (stimulus) in the internal or external environment that attempts to restore homeostasis; passes over a reflex arc.

Reflex arc The most basic conduction pathway through the nervous system, connecting a receptor and an effector and consisting of a receptor, a sensory neuron, a center in the central nervous system for a synapse, a motor neuron, and an effector.

Refraction (rē-FRAK-shun) The bending of light as it passes from one medium to another.

Refractory (re-FRAK-to-rē) ***period*** A time during which an excitable cell cannot respond to a stimulus that is usually adequate to evoke an action potential.

Regeneration (rē-jen'-er-Ā-shun) The natural renewal of a structure.

Regimen (REJ-i-men) A strictly regulated scheme of diet, exercise, or activity designed to achieve certain ends.

Regional anatomy The division of anatomy dealing with a specific region of the body, such as the head, neck, chest, or abdomen.

Regulating factor Chemical secretion of the hypothalamus whose structure is unknown that can either stimulate or inhibit secretion of hormones of the adenohypophysis (anterior pituitary).

Regulating hormone Chemical secretion of the hypothalamus whose structure is known that can either stimulate or inhibit secretion of hormones of the adenohypophysis (anterior pituitary).

Regurgitation (rē-gur'-ji-TĀ-shun) Return of solids or fluids to the mouth from the stomach; flowing backward of blood through incompletely closed heart valves.

Relapse (RĒ-laps) The return of a disease weeks or months after its apparent cessation.

Relaxin (RLX) A female hormone produced by the ovaries that relaxes the symphysis pubis and helps dilate the uterine cervix to facilitate delivery.

Remodeling Replacement of old bone by new bone tissue.

Renal (RĒ-nal) Pertaining to the kidney.

Renal corpuscle (KOR-pus'-l) A glomerular (Bowman's) capsule and its enclosed glomerulus.

Renal erythropoietic (ē-rith'-rō-poy-Ē-tik) ***factor*** An enzyme released by the kidneys and liver in response to hypoxia that acts on a plasma protein to bring about the production of erythropoietin, a hormone that stimulates red blood cell production.

Renal failure Inability of the kidneys to function properly, due to abrupt failure (acute) or progressive failure (chronic).

Renal pelvis A cavity in the center of the kidney formed by the expanded, proximal portion of the ureter, lying within the kidney, and into which the major calyces open.

Renal pyramid A triangular structure in the renal medulla composed of the straight segments of renal tubules.

Renin (RĒ-nin) An enzyme released by the kidney into the plasma where it converts angiotensinogen into angiotensin I.

Renin-angiotensin (an'-jē-ō-TEN-sin) ***pathway*** A mechanism for the control of aldosterone secretion by angiotensin II, initiated by the secretion of renin by the kidney in response to low blood pressure.

Reproduction (rē'-prō-DUK-shun) Either the formation of new cells for growth, repair, or replacement, or the production of a new individual.

Reproductive cell division Type of cell division in which sperm and egg cells are produced; consists of meiosis and cytokinesis.

Residual (re-ZID-yoo-al) ***volume*** The volume of air still contained in the lungs after a maximal expiration; about 1,200 ml.

Resistance Ability to ward off disease. The hindrance encountered by an electrical charge as it moves through a substance from one point to another. The hindrance encountered by blood as it flows through the vascular system or by air through respiratory passageways.

Respiration (res-pi-RĀ-shun) Overall exchange of gases between the atmosphere, blood, and body cells consisting of pulmonary ventilation, external respiration, and internal respiration.

Respirator (RES-pi-rā'-tor) An apparatus fitted to a mask over the nose and mouth, or hooked directly to an endotracheal or tracheotomy tube, that is used to assist or support ventilation or to provide nebulized medication to the air passages under positive pressure.

Respiratory center Neurons in the reticular formation of the brain stem that regulate the rate of respiration.

Respiratory distress syndrome (RDS) of the newborn A disease of newborn infants, especially premature ones, in which insufficient amounts of surfactant are produced and breathing is labored. Also called ***hyaline*** (HĪ-a-lin) ***membrane disease (HMD).***

Respiratory failure Condition in which the respiratory system cannot supply sufficient oxygen to maintain metabolism or eliminate enough carbon dioxide to prevent respiratory acidosis.

Resting membrane potential The voltage that exists between the inside and outside of a cell membrane when the cell is not responding to a stimulus; about −70 to −90 mV, with the inside of the cell negative.

Resuscitation (rē-sus'-i-TĀ-shun) Act of bringing a person back to full consciousness.

Retention (rē-TEN-shun) A failure to void urine due to obstruction, nervous contraction of the urethra, or absence of sensation of desire to urinate.

Rete (RĒ-tē) ***testis*** The network of ducts in the testes.

Reticular (re-TIK-yoo-lar) ***activating system (RAS)*** An extensive network of branched nerve cells running through the core of the brain stem. When these cells are activated, a generalized alert or arousal behavior results.

Reticular formation A network of small groups of nerve cells scattered among bundles of fibers beginning in the medulla as a continuation of the spinal cord and extending upward through the central part of the brain stem.

Reticulocyte (re-TIK-yoo-lō-sīt) An immature red blood cell.

Reticulocyte (re-TIK-yoo-lō-sīt) ***count*** Examination of a stained sample of blood to determine the percentage of reticulocytes in the total number of red blood cells; used to evaluate rate of erythropoiesis and monitor treatment for anemia.

Reticulum (re-TIK-yoo-lum) A network.

Retina (RET-i-na) The inner coat of the eyeball, lying only in the posterior portion of the eye and consisting of nervous tissue and a pigmented layer comprised of epithelial cells lying in contact with the choroid. Also called the ***nervous tunic*** (TOO-nik).

Retinal (RE'-ti-nal) The pigment portion of the photopigment rhodopsin. Also called ***visual yellow.***

Retraction (rē-TRAK-shun) The movement of a protracted part of the body backward on a plane parallel to the ground, as in pulling the lower jaw back in line with the upper jaw.

Retroflexion (re-trō-FLEK-shun) A malposition of the uterus in which it is tilted posteriorly.

Retrograde degeneration (RE-trō-grād dē-jen-er-Ā-shun) Changes that occur in the proximal portion of a damaged axon only as far as the first neurofibral node (node of Ranvier); similar to changes that occur during Wallerian degeneration.

Retroperitoneal (re'-trō-per-i-tō-NĒ-al) External to the peritoneal lining of the abdominal cavity.

Rheumatism (ROO-ma-tizm') Any painful state of the supporting structures of the body—bones, ligaments, joints, tendons, or muscles.

RH factor An inherited agglutinogen (antigen) on the surface of red blood cells.

Rhinology (rī-NOL-ō-jē) The study of the nose and its disorders.

Rhinoplasty (RĪ-nō-plas'-tē) Surgical procedure in which the structure of the external nose is altered.

Rhodopsin (rō-DOP-sin) A photopigment in rods of the retina, consisting of a protein scotopsin plus retinal, that is sensitive to low levels of illumination. Also called ***visual purple.***

Ribonucleic (rī'-bō-nyoo-KLĒ-ik) ***acid (RNA)*** A single-stranded nucleic acid constructed of nucleotides consisting of one of four possible nitrogenous bases (adenine, cytosine, guanine, or uracil), ribose, and a phosphate group; three types are messenger RNA (mRNA), transfer RNA (tRNA), and ribosomal RNA (rRNA), each of which cooperates with DNA for protein synthesis.

Ribosome (RĪ-bō-sōm) An organelle in the cytoplasm of cells, composed of ribosomal RNA and ribosomal proteins, that synthesizes proteins; nicknamed the "protein factory."

Rickets (RIK-ets) Condition affecting children characterized by soft and deformed bones resulting from inadequate calcium metabolism due to a vitamin D deficiency.

Right heart (atrial) reflex A reflex concerned with maintaining normal venous blood pressure.

Right lymphatic (lim-FAT-ik) ***duct*** A vessel of the lymphatic system that drains lymph from the upper right side of the body and empties it into the right subclavian vein.

Rigidity (ri-JID-i-tē) Hypertonia characterized by increased muscle tone, but reflexes are not affected.

Rigor mortis State of partial contraction of muscles following death due to lack of ATP that causes cross bridges of thick myofilaments to remain attached to thin myofilaments, thus preventing relaxation.

Rod A visual receptor in the retina of the eye that is specialized for vision in dim light.

Roentgen (RENT-gen) The international unit of radiation; a standard quantity of x or gamma radiation.

Roentgenogram (RENT-gen-ō-gram) A photographic image produced by x rays.

Root canal A narrow extension of the pulp cavity lying within the root of a tooth.

Root of penis Attached portion of penis that consists of the bulb and crura.

Rotation (rō-TĀ-shun) Moving a bone around its own axis, with no other movement.

Round ligament (LIG-a-ment) A band of fibrous connective tissue enclosed between the folds of the broad ligament of the uterus, emerging from a point on the uterus just below the uterine (Fallopian) tube, extending laterally along the pelvic wall, and penetrating the abdominal wall through the deep inguinal ring to end in the labia majora.

Round window A small opening between the middle and inner ear, directly below the oval window, covered by the secondary tympanic membrane. Also called the ***fenestra cochlea*** (fe-NES-tra KŌK-lē-a).

Rugae (ROO-jē) Large folds in the mucosa of an empty hollow organ, such as the stomach and vagina.

Saccule (SAK-yool) The lower and smaller of the two chambers in the membranous labyrinth inside the vestibule of the inner ear containing a receptor organ for static equilibrium.

Sacral hiatus (hi-Ā-tus) Inferior entrance to the vertebral canal formed when the laminae of the fifth sacral vertebra (and sometimes fourth) fail to meet.

Sacral plexus (PLEK-sus) A network formed by the anterior branches of spinal nerves L4 through S3.

Sacral promontory (PROM-on-tor'-ē) The superior surface of the body of the first sacral vertebra that projects anteriorly into the pelvic cavity; a line from the sacral promontory to the superior border of the symphysis pubis divides the abdominal and pelvic cavities.

Saddle joint A synovial joint in which the articular surface of one bone is saddle shaped and the articular surface of the other bone is shaped like a rider sitting in the saddle, as in the joint between the trapezium and the metacarpal of the thumb. Also called a ***sellaris*** (sel-LA-ris) ***joint.***

Sagittal (SAJ-i-tal) ***plane*** A vertical plane that divides the body or organs into left and right portions. Such a plane may be ***midsagittal (median),*** in which the divisions are equal, or ***parasagittal,*** in which the divisions are unequal.

Saliva (sa-LĪ-va) A clear, alkaline, somewhat viscous secretion produced by the three pairs of salivary glands; contains various salts, mucin, lysozyme, and salivary amylase.

Salivary amylase (SAL-i-ver-ē AM-i-lās) An enzyme in saliva that initiates the chemical breakdown of starch, mostly in the mouth.

Salivary gland One of three pairs of glands that lie outside the mouth and pour their secretory product (called saliva) into ducts that empty into the oral cavity; the parotid, submandibular, and sublingual glands.

Salpingitis (sal'-pin-JĪ-tis) Inflammation of the uterine (Fallopian) or auditory (Eustachian) tube.

Sarcolemma (sar'-kō-LEM-ma) The cell membrane of a muscle fiber (cell), especially of a skeletal muscle fiber.

Sarcoma (sar-KŌ-ma) A connective tissue tumor, often highly malignant.

Sarcomere (SAR-kō-mēr) A contractile unit in a striated muscle fiber (cell) extending from one Z line to the next Z line.

Sarcoplasm (SAR-kō-plazm) The cytoplasm of a muscle fiber (cell).

Sarcoplasmic reticulum (sar'-kō-PLAZ-mik re-TIK-yoo-lum) A network of saccules and tubes surrounding myofibrils of a muscle fiber (cell), comparable to endoplasmic reticulum; functions to reabsorb calcium ions during relaxation and to release them to cause contraction.

Satiety (sa-TĪ-e-tē) Fullness or gratification, as of hunger or thirst.

Satiety center A collection of nerve cells located in the ventromedial nuclei of the hypothalamus that, when stimulated, brings about the cessation of eating.

Saturated fat A fat that contains no double bonds between any of its carbon atoms; all are single bonds and all carbon atoms are bonded to the maximum number of hydrogen atoms; found naturally in animal foods such as meat, milk, milk products, and eggs.

Scala tympani (SKA-la TIM-pan-ē) The lower spiral-shaped channel of the bony cochlea, filled with perilymph.

Scala vestibuli (ves-TIB-yoo-lē) The upper spiral-shaped channel of the bony cochlea, filled with perilymph.

Schwann cell *See* ***Neurolemmocyte.***

Sciatica (sī-AT-i-ka) Inflammation and pain along the sciatic nerve; felt at the back of the thigh running down the inside of the leg.

Sclera (SKLE-ra) The white coat of fibrous tissue that forms the outer protective covering over the eyeball except in the most anterior portion; the posterior portion of the fibrous tunic.

Scleral venous sinus A circular venous sinus located at the junction of the sclera and the cornea through which aqueous humor drains from the anterior chamber of the eyeball into the blood. Also called the ***canal of Schlemm*** (SHLEM).

Sclerosis (skle-RŌ-sis) A hardening with loss of elasticity of tissues.

Scoliosis (skō'-lē-Ō-sis) An abnormal lateral curvature from the normal vertical line of the backbone.

Scotoma (skō-TŌ-ma) An area of depressed or lost vision within the visual field.

Scotopsin (skō-TOP-sin) The protein portion of the visual pigment rhodopsin found in rods of the retina.

Scrotum (SKRŌ-tum) A skin-covered pouch that contains the testes and their accessory structures.

Sebaceous (se-BĀ-shus) Secreting oil.

Sebaceous (se-BĀ-shus) ***gland*** An exocrine gland in the dermis of the skin, almost always associated with a hair follicle, that secretes sebum. Also called an ***oil gland.***

Sebum (SĒ-bum) Secretion of sebaceous (oil) glands.

Secondary sex characteristic A feature characteristic of the male or female body that develops at puberty under the stimulation of sex hormones but is not directly involved in sexual reproduction, such as distribution of body hair, voice pitch, body shape, and muscle development.

Secretion (se-KRĒ-shun) Production and release from a gland cell of a fluid, especially a functionally useful product as opposed to a waste product.

Selectively permeable membrane A membrane that permits the passage of certain substances, but restricts the passage of others. Also called a ***semipermeable*** (sem'-ē-PER-mē-a-bl) ***membrane.***

Sella turcica (SEL-a TUR-si-ka) A depression on the superior surface of the sphenoid bone that houses the pituitary gland.

Semen (SĒ-men) A fluid discharged at ejaculation by a male that consists of a mixture of spermatozoa and the secretions of the seminal vesicles, prostate gland, and bulbourethral (Cowper's) glands. Also called ***seminal*** (SEM-i-nal) ***fluid.***

Semicircular canals Three bony channels (anterior, posterior, lateral), filled with perilymph, in which lie the membranous semicircular canals filled with endolymph. They contain receptors for equilibrium.

Semicircular ducts The membranous semicircular canals filled with endolymph and floating in the perilymph of the bony semicircular canals. They contain cristae that are concerned with dynamic equilibrium.

Semiconservative Replication of DNA in which one complete strand is from an original (conserved) molecule of DNA and one is newly produced.

Semilunar (sem'-ē-LOO-nar) ***valve*** A valve guarding the entrance into the aorta or the pulmonary trunk from a ventricle of the heart.

Seminal vesicle (SEM-i-nal VES-i-kul) One of a pair of convoluted, pouchlike structures, lying posterior and inferior to the urinary bladder and anterior to the rectum, that secrete a component of semen into the ejaculatory ducts.

Seminiferous tubule (sem'-i-NI-fer-us TOO-byool) A tightly coiled duct, located in a lobule of the testis, where spermatozoa are produced.

Senescence (se-NES-ens) The process of growing old; the period of old age.

Senile macular (MAK-yoo-lar) ***degeneration (SMD)*** A disease in which blood vessels grow over the macula lutea.

Senility (se-NIL-i-tē) A loss of mental or physical ability due to old age.

Sensation A state of awareness of external or internal conditions of the body.

Sensory area A region of the cerebral cortex concerned with the interpretation of sensory impulses.

Sepsis (SEP-sis) A morbid condition that results from the presence in the blood or other body tissues of pathogenic bacteria and their products.

Septal defect An opening in the septum (interatrial or interventricular) between the left and right sides of the heart.

Septicemia (sep'-ti-SĒ-mē-a) Toxins or disease-causing bacteria in blood. Also called ***"blood poisoning."***

Septum (SEP-tum) A wall dividing two cavities.

Serosa (ser-Ō-sa) Any serous membrane. The outermost layer of an organ formed by a serous membrane. The membrane that lines the pleural, pericardial, and peritoneal cavities.

Serous (SIR-us) ***membrane*** A membrane that lines a body cavity that does not open to the exterior. Also called the ***serosa*** (se-RŌ-sa).

Serum Plasma minus its clotting proteins.

Serum enzyme studies Evaluation of levels of certain enzymes in blood (creatine phosphokinase, serum glutamic oxaloacetic transaminase, lactic dehydrogenase) to diagnose and monitor heart attacks.

Sesamoid bones (SES-a-moyd) Small bones usually found in tendons.

Sex chromosomes The twenty-third pair of chromosomes, designated X and Y, which determine the genetic sex of an individual; in males, the pair is XY; in females, XX.

Sexual intercourse The insertion of the erect penis of a male into the vagina of a female. Also called ***coitus*** (KŌ-i-tus) or ***copulation.***

Sexually transmitted disease (STD) General term for any of a large number of diseases spread by sexual contact. Also called a ***venereal disease (VD).***

Sheath of Schwann *See* ***Neurolemma.***

Shingles Acute infection of the peripheral nervous system caused by a virus.

Shinsplints Soreness or pain along the tibia probably caused by inflammation of the periosteum brought on by repeated tugging of the muscles and tendons attached to the periosteum. Also called ***tibia stress syndrome.***

Shivering Involuntary contraction of a muscle that generates heat.

Shock Failure of the cardiovascular system to deliver adequate amounts of oxygen and nutrients to meet the metabolic needs of the body due to inadequate cardiac output. It is characterized by hypotension; clammy, cool, and pale skin; sweating; reduced urine formation; altered mental state; acidosis; tachycardia; weak, rapid pulse; and thirst. Types include hypovolemic, cardiogenic, obstructive, neurogenic, and septic.

Shoulder A synovial or diarthrotic joint where the humerus joins the scapula.

Sigmoid colon (SIG-moyd KŌ-lon) The S-shaped portion of the large intestine that begins at the level of the left iliac crest, projects inward to the midline, and terminates at the rectum at about the level of the third sacral vertebra.

Sigmoidoscopy (sig'-moy-DOS-kō-pē) Visualization of the anal canal, rectum, and colon to screen for colorectal cancer, collect biopsy samples, remove polyps, gather specimens for culture, and photograph the intestinal mucosa.

Sign Any objective evidence of disease that can be observed or measured such as a lesion, swelling, or fever.

Sinoatrial (si-nō-Ā-trē-al) ***(SA) node*** A compact mass of cardiac muscle fibers (cells) specialized for conduction, located in the right atrium beneath the opening of the superior vena cava. Also called the ***sinuatrial node*** or ***pacemaker.***

Sinus (SĪ-nus) A hollow in a bone (paranasal sinus) or other tissue; a channel for blood (vascular sinus); any cavity having a narrow opening.

Sinusitis (sīn-yoo-SĪT-is) Inflammation of the mucous membrane of a paranasal sinus.
Sinusoid (SĪN-yoo-soyd) A microscopic space or passage for blood in certain organs such as the liver or spleen.
Skeletal muscle An organ specialized for contraction, composed of striated muscle fibers (cells), supported by connective tissue, attached to a bone by a tendon or an aponeurosis, and stimulated by somatic efferent neurons.
Skene's gland *See* ***Paraurethral gland.***
Skull The skeleton of the head consisting of the cranial and facial bones.
Sliding-filament theory The most commonly accepted explanation for muscle contraction in which actin and myosin myofilaments move into interdigitation with each other, decreasing the length of the sarcomeres.
Small intestine A long tube of the gastrointestinal tract that begins at the pyloric sphincter of the stomach, coils through the central and lower part of the abdominal cavity, and ends at the large intestine; divided into three segments: duodenum, jejunum, and ileum.
Smooth muscle An organ specialized for contraction, composed of smooth muscle fibers (cells), located in the walls of hollow internal structures, and innervated by a visceral efferent neuron.
Snellen (SNEL-en) ***test*** Test used to evaluate any problem or changes in vision by measuring visual acuity (sharpness).
Sodium-potassium pump An active transport system located in the cell membrane that transports sodium ions out of the cell and potassium ions into the cell at the expense of cellular ATP. It functions to keep the ionic concentrations of these elements at physiological levels.
Soft palate (PAL-at) The posterior portion of the roof of the mouth, extending posteriorly from the palatine bones and ending at the uvula. It is a muscular partition lined with mucous membrane.
Solution A homogeneous molecular or ionic dispersion of one or more substances (solutes) in a usually liquid-dissolving medium (solvent).
Somatic cell division Type of cell division in which a single starting cell (parent cell) duplicates itself to produce two identical cells (daughter cells); consists of mitosis and cytokinesis.
Somatic (sō-MAT-ik) ***nervous system (SNS)*** The portion of the peripheral nervous system made up of the somatic efferent fibers that run between the central nervous system and the skeletal muscles and skin.
Somatomedin (sō'-ma-tō-MĒ-din) Small protein produced by the liver in response to stimulation by human growth hormone (hGH) that mediates most of the effects of human growth hormone.
Somesthetic (sō'-mes-THET-ik) Pertaining to sensations and sensory structures of the body.
Somite (SŌ-mīt) Block of mesodermal cells in a developing embryo that is distinguished into a myotome (which forms most of the skeletal muscles), dermatome (which forms connective tissues), and sclerotome (which forms the vertebrae).
Spasm (spazm) A sudden, involuntary contraction of large groups of muscles.
Spastic (SPAS-tik) An increase in muscle tone (stiffness) associated with an increase in tendon reflexes and abnormal reflexes (Babinski sign).
Spasticity (spas-TIS-i-tē) Hypertonia characterized by increased muscle tone, increased tendon reflexes, and pathological reflexes (Babinski sign).
Spermatic (sper-MAT-ik) ***cord*** A supporting structure of the male reproductive system, extending from a testis to the deep inguinal ring, that includes the ductus (vas) deferens, arteries, veins, lymphatics, nerves, cremaster muscle, and connective tissue.
Spermatogenesis (sper'-ma-tō-JEN-e-sis) The formation and development of spermatozoa in the seminiferous tubules of the testes.
Spermatozoon (sper'-ma-tō-ZŌ-on) A mature sperm cell.
Spermicide (SPER-mi-sīd') An agent that kills spermatozoa.
Spermiogenesis (sper'-mē-ō-JEN-e-sis) The maturation of spermatids into spermatozoa.
Sphincter (SFINGK-ter) A circular muscle constricting an orifice.
Sphincter of Oddi *See* ***Sphincter of the hepatopancreatic ampulla.***
Sphincter of the hepatopancreatic ampulla A circular muscle at the opening of the common bile and main pancreatic ducts in the duodenum. Also called the ***sphincter of Oddi*** (OD-ē).
Sphygmomanometer (sfig'-mō-ma-NOM-e-ter) An instrument for measuring arterial blood pressure.
Spina bifida (SPĪ-na BIF-i-da) A congenital defect of the vertebral column in which the halves of the neural arch of a vertebra fail to fuse in the midline.
Spinal (SPĪ-nal) ***cord*** A mass of nerve tissue located in the vertebral canal from which 31 pairs of spinal nerves originate.
Spinal nerve One of the 31 pairs of nerves that originate on the spinal cord from posterior and anterior roots.
Spinal shock A period of time, from several days to several weeks, following transection of the spinal cord and characterized by the abolition of all reflex activity.
Spinal (lumbar) tap (puncture) Withdrawal of some of the cerebrospinal fluid from the subarachnoid space in the lumbar region for diagnostic purposes, introduction of various substances, and evaluation of the effects of treatment.
Spinous (SPĪ-nus) ***process*** A sharp or thornlike process or projection. Also called a ***spine.*** A sharp ridge running diagonally across the posterior surface of the scapula.
Spiral organ The organ of hearing, consisting of supporting cells and hair cells that rest on the basilar membrane and extend into the endolymph of the cochlear duct. Also called the ***organ of Corti*** (KOR-tē).
Spirometer (spī-ROM-e-ter) An apparatus used to measure air capacity of the lungs.
Splanchnic (SPLANK-nik) Pertaining to the viscera.
Spleen (SPLĒN) Large mass of lymphatic tissue between the fundus of the stomach and the diaphragm that functions in phagocytosis, production of lymphocytes, and blood storage.
Sprain Forcible wrenching or twisting of a joint with partial rupture or other injury to its attachments without dislocation.
Sputum (SPYOO-tum) Substance ejected from the mouth containing saliva and mucus.
Squamous (SKWĀ-mus) Scalelike.
Starling's law of the capillaries The movement of fluid between plasma and interstitial fluid is in a state of near equilibrium at the arterial and venous ends of a capillary; that is, filtered fluid and absorbed fluid plus that returned to the lymphatic system are nearly equal.
Starling's law of the heart The force of muscular contraction is determined by the length of the cardiac muscle fibers (cells); the greater the length of stretched fibers, the stronger the contraction.
Starvation (star-VĀ-shun) The loss of energy stores in the

form of glycogen, fats, and proteins due to inadequate intake of nutrients or inability to digest, absorb, or metabolize ingested nutrients.

Stasis (STĀ-sis) Stagnation or halt of normal flow of fluids, as blood, urine, or of the intestinal mechanism.

Static equilibrium (ē-kwi-LIB-rē-um) The maintenance of posture in response to changes in the orientation of the body, mainly the head, relative to the ground.

Stellate reticuloendothelial (STEL-āte re-tik'-yoo-lō-en'-dō-THĒ-lē-al) ***cell*** Phagocytic cell that lines a sinusoid of the liver. Also called a ***Kupffer's*** (KOOP-ferz) ***cell.***

Stenosis (sten-Ō-sis) An abnormal narrowing or constriction of a duct or opening.

Stereocilia (ste'-rē-ō-SIL-ē-a) Groups of extremely long, slender, nonmotile microvilli projecting from epithelial cells lining the epididymis.

Stereognosis (ste'-rē-og-NŌ-sis) The ability to recognize the size, shape, and texture of an object by touch.

Sterile (STE-ril) Free from any living microorganisms. Unable to conceive or produce offspring.

Sterilization (ster'-i-li-ZĀ-shun) Elimination of all living microorganisms. The rendering of an individual incapable of reproduction (e.g., castration, vasectomy, hysterectomy).

Sternal puncture Introduction of a wide-bore needle into the marrow cavity of the sternum for aspiration of a sample of red bone marrow.

Stimulus Any change in the environment capable of altering the membrane potential.

Stomach The J-shaped enlargement of the gastrointestinal tract directly under the diaphragm in the epigastric, umbilical, and left hypochondriac regions of the abdomen, between the esophagus and small intestine.

Strabismus (stra-BIZ-mus) A condition in which the visual axes of the two eyes differ, so that they do not fix on the same object.

Straight tubule (TOO-byool) A duct in a testis leading from a convoluted seminiferous tubule to the rete testis.

Stratum (STRĀ-tum) A layer.

Stratum basalis (STRĀ-tum ba-SAL-is) The outer layer of the endometrium, next to the myometrium, that is maintained during menstruation and gestation and produces a new functionalis following menstruation or parturition.

Stratum functionalis (funk'-shun-AL-is) The inner layer of the endometrium, the layer next to the uterine cavity, that is shed during menstruation and that forms the maternal portion of the placenta during gestation.

Stressor A stress that is extreme, unusual, or long-lasting and triggers the general adaptation syndrome.

Stretch receptor Receptor in the walls of bronchi, bronchioles, and lungs that sends impulses to the respiratory center that prevents overinflation of the lungs.

Stretch reflex A monosynaptic reflex triggered by a sudden stretch of a muscle and ending with a contraction of that same muscle. Also called a ***tendon jerk.***

Stricture (STRIK-cher) A local constriction of a tubular structure.

Stroke volume The volume of blood ejected by either ventricle in one systole; about 70 ml.

Stroma (STRŌ-ma) The tissue that forms the ground substance, foundation, or framework of an organ, as opposed to its functional parts.

Stupor (STOO-por) Unresponsiveness from which a patient can be aroused only briefly and by vigorous and repeated stimulation.

Subarachnoid (sub'-a-RAK-noyd) ***space*** A space between the arachnoid and the pia mater that surrounds the brain and spinal cord and through which cerebrospinal fluid circulates.

Subcutaneous (sub'-kyoo-TĀ-nē-us) Beneath the skin. Also called ***hypodermic*** (hī-pō-DER-mik).

Subcutaneous layer A continuous sheet of loose connective tissue and adipose tissue between the dermis of the skin and the deep fascia of the muscles. Also called the ***superficial fascia*** (FASH-ē-a).

Subdural (sub-DOO-ral) **space** A space between the dura mater and the arachnoid of the brain and spinal cord that contains a small amount of fluid.

Sublingual (sub-LING-gwal) ***gland*** One of a pair of salivary glands situated in the floor of the mouth under the mucous membrane and to the side of the lingual frenulum, with a duct (Rivinus's) that opens into the floor of the mouth.

Submandibular (sub'-man-DIB-yoo-lar) ***gland*** One of a pair of salivary glands found beneath the base of the tongue under the mucous membrane in the posterior part of the floor of the mouth, posterior to the sublingual glands, with a duct (Wharton's) situated to the side of the lingual frenulum. Also called the ***submaxillary*** (sub'-MAK-si-ler-ē) ***gland.***

Submucosa (sub-myoo-KŌ-sa) A layer of connective tissue located beneath a mucous membrane, as in the gastrointestinal tract or the urinary bladder; the submucosa connects the mucosa to the muscularis tunic.

Submucosal plexus A network of autonomic nerve fibers located in the outer portion of the submucous layer of the small intestine. Also called the ***plexus of Meissner*** (MĪS-ner).

Subserous fascia (sub-SE-rus FASH-ē-a) A layer of connective tissue internal to the deep fascia, lying between the deep fascia and the serous membrane that lines the body cavities.

Substrate A substance with which an enzyme reacts.

Sudoriferous (soo'-dor-IF-er-us) ***gland*** An apocrine or eccrine exocrine gland in the dermis or subcutaneous layer that produces perspiration. Also called a ***sweat gland.***

Sulcus (SUL-kus) A groove or depression between parts, especially between the convolutions of the brain. *Plural,* ***sulci*** (SUL-sē).

Summation (sum-MĀ-shun) The algebraic addition of the excitatory and inhibitory effects of many stimuli applied to a nerve cell body. The increased strength of muscle contraction that results when stimuli follow in rapid succession.

Superficial (soo'-per-FISH-al) Located on or near the surface of the body.

Superficial fascia (FASH-ē-a) A continuous sheet of fibrous connective tissue between the dermis of the skin and the deep fascia of the muscles. Also called ***subcutaneous*** (sub'-kyoo-TĀ-nē-us) ***layer.***

Superficial inguinal (IN-gwi-nal) ***ring*** A triangular opening in the aponeurosis of the external oblique muscle that represents the termination of the inguinal canal.

Superior (soo-PĒR-ē-or) Toward the head or upper part of a structure. Also called ***cephalad*** (SEF-a-lad) or ***craniad.***

Superior vena cava (VĒ-na CĀ-va) ***(SVC)*** Large vein that collects blood from parts of the body superior to the heart and returns it to the right atrium.

Supination (soo-pī-NĀ-shun) A movement of the forearm in which the palm of the hand is turned anteriorly or superiorly.

Suppuration (sup'-yoo-RĀ-shun) Pus formation and discharge.

Surface anatomy The study of the structures that can be identified from the outside of the body.

Surfactant (sur-FAK-tant) A phospholipid substance produced by the lungs that decreases surface tension.

Susceptibility (sus-sep'-ti-BIL-i-tē) Lack of resistance of a body to the deleterious or other effects of an agent such as pathogenic microorganisms.

Suspensory ligament (sus-PEN-so-rē LIG-a-ment) A fold of peritoneum extending laterally from the surface of the ovary to the pelvic wall.

Sustentacular (sus'-ten-TAK-yoo-lar) ***cell*** A supporting cell of seminiferous tubules that produces secretions for supplying nutrients to spermatozoa and the hormone inhibin. Also called a ***Sertoli*** (ser-TŌ-lē) ***cell.***

Sutural (SOO-cher-al) ***bone*** A small bone located within a suture between certain cranial bones. Also called ***Wormian*** (WER-mē-an) ***bone.***

Suture (SOO-cher) An immovable fibrous joint in the skull where bone surfaces are closely united.

Sympathetic (sim'-pa-THET-ik) ***division*** One of the two subdivisions of the autonomic nervous system, having cell bodies of preganglionic neurons in the lateral gray columns of the thoracic segment and first two or three lumbar segments of the spinal cord; primarily concerned with processes involving the expenditure of energy. Also called the ***thoracolumbar*** (thō'-ra-kō-LUM-bar) ***division.***

Sympathetic trunk ganglion (GANG-glē-on) A cluster of cell bodies of postganglionic sympathetic neurons lateral to the vertebral column, close to the body of a vertebra. These ganglia extend downward through the neck, thorax, and abdomen to the coccyx on both sides of the vertebral column and are connected to one another to form a chain on each side of the vertebral column. Also called ***lateral,*** or ***sympathetic, chain*** or ***vertebral chain ganglia.***

Sympathomimetic (sim'-pa-thō-mi-MET-ik) Producing effects that mimic those brought about by the sympathetic division of the autonomic nervous system.

Symphysis (SIM-fi-sis) A line of union. A slightly movable cartilaginous joint such as the symphysis pubis between the anterior surfaces of the coxal (hip) bones.

Symphysis pubis (PYOO-bis) A slightly movable cartilaginous joint between the anterior surfaces of the coxal (hip) bones.

Symptom (SIMP-tum) A subjective change in body function not apparent to an observer, such as fever or nausea, that indicates the presence of a disease or disorder of the body.

Synapse (SIN-aps) The junction between the process of two adjacent neurons; the place where the activity of one neuron affects the activity of another; may be electrical or chemical.

Synapsis (sin-AP-sis) The pairing of homologous chromosomes during prophase I of meiosis.

Synaptic (sin-AP-tik) ***cleft*** The narrow gap that separates the axon terminal of one nerve cell from another nerve cell or muscle fiber (cell) and across which a neurotransmitter diffuses to affect the postsynaptic cell.

Synaptic delay The length of time between the arrival of the action potential at the axon terminal and the membrane potential (IPSP or EPSP) change on the postsynaptic membrane; usually about 0.5 msec.

Synaptic end bulb Expanded distal end of an axon terminal that contains synaptic vesicles. Also called a ***synaptic knob*** or ***end foot.***

Synaptic gutter Invaginated portion of a sarcolemma under an axon terminal. Also called a ***synaptic trough*** (TROF).

Synaptic vesicle Membrane-enclosed sac in a synaptic end bulb that stores neurotransmitters.

Synarthrosis (sin'-ar-THRŌ-sis) An immovable joint.

Synchondrosis (sin'-kon-DRŌ-sis) A cartilaginous joint in which the connecting material is hyaline cartilage.

Syncope (SIN-kō-pē) Faint; a sudden temporary loss of consciousness associated with loss of postural tone and followed by spontaneous recovery; most commonly caused by cerebral ischemia.

Syndesmosis (sin'-dez-MŌ-sis) A fibrous joint in which articulating bones are united by dense fibrous tissue.

Syndrome (SIN-drōm) A group of signs and symptoms that occur together in a pattern that is characteristic of a particular disease or abnormal condition.

Syneresis (si-NER-e-sis) The process of clot retraction.

Synergist (SIN-er-jist) A muscle that assists the prime mover by reducing undesired action or unnecessary movement.

Synergistic (syn-er-GIS-tik) ***effect*** A hormonal interaction in which the effects of two or more hormones complement each other so that the target cell responds to the sum of the hormones involved. An example is the combined actions of estrogens, progesterone (PROG), prolactin (PRL), and oxytocin (OT) necessary for lactation.

Synostosis (sin'-os-TŌ-sis) A joint in which the dense fibrous connective tissue that unites bones at a suture has been replaced by bone, resulting in a complete fusion across the suture line.

Synovial (si-NŌ-vē-al) ***cavity*** The space between the articulating bones of a synovial (diarthrotic) joint, filled with synovial fluid. Also called a ***joint cavity.***

Synovial fluid Secretion of synovial membranes that lubricates joints and nourishes articular cartilage.

Synovial joint A fully movable or diarthrotic joint in which a synovial (joint) cavity is present between the two articulating bones.

Synovial membrane The inner of the two layers of the articular capsule of a synovial joint, composed of loose connective tissue that secretes synovial fluid into the synovial (joint) cavity.

Syphilis (SIF-i-lis) A sexually transmitted disease caused by the bacterium *Treponema pallidum.*

System An association of organs that have a common function.

Systemic (sis-TEM-ik) Affecting the whole body; generalized.

Systemic anatomy The study of particular systems of the body, such as the skeletal, muscular, nervous, cardiovascular, or urinary systems.

Systemic circulation The routes through which oxygenated blood flows from the left ventricle through the aorta to all the organs of the body and deoxygenated blood returns to the right atrium.

Systemic lupus erythematosus (er-i-them-a-TŌ-sus) ***(SLE)*** An autoimmune, inflammatory disease that may affect every tissue of the body.

Systole (SIS-tō-lē) In the cardiac cycle, the phase of contraction of the heart muscle, especially of the ventricles.

Systolic (sis-TO-lik) ***blood pressure*** The force exerted by blood on arterial walls during ventricular contraction; the highest pressure measured in the large arteries, about 120 mm Hg under normal conditions for a young, adult male.

Tachycardia (tak'-i-KAR-dē-a) A rapid heartbeat or pulse rate.

Tactile (TAK-tīl) Pertaining to the sense of touch.

Tactile disc Modified epidermal cell in the startum basale of hairless skin that functions as a cutaneous receptor for discriminative touch. Also called a ***Merkel's*** (MER-kelz) ***disc.***

Taenia coli (TĒ-nē-a KŌ-lī) One of three flat bands of thickened, longitudinal muscles running the length of the large intestine.

Target cell A cell whose activity is affected by a particular hormone.

Tarsal gland Sebaceous (oil) gland that opens on the edge of each eyelid. Also called a ***Meibomian*** (mī-BŌ-mē-an) ***gland.***

Tarsal plate A thin, elongated sheet of connective tissue, one in each eyelid, giving the eyelid form and support. The aponeurosis of the levator palpebrae superioris is attached to the tarsal plate of the superior eyelid.

Tarsus (TAR-sus) A collective term for the seven bones of the ankle.

Tay-Sachs (TĀ SAKS) ***disease*** Inherited, progressive neuronal degeneration of the central nervous system due to a deficient lysosomal enzyme that causes excessive accumulations of a lipid called ganglioside.

T cell A lymphocyte that can differentiate into one of six kinds of cells—killer, helper, suppressor, memory, amplifier, or delayed hypersensitivity—all of which function in cellular immunity.

Tectorial (tek-TŌ-rē-al) ***membrane*** A gelatinous membrane projecting over and in contact with the hair cells of the spiral organ (organ of Corti) in the cochlear duct.

Telophase (TEL-ō-fāz) The final stage of mitosis in which the daughter nuclei become established.

Temporomandibular joint (TMJ) syndrome A disorder of the temporomandibular joint (TMJ) characterized by dull pain around the ear, tenderness of jaw muscles, a clicking or popping noise when opening or closing the mouth, limited or abnormal opening of the mouth, headache, tooth sensitivity, and abnormal wearing of the teeth.

Tendon (TEN-don) A white fibrous cord of dense, regularly arranged connective tissue that attaches muscle to bone.

Tendon organ A proprioceptive receptor, sensitive to changes in muscle tension and force of contraction, found chiefly near the junction of tendons and muscles. Also called a ***Golgi*** (GOL-jē) ***tendon organ.***

Tendon reflex A polysynaptic, ipsilateral reflex that is designed to protect tendons and their associated muscles from damage that might be brought about by excessive tension. The receptors involved are called tendon organs (Golgi tendon organs).

Tenosynovitis (ten′-ō-sin-ō-VĪ-tis) Inflammation of a tendon sheath and synovial membrane at a joint.

Tentorium cerebelli (ten-TŌ-rē-um ser′-e-BEL-ē) A transverse shelf of dura mater that forms a partition between the occipital lobe of the cerebral hemispheres and the cerebellum and that covers the cerebellum.

Teratogen (TER-a-tō-jen) Any agent or factor that causes physical defects in a developing embryo.

Terminal ganglion (TER-min-al GANG-lē-on) A cluster of cell bodies of postganglionic parasympathetic neurons either lying very close to the visceral effectors or located within the walls of the visceral effectors supplied by the postganglionic fibers.

Testis (TES-tis) Male gonad that produces sperm and the hormones testosterone and inhibin. Also called a ***testicle.***

Testosterone (tes-TOS-te-rōn) A male sex hormone (androgen) secreted by interstitial endocrinocytes (cells of Leydig) of a mature testis; controls the growth and development of male sex organs, secondary sex characteristics, spermatozoa, and body growth.

Tetanus (TET-a-nus) An infectious disease caused by the toxin of *Clostridium tetani,* characterized by tonic muscle spasms and exaggerated reflexes, lockjaw, and arching of the back. A smooth, sustained contraction produced by a series of very rapid stimuli to a muscle.

Tetany (TET-a-nē) A nervous condition caused by hypoparathyroidism and characterized by intermittent or continuous tonic muscular contractions of the extremities.

Tetralogy of Fallot (tet-RAL-ō-jē of fal-Ō) A combination of four congenital heart defects: (1) constricted pulmonary semilunar valve, (2) interventricular septal opening, (3) emergence of aorta from both ventricles instead of from the left only, and (4) enlarged right ventricle.

Thalamus (THAL-a-mus) A large, oval structure located above the midbrain, consisting of two masses of gray matter covered by a thin layer of white matter.

Thalassemia (thal′-a-SĒ-mē-a) A group of hereditary hemolytic anemias.

Thallium (THAL-ē-um) ***imaging*** Diagnostic procedure used to evalute blood flow through coronary arteries, cardiac disorders, and effectiveness of drug therapy; thallium concentrates in healthy myocardial tissue.

Therapy (THER-a-pē) The treatment of a disease or disorder.

Thermoreceptor (THER-mō-rē-sep-tor) Receptor that detects changes in temperature.

Thigh The portion of the lower extremity between the hip and the knee.

Third ventricle (VEN-tri-kul) A slitlike cavity between the right and left halves of the thalamus and between the lateral ventricles.

Thoracic (thō-RAS-ik) ***cavity*** Superior component of the ventral body cavity that contains two pleural cavities, the mediastinum, and the pericardial cavity.

Thoracic duct A lymphatic vessel that begins as a dilation called the cisterna chyli, receives lymph from the left side of the head, neck, and chest, the left arm, and the entire body below the ribs, and empties into the left subclavian vein. Also called the ***left lymphatic*** (lim-FAT-ik) ***duct.***

Thoracolumbar (thō′-ra-kō-LUM-bar) ***outflow*** The fibers of the sympathetic preganglionic neurons, which have their cell bodies in the lateral gray columns of the thoracic segment and first two or three lumbar segments of the spinal cord.

Thorax (THŌ-raks) The chest.

Threshold potential The membrane voltage that must be reached in order to trigger an action potential (nerve impulse).

Threshold stimulus Any stimulus strong enough to initiate an action potential (nerve impulse). Also called a ***liminal*** (LIM-i-nal) ***stimulus.***

Thrombin (THROM-bin) The active enzyme formed from prothrombin that acts to convert fibrinogen to fibrin.

Thrombocyte (THROM-bō-sīt) A fragment of cytoplasm enclosed in a cell membrane and lacking a nucleus; found in the circulating blood; plays a role in blood clotting. Also called a ***platelet*** (PLĀT-let).

Thrombolytic (throm-bō-LIT-ik) ***agent*** Chemical substance injected into the body to dissolve blood clots and restore circulation; mechanism of action is direct or indirect activation of plasminogen; examples include tissue plasminogen activator (t-PA), streptokinase, and urokinase.

Thrombophlebitis (throm′-bo-fle-BĪ-tis) A disorder in which inflammation of the wall of a vein is followed by the formation of a blood clot (thrombus).

Thrombosis (throm-BŌ-sis) The formation of a clot in an unbroken blood vessel, usually a vein.

Thrombus A clot formed in an unbroken blood vessel, usually a vein.

Thymectomy (thī-MEK-tō-mē) Surgical removal of the thymus gland.

Thymus (THĪ-mus) ***gland*** A bilobed organ, located in the upper mediastinum posterior to the sternum and between the lungs, that plays a role in the immune mechanism of the body.

Thyroglobulin (thī-rō-GLŌ-byoo-lin) ***(TGB)*** A large glycoprotein molecule secreted by follicle cells of the thyroid gland in which iodine is combined with tyrosine to form thyroid hormones.

Thyroid cartilage (THĪ-royd KAR-ti-lij) The largest single cartilage of the larynx, consisting of two fused plates that form the anterior wall of the larynx. Also called the ***Adam's apple.***

Thyroid colloid (KOL-loyd) A complex in thyroid follicles consisting of thyroglobulin and stored thyroid hormones.

Thyroid follicle (FOL-i-kul) Spherical sac that forms the parenchyma of the thyroid gland and consists of follicular cells that produce thyroxine (T_4) and triiodothyronine (T_3) and parafollicular cells that produce calcitonin (CT).

Thyroid function tests Tests used to evaluate a swelling or lump in the thyroid gland, ascertain symptoms of abnormal thyroxine levels, monitor responses of thyroid diseases to therapy, and screen newborns for cretinism. Examples are the radioiodine uptake (RAIU), serum T_4 concentration, and serum T_3 concentration tests.

Thyroid gland An endocrine gland with right and left lateral lobes on either side of the trachea connected by an isthmus located in front of the trachea just below the cricoid cartilage.

Thyroid-stimulating hormone (TSH) A hormone secreted by the adenohypophysis (anterior lobe) of the pituitary gland that stimulates the synthesis and secretion of hormones produced by the thyroid gland.

Thyroxine (thī-ROK-sēn) ***(T_4)*** A hormone secreted by the thyroid that regulates organic metabolism, growth and development, and the activity of the nervous system.

Tic Spasmodic twitching made involuntarily by muscles that are ordinarily under voluntary control.

Tidal volume The volume of air breathed in and out in any one breath; about 500 ml in quiet, resting conditions.

Tinnitus (ti-NĪ-tus) A ringing, roaring, or clicking in the ears.

Tissue A group of similar cells and their intercellular substance joined together to perform a specific function.

Tissue factor (TF) A factor, or collection of factors, whose appearance initiates the blood clotting process. Also called ***thromboplastin*** (throm-bō-PLAS-tin).

Tissue macrophage system General term that refers to wandering and fixed macrophages.

Tissue plasminogen activator (t-PA) An enzyme that dissolves small blood clots by initiating a process that converts plasminogen to plasmin, which degrades the fibrin of a clot.

Tissue rejection Phenomenon by which the body recognizes the protein (HLA antigens) in transplanted tissues or organs as foreign and produces antibodies against them.

Tongue A large skeletal muscle covered by a mucous membrane located on the floor of the oral cavity.

Tonometry (tō-NOM-e-trē) A test used to measure intraocular pressure to screen for glaucoma.

Tonsil (TON-sil) A multiple aggregation of large lymphatic nodules embedded in mucous membrane.

Topical (TOP-i-kal) Applied to the surface rather than ingested or injected.

Torn cartilage A tearing of an articular disk in the knee.

Torpor (TOR-por) State of lethargy and sluggishness that precedes stupor, which precedes semicoma, which precedes coma.

Total lung capacity The sum of tidal volume, inspiratory reserve volume, expiratory reserve volume, and residual volume; about 6,000 ml.

Toxic (TOK-sik) Pertaining to poison; poisonous.

Toxic shock syndrome (TSS) A disease caused by the bacterium *Staphylococcus aureus,* occurring among menstruating females who use tampons and characterized by high fever, sore throat, headache, fatigue, irritability, and abdominal pain.

Trabecula (tra-BEK-yoo-la) Irregular latticework of thin plate of spongy bone. Fibrous cord of connective tissue serving as supporting fiber by forming a septum extending into an organ from its wall or capsule. *Plural,* ***trabeculae*** (tra-BEK-yoo-lē).

Trabeculae carneae (tra-BEK-yoo-lē KAR-nē-ē) Ridges and folds of the myocardium in the ventricles.

Trachea (TRĀ-kē-a) Tubular air passageway extending from the larynx to the fifth thoracic vertebra. Also called the ***windpipe.***

Tracheostomy (trā-kē-OS-tō-mē) Creation of an opening into the trachea through the neck (below the cricoid cartilage), with insertion of a tube to faciliate passage of air or evacuation of secretions.

Trachoma (tra-KŌ-ma) A chronic infectious disease of the conjunctiva and cornea of the eye caused by *Chlamydia trachomatis.*

Tract A bundle of nerve fibers in the central nervous system.

Transfusion (trans-FYOO-shun) Transfer of whole blood, blood components, or bone marrow directly into the bloodstream.

Transient ischemic (is-KĒ-mik) ***attack (TIA)*** Episode of temporary focal, nonconvulsive cerebral dysfunction caused by interference of the blood supply to the brain.

Transplantation (trans-plan-TĀ-shun) The replacement of injured or diseased tissues or organs with natural ones.

Transvaginal oocyte retrieval Procedure in which aspirated secondary oocytes are combined with a solution containing sperm outside the body and then the fertilized ova are implanted in the uterus.

Transverse colon (trans-VERS KŌ-lon) The portion of the large intestine extending across the abdomen from right colic (hepatic) flexure to the left colic (splenic) flexure.

Transverse fissure (FISH-er) The deep cleft that separates the cerebrum from the cerebellum.

Transverse tubules (TOO-byools) ***(T tubules)*** Minute, cylindrical invaginations of the muscle fiber (cell) membrane that carry the muscle action potentials deep into the muscle fiber.

Trauma (TRAW-ma) An injury, either a physical wound or psychic disorder, caused by an external agent or force, such as a physical blow or emotional shock; the agent or force that causes the injury.

Traveler's diarrhea Infectious disease of the gastrointestinal tract that results in loose, urgent bowel movements, cramping, abdominal pain, malaise, nausea, and occasionally fever and dehydration. It is acquired through ingestion of food or water that has become contaminated with fecal material containing mostly bacteria (especially *Escherichia coli*).

Also called ***Montezuma's revenge, turista,*** and ***Tut's tummy.***

Tremor (TREM-or) Rhythmic, involuntary, purposeless contraction of opposing muscle groups.

Treppe (TREP-ē) The gradual increase in the amount of contraction by a muscle caused by rapid, repeated stimuli of the same strength.

Triad (TRĪ-ad) A complex of three units in a muscle fiber (cell) composed of a transverse tubule and the segments of sarcoplasmic reticulum on both sides of it.

Tricuspid (trī-KUS-pid) ***valve*** Artrioventricular (AV) valve on the right side of the heart.

Trigeminal neuralgia (trī-JEM-i-nal noo-RAL-jē-a) Pain in one or more of the branches of the trigeminal (V) nerve. Also called ***tic douloureux*** (doo-loo-ROO).

Trigone (TRĪ-gon) A triangular area at the base of the urinary bladder.

Triiodothyronine (trī-ī-od-ō-THĪ-rō-nēn) (***T_3***) A hormone produced by the thyroid gland that regulates organic metabolism, growth and development, and the activity of the nervous system.

Trochlea (TROK-lē-a) A pulleylike surface.

Trophoblast (TRŌF-ō-blast) The outer covering of cells of the blastocyst.

Tropic (TRŌ-pik) ***hormone*** A hormone whose target is another endocrine gland.

Trunk The part of the body to which the upper and lower extremities are attached.

Tubal ligation (lī-GĀ-shun) A sterilization procedure in which the uterine (Fallopian) tubes are tied and cut.

Tuberculosis (too-berk-yoo-LŌ-sis) An infection of the lungs and pleurae caused by *Mycobacterium tuberculosis* resulting in destruction of lung tissue and its replacement by fibrous connective tissue.

Tubular transport maximum (Tm) The maximum amount of a substance that can be reabsorbed by renal tubules under any condition.

Tubular reabsorption The movement of filtrate from renal tubules back into blood in response to the body's specific needs.

Tubular secretion The movement of substances in blood back into filtrate in response to the body's specific needs.

Tumor (TOO-mor) A growth of excess tissue due to an unusually rapid division of cells.

Tunica albuginea (TOO-ni-ka al'-byoo-JIN-ē-a) A dense layer of white fibrous tissue covering a testis or deep to the surface of an ovary.

Tunica externa (eks-TER-na) The outer coat of an artery or vein, composed mostly of elastic and collagenous fibers. Also called the ***adventitia.***

Tunica interna (in-TER-na) The inner coat of an artery or vein, consisting of a lining of endothelium, basement membrane, and internal elastic lamina. Also called the ***tunica intima*** (IN-ti-ma).

Tunica media (MĒ-dē-a) The middle coat of an artery or vein, composed of smooth muscle and elastic fibers.

T wave The deflection wave of an electrocardiogram that records ventricular repolarization (relaxation).

Twitch Rapid, jerky contraction of a musle in response to a single stimulus.

Tympanic antrum (tim-PAN-ik AN-trum) An air space in the posterior wall of the middle ear that leads into the mastoid air cells or sinus.

Tympanic (tim-PAN-ik) ***membrane*** A thin, semitransparent partition of fibrous connective tissue between the external auditory meatus and the middle ear. Also called the ***eardrum.***

Type II cutaneous mechanoreceptor A receptor embedded deeply in the dermis and deeper tissues that detects heavy and continuous touch sensations. Also called an ***end organ of Ruffini.***

Ulcer (UL-ser) An open lesion of the skin or a mucous membrane of the body with loss of substance and necrosis of the tissue.

Ultrasound (US) Medical imaging technique that utilizes high-frequency sound waves to produce an image called a ***sonogram.***

Umbilical (um-BIL-i-kal) Pertaining to the umbilicus or navel.

Umbilical (um-BIL-i-kal) ***cord*** The long, ropelike structure, containing the umbilical arteries and vein that connect the fetus to the placenta.

Umbilicus (um-BIL-i-kus or um-bil-Ī-kus) A small scar on the abdomen that marks the former attachment of the umbilical cord to the fetus. Also called the ***navel.***

Upper extremity The appendage attached at the shoulder girdle, consisting of the arm, forearm, wrist, hand, and fingers.

Up-regulation Phenomenon in which there is an increase in the number of receptors in response to a deficiency of a hormone or neurotransmitter.

Uremia (yoo-RĒ-mē-a) Accumulation of toxic levels of urea and other nitrogenous waste products in the blood, usually resulting from severe kidney malfunction.

Ureter (YOO-re-ter) One of two tubes that connect the kidney with the urinary bladder.

Urethra (yoo-RĒ-thra) The duct from the urinary bladder to the exterior of the body that conveys urine in females and urine and semen in males.

Urinalysis The physical, chemical, and microscopic analysis or examination of urine.

Urinary (YOO-ri-ner-ē) ***bladder*** A hollow, muscular organ situated in the pelvic cavity posterior to the symphysis pubis.

Urinary tract infection (UTI) An infection of a part of the urinary tract or the presence of large numbers of microbes in urine.

Urine The fluid produced by the kidneys that contains wastes or excess materials and is excreted from the body through the urethra.

Urobilinogenuria The presence of urobilinogen in urine.

Urogenital (yoo'-rō-JEN-i-tal) ***triangle*** The region of the pelvic floor below the symphysis pubis, bounded by the symphysis pubis and the ischial tuberosities and containing the external genitalia.

Urology (yoo-ROL-ō-jē) The specialized branch of medicine that deals with the structure, function, and diseases of the male and female urinary systems and the male reproductive system.

Urticaria (ur'-ti-KĀ-rē-a) A skin reaction to certain foods, drugs, or other substances to which a person may be allergic; hives.

Uterine (YOO-ter-in) ***tube*** Duct that transports ova from the ovary to the uterus. Also called the ***Fallopian*** (fal-LŌ-pē-an) ***tube*** or ***oviduct.***

Uterosacral ligament (yoo'-ter-ō-SĀ-kral LIG-a-ment) A fibrous band of tissue extending from the cervix of the uterus laterally to attach to the sacrum.

Uterovesical (yoo'-ter-ō-VES-ī-kal) ***pouch*** A shallow pouch formed by the reflection of the peritoneum from the anterior surface of the uterus, at the junction of the cervix and the body, to the posterior surface of the urinary bladder.

Uterus (YOO-te-rus) The hollow, muscular organ in females that is the site of menstruation, implantation, development of the fetus, and labor. Also called the ***womb.***

Utricle (YOO-tri-kul) The larger of the two divisions of the membranous labyrinth located inside the vestibule of the inner ear, containing a receptor organ for static equilibrium.

Uvea (YOO-vē-a) The three structures that together make up the vascular tunic of the eye.

Uvula (YOO-vyoo-la) A soft, fleshy mass, especially the V-shaped pendant part, descending from the soft palate.

Vacuole (VAK-you-ōl) Membrane-bound organelle that, in animal cells, frequently functions in temporary storage or transportation.

Vagina (va-JĪ-na) A muscular, tubular organ that leads from the uterus to the vestibule, situated between the urinary bladder and the rectum of the female.

Valvular stenosis (VAL-vyoo-lar STEN-Ō-sis) A narrowing of heart valve, usually the bicuspid (mitral) valve.

Varicocele (VAR-i-kō-sēl) A twisted vein; especially, the accumulation of blood in the veins of the spermatic cord.

Varicose (VAR-i-kōs) Pertaining to an unnatural swelling, as in the case of a varicose vein.

Vas A vessel or duct.

Vasa recta (REK-ta) Extensions of the efferent arteriole of a juxtaglomerular nephron that run alongside the loop of the nephron (Henle) in the medullary region.

Vasa vasorum (VĀ-sa va-SŌ-rum) Blood vessels that supply nutrients to the larger arteries and veins.

Vascular (VAS-kyoo-lar) Pertaining to or containing many blood vessels.

Vascular spasm Contraction of the smooth muscle in the wall of a damaged blood vessel to prevent blood loss.

Vascular tunic (TOO-nik) The middle layer of the eyeball, composed of the choroid, ciliary body, and iris. Also called the ***uvea*** (YOO-vē-a).

Vascular (venous) sinus A vein with a thin endothelial wall that lacks a tunica media and externa and is supported by surrounding tissue.

Vasectomy (va-SEK-tō-mē) A means of sterilization of males in which a portion of each ductus (vas) deferens is removed.

Vasoconstriction (vāz-ō-kon-STRIK-shun) A decrease in the size of the lumen of a blood vessel caused by contraction of the smooth muscle in the wall of the vessel.

Vasodilation (vās'-ō-DĪ-lā-shun) An increase in the size of the lumen of a blood vessel caused by relaxation of the smooth muscle in the wall of the vessel.

Vasomotion (vāz-ō-MŌ-shun) Intermittent contraction and relaxation of the smooth muscle of the metarterioles and precapillary sphincters that result in an intermittent blood flow.

Vasomotor (vā-sō-MŌ-tor) ***center*** A cluster of neurons in the medulla that controls the diameter of blood vessels, especially arteries.

Vein A blood vessel that conveys blood from tissues back to the heart.

Vena cava (VĒ-na KĀ-va) One of two large veins that open into the right atrium, returning to the heart all of the deoxygenated blood from the systemic circulation except from the coronary circulation.

Venesection (vēn'-e-SEK-shun) Opening of a vein for withdrawal of blood.

Ventral (VEN-tral) Pertaining to the anterior or front side of the body; opposite of dorsal.

Ventral body cavity Cavity near the ventral aspect of the body that contains viscera and consists of a superior thoracic cavity and an inferior abdominopelvic cavity.

Ventral ramus (RĀ-mus) The anterior branch of a spinal nerve, containing sensory and motor fibers to the muscles and skin of the anterior surface of the head, neck, trunk, and the extremities.

Ventricle (VEN-tri-kul) A cavity in the brain or an inferior chamber of the heart.

Ventricular fibrillation (ven-TRIK-yoo-lar fib-ri-LĀ-shun) Asynchronous ventricular contractions that result in cardiovascular failure.

Venule (VEN-yool) A small vein that collects blood from capillaries and delivers it to a vein.

Vermiform appendix (VER-mi-form a-PEN-diks) A twisted, coiled tube attached to the cecum.

Vermilion (ver-MIL-yon) The area of the mouth where the skin on the outside meets the mucous membrane on the inside.

Vermis (VER-mis) The central constricted area of the cerebellum that separates the two cerebellar hemispheres.

Vertebral (VER-te-bral) ***canal*** A cavity within the vertebral column formed by the vertebral foramina of all the vertebrae and containing the spinal cord. Also called the ***spinal canal.***

Vertebral column The 26 vertebrae; encloses and protects the spinal cord and serves as a point of attachment for the ribs and back muscles. Also called the ***spine, spinal column,*** or ***backbone.***

Vertigo (VER-ti-go) Sensation of spinning or movement.

Vesicle (VES-i-kul) A small bladder or sac containing liquid.

Vesicular ovarian follicle A relatively large, fluid-filled follicle containing an immature ovum and its surrounding tissues that secretes estrogens. Also called a ***Graafian*** (GRAF-ē-an) ***follicle.***

Vestibular (ves-TIB-yoo-lar) ***membrane*** The membrane that separates the cochlear duct from the scala vestibuli.

Vestibule (VES-ti-byool) A small space or cavity at the beginning of a canal, especially the inner ear, larynx, mouth, nose, and vagina.

Villus (VIL-lus) A projection of the intestinal mucosal cells containing connective tissue, blood vessels, and a lymphatic vessel; functions in the absorption of the end products of digestion. *Plural,* ***villi*** (VIL-Ī).

Viscera (VIS-er-a) The organs inside the ventral body cavity. *Singular,* ***viscus*** (VIS-kus).

Visceral (VIS-er-al) Pertaining to the organs or to the covering of an organ.

Visceral effector (e-FEK-tor) Cardiac muscle, smooth muscle, and glandular epithelium.

Visceral muscle An organ specialized for contraction, composed of smooth muscle fibers (cells), located in the walls of hollow internal structures, and stimulated by visceral efferent neurons.

Visceral pleura (PLOO-ra) The inner layer of the serous membrane that covers the lungs.

Visceroceptor (vis'-er-ō-SEP-tor) Receptor that provides information about the body's internal environment.

Viscosity (vis-KOS-i-tē) The state of being sticky or thick.

Vital capacity The sum of inspiratory reserve volume, tidal volume, and expiratory reserve volume; about 4,800 ml.

Vital signs Signs necessary to life that include temperature (T), pulse (P), respiratory rate (RR), and blood pressure (BP).

Vitamin An organic molecule necessary in trace amounts that acts as a catalyst in normal metabolic processes in the body.

Vitiligo (vit-i-LĪ-go) Patchy, white spots on the skin due to partial or complete loss of melanocytes.

Vitreous (VIT-rē-us) ***body*** A soft, jellylike substance that fills the vitreous chambers of the eyeball, lying between the lens and the retina.

Vocal folds Pair of mucous membrane folds below the ventricular folds that function in voice production. Also called ***true vocal cords.***

Volkmann's canal *See* ***Perforating canal.***

Voltage-sensitive channel An ion channel in a plasma membrane composed of integral proteins that functions like a gate to permit or restrict the movement of ions in response to the voltage state of the membrane.

Vomiting Forcible expulsion of the contents of the upper gastrointestional tract through the mouth.

Vulva (VUL-va) Collective designation for the external genitalia of the female. Also called the ***pudendum*** (poo-DEN-dum).

Wallerian (wal-LE-rē-an) ***degeneration*** Degeneration of the portion of the axon and myelin sheath of a neuron distal to the site of injury.

Wandering macrophage (MAK-rō-fāj) Phagocytic cell that develops from a monocyte, leaves the blood, and migrates to infected tissues.

Wart Generally benign tumor of epithelial skin cells caused by a virus.

Wave summation (sum-MĀ-shun) The algebraic addition of the excitatory and inhibitory effects of many stimuli applied to a nerve cell body. The increased strength of muscle contraction that results when stimuli follow in rapid succession. Also called ***temporal summation.***

Wheal (hwēl) Elevated lesion of the skin.

White matter Aggregations or bundles of myelinated axons located in the brain and spinal cord.

White matter tract The treelike appearance of the white matter of the cerebellum when seen in midsagittal section. Also called ***arbor vitae*** (AR-bōr VĒ-te). A series of branching ridges within the cervix of the uterus.

White pulp The portion of the spleen composed of lymphatic tissue, mostly lymphocytes, arranged around central arteries; in some areas of white pulp, lymphocytes are thickened into lymphatic nodules called splenic nodules (Malpighian corpuscles).

White ramus communicans (RĀ-mus ko-MYOO-ni-kans) The portion of a preganglionic sympathetic nerve fiber that branches away from the anterior ramus of a spinal nerve to enter the nearest sympathetic trunk ganglion.

Wormian bone *See* ***Sutural bone.***

Xiphoid (ZĪ-foyd) Sword-shaped. The lowest portion of the sternum.

Yolk sac An extraembryonic membrane that connects with the midgut during early embryonic development, but is nonfunctional in humans.

Zona fasciculata (ZŌ-na fa-sik'-yoo-LA-ta) The middle zone of the adrenal cortex that consists of cells arranged in long, straight cords and that secretes glucocorticoid hormones.

Zona glomerulosa (glo-mer'-yoo-LŌ-sa) The outer zone of the adrenal cortex, directly under the connective tissue covering, that consists of cells arranged in arched loops or round balls and that secretes mineralocorticoid hormones.

Zona pellucida (pe-LOO-si-da) Gelatinous glycoprotein layer internal to the corona radiata that surrounds a secondary oocyte.

Zona reticularis (ret-ik'-yoo-LAR-is) The inner zone of the adrenal cortex, consisting of cords of branching cells that secrete sex hormones, chiefly androgens.

Zygote (ZĪ-gōt) The single cell resulting from the union of a male and female gamete; the fertilized ovum.

Zymogenic (zī'-mō-JEN-ik) ***cell*** One of the cells of a gastric gland that secretes the principal gastric enzyme precursor, pepsinogen. Also called a ***peptic cell.***

Index

Note: Page numbers followed by the letter E indicate terms to be found in exhibits.

EPONYMS USED IN THIS TEXT

Eponymous terms are those named after a person. In general, eponyms should be avoided where possible, since they are totally nondescriptive, often vague, and do not necessarily indicate that the person whose name is used actually contributed anything very original. However, since eponyms are still in frequent use, this glossary has been prepared to indicate which current terms have been used to replace eponyms in this book. In the body of the text eponyms are cited in parentheses, immediately following the current terms where they are used for the first time in a chapter or later in the book. In addition, although eponyms are included in the index, they have been cross-referenced to their current terminology.

EPONYM	CURRENT TERMINOLOGY
Achilles tendon	calcaneal tendon
Adam's apple	thyroid cartilage
ampulla of Vater (VA-ter)	hepatopancreatic ampulla
Bartholin's (BAR-tō-linz) gland	greater vestibular gland
Billroth's (BIL-rōtz) cord	splenic cord
Bowman's (BŌ-manz) capsule	glomerular capsule
Bowman's (BŌ-manz) gland	olfactory gland
Broca's (BRŌ-kaz) area	motor speech area
Brunner's (BRUN-erz) gland	duodenal gland
bundle of His (HISS)	atrioventricular (AV) bundle
canal of Schlemm (SHLEM)	scleral venous sinus
circle of Willis (WIL-is)	cerebral arterial circle
Cooper's (KOO-perz) ligament	suspensory ligament of the breast
Cowper's (KOW-perz) gland	bulbourethral gland
crypt of Lieberkühn (LĒ-ber-kyoon)	intestinal gland
duct of Rivinus (re-VĒ-nus)	lesser sublingual duct
duct of Santorini (san′-tō-RĒ-nē)	accessory duct
duct of Wirsung (VĒR-sung)	pancreatic duct
end organ of Ruffini (roo-FĒ-nē)	type II cutaneous mechanoreceptor
Eustachian (yoo-STĀ-kē-an) tube	auditory tube
Fallopian (fal-LŌ-pē-an) tube	uterine tube
gland of Littré (LĒ-tra)	urethral gland
gland of Zeis (ZĪS)	sebaceous ciliary gland
Golgi (GOL-jē) tendon organ	tendon organ
Graafian (GRAF-ē-an) follicle	vesicular ovarian follicle
Granstein (GRAN-stēn) cell	nonpigmented granular dendrocyte
Hassall's (HAS-alz) corpuscle	thymic corpuscle
Haversian (ha-VĒR-shun) canal	central canal
Haversian (ha-VĒR-shun) system	osteon
Heimlich (HĪM-lik) maneuver	abdominal thrust maneuver
interstitial cell of Leydig (LĪ-dig)	interstitial endocrinocyte
islet of Langerhans (LANG-er-hanz)	pancreatic islet
Kupffer's (KOOP-ferz) cells	stellate reticuloendothelial cell
Langerhans (LANG-er-hanz) cell	nonpigmented granular dendrocyte
loop of Henle (HEN-lē)	loop of the nephron
Luschka's (LUSH-kaz) aperture	lateral aperture
Magendie's (ma-JEN-dēz) aperture	median aperture
Malpighian (mal-PIG-ē-an) corpuscle	splenic nodule
Meibomian (mi-BŌ-mē-an) gland	tarsal gland
Meissner's (MĪS-nerz) corpuscle	corpuscle of touch
Merkel's (MER-kelz) disc	tactile disc
Müller's (MIL-erz) duct	paramesonephric duct
Nissl (NIS-l) bodies	chromatophilic substance
node of Ranvier (ron-VĒ-ā)	neurofibral node
organ of Corti (KOR-tē)	spiral organ
Pacinian (pa-SIN-ē-an) corpuscle	lamellated corpuscle
Peyer's (PĪ-erz) patch	aggregated lymphatic follicle
plexus of Auerbach (OW-er-bak)	myenteric plexus
plexus of Meissner (MĪS-ner)	submucosal plexus
pouch of Douglas	rectouterine pouch
Purkinje (pur-KIN-jē) fiber	conduction myofiber
Rathke's (rath-KĒZ) pouch	hypophyseal pouch
Schwann (SCHVON) cell	neurolemmocyte
Sertoli (ser-TŌ-lē) cell	sustentacular cell
Skene's (SKĒNZ) gland	paraurethral gland
sphincter of Oddi (OD-dē)	sphincter of the hepatopancreatic ampulla
Stensen's (STEN-senz) duct	parotid duct
Volkmann's (FŌLK-manz) canal	perforating canal
Wernicke's (VER-ni-kēz) area	auditory association area
Wharton's (HWAR-tunz) duct	submandibular duct
Wharton's (HWAR-tunz) jelly	mucous connective tissue
Wormian (WER-mē-an) bone	sutural bone